ASTRONOMY AND ASTROPHYSICS ABSTRACTS

A Publication of the Astronomisches Rechen-Institut Heidelberg
Produced in Cooperation
with the Fachinformationszentrum Karlsruhe

Astronomy and Astrophysics Abstracts is Prepared
Under the Auspices of the International Astronomical Union

Volume 55 A
Literature 1992, Part 1

Edited by
G. Burkhardt U. Esser H. Hefele I. Heinrich W. Hofmann
D. Krahn V. R. Matas L. D. Schmadel R. Wielen G. Zech

Springer-Verlag Berlin Heidelberg GmbH

Astronomisches Rechen-Institut, Mönchhofstraße 12–14,
W-6900 Heidelberg 1, Fed. Rep. of Germany
Telex: 461 336 ARIHD D
Telefax: 0 62 21-40 52 97
Telephone: 0 62 21-40 50
Director: Prof. Dr. Roland Wielen

Astronomy and Astrophysics Abstracts
Department Head: Dr. Lutz D. Schmadel
Editors-in-Chief: Inge Heinrich, Dr. Lutz D. Schmadel

ISBN 978-3-662-12381-2   ISBN 978-3-662-12379-9 (eBook)
DOI 10.1007/978-3-662-12379-9

Library of Congress Catalog Card Number 72-104650.
Media conversion: Daten- und Lichtsatz-Service, Würzburg.
55/3140-5 4 3 2 1 0 – Printed on acid-free paper

Preface

Astronomy and Astrophysics Abstracts aims to present a comprehensive documentation of the literature concerning all aspects of astronomy, astrophysics, and their neighbouring fields. It is devoted to the recording, summarizing, and indexing of the relevant publications throughout the world. *Astronomy and Astrophysics Abstracts* is prepared by a special department of the *Astronomisches Rechen-Institut* in close cooperation with the *Fachinformationszentrum Karlsruhe*. The work is done under the auspices of the *International Astronomical Union*.

Volume 55 records literature published in the first half of 1992. Some older documents which we received late and which are not surveyed in earlier volumes are included too. The input work was shared between the Fachinformationszentrum Karlsruhe and the Astronomisches Rechen-Institut. We thank our Karlsruhe colleagues Ms. Eike Hellmann and Mr. Wolfgang Lück for their willing and effective cooperation. We acknowledge with thanks contributions from our colleagues all over the world. Last but not least we express our gratitude to all the organizations, observatories, and publishers that provide us with complimentary copies of their publications.

The recording process and corrections were carried out by our technical staff members Ms. Helga Ballmann, Ms. Monika Kohl, and Ms. Sylvia Matyssek. It is a pleasure to thank them all for their diligence.

Heidelberg, December 1992 **The Editors**

Contents

PART B

Stars

Interstellar Matter, Nebulae

Radio Sources, X-ray Sources, Cosmic Rays

Stellar Systems, Galaxy, Extragalactic Objects, Cosmology

Introduction

Astronomical Bibliographies

Astronomy and Astrophysics Abstracts started documentation and abstracting work in 1969 as the direct successor of the Astronomischer Jahresbericht. For information on astronomical literature before this date consultation of one of the following bibliographies is suggested:

(1) J. J. de Lalande, Bibliographie Astronomique, Paris 1803 (this work covers the time from 480 B. C. to the year 1803, VIII + 966 pages).

(2) J. C. Houzeau, A. Lancaster, Bibliographie générale de l'astronomie, Volume I (in two parts), Bruxelles 1887, 1889, Volume II, Bruxelles 1882. The complete title of Volume II is "Bibliographie générale de l'astronomie ou catalogue méthodique des ouvrages, des mémoires et des observations astronomiques, publiés depuis l'origine de l'imprimerie jusqu'en 1880". A new edition of these volumes was prepared by D. W. Dewhirst in 1964.

(3) Bibliography of Astronomy, 1881–1898. The literature of this period was recorded on standard slips by the Observatoire Royal de Belgique. From the material (some 52,000 items) a microfilm version was produced by University Microfilms Limited, Tylers Green, High Wycombe, Buckinghamshire, England, in 1970.

(4) Astronomischer Jahresbericht, 1899 gegründet von Walter Wislicenus, herausgegeben vom Astronomischen Rechen-Institut in Heidelberg (formerly in Berlin), Verlag W. de Gruyter, Berlin. For the period from 1899 to 1968 sixty-eight volumes were published, each of which, in general, covers the literature of one year.

Concept of *Astronomy and Astrophysics Abstracts*

This abstracting service aims to present a comprehensive documentation of the literature in all fields of astronomy and astrophysics and their border fields. It appears in semi-annual volumes. Two of these volumes cover the literature of one calendar year. Every effort will be made to ensure that the average time interval between the receiving date of the original documents and publication of the abstracts will not exceed eight months. This time interval is near to that achieved by monthly abstracting journals, compared to which our system of accumulation of information over six months offers the advantage of greater convenience for the user.

The main characteristics of the concept of *Astronomy and Astrophysics Abstracts* may be summarized as follows:

(1) The subdivision of astronomy and its border fields into subject categories is facilitated by the fact that the astronomical objects appear to be particularly well suited for the formation of categories. It may be assumed that such subdivisions can be maintained for a long period. Experience shows, however, that progress in research might imply minor changes in the classification scheme.

(2) Each paper has been classified into one of 106 numbered subject categories and given a serial number within the category. In this way each item is numbered by six figures: the first three indicate the number of the category, the following three the serial number within the category. Reference to an abstract in Volume 1 is indicated by "01" before the number of the category; for example: 01.074.028, denotes Volume 1, category 074, abstract 028. A paper might be classified into more than one

category. In this case, its abstract is placed only in one category, whereas in the other categories only cross references are given. These are listed at the end of each category.

(3) Authors' abstracts are used whenever possible. Popular articles are not abstracted.

(4) If possible, titles of papers and abstracts are given in English. A special reference is made to titles which we have not taken in the original language.

The whole material was recorded by means of a local AT network. All text recording programs and other data processing software were developed by Multicom GmbH, Gröbenzell, F. R. Germany and by our staff members as well. The index computations were carried out on the IBM 3090-180 computer of the University of Heidelberg.

Classification Systems

The two most common and widely used classification systems in astronomy and astrophysics are given by Class 9 of the revised edition of the International Classification System for Physics, published by the International Council of Scientific Unions Abstracting Board (Second edition 1978. ICSU-AB, 17 Rue Mirabeau, 75017 Paris, France, ISSN 0305-9618), and the *Astronomy and Astrophysics Abstracts* classification. In order to facilitate literature searches, we introduce a concordance relation between these two very different systems. This solution is only a unilateral one. Starting from the third hierarchical level of the PHYS-Classification Scheme 1987, the appropriate *Astronomy and Astrophysics Abstracts* chapter numbers are listed. This cannot imply an identical content of the respective chapters in both systems. In many cases there is only a rather partial concordance, and therefore the *Astronomy and Astrophysics Abstracts* numbers are enclosed in parentheses.

Transliteration Scheme for the Russian Alphabet

The transliteration of the Russian alphabet in use in *Astronomy and Astrophysics Abstracts* is presented here.

А	а	a	П	п	p
Б	б	b	Р	р	r
В	в	v	С	с	s
Г	г	g	Т	т	t
Д	д	d	У	у	u
Е	е	e	Ф	ф	f
Ё	ё	e	Х	х	kh
Ж	ж	zh	Ц	ц	ts
З	з	z	Ч	ч	ch
И	и	i	Ш	ш	sh
Й	й	j	Щ	щ	shch
К	к	k	Ы	ы	y
Л	л	l	Ь	ь	'
М	м	m	Э	э	eh
Н	н	n	Ю	ю	yu
О	о	o	Я	я	ya

This transliteration was recommended by the Abstracting Board of the International Council of Scientific Unions in 1969. It corresponds essentially to the transliteration proposed by the Academy of Sciences, Moscow. In this case the letters can be read and printed by usual data processing machines. If the names of Russian authors in the literature are transliterated in a different scheme, we present the names as they are given in the references cited and in addition in brackets according to our transliteration table.

Sources of Information

The majority of sources of information for this volume is given in category **001 Periodicals** and in category **008 Observatories, Institutes.** It may be noted that the titles of the periodicals are given in the original languages, and that Russian titles have been transliterated applying the transliteration scheme given above. Category 008 records publication series of observatories and astronomical institutes. Titles of the periodicals have been given following the recommendations of the "International List of Periodical Title Word Abbreviations" and its additions (see also **Abbreviations**). In most cases they permit recognition of the full title without recourse to the key in category 001. If other secondary sources have been consulted, we cite these papers and give reference to the respective services.

The total number of papers (some do not give names of authors) recorded in this volume amounts to 10,063.

Author, Subject, and Object Indexes

The subject category and the serial number have been used as a reference in all three indexes. These references are more precise than page references and offer considerable advantages in indexing by means of computers.

The author index of this volume contains 13,723 names.

We consider the subject index as an approximation to an optimal index covering all fields of astronomy and astrophysics. At present, the *Astronomy and Astrophysics Abstracts Vocabulary* 30.92, containing 2,447 key words, is in use. This is done not only for the users' convenience, but also with the intention to propose the use of special key words to authors and publishers.

While each volume is scheduled to contain an author index, a subject index and an object index, the magnetic tapes containing the index information will be used to produce separate index volumes (authors, subjects, and objects) at intervals of five years.

Beginning with Vol. 39 *Astronomy and Astrophysics Abstracts* will additionally contain an object index providing a further key to the documents abstracted. For detailed information concerning selection, standardization, sorting, etc, see introduction to the object index.

The sorting program for the indexes is based on the IBM SORT/MERGE Program. This program sorts blank before hyphen and before letters. Apostrophes are ignored by a special routine.

The users are requested to inform us on spelling errors within the author index in order to assist us in eliminating mistakes in future cumulative indexes.

Information Retrieval

Most of the information given in this volume of *Astronomy and Astrophysics Abstracts* is incorporated into the data base PHYS of the Fachinformationszentrum Karlsruhe. PHYS is on-line retrievable via STN International. Inquiries about PHYS should be directly addressed to:

STN International
c/o Fachinformationszentrum Karlsruhe
P.O. Box 2465
D-7500 Karlsruhe 1
F. R. Germany

Telex: 17 724 710 +
Telefax: 0 72 47-80 86 66
Teletex: 72 47 10-FIZKA
Telephone: 0 72 47-80 85 55

The Astronomisches Rechen-Institut Heidelberg does not distribute the astronomical bibliography in any on-line or off-line form.

Astronomy and Astrophysics Abstracts is prepared at the *Astronomisches Rechen-Institut*, Heidelberg under the auspices of the *International Astronomical Union* on a non-profit basis. The editors urge publishers of literature related to astronomy and astrophysics to provide our service in due time with complimentary copies of their material.

Publications should be mailed to:

Astronomy and Astrophysics Abstracts
Astronomisches Rechen-Institut
Moenchhofstrasse 12–14
W-6900 Heidelberg 1
Fed. Rep. of Germany
Telex: 461 336 ARIHD D
Telefax: 0 62 21-40 52 97
Telephone: 0 62 21-4050

Concordance Relation

between the PHYS-Classification Scheme 1987
and the *Astronomy and Astrophysics Abstracts* Classification Scheme

PHYS Classification Scheme	AAA Classification Scheme
0 General	
0100 Communication, education, history, and philosophy	
0130 Physics literature and publications	002, 003, 012, 046
0140 Education	014
0150 Educational aids	014
0160 Biographical, historical, and personal notes	004–007
0190 Other topics of general interest	015
9 Geophysics, Astronomy, and Astrophysics	
9100 Solid Earth physics	
9110 Geodesy and gravity	044, 045, 081
9125 Geomagnetism and paleomagnetism; geoelectricity	084 (081)
9135 Earth's interior structure and properties	081
9190 Other topics in solid Earth physics	081
9400 Aeronomy	
9410 Physics of the neutral atmosphere	082 (063, 084)
9420 Physics of the ionosphere	083 (062, 084)
9430 Physics of the magnetosphere	084 (062, 106)
9440 Cosmic-ray interactions with the Earth	078, 144 (085, 105)
9500 Fundamental astronomy and astrophysics, instrumentation, techniques, and astronomical observations	
9510 Fundamental astronomy	041–043 (052, 079, 095, 096)
9520 Historical and archaeoastronomy	004

PHYS Classification Scheme	AAA Classification Scheme
9530 Fundamental aspects of astrophysics	022, 061–063, 066
9540 Artificial Earth satellites	051, 053
9545 Observatories	008, 009
9555 Astronomical and space-research instrumentation	032–035 (031, 036, 053)
9575 Observation and reduction techniques	036 (021)
9580 Catalogues, atlases, etc.	002 (046)

9600   Solar system

9610 General, solar nebula, and cosmogony	107 (091)
9620 Moon	094
9630 Planets and satellites (excluding the Moon)	091–093, 097, 099–101
9640 Cosmic rays	078, 144 (106)
9650 Interplanetary space	074, 098, 102–106
9660 Solar physics	071–080

9700   Stars

9710 Stellar characteristics and properties	064, 065, 111–116, 131, (061, 062, 063)
9720 Normal stars (by class); general or individual	113–116, 121, 126
9730 Variable and peculiar stars (including novae)	112, 116, 117, 122–124
9760 Late stages of stellar evolution (including black holes)	067, 125, 126 (065)
9780 Binary and multiple stars (including extrasolar planetary systems)	117–120

9800   Stellar systems; galactic and extragalactic objects and systems; the Universe

9810 Stellar dynamics	151
9820 Stellar clusters and associations	152–154
9840 Interstellar matter and nebulae	125, 131–134
9850 Galaxies, extragalactic objects and systems	155–160 (066, 161)
9870 Other objects and background radiations of unknown origin or distances	133, 141–144, 161
9880 Cosmology	161 (061, 066)

Abbreviations

Abbreviations used in *Astronomy and Astrophysics Abstracts* are primarily based on the 'International List of Periodical Title Word Abbreviations', prepared for the UNISIST/ICSU-AB Working Group on Bibliographic Descriptions (1970).

A.A.B.	Associazione Astrofili Bolognesi
Aarg.	Aargang
AAS	American Astronomical Society
AAVSO	American Association of Variable Star Observers
Abh.	Abhandlung –
Abstr.	Abstract –
Abt.	Abteilung
Acad.	Academi –, Academy
Accad.	Accademi –
Act.	Active, Activit –
Adm.	Administr –
Adv.	Advanc –
Aehron.	Aehronomi –
Aeron.	Aeronom –
Aeronaut.	Aeronauti –
Aerosp.	Aerospace
Afr.	Africa –
AG	Astronomische Gesellschaft
AIAA	American Institute of Aeronautics and Astronautics
AJB	Astronomischer Jahresbericht
Akad.	Akadem –
AL	Alabama
Alm.	Almanac –
Am.	America –, Amerika –, Amerique –
Amat.	Amateur –
Amst.	Amsterdam
An.	Anais, Anale –, Anali –, Anals
Anal.	Analis –, Analit –, Analys –, Analyt –
Angew.	Angewandt –
Ann.	Annaes, Annal –
Annu.	Annu –
Anst.	Anstalt
Anu.	Anual –, Anuar –
Anz.	Anzeiger
Appl.	Applied
AR	Arkansas
Arb.	Arbeit
Arch.	Archiv –
Årg.	Årgang
Argent	Argentin –
Ark.	Arkiv –
Arkh.	Arkhiv –
Artif.	Artifici –
ASA	Astronomical Society of Australia
Asoc.	Asocia –
ASP	Astronomical Society of the Pacific
ASSA	Astronomical Society of Southern Africa
Assem.	Assembl –
Assoc.	Associ –
Assoz.	Assozi –
Astrofis.	Astrofisic –
Astrofiz.	Astrofizi –
Astrometr.	Astrometr –
Astron.	Astronom –
Astronaut.	Astronauti –, Astronauty –
Astrophys.	Astrophys –
ASV	Astronomical Society of Victoria
ASWA	Astronomical Society of Western Australia
At.	Atom –

Atmos.	Atmosf –, Atmosph –
Aust.	Australi –
AZ	Arizona
BAA	British Astronomical Association
Barc.	Barcelona
Bayer.	Bayerisch –
Beitr.	Beitrag, Beiträge
Belg.	Belge –, Belgi –
Beob.	Beobacht –
Beogr.	Beograd –
Ber.	Bericht –
Bibl.	Bibliot –
Bibliogr.	Bibliograf –, Bibliograph –
BIH	Bureau International de l'Heure
Bimest.	Bimestr –
Bl.	Blatt, Blätter
Bol.	Boletin
Boll.	Bolletino
Br.	British
Bras.	Brasil –
Brun.	Brunens –
Bruss.	Brussel, s
Brux.	Bruxelles
Bul.	Buleten –, Buletin –, Bulten
Bulg.	Bulgar –
Bull.	Bulletin –, Bullettino
Bur.	Bureau –
Byul.	Byuleten –, Byuletin –
Byull.	Byulleten –
C. R.	Comptes Rendus
CA	California
Cah.	Cahier –
Camb.	Cambridge
Can.	Canadi –, Canada
Carol.	Carolina –
Cas.	Casopis
Cat.	Catalog –
Celest.	Celestial
Cent.	Center, Central, Centrale, Centrally, Centre
Cercet.	Cercetari
Cesk.	Ceskoslov –
Chem.	Chemi –
Chim.	Chimi –
Chin.	Chinese
Chron.	Chronic –, Chronik, Chronique
Chronom.	Chronometr –
Cie.	Compagnie
Cienc.	Ciencia –
Cient.	Cientific –
Circ.	Circolar –, Circolo, Circolaire –, Circular –, Circulo
Cirk.	Cirkulaer –
Cl.	Clasa, Classe –
CO	Colorado
Co.	Companies, Company
Coll.	College
Collect.	Collect –
Colloq.	Colloqui –

Comet.	Cometary
Commentat.	Commentat–
Commun.	Communica–
Comput.	Computation, Computer–, Computing
Comun.	Comunica–
Conf.	Conferen–
Congr.	Congres–
Contract.	Contract–
Contrib.	Contribu–
Copenh.	Copenhagen
Cosmochim.	Cosmochimi–
COSPAR	Committee on Space Research
Crystallogr.	Crystallograph–
CSIRO	Commonwealth Scientific and Industrial Research Organization
CT	Connecticut
Cult.	Cultur–, Cultuur
Curr.	Current
Czech.	Czechoslovak–
DC	District of Columbia
DE	Delaware
Dep.	Departament, Département, Department
Dev.	Development–, Développement–
Dig.	Digest
Dir.	Director–
Diss.	Disserta–
Div.	Divis–
Doc.	Document–
Dok.	Dokument–
Dokl.	Doklad–
Dom.	Dominion
Dtsch.	Deutsch
Ed.	Edit–
Edinb.	Edinburgh
Ehksp.	Ehksperiment–
Eidg.	Eidgenössisch–
Eksp.	Eksperiment–
Electron.	Electroni–
Eng.	Engineer–
Environ.	Environment–
Equip.	Equipement, Equipment
Ergeb.	Ergebnis–
ESA	European Space Agency
ESO	European Southern Observatory
Espec.	Especial–
ESRO	European Space Research Organization
Eur.	Europ–
Eval.	Evaluation–
Exp.	Experiment–
Extraterr.	Extraterrestr–
Fac.	Facolt–, Faculd–, Facult–
Fak.	Fakult
Fasc.	Fascicul–
Fenn.	Fenni–
Finn.	Finni–
Fis.	Fisic–, Fisik–
Fiz.	Fizic–, Fizik–, Fizyk–
FL	Florida
Fluid.	Fluidi–
Fond.	Fondation–, Fondazione
Fortschr.	Fortschritt–
Fotogr.	Fotograf–
Found.	Foundation–
Fr.	Français–
Freq.	Frequen–
Fundam.	Fundamenta–
Fys.	Fysik–, Fysisch, Fysisk–
Fyz.	Fyzik–
G.	Giornale
GA	Georgia
Gaz.	Gazeta, Gazette
Gazz.	Gazzetta
Gen.	General
Geochem.	Geochem–
Geochim.	Geochim–
Geod.	Geodaes–, Geodaet–, Geodes–, Geodet–, Geodez–
Geofis.	Geofis–
Geofiz.	Geofiz–
Geofys.	Geofys–
Geogr.	Geograf–, Geograph–
Geokhim.	Geokhim–
Geol.	Geolog–, Geolosk–
Geomagn.	Geomagneti–
Geophys.	Geophys–
Ges.	Gesellschaft
Gesch.	Geschichte
Gl.	Glavno–
Glas.	Glasnik
Gos.	Gosudarst–
Gov.	Government–
Grenzgeb.	Grenzgebiet–
GSFC	Goddard Space Flight Center
H.M.	Her Majesty's, His Majesty's
Hamb.	Hamburg
Handb.	Handbook, Handbuch
Heidelb.	Heidelberg–
Helv.	Helveti–
Her.	Herald–
HI	Hawaii
Hist.	History
Hochsch.	Hochschule
Hoegsk.	Hoegskol–
HR-diagram	Hertzsprung-Russell diagram
Hung.	Hungar–
Hydrogr.	Hydrograf–, Hydrograph–
IA	Iowa
IAF	International Astronautical Federation
IAU	International Astronomical Union
IBM	International Business Machines Corporation
ICSTI	International Council for Scientific and Technical Information
ICSU	International Council of Scientific Unions
ID	Idaho
IEEE	Institute of Electrical and Electronics Engineers
IL	Illinois
IN	Indiana
Inc.	Incorporated
Ind.	Industr–
Inf.	Informat–, Informaz–, Informe–
Ing.	Ingenieur
INIS	International Nuclear Information System
INSPEC	International Information Services for the Physics and Engineering Communities
Inst.	Institut–, Instytut–
Instrum.	Instrument–
Int.	International, Internazional–
Intell.	Intelligenc–
Inter.	Intérieur–, Interior
Interplanet.	Interplanetary
Intez.	Intezet–
Invest.	Investiga–
Ionos.	Ionosfer–, Ionospher–
Ir.	Irish
Iskusstv.	Iskusstvenn–
Isr.	Israel–
Issled.	Issledovan–
Ist.	Istitut
Ital.	Itali–
Izd.	Izdatel–
Izv.	Izvesti–

J.	Joernaal–, Jornal–, Journal–
Jaarb.	Jaarboek–
Jahrb.	Jahrbuch, Jahrbücher
Jahresber.	Jahresbericht–
Jahresschr.	Jahresschrift
Jahrg.	Jahrgang
JPL	Jet Propulsion Laboratory
Jpn.	Japan–
K.	Königlich–, Koninkljik–, Kunglig–
Kartogr.	Kartograf–
Kernforsch.	Kernforschung
Kernphys.	Kernphysik–
Khem.	Khemyi–
Khim.	Khimi–
Kim.	Kimija–, Kimya
Kl.	Klass–
Kolloq.	Kolloquium–
Komet.	Kometnyj
Komm.	Kommission–
Konf.	Konfer–
Kongr.	Kongress
Kosm.	Kosmich–
Kosmog.	Kosmogon–
Kozp.	Kozponti
KPNO	Kitt Peak National Observatory
KS	Kansas
KY	Kentucky
LA	Louisiana
Lab.	Laborato–
LEST	Large European Solar Telescope
Lett.	Letter–, Lettra, Lettre
Libr.	Librair–, Librar–
MA	Massachusetts
Madr.	Madrid
Mag.	Magasin, Magazin–
Magn.	Magneti–, Magnitn–
Mar.	Marin–
Mat.	Matemaat–, Matemat–
Mater.	Material–
Math.	Mathemat–
MD	Maryland
ME	Maine
Meas.	Measur–
Mec.	Mecani–
Mech.	Mechani–
Medd.	Meddelande–, Meddelelse
Meded.	Mededeeling, Mededeling–
Mekh.	Mekhani–
Mem.	Memento–, Memoir–, Memori–, Memory–, Memuary
Memo.	Memorand–
Mens.	Mensile, Mensual–, Mensuel–
Messtech.	Messtechni–
Meteorol.	Meteorolog–
Mex.	Mexic–
MI	Michigan
Micromec.	Micromecaniq–
Miner.	Mineral, Minerale–, Minerali–
Mineral.	Mineralog–
MIT	Massachusetts Institute of Technology
Mitt.	Mitteilung–
MN	Minnesota
MO	Missouri
Mod.	Modern–
Mol.	Molecul–, Molekul–
Mon.	Monat, Monatlich–, Month–
Monogr.	Monograph–
MPI	Max-Planck-Institut
MS	Mississippi
MT	Montana
Mt.	Mount

Muench.	Muenchen
Mus.	Museum
N.Z.	New Zealand
Nablyud.	Nablyudeni–
Nac.	Nacion–
Nachr.	Nachricht–
NASA	National Aeronautics and Space Administration
Nat.	Natur–
Natl.	National–
Naturforsch.	Naturforsch–
Naturwiss.	Naturwissenschaft–
Natuurkd.	Natuurkunde
Nauchn.	Nauchny–
Nauk.	Nauka, Naukite, Naukov–, Naukow–
Naut.	Nautic–
Nav.	Naval–
Navig.	Navigat–
Naz.	Nazion–
NC	North Carolina
ND	North Dakota
NE	Nebraska
Ned.	Nederland–
Newsl.	Newsletter–
NH	New Hampshire
NJ	New Jersey
NM	New Mexico
Not.	Notationes, Notic–, Notise, Notizi–
Nouv.	Nouveau–, Nouvell–
Nov.	Novoe
Nucl.	Nucléaire–, Nuclear–, Nucl–
Nukl.	Nukle–
Numer.	Numeri–
NV	Nevada
NY	New York
O-va	Obshchestva
O-vo	Obshchestvo
Obs.	Observ–
Österr.	Österreich–
Off.	Offic–
OH	Ohio
OK	Oklahoma
Opt.	Optic–, Optik–, Optique
OR	Oregon
Oss.	Osserva–
PA	Pennsylvania
Pac.	Pacific
Paleontol.	Paleontolog–
Pap.	Paper–, Papier
Part.	Particle
Pekin.	Pekinens–
Perem.	Peremen–
Period.	Periodi–
Petrol.	Petrolog–
Philos.	Philosoph–
Photogr.	Photograf–, Photograph–
Photogramm.	Photogrammetr–
Photom.	Photometr–
Phys.	Physic–, Physik–, Physique–, Physisch–
Pict.	Picture–
Planet.	Planetary
Pol.	Polish, Polon–
Pr.	Prac–
Prelim.	Prelimin–
Prepr.	Preprint
Prib.	Pribor–
Prikl.	Prikladn–
Prilozh.	Prilozhen–
Prir.	Prirodn–
Prirodoved.	Prirodoved–

Probl.	Problem –		**Sitzungsber.**	Sitzungsbericht –
Proc.	Proceedings		**Skr.**	Skrift –
Prod.	Prodott –, Produc –, Produkt,		**Smithson.**	Smithsonian
Prog.	Progres –		**Soc.**	Sociedad –, Societ –
Propag.	Propagation		**Sol.**	Solar
Prospect.	Prospecting		**Soln.**	Solnechn –
Prov.	Provinc –, Provints –, Provinz –		**Sonderdr.**	Sonderdruck –
Pubbl.	Pubblicazion –		**Soobshch.**	Soobshchen –
Publ.	Publicac –, Publicas –, Publicat –,		**South.**	Southern
	Publikas –, Publikat –		**Spacecr.**	Spacecraft
			Spat.	Spatial –
			Spec.	Special –
Q.	Quarterly		**Spectrosc.**	Spectroscop –
Quant.	Quantit –		**Spectrosk.**	Spectroskop –
			Spets.	Spetsial –
			Spez.	Spezial –, Speziell –
R.	Royal		**SSR**	Sovetskaya Sotsialisticheskaya
Radiat.	Radiati –			Respublika
Radioact.	Radioactiv –, Radioaktiv –		**SSSR**	Soyuz Sovetskikh Sotsialisticheskikh
Radioisot.	Radioisotop –			Respublik
Rap.	Raport –		**St.**	Saint –, Sankt –, Sant –
Rapp.	Rapport –		**– St.**	– Straße, Street
RAS	Royal Astronomical Society		**Stand.**	Standard –, Standart –
Rec.	Record –		**Sternw.**	Sternwarte –
Rech.	Recherche –		**Stiint.**	Stiintific –
Ref.	Referat –, Reference –, Referieren		**Stn.**	Station, Stazione
Relat.	Related, Relation –		**Stud.**	Studia, Studie –, Studii
Relativ.	Relativit –		**Supl.**	Suplement –, Supliment –
Rend.	Rendicont –		**Suppl.**	Supplement –
Rep.	Report –		**Surv.**	Survey –
Repr.	Reprint –		**Syd.**	Sydney
Repub.	Republi –		**Symp.**	Sympos –, Sympoz –
Res.	Research –		**Syst.**	System –
Result.	Resultad –, Resultat –		**Sz.**	Szemle
Rev.	Review –, Revisio, Revista, Revue –			
Rezul't.	Rezul'tat –			
RI	Rhode Island		**Teach.**	Teacher –, Teaching
Ric.	Ricerca, Ricerche		**Tec.**	Tecni –
Riv.	Rivist –		**Tech.**	Techni –
Roum.	Roumain –		**Technol.**	Technolog –
Rundsch.	Rundschau		**Tecnol.**	Tecnolog –
			Teh.	Tehnic –, Tehnika, Tehnisk –
			Tehnol.	Tehnolog –, Tehnolosk –
S. Afr.	South Africa		**Tek.**	Tekni –
SAF	Société Astronomique de France		**Tekh.**	Tekhni –
SAI	Società Astronomica Italiana		**Tekhnol.**	Tekhnolog –
Samml.	Sammlung –		**Teknol.**	Teknolog –
SAO	Smithsonian Astrophysical Observatory		**Telesc.**	Telescop –
SAS	Société Astronomique de Suisse		**Telev.**	Television –
Satell.	Satellite		**Teor.**	Teoret –, Teori –
Sb.	Sbornik –		**Terr.**	Terrestr –
SC	South Carolina		**Test.**	Testing
Scand.	Scandinavi –		**TH**	Technische Hochschule
Sch.	Schul –		**Theor.**	Theoret –, Theori –
Schr.	Schrift –		**Tidschr.**	Tidschrift –
Schriftenr.	Schriftenreihe		**Tidskr.**	Tidskrift –
Schweiz.	Schweizer –		**Tidsskr.**	Tidsskrift –
Sci.	Scienc –, Scient –, Scienz –		**TN**	Tennessee
Scr.	Scripta, Scritt –		**Top.**	Topic –
SD	South Dakota		**Torun.**	Torunensis
Secc.	Seccion –		**Tr.**	Trudy
Sect.	Secti –		**Trans.**	Transactions, Transazione
Sekc.	Sekci –, Sekcj –		**Tsentr.**	Tsentral –
Sekt.	Sektion –, Sektor –		**Tsirk.**	Tsirkulyar –
Sekts.	Sektsi –		**TU**	Technical University
Sel.	Seleccion –, Select –, Selek –, Selezione		**TX**	Texas
Selsk.	Selskab –, Selskap –			
Semin.	Séminair –, Seminar –			
Sep.	Separat –		**Uch.**	Uchen –
Ser.	Seria –, Serie –, Seriya		**Uchebn.**	Uchebn –
Serv.	Servic –, Serviz –		**UK**	United Kingdom
Sess.	Sessi –		**Umsch.**	Umschau
Signal.	Signalétique –		**UN**	United Nations
Simp.	Simpoz –		**Univ.**	Universidad –, Universit –, Univerzitet –
Sin.	Sinica		**Ups.**	Upsaliens –

US	United States	Vyp.	Vypusk —
USA	United States of America	Vyssh.	Vyssh —
USSR	Union of Soviet Socialist Republics	Vyzk.	Vyzkum —
UT	Utah		
Utr.	Utrecht	WA	Washington
		West.	Western
VA	Virginia	Wet.	Wetenschap —, Wetenskap —
Var.	Various	WI	Wisconsin
Ver.	Verein —, Verenig —	Wiss.	Wissenschaft —
Veränderl.	Veränderlich —	WV	West Virginia
Verh.	Verhandl —	WY	Wyoming
Vermess.	Vermessung —		
Vermessungswes.	Vermessungswesen	Yad.	Yadern —
Veröff.	Veröffentlich —		
Vesn.	Vesnik	Z.	Zeitschrift —
Vestn.	Vestnik	ZA	Zero Age
Vetensk.	Vetenskap —	ZAED	Zentralstelle für Atomkernenergie-Dokumentation
Vgl.	Vergleich —		
Vidensk.	Videnskab —, Videnskap	Zap.	Zapisk —, Zapyisk —
Vierteljahresschr.	Vierteljahresschrift —	Zaved.	Zaveden —
Vierteljahrsschr.	Vierteljahrsschrift	Zent.	Zentral
VLB	Very Long Baseline	Zentralbl.	Zentralblatt
Volcanol.	Volcanolog —	Zesz.	Zeszyt
Vopr.	Vopros —	Zh.	Zhurnal —
Vortr.	Vorträge	Zirk.	Zirkular
Vses.	Vsesoyuzn —		
VT	Vermont		

Periodicals, Proceedings, Books, Activities

001 Periodicals

AAO Newsl.
AAO Newsletter. Anglo–Australian Observatory, PO Box 296, Epping, NSW 2121 (Australia). Quarterly.

AAVSO Monogr.
AAVSO Monograph. American Association of Variable Star Observers, 25 Birch Street, Cambridge, MA 02138 (USA). Irregularly.

Acta Astron.
Acta Astronomica. Polska Akademia Nauk, Komitet Astronomii. Państwowe Wydawnictwo Naukowe, Warszawa–Kraków (Poland). Quarterly. Subscription address: Ars Polona, 00–068 Warszawa, Krakowskie Przedmieście 7, Poland. ISSN 0001–5237, AASWAM.

Acta Astron. Sin.
Acta Astronomica Sinica. Purple Mountain Observatory, Academia Sinica, Nanjing (China). Quarterly. Selected English translation in Chinese Astronomy and Astrophysics. ISSN 0001–5245, TIWHAO.

Acta Astrophys. Sin.
Acta Astrophysica Sinica. Academia Sinica. Beijing Astronomical Obervatory, Beijing (China). Subscription address: Guozi Shudian, PO Box 399, Beijing, China. Selected English translation in Chinese Astronomy and Astrophysics. ISSN 0253–2379.

Acta Cosmologica
Acta Cosmologica. Zeszyty Naukowe Uniwersytetu Jagiellońskiego. Państwowe Wydawnictwo Naukowe, Warszawa–Kraków (Poland). ISSN 0137–2386.

Acta Geophys. Sin.
Acta Geophysica Sinica. Department of Geophysical Research, Academia Sinica, Beijing (China). Bimonthly. Subscription address: Guozi Shudian, PO Box 399, Beijing, China. ISSN 0001–5733, AHGSBY.

Acta Opt. Sin.
Acta Optica Sinica. Monthly. ISSN 0253–2239.

Acta Phys. Hung.
Acta Physica Hungarica. Akademiai Kiado, Publishing House of the Hungarian Academy of Sciences, H–1054 Budapest, Alkotmany U. 21 (Hungary). Quarterly. ISSN 0001–6705, APHUE2.

Acta Phys. Pol., A
Acta Physica Polonica, Series A (General Physics, Atomic and Molecular Spectroscopy, Applied Physics). Polska Akademia Nauk (PAN), Warsaw (Poland). Bimonthly. ISSN 0587–4246, ATPLB6.

Acta Phys. Pol., B
Acta Physica Polonica, Series B (Elementary Particle Physics, Nuclear Physics, Theory of Relativity, Field Theory). Polska Akademia Nauk (PAN), Warsaw (Poland). Monthly. ISSN 0587–4254, APOBBB.

Acta Phys. Sin.
Acta Physica Sinica. Academia Sinica, Beijing (China). Monthly. ISSN 0372–736X, WLHPAR.

Acta Phys. Slovaca
Acta Physica Slovaca. VEDA Publishing House of the Slovak Academy of Sciences, 89530 Bratislava, Klemensova 19 (Czechoslovakia). Bimonthly. ISSN 0323–0465, APSVCO.

Adv. Space Res.
Advances in Space Research. Pergamon Press, Oxford (UK). Irregularly. The official journal of the Committee on Space Research (COSPAR). ISSN 0273–1177, ASRSDW.

Aerospace Res. Bulg.
Aerospace Research in Bulgaria. Space Research Institute, Bulgarian Academy of Sciences. Sofia (Bulgaria). Aerokosmicheski Izsledvaniya v B'lgariya. Irregularly. ISSN 0861–1432.

AIP Conf. Proc.
AIP Conference Proceedings. American Institute of Physics, New York, NY (USA). Irregularly. ISSN 0094–243X, APCPCS.

Am. Assoc. Variable Star Obs. Circ.
American Association of Variable Star Observers Circular. American Association of Variable Star Observers, Cambridge, MA (USA). Irregularly. Subscription address: AAVSO, C.E. Scovil, Stamford Observatory, 39 Scofieldtown Rd., Stamford, CT 06903, USA. ISSN 0197–2979, AACID5.

Am. J. Phys.
American Journal of Physics. American Institue of Physics, New York, NY (USA). Monthly. ISSN 0002–9505, AJPIAS.

Ann. Geophys.
Annales Geophysicae (Upper Atmosphere and Space Sciences). European Geophysical Society. Gauthier–Villars, Montrouge (France). Bimonthly.

Ann. Inst. Henri Poincaré, Phys. Théor.
Annales de l'Institut Henri Poincaré, Physique Théorique. Gauthier–Villars, Montrouge (France). Quarterly. ISSN 0246–0211, AHPAAO.

Ann. Isr. Phys. Soc.
Annals of the Israel Physical Society. Department of Physics, Bar–Ilan University, Ramat–Gan, Israel. Adam Hilger, Bristol (UK). Irregularly. ISSN 0309–8710, AIPSDK.

Ann. N.Y. Acad. Sci.
Annals of the New York Academy of Sciences. The New York Academy of Sciences, New York, NY (USA). Irregularly. ISSN 0077–8923, ANYAA9.

Ann. Phys. (N.Y.)
Annals of Physics (New York). Academic Press, New York, NY (USA). 15 issues per year. ISSN 0003–4916, APNYA6.

Ann. Phys. (Paris)
Annales de Physique (Paris). Masson, Paris (France). Bimonthly. ISSN 0003–4169, ANPHAJ.

Ann. Shanghai Obs., Acad. Sin.
Annals of Shanghai Observatory, Academia Sinica. Shanghai Scientific and Technical Publishers, Shanghai, Rei Jing Er Street 450 (China). Irregularly.

Annu. Rep. BIPM Time Sect.
Annual Report of the BIPM Time Section. Bureau International des Poids et Mesures (BIPM), Section du temps, Pavillon de Breteuil, F–92312 Sèvres Cedex (France). Annual. ISSN 1016–6114.

Annu. Rev. Astron. Astrophys.
Annual Review of Astronomy and Astrophysics. Annual Reviews Inc., 4139 El Camino Way, Palo Alto, CA 94303–0897 (USA). Annual. ISSN 0066–4146, ARAAAJ.

Annu. Rev. Earth Planet. Sci.
Annual Review of Earth and Planetary Sciences. Annual Reviews Inc., 4139 El Camino Way, Palo Alto, CA 94303–0897 (USA). Annual. ISSN 0084–6597, AREPLR.

Appl. Opt.
Applied Optics. Optical Society of America. American Institute of Physics, New York, NY (USA). Semimonthly. ISSN 0003–6935, APOPAI.

Appl. Phys. Lett.
Applied Physics Letters. American Institute of Physics, New York, NY (USA). Semimonthly. ISSN 0003–6951, APPLAB.

Appl. Phys., B
Applied Physics, B (Photophysics and Laser Chemistry). Springer, Berlin (Germany). Monthly. ISSN 0721–7269, APPCDL.

Appl. Spectrosc.
Applied Spectroscopy. Society for Applied Spectroscopy. Williams and Wilkins, Baltimore, MD (USA). Bimonthly. ISSN 0003–7028, APSPA4.

Arch. Hist. Exact Sci.
Archive for History of Exact Sciences. Springer, Berlin (Germany). ISSN 0003–9519, AHESAN.

Arch. Ration. Mech. Anal.
Archive for Rational Mechanics and Analysis. Springer, Berlin (Germany, F.R.). Monthly. ISSN 0003–9527, AVRMAW.

Archaeoastronomy (U.K.)
Archaeoastronomy. Science History Publications Ltd., Halfpenny Furze, Mill Lane, Chalfont St Giles, Bucks HP8 4NR (UK). Annual or semiannual. Supplement to Journal for the History of Astronomy. ISSN 0142–7253.

Astrofizika
Astrofizika. Izdatel'stvo Akademii Nauk Armyanskoj SSR, Erevan (Armenia). Bimonthly. English translation in Astrophysics. ISSN 0571–7132, ASTKBG.

Astron. Astrophys.
Astronomy and Astrophysics. Springer, Berlin (Germany). Semimonthly. ISSN 0004–6361, AAEJAF.

Astron. Astrophys. Rev.
The Astronomy and Astrophysics Review. Springer, Berlin (Germany). Irregularly. ISSN 0935–4956, AASREB.

Astron. Astrophys. Trans.
Astronomical and Astrophysical Transactions. Gordon and Breach, Philadelphia, PA (USA). The Journal of the Soviet Astronomical Society. ISSN 1055–6796, AATREG.

Astron. Astrophys., Suppl. Ser.
Astronomy and Astrophysics, Supplement Series. Les Editions de Physique, Z.I. de Courtabœuf, B.P. 112, F–91944 Les Ulis Cedex (France). Monthly. ISSN 0365–0138, AAESBO.

Astron. Her.
The Astronomical Herald. Astronomical Society of Japan, Tokyo (Japan). Monthly. ISSN 0374–2466, TGEPAC.

Astron. J.
The Astronomical Journal. American Astronomical Society. American Institute of Physics, New York, NY (USA). Monthly. ISSN 0004–6256, ANJOAA.

Astron. Nachr.
Astronomische Nachrichten. Akademie Verl., Berlin (Germany). Bimonthly. ISSN 0004–6337, ASNAAN.

Astron. Now
Astronomy Now. Intra Press, London (UK). Monthly. ISSN 0951–9726.

Astron. Raumfahrt
Astronomie und Raumfahrt. Kulturbund der DDR, Zentrale Kommission Astronomie und Raumfahrt, Berlin (German Democratic Republic). Bimonthly. Available from Zeitungsvertriebsamt, Abt. Export, DDR–1004 Berlin, Straße der Pariser Kommune 3–4, German Democratic Republic. ISSN 0587–565X.

Astron. Rechen–Inst. Heidelb., Mitt., Ser. A
Astronomisches Rechen–Institut Heidelberg, Mitteilungen, Serie A. Astronomisches Rechen–Institut, Mönchhofstraße 12–14, D–6900 Heidelberg 1 (Germany). Irregularly. ISSN 0440–5951, ARIMBS.

Astron. Rechen–Inst. Heidelb., Mitt., Ser. B
Astronomisches Rechen–Institut Heidelberg, Mitteilungen, Serie B. Astronomisches Rechen–Institut, Mönchhofstraße 12–14, D–6900 Heidelberg 1 (Germany). Irregularly.

Astron. Sch.
Astronomie in der Schule. Friedrich Verlag, Velber (Germany). Bimonthly. ISSN 0004–6310.

Astron. Tidsskr.
Astronomisk Tidsskrift. Astronomisk Selskab, København (Denmark). Quarterly. ISSN 0004–6345, ANTKBF.

Astron. Tsirk.
Astronomicheskij Tsirkulyar. Institut Astronomii, Rossijskaya Akademiya Nauk. Byuro Astronomicheskikh Soobshchenij, Moskva (Russia). Monthly. ISSN 0236–2457.

Astron. Vestn.
Astronomicheskij Vestnik. Issledovaniya Solnechnoj Sistemy. Rossijskaya Akademiya Nauk. Nauka, Moskva (Russia). Bimonthly. English translation in Solar System Research. ISSN 0320–930X, ASVEA7.

Astron. Zh.
Astronomicheskij Zhurnal. Rossijskaya Akademiya Nauk. Izdatel'stvo Nauka, Moskva (Russia). Bimonthly. English translation in Soviet Astronomy. ISSN 0004–6299, ASZHA2.

Astronomia UAI
Astronomia UAI. Unione Astrofili Italiani. Università degli Studi di Padova, Dipartimento di Astronomia, Vicolo dell'Osservatorio, 5, I–35122 Padova (Italy). Bimonthly. Periodico bimestrale dell' Unione Astrofili Italiani. ISSN 0392–2308.

Astronomie
L'Astronomie. Société Astronomique de France, 3, rue Beethoven, F–75016 Paris (France). Monthly. ISSN 0004–6302, ATROAO.

Astronomy
Astronomy. Kalmbach Publishing Co., 1027 N. 7th Street, Milwaukee, WI 53233 (USA). Monthly. ISSN 0091–6358, ASTRD5.

Astrophys. Invest.
Astrophysical Investigations. Department of Astronomy and National Astronomical Observatory, Bulgarian Academy of Sciences, Sofia (Bulgaria). Irregularly. Formerly entitled Astrofizicheskie Issledovaniya. ISSN 0861–265X.

Astrophys. J.
The Astrophysical Journal. American Astronomical Society. University of Chicago Press, Chicago, IL (USA). Semimonthly. ISSN 0004–637X, ASJOAB.

Astrophys. J., Lett.
The Astrophysical Journal, Letters. American Astronomical Society. University of Chicago Press, Chicago, IL (USA). Semimonthly. ISSN 0004–637X, ASJOAB.

Astrophys. J., Suppl. Ser.
The Astrophysical Journal, Supplement Series. American Astronomical Society. University of Chicago Press, Chicago, IL (USA). Monthly. ISSN 0067–0049, APJSA2.

Astrophys. Lett. Commun.
Astrophysical Letters and Communications. Gordon and Breach, New York, NY (USA). ISSN 0888–6512, ALECE7.

Astrophys. Space Sci.
Astrophysics and Space Science. Kluwer Academic Publ., Dordrecht (Netherlands). Semimonthly. ISSN 0004–640X, APSSBE.

Astrophys. Space Sci. Libr.
Astrophysics and Space Science Library. Kluwer Academic Publ., Dordrecht (Netherlands).

Astrophysics
Astrophysics. Consultants Bureau, New York, NY (USA). Bimonthly. A cover–to–cover translation of Astrofizika. ISSN 0571–7256, ATPYAA.

Atti Accad. Naz. Lincei, Ser. 9, Rend. Lincei: Sci. Fis. Nat.
Atti della Accademia Nazionale dei Lincei, Serie 9, Rendiconti Lincei: Scienze Fisiche e Naturali. Accademia Nazionale dei Lincei, Roma (Italy). 4 issues per year. Formerly entitled: Atti Accad. Naz. Lincei, Ser. Ottava, Rend.

Atti Accad. Naz. Lincei, Ser. 9, Rend. Lincei: Suppl.
Atti della Accademia Nazionale dei Lincei, Serie 9, Rendiconti Lincei: Supplemento. Accademia Nazionale dei Lincei, Roma (Italy).

Aust. J. Astron.
Australian Journal of Astronomy. Astral Press, Wembley, W.A. 6014 (Australia). Semiannual. ISSN 0814–5628.

Aust. J. Phys.
Australian Journal of Physics. Commonwealth Scientific and Industrial Research Organization (CSIRO), 314 Albert Street, East Melbourne, Victoria 3002 (Australia). Bimonthly. ISSN 0004–9506, AUJPAS.

BAV Mitt.
BAV Mitteilungen. Bundesdeutsche Arbeitsgemeinschaft für Veränderliche Sterne e.V., Munsterdamm 90, D–1000 Berlin 41 (Germany).

BAV Rundbrief
BAV Rundbrief. Bundesdeutsche Arbeitsgemeinschaft für Veränderliche Sterne e.V., Munsterdamm 90, D–1000 Berlin 41 (Germany). Quarterly. ISSN 0405–5497.

BBSAG Bull.
Bedeckungsveränderlichen Beobachter der Schweizerischen Astronomischen Gesellschaft Bulletin. Subscription address: N. Hasler, Huzlenstraße 3, CH–8604 Volketswil, Switzerland.

Be Star Newsl.
Be Star Newsletter. European Southern Observatory, Karl–Schwarzschild–Straße 2, D–8046 Garching (Germany, F.R.). ISSN 0296–3140.

Bibliogr. Program Notes Close Bin.
Bibliography and Program Notes on Close Binaries. International Astronomical Union. Department of Geoscience, National Defence Academy, Yokosuka (Japan). Irregularly. ISSN 1017–0634.

Bild Wiss.
Bild der Wissenschaft. Deutsche Verlagsanstalt, Stuttgart (Germany). Monthly. ISSN 0006–2375, BIWIAX.

BIPM Circ. T
Bureau International des Poids et Mesures (BIPM), Circular T. Bureau International des Poids et Mesures (BIPM), Section du temps, Pavillion de Breteuil, F–92312 Sèvres CEDEX (France). Monthly. ISSN 1143–1393.

Bol. Obs. Astron. Quito
Boletin del Observatorio Astronomico de Quito. Escuela Politecnica Nacional, Observatorio Astronomico de Quito, Quito (Ecuador). Irregularly.

Boyden Obs., Occas. Publ.
Boyden Observatory, Occasional Publication. Boyden Observatory, Institute and Department of Astronomy, University of the Orange Free State, PO Box 339, Bloemfontein 9300, (South Africa). Irregularly.

Bull. Am. Astron. Soc.
Bulletin of the American Astronomical Society. American Astronomical Society. American Institute of Physics, New York, NY (USA). Quarterly. ISSN 0002–7537, ASLBBH.

Bull. Assoc. Fr. Obs. Etoiles Variables
Bulletin de l'Association Française des Observateurs d'Etoiles Variables. Association Française des Observateurs d'Etoiles Variables, Observatoire de Strasbourg, 11, rue de l'Université, F–67000 Strasbourg (France). Quarterly. ISSN 0153–9949.

Bull. Crimean Astrophys. Obs.
Bulletin of the Crimean Astrophysical Observatory. Allerton Press, New York, NY (USA). Annual. English translation of Izvestiya Krymskoj Astrofizicheskoj Observatorii. ISSN 0190–2717, BCAOD4.

Bull. Inf. Cent. Données Astron. Strasb.
Bulletin d'Information du Centre de Données Astronomiques de Strasbourg. Observatoire Astronomique de Strasbourg, 11, rue de l'Université, F–67000 Strasbourg (France). Semiannual. Formerly entitled Bulletin d'Information du Centre de Données Stellaires. ISSN 0242–6536, BICSDY.

Bull. Soc. Fribourgeoise Sci. Nat.
Bulletin de la Société Fribourgeoise des Sciences Naturelles. Editions Universitaires, Fribourg (Switzerland).

Bull. Spec. Astrophys. Obs. – North Caucasus
Bulletin of the Special Astrophysical Observatory – North Caucasus. Allerton Press, New York, NY (USA). Irregularly. A cover-to-cover translation of Astrofizicheskie Issledovaniya – Izvestiya Spetsial'noj Astrofizicheskoj Observatorii. ISSN 0190–2709, BSACDC.

C. R. Acad. Sci., Sér. II
Comptes Rendus de l'Académie des Sciences, Série II: Mécanique, Physique, Chimie, Sciences de la Terre et de l'Univers. Académie des Sciences, Paris (France). Gauthier–Villars, Montrouge (France). Weekly. ISSN 0764–4450, CRAMED.

Can. J. Phys.
Canadian Journal of Physics. National Research Council of Canada, Ottawa, K1A 0R6, Ontario, (Canada). Monthly. ISSN 0008–4204, CJPHAD.

Celest. Mech. Dyn. Astron.
Celestial Mechanics and Dynamical Astronomy. Kluwer Academic Publ., Dordrecht (Netherlands). Quarterly. ISSN 0923–2958, CLMCAV.

Chem. Phys. Lett.
Chemical Physics Letters. North–Holland, Amsterdam (Netherlands). Semimonthly. ISSN 0009–2614, CHPLBC.

Chin. Astron. Astrophys.
Chinese Astronomy and Astrophysics. Pergamon Press, Oxford (UK). Quarterly. A selected translation from the current issues of Acta Astronomica Sinica, Acta Astrophysica Sinica, and Chinese Journal of Space Sciences. ISSN 0275–1062.

Chin. Phys.
Chinese Physics. American Institute of Physics, New York, NY (USA). Quarterly. English translation of Wuli Xuebao. ISSN 0273–429X, CHPHD2.

Chin. Phys. Lett.
Chinese Physics Letters. Science Press, 137 Chaoyangmennei Street, Beijing (China). Monthly. ISSN 0256–307X, CPLEEU.

Chin. Sci. Bull.
Chinese Science Bulletin.

Ciel
Le Ciel. Société Astronomique de Liège, B–4200 Cointe–Liège (Belgium). Monthly. Bulletin de la Société Astronomique de Liège. ISSN 0771–3010.

Ciel Terre
Ciel et Terre. Société Royale Belge d'Astronomie, de Météorologie et de Physique du Globe, Avenue Circulaire 3, 1180 Bruxelles (Belgium). Bimonthly. Bulletin de la Société Royale Belge d'Astronomie, de Météorologie et de Physique du Globe. ISSN 0009–6709, CIELAV.

Circ. Czech. Obs., Time Latitude
Circular of the Czechoslovak Observatories, Time and Latitude. Centre of Scientific Information, Astronomical Institute of the Czechoslovak Academy of Sciences, 251 65 Ondřejov (Czechoslovakia). Quarterly. ISSN 1210–0463.

Circ. Inf. – IAU Comm. Etoiles Doubles
Circulaire d'Information. Union Astronomique Internationale, Commission des Etoiles Doubles. Observatoire de la Côte d'Azur, B.P. 139, F–06003 Nice (France). 3 Nos. per year.

Classical Quantum Gravity
Classical and Quantum Gravity. Institute of Physics, London (UK). Bimonthly. ISSN 0264–9381, CQGRDG.

Comments Astrophys.
Comments on Astrophysics. Comments on Modern Physics: Part C. Gordon and Breach, New York, NY (USA). Bimonthly. ISSN 0146–2970, COASB7.

Commun. Math. Phys.
Communications in Mathematical Physics. Springer, Berlin (Germany). Semimonthly. ISSN 0010–3616, CMPHAY.

Comput. Phys.
Computers in Physics. American Institute of Physics, New York, N.Y. (USA). Bimonthly. ISSN 0894–1866, CPHYE2.

Comput. Phys. Commun.
Computer Physics Communications. North–Holland, Amsterdam (Netherlands). Monthly. ISSN 0010–4655, CPHCBZ.

Contemp. Phys.
Contemporary Physics. Taylor and Francis, London (UK). Bimonthly. ISSN 0010–7514, CTPHAF.

Contrib. Atmos. Phys.
Contributions to Atmospheric Physics. Vieweg, Wiesbaden (Germany). Quarterly. ISSN 0005–8173, BPYAAY.

Contrib. Dep. Astron., Kyoto Univ.
Contributions from the Department of Astronomy, Kyoto University. Department of Astronomy, Kyoto University, Kyoto (Japan). Irregularly.

Contrib. Inst. Copérnico (Buenos Aires)
Contribuciones del Instituto Copérnico (Buenos Aires). Instituto Copérnico, Buenos Aires (Argentina). Monthly.

Contrib. Nicholas Copernicus Obs. Planetarium Brno
Contributions of the Nicholas Copernicus Observatory and Planetarium in Brno. Nicholas Copernicus Observatory and Planetarium Brno (Czechoslovakia). Irregularly. ISSN 0862–173X.

Cosmic Res.
Cosmic Research. Consultants Bureau, New York, NY (USA). Bimonthly. A cover-to-cover translation of Kosmicheskie Issledovaniya. ISSN 0010–9525, CSCRA7.

Data Rep. Hydrogr. Obs., Ser. Astron. Geod.
Data Report of Hydrographic Observations, Series of Astronomy and Geodesy. Hydrographic Department of Japan, Tsukiji-5, Chuo-ku, Tokyo 104 (Japan). Annual. ISSN 0287–2633.

Dtsch. Geod. Komm. Bayer. Akad. Wiss., Reihe A
Deutsche Geodätische Kommission bei der Bayerischen Akademie der Wissenschaften, Reihe A: Theoretische Geodäsie. Beck, München (Germany). Irregularly. ISSN 0938–2836.

Earth Planet. Sci. Lett.
Earth and Planetary Science Letters. Elsevier, Amsterdam (Netherlands). Irregularly. ISSN 0012–821X, EPSLA2.

Earth Space
Earth in Space. American Geophysical Union, Washington, DC (USA). 9 issues per year (monthly Sep – May). A magazine for teachers and students of science. ISSN 1040–3124.

Earth, Moon, Planets
Earth, Moon, and Planets. Kluwer Academic Publ., Dordrecht (Netherlands). Monthly. ISSN 0167–9295, EMPLD3.

ESA Bull.
ESA Bulletin. ESA Publications Division, c/o ESTEC, Noordwijk (Netherlands). Quarterly. ISSN 0376–4265, ESABD8.

ESA IUE Newsl.
ESA IUE Newsletter. The ESA IUE Observatory, Apartado 54065, 28080 Madrid (Spain). ISSN 1011–0100.

ESA J.
ESA Journal. ESA Publications Division, ESTEC, PO Box 299, 2200 AG Noordwijk (Netherlands). Quarterly. ISSN 0379–2285, ESAJDW.

ESO Sci. Rep.
ESO Scientific Report. European Southern Observatory, Karl–Schwarzschild–Straße 2, D–8046 Garching bei München (Germany). Irregularly.

Europhys. Lett.
Europhysics Letters. European Physical Society, PO Box 69, CH–1213 Petit–Lancy 2 (Switzerland). Semimonthly. ISSN 0302–072X, EULEEJ.

Europhys. News
Europhysics News. European Physical Society, PO Box 69, CH–1213 Petit–Lancy 2 (Switzerland). Monthly. ISSN 0531–7479, EUPNAS.

Found. Phys.
Foundations of Physics. Plenum Publishing Corporation, New York, NY (USA). Monthly. ISSN 0015–9018, FNDPA4.

Found. Phys. Lett.
Foundations of Physics Letters. Plenum Press, New York, NY (USA). Bimonthly. ISSN 0894–9875, FPLEET.

G. Astron.
Giornale di Astronomia. Società Astronomica Italiana, Largo E. Fermi, 5–50125 Firenze (Italy). Quarterly. ISSN 0390–1106.

Gemini
Gemini. Newsletter of the Royal Greenwich Observatory. Royal Greenwich Observatory, Madingley Road, Cambridge, CB3 0EZ (UK). Quarterly. ISSN 0960–670X.

Gen. Relativ. Gravitation
General Relativity and Gravitation. Plenum Publishing Corporation, New York, NY (USA). Monthly. ISSN 0001–7701, GRGVA8.

Geochim. Cosmochim. Acta
Geochimica et Cosmochimica Acta. Pergamon Press, New York (USA). Monthly. Journal of the Geochemical Society and the Meteoritical Society. ISSN 0016–7037, GCACAK.

Geomagn. Aehron.
Geomagnetizm i Aehronomiya. Rossijskaya Akademiya Nauk. Izdatel'stvo Nauka, Moskva (Russia). Bimonthly. English translation in Geomagnetism and Aeronomy. ISSN 0016–7940, GEAEA6.

Geomagn. Aeron.
Geomagnetism and Aeronomy. American Geophysical Union, 2000 Florida Avenue, N.W., Washington, DC 20009 (USA). Bimonthly. Cover–to–cover translation of Geomagnetizm i Aehronomiya. ISSN 0016–7932, GMARAX.

Geophys. Astrophys. Fluid Dyn.
Geophysical and Astrophysical Fluid Dynamics. Gordon and Breach, New York, NY (USA). Irregularly. ISSN 0309–1929, GAFDD3.

Geophys. J. Int.
Geophysical Journal International. Royal Astronomical Society, Deutsche Geophysikalische Gesellschaft, European Geophysical Society. Blackwell, Oxford (UK). Monthly. ISSN 0955–419X.

Geophys. Res. Lett.
Geophysical Research Letters. American Geophysical Union, Washington, DC (USA). Monthly. ISSN 0094–8276, GPRLAJ.

Geophysics
Geophysics. Society of Exploration Geophysicists, PO Box 3098, Tulsa, OK 74101 (USA). Monthly. ISSN 0016–8033, GPYSA7.

Gnomonica Nova
Gnomonica Nova. Andreas Verdun, Astronomisches Institut der Universität Bern, Bern (Switzerland). Annual.

Heavens
The Heavens. Oriental Astronomical Association, Ōtsu–shi, Shiga–ken (Japan). Monthly. In Japanese. ISSN 0287–6906.

Helsinki Univ. Technol., Metsähovi Radio Res. Stn., Rep. Ser. A
Helsinki University of Technology, Metsähovi Radio Research Station, Report Series A. Metsähovi Radio Research Station, Kylmälä (Finland). ISSN 0783–8751.

High Energ. Phys. Nucl. Phys.
High Energy Physics and Nuclear Physics. Science Press, Beijing (China). Bimonthly. ISSN 0254–3052, KNWLD9.

Hvar Obs. Bull.
Hvar Observatory Bulletin. Hvar Observatory, Faculty of Geodesy, Zagreb (Croatia). Annual. ISSN 0351–2651, HOBUD7.

Hyperfine Interact.
Hyperfine Interactions. Baltzer, Basel (Switzerland). Monthly. ISSN 0304–3843, HYINDN.

I.A.P.P.P. Commun.
International Amateur–Professional Photoelectric Photometry Communication. Dyer Observatory, Vanderbilt University, Nashville, TN 37235 (USA). Quarterly. ISSN 0886–6961.

IAU Circ.
International Astronomical Union, Circular. International Astronomical Union. Central Bureau for Astronomical Telegrams, Smithsonian Astrophysical Observatory, 60 Garden Street, Cambridge, MA 02138 (USA). Irregularly. ISSN 0081–0304, IANUAB.

Icarus
Icarus. Academic Press, 1 East First Street, Duluth, MN 55802 (USA). Monthly. ISSN 0019–1035, ICRSA5.

IERS Bull., A
IERS Bulletin. A. International Earth Rotation Service, U.S. Naval Observatory, Washington, DC (USA). Weekly. ISSN 1048–6127.

IERS Bull., B
IERS Bulletin. B. International Earth Rotation Service, Central Bureau of IERS, Observatoire de Paris, Paris (France). Monthly.

IERS Tech. Note
IERS (International Earth Rotation Service) Technical Note. Central Bureau of IERS, Observatoire de Paris, Paris (France). Irregularly.

Inf. Bull. Variable Stars
Information Bulletin on Variable Stars. Commission 27 of the IAU. Konkoly Observatory, Budapest (Hungary). Irregularly. ISSN 0374–0676.

Int. Astron. Union Colloq.
International Astronomical Union Colloquium. Various publishers.

Int. Astron. Union Symp.
International Astronomical Union Symposium. Kluwer Academic Publ., Dordrecht (Netherlands).

Int. Comet Q.
The International Comet Quarterly. Smithsonian Astrophysical Observatory, 60 Garden Street, Cambridge, MA 02138, (USA). Quarterly. ISSN 0736–6922, ICOQDL.

Int. J. Mod. Phys. A
International Journal of Modern Physics A. World Scientific, Singapore (Singapore). 28 issues per year. ISSN 0217–751X.

Inverse Probl.
Inverse Problems. Institute of Physics, Bristol (UK). Quarterly. ISSN 0266–5611, INPEEY.

Ir. Astron. J.
The Irish Astronomical Journal. Armagh Observatory, Armagh BT61 9DG (UK). Semiannual. ISSN 0021–1052, IRAJAW.

IRIS Bull. A
IRIS Bulletin A. Subcommission International Radio Interferometric Surveying (IRIS) Steering Committee. National Geodetic Survey, National Oceanic and Atmospheric Administration, N/CG114, Rockville, MD 20852 (USA). Monthly. ISSN 1016–8648.

Isis
Isis. Journal of the History of Science Society. University of Chicago Press, Chicago, IL (USA). Editorial Office: Department of the History of Medicine, University of Wisconsin, Madison, WI (USA). Quarterly. An international review devoted to the history of science and its cultural influences. ISSN 0021–1753.

Izv. Acad. Sci. USSR, Phys. Solid Earth
Izvestiya Academy of Sciences USSR, Physics of the Solid Earth. American Geophysical Union (AGU), Washington, DC (USA). Monthly. Cover-to-cover translation of Izv. Akad. Nauk SSSR, Fiz. Zemli. ISSN 0001–4354, IPSEBQ.

J. Acoust. Soc. Am.
Journal of the Acoustical Society of America. American Institute of Physics, New York, N.Y. (USA). Monthly. ISSN 0001–4966, JASMAN.

J. Am. Assoc. Variable Star Obs.
The Journal of the American Association of Variable Star Observers. The American Association of Variable Star Observers, 25 Birch Street, Cambridge, MA 02138 (USA). Quarterly. ISSN 0271–9053, JAAODA.

J. Appl. Phys.
Journal of Applied Physics. American Institute of Physics, New York, NY (USA). Semimonthly. ISSN 0021–8979, JAPIAU.

J. Astron. Fr.
Journal des Astronomes Français. Société Française des Spécialistes d'Astronomie. Edité à l'Observatoire de Midi-Pyrénées, 14, avenue Ed. Belin, F–31400 Toulouse (France). Bulletin de la Société Française des Spécialistes d'Astronomie. ISSN 0181–9429.

J. Astrophys. Astron.
Journal of Astrophysics and Astronomy. Indian Academy of Sciences, PO Box 8005, Bangalore 560080 (India). Quarterly. ISSN 0250–6335, JASRD7.

J. Atmos. Sci.
Journal of the Atmospheric Sciences. American Meteorological Society, 45 Beacon Street, Bosten, MA 02108 (USA). Semimonthly. ISSN 0022–4928, JAHSAK.

J. Atmos. Terr. Phys.
Journal of Atmospheric and Terrestrial Physics. Pergamon Press, Oxford (UK). Bimonthly. ISSN 0021–9169, JATPA3.

J. Beijing Normal Univ. Nat. Sci.
Journal of the Beijing Normal University Natural Sciences. Quarterly. In Chinese.

J. Br. Astron. Assoc.
Journal of the British Astronomical Association. The British Astronomical Association, Burlington House, Picadilly, London W1V 9AG (UK). Bimonthly. ISSN 0007–0297, JBAAA6.

J. Br. Interplanet. Soc.
Journal of the British Interplanetary Society. The British Interplanetary Society, 27/29 South Lambeth Road, London SW8 1SZ (UK). Monthly. ISSN 0007–084X, JBISAW.

J. Chem. Phys.
Journal of Chemical Physics. American Institute of Physics, New York, NY (USA). Monthly. ISSN 0021–9606, JCPSA6.

J. Fluid Mech.
Journal of Fluid Mechanics. Cambridge University Press, London (UK). Monthly. ISSN 0022–1120, JFLSA7.

J. Geophys. Res.
Journal of Geophysical Research, Part A: Space Physics. American Geophysical Union, Washington, DC (USA). Monthly. ISSN 0148–0227, JGREA2.

J. Geophys. Res.
Journal of Geophysical Research, Part B: Solid Planets. American Geophysical Union, Washington, DC (USA). Monthly. ISSN 0148–0227, JJGBDO.

J. Geophys. Res.
Journal of Geophysical Research, Part C: Oceans and Atmospheres. American Geophysical Union, Washington, DC (USA). Monthly. ISSN 0148–0227, JJGADR.

J. Geophys. Res.
Journal of Geophysical Research, Part D: Atmospheres. American Geophysical Union, Washington, DC (USA). Monthly. ISSN 0148–0227, JGREA2.

J. Geophys. Res.
Journal of Geophysical Research, Part E: Planets. American Geophysical Union, Washington, DC (USA). Monthly. ISSN 0148–0227, JGREA2.

J. Hist. Astron.
Journal for the History of Astronomy. Science History Publications Ltd., Halfpenny Furze, Mill Lane, Chalfont St Giles, Buckinghamshire, HP8 4NR (UK). Quarterly. ISSN 0021–8286, JHSAA2.

J. Magn. Magn. Mater.
Journal of Magnetism and Magnetic Materials. North–Holland, Amsterdam (Netherlands) Monthly. ISSN 0304–8853, JMMMDC.

J. Math. Phys.
Journal of Mathematical Physics. American Institute of Physics, New York, NY (USA). Monthly. ISSN 0022–2488, JMAPAQ.

J. Mod. Opt.
Journal of Modern Optics. Taylor and Francis, London (UK). Monthly. Formerly: Optica Acta. ISSN 0950–0340.

J. Opt. (Paris)
Journal of Optics. Masson, Paris (France). Bimonthly. ISSN 0150–536X, JOOPDB.

J. Opt. Soc. Am. A
Journal of the Optical Society of America A. (Optics and Image Science). Optical Society of America. American Institute of Physics, New York, NY (USA) 13 issues per year. Formerly: Journal of the Optical Society of America. ISSN 0740–3232, JOAOD6.

J. Opt. Soc. Am. B
Journal of the Optical Society of America B (Optical Physics). Optical Society of America. American Institute of Physics, New York, NY (USA). 13 issues per year. Formerly: Journal of the Optical Society of America. ISSN 0740–3224, JOBPDE.

J. Phys. A
Journal of Physics A (Mathematical and General Physics). Institute of Physics, London (UK). 18 issues per year. ISSN 0305–4470, JPHAC5.

J. Phys. B
Journal of Physics B (Atomic, Molecular and Optical Physics). Institute of Physics, London (UK). Semimonthly. ISSN 0935–4075, JPAPEH.

J. Phys. Chem. Ref. Data
Journal of Physical and Chemical Reference Data. American Chemical Society, 1155 16th Street, N.W., Washington, DC 20036 (USA). Quarterly. ISSN 0047–2689, JPCRBU.

J. Phys. D
Journal of Physics D (Applied Physics). Institute of Physics, London (UK). Monthly. ISSN 0022–3727, JPAPBE.

J. Phys. G
Journal of Physics G (Nuclear Physics). Institute of Physics, London (UK). Monthly. ISSN 0305–4616, JPHGBM.

J. Phys. Soc. Jpn.
Journal of the Physical Society of Japan. Physical Society of Japan. Room 211, Kikai Shinko Building, 3–5–8 Shiba Koen, Minato–ku, Tokyo 105 (Japan) Monthly. ISSN 0031–9015, JUPSAU.

J. Plasma Phys.
Journal of Plasma Physics. Cambridge University Press, London (UK) Bimonthly. ISSN 0022–3778, JPLPBZ.

J. Quant. Spectrosc. Radiat. Transfer
Journal of Quantitative Spectroscopy and Radiative Transfer. Pergamon Press, Oxford (UK). Monthly. ISSN 0022–4073, JQSRAE.

J. R. Astron. Soc. Can.
The Journal of the Royal Astronomical Society of Canada. The Royal Astronomical Society of Canada, 136 Dupont Street, Toronto, Ontario M5R 1V2 (Canada). Bimonthly. ISSN 0035–872X, JRASA2.

J. Vac. Sci. Technol., A
Journal of Vacuum Science and Technology, A: Vacuums Surfaces and Films. American Institute of Physics, New York, NY (USA). Bimonthly. ISSN 0734–2101, JVTAD6.

J. Volcanol. Geotherm. Res.
Journal of Volcanology and Geothermal Research. Elsevier, Amsterdam (Netherlands). 16 issues per year. ISSN 0377–0273.

JETP Lett.
JETP Letters. American Institute of Physics, New York, NY (USA). Monthly. Cover–to–cover translation of Pis'ma v Zhurnal Ehksperimental'noj i Teoreticheskoj Fiziki. ISSN 0021–3640, JTPLA2.

Johns Hopkins APL Tech. Dig.
Johns Hopkins APL Technical Digest. The Johns Hopkins University Applied Physics Laboratory, Johns Hopkins Road, Laurel, Md. 20707 (USA). Quarterly. ISSN 0270–5214, APLDAZ.

Jpn. J. Appl. Phys., Part 1
Japanese Journal of Applied Physics, Part 1 (Regular Papers and Short Notes). Publication Office, Daini Toyokaiji Building, 24–8, Shinbashi 4–chome, Minato–ku, Tokyo 105 (Japan). Monthly. ISSN 0021–2922, JAPNDE.

Kinematics Phys. Celest. Bodies
Kinematics and Physics of Celestial Bodies. Allerton Press, New York, NY (USA). A cover–to–cover translation of Kinematika i Fizika Nebesnykh Tel. ISSN 0884–5913.

Kinematika Fiz. Nebesn. Tel
Kinematika i Fizika Nebesnykh Tel. Akademiya Nauk Ukrainy. Otdelenie Fiziki i Astronomii. Glavnaya Astronomicheskaya Observatoriya Akademii Nauk Ukrainy. Naukova Dumka, Kiev (Ukraine). Bimonthly. English translation in Kinematics and Physics of Celestial Bodies. ISSN 0233–7665.

Kodaikanal Obs. Bull.
Kodaikanal Observatory Bulletins. Indian Institute of Astrophysics, Bangalore 560 034 (India). Irregularly.

Kosm. Issled.
Kosmicheskie Issledovaniya. Rossijskaya Akademiya Nauk. Nauka, Moskva (Russia). Bimonthly. English translation in Cosmic Research. ISSN 0023–4206.

Kozmos
Kozmos. Obzor, Bratislava (Czechoslovakia). Monthly. Slovak popular astronomical journal. ISSN 0323–049X.

KPM
Kometen Planetoiden Meteore. 3 issues per year. Subscription address: Michael Möller, Steiluferallee 7, D–2408 Timmendorfer Strand (Germany). ISSN 0930–102X.

LEST Found., Tech. Rep.
LEST Foundation, Technical Report. Institute of Theoretical Astrophysics, University of Oslo, Oslo (Norway). Irregularly. ISSN 0800–7780.

Lett. Math. Phys.
Letters in Mathematical Physics. Kluwer, Dordrecht (Netherlands). Quarterly. ISSN 0377–9017, LMPDHQ.

Lick Obs. Bull.
Lick Observatory Bulletin. Board of Studies in Astronomy and Astrophysics, University of California at Santa Cruz, Santa Cruz, CA 95064 (USA). Irregularly. ISSN 0075–9317.

Manuscr. Geod.
Manuscripta Geodaetica. Springer, Berlin (Germany). Bimonthly. ISSN 0340–8825, MANGEH.

Mater. Res. Bull.
Materials Research Bulletin. Pergamon Press, Elmsford, NY (USA). Monthly. ISSN 0025–5408, MRBUAC.

Meas. Sci. Technol.
Measurement Science and Technology. Institute of Physics (IOP), London (UK). Monthly. Formerly: J. Phys. E (London). Sci. Instrum. ISSN 0957–0233, MSTCEP.

Mem. Soc. Astron. Ital.
Memorie della Società Astronomica Italiana. Osservatorio Astrofisica di Arcetri, Largo Fermi, 5, I–50125 Firenze (Italy). Quarterly. ISSN 0037–8720, MSATAB.

Mercury
Mercury. Astronomical Society of the Pacific, 390 Ashton Avenue, San Francisco, CA 94112 (USA). Bimonthly. The Journal of the Astronomical Society of the Pacific. ISSN 0047–6773, MRCYAT.

Messenger
The Messenger – El Mensajero. European Southern Observatory, Karl–Schwarzschild–Straße 2, D–8046 Garching bei München (Germany). Quarterly. ISSN 0722–6691, MESSE4.

Meteoritics
Meteoritics. Meteoritical Society at the Institute of Geophysics and Planetary Physics. University of California, Los Angeles, CA 90024–1567 (USA). Quarterly. ISSN 0026–1114, MERTAW.

Meteorol. Atmos. Phys.
Meteorology and Atmospheric Physics. Springer, Wien (Austria). Irregularly. ISSN 0177–8971, MAPHEU.

Minor Planet Bull.
The Minor Planet Bulletin. Association of Lunar and Planetary Observers. Editorial Office: R. P. Binzel, MIT 54–426, Cambridge, MA 02139 (USA). Quarterly. Bulletin of the Minor Planet Section of the Association of Lunar and Planetary Observers. ISSN 1052–8091.

Minor Planet Circ.
The Minor Planet Circulars/Minor Planets and Comets. International Astronomical Union. Minor Planet Center, Smithsonian Astrophysical Observatory, 60 Garden Street, Cambridge, MA 02138, (USA). Published usually at the date of full moon. ISSN 0736–6884, MPCIB2.

Mitt. Astron. Ges.
Mitteilungen der Astronomischen Gesellschaft. Astronomische Gesellschaft, Hamburg (Germany, F.R.). Available from H.–U. Keller, Planetarium Stuttgart, Neckarstraße 47, D–7000 Stuttgart 1 (Germany, F.R.). ISSN 0172–5483.

Mitt. Satell.–Beobachtungsstn. Zimmerwald
Mitteilungen der Satelliten–Beobachtungsstation Zimmerwald. Universität Bern, Bern (Switzerland). Irregularly. ISSN 1017–0413.

Mitt. Sternw. Sonneberg
Mitteilungen der Sternwarte zu Sonneberg. Sternwarte Sonneberg, Sonneberg (Germany). Irregularly. ISSN 0373–6202.

Mod. Phys. Lett. A
Modern Physics Letters A. World Scientific, Singapore (Singapore). 32 issues per year. ISSN 0217–7323.

Mon. Not. R. Astron. Soc.
Monthly Notices of the Royal Astronomical Society. Royal Astronomical Society. Blackwell, London (UK). Semimonthly. ISSN 0035–8711, MNRAA4.

Mon. Notes Astron. Soc. S. Afr.
Monthly Notes of the Astronomical Society of Southern Africa. South African Astronomical Observatory, PO Box 9, Observatory, 7935 Cape (South Africa). Bimonthly. ISSN 0024–8266, MASAAK.

Nachr. Olbers–Ges. Bremen
Nachrichten der Olbers–Gesellschaft Bremen. Olbers–Gesellschaft Bremen e.V., Bremen (Germany). Quarterly.

Natl. Astron. Obs. (Jpn.), Repr.
National Astronomical Observatory (Japan), Reprint. National Astronomical Observatory, Mitaka, Tokyo 181 (Japan). Formerly entitled Tokyo Astron. Obs. Repr. ISSN 0915–0021.

Natl. Geogr.
National Geographic. National Geographic Society, Washington, DC (USA). Monthly. ISSN 0027–9358.

NATO ASI Ser., Ser. C, Math. Phys. Sci.
NATO Advanced Science Institutes Series. Series C: Mathematical and Physical Sciences. Kluwer Academic Publ., Dordrecht (Netherlands). Irregularly.

Nature
Nature. Macmillan Magazines Ltd., London (UK). Weekly. Subscription address: Nature, Subscription Dept., Brunel Road, Basingstoke, Hants RG21 2XS, UK. ISSN 0028–0836, NATUAS.

Naturwissenschaften
Die Naturwissenschaften. Springer, Heidelberg (Germany). Monthly. ISSN 0028–1042, NATWAY.

News Lett. Astron. Soc. N.Y.
News Letter of the Astronomical Society of New York. Astronomical Society of New York, 1125 Oxford Place, Schenectady, NY 12308 (USA). Semiannual. ISSN 0148–9992.

Nucl. Instrum. Methods Phys. Res., Sect. A
Nuclear Instruments and Methods in Physics Research, Section A (Accelerators, Spectrometers, Detectors and Associated Equipment). North–Holland, Amsterdam (Netherlands). Bimonthly. Formerly: Nuclear Instruments and Methods in Physics Research. ISSN 0168–9002, NIMRD9.

Nucl. Instrum. Methods Phys. Res., Sect. B
Nuclear Instruments and Methods in Physics Research, Section B (Beam Interactions with Materials and Atoms). North–Holland, Amsterdam (Netherlands). Monthly. ISSN 0168–583X, NIMAER.

Nucl. Phys. A
Nuclear Physics A. North–Holland, Amsterdam (Netherlands). Weekly. ISSN 0375–9474, NUPABL.

Nucl. Phys. B, Part. Phys.
Nuclear Physics B, Particle Physics. North–Holland, Amsterdam (Netherlands). Weekly. Formerly: Nuclear Physics, B. ISSN 0550–3213, NUPBBO.

Nucl. Phys. B, Proc. Suppl.
Nuclear Physics B, Proceedings Supplements. North–Holland, Amsterdam (Netherlands). Irregularly. ISSN 0920–5632, NPBSE7.

Nuovo Cimento A
Il Nuovo Cimento della Società Italiana di Fisica A (Nuclei, Particles and Fields). Società Italiana di Fisica. Editrice Compositori, viale XII Giugno, 1, I–40124 Bologna (Italy). 20 issues per year. ISSN 0369–3546, NIFAAM.

Nuovo Cimento B
Il Nuovo Cimento della Società Italiana di Fisica B (General Physics, Relativity, Astronomy and Plasmas). Società Italiana di Fisica. Editrice Compositori, viale XII Giugno, 1, I–40124 Bologna (Italy). Monthly. ISSN 0369–3554, NIFBAP.

Nuovo Cimento C
Il Nuovo Cimento della Società Italiana di Fisica C (Geophysics and Space Physics). Società Italiana di Fisica. Editrice Compositori, viale XII Giugno, 1, I–40124 Bologna (Italy). Bimonthly. ISSN 0390–5551, NIFCAS.

Nuovo Cimento D
Il Nuovo Cimento della Società Italiana di Fisica D (Condensed Matter, Atomic, Molecular and Chemical Physics, Biophysics). Società Italiana di Fisica. Editrice Compositori, viale XII Giugno, 1, I–40124 Bologna (Italy). Bimonthly. ISSN 0392–6737, NCSDDN.

Obs. Trav.
Observations et Travaux. Société Astronomique de France, 3, rue Beethoven, F–75016 Paris (France). Quarterly. ISSN 0769–0878.

Observatory
The Observatory. Rutherford Appleton Laboratory, Chilton, Didcot, Oxon, OX11 0QX (UK). Bimonthly. ISSN 0029–7704, OBSEAR.

Occultation Newsl.
Occultation Newsletter. International Occultation Timing Association (IOTA), 6N106 White Lake Lane, St. Charles, IL 60175 (USA). Quarterly. ISSN 0737–6766.

Opt. Commun.
Optics Communications. North–Holland, Amsterdam (Netherlands) Biweekly. ISSN 0030–4018, OPCOB8.

Opt. Eng.
Optical Engineering. Society of Photo–Optical Instrumentation Engineers, Bellingham, WA (USA). Monthly. ISSN 0091–3286, OPEGAR.

Opt. Laser Technol.
Optics and Laser Technology. IPC Science and Technology Press, Guildford (UK). Bimonthly. ISSN 0030–3992, OLTCAS.

Opt. Lett.
Optics Letters. American Institute of Physics, New York, NY (USA). Monthly. ISSN 0146–9592, OPLEDP.

Opt. Photonics News
Optics and Photonics News. Optical Society of America, Washington, DC (USA). Monthly. ISSN 1047–6938, OPPHRI.

Opt. Spectrosc.
Optics and Spectroscopy. American Institute of Physics, New York, NY (USA). Monthly. Cover–to–cover translation of Optica i Spektroskopiya. ISSN 0030–400X, OPSUA3.

Origins Life Evol. Biosphere
Origins of Life and Evolution of the Biosphere. Kluwer Academic Publ., Dordrecht (Netherlands). Bimonthly. The Journal of the International Society for the Study of the Origin of Life. ISSN 0169–6149, OGLFAU.

Orion
Orion. Zeitschrift der Schweizerischen Astronomischen Gesellschaft. Schweizerische Astronomische Gesellschaft, Chemin du Marais–Long, 10, CH–1217 Meyrin (Switzerland). Bimonthly. ISSN 0030–557X, ORIOAH.

Phys. Bl.
Physikalische Blätter. Physik–Verlag, Postfach 1260/1280, D–6940 Weinheim (Germany). Monthly. ISSN 0031–9279, PHBLAG.

Phys. Chem. Miner.
Physics and Chemistry of Minerals. Springer, Berlin (Germany). Irregularly. ISSN 0342–1791, PCMIDU.

Phys. Earth Planet. Inter.
Physics of the Earth and Planetary Interiors. Elsevier, Amsterdam (Netherlands). Quarterly. ISSN 0031–9201, PEPIAM.

Phys. Fluids, A
Physics of Fluids, A. American Institute of Physics, New York, NY (USA). Monthly. ISSN 0031–9171, PFLDAS.

Phys. Fluids, B
Physics of Fluids, B. American Institute of Physics, New York, NY (USA). Monthly. ISSN 0031–9171, PFLDAS.

Phys. Lett. A
Physics Letters A (General, Atomic, and Solid State Physics). North–Holland, Amsterdam (Netherlands). Weekly. ISSN 0375–9601, PYLAAG.

Phys. Lett. B
Physics Letters B (Nuclear, Elementary Particle, and High–Energy Physics). North–Holland, Amsterdam (Netherlands). Weekly. ISSN 0370–2693, PYLBAJ.

Phys. Rep.
Physics Reports. North–Holland, Amsterdam (Netherlands). Weekly. Review Section of Physics Letters, Section C. ISSN 0370–1573, PRPLCM.

Phys. Rev. A
Physical Review A (General Physics). American Physical Society. American Institute of Physics, New York, NY (USA). Biweekly. ISSN 0556–2791, PLRAAN.

Phys. Rev. B
Physical Review B (Condensed Matter). American Physical Society. American Institute of Physics, New York, NY (USA). 36 issues per year. ISSN 0556–2805, PRBMDO.

Phys. Rev. C
Physical Review C (Nuclear Physics). American Physical Society. American Institute of Physics, New York, NY (USA). Bimonthly. ISSN 0556–2813, PRVCAN.

Phys. Rev. D
Physical Review D (Particles and Fields). American Physical Society. American Institute of Physics, New York, NY (USA). Biweekly. ISSN 0556–2821, PRVDAQ.

Phys. Rev. Lett.
Physical Review Letters. American Physical Society. American Institute of Physics, New York, NY (USA). Weekly. ISSN 0031–9007, PRLTAO.

Phys. Scr.
Physica Scripta. The Royal Swedish Academy of Sciences, Box 50005, S–10405 Stockholm, (Sweden). Monthly. ISSN 0031–8949, PHSTBO.

Phys. Scr. T
Physica Scripta T. The Royal Swedish Academy of Sciences, Box 50005, S–10405 Stockholm (Sweden). Irregularly. ISSN 0281–1847, PHSTBO.

Phys. Teach.
Physics Teacher. American Institute of Physics, New York, NY (USA). 9 issues per year. ISSN 0031–921X, PHTEAH.

Phys. Today
Physics Today. American Institute of Physics, New York, NY (USA). Monthly. ISSN 0031–9228, PHTOAD.

Phys. Unserer Zeit
Physik in Unserer Zeit. Verlag Chemie, Postfach 1260/1280, D–6940 Weinheim (Germany) Bimonthly. ISSN 0031–9252, PHUZAH.

Physica A
Physica A (Theoretical and Statistical Physics). North–Holland, Amsterdam (Netherlands). Monthly. ISSN 0378–4371, PHYADX.

Pis'ma Astron. Zh.
Pis'ma v Astronomicheskij Zhurnal. Rossijskaya Akademiya Nauk. Nauka, Moskva (Russia). Monthly. English translation in Soviet Astronomy Letters. ISSN 0320–0108, PAZHDA.

Planet. Space Sci.
Planetary and Space Science. Pergamon Press, Oxford (UK). Monthly. ISSN 0032–0633, PLASSE.

Plasma Phys. Controlled Fusion
Plasma Physics and Controlled Fusion. Pergamon Press, Oxford (UK). Monthly. ISSN 0741–3335, PLPHBZ.

Postepy Astron.
Postepy Astronomii. Czasopismo Póswiecone Upowszechnianiu Wiedzy Astronomicznej. Polskie Towarzystwo Astronomiczne, Warszawa. Pánstwowe Wydawnictwo Naukowe, Warszawa (Poland). Quarterly., PYAIAJ.

Priroda
Priroda. Nauka, Moskva (Russia). Monthly. ISSN 0032–874X.

Proc. Astron. Soc. Aust.
Proceedings of the Astronomical Society of Australia. CSIRO Division of Radiophysics, Box 76, P.O., Epping, N.S.W. 2121 (Australia). Semiannual. ISSN 0066–9997, AAUPBC.

Proc. IEEE
Proceedings of the IEEE. The Institute of Electrical and Electronics Engineers, New York, NY (USA). Monthly. Subscriptions: 445 Hoes Lane, Piscataway, NJ 08854–4150, USA. ISSN 0018–9219.

Prog. Astron.
Progress in Astronomy. Knowledge Press, 650 Gubei Road, Shanghai (China). Quarterly. Distributed by Shanghai Distribution Office of Xinhua Bookstore. ISSN 1000–8349.

Prog. Theor. Phys.
Progress of Theoretical Physics. Kyoto Univ. (Japan). Research Inst. for Fundamental Physics. Monthly. ISSN 0033–068X, PTPKAV.

Publ. Astron. Soc. Jpn.
Publications of the Astronomical Society of Japan. Astronomical Society of Japan. National Astronomical Observatory, Mitaka–shi, Tokyo 181 (Japan). Bimonthly. ISSN 0004–6264, PASJA.

Publ. Astron. Soc. Pac.
Publications of the Astronomical Society of the Pacific. American Institute of Physics, New York, NY (USA). Astronomical Society of the Pacific, 390 Ashton Avenue, San Francisco, CA 94122 (USA). Monthly. ISSN 0004–6280, PASPAU.

Publ. Beijing Astron. Obs.
Publications of the Beijing Astronomical Observatory. Beijing Astronomical Observatory, Academica Sinica, Beijing (China). Irregularly.

Publ. Istanbul Univ. Obs.
Publications of the Istanbul University Observatory. Istanbul Üniversitesi, Fen Fakültesi Basimevi, Istanbul (Turkey). Irregularly.

Publ. Korean Astron. Soc.
Publications of the Korean Astronomical Society. Korean Astronomical Society, Seoul National University, Seoul (Republic of Korea). Irregularly.

Publ. Natl. Astron. Obs. Jpn.
Publications of the National Astronomical Observatory of Japan. National Astronomical Observatory of Japan, Mitaka, Tokyo (Japan). Irregularly. ISSN 0915–3640.

Publ. Obs. Astron. Strasbourg, Sér. Astron. Sci. Hum.
Publication de l'Observatoire Astronomique de Strasbourg, Série "Astronomie et Science Humaines". Observatoire de Strasbourg, 11, rue de l'Université, F–67000 Strasbourg (France). Irregularly. ISSN 0989–6236.

Publ. Obs. Ebro, Misc.
Publicaciones del Observatorio del Ebro, Miscelánea. Consejo Superior de Investigaciones Cientificas, Roquetes (Spain). Irregularly.

Publ. Purple Mt. Obs.
Publications of Purple Mountain Observatory. Purple Mountain Observatory, Academia Sinica, Nanking (China). Quarterly. ISSN 1000–3681.

Publ. Shaanxi Astron. Obs.
Publications of the Shaanxi Astronomical Observatory. Shaanxi Astronomical Observatory, Academia Sinica, Lintong, Xian (China). Irregularly. ISSN 1001–1544.

Publ. Spéc. Cent. Données Stellaires
Publication Spéciale du Centre de Données Stellaires. Observatoire Astronomique, 11, rue de l'Université, F–67000 Strasbourg (France). Irregularly. ISSN 0764–9614.

Publ. U.S. Nav. Obs., Second Ser.
Publications of the United States Naval Observatory, Second Series. U.S. Government Printing Office, Washington, D.C. 20402 (USA). Irregularly. ISSN 0083–2448.

Publ. Yunnan Obs.
Publications of Yunnan Observatory. Yunnan Observatory, Academia Sinica, PO Box 110, Kunming, Yunnan Province (China). Quarterly.

Pure Appl. Geophys.
Pure and Applied Geophysics. Birkhäuser, Basel (Switzerland). Bimonthly. ISSN 0033–4553, PAGYAV.

Q. Bull. Sol. Act.
Quarterly Bulletin on Solar Activity. International Astronomical Union. National Astronomical Observatory of Japan, Mitaka, Tokyo (Japan). Quarterly. ISSN 0373–546X.

Q. J. R. Astron. Soc.
The Quarterly Journal of the Royal Astronomical Society. Royal Astronomical Society. Blackwell, Oxford (UK). Quarterly. ISSN 0035–8738, QJRAAK.

Q. J. R. Meteorol. Soc.
Quarterly Journal of the Royal Meteorological Society. Royal Meteorological Society, Bracknell (UK). Quarterly. ISSN 0035–9009, QJRMAM.

Radiant
Radiant. Dutch Meteor Society, Lederkarper 4, 2318 NB Leiden (Netherlands). Bimonthly. Journal of the Dutch Meteor Society. ISSN 0925–8566.

Recherche
Recherche. Société d'Editions Scientifiques, 57, rue de Seine, F–75006 Paris (France). 11 issues per year. ISSN 0029–5671, RCCHBV.

Rep. Math. Phys.
Reports on Mathemtical Physics. Pergamon Press, Oxford (UK). Bimonthly. ISSN 0034–4877, RMHPBE.

Rep. Natl. Astron. Obs. Jpn.
Report of the National Astronomical Observatory of Japan. National Astronomical Observatory of Japan, Mitaka, Tokyo (Japan). ISSN 0915–6321.

Rep. Prog. Phys.
Reports on Progress in Physics. Institute of Physics, Bristol (UK). Monthly. ISSN 0034–4885, RPPHAG.

Rev. Mex. Astron. Astrofis.
Revista Mexicana de Astronomia y Astrofisica. Instituto de Astronomia, Universidad Nacional Autónoma de México, Apartado Postal 70–264, México 20, D.F. (México). Semiannual. ISSN 0185–1101, RMAAD4.

Rev. Mex. Fis.
Revista Mexicana de Fisica. Sociedad Mexicana de Fisica, Apartado Postal 20–364, México 20, D.F. (México). Quarterly. ISSN 0035–001X, RMXFAT.

Rev. Mod. Astron.
Reviews in Modern Astronomy. Springer–Verlag, Berlin (Germany). Irregularly. Series of publications of the Astronomische Gesellschaft (AG).

Rev. Sci. Instrum.
Review of Scientific Instruments. American Institute of Physics, New York, NY (USA). Monthly. ISSN 0034–6748, RSINAK.

Riv. Nuovo Cimento
La Rivista del Nuovo Cimento della Società Italiana di Fisica. Società Italiana di Fisica. Editrice Compositori, viale XII Giugno 1, I–40124 Bologna (Italy). Monthly. ISSN 0393–697X, RNUCAC.

Rom. Astron. J.
Romanian Astronomical Journal. Editura Academiei Române, Bucuresti (Romania). Biannually. ISSN 1210–5168.

Říše hvězd
Říše hvězd. Panorama, Praha (Czechoslovakia). Monthly. Czech popular astronomical journal. ISSN 0035–5550, RIHVA3.

SAAO Newsl.
SAAO Newsletter. South African Astronomical Observatory, PO Box 9, Observatory 7935, Cape (South Africa).

Sci. Am.
Scientific American. Scientific American Inc., 415 Madison Avenue, New York, NY 10017 (USA). Monthly. ISSN 0036–8733, SCAMAC.

Sci. China, Ser. A
Science in China, Series A: Mathematics, Physics, Astronomy & Technological Sciences. Science Press, Beijing (China). Monthly. (Scientia Sinica). ISSN 0253–5831, SSINAV.

Sci. Rep. Tôhoku Univ., Eighth Ser.
The Science Reports of the Tôhoku University, Eighth Series (Physics and Astronomy). Faculty of Science, Tôhoku University, Sendai (Japan). Irregularly. ISSN 0388–5607.

Science
Science. American Association for the Advancement of Science, 1333 H Street, NW, Washington, DC 20005 (USA). Weekly. ISSN 0036–8075, SCIEAS.

Scientometrics
Scientometrics. Elsevier, Amsterdam (Netherlands).

Sendai Astron. Rap.
Sendai Astronomiaj Raportoj. Faculty of Science, Tôhoku University, Sendai 980 (Japan). Irregularly. ISSN 0386–0817.

SID Tech. Bull.
SID Technical Bulletin. The American Association of Variable Star Observers. A. J. Stokes, PO Box 398, Hudson, OH 44236 (USA). Quarterly.

Sidereal Times
The Sidereal Times. Astronomical Society of Western Australia, PO Box 421, Subiaco, W.A. 6008 (Australia).

Sitzungsber., Österr. Akad. Wiss., Math.–Naturwiss. Kl., Abt. II
Sitzungsberichte, Österreichische Akademie der Wissenschaften, Mathematisch–Naturwissenschaftliche Klasse, Abteilung II (Mathematische, Physikalische und Technische Wissenschaften). Springer, Wien (Austria). Irregularly. ISSN 0029–8816.

Sky Telesc.
Sky and Telescope. Sky Publ., Cambridge, MA (USA). Monthly. ISSN 0037–6604, SKTEA3.

Soc. Ital. Fis., Atti Conf.
Società Italiana di Fisica, Atti di Conferenze. Società Italiana di Fisica. Editrice Compositori, Bologna (Italy). Irregularly. English title: Italian Physical Society, Conference Proceedings.

Sol. Bull. (AAVSO)
Solar Bulletin (American Association of Variable Star Observers). The American Association of Variable Star Observers – Solar Division. Peter O. Taylor, PO Box 5685, Athens, GA 30604 (USA). Monthly. ISSN 0251–8480.

Sol. Phys.
Solar Physics. Kluwer Academic Publ., Dordrecht (Netherlands). Irregularly. ISSN 0038–0938, SLPHAX.

Sol. Syst. Res.
Solar System Research. Consultants Bureau, New York, NY (USA). Quarterly. Cover–to–cover translation of Astronomicheskij Vestnik. ISSN 0038–0946, SSYRAL.

Sonne
Sonne. Wilhelm–Foerster–Sternwarte, Munsterdamm 90, D–1000 Berlin 41 (Germany). Quarterly. Mitteilungsblatt der Amateursonnenbeobachter. ISSN 0721–0094.

South. Stars
Southern Stars. Royal Astronomical Society of New Zealand, PO Box 3181, Wellington, (New Zealand). Quarterly. ISSN 0049–1640, SNSRBP.

Sov. Astron.
Soviet Astronomy. American Institute of Physics, New York, NY (USA). Bimonthly. Cover–to–cover translation of Astronomicheskij Zhurnal. ISSN 0038–5301, SAAJAN.

Sov. Astron. Lett.
Soviet Astronomy Letters. American Institute of Physics, New York, NY (USA). Bimonthly. Cover–to–cover translation of Pis'ma v Astronomicheskij Zhurnal. ISSN 0360–0327, SALEDW.

Sov. J. Nucl. Phys.
Soviet Journal of Nuclear Physics. American Institute of Physics, New York, NY (USA). Bimonthly. A cover–to–cover translation of Yadernaya Fizika. ISSN 0038–5506, SJNCAS.

Sov. J. Opt. Technol.
Soviet Journal of Optical Technology. American Institute of Physics, New York, NY (USA). Monthly. Cover–to–cover translation of Optiko–Mekhanicheskaya Promyshlennost'. ISSN 0038–5514, SJOTBH.

Sov. J. Quantum Electron.
Soviet Journal of Quantum Electronics. American Inst. of Physics (AIP), New York, NY (USA). Monthly. Cover–to–cover translation of Kvantovaya Ehlektronika (Moskva). ISSN 0049–1748, SJQEAF.

Sov. Phys. – Dokl.
Soviet Physics – Doklady. American Institute of Physics, New York, NY (USA). Monthly. Cover–to–cover translation of Doklady Akademii Nauk SSSR. ISSN 0038–5689, SPHDAO.

Sov. Phys. – JETP
Soviet Physics – JETP. American Institute of Physics, New York, NY (USA). Monthly. Cover–to–cover translation of Zhurnal Ehksperimental'noj i Teoreticheskoj Fiziki. ISSN 0038–5646, SPHJAR.

Sov. Phys. – Usp.
Soviet Physics – Uspekhi. American Institute of Physics, New York, NY (USA). Monthly. Cover–to–cover translation of Uspekhi Fizicheskikh Nauk. ISSN 0038–5670, SOPUAP.

Space Sci. Rev.
Space Science Reviews. Kluwer Academic Publ., Dordrecht (Netherlands). Quarterly. ISSN 0038–6308, SPSRA4.

Space Telesc. Sci. Inst., Newsl.
Space Telescope Science Institute, Newsletter. Space Telescope Science Institute, 3700 San Martin Drive, Baltimore, MD 21218 (USA). Quarterly. ISSN 1055–6524.

Spaceflight
Spaceflight. The British Interplanetary Society Ltd., 27/29 South Lambeth Road, London SW8 1SZ (UK). Monthly. The international magazine of space and astronautics. ISSN 0038–6340, SPFLAN.

Spektrum Wiss.
Spektrum der Wissenschaft. VCH Verlagsgesellschaft, Weinheim (Germany). Monthly. ISSN 0170–2971, SPWIDY.

ST–ECF Newsl.
Space Telescope–European Coordinating Facility Newsletter. Space Telescope–European Coordinating Facility, European Southern Observatory, Karl–Schwarzschild–Str. 2, D–8046 Garching bei München (Germany). ISSN 0258–8080.

Sterne
Die Sterne. Barth, Leipzig (Germany). Bimonthly. ISSN 0039–1255, STNEA2.

Sterne Weltraum
Sterne und Weltraum. Verlag Sterne und Weltraum Dr. Vehrenberg GmbH, Portiastraße 10, D–8000 München 90 (Germany). Monthly. ISSN 0039–1263, STUWAN.

Sternenbote
Der Sternenbote. Astronomisches Büro, Hasenwartgasse 32, A–1238 Wien (Austria). Monthly. Österreichische Astronomische Monatsschrift. ISSN 0039–1271.

Strolling Astron.
The Strolling Astronomer. The Journal of the Association of Lunar and Planetary Observers. A.L.P.O., PO Box 16131, San Francisco CA 94116 (USA). ISSN 0039–2502, STASAD.

Stud. Geophys. Geod.
Studia Geophysica et Geodaetica. Geophysical Institute of the Czechoslovak Academy of Sciences. Academia, Prague (Czechoslovakia).

Surf. Sci.
Surface Science. North–Holland, Amsterdam (Netherlands). 48 issues per year. A journal devoted to the physics and chemistry of interfaces. ISSN 0039–6028, SUSCAS.

Tectonophysics
Tectonophysics. Elsevier, Amsterdam (Netherlands). Irregularly. ISSN 0040–1951, TCTOAM.

Telesc. Naz. Galileo, Tech. Rep.
Telescopio Nazionale Galileo, Technical Report. Osservatorio Astronomico di Padova–Asiago, Asiago (Italy). Irregularly.

Teor. Mat. Fiz.
Teoreticheskaya i Matematicheskaya Fizika. AN, Moscow (Russia). Inst. Fiziki. 3 issues per year. English translation in Theoretical and Mathematical Physics. ISSN 0564–6162, TMFZAL.

Time Serv. Announcement, Ser. 14
Time Service Announcement, Series 14. U.S. Naval Observatory, Washington, DC 20392 (USA). Irregularly.

Tr. Gos. Astron. Inst. im. Shternberg
Trudy Gosudarstvennogo Astronomicheskogo Instituta im. P. K. Shternberga. Izdatel'stvo Moskovskogo Universiteta, Moskva (Russia). Irregularly. ISSN 0371–6791.

Trans. Int. Astron. Union
Transactions of the International Astronomical Union. Kluwer Academic Publ., Dordrecht (Netherlands). ISSN 0251–107X.

Tuorla Obs. Rep., Informo
Tuorla Observatory Reports, Informo. Tuorla Observatory, University of Turku, Turku (Finland). Irregularly. ISSN 0789–6719.

Ukr. Fiz. Zh.
Ukrainskij Fizicheskij Zhurnal. Naukova Dumka, Kiev (Ukrainian SSR). Monthly. ISSN 0503–1265, UFIZAW.

Universe Classroom
The Universe in the Classroom. Astronomical Society of the Pacific, 390 Ashton Ave., San Francisco, CA 94112 (USA). A newsletter on teaching astronomy.

Universo
Universo. Liga Ibero–Americana de Astronomía (LIADA), Apartado 700, Mérida 5101–A (Venezuela). 3 issues per year. ISSN 0012–9820.

Urania
Urania. Zakład Narodowy imienia Ossolińskich, Wydawnictwo Polskiej Akademii Nauk, Kraków (Poland). Monthly. Miesiecznik Polskiego Towarzystwa Miłośników Astronomii. ISSN 0042–0794.

U.S. Nav. Obs., Ser. 4
United States Naval Observatory, Series 4, Daily time differences. U.S. Naval Observatory, Washington D.C. 20392 (USA). Weekly.

Uttar Pradesh State Obs., Repr.
Uttar Pradesh State Observatory, Reprint. Uttar Pradesh State Observatory, Manora Peak, Naini Tal (India). Irregularly.

Veröff. – Astron. Rechen–Inst. Heidelb.
Veröffentlichungen – Astronomisches Rechen–Institut Heidelberg. Braun, Karlsruhe (Germany). Irregularly.

Vesmír
Vesmír. Academia, Prague (Czechoslovakia). Monthly. Natural scientific journal. ISSN 0042–4544.

Visn. Kiiv. Univ., Fiz.–Mat. Nauki, Astron.
Visnik Kiivskogo Universitetu, Fiziko–Matematichni Nauki, Astronomiya. Libid', Kiiv (Ukraine). Annual. Formerly entitled Vestn. Kiev. Univ., Astron. ISSN 0868–510X.

Vistas Astron.
Vistas in Astronomy. Pergamon Press, Oxford (UK). Quarterly. ISSN 0083–6656, VASTA6.

WGN
WGN. Werkgroepnieuws – meteoren. The International Meteor Organization. P. Roggemans, Pijnboomstraat 25, B–2800 Mechelen (Belgium). Bimonthly. The Journal of the International Meteor Organization. ISSN 1016–3115.

Yamamoto Circ.
Yamamoto Circular. Yamamoto Observatory, 289 Kamitanakami–Kiryutyo, Otu, Sigaken, 520–21 (Japan). Irregularly.

Z. Naturforsch., A
Zeitschrift für Naturforschung, Section A. A Journal of Physical Sciences. Verlag der Zeitschrift für Naturforschung, PO Box 2645, D–7400 Tübingen (Germany). Monthly. ISSN 0932–0784, ZNASEI.

Z. Phys., B
Zeitschrift für Physik, B (Condensed Matter). Springer, Berlin (Germany). Monthly. ISSN 0722–3277, ZPCMDN.

Z. Phys., C
Zeitschrift für Physik, C (Particles and Fields). Springer, Berlin (Germany). Monthly. ISSN 0170–9739, ZPCFD2.

Zeiss Inf. Jenaer Rundsch.
Zeiss Information mit Jenaer Rundschau. Carl Zeiss, Oberkochen und Carl Zeiss Jena GmbH, Jena (Germany). Irregularly. Formed by merger of Zeiss Inf. and Jenaer Rundsch. ISSN 0941–7559.

Zenit
Zenit. Stichting De Koepel, Zonnenburg 2, 3512 Utrecht (Netherlands). Monthly. Populair–wetenschappelijk maandblad over sterrenkunde, weerkunde, ruimtevaart, ruimteonderzoek en aanverwante wetenschappen en technieken. ISSN 0165–0211.

Zh. Ehksp. Teor. Fiz.
Zhurnal Ehksperimental'noj i Teoreticheskoj Fiziki. Rossijskaya Akademiya Nauk, Leninskij Prosp. 14, Moskva (Russia). Bimonthly. English translation in Soviet Physics – JETP. ISSN 0044–4510, ZETFA7.

Journals Abstracted Completely

A selected number of journals listed in category 001 (periodicals) are central to the subject scope of *Astronomy and Astrophysics Abstracts*. Depending on their relevance, almost all papers of the journals listed below are abstracted in our service.

Acta Astron.
Acta Astron. Sin.
Acta Astrophys. Sin.
Acta Cosmologica
Am. Assoc. Variable Star Obs. Bull.
Annu. Rev. Astron. Astrophys.
Archaeoastronomy (U.K.)
Astrofizika
Astron. Astrophys.
Astron. Astrophys. Rev.
Astron. Astrophys., Suppl. Ser.
Astron. Ges., Abstr. Ser.
Astron. J.
Astron. Nachr.
Astron. Tidsskr.
Astron. Tsirk.
Astron. Vestn.
Astron. Zh.
Astronomia UAI
Astronomie
Astrophys. J.
Astrophys. J., Lett.
Astrophys. J., Suppl. Ser.
Astrophys. Lett. Commun.
Astrophys. Space Sci.
Aust. J. Astron.

BAV Rundbrief
BBSAG Bull.
Bol. Asoc. Argent. Astron.
Br. Astron. Assoc. Circ.
Bull. Am. Astron. Soc.
Bull. Assoc. Fr. Obs. Etoiles Variables
Bull. Astron. Soc. India
Bull. Inf. Cent. Données Astron. Strasb.

Celest. Mech. Dyn. Astron.
Chin. Astron. Astrophys.
Ciel
Ciel Terre
Circ. Inf.
Comments Astrophys.

Earth, Moon, Planets
Exp. Astron.

Fundam. Cosmic Phys.

G. Astron.
Gen. Relativ. Gravitation
GEOS Circ.

I.A.P.P.P. Commun.
IAU Circ.
Icarus
Inf. Bull. Variable Stars
Int. Comet Q.
Ir. Astron. J.
Issled. Solntsa Krasnykh Zvezd
Itogi Nauki Tekh., Ser. Astron.

J. Am. Assoc. Variable Star Obs.
J. Astrophys. Astron.
J. Br. Astron. Assoc.
J. Hist. Astron.
J. Korean Astron. Soc.
J. R. Astron. Soc. Can.

Kinematika Fiz. Nebesn. Tel
Komet. Tsirk.
Komety Meteory

Mem. Soc. Astron. Ital.
Mercury
Messenger
Meteoritics
Meteoritika
Minor Planet Bull.
Minor Planet Circ.
Mitt. Astron. Ges.
Mon. Not. R. Astron. Soc.
Mon. Notes Astron. Soc. S.Afr.

Nablyud. Iskusstv. Nebesn. Tel
Nauchn. Inf.
News Lett. Astron. Soc. N.Y.

Observatory
Occultation Newsl.
Orion
Orione

Perem. Zvezdy
Perem. Zvezdy, Prilozh.
Pis'ma Astron. Zh.
Postępy Astron.
Probl. Kosm. Fiz.
Proc. Astron. Soc. Aust.
Prog. Astron.
Publ. Astron. Soc. Jpn.
Publ. Astron. Soc. Pac.
Publ. Variable Star Sect., R. Astron. Soc. N.Z.

Q. J. R. Astron. Soc.

Rev. Astron.
Rev. Mex. Astron. Astrofis.
Rev. Mod. Astron.

Sky Telesc.
Sol. Phys.
Soln. Dannye Byull.
South. Stars
Sov. Sci. Rev., Sect. E
Space Sci. Rev.
Sterne
Sterne Weltraum
Strolling Astron.

Trans. IAU

Vasiona
Visn. Kiiv. Univ., Fiz.-Mat. Nauki, Astron.
Vistas Astron.

Yamamoto Circ.

Zenit
Zvaigžnota Debess

Publications of Observatories and Astronomical Institutes

Reports, communications, publications, and numbered series of reprints of observatories and astronomical institutes (listed in category 001) which are scanned completely in our service are listed below.

AAO Newsl.
Abastumanskaya Astrofiz. Obs., Byull.
Anglo-Aust. Telesc., Annu. Rep.
Ann. Shanghai Obs., Acad. Sin.
Annu. Rep. BIPM Time Sect.
Annu. Rep. IERS
Astrofiz. Issled. Izv. Spets. Astrofiz. Obs.
Astron. Bull., Carter Obs.
Astron. Inst. Univ. Brno, Contrib.
Astron. Inst. Univ. Brno, Publ.
Astron. Pap.
Astron. Rechen-Inst. Heidelb., Mitt., Ser. A
Astron. Rechen-Inst. Heidelb., Mitt., Ser. B

BAV Mitt.
BIPM Circ. T
Bol. Astron. Obs. Madr.
Bol. Obs. Astron. Quito
Bol. Obs. Ebro
Bol. R. Inst. Obs. Armada
Boyden Obs., Occas. Publ.
Boyden Obs., Repr.
Bull. Astron.
Bull. Crimean Astrophys. Obs.
Bull. Inf. Cent. Données Astron. Strasb.
Bull. Obs. Astron. Belgr.
Bull. Spec. Astrophys. Obs. – North Caucasus
Byull. Inst. Astrofiz.
Byull. Inst. Teor. Astron.

Carter Obs., Repr. Ser.
Cartes Synoptiques
Cent. Astron. Sci. Spat., Obs. Sol.
Circ. Czech. Obs., Time Latitude
Circ. Stn. Astron. Int. Latitudine, Carloforte-Cagliari, Ser. A
Commun. Konkoly Obs.
Commun. Univ. Lond. Obs.
Contrib. Astron. Obs. Skalnaté Pleso
Contrib. Bosscha Obs.
Contrib. Dept. Astron., Univ. Tokyo
Contrib. Inst. Argent. Radioastron.
Contrib. Nicholas Copernicus Obs. Planetarium Brno
Contrib. Nizamiah Japal-Rangapur Obs.
Contrib. Van Vleck Obs.

Data Rep. Hydrogr. Obs., Ser. Astron. Geod.
Dept. Radio Space Sci., Onsala Space Obs., Res. Rep.
Dtsch. Hydrogr. Inst. Hamb., Jahresber.

ESO Annu. Rep.
ESO Conf. Workshop Proc.
ESO Sci. Rep.
ESO Tech. Rep.

Gemini

Hvar Obs. Bull.

IERS Bull. A
IERS Bull. B
IERS Tech. Notes
Inst. Astrofis. Canarias, Mem.
Inst. Astron. Astrophys. Tech. Univ. Berlin, Mitt.
Inst. Astron. Geod., Univ. Madr., Publ.
Inst. Astron., Univ. Camb., Annu. Rep.
Inst. Astrophys., Univ. Liège, Annu. Rep.
Izv. Astron. Ehngel'gardt. Obs.
Izv. Glav. Astron. Obs. Pulkovo
Izv. Krym. Astrofiz. Obs.

Kapteyn Astron. Inst., Annu. Rep.
Kodaikanal Obs. Bull.

Latitude Circ.
Lick Obs. Bull.

Max-Planck-Inst. Kernphys., Jahresber.
Messenger
Mitt. Geod. Inst. Rheinischen Friedrich-Wilhelms-Univ. Bonn
Mitt. Karl-Schwarzschild Obs. Tautenburg
Mitt. Lohrmann-Obs. Tech. Univ. Dresden
Mitt. Satell.-Beobachtungsstn. Zimmerwald
Mitt. Sonnenobs. Kanzelhöhe
Mitt. Sternw. Sonneberg
Mitt. Universitätssternw. Graz
Mitt. Veränderliche Sterne
Mitt. Zentralinst. Phys. Erde
MPE Rep.
MPE Repr.

Natl. Astron. Obs. (Jpn.), Repr.
Natl. Radio Astron. Obs., Repr., Ser. A
Natl. Radio Astron. Obs., Repr., Ser. B
Nizamiah Rangapur Obs. Dept. Astron. Osmania Univ., Repr.
NRAO Workshops

Obs. Astron. Córdoba, Tirada Aparte
Obs. R. Belg., Commun., Sér. A
Obs. R. Belg., Commun., Sér. B
Occas. Rep. R. Obs., Edinb.
Oss. Astrofis. Catania, Pubbl.

Pubbl. Stn. Astron. Int. Latitudine Carloforte-Cagliari,
 Nuov. Ser.
Publ. Astron. Dept. Eötvös Univ.
Publ. Astron. Inst. Czech. Acad. Sci.
Publ. Astron. Opservatorije Beogr.
Publ. Astrophys. Obs. Potsdam
Publ. Beijing Astron. Obs.
Publ. Debrecen Heliophys. Obs.
Publ. Debrecen Heliophys. Obs., Heliogr. Ser.
Publ. Dep. Astron. Univ. Beogr.
Publ. Dom. Astrophys. Obs.
Publ. Ege Univ. Obs.
Publ. Istanbul Univ. Obs.
Publ. Natl. Astron. Obs. Jpn.
Publ. Obs. Astron. Strasbourg, Sér. Astron. Sci. Hum.
Publ. Obs. Ebro
Publ. Purple Mt. Obs.
Publ. Shaanxi Astron. Obs.
Publ. Spéc. Cent. Données Stellaires
Publ. U.S. Nav. Obs., Second Ser.
Publ. Warner Swasey Obs.
Publ. Yunnan Obs.

Q. Bull. Sol. Act.

R. Greenwich Obs., Bull.
R. Obs. Edinb., Res. Facilities
Rep. Natl. Astron. Obs. Jpn.
Rep. Ser., Dept. Phys. Sci., Univ. Turku
Rutherford Appleton Lab., Rep.

SAAO Newsl.
S.Afr. Astron. Obs., Annu. Rep.
S.Afr. Astron. Obs., Circ.
Sendai Astron. Rap.
Sol. Maps Act.
Soobshch. Byurakan. Obs.
Soobshch. Spets. Astrofiz. Obs.

Space Telesc. Sci. Inst., Newsl.
ST-ECF Newsl.

Tartu Astrofüüs. Obs. Publ.
Tartu Astrofüüs. Obs. Teated
Telesc. Naz. Galileo, Tech. Rep. (Asiago)
Theor. Pap.
Time Freq. Serv., Bull.
Time Serv. Bull.
Tr. Astrofiz. Inst. (Alma-Ata)
Tr. Astron. Obs. (Leningrad)
Tr. Glav. Astron. Obs. Pulkovo, Ser. 2
Tr. Gos. Astron. Inst. im. Shternberga
Tr. Inst. Teor. Astron., Leningrad
Tr. Kazan. Gorod. Astron. Obs.
Tr. Tashkent. Astron. Obs.
Tsirk. Astron. Inst. (Tashkent)
Tsirk. Astron. Obs. L'vov
Tsirk. Shemakh. Astrofiz. Obs.
Tuorla Obs. Rep., Informo

Univ. Tex., Monogr. Astron.
Univ. Tex., Publ. Astron.
Upps. Astron. Obs. Ann.
Upps. Astron. Obs., Rep.
U.S. Nav. Obs., Circ.
U.S. Nav. Obs., Time Serv. Publ., Ser. 4
U.S. Nav. Obs., Time Serv. Publ., Ser. 14
Uttar Pradesh State Obs., Repr.

Vatican Obs. Publ.
Vatican Obs. Publ., Spec. Ser., Studi Galileiani
Veröff. Archenhold-Sternw. Berlin-Treptow
Veröff.-Astron. Rechen-Inst. Heidelb.
Veröff. Remeis-Sternw. Bamberg
Veröff. Sternw. Pulsnitz
Veröff. Sternw. Sonneberg
Veröff. Wilhelm-Foerster-Sternw.
Vilniaus Astron. Obs. Biul.

002 Bibliographical Publications, Documentation, Catalogues, Data Bases

002.001 Dictionary of minor planet names.
L. D. Schmadel.
Springer, Berlin (Germany). 696 p. (1992). ISBN 3–540–54384–8.
Price DM 98.00. ISBN 0–387–54384–8 (USA). Price US$ 59.00.

The book contains the names, their meanings, and discovery circumstances of all minor planets numbered until the end of 1991. The following informations are given for the first 5012 numbered objects: definitive number, name (or provisional designation), provisional designation, time of discovery, discoverer(s), place of discovery, data on possible independent discoveries, naming citation and possible cross references, source of the naming information, and – if existent – name(s) of the naming proposers, biographical and bibliographical data. The catalogue comprises naming explanations for all 3957 planets that obtained a name so far. Ranking lists, names classifications, compilations of special type numbered minor planets and a comprehensive index of names are provided in appendices.

002.002 CPC 2 – the second cape photographic catalog. II. Conventional plate adjustment and catalog construction.
N. Zacharias, C. de Vegt, W. Nicholson, M. J. Penston.
Astron. Astrophys., Vol. 254, No. 1/2, p. 397 – 421 (Feb 1992).

The Second Cape Photographic Catalog CPC 2, containing positions and visual magnitudes of 276131 stars in the approximate magnitude range $V = 6.5$–10.5 is the first modern photographic catalog project on the southern hemisphere with an optimal fourfold plate overlap pattern and an astrometrically optimized spectral bandpass in the yellow–red region (530–640 nm). During the period 1962–1972 a total number of 5820 plates has been taken at Cape Observatory with a newly designed 4–elements lens ($F{:}10$, $f = 2000$ mm, $4^\circ\!.1 \times 4^\circ\!.1$ field, scale = $100''\mathrm{mm}^{-1}$). All plates have two 3 min exposures, shifted by about $50''$ in declination. Plate measurement has been accomplished on the GALAXY astrometric measuring machine at RGO Herstmonceux. As a joint effort of RGO and Hamburg Observatories the astrometric data reduction and catalog construction is being performed at Hamburg Observatory whereas the photographic photometry which provided visual magnitudes for all program stars has been carried out entirely at RGO by M. Penston.

002.003 A catalog of co–added IRAS fluxes of Orion Population stars.
W. B. Weaver, G. Jones.
Astrophys. J., Suppl. Ser., Vol. 78, No. 1, p. 239 – 266 (Jan 1992).

A catalog of co–added IRAS fluxes for the pre–main–sequence objects in the Herbig–Bell catalog (HBC) is presented. This catalog doubles the number of HBC stars with detected IRAS fluxes and provides improved flux values for the previously known sources. Noise levels are given for all HBC fields in each band, permitting upper limits to be estimated for all undetected sources.

002.004 Updated catalogue of DC dwarfs (1987 variant).
G. M. Beskin, S. N. Mitronova.
Bull. Spec. Astrophys. Obs. – North Caucasus, Vol. 31, p. 33 – 76 (1991). English translation of Astrofiz. Issled. Izv. Spets. Astrofiz. Obs., Tom 31, p. 36 – 79 (1991).

A list of 123 objects having noticeable proper motion, the spectra of which have no lines, is compiled on the basis of published data. Their coordinates and kinematic, photometric, and spectral characteristics are given. An analysis of the parameters of DC dwarfs from Nesterenko and Chuprina's list (1976) for which lines were found showed that DC dwarfs for the most part are white dwarfs of various spectral classes with weakened lines and not cold DB dwarfs with spectra deviating from purely black–body ones and which, possibly, are variable

on time scale of months–years. These objects are of interest as probable black holes and therefore are priority candidates for searching for superrapid brightness variability in them.

002.005 Uranografie: de cartografie van de sterrenhemel.
W. Tirion.
Zenit, Jaarg. 19, Nr. 4, p. 156 – 160 (Apr 1992).

002.006 Philip's Color Star Atlas. Epoch 2000.
J. Cox, R. Monkhouse.
Philip, London (UK). 40 p. (1991). ISBN 0–913135–08–9. Price £ 19.95.

This atlas charts the stars of the northern and southern hemispheres. The computer–plotted maps use color to show spectral type, and special symbols to indicate dwarf stars, giants, and supergiants, as well as variable and multiple stars. All stars down to magnitude 6.75 are included, as well as fainter star clusters, nebulae, and galaxies. A concise text explains the theory of color in stars and how to perceive it.

002.007 Compiled list of clusters of galaxies with measured redshifts.
V. S. Lebedev, I. A. Lebedeva.
Bull. Spec. Astrophys. Obs. – North Caucasus, Vol. 31, p. 88 – 125 (1991). English translation of Astrofiz. Issled. Izv. Spets. Astrofiz. Obs., Tom 31, p. 91 – 127 (1991).

The data on measured radial velocities are given for 1943 clusters of galaxies.

002.008 Atlas zvezdnogo neba. (The star sky atlas.)
V. K. Abalakin (ed.).
Akademiya Nauk SSSR, Moskva (USSR). 80 p. (1991). With 20 charts. Price 50 Rbl.

This atlas consists of 20 star charts together with a stellar catalogue. A booklet with explanations to the star sky atlas and to the stellar catalogue is included. The charts of the atlas contain stars to visual magnitude 6.5. The total number of stars is ~8,500. The star charts also contain star clusters, nebulae and galaxies. The atlas is referred to 2000.0 equinox.

002.009 What fraction of literature references are incorrect?
H. A. Abt.
Publ. Astron. Soc. Pac., Vol. 104, No. 673, p. 235 – 236 (Mar 1992).

From a systematic study of 1009 references in *The Astrophysical Journal* it was found that 12.2% had errors. Only 0.4% of the referenced papers could not be found at all; another 3.0% were found by searching in volume, annual, and five–year indices and the remaining correctly and incorrectly referenced papers were found right away. Another 8.3% of the references have errors in the first authors' names or in the journal names, and volume and page numbers such that they could be misplaced in the *Science Citation Index* (SCI). However, the compilers of SCI match all citations against a computerized file of the source papers and correct some of the citations, so that only 3.6% of the citations are missing or displaced in SCI.

002.010 A list of highly reddened stars likely to be distant luminous stars.
C. B. Stephenson.
Astron. J., Vol. 103, No. 1, p. 263 – 266 (Jan 1992). Current Physics Microform No.: 9202A2173. With plates 24 – 38.

A list of highly reddened stars found in a photographic – infrared objective prism survey is presented. Many of these stars appear to have (R–I) color indices well in excess of 4–5 mag, yet their spectra do not exhibit molecular absorption bands on the spectral plates, so any intrinsic contribution to the redness should be small. Nor do most of the stars occur in the IRAS Point

Source Catalog, so the reddening is unlikely to be circumstellar. The authors infer that they are reddened by interstellar absorption, hence they are generally at great distances and must be quite luminous in order to be detected at all.

002.011 Das Plattenarchiv der Sternwarte Sonneberg und die Sonneberger Himmelsüberwachung.
H.-J. Bräuer, B. Fuhrmann.
Sterne, Band 68, Heft 1, p. 19 – 33 (1992).

002.012 Der Sonneberger "Bibliographische Katalog der veränderlichen Sterne".
S. Rößiger.
Sterne, Band 68, Heft 1, p. 43 – 46 (1992).

002.013 A catalogue of meteor showers in medieval Arab chronicles.
W. S. Rada, F. R. Stephenson.
Q. J. R. Astron. Soc., Vol. 33, No. 1, p. 5 – 16 (Mar 1992).

002.014 Das Plattenarchiv der Sternwarte Sonneberg und die Sonneberger Himmelsüberwachung.
H.-J. Bräuer, B. Fuhrmann.
Sternenbote, Jahrg. 35, Nr. 4, p. 78 – 87 (1992).

002.015 IUE Bibliography Index (1978 – 1990).
ESA IUE Newsl., No. 39, p. 5 – 122 (Jan 1992).

002.016 Spectroscopic investigations of objects of the Second Byurakan Survey. Stellar objects. IV.
D. A. Stepanyan, V. A. Lipovetskij, V. O. Chavushyan, L. K. Erastova, A. I. Shapovalova.
Astrophysics, Vol. 34, No. 1, p. 1 – 6 (Jan – Feb 1991). English translation of Astrofizika, Tom 34, Vyp. 1, p. 5 – 12 (Feb 1991).
Data are given on 26 quasistellar objects found during the Second Byurakan Spectral Sky Survey in a region of the northern sky bounded by the coordinates $8^h < \alpha < 17^h$ and $+49° < \delta < 61°$. The main parameters of the emission lines and other data for all quasistellar objects are given, together with scans of the majority of them.

002.017 The La Palma data archive.
E. Zuiderwijk, P. Meikle.
Gemini, No. 36, p. 12 – 13 (Jun 1992).

002.018 First Byurakan Spectral Sky Survey. Blue stellar objects. IV. The region $+41° \leqslant \delta \leqslant 45°$.
G. V. Abramyan, V. A. Lipovetskij, A. M. Mikaelyan, D. A. Stepanyan.
Astrophysics, Vol. 34, No. 1, p. 7 – 15 (Jan – Feb 1991). English translation of Astrofizika, Tom 34, Vyp. 1, p. 13 – 20 (Feb 1991).
The fourth list of blue stellar objects of the second part of the First Byurakan Spectral Sky Survey is given. The objects are situated in the region $+41° \leqslant \delta \leqslant +45°$, $13^h30^m \leqslant \alpha \leqslant 19^h10^m$ and $21^h50^m \leqslant \alpha \leqslant 24^h00^m$. The list contains data on 106 stellar objects, of which 64 have been found for the first time. A preliminary classification of the objects is given. The preliminary classification by the authors is compared with the generally accepted spectral classification, and subsamples fort searches for QSOs, white dwarfs, and subdwarfs are identified.

002.019 Bibliography on Be Stars.
A. M. Hubert, J. Jugaku, P. Koubský, G. J. Peters, M. Ruusalepp, A. Slettebak.
Be Star Newsl., No. 25, p. 20 – 22 (Feb 1992).

002.020 The astronomy and astrophysics encyclopedia.
S. P. Maran (ed.).
Cambridge University Press, Cambridge (UK). 1031 p. (1992). ISBN 0-521-41744-9. Price £ 60.00. Van Nostrand Reinhold, New York, NY (USA), ISBN 0-442-26364-3, Price US$ 89.95.
This volume features more than 400 articles on today's hottest research subjects – from accretion to zodiacal light. The articles include in-depth coverage of basic theory, the state of current research, and a forecast of future scientific investigation. And they incorporate the latest findings from today's technologically advanced observatories, satellites, and space probes. Exclusively focused on astronomy and astrophysics, including space science, this volume offers in-depth, comprehensive coverage of subjects. It features a complete range of topics, presented entirely by recognized experts in astronomy and astrophysics, plus thorough bibliographies of other sources in the field.

002.021 The AGK3U: an updated version of the AGK3.
B. Bucciarelli, D. Daou, M. G. Lattanzi, L. G. Taff.
Astron. J., Vol. 103, No. 5, p. 1689 – 1700 (May 1992). Current Physics Microform No.: 9205G0263.
The authors present the details of construction of an updated version of the AGK3. Unlike the builders of the Positions and Proper Motions (PPM) catalog, they have gone forward in time to gain new positions to use to improve the AGK3 proper motions. The source of these new observations is the Palomar "Quick V" survey made for the construction of the Hubble Space Telescope Guide Star Catalog. The authors have been able to recover individual equatorial coordinates from a Schmidt plate with a typical standard deviation about the mean of $0\rlap{.}''33$. Moreover, the means (i.e., the systematic errors) are at the $0\rlap{.}''01$ level. These realistic, external errors compare favorably with the reduction of the AGK3 astrograph material ($0\rlap{.}''21$ per equatorial coordinate). The authors believe that they have proven that the astrometric potential of large–scale Schmidt plates can be realized via the subplate technique and then successfully integrated with the results of classical photographic astrometry. The authors' AGK3U positions have a mean error of $0\rlap{.}''167$ at an average epoch of 1950.62. The new proper motions have a total (i.e., two–dimensional) formal mean error of $0\rlap{.}''82$/cy. The authors present absolute comparisons via the GC. As viewed through the GC, the mean error of an AGK3U star's total proper motion is $\sim 1\rlap{.}''6$/cy. The AGK3U has been deposited at the National Space Science Data Center and at the Strasbourg Data Center.

002.022 Publication rates and trends in international collaborations for astronomers in developing countries, Eastern European countries, and the former Soviet Union.
J. C. White II.
Publ. Astron. Soc. Pac., Vol. 104, No. 676, p. 472 – 476 (Jun 1992).
The author surveyed two major astronomical journals for the 30 yr period 1960 – 1989 looking for papers with either principal or co–authors from developing countries, formerly communist Eastern European countries, and the former Soviet Union. The number of papers with authors from these areas has increased during the period surveyed, but the percentage of papers with such authors is less than 10% of the total number of papers over the period. The number of papers by collaborations between astronomers from developing countries and industrialized countries was found to be approximately the same as the number of papers by astronomers from developing countries only. Astronomers from Eastern European countries and the former Soviet Union, however, were found to either prefer or require collaborations with Western astronomers.

002.023 Star and star cluster spectral libraries.
E. Bica.
IAU Symposium No. 149: The stellar populations of galaxies, p. 215 – 224 (1992). – See Abstr. 012.007 for the main entry.
This paper reviews spectral libraries of stars and star clusters, together with their applications to population synthesis. The problem of abundance calibrations for metal rich populations is also addressed, in particular index definitions and non–solar CNO/Fe ratios. A stellar population data bank would be important to accelerate progress in the field and would optimize the use of future telescope time.

002.024 A new catalog of 30,239 1.4 GHz sources.
R. L. White, R. H. Becker.
Astrophys. J., Suppl. Ser., Vol. 79, No. 2, p. 331 – 467
(Apr 1992).
Using the Green Bank 1.4 GHz Northern Sky Survey, the
authors have generated a catalog of 30,239 sources over the
declination range of –5° to 82° to a flux density limit of 100 mJy.
Spectral indices were found for over 90% of these sources, based
on flux densities at 0.365 or 4.85 GHz taken from other catalogs.
Flux densities at all three frequencies are available for 20,000
sources. The spectral properties of 1.4 GHz sources are discussed
along with the selection effects induced by the idiosyncracies of
the various catalogs.

**002.025 A new catalogue of semidetached Algol–type binaries
with well determined absolute dimensions.**
J. M. García, A. Giménez.
IAU Symposium No. 151: Evolutionary processes in interacting
binary stars, p. 295 – 298 (1992). – See Abstr. 012.008 for the
main entry.
Data on a selected sample of semidetached Algol–type
eclipsing binaries with well–known absolute dimensions have
been collected. The strict criteria of selection adopted are given
together with the provisional form of the catalogue. This includes
besides the absolute dimensions other relevant parameters that
may be of interest.

002.026 Catalogue of photoelectric (UBV) star magnitudes.
V. P. Ryl'kov, N. D. Kostyuk, K. A. Kandaurova.
Kinematics Phys. Celest. Bodies, Vol. 7, No. 3, p. 86 – 89 (1991).
English translation of Kinematika Fiz. Nebesn. Tel, Tom 7,
No. 3, p. 92 – 95 (1991).
The summary S2122 catalogue of photoelectric measurements
in the UBV system is identified with catalogues that contain the
coordinates and spectral classes of the stars. Equatorial coordi-
nates at the equinox B1950.0 were obtained for 43,027 stars,
including 39,005 for which HD spectral classes were also found.

**002.027 The astrometric data bank of the Pulkovo Observatory
and some examples of its use.**
K. A. Kandaurova, E. V. Khrutskaya.
Kinematika Fiz. Nebesn. Tel, Tom 8, No. 1, p. 89 – 96
(Jan–Feb 1992). In Russian. English translation in Kinematics
Phys. Celest. Bodies, Vol. 8, No. 1.
Astrometric information on the Pulkovo Observatory's data
bank is given. Possibilities of applying these informations are
demonstrated.

002.028 Secondary spectrophotometric standards.
I. N. Glushneva, A. V. Kharitonov, L. N. Knyazeva,
V. I. Shenavrin.
Astron. Astrophys., Suppl. Ser., Vol. 92, No. 1, p. 1 – 29
(Jan 1992).
Energy distribution data on 238 secondary standard stars are
presented in the range 3200 – 7600 Å with 50 Å steps. These stars
are common to the Catalogue of the Sternberg State Astronomical Institute and the Fessenkov Astrophysical Institute. For these
stars, the differences between spectral energy distribution data of
the two catalogues do not exceed 5%, while the mean internal
accuracy of both catalogues data in this range are about 3.5%.
For 99 stars energy distribution data in the near infrared (6000 –
10800 Å) obtained at the Sternberg State Astronomical Institute
are also presented.

**002.029 The comet light curve atlas. (The comet light curve
catalogue/atlas. III. The atlas).**
L. Kamél.
Astron. Astrophys., Suppl. Ser., Vol. 92, No. 1, p. 85 – 149
(Jan 1992).
The comet light curve atlas is a compilation of light curves of
periodic comets, covering the period 1899 – 1989, except for
P/Encke, which also has the period 1832 – 1898 included. The
magnitudes have been taken from a compilation of a large
number of observations. The different corrections applied to the

magnitudes are described. These corrections are colour corrections, a delta effect correction, telescope aperture corrections and
corrections for sky conditions.

**002.030 Slit spectra of galaxies of the Second Byurakan
Survey. IV.**
D. A. Stepanyan, V. A. Lipovetskij, L. K. Erastova,
A. I. Shapovalova.
Astrophysics, Vol. 34, No. 2, p. 99 – 103 (Mar – Apr 1991).
English translation of Astrofizika, Tom 34, Vyp. 2, p. 205 – 211
(Apr 1991).
The results are given of spectroscopic observations of 32
galaxies in the Second Byurakan Survey. The observations were
made with the 6–m telescope of the Special Astrophysical
Observatory of the USSR Academy of Sciences and the 2.6–m
telescope of the Byurakan Astrophysical Observatory. Emission
lines have been found in the spectra of the majority of the
galaxies. The red shifts and luminosities of all the galaxies have
been determined. It is shown that SBS 1212 + 558 and
SBS 1213 + 549 A are type 1 Seyfert galaxies and that
SBS 1220 + 554 is a possible Seyfert. The galaxy SBS 0948 + 532
has been classified as BCDG.

**002.031 Investigation of photometric systems of the CdC and
AGK catalogues and their derivatives.**
L. K. Pakulyak, V. P. Trofimyuk, N. V. Kharchenko.
Kinematika Fiz. Nebesn. Tel, Tom 8, No. 2, p. 65 – 70
(Mar–Apr 1992). In Russian. English translation in Kinematics
Phys. Celest. Bodies, Vol. 8, No. 2.
Systematic (up to 0.5 – 1^{m}0) and random (up to ± 0^{m}5) errors
of estimates of brightness for stars in the CdC and AGK
catalogues are determined on the basis of comparison with
photoelectric stellar magnitudes in the B system. When using
stellar magnitudes from these catalogues, it is necessary to rid
them of the scale errors and zero–point errors, and convert from
the diameters of stellar images (CdC) or m_{pg} (AGK) to
B magnitudes. Values of the relation coefficients for the stellar
magnitudes m_{pg} of the AGK3 catalogue and B magnitudes are
given.

002.032 The "Fraunhofer Solar Spectrum" data bank.
A. S. Gadun, M. M. Sosonkina, V. A. Sheminova.
Kinematika Fiz. Nebesn. Tel, Tom 8, No. 2, p. 80 – 82
(Mar–Apr 1992). In Russian. English translation in Kinematics
Phys. Celest. Bodies, Vol. 8, No. 2.
A relational data base model has been developed
for the IBM PC XT/AT–type computers managed by
MS–DOS/PC–DOS of the versions 3.0 and later. The data base
includes the following characteristics of the solar spectrum
Fraunhofer lines: the wavelength, chemical element symbol and
its ionization state, atomic transition, Lande factors, excitation
potential of the lower level, central depth, equivalent width,
oscillator strengths, height of formation. The program of data
manipulation provides a user with ample opportunities in
sampling lines. At present the data bank contains information
about 662 unblended lines in the solar spectrum belonging to
Fe I, Fe II, Ni I, Sc I, Sc II, Ti I, Ti II, V I, V II, Cr I, Cr II, Y I,
Y II, Zr I, Zr II.

002.033 Solar–terrestrial models and application software.
D. Bilitza.
Planet. Space Sci., Vol. 40, No. 4, p. 541 – 579 (Apr 1992).
Models describing the different regions of the solar–terrestrial
environment are of great importance for a wide spectrum of
scientific, engineering and educational applications . In an effort
to assist model users, more than 80 models and application
programs are described in this paper. For each entry the
information provided includes model author(s), availability, brief
description, and references. These models describe parameters in
Earth's ionosphere, atmosphere, and magnetosphere, in the solar
wind, and in the atmosphere and ionosphere of planet Venus.
Almost all of the listed entries are empirical models i.e. models
based primarily on measurements, rather than theoretically
obtained values. All models use some type of solar–terrestrial

index to describe parameter variations with solar cycle and magnetic activity. The definition and the availability of these indices is described in a separate chapter. Finally solar–terrestrial coordinate systems are explained and software for coordinate conversion between different systems is listed.

002.034 The spectrum of the VV Cephei star KQ Puppis (Boss 1985). II. Atlas of the optical and ultraviolet spectrum.
A. Altamore, C. Rossi, R. Viotti, G. B. Baratta.
Astron. Astrophys., Suppl. Ser., Vol. 92, No. 4, p. 685 – 719 (Mar 1992).
    The authors present an atlas of the high resolution spectrum of the VV Cep M1–2Iab + Be variable KQ Pup (Boss 1985). The atlas is based on ultraviolet observations obtained in 1979 with the International Ultraviolet Explorer covering the region from 1223 Å to 3228 Å, and on coudé plates collected in 1969 at the Dominion Astrophysical Observatory, Canada, and in 1979 and 1983 at the European Southern Observatory, Chile, covering the blue to the optical UV. The equivalent widths and radial velocities of the main features present in the optical spectra are given. Additional CAT/CES observations of Hα taken in February 1984 are shown. The atlas should be a basis for future investigation on luminous cool variables. The digitized spectra are available upon request.

002.035 The far–infrared properties of the CfA galaxy sample. I. The catalog.
T. X. Thuan, M. Sauvage.
Astron. Astrophys., Suppl. Ser., Vol. 92, No. 4, p. 749 – 839 (Mar 1992).
    The authors present IRAS flux densities for all galaxies in the Center for Astrophysics (CfA) magnitude–limited sample ($m_B \leqslant 14.5$) detected in the IRAS Faint Source Survey (FSS), a total of 1544 galaxies. The FSS is an attempt to reach lower sensitivity limits than the Point Source Catalog (PSC) in the IRAS data by generalizing the coadding method to the whole sky. The detection rate in the FSS is slightly larger than in the PSC for the long wavelength 60 and 100 μm bands, but improves substantially (by a factor of ~ 3 or more) for the short wavelength 12 and 25 μm bands. 63% of all 2445 CfA galaxies were detected in at least one IRAS band in the FSS, and, compared to the PSC, the authors have added $\sim 50\%$ more flux densities, with the additions coming essentially all in the short wavelength range. This optically selected sample consists of galaxies which are, on average, much less infrared–active than galaxies in infrared–selected samples. It possesses accurate and complete redshift, morphological and magnitude information, along with observations at other wavelengths, and forms the basis for studies of the far–infrared properties of optically selected galaxies in a forthcoming series of papers.

002.036 uvbyβ observations of 528 type B stars with V between the 8th and 9th magnitude.
J. Knude.
Astron. Astrophys., Suppl. Ser., Vol. 92, No. 4, p. 841 – 861 (Mar 1992).
    uvbyβ measurements of 528 type B stars are presented. The stars were selected from the SAO Catalog fulfilling two criteria: the spectral types in the range B3 – B5 and m_v between the 8th and the 9th magnitude. The observations were mainly undertaken to form a new homogeneous basis for a study of the spatial frequency of optical reflection nebulae and infrared emission clouds. Particularly the b–y versus m_1 and u–b versus b–y diagrams seem useful and may probably be used to derive reddening ratios. The data are compared to those already published and average differences are given.

002.037 Erratum: "The interstellar lines catalogue" [Astron. Astrophys., Suppl. Ser., Vol. 89, No. 3, p. 469 – 527 (Sep 1991)].
B. Garcia.
Astron. Astrophys., Suppl. Ser., Vol. 92, No. 4, p. 885 – 887 (Mar 1992). See 54.002.027.

002.038 UBVRI–JHKL photometric catalogue of symbiotic stars.
U. Munari, B. F. Yudin, O. G. Taranova, G. Massone, F. Marang, G. Roberts, H. Winkler, P. A. Whitelock.
Astron. Astrophys., Suppl. Ser., Vol. 93, No. 2, p. 383 – 390 (May 1992).
    The authors present the results of an optical and infrared photometric survey of a large sample of symbiotic stars: 93 objects out of the approximately 150 known class members. A sample of 78, mostly southern, symbiotic stars was observed during 1990 at ESO and SAAO. Of these, 63 were contemporaneously observed in $UBVR_CI_C$ from La Silla and in JHKL from Sutherland. A further 15 stars were observed only in the optical or in the infrared. Twenty northern symbiotic stars have been contemporaneously observed in $UBVR_J$ and JHKL at the Crimean Astrophysical Observatory of the Shternberg Astronomical Institute. Emphasis has been placed on: (1) nearly simultaneous optical and IR observations, to avoid any biasing from the long term variability of symbiotic stars; (2) the observation of a large sample in fixed and well known photometric systems; the strong emission line spectrum of these objects prevents a meaningful comparison of photometric observations made at different sites with different photometers and using different sets of standard stars.

002.039 Fifth Fundamental Catalogue (FK5). Part II. The FK5 extension – new fundamental stars.
W. Fricke, H. Schwan, T. Corbin, U. Bastian, R. Bien, C. Cole, E. Jackson, R. Jährling, H. Jahreiß, T. Lederle, S. Röser (comps.).
Veröff. – Astron. Rechen–Inst. Heidelb., No. 33.
Astronomisches Rechen–Institut, Heidelberg (Germany), Braun, Karlsruhe (Germany). 143 p. (1991).
    The FK5 extension gives mean positions and proper motions for 3117 new fundamental stars extending the fundamental system to about apparent visual magnitude 9.5. The new fundamental stars were selected from the FK4 Supplement and the IRS list. The catalogue has been constructed in co–operation between the Astronomisches Rechen–Institut, Heidelberg and the U.S. Naval Observatory, Washington, DC.

002.040 Astrophysical supplement to the general catalogue of geodetic stars.
K. A. Kandaurova, E. V. Khrutskaya.
Astron. Tsirk., No. 1550, p. 39 – 40 (Sep – Oct 1991). In Russian.
    The astrophysical data for stars entering the general catalogue of positions and proper motions of 4949 geodetic stars from $+90°$ to $-90°$ are chosen from available catalogues. The degree of availability of astrophysical information for bright stars (to $6^m.1$) has been estimated.

002.041 The Hipparcos Input Catalogue. I. Star selection.
C. Turon, A. Gómez, F. Crifo, M. Crézé, M. A. C. Perryman, D. Morin, F. Arenou, B. Nicolet, M. Chareton, D. Egret.
Astron. Astrophys., Vol. 258, No. 1, p. 74 – 81 (May 1992).
    The Hipparcos Input Catalogue has been compiled, over the period 1982–1991, as the definitive observing catalogue for the European Space Agency's Hipparcos satellite, launched on 8 August 1989. It contains the most up–to–date, comprehensive and homogeneous information on the 118000 stars being observed by Hipparcos. The present paper deals with the stellar content of the catalogue and the way it was constructed.

002.042 The Hipparcos Input Catalogue. II. Astrometric data.
H. Jahreiß, Y. Réquième, A. N. Argue, J. Dommanget, M. Rousseau, T. Lederle, R. S. Le Poole, J. M. Mazurier, L. V. Morrison, O. Nys, M. J. Penston, J. P. Périé, L. Prévot, H. J. Tucholke, C. de Vegt.
Astron. Astrophys., Vol. 258, No. 1, p. 82 – 87 (May 1992).
    A positional accuracy of 1.5 arcsec rms at the epoch of observation was required for the stars proposed for observation by the Hipparcos satellite. How this was achieved by collecting and investigating already available positions and proper motions

in a special compilation catalogue, carrying out new ground–based measurements, and a detailed analysis in combination with new observations of double and multiple stars is described in this paper. Special emphasis is put on how to achieve a link of the resulting Hipparcos catalogue to an inertial system.

002.043   The Hipparcos Input Catalogue. III. Photometry.
M. Grenon, M. Mermilliod, J. C. Mermilliod.
Astron. Astrophys., Vol. 258, No. 1, p. 88 – 93 (May 1992).

Photometric information about stars proposed for observation by the astrometric satellite Hipparcos is used for several purposes. Instrumental colours and magnitudes served as an input to the definition of the observing programme (e.g. in defining the magnitude–limited survey), in optimising the observations by allocating observing time according to the star's magnitude and for the calibrations, on–orbit, of the satellite detection chains (main mission and Tycho). In cases where existing magnitudes data were inadequate or missing new ground–based observations were required. The work undertaken in the INCA consortium in this area is described here.

002.044   An astrometric catalogue of southern and equatorial dwarf novae.
A. Bruch, J. Meijer, M. Naumann, T. Schimpke, R. Ungruhe, N. Vogt.
Astron. Astrophys., Suppl. Ser., Vol. 93, No. 3, p. 463 – 468 (Jun 1992).

An astrometric catalogue of 117 southern and elquatorial dwarf novae is presented, complementing earlier work for the northern sky. The mean internal accuracy of the coordinates if of the order of $\pm 0\overset{s}{.}025$ in $\alpha \cos \delta$ and $\pm 0\overset{"}{.}3$ in δ.

002.045   Reliability of the Hipparcos Input Catalogue tested by the "First look".
F. Crifo, A. Gómez, F. Arenou, D. Morin, H. Schrijver.
Astron. Astrophys., Vol. 258, No. 1, p. 116 – 118 (May 1992).

The reliability of the Input Catalogue is one of the keys the success of the Hipparcos mission. As the Input Catalogue is catalogue made of very heterogeneous sources, many systematic ability tests were conducted to eliminate errors on identifications, condition magnitude, etc. In March 1991 half of the Input Catalogue had already been checked by the "First look" facility developed by the FAST consortium at Utrecht. It is shown that less than 0.1% of the 118000 stars of the observing program will not be detected.

002.046   The Tycho Input Catalogue. Cross–matching the Guide Star Catalog with the Hipparcos INCA Data Base.
D. Egret, P. Didelon, B. J. McLean, J. L. Russell, C. Turon.
Astron. Astrophys., Vol. 258, No. 1, p. 217 – 222 (May 1992).

A Tycho Input Catalogue of three million stars brighter than $V = 12.1$ has been produced for the needs of the Tycho data analysis. This catalogue results from the cross–matching of a subset of the Hubble Space Telescope Guide Star Catalog with the Hipparcos NCA Data Base. The cross–matching procedure is explained and statistics about the magnitude distribution and errors of the resulting catalogue are given.

002.047   Recommended rest frequencies for observed interstellar molecular microwave transitions – 1991 revision.
F. J. Lovas.
J. Phys. Chem. Ref. Data, Vol. 21, No. 2, p. 181 – 272 (Mar 1992). Current Physics Microform No.: 9209E0001.

Critically evaluated transition frequencies for the molecular transitions detected in interstellar and circumstellar clouds are presented. The tabulated transitions are recommended for reference in future astronomical observations in the microwave and millimeter wavelength regions. The transition frequencies have been selected through a critical examination and analysis of the laboratory spectral data obtained from the literature. The information tabulated includes the species identity, transition frequency, uncertainty, and quantum state labels. In addition, representative line antenna temperatures are listed for a typical astronomical source for each transition as a convenience to users,

and the references are cited for the laboratory and astronomical literature which have been employed.

002.048   The astrometric standard region in the Cyg constellation: catalog of 2197 stars.
E. V. Rel'ke.
Astron. Tsirk., No. 1551, p. 36 – 37 (Nov – Dec 1991). In Russian.

A catalogue of 2197 star positions from $6^m – 7^m$ to $20^m – 22^m$ was obtained in the Cyg constellation.

002.049   Compact radio sources near the Galactic plane.
D. J. Helfand, S. Zoonematkermani, R. H. Becker, R. L. White.
Astrophys. J., Suppl. Ser., Vol. 80, No. 1, p. 211 – 255 (May 1992).

The results of an extension of the 20 cm Galactic plane survey reported by Zoonematkermani and coworkers to Galactic latitudes of $\pm 1\overset{\circ}{.}8$ over the central region of the Milky Way ($-10° \leqslant l \leqslant 40°$) are reported. The authors catalog 1457 new discrete radio sources down to flux densities of $\lesssim 5$ mJy. The sample is 95% complete at 20 mJy. A detailed comparison of all radio sources from the complete survey in this longitude range with the IRAS Point Source Catalog provides classification for 13% of the objects, including 159 compact H II regions, and nearly 100 planetary nebulae, of which more than 70 are newly identified. The authors comment briefly on the identity of the remaining radio sources and on the importance of their 6 cm survey, currently underway, in defining the radio source population of the Galacy.

002.050   The EINSTEIN Slew Survey.
M. Elvis, D. Plummer, J. Schachter, G. Fabbiano.
Astrophys. J., Suppl. Ser., Vol. 80, No. 1, p. 257 – 303 (May 1992).

A catalog of 819 sources detected in the Einstein IPC Slew Survey of the X–ray sky is presented; 313 of the sources were not previously known as X–ray sources. Typical count rates are 0.1 IPC count s^{-1}, roughly equivalent to a flux of 3×10^{-12} ergs $cm^{-1} s^{-1}$. The sources have positional uncertainties of $1\overset{'}{.}2$ (90% confidence) radius, based on a subset of 452 sources identified with previously known pointlike X–ray sources. Identifications based on a number of existing catalogs of X–ray and optical objects are proposed for 637 of the sources, 78% of the survey, (within a 3' error radius) including 133 identifications of new X–ray sources. A public identification data base for the Slew Survey sources will be maintained at CfA, and contributions to this data base are invited.

002.051   The Strasbourg–ESO Catalogue of Galactic Planetary Nebulae. Parts I, II.
A. Acker, J. Marcout, F. Ochsenbein, B. Stenholm, R. Tylenda, C. Schohn.
European Southern Observatory, Garching (Germany). 1047 p. (1992). ISBN 3–923524–41–2.

The selection of the objects populating this catalogue is mainly based on an observational programme, and a verification programme consisting of spectroscopic observations for all suspected planetary nebulae, assuming that a planetary nebula has a spectral signature which can easy be recognised. For the present catalogue, a list of 1820 objects, each of them called at least once a planetary nebula, have been inspected; 1143 of them have been classified as true or probable planetary nebulae; 347 objects, which status is still unclear, were classified among the "possible" planetary nebulae. Finally, 330 objects have been rejected.
Part I: A. Explanation of the catalogue. B. Tables. C. References of papers containing 20 objects or more. D. Finding charts.
Part II: The catalogue.

002.052 **The "visibility" of West European astronomical research.**
C. Jaschek.
Scientometrics, Vol. 23, No. 3, p. 377 – 393 (1992).

Publications and citations of five West European astronomical communities (Switzerland, Sweden, GFR, France and Spain) are compared. A large proportion of astronomers are sparsely cited or not cited at all, a fact which shows that estimations of the number of scientists based upon citation statistics are underestimates. It is found that publication rates are similar but citation rates very dissimilar in the five countries. No clear explanation of these differences is found, except for Spain. A plea is made to use citation statistics rather than publication statistics for evaluation.

002.053 **The role of spectroscopic archives in binary research.**
R. Viotti, M. Friedjung.
13. Journée de Strasbourg: Stades avancés dans l'évolution des étoiles binaires serrées, p. 56 – 57 (1992). – See Abstr. 012.034 for the main entry.

002.054 **IAU archives of unpublished observations of variable stars.**
E. G. Schmidt.
Inf. Bull. Variable Stars, No. 3733.
International Astronomical Union. Commission 27; Konkoly Obs., Budapest (Hungary). 1 p. (29 May 1992).

002.055 **The Hipparcos Input Catalogue. Volumes 1 – 7.**
C. Turon, M. Crézé, D. Egret, A. Gómez,
M. Grenon, H. Jahreiß, Y. Réquième, A. N. Argue,
A. Bec–Borsenberger, J. Dommanget, M. O. Mennessier,
F. Arenou, M. Chareton, F. Crifo, J. C. Mermilliod, D. Morin,
B. Nicolet, O. Nys, L. Prévot, M. Rousseau,
M. A. C. Perryman, J. E. Arlot, A. Baglin, D. Barthès,
M. O. Baylac, P. Brosche, M. Burnet, J. Delhaye, C. Dettbarn,
M. Erbach, F. Figueras, W. Fricke, L. Helmer, P. Hemenway,
C. Jordi, P. Lampens, T. Lederle, J. Lub, J. Manfroid,
J. A. Mattei, J. M. Mazurier, M. Mermilliod, L. V. Morisson,
C. A. Murray, E. Oblak, J. P. Périé, B. Pernier, R. S. Le Poole,
L. Quijano, M. Rapaport, A. Sellier, J. Torra, H.–J. Tucholke,
C. de Vegt, E. Høg, J. Kovalevsky, F. van Leeuwen,
L. Lindegren, A. Schütz, H. Schrijver.
European Space Agency, Paris (France), ESA–SP—1136. 3211 p. (Mar 1992). ISBN 92–9092–120. Price Dfl. 180.00.
Vol. 1 – 5: The Hipparcos Input Catalogue.
Vol. 6: Annex 1. Double and multiple stars.
Vol. 7: Annex 2. The atlas of identification charts for faint stars.
Annex 3. Identification charts for stars in galactic open clusters.
Annex 4. Identification charts for stars in the Magellanic Clouds.
The Hipparcos Input Catalogue was constructed as the observing programme for the European Space Agency's Hipparcos astrometry mission. The requirements of the project in terms of completeness, sky coverage, astrometric and photometric accuracy, as well as the necessary optimisation of the scientific impact, resulted in an extended effort to compile and homogenize existing data, to clarify sources and identifications and, where needed, to collect new data matching the required accuracy. This has resulted in an unprecedented catalogue of stellar data including up–to–date information on positions, proper motions, magnitudes and colours, and (whenever available) spectral types, radial velocities, multiplicity and variability information. The catalogue is complete to well–defined magnitude limits, and includes a substantial sampling of the most important stellar categories present in the solar neighborhood beyond these limits. The magnitude limits vary from 7.3 to 9 mag as a function of galactic latitude and spectral type, and there are no stars fainter than about V = 13 mag.

002.056 **Astrochemistry library with artificial intelligence for quality control.**
S. S. Prasad, P. Gangopadhaya.
IAU Symposium No. 150: Astrochemistry of cosmic phenomena, p. 19 – 20 (1992). – See Abstr. 012.029 for the main entry.

Libraries of reactions used in astrochemistry modeling have seen an explosive increase in size in recent years. Their quality control by manual effort is almost impossible. Expert systems with artificial intelligence are now needed to ensure the quality of large scale astrochemistry libraries.

002.057 **Cyclopaedia of telescope makers.**
A. D. Andrews.
Ir. Astron. J., Vol. 20, No. 3, p. 102 – 183 (Mar 1992).

The Cyclopaedia consists of short biographical notes on telescope–makers, telescope–opticians, and telescope–engineers, and includes contemporary astronomers, natural philosophers, and a few surveyors, and wholesalers and retailers of optical scientific instruments of particular influence in their time. The names, places of work, family business connections, business associations with astronomers, and a few historical references, are listed alphabetically together with an outline of the type of constructional work carried out and brief details of some notable telescopes. The period covered is roughly from 1600 to 1970.

002.058 **Catalogue of WBVR–magnitudes of bright stars of the northern sky.**
V. G. Kornilov, I. M. Volkov, A. I. Zakharov, V. S. Kozyreva,
L. N. Kornilova, A. N. Krutyakov, A. V. Krylov,
A. V. Kusakin, S. E. Leont'ev, A. V. Mironov,
V. G. Moshkalev, T. M. Pogrosheva, V. N. Sementsov,
Kh. F. Khaliullin.
Tr. Gos. Astron. Inst. im. Shternberga, Tom 63, p. 1 – 400 (1991). ISBN 5–211–02446–X. Price 7 Rbl. 30 Kop.

This publication contains the catalogue of photoelectric magnitudes of bright stars of the northern sky in the WBVR four–colour photometric system. For 13586 objects (stars and multiple systems) the V–magnitudes and W–B, B–V, and V–R colour indices are given, which were obtained from original observations made during the Tien Shan expedition of the Sternberg Institute in the years 1985 to 1988. The catalogue contains practically all stars brighter than 7^m2 with declination greater than $-14°$. Further the authors give informations on the observational instruments, the photometric system of the catalogue, on the duplicity and variability for a considerable part of stars of the catalogue.

002.059 **Moons of the solar system. An illustrated encyclopedia.**
J. Stewart.
McFarland, Jefferson, NC (USA). 260 p. (1991). ISBN 0–88950–568–6. Price US$ 45.00.
Contents: Part I: Luna, Earth's moon.
Lunar data. Luna A – Z. Lunar bibliography.
Part II: Photographs.
Checklist of moons depicted.
Part III: The other moons of the solar system.
Pronunciation guide. The other moons A – Z.
Part IV: Appendices.
1. The moons planet by planet. 2. Order of discovery. 3. Order of size. 4. Order of visible magnitudes. 5. Moons with an atmosphere. 6. The early discoverers. 7. Relevant space missions.

002.060 **A variable stars analysis computer program archive.**
T. Banks, F. Jansen.
Astrophys. Space Sci., Vol. 190, No. 1, p. 155 – 158 (Apr 1992). Letter–to–the–editor.

A library site for the storage of computer programs used in the analysis of variable stars has recently been set up. It is directly accessible across the computer networks, using standard procedures. Currently only a few programs for the analysis of the light curves of eclipsing binaries are stored at the site, although it is hoped that as the library becomes more well known, more investigators will deposit copies of their programs in it. This would result in a library where interested researchers could obtain the latest versions of analysis techniques from a comprehensive listing. Instructions for access to the site are also discussed.

002.061   **Guide to the history of science. 8th edition.**
P. T. Carroll (ed.).
Rensselaer Polytech. Inst., Troy, NY (USA). Dept. of Sci.
Technol. Stud.; Pennsylvania Univ., Philadelphia (USA). Dept.
of History and Sociology of Science, History of Science Society
Publ. Office, University of Chicago Press, Chicago, IL (USA).
315 p. (1992). ISBN 0–934235–20–1.
Contents: 1. Preface. 2. The History of Science Society.
3. Statutes of the Society. 4. Directory of members. 5. Guide to
the profession. 6. Indexes.

002.062   **Catalog of galaxies behind the Milky Way. Vol. 1:**
l = 210 to 230 degrees.
M. Saitō, H. Ohtani, A. Asomuna, N. Kashikawa, T. Maki,
S. Nishida, T. Watanabe.
Contrib. Dep. Astron., Kyoto Univ., No. 300.
Kyoto University (Japan). Dept. of Astronomy. 72 p. (1991).
A systematic galaxy search has been carried out by means of 18
film copies of the UK Schmidt Southern Infrared Atlas on the
Milky Way between *l* = 209° and 234° at *b* = 0° covering about
500 square degrees. The authors have detected galaxies larger
than 0.1 mm (6.7 arcsec) in diameter on film copy by visual
inspection, the limiting size of which is about twice that of the
faintest stellar images. A total of 2411 galaxies were detected. A
catalog of the detected galaxies was made regarding position,
size, galaxy type, and cross identification. The material and
procedure of search are described as well as the detectability of
galaxies in Publ. Astron. Soc. Jpn., Vol. 42, p. 603 – 624 (1990)
and appended before this catalogue. The parameters of the
galaxies are also explained in this article. Cross identifications
with other catalogs are shown in the last column only for the
names of catalogs and the numbers in these catalogs are listed.

002.063   **Catalog of galaxies behind the Milky Way. Vol. 2:**
l = 230 to 250 degrees.
M. Saitō, H. Ohtani, A. Baba, N. Hotta, S. Kameno,
S. Kurosu, K. Nakada, T. Takata.
Contrib. Dep. Astron., Kyoto Univ., No. 300.
Kyoto University (Japan). Dept. of Astronomy. 123 p. (1991).
A systematic search for galaxies behind the Milky Way by
means of film–copies of the UK Schmidt Southern Infrared Atlas
at an effective wavelength of 790 nm has been made. This search
was carried out by visual inspection of 14 films between *l* = 232°
and 250° at *b* = 0° covering 411 square degrees; a total of 4633
galaxies were detected. A catalog of the detected galaxies was
made regarding position, size, galaxy type, and cross identification. The material and procedure of search are described as well
as the detectability of galaxies in Publ. Astron. Soc. Jpn., Vol. 43,
p. 449 – 468 (1991) and appended before this catalogue. Cross
identifications with other catalogs are shown in the last column
only for the names of catalogs and the numbers in these catalogs
are listed.

002.064   **Women in astronomy: a bibliography.**
A. Fraknoi, R. Freitag.
Mercury, Vol. 21, No. 1, p. 46 – 47 (Jan – Feb 1992).

002.065   **RRS2 – the list of reference stars in the field around 238
extragalactic radio sources.**
V. V. Tel'nyuk–Adamchuk, A. A. Molotaj.
Visn. Kiiv. Univ., Fiz.–Mat. Nauki, Astron., Vip. 3, p. 53 – 56
(1992). In Ukrainian.
A motivation has been given briefly for works on purposeful
formation of special reference systems in fields with a compact
extragalactic radio/optical sources. The description is presented
of a broadened list of meridian reference stars in the fields around
the radio sources. Prospects for the observations of these stars are
discussed. The RRS2 list contains 2575 stars in the vicinity of 238
radio sources, the fields –2°, magnitudes – mainly up to 9ᵐ0.
Bright radiostars have been included into this list. RRS2 star
distributions over the coordinates, magnitudes, spectra and
distances to the sources are presented.

002.066   **The future of astronomy publications: electronic publish-
ing?**
J. Lequeux.
Messenger, No. 67, p. 58 – 61 (Mar 1992).

002.067   **Astronomy acknowledgements index 1991.**
D. A. Verner.
Messenger, No. 67, p. 61 – 62 (Mar 1992).

002.068   **An X–ray catalog and atlas of galaxies.**
G. Fabbiano, D.–W. Kim, G. Trinchieri.
Astrophys. J., Suppl. Ser., Vol. 80, No. 2, p. 531 – 644 (Jun 1992).
The authors present an X–ray catalog and atlas of galaxies
observed with the Einstein Observatory imaging instruments
(IPC and HRI). The catalog comprises 493 galaxies, including
targets of pointed observations, and RSA or RC2 galaxies
serendipitously included in Einstein fields. A total of 450 of these
galaxies were imaged well within the instrumental fields, resulting
in 238 detections and 212 upper limits (3σ). The other galaxies
were either at the edge of the visible field of view or confused with
other X–ray sources. For these the authors also give a rough
measure of their X–ray emission. The atlas shows X–ray contour
maps of detected galaxies superposed on optical photographs
and gives azimuthally averaged surface brightness profiles of
galaxies detected with high signal–to–noise ratio.

002.069   **The present state of the compilation of the General
Catalogue of Variable Stars (GCVS).**
N. Samus.
1. European Meeting of the American Association of Variable
Star Observers (AAVSO): International Cooperation and Coordination in Variable Star Research, p. 52 – 54 (1992). – See Abstr.
012.048 for the main entry.

002.070   **The extragalactic variables in the fourth edition of the
GCVS.**
N. A. Lipunova.
1. European Meeting of the American Association of Variable
Star Observers (AAVSO): International Cooperation and Coor-
dination in Variable Star Research, p. 55 – 59 (1992). – See Abstr.
012.048 for the main entry.
The data on the extragalactic variables will be the first
systematic catalogue of extragalactic variables of its type. It will
contain the data on variables in the nearby galaxies, namely the
Large and Small Magellanic Clouds, the Andromeda Nebula
(M31), the Triangulum Nebula (M33), the dwarf galaxies Draco,
Ursa Minor, Sculptor, etc.

002.071   **Carlsberg Meridian Catalogue, La Palma. Number 6.
Observations of positions of stars and planets 1990.**
Copenhagen Univ. (Denmark). Astronomisk Observatorium;
Royal Greenwich Observatory, Cambridge (UK); Real Instituto
y Observatorio de la Armada, San Fernando, Cadiz (Spain).
252 p. (1992). ISBN 84–7469–052–8.
This catalogue (CMC6) contains positions and magnitudes of
12771 stars north of declination –45°, 11555 proper motions, and
1368 positions and magnitudes of 69 solar system objects
obtained with the Carlsberg Automatic Meridian Circle on
La Palma during the year 1990. The positions of the stars are for
the epoch of observation and the equinox J2000.0, and are in a
system close to that of the FK5. The limiting magnitude is
m_v = 14.8. The catalogue mainly comprises the following stars:
4143 faint reference stars in a global net; 3162 International
Reference Stars; 1040 faint reference stars in the fields of about
200 benchmark radio sources; 889 A5–G0 stars near the north
galactic pole; 535 stars from an unbiased sample of K–M dwarfs;
and 751 stars in nearby O–B associations. The catalogue also
contains observations of Callisto, Titan, Iapetus, Uranus, Neptune, Pluto and 63 minor planets. The mean error of a catalogue
position in the zenith is 0."08 in right ascension and declination,
and 0."04 in magnitude. The mean error of the proper motions
derived by combining the position in this catalogue with those at
earlier epochs, is typically 0."002 per year. Cross–references are

given to DM, AGK, SAO, HD and the double star catalogues ADS and WDS.

002.072   A catalogue of variable components in visual double and multiple stars.
P. Lampens.
1. European Meeting of the American Association of Variable Star Observers (AAVSO): International Cooperation and Coordination in Variable Star Research, p. 60 – 63 (1992). – See Abstr. 012.048 for the main entry.
The various reasons for a compilation of a "Catalogue of Variable Components in Visual Double and Multiple Stars" are described. The present status is given and a first version is announced.

002.073   Computer networks in variable star activities.
V. Mäkelä, A. Kellomäki.
1. European Meeting of the American Association of Variable Star Observers (AAVSO): International Cooperation and Coordination in Variable Star Research, p. 156 – 158 (1992). – See Abstr. 012.048 for the main entry.
The authors discuss some examples of possible and already existing ways of using computer networks.

002.074   Determination of star coordinates and proper motion from FON program plates.
A. I. Yatsenko.
Kinematics Phys. Celest. Bodies, Vol. 7, No. 4, p. 46 – 49 (1991). English translation of Kinematika Fiz. Nebesn. Tel, Tom 7, No. 4, p. 56 – 60 (1991).
The procedure used to reduce measurements on plates from the northern sky photographic survey program to determine star positions and proper motions is described. The systematic position and proper-motion differences AGK3–PPM are investigated.

002.075   Proper motion catalogue of stars in the region of the North Galactic Pole.
L. K. Pakulyak.
Kinematics Phys. Celest. Bodies, Vol. 7, No. 4, p. 50 – 55 (1991). English translation of Kinematika Fiz. Nebesn. Tel, Tom 7, No. 4, p. 61 – 66 (1991).
A catalogue of 3304 stars in the region of the North Galactic Pole was prepared on the basis of reduction of 43 plates of the Oxford and Paris zones of the Astrographic Catalogue. The position errors amounted to 0.2″ in both coordinates, and the magnitude–estimation errors to 0.1 – 0.4^m. The proper motions of 1561 stars were derived from a previously prepared astrometric standard catalogue for the same region, with an epoch difference of about 100 years. The proper-motion errors amount to 0.003″/yr in both coordinates. The catalogue was investigated and corrected for the errors of the magnitude equation in the proper motions (by two procedures).

002.076   The second HST Guide Star photometric catalog.
M. Postman, L. Siciliano, M. Shara, D. Rehner, N. Brosch, C. Sturch, B. Bucciarelli, C. Lopez.
2. Conference on Digitised Optical Sky Surveys, p. 61 – 63 (1992). – See Abstr. 012.059 for the main entry.

002.077   The COSMOS/UKST catalog of the southern sky.
D. J. Yentis, R. G. Cruddace, H. Gursky, B. V. Stuart, J. F. Wallin, H. T. MacGillivray, C. A. Collins.
2. Conference on Digitised Optical Sky Surveys, p. 67 – 75 (1992). – See Abstr. 012.059 for the main entry.
The authors have prepared an "object" catalog of the Southern sky south of + 2.5 degrees declination from COSMOS scans of the IIIa–J and Short Red surveys taken with the UK Schmidt Telescope. The catalog consists of upwards of 500 million objects down to the limit of the plates. A version of the catalog in compact form is available for distribution, and a database management system has been produced which allows rapid access to any part of the catalog. The authors also describe the creation of a catalog of clusters of galaxies derived from the galaxies

appearing in the object catalog. The cluster catalog contains some 70000 candidate clusters in the Southern hemisphere and forms the basis for further major programs of follow–up study.

002.078   The automated plate scanner catalog of the Palomar Sky Survey.
R. L. Pennington, R. M. Humphreys, S. C. Odewahn, W. A. Zumach, E. B. Stockwell.
2. Conference on Digitised Optical Sky Surveys, p. 77 – 85 (1992). – See Abstr. 012.059 for the main entry.
The automated plate scanner (APS) is used to scan 936 pairs of plates comprising the Palomar Observatory – National Geographic Sky Survey. The resultant dataset will be used to produce a catalogue of a billion stellar objects and several million galaxies. This catalogue will be publicly available (also an online version over the Internet).

002.079   MOGADOC – a personal computer database for atmospheric and interstellar molecules in microwave spectroscopy and radio astronomy.
J. Vogt.
Rev. Mex. Astron. Astrofís., Vol. 23, p. 119 – 125 (Mar 1992). – See Abstr. 012.054 for the main entry.
MOGADOC is a comprehensive database for gas–phase molecules, investigated by microwave spectroscopy, radio astronomy and electron diffraction. It contains data on electrical, magnetic, dynamical and spectroscopic properties of inorganic, organic and organometallic compounds in the gas phase. As a special feature the in–house database, which can be run on a personal computer by means of the well known Messenger retrieval language, contains numerical data sets for structural parameters such as internuclear distances and bond angles.

002.080   The Sonneberg plate archive.
H.–J. Bräuer, B. Fuhrmann.
Messenger, No. 68, p. 24 – 26 (Jun 1992).

002.081   Updating the AGK3 reference star catalog with astrometrically reduced Schmidt plates.
L. G. Taff, M. G. Lattanzi, B. Bucciarelli, D. Daou.
2. Conference on Digitised Optical Sky Surveys, p. 185 – 191 (1992). – See Abstr. 012.059 for the main entry.

002.082   The Edinburgh/Durham Southern Galaxy Catalogue.
R. C. Nicol, C. A. Collins, L. Guzzo, S. Lumsden.
2. Conference on Digitised Optical Sky Surveys, p. 335 – 344 (1992). – See Abstr. 012.059 for the main entry.

002.083   Les moitiés masculines et féminines du ciel: astronomie de quelques tribus Guyanaises. Bibliographie (Volume 5).
E. Magana–Torres.
Publ. Obs. Astron. Strasbourg, Sér. Astron. Sci. Hum., No. 6, p. 75 – 77 (1992).

002.084   Asteroid photometric database.
M. T. Capria, A. Barucci, M. Dahlgren, M. Fulchignoni, C.–I. Lagerkvist, P. Magnusson.
Minor Planet Bull., Vol. 19, No. 2, p. 12 – 13 (Apr – Jun 1992).
The authors started to collect in 1984 all published asteroid lightcurves with the goals (1) to create and maintain a digital database, and (2) to publish the data in a consistent way. They have published two catalogues and the third is in press.

002.085   A ∼ 700 galaxy magnitude–limited redshift survey near the SGP.
Q. A. Parker.
2. Conference on Digitised Optical Sky Surveys, p. 383 – 387 (1992). – See Abstr. 012.059 for the main entry.
A description of a magnitude–limited redshift survey covering 7 adjacent UKST fields is presented together with some preliminary results. The survey is ∼91% complete to $b_j \leqslant 16.5$ but does extend to $b_j \leqslant 16.8$. The survey of ∼ 700 redshifts will form one of the largest, independent, contiguous, well sampled

and most homogeneous redshift catalogues ever compiled. The special features of this catalogue are its coverage (~ 200 deg^2), mean sampling depth (~ 125 h^{-1}Mpc) and field–to–fild photometric accuracy ($\Delta b_j \pm 0.05$). Morphological classifications have been determined for all 937 galaxies in the survey fields down to $b_j \leqslant 16.8$ and COSMOS image parameters are also available. This extensive catalogue should prove an extremely valuable resource for many applications, but particularly in characterising the precise form and nature of the features in the large–scale galaxy distribution with which any theoretical model must be necessarily reconciled.

002.086   The Einstein Slew Survey catalog.
    J. F. Schachter, M. Elvis, D. Plummer, G. Fabbiano, J. Huchra.
2. Conference on Digitised Optical Sky Surveys, p. 441 – 451 (1992). – See Abstr. 012.059 for the main entry.
    The Slew Survey catalog contains 1075 bright X–ray sources, including 557 objects with no previous X–ray detection. Two–thirds of the survey has been identified with counterparts of known optical type. Source samples to date provide insight on low–luminosity AGN and clusters. The remainder, which contains many uncatalogued BL Lacs and clusters, will be identified by using digitized photographic plates.

002.087   Optical identification of sources in the IRAS Faint Sources database.
R. D. Wolstencroft, H. T. MacGillivray, C. J. Lonsdale, T. Conrow, D. J. Yentis, J. F. Wallin, G. Hau.
2. Conference on Digitised Optical Sky Surveys, p. 471 – 483 (1992). – See Abstr. 012.059 for the main entry.

002.088   The 1984 – 1987 Solar Maximum Mission event list.
    B. R. Dennis, J. P. Licata, J. J. Nelson, A. K. Tolbert.
National Aeronautics and Space Administration, Washington, DC (USA); National Aeronautics and Space Administration, Greenbelt, MD (USA). Goddard Space Flight Center, Lab. for Astronomy and Solar Physics, NASA–TM—4342. 288 p. (Feb 1992).
    This document contains information on solar burst and transient activity observed by the Solar Maximum Mission (SMM) during 1984 through 1987 pointed observations. Data from the following SMM experiments are included: 1) Gamma Ray Spectrometer, 2) Hard X–Ray Burst Spectrometer, 3) Flat Crystal Spectrometer, 4) Bent Crystal Spectrometer, 5) Ultraviolet Spectrometer Polarimeter, and 6) Coronagraph/Polarimeter. Correlative optical, radio, and Geostationary Operational Environmental Satellite (GOES) X–ray data are also presented. Where possible, bursts or transients observed in the various wavelengths were grouped into discrete flare events identified by unique event numbers. Each event carries a qualifier denoting the quality or completeness of the observations. Spacecraft pointing coordinates and flare site angular displacement values from Sun center are also included.

002.089   The 1988 Solar Maximum Mission event list.
    B. R. Dennis, J. P. Licata, A. K. Tolbert.
National Aeronautics and Space Administration, Washington, DC (USA); National Aeronautics and Space Administration, Greenbelt, MD (USA). Goddard Space Flight Center, Lab. for Astronomy and Solar Physics, NASA–TM—4343. 353 p. (Feb 1992).
    This document contains information on solar burst and transient activity observed by the Solar Maximum Mission (SMM) during 1988 pointed observations. Data from the following SMM experiments are included: 1) Gamma Ray Spectrometer, 2) Hard X–Ray Burst Spectrometer, 3) Flat Crystal Spectrometer, 4) Bent Crystal Spectrometer, 5) Ultraviolet Spectrometer Polarimeter, and 6) Coronagraph/Polarimeter. Correlative optical, radio, and Geostationary Operational Environmental Satellite (GOES) X–ray data are also presented. Where possible, bursts or transients observed in the various wavelengths were grouped into discrete flare events identified by unique event numbers. Each event carries a qualifier denoting the

quality or completeness of the observations. Spacecraft pointing coordinates and flare site angular displacement values from Sun center are also included.

002.090   The 1989 Solar Maximum Mission event list.
    B. R. Dennis, J. P. Licata, A. K. Tolbert.
National Aeronautics and Space Administration, Washington, DC (USA); National Aeronautics and Space Administration, Greenbelt, MD (USA). Goddard Space Flight Center, Lab. for Astronomy and Solar Physics, NASA–TM—4344. 610 p. (Mar 1992).
    This document contains information on solar burst and transient activity observed by the Solar Maximum Mission (SMM) during 1989 pointed observations. Data from the following SMM experiments are included: 1) Gamma Ray Spectrometer, 2) Hard X–Ray Burst Spectrometer, 3) Flat Crystal Spectrometer, 4) Bent Crystal Spectrometer, 5) Ultraviolet Spectrometer Polarimeter, and 6) Coronagraph/Polarimeter. Correlative optical, radio, and Geostationary Operational Environmental Satellite (GOES) X–ray data are also presented. Where possible, bursts or transients observed in the various wavelengths were grouped into discrete flare events identified by unique event numbers. Each event carries a qualifier denoting the quality or completeness of the observations. Spacecraft pointing coordinates and flare site angular displacement values from Sun center are also included.

002.091   Star catalogs: how good are they really.
    L. G. Taff, B. Bucciarelli, M. G. Lattanzi, D. Daou, W. Jin.
Bull. Am. Astron. Soc., Vol. 23, No. 3, p. 1258 (1991). Abstract. – See Abstr. 012.065 for the main entry.

002.092   An automated information system for the Lunar Nomenclature data base processing.
V. V. Shevchenko, S. G. Pugacheva, K. I. Dekhtyareva, T. P. Skobeleva.
Sol. Syst. Res., Vol. 25, No. 5, p. 435 – 445 (Mar 1992). English translation of Astron. Vestn., Tom 25, No 5, p. 578 – 592 (1991).
    The data base of Lunar Nomenclature contains a set of data on toponymy, morphology, selenographic coordinates, and dimensions of lunar formations. The total number of lunar objects in the data base is 1920. Terms used for lunar names have been approved by ILAU. The Lunar Nomenclature data base is constructed as a logical relational structure. The paper describes a statistical sampling of data produced by an automated control program. The information has been classified according to the following parameters: lunar relief type, origin, name assignment year, and bibiliographic data on famous scientists whose names are immortalized in the names of lunar objects. A map of the density of named craters of various diameters has been plotted. An analysis of the name "loads" of lunar craters has been performed for map sheets with scales of 1:5,000,000; 1:1,000,000; and 1:250,000.

002.093   Bibliography (1990 Part 1).
    E. B. Carling, Z. Kopal, S. H. Tellier (eds.).
Earth, Moon, Planets, Vol. 57, No. 2, p. 139 – 189 (May 1992).
    Contents: 1. Moon. 2. Planets. 3. Jupiter. 4. Satellites of Jupiter. 5. Mars. 6. Satellites of Mars. 8. Neptune. 10. Pluto. 12. Saturn. 13. Satellites of Saturn.

002.094   The IAU Meteor Data Center in Lund.
    B. A. Lindblad.
IAU Colloquium No. 126: Origin and evolution of interplanetary dust, p. 311 – 314 (1991). – See Abstr. 012.068 for the main entry.
    The purpose of the IAU Meteor Data Center in Lund is to archive, document and disseminate information on meteoroid orbits. At present some 6000 photographic double–station orbits and 60000 radio determined orbits are archived.

002.095 **Atlas de spectres stellaires. Standards de classification MK – binaires spectroscopiques – étoiles particulières.** *(Atlas of stellar spectra. MK standards – spectroscopic binaries – peculiar stars.)*
N. Ginestet, J. M. Carquillat, M. Jaschek, C. Jaschek, A. Pédoussaut, J. Rochette.
Observatoire Midi–Pyrénées, Toulouse (France); Observatoire Astronomique, Strasbourg (France), Centre Régional de Documentation Pédagogique, Toulouse (France). 61 p. (1992). With 75 sheets. Price FF 380.00.

Chapters I, II and III are devoted to the description of the atlas; the first concerns the MK standards, the second the spectroscopic binaries and the third the peculiar spectra. All stars are listed in a table in order of increasing HD number. The ten columns of this table provide successively: HD number; name or HR number; the spectral classification for MK standards; the sheet number on which figures the spectrum (or spectra) of the object; the equatorial coordinates for equinox 2000; the V magnitude; the color index B–V; the projected rotational velocity V sin i; informations regarding binarity and finally notes in which the authors provide their classification for the non-standard stars or more detailed classifications for certain MK standards. For some stars, remarks re the spectral peculiarities are given after the table. In Chapter IV the authors provide a summary of the technical data re the spectrograph BS Cass as well as a plan of the instrument. Chapter V finally provides the criteria used for classification. The sheets constituting the atlas were made from positive copies of the film, enlarged ten times. The spectral range covered is approximately: $\lambda 3800 - 4700$ Å for stars of types O to G, $\lambda 3900 - 4800$ Å for stars of types K and M. On each sheet the following items were noted: name of the star or its HD number, spectral type, luminosity class, identification of the most important lines used for the classification.

002.096 **StarBriefs. A dictionary of abbreviations, acronyms, and symbols in astronomy, space sciences, and related fields.**
A. Heck.
Publ. Spéc. Cent. Données Stellaires, No. 19.
Observatoire Astronomique, Strasbourg (France). 694 p. (Feb 1992). ISBN 2-908064-17-0. Price FF 250.00, US$ 50.00.

This list gathers presently about 50,000 abbreviations, acronyms and symbols with sections devoted to Greek letters, mathematical symbols, special signs and characters, as well as to entries with a numerical beginning. Besides astronomy and space sciences, related fields such as aeronautics, aeronomy, astronautics, atmospheric sciences, chemistry, communications, computer sciences, data processing, education, electronics, energetics, engineering, environment, geodesy, geophysics, information retrieval, management, mathematics, meteorology, optics, physics, remote sensing, and so on, are also covered when justified. Abbreviations, acronyms and symbols in common use and/or of general interest have also been included when appropriate. The travelling scientist has not been forgotten and humor is not quite absent from this publication either.

002.097 **Desktop publishing in astronomy & space sciences: an introduction to the colloquium.**
A. Heck.
Colloquium on Desktop Publishing in Astronomy and Space Sciences, p. 3 – 9 (1992). – See Abstr. 012.053 for the main entry.

A background is set up for the colloquium. A few comments and worries are expressed. Some challenges are pointed out.

002.098 **SGML in the real world.**
E. van Herwijnen.
Colloquium on Desktop Publishing in Astronomy and Space Sciences, p. 11 – 20 (1992). – See Abstr. 012.053 for the main entry.

SGML, the Standard Generalized Markup Language, is the ISO (International Organization for Standardization) standard for text representation. The SGML format is independent of a particular text formatter and of any kind of processing. It makes documents portable and enables them to be processed in many different ways. Although the advantages of SGML are apparent,

it is not obvious how to set up a working SGML system in the real world. The author reports on some concrete realizations of SGML.

002.099 **T$_E$X, an universal formatter?**
P. Louarn.
Colloquium on Desktop Publishing in Astronomy and Space Sciences, p. 21 – 30 (1992). – See Abstr. 012.053 for the main entry.

After a brief talk on T$_E$X capabilities, the author presents T$_E$X's fellow tools: the T$_E$X ware, METAFONT and MF ware, and the major macros packages, LAT$_E$X and AMST$_E$X. He also presents links between T$_E$X and other products, and lays special stress on standards (SGML, FOSI, PostScript...).

002.100 **WordPerfect 5.1 does it all.**
R. E. White.
Colloquium on Desktop Publishing in Astronomy and Space Sciences, p. 31 – 45 (1992). – See Abstr. 012.053 for the main entry.

The various kinds of T$_E$X–type wordprocessing routines require definition of many different macros at the head of each document before it can be printed–out in any desired format. Mathematical equations and tables of data have been especially difficult to handle in such word–processing packages. With the use of WordPerfect 5.1 (WP5.1), a commercial word–processing program, both text–, equation–, and tabular–formats are swiftly and easily set–up for printing. Text may be laid–down in multiple column formats, with "boxes" included for graphics or for pictorial material. A number of different examples of the flexibility of format and font possible with WP5.1 are presented.

002.101 **American on–line publication tests and long–range plans.**
H. A. Abt.
Colloquium on Desktop Publishing in Astronomy and Space Sciences, p. 47 – 53 (1992). – See Abstr. 012.053 for the main entry.

The author explores the problems involved in the computer–readable submission of manuscripts, on–line publication, and electronic literature searches. The main problems are the transmission of halftones, funding of publication, and copyright. A pilot study is planned for late in 1991 to read optically five years of about four astronomical journals and to operate literature searches over a network.

002.102 **Results of a DeskTop Publishing Survey.**
A. Heck.
Colloquium on Desktop Publishing in Astronomy and Space Sciences, p. 55 – 64 (1992). – See Abstr. 012.053 for the main entry.

A DeskTop Publishing Survey has been run both by normal mail addressed to institutions and by electronic messages sent to institutions and individuals. The returned questionnaires concern about 3,700 persons from more than 160 institutions in 23 countries. The results of the survey are described and point to a majoritarian use of T$_E$X and associated packages on machines approximately equally distributed among PCs and "compatible" machines, DEC computers and SUN stations. This rough tendency was nuanced by the discussion presented in the paper and by the comments reported from the questionnaires.

002.103 **Desktop publishing applied to astronomical books and journals.**
S. Mitton.
Colloquium on Desktop Publishing in Astronomy and Space Sciences, p. 67 – 76 (1992). – See Abstr. 012.053 for the main entry.

Desktop publishing has important applications in astronomy publishing because the market for professional astronomy publications is intrinsically small, and therefore cost savings in prior to manufacturing are significant. Many commercial word processing applications are available for scientists to use. Experience has been that authors using TeX or its derivatives are frequently able to produce material of a high quality. The principle benefits of using TeX are availability, low cost, flexibility, and pleasing results.

002.104   The author as typesetter: the publisher's problem.
G. Kiers.
Colloquium on Desktop Publishing in Astronomy and Space Sciences, p. 77 – 83 (1992). – See Abstr. 012.053 for the main entry.
The incredibly quick developments in computer hard– and software of the past years make it easier for publishers to produce good looking pages, while authors can be published much quicker. Authors and publishers face one distinct problem: transferring knowledge of style. Style is one of the keys to the readability of text. Knowledge of style can expected to be found in a publishing house. However, this knowledge should be copied to authors in order to benefit from the possibilities of new hard– and software to improve the readability of text, the shared goal of authors, publishers and readers.

002.105   Ten years of the L. Davis Press.
A. G. D. Philip.
Colloquium on Desktop Publishing in Astronomy and Space Sciences, p. 85 – 90 (1992). – See Abstr. 012.053 for the main entry.
The L. Davis Press has existed during a period of rapid change in the way the proceedings of meetings are prepared and published. This article outlines the methods used in publishing the proceedings of astronomical meetings starting in 1972 (pre-press), continuing with the hardware and software currently used by the press and ending with some recommendations for publishing proceeedings in the future.

002.106   The experience of astronomy journals with desktop publishing, and the long–term future.
J. Lequeux.
Colloquium on Desktop Publishing in Astronomy and Space Sciences, p. 93 – 101 (1992). – See Abstr. 012.053 for the main entry.
The experience of astronomy journals with papers prepared electronically by the authors is reviewed on the basis of personal experience and of an enquiry. The future for journal publications in electronic form is discussed. Two main types of solutions can be envisaged: e–mail journals and journals distributed on a digital support: their respective advantages and inconveniences are presented.

002.107   Future plans for DTP (*Desk Top Publishing*) in Monthly Notices of the Royal Astronomical Society.
J. Shakeshaft.
Colloquium on Desktop Publishing in Astronomy and Space Sciences, p. 103 – 105 (1992). – See Abstr. 012.053 for the main entry.

002.108   Experiences of the journal Celestial Mechanics with T_EX and LaT_EX.
J. Henrard.
Colloquium on Desktop Publishing in Astronomy and Space Sciences, p. 107 – 111 (1992). – See Abstr. 012.053 for the main entry.
The journal Celestial Mechanics is now formatted in LaT_EX with the help of an ad–hoc style, the Spacekap.sty, developed by the Kluwer Academic Publishers. This new policy was preceeded by an experiment in which authors were invited to submit camera–ready papers formatted in T_EX or LaT_EX.

002.109   Bittersweet experiences during 6 years using T_EX and LAT_EX.
P. Bartholdi.
Colloquium on Desktop Publishing in Astronomy and Space Sciences, p. 113 – 116 (1992). – See Abstr. 012.053 for the main entry.

002.110   Assistance in writing proposals.
E. R. Deul.
Colloquium on Desktop Publishing in Astronomy and Space Sciences, p. 117 – 122 (1992). – See Abstr. 012.053 for the main entry.
A system to assist the proposal writer with filling out the proposal forms is described. The software is based on the UNIX environment, including the X–Windows user–interface, and the T_EX text formatting and processing system. Although the system is still under development, it finds current application for some Dutch proposal forms.

002.111   From compuscripts to intelligent information retrieval.
A. Heck.
Colloquium on Desktop Publishing in Astronomy and Space Sciences, p. 125 – 127 (1992). – See Abstr. 012.053 for the main entry.
The impact of desktop and electronic publishing on intelligent information retrieval is considered through a few prospective thoughts based on the development in the recent years of new communication tools in science and new concepts such as knowledge bases, leading unevitably at middle or long term to new behaviors towards information access.

002.112   Publishing, archiving and retrieving astronomical data.
M. Crézé.
Colloquium on Desktop Publishing in Astronomy and Space Sciences, p. 129 – 131 (1992). – See Abstr. 012.053 for the main entry.
Most published papers are now at some stage in computer readable form. There are editing software packages which may help organizing the published material for efficient retrieval of information. There is strong need for upgrading the efficiency of the processes which feed the data bases. All three circumstances work towards a coordination of the efforts of editors and data center people to capture and structure the data and information.

002.113   Databases and T_EX.
G. Weil.
Colloquium on Desktop Publishing in Astronomy and Space Sciences, p. 133 – 136 (1992). – See Abstr. 012.053 for the main entry.
This paper describes the structure of a database to be querried easily and to be printed in a very legible way.

002.114   Automatic production of star charts with LAT_EX.
D. Mégevand.
Colloquium on Desktop Publishing in Astronomy and Space Sciences, p. 137 – 148 (1992). – See Abstr. 012.053 for the main entry.
The author has developed two different applications involving LaT_EX in the production of observation or identification charts. Both use the same technique to draw the charts: A first program scans the star database to extract all the objects within a certain angular distance from the selected star. The coordinates and visual magnitude of the stars are then embedded in LAT_EX macros. The production of the charts is then realized by an appropriate LAT_EX program.

002.115   STELAR: study of electronic literature for astronomical research. A status report.
M. E. Van Steenberg.
Colloquium on Desktop Publishing in Astronomy and Space Sciences, p. 149 – 153 (1992). – See Abstr. 012.053 for the main entry.
A pilot project was started to study the technical and practical aspects of placing refereed scientific literature on–line for access by the astronomical community. This access would include graphics and full text retrieval capabilities. The current plan is to create bitmaps of five years of the ApJ, AJ, ApJS, and PASP, and to place them on–line with their corresponding text abstracts. This will allow a series of evaluations by the astronomical community as to what is truly needed as well as simply what works. Finally, the goal of STELAR is to explore the use of

electronic means for improving access to scientific literature, using astronomical publications to evaluate distribution, search, and retrieval techniques for full text and graphics display.

002.116 New microcomputer technologies to facilitate collaboration and publication.
J. O. Breen.
Colloquium on Desktop Publishing in Astronomy and Space Sciences, p. 157 – 164 (1992). – See Abstr. 012.053 for the main entry.
As microcomputer operating systems gain more advanced networking and communication facilities, new software applications are becoming available which will fundamentally change the nature of desktop publishing. The desktop microcomputer has rapidly evolved from a replacement for an office typewriter into a powerful tool capable not only of producing publication-quality documents, but also of collecting, storing, and analyzing the data on which these documents are based. Applications based on new technologies will allow researchers to spend less time managing revisions of their documents, and more time concentrating on their research.

002.117 Designing with TyPe.
J. Godfrey Jr.
Colloquium on Desktop Publishing in Astronomy and Space Sciences, p. 165 – 170 (1992). – See Abstr. 012.053 for the main entry.
Scientific pages are often difficult to read. One reason is obvious; the reader must put forth an effort to understand the material. Statistics and formulas are not easily digested like a magazine article or novel. There are, however, some rectifiable difficulties. Lines of scientific pages are often very crowded. Applying good typographic techniques can help you to correct these kinds of problems and bring life to publications. That is not to say the effectiveness of a page is hinged on its appearance, although appearance is very influential, however, good typography simply helps readers comprehend the material.

002.118 LAD$_{IE}$S and LAT$_E$X.
T. A. Jurriens.
Colloquium on Desktop Publishing in Astronomy and Space Sciences, p. 171 – 175 (1992). – See Abstr. 012.053 for the main entry.
This paper describes the use of LAT$_E$X by non–astronomers at the Kapteyn Institute. Although the general opinion is that it is too complicated for non–scientists, experience in Groningen proves otherwise.

002.119 AST$_E$X: a software environment on PC for creating multi–author scientific books.
M. Lavaud.
Colloquium on Desktop Publishing in Astronomy and Space Sciences, p. 177 – 186 (1992). – See Abstr. 012.053 for the main entry.
The program AST$_E$X is designed to ease the creation of high-quality scientific books by several authors, on low–cost systems. It runs on any PC, under the control of Framework 3. It is multilingual. AST$_E$X provides, on a plain PC, an integrated and customizable multiwindow environment for everyday scientific work, that is as comfortable to use as the one of the workstation. If the PC is connected to a workstation, AST$_E$X transforms it into a machine virtually as powerful as the workstation itself, for most scientific tasks. This provides a very low–cost alternative to color workstations or X–Window terminals, while preserving the investment in PC hardware and software, and allowing to enter progressively and painlessly into the Unix world.

002.120 How to use T$_E$X macro A$_P$T$_E$X Ver. 2.0. Reference sort and figure spacing.
M. Nagasawa.
Colloquium on Desktop Publishing in Astronomy and Space Sciences, p. 187 – 193 (1992). – See Abstr. 012.053 for the main entry.
The author presents the new T$_E$X command macro named A$_P$T$_E$X which is an extended version of PHYZZX and dedicated

to Astronomy and Physics Society. A$_P$T$_E$X has some powerful macros as follows: (1) If you select the proper journal type, alphabetical reference ordering becomes possible. The arguments in reference macros need not be changed at all. (2) Chapter heading, section heading, equation numbering etc. are converted automatically to fit the selected journal type. (3) Various figure spacing commands are prepared for camera–ready format. (4) Graphic tool is available to draw graphs in dvi file.

002.121 ST ScI information services.
P. Reppert.
Colloquium on Desktop Publishing in Astronomy and Space Sciences, p. 195 – 201 (1992). – See Abstr. 012.053 for the main entry.
The Space Telescope Science Institute Newsletter has grown from a brief four pages to an average 20 – 30 pages with a circulation of about 2900 in three years. The target readership is anyone interested in submitting a proposal, anyone with an active observing program, students and educators. The Newsletter presents highlights of recent observations, information on the status of the six scientific instruments, proposal and software news, and miscellaneous articles of interest to readers.

002.122 A case for LAT$_E$X use in astronomy publishing.
A. K. Tzioumis.
Colloquium on Desktop Publishing in Astronomy and Space Sciences, p. 203 – 206 (1992). – See Abstr. 012.053 for the main entry.
T$_E$X and LAT$_E$X offer clear advantages over other DTP packages in astronomy publishing. LAT$_E$X provides a "standard" macro package which can be used to produce the publishing styles required by the different journals, both for text and references. A case is made here, on behalf of the most of the astronomy community, for LAT$_E$X to be adopted as the future standard in astronomy publishing.

002.123 An example of author guidelines for the preparation of scientific colloquium proceedings usig logical formatting with T$_E$X.
L. Siebenmann.
Colloquium on Desktop Publishing in Astronomy and Space Sciences, p. 211 – 221 (1992). – See Abstr. 012.053 for the main entry.
Many editors and publishers are currently perfecting mechanisms for printing and publishing articles prepared originally by scientists using one or another flavor of T$_E$X. Since this involves authors more intimately in the publication of articles than did classical typesetting, author guidelines now have an enhanced role to play. This note presents sample guidelines based on Plain T$_E$X or AMS–T$_E$X. They emphasize what is called logical formatting – which signals logical structures and tends to bypass geometrical prescription of page layout.

002.124 Desktop publishing in astronomy & space sciences: closing comments.
H. A. Abt.
Colloquium on Desktop Publishing in Astronomy and Space Sciences, p. 223 (1992). – See Abstr. 012.053 for the main entry.

002.125 Application of the PPM catalogue in the reduction of the Carte du Ciel plates.
I. E. Tselishchev.
Kinematika Fiz. Nebesn. Tel, Tom 8, No. 3, p. 92 – 96 (May–Jun 1992). In Russian. English translation in Kinematics Phys. Celest. Bodies, Vol. 8, No. 3.
35 plates of the Paris zone and 10 plates of the Bordeaux zone of the Carte du Ciel were re–reduced in the framework of the construction of the wide angle astronomical standard around Praesepe open cluster. The PPM catalogue is used as a reference frame. The reduction procedure is described and the results are analysed. The internal error of the catalogue star positions is $0.17''$ in both coordinates. The catalogue obtained is compared with the Fresneau CdC catalogue.

002.126   A list of some corrections to Zwicky's *Catalogue of Galaxies and Clusters of Galaxies.*
M. J. Thomson.
Q. J. R. Astron. Soc., Vol. 33, No. 2, p. 59 – 81 (Jun 1992).

002.127   Astronomy and Astrophysics Abstracts.
   Vol. 53 A, B. Literature 1991, Part 1.
G. Burkhardt, U. Esser, H. Hefele, I. Heinrich, W. Hofmann, D. Krahn, V. R. Matas, L. D. Schmadel, R. Wielen, G. Zech (eds.).
Springer, Berlin (Germany). 1635 p. (1992). ISBN 3–540–55314–2. Price DM 498.00 [Subscription Price DM 398.40]. ISBN 0–387–55314–2 (USA), Price US$ 349.00 [Subscription Price US$ 279.20].
   A publication of the Astronomisches Rechen–Institut Heidelberg produced in cooperation with the Fachinformationszentrum Karlsruhe.

002.128   Asteroid Photometric Catalogue. Second update.
   C.–I. Lagerkvist, M. A. Barucci, M. T. Capria, M. Dahlgren, A. Erikson, M. Fulchignoni, P. Magnusson.
Astronomiska Observatoriet, Uppsala (Sweden). 180 p. (1992). ISBN 91–506–0895–9. Price SEK 500.00.
   The original version of the Asteroid Photometric Catalogue (Lagerkvist et al., 1987) and the first update (Lagerkvist et al., 1988) contained about 3500 lightcurves of more than 500 asteroids. This updating of the catalogue contains almost 1000 lightcurves of 158 asteroids published during 1988 – 1990.

002.129   Catalogue of proper motions in NGC 752.
   I. Platais.
Bull. Inf. Cent. Données Astron. Strasb., No. 40, p. 5 – 6 (Jan 1992).

002.130   Catalogue of proper motions in NGC 7209.
   I. Platais.
Bull. Inf. Cent. Données Astron. Strasb., No. 40, p. 7 – 8 (Jan 1992).

002.131   Second general catalogue of stars observed with the photoelectric astrolabes (GCPA2)
Lu Lizhi.
Bull. Inf. Cent. Données Astron. Strasb., No. 40, p. 9 (Jan 1992).

002.132   Fourth preliminary catalogue of stars observed with the photoelectric astrolabe (PACP4).
Lu Lizhi.
Bull. Inf. Cent. Données Astron. Strasb., No. 40, p. 11 (Jan 1992).

002.133   Machine–readable version of the Parenago catalogue of stars in the area of the Orion nebula.
O. Yu. Malkov.
Bull. Inf. Cent. Données Astron. Strasb., No. 40, p. 13 – 14 (Jan 1992).

002.134   General catalogue of variable stars, fourth edition, Vol. IV.
P. N. Kholopov, N. N. Samus', O. V. Durlevich, E. V. Kazarovets, N. N. Kireeva, T. M. Tsvetkova.
Bull. Inf. Cent. Données Astron. Strasb., No. 40, p. 15 – 17 (Jan 1992).

002.135   Fourth catalogue of Am stars with known spectral types.
   B. Hauck.
Bull. Inf. Cent. Données Astron. Strasb., No. 40, p. 19 – 21 (Jan 1992).

002.136   A general catalogue of Ap and Am stars.
   P. Renson.
Bull. Inf. Cent. Données Astron. Strasb., No. 40, p. 23 – 29 (Jan 1992).

002.137   Bibliographic catalogue of variable stars (BCVS): second supplement.
S. Rössiger.
Bull. Inf. Cent. Données Astron. Strasb., No. 40, p. 39 (Jan 1992).

002.138   A list of faint photometric sequences (1980 – 1990) based on photoelectric and CCD measurements.
B.–M. Ritzmann.
Bull. Inf. Cent. Données Astron. Strasb., No. 40, p. 41 (Jan 1992).

002.139   A bibliography of radio surveys that include maps or atlases, 1955 – 1989.
E. C. Campbell.
Bull. Inf. Cent. Données Astron. Strasb., No. 40, p. 43 – 70 (Jan 1992).

002.140   Some revisions to the Bright Star Catalogue and its supplement.
A. R. Peters, D. E. Hoffleit.
Bull. Inf. Cent. Données Astron. Strasb., No. 40, p. 71 – 93 (Jan 1992).

002.141   The database for stars in open clusters. II. A progress report on the introduction of new data.
J.–C. Mermilliod.
Bull. Inf. Cent. Données Astron. Strasb., No. 40, p. 115 – 120 (Jan 1992).
   The development of the database for stars in open clusters has been continued. 238793 new data have been entered in the database during the last three years concerning 212128 stars. The emphasis has been put on coordinates, and (x,y) positions, but a large number of UBV photographic and CCD data have also been added. The command soft–ware and graphics have been improved.

002.142   Building the astronomical semantic network for ESIS.
   D. Egret, S. G. Ansari, L. Denizman, A. Preite–Martinez.
Bull. Inf. Cent. Données Astron. Strasb., No. 40, p. 121 – 126 (Jan 1992).
   The European Space Information System (ESIS) is a project of the European Space Agency to integrate space mission archives, scientific databases and other archive services. The authors present an overview of the astronomical semantic network, an integrating layer tailored to astronomical needs, and point out what methods were used to achieve this.

002.143   Electronic publishing and intelligent information retrieval.
A. Heck.
Bull. Inf. Cent. Données Astron. Strasb., No. 40, p. 127 – 131 (Jan 1992).

002.144   Introducing the Star's Family.
   A. Heck.
Bull. Inf. Cent. Données Astron. Strasb., No. 40, p. 139 – 141 (Jan 1992).

002.145   SIMBAD: new features (1).
   D. Egret.
Bull. Inf. Cent. Données Astron. Strasb., No. 40, p. 147 – 150 (Jan 1992).
   This paper is the first one of a series which aims at describing the current developments of the SIMBAD database. In this first issue the author presents some new features which have been developed recently, and are therefore not documented in the User's Guide.

002.146   New catalogues and files available at the CDS.
   F. Ochsenbein.
Bull. Inf. Cent. Données Astron. Strasb., No. 40, p. 151 – 158 (Jan 1992).

002.147   **Addendum: Physical data of the fundamental stars [Publ. Beijing Astron. Obs., No. 16, p. 18 – 30 (Dec 1990)].**
Luo Dingjiang, Zhang Baocai.
Publ. Beijing Astron. Obs., No. 18, p. 92 – 106 (Oct 1991). See Abstr. 53.002.130.

002.148   **American Astronomical Society 1993 Membership Directory.**
American Astronomical Society (AAS), AAS Executive Office, Washington, DC (USA). 224 p. (1992). Price US$ 12.00 (USA), US$ 20.00 (overseas).

002.149   **La communication en astronomie.**
A. Nath.
Orion, Jahrg. 50, Nr. 248, p. 66 – 70 (Apr 1992).

002.150   **Sur la visibilité de la recherche (suite et fin).**
C. Jaschek.
J. Astron. Fr., No. 41, p. 23 – 24 (Feb 1992).

002.151   **Kiso survey for ultraviolet–excess galaxies. XV.**
B. Takase, N. Miyauchi–Isobe.
Publ. Natl. Astron. Obs. Jpn., Vol. 2, No. 3, p. 399 – 429 (1992).
Presented here are the fifteenth list and identification charts of the ultraviolet–excess galaxies which have been detected on the multi–color plates taken with the Kiso Schmidt telescope for 10 survey fields. In the sky area of some 300 square degrees 544 objects are catalogued down to the photographic magnitude of about 17.5.

002.152   **The Tokyo PMC catalog 88: catalog of positions of 3800 stars observed in 1988 and planetary positions observed in 1986 to 1988 with Tokyo Photoelectric Meridian Circle.**
M. Yoshizawa, S. Suzuki, M. Sôma.
Publ. Natl. Astron. Obs. Jpn., Vol. 2, No. 3, p. 475 – 546 (1992).
The fourth annual catalog of the Tokyo PMC is presented for 3800 stars (Part I and Part II) which had been observed at least two times in the 1988 period, that is, from January 1, 1988 to December 31, 1988. The positions of the stars given in the catalog are those at the mean epoch of observations. The coordinates of the catalog are based on the FK5 system, and are referred to the equinox and equator of J2000.0. The positions of five major and nine minor planets are also given (Part III) for all individual observations made in 1986 to 1988.

002.153   **Bibliography.**
E. B. Carling, Z. Kopal, S. H. Tellier (eds.).
Earth, Moon, Planets, Vol. 57, No. 3, p. 231 – 284 (Jun 1992).
Contents: 14. Uranus. 15. Satellites of Uranus. 16. Venus. 18. Space exploration. 19. Asteroids. 20. Comets. 21. Meteorites. 22. Miscellaneous.

002.154   **The Case Low–Dispersion Northern Sky Survey. XIII. A region in central Boötes and Corona Borealis.**
C. B. Stephenson, P. Pesch, D. J. MacConnell.
Warner and Swasey Observatory, Cleveland, OH (USA). vp. (1991).
Positions, estimated magnitudes, and finding charts (when necessary) are provided for 138 blue and/or emission–line galaxies, 4 H II regions in 2 galaxies, 83 unresolved blue and/or emission–line objects, including QSO candidates, and 36 known and suspected blue stars in a ~ 100 deg^2 region in central Boötes and Corona Borealis encompassed primarily in the region $14^h08^m <$ R.A. $< 16^h10^m$ and $+29°30' <$ DEC $< +33°30'$ (1950). The objects, whose blue magnitudes are mostly within the range 15 – 18, were identified on low–dispersion objective–prism plates taken with the Burrell Schmidt telescope at Kitt Peak.

002.155   **Composition and analysis of the GZ listing from the NRAO survey.**
V. I. Zhuravlev, M. G. Larionov.
Astron. Zh., Tom 69, Vyp. 3, p. 449 – 460 (May – Jun 1992). In Russian. English translation in Sov. Astron., Vol. 36, No. 3 (May – Jun 1992).
From the continuum radio maps of the northern sky survey, carried out at 4,85 GHz at NRAO 91–m telescope, the version of the GZ catalogue was obtained, containing 89 974 discrete radio sources stronger than 20 mJy. The comparison was made between the sources of the GZ survey with the Z2 Zelenchuk catalogue and other surveys at different frequencies. The coordinates precision and the confusion for Z2 survey were estimated. The information on the distribution of spectral indices for a number of catalogues relative to the GZ listing was obtained.

002.156   **The Einstein Observatory data bases.**
G. Fabbiano.
Symposium on Imaging X–Ray Astronomy – a Decade of Einstein Observatory Achievements, p. 307 (1990). Abstract. – See Abstr. 012.097 for the main entry.

002.157   **The Einstein Observatory Source Catalog.**
D. E. Harris, W. Forman, I. M. Gioia, J. A. Hale, F. R. Harnden Jr., C. Jones, T. Maccacaro, J. D. McSweeney, F. A. Primini, J. Schwarz, H. D. Tananbaum.
Symposium on Imaging X–Ray Astronomy – a Decade of Einstein Observatory Achievements, p. 309 – 311 (1990). – See Abstr. 012.097 for the main entry.
Contents: The IPC Source Catalog. A multi–volume set. Description of a catalog page. The catalog in FITS format. The on–line version of the catalog.

002.158   **The Einstein Observatory Stellar X–ray Database.**
F. R. Harnden Jr., S. Sciortino, G. Micela, A. Maggio, G. S. Vaiana, J. H. M. M. Schmitt.
Symposium on Imaging X–Ray Astronomy – a Decade of Einstein Observatory Achievements, p. 313 – 317 (1990). – See Abstr. 012.097 for the main entry.
The authors present the motivation for and methodology followed in constructing the Einstin Observatory Stellar X–ray Database from a uniform analysis of nearly 4000 Imaging Proportional Counter fields obtained during the life of this mission. This project has been implemented using the INGRES™ database system, so that statistical analyses of the properties of detected X–ray sources are relatively easy and flexibly accomplished.

002.159   **The Einstein normal galaxy database.**
G. Fabbiano, G. Trinchieri, S. Hazelton, D.–W. Kim.
Symposium on Imaging X–Ray Astronomy – a Decade of Einstein Observatory Achievements, p. 323 – 326 (1990). – See Abstr. 012.097 for the main entry.
An X–ray catalog and a data base for normal galaxies surveyed with the Einstin Observatory are under construction and will be made available for use by the astronomical community at large. The X–ray database contains information for all the galaxies including detections and upper limits. The data reduction method is described.

002.160   **The Einstein database of optically and radio selected quasars.**
B. J. Wilkes, H. Tananbaum, Y. Avni, M. S. Oey, D. M. Worrall.
Symposium on Imaging X–Ray Astronomy – a Decade of Einstein Observatory Achievements, p. 327 – 330 (1990). – See Abstr. 012.097 for the main entry.
The Einstein quasar database will provide soft X–ray count rates or upper limits, fluxes, and luminosities for all previously known, optically– or radio–selected quasars observed with the Imaging Proportional Counter aboard the Einstein observatory. Data for all the quasars will be available in published hardcopy,

on–line via computer network, and in a form suitable for distribution (e.g. tape, CDROM).

002.161   The Einstein all–sky slew survey data base.
M. Elvis, D. Plummer, G. Fabbiano.
Symposium on Imaging X–Ray Astronomy – a Decade of Einstein Observatory Achievements, p. 331 – 332 (1990). Abstract. Abstract. – See Abstr. 012.097 for the main entry.

002.162   The Einstein Extended Medium Sensitivity Survey database.
I. M. Gioia, T. Maccacaro, S. L. Morris, R. E. Schild, J. T. Stocke, A. Wolter.
Symposium on Imaging X–Ray Astronomy – a Decade of Einstein Observatory Achievements, p. 333 – 336 (1990). – See Abstr. 012.097 for the main entry.
Contents: The survey. Selection Criteria. Optical work. Catalogs and literature search. The catalog.

002.163   The Einstein Observatory Deep Survey database.
F. A. Primini.
Symposium on Imaging X–Ray Astronomy – a Decade of Einstein Observatory Achievements, p. 337 – 338 (1990). – See Abstr. 012.097 for the main entry.
Contents: Introduction. Data Analysis. Data Products. Using the database.

Astrophotography with the Schmidt telescope.
See Abstr. 003.012.

Astrophysical data. Planets and stars.
See Abstr. 003.017.

Desktop publishing in astronomy & space sciences. Proceedings.
See Abstr. 012.053.

Digitised optical sky surveys. Proceedings.
See Abstr. 012.059.

Solar–terrestrial models at the National Space Science Data Center.
See Abstr. 013.027.

AAVSO and variable star observing.
See Abstr. 013.051.

The study of short–period variable stars through international cooperation.
See Abstr. 013.060.

Current and future programmes with the UK Schmidt Telescope.
See Abstr. 013.073.

Digitised sky surveys at the Space Telescope Science Institute.
See Abstr. 013.079.

The Opacity Project – the TOPBASE atomic database.
See Abstr. 013.081.

Archiving data: a must, but how?
See Abstr. 013.117.

Investigators of resources and distribution for astronomical literature in Academia Sinica.
See Abstr. 013.131.

The IUE Final Archive – III.
See Abstr. 036.029.

A new approach to point–pattern matching.
See Abstr. 036.053.

Early improvements to the Hipparcos Input Catalogue through the accumulation of data from the satellite. Including the NDAC attitude reconstruction description.
See Abstr. 036.096.

Tycho star recognition.
See Abstr. 036.104.

Tycho astrometry calibration.
See Abstr. 036.105.

Tycho photometry calibration and first results.
See Abstr. 036.107.

Optical searches for ROSAT X–ray cluster sources.
See Abstr. 036.115.

Developments in compression techniques for COSMOS/Super-COSMOS data.
See Abstr. 036.152.

A comparison of star/galaxy classification approaches on digitised POSS–II plates.
See Abstr. 036.160.

Making an astrometric standard in the region of Cygnus.
See Abstr. 041.005.

Results of observations made with the seven–inch transit circle 1967 – 1973.
Observations of the Moon and minor planets.
Catalog of 23,001 stars for 1950.0.
Comparisons with FK4.
See Abstr. 041.008.

Catalog–to–catalog reductions: results for the FK, N30, and GC catalogs.
See Abstr. 041.013.

The Hipparcos observing programme: preparation of the input catalogue.
See Abstr. 051.059.

Hipparcos data reduction – construction of the Hipparcos star catalogue.
See Abstr. 051.060.

Early scientific results from Hipparcos and future expectations.
See Abstr. 051.061.

A catalogue of the observations of the mutual phenomena of the Galilean satellites of Jupiter made in 1985 during the PHEMU85 campaign.
See Abstr. 099.027.

PosDat – the positional meteor database of the IMO.
See Abstr. 104.030.

Radial velocities from objective prism plates.
See Abstr. 111.016.

UBV corrections for bright stars in SIMBAD.
See Abstr. 113.018.

Spectroscopy and spectral types for 387 stellar objects from the Large, Bright QSO Survey.
See Abstr. 114.016.

A master catalogue of equivalent widths of the interstellar 217 nm band.
See Abstr. 114.058.

UBVRI distances and metallicities for a sample of late–type HIPPARCOS stars.
See Abstr. 115.008.

Speckle observations of visual and spectroscopic binaries. III.
See Abstr. 120.030.

Photoelectric observations of Cepheids in 1991.
See Abstr. 122.053.

Photoelectric observations of Cepheids in 1991.
See Abstr. 122.108.

The American Association of Variable Star Observers (AAVSO) photoelectric photometry archive.
See Abstr. 123.019.

Einstein observations of supernova remnants.
See Abstr. 125.080.

Galactic worms. I. Catalog of worm candidates.
See Abstr. 131.103.

Low mass star formation in Orion.
See Abstr. 131.113.

Interferometer phase calibration sources. I. The region $35° \leqslant \delta \leqslant 75°$.
See Abstr. 141.002.

The spectral archive of cosmic X–ray sources observed by the Einstein Observatory focal plane crystal spectrometer.
See Abstr. 142.018.

Liste des étoiles Ap et Am dans les amas ouverts (édition révisée).
See Abstr. 153.051.

Separation of halo and thick disc stars in two catalogs.
See Abstr. 155.057.

Morphological and spatial distributions of high velocity molecular outflows.
See Abstr. 155.083.

Carbon stars in the Small Magellanic Cloud: positions, finding charts and spectrophotometry.
See Abstr. 156.023.

New planetary nebulae in the Large Magellanic Cloud.
See Abstr. 156.029.

A catalog of low surface brightness galaxies. List II.
See Abstr. 157.071.

A near–infrared imaging survey of interacting galaxies: the small angular–size Arp systems.
See Abstr. 157.168.

A spectrophotometric atlas of galaxies.
See Abstr. 157.169.

Nonstellar objects in M33.
See Abstr. 157.283.

Normal galaxies. IUE–ULDA Access Guide Vol. 3. International Ultraviolet Explorer – Uniform Low Dispersion Archive.
See Abstr. 157.286.

Abell clusters of galaxies: catalogs and large–scale structures.
See Abstr. 160.063.

The APM bright galaxy surveys.
See Abstr. 160.108.

The APM cluster redshift survey.
See Abstr. 160.109.

The ROE/NRL cluster catalog: I. Correlation with southern Abell clusters.
See Abstr. 160.111.

The ROE/NRL cluster catalog: II. Corelation with serendipitous Einstein X–ray sources.
See Abstr. 160.112.

The model of two points correlation functions for the determination of the large scale structure of the Universe (I).
See Abstr. 161.509.

003 Books

003.001   **The early universe. Facts and fiction.**
G. Börner.
Texts Monogr. Phys.
Springer, Berlin (Germany). 2. enl. ed. 478 p. (1992). ISBN 3–540–54656–1. Price DM 88.00. ISBN 0–387–54656–1 (USA).

This book (for the first edition 1988 see 46.003.028) is concerned with the connections which have been developed in recent years between particle physics and cosmology. The standard big bang model and modifications introduced by the fundamental interactions of elementary particles, such as inflation baryon synthesis, are described. The author presents an up–to–date picture of this rapidly changing field, taking care to sort out speculations from well–established results.
Contents:
Part I. The standard big–bang model.
1. The cosmological models. 2. Facts – observations of cosmological significance. 3. Thermodynamics of the early universe in the classical hot–big–bang picture. 4. Can the standard model be verified experimentally?
Part II. Particle physics and cosmology.
5. Gauge theories and the standard model. 6. Grand unification schemes. 7. Relic particles from the early universe. 8. Baryon synthesis. 9. The inflationary universe.

Part III. Dark matter and galaxy formation.
10. Typical scales – from observation and theory. 11. The evolution of small perturbations. 12. Computer simulations and the large–scale structure.
Appendix. Recent developments.

003.002   **Spacetime and gravitation.**
W. Kopczyński, A. Trautman.
Wiley, Chichester (UK). 176 p. (1992). ISBN 0–471–92186–6. Price US$ 62.95.Revised translation of "Czasoprzestrzeń i grawitacja", Warszawa (Poland), Państwowe Wydawnictwo Naukowe, 1984.

This book is a revised translation of the Polish original "Czasoprzestrzeń i grawitacja", Warszawa (Poland), Państwowe Wydawnictwo Naukowe, 1984.
Ideas about space and time are at the root of one's understanding of nature, both at the intuitive level of everyday experience and in the framework of sophisticated physical theories. These ideas have led to the development of geometry and its applications to physics. The contemporary physical theory of space and time, including its extention to the phenomena of gravitation, is Einstein's theory of relativity. The book is a short introduction to

this theory. A great deal of emphasis is given to the geometrical aspects of relativity theory and its comparison with the Newtonian view of the world. There are short chapters on the origins of Einstein's theory, gravitational waves, cosmology, spinors and the Einstein–Cartan theory.

003.003 Knaurs moderne Astronomie.
Das Standardwerk völlig neu bearbeitet.
H. J. Störig.
Droemer Knaur, München (Germany). 3. rev. ed. 310 p. (1992). ISBN 3–426–26462–5. Price DM 48.00.

This elementary textbook covers the following topics: cosmic dimensions, the sun as a star, the solar system, basic knowledge about stars, binary and variable stars, the cosmic windows, birth and death of stars, the Milky Way, extragalactic systems and cosmology.
Contents: 1. Weißt du, wieviel Sterne stehen... 2. Die Sonne als Durchschnittsstern. 3. Das Sonnensystem im Überblick. 4. Grundwissen über Sterne. 5. Doppelsterne und Veränderliche. 6. Fenster zum Weltall. 7. Geburt und Tod der Sterne. 8. Die Milchstraße. 9. Extragalaktische Systeme. 10. Das Weltganze in Raum und Zeit.

003.004 Astronomical masers.
M. Elitzur.
Astrophys. Space Sci. Libr., Vol. 170.
Kluwer, Dordrecht (Netherlands). 365 p. (1992). ISBN 0–7923–1217–1. Price Dfl. 75.00, US$ 45.00, £ 26.00.

One of the most spectacular discoveries of molecular astronomy has been the detection of maser emission. The same radiation that is generated in the laboratory only with elaborate, special equipment occurs naturally in interstellar space. By a fortunate coincidence maser radiation is generated in both star forming regions and the envelopes of late–type stars. Maser emission has also been detected in external galaxies. This book provides an extensive coverage of the interstellar maser phenomenon. A precondition for maser action is departure from thermal equilibrium. The book therefore starts with a detailed coverage of the basic background concepts required for an understanding of line formation and radiative transfer. It goes on to describe the theoretical and phenomenological aspects of interstellar masers, their formation sites and the inversion mechanisms.
Contents: 1. Introduction. 2. Basic background concepts. 3. Astronomical maser radiation. 4. Maser theory. 5. Effects of geometry. 6. Polarization. 7. Pumping. 8. The environment of astronomical masers. 9. OH masers. 10. H_2O masers. 11. SiO masers. 12. Other masers. 13. Extragalactic masers. 14. Masers as astronomical tools.

003.005 Totality. Eclipses of the Sun.
M. Littmannn, K. Willcox.
University of Hawaii Press, Honolulu, HI (USA). 240 p. (1991). ISBN 0–8248–1371–5. Price US$ 10.47.

This book provides the reader a historical account of eclipses from the earliest times through recent scientific research together with a clear description of the mechanics of eclipses and their place in modern solar research.
Contents: 1. The experience of totality. 2. The great celestial cover–up. 3. A quest to understand. 4. Eclipses in mythology. 5. Strange behavior of man and beast. 6. Anatomy of the Sun. 7. Lessons from eclipses. 8. A magic shadow show. 9. Modern scientific uses for eclipses. 10. Observing a total eclipse. 11. Observing safely. 12. Eclipse photography. 13. The pedigree of an eclipse. 14. The eclipse of July 11, 1991. Appendix A: Coming attractions – total eclipses, 1990 – 2052 – total eclipses and annular eclipses for the United States, 1990 – 2052. Appendix B: Maps of eclipse paths, 1986 – 2052. Appendix C: Chronology of discoveries about the Sun from eclipses.

003.006 Der Himmel auf Erden. Die Welt der Planetarien.
L. Meier.
Barth, Leipzig (Germany). 159 p. (1992). ISBN 3–335–00279–2. Price DM 68.00.
Contents: 1. Himmelsmaschinen aus zwei Jahrtausenden. 2. Erfindung und Entwicklung des Projektionsplanetariums.

3. Astronomie im Spiegel der Planetariumstechnik. 4. Planetarien im Einsatz.

003.007 Bau und Physik der Galaxis.
H. Scheffler, H. Elsässer.
BI–Wissenschaftsverlag, Mannheim (Germany). 2., überarb. Aufl. 568 p. (1992). ISBN 3–411–14402–5. Price DM 98.00.

This book is the second revised and enlarged edition of the 1982 version (see 32.003.108).
Contents: Einleitender Überblick. 1. Positionen, Bewegungen und Entfernungen der Sterne – Begriffe und Methoden. 2. Aufbau und Kinematik des Sternsystems. 3. Interstellare Phänomene. 4. Physik der interstellaren Materie. 5. Dynamik der Galaxis.

003.008 Das Echo des Urknalls. Kernfragen der modernen Kosmologie.
D. Overbye.
Droemer Knaur, München (Germany). 544 p. (1991). ISBN 3–426–26267–3. Price 48.00.

This book is a German translation of the American original *"Lonely hearts of the cosmos"*.

003.009 Gauge gravitation theory.
G. Sardanashvili, O. Zakharov.
World Scientific, Singapore (Singapore). 129 p. (1992). ISBN 981–02–0799–9. Price £ 22.00.

The gauge theory as presented by the authors incorporates Einstein's gravity into the universal picture of fundamental interactions and clarifies its physical nature as a Higgs field. A key point advanced here is the spontaneous breakdown of space–time symmetries according to the equivalence principle.

003.010 Special relativity.
N. M. J. Woodhouse.
Lect. Notes Phys., New Ser. m, Monogr., Vol. 6.
Springer, Berlin (Germany). 94 p. (1992). ISBN 3–540–55049–6. Price DM 54.00. ISBN 0–387–55049–6 (USA).

The course on which this book is based was given to students with a strong mathematical background who already had a good grounding in classical mathematical physics, but who had not yet met relativity. The emphasis is on the use of coordinate–free and tensorial methods. The author has tried to avoid the traditional arguments based on the standard Lorentz transformation and to encourage students to look at problems from a four–dimensional point of view.

003.011 Karl Schwarzschild. Gesammelte Werke. Band 1. *(Karl Schwarzschild. Collected works. Volume 1.)*
H.–H. Voigt (ed.).
Springer, Berlin (Germany). 516 p. (1992). ISBN 3–540–52455–X. Price DM 278.00. ISBN 0–387–52455–X (USA).

This volume is the first part of a three–volume collection of research papers by Karl Schwarzschild. The complete collection will comprise a total of 119 papers, organized in ten categories. Karl Schwarzschild was one of the most important German astronomers and was, in addition, one of the founders of modern astrophysics. His personal papers are held at the Niedersächsische Staats– und Universitätsbibliothek in Göttingen, and an extensive collection of reprints can be found in the Observatory in Göttingen. The material presented in these volumes is restricted intentionally to Schwarzschild's published works.
Contents: Biography of Karl Schwarzschild (1873 – 1916).
1. Celestial mechanics. Commentary by R. Dvorak.
2. Sun and stellar atmospheres. Commentary by W. Mattig.
3. Cometary tails. Commentary by K. Jockers.
4. Structure, kinematics and dynamics of stellar systems. Commentary by R. Wielen.

003.012 Astrophotography with the Schmidt telescope.
S. Marx, W. Pfau.
Cambridge University Press, Cambridge (UK). 174 p. (1992). ISBN 0–521–39549–6. Price £ 30.00, US$ 59.95.

This book is an English translation by P. Lamble of the German original *"Himmelsfotographie mit Schmidt Teleskopen"* published in 1990 (see 51.003.072).

It reviews the history and achievements of the Schmidt telescope. It includes a portfolio of 43 photographic plates produced by large Schmidt telescopes, in particular the Karl Schwarzschild Observatory, Tautenburg. The book contains photographs of galactic nebulae of different species, galactic and globular clusters of stars, stellar fields, external galaxies and comets.
Contents: 1. Astronomy as a natural science. 2. The construction of astronomical telescopes. 3. The Schmidt telescope. 4. Astronomical applications of Schmidt telescopes. 5. The life and work of Bernhard Schmidt. 6. Astronomical objects in words and pictures.

003.013   **The physical basis of the direction of time.**
H.–D. Zeh.
Springer, Berlin (Germany). 2. ed. 198 p. (1992). ISBN 3–540–54884–X. Price DM 68.00. ISBN 0–387–54884–X (USA).
The author investigates the most important classes of physical phenomena that characterize the arrow of time, discussing their interrelations as well as striving to uncover a cosmological common root of the phenomena, such as the time–independent wave function of the universe. The description of irreversible phenomena is shown to be fundamentally "observer–related".

003.014   **Belye i chernye dyry vo Vselennoj. (White and black holes in the Universe.)**
A. P. Trofimenko.
Neobychnoe v Obychnom.
Universitetskoe, Minsk (Belarus). 174 p. (1991). ISBN 5–7855–0193–7. Price 3 Rbl.
This book is the first one in which not only black holes but also white and grey holes are considered. The author's original interpretation of white and black holes is presented, which is founded on many–world representations in the many–dimensional Universe. Possible remnants of white and grey holes explosions can be quasars, active galactic nuclei, rays of super–high energy, cosmic voids and other non–stationary objects. White holes are thought to be principally new and most powerful energy sources in the Universe. Another version of black bodies formation is proposed according to which they are relics of grey holes. Two new mechanisms of chemical elements formation in thermonuclear reactions in white holes and magmatic cells, where matter is heated to high temperatures by black hole radiation, are pointed out.

003.015   **Stars.**
J. B. Kaler.
Sci. Am. Libr. Ser., No. 39.
Scientific American Library, New York, NY (USA). 280 p. (1992). ISBN 0–7167–5033–3. Price US$ 32.95, £ 17.95.
The author explores the nature of stars, describing our current knowledge of their origin, variety, distribution, composition, and distinctive histories. He demonstrates that stars are the key to our comprehension of how the universe evolved – that the development, death, and birth of stars is intimately associated with our own origins and continued existence. From the calculations of Eratosthenes and Aristarchus to recent estimates of galactic motion and theories about dark matter, the book charts the development of the science of astronomy.
Contents: 1. From ancient wonders. 2. The tools of discovery. 3. The discovery of reality. 4. To build a star. 5. Coming of age. 6. Catastrophe. 7. First light of day.

003.016   **Introduction to stellar astrophysics. Volume 3. Stellar structure and evolution.**
E. Böhm–Vitense.
Cambridge University Press, Cambridge (UK). 300 p. (1992). ISBN 0–521–34404–2. Price £ 40.00, US$ 64.95 (cloth). ISBN 0–521–34871–4, Price £ 13.95, US$ 27.95 (paper).
This text discusses the internal structure and the evolution of stars. It emphasises the basic physics governing stellar structure and the basic ideas on which our understanding of stellar structure is founded. The book also provides a comprehensive discussion of stellar evolution. Careful comparison is made between theory and observation, and the author has thus provided a lucid and balanced introductory text for the student. For Volume 1 in this series *Basic stellar observations and data* see 49.003.091, Volume 2: *Stellar atmospheres* see 49.003.094. The present volume is the final one in this short series of books which together provide a modern, complete and authoritative account of our present knowledge of the stars.
Contents: 1. Introduction. 2. Hydrostatic equilibrium. 3. Thermal equilibrium. 4. The opacities. 5. Convective instability. 6. Theory of convective energy transport. 7. Depths of the outer convection zones. 8. Energy generation in stars. 9. Basic stellar structure equations. 10. Homologous stars in radiative equilibrium. 11. Influence of convection zones on stellar structure. 12. Calculation of stellar models. 13. Models for main sequence stars. 14. Evolution of low mass stars. 15. Evolution of massive stars. 16. Late stages of stellar evolution. 17. Observational tests of stellar evolution theory. 18. Pulsating stars. 19. The Cepheid mass problem. 20. Star formation.

003.017   **Astrophysical data. Planets and stars.**
K. R. Lang.
Springer, New York, NY (USA). 947 p. (1992). ISBN 0–387–97109–2. ISBN 3–540–97109–2, Price DM 118.00.
This volume collects an enormous amount of reference information ranging from fundamental physical constants to orbital data for selected asteroids, comets, meteor streams, and satellites. The chapters on the Earth and planets include, for example, data on magnetic fields, masses, sizes, rotation, orbits, and rings. Other chapters cover the Sun, stars in general, bright and nearby stars. Wolf–Rayet stars, magnetic stars and active stars. Larger objects covered include star clusters and associations, regions of star formation, emission nebulae, reflection nebulae, planetary nebulae, and supernova remnants. Also included are data on white dwarf stars, pulsars, candidate black holes, interacting binary systems, symbiotic stars, supernovae, and X–ray and gamma–ray sources.

003.018   **High energy astrophysics. Volume 1. Particles, photons and their detection.**
M. S. Longair.
Cambridge University Press, Cambridge (UK). 2. ed. 435 p. (1992). ISBN 0–521–38374–9. Price £ 45.00, US$ 69.95 (cloth). ISBN 0–521–38773–6, Price £ 16.95, US$ 34.95 (paper).
High energy astrophysics is one of the most exciting areas of contemporary astronomy, covering the most energetic phenomena in the universe. The highly acclaimed first edition of the author's book immediately established itself as an essential textbook on high energy astrophysics. In this complete revision, the subject matter has expanded to the point where two volumes are desirable. In the first a thorough treatment is given of the physical processes that govern the behaviour of particles in astrophysical environments such as interstellar gas, neutron stars, and black holes. Special emphasis is placed on how observations are made in high energy astrophysics and the limitations imposed on them. The tools of the astronomer and high energy astrophysicist are introduced in the context of specific astronomical problems. The material in Volume 1 leads to a study of all kinds of high energy phenomena in the Galaxy and the universe, given in the second volume.

003.019   **Introduction to Hamiltonian dynamical systems and the N–body problem.**
K. R. Meyer, G. R. Hall.
Appl. Math. Sci., Vol. 90.
Springer, New York, NY (USA). 304 p. (1992). ISBN 0–387–97637–X. ISBN 3–540–97637–X, Price DM 98.00.
This book develops the basic theory of Hamiltonian differential equations from a dynamical systems point of view. That is, the solutions of the differential equations are thought of as curves in a phase space and it is the geometry of these curves that is the important object of study. The analytic underpinnings of the subject are developed in detail. The last chapter on twist maps has a more geometric flavor. The main example developed in the text is the classical N–body problem, i.e., the Hamiltonian system of differential equations which describes the motion of N point

masses moving under the influence of their mutual gravitational attraction. Many of the general concepts are applied to this example. But this is not a book about the N–body problem for its own sake. The N–body problem is a subject in its own right which would require a sizable volume of its own. Very few of the special results which only apply to the N–body problem are given.

003.020   Physics of the plasma universe.
A. L. Peratt.
Springer, New York, NY (USA). 384 p, (1992). ISBN 0–387–97575–6. ISBN 3–540–97575–6, Price DM 148.00.
The explosive growth in plasma physics has been concentrated mainly in two research areas: work directed toward controlled nuclear fusion, and work in space physics. This book addresses the growing need to apply these complementary discoveries to astrophysics. Today many scientists recognize plasma as the key element to understanding the generation of magnetic fields in planets, stars, and galaxies; the behavior of stellar atmospheres and the interstellar medium; the acceleration and transport of cosmic rays; and many other phenomena occurring in radio galaxies, galactic nuclei, quasars, and so forth. These phenomena have only recently been recognized as a unified discipline. The material in this book addresses the known properties of matter in the plasma state and addresses topics in contemporary astrophysics such as magnetism in plasma; electric fields in space plasmas; double layers, synchrotron radiation, and energy transport in plasmas; and particle–in–cell simulation of astrophysical plasmas. Examples of specific problems are included, as well as numerous useful illustrations and appendixes that discuss transmission line concepts in space plasmas, the polarization properties of plasma waves, and dusty and grain plasmas.

003.021   Dust in the galactic environment.
D. C. B. Whittet.
Grad. Ser. Astron.
Institute of Physics, Bristol (UK). 304 p. (1992). ISBN 0–7503–0204–6. Price £ 47.50 (cloth). ISBN 0–7503–0209–7, Price £ 22.50 (paper).
Dust is a ubiquitous feature of the cosmos, impinging directly or indirectly on most fields of modern astronomy. This book covers methods of investigation, important results and their significance, and suggestions for promising avenues of future research. The text is divided into eight chapters, the first of which provides a historical perspective for current research and a review of interstellar environments. The observed properties of interstellar grains are considered in Chapters 2 to 6, beginning, in Chapter 2, with their influence on gas phase abundances. Chapters 3, 4 and 5 examine the extinction, polarization and spectral absorption characteristics of the grains, respectively, and Chapter 6 considers continuum and line emission. In Chapter 7, the author discusses the origin and evolution of the dust, tracing its lifecycle in a succession of environments from circumstellar shells to diffuse interstellar clouds, molecular clouds, protostars and protoplanetary disks. The final chapter reviews progress towards a unified model for interstellar grains.

003.022   Cuno Hoffmeister. Festschrift zum 100. Geburtstag.
S. Marx (ed.).
Barth, Leipzig (Germany). 137 p. (1992). ISBN 3–335–00282–2. Price DM 88.00.
The individual contributions within the subject scope of AAA are included in their corresponding categories – see abstracts 005.014, 005.015, 009.005.

003.023   Himmelsmechanik. Band I: Grundlagen, Determinierung.
M. Schneider.
BI–Wissenschaftsverlag, Mannheim (Germany). 3., neu bearb. Aufl. 670 p. (1992). ISBN 3–411–15223–0. Price DM 78.00.
This volume is the first in a now three volume set of the book *Himmelsmechanik*, published in 1979. This third edition is a fully new and enlarged version.
Contents: Einführung.
Teil I: Himmelsmechanik in der Newtonschen Raumzeit.

Grundlagen: A. Strukturen der Newtonschen Raumzeit. B. Bilanzgleichungen, Materialtheorie, Gravitationswechselwirkung. C. Alternative Fassung von Bewegungsproblemen. Determinierung: A. Determinierung durch Anfangswerte. B. Determinierung durch zeitliche Randwerte. C. Lösungsverhalten in nichtlinearen dynamischen Systemen. Anhänge.

003.024   Mars. Unser geheimnisvoller Nachbar. Vom antiken Mythos zur bemannten Mission.
J. N. Wilford.
Birkhäuser, Basel (Switzerland). 288 p. (1992). ISBN 3–7643–2643–3. Price DM 68.00.
This book is a German translation, by D. Gerstner, S. Khan, of the American original *"Mars beckons"*, published in 1990.
Contents: 1. Verlockung einer anderen Welt. 2. Lowells Mars. 3. Die Marsmenschen kommen! 4. Erste Begegnungen. 5. Die Marsmonde. 6. Viking und die Suche nach Leben auf dem Mars. 7. Erkenntnisse und Fragen – eine Bestandsaufnahme. 8. Die Russen sind bereit. 9. Die Unentschlossenheit der Amerikaner. 10. Gemeinsam zum Mars. 11. Aufbruchstimmung. 12. Wege zum Mars. 13. Die künftigen Marsbewohner. 14. Der Mars – unsere Bestimmung?

003.025   Space sailing.
J. L. Wright.
Gordon and Breach, Philadelphia, PA (USA). 271 p. (1992). ISBN 2–88124–803–9. Price US$ 68.00, £ 40.00. ISBN 2–88124–842–X (paper).
This book offers a clear, concise and fascinating insight into the concepts behind solar sails, covering all aspects of the technology from the basic principles of the solar sail to proposed trajectories for travel into various planetary orbits. The author shows that, in principle, sailing ships are practical, low–cost concepts. Also detailed is the proposed use of powerful lasers to drive sailing ships on interstellar journeys.

003.026   Solitary waves in plasmas and in the atmosphere.
V. Petviashvili, O. Pokhotelov.
Gordon and Breach, Philadelphia, PA (USA). 262 p. (1992). ISBN 2–88124–787–3. Price US$ 90.00, £ 50.00.
This book is an English of the Russian original *"Uedinennye volny v plazme i atmosfere"*. The authors introduce the theory of highly nonlinear phenomena in plasmas and in the atmosphere, and then study the development of these phenomena under the influence of various characteristics of the surrounding media. A great deal of attention is devoted to recent progress in stability studies using the Lyapunov method.

003.027   My universe. Selected papers of Ya. B. Zel'dovich.
B. Ya. Zel'dovich, M. V. Sazhin (eds.).
Harwood Academic Publ., Chur (Switzerland). 267 p. (1992). ISBN 3–7186–5004–5. Price US$ 95.00, £ 52.00. ISBN 3–7186–5238–2, Price US$ 30.00, £ 19.00 (paper).
This book is a selection of papers summarizing the contribution to modern cosmology of the Soviet school, led by the late Ya. B. Zel'dovich. It is translated from the Russian by M. V. Sazhin. The first three articles present the underlying theories of cosmology. These papers were originally published in Russian in the journal Priroda, Nos. 9 (1982), 2 (1983) 8 (1984). The remaining five contributions are review papers offering a deeper understanding of the structure of the universe.
Contents: Preface. Introduction.
Part I: 1. Modern cosmology (*Ya. B. Zel'dovich*). 2. Why the universe is expanding (*Ya. B. Zel'dovich*). 3. Matter and antimatter in the universe (*A. D. Dolgov, Ya. B. Zel'dovich*).
Part II: 4. Cosmological field theory for observational astronomers (*Ya. B. Zel'dovich*). 5. Spontaneous creation of the universe (*A. A. Starobinskij, Ya. B. Zel'dovich*). 6. Gravitational waves in cosmology: sources and detection (*M. V. Sazhin, Ya. B. Zel'dovich*). 7. Intergalactic gas in clusters of galaxies, the microwave background, and cosmology (*R. A. Syunyaev, Ya. B. Zel'dovich*). 8. Structure of the universe (*Ya. B. Zel'dovich*).

003.028 **The observation and analysis of stellar photospheres.**
D. F. Gray.
Camb. Astrophys. Ser., Vol. 20.
Cambridge University Press, Cambridge (UK). 2. ed. 469 p. (1992). ISBN 0–521–40320–0. Price £ 50.00, ISBN 0–521–40868–7, Price £ 25.00 (paper).

The starlight we see comes from the outer layers of a star, from the region known as the photosphere. Most of what we know about stars is learned by studying the light from the photosphere. This book describes the equipment, observational techniques and analysis used in the investigation of stellar photospheres. The opening chapters describe the basic tools, such as spectrographs and light detectors, as well as the physics of radiative transfer and the construction of models. Next the author introduces the measurement and modelling of the continuum spectrum. This is followed by the study of spectral line radiation. The final chapters explain how these techniques enable astronomers to deduce valuable information on basic properties of stars. For example, temperature, radius, surface gravity, chemical composition, rotation rate, and velocity fields can be derived from stellar spectroscopy.

003.029 **Monde und Ringe.**
Redaktion der Time–Life Bücher.
Reise durch das Universum.
Time–Life Bücher, Amsterdam (Netherlands). Authoriz. German ed. 144 p. (1991). ISBN 90–6182–475–3. Price DM 46.00.

Contents: 1. Der Begleiter der Erde. Eine Chronik des Mondes. 2. Im Reich des Mars und Jupiter. Satelliten–Zählung. 3. Die geheimnisvolle Welt der Ringe. Bewegte Systeme.

003.030 **Strukturen des Universums.**
Redaktion der Time–Life Bücher.
Reise durch das Universum.
Time–Life Bücher, Amsterdam (Netherlands). Authoriz. German ed. 144 p. (1992). ISBN 90–6182–476–1. Price DM 46.00.

Contents: 1. Wege zur Vereinheitlichung. Der Ursprung der Materie. 2. Die Suche nach der Quanten–Gravitation. Ein Superstring–Universum. 3. Über die Bedeutung der Zeit. Das Rätsel der Zeitrichtung.

003.031 **Far encounter. The Neptune system.**
E. Burgess.
Columbia University Press, New York, NY (USA). 164 p. (1991). ISBN 0–231–07412–3. Price US$ 34.95.

This book describes the encounter of Voyager 2 with the Neptune system. The author describes the surprisingly active planet, Neptune, its rings, its large satellite Triton, its smaller satellites, and other aspects of the planet. He details Neptune's complex atmospheric structure as well as other unexpected structures and characteristics including the hypothesized internal structure and its gravitational field. A chapter on Triton reveals that it is the only large satellite travelling in a retrograde orbit around its primary and has a surprising pink and blue surface. The author investigates Pluto and its satellite Charon. The book presents over 35 Voyager 2 mission photos from NASA.

003.032 **Das neue Bild vom Sonnensystem.**
H.–M. Hahn.
Franckh–Kosmos, Stuttgart (Germany). 142 p. (1992). ISBN 3–440–06368–2.

Contents: 1. Prolog. 2. Atmosphären. 3. Krater. 4. Vulkane. 5. Innenleben. 6. Monde und Ringe. 7. Kleinkörper. 8. Wasser. 9. Epilog.

003.033 **Spacetime physics. Introduction to special relativity.**
E. F. Taylor, J. A. Wheeler.
Freeman, New York, NY (USA). 2. ed. 319 p. (1992). ISBN 0–7167–2326–3. Price US$ 29.95.

In the 26 years since its initial publication, *Spacetime physics* has remained the most informative and accessible introduction to modern relativity theory. The second edition examines the principles of relativity in the same clear style. Thoroughly revised and updated, the book features an expanded chapter on general relativity and touches on the major aspects of the "new physics", such as – microgravity – collider accelerators – satellite probes – neutrino detectors – radioastronomy – and pulsars.

003.034 **Rendezvous in space. The science of comets.**
J. C. Brandt, R. D. Chapman.
Freeman, New York, NY (USA). 300 p. (1992). ISBN 0–7167–2175–9. Price £ 16.95.

The return of Halley's comet has historically served to reinvigorate comet research. This book conveys our current understanding of comets through an engaging blend of science and history. The primary focus is on the milestones of the modern era of comet research (from 1950 to the present), including Whipple's "dirty snowball" model of the comet nucleus; the innovative probe of comet Giacobini–Zinner; and the remarkable work of "Halley's Armada" – the fleet of spacecraft that observed the comet up close. One will find tips for optimum comet watching, computer programs for plotting comet orbits and positions, instructions for reporting a new comet, a full color photo set and sources for keeping up with on–going and future comet studies.

003.035 **ESO's early history. The European Southern Observatory from concept to reality.**
A. Blaauw.
European Southern Observatory, Garching b. München (Germany). 283 p. (1991). ISBN 3–923524–40–4.

This book tells the early story of the European Southern Observatory. It begins in the early 1950's when leading European astronomers initiated the project and started to search for the best possible observatory site under the southern sky. In 1962, ESO was established by an international convention and a few years later a remote mountain top in the Chilean Atacama desert, La Silla, was acquired. It took another decade to transform this site into the world's largest optical observatory, now serving more than 2000 astronomers in eight member countries. The story of ESO is that of a highly successful European integration in a fundamental field of science, providing European scientists with modern facilities for front–line investigations beyond the capacities of the individual member states.

003.036 **The secret of the universe.**
I. Asimov.
Oxford University Press, Oxford (UK). 254 p. (1992). ISBN 0–19–286144–1. Price £ 6.99.

The author turns his attention to such questions as: How near is the nearest star? How heavy is the Sun? How does the Doppler effect work? and countless others. In addition, he provides an explanation of how mankind first became engaged in business and commerce, and advances his own unique theory on the secret of the universe.

003.037 **An introduction to astrophysical hydrodynamics.**
S. N. Shore.
Academic Press, San Diego, CA (USA). 468 p. (1992). ISBN 0–12–640670–7. Price US$ 49.95, £ 33.00.

Contents: 1. The equations of fluid motion. 2. Viscosity and diffusion. 3. Vorticity and rotation. 4. Shocks. 5. Similarity methods. 6. Magnetic fields in astrophysics. 7. Turbulence. 8. Outflows and accretion. 9. Instabilities. 10. Diagnosis of astrophysical flows.

003.038 **The protection of astronomical and geophysical sites.**
J. Kovalevsky (ed.).
NATO–Committee on the Challenges of Modern Society, Pilot Study, No. 189.
Editions Frontières, Gif-sur-Yvette (France). 216 p. (1992). ISBN 2–86332–109–9. Price FF 290.00, US$ 48.00.

The problem of protecting astronomical and geophysical observatory sites against environmental pollution is long standing and there are, in a number of places, local arrangements to limit such interference. More global guide–lines for specific nuisances have been published. NATO's Committee on the Challenges of Modern Society established a group for a Pilot

Study on the protection of astronomical and geophysical observatory sites. Several meetings of the study group were held between 1986 and 1988. The chapters of this book deal with each technical aspect linked with the object of the study.
The individual contributions within the subject scope of AAA are included in their corresponding categories – see abstracts 013.028, 013.029, 013.030, 013.031, 013.032, 013.033, 013.034.

003.039   **Low mass star formation in southern molecular clouds.**
B. Reipurth (ed.).
ESO Sci. Rep., No. 11(Nov 1991).
European Southern Observatory, Garching (Germany). 207 p. (1992).
The individual contributions within the subject scope of AAA are included in their corresponding categories – see abstracts 131.113, 131.114, 131.115, 131.116, 131.117, 131.118, 131.119, 131.120, 131.121, 131.122.

003.040   **Die Ordnung des Universums. Eine Einführung in die Astronomie.**
K. Rohlfs.
Birkhäuser, Basel (Switzerland). 316 p. (1992). ISBN 3-7643-2706-5. Price DM 58.00.
Contents: 1. Astronomie in der Welt von heute. 2. Der Stoff, aus dem der Himmel ist. 3. Aufbau und Entwicklung der Sterne. 4. Zwischen den Sternen. 5. Die Milchstraße als Sternsystem. 6. Bausteine der Welt im Großen. 7. Kosmologie als Naturwissenschaft. 8. Das Planetensystem. 9. Astrophysik als Naturwissenschaft.

003.041   **Accretion power in astrophysics.**
J. Frank, A. King, D. Raine.
Camb. Astrophys. Ser., Vol. 21.
Cambridge University Press, Cambridge (UK). 2. ed. 310 p. (1992). ISBN 0-521-40863-6. Price £ 19.95, US$ 37.95. ISBN 0-521-40306-5, Price £ 50.00, US$ 79.95 (cloth).
Accretion is recognized as a phenomenon of fundamental importance in astrophysics. This book examines accretion as a source of energy in binary star systems containing compact objects and in active galactic nuclei. The authors assume a basic knowledge of physics in order to describe the physical processes at work in accretion discs. The first three chapters explain why accretion is a source of energy, and then present the gas dynamics and plasma concepts necessary for astrophysical applications. The next three chapters then develop accretion in stellar systems, including accretion onto compact objects. Three further chapters give extensive treatment of accretion in active galactic nuclei, and the concluding chapter describes thick accretion discs. The second edition is a complete revision of the earlier account in 1985 (see 39.003.012). In particular it gives much greater attention to active galaxies and quasars, where the accretion model is now accepted as the central energy source.

003.042   **Causality electromagnetic induction and gravitation. A different approach to the theory of electromagnetic and gravitional fields.**
O. D. Jefimenko.
Electret Scientific Co., Star City, WV (USA). 192 p. (1992). ISBN 0-917406-09-5.
There are two important theories in classical physics, which have not been properly developed to their logical and mathematical conclusion. They are the Faraday–Maxwell theory of electromagnetic induction and Newton's theory of gravitation. Electromagnetic induction is one of the most important physical phenomena. Any misinterpretation or misrepresentation of this phenomenon may weaken the entire electromagnetic theory and may have undesirable practical consequences. Newton's theory of gravitation is the basic working theory of astronomers and other scientists dealing with space exploration and celestial mechanics. Therefore this theory must also be as accurate and complete as possible. What is more, one cannot really judge the significance and value of alternative theories of gravitation without a thorough understanding of all the peculiarities and consequences of Newton's gravitational theory in its most general form.

003.043   **Atomic and molecular spectroscopy. Basic aspects and practical applications.**
S. Svanberg.
Springer Ser. At. Plasmas., Vol. 6.
Springer, Berlin (Germany). 2. ed. 418 p. (1992). ISBN 3-540-55243-X.
This book is a wide–ranging review of modern spectroscopic techniques such as X–ray, photoelectron, optical and laser spectroscopy, and radiofrequency and microwave techniques. On the fundamental side it focuses on physical principles and the impact of spectroscopy on our understanding of the building blocks of matter, while in the area of applications particular attention is given to those in chemical analysis, photochemistry, surface characterization, environmental and medical diagnostics, remote sensing and astrophysics.

003.044   **Dynamics of the Standard Model.**
J. F. Donoghue, E. Golowich, B. R. Holstein.
Camb. Monogr. Part. Phys. Nucl. Phys. Cosmol., Vol. 2.
Cambridge University Press, Cambridge (UK). 558 p. (1992). ISBN 0-521-36288-1.
This book describes the practical techniques for connecting the phenomenology of particle physics with the accepted modern theory known as the "Standard Model". The Standard Model of elementary particle interactions is the outstanding achievement of the past forty years of experimental and theoretical activity in particle physics. This book gives a detailed account of the Standard Model, focussing on the techniques by which the model can produce information about real observed phenomena. The text opens with a pedagogic account of the theory of the Standard Model. Introductions to the essential calculational techniques needed, including effective Lagrangian techniques and path integral methods, are inlcuded. The major part of the text is concerned with the use of the Standard Model in the calculation of physical properties of particles. Rigorous and reliable methods (radiative corrections and nonperturbative techniques based on symmetries and anomalies) are emphasized, but other useful models (such as the quark and Skyrme models) are also described.

003.045   **Understanding catastrophe.**
J. Bourriau (ed.).
The Darwin College Lectures, 1990.
Cambridge University Press, Cambridge (UK). 217 p. (1992). ISBN 0-521-41324-9. Price £ 17.95, US$ 29.95.
This book tries multi–disciplinary communication understanding catastrophe.
Contents: Introduction: Understanding catastrophe (*G. Lloyd*). 1. Supernovae and stellar catastrophe (*R. P. Kirshner*). 2. The extinction of the dinosaurs (*W. Alvarez, F. Asaro*). 3. Darwin and catastrophism (*M. Rudwick*). 4. Evolution and catastrophe theory (*C. Zeeman*). 5. Earthquakes (*C. Vita–Finzi*). 6. Storms and cyclones (*N. Cook*). 7. Famine in history (*P. Garnsey*). 8. The case of consumption (*R. Porter*).

003.046   **Die Geschichte der Astronomischen Gesellschaft gegründet in Lilienthal am 20. September 1800. Die ersten 63 Jahre ihres Bestehens von 1800 bis 1863.**
D. Gerdes.
Heimatverein, Lilienthal (Germany). 128 p. (1990). ISBN 3-927723-06-1.

003.047   **Venus geology, geochemistry, and geophysics. Research results from the USSR.**
V. L. Barsukov, A. T. Basilevsky (*A. T. Bazilevskij*), V. P. Volkov, V. N. Zharkov (eds.).
University of Arizona Press, Tucson, AZ (USA). 436 p. (1992). ISBN 0-8165-1222-1. Price US$ 75.00.
The Venera 15/16 mission provided geophysically understandable radar images of a major portion of the Venus surface. With the background of data from previous and later missions this book summarizes the available data on Venus geology, geochemistry, and geophysics.
The individual contributions within the subject scope of AAA are included in their corresponding categories – see abstracts

093.024, 093.025, 093.026, 093.027, 093.028, 093.029, 093.030, 093.031, 093.032, 093.033, 093.034, 093.035, 093.036, 093.037, 093.038, 093.039, 093.040, 093.041, 093.042, 093.043, 093.044, 093.045, 093.046, 093.047.

003.048   Astrophysics on the threshold of the 21st century.
N. S. Kardashev (ed.).
Gordon and Breach, Philadelphia, PA (USA). 389 p. (1992). ISBN 2–88124–817–9. Price US$ 165.00.
Translated from the Russian by D. F. Smith.
This book is dedicated to Joseph Samuilovich Shklovsky and Solomon Borisovich Pikel'ner who were responsible for some remarkable developments in 20th–century astrophysics. The book covers a wide range of fundamental problems in modern astrophysics (solar and solar system physics, physics of different scale astronomical objects, evolution of the universe and the search for extraterrestrial intelligence). It also contains prospects for the progress of astronomy in the 21st century and some reminiscences about the history of science, particularly in the Soviet Union.
The individual contributions within the subject scope of AAA are included in their corresponding categories – see abstracts 005.022, 005.023, 005.024, 005.026, 013.048, 013.049, 013.065, 013.066, 013.067, 013.068, 013.069, 013.070, 066.223, 080.060, 091.041, 125.061, 131.248, 131.249, 151.071, 155.137, 159.090, 161.395.

003.049   Conceptual problems of quantum gravity.
A. Ashtekar, J. Stachel (eds.).
Einstein Stud., Vol. 2.
Birkhäuser, Boston, MA (USA). 617 p. (1991). ISBN 0–8176–3443–6. ISBN 3–7643–3443–6, Price DM 238.00. Based on the Proceedings of the 1988 Osgood Hill Conference, North Andover, MA (USA), 15 – 19 May 1988.
Contents: 1. Quantum mechanics, measurement, and the universe. 2. The issue of time in quantum gravity. 3. Strings and gravity. 4. Approaches to the quantization of gravity. 5. Role of topology and black holes in quantum gravity.
The individual contributions within the subject scope of AAA are included in their corresponding categories – see abstracts 061.143, 061.144, 061.145, 061.146, 066.290, 066.291, 066.294, 067.228, 067.229.

003.050   Studies in the history of general relativity.
J. Eisenstaedt, A. J. Kox (eds.).
Einstein Stud., Vol. 3.
Birkhäuser, Boston, MA (USA). 480 p. (1992). ISBN 0–8176–3479–7. Based on the Proceedings of the 2. International Conference on the History of General Relativity, Luminy (France), 1988.
This volume is based on the proceedings of the 2. International Conference on the History of General Relativity, Luminy (France), fall 1988. Several of the papers presented are devoted to the reception and development of the general theory of relativity around the world and on the institutional history of general relativity. The discussion of conceptual issues in general relativity constitutes a second issue of this book. The papers in this category concentrate on various themes, concepts and principles that play a role in Einstein's theory of gravitation. A third set of papers deals with more technical aspects of general relativity. The topics range from the pre–history of general relativity to some quite recent issues in general relativity and even extend to the history of relativistic cosmology.

003.051   Recent advances in general relativity. Essays in honor of Ted Newman.
A. I. Janis, J. R. Porter (eds.).
Einstein Stud., Vol. 4.
Birkhäuser, Boston, MA (USA). 278 p. (1992). ISBN 0–8176–3541–6.
This volume is based on the Conference on Recent Advances in General Relativity, Pittsburgh, PA (USA), 3 – 5 May 1990. Classical general relativity, astrophysics, and quantum gravity have increasingly influenced one another in recent years. Black holes have moved to the forefront of astrophysical research, and pulsars have provided the first observational evidence for the

validity of Einstein's prediction of the rate at which gravitational waves would carry energy from a system. New mathematical techniques, introduced in classical relativity, are now being applied in the study of quantum gravity. This collection of high–level review papers with extensive bibliographies surveys these interacting fields of research.
The individual contributions within the subject scope of AAA are included in their corresponding categories – see abstracts 066.304, 066.305, 066.306, 066.307, 066.308, 066.309, 066.310, 067.232.

003.052   Plasmaphysik im Sonnensystem.
K.–H. Glaßmeier, M. Scholer.
BI–Wissenschaftsverlag, Mannheim (Germany). 378 p. (1991). ISBN 3–411–15151–X. Price DM 46.00.
Contents: 1. Die Sonne (*W. Deinzer*). 2. Der Sonnenwind (*R. Schwenn*). 3. Die kosmische Strahlung im Sonnensystem (*G. Wibberenz*). 4. Stoßwellen in stoßfreien Plasmen (*M. Scholer*). 5. Die Erdmagnetosphäre (*W. Baumjohann*). 6. Die Magnetopause der Erdmagnetosphäre (*M. Scholer*). 7. Grundprozesse magnetosphärischer Aktivität (*H. Wiechen, K. Schindler*). 8. Polarlichter (*K. Schlegel*). 9. Die Magnetosphären anderer Planeten im Sonnensystem (*F. M. Neubauer*). 10. Eigenschwingungen planetarer Magnetosphären (*K.–H. Glaßmeier*). 11. Ausbreitung von MHD–Wellen unter inhomogenen Bedingungen (*F. Krummheuer*). 12. Plasmawellen und Welle–Teilchen–Wechselwirkungen (*E. Marsch*). 13. Der Plasmazustand der Atmosphäre (*H. Volland*). 14. Staub–Plasma–Wechselwirkungen (*C. K. Goertz*). 15. Grundlagen der numerischen Plasmasimulation (*J. Raeder*).

003.053   Unsere Sonne – ein rätselhafter Stern? Erkenntnisse und Spekulationen der Astrophysik.
J. Gribbin.
Birkhäuser, Basel (Switzerland). 290 p. (1992). ISBN 3–7643–2683–2. Price DM 58.00.
This book is a German translation, by A. Ehlers, of the English original "*Blinded by the light. The secret life of the Sun*", published in 1991.
Contents: 1. Vorgeschichte. 2. Quelle gewaltiger Energien. 3. Im Innern der Sonne. 4. Zuwenig Geister. 5. Ein verrückter Gedanke. 6. Die atmende Sonne. 7. Die zitternde Sonne. 8. Das Große und das Kleine. 9. Der Beitrag der Supernova.

003.054   Astrophysics of neutron stars.
V. M. Lipunov.
Astron. Astrophys. Libr.
Springer, Berlin (Germany). 335 p. (1992). ISBN 3–540–53568–3. Price DM 168.00. ISBN 0–387–53568–3 (USA).
This book is an English translation, by R. S. Wadhwa, of the Russian original "*Astrofizika nejtronnykh zvezd*" published in 1987 (see 45.003.106).
The reader will find a description of the current state of experimental and theoretical investigations of neutron stars in all their manifestations: radio pulsars, X–ray pulsars, X–ray bursters, transient X–ray sources, and so on. The approach adopted stresses the idea that the astrophysical properties of a neutron star are determined mainly by its interaction with its surroundings, that is, the interaction of the intense intrinsic magnetic field of the neutron star with an accreting plasma. Thus the book treats topics in plasma physics and magnetohydrodynamics.
Contents: 1. Theoretical and observational principles of the astrophysics of neutron stars. 2. Structure of neutron stars. 3. Fluid dynamics of accretion. 4. Classification of neutron stars. 5. Boundaries, magnetospheres of slowly rotating neutron stars. 6. Accreting neutron stars. 7. The "propeller" regime. 8. Ejecting stars. 9. Supercritical regimes. 10. Stars with an anomalously low value of gravimagnetic parameter. 11. Evolution of stars.

003.055   Der Einstein–Turm. Erwin F. Freundlich und die Relativitätstheorie – Ansätze zu einer "dichten Beschreibung" von institutionellen, biographischen und theoriengeschichtlichen Aspekten.
K. Hentschel.
Spektrum, Akademischer Verlag, Heidelberg (Germany). 192 p. (1992). ISBN 3–86025–025–6. Price DM 38.00.

003.056 **Black holes.**
J.-P. Luminet.
Cambridge University Press, Cambridge (UK). 331 p. (1992). ISBN 0–521–40906–3. Price £ 10.95. ISBN 0–521–40029–5, Price £ 30.00.
This book is an English translation, by A. Bullough and A. King, of the French original "*Les trous noirs*", published in 1987 (see 45.003.109).
In answering questions as "Are black holes purely hypothetical objects from the theory of relativity or are they an observable reality?" the author takes the reader on a fabulous voyage through space and time. He explains how stars are born, light up and die. He takes us into the strange world of supernovae, X–ray stars and quasars.

003.057 **Stellar astrophysics.**
R. J. Tayler (ed.).
Grad. Ser. Astron.
Institute of Physics, Bristol (UK). 368 p. (1992). ISBN 0–7503–0200–3. Price £ 37.50.
This volume contains a collection of articles on stellar astrophysics which have appeared in *Reports on Progress in Physics* in the past few years.
Contents: 1. s–process nucleosynthesis – nuclear physics and the classical model (1989, see 50.061.035) (*F. Käppeler, H. Beer, K. Wisshak*). 2. Physics of white dwarf stars (1990, see 52.065.017) (*D. Koester, G. Chanmugam*). 3. Type I supernovae (1990, see 52.125.062) (*J. C. Wheeler, R. P. Harkness*). 4. The supernova 1987A in the Large Magellanic Cloud (1989, see 50.125.146) (*W. Hillebrandt, P. Höflich*). 5. Neutrino astronomy (1992, see 55.061.089) (*Y. Totsuka*). The work on neutrino astronomy hints at where the future may lie. In addition to the importance of neutrino astronomy to stellar evolution, it provides clues on the origin of cosmic rays, cosmology and grand unified theories.

003.058 **Diffraction effects in semiclassical scattering.**
H. M. Nussenzveig.
Montroll Meml. Lect. Ser. Math. Phys., Vol. 1.
Cambridge University Press, Cambridge (UK). 251 p. (1992). ISBN 0–521–38318–8.
Critical effects in semiclassical scattering, in which the standard approximations break down, are associated with forward peaking, rainbows, glories, orbiting and resonances. Besides giving rise to beautiful optical effects in the atmosphere, critical effects have important applications in many areas of physics. However, their interpretation and accurate treatment is difficult. This book deals with the theory of these critical effects. After a preliminary chapter in which the problem of critical effects is posed, the next three chapters on coronae, rainbows and glories are written so as to be accessible to a broader audience. The main part of the book then describes the results obtained from the application of complex angular momentum techniques to scattering by homogeneous spheres. These techniques lead to practically usable asymptotic approximations, and to new physical insights into critical effects. A new conceptual picture of diffraction, regarded as a tunneling effect, emerges. The final two chapters contain brief descriptions of applications to a broad range of fields, including linear and nonlinear optics, radiative transfer, astronomy, acoustics, seismology, and atomic, nuclear and particle physics.

003.059 **Observational astrophysics.**
R. E. White (ed.).
Grad. Ser. Astron.
Institute of Physics, Bristol (UK). 366 p. (1992). ISBN 0–7503–0201–1. Price £ 37.50.
This volume contains a collection of articles on recent results of a number of currently active fields of observational research which have appeared in Reports on Progress in Physics in the past few years.
Contents: 1. Planetary nebulae (1990, see 52.134.062) (*M. Peimbert*). 2. Compact maser sources (1989, see 50.131.070) (*R. J. Cohen*). 3. Intergalactic matter (1991, see 54.160.008)

(*A. C. Fabian, X. Barcons*). 4. Active galactic nuclei (1991, see 53.158.022) (*D. E. Osterbrock*). 5. The detection of high–redshift quasars (1990, see 51.159.017) (*S. J. Warren, P. C. Hewett*). 6. Large–scale motions in the universe: a review (1990, see 51.161.222) (*D. Burstein*).

003.060 **Fernrohre und ihre Meister.**
R. Riekher.
Verlag Technik, Berlin (Germany). 2.stark bearb. Aufl. 443 p. (1990). ISBN 3–341–00791–1. Price DM 56.00.
Contents: 1. Wegbereiter. 2. Die ersten Linsenfernrohre. 3. Die nichtachromatischen Linsenfernrohre. 4. Die ersten Spiegelfernrohre. 5. Die ersten Achromate. 6. Friedrich Wilhelm Herschel. 7. Joseph Fraunhofer und seine Zeit. 8. Die großen Metallspiegelteleskope. 9. Die Epoche der großen Refraktoren. 10. Die Spiegelteleskope in der zweiten Hälfte des 19. Jahrhunderts. 11. Fernrohre mit optischen Besonderheiten. 12. Die Handfernrohre nach 1800. 13. Fernrohre mit konstruktiven Besonderheiten. 14. Sonnenbeobachtungsanlagen und Turmteleskope. 15. Der Siegeszug der Spiegelteleskope. 16. Das 5–m–Spiegelteleskop auf dem Mount Palomar. 17. Die komafreien Spiegelsysteme. 18. Bernhard Schmidt und sein Spiegelsystem. 19. Maksutov und die Meniskusteleskope. 20. Das 2–m–Universal–Spiegelteleskop des Karl–Schwarzschild–Observatoriums und seine Jenaer Nachfolger. 21. Die Spiegelteleskope der 3,5– bis 4–m–Klasse. 22. Das 6–m–Spiegelteleskop im Kaukasus. 23. Das Multiple–Mirror–Teleskop. 24. Teleskope für den Weltraum. 25. Eine neue Generation von Teleskopen für die Zukunft.

003.061 **The Galileo mission.**
C. T. Russell (ed.).
Kluwer Academic Publ., Dordrecht (Netherlands). 622 p. (1992). ISBN 0–7923–1719–X. Price Dfl. 390.00, US$ 225.00, £ 134.00.
Reprinted from Space Science Reviews, Vol. 60, Nos. 1–4 (May 1992).
The articles in this volume are a document of the Galileo mission to Jupiter. The mission overview is the first article; the second is a description of the design of the very complex spacecraft trajectory in relation to the scientific objects. Subsequent articles describe the various investigations planned by the scientific groups. These are divided in three groups. The probe articles are presented first. Then the "magnetospheric" instruments which provide local measurements of the Jovian system are presented. Finally the remote sensing instruments and the radio science investigations are presented.

003.062 **Die Lilienthaler Sternwarte 1781 bis 1818. Machinae Coelestes Lilienthalienses: Die Instrumente. Eine zeitgeschichtliche Dokumentation.**
D. Gerdes.
Heimatverein, Lilienthal (Germany), Simmering, Lilienthal (Germany). 297 p. (1991). ISBN 3–927723–09–6. Price DM 39.00.

003.063 **Introducing Einstein's relativity.**
R. d'Inverno.
Clarendon Press, Oxford (UK). 394 p. (1992). ISBN 0–19–859686–3. Price £ 22.50. ISBN 0–19–859653–7 (cloth).
There is little doubt that Einstein's theory of relativity captures the imagination. Not only has it radically altered the way we view the universe, but the theory has a considerable number of surprises in store. This is especially so in the three main topics of current interest that this book reaches, namely: black holes, gravitational waves, and cosmology. The main aim of this textbook is to provide students with a sound mathematical introduction coupled to an understanding of the physical insights needed to explore the subject. Indeed, the book follows Einstein in that it introduces the theory very much from a physical point of view. After introducing the special theory of relativity, the basic field equations of gravitation are derived and discussed carefully as a prelude to first solving them in simple cases and then exploring the three main areas of application. The book includes numerous illustrative diagrams and exercises (of varying

degrees of difficulty), and as a result this book makes an excellent course for any student coming to the subject for the first time.

003.064   **Basics of interferometry.**
P. Hariharan.
Academic Press, Boston, MA (USA). 230 p. (1992). ISBN 0–12–325218–0. Price US$ 39.95.
This book is an introduction to the use of interferometric techniques for precision measurements in science and engineering.
Contents: 1. Introduction. 2. Interference: a primer. 3. Two-beam interferometers. 4. Light sources. 5. Multiple–beam interference. 6. The laser as a light source. 7. Detectors. 8. Measurements of length. 9. Optical testing. 10. Digital techniques. 11. Macro– and micro–interferometry. 12. Holographic and speckle interferometry. 13. Interferometric sensors. 14. Interference spectroscopy. 15. Fourier–transform spectroscopy. 16. Choosing an interferometer.

003.065   **Fernrohr–Selbstbau. Fenster ins Weltall. Praktische Anleitungen für Freunde des Sternenhimmels und solche, die es werden wollen.**
H. Oberndorfer.
Verlag Sterne und Weltraum Dr. Vehrenberg, München (Germany). 7. überarb. Aufl. 223 p. (1992). ISBN 3–87973–909–9.

003.066   **Theoretical foundations of cosmology. Introduction to the global structure of space–time.**
M. Heller.
World Scientific, Singapore (Singapore). 155 p. (1992). ISBN 981–02–0756–5. Price £ 20.00.
The aim of the present book is to discuss theoretical foundations of cosmology. One of its principal goals is to study the assumptions upon which relativistic cosmology is founded, and, first of all, those assumptions that are implicit in the mathematical structure of cosmology. The organisation of the material is the following. In the first two chapters foundations and assumptions of the contemporary theory of space–time are discussed. The third chapter is a concise repetition of the fibre bundle method, one of the main theoretical tools used in the book. With the help of it the general theory of relativity is presented in the fourth chapter. The method, on the one hand, makes manifest similarities of this theory with gauge theories of contemporary physics and, on the other hand, stresses the peculiarity of Einstein's theory of gravity as a theory of the frame bundle over space–time. The fifth chapter occupies the central place in the book. It poses, to use Einstein's expression, "the cosmological problem", and explicitly formulates the leading idea of the entire book. Although no single formula appears in this chapter, it is by no means a popular exposition: its full understanding presupposes a considerable mathematical and cosmological culture by the reader. Analyses caried out in this chapter (and a program it outlines) are applied to the simplest world models, the so–called Robertson–Walker cosmology, in the sixth chapter.

003.067   **Hundert Jahre Astronomie an der Leopold–Franzens–Universität Innsbruck (1892 – 1992).**
Uni Innsbruck 1669 – 2000 Retrospektiven.
Innsbruck Univ. (Austria). Inst. für Astronomie; Innsbruck Univ. (Austria). Universitätsarchiv. 116 p. (1992).
The individual contributions within the subject scope of AAA are included in their corresponding categories – see abstracts 004.084, 009.025.

003.068   **Chaotic processes in the geological sciences.**
D. A. Yuen (ed.).
IMA Vol. Math. its Appl., Vol. 41.
Springer, New York, NY (USA). 317 p. (1992). ISBN 0–387–97789–9.
This IMA Volume in Mathematics and its Applications is based on the proceedings of a workshop which was an integral part of the 1989 – 90 IMA program on "Dynamical systems and their applications". The workshop was intended for scientific exchanges between earth scientists and mathematical researchers, especially with experts in dynamical systems.

003.069   **Advances in solar–terrestrial science of China.**
Hu Wenrui, Zhang Bairong, Lu Daren (eds.).
Science Press, Beijing (China). 315 p. (1992). ISBN 7–03–003037–0. Price US$ 78.50.
The articles summarize and review the progress of solar–terrestrial sciences in China, especially the contributions over the last five years.
Contents: 1. Generality. 2. Solar processes. 3. Interplanetary and magnetospheric processes. 4. Ionospheric processes. 5. Atmospheric processes. 6. Synthetic analysis.
The individual contributions within the subject scope of AAA are included in their corresponding categories – see abstracts 004.076, 013.123, 013.124, 013.125, 013.126, 062.204, 072.053, 072.054, 073.077, 074.064, 075.036, 075.037, 076.023, 077.039, 082.073, 083.039, 083.040, 083.041, 083.042, 084.073, 084.074, 085.016, 085.017, 085.018, 085.019, 106.090, 144.073.

003.070   **Das geozentrische Weltbild. Astronomie, Geographie und Mathematik der Griechen.**
A. Szabó.
dtv, No. 4490.
Deutscher Taschenbuch Verlag, München (Germany). 378 p. (1992). ISBN 3–423–04490–X. Price DM 26.80.
Contents: 1. Einleitung. 2. Der wissenschaftshistorische Rahmen. 3. Die Erde im Weltall. 4. Die geographische Breite. 5. Die Polhöhe. 6. Weltbild und Zeitmessung. 7. Astronomie und Mathematik.

003.071   **Relativity theory. Concepts and basic principles.**
A. Harpaz.
Jones and Bartlett Publ., Boston, MA (USA). 232 p. (1992). ISBN 0–86720–220–3. Price US$ 24.00.
This book intends to explain the principles of the general theory of relativity (GTR) and can serve as a textbook for an introductory course.
Contents: 1. The theory of relativity. 2. Mathematical introduction. 3. The metric tensor. 4. Space dependent metric. 5. Four dimensional space. 6. The principles of GTR. 7. Einstein's equations. 8. Schwarzschild's solution. 9. Cosmological solutions. 10. Relativistic astrophysics phenomena. 11. Epilogue.

003.072   **Worlds in the sky. Planetary discovery from earliest times through Voyager and Magellan.**
W. Sheehan.
University of Arizona Press, Tucson, AZ (USA). 258 p. (1992). ISBN 0–8165–1290–6. Price US$ 35.00. ISBN 0–8165–1308–2 (paper).
The author gives a scientific history together with anecdotes surrounding planetary discoveries, and some personal reflections. He describes how we arrived at our current understanding of the Moon and the planets and shows how certain individuals in history shaped the knowledge about the solar system.

003.073   **Annual Review of Astronomy and Astrophysics. Volume 30.**
G. Burbidge, D. Layzer, A. Sandage (eds.).
Annu. Rev. Astron. Astrophys., Vol. 30, 779 p. (1992). ISBN 0–8243–0930–8. Price US$ 57.00 (USA, Canada), US$ 62.00 (elsewhere).

003.074   **Bildatlas des Kosmos.**
Redaktion der Time–Life Bücher.
Reise durch das Universum.
Time–Life Bücher, Amsterdam (Netherlands). Authoriz. German ed. 144 p. (1992). ISBN 90–6182–477–X. Price DM 46.00.
Contents: 1. Das Sonnensystem. Ein Reigen der Planeten. 2. Der Sternhimmel. Glanzpunkte am Firmament. 3. Die Vielfalt der Galaxien. Ein Teleskop–Führer.

003.075   **Biographien bedeutender Astronomen. Eine Sammlung von Biographien.**
D. B. Herrmann (ed.).
Volk und Wissen Verlag, Berlin (Germany). 159 p. (1991). ISBN 3–06–082502–5. Price DM 24.90.
This collection of articles contains 27 biographies of astronomers from Claudius Ptolemäus to Cuno Hoffmeister.

003.076   **Blick in das kalte Weltall. Protosterne, Staubscheiben und schwarze Löcher.**
P. G. Mezger.
Deutsche Verlags–Anstalt, Stuttgart (Germany). 320 p. (1992). ISBN 3–421–02765–X. Price DM 58.00.
Contents: 1. Radioastronomie: Ein neues Fenster ins Weltall öffnet sich.
Die Entwicklung der Radioastronomie. Vom Anfang der Welt: Der Urknall. Die Milchstraße und ihr infrarotes Bild. Sterne und interstellare Materie. Die Suche nach den Protosternen. Der Zentralbereich der Milchstraße. Im Zentrum der Milchstraße: Ein schwarzes Loch mit einer Akkretionsscheibe?
2. Der Wissenschaftsbetrieb.

003.077   **Supernovae and stellar wind in the interstellar medium.**
T. A. Lozinskaya.
American Institute of Physics, New York, NY (USA). 478 p. (1992). ISBN 0–88318–659–4. Price £ 83.25.
This book is an English translation, by M. Damashek, of the Russian original "*Sverkhnovye zvezdy i zvezdnyj veter*" published in 1986 (see 42.003.040).
The author summarizes our present understanding of supernovae and their far–reaching effects within the interstellar medium. She demonstrates how interstellar gas, the gas ejected in the explosions themselves, and the gas emitted as stellar wind by the progenitor all interact to reshape the physical state of the interstellar medium. She goes on to explore different types of supernovae and the evolution of supernova remnants, and includes new information derived from the explosion of supernova 1987A.

003.078   **Der französische Revolutionskalender (1792 – 1805). Planung, Durchführung und Scheitern einer politischen Zeitrechnung.**
M. Meinzer.
Ancien Régime, Aufklärung und Revolution, Band 20.
Oldenbourg Verlag, München (Germany). 307 p. (1992). ISBN 3–486–55791–2. Price DM 128.00. Zugl.: Bielefeld, Univ., Diss., 1986.

003.079   **Sternwarten in Bildern. Architektur und Geschichte der Sternwarten von den Anfängen bis ca. 1950.**
P. Müller.
Springer, Berlin (Germany). 265 p. (1992). ISBN 3–540–52771–0. Price DM 98.00. ISBN 0–387–52771–0 (USA).

003.080   **The book of nature.**
O. Pedersen.
Vatican Obs. Publ.
Vatican Observatory, Città del Vaticano (Vatican City State). 98 p. (1992). ISBN 0–268–00690–3. Distributed: University of Notre Dame Press, Notre Dame, IN (USA). Libreria Editrice Vaticana, Città del Vaticano (Vatican City State) (in Italy and Vatican City State).

003.081   **Satellites of the outer planets. Worlds in their own right.**
D. A. Rothery.
Clarendon Press, Oxford (UK). 221 p. (1992). ISBN 0–19–854289–5. Price £ 40.00 (cloth). ISBN 0–19–854290–9, Price £ 19.50 (paper).
The author presents geological investigations of the major satellites of the outer planets (from Jupiter to Neptune) and shows that each one is a distinctive member of a family of worlds. He draws attention to the similarities and differences between them and discusses in particular how tectonic and volcanic processes have shaped their rigid outer layers, driven by heat

from within. The text is illustrated with close–up images from the Voyager space probes and explanatory drawings.

003.082   **The physics of astrophysics. Volume II: Gas dynamics.**
F. H. Shu.
University Science Books, Mill Valley, CA (USA). 493 p. (1992). ISBN 0–935702–65–2. Price £ 36.95.
Volume II is a self–contained textbook, and is not dependent on Volume I (see 53.003.096). It can be used as the text for a separate, one–semester course on its subject matter, which includes the interactions of matter and radiation, and electromagnetic fields of macroscopic scale in both the strongly collisional and collisionless regimes. It covers such fields as single–fluid theory, including radiative processes; waves, shocks, and fronts; magnetohydrodynamics and plasma physics; as well as their applications to such topics as self–gravitating spherical masses, accretion disks, spiral density waves, star formation, and dynamo theory.

003.083   **Physical processes in solar flares.**
B. V. Somov.
Astrophys. Space Sci. Libr., Vol. 172.
Kluwer Academic Publ., Dordrecht (Netherlands). 257 p. (1992). ISBN 0–7923–1261–9. Price Dfl. 180.00, US$ 99.00, £ 61.00.
This book provides three stages in the solution of the solar flare problem. Chapter one describes the connection between observational data and theoretical concepts, where it is stressed that next to investigating flares, the related non–stationary large–scale phenomena must be studied as well. The second chapter deals with secondary physical processes, in particular the study of high–temperature plasma dynamics during impulsive heating. The last chapter presents a model built on the knowledge of the two previous chapters and it constructs a theory of non–neutral turbulent current sheets.

003.084   **Multicolor stellar photometry.**
V. Straižys.
Pachart Astron. Astrophys. Ser., Vol. 15.
Pachart Publ. House, Tucson, AZ (USA). 584 p. (1992). ISBN 0–88126–029–0. Price US$ 63.00.
In this book one can find detailed descriptions of almost all photometric systems with their advantages and drawbacks. The ample references facilitate finding primary sources on each system. The last section of the book describes the Vilnius system in detail for its practical use. It can be regarded as a valuable generic system to classify stars of almost all types – even in the most distant and dusty parts of the Galaxy.

003.085   **The fullness of space. Nebulae, stardust, and the interstellar medium.**
C. G. Wynn–Williams.
Cambridge University Press, Cambridge (UK). 217 p. (1992). ISBN 0–521–42638–3. Price £ 15.95, US$ 29.95 (paper). ISBN 0–521–35591–5, Price £ 35.00, US$ 65.00 (cloth).
This book is a comprehensive account of what astronomers have learned about interstellar matter – where it comes from, what it is made of, and how it collects together to form new stars and planets. It is illustrated with photographs and computer–generated images of nebulae, dust clouds and galaxies. The text is non–technical.

003.086   **Karl Schwarzschild. Gesammelte Werke. Band 2. (*Karl Schwarzschild. Collected works. Volume 2.*)**
H.–H. Voigt (ed.).
Springer, Berlin (Germany). 558 p. (1992). ISBN 3–540–52456–8. Price DM 298.00. ISBN 0–387–52456–8 (USA).
This volume is the second part of a three–volume collection of research papers by Karl Schwarzschild. The complete collection will comprise a total of 119 papers, organized into ten categories. (For Vol. 1 see 003.011).
Contents: 5. Astronomical positioning. Commentary by G. Seeber.

6. Photographic photometry. Commentary by U. Haug.
7. Measuring techniques, binary stars, variable stars and spectroscopy. Commentary by E. H. Geyer.

003.087   **Reviews in Modern Astronomy 5: Variabilities in stars and galaxies.**
G. Klare (ed.).
Springer, Berlin (Germany). 281 p. (1992). ISBN 3–540–55523–4. ISBN 0–387–55523–4 (USA).
The fifth yearbook of the Astronomische Gesellschaft (AG) contains eighteen invited reviews and highlight contributions presented at the Spring Meeting of the AG, Bamberg, Germany in April 1991. The contributions deal with time–dependent phenomena of a wide range of astronomical objects, both stellar (for example, galactic cataclysmic variables, symbiotic stars, luminous blue variables, and novae) and extragalactic (for example, active galactic nuclei and quasars). Different physical mechanisms causing the variability, including pulsation, accretion, and eruptive events connected with mass loss, are discussed. The individual contributions within the subject scope of AAA are included in their corresponding categories – see abstracts 064.088, 065.114, 065.116, 116.076, 117.274, 117.275, 122.141, 122.153, 122.154, 124.009, 124.010, 142.052, 155.152, 158.277, 158.278, 158.279, 158.280, 161.501.

003.088   **Gravitational lenses.**
P. Schneider, J. Ehlers, E. E. Falco.
Astron. Astrophys. Libr.
Springer, Berlin (Germany). 573 p. (1992). ISBN 3–540–97070–3. Price DM 128.00. ISBN 0–387–97070–3 (USA).
This is the first systematic presentation of gravitational lens theory as developed from first principles. Beginning with simple models and basic properties of the lens mapping, the book goes on to consider such topics as microlensing and the prominent role played lens statistics in the interpretation of high–redshift objects. The relevant basics of catastrophe theory are derived and techniques for numerical treatment of gravitational lensing are listed. On the observational side, details are summarized of several known multiple QSOs, radio rings and luminous arcs, together with the difficulties of the observation and verification of lens systems. The potential role of gravitational lenses as astronomical tools ("natural telescopes") is pointed out. This book can be viewed as both an advanced textbook and a research monograph.

003.089   **Carl Friedrich Gauß. Zwölf Kapitel aus seinem Leben. Von Ludwig Hänselmann.**
Braunschweig 1878.
H. Michling (comp.).
Gauß–Gesellschaft e.V., Göttingen (Germany). Reprint 106 p. (1992). Herstellung: Goltze, Göttingen (Germany).

003.090   **Images of the universe.**
C. Stott (ed.).
Cambridge University Press, Cambridge (UK). 245 p. (1991). ISBN 0–521–39178–4.   Price £ 35.00 (cloth).   ISBN 0–521–42419–4, Price £ 15.95 (paper).
The individual contributions within the subject scope of AAA are included in their corresponding categories – see abstracts 013.136, 065.117, 071.025, 091.057, 099.109, 100.073, 101.093, 102.057, 122.155, 125.071, 126.116, 155.153, 158.283, 161.503, 161.504.

003.091   **Physics of massive neutrinos.**
F. Boehm, P. Vogel.
Cambridge University Press, Cambridge (UK). 2. ed. 129 p. (1992). ISBN 0–521–42849–1. Price £ 15.95, US$ 27.95 (paper). ISBN 0–521–41824–0, Price £ 40.00, US$ 69.95 (cloth).
This book, while describing all aspects of neutrino physics, focuses on what we know and may hope to know about the mass of the neutrino and its particle–antiparticle symmetry. Topics include neutrino mixing, neutrino decay, neutrino oscillations, double beta decay, solar neutrinos, supernova neutrinos and related issues. The authors stress the physical concepts, and

discuss both theoretical and experimental techniques. This second edition (the first edition was published in 1987, see 45.003.039) is completely up to date and differs from the first in that it contains an expanded coverage of new experimental results and recent theoretical advances. Since publication of the first edition, many issues that were at that time unresolved, such as tritium beta decay and reactor neutrino oscillations, have been clarified and these are discussed in the new edition. Also included is an expanded coverage of solar and supernova neutrinos.

003.092   **Annual Review of Earth and Planetary Sciences. Volume 19.**
G. W. Wetherill, A. L. Albee, K. C. Burke (eds.).
Annu. Rev. Earth Planet. Sci., Vol. 19, 492 p. (1991). ISBN 0–8243–2019–0. Price US$ 55.00 (USA, Canada), US$ 60.00 (elsewhere).

003.093   **Annual Review of Earth and Planetary Sciences. Volume 20.**
G. W. Wetherill, A. L. Albee, K. C. Burke (eds.).
Annu. Rev. Earth Planet. Sci., Vol. 20, 641 p. (1992). ISBN 0–8243–2020–4. Price US$ 59.00 (USA, Canada), US$ 64.00 (elsewhere).

003.094   **Die neue Kosmologie. Von Dunkelmaterie, GUTs und Superhaufen.**
J. Cornell (ed.).
Birkhäuser, Basel (Switzerland). 239 p. (1991). ISBN 3–7643–2516–X. Price DM 58.00.
This book is a German translation, by M. Röser, of the American original "Bubbles, voids, and bumps in time: the new cosmology", published 1989 (see 49.003.045).
Contents: 1. Die Entdeckung des Weltalls: Eine Einleitung (A. P. Lightman). 2. Die Vermessung des Weltalls: Rotverschiebungen und Standardkerzen (R. P. Kirshner). 3. Die Kartierung des Weltalls: Scheiben und Blasen (M. J. Geller). 4. Die Bestimmung der Masse im Weltall: Dunkelmaterie und fehlende Masse (V. C. Rubin). 5. Der Beginn des Weltalls: Urknall und kosmische Inflation (A. H. Guth). 6. Die Ausweitung des Weltalls: Das Weltraumteleskop und Perspektiven für die nächsten 20 Jahre (J. E. Gunn).

003.095   **The glorious constellations. History and mythology.**
G. M. Sesti.
Abrams, New York, NY (USA). 495 p. (1991). ISBN 0–8109–3355–1. With 48 plates. Price US$ 125.00.
This book is an English translation, by K. H. Ford, of the Italian original "Le dimore del cielo".
It tells the tale of the nighttime sky by concentrating on forty-eight ancient constellations and the Milky Way. Each is presented with a description of the lore surrounding it in various cultures – Sumerian, Egyptian, Phoenician, Assyrian, Greek, Hebrew – augmented when possible by corresponding legends from India, China, Australia, and the Americas. Supplementing the thorough text are forty-eight color-plates reproducing the astronomical fresco in the Villa Farnese in Caprarola, Italy. A far-reaching yet concise and easy–to–understand history of astronomy sets the scene. These pages, as well as those discussing each constellation, are illustrated with a variety of black–and–white images.

003.096   **Relativity on curved manifolds.**
F. de Felice, C. J. S. Clarke.
Camb. Monogr. Math. Phys.
Cambridge University Press, Cambridge (UK). 459 p. (1992). ISBN 0–521–42908–0. Price £ 19.95, US$ 37.95.
Paperback reprint of the textbook first published in 1990 as hardback edition (see 52.003.083).

003.097 The detection of gravitational waves.
D. G. Blair (ed.).
Cambridge University Press, Cambridge (UK). 505 p. (1991). ISBN 0–521–35278–9. Price £ 95.00, US$ 125.00.
The individual contributions within the subject scope of AAA are included in their corresponding categories – see abstracts 022.231, 034.112, 034.114, 034.115, 034.116, 034.117, 034.118, 034.119, 034.120, 034.121, 034.122, 034.123, 036.202, 036.203, 036.204, 066.292, 066.293.

003.098 Development of large–scale structure in the universe.
J. P. Ostriker.
Lezioni Fermiane, 1988.
Accademia Nazionale dei Lincei, Rome (Italy); Scuola Normale Superiore, Pisa (Italy), Cambridge University Press, Cambridge (UK). 74 p. (1991). ISBN 0–521–42361–9. Price £ 9.95, US$ 19.95.
Contents: 1. Overview of elementary observational results: inhomogeneities and velocity perturbations; support for homogeneous isotropic model universes. 2. Homogeneous isotropic models: Friedmann–Lemaître cosmological models; cosmological structure and the quest for the numbers H_0 and Ω_0. 3. Perturbations in the universe: observed correlation functions; perturbations on standard cosmologies; the standard model for

growth of structures. 4. Non–linear isolated perturbations: classical positive energy perturbations; cosmological positive energy disturbances; stability of expanding shells; cosmological consequences of superconducting strings; negative energy perturbations – infall and accretion.

003.099 Beams and jets in astrophysics.
P. A. Hughes (ed.).
Camb. Astrophys. Ser., Vol. 19.
Cambridge University Press, Cambridge (UK). 595 p. (1991). ISBN 0–521–34025–X. Price £ 55.00 (cloth). ISBN 0–521–33576–0, Price £ 19.95 (paper).
The material covered by this book includes a description and classification of parsec– and kiloparsec–scale extragalactic jets and of Galactic jets; basic jet physics and the derivation of jet parameters; the theories of jet propagation and stability; the processes which accelerate particles and generate magnetic fields; the mechanisms by which jets radiate; hydrodynamics simulations; and the engines which generate and collimate the flows.
The individual contributions within the subject scope of AAA are included in their corresponding categories – see abstracts 062.137, 062.242, 063.150, 155.159, 158.302, 158.303, 158.304, 158.305, 158.306, 158.307.

004 History of Astronomy

004.001 Galileo revisited.
F. R. Hickey.
Phys. Teach., Vol. 30, No. 2, p. 103 – 104 (Feb 1992). Abstract. Current Physics Microform No.: 9203G1679.

004.002 The coins of Antioch.
M. R. Molnar.
Sky Telesc., Vol. 83, No. 1, p. 37 – 39 (Jan 1992).
Coins of imperial Rome could shed new light on the Star of Bethlehem.

004.003 Midwinter sunrise at El Karnak.
R. L. Reese.
Sky Telesc., Vol. 83, No. 3, p. 276 – 278 (Mar 1992).
A lone observer witnesses the first sunrise of winter from an ancient Egyptian temple.

004.004 Die Heiligen Drei Könige und ihr Stern.
W. Seggewiß.
Sterne Weltraum, Jahrg. 31, Nr. 1, p. 14 – 19 (Jan 1992).

004.005 Sir William Herschel's notebooks: abstracts of solar observations.
D. V. Hoyt, K. H. Schatten.
Astrophys. J., Suppl. Ser., Vol. 78, No. 1, p. 301 – 340 (Jan 1992).
Sir William Herschel observed the Sun from 1779 to 1818 with most of his observations made between 1799 and 1806. These observations are interesting both because other solar observers at this time were not very active and because solar behavior was quite unusual. Solar activity was nearly dormant and sunspot periods were very long. In this paper the authors provide an introduction to the background of Herschel's notebooks and the historical context within which his observations were made. The observations have relevance in reconstructing solar behavior, as discussed in a separate analysis paper by Hoyt and Schatten, and in understanding active features on the Sun such as faculae. The text of Herschel's notebooks with modern terms used throughout is presented here. The complete text has not previously been published and is not easily accessible to scholars. A glossary

explaining the terminology and biographical data of several contemporaries mentioned in the notes are included.

004.006 The English quadrant in Europe: instruments and the growth of consensus in practical astronomy.
J. A. Bennett.
J. Hist. Astron., Vol. 23, Part 1, p. 1 – 14 (Feb 1992).

004.007 The "Notes on al–Bitrûji" attributed to Regiomontanus: second thoughts.
M. H. Shank.
J. Hist. Astron., Vol. 23, Part 1, p. 15 – 30 (Feb 1992).

004.008 Astronomical records in the *Ch'un–ch'iu* chronicle.
F. R. Stephenson, K. K. C. Yau.
J. Hist. Astron., Vol. 23, Part 1, p. 31 – 51 (Feb 1992).

004.009 Walter S. Adams and the imposed settlement between Edwin Hubble and Adriaan van Maanen.
N. S. Hetherington, R. S. Brashear.
J. Hist. Astron., Vol. 23, Part 1, p. 53 – 56 (Feb 1992).

004.010 Orientations of megalithic sepulchres in Salamanca, Spain.
M. Hoskin.
J. Hist. Astron., Vol. 23, Part 1, p. 57 – 60 (Feb 1992).

004.011 James Gregory and the reflecting telescope.
A. D. C. Simpson.
J. Hist. Astron., Vol. 23, Part 2, p. 77 – 92 (May 1992).

004.012 Seeing red: observations of colour in Jupiter's Equatorial Zone on the eve of the modern discovery of the Great Red Spot.
T. A. Hockey.
J. Hist. Astron., Vol. 23, Part 2, p. 93 – 105 (May 1992).

004.013   **The orientations of the temples of Malta.**
G. Foderà Serio, M. Hoskin, F. Ventura.
J. Hist. Astron., Vol. 23, Part 2, p. 107 – 119 (May 1992).

004.014   **The 1816 solar eclipse and comet 1811 I in John Linnell's astronomical album.**
R. J. M. Olson, J. M. Pasachoff.
J. Hist. Astron., Vol. 23, Part 2, p. 121 – 133 (May 1992).

004.015   **Enigma of Sirius' color.**
J.–M. Bonnet–Bidaud, C. Gry.
Recherche, Vol. 23, No. 239, p. 105 – 107 (Jan 1992). In French.

004.016   **Europäische Astronomen in Indien. Zum Ausbleiben der kopernikanischen Wende im 18. Jahrhundert.**
B. du Mont.
Sterne Weltraum, Jahrg. 31, Nr. 4, p. 233 – 237 (Apr 1992).

004.017   **210 years since the discovery of Uranus.**
L. Lenža.
Říše hvězd, Vol. 72, No. 5, p. 86 – 87 (May 1991). In Czech.

004.018   **Johann Kepler, Brandenburg army in Bohemia and Nova Geminorum 1283.**
A. Hadravová, P. Hadrava.
Vesmír, Vol. 70, No. 1, p. 44 – 47 (Jan 1991). In Czech.

004.019   **When have been the Mayan calendar started?**
B. Böhm, V. Böhm.
Vesmír, Vol. 70, No. 2, p. 98 – 105 (Feb 1991). In Czech.

004.020   **Cosmographical mystery of contemporaneity. I.**
P. Hadrava.
Vesmír, Vol. 71, No. 5, p. 263 – 265 (May 1992). In Czech.

004.021   **Cosmographical mystery of contemporaneity. II.**
P. Hadrava.
Vesmír, Vol. 71, No. 6, p. 337 – 338 (Jun 1992). In Czech.

004.022   **Aus den Anfängen der lichtelektrischen Sternphotometrie.**
H. Schmidt.
BAV Rundbrief, Jahrg. 41, Nr. 2, p. 72 – 79 (1992).

004.023   **A propos du zodiaque....**
P. Barthel.
Obs. Trav., No. 30, p. 33 – 37 (1992).

004.024   **Historical review of a long–overlooked paper by R. A. Daly concerning the origin and early history of the Moon.**
R. B. Baldwin, D. E. Wilhelms.
J. Geophys. Res., Vol. 97, No. E3, p. 3837 – 3843 (25 Mar 1992).
In 1946 the great geologist R. A. Daly published an important paper in which he discussed a great many problems concerning the Moon and its features and origin. His paper was almost completely ignored by the scientists of the day and was "lost" for nearly half a century. The present paper marks an attempt to outline Daly's contributions to the interpretation of these lunar problems, in particular the origin of the Moon.

004.025   **The 1095 AD meteor event as described in the Anglo Saxon Chronicle.**
E. G. Mardon, A. A. Mardon.
54. Annual Meeting of the Meteoritical Society, p. 147 (1991). Abstract. – See Abstr. 012.010 for the main entry.

004.026   **The history of the Meteorological and Astronomical Observatory on the Pop Ivan Mountain – post scriptum.**
J. M. Kreiner.
Urania, Rok 63, Nr. 2, p. 49 – 53 (Feb 1992). In Polish.

004.027   **Die Anfangsjahre der Radioastronomie in Deutschland.**
K. Rohlfs.
Sterne, Band 68, Heft 2, p. 80 – 93 (1992).

004.028   **Eine Medaille von der größten Kometenerscheinung des 17. Jahrhunderts.**
F. Lehmann.
Sterne, Band 68, Heft 2, p. 113 – 114 (1992).

004.029   **Newton's solution of the one–body problem.**
B. Pourciau.
Arch. Hist. Exact Sci., Vol. 44, No. 2, p. 125 – 146 (25 Jun 1992).

004.030   **The moon–test in Newton's *Principia*: accuracy of inverse–square law of universal gravitation.**
S. Aoki.
Arch. Hist. Exact Sci., Vol. 44, No. 2, p. 145 – 190 (25 Jun 1992).

004.031   **A study of astrolabes.**
R. K. E. Torode.
J. Br. Astron. Assoc., Vol. 102, No. 1, p. 25 – 30 (Feb 1992).
The rete of a planispheric astrolabe has pointers representing stars. From their positions these stars can usually be identified. The positions can then be compared with modern positions of the same stars to find the amount of precessional movement and hence estimate the age of the instrument.

004.032   **The two comets of August AD 1165.**
D. Wright.
J. Br. Astron. Assoc., Vol. 102, No. 1, p. 39 (Feb 1992).

004.033   **The history of the tropical year.**
J. Meeus, D. Savoie.
J. Br. Astron. Assoc., Vol. 102, No. 1, p. 40 – 42 (Feb 1992).

004.034   **The birth of radio astronomy.**
R. Wearner.
Astronomy, Vol. 20, No. 6, p. 46 – 49 (Jun 1992).

004.035   **Carter Observatory's 9–inch refractor: the Crossley connection.**
F. P. Andrews, E. Budding.
South. Stars, Vol. 34, No. 6, p. 358 – 366 (Mar 1992).
Some of the historical background of Carter Observatory's main refractor at Kelburn is reviewed.

004.036   **A brief history of the Crossley 36–inch reflector.**
F. Dyson.
South. Stars, Vol. 34, No. 6, p. 367 – 372 (Mar 1992).
The story of the more well–known 36–inch telescope associated with Edward Crossley is revisited.

004.037   **Behind the "red Sirius" myth.**
R. Ceragioli.
Sky Telesc., Vol. 83, No. 6, p. 613 – 615 (Jun 1992).
Why did certain ancient writers call the white stars Sirius red? The answer has nothing to do with astrophysics.

004.038   **Representing the heavens: Galileo and visual astronomy.**
M. G. Winkler, A. Van Helden.
Isis, Vol. 83, No. 2, p. 195 – 217 (Jun 1992).
The authors present the following conclusion. Galileo was not alone in his ambivalent attitude toward visual communication in astronomy. His attitude was shared by his contemporaries. In fact, the use of visual evidence is surprisingly rare until after 1640. And when astronomers finally began using pictorial evidence, they did so with an explicit commitment to representing the heavens faithfully and accurately. Although Francesco Fontana was the first to publish an astronomical book in which pictorial information was central, it is in the work of Johannes Hevelius (1611 – 1687), a university trained brewer in the Polish city of Gdansk, that we see the new visual dimension of telescopic astronomy best exemplified. Hevelius's *Selenographia sive lunae descriptio* of 1647 contained figures of forty different lunar

phases, four views of the full moon, eighty–three diagrams, and several illustrations of his equipment and the appearances of other heavenly bodies. What is even more interesting, Hevelius made his own telescopes and, he himself engraved virtually every illustration – diagram or picture – in the book, thus combining the roles of the natural philosopher and the lowly artisan. Hevelius's approach to representing the heavens was so different from Galileo's that he utterly misunderstood the purpose behind the views of the moon shown in *Sidereus nuncius*. Such a completely wrongheaded judgment of Galileo's instruments and his ability as an observer and draftsman shows just how different the worlds of these two men were. When, within the range of media available to them, Hevelius and others chose to make the visual component central in communicating their observations, astronomy became a visual science.

004.039   Representing the Earth's shape: the polemics surrounding Maupertuis's expedition to Lapland.
M. Terrall.
Isis, Vol. 83, No. 2, p. 218 – 237 (Jun 1992).

Historical accounts of quantification in the physical sciences in the eighteenth century have often been described as a straightforward series of steps in a process of maturation, as instruments and standards advanced in precision. This paper calls into question the self–evidence nature of precision by investigating the production and uses of measurements. In the case of the dispute over the shape of the Earth, centered in Paris in the 1730s, the precision of measurements was a matter to be interpreted, attacked, defended, and represented. The whole messy business, undertaken by the participants to win consensus from their contemporaries, took place in the context of academic politics and the intellectual fashions of the salons and the court. All parties to the dispute claimed to be drawing on precision measurements; evaluating precision turned out to require the use of a range of intellectual, mathematical, instrumental, political, and textual resources. The alleged precision was then used to construct and defend rival scientific programs and practices.

004.040   Annals of scientific publishing: Johannes Petreius's letter to Rheticus.
N. M. Swerdlow.
Isis, Vol. 83, No. 2, p. 270 – 274 (Jun 1992).

In the early 16th century Nuremberg had become the center of scholarly publishing in Germany, and Petreius was then its most distinguished printer. His letter has a naive humor and charm, particularly in its timeless commonplaces about riches and learning, its good humanist (and Lutheran) dismissal of scholasticism, and its pride in the culture and learning of Nuremberg. It is here translated in its entirety. While known slightly for its remarks on Copernicus, it is also of interest for its no doubt deliberate depiction of a discerning publisher, encouraging and rewarding learning, and always looking out to publish worthy books by learned men. Indeed, even today Petreius, in his own way a predecessor of Breitkopf, Teubner, and Springer, provides and admirable model of the scholarly and scientific publisher, promoting higher standards and willing to take more risks, not the least of them financial, than most modern publishers.

004.041   Astronomical use of pinhole images in William of Saint–Cloud's *Almanach Planetarum* (1292).
J. L. Mancha.
Arch.   Hist.   Exact   Sci.,   Vol. 43,   No. 4,   p. 275 – 298 (21 May 1992).

004.042   Galilée et la naissance de la science moderne.
Y. De Rop.
Ciel, Vol. 54, p. 128 – 137 (Apr 1992).

004.043   Urania's heritage: a historical introduction to women in astronomy.
A. K. Dobson, K. Bracher.
Mercury, Vol. 21, No. 1, p. 4 – 15 (Jan – Feb 1992).

004.044   Aboriginal astronomy.
R. D. Haynes.
Aust. J. Astron., Vol. 4, No. 3, p. 127 – 140 (Apr 1992).

The accuracy and purpose of astronomical observations by the Australian Aborigines are discussed. The predictive and moral functions inherent in the legends associated with particular celestial objects are outlined and compared with the purpose of western science.

004.045   The last great speculum: the 48–inch Great Melbourne Telescope.
J. Perdrix.
Aust. J. Astron., Vol. 4, No. 3, p. 149 – 163 (Apr 1992).

The injection of gold royalties into the Treasury of the Colony of Victoria, and Herschel's observations of nebulae and double stars in southern skies were the catalysts for the Great Melbourne Telescope. The conception, gestation, birth, life, and death are discussed. Was its failure due to design and materials of construction or colonial incompetence? The inability to take photographs of greater than two minutes exposure restricted its usefulness.

004.046   The meteorite of Ensisheim: 1492 to 1992.
U. B. Marvin.
Meteoritics, Vol. 27, No. 1, p. 28 – 72 (Mar 1992).

On Nov 7, 1492, a 127–kg stony meteorite fell at Ensisheim in Alsace after a fireball explosion that was heard for a distance of 150 km over the upper Rhineland. Today a 56–kg specimen of the stone, an LL6 chondrite with large patches of fusion crust, remains on display in the Hotel de Ville at Ensisheim. This was the earliest witnessed meteorite fall in the West from which pieces are preserved. In nearby Basel, broadsheets were printed within weeks bearing the story in Latin and German verses by the eminent poet, Sebastian Brant, who turned the sheets into propaganda tracts by claiming the stone as a portent of victory and admonishing Maximilian to make war on the French without delay. Maximilian declared the stone to be a sign of divine favor and ordered it to be preserved in the Ensisheim parish church. Through centuries of battle and political changes, the stone remained in the church until 1793 when French revolutionaries transferred it to a new National Museum in Colmar. There, many pieces were taken for chemical analyses during the birth of the meteoritics at the turn of the 19th century. This paper traces the history of the stone itself and people's responses to it through the 500 years since the fall at Ensisheim.

004.047   The man who found a city in the Moon.
R. Baum.
J. Br. Astron. Assoc., Vol. 102, No. 2, p. 157, 159 (Jun 1992).

004.048   Lancaster's lost observatory.
P. Wade.
J. Br. Astron. Assoc., Vol. 102, No. 2, p. 160 – 162 (Jun 1992).

004.049   Lord Crawford's Observatory at Dun Echt, 1872 – 1892.
H. A. Brück.
Vistas Astron., Vol. 35, Part 1, p. 81 – 138 (Jan 1992).

004.050   Ground–based planetary science at Lick Observatory 1888 – 1938.
D. E. Osterbrock.
Bull. Am. Astron. Soc., Vol. 23, No. 3, p. 1202 (1991). Abstract. – See Abstr. 012.037 for the main entry.

004.051   Freezing fire: measuring planetary heat, 1900 – 1930.
R. S. Brashear.
Bull. Am. Astron. Soc., Vol. 23, No. 3, p. 1202 (1991). Abstract. – See Abstr. 012.037 for the main entry.

004.052 **Yale and USNO cooperation especially in the Brouwer and Clemence era.**
D. Hoffleit.
Comments Astrophys., Vol. 16, No. 1, p. 17 – 30 (Feb 1992).
The United States Naval Observatory (USNO) and Yale University Observatory have always had many research interests in common and often worked in collaboration. More Yale graduates found employment at USNO than at any other institution, including Yale. Several USNO astronomers upon retirement transferred to Yale in order to collaborate more effectively with Yale astronomers. Finally, from 1947 until Brouwer's death, ONR awarded Yale a contract for collaborative research among Yale, USNO and the IBM Watson Scientific Computing Laboratory in New York. The director of the latter, Wallace Eckert, a Yale Ph.D., had previously been the director of the U.S. Nautical Almanac. Under the ONR contract over twenty important papers were published in the "Papers of the American Ephemeris" and "Nautical Almanac".

004.053 **Die Entdeckung des Planeten Neptun an Enckes 55. Geburtstag. Vorgeschichte, Ablauf und Wirkung einer "geplanten" Entdeckung.**
J. Hamel.
Sterne, Band 68, Heft 3, p. 161 – 174 (1992).

004.054 **Humboldts Bemühungen um die Gründung einer neuen Sternwarte in Gotha. Ein neu entdeckter Briefwechsel, der auf eine Mitwirkung Enckes hinweist.**
M. Strumpf, O. Schwarz.
Sterne, Band 68, Heft 3, p. 175 – 178 (1992).

004.055 **Le cycle lunaire et sa signification chez les Indiens Mexicains.**
U. Köhler.
Publ. Obs. Astron. Strasbourg, Sér. Astron. Sci. Hum., No. 6, p. 1 – 13 (1992).

004.056 **Les mégalithes de Bretagne et les théories astronomiques. Cent ans d'interrogations.**
J. Briard.
Publ. Obs. Astron. Strasbourg, Sér. Astron. Sci. Hum., No. 6, p. 15 – 33 (1992).

004.057 **Sökandet efter rymdens krökning.**
A. Sandage.
Astron. Tidsskr., Årg. 25, Nr. 1, p. 4 – 23 (Mar 1992).

004.058 **Stjernebildene.**
T. S. Ringnes.
Astron. Tidsskr., Årg. 25, Nr. 2, p. 63 – 77 (Jun 1992).

004.059 **Préoccupations cosmologiques et astronomiques dans les travaux de l'ecole Française d'ethnologie dans la boucle du Niger.**
P. Erny.
Publ. Obs. Astron. Strasbourg, Sér. Astron. Sci. Hum., No. 6, p. 53 – 74 (1992).

004.060 **La supernova del 1054 osservata in Occidente.**
G. Lupato.
Astronomia UAI, N. 1 – 2, p. 17 – 24 (Jan – Apr 1992).

004.061 **The North Mull Project (3): prominent hill summits and their astronomical potential.**
C. L. N. Ruggles, R. D. Martlew.
Archaeoastronomy (U.K.), No. 17, p. S1 – S13 (1992).

004.062 **Orientations of the sesi of Pantelleria.**
S. Tusa, G. Foderà Serio, M. Hoskin.
Archaeoastronomy (U.K.), No. 17, p. S15 – S20 (1992).

004.063 **Orientations of tombs in the Late–Minoan cemetery at Armenoi, Crete.**
M. Papathanassiou, M. Hoskin, H. Papadopoulou.
Archaeoastronomy (U.K.), No. 17, p. S43 – S55 (1992).

004.064 **Two observations by Galileo of unidentified objects.**
E. M. Standish, D. A. Pierce, A. M. Nobili.
Bull. Am. Astron. Soc., Vol. 23, No. 3, p. 1258 (1991). Abstract. – See Abstr. 012.065 for the main entry.

004.065 **Quelques occultations chez Képler.**
S. De Meis, J. Meeus.
Astronomie, p. 1 – 5 (Jan 1992).

004.066 **Passage de Mercure devant le Soleil, observé par Gassendi (1592 – 1655), le 7 novembre 1631.**
S. Dumont, J. Meeus, M. Anstett.
Astronomie, p. 5 – 7 (Apr 1992).

004.067 **Zum "Stern von Bethlehem".**
A. Kunert.
Astron. Sch., Jahrg. 28, Heft 6, p. 22 – 24 (Dec 1991).

004.068 **Peter Apian und seine volkstümliche Darstellung der Polfindung.**
J. Hamel.
Astron. Sch., Jahrg. 28, Heft 6, p. 25 (Dec 1991).

004.069 **Das "Instrument Buch" von Peter Apian (1495 – 1552).**
J. Hamel.
Astron. Raumfahrt, Jahrg. 29, Heft 2, p. 8 – 11 (1991).

004.070 **The AGN paradigm: historical highlights.**
V. Trimble.
AIP Conf. Proc., No. 254, p. 647 – 656 (1 May 1992). – See Abstr. 012.057 for the main entry.
The observations and ideas that most of us feel are critical to our present understanding of AGNs are the outcome of winnowing processes operating on a much larger initial supply. Some of those processes are described, together with a few incidents whose importance has only gradually become clear. Other items touched upon include the life of Carl Seyfert, the state of the astronomical world in 1960, and citation histories of some key papers.

004.071 **Archaeoastronomy of the Old Stone Tower, Newport, RI.**
W. S. Penhallow, M. J. Brennan.
J. Am. Assoc. Variable Star Obs., Vol. 20, No. 1, p. 109 – 110 (1991). Abstract. – See Abstr. 012.066 for the main entry.

004.072 **Shapley's model for Cepheids.**
B. L. Welther.
J. Am. Assoc. Variable Star Obs., Vol. 20, No. 1, p. 111 (1991). Abstract. – See Abstr. 012.066 for the main entry.

004.073 **The birth of electronic astronomy.**
M. T. Svec.
Sky Telesc., Vol. 83, No. 5, p. 496 – 499 (May 1992).
The modern astronomer's light detectors, from phototubes to CCD's, all trace their lineage to crude devices developed by Joel Stebbins some 80 years ago.

004.074 **A re–investigation of the "double dawn" event recorded in the Bamboo Annals.**
F. R. Stephenson.
Q. J. R. Astron. Soc., Vol. 33, No. 2, p. 91 – 98 (Jun 1992).
An allusion to a "double dawn" phenomenon in an ancient Chinese chronicle, which has been identified as caused by a sunrise eclipse occuring in 899 BC, is discussed. This event has been regarded as of considerable importance in the investigation of Earth's past rotation. It is shown that an eclipse interpretation is implausible, not least because the eclipse in question was only annular.

004.075   Astronomical and entoptic phenomena.
F. Thackeray, P. Knox–Shaw.
Mon. Notes Astron. Soc. S. Afr., Vol. 51, Nos. 1/2, p. 6 – 12 (Feb 1992).
   Linguistic and textural evidence from various sources are used to explore a suggestion that comets, meteorites and fireballs were conceptually associated with "entoptic" imagery of the kind perceived in states of trance. Such associations are likely to have been held not only by peoples of southern Africa, but also by others in Eurasia and America in recent periods of human history.

004.076   Historical reverse of solar activities.
Xu Zhentao.
Advances in solar–terrestrial science of China, p. 258 – 267 (1992). – See Abstr. 003.069 for the main entry.
   The author reviews the general progress of investigations on the history of solar activity. Although the 11–year cycle of solar activity has been confirmed since Wolf introduced the relative sunspot number and established the 11–year cycle as one of the basic features of solar activity, many scientists have been interested in the historical behaviour of solar activity, especially before the invention of the telescope. "The Maunder Minimum", named by Eddy has given strong stimulation to this research. The author briefly introduces and comments on recent achievements by Chinese and foreign scientists. Some questions and prospects in the field are presented.

004.077   A study on the calculation of true new moon and almanac in Xuanming calendar.
Zhang Pei–yu, Wang Gui–fen, Chen yue–ying, Lu Xiu–hua.
Publ. Purple Mt. Obs., Vol. 11, No. 2, p. 121 – 155 (Jun 1992). In Chinese.

004.078   A brief history of the Minor Planet Center.
C. M. Bardwell.
Publ. Purple Mt. Obs., Vol. 11, No. 2, p. 156 – 172 (Jun 1992).

004.079   Tejo pre–historical engravings – astronomical aspects involved.
I. D. Simões.
Coimbra Univ. (Portugal). Observatorio Astronomico. 21 p. (1992). Given at Ateneu de Coimbra, 12 Dec 1991.
   Contents: 1. Man is also an astronomical beeing. 2. Pre–historical economy. 3. Power and survival. 4. Survival and calendar. 5. Society, science and technical knowledge.

004.080   La natura delle nubulose a spirale. I. La Via Lattea da Galileo a Herschel e Kapteyn.
R. Gallino, C. Lento.
G. Astron., Vol. 18, N. 1, p. 2 – 11 (Mar 1992).

004.081   La natura delle nebulose a spirale. II. Dall'Universo di Shapley alla recessione delle galassie di Hubble.
R. Gallino, C. Lento.
G. Astron., Vol. 18, N. 2, p. 7 – 22 (Jun 1992).

004.082   A report on mediaeval calendars discovered at Syogonji–temple.
T. Kanda, S. Ito, Y. Okada.
Rep. Natl. Astron. Obs. Jpn., Vol. 1, No. 3, p. 229 – 292 (1992). In Japanese.
   Three complete sets of Kanagoyomi–calendar from Kôei–4 (1345) to Jôwa–3 (1347) of the Nanbokucho era were discovered at Shogonji–temple in Tochigi Prefecture. This report presents a preliminary analysis on these calendars, i.e. how to read them.

004.083   Spiru Haret's theorem.
A. Pál.
Rom. Astron. J., Vol. 1, No. 1 – 2, p. 5 – 11 (1991).
   The present paper, dedicated to the 140th anniversary of the birthday of the Romanian mathematician, mechanicist, and astronomer Spiru Haret (1851 – 1912), is an attempt to state as a theorem his famous result concerning the well–known problem

on the invariability of the major axes of planetary orbits, related to the stability of the solar–planetary system.

004.084   Die Astronomie an der Universität Innsbruck (1888/92 – 1929).
G. Oberkofler, P. Goller.
Hundert Jahre Astronomie an der Leopold–Franzens–Universität Innsbruck (1892 – 1992), p. 5 – 103 (1992). – See Abstr. 003.067 for the main entry.
   Contents: 1. Schwieriger Start. Das astronomische Vorlesungsprogramm vor 1914. 2. Die Sternwarte. Aus den Jahresberichten von Adalbert Prey. 3. Joseph Hepperger. Ein Tiroler Astronom an den Universitäten Graz und Wien. Zwei Dokumente. 4. Faksimiles. Aus den Arbeiten von Eduard Haerdtl, Egon von Oppolzer, Adalbert Prey und Arthur Scheller.

Dictionary of minor planet names.
See Abstr. 002.001.

Uranografie: de cartografie van de sterrenhemel.
See Abstr. 002.005.

Guide to the history of science. 8th edition.
See Abstr. 002.061.

Les moitiés masculines et féminines du ciel: astronomie de quelques tribus Guyanaises. Bibliographie (Volume 5).
See Abstr. 002.083.

Der Himmel auf Erden. Die Welt der Planetarien.
See Abstr. 003.006.

Cuno Hoffmeister. Festschrift zum 100. Geburtstag.
See Abstr. 003.022.

ESO's early history. The European Southern Observatory from concept to reality.
See Abstr. 003.035.

Die Geschichte der Astronomischen Gesellschaft gegründet in Lilienthal am 20. September 1800. Die ersten 63 Jahre ihres Bestehens von 1800 bis 1863.
See Abstr. 003.046.

Studies in the history of general relativity.
See Abstr. 003.050.

Der Einstein–Turm. Erwin F. Freundlich und die Relativitätstheorie – Ansätze zu einer "dichten Beschreibung" von institutionellen, biographischen und theoriengeschichtlichen Aspekten.
See Abstr. 003.055.

Fernrohre und ihre Meister.
See Abstr. 003.060.

Die Lilienthaler Sternwarte 1781 bis 1818. Machinae Coelestes Lilienthalienses: Die Instrumente. Eine zeitgeschichtliche Dokumentation.
See Abstr. 003.062.

Hundert Jahre Astronomie an der Leopold–Franzens–Universität Innsbruck (1892 – 1992).
See Abstr. 003.067.

Das geozentrische Weltbild. Astronomie, Geographie und Mathematik der Griechen.
See Abstr. 003.070.

The glorious constellations. History and mythology.
See Abstr. 003.095.

Was Leonard Digges de uitvinder van de spiegeltelescoop?
See Abstr. 005.002.

W. G. B. Baumann und sein "Multiplications–Kreis".
See Abstr. 005.008.

G. B. Airy and J. C. Adams: like poles repel.
See Abstr. 005.011.

Bartholomaeus Scultetus: Astronom und Kalenderautor des 16.
Jahrhunderts (14. Mai 1540 bis 21. Juni 1614).
See Abstr. 005.033.

Zur Geschichte der Sternwarte Sonneberg.
See Abstr. 009.005.

Major advances in astronomy since 1890.
See Abstr. 013.136.

De geschiedenis van de astrofotografie.
See Abstr. 036.016.

The introduction of the cosmological constant.
See Abstr. 066.271.

New information on solar activity, 1779 – 1818, from Sir William
Herschel's unpublished notebooks.
See Abstr. 072.005.

A fresh analysis of some recent data on atmospheric refraction near
the horizon with implications in archaeoastronomy.
See Abstr. 082.065.

Variable stars: a historical perspective.
See Abstr. 122.089.

005 Biography

005.001   Hermann Strebel. Lichttherapeut und Sonnenforscher.
F. Litten.
Sterne Weltraum, Jahrg. 31, Nr. 3, p. 154 – 157 (Mar 1992).

005.002   Was Leonard Digges de uitvinder van de spiegelteles-
coop?
B. Ernst.
Zenit, Jaarg. 19, Nr. 5, p. 214 – 215 (May 1992).

005.003   H. C. Schumacher: astronoom onder opstandelingen.
G. Beekman.
Zenit, Jaarg. 19, Nr. 6, p. 267 – 269 (Jun 1992).

005.004   Mirror maker.
B. Marriott.
Astron. Now, Vol. 6, No. 2, p. 18 – 20 (Feb 1992).
Despite early attempts to deposit metal films on speculum–
metal mirrors and glass, it was not until the mid–nineteenth
century that a surface fine enough for astronomical purposes
could be achieved. In England, one of the first to produce silver–
on–glass mirrors was George Henry With.

005.005   Zum 100. Geburtstag von Cuno Hoffmeister.
W. C. Seitter.
Sterne, Band 68, Heft 1, p. 3 – 5 (1992).

005.006   Cuno Hoffmeister (1892 – 1968). Leben und Werk.
W. Götz.
Sterne, Band 68, Heft 1, p. 6 – 18 (1992).

005.007   Cuno Hoffmeister und die Planetoiden.
F. Börngen.
Sterne, Band 68, Heft 1, p. 47 – 51 (1992).

005.008   W. G. B. Baumann und sein "Multiplications–Kreis".
H. Schmidt.
Sterne Weltraum, Jahrg. 31, Nr. 4, p. 228 – 229 (Apr 1992).

005.009   Galilei – 350th anniversary of his death.
J. Dvořák, L. Křivský.
Vesmír, Vol. 71, No. 6, p. 332 – 334 (Jun 1992). In Czech.

005.010   Cuno Hoffmeister – Leben und Werk. Zu seinem
100. Geburtstag.
S. Marx.
Astron. Sch., Jahrg. 29, Heft 7, p. 10 – 11 (Feb 1992).

005.011   G. B. Airy and J. C. Adams: like poles repel.
A. Perkins.
Gemini, No. 36, p. 29 – 34 (Jun 1992).

005.012   Microbiografias: Ptolomeo, Claudio.
H. Dávila S.
Bol. Obs. Astron. Quito, p. 85 – 89 (Jan 1992).

005.013   Yakov   Borisovich   Zel'dovich   (8 March   1914 –
2 December 1987).
Priroda, No. 2, p. 84 – 111 (Feb 1992). In Russian.
Reminiscences of his friends and colleagues. 1. How Zel'do-
vich was "discovered" (L. A. Sena). 2. He was born a leader
(L. P. Feoktistov). 3. Knight of sciences (Yu. N. Smirnov).
4. Towards universal weak interaction (S. S. Gershtejn). 5. Yakov
Borisovich and mathematics (V. I. Arnol'd). 6. Ten years with
Zel'dovich (A. D. Dolgov).

005.014   Der Lebensweg Cuno Hoffmeisters.
W. Götz.
Cuno Hoffmeister. Festschrift zum 100. Geburtstag, p. 11 – 49
(1992). – See Abstr. 003.022 for the main entry.

005.015   Meine Begegnung mit Cuno Hoffmeister.
A. Jensch, R, Kippenhahn, W. Deinzer, J. Schubart,
E. H. Geyer, M. Eichhorn, H. Busch.
Cuno Hoffmeister. Festschrift zum 100. Geburtstag, p. 83 – 137
(1992). – See Abstr. 003.022 for the main entry.

005.016   Harold Jeffreys. Commemorazione tenuta nella seduta
del 15 dicembre 1990.
A. H. Cook.
Atti Accad. Naz. Lincei, Ser. 9, Rend. Lincei: Suppl., Vol. 2,
p. 35 – 58 (1991).

005.017   The family background of Lady Huggins (Margaret
Lindsay Murray).
M. T. Brück, I. Elliott.
Ir. Astron. J., Vol. 20, No. 3, p. 210 – 211 (Mar 1992).

005.018   **Some glimpses from my career.**
D. Hoffleit.
Mercury, Vol. 21, No. 1, p. 16 – 18 (Jan – Feb 1992).

005.019   **One woman's journey.**
A. M. Boesgaard.
Mercury, Vol. 21, No. 1, p. 19 – 22, 37 (Jan – Feb 1992).

005.020   **Wallace Campbell: the twelfth Bruce medalist.**
J. S. Tenn.
Mercury, Vol. 21, No. 2, p. 62 – 63, 75 (Mar – Apr 1992).

005.021   **George Ellery Hale: the thirteenth Bruce medalist.**
J. S. Tenn.
Mercury, Vol. 21, No. 3, p. 94 – 96, 110 (May – Jun 1992).

005.022   **Two great astrophysicists: some personal reflections.**
H. C. van de Hulst.
Astrophysics on the threshold of the 21st century, p. 1 – 5 (1992).
– See Abstr. 003.048 for the main entry.
The author presents some personal reflections on Pikel'ner and Shklovsky.

005.023   **The Shklovsky phenomenon.**
N. S. Kardashev, L. S. Marochnik.
Astrophysics on the threshold of the 21st century, p. 7 – 24 (1992). – See Abstr. 003.048 for the main entry.
Translated from the Russian Priroda, 1986.

005.024   **Words about Pikel'ner.**
L. S. Marochnik.
Astrophysics on the threshold of the 21st century, p. 25 – 35 (1992). – See Abstr. 003.048 for the main entry.

005.025   **Some little known variable star astronomers from around the world.**
T. R. Williams.
1. European Meeting of the American Association of Variable Star Observers (AAVSO): International Cooperation and Coordination in Variable Star Research, p. 21 – 35 (1992). – See Abstr. 012.048 for the main entry.
Amateur astronomy is rich with opportunities for individuals with limited resources to make substantive scientific contributions. Noteworthy contributions have been made over the years, and around the world, by thousands of individuals who have received little notice for their efforts. This paper is intended to highlight a few outstanding amateur contributors to variable star astronomy.

005.026   **The last love.**
I.L. Rosenthal (*I. L. Rozental'*).
Astrophysics on the threshold of the 21st century, p. 295 – 302 (1992). – See Abstr. 003.048 for the main entry.
Probability of the formation of a metagalaxy.

005.027   **Vom Artillerieleutnant zum Sternwartendirektor. Enckes Gothaer Jahre 1816 – 1825.**
M. Strumpf.
Sterne, Band 68, Heft 3, p. 137 – 142 (1992).

005.028   **Sternwartendirektor, Universitätslehrer und Akademiesekretär. Enckes berufliches Wirken in Berlin.**
K. H. Tiemann.
Sterne, Band 68, Heft 3, p. 143 – 160 (1992).

005.029   **Erinnerung an bedeutende Astronomen im Schuljahr 1992/93.**
D. Fürst.
Astron. Sch., Jahrg. 29, Heft 9, p. 12 – 14 (Jun 1992).
Contents: 350. Geburtstag Isaac Newtons. 100. Geburtstag von Walter Baade. 200. Geburtstag von Karl Hencke. 200. Geburtstag von Friedrich Georg Wilhelm Struve. 450. Todestag von Nicolaus Copernicus.

005.030   **L'astronomo Ruggero Boscovich.**
G. Nibaldi.
Astronomia UAI, N. 3, p. 13 – 14 (May – Jun 1992).
The author sketches out a short biography of the Croat astronomer Ruggero Boscovich (1711 – 1787), founder and director of Brera Observatory.

005.031   **1992: un cent–cinquantenaire.**
J. Pernet.
Astronomie, p. 9 – 12 (Feb 1992).
150th anniversary of Camille Flammarion.

005.032   **Phoebe Haas – an AAVSO volunteer.**
T. R. Williams.
J. Am. Assoc. Variable Star Obs., Vol. 20, No. 1, p. 18 – 22 (1991).
When the AAVSO first separated from Harvard College Observatory in 1953, Phoebe Waterman Haas quickly volunteered to assist the AAVSO Director. For over ten years Haas calculated the five and ten day means for light curves of southern variable stars, an involvement that partially fulfilled her early aspirations for a career in astronomy.

005.033   **Bartholomaeus Scultetus: Astronom und Kalenderautor des 16. Jahrhunderts (14. Mai 1540 bis 21. Juni 1614).**
J. Helfricht.
Astron. Raumfahrt, Jahrg. 29, Heft 1, p. 22 – 25 (1991).

005.034   **My life.**
P. G. Bergmann.
Symposium on Gravitation and Modern Cosmology – the Cosmological Constant Problem, p. 1 – 4 (1991). – See Abstr. 012.076 for the main entry.

005.035   **Maria Mitchell's haunting legacy.**
J. Opalko.
Sky Telesc., Vol. 83, No. 5, p. 505 – 507 (May 1992).

005.036   **From steam to stars to the early universe.**
W. A. Fowler.
Annu. Rev. Astron. Astrophys., Vol. 30, p. 1 – 9 (1992).

005.037   **Vera Rubin: an unconventional career.**
S. Stephens.
Mercury, Vol. 21, No. 1, p. 38 – 45 (Jan – Feb 1992).

005.038   **Adriaan Blaauw: zestig jaar sterrenkunde.**
E. Echternach.
Zenit, Jaarg. 19, Nr. 5, p. 204 – 208 (May 1992).

005.039   **Galileo y la inquisición.**
M. Navarro.
Universo, Vol. 12, No. 36, p. 12 – 14 (Feb – Apr 1992).

005.040   **Gheorghe Bratu (1881 – 1941).**
A. Pál, V. Mioc.
Rom. Astron. J., Vol. 1, No. 1 – 2, p. 118 – 119 (1991).

Cyclopaedia of telescope makers.
See Abstr. 002.057.

Das Echo des Urknalls. Kernfragen der modernen Kosmologie.
See Abstr. 003.008.

Karl Schwarzschild. Collected works. Volume 1.
See Abstr. 003.011.

Cuno Hoffmeister. Festschrift zum 100. Geburtstag.
See Abstr. 003.022.

My universe. Selected papers of Ya. B. Zel'dovich.
See Abstr. 003.027.

Astrophysics on the threshold of the 21st century.
See Abstr. 003.048.

Der Einstein–Turm. Erwin F. Freundlich und die Relativitätstheorie – Ansätze zu einer "dichten Beschreibung" von institutionellen, biographischen und theoriengeschichtlichen Aspekten.
See Abstr. 003.055.

Fernrohre und ihre Meister.
See Abstr. 003.060.

Hundert Jahre Astronomie an der Leopold–Franzens–Universität Innsbruck (1892 – 1992).
See Abstr. 003.067.

Biographien bedeutender Astronomen. Eine Sammlung von Biographien.
See Abstr. 003.075.

Karl Schwarzschild. Collected works. Volume 2.
See Abstr. 003.086.

Carl Friedrich Gauß. Zwölf Kapitel aus seinem Leben. Von Ludwig Hänselmann. Braunschweig 1878.
See Abstr. 003.089.

Galileo revisited.
See Abstr. 004.001.

Walter S. Adams and the imposed settlement between Edwin Hubble and Adriaan van Maanen.
See Abstr. 004.009.

James Gregory and the reflecting telescope.
See Abstr. 004.011.

Galilée et la naissance de la science moderne.
See Abstr. 004.042.

Urania's heritage: a historical introduction to women in astronomy.
See Abstr. 004.043.

Joseph Shklovsky and X–ray astronomy.
See Abstr. 013.066.

Hiroko Suzuki and chemistry in dark clouds.
See Abstr. 131.176.

006 Personal Notes

006.001   Edward Anders received the 1991 Gerard Kuiper Prize of the American Astronomical Society.
Phys. Today, Vol. 45, No. 2, p. 117 – 118 (Feb 1992).

006.002   Richard Binzel received the 1991 Harold C. Urey Prize of the American Astronomical Society.
Phys. Today, Vol. 45, No. 2, p. 118 (Feb 1992).

006.003   Leonard Medal citation for Donald D. Clayton.
G. W. Lugmair.
Meteoritics, Vol. 27, No. 1, p. 4 (Mar 1992).

006.004   Address by Professor R. D. Davies on the presentation of the Gold Medal of the Society to Professor Vitaly L. Ginzburg on Thursday 1991 October 24 in Moscow.
R. D. Davies.
Q. J. R. Astron. Soc., Vol. 33, No. 2, p. 43 – 44 (Jun 1992).

006.005   Zdeněk Kopal – honorary freeman of the town Litomyšl.
J. Grygar.
Vesmír, Vol. 70, No. 9, p. 491 (Sep 1991). In Czech.

006.006   Barringer Medal citation for Victor L. Masaitis.
R. S. Dietz.
Meteoritics, Vol. 27, No. 1, p. 19 – 20 (Mar 1992).

006.007   Carl Sagan received the 1991 Harold Masursky Award of the American Astronomical Society.
Phys. Today, Vol. 45, No. 2, p. 118 (Feb 1992).

006.008   Martin Schwarzschild received the 1991 Dirk Brouwer Award of the American Astronomical Society.
Phys. Today, Vol. 45, No. 2, p. 117 (Feb 1992).

006.009   Juan Vernet received the Sarton Medal.
T. F. Glick.
Isis, Vol. 83, No. 2, p. 284 – 286 (Jun 1992).

007 Obituaries

007.001   **Isaac Asimov (1920 – 1992).**
C. Sagan.
Nature, Vol. 357, No. 6374, p. 113 (14 May 1992).

007.002   **Alan H. Barrett died 3 July 1991.**
P. C. Myers, J. M. Moran, P. T. P. Ho.
Phys. Today, Vol. 45, No. 6, p. 97 – 98 (Jun 1992).

007.003   **Angelo Bernasconi (1911 – 1990).**
L. Pansecchi.
Universo, Vol. 12, No. 36, p. 51 (Feb – Apr 1992).

007.004   **Memorial for James Maxime DuPont (1912 – 1991).**
E. J. Olsen.
Meteoritics, Vol. 27, No. 1, p. 105 (Mar 1992).

007.005   **Tord Elvius, 30 August 1915 – 6 March 1992.**
G. Larsson–Leander.
Astron. Tidsskr., Årg. 256, Nr. 2, p. 78 – 80 (Jun 1992). In Swedish.

007.006   **Memorial for Louis H. Fuchs (1915 – 1991).**
E. J. Olsen.
Meteoritics, Vol. 27, No. 1, p. 106 (Mar 1992).

007.007   **R. H. Giese (1931 – 1988).**
H. Fechtig, C. Leinert.
IAU Colloquium No. 126: Origin and evolution of interplanetary dust, p. xxiii – xxiv (1991).

007.008   **Fritz Hinderer (24. September 1912 – 1991).**
I. Bues.
Mitt. Astron. Ges., Nr. 75, p. 5 – 7 (1992).

007.009   **M. Huruhata (1912 – 1988).**
H. Tanabe.
IAU Colloquium No. 126: Origin and evolution of interplanetary dust, p. xxv – xxvi (1991).

007.010   **Memorial for Glenn I Huss (10 May 1921 – 28 Sep 1991).**
R. Hutchison, C. F. Lewis.
Meteoritics, Vol. 27, No. 1, p. 107 – 108 (Mar 1992).

007.011   **Sergei Leonidovitch Mandelshtam (1910 – 1990) in memoriam.**
W. I. Axford.
Sol. Phys., Vol. 138, No. 1, p. iv – vi (Mar 1992).

007.012   **Peter Mackenzie Millman (1906 – 1990).**
B. A. McIntosh.
IAU Colloquium No. 126: Origin and evolution of interplanetary dust, p. xxvii (1991).

007.013   **Syotaro Miyamoto (1912 – 1992).**
Yamamoto Circ., No. 2185, p. 2 (23 May 1992).

007.014   **Vishnu Vasudev Narlikar (26 September 1908 – 1 April 1991).**
P. C. Vaidya.
Q. J. R. Astron. Soc., Vol. 33, No. 1, p. 33 – 34 (Mar 1992).

007.015   **Jean–Luc Nieto, 1950 – 5 January 1992.**
E. Davoust.
Messenger, No. 67, p. 48 (Mar 1992).

007.016   **Karl Pilowski (24. August 1905 – 26. Dezember 1991).**
G. Seeber.
Mitt. Astron. Ges., Nr. 75, p. 9 – 10 (1992).

007.017   **Walter Orr Roberts (20 August 1915 – 12 March 1990).**
J. Firor.
Q. J. R. Astron. Soc., Vol. 33, No. 1, p. 35 – 37 (Mar 1992).

007.018   **In memoriam: Georges Roland (8 June 1922 – 26 October 1991).**
J. Dommanget.
Ciel Terre, Vol. 108, No. 1, p. 23 (Jan–Feb 1992).

007.019   **Harlan J. Smith, 25 August 1924 – 17 October 1991.**
T. G. Barnes III.
J. Astrophys. Astron., Vol. 13, No. 1, p. 145 – 150 (Mar 1992).

007.020   **Ricordo di Giuseppe Salvatore Vaiana.**
S. Serio.
G. Astron., Vol. 18, N. 2, p. 3 – 6 (Jun 1992).
Giuseppe Salvatore Vaiana died 25 August 1991.

007.021   **A. J. J. van Woerkom (3 October 1915 – 8 July 1991).**
R. L. Duncombe, M. S. Davis.
Celest. Mech. Dyn. Astron., Vol. 53, No. 1, p. 1 (1992).

007.022   **Hans Vehrenberg (6. März 1910 – 2. Oktober 1991).**
K. Schaifers.
Mitt. Astron. Ges., Nr. 75, p. 11 – 12 (1992).

008 Publications of Observatories, Institutes

Reports, communications, and publications of observatories and astronomical institutes are recorded in this section; included are numbered series of reprints. Whenever possible, the numbers of the abstracts referring to the publications are given. The places of observatories and institutes are listed in alphabetical order. If only the formal name of an observatory or institute is known, its place can be found in the following index list.

Aerospace Corporation — **Los Angeles, CA (USA)**
Allegheny Observatory — **Pittsburgh, PA (USA)**
Anglo-Australian Observatory — **Epping (Australia)**
Arcetri Astrophysical Observatory — **Florence (Italy)**
Archenhold Observatory — **Berlin (Germany)**
Argentine Institute of Radio Astronomy — **Villa Elisa (Argentina)**
Arizona State University — **Tempe, AZ (USA)**
Astronomical Institute of the Czechoslovak Academy
    of Sciences — **Ondřejov (Czechoslovakia)**
Astronomisches Rechen-Institut — **Heidelberg (Germany)**
AT&T Bell Laboratories — **Holmdel, NJ (USA)**

Babelsberg Observatory — **Potsdam (Germany)**
Bartol Research Foundation — **Newark, DE (USA)**
Battelle Memorial Institute — **Richland, WA (USA)**
Behlen Observatory — **Lincoln, NB (USA)**
Bergedorf Observatory — **Hamburg (Germany)**
Blindern Institute of Theoretical Astrophysics — **Oslo (Norway)**
Bosscha Observatory — **Lembang (Indonesia)**
Boyden Observatory — **Bloemfontein (South Africa)**
Braeside Observatory — **Flagstaff, AZ (USA)**
Bundesamt für Seeschiffahrt und Hydrographie — **Hamburg (Germany)**
Bureau Central de l'IERS — **Paris (France)**
Bureau International des Poids et Mesures, Time Section — **Sèvres (France)**

C. E. Kenneth Mees Observatory — **Rochester, NY (USA)**
California Institute of Technology — **Pasadena, CA (USA)**
Cantonal Observatory — **Neuchâtel (Switzerland)**
Capodimonte Astronomical Observatory — **Naples (Italy)**
Carlsberg Automatic Meridian Circle — **La Palma (Spain)**
Carnegie Institution of Washington, The Observatories — **Pasadena, CA (USA)**
Carter Observatory — **Wellington (New Zealand)**
Center for Astrophysics — **Cambridge, MA (USA)**
Center for High Angular Resolution
    Astronomy, Georgia State University — **Atlanta, GA (USA)**
Centre de Données Astronomiques de Strasbourg — **Strasbourg (France)**
Centro Astronomico Hispano-Aleman — **Calar Alto (Spain)**
Centro de Investigación de Astronomía (CIDA) — **Mérida (Venezuela)**
Cerro Calan National Astronomical Observatory — **Santiago (Chile)**
Cerro Tololo Interamerican Observatory — **La Serena (Chile)**
Charles University Astronomical Institute — **Prague (Czechoslovakia)**
Climenhaga Observatory — **Victoria (Canada)**
Clyde W. Tombaugh Observatory — **Lawrence, KS (USA)**
Cointe Observatory — **Liège (Belgium)**
Collurania Astronomical Observatory — **Teramo (Italy)**
Computer Sciences Corporation — **Beltsville, MD (USA)**
Cornell University — **Ithaca, NY (USA)**

David Dunlap Observatory — **Richmond Hill (Canada)**
Dearborn Observatory — **Evanston, IL (USA)**
Department of Astronomy and Space Sciences,
    Punjabi University — **Patiala (India)**
Department of Mathematics, University of Poona — **Poona (India)**
Dominion Astrophysical Observatory — **Victoria (Canada)**
Dominion Radio Astrophysical Observatory — **Penticton (Canada)**
Dudley Observatory — **Schenectady, NY (USA)**
Dunsink Observatory — **Dublin (Ireland)**
Dyer Observatory — **Nashville, TN (USA)**

Ebro Observatory — **Roquetas (Spain)**
Effelsberg Radio Observatory — **Bonn (Germany)**
Ege University Observatory — **Izmir (Turkey)**
Electronics Research Laboratory — **Göteborg (Sweden)**
Engelhardt Observatory — **Kazan' (Russia)**
Erwin W. Fick Observatory — **Ames, IA (USA)**

European Southern Observatory   **Garching** (Germany)
European Southern Observatory   **La Silla** (Chile)

Felix Aguilar Observatory   **San Juan** (Argentina)
Figl Observatory   **Vienna** (Austria)
Five College Astronomy Department   **Amherst,** MA (USA)
Floirac Observatory   **Bordeaux** (France)
Florida State University Radio Observatory   **Tallahassee,** FL (USA)
Flower and Cook Observatory   **Malvern,** PA (USA)
Franko State University Astronomical Observatory   **L'vov** (Ukraine)

Georgia State University   **Atlanta,** GA (USA)
Goddard Space Flight Center   **Greenbelt,** MD (USA)
Goethe Link Observatory   **Bloomington,** IN (USA)
Griffith Observatory   **Los Angeles,** CA (USA)

Hartebeesthoek Radio Astronomy Observatory   **Johannesburg** (South Africa)
Haute Provence Observatory   **Saint Michel l'Observatoire** (France)
Haystack Observatory   **Westford,** MA (USA)
Heinrich-Hertz-Institut   **Berlin** (Germany)
Heliophysical Observatory   **Debrecen** (Hungary)
Herzberg Institute of Astrophysics   **Ottawa** (Canada)
High Altitude Observatory   **Boulder,** CO (USA)
H.M. Nautical Almanac Office   **Cambridge,** (UK)
Hoher List Observatory   **Bonn** (Germany)
Hopkins Observatory   **Williamstown,** MA (USA)
Hvar Observatory   **Zagreb** (Croatia)

IBM Thomas J. Watson Research Center   **Yorktown Heights,** NY (USA)
Indian Institute of Astrophysics   **Bangalore** (India)
Infrared Telescope Facility   **Honolulu,** HI (USA)
Institut Astrofiziki   **Dushanbe** (Tajikistan)
Instituto de Astrofisica de Canarias   **La Laguna** (Spain)
International Earth Rotation Service   **Paris** (France)
International Latitude Observatory   **Carloforte-Cagliari** (Italy)
IUE Observatory   **Villafranca** (Spain)

James Clerk Maxwell Telescope   **Hilo,** HI (USA)
Jet Propulsion Laboratory   **Pasadena,** CA (USA)
Jodrell Bank Radio Observatory   **Manchester** (UK)
Judson B. Coit Observatory   **Boston,** MA (USA)

Kandilli Observatory   **Istanbul** (Turkey)
Kanzelhöhe Solar Observatory   **Graz** (Austria)
Kapteyn Astronomical Laboratory   **Groningen** (Netherlands)
Kiepenheuer Institut   **Freiburg** (Germany)
Kiso Observatory   **Tokyo** (Japan)
Kitt Peak National Observatory   **Tucson,** AZ (USA)
Kodaikanal Observatory   **Bangalore** (India)
Königstuhl Observatory   **Heidelberg** (Germany)
Konkoly Observatory   **Budapest** (Hungary)
Korean National Astronomical Observatory   **Seoul** (Korea)
Kwasan and Hida Observatories   **Kyoto** (Japan)

Laboratory for High Energy Astrophysics   **Greenbelt,** MD (USA)
Las Campanas Observatory   **Pasadena,** CA (USA)
Latitude Station of the Polish Academy of Sciences   **Borowiec** (Poland)
Lawrence Livermore National Laboratory   **Livermore,** CA (USA)
Leander McCormick Observatory   **Charlottesville,** VA (USA)
Lick Observatory   **Santa Cruz,** CA (USA)
Lindheimer Astronomical Research Center   **Evanston,** IL (USA)
Lockheed Palo Alto Research Laboratory   **Palo Alto,** CA (USA)
Lockheed Solar Observatory   **Saugus,** CA (USA)
Lohrmann Observatory   **Dresden** (Germany)
Louisiana State University Observatory   **Baton Rouge,** LA (USA)
Lowell Observatory   **Flagstaff,** AZ (USA)
Lunar and Planetary Laboratory   **Tucson,** AZ (USA)

Main Astronomical Observatory of the USSR Academy
   of Sciences   **Pulkovo** (Russia)
Max-Planck-Institut für Astronomie   **Calar Alto** (Spain)
Max-Planck-Institut für Astronomie   **Heidelberg** (Germany)
Max-Planck-Institut für Kernphysik   **Heidelberg** (Germany)
Max-Planck-Institut für Physik und Astrophysik,
   Institut für Astrophysik   **Garching** (Germany)

Max-Planck-Institut für Physik und Astrophysik, Institut für Extraterrestrische Physik	Garching (Germany)
Max-Planck-Institut für Radioastronomie	Bonn (Germany)
McDonald Observatory	Fort Davis, TX (USA)
McDonnell Center for the Space Sciences	St. Louis, MO (USA)
Meudon Observatory	Paris (France)
Michigan State University Observatory	East Lansing, MI (USA)
Millstone Hill Radar Observatory	Westford, MA (USA)
Minor Planet Center	Cambridge, MA (USA)
Molonglo Radio Observatory	Sydney (Australia)
Monterey Institute for Research in Astronomy	Carmel Valley, CA (USA)
Mount Hamilton	Santa Cruz, CA (USA)
Mount John University Observatory	Lake Tekapo (New Zealand)
Mount Palomar Observatory	Pasadena, CA (USA)
Mount Stromlo Observatory	Canberra (Australia)
Mount Wilson Observatory	Pasadena, CA (USA)
Mullard Radio Astronomy Observatory	Cambridge (UK)
Mullard Space Science Laboratory	London (UK)
Multiple Mirror Telescope Observatory	Tucson, AZ (USA)
N. Copernicus Astronomical Center	Warsaw (Poland)
N. Copernicus University Observatory	Torun (Poland)
NASA Headquarters	Washington, DC (USA)
National Astronomical Observatory	Tokyo (Japan)
National Astronomy and Ionosphere Center	Ithaca, NY (USA)
National Bureau of Standards	Washington, DC (USA)
National Radio Astronomy Observatory	Charlottesville, VA (USA)
National Radio Astronomy Observatory	Green Bank, WV (USA)
National Radio Astronomy Observatory	Socorro, NM (USA)
National Solar Observatory	Sunspot, NM (USA)
New Mexico State University Observatory	Las Cruces, NM (USA)
Nicholas Copernicus Observatory	Brno (Czechoslovakia)
Nizamiah and Japal-Rangapur Observatories	Hyderabad (India)
Nuffield Radio Astronomy Laboratories	Manchester (UK)
Observatorio del Ebro	Roquetas (Spain)
Ole Roemer Observatory	Aarhus (Denmark)
Onsala Space Observatory	Göteborg (Sweden)
Osservatorio Astronomico di Brera	Milano-Merate (Italy)
Owens Valley Radio Observatory	Big Pine, CA (USA)
Pennsylvania State University	University Park, PA (USA)
Perkins Observatory	Delaware, OH (USA)
Perth Observatory	Bickley (Australia)
Pic du Midi Observatory	Toulouse (France)
Pino Torinese Observatory	Turin (Italy)
Planetary Science Institute	Tucson, AZ (USA)
Purple Mountain Observatory	Nanking (China)
Raman Research Institute	Bangalore (India)
Rattlesnake Mountain Observatory	Richland, WA (USA)
Real Instituto y Observatorio de la Armada	San Fernando (Spain)
Remeis Observatory	Bamberg (Germany)
Rensselaer Observatory	Troy, NY (USA)
Ritter Astrophysical Research Center	Toledo, OH (USA)
Rosemary Hill Observatory	Bronson, FL (USA)
Rothney Astrophysical Observatory	Calgary (Canada)
Royal Greenwich Observatory	Cambridge (UK)
Royal Observatory	Edinburgh (UK)
Royal Observatory of Belgium	Uccle (Belgium)
Rutgers University	Piscataway, NJ (USA)
Rutherford Appleton Laboratory	Chilton (UK)
Sagamore Hill Radio Observatory	Hanscom, MA (USA)
Saltsjöbaden Observatory	Stockholm (Sweden)
San Vittore Observatory	Bologna (Italy)
Satelliten-Beobachtungsstation	Zimmerwald (Switzerland)
Shaanxi Astronomical Observatory	Lintong (China)
Smithsonian Astrophysical Observatory	Cambridge, MA (USA)
Sonnenborgh Observatory	Utrecht (Netherlands)
Sonoma State University	Rohnert Park, CA (USA)
South African Astronomical Observatory	Cape Town (South Africa)
Space Telescope Science Institute	Baltimore, MD (USA)
Special Astrophysical Observatory	Zelenchukskaya (Russia)
Sproul Observatory	Swarthmore, PA (USA)

Stanford Center for Radar Astronomy	**Menlo Park,** CA (USA)
Stellar Data Center	**Strasbourg** (France)
Sternberg State Astronomical Institute	**Moscow** (Russia)
Steward Observatory	**Tucson,** AZ (USA)
Stockert Radio Observatory	**Bonn** (Germany)
Struve Astrophysical Observatory	**Tartu** (Estonia)
Table Mountain Observatory	**Wrightwood,** CA (USA)
Tata Institute of Fundamental Research	**Bombay** (India)
Thüringer Landessternwarte	**Tautenburg** (Germany)
Tôhoku University Observatory	**Sendai** (Japan)
United Kingdom Infrared Telescope (UKIRT)	**Hilo,** HI (USA)
University of Sussex	**Brighton** (UK)
U.S. Naval Observatory	**Washington,** DC (USA)
UT Radio Astronomy Observatory	**Marfa,** TX (USA)
Uttar Pradesh State Observatory	**Naini Tal** (India)
Van Vleck Observatory	**Middletown,** CT (USA)
Vassar College Observatory	**Poughkeepsie,** NY (USA)
Vatican Observatory	**Castel Gandolfo** (Vatican City)
W. M. Keck Observatory	**Berkeley,** CA (USA)
W. M. Keck Observatory	**Kamuela,** HI (USA)
Warner and Swasey Observatory	**Cleveland,** OH (USA)
Washburn Observatory	**Madison,** WI (USA)
Wesleyan Radio Observatory	**Delaware,** OH (USA)
Western Ontario University Observatory	**London** (Canada)
Whipple Observatory (MMT)	**Tucson,** AZ (USA)
Wilhelm Foerster Observatory	**Berlin** (Germany)
William Marsh Rice University, Department of Space Physics and Astronomy	**Houston,** TX (USA)
Wyoming Infrared Observatory	**Laramie,** WY (USA)
Yale Astronomy Department	**New Haven,** CT (USA)
Yale University Astronomical Observatory	**New Haven,** CT (USA)
Yerkes Observatory	**Chicago,** IL (USA)
Yunnang Obs.	**Kunming** (China)
Zentralinstitut für Physik der Erde	**Potsdam** (Germany)

Asiago (Italy)

008.001   **TNG (Telescopio Nazionale Galileo)–Newsletter.**
No. 1 (Jan 1992).

008.002   **Telescopio Nazionale Galileo, Technical Report.**
Nos. 10 (55.031.076), 11 (55.034.108).

Baltimore, MD (USA)

008.003   **Space Telescope Science Institute, Newsletter.**
Vol. 9, No. 1 (Mar 1992).

Bamberg (Germany)

008.004   **Dr. Remeis–Sternwarte Bamberg, Astronomisches Institut der Universität Erlangen–Nürnberg. Jahresbericht für 1991.**
I. Bues.
Mitt. Astron. Ges., Nr. 75, p. 15 – 22 (1992).

Bangalore (India)

008.005   **Indian Institute of Astrophysics. Annual report 1989 – 90.**
M. Parthasarathy, S. S. Hasan.
Indian Inst. of Astrophysics, Bangalore, 70 p. (1991).

008.006   **Indian Institute of Astrophysics. Annual report 1990 – 91.**
M. Parthasarathy, S. S. Hasan, R. Srinivasan.
Indian Inst. of Astrophysics, Bangalore, 112 p. (1992).

008.007   **Kodaikanal Observatory Bulletins.**
Vol. 11 (1991) (55.012.072).

Basel (Switzerland)

008.008   **Astronomisches Institut der Universität Basel. Jahresbericht für 1991.**
G. A. Tammann.
Mitt. Astron. Ges., Nr. 75, p. 23 – 29 (1992).

Berlin (Germany)

008.009   **Institut für Astronomie und Astrophysik der Technischen Universität Berlin. Jahresbericht für 1991.**
E. Sedlmayr.
Mitt. Astron. Ges., Nr. 75, p. 31 – 38 (1992).

Bochum (Germany)

008.010   **Astronomisches Institut der Ruhr–Universität. Jahresbericht für 1991.**
T. Schmidt–Kaler.
Mitt. Astron. Ges., Nr. 75, p. 39 – 44 (1992).

008.011   **Institut für Theoretische Physik, Lehrstuhl IV der Ruhr–Universität Bochum. Jahresbericht für 1991.**
K. Schindler.
Mitt. Astron. Ges., Nr. 75, p. 45 – 46 (1992).

Bonn (Germany)

008.012   **Astronomische Institute der Universität Bonn. I. Sternwarte mit Observatorium Hoher List. Jahresbericht für 1991.**
K. S. de Boer.
Mitt. Astron. Ges., Nr. 75, p. 47 – 57 (1992).

008.013   **Astronomische Institute der Universität Bonn. II. Radioastronomisches Institut der Universität Bonn. Jahresbericht für 1991.**
U. Mebold.
Mitt. Astron. Ges., Nr. 75, p. 59 – 66 (1992).

008.014   **Astronomische Institute der Universität Bonn. III. Institut für Astrophysik und Extraterrestrische Forschung. Jahresbericht für 1991.**
M. Römer.
Mitt. Astron. Ges., Nr. 75, p. 67 – 74 (1992).

008.015   **Max–Planck–Institut für Radioastronomie. Jahresbericht für 1991.**
R. Schwartz.
Mitt. Astron. Ges., Nr. 75, p. 75 – 92 (1992).

008.016   **Max–Planck–Institut für Radioastronomie.**
Max–Planck–Gesellschaft, Berichte und Mitteilungen, No. 4, 104 p. (1992).
Contents: 1. Geschichte des Instituts. 2. Radioteleskope. 3. Internationale Zusammenarbeit. 4. Aus dem Forschungsprogramm: Radiokarten des Himmels. Interstellare Materie, Sternentstehung und das Zentrum der Milchstraße. Wasserstoff und Moleküle in Galaxien. Magnetfelder in Spiralgalaxien. Aktive galaktische Kerne. Neue hochauflösende Abbildungsmethoden für die optische Astronomie. Radiogalaxien und der Sunyaev–Zeldovich–Effekt.

Brno (Czechoslovakia)

008.017   **Contributions of the Nicholas Copernicus Observatory and Planetarium in Brno.**
No. 30 (1992).

Buenos Aires (Argentina)

008.018   **Contribuciones del Instituto Copérnico (Buenos Aires).**
Vol. 1, Nos. 1 (55.013.129), 2 (55.013.130), 3 (55.113.020).

Cambridge (UK)

008.019   **Gemini. Newsletter of the Royal Greenwich Observatory.**
Nos. 35 (Mar 1992), 36 (Jun 1992).

Cambridge, MA (USA)

008.020   **The Minor Planet Circulars/Minor Planets and Comets.**
Nos. 19347 – 20366 (Jan – Jun 1992).

008.021   **International Astronomical Union, Circular.**
Nos. 5420 – 5551 (Jan – Jun 1992).

Cape Town (South Africa)

008.022   **The South African Astronomical Observatory. Report for the year ending 1991 December 31.**
M. W. Feast.
South African Astronomical Observatory, Cape Town, 47 p. (1992). ISBN 0–9583227–1–6. ISSN 0250–0671.

008.023   **SAAO Newsletter.**
No. 19 (Apr 1992).

008.024   **South African Astronomical Observatory. Report for the year ending 1990 December 31.**
M. W. Feast.
Q. J. R. Astron. Soc., Vol. 33, No. 2, p. 111 – 151 (Jun 1992).

Coimbra (Portugal)

008.025   **Longitudinal position of sunspots and chromospheric filaments.**
Vol. 1, Fasc. 6 – 8 (1990) (55.072.033 – 55.072.035).

Cologne (Germany)

008.026   **I. Physikalisches Institut der Universität zu Köln. Jahresbericht für 1991.**
G. Winnewisser.
Mitt. Astron. Ges., Nr. 75, p. 305 – 310 (1992).

Crimea (Ukraine)

008.027   **Bulletin of the Crimean Astrophysical Observatory.**
Vol. 82 (1992).

Dresden (Germany)

008.028   **Lohrmann–Observatorium, Lehrstuhl für Astronomie im Institut für Planetare Geodäsie der Technischen Universität Dresden. Jahresbericht für 1991.**
K.–G. Steinert.
Mitt. Astron. Ges., Nr. 75, p. 93 – 95 (1992).

Epping (Australia)

008.029   **AAO Newsletter.**
Nos. 60 (Jan 1992), 61 (Apr 1992).

008.030   **Anglo–Australian Observatory. Report for the year 1990 July 1 – 1991 June 30.**
R. D. Cannon, D. A. Allen.
Q. J. R. Astron. Soc., Vol. 33, No. 2, p. 99 – 109 (Jun 1992).

Frankfurt (Germany)

008.031   **Institut für Theoretische Physik (Astrophysik). Jahresbericht für 1991.**
W. H. Kegel.
Mitt. Astron. Ges., Nr. 75, p. 97 – 98 (1992).

Freiburg (Germany)

008.032   **Kiepenheuer–Institut für Sonnenphysik. Jahresbericht für 1991.**
E. H. Schröter.
Mitt. Astron. Ges., Nr. 75, p. 99 – 117 (1992).

Garching (Germany)

008.033   **Space Telescope–European Coordinating Facility Newsletter.**
No. 17 (Feb 1992).

008.034   **European Southern Observatory. Annual report 1991.**
H. van der Laan.
European Southern Observatory, Garching (Germany), 131 p. (1992). In English, French and German. ISSN 0531–4496.

008.035   **MPE Reports.**
Nos. 227 (55.012.020), 232 (55.117.250), 233 (55.084.059), 236 (55.036.148), 237 (55.034.073), 238 (55.161.400), 239 (55.034.074).

008.036   **ESO Scientific Report.**
No. 11 (55.003.039).

008.037   **Max–Planck–Institut für Astrophysik. Jahresbericht für 1991.**
W. Hillebrandt.
Mitt. Astron. Ges., Nr. 75, p. 119 – 146 (1992).

008.038   **Max–Planck–Institut für Extraterrestrische Physik. Jahresbericht für 1991.**
G. Morfill.
Mitt. Astron. Ges., Nr. 75, p. 147 – 163 (1992).

008.039   **The Messenger – El Mensajero.**
Nos. 67 (Mar 1992), 68 (Jun 1992).

008.040   **Max–Planck–Institut für Physik und Astrophysik, Institut für Astrophysik. Proceedings.**
MPA/P—6 (55.012.101).

Göttingen (Germany)

008.041   **Universitäts–Sternwarte. Jahresbericht für 1991.**
F. Kneer.
Mitt. Astron. Ges., Nr. 75, p. 165 – 175 (1992).

008.042   **Arbeitsgemeinschaft Deutsches Großteleskop (AG–DGT). Jahresbericht für 1991.**
K. J. Fricke.
Mitt. Astron. Ges., Nr. 75, p. 177 – 178 (1992).

Graz (Austria)

008.043   **Institut für Astronomie der Universität Graz, Institut für Astronomie (Universitätssternwarte), Observatorium Lustbühel, Sonnenobservatorium Kanzelhöhe. Jahresbericht für 1991.**
H. Haupt.
Mitt. Astron. Ges., Nr. 75, p. 179 – 185 (1992).

Groningen (Netherlands)

008.044   **Kapteyn Astronomical Institute, Department of Astronomy, University of Groningen. Annual report 1991.**
P. van der Kruit.
Rijksuniversiteit Groningen (Netherlands). Kapteyn Instituut, 110 p. (1992). ISSN 0925–5826.

Hamburg (Germany)

008.045   **Hamburger Sternwarte. Jahresbericht für 1991.**
D. Reimers.
Mitt. Astron. Ges., Nr. 75, p. 187 – 193 (1992).

008.046   **Bundesamt für Seeschiffahrt und Hydrographie. Jahresbericht 1991.**
P. Ehlers.
Bundesamt für Seeschiffahrt und Hydrographie, Hamburg (Germany), 223 p. (1992).

Hannover (Germany)

008.047   **Astronomische Station am Institut für Erdmessung der Universität. Jahresbericht für 1991.**
G. Seeber.
Mitt. Astron. Ges., Nr. 75, p. 195 – 198 (1992).

Heidelberg (Germany)

008.048   **Max–Planck–Institut für Kernphysik Heidelberg. Jahresbericht 1991.**
H. V. Klapdor–Kleingrothaus, J. Kiko.
Max–Planck–Institut für Kernphysik, Heidelberg (Germany), 285 p. (1992).

008.049   **Astronomische Grundlagen für den Kalender 1994.**
(55.046.006).

008.050   **Veröffentlichungen – Astronomisches Rechen–Institut Heidelberg.**
No. 33 (1991) (55.002.039).

008.051   **Astronomisches Rechen–Institut. Jahresbericht für 1991.**
R. Wielen.
Mitt. Astron. Ges., Nr. 75, p. 199 – 214 (1992).

008.052   **Institut für Theoretische Astrophysik. Jahresbericht für 1991.**
W. M. Tscharnuter.
Mitt. Astron. Ges., Nr. 75, p. 215 – 224 (1992).

008.053   **Landessternwarte. Jahresbericht für 1991.**
I. Appenzeller.
Mitt. Astron. Ges., Nr. 75, p. 225 – 245 (1992).

008.054   **Max–Planck–Institut für Astronomie. Jahresbericht für 1991.**
H. Elsässer.
Mitt. Astron. Ges., Nr. 75, p. 247 – 280 (1992).

008.055   **Astronomisches Rechen–Institut Heidelberg, Mitteilungen, Serie A.**
Nr. 233 (55.041.016), 234 (55.151.023).

008.056   **Astronomisches Rechen–Institut Heidelberg, Mitteilungen, Serie B.**
Nr. 179 (55.036.089), 180 (55.002.042), 181 (55.036.102), 182 (55.036.104), 183 (55.036.105).

008.057   **Astronomy and Astrophysics Abstracts.**
Vol. 53 (55.002.127).

Helsinki (Finland)

008.058   **Helsinki University of Technology, Metsähovi Radio Research Station, Report Series A.**
Nos. 10 (1992) (55.077.021), 11 (1992) (55.077.022), 12 (55.077.030).

Hilo, HI (USA)

008.059   **United Kingdom Infrared Telescope. Annual report 1991.**
Royal Observatory, Edinburgh (UK); Joint Astronomy Center, Hilo, HI (USA), 53 p. (1992).

Houston, TX (USA)

008.060   **Lunar and Planetary Institute Technical Report.**
LPI–TR—91–01 (55.012.026), 91–02 (55.012.027), 91–03 (55.012.028), 92–01 (55.012.030), 92–02 (55.012.031), 92–03 (55.012.032).

Innsbruck (Austria)

008.061   **Institut für Astronomie der Leopold–Franzens–Universität. Jahresbericht für 1991.**
J. Pfleiderer.
Mitt. Astron. Ges., Nr. 75, p. 281 – 284 (1992).

Istanbul (Turkey)

008.062   **Publications of the Istanbul University Observatory.**
Nos. 157 (55.072.048), 158 (55.155.147).

Jena (Germany)

008.063   **Astrophysikalisches Institut und Universitäts–Sternwarte Jena. Jahresbericht für 1991.**
W. Pfau.
Mitt. Astron. Ges., Nr. 75, p. 285 – 289 (1992).

Kiel (Germany)

008.064   **Institut für Theoretische Physik und Sternwarte der Universität Kiel. Jahresbericht für 1991.**
G. Hensler.
Mitt. Astron. Ges., Nr. 75, p. 291 – 300 (1992).

008.065   **Institut für Reine und Angewandte Kernphysik, Abteilung Mathematische Physik. Jahresbericht für 1991.**
K. O. Thielheim.
Mitt. Astron. Ges., Nr. 75, p. 301 – 304 (1992).

Kunming (China)

008.066   **Publications of Yunnan Observatory.**
No. 1 (1992).

Kyoto (Japan)

008.067   **Contributions from the Department of Astronomy, Kyoto University.**
No. 300 (1991) (55.002.062, 55.002.063).

La Laguna (Spain)

008.068   **International Scientific Committee. Annual report 1991.**
(*Comite Cientifico Internacional. Informe anual 1991*).
Instituto de Astrofisica de Canarias, La Laguna (Spain), Observatorio del Roque de los Muchachos, Observatorio del Teide, 52 p. (1992). In English and Spanish.

La Palma (Spain)

008.069   **Carlsberg Meridian Catalogue, La Palma.**
No. 6 (55.002.071).

Lake Tekapo (New Zealand)

008.070   **Astronomy at the University of Canterbury, Physics Department and Mt John University Observatory. Annual report 1990.**
P. L. Cottrell, J. B. Hearnshaw.
South. Stars, Vol. 34, No. 7, p. 407 – 411 (Jun 1992).

008.071   **Astronomy at the University of Canterbury, Physics Department and Mt John University Observatory. Annual report 1991.**
J. B. Hearnshaw.
South. Stars, Vol. 34, No. 7, p. 412 – 417 (Jun 1992).

Locarno (Switzerland)

008.072   **Istituto Ricerche Solari Locarno. Jahresbericht für 1991.**
M. Bianda.
Mitt. Astron. Ges., Nr. 75, p. 311 – 312 (1992).

Moscow (Russia)

008.073   **Trudy Gosudarstvennogo Astronomicheskogo Instituta im. P. K. Shternberga.**
Tom 63 (1991) (55.002.058).

Münster (Germany)

008.074   **Astronomisches Institut der Westfälischen Wilhelms–Universität. Jahresbericht für 1991.**
W. C. Seitter.
Mitt. Astron. Ges., Nr. 75, p. 323 – 335 (1992).

Munich (Germany)

008.075   **Institut für Astronomie und Astrophysik der Universität München, Universitäts–Sternwarte. Jahresbericht für 1991.**
R.–P. Kudritzki.
Mitt. Astron. Ges., Nr. 75, p. 313 – 322 (1992).

Naini Tal (India)

008.076   **Uttar Pradesh State Observatory, Reprint.**
Nos. 409 (51.073.135), 410 (46.071.005), 411 (51.103.038), 412 (49.153.017), 413 (51.119.056), 414 (51.119.054), 415 (51.119.055), 416 (51.112.146), 417 (51.117.289), 418 (51.153.017), 419 (51.119.065), 420 (49.103.191), 421 (55.112.116), 422 (55.103.143), 423 (52.131.118), 424 (51.103.068), 425 (52.155.008), 426 (50.156.005), 427 (52.071.033), 428 (53.022.002), 429 (54.153.044), 430 (54.153.045), 431 (54.153.060), 432 (55.031.079), 433 (52.071.049), 434 (53.119.090), 435 (53.119.032).

Nanking (China)

008.077   **Publications of Purple Mountain Observatory.**
Vol. 10, No. 3 (Sep 1991).
Vol. 11, Nos. 1 (Mar 1992), 2 (Jun 1992).

Ondřejov (Czechoslovakia)

008.078   **Circular of the Czechoslovak Observatories, Time and Latitude.**
Jul – Sep 1991 (55.044.008).

Paris (France)

008.079   **IERS Bulletin. B.**
Nos. 47 – 52 (Jan – Jun 1992) (55.044.010).

008.080   **IERS (International Earth Rotation Service) Technical Note.**
No. 10 (Jan 1992) (55.043.009).

Peking (China)

008.081   **Publications of the Beijing Astronomical Observatory.**
No. 18 (Oct 1991).

Poona (India)

008.082   **Khagol. The IUCAA Bulletin.**
Nos. 9 (Jan 1992), 10 (Apr 1992).

Potsdam (Germany)

008.083   **Zentralinstitut für Astrophysik: Sternwarte Babelsberg, Sternwarte Sonneberg, Karl–Schwarzschild–Observatorium Tautenburg, Sonnenobservatorium Einsteinturm, Observatorium für solare Radioastronomie Tremsdorf. Jahresbericht für 1991.**
K.–H. Rädler.
Mitt. Astron. Ges., Nr. 75, p. 337 – 359 (1992).

Quito (Ecuador)

008.084   **Boletin del Observatorio Astronomico de Quito.**
Jan 1992.

Roquetas (Spain)

008.085   **Publicaciones del Observatorio del Ebro, Memoria.**
No. 15 (1992).

Saint Michel l'Observatoire (France)

008.086   **La Lettre de l'OHP.**
Observatoire de Haute–Provence, St. Michel l'Observatoire (France)
Nos. 8 (Feb 1992), 9 (Jun 1992).

008.087   **Observatoire de Haute Provence. Rapport d'activité, exercice 1991.**
P. Véron, C. Chevalier, S. Guyot.
Centre National de la Recherche Scientifique, Observatoire de Haute Provence, St. Michel l'Observatoire (France), 81 p. (1992). ISSN 0750–6650.

Santa Cruz, CA (USA)

008.088   **Lick Observatory Bulletin.**
Nos. 1180 (53.121.005), 1187 (54.154.102), 1188 (54.158.155), 1189 (54.101.053), 1190 (54.158.085), 1191 (54.151.111), 1192 (54.116.081), 1193 (54.153.053), 1195 (55.063.016), 1196 (55.151.015), 1197 (55.151.016), 1198 (54.154.124), 1199 (55.111.003), 1200 (54.155.086), 1201 (55.153.010), 1202 (55.158.097), 1203 (55.132.020), 1204 (55.117.162), 1205 (55.082.010), 1206 (55.065.076), 1208 (55.080.026), 1209 (55.154.047), 1210 (55.117.172), 1211 (55.157.034), 1214 (55.062.087).

Sendai (Japan)

008.089   **Sendai Astronomiaj Raportoj.**
Nos. 371 (53.065.124), 374 (54.065.009), 377 (53.151.094), 378 (53.065.130), 379 (54.062.181), 380 (54.065.112).

Sèvres (France)

008.090   **Bureau International des Poids et Mesures (BIPM), Circular T.**
Nos. 47 – 52 (Jan – Jun 1992) (55.044.009).

008.091   **Annual Report of the BIPM Time Section.**
Vol. 4 (1991) (55.044.013).

Shanghai (China)

008.092   **Annals of Shanghai Observatory, Academia Sinica.**
No. 13 (1992).

Sonneberg (Germany)

008.093   **Mitteilungen der Sternwarte zu Sonneberg.**
Nr. 98: Sterne, Band 68, Heft 1 (1992) (55.005.005, 55.005.006, 55.002.011, 55.013.011, 55.002.012, 55.005.007).

Strasbourg (France)

008.094   Publication de l'Observatoire Astronomique de Strasbourg, Série "Astronomie et Science Humaines".
No. 6 (1992).

008.095   Publication Spéciale du Centre de Données Stellaires.
No. 19 (55.002.096).

008.096   Bulletin d'Information du Centre de Données Astronomiques de Strasbourg.
No. 40 (Jan 1992).

Tokyo (Japan)

008.097   Data Report of Hydrographic Observations, Series of Astronomy and Geodesy.
No. 26 (Mar 1992) (55.096.010).

008.098   Quarterly Bulletin on Solar Activity.
Vol. 31   (1989),   Part I   (55.072.041),   Part II (55.075.021), Part IV (55.074.046).

008.099   Publications of the National Astronomical Observatory of Japan.
Vol. 2, No. 3 (1992).

008.100   Report of the National Astronomical Observatory of Japan.
Vol. 1, No. 3 (1992).

008.101   National Astronomical Observatory (Japan), Reprint.
Nos. 117 (55.073.063), 131 (55.080.015), 133 (55.067.055), 134 (55.155.015), 135 (55.073.062), 137 (55.161.105), 138 (55.131.045), 139 (55.101.089), 140 (55.098.142), 141 (55.103.067), 142 (55.161.235), 143 (55.034.125), 144 (55.067.075), 145 (55.062.057), 146 (55.155.119), 147 (55.107.019), 149 (55.131.238), 152 (55.159.125), 154 (55.131.278).

Tübingen (Germany)

008.102   Universität Tübingen, Astronomisches Institut. Jahresbericht für 1991.
R. Staubert.
Mitt. Astron. Ges., Nr. 75, p. 361 – 372 (1992).

008.103   Universität Tübingen, Institut für Theoretische Astrophysik. Jahresbericht für 1991.
H. Ruder.
Mitt. Astron. Ges., Nr. 75, p. 373 – 388 (1992).

Turku (Finland)

008.104   Tuorla Observatory. Annual report 1991.
P. Teerikorpi.
Tuorla Obs. Rep., Informo, No. 165, 40 p. (1992).

008.105   Tuorla Observatory Reports, Informo.
Nos. 165 (55.008.104), 166 (55.012.033).

Vienna (Austria)

008.106   Institut für Astronomie der Universität Wien. Jahresbericht für 1991.
P. Jackson.
Mitt. Astron. Ges., Nr. 75, p. 389 – 402 (1992).

008.107   Institut für Mathematik der Universität, Extraordinariat für Angewandte Mathematik mit besonderer Berücksichtigung der Astrophysik. Jahresbericht für 1991.
H. Muthsam.
Mitt. Astron. Ges., Nr. 75, p. 403 – 404 (1992).

Washington, DC (USA)

008.108   Time Service Announcement, Series 14.
No. 52 (55.044.005).

008.109   United States Naval Observatory, Series 4, Daily time differences.
Nos. 1300 – 1325 (Jan – Jun 1992) (55.044.006).

008.110   IERS Bulletin. A.
Vol. 5, Nos. 1 – 26 (Jan – Jun 1992) (55.044.007).

008.111   Publications of the United States Naval Observatory, Second Series.
Vol. 26, Part 2 (1992) (55.041.008).

Würzburg (Germany)

008.112   Institut für Astronomie and Astrophysik, Lehrstuhl Astronomie. Jahresbericht für 1991.
F.–L. Deubner.
Mitt. Astron. Ges., Nr. 75, p. 405 – 409 (1992).

Zagreb (Croatia)

008.113   Hvar Observatory Bulletin.
Vol. 15, No. 1 (1991).

Zelenchukskaya (Russia)

008.114   Bulletin of the Special Astrophysical Observatory – North Caucasus.
Vol. 31 (1991).

Zimmerwald (Switzerland)

008.115   Mitteilungen der Satelliten–Beobachtungsstation Zimmerwald.
Nr. 27 (55.031.020).

008.116   Satelliten–Beobachtungsstation Zimmerwald, Bericht.
Nr. 21 (55.044.019).

Zurich (Switzerland)

008.117   Institut für Astronomie. Jahresbericht für 1991.
J. O. Stenflo.
Mitt. Astron. Ges., Nr. 75, p. 411 – 419 (1992).

009 Notes on Observatories, Planetaria, Exhibitions

009.001 Planetaria – the American way.
A. Rükl.
Říše hvězd, Vol. 72, No. 3, p. 41 – 43 (Mar 1991). In Czech.

009.002 Visiting European Southern Observatory.
L. Kohoutek.
Říše hvězd, Vol. 72, No. 7, p. 128 – 132 (Jul 1991). In Czech.

009.003 Hundred and one years of the Dr. Remeis Observatory.
V. Vanýsek.
Říše hvězd, Vol. 72, No. 9, p. 174 – 177 (Sep 1991). In Czech.

009.004 Report from the Canary Islands.
M. Sobotka.
Říše hvězd, Vol. 72, No. 11, p. 209 – 210 (Nov 1991). In Czech.

009.005 Zur Geschichte der Sternwarte Sonneberg.
G. A. Richter.
Cuno Hoffmeister. Festschrift zum 100. Geburtstag, p. 51 – 81 (1992). – See Abstr. 003.022 for the main entry.

009.006 Fifteen years of the activity of the Space Research Centre.
K. Ziołkowski.
Urania, Rok 63, Nr. 2, p. 38 – 49 (Feb 1992). In Polish.

009.007 A public observatory trying at fresh activities. Nishi Harima Astronomical Observatory.
T. Kuřoda.
Astron. Her., Vol. 85, No. 4, p. 155 – 161 (1992). In Japanese.

009.008 Three nights on Kitt Peak.
D. Bruning.
Astronomy, Vol. 20, No. 4, p. 38 – 43 (Apr 1992).

009.009 Der Himmel steht Kopf – zu Gast bei ESO auf La Silla.
R. Luthardt.
Ahnerts Kalender für Sternfreunde 1993.
p. 175 – 176 (1992). With plates 35, 49 – 55. – See Abstr. 046.012 for the main entry.

009.010 LEP, the Laboratory for Electrostrong Physics, one year later.
U. Amaldi.
15. Texas Symposium on Relativistic Astrophysics and 4. ESO–CERN Symposium, p. 244 – 261 (1991). – See Abstr. 012.060 for the main entry.
Contents: 1. Introduction. 2. Around the Z–peak. 3. Search for new particles. 4. Measurements of the strong coupling constant. 5. A LEP view of grand unified theories. 6. Conclusions.

009.011 The objectives, construction, and many uses of the Drew Observatory.
J. C. Dawson Jr.
J. Am. Assoc. Variable Star Obs., Vol. 20, No. 1, p. 108 (1991). Abstract. – See Abstr. 012.066 for the main entry.

009.012 A geological description of Cerro Paranal or another insight into the "perfect site for astronomy".
F. Bourlon.
Messenger, No. 67, p. 4 – 9 (Mar 1992).

009.013 Observatory profiles from the CSFR.
P. Ambrož, A. Kučera.
Annual report, 1991.
Joint Organization for Solar Observations (JOSO). p. 27 – 36 (1992).
Contents: 1. Ondřejov Observatory. 2. Tatranská Lomnica Observatory.

009.014 Potchefstroom University for C.H.E.: Cosmic Ray Research Unit (Astrophysics Division). Report for 1991.
Mon. Notes Astron. Soc. S. Afr., Vol. 51, Nos. 3/4, p. 23 – 26 (Apr 1992).

009.015 Rhodes University. Department of Physics and Electronics. Report for 1991.
Mon. Notes Astron. Soc. S. Afr., Vol. 51, Nos. 3/4, p. 27 – 28 (Apr 1992).

009.016 Observations and researches at the Huairou Solar Observing Station in 1989.
Zhang Hongqi, Ai Guoxiang, Ming Changrong.
Publ. Beijing Astron. Obs., No. 18, p. 62 – 65 (Oct 1991).

009.017 The New Zealand Astronomy Centre.
G. van Dijk.
South. Stars, Vol. 34, No. 7, p. 403 – 406 (Jun 1992).

009.018 Algol–observatoriet på Frederiksberg.
P. B. Darnell.
Astron. Tidsskr., Årg. 25, Nr. 1, p. 24 – 26 (Mar 1992).

009.019 Die Urania–Sternwarte Zürich. Ein Besucherbericht.
D. B. Herrmann.
Astron. Sch., Jahrg. 29, Heft 9, p. 26 – 27 (Jun 1992). In German.

009.020 ESO exhibitions in Chile – a tremendous success.
P. Bouchet, A. Cabillic, C. Madsen.
Messenger, No. 68, p. 18 – 20 (Jun 1992).

009.021 A most impressive astronomy exhibition.
R. West.
Messenger, No. 68, p. 21 – 22 (Jun 1992).
A new 1000 m² astronomy exhibition in the Deutsches Museum, Munich (Germany).

009.022 A planetarium and observatory complex for Sydney.
A. E. Vaughan.
Proc. Astron. Soc. Aust., Vol. 9, No. 2, p. 338 – 339 (1991). – See Abstr. 012.090 for the main entry.
This paper describes the plans for a planetarium and observatory at Sydney's Macquarie University.

009.023 CCD observations from the centre of a city: use in astronomical education.
N. R. Lomb.
Proc. Astron. Soc. Aust., Vol. 9, No. 2, p. 340 – 341 (1991). – See Abstr. 012.090 for the main entry.
One of the two main telescopes at Sydney Observatory is a computer–controlled 35.6–cm Celestron. It is used with a television–type system. There is an integrating CCD camera available as well as an ordinary TV camera for bright objects. The images from the cameras can be shown in the Observatory's lecture room. They are first put through an image processor which allows constant enhancement, spatial filtering and the recording of images for use on cloudy nights. Preliminary results are encouraging, and show the value of the system, even in Sydney's highly light–polluted sky.

009.024 Tremsdorf solar radio astronomy observatory – the scientific programme and an interesting observation.
H. Aurass.
International Workshop on Reconnection in Space Plasma, p. 139 – 144 (Jan 1989). – See Abstr. 012.099 for the main entry.

009.025 **Das Institut für Astronomie an der Leopold–Franzens–Universität Innsbruck in der Gegenwart. Aktuelle Forschungsschwerpunkte.**
J. Pfleiderer.
Hundert Jahre Astronomie an der Leopold–Franzens–Universität Innsbruck (1892 – 1992), p. 105 – 116 (1992). – See Abstr. 003.067 for the main entry.

009.026 **Nuevo observatorio en Sonora, Mexico.**
A. Sánchez Ibarra.
Universo, Vol. 12, No. 36, p. 21 (Feb – Apr 1992).

Der Himmel auf Erden. Die Welt der Planetarien.
See Abstr. 003.006.

ESO's early history. The European Southern Observatory from concept to reality.
See Abstr. 003.035.

The protection of astronomical and geophysical sites.
See Abstr. 003.038.

Der Einstein–Turm. Erwin F. Freundlich und die Relativitätstheorie – Ansätze zu einer "dichten Beschreibung" von institutionellen, biographischen und theoriengeschichtlichen Aspekten.
See Abstr. 003.055.

Die Lilienthaler Sternwarte 1781 bis 1818. Machinae Coelestes Lilienthalienses: Die Instrumente. Eine zeitgeschichtliche Dokumentation.
See Abstr. 003.062.

Sternwarten in Bildern. Architektur und Geschichte der Sternwarten von den Anfängen bis ca. 1950.
See Abstr. 003.079.

Carter Observatory's 9–inch refractor: the Crossley connection.
See Abstr. 004.035.

A brief history of the Crossley 36–inch reflector.
See Abstr. 004.036.

Lord Crawford's Observatory at Dun Echt, 1872 – 1892.
See Abstr. 004.049.

Yale and USNO cooperation especially in the Brouwer and Clemence era.
See Abstr. 004.052.

A brief history of the Minor Planet Center.
See Abstr. 004.078.

Protection of observatories: the legal avenues.
See Abstr. 013.034.

Ten years monitoring of blazars at Metsähovi.
See Abstr. 013.122.

The CCD camera in the Suhora Observatory.
See Abstr. 034.049.

A history of laser ranging at McDonald Observatory.
See Abstr. 045.005.

Observing conditions at the Wise Observatory.
See Abstr. 082.014.

The influence of the Pinatubo eruption on the atmospheric extinction at La Silla.
See Abstr. 082.063.

010 Societies, Associations, Organizations

American Association of Variable Star Observers (AAVSO)

010.001 **Solar Bulletin (American Association of Variable Star Observers).**
Vol. 48, Nos. 1 – 6 (Jan – Jun 1992).

010.002 **American Association of Variable Star Observers Circular.**
Nos. 255 – 260 (Jan – Jun 1992).

010.003 **SID Technical Bulletin.**
Vol. 3, Nos. 1 (Jan 1992), 2 (Apr 1992).

010.004 **The Journal of the American Association of Variable Star Observers.**
Vol. 20, No. 1 (1991).

010.005 **Meetings, activities, and committee reports of the AAVSO.**
J. Am. Assoc. Variable Star Obs., Vol. 20, No. 1, p. 112 – 127 (1991).

010.006 **AAVSO Eclipsing Binary Bulletin.**
M. E. Baldwin.
AAVSO Eclipsing Binary Bull., No. 48, p. 1 – 2 (Jan 1992).
Concerning three projects which involve the stars V342 Aql, TU Mon and EE Cep.

American Astronomical Society (AAS)

010.007 **Bulletin of the American Astronomical Society.**
Vol. 23, No. 3 (1991) (012.037, 012.065, 012.067).

010.008 **1990 – 91 annual report of the American Astronomical Society.**
Bull. Am. Astron. Soc., Vol. 23, No. 3, p. 1081 – 1099 (1991).

010.009 **Some historical notes on the DPS/AAS.**
Bull. Am. Astron. Soc., Vol. 23, No. 3, p. 1245 – 1246 (1991).

010.010 **Minutes of the Division for Planetary Sciences Annual Business Meeting.**
L. A. Lebofsky.
Bull. Am. Astron. Soc., Vol. 23, No. 3, p. 1247 – 1251 (1991).

010.011 **1993 Membership Directory.**
See Abstr. 002.148.

Association Française des Observateurs d'Etoiles Variables

010.012   **Bulletin de l'Association Française des Observateurs d'Etoiles Variables.**
Nos. 59, 60 (1992).

010.013   **La vie de l'Association.**
E. Schweitzer.
Bull. Assoc. Fr. Obs. Etoiles Variables, No. 59, p. 18 (1992).
Further information is given in No. 60, p. 25 (1992).

010.014   **Activité de l'AFOEV en 1991 (1ère partie).**
E. Schweitzer.
Bull. Assoc. Fr. Obs. Etoiles Variables, No. 60, p. 1 – 14 (1992).
Contents: 1. Activité générale. 2. Variables du type Mira.
3. Variables semi–régulières. 4. Variables du type RV Tauri.
5. Variables à éclipses.

Association of Lunar and Planetary Observers (A.L.P.O.)

010.015   **The Strolling Astronomer. The Journal of the Association of Lunar and Planetary Observers.**
Vol. 36, No. 1 (Mar 1992).

010.016   **The Minor Planet Bulletin.**
Vol. 19, Nos. 1 (Jan – Mar 1992), 2 (Apr – Jun 1992).

Astronomical Society of Australia (ASA)

010.017   **Proceedings of the Astronomical Society of Australia.**
Vol. 9, No. 2 (1991) (012.090).

Astronomical Society of Japan

010.018   **The Astronomical Herald.**
Vol. 85, Nos. 1 – 6 (1992). In Japanese.

010.019   **Publications of the Astronomical Society of Japan.**
Vol. 44, Nos. 1 – 3 (1992).

Astronomical Society of New York

010.020   **News Letter of the Astronomical Society of New York.**
Vol. 4, No. 1 (Feb 1992).

Astronomical Society of Southern Africa (ASSA)

010.021   **Monthly Notes of the Astronomical Society of Southern Africa.**
Vol. 51, Nos. 1/2 (Feb 1992), 3/4 (Apr 1992), 5/6 (Jun 1992).

Astronomical Society of the Pacific (ASP)

010.022   **Publications of the Astronomical Society of the Pacific.**
Vol. 104, Nos. 671 – 676 (Jan – Jun 1992).

010.023   **Mercury.**
Vol. 21, Nos. 1 (Jan – Feb 1992), 2 (Mar – Apr 1992), 3 (May – Jun 1992).

010.024   **The Universe in the Classroom.**
No. 20 (Spr 1992).

Astronomical Society of Western Australia (ASWA)

010.025   **The Sidereal Times.**
Vol. 42, Nos. 4 (Feb – Mar 1992), 5 (Apr – May 1992), 6 (Jun – Jul 1992).

Astronomische Gesellschaft (AG)

010.026   **Mitteilungen der Astronomischen Gesellschaft.**
Nr. 75 (1992) (Jahresberichte Astronomischer Institute für 1991).

010.027   **Die Geschichte der Astronomischen Gesellschaft gegründet in Lilienthal am 20. September 1800. Die ersten 63 Jahre ihres Bestehens von 1800 bis 1863.**
See Abstr. 003.046.

010.028   **Reviews in Modern Astronomy.**
Vol. 5 (1992) (003.087).

British Astronomical Association (BAA)

010.029   **The British Astronomical Association Newsletter.**
Nos. 52 (Feb 1992), 53 (Apr 1992), 54 (Jun 1992).

010.030   **Journal of the British Astronomical Association.**
Vol. 102, No. 1 (Feb 1992).

010.031   **Meeting reports.**
J. Br. Astron. Assoc., Vol. 102, No. 1, p. 63 – 67 (Feb 1992).
Further information is given in Vol. 103, No. 2, p. 176 – 182 (Jun 1992).

British Interplanetary Society (BIS)

010.032   **Spaceflight.**
Vol. 34, Nos. 1 – 6 (Jan – Jun 1992).

010.033   **Journal of the British Interplanetary Society.**
Vol. 45, Nos. 1 – 6 (Jan – Jun 1992).

Bundesdeutsche Arbeitsgemeinschaft für Veränderliche Sterne (BAV)

010.034   **BAV Rundbrief.**
41. Jahrg., Nr. 1, 2 (1992).

010.035   **BAV Blätter.**
Nr. 11 (1992).

010.036   **BAV Mitteilungen.**
Nr. 60 (1992) (55.123.027).

European Space Agency (ESA)

010.037   **ESA Bulletin.**
Nos. 69 (Feb 1992), 70 (May 1992).

010.038   **ESA Special Publication.**
ESA–SP—285(Vol. 2)   (012.099),   ESA–SP—317 (012.088), ESA–SP—328 (012.093), ESA–SP—329 (012.082), ESA–SP—330 (012.071), ESA–SP—1152 (157.286), ESA–SP—1136 (002.055).

010.039   **ESA Journal.**
Vol. 16, Nos. 1, 2 (1992).

010.040   **ESA IUE Newsletter.**
Nos. 39 (Jan 1992), 40 (Mar 1992).

International Amateur–Professional Photoelectric Photometry

010.041   **International Amateur–Professional Photoelectric Photometry Communication.**
No. 47 (Mar – May 1992).

International Astronomical Union (IAU)

010.042   **The Minor Planet Circulars/Minor Planets and Comets.**
Nos. 19347 – 20366 (Jan – Jun 1992).

010.043   **IAU Information Bulletin 67.**
J. Bergeron.
Kluwer Academic Publ., Dordrecht (Netherlands), 70 p. (Jan 1992).
Contents: 1. General Assemblies. 2. Executive Committees. 3. Commission matters. 4. International organisations. 5. IAU meetings. 6. Other scientific meetings. 7. IAU publications. 8. Other publications received. 9. Membership. 10. Other matters.

010.044   **International Astronomical Union Symposium.**
Nos. 149 (012.007), 150 (012.029), 151 (012.008), 152 (012.025).

010.045   **International Astronomical Union Colloquium.**
Nos. 126 (012.068), 128 (012.022), 133 (012.096).

010.046   **Information Bulletin on Variable Stars. Commission 27 of the IAU.**
Nos. 3693 – 3741 (Jan – Jun 1992).

010.047   **Circulaire d'Information. Union Astronomique Internationale, Commission des Etoiles Doubles.**
Nos. 116 (Feb 1992), 117 (Jun 1992).

010.048   **International Astronomical Union, Circular.**
Nos. 5420 – 5551 (Jan – Jun 1992).

010.049   **Transactions of the International Astronomical Union.**
Vol. 21B (1992) (XXIst General Assembly, Buenos Aires 1991, see Abstr. 012.050).

International Occultation Timing Association (IOTA)

010.050   **Occultation Newsletter.**
Vol. 5, No. 7 (May 1992).

Joint Organization for Solar Observations (JOSO)

010.051   **Annual report 1991.**
A. von Alvensleben.
Kiepenheuer–Institut für Sonnenphysik, Freiburg (Germany), 97 p. (1992).
Contents: 1. Introduction. 2. Report for 1991 of the LEST Foundation Council. 3. 23. Meeting of the JOSO Board. 4. Observatory profiles from the CSFR. 5. Solar physics at Potsdam. 6. Proceedings of the Meeting of Working Group 6. 7. Publications issued by or related to JOSO in 1991.

LEST Foundation

010.052   **LEST Foundation, Technical Report.**
Nos. 51 (55.034.033), 52 (55.082.017), 53 (55.034.034), 54 (55.034.035), 55 (55.032.016).

010.053   **LEST Foundation. Annual report 1991.**
Ø. Hauge.
Oslo Univ. (Norway). Inst. for Teoretisk Astrofysikk, 39 p. (1992).

Liga Ibero–Americana de Astronomía (LIADA)

010.054   **Universo.**
Vol. 12, No. 36 (Feb – Apr 1992).

Meteoritical Society

010.055   **Geochimica et Cosmochimica Acta.**
Vol. 56, Nos. 1 – 6 (Jan – Jun 1992).

010.056   **Meteoritics.**
Vol. 26, No. 4 (Dec 1991) (012.010).
Vol. 27, Nos. 1 (Mar 1992), 2 (Jun 1992).

Nantucket Maria Mitchell Association

010.057   **Eighty–ninth Annual Report of The Nantucket Maria Mitchell Association for the year ending December 31, 1990.**
R. K. Noyes.
The Nantucket Maria Mitchell Association, 2 Vestal Street, Nantucket, MA 02554 (USA), vp. (1991).

Oriental Astronomical Association

010.058   **The Heavens.**
Vol. 73, Nos. 1 – 6 (Jan – Jun 1992). In Japanese.

Royal Astronomical Society (RAS)

010.059   **The Quarterly Journal of the Royal Astronomical Society.**
Vol. 33, Nos. 1 (Mar 1992), 2 (Jun 1992).

010.060   **Reports of meetings.**
Q. J. R. Astron. Soc., Vol. 33, No. 1, p. 39 – 40 (Mar 1992).

010.061   **Meetings of the Society.**
Observatory, Vol. 112, No. 1108, p. 81 – 98 (Jun 1992).

010.062   **Summary of the RAS Specialist Discussion Meeting on Galactic and Extragalactic Magnetic Fields, London (UK), 8 Feb 1991.**
L. Mestel, A. W. Wolfendale.
Observatory, Vol. 112, No. 1108, p. 99 – 104 (Jun 1992).

010.063   **Monthly Notices of the Royal Astronomical Society.**
Vol. 254, Nos. 1 – 4, Vol. 255, Nos. 1 – 4, Vol. 256, Nos. 1 – 4 (Jun – Dec 1992).

010.064   **Royal Astronomical Society Newsletter.**
No. 22 (Mar 1992).

Royal Astronomical Society of Canada

010.065 **Royal Astronomical Society of Canada Bulletin.**
Vol. 2, Nos. 1 (Feb 1992), 2 (Apr 1992).

010.066 **The Journal of the Royal Astronomical Society of Canada.**
Vol. 86, Nos. 1 (Feb 1992), 2 (Apr 1992), 3 (Jun 1992).

010.067 **Observer's Handbook 1993.**
See Abstr. 046.029.

Royal Astronomical Society of New Zealand

010.068 **Southern Stars.**
Vol. 34, Nos. 6 (Mar 1992), 7 (Jun 1992).

Schweizerische Astronomische Gesellschaft (SAG)

010.069 **Bedeckungsveränderlichen Beobachter der Schweizerischen Astronomischen Gesellschaft Bulletin.**
Nos. 99 (15 Jan 1992), 100 (15 May 1992).

010.070 **Orion. Zeitschrift der Schweizerischen Astronomischen Gesellschaft.**
50. Jahrg., Nr. 248 (Feb 1992), 249 (Apr 1992), 250 (Jun 1992).

010.071 **Mitteilungen.**
Orion, Jahrg. 50, Nr. 248, p. 25 – 28 (Feb 1992).
Further information is given in Nr. 249, p. 71 – 78 (Apr 1992), Nr. 250, p. 115 – 118 (Jun 1992).

Società Astronomica Italiana (S.A.It.)

010.072 **Memorie della Società Astronomica Italiana.**
Vol. 62, No. 4 (1991) (012.042).
Vol. 63, Nos. 1 (1992) (012.087), 2 (1992) (012.084).

010.073 **Giornale di Astronomia.**
Vol. 18, N. 1 (Mar 1992), 2 (Jun 1992).

Société Astronomique de France

010.074 **Observations et Travaux.**
Nos. 29, 30 (1992).

010.075 **L'Astronomie.**
Janvier – juin (1992).

Société Astronomique de Liège

010.076 **Le Ciel.**
Vol. 54, janvier – juin (1992), Numéro spécial (046.010).

Société Française des Spécialistes d'Astronomie (SFSA)

010.077 **Journal des Astronomes Français.**
No. 41 (Feb 1992).

Société Royale Belge d'Astronomie

010.078 **Ciel et Terre.**
Vol. 108, Nos. 1 (Jan – Feb 1992), 2 (Mar – Apr 1992), 3 (May – Jun 1992).

Vereinigung der Sternfreunde e.V. (VdS)

010.079 **VdS Nachrichten für Sternfreunde.**
Sterne Weltraum, Jahrg. 31, Nr. 1, p. 62 – 66 (Jan 1992).
Further information is given in Nr. 2, p. 132 – 135 (Feb 1992), Nr. 3, p. 200 – 203 (Mar 1992), Nr. 4, p. 268 – 271 (Apr 1992), Nr. 5, p. 344 – 346 (May 1992), Nr. 6, p. 420 – 422 (Jun 1992).

010.080 **Sonne.**
Jahrg. 16, Nr. 61 (Mar 1992).

011 Reports on Colloquia, Congresses, Meetings, Symposia, Expeditions

011.001 **First European Meeting of the A.A.V.S.O.**
M. Wolf.
Říše hvězd, Vol. 72, No. 1, p. 2 – 3 (Jan 1991). In Czech.

011.002 **Meeting of European astronomers in Davos.**
V. Vanýsek.
Říše hvězd, Vol. 72, No. 4, p. 71 – 73 (Apr 1991). In Czech.

011.003 **The evolution of interstellar matter and dynamics of galaxies (CTS Workshop No. 1).**
J. Palouš.
Říše hvězd, Vol. 72, No. 10, p. 194 – 195 (Oct 1991). In Czech.

011.004 **Summary of the RAS Specialist Meeting on Atomic and Molecular Data for Astrophysics, held at the Scientific Societies' Lecture Theatre, 8 Mar 1991.**
K. L. Bell, P. L. Dufton.
Observatory, Vol. 112, No. 1106, p. 1 – 4 (Feb 1992).

011.005 **Summary of the RAS Discussion Meeting on Solar–System Dynamics and Planet X, held at the Geological Society Lecture Theatre, Burlington House, 8 Nov 1991.**
L. V. Morrison.
Observatory, Vol. 112, No. 1107, p. 37 – 40 (Apr 1992).

011.006   **A colloquium of meteor observers in Kirov, 22 – 24 April 1991.**
M. V. Gorshechnikov, R. L. Khotinok.
Astron. Vestn., Tom 26, No. 2, p. 126 (Mar – Apr 1992). In Russian. English translation in Solar Syst. Res., Vol. 26, No. 2.

011.007   **Conference of the problem group "Radiation and structure of the Sun", held in L'vov, 23 – 25 September 1991.**
Eh. G. Gurtovenko, R. I. Kostyk.
Kinematika Fiz. Nebesn. Tel, Tom 8, No. 2, p. 91 (Mar–Apr 1992). In Russian.

011.008   **Information on the conferene of the Working Group for "Studies of Long–Period Variables", held in Pulkovo, 16 – 17 September 1991.**
V. S. Strel'nitskij.
Astron. Tsirk., No. 1551, p. 40 (Nov – Dec 1991). In Russian.

011.009   **Stardust memories.**
I. P. Wright.
Nature, Vol. 356, No. 6370, p. 567 – 568 (16 Apr 1992).
The author reports on a discussion at the 23rd Lunar and Planetary Science Conference, Houston, TX (USA), 16–20 Mar 1992, concerning SiC grains, embedded in meteorites which turned out to be "stardust" and on the consequences on stellar nuclear burning processes.

011.010   **Planetary geology: Killer acid at the K/T boundary.**
W. B. McKinnon.
Nature, Vol. 357, No. 6373, p. 15 – 16 (7 May 1992).
The author reports on a special session at the 23rd Lunar and Planetary Science Conference, Houston, TX, USA, 16 – 20 Mar 1992. In this session the impact of a massive asteroid or comet at the end of the Cretaceous period, which should be responsible for the extinction of the dinosaurs, was discussed.

011.011   **Stellar evolution: Stars arriving two by two.**
C. Clarke.
Nature, Vol. 357, No. 6375, p. 197 – 198 (21 May 1992).
The author reports on the IAU Colloquium No. 135: "Complementary approaches to double and multiple star research", Pine Mountain, GA (USA), 5 – 10 Apr 1992.

011.012   **Dateline Atlanta (The 1992 Meeting of the American Astronomical Society).**
S. Stephens.
Mercury, Vol. 21, No. 2, p. 64 – 69, 75 (Mar – Apr 1992).

011.013   **The 4th International Space University: an International Mars Mission.**
B. Clark.
Bull. Am. Astron. Soc., Vol. 23, No. 3, p. 1190 – 1191 (1991). Abstract. – See Abstr. 012.037 for the main entry.

011.014   **The DPS Teachers Workshop.**
M. S. Hanner.
Bull. Am. Astron. Soc., Vol. 23, No. 3, p. 1191 (1991). Abstract. – See Abstr. 012.037 for the main entry.

011.015   **Kosmologie und Relativitätstheorie im Aufschwung. Bericht vom 10. Seminar über Relativistische Astrophysik und Gravitation in Potsdam.**
G. Börner.
Phys. Bl., Jahrg. 48, Heft 1, p. 39 – 40 (Jan 1992). Report on a conference, held at Potsdam, F.R. Germany, 21 –25 October 1991.

011.016   **Particle astronomy minisymposium: introductory remarks.**
P. F. Smith, B. Sadoulet.
15. Texas Symposium on Relativistic Astrophysics and 4. ESO–CERN Symposium, p. 323 – 325 (1991). – See Abstr. 012.060 for the main entry.
The objective of the papers in this part of the Texas/ESO–CERN Symposium is to review the status and future prospects for the detection of all types of particles incident (or hypothetically incident) upon the earth.

011.017   **Summary: miniworkshops on space–based astrophysics.**
H. J. Smith.
15. Texas Symposium on Relativistic Astrophysics and 4. ESO–CERN Symposium, p. 628 – 634 (1991). – See Abstr. 012.060 for the main entry.

011.018   **Conference: The scientific problems of creating a Lunar base. Moscow, 5 – 8 Feb 1991.**
V. V. Shevchenko.
Astron. Astrophys. Trans., Vol. 1, No. 1, p. 77 – 79 (1991).

011.019   **Conference: "Analytical celestial mechanics", Kiev, 1 – 3 Oct 1991.**
K. V. Kholshevnikov.
Kinematika Fiz. Nebesn. Tel, Tom 8, No. 3, p. 109 – 111 (May–Jun 1992). In Russian. English translation in Kinematics Phys. Celest. Bodies, Vol. 8, No. 3.

011.020   **The 1992 A.S.P. Meeting in Wisconsin.**
A. Fraknoi.
Mercury, Vol. 21, No. 2, p. 72 – 74 (Mar – Apr 1992).

011.021   **Objective–prism and other surveys. A meeting held at Van Vleck Observatory.**
A. G. D. Philip.
News Lett. Astron. Soc. N.Y., Vol. 4, No. 1, p. 26 – 28 (Feb 1992). – See Abstr. 012.078 for the main entry.
The meeting was held at the Wesleyan University in Middletown, CT (USA), 9 – 11 May 1991.

011.022   **The research seminars of the Astronomical Observatory Cluj–Napoca.**
A. Pál, V. Ureche.
Rom. Astron. J., Vol. 1, No. 1 – 2, p. 116 – 117 (1991).

011.023   **Scientific session, Bucharest (Romania), 8 – 9 Nov 1990.**
M. Stavinschi.
Rom. Astron. J., Vol. 1, No. 1 – 2, p. 117 – 118 (1991).

011.024   **Conference report: IAU Symposium No. 150 on "Astrochemistry of Cosmic Phenomena" (5 – 9 August 1991, Campos do Jordão, Brazil).**
J. M. Vrtilek.
Comments Astrophys., Vol. 16, No. 2, p. 87 – 113 (May 1992).
Contents: 1. Introduction. 2. Basic studies. 3. Early universe. 4. External galaxies. 5. Diffuse, translucent, and high–latitude clouds. 6. Quiescent clouds and regions of low mass star formation. 7. Regions of high mass star formation. 8. Chemistry of interface regions. 9. Near–stellar environments. 10. Solar system. 11. Panel discussions.

011.025   **A CCD Workshop in Germany.**
W. Quester.
I.A.P.P.P. Commun., No. 47, p. 17 – 18 (Mar – May 1992).
Report on a workshop, held Nov 9, 1991 in Erlangen (Germany).

011.026   **Chronicle (Report on): All–Union Conference of the Working Group "Stellar Atmospheres", held at the Institute of Astrophysics and Physics of the Atmosphere of the Academy of Sciences of Estonia, 28 – 30 May 1991.**
N. A. Sakhibullin.
Astron. Zh., Tom 69, Vyp. 3, p. 663 – 668 (May – Jun 1992). In Russian. English translation in Sov. Astron., Vol. 36, No. 3 (May – Jun 1992).

Bigger telescopes and better instrumentation: report on the 1992 ESO conference.
See Abstr. 013.082.

The emerging picture of eruptive solar flares.
See Abstr. 073.122.

The pulsar radio–frequency emission problem.
See Abstr. 126.110.

012 Proceedings of Colloquia, Congresses, Meetings, Symposia

012.001 New windows to the universe. Volumes I, II. Invited review papers and general lectures. 11. European Regional Astronomy Meeting of the IAU: New windows to the universe, Tenerife, Canary Islands (Spain), 3 – 9 Jul 1989.
F. Sanchez, M. Vazquez (eds.).
Cambridge University Press, Cambridge (UK). 1158 p. (1990). ISBN 0–521–38429–X. ISBN 0–521–40140–2. Price £ 90.00, US$ 105.00. The contributed papers are published in Astrophys. Space Sci., Vols. 169 – 171 (Jul – Sep 1990). See 52.012.030.
Contents:
Vol. I: The Sun and solar–like stars. Stellar structure and evolution. Astronomical instrumentation. General lectures.
Vol. II: Structure and evolution of galaxies. Active galaxies and cosmology. Interstellar and intergalactic medium.
The individual contributions within the subject scope of AAA are included in their corresponding categories – see abstracts 013.001, 013.002, 013.003, 013.004, 013.005, 034.002, 035.007, 036.008, 036.009, 063.001, 064.002, 064.003, 065.004, 065.006, 065.007, 065.008, 072.001, 073.003, 074.001, 080.003, 082.004, 112.005, 112.009, 112.012, 114.003, 114.004, 114.005, 116.001, 116.002, 118.002, 121.003, 125.003, 125.006, 125.090, 131.007, 131.008, 131.009, 131.010, 131.011, 131.012, 154.002, 155.005, 157.008, 157.009, 157.010, 157.011, 157.012, 157.013, 157.014, 158.010, 158.011, 158.012, 158.013, 158.014, 158.015, 158.016, 158.017, 159.007, 161.013, 161.023, 161.024.

012.002 Quantum chaos – quantum measurement. Proceedings. NATO Advanced Research Workshop on Quantum Chaos – Theory and Experiment, Copenhagen (Denmark), 28 May – 1 Jun 1991.
P. Cvitanović, I. Percival, A. Wirzba (eds.).
NATO ASI Ser., Ser. C, Math. Phys. Sci., Vol. 357.
Kluwer Academic Publ., Dordrecht (Netherlands). 341 p. (1992). ISBN 0–7923–1599–5.
The objective of the workshop was to bring together quantum chaos theorists and experimentalists with the aim of improving our understanding of the physics of quantum systems whose classical limit is chaotic.
The individual contributions within the subject scope of AAA are included in their corresponding categories – see abstracts 161.022.

012.003 Gamma–ray line astrophysics. Proceedings. International Symposium on Gamma–Ray Line Astrophysics, Paris–Saclay (France), 10 – 13 Dec 1990.
P. Durouchoux, N. Prantzos (eds.).
AIP Conf. Proc., No. 232.
American Institute of Physics, New York, NY (USA). 541 p. (1991). ISBN 0–88318–875–9. Price US$ 95.00.
The scientific program of this symposium covered all topics relevant to the spectroscopy of astrophysical gamma–ray line sources, whether detected or potentially interesting: the galactic center and plane, solar flares, compact objects (e.g., neutron stars and gamma–ray bursts), stellar explosions (e.g., novae, supernovae, and SN1987A in particular), quiescently burning stars (like red giants and Wolf–Rayet stars) and extragalactic sources.
The individual contributions within the subject scope of AAA are included in their corresponding categories – see abstracts 035.010, 035.011, 035.012, 035.013, 035.014, 035.015, 035.016, 035.017, 035.018, 035.019, 061.034, 061.035, 061.036, 061.037, 063.013, 067.025, 067.026, 073.008, 076.002, 080.022, 125.011, 125.012, 125.013, 125.014, 125.096, 125.097, 125.098, 125.099, 125.100, 125.101, 125.102, 126.004, 126.005, 142.004, 142.005, 142.006, 142.007, 142.008, 142.010, 142.011, 142.012, 142.013, 142.014, 143.008, 143.009, 143.010, 143.011, 143.012, 143.013, 143.014, 143.015, 143.016, 143.017, 143.018, 143.019, 143.020, 143.021, 143.022, 143.023, 143.024, 143.025, 143.026, 143.028, 143.029, 143.030, 144.014.

012.004 Atomic physics 12. Proceedings. 12. International Conference on Atomic Physics (ICAP–12), Ann Arbor, MI (USA), 29 Jul – 3 Aug 1990.
J. C. Zorn, R. R. Lewis (eds.).
AIP Conf. Proc., No. 233.
American Institute of Physics, New York, NY (USA). 645 p. (1991). ISBN 0–88318–811–2.
The sessions of this conference covered a broad range of fundamental questions. The stopping and trapping of neutral atoms represents an important new direction for atomic physics. This work is represented by reports of experiments done at five leading laboratories and by summaries of theoretical work on ultraslow collisions and on light–induced drift. Experiments on trapped ions are also reported. There are reports of the Casimir effect seen in spectroscopy and in beam deflection experiments. Spectroscopy is represented by a number of reports including a precision measurement of the 1s–2s interval in hydrogen, and measurements of the Lamb shift and the isotope shift in helium.
The individual contributions within the subject scope of AAA are included in their corresponding categories – see abstracts 066.035, 067.032.

012.005 Bioastronomy. The search for extraterrestrial life – the exploration broadens. Proceedings. 3. Symposium International de Bioastronomie, Lanslevillard, Val Cenis (France), 18 – 23 Jun 1990.
J. Heidmann, M. J. Klein (eds.).
Lect. Notes Phys., Vol. 390.
Springer, Berlin (Germany). 413 p. (1991). ISBN 3–540–54752–5.
The scientific program was organized to maintain a broad perspective because the previous meetings had shown that the interaction of scientists from various disciplines was productive and stimulating. Consequently, the symposium was open to scientists from a variety of related fields: geology, climatology, biochemistry, origins of life, paleontology, ecology, neurology, sociology, intelligence and civilizations. The program was organized into sessions dealing with the five main stages of bioastronomy and a sixth session to encourage wider interdisciplinary connections.
The individual contributions within the subject scope of AAA are included in their corresponding categories – see abstracts 013.026, 015.014, 015.015, 015.016, 015.017, 015.018, 015.019, 015.020, 015.021, 015.022, 015.023, 015.024, 015.025, 015.026, 015.027, 015.028, 015.029, 015.030, 015.031, 015.032, 015.033, 015.034, 015.035, 015.036, 015.037, 015.038, 015.039, 036.109, 036.110, 065.061, 097.027, 097.028, 097.029, 100.011, 100.012, 100.013, 100.014, 102.016, 105.204, 107.011, 107.012, 112.066, 114.041, 118.020, 118.021, 118.022, 118.023, 118.024, 121.032, 131.096, 131.097, 131.098, 131.099, 131.100, 131.101, 131.102, 153.027, 155.112, 161.275.

012.006 Proceedings of the Fourth Asia Pacific Physics Conference. 4. Asia Pacific Physics Conference, Seoul (Republic of Korea), 13 – 17 Aug 1990.
S. H. Ahn, S. H. Choh, I. T. Cheon, C. Lee (eds.).
World Scientific, Singapore (Singapore). 2272 p. (1991). ISBN 981–02–0538–4. Published in two volumes.
Contents: 1. Particle physics. 2. Nuclear physics. 3. Condensed matter physics. 4. Thermal and statistical physics. 5. Plasma physics. 6. Optics and quantum electronics. 7. Atomic and molecular physics. 8. Applied physics. 9. Accelerator physics. 10. Physics general.
The individual contributions within the subject scope of AAA are included in their corresponding categories – see abstracts 061.040, 062.071, 063.065, 066.094, 067.086, 077.015, 125.104, 144.015, 161.091.

012.007   The stellar populations of galaxies. Proceedings. IAU Symposium No. 149: The stellar populations of galaxies, Angra dos Reis (Brazil), 5 – 9 Aug 1991.
B. Barbuy, A. Renzini (eds.).
Int. Astron. Union Symp., No. 149.
Kluwer Academic Publ., Dordrecht (Netherlands). 543 p. (1992).
ISBN 0–7923–1698–3. Price Dfl. 240.00.

The symposium covered the whole subject of stellar populations from the solar neighborhood to the most distant radiogalaxies. The meeting was structured in three phases. 1. Focussing on our own Galaxy, the main reviews are devoted to the stellar populations in the Galactic bulge, halo, disk and spiral arms, then touching upon their age, composition, and kinematics. 2. Nearby galaxies that observations are able to resolve into individual stars, such as the Magellanic Clouds, the dwarf spheroidals, Andromeda, and the other members of the Local Group. 3. Distant, non–resolved galaxies, whose stellar populations can only be studied in integrated light. Here the main reviews were devoted to low, intermediate, and high–redshift galaxies, as well as to the ongoing search for primeval galaxies. Through the whole Symposium theoretical papers innervated the complex wealth of observational evidences being provided, thus addressing topics such as the dynamics of the stellar populations in the Galaxy, stellar evolution, stellar spectra and model atmospheres as main ingredients in the construction of the population sythesis tools which are necessary to study unresolved galaxies, etc. The question of how galaxies formed, evolved, and acquired their present morphology has percurred the whole meeting.
The individual contributions within the subject scope of AAA are included in their corresponding categories – see abstracts 002.023, 022.041, 064.027, 064.029, 065.033, 065.034, 114.020, 114.021, 114.022, 114.030, 114.032, 115.008, 115.009, 115.010, 121.018, 122.033, 122.039, 122.040, 124.027, 125.029, 131.058, 134.012, 134.014, 134.017, 134.018, 134.020, 141.001, 142.020, 151.023, 151.024, 151.028, 154.026, 154.027, 154.029, 154.030, 154.031, 154.033, 154.034, 154.035, 154.036, 154.037, 155.036, 155.037, 155.038, 155.039, 155.040, 155.041, 155.042, 155.043, 155.044, 155.046, 155.047, 155.048, 155.049, 155.050, 155.051, 155.052, 155.053, 155.055, 155.056, 155.057, 155.058, 155.059, 155.060, 155.062, 155.063, 155.064, 155.065, 155.066, 155.067, 155.068, 155.069, 155.070, 155.071, 155.072, 155.073, 155.074, 155.075, 155.076, 155.077, 155.078, 155.079, 155.080, 155.081, 155.082, 155.083, 155.084, 155.085, 155.086, 155.087, 156.011, 156.012, 156.013, 156.014, 156.015, 156.016, 156.017, 156.018, 156.019, 156.020, 156.022, 156.023, 156.025, 157.099, 157.102, 157.103, 157.104, 157.105, 157.106, 157.107, 157.108, 157.109, 157.110, 157.111, 157.112, 157.113, 157.114, 157.116, 157.117, 157.118, 157.119, 157.120, 157.127, 157.128, 157.129, 157.137, 157.138, 157.139, 157.140, 157.141, 157.142, 157.143, 157.144, 157.145, 157.146, 157.147, 157.148, 157.149, 157.152, 157.153, 157.154, 157.155, 157.156, 157.157, 157.158, 157.159, 157.161, 157.162, 157.163, 157.164, 157.170, 157.171, 157.172, 157.173, 157.174, 157.175, 157.176, 157.177, 157.178, 157.179, 157.180, 157.181, 157.182, 157.183, 158.059, 158.066, 158.068, 158.069, 158.070, 159.040, 160.028, 160.030, 160.034, 160.035, 160.036, 160.037, 160.046, 161.120, 161.121, 161.122, 161.127, 161.128, 161.131.

012.008   Evolutionary processes in interacting binary stars. Proceedings. IAU Symposium No. 151: Evolutionary processes in interacting binary stars, Córdoba (Argentina), 5 – 9 Aug 1991.
Y. Kondo, R. F. Sisteró, R. S. Polidan (eds.).
Int. Astron. Union Symp., No. 151.
Kluwer Academic Publ., Dordrecht (Netherlands). 555 p. (1992).
ISBN 0–7923–1731–9. Price Dfl. 246.00.

The IAU Symposium No. 151 is dedicated to Jorge Sahade, who has served as President of IAU Commission 42 (Close Binary Stars) and a Vice President and later President of the International Astronomical Union. The primary aim of this conference was to review and evaluate our current understanding of the evolutionary processes in a wide variety of interacting binary stars from birth to death. Subjects included the formation of binaries, mass flow and transfer, accretion processes, and

binaries with collapsed components, such as novae, X–ray binaries, and binary pulsars.
The individual contributions within the subject scope of AAA are included in their corresponding categories – see abstracts 002.025, 064.031, 064.032, 064.037, 115.011, 117.059, 117.063, 117.064, 117.065, 117.066, 117.067, 117.068, 117.069, 117.070, 117.072, 117.074, 117.075, 117.079, 117.082, 117.083, 117.084, 117.085, 117.086, 117.087, 117.088, 117.091, 117.092, 117.093, 117.094, 117.095, 117.096, 117.097, 117.099, 117.100, 117.101, 117.102, 117.103, 117.104, 117.105, 117.106, 117.107, 117.108, 117.109, 117.110, 117.111, 117.112, 117.113, 117.114, 117.116, 117.117, 117.118, 117.119, 117.120, 117.121, 117.122, 117.123, 117.124, 117.125, 117.126, 117.127, 117.128, 117.129, 117.130, 117.133, 117.134, 117.136, 117.137, 117.138, 117.139, 119.013, 119.014, 119.015, 119.016, 119.017, 119.018, 119.019, 119.020, 119.021, 119.022, 119.024, 119.025, 120.015, 122.042, 125.031, 134.021, 153.019, 153.020, 153.021.

012.009   The infrared and submillimetre sky after COBE. Proceedings. NATO Advanced Study Institute on The Infrared and Submillimetre Sky after COBE, Les Houches (France), 20 – 30 Mar 1991.
M. Signore, C. Dupraz (eds.).
NATO ASI Ser., Ser. C, Math. Phys. Sci., Vol. 359.
Kluwer Academic Publ., Dordrecht (Netherlands). 488 p. (1992).
ISBN 0–7923–1602–9. Price Dfl. 235.00, US$ 144.00, £ 82.00.

This volume gives an up–to–date account of the cosmic microwave background, the cosmic infrared background (if any), and the infrared emission of the Galaxy, after the early results from COBE (Cosmic Background Explorer). It represents a complete coverage of our present knowledge and understanding of: the early Universe, large–scale structure, dust in galaxies, infrared to submillimetre backgrounds, CMB anisotropies, complementary observations and instrumentation problems, etc.
The individual contributions within the subject scope of AAA are included in their corresponding categories – see abstracts 022.060, 036.066, 036.067, 051.016, 051.017, 155.093, 158.085, 161.160, 161.161, 161.162, 161.163, 161.164, 161.165, 161.166, 161.167, 161.168, 161.169, 161.170, 161.171, 161.172, 161.173, 161.174, 161.175, 161.176, 161.178, 161.179.

012.010   Abstracts for the 54th Annual Meeting of the Meteoritical Society. 54. Annual Meeting of the Meteoritical Society, Monterey, CA (USA), 21 – 26 Jul 1991.
LPI Contrib., No. 766.
Lunar and Planetary Inst., Houston, TX (USA). 321 p. (1991).
The abstracts of this Annual Meeting are also published in Meteoritics, Vol. 26, No. 4, p. 311 – 414 (Dec 1991).
The individual contributions within the subject scope of AAA are included in their corresponding categories – see abstracts 004.025, 022.063, 022.064, 022.066, 022.070, 022.071, 064.044, 094.013, 094.014, 094.015, 094.016, 094.017, 094.018, 094.019, 094.020, 094.021, 097.021, 097.022, 098.046, 098.047, 098.048, 098.049, 098.051, 098.052, 098.053, 098.054, 098.055, 098.056, 102.009, 103.115, 104.055, 104.056, 104.057, 105.021, 105.022, 105.023, 105.024, 105.025, 105.026, 105.027, 105.028, 105.029, 105.030, 105.031, 105.032, 105.033, 105.034, 105.035, 105.036, 105.037, 105.038, 105.039, 105.040, 105.041, 105.042, 105.043, 105.044, 105.045, 105.046, 105.047, 105.048, 105.049, 105.050, 105.051, 105.052, 105.053, 105.054, 105.055, 105.056, 105.057, 105.058, 105.059, 105.060, 105.061, 105.062, 105.063, 105.064, 105.065, 105.066, 105.067, 105.068, 105.069, 105.070, 105.071, 105.072, 105.073, 105.074, 105.075, 105.076, 105.077, 105.078, 105.080, 105.081, 105.082, 105.083, 105.084, 105.085, 105.086, 105.087, 105.088, 105.089, 105.090, 105.091, 105.092, 105.093, 105.094, 105.095, 105.096, 105.097, 105.098, 105.099, 105.100, 105.101, 105.102, 105.103, 105.104, 105.105, 105.106, 105.107, 105.108, 105.109, 105.110, 105.111, 105.112, 105.113, 105.114, 105.115, 105.116, 105.117, 105.118, 105.119, 105.120, 105.121, 105.122, 105.123, 105.124, 105.125, 105.126, 105.127, 105.128, 105.129, 105.130, 105.131, 105.132, 105.133, 105.134, 105.135, 105.136, 105.137, 105.138, 105.139, 105.140, 105.141, 105.142, 105.143, 105.144, 105.145, 105.146, 105.147, 105.148, 105.149, 105.150, 105.151, 105.152, 105.153, 105.154, 105.155, 105.156,

105.157, 105.158, 105.159, 105.160, 105.161, 105.162, 105.163, 105.164, 105.165, 105.166, 105.167, 105.168, 105.169, 105.170, 105.171, 105.172, 105.173, 105.174, 105.175, 105.176, 105.177, 105.178, 105.179, 105.180, 105.181, 105.182, 105.183, 105.184, 105.185, 105.186, 105.187, 105.188, 105.189, 105.190, 105.191, 105.192, 106.019, 106.020, 106.021, 106.022, 106.024, 107.005, 107.006, 107.007, 107.008, 107.009, 107.010, 112.049, 112.052, 114.037, 125.033, 125.038, 131.071.

012.011   **Current developments in optical engineering 4. Proceedings.** SPIE Conference on Current Developments in Optical Engineering IV as Part of SPIE's International Symposium on Optical and Optoelectronics Applied Science and Engineering, San Diego, CA (USA), 9 – 10 Jul 1990.
R. E. Fischer, W. J. Smith (eds.).
SPIE Proc. Ser., Vol. 1334.
SPIE, Bellingham, WA (USA). 251 p. (1990). ISBN 0–8194–0395–4.
These proceedings includes timely papers on basic optical design and engineering, such as "Optomechanical laboratory hardware: organization, terminology, and comments" and "Precision dual positioning system". The session on Optical fabrication and testing covers contemporary subjects with papers such as "Adaptive optics wavefront corrector using addressable liquid crystal variable retarders", "New process for manufacturing arrays of microlenses on a wide range of substrates", and others. There was a lot of interest in waveguide and diffractive optics as well as in microlithography.
The individual contributions within the subject scope of AAA are included in their corresponding categories – see abstracts 031.024, 031.025, 032.011, 034.046, 034.047.

012.012   **Infrared technology 16. Proceedings.** SPIE Conference on Infrared Technology XVI as Part of SPIE's International Symposium on Optical and Optoelectronic Applied Science and Engineering, San Diego, CA (USA), 11 – 13 Jul 1990.
I. J. Spiro (ed.).
SPIE Proc. Ser., Vol. 1341.
SPIE, Bellingham, WA (USA). 477 p. (1990). ISBN 0–8194–0402–0.
This infrared technology series of conference proceedings includes all aspects of IR technology, radiative properties of sources, atmospherics, and civilian and military applications.
The individual contributions within the subject scope of AAA are included in their corresponding categories – see abstracts 013.021, 032.012, 034.048, 035.049, 035.050.

012.013   **The Taylor Colloquium: Origin and evolution of planetary crusts,** Canberra (Australia), 1 – 2 Oct 1990.
S. M. McLennan, R. L. Rudnick (eds.).
Geochim. Cosmochim. Acta, Vol. 56, No. 3, p. 869 – 1064 (Mar 1992).
In recognition of Stuart Ross Taylor on the occasion of his sixty-fifth birthday and retirement from the Research School of Earth Science.
The individual contributions within the subject scope of AAA are included in their corresponding categories – see abstracts 081.013, 081.014, 094.022, 094.023.

012.014   **Space dust and debris. Proceedings.** Topical Meeting of the COSPAR Interdisciplinary Scientific Commission B (Meetings B2, B3 and B5) of the COSPAR 28. Plenary Meeting, The Hague (Netherlands), 25 Jun – 6 Jul 1990.
D. J. Kessler, J. C. Zarnecki, D. L. Matson (eds.).
Adv. Space Res., Vol. 11, No. 12, 207 p. (1991). ISBN 0–08–041842–2.
Contents: 1. Orbital debris. 2. In situ measurements and laboratory analysis of space dust particles. 3. Comparative studies of comets, asteroids, and dust.
The individual contributions within the subject scope of AAA are included in their corresponding categories – see abstracts 022.073, 098.058, 098.059, 098.060, 098.061, 102.011, 103.017, 103.018, 103.116, 104.059, 106.029, 106.030, 106.031, 106.032.

012.015   **Life sciences and space research XXIV(1): Gravitational biology. Proceedings.** Symposia 10 and 13 and Topical Meeting of the COSPAR Interdisciplinary Scientific Commission F (Meetings F1 and F2) of the COSPAR 28. Plenary Meeting, The Hague (Netherlands), 25 Jun – 6 Jul 1990.
R. S. Young, A. Cogoli, H. Planel, G. A. Ubbels, A. Sievers, H. Oser, G. Horneck, H. Wagner (eds.).
Adv. Space Res., Vol. 12, No. 1, 411 p. (1992). ISBN 0–08–041843–0.
Contents: 1. Biological response to gravity. 2. Human factors with respect to manned missions to Mars. 3. Human performance in space. 4. Gravitational effects in bio–processing and bio–separation.

012.016   **Life sciences and space research XXIV(2): Radiation biology. Proceedings.** Topical Meeting of the COSPAR Interdisciplinary Scientific Commission F (Meetings F3, F4, F5, F6 and F1) of the COSPAR 28. Plenary Meeting, The Hague (Netherlands), 25 Jun – 6 Jul 1990.
G. Kraft, A. B. Cox, J. R. Maisin, E. J. Ainsworth, G. Reitz, G. Horneck (eds.).
Adv. Space Res., Vol. 12, No. 2/3, 475 p. (1992). ISBN 0–08–041844–9.
Contents: 1. Heavy–ion effects in genetically relevant cellular structures. 2. Combined effects: radiation, microgravity, trauma and other factors. 3. Physical and chemical protection against ionizing radiation. 4. Radiation risk assessment for manned spaceflight. 5. Radiation protection aspects related to human missions to Mars.
The individual contributions within the subject scope of AAA are included in their corresponding categories – see abstracts 072.027, 073.028, 076.009, 078.007, 078.008, 078.009, 106.033, 144.023.

012.017   **Evolutionary trends in the physical sciences. Proceedings.** Yoshio Nishina Centennial Symposium: Evolutionary trends in the physical sciences, Tokyo (Japan), 5 – 7 Dec 1990.
M. Suzuki, R. Kubo (eds.).
Springer Proc. Phys., Vol. 57.
Springer, Berlin (Germany). 251 p. (1991). ISBN 3–540–54568–9.
The contributions to this volume include recollections of Nishina's life and work as well as reviews of current work in fields in which contemporary physicists are most actively engaged. The scope is extremely broad, ranging from nuclear and high–energy physics to astrophysics, solid state physics and even molecular biology, reflecting current developments and delineating future potential.
The individual contributions within the subject scope of AAA are included in their corresponding categories – see abstracts 013.020.

012.018   **The Compton Observatory Science Workshop. Proceedings.** Compton Observatory Science Workshop, Annapolis, MD (USA), 23 – 25 Sep 1991.
C. R. Shrader, N. Gehrels, B. Dennis (eds.).
National Aeronautics and Space Administration, Washington, DC (USA); National Aeronautics and Space Administration, Greenbelt, MD (USA). Goddard Space Flight Center, NASA–CP—3137. 566 p. (Feb 1992).
The primary purpose of the workshop was to provide a forum for the exchange of ideas and information among scientists with interests in various areas of high–energy astrophysics, with emphasis on the scientific capabilities of the Compton Observatory. At the time of the workshop, the observatory had completed 5 months of successful in–orbit operations. Early scientific results, as well as reports on in–flight instrument performance and calibrations were presented. Additional reports describing guest investigator data products, analysis techniques and associated software as well as a report on the Phase–2 NASA Research Announcement were presented. Theoretical papers, as well as papers describing results from other experiments were also presented. Scientific topics included active galaxies, cosmic gamma–ray bursts, solar physics, pulsars, novae, supernovae,

galactic binary sources and diffuse galactic and extragalactic emission.

The individual contributions within the subject scope of AAA are included in their corresponding categories – see abstracts 013.022, 021.007, 034.050, 035.046, 035.047, 035.048, 035.051, 035.052, 036.078, 036.079, 036.080, 036.081, 036.082, 036.083, 036.084, 036.085, 036.086, 067.093, 067.094, 072.030, 073.030, 073.031, 073.032, 073.034, 073.035, 073.036, 073.037, 073.038, 073.039, 073.040, 075.014, 076.010, 076.012, 077.016, 080.044, 117.145, 117.146, 117.148, 117.149, 117.151, 121.030, 124.004, 125.039, 125.040, 126.027, 126.028, 126.029, 126.030, 126.031, 126.032, 142.024, 142.026, 143.042, 143.043, 143.044, 143.045, 143.046, 143.047, 143.048, 143.050, 143.051, 143.052, 143.053, 143.054, 143.055, 143.056, 155.106, 155.107, 155.108, 158.102, 158.103, 158.104, 158.105, 158.106, 158.107, 159.052, 160.065.

012.019   **Gravitation. A Banff Summer Institute. Banff Summer Institute on Gravitation, Banff (Canada), 12 – 25 Aug 1990.**
R. Mann, P. Wesson (eds.).
World Scientific, Singapore (Singapore). 663 p. (1991). ISBN 981–02–0751–4. Price £ 50.00.

The 1990 Banff Summer Institute on Gravitation was designed to provide a forum for both beginning and established researchers in gravitation and cosmology to address issues such as these. Frontier issues in cosmology, quantum gravity and tests of gravitational theory were discussed.

The individual contributions within the subject scope of AAA are included in their corresponding categories – see abstracts 066.104, 066.105, 066.106, 066.107, 066.108, 066.109, 066.110, 066.111, 066.112, 066.113, 066.114, 066.115, 066.116, 066.117, 066.118, 066.119, 066.120, 066.121, 066.122, 066.123, 161.224, 161.225, 161.226, 161.227, 161.228, 161.229, 161.230, 161.231, 161.232.

012.020   **Traces of the primordial structure in the universe. Proceedings. Workshop on Traces of the Primordial Structure in the Universe, Schloß Ringberg, Tegernsee (Germany), 1 – 5 Oct 1990.**
H. Böhringer, R. A. Treumann (eds.).
Max–Planck–Institut für Physik und Astrophysik, Garching (Germany). Inst. für Extraterrestrische Physik, MPE—227. 221 p. (May 1991). ISSN 0178–0719.

This workshop was intended to provide a forum for discussing some of the relevant problems in detecting the large structure of the universe, the confidence in the structures found, and of the theories of structure formation. The main emphasis of the workshop was to confront theory with observations. The successful launch of the German X–ray observatory ROSAT on June 1, 1990 and the start of the first all–sky X–ray survey using an imaging telescope provided reason enough for organizing this meeting in time to assemble some expertise about the role the X–ray survey could play in the study of the large scale structure as well as to discuss collaborative work on this topic. The present proceedings collect the main contributions to the workshop.

The individual contributions within the subject scope of AAA are included in their corresponding categories – see abstracts 036.115, 142.027, 160.061, 160.063, 160.064, 160.066, 160.067, 160.069, 160.070, 160.071, 160.072, 160.073, 160.074, 161.221, 161.222, 161.223, 161.237, 161.238, 161.239, 161.240.

012.021   **9th Italian Conference on General Relativity and Gravitational Physics. Proceedings. 9. Italian Conference on General Relativity and Gravitational Physics, Capri (Italy), 25 – 28 Sep 1990.**
R. Cianci, R. de Ritis, M. Francaviglia, G. Marmo, C. Rubano, P. Scudellaro (eds.).
World Scientific, Singapore (Singapore). 725 p. (1991). ISBN 981–02–0765–4. Price £ 50.00.

This conference was preceded by a special meeting on 24 Sep 1990 to celebrate Peter G. Bergmann's 75th birthday. The invited lectures of both meetings where held in plenary sessions. For the conference the contributed papers where subdivided into four workshops according to the following main themes: A. General relativity and classical theories of gravitation. B. Relativistic astrophysics and cosmology. C. Experimental and observational gravitation. D. Supergravity and quantum gravity. These proceedings contain the invited talks presented both at P. G. Bergmann's Celebration and at the conference, as well as most of the contributed talks to the latter.

The individual contributions within the subject scope of AAA are included in their corresponding categories – see abstracts 022.080, 022.081, 022.082, 022.083, 022.084, 022.085, 022.086, 022.087, 034.052, 065.056, 066.125, 066.126, 066.127, 066.128, 066.129, 066.130, 066.131, 066.132, 066.133, 066.134, 066.135, 066.136, 066.137, 066.138, 066.139, 067.096, 067.097, 117.158, 125.043, 151.045, 157.215, 158.110, 160.077, 161.243, 161.244, 161.245, 161.246, 161.247, 161.248, 161.249, 161.250, 161.251, 161.252, 161.253, 161.254.

012.022   **The magnetospheric structure and emission mechanisms of radio pulsars. Proceedings. IAU Colloquium No. 128: The magnetospheric structure and emission mechanisms of radio pulsars, Łagów (Poland), 17 – 23 Jun 1990.**
T. H. Hankins, J. M. Rankin, J. A. Gil (eds.).
Int. Astron. Union Colloq., No. 128.
Pedagogical University Press, Zielona Góra (Poland). 446 p. (1992). ISBN 83–00–03596–6. Price US$ 40.00.
Contents: 1. Structure of the magnetic field. 2. Evolution of the magnetic field. 3. Magnetospheric models. 4. Form and spectra of emission beams. 5. X–ray, γ–ray and millisecond pulsars. 6. Polar cap theories. 7. Polar cap phenomena. 8. Coherence phenomena. 9. Polarization phenomena. 10. What are the priorities for future work?

The individual contributions within the subject scope of AAA are included in their corresponding categories – see abstracts 015.013, 062.083, 062.086, 063.068, 063.069, 063.070, 063.071, 063.072, 063.073, 063.074, 063.075, 063.076, 063.077, 063.078, 063.079, 063.082, 063.083, 063.084, 063.085, 063.086, 063.087, 063.088, 067.100, 067.102, 067.103, 067.105, 067.106, 067.107, 067.108, 067.109, 067.110, 067.111, 067.112, 067.114, 067.116, 067.117, 067.118, 067.119, 067.126, 067.127, 067.128, 067.129, 067.131, 067.132, 125.126, 126.034, 126.035, 126.036, 126.037, 126.038, 126.039, 126.040, 126.041, 126.042, 126.043, 126.044, 126.045, 126.046, 126.048, 126.049, 126.050, 126.051, 126.053, 126.054, 126.055, 126.056, 126.057, 126.058, 126.059, 126.060, 126.061, 126.062, 126.063, 126.064, 126.065, 126.066, 126.067, 126.068, 126.069, 126.070, 126.071, 126.072, 126.073, 126.074, 126.075, 126.076, 126.077, 126.078, 126.079, 126.080, 126.081, 126.082, 126.083, 126.084, 126.085, 126.086, 126.087, 126.088, 126.089, 126.090, 126.092, 126.093.

012.023   **Dusty plasmas. Proceedings. Special session during the 4. week of the Spring College on Plasma Physics, Trieste (Italy), 14 – 21 Jun 1991.**
P. K. Shukla, U. de Angelis, L. Stenflo (eds.).
Phys. Scr., Vol. 45, No. 5, p. 463 – 544 (May 1992).
The individual contributions within the subject scope of AAA are included in their corresponding categories – see abstracts 022.088, 022.089, 022.090, 091.019, 091.020, 131.093.

012.024   **7. Quadrennial Symposium on Solar–Terrestrial Physics (SCOSTEP-7), The Hague (Netherlands), 25 – 30 Jun 1990.**
M. J. Rycroft (ed.).
J. Atmos. Terr. Phys., Vol. 53, No. 11/12, p. 993 – 1211 (Nov–Dec 1991).
The individual contributions within the subject scope of AAA are included in their corresponding categories – see abstracts 013.027, 022.096, 062.090, 062.091, 072.031, 076.015, 076.016, 076.017, 082.046, 082.047, 083.015, 083.016, 083.017, 083.018, 084.034, 084.035, 084.036, 084.037, 084.038, 085.008, 106.038, 106.039, 106.040.

012.025   Chaos, resonance and collective dynamical phenomena in the solar system. Proceedings. IAU Symposium No. 152: Chaos, resonance and collective dynamical phenomena in the solar system, Angra dos Reis (Brazil), 15 – 19 Jul 1991.
S. Ferraz–Mello (ed.).
Int. Astron. Union Symp., No. 152.
Kluwer Academic Publ., Dordrecht (Netherlands). 428 p. (1992).
ISBN 0–7923–1781–5. Price Dfl. 190.00.
Contents: 1. The planetary system. 2. Planetary rings. 3. The asteroidal belt. 4. Planetary satellites. 5. Comets. 6. Meteors, zodiacal cloud, nebulae. 7. Dynamical systems, maps, integrators.
The individual contributions within the subject scope of AAA are included in their corresponding categories – see abstracts 021.008, 042.017, 042.018, 042.019, 042.021, 042.022, 042.023, 042.024, 042.025, 042.026, 042.027, 042.028, 042.029, 042.030, 042.031, 042.032, 042.033, 042.034, 042.035, 042.036, 042.037, 042.038, 042.039, 042.040, 042.041, 042.042, 042.043, 042.044, 042.045, 042.046, 042.047, 042.048, 042.049, 042.050, 042.051, 042.052, 042.053, 042.054, 042.055, 062.098, 091.021, 098.070, 098.073, 100.015, 100.016, 102.017, 102.018, 102.019, 102.020, 102.021, 102.022, 103.019, 104.067, 104.068, 104.069, 104.070, 106.041, 106.042.

012.026   Production and uses of simulated lunar materials. Workshop on Production and Uses of Simulated Lunar Materials, Houston, TX (USA), 25 – 27 Sep 1989.
D. S. McKay, J. D. Blacic (eds.).
Lunar and Planetary Inst., Houston, TX (USA), LPI–TR—91–01. 88 p. (1991).
The workshop was convened to define the need for simulated lunar materials and examine related issues in support of extended space exploration and development. Lunar samples cannot be sacrificed in sufficient quantity to test lunar resource utilization processes adequately. Hence, the workshop focused on a detailed examination of the variety of potential simulants and the methods for their production. The workshop participants also touched on policy issues concerned with making and distributing simulants. This report will be useful as a guide to potential producers and users of simulated lunar materials.

012.027   Scientific rationale and requirements for a global seismic network on Mars. Report of a workshop. Workshop on the Scientific Rationale and Requirements for a Global Seismic Network on Mars, Morro Bay, CA (USA), 7 – 9 May 1990.
S. C. Solomon, D. L. Anderson, W. B. Banerdt, R. G. Butler, P. M. Davis, F. K. Duennebier, Y. Nakamura, E. A. Okal, R. J. Phillips.
Lunar and Planetary Inst., Houston, TX (USA), LPI–TR—91–02. 57 p. (1991).
Following a brief overview of the mission concepts for a Mars Global Network Mission as of the time of the workshop, the authors present the principal scientific objectives to be achieved by a Mars seismic network. They review the lessons for extraterrestrial seismology gained from experience to date on the Moon and on Mars. They discuss how particular types of seismic waves will provide the most useful information to address each of the scientific objectives, and this discussion provides the basis for a strategy for station siting. Finally, the authors define the necessary technical requirements for the seismic stations.

012.028   Mare volcanism and basalt petrogenesis: "Astounding Fundamental Concepts (AFC)" developed over the last fifteen years. Abstracts of presented papers. Workshop on Mare Volcanism and Basalt Petrogenesis: "Astounding Fundamental Concepts (AFC)" Developed Over the Last Fifteen Years, Dallas, TX (USA), 27 – 28 Oct 1990.
L. A. Taylor, J. Longhi (eds.).
Lunar and Planetary Inst., Houston, TX (USA), LPI–TR—91–03. 79 p. (1991).
Contents: 1. Geological setting. 2. Magma evolution and source regions. 3. Magma source and ascent processes. 4. History of volcanism.

The individual contributions within the subject scope of AAA are included in their corresponding categories – see abstracts 094.026, 094.027, 094.028, 094.029, 094.030, 094.031, 094.032, 094.033, 094.034, 094.035, 094.036, 094.037, 094.038, 094.039, 094.040, 094.041, 094.042, 094.043, 094.044, 094.045, 094.046, 094.047, 094.048, 094.049, 105.206.

012.029   Astrochemistry of cosmic phenomena. Proceedings. IAU Symposium No. 150: Astrochemistry of cosmic phenomena, São Paulo (Brazil), 5 – 9 Aug 1991.
P. D. Singh (ed.).
Int. Astron. Union Symp., No. 150.
Kluwer Academic Publ., Dordrecht (Netherlands). 535 p. (1992).
ISBN 0–7923–1824–2. Price Dfl. 240.00.
This symposium had a wide ranging discussion of the chemistry of astronomical environments with an emphasis on the description of molecular processes that critically influence the nature and evolution of astronomical objects and the identification of specific observations that directly address significant astronomical questions. The subject areas of the symposium included atomic and molecular processes at low and high temperatures and photon interactions, the chemical structure of molecular clouds in the Milky Way and in external galaxies, the chemistry of outflows and their interactions with the interstellar medium, the chemical connections between the interstellar medium and the solar system and pregalactic chemistry.
The individual contributions within the subject scope of AAA are included in their corresponding categories – see abstracts 002.056, 022.105, 022.106, 022.107, 022.108, 022.109, 022.112, 022.113, 022.114, 022.115, 022.116, 022.117, 022.118, 022.121, 022.122, 022.129, 022.156, 061.099, 062.105, 062.108, 064.060, 071.014, 091.025, 098.109, 099.072, 103.026, 103.027, 103.028, 103.029, 103.030, 103.031, 103.032, 103.158, 107.020, 107.021, 112.077, 112.078, 112.079, 112.080, 112.081, 112.082, 117.207, 121.043, 121.044, 121.045, 124.005, 125.131, 131.123, 131.129, 131.130, 131.132, 131.133, 131.135, 131.136, 131.137, 131.138, 131.139, 131.140, 131.141, 131.142, 131.143, 131.144, 131.145, 131.146, 131.147, 131.148, 131.149, 131.151, 131.152, 131.153, 131.154, 131.155, 131.156, 131.157, 131.158, 131.159, 131.160, 131.161, 131.162, 131.163, 131.164, 131.165, 131.166, 131.172, 131.173, 131.174, 131.175, 131.177, 131.178, 131.181, 131.182, 131.184, 131.187, 131.188, 131.189, 131.190, 131.192, 131.195, 131.196, 131.198, 131.201, 131.209, 131.210, 131.212, 131.215, 134.033, 134.034, 134.035, 155.126, 155.127, 156.035, 157.239, 157.240, 157.241, 157.242, 161.293, 161.295.

012.030   Towards Other Planetary Systems (TOPS): A Technology Needs Identification Workshop. TOPS: A Technology Needs Identification Workshop, Houston, TX (USA), 22 – 24 Apr 1991.
D. C. Black, K. Nishioka (eds.).
Lunar and Planetary Inst., Houston, TX (USA), LPI–TR—92–01. 28 p. (1992).
The purpose of this workshop was to identify and document key technology issues that are associated with the TOPS (Toward Other Planetary Systems) program in general, and with some of the candidate observational facilities specifically. An effort was made to define what the current state of the art is in each area, and to forecast technology trends or studies that will be relevant to the development of TOPS instrumentation. The workshop was structured along four major technology theme areas, viz., optics, metrology, structures, and detectors.

012.031   Mars surface and atmosphere through time (MSATT). Abstracts of presented papers. Workshop on the Martian Surface and Atmosphere Through Time, Boulder, CO (USA), 23 – 25 Sep 1991.
R. M. Haberle, B. M. Jakosky (eds.).
Lunar and Planetary Inst., Houston, TX (USA), LPI–TR—92–02. 199 p. (1992).
The purpose of the workshop was to begin to explore the interdisciplinary nature of, and to determine the relationships between, various aspects of Mars science that involve the geological and chemical evolution of its surface, the structure and

dynamics of its atmosphere, interactions between the surface and atmosphere, and the present and past states of its volatile endowment and climate system.
The individual contributions within the subject scope of AAA are included in their corresponding categories – see abstracts 022.102, 022.103, 022.104, 051.025, 051.026, 051.027, 051.028, 097.030, 097.031, 097.032, 097.033, 097.034, 097.035, 097.036, 097.037, 097.038, 097.039, 097.040, 097.041, 097.042, 097.043, 097.044, 097.045, 097.046, 097.047, 097.048, 097.049, 097.050, 097.051, 097.052, 097.053, 097.054, 097.055, 097.056, 097.057, 097.058, 097.059, 097.060, 097.061, 097.062, 097.063, 097.064, 097.065, 097.066, 097.067, 097.068, 097.069, 097.070, 097.071, 097.072, 097.073, 097.074, 097.075, 097.076, 097.077, 097.078, 097.079, 097.080, 097.081, 097.082, 097.083, 097.084, 097.085, 097.086, 097.087, 097.088, 097.089, 097.090, 097.091, 097.092, 097.093, 097.094, 097.095, 097.096, 097.097, 097.098, 097.099, 097.100, 097.101, 097.102, 097.103, 097.104, 097.105, 097.106, 097.107, 097.108, 097.109, 097.110, 097.111, 097.112, 105.208, 105.220, 105.221.

012.032   Physics and chemistry of magma oceans from 1 bar to 4 Mbar. Abstracts of presented papers. Workshop on the Physics and Chemistry of Magma Oceans from 1 bar to 4 Mbar, Burlingame, CA (USA), 6 – 8 Dec 1991.
C. B. Agee, J. Longhi (eds.).
Lunar and Planetary Inst., Houston, TX (USA), LPI–TR—92–03. 89 p. (1992).
Evidence for the existence of magma oceans was discussed in great detail, and among the many new items introduced were high–pressure phase equilibrium experiments, calculations of depth of impact–produced melting, models incorporating crystal growth rates with degree of crystallinity and convection, and models of hard turbulent convection.
The individual contributions within the subject scope of AAA are included in their corresponding categories – see abstracts 022.133, 081.029, 081.030, 081.031, 081.032, 081.033, 081.034, 081.035, 081.036, 091.028, 091.029, 094.054, 094.055, 094.056, 094.057, 094.058, 094.059, 097.113, 105.222, 107.014, 107.015, 107.016, 107.017, 107.018.

012.033   Proceedings of the Finnish Astronomical Society 1991. Annual meeting of the Finnish Astronomical Society, Tuorla (Finland), 9 – 10 May 1991.
M. Lainela (ed.).
Tuorla Obs. Rep., Informo, No. 166.
Turku Univ. (Finland). Tuorla Observatory. 48 p. (Apr 1992).
The individual contributions within the subject scope of AAA are included in their corresponding categories – see abstracts 033.007, 093.020, 113.013, 131.168, 131.169, 151.060, 158.123, 158.124, 160.093, 161.322.

012.034   Stades avancés dans l'évolution des étoiles binaires serrées. Comptes rendus. 13. Journée de Strasbourg: Stades avancés dans l'évolution des étoiles binaires serrées, Strasbourg (France), 23 May 1991.
G. Jasniewicz (ed.).
Observatoire Astronomique, Strasbourg (France). 60 p. (1992).
The individual contributions within the subject scope of AAA are included in their corresponding categories – see abstracts 002.053, 065.071, 117.176, 117.177, 117.178, 117.179, 134.031, 155.123.

012.035 Z^0 physics. Cargèse 1990. NATO Advanced Study Institute: Physique du Z^0, Cargèse (France), 13 – 25 Aug 1990.
M. Lévy, J. L. Basdevant, M. Jacob, D. Speiser, J. Weyers, R. Gastmans (eds.).
NATO ASI Ser., Ser. B, Phys., Vol. 261.
Plenum Press, New York, NY (USA). 538 p. (1991). ISBN 0–306–43934–4.
The main theme of the school was Z^0–physics, with particular emphasis on the way the experiments at LEP are analyzed.
The individual contributions within the subject scope of AAA are included in their corresponding categories – see abstracts 161.288.

012.036   Capture gamma–ray spectroscopy. Proceedings. 7. International Symposium on Capture Gamma–Ray Spectroscopy and Related Topics (ISCGRSART–7), Pacific Grove, CA (USA), 14 – 19 Oct 1990.
R. W. Hoff (ed.).
AIP Conf. Proc., No. 238.
American Institute of Physics, New York, NY (USA). 1051 p. (1991). ISBN 0–88318–830–9.
Topics discussed during the symposium were the following: neutron and proton capture (high–resolution gamma spectroscopy, short lifetimes, gamma–ray production, reaction mechanisms), nuclear structure (collective and single–particle phenomena, superdeformation, proton–neutron interactions), statistical properties of nuclear levels (chaotic behavior), nuclear astrophysics (nucleosynthesis, chronometry), fundamental physics with neutrons (neutron lifetime), new facilities, new instrumentation, new neutron sources, and applications (elemental mapping of planetary surfaces, analysis of near–surface solid-state reactions).
The individual contributions within the subject scope of AAA are included in their corresponding categories – see abstracts 061.093, 061.094, 061.095, 061.096, 061.097, 061.098, 091.024.

012.037   23. Annual Meeting of the Division for Planetary Sciences (DPS) of the American Astronomical Society (AAS), Palo Alto, CA (USA), 4 – 8 Nov 1991. Abstracts of presented papers.
Bull. Am. Astron. Soc., Vol. 23, No. 3, p. 1101 – 1244 (1991).
The individual contributions within the subject scope of AAA are included in their corresponding categories – see abstracts 004.050, 004.051, 011.013, 011.014, 014.046, 014.047, 014.048, 014.049, 014.050, 015.042, 015.043, 015.048, 022.123, 022.124, 022.125, 022.126, 022.127, 022.128, 022.130, 022.131, 022.134, 022.135, 022.136, 022.170, 022.171, 022.172, 022.176, 022.178, 022.179, 022.180, 022.181, 034.075, 036.123, 036.141, 036.142, 036.143, 036.147, 042.060, 042.080, 042.081, 051.029, 051.034, 063.098, 063.116, 081.037, 081.040, 082.052, 091.026, 091.027, 091.030, 091.031, 091.032, 091.033, 091.034, 091.035, 091.042, 091.043, 091.044, 091.045, 091.048, 092.005, 092.006, 092.007, 092.008, 092.009, 092.010, 092.011, 092.012, 092.013, 093.049, 093.050, 093.052, 093.053, 093.054, 093.055, 093.056, 093.057, 093.058, 093.059, 093.060, 093.061, 093.062, 093.063, 093.064, 093.065, 093.066, 093.067, 093.068, 093.069, 093.070, 093.071, 093.072, 093.073, 093.074, 093.075, 093.076, 093.077, 093.078, 093.079, 093.080, 093.081, 093.082, 093.083, 093.084, 093.085, 093.086, 093.087, 093.088, 093.089, 093.090, 093.091, 094.065, 094.066, 094.067, 094.068, 094.069, 094.070, 094.071, 094.072, 094.073, 094.074, 094.075, 094.076, 094.077, 096.013, 097.114, 097.115, 097.116, 097.117, 097.118, 097.119, 097.120, 097.121, 097.122, 097.123, 097.124, 097.125, 097.126, 097.127, 097.128, 097.129, 097.130, 097.131, 097.135, 097.136, 097.137, 097.138, 097.139, 097.146, 097.147, 097.148, 097.149, 097.150, 097.151, 097.152, 097.153, 097.154, 097.155, 097.156, 097.157, 097.158, 097.159, 097.160, 097.161, 097.162, 097.163, 097.164, 097.165, 097.166, 097.167, 097.168, 097.169, 097.170, 097.171, 097.172, 098.074, 098.075, 098.076, 098.077, 098.078, 098.079, 098.080, 098.081, 098.082, 098.083, 098.084, 098.085, 098.086, 098.087, 098.088, 098.089, 098.090, 098.091, 098.092, 098.093, 098.094, 098.095, 098.096, 098.099, 098.100, 098.101, 098.102, 098.103, 098.104, 098.105, 098.106, 098.107, 098.108, 098.143, 099.035, 099.036, 099.037, 099.038, 099.039, 099.040, 099.041, 099.042, 099.043, 099.044, 099.045, 099.046, 099.047, 099.048, 099.049, 099.050, 099.051, 099.052, 099.053, 099.054, 099.055, 099.056, 099.057, 099.058, 099.059, 099.060, 099.061, 099.062, 099.063, 099.064, 099.065, 099.066, 099.068, 099.069, 099.070, 099.071, 099.076, 099.077, 099.078, 099.079, 099.080, 099.081, 099.082, 099.083, 099.084, 099.085, 099.086, 099.087, 099.088, 099.089, 099.090, 099.091, 099.092, 099.093, 100.017, 100.018, 100.019, 100.020, 100.021, 100.022, 100.023, 100.024, 100.025, 100.026, 100.027, 100.029, 100.030, 100.031, 100.032, 100.033, 100.034, 100.035, 100.036, 100.037, 100.038, 100.039, 100.040, 100.041, 100.042, 100.043, 100.044, 100.045, 100.046, 100.047, 100.048, 100.049, 100.050, 100.051, 100.052, 100.053, 100.054, 100.055, 100.056, 100.057, 100.058, 100.059, 100.060, 100.061, 100.062, 100.063, 100.064, 101.028, 101.029, 101.030, 101.031, 101.032,

101.033, 101.034, 101.035, 101.036, 101.037, 101.038, 101.039, 101.042, 101.043, 101.044, 101.045, 101.046, 101.047, 101.048, 101.049, 101.050, 101.051, 101.052, 101.053, 101.054, 101.055, 101.062, 101.063, 101.064, 101.065, 101.066, 101.067, 101.068, 101.069, 101.071, 101.072, 101.073, 101.074, 101.075, 101.076, 101.077, 101.078, 101.079, 101.080, 101.081, 101.082, 101.083, 101.084, 101.085, 101.086, 101.087, 101.088, 102.024, 102.026, 102.027, 102.028, 102.029, 102.030, 102.031, 102.032, 102.033, 102.034, 102.035, 102.036, 102.037, 102.038, 102.039, 102.040, 103.020, 103.022, 103.023, 103.024, 103.025, 103.046, 103.048, 103.052, 103.053, 103.054, 103.055, 103.056, 103.057, 103.065, 103.066, 103.120, 103.121, 103.122, 103.123, 103.124, 103.125, 103.126, 103.127, 103.147, 105.215, 105.216, 105.223, 106.044, 106.045, 106.046, 106.047, 106.048, 106.049, 106.050, 107.024, 107.025, 107.026, 107.027, 107.028, 107.029, 107.030, 107.031, 107.032, 107.033, 107.034, 107.035, 107.036, 107.037, 107.038, 107.039, 107.040, 107.041, 107.042, 107.043, 112.098, 118.031, 118.032, 118.033, 121.057, 131.253.

012.038   Meeting held by the Astronomical Science Group of Ireland, Cork (Ireland), 14 Sep 1990. Proceedings.
Ir. Astron. J., Vol. 20, No. 3, p. 184 – 206 (Mar 1992).
The individual contributions within the subject scope of AAA are included in their corresponding categories – see abstracts 035.065, 036.122, 064.059, 066.181, 073.055, 132.026.

012.039   Chemistry and spectroscopy of interstellar molecules. Proceedings. Symposium on Chemistry and Spectroscopy of Interstellar Molecules, Honolulu, HI (USA), Dec 1989.
D. K. Bohme, E. Herbst, N. Kaifu, S. Saito (eds.).
University of Tokyo Press, Tokyo (Japan). 291 p. (1992). ISBN 4–13–068205–9. ISBN 0–86008–465–5. Price US$ 100.00.
This symposium was dedicated to Hiroko Suzuki.
Contents: 1. Observational astronomy. 2. Laboratory spectroscopy. 3. Chemical kinetics and theory. 4. Modelling.
The individual contributions within the subject scope of AAA are included in their corresponding categories – see abstracts 013.035, 013.036, 022.137, 022.138, 022.139, 022.140, 022.141, 022.142, 022.143, 022.144, 022.145, 022.146, 022.148, 022.149, 022.150, 022.151, 022.152, 022.153, 022.154, 022.155, 022.157, 022.158, 022.159, 064.061, 125.132, 131.176, 131.180, 131.183, 131.185, 131.191, 131.193, 131.194, 131.197, 131.199, 131.200, 131.202, 131.203, 131.204, 131.205, 131.206, 131.207, 131.208, 131.211, 131.213, 131.216, 131.217, 131.218, 131.219, 131.220, 131.221, 131.222, 131.223.

012.040   Frontier objects in astrophysics and particle physics. Proceedings. Vulcano Workshop 1990: Frontier objects in astrophysics and particle physics, Vulcano (Italy), 21 – 25 May 1990.
F. Giovannelli, G. Mannocchi (eds.).
Soc. Ital. Fis., Atti Conf., Vol. 28.
Editrice Compositori, Bologna (Italy). 566 p. (1991). ISBN 88–7794–038–7. Price L. 110000.
Contents: 1. Physics of accreting matter onto collapsed objects. 2. Cosmic rays: acceleration mechanisms. 3. The state of the art in detection. 4. Neutrino and γ–ray astronomy. 5. Gravitational waves. 6. Concluding remarks.
The individual contributions within the subject scope of AAA are included in their corresponding categories – see abstracts 013.040, 013.041, 013.042, 013.043, 013.044, 013.045, 022.164, 022.165, 034.057, 034.058, 034.059, 034.060, 034.061, 034.062, 034.063, 035.070, 061.107, 062.116, 062.118, 063.100, 063.101, 064.062, 064.063, 067.146, 067.147, 067.148, 117.208, 117.209, 117.210, 117.211, 117.212, 117.213, 117.214, 117.215, 142.035, 143.072, 143.073, 143.074, 144.038, 144.040, 144.041, 144.042, 144.043, 144.044, 144.045, 144.046, 144.047, 144.048, 158.126, 160.094, 161.350.

012.041   Extragalactic radio sources – from beams to jets. Proceedings. 7. IAP Meeting: Extragalactic radio sources – from beams to jets, Paris (France), 2 – 5 Jul 1991.
J. Roland, H. Sol, G. Pelletier (eds.).
Cambridge University Press, Cambridge (UK). 388 p. (1992). ISBN 0–521–41602–7. Price £ 40.00, US$ 69.95.
Contents: 1. Physical conditions around the central engine. 2. Beams and jets observed on VLBI scale. 3. Production and propagation of beams and jets. 4. The large scale structure as a guide to the compact source properties.
The individual contributions within the subject scope of AAA are included in their corresponding categories – see abstracts 036.133, 062.133, 062.134, 062.135, 062.136, 062.139, 062.140, 062.141, 062.142, 062.143, 062.144, 062.145, 062.146, 062.147, 062.149, 062.150, 062.151, 062.152, 062.153, 063.104, 067.157, 067.158, 067.159, 067.160, 143.075, 155.133, 158.132, 158.133, 158.134, 158.135, 158.136, 158.137, 158.138, 158.139, 158.140, 158.143, 158.144, 158.145, 158.146, 158.147, 158.148, 158.149, 158.150, 158.151, 158.152, 158.153, 158.162, 158.163, 158.165, 158.166, 158.167, 159.070, 159.071, 159.072, 159.074, 159.075, 159.076, 159.077, 159.082.

012.042   Young star clusters and early stellar evolution. Proceedings. Vulcano Workshop on Young Star Clusters and Early Stellar Evolution, Vulcano (Italy), 16 – 20 Sep 1991.
F. Palla, P. Persi, H. Zinnecker (eds.).
Mem. Soc. Astron. Ital., Vol. 62, No. 4, p. 705 – 976 (1991).
Contents: 1. Embedded clusters and luminosity functions. 2. Molecular cores and star formation. 3. Young open clusters.
The individual contributions within the subject scope of AAA are included in their corresponding categories – see abstracts 064.064, 064.065, 065.083, 111.014, 114.050, 121.046, 121.047, 131.224, 131.225, 131.226, 131.227, 131.228, 131.229, 131.230, 131.231, 131.232, 131.233, 131.234, 131.235, 131.236, 131.237, 133.008, 153.028, 153.029, 153.030, 153.031, 153.032, 153.033, 153.034, 153.035, 153.036, 153.037, 153.038, 153.039, 153.040, 154.051, 157.262, 157.263.

012.043   Nuclear data for science and technology. Proceedings. International Conference on Nuclear Data for Science and Technology, Jülich (Germany), 13 – 17 May 1991.
S. M. Qaim (ed.).
Res. Rep. Phys.
Springer, Berlin (Germany). 1060 p. (1992). ISBN 3–540–55100–X.
This book is the most up–to–date and complete work in the field of nuclear data and an essential reference material for scientists working in this area. It includes a wide range of interdisciplinary topics dealing with measurement, calculation, evaluation and application of nuclear data in various fields. Both energy and non–energy related data are treated, with major emphasis on application oriented numerical data. Nuclear theories relevant to the understanding of data are also covered. A special feature of these proceedings is that all the major nuclear data files are described.
The individual contributions within the subject scope of AAA are included in their corresponding categories – see abstracts 061.112, 061.113, 061.114.

012.044   Quarks, symmetries and strings. Proceedings. Symposium on Quarks, Symmetries and Strings in Honor of Bunji Sakita's 60th Birthday, New York, NY (USA), 1 – 2 Oct 1990.
M. Kaku, A. Jevicki, K. Kikkawa (eds.).
World Scientific, Singapore (Singapore). 428 p. (1991). ISBN 981–02–0526–0.
Contents: 1. Quarks. 2. String theory. 3. Methods of field theory. 4. Statistical methods.
The individual contributions within the subject scope of AAA are included in their corresponding categories – see abstracts 061.115, 061.116, 080.055.

012.045   Solar observations: techniques and interpretation. Proceedings. 1. Canary Islands Winter School of Astrophysics: Solar observations – techniques and interpretation, La Laguna, Tenerife (Spain), 23 Oct – 3 Nov 1989.
F. Sánchez, M. Collados, M. Vázquez (eds.).
Cambridge University Press, Cambridge (UK). 258 p. (1992). ISBN 0–521–40251–4.
This book presents four detailed analyses on some of the most important aspects of the Sun, going from solar interior to the chromosphere. The authors discuss high spatial resolution

techniques, magnetic field measurements, solar post–focus instrumentation, and the dynamics of the solar atmosphere.

The individual contributions within the subject scope of AAA are included in their corresponding categories – see abstracts 034.070, 036.138, 036.139, 080.064.

012.046 Lunar and planetary science, Volume 22. Proceedings. 22. Lunar and Planetary Science Conference, Houston, TX (USA), 18 – 22 Mar 1991.
G. Ryder, V. L. Sharpton (eds.).
Proc. Lunar Planet. Sci., Vol. 22.
Lunar and Planetary Inst., Houston, TX (USA). 480 p. (1992). ISBN 0–942862–06–6. Price US$ 25.00.

Contents: 1. Venus: review paper. 2. The surface of Mars. 3. The Cretaceous–Tertiary boundary and impact processes. 4. The interplanetary medium then and now. 5. Lunar surface characterization and processes. 6. Basic magmatic rocks and processes. 7. Laboratory precautions. 8. Errata.

The individual contributions within the subject scope of AAA are included in their corresponding categories – see abstracts 022.184, 022.185, 022.191, 093.092, 094.078, 094.079, 094.080, 094.081, 094.082, 094.083, 094.084, 094.086, 094.087, 094.088, 094.089, 094.090, 094.091, 094.092, 094.093, 097.176, 097.177, 097.178, 097.179, 097.180, 097.181, 105.248, 105.249, 105.250, 105.251, 105.252, 105.253, 105.255, 106.060, 106.061, 106.062, 106.063, 106.064, 107.044.

012.047 Intersections between particle and nuclear physics. Proceedings. 4. Conference on the Intersections Between Particle and Nuclear Physics (CIPANP–4), Tucson, AZ (USA), 23 – 29 May 1991.
W. T. H. van Oers (ed.).
AIP Conf. Proc., No. 243.
American Institute of Physics, New York, NY (USA). 1179 p. (1992). ISBN 0–88318–950–X.

The quest for physics beyond the Standard Model of quarks and leptons is to be pursued along two complementary directions. The first direction is to search for new physica along the energy frontier with the proposed high energy accelerators. The second direction is to test the Standard Model through precision experiments requiring intense beams of mesons containing strangeness, charm or beauty. This was in part concerned with such precision experiments and their interpretation. Extremely high levels of precision are being obtained in studies of rare decays. Intriguing results were obtained in proton–antiproton annihilation experiments in terms of exotic mesons. The current neutrino problem was discussed at length. The search for manifestations of a quark–gluon plasma in relativistic heavy ion collisions continues. The question about the spin contents of the nucleon remain unresolved.

The individual contributions within the subject scope of AAA are included in their corresponding categories – see abstracts 034.069, 061.120, 080.061, 080.062, 080.063, 155.136.

012.048 Variable star research: an international perspective. Proceedings. 1. European Meeting of the American Association of Variable Star Observers (AAVSO): International Cooperation and Coordination in Variable Star Research, Brussels (Belgium), 24 – 28 Jul 1990.
J. R. Percy, J. A. Mattei, C. Sterken (eds.).
Cambridge University Press, Cambridge (UK). 343 p. (1992). ISBN 0–521–40469–X. Price £ 35.00, US$ 49.95.

This book presents results from the First European Meeting of the American Association of Variable Star Observers which was attended by professional and amateur astronomers. The authors cover such topics as the history of variable star research, the contributions of amateurs, current understanding of the major classes of variable stars, and prospects for amateur–professional collaboration in the future. Visual, photographic, photoelectric and CCD observing techniques are discussed in a practical way. Throughout the book there are many suggestions for research projects that can be undertaken with modest equipment under average observing conditions.

Further papers of this conference are published in J. Am. Assoc. Variable Star Obs., Vol. 19, Nos. 1, 2 (1990), see Abstr. 53.012.103.

The individual contributions within the subject scope of AAA are included in their corresponding categories – see abstracts 002.069, 002.070, 002.072, 002.073, 005.025, 013.050, 013.051, 013.052, 013.053, 013.054, 013.055, 013.056, 013.057, 013.058, 013.059, 013.060, 013.061, 013.062, 013.063, 013.064, 034.064, 098.139, 113.016, 117.241, 117.242, 117.243, 117.244, 122.089, 122.090, 122.091, 122.092, 122.093, 122.094, 122.095, 122.096, 122.097, 122.098, 122.099, 122.100.

012.049 ISPP–11. Physics of charged bodies in space plasmas. Proceedings. Workshop on Physics of Charged Bodies in Space Plasmas, Varenna (Italy), 23 – 27 Sep 1991.
M. Dobrowolny, E. Sindoni (eds.).
Editrice Compositori, Bologna (Italy). 369 p. (1992). ISBN 88–7794–047–6. Price Lit 80.000.

The main motivation of the workshop was that of assess and understand the problems of active and passive charging phenomena for space platforms. This includes results from space experiments, laboratory simulations, numerical simulations and theoretical studies. Topics areas covered in the meeting were: wake effects, responses to natural charging, methods for neutralizing charged platforms (through ion and electron sources and plasma contactors), charging in active experiments, through electron or ion beams (including interaction of the ejected beam with the environment), high voltage phenomena, radiation from moving charged bodies, global phenomena induced by current closure.

012.050 Proceedings of the Twenty–first General Assembly. 21. General Assembly of the IAU, Buenos Aires (Argentina), 23 Jul – 1 Aug 1991.
J. Bergeron (ed.).
Trans. Int. Astron. Union, Vol. 21B.
Kluwer Academic Publ., Dordrecht (Netherlands). 947 p. (1992). ISBN 0–7923–1914–1. Price Dfl. 390.00.

This volume summarizes the work of the XXIst General Assembly of the IAU. The discourses given during the Inaugural and Closing Ceremonies are reproduced in Chapters I & III respectively. The proceedings of the two sessions of the General Assembly will be found in Chapter II, which includes the Resolutions, the report of the Finance Committee and the Accounts and other aspects of the administration of the Union. Together with the report of the Executive Committee for this last triennium (Chapter IV), they provide the permanent record for the Union in the period 1988 – 1991. This volume also contains the Commission reports from Buenos Aires compiled by the Presidents of Commissions (Chapter V). The Statutes, By–Laws and a few working rules of the Union are published in Chapter VI. Finally, Chapter VII contains the list of countries adhering to the Union and the alphabetical, geographical and commission membership lists of over 7350 individual IAU members. The IAU still appears to be unique among the scientific Unions in maintaining this category of individual membership which contributes in a crucial way to the spirit and the aims of the Union.

012.051 Infrared technology XVII. Dedicated to the memory of Irving J. Spiro. Proceedings. SPIE Conference on Infrared Technology XVII – Dedicated to the Memory of Irving J. Spiro as Part of SPIE's International Symposium on Optical Applied Science and Engineering, San Diego, CA (USA), 22 – 26 Jul 1991.
B. F. Andresen, M. S. Scholl, I. J. Spiro (eds.).
SPIE Proc. Ser., Vol. 1540.
SPIE, Bellingham, WA (USA). 809 p. (1991). ISBN 0–8194–0668–6.

Infrared–imaging technology is evolving at a rapid pace. At this conference there were, for the first time, five different focal plane technologies reported on, all of which have the required performance parameters needed to support staring sensor development. These technologies comprise PtSi, InSb, HgCdTe, GaAs/AlGaAs, and GeSi/Si. Infrared imagery was demonstrated

having higher resolution than current broadcast television. Also discussed were optical designs for imagers using hybrid diffractive/refractive elements as replacements for conventional elements.
The individual contributions within the subject scope of AAA are included in their corresponding categories – see abstracts 031.047, 034.071, 034.072, 035.094, 035.095, 035.096, 035.097, 035.098, 035.099, 035.100, 035.101, 035.102, 035.103, 035.104, 035.105, 035.106, 035.107, 035.108, 035.109.

012.052 **Active and adaptive optical components. Proceedings. SPIE Conference on Active and Adaptive Optical Components as Part of SPIE's International Symposium on Optical Applied Science and Engineering, San Diego, CA (USA), 24 – 26 Jul 1991.**
M. A. Ealey (ed.).
SPIE Proc. Ser., Vol. 1543.
SPIE, Bellingham, WA (USA). 524 p. (1992). ISBN 0–8194–0671–6.
This conference included six sessions plus a poster session. Session 1, *Historical perspective and future trends*, was devoted to tracing the roots of adaptive optical components. Session 2, *Deformable mirrors*, provided an overview of wavefront corrector devices. Session 3, *Large active mirrors*, discussed recent developments in manufacturing large-aperture optical mirrors for ground-based and space-based operation. The Keck telescope segments are discussed in terms of schedule and process enhancements. A novel design to make large-aperture mirrors using small modular segments is described, including the novel inductive sensor that is used to phase the segments. Session 4, *Fast-steering mirrors*, began with an overview of critical specifications in terms of system requirements. Tilt mirrors developed for the European Southern Observatory and the University of Hawaii were reviewed in terms of design configuration and performance results. A novel design that features a magnetic suspension system was described, with emphasis placed on its potential to eliminate failure associated with mechanical flexures. Session 6, *Actuators and controls*, discussed the newly developed solid state materials that have led to the production of more precise and accurate deformable mirrors.
The individual contributions within the subject scope of AAA are included in their corresponding categories – see abstracts 031.028, 031.029, 031.030, 031.031, 031.032, 031.033, 031.034, 031.035, 031.036, 031.037, 031.038, 031.039, 031.040, 031.041, 031.042, 031.043, 031.044, 031.045, 031.046, 036.144, 036.145, 036.146.

012.053 **Desktop publishing in astronomy & space sciences. Proceedings. Colloquium on Desktop Publishing in Astronomy and Space Sciences, Strasbourg (France), 1 – 3 Oct 1991.**
A. Heck (ed.).
World Scientific, Singapore (Singapore). 251 p. (1992). ISBN 981–02–0915–0. Price £ 38.00, US$ 58.00.
At the first session of this colloquium introductory papers on packages and publication polices were presented. Then three thematic sessions: *publishers, scientific editors, electronic publishing and intelligent information retrieval* followed. The last session was devoted to contributed papers.
The individual contributions within the subject scope of AAA are included in their corresponding categories – see abstracts 002.097, 002.098, 002.099, 002.100, 002.101, 002.102, 002.103, 002.104, 002.105, 002.106, 002.107, 002.108, 002.109, 002.110, 002.111, 002.112, 002.113, 002.114, 002.115, 002.116, 002.117, 002.118, 002.119, 002.120, 002.121, 002.122, 002.123, 002.124.

012.054 **Astrophysical opacities. Proceedings. Workshop on Astrophysical Opacities (WAO), Caracas (Venezuela), 15 – 19 Jul 1991.**
A. E. Lynas–Gray, C. Mendoza, C. J. Zeippen (eds.).
Rev. Mex. Astron. Astrofis., Vol. 23, special issue, 259 p. (Mar 1992).
Contents: 1. Atomic and molecular data. 2. Equation of state. 3. Opacities. 4. Astronomical problems.
The individual contributions within the subject scope of AAA are included in their corresponding categories – see abstracts

002.079, 013.078, 013.081, 013.084, 013.085, 013.086, 022.186, 022.187, 022.188, 022.189, 022.190, 022.192, 022.193, 022.194, 061.122, 062.178, 063.119, 063.120, 063.121, 063.122, 063.123, 063.124, 063.125, 063.126, 063.127, 063.128, 064.075, 064.076, 065.101.

012.055 **Interrelations between physics and dynamics for minor bodies in the solar system. Proceedings. 15. Ecole de Printemps d'Astrophysique de Goutelas: Interrelations between physics and dynamics for minor bodies in the solar system, Château de Goutelas (France), 29 Apr – 4 May 1991.**
D. Benest, C. Froeschlé (eds.).
Editions Frontières, Gif–sur–Yvette (France). 651 p. (1992). ISBN 2–86332–112–9. Price FF 420.00, US$ 75.00.
Contents: 1. Asteroids. 2. Comets. 3. Meteors. 4. Rings.
The individual contributions within the subject scope of AAA are included in their corresponding categories – see abstracts 021.021, 022.195, 022.196, 042.082, 042.083, 042.084, 042.085, 042.086, 042.087, 063.130, 098.153, 098.154, 102.049, 102.050, 104.083, 106.065.

012.056 **General relativity and relativistic astrophysics. Proceedings. 4. Canadian Conference on General Relativity and Relativistic Astrophysics, Winnipeg (Canada), 16 – 18 May 1991.**
G. Kunstatter, D. E. Vincent, J. G. Williams (eds.).
World Scientific, Singapore (Singapore). 385 p. (1992). ISBN 981–02–0965–7. Price £ 44.00.
The individual contributions within the subject scope of AAA are included in their corresponding categories – see abstracts 022.198, 022.199, 061.123, 063.129, 067.172, 067.173, 161.403, 161.405, 161.406, 161.407, 161.408, 161.409, 161.410, 161.411, 161.412, 161.413, 161.414, 161.415, 161.416, 161.417.

012.057 **Testing the AGN paradigm. Proceedings. 2. Annual Mid–October Topical Astrophysics Conference: Testing the AGN paradigm, College Park, MD (USA), 13 – 16 Oct 1991.**
S. S. Holt, S. G. Neff, C. M. Urry (eds.).
AIP Conf. Proc., No. 254.
American Institute of Physics, New York, NY (USA). 718 p. (1992). ISBN 1–56396–009–5. Price US$ 99.00.
Contents: 1. Introduction. 2. Black holes. 3. Accretion disks. 4. High energy continuum. 5. Jets. 6. AGN geometry. 7. Broad line regions. 8. Beyond the broad line region.
The individual contributions within the subject scope of AAA are included in their corresponding categories – see abstracts 004.070, 061.130, 062.189, 062.190, 062.191, 062.192, 062.193, 063.134, 063.135, 063.136, 063.137, 063.138, 063.140, 064.077, 064.078, 067.174, 067.175, 067.176, 067.177, 067.178, 067.184, 067.185, 067.186, 067.188, 067.189, 067.191, 067.192, 067.196, 067.197, 067.198, 067.199, 067.200, 067.207, 117.255, 143.088, 151.075, 155.143, 155.144, 157.277, 157.278, 157.279, 157.281, 157.294, 157.295, 157.296, 157.297, 158.179, 158.180, 158.181, 158.182, 158.183, 158.184, 158.185, 158.187, 158.189, 158.190, 158.191, 158.192, 158.193, 158.194, 158.195, 158.196, 158.197, 158.198, 158.199, 158.200, 158.201, 158.202, 158.203, 158.204, 158.205, 158.206, 158.207, 158.208, 158.209, 158.210, 158.211, 158.212, 158.213, 158.214, 158.215, 158.216, 158.217, 158.218, 158.219, 158.220, 158.221, 158.222, 158.223, 158.224, 158.225, 158.226, 158.227, 158.228, 158.229, 158.230, 158.231, 158.232, 158.233, 158.234, 158.235, 158.236, 158.237, 158.238, 159.094, 159.100, 159.102, 159.103, 159.104, 159.105, 159.110, 159.111, 159.112, 159.113, 159.114, 159.115, 159.116, 159.118, 161.418, 161.445, 161.457.

012.058 **Les Journées Relativistes, Cargèse (France), 1 – 4 May 1991.**
A. Folacci, B. P. Jensen (eds.).
Classical Quantum Gravity, Vol. 9, Suppl., 213 p. (1992).
The individual contributions within the subject scope of AAA are included in their corresponding categories – see abstracts 066.234, 066.263, 066.264, 066.276, 066.277, 066.278, 066.279, 066.280, 066.281, 067.215, 067.216.

012.059 **Digitised optical sky surveys. Proceedings. 2. Conference on Digitised Optical Sky Surveys, Edinburgh (UK), 18 – 21 Jun 1991.**
H. T. MacGillivray, E. B. Thomson (eds.).
Astrophys. Space Sci. Libr., Vol. 174.
Kluwer Academic Publ., Dordrecht (Netherlands). 542 p. (1992). ISBN 0–7923–1642–8. Price Dfl. 250.00, US$ 149.00, £ 87.00.
Contents: 1. Sky surveys and calibration programmes. 2. Digitisation programmes – current and planned. 3. Developments in techniques – compression, astrometry, photometry, and object classification. 4. Astronomy from large scale digitised surveys. 5. Optical identification programmes using digitised surveys.
The individual contributions within the subject scope of AAA are included in their corresponding categories – see abstracts 002.076, 002.077, 002.078, 002.081, 002.082, 002.085, 002.086, 002.087, 013.071, 013.072, 013.073, 013.074, 013.075, 013.076, 013.077, 013.079, 013.080, 013.087, 013.088, 013.089, 013.090, 013.091, 013.092, 034.076, 034.077, 034.078, 034.079, 034.080, 036.149, 036.150, 036.151, 036.152, 036.154, 036.155, 036.156, 036.157, 036.158, 036.159, 036.160, 036.161, 036.163, 117.256, 153.045, 154.057, 155.141, 155.142, 156.040, 157.280, 159.095, 159.096, 159.097, 159.098, 160.106, 160.107, 160.108, 160.109, 160.110, 160.111, 160.112, 160.113, 161.421.

012.060 **Texas/ESO–CERN Symposium on Relativistic Astrophysics, Cosmology, and Fundamental Physics. Proceedings. 15. Texas Symposium on Relativistic Astrophysics and 4. ESO–CERN Symposium, Brighton (UK), 16 – 21 Dec 1990.**
J. D. Barrow, L. Mestel, P. A. Thomas (eds.).
Ann. N.Y. Acad. Sci., Vol. 647.
New York Academy of Sciences, New York, NY (USA). 854 p. (1991). ISBN 0–89766–707–7. Price US$ 190.00. ISBN 0–89766–708–5 (paper).
This symposium was dedicated to Professor Léon van Hove (1924 – 1990).
Contents: 1. Plenary talks. 2. Particle astronomy minisymposium. 3. Neutron stars and black hole astrophysics minisymposium. 4. Lunar and space-based astrophysics minisymposium. 5. Large-scale structure and galaxy formation minisymposium.
The individual contributions within the subject scope of AAA are included in their corresponding categories – see abstracts 009.010, 011.016, 011.017, 034.086, 034.089, 034.090, 034.091, 034.126, 036.164, 036.165, 036.166, 036.207, 051.035, 051.036, 051.046, 051.047, 061.124, 061.125, 061.126, 061.128, 061.129, 066.255, 066.260, 067.182, 067.183, 067.190, 067.193, 067.194, 067.195, 067.202, 067.203, 067.205, 080.069, 080.073, 080.075, 080.078, 080.080, 125.139, 125.140, 126.110, 126.111, 142.050, 142.063, 143.082, 143.083, 143.084, 143.085, 143.086, 144.062, 144.063, 144.068, 157.282, 159.099, 159.101, 160.115, 160.116, 161.422, 161.424, 161.425, 161.427, 161.428, 161.430, 161.446, 161.447, 161.448, 161.449, 161.450, 161.451, 161.452, 161.454, 161.455, 161.456, 161.458, 161.459, 161.460, 161.461, 161.462, 161.463, 161.464, 161.465, 161.466, 161.467, 161.468, 161.471.

012.061 **Advances in theoretical physics. Proceedings. Italo–Soviet Workshop on Advances in Theoretical Physics, Vietri sul Mare (Italy), 23 – 28 Oct 1990.**
E. R. Caianiello (ed.).
World Scientific, Singapore (Singapore). 225 p. (1991). ISBN 981–02–0717–4.
The individual contributions within the subject scope of AAA are included in their corresponding categories – see abstracts 022.200, 161.419.

012.062 **Structure of hadrons and hadronic matter. Proceedings. International Summer School on the Structure of Hadrons and Hadronic Matter – NATO Advanced Study Institute, Dronten (Netherlands), 5 – 18 Aug 1990.**
O. Scholten, J. H. Koch (eds.).
World Scientific, Singapore (Singapore). 360 p. (1991). ISBN 981–02–0590–2.
The lectures at the summer school were focussed on the dynamics and structure of hadronic systems. This theme was examined from various perspectives. For nuclear matter close to normal densities and for relatively low excitation energies, a description in terms of nucleon degrees of freedom is appropriate. As the density increases, but in some case already under normal conditions, relativistic effects become important and a relativistic approach is necessary. For the description of heavy ion scattering at high energies or to understand the dynamics governing neutron stars, one must explicitly take into account also the non-nucleon degrees of freedom.
The individual contributions within the subject scope of AAA are included in their corresponding categories – see abstracts 067.179.

012.063 **Testing the standard model. Proceedings. Theoretical Advanced Study Institute (TASI) in Elementary Particle Physics: Testing the standard model, Boulder, CO (USA), 3 – 27 Jun 1990.**
M. Cvetič, P. Langacker (eds.).
World Scientific, Singapore (Singapore). 916 p. (1991). ISBN 981–02–0314–4.
The school explored the theory of the standard model (both electroweak and QCD), possible physics beyong the standard model (especially grand unification, supersymmetry, and superstrings) and the experimental tests of the standard model and searches for new physics.
The individual contributions within the subject scope of AAA are included in their corresponding categories – see abstracts 080.074, 161.429.

012.064 **Basic space science. Proceedings. Workshop on Basic Space Science, Bangalore (India), 30 Apr – 3 May 1991.**
H. J. Haubold, R. K. Khanna (eds.).
AIP Conf. Proc., No. 245.
American Institute of Physics, New York, NY (USA). 350 p. (1992). ISBN 0–88318–951–8. Price US$ 95.00.
This workshop *"Basic space science for the benefit of developing countries"* was devoted to the study of international cooperation on basic space science, basic space science in developing countries, solar–terrestrial interaction, solar system science and space astronomy and astrophysics. Two introductory addresses were presented: 1. United Nations address (*N. Jasentuliyana*). 2. Keynote address: Importance of basic space science for developing countries (*U. R. Rao*).
The individual contributions within the subject scope of AAA are included in their corresponding categories – see abstracts 013.093, 013.094, 013.095, 013.096, 013.097, 013.098, 015.051, 021.022, 051.037, 051.038, 062.183, 073.068, 076.020, 080.076, 080.077, 085.014, 085.015, 161.432.

012.065 **22. Meeting of the American Astronomical Society Division on Dynamical Astronomy, Key Biscayne, FL (USA), 16 – 17 May 1991. Abstracts of presented papers.**
Bull. Am. Astron. Soc., Vol. 23, No. 3, p. 1253 – 1260 (1991).
The individual contributions within the subject scope of AAA are included in their corresponding categories – see abstracts 002.091, 004.064, 021.024, 034.088, 035.053, 035.113, 036.167, 042.090, 043.007, 045.005, 051.039, 066.262, 098.163, 098.164, 098.165, 098.166, 099.104, 099.105, 100.071, 101.091, 151.076, 151.077, 157.285.

012.066 **80. Spring Meeting of the AAVSO, Charlestown, RI (USA), 10 – 12 May 1991. Abstracts of papers presented.**
J. Am. Assoc. Variable Star Obs., Vol. 20, No. 1, p. 108 – 111 (1991).
The individual contributions within the subject scope of AAA are included in their corresponding categories – see abstracts 004.071, 004.072, 009.011, 013.116, 014.071, 033.014, 122.130, 123.026, 131.271.

012.067 **178. Meeting of the American Astronomical Society (AAS), Seattle, WA (USA), 26 – 30 May 1991. Late-paper abstracts.**
Bull. Am. Astron. Soc., Vol. 23, No. 3, p. 1261 – 1271 (1991).
The individual contributions within the subject scope of AAA are included in their corresponding categories – see abstracts 013.100, 014.060, 014.061, 034.092, 034.093, 035.114, 051.040, 062.188, 064.079, 064.080, 065.103, 067.201, 073.029, 074.058, 076.021, 112.101, 116.034, 117.260, 119.054, 121.059, 121.060, 122.107, 125.141, 126.112, 131.265, 131.266, 131.267, 132.028, 134.042, 152.010, 153.046, 153.047, 157.287, 157.288, 157.289, 157.290, 157.291, 157.292, 159.108, 159,109, 160.114, 161.442.

012.068 **Origin and evolution of interplanetary dust. Proceedings. IAU Colloquium No. 126: Origin and evolution of interplanetary dust, Kyoto (Japan), 27 – 30 Aug 1990.**
A. C. Levasseur–Regourd, H. Hasegawa (eds.).
Astrophys. Space Sci. Libr., Vol. 173.
Kluwer Academic Publ., Dordrecht (Netherlands). 480 p. (1991). ISBN 0–7923–1365–8. Price Dfl. 240.00, US$ 129.00, UK£ 82.00.
The book begins with investigations of interplanetary dust by space and Earth environmental studies (Part I), by physical and chemical analysis (Part II), and by zodiacal light and optical studies (Part III). Topics related to cometary dust (Part IV), meteoroids and meteor streams (Part V), and circumplanetary dust, are then presented. Finally, the origin of interplanetary dust (Part VII) is tracked back to comets or asteroids and to interstellar or circumstellar dust. A summary demonstrates that interplanetary dust studies are thriving and may provide a clearer understanding of the formation of the solar system.
The individual contributions within the subject scope of AAA are included in their corresponding categories – see abstracts 002.094, 013.107, 022.203, 022.204, 022.205, 022.206, 022.207, 022.208, 022.209, 022.210, 022.211, 022.212, 022.213, 022.214, 022.215, 022.216, 022.217, 022.218, 022.219, 022.220, 022.221, 022.222, 022.223, 022.224, 022.225, 022.226, 035.115, 035.116, 051.041, 051.042, 051.043, 051.044, 051.045, 051.048, 063.141, 063.142, 074.060, 082.067, 091.054, 100.072, 103.037, 103.038, 103.039, 103.040, 103.041, 103.042, 103.091, 103.092, 103.135, 103.136, 103.137, 103.138, 103.139, 104.085, 104.086, 104.087, 104.088, 104.089, 104.090, 104.091, 104.092, 105.259, 106.035, 106.066, 106.067, 106.068, 106.069, 106.070, 106.071, 106.072, 106.073, 106.074, 106.075, 106.076, 106.077, 106.078, 106.079, 106.080, 106.081, 106.082, 106.083, 106.084, 106.085, 106.086, 106.087, 106.088, 107.049, 112.104, 131.269, 131.270, 144.071.

012.069 **Supercomputing astronomy and astrophysics. Proceedings. Symposium of Supercomputing Astronomy and Astrophysics in Japan, Mitaka, Tokyo (Japan), 24 – 25 Dec 1990.**
S. M. Miyama, M. Nagasawa (eds.).
NAO–TAP Rep., 1991.
National Astronomical Observatory, Mitaka, Tokyo (Japan). Division of Theoretical Astrophysics. 394 p. (1991).
The main purpose of this workshop was focused on the theoretical activities in computational astronomy and astrophysics in Japan. The various results in subjects of hydrodynamics, magnetohydrodynamics, plasma physics, gravitating systems, radiative transfer, high–energy astrophysics, general relativity are also presented.
The individual contributions within the subject scope of AAA are included in their corresponding categories – see abstracts 021.025, 021.026, 021.027, 021.028, 021.029, 021.030, 021.031, 021.032, 061.131, 062.194, 062.195, 062.196, 062.197, 062.198, 062.199, 062.200, 062.201, 062.202, 065.105, 067.209, 067.210, 074.061, 107.048, 161.472, 161.473, 161.474, 161.476.

012.070 **Journées Scientifiques de la S.F.S.A., Montpellier (France), 12 – 14 Nov 1990.**
J. Astron. Fr., No. 41, p. 3 – 19 (Feb 1992).
The individual contributions within the subject scope of AAA are included in their corresponding categories – see abstracts 013.133, 061.137, 061.138, 065.111, 097.189, 097.190, 098.177, 112.107, 117.273, 122.151, 122.152, 122.161, 157.307, 159.124, 160.120, 160.121, 161.502.

012.071 **Cluster dayside polar cusp. Planning and coordination of measurements from Cluster, ground stations, balloons and rockets in the dayside polar–cusp region. Proceedings. International Workshop on Cluster Dayside Polar Cusp, Longyearbyen, Spitzbergen, Svalbard (Norway), 16 – 19 Sep 1991.**
C. I. Barron (ed.).
European Space Agency, Paris (France), ESA–SP—330. 225 p. (Dec 1991).
Contents: 1. Present research programmes in Svalbard. 2. Other programmes for cusp studies. 3. Satellite projects. 4. Radar studies of the cusp – future plans. 5. Polar cusp problems. 6. Needs and requirements. 7. Present status of observation techniques. 8. Need for balloon and rocket campaigns. 9. Data systems, future meetings and outlook.
The individual contributions within the subject scope of AAA are included in their corresponding categories – see abstracts 013.108, 013.109, 013.110, 013.111, 013.112, 013.113, 013.114, 013.115, 036.174, 036.175, 036.176, 051.049, 051.050, 083.034, 083.035, 083.036, 083.037, 084.066, 084.067, 084.068, 084.069, 084.070, 084.071.

012.072 **18. Optical Society of India Symposium on Optical Science and Engineering, Bangalore (India), 21 – 23 Mar 1990.**
A. K. Saxena, A. Vagiswari (eds.).
Kodaikanal Obs. Bull., Vol. 11, 127 p. (1991).
The individual contributions within the subject scope of AAA are included in their corresponding categories – see abstracts 031.074, 034.098, 034.099, 063.143, 082.069.

012.073 **Activities and ejecta of supergiant stars. Proceedings. Workshop on the Activities and Ejecta of Supergiant Stars, Sendai (Japan), 22 Jan 1992.**
M. Takeuti (ed.).
Sci. Rep. Tôhoku Univ., Eighth Ser., Vol. 12, Nos. 2/3, p. 111 – 165 (Jan 1992).
The purpose of this workshop was to discuss the observational facts and theoretical results on the circumstellar environment of proto–planetary nebulae stars and on the nature of the stars themselves.
The individual contributions within the subject scope of AAA are included in their corresponding categories – see abstracts 064.082, 065.106, 112.106, 114.057, 122.131.

012.074 **Active and adaptive optical systems. Proceedings. SPIE Conference on Active and Adaptive Optical Systems as Part of SPIE's International Symposium on Optical Applied Science and Engineering, San Diego, CA (USA), 22 – 24 Jul 1991.**
M. A. Ealey (ed.).
SPIE Proc. Ser., Vol. 1542.
SPIE, Bellingham, WA (USA). 566 p. (1991). ISBN 0–8194–0670–8.
This conference covered a range of topics, including a historical perspective of component and systems developments, an overview of currently available hardware, and a discussion of future development needs. The papers presented in these proceedings fall into the following areas: historical perspective tracing the beginnings of the adaptive optical methodology, including landmark components and systems, systems considerations for both laser beam control and astronomical imaging applications, deformable and segmented wavefront corrector design, fabrication, and performance evaluation in terms of wavefront correctability, wavefront sensor developments in terms of hardware implementation and theoretical performance limitations, actuators and their electronic control systems performance, including active materials response, stability, set–point accuracy, and the effects of hysteresis on system performance, fast–steering mirror design, fabrication, and performance evaluation for line–of–sight stabilization, precision pointing, and tilt compensation, large, active primary mirrors, including fabrication and test techniques, active structures for active vibration damping and isolation for new–generation space–based optical systems, European Space Observatory and NASA telescope programs progress in adaptive systems and active components.

The individual contributions within the subject scope of AAA are included in their corresponding categories – see abstracts 031.052, 031.053, 031.054, 031.055, 031.056, 031.057, 031.058, 031.059, 031.060, 031.061, 031.063, 031.064, 031.065, 031.066, 031.067, 031.068, 031.069, 031.070, 031.071, 031.072, 031.073, 032.018, 032.019, 032.020, 032.021, 034.094, 034.095, 034.096, 035.117, 035.118, 035.119, 036.177, 036.178, 036.179, 036.183, 036.184, 036.185, 036.186.

012.075 **Planetary sciences. American and Soviet research. Proceedings. US–USSR Workshop on Planetary Sciences, Moscow (USSR), 2 – 6 Jan 1989.**
T. M. Donahue, K. K. Trivers, D. M. Abramson (eds.).
National Academy Press, Washington, DC (USA). 302 p. (1991). ISBN 0–309–04333–6. Price US$ 22.00.
The purpose of the workshop was to examine the current state of the theoretical understanding of how the planets were formed and how they evolved to their present state.
The individual contributions within the subject scope of AAA are included in their corresponding categories – see abstracts 013.121, 081.041, 082.070, 093.098, 093.099, 102.053, 102.054, 106.089, 107.050, 107.051, 107.052, 107.053, 107.054, 107.055, 107.056, 107.057, 107.058, 107.059, 107.060, 121.061.

012.076 **Gravitation and modern cosmology. The cosmological constant problem. Volume in honor of Peter Gabriel Bergmann's 75th birthday. Proceedings. Symposium on Gravitation and Modern Cosmology – the Cosmological Constant Problem, Erice (Italy), 17 – 20 Sep 1990.**
A. Zichichi, V. de Sabbata, N. Sánchez (eds.).
Ettore Majorana Int. Sci. Ser., Phys. Sci., Vol. 56.
Plenum Press, New York, NY (USA). 241 p. (1991). ISBN 0–306–44054–7. Price US$ 79.50 (USA, Canada), US$ 95.40 (elsewhere).
The reader will find in this volume an updated version of different approaches to the cosmological constant problem, as well as contributions on classical and quantum cosmology, cosmic and quantum strings, classical general relativity, and gravitational radiation and its experimental search.
The individual contributions within the subject scope of AAA are included in their corresponding categories – see abstracts 005.034, 036.180, 066.266, 066.267, 066.268, 066.269, 066.270, 066.271, 066.273, 067.211, 161.374, 161.475, 161.477, 161.478, 161.479, 161.480, 161.481, 161.482, 161.484, 161.485.

012.077 **Variability of blazars. Conference in honour of the 100th anniversary of the birth of Yrjö Väisälä. Proceedings. Conference on the Variability of Blazars, Turku (Finland), 6 – 10 Jan 1991.**
E. Valtaoja, M. Valtonen (eds.).
Cambridge University Press, Cambridge (UK). 478 p. (1992). ISBN 0–521–41351–6. Price £ 40.00, US$ 64.95.
In this book a complete summary of the observations of blazars and the theoretical interpretation are given. A comprehensive listing of confirmed and candidate objects is included. Mechanisms in which the variability can arise from shocks and relativistic jets are discussed. There are at least four different answers given to the question: what is a blazar? This book is a complete overview of the violent activity observed in these extreme active galactic nuclei.
The individual contributions within the subject scope of AAA are included in their corresponding categories – see abstracts 013.122, 062.205, 066.272, 067.224, 158.240, 158.241, 158.242, 158.243, 158.244, 158.245, 158.246, 158.247, 158.248, 158.249, 158.250, 158.251, 158.252, 158.253, 158.254, 158.255, 158.256, 158.257, 158.258, 158.259, 158.260, 158.261, 158.262, 158.263, 158.264, 158.265, 158.268, 158.269, 158.270, 158.271, 158.272, 158.273, 158.274, 158.275, 158.276, 158.281, 158.282, 158.284, 158.285, 158.286, 158.287, 158.288, 158.289, 158.290, 158.291, 158.292, 158.293, 158.294, 159.119, 159.120, 159.121, 159.122, 159.123, 159.127, 159.128, 159.129.

012.078 **Fall meeting of the Astronomical Society of New York, Schenectady, NY (USA), 2 Nov 1991. Abstracts of papers presented.**
A. G. D. Philip (ed.).
News Lett. Astron. Soc. N.Y., Vol. 4, No. 1, 53 p. (Feb 1992).
The individual contributions within the subject scope of AAA are included in their corresponding categories – see abstracts 011.021, 014.076, 014.078, 022.229, 062.214, 064.090, 067.225, 073.081, 082.076, 099.110, 101.094, 131.280, 131.281, 133.010, 156.049.

012.079 **Galaxies and the universe. Proceedings. Symposium on Galaxies and the Universe, Seoul (Republic of Korea), 5 Mar 1992.**
S.–W. Lee (ed.).
Publ. Korean Astron. Soc., Vol. 7, No. 1, 148 p. (1992).
The individual contributions within the subject scope of AAA are included in their corresponding categories – see abstracts 066.282, 067.217, 067.218, 157.298, 160.117, 161.488, 161.489, 161.490, 161.491, 161.492, 161.493, 161.494, 161.495.

012.080 **Coordination of SOHO and ground–based observations. Proceedings. Meeting of JOSO Working Group 6: Coordination of SOHO and ground–based observations, Berlin (Germany), 23 Oct 1991.**
Annual report, 1991.
Joint Organization for Solar Observations (JOSO). p. 45 – 85 (1992).
The presentations included a description of the SOHO experiments and operations, and the description of a number of ground–based facilities both in Europe and in the USA. Several presentations have addressed the practical aspects of how to make best uses of the coordinated ground–based and space observations.

012.081 **Particles, quantum groups, high Tc, phase transitions and all that. Proceedings. 1. Yanbian International Workshop on Modern Physics: Particles, quantum groups, high Tc, phase transitions and all that, Yanji (People's Republic of China), 15 – 18 Jul 1990.**
K. Kang, C. W. Kim (eds.).
World Scientific, Singapore (Singapore). 394 p. (1991). ISBN 981–0205–75–9.
The individual contributions within the subject scope of AAA are included in their corresponding categories – see abstracts 061.132, 061.133, 067.213, 067.214, 161.486.

012.082 **The Solid–Earth Mission ARISTOTELES. Proceedings. International Workshop on The Solid–Earth Mission ARISTOTELES, Anacapri (Italy), 23 – 24 Sep 1991.**
C. Mattok (ed.).
European Space Agency, Paris (France), ESA–SP—329. 143 p. (Dec 1991). ISBN 92–9092–158–7.
The individual contributions within the subject scope of AAA are included in their corresponding categories – see abstracts 045.006, 051.051, 051.052, 051.053, 081.042.

012.083 **Comets and the origin and evolution of life. Proceedings. Conference on Comets and the Origin and Evolution of Life, Eau Claire, WI (USA), 30 Sep – 2 Oct 1991.**
P. J. Thomas (ed.).
Origins Life Evol. Biosphere, Vol. 21, Nos. 5 – 6, p. 265 – 434 (1991 – 1992).
The scope of the conference was to consider the role of comets in the origin and evolution of life, in light of new findings about the chemical nature of Comet Halley, the study of interplanetary dust particles, an improved understanding of plausible mechanisms of organic synthesis in meteorites and comets, progress in numerical simulations of cometary orbital evolution, and models of comet impacts on the Earth.
The individual contributions within the subject scope of AAA are included in their corresponding categories – see abstracts

091.056, 102.060, 105.261, 105.262, 107.061, 107.062, 107.063, 107.064, 107.065, 107.066, 131.277.

012.084   **New results on standard candles. Proceedings. Trani Workshop on New Results on Standard Candles, Trani, Bari (Italy), 26 – 30 Aug 1991.**
F. Caputo (ed.).
Mem. Soc. Astron. Ital., Vol. 63, No. 2, p. 223 – 532 (1992).
The individual contributions within the subject scope of AAA are included in their corresponding categories – see abstracts 115.019, 115.020, 115.021, 115.022, 115.023, 122.133, 122.134, 122.135, 122.136, 122.137, 122.138, 122.139, 122.140, 125.066, 125.067, 125.068, 153.049, 153.050, 154.058, 154.059, 156.042, 157.299, 157.300, 157.303, 160.119.

012.085   **Surface inhomogeneities on late–type stars. Proceedings. Armagh Observatory Bicentenary Colloquium on Surface Inhomogeneities on Late–Type Stars, Armagh (UK), 24 – 27 Jul 1990.**
P. B. Byrne, D. J. Mullan (eds.).
Lect. Notes Phys., Vol. 397.
Springer, Berlin (Germany). 371 p. (1992). ISBN 3–540–55310–X. Price DM 98.00. ISBN 0–387–55310–X (USA).
The individual contributions within the subject scope of AAA are included in their corresponding categories – see abstracts 013.128, 032.024, 064.083, 064.084, 064.085, 064.086, 064.087, 065.108, 072.056, 072.057, 072.059, 080.088, 114.059, 116.036, 116.037, 116.038, 116.039, 116.042, 116.043, 116.044, 116.045, 116.046, 116.047, 116.048, 116.049, 116.050, 116.051, 116.052, 116.053, 116.054, 116.055, 116.056, 116.057, 116.058, 116.059, 116.060, 116.061, 116.062, 116.063, 116.064, 116.065, 116.066, 116.067, 116.068, 116.069, 116.070, 116.071, 116.072, 116.073, 116.074, 116.075, 117.266, 117.269, 117.270, 117.271, 117.272, 117.277, 117.283, 120.029, 121.062, 122.148, 122.149, 122.150.

012.086   **Mare volcanism and basalt petrogenesis. Papers presented at a workshop at the Annual Meeting of the Geological Society of America, Dallas, TX (USA), 27 – 28 Oct 1990.**
L. A. Taylor, J. Longhi (eds.).
Geochim. Cosmochim. Acta, Vol. 56, No. 6, p. 2153 – 2265 (Jun 1992).
These five review papers are not simply reviews of the literature, in addition, they use the entire data base generated over the last 20 some years to evaluate old and new concepts and to synthesize and generate today's best expression of our understanding of mare basalts, their origins, petrologic evolutions, emplacements, and occurrences upon the surface of the Moon.
The individual contributions within the subject scope of AAA are included in their corresponding categories – see abstracts 094.096, 094.097, 094.098, 094.099, 094.100.

012.087   **Star clusters and stellar evolution. Proceedings. Teramo Workshop on Star Clusters and Stellar Evolution, Teramo (Italy), 18 – 20 Sep 1991.**
E. Brocato, F. R. Ferraro, G. Piotto (eds.).
Mem. Soc. Astron. Ital., Vol. 63, No. 1, p. 1 – 217 (1992).
Contents: 1. Recent developments in the stellar evolution theory. 2. Variable stars. 3. Stellar photometry of clusters. 4. Spectroscopy.
The individual contributions within the subject scope of AAA are included in their corresponding categories – see abstracts 036.193, 036.194, 061.135, 061.136, 065.110, 114.060, 115.024, 115.025, 122.143, 122.144, 122.145, 122.146, 122.147, 153.052, 153.053, 153.054, 153.056, 153.057, 154.060, 154.062, 154.063, 154.064, 154.065, 154.066, 154.067, 154.068, 154.070, 154.071, 154.072, 156.043, 156.044, 156.045, 156.046, 156.047, 156.048, 157.304, 157.305.

012.088   **Proceedings of the 10th ESA Symposium on European Rocket and Balloon Programmes and Related Research. 10. ESA Symposium on European Rocket and Balloon Programmes and Related Research, Mandelieu (France), 27 – 31 May 1991.**
B. Kaldeich (ed.).
European Space Agency, Paris (France), ESA–SP—317. 461 p. (Nov 1991). ISBN 92–9092–110–2.
This volume contains 70 papers covering the topics ionosphere and magnetosphere, mesosphere, microgravity, new instrumentation and technology, astronomy and DYANA project.
The individual contributions within the subject scope of AAA are included in their corresponding categories – see abstracts 035.120, 035.121, 035.122, 035.126, 035.127, 051.062, 071.027, 082.074, 082.075, 082.079, 082.080, 083.043, 106.091.

012.089   **Predictability, stability, and chaos in N–body dynamical systems. Proceedings. NATO Advanced Study Institute on Predictability, Stability, and Chaos in N–Body Dynamical Systems, Cortina d'Ampezzo (Italy), 6 – 17 Aug 1990.**
A. E. Roy (ed.).
NATO ASI Ser., Ser. B, Phys., Vol. 272.
Plenum Press, New York, NY (USA). 613 p. (1991). ISBN 0–306–44034–2. Price US$ 135.00 (USA, Canada), US$ 162.00 (elsewhere).
Recent progress in the study of the dynamics of N–body systems, both point–mass and non–point–mass, has led to the realization that the influence of chaos in dynamical systems is of supreme importance in understanding their behaviour. The present Institute therefore had as its main goal the examination of the relationships between predictability, stability and chaos in N–body dynamical systems. Its main topics included: 1. Aspects of chaos. 2. Dynamics of asteroids, comets and meteors. 3. Dynamics of natural and artificial satellites. 4. The three–body problem. 5. Selected topics in dynamics.
The individual contributions within the subject scope of AAA are included in their corresponding categories – see abstracts 042.093, 042.094, 042.095, 042.096, 042.097, 042.098, 042.099, 042.100, 042.101, 042.102, 042.103, 042.104, 042.105, 042.106, 042.107, 042.108, 042.109, 042.110, 042.111, 042.112, 042.113, 042.114, 042.115, 042.116, 042.117, 042.119, 042.120, 042.121, 042.122, 042.123, 042.124, 042.125, 042.126, 042.127, 042.129, 042.130, 042.131, 042.132, 052.032, 052.033, 052.034, 052.035, 052.036, 052.037, 091.058, 091.059, 094.101, 094.102, 094.103, 102.056, 104.096, 151.082, 151.083.

012.090   **5. Asian–Pacific Regional Astronomy Meeting of the IAU, Sydney (Australia), 16 – 20 Jul 1990. Proceedings.**
M. C. B. Ashley, J. L. Caswell, W. J. Couch, R. W. Hunstead, K. M. Proust (eds.).
Proc. Astron. Soc. Aust., Vol. 9, No. 2, 153 p. (1991).
There are 10 invited and 52 contributed papers in this second and concluding part of the Proceedings – for the first part see Abstr. 54.012.078.
The individual contributions within the subject scope of AAA are included in their corresponding categories – see abstracts 009.022, 009.023, 013.140, 015.052, 022.232, 034.124, 035.128, 051.067, 062.220, 062.221, 063.149, 066.287, 067.227, 072.067, 073.080, 073.085, 073.123, 073.125, 080.094, 082.081, 091.065, 092.014, 098.181, 102.058, 107.067, 107.068, 112.112, 112.115, 114.094, 117.279, 117.280, 117.281, 117.282, 119.063, 120.028, 121.065, 121.066, 121.067, 122.164, 122.165, 122.166, 124.054, 125.072, 125.073, 131.282, 131.283, 134.047, 134.048, 141.011, 153.055, 155.155, 155.156, 155.157, 157.245, 157.310, 157.311, 157.312, 157.313, 158.295, 158.296, 158.297, 159.130.

012.091   **The Sun: a laboratory for astrophysics. Proceedings. NATO Advanced Study Institute on the Sun: a Laboratory for Astrophysics, Crieff (UK), 16 – 29 Jun 1991.**
J. T. Schmelz, J. C. Brown (eds.).
NATO ASI Ser., Ser. C, Math. Phys. Sci., Vol. 373.
Kluwer Academic Publ., Dordrecht (Netherlands). 630 p. (1992). ISBN 0–7923–1811–0. Price Dfl. 395.00, US$ 234.00, £ 139.00.
Part I covers the solar interior from various interwoven viewpoints, reflecting the diversity both of the astrophysics

involved and of the individual lecturer's approaches. After details of the solar structure models and their helioseismic exploration, the processes of convection and dynamo action are presented alongside the basics of MHD theory. Part II describes solar and stellar photospheres, chromospheres, and coronae, radio and X-ray diagnostics for these, plus sunspot phenomena and the modelling of winds and outflows. Part III deals with solar instrumentation with details of methodology for all frequencies from radio to gamma-ray, giving both a historical perspective and the future outlook for space- and ground-based solar instruments. Part IV addresses solar and stellar activity, covering observational and theoretical aspects of flares on the Sun and other stars, and related magnetised plasma phenomena in accretion disks.

The individual contributions within the subject scope of AAA are included in their corresponding categories – see abstracts 013.138, 032.028, 033.023, 035.123, 035.124, 035.125, 062.215, 062.216, 062.217, 062.218, 062.219, 071.026, 072.062, 072.063, 073.082, 073.083, 073.084, 074.065, 075.040, 076.024, 077.041, 080.089, 080.091, 080.092, 080.093, 112.109, 122.157.

012.092 Cosmology and elementary particles. Proceedings. 2. Winter School of Physics: Cosmology and elementary particles, Rio Piedras, Puerto Rico, PR (USA), 27 Mar – 5 Apr 1991.
D. R. Altschuler, J. F. Nieves, J. Ponce de Leon, M. R. Ubriaco (eds.).
World Scientific, Singapore (Singapore). 397 p. (1992). ISBN 981–02–0808–1. Price £ 53.00.
The individual contributions within the subject scope of AAA are included in their corresponding categories – see abstracts 061.142, 066.288, 126.117, 161.402, 161.505, 161.506, 161.507, 161.508.

012.093 Radars and lidars in Earth and planetary sciences. Proceedings. International Symposium on Radars and Lidars in Earth and Planetary Sciences, Cannes (France), 2 – 4 Sep 1991.
T. D. Guyenne, J. J. Hunt (eds.).
European Space Agency, Paris (France), ESA–SP—328. 153 p. (Dec 1991). ISBN 92–9092–154–4.
The individual contributions within the subject scope of AAA are included in their corresponding categories – see abstracts 035.129, 035.130, 035.131, 035.132, 035.133, 051.063, 051.064, 051.065, 053.006, 091.061, 091.062, 093.102, 093.103, 098.178.

012.094 Frontier topics in nuclear and astrophysics – graduate lectures. 22. Masurian Lakes Summer School on Nuclear Physics: Frontier topics in nuclear and astrophysics, Piaski (Poland), 26 Aug – 5 Sep 1991.
Z. Sujkowski, G. Szeflińska (eds.).
Institute of Physics, Bristol (UK). 402 p. (1992). ISBN 0–7503–0172–4. Price £ 24.50.
The subjects covered are the properties of hot, dense nuclear matter in the laboratory and in stellar objects, nuclear reactions of astrophysical interest and nucleosynthesis. These include heavy-ion reaction dynamics, high-spin physics, radioactive beams, neutron stars, supernovae and solar neutrinos. Experimental techniques, both established and emerging and theoretical developments are included.
The individual contributions within the subject scope of AAA are included in their corresponding categories – see abstracts 061.140, 061.141, 067.226.

012.095 The atmospheres of early-type stars. Proceedings. Workshop on the Atmospheres of Early-Type Stars, Kiel (Germany), 18 – 20 Sep 1991.
U. Heber, C. S. Jeffery (eds.).
Lect. Notes Phys., Vol. 401.
Springer, Berlin (Germany). 469 p. (1992). ISBN 3–540–55256–1. Price DM 118.00. ISBN 0–387–55256–1 (USA).
The broad divisions into which the subject of hot-star atmospheres naturally fall have been followed in these proceedings. Thus the first two sections deal with the analyses of O, B and

A stars in our own and other galaxies, in which methods similar to those first developed by Unsöld for τ Sco have been employed. The more complex situations created by the presence of mass outflows, magnetic fields, diffusion and pulsation are described in the following two sections. These first four sections deal mainly with upper main–sequence stars. The large variety of evolved low–mass stars with early–type spectra, from blue horizontal branch stars to PG 1159 stars and white dwarfs are discussed in section five. Two final sections deal with the derivation of accurate atomic data, and the modern methods used in radiative transfer calculations, both vital ingredients for the modelling of stellar spectra.
The individual contributions within the subject scope of AAA are included in their corresponding categories – see abstracts 062.226, 064.091, 064.092, 064.093, 064.094, 064.095, 064.096, 064.097, 064.098, 064.099, 064.100, 064.103, 064.104, 064.105, 064.106, 064.107, 064.108, 064.109, 064.110, 064.111, 064.112, 064.113, 064.114, 064.115, 064.116, 064.117, 064.118, 064.119, 065.118, 112.108, 112.110, 112.111, 112.113, 112.114, 113.021, 114.061, 114.062, 114.063, 114.064, 114.065, 114.066, 114.067, 114.068, 114.069, 114.070, 114.071, 114.072, 114.074, 114.075, 114.076, 114.077, 114.078, 114.079, 114.080, 114.081, 114.082, 114.083, 114.084, 114.085, 114.086, 114.087, 114.088, 114.089, 114.090, 114.091, 114.092, 114.093, 115.026, 116.006, 116.077, 116.078, 117.276, 122.158, 122.159, 122.160, 126.118, 126.119, 126.120, 126.121, 126.122, 126.123, 126.124, 134.044, 134.045, 134.046, 154.061, 154.069, 155.154.

012.096 Eruptive solar flares. Proceedings. IAU Colloquium No. 133: Eruptive solar flares, Iguazú (Argentina), 2 – 6 Aug 1991.
Z. Švestka, B. V. Jackson, M. E. Machado (eds.).
Lect. Notes Phys., Vol. 399.
Springer, Berlin (Germany). 423 p. (1992). ISBN 3–540–55246–4. Price DM 108.00. ISBN 0–387–55246–4 (USA).
The scientific program was divided into 10 sessions. Session 1 was concerned with the build-up and triggering of eruptive flares. Sessions 2 and 3 centered on filament eruption, field opening and successive reconnection, and energy release and transport. In sessions 4 and 5 the flare impulsive phase and the formation of loops were discussed. Session 6 was concerned with large-scale coronal structures associated with eruptive flares in X-rays and radio waves. A discussion about mass ejections from flares in session 6 was continued in session 7. Session 8 was concerned with the terrestrial response to eruptive solar flares. The first two invited talks of session 9 summarized our knowledge of stellar flares while the rest of this session as well as session 10 were devoted to the presentation of plans for related observations, studies, and cooperative projects in the future. These projects included a discussion about Solar-A (Yohkoh, launched three weeks after this colloquium). The posters were divided into two sets and were briefly discussed in sessions 5 and 8.
The individual contributions within the subject scope of AAA are included in their corresponding categories – see abstracts 033.024, 033.025, 035.134, 051.066, 062.223, 062.224, 062.225, 064.101, 064.102, 072.065, 073.086, 073.087, 073.088, 073.089, 073.090, 073.091, 073.092, 073.093, 073.094, 073.095, 073.096, 073.097, 073.098, 073.099, 073.100, 073.101, 073.102, 073.104, 073.105, 073.106, 073.107, 073.108, 073.109, 073.110, 073.111, 073.112, 073.113, 073.114, 073.115, 073.116, 073.117, 073.118, 073.119, 073.120, 073.121, 073.122, 074.066, 074.067, 074.068, 074.069, 074.070, 074.071, 074.072, 074.073, 074.074, 074.075, 075.041, 077.042, 077.043, 077.044, 077.045, 077.046, 077.047, 078.020, 078.021, 084.077, 084.078, 084.079, 116.079.

012.097 Imaging X-ray astronomy. A decade of Einstein Observatory achievements. Symposium on Imaging X-Ray Astronomy – a Decade of Einstein Observatory Achievements, Cambridge, MA (USA), 13 – 15 Nov 1988.
M. Elvis (ed.).
Cambridge University Press, Cambridge (UK). 359 p. (1990). ISBN 0–521–38105–3. Price £ 30.00, US$ 49.50.
The individual contributions within the subject scope of AAA are included in their corresponding categories – see abstracts

002.156, 002.157, 002.158, 002.159, 002.160, 002.161, 002.162, 002.163, 013.017, 035.135, 036.206, 064.120, 116.080, 116.081, 117.284, 121.068, 125.074, 125.075, 125.080, 134.050, 142.053, 142.054, 142.055, 142.056, 142.057, 142.058, 142.059, 142.060, 142.061, 142.062, 158.298, 158.299, 158.300, 158.301, 161.521.

012.098 **Particles, strings and cosmology. Proceedings.
    1. International Symposium on Particles, Strings and Cosmology, Boston, MA (USA), 27 – 31 Mar 1990.**
P. Nath, S. Reucroft (eds.).
World Scientific, Singapore (Singapore). 702 p. (1991). ISBN 981–02–0392–6. Price £ 64.00.
Contents: 1. New results from accelerators. 2. Non–accelerator physics and neutrino physics. 3. Particle physics phenomenology. 4. Beyond the standard model. 5. Superstring phenomenology. 6. String theory and conformal field theory. 7. Astrophysics and cosmic strings. 8. Quantum gravity. 9. Quantum cosmology.
The individual contributions within the subject scope of AAA are included in their corresponding categories – see abstracts 022.233, 061.147, 061.148, 061.150, 066.295, 066.296, 066.298, 066.299, 066.300, 066.301, 066.302, 066.303, 161.525, 161.526, 161.528, 161.534.

012.099 **Reconnection in space plasma. Proceedings. International Workshop on Reconnection in Space Plasma, Potsdam (Germany), 5 – 9 Sep 1988.**
T. D. Guyenne, J. J. Hunt (eds.).
European Space Agency, Paris (France), ESA–SP—285(Vol. 2). 333 p. (Jan 1989). Price Dfl. 80.00, US$ 45.00.
Contents: 1. General theory of reconnection. 2. Solar reconnection, theory and observations. 3. Laboratory reconnection 4. Reconnection at and in magnetospheres. 5. Astrophysical reconnection.
The individual contributions within the subject scope of AAA are included in their corresponding categories – see abstracts 009.024, 022.235, 022.236, 022.237, 022.238, 062.209, 062.227, 062.228, 062.229, 062.230, 062.231, 062.232, 062.233, 062.234, 062.235, 062.236, 062.237, 062.238, 062.239, 062.240, 064.121, 064.122, 072.068, 072.069, 073.128, 073.129, 073.130, 073.131, 073.132, 074.076, 074.077, 074.078, 074.079, 074.083, 074.085, 075.042, 075.043, 075.044, 075.045, 077.049, 077.050, 084.081, 084.082, 084.083, 084.084, 084.085, 084.086, 084.087, 084.088, 084.089, 084.090, 084.091, 091.066, 106.114, 122.167.

012.100 **Physics of the outer heliosphere. Proceedings.
    1. COSPAR Colloquium: Physics of the outer heliosphere, Warsaw (Poland), 19 – 22 Sep 1989.**
S. Grzedzielski, D. E. Page (eds.).
COSPAR Colloq. Ser., Vol. 1.
Pergamon Press, Oxford (UK). 418 p. (1990). ISBN 0–08–040780–3. Price US$ 80.00.
Contents: 1. Opening session: Introductory lecture – the heliosphere. 2. Session I: Spectroscopic data on the local interstellar medium and the related XUV radiation background. 3. Session II: Solar UV backscatter on neutral galactic gases. 4. Session III: Entry and dynamics of galactic and anomalous cosmic rays in the heliosphere. 5. Session IV: Distant solar wind plasma, magnetic field and solar energetic particles. 6. Special session: Deep space missions. 7. Poster session.
The individual contributions within the subject scope of AAA are included in their corresponding categories – see abstracts 036.205, 036.208, 036.209, 051.068, 051.069, 062.222, 062.241, 064.123, 072.070, 073.133, 074.080, 074.081, 074.082, 074.084, 074.086, 074.087, 074.088, 074.090, 074.091, 102.059, 106.092, 106.093, 106.094, 106.095, 106.096, 106.097, 106.098, 106.099, 106.100, 106.101, 106.102, 106.103, 106.104, 106.105, 106.106, 106.107, 106.108, 106.109, 106.110, 106.111, 106.112, 106.113, 106.115, 106.116, 106.117, 106.118, 106.119, 131.284, 131.285, 131.286, 131.287, 131.288, 144.074, 144.075, 144.076, 144.077, 144.078, 144.079, 144.080, 144.081, 144.082, 144.084, 144.086.

012.101 **Methods in computational molecular physics. Lecture notes. Advanced NATO Study Institute on Methods in Computational Molecular Physics, Bad Windsheim (Germany), 22 Jul – 2 Aug 1991.**
Max–Planck–Institut für Physik und Astrophysik, Garching (Germany). Inst. für Astrophysik, MPA/P—6. 568 p. (Sep 1991).
This Advanced Study Institute seeks to bridge the gap which exists between the presentation of molecular electronic structure theory in contemporary monographs and the realization of the sophisticated computational algorithms required for their practical application.

012.102 **Trends in astroparticle physics. Proceedings.
    1. International Trends in Astroparticle Physics Conference together with the SuperNova Watch Workshop, Santa Monica, CA (USA), 28 Nov – 1 Dec 1990.**
D. B. Cline, R. Peccei (eds.).
World Scientific, Singapore (Singapore). 636 p. (1992). ISBN 981–02–0825–1. Price £ 67.00.
The Trends meeting covered the interface between elementary particles and the universe. Topics covered range from massive neutrinos, axions, WIMPS, large–scale structure of the universe, COBE results to the early universe particle physics at the GUT and Planck scales. The SuperNova Watch is a world wide search for neutrino bursts from a galactic supernova as well as the automated Berkeley search for the optical pulse from the supernova. Detection of supernova pulses would be the ultimate neutrino laboratory as well as advancing our understanding of nuclear physics and the search for new particles such as axions.
The individual contributions within the subject scope of AAA are included in their corresponding categories – see abstracts 022.234, 022.239, 022.240, 022.241, 022.242, 022.243, 022.244, 022.245, 022.246, 022.247, 022.248, 022.249, 022.250, 022.251, 022.252, 061.127, 061.149, 061.151, 065.109, 065.113, 065.119, 067.231, 080.095, 125.077, 125.078, 125.079, 144.083, 144.085, 155.158, 160.125, 161.431, 161.512, 161.513, 161.514, 161.515, 161.516, 161.517, 161.518, 161.519, 161.520, 161.522, 161.523, 161.524, 161.527, 161.529, 161.530, 161.531, 161.532, 161.533, 161.535, 161.536, 161.537.

013 Reports on Astronomy in Various Countries and Particular Fields

013.001 **The LEST project.**
O. Engvold.
11. European Regional Astronomy Meeting of the IAU: New windows to the universe, Vol. 1, p. 451 – 461 (1990). – See Abstr. 012.001 for the main entry.
LEST is a most ambitious ground based solar telescope program. The multi–national, non–profit organization behind the project, the LEST Foundation, presently counts 9 member countries. The construction of LEST could start in 1992 and the telescope may be ready for first light in 1995. The current design is based on a 2.4 m and the diffraction limit of the modified Gregorian system is 0.05 arcsec at λ 5,000 Å.

013.002 **The instrumentation plan for the Very Large Telescope of the European Southern Observatory.**
S. D'Odorico.
11. European Regional Astronomy Meeting of the IAU: New windows to the universe, Vol. 1, p. 463 – 464 (1990). Abstract. – See Abstr. 012.001 for the main entry.

013.003 **Ground–based European astronomical projects.**
A. Boksenberg.
11. European Regional Astronomy Meeting of the IAU: New windows to the universe, Vol. 1, p. 521 – 533 (1990). – See Abstr. 012.001 for the main entry.

013.004 **ESA astronomical projects.**
M. C. E. Huber.
11. European Regional Astronomy Meeting of the IAU: New windows to the universe, Vol. 1, p. 535 – 549 (1990). – See Abstr. 012.001 for the main entry.

013.005 **Prospects of the development of ground–based and space astronomy in the USSR.**
A. A. Boyarchuk.
11. European Regional Astronomy Meeting of the IAU: New windows to the universe, Vol. 1, p. 551 – 564 (1990). – See Abstr. 012.001 for the main entry.

013.006 **Planetary astronomy in the 1990's.**
D. Morrison.
Sky Telesc., Vol. 83, No. 2, p. 151 – 156 (Feb 1992).
Traditional astronomers, armed with new equipment, will have many opportunities for solar system discoveries in the coming decade.

013.007 **Astrophysics in 1991.**
V. Trimble.
Publ. Astron. Soc. Pac., Vol. 104, No. 671, p. 1 – 14 (Jan 1992). Invited review paper.
This review attempts to shine a narrow and not unprejudiced beam of light on areas of astrophysics where the author believes that something interesting happened in 1991. Length scales range from the kilometers of comets to the gigaparsecs of cosmology, and time scales from milliseconds upward. Eight self–contained sections discuss the surface of Venus, late activity in Comet Halley, convective overshoot, blue stragglers, new examples and kinds of X–ray binaries and recycled pulsars, hard X and γ rays from the galactic–center region, microlensing, and biased cold dark matter. The last two sections contain very short descriptions, with references, of a number of items that did not fit into the main sections but seemed a shame to leave out.

013.008 **The NASA airborne astronomy program: a perspective on its contributions to science, technology, and education.**
H. P. Larson.
Publ. Astron. Soc. Pac., Vol. 104, No. 672, p. 146 – 153 (Feb 1992).
The scientific, educational, and instrumental contributions from NASA's airborne observatories are deduced from the program's publication record (789 citations, excluding abstracts, involving 580 authors at 128 institutions in the United States and abroad between 1967 – 1990).

013.009 **Golden age of the earth–based astronomy.**
P. Koubský.
Kozmos, Vol. 22, No. 1, p. 16 – 21 (Jan 1991). In Czech.

013.010 **The Czech and Slovak astronomy: the year 0.**
E. Gindl.
Kozmos, Vol. 22, No. 6, p. 186 – 189 (Nov 1991). In Slovak.

013.011 **Der Felderplan der Sternwarte Sonneberg.**
G. A. Richter.
Sterne, Band 68, Heft 1, p. 34 – 42 (1992).

013.012 **Highlights of astronomy in the year 1990.**
J. Grygar.
Říše hvězd, Vol. 72, No. 4, p. 65 – 69 (Apr 1991). In Czech.
Further information is given in Nos. 5, p. 81 – 88 (May 1991), 6, p. 105 – 111 (Jun 1991), 7, p. 121 – 124 (Jul 1991), 8, p. 145 – 151 (Aug 1991).

013.013 **Highlights of astronomy 1991.**
J. Grygar.
Říše hvězd, Vol. 73, No. 4/5, p. 51 – 70 (Apr – May 1992). In Czech.

013.014 **LIGO: The Laser Interferometer Gravitational–Wave Observatory.**
A. Abramovici, W. E. Althouse, R. W. P. Drever, Y. Gürsel, S. Kawamura, F. J. Raab, D. Shoemaker, L. Sievers, R. E. Spero, K. S. Thorne, R. E. Vogt, R. Weiss, S. E. Whitcomb, M. E. Zucker.
Science, Vol. 256, No. 5055, p. 325 – 333 (17 Apr 1992).
The goal of the Laser Interferometer Gravitational–Wave Observatory (LIGO) project is to detect and study astrophsical gravitational waves and use data from them for research in physics and astronomy. LIGO will support studies concerning the nature and nonlinear dynamics of gravity, the structures of black holes, and the equation of state of nuclear matter. It will also measure the masses, birth rates, collisions, and distributions of black holes and neutron stars in the universe and probe the cores of supernovae and the very early universe.

013.015 **European astronomy.**
M. Rees.
Science, Vol. 256, No. 5056, p. 485 – 487 (24 Apr 1992).

013.016 **Astronomie zur Jahrtausendwende.**
H.–H. Voigt.
Astron. Sch., Jahrg. 29, Heft 7, p. 4 – 9 (Feb 1992).

013.017 **Personal recollections on the origins of the Einstein Observatory.**
B. Rossi, H. Gursky, E. Boldt, G. W. Clark, R. Giacconi.
Symposium on Imaging X–Ray Astronomy – a Decade of Einstein Observatory Achievements, p. 3 – 12 (1990). – See Abstr. 012.097 for the main entry.

013.018   The role of solar system photometry.
F. J. Melillo.
I.A.P.P.P. Commun., No. 47, p. 21 – 24 (Mar – May 1992).

013.019   Towards the archiving of astronomical spectra.
R. E. Griffin.
Gemini, No. 36, p. 10 – 11 (Jun 1992).

013.020   How has space astrophysics expanded the horizon of physics?
M. Oda.
Yoshio Nishina Centennial Symposium: Evolutionary trends in the physical sciences, p. 113 – 126 (1991). – See Abstr. 012.017 for the main entry.
Serendipitous discoveries in the 1960s drastically expanded the horizon of physics. In this paper, the evolution of major problems in astrophysics since the 1950s is analysed by decades. Several questions as of 1990 are raised. Progress of the understanding in a couple of specific subjects, i.e. the neutron star and the black hole, are briefly described. Future directions are discussed.

013.021   Advances in IR technology at Paris Observatory.
F. Lacombe, M. Combes, P. Léna, F. Rigaut,
D. Rouan, E. Tessier, D. Tiphène.
SPIE Conference on Infrared Technology XVI as Part of SPIE's International Symposium on Optical and Optoelectronic Applied Science and Engineering, p. 187 – 192 (1990). – See Abstr. 012.012 for the main entry.
For more than twenty years, the scientific activity of the spatial research department of the Paris Observatory has been centered on ground based and spatial infrared astronomy together with radio and plasma astrophysics. This paper summarizes the most recent projects that the authors have participated to, or led, in collaboration with Canadian, Soviet and various European institutes. From Canada France Hawaii Observatory to the European Southern Observatory's Very Large Telescope, or Infrared Space Observatory, recent and future IR cameras and now existing Adaptive Optics are briefly described.

013.022   The Phase 2 NRA.
E. Chipman.
Compton Observatory Science Workshop, p. 164 – 170 (Feb 1992). – See Abstr. 012.018 for the main entry.
The author presents points of special interest to potential proposers for the Compton Observatory Phase 2 Guest Investigator Program.

013.023   Radioastronomy on the beginning of the 21st century.
M. Łysik.
Urania, Rok 63, Nr. 6, p. 179 – 183 (Jun 1992). In Polish.

013.024   Die 21. Generalversammlung der Internationalen Astronomischen Union.
S. Klose.
Sterne, Band 68, Heft 2, p. 100 – 112 (1992).

013.025   Planetary influences on electrical engineering.
R. N. Bracewell.
Proc. IEEE, Vol. 80, No. 2, p. 230 – 237 (Feb 1992).
This paper sketches some of the interplay between researches into the behavior of planets and the development of electrical science and engineering. It begins with Ptolemy, continues through the period of Newton and Maxwell, and touches on some of the most recent advances in radio science.

013.026   The ESO microvariability key program and the detection of extrasolar planets and brown dwarfs.
N. Vogt.
3. Symposium International de Bioastronomie, p. 31 – 32 (1991). – See Abstr. 012.005 for the main entry.

013.027   Solar–terrestrial models at the National Space Science Data Center.
D. Bilitza.
J. Atmos. Terr. Phys., Vol. 53, No. 11/12, p. 1207 – 1211 (Nov–Dec 1991). – See Abstr. 012.024 for the main entry.
The National Space Science Data Center (NSSDC) and World Data Center A for Rockets and Satellites (WDC–A–R and S) has a long record of participation in the worldwide efforts to establish and improve empirical models for the different regions of the solar–terrestrial environment. The center maintains a unique archive of solar–terrestrial models and related applications software, described in a recently published models catalog. The software packages are distributed on tape, diskette, and on–line on the Space Physics Analysis Network (SPAN). Four of the most frequently requested models (IRI, MSIS/CIRA, IGRF, AE–8/AP–8) can also be accessed and run on the NSSDC Online Documentation and Information Service (NODIS) account, which can be reached from any SPAN node.

013.028   The protection of astronomical and geophysical sites: general introduction.
J. Kovalevsky.
The protection of astronomical and geophysical sites, p. 1 – 29 (1992). – See Abstr. 003.038 for the main entry.
Contents: 1. Scientific context. 2. Description of the nuisances. 3. How to reduce the nuisances. 4. Conclusions and recommendations. 5. General bibliography.

013.029   Light pollution.
D. L. Crawford.
The protection of astronomical and geophysical sites, p. 31 – 78 (1992). – See Abstr. 003.038 for the main entry.
Contents: 1. Introduction. 2. The problem of light pollution. 3. Other adverse effects of outdoor lighting. 4. Sky glow and astronomy. 5. Outdoor lighting. 6. Solutions to light pollution. 7. Implementation of the solutions. 8. Concerns about lighting controls. 9. Help! 10. Summary and conclusions.

013.030   Radio–interference.
T. Gergely, H. C. Kahlmann.
The protection of astronomical and geophysical sites, p. 79 – 110 (1992). – See Abstr. 003.038 for the main entry.
Contents: 1. Summary. 2. Radio astronomy. 3. Interferences in radio–astronomy. 4. The international regulatory situation. 5. Present day congestion. 6. Observability. 7. International cooperation. 8. The future. 9. Conclusions and recommendations. 10. References. 11. Annex: Technical recommendations of the International Astronomical Union.

013.031   Millimeter band radio astronomy.
P. Encrenaz, J. Kovalevsky.
The protection of astronomical and geophysical sites, p. 111 – 122 (1992). – See Abstr. 003.038 for the main entry.
Contents: 1. Overview.   2. Scientific   background. 3. Atmospheric limitations and studies. 4. Major nuisances and protection policy. 5. References. 6. Annex: some frequencies to be protected.

013.032   Pollution of geophysical sites.
F. Barlier, J. Kovalevsky.
The protection of astronomical and geophysical sites, p. 123 – 141 (1992). – See Abstr. 003.038 for the main entry.
Contents: 1. General introduction. 2. Light and electromagnetic radiations. 3. Heat production. 4. Magnetic field measurements. 5. Vibrations. 6. Chemical composition of the atmosphere. 7. Human degradation (automatic stations).

013.033   Satellites, space debris, aircraft and astronomy.
J. Kovalevsky.
The protection of astronomical and geophysical sites, p. 143 – 158 (1992). – See Abstr. 003.038 for the main entry.
Contents: 1. Satellites and space debris. 2. Space objects as nuisances for ground–based astronomy. 3. Very large space

objects. 4. Space astronomy and space debris. 5. Aircraft hazards. 6. Bibliography. 7. International Union resolution.

013.034   Protection of observatories: the legal avenues.
P. Murdin.
The protection of astronomical and geophysical sites, p. 159 – 207 (1992). – See Abstr. 003.038 for the main entry.
   Contents: 1. General comments. 2. Examples in United States. 3. Site protection in Australia. 4. Other countries. 5. Radio protection. 6. Conclusions.
Annexes: 1. City of San Diego outdoor lighting control ordinance. 2. Tucson and Pima County Arizona outdoor lighting control ordinance. 3. Orana environmental plan No. 1, Siding Spring. 4. Report and recommendations of Commission 50 of the International Astronomical Union. 5. Statement of the Commission Internationale de l'Eclairage concerning protections of sites of astronomical observations. 6. Law of the Kingdom of Spain on the protection of the Canarian Island observations.

013.035   New development in radio astronomy in the submillimeter–wave region.
H. E. Matthews.
Symposium on Chemistry and Spectroscopy of Interstellar Molecules, p. 81 – 86 (1992). – See Abstr. 012.039 for the main entry.
   Submillimeter–wave astronomy is a technically demanding discipline which has seen the recent construction of several large ground–based facilities and major advances in receiver technology. The status of the field is reviewed with attention being paid to the contributions to astrophysics and to the major difficulties facing the observer at these wavelengths. The results of surveys for molecular species, and observations of dense, warm gas and of cold dust are discussed. Both continuum and spectral lines have great potential at these wavelengths particularly for the investigation of protostellar objects, and of the late stages of stellar evolution. Prospects for the future development of submillimeter astronomy are bright; the next years should see a wealth of new information obtained in this wavelength range.

013.036   Precise position measurements of molecular maser sources (a plan).
M. Morimoto.
Symposium on Chemistry and Spectroscopy of Interstellar Molecules, p. 99 – 100 (1992). – See Abstr. 012.039 for the main entry.
   VLBI observations of molecular line maser sources are in progress. They will provide accurate positions of maser spots and so physical and dynamical conditions of extended envelopes of late type variable stars.

013.037   Space plasma physics at the Applied Physics Laboratory over the past half–century.
T. A. Potemra.
Johns Hopkins APL Tech. Dig., Vol. 13, No. 1, p. 182 – 199 (Jan–Mar 1992).
   Exploration of our planet's space plasma environment began with the use of ground–based radio propagation experiments and with instruments carried to high altitudes by balloons and rockets. During the past half–century, APL's space plasma physics activities have evolved in a major and multifaceted effort involving the observation and study of phenomena from the Earth's atmosphere to the Sun and beyond to the outer planets and heliosphere.

013.038   The presence and future of the Norwegian program of investigations on the Spitzbergen archipelago in the field of cosmic physics.
A. Egeland.
Geomagn. Aeron., Vol. 31, No. 1, p. 9 – 13 (Aug 1991). English translation of Geomagn. Aehron., Tom 31, No. 1, p. 13 – 19 (1991).
   The basic positions of the Norwegian program of investigations on the Spitzbergen archipelago in the field of cosmic physics within the framework of the international program of modeling

of near–earth space (Geospace Environment Modeling–GEM) proposed by American scientists, which in turn is a component part of the international project Solar–Terrestrial Energy Program, are outlined. It is shown that Spitzbergen is a unique site for making observations of geomagnetic and auroral phenomena in the polar cusp.

013.039   Women in astronomy: a sampler of issues and ideas.
N. Barlow, F. A. Cordova, J. Bahcall, J. Price, K. Eastwood, N. Bahcall, G. Clayton, J. Lutz, J. Bell Burnell, D. Hunter, V. Rubin, L. McFadden, S. Faber, G. Knapp, E. M. Alvarez del Castillo, V. Trimble.
Mercury, Vol. 21, No. 1, p. 27 – 36 (Jan – Feb 1992).

013.040   Correlation between the Maryland and Rome gravitational wave detectors and the IMB detector during SN1987A.
P. Astone, M. Bassan, P. Bonifazi, M. G. Castellano, E. Coccia, C. Cosmelli, S. Frasca, I. Modena, E. Majorana, D. Gretz, G. V. Pallottino, G. Pizzella, P. Rapagnani, F. Ricci, M. Visco, J. Weber, G. Wilmot.
Vulcano Workshop 1990: Frontier objects in astrophysics and particle physics, p. 487 – 496 (1991). – See Abstr. 012.040 for the main entry.
   The authors present the results they found when investigating possible correlations of MGW and RGW with the IMB detector.

013.041   Coincidences among Mont Blanc, Kamiokande, Baksan, IMB, Frejus, Homestake and Plateau Rosa detectors during SN1987A.
M. Aglietta, G. Badino, G. Bologna, C. Castagnoli, A. Castellina, W. Fulgione, P. Galeotti, O. Saavedra, G. Trinchero, S. Vernetto.
Vulcano Workshop 1990: Frontier objects in astrophysics and particle physics, p. 497 – 504 (1991). – See Abstr. 012.040 for the main entry.
   Correlation analysis has been performed among the events recorded during the occurrence of SN1987A by the underground detectors of Mont Blanc, Kamiokande, Baksan, IMB, Frejus, Homestake and the EAS array of Plateau Rosa. A significative excess of coincidence exists between low energy events of Mont Blanc and IMB muons, in the time interval from 1:45 to 3:45 UT on Feb. 23, 1987.

013.042   What effect has been detected by underground detectors and gravitational wave antennas a few hours before SN1987A observation?
E. N. Alexeyev (*E. N. Alekseev*), L. N. Alexeyeva (*L. N. Alekseeva*), I. M. Kogaj, I. V. Krivoshejna, V. Ya. Poddubnyj, V. N. Zakidyshev.
Vulcano Workshop 1990: Frontier objects in astrophysics and particle physics, p. 505 – 513 (1991). – See Abstr. 012.040 for the main entry.
   Data recorded by the LSD underground detector, by the Baksan underground telescope, by the IMB underground detector and by the Rome and the Maryland gravitational wave antennas have been studied to look for the temporal correlations between them. The bulk effect was found to occur within one–hour–period of 1:45 – 2:45 UT on February 23, 1987. Discussion on some properties of the effect is presented.

013.043   Can astrophysics rescue particle physics from the standard model impasse?
A. Masiero.
Vulcano Workshop 1990: Frontier objects in astrophysics and particle physics, p. 529 – 542 (1991). – See Abstr. 012.040 for the main entry.
   The solar neutrino and the dark matter problems have the potentiality to shake our belief in (at least) one of the following standard models (SM): the SM of electroweak interactions, the SM of the Sun and the SM of the early Universe. Given that the SM of particle physics is perfectly unscathed after the first results of Tevatron and LEP I. The author considers the chances that the astroparticle joint venture has to provide us guidelines in the

search for new physics. He discusses some aspects of the determination of the number of neutrino species, of the solar neutrino problem and the dark matter mystery.

013.044   Comments on some aspects of experimental particle astrophysics.
G. Giacomelli.
Vulcano Workshop 1990: Frontier objects in astrophysics and particle physics, p. 547 – 548 (1991). – See Abstr. 012.040 for the main entry.

013.045   Some important needs in current experiments on air showers and point sources.
A. M. Hillas.
Vulcano Workshop 1990: Frontier objects in astrophysics and particle physics, p. 549 – 551 (1991). – See Abstr. 012.040 for the main entry.

013.046   Impact craters: are they useful?
V. L. Masaitis.
Meteoritics, Vol. 27, No. 1, p. 21 – 27 (Mar 1992). The Barringer Medal address, presented 25 Jul 1991 at Monterey, CA (USA).
Terrestrial impact craters are important geological and geomorphological objects that are significant not only for scientific research but for industrial and commercial purposes. The structures may contain commercial minerals produced directly by thermodynamic transformation of target rocks (including primary forming ores) controlled by some morphological, structural or lighological factors and exposed in the crater. Iron and uranium ores, nonferrous metals, diamonds, coals, oil shales, hydrocarbons, mineral waters and other raw materials occur in impact craters. Impact morphostructures may be used for underground storage of gases or liquid waste material. Surface craters may serve a sreservoirs for hydropower. These ring structures may be of value to society in other ways. Scientific investigation of them is especially important in comparative planetology, terrestrial geology and in other divisions of the natural sciences.

013.047   Canadian Arctic Meteorite Project (CAMP): 1990.
R. G. Cresswell, R. K. Herd.
Meteoritics, Vol. 27, No. 1, p. 81 – 85 (Mar 1992).
The Devon Ice Cap, Northwest Territories, has been targeted for searches for extra–terrestrial material in the Canadian Arctic. Of three expeditions (1981, 1986, 1990), only the last met with weather conditions favourable for meteorite reconnaissance. Surveys were carried out on the ice cap margin. In addition, outwash streams and plains were covered, in the hope of discerning meteoritic material lying on the Lower Palaeozoic bedrock. No meteorites were recovered, though a number of pseudotaks were identified as potential searching grounds for future expeditions.

013.048   Reflections on the Soviet–American VLBI program.
K. I. Kellermann.
Astrophysics on the threshold of the 21st century, p. 37 – 51 (1992). – See Abstr. 003.048 for the main entry.

013.049   Decametric radioastronomy.
S. Ya. Braude.
Astrophysics on the threshold of the 21st century, p. 81 – 102 (1992). – See Abstr. 003.048 for the main entry.
The main results in experimental radio astronomy obtained at the Radioastronomy Institute of the Ukrainian SSR Academy of Sciences from 1958 to 1989 are presented. The decameter radio telescope and interferometers used, UTR–2 and URAN, are briefly described. Data at these wavelengths radiated by both galactic and metagalactic objects are presented, namely: recombination lines, the Sun, H II regions, supernova remnants (SNR), pulsars, radio galaxies, and quasars. Some information on a discrete source catalogue and the cosmological conclusions obtained from a source count are presented.

013.050   International cooperation and coordination in astronomical research.
A. Blaauw.
1. European Meeting of the American Association of Variable Star Observers (AAVSO): International Cooperation and Coordination in Variable Star Research, p. 3 – 7 (1992). – See Abstr. 012.048 for the main entry.
The author looks at a few of the great joint undertakings in the past; the "Carte du Ciel" – Astrographic Catalogue, the Plan of Selected Areas, the International Astronomical Union.

013.051   AAVSO and variable star observing.
J. A. Mattei.
1. European Meeting of the American Association of Variable Star Observers (AAVSO): International Cooperation and Coordination in Variable Star Research, p. 36 – 49 (1992). – See Abstr. 012.048 for the main entry.
The American Association of Variable Star Observers (AAVSO) was founded in 1911 at Harvard College Observatory to coordinate variable star observations made largely by amateur astronomers and to make them available to professional astronomers. The archives of the AAVSO currently contain over 6.5 million visual observations and about 5,000 photoelectric ones. Approximately 550 observers from around the world send in between 240,000 and 265,000 observations yearly. They are then converted into computer–readable form and processed using computer systems at AAVSO Headquarters. These observations are then added to the data files for each star, and the corresponding computer–generated light curves are brought up to date.

013.052   International Astronomical Union Commission 27: variable stars.
M. Breger.
1. European Meeting of the American Association of Variable Star Observers (AAVSO): International Cooperation and Coordination in Variable Star Research, p. 50 – 51 (1992). – See Abstr. 012.048 for the main entry.
The Commission 27 of the IAU provides several services to the astronomical community: (1) The publication of the Information Bulletin on Variable Stars. (2) The commission maintains the Archives of Unpublished Photoelectric Observations of Variable Stars. (3) The commission sponsors the important General Catalog of Variable Stars. (4) Every three years the commission issues a triannual report on the new developments in the field of variable stars.

013.053   Coordination of visual observing programs.
J. A. Mattei.
1. European Meeting of the American Association of Variable Star Observers (AAVSO): International Cooperation and Coordination in Variable Star Research, p. 67 – 74 (1992). – See Abstr. 012.048 for the main entry.
The observing program of the AAVSO is described and the cooperation with other data banks is outlined.

013.054   On the homogeneity of visual photometry.
C. Sterken, J. Manfroid.
1. European Meeting of the American Association of Variable Star Observers (AAVSO): International Cooperation and Coordination in Variable Star Research, p. 75 – 87 (1992). – See Abstr. 012.048 for the main entry.
We are in an era where visual photometry is under question. The greatest asset of existing visual data is that they have been collected over a very long time baseline (in principle over thousands of years). But there are reasons for which modern visual estimates are not comparable to ancient ones.

013.055   Visual searching for supernovae.
R. Evans.
1. European Meeting of the American Association of Variable Star Observers (AAVSO): International Cooperation and Coordination in Variable Star Research, p. 88 – 92 (1992). – See Abstr. 012.048 for the main entry.
Visual searching provides the best and cheapest opportunities for amateurs to be involved in the business of supernova hunting.

013.056 Photoelectric photometry of variable stars.
D. S. Hall.
1. European Meeting of the American Association of Variable Star Observers (AAVSO): International Cooperation and Coordination in Variable Star Research, p. 95 – 108 (1992). – See Abstr. 012.048 for the main entry.
Photoelectric photometry has two major advantages over photometry which uses the photographic emulsion or the human eye as a detector. First, it is capable of greater precision. Second, it is linear.

013.057 Robotic telescopes for photometry.
J. Baruch.
1. European Meeting of the American Association of Variable Star Observers (AAVSO): International Cooperation and Coordination in Variable Star Research, p. 109 – 116 (1992). – See Abstr. 012.048 for the main entry.
There are new areas of astronomy which are being opened up by the development of robotic telescope: parallel multi–waveband observing and long–period and continuous observing. The program and development of a robotic telescope is described. Automated data reduction and community confidence are discussed.

013.058 Multichannel photometry for amateur astronomy groups.
J. M. Le Contel, E. N. Walker.
1. European Meeting of the American Association of Variable Star Observers (AAVSO): International Cooperation and Coordination in Variable Star Research, p. 117 – 121 (1992). – See Abstr. 012.048 for the main entry.
Comparative photometers and the possibility to make them available for amateurs are discussed.

013.059 Extragalactic photometer in the transition era between photoelectric and CCD photometry.
G. Longo, G. Busarello, C. Sterken.
1. European Meeting of the American Association of Variable Star Observers (AAVSO): International Cooperation and Coordination in Variable Star Research, p. 126 – 134 (1992). – See Abstr. 012.048 for the main entry.

013.060 The study of short–period variable stars through international cooperation.
M. Breger.
1. European Meeting of the American Association of Variable Star Observers (AAVSO): International Cooperation and Coordination in Variable Star Research, p. 171 – 184 (1992). – See Abstr. 012.048 for the main entry.
Contents: 1. Introduction. 2. The international observatory network. 3. How to organize successful multisite campaigns. 4. Comparison stars. 5. Observing sequence. 6. Filters used. 7. Reduction procedures. 8. Relative zero–points of the different observers. 9. Techniques of period findings. 10. Fourier analysis. 11. The LSR method.

013.061 International cooperation for coordinated studies of Mira variables.
M. Karovska.
1. European Meeting of the American Association of Variable Star Observers (AAVSO): International Cooperation and Coordination in Variable Star Research, p. 255 – 258 (1992). – See Abstr. 012.048 for the main entry.

013.062 The benefit of amateur observations for research in dwarf novae.
C. la Dous.
1. European Meeting of the American Association of Variable Star Observers (AAVSO): International Cooperation and Coordination in Variable Star Research, p. 279 – 289 (1992). – See Abstr. 012.048 for the main entry.
The author concentrates on three main aspects: on the analysis of outburst light curves, in the UV delay, and, finally she presents one sort of observation that has not generally been carried out by

amateurs so far, but which might be just at the limit of possibilities: orbital photometry of dwarf novae.

013.063 Coordination of multiwavelength groundbased and satellite observations of novae in outburst.
S. N. Shore.
1. European Meeting of the American Association of Variable Star Observers (AAVSO): International Cooperation and Coordination in Variable Star Research, p. 290 – 301 (1992). – See Abstr. 012.048 for the main entry.
The author reviews groundbased and satellite (UV) observations, circulars and data bases, contributions from astronomers and amateurs to understand the phenomenon of outbursts in novae.

013.064 Variable star research: an international perspective. Concluding remarks.
J. R. Percy.
1. European Meeting of the American Association of Variable Star Observers (AAVSO): International Cooperation and Coordination in Variable Star Research, p. 321 – 325 (1992). – See Abstr. 012.048 for the main entry.
Amateur–professional cooperation has also been a prominent feature of this meeting. The review papers have provided numerous examples of how amateurs contribute to variable star astronomy, and it is not surprising that the demand for AAVSO data and services has increased by a factor of ten in the last two decades.

013.065 Recombination radio lines.
R. L. Sorochenko.
Astrophysics on the threshold of the 21st century, p. 131 – 150 (1992). – See Abstr. 003.048 for the main entry.
Recombination radio lines (RRL) detected more than 25 years ago become an effective tool for astrophysical studies. Presently they are observed in a wide range from 1 mm to 20 m; radio lines are recorded from levels of excited atoms up to n = 768. They were used to determine the temperature, density, and velocity of internal motions in H II regions, to obtain the distribution of ionized hydrogen and the helium abundance in the Galaxy, to estimate the intensities of galactic cosmic rays, etc. RRL have begun to be used to study extragalactic objects.

013.066 Joseph Shklovsky and X–ray astronomy.
H. Friedman.
Astrophysics on the threshold of the 21st century, p. 245 – 252 (1992). – See Abstr. 003.048 for the main entry.
Contents: Introduction. Solar X–ray Astronomy. Galactic X–ray astronomy. Some thoughts for the future. X–ray source studies with a very large area detector.

013.067 Searching for planetary systems.
B. F. Burke.
Astrophysics on the threshold of the 21st century, p. 303 – 313 (1992). – See Abstr. 003.048 for the main entry.
Contents: 1. Fermi's question. 2. To find a planet. 3. To see a planet. 4. One step beyond. 5. The search for life. 6. L'envoi.

013.068 Space radiointerferometry and gravitational waves.
V. B. Braginsky (*V. B. Braginskij*), N. S. Kardashev, I. D. Novikov, A. G. Palnarev (*A. G. Polnarev*).
Astrophysics on the threshold of the 21st century, p. 315 – 330 (1992). – See Abstr. 003.048 for the main entry.
The problem to which the article is devoted lies at the intersection of radioastronomy, the theory of the Early Universe and gravitational–wave astronomy.

013.069 Radio astronomy of the next century.
Y. N. Pariiskii (*Yu. N. Parijskij*).
Astrophysics on the threshold of the 21st century, p. 331 – 355 (1992). – See Abstr. 003.048 for the main entry.
Contents: 1. General trends in radio astronomy instrumentation. 2. Natural limits of radio astronomy. 3. Giant projects (short summary). 4. RATAN–600 and the next century.

013.070 On astronomy for the twenty–first century.
N. S. Kardashev.
Astrophysics on the threshold of the 21st century, p. 357 – 376 (1992). – See Abstr. 003.048 for the main entry.
Contents: 1. The contribution of the twentieth century to astronomy and physics. 2. Possible development of telescopes and methods of atronomical research. 3. Conclusion: the main problems of astronomy.

013.071 Welcoming address: Two thousand years of optical sky surveys.
P. G. Murdin.
2. Conference on Digitised Optical Sky Surveys, p. 1 – 2 (1992). – See Abstr. 012.059 for the main entry.

013.072 Digitised optical sky surveys: introduction.
V. C. Reddish.
2. Conference on Digitised Optical Sky Surveys, p. 3 – 8 (1992). – See Abstr. 012.059 for the main entry.

013.073 Current and future programmes with the UK Schmidt Telescope.
D. H. Morgan, S. B. Tritton, A. Savage, M. Hartley, R. D. Cannon.
2. Conference on Digitised Optical Sky Surveys, p. 11 – 22 (1992). – See Abstr. 012.059 for the main entry.
The status of the major sky surveys and sky atlases is reviewed. The new second epoch southern sky survey being taken with the UK 1.2 m Schmidt Telescope (UKST) is described. Trends in the astronomical use of the UKST are presented, and an outline of possible future work is given.

013.074 The Chinese Large Schmidt Telescope project.
J.–S. Chen.
2. Conference on Digitised Optical Sky Surveys, p. 35 – 39 (1992). – See Abstr. 012.059 for the main entry.
The idea of building the world's largest Schmidt telescope in China has been discussed for several years among the Chinese astronomical community. The basic point to motivate such a facility is the following: China, as a developing country, has only modest means for the development of astrophysics, including the capabilities of financial support, technical feasibility, and site quality. Chinese astronomers should determine a strategical direction in which they can make, for the coming decades, important contributions to the development of world astronomy. They hope that such a facility would be a unique one, so that it would become a very real attraction for other astronomers of the world.

013.075 The deep 2–micron survey of the southern sky.
A. Blanchard, N. Epchtein.
2. Conference on Digitised Optical Sky Surveys, p. 41 – 42 (1992). – See Abstr. 012.059 for the main entry.
The authors envisage a digital sky survey in the near infrared. The purpose of DENIS (Deep Near Infrared Survey of the Southern Sky) is to provide an all–sky survey in the J (1.25 μm) and K (2.2 μm) bands with a pixel size of 3″. An optical band, I (0.9 μm), with a better spatial resolution will be also included. The survey will be done on the 1 m ESO telescope, with a camera using 256x256 Rockwell arrays. The limiting magnitude in the K–band should be 14.5. The importance of such surveys for infrared astronomy will be comparable to Schmidt surveys in the optical range.

013.076 Next generation optical sky surveys.
M. J. Irwin.
2. Conference on Digitised Optical Sky Surveys, p. 43 – 52 (1992). – See Abstr. 012.059 for the main entry.
By building a prime focus CCD array camera system capable of taking advantage of the full 40 arcmin diameter field of the 2.5 m Isaac Newton telescope on La Palma, a dramatic advance in optical sky surveys would be possible. With such a system, a deep CCD–based sky survey in several optical passbands over several hundred square degrees could be achieved using an existing medium sized telescope at modest cost. The key technology and expertise in processing sky surveys already exist and the scientific returns would be enormous: ranging from Solar System studies, such as searches for primordial comets, out to searches for the most distant objects in the Universe, such as high redshift quasars and primordial galaxies.

013.077 Photometric calibration of the southern sky surveys.
S. J. Maddox, W. J. Sutherland.
2. Conference on Digitised Optical Sky Surveys, p. 53 – 60 (1992). – See Abstr. 012.059 for the main entry.
The authors describe a project proposed by a large consortium of UK and Australian surveyors with the aim of obtaining accurate B and R photometric calibrations for all 270 UK Schmidt fields in the South Galactic Cap. This will have major benefits for all aspects of survey astronomy, including studies of the large–scale structure of the universe, galaxy evolution, quasar surveys, galactic structure, rare–object searches, and optical identifications of objects in surveys at other wavebands, such as the ROSAT, MIT/Parkes, and IRAS surveys.

013.078 The Opacity Project – computation of atomic data.
M. J. Seaton, C. J. Zeippen, J. A. Tully, A. K. Pradhan, C. Mendoza, A. Hibbert, K. A. Berrington.
Rev. Mex. Astron. Astrofis., Vol. 23, p. 19 – 43 (Mar 1992). – See Abstr. 012.054 for the main entry.
A general description is given of the methods used by participants in the international Opacity Project to produce massive sets of accurate radiative atomic data, followed by some illustrative examples of results obtained.

013.079 Digitised sky surveys at the Space Telescope Science Institute.
B. M. Lasker.
2. Conference on Digitised Optical Sky Surveys, p. 87 – 94 (1992). – See Abstr. 012.059 for the main entry.
Space Telescope Science Institute activity in the general area of 'Digitised Optical Sky Surveys' is directed, on the operational side, to supporting the use of Hubble Space Telescope by providing guide stars for pointing and calibrated images for observation–planning, and, on the scientific side, to applying the same resources to selected programs in community service and astrophysics.

013.080 A colour/proper motion study of the southern sky – a planned digitisation programme on SuperCOSMOS.
M. R. S. Hawkins.
2. Conference on Digitised Optical Sky Surveys, p. 147 – 150 (1992). – See Abstr. 012.059 for the main entry.

013.081 The Opacity Project – the TOPBASE atomic database.
W. Cunto, C. Mendoza.
Rev. Mex. Astron. Astrofis., Vol. 23, p. 107 – 118 (Mar 1992). – See Abstr. 012.054 for the main entry.
TOPBASE is the first prototype in the development of a tailored DBMS for the efficient manipulation of the large volume of atomic data that resulted from the "Opacity Project". The authors describe the details of the adopted conceptual approach, its physical design and query language. A few examples are included to illustrate its present search, data manipulation and graphic capabilities.

013.082 Bigger telescopes and better instrumentation: report on the 1992 ESO conference.
M.–H. Ulrich.
Messenger, No. 68, p. 1 – 6 (Jun 1992).
Report of the conference "Progress in telescope and instrumentation technologies", Garching (Germany), 27 – 30 April 1992.

013.083 Astronomical observations in 2001.
D. Alloin, T. Le Bertre.
Messenger, No. 68, p. 22 – 24 (Jun 1992).

013.084   The Opacity Project – equation of state.
      D. Mihalas.
Rev. Mex. Astron. Astrofis., Vol. 23, p. 127 – 132 (Mar 1992). –
See Abstr. 012.054 for the main entry.
      The equation of state used in the opacity calculations of the
Opacity Project is described briefly.

013.085   The Opacity Project – results for opacities.
      Yu Yan.
Rev. Mex. Astron. Astrofis., Vol. 23, p. 171 – 179 (Mar 1992). –
See Abstr. 012.054 for the main entry.
      Methods and results of the Opacity Project are discussed. An
emerging opacity library for arbirary compositions is presented.

013.086   The Opacity Project – a post–script.
      M. J. Seaton.
Rev. Mex. Astron. Astrofis., Vol. 23, p. 180 (Mar 1992). – See
Abstr. 012.054 for the main entry.
      New Opacity Project results of January 1992 are presented.

013.087   Asteroid searches from UKST material.
      K. S. Russell, E. Bowell, S. J. Bus, B. Skiff.
2. Conference on Digitised Optical Sky Surveys, p. 237 – 244
(1992). – See Abstr. 012.059 for the main entry.
      Contents: 1. Introduction. 2. UCAS. 3. LUCAS. 4. Reduction
techniques. 5. Jupiter Trojans. 6. Isolated plates. 7. Conclusions.

013.088   The Muenster Redshift Project – MRSP.
      W. C. Seitter.
2. Conference on Digitised Optical Sky Surveys, p. 367 – 381
(1992). – See Abstr. 012.059 for the main entry.
      MRSP is the attempt to provide a three–dimensional survey of
galaxies and quasars to be employed as a uniform data set for
studying strongly interdependent cosmological properties: large–
scale structure and larger–scale homogeneity and isotropy,
parameters of the Friedmann–Lemaître universe, and predictions
from hierarchical and inflationary models. So far, the survey
covers about 360 square degrees. Data from the first 180 square
degrees were used to develop methods and to obtain preliminary
results, presented here to illustrate the above applications.

013.089   The Durham/UKST Galaxy Redshift Survey.
      A. Broadbent, D. Hale–Sutton, T. Shanks, R. Fong,
A. P. Oates, F. G. Watson, C. A. Collins, H. T. MacGillivray,
Q. A. Parker, R. C. Nichol.
2. Conference on Digitised Optical Sky Surveys, p. 389 – 395
(1992). – See Abstr. 012.059 for the main entry.
      The authors are currently engaged in a long term project to
make a redshift survey of ~ 4000 galaxies with $b_j \leqslant 16\overset{m}{.}75$. When
complete, they will have mapped a continuous volume of
$\sim 4 \times 10^6 \mathrm{Mpc}^3$ and to a depth $\sim 300\ h^{-1}\mathrm{Mpc}$ in a region around
the South Galactic Pole.

**013.090   The ROSAT all–sky survey and first identifications of
      X–ray sources using digitised optical plates.**
W. Voges.
2. Conference on Digitised Optical Sky Surveys, p. 453 – 463
(1992). – See Abstr. 012.059 for the main entry.
      The ROSAT X–ray astronomy satellite has completed the first
all–sky X–ray and XUV survey with imaging telescopes. About
50000 new X–ray and 1500 new XUV sources were detected. The
identification of these sources using digitised optical sky surveys
has begun.

**013.091   Optical identification of the new PARKES/MIT/NRAO
      radio sources from digitised sky survey data.**
A. Savage, A. E. Wright.
2. Conference on Digitised Optical Sky Surveys, p. 485 – 489
(1992). – See Abstr. 012.059 for the main entry.
      In 1990 the southern sky was surveyed at Parkes using a multi-
beam receiver at frequencies of 843 and 4850 MHz. The surveys
cover the declination range between + 10 and –90 degrees and are
complete in right ascension, an area of 7.30 steradians. Preliminary analysis of the 5 GHz data indicates a flux limit of about

30 mJy. The authors estimate to find some 90000 new sources
above this limit, or some 90 radio sources per Schmidt field. Such
source densities, combined with accurate positions from the
Australia Telescope, means that this survey will be ideally suited
to an automated identification procedure using the digitised sky
survey data from the UK Measuring Machines and Schmidt
Telescope plate material.

013.092   Conference overview and concluding remarks.
      R. D. Cannon.
2. Conference on Digitised Optical Sky Surveys, p. 499 – 508
(1992). – See Abstr. 012.059 for the main entry.

**013.093   The current status and future focus of basic space science
      in India.**
K. Kasturirangan.
AIP Conf. Proc., No. 245: Workshop on Basic Space Science,
p. 23 – 35 (1992). – See Abstr. 012.064 for the main entry.
      A brief overview of the Indian basic space science programme
with special emphasis on the aeronomy and astronomy related
research activities is presented.

013.094   Astronomical research in Japan.
      Y. Kozai.
AIP Conf. Proc., No. 245: Workshop on Basic Space Science,
p. 36 – 39 (1992). – See Abstr. 012.064 for the main entry.
      The author describes a brief history of astronomical research in
Japan, research institutes and observatories as well as their
facilities and their future plans, particularly, those for the
National Astronomical Observatory and the Institute for Space
and Astronautical Sciences. Also it is mentioned how their
facilities can be used and what are available by astronomers not
directly associated with these institutes.

**013.095   International co–operation in basic sciences research, its
      significance for developing countries: a personal experi-
      ence.**
A. M. Mathai.
AIP Conf. Proc., No. 245: Workshop on Basic Space Science,
p. 40 – 48 (1992). – See Abstr. 012.064 for the main entry.
      The author speaks on some of his personal experiences in
international collaborations and on the need for collaborative
research in basic sciences, especially with developing countries.
After discussing some general aspects he concentrates on his
collaborative work in space sciences.

**013.096   Space research: developing university level curricula and
      research projects.**
H. S. Gurm.
AIP Conf. Proc., No. 245: Workshop on Basic Space Science,
p. 49 – 56 (1992). – See Abstr. 012.064 for the main entry.
      Challenge of space demands an integration of diverse and
varied fields in order to develop a new discipline of teaching at the
university level. Interaction of education and space research has
mutual benefit. Impact, characteristics of space research and its
suitability as an academic pursuit, prompt to generate education
system and its infrastructure. Manpower requirements are
estimated to the order of 100,000 persons besides a high potential
of employment in allied fields. Various curricula and research
projects are identified. The universities of the developing world
do not have large scale infrastructure and a culture for innovative
experimentation. Their problems and options have been addressed to. An interactional centre for space studies is proposed
to redress the imbalance and enhance the involvement of the
developing world.

**013.097   The role of space–based observations in astrophysical
      research.**
J. Grygar.
AIP Conf. Proc., No. 245: Workshop on Basic Space Science,
p. 219 – 229 (1992). – See Abstr. 012.064 for the main entry.
      Since the Earth's atmosphere is transparent to electromagnetic
radiation only in few rather narrow windows, space–based
observations are essential for understanding the nature of various

astronomical objects. Present status of research in different portions of the electromagnetic spectrum is briefly reviewed and some prospects of the development in the future are mentioned. Although space observations are costly, their impact on astrophysics is so substantial that nobody can ignore this area of research. As a consequence, the very nature of astrophysical research is subject to profound change.

013.098   Neutrino and gravitational wave astronomy: grand visions.
H. J. Haubold.
AIP Conf. Proc., No. 245: Workshop on Basic Space Science, p. 230 – 249 (1992). – See Abstr. 012.064 for the main entry.
Three astronomies seem to exist simultaneously in the future to expand the horizon of astronomical observation: photon astronomy, neutrino astronomy, and gravitational wave astronomy. Sources of the radiation, their spectra, and ground–based and space–based observations for photons, neutrinos, and gravitational waves are briefly reviewed.

013.099   Un bilan de la recherche astronomique en France entre 1987 et 1991.
P. de La Cotardière.
Astronomie, p. 9 – 11 (Mar 1992).

013.100   The Soviet Astronomical Society (SAS).
N. G. Bochkarev.
Bull. Am. Astron. Soc., Vol. 23, No. 3, p. 1269 (1991). Abstract. – See Abstr. 012.067 for the main entry.

013.101   Focus: ultra–violet astronomy. Introduction.
Astron. Now, Vol. 6, No. 6, p. 37 (Jun 1992). Special issue: "Focus: ultra–violet astronomy".

013.102   Map of the sky in ultra–violet.
S. Mitton.
Astron. Now, Vol. 6, No. 6, p. 38 – 40 (Jun 1992). Special issue: "Focus: ultra–violet astronomy".

013.103   A vital window on the Universe.
D. Stickland.
Astron. Now, Vol. 6, No. 6, p. 41 – 43 (Jun 1992). Special issue: "Focus: ultra–violet astronomy".
Over the past fourteen years the International Ultraviolet Explorer has studied the Universe at ultra–violet wavelengths. Observations in this "invisible colour" give a rare insight into stars, comets, and the interstellar medium.

013.104   A look inside the bubble.
B. Welsh.
Astron. Now, Vol. 6, No. 6, p. 44 – 46 (Jun 1992). Special issue: "Focus: ultra–violet astronomy".
The Extreme Ultraviolet Explorer is a NASA probe due for launch on 28 May 1992. It will explore the entire extreme ultra–violet waveband and give astronomers a new insight into the bubble we call space.

013.105   Ultra–violet in the extreme.
M. Barstow.
Astron. Now, Vol. 6, No. 6, p. 47 – 49 (Jun 1992). Special issue: "Focus: ultra–violet astronomy".
What techniques do astronomers use in the EUV and what has been discovered in a first EUV survey of the sky, carried out by the UK's ROSAT Wide Field Camera?

013.106   Man behind the mission.
N. Booth.
Astron. Now, Vol. 6, No. 6, p. 50 – 52 (Jun 1992). Special issue: "Focus: ultra–violet astronomy".
An interview with Professor Sir Robert Wilson, the prime mover behind IUE.

013.107   The International Meteor Organization.
M. Gyssens, A. Knöfel, J. Rendtel, P. Roggemans.
IAU Colloquium No. 126: Origin and evolution of interplanetary dust, p. 335 – 338 (1991). – See Abstr. 012.068 for the main entry.
Founded in 1988, the International Meteor Organization (IMO) is an international scientific non–profit association with members all over the world. The IMO was created in response to an ever growing need for international cooperation of amateur work. As such, the main objectives of the IMO are to encourage, support and coordinate meteor observing, to improve the quality of amateur observations, to make global analyses of observations received world–wide, to develop contacts between amateurs and professionals, and to disseminate the observations and results obtained to other amateurs and to the professional community.

013.108   The Norwegian upper atmosphere programme in Svalbard.
J. A. Holtet, A. Brekke.
International Workshop on Cluster Dayside Polar Cusp, p. 5 – 9 (Dec 1991). – See Abstr. 012.071 for the main entry.
Svalbard is the Arctic archipelago between 10° and 35° East and 74° and 81° North. The geomagnetic latitudes spans from approximately 71° to 77° north. The high geographic latitude at a geomagnetic latitude of 75°, makes Svalbard an ideal place for studies of dayside aurora and processes in the magnetospheric boundary regions. Conjugacy to stations in Antarctica makes Svalbard even more attractive. The observatories at Ny–Ålesund and Longyearbyen are the master stations in a network which also includes field stations at Hopen, Hornsund, Bøornøya and Jan Mayen. The core instruments are photometers, optical imagers, magnetometers and riometers. Instruments as spectrometers and ULF/ELF/VLF receivers add to the data sets. The observations form the basis for various studies of solar wind/magnetosphere coupling, electrodynamics, transient events and other magnetospheric and boundary layer phenomena. International cooperation is essential in the programme.

013.109   U.S. ground–based space research programs in Svalbard.
C. Deerh, R. W. Smith.
International Workshop on Cluster Dayside Polar Cusp, p. 11 – 14 (Dec 1991). – See Abstr. 012.071 for the main entry.
U.S. participation on ground–based space research programs in Svalbard has mainly been in cooperation with Norwegian groups. Coordination of instruments, personnel, and observing time and place has proceeded on an informal, but effective basis. Most studies are tied to the CEDAR, GEM, and STEP programs and involve optical observations of the aurora and airglow which takes advantage of the 24–hour darkness, but there are other polar cap studies as well. The Auroral Station at Longyearbyen is central to the American effort and would provide a unique base for ground–based observations associated with the CLUSTER mission.

013.110   Research programmes at Svalbard.
T. Oguti.
International Workshop on Cluster Dayside Polar Cusp, p. 15 – 18 (Dec 1991). – See Abstr. 012.071 for the main entry.
Research programmes of aurorae, ionospheric absorption, electric and magnetic fields at Svalbard for the study of polar cusp dynamics, in connection with boundary layer processes, are briefly described. The necessity for networks of low light level TV cameras, imaging riometers and magnetic stations is emphasized in conjunction with an incoherent scatter radar. These networks would provide data of fine space–time resolution and serve as a reference frame for satellite measurements. Significant conjugates studies could then be made between polar ionispheric and magnetospheric phenomena.

013.111   The Polar Geophysical Institute program of the theoretical and experimental investigation on cusp.
V. G. Pivovarov.
International Workshop on Cluster Dayside Polar Cusp, p. 19 – 22 (Dec 1991). – See Abstr. 012.071 for the main entry.
A theoretical and experimental program for studies of the electro–dynamical structure of the cusp is discussed. The

program is part of a general program – "Generation of Electrical Fields and Currents in the Ionosphere and Magnetosphere". Some tasks are dicussed step by step within the framework of the program. The first of them are those on calculation of the magnetic barrier near the magnetospause, geomagnetic field inside of the magnetosphere, electrical, field and fluxes in the near Earth space and others. Results of the research can be used during studying the cusp problem. Theoretical and experimental investigations are conducted by scientists from the Polar Geophysical Institute and groups of researchers from other institutes of the USSR Academy of Sciences.

013.112   Cusp and cleft studies in Greenland.
E. Friis–Christensen.
International Workshop on Cluster Dayside Polar Cusp, p. 27 – 29 (Dec 1991). – See Abstr. 012.071 for the main entry.
The Danish Meteorological Institute (DMI) has a long tradition of performing geophysical observations in Greenland. The permanent observatory at Godhavn was established in 1926 as the first permanent observatory poleward of the auroral zone. DMI has also traditionally taken an active part in international observational programs in the Arctic regions. Today the Institute is the only Danish institution carrying on ground–based upper atmosphere and magnetosphere studies on a permanent basis. This is done by an extensive network of geophysical observatories and stations on the west– and east coast of Greenland. The network forms the basis for a number of co–operative studies with foreign research institutions. This paper briefly outlines the Danish plans for the coming years with special emphasis on STEP and CLUSTER releated activities.

013.113   Plans for the EISCAT Svalbard radar.
J. Röttger.
International Workshop on Cluster Dayside Polar Cusp, p. 77 – 80 (Dec 1991). – See Abstr. 012.071 for the main entry.
The EISCAT Scientific Association operates incoherent scatter radars in northern Europe for studies of the auroral ionosphere and upper atmosphere. In order to investigate phenomena in the cusp/cleft region and the polar cap it was decided to plan for an extension of the EISCAT facilities by constructing an additional radar on Svalbard. The scientific rationale behind these plans and an outline of the possible instrumental configuration is briefly described.

013.114   The need of coordinated observations in space and from ground.
B. Hultqvist.
International Workshop on Cluster Dayside Polar Cusp, p. 123 – 127 (Dec 1991). – See Abstr. 012.071 for the main entry.
The different roles of ground based and spacecraft measurements is reviewed. The importance of the groundbased measurements also for the satellite projects is discussed. The groundbased global observatory networks provide necessary reference systems for the "front line" measurements on board satellites. Investigations of dynamic phenomena on a global scale depend primarily on the ground based networks, which provide continuity in space and time, but satellites contribute local observations of great importance for the progress of the research. The need for coordination of many kinds of measurements characterizes the solar–terrestrial physics field. The main international coordinated program in the 90ies is STEP (the Solar–Terrestrial Energy Program).

013.115   Requirements for coordinated Cluster and ground–based observations of the cusp.
M. Lockwood.
International Workshop on Cluster Dayside Polar Cusp, p. 203 – 207 (Dec 1991). – See Abstr. 012.071 for the main entry.
The spatial and temporal resolutions and the field–of–view required for ground–based observations of the cusp are discussed by considering a typical example of a dayside auroral transient and plasma flow burst. Systems meeting such specifications would also be ideal for studying travelling convection vortices

and would undoubtedly reveal new spatial structure and temporal behaviour of the cusp region. The implications of the CLUSTER orbit characteristics for coordinated cusp observations with ground–based instruments are discussed.

013.116   Radial velocities and photometry: FS Comae Berenices, Epsilon Aurigae, Miras, and the search for low–mass stellar companions.
R. Stefanik.
J. Am. Assoc. Variable Star Obs., Vol. 20, No. 1, p. 110 (1991). Abstract. – See Abstr. 012.066 for the main entry.

013.117   Archiving data: a must, but how?
E. Griffin.
ESA IUE Newsl., No. 40, p. 16 – 18 (Mar 1992).

013.118   A galaxy redshift survey in the south galactic pole region.
G. Vettolani, J. M. Alimi, C. Balkowski, C. Blanchard, A. Cappi, V. Cayatte, G. Chincarini, C. Collins, P. Felenbok, L. Guzzo, D. Maccagni, H. MacGillivray, S. Maurogordato, R. Merighi, M. Mignoli, D. Proust, M. Ramella, R. Scaramella, G. M. Stirpe, G. Zamorani, E. Zucca.
Messenger, No. 67, p. 26 – 29 (Mar 1992).
Profile of a key programme.

013.119   Solar physics at Potsdam: vector magnetic field measurements, diagnostics and modelling of sunspot structure and dynamics.
A. Hofmann, J. Staude.
Annual report, 1991.
Joint Organization for Solar Observations (JOSO). p. 37 – 44 (1992).
The authors outline the main topics of solar research in the Astrophysical Institute Potsdam in general and in the field of optical solar research in particular. In the Solar Observatory "Einsteinturm" the interest is focused on the measurement of the magnetic field structure in active regions, especially in sunspots, by means of a double vector magnetograph and on modelling of related phenomena.

013.120   Gamma–ray astronomy.
V. A. Dogiel.
Contemp. Phys., Vol. 33, No. 2, p. 91 – 109 (Mar–Apr 1992).
Gamma–rays are located in the hard electromagnetic wave band. The range of measured energies is restricted to the interval between about 10^5eV and 10^{16}eV (the latter is the highest energy of detected particles which are supposed to be gamma–ray photons). Gamma–ray astronomy is relatively young – approximately 25 years old. So only a few space experiments have been performed in this energy range. New perspectives opened since the launch of the Gamma–Ray Observatory will provide the basis for experiments over the coming years. In this paper the problems and achievements of gamma–ray astronomy are described.

013.121   Progress in extra–solar planet detection.
R. A. Brown.
US–USSR Workshop on Planetary Sciences, p. 270 – 287 (1991). – See Abstr. 012.075 for the main entry.
The solar system's existence poses this fundamental question: Are planetary systems a common by–product of star formation? One supporting argument is that flattened disks appear to be abundant around pre–main sequence stars (Strom et al. 1988). Perhaps the planetary orbits in the solar system preserve the form of such a disk that existed around the young Sun. Such heuristic evidence notwithstanding, real progress on the general question requires determining the frequency of occurrence of extra–solar planetary systems and measuring their characteristics (Black 1980). At the current time (the beginning of 1989) no investigator has announced an extra–solar planet detection that is unqualified or that has been generally accepted as such. This paper reviews progress to date. Three investigator groups claim to have found evidence for smaller bodies, perhaps planets, by studying perturbations in star motions. Those observations are instructive

about the specific strengths and weaknesses of indirect techniques for detecting planets with various masses and orbits.

013.122 Ten years monitoring of blazars at Metsähovi.
H. Teräsranta, M. Tornikoski, K. Karlamaa,
E. Valtaoja, S. Urpo, M. Lainela, J. Kotilainen, S. Wirén,
S. Laine, K. Nilsson, A. Lähteenmäki, R. Korpi, M. Valtonen.
Conference on the Variability of Blazars, p. 159 – 166 (1992). – See Abstr. 012.077 for the main entry.

The authors give an overall presentation of the monitoring programme of extragalactic sources at Metsähovi Radio Research Station. The monitoring has now lasted for 10 years and most of the 12000 observations are at 22 and 37 GHz, thus giving the largest existing database at these frequencies.

013.123 The program on global character research of solar–terrestrial system in the maximum period of the 22nd solar cycle.
Hu Wenrui, Zhang Bairong.
Advances in solar–terrestrial science of China, p. 1 – 8 (1992). – See Abstr. 003.069 for the main entry.

The program of the global character research of the solar-terrestrial system in the 22nd solar cycle organized by the Chinese Academy of Sciences in the period 1987 – 1991 is introduced and the research activities of solar–terrestrial sciences in the 1990's in China are discussed.

013.124 Coordinated observations and measurements of solar-terrestrial system.
Zhang Bairong, Luo Baorong, Zhang Qin, Hou Shuming.
Advances in solar–terrestrial science of China, p. 9 – 14 (1992). – See Abstr. 003.069 for the main entry.

Coordinated observations and measurements of important events in the solar–terrestrial system have been organized as a prominent part of the program of the global character of solar-terrestrial systems in the maximum period of the 22nd solar cycle promoted by the Chinese Academy of Sciences. An overview is given in the present paper.

013.125 The progress of solar–terrestrial sciences in China.
Hu Wenrui, Fu Zhufeng.
Advances in solar–terrestrial science of China, p. 15 – 24 (1992). – See Abstr. 003.069 for the main entry.

The present paper gives an overview of the progress made in the solar–terrestrial sciences in China, mainly deverloped since the early 1960's. The primary capabilities of space exploration and the bases of synthetic ground facilities are discussed, and current theoretical and applied studies are presented.

013.126 China meridian chain of magnetometers and the global electric current system.
Xu Wenyao.
Advances in solar–terrestrial science of China, p. 140 – 147 (1992). – See Abstr. 003.069 for the main entry.

The China Meridian Chain of Magnetometers (CMCM) is briefly introduced. The scientific purpose of this chain is to provide ground–based data for studying (1) the coupling of the solar wind, magnetosphere, ionosphere and neutral atmosphere, (2) the behaviour and the role of the mid–low latitude ionosphere and magnetosphere in energy coupling processes, and (3) large–scale magnetospheric–ionospheric current systems at midlow latitudes during solar–terrestrial events. On the basis of the magnetic data recorded at the CMCM stations during major solar–terrestrial events in 1988 – 1989, large–scale current systems are constructed. The contributions from the ring currents and field–aligned currents to the magnetic disturbances are recognized.

013.127 Activities of research work related to Earth rotation in China from January 1988 to June 1990.
Jin Wenjing, Zhang Chengzhi, Li Zhisen, Wang Zhengming.
Publ. Beijing Astron. Obs., No. 18, p. 66 – 71 (Oct 1991).

013.128 Multi–site spectroscopic networks for the study of late–type stars.
B. H. Foing.
Armagh Observatory Bicentenary Colloquium on Surface Inhomogeneities on Late–Type Stars, p. 224 – 228 (1992). – See Abstr. 012.085 for the main entry.

Contents: 1. The need for multi–site spectroscopy. 2. The MUSICOS project. 3. Requirements for asteroseismology and stellar activity. 4. Technical design of the MUSICOS instrument. 5. The MUSICOS December 1989 observing campaign. 6. MUSICOS project organization and collaborations. 7. MUSICOS planning and perspectives.

013.129 Prospects for robotic observatories and stellar seismology in Latin America.
J. R. García.
Contrib. Inst. Copérnico (Buenos Aires), Vol. 1, No. 1.
Instituto Copérnico, Buenos Aires (Argentina). 3 p. (Jan 1992). Invited talk to the Symposium on Robotic Observatories, Boston, MA (USA), 13 – 15 Jul 1990.

013.130 Iniciación a la investigación científica.
J. R. García, R. Cebral, E. R. Scoccimarro,
P. S. Whanon, M. Zimmermann.
Contrib. Inst. Copérnico (Buenos Aires), Vol. 1, No. 2.
Instituto Copérnico, Buenos Aires (Argentina). 2 p. (Feb 1992). Presentado al Coloquio Internacional "La cultura astronómica en la sociedad moderna", Montevideo (Uruguay), 18 – 22 Jul 1991.

013.131 Investigators of resources and distribution for astronomical literature in Academia Sinica.
Pang Xiaojun, Li Zhifang, Xu Mianqin, Ke Darong.
Ann. Shanghai Obs., Acad. Sin., No. 13, p. 203 – 217 (1992). In Chinese.

013.132 Geophysics news 1991.
Earth Space, Vol. 4, No. 5, p. 5 – 14 (Jan 1992).
A look back at some of the most important events and developments in geophysics during 1991, from the first detailed views of the surface of Venus to the whereabouts of the continents 500 million years ago.

013.133 Les astronomes amateurs et les variables à longue période.
M.–O. Mennessier.
J. Astron. Fr., No. 41, p. 14 (Feb 1992). – See Abstr. 012.070 for the main entry.

013.134 La Suisse et l'espace.
S. Berthet.
Orion, Jahrg. 50, Nr. 248, p. 10 – 15 (Feb 1992).

013.135 On the evolving role of computational astrophysics.
P. Charbonneau.
J. R. Astron. Soc. Can., Vol. 86, No. 1, p. 31 – 53 (Feb 1992).
The aim of this paper is to illustrate, with the help of two specific examples, a change currently taking place with regard to our perception of the role and scope of computational astrophysics in the broader context of astronomical and astrophysical research. Current supercomputers and numerical techniques are now powerful enough to allow a detailed, accurate, and reliable multidimensional modelling of an increasing number of astrophysical systems. Furthermore, modern graphics and visualization tools allow the study of simulation results in an intuitively appealing format, such as video animation. This makes the work of a computational astrophysicist increasingly similar, conceptually, to that of a "classical" experimentalist working in another subfield of physics.

013.136   Major advances in astronomy since 1890.
C. Ronan.
Images of the universe, p. 1 – 19 (1991). – See Abstr. 003.090 for the main entry.

Over the last one hundred years, astronomy has seen an advance unprecedented during any previous century. This has been due primarily to two factors. In the first place there has been a vast growth in ways of observing the universe. Not only have telescope apertures increased in size, allowing ever dimmer objects to be discerned and so enabling astronomers to penetrate much deeper into space, but also totally new techiques have made it possible to examine hitherto unavailable evidence. The second significant change which has come about in astronomy is due to the increasingly effective interplay between astronomy and physics – particularly particle physics. Admittedly this last began more than a century ago, but only within the last hundred years has it been possible for these results to yield their true significance.

013.137   Entrevista: Manuel Peimbert.
I. Ferrin.
Universo, Vol. 12, No. 36, p. 5 – 8 (Feb – Apr 1992).
This paper presents the text of an interview given by Prof. Manuel Peimbert to the author.

013.138   The Sun: a laboratory for astrophysics. Opening address.
J. C. Brown.
NATO Advanced Study Institute on the Sun: a Laboratory for Astrophysics, p. 1 – 7 (1992). – See Abstr. 012.091 for the main entry.
The author gives a brief introduction to Scotland and the Scots, a short historical overview of solar science in Scotland, and the current state of astronomical research in Scotland.

013.139   Airglow observation program carried out by Tokyo Astronomical Observatory.
H. Tanabe, A. Takechi, A. Miyashita, K. Tanaka.
Rep. Natl. Astron. Obs. Jpn., Vol. 1, No. 3, p. 309 – 347 (1992). In Japanese.
This is a record of more than 30 years' history of an airglow observation program, which was terminated in March 1990, carried out by a group of Tokyo/National Astronomical Observatory. The instruments and methods of observation used at each station and results obtained with them are explained. In addition, reports on temporal airglow observations and a report on rocket observations are given. Finally, functions and activities of "World Data Center C2 for Airglow" are mentioned.

013.140   Space astronomy in India.
P. C. Agrawal.
Proc. Astron. Soc. Aust., Vol. 9, No. 2, p. 229 – 233 (1991). – See Abstr. 012.090 for the main entry.
Astronomical observations from space–borne instruments are carried out in India in the areas of infrared, X–ray and gamma–ray astronomy. This paper briefly describes the facilities available in India for conducting experiments in space astronomy using balloons, rockets and satellites. It briefly reviews the important results obtained by Indian astronomers from observations made in India with the balloon, rocket and satellite experiments. The present status of research in different disciplines of space astronomy is discussed.

013.141   Report of the IAU/IAG/COSPAR Working Group on Cartographic Coordinates and Rotational Elements of the Planets and Satellites: 1991.
M. E. Davies, V. K. Abalakin, A. Brahic, M. Burša, B. H. Chovitz, J. H. Lieske, P. K. Seidelmann, A. T. Sinclair, Y. S. Tjuflin (*Yu. S. Tyuflin*).
Celest. Mech. Dyn. Astron., Vol. 53, No. 4, p. 377 – 397 (1992).
Every three years the IAU/IAG/COSPAR Working Group on Cartographic Coordinates and Rotational Elements of the Planets and Satellites revises tables giving the directions of the north poles of rotation and the prime meridians of the planets

and satellites. Also presented are revised tables giving their sizes and shapes.

013.142   Astronomical Resarch in Romania.
M. Stavinschi.
Rom. Astron. J., Vol. 1, No. 1 – 2, p. 115 – 116 (1991).

The protection of astronomical and geophysical sites.
See Abstr. 003.038.

Astrophysics on the threshold of the 21st century.
See Abstr. 003.048.

A brief history of the Minor Planet Center.
See Abstr. 004.078.

A CCD Workshop in Germany.
See Abstr. 011.025.

Chronicle (Report on): All–Union Conference of the Working Group "Stellar Atmospheres", held at the Institute of Astrophysics and Physics of the Atmosphere of the Academy of Sciences of Estonia, 28 – 30 May 1991.
See Abstr. 011.026.

Variable star research: an international perspective. Proceedings.
See Abstr. 012.048.

Digitised optical sky surveys. Proceedings.
See Abstr. 012.059.

Coordination of SOHO and ground–based observations. Proceedings.
See Abstr. 012.080.

Nobeyama radioheliograph.
See Abstr. 033.024.

Decimeter high resolution solar radio spectroscope.
See Abstr. 033.025.

Particle astrophysics on the moon.
See Abstr. 051.046.

Astrophysics from the moon.
See Abstr. 051.047.

ESOC's role in routine Hipparcos operations.
See Abstr. 051.057.

Sources of gravitational waves and prospects for their detection.
See Abstr. 066.308.

L'esperimento Gallex del Gran Sasso ed il problema dei neutrini solari.
See Abstr. 080.090.

Spectroscopy of young open clusters.
See Abstr. 153.053.

A programme on young open clusters: the combined use of astrometry, photometry and spectroscopy for membership, CM diagrams and internal kinematics.
See Abstr. 153.054.

Early studies of globular clusters in Italy.
See Abstr. 154.060.

Results of radio astronomy obtained by Japan–US VLBI experiment in CDP.
See Abstr. 159.028.

A homogeneous bright quasar survey.
See Abstr. 159.095.

The Edinburgh multi–colour survey for quasars.
See Abstr. 159.096.

APM surveys for high redshift quasars.
See Abstr. 159.097.

The APM bright galaxy surveys.
See Abstr. 160.108.

An automated search for compact groups of galaxies.
See Abstr. 160.113.

Extended structure in the 3–D galaxy distribution at the Galactic poles.
See Abstr. 161.421.

014 Teaching in Astronomy

014.001 **Counting distant radio sources to determine the overall curvature of space.**
C. A. Eckroth.
Phys. Teach., Vol. 30, No. 2, p. 92 – 93 (Feb 1992). Abstract.
Current Physics Microform No.: 9203G1668.

014.002 **Sensing the rotation of the Earth.**
H. R. Crane.
Phys. Teach., Vol. 30, No. 2, p. 111 (Feb 1992). Abstract. Current Physics Microform No.: 9203G1687.

014.003 **Science in an imaginary sky.**
G. Mumford.
Sky Telesc., Vol. 83, No. 2, p. 146 – 148 (Feb 1992).
The computer screen offers a new window on the universe for students using a revolutionary observatory simulator.

014.004 **Astronomical computing: the green flash.**
B. E. Schaefer.
Sky Telesc., Vol. 83, No. 2, p. 200 – 203 (Feb 1992).

014.005 **Bildverarbeitung auf dem Personal Computer. Teil 1: Technik.**
H. Hilbrecht.
Sterne Weltraum, Jahrg. 31, Nr. 2, p. 128 – 131 (Feb 1992).

014.006 **Planetenkartographie.**
M. Hauss.
Sterne Weltraum, Jahrg. 31, Nr. 3, p. 172 – 174 (Mar 1992).

014.007 **Bildverarbeitung auf dem Personal Computer. Teil 2: Beispiele für die Anwendung.**
H. Hilbrecht.
Sterne Weltraum, Jahrg. 31, Nr. 3, p. 196 – 199 (Mar 1992).

014.008 **The Milky Way and other galaxies.**
I. Nicolson.
Astron. Now, Vol. 6, No. 1, p. 16 – 17 (Jan 1992).
Astronomy for absolute beginners.

014.009 **Six inches of city skies.**
G. McNamera.
Astron. Now, Vol. 6, No. 1, p. 19 – 24 (Jan 1992).
Living in a city and pursuing astronomy need not be mutually exclusive.

014.010 **Spacecraft, orbits and slingshots.**
I. Nicolson.
Astron. Now, Vol. 6, No. 3, p. 26 – 27 (Mar 1992).
Astronomy for absolute beginners.
For the second part of this article see No. 4, p. 26 – 27 (Apr 1992).

014.011 **A new window on the sky.**
G. Carr.
Astron. Now, Vol. 6, No. 1, p. 50 – 51 (Jan 1992).

014.012 **Nebulae and interstellar matter.**
I. Nicolson.
Astron. Now, Vol. 6, No. 2, p. 16 – 17 (Feb 1992).
Astronomy for absolute beginners.

014.013 **A point–and–shoot planetarium.**
G. Carr.
Astron. Now, Vol. 6, No. 2, p. 48 – 49 (Feb 1992).

014.014 **The Messier challenge.**
B. Ewen–Smith.
Astron. Now, Vol. 6, No. 3, p. 20 – 24 (Mar 1992).
Most people have seen a few Messier objects, but how many have seen over a hundred? Here's a step–by–step guide for those who want to rise to the challenge.

014.015 **Focus: Jupiter. Introduction.**
Astron. Now, Vol. 6, No. 3, p. 37 – 40 (Mar 1992).
Special issue: "Focus: Jupiter".

014.016 **Jupiter – a failed star?**
P. Cattermole.
Astron. Now, Vol. 6, No. 3, p. 52 (Mar 1992). Special issue: "Focus: Jupiter".

014.017 **Focus: cosmic distance ladder. Introduction.**
J. Gribbin.
Astron. Now, Vol. 6, No. 4, p. 37 – 40 (Apr 1992). Special issue: "Focus: cosmic distance ladder".

014.018 **Sizing up the solar system.**
T. Lyster.
Astron. Now, Vol. 6, No. 4, p. 41 – 42 (Apr 1992). Special issue: "Focus: cosmic distance ladder".
The author shortly describes the methods used to determine the size of the Solar System: radar astronomy, triangulation, and Kepler's laws.

014.019 **How far the star?**
M. Penston.
Astron. Now, Vol. 6, No. 4, p. 43 – 45 (Apr 1992). Special issue:
"Focus: cosmic distance ladder".
Hipparcos will make a valuable contribution to our knowledge
of the structure and dynamics of our Galaxy by measuring to
unprecedented accuracy the positions, parallaxes, and proper
motions of 120000 stars.

014.020 **From here to infinity.**
C. Kitchin.
Astron. Now, Vol. 6, No. 4, p. 46 – 50 (Apr 1992). Special issue:
"Focus: cosmic distance ladder".
Measuring distances outside the Milky Way Galaxy.

014.021 **Through the eyepiece.**
T. Cave.
Astron. Now, Vol. 6, No. 4, p. 52 – 53 (Apr 1992).
The author looks at the many different varieties of lens.
In the second part (No. 5, p. 52 – 53 (May 1992)) he explores the
strengths and weaknesses of particular design.

014.022 **Colliding galaxies.**
N. Henbest.
Astron. Now, Vol. 6, No. 5, p. 16 – 18 (May 1992).
If galaxies pass too close to one another, the results will be
dynamic and possibly catastrophic.

014.023 **Hot Big Bang.**
I. Nicolson.
Astron. Now, Vol. 6, No. 5, p. 26 – 27 (May 1992).
Astronomy for absolute beginners.
The Big Bang theory suggests that the universe is expanding
because it erupted from a hot dense state some 10 to 20 billion
years ago.

014.024 **Focus: professional amateurs. Introduction.**
P. Murdin.
Astron. Now, Vol. 6, No. 5, p. 37 – 40 (May 1992). Special issue:
"Focus: professional amateurs".

014.025 **The search for nova.**
J. Fletcher.
Astron. Now, Vol. 6, No. 5, p. 41 – 42 (May 1992). Special issue:
"Focus: professional amateurs".
Novae, or exploding stars at the end of their evolutionary cycle
are often discovered by keen eyed amateurs who constantly scan
the skies for a stellar interloper.

014.026 **The new technologies.**
M. Gavin.
Astron. Now, Vol. 6, No. 5, p. 43 – 45 (May 1992). Special issue:
"Focus: professional amateurs".
The electronics revolution has armed amateur astronomers
with a vast array of new detectors.

014.027 **At the eyepiece.**
P. Moore.
Astron. Now, Vol. 6, No. 5, p. 46 – 48 (May 1992). Special issue:
"Focus: professional amateurs".
Although electronic imaging can produce stunning results
quickly, you can't beat the sheer fun of optical observing.

014.028 **Ein computergesteuertes Schulplanetarium im Selbstbau.**
A. Gruppe.
Sterne Weltraum, Jahrg. 31, Nr. 4, p. 238 – 240 (Apr 1992).

014.029 **An exercise on the altitude of the noon sun.**
J. Hickman, H. Kruglak.
Phys. Teach., Vol. 30, No. 4, p. 236 – 237 (Apr 1992). Abstract.
Current Physics Microform No.: 9205H2190.

014.030 **Astronomische Bildung für alle. Eine Herausforderung für die Fachdidaktik.**
W. Winnenburg.
Astron. Sch., Jahrg. 29, Heft 7, p. 12 – 15 (Feb 1992).

014.031 **Astronomie an Berliner Schulen. Situation – Tendenzen – Hoffnungen.**
P. Kriesel.
Astron. Sch., Jahrg. 29, Heft 7, p. 22 – 23 (Feb 1992).

014.032 **Dem Kosmos ein Stück näher.**
M. Stark.
Astron. Sch., Jahrg. 29, Heft 8, p. 4 – 5 (Apr 1992).

014.033 **Der Mond ist aufgegangen. Kinder beobachten den Erdtrabanten.**
A. Anhalt, W. Winnenburg.
Astron. Sch., Jahrg. 29, Heft 8, p. 15 – 16, 21 (Apr 1992).

014.034 **Lernhilfen für den Astronomieunterricht.**
K. Lindner.
Astron. Sch., Jahrg. 29, Heft 8, p. 24 – 25 (Apr 1992).

014.035 **Die Welt durch Vergleiche erschließen. Planetenmodelle.**
A. Christian, W. Winnenburg.
Astron. Sch., Jahrg. 29, Heft 8, p. 28 – 30 (Apr 1992).

014.036 **Spring galaxies.**
B. Abrams.
Astron. Now, Vol. 6, No. 4, p. 22 – 24 (Apr 1992).
Spring skies present an opportunity to observe and enjoy some
fine galaxies.

014.037 **Measuring the distance of stars in the TS–24 sky.**
D. Earley.
I.A.P.P.P. Commun., No. 47, p. 31 – 32 (Mar – May 1992).

014.038 **Projet d'action éducative: surveillance de l'activité solaire au cours du maximum 1990 – 1991.**
J. L. Delon.
Obs. Trav., No. 30, p. 24 – 32 (1992).

014.039 **Daylight savings.**
J. M. Pasachoff.
Phys. Teach., Vol. 29, No. 2, p. 71 (Feb 1991). Abstract.

014.040 **Die totale Sonnenfinsternis vom 11. Juli 1991 – Beobachtungsort San Pedrito.**
T. Marold.
Sterne, Band 68, Heft 2, p. 126 – 131 (1992).

014.041 **Give your camera a piggyback ride.**
J. Sanford.
Astronomy, Vol. 20, No. 1, p. 76 – 81 (Jan 1992).

014.042 **Seeing the most on Jupiter.**
J. Olivarez.
Astronomy, Vol. 20, No. 3, p. 85 – 87 (Mar 1992).

014.043 **Polar aligning your telescope.**
M. Porcellino.
Astronomy, Vol. 20, No. 5, p. 68 – 72 (May 1992).

014.044 **The design and construction of a run–off roof observatory.**
M. J. Ropelewski.
J. Br. Astron. Assoc., Vol. 102, No. 3, p. 132 – 134 (Jun 1992).
This paper describes the construction of a low–cost, run–off
roof observatory, which was designed to house the author's
25–cm Newtonian reflector.

014.045   **A new lunar occultation package for microcomputers.**
G. E. Taylor.
J. Br. Astron. Assoc., Vol. 102, No. 3, p. 158 – 159 (Jun 1992).
   This report is a brief description of a suite of programs designed to enable occultation observers to predict all lunar occultations of SAO stars during one night. The computer time will normally be only a few minutes.

014.046   **Hands–on astronomy in the middle and secondary school.**
L. M. French.
Bull. Am. Astron. Soc., Vol. 23, No. 3, p. 1190 (1991). Abstract. – See Abstr. 012.037 for the main entry.

014.047   **Getting girls into America's science education agenda.**
N. G. Barlow.
Bull. Am. Astron. Soc., Vol. 23, No. 3, p. 1190 (1991). Abstract. – See Abstr. 012.037 for the main entry.

014.048   **Resources for teaching about the planets: slides, videos, software and a teachers' resource notebook.**
A. Fraknoi.
Bull. Am. Astron. Soc., Vol. 23, No. 3, p. 1191 (1991). Abstract. – See Abstr. 012.037 for the main entry.

014.049   **Astronomy–related teacher inservice training.**
L. A. Lebofsky, N. R. Lebofsky.
Bull. Am. Astron. Soc., Vol. 23, No. 3, p. 1191 (1991). Abstract. – See Abstr. 012.037 for the main entry.

014.050   **Ways to become personally involved in education.**
M. L. Nelson, L. A. Lebofsky.
Bull. Am. Astron. Soc., Vol. 23, No. 3, p. 1191 (1991). Abstract. – See Abstr. 012.037 for the main entry.

014.051   **Das Riesenstadium der Sterne.**
H. Zimmermann.
Astron. Sch., Jahrg. 29, Heft 9, p. 4 – 6 (Jun 1992).

014.052   **Eine leuchtende Sternkarte im Klassenzimmer.**
R. Szostak.
Astron. Sch., Jahrg. 29, Heft 9, p. 15 – 21 (Jun 1992).

014.053   **Feuerwerk aus dem All.**
K. Düber, S. Molau, M. Nitschke, D. Przewozny.
Astron. Sch., Jahrg. 29, Heft 9, p. 24 – 25 (Jun 1992).

014.054   **Le stelle doppie (seconda parte).**
P. Tempesti.
Astronomia UAI, N. 1 – 2, p. 3 – 8 (Jan – Apr 1992).
   In the second part of this article on double stars, the author explains how astronomers can deduce the absolute orbital motion and the mass of stars in a double system. He examines also such classical cases as Krueger 80 and Sirius.

014.055   **Acquisizione digitale ed elaborazione numerica di immagini planetarie (II).**
A. Leo, G. Quarra Sacco, D. Sarocchi.
Astronomia UAI, N. 1 – 2, p. 9 – 16 (Jan – Apr 1992).
   The problem of data reduction, enhancement and analysis of CCD images are examined.

014.056   **Beobachtungen der Sonnenbahn durch Schüler.**
J. Alean.
Sterne Weltraum, Jahrg. 31, Nr. 5, p. 306 – 309 (May 1992).

014.057   **Astronomie mit dem Atari ST.**
D. Roth.
Sterne Weltraum, Jahrg. 31, Nr. 5, p. 338 – 339 (May 1992).

014.058   **Le stelle doppie (terza parte).**
P. Tempesti.
Astronomia UAI, N. 3, p. 5 – 12 (May – Jun 1992).
   The author examines the argument of spectroscopic double stars and binaries at eclipse. He finally examines the multiple systems and the origin of double stars.

014.059   **Construzione e impiego di uno spettrografo.**
I. Dalmeri, G. Favero, G. Favero.
Astronomia UAI, N. 3, p. 23 – 27 (May – Jun 1992).

014.060   **Teaching physics and astronomy to disadvantaged college students.**
H. J. Augensen.
Bull. Am. Astron. Soc., Vol. 23, No. 3, p. 1263 (1991). Abstract. – See Abstr. 012.067 for the main entry.

014.061   **Introductory astronomy: fundamentals and tools.**
R. Ruotsalainen.
Bull. Am. Astron. Soc., Vol. 23, No. 3, p. 1263 (1991). Abstract. – See Abstr. 012.067 for the main entry.

014.062   **Will the Universe expand for ever?**
I. Nicolson.
Astron. Now, Vol. 6, No. 6, p. 26 – 27 (Jun 1992).
   Astronomy for absolute beginners.

014.063   **Variable star observations in an introductory astronomy course.**
C. M. Gaskell.
J. Am. Assoc. Variable Star Obs., Vol. 20, No. 1, p. 41 – 50 (1991).
   Criteria for the choice of stars used in an introductory astronomy course are discussed and a list of suitable bright variable stars is given. Consideration is given to the constraints of the academic calendar, student interest, and the probable sky conditions under which the students will have to observe.

014.064   **Astronomieunterricht in Mecklenburg–Vorpommern.**
M. Schukowski.
Astron. Sch., Jahrg. 28, Heft 6, p. 12 – 14 (Dec 1991).

014.065   **Prüfung der Leistungsfähigkeit des Schulfernrohres.**
W. Knobel.
Astron. Sch., Jahrg. 28, Heft 6, p. 14 – 15 (Dec 1991).

014.066   **Die Sterne auf die Erde holen.**
S. von der Weiden.
Astron. Sch., Jahrg. 28, Heft 6, p. 33 – 34 (Dec 1991).

014.067   **Mars – erkundet und interpretiert.**
K. Hiller.
Astron. Raumfahrt , Jahrg. 29, Heft 2, p. 20 – 22 (1991).

014.068   **Planetenphotographie mit extremen Brennweiten.**
L. Stephan.
Astron. Raumfahrt , Jahrg. 29, Heft 2, p. 27 – 29 (1991).

014.069   **Video in der Planetenastronomie.**
G. Dittié.
Astron. Raumfahrt , Jahrg. 29, Heft 2, p. 30 – 33 (1991).

014.070   **The search for extraterrestrial intelligence.**
Universe Classroom, No. 20, p. 1 – 4 (Spr 1992).

014.071   **Some variable star research projects for undergraduate college students.**
F. R. West.
J. Am. Assoc. Variable Star Obs., Vol. 20, No. 1, p. 111 (1991). Abstract. – See Abstr. 012.066 for the main entry.

014.072   **Ricerca amatoriale di supernovae extragalattiche.**
M. Villi, G. Cortini.
Astronomia UAI, N. 3, p. 31 – 35 (May – Jun 1992).

014.073 **Introduzione all'osservazione visuale delle nebulose planetarie.**
A. Bertoglio, P. Tanga.
Astronomia UAI, N. 3, p. 36 – 38 (May – Jun 1992).

014.074 **Per una didattica "debole" l'astronomia come traccia pluridisciplinare.**
F. Romano.
G. Astron., Vol. 18, N. 1, p. 20 – 23 (Mar 1992).

014.075 **L'astronomia nelle proposte della Commissione Brocca per i programmi dei trienni della scuola secondaria superiore.**
L. E. Abati.
G. Astron., Vol. 18, N. 2, p. 34 – 42 (Jun 1992).

014.076 **Computers in introductory astronomy: who needs them?**
D. D. Meisel, K. F. Kinsey.
News Lett. Astron. Soc. N.Y., Vol. 4, No. 1, p. 9 (Feb 1992).
Abstract. – See Abstr. 012.078 for the main entry.

014.077 **Astrofotografie als Anfänger.**
P. Frauenfelder.
Orion, Jahrg. 50, Nr. 248, p. 123 – 127 (Jun 1992).

014.078 **CCDs for everyman.**
J. Stull.
News Lett. Astron. Soc. N.Y., Vol. 4, No. 1, p. 10 (Feb 1992).
Abstract. – See Abstr. 012.078 for the main entry.

014.079 **O uso de cameras de video portateis CCD na astronomia.**
Universo, Vol. 12, No. 36, p. 22 – 26 (Feb – Apr 1992).

014.080 **Magnitud limite de un telescopio.**
J. Luna Tirado.
Universo, Vol. 12, No. 36, p. 38 – 40 (Feb – Apr 1992).

Fernrohr–Selbstbau. Fenster ins Weltall. Praktische Anleitungen für Freunde des Sternenhimmels und solche, die es werden wollen.
See Abstr. 003.065.

CCD observations from the centre of a city: use in astronomical education.
See Abstr. 009.023.

The 4th International Space University: an International Mars Mission.
See Abstr. 011.013.

The DPS Teachers Workshop.
See Abstr. 011.014.

Space research: developing university level curricula and research projects.
See Abstr. 013.096.

Die Sonnenuhr – Bastelobjekt in der pädagogischen Arbeit.
See Abstr. 034.022.

Infrared and the search for extrasolar planets.
See Abstr. 035.098.

015 Miscellanea (Philosophical Aspects, Extraterrestrial Life, etc.)

015.001 **What's wrong with a gibbous Moon?**
W. Livingston.
Sky Telesc., Vol. 83, No. 2, p. 159 – 160 (Feb 1992).

015.002 **De speurtocht naar kosmische levenstekens begint.**
S. Shostak.
Zenit, Jaarg. 19, Nr. 2, p. 66 – 68 (Feb 1992).

015.003 **What's up, popular astronomy?**
R. Piffl.
Kozmos, Vol. 22, No. 5, p. 162 – 164 (Oct 1991). In Slovak.

015.004 **Overhearing the extraterrestrials at Harvard.**
J. Grygar.
Říše hvězd, Vol. 72, No. 1, p. 11 – 12 (Jan 1991). In Czech.

015.005 **From the history of the Bulletin of the Astronomical Institutes of Czechoslovakia.**
M. Kopecký.
Říše hvězd, Vol. 73, No. 1, p. 6 (Jan 1992). In Czech.

015.006 **Die Sonne im Spiegel der Wissenschaft.**
I. Mügge.
Astron. Sch., Jahrg. 29, Heft 8, p. 6 – 7 (Apr 1992).

015.007 **SETI, a long way to go.**
H. Hirabayashi.
Astron. Her., Vol. 85, No. 6, p. 252 – 256 (1992). In Japanese.

015.008 **The anthropomorphic fallacy.**
E. J. Coffey.
J. Br. Interplanet. Soc., Vol. 45, No. 1, p. 23 – 29 (Jan 1992).
Special issue: Exobiology. S. Santoli (ed.).
Central to human creativity is the projection of human qualities upon the external world. On the other hand, so subtle is the latter's mode of operation and so all–persuasive is human nature, that it persistently generates a mode of thought which, although fallacious, exerts a tremendous hold over us. Several examples – schemas, technologies, stories, evolutionary "ladder", contingency – reveal the impact of the resulting anthropomorphic fallacy. It casts serious doubts upon any presumptions as to what extraterrestrials should be like.

015.009 **What makes a planet habitable, and how to search for habitable planets in other solar systems.**
M. D. Papagiannis.
J. Br. Interplanet. Soc., Vol. 45, No. 6, p. 227 – 230 (Jun 1992).
Special issue: Exobiology. S. Santoli (ed.).

015.010 **Jules Verne et l'astronomie.**
P. Bacchus.
Obs. Trav., No. 29, p. 3 – 19 (1992).

015.011 **Endogenous production, exogenous delivery and impact–shock synthesis of organic molecules: an inventory for the origins of life.**
C. Chyba, C. Sagan.
Nature, Vol. 355, No. 6356, p. 125 – 132 (9 Jan 1992).
Sources of organic molecules on the early Earth divide into three categories: delivery by extraterrestrial objects; organic

synthesis driven by impact shocks; and organic synthesis by other energy sources. Estimates of these sources for plausible end-member oxidation states of the early terrestrial atmosphere suggest that the heavy bombardment before 3.5 Gyr ago either produced or delivered quantities of organics comparable to those produced by other energy sources.

015.012   **Kontakte mit außerirdischen Intelligenzen.**
H.–J. Treder.
Sterne, Band 68, Heft 2, p. 98 – 99 (1992).

015.013   **The historical development of polar–cap emission models.**
D. G. Lominadze.
IAU Colloquium No. 128: The magnetospheric structure and emission mechanisms of radio pulsars, p. 53 – 55 (1992). – See Abstr. 012.022 for the main entry.
As part of the introduction to the session "magnetospheric models" the author presented a penetrating summary of the historical development of polar cap emission models. The remarks were complete, incisive, and very amusing. Here only the transparencies used are presented.

015.014   **The habitability of Mars–like planets around main sequence stars.**
W. L. Davis, L. R. Doyle, D. E. Backman, C. P. McKay.
3. Symposium International de Bioastronomie, p. 55 – 61 (1991). – See Abstr. 012.005 for the main entry.
The authors have developed a model to investigate the duration of conditions necessary for the origin of life on a Mars–like planet. They investigate various star–planet distances and a range of stellar spectral types in order to provide guidance for the SETI. Planets suitable for the origin of life if they contain liquid water habitats for time periods comparable to the maximum time required for the origin of life on Earth are considered. A Mars–like planet will differ from an Earth–like planet primarily because it will have no plate tectonic activity and therefore no long–term recycling of CO_2. It is found that there is sufficient time for the origin of life on Mars–like planets around F, G, K, and M type stars.

015.015   **Geophysiology and habitable zones around Sun–like stars.**
D. W. Schwartzman, T. Volk.
3. Symposium International de Bioastronomie, p. 155 – 162 (1991). – See Abstr. 012.005 for the main entry.

015.016   **Exobiological habitats: an overview.**
L. R. Doyle, C. P. McKay.
3. Symposium International de Bioastronomie, p. 163 – 172 (1991). – See Abstr. 012.005 for the main entry.
Exobiological habitats, defined here as a long–term (-10^9y) liquid water environment with available biogenic materials, can be examined from both a stellar and a planetary perspective. From the stellar perspective the solar system's galactic orbit may be fortuitous "straddling" galactic spiral arms where stellar activity is high. However, interstellar molecular clouds, possibly enhanced in spiral arms, may also be a substantial source of organics to the protoplanetary disc. Expected ecospheres around various stellar spectral types can also be examined, in light of new atmospheric greenhouse models, to re–define the evolution of liquid water habitats with stellar evolution. From the planetary perspective of the solar system, perhaps four exobiological habitats have existed. In addition to Earth, the authors examine primordial Mars (early heating via greenhouse gases), the moon Europa (tidal heating via the orbital eccentricity), and possible large cometary nuclei (radioactive heating from ^{26}Al).

015.017   **Habitable zones for Earth–like planets around main sequence stars.**
D. P. Whitmire, R. T. Reynolds, J. F. Kasting.
3. Symposium International de Bioastronomie, p. 173 – 178 (1991). – See Abstr. 012.005 for the main entry.
As stars evolve and brighten, the radial zone within which liquid water can exist at the surface of an Earth–like planet expands outward. Using a new planetary climate model the authors have calculated the evolution of these habitable zones around several main sequence stars of masses between 0.50 and 1.25 $M_\odot$. This evolution is presented in the form of a habitability continuum diagram for each star. The authors also give results for post main sequence evolution of the habitable zone around a 1.0 $M_\odot$ star. Preliminary results indicate that the range of planetary radii for which liquid water can exist for >4.5 Byr is considerably broader than previously calculated.

015.018   **The impact of technology on SETI.**
M. J. Klein, S. Gulkis.
3. Symposium International de Bioastronomie, p. 203 – 209 (1991). – See Abstr. 012.005 for the main entry.

015.019   **The SERENDIP II SETI project: observations and RFI analysis.**
C. Donnelly, S. Bowyer, W. Herrick, D. Werthimer, M. Lampton, T. Hiatt.
3. Symposium International de Bioastronomie, p. 223 – 228 (1991). – See Abstr. 012.005 for the main entry.

015.020   **SETI: on the telescope and on the drawing board.**
J. Tarter, M. J. Klein.
3. Symposium International de Bioastronomie, p. 229 – 235 (1991). – See Abstr. 012.005 for the main entry.
In 1992, NASA will initiate the observational phase of its SETI Microwave Observing Project that will continue to the end of the century. This paper discusses the searches as well as the strategies and technologies that have been suggested for the next generation of searching.

015.021   **Project of ETI signal search at the wavelength 1.47 mm.**
S. F. Likhachev, G. M. Rudnitskij.
3. Symposium International de Bioastronomie, p. 236 – 239 (1991). – See Abstr. 012.005 for the main entry.

015.022   **A SETI search technique: monitor stars to which we have sent signals.**
P. B. Boyce, L. H. Wasserman.
3. Symposium International de Bioastronomie, p. 240 – 243 (1991). – See Abstr. 012.005 for the main entry.
After the appropriate light travel time, one should monitor stars which have been illuminated by earth's high power radar systems. One knows the position of the star, the frequency and the time after which a reply, if sent, could arrive. The problem of computing which stars have been illuminated by the Arecibo planetary radar has been solved by using occultation predictions. It is possible that four stars have been illuminated. Because of planetary motion, radar signals beamed toward any given star have a maximum duration of a few hours, making verification difficult. In order for the searches to be sensitive to an alien civilization's radar signals, immediate verification of any suspicious signal is essential.

015.023   **Radio search for alien space probes.**
A. V. Arkhipov.
3. Symposium International de Bioastronomie, p. 244 – 246 (1991). – See Abstr. 012.005 for the main entry.
It is shown that the search for alien space probes by occasional interceptions of their radio communication beams appears to be a promising task not beyong the ability of amateur radio astronomers and all–sky monitoring systems.

015.024　**Karhunen–Loève versus Fourier transform for SETI.**
C. Maccone.
3. Symposium International de Bioastronomie, p. 247 – 253 (1991). – See Abstr. 012.005 for the main entry.

015.025　**On the strategy of SETI.**
L. N. Filippova, N. S. Kardashev, S. F. Likhachev, V. S. Strelnitskij.
3. Symposium International de Bioastronomie, p. 254 – 258 (1991). – See Abstr. 012.005 for the main entry.
Arguments in favor of narrow–beamed SETI signals transmitted by "civilization–senders" (CS) directly towards target stars are revealed. This consideration gives one, as a "civilization-receiver" (CR), the grounds to search near the ecliptic and around the moment when the candidate CS is in astronomical opposition with the Sun. A list of 29 candidate stars, with their dates of opposition, are presented for desirable international patrol observations.

015.026　**Pan–galactic pulse periods and the pulse window for SETI.**
W. T. Sullivan III.
3. Symposium International de Bioastronomie, p. 259 – 268 (1991). – See Abstr. 012.005 for the main entry.
It is argued that more attention should be paid in SETI programs to the possibility of finding rationalized, preferred pulse periodicities, in the same sense that many have argued for preferred frequencies. Within the range of detectable pulse periods, which is from 10^{-6} to 10^5sec, the Pulse Window, from ~ 0.1 to ~ 3.0 sec and defined by the histogram of observed pulsar periods, is suggested as a natural galactic communications channel for pulse–like signals.

015.027　**Cosmic background radiation limits for SETI.**
S. Gulkis.
3. Symposium International de Bioastronomie, p. 269 (1991). Abstract. – See Abstr. 012.005 for the main entry.

015.028　**VLBI and interstellar scattering tests for SETI signals.**
V. Slysh.
3. Symposium International de Bioastronomie, p. 270 (1991). Abstract. – See Abstr. 012.005 for the main entry.

015.029　**A test for the interstellar contact channel hypothesis in SETI.**
D. G. Blair, A. Williams, R. Norris, K. J. Wellington, A. Wright.
3. Symposium International de Bioastronomie, p. 271 – 279 (1991). – See Abstr. 012.005 for the main entry.
This paper describes a SETI research programme which commenced in 1990 in Australia. The project is intended to be a specific test for the existence of definable interstellar contact channels. The rationale for the choice of interstellar contact frequency is described.

015.030　**The SETI program of the Planetary Society.**
T. R. McDonough.
3. Symposium International de Bioastronomie, p. 280 – 282 (1991). – See Abstr. 012.005 for the main entry.

015.031　**The potential contribution of the Northern Cross radio-telescope to the SETI program.**
C. Bortolotti, A. Cattani, A. Maccaferri, S. Montebugnoli, M. Roma, N. d'Amico, G. Grueff.
3. Symposium International de Bioastronomie, p. 285 – 288 (1991). – See Abstr. 012.005 for the main entry.

015.032　**A proposal for a SETI global network.**
J. Heidmann.
3. Symposium International de Bioastronomie, p. 289 – 291 (1991). – See Abstr. 012.005 for the main entry.

015.033　**A search for Dyson spheres around late–type stars in the IRAS catalog.**
J. Jugaku, S. Nishimura.
3. Symposium International de Bioastronomie, p. 295 – 298 (1991). – See Abstr. 012.005 for the main entry.

015.034　**Gravitational, plasma, and black–hole lenses for interstellar communications.**
V. R. Eshleman.
3. Symposium International de Bioastronomie, p. 299 (1991). Abstract. – See Abstr. 012.005 for the main entry.

015.035　**SETI through the gamma–ray window: a search for interstellar spacecraft.**
M. J. Harris.
3. Symposium International de Bioastronomie, p. 300 – 305 (1991). – See Abstr. 012.005 for the main entry.
Consideration of the Fermi Paradox leads to the conclusion that SETI strategies should focus on the detection of interstellar spacecraft. Although current concepts for such spacecraft are highly speculative, two very general distinctive features can be identified. First, autonomous propulsion on rocket principles is limited to a nuclear fusion or antimatter annihilation power source, both of which will emit γ–rays. Second, interstellar spacecraft are identified most readily by their very higher proper motions. The strategy is therefore to search for γ–ray sources with large proper motions. A variant of this method involves searching for transient sources which fall along a straight line in space. The limitations due to the poor measurement of γ–ray source positions are emphasized.

015.036　**Criteria of artificiality in SETI.**
V. V. Rubtsov.
3. Symposium International de Bioastronomie, p. 306 – 310 (1991). – See Abstr. 012.005 for the main entry.

015.037　**SETI searches with the 70 m SUFFA radio telescope.**
N. S. Kardashev.
3. Symposium International de Bioastronomie, p. 344 – 345 (1991). – See Abstr. 012.005 for the main entry.

015.038　**Strategy of the mutual search for civilizations by means of probes.**
U. N. Zakirov.
3. Symposium International de Bioastronomie, p. 346 (1991). Abstract. – See Abstr. 012.005 for the main entry.

015.039　**From the physical world to the biological Universe: historical developments underlying SETI.**
S. J. Dick.
3. Symposium International de Bioastronomie, p. 356 – 363 (1991). – See Abstr. 012.005 for the main entry.
More than thirty years ago the French historian of science Alexandre Koyré (1957) wrote his classic volume, "From the Closed World to the Infinite Universe", in which he argued that a fundamental shift in world view had taken place in 17th century cosmmology. Between Nicholas of Cusa in the fifteenth century and Newton and Leibniz in the seventeenth, he found that the very terms in which humans thought about their universe hat changed. These changes he characterized broadly as the destruction of the closed finite cosmos and the geometrization of space. The author argues that the SETI endeavor represents a test for a similar fundamental shift in cosmological world view, from the physical world to the biological universe.

015.040　**Chemical studies on the existence of extraterrestrial life.**
C. Ponnamperuma, Y. Honda, R. Navarro–González.
J. Br. Interplanet. Soc., Vol. 45, No. 6, p. 241 – 249 (Jun 1992). Special issue: Exobiology. S. Santoli (ed.).

015.041　**Life around a larger sun.**
N. F. Comins.
Astronomy, Vol. 20, No. 5, p. 50 – 55 (May 1992).

015.042   Comets and meteorites were a minor source of prebiotic organic compounds on the early Earth.
S. L. Miller.
Bull. Am. Astron. Soc., Vol. 23, No. 3, p. 1172 (1991). Abstract. – See Abstr. 012.037 for the main entry.

015.043   Sources of organic material for the origin of life on Earth.
C. F. Chyba.
Bull. Am. Astron. Soc., Vol. 23, No. 3, p. 1172 (1991). Abstract. – See Abstr. 012.037 for the main entry.

015.044   Tunnels through time.
B. Parker.
Astronomy, Vol. 20, No. 6, p. 28 – 35 (Jun 1992).

015.045   Astronomical light pollution by artificial earth satellites.
E. Fosbury, A. Turtle, M. Black.
Messenger, No. 67, p. 53 – 56 (Mar 1992).

015.046   On the life expectancy of astronomers.
D. B. Herrmann.
Messenger, No. 67, p. 62 – 63 (Mar 1992).

015.047   The process of discovery: supernovae, comets, and extraterrestrial life.
R. F. Garrison.
Vistas Astron., Vol. 35, Part 1, p. 73 – 80 (Jan 1992).
   The three topics in the subtitle have been the subjects of many of the public lectures the author has given during the past twenty years at the University of Toronto. The Jacob Bronowski Memorial Lecture (Univ. Toronto, 9 Nov. 1988) gives occasion for some philosophical thoughts on the process of discovery in science.

015.048   Astrophysical constraints on planetary environments.
L. Doyle, W. Davis, C. McKay, R. Reynolds, D. Whitmire, J. Matese.
Bull. Am. Astron. Soc., Vol. 23, No. 3, p. 1234 (1991). Abstract. – See Abstr. 012.037 for the main entry.

015.049   Dans le procès de l'astrologie, le rationalisme est–il tout à fait rationnel?
C. Maillard.
Publ. Obs. Astron. Strasbourg, Sér. Astron. Sci. Hum., No. 6, p. 35 – 48 (1992).

015.050   Commentaire sur l'exposé de M. Maillard.
H. Andrillat.
Publ. Obs. Astron. Strasbourg, Sér. Astron. Sci. Hum., No. 6, p. 49 – 52 (1992).

015.051   Search for life beyond Earth.
R. K. Khanna, C. Ponnamperuma, R. Navarro–González.
AIP Conf. Proc., No. 245: Workshop on Basic Space Science, p. 279 – 290 (1992). – See Abstr. 012.064 for the main entry.
   A brief account of our understanding of prebiotic conditions which may have existed for the origin of life on Earth are presented. The production of key molecules, HCN and HCHO, from less reducing atmospheric conditions is not extremely efficient. Their incorporation into the Earth from cometary impacts needs to be given serious consideration.

015.052   Light–pollution prevention ordinance in the town of Bisei.
S. Isobe, N. Sugihara.
Proc. Astron. Soc. Aust., Vol. 9, No. 2, p. 336 – 337 (1991). – See Abstr. 012.090 for the main entry.
   The authors discuss the light–pollution prevention ordinance passed by the Japanese town of Bisei in 1989.

Understanding catastrophe.
See Abstr. 003.045.

The book of nature.
See Abstr. 003.080.

Bioastronomy. The search for extraterrestrial life – the exploration broadens. Proceedings.
See Abstr. 012.005.

Searching for planetary systems.
See Abstr. 013.067.

The search for extraterrestrial intelligence.
See Abstr. 014.070.

Hydrogen cyanide polymerization: a preferred cosmochemical pathway.
See Abstr. 022.101.

The anthropic principle and nucleosynthesis.
See Abstr. 061.039.

Bursts of star formation in the local galactic disk and their implications for the origin and evolution of life around the Sun and nearby stars.
See Abstr. 155.112.

Applied Mathematics, Physics

021 Mathematical Papers Related to Astronomy and Astrophysics, Computing

021.001 A search for a quantitative comparison of galaxy clustering algorithms.
A. M. Garcia, V. Morenas, G. Paturel.
Astron. Astrophys., Vol. 253, No. 1, p. 74 – 76 (Jan 1992).
Several algorithms exist to generate groups or clusters of galaxies from a sample of individual objects. The authors propose an estimate to compare the results obtained in different ways. This estimate is expressed as a mathematical distance between two classifications. One example is given to study the group–membership stability in each of the two cases: (i) when a single algorithm is applied with different parameters (ii) when two different algorithms are applied.

021.002 On an alternative method for approximate solution of linear stochastic differential equations.
J. Stahlberg.
Astron. Nachr., Vol. 313, No. 1, p. 53 – 56 (1992).
An approximate method for solving formal linear stochastic differential equations of first order is proposed. On the basis of the Reynold averaging technique, the stochastic differential equation is transformed into an infinite hierarchical system. This infinite system is cut off in such a way that uncorrelated (totally randomly) and totally correlated stochastic processes are exactly included. By this the applicability of the Reynold method is extended.

021.003 Minimax property of a discrete Kalman filter for imprecisely specified variance matrices.
A. I. Rusakov, I. K. Kokhanenko.
Cosmic Res., Vol. 29, No. 2, p. 169 – 174 (Sep 1991). English translation of Kosm. Issled., Tom 29, Vyp. 2, p. 194 – 200 (1991).
The deviation of the nominal variance matrices used in the discrete Kalman filter algorithm from the true variance matrices lowers the accuracy of estimation of the state vector. It is shown that when the actual variance matrices are majorized by standard matrices, the use of the latter for filtering will provide the best guaranteed value of the quadratic performance index among all linear estimators. A relation is derived as a characteristic of the reduction in the index of the error of estimation as a result of the transition to guaranteed variance matrices. The results of calculations are given in connection with the identification of the parameters of motion of an artificial earth satellite in a near–circular orbit.

021.004 Special–purpose computers in astronomy and physics – how to get maximum results with minimum effort.
J. Makino.
Astron. Her., Vol. 85, No. 4, p. 142 – 147 (1992). In Japanese.

021.005 Some representations of unified Voigt functions $\Omega^\mu_{\eta,v,\lambda}(x,y)$.
A. Siddiqui, S. M. Uppal.
Astrophys. Space Sci., Vol. 188, No. 1, p. 1 – 8 (Feb 1992).
This paper aims at some representation of unified Voigt functions which play a rather important role in several diverse fields of physics such as astrophysical spectroscopy and the theory of neutron reactions. The authors derive several representations of $\Omega^\mu_{\eta,v,\lambda}(x,y)$ in terms of series and integrals which are especially useful in situations when the parameters μ, η, v, and λ take on particular values.

021.006 A smoothing scheme for reducing field anisotropy in N–body simulation.
W. Y. Chau, S. S. Mak, Y. W. Law, K. L. Chan.
Astrophys. Space Sci., Vol. 188, No. 1, p. 159 – 164 (Feb 1992).
A method is proposed to reduce the field anisotropy in N–body calculations arising from the discretization of Poisson's equation in a particle–mesh scheme. Specifically, the mass is split (in addition to that employed in the mass assignment algorithm) over neighbouring points to create compensating higher–order multipoles. Improvement by factor 2 has been observed in the field calculations, and spurious geometric effects in the example of a 4000 particle free–fall collapse have been reduced.

021.007 Gamma–ray burst astrometry II: numerical tests.
L. G. Taff, S. T. Holfeltz.
Compton Observatory Science Workshop, p. 301 – 308 (Feb 1992). – See Abstr. 012.018 for the main entry.
Since the announcement of the discovery of sources of bursts of gamma–ray radiation in 1973, many more reports of such bursts have been published. Numerous artificial satellites have been equipped with gamma–ray detectors including GRO. Unfortunately, one has made almost no progress in identifying the sources of this high energy radiation. Only one visible counterpart is known. The authors suspect that this is a consequence of the methods currently used to define gamma–ray burst source "error boxes". An alternative procedure was proposed in 1988 by Taff. In this paper the authors report on Monte Carlo simulations of the efficacy of this technique using realistic burst timing uncertainties and satellite location errors as well as a variety of satellite constellations. The results clearly show that an arc minute prediction of a unique burst location is routinely obtainable once there are at least two interplanetary detectors.

021.008 Symplectic integrators for Hamiltonian systems: basic theory.
H. Yoshida.
IAU Symposium No. 152: Chaos, resonance and collective dynamical phenomena in the solar system, p. 407 – 411 (1992). – See Abstr. 012.025 for the main entry.
Symplectic integrators are numerical integration methods for Hamiltonian systems, which conserves the symplectic 2–form exactly. With use of symplectic integrators there is no secular increase in the error of the energy because of the existence of a conserved quantity closed to the original Hamiltonian. Higher order symplectic integrators are obtained by a composition of 2nd order ones.

021.009   From photons to hadrons to galaxies – how to analyze the texture of matter distributions.
P. Carruthers.
Acta Phys. Pol., B, Vol. 22, No. 11–12, p. 931 – 953 (Nov–Dec 1991). Paper presented at the 31. Cracow School of Theoretical Physics: Particle physics at high energy, Zakopane (Poland), 4 – 14 Jun 1991.

The author discusses the phenomenology of point matter distributions, using examples of photoelectron count arrival times, galaxy distributions and multihadron production at high energies. Count distributions in phase space cells, multiplicity moments and correlation functions are discussed along with their interconnections. The possible description of higher order cumulant correlations (both for hadrons and galaxies) by linked two–particle cumulants and negative binomial coefficients, is reviewed.

021.010   Estimation of the number of extrema of a spherical harmonic.
Eh. D. Kuznetsov, K. V. Kholshevnikov.
Astron. Zh., Tom 69, Vyp. 2, p. 439 – 442 (Mar–Apr 1992). In Russian. English translation in Sov. Astron., Vol. 36, No. 2 (1992).

The maximum possible number of stationary points $2n^2-2n+2$ and extrema n^2-n+2 for restriction of a polynomial of degree n to the unit sphere is obtained. The same is valid for a spherical harmonic of degree n.

021.011   Estimation of the number of extrema of a spherical harmonic.
Eh. D. Kuznetsov, K. V. Kholshevnikov.
Sov. Astron., Vol. 36, No. 2, p. 220 – 222 (Mar 1992). Current Physics Microform No.: 9208X2056. English translation of Astron. Zh., Tom 69, Vyp. 2, p. 439 – 442 (1992).

The authors find the maximum possible number of stationary points $(2n^2-2n+2)$ and extrema (n^2-n+2) when a polynomial of degree n is restricted to the unit sphere. The same numbers hold for a spherical harmonic of degree n.

021.012   On some generalizations of the manifold concept.
M. Heller, P. Multarzyński, W. Sasin, Z. Żekanowski.
Acta Cosmologica, Fasc. 18, p. 31 – 44 (1992).

Some generalizations of the smooth manifold concept, currently known in the mathematical literature, are critically reviewed. Mutual dependencies between these concepts are analysed, and the results expressed in the form of propositions.

021.013   Numerical simulators of stability in astrophysical problems.
J. N. Tokis.
Astrophys. Space Sci., Vol. 191, No. 1, p. 131 – 135 (May 1992).

The aim of this paper is to discuss systematically two numerical methods that can play a significant role of "numerical simulators of stability" in the corresponding astrophysical problems to which they are applied.

021.014   Time series analysis in astronomy: an application to quasar variability studies.
R. Vio, S. Cristiani, O. Lessi, A. Provenzale.
Astrophys. J., Vol. 391, No. 2, p. 518 – 530 (1 Jun 1992).

Many astrophysical objects can be associated with continuous non–linear stochastic systems. Up to now, however, the signals coming from these systems have been analyzed mainly through linear approaches, such as the power spectrum (PS) and the structure function (SF) techniques, frequently with controversial results. In this paper the authors show that in the study of astronomical time series the PS (even in the maximum entropy or CLEAN version) and the SF techniques are often of little use or even misleading, since they do not take into account all the information contained in the data. New techniques, such as the bispectrum and multifractal analyses, are necessary to gain further insight into the dynamics of these systems. The effects of the discrete sampling of the signal are also considered, and the evolution of a single system and that of a multiple subunit system

are distinguished. The approach based on phase–space reconstruction algorithms and on dimension and entropy estimators as derived from the theory of nonlinear dynamical systems is discussed as well. As example, an analysis of the X–ray light curve of the BL Lac object PKS 2155–304 and of the optical light curve of the quasar 3C 345 is presented.

021.015   Some unified presentations of the Voigt functions.
H. M. Srivastava, Chen Mingpo.
Astrophys. Space Sci., Vol. 192, No. 1, p. 63 – 74 (Jun 1992).

The principal object of this note is to provide a natural further step toward the unified presentations of the Voigt functions $K(x,y)$ and $L(x,y)$ which play a rather important role in such diverse fields of physics as astrophysical spectroscopy and the theory of neutron reactions. Explicit representations for these functions, given in terms of some relatively more familiar special functions of one and two variables, are potentially useful in finding many other needed (numerical or analytical) properties of the Voigt functions. Several erroneous recent contributions to the theory of Voigt functions, including (for example) the main result of A. Siddiqui (1990), are also corrected here.

021.016   On the periodic solutions of a particular Lamé equation.
M. Amar.
Celest. Mech. Dyn. Astron., Vol. 52, No. 4, p. 397 – 406 (1991).

The author considers the Hill's equation: $d^2\xi/dt^2 + m(m+1)/2\,C^2(t)\xi = 0$, where $C(t) = Cn(t,1/\sqrt{2})$ is the elliptic function of Jacobi and m a given real number. It is a particular case of the Lamé equation. By a change of variable it is transformed to the Ince equation: $(1+a\cos(2\Phi))y'' + b\sin(2\Phi)\,\dot{y} + (c+d\cos(2\Phi))y = 0$ where $a = -b = 1/3$, $c = d = m(m+1)/3$. In the neighourhood of the poles, the author gives the expression of the solutions. The periodic solutions of the former equation correspond to the periodic solutions of the latter equation. Magnus and Winkler give a theory of their existence. By comparing these results to those of the author's study in the case of the Hill's equation, it is possible to find the development in Fourier series of periodic solutions in function of the variable Φ and to deduce the development of solutions of the above equation in function of $C(t)$.

021.017   Numerical integration methods for orbital motion.
O. Montenbruck.
Celest. Mech. Dyn. Astron., Vol. 53, No. 1, p. 59 – 69 (1992).

The present report compares Runge–Kutta, multistep and extrapolation methods for the numerical integration of ordinary differential equations and assesses their usefulness for orbit computations of solar system bodies or artificial satellites. The scope of earlier studies is extended by including various methods that have been developed only recently. Several performance tests reveal that modern single– and multistep methods can be similarly efficient over a wide range of eccentricities. Multistep methods are still preferable, however, for ephemeris predictions with a large number of dense output points.

021.018   The problem on step–size change of symplectic integrator.
Zhao Zhangyin, Liu Lin.
Publ. Purple Mt. Obs., Vol. 10, No. 3, p. 196 – 203 (Sep 1991). In Chinese.

The comparison between symplectic difference schemes and Runge–Kutta difference schemes is done and a variable step–size sympelctic integrator, which is similar to RKF algorithm, is presented. The advantage in efficiency of this algorithm is verified by practical computations.

021.019   On a Hermite integrator with Ahmad–Cohen scheme for gravitational many–body problems.
J. Makino, S. J. Aarseth.
Publ. Astron. Soc. Jpn., Vol. 44, No. 2, p. 141 – 151 (1992).

The authors describe the implementation of the Ahmad–Cohen scheme based on a fourth–order Hermite integrator. With the fourth–order Hermite scheme, the authors calculate the force and the time derivative of the force analytically, and construct a third–order interpolation polynomian using two points in time. Compared with the standard scheme which is widely used, it allows a longer stepsize for the same accuracy, and the program is much simpler. In the case of the Ahmad–Cohen scheme, which uses different stepsizes for the forces from neighboring particles and that from distant particles, the difference in the programming complexity is even larger, since the Hermite scheme does not require corrections of the higher order divided differences for the forces from distant particles. On scalar computers the Hermite schemes are marginally faster than the standard scheme for the same level of accuracy, both with and without the Ahmad–Cohen scheme. On vector machines or special–purpose hardware, such as GRAPE, the Hermite scheme would be significantly faster since the number of scalar operation is much smaller. The gain in computing speed using the Ahmad–Cohen scheme is $(N/3.8)^{1/4}$ for both the standard and Hermite schemes, where N is the total number of particles. However, this gain can be significantly smaller on vector or parallel machines.

021.020   Orbital evolution of dust particles from comets and asteroids.
A. A. Jackson, H. A. Zook.
Icarus, Vol. 97, No. 1, p. 70 – 84 (May 1992).

In a computer simulation, dust grains of radius 10, 30, and 100 μm were released at perihelion passage from each of 35 different celestial bodies: 15 main belt asteroids, 15 short period comets with perihelion greater than 1 AU, and 5 short period comets with perihelion less than 1 AU. The evolving orbit of each of the 105 released dust grains was then continuously computed until the orbit aphelion passed inside of 0.387 AU, or the dust grain had been ejected from the Solar System. The forces due to the gravity of the Sun and the planets as well as radiation pressure, Poynting–Robertson drag, and solar wind drag were all included in these numerical simulations. It is found that when dust grains evolve to intersection with the Earth's orbit, they nearly always retain orbital characteristics indicative of their origins. The results mean that accurate trajectory measurements of meteoroids collected with a near–Earth space platform would make it possible to distinguish asteroidal grains from cometary grains.

021.021   Monte Carlo methods.
J.–M. Petit.
15. Ecole de Printemps d'Astrophysique de Goutelas: Interrelations between physics and dynamics for minor bodies in the solar system, p. 599 – 629 (1992). – See Abstr. 012.055 for the main entry.

Contents: 1. Introduction. 2. Globular clusters. 3. Philosophy of the Monte Carlo approach. 4. Detailed description (Superstars. Initial conditions. Potential. Time step. Encounters. New positions. Selection of next pair). 5. Planetary rings (Selection of interacting particles. Dimensionless variables. Outcome of an interaction. Algorithm). 6. Fokker–Planck equation (Derivation of the equation. Asymptotic solution for identical particles).

021.022   Basic research in mathematical and space sciences.
A. M. Mathai.
AIP Conf. Proc., No. 245: Workshop on Basic Space Science, p. 201 – 205 (1992). – See Abstr. 012.064 for the main entry.

Contents: 1. Introduction. 2. G–function and the study of nuclear reaction rates. 3. Generalized special functions and stellar models. 4. Conclusion.

021.023   The solution topology of the Eddington quartic–like equation $Y = \alpha X^4 + X - 1$.
M. Beech.
Astrophys. Space Sci., Vol. 192, No. 2, p. 329 – 334 (Jun 1992). Letter–to–the–editor.

The root structure of the Eddington quartic equation is examined in a general manner. The complex and real root topology is annotated, and an explicit expression for the real positive root is derived.

021.024   Sharpening Occam's Razor on a Bayesian strop.
W. H. Jefferys, J. O. Berger.
Bull. Am. Astron. Soc., Vol. 23, No. 3, p. 1259 (1991). Abstract. – See Abstr. 012.065 for the main entry.

021.025   Memorandum on basic upwind schemes for compressible flows.
K. Fujii.
Symposium of Supercomputing Astronomy and Astrophysics in Japan, p. 1 – 9 (1991). – See Abstr. 012.069 for the main entry.

021.026   Numerical simulations of two–dimensional and three–dimensional accretion flows.
T. Matsuda, T. Ishii, N. Sekino, K. Sawada, E. Shima, M. Livio, U. Anzer.
Symposium of Supercomputing Astronomy and Astrophysics in Japan, p. 11 – 25 (1991). With 2 plates. – See Abstr. 012.069 for the main entry.

Numerical simulations of two–dimensional (2D) as well as three–dimensional (3D) accretion flows on to a gravitating compact object from a uniform flow at far upstream are performed by solving Eulerian equations. The authors assume that gas is adiabatic except at shocks. 2D flows exhibit a "flip-flop instability" and the shock cone oscillates aperiodically, if the central accreting body is small. If the central body is enlarged at a some instance in the flip–flopping flow, then the accretion shock shows a rather periodic oscillation similar to the von Karman vortex street. In the case of 3D flow, it is found that the shock cone is much more robust than in 2D, and the flip–flop instability takes a different, probably less violent form in 3D. The causes of the instabilities are discussed.

021.027   Smoothed particle rendering for fluid visualization – three–dimensional accretion disk and jet formation.
M. Nagasawa, T. Matsuda, K. Kuwahara.
Symposium of Supercomputing Astronomy and Astrophysics in Japan, p. 27 – 42 (1991). – See Abstr. 012.069 for the main entry.

The smoothed particle method is applied to the radiative transfer problem for the direct visualization of 3–D scalar fields. The Smoothed Particle Rendering (SPR) integrates the ray equation through the opaque medium and calculates the global contribution of scattered light. The opacity represents the density scalar. The emissivity and the flux direction are derived by the temperature field and its gradient. This method has some common features with Voxel Volume Rendering and Radiosity Method and is applicable both to grid data and to particle configurations. The validity of SPR indicates the possibility to simulate the radiation hydrodynamics. The authors present full the three–dimensional hydrodynamic calculations of gas flow around a compact object by Smoothed Particle Hydrodynamics. In order to investigate the tidal effect, Roche–lobe overflow in a semi–detached close binary system is considered.

021.028   Unrivaled monster TTTP: a computer system composed of massively parallel specialized units.
T. Ebisuzaki, J. Makino, S. K. Okumura, Y. Chikada.
Symposium of Supercomputing Astronomy and Astrophysics in Japan, p. 233 – 235 (1991). – See Abstr. 012.069 for the main entry.

The authors propose a computer system composed of massively parallel specialized units as a next generation architecture of a computer for astronomical use. A massively parallelized machine can achieve very high performance when it is specialized to a single problem. By combining specialized machines, the authors construct a high–speed multi–purpose machine. TTTP aims at: Tera flops: computational speed; Tera bytes:main storage; Tera bit/s: communication speed within the machine; Peta bit: external storage.

021.029   DREAM: a special–purpose computer for large scale mesh calculations.
J. Makino, T. Ebisuzaki, S. Okumura.
Symposium of Supercomputing Astronomy and Astrophysics in Japan, p. 237 – 246 (1991). – See Abstr. 012.069 for the main entry.

The authors present the concept of DREAM (Disk REsouce Array Machine), a computer designed for large scale simulations using finite–difference grids. The DREAM system uses an array of magnetic disks as the main memory, instead of silicon solid–state memory chips. Since the price per bit of a magnetic disk is about 1/100 of that of a silicon memory, it is possible to construct a machine with 100 times larger memory for the same cost. The access time of a disk is about 10^5 time longer than that of a silicon memory. In the case of finite–difference calculations using regular grids, however, this access time becomes negligible since the memory is accessed only in long vectors. The authors also discuss the reliability of the hardware and the possibility to apply DREAM to other fields such as the reduction of the large observational data.

021.030   A universal solver both for compressible and incompressible fluid by cubic interpolation.
T. Yabe.
Symposium of Supercomputing Astronomy and Astrophysics in Japan, p. 247 – 265 (1991). – See Abstr. 012.069 for the main entry.

A universal numerical solver commonly usable for compressible and incompressible fluid is proposed. The method approaches the MAC algorithm at very high sound speed and continuously approaches the algorithm for compressible fluid with decreasing sound speed. Advection term is treated by the CIP algorithm which was previously proposed. A single program is applied to one– and two–dimensional shock–tube problems, and two–dimensional liquid flow inside a cavity at high–Reynolds number. The program is used to examine the Rayleigh–Taylor instability in two and three dimensions as application to inertial confinement fusion and supernova explosions. The mushroom structure owing to the Kelvin–Helmholtz instability in three dimensions is much smaller and hence the nonlinear growth of the R–T instability is faster in three dimensions.

021.031   Integration of the collisionless Boltzmann equation – the tracing back method.
T. Fujiwara.
Symposium of Supercomputing Astronomy and Astrophysics in Japan, p. 327 – 331 (1991). – See Abstr. 012.069 for the main entry.

A numerical method for studying collisionless stellar systems has been examined. The method exploits Liouville's theorem directly: the trajetories of phase particles are traced back to their initial positions to determine the evolution of the distribution function in phase space. Numerical results are presented, and the advantages and disadvantages of the method are discussed.

021.032   Simulations of globular cluster evolution.
S. Inagaki.
Symposium of Supercomputing Astronomy and Astrophysics in Japan, p. 333 – 349 (1991). – See Abstr. 012.069 for the main entry.

In this paper the author reviews numerical methods for the simulations of dynamical evolution of globular clusters and next he reports some recent results of simulations.

021.033   A FORTRAN program for multiple–period analysis.
Hao Jinxin.
Publ. Beijing Astron. Obs., No. 18, p. 35 – 42 (Oct 1991).

021.034   Length of arc as independent argument for highly eccentric orbits.
E. V. Brumberg.
Celest. Mech. Dyn. Astron., Vol. 53, No. 4, p. 323 – 328 (1992).

For analytic step regulation in numerical integration of highly eccentric orbits it is proposed to use the orbital arc length of a moving particle as independent argument.

021.035   On the efficiency of Runge–Kutta–Nystrom methods with interpolants for solving equations of the form $y'' = f(t,y,y')$ over short timespans.
C. Tsitouras, G. Papageorgiou, T. Kalvouridis.
Celest. Mech. Dyn. Astron., Vol. 53, No. 4, p. 329 – 346 (1992).

Runge–Kutta–Nystrom codes for the solution of the initial value problem for the general second order differential system have been developed recently, although the methodology on which they are based was known many years ago. In this paper the authors try to examine the efficiency of several known general Runge–Kutta–Nystrom methods by posing some criteria of cost and accuracy. These methods supplied with the corresponding interpolants, have been applied to some problems of Celestial Dynamics. The results obtained show that these codes have a good response in the approximation of the solution of these problems.

021.036   Increased accuracy of computations in the main satellite problem through linearization methods.
J.–M. Ferrándiz, M.–E. Sansaturio, J. R. Pojman.
Celest. Mech. Dyn. Astron., Vol. 53, No. 4, p. 347 – 363 (1992).

The set of canonical redundant variables previously introduced by the first author is derived from Cartesian coordinates in a simplified form which allows the reduction of the Kepler problem to four harmonic oscillators with unit frequency. The coordinates are defined to be the direction cosines of the position of the particle along with the inverse of its distance. True anomaly is the new independent variable. The behavior of this new transformation is studied when applied to the numerical integrations of the main problem in satellite theory. In particular, computation time and accuracy of orbits in the new variables are compared with those in K–S and Cartesian variables. It is noteworthy that for high eccentricities the new variables require the least computation time for comparable accuracy, regardless of the integration scheme.

021.037   Stochasticity of motions and the structure of directions field in the Henon – Heiles model.
T. A. Agekyan, A. A. Myullyari, V. V. Orlov.
Astron. Zh., Tom 69, Vyp. 3, p. 469 – 478 (May – Jun 1992). In Russian. English translation in Sov. Astron., Vol. 36, No. 3 (May – Jun 1992).

A transition from regularity to stochasticity in the Hénon – Heiles (1964) model is investigated. A new method for study of the stochasticity origin and development is suggested. The method is based on constructing the contours of the orbit and folds of directions fields; it is more informative and obvious in comparison with the classical methods using the Poincaré maps and the Lyapunov characteristic numbers. Scanning the region of initial conditions in the Hénon – Heiles model allows to outline some zones of the stochasticity and regularity. As for the ordered and semi–stochastic orbits, the contours of the orbit and folds are smooth. The origin of the stochasticity is connected with an appearance of the first folds in the directions field. The growth of stochasticity is caused by an increase of the size and multiplicity of the folds. The Hausdorff dimension of the set of points at the orbit and folds contours is suggested as a stochasticity measure.

Introduction to Hamiltonian dynamical systems and the N–body problem.
See Abstr. 003.019.

Supercomputing astronomy and astrophysics. Proceedings.
See Abstr. 012.069.

Astrophysical use of the principal component analysis of imperfect data.
See Abstr. 036.058.

A topological / geometrical approach to the study of astrophysical maps.
See Abstr. 036.059.

Revision of the Theil fitting method.
See Abstr. 036.120.

Collisional simulations of isolated Lindblad resonances.
See Abstr. 042.026.

Numerical simulations of dense collisional systems with extended distribution of particle sizes.
See Abstr. 042.027.

A collisional and self–gravitational model to simulate numerically the dynamics of planetary disks.
See Abstr. 042.028.

New methods for long–time numerical integration of planetary orbits.
See Abstr. 042.055.

On basic families of three–dimensional periodic orbits of three massive bodies and their stability.
See Abstr. 042.063.

A note on the Hori–Lie perturbation technique.
See Abstr. 042.064.

Expression of the initial Poincaré canonical variables as functions of the new in a ninth order J–S theory.
See Abstr. 042.065.

Conformal geometry of the Kepler orbit space.
See Abstr. 042.066.

On a question of A. Wintner about Jacobi coordinates in the three–body problem.
See Abstr. 042.067.

Coordinates for perturbed Keplerian systems with axial symmetry.
See Abstr. 042.069.

On a class of variational equations transformable to the Gauss hypergeometric equation.
See Abstr. 042.071.

Analytical expansions of torque–free motions for short and long axis modes.
See Abstr. 042.128.

Motion of artifical satellites in the set of Eulerian redundant parameters. III.
See Abstr. 052.013.

ZEUS–2D: a radiation magnetohydrodynamics code for astrophysical flows in two space dimensions. I. The hydrodynamic algorithms and tests.
See Abstr. 062.154.

ZEUS–2D: a radiation magnetohydrodynamics code for astrophysical flows in two space dimensions. II. The magnetohydrodynamic algorithms and tests.
See Abstr. 062.155.

ZEUS–2D: a radiation magnetohydrodynamics code for astrophysical flows in two space dimensions. III. The radiation hydrodynamic algorithms and tests.
See Abstr. 062.156.

Mixing in ejecta of supernovae. I. General properties of 2–D Rayleigh–Taylor instabilities and mixing depth in ejecta of supernovae.
See Abstr. 062.194.

The elimination of the 1:2 critical terms of a first order theory of Uranus perturbed by Neptune through Von Zeipel method.
See Abstr. 101.057.

A planet in the inner clearing zone of β–Pic disk?
See Abstr. 112.098.

Comoving frame calculations for λ–Cephei.
See Abstr. 112.111.

A general analytical solution to the problem of Malmquist bias due to lognormal distance errors.
See Abstr. 160.096.

On the error estimates of correlation functions.
See Abstr. 160.105.

022 Physical Papers Related to Astronomy and Astrophysics

022.001   **The millimeter–wave spectrum of the CaOH radical (X $^2\Sigma^+$).**
L. M. Ziurys, W. L. Barclay Jr., M. A. Anderson.
Astrophys. J., Lett., Vol. 384, No. 2, p. L63 – L66 (10 Jan 1992).
  The pure rotational spectrum of the calcium hydroxide radical (CaOH) in its X $^2\Sigma^+$ ($v = 0$) ground state has been observed in the laboratory using millimeter / submillimeter direct absorption spectroscopy. CaOH was generated by the reaction of hydrogen peroxide with calcium vapor. Eleven rotational transitions of the species were measured in the range 80 – 320 GHz to an accuracy of ± 50 kHz. The spin–rotation splitting was readily observed in these data. The magnetic hyperfine structure due to the proton spin, however, could no be resolved. The rotational and spin–rotation constants of CaOH were determined from a nonlinear least–squares fit to the data, using a $^2\Sigma$ Hamiltonian. These laboratory measurements allow for an in–depth radio astronomical search for CaOH in interstellar clouds and circumstellar envelopes of late–type stars.

022.002   **Spectral lines in the beryllium sequence.**
W. B. Eissner, J. A. Tully.
Astron. Astrophys., Vol. 253, No. 2, p. 625 – 631 (Jan 1992).
  The Einstein A coefficients which Nussbaumer calculated two decades ago for 10 Be–like ions are compared with recent data from the Opacity Project. The overall agreement is good except for the 2–electron transitions $2p^2\ ^{1,3}L^e \leftarrow 2s\ 3p\ ^{1,3}p^0$. A careful analysis of the early work allows to elucidate this discrepancy. The authors also demonstrate the increasing influence of relativistic effects as one moves up the iso–electronic sequence towards iron. They correct the published A values that are in

error and present new results for some other transitions in the sequence.

022.003   Stark shifts of singly–ionized nitrogen spectral lines.
S. Djeniže, A. Srećković, J. Labat.
Astron. Astrophys., Vol. 253, No. 2, p. 632 – 634 (Jan 1992).
Stark shifts and widths measurements of fifteen N II spectral lines (of 3s–3p, 3p–3d and 3d–4f transition arrays) have been performed at the electron density of $1.6 \times 10^{23} m^{-3}$ and for the electron temperature of 31000K. Stark parameters have been measured in a linear pulsed arc operating in nitrogen.

022.004   The fundamental constants of physics and spectroscopy.
B. W. Petley.
Phys. Scr. T, Vol. T40, p. 5 – 14 (1992). Paper presented at the 23. Conference of the European Group for Atomic Spectroscopy (EGAS), Toruń (Poland), 9 – 12 Jul 1991. S. Łegowski (ed.). ISBN 91–87308–84–3.
The paper discusses some of the recent measurements of the fundamental physical constants and their "constancy" (particularly of those measurements relating to Lamb shifts, the fine structure constant, the Planck constant, the Rydberg constant, and the speed of light), and their implications for physics and spectroscopy.

022.005   Shock–induced transformations in the system NaAl-SiO$_4$–SiO$_2$: a new interpretation.
T. Sekine, T. J. Ahrens.
Phys. Chem. Miner., Vol. 18, No. 6, p. 359 – 364 (Feb 1992).
New internally consistent interpretations of the phases represented by the high pressure phase shock wave data for an albite-rich rock, jadeite, and nepheline in the system $NaAlSiO_4$–SiO_2, are obtained using the results of static high pressure investigations, and the recent discovery of the hollandite phase in a shocked meteorite. The authors conclude that nepheline transforms directly to the calcium ferrite structure, whereas albite transforms possibly to the hollandite structure.

022.006   Level populations for Fe III applicable to astrophysical plasmas and a comparison with planetary nebula observations.
F. P. Keenan, K. A. Berrington, P. G. Burke, C. J. Zeippen, M. Le Dourneuf, R. E. S. Clegg.
Astrophys. J., Vol. 384, No. 1, p. 385 – 389 (1 Jan 1992).
Recent **R**–matrix calculations of electron impact excitation rates in Fe III, which are significantly different from the earlier results of Garstang et al., are used to derive relative populations for the 17 fine–structure levels in the ^5D, ^3P, ^3H, ^3F, and ^3G states of the $3d^6$ configuration. Populations are presented for a wide range of electron temperatures (T_e = 5000 – 20,000K) and densities (N_e = 10^2– $10^9 cm^{-3}$) applicable to astrophysical plasmas. A comparison of theoretical emission–line ratios generated using these results with observational data for the planetary nebulae DDDM–1, Vy 2–2, and NGC 7027 reveals general agreement between theory and observation, with discrepancies that average only 10%. In addition, the present calculations remove the disagreement found between theory and observations for the I(4881 Å)/I(4658 Å) line intensity ratio in DDDM–1 when the theoretical ratios of Garstang et al. are adopted.

022.007   Hydrogen molecules and chains in a superstrong magnetic field.
Lai Dong, E. E. Salpeter, S. L. Shapiro.
Phys. Rev. A, Vol. 45, No. 7, Part B, p. 4832 – 4847 (1 Apr 1992). Current Physics Microform No.: 9205G1466.
The authors study the electronic structures of hydrogen polymolecules H_n (n = 2, 3, 4,...) in a superstrong magnetic field ($B \gtrsim 10^{12}$G) typically found on the surface of a neutron star. Simple analytical scaling relations for several limiting cases are derived. The authors numerically calculate the binding energies of H_n molecules for various magnetic–field strengths. For a given magnetic–field strength, the binding energy per atom in the H_n molecule is found to approach a constant value as n increases. They also consider the structure of negative H ions in a high

magnetic field. For $B \sim 10^{12}$G the dissociation energy of an atom in a hydrogen chain and the ionization potential of H^- are smaller than the ionization potential of neutral atomic hydrogen.

022.008   Mössbauer spectroscopy on the surface of Mars. Why?
J. M. Knudsen, M. B. Madsen, M. Olsen, L. Vistisen, C. B. Koch, S. Mørup, E. Kankeleit, G. Klingelhöfer, E. N. Evlanov, V. N. Khromov, L. M. Mukhin, O. F. Prilutskij, B. Zubkov, G. V. Smirnov, J. Juchniewicz.
Hyperfine Interact., Vol. 68, No. 1–4, p. 83 – 94 (Apr 1992). Paper presented at the International Conference on the Applications of the Mössbauer Effect (ICAME), Nanjing (People's Republic of China), 16 – 20 Sep 1991.
A Mössbauer spectrometer is included in the preliminary payload of a rover to be placed on the surface of Mars in the Soviet to the planet in 1996. In connection with the American planetary program it has also been suggested to construct a Mössbauer spectrometer to be landed on Mars. The objective is to study the iron compounds of the Martian soil and rocks by backscattering Mössbauer spectroscopy. The paper describes the significance of the element iron in the study of the evolution of the planetary system and what we might expect to learn from Mössbauer spectroscopy of the surface materials of Mars. The study of Mars is expected to expand substantially in the coming decades, probably culminating with a manned flight to the planet. The international Mössbauer community may contribute significantly to the preparation of these events.

022.009   Laboratory spectra of gas–phase coronene at elevated temperatures.
J. Kurtz.
Astron. Astrophys., Vol. 255, No. 1/2, p. L1 – L4 (Feb 1992). Letter–to–the–editor.
Gas–phase infrared spectra of coronene from 4000 to 400 cm^{-1} are presented in emission and absorption at temperatures from 350 to 450°C. Peak positions are compared to KBr pellet data and to a previously published neon matrix–isolated coronene spectrum. Relative feature strengths are analyzed and used to estimate a temperature of interstellar PAHs assuming thermal emission.

022.010   SO$_2$ absorption cross–section measurements from 197 nm to 240 nm.
R. D. Martinez, J. A. Joens.
Geophys. Res. Lett., Vol. 19, No. 3, p. 277 – 280 (7 Feb 1992).
Absorption cross–sections are reported for SO_2 for wavelengths between 197 nm and 240 nm at a temperature of 300K and a spectral bandwidth of 0.10 nm. The results are in good agreement with those previously given by Warneck et al. (1964) if a correction is applied to their wavelength scale. Differences between the present results and the high resolution cross–sections reported by Freeman et al. (1984) at 213K, degraded to a resolution of 0.10 nm, can be attributed to changes in the spectrum with temperature.

022.011   Radiative lifetimes in B I using ultraviolet and vacuum–ultraviolet laser–induced fluorescence.
T. R. O'Brian, J. E. Lawler.
Astron. Astrophys., Vol. 255, No. 1/2, p. 420 – 426 (Feb 1992).
Radiative lifetimes of the 8 lowest even parity levels in the doublet system of B I are measured using time–resolved laser–induced fluorescence in the ultraviolet and vacuum ultraviolet on an atomic beam of boron. The accurate ($\pm 5\%$) lifetimes provide a base for improved determination of absolute transition probabilities in B I. The techniques described are broadly applicable to measurement of lifetimes of levels with transitions in the visible, ultraviolet and vacuum ultraviolet in almost any element.

022.012   Partition functions and equilibrium constants for H$_3$$^+$ and H$_2$D$^+$.
K. S. Sidhu, S. Miller, J. Tennyson.
Astron. Astrophys., Vol. 255, No. 1/2, p. 453 – 456 (Feb 1992).
Partition functions for H_3^+ and H_2D^+ are calculated by explicit summation of ab initio rotation–vibration energy levels

for temperatures up to 2800K. Estimates of the errors are given. A previously proposed high temperature approximation to the nuclear spin statistics of H_3^+ is tested and found to be reliable. Equilibrium constants are given as a function of temperature for the main H_3^+ forming reaction $H_2 + H_2^+ \rightarrow H_3^+ + H$ and reactions responsible for deuterium fractionation $H_3^+ + D \rightarrow H_2D^+ + H$ and $H_3^+ + HD \rightarrow H_2D^+ + H_2$. Comparisons are made with data previously used for modelling.

022.013   **Oscillator strengths and branching ratios of transitions between low–lying levels in the barium II spectrum.**
M. D. Davidson, L. C. Snoek, H. Volten, A. Dönszelmann.
Astron. Astrophys., Vol. 255, No. 1/2, p. 457 – 458 (Feb 1992).
The branching ratios of transitions between the $6s^2S_{1/2}$, $7s^2S_{1/2}$, $6p^2P_{1/2,3/2}$, $5d^2D_{3/2,5/2}$ and $6d^2D_{3/2,5/2}$ levels in the spectrum of Ba II have been measured. The accuracy of oscillator strengths have been improved in respect to literature values.

022.014   **The possibility of observation of the Lα–line of positronium (e^+e^-) from astronomical objects.**
V. V. Burdyuzha, V. L. Kauts, N. P. Yudin.
Astron. Astrophys., Vol. 255, No. 1/2, p. 459 – 461 (Feb 1992).
The authors analyze from a physical point of view the possibility of observation of the Lα–line of the positronium (Ps). In a broad range of temperatures the processes of recombination to states of Ps with different quantum number nl^- and collisions of Ps with electrons and protons are examined. It is shown that the electron density in annihilation sources can be obtained from the intensity of the Lα–line of positronium.

022.015   **Observation of dust shedding from material bodies in a plasma.**
T. E. Sheridan, J. Goree, Y. T. Chiu, R. L. Rairden, J. A. Kiessling.
J. Geophys. Res., Vol. 97, No. A3, p. 2935 – 2942 (1 Mar 1992).
Exposure to a space plasma can cause a dusty body, such as a spacecraft or a boulder in Saturn's rings, to release dust into its environment. This is demonstrated in a laboratory experiment with an aluminum sphere covered with micrometer–sized dust grains. The sphere was rotating and electrically floating like an object space. Laser light scattering was used to detect dust falling from the body. When a low–temperature nitrogen plasma was turned on, rapid dust shedding was observed, and when it was turned off, the shedding stopped. The rate of shedding increases with plasma density. The dust is not all released the instant the plasma is turned on but rather takes place over an extended period of time, with individual grains jumping off at random intervals with a certain probability per unit time.

022.016   **Al II emission–line strengths in low–density astrophysical plasmas.**
F. P. Keenan, L. K. Harra, K. M. Aggarwal, W. A. Feibelman.
Astrophys. J., Vol. 385, No. 1, p. 375 – 377 (20 Jan 1992).
Theoretical Al II emission line ratios, determined using electron impact excitation rates calculated with the **R**–matrix code, are presented for the ratio $I(3s^2\ {}^1S - 3s3p\ {}^3P_2)/I(3s^2\ {}^1S - 3s3p\ {}^3P_1) = I(2660\ \text{Å})/I(2669\ \text{Å})$. This ratio is a useful electron density diagnostic for $N_e \geqslant 10^2 \text{cm}^{-3}$. Its use is illustrated for the planetary nebula NGC 7027 and the symbiotic star RR Tel.

022.017   **Temperature dependence of infrared bands produced by polycyclic aromatic hydrocarbons.**
L. Colangeli, V. Mennella, E. Bussoletti.
Astrophys. J., Vol. 385, No. 2, p. 577 – 584 (1 Feb 1992).
The behavior of IR absorption bands as a function of temperature has been examined systematically in the laboratory for three representative polycyclic aromatic hydrocarbon molecules: coronene, chrysene, and 1–methylcoronene. A careful description of both intensity and profile measured for most of the bands is reported. A tentative interpretation of the observed variations is given in terms of extramolecular effects produced by the anharmonicity of the vibrational energy levels as a function of temperature. These new laboratory data provide an accurate description of the optical properties for representative molecules often used to account for the so–called unidentified infrared bands emitted by astronomical sources.

022.018   **Measurement of the ^{8}Li(α, n)^{11}B reaction cross section at energies of astrophysical interest.**
R. N. Boyd, I. Tanihata, N. Inabe, T. Kubo, T. Nakagawa, T. Suzuki, M. Yonokura, X. X. Bai, K. Kimura, S. Kubono, S. Shimoura, H. S. Xu, D. Hirata.
Phys. Rev. Lett., Vol. 68, No. 9, p. 1283 – 1286 (2 Mar 1992).
Current Physics Microform No.: 9205C0033.
The cross section for the ^{8}Li(α, n)^{11}B reaction, which is crucial to predictions of primordial nucleosynthesis in inhomogeneous models, has been measured using the radioactive–beam facility of the Institute for Physical and Chemical Research (RIKEN). The reaction cross section to all allowed ^{11}B states was found to be larger than that to just the ^{11}B ground state by about a factor of 5.

022.019   **Accurate far–infrared rotational frequencies of carbon monoxide.**
T. D. Varberg, K. M. Evenson.
Astrophys. J., Vol. 385, No. 2, p. 763 – 765 (1 Feb 1992).
High–resolution measurements of the pure rotational absorption spectrum of CO in its ground state are reported for the range $J'' = 5 - 37$. A least–squares fit to this data set, augmented by previous microwave measurements of the $J'' = 0 - 4$ rotational transitions by other workers, determined the following accurate values for the molecular constants (1σ errors of the last digits in parentheses): $B_0 = 57635.96826(12)$ MHz, $D_0 = 0.18350552(46)$ MHz, $H_0 = 1.7249(59) \times 10^{-7}$ MHz, and $L_0 = -3.1(23) \times 10^{-13}$ MHz. A table of calculated CO rotational frequencies is given for the range $J'' = 0 - 45$; these frequencies are accurate to $\leqslant 10$ kHz (2σ) for $J'' \leqslant 28$.

022.020   **Chaos studied beyond the limits of thermodynamics.**
P. Andrle.
Říše hvězd, Vol. 73, No. 1, p. 4 – 5 (Jan 1992). In Czech.

022.021   **The rotational spectrum of the carbon chain radical HCCCO.**
A. L. Cooksy, J. K. G. Watson, C. A. Gottlieb, P. Thaddeus.
Astrophys. J., Lett., Vol. 386, No. 1, p. L27 – L30 (10 Feb 1992).
The bent chain radical HCCCO has been identified from laboratory measurement of 248 rotational transitions between 81 and 400 GHz. Because two molecules with the same heavy atom backbone, CCCO and HCCCHO, have already been observed in TMC–1, HCCCO is a good candidate for detection in molecular clouds and circumstellar envelopes. Rotational transition frequencies for the $K_a = 0$ manifold are tabulated, as well as precise values for the rotational, centrifugal distortion, and spin–rotation constants, which allow calculation of the entire rotational spectrum into the far–IR.

022.022   **Spectrum of particles' size formed in the course of meteorites ablation under model conditions.**
V. A. Bronshtehn, V. N. Zelenin, S. G. Mikheenko.
Astron. Vestn., Tom 26, No. 1, p. 72 – 76 (Jan – Feb 1992). In Russian. English translation in Solar Syst. Res., Vol. 26, No. 1.
The specimens of stony and iron meteorites as well as those of steel and basalt were subjected to action of a hot gas flow in an electric arc plasmotron (temperature of the mixture airnitrogen 4500K, heat inflow on the specimen 10 MWt/m², pressure 10^5Pa). The distribution of sizes of melted particles detached from the meteor body is well–approximated by the logarithmic normal law with two maxima corresponding to the following values of the aerodynamical mass median diameter: $2 - 4$ and $0.4 - 1.2\ \mu$. The first is due to small particles formed by boundary effects in process of fragmentation of melted flow, the second – to the condensated particles. The mass loss rates given are in good coincidence with theoretical results of Bronshtehn.

022.023 Absolute differential and integral electron excitation cross sections for atomic nitrogen. 2. The $^4S^0 \rightarrow 2p^4\ ^4P$ ($\lambda1135$ Å) transition from 30 to 100 eV.
J. P. Doering, L. Goembel.
J. Geophys. Res., Vol. 97, No. A4, p. 4295 – 4298 (1 Apr 1992).

The absolute direct differential and integral electron excitation cross sections for the atomic nitrogen $^4S^0 \rightarrow 2p^4\ ^4P$ ($\lambda1135$ Å) transition have been measured at incident energies of 30, 50, and 100 eV. The differential cross sections measured versus scattering angle were integrated to give the integral cross section (ICS) as a function of incident energy. The ICS has a maximum value of $0.43 \times 10^{-16} cm^2$ ($\pm 30\%$) near 50 eV. Above 50 eV, the direct excitation cross section is within 20% of the fluorescence cross section of Stone and Zipf (1973). At 30 eV, the direct cross section is $\sim 1/10$ of the fluorescence cross section. These results show that as for the $\lambda1200$–Å N I transition cross section, there is a strong cascade contribution to the fluorescence cross section from forbidden transitions to higher states excited by low–energy electron impact near threshold; but above 50 eV, unlike the $\lambda1200$–Å N I transition cross section, there is only a small cascade contribution. The optical oscillator strength of the N I ($\lambda1135$ Å) multiplet obtained from the present data is 0.1 ± 0.005.

022.024 A measurement of the $^{14}C(n, \gamma)^{15}C$ cross section at a stellar temperature of kT = 23.3 keV.
H. Beer, M. Wiescher, F. Käppeler, J. Görres, P. E. Koehler.
Astrophys. J., Vol. 387, No. 1, p. 258 – 262 (1 Mar 1992).

The capture cross section of ^{14}C has been determined by a fast cyclic activation technique. The measurements were carried out at the Karlsruhe 3.75 MV pulsed Van de Graaf accelerator using the $^7Li(p, n)$ reaction close to the reaction threshold to generate neutrons with a distribution resembling a Maxwell spectrum of kT = 25 keV. The activation sample consisted of 605 ± 15 mg carbon powder enriched in ^{14}C by 89%. The ^{14}C capture cross section was measured relative to the ^{197}Au standard cross section. The activity produced by neutron captures in the ^{14}C sample was counted with a high–resolution HPGe detector via the characteristic 5297.79 keV ^{15}C γ-ray line. The ^{14}C capture cross section at kT = 23.3 keV was found to be (1.72 ± 0.43) μbarn.

022.025 Radiative association of N and O atoms at low temperatures.
Y. Sun, A. Dalgarno.
J. Geophys. Res., Vol. 97, No. A5, p. 6537 – 6539 (1 May 1992).

The rate coefficients for the radiative association of $N(^4S)$ and $O(^3P)$ atoms are calculated for temperatures between 1K and 2000K. Below 200K the effects of the interaction between the $B^2\Pi$ and $C^2\Pi$ states have to be included. Experimental spectroscopic data are analyzed to determine the mixing of the states and to infer the radiative and predissociation lifetimes for the quasi–bound rotational levels through which the radiative association occurs at low temperatures. At 20K the rate coefficient is $6.5 \times 10^{-17} cm^3 s^{-1}$, and at 200K it is $2.3 \times 10^{-17} cm^3 s^{-1}$.

022.026 Thermal detection of dark matter.
E. Fiorini.
Atti Accad. Naz. Lincei, Ser. 9, Rend. Lincei: Sci. Fis. Nat., Vol. 3, Fasc. 1, p. 11 – 22 (1992).

Bolometers operating at low temperature are discussed as detectors of heavy particles which have been recently considered as constituents of cosmic dark matter. The study of the heat leak in a cascade refrigerator operating underground is suggested for the indirect detection of weak interacting massive particles.

022.027 The oscillator strength of the Si II $3s^23p\ ^2P - 3s3p^2\ ^2D$ multiplet and the interstellar abundance of silicon.
P. L. Dufton, F. P. Keenan, A. Hibbert, P. C. Ojha, R. P. Stafford.
Astrophys. J., Vol. 387, No. 1, p. 414 – 416 (1 Mar 1992).

Spontaneous radiative rates, calculated using sophisticated configuration interaction wavefunctions, are presented for Si II resonance transitions in the $3s^23p\ ^2P_J - 3s3p^2\ ^2D_{J'}$ multiplet. For the J = 1/2 to J' = 3/2 transition, an oscillator strength of 0.0020

(with an estimated uncertainty of 25%) is deduced. This value is significantly lower than those found in some previous studies; the reasons for this and in particular the crucial role of the energy splitting of the $3s3p^2\ ^2D$ and $3s^23d\ ^2D$ levels are investigated. The implication of these new atomic data for gas phase silicon abundances in the interstellar medium is briefly discussed.

022.028 Potential energy curves and dissociation energies of NbO, SiC, CP, PH^+, SiF^+, and NH^+.
R. R. Reddy, T. V. R. Rao, R. Viswanath.
Astrophys. Space Sci., Vol. 189, No. 1, p. 29 – 38 (Mar 1992).

The potential energy curves for the electronic ground states of astrophysically important NbO, SiC, CP, PH^+, SiF^+, and NH^+ molecules are constructed by the RKRV method. The dissociation energies are determined by curve–fitting techniques using the five–parameter Hulburt-Hirschfelder function. The estimated dissociation energies are 7.86 ± 0.16, 3.66 ± 0.09, 5.12 ± 0.12, 3.08 ± 0.09, 6.46 ± 0.14, and 3.02 ± 0.09 eV for NbO, SiC, CP, PH^+, SiF^+, and NH^+, respectively. The estimated D_0 values are in reasonably good agreement with literature values. If one utilizes D_0 values of PH^+, SiF^+, and NH^+, ionization potentials for PH, SiF, and NH are derived. The ionization potentials are 10.12, 7.13, and 13.66 eV, respectively, for PH, SiF, and NH. Dissociation energies for the above molecules are also estimated by use of the Birge–Sponer extrapolation and Hildenbrand and Murad methods.

022.029 Clock synchronization and isotropy of the one–way speed of light.
C. M. Will.
Phys. Rev. D, Vol. 45, No. 2, p. 403 – 411 (15 Jan 1992). Current Physics Microform No.: 9203G0937.

Experimental tests of the isotropy of the speed of light using one–way propagation are analyzed using a test theory of special relativity. It is shown that, when properly expressed in terms of measurable quantities, the results of such experiments are independent of the method of global synchronization of clocks. Experiments analyzed include a Jet Propulsion Laboratory time-of–flight measurement, a resonant two–photon absorption experiment, the SAO – NASA 1976 rocket gravitational redshift experiment, and Mössbauer rotor experiments. If the characteristic anisotropy is proportional to αw, where w is the velocity of the Earth relative to the cosmic background radiation, the best bound on α from these experiments is $|\alpha| < 9 \times 10^{-8}$.

022.030 Atmospheric effects on cratering efficiency.
P. H. Schultz.
J. Geophys. Res., Vol. 97, No. E1, p. 975 – 1005 (25 Jan 1992).

Laboratory experiments permit quantifying the effects of an atmosphere on cratering efficiency by hypervelocity impacts. Three separable processes have been identified: ambient atmospheric pressure, aerodynamic drag, and projectile–atmosphere interactions. The present paper re–examines the possible role of the atmosphere for crater scaling by varying not only pressure but also ejecta size and atmospheric density.

022.031 Secondary electron yields of solar system ices.
D. M. Suszcynsky, J. E. Borovsky, C. K. Goertz.
J. Geophys. Res., Vol. 97, No. E2, p. 2611 – 2619 (25 Feb 1992).

The secondary electron yields of H_2O, CO_2, NH_3 (ammonia), and CH_3OH (methanol) ices have been measured as a function of electron beam energy in the 2– to 30–keV energy range. The ices were produced on a liquid–nitrogen–cooled cold finger and transferred under vacuum to a scanning electron microscope where the yield measurements were made. The imaging capabilities of the scanning electron microscope provide a means of correlating the yield measurements with the morphology of the ices and are also used to monitor charging effects. The yields were determined by measuring the amplified current from a secondary electron detector and calibrating this current signal with the amplified current signal from samples of metals with known secondary electron yields. Each of the measured yields is found to decrease with an increase in energy in the 2– to 30–keV range. Estimates are given for the maximum secondary electron yield

Y_{max} of each ice and the energy at which this maximum yield occurs. Implications for the charging of solar system ice grains are discussed.

022.032 Adsorption of CO on oxide and water ice surfaces: implications for the Martian atmosphere.
M.–T. Leu, J. E. Blamont, A. D. Anbar, L. F. Keyser, S. P. Sander.
J. Geophys. Res., Vol. 97, No. E2, p. 2621 – 2627 (25 Feb 1992).
The adsorption of carbon monoxide (CO) on water ice and on the oxides Fe_2O_3, Fe_3O_4, Al_2O_3, SiO_2, CaO, MgO, and TiO_2 (rutile and anatase) has been investigated in a flow reactor. A mass spectrometer was employed as a detector to monitor the temporal concentrations of CO. The authors have measured adsorption coefficients as large as 1×10^{-4} for CO on TiO_2 solids in helium at 196K. The fractional surface coverage for CO on TiO_2 solids in helium was also determined to be approximately 10% at 196K. The upper limits of the fractional surface coverage for the other oxides (Fe_2O_3, Fe_3O_4, Al_2O_3, SiO_2, CaO, and MgO) and water ice were also measured to be less than 1%. The implications for the stability of CO_2 in the Martian atmosphere and the "CO hole" observed by the Phobos/ISM (infrared spectrometer) experiment are discussed.

022.033 Simple algorithms for remote determination of mineral abundances and particle sizes from reflectance spectra.
P. E. Johnson, M. O. Smith, J. B. Adams.
J. Geophys. Res., Vol. 97, No. E2, p. 2649 – 2657 (25 Feb 1992).
Simple algorithms for quantitatively modeling the reflectance spectra of mineral particulates are tested. Although more sophisticated models exist, these algorithms are particularly suited for remotely sensed data, where little or no opportunity exists to independently measure reflectance versus particle size and phase function. Previously, the authors introduced this method in the analysis of the directional–hemispherical reflectance spectra of binary mineral mixtures containing a single particle size distribution. In this study the technique is extended to multicomponent mixtures, various size separates, and spectra with differing illumination/viewing geometries. It is found that the theoretical calculations and measured data agree to nearly the level of experimental error. This method is also used to determine the threshold abundance at which a mineral can be detected when mixed with another mineral.

022.034 Quantitative subpixel spectral detection of targets in multispectral images.
D. E. Sabol Jr., J. B. Adams, M. O. Smith.
J. Geophys. Res., Vol. 97, No. E2, p. 2659 – 2672 (25 Feb 1992).
Spectral mixture analysis was used to determine threshold detection limits of target materials in the presence of background materials within the field of view under various simulated but realistic compositional, instrumental, and topographical conditions. Detection thresholds were determined for the cases where the target is detected as (1) a component of a spectral mixture (continuum threshold analysis) and (2) residuals (residual threshold analysis). In continuum threshold analysis, the target was included as a component during unmixing thereby permitting evaluation of target detectability. In residual threshold analysis, the unmodeled target was detected as wavelength–dependent deviations of the spectral mixture (target included) from the predicted spectrum (mixtures of the modeled background spectra). High resolution laboratory spectra were used to test the "best case" for target detection in spectral mixtures. Data quality was then decreased to simulate the effects of various imaging instruments (spectral sampling and noise) and changes in lighting geometry.

022.035 Einstein coefficients for rotational lines of the (0, 0) band of the NO $A^2\Sigma^+ – X^2\Pi$ system.
J. R. Reisel, C. D. Carter, N. M. Laurendeau.
J. Quant. Spectrosc. Radiat. Transfer, Vol. 47, No. 1, p. 43 – 54 (Jan 1992).
The authors present a summary of the spectroscopic equations necessary for prediction of the molecular transition energies and

the Einstein A and B coefficients for rovibronic lines of the $\gamma(0, 0)$ band of nitric oxide (NO). The calculated molecular transition energies are all within $0.57\,cm^{-1}$ of published experimental values; in addition, over 95% of the calculated energies give agreement with measured results within $0.25\,cm^{-1}$. Einstein $A_{J'J''}$ and $B_{J'J''}$ coefficients are calculated from the band A_{00} value and the known Hönl–London factors and are tabulated for individual rovibronic transitions in the NO $A^2\Sigma^+ – X^2\Pi(0, 0)$ band.

022.036 On the calculation of bound–bound, bound–free and free–free dipole transitions in a non–hydrogenic atom.
L. G. D'yachkov, P. M. Pankratov.
J. Quant. Spectrosc. Radiat. Transfer, Vol. 47, No. 1, p. 75 – 79 (Jan 1992).
The semi–classical expressions for the radial integrals of bound–bound, bound–free and free–free dipole transitions in a non–hydrogenic atom or positive ion are obtained. Their region of validity is restricted to the Coulomb approximation and an approximation analogous to that made by Kramers $(Z\omega/|E|^{3/2} \gg 1)$. The well–known Kramers formula for the bremsstrahlung cross section in the pure Coulomb potential is generalized to the case when a short–range potential is added. Comparisons are made with more exact calculations.

022.037 Collisional broadening of rotational lines in the simulated Raman pentad Q–branch of CD_4.
G. Millot, B. Lavorel, J. I. Steinfeld.
J. Quant. Spectrosc. Radiat. Transfer, Vol. 47, No. 2, p. 81 – 90 (Feb 1992).
Self–and argon–broadening coefficients are reported for a number of Raman Q–branch transitions in the v_1 and $v_2 + v_4$ bands of $^{12}CD_4$ at room temperature (296K). The coefficients display a variation with j and with C^n (symmetry species A, E, F) that is essentially independent of collision partner and which is similar to the j– and C^n–dependence found in previous measurements of the i.r. line–broadening coefficients. The rotationally inelastic collision rates previously measured by Foy et al. for $^{13}CD_4(v_4 = 0, 1)$ in collision with $^{13}CD_4$ or Ar account for only a part of the Raman broadening rate, suggesting possibly significant contributions to the linewidths from efficient $V–V$ transfer or elastic dephasing collisions.

022.038 Application of the Stogryn–Hirschfelder treatment of weak dimers to planetary atmospheres.
Z. Slanina, K. Fox, S. J. Kim.
J. Quant. Spectrosc. Radiat. Transfer, Vol. 47, No. 2, p. 91 – 94 (Feb 1992).
The thermodynamics of carbon dioxide dimerization is treated as an example of the evaluation of weak dimer populations in planetary atmospheres (e.g., Mars and Venus). Two approaches considered are the Stogryn–Hirschfelder treatment using the Lennard–Jones interaction and calculations based on recent quantum chemical data. Several improvements of these treatments are developed. It is indicated that carbon dioxide dimers may be less or more abundant at the surfaces of Mars or Venus, respectively, than is suggested by previous calculations.

022.039 Reinvestigation of some of the autoionizing levels in the spectrum of Cu I.
P. M. R. Rao, S. Padmanabhan, G. Krishnamurty, B. N. R. Sekhar.
J. Quant. Spectrosc. Radiat. Transfer, Vol. 47, No. 2, p. 113 – 119 (Feb 1992).
The emission spectrum of Cu I was generated in a 10 A d.c. arc and photographed in the second and third orders of a 3.4 m Jarrell–Ash spectrograph using a 1200 grooves/mm grating. The spectral lines involving the autoionizing levels $5s'^4D$, $5s'^2D$ and $5s''^2D$ arising from the $3d^94s5s$ configuration have been reinvestigated to obtain comprehensive data on the half widths of all of the diffuse lines.

022.040   Intensities and electronic transition strengths of seven T_{e2} visible and i.r. band systems.
R. S. Ferber, Ya. A. Harya, A. V. Stolyarov.
J. Quant. Spectrosc. Radiat. Transfer, Vol. 47, No. 2, p. 143 – 158 (Feb 1992).

022.041   Synthetic Mg_1, Mg_2 and Mgb indices.
M. Erdelyi–Mendes, B. Barbuy, A. Milone.
IAU Symposium No. 149: The stellar populations of galaxies, p. 415 (1992). – See Abstr. 012.007 for the main entry.

022.042   Regularities in experimental Stark shifts.
W. L. Wiese, N. Konjević.
J. Quant. Spectrosc. Radiat. Transfer, Vol. 47, No. 3, p. 185 – 200 (Mar 1992).
The authors have examined regularities in plasma–produced line shifts (Stark shifts) by a comprehensive analysis of literature data. Since the shifts are the result of atomic collision processes, regularities are expected from general atomic structure considerations. Specificlly, systematic behavior should occur for spectral series and for corresponding transitions in homologous atoms and isoelectronic ions. Also, Stark shifts should be similar for lines within multiplets and, to a lesser degree, within supermultiplets and transition arrays. Numerous examples show conclusively that the measured data exhibit these predicted regularities. When pronounced irregularities occur, they are readily explainable in terms of special circumstances in the atomic structure.

022.043   Infrared emission spectra of benzene and naphthalene: implications for the interstellar polycyclic aromatic hydrocarbon hypothesis.
J. D. Brenner, J. R. Barker.
Astrophys. J., Lett., Vol. 388, No. 1, p. L39 – L43 (20 Mar 1992).
Emissions from the fundamental region (~ 3050 cm^{-1}, 3.3 μm) and first overtone (~ 6000 cm^{-1}, 1.7 μm) of the C–H streching modes in the small aromatic hydrocarbons benzene (C_6H_6) and naphthalene ($C_{10}H_8$) were observed following ultraviolet laser excitation. These two molecules respectively represent the "prototype" and smallest members of the polycyclic aromatic hydrocarbon (PAH) chemical family, proposed as likely carriers of a set of infrared features widely observed in various dust–containing astronomical sources. Wavelength – and time – resolved emission spectra in the 3050 cm^{-1} region show contributions from anharmonically shifted $\Delta v = -1$ transitions originating in $v = 1$, 2, and 3 in benzene and $v = 1$ and 3 in naphthalene, as well as underlying continuum emission. The emission transition frequencies differ significantly from the corresponding absorption frequencies for both molecules, and depend on the vibrational energy of the emitter. Thus, unambiguous identification of the interstellar emitters cannot be established just on the basis of matching the emission frequencies to laboratory absorption spectra. The laboratory emission spectra were used to calculate the spectra expected under conditions prevailing in the interstellar medium. The overtone emission measurements were used to predict the intensity of a weak 5980 cm^{-1} overtone feature which should be present in the interstellar spectra according to the PAH hypothesis.

022.044   Tunable diode laser measurements on the 951.7393 cm^{-1} line of $^{12}C_2H_4$ at planetary atmospheric temperatures.
J. F. Brannon Jr., P. Varanasi.
J. Quant. Spectrosc. Radiat. Transfer, Vol. 47, No. 4, p. 237 – 242 (Apr 1992).
The absolute intensity and collision–broadened half–widths of the $5_{0,5} \rightarrow 5_{1,5}$ transition at 951.7393 cm^{-1} in the v_7–fundamental band of $^{12}C_2H_4$ have been measured at 152, 202, 252, and 295K with the Doppler–limited spectral resolution ($\sim 10^{-4}$cm^{-1}) of a tunable diode laser spectrometer. The temperature dependence of the half–width has been determined for line broadening by the planetary atmospheric gases He, H_2, and N_2.

022.045   Recombination line intensities for hydrogenic ions. III. Effects of finite optical depth and dust.
D. G. Hummer, P. J. Storey.
Mon. Not. R. Astron. Soc., Vol. 254, No. 2, p. 277 – 290 (15 Jan 1992).
The authors explore systematically the effect on the recombination spectrum of hydrogen arising from: (1) finite optical thickness in the Lyman lines; (2) the overlapping of Lyman lines near the series limit; (3) the absorption of Lyman lines by dust or photoionization, and (4) the long–wave radiation emitted by dust. Full account is taken of electron and heavy particle collisions in redistributing energy and angular momentum. The authors find that each of these deviations from the classical Case B leads to observable effects, and that dust influences the recombination spectrum in characteristic ways that may make possible new observational constraints on dust properties in nebulosities. On the basis of these calculations the authors believe the uncertainty in the determination of the helium–to–hydrogen abundance ratio in the Universe may be larger than currently claimed.

022.046   Site–site Lennard–Jones potential parameters for N_2, O_2, H_2, CO and CO_2.
J.–P. Bouanich.
J. Quant. Spectrosc. Radiat. Transfer, Vol. 47, No. 4, p. 243 – 250 (Apr 1992).
The classical second–virial coefficients $B(T)$ for identical linear molecules have been calculated exactly by using a potential model that includes site–site Lennard–Jones 12–6 interactions with added dipole and quadrupole interactions. By fitting selected experimental values of $B(T)$ and using up to three different plausible quadrupole moments of the molecules, the author has determined site–site Lennard–Jones parameters for N_2, O_2, H_2, CO and CO_2. He has also considered, in addition to electrostatic interactions, a limited spherical harmonics expansion of the site–site potential. By using at short range only the isotropic part of this expansion, the author has derived parameters that are close for N_2, O_2 and H_2 to the exact site–site parameters. The validity of this approximation in the truncated site–site potential is more questionabe for CO and CO_2.

022.047   Extremely low thermal conductivity of amorphous ice: relevance to comet evolution.
A. Kouchi, J. M. Greenberg, T. Yamamoto, T. Mukai.
Astrophys. J., Lett., Vol. 388, No. 2, p. L73 – L76 (1 Apr 1992). With plate L1.
The thermal conductivity of very slowly deposited amorphous ice, derived from laboratory experiments, is shown to be a factor of 10^{-4} to 10^{-5} less than hitherto estimated. Using the exceedingly low value of the thermal conductivity of comets deduced from the amorphous ice properties leads to the expectation that internal heating of comets is negligible below the outer several tens of centimeters.

022.048   Intensity and linewidth measurements in the 13.7 μm fundamental bands of $^{12}C_2H_2$ and $^{12}C^{13}CH_2$ at planetary atmospheric temperatures.
P. Varanasi.
J. Quant. Spectrosc. Radiat. Transfer, Vol. 47, No. 4, p. 263 – 274 (Apr 1992).
The absolute intensities and collision–broadened half–widths of several lines in the P–, Q–, and R–branches of the v_5–fundamental band of $^{12}C_2H_2$ have been measured at various temperatures between 147 and 295K employing the Doppler–limited spectral resolution ($\sim 10^{-4}$cm^{-1}) of a tunable diode laser spectrometer. The absolute intensities of $R(5)$, $R(7)$, $R(9)$, and $R(20)$ in the same fundamental belonging to $^{12}C^{13}CH_2$ have also been measured at 294K. The temperature dependence of the collision–broadened half–width has been determined for some of the lines broadened by planetary atmospheric gases, namely, He, Ar, H_2, and N_2. Four self–broadened linewidths of $^{12}C_2H_2$, as well as three H_2–broadened linewidths and a self–broadened half–width of $^{12}C^{13}CH_2$, have also been retrieved from the measurements.

022.049   Laboratory studies of water vapor absorption in the atmospheric window at 213 GHz.
M. Godon, J. Carlier, A. Bauer.
J. Quant. Spectrosc. Radiat. Transfer, Vol. 47, No. 4, p. 275 – 285 (Apr 1992).
Absolute absorption rates of water vapor have been measured in the atmospheric window between the rotational lines at 183 and 321 – 325 GHz. Measurements have been carried out for pure water vapor and mixtures with N_2 at atmospheric pressure. Pressure and temperature dependences are compared with models involving different lineshapes and different types of continua.

022.050   Stark–broadening parameters of ionized mercury spectral lines of astrophysical interest.
M. S. Dimitrijević.
J. Quant. Spectrosc. Radiat. Transfer, Vol. 47, No. 5, p. 315 – 318 (May 1992).
Using a semiclassical approach, the author has calculated electron–, proton–, and ionized– helium–impact line widths and shifts for seven ionized mercury lines observed in the spectra of Mn and magnetic stars, as well as in laboratory plasmas.

022.051   The optically thick C III spectrum. I. Term populations and multiplet intensities at lower optical depths.
A. K. Bhatia, S. O. Kastner.
Astrophys. J., Suppl. Ser., Vol. 79, No. 1, p. 139 – 156 (Mar 1992).
The C III spectrum is studied quantitatively under both optically thin and optically thick conditions, yielding term populations and line/multiplet intensities for column lengths from zero to $10^{18} cm^{-2}$. The role of escape probabilities and line profiles in the calculation is discussed. It is shown that use of the fully integrated escape factor, rather than the more appropriate monodirectional escape probability, can lead to appreciable errors in calculated intensities. The results for populations and intensities are used to identify two unassigned featues in the solar EUV spectrum of Vernazza and Reeves as C III multiplets, and to establish that an unidentified infrared solar feature at 8500.32 Å, seen in both absorption (Fraunhofer) and emission (chromospheric) spectra, is the C III transition 2s3s(^1S) – 2s3p(^1P). In addition, the authors tabulate Voigt parameters for the C III lines and multiplets, obtained by a modified semiclassical method.

022.052   Quantitative photoabsorption and fluorescence spectroscopy of SO_2 at 188–231 and 278.7–320 nm.
S. M. Ahmed, V. Kumar.
J. Quant. Spectrosc. Radiat. Transfer, Vol. 47, No. 5, p. 359 – 373 (May 1992).
Absolute photoabsorption and relative fluorescence cross sections for SO_2 have been measured in the 188–231 and 278.7–320 nm regions using an argon mini–arc light source. The absorption cross sections have been measured with an accuracy of $\pm 3.1\%$; the most probable error for the fluorescence cross section is $\pm 4.6\%$. The fluorescence quantum yields for SO_2 have also been obtained in the two spectral regions.

022.053   Schrödinger's radial equation: solution by extrapolation.
D. Goorvitch, D. C. Galant.
J. Quant. Spectrosc. Radiat. Transfer, Vol. 47, No. 5, p. 391 – 399 (May 1992).
Combining an appropriate finite difference method with iterative extrapolation to the limit results in a simple, highly accurate, numerical method for solving a one–dimensional Schrödinger's equation appropriate for a diatomic molecule. This numerical procedure has several distinct advantages over the more conventional methods such as Numerov's method or the method of finite differences without extrapolation. The authors demonstrate the advantages of the present algorithm by solving Schrödinger's equation for (1) a Morse potential function appropriate for HCl and (2) a numerically derived Rydberg–Klein–Rees potential function for the $X^1\Sigma^+$ state of CO. A direct

comparison of the results for the $X^1\Sigma^+$ state of CO is made with results obtained using Numerov's method.

022.054   Quantum–defect studies of transitions in the diffuse spectral series of the potassium isoelectronic sequence.
C. Lavin, C. Barrientos, I. Martín.
J. Quant. Spectrosc. Radiat. Transfer, Vol. 47, No. 5, p. 411 – 419 (May 1992).
Theoretical oscillator strengths are reported for transitions in the diffuse spectral series of some members of the potassium isoelectronic sequence (K I–Cr VI). The calculations have been performed with the Quantum Defect Orbital (QDO) method. A core–polarization correction to the dipole transition moment has also been included in the formalism.

022.055   The solution of coupled Schrödinger equations using an extrapolation method.
D. Goorvitch, D. C. Galant.
J. Quant. Spectrosc. Radiat. Transfer, Vol. 47, No. 6, p. 505 – 513 (Jun 1992).
The authors apply extrapolation to the limit in a finite–difference method to solve a system of coupled Schrödinger equations. This combination results in a method that only requires knowledge of the potential energy functions for the system. This numerical procedure has several distinct advantages over the more conventional methods such as Numerov's method or the method of finite differences without extrapolation. The authors solve the coupled Schrödinger equation for the $X^2\Pi$ state of OH. The algorithm results in term values that agree with experimentally–derived values within 6 parts in 10^4. The calculated wavefunctions are compared indirectly through experimentally–derived rotational constants and are found to be accurate to better than 3 parts in 10^3. A comparison of the authors' method with another numerical method shows results agreeing within 1 part in 10^4.

022.056   The effect of H_2O gas on volatilities of planet–forming major elements: I. Experimental determination of thermodynamic properties of Ca–, Al–, and Si–hydroxide gas molecules and its application to the solar nebula.
A. Hashimoto.
Geochim. Cosmochim. Acta, Vol. 56, No. 1, p. 511 – 532 (Jan 1992).
This paper reports experimental determinations of the heat of formation and entropy for the hydroxide species which are expected to be most abundant under the physical conditions in the primordial solar nebula that current astrophysical theories assume. It also reports the discovery of Si–tetrahydroxide, the first stable Si–hydroxide gas molecules to be found. These data (along with literature thermodynamic data) are used to calculate the relative abundances of M, MO_x, and $M(OH)_n$ gas speciesand relative volatilities of the five major elements for ranges of temperature, total pressure, and H/O abundance ratio corresponding to plausible ranges of physical conditions in the nebula.

022.057   Multipump and quasistroboscopic back–action evasion measurements for resonant–bar gravitational wave antennas.
L. E. Marchese, M. F. Bocko, R. Onofrio.
Phys. Rev. D, Vol. 45, No. 6, p. 1869 – 1877 (15 Mar 1992).
Current Physics Microform No.: 9204H1255.
A generalization of the back–action evasion (BAE) measurement technique, which is called multipump back–action evasion, has been demonstrated with a parametric electromechanical transducer similar to those used at resonant–bar gravitational wave detectors. The benefit of a BAE measurement is that the fluctuating back–action force of the transducer readout circuit acting on a test mass is squeezed, i.e., reduced in one of the quadrature phases of the test mass, thus improving the sensitivity for the detection of weak forces. The multipump BAE technique may be used to further improve the sensitivity of resonant–bar gravitational wave antennas.

022.058 Simulation and alteration for amorphous silicates with very broad bands in infrared spectra.
C. Koike, A. Tsuchiyama.
Mon. Not. R. Astron. Soc., Vol. 255, No. 2, p. 248 – 254 (15 Mar 1992).

Amorphous silicates were synthesized by evaporation of olivine and pyroxene. It is found that amorphous condensates of olivine have nearly the same composition as source olivine. They show two very broad bands at wavelengths λ of 10–11 and 18–19 μm in their infrared spectra. The authors have found that the 18-μm band is very unstable, that is, it easily shifts no longer wavelength by heating or hydration. On the other hand, amorphous condensates of pyroxene were enriched in Si compared with source pyroxene and showed two bands at about λ = 9.5 and 22–23 μm.

022.059 The photolysis of NH_3 in the presence of substituted acetylenes: a possible source of oligomers and HCN on Jupiter.
J. P. Ferris, R. R. Jacobson, J. C. Guillemin.
Icarus, Vol. 95, No. 1, p. 54 – 59 (Jan 1992).

Photolysis of NH_3 in the presence of propyne yields dimethylketazine (4) as the main product along with dimethylketimine, isopropylamine, and propioazine (7). Quantum yield and percentage conversion to products are reported. These studies show that acetylenic hydrocarbons formed by the photolysis of methane in the stratosphere of Jupiter may react with radicals formed by NH_3 photolysis to give nonvolatile yellow–brown polymers, dialkylazines, alkylnitriles, and eventually HCN. This scenario accounts for the observation of both HCN and chromophores on Jupiter.

022.060 Strings, gravity, and the constants of nature.
G. Veneziano.
NATO Advanced Study Institute on The Infrared and Submillimetre Sky after COBE, p. 15 – 33 (1992). – See Abstr. 012.009 for the main entry.

After discussing some non–trivial properties of classical string motions in gravitational backgrounds, the author turns to the standard model and, in particular, to its (usually understated) shortcomings. He then discusses how quantum strings might resolve the SM difficulties by providing a finite theory of all interactions, including gravity. An outcome of this optimistic scenario is the calculability–in–principle of the fundamental constants of Nature, which simply appear, in string theory, as vacuum parameters.

022.061 Theoretical study of the deexcitation of C_2 in collisions with helium.
J. M. Robbe, H. Lavendy, D. Lemoine, B. Pouilly.
Astron. Astrophys., Vol. 256, No. 2, p. 679 – 682 (Mar 1992).

Rotational deexcitation cross–sections are calculated for C_2 ground state in collisions with He in an IOS treatment after the relevant ab initio potential surfaces have been determined. The corresponding rate constants are then determined for temperatures relevant to interstellar clouds.

022.062 Improved determination of the ground state molecular constants of the CS^+ cation to aid possible astrophysical detection.
M. Horani, M. Vervloet.
Astron. Astrophys., Vol. 256, No. 2, p. 683 – 685 (Mar 1992).

The electronic spectrum, $A^2\Pi_i$–$X^2\Sigma^+$ transition, of the CS^+ radical ion has been recorded by Fourier Transform Spectroscopy between 14000 and 5800 cm^{-1}. Precise molecular constants have been derived for the ground state $X^2\Sigma^+$ in order to aid detection of the CS^+ molecular ion in astrophysical objects. From this new set of molecular constants, the frequencies of some CS^+ lines of astrophysical interest are calculated; they are significantly different from an earlier prediction (Quarta and Singh 1981).

022.063 Meteoritics and the origins of atomic nuclei.
D. D. Clayton.
54. Annual Meeting of the Meteoritical Society, p. 47 (1991). Abstract. – See Abstr. 012.010 for the main entry.

022.064 Two–step laser mass spectrometry: analysis at high spatial resolution of cosmochemical samples.
S. J. Clemett, L. J. Kovalenko, C. R. Maechling, R. N. Zare.
54. Annual Meeting of the Meteoritical Society, p. 48 (1991). Abstract. – See Abstr. 012.010 for the main entry.

022.065 Erratum: "Electron collisional rates for atomic hydrogen, revisited" [Astron. Astrophys., Vol. 247, No. 2, p. 580 – 583 (Jul 1991)].
E. S. Chang, E. H. Avrett, R. Loeser.
Astron. Astrophys., Vol. 256, No. 2, p. 724 (Mar 1992). See Abstr. 54.022.032.

022.066 Simulation of the interaction of galactic protons with meteoroids: on the production of ^7Be, ^{10}Be and ^{22}Na in an artificial meteoroid irradiated isotropically with 1.6 GeV protons.
U. Herpers, R. Rösel, R. Michel, M. Lüpke, D. Filges, P. Dragovitsch, W. Wölfli, B. Dittrich, H. J. Hofmann.
54. Annual Meeting of the Meteoritical Society, p. 89 (1991). Abstract. – See Abstr. 012.010 for the main entry.

022.067 The comet–tail (A–X) system of CO^+: precise molecular constants of its $X^2\Sigma^+$, $A^2\Pi_i$, and $B^2\Sigma^+$ states.
C. Haridass, C. V. V. Prasad, S. P. Reddy.
Astrophys. J., Vol. 388, No. 2, p. 669 – 677 (1 Apr 1992). With plate 6.

The comet–tail ($A^2\Pi_i$ – $X^2\Sigma^+$) system of the molecular ion $^{12}C^{16}O^+$, excited in the cathode glow of a hollow–cathode discharge tube of special design, was photographed in the spectral region 3400 – 8500 Å. Of the 17 observed bands with $v' = 0$–8 and $v'' = 0$–4, the rotational structure of 12 bands with $v' = 0$–4 and six with $v'' = 0$–4 was analyzed. An effective Hamiltonian was used to obtain the molecular constants of the individual bands from their wavenumber data, and the constants for A and X states were then determined by merging the molecular constants thus obtained by the method of correlated least–squares fit. Recent experimental data of $^{12}C^{16}O^+$ available for the first negative (B $^2\Sigma^+$ – X $^2\Sigma^+$) and the Baldet–Johnson (B $^2\Sigma^+$ – A $^2\Pi_i$) systems and for the infrared spectrum involving $v = 0$ and 1, and the microwave spectra involving $v = 0$, 1, and 2, of its ground state X, were reanalyzed using this effective Hamiltonian. Finally, a precise set of molecular constants for the X, A, and B states of $^{12}C^{16}O^+$ were determined. The astrophysical importance of the spectra of CO^+ is briefly outlined.

022.068 Circular polarization as an instrument for investigation of surfaces of atmosphereless celestial bodies. I. Laboratory measurements of highly absorptive substances.
V. S. Degtyarev, L. O. Kolokolova.
Kinematika Fiz. Nebesn. Tel, Tom 8, No. 2, p. 3 – 7 (Mar–Apr 1992). In Russian. English translation in Kinematics Phys. Celest. Bodies, Vol. 8, No. 2.

Phase dependences of the circular polarization for light scattered from surfaces formed by layers of tiny powder particles were obtained with a high–precision Stokes polarimeter. Powders of nickel, iron and graphite which have large absorptive indices were used, the particles having similar sizes, but differing in shape. The measurements showed that highly absorptive substances demonstrate large values of circular polarization (more than 1%), especially for large phase angles. The trend of the circular polarization as a function of the phase angle turned out to depend strongly on structure characteristics of a surface. For a porous layer of particles, the absolute value of circular polarization increases slowly up to angles of $\alpha \approx 150°$ and then decreases down to zero, retaining their sign at all phase angles. Measurements for a pressed layer of particles (graphite) showed that the phase curve demonstrates negative values of circular polarization at first, and then becomes positive reaching its maximum at

$\alpha \approx 130°$. Hence, the circular polarization of light scattered by the surfaces of atmosphereless celestial bodies, by asteroids in particular, can give information about the presence of metals in the matter of their surface layers and about the presence of regolith on the surface of metalliferous objects.

022.069   Circular polarization as an instrument for investigation of surfaces of atmosphereless celestial bodies. II. Theoretical simulation of light scattering by rough surfaces.
V. S. Degtyarev, L. O. Kolokolova.
Kinematika Fiz. Nebesn. Tel, Tom 8, No. 2, p. 8 – 14, 88 (Mar–Apr 1992). In Russian. English translation in Kinematics Phys. Celest. Bodies, Vol. 8, No. 2.

A model that allows to calculate the Stokes vector for light scattered by a rough surface is described. The surface is considered as a set of microfacets that reflect light in accordance with the Fresnel laws and obey a certain distribution over the inclination. The calculations are based on the Mueller 4x4 matrix formalism and take account of multiple reflections of light from surface microfacets. The model has been used for interpretation of the experimental phase dependences of the circular polarization for absorptive substances which were described in Part I (see Abstr. 022.068) of this work. Theoretical phase dependences for iron, nickel and graphite agree well with the observed ones and show the same modifications when optical and structural properties of a surface change. The calculations confirm an earlier conclusion that the circular polarization can be used for determination of metals on the surfaces of atmosphereless celestial bodies and for detection of regolith on the surface of metaliferous objects.

022.070   Hot shock experiments: simulation of an important process in the early solar system and in multi–ring cratering.
F. Langenhorst, A. Deutsch.
54. Annual Meeting of the Meteoritical Society, p. 127 (1991). Abstract. – See Abstr. 012.010 for the main entry.

022.071   Simulation of the interaction of galactic protons with meteoroids: isotropic irradiation of an artificial meteoroid with 1.6 GeV protons.
54. Annual Meeting of the Meteoritical Society, p. 156 (1991). Abstract. – See Abstr. 012.010 for the main entry.

022.072   High resolution absorption cross sections in the transmission window region of the Schumann–Runge bands and Herzberg continuum of O_2.
K. Yoshino, J. R. Esmond, A. S.–C. Cheung, D. E. Freeman, W. H. Parkinson.
Planet. Space Sci., Vol. 40, No. 2/3, p. 185 – 192 (Feb–Mar 1992).

The absorption cross sections of the Schumann–Runge bands in the window region between the rotational lines have been measured in the wavelength region 180-195 nm. The measurements have been done with many different pressures of oxygen, 2.5–760 torr, so that the pressure–dependent absorption can be separated from the main cross sections. The published cross sections in the window region are superseded by the present cross sections. The combined cross sections are presented graphically in the paper and are available at wavenumber intervals of ~ 0.1 cm^{-1} as numerical compilations stored on magnetic tape, from the National Space Science Data Center, NASA/Goddard Space Flight Center, Greenbelt MD 20771, U.S.A. The Herzberg continuum cross sections are derived after subtracting calculated contributions from the Schumann–Runge bands and are significantly smaller than any previous measurements.

022.073   Dust emission phenomena of cometary analogues.
H. Kochan, W. Koerver.
Adv. Space Res., Vol. 11, No. 12, p. 161 – 174 (1991). – See Abstr. 012.014 for the main entry.

Experiments with cometary analogues are reported, which were performed in a small vacuum chamber and in the Space Simulator. The results apply to the emission processed of ice-dust grains from the surface of mineral–ice mixtures during insolation. The emission phenomena were observed with CCD-video–cameras equipped with a magnifying (10 ×) macro–lens. The pictures show that the ice–dust agglomerates carry tiny ice particles on their surfaces. These may play the role of interlinking bonds between one grain and its neighbors. In the enlarged view it can be seen that the ice–dust grains do not erupt with high velocity out of the surface. The upstreaming gas jet originating from the sublimation of the volatile components first erodes the interlinking bonds between the particles. When the drag force exerted on the particles exceeds the force of gravity, the particles lift off. In most of the observed cases this initial lift off velocity is very small.

022.074   Intensities and transition probabilities for selected Dy I and Dy II lines emitted from a ferroelectric plasma source.
J. Kusz.
Astron. Astrophys., Suppl. Ser., Vol. 92, No. 3, p. 517 – 532 (Feb 1992).

An argon–dysprosium plasma was generated at atmospheric pressure between a ceramic ferroelectric plate and a dysprosium plate. The system of plates was connected by an acoustic frequency power supply. The plasma radiation was analyzed in the spectral range from 2500 to 6700 Å by using a grating spectrograph with a linear dispersion of about 1 mm/Å, adapted for photoelectric measurements. The emission spectrum of dysprosium was recorded and the intensities of more than 500 Dy I and Dy II lines were measured. For 133 Dy I–lines and 228 Dy II–lines the transition probabilities, absorption oscillator strengths and log $(g_i f_{ik})$ were determined. The obtained values were compared with the literature data.

022.075   Solar–neutrino neutral–current detection methods in the Sudbury Neutrino Observatory.
C. K. Hargrove, D. J. Paterson.
Can. J. Phys., Vol. 69, No. 11, p. 1309 – 1316 (Nov 1991).

The Sudbury Neutrino Observatory will study the solar-neutrino problem through the detection of charged–current (CC), neutral–current (NC), and elastic–scattering (ES) interactions of solar neutrinos with heavy water. The measurement of the NC rate relative to the CC rate provides a nearly model-independent method of observing neutrino oscillations. The interaction rate in the original design is measured by observing Čerenkov light from showers produced by neutron–capture γ rays from the capture of the NC neutrons by a selected additive to the heavy water. These signals overlap the CC and ES signals, so that the measurement of the NC rate requires the subtraction of two signals obtained at different times. This paper describes the authors' investigation of an alternate detection method in which the thermalized neutrons are captured by (n, α) or (n, p) reactions on light nuclei. The resulting charged–particle products are uniquely detected by scintillators or proportional counters, completely separating this NC signal from the CC and ES Čerenkov signals, thus simplifying its measurement, improving its significance, and allowing observation of otherwise unobservable short–term NC fluctuations.

022.076   Experimental simulation of Martian neutron leakage spectra.
D. M. Drake, S. A. Wender, R. O. Nelson, D. A. Clark, M. Drosg, W. Amian, J. Brückner, P. A. J. Englert.
Nucl. Instrum. Methods Phys. Res., Sect. A, Vol. 309, No. 3, p. 575 – 580 (15 Nov 1991).

The LAMPF 800 MeV proton beam and a container of sand, mixed to resemble Martian soil, were used to simulate the neutron leakage spectrum of Mars. The neutron spectra were measured by time–of–flight techniques and compared with Monte Carlo calculations.

022.077   Synthetic Mg_1, Mg_2 and Mgb indices: relative intensities of molecular bands as a function of stellar parameters.
B. Barbuy, M. Erdelyi–Mendes, A. Milone.
Astron. Astrophys., Suppl. Ser., Vol. 93, No. 2, p. 235 – 246 (May 1992).

Synthetic fluxes were computed in the wavelength region $\lambda\lambda 490 - 530$ nm, in order to study the behaviour of the Mg indices: Mgb, Mg_1, Mg_2 and the DDO "51" filter bandpasses, and of its main molecular constituents: the MgH, C_2, CN and TiO molecular bands, as a function of the stellar parameters effective temperature T_{eff}, gravity log g, and metallicity [M/H]. The aim of this work is to verify the validity of use of the Mg indices as metallicity indicators: the authors conclude that Mg_2 and Mgb indices are adequate metallicity indicators, being practically independent of temperature and gravity.

022.078   Stark broadening of spectral lines of multicharged ions of astrophysical interest. III. O VI lines.
M. S. Dimitrijević, S. Sahal–Bréchot.
Astron. Astrophys., Suppl. Ser., Vol. 93, No. 2, p. 359 – 371 (May 1992).
Using a semiclassical approach, the authors have calculated electron–, proton–, and ionized helium–impact line widths and shifts for 30 O VI multiplets. This comprehensive set of data has been used for the investigation of Stark broadening parameter regularities within spectral series.

022.079   Improved calculations for the C III $\lambda\lambda$1907, 1909 and Si III $\lambda\lambda$1883, 1892 electron density sensitive emission–line ratios, and a comparison with IUE observations.
F. P. Keenan, W. A. Feibelman, K. A. Berrington.
Astrophys. J., Vol. 389, No. 1, p. 443 – 446 (10 Apr 1992).
Theoretical electron density sensitive emission–line ratios are presented for $R_1 = F(2s^2\ ^1S - 2s2p\ ^3P_2) / F(2s^2\ ^1S - 2s2p\ ^3P_1) = F(1907\ \text{Å}) / F(1909\ \text{Å})$ in C III and $R_2 = F(3s^2\ ^1S - 3s3p\ ^3P_2) / F(3s^2\ ^1S - 3s3p\ ^3P_1) = F(1883\ \text{Å}) / F(1892\ \text{Å})$ in Si III. These are significantly different from those deduced by previous authors, principally due to adoption of improved electron excitation rates (for C III) and A values (for Si III) in the present analysis. The observed values of R_1 and R_2 for several planetary nebulae and a symbiotic star, measured from high–resolution spectra obtained with the IUE satellite, lead to electron densities that are compatible, and are also in good agreement with those deduced from line ratios in other species. This provides observational support for the accuracy of the atomic data adopted in the present calculations.

022.080   Present and future of the Rome gravitational wave experiment.
F. Ricci.
9. Italian Conference on General Relativity and Gravitational Physics, p. 375 – 388 (1991). – See Abstr. 012.021 for the main entry.

022.081   Approaching the dc SQUID limit for a conventional cryogenic gravitational radiation detector.
C. Cosmelli.
9. Italian Conference on General Relativity and Gravitational Physics, p. 557 – 561 (1991). – See Abstr. 012.021 for the main entry.

022.082   Progress report on the development of the Gyromagnetic Electron Gyroscope.
P. Falferi, M. Cerdonio, R. Macchietto, G. A. Prodi, S. Vitale.
9. Italian Conference on General Relativity and Gravitational Physics, p. 573 – 576 (1991). – See Abstr. 012.021 for the main entry.
The authors are developing a gyroscope based on the phenomenon of magnetization by rotation, the Gyromagnetic Electron Gyroscope. The ferromagnetic core of the prototype under assembly is made by many rods whose magnetization is read in parallel by a rf SQUID via a superconducting transformer. A dominant noise source, which limits severely the performances of the gyroscope, is the thermal magnetic noise due to the ferromagnet. They have identified the parameters which describe this noise.

022.083   Spectral analysis for gravitational antennas.
S. Frasca, F. R. Mariani.
9. Italian Conference on General Relativity and Gravitational Physics, p. 577 – 581 (1991). – See Abstr. 012.021 for the main entry.

022.084   Fast estimation of the noise of a graviational wave antenna.
P. Astone, P. Bonifazi, G. V. Pallottino.
9. Italian Conference on General Relativity and Gravitational Physics, p. 582 – 588 (1991). – See Abstr. 012.021 for the main entry.
The authors report on the estimation of the noise of a gravitational wave antenna. The aim is to reduce the measurement time required to obtain accurate estimates. They discuss a solution based on the Wiener–Kolmogoroff optimum filter and present experimental results on the noise of a gravitational wave detector, that show that the variance of a narrowband process can be determined in a time smaller than its correlation time.

022.085   Fabry–Perot resonators with oscillating mirrors for interferometric GW antennas.
S. Solimeno, F. Barone, L. Di Fiore, L. Milano, G. Russo.
9. Italian Conference on General Relativity and Gravitational Physics, p. 603 – 607 (1991). – See Abstr. 012.021 for the main entry.

022.086   Report on the operation of the 389 kg cryogenic gravitational wave antenna ALTAIR at I.F.S.I.
P. Bonifazi, M. G. Castellano, V. Loschiavo, M. Visco.
9. Italian Conference on General Relativity and Gravitational Physics, p. 624 – 629 (1991). – See Abstr. 012.021 for the main entry.

022.087   Low noise ultra high frequency R.F.–SQUID.
A. Cavalleri, M. Cerdonio, G. Fontana, G. Jung, R. Mezzena, S. Vitale, J. P. Zendri.
9. Italian Conference on General Relativity and Gravitational Physics, p. 630 – 633 (1991). – See Abstr. 012.021 for the main entry.
SQUID devices are used as low noise amplifiers in the instruments chain necessary to detect gravitational waves by Weber's bars. The authors present the characteristic of a two holes R.F.–SQUID, operating at 400 MHz frequency.

022.088   Dusty plasmas.
T. G. Northrop.
Phys. Scr., Vol. 45, No. 5, p. 475 – 490 (May 1992). – See Abstr. 012.023 for the main entry.
Dust grains immersed in plasma become charged. The charge is determined by the plasma characteristics, by secondary and photoemission from the grain, by grain velocity, and at any given instant by the past time history of the charging currents. This charge affects the Coulomb drag on a grain moving through the plasma. It affects the motion of the grain in an electromagnetic field of a planetary magnetosphere, and it is involved in the formation of the spokes in Saturn's rings and in the erosion of the rings by micrometeorites. And finally it affects the coagulation rate of dust into larger bodies.

022.089   Low–frequency modes in dusty plasmas.
P. K. Shukla.
Phys. Scr., Vol. 45, No. 5, p. 504 – 507 (May 1992). – See Abstr. 012.023 for the main entry.
Linear properties of low–frequency electrostatic and electromagnetic modes in dusty plasmas are studied. The wave spectra in the presence of static and non-static charged dust grains are presented. The relevance of this investigation to astrophysical and cometary plasmas is pointed out.

022.090 **Two–stream instabilities in unmagnetized dusty plasmas.**
R. Bharuthram, H. Saleem, P. K. Shukla.
Phys. Scr., Vol. 45, No. 5, p. 512 – 514 (May 1992). – See Abstr. 012.023 for the main entry.
Two–stream instabilities in an unmagnetized multi–species dusty plasma are investigated. Four different plasma models are considered. The effect of the dust particles on the instability growth rate and the threshold drift velocity for excitation of the instability are examined.

022.091 **The atomic oxygen $^3P \to {}^1D$ electron–excitation cross section near threshold.**
J. P. Doering.
Geophys. Res. Lett., Vol. 19, No. 5, p. 449 – 451 (3 Mar 1992).
Preliminary results of a new measurement of the differential electron excitation cross section of the O I ($^3P \to {}^1D$) transition near threshold confirm the presence of a large peak in the inelastic cross section at 5 to 6 eV. The new results are compared with previous measurements and theoretical calculations. The near–threshold shape of the O I ($^3P \to {}^1D$) cross section has interesting implications for the aeronomy of the thermosphere.

022.092 **Hyperfine structure of the $2s\ ^2S$ level of Li–like multicharged ions.**
M. B. Shabaeva, V. M. Shabaev.
Phys. Lett. A, Vol. 165, No. 1, p. 72 – 78 (4 May 1992).
All corrections of relative order $(1/Z)\,(\alpha Z)^2$ to the hyperfine structure for the $2s\ ^2S$ level of a lithium–like multicharged ion are calculated. $\lambda = 0.3071(3)$cm is obtained for the wavelength of the transition between the hyperfine structure components for the $2s\ ^2S$ level of $^{57}Fe^{23+}$, which is important for astrophysics researches. A possibility for a verification of the quantum electrodynamics effects in the investigations of the hyperfine structure of the lithium–like multicharged ions is noted.

022.093 **Rate coefficients for the excitation of infrared and ultraviolet lines in C II, N III, and O IV.**
R. D. Blum, A. K. Pradhan.
Astrophys. J., Suppl. Ser., Vol. 80, No. 1, p. 425 – 452 (May 1992).
Electron impact excitation of boron–like C, N, and O is the primary mechanism for the formation of a number of IR and UV emission lines that are useful density and temperature diagnostics for a variety of astrophysical sources. New and improved collision strengths and Maxwellian–averaged rate coefficients for average temperatures between 1000K and 40,000K are presented for all the prominent transitions in the spectra of C II, N III, and O IV. The collision strengths show extensive autoionization structures that are delineated in detail and which enhance the rate coefficients for several transitions by a considerable amount. Particular attention is directed toward the fine–structure IR transition $^2P^0_{1/2} - {}^2P^0_{3/2}$ and the dipole–allowed and intercombination UV transitions $^2P^0_{1/2,3/2} - {}^2D^2_{3/2,5/2}$, $^2S^2_{1/2} - {}^2P_{1/2,3/2}$, and $^2P^0_{1/2,3/2} - {}^4P_{1/2,3/2,5/2}$, respectively, in the three ions. Maxwellian–averaged collision strengths are calculated for all possible fine–structure transitions among the states included in the eigenfunction expansion of the target ion.

022.094 **Collision strengths and excitation rate coefficients for transitions in Ca XV.**
K. M. Aggarwal.
Astrophys. J., Suppl. Ser., Vol. 80, No. 1, p. 453 – 471 (May 1992).
Collision strengths among fine–structure levels of the $1s^2 2s^2 2p^2$, $1s^2 2s 2p^3$, and $1s^2 2p^4$ configurations of Ca XV have been computed in the j–j coupling scheme using the Dirac–Fock R–matrix program. All partial waves with $J \leqslant 29/2$ are included, and collision strengths have been computed on a dense energy grid in order to elucidate the resonances. The results are tabulated in a wide range below 200 Ry. Contributions of the relativistic effects are assessed and comparisons are made with the earlier available collision strengths. Agreement with the available values, at electron energies above thresholds, is good for most of the transitions, but striking differences (of up to an order of magnitude) are noticed for some of the transitions. Effective collision strengths are also obtained after integrating collision strengths over a Maxwellian distribution of electron velocities, and these are tabulated at a number of temperatures below 10^7K. Comparisons are made with the earlier effective collision strengths obtained with the R–matrix method.

022.095 **Excitation rate coefficients for transitions among the $n = 1, 2$, and 3 levels of He$^+$.**
K. M. Aggarwal, J. Callaway, A. E. Kingston, K. Unnikrishnan.
Astrophys. J., Suppl. Ser., Vol. 80, No. 1, p. 473 – 477 (May 1992).
Excitation rate coefficients for transitions among the $n = 1, 2$, and 3 levels of He$^+$ are reported in the temperature range of 5×10^3K – 5×10^5K. These are based on the best available R–matrix collision strengths of Aggarwal et al. in the threshold energy region including resonances, and of the pseudostate close coupling calculations of Unnikrishnan, Callaway, and Oza at energies above thresholds. Comparisons with the earlier available effective collision strengths are made and differences are discussed. The present results are believed to be the best available in the entire electron temperature range, although the differences with the earlier published values of Aggarwal et al. are not more than 15% at temperatures below 10^5K, except for the 1s – 3p transition for which they differ up to 25%.

022.096 **Optical manifestation of microbursts of electron fluxes.**
S. Yu. Ermilov, A. V. Mikhalev.
J. Atmos. Terr. Phys., Vol. 53, No. 11/12, p. 1157 – 1160 (Nov–Dec 1991). – See Abstr. 012.024 for the main entry.
The authors consider the behaviour of populations of the levels ^1S and ^1D of neutral oxygen [OI] at ionospheric heights when microbursts of electron fluxes impulsively excite the excitation mechanism. The authors also discuss the possibility that microbursts of electron fluxes can manifest themselves in the form of optical flashes at 557.7 and 630.0 nm.

022.097 **Emission lines from O IV as a plasma diagnostic.**
B. N. Dwivedi, A. K. Gupta.
Sol. Phys., Vol. 138, No. 2, p. 283 – 290 (Apr 1992).
Using several density diagnostic O IV theoretical line ratios and corresponding observed values for the same source by Sandlin et al. (1984) and Sandlin, Brueckner, and Tousey (1977), the authors find that an emitting region has a multidensity structure. They discuss several other line ratios for density measurement in sunspots, active regions, and flares.

022.098 **Effect of temperature on shock metamorphism of single–crystal quartz.**
F. Langenhorst, A. Deutsch, D. Stöffler, U. Hornemann.
Nature, Vol. 356, No. 6369, p. 507 – 509 (9 Apr 1992). Letter–to–the–editor.
Features characteristic of shock metamorphism in target rocks are the main diagnostic tool for recognizing impact phenomena on the Earth and other planetary bodies, and experimentally calibrated shock effects in silicate minerals have been important in elucidating the pressure histories of these rocks. High–temperature shock metamorphism must also have been of great importance in the collision history of meteorite parent bodies in the early solar system. The authors report the results of shock experiments on single–crystal quartz heated to 630°C, which show that the physical properties of shocked quartz depend strongly on the pre–shock temperature. They conclude that existing shock–wave barometers cannot be applied to high–temperature target rocks, and that new barometers independent of pre–shock temperature will be required to understand the shock pressure history of terrestrial and planetary impact formations.

022.099   Polynomial coefficients for calculating O_2 Schumann–Runge cross sections at 0.5 cm^{-1} resolution.
K. Minschwaner, G. P. Anderson, L. A. Hall, K. Yoshino.
J. Geophys. Res., Vol. 97, No. D9, p. 10103 – 10108 (20 Jun 1992).
The authors have fitted O_2 cross sections from 49,000 and 57,000 cm^{-1} with temperature dependent polynomial expressions, providing an accurate and efficient means of determining Schumann–Runge band cross sections for temperatures $130 < T < 500$K. The least squares fits were carried out on a 0.5 cm^{-1} spectral grid, using cross sections obtained from a Schumann–Runge line–by–line model that incorporates the most recent spectroscopic data. The O_2 cross sections do not include the underlying Herzberg continuum, but they do contain contributions from the temperature dependent Schumann–Runge continuum. The cross sections are suitable for use in ultra–violet transmission calculations at high spectral resolution. They should also prove useful for updating existing parameterizations of ultraviolet transmission and O_2 photolysis.

022.100   The 2140 cm^{-1} band of frozen CO in ion–irradiated and unirradiated mixtures with methanol and water.
M. E. Palumbo, G. Strazzulla.
Astron. Astrophys., Vol. 259, No. 1, p. L12 – L14 (Jun 1992).
Letter–to–the–editor.
The authors present laboratory data on the shape of the 4.67 μm band of frozen CO in mixtures with CH_3OH and $CH_3OH + H_2O$. The mixtures have been produced both by depositing CO together with CH_3OH (or $CH_3OH + H_2O$) and by depositing only CH_3OH (or $CH_3OH + H_2O$) and and producing CO by ion irradiation. The authors find that the mixtures here studied are good candidates to reproduce the band observed in astronomical sources. The effect of producing CO by ion irradiation is that the band results to be broader. The same is observed for the narrow band of irradiated pure CO. Such a broadening might help, under circumstances to be investigated in detail, to better reproduce the astronomical spectra.

022.101   Hydrogen cyanide polymerization: a preferred cosmochemical pathway.
C. N. Matthews.
J. Br. Interplanet. Soc., Vol. 45, No. 1, p. 43 – 48 (Jan 1992).
Special issue: Exobiology. S. Santoli (ed.).

022.102   Simulations of surface winds at the Viking Lander sites using a one–level model.
A. F. C. Bridger, R. M. Haberle.
Workshop on the Martian Surface and Atmosphere Through Time, p. 24 (1992). Abstract. – See Abstr. 012.031 for the main entry.

022.103   Simulations of surface winds at the Viking Lander sites.
A. F. C. Bridger, R. M. Haberle.
Workshop on the Martian Surface and Atmosphere Through Time, p. 25 (1992). Abstract. – See Abstr. 012.031 for the main entry.

022.104   Martian surface simulations.
R. W. Gaskell.
Workshop on the Martian Surface and Atmosphere Through Time, p. 57 – 58 (1992). Abstract. – See Abstr. 012.031 for the main entry.

022.105   Chemical reactions in astrochemistry.
B. R. Rowe.
IAU Symposium No. 150: Astrochemistry of cosmic phenomena, p. 7 – 12 (1992). – See Abstr. 012.029 for the main entry.
This paper is devoted to chemistry in the gas phase dealing with ion–molecule reactions at extremely low temperature. The experimental techniques that have been used in this field are shortly presented and the reactions that have been studied using the CRESU(S) method reviewed. The most recent measurements concerning dissociative recombination are discussed, including

studies of branching ratio and new determination of the rate coefficient for H_3^+ ions.

022.106   Neutral–neutral reactions: a new CRESU study.
C. Rebrion, A. Defrance, J. L. Queffelec, B. R. Rowe, D. Travers.
IAU Symposium No. 150: Astrochemistry of cosmic phenomena, p. 13 – 14 (1992). – See Abstr. 012.029 for the main entry.
A new CRESU experiment devoted to neutral–neutral chemistry at low temperature is being built. The first reactions to be studied are $N + NO$ and $CN + O_2$.

022.107   ISO and laboratory astrophysics.
L. J. Allamandola.
IAU Symposium No. 150: Astrochemistry of cosmic phenomena, p. 15 – 16 (1992). – See Abstr. 012.029 for the main entry.

022.108   Electrostatic fragmentation of dust particles in laboratory.
J. Svestka, E. Grün.
IAU Symposium No. 150: Astrochemistry of cosmic phenomena, p. 17 – 18 (1992). – See Abstr. 012.029 for the main entry.
The electrostatic fragmentation of dust particles is one of important destruction mechanisms for cosmic dust particles which leads to conversion of the solid phase of interstellar and interplanetary matter to the gas phase. Experimental laboratory work on simulation of the electrostatic fragmentation was started with loosely bound Al_2O_3 particles of 1 to 10 micrometers size.

022.109   Contribution of PAHs to the interstellar extinction curve.
C. Joblin, A. Leger, P. Martin, D. Defourneau.
IAU Symposium No. 150: Astrochemistry of cosmic phenomena, p. 21 – 22 (1992). – See Abstr. 012.029 for the main entry.
Absorption spectra of some gaseous PAHs, either pure species or natural mixtures, have been obtained in the VUV–visible region and compared to the interstellar extinction curve. The assumption that free PAHs are ubiquitous in the ISM cannot be rejected by incompatibility between the interstellar extinction curve and the absorption spectra of such molecules. PAHs absorb in the FUV rise and may give an important contribution to the bump at 2200 Å. The authors have derived that about 15% of the cosmic carbon is involved in these molecules.

022.110   Search for the 4430 Å DIB in the spectra of coronene cation and neutral ovalene.
P. Ehrenfreund, L. d'Hendecourt, L. Verstraete, A. Léger, W. Schmidt, D. Defourneau.
Astron. Astrophys., Vol. 259, No. 1, p. 257 – 264 (Jun 1992).
Polycyclic aromatic hydrocarbons (PAHs) have been proposed as candidates to explain the diffuse interstellar bands (DIBs). Among the great variety of these species, two seemed of special interest: the coronene cation and neutral ovalene, because solution spectra reported in the literature show that they possess transitions near the strong 4430 Å DIB. The authors have performed laboratory measurements, using rare gas matrix isolation techniques and UV photolysis to provide a data base with the spectra of these two species and their ions, almost free from environmental shifts and broadening. Bands are found at 4590 Å for the coronene cation and 4305 Å for neutral ovalene. The poor agreement of the visible spectra of the coronene cation and the neutral and ionized ovalene with current astronomical observations excludes these species as carriers of already known DIBs. Conversely, these laboratory data can be used to search for the presence of these two PAHs in the spectra of interstellar objects. They provide the first data on coronene and ovalene neutrals and cations in neon matrices with a direct relevance to the astrophysical context. The authors further discuss the possibility that other PAH molecules could be carriers of broad DIBs, considering the widths and strengths of measured bands for coronene and ovalene.

022.111 Thermal emission spectra from polycyclic aromatic hydrocarbon molecules of astrophysical interest.
V. Mennella, L. Colangeli, E. Bussoletti.
Astrophys. J., Lett., Vol. 391, No. 1, p. L45 – L48 (20 May 1992).

Polycyclic aromatic hydrocarbon (PAH) molecules are among the major candidates that have been proposed in the past to explain a class of infrared bands, the so–called unidentified infrared bands detected in emission toward a wide variety of Galactic sources rich in ultraviolet flux. Up to now, laboratory measurements on representative PAH molecules have been mostly performed in absorption. Here, the authors present infrared thermal emission spectra of three PAHs, coronene, chrysene, and 1–methylcoronene, obtained at a temperature of 240°C. These data may help clarify open questions concerning both the presence of polycyclic aromatic hydrocarbons and the actual emission mechanisms active in space.

022.112 Cumulene carbenes in space and in the laboratory.
J. M. Vrtilek, C. A. Gottlieb, T. C. Killian,
P. Thaddeus, J. Cernicharo, M. Guélin, G. Paubert.
IAU Symposium No. 150: Astrochemistry of cosmic phenomena, p. 23 – 24 (1992). – See Abstr. 012.029 for the main entry.

Astronomical searches for H_2CCC and H_2CCCC, based on frequencies from laboratory identifications, have resulted in detections toward TMC–1 and IRC +10216. These new interstellar species are possibly the first of a new family of highly polar carbon chains; they are only the second and third carbenes (carbon molecules with two nonbonded electrons) known in space.

022.113 UV–visible and near IR absorption characteristics of interstellar PAHs. I. $C_{10}H_8{}^+$.
IAU Symposium No. 150: Astrochemistry of cosmic phenomena, p. 25 – 26 (1992). – See Abstr. 012.029 for the main entry.

The authors have initiated a systematic and detailed study of the spectroscopy of neutral and ionized polycyclic aromatic hydrocarbons (PAHs). They report the results obtained for the smallest PAH ($C_{10}H_8$) and discuss their astrophysical applications.

022.114 Infrared spectroscopy of interstellar and solar system ice analogs: measurement of optical constants.
D. M. Hudgins, S. A. Sandford, A. G. G. M. Tielens,
L. J. Allamandola.
IAU Symposium No. 150: Astrochemistry of cosmic phenomena, p. 27 – 28 (1992). – See Abstr. 012.029 for the main entry.

Laboratory spectra through the mid–infrared have been used to calculate the optical constants (n and k) for a variety of pure and mixed molecular ices.

022.115 Formation of organic molecules by formaldehyde reactions in astrophysical ices at very low temperatures.
W. A. Schutte, L. J. Allamandola, S. A. Sandford.
IAU Symposium No. 150: Astrochemistry of cosmic phenomena, p. 29 – 30 (1992). – See Abstr. 012.029 for the main entry.

Warm–up of astrophysical ice analogues containing formaldehyde produced organic residues in large abundances. It is argued that formaldehyde reactions at very low temperatures could be an important source of interstellar and cometary organic molecules.

022.116 Computational chemistry approach to space chemistry.
Y. Ellinger.
IAU Symposium No. 150: Astrochemistry of cosmic phenomena, p. 31 – 38 (1992). – See Abstr. 012.029 for the main entry.

This review paper presents the results of state of the art Quantum Chemistry calculations in the field of astrochemistry. It provides selected examples to illustrate the possible contribution of molecular orbital theories to solving a number of problems of astrophysical interest ranging from identification of new molecules to IR emission analysis and rate constants determinations.

022.117 Metamorphism of cosmic dust: diagnostic infrared signatures.
J. A. Nuth III.
IAU Symposium No. 150: Astrochemistry of cosmic phenomena, p. 39 – 40 (1992). – See Abstr. 012.029 for the main entry.

022.118 Astropysical problems involving carbon re–appraised.
J. P. Hare, H. W. Kroto.
IAU Symposium No. 150: Astrochemistry of cosmic phenomena, p. 47 – 54 (1992). – See Abstr. 012.029 for the main entry.

The article contains a brief account of the processes responsible for the synthesis of carbon in stars and its dissemination throughout the Galaxy as this information is deemed necessary to gain an intrinsic understanding of the amazing role carbon plays in nature.

022.119 Iron and chromium absorption bands in the spectra of terrestrial pyroxenes: application to mineralogic remote sensing of asteroid surfaces.
D. I. Shestopalov, L. F. Golubeva, M. N. Taran,
V. M. Khomenko.
Sol. Syst. Res., Vol. 25, No. 4, p. 332 – 340 (Jan 1992). English translation of Astron. Vestn., Tom 25, No. 4, p. 442 – 452 (1991).

The dependence of Fe^{2+} and Cr^{3+} absorption bands position in the spectra of terrestrial pyroxenes upon temperature has been studied. The study found a shortwave shift (~ 10 nm) of 0.95 μm Fe^{2+} band and the absence of the Cr^{3+} bands shift in the interval from room temperature to the nitrogen temperature. The possibility has been demonstrated of remote diagnosing of pyroxene composition by the position of Fe^{2+} absorption band center near 0.505 μm. The chemical compositon of pyroxenes in several light asteroids has been estimated. Calcium– and iron–rich pyroxenes seem to be more widespread on the surface of S–asteroids than in achondrites and ordinary chrondrites. The existence of chromium–containing pyroxenes in the main belt appears likely.

022.120 Reaction rates of C^+ with OH at low interstellar temperatures.
M. L. Dubernet, M. Gargaud, R. McCarroll.
Astron. Astrophys., Vol. 259, No. 1, p. 373 – 376 (Jun 1992).

The adiabatic rotational state method is used to investigate the reaction of C^+ ions with the OH radical for the formation of CO and CO^+ at low interstellar temperatures. The OH molecule being in an $X^2\Pi$ state, full account has been taken of the coupling of the electronic and rotational angular momentum in treating the collision dynamics of the ion–molecule system. The potential energies of the adiabatic rotational states correlated to the lowest rotational state mainfolds of the $^2\Pi_{3/2}$ and $^2\Pi_{1/2}$ electronic states of OH are first obtained. The adiabatic capture model is subsequently used to determine the reactive rate constants for temperatures in the range from 1 to 200K.

022.121 Search for the 4430 Å DIB in the spectrum of coronene.
P. Ehrenfreund, L. d'Hendecourt, L. Verstraete,
A. Leger, W. Schmidt.
IAU Symposium No. 150: Astrochemistry of cosmic phenomena, p. 135 – 136 (1992). – See Abstr. 012.029 for the main entry.

Polycyclic aromatic hydrocarbon (PAH) molecules have been proposed as candidates to explain the diffuse interstellar bands (DIBs). The authors have performed laboratory measurements of coronene, using rare gas matrix isolation techniques and UV photolysis, to search for a possible identification of the 4430 Å DIB.

022.122 Theoretical modelling of the infrared fluorescence by interstellar polycyclic aromatic hydrocarbons.
W. A. Schutte, A. G. G. M. Tielens, L. J. Allamandola.
IAU Symposium No. 150: Astrochemistry of cosmic phenomena, p. 137 – 138 (1992). – See Abstr. 012.029 for the main entry.

The authors investigated the relation between the PAH size distribution and the observed emission spectrum by theoretical modelling. Substantial differences between the IR properties of interstellar and laboratory PAHs are found, possibly resuting from ionization.

022.123   Quasi–random narrow band model fits to 1.6 – 2.5 μm laboratory methane spectra and application to Jupiter.
K. H. Baines, R. A. West, L. P. Giver, F. Moreno,
T. W. Momary.
Bull. Am. Astron. Soc., Vol. 23, No. 3, p. 1133 (1991). Abstract. – See Abstr. 012.037 for the main entry.

022.124   Computational model of the collision induced absorption spectra of H_2–He pairs in the fundamental band.
A. Borysow.
Bull. Am. Astron. Soc., Vol. 23, No. 3, p. 1133 – 1134 (1991). Abstract. – See Abstr. 012.037 for the main entry.

022.125   Laboratory studies of radiation chemistry in the Jovian atmosphere.
G. D. McDonald, W. R. Thompson, C. Sagan.
Bull. Am. Astron. Soc., Vol. 23, No. 3, p. 1136 (1991). Abstract. – See Abstr. 012.037 for the main entry.

022.126   Density shifts and line strengths for 4–0 and 5–0 quadrupole transitions in molecular hydrogen: implications for the spectra of the Jovian planets.
M. E. Mickelson, L. E. Larson, D. W. Ferguson, K. N. Rao.
Bull. Am. Astron. Soc., Vol. 23, No. 3, p. 1136 (1991). Abstract. – See Abstr. 012.037 for the main entry.

022.127   Measurements of line strengths and pressure broadening coefficients of selected lines of the 6190 band of methane at 77, 180, and 297 Kelvin.
W. K. Wells, D. M. Hunten, J. Lunine, P. V. Cvijin,
G. H. Atkinson.
Bull. Am. Astron. Soc., Vol. 23, No. 3, p. 1137 (1991). Abstract. – See Abstr. 012.037 for the main entry.

022.128   Laboratory study of the opposition effect.
B. W. Hapke, R. M. Nelson, W. D. Smythe,
V. Gharakanian, L. J. Horn, A. L. Lane.
Bull. Am. Astron. Soc., Vol. 23, No. 3, p. 1139 (1991). Abstract. – See Abstr. 012.037 for the main entry.

022.129   The formation of deuterated molecules in dense clouds.
T. J. Millar.
IAU Symposium No. 150: Astrochemistry of cosmic phenomena, p. 211 – 215 (1992). – See Abstr. 012.029 for the main entry.
The author lists the deuterated species detected so far, together with an estimate of their abundance ratios relative to their hydrogen bearing parent for a variety of astronomical regions, discusses the basic chemical processes which fractionate deuterium together with some simple estimates of the degree of fractionation which can result from these processes. He then discusses in detail the deuteration of hydrocarbon molecules and shows that the effects of chemical reactions which can recycle parent species can appreciably affect estimates of fractionation. Finally he discusses how structural effects may inhibit fractionation in the cyanopolyynes and related species.

022.130   Optical properties of dust aggregates.
T. Kozasa, J. Blum, T. Mukai.
Bull. Am. Astron. Soc., Vol. 23, No. 3, p. 1143 (1991). Abstract. – See Abstr. 012.037 for the main entry.

022.131   Light scattering by a randomly oriented cluster of spheres.
K. Muinonen, K. Lumme.
Bull. Am. Astron. Soc., Vol. 23, No. 3, p. 1144 (1991). Abstract. – See Abstr. 012.037 for the main entry.

022.132   Formaldehyde reactions in dark clouds.
A. D. Sen, V. G. Anicich, S. R. Federman.
Astrophys. J., Vol. 391, No. 1, p. 141 – 143 (20 May 1992).
The low–pressure reactions of formaldehyde (H_2CO) with D^+, D_2^+, D_3^+, and He^+ have been studied by the ion cyclotron resonance technique. These reactions are potential loss processes for formaldehyde in cores of dark interstellar clouds. The deuterated reactants, which are easier to study experimentally, represent direct analogs for protons. Rate coefficients and branching ratios of product channels have been measured. Charge transfer is observed to be the dominant reaction of H_2CO with D^+, D_2^+, and He^+ ions. Only the D_3^+ reaction exhibits a proton transfer channel. All reactions proceed at rate coefficients near the collision limit. Proton–deuteron exchange reactions were found to be inefficient processes in the formaldehyde system.

022.133   Evidences for the terrestrial magma ocean from high–pressure melting experiments.
E. Takahashi.
Workshop on the Physics and Chemistry of Magma Oceans from 1 bar to 4 Mbar, p. 56 – 57 (1992). Abstract. – See Abstr. 012.032 for the main entry.

022.134   An expanded program of laboratory measurements on ammonia's microwave absorption spectrum.
T. R. Spilker.
Bull. Am. Astron. Soc., Vol. 23, No. 3, p. 1155 (1991). Abstract. – See Abstr. 012.037 for the main entry.

022.135   Laboratory reflectance in the UV/VIS of ion bombarded ices: application to the Jovian satellites.
N. J. Sack, R. E. Johnson, R. A. Baragiola.
Bull. Am. Astron. Soc., Vol. 23, No. 3, p. 1170 – 1171 (1991). Abstract. – See Abstr. 012.037 for the main entry.

022.136   The spectral absorption of CO_2 ice in the ultraviolet, visible, and near–infrared.
G. B. Hansen.
Bull. Am. Astron. Soc., Vol. 23, No. 3, p. 1176 (1991). Abstract. – See Abstr. 012.037 for the main entry.

022.137   Microwave spectroscopy of transient interstellar molecules.
S. Saito.
Symposium on Chemistry and Spectroscopy of Interstellar Molecules, p. 103 – 109 (1992). – See Abstr. 012.039 for the main entry.
Laboratory microwave spectroscopy of transient molecules is discussed from an astronomical point of view. A high sensitivity millimeter and submillimeter wave spectrometer has been developed and used to detect more than twenty transient molecules. Seven transient species among the molecules studied in the laboratory have been identified as new interstellar molecules. Finally, the discussion is directed at the spectroscopically and astronomically important aspects of the author's latest study on the CH_2CN radical.

022.138   Microwave spectrum of CCS radical in vibrationally excited states.
M. Tanimoto, S. Saito, S. Yamamoto, K. Kawaguchi.
Symposium on Chemistry and Spectroscopy of Interstellar Molecules, p. 111 – 112 (1992). – See Abstr. 012.039 for the main entry.
Microwave spectra of the interstellar molecule CCS have been observed in three vibrationally excited states $v_1 = 1$, $v_2 = 1$ and $v_3 = 1$ to obtain information on the vibrational frequencies.

022.139   Vibrationally excited carbon–chain molecules in the laboratory and in space; C_3N and C_4H.
H. Mikami.
Symposium on Chemistry and Spectroscopy of Interstellar Molecules, p. 113 – 115 (1992). – See Abstr. 012.039 for the main entry.

022.140 **The structure and chemistry of the carbon–chain molecule, C₃H.**
M. Kanada.
Symposium on Chemistry and Spectroscopy of Interstellar Molecules, p. 117 – 120 (1992). – See Abstr. 012.039 for the main entry.
The rotational spectra of three ^{13}C isotopic species of the C_3H radical were studied with laboratory microwave spectroscopy.

022.141 **The dissociative recombination of H₃⁺ with electrons is not very slow.**
T. Amano.
Symposium on Chemistry and Spectroscopy of Interstellar Molecules, p. 121 – 126 (1992). – See Abstr. 012.039 for the main entry.
The dissociative recombination rate coefficients for H_3^+ in the ground vibrational state is measured for the five rotational levels by monitoring the decay of the infrared absorption signals as a function of time.

022.142 **Infrared spectroscopy of transient species.**
K. Kawaguchi.
Symposium on Chemistry and Spectroscopy of Interstellar Molecules, p. 127 – 132 (1992). – See Abstr. 012.039 for the main entry.
High–resolution infrared spectroscopy of transient species such as free radicals and molecular ions is discussed with reference to astrochemical interest. The infrared diode laser and Fourier transform studies of the C_3 molecule are described in connection with its detection in IRC + 10216; possibility of the observation of far infrared transitions in molecular clouds is discussed.

022.143 **Interstellar polycyclic aromatic hydrocarbons.**
L. J. Allamandola.
Symposium on Chemistry and Spectroscopy of Interstellar Molecules, p. 133 – 140 (1992). – See Abstr. 012.039 for the main entry.
The infrared evidence which supports the PAH hypothesis is briefly summarized. Rather than presenting a general discussion of the assignments, this paper focuses on the spectroscopic constraints placed on the molecular shape, size and structure of interstellar PAHs. It is pointed out that ionized, gas phase, symmetric PAHs containing 20 – 50 carbon atoms are the dominant contributors to the narrow interstellar infrared emission bands while larger species contribute to the broader components. These apparently ubiquitous, complex, organic ring molecules are more abundant than the other interstellar polyatomic molecules known. This points to a unique chemical history of the interstellar PAHs.

022.144 **IR emission from PAH molecules and interstellar dust.**
W. W. Duley.
Symposium on Chemistry and Spectroscopy of Interstellar Molecules, p. 141 – 144 (1992). – See Abstr. 012.039 for the main entry.
The population of vibrational levels in PAH emitters is examined using a simple optical pumping model. It is shown that significant pumping effects occur for vibrational states with energies $\leqslant 200$ cm^{-1}. The long radiative lifetime of these states implies that they may exhibit enhanced or inverted population relative to lower levels. The apparent absence of strong discrete PAH emission at $\lambda > 15 \mu m$ suggests the existence of a hitherto unrecognized quenching mechanism and may provide useful insight into the physical state of these emitters.

022.145 **The time dependent structure of carbonaceous grains and mantles in the interstellar medium.**
A. P. Jones, W. W. Duley, D. A. Williams.
Symposium on Chemistry and Spectroscopy of Interstellar Molecules, p. 145 – 147 (1992). – See Abstr. 012.039 for the main entry.
Interstellar HAC likely contains two major carbon phases; graphitic and polymeric/diamond–like. This structure is the result of carbon species accreting onto interstellar grains as hydrogen–rich polymeric and/or diamond–like materials which are subsequently partially photoprocessed into a graphitic type of material. It is shown that a HAC model for interstellar carbon grains can explain the IR emission features (UIR bands) in terms of small linear PAHs that are an intimate component of interstellar HAC.

022.146 **Growing silicon–bearing and PAH molecules with ion/molecule reactions.**
D. K. Bohme.
Symposium on Chemistry and Spectroscopy of Interstellar Molecules, p. 155 – 160 (1992). – See Abstr. 012.039 for the main entry.
Results of recent gas–phase laboratory studies are highlighted which provide insight into of the formation of silicon–bearing molecules with chemistry initiated by ground–state atomic silicon ions, the formation of naphthalene ions from benzene ions, and the influence of naphthalene on the intrinsic chemistry of atomic silicon ions.

022.147 **On the dynamical problem of a generalized thermoelastic granular infinite cylinder under initial stress.**
A. M. Elnaggar.
Astrophys. Space Sci., Vol. 190, No. 2, p. 177 – 190 (Apr 1992).
The object of the present paper is to investigate the influence of initial stress on the waves propagation in a generalized thermoelastic granular medium subjected to the boundary conditions that the outer surface is traction free. In addition, it is subjected to temperature boundary conditions. The wave velocity equation for the generalized thermoelastic granular medium Rayleigh wave under the influence of initial stress has been obtained. The classical result has been derived as a limiting case similar to that obtained by Ewing et al. (1957).

022.148 **The yield of H atoms through dissociative recombination of polyatomic ions.**
B. R. Rowe, A. Canosa, J. C. Gomet, J. L. Queffelec, M. Morlais.
Symposium on Chemistry and Spectroscopy of Interstellar Molecules, p. 161 – 166 (1992). – See Abstr. 012.039 for the main entry.
The yield of H atoms through dissociative recombination of several polyatomic ions with electrons has been studied using a plasma flow tube experiment. It was found that CH_5^+, H_3O^+, N_2OH^+, NH_4^+, HCO_2^+ produce about one H atom per dissociative recombination. On the other hand, a low branching ratio was obtained concerning $OCSH^+$ and H_3S^+. Different precursor ions were used and a good agreement was found between all measurements.

022.149 **The CRESUS apparatus: a tool for studying mass–selected ion–molecule reactions at interstellar temperatures.**
J. B. Marquette, B. R. Rowe, C. Rebrion.
Symposium on Chemistry and Spectroscopy of Interstellar Molecules, p. 167 – 172 (1992). – See Abstr. 012.039 for the main entry.
The CRESUS (Cinétique de Réactions en Ecoulement Supersonique Uniforme avec Sélection) experiment allows rate coefficients of reactions of mass–selected ions with molecules to be determined down to 20K. The result is compared to other experimental studies and discussed in the light of association reaction theories including the related interstellar radiative association processes.

022.150 **CRESU measurements of ion–molecules reactions down to 20K.**
C. Rebrion, J. B. Marquette, B. R. Rowe.
Symposium on Chemistry and Spectroscopy of Interstellar Molecules, p. 173 – 178 (1992). – See Abstr. 012.039 for the main entry.
The few experimental techniques allowing measurements of ion–molecule reaction rate coefficients at very low temperatures

are reviewed. For numerous reactions, capture theories which consider only the long–range part of the intermolecular potential are in satisfying agreement with the experimental data. Two of them, having an analytical expression, can be used for straight-forward calculations. Special attention has been paid to the influence of rotational energy.

022.151   **Ab initio study of the potential energy surfaces for the radiative association reaction $C^+ + H_2 \rightarrow CH_2^+ + h\nu$.**
A. Ozeki, S. Iwata.
Symposium on Chemistry and Spectroscopy of Interstellar Molecules, p. 179 – 181 (1992). – See Abstr. 012.039 for the main entry.

022.152   **Time–of–flight analysis of $C_3H_3^+$ ions produced by the reactions $C_3H_4^+ + C_3H_4$.**
T. Ogata, S. Suzuki, I. Koyano.
Symposium on Chemistry and Spectroscopy of Interstellar Molecules, p. 183 – 186 (1992). – See Abstr. 012.039 for the main entry.
This paper reports the results from the CP^+ and AL^+ reactions.

022.153   **Laboratory studies of possible interstellar isomeric ions.**
C. G. Freeman, M. J. McEwan.
Symposium on Chemistry and Spectroscopy of Interstellar Molecules, p. 187 – 191 (1992). – See Abstr. 012.039 for the main entry.
The authors have used the selected ion flow tube technique to examine the behaviour of four isomeric pairs of ions that are relevant to chemical processes occurring in interstellar clouds: CCN^+ and CNC^+; HOC^+ and HCO^+; HCN^+ and HNC^+; $CCCN^+$ and c–C_3N^+.

022.154   **Fast neutral reactions in cold interstellar clouds.**
M. M. Graff.
Symposium on Chemistry and Spectroscopy of Interstellar Molecules, p. 193 – 195 (1992). – See Abstr. 012.039 for the main entry.
The dynamics of exothermic neutral reactions between radical species are examined, with particular attention to reactivity at the very low energies characteristic of cold interstellar clouds. The reaction systems $O + OH$ and $O + CH$ and $C + CH$ and $C + OH$ are examined, and upper limits of rate constants for these reactions have been estimated. General predictions are made for other reactions systems. Implications for interstellar chemistry are discussed.

022.155   **Rate constants of ion–molecule reactions: importance of long–range forces.**
K. Sakimoto.
Symposium on Chemistry and Spectroscopy of Interstellar Molecules, p. 197 – 202 (1992). – See Abstr. 012.039 for the main entry.
The author summarizes recent theoretical work on low-temperature ion–molecule reactions. Particular emphasis is laid on the importance of long–range interaction. The low tempera-ture limit of the rate constant and quantum mechanical effects are also discussed.

022.156   **Type II clathrate hydrate formation in cometary ice analogs in vacuo.**
D. F. Blake, L. Allamandola, S. Sandford, D. Hudgins, F. Freund.
IAU Symposium No. 150: Astrochemistry of cosmic phenomena, p. 437 – 438 (1992). – See Abstr. 012.029 for the main entry.

022.157   **Ab initio study of the $NH_3^+ + H_2$ reaction.**
J. D. DeFrees, D. Talbi, F. Pauzat, W. Koch, A. D. McLean.
Symposium on Chemistry and Spectroscopy of Interstellar Molecules, p. 203 – 209 (1992). – See Abstr. 012.039 for the main entry.
The unusual dependence of the rate of this reaction on temperature has been explained via a two–step reaction mecha-nism. An initio molecular orbital theory has been used to confirm the qualitative features of this hypothesis.

022.158   **Ion–molecule radiative association.**
A. D. Sen, V. G. Anicich, M. J. McEwan.
Symposium on Chemistry and Spectroscopy of Interstellar Molecules, p. 211 – 213 (1992). – See Abstr. 012.039 for the main entry.
Results of a number of ion–molecule association reactions are presented. Bimolecular and termolecular association rate coeffi-cients have been measured for a number of ion–molecule system of relevance in interstellar cloud chemistry by ion cyclotron resonance spectrometry.

022.159   **Gas–phase reactions of $SiC_6H_6^+$ and $SiC_{10}H_8^+$ with simple molecules: a possible role for polycyclic aromatic hydrocarbons in the synthesis of interstellar molecules.**
S. Wlodek, H. Wincel, D. K. Bohme.
Symposium on Chemistry and Spectroscopy of Interstellar Molecules, p. 215 – 217 (1992). – See Abstr. 012.039 for the main entry.

022.160   **L–shell X–ray opacity of many–electron atoms.**
Y.–D. Jung, R. J. Gould.
Astrophys. J., Vol. 391, No. 1, p. 403 – 408 (20 May 1992).
A formulation is developed for obtaining the L–shell photo-electric cross section (σ_L) for many–electron atomic systems. A scaling law is suggested for the evaluation of the contribution to σ_L from 2s and 2p ejection from a general system in any ionization state. The formula is expressed in terms of the effective nuclear charge (Z_{2s} and Z_{2p}) seen by the corresponding bound electrons, with Z_{2s} and Z_{2p} obtained from improved screening constants derived in a recent investigation. A comparison is made of Hartree–Fock and hydrogenic expression for σ_L with experi-mental results for atomic neon at X–ray photon energies. The special case of σ_L for atomic Fe is considered in more detail, since the L threshold is at soft X–ray energies around 0.8 keV. From the results in this work the X–ray opacity contribution from L–shell photoejection can be evaluated easily for any condition in the interstellar gas or in discrete X–ray sources.

022.161   **Charged particle motion in the field of a massive charged filament.**
S. L. Parnovskij.
Visn. Kiiv. Univ., Fiz.–Mat. Nauki, Astron., Vip. 3, p. 17 – 21 (1992). In Ukrainian.
The motion of a charged test particle in the field of an infinitely long massive charged filament is investigated. It is shown that an electric repulsion cannot prevent its fall on the filament. Hence it follows that Coulomb forces cannot decisively prevent the formation of naked linear singularity by collapse of charged matter.

022.162   **Equilibrium constants of the molecular ion H_3^+.**
V. P. Gaur, M. C. Pande, S. Chandra.
Astrophys. Space Sci., Vol. 191, No. 1, p. 147 – 149 (May 1992). Letter–to–the–editor.
The H_3^+ molecular ion plays an important role in the chemistry of astronomical objects as it protonates the neutral species. The authors have recently calculated the partition functions of H_3^+ which may be used to compute the equilibrium constants for the chemical reaction $H_2 + H_2^+ \rightarrow H_3^+ + H$. In this short communication the equilibrium constants for the temperature range from 500 to 8000K are calculated. The results are also presented in the polynomial form.

022.163 **Theoretical problems on gravitational–wave detectors.**
V. Faraoni.
Nuovo Cimento B, Vol. 107, No. 6, p. 631 – 642 (Jun 1992).
The author investigates the physical meaning of the observes (defined in a geometrical coordinate–independent way) associated with the TT gauge and other ad hoc coordinate systems that are often used in the description of gravitational–wave detectors. TT observers undergo very complicated motions when seen by a freely falling observer; for most (but not all) kinds of detectors with size l much smaller than the wavelength λ of the radiation to be detected, Fermi normal coordinates are the best. The author points out the difficulty of giving a clear, geometrical description of the family of fundamental observers associated with the detector.

022.164 **Monte Carlo simulation of photon induced air showers.**
B. D'Ettorre Piazzoli, G. Di Sciascio.
Vulcano Workshop 1990: Frontier objects in astrophysics and particle physics, p. 443 – 468 (1991). – See Abstr. 012.040 for the main entry.
The EPAS code (Electron Photon induced Air Showers) is a three dimensional Monte Carlo simulation developed to study the properties of extensive air showers generated by the interaction of high energy photons (or electrons) in the atmosphere. Results of the present simulation concern the longitudinal, lateral, temporal and angular distributions of electrons in atmospheric cascades initiated by photons of energies up to 100 TeV.

022.165 **Gravity wave astronomy.**
F. Fuligni, V. Iafolla.
Vulcano Workshop 1990: Frontier objects in astrophysics and particle physics, p. 471 – 486 (1991). – See Abstr. 012.040 for the main entry.
Contents: 1. Introduction. 2. Description of gravitational waves. 3. The role of gravity waves in astrophysics. 4. Sources of gravitational radiation. 5. Detection of gravitational radiation.

022.166 **Reflectivity (visible and near IR), Mössbauer, static magnetic, and X ray diffraction properties of aluminum–substituted hematites.**
R. V. Morris, D. G. Schulze, H. V. Lauer, D. G. Agresti, T. D. Shelfer.
J. Geophys. Res., Vol. 97, No. E6, p. 10257 – 10266 (25 Jun 1992).
Because aluminum substitution for iron occurs in polymorphs of Fe_2O_3 and FeOOH in terrestrial (and by inference Martian) environments, it is important for the mineralogical remote sensing of both planets at visible and near–IR wavelengths to know the effects of Al substitution on their reflectivity spectra. Diffuse reflectivity (350–2200 nm), Mössbauer, static magnetic, and X ray diffraction data are reported for a series of aluminum–substituted hematites α–$(Fe,Al)_2O_3$ for compositions having values of Al_s (mole ratio $Al/(Al+Fe)$) up to 0.61. On the basis of Martian spectral data, the range in Al_s for Martian hematites is $0 < Al_s < 0.19$.

022.167 **Application of ion irradiation experiments to planetary surfaces in the outer solar system.**
G. Strazzulla, G. A. Baratta, G. Leto, M. E. Palumbo.
Earth, Moon, Planets, Vol. 56, No. 1, p. 35 – 45 (Jan 1992).
Experimental results on the interaction between fast bombarding ions and solid targets simulating satellite surfaces in the Outer Solar System are reviewed. Applications to Jovian, Saturnian, Uranian, Neptunian, and Plutonian systems suggest the important role played by cosmic and magnetospheric ions in eroding material, in redistributing it on the surfaces of some objects, and in producing either thin or thick mantles of dark organics.

022.168 **Electron energy loss spectra of polycyclic aromatic hydrocarbons.**
J. W. Keller, M. A. Coplan, R. Goruganthu.
Astrophys. J., Vol. 391, No. 2, p. 872 – 875 (1 Jun 1992).
Polycyclic aromatic hydrocarbons (PAHs) have been tentatively identified as the source of infrared emission bands seen in reflection nebulae and other astronomical objects. PAHs may also play a role in other astronomical phenomena, but concrete identification of these systems await further understanding of the properties of PAHs as a class of molecules. The authors report on a survey of the electron energy loss spectroscopy of gas phase PAH molecules consisting of up to seven rings. The study is limited to the more thermodynamically stable pericondensed systems. Absorption profiles (proportional to the oscillator strengths) from the visible to the soft X–ray region near 30 eV are presented.

022.169 **Submillimeter spectrum of low–temperature hydrogen: the pure translational band of H_2 and the R(0) line of HD.**
E. H. Wishnow, I. Ozier, H. P. Gush.
Astrophys. J., Lett., Vol. 392, No. 1, p. L43 – L46 (10 Jun 1992).
The pure translational absorption spectrum of hydrogen gas has been measured from 20 to 100 cm^{-1} at temperatures in the range 24K – 37K. Superposed on the broad translational band is the sharp rotational line R(0) of HD, present in the gas in natural abundance. From the intensity of the line, the electric dipole moment is determined to be $(0.81 \pm 0.05) \times 10^{-3}$ debye. The measurements are of relevance to the infrared opacity and to the D/H ratio of the atmospheres of the giant planets.

022.170 **Modelling particle size effects on the emissivity spectra of minerals in the thermal infrared.**
J. E. Moersch, P. R. Christensen.
Bull. Am. Astron. Soc., Vol. 23, No. 3, p. 1183 (1991). Abstract. – See Abstr. 012.037 for the main entry.

022.171 **Optical properties of tholin from H_2O/C_2H_6(6:1) ice, and comparison with Titan tholin, kerogen and meteoritic organics.**
B. N. Khare, W. R. Thompson, L. Cheng, C. Sagan, C. Meisse, E. T. Arakawa, C. N. Matthews.
Bull. Am. Astron. Soc., Vol. 23, No. 3, p. 1186 (1991). Abstract. – See Abstr. 012.037 for the main entry.

022.172 **Optical constants of poly–HCN for astronomical applications.**
C. Meisse, E. T. Arakawa, B. N. Khare, W. R. Thompson, C. Sagan.
Bull. Am. Astron. Soc., Vol. 23, No. 3, p. 1189 (1991). Abstract. – See Abstr. 012.037 for the main entry.

022.173 **$O(^1S)$ and $O(^1D)$ quantum yields from rocket measurements of electron densities and 557.7 and 630.0 nm emissions in the nocturnal F–region.**
J. H. A. Sobral, H. Takahashi, M. A. Abdu, P. Muralikrishna, Y. Sahai, C. J. Zamlutti.
Planet. Space Sci., Vol. 40, No. 5, p. 607 – 619 (May 1992).
The quantum yields $f(^1D)$ and $f(^1S)$ of the atomic oxygen excited states $O(^1D)$ and $O(^1S)$, respectively, from dissociative recombination in the nocturnal F–region are determined utilizing rocket airglow (O I 630 nm and O I 557.7 nm) and electron density data obtained in two experiments that were carried out at Natal (geogr. 5.8°S, 35.2°W), Brazil, on 11 December 1985 and 31 October 1986. Using the 557.7 and 630 nm airglow data from the second experiment Takahashi et al. calculated the ratio $f(^1S)/f(^1D)$. From a different approach the authors utilize in this paper the airglow volumetric emission rate and electron density to calculate $f(^1S)$ and $f(^1D)$, individually, which are compared with the previously published results from in situ and laboratory measurements. The results show $f(^1S)$ to be height dependent. In the second experiment it varied from 1.96×10^{-2} at 190 km to 1.33×10^{-1} at 315 km with an average value of 0.053. The average value for the first experiment was 0.041. The $f(^1D)$ magnitude was also found to vary significantly with altitude, its value increasing both upwards and downwards from 250 km where it attains a minimum of about 0.77.

022.174   Vacuum ultraviolet extinction measurements on cosmic dust analog carbon grains.
L. Colangeli, A. Blanco, S. Fonti, E. Bussoletti.
Astrophys. J., Vol. 392, No. 1, p. 284 – 288 (10 Jun 1992).

The authors present the results of new systematic spectroscopic analyses performed in the vacuum ultraviolet spectral range (110 – 300 nm) on amorphous carbon grains, produced under various and controlled physical conditions and with two different production methods. The results confirm that particles similar to those obtained in the laboratory might reproduce the extinction observed toward a class of stellar sources characterized by an absorption "bump" at around 240 nm. The 220 nm interstellar hump is not reproduced by laboratory particles. However, a form of the same carbonaceous grains, partially modified in its chemical, structural, and/or morphological properties, may be used to simulate the interstellar extinction curve.

022.175   N(^{2}D) + O$_2$: a source of thermospheric 6300 Å emission?
R. Link, P. K. Swaminathan.
Planet. Space Sci., Vol. 40, No. 5, p. 699 – 705 (May 1992).

The aeronomic and laboratory evidence concerning the efficiency of the reaction N(^{2}D) + O$_2$ → NO(v) + O(^{1}D) in producing thermospheric 6300 Å emission is reviewed. Dayglow and nightside auroral measurements do not offer unambiguous evidence concerning the need for an additional source. Measurements of the temporal variation and lifetime of the 6300 and 5200 Å emissions in the polar cap and cleft aurora are not consistent with the proposed mechanism. Through consideration of reaction dynamics, the authors conclude that the laboratory inference of a large O(^{1}D) yield at cryogenic temperatures is not in conflict with reports of a low O(^{1}D) yield at thermospheric temperatures.

022.176   Laboratory measurements of weak carbon dioxide bands relevant to Venus' nightside emission spectrum at 2.2 microns.
L. P. Giver, C. Chakerian Jr., J. B. Pollack, R. B. Wattson.
Bull. Am. Astron. Soc., Vol. 23, No. 3, p. 1192 – 1193 (1991). Abstract. – See Abstr. 012.037 for the main entry.

022.177   Study of new fundamental forces in a microgravity environment.
R. J. Slobodrian.
Classical Quantum Gravity, Vol. 9, No. 4, p. 1115 – 1119 (Apr 1992).

A technique based on the mutual interaction of masses in a microgravity environment is described. There are no extraneous materials in the system except the masses. This technique may provide a very precise comparison of the interaction of test bodies differing in baryonic content.

022.178   Laboratory and theoretical studies of thermal emission spectroscopy.
P. G. Lucey, B. C. Bruno, B. G. Henderson, B. M. Jakosky, C. E. Randall.
Bull. Am. Astron. Soc., Vol. 23, No. 3, p. 1199 (1991). Abstract. – See Abstr. 012.037 for the main entry.

022.179   Structure, viscosity, and changes of silicate melts at impact sites.
P. Jakeš, S. Sen, K. Matsuishi.
Bull. Am. Astron. Soc., Vol. 23, No. 3, p. 1201 (1991). Abstract. – See Abstr. 012.037 for the main entry.

022.180   The absorption coefficient of nitrogen with application to Triton.
J. R. Green, R. H. Brown, V. Anicich, D. P. Cruikshank.
Bull. Am. Astron. Soc., Vol. 23, No. 3, p. 1208 (1991). Abstract. – See Abstr. 012.037 for the main entry.

022.181   The adsorption of HO$_x$ on surfaces: implications for the stability of CO$_2$ in the atmosphere of Mars.
A. D. Anbar, M. T. Leu, Y. L. Yung.
Bull. Am. Astron. Soc., Vol. 23, No. 3, p. 1212 (1991). Abstract. – See Abstr. 012.037 for the main entry.

022.182   Highly excited level population of carbon.
V. O. Ponomarev, R. L. Sorochenko.
Pis'ma Astron. Zh., Tom 18, No. 6, p. 541 – 546 (Jun 1992). In Russian. English translation in Sov. Astron. Lett., Vol. 18, No. 3.

The carbon departure coefficients b_n and β_n are calculated for typical conditions of molecular clouds; radiationless electron capture with simultaneous excitation of the fine structure of the core and collisional stabilization of the arising highly excited levels were taken into account. Relative populations of the fine structure levels of the core are derived from the clouds characteristics.

022.183   New model of collision–induced infrared absorption spectra of H$_2$–He pairs in the 2 – 2.5 μm range at temperatures from 20 to 300K: an update.
A. Borysow.
Icarus, Vol. 96, No. 2, p. 169 – 175 (Apr 1992).

A revised numerical method to calculate the collision–induced absorption rotovibrational (RV CIA) spectra of H$_2$–He pairs in the fundamental band of hydrogen, at temperatures from 20 to 300K is presented. The paper corrects few inaccuracies which exist in the previous modeling of the H$_2$–He absorption in this frequency range. New computations account for the previously neglected J–dependence of the vibrational matrix elements of the induced dipoles. Model spectra are presented which reproduce closely the latest quantum mechanical results and are found to agree with most experimental CIA coefficients of H$_2$–He to within a few percent. The numerical model uses simple analytical model lineshapes and can generate the RV CIA spectra of H$_2$–He extremely efficiently.

022.184   NMR spectroscopy of experimentally shocked quartz and plagioclase feldspar powders.
R. T. Cygan, M. B. Boslough, R. J. Kirkpatrick.
22. Lunar and Planetary Science Conference, p. 127 – 136 (1992). – See Abstr. 012.046 for the main entry.

Magic–angle spinning nuclear magnetic resonance (MAS NMR) spectroscopy is described as diagnostic method for examining shock features in minerals.

022.185   Effects of grain size and shape in modeling reflectance spectra of mineral mixtures.
T. Hiroi, C. M. Pieters.
22. Lunar and Planetary Science Conference, p. 313 – 325 (1992). – See Abstr. 012.046 for the main entry.

The effects of grain size and shape in modeling reflectance spectra of mineral mixtures have been investigated using a simple model called the "isograin model". This model treats reflectance from a particulate surface as a series of grain interactions that are coupled to the optical constants of mineral constituents. The validity of the model is demonstrated.

022.186   Calculation of detailed atomic data using parametric potentials.
C. A. Iglesias, F. J. Rogers, B. G. Wilson.
Rev. Mex. Astron. Astrofis., Vol. 23, p. 9 – 18 (Mar 1992). – See Abstr. 012.054 for the main entry.

Parametric potentials consisting of a sum of Yukawa terms plus a long–ranged Coulomb tail can provide atomic data of accuracy comparable to single–configuration, self–consistent field calculations with relativistic corrections. The method is used to generate prefitted effective potentials for isoelectronic sequences up to zinc, not only for valence electrons but also for multiply excited configurations as well as inner–core excitations. Comparisons to experimental and other theoretical calculations are presented.

022.187 **Atomic and molecular data for opacity calculations.**
R. L. Kurucz.
Rev. Mex. Astron. Astrofis., Vol. 23, p. 45 – 48 (Mar 1992). – See Abstr. 012.054 for the main entry.
The author is attempting to produce line lists for all atoms and diatomic molecules that are important in stars. He collects all published data on spectrum analysis and oscillator strengths. The author computes the energy levels, wavelengths, gf values, and damping constants that are not available from the literature. Line lists have been computed for diatomic molecules H_2, CH, NH, OH, MgH, SiH, CN, C_2, CO, SiO, and TiO, and for the iron group atoms Ca I–IX to Ni I–IX. These lists total 58 million lines. These calculations are being revised as new laboratory data become available. The work is being extended to other diatomic molecules, to lighter and heavier elements, and to higher stages of ionization.

022.188 **Molecular opacity data for stellar atmospheres.**
U. G. Jørgensen.
Rev. Mex. Astron. Astrofis., Vol. 23, p. 49 – 62 (Mar 1992). – See Abstr. 012.054 for the main entry.
The author reviews the molecular opacity data that are available from various sources, and compares the methods that have been used in computing them. Line strengths, excitation energies and frequencies for 22 million lines of 20 different diatomic molecules and 35 million lines of 4 polyatomic molecules exist in astrophysical data bases, which is probably far from sufficient for interpretation of the accumulating amount of satellite data especially in the infrared. Also, data are lacking for good model atmospheric computation, and the quality and effect of a great fraction of the data on the weak lines have never been analysed. Practically nothing is known about the opacity of the bigger molecules that are observed or suspected in the circumstellar envelopes, and their possible presence in the upper photosphere may substantially alter our picture of cool stellar atmospheres. The author reviews the recent progress in the understanding of this new field and its possible importance for stellar atmospheres.

022.189 **Calculation of transition frequencies and line strengths of water for cool star opacities.**
S. Miller, J. Tennyson, J. Fernley.
Rev. Mex. Astron. Astrofis., Vol. 23, p. 63 – 70 (Mar 1992). – See Abstr. 012.054 for the main entry.
First principles calculations for water, using a number of electronic potential surfaces, are presented as a first step towards the computation of an accurate water opacity for cool stars such as M dwarfs.

022.190 **Molecular data from solar spectroscopy.**
N. Grevesse, A. J. Sauval.
Rev. Mex. Astron. Astrofis., Vol. 23, p. 71 – 77 (Mar 1992). – See Abstr. 012.054 for the main entry.
The authors show through a few examples how the analysis of molecular transitions present in the solar visible and infrared spectrum can be used to refine our knowledge of the molecular constants and to test the accuracy of available molecular data like transition probabilities and dissociation energies for a few diatomic molecules.

022.191 **Xylan: a potential contaminant for lunar samples and antarctic meteorites.**
I. P. Wright, S. S. Russell, S. R. Boyd, C. Meyer, C. T. Pillinger.
22. Lunar and Planetary Science Conference, p. 449 – 458 (1992). – See Abstr. 012.046 for the main entry.
A proprietary lubricant paint, known as Xylan, has in the past been used to coat screw threads in the dry-N, sample–processing cabinets at the Planetary Materials Branch, NASA Johnson Space Center. This material therefore has to be considered a potential contaminant for those lunar samples and Antarctic meteorites that have been processed at JSC.

022.192 **What can molecular spectroscopy tell us about hot bands?**
G. Graner.
Rev. Mex. Astron. Astrofis., Vol. 23, p. 79 – 89 (Mar 1992). – See Abstr. 012.054 for the main entry.
After a definition of hot bands, the author studies the case of CO and shows that, both for frequencies and for intensities, the zeroth order model is clearly inadequate. The case of polyatomic molecules is more complex mainly due to resonances. The examples of carbon dioxide and water are taken to show the importance of hot bands at high temperature.

022.193 **Relativistic free–free Gaunt factors for high–temperature stellar plasmas.**
N. Itoh.
Rev. Mex. Astron. Astrofis., Vol. 23, p. 91 – 93 (Mar 1992). – See Abstr. 012.054 for the main entry.
The free–free Gaunt factor of the dense high–temperature stellar plasma is calculated by using the accurate relativistic cross section and is compared with the Gaunt factor derived by using Sommerfeld's exact nonrelativistic cross section. A wide range of electron degeneracy is accurately taken into account. Significant deviations are found for high–temperature cases.

022.194 **Local–scaling density functional theory: prospects for applications to the electronic structure of atoms.**
E. V. Ludeña, E. S. Kryachko.
Rev. Mex. Astron. Astrofis., Vol. 23, p. 95 – 106 (Mar 1992). – See Abstr. 012.054 for the main entry.
The Hohenberg–Kohn version of density functional theory is critically reviewed. Some of its shortcomings having to do with atomic multiplet representation, calculation of excited states and N–representability of the functional and of the one–particle density are discussed. An alternative version of density functional theory based on local–scaling transformations, where the above shortcomings are solved, is presented. Some prospective applications of this novel theory to the calculation of electronic structure of atoms is discussed.

022.195 **The physics and dynamics of charged dust grains.**
P. Lamy.
15. Ecole de Printemps d'Astrophysique de Goutelas: Interrelations between physics and dynamics for minor bodies in the solar system, p. 369 – 400 (1992). – See Abstr. 012.055 for the main entry.
Contents: 1. The equilibrium potential of a dust grain. 2. The physical processes determining the charge of a dust grain (Photoemission. Secondary electron emission by electron impact. Secondary electron emission by ion impact. Thermoionic emission. The case of porous grains). 3. Results for the potential of interplanetary dust grains. 4. Forces acting on interplanetary dust (The Coulomb or plasma drag. The Lorentz and convective forces. Diffusion of Keplerian orbits by electromagnetic scattering). 5. Other applications (Cometary dust grains. Circumplanetary dust grains).

022.196 **Boltzmann and Fokker–Planck equations.**
F. Pereira Gama.
15. Ecole de Printemps d'Astrophysique de Goutelas: Interrelations between physics and dynamics for minor bodies in the solar system, p. 587 – 598 (1992). – See Abstr. 012.055 for the main entry.
Contents: 1. Introduction. 2. Boltzmann transport equation (Absence of collision. Presence of collisions). 3. Fokker–Planck equation (Local equation. Global equation). 4. Conclusion.

022.197 **New measurements with a torsion pendulum during the solar eclipse.**
T. Kuusela.
Gen. Relativ. Gravitation, Vol. 24, No. 5, p. 543 – 550 (May 1992).
During the solar eclipse of 11 July 1991 in Mexico the period of a torsion pendulum was measured in order to reexamine possible anomalies observed in previous experiments of this kind. An

upper limit for the relative change of the pendulum's period during the eclipse of 2.0×10^{-6} (90% confidence limit) was derived. This result is similar to a previous one obtained during the eclipse in 1990 in Finland, when the Sun was much lower in the horizon. However, two small but distinct shifts were observed in the horizontal position of the pendulum wire, which were well correlated with the beginning and the end of the eclipse.

022.198   **Time and interpretations of quantum gravity.**
K. V. Kuchař.
4. Canadian Conference on General Relativity and Relativistic Astrophysics, p. 211 – 314 (1992). – See Abstr. 012.056 for the main entry.
In canonical quantization of gravity, the state functional does not seem to depend on time. This hampers the physical interpretation of quantum gravity. The author critically examines ten major attempts to circumvent this problem and discusses their shortcomings.

022.199   **Derivation of the spectral energy density in S^3.**
W. M. Stuckey, G. Bambakidis.
4. Canadian Conference on General Relativity and Relativistic Astrophysics, p. 347 – 349 (1992). – See Abstr. 012.056 for the main entry.
The spectral energy density as a function of temperature is derived for a photon gas in S^3. The photons are assumed to obey Bose–Einstein statistics. The frequency spectrum of the photon gas is obtained from the eigenvalues of the Laplacian on S^3. The result is precisely that of a like photon gas in R^3. Therefore, as widely accepted, one cannot use the blackbody distribution of the cosmic microwave background to identify our universe as spatially open or closed.

022.200   **Some consequences of phase–space QM: Urfelder, maximal acceleration.**
E. R. Caianiello.
Italo–Soviet Workshop on Advances in Theoretical Physics, p. 58 – 67 (1991). – See Abstr. 012.061 for the main entry.
The author discusses the "natural" appearence of Cartan's triality, and hence of supersymmetry, in the 8–dimensional space, together with the realization that space and time themselves may derive from a hidden structure made of "simple spinors" playing the role of "Urfelder", for which "left (or right)–handedness" would be a "primitive" feature; "strings" might easily follow from the new factorization properties of Dirac–like equations. He discusses a new concept, that of "maximal acceleration" for any extended object (the standard dimensionless 4–dimensional "material point" is exceptional in this context); additional independent proofs of its necessity have been found afterwards by several authors; its cosmological implications are a subject of investigation.

022.201   **Neutral Ti line oscillator strengths.**
D. A. Vakulenko, I. S. Savanov.
Bull. Crimean Astrophys. Obs., Vol. 82 , p. 78 – 98 (1992). English translation of Izv. Krym. Astrofiz. Obs., Tom 82, p. 87 – 108 (1990).
Published oscillator strengths for neutral titanium lines are critically surveyed as reduced to the scale of the precision Oxford measurements by Blackwell et al. The final values for the reduced and adopted gf for 795 neutral titanium lines are given. In that system, the Ti content in the solar atmosphere is in the range $\log \varepsilon$ (Ti) = 5.07 – 5.09 with an error of 0.04 dex.

022.202   **The role of volume scattering in reducing spectral contrast of reststrahlen bands in spectra of powdered minerals.**
J. W. Salisbury, A. Wald.
Icarus, Vol. 96, No. 1, p. 121 – 128 (Mar 1992).
As particle size decreases, the spectral contrast of reststrahlen bands also decreases. It has been generally agreed that this loss of spectral contrast is due to the increased porosity associated with fine particle size, resulting in formation of photon traps. That is, that the pores acted like small black bodies. However, the authors

show here that the reststrahlen bands change in shape as well as intensity, and this change in shape can only be explained by the occurrence of substantial volume scattering at fine particle size, rather than by photon trapping. It appears that the role of porosity is to physically separate $1 – 5–\mu m$–diameter particles that are optically thin. When such particles are separated by more than a wavelength, they scatter independently as optically thin, volume–scattering particles. When packed closely together, however, they scatter coherently as if they were large, optically thick, surface–scattering particles. Thus, the loss of spectral contrast of reststrahlen bands for fine particle size materials appears to be due directly to particle size and only indirectly (but critically) to porosity. The practical implication of this finding for remote sensing of the Earth, Moon, Mercury, Mars, and the asteroids is that the spectral features displayed by some particulate materials may be substantially changed from those seen in spectra of solids, requiring the use of a separate spectral search library to identify component minerals.

022.203   **The effect of total pressure on vaporization of alkalis from partially molten chondritic material.**
T. Shimaoka, N. Nakamura.
IAU Colloquium No. 126: Origin and evolution of interplanetary dust, p. 79 – 82 (1991). – See Abstr. 012.068 for the main entry.
In order to examine the effect of total pressure on vaporization of alkalis (Na, K, Rb) from a partially molten chondritic material, heating experiments were carried out under various He gas pressures ($\sim 10^{-5} – \sim 10^{-1}$torr) at 1300°C. The rate of vaporization decreased in the order of Na > K > Rb with the increasing of the pressure, and reached a minimum at $\sim 10^{-1}$torr.

022.204   **Condensation experiments of Mg–silicate minerals.**
A. Tsuchiyama.
IAU Colloquium No. 126: Origin and evolution of interplanetary dust, p. 83 – 86 (1991). – See Abstr. 012.068 for the main entry.
Condensation experiments were performed in the simple but most fundamental system Mg–Si–O–H with forsterite vaporization source. At temperatures above about 1000°C, euhedral crystals of forsterite (Mg_2SiO_4) of a few μm were formed. These crystals are similar to olivines in Allende matrix. At temperatures below about 1000°C, whiskers of forsterite and enstatite ($MgSiO_3$) were formed by vapor–liquid–solid growth mechnism. These whiskers are different from enstatite whiskers in interplanetary dust, which were probably formed at small super coolings.

022.205   **Ultraviolet–induced amorphization of cubic ice and its implication for the evolution of ice grains.**
A. Kouchi, T. Kuroda.
IAU Colloquium No. 126: Origin and evolution of interplanetary dust, p. 87 – 90 (1991). – See Abstr. 012.068 for the main entry.
The authors found that cubic ice is transformed below 70K to amorphous ice by ultraviolet irradiation, whereas no change in structure is observed at temperatures above 70K, regardless of the irradiation time. Experimental results can be interpreted by theoretical consideration of nucleation and growth of cubic ice in amorphous ice. The authors also discuss the evolution of ice grains in space on the basis of the experimental results.

022.206   **Simulation in laboratory of solid grains present in space.**
L. Colangeli, E. Bussoletti, V. Mennella.
IAU Colloquium No. 126: Origin and evolution of interplanetary dust, p. 91 – 94 (1991). – See Abstr. 012.068 for the main entry.
Laboratory data on cosmic dust analogue materials are compared with recent results obtained by means of spectroscopy and mass spectrometry on cometary dust, meteorites and interplanetary dust. Their actual chemical and physical properties can be further clarified, as well as possible links with interstellar dust.

022.207   The infrared spectra of synthesized amorphous silicates with compositions of olivine pyroxene.
C. Koike, A. Tsuchiyama.
IAU Colloquium No. 126: Origin and evolution of interplanetary dust, p. 95 – 98 (1991). – See Abstr. 012.068 for the main entry.

Amorphous olivines synthesized by evaporation method show two very broad bands at $10 - 11\,\mu m$ and $17.5 - 19\,\mu m$, which resemble the spectra of symbiotic stars. On the other hand, amorphous pyroxenes produced by the same method show two broad bands at $9.5 - 10.3\,\mu m$ and $20 - 22\,\mu m$, which are narrower than that of amorphous olivine. The features of amorphous olivine were easily altered by heating or hydration, and the peak wavelength of $18\,\mu m$ band was easily shifted to longer wavelengths.

022.208   Optical constants of kerogen from 0.15 to 40 μm: comparision with meteoritic organics.
B. N. Khare, W. R. Thompson, C. Sagan, E. T. Arakawa,
C. Meisse, I. Gilmour.
IAU Colloquium No. 126: Origin and evolution of interplanetary dust, p. 99 – 101 (1991). – See Abstr. 012.068 for the main entry.

A vacuum evaporation technique has been used to produce thin, optical quality films of samples of type II kerogen and of insoluble organic residue from the Murchison meteorite. Using these films, optical constants have been measured form 0.15 to $40\,\mu m$ for kerogen, and from 2.5 to $40\,\mu m$ for the Murchison residue. The infrared absorption properties of these materials show many similarities, although Murchison residue is more opaque throughout the infrared than is kerogen, and shows no distinct aliphatic absorptions.

022.209   Optical constants of basaltic glass from 0.0173 to 50 μm.
E. T. Arakawa, D. W. Young, J. M. Zhang,
P. C. Eklund, B. N. Khare, W. R. Thompson, C. Sagan.
IAU Colloquium No. 126: Origin and evolution of interplanetary dust, p. 102 – 104 (1991). – See Abstr. 012.068 for the main entry.

The authors have revised the former measurements for basaltic glass and extended them into the extreme UV to $0.0173\,\mu m$.

022.210   Noble metal enrichments in cosmic spherules.
K. Nogami, K. Misawa, R. Omori, M. Jianguo,
K. Yamakoshi.
IAU Colloquium No. 126: Origin and evolution of interplanetary dust, p. 105 – 108 (1991). – See Abstr. 012.068 for the main entry.

Studies on relationships of chemical compositions between fusion crust and nucleus in iron spherules are reported. More than 10% of the iron spherules which were picked out from deep sea sediment, have cores and crusts. The authors were able to divide three of them into cores and crusts. Each cores and crusts were analyzed individually by INAA. The core mainly consists of iron and nickel. Other trace elements, especially noble metal Au and Ir were concentrated in the core. The mechanism of core formation in the iron spherules shows us the origin of them.

022.211   Studies on isotopic ratios of osmium and iridium in cosmic spherules using instrumental neutron activation analysis.
K. Yamakoshi, K.'i. Nogami.
IAU Colloquium No. 126: Origin and evolution of interplanetary dust, p. 109 – 112 (1991). – See Abstr. 012.068 for the main entry.

Isotopic ratios of the elements having high condensation temperatures, such as Os and Ir, in cosmic spherules are examined using instrumental neutron activation analysis.

022.212   Structures of amorphous silicate dusts simulated by molecular dynamics method.
A. Tsuchiyama, K. Kawamura.
IAU Colloquium No. 126: Origin and evolution of interplanetary dust, p. 113 – 116 (1991). – See Abstr. 012.068 for the main entry.

Atomic structures of amorphous silicate dusts with $MgSiO_3$ composition were simulated by molecular dynamics method as a function of the dust density based on the assumption that the density corresponds to cooling rate of dust formation.

022.213   Astrophysical interesting compound grains produced by a gas evaporation method.
C. Kaito, Y. Saito.
IAU Colloquium No. 126: Origin and evolution of interplanetary dust, p. 117 – 120 (1991). – See Abstr. 012.068 for the main entry.

Production methods of Fe_3O_4 grain and MgS grain have been introduced. Fe_3O_4 grain was produced in an Ar gas pressure range of 25 to 100 torr by evaporating FeO powder as result of oxidation of Fe grains. MgS grain produced by the reaction of Mg and S vapors grew in the coagulation of tiny cubic sulfide.

022.214   Measurement of far–infrared absorption for amorphous silicates between 27 and 400 μm.
C. Koike, H. Shibai.
IAU Colloquium No. 126: Origin and evolution of interplanetary dust, p. 121 – 124 (1991). – See Abstr. 012.068 for the main entry.

The far–infrared extinction of various silicates was measured in the $27 - 400\,\mu m$ range of wavelength. There is no distinct absorption band in the far–infrared region and the extinction decreases in proportion to $\lambda^{-1.4}$.

022.215   Laboratory spectra of amorphous and crystalline olivine: an application to comet Halley IR spectrum.
A. Blanco, V. Orofino, E. Bussoletti, S. Fonti, L. Colangeli,
J. R. Stephens.
IAU Colloquium No. 126: Origin and evolution of interplanetary dust, p. 125 – 128 (1991). – See Abstr. 012.068 for the main entry.

The authors present the infrared spectra of three different types of olivine grains: crystalline, amorphous and synthetic (also amorphous). While the first and second sample derive from the same natural mineral, the third one has been prepared in the laboratory according to the relative cosmic abundances of the elements. The experimental data is used to fit the emission feature observed in the comet Halley spectrum between 8 and $13\,\mu m$. Satisfactory results are obtained by using synthetic olivine mixed with a small amount (5%) of crystalline grains.

022.216   Chemical composition of an emanation from comets: identification of the 3 micron comet feature.
A. Sakata, S. Wada, A. T. Tokunaga.
IAU Colloquium No. 126: Origin and evolution of interplanetary dust, p. 241 – 244 (1991). – See Abstr. 012.068 for the main entry.

Recent high resolution observations of comets revealed a detailed spectral shape of the $3.4\,\mu m$ feature. The authors measured IR spectra of simple 14 hydrocarbon molecules and made "synthesized comet spectrum". Peak wavelength and spectral shape of the synthesized spectrum are well in agreement with the observed comet features.

022.217   Ice particle emission from cometary analogues.
H. Kohl, E. Grün.
IAU Colloquium No. 126: Origin and evolution of interplanetary dust, p. 257 – 260 (1991). – See Abstr. 012.068 for the main entry.

Dust particles originating from comets are an important constituent of the interplanetary dust regime. In order to study the ejection mechanisms from the cometary nucleus surface simulation experiments in the laboratory have been performed. Samples consisting of water ice, carbon dioxide ice and dust grains have been studied when they are irradiated by artificial sunlight within a cooled vacuum system. It has been shown that particle emission is extremely dependent on the initial composition of the samples. For samples with a distinct amount of non–volatile, mineral particles the formation of a dust mantle and, as a consequence, rapid decrease of particle ejection has been observed.

022.218   Polyoxymethylene in cometary dust: laboratory tests.
D. C. Boice, D. W. Naegeli, W. F. Huebner.
IAU Colloquium No. 126: Origin and evolution of interplanetary dust, p. 265 – 268 (1991). – See Abstr. 012.068 for the main entry.

The authors have investigated the stability of gas–phase formaldehyde oligomers and its implications for cometary science. The experiments complement previous experiments using a formaldehyde–water solution and indicate that formaldehyde oligomers are stable in the gas phase up to at least 6 monomeric units in length. Methanol is important in the end–capping process of the oligomers, leading to increased stability and a richer mass spectrum when compared to the formaldehyde–water solution. The results are consistent with mass spectra obtained by the Giotto PICCA instrument exhibiting alternating 14 – 16 amu mass peaks.

022.219   Penetration of hypervelocity projectiles into low density materials.
A. Fujiwara, T. Kadono, A. Nakamura, T. Ishibashi, N. Fujii.
IAU Colloquium No. 126: Origin and evolution of interplanetary dust, p. 281 – 284 (1991). – See Abstr. 012.068 for the main entry.
Spherical nylon projectiles of 7 mm diameter and up–to 4 km/s velocities were penetrated into three types of targets; aluminum multisheet stacks, foamed polystyrene, and 1–atm air. Penetration depth, recovery rate of the projectiles were determined as a function of projectile velocity and targe density, and a new type of dust collector is proposed.

022.220   Catastrophic disruption of solid bodies by collision – experimental approach.
A. Fujiwara.
IAU Colloquium No. 126: Origin and evolution of interplanetary dust, p. 361 – 366 (1991). – See Abstr. 012.068 for the main entry.
The results on mass, velocity, shape, and rotation of the fragments produced in the disruption experiments by impact are surveyed. Some future works are also suggested.

022.221   Methods, difficulties, and first results in laboratory simulation of cosmic dust electric charging.
J. Svestka, E. Grün.
IAU Colloquium No. 126: Origin and evolution of interplanetary dust, p. 367 – 370 (1991). – See Abstr. 012.068 for the main entry.
Particles of radii 0.2 to 3 μm and of different materials were suspended in an electrodynamic quadrupole inside a vacuum chamber and exposed to beams of electrons and ions of energies up to 20 keV and 5 keV, respectively, with the aim to simulate electric charging of cosmic dust particles.

022.222   Electrostatic fragmentation of irregularly shaped particles.
T. Mukai.
IAU Colloquium No. 126: Origin and evolution of interplanetary dust, p. 371 – 374 (1991). – See Abstr. 012.068 for the main entry.
An enhancement of the electrostatic stress forces due to the charge concentration on a position with small radius of curvature on the surface of irregularly shaped particle causes the fragmentation of fluffy particle more than expected for a spherical particle. This mechanism may act to produce "dust clusters" as detected in comet P/Halley by Simpson et al.(1987) and also "dust swarms" near the Earth as reported in Fechtig (1982).

022.223   Plasma emission from high velocity impacts of microparticles onto water ice.
R. Timmermann, E. Grün.
IAU Colloquium No. 126: Origin and evolution of interplanetary dust, p. 375 – 378 (1991). – See Abstr. 012.068 for the main entry.
Collisions of icy objects play a major role in the outer solar system. The purpose of this investigation is the experimental study of plasma production by dust impacts on icy surfaces. Impact speeds ranged from 3 to 60 km/s. It was found that the dominant ion species which were released are both positive and negative water clusters. The impact charge yield from icy surfaces is approximately a factor 100 below that from previously studied gold surfaces.

022.224   Velocity distribution of fragments in collisional breakup.
A. Nakamura, A. Fujiwara.
IAU Colloquium No. 126: Origin and evolution of interplanetary dust, p. 379 – 382 (1991). – See Abstr. 012.068 for the main entry.
One of the key outcomes of collisional disruptions to the dust system is the velocity distribution of fragments. A series of laboratory impact experiments were carried out to obtain the

mass–velocity and the position–velocity relation of the fragments by taking movie films, and films for tow impacts were completely analyzed.

022.225   Jets of fragments from catastrophic break–up and their astrophysical implications.
G. Martelli, P. Rothwell, P. N. Smith, I. Giblin, J. Martinsson, E. Ducrocq, M. Wettstein, M. di Martino, P. Farinella.
IAU Colloquium No. 126: Origin and evolution of interplanetary dust, p. 383 – 386 (1991). – See Abstr. 012.068 for the main entry.
The authors present some preliminary results of a series of catastrophic break–up experiments carried out in the open, against targets of natural and artificial rock, with and without a harder core. For the first time in this kind of experiments, evidence was found of collimated jets, i.e. the ejection of a statistically significant number of fragments all closely aligned about some preferential planes. Moreover, the presence of some groups of fragments lying close to each other on the ground was also detected.

022.226   Laboratory studies of grain mantles in interstellar space.
C. X. Mendoza–Gomez, J. M. Greenberg.
IAU Colloquium No. 126: Origin and evolution of interplanetary dust, p. 437 – 440 (1991). – See Abstr. 012.068 for the main entry.
The author simulate the most relevant conditions in interstellar space in order to follow the chemical and physical evolution of interstellar organic grain mantles.

022.227   A Galilean experiment using a holographic technique.
R. Takahashi, A. Yamaguchi, S. Tanaka.
Sci. Rep. Tôhoku Univ., Eighth Ser., Vol. 12, Nos. 2/3, p. 167 – 176 (Jan 1992).
The authors have done a Galilean experiment, making use of a holographic technique. The angular accelerations of four freely falling objects were measured in an experiment at the precision of about 10^{-4}rad/s^2. From this measurement, the acceleration differences for two compound pairs of an aluminum–copper (Al–Cu) test object and an aluminum–stainless steel (Al–SUS) object in the Earth's gravitational field were obtained at the 10^{-6} level. The results show non–zero acceleration differences for the objects.

022.228   Partition coefficients for iron between plagioclase and basalt as a function of oxygen fugacity: implications for Archean and lunar anorthosites.
W. C. Phinney.
Geochim. Cosmochim. Acta, Vol. 56, No. 5, p. 1885 – 1895 (May 1992).

022.229   Laboratory study of early solar nebula condensed object analogs.
D. Peak.
News Lett. Astron. Soc. N.Y., Vol. 4, No. 1, p. 7 – 8 (Feb 1992). Abstract. – See Abstr. 012.078 for the main entry.

022.230   Shock modification and chemistry and planetary geologic processes.
M. B. Boslough.
Annu. Rev. Earth Planet. Sci., Vol. 19, p. 101 – 130 (1991).
The purpose of this paper is to bring the rapid advances on shock processing of materials to the attention of Earth scientists, and to put these advances in the context of planetary geologic processes. As a case study, the surface of Mars is suggested as a place where conditions are optimal for shock processing to be a dominant factor. The various mechanisms of shock modification, activation, synthesis and decomposition are all proposed as major contributors to the evolution of chemical, mineralogical, and physical properties of the Martian regolith.

022.231 **Internal friction in high *Q* materials.**
J. Ferreirinho.
The detection of gravitational waves, p. 116 – 168 (1991). – See Abstr. 003.097 for the main entry.
Contents: 1. Introduction. 2. Anelastic relaxation: The anelastic model; Thermal activation. 3. Anelastic relaxation mechanisms in crystalline solids: Outline; Relaxation mechanisms in a perfect crystal; Defect relaxation mechanisms. 4. Measured internal friction in niobium and other high *Q* materials: Internal friction in polycrystalline niobium; Aluminium alloys; Sapphire; Quartz; Silicon. 5. Summary and comparison of relevant properties of high *Q* materials: Covalently bonded materials – sapphire, quartz and silicon; Suggestions for further work on the other bcc transition metals.

022.232 **Studies on molecules of astrophysical interest.**
M. Singh, J. P. Chaturvedi.
Proc. Astron. Soc. Aust., Vol. 9, No. 2, p. 328 (1991). Extended abstract. – See Abstr. 012.090 for the main entry.

022.233 **Searches for dark matter particles.**
B. Sadoulet.
1. International Symposium on Particles, Strings and Cosmology, p. 147 – 184 (1991). – See Abstr. 012.098 for the main entry.
This paper reviews the present searches for the weakly interacting massive particles, which may constitute dark matter. The paper discusses in detail the experimental challenges of such an endeavor. In particular, recent results obtained with cryogenic detectors are considered.

022.234 **A proposed search for dark–matter axions in the 0.6 – 16 μeV range.**
K. van Bibber, P. Sikivie, N. S. Sullivan, D. B. Tanner, M. S. Turner, D. M. Moltz.
1. International Trends in Astroparticle Physics Conference together with the SuperNova Watch Workshop, p. 154 – 166 (1992). – See Abstr. 012.102 for the main entry.
A proposed experiment is described to search for dark–matter axions in the mass range 0.6 – 16 μeV. The method is based on the Primakoff conversion of axions into monochromatic microwave photons inside a tunable microwave cavity in a large volume high field magnet, as described by Sikivie (1983). This proposal capitalizes on the availability of two Axicell magnets from the decommissioned MFTF–B fusion machine at LLNL. Assuming a local dark–matter density in axions of $\varrho = 0.3$ GeV/cm^3, the axion would be found or ruled out at the 97% c.l. in the above mass range in 48 months.

022.235 **Flare–type plasma processes and explosive disruption of pinch current sheets.**
A. G. Frank.
International Workshop on Reconnection in Space Plasma, p. 185 – 192 (Jan 1989). – See Abstr. 012.099 for the main entry.
The results of laboratory experiments with plane pinch current sheets are presented.

022.236 **Current sheet structure at different stages of magnetic reconnection.**
S. Yu. Bogdanov, A. G. Frank, V. S. Markov.
International Workshop on Reconnection in Space Plasma, p. 193 – 198 (Jan 1989). – See Abstr. 012.099 for the main entry.
The structure of the magnetic field of the plane pinch current sheet (CS) was experimentally investigated. The current distributions in different cross sections of the CS were obtained. The structure of the current region was found to be rather complicated and essentially changed in the course of the CS evolution.

022.237 **Impulsive heating, plasma turbulence and magnetic reconnection inside current sheets.**
A. G. Frank, N. P. Kyrie, V. S. Markov.
International Workshop on Reconnection in Space Plasma, p. 199 – 202 (Jan 1989). – See Abstr. 012.099 for the main entry.
Results are presented about the evolution of electron and ion temperatures in the current sheet.

022.238 **Spontaneous and forced reconnection in a magnetic type–X structure.**
A. T. Altyntsev, N. V. Lebedev, N. A. Strokin.
International Workshop on Reconnection in Space Plasma, p. 207 – 213 (Jan 1989). – See Abstr. 012.099 for the main entry.
An experimental study is made of the evolution of a magnetic type–X structure which is produced on a theta–pinch device. It is found that during the initial stage the reconnection process is proceeding in a spontaneous fashion, with periodic restructurings of the current sheet and abrupt accelerations of ions. In the subsequent stage it is stationary in character, with the reconnection rate being in agreement with estimates obtained in terms of the forced reconnection Sweet–Parker models.

022.239 **Prospects for relic neutrino detection.**
P. F. Smith.
1. International Trends in Astroparticle Physics Conference together with the SuperNova Watch Workshop, p. 311 – 325 (1992). – See Abstr. 012.102 for the main entry.
The aim of this paper is to illustrate the difficulties of neutrino background detection by summarizing six detection ideas which have been previously considered, indicating in each case the problems which have prevented the idea being developed into an experimental proposal.

022.240 **"Long–term" neutrino flux integrations.**
W. C. Haxton.
1. International Trends in Astroparticle Physics Conference together with the SuperNova Watch Workshop, p. 369 – 377 (1992). – See Abstr. 012.102 for the main entry.
The standard solar model predicts that the sun's luminosity has increased by 40% over the past 5 Gyrs of main–sequence burning, reflecting the evolving chemistry of the solar core. This increase is accompanied by an exponential growth in the ^{8}B neutrino flux, with a doubling time of 0.85 Gyr. The author describes an unusual nuclear system that, in principle, could yield a quantitative terrestrial record of these past changes, and he discusses some of the practical obstacles to reading this record. The author also argues that there exists a "twin" of the ^{37}Cl solar neutrino experiment that could be mounted with modest effort.

022.241 **Study of mixing of solar neutrinos with a 1000 ton ICARUS detector.**
M. Cheng, D. Cline, J. Park, M. Zhou.
1. International Trends in Astroparticle Physics Conference together with the SuperNova Watch Workshop, p. 393 – 404 (1992). – See Abstr. 012.102 for the main entry.
The authors review recent measurements of Solar neutrinos with the view towards the goals of 1000 ton Liquid Argon detector for the Gran Sasso laboratory (ICARUS). The technique allows the study of both elastic scattering on electrons and inverse β decay and is thus independent of calculations of solar neutrino flux. The authors show how the flux of neutrinos (ν_μ, ν_τ) that could arise from neutrino mixing can be inferred from these two measurements for a large range of parameters that are allowed by current measurements.

022.242 **Greenland '90: a first step toward using the polar ice cap as a Cherenkov detector.**
S. W. Barwick, F. Halzen, D. Lowder, T. Miller, R. Morse, P. B. Price, A. Westphal.
1. International Trends in Astroparticle Physics Conference together with the SuperNova Watch Workshop, p. 413 – 417 (1992). – See Abstr. 012.102 for the main entry.

022.243 **A double beta decay experiment using bolometric detectors.**
A. Alessandrello, C. Brofferio, D. Camin, O. Cremonesi, E. Fiorini, G. Gervasio, A. Giuliani, F. Passoni, F. Pavan, G. Pessina, E. Previtali, L. Zanotti.
1. International Trends in Astroparticle Physics Conference together with the SuperNova Watch Workshop, p. 418 – 424 (1992). – See Abstr. 012.102 for the main entry.
A new way to search for double beta decay and heavy dark matter candidates consists in using large mass (100 g or more)

cryogenic particle detectors. The operation of such large detectors with good energy resolution requires the optimization of the system at very low temperature, $20-30$ mK typically, in order to achieve low heat capacity. The authors report on several tests that have been performed on double beta candidate materials, with the aim of selecting the most promising ones to detect this rare process. Among them the best results have been obtained with a 6 g crystal of TeO_2. Preliminary background measurements with crystals of this compound will take place in the Gran Sasso Laboratory, with an especially built low–activity cryostat.

022.244   A cosmic axion experiment.
C. Hagmann, P. Sikivie, N. S. Sullivan, D. B. Tanner.
1. International Trends in Astroparticle Physics Conference together with the SuperNova Watch Workshop, p. 425 – 434 (1992). – See Abstr. 012.102 for the main entry.

The authors report on the status of a pilot experiment to search for cosmic axions. The detector utilizes the electromagnetic coupling of the axion by stimulating the decay of halo axions into photons inside a microwave cavity. A narrow range of axion masses m_a was explored by sweeping the cavity resonance frequency with a tuning rod. The negative result of the search puts a constraint on the axion–photon coupling assuming our halo is made of axions.

022.245   Searching for WIMPs with mica.
D. P. Snowden–Ifft, Y. D. He, P. B. Price.
1. International Trends in Astroparticle Physics Conference together with the SuperNova Watch Workshop, p. 440 – 448 (1992). – See Abstr. 012.102 for the main entry.

The authors propose two new ideas which use mica to search for WIMPs in our galaxy. The first sets an upper limit on WIMP elastic scattering cross sections by utilizing the long integration time associated with natural mica. The second, using annealed mica, not only limits WIMP cross sections but also provides a strong signature for WIMPs, allowing the possibility of detection. Both ideas can, potentially, set better limits than existing experiments.

022.246   The possibility of radiogeochemical limits on stellar collapse rates in the Galaxy.
C. W. Johnson.
1. International Trends in Astroparticle Physics Conference together with the SuperNova Watch Workshop, p. 526 – 532 (1992). – See Abstr. 012.102 for the main entry.

The author discusses the theoretical possibility of extracting non–trivial limits on stellar collapses in our galaxy by measuring the amount of ^{97}Tc in deeply buried molybdenum ore produced by solar and supernova neutrinos. The solar neutrino contribution is constrained by experiment and here is calculated in the famework of the nonadiabatic Mikheyev–Smirnov–Wolfenstein "solution" to the solar neutrino puzzle. A singinificant "poison" to a measure of the stellar collapse rate is ^{97}Tc produced by solar neutrinos on ^{97}Mo; since the nuclear structure of the latter is still unknown, a definitive answer on the practicalities of this experiment cannot yet be given.

022.247   A large low energy neutrino detector for oscillations and supernovae watch.
F. Boehm.
1. International Trends in Astroparticle Physics Conference together with the SuperNova Watch Workshop, p. 533 – 541 (1992). – See Abstr. 012.102 for the main entry.

The author describes a large, low energy, low background, Gd–located liquid scintillation detector with a fiducial volume of 1000 tons. Installed at a distance of $10-15$ km from the San Onofre power reactors (7.5 GW), it will be capable of exploring neutrino oscillations via $\bar{\nu}_e$ disappearance down to $\Delta m^2 = 10^{-4} eV^2$ and mixing angles down to $\sin^2 2\theta \approx 0.1$. The detector will also serve as a supernova neutrino observatory.

022.248   The LVD supernova detector.
E. S. Hafen.
1. International Trends in Astroparticle Physics Conference together with the SuperNova Watch Workshop, p. 542 – 553 (1992). – See Abstr. 012.102 for the main entry.

The LVD is a large underground detector with powerfull detection capabilities for a wide variety of particle astrophysics topics. The LVD detector is fully active, with tracking, timing, and energy deposition measurement interspersed throughout its volume. It is thus an especially useful tool for studying neutrinos and antineutrinos over a wide range of energies. The capabilities below 100 MeV are discussed.

022.249   Expected supernova neutrino signal in IMB–3 detector.
D. Kiełczewska.
1. International Trends in Astroparticle Physics Conference together with the SuperNova Watch Workshop, p. 554 – 570 (1992). – See Abstr. 012.102 for the main entry.

The IMB–3 detector has been upgraded since February 1987 and is now even better prepared for detection of a gravitational collapse. As the largest existing detector, which can measure both the electron energies and angles, it would provide unique information about muon and tau neutrino properties. Thanks to the kinematics of neutrino scattering on electrons their signal could be enhanced and the direction of a source could be indicated. Observation of hundreds of electron antineutrinos would allow precise measurement of their energy spectra and the burst time evolution, which would lead to a better understanding of processes occuring in the protoneutron star.

022.250   Sudbury Neutrino Observatory.
G. T. Ewan.
1. International Trends in Astroparticle Physics Conference together with the SuperNova Watch Workshop, p. 571 – 580 (1992). – See Abstr. 012.102 for the main entry.

The Sudbury Neutrino Observatory (SNO) detector is a 1000 tonne heavy water (D_2O) Čerenkov detector designed to study neutrinos from the sun and other astrophysical sources. The use of heavy water allows both electron neutrinos and all types of neutrinos to be observed by three complementary reactions. The detector will be sensitive to the electron neutrino flux and energy spectrum shape and to the total neutrino flux irrespective of neutrino type. These measurements will provide information on both vacuum neutrino oscillations and matter enhanced oscillations, the MSW effect. In the event of a supernova it will be very sensitive to muon and tau neutrinos as well as the electron neutrinos emitted in the initial burst enabling sensitive mass measurements as well as providing details of the physics of stellar collapse.

022.251   Extra galactic supernova detector, neutrino mass and the supernova watch.
D. B. Cline.
1. International Trends in Astroparticle Physics Conference together with the SuperNova Watch Workshop, p. 590 – 611 (1992). – See Abstr. 012.102 for the main entry.

The detection of finite neutrino mass above a few eV would have profound consequences on cosmology and particle physics. Supernova neutrino bursts provide the most reliable technique to detect finite neutrino mass. The author reports on techniques to detect extra–galactic supernova bursts that would be sensitive to $\sim$eV neutrino mass and galactic supernova detection (SuperNova Watch) that is sensitive to the $20-50$ eV mass range. A brief remark is made on a real–time supernova network being discussed.

022.252   Supernova detectors based on coherent nuclear recoil.
P. F. Smith, J. D. Lewin.
1. International Trends in Astroparticle Physics Conference together with the SuperNova Watch Workshop, p. 581 – 589 (1992). – See Abstr. 012.102 for the main entry.

Recommended rest frequencies for observed interstellar molecular microwave transitions – 1991 revision.
See Abstr. 002.047.

Astrochemistry library with artificial intelligence for quality control.
See Abstr. 002.056.

MOGADOC – a personal computer database for atmospheric and interstellar molecules in microwave spectroscopy and radio astronomy.
See Abstr. 002.079.

The physical basis of the direction of time.
See Abstr. 003.013.

Atomic and molecular spectroscopy. Basic aspects and practical applications.
See Abstr. 003.043.

LEP, the Laboratory for Electrostrong Physics, one year later.
See Abstr. 009.010.

Production and uses of simulated lunar materials.
See Abstr. 012.026.

Chemistry and spectroscopy of interstellar molecules. Proceedings.
See Abstr. 012.039.

Astrophysical opacities. Proceedings. Workshop on Astrophysical Opacities (WAO), Caracas (Venezuela), 15 – 19 Jul 1991.
See Abstr. 012.054.

Reconnection in space plasma. Proceedings.
See Abstr. 012.099.

Methods in computational molecular physics. Lecture notes.
See Abstr. 012.101.

The Opacity Project – computation of atomic data.
See Abstr. 013.078.

The Opacity Project – the TOPBASE atomic database.
See Abstr. 013.081.

The Opacity Project – a post–script.
See Abstr. 013.086.

On the analysis of collision strengths and rate coefficients.
See Abstr. 062.012.

Plasma diagnostics based on self–reversed lines. I. Model calculation and application to argon arc measurements in the near–infrared and vacuum ultraviolet regions.
See Abstr. 062.050.

Plasma diagnostics based on self–reversed lines. II. Application to nitrogen, carbon and oxygen arc measurements in the vacuum ultraviolet.
See Abstr. 062.051.

Review of the CIV phenomenon.
See Abstr. 062.070.

Analysis of McCoyd's mechanism of the negative polarization of light scattered by atmosphereless celestial bodies.
See Abstr. 063.022.

Molecular opacity and stellar structure.
See Abstr. 063.124.

Influence of ion–atom collisions on the absorption of radiation in white dwarfs.
See Abstr. 064.030.

The opacity project – a review.
See Abstr. 064.104.

Stark broadening parameters for spectral lines of multicharged ions in stellar atmospheres: C IV, N V, O VI lines and regularities within an isoelectronic sequence.
See Abstr. 064.107.

Accelerated Lambda Iteration.
See Abstr. 064.109.

Supernova 1987A gravitational wave antenna observations, cross sections, correlations with six elementary particle detectors, and resolution of past controversies.
See Abstr. 066.309.

On Joseph Weber's new cross section for resonant–bar gravitational wave detectors.
See Abstr. 066.310.

Lifetimes in Fe II and the solar abundance of iron.
See Abstr. 071.012.

Theoretical analysis of the Fe XVIII X–ray spectrum and application to solar coronal observations.
See Abstr. 074.045.

Fe XVIII emission–line intensities in the Sun.
See Abstr. 076.003.

C IV line ratios in the Sun.
See Abstr. 076.004.

Mg IX line ratios in the Sun.
See Abstr. 076.006.

Macroscopic fluctuations, Sun–Earth links, and methodological aspects of excat measurements.
See Abstr. 085.013.

Iron Mössbauer spectroscopy: superparamagnetism in hydrothermal vents and the search for evidence of past life on Mars.
See Abstr. 097.030.

Extraction of ^{4}He from IDPs by step–heating.
See Abstr. 106.024.

Physics of the rotation of a PAH molecule in interstellar environments.
See Abstr. 131.003.

The chemistry of H_2D^+ in cold clouds.
See Abstr. 131.109.

The charge state in a highly dense molecular cloud.
See Abstr. 131.214.

Astronomical Instruments and Techniques

031 Astronomical Optics

031.001 Some infrared materials for the Cooke triplet design for the 3 – 5 μm spectral region: a comparison.
K. D. Sharma.
Appl. Opt., Vol. 31, No. 1, p. 101 – 105 (1 Jan 1992). Current Physics Microform No.: 9404C0091.

The suitability of zinc sulfide versus germanium for the middle negative lens of the Cooke triplet is studied. For this purpose two designs based on the Cooke triplet configuration are developed with a focal length of 100 mm and a relative aperture of f/2 to cover a total field of 14° for use in the 3 – 5 μm region of the spectrum. For the outer positive lenses both designs use silicon. For the middle negative lens one design uses zinc sulfide and the other uses germanium. The performances of the two designs are compared. It is found that the design with the zinc sulfide negative lens performs better in monochromatic applications, and the design with a germanium negative lens performs better in polychromatic applications.

031.002 Delayed elastic effects in Zerodur at room temperature.
J. W. Pepi.
Appl. Opt., Vol. 31, No. 1, p. 115 – 119 (1 Jan 1992). Current Physics Microform No.: 9404C0105.

Continuous testing at room temperature of large optics made of Zerodur has revealed a delayed elastic effect under low stress levels during both load and recovery after removal. Using a high-performance mechanical profilometer, a delayed strain of the order of 1% is realized over a period of a few weeks. The time-dependent phenomenon is elastic and reversible, but must be accounted for in various applications of optical design.

031.003 Distortion–adjusting optical elements.
D. J. Reiley, R. A. Chipman.
Appl. Opt., Vol. 31, No. 13, p. 2188 – 2193 (1 May 1992). Current Physics Microform No.: 9207B0020.

Distortion can be corrected in an image by placing a fourth-order aspheric optical element near the image plane. Moving the aspheric surface longitudinally changes the amount of distortion that is added by the aspheric surface without changing the paraxial image. Third–order astigmatism limits the performance of distortion correctors and may be eliminated by adding another fourth–order aspheric surface. Example elements were fabricated by diamond turning and were shown to introduce distortion without significantly degrading image quality. Three arrangements of distortion correctors are discussed: a single–element planoaspheric arrangement, an antisymmetric two–element arrangement, and a biaspheric arrangement in which distortion is not adjustable.

031.004 Evaluation of optical aberrations in point images.
J. S. Loomis.
Appl. Opt., Vol. 31, No. 13, p. 2211 – 2222 (1 May 1992). Current Physics Microform No.: 9207B0043.

Merit figures that are based on ray aberrations give different results from those based on wave–front aberrations. Zernike polynomials and a set of orthogonal polynomials that are based on the wave–front slope are used to study the transition from physical optics to geometrical optics. A knife–edge energy distribution is used to evaluate point images that are derived from both physical optics and geometrical optics. A direct comparison of geometrical and physical aberrations through tenth order is made.

031.005 Modeling diffraction efficiency effects when designing hybrid diffractive lens systems.
C. Londoño, P. P. Clark.
Appl. Opt., Vol. 31, No. 13, p. 2248 – 2252 (1 May 1992). Current Physics Microform No.: 9207B0080.

The authors investigated the design of two broadband hybrid diffractive–refractive optical systems, a landscape lens, and a Schmidt telescope. The systems were achromatized by using the characteristically large negative dispersion of kinoforms. In the scalar wave regime kinoforms can approach 100% efficiency but only for one object point and wavelength. The authors evaluated polychromatic image quality, accounting for diffraction efficiency, by constructing weighted geometric point–spread functions from several diffracted orders and then calculating modulation transfer functions. The MTF's of the hybrid achromats were improved at high spatial frequencies but were reduced at low frequencies because of diffraction into nondesign orders.

031.006 Wide–field f/3.5 Rosin camera.
T. M. Brown.
Appl. Opt., Vol. 31, No. 13, p. 2314 – 2316 (1 May 1992). Current Physics Microform No.: 9207B0146.

Modification of Rosin's (1961) simple design yields an f/3.5 camera that gives 2.5″ resolution over a flat field that is 5° in diameter.

031.007 Ronchi and Hartmann tests with the same mathematical theory.
A. Cordero–Davila, A. Cornejo–Rodriguez, O. Cardona–Nuñez.
Appl. Opt., Vol. 31, No. 13, p. 2370 – 2376 (1 May 1992). Current Physics Microform No.: 9207B0202.

A common mathematical model is established for the Ronchi and Hartmann tests and for interpretation of the Ronchigrams as level curves of the components of the transversal aberrations. With the same point of view, a Hartmanngram is regarded as two 90° crossed null Ronchi gratings. A simple and direct method is also developed for calculating Ronchigrams for the cases of centered and off–axis conic sections with the point light source at any location.

031.008 Optimization of partial adaptive optics.
J. M. Beckers.
Appl. Opt., Vol. 31, No. 4, p. 424 – 425 (1 Feb 1992). Current Physics Microform No.: 9205E0478.

Astronomical applications suggest wave–front control algorithms that maximize the fractional area of the pupil over which the wave–front distortion is smaller than an eighth of a wave.

031.009 Design for an all–reflection Michelson interferometer.
D. D. Cleary, J. W. Nichols, D. S. Davis.
Appl. Opt., Vol. 31, No. 4, p. 433 – 435 (1 Feb 1992). Current Physics Microform No.: 9205E0487.

The authors present a new design for an all–reflection Michelson interferometer that uses a concave spherical grating in an off–plane Rowland circle configuration.

031.010 **Calculation of wave–front–tilt correlations associated with atmospheric turbulence.**
G. A. Chanan.
J. Opt. Soc. Am. A, Vol. 9, No. 2, p. 298 – 301 (Feb 1992).
Current Physics Microform No.: 9205E2026.
Certain practical problems in astronomical optics require a detailed knowledge of aperture–to–aperture correlations in the instantaneous tilt of a wave–front distorted by atmospheric turbulence. The corresponding correlation coefficients are calculated by Fourier analysis. The results are expressed as one–dimensional integrals (over k space), which are equivalent to, but possess computational advantages over, the two–dimensional integrals (over aperture space) that appear in the literature. In addition, the angular dependence of the tilt correlation in these expressions is simple and explicit. Tables of these correlation coefficients, as well as analytic forms for the asymptotic values corresponding to large and small separations between the two apertures, are presented.

031.011 **Prediction of atmospherically induced wave–front degradations.**
M. B. Jorgenson, G. J. M. Aitken.
Opt. Lett., Vol. 17, No. 7, p. 466 – 468 (1 Apr 1992). Current Physics Microform No.: 9205F2026.
Accurate prediction of the short–term future behavior of atmospherically distorted wave fronts would permit the elimination of delays inherent in current adaptive–optics systems. It is shown by using astronomical image data that atmospherically induced wave–front distortions as represented by time series of wave–front tips and tilts measured in the visible and piston values measured in the infrared are predictable to a degree that would be useful in an adaptive–optics system. Adaptive linear predictors as well as predictors based on the back– propagation neural network are employed in this study.

031.012 **Exact expressions for thermal deformations of cooled mirrors under axisymmetric illumination.**
D. Yu. Tarutin, V. V. Kharitonov.
Sov. J. Opt. Technol., Vol. 58, No. 12, p. 766 – 768 (Dec 1991). Current Physics Microform No.: 9207X0064.
Exact expressions are obtained for calculating the steady–state profile of thermally deformed surfaces of disk and annular mirrors cooled from the opposite face in the presence of heating by beams having different heat–loading profiles.

031.013 **De atmosfeer overwonnen.**
G. Beekman.
Zenit, Jaarg. 19, Nr. 2, p. 60 – 63 (Feb 1992).

031.014 **Simple method how to manufacture eyepiece lenses.**
I. Šolc.
Říše hvězd, Vol. 72, No. 7, p. 132 – 134 (Jul 1991). In Czech.

031.015 **Sur la réduction des aberrations d'un miroir de télescope.**
E. Soulie.
Obs. Trav., No. 29, p. 31 – 38 (1992).

031.016 **Optical fibers in astronomical instruments.**
W. D. Heacox, P. Connes.
Astron. Astrophys. Rev., Vol. 3, No. 3–4, p. 169 – 199 (Apr 1992).
This review is of current and projected applications of optical fibers to observational astronomy. The intent is to provide astronomers with a broad perspective on the subject, with the hope of encouraging productive use of optical fibers in the design of new instrumentation. The unique characteristics of fibers have been (or soon well be) exploited to advantage in several areas of astronomical instrumentation, including multiplexers for multi–object spectrographs, remote optical feeds for spectrographs and photometers, coherent beam recombiners for optical interferometry, and many miscellaneous applications. The authors discuss the most important such applications in detail, with reference to operational instruments wherever possible, and with emphasis on

the optical properties of fibers and the engineering considerations encountered in their application to observational astronomy.

031.017 **New contra old wavefront measurement concepts for interferometric optical testing.**
R. Jóźwicki, M. Kujawińska, L. Sałbut.
Opt. Eng., Vol. 31, No. 3, p. 422 – 433 (Mar 1992).
Three new and modified wavefront analysis concepts for interferometric optical testing are presented, specifically a moiré fringes version of a temporal phase–shifting method and spatial–carrier phase–shifting and Fourier transform methods. All of these techniques require working with a finite fringe observation field in the interferometer, therefore additional analysis of the imaging optics is necessary. The impact of the interferometer construction and the method of analysis on the measurement error is discussed. A detailed experimental and theoretical comparison of the techniques is presented to fulfill a wide range of user requirements.

031.018 **Image plane tilt in optical systems.**
J. M. Sasian.
Opt. Eng., Vol. 31, No. 3, p. 527 – 532 (Mar 1992).
Mathematical expressions are presented to relate object and image plane tilts in nonaxially symmetrical optical systems. Examples that illustrate the use of these expressions are given. Keystone distortion is described and quantified as a function of image plane tilt.

031.019 **Determination of the coordinates of the Hartmann spots on a photographic plate.**
M. A. Dubinovskij, V. V. Kuznetsov.
Sov. J. Opt. Technol., Vol. 59, No. 1, p. 26 – 29 (Jan 1992). Current Physics Microform No.: 9207X1980.
A mathematical model is constructed of the formation of Hartmann spots, and a technique is described for finding the coordinates of the spots on a photographic plate with enhanced accuracy. Numerical results of measurements of Hartmann photographs obtained during certification of the main mirror of the AZT–22 telescope at LOMO are presented.

031.020 **Der Schmidtspiegel.**
M. Schürer.
Mitt. Satell.–Beobachtungsstn. Zimmerwald, Nr. 27.
Satelliten–Beobachtungsstation Zimmerwald (Switzerland), Universität Bern, Bern (Switzerland). 13 p. (1991).
Contents: 1. Theorie der Korrektionsplatte. 2. Das Schleifen der Korrektionsplatte. 3. Die Toleranzen für den Zusammenbau des Schmidtspiegels und die Kollimationen. 4. Der Zimmerwalder Schmidtspiegel.

031.021 **Solar imaging with a segmented adaptive mirror.**
D. S. Acton, R. C. Smithson.
Appl. Opt., Vol. 31, No. 16, p. 3161 – 3169 (1 Jun 1992). Current Physics Microform No.: 9208D1882.
A 19–segment adaptive–mirror system is currently being used on the Sacramento Peak 76–cm Tower Telescope to remove wave–front distortions resulting from atmospheric turbulence. The system has proven to be capable of substantially improving the quality of an image, at times achieving 0.33–arcsec resolution in visible wavelengths under 1 – 3–arcsec seeing conditions. An improvement in resolution seems to occur across a large field of view that is, at times, 30″ in diameter.

031.022 **Redoubling spectral resolution.**
C. G. Wynne.
Mon. Not. R. Astron. Soc., Vol. 254, No. 1, p. 7P – 10P (1 Jan 1992).
Using two immersed gratings in tandem, a spectrograph can be designed of size not much larger than a simple conventional Cassegrain spectrograph, with the same advantages of long accessible free spectral range, long slit, and multi–object capability, and in addition with spectral resolutions at least as high as are provided by an echelle spectrograph at the same slit width.

031.023 Making stars to see stars: DOD adaptive optics work is declassified.
G. P. Collins.
Phys. Today, Vol. 45, No. 2, p. 17 – 21 (Feb 1992).

031.024 Evaluation of optical systems designed using the "concentric–aplanatic" method.
A. Deslis.
SPIE Conference on Current Developments in Optical Engineering IV as Part of SPIE's International Symposium on Optical and Optoelectronics Applied Science and Engineering, p. 18 – 30 (1990). – See Abstr. 012.011 for the main entry.
The "concentric–aplanatic" method used for lens design is described in many papers. While many shapes and forms of these lenses have been shown, performance evaluation of these lenses has never been published. How good is the method? Can one design optical systems using this technique that have acceptable image quality? This paper deals step by step with these questions, both from first principles and by the basis of third order theory. Performance evaluations are given for the lenses designed.

031.025 What is wrong in extended source adaptive optics?
M. Abitbol, N. Ben–Yosef.
SPIE Conference on Current Developments in Optical Engineering IV as Part of SPIE's International Symposium on Optical and Optoelectronics Applied Science and Engineering, p. 98 – 109 (1990). – See Abstr. 012.011 for the main entry.
A novel method for the simulation of atmospheric turbulence degraded wavefronts is shown. The computer generated wavefronts contain all the statistical and spatial properties predicted for Kolmogorov type turbulence. The technique enables the generation of wavefronts from two separated point sources while maintaining the correct correlation. The techniques was used to check the effects of anisoplanitism on the performance of wavefront compensation by adaptive optics using the modal correction. The dependance on the order of correction was estimated as well. It is shown that for strong turbulence, for extended sources, there is no benefit of correction beyond the third order (tilt).

031.026 First results of an on–line adaptive optics system with atmospheric wavefront sensing by an artificial neural network.
M. Lloyd–Hart, P. Wizinowich, B. McLeod, D. Wittman, D. Colucci, R. Dekany, D. McCarthy, J. R. P. Angel, D. Sandler.
Astrophys. J., Lett., Vol. 390, No. 1, p. L41 – L44 (1 May 1992). With plates L6 – L8.
The authors report the first results from an adaptive optics system operating on–line at the telescope with the wavefront aberration sensed by a trained artificial neural network. Star images were formed at 2.2 μm wavelength by two coherently phased apertures of the Multiple Mirror Telescope, and analyzed by the neural net. The net derives wavefront parameters in a few milliseconds, and the system performance is fast enough that the aberration is nearly frozen during the time needed to make a correction. With the servo loop in operation, the corrected image shows significant power at the diffraction limit of 0″.1. The authors discuss some areas of scientific interest which will benefit from the application of adaptive wavefront correction.

031.027 Liquid mirror surface aberrations. I. Wavefront analysis.
B. K. Gibson, P. Hickson.
Astrophys. J., Vol. 391, No. 1, p. 409 – 417 (20 May 1992).
Five sources of wavefront aberration for rotating liquid mirrors are investigated: the Earth's curvature, the Coriolis force, a gravitational field gradient perpendicular to the mirror surface, a symmetry axis misalignment, and lunar tides. Coriolis effects provide the single largest perturbation, introducing correctable astigmatic and pseudo–comatic images of up to 10″ in diameter for a typical 10 m – class mirror. The curvature of the Earth, a gravitational field gradient, and lunar tides all have negligible

effects on the surface of the liquid mirror. Small axis misalignments do not introduce any wavefront aberration.

031.028 Cooled deformable mirror.
D. Kittell, C. La Fiandra.
SPIE Conference on Active and Adaptive Optical Components as Part of SPIE's International Symposium on Optical Applied Science and Engineering, p. 101 – 106 (1992). – See Abstr. 012.052 for the main entry.
This technical paper describes a new unique form of deformable mirror that was engineered, designed, and built at Hughes Danbury Optical Systems. This paper discusses the current state of the art, the requirements, configuration and special features such as the choice of materials for the mirror, interchangeable actuators, cooling technique and fabrication details.

031.029 Active and adaptive optical components: the technology and future trends.
M. A. Ealey.
SPIE Conference on Active and Adaptive Optical Components as Part of SPIE's International Symposium on Optical Applied Science and Engineering, p. 2 – 34 (1992). – See Abstr. 012.052 for the main entry.
Though the exact origins of active and adaptive optics are unknown, recent history can be traced by the development of components. Development of active and adaptive optical components began in the early 1970's with two areas of emphasis: 1) compensated imaging and 2) laser beam propagation. Typical for defense driven research and development programs, much of the evolutionary successes and pitfalls will forever remain undisclosed. Recent declassification of adaptive optics technology including laser guide star use has opened the way for exploitation by astronomers and others of the component development.

031.030 Deformable mirrors: design fundamentals, key performance specifications, and parametric trades.
M. A. Ealey, J. A. Wellman.
SPIE Conference on Active and Adaptive Optical Components as Part of SPIE's International Symposium on Optical Applied Science and Engineering, p. 36 – 51 (1992). – See Abstr. 012.052 for the main entry.
A wide variety of deformable mirror structures have been studied for wavefront correction since the advent of adaptive optics nearly two decades ago. These structures generally fall into two categories: 1) segmented facesheet and 2) continuous facesheet. In addition there are two methods of correction: 1) zonal control and 2) modal control. The basic mirror types are discussed and analyzed in terms of wavefront correction capabilities. Curve fitting characteristics are explained in terms of the optical influence function and mirror meshing functions. The continuous facesheet deformable mirror is used as a model to develop basic design equations which are used for parametric trades.

031.031 A low voltage electrodistortive mirror system for wavefront control.
W. G. Thorburn, L. Kaplan.
SPIE Conference on Active and Adaptive Optical Components as Part of SPIE's International Symposium on Optical Applied Science and Engineering, p. 52 – 63 (1992). – See Abstr. 012.052 for the main entry.
A computerized closed loop wavefront control system has been developed. This system uses a 97 channel deformable mirror to provide accurate wavefront control. The continuous facesheet mirror has a 63 mm aperture and drivers which address each of the 97 low voltage electrostrictive actuators independently. The computerized wavefront control system is capable of accurately measuring the output wavefront and generating commands to produce the desired shape. Closed loop control has been demonstrated to produce various mirror surfaces with an

accuracy of 0.020 microns rms. Data are presented showing the control capability.

031.032 High bandwidth, long stroke segmented mirror for atmospheric compensation.
B. Hulburd, T. Barrett, L. Cuellar, D. Sandler.
SPIE Conference on Active and Adaptive Optical Components as Part of SPIE's International Symposium on Optical Applied Science and Engineering, p. 64 – 75 (1992). – See Abstr. 012.052 for the main entry.

Segmented adaptive optic mirrors have been developed, fabricated and demonstrated in real time atmospheric compensation systems. Until recently, most segmented adaptive optic mirrors have been designed for single wavelength applications and have not required more than 1.5 microns of surface motion since absolute phasing of the surface is not required for very narrow bandwidth compensation. Requirements for astronomical and imaging systems have required the design and fabrication of long stroke (6 – 10 micron) segmented mirrors capable of absolute phasing of the segments, optical response from 0.4 to 3.5 microns and bandwidths above 2.5 kHz.

031.033 The NSO/Sac Peak continuous–face–plate adaptive mirror.
R. B. Dunn, G. W. Streander, W. Hull, L. Wilkins.
SPIE Conference on Active and Adaptive Optical Components as Part of SPIE's International Symposium on Optical Applied Science and Engineering, p. 88 – 100 (1992). – See Abstr. 012.052 for the main entry.

The National Solar Observatory is constructing a continuous–face–plate mirror with 61 actuators. The mirror, which has a clear aperture of 218 mm, features a detachable face plate and replaceable actuators that are servoed to maintain a position measured by capacitors which are within the actuators themselves. The servos have a bandwidth of 1 kHz (–3 db). At Sacramento Peak the mirror will be used with the 76–cm aperture Vacuum Tower Telescope to observe small details on the Sun.

031.034 Adaptive optics wavefront corrector using addressable liquid crystal retarders: Part 2.
D. Bonaccini, G. Brusa, S. Esposito, P. Salinari, P. Stefanini, V. Biliotti.
SPIE Conference on Active and Adaptive Optical Components as Part of SPIE's International Symposium on Optical Applied Science and Engineering, p. 133 – 143 (1992). – See Abstr. 012.052 for the main entry.

The authors present recent results of their investigations into the feasibility of a wavefront corrector for adaptive optics, using nematic liquid crystal material. In particular, they are aiming at a small, flexible adaptive optics module, to be located at the focal plane of existing and new generation telescopes in order to improve imaging quality. The authors address some of the specific problems investigated, concerning the theoretical behaviour of the LC corrector, its comparison with other state–of–the–art wavefront correctors, modal and zonal, the speed and the slew rate of the actuators, and a real time capacitive servo loop scheme controlling the actuator retardance.

031.035 Large Active Mirror Program (LAMP).
R. L. Plante.
SPIE Conference on Active and Adaptive Optical Components as Part of SPIE's International Symposium on Optical Applied Science and Engineering, p. 146 – 160 (1992). – See Abstr. 012.052 for the main entry.

In the fall of 1988 the construction and testing of a 4–meter–diameter, segmented, active mirror was completed. Developed to demonstrate the technology for 10–meter–class mirrors for beam expanders in high energy laser systems, diffraction–limited performance at the design wavelength was achieved.

031.036 Applications of stress polishing techniques as developed for the Keck Observatory primary mirror fabrication.
K. W. Johnson.
SPIE Conference on Active and Adaptive Optical Components as Part of SPIE's International Symposium on Optical Applied Science and Engineering, p. 161 – 164 (1992). – See Abstr. 012.052 for the main entry.

The stress polishing technique is a powerful tool for fabrication of off–axis mirror profiles. The inherent smoothness that can be achieved from polishing spheres can now be applied to asymmetric profiles. This smoothness combined with the need for the mirror to be thin and flexible can be coupled to active mounting systems which readily correct low order shapes to produce a near ideal system. This technique will allow the fabrication of complete off–axis telescopes within the same basic cost range as on–axis systems, if not for less. The strength of this statement comes from the fact that both the primary and secondary mirror can be fabricated and mounted using the same techniques, affording less stringent global profile requirements during fabrication. The components are literally bent into final shape with minimal high spacial residual error in the telescope, making the system function at optimum.

031.037 PAMELA: control of a segmented mirror via wavefront tilt and segment piston sensing.
A. D. Gleckler, D. J. Markason, G. H. Ames.
SPIE Conference on Active and Adaptive Optical Components as Part of SPIE's International Symposium on Optical Applied Science and Engineering, p. 176 – 189 (1992). – See Abstr. 012.052 for the main entry.

A hardware demonstration of segmented mirror systems for adaptive optics is described. The basis of the PAMELA™ concept (Phased Array Mirror, Extendible Large Aperture) is that large adaptive mirrors can be fabricated from many small segments by utilizing edge–sensors, which measure the piston error between segments. The authors have investigated the interaction between the piston and tilt control loops which direct the motion of individual segments.

031.038 Final surface error correction of an off–axis aspheric petal by ion figuring.
L. N. Allen, J. J. Hannon, R. W. Wambach.
SPIE Conference on Active and Adaptive Optical Components as Part of SPIE's International Symposium on Optical Applied Science and Engineering, p. 190 – 200 (1992). – See Abstr. 012.052 for the main entry.

The final surface figure error correction of a 1.3 m ULE™ frit–bonded, ultra–lightweight, off–axis primary mirror petal was successfully completed using the ion figuring process. The petal was a concave aspheric optical element. Ion figuring is an optical fabrication method that provides highly controlled error correction of previously polished surfaces using a directed, inert and neutralized ion beam to physically sputter material from the optic surface. The surface figure error of the petal following conventional polishing was 5.02 λ p–v, 0.62 λ rms, and was improved to 0.17 λ p–v, 0.015 λ rms in four process (test–ion figure) iterations (λ = 632.8 nm). The benefits of ion figuring a complex shaped optic using multiple figuring–testing iterations were clearly demonstrated.

031.039 Specification of fine–steering mirrors for line–of–sight stabilization systems.
L. M. Germann.
SPIE Conference on Active and Adaptive Optical Components as Part of SPIE's International Symposium on Optical Applied Science and Engineering, p. 202 – 212 (1992). – See Abstr. 012.052 for the main entry.

Fine–steering mirrors (FSM) are being used in a greater variety of optical systems than ever before. The performance requirements for these systems vary just as widely. It is important for optical system designers, space–based–observation instrument principal investigators, astronomers, tactical weapons seeker designers, and others to understand the capabilities of FSM. This paper introduces the critical and secondary performance and environmental parameters that make up an FSM specification. Also discussed are situations when the various parameters become important.

031.040 Design of a high–bandwidth steering mirror for space–based optical communications.

G. C. Loney.
SPIE Conference on Active and Adaptive Optical Components as Part of SPIE's International Symposium on Optical Applied Science and Engineering, p. 225 – 235 (1992). – See Abstr. 012.052 for the main entry.

A space–based optical communications experiment requires a fast steering mirror as part of its spatial pointing, tracking and acquisition system. The High Bandwidth Steering Mirror version C, has been designed, built and tested. This device steers a small–aperture mirror of 6 mm about two axes, through an operating range of 25 mrad and a small–signal closed–loop bandwidth up to 2 kHz. A description of the functional requirements, design and assembly, and analytical methods used is presented. Key results from performance and environmental testing are shown.

031.041 Latest experience in design of piezoelectric driven fine steering mirrors.

H. Marth, M. Donat, C. Pohlhammer.
SPIE Conference on Active and Adaptive Optical Components as Part of SPIE's International Symposium on Optical Applied Science and Engineering, p. 248 – 261 (1992). – See Abstr. 012.052 for the main entry.

The European Space Organization requested a system to compensate for atmospherically induced image jitter in astronomical telescopes. The product is a sophisticated adaptive optic system using closed loop piezoelectric actuators and momentum compensation to significantly improve telescope resolution during long integrations by correcting for image jitter in real time. Optimizing the design of this system involved solving several interdependent problems including: 1. Selection of the motion system. 2. Arrangement of the pivot points and actuators. 3. Momentum compensation. 4. Selection of the sensor system. This paper presents the trade–offs leading to final design of the system.

031.042 Magnetostrictive actuators in optical design.

V. V. Apollonov, V. I. Aksinin, S. A. Chetkin, V. V. Kijko, S. V. Muraviev, G. V. Vdovin.
SPIE Conference on Active and Adaptive Optical Components as Part of SPIE's International Symposium on Optical Applied Science and Engineering, p. 313 – 324 (1992). – See Abstr. 012.052 for the main entry.

A number of the important applications of high precision actuators are related with laser technique and technology, optics, especially with large optical devices for the ground and space based astronomy. These actuators can be used in adaptive optics as the displacement and force actuators for correction of large optics tilts.

031.043 Real–time Hartmann sensor for phase–conjugated adaptive optical system.

V. V. Apollonov, S. A. Chetkin, V. V. Kijko, S. V. Muraviev, G. V. Vdovin.
SPIE Conference on Active and Adaptive Optical Components as Part of SPIE's International Symposium on Optical Applied Science and Engineering, p. 325 – 336 (1992). – See Abstr. 012.052 for the main entry.

031.044 Wide field of view adaptive optics.

A. J. Jankevics, A. Wirth.
SPIE Conference on Active and Adaptive Optical Components as Part of SPIE's International Symposium on Optical Applied Science and Engineering, p. 438 – 448 (1992). – See Abstr. 012.052 for the main entry.

One of the most significant limitations to conventional atmospheric compensation systems is their very restricted field of view (FOV), generally equal to an isoplanatic patch size. A wavefront sensing and compensation concept is proposed that should allow the FOV to be increased in size by factors of ten or more. The kernel of the idea is to use wavefront measurements in several (~ 9) directions separated by 100 – 200 μrad to deduce an estimate of the three dimensional Optical Path Difference (OPD) distribution in the atmosphere. Initial indications are that the FOV may be increased to 500 μrad for a 3.5 m telescope operating at 0.8 μm.

031.045 High precision, wide dynamic range WCE wavefront sensor.

B. A. Horwitz.
SPIE Conference on Active and Adaptive Optical Components as Part of SPIE's International Symposium on Optical Applied Science and Engineering, p. 449 – 459 (1992). – See Abstr. 012.052 for the main entry.

A temporal carrier frequency shearing interferometer was designed and fabricated to support the Space Shuttle based Wavefront Control Experiment. This platform and the intended experiments placed many constraints (weight, size, power, sensitivity, etc.) on the sensor design process. The sensor resulting from this design meets all requirements by using a rotating crossed log spiral grating and PMT detectors to achieve both a large tilt range and high precision. A measurement dynamic range of 1340:1 is obtained at 3875 wavefronts per second and the sensor is shot noise limited over 5 decades of signal.

031.046 MARTINI: sensing and control system design.

A. P. Doel, C. N. Dunlop, J. V. Major, R. M. Myers, R. M. Sharples.
SPIE Conference on Active and Adaptive Optical Components as Part of SPIE's International Symposium on Optical Applied Science and Engineering, p. 472 – 478 (1992). – See Abstr. 012.052 for the main entry.

MARTINI (Multiple Aperture Real Time Image Normalisation Instrument) is an astronomical adaptive optics system for visible imaging and spectroscopic feedthrough at the 4.2 m William Herschel Telescope on La Palma. It consists of a six–subaperture, tip–tilt–piston, segmented mirror device and uses $4r_0$ aperture–matching to provide optimum slope removal in zones large enough for operation in the visible and with reference objects fainter than $V = 13^m$. This limit is achieved by optimizing the use of reference light, by analysing the information from a photon counting wavefront sensor using a non–framing (ie. irregular sampling) infinite impulse response filter for estimation and prediction of the wavefront slopes. The value of this approach is discussed along with its extension to higher–order correction schemes.

031.047 Stray light issues for background–limited infrared telescope operation.

M. S. Scholl, J. W. Scholl.
SPIE Conference on Infrared Technology XVII – Dedicated to the Memory of Irving J. Spiro as Part of SPIE's International Symposium on Optical Applied Science and Engineering, p. 109 – 118 (1991). – See Abstr. 012.051 for the main entry.

An infrared telescope needs to be designed in such a way that the telescope itself contributes a minimal amount of noise to the detector. When, under the most favorable conditions, the telescope is looking for faint sources in the 3K background, it is important that it contributes appreciably less than the outside background. There are two ways of achieving this goal. The first one is to decrease the temperature of the objects seen by the detector. The second one is to decrease the configuration factor for the high temperature telescope components. The analytical expressions have been obtained to graph the photon radiance in a band as a function of temperature. Also, the derivative of the photon radiance in a band with respect to temperature has been obtained. The temperature dependence of photon radiance and its derivative are applied to a background–limited telescope to determine the temperature tolerances.

031.048 **Optimal grazing incidence optics and its application to wide-field X-ray imaging.**
C. J. Burrows, R. Burg, R. Giacconi.
Astrophys. J., Vol. 392, No. 2, p. 760 – 765 (20 Jun 1992).
The authors discuss a class of high–resolution, efficient, and wide–field grazing incidence optics. Although such designs have been discussed in the literature before, they have rarely been considered for a practical application. The authors have developed optical designs to search efficiently for distant X–ray clusters. They wish to resolve clusters of galaxies with an angular diameter of ~ 5″ over a field of ~ 1°, so that a comprehensive deep search is possible with a payload of 1/10 the linear dimensions of AXAF. No design of a standard type is capable of meeting this requirement. Wolter – Schwarzschild and parabola – hyperbola designs yield a useful field that is many times smaller than needed because they suffer from unacceptably large off–axis aberration, even though they yield perfect images on–axis. By dropping the requirement for perfect on–axis imagery, searching within a suitably general class of telescope design, and optimizing a quantity directly related to the scientific requirement, the authors have been able to show that satisfactory designs do exist. The resulting telescope is shown to be no more difficult to fabricate than existing mirrors and that it can be nested.

031.049 **A high–aperture mirror–lens optical system.**
G. M. Popov.
Bull. Crimean Astrophys. Obs., Vol. 82 , p. 173 – 175 (1992). English translation of Izv. Krym. Astrofiz. Obs., Tom 82, p. 189 – 192 (1990).
High–aperture Cassegrain–type systems are considered in which there is a two–lens corrector and a field lens in each case. There is a relationship between the aperture and the refractive index. The system can have a numerical aperture of up to 1:1.25 – 1:1.5 with a field of view of 16° and a convenient external position for the focal phase.   .

031.050 **Mirror for optical telescopes.**
M. M. Miroshnikov, S. V. Ljubarsky, Y. P. Khimich.
Opt. Eng., Vol. 31, No. 4, p. 701 – 710 (Apr 1992).
From the viewpoint of up–to–date requirements, fundamental scientific and technonolgical problems arising in the design and manufacture of optical reflecting telescopes are formulated as compared to those of the refracting systems. Nontraditional materials, such as beryllium, silicon, and silicon carbide, are shown to be capable of successful competition with traditional optical materials for the telescope mirror. A brief overview is given of the development of beryllium–based mirrors.

031.051 **Determining the requirements for the thermal stabilization system of the mirrors of large telescopes.**
V. V. Barantsev, A. A. Kamenskaya, V. G. Parfenov, A. M. Savitskij.
Sov. J. Opt. Technol., Vol. 59, No. 5, p. 267 – 270 (May 1992). Current Physics Microform No.: 9211X0011.
A method is proposed for determining the requirements for the thermal stabilization system of thin adaptive mirrors of large telescopes, taking into account the amplitude of the dynamic error of the temperature of the thermostatically controlled enclosure and the parameters that characterize the physical and mechanical properties of the mirror material as well as its geometrical dimensions.

031.052 **Fitting capability of deformable mirror.**
Jiang Wenhan, Ling Ning, Rao Xuejun, Shi Fan.
SPIE Conference on Active and Adaptive Optical Systems as Part of SPIE's International Symposium on Optical Applied Science and Engineering, p. 130 – 137 (1991). – See Abstr. 012.074 for the main entry.
Deformable mirror is the key element for adaptive optical wavefront correction. The number of actuators decides the complexity and cost of adaptive optical system. In this paper computer simulations of wavefront error for fitting different Zernike terms by deformable mirror with different number of actuators are presented. The arrangement of actuator and the influence function of mirror are discussed in respect of fitting error. The minimum number of actuators for fitting different Zernike orders of wavefront are given. Some optical experiments of fitting capability have been done with 19 and 37–element deformable mirrors and a Zygo interferometer.

031.053 **Adaptive optics using curvature sensing.**
F. F. Forbes, N. Roddier.
SPIE Conference on Active and Adaptive Optical Systems as Part of SPIE's International Symposium on Optical Applied Science and Engineering, p. 140 – 147 (1991). – See Abstr. 012.074 for the main entry.
An adaptive optics scheme is proposed which uses curvature sensing, a bimorph PZT deformable mirror, a simple extrafocal image splitter and a sensitive, low noise Solid State Photomultiplier array detector. With curvature sensing the incoming wavefront is described in terms of curvature which can be directly mapped point–by–point to a driven deformable segmented mirror. While other methods such as Hartmann, require time and cost consuming computations, the curvature mirror can be driven in near real–time taking full advantage of the 11 kHz bandwidth of the bimorph mirror. The SSPM is easily configured to any array pattern without loss of performance. The arrangement is particularly studied to large telescopes in the 8 m class for which the wavefront curvature sensor can provide additional useful information to correct for collimation and structural variations as well as the rapid seeing correction.

031.054 **Neural network adaptive optics for the Multiple Mirror Telescope.**
P. Wizinowich, M. Lloyd–Hart, B. McLeod, D. Colucci, R. Dekany, D. Wittman, R. Angel, D. McCarthy, B. Hulburd, D. Sandler.
SPIE Conference on Active and Adaptive Optical Systems as Part of SPIE's International Symposium on Optical Applied Science and Engineering, p. 148 – 158 (1991). – See Abstr. 012.074 for the main entry.
The Multiple Mirror Telescope (MMT) consists of six co–mounted 1.8 m telescopes from which the light is brought to a combined coherent focus. Atmospheric turbulence spoils the MMT diffraction–limited beam profile, which would otherwise have a central peak of 0.06″ full width at half maximum, at 2 μm wavelength. At this wavelength adaptive correction of the tilt and path difference of each telescope beam is sufficient to recover diffraction–limited angular resolution. Computer simulations have shown that these tilts and pistons can be derived by an artificial neutral network, given only a simultaneous pair of in–focus and out–of–focus images of a reference star formed at the combined focus of all the array elements. The authors describe an adaptive optics system, based on this approach, which they have developed for the MMT.

031.055 **Solar astronomy with a 19–segment adaptive mirror.**
D. S. Acton, R. C. Smithson.
SPIE Conference on Active and Adaptive Optical Systems as Part of SPIE's International Symposium on Optical Applied Science and Engineering, p. 159 – 164 (1991). – See Abstr. 012.074 for the main entry.
A 19–segment adaptive mirror system for use in solar astronomy has been developed and operated on the Sacramento Peak Tower Telescope. The system has proven itself to be capable of improving the quality of an image, at times achieving 1/3 arcsecond resolution in 1 – 3 arcsecond seeing conditions.

031.056 **The Johns Hopkins adaptive optics coronagraph.**
M. Clampin, S. T. Durrance, D. A. Golimowski, R. H. Barkhouser.
SPIE Conference on Active and Adaptive Optical Systems as Part of SPIE's International Symposium on Optical Applied Science and Engineering, p. 165 – 174 (1991). – See Abstr. 012.074 for the main entry.
The Johns Hopkins University is developing a stellar coronagraph which will use adaptive optics to achieve nearly diffraction–limited imaging at optical wavelengths with 1 – 2 meter class

telescopes. The first phase of development, the incorporation of an image motion compensation system into the coronagraph, is complete. Performance tests have resulted in a factor of 2 gain in image resolution, corresponding to the maximum grain predicted by theory. The next phase of development involves the construction of an electrostatically deformable membrane mirror and a wavefront curvature sensor for the removal of higher order aberrations. A membrane mirror with 91 actuators has been built for laboratory testing. Integration of the adaptive mirrors, higher speed wavefront sensor, and control processor is forthcoming.

031.057   **Active optics system for a 3.5–meter structured mirror.**
L. Stepp, N. Roddier, D. Dryden, Cho Myung.
SPIE Conference on Active and Adaptive Optical Systems as Part of SPIE's International Symposium on Optical Applied Science and Engineering, p. 175 – 185 (1991). – See Abstr. 012.074 for the main entry.
An active optics system for a 3.5–meter f/1.75 borosilicate honeycomb mirror has been designed and built. The system hardware and software are described, and preliminary test results are presented that demonstrate the structured mirror responds well to the active optics control. Plans for extensive further testing are described. The results of the testing will guide a redesign of the system, before installation of the second-generation system in the WIYN Telscope, to be built on Kitt Peak in Arizona.

031.058   **Alignment and focus control of a telescope using image sharpening.**
P. A. Jones.
SPIE Conference on Active and Adaptive Optical Systems as Part of SPIE's International Symposium on Optical Applied Science and Engineering, p. 194 – 204 (1991). – See Abstr. 012.074 for the main entry.
Two alternative methods have traditionally been used to maintain telescope alignment and focus. The direct method measures rigid body changes (decentration, tilt and despace) using internal sensors and predicts the resulting performance degradation. The indirect method measures the performance degradation with a wavefront sensor and predicts the rigid body changes. An alternative indirect method uses an image sharpening merit function value derived from the image irradiance data. The image sharpening function has the property that it achieves its maximum value when the image has no residual aberrations. A comparison of the direct and indirect methods and recent experimental results of the image sharpening method are presented.

031.059   **Comparisons of deformable mirror models and influence functions.**
H. R. Hiddleston, D. D. Lyman, E. L. Schafer.
SPIE Conference on Active and Adaptive Optical Systems as Part of SPIE's International Symposium on Optical Applied Science and Engineering, p. 20 – 33 (1991). – See Abstr. 012.074 for the main entry.

031.060   **The University of Hawaii adaptive system. I. General approach.**
F. Roddier, J. E. Graves, D. McKenna, M. Northcott.
SPIE Conference on Active and Adaptive Optical Systems as Part of SPIE's International Symposium on Optical Applied Science and Engineering, p. 248 – 253 (1991). – See Abstr. 012.074 for the main entry.
An innovative adaptive optics system is being developed at the University of Hawaii to sharpen images produced by telescopes on Mauna Kea.

031.061   **The University of Hawaii adaptive optics system. II. Computer simulation.**
M. Northcott.
SPIE Conference on Active and Adaptive Optical Systems as Part of SPIE's International Symposium on Optical Applied Science and Engineering, p. 254 – 261 (1991). – See Abstr. 012.074 for the main entry.
The author has proposed and is building a low order adaptive optics system based upon curvature sensing and a bimorph

mirror. He describes a computer simulation of this adaptive optics system and some of the results obtained from it. The simulation has been used to help refine the design of the system, and to evaluate its performance limits.

031.062   **Optical performance of large ground–based telescopes.**
P. Dierickx.
J. Mod. Opt., Vol. 39, No. 3, p. 569 – 588 (Mar 1992).
Images taken with ground–based telescopes are dominated by atmospheric seeing. Analytical expressions of long–exposure optical functions, namely the modulation transfer function, point spread function and encircled energy are established, under the assumption that dome and telescope seeing are brought to negligible values, and that the diameter of the telescope is larger than the atmospheric coherence length. The influence of guiding errors and axisymmetrical telescope aberrations is also assessed, and a definition of optical quality is proposed. The results are generalized and the optical performance of a ground–based telescope is expressed in terms of effective diameter and signal–to–noise ratio.

031.063   **Latest developments of active optics of the ESO NTT and the implications for the ESO VLT.**
L. Noethe, G. Andreoni, F. Franza, P. Giordano, F. Merkle, R. N. Wilson.
SPIE Conference on Active and Adaptive Optical Systems as Part of SPIE's International Symposium on Optical Applied Science and Engineering, p. 293 – 296 (1991). – See Abstr. 012.074 for the main entry.
The latest developments of active optics of the ESO NTT include the reduction of friction in the lateral supports of the primary mirror and in the positioning system of the secondary mirror. The most important remaining problem is the local air condition. The implications for the ESO VLT and the latest developments in the design of its active optics are discussed.

031.064   **The Come–On–Plus project: an upgrade of the Come–On adaptive optics prototype system.**
E. Gendron, J. G. Cuby, F. Rigaut, P. Lena, J. C. Fontanella, G. Rousset, J. P. Gaffard, C. Boyer, J. C. Richard, M. Vittot, F. Merkle, N. Hubin.
SPIE Conference on Active and Adaptive Optical Systems as Part of SPIE's International Symposium on Optical Applied Science and Engineering, p. 298 – 307 (1991). – See Abstr. 012.074 for the main entry.
This paper is a presentation of the Come–On–Plus adaptive optics system, based on the Come–On prototype. Come–On–Plus will be set up in 1992 on the ESO 3.6 m telescope in La Silla (Chile). It is an upgrade of the Come–On instrument, with a 52 actuator deformable mirror, and 30 Hz correction bandwidth. But the main improvement concerns the wavefront sensing, designed in this instrument for astronomical applications, with a high detectivity wavefront sensor and a specific mirror control algorithm. This system is planned for routine astronomical observing as well as providing design parameters for the adaptive optics system of the ESO Very Large Telescope (VLT).

031.065   **Adaptive optics – a progress review.**
J. W. Hardy.
SPIE Conference on Active and Adaptive Optical Systems as Part of SPIE's International Symposium on Optical Applied Science and Engineering, p. 2 – 17 (1991). – See Abstr. 012.074 for the main entry.
Active and adaptive optics technology has emerged from the laboratory and is being applied to improve the performance of optical imaging and laser systems. In the last few years, development of both systems and components has accelerated. Many new concepts and devices have appeared, among which are high–performance deformable mirrors, new types of wavefront sensors, and more sophisticated wavefront processing algorithms. Current developments in adaptive optics for ground–based astronomy include the use of IR wavelengths, partial wavefront compensation using natural guide stars, and the use of

laser guide stars to allow all–sky coverage with full compensation at visible wavelengths.

031.066   **MARTINI: system operation and astronomical performance.**
A. P. Doel, C. N. Dunlop, J. V. Major, R. M. Myers, R. M. Sharples.
SPIE Conference on Active and Adaptive Optical Systems as Part of SPIE's International Symposium on Optical Applied Science and Engineering, p. 319 – 326 (1991). – See Abstr. 012.074 for the main entry.
The MARTINI adaptive optics system has been engineered for regular astronomical observations on the 4.2 m William Herschel Telescope on La Palma. The design specifications for such a general purpose image–sharpening device are discussed. Examples of the imaging performance achieved during development of MARTINI are presented, together with results from visible-region astronomical imaging programmes. The prospects for spectroscopic observations with this class of system are outlined.

031.067   **Adaptive optical transfer function modeling.**
J. P. Gaffard, G. Ledanois.
SPIE Conference on Active and Adaptive Optical Systems as Part of SPIE's International Symposium on Optical Applied Science and Engineering, p. 34 – 45 (1991). – See Abstr. 012.074 for the main entry.
The authors revise their model described in an earlier paper to introduce modal control and take into account the effects of time lags and/or angular depointings on the corrected Optical Transfer Function (O.T.F.). These lags represent the time needed to perform the computations between phase measurements and corrections, and the angular depointings define the angular field where the corrections are valuables. The model permits an evaluation of the O.T.F. decay in respect to the Fried diameter and for different time lags for an Adaptive Mirror corresponding to the Come–On project specifications.

031.068   **The MIT multipoint alignment testbed: technology development for optical interferometry.**
G. H. Blackwood, R. N. Jacques, D. W. Miller.
SPIE Conference on Active and Adaptive Optical Systems as Part of SPIE's International Symposium on Optical Applied Science and Engineering, p. 371 – 391 (1991). – See Abstr. 012.074 for the main entry.
A class of proposed space–based astronomical missions requiring large baselines and precision alignment can benefit from the application of Controlled Structures Technology. One candidate mission, that of a 35 m baseline orbiting optical interferometer, is studied as a focus mission for a testbed for controlled structures research. Interferometry science requirements are investigated and used to design a laboratory testbed which captures the essential architecture, physics and performance requirements of a full scale instrument. The testbed and research program are discussed in terms of controlled structures design and in terms of the expected benefits to the optical engineering and science communities.

031.069   **Implementation issue in the control of a flexible mirror testbed.**
E. H. Anderson, J. P. How.
SPIE Conference on Active and Adaptive Optical Systems as Part of SPIE's International Symposium on Optical Applied Science and Engineering, p. 392 – 405 (1991). – See Abstr. 012.074 for the main entry.
The goal is to develop and demonstrate the potential benefits of applying the controlled structures technology (CST) approach to active and adaptive optics. Two testbeds are envisioned for the experimental work. CST techniques are demonstrated experimentally on the first, a simple deformable mirror testbed incorporating piezoelectric sensors and actuators and a single optical displacement sensor. Implementation of passive damping augmentation, local control and global high authority control is demonstrated. Some of the key issue in the approach to the

control of larger interferometers or lightweight Cassegrain telescopes are discussed.

031.070   **Performance tests of a 1500 degree–of–freedom adaptive optics system for atmospheric compensation.**
L. Cuellar, P. Johnson, D. G. Sandler.
SPIE Conference on Active and Adaptive Optical Systems as Part of SPIE's International Symposium on Optical Applied Science and Engineering, p. 468 – 476 (1991). – See Abstr. 012.074 for the main entry.
Results from a benchtop experiment to demonstrate phase compensation using a 512 segment, 1500 degree–of–freedom adaptive optic system are presented. Atmospheric phase distortion is simulated by a static Kolmogorov spectrum aberration plate with r_0 equal to the subaperture size. The phase gradients are measured using a Poisson–limited, self–referenced shearing interferometer which operated at two distinct shear lengths. A parallel processor is then employed utilizing a sparse matrix multiply to reconstruct the phase front in realtime. The performance of the compensation was determined by measuring the normalized half λ/D intensity ratio in the Fourier transform plane. Corrections to a Strehl ratio of 0.55 were performed, consistent with the measured sensitivity of the system.

031.071   **Adaptive optics, transfer loops modeling.**
C. Boyer, J. P. Gaffard.
SPIE Conference on Active and Adaptive Optical Systems as Part of SPIE's International Symposium on Optical Applied Science and Engineering, p. 46 – 61 (1991). – See Abstr. 012.074 for the main entry.
An adaptive optical system dedicated to high resolution imaging can be modelized in terms of transfer loops. This model permits to estimate the response of such a system to time varying wavefront perturbation. In the present paper a block diagram describing the closed loop of the adaptive optics of the Come–On project is given and a theoretical expression termed in Z transform is found in the case of a nodal approach. In a second step, identification methods are used to determine the best parameter values of these nodal models. Finally the responses of these models to known perturbations are compared with the experimental data recorded during the experiments of Come–On. Same analysis are made for modal models. In the two cases the results are found to be in good agreement.

031.072   **Optical figure testing of prototype mirrors for JPL's Precision Segmented Reflector (PSR) program.**
E. B. Hochberg.
SPIE Conference on Active and Adaptive Optical Systems as Part of SPIE's International Symposium on Optical Applied Science and Engineering, p. 511 – 522 (1991). – See Abstr. 012.074 for the main entry.
JPL's Precision Segmented Reflector (PSR) program is developing enabling technologies for large space telescopes employing segmented optics, and in particular, lightweight, thermally stable mirrors for telescopes operating at sub–millimeter wavelengths. This paper describes JPL's vacuum cryointerferometric optical test facility. The paper discusses performance and limitations of the optical metrology hardware and software components. Representative test results on prototype one meter–class composite mirrors being developed for PSR and related ground–based programs are also presented.

031.073   **Imaging performance analysis of adaptive optical telescopes using laser guide stars.**
B. M. Welsh.
SPIE Conference on Active and Adaptive Optical Systems as Part of SPIE's International Symposium on Optical Applied Science and Engineering, p. 88 – 99 (1991). Also published in Appl. Opt., Vol. 30, No. 34, p. 5021–5030 (1 Dec 1991). – See Abstr. 012.074 for the main entry.
The use of laser guide stars in conjunction with adaptive optical telescopes offers the possibility of nearly diffraction

limited imaging performance from large, ground–based telescopes. The author investigates the expected imaging performance of an adaptive telescope using laser guide stars created in the mesospheric sodium (Na) layer. A two to three meter class telescope is analyzed for the case of a single, on axis guide star at an altitude of 92 km. The results of the investigation indicate that a 3 m adaptive telescope using a single Na guide star is capable of achieving a Strehl ratio of 0.57 and an angular resolution nearly matching that of diffraction limited performance (0.05″). This performance is achieved assuming $r_0 = 20$ cm and a 5 watt laser is used to create the guide star.

031.074   An aspheric grin ray trace program for thermal analysis of optical systems.
D. K. Sharma, D. Shyam, G. Sujatha.
Kodaikanal Obs. Bull., Vol. 11, p. 39 – 46 (1991). – See Abstr. 012.072 for the main entry.
   The authors describe a program which can ray trace through systems bounded by spherical/aspheric surfaces together with the option to trace through media having graded index profile (GRIN) for one or more components. This program is useful for analysing the performance of lenses subjected to thermal gradients.

031.075   Test of spherical surface with small curvature and large aperture.
Zheng Gui–tang.
Publ. Purple Mt. Obs., Vol. 11, No. 1, p. 50 – 53 (Mar 1992). In Chinese.

031.076   IDL SH. A package of Shack–Hartmann data reduction under IDL environment.
R. Ragazzoni.
Telesc. Naz. Galileo, Tech. Rep., No. 10.
Osservatorio Astronomico di Padova–Asiago, Asiago (Italy). 30 p. (Feb 1992).
   This document describes the software produced in the Active Optics Group of the Telescopio Nazionale Galileo (TNG) for wavefront sensing using a Shack–Hartmann Wavefront Sensor. Such a software was developed under IDL, using occasional codes written in other languages and linked in a transparent way to the IDL environment.

031.077   Performance of the microlens array.
            Y. Torii, A. Miyashita, M. Nakagiri, H. Ando,
Y. Yamashita.
Rep. Natl. Astron. Obs. Jpn., Vol. 1, No. 3, p. 213 – 220 (1992). In Japanese.
   Several sets of plane microlens array were manufactured by an ion exchange method. Optical experiments are carried out, in order to evaluate these microlens arrays. It turned out that the Fresnel diffraction has an important role in understanding the lens quality. Positions of several maxima and minima of the Fresnel diffraction pattern were measured. Fresnel diffraction patterns are also calculated. Comparing these two, the authors try to evaluate the focal length of the microlens array.

031.078   Application of diffraction theory to telescope optics.
            A. Miyashita, H. Ando.
Publ. Natl. Astron. Obs. Jpn., Vol. 2, No. 3, p. 385 – 397 (1992).
   Based on the diffraction theory derived from the principle of Huygens–Fresnel, the realistic energy distribution (or Point Spread Function) at the focal plane of the telescope optics has been calculated. This method is very useful to make a final answer or a final acceptance check to the optical performance of telescope optics. The authors give some applications of this calculation to the telescope optics.

031.079   A concept to increase thickness uniformity and to reduce pinholes in vacuum aluminized coatings.
T. D. Padalia.
Uttar Pradesh State Obs., Repr., No. 432.
Uttar Pradesh State Observatory, Naini Tal (India). 6 p. (1992).

Basics of interferometry.
See Abstr. 003.064.

Current developments in optical engineering 4. Proceedings.
See Abstr. 012.011.

Active and adaptive optical components. Proceedings.
See Abstr. 012.052.

Active and adaptive optical systems. Proceedings.
See Abstr. 012.074.

The LEST project.
See Abstr. 013.001.

Through the eyepiece.
See Abstr. 014.021.

Temperatures of the LEST tube.
See Abstr. 032.016.

Primary mirror control system for the GALILEO telescope.
See Abstr. 032.018.

Moving M2 mirror without pointing offset.
See Abstr. 032.019.

Progress report on a five–axis fast guiding secondary for the University of Hawaii 2.2–m telescope.
See Abstr. 032.020.

Adaptive optics for the European Very Large Telescope.
See Abstr. 032.021.

Optical design of two spectrographs for the Canada–France–Hawaii telescope.
See Abstr. 034.001.

Flexures of conventional Cassegrain–fed spectrographs.
See Abstr. 034.013.

A near–infrared camera for Las Campanas Observatory.
See Abstr. 034.017.

A possible correlation tracker for LEST.
See Abstr. 034.033.

The limitation of wavefront tilt sensor in adaptive astronomical telescope.
See Abstr. 034.046.

Adaptive optics wavefront corrector using addressable liquid crystal retarders.
See Abstr. 034.047.

Solar feature correlation tracker.
See Abstr. 034.094.

The University of Hawaii adaptive optics system: III. The wavefront curvature sensor.
See Abstr. 034.096.

Active control of spectrum drifts in spectrographs.
See Abstr. 034.108.

Aberration–corrected aspheric grating designs for the Lyman/Far–Ultraviolet Spectroscopic Explorer high–resolution spectrograph: a comparison.
See Abstr. 035.004.

The case for liquid mirrors in orbiting telescopes.
See Abstr. 035.074.

SIRTF stray light analysis.
See Abstr. 035.106.

Analysis and testing of a soft actuation system for segmented reflector articulation and isolation.
See Abstr. 035.117.

Need for active structures in future large IR and sub–mm telescopes.
See Abstr. 035.118.

A deformable mirror concept for adaptice optics in space.
See Abstr. 035.119.

Real time wavefront reconstruction for a 512 subaperture adaptive optical system.
See Abstr. 036.144.

Wavefront sensing in imaging through the atmosphere: a detector strategy.
See Abstr. 036.145.

Recovery of atmospheric phase distortion from stellar images using an artificial neural network.
See Abstr. 036.146.

Estimate of the expected number of stars within the field of view of an astronomical instrument.
See Abstr. 036.171.

Anisoplanatism and use of laser guide stars.
See Abstr. 036.177.

Laser guide stars for adaptive optics systems: Rayleigh scattering experiments.
See Abstr. 036.178.

Algorithms for wavefront reconstruction out of curvature sensing data.
See Abstr. 036.179.

Adaptive optics system tests at the ESO 3.6–m telescope.
See Abstr. 036.183.

Partially compensated speckle imaging: Fourier phase spectrum estimation.
See Abstr. 036.184.

Measuring phase errors of an array or segmented mirror with a single far–field intensity distribution.
See Abstr. 036.185.

Atmospheric turbulence sensing for a multiconjugate adaptive optics system.
See Abstr. 036.186.

Atmospheric fluctuations: empirical structure functions and projected performance of future instruments.
See Abstr. 082.062.

032 Astronomical Instruments

032.001 Low–cost, high–resolution, single–structure array telescopes for imaging of low–Earth–orbit satellites.
N. A. Massie, Y. Oster, G. Poe, L. Seppala, M. Shao.
Appl. Opt., Vol. 31, No. 4, p. 447 – 456 (1 Feb 1992). Current Physics Microform No.: 9205E0501.

Telescopes that are designed for the unconventional imaging of near–Earth satellites must follow unique design rules. The costs must be reduced substantially over those of the conventional telescope designs, and the design must accommodate a technique to circumvent atmospheric distortion of the image. Apertures of 12 m and more along with altitude–altitude mounts that provide high tracking rates are required. A novel design for such a telescope, optimized for speckle imaging, has been generated. Its mount closely resembles a radar mount, and it does not use the conventional dome. Costs for this design are projected to be considerably lower than those for the conventional designs. Results of a design study are presented with details of the electro–optical and optical designs.

032.002 The quest for high resolution.
J. Davis.
Sky Telesc., Vol. 83, No. 1, p. 29 – 33 (Jan 1992).

Astronomers always seek ways to see the universe in sharper detail. New developments in optical interferometry promise major advances.

032.003 Beobachtungen mit dem Vakuum–Turm–Teleskop auf Teneriffa.
W. Schmidt.
Sterne Weltraum, Jahrg. 31, Nr. 3, p. 167 – 171 (Mar 1992).

032.004 Ein neuartiges Schiefspiegler–System.
E. Herrig.
Sterne Weltraum, Jahrg. 31, Nr. 3, p. 193 – 195 (Mar 1992).

032.005 Het astronomisch paradijs.
G. Schilling.
Zenit, Jaarg. 19, Nr. 2, p. 52 – 59 (Feb 1992).

032.006 Das Houghton–Teleskop. Ein idealer Kompromiß?
H. G. J. Rutten, M. A. M. van Venrooij.
Sterne Weltraum, Jahrg. 31, Nr. 4, p. 264 – 266 (Apr 1992).

032.007 The largest telescope of the world will be a twin.
M. Plavec.
Říše hvězd, Vol. 72, No. 9, p. 167 – 169 (Sep 1991). In Czech.

032.008 The largest astronomical telescope on the way to desert Atacama.
S. Štefl.
Vesmír, Vol. 70, No. 9, p. 497 – 503 (Sep 1991). In Czech.

032.009 Automated instrument complex for determining positions of natural and artificial celestial bodies.
D. P. Duma, Yu. N. Ivashchenko, N. I. Laptienko, M. A. Mel'nikov.
Kinematics Phys. Celest. Bodies, Vol. 7, No. 3, p. 28 – 31 (1991). English translation of Kinematika Fiz. Nebesn. Tel, Tom 7, No. 3, p. 29 – 32 (1991).

An instrument complex for position observations of natural and artificial celestial bodies having various magnitudes and apparent speeds relative to the stars varying from zero to several hundred arcseconds per second of time is described. The complex consists of a telescope, special cameras with radiation detectors to track the movement of the object in the focal plane of the telescope, logic and control apparatus, a minicomputer, a self–contained time service, and devices for registration of information on the observations. The apparatus is in use at the GAO

(Main Astronomical Observatory), AS UkrSSR for determining positions of high–orbit satellites.

032.010 An all–latitude automated astrolabe.
V. A. Vasil'ev, V. M. Zinenko, L. B. Kogan,
V. G. Peshekhonov, V. F. Savik, S. K. Romanenko.
Kinematics Phys. Celest. Bodies, Vol. 7, No. 3, p. 32 – 34 (1991).
English translation of Kinematika Fiz. Nebesn. Tel, Tom 7, No. 3, p. 33 – 36 (1991).
A computer–controlled automatic prism astrolabe with television camera and processor, mirror and mirror–angle sensors, and horizon and azimuth drives is described. Using a modified equal-altitude method, the instrument can be used to determine the astronomical coordinates of a point on the surface of the Earth at any latitude from observations of 30 – 40 stars made during the course of an hour, with an rms error of about 0.3″.

032.011 Design consideration of Chinese long baseline stellar interferometer.
Zhao Peiqian, Zhou Bifang.
SPIE Conference on Current Developments in Optical Engineering IV as Part of SPIE's International Symposium on Optical and Optoelectronics Applied Science and Engineering, p. 129 – 138 (1990). – See Abstr. 012.011 for the main entry.
China is planning to build a small aperture long baseline stellar optical interferometer for three purposes. One purpose is to clear up some problems before building a large and expensive interferometer. The second is to measure angular diameter of star and angular distance of double stars. The third is to modify it for astrometric use in the future. The design tolerances and specifications of the interferometer are discussed based on the astronomical requirements.

032.012 Infrared interferometry at Observatoire de la Côte d'Azur, France.
Y. Rabbia, D. Mékarnia, J. Gay.
SPIE Conference on Infrared Technology XVI as Part of SPIE's International Symposium on Optical and Optoelectronic Applied Science and Engineering, p. 172 – 182 (1990). – See Abstr. 012.012 for the main entry.
Principle and up–to–date instrumentation of the infrared interferometer (2 1 m–telescopes, EW 1 5m baseline) at Observatoire de la Côte d'Azur are described. This interferometer is designed for multiwavelengths observations within given spectral ranges located in K,L,M and N photometric bands (respectively centered at $\lambda\lambda$: 2.2 μm, 3.5 μm, 5 μm and 10 μm) and is an example of realisation of double spatio–spectral interferometry in the diluted–aperture long baseline scheme. A preliminary result on α Ori, from initial tests using a former simplified design in the N band is reported, which suggests to add a morphological component to the currently adopted brightness spatial distribution of this source.

032.013 Keck's telescope.
S. R. Brzostkiewicz.
Urania, Rok 63, Nr. 5, p. 144 – 148 (May 1992). In Polish.

032.014 Galaxies through a red giant.
W. C. Keel.
Sky Telesc., Vol. 83, No. 6, p. 626 – 627, 630 – 632 (Jun 1992).

032.015 An approximate calculation for supporting a medium–sized mirror mounted on the table of polishing machine.
Cai Biyun.
Publ. Purple Mt. Obs., Vol. 10, No. 3, p. 254 – 262 (Sep 1991). In Chinese.
A nine pads support system for a medium–sized mirror is studied. The formulae and calculated positions of the supporting pads are derived for plane concavé and convex mirrors, with hole or without hole. A set of related programms is presented.

032.016 Temperatures of the LEST tube.
E. Nielsen.
LEST Found., Tech. Rep., No. 55.
Oslo Univ. (Norway). Inst. for Teoretisk Astrofysikk. 64 p. (1992).
This report presents a mathematical model and a simulation of the thermal performance of the LEST tube.

032.017 The dust war.
A. Gilliotte, P. Giordano, A. Torrejon.
Messenger, No. 68, p. 46 – 48 (Jun 1992).

032.018 Primary mirror control system for the GALILEO telescope.
F. Bortoletto, A. Baruffolo, C. Bonoli, M. d'Alessandro,
D. Fantinel, R. Ragazzoni, L. Salvadori, G. Giudici, P. Vanini.
SPIE Conference on Active and Adaptive Optical Systems as Part of SPIE's International Symposium on Optical Applied Science and Engineering, p. 225 – 235 (1991). – See Abstr. 012.074 for the main entry.
The Italian GALILEO telescope (TNG) is in an advanced phase of construction. Among the various new technical aspects of this telescope the active optics system is now receiving special consideration. In particular, the optical and informatic groups are considering the definition of the control environment dedicated to the active–optics. A solution based on an array of interconnected 16 bit transputers is described with the main requirements for the inter–communication and monitoring software.

032.019 Moving M2 mirror without pointing offset.
R. Ragazzoni, F. Bortoletto.
SPIE Conference on Active and Adaptive Optical Systems as Part of SPIE's International Symposium on Optical Applied Science and Engineering, p. 236 – 246 (1991). – See Abstr. 012.074 for the main entry.
New telescopes, using active optics features and high quality tracking capabilities, require a high precision movement of the secondary mirror M2 during exposures. Moving this mirror to introduce an amount of decentering coma is one of the tasks of active optics. The authors show that this target is accomplished with high accuracy rotating the mirror around a point located near, but not excactly at the center of curvature of M2. Ray tracing results are compared to analytical ones in the case of the Italian National Galileo telescope, that will be equipped with a high precision M2 driving device; the close matching with the analytical calculations is demonstrated.

032.020 Progress report on a five–axis fast guiding secondary for the University of Hawaii 2.2–m telescope.
C. P. Cavedoni, J. E. Graves, A. J. Pickles.
SPIE Conference on Active and Adaptive Optical Systems as Part of SPIE's International Symposium on Optical Applied Science and Engineering, p. 273 – 282 (1991). – See Abstr. 012.074 for the main entry.
A telescope aperture of 2.2–m on Mauna Kea that routinely experiences $d/r_0 = 4$ in the near infrared can achieve a factor of 2 gain in angular resolution by tip–tilt correction of atmospheric–induced wavefront errors. To utilize the gains possible from tip–tilt correction, collimation errors and focus errors must also be removed. For its 2.2–m f/31 telescope, the University of Hawaii is in the process of implementing a five–axis fast guiding secondary consisting of a fast steering mirror platform and slow remote detilt, decenter, and despace collimation and focus drives. The near–term goal is to implement closed–loop tip–tilt image motion correction with open–loop collimation and focus control. The long–term goal is to add closed–loop collimation and focus control. This paper documents the progress to date on the fast steering mirror platform and its spider support structure.

032.021   **Adaptive optics for the European Very Large Telescope.**
F. Merkle, N. Hubin.
SPIE Conference on Active and Adaptive Optical Systems as Part of SPIE's International Symposium on Optical Applied Science and Engineering, p. 283 – 292 (1991). – See Abstr. 012.074 for the main entry.

Adaptive optics is one of the main features of the Very Large Telescope (VLT) of the European Southern Observatory (ESO) – an array of four 8 meter telescopes. These telescopes can be operated individually, in an incoherent and in a coherent interferometric beam combination mode. Each telescope will be equipped with adaptive optics systems for real–time correction of atmospheric turbulence effects. First results with a prototype system developed for the VLT demonstrated the feasibility and the significant grain of this technology for astronomical imaging. This paper describes the VLT adaptive optics system and its implementation program.

032.022   **An astrometric test for the 40 cm refractor at Zô–Sè.**
Mao Yaqing.
Ann. Shanghai Obs., Acad. Sin., No. 13, p. 84 – 89 (1992). In Chinese.

Since 1900 the double astrograph at Zô–Sé Station of Shanghai Observatory, with aperture of 40 cm and photographic and visual focal lengths of 6.90 m and 7.14 m respectively, has been in use for more than 90 years, with which some seven thousand plates have been taken. An astrometric test for this refractor has been done in order to examine the astrometric capability of the telescope in the present stage. It is concluded that the 40 cm refractor still satisfies the requirements of accurate astrometry.

032.023   **The adjustment of polar axis and the measurement of tracking error for 1.56 m telescope.**
Qian Bochen.
Ann. Shanghai Obs., Acad. Sin., No. 13, p. 167 – 169 (1992). In Chinese.

032.024   **Proposed upgrade of the McMath Solar/Stellar Telescope to a 4 m aperture.**
M. S. Giampapa, W. C. Livingston, D. R. Rabin.
Armagh Observatory Bicentenary Colloquium on Surface Inhomogeneities on Late–Type Stars, p. 279 – 280 (1992). – See Abstr. 012.085 for the main entry.

032.025   **Ein computergesteuertes Dobsonian–Teleskop.**
A. Kunzmann.
Orion, Jahrg. 50, Nr. 248, p. 37 – 38 (Feb 1992).

032.026   **Zehn Meter. Das Keck–Teleskop auf Hawaii kurz vor der Vollendung.**
J. Alean.
Orion, Jahrg. 50, Nr. 249, p. 86 – 90 (Apr 1992).

032.027   **Water tunnel test of telescope enclosure models.**
A. Miyashita, H. Ando, S. Shindo, K. Sakata.
Rep. Natl. Astron. Obs. Jpn., Vol. 1, No. 3, p. 293 – 300 (1992). In Japanese.

The Japan National Large Telescope (JNLT) Project group initially adopted a conventional hemispherical dome as the telescope enclosure design, but recently they have considered alternative shapes. A water tunnel test has been carried out at the National Aeronautical Laboratory. Three different enclosure styles have been modeled and tested in this setup, with several variations in the venting arrangement. The authors describe the experimental results and conclusions.

032.028   **Solar optical instrumentation.**
H. Zirin.
NATO Advanced Study Institute on the Sun: a Laboratory for Astrophysics, p. 379 – 393 (1992). – See Abstr. 012.091 for the main entry.

The author discusses the various aspects of specialized instrumentation for solar observation. Telescopes are designed to minimize heating and turbulence. Because two–dimensional images are so important, two–dimensional monochromators are particularly important. Devices for magnetic field measurement and coronal observation are described, and the resolution of the various instruments is discussed.

032.029   **Performance of SUBARU telescope in 0.3 – 30 microns.**
S. S. Hayashi, S. Okamura, H. Shibai.
Publ. Natl. Astron. Obs. Jpn., Vol. 2, No. 3, p. 547 – 550 (1992).

One of the remarkable features of SUBARU telescope, a Japanese National Large Telescope of 8.2 m diameter, to be put on Mauna Kea, Hawaii, is its wide coverage over a spectral range of 0.3 to 30 microns, from near–ultraviolet to the middle infrared wavelengths. Sensitvity of the observations with this telescope is examined and compared with virtual telescopes with different structure and located in different environments. As expected, SUBARU is characterized by high sensitivity in the high dispersion spectroscopy mode. The performance in optical–near infrared regime is approaching the optimum of the detecting technology. It is important for the telescope design to suppress the thermal background for the observations at middle infrared wavelengths.

Fernrohre und ihre Meister.
See Abstr. 003.060.

Fernrohr–Selbstbau. Fenster ins Weltall. Praktische Anleitungen für Freunde des Sternenhimmels und solche, die es werden wollen.
See Abstr. 003.065.

Carter Observatory's 9–inch refractor: the Crossley connection.
See Abstr. 004.035.

A brief history of the Crossley 36–inch reflector.
See Abstr. 004.036.

The last great speculum: the 48–inch Great Melbourne Telescope.
See Abstr. 004.045.

New windows to the universe. Volumes I, II. Invited review papers and general lectures.
See Abstr. 012.001.

The LEST project.
See Abstr. 013.001.

The instrumentation plan for the Very Large Telescope of the European Southern Observatory.
See Abstr. 013.002.

Ground–based European astronomical projects.
See Abstr. 013.003.

Prospects of the development of ground–based and space astronomy in the USSR.
See Abstr. 013.005.

Advances in IR technology at Paris Observatory.
See Abstr. 013.021.

Robotic telescopes for photometry.
See Abstr. 013.057.

The Chinese Large Schmidt Telescope project.
See Abstr. 013.074.

Bigger telescopes and better instrumentation: report on the 1992 ESO conference.
See Abstr. 013.082.

Polar aligning your telescope.
See Abstr. 014.043.

Solar imaging with a segmented adaptive mirror.
See Abstr. 031.021.

First results of an on–line adaptive optics system with atmospheric wavefront sensing by an artificial neural network.
See Abstr. 031.026.

The NSO/Sac Peak continuous–face–plate adaptive mirror.
See Abstr. 031.033.

Applications of stress polishing techniques as developed for the Keck Observatory primary mirror fabrication.
See Abstr. 031.036.

MARTINI: sensing and control system design.
See Abstr. 031.046.

Neural network adaptive optics for the Multiple Mirror Telescope.
See Abstr. 031.054.

Solar astronomy with a 19–segment adaptive mirror.
See Abstr. 031.055.

The Johns Hopkins adaptive optics coronagraph.
See Abstr. 031.056.

Active optics system for a 3.5–meter structured mirror.
See Abstr. 031.057.

Latest developments of active optics of the ESO NTT and the implications for the ESO VLT.
See Abstr. 031.063.

The Come–On–Plus project: an upgrade of the Come–On adaptive optics prototype system.
See Abstr. 031.064.

Application of diffraction theory to telescope optics.
See Abstr. 031.078.

A coronagraph for COME–ON, the adaptive optics VLT proto-type.
See Abstr. 034.066.

The case for liquid mirrors in orbiting telescopes.
See Abstr. 035.074.

Interferometry with large optical telescopes.
See Abstr. 036.009.

Adaptive optics system tests at the ESO 3.6–m telescope.
See Abstr. 036.183.

Atmospheric fluctuations: empirical structure functions and projected performance of future instruments.
See Abstr. 082.062.

Automated photometric telescopes in the study of active stars.
See Abstr. 116.054.

Near–term prospects for extra–solar planet detection: the Astrometric Imaging Telescope.
See Abstr. 118.024.

033 Radio Telescopes and Equipment

033.001   A quick method of checking the pointing of the antenna pattern of the RATAN–600 telescope with respect to the angle of elevation.
O. A. Golubchina.
Bull. Spec. Astrophys. Obs. – North Caucasus, Vol. 31, p. 141 – 145 (1991). English translation of Astrofiz. Issled. Izv. Spets. Astrofiz. Obs., Tom 31, p. 144 – 148 (1991).

A quick method is proposed for checking the pointing of the antenna pattern of the RATAN–600 radio telescope with respect to the angle of elevation. This method makes it possible to determine the position of the center of the antenna pattern with respect to the angle of elevation with an accuracy greater than 1' from one recording of radio emission of the Sun whose azimuth at the time of observation is not equal to zero. The results of an experiment are given. The quick check method can be used for determining radio refraction at low observation angles, which is also confirmed experimentally.

033.002   A fast noise–immunity estimation of the noise–track level of a radiometer by means of the absolute median deviation.
B. L. Erukhimov.
Bull. Spec. Astrophys. Obs. – North Caucasus, Vol. 31, p. 146 – 149 (1991). English translation of Astrofiz. Issled. Izv. Spets. Astrofiz. Obs., Tom 31, p. 149 – 152 (1991).

A real–time method of noise–immunity estimation of the σ level of a radiometer is examined. A known algorithm of the absolute median deviation is used. A comparison of the proposed method with the classical one on noise records contaminated by noise typical for radio astronomical observations and the efficiency of the estimate used are given.

033.003   Accuracy of coordinate measurements on the RATAN–600 radio telescope.
M. G. Mingaliev, V. N. Chernenkov.
Bull. Spec. Astrophys. Obs. – North Caucasus, Vol. 31, p. 150 – 159 (1991). English translation of Astrofiz. Issled. Izv. Spets. Astrofiz. Obs., Tom 31, p. 153 – 162 (1991).

The main causes limiting the coordinate accuracy during observations on the RATAN–600 radio telescope are analyzed. The main contribution to the error of determining right ascension is made by the error due to the accuracy of setting the primary horns in the electrical axis of the "primary mirror + secondary mirror" system. The accuracy of a single measurement is about 4". The use of additional methods of checking the orientation of the electrical axis and primary horns makes it possible to substantially increase the accuracy. The accuracy of one determination of right ascension is within 0.3"–0.5". An estimate of the contribution of the atmosphere to the coordinate accuracy during observations on the RATAN–600 is also given.

033.004   Perspectives of advance of RATAN–600 solar observations.
V. M. Bogod.
Astron. Nachr., Vol. 313, No. 2, p. 97 – 100 (1992).

The potential possibilities of RATAN–600 for solar studies are considered which are going on to be realized during the present maximum of solar activity. Different methods of an improvement of the characteristics of the antenna–receiver system are analyzed. These are: realization of tracking the Sun during about 4 hours for a day and attaining temporal resolution up to some msec; realization of radio–heliographic observations; realization of a full spectral coverage of the radio telescope from 1.5 to 18 GHz with a frequency resolution up to 1–2 MHz; combination of all possibilities mentioned above in an adaptable solar observing complex.

033.005 **Lense antennas for astronomy and geodesy. Beyond limits of reflectors.**
Y. Shiratori.
Astron. Her., Vol. 85, No. 1, p. 13 – 18 (1992). In Japanese.

033.006 **10–meter radiotelescope – VLBI network element.**
V. Dhawan, N. V. G. Sarma, R. Ganesh,
L. I. Matveenko, L. R. Kogan, A. P. Molodyanu, D. Graham.
Pis'ma Astron. Zh., Tom 18, No. 4, p. 391 – 400 (Apr 1992). In Russian. English translation in Sov. Astron. Lett., Vol. 18, No. 2.
A 10–m radiotelescope of Raman Research Institute, Bangalore, India, had been equipped by a 22 GHz low noise receiver, a coherent local oscillator, a hydrogen frequency standard, and MK–2 terminal. The effective area of antenna is equal to $A_{eff} = 50\ m^2$ and the system noise temperature is $T_{sys} = 60$ K. VLBI observations of H_2O maser sources had been made in May 1990 at Bangalore – EVN VLBI network. The angular sizes of compact components and coordinates of the radio telescope had been determined.

033.007 **Automation of radioastronomical observing programs in Metsähovi.**
M. Tornikoski.
Annual meeting of the Finnish Astronomical Society, p. 27 – 31 (Apr 1992). – See Abstr. 012.033 for the main entry.
Automation of the quasar and solar observing programs used at the Metsähovi Radio Research Station is discussed. The new solar observing system SMART is a knowledge system which is able to perform (semi–)automatic observations using heuristic rules and parameters as "intelligence". The new quasar observing program CONTOBS performs routine observations but is striving for a user–friendly and effective way of observing.

033.008 **75 – 115 GHz phase–locked gunn oscillator.**
Xu Zhicai, Chen Shanhuai, Xiao Kechen,
Shi Shengcai, Yan Guipan, Lue Xun, Din Zhugao, Li Min.
Publ. Purple Mt. Obs., Vol. 10, No. 3, p. 229 – 236 (Sep 1991). In Chinese.
A W–band phase–locked Gunn oscillator has been developed for using as a local oscillator for Purple Mountain Observatory 13.7 m millimeter–wave radio telescope. Two kinds of Gunn oscillators and the harmonic mixer of the PLO system are constructed. The operating range is 40 GHz and the output power is greater than 5 mW over its operating range. The output signal of servo amplifier in PLO controls the bias voltage of Gunn oscillator to phase lock the Gunn oscillator over more than 100 MHz tuning range.

033.009 **An improved W–band wide band Gunn oscillator.**
Lue Xun.
Publ. Purple Mt. Obs., Vol. 10, No. 3, p. 237 – 240 (Sep 1991). In Chinese.
An improved second–harmonic Gunn oscillator with new design has been built. It has the features of simple configuration, easily processing and working stably. Its operating frequency covers wide band with a maximum output power of 23 mW and more than 450 MHz bias tuning bandwidth. After phase–locking, a locking bandwidth of more than 800 MHz is achieved.

033.010 **3 mm–band harmonic mixer.**
Shi Shengcai, Chen Shanhuai.
Publ. Purple Mt. Obs., Vol. 10, No. 3, p. 241 – 246 (Sep 1991). In Chinese.
This paper describes the design and performance of 3 mm–band harmonic mixer in PLO (phase–locking oscillator) system of Purple Mountain Observatory 13.7 m millmeter radio telescope.

033.011 **A high performance, broad–band mechanical tunable 100 – 115 GHz Gunn VCO.**
Xiao Kecheng.
Publ. Purple Mt. Obs., Vol. 10, No. 3, p. 247 – 253 (Sep 1991). In Chinese.
A high performance, W–band 2nd harmonic Gunn oscillator has been developed. Using a common waveguide cavity, designed for both the fundamental and the 2nd harmonic frequency, this oscillator is easily backshort and varactor tunable. The VCO has high frequency, great output power, and pure spectrum. It has been used in 2.6 mm wave bend phase–lock local system of large millimeter wave radio telescope of Purple Mountain Observatory.

033.012 **350 GHz SIS receiver installed at SEST.**
N. Whyborn, L.–A. Nyman, W. Wild, G. Delgado.
Messenger, No. 68, p. 45 (Jun 1992).

033.013 **RT–10: a new element in the VLBI network.**
V. Dhawan, N. V. G. Sarma, R. Ganesh,
L. I. Matveenko, L. R. Kogan, A. P. Molodyanu, D. Graham.
Sov. Astron. Lett., Vol. 18, No. 2, p. 149 – 152 (Mar 1992). Current Physics Microform No.: 9211X0788. English translation of Pis'ma Astron. Zh., Tom 18, No. 4, p. 391 – 400 (1992).
The 10–meter radio telescope at the Raman Research Institute in Bangalore (India) has been fitted out with a low–noise 22–GHz receiver, coherent local oscillator, hydrogen frequency standard, and a Mark II recording system. The effective aperture of the antenna is $A_{eff} = 50\ m^2$ and the system noise temperature is $T_{ns} = 60$ K. Interferometric observations of H_2O maser sources were carried out in May 1990 using the Bangalore telescope in collaboration with the European VLBI network, yielding the angular sizes of the compact components of the maser sources and the coordinates of the radio telescope.

033.014 **A low frequency tunable gravitational wave antenna.**
C. Hossfield.
J. Am. Assoc. Variable Star Obs., Vol. 20, No. 1, p. 108 (1991). Abstract. – See Abstr. 012.066 for the main entry.

033.015 **Development of a reflection dish antenna of glass fiber reinforced plastic.**
Du Hong, Yang Kaiping.
Publ. Yunnan Obs., No. 1, p. 69 – 72 (1992). In Chinese.
An antenna of glass fiber reinforced plastic is introduced and the application of the technology of glass fiber reinforced plastic to the manufacture of reflection dish antenna is also proposed. The tested results show that the antenna meets the requirements of the design and utilization, and its performance is good.

033.016 **Data transmission between the control computer and data processing computer of PMO 13.7 m radio telescope.**
Zhou Xue–ya, Han Fu, Lu Jing.
Publ. Purple Mt. Obs., Vol. 11, No. 1, p. 54 – 60 (Mar 1992). In Chinese.
For efficiency of observation and convenience to observers, data transmission between the control computer and data processing computer of the telescope has been realized and make the real–time data processing possible. In this paper, the work on the hardware interface and programming is presented. Examples of data transmission and real–time data processing are also given.

033.017 **327–MHz low noise front end.**
Xu Xiang, Chen Hongsheng.
Publ. Beijing Astron. Obs., No. 18, p. 81 – 84 (Oct 1991).

033.018 **A solar decimeter radio dynamic spectrometer.**
Ji Huirong, Jin Shenzheng, Fu Qijun.
Publ. Beijing Astron. Obs., No. 18, p. 85 – 88 (Oct 1991).
A decimeter solar dynamic spectrometer with high time and high spectral resolutions is described. The instrument can be used to observe the circular polarization spectra of the microwave

type III bursts and the spikes with fine structure in time and frequency.

033.019   A sweeping local oscillator system for pulsar observations.
A. A. Deshpande.
J. Astrophys. Astron., Vol. 13, No. 2, p. 167 – 173 (Jun 1992).
The authors discusses the design details of an inexpensive programmable sweeping local oscillator system (SLOS) built for use in a "swept frequency dedispersion scheme" for pulsar observations. A useful extension of the basic Divide–and–Add algorithm for frequency synthesis is developed for this purpose. An SLOS based on this design has been built and used for high time–resolution observations of pulsars at low radio–frequencies.

033.020   The mathematical model for calculating effective section area of space VLBI antenna surface.
Zheng Yong, Qian Zhihan.
Ann. Shanghai Obs., Acad. Sin., No. 13, p. 140 – 144 (1992). In Chinese.
The authors have derived the mathematical model for calculating effective section area of space VLBI antenna surface. This model will be used to calculate A/m ratio of non–gravitational perturbations, which include solar radiation pressure, atmospheric drag and earth albedo radiation pressure, etc., needed for precise determination of space VLBI station orbit.

033.021   A modified scheme on raising the most slew rate for the 25 m radio telescope of Shanghai Observatory.
Wang Husheng.
Ann. Shanghai Obs., Acad. Sin., No. 13, p. 181 – 187 (1992). In Chinese.

033.022   First fringe with the Waseda FFT radio telescope.
J. Nakajima, E. Otobe, K. Nishibori, N. Watanabe, K. Asuma, T. Daishido.
Publ. Astron. Soc. Jpn., Vol. 44, No. 3, p. L35 – L38 (1992).
A sixty–four–element FFT (Fast Fourier Transform) type radio interferometer has been constructed at Waseda University. Using 2 elements, the authors have started fringe observations of radio sources. During observations of 1991 November 10 – 20, they detected the fringes of Cas A and Tau A. The performance of the radio telescope, which includes backend digital processors, have been confirmed by switching observations. Each element is a 2.4–m diameter Cassegrain antenna with a receiver of 200K noise temperature. The purpose of the telescope is to survey transient radio sources.

033.023   Solar radio instrumentation.
G. J. Hurford.
NATO Advanced Study Institute on the Sun: a Laboratory for Astrophysics, p. 411 – 422 (1992). – See Abstr. 012.091 for the main entry.
The definitions of basic quantities such as flux density, brightness temperature and optical depth are reviewed in the context of solar radio astronomy. Current observational techniques and hardware are discussed in terms of the spatial and spectral resolution required for solar observations. The fundamentals of interferometry are outlined with consideration given to the limitations of interferometric imaging.

033.024   Nobeyama radioheliograph.
S. Enome.
IAU Colloquium No. 133: Eruptive solar flares, p. 314 – 317 (1992). – See Abstr. 012.096 for the main entry.
A new dedicated radioheliograph, or a tee–shaped interferometer, is under construction at Nobeyama by the solar radio astronomy group during fiscal years 1990 and 1991. It will be used to obtain images of active region magnetic fields during solar flares by observing microwave radiation through the mechanism of gyro–synchrotron emission. The observation frequency is 17 GHz, the spatial and time resolutions are 10 arcsec and 50 msec, respectively. It is scheduled to be completed by March 1992, the end of the fiscal year 1991 and to go into full–scale routine observations in the summer of 1992. The major performance characteristics, the current status of construction and expected programs of observations are briefly described.

033.025   Decimeter high resolution solar radio spectroscope.
H. S. Sawant, J. H. A. Sobral, J. A. C. F. Neri, F. C. R. Fernandes, R. R. Rosa, J. R. Cecatto, D. Martinazzo.
IAU Colloquium No. 133: Eruptive solar flares, p. 318 – 321 (1992). – See Abstr. 012.096 for the main entry.
High sensitivity (8–0.8 s.f.u.), high time (100 ms), frequency (300–3000 kHz) resolution decimeter spectroscope operating at present in the frequency range of (1600 ± 50) MHz has been put into operation in São José dos Campos at the National Space Research Institute, Brazil. Finally, this spectroscope will be operated over a frequency range of 200–2500 MHz. However, in high resolution mode, it will be operating over a frequency range of $(f_s \pm \Delta f)$, (Δf will be selected in the range of 50–500 MHz or as desirable), where f_s is the frequency selected in the entire range of band. At the same time, the entire frequency band will be monitored with low resolution mode. Initial observations of this spectroscope at (1600 ± 50) MHz, the details of spectroscope and future development plans (up to 1992) are presented.

Die Anfangsjahre der Radioastronomie in Deutschland.
See Abstr. 004.027.

Radio–interference.
See Abstr. 013.030.

Millimeter band radio astronomy.
See Abstr. 013.031.

New development in radio astronomy in the submillimeter–wave region.
See Abstr. 013.035.

Decametric radioastronomy.
See Abstr. 013.049.

Radio astronomy of the next century.
See Abstr. 013.069.

On astronomy for the twenty–first century.
See Abstr. 013.070.

The potential contribution of the Northern Cross radiotelescope to the SETI program.
See Abstr. 015.031.

SETI searches with the 70 m SUFFA radio telescope.
See Abstr. 015.037.

High spatial resolution techniques.
See Abstr. 036.138.

Gain measurement of small–sized antenna by means of radioastronomical methods.
See Abstr. 036.192.

VLA/Goldstone Planetary Radar results.
See Abstr. 091.061.

Molecular line observations in the 85 – 90 GHz band using a maser receiver at the Crimean Astrophysical Observatory 22–meter radio telescope.
See Abstr. 131.031.

Geometry of star forming regions and magnetic fields from polarimetry. 2D infrared array detectors and millimeter–wave telescopes.
See Abstr. 131.032.

An improved method to derive H I absorption spectra through the disks of galaxies: application to the Sculptor Group.
See Abstr. 157.045.

034 Auxiliary Instrumentation, Photographic Materials, Clocks

034.001 Optical design of two spectrographs for the Canada–France–Hawaii telescope.
C. L. Morbey.
Appl. Opt., Vol. 31, No. 13, p. 2291 – 2300 (1 May 1992). Current Physics Microform No.: 9207B0123.
Optical designs of two new spectrographs for the Canada–France–Hawaii telescope Cassegrain focus are described. Also given is a summary of the design procedure using the Dominion Astrophysical Observatory optical design code OPTESA (optical system optimization by educated simulated annealing). The f/2.8 multiobject spectrograph has a field of view of 10′, whereas the f/10 subsecond of arc imaging spectrograph has a field of view of 3′.

034.002 CCDs for the 1990s.
P. R. Jorden.
11. European Regional Astronomy Meeting of the IAU: New windows to the universe, Vol. 1, p. 465 – 481 (1990). – See Abstr. 012.001 for the main entry.

034.003 Testing the FEU–165 photomultiplier.
M. R. Ainbund, E. A. Vitrichenko, F. A. Maslyukov, G. A. Men'shikov.
Sov. J. Opt. Technol., Vol. 58, No. 12, p. 777 – 779 (Dec 1991). Current Physics Microform No.: 9207X0075.
The results of tests of the new FEU–165 photomultiplier employing multiplication in microchannel plates in the photon-counting regime are presented. The photomultiplier is proposed for use in a star sensor, a prototype of which has undergone successful tests in an observatory.

034.004 Astrophotographie im Mittelformat.
P. Keller, G. Schmidbauer.
Sterne Weltraum, Jahrg. 31, Nr. 1, p. 43 – 44, 46 – 47 (Jan 1992).

034.005 Teleskop–Nachführung aus der Sicht des Elektronik–Entwicklers.
M. Rök–Ramirez.
Sterne Weltraum, Jahrg. 31, Nr. 1, p. 48 – 51 (Jan 1992).

034.006 Neue Entwicklungen in der Speckle– und Stellar–Interferometrie.
D. Fischer.
Sterne Weltraum, Jahrg. 31, Nr. 3, p. 161 – 166 (Mar 1992).

034.007 MWP–2: een prima ontwikkelaar voor TP 2415.
J. Sussenbach, J. Koet.
Zenit, Jaarg. 19, Nr. 3, p. 136 – 137 (Mar 1992).

034.008 Comparative characteristics of the coude spectrograph of the Shemakha 2–m telescope and main stellar spectrograph of the 6–m telescope.
A. Kh. Rzaev, E. L. Chentsov.
Bull. Spec. Astrophys. Obs. – North Caucasus, Vol. 31, p. 132 – 140 (1991). English translation of Astrofiz. Issled. Izv. Spets. Astrofiz. Obs., Tom 31, p. 134 – 143 (1991).
The position and photometric characteristics, reaction curves, and penetrating power of the coude spectrograph of the 2–m telescope of the Shemakha Astrophysical Observatory, Azerbaijan Academy of Sciences, and the Main Stellar Spectrograph of the 6–m telescope of the Special Astrophysical Observatory, USSR Academy of Sciences, are given and compared.

034.009 Improved band filters for a millimeter photometer.
G. A. Chuntonov, I. A. Maslov, V. A. Soglasnova.
Bull. Spec. Astrophys. Obs. – North Caucasus, Vol. 31, p. 160 – 161 (1991). English translation of Astrofiz. Issled. Izv. Spets. Astrofiz. Obs., Tom 31, p. 163 – 164 (1991).
Filters for studying the Syunyaev–Zel'dovich effect in the near–millimeter range have been developed and built.

034.010 Digicon spectrophotometer description of photodetectors and investigation of their characteristics.
V. S. Rylov.
Bull. Spec. Astrophys. Obs. – North Caucasus, Vol. 31, p. 162 – 176 (1991). English translation of Astrofiz. Issled. Izv. Spets. Astrofiz. Obs., Tom 31, p. 165 – 179 (1991).
The description and characteristics of all elements composing a digicon spectrophotometer (DSP–40) are given. The DSP–40 is structurally attached to the standard fast spectrograph SP–160, which is intended for the 6–m telescope, or to the universal spectrograph UAGS (Germany), in connection with which particular information is given about them in combination with the 40–channel photon detector – the digicon – and its characteristics. Methods of focusing the digicon in the DSP–40 system, setting the discrimination threshold, identifying the comparison spectrum and calculating the wavelength scale, and one of the possible methods of primary processing of the data obtained, including the reaction curve of the entire system, are examined. The results of testing the DSP–40 under laboratory conditions and on the telescope are given in conclusion. The data show the high efficiency of the DSP–40 in recording faint luminous fluxes. The temporal stability of the digicon at various levels of luminous fluxes was studied under laboratory conditions. The author considers that the 40–channel digicon can find most effective use in spectrophotometry and spectropolarimetry when studying faint objects.

034.011 Kleurrijk heelel.
D. Malin.
Zenit, Jaarg. 19, Nr. 5, p. 196 – 203 (May 1992).

034.012 Het fotograferen van extragalactische sterrenstelsels.
J. S. Sussenbach.
Zenit, Jaarg. 19, Nr. 5, p. 220 – 223 (May 1992).

034.013 Flexures of conventional Cassegrain–fed spectrographs.
U. Munari, M. G. Lattanzi.
Publ. Astron. Soc. Pac., Vol. 104, No. 672, p. 121 – 126 (Feb 1992). Current Physics Microform No.: 9203B2038.
The authors have derived the flexure patterns of two classical Cassegrain–fed spectrographs, by cross–correlating CCD spectra. After a detailed description of the method used to derive the flexure models, emphasis is given to the fact that extreme care must be used in compensating for those flexures when high precision is sought in wavelength–calibration and radial–velocity work involving instrumentation similar to that investigated here. If not fully accounted for, the flexures can introduce spurious shifts up to one hundred times the typical error of the cross–correlation technique usually used to compare the spectra. With the aid of the models it is now possible to remove most of the flexure biases from spectra taken with both spectrographs of the Asiago Astrophysical Observatory over a large fraction of the sky accessible to the telescope. This will ensure high consistency of the radial–velocity measurements derived with the cross–correlation technique. A brief discussion on possible ways to monitor in real time the flexure pattern of a telescope + spectrograph combination is also given.

034.014 Astronomical observations at 10 and 20 μm with the NRL infrared camera.
S. Odenwald, K. Shivanandan, H. A. Thronson Jr.
Publ. Astron. Soc. Pac., Vol. 104, No. 672, p. 127 – 139 (Feb 1992). Current Physics Microform No.: 9203B2044.
An infrared camera for astronomical research has been developed at the Naval Research Laboratory, Center for Advanced Space Sensing. A series of observations in 1990 December at the 2.3–m Wyoming Infrared Observatory (WIRO) yielded new astronomical images of the star forming regions M42/IRc 1, NGC 2264/IRS1 and NGC 7538/IRS1 at 12 and 20 μm, in addition to images of a variety of stellar calibrators.

This camera system (IRCAM) achieves its peak sensitivity between 18 – 22 μm, making it well matched for observing in the 20 μm atmospheric window. Substantial sensitivity is also available at 10 μm so that near–simultaneous observations are feasible through both windows. IRCAM employs four, ($\Delta\lambda \approx 1.0\ \mu$m) filters at 8, 10, 12, and 12.4 μm; a broadband filter: 18 – 24 μm, as well as standard K, L, and M filters. The array field of view is 20″ × 100″ with 1″.6 pixels surrounded by 0″.4 dead space, allowing diffraction–limited observing at 20 μm with a 2.3–m telescope. In this paper, the authors describe the camera system, and present the results of the astronomical observations conducted at the Wyoming Infrared Observatory.

034.015   Choosing your film.
B. Rose.
Astron. Now, Vol. 6, No. 5, p. 20 – 24 (May 1992).
A guide to films for 35 mm amateur astrophotography.

034.016   A 5 – 18 μm array camera for high–background astronomical imaging.
D. Y. Gezari, W. C. Folz, L. A. Woods, F. Varosi.
Publ. Astron. Soc. Pac., Vol. 104, No. 673, p. 191 – 203 (Mar 1992).
A new infrared array camera system using a Hughes/SBRC 58 × 62 pixel hybrid Si:Ga array detector has been successfully applied to high–background 5 – 18 μm astronomical imaging observations. The electronic and optical design of the camera, its photometric characteristics, examples of observational results, and techniques for successful array imaging in a high–background astronomical application are discussed.

034.017   A near–infrared camera for Las Campanas Observatory.
S. E. Persson, S. C. West, D. M. Carr, A. Sivaramakrishnan, D. C. Murphy.
Publ. Astron. Soc. Pac., Vol. 104, No. 673, p. 204 – 214 (Mar 1992). Current Physics Microform No.: 9204G1308.
The optical, mechanical, and electronic designs of a multipurpose near–infrared (1.1 – 2.5 μm) camera are described. The camera is used on both the 2.5–m duPont and 1–m Swope telescopes. A refractive all–spherical achromatic reimager incorporates three confocal discrete zooms which image onto a Rockwell HgCdTe (NICMOS2) 128 × 128 detector array. Although the instrument was designed primarily for imaging, the authors have included two experimental modes for future use – a low–resolution grism and a coronagraph. The optical design choices and procedures are described in some detail. The instrument is remotely controlled. Electronic designs suitable for use with a NICMOS2 array are given. The data are acquired and inspected using IRAF routines running under the UNIX multitasking environment. Initial performance results on camera optical quality and system sensitivity are given.

034.018   A modular dewar design and detector mounting strategy for large–format, astronomical CCD mosaics.
G. A. Luppino, K. R. Miller.
Publ. Astron. Soc. Pac., Vol. 104, No. 673, p. 215 – 222 (Mar 1992). Current Physics Microform No.: 9204G1319.
The authors present their mechanical design for a very–large–format, 4096 × 4096 pixel charge–coupled device (CCD) mosaic constructed from a 2 × 2 array of two–edge–abuttable, 15 μm pixel 2048 × 2048 CCDs. They describe the advanced manufacturing techniques used to fabricate their package and mounting structure so that the assembled mosaic is flat ($\pm$ 1 pixel), and the rows and columns of the individual quadrants are aligned. They describe their design for a modular dewar built to house the 4096 × 4096 CCD mosaic.

034.019   Ein Detektorsystem zum Nachweis der e/γ–Komponente großer Luftschauer im UHE–Bereich.
G. Völker.
Diss., Kernforschungszentrum Karlsruhe GmbH (Germany). Inst. für Kernphysik; Karlsruhe Univ. (T.H.) (Germany). Fakultät für Physik. KFK—4983, 109 p. (Jan 1992).
To learn more about the origin and the composition of cosmic rays at high energies, the detectorsystem KASCADE, now under construction, was proposed in the Kernforschungszentrum Karlsruhe. KASCADE will measure simultaneously many parameters of extensive air showers, which are produced by cosmic ray particles in the atmosphere. An important part of this experimental setup is a detector array for the electromagnetic and the muonic component of the showers. With these informations characteristic parameters for each shower can be reconstructed. For the detection of electrons and photons, a liquid scintillation detector, which has a large effective area of 0.78 m² as well as a very good time resolution of σ_t = 0.77 ns and an energy resolution of σ_E = 25%/$\sqrt{E}$ at an energy loss of 7.8 MeV, was developed in extensive prototype studies. With 64 of these detectors a prototyp array was built, which corresponds to 5% of the planned array.

034.020   Turning a keen eye on the stars.
M. Bartusiak.
Science, Vol. 256, No. 5055, p. 316 – 317 (17 Apr 1992).
By collecting starlight with widely spaced mirrors, astronomers are mapping the heavens with superlative precision. Soon they'll be turning those maps into images.

034.021   Sonnenuhr und Sonnenkompass.
H. Lichtenegger.
Sternenbote, Jahrg. 35, Nr. 1, p. 2 – 8 (1992).

034.022   Die Sonnenuhr – Bastelobjekt in der pädagogischen Arbeit.
A. Zenkert.
Astron. Sch., Jahrg. 29, Heft 8, p. 10 – 14 (Apr 1992).

034.023   Eine Rarität – Die astronomische Uhr zu Rostock.
M. Schukowski.
Astron. Sch., Jahrg. 29, Heft 8, p. 31 – 32 (Apr 1992).

034.024   The physics of astronomical infrared detectors.
J. D. Patterson, W. A. Gobba.
I.A.P.P.P. Commun., No. 47, p. 1 – 16 (Mar – May 1992).

034.025   Sensitivity of ING (*Isaac Newton Group*) La Palma instruments.
R. Clegg, D. Carter, C. Benn.
Gemini, No. 35, p. 16 – 19 (Mar 1992).

034.026   Charge–coupled devices: frame adding as an alternative to long integration times and cooling.
M. O'Malley, E. O'Mongain.
Opt. Eng., Vol. 31, No. 3, p. 522 – 526 (Mar 1992).
CCDs used for low light level applications require a long integration time and/or cooling to have acceptable SNR. Standard commercially available CCD cameras (e.g. for TV applications) are used with a relatively short integration time ($\approx$1/60 s per field) and are operated at room temperature. Consequently, these cameras will have poor SNR if used for low light level applications. The addition of successive frames from a standard CCD camera for the improvement of the SNR is considered. The authors show, using a simple model, that a range of signal levels (as low as a few millilux) exists for which frame addition is a viable alternative to long integration times and cooling.

034.027   Ein Photometerkopf nach der Zweiblendenmethode (I).
R. Gröbel.
BAV Rundbrief, Jahrg. 41, Nr. 2, p. 80 – 88 (1992).

034.028   LDSS–2 pushes back the frontiers on the WHT.
J. Allington–Smith, R. Ellis, K. Glazebrook, G. Shaw, N. Tanvir, J. Webster, M. Breare, D. Gellatly, P. Jorden, J. MacLean, P. Oates, P. Taylor, S. Worswick, K. Taylor.
Gemini, No. 36, p. 4 – 7 (Jun 1992).

034.029   **Een batterij Canon camera's geautomatiseerd.**
H. Betlem.
Radiant, Jaarg. 14, Nr. 2, p. 34 – 37 (Apr 1992).

034.030   **Even more large CCDs at the ING telescopes.**
P. Oates, P. Jorden.
Gemini, No. 36, p. 23 – 25 (Jun 1992).

034.031   **Millisecond time resolution with the Kitt Peak Photon–Counting Array.**
N. A. Sharp.
Publ. Astron. Soc. Pac., Vol. 104, No. 674, p. 263 – 269 (Apr 1992).
The Kitt Peak Photon–Counting Array (KPCA) has been modified to provide access to individual frames, with the accompanying short frame time per exposure of between 2 and 5 ms. These changes are described and the instruments performance is illustrated. Imaging of the Crab pulsar with millisecond time resolution is presented as an astronomical example. Poor observing conditions precluded the photometric quality which would normally be achieved with a photon counter.

034.032   **A simple visual Cassegrain CCD camera for the Wyoming Infrared Observatory.**
S. D. Gillam, P. E. Johnson, M. Smith.
Publ. Astron. Soc. Pac., Vol. 104, No. 674, p. 278 – 284 (Apr 1992). Current Physics Microform No.: 9205C2234.
A visual Cassegrain CCD camera system, called the C^3, has been constructed for use at the 2.3–m telescope at the Wyoming Infrared Observatory. It is optically, mechanically, and cryogenically simple. The instrument is based upon a Texas Instruments TI–4849 (584 × 390) virtual–phase CCD and an Hitachi 63701 microprocessor. It is controlled remotely from an IBM PC XT microcomputer. The system will include an Ethernet connection to another computer which will be used as an in situ image reduction facility. The design and photometric performance of the instrument are described, and an image taken using the C^3 is presented. Further developments of the system are also discussed.

034.033   **A possible correlation tracker for LEST.**
J. A. Bonet, C. Martin, E. Ballesteros, F. J. Fuentes, F. Lorenzo, A. Manescau, L. F. Rodriguez, A. Zadrozny.
LEST Found., Tech. Rep., No. 51.
Oslo Univ. (Norway). Inst. for Teoretisk Astrofysikk. 23 p. (1992).
This document presents a proposal for the design of the LEST correlation tracker prototype. The proposed correlator, based on the absolute differences algorithm, is built around the LSI L64720 (Video Motion Estimation Processor) chip. The performances of the tracker system are established according to the LEST specifications and the experience derived from similar programmes. The design for the optomechanical support system to test the correlation tracker is also included.

034.034   **Zürich Imaging Stokes Polarimeter – ZIMPOL I. Design review.**
C. U. Keller, F. Aebersold, U. Egger, H. P. Povel, P. Steiner, J. O. Stenflo.
LEST Found., Tech. Rep., No. 53.
Oslo Univ. (Norway). Inst. for Teoretisk Astrofysikk. 82 p. (1992).
This document describes the design of ZIMPOL I, the first Zürich Imaging Stokes Polarimeter. This solar vector polarimeter will mainly be used for observations of the solar magnetic field at high spatial and/or spectral resolution. A brief overview of ZIMPOL I is given in the preface. The scientific requirements are then specified in detail. They lead to an instrument concept which consists of several parts: the optical system, the camera system, the real–time image processing system, and the graphical user interface. Data reduction and analysis of observations recorded with this polarimeter are also dealt with in detail. Prototypes of the modulator package and the CCD camera have been tested at various observatories. Results from these tests are presented and discussed.

034.035   **Demodulation of all four Stokes parameters with a single CCD – ZIMPOL II. Conceptual design.**
J. O. Stenflo, C. U. Keller, H. P. Povel.
LEST Found., Tech. Rep., No. 54.
Oslo Univ. (Norway). Inst. for Teoretisk Astrofysikk. 14 p. (1992).
It is shown how it is possible to simultaneously record images of all four Stokes parameters with a single CCD detector chip when fast (50 kHz), piezoelastic modulation of the polarization state is used. As the four image planes use the identical pixels of the CCD, all gain–table or flat–field effects vanish when forming the fractional polarization images. For each group of four pixel rows, one row collects the photons, while the other three are used for fast buffer storage. There are no light losses caused by masking of the pixel rows used for buffer storage, sinced a microlens array collects all the photons and directs them to the unmasked pixel rows. The efficiency of the system for simultaneous recording of all four Stokes parameters is six times greater than that of ZIMPOL I, the first generation of the Zürich Imaging Stokes Polarimeter, since no beam splitter with three separate CCD cameras is needed and no significant light losses occur at the masked pixel rows. The theoretically possible efficiency limit is thereby practically reached. The system is planned to be developed as ZIMPOL II, the second generation of the Zürich Imaging Stokes Polarimeter.

034.036   **The University of Hawaii NICMOS–3 Near–Infrared Camera.**
K.–W. Hodapp, J. Rayner, E. Irwin.
Publ. Astron. Soc. Pac., Vol. 104, No. 676, p. 441 – 451 (Jun 1992). Current Physics Microform No.: 9208B0697.
The University of Hawaii Near–Infrared Camera is equipped with a NICMOS–3 HgCdTe detector array sensitive from ∼ 1 to 2.5 µm and produced by the Rockwell International Science Center for the Near–Infrared Camera and Multi–Object Spectrometer Project (NICMOS). In the authors' camera, the NICMOS–3 array operates with 53 electrons readnoise in a double–correlated sampling mode. A dark current below 1 electron per second at an operating temperature of 60K has been achieved. The device works linearly within 1% up to 250000 electrons at 1.0 V bias. Techniques for further reduction of readnoise and dark current are discussed. The only significant remaining problem is a residual excess dark current, remaining from previous exposure of the device.

034.037   **Application of the intensified CCD to airglow and auroral measurements.**
A. L. Broadfoot, B. R. Sandel.
Appl. Opt., Vol. 31, No. 16, p. 3097 – 3108 (1 Jun 1992). Current Physics Microform No.: 9208D1818.
New detector technology exemplified by advanced CCD and intensified CCD (ICCD) systems have important advantages for both spectrographic and imaging research. However, to realize the full potential of this new technology, one must consider the detector and the optical system as a whole. It is frequently not enough to simply substitute an ICCD for an earlier detector; rather, to achieve optimum results, the optics must be adapted to the specific detector. Properly designed airglow spectrographs based on the ICCD detector offer the advantages of high throughput over a broad spectral range, precise wavelength stability, low noise, and compactness. Imagers having the wide field and the high sensitivity needed for airglow research are practical as well.

034.038   **Narrow–band tunable filter for solar observations.**
E. S. Kulagin.
Kinematika Fiz. Nebesn. Tel, Tom 8, No. 1, p. 24 – 35 (Jan–Feb 1992). In Russian. English translation in Kinematics Phys. Celest. Bodies, Vol. 8, No. 1.
A scheme of a tunable solar filter is described and analysed. As a preliminary monochromator, a double monochromator with subtraction of dispersions is used, and the final band of transmission is formed by the Fabry–Perot tunable interferometer. Basic parameters of the working model of the filter and

samples of filtergrams obtained on the horizontal solar telescope are given.

034.039 Laboratory polarimeter for measurement of the Stokes vector of light scattered from surfaces.
V. S. Degtyarev, N. V. Karpov, L. O. Kolokolova.
Kinematika Fiz. Nebesn. Tel, Tom 8, No. 2, p. 83 – 88 (Mar–Apr 1992). In Russian. English translation in Kinematics Phys. Celest. Bodies, Vol. 8, No. 2.

The laboratory polarimeter constructed at the Main Astronomical Observatory, Academy of Sciences, Ukraine, is described. It measures all four Stokes parameters of light scattered from surfaces. The block diagram of the polarimeter is given, and the mathematical algorithm of its operation is described. The polarimeter is linked with a minicomputer, so that the data collection and processing are carried out automatically and are realized in the dialogue mode. Characteristics of the polarimeter are listed. The source polarization for the Stokes parameters divided by intensity does not exceed 0.01%. The polarimeter is used for investigation of diagnostic abilities of the Stokes vector in remote sensing of atmosphereless celestial bodies.

034.040 Estimation of photometric characteristics of the PARSEC automatic measuring system.
V. I. Voroshilov, L. K. Pakulyak, T. P. Sergeeva.
Kinematika Fiz. Nebesn. Tel, Tom 8, No. 2, p. 89 – 90, 95 (Mar–Apr 1992). In Russian. English translation in Kinematics Phys. Celest. Bodies, Vol. 8, No. 2.

Photographic photometry at photoelectric standard stars in the region of the galactic cluster NGC 188 was carried out using the PARSEC, "Askoris" and MF–2 photometers. Comparison of the results shows that measurements made with PARSEC have the smallest errors.

034.041 Measurement of energy transfer in a five–mode gravitational wave bar detector.
Pang Yi, J. P. Richard.
Rev. Sci. Instrum., Vol. 63, No. 1, p. 56 – 63 (Jan 1992).

A 1400–kg five–mode bar detector has been constructed and tested at room temperature. Detailed calculations and computer simulations have been performed to optimize the design and study the dynamics of the system. The tests have confirmed the analytical prediction that a large fraction (67% in this experiment) of the energy deposited into the bar is transferred to the last resonator. The rise time of the energy at the last resonator is short (≈ 2 ms), and corresponds to an effective detector bandwidth of ≈ 500 Hz.

034.042 A new–type antenna for continuous gravitational radiation.
T. Suzuki, N. Akasaka, Y. Ogawa, N. Kudo, K. Morimoto.
Rev. Sci. Instrum., Vol. 63, No. 3, p. 1880 – 1883 (Mar 1992).

A new–type disk antenna has been developed to search for continuous gravitational waves emitted from millisecond and submillisecond pulsars. The antenna not only has a wide tunable range of the eigenfrequency that covers down to almost half of the original frequency of the quadrupole mode, but also is easily tuned to an objective frequency with an accuracy of 4×10^{-5} at 4.2K. The mechanical quality factor has reached 3.0×10^{7} at 4.2K in an antenna made of Al5056.

034.043 Automatic wide–range scanning and calibration over 220–740 nm using a dye laser with a rapid cell exchanger.
Y. Oki, E. Tashiro, M. Maeda, C. Honda, Y. Hasegawa, H. Futami, J. Izumi, K. Matsuda.
Rev. Sci. Instrum., Vol. 63, No. 5, p. 2927 – 2931 (May 1992).

A pulsed dye laser system fully controlled by a microcomputer has been developed as a tunable light source for analytical spectroscopy. The system has 13 dyes with a rapid replacement mechanism and a frequency doubler inside, and any desired wavelength from 220 to 740 nm can be instantaneously generated by the aid of the microcomputer control. The absolute wavelength can be automatically calibrated with a newly developed optogalvanic wavemeter with an accuracy of ± 1 cm^{-1} over the whole spectral range.

034.044 Catching the wave.
R. Ruthen.
Sci. Am., Vol. 266, No. 3, p. 72 – 81 (Mar 1992).

The gravitational waves that ripple the fabric of space have never been conclusively observed. A team of U.S. scientists hopes by the end of the decade to be the first to build a device that will detect these extremely weak undulations. If they succeed, their unique telescope may also illuminate black holes and detect unknown cosmic structures invisible in the electromagnetic spectrum.

034.045 The CYGNUS extensive air–shower experiment.
D. E. Alexandreas, R. C. Allen, S. D. Biller, R. S. Delay, G. M. Dion, X. Q. Lu, P. R. Vishwanath, G. B. Yodh, D. Berley, C. Y. Chang, B. L. Dingus, J. A. Goodman, T. J. Haines, S. Gupta, D. A. Krakauer, M. J. Stark, R. L. Talaga, R. L. Burman, K. Butterfield, R. Cady, C. M. Hoffman, J. Lloyd–Evans, D. E. Nagle, M. E. Potter, V. D. Sandberg, C. Sinnis, S. Stanislaus, T. N. Thompson, C. A. Wilkinson, W. Zhang, R. W. Ellsworth.
Nucl. Instrum. Methods Phys. Res., Sect. A, Vol. 311, No. 1/2, p. 350 – 367 (1 Jan 1992).

The CYGNUS extensive air–shower experiment is described. The design criteria, construction and operation details, and performance characteristics are presented. A discussion of the data analysis techniques is given. Finally, several enhancements and improvements in the apparatus are described.

034.046 The limitation of wavefront tilt sensor in adaptive astronomical telescope.
Cao Genrui, Yu Xin, Gu Ruowei, Zhou Renzhong.
SPIE Conference on Current Developments in Optical Engineering IV as Part of SPIE's International Symposium on Optical and Optoelectronics Applied Science and Engineering, p. 139 – 145 (1990). – See Abstr. 012.011 for the main entry.

The limitations of a Hartmann type wavefront sensor are theoretically and experimentally studied based on photon counting techniques. An experimental set–up for simulation of single subaperture adaptive astronomical telescope is described. The results of experiments and analyses show that a star of the 8th magnitude can be dealt with by this kind of wavefront sensors within 1 ms of time and with $\lambda/10$ of wavefront distortion detection sensitivity. As for calibration of experiments, an algorithm related to luminous flux at the entrance pupil of telescope and its corresponding photo–electron rate detected by the wavefront sensor is presented and verified by experiments.

034.047 Adaptive optics wavefront corrector using addressable liquid crystal retarders.
D. Bonaccini, G. Brusa, S. Esposito, P. Salinari, P. Stefanini.
SPIE Conference on Current Developments in Optical Engineering IV as Part of SPIE's International Symposium on Optical and Optoelectronics Applied Science and Engineering, p. 89 – 97 (1990). – See Abstr. 012.011 for the main entry.

For the 8m astronomical telescopes which will see first light in the '90s, interest in adaptive optics correction of the wavefront degraded by atmospheric turbulence is very great. A wavefront corrector is described in this paper, in which the authors present an approach based on liquid crystal material. Some of the properties in favor of using this technique are analyzed, together with some experimental results on a prototype. The great advantage of the liquid crystal approach is that a large number of actuators can be obtained in a 1 – 2 inch device, at a cost orders of magnitude lower than the conventional mirror–piezoactuator device.

034.048   Applications of infrared bidimensional devices in astronomy.

J. L. Monin, M. Caes, J. P. Chatard, P. Nicolas, R. Boch.

SPIE Conference on Infrared Technology XVI as Part of SPIE's International Symposium on Optical and Optoelectronic Applied Science and Engineering, p. 202–213 (1990). – See Abstr. 012.012 for the main entry.

The characteristics of some bidimensional infrared devices are reviewed within the framework of astronomical observations. Some recent laboratory results and infrared astronomical observations are presented.

034.049   The CCD camera in the Suhora Observatory.

J. Krzesiński.

Urania, Rok 63, Nr. 1, p. 5–7 (Jan 1992). In Polish.

034.050   Status of the Whipple Observatory Čerenkov air shower imaging telescope array.

C. W. Akerlof, M. F. Cawley, D. F. Fegan, S. Fennell, S. Freeman, D. Frishman, K. Harris, A. M. Hillas, D. Jennings, R. C. Lamb, M. A. Lawrence, D. A. Lewis, D. I. Meyer, M. Punch, P. T. Reynolds, M. S. Schubnell, T. C. Weekes.

Compton Observatory Science Workshop, p. 406–411 (Feb 1992). Abstract. – See Abstr. 012.018 for the main entry.

034.051   A new instrument for high resolution, two–dimensional solar spectroscopy.

C. Bendlin, R. Volkmer, F. Kneer.

Astron. Astrophys., Vol. 257, No. 2, p. 817–823 (Apr 1992).

A two–dimensional spectrometer suitable for solar observations of high spatial, spectral, and temporal resolution was built and successfully tested with the German Vacuum Tower Telescope at the Observatorio del Teide/Tenerife. Using a universal birefringent filter and a Fabry–Perot interferometer (FPI), narrow–band ($\lesssim 30$ mÅ) filtergrams can be obtained. With this instrument, it is possible to scan through a Fraunhofer line at an appropriate number of wavelength settings within a few seconds. Here, fast scanning is accomplished by tuning only the FPI. The images are taken by two CCDs. One of them is coupled with an image intensifier to achieve short integration times. Due to the instrument's design, the adequate field of view has a diameter of about 20–40 arsec. The spectrometer can be tuned to virtually any wavelength in the visible spectrum.

034.052   Automatic control of a Michelson interferometer.

F. Barone, L. Di Fiore, L. Milano, G. Russo.

9. Italian Conference on General Relativity and Gravitational Physics, p. 562–572 (1991). – See Abstr. 012.021 for the main entry.

The authors present the current status of the Automatic Control System for mirror alignment of the VIRGO antenna. The performed experiments are very important, if they are interpreted in the framework of the project of the final control system of the whole antenna. Two important results have been achieved: 1) the authors have demonstrated that the mechanical modulation technique can be used in mirror alignment; 2) they have proved the feasibility of the implementation of this subsystem by means of a computer aided system, more flexible than an analogic control.

034.053   Visible and infrared wavefront sensing for astronomical adaptive optics.

F. Rigaut, J. G. Cuby, M. Caes, J. L. Monin, M. Vittot, J. C. Richard, G. Rousset, P. Léna.

Astron. Astrophys., Vol. 259, No. 2, p. L57–L60 (Jun 1992). Letter–to–the–editor.

The applicability of adaptive optics to astronomical standard observing critically depends on the ability to provide with sufficient sensitivity a reference measurement of the wave–fronts distorted by the atmosphere. The authors describe successful operation of the VLT Adaptive Optics Prototype, called COMEON, with two reference channels: a visible one, quantum noise limited, and an infrared one. They define the adaptive optics

limiting magnitudes $(m_\lambda)_{lim}$ for closed–loop operation of a given system with a given turbulence and demonstrate on a 3.6–m telescope the lower limits $(m_R)_{lim} = 13$ and $(m_K)_{lim} = 1.6$. Optimization of the system transmission over the current prototype should soon lead to $(m_R)_{lim} = 14$ to 16 and $(m_K)_{lim} \geqslant 6$.

034.054   Magnetic and radio detection of aurorae.

D. J. Smillie.

J. Br. Astron. Assoc., Vol. 102, No. 1, p. 16–20 (Feb 1992).

A clear sky during the hours of darkness seldom occurs at the author's location. The prospect of seeing a visual aurora is almost zero, due to cloud and sodium street lighting. However, other alternative methods of aurorae detection are available for those unlucky enough not to have a clear night sky. These alternatives include magnetometry, which monitors changes in the Earth's magnetic field, and the recording of radio signals whose reception is modified by the presence of ionospheric disturbances and aurorae.

034.055   MUSICOS: a fiber–fed spectrograph for multi–site observations.

J. Baudrand, T. Böhm.

Astron. Astrophys., Vol. 259, No. 2, p. 711–719 (Jun 1992).

This paper describes a fiber fed echelle spectrograph designed and constructed at the Paris–Meudon Observatory. This instrument is dedicated to the MUSICOS project, whose purpose is to organize multisite continuous spectroscopic observations. The optical scheme is described in some detail, as well as the data reduction software. The authors give a review of the spectrograph main performances which were obtained during laboratory tests and in real observational conditions at the 2 m diameter telescope at Pic du Midi.

034.056   Le développement d'un film noir et blanc.

L. Spede.

Ciel, Vol. 54, p. 175–178 (Jun 1992).

034.057   A status report of the MINI experiment.

G. Iaselli, F. D'Aquino, N. Mirizzi, S. Nuzzo, A. Ranieri, F. Romano, A. Rossi, P. Bernardini, P. Pistilli, J. Beman, M. Lawrence, J. Lloyd–Evans, R. Reid, A. Watson, M. Ambrosio, G. C. Barbarino, B. Bartoli, D. Campana, J. W. Elbert, F. Guarino, M. Jacovacci, G. Osteria, V. Silvestrini, R. Buccheri, O. Catalano, S. del Sordo, J. Linsley, L. Scarsi, G. Bressi, M. Cambiaghi, A. Lanza, S. Ratti, G. Auriemma, M. Bonori, A. Capone, G. D'Agostini, D. de Pedis, M. de Vincenzi, P. Lipari, F. Massa, M. Mattioli, A. Nigro, G. Piredda, D. Zanello, R. Cardarelli, R. Santonico, L. de Cesare, G. Grella, M. Guida, F. Mancini, G. Marini, G. Romano, G. Vitiello.

Vulcano Workshop 1990: Frontier objects in astrophysics and particle physics, p. 337–348 (1991). – See Abstr. 012.040 for the main entry.

A telescope to detect and to track cosmic muons has been built by the MINI collaboration in the laboratory of the Physics Department of the University of Bari. The telescope has been partially equipped using gaseous detectors developed with a recently proposed technique (RPC). To debug the hardware a first technical run was performed around the middle of May 1990. Physics runs will start during next summer. The assembly of the telescope will be completed before the end of the year. A short presentation of the telescope layout and a preliminary report of its behaviour in this very early set up phase are given.

034.058   Status and performance of the Chicago Air Shower Array and related experiments.

B. J. Newport.

Vulcano Workshop 1990: Frontier objects in astrophysics and particle physics, p. 387–394 (1991). – See Abstr. 012.040 for the main entry.

The partially completed Chicago Air Shower Array (CASA) is an instrument whose principal goal is the unambiguous detection of astrophysical point sources of particles with energies $\gtrsim 100$ TeV. CASA operates in conjunction with a very large muon detector built by the University of Michigan, and with several University of Utah experiments which use optical shower detection techniques. This paper describes the status of these experiments, presents some measurements of the performance of CASA, and provides schedules for the completion of the various detectors and the production phase of the data analysis system.

034.059   High sensitivity UHE gamma–ray Cherenkov telescope with time delay technique.
R. A. Antonov, I. P. Ivanenko, S. Karakuła, W. Tkaczyk.
Vulcano Workshop 1990: Frontier objects in astrophysics and particle physics, p. 395 – 400 (1991). – See Abstr. 012.040 for the main entry.
The authors compare the main parameters of Cherenkov telescopes.

034.060   Calculation of the upward–going muon background for a surface neutrino detector.
J. W. Elbert, M. Iacovacci, V. Silvestrini.
Vulcano Workshop 1990: Frontier objects in astrophysics and particle physics, p. 417 – 422 (1991). – See Abstr. 012.040 for the main entry.
The authors have calculated the background in a neutrino detector on the Earth's surface due to atmospheric muons which penetrate underground, scatter, and emerge traveling upwards.

034.061   CLUE: Cerenkov light ultraviolet experiment. Preliminary results and future plans.
L. Peruzzo, G. Sartori, F. Bedeschi, E. Bertolucci, M. Mariotti, A. Menzione, L. Ristori, A. Scribano, A. Stefanini, F. Zetti, B. Bartoli, M. Budinich, F. Liello.
Vulcano Workshop 1990: Frontier objects in astrophysics and particle physics, p. 423 – 430 (1991). – See Abstr. 012.040 for the main entry.
The results of the test of a prototype apparatus aimed to detect the ultraviolet Cerenkov light in the wavelength range 1900 – 2300 Å are presented.

034.062   First results from MACRO an underground detector in the Gran Sasso laboratory.
P. Lipari.
Vulcano Workshop 1990: Frontier objects in astrophysics and particle physics, p. 431 – 442 (1991). – See Abstr. 012.040 for the main entry.
The MACRO detector is presently under construction in the underground Gran Sasso laboratory. The detector is dedicated to the search of rare phenomena in the cosmic radiation. Some results obtained in this first run are described.

034.063   Ultralow temperatures gravitational wave detectors.
E. Coccia.
Vulcano Workshop 1990: Frontier objects in astrophysics and particle physics, p. 515 – 526 (1991). – See Abstr. 012.040 for the main entry.
Results are reported of the feasibility study for large milliKelvin temperature resonant gravitational wave antennae. The aim of this third generation of resonant detectors is to observe gravitational collapses occurring at distances of the order of 10 Mpc, approaching the Virgo cluster of galaxies, at a rate of several per year.

034.064   The Joint European Amateur Photometer: an update.
E. N. Walker.
1. European Meeting of the American Association of Variable Star Observers (AAVSO): International Cooperation and Coordination in Variable Star Research, p. 122 – 125 (1992). – See Abstr. 012.048 for the main entry.
If the potential for serious research that exists among amateur astronomers was to be realised, then it was necessary to develop a robust photoelectric photometer. The authors gives a brief description of what has become known as the JEAP (The Joint European Amateur Photometer), and of the design philosophy behind many of its features.

034.065   FORS – the focal reducer for the VLT.
I. Appenzeller, G. Rupprecht.
Messenger, No. 67, p. 18 – 21 (Mar 1992).

034.066   A coronagraph for COME–ON, the adaptive optics VLT prototype.
F. Malbet.
Messenger, No. 67, p. 46 – 48 (Mar 1992).

034.067   A new cross disperser for CASPEC.
L. Pasquini, G. Rupprecht, A. Gilliotte, J.–L. Lizon.
Messenger, No. 67, p. 50 – 51 (Mar 1992).

034.068   Compact fibre–optic format changers for a multislit echelle spectrometer – initial results on the Dumbbell nebula (NGC 6853).
J. Meaburn, P. E. Christopoulou, C. D. Goudis.
Mon. Not. R. Astron. Soc., Vol. 256, No. 1, p. 97 – 102 (1 May 1992).
Eight compact fibre–optic format changers have been manufactured. These convert the shapes of line–emission sources, imaged in the focal plane of the 4.2–m William Herschel telescope, into three or five long entrance slits for the Manchester echelle spectrometer. The construction and performance of one such array, with 169 fibres feeding three parallel slits, is described in detail. Improvements which will enhance its performance by at least a factor of 2 are suggested, as is the best way of spectrally calibrating these data. A gas–spaced, optically contacted Fabry–Perot which produces white–light (Edser–Butler) on–axis fringes has been manufactured for this purpose. The initial use of the 169–fibre array on the core of the Dumbbell planetary nebula has revealed the presence of four separate velocity components in the [O III] 5007 Å profiles. The existence of an inner, highly ionized shell, expanding radially at 12 km s^{-1} and contained within an outer one expanding at 31 km s^{-1}, is implied.

034.069   DUMAND–II progress report.
R. J. Wilkes.
AIP Conf. Proc., No. 243, p. 1086 – 1088 (1 Jan 1992). Current Physics Microform No.: 9203C1934. – See Abstr. 012.047 for the main entry.
The design, scientific goals, and capabilities of the DUMAND II detector system are described. Construction was authorized by DOE in 1990, and development of various detector subsystems is under way. Current plans include deployment of the shore cable, junction box and three strings of optical detector modules in 1992, with expansion to the full 9–string configuration about one year later.

034.070   Solar post–focus instrumentation.
H. Wöhl.
1. Canary Islands Winter School of Astrophysics: Solar observations – techniques and interpretation, p. 145 – 178 (1992). – See Abstr. 012.045 for the main entry.
The following lecture includes post–focus instrumentation for solar observations. The wavelength region covered is mainly the visible, although most of the equipment will be similar for observations in the near ultraviolet or near infrared. Complex instrumentation as well as detectors used in solar observatories are described.
Contents: (1) General remarks. (2) Locations of post–focus instrumentation. (3) Post–focus imaging. (4) Spectrometers: prismspectrometer; plane grating spectrometer; concave grating spectrometer; Fourier transform spectrometer; Fabry–Perot spectrometer; pinhole photometer. (5) Detectors for spectrometers: photographic film; photomultiplier; linear photodiode array; two–dimensional photodiode array; image intensifier. (6) Special instruments at spectrometer foci: spectrum scanner; Doppler compensator; spectroheliograph. (7) Instrument control. (8) Data collection. (9) Data reduction.

034.071   High–performance 256 × 256 InSb FPA for astronomy.
A. Hoffman, D. Randall.
SPIE Conference on Infrared Technology XVII – Dedicated to the Memory of Irving J. Spiro as Part of SPIE's International Symposium on Optical Applied Science and Engineering, p. 297 – 302 (1991). – See Abstr. 012.051 for the main entry.
This paper describes a high–performance, large–format staring array. A 256 × 256 element indium antimonide (InSb) array has been developed for low–background commercial applications. The high sensitivity of the array makes it useful for space surveillance as well as astronomy. The array responds over a broad range of wavelengths, 1 μm to 5 μm, with 90% peak detector quantum efficiency and nearly 100% fill factor.

034.072   Development of a low pass far infrared filter for lunar observer horizon sensor application.
S. Mobasser, L. Horwitz, O'D. Griffith.
SPIE Conference on Infrared Technology XVII – Dedicated to the Memory of Irving J. Spiro as Part of SPIE's International Symposium on Optical Applied Science and Engineering, p. 764 – 774 (1991). – See Abstr. 012.051 for the main entry.
A study was conducted to 1) determine the feasibility of design and fabrication of a low pass filter with a relatively sharp cut–on at higher wavelengths (i.e. 30 – 40 μm) using metallic mesh technique, and to 2) investigate whether the combination of this filter and a suitable IR detector, as a part of a Lunar Observer horizon sensor, is capable of detecting radiation emanating from two blackbody sources kept at tempeatures simulating space and the surface temperature of dark or lit sides of the moon.

034.073   Kryogener Shutter für die NIR–Kamera am VLT.
F.–J. Schöniger.
Max–Planck–Institut für Physik und Astrophysik, Garching (Germany). Inst. für Extraterrestrische Physik, MPE—237. 94 p. (Mar 1992). ISSN 0178–0719.
Contents: 1. Einleitung. 2. Der Einfluß der Erdatmosphäre auf das astronomische Bild. 3. Das Very Large Telescope (VLT) der ESO. 4. Der kryogene Shutter.

034.074   Die räumliche Verteilung von Nahinfrarotlinienstrahlung in galaktischen Kernen gemessen mit dem abbildenden Fabry–Pérot–Spektrometer FAST.
V. Rotaciuc.
Diss. (Dr.rer.nat.), München Univ. (Germany). Fakultät für Physik; Max–Planck–Institut für Physik und Astrophysik, Garching (Germany). Inst. für Extraterrestrische Physik. MPE—239, 115 p. (Jun 1992). ISSN 0178–0719.
Contents: 1. Einleitung. 2. FAST – das abbildende Nahinfrarotspektrometer.   3. Astrophysikalische   Grundlagen. 4. NGC 1068. 5. Das galaktische Zentrum.

034.075   Polaris II, a polarimetric imaging spectrometer.
J. T. Bergstralh, D. Glenar, R. Zimmerman, B. Saif, W. H. Smith.
Bull. Am. Astron. Soc., Vol. 23, No. 3, p. 1234 – 1235 (1991). Abstract. – See Abstr. 012.037 for the main entry.

034.076   The use of Eastman Kodak 4415 film in the UK Schmidt Telescope.
K. S. Russell, D. F. Malin, A. Savage, M. Hartley, Q. A. Parker.
2. Conference on Digitised Optical Sky Surveys, p. 23 – 33 (1992). – See Abstr. 012.059 for the main entry.
The authors report experiments with 4415 film in the UK Schmidt telescope which may enable to use a much wider range of sensitized materials than was previously possible, and to exploit the superior imaging properties of modern materials that are only available on film.

034.077   The MAMA facility: a survey of scientific programmes.
J. Guibert.
2. Conference on Digitised Optical Sky Surveys, p. 103 – 108 (1992). – See Abstr. 012.059 for the main entry.
MAMA (machine automatique à mésurer pour l'astronomie) is a fast and accurate multichannel microdensitometer developed and operated by INSU and located at the Observatoire de Paris. MAMA is widely used for the identification, astrometry and photometry of optical counterparts of sources detected at other wavelength.

034.078   ASTROSCAN II: the next Leiden microdensitometer.
E. R. Deul.
2. Conference on Digitised Optical Sky Surveys, p. 109 – 114 (1992). – See Abstr. 012.059 for the main entry.

034.079   A new astrometric measuring machine: design and astronomical programmes.
C. de Vegt, L. Winter, N. Zacharias.
2. Conference on Digitised Optical Sky Surveys, p. 115 – 121 (1992). – See Abstr. 012.059 for the main entry.
The design of a new type of astrometric measuring machine, recently installed at Hamburg Observatory is described. The measuring system consists of a very compact high–precision, air-bearing, granite x–y measuring table; the photoplate is digitised in a frame-by-frame mode using the VIDEK–KODAK MEGAPLUS CCD–camera which provides a frame size of 1035x1320 square pixels of 6.8 × 6.8 micrometers at unit magnification. The machine design is based on a fully modular concept; machine operation and camera data are handled by independent computer systems. The maximum measuring area is about 270 × 270 mm², typical astrograph plates (measuring area $\approx$ 220 × 220 mm²) can be digitised in about 30 minutes. First measuring programs will concentrate on the Hipparcos ground-based extra–galactic reference link and various catalogues, in particular remeasurement of AC zones and the AGK2 plates.

034.080   SuperCOSMOS.
L. Miller, W. Cormack, M. Paterson, S. Beard, L. Lawrence.
2. Conference on Digitised Optical Sky Surveys, p. 133 – 139 (1992). – See Abstr. 012.059 for the main entry.
The authors describe the design of and progress on a new, highly accurate machine for digitising photographic plates which is under construction at Edinburgh.

034.081   Ein Detektorsystem zum Nachweis von Myonen in ausgedehnten Luftschauern für das KASCADE–Projekt. (*A detectorsystem to measure muons in extensive air showers for the KASCADE–project.*)
W. Kriegleder.
Diss., Kernforschungszentrum Karlsruhe GmbH (Germany); Karlsruhe Univ. (T.H.) (Germany). Fakultät für Physik. KfK—5023, 115 p. (Apr 1992). ISSN 0303–4003.
The KASCADE–experiment aims to measure the hadronic, the electromagnetic and the muonic component of an extensive air shower simultaneously with high precision. Energy, direction and nature of the primary cosmic particle has been determined, which has initiated the particle cascade by a collision with a nucleus in the higher atmosphere. The determinaton of the primary cosmic ray energy spectrum, the analysis of the chemical composition at different energies and the search for point sources of particles with ultra high energies are the main goals of the experiment. With the help of these data, different models concerning the origin of ultra high energy cosmic ray particles can be tested. To measure the muonic component, a detector has been developed consisting of large scintillator sheets read out by wave lengths shifter bars at all edges.

034.082   Die neuen Instrumente für das VLT.
R. Lenzen.
Sterne Weltraum, Jahrg. 31, Nr. 5, p. 290 – 291 (May 1992).

034.083   Vier Millionen Bildelemente. CCD–Kamera–Entwicklung an der Universität Bonn.
K. Reif.
Sterne Weltraum, Jahrg. 31, Nr. 5, p. 300 – 303 (May 1992).

034.084   Un metodo semplice e preciso per disegnare meridiane verticali declinanti.
N. Severino.
Astronomia UAI, N. 3, p. 28 – 30 (May – Jun 1992).
The author presents a simple method to plan vertical declining sundials.

034.085   Total internal reflection phase shifters for astronomical polarimeters.
A. V. Bruns, O. P. Gollandskij.
Bull. Crimean Astrophys. Obs., Vol. 82 , p. 169 – 172 (1992). English translation of Izv. Krym. Astrofiz. Obs., Tom 82, p. 185 – 188 (1990).
Optical parameters are considered for two types of achromatic quarter–wave phase shifters based on total internal reflection: with parallel input and output rays (in particular, a Fresnel rhomb) and with deflected output beam (in particular, a Mooney rhomb). The second type produces much less discrepancy between the phase shifts for the axial ray and the peripheral ones, so the latter is used in high-aperture polarimeters.

034.086   Extragalactic supernova detector, neutrino mass, and the Supernova Watch.
D. B. Cline.
15. Texas Symposium on Relativistic Astrophysics and 4. ESO–CERN Symposium, p. 413 – 424 (1991). – See Abstr. 012.060 for the main entry.
The author lists some of the supernova detectors operating, being constructed, or being planned around the world. This is an impressive array of detectors. The basic idea of a Supernova Watch network is shown. A real–time Supernova Watch discussion group has been formed following the UCLA meeting.

034.087   Abnormalities of the time comparisons of atomic clocks during the solar eclipses.
S. W. Zhou, B. J. Huang.
Nuovo Cimento C, Vol. 15, N. 2, p. 133 – 137 (Mar–Apr 1992).
The authors have investigated the time comparison data of the atomic clocks of some stations and have found that the solar eclipse has influence on the time comparison of atomic clocks.

034.088   New detector systems for dynamical astronomy.
D. G. Monet.
Bull. Am. Astron. Soc., Vol. 23, No. 3, p. 1256 (1991). Abstract. – See Abstr. 012.065 for the main entry.

034.089   Thermal detection of dark matter.
E. Fiorini.
15. Texas Symposium on Relativistic Astrophysics and 4. ESO–CERN Symposium, p. 446 – 456 (1991). – See Abstr. 012.060 for the main entry.
Contents: 1. Introduction. 2. Thermal detectors. 3. Underground cryogenics. 4. Conclusions.

034.090   Recent results on cryogenic detector developments.
F. von Feilitzsch.
15. Texas Symposium on Relativistic Astrophysics and 4. ESO–CERN Symposium, p. 457 – 463 (1991). – See Abstr. 012.060 for the main entry.
In this paper the author discusses briefly developments based upon phonon excitations and the detection of quasi particles in superconductors.

034.091   Neutrino astronomy with large Cerenkov detectors.
J. G. Learned.
15. Texas Symposium on Relativistic Astrophysics and 4. ESO–CERN Symposium, p. 464 – 482 (1991). – See Abstr. 012.060 for the main entry.
In this paper the author summarizes the present status of experiments and prospects for future high–energy neutrino astrophysics endeavors employing Cerenkov radiation detection. Roughly six third–generation detectors (in the $> 10000 \ m^2$ class) are in various stages of proposal, test, or construction, some of which will come to operation by the mid–1990s. The author describes the DUMAND II detector, now in construction for deployment in Hawaii for operation beginning in late 1993, in some detail. He also briefly discusses several novel detection techniques, particularly for application in the Antarctic, and examines prospects for the future. For the present it seems that water (or ice) Cerenkov detectors employing photomultipliers remain the most cost effective means to reach the $1 \ km^2$ sizes needed for neutrino astronomy in the next generation.

034.092   An astronomical camera for imaging at mid–infrared wavelengths.
R. K. Pina, B. Jones, R. C. Puetter.
Bull. Am. Astron. Soc., Vol. 23, No. 3, p. 1268 – 1269 (1991). Abstract. – See Abstr. 012.067 for the main entry.

034.093   Preliminary report on the "lattice gate" CCD.
W. V. Schempp.
Bull. Am. Astron. Soc., Vol. 23, No. 3, p. 1269 (1991). Abstract. – See Abstr. 012.067 for the main entry.

034.094   Solar feature correlation tracker.
T. Rimmele, O. von der Lühe, P. H. Wiborg, A. L. Widener, R. B. Dunn, G. Spence.
SPIE Conference on Active and Adaptive Optical Systems as Part of SPIE's International Symposium on Optical Applied Science and Engineering, p. 186 – 193 (1991). – See Abstr. 012.074 for the main entry.
The authors present a tracking system that stabilizes atmospheric and instrumental image motion at the vacuum tower telescopes of the National Solar Observatory at Sacramento Peak and the Kiepenheuer Institut für Sonnenphysik at Tenerife. A matrix diode array rapidly scans the scene of interest, usually with a field of 5 arcsec. Images are cross–correlated in real time with a previously recorded reference image of the same area. Reference pictures are updated every 30 s. Recent performance tests show that the residual image motion in the tracked image is 0.05 arcsec rms compared to a typical 0.5 arcsec rms for the untracked image. The correlation tracker also includes a seeing monitor providing a relative seeing measure at a two millisecond rate, which can be used for frame selection and shutter control.

034.095   Prototype high speed optical delay line for stellar interferometry.
M. M. Colavita, B. E. Hines, M. Shao, G. J. Klose, B. V. Gibson.
SPIE Conference on Active and Adaptive Optical Systems as Part of SPIE's International Symposium on Optical Applied Science and Engineering, p. 205 – 212 (1991). – See Abstr. 012.074 for the main entry.
The long baselines of the next–generation ground–based optical stellar interferometers require optical delay lines which can maintain nanometer–level pathlength accuracy while moving at high speeds. JPL is currently designing delay lines to meet these requirements for the NRL Big Optical Array and the USNO Astrometric Interferometer, successors to the Mt. Wilson Mark III interferometer. The design is discussed. The delay line is fully programmable in position and velocity, and the system is controlled with 4 cascaded software feedback loops implemented with a VME–based control system. Preliminary performance is a jitter in any 5 ms window of less than 10 nm rms for delay rates of up to 28 mm/s; total jitter is less than 10 nm rms for delay rates up to 20 mm/s.

034.096   **The University of Hawaii adaptive optics system: III. The wavefront curvature sensor.**
J. E. Graves, D. L. McKenna.
SPIE Conference on Active and Adaptive Optical Systems as Part of SPIE's International Symposium on Optical Applied Science and Engineering, p. 262 – 272 (1991). – See Abstr. 012.074 for the main entry.
A wavefront curvature sensor suitable for adaptive optics has been developed at the University of Hawaii. This paper describes a curvature sensor that can rapidly scan between extra focus images with a vibrating membrane acting as a variable curvature active optical element at 7 – 20 kHz. This optical system is coupled to an array detector, of a special design, that is computer optimized to match the response of a bimorph adaptive mirror. Recent laboratory results will show wavefront comparisons with other more established technique for curvature sensing and closed loop performance test with a monolithic piezo deformable mirror with 21 actuators. Results will also be given showing tip/tilt corrections achieved on Procyon with this wavefront sensor at the Canada France Hawaii 3.5 m Telescope on Mauna Kea during April 1991.

034.097   **The NASA/NSO Spectromagnetograph.**
H. P. Jones, T. L. Duvall Jr., J. W. Harvey, C. T. Mahaffey, J. D. Schwitters, J. E. Simmons.
Sol. Phys., Vol. 139, No. 2, p. 211 – 232 (Jun 1992).
The NASA/NSO Spectromagnetograph is a new focal plane instrument for the National Solar Observatory/Kitt Peak Vacuum Telescope which features real–time digital analysis of long–slit spectra formed on a two–dimensional CCD detector. The instrument is placed at an exit port of a Littrow spectrograph and uses an existing modulator of circular polarization. Commercial video processing boards are used to digitize the spectral images at video rates and to separate, accumulate, and buffer the spectra in the two polarization states. An attached processor removes fixed–pattern bias and gain from the spectra in cadence with spatial scanning of the image across the entrance slit. The data control computer performs position and width analysis of the line profiles as they are acquired and records line–of–sight magnetic field, Doppler shift, and other computed parameters. The observer controls the instrument through windowed processes on a data control console using a keyboard and mouse. Early observations made with the spectromagnetograph are presented and plans for future development are discussed.

034.098   **Intensified CCD camera based remote guiding unit for VBT and observation of speckles with ICCD.**
V. Chinnappan, S. K. Saha, Faseehana.
Kodaikanal Obs. Bull., Vol. 11, p. 87 – 92 (1991). – See Abstr. 012.072 for the main entry.
For successful observations in any telescope, guiding becomes important. On the 2.34 m telescope at the Vainu Bappu Observatory, Kavalur, a remote guiding system is assembled using an intensified CCD camera and a PC based image processing system. It is found that the camera reaches up to 14th magnitude at dark nights. The PC based image processing system eliminates noise in the picture and helps in integration of faint images, contrast enhancement and other related function. Though not ideally suited for speckle observations, the camera is used to record speckles at the Vainu Bappu Telescope (VBT). Requirements for speckle observations are included.

034.099   **Fast photometry using CCD.**
R. Srinivasan, S. Murali Shankar, R. Rajamohan.
Kodaikanal Obs. Bull., Vol. 11, p. 93 – 97 (1991). – See Abstr. 012.072 for the main entry.

034.100   **The spectral characteristics of the No. 2 thin CCD of the Yunnan Observatory.**
Qin Songnia, Gao Cai.
Publ. Yunnan Obs., No. 1, p. 15 – 19 (1992). In Chinese.
The spectral characteristics and measuring methods of the No. 1 and No. 2 thin CCDs of the Yunnan Observatory are given. The measured results show that the No. 2 thin CCD

system has spectral response within the wavelength range of 3300 Å to 11000 Å. The observed results show that the efficiency of the No. 2 CCD is 1.5 or 3 magnitudes higher than that of the photographic dry plate of Kodak IIaO or 103 a–F.

034.101   **Polarization characteristics of the Jensch coelostat.**
M. L. Demidov.
Kinematics Phys. Celest. Bodies, Vol. 7, No. 6, p. 32 – 39 (1991). English translation of Kinematika Fiz. Nebesn. Tel., Tom 7, No. 6, p. 62 – 70 (Nov–Dec 1991).
The polarization characteristics of Jensch coelostats are investigated with the coelostat of the Sayan Observatory service-forecast solar telescope (STOP) as an example. The theory of coelostats of this design is examined briefly, and the basic relationships needed for practical calculation of Stokes–parameter transformations on reflection of light from the mirrors a functions of the Sun's declination and hour angle are presented. Note is taken of the highly important dependence of the instrumental polarization (IP) on the coordinates of the Sun and on the latitude of the coelostat station. At the same time, the IP depends quite weakly on the errors of adjustment of the coelostat. The results of theoretical IP calculations for Jensch coelostats are illustrated with actual observations made at the STOP.

034.102   **Hinweise zur Konstruktion von Präzisions–Sonnenuhren.**
A. Verdun.
Gnomonica Nova, Nr. 1, p. 1 – 20 (Sep 1990).

034.103   **Äquatoriale Rotationshohlkörper–Sonnenuhren mit peripher–integriertem Lichtzeiger–System.**
A. Verdun.
Gnomonica Nova, Nr. 2, p. 1 – 18 (Aug 1991).

034.104   **The birefringent filter for solar fine structure telescope.**
Hua Jiajun, Wang Yanan, Gu Zhenlei, He Fengbao, Liu Guanqun.
Publ. Beijing Astron. Obs., No. 18, p. 89 – 91 (Oct 1991).

034.105   **A high speed photometer in the optical region for lunar occultation studies.**
T. Chandrasekhar, N. M. Ashok, S. Ragland.
J. Astrophys. Astron., Vol. 13, No. 2, p. 195 – 207 (Jun 1992).
High speed photometry during the lunar occultation of a stellar system provides an effective means of achieving high angular resolution in one dimension at the sub arc second level which is well suited for resolving close binary projected separations in the range of 10 – 100 milliarcseconds. An optical fast photometer designed for such a purpose is described and some results from the initial observations taken with the system including the resolution of a projected separation of 55 milliarcsecond in one binary system are detailed.

034.106   **Die Weltkarte als astrolabische Sonnenuhr (4). (Stereographische Transversalprojektionen).**
H. Sigmund.
Schriften des historisch–wissenschaftlichen Fachkreises "Freunde alter Uhren", Band 31.
Deutsche Gesellschaft für Chronometrie. Historisch–wissenschaftlicher Fachkreis. p. 149 – 160 (1992). ISBN 3–923422–09–1.

034.107   **The efficiency of photon–detecting and high efficient photon–counting detecting system.**
Tan Zheng.
Publ. Purple Mt. Obs., Vol. 11, No. 2, p. 111 – 120 (Jun 1992). In Chinese.
The detecting efficiency should be regarded as an important performance of a photon–detecting system. Many photon-detecting systems show obviously lower detecting efficiency than that expected, due to the leakage. In this paper the author has investigated the reason of count leakage, proposed some methods to eliminate it, and furthermore, developed a new high efficient photon–counting system with detecting efficiency over 95%.

034.108   Active control of spectrum drifts in spectrographs.
R. Bhatia, A. Ciani.
Telesc. Naz. Galileo, Tech. Rep., No. 11.
Osservatorio Astronomico di Padova–Asiago, Asiago (Italy).
8 p. (Feb 1992).
Astronomical spectrographs mounted at the telescope suffer from drifts of the spectrum on the detector due to flexure effects, as well as changes in focus due to temperature effects, while those mounted in the Coudé room suffer from high frequency vibrations. The authors propose a system to measure and correct these effects, which improves substantially the efficiency, and reduces the weight and cost of the spectrographs.

034.109   The development of a frequency synthesizer used for the hydrogen maser.
Sheng Jiliang.
Ann. Shanghai Obs., Acad. Sin., No. 13, p. 188 – 192 (1992). In Chinese.

034.110   Improvements to the PDS microdensitometer of the Dominion Astrophysical Observatory.
J. R. Stilburn, P. B. Stetson, W. A. Fisher.
J. R. Astron. Soc. Can., Vol. 86, No. 3, p. 140 – 152 (Jun 1992).
Modifications to improve the photometric response, positional accuracy, resolution, and reliability of a Photometrics Data Systems (PDS) model 1010 microdensitometer are described. The modifications resulted in a greatly improved utility of the machine in the areas of stellar photometry, astrometry, and measurement of stellar spectra.

034.111   Fourier spectrometer – III.
W. Tanaka, T. Okada, Y. Yamashita.
Rep. Natl. Astron. Obs. Jpn., Vol. 1, No. 3, p. 301 – 307 (1992). In Japanese.
It is investigated how to calibrate the wave numbers of spectral lines observed with the Fourier spectrometer located at the coudé focus of the 188 cm telescope of the Okayama Astrophysical Observatory. Also its spectroscopic sensitivity characteristics are examined.

034.112   Gravitational wave detectors.
D. G. Blair, D. E. McClelland, H.–A. Bachor, R. J. Sandeman.
The detection of gravitational waves, p. 43 – 70 (1991). – See Abstr. 003.097 for the main entry.
Contents: 1. Introduction. 2. Resonant–bar antennas. 3. Noise contributions to resonant bars: Brownian motion; Series noise; Back–action noise. 4. Problems and progress with resonant bars: The acoustic–loss problem; The impedance–matching problem; The transducer problem; The quantum–limit problem. 5. Electromagnetic detectors. 6. The Michelson laser interferometer: Fundamental constraints. 7. Michelson interferometer designs: Multi–pass Michelson; The Fabry–Perot Michelson; The locked double Fabry–Perot interferometer. 8. Conclusion.

034.113   Ortsfeste Sonnenuhren im Kanton Freiburg.
F. Mäder.
Bull. Soc. Fribourgeoise Sci. Nat., Vol. 80, Fasc. 1/2, p. 121 – 158 (1991).
The author localised and described a total of 33 sundials with plane faces on the territory of the canton of Fribourg (Switzerland). The oldest historically proven and still existing sundial dates from 1541, the most recent from 1990.

034.114   Resonant–bar detectors.
D. G. Blair.
The detection of gravitational waves, p. 73 – 99 (1991). – See Abstr. 003.097 for the main entry.
Contents: 1. Introduction. 2. Intrinsic noise in resonant–mass antennas. 3. The signal–to–noise ratio. 4. Introduction to transducers. 5. Antenna materials. 6. Antenna suspension and isolation systems: Cable suspension; Magnetic levitation; 4–cables; Four–point suspension; Nodal point suspension; Vibration isolation at room temperature. 7. Excess noise and multiple

antenna correlation. 8. Quantum non–demolition and back–action evasion.

034.115   Gravity wave dewars.
W. O. Hamilton.
The detection of gravitational waves, p. 100 – 115 (1991). – See Abstr. 003.097 for the main entry.
Contents: 1. Introduction. 2. Thermodynamic considerations. 3. Mechanical considerations. 4. Practical aspects: Pump out time; Cooldown time; Recovery from accidents.

034.116   Motion amplifiers and passive transducers.
J.–P. Richard, W. M. Folkner.
The detection of gravitational waves, p. 169 – 185 (1991). – See Abstr. 003.097 for the main entry.
Contents: 1. Introduction. 2. Multi–mode system analysis: Two–mode systems; Three–mode systems; Generalization to n–mode systems. 3. Passive transducers and associated amplifiers: Capacitance transducer coupled to FET; Inductance modulation transducer coupled to a SQUID; Capacitive transducer coupled to a SQUID amplifier. 4. Analysis of multi–mode systems: Signal–to–noise ratio with the optimum filter; Analysis of a three–mode system; Analysis of a five–mode system.

034.117   Parametric transducers.
P. J. Veitch.
The detection of gravitational waves, p. 186 – 225 (1991). – See Abstr. 003.097 for the main entry.
Contents: 1. Introduction. 2. The Manley–Rowe equations. 3. Impedance matrix description. 4. Modification on the antenna's frequency and acoustic quality factor by the transducer. 5. Calculation of the transducer sensitivity and noise characteristics using the equivalent electrical circuit: Transducer sensitivity; Calibration of the transducer; Modification of pump noise by the transducer; Power dissipated in the transducer; Nyquist noise produced by the transducer resonant circuit. 6. Noise analysis: general comments: Description of the phase bridge. 7. Practical implementation of parametric transducers: The UWA transducer; The Tokyo transducer; The Moscow transducer; The LSU transducer. 8. Conclusion.

034.118   Detection of continuous waves.
K. Tsubono.
The detection of gravitational waves, p. 226 – 242 (1991). – See Abstr. 003.097 for the main entry.
Contents: 1. Antenna properties. 2. Frequency tuning. 3. Cold damping. 4. Detector sensitivity.

034.119   A Michelson interferometer using delay lines.
W. Winkler.
The detection of gravitational waves, p. 269 – 305 (1991). – See Abstr. 003.097 for the main entry.
Contents: 1. Principle of measurement. 2. Sensitivity limits. 3. The optical delay line: A laser beam in an optical delay line; Imperfect spherical mirrors; Mirror size; Misalignment and path length variations. 4. Mechanical noise. 5. Thermal mechanical noise. 6. Laser noise and a Michelson interferometer with delay lines: Power fluctuations; Frequency noise; Instabilities in beam geometry. 7. Scattered light: Amplitudes of scattered light interfering with the main beam; Scattered light and spurious interferometer signals. 8. Multi–mirror delay line. 9. Sensitivity of prototype experiments. 10. Conclusion.

034.120   Fabry–Perot gravity–wave detectors.
R. W. P. Drever.
The detection of gravitational waves, p. 306 – 328 (1991). – See Abstr. 003.097 for the main entry.
Contents: 1. Introduction. 2. Principle of basic interferometer. 3. Enhancement of sensitivity by light recycling. 4. Resonant recycling and dual recycling: Resonant recycling; Dual recycling; Resonant recycling interferometers in general. 5. Other techniques for achieving high sensitivity: Use of squeezed light techniques; Use of auxiliary interferometers to reduce seismic noise. 6. Experimental strategies with Fabry–Perot systems: Use

of interferometers of different length; Concurrent operation of interferometers for different purposes; Detector and vacuum system arrangements to facilitate efficient experiments. 7. Some practical issues: Mode cleaners and fibre filters; Beam heating effects in mirrors and other components, and techniques for reducing it. 8. Conclusion.

034.121   The stabilisation of lasers for interferometric gravitational wave detectors.
J. Hough, H. Ward, G. A. Kerr, N. L. Mackenzie, B. J. Meers, G. P. Newton, D. I. Robertson, R. Schilling.
The detection of gravitational waves, p. 329 – 352 (1991). – See Abstr. 003.097 for the main entry.
Contents: 1. Introduction. 2. Laser frequency stability: Delay line systems; Cavity systems; Laser frequency stabilisation; Optical cavity as a frequency discriminator; Transducers for laser frequency control; Feedback amplifying system; Design of the servo system; Typical performance of such a system; Current developments; Future prospects. 3. Laser beam geometry stabilisation: Passive suppression of geometry fluctuations; Active control of beam pointing. 4. Laser intensity stabilisation: Fringe detection process; Radiation pressure effects; Methods of intensity stabilisation. 5. Conclusion.

034.122   Vibration isolation for the test masses in interferometric gravitational wave detectors.
N. A. Robertson.
The detection of gravitational waves, p. 353 – 368 (1991). – See Abstr. 003.097 for the main entry.
Contents: 1. Introduction: Why good broadband seismic isolation is an essential design feature for laser interferometric antennas; The spectrum of seismic noise. 2. Methods of isolation: Passive techniques; Active techniques. 3. Conclusions.

034.123   Advanced techniques: recycling and squeezing.
A. Brillet, J. Gea–Banacloche, G. Leuchs, C. N. Man, J. Y. Vinet.
The detection of gravitational waves, p. 369 – 405 (1991). – See Abstr. 003.097 for the main entry.
Contents: 1. Introduction to recycling. 2. Theory of recycling interferometers: Optics in a weakly modulating medium; Standard recycling; Numerical estimations. 3. Experimental results: Internal modulation; External modulation. 4. Recycling: the current status. 5. Use of squeezed states in interferometric gravitational–wave detectors. 6. The principles of noise reduction using squeezed states. 7. Squeezed states for non–ideal interferometers. 8. Squeezing and light recycling.

034.124   A cloud detector for automated telescopes.
M. C. B. Ashley, J. S. Jurcevic.
Proc. Astron. Soc. Aust., Vol. 9, No. 2, p. 334 – 335 (1991). – See Abstr. 012.090 for the main entry.
An instrument is described that can detect clouds at night–time by sensing their infrared emission. The device can readily detect clouds that are difficult to see with the unaided eye on a moon–lit night. It can be used to provide an indication of how photometric the conditions are, to terminate exposures when cloud forms, and to close the dome when conditions become unsuitable for observing. The detector also has applications as an astronomical site–surveying instrument.

034.125   A 16 channel array infrared prism spectro–polarimeter for astronomical observation.
H. Shiba, H. Takami, S. Sato.
Natl. Astron. Obs. (Jpn.), Repr., No. 143, vp. (1992). In Japanese.
The authors describe the construction of a 16 channel, near infrared spectro–polarimeter capable of simultaneous measurement of polarimetry from 0.9 to 2.5 μm with low spectral resolution and present the polarimetric data obtained with it.

034.126   Low–temperature extragalactic supernova neutrino detector.
L. Stodolsky.
15. Texas Symposium on Relativistic Astrophysics and 4. ESO–CERN Symposium, p. 405 – 412 (1991). – See Abstr. 012.060 for the main entry.

Astrophotography with the Schmidt telescope.
See Abstr. 003.012.

High energy astrophysics. Volume 1. Particles, photons and their detection.
See Abstr. 003.018.

Basics of interferometry.
See Abstr. 003.064.

Infrared technology 16. Proceedings.
See Abstr. 012.012.

Towards Other Planetary Systems (TOPS): A Technology Needs Identification Workshop.
See Abstr. 012.030.

18. Optical Society of India Symposium on Optical Science and Engineering, Bangalore (India), 21 – 23 Mar 1990.
See Abstr. 012.072.

The instrumentation plan for the Very Large Telescope of the European Southern Observatory.
See Abstr. 013.002.

Advances in IR technology at Paris Observatory.
See Abstr. 013.021.

Some important needs in current experiments on air showers and point sources.
See Abstr. 013.045.

Multichannel photometry for amateur astronomy groups.
See Abstr. 013.058.

Extragalactic photometer in the transition era between photoelectric and CCD photometry.
See Abstr. 013.059.

Next generation optical sky surveys.
See Abstr. 013.076.

Through the eyepiece.
See Abstr. 014.021.

The new technologies.
See Abstr. 014.026.

Construzione e impiego di uno spettrografo.
See Abstr. 014.059.

O uso de cameras de video portateis CCD na astronomia.
See Abstr. 014.079.

Gravity wave astronomy.
See Abstr. 022.165.

Design for an all–reflection Michelson interferometer.
See Abstr. 031.009.

Optical fibers in astronomical instruments.
See Abstr. 031.016.

Adaptive optics using curvature sensing.
See Abstr. 031.053.

MARTINI: system operation and astronomical performance.
See Abstr. 031.066.

Automated instrument complex for determining positions of natural and artificial celestial bodies.
See Abstr. 032.009.

An all–latitude automated astrolabe.
See Abstr. 032.010.

Progress report on a five–axis fast guiding secondary for the University of Hawaii 2.2–m telescope.
See Abstr. 032.020.

Solar optical instrumentation.
See Abstr. 032.028.

Panchromatic spectrograph with supporting monochromatic imagers.
See Abstr. 035.027.

A bolometric millimeter–wave system for observations of anisotropy in the cosmic microwave background radiation on medium angular scales.
See Abstr. 035.028.

French activity in infrared astronomy from stratospheric balloons.
See Abstr. 035.049.

Long–wave infrared (LWIR) detectors based on III–V materials.
See Abstr. 035.095.

Solar ultraviolet instrumentation.
See Abstr. 035.123.

Soft X–ray instrumentation.
See Abstr. 035.124.

New trends in ground–based astronomy: fast real time processing needed.
See Abstr. 036.008.

The object–image relationship in Michelson stellar interferometry.
See Abstr. 036.011.

De geschiedenis van de astrofotografie.
See Abstr. 036.016.

Het fotograferen van emissienevels.
See Abstr. 036.017.

Observations with the scanner of the 6–m telescope at Nasmyth–1 focus and automated software for reduction of spectra.
See Abstr. 036.019.

Astrophotographie von Deep–Sky–Objekten mit Bildverstärker.
See Abstr. 036.028.

TVSTAR 2.0 interface software for the STARLIGHT–1 photometer.
See Abstr. 036.034.

Planetary imaging with a small CCD camera.
See Abstr. 036.049.

High–fidelity copying of large–area astronomical plates: principles and some practical results.
See Abstr. 036.064.

Gain calibration of CCD systems at VBO.
See Abstr. 036.137.

Magnetic field measurements.
See Abstr. 036.139.

Wavefront sensing in imaging through the atmosphere: a detector strategy.
See Abstr. 036.145.

Hardware and software aspects of CCD camera–based astrometric plate measurements.
See Abstr. 036.150.

Real–time data processing for SuperCOSMOS.
See Abstr. 036.151.

Developments in compression techniques for COSMOS/Super-COSMOS data.
See Abstr. 036.152.

An automated image measuring system.
See Abstr. 036.157.

Past and future dark matter experiments.
See Abstr. 036.164.

Direct detection of solar neutrinos.
See Abstr. 036.166.

Experimental search of gravitational waves.
See Abstr. 036.180.

Two dimensional photon–counting techniques.
See Abstr. 036.190.

Use of photon counts in the photoelectric astrolabe (Type–II).
See Abstr. 036.197.

Data analysis and algorithms for gravitational wave antennas.
See Abstr. 036.202.

Data processing, analysis and storage for interferometric antennas.
See Abstr. 036.203.

Source location determination for gravitational wave pulses with a network of four earth–based laser–interferometric detectors.
See Abstr. 066.137.

The one–way propagation of light near the surface of the Earth in metric theories of gravity.
See Abstr. 066.145.

A CAMAC–MERA 60 data–acquisition system applied to solar spectra and maps in the He I 10830 Å line.
See Abstr. 071.021.

Night sky brightness from visual observations. II. A visual photometer.
See Abstr. 082.076.

Stellar magnetic–field measurement with a double–beam polarimeter fitted with a photoelastic modulator.
See Abstr. 116.033.

Extensive air shower arrays.
See Abstr. 144.044.

Experiment OMEGA (observation of multiple particle production, exotic interactions and gamma–ray air shower) at Mt. Chacaltaya laboratory (5200 m.a.s.l.).
See Abstr. 144.046.

Experimental search for direct detection of WIMPs.
See Abstr. 144.047.

Search for 100 TeV gamma rays with an air shower array at the South Pole.
See Abstr. 144.048.

Deep Hα survey of the Milky Way. I. Instrument description and detection of a distant H II region.
See Abstr. 155.099.

Emission–line objects in the Large Magellanic Cloud. Observations obtained with the FLAIR system.
See Abstr. 156.033.

035 Space Instrumentation

035.001 Bragg diffraction technique for the concentration of hard x–rays for space astronomy.
P. de Chiara, F. Frontera.
Appl. Opt., Vol. 31, No. 10, p. 1361 – 1369 (1 Apr 1992). Current Physics Microform No.: 9206B1731.
Results of an investigation that was devoted to determining whether the Bragg diffraction technique can be used for the concentration of hard x rays (< keV) are reported. The final goal was to develop a hard x–ray concentrator for space astronomy. General formulas for the reflecting power of the mirrors are given, and criteria for optimizing the integrated reflectivity are derived. An application of these optimization criteria is presented to evaluate the performances that can be expected of a particular configuration of the Bragg concentrator for space astronomy.

035.002 Polarization aberration analysis of the advanced X–ray astrophysics facility telescope assembly.
R. A. Chipman, D. M. Brown, J. P. McGuire.
Appl. Opt., Vol. 31, No. 13, p. 2301 – 2313 (1 May 1992). Current Physics Microform No.: 9207B0133.
The advanced X–ray astrophysics facility (AXAF) telescope consists of six concentric paraboloid–hyperboloid pairs of mirrors that operate near grazing incidence. Because of the substantial polarization effects at large angles of incidence there has been concern regarding the feasibility of doing polarimetry near the telescope focal plane. Polarization aberration functions are used to calculate the effect of instrumental polarization on the transmitted wave front and the polarization state that is due to the primary mirror and the telescope assembly.

035.003 Three–element stressed Ge:Ga photoconductor array for the infrared telescope in space.
N. Hiromoto, T. Itabe, H. Shibai, H. Matsuhara, T. Nakagawa, H. Okuda.
Appl. Opt., Vol. 31, No. 4, p. 460 – 465 (1 Feb 1992). Current Physics Microform No.: 9205E0514.
A stressed Ge:Ga photoconductor array with three elements applied to the Infrared Telescope in Space satellite was fabricated and tested in experiments at 2.0K in very low–photon–influx conditions ($\sim 10^5$ photons/s). Stress was applied to three Ge:Ga detectors in a series by a stable and compact stressing apparatus by using cone–disk springs. The cutoff wavelength was $\sim 180 \, \mu m$. Responsivity was $\sim 100 \, A/W$, and the product of quantum efficiency and photoconductive gain, was ~ 1 with a chopping frequency of 2 Hz. The noise equivalent power was $< 5 \times 10^{-18} W/Hz^{1/2}$ when low–noise transimpedance amplifiers were used.

035.004 Aberration–corrected aspheric grating designs for the Lyman/Far–Ultraviolet Spectroscopic Explorer high–resolution spectrograph: a comparison.
C. Trout, D. Content, P. Davila.
Appl. Opt., Vol. 31, No. 7, p. 943 – 948 (1 Mar 1992). Current Physics Microform No.: 9206B1313.
Two approaches to reducing the optical aberrations of concave diffraction gratings have been studied to obtain candidate grating designs for the Lyman/Far–Ultraviolet Spectroscopic Explorer mission. The first approach involves shaping the grating substrate while using straight and equally spaced grooves. The second approach involves using a gating substrate with a relatively simple figure and holographically controlling the groove curvature and spacing. The authors analyze and compare specific designs derived from both approaches.

035.005 Aberration–corrected aspheric gratings for far–ultraviolet spectrographs: holographic approach.
P. Davila, D. Content, C. Trout.
Appl. Opt., Vol. 31, No. 7, p. 949 – 954 (1 Mar 1992). Current Physics Microform No.: 9206B1319.
Two approaches to reducing the optical aberrations of concave diffraction gratings have been studied: holographically controlling the groove curvature and spacing, and shaping the optical substrate while keeping the grooves straight and equally spaced. The authors develop the theory of ellipsoidal holographic diffraction gratings and apply this theory to Rowland circle spectrographs. They show that ellipsoidal holographic gratings used in second order can yield high spectral resolutions across the spectral band. These gratings may be suitable for use in far–ultraviolet Rowland spectrographs with small bandpasses.

035.006 Characterization of X–ray transmission gratings.
H. Lochbihler, P. Predehl.
Appl. Opt., Vol. 31, No. 7, p. 964 – 971 (1 Mar 1992). Current Physics Microform No.: 9206B1334.
Self–supporting transmission gratings with periods of 1 μm or below are used in combination with grazing–incidence telescopes in celestial X–ray astronomy. They can be produced with sizes up to only a few cm^2; therefore, several hundreds or even thousands of individual elements are needed in order to cover the aperture of a telescope. This large number leads to the problem of characterization of the gratings regarding their X–ray performance. The authors demonstrate that spectrometry in the resonance domain using H polarization is a suitable method for the determination of the grating wire profile and deviations of the grating surface from a plane.

035.007 The ISO instrumentation and expected performances.
T. de Graauw.
11. European Regional Astronomy Meeting of the IAU: New windows to the universe, Vol. 1, p. 497 – 505 (1990). – See Abstr. 012.001 for the main entry.
With the instrumentation of the Infrared Space Observatory, in an advanced development stage, to be launched mid 1993, it is now possible to make a more reliable estimate of the scientific performance of these instruments. In this contribution the author gives a brief outline of the instruments and their expected sensitivities.

035.008 Aftermath of the Hubble disaster and its effect on the optical industry.
R. R. Shannon.
Opt. Photonics News, Vol. 3, No. 2, p. 35 (Feb 1992). Abstract. Current Physics Microform No.: 9202D2228.

035.009   Telescope of tomorrow.
F. Gammie.
Astron. Now, Vol. 6, No. 2, p. 33 – 37 (Feb 1992).
Concerning NASA's Compton Observatory or Gamma–Ray Observatory.

035.010   The INTEGRAL mission.
C. Winkler.
AIP Conf. Proc., No. 232, p. 483 – 488 (1 Aug 1991). Current Physics Microform No.: 9110A1179. – See Abstr. 012.003 for the main entry.
The International Gamma-Ray Astrophysics Laboratory (INTEGRAL) is a candidate for the next ESA medium–size scientific mission to be launched in 2000. INTEGRAL addresses the fine spectroscopy and accurate positioning of celestial, gamma–ray sources in the energy range 15 keV to MeV. A joint ESA–NASA assessment study has been completed recently. The slope of this study covered the mission's scientific prospects; the overall payload concept; payload accommodation and system analysis and an investigation of possible mission scenarios within the framework of an international collaboration.

035.011   The imaging and spectroscopy capabilities of the INTE-
GRAL imager: Monte–Carlo simulation results.
G. Malaguti, A. J. Bird, E. Caroli, A. J. Dean, G. Di Cocco,
B. M. Swinyard, G. E. Villa.
AIP Conf. Proc., No. 232, p. 489 – 491 (1 Aug 1991). Current Physics Microform No.: 9110A1185. – See Abstr. 012.003 for the main entry.
For the gamma–ray satellite INTEGRAL, A 3–D position–sensitive imaging detector has been proposed. The instrument is based on a two layer CsI(Tl) device coupled with a coded aperture mask placed at ~4 m from the detector. This particular configuration of discrete detection elements, together with the 3–D position sensitivity will ensure good imaging capabilities throughout a large energy range from 40 – 50 keV up to ~ 20 – 30 MeV. The imager will also have spectroscopy capability in the nuclear line region. The authors present the first Monte–Carlo simulation results regarding the reconstruction of the shadowgram of the mask pattern. A simulated image of a 511 keV Galactic Centre point source is presented for both the low and the high state.

035.012   SIGMA: A soft gamma–ray imaging telescope in–flight
performances.
P. Mandrou, J. P. Chabaub, E. Ehanno, J. Lande, M. Niel,
J. P. Roques, G. Rouaix, P. Souleille, J. Paul, J. Ballet,
M. Cantin, B. Cordier, A. Goldwurm, P. Laurent, F. Lebrun,
J. P. Leray.
AIP Conf. Proc., No. 232, p. 492 – 494 (1 Aug 1991). Current Physics Microform No.: 9110A1188. – See Abstr. 012.003 for the main entry.
The low energy gamma–ray telescope SIGMA, aboard the GRANAT observatory, is now entering in his second year of life. It has provided during the last year, high angular resolution images of the sky in the energy domain 35 keV – 1.3 MeV. The data accumulated in different background conditions on series of astrophysical sites allow to present the results on in–flight performances of this instrument.

035.013   Calculation of the gamma–ray line sensitivity of the
BATSE large area detectors on GRO.
G. N. Pendleton, W. S. Paciesas, G. J. Fishman, R. B. Wilson,
C. A. Meegan.
AIP Conf. Proc., No. 232, p. 495 – 497 (1 Aug 1991). Current Physics Microform No.: 9110A1191. – See Abstr. 012.003 for the main entry.
The gamma–ray line sensitivity of the BATSE detectors depends primarily upon the characteristics of the detectors' response and the nature of the backgrounds observed by the instruments. Using the example of the detectability of prompt annihilation radiation from nearby novae, the detector response characteristics and data analysis tools relevant to gamma–ray line and continuum sensitivity determination will be presented.

Sensitivities to gamma–ray lines at various energies will be presented.

035.014   MART–LIME: the spectrosocpic hard x–ray imager for
spectrum–X–gamma.
P. Ubertini, A. Bazzano, R. Sunyaev, N. Yamburienco.
AIP Conf. Proc., No. 232, p. 498 – 500 (1 Aug 1991). Current Physics Microform No.: 9110A1194. – See Abstr. 012.003 for the main entry.
A high spatial resolution spectroscopic proportional counter has been developed. The features of this instrument are discussed in some detail. It will provide sub–arcmin spatially resolved images in the energy range 5 – 150 keV with good spectral resolution.

035.015   Advanced Ge detectors for gamma–ray astronomy.
L. S. Varnell.
AIP Conf. Proc., No. 232, p. 501 – 505 (1 Aug 1991). Current Physics Microform No.: 9110A1197. – See Abstr. 012.003 for the main entry.
Future observations of cosmic sources of gamma–ray lines will require instruments with sensitivities of the order of 10^{-6} photons/cm^2–s. To achieve this sensitivity, external background can be reduced by placing the Ge detector array inside a thick active shield, but the internal background produced by beta decay in the detector must be rejected by the Ge detector itself. First results from laboratory measurements are discussed.

035.016   Germanium detector vacuum encapsulation.
N. W. Madden, D. F. Malone, R. H. Pehl, C. P. Cork,
P. N. Luke, D. A. Landis, M. J. Pollard.
AIP Conf. Proc., No. 232, p. 506 – 508 (1 Aug 1991). Current Physics Microform No.: 9110A1202. – See Abstr. 012.003 for the main entry.
The encapsulation of germanium detectors has been a long sought after goal. The authors have begun to develop encapsulation technology that should significantly improve the viability of germanium gamma–ray detectors for a number of important applications. A specialized vacuum chamber has been constructed in which the detector and the encapsulating module are processed in high vacuum. Very high vacuum conductance is achieved within the valveless encapsulating module. The detector module is then sealed without breaking the chamber vacuum. The details of the vacuum chamber, valveless module, processing, and sealing method are presented in the paper.

035.017   Electronic considerations for externally segmented ger-
manium detectors.
N. W. Madden, D. A. Landis, F. S. Goulding, R. H. Pehl,
C. P. Cork, P. N. Luke, D. F. Malone, M. J. Pollard.
AIP Conf. Proc., No. 232, p. 509 – 511 (1 Aug 1991). Current Physics Microform No.: 9110A1205. – See Abstr. 012.003 for the main entry.
The dominant background source for germanium gamma ray detector spectrometers used for some astrophysics observations is internal β decay. Externally segmented germanium gamma ray coaxial detectors can identify β decay by localizing the event. Energetic gamma rays interact in the germanium detector by multiple Compton interactions while β decay is a local process. In order to recognize the difference between gamma rays and β decay events the external electrode (outside of detector) is electrically partitioned. The instrumentation of these external segments and the consequence with respect to the spectrometer energy signal is examined.

035.018   Random charge pulser: A diagnostic tool.
D. A. Landis, N. W. Madden, F. S. Goulding.
AIP Conf. Proc., No. 232, p. 512 – 514 (1 Aug 1991). Current Physics Microform No.: 9110A1208. – See Abstr. 012.003 for the main entry.
Electronic instrumentation used in conjunction with germanium gamma–ray detectors for astrophysics observations must be tested under conditions similar to those encountered during

actual operations. Simulation of cosmic–rays, gamma–ray bursts and solar flares is necessary to validate the performance of the entire spectroscopy chain. A random charge pulser used in conjunction with other diagnostic pulsers, and the techniques developed to verify the performance with and without a germanium gamma–ray detector are described.

035.019   **Optimisation of field of view size in a square element coded aperture imaging system incorporating a circular detector.**
K. Byard.
AIP Conf. Proc., No. 232, p. 515 – 517 (1 Aug 1991). Current Physics Microform No.: 9110A1211. – See Abstr. 012.003 for the main entry.
A new coded aperture detector design approach is discussed, whereby square element coded aperture unit patterns are a mosaiced onto a circular detector with a view to maximising the are occupied by the unit pattern, and hence the size of the field of view. The results for these optimum patterns are presented for all URA and MURA unit patterns of order v < 400.

035.020   **An optimised coded aperture imaging system.**
K. Byard.
Nucl. Instrum. Methods Phys. Res., Sect. A, Vol. 313, No. 1/2, p. 283 – 289 (1 Mar 1992).
A new coded aperture system design for use with circular detectors is discussed, whereby square element coded aperture unit patterns are mosaiced onto a circular detector with a view to maximising the size of the unambiguous field of view for any given detector radius. The results for these optimum configurations are presented for all 50% transparency URA and MURA unit patterns of order v < 600. Improvements in the FOV size over standard rectangular mosaicing are presented with the available increase being found to be around 23% for large value of v.

035.021   **Hubble Space Telescope – first 500 days.**
P. Koubský.
Říše hvězd, Vol. 72, No. 11, p. 201 – 204 (Nov 1991). In Czech.

035.022   **Influence of the solar and terrestrial magnetic field on the flux particle monitor.**
E. Solano.
ESA IUE Newsl., No. 40, p. 19 – 29 (Mar 1992).
Fourier analysis of the centimetric solar radiation and of the daily peak radiation measured by the Flux Particle Monitor on board the IUE has been performed in order to establish relations between the periodic variations of the Solar activity and the incident radiation on the satellite.

035.023   **An EUV imaging spectrograph for high–resolution observations of the solar corona.**
W. M. Neupert, G. L. Epstein, R. J. Thomas, W. T. Thompson.
Sol. Phys., Vol. 137, No. 1, p. 87 – 104 (Jan 1992). With a correction in Vol. 139, No. 1, p. 209 – 210 (May !992).
An extreme ultraviolet (EUV) imaging spectrograph for the wavelength range from 235 to 450 Å has been developed and used for high resolution observations of the Sun. The instrument incorporates a glancing incidence Wolter Type II Telescope and a near–normal incidence toroidal grating spectrograph to achieve near–stigmatic performance over this spectral range. The design of the spectrograph entrance aperture enables both stigmatic spectra with spectral resolution adequate to observe emission line profiles and spectroheliograms of restricted portions of the Sun to be obtained concurrently. The authors describe the design and performance of the instrument and provide an overview of results obtained during a sounding rocket flight on May 5, 1989.

035.024   **Resolution and noise properties of the Goddard High–Resolution Spectrograph.**
R. L. Gilliland, S. L. Morris, R. J. Weymann, D. C. Ebbets, D. J. Lindler.
Publ. Astron. Soc. Pac., Vol. 104, No. 675, p. 367 – 382 (May 1992). Current Physics Microform No.: 9206D2121.
The Goddard High–Resolution Spectrograph (GHRS) on the Hubble Space Telescope (HST) provides an unprecedented capability for ultraviolet spectroscopy of a wide range of astronomical targets. Unfortunately, the presence of spherical aberration in the primary mirror of HST complicates issues of both observing strategy and data analysis procedures, since the throughput and resolving–power losses of the two available GHRS apertures are often offsetting. The authors discuss several characteristics of GHRS observations, and data–analysis procedures, including deconvolution, with a particular emphasis on the detection of weak absorption lines in low–signal–to–noise observations. They find that observations with the GHRS large aperture generally provide greater sensitivity for the detection of weak, isolated spectral features than does use of the small aperture for equal observing times. Analysis of the UV spectrum of 3C 273 serves as the motivation for this study.

035.025   **An analysis of the Hubble Space Telescope fine guidance sensor Fine Lock mode.**
L. G. Taff.
Publ. Astron. Soc. Pac., Vol. 104, No. 676, p. 452 – 466 (Jun 1992). Current Physics Microform No.: 9208B0708.
There are two guiding modes of the Hubble Space Telescope used for the acquisition of astronomical data by one of its six scientific instruments. The more precise one is called Fine Lock. Command and control problems in the on–board electronics, compounded by the aberrations in the main optics, has limited routinely successful Fine Lock to brighter stars, V < 13.0 mag, instead of fulfilling its prelaunch goal of V = 14.5 mag. In this paper the author reports on the only realistic or extensive simulations of the Fine Lock guidance mode. The theoretical analysis underlying the Monte Carlo experiments and the numerical computations show that the control electronics have significant limitations and how to adjust the various control parameters in an attempt to extend Fine Lock guiding performance back to V = 14.0 mag.

035.026   **Possible magnetic experiments on the surface of Mars.**
O. V. Nielsen, T. Johansson, J. M. Knudsen, F. Primdahl.
J. Geophys. Res., Vol. 97, No. E1, p. 1037 – 1044 (25 Jan 1992).
It is proposed that important magnetic properties of the surface materials on Mars be measured by a simple instrument comprising a flux gate magnetometer and a magnetizing coil. The paper describes the basic construction principles and demonstrates the instrument's usability on a few materials that are expected to be similar to those on Mars. Very small traces of ferromagnetic materials can be detected by the measuring technique. In addition to materials studies, the instrument is suitable for magnetospheric, paleomagnetic, and magnetotelluric sounding measurements.

035.027   **Panchromatic spectrograph with supporting monochromatic imagers.**
A. L. Broadfoot, B. R. Sandel, D. Knecht, R. Viereck, E. Murad.
Appl. Opt., Vol. 31, No. 16, p. 3083 – 3096 (1 Jun 1992). Current Physics Microform No.: 9208D1804.
The Arizona Imager/Spectrograph is a set of imaging spectrographs and two–dimensional imagers for space flight. Nine nearly identical spectrographs record wavelengths from 114 to 1090 nm with a resolution of 0.5 – 1.3 nm. The spatial resolution along the slit is electronically selectable and can reach 192 elements. Twelve passband imagers cover wavelengths in the 160 – 900 nm range and have fields of view from 2° to 21°. The fields of view of the spectrographs and imagers are coaligned, and all spectra and images can be exposed simultaneously. A scan platform can rotate the sensor head about two orthogonal axes.

The Arizona imager/spectrograph is designed for investigations of the interaction between the Space Shuttle and its environment. It is scheduled for flight on a Shuttle subsatellite.

035.028   **A bolometric millimeter–wave system for observations of anisotropy in the cosmic microwave background radiation on medium angular scales.**
M. L. Fischer, D. C. Alsop, E. S. Cheng, A. C. Clapp, D. A. Cottingham, J. O. Gundersen, T. C. Koch, E. Kreysa, P. R. Meinhold, A. E. Lange, P. M. Lubin, P. L. Richards, G. F. Smoot.
Astrophys. J., Vol. 388, No. 2, p. 242–252 (1 Apr 1992).
The authors report the performance of a bolometric system designed to measure the anisotropy of the cosmic microwave background (CMB) radiation on angular scales from $0°.3$ to $3°$. The system represents a collaborative effort combining a low-background 1 m diameter balloon–borne telescope with new multimode feed optics, a beam modulation mechanism with high stability, and a four–channel bolometric receiver with passbands centered near frequencies of 3 (90), 6 (180), 9 (270), and 12 (360) cm^{-1} (GHz). The telescope has been flown three times with the bolometric receiver and has demonstrated detector noise limited performance capable of reaching sensitivity levels of $\Delta T/T_{CMB} \approx 10^{-5}$ with detectors operated at T = 0.3K.

035.029   **Compton Observatory data deepen the gamma ray burster mystery.**
B. Schwarzschild.
Phys. Today, Vol. 45, No. 2, p. 21–24 (Feb 1992).

035.030   **The magnetic field investigation on the Ulysses mission: instrumentation and preliminary scientific results.**
A. Balogh, T. J. Beek, R. J. Forsyth, P. C. Hedgecock, R. J. Marquedant, E. J. Smith, D. J. Southwood, B. T. Tsurutani.
Astron. Astrophys., Suppl. Ser., Vol. 92, No. 2, p. 221–236 (Jan 1992). Special issue: Ulysses instruments.
A fundamental feature of the heliosphere is the three dimensional structure of the interplanetary magnetic field. The magnetic field investigation on Ulysses, the first space probe to explore the out–of–ecliptic and polar heliosphere, aims at determining the large scale features and gradients of the field, as well as the heliolatitude dependence of interplanetary phenomena so far only observed near the ecliptic plane. The Ulysses magnetometer uses two sensors, one a Vector Helium Magnetometer, the other a Fluxgate Magnetometer. Onboard data processing yields measurements of the magnetic field vector with a time resolution up to 2 vectors/second and a sensitivity of about 10 pT. Since the switch–on of the instrument in flight on 25 October 1990, a steady stream of observations have been made, indicating that at this phase of the solar cycle the field is generally disturbed: several shock waves and a large number of discontinuities have been observed, as well as several periods with apparently intense wave activity. The paper gives a brief summary of the scientific objectives of the investigation, followed by a detailed description of the instrument and its characteristics. Examples of wave bursts, interplanetary shocks and crossings of the heliospheric current sheet are given to illustrate the observations made with the instrument.

035.031   **The Ulysses solar wind plasma experiment.**
S. J. Bame, D. J. McComas, B. L. Barraclough, J. L. Phillips, K. J. Sofaly, J. C. Chavez, B. E. Goldstein, R. K. Sakurai.
Astron. Astrophys., Suppl. Ser., Vol. 92, No. 2, p. 237–265 (Jan 1992). Special issue: Ulysses instruments.
The Solar Wind Plasma Experiment on Ulysses is accurately characterizing the bulk flow and internal state conditions of the interplanetary plasma in three dimensions on the way out to Jupiter. These observations will continue over the full range of heliocentric distances and heliographic latitudes reached by the probe after its encounter with Jupiter and consequent deflection out of the ecliptic plane. Solar wind electrons and ions are measured simultaneously with independent curved–plate electrostatic analyzers equipped with multiple Channel Electron Multipliers (CEMs). The CEMs are arranged to detect particles at chosen polar angles from the spacecraft spin axis; resolution in spacecraft azimuth is obtained by timing measurements with the spacecraft Sun clock as the spacecraft spins. Electrons with central energies extending from 0.86 eV to 814 eV are detected at seven polar angles and various combinations of azimuth angle to cover the unit sphere comprehensively, so as to enable computation of the pertinent electron velocity distribution parameters. As the average electron flux level changes with heliocentric distance, command control of the CEM counting intervals is used to extend the dynamic range. Ions are detected between 255 eV/q and 34.4 keV/q using appropriate subsets of 16 CEMs at spin angles designed to provide matrices of counts as a function of energy per charge, azimuth angle, and polar angle centered on the average direction of solar–wind flow.

035.032   **The Solar Wind Ion Composition Spectrometer.**
G. Gloeckler, J. Geiss, H. Balsiger, P. Bedini, J. C. Cain, J. Fischer, L. A. Fisk, A. B. Galvin, F. Gliem, D. C. Hamilton, J. V. Hollweg, F. M. Ipavich, R. Joos, S. Livi, R. Lundgren, U. Mall, J. F. McKenzie, K. W. Ogilvie, F. Ottens, W. Rieck, E. O. Tums, R. von Steiger, W. Weiss, B. Wilken.
Astron. Astrophys., Suppl. Ser., Vol. 92, No. 2, p. 267–289 (Jan 1992). Special issue: Ulysses instruments.
The Solar Wind Ion Composition Spectrometer (SWICS) on Ulysses is designed to determine uniquely the elemental and ionic–charge composition, and the temperatures and mean speeds of all major solar–wind ions, from H through Fe, at solar wind speeds ranging from 175 km/s (protons) to 1280 km/s (Fe^{8+}). The instrument, which covers an energy per charge range from 0.16 to 59.6 keV/e in $\sim$13 min, combines an electrostatic analyzer with post–acceleration, followed by a time–of–flight and energy measurement. The measurements made by SWICS will have an impact on many areas of solar and heliospheric physics, in particular providing essential and unique information on: (i) conditions and processes in the region of the corona where the solar wind is accelerated; (ii) the location of the source regions of the solar wind in the corona; (iii) coronal heating processes; (iv) the extent and causes of variations in the composition of the solar atmosphere; (v) plasma processes in the solar wind; (vi) the acceleration of energetic particles in the solar wind; (vii) the thermalization and acceleration of interstellar ions in the solar wind, and their composition; and (viii) the composition, charge states and behavior of the plasma in various regions of the Jovian magnetosphere.

035.033   **The unified radio and plasma wave investigation.**
R. G. Stone, J. L. Bougeret, J. Caldwell, P. Canu, Y. de Conchy, N. Cornilleau–Wehrlin, M. D. Desch, J. Fainberg, K. Goetz, M. L. Goldstein, C. C. Harvey, S. Hoang, R. Howard, M. L. Kaiser, P. J. Kellogg, B. Klein, R. Knoll, A. Lecacheux, D. Lengyel–Frey, R. J. MacDowall, R. Manning, C. A. Meetre, A. Meyer, N. Monge, S. Monson, G. Nicol, M. J. Reiner, J. L. Steinberg, E. Torres, C. de Villeday, F. Wouters, P. Zarka.
Astron. Astrophys., Suppl. Ser., Vol. 92, No. 2, p. 291–316 (Jan 1992). Special issue: Ulysses instruments.
The scientific objectives of the Ulysses Unified Radio and Plasma wave (URAP) experiment are twofold: 1) the determination of the direction, angular size, and polarization of radio sources for remote sensing of the heliosphere and the Jovian magnetosphere and 2) the detailed study of local wave phenomena, which determine the transport coefficients of the ambient plasma. The tracking of solar radio bursts, for example, can provide three dimensional "snapshots" of the large scale magnetic field configuration along which the solar exciter particles propagate. URAP observations of Jovian radio emissions should greatly improve the determination of source locations and consequently our understanding of the generation mechanism(s) of planetary radio emissions. The study of observed wave–particle interactions will improve our understanding of the processes that occur in the solar wind and at Jupiter and of radio

wave generation. A brief discussion of the scientific goals of the experiment is followed by a comprehensive description of the instrument. The URAP sensors consist of a 72.5 m electric field antenna in the spin plane, a 7.5-m electric field monopole along the spin axis and a pair of orthogonal search coil magnetic antennas. The various receivers, designed to encompass specific needs of the investigation, cover the frequency range from DC to 1 MHz. A relaxation sounder provides very accurate electron density measurements. Radio and plasma wave observations are shown to demonstrate the capabilities and limitations of the URAP instruments: radio observations include solar bursts, auroral kilometric radiation, and Jovian bursts; plasma waves include Langmuir waves, ion acoustic–like noise, and whistlers.

035.034   The ULYSSES energetic particle composition experiment EPAC.
E. Keppler, J. B. Blake, D. Hovestadt, A. Korth, J. Quenby, G. Umlauft, J. Woch.
Astron. Astrophys., Suppl. Ser., Vol. 92, No. 2, p. 317 – 331 (Jan 1992). Special issue: Ulysses instruments.

The purpose of this paper is to briefly describe the EPAC investigation aboard the ULYSSES spacecraft. The EPAC sensor will measure the fluxes, angular distributions, energy spectra, and composition of ions in the enrgy range from 300 keV/nucleon to 25 MeV/nucleon. Some measurements are shown obtained after turn–on in interplanetary space.

035.035   The interstellar neutral–gas experiment on ULYSSES.
M. Witte, H. Rosenbauer, E. Keppler, H. Fahr, P. Hemmerich, H. Lauche, A. Loidl, R. Zwick.
Astron. Astrophys., Suppl. Ser., Vol. 92, No. 2, p. 333 – 348 (Jan 1992). Special issue: Ulysses instruments.

The properties (density, bulk velocity relative to the solar system, and temperature) of the local interstellar gas, represented by neutral helium penetrating the heliosphere, will be measured in–situ for the first time by the ULYSSES GAS instrument. By employing the solar gravitational field as a natural velocity analyser, the bulk velocity relative to the solar system and temperature of the gas can be derived from the angular distributions of the particles measured in at least two widely separated points in the heliosphere. The gas density can be determined if a composition corresponding to cosmic abundances is assumed. The neutral particles are detected via the secondary electrons or ions which are emitted upon particle impact from a freshly deposited lithium–fluoride (LiF) layer. The physical principles and assumptions on which the experiment is based, the main technical features of the instrument, and first measurements in space are briefly described.

035.036   Heliosphere Instrument for Spectra, Composition and Anisotropy at Low Energies.
L. J. Lanzerotti, R. E. Gold, K. A. Anderson, T. P. Armstrong, R. P. Lin, S. M. Krimigis, M. Pick, E. C. Roelof, E. T. Sarris, G. M. Simnett, W. E. Frain.
Astron. Astrophys., Suppl. Ser., Vol. 92, No. 2, p. 349 – 363 (Jan 1992). Special issue: Ulysses instruments.

The Heliosphere Instrument for Spectra, Composition, and Anisotropy at Low Energies (HI–SCALE) is designed to make measurements of interplanetary ions and electrons throughout the entire Ulysses mission. The ions ($E_i \gtrsim 50$ keV) and electrons ($E_e \gtrsim 30$ keV) are identified uniquely and detected by five separate solid–state detector telescopes that are oriented to give nearly complete pitch–angle coverage (i.e., coverage of essentially 4π ster) from the spinning spacecraft. Ion elemental abundances are determined by a ΔE vs E telescope using a thin (5 μm) front solid state detector element in a three–element telescope. Experiment operation is controlled by a microprocessor–based data system. Inflight calibration is provided by radioactive sources mounted on telescope covers which can be closed for calibration purposes and for radiation protection during the course of the mission. Ion and electron spectral information is determined using both broad–energy–range rate channels and a 32 channel pulse–height analyser (channels spaced logarithmically) for more detailed spectra. The instrument weighs 5.775 kg and uses 4.0 W

of power. Some initial in–ecliptic measurements are presented which demonstrate the features of the instrument.

035.037   The Ulysses Cosmic Ray and Solar Particle Investigation.
J. A. Simpson, J. D. Anglin, A. Balogh, M. Bercovitch, J. M. Bouman, E. E. Budzinski, J. R. Burrows, R. Carvell, J. J. Connell, R. Ducros, P. Ferrando, J. Firth, M. Garcia–Munoz, J. Henrion, R. J. Hynds, B. Iwers, R. Jacquet, H. Kunow, G. Lentz, R. G. Marsden, R. B. McKibben, R. Mueller–Mellin, D. E. Page, M. Perkins, A. Raviart, T. R. Sanderson, H. Sierks, L. Treguer, A. J. Tuzzolino, K.–P. Wenzel, G. Wibberenz.
Astron. Astrophys., Suppl. 92, No. 2, p. 365 – 399 (Jan 1992). Special issue: Ulysses instruments.

A special high flux telescope provides measurements of protons and heavier particles ~0.2 to ~36 MeV with high azimuthal resolution. The program is called "Cosmic Ray and Solar Particle Investigation" or COSPIN. Examples of the COSPIN scientific goals include: (i) for energetic charge particles of solar origin, to determine the role of coronal magnetic fields in their acceleration and propagation and to search for the origin of the enrichment of ^{3}He and Fe nuclei observed in some solar particle events; (ii) using galactic cosmic radiation measurements, to explore the likely reduction or elimination of solar modulation in polar regions relative to the equator, to search for the origin of the anomalous nuclear component, and to determine the nucleosynthetic origins of nuclei at lowest measurable energies; (iii) for energetic nuclei and electrons of interplanetary origin, to study the three–dimensional character of traveling shocks, CIRs and their associated charged particle acceleration, and; (iv) as a secondary scientific objective at Jupiter encounter, to characterize the energetic charged particle populations during the first traversal of the dusk side of the Jovian magnetosphere and to search for the mechanism producing the ~10 hour "clock" variation of Jovian electrons in the interplanetary medium. The authors give a detailed description of the instruments, their pre-launch performance and a sampling of preliminary results based on the data so far available. The preliminary results confirm that the instruments are functioning properly.

035.038   The solar X–ray/cosmic gamma–ray burst experiment aboard Ulysses.
K. Hurley, M. Sommer, J.–L. Atteia, M. Boer, T. Cline, F. Cotin, J.–C. Henoux, S. Kane, P. Lowes, M. Niel, J. Van Rooijen, G. Vedrenne.
Astron. Astrophys., Suppl. Ser., Vol. 92, No. 2, p. 401 – 410 (Jan 1992). Special issue: Ulysses instruments.

The authors describe the scientific objectives of the Ulysses solar X–ray/cosmic gamma–ray burst experiment, and the unique features of the Ulysses mission which will help to achieve them. After a discussion of the special design constraints imposed by the mission, they describe the sensor systems, consisting of two CsI scintillators and two Si surface barrier detectors covering the energy range 5 keV – 150 keV. The operating modes and inflight performance are also given.

035.039   The Ulysses dust experiment.
E. Grün, H. Fechtig, R. H. Giese, J. Kissel, D. Linkert, D. Maas, J. A. M. McDonnell, G. E. Morfill, G. Schwehm, H. A. Zook.
Astron. Astrophys., Suppl. Ser., Vol. 92, No. 2, p. 411 – 423 (Jan 1992). Special issue: Ulysses instruments.

The Ulysses dust experiment is intended to provide direct observations of dust grains with masses between 10^{-16}g and 10^{-6}g in interplanetary space, to investigate their physical and dynamical properties as functions of heliocentric distance and ecliptic latitude. Of special interest is the question what portion is provided by comets, asteroids and interstellar particles. The investigation is performed with an instrument that measures the mass, speed, flight direction and electric charge of individual dust particles. It is a multicoincidence detector with a mass sensitivity 10^6 times higher than that of previous in–situ experiments which measured dust in the outer solar system. On 27th October 1990 the instrument was switched–on. The instrument was configured to flight conditions and science data collection started immediately. In the period to 13th January 1991 at least 44 dust impacts have been recorded. Flux values are given covering the heliocentric distance range from 1.04 to 1.7 AU.

035.040 The coronal–sounding experiment.
M. K. Bird, S. W. Asmar, J. P. Brenkle, P. Edenhofer, M. Pätzold, H. Volland.
Astron. Astrophys., Suppl. Ser., Vol. 92, No. 2, p. 425 – 430 (Jan 1992). Special issue: Ulysses instruments.

The main science objective of the Ulysses Solar Corona Experiment is to derive the plasma parameters of the solar atmosphere using established coronal–sounding techniques. Applying appropriate model assumptions, the 3–D electron density distribution will be determined from dual–frequency ranging and Doppler measurements recorded at the NASA Deep Space Network during the solar conjunctions. Multi–station observations will be used to derive the plasma bulk velocity at solar distances where the solar wind is expected to undergo its greatest acceleration. As a secondary objective profiling from the favorable geometry during Jupiter encounter, radio–sounding measurements will yield a unique cross–scan of the electron density in the Io Plasma Torus.

035.041 The gravitational wave experiment.
B. Bertotti, R. Ambrosini, S. W. Asmar, J. P. Brenkle, G. Comoretto, G. Giampieri, L. Iess, A. Messeri, H. D. Wahlquist.
Astron. Astrophys., Suppl. Ser., Vol. 92, No. 2, p. 431 – 440 (Jan 1992). Special issue: Ulysses instruments.

Since the optimum size of a gravitational wave detector is the wave length, interplanetary dimensions are needed for the mHz band of interest. Doppler tracking of Ulysses will provide the most sensitive attempt to date at the detection of gravitational waves in the low frequency band. The driving noise source is the fluctuations in the refractive index of interplanetary plasma. This dictates the timing of the experiment to be near solar opposition and sets the target accuracy for the fractional frequency change at 3.0×10^{-14} for integration times of the order of 1000 sec. The instrumentation utilized by the experiment is distributed between the radio systems on the spacecraft and the seven participating ground stations of the Deep Space Network and Medicina. Preliminary analysis is available of the measurements taken during the Ulysses first opposition test.

035.042 A rotating tomographic imager for solar extreme–ultraviolet / soft X–ray emission.
J. M. Davila, W. T. Thompson.
Astrophys. J., Lett., Vol. 389, No. 2, p. L91 – L93 (20 Apr 1992). With plates L10 – L12.

The authors present a concept for a high–resolution EUV/soft X–ray imager that has much in common with the medical imaging procedure of tomography. The resulting instrument is compatible with a simpler, less costly spin–axis – stabilized spacecraft. To demonstrate the fidelity of the reconstruction procedure, the authors simulate the observation and reconstruction process and compare the results with the original image.

035.043 A SPAN MCP detector for the SOHO Coronal Diagnostic Spectrometer.
A. A. Breeveld, M. L. Edgar, A. Smith, J. S. Lapington, P. D. Thomas.
Rev. Sci. Instrum., Vol. 63, No. 1, p. 673 – 676 (Jan 1992). Paper presented at the 4. International Conference on Synchrotron Radiation Instrumentation (SRI–4), Chester (UK), 15–19 Jul 1991.

Solar and Heliospheric Observatory (SOHO) is a European Space Agency project scheduled for launch in 1995. Part of its payload will be a coronal diagnostic spectrometer that includes a grazing incidence spectrometer (GIS) with four detectors set at nonoverlapping wavelength ranges between 15.5 and 78.7 nm. The UV detectors are being developed at MSSL and consist of microchannel plate (MCP) stacks with one–dimensional SPAN

anode readout. The dimensions of the sensitive area of each detector is 50×16 mm and it is required that 47–μm FWHM positional resolution is obtained in the 50 mm direction together with a throughput of 100,000 counts/s (random). The design of the detector system and readout electronics is described. A Monte Carlo PC based simulator has been developed to support the research program and permit an early optimization of various parameters.

035.044 Cooled submillimeter Fourier transform spectrometer flown on a rocket.
H. P. Gush, M. Halpern.
Rev. Sci. Instrum., Vol. 63, No. 6, p. 3249 – 3260 (Jun 1992).

A detailed description is given of a liquid helium cooled Fourier transform spectrometer recently flown on a rocket. The instrument was used to make new precise measurements of the spectrum of the cosmic background radiation in the wave number range 2–30 cm^{-1}, as reported by H. P. Gush, M. Halpern, and E. Wishnow (1990).

035.045 The white–light coronograph for KORONAS–I.
J. Buzaši, L. Klocok, M. Rybanský.
Astron. Astrophys., Vol. 257, No. 2, p. L7 – L8 (Apr 1992). Letter–to–the–editor.

The design of the white–light coronograph for and the scientific goals of the experiment of the "KORONAS–I" satellite are described. The observations in the solar corona will range from 2.2 to 10.0 solar radii, the resolving power is 76″. The observations will begin in September 1992.

035.046 Operation and performance of the OSSE instrument.
R. A. Cameron, J. D. Kurfess, W. N. Johnson, R. L. Kinzer, R. A. Kroeger, M. D. Leising, R. J. Murphy, G. H. Share, M. S. Strickman, J. E. Grove, G. V. Jung, D. A. Grabelsky, S. M. Matz, W. R. Purcell, M. P. Ulmer.
Compton Observatory Science Workshop, p. 3 – 14 (Feb 1992). – See Abstr. 012.018 for the main entry.

The Oriented Scintillation Spectrometer Experiment (OSSE) on the Compton Gamma Ray Observatory is described. An overview of the operation and control of the instrument is given, together with a discussion of typical observing strategies used with OSSE and basic data types produced by the instrument. Some performance measures for the instrument are presented, obtained from pre–launch and in–flight data. These include observing statistics, continuum and line sensitivity, and detector effective area and gain stability.

035.047 The BATSE experiment on the Compton Gamma Ray Observatory: status and some early results.
G. J. Fishman, C. A. Meegan, R. B. Wilson, W. S. Paciesas, G. N. Pendleton.
Compton Observatory Science Workshop, p. 26 – 34 (Feb 1992). – See Abstr. 012.018 for the main entry.

The Burst and Transient Source Experiment (BATSE) on the Compton Gamma Ray Observatory is a sensitive all–sky detector system. It consists of eight uncollimated detectors at the corners of the spacecraft which have a total energy range of 15 eV to 100 Mev. The primary objective of BATSE is the detection, location and study of gamma–ray bursts and other transient sources. The experiment is now in full operation, detecting about one gamma–ray burst per day. A brief description of the on–orbit performance of BATSE is presented, along with examples of early results from some of the gamma–ray bursts observed.

035.048 COMPTEL: instrument description and performance.
J. W. den Herder, H. Aarts, K. Bennett, H. de Boer, M. Busetta, W. Collmar, A. Connors, R. Diehl, W. Hermsen, J. Ryan, M. Kippen, L. Kuiper, G. Lichti, J. Lockwood, J. Macri, M. McConnell, D. Morris, R. Much, V. Schoenfelder, G. Stacy, H. Steinle, A. Strong, B. Swanenburg, B. G. Taylor, M. Varendorff, C. de Vries, C. Winkler.
Compton Observatory Science Workshop, p. 85 – 94 (Feb 1992). – See Abstr. 012.018 for the main entry.

The imaging Compton telescope, COMPTEL, is one of the four gamma–ray detectors onboard of the Compton Gamma–Ray Observatory (GRO). COMPTEL is sensitive to gamma–rays from 800 keV to 30 MeV with a field of view of approximately 1 sr. Its angular resolution ranges between 1° and 2° depending on the energy and incident angle. The energy resolution of better than 10% FWHM enables COMPTEL to provide spectral resolution in the regime of astrophysical nuclear lines. In its telescope mode COMPTEL is able to study a wide variety of objects, pointlike as well as extended in space. In single detector mode COMPTEL uses two of its detectors to study the temporal spectral evolution of strong gamma–ray bursts or transients.

035.049 French activity in infrared astronomy from stratospheric balloons.
J. M. Lamarre, F. Pajot, M. Giard, G. Serra, P. Encrenaz.
SPIE Conference on Infrared Technology XVI as Part of SPIE's International Symposium on Optical and Optoelectronic Applied Science and Engineering, p. 183 – 186 (1990). – See Abstr. 012.012 for the main entry.

The infrared and far–infrared ranges give informations on cold astronomical objects, such as molecular clouds in the interstellar medium. For more than fifteen years, French astronomers have been using stratospheric balloons to carry telescopes and focal plane instruments up to altitudes where the atmosphere is transparent in this wavelength range. The experiment AROME has measured in the diffuse galactic emission the 3.3 μm feature attributed to polycyclic aromatic hydrocarbons (PAHs). The experimental set–up and the results are described. The main balloon project, Pronaos, consists of a 2 m telescope with two focal plane instruments: a submillimeter photometer dedicated to the measurement of very faint sources, and a high resolution heterodyne spectrometer that will measure water vapor and other species not observable from the ground. Many technical developments made for Pronaos will be reusable in future submillimeter satellites.

035.050 Recent developments on ISOCAM long wavelength channel detector.
P. Mottier, C. Lucas, M. Ravetto, P. Agnèse.
SPIE Conference on Infrared Technology XVI as Part of SPIE's International Symposium on Optical and Optoelectronic Applied Science and Engineering, p. 368 – 374 (1990). – See Abstr. 012.012 for the main entry.

Since 1984 the Infrared Laboratory (LIR) has developed an infrared sensor as a part of the Infrared Space Observatory (ISO) focal plane, for sky observation in the 4 – 17 μm wavelength band, where the very low involved photon fluxes imply a low readout noise and liquid helium operating temperature. The flight model sensor will be delivered this year. All electro–optical specifications are already met, and simulations of space environment took place successfully. In this paper last major results are presented.

035.051 Neutron induced background in the COMPTEL detector on the Gamma Ray Observatory.
D. J. Morris, H. Aarts, K. Bennett, M. Busetta, R. Byrd, W. Collmar, A. Connors, R. Diehl, G. Eymann, C. Foster, J. W. den Herder, W. Hermsen, M. Kippen, L. Kuiper, J. Lockwood, J. Macri, M. McConnell, R. Much, J. Ryan, V. Schönfelder, G. Simpson, M. Snelling, G. Stacy, H. Steinle, A. Strong, B. Swanenburg, B. G. Taylor, T. Taddeucci, M. Varendorff, C. de Vries, C. Winkler.
Compton Observatory Science Workshop, p. 102 – 108 (Feb 1992). – See Abstr. 012.018 for the main entry.

Interactions of neutrons in a prototype of the Comptel gamma–ray detector for the Gamma Ray Observatory were studied both to determine Comptel's sensitivity as a neutron telescope and to estimate the gamma–ray background resulting from neutron interactions. The measurements showed that the gamma–ray background from neutron interactions is greater than previously expected. It was thought that most such events would be due to interactions in the upper (D1) detector modules

of Comptel and could be distinguished by pulse–shape discrimination. In order to assess the significance of this background the flux of neutrons in orbit has been estimated based on observed events with neutron pulse–shape signature in D1. The strength of the neutron–induced background is estimated. This is compared with the rate expected from the isotropic cosmic gamma–ray flux.

035.052 The EGRET High Energy Gamma Ray Telescope.
R. C. Hartman, D. L. Bertsch, C. E. Fichtel, S. D. Hunter, G. Kanbach, D. A. Kniffen, P. W. Kwok, Y. C. Lin, J. R. Mattox, H. A. Mayer–Hasselwander, P. F. Michelson, C. von Montigny, P. L. Nolan, K. Pinkau, H. Rothermel, E. Schneid, M. Sommer, P. Sreekumar, D. J. Thompson.
Compton Observatory Science Workshop, p. 116 – 125 (Feb 1992). – See Abstr. 012.018 for the main entry.

The Energetic Gamma Ray Experiment Telescope on the Compton Gamma Ray Observatory is sensitive in the energy range from about 20 MeV to about 30,000 MeV. Electron–positron pair production by incident gamma ray photons is utilized as the detection mechanism. The Compton Observatory was placed in orbit on Apr 6, 1991; EGRET completed activation and on–orbit testing and calibration on May 16, 1991, at which time it began a 15–month full sky survey.

035.053 Status of Hubble Space Telescope astrometry.
W. H. Jefferys.
Bull. Am. Astron. Soc., Vol. 23, No. 3, p. 1257 (1991). Abstract. – See Abstr. 012.065 for the main entry.

035.054 A Fourier–Bessel telescope for hard X–ray astronomy.
D. Cardini, J. M. Poulsen, E. Costa, D. Dal Fiume, A. Emanuele, F. Frontera, A. Basili, T. Franceschini, M. Frutti, G. Landini, S. Silvestri.
Astron. Astrophys., Vol. 257, No. 2, p. 824 – 830 (Apr 1992).

A hard X–ray (20–300 keV) Fourier–Bessel imaging telescope (FOBET) devoted to observations of celestial X–ray sources from stratospheric balloons is described. The imaging capabilities are obtained by the use of 16 rotating modulation collimators, each with different pitch angle, mounted above the field collimators of the large area hard X–ray experiment LAPEX. The detection plane of this experiment is made of 16 Na I(Tl)/Cs I(Na) phoswich detectors. The theory of the Fourier–Bessel transform image is reviewed and the optical properties of the imaging device are determined. Also the effects of the gondola azimuth instability are studied. The expected angular resolution of the telescope is 22 arcmin. Its expected 3σ sensitivity in the 20–200 keV energy band corresponds to 10 mCrab for an on–axis source observed for 10^4s at a residual atmospheric depth of 3 g cm^{-2}.

035.055 Geometrical calibration and assessment of the stability of the Hipparcos payload.
H. Schrijver, H. van der Marel.
Astron. Astrophys., Vol. 258, No. 1, p. 31 – 34 (May 1992).

The various geometrical calibrations of the Hipparcos telescope and detection system are described. It is shown that these calibrations can be executed with the required precision of better than a milliarcsecond. Moreover, it is shown that variations in the parameters describing the calibration results are very slow, yielding the picture of a very stable instrument. With the exception of two short time periods where the thermal control of the instrument was not performing correctly, the basic angle is varying at a rate smaller than 2 milliarcsec/month.

035.056 Geometrical stability and evolution of the Hipparcos telescope.
L. Lindegren, R. S. Le Poole, M. A. C. Perryman, C. Petersen.
Astron. Astrophys., Vol. 258, No. 1, p. 35 – 40 (May 1992).

A simple geometrical model of the Hipparcos telescope is described whereby the relative positions and orientations of critical optical elements can be derived from the field–to–grid transformation determined as part of the normal data processing. The instrument is very stable on time scales of hours, but secular drifts of the order of $0.1 \mu m\, d^{-1}$ are observed and there is evidence for a progressive deformation of the mirrors which may eventually also affect the image quality.

035.057   Science applications of the Mars Observer gamma ray spectrometer.
W. V. Boynton, J. I. Trombka, W. C. Feldman, J. R. Arnold, P. A. J. Englert, A. E. Metzger, R. C. Reedy, S. W. Squyres, H. Wänke, S. H. Bailey, J. Brückner, J. L. Callas, D. M. Drake, P. Duke, L. G. Evans, E. L. Haines, F. C. McCloskey, H. Mills, C. Shinohara, R. Starr.
J. Geophys. Res., Vol. 97, No. E5, p. 7681 – 7698 (25 May 1992).
The Mars Observer gamma ray spectrometer will return data related to the elemental composition of Mars. The instrument has both a gamma ray spectrometer and several neutron detectors. The gamma ray spectrometer will return a spectrum nominally every 20 s from Mars permitting a map of the elemental abundances to be made. The gamma rays are emitted from nuclei involved in radioactive decay, from nuclei formed by capture of a thermal neutron, and from nuclei put in an excited state by a fast-neutron interaction. The gamma rays come from an average depth of the order of a few tens of centimeters. The spectrum will show sharp emission lines whose intensity determines the concentration of the element and whose energy identifies the element. The neutron detectors, using the fact that the orbital velocity of the Mars Observer spacecraft is similar to the velocity of thermal neutrons, determine both the thermal and epithermal neutron flux. These parameters are particularly sensitive to the concentration of hydrogen in the upper meter of the surface. By combining the results from both techniques it is possible to map the depth dependence of hydrogen in the upper meter as well. These data permit a variety of Martian geoscience problems to be addressed including the crust and middle composition, weathering processes, volcanism, and the volatile reservoirs and processes. In addition, the instrument is also sensitive to gamma ray and particle fluxes from non-Martian sources and will be able to address problems of astrophysical interest including gamma ray bursts, the extragalactic background, and solar processes.

035.058   Mars Observer camera.
M. C. Malin, G. E. Danielson, A. P. Ingersoll, H. Masursky, J. Veverka, M. A. Ravine, T. A. Soulanille.
J. Geophys. Res., Vol. 97, No. E5, p. 7699 – 7718 (25 May 1992).
The Mars Observer camera (MOC) is a three-component system (one narrow-angle and two wide-angle cameras) designed to take high spatial resolution pictures of the surface of Mars and to obtain lower spatial resolution, synoptic coverage of the planet's surface and atmosphere. The cameras are based on the "push broom" technique; that is, they do not take "frames" but rather build pictures, one line at a time, as the spacecraft moves around the planet in its orbit. The Mars imaging capabilities of the MOC components are discussed in detail.

035.059   Thermal emission spectrometer experiment: Mars Observer mission.
P. R. Christensen, D. L. Anderson, S. C. Chase, R. N. Clark, H. H. Kieffer, M. C. Malin, J. C. Pearl, J. Carpenter, N. Bandiera, F. G. Brown, S. Silverman.
J. Geophys. Res., Vol. 97, No. E5, p. 7719 – 7734 (25 May 1992).
Thermal infrared spectral measurements will be made of the surface and atmosphere of Mars by the thermal emission spectrometer (TES) on board Mars Observer. By using these observations the composition of the surface rocks, minerals, and condensates will be determined and mapped. In addition, the composition and distribution of atmospheric dust and condensate clouds, together with temperature profiles of the CO_2 atmosphere, will be determined. Broadband solar reflectance and thermal emittance measurements will also be made to determine the energy balance in the polar regions and to map the thermophysical properties of the surface. The instrument consists of three subsections: a Michelson interferometer, a solar reflectance sensor, and a broadband radiance sensor. The properties and scientific goals of TES are discussed in detail.

035.060   Atmosphere and climate studies of Mars using the Mars Observer pressure modulator infrared radiometer.
D. J. McCleese, R. D. Haskins, J. T. Schofield, R. W. Zurek, C. B. Leovy, D. A. Paige, F. W. Taylor.
J. Geophys. Res., Vol. 97, No. E5, p. 7735 – 7757 (25 May 1992).
Studies of the climate and atmosphere of Mars are limited at present by a lack of meteorological data having systematic global coverage with good horizontal and vertical resolution. The Mars Observer spacecraft in a low, nearly circular, polar orbit will provide an excellent platform for acquiring the data needed to advance significantly our understanding of the Martian atmosphere and its remarkable variability. The Mars Observer pressure modulator infrared radiometer (PMIRR) is a nine-channel limb and nadir scanning atmospheric sounder which will observe the atmosphere of Mars globally from 0 to 80 km for a full Martian year. PMIRR employs narrow-band radiometric channels and two pressure modulation cells to measure atmospheric and surface emission in the thermal infrared; a visible channel (0.39–$4.7 \mu m$) is used to measure solar radiation reflected from the atmosphere and surface. Vertical profiles of atmospheric temperature, the infrared extinction of dust suspended in the atmosphere, atmospheric water vapor, and condensate hazes will be retrieved from infrared measurements having a vertical resolution of 5 km, which is half an atmospheric scale height. PMIRR infrared and visible measurements will be combined to determine the radiative balance of the polar regions, where a sizeable fraction of the global atmospheric mass annually condenses onto and sublimes from the surface. Derived meteorological fields, including diabatic heating and cooling and the vertical variation of horizontal winds, will be computed from globally mapped fields retrieved from PMIRR data.

035.061   The Mars Observer laser altimeter investigation.
M. T. Zuber, D. E. Smith, S. C. Solomon, D. O. Muhleman, J. W. Head, J. B. Garvin, J. B. Abshire, J. L. Bufton.
J. Geophys. Res., Vol. 97, No. E5, p. 7781 – 7798 (25 May 1992).
The primary objective of the Mars Observer laser altimeter (MOLA) investigation is to determine globally the topography of Mars at a level suitable for addressing problems in geology and geophysics. Secondary objectives are to characterize the 1064-nm wavelength surface reflectivity of Mars to contribute to analyses of global surface mineralogy and seasonal albedo changes, to assist in addressing problems in atmospheric circulation, and to provide geodetic control and topographic context for the assessment of possible future Mars landing sites. The principal components of MOLA are a diode-pumped, neodymium-doped yttrium aluminum garnet laser transmitter that emits 1064-nm wavelength laser pulses, a 0.5-m-diameter telescope, a silicon avalanche photodiode detector, and a time interval unit with 10-ns resolution. MOLA will provide measurements of the topography of Mars within approximately 160-m footprints and a center-to-center along-track footprint spacing of 300 m along the Mars Observer subspacecraft ground track. The elevation measurements will be quantized with 1.5 m vertical resolution before correction for orbit- and pointing-induced errors. MOLA profiles will be assembled into a global $0.2° \times 0.2°$ grid that will be referenced to Mars's center of mass with an absolute accuracy of approximately 30 m. Other data products will include a global grid of topographic gradients, corrected individual profiles, and a global $0.2° \times 0.2°$ grid of 1064-nm surface reflectivity.

035.062   Mars Observer magnetic fields investigation.
M. H. Acuña, J. E. P. Connerney, P. Wasilewski, R. P. Lin, K. A. Anderson, C. W. Carlson, J. McFadden, D. W. Curtis, H. Réme, A. Cros, J. L. Médale, J. A. Sauvaud, C. d'Uston, S. J. Bauer, P. Cloutier, M. Mayhew, N. F. Ness.
J. Geophys. Res., Vol. 97, No. E5, p. 7799 – 7814 (25 May 1992).
The Mars Observer magnetic fields investigation will provide fast vector measurements of the Martian magnetic field over a wide dynamic range. The fundamental objectives of this investigation are (1) to establish the nature of the magnetic field of Mars, (2) to develop appropriate models for its representation, which take into account the internal sources of magnetism and

the effects of the interaction with the solar wind, and (3) to map the Martian crustal remanent field of a resolution consistent with the Mars Observer orbit altitude and ground track separation. The basic instrumentation complement implemented for this mission is a synergistic combination of a dual, triaxial, flux gate magnetometer system and an electron reflectometer with sensors mounted on a spacecraft boom. The dual magnetometer system allows the real–time estimation and correction of spacecraft–generated fields, while the electron reflectometer provides remote magnetic field sensing capabilities. These instruments have an extensive spaceflight heritage, and similar version of the same have been flown in numerous missions like Voyager, Magsat, International Solar Polar mission, Giotto, Active Magnetospheric Particle Tracer Explorers, and Global Geospace Science. Depending on the telemetry rate supported, a minimum of 2–16 vector samples per second will be acquired. The instrument is microprocessor controlled, can be partially reprogrammed in flight, and supports the packet telemetry protocol implemented for Mars Observer.

035.063   **Characterizing the Hubble Space Telescope using retrieval alogorithms.**
J. R. Fienup.
Opt. Photonics News, Vol. 2, No. 12, p. 41 – 42 (Dec 1991).
Abstract. Current Physics Microform No.: 9202B1047.

035.064   **Mössbauer backscattering spectrometer for mineralogical analysis of the Mars surface.**
G. Klingelhöfer, J. Foh, P. Held, H. Jäger, E. Kankeleit, R. Teucher.
Hyperfine Interact., Vol. 71, No. 1–4, p. 1449 – 1452 (Apr 1992).
Paper presented at the International Conference on the Applications of the Mössbauer Effect (ICAME'91), Nanjing (China), 16 – 20 Sep 1991.
A Mössbauer spectrometer for the mineralogical analysis of the Mars surface is under development. This instrument will be installed on a Mars–Rover, included in the Soviet Union Mars–94/96 Mars mission. Due to power and mass restrictions the electromechanical drive and the electronic components have been extremely miniaturized in comparison to standard systems. Solid state detectors (PIN–diodes) are used for γ– and X–ray detection. The whole spectrometer is controlled by a microprocessor (transputer). An additional application as X–ray fluorescence spectrometer is proposed.

035.065   **The status of the HST.**
S. Jörsäter.
Ir. Astron. J., Vol. 20, No. 3, p. 198 – 200 (Mar 1992). – See Abstr. 012.038 for the main entry.

035.066   **Hubble illuminates the universe.**
S. P. Maran.
Sky Telesc., Vol. 83, No. 6, p. 619 – 625 (Jun 1992).
After two eventful years in orbit, the Hubble Space Telescope has become a cornucopia of astronomical discovery.

035.067   **Radiometric calibration of solar space telescopes – the development of a vacuum–ultraviolet transfer source standard.**
J. Hollandt, M. Kühne, M. C. E. Huber.
ESA Bull., No. 69, p. 78 – 89 (Feb 1992).
The development of a source standard for the vacuum–ultraviolet spectral region will facilitate the laboratory calibration and radiometric intercomparison of the coronal telescopes to be flown on ESA's Solar and Heliospheric Observatory (Soho) spacecraft. Radiometric intercomparison of Soho's instruments on the ground, strict attention to cleanliness, and in–orbit intercomparisons are providing the means for vastly improved solar radiometry.

035.068   **Giotto – the second encounter.**
G. Schwehm, M. G. Grensemann.
ESA Bull., No. 70, p. 54 – 59 (May 1992).
On 4 May 1992 the Giotto spacecraft was successfully reactivated after its second period of hibernation and is currently being prepared for its encounter with Comet P/Grigg–Skjellerup on 10 July 1992. The elements of the scientific payload that survived the earlier encounter with Comet Halley are expected to make a further important contribution to our understanding of cometary phenomena.

035.069   **COBE Differential Microwave Radiometers: calibration techniques.**
C. L. Bennett, G. F. Smoot, M. Janssen, S. Gulkis, A. Kogut, G. Hinshaw, C. Backus, M. G. Hauser, J. C. Mather, L. Rokke, L. Tenorio, R. Weiss, D. T. Wilkinson, E. L. Wright, G. De Amici, N. W. Boggess, E. S. Cheng, P. D. Jackson, P. Keegstra, T. Kelsall, R. Kummerer, C. Lineweaver, S. H. Moseley, T. L. Murdock, J. Santana, R. A. Shafer, R. F. Silverberg.
Astrophys. J., Vol. 391, No. 2, p. 466 – 482 (1 Jun 1992).
The COBE spacecraft was launched 1989 November 18 carrying three scientific instruments into Earth orbit for studies of cosmology. One of these instruments, the Differential Microwave Radiometer (DMR), is designed to measure the large–angular–scale temperature anisotropy of the cosmic microwave background radiation at three frequencies (31.5, 53, and 90 GHz). In this paper the authors present three methods used to calibrate the DMR. First, the signal difference between beam–filling hot and cold targets observed on the ground provides a primary calibration that is transferred to space by noise sources internal to the instrument. Second, the Moon is used in flight as an external calibration source. Third, the signal arising from the Doppler effect due to the Earth's motion around the barycenter of the solar system is used as an external calibration source. Preliminary analysis of the external source calibration techniques confirms the accuracy of the currently more precise ground–based calibration. Assuming the noise source behavior did not change from the ground–based calibration to flight, the authors derive a 0.1% – 0.4% relative and 0.7% – 2.5% absolute calibration uncertainty, depending on the radiometer channel.

035.070   **Electrons, antiprotons and positrons from the 89 MASS flight.**
P. Spillantini.
Vulcano Workshop 1990: Frontier objects in astrophysics and particle physics, p. 383 – 386 (1991). – See Abstr. 012.040 for the main entry.
Performance, first data taking and launch program of the MASS instrument are discussed.

035.071   **Extraterrestrial Mössbauer spectrometry.**
D. G. Agresti, E. L. Wills, T. D. Shelfer, M. M. Pimperl, Shen Minghung, R. V. Morris, B. C. Clark, B. D. Ramsey.
Hyperfine Interact., Vol. 72, No. 1–3, p. 285 – 298 (May 1992).
Paper presented at the International Workshop in honor of Professor Stanley S. Hanna: Nuclear Zeeman effect and recent advances in Mössbauer spectroscopy, Stanford, CA (USA), 15 Mar 1991.
The authors describe a combined backscatter Mössbauer spectrometer and X–ray fluorescence analyzer (BaMS/XRF) instrument suitable for planetary missions to the surfaces of Mars (MESUR Program), the Moon, asteroids, or other solid–solar–system objects. The BaMS/XRF instrument is designed to be capable of concurrent analysis of a sample for its elemental abundances (XRF) and for the mineralogy of its iron–bearing phases (BaMS) without any sample preparation.

035.072 **Measurement of the magnetic field vector from a rotating spacecraft.**
M. K. Trubetskov, E. G. Eroshenko, I. P. Lyannaya,
A. A. Ruzmajkin, D. D. Sokolov, V. A. Styazhkin,
A. M. Shukurov.
Cosmic Res., Vol. 29, No. 4, p. 513 – 518 (Jan 1992). English translation of Kosm. Issled., Tom 29, Vyp. 4, p. 597 – 603 (1991).

A method is developed for reconstructing the magnetic field components measured by a vector magnetometer aboard a spacecraft that rotates at an angular velocity not known beforehand. The magnetic data is used to establish the angular velocity of rotation with high accuracy, whereupon the magnetic field components are found from the initial orientation of the vehicle. The method is used to process magnetic field measurements in the vicinity of Mars from the Phobos–2 spacecraft. The angular velocity is determined to an accuracy of better than 0.1%.

035.073 **The Hopkins Ultraviolet Telescope: performance and calibration during the Astro–1 mission.**
A. F. Davidsen, K. S. Long, S. T. Durrance, W. P. Blair,
C. W. Bowers, S. J. Conard, P. D. Feldman, H. C. Ferguson,
G. H. Fountain, R. A. Kimble, G. A. Kriss, H. W. Moos,
K. A. Potocki.
Astrophys. J., Vol. 392, No. 1, p. 264 – 271 (10 Jun 1992).

The Hopkins Ultraviolet Telescope (HUT) was flown aboard the Space Shuttle Columbia on the Astro–1 mission 1990 December 2 – 11. Spectrophotometric observations of 77 astronomical sources were made throughout the far–ultraviolet (912 – 1850 Å) at a resolution of ~ 3 Å, and, for a small number of sources, in the extreme ultraviolet (415 – 912 Å) beyond the Lyman limit at a resolution of ~ 1.5 Å. The objects observed include quasars, galaxy clusters, active and normal galaxies, cataclysmic variables, globular clusters, supernova remnants, planetary nebulae, white dwarfs, Wolf–Rayet stars, Be stars, cool stars with active coronae, comet Levy (1990c), Jupiter, and Io. HUT has provided the first spectrophotometry in the sub–Lyman–α region for most of these sources. In this paper the authors describe the HUT instrument and its performance in orbit. They also present a HUT observation of the DA white dwarf G191–B2B and derive the photometric and calibration curve for the instrument from a comparison of the observation with a model stellar atmosphere. The sensitivity of the instrument is found to reach a maximum at 1050 Å.

035.074 **The case for liquid mirrors in orbiting telescopes.**
E. F. Borra.
Astrophys. J., Vol. 392, No. 1, p. 375 – 383 (10 Jun 1992).

This paper considers the characteristics of very large (10 – 1000 m diameter) orbiting liquid mirror telescopes (OLMTs) continuously accelerated by the radiation pressure from the Sun on solar sails. OLMTs in halo orbits could be used for pointed observations. Orbit patching would allow access to different regions of the sky. Several practical considerations and issues are discussed.

035.075 **Energetic Particles Investigation (EPI).**
H. M. Fischer, J. D. Mihalov, L. J. Lanzerotti,
G. Wibberenz, K. Rinnert, F. O. Gliem, J. Bach.
Space Sci. Rev., Vol. 60, No. 1–4, p. 79 – 90 (May 1992). Special issue: The Galileo Mission. C. T. Russell (ed.).

The Energetic Particles Investigation (EPI) instrument operates during the pre–entry phase of the Galileo Probe. The major science objective is to study the energetic particle population in the innermost regions of the Jovian magnetosphere – within 4 radii of the cloud tops – and into the upper atmosphere. To achieve these objectives the EPI instrument will make omnidirectional measurements of four different particle species – electrons, protons, alpha–particles, and heavy ions (Z > 2). Intensity profiles with a spatial resolution of about 0.02 Jupiter radii will be recorded. Three different energy range channels are allocated to both electrons and protons to provide a rough estimate of the spectral index of the energy spectra. In addition to the omnidirectional measurements, sectored data will be obtained for certain energy range electrons, protons, and alpha–particles to determine directional anisotropies and particle pitch angle distributions. The design and the properties of the EPI instrument are described in detail.

035.076 **The lightning and Radio Emission Detector (LRD) instrument.**
L. J. Lanzerotti, K. Rinnert, G. Dehmel, F. O. Gliem,
E. P. Krider, M. A. Uman, G. Umlauft, J. Bach.
Space Sci. Rev., Vol. 60, No. 1–4, p. 91 – 109 (May 1992). Special issue: The Galileo Mission. C. T. Russell (ed.).

The Lightning and Radio Emission Detector (LRD) instrument will be carried by the Galileo Probe into Jupiter's atmosphere. The LRD will verify the existence of lightning in the atmosphere and will determine the details of many of its basic characteristics. The instrument, operated in its magnetospheric mode at distances of about 5, 4, 3, and 2 planetary radii from Jupiter's center, will also measure the radio frequency (RF) noise spectrum in Jupiter's magnetosphere. The LRD instrument is composed of a ferrite–core radio frequency antenna (~ 100 Hz to ~ 100 kHz) and two photodiodes mounted behind individual fisheye lenses. The output of the RF antenna is analyzed both separately and in coincidence with the optical signals from the photodiodes. The RF antenna provides data both in the frequency domain (with three narrow–band channels, primarily for deducing the physical properties of distant lightning) and in the time domain with a priority scheme (primarily for determining from individual RF waveforms the physical properties of closeby–lightning).

035.077 **Galileo Probe Mass Spectrometer experiment.**
H. B. Niemann, D. N. Harpold, S. K. Atreya,
G. R. Carignan, D. M. Hunten, T. C. Owen.
Space Sci. Rev., Vol. 60, No. 1–4, p. 111 – 142 (May 1992). Special issue: The Galileo Mission. C. T. Russell (ed.).

The Galileo Probe Mass Spectrometer (GPMS) is a Probe instrument designed to measure the chemical and isotopic composition including vertical variations of the constituents in the atmosphere of Jupiter. The measurement will be performed by in situ sampling of the ambient atmosphere in the pressure range from approximately 150 mbar to 20 bar. In addition batch sampling will be performed for noble gas composition measurement and isotopic ratio determination and for sensitivity enhancement of non–reactive trace gases. The instrument consists of a gas sampling system which is connected to quadrupole mass analyzer for molecular weight analysis. In addition two sample enrichment cells and one noble gas analysis cell are part of the sampling system. The mass range of the quadrupole analyzer is from 2 amu to 150 amu. The maximum dynamic range is 10^8. The detector threshold ranges from 10 ppmv for H_2O to 1 ppbv for Kr and Xe. It is dependent on instrument background and ambient gas composition because of spectral interference. The threshold values are lowered through sample enrichment by a factor of 100 to 500 for stable hydrocarbons and by a factor of 10 for noble gases. The instrument follows a sampling sequence of 8192 steps and a sampling rate of two steps per second. The measurement period lasts appropriately 60 min through the nominal pressure and altitude range.

035.078 **Retrieval of a wind profile from the Galileo Probe telemetry signal.**
J. B. Pollack, D. H. Atkinson, A. Seiff, J. D. Anderson.
Space Sci. Rev., Vol. 60, No. 1–4, p. 143 – 178 (May 1992). Special issue: The Galileo Mission. C. T. Russell (ed.).

Ultrastable oscillators onboard the Galileo Probe and Orbiter will permit very accurate determinations of the frequency of the Probe's telemetry signal as the Probe descends from a pressure level of several hundred mb to a level of about 20 bars. Analysis of the time–varying frequency can provide, in principle, a unique and important definition of the vertical profile of the zonal wind speed in the Jovian atmosphere. In this paper, the authors develop a protocol for retrieving the zonal wind profile from the Doppler shift of the measured frequency; assess the impact of a wide range of error sources on the accuracy of the retrieved wind profile; and perform a number of simulations to illustrate the technique and to assess the likely accuracy of the retrieval.

035.079 Galileo Probe Nephelometer Experiment.
B. Ragent, C. A. Privette, P. Avrin, J. G. Waring, C. E. Carlston, T. C. D. Knight, J. P. Martin.
Space Sci. Rev., Vol. 60, No. 1–4, p. 179 – 201 (May 1992).
Special issue: The Galileo Mission. C. T. Russell (ed.).
The objective of the Nephelometer Experiment aboard the Probe of the Galileo mission is to explore the vertical structure and microphysical properties of the clouds and hazes in the atmosphere of Jupiter along the descent trajectory of the Probe (nominally from 0.1 to >10 bars). The measurements, to be obtained at least every kilometer of the Probe descent, will provide the bases for inferences of mean particle sizes, particle number densities (and hence, opacities, mass densities, and columnar mass loading) and, for non–highly absorbing particles, for distinguishing between solid and liquid particles. These quantities, especially the location of the cloud bases, together with other quantities derived from this and other experiments aboard the Probe, will not only yield strong evidence for the composition of the particles, but, using thermochemical models, for species abundances as well. The measurements in the upper troposphere will provide "ground truth" data for correlation with remote sensing instruments aboard the Galileo Orbiter vehicle. The instrument is carefully designed and calibrated to measure the light scattering properties of the particulate clouds and hazes at scattering angles of 5,8°, 16°, 40°, 70°, and 178°. The measurement sensitivity and accuracy is such that useful estimates of mean particle radii in the range from about 0.2 to 20 μ can be inferred. The instrument will detect the presence of typical cloud particles with radii of about 1.0 μ, or larger, at concentrations of less than 1 cm^3.

035.080 The Galileo Probe Atmosphere Structure Instrument.
A. Seiff, T. C. D. Knight.
Space Sci. Rev., Vol. 60, No. 1–4, p. 203 – 232 (May 1992).
Special issue: The Galileo Mission. C. T. Russell (ed.).
The Galileo Probe Atmosphere Structure Instrument will make in–situ measurements of the temperature and pressure profiles of the atmosphere of Jupiter, starting at about 10^{-10}bar level, when the Probe enters the upper atmosphere at a velocity of 48 km s^{-1}, and continuing through its parachute descent to the 16 bar level. The data should make possible a number of inferences relative to atmospheric and cloud physical processes, cloud location and internal state, and dynamics of the atmosphere. For example, atmospheric stability should be defined, from which the convective or stratified nature of the atmosphere at levels surveyed should be determined and characterized, as well as the presence of turbulence and/or gravity waves. Because this is a rare opportunity, sensors have been selected and evaluated with great care, making use of prior experience at Mars and Venus, but with an eye to special problems which could arise in the Jupiter environment. The temperature sensors are similar to those used on Pioneer Venus; pressure sensors are similar to those used in the Atmosphere Structure Experiment during descent of the Viking Landers (and by the Meteorology Experiment after landing on the surface); the accelerometers are a miniaturized version of the Viking accelerometers. The microprocessor controlled experiment electronics serve multiple functions, including the sequencing of experiment operation in three modes and performing some on–board data processing and data compression.

035.081 Galileo Net Flux Radiometer Experiment.
L. A. Sromovsky, F. A. Best, H. E. Revercomb, J. Hayden.
Space Sci. Rev., Vol. 60, No. 1–4, p. 233 – 262 (May 1992).
Special issue: The Galileo Mission. C. T. Russell (ed.).
The Galileo Net Flux Radiometer (NRF) is a Probe instrument designed to measure the vertical profile of upward and net radiation fluxes in five spectral bands spanning the range from solar to far infrared wavelengths. These unique measurements within Jupiter's atmosphere, from which radiative heating and cooling profiles will be derived, will contribute to our understanding of Jovian atmospheric dynamics, to the detection of cloud layers and determination of their opacities, and to the estimation of water vapor abundance. The NFR uses an array of pyroelectric detectors and individual bandpass filters in a sealed detector package. The detector package and optics rotate as a unit to provide chopping between views of upward and downward radiation fluxes. This arrangement makes possible the measurement of small net fluxes in the presence of large ambient fluxes. A microprocessor–controlled electronics package handles instrument operation.

035.082 The Jupiter Helium Interferometer Experiment on the Galileo entry probe.
U. von Zahn, D. M. Hunten.
Space Sci. Rev., Vol. 60, No. 1–4, p. 263 – 281 (May 1992).
Special issue: The Galileo Mission. C. T. Russell (ed.).
The authors discuss the scientific objective, instrument design, and calibration of a miniaturized Jamin–Mascart interferometer which is to perform an accurate measurement of the refractive index of the Jovian atmosphere in the pressure range 2.5 to 10 bar. The instrument is to perform this measurement in December 1995 aboard the entry probe of the NASA Galileo spacecraft. From the data obtained the mole fraction of helium in the atmosphere of Jupiter is to be calculated with an estimated uncertainty of ± 0.0015. The instrument has a total mass of 1.4 kg and consumes 0.9 W of electrical power.

035.083 The plasma instrumentation for the Galileo Mission.
L. A. Frank, K. L. Ackerson, J. A. Lee, M. R. English, G. L. Pickett.
Space Sci. Rev., Vol. 60, No. 1–4, p. 283 – 307 (May 1992).
Special issue: The Galileo Mission. C. T. Russell (ed.).
The plasma instrumentation (PLS) for the Galileo Mission comprises a nested set of four spherical–plate electrostatic analyzers and three miniature, magnetic mass spectrometers. The three–dimensional velocity distributions of positive ions and electrons, separately, are determined for the energy–per–unit charge (E/Q) range of 0.9 w V to 52 kV. A large fraction of the 4π–steradian solid angle for charged particle velocity vectors is sampled by means of the fan–shaped field–of–view of 160°, multiple sensors, and the rotation of the spacecraft spinning section. The fields–of–view of the three mass spectrometers are respectively directed perpendicular and nearly parallel and anti–parallel to the spin axis of the spacecraft. These mass spectrometers are used to identify the composition of the positive ion plasmas, e.g., H^+, O^+, Na^+, and S^+, in the Jovian magnetosphere. The energy range of these three mass spectrometers is dependent upon the species. The maximum temporal resolutions of the instrument for determining the energy (E/Q) spectra of charged particles and mass (M/Q) composition of positive ion plasmas are 0.5 s. Because the instrument is specifically designed for measurements in the environs of Jupiter with the advantages of previous surveys with the Voyager spacecraft, first determinations of many plasma phenomena can be expected.

035.084 The Galileo heavy element monitor.
T. L. Garrard, N. Gehrels, E. C. Stone.
Space Sci. Rev., Vol. 60, No. 1–4, p. 305 – 315 (May 1992).
Special issue: The Galileo Mission. C. T. Russell (ed.).
The Heavy Ion Counter on the Galileo spacecraft will monitor energetic heavy nuclei of the elements from C to Ni, with energies from ~6 to ~200 MeV nucl^{-1}. The instrument will provide measurements of trapped heavy ions in the Jovian magnetosphere, including those high–energy heavy ions with the potential for affecting the operation of the spacecraft electronic circuitry. The authors describe the instrument, which is a modified version of the Voyager CRS instrument.

035.085 The Galileo Dust Detector.
    E. Grün, H. Fechtig, M. S. Hanner, J. Kissel,
B.-A. Lindblad, D. Linkert, D. Maas, G. E. Morfill,
H. A. Zook.
Space Sci. Rev., Vol. 60, No. 1–4, p. 317 – 340 (May 1992).
Special issue: The Galileo Mission. C. T. Russell (ed.).
    The Galileo Dust Detector is intended to provide direct observations of dust grains with masses between 10^{-19} and 10^{-9}kg in interplanetary space and in the Jovian system, to investigate their physical and dynamical properties as functions of the distances to the Sun, to Jupiter and to its satellites, to study its interaction with the Galilean satellites and the Jovian magnetosphere. Surface phenomena of the satellites (like albedo variations), which might be effects of meteoroid impacts will be compared with the dust environment. Electric charges of particulate matter in the magnetosphere and its consequences will be studied; e.g., the effects of the magnetic field on the trajectories of dust particles and fragmentation of particles due to electrostatic disruption. The investigation is performed with an instrument that measures the mass, speed, flight direction and electric charge of individual dust particles. It is a multicoincidence detector with a mass sensitivity 10^6 times higher than that of previous in–situ experiments which measured dust in the outer solar system. On December 29, 1989 the instrument was switched–on. After the instrument had been configured to flight conditions cruise science data collection started immediately. In the period to May 18, 1990 at least 168 dust impacts have been recorded. For 81 of these dust grains masses and impact speeds have been determined. First flux values are given.

035.086 The Galileo plasma wave investigation.
    D. A. Gurnett, W. S. Kurth, R. R. Shaw, A. Roux,
R. Gendrin, C. F. Kennel, F. L. Scarf, S. D. Shawhan.
Space Sci. Rev., Vol. 60, No. 1–4, p. 341 – 355 (May 1992).
Special issue: The Galileo Mission. C. T. Russell (ed.).
    The purpose of the Galileo plasma wave investigation is to study plasma waves and radio emissions in the magnetosphere of Jupiter. The plasma wave instrument uses an electric dipole antenna to detect electric fields, and two search coil magnetic antennas to detect magnetic fields. The frequency range covered is 5 Hz to 5.6 MHz for electric fields and 5 Hz to 160 kHz for magnetic fields. Low time–resolution survey spectrums are provided by three on–board spectrum analyzers. In the normal mode of operation the frequency resolution is about 10%, and the time resolution for a complete set of electric and magnetic field measurements is 37.33 s. High time–resolution spectrums are provided by a wideband receiver. The wideband receiver provides waveform measurements over bandwidths of 1, 10 and 80 kHz. Compared to previous measurements at Jupiter this instrument has several new capabilities. These new capabilities include (1) both electric and magnetic field measurements to distinguish electrostatic and electromagnetic waves, (2) direction finding measurements to determine source locations, and (3) increased bandwith for the wideband measurements.

035.087 The Galileo magnetic field investigation.
    M. G. Kivelson, K. K. Khurana, J. D. Means,
C. T. Russell, R. C. Snare.
Space Sci. Rev., Vol. 60, No. 1–4, p. 357 – 383 (May 1992).
Special issue: The Galileo Mission. C. T. Russell (ed.).
    The Galileo Orbiter carries a complement of fields and particles instruments designed to provide data needed to shed light on the structure and dynamical variations of the Jovian magnetosphere. Many questions remain regarding the temporal and spatial properties of the magnetospheric magnetic field, how the magnetic field maintains corotation of the embedded plasma and the circumstances under which corotation breaks down, the nature of magnetic perturbations that transport plasma across magnetic shells in different parts of the system, and the electromagnetic properties of the Jovian moons and how they interact with the magnetospheric plasma. Critical to answering these closely related questions are measurements of the dc and low–frequency magnetic field. The Galileo Orbiter carries a

fluxgate magnetometer designed to provide the sensitive measurements required for this purpose. In this paper, the magnetometer is described. The instrument has two–boom–mounted, three–axis sensor assemblies. Flipper mechanisms are included in each sensor assembly for the purpose of offset calibration. The microprocessor controlled data handling system produces calibrated despun data that can be used directly without further processing. A memory system stores data for those periods when the spacecraft telemetry is not active. This memory system can also be used for storing high time–resolution snapshots of data.

035.088 The Galileo Energetic Particles Detector.
    D. J. Williams, R. W. McEntire, S. Jaskulek,
B. Wilken.
Space Sci. Rev., Vol. 60, No. 1–4, p. 385 – 412 (May 1992).
Special issue: The Galileo Mission. C. T. Russell (ed.).
    Amongst its complement of particles and fields instruments, the Galileo spacecraft carries an Energetic Particles Detector (EPD) designed to measure the characteristics of particle populations important in determining the size, shape, and dynamics of the Jovian magnetosphere. To do this the EPD provides 4π angular coverage and spectral measurements for $Z \geqslant 1$ ions from 20 keV to 55 MeV, for electrons from 15 keV to >11 MeV, and for the elemental species helium through iron from approximately $10\,keV\,nucl^{-1}$ to $15\,MeV\,nucl^{-1}$. Two bi–directional telescopes, mounted on a stepping platform, employ magnetic deflection, energy loss versus energy, and time–of–flight techniques to provide 64 rate channels and pulse height analysis of priority selected events. The EPD data system provides a large number of possible operational modes from which a small number will be selected to optimize data collection during the many encounter and cruise phases of the mission. The EPD has demonstrated its operational flexibility throughout the long evolution of the Galileo program by readily accommodating a variety of secondary mission objectives occasioned by the changing mission profile, such as the Venus flyby and the Earth 1 and 2 encounters. To date the EPD performance in flight has been nominal. In this paper the authors describe the instrument and its operation.

035.089 The Galileo Solid–State Imaging experiment.
    M. J. S. Belton, K. P. Klaasen, M. C. Clary,
J. L. Anderson, C. D. Anger, M. H. Carr, C. R. Chapman,
M. E. Davies, R. Greeley, D. Anderson, L. K. Bolef,
T. E. Townsend, R. Greenberg, J. W. Head III, G. Neukum,
C. B. Pilcher, J. Veverka, P. J. Gierasch, F. P. Fanale,
A. P. Ingersoll, H. Masursky, D. Morrison, J. B. Pollack.
Space Sci. Rev., Vol. 60, No. 1–4, p. 413 – 455 (May 1992).
Special issue: The Galileo Mission. C. T. Russell (ed.).
    The Solid State Imaging (SSI) experiment on the Galileo Orbiter spacecraft utilizes a high–resolution (1500 mm focal length) television camera with an 800x800 pixel virtual–phase, charge–coupled detector. It is designed to return images of Jupiter and its satellites that are characterized by a combination of sensitivity levels, spatial resolution, geometric fiedelity, and spectral range unmatched by imaging data obtained previously. The spectral range extends from approximately 375 to 1100 nm and only in the near ultra–violet region (~ 350 nm) is the spectral coverage reduced from previous missions. The camera is approximately 100 times more sensitive than those used in the Voyager mission, and, because of the nature of the satellite encounters, will produce images with approximately 100 times the ground resolution (i.e., $\sim 50\,m\,lp^{-1}$) on the Galilean satellites. The authors describe aspects of the detector including its sensitivity to energetic particle radiation and how the requirements for a large full–well capacity and long–term stability in operating voltages led to the choice of the virtual phase chip. They describe the performance of the system as determined by ground calibration and the improvements that have been made to the telescope to reduce the scattered light reaching the detector. Information "preserving" and "non–preserving" on–board data compression capabilities are outlined. A special "summation" mode, designed for use deep in the Jovian radiation belts, near Io, is also described. The authors discuss the measurement objectives of the

SSI experiment in the Jupiter system and emphasize their relationships to those of other experiments in the Galileo project.

035.090 **Near–Infrared Mapping Spectrometer experiment on Galileo.**
R. W. Carlson, P. R. Weissman, W. D. Smythe, J. C. Mahoney.
Space Sci. Rev., Vol. 60, No. 1–4, p. 457 – 502 (May 1992). Special issue: The Galileo Mission. C. T. Russell (ed.).

The Galileo Near–Infrared Mapping Spectrometer (NIMS) is a combination of imaging and spectroscopic methods. Simultaneous use of these two methods yields a powerful combination, far greater than when used individually. The NIMS experiment will investigate Jupiter and the Galilean satellites during the two year orbital operation period, commencing December 1995. Prior to that, Galileo will have flown past Venus, the Earth/Moon system (twice), and two asteroids; obtaining scientific measurements for all of these objects. The NIMS instrument covers the spectral range 0.7 to 5.2 μ, which includes the reflected–sunlight and thermal–radiation regimes for many solar system objects. This spectral region contains diagnostic spectral signatures, arising from molecular vibrational transitions (and some electronic transitions) of both solid and gaseous species. Imaging is performed by a combination of one–dimensional instrument spatial scanning, coupled with orthogonal spacecraft scanplatform motion, yielding two–dimensional images for each of the NMS wavelenghts. The instrument consists of a telescope, with one dimension of spatial scanning, and a diffraction grating spectrometer. The detectors consist of an array of indium antimonid and silicon photovoltaic diodes, contained within a focal–plane assembly, and cooled to cryogenic temperatures using a radiative cooler. Spectral and spatial scanning is accomplished by electro–mechanical devices, with motions executed using commandable instrument modes. Particular attention was given to the thermal and contamination aspects of the Galileo spacecraft.

035.091 **Galileo ultraviolet spectrometer experiment.**
C. W. Hord, W. E. McClintock, A. I. F. Stewart,
C. A. Barth, L. W. Esposito, G. E. Thomas, B. R. Sandel,
D. M. Hunten, A. L. Broadfoot, D. E. Shemansky, J. M. Ajello,
A. L. Lane, R. A. West.
Space Sci. Rev., Vol. 60, No. 1–4, p. 503 – 530 (May 1992). Special issue: The Galileo Mission. C. T. Russell (ed.).

The Galileo ultraviolet spectrometer experiment uses data obtained by the Ultraviolet Spectrometer (UVS) mounted on the pointed orbiter scan platform and from the Extreme Ultraviolet Spectrometer (EUVS) mounted on the spinning part of the orbiter with the field of view perpendicular to the spin axis. The UVS is a Ebert–Fastie design that covers the range 113–432 nm with a wavelength resolution of 0.7 nm below 190 and 1.3 nm at longer wavelengths. The UVS spatial resolution is 0.4 deg × 0.1 deg for illuminated disc observations and 1 deg × 0.1 deg for limb geometries. The EUVS is a Voyager design objective grating spectrometer, modified to cover the wavelength range from 54 to 128 nm with wavlength resolution 3.5 nm for extended sources and 1.5 nm for point sources and spatial resolution of 0.87 deg × 0.17 deg. The EUVS instrument will follow up on the many Voyager UVS discoveries, particularly the sulfur and oxygen ion emissions in the Io torus and molecular and atomic hydrogen auroral and airglow emissions from Jupiter. The UVS will obtain spectra of emission, absorption, and scattering features in the unexplored, by spacecraft, 170–432 nm wavelength region. The UVS and EUVS instruments will provide a powerful instrument complement to investigate volatile escape and surface composition of the Galilean satellites, the Io plasma torus, micro– and macro–properties of the Jupiter clouds, and the composition structure and evolution of the Jupiter upper atmosphere.

035.092 **Galileo Photopolarimeter/Radiometer experiment.**
E. E. Russell, F. G. Brown, R. A. Chandos,
W. C. Fincher, L. F. Kubel, A. A. Lacis, L. D. Travis.
Space Sci. Rev., Vol. 60, No. 1–4, p. 531 – 563 (May 1992). Special issue: The Galileo Mission. C. T. Russell (ed.).

The Photopolarimeter/Radiometer (PPR) is a remote sensing instrument on the Galileo Orbiter designed to measure the degree of linear polarization and the intensity of reflected sunlight in ten spectral channels between 410 and 945 nm to determine the physical properties of Jovian clouds and aerosols, and to characterize the texture and microstructure of satellite surfaces. The PPR also measures thermal radiation in five spectral bands between 15 and 100 μm to sense the upper tropospheric temperature structure. Two additional channels which measure spectrally integrated solar and solar plus thermal radiation are used to determine the planetary radiation budget components. The PPR photopolarimetric measurements utilize previously flown technology for high–precision polarimetry using a calcite Wollaston prism and two silicon photodiodes to enable simultaneous detection of the two orthogonal polarization components. The PPR radiometry measurements are made with a lithium tantalate pyroelectric detector utilizing a unique arrangement of radiometric stops and a scene/space chopper blade to enable a warm instrument to sense accurately the much colder scene temperatures.

035.093 **Galileo radio science investigations.**
H. T. Howard, V. R. Eshleman, D. P. Hinson,
A. J. Kliore, G. F. Lindal, R. Woo, M. K. Bird, H. Volland,
P. Edenhofer, M. Pätzold, H. Porsche.
Space Sci. Rev., Vol. 60, No. 1–4, p. 565 – 590 (May 1992). Special issue: The Galileo Mission. C. T. Russell (ed.).

The radio science investigations planned for Galileo's 6–year flight to and 2–year orbit of Jupiter use as their instrument the dual–frequency radio system on the spacecraft operating in conjunction with various US and German tracking stations on Earth. The planned radio propagation experiments are based on measurements of absolute and differential propagation time delay, differential phase delay, Doppler shift, signal strength, and polarization. These measurements will be used to study: the atmospheric and ionospheric structure, constituents, and dynamics of Jupiter; the magnetic field of Jupiter; the diameter of Io, its ionospheric structure, and the distribution of plasma in the Io torus; the diameters of the other Galilean satellites, certain properties of their surfaces, and possibly their atmospheres and ionospheres; and the plasma dynamics and magnetic field of the solar corona. The spacecraft system used for these investigations is based on Voyager heritage but with several important additions and modifications that provide linear rather than circular polarization on the S–band downlink signal, the capability to receive X–band uplink signals, and a differential downlink ranging mode. Collaboration between the investigators and the spacecraft communications engineers has resulted in the first highly–stable, dual–frequency, spacecraft radio system suitable for simultaneous measurements of all the parameters normally attributed to radio waves.

035.094 **Materials technology for SIRTF.**
D. Coulter, B. Dolgin, R. Rainen, T. O'Donnell.
SPIE Conference on Infrared Technology XVII – Dedicated to the Memory of Irving J. Spiro as Part of SPIE's International Symposium on Optical Applied Science and Engineering, p. 119 – 126 (1991). – See Abstr. 012.051 for the main entry.

035.095 **Long–wave infrared (LWIR) detectors based on III–V materials.**
J. Maserjian.
SPIE Conference on Infrared Technology XVII – Dedicated to the Memory of Irving J. Spiro as Part of SPIE's International Symposium on Optical Applied Science and Engineering, p. 127 – 134 (1991). – See Abstr. 012.051 for the main entry.

Future NASA missions for Earth observation and planetary science require large photovoltaic detector arrays with high performance in the long wavelength region to 18 μm and at operating temperatures above 65K where single–cycle long–life cryocoolers are being developed. Advanced growth techniques (e.g., MBE and MOCVD) of column III–V semiconductors have opened opportunities for engineering new detector materials and device structures. The technical approaches under investigaton include: quantum well infrared photodetectors, heterojunction internal photoemission photodetectors, type–II strained layer superlattices, and nipi doping superlattices. Each of these options are briefly described with some of their pros and cons.

035.096 Space infrared telescope facility science instruments overview.
M. Bothwell.
SPIE Conference on Infrared Technology XVII – Dedicated to the Memory of Irving J. Spiro as Part of SPIE's International Symposium on Optical Applied Science and Engineering, p. 15 – 26 (1991). – See Abstr. 012.051 for the main entry.

The Space Infrared Telescope Facility (SIRTF) will contain three cryogenically cooled infrared instruments: the Infrared Array Camera, the Infrared Spectrograph, and the Multiband Infrared Photometer for SIRTF. These instruments are sensitive to infrared radiation in the 1.8 – 1,200 micrometer range. This paper discusses the three instruments' functional requirements and their accommodation in the SIRTF telescope system.

035.097 Current instrument status of the Airborne Visible/Infrared Imaging Spectrometer (AVIRIS).
M. L. Eastwood, C. M. Sarture, T. G. Chrien, R. O. Green, W. M. Porter.
SPIE Conference on Infrared Technology XVII – Dedicated to the Memory of Irving J. Spiro as Part of SPIE's International Symposium on Optical Applied Science and Engineering, p. 164 – 175 (1991). – See Abstr. 012.051 for the main entry.

The Airborne Visible/Infrared Imaging Spectrometer (AVIRIS) has been upgraded a number of times since its debut in 1987. This paper describes these improvements and is meant to serve as a reference for scientists working with AVIRIS data.

035.098 Infrared and the search for extrasolar planets.
A. B. Meinel, M. P. Meinel.
SPIE Conference on Infrared Technology XVII – Dedicated to the Memory of Irving J. Spiro as Part of SPIE's International Symposium on Optical Applied Science and Engineering, p. 196 – 201 (1991). – See Abstr. 012.051 for the main entry.

Search for evidence concerning the existence of extrasolar planets will involve both indirect detection as well as direct (imaging). Indirect detection may be possible using ground based instrumentation on the Keck telescope. Imaging probably will require an orbiting system. Characterizing other planets for complex molecules will require a large orbiting or lunar-based telescope or interferometer. Cryogenic infrared techniques appear to be necessary. Planning for a NASA ground and space-based program, Toward Other Planet Systems (TOPS), is proceeding.

035.099 The Pressure Modulator Infrared Radiometer (PMIRR) optical system alignment and performance.
M. P. Crisp, S. A. Macenka.
SPIE Conference on Infrared Technology XVII – Dedicated to the Memory of Irving J. Spiro as Part of SPIE's International Symposium on Optical Applied Science and Engineering, p. 213 – 218 (1991). – See Abstr. 012.051 for the main entry.

The alignment and performance of the optical system for the Pressure Modulator Infrared Radiometer (PMIRR) are described. This limb and nadir scanning instrument will be used for remote sounding of the Martian atmosphere and will be launched on Mars Observer in 1992. The instrument has nine channels distributed over the wavelength range 0.3 to 50 microns and has two pressure modulator cells for water vapor and carbon dioxide.

035.100 Infrared focal plane design for the Comet Rendezvous/Asteroid Flyby and Cassini Visible and Infrared Mapping Spectrometers.
C. Staller, C. Niblack, T. Evans, M. Blessinger, A. Westrick.
SPIE Conference on Infrared Technology XVII – Dedicated to the Memory of Irving J. Spiro as Part of SPIE's International Symposium on Optical Applied Science and Engineering, p. 219 – 230 (1991). – See Abstr. 012.051 for the main entry.

A focal plane assembly combining hybrid electronic components with passive optical components within a single hermetically sealed package has been designed to meet the performance requirements imposed by the Comet Rendezvous/Asteroid Flyby and Cassini Visible and Infrared Mapping Spectrometers. A single line array of 256 InSb photodiodes, accessed by two 1x128 multiplexers, provides continuous spectral coverage from 0.85 to 5.1 μm. Intrinsic field-of-view apertures and a unique order sorting filter require critical optical alignment within the hybrid. FPA performance requirements, design approach, and critical issues are discussed.

035.101 Mission design for the Space Infrared Telescope Facility (SIRTF).
J. H. Kwok, M. G. Osmolovsky.
SPIE Conference on Infrared Technology XVII – Dedicated to the Memory of Irving J. Spiro as Part of SPIE's International Symposium on Optical Applied Science and Engineering, p. 27 – 37 (1991). – See Abstr. 012.051 for the main entry.

The Space Infrared Telescope Facility (SIRTF) is the fourth in NASA's series of Great Observatories. It will feature a one-meter class cryogenically cooled telescope. It is planned for a NASA fiscal start for the development phase in 1994 with a launch in about 2001. The operational orbit will be circular at an altitude of about 100,000 km. The planned mission lifetime is 5 years. This paper addresses the rationale in the selection of the high altitude orbit, the performance of the launch vehicle in delivering the observatory to orbit, other orbit options, and the planned observational modes and capabilities of the observatory. The paper also addresses the viewing geometry and viewing constraints affecting science observation, telescope aperture shade design, and spacecraft solar-panel and communication design.

035.102 Objectives for the Space Infrared Telescope Facility.
R. J. Spehaslki, M. W. Werner.
SPIE Conference on Infrared Technology XVII – Dedicated to the Memory of Irving J. Spiro as Part of SPIE's International Symposium on Optical Applied Science and Engineering, p. 2 – 14 (1991). – See Abstr. 012.051 for the main entry.

SIRTF – the Space Infrared Telescope Facility – is a one-meter-class, liquid-helium-cooled, Earth-orbiting astronomical observatory that will be the infrared component of NASA's family of Great Observatories. SIRTF will investigate numerous scientific areas including formation and evolution of galaxies, stars, and other solar systems; supernovae; phenomena in our own solar system; and, undoubtedly, topics that are outside today's scientific domain. SIRTF's three instruments will permit imaging at all infrared wavelengths from 1.8 to 1200 μm and spectroscopy from 2.5 to 200 μm.

035.103 Ground systems and operations concepts for the Space Infrared Telescope Facility (SIRTF).
R. B. Miller.
SPIE Conference on Infrared Technology XVII – Dedicated to the Memory of Irving J. Spiro as Part of SPIE's International Symposium on Optical Applied Science and Engineering, p. 38 – 46 (1991). – See Abstr. 012.051 for the main entry.

The Space Infrared Telescope Facility (SIRTF) presents a significant challenge in the operational phase of the mission. A guaranteed time program, requests from about 200 guest observer teams per year, and observatory "maintenance" must be integrated reliably. The five-year lifetime due to cryogen boil-off means that the ground system must be fully operational at launch and must operate with an efficiency and timeliness rarely achieved in previous space missions. The operations concepts and implementation strategy for the SIRTF Science and Mission Operations is the topic of this paper.

035.104 SIRTF focal plane technologies.
R. W. Capps, M. Bothwell.
SPIE Conference on Infrared Technology XVII – Dedicated to the Memory of Irving J. Spiro as Part of SPIE's International Symposium on Optical Applied Science and Engineering, p. 47 – 50 (1991). – See Abstr. 012.051 for the main entry.

The Space Infrared Telescope Facility (SIRTF) will have three science instruments, the Infrared Array Camera which will obtain multispectral images between 1.8 μm and 26 μm, the Infrared

Spectrometer which is a set of two dispersive spectrometers covering the wavelength range between 2.5 μm and 200 μm, and the Multiband Imaging Photometer for SIRTF which is a general purpose photometric instrument which operates between 30 μm and 1,200 μm.

035.105 Space Infrared Telescope Facility (SIRTF) telescope overview.
H. Schember, P. Manhart, C. Guiar, J. H. Stevens.
SPIE Conference on Infrared Technology XVII – Dedicated to the Memory of Irving J. Spiro as Part of SPIE's International Symposium on Optical Applied Science and Engineering, p. 51 – 62 (1991). – See Abstr. 012.051 for the main entry.

The Space Infrared Telescope Facility (SIRTF) is a space-borne astronomical observatory which is intended to complete the wavelength coverage of the series of NASA spacecraft designated Great Observatories. This paper discusses a candidate design for the SIRTF telescope, encompassing optics, cryostat and instrument accomodation. The telescope optics employ a baffled Ritchey–Crétien Cassegrain system with a one-meter class primary mirror, and active secondary mirror, and a stationary facetted tertiary mirror. The optics are embedded in a large superfluid Helium cryostat designed to maintain the entire telescope–instrument system at extremely cold temperatures, below 3K in most instances.

035.106 SIRTF stray light analysis.
D. G. Elliott, A. S. Dinger.
SPIE Conference on Infrared Technology XVII – Dedicated to the Memory of Irving J. Spiro as Part of SPIE's International Symposium on Optical Applied Science and Engineering, p. 63 – 67 (1991). – See Abstr. 012.051 for the main entry.

The Space Infrared Telescope Facility (SIRTF) is a 1-meter crogenic infrared telescope. Stray light is kept below the natural background by restrictions on Sun, Earth, and Moon off–axis angles; by conservative baffle design; by the use of advanced diffuse black coatings; and by superfluid helium cooling. The aperture stop is located at the primary mirror rather than at the secondary mirror to increase the aperture and reduce the central obscuration. Stray light from off–axis sources is greater with the aperture stop at the primary than with the aperture stop at the secondary, but the modulation of the signal produced by tilting of the secondary mirror for chopping is less. Stray light from telescope thermal emission is lower with the aperture stop at the primary.

035.107 The Space Infrared Telescope Facility structural design requirements.
P. D. MacNeal, M. C. Lou, Chen Gunshing.
SPIE Conference on Infrared Technology XVII – Dedicated to the Memory of Irving J. Spiro as Part of SPIE's International Symposium on Optical Applied Science and Engineering, p. 68 – 85 (1991). – See Abstr. 012.051 for the main entry.

Structural design requirements have been derived based on the stated mission objectives. To illustrate how the structural design requirements can be met, a point design of the SIRTF flight description of the key features of this point design, along with pertinent modeling and analysis results, are discussed in this paper.

035.108 Space Infrared Telescope Facility cryogenic and optical technology.
P. Mason, T. Kiceniuk, J. Plamondon, W. Petrick.
SPIE Conference on Infrared Technology XVII – Dedicated to the Memory of Irving J. Spiro as Part of SPIE's International Symposium on Optical Applied Science and Engineering, p. 88 – 96 (1991). – See Abstr. 012.051 for the main entry.

SIRTF will require new liquid helium cryogenics and optical technology at liquid helium temperatures to meet the scientific requirements. In particular, it will require a helium cryogenic system operating at 1.25K with a lifetime of 5 years. The optical system will require a 1 meter mirror operating at 2K which is diffraction limited at 3 μm. This paper describes the advance which will be needed and the approaches to be taken.

035.109 Thermal systems analysis for the Space Infrared Telescope Facility dewar.
P. Bhandari, S. W. Petrick, H. Schember.
SPIE Conference on Infrared Technology XVII – Dedicated to the Memory of Irving J. Spiro as Part of SPIE's International Symposium on Optical Applied Science and Engineering, p. 97 – 108 (1991). – See Abstr. 012.051 for the main entry.

The Space Infrared Telescope Facility (SIRTF) is a 1 m class cryogenically cooled observatory for infrared astronomy. It is 100 to 10,000 times more sensitive than the Infrared Astronomical Satellite which completed a successful 10 month mission in 1983. This paper discusses the analysis, thermal design, and predicted performance of the current cryogenic system.

035.110 The space coronograph for "KORONAS–I" satellite.
J. Buzashy, L. Klocok, M. Rybansky.
Pis'ma Astron. Zh., Tom 18, No. 6, p. 537 – 540 (Jun 1992). In Russian. English translation in Sov. Astron. Lett., Vol. 18, No. 3.

The authors describe the white–light coronograph construction and its mission for the satellite "KORONAS–I". The space resolution of the apparatus is 76″, the range in solar corona observations is from 2.2 up to 10.0 $R_\odot$.

035.111 Early results from the Hubble Space Telescope.
E. C. Chaisson.
Sci. Am., Vol. 226, No. 6, p. 18 – 25 (Jun 1992).

035.112 Scintillation gamma–ray spectrometer for determining the element composition of the rocks on Mars from the Phobos spacecraft.
Yu. A. Surkov, L. P. Moskaleva, A. G. Mityugov, V. P. Kharyukova, S. E. Zajtseva, G. G. Smirnov, O. P. Shcheglov, V. L. Gimadov, V. N. Rasputnyj, L. N. Myasnikova, S. S. Bulychev.
Cosmic Res., Vol. 29, No. 6, p. 801 – 810 (May 1992). English translation of Kosm. Issled., Tom 29, Vyp. 6, p. 933 – 943 (1991).

A scintiallation gamma–ray spectrometer designed to determine the element composition of Martian rocks was included in the scientific equipment of the Phobos spacecraft. Approximately 10 background spectra on the Earth to Mars flight path and about 80 spectra of Martian rocks in the equatorial region of Mars were measured during the spacecraft's operation. The gamma–ray spectrometer consisted of a detection assembly based on a CsI(Tl) crystal of 100×100 mm size, a pulse amplitude analyzer, and of an assembly for recording X–ray flashes and hard radiation from solar flares. A set of preflight tests of the equipment was conducted during preparation of the equipment.

035.113 The scientific justification for an FGS II.
L. G. Taff, M. G. Lattanzi.
Bull. Am. Astron. Soc., Vol. 23, No. 3, p. 1257 (1991). Abstract. – See Abstr. 012.065 for the main entry.

035.114 Collimating the Hubble Space Telescope with the Fine Guidance Sensors.
D. Story, E. Nelan, B. McArthur, W. Jefferys, C. Ftaclas, R. Basedow, H. J. Wood.
Bull. Am. Astron. Soc., Vol. 23, No. 3, p. 1269 (1991). Abstract. – See Abstr. 012.067 for the main entry.

035.115 A balloon–borne detector for stratospheric cosmic dust detection.
Wan Gucun, Ouyang Ziyuan, Xu Yiwen, Wu Xiguang.
IAU Colloquium No. 126: Origin and evolution of interplanetary dust, p. 33 – 36 (1991). – See Abstr. 012.068 for the main entry.

This paper introduces a kind of cosmic dust collecting technique and describes in detail the structure of the collector, the balloon–basket system and concerning experimental skill.

035.116   The Munich Dust Counter – a cosmic dust experiment on board of the MUSES–A mission of Japan.
E. Igenbergs, A. Hüdepohl, K. Uesugi, T. Hayashi, H. Svedhem, H. Iglseder, G. Koller, A. Glasmachers, E. Grün, G. Schwehm, H. Mizutani, T. Yamamoto, A. Fujimura, N. Ishii, H. Araki, K. Yamakoshi, K. Nogami.
IAU Colloquium No. 126: Origin and evolution of interplanetary dust, p. 45 – 48 (1991). – See Abstr. 012.068 for the main entry.

The Munich Dust Counter (MDC) is a scientific experiment on board of the MUSES–A mission of Japan. It is an impact ionization detector designed to determine mass and velocity of cosmic dust. A short overview over the MUSES–A mission is given to show the measurement situation of the MDC experiment. The measurement principle of the instrument together with a discussion of the scientific objectives and the design of the experiment is summarized.

035.117   Analysis and testing of a soft actuation system for segmented reflector articulation and isolation.
L. Jandura, M. L. Agronin.
SPIE Conference on Active and Adaptive Optical Systems as Part of SPIE's International Symposium on Optical Applied Science and Engineering, p. 213 – 224 (1991). – See Abstr. 012.074 for the main entry.

Segmented reflectors have been proposed for space–based applications such as optical communication and large–diameter telescopes. An actuation system for mirrors in a space–based segmented mirror array has been developed. The actuation system, called the Articulated Panel Module (APM), articulates a mirror panel in 3 degrees of freedom in the submicron regime, isolates the panel from structural motion, and simplifies space assembly of the mirrors to the reflector backup truss. A breadboard of the APM has been built and is described. Three–axis modeling, analysis, and testing of the breadboard is discussed.

035.118   Need for active structures in future large IR and sub–mm telescopes.
D. Rapp.
SPIE Conference on Active and Adaptive Optical Systems as Part of SPIE's International Symposium on Optical Applied Science and Engineering, p. 328 – 358 (1991). – See Abstr. 012.074 for the main entry.

This paper discusses the potential of active structures technology for future IR and sub–mm telescopes deployed in space. There is a need for a space IR telescope with an aperture of at least 8 meters, which operates at $\sim 50K$ or lower, with a surface figure precision better than 0.5 microns. Active structures technology can produce lightweight structures with superior stability and precision through the use of feedback control on actuation and sensing devices embedded with the structure itself. Active structures will find application in truss support structures and in deformable reflector panels for IR and sub–mm telescopes.

035.119   A deformable mirror concept for adaptice optics in space.
C. P. Kuo.
SPIE Conference on Active and Adaptive Optical Systems as Part of SPIE's International Symposium on Optical Applied Science and Engineering, p. 420 – 431 (1991). – See Abstr. 012.074 for the main entry.

A concept for correcting long wave low order distortions of a lightweight composite mirror for space applications is described. One of the attractive features of this concept for space applications is that a backup structure is not required. The actuation system consists of piezoelectric elements, attached directly to the back of the mirror surface. The system is self balancing. This paper describes the test results for a one–half meter curved composite reflector.

035.120   Balloon–borne solar occultation Fourier transform spectrometry for measurements of stratospheric trace species.
C. Camy–Peyret, J. M. Flaud, A. Perrin.
10. ESA Symposium on European Rocket and Balloon Programmes and Related Research, p. 179 – 186 (Nov 1991). – See Abstr. 012.088 for the main entry.

The LPMA (Limb Profile Monitor of the Atmosphere) instrument is a balloon–borne Fourier transform interferometer based on a commercial BOMEN DA2 design. Its present operating capabilities cover the range 2.5 μm to 14 μm using liquid nitrogen cooled HgCdTe or InSb detectors. The maximum optical path difference achievable by the moving mirror is 50 cm leading to a theoretical apodized resolution of $0.020 \, cm^{-1}$. Technical details on the different subunits of the payload are presented together with their general accomodation into a medium weight gondola (~ 400 kg). Scientific results of the last successful flights are discussed.

035.121   Remote sensing of trace gases with a balloon borne version of the Michelson interferometer for passive atmospheric sounding (MIPAS).
H. Oelhaf, T. v. Clarmann, H. Fischer, F. Friedl–Vallon, C. Fritzschi, C. Piesch, M. Seefeldner, F. Fergg, D. Rabus, W. Völker.
10. ESA Symposium on European Rocket and Balloon Programmes and Related Research, p. 207 – 213 (Nov 1991). – See Abstr. 012.088 for the main entry.

A novel cryogenic Fourier transform spectrometer (FTS) has been developed for limb emission measurements in the mid IR–region from balloon–borne platforms. The FTS is a rapid scanning interferometer using a modified Michelson arrangement achieving a spectral resolution of $0.04 \, cm^{-1}$. Two balloon flights were undertaken up to now from Aire sur L'Adour. During the first one, in May 1989, the spectral region $690–960 \, cm^{-1}$ was covered, the second flight took place in May 1990 with an additional filter region from 1160 to $1400 \, cm^{-1}$. Limb emission spectra were collected from 33 km and 38 km floating altitudes covering tangent heights between the middle troposphere and the float altitude. The trace gases CO_2, H_2O, O_3, CH_4, N_2O, HNO_3, N_2O_5, $ClONO_2$, CF_2Cl_2, $CFCl_3$, CHF_2Cl, CCl_4 and C_2H_6 have been identified up to now in the measured spectra. These gase play an important role on the ozone chemistry and/or climate system.

035.122   A light UV–visible spectrometer for atmospheric composition measurements by solar occultation.
J. P. Pommereau, J. Piquard, F. Goutail.
10. ESA Symposium on European Rocket and Balloon Programmes and Related Research, p. 215 – 218 (Nov 1991). – See Abstr. 012.088 for the main entry.

A light UV–Visible spectrometer for investigating the chemistry of the atmosphere by solar occultation at sunset and sunrise, has been designed. The microprocessor controlled instrument is able to perform measurements automatically without remote control and to transmit the compressed spectrometric data even with a low rate telemetry required for long distance transmission. The spectrometer will be used 1991 and 1992 on–board simplified and unexpensive balloons for regular soundings of the stratosphere and the upper troposphere at midlatitudes, and for the study of perturbed heterogeneous chemistry in the winter arctic stratosphere. The same instruments will be flown later in 1993 on–board long duration Infra Red Montgolfier balloons for investigating the troposphere stratosphere exchange in the tropical regions around the Inter–Tropical Convergence Zone.

035.123   Solar ultraviolet instrumentation.
J. B. Gurman.
NATO Advanced Study Institute on the Sun: a Laboratory for Astrophysics, p. 395 – 410 (1992). – See Abstr. 012.091 for the main entry.

The ultraviolet is the only waveband in which the entire outer solar atmosphere, from chromosphere to corona, can be accessed. After examining the types of ultraviolet observations necessary to determine morphology and the state variables of the atmospheric plasma, the author reviews briefly some of the solar ultraviolet instrumentation of the last two decades. Among recent developments of note are changing detector technologies, multilayer coatings for high–resolution imaging at short wavelengths, and "solar–blind" optics.

035.124   Soft X–ray instrumentation.
A. H. Gabriel.
NATO Advanced Study Institute on the Sun: a Laboratory for Astrophysics, p. 423 – 434 (1992). – See Abstr. 012.091 for the main entry.
The components of an X–ray spectroscopy configuration are identified by their function. Many physical components fulfil more than one of these functions. The author reviews briefly the principles underlying a number of X–ray components, with emphasis on those of use in the soft X–ray region.

035.125   X–ray instrumentation.
G. J. Hurford.
NATO Advanced Study Institute on the Sun: a Laboratory for Astrophysics, p. 435 – 445 (1992). – See Abstr. 012.091 for the main entry.
The observational characteristics of solar hard X–ray and gamma–ray emission above 10 keV are reviewed to identify the sensitivity, spectral and spatial resolution desired for the corresponding X–ray instrumentation. The operation principles of non–imaging instrumentation are reviewed and their limitations discussed. Techniques for achieving high–sensitivity, high–spatial resolution hard X–ray images are outlined with emphasis on Fourier transform imaging.

035.126   Contamination of terrestrial EUV observations by energetic particles.
G. Lay, H. U. Nass, H. J. Fahr.
10. ESA Symposium on European Rocket and Balloon Programmes and Related Research, p. 339 – 342 (Nov 1991). – See Abstr. 012.088 for the main entry.
On September 3, 1988 at 14.10 UT the payload INTERZODIAK II was carried on board a SKYLARK 12 from Natal, Brazil, to an apogee of 857 km. This mission was aimed at the observation of interplanetary and geocoronal EUV resonance radiation, especially at 58.4 nm (He) and 121.6 nm (H). During the 16 minute flight an unexpectedly high background signal was registered. A first analysis already showed that this signal could not be explained as resonance radiation. A Japanese mission launched in 1970 from Japan registered a comparably intense signal, however at much higher altitudes. The comparison with these data and with some earlier results obtained at Natal in 1967 allowed to identify the background signal as being due to high energetic particles trapped in the magnetic field of the earth.

035.127   Observation of the solar Lyman–α line.
H. U. Nass, G. Lay, H. J. Fahr.
10. ESA Symposium on European Rocket and Balloon Programmes and Related Research, p. 343 – 347 (Nov 1991). – See Abstr. 012.088 for the main entry.
On August 23, 1989 at 18.05 UT the payload SOLLY was carried on board a BLACK BRANT IX rocket from White Sands Missile Range/USA to an apogee of about 330 km. The experiment SOLLY consists mainly of a hydrogen cell with a channeltron mounted sideways of the main optical axis. By only measuring the scattered Lyman–α photons, it is possible to study in detail the core region of the solar Lyman–α line. The experiment SOLLY was already flown on October 24, 1988 for the first time. The data sets of this flight showed a very high background signal and in order to improve the signal/noise ratio in the next flight, a Lyman–α interference filter was mounted in front of the channeltron detector. The effect of this filter will be presented. Up to now the data sets were evaluated by simulating the photon flux by means of a Monte–Carlo technique, which is a very time consuming computer method. An analytical approximation was evaluated, and a comparison of both methods will be given.

035.128   The ISO instrumentation and expected performances.
T. de Graauw.
Proc. Astron. Soc. Aust., Vol. 9, No. 2, p. 212 – 214 (1991). – See Abstr. 012.090 for the main entry.
In this paper the author describes the instrumentation for the Infrared Space Observatory to be launched in 1993.

035.129   Mars 96 subsurface radar.
Y. Barbin, W. Kofman, M. Elkine, M. Finkelstein, V. Glotov, V. Zolotarev.
International Symposium on Radars and Lidars in Earth and Planetary Sciences, p. 51 – 58 (Dec 1991). – See Abstr. 012.093 for the main entry.
The Mars 96 International Scientific mission is to launch an aerostat that will drift in the martian atmosphere for ten days. The stabilizing element of the aerostat (guiderope) will be dragged on the martian surface every night. A ground penetrating radar will be installed within the guiderope. Its external surface will act as a transmit and receive antenna. A full scale model has been built and tested on different soils and glaciers. Further experiments will be performed to test the full specifications: radar potential and data processing could yield, on planet Mars, a penetrating depth down to 2.5 km with 30 m resolution. In this paper, the authors describe the main technical features of the radar, examine its implementation into the guiderope, and present some experimental results.

035.130   Definition of a L–band SAR for a Mars Rover mission.
A. Lifermann, E. Thouvenot.
International Symposium on Radars and Lidars in Earth and Planetary Sciences, p. 59 – 63 (Dec 1991). – See Abstr. 012.093 for the main entry.
In the frame of the Automatic Planetary Rover Program under study at CNES a radar with surface and subsurface imagery capabilities is proposed both for scientific applications (30–100 m resolution imagery with typical penetration of 10 m) and as additional support to the rover planification through the potential information brought about terrain roughness at relevant scale. The SAR frequency should be selected in the range 500 MHz to 1 GHz. The design of the instrument within the mission constraints is considered in terms of data rate transmission, power and antenna size.

035.131   The Italian involvement in Cassini Radar.
F. Nirchio, B. Pernice, L. Borgarelli, C. Dionisio.
International Symposium on Radars and Lidars in Earth and Planetary Sciences, p. 79 – 81 (Dec 1991). – See Abstr. 012.093 for the main entry.
The paper describes the Radio Frequency Electronic Subsystem of the Cassini Radar, a key instrument in this interplanetary exploration. It will be provided by the Italian Space Agency which contracted its development to Alenia Spazio.

035.132   Laser sensors for planetary research – ESA instrument development.
R. Flatscher, M. Hueber.
International Symposium on Radars and Lidars in Earth and Planetary Sciences, p. 109 – 116 (Dec 1991). – See Abstr. 012.093 for the main entry.
This paper reviews the requirements of exploratory missions to celestial bodies and translates them into the specific laser sensor requirements and constraints. Typical equipment is reviewed and characterized. Possible laser instrumentation concepts are defined. A comparison of the performance of these configurations is provided in parametric form.

035.133   European Space Agency. Lidar Technology Programme.
M. F. Hueber.
International Symposium on Radars and Lidars in Earth and Planetary Sciences, p. 117 – 121 (Dec 1991). – See Abstr. 012.093 for the main entry.
This paper describes the preparatory efforts of the European Space Agency (ESA) in the area of laser remote sensing. Main

emphasis in technology development is placed on the achievement of efficient and powerful laser sources with the required levels of beam quality, reliability, and lifetime. Activities include the development of flash–lamp and diode–pumped Nd:YAG lasers for backscatter lidar, CO_2 lasers for Doppler wind lidar, and rare–earth solid–state laser activities for Doppler wind lidar and DIAL applications. At instrument level, priority is given to the development of an atmospheric backscatter lidar, termed "ATLID", a promising first candidate for early space implementation. The paper describes ESA's preparatory steps towards a space–qualifiable instrument demonstrator of ATLID, and discusses results from related user studies and campaigns.

035.134   Considerations of a solar mass ejection imager in a low–earth orbit.
B. V. Jackson, D. F. Webb, R. C. Altrock, R. Gold.
IAU Colloquium No. 133: Eruptive solar flares, p. 322 – 328 (1992). – See Abstr. 012.096 for the main entry.

The authors are designing an imager capable of observing the Thomson scattering signal from transient, diffuse features in the heliosphere. The imager is expected to trace these features, which include coronal mass ejections, co–rotating structures and shock waves, to elongations greater than 90° from the Sun from a spacecraft in an ≈ 800 km Earth orbit. The predecessor of this instrument was the zodiacal–light photometer experiment on the HELIOS spacecraft which demonstrated the capability of remotely imaging transient heliospheric structures. The HELIOS photometers have shown it possible to image mass ejections, co–rotating structures and the density enhancements behind shock waves. The second–generation imager the authors are designing, would have far higher spatial resolution enabling to make a more complete description of these features from the Sun to 1 AU. In addition, an imager at Earth could allow up to three days warning of the arrival of a scalar mass ejection.

035.135   From Einstein to AXAF.
H. Tananbaum.
Symposium on Imaging X–Ray Astronomy – a Decade of Einstein Observatory Achievements, p. 15 – 37 (1990). – See Abstr. 012.097 for the main entry.
Contents: Introduction. The Einstein Observatory (Stars, supernova remnants, normal galaxies, active galactic nuclei and quasars, hot gas in galaxies and clusters of galaxies), AXAF.

The Einstein Observatory data bases.
See Abstr. 002.156.

The Galileo mission.
See Abstr. 003.061.

The Compton Observatory Science Workshop. Proceedings.
See Abstr. 012.018.

Towards Other Planetary Systems (TOPS): A Technology Needs Identification Workshop.
See Abstr. 012.030.

Proceedings of the 10th ESA Symposium on European Rocket and Balloon Programmes and Related Research.
See Abstr. 012.088.

Radars and lidars in Earth and planetary sciences. Proceedings.
See Abstr. 012.093.

Personal recollections on the origins of the Einstein Observatory.
See Abstr. 013.017.

Ultra–violet in the extreme.
See Abstr. 013.105.

High precision, wide dynamic range WCE wavefront sensor.
See Abstr. 031.045.

Optimal grazing incidence optics and its application to wide–field X–ray imaging.
See Abstr. 031.048.

The MIT multipoint alignment testbed: technology development for optical interferometry.
See Abstr. 031.068.

Optical figure testing of prototype mirrors for JPL's Precision Segmented Reflector (PSR) program.
See Abstr. 031.072.

A calculation of confusion noise due to infrared cirrus.
See Abstr. 036.042.

OSSE spectral analysis techniques.
See Abstr. 036.078.

Data analysis of the COMPTEL instrument on the NASA Gamma Ray Observatory.
See Abstr. 036.082.

High spatial resolution techniques.
See Abstr. 036.138.

Long–baseline optical and infrared stellar interferometry.
See Abstr. 036.189.

Integrity of HRI images.
See Abstr. 036.206.

An overview of the cosmic background explorer (COBE) and its observations: new sky maps of the early universe.
See Abstr. 051.016.

In–orbit performance of the Hipparcos astrometry satellite.
See Abstr. 051.021.

Mars Observer mission.
See Abstr. 051.022.

Marsnet surface and atmosphere investigations.
See Abstr. 051.025.

Space Science Reviews volume on Galileo Mission overview.
See Abstr. 051.031.

ROSAT: early results.
See Abstr. 051.035.

Early results from the Cosmic Background Explorer (COBE).
See Abstr. 051.038.

Future observation of the F–corona with the LASCO coronograph space experiment.
See Abstr. 051.048.

X–ray astronomy missions.
See Abstr. 051.054.

F–corona–experiment: requirements for remote sensing of inter-planetary dust.
See Abstr. 051.062.

Magellan mission description and radar system.
See Abstr. 051.063.

ASTRA: altimetry and sounding of Titan with a radar on a descending craft.
See Abstr. 051.065.

Ulysses: a status report.
See Abstr. 051.069.

High resolution extreme ultraviolet (EUV) studies of the Sun.
See Abstr. 076.020.

Preliminary results of a balloon flight of the solar disk sextant.
See Abstr. 080.042.

Elemental mapping of planetary surfaces using gamma–ray spectroscopy.
See Abstr. 091.024.

Bistatic radar studies of the planets.
See Abstr. 091.062.

Observations of Mars using Hubble Space Telescope.
See Abstr. 097.063.

The problem of the oxygen presence in the heliosphere and observational implications.
See Abstr. 106.091.

Rocket observation of the near–infrared spectrum of the sky.
See Abstr. 133.009.

X–ray astronomy beyond AXAF and XMM.
See Abstr. 142.059.

The Snapshot Survey: a search for gravitationally lensed quasars with the Hubble Space Telescope.
See Abstr. 159.036.

COBE measures anisotropy in cosmic microwave background radiation.
See Abstr. 161.191.

The Cosmic Background Radiation anisotropy program at UCSB.
See Abstr. 161.530.

MAX, a millimeter–wave anisotropy experiment to search for anisotropy in the cosmic background radiation on intermediate angular scales.
See Abstr. 161.531.

036 Methods of Observation and Reduction, Data Processing

036.001 On the relationship between conventional and overlap reduction techniques in positional astronomy.
P. Benevides–Soares, R. Teixeira.
Astron. Astrophys., Vol. 253, No. 1, p. 307 – 310 (Jan 1992).

The authors show that the familiar night–to–night or plate–to–plate reduction of astrometric observations, if iterated after removal of mean star residuals, is equivalent to the more sophisticated overlap reduction. As an example, they consider in detail the reduction of a meridian circle differential program in right ascension. It is shown that the iteration is convergent even in the case, where the Jacobian matrix is rank deficient. Practical means to deal with the indeterminacy which arises from the singularity are outlined.

036.002 The Cramér–Rao lower bound and stellar photometry with aberrated HST images.
P. Jakobsen, P. Greenfield, R. Jedrzejewski.
Astron. Astrophys., Vol. 253, No. 1, p. 329 – 332 (Jan 1992).

The authors apply the Cramér–Rao lower bound theorem of statistics to the problem of extracting two–dimensional stellar photometry from aberrated Hubble Space Telescope camera images. The theorem is used to derive a simple and general expression for the best possible photometric precision that can be derived from a given image – regardless of what image processing or deconvolution techniques are applied to the data. This hard limit, which reflects the underlying photon statistics of the overlapping stellar images, is quite universal in nature and valid for arbitrary point spread function and any degree of crowding. The form of the expression is also such that it can be used to estimate a minimum statistical photometric error directly from the raw data once the positions of the stars in the image are known. As a specific example, the authors use the Cramér–Rao lower bound to quantify the irrecoverable loss in sensitivity introduced by the HST spherical aberration in the case of the Faint Object Camera.

036.003 Contrast of the vibration fringes in time–averaged electronic speckle–pattern interferometry: effect of speckle averaging.
C. Joenathan, B. M. Khorana.
Appl. Opt., Vol. 31, No. 11, p. 1863 – 1870 (10 Apr 1992). Current Physics Microform No.: 9206C0199.

This paper presents a detailed investigation of the effect of speckle averaging in electronic speckle–pattern interferometric fringes. The theory states that the contrast of the resultant smoothed fringes increases as the number of frames is increased. It is shown that the contrast of the fringes is optimized with a limited number of superpositions, and further addition results in the reduction of speckle noise with the contrast remaining almost the same. The contrast of the fringes obtained with π and $\pi/2$ phase–stepping methods with speckle averaging is also discussed. Both theoretical and experimental results are presented in this paper.

036.004 Restoration of images degraded by atmospheric turbulence and detection noise.
Shi Xiaotian, R. K. Ward.
J. Opt. Soc. Am. A, Vol. 9, No. 3, p. 364 – 370 (Mar 1992). Current Physics Microform No.: 9206D1788.

The restoration of images blurred by atmospheric turbulence and contaminated by additive–signal–independent noise is investigated. A series of noisy short–exposure images and the a priori knowledge of the time–averaged autocorrelation function of the optical transfer function are assumed to be known. Three kinds of filter are discussed. These are based on the speckle interferometry technique, the Wiener criterion, and an ad hoc scheme.

036.005 Estimation of binary star parameters by model fitting the bispectrum phase.
A. Glindemann, R. G. Lane, J. C. Dainty.
J. Opt. Soc. Am. A, Vol. 9, No. 4, p. 543 – 548 (Apr 1992). Current Physics Microform No.: 9206D1967.

The analysis of binary stars has to date been one of the major successes of speckle interferometry. A new technique for estimating the parameters of a binary star is presented. Unlike earlier methods, the system does not require the measurement of a reference star to compensate for the speckle transfer function. The algorithm relies on model fitting to the bispectrum phase and can obtain the separation, position angle, and relative brightness of the two components.

036.006 Image–position error associated with a focal plane array.
D. Down.
J. Opt. Soc. Am. A, Vol. 9, No. 5, p. 700 – 707 (May 1992). Current Physics Microform No.: 9207A0820.

Assuming Poisson statistics for object and background image shot noise and detector dark noise, exact expressions are derived for the mean and the standard deviation (called the noise equivalent angle of a noise–subtracted centroid image–position estimator assumed to use the output of a focal plane array. The derived expressions are examined to determine (1) the dependence of the noise equivalent angle on the signal and the noise, (2) the relative performance of a noise–subtracted versus a non-noise–subtracted estimator, and (3) the optimum pixel–to–spot–size ratio.

036.007 Model–independent mapping by optical aperture synthesis: basic principles and computer simulation.
P. Cruzalèbes, G. Schumacher, J. L. Starck.
J. Opt. Soc. Am. A, Vol. 9, No. 5, p. 708 – 724 (May 1992). Current Physics Microform No.: 9207A0828.
   The basic concepts of optical interferometric imaging through the atmosphere at a low light level are applied to the case of the Calern High Angular Resolution Optical Network (CHARON) stellar interferometric array. The numerical simulation that was implemented to create the interferometric data is presented. The processing algorithms used to process the raw data, to extract the object parameters, and to restore the initial map are pointed out. The multiresolution approach provides an objective way of analyzing the reconstruction procedure. Reconstructed maps under different conditions of brightness and turbulence are shown and discussed. The advantages and the drawbacks of the different steps of the computer simulation are analyzed.

036.008 New trends in ground–based astronomy: fast real time processing needed.
J. P. Picat.
11. European Regional Astronomy Meeting of the IAU: New windows to the universe, Vol. 1, p. 483 – 496 (1990). – See Abstr. 012.001 for the main entry.
   The next decade will be of major importance for ground–based astronomy development with the coming of very efficient detectors, of Very Large New Technology Telescopes and of new observing techniques. The major result will be a tremendous increase in the information to be processed which calls for developing new real time processing facilities. This need is shown and discussed through some examples already implemented on present telescopes.

036.009 Interferometry with large optical telescopes.
P. J. Lena.
11. European Regional Astronomy Meeting of the IAU: New windows to the universe, Vol. 1, p. 507 – 514 (1990). – See Abstr. 012.001 for the main entry.

036.010 How to monitor optimum exposure times for high resolution imaging modes?
B. Lopez.
Astron. Astrophys., Vol. 253, No. 2, p. 635 – 640 (Jan 1992).
   Optimum exposure times for high resolution imaging modes are functions of the distribution of wind and turbulence in the Earth's atmosphere. The theory which is developed here shows that the velocity of the tilt for the angle of arrival of light at the ground level, or equivalently the observed image motion velocity, is also statistically related to turbulent atmosphere motions. The method of differential image motion which is actually used for monitoring the "Fried parameter", can be extended to one new differential image motion velocity method. It can therefore be used as a diagnostic to estimate exposure times needed to freeze the drastic effect of the wavefront corrugation evolution in high angular resolution imagery. Technically, this requires a growth of the sampling rate of ESO Differential Image Motion Monitor.

036.011 The object–image relationship in Michelson stellar interferometry.
M. Tallon, I. Tallon–Bosc.
Astron. Astrophys., Vol. 253, No. 2, p. 641 – 645 (Jan 1992).
   The authors describe the imaging capabilities of a Michelson stellar interferometer compared to a Fizeau stellar (or Young) interferometer. The re-arrangement of the entrance pupil in the Michelson interferometer prevents one from describing the

object–image transformation by a convolution operation. In order to reach a wide field–of–view, the only known solution is to make the Michelson interferometer optically equivalent to a Fizeau interferometer and so to lose specific advantages like the possible re–arrangement of the pupil. For incoming monochromatic light, it is shown that in fact, the only effect of the pupil re–arrangement is a shift of the high spatial frequency components of the object spectrum. This shift can be easily corrected, thus yielding the same relationship as the Fizeau interferometer. As a consequence, full imaging of an extended object with a Michelson stellar interferometer is possible, without the constraints of a Fizeau configuration. Emphasis is also given to the necessity, for any interferometer, to have a good resolution in the Fourier plane for getting a wide field–of–view. These results concern particularly the arrays of large apertures like the Very Large Telescope Interferometer.

036.012 Speckle interferometry: noise reduction by correlation fringe averaging.
J. M. Huntley, L. Benckert.
Appl. Opt., Vol. 31, No. 14, p. 2412 – 2414 (10 May 1992). Current Physics Microform No.: 9207B0244.
   A method for noise reduction in double–exposure speckle interferometry is proposed, based on averaging independent spatially filtered correlation fringe patterns.

036.013 Unconventional astronomical imaging.
M. C. Roggemann, D. W. Tyler.
Opt. Photonics News, Vol. 3, No. 3, p. 16 – 21 (Mar 1992). Current Physics Microform No.: 9204G1358.

036.014 The OH airglow spectrum: a calibration source for infrared spectrometers.
E. Oliva, L. Origlia.
Astron. Astrophys., Vol. 254, No. 1/2, p. 466 – 471 (Feb 1992).
   A finding list of OH sky lines suitable for the wavelength calibration of infrared spectrometers is presented. The selection of the sky lines which are sufficiently bright and isolated was made on the basis of long slit infrared spectra collected at ESO. The authors also briefly discuss how this list could be used to choose the parameters of narrow band filters where it is of primary importance to minimize the contribution of strong sky lines to the background seen through the filter.

036.015 Binary star studies with small telescopes: photometry of close visual systems.
J. S. B. Dick, M. N. Devaney, R. W. Argyle, D. H. P. Jones.
Meas. Sci. Technol., Vol. 3, No. 2, p. 184 – 187 (Feb 1992).
   The astrophysical study of close visual binary stars presents many problems due to the degradation in spatial information by atmospheric turbulence. An image scanner, part of the instrumentation for the 1 m telescope at La Palma, is described which allows photometry of multiple stellar systems to be undertaken. The instrument is optimized for correct sampling of the atmospheric turbulence and uses a photomultiplier detector linked to a high–speed photon counting system.

036.016 De geschiedenis van de astrofotografie.
D. Malin.
Zenit, Jaarg. 19, Nr. 1, p. 4 – 11 (Jan 1992).

036.017 Het fotograferen van emissienevels.
J. S. Sussenbach.
Zenit, Jaarg. 19, Nr. 1, p. 12 – 15 (Jan 1992).

036.018 Astrofotografie met eenvoudige middelen... niet alleen voor beginners.
S. Klein.
Zenit, Jaarg. 19, Nr. 2, p. 86 – 93 (Feb 1992).

036.019   Observations with the scanner of the 6–m telescope at Nasmyth–1 focus and automated software for reduction of spectra.
V. L. Afanas'ev, V. A. Lipovetskij, V. P. Mikhajlov, E. A. Nazarov, A. I. Shapovalova.
Bull. Spec. Astrophys. Obs. – North Caucasus, Vol. 31, p. 126 – 131 (1991). English translation of Astrofiz. Issled. Izv. Spets. Astrofiz. Obs., Tom 31, p. 128 – 133 (1991).
The hardware of the 1000–channel TV scanner of the 6–m telescope mounted on the SP–124 spectrograph at the Nasmyth focus is described. The software enabling complete automation of observations and reduction of the scanner spectra is described briefly.

036.020   Processing digital spectroscopic data: the point–spread function and second–order scaling errors.
A. Young, L. Rottler.
Publ. Astron. Soc. Pac., Vol. 104, No. 671, p. 71 – 75 (Jan 1992). Current Physics Microform No.: 9202F2097.
The authors discuss the analysis of some peculiarities found in semi–raw spectroscopic CCD data, and demonstrate that they result from effects that correlate with the zenith angle at the time of observation. They present evidence and arguments that suggest that the ultimate cause of such effects are alterations of the point–spread function of the stellar image that is delivered to the entrance aperture of the spectrograph by the combination of the telescope optics and the terrestrial atmosphere along different sight paths. The coping strategy for this problem leads to a scale–normalization procedure akin to fitting a (pseudo) continuum to stellar spectra. Ideally, such a procedure should produce stellar spectra that can be directly compared numerically for the purpose of detecting and measuring small variations of spectral features. However, small subjective (personal) errors in any such procedure introduce small systematic errors that can raise the detection threshold for small variations above the noise level of the data; or even produce chimerical variations that can be misinterpreted as being real. The authors discuss and demonstrate an objective method for detecting such systematic effects and eliminating them, so as to achieve the optimum detection capability that the signal–to–noise ratio of the data will permit.

036.021   Statistical analysis and separation of systematic errors in astrometric data.
M. L. Bougeard.
Astron. Astrophys., Vol. 255, No. 1/2, p. 388 – 400 (Feb 1992).
This study concerns the combined effects in temporal variation, magnitude, and colour that are likely to affect the results of astrometric observations. A statistical analysis is performed on the Paris astrolabe data from two test periods each with a different instrumental set–up. The results are obtained using the approach described by Bougeard (1987, 1988), with slight modifications.

036.022   Pupil plane interferometry in the near infrared. I. Methodology of observation and first results.
J.–M. Mariotti, J.–L. Monin, P. Ghez, C. Perrier, A. Zadrozny.
Astron. Astrophys., Vol. 255, No. 1/2, p. 462 – 476 (Feb 1992).
The authors describe an experiment aimed at diffraction–limited imaging of astronomical sources in the 2–5 μm range. It is based on a rotation shearing interferometer that extracts the complex degree of coherence of the wavefront directly in the pupil plane. Interference fringes are produced by a Michelson interferometer in the image of the telescope pupil. They are scanned through the zero optical path difference faster than the deformations produced by atmospheric turbulence. The two–dimensional interferograms are recorded with an infrared camera located at the recombined pupil image. For each spatial frequency, the modulus and the phase of the Fourier transform of the object intensity distribution are derived from these interferograms. The main advantage of this technique is its high constant transfer function that makes it independent of seeing variations and instrumental aberrations. The authors describe the experimental set–up and discuss some modelbased simulation results which illustrate the operation of the interferometer. They present astronomical data recently obtained at the 4.20 m William Herschel Telescope of the Royal Greenwich Observatory in La Palma, and derive the performances of rotation shearing interferometry for diffraction–limited observations of infrared astronomical objects.

036.023   On the correction of stellar spectra for the loss of radiation during its passage through the Earth's atmosphere and through the spectrograph–slit.
M. A. Fluks, P. S. Thé.
Astron. Astrophys., Vol. 255, No. 1/2, p. 477 – 489 (Feb 1992).
In the present paper the results of a study of the depletion of stellar radiation during its passage through the atmosphere and the blocking of part of it by the spectrograph–slit are reported. The spectral range is $375 < \lambda[nm] < 900$. The atmospheric scattering plus absorption at the ESO is given. A multi–disciplinary approach comprising astrometry, astronomical spectroscopy, advanced integral calculus, meteorology and atmospheric optics is followed for solving the transmittance problem at the spectrograph–slit.

036.024   Monitoring meteors.
J. Rowlands.
Astron. Now, Vol. 6, No. 5, p. 49 (May 1992). Special issue: "Focus: professional amateurs".

036.025   Photometric calibration of NGS/POSS and ESO/SRC plates using the NOAO PDS measuring engine. I. Stellar photometry.
R. M. Cutri, F. J. Low, K. B. Marvel.
Publ. Astron. Soc. Pac., Vol. 104, No. 673, p. 223 – 234 (Mar 1992). Current Physics Microform No.: 9204G1327.
The PDS/Monet measuring engine at NOAO was used to obtain photometry of nearly 10,000 stars on the NGS/POSS and 2000 stars on the ESO/SRC Survey glass plates. These measurements have been used to show that global transformation functions exist that allow calibration of stellar photometry from any blue or red plate to equivalent Johnson B and Cousins R photoelectric magnitudes. The authors have characterized the four transformation functions appropriate for the POSS O and E and ESO/SRC J and R plates, and found that within the measurement uncertainties they vary from plate to plate only by photometric zero–point offsets. A method is described to correct for the zero–point shifts and to obtain calibrated B and R photometry of stellar sources to an average accuracy of $0.3 – 0.4$ mag within the range $8 \leqslant R \leqslant 19.5$ for red plates in both surveys, $9 \leqslant B \leqslant 20.5$ on POSS blue plates, and $10 \leqslant B \leqslant 20.5$ on ESO/SRC blue plates.

036.026   Automated star/galaxy discrimination with neural networks.
S. C. Odewahn, E. B. Stockwell, R. L. Pennington, R. M. Humphreys, W. A. Zumach.
Astron. J., Vol. 103, No. 1, p. 318 – 331 (Jan 1992). Current Physics Microform No.: 9202A2228.
The authors discuss progress in the development of automatic star/galaxy discriminators for processing images generated by the University of Minnesota Automated Plate Scanner (APS) for cataloging the first epoch Palomar Sky Survey. Classifications are based on 14 image parameters computed for each object detected by the APS operating in a threshold densitometry mode. It is shown that a number of parameter spaces formed with these vector elements are effective in separating a sample into the two basic populations of stellar and nonstellar objects. An artificial intelligence technique known as a neural network is employed to perform the image classification. The authors have experimented with a simple linear classifier known as a perceptron, as well as with a more sophisticated backpropagation neural network with the result that they are able to attain classification success rates of 99% for galaxy images with $B \leqslant 18.5$ and above 95% for the magnitude range $18.5 \leqslant B \leqslant 19.5$. Simple numerical experiments have been conducted in an effort to illustrate the robust nature of this method as well as to isolate the most significant

image parameters used by the networks in distinguishing image class.

036.027   A filter for deep near–infrared imaging.
R. J. Wainscoat, L. L. Cowie.
Astron. J., Vol. 103, No. 1, p. 332 – 337 (Jan 1992). Current Physics Microform No.: 9202A2242.
The K passband (central wavelength 2.2 μm, FWHM 0.4 μm) is the longest wavelength standard near–infrared passband through which deep ground based imaging is possible. Thermal emission from telescope, instrument, and sky limits the depth to which such imaging can reach by producing strongly temperature dependent backgrounds in the range 11–13.5 mag arcsec^{-2}. The authors describe how a passband, which they denote as K', located slightly shortward of the standard K passband (central wavelength 2.1 μm), yet still within the same atmospheric window, leads to a significantly lower thermal component of the background, reducing the background surface brightness by up to 0.9 mag arcsec^{-2}, and thereby allowing deeper imaging to be obtained in the same integration time. The photometric differences between the K' filter and the standard K filter are discussed.

036.028   Astrophotographie von Deep–Sky–Objekten mit Bildverstärker.
P. Winter, K. Völkel.
Sterne Weltraum, Jahrg. 31, Nr. 4, p. 252 – 253 (Apr 1992).

036.029   The IUE Final Archive – III.
A. Talavera, J. D. Ponz.
ESA IUE Newsl., No. 40, p. 14 – 15 (Mar 1992).

036.030   Automatic determination of the astronomical azimuth by observing a celestial body using the electronic theodolite Kern E2 and the laptop computer Toshiba T1600.
N. Solarič, D. Špoljarić, M. Vresk, I. Skender.
Hvar Obs. Bull., Vol. 15, No. 1, p. 35 – 43 (1991).
Automatic determination of the grid azimuth by the Sun and stars using the electronical theodolite Leica Kern E2 and laptop computer Toshiba T1600. 1. An astronomical almanac for either Sun or for the stars coordinates is not necessary. 2. Laptop Toshiba T1600 is also used for time measuring. 3. It is pointed to the target mark up to 10 times in the same telescope position, and Toshiba T1600 will remove automatically measurements with a large error. The same procedure is repeated in the observations of a celestial body. 4. The observation can start either in the direct or reverse telescope position. Then it is checked at the Toshiba T1600 if the telescope was in the appropriate position. 5. After each of the observing sets, the mean value of the grid azimuth is obtained immediately from all the sets observed so far together with the standard deviation. 6. First test measurements indicate that an approximate 20–minute observation is required to obtain the mean grid azimuth with a standard deviation smaller than 0.5".

036.031   Bayesian deconvolution in optical astronomy.
R. Molina, A. del Olmo, J. Perea, B. D. Ripley.
Astron. J., Vol. 103, No. 2, p. 666 – 675 (Feb 1992). Current Physics Microform No.: 9202D1996. With plate 48.
In this work the authors use Bayesian methods and spatial stochastic processes in the deconvolution of images of galaxies. Under very simple but realistic prior assumptions about the true underlying image of a galaxy the Bayesian framework is put to work. The method is tested in CCD images of extragalactic objects of different morphological types and an analysis of the deconvolutions obtained is performed emphasizing the comparison with other observational results.

036.032   Diffraction–limited imaging with ground–based telescopes.
T. Nakajima.
Astron. Her., Vol. 85, No. 3, p. 98 – 103 (1992). In Japanese.

036.033   A full function photoelectric data gathering system.
T. Arnold.
I.A.P.P.P. Commun., No. 47, p. 25 – 27 (Mar – May 1992).

036.034   TVSTAR 2.0 interface software for the STARLIGHT–1 photometer.
R. C. Wolpert, J. B. Gunn.
I.A.P.P.P. Commun., No. 47, p. 28 – 30 (Mar – May 1992).

036.035   The JKT scanning slits: investigating binary stars.
J. Dick, N. Devaney, B. Argyle, D. Jones.
Gemini, No. 35, p. 3 (Mar 1992).

036.036   SIGNAL.
C. Benn.
Gemini, No. 35, p. 20 (Mar 1992).

036.037   Bestimmung des Wilson–Effektes an H–Flecken.
H. Joppich.
Sonne, Jahrg. 16, Nr. 61, p. 21 – 23 (Mar 1992).

036.038   Auswertung von Lichtkurven Bedeckungsveränderlicher Sterne.
S. Kohle.
BAV Rundbrief, Jahrg. 41, Nr. 2, p. 89 – 107 (1992).

036.039   Astronomy and personal computers.
J. B. Dunham.
Occultation Newsl., Vol. 5, No. 7, p. 170 – 171 (May 1992).

036.040   Reducing graze observations with a personal computer.
R. Giller.
Occultation Newsl., Vol. 5, No. 7, p. 179 – 180 (May 1992).

036.041   HIPPARCOS: one year gone, a few more to go.
F. van Leeuwen, M. Penston, D. W. Evans, N. Ramamani, C. Petersen, E. Hög, L. Lindegren, S. Söderhjelm.
Gemini, No. 36, p. 14 – 18 (Jun 1992).

036.042   A calculation of confusion noise due to infrared cirrus.
T. N. Gautier III, F. Boulanger, M. Pérault, J. L. Puget.
Astron. J., Vol. 103, No. 4, p. 1313 – 1324 (Apr 1992). Current Physics Microform No.: 9204O0267.
General expressions are developed for the statistical errors to be expected in photometric measurements due to confusion in a background of fluctuating surface brightness. Backgrounds actually observed in the far infrared by the IRAS satellite are used to calculate tables of these error expressions for two simple measurement techniques. The confusion noise limited sensitivities for NASA's planned Space Infrared Telescope Facility and the European Space Agency's Infrared Space Observatory are estimated at a wavelength of 100 μm from these tables.

036.043   Absolute quadrant determinations from speckle observations of binary stars.
W. G. Bagnuolo Jr., B. D. Mason, D. J. Barry, W. I. Hartkopf, H. A. McAlister.
Astron. J., Vol. 103, No. 4, p. 1399 – 1407 (Apr 1992). Current Physics Microform No.: 9204G0353.
Reduction of speckle data obtained for binary stars is typically carried out using power spectrum or, equivalently, autocorrelation methods. An especially powerful algorithm from which accurate differential astrometry can be obtained is the vector–autocorrelation technique. While such methods are highly suited to extracting astrometric information from very large volumes of speckle data in near real time, they inherently introduce a 180° ambiguity in the position angle measurement. The authors briefly describe why this can be a problem in determining the orbital motions in binaries, present a new algorithm which maintains most of the simplicity of vector autocorrelation while removing the quadrant ambiguity, and provide results of new absolute quadrant determinations for 66 binary star systems first resolved

by the long–term GSU/CHARA speckle program carried out at the National Optical Astronomy Observatories. Finally, the authors use this new algorithm to eliminate the period ambiguity in the orbit of the close visual binary ADS 9744.

036.044 Possibilities of RDS in meteor back–scatter.
C. Steyaert.
WGN, Vol. 20, No. 1, p. 51 – 54 (Feb 1992).

036.045 The software "Radiant"
R. Arlt.
WGN, Vol. 20, No. 2, p. 62 – 69 (Apr 1992).
The main algorithms underlying the program *Radiant* are described. Backward tracings and probability functions are used to determine radiants. First experiences and methods to estimate the reliability of the results obtained by the software are discussed.

036.046 Precision of telescopic meteor recordings: plotting errors and recording probability.
P. Pravec, J. Boček.
WGN, Vol. 20, No. 2, p. 70 – 83 (Apr 1992).
Forty meteors of magnitudes between 4 and 8.5 were observed both telescopically and by means of a TV–camera during the 1991 Perseid campaign at the Ondřejov Observatory. Using standard telescopes for telescopic observations of meteors, 152 individual telescopic recordings of those 40 meteors were obtained. An analysis of the precision of these telescopic recordings is presented. No systematic deviations were found.

036.047 Maximum likelihood image restoration. V. Incoherent fluxes.
V. Yu. Terebizh, O. K. Cherbunina, Yu. G. Cherbunin, V. V. Biryukov.
Astrophysics, Vol. 34, No. 1, p. 56 – 62 (Jan – Feb 1991). English translation of Astrofizika, Tom 34, Vyp. 1, p. 91 – 100 (Feb 1991).
The general scheme of maximum likelihood image restoration is particularized to the case when there is a priori information about the Poisson distribution of events stored during the exposure time. In this case, the restoring algorithm is significantly simpler, but a real gain due to such information is appreciable only for very faint objects.

036.048 Erratum: "Maximum likelihood image restoration. III. Algorithm. One–dimensional test cases" [Astrofizika, Tom 33, Vyp. 2, p. 305 – 315 (Oct 1990) and Astrophysics, Vol. 33, No. 2, p. 475 – 482 (Sep – Oct 1990)].
V. Yu. Terebizh, V. V. Biryukov.
Astrophysics, Vol. 34, No. 1, p. 63 (Jan – Feb 1991). See abstracts 53.036.263, 54.036.076.

036.049 Planetary imaging with a small CCD camera.
D. Parker, R. Berry.
Strolling Astron., Vol. 36, No. 1, p. 1 – 8 (Mar 1992).
CCD images rival the human eye, capturing diffraction–limited planetary detail with short exposure times. The authors present their first six months' experience with a Lynxx PC CCD camera and outline the potential that CCD imaging offers amateur planetary observers.

036.050 Precision radial velocities with an iodine absorption cell.
G. W. Marcy, R. P. Butler.
Publ. Astron. Soc. Pac., Vol. 104, No. 674, p. 270 – 277 (Apr 1992). Current Physics Microform No.: 9205C2226.
The authors have used gaseous iodine for generating reference absorption lines in stellar spectra taken at high resolution. A major advance involves the use of a fast echelle spectrograph and a CCD which acquires the near–ultraviolet, the visible, and the near–infrared spectrum in a single exposure. The superimposed iodine lines provide both a highly precise wavelength scale and a specification of the spectrograph PSF in situ over the entire echelle format. Test observations of three solar–type stars exhibit a velocity scatter of less than 25 m s^{-1} over a 1–yr duration, and

only 1/5 of the available spectrum has been employed in the analysis to date. Velocity precision of 50 m s^{-1} can be achieved for magnitude V $= 12$ in 1 h exposures on a 3–m telescope. The authors discuss an on–going project to detect brown–dwarf and planetary companions to F–, G–, K–, and M–type main–sequence stars, designed to complement other efforts. The current velocity precision permits detection of companions with masses as low as 3 M_{Jup} located up to 5 AU from the star. They also discuss the use of precision velocities in revising Cepheid distances.

036.051 The Nyquist criterion in CCD photometry for surface brightness.
R. L. Wildey.
Publ. Astron. Soc. Pac., Vol. 104, No. 674, p. 285 – 289 (Apr 1992). Current Physics Microform No.: 9205C2241.
Astronomers routinely violate the directive to sample surface brightness with at least twice the frequency of the highest spatial frequency of the Fourier transform of the continuous image, when doing direct CCD imaging. It is reasonably speculated that this practice is rationalized on the basis that the CCD does not actually sample the surface brightness at periodic intervals, but instead integrates the surface brightness over contiguous regions (the CCD pixels). It is herein derived that this mode of sampling changes the form of aliasing error, but the aliasing error is nevertheless present when undersampling occurs. The very nature of the error betrays the possibility of detecting its presence, a priori. It would be of value to develop an active optical apodizer to accommodate a given CCD, in terms of the Nyquist criterion, without the need to abandon either the full light–gathering power of the telescope or the plate scale at the chosen observing station.

036.052 On blackbody behavior and the transformation from instrumental to standard magnitudes and color indices.
R. L. Wildey.
Publ. Astron. Soc. Pac., Vol. 104, No. 674, p. 290 – 300 (Apr 1992). Current Physics Microform No.: 9205C2246.
A first–order error analysis in the application of standard transformation techniques connecting instrumental to standard magnitude/color–index systems has been performed in which terms for departures from blackbody behavior and departures of the latter from the Wien approximation are separated. A best case is taken in the sense that confoundment by interstellar reddening or reflection from planetary surfaces is omitted, as well as the effects of finite spatial bandwidth. The results are presented in a form easily delineable with respect to departures from unity in the coefficients of the color equations. What has been determined can therefore be summarized as the error resulting from a mismatch of effective wavelength, at a spectral resolution of nominally 200 Å.

036.053 A new approach to point–pattern matching.
F. Murtagh.
Publ. Astron. Soc. Pac., Vol. 104, No. 674, p. 301 – 307 (Apr 1992). Current Physics Microform No.: 9205C2257.
The author describes a new algorithm for matching star lists, given by their two–dimensional coordinates. Such matching should be unaffected by translation, rotation, rescaling, random perturbations, and some random additions and deletions of coordinate couples in one list relative to another. The first phase of the algorithm is based on a characterization of a set of coordinate couples, relative to each individual coordinate couple. In the second phase of the algorithm, the matching of stars in different lists is based on proximity of feature vectors associated with coordinate couples in the two lists. The order of magnitude computational complexity of the overall algorithm is n^2 for O(n) coordinate couples in the coordinate lists.

036.054 Automated morphological classification of faint galaxies.
G. Spiekermann.
Astron. J., Vol. 103, No. 6, p. 2102 – 2110 (Jun 1992). Current Physics Microform No.: 9207A0382.
A fully automated morphological classification system for faint galaxies is described. By means of fuzzy algebra and the

subsequent application of heuristic methods, five morphological classes of galaxies, corresponding to Hubble types, are determined in the range $14 < m \leqslant 19$. The application of the classification program to 16 ESO–SERC fields near the South Galactic Pole leads to a homogeneous catalogue of more than 100,000 galaxies. From these data, the morphological type mixture is determined to be E:S0:S/Ir = 14:21:65 (%), with slight variations between areas with different degrees of clustering. The two–point correlation functions of the different morphological classes show the decreasing tendency for clustering from early to late type galaxies.

036.055   High–precision time–resolved CCD photometry.

H. Kjeldsen, S. Frandsen.
Publ. Astron. Soc. Pac., Vol. 104, No. 676, p. 413 – 434 (Jun 1992). Current Physics Microform No.: 9208B0669.

This paper describes some of the problems related to high–precision photometry, especially focusing on CCD photometry both on focused stars and defocused field stars. The authors show that atmospheric transparency variations and instrumental drift are the main noise sources at low frequencies and that CCD photometry is the best way of doing differential measurements in order to detect oscillations at low frequencies. They also describe a special reduction program, multiobject multiframe photometric package, which can reduce time–resolved photometry in half–crowded fields. It has been developed to limit the photometric noise and runs fully automatically, with only a few interactive steps. The authors compare the theoretical description of noise and problems in high–precision photometry with CCD time–series observations in a noncrowded test field observed with the Danish 1.5–m telescope at ESO in 1990. In addition, they describe observations of defocused double stars observed with the Nordic Optical Telescope in 1991.

036.056   CCD ensemble photometry on an inhomogeneous set of exposures.

R. K. Honeycutt.
Publ. Astron. Soc. Pac., Vol. 104, No. 676, p. 435 – 440 (Jun 1992). Current Physics Microform No.: 9208B0691.

A method is described to perform ensemble stellar photometry on a series of CCD images for which the number and identity of the comparison stars vary throughout the set of exposures. The technique is particularly useful for the production of light curves of variable stars from long–term photometric programs where the inhomogeneity of the data set can be large. The linear least–squares solution is derived, paying particular attention to the evaluation of errors. Examples are presented of the use of two interactive computer programs which implement this technique.

036.057   A study of polynomial models for astrographic plate reductions.

G. Vieira, M. Assafin, R. Vieira Martins.
Publ. Astron. Soc. Pac., Vol. 104, No. 676, p. 467 – 471 (Jun 1992). Current Physics Microform No.: 9208B0723.

Polynomial adjustments for the reduction of plates taken with the astrograph at the Observatório Municipal de Campinas, São Paulo, Brazil, were investigated. A statistical analysis was performed on the vectorial residual maps derived from the reductions of 11 photographic plates. The results have shown the second degree model to be inadequate for the reductions. A radial distortion effect in the $7° \times 7°$ field was detected. The fifth degree radial distortion term was found to be not significant. The second degree plus a term of radial distortion of the third degree and the third degree polynomial models give residuals of the order of $0\rlap{.}''30$, compatible with the errors of the catalogues used (Perth 70 and PPM South, preliminary version). However, there is evidence that the last model gives slightly better results than the former.

036.058   Astrophysical use of the principal component analysis of imperfect data.

W. Unno, M. Yuasa.
Astrophys. Space Sci., Vol. 189, No. 2, p. 271 – 278 (Mar 1992).

In natural historical studies, no uniform accuracy of observational data can generally be expected for different quantities and for different samples. Principal component analysis which is the basic mathematical method of natural history is generalized in such a way that maximum information can be drawn from any given set of imperfect data. The principle of the generalization is to supplement data to the given set of data so that the combined system does attain the maximum probability distribution. The utility of the method in a few astrophysical and other applications are discussed briefly.

036.059   A topological / geometrical approach to the study of astrophysical maps.

F. C. Adams.
Astrophys. J., Vol. 387, No. 2, p. 572 – 590 (10 Mar 1992).

This paper discusses topological and geometrical methods for analyzing astrophysical maps, ranging from emission maps of molecular clouds to surveys of the large–scale structure of the universe. The author constructs a formalism which considers the collection of all maps of a given type as a metric space; he then constructs a set of distance functions (pseudometrics), which measure the difference between any two astrophysical maps and thereby topologize the space of maps. The author uses these pseudometrics to construct a procedure for ordering the space of maps. The results of this work provide a formalism for analyzing and classifying astrophysical maps.

036.060   Dual beam mapping with a maximum entropy algorithm.

J. S. Richer.
Mon. Not. R. Astron. Soc., Vol. 254, No. 1, p. 165 – 176 (1 Jan 1992).

The problem of making images of the sky from dual beam (or differential) maps is a linear inversion problem in which the data are related to the unknown sky brightness distribution via a convolution with a point spread function which in general varies across the map. A solution to this problem based on the maximum entropy method has been implemented and tested, and is described in detail with particular reference to data obtained at the James Clerk Maxwell Telescope.

036.061   The intrinsic scatter in a linear regression with errors in the variables.

C. Koen.
Mon. Not. R. Astron. Soc., Vol. 254, No. 3, p. 383 – 388 (1 Feb 1992).

Under certain circumstances the slope of a linear regression of two variables measured with error may be accurately determined if an auxiliary variable is available. Equating the result to a second estimator based on error variances then allows limits to be set on the error dispersions and on the intrinsic scatter in the linear relation. The results are illustrated with simulated data and applied to a two–colour diagram based on observations of Cepheids.

036.062   Photometric calibration of the International Ultraviolet Explorer (IUE) at low resolution: the LWP camera.

A. Cassatella, R. Gonzalez–Riestra, C. Imhoff, N. Oliversen, C. Lloyd.
Astron. Astrophys., Vol. 256, No. 1, p. 309 – 320 (Mar 1992).

The authors present and discuss the low resolution absolute calibration for the LWP camera of the International Ultraviolet Explorer (IUE). The calibration, implemented in routine processing of LWP low resolution images on December 22, 1987 at both GSFC and VILSPA IUE ground stations, is based on the last set of Intensity Transfer Function (ITF2) obtained in September 1984 and on a substantially larger amount of IUE observations of standard stars than used for the preliminary calibration by Cassatella and Harris (1983) based on the earlier ITF (ITF1). The new ITF and the associated calibration provide upgraded performances in flux accuracy, signal–to–noise ratio and linearity with respect to the earlier LWP ITF1 which was implemented in processing LWP data obtained in the period from October 16, 1983 until December 21, 1987.

036.063   Pearson type VII distribution of errors of laser satellite observations.
I. V. Dzhun'.
Kinematics Phys. Celest. Bodies, Vol. 7, No. 3, p. 76 – 85 (1991). English translation of Kinematika Fiz. Nebesn. Tel, Tom 7, No. 3, p. 82 – 91 (1991).
The distribution of the difference O–C between observed and calculated satellite distances obtained on execution of the short MERIT program is studied. It is shown that the distribution of these differences is described better by the curve of a Pearson type VII distribution than by the normal law. This can be explained by fluctuations in the accuracy of the observations due to instability of the metrological situation. It is shown that the parameter m of the Pearson distribution varies from 2.7 to ∞ (normal law).

036.064   High–fidelity copying of large–area astronomical plates: principles and some practical results.
M. P. van Haarlem, R. S. Le Poole, P. Katgert, S. Tritton.
Mon. Not. R. Astron. Soc., Vol. 255, No. 2, p. 295 – 307 (15 Mar 1992).
The authors report on an investigation into the achieved fidelity of copies of a large Schmidt telescope plate. Atlas quality copies both on glass and film and the original are compared. Two glass copies made through different intermediate positives are used to look at the fidelity of the separate positives. The authors confirm the general belief that glass copies are capable of impeccable photogrammatic quality. However, film copies are shown to achieve essentially the same astrometric fidelity. The general feeling that for photometric applications film copies should be equivalent to glass ones is not confirmed. The authors find that glass copies are generally better in densitometry.

036.065   Tomographic imaging of late–type stars from spectro-scopic and photometric rotational modulation. I. Principle and mathematical formulation of the method.
S. Jankov, B. H. Foing.
Astron. Astrophys., Vol. 256, No. 2, p. 533 – 550 (Mar 1992).
The authors describe the scientific interest, context and methods for flux rotational modulation imaging of late–type stars. Then they treat the problem of indirect stellar imaging from projections to recover the specific intensities on the surface of a star, from its integrals (fluxes) over the parts of the stellar surface observed at different angles of view. The full mathematical formulation of the problem is given in terms of matricial formalism. Finally, the principles of an optimal strategy of observations for Doppler imaging are discussed. The authors use their indirect imaging code built on the basis of developed approach to reconstruct an input image, from a series of generated noisy spectra, using maximum entropy technique, to show the performances and intrinsic limitations of the method.

036.066   Fundamental and practical limits to the sensitivity of submillimeter astronomical observations.
J. M. Lamarre.
NATO Advanced Study Institute on The Infrared and Submilli-metre Sky after COBE, p. 409 – 421 (1992). – See Abstr. 012.009 for the main entry.
Unwanted radiation emitted by sources in the foreground and immediate environment of the detectors is much more important than what astronomers want to observe. The success or the failure of submillimeter observations depends mainly on the techniques used to remove this parasitic radiation and its fluctuations such as photon noise. The author analyses with some detail the formation of photon noise in multimode instruments and shows that the usual simplified equations may give signifi-cantly wrong results in practical cases. Other sources of fluctuations, such as the atmospheric noise, and the way used to partly remove them are also described.

036.067   Observations at (sub)millimetre wavelengths: effect of atmosphere and telescope sidelobes.
· M. Guélin.
NATO Advanced Study Institute on The Infrared and Submilli-metre Sky after COBE, p. 423 – 443 (1992). – See Abstr. 012.009 for the main entry.
The author focusses on two topics of special importance for millimetre–wave measurements requiring accuracy (e.g. line intensity ratios, background inhomogeneities, ...). The first deals with atmospheric absorption and related problems; the second with telescope sidelobes.

036.068   Possibility of outer space X–ray diffractometry.
V. G. Feklichev, U. V. Khangil'din.
Sol. Syst. Res., Vol. 25, No. 1, p. 46 – 48 (Jul 1991). English translation of Astron. Vestn., Tom 25, No. 1, p. 61 – 64 (1991).
The possibility of remote mineral composition diagnostics of space objects by the powder X–ray diffractometry method is demonstrated. Equipment variants are evaluated.

036.069   Photon bias compensation in triple correlation imaging and observation of R 136.
E. Pehlemann, K.–H. Hofmann, G. Weigelt.
Astron. Astrophys., Vol. 256, No. 2, p. 701 – 714 (Mar 1992).
The authors report diffraction–limited speckle masking obser-vations of the central object R 136 in the 30 Doradus Nebula, and describe a new method for the compensation of the 4–dimensional photon bias terms in the average bispectrum. More than 40 stellar components are present in the $4\overset{''}{.}9 \times 4\overset{''}{.}9$ field of view of their reconstructed images. The closest binaries found have separations between $0\overset{''}{.}03$ and $0\overset{''}{.}05$. Because of the large number of photon events in the speckle interferograms, it was not possible to apply photon counting techniques. Therefore, the extended photon events caused various frequency–dependent bias terms in the average image bispectrum. To overcome this influence of photon noise, the authors have developed a new technique for the compensation of the 4–dimensional photon bias terms in the bispectrum. They describe the theory of the method and demonstrate its feasibility by various image reconstruction experiments.

036.070   Optical aperture synthesis and wave–front sensing.
M. Tallon, I. Tallon–Bosc.
Astron. Astrophys., Vol. 256, No. 2, p. 715 – 722 (Mar 1992).
To apply phase closure with an array of optical telescopes, each aperture must be phased. This sets an upper limit on the size of each aperture equal to the coherence area of the wave fronts (if no adaptive optics is used). By recording wave–front sensing on each aperture and by post–processing, it is possible to apply phase closure algorithms with apertures sustaining an arbitrary high number of coherence areas, as far as the combination remains coherent. This procedure increases substantially the signal–to–noise ratio in the visible and stands as a complement when adaptive optics only provides a partial correction of wave–fronts.

036.071   Time variability studies with photon–counting imaging detectors. I. A maximum likelihood technique.
S. Sciortino, G. Micela.
Astrophys. J., Vol. 388, No. 2, p. 595 – 602 (1 Apr 1992).
The authors present a new method for characterizing the variability properties of a preselected class of celestial objects which are repeatedly observed with the same photon–counting imaging detector. Instrument–specific characteristics, such as dectector instrument point response function and mirror system off–axis response, are separated and taken into account in the form of input parameters. All collected source data, both in the form of detections and upper bounds at putative source positions, are retained and fully used in determining the maximum likelihood (ML) estimator of the intrinsic source count rate. For each distinct observation one can determine, at the chosen significane level based on Poisson statistics, whether the observed flux (or upper bound) undergoes a statistically signifi-cant variation with respect to the derived ML flux estimator, and one can evaluate the variation amplitude, or an upper bound for

those cases where significant variations have not been detected. The authors illustrate the capabilities of the proposed technique, applying it to the samples of the Pleiades and the Hyades stellar X–ray sources surveyed with the IPC of the Einstein Observatory.

036.072   The GDDSYN light curve synthesis method.
P. D. Hendry, S. W. Mochnacki.
Astrophys. J., Vol. 388, No. 2, p. 603 – 613 (1 Apr 1992).

The authors present a close binary system light curve synthesis algorithm called geodesic distribution binary synthesis (GDDSYN), which uses a novel surface element distribution and partial visibility computational scheme for the Roche model. The surface element distribution, based on the geodesic sphere, is well suited to maximum entropy imaging of surface features on eclipsing binaries. It allows the fitting of spot distributions which are less sensitive to observational errors than schemes which have too many low weight surface elements. The algorithm for computation of surface element visibility is based on techniques used in computer graphics, and is shown to be more accurate and faster than the Wilson – Devinney (WD) code. Light curves produced with GDDSYN are at least as accurate as those of the WD code, and in addition the light curve of each individual surface element is accurately computed to form the kernel needed for surface feature imaging. Examples and numerical tests have been computed.

036.073   Determination of the equatorial velocity and inclination of the equatorial plane of an astrophysical object in space.
V. O. Gladyshev.
Astrophysics, Vol. 34, No. 2, p. 111 – 114 (Mar – Apr 1991). English translation of Astrofizika, Tom 34, Vyp. 2, p. 227 – 232 (Apr 1991).

Expressions relating the equatorial velocity V_e and the angle of inclination φ of the equatorial plane of an astrophysical object in space to variations in the profiles of spectral lines are discussed. The results of the analysis are discussed from the point of view of the application of the methods of optical spectrometry.

036.074   Maximum likelihood image restoration. VI. The Cramér–Rao limit of restoration efficiency.
V. Yu. Terebizh, V. V. Biryukov.
Astrophysics, Vol. 34, No. 2, p. 114 – 123 (Mar – Apr 1991). English translation of Astrofizika, Tom 34, Vyp. 2, p. 233 – 248 (Apr 1991).

The presence of intrinsic (radiative) noise in radiation necessarily reduces the problem of restoration of images that are blurred and have noise added to the statistical problem of estimating many unknown parameters – the intensities of the original in the pixels. Moreover, irrespective of the method there exists a theoretical limit to the efficiency of restoration, which is determined by the Cramér-Rao theorem. General considerations, and also numerical and analytical examples show that restoration by means of the maximum likelihood image restoration method described achieves the maximal efficiency already at an intensity of the luminous image that is comparable with the mean level of the external noise. Computational relations that make it possible to find the natural accuracy of image restoration (the error corridor) are given.

036.075   Improvements to photometry. V. High–order moments in transformation theory.
A. T. Young.
Astron. Astrophys., Vol. 257, No. 1, p. 366 – 388 (Apr 1992).

Photometric transformation theory is re-examined to determine the importance of high–order terms that have previously been neglected. These terms depend on high–order central moments of the passband response profiles, as functions of

wavelength, so simple rules are developed for estimating these moments.

036.076   A method for characterizing transient ionospheric disturbances using a large radiotelescope array.
A. R. Jacobson, W. C. Erickson.
Astron. Astrophys., Vol. 257, No. 1, p. 401 – 409 (Apr 1992).

Radio interferometers have been previously used for study of wave–like disturbances in the overhead ionosphere. However, all such studies involved baselines ($\leqslant 3$ km) which were very short compared to the disturbance wavelength, so that meaningful lag information was unavailable, and hence the waves' phase velocity usually could not be measured. Recently the authors have employed the VLA radiotelescope at 90 cm radio wavelength to study ionospheric waves possessing disturbance wavelengths > 20 km, using baselines which extend to 35 km. They have developed an analysis algorithm which fits spatially monochromatic plane waves for each angular frequency in the visibility–phase data. This paper examines (1) analysis procedures appropriate to long–baseline interferometers, and (2) the basis for dynamic phase compensation using continual wave–parameter fits.

036.077   Physical implementation of an antimask in URA based coded mask systems.
U. B. Jayanthi, J. Braga.
Nucl. Instrum. Methods Phys. Res., Sect. A, Vol. 310, No. 3, p. 685 – 689 (15 Dec 1991).

X– and gamma–ray astronomy experiments which employ rectangular URA coded masks alone show artifacts in the images reconstructed due to nonuniform background levels in the detector plane. The employment of a separate antimask in addition to the mask in observations is useful to eliminate this problem. The authors propose a method to implement the antimask with the same mask, utilizing the antisymmetric properties in the mask pattern, thereby avoiding the need for a separate antimask in an experiment. Simulations performed with this mask–antimask system are presented to show its advantages.

036.078   OSSE spectral analysis techniques.
W. R. Purcell, K. M. Brown, D. A. Grabelsky, W. N. Johnson, G. V. Jung, R. L. Kinzer, R. A. Kroeger, J. D. Kurfess, S. M. Matz, M. S. Strickman, M. P. Ulmer.
Compton Observatory Science Workshop, p. 15 – 25 (Feb 1992).
– See Abstr. 012.018 for the main entry.

Analysis of spectra from the Oriented Scintillation Spectrometer Experiment (OSSE) is complicated because of the typically low signal–to–noise and the large background variability. The OSSE instrument was designed to address these difficulties by periodically offset–pointing the detectors from the source to perform background measurements. These background measurements are used to estimate the background during each of the source observations. The resulting background–subtracted spectra can then be accumulated and fitted for spectral lines and/or continua. Data selection based on various environmental parameters can be performed at several stages during the analysis procedure. A brief description of the major steps in the OSSE spectral analysis process will be described, including a discussion of the OSSE background spectrum and examples of several observation strategies.

036.079   Long–term source monitoring with BATSE.
R. B. Wilson, B. A. Harmon, M. H. Finger, G. J. Fishman, C. A. Meegan, W. S. Paciesas.
Compton Observatory Science Workshop, p. 35 – 46 (Feb 1992).
– See Abstr. 012.018 for the main entry.

The uncollimated BATSE large area detectors (LADs) are well suited to nearly continuous monitoring of the stronger hard X–ray sources, by use of Earth occultation for non–pulsed sources, and time series analysis for pulsars. An overview of the analysis techniques presently being applied to the data are discussed, including representative observations of the Crab Nebula, Crab pulsar, and summaries of the sources detected to date. Results of a search for variability in the Crab Pulsar pulse profile are presented.

036.080 Preliminary calibration results for the BATSE instrument on CGRO.
G. N. Pendleton, W. S. Paciesas, G. J. Fishman, R. B. Wilson, C. A. Meegan, F. E. Roberts, J. P. Lestrade, J. M. Horack, M. N. Brock, M. D. Flickinger.
Compton Observatory Science Workshop, p. 47 – 52 (Feb 1992). – See Abstr. 012.018 for the main entry.
Preliminary results pertaining to spectral reconstruction using BATSE Large Area Detector measurements of solar flares will be presented. These solar flare measurements are currently being used to fine tune the calibration of the data analysis software. The current status of the stability of spectral analysis given the systematic errors present in burst location at the time of the writing of this paper discussed. A brief description of enhancements to the input data for the atmospheric scattering algorithm that will be implemented in the data analysis software is presented.

036.081 BATSE spectroscopy analysis system.
B. E. Schaefer, S. Bansal, A. Basu, P. Brisco, T. L. Cline, E. Friend, N. Laubenthal, E. S. Panduranga, N. Parker, B. Rust, T. Sheets, B. J. Teegarden, G. J. Fishman, C. A. Meegan, R. B. Wilson, W. S. Paciesas, G. Pendleton, J. L. Matteson.
Compton Observatory Science Workshop, p. 53 – 59 (Feb 1992). – See Abstr. 012.018 for the main entry.
The BATSE Spectroscopy Analysis System (BSAS) is the software system which is the primary tool for analysis of spectral data from BATSE. As such, one needs to know its basic properties and capabilities. This paper describes the characteristics of the BATSE spectroscopy detectors and the BSAS.

036.082 Data analysis of the COMPTEL instrument on the NASA Gamma Ray Observatory.
R. Diehl, K. Bennett, W. Collmar, A. Connors, J. W. den Herder, W. Hermsen, G. G. Lichti, J. A. Lockwood, J. Macri, M. McConnell, D. Morris, J. Ryan, V. Schönfelder, H. Steinle, A. W. Strong, B. N. Swanenburg, C. de Vries, C. Winkler.
Compton Observatory Science Workshop, p. 95 – 101 (Feb 1992). – See Abstr. 012.018 for the main entry.
The imaging Compton telescope COMPTEL on the Gamma Ray Observatory is a wide field–of–view instrument. The coincidence measurement technique in two scintillation detector layers requires specific analysis methods. Detector events are analyzed in a multi–dimensional dataspace using a gamma–ray sky hypothesis convolved with the point spread function of the instrument in this dataspace. Background suppression and analysis techniques have important implications on the gamma ray source results for this background limited telescope. The COMPTEL collaboration applies a software system of analysis utilities, organized around a database management system.

036.083 The EGRET data products.
J. R. Mattox, D. L. Bertsch, C. E. Fichtel, R. C. Hartman, S. D. Hunter, G. Kanbach, D. A. Kniffen, P. W. Kwok, Y. C. Lin, H. A. Mayer–Hasselwander, P. F. Michelson, C. von Montigny, P. L. Nolan, K. Pinkau, H. D. Radecke, H. Rothermel, E. Schneid, M. Sommer, P. Sreekumar, D. J. Thompson.
Compton Observatory Science Workshop, p. 126 – 136 (Feb 1992). – See Abstr. 012.018 for the main entry.
The authors describe the EGRET data products which they anticipate will suffice for virtually all guest and archival investigations. The production process, content, availability, format, and the associated software of each product is described. This paper also supplies sufficient detail for the archival

researcher to do analysis which is not supported by extant software.

036.084 Neural network classification of "questonable" EGRET events.
C. A. Meetre, J. P. Norris.
Compton Observatory Science Workshop, p. 137 – 144 (Feb 1992). – See Abstr. 012.018 for the main entry.
High–energy gamma rays (> 20 MeV) pair producing in the spark chamber of EGRET give rise to a characteristic but highly variable 3–D locus of spark sites, which must be processed to decide whether the event is to be included in the database. A significant fraction ($\sim 15\%$: 10^4events/day) of the candidate events cannot be categorized (accept/reject) by an automated rule–based procedure, and must be examined and classified manually by a team of expert analysts.The authors describe a feed–forward, back–propagation neural network approach to the classification of the questionable events. The algorithm computes a set of coefficients using representative exemplars drawn from the preclassified set of questionable events. These coefficients map a given input event into a decision vector that, ideally, describes the correct disposition of the event. The net's accuracy is then tested using a different subset of preclassified events. Preliminary results demonstrate the net's ability to correctly classify a large proportion of the events for some categories of questionables.

036.085 Gamma ray pulsar analysis from photon probability maps.
L. E. Brown, D. D. Clayton, D. H. Hartmann.
Compton Observatory Science Workshop, p. 267 – 272 (Feb 1992). – See Abstr. 012.018 for the main entry.
The authors present a new method of analyzing skymap–type γ–ray data. Each photon event is replaced by a probability distribution on the sky corresponding to the observing instrument's point spread function. The skymap produced by this process may be useful for source detection or identification. Most important, the use of these photon weights for pulsar analysis promises significant improvement over traditional techniques.

036.086 Ulysses/BATSE observations of cosmic gamma–ray bursts.
K. Hurley, M. Boer, M. Sommer, G. Fishman, C. Meegan, W. Paciesas, R. Wilson, C. Kouveliotou, T. Cline.
Compton Observatory Science Workshop, p. 288 – 292 (Feb 1992). – See Abstr. 012.018 for the main entry.
The gamma–ray burst detector aboard the ESA–NASA Ulysses spacecraft has detected numerous gamma–bursts in conjunction with the BATSE experiment aboard the Compton Observatory. The authors present initial results on burst locations for three events (Apr 21, May 2, and May 3 1991) obtained by arrival time analysis, and compare them with the BATSE locations. The arrival time analysis annuli have typical widths of $5'$. This preliminary analysis indicates that both experiments are likely to have unresolved systematic errors, but that further work will improve the location accuracy substantially.

036.087 Calibration of radioastronomical observations in the presence of a radome.
Z. Abraham, F. Kokubun.
Astron. Astrophys., Vol. 257, No. 2, p. 831 – 834 (Apr 1992).
The authors studied the effects of a wet radome on the determination of the atmospheric opacity and the calibration of radioastronomical observations. They derived an algorithm, based on the measurements of a noise source and of a room temperature load, that compensates the atmospheric absorption even in the presence of a wet radome. The algorithm was tested with observations of Jupiter at 22 and 30 GHz under very different weather conditions and proved to be satisfactory. This result is very important in the evaluation of the reliability of long–term variability studies, where the observations are made during part of the year at night or early morning, when the radome is likely to be wet.

036.088　Photometric techniques for close binary and multiple systems.
M. N. Devaney.
Astron. Astrophys., Vol. 257, No. 2, p. 835 – 843 (Apr 1992).

Techniques for the photometry of close binaries are reviewed and developed in the context of seeing theory. Different techniques are appropriate depending upon whether the binary separation is greater than the seeing–limited resolution ($\sim 1''$), in the range $0\rlap{.}''4$–$1\rlap{.}''0$, or close to the diffraction limit of the telescope being used. These techniques may also be useful for improving the photometry of crowded fields.

036.089　The FAST Hipparcos Data Reducation Consortium: overview of the reduction software.
J. Kovalevsky, J. L. Falin, J. L. Pieplu, P. L. Bernacca, F. Donati, M. Froeschlé, I. Galligani, F. Mignard, B. Morando, M. A. C. Perryman, H. Schrijver, D. T. van Daalen, H. van der Marel, M. Villenave, H. G. Walter, M. Badiali, L. Borriello, W. N. Brouw, E. Canuto, A. Guerry, R. Hering, C. Huc, D. Iorio–Fili, P. Lacroute, M. Lattanzi, R. S. Le Poole, F. P. Murgolo, R. A. Preston, S. Röser, F. Sansò, R. Wielen, P. Belforte, H. H. Bernstein, B. Bucciarelli, D. Cardini, A. Emanuele, B. Fassino, H. Lenhart, J. F. Lestrade, G. Prezioso, T. Tommasini Montanari.
Astron. Astrophys., Vol. 258, No. 1, p. 7 – 17 (May 1992).

This paper gives the main features of the data analysis methods adopted by FAST in the calibration and processing of the Hipparcos data. The structure of the data reduction software is described and the actual processing and evaluation is presented.

036.090　The NDAC Hipparcos data analysis consortium. Overview of the reduction methods.
L. Lindegren, E. Høg, F. van Leeuwen, C. A. Murray, D. W. Evans, M. J. Penston, M. A. C. Perryman, C. Petersen, N. Ramamani, M. A. J. Snijders, S. Söderhjelm, G. K. Andreasen, A. M. Cruise, N. Elton, N. Lund, K. Poder.
Astron. Astrophys., Vol. 258, No. 1, p. 18 – 30 (May 1992).

This paper gives an overview of the assumptions and algorithms adopted by the Northern Data Analysis Consortium (NDAC) for the reduction of data from the space astrometry satellite Hipparcos. A fairly detailed account is given of the main steps of the reduction, viz., the attitude determination, analysis of image dissector tube data, great–circle reductions and sphere solution. Of these, only aspects of the attitude determination have required substantial revision due to the satellite's failure to reach geostationary orbit. A brief outline is given of the instrument calibrations and the analysis of non–single stars.

036.091　Attitude determination in the Hipparcos revised mission.
F. Donati, M. Froeschlé, J. L. Falin, E. Canuto, J. Kovalevsky.
Astron. Astrophys., Vol. 258, No. 1, p. 41 – 45 (May 1992).

An accurate knowledge of the attitude as determined by the observation of star transits through the star mapper is an essential factor in the final accuracy of the Hipparcos reductions. In comparison with the nominal mission, several features tend to degrade the precision of the attitude determination: an irregular increase of the background noise, an increased number of data recovery interruptions, and large domain of variation of the gravity gradient torque. It is shown in this paper how these difficulties have been overcome. A precision assessment of the attitude determination is given for the first treatment showing that the effects of the remaining errors (30 to 100 milli–arcsec rms) on the reduction is within acceptable limits. In addition, the software permits an improvement of the along–scan positions of the observed stars. The mean precision of the Input Catalogue can be assessed from the results: 0.25 arcsec rms.

036.092　Method of comparison between determinations of the Hipparcos attitude.
F. Donati, G. Sechi.
Astron. Astrophys., Vol. 258, No. 1, p. 46 – 52 (May 1992).

Attitude comparison is the way adopted to assess the quality of the attitude estimates obtained by the Hipparcos data reduction Consortia (FAST and NDAC). The comparison method is described and illustrated by results obtained in a typical FAST–NDAC comparison developed during the attitude determination assessment phase.

036.093　Modelling the torques affecting the Hipparcos satellite.
F. van Leeuwen, M. J. Penston, M. A. C. Perryman, D. W. Evans, N. Ramamani.
Astron. Astrophys., Vol. 258, No. 1, p. 53 – 59 (May 1992).

This paper presents a description of the methods used to derive a model of the torques working on the Hipparcos satellite and results based on 12 weeks of data spread over 14 months.

036.094　Hipparcos great–circle reduction. Theory, results and intercomparisons.
H. van der Marel, C. Petersen.
Astron. Astrophys., Vol. 258, No. 1, p. 60 – 69 (May 1992).

In this paper some of the results of the great–circle reduction in the two data reduction consortia (NDAC and FAST) are presented. It is shown that the root–mean–square (rms) error of the great–circle abscissa of stars brighter than 10 mag is between 3–5 milli–arcsec. This is as was expected before launch, even in spite of the fact that the data–sets are shorter in the revised mission. There is a large positive correlation between the abscissa of stars separated by a multiple of the basic angle. Moreover, the great–circle results are affected by systematic errors at the beginning of the reductions. This is confirmed by intercomparisons between the two consortia. Finally, it is shown that the along–scan attitude of the Hipparcos satellite can be described with a few hundred parameters with milli–arcsec accuracy.

036.095　Comparison of Hipparcos results obtained on different dates on the same great circle.
H. Schrijver.
Astron. Astrophys., Vol. 258, No. 1, p. 70 – 73 (May 1992).

The results on great circles that have been scanned by Hipparcos on dates separated by several months are compared. The differences in the abscissae, after correction for known proper motions and parallaxes show a noise and a systematic effect that correspond to expected inaccuracies in proper motions and parallaxes. Instrumental terms related to the position of the sun are not significantly detected, with the possible exception of a first harmonic term. If the systematic effect is entirely due to parallax, the mean value of the parallax in the observed sets of stars would be of the order of 5 milli–arcsec.

036.096　Early improvements to the Hipparcos Input Catalogue through the accumulation of data from the satellite. Including the NDAC attitude reconstruction description.
F. van Leeuwen, D. W. Evans, L. Lindegren, M. J. Penston, N. Ramamani.
Astron. Astrophys., Vol. 258, No. 1, p. 119 – 124 (May 1992).

The authors present a description of ways in which the Northern Data Analysis Consortium (NDAC) accumulates positional and photometric data as obtained in the early stages of the reduction of data from the Hipparcos mission. An outline of the closely associated attitude reconstruction process is also presented.

036.097　Hipparcos photometry: FAST main mission reduction.
F. Mignard, M. Froeschlé, J. L. Falin.
Astron. Astrophys., Vol. 258, No. 1, p. 142 – 148 (May 1992).

The authors describe a photometric analysis of the Hipparcos data collected with the main detector. The overall principles of the photometric reduction are presented along with the first results. Both results at the transit level and over longer time scales are considered. It is shown that for a non–variable star of magnitude 8.5 an accuracy as good as 0.002 mag will be achieved at the mission completion.

036.098   **Hipparcos photometry: NDAC reductions.**
D. W. Evans, F. van Leeuwen, M. J. Penston,
N. Ramamani, E. Høg.
Astron. Astrophys., Vol. 258, No. 1, p. 149 – 156 (May 1992).
The Hipparcos mission will provide astrometry for about
120000 stars. It will also give very accurate photometry for these
stars. The resulting data base will by far surpass any existing
photometric catalogue, both in accuracy and number of stars
measured. This paper describes the calibrations and initial
reductions of the photometry carried out by the Northern Data
Analysis Consortium (NDAC), one of the two data reduction
groups analysing the data.

036.099   **Detection and measurement of double stars with the**
Hipparcos satellite: NDAC reductions.
S. Söderhjelm, D. W. Evans, F. van Leeuwen, L. Lindegren.
Astron. Astrophys., Vol. 258, No. 1, p. 157 – 164 (May 1992).
The Hipparcos satellite has a predefined observing program
given by an Input Catalogue with 118000 entries. This includes
about 10000 members of known double or multiple systems (with
separations above 0.1 arcsec and magnitude differences below
4–5), which will be observably non–single and thus need a more
complex reduction procedure. Also, a similar number of objects
showing deviations from a single–star observation model will be
included in these "double star" reductions, and a few thousand
hitherto unknown doubles may thus be discovered. This paper
describes the methods used by the Northern Data Analysis
Consortium (NDAC) for detecting suspected non–singles and for
determining the photometric and astrometric parameters for the
non–single objects. A key feature of these methods is that they
use the standard reduction results for calibrating and collecting
all the observations of a specific object. Only after this derivation
of individual "Case History Files" is completed, the astrometric
parameters are determined for one object at a time by one of
several solution programs. The solutions use all the available
information in one step, and all double star parameters are in the
absolute reference system defined by the standard Hipparcos
reductions. Some early results derived from "provisonal" data
tapes are given which illustrate and validate the main steps of the
processing.

036.100   **Hipparcos double star recognition and processing within**
the FAST consortium.
F. Mignard, M. Froeschlé, M. Badiali, D. Cardini,
A. Emanuele, J. L. Falin, J. Kovalevsky.
Astron. Astrophys., Vol. 258, No. 1, p. 165 – 172 (May 1992).
The Hipparcos observations are potentially a source of
discovery of new systems of double and multiple stars. Procedures have been developed to recognize in the signal parameters
the multiplicity of the stellar objects. The principles applied to the
recognition of double stars are presented in this paper along with
statistical tests. Results based on real data indicate that about 600
new double stars will be discovered at the mission completion.
The authors also describe the main algorithms which allow the
determination of the photometric and astrometric parameters of
double stars.

036.101   **The treatment of Hipparcos observations of some**
peculiar double stars: anomalous cases.
R. Pannunzio, A. Spagna, M. G. Lattanzi, R. Morbidelli,
M. Sarasso.
Astron. Astrophys., Vol. 258, No. 1, p. 173 – 176 (May 1992).
The authors present an investigation on a few different types of
double stars which, when observed by the Hipparcos satellite, will
probably produce anomalous signals. Specifically, for these
objects the automatic data reduction chain set up by the FAST
Consortium will usually give inconsistent results. Thus, some new
mathematical models have been considered in order to obtain
reliable parameters of such anomalous pairs, quite interesting
from the astrophysical point of view. Those models are based on
an attempt to systematically classify which systems are to be
considered peculiar within the Hipparcos mission. As recognized
in the past, among those double stars giving anomalous signals
the authors find pairs with fast relative motion (rectilinear,

parallactic and orbital), doubles with variable components, and
multiple stars.

036.102   **Tycho data analysis. Overview of the adopted reduction**
software and first results.
E. Høg, U. Bastian, D. Egret, M. Grewing, J. L. Halbwachs,
A. Wicenec, G. Bässgen, P. L. Bernacca, F. Donati,
J. Kovalevsky, F. van Leeuwen, L. Lindegren, H. Pedersen,
M. A. C. Perryman, C. Petersen, D. Scales, M. A. J. Snijders,
P. R. Wesselius.
Astron. Astrophys., Vol. 258, No. 1, p. 177 – 185 (May 1992).
In the Tycho project astrometric and photometric data of
about 1000000 stars to a limit of $B = 12$ mag will be derived. The
brightest 500000 stars will obtain magnitudes in two colours B_T
and V_T, the fainter 500000 usually only one broad–band
magnitude T. There is confidence that an accuracy will be
achieved about 0.03 arcsec for positions and 0.03 mag for
magnitudes, at $B = 10.5$ mag. The results are obtained by
appropriate treatment of the continuous data records generated
by the Hipparcos satellite's star mapper (which provide simultaneous measurements in two spectral channels), based on predicted star transits using its own "Tycho Input Cataloue". The
data treatment is carried out by the Tycho Data Analysis
Consortium (TDAC), using calibration and satellite attitude
information from the two Hipparcos data reduction consortia.
An overview of the data reductions is given. Raw observation
data and their numerical treatment are described. This assessment is based on a small data set, i.e. 12 weeks spread over the
first 8 months of the Tycho "revised mission" where the satellite
is in the elliptical transfer orbit, instead of the intended
geostationary orbit.

036.103   **Tycho transit detection.**
G. Bässgen, A. Wicenec, G. K. Andreasen, E. Høg,
K. Wagner, P. Wesselius.
Astron. Astrophys., Vol. 258, No. 1, p. 186 – 192 (May 1992).
By means of some extracted parts of the raw countrates of the
Tycho photomultipliers the method of detecting signals of stars
and estimating their amplitudes and transit positions is described.
The elimination of disturbancies due to spikes with the help of
sophisiticated folding and background determination methods is
presented.

036.104   **Tycho star recognition.**
J. L. Halbwachs, E. Høg, U. Bastian, P. C. Hansen,
P. Schwekendiek.
Astron. Astrophys., Vol. 258, No. 1, p. 193 – 200 (May 1992).
The observations of the first year of mission of the Tycho
program will be used for revising the Tycho Input Catalogue
(TIC). The Tycho Input Catalogue Revision essentially defines
the list of objects in the final Tycho output catalogues. This paper
describes the mathematical and practical details of this revision
process. The stars will be recognized with three different
processes, according to their distances from the positions in the
TIC. The main process concerns the stars closer than 6 arcsec to
the TIC positions; stars with separations between 6 and 20 arcsec
are recognized too, but the threshold in detection is slightly
brighter than in the main process. Stars absent from the input
catalogue could also be recognized, but with an even higher
threshold in detection. An assessment based on about 85 h of
actual Hipparcos observations is presented. It points to a Tycho
Input Catalogue Revision containing about 1 million stars.

036.105   **Tycho astrometry calibration.**
E. Høg, U. Bastian, P. C. Hansen, F. van Leeuwen,
L. Lindegren, H. Pedersen, A. B. Saust, P. Schwekendiek,
K. Wagner.
Astron. Astrophys., Vol. 258, No. 1, p. 201 – 205 (May 1992).
The Hipparcos satellite's star mapper gives photon counts in
two spectral channels simultaneously. The transit times and the
signal amplitudes for each star across two groups of four slits are
derived and used for astrometry and photometry, respectively,
and this constitutes the Tycho project. The present paper
describes the Tycho astrometric data processing, leading from the

transit times to a geometric calibration of the slit system and, finally, to the astrometric parameters of the Tycho stars. A small part, mainly three weeks of continuous Tycho observations has been used to study the accuracy of the star mapper calibration.

036.106   Tycho background determination and monitoring.
A. Wicenec, G. Bässgen.
Astron. Astrophys., Vol. 258, No. 1, p. 206 – 210 (May 1992).

The method of background determination in the Tycho photon counts is given. By using a part of the data from the first 8 months periodical and aperiodical behaviour of the background is studied and statistics on the usability of Tycho data are given.

036.107   Tycho photometry calibration and first results.
D. R. Scales, M. A. J. Snijders, G. K. Andreasen, M. Grenon, M. Grewing, E. Høg, F. van Leeuwen, L. Lindegren, H. Mauder.
Astron. Astrophys., Vol. 258, No. 1, p. 211 – 216 (May 1992).

The data reduction methods used for the Tycho photometry are presented and the software package to implement these methods is described. Results derived from a preliminary reduction of eight weeks of data using these methods are presented. A critical discussion is presented of the sources of photometric errors and the expected and the achieved photometric accuracy.

036.108   Statistical analysis of data in planetary observations.
S. G. Valeev.
Sol. Syst. Res., Vol. 25, No. 2, p. 182 – 187 (Sep 1991). English translation of Astron. Vestn., Vol. 25, No. 2, p. 245 – 252 (1991).

In the processing of planetary observations, certain incorrect mathematical models are often applied. The accuracy of data processing can be improved by using the capabilities of applied statistics and computer technology on the basis of the optimal model principle. Hypsometric data for the Moon are processed as a brief illustration of the main limitations of standard mathematical models: 1) redundancy in the number of terms included in the model; 2) overloading with duplicate (interdependent) terms; and 3) violation of certain other assumptions of the least squares method. The chief results of years of observations are briefly summarized.

036.109   On the feasibility of extra–solar planetary detection at very low radio frequencies.
A. Lecacheux.
3. Symposium International de Bioastronomie, p. 21 – 30 (1991). – See Abstr. 012.005 for the main entry.

The search for extra–solar planetary systems remains an important, continuing, astrophysical problem as well as a primary step in the quest for knowledge of the existence of life in the universe. In addition to the methods currently in use for detecting planets, there is the possibility of revealing the existence of planets in other stellar systems by observing their natural, non–thermal low–frequency radiation. Five planets in the solar system, including the Earth, are known to produce low–frequency radiations in the kilometer to decameter wavelength range. The intensity of the planetary radiation is high enough to be detectable, in a reasonable range of distance, with a large but feasible low–frequency radio telescope, operated on the ground, on the Moon or in space. Taking into account the observed properties of the planetary radio emissions in the solar system, and the scaling rules which can be inferred from the plasma physics, the author discusses the feasibility of such a detection.

036.110   A proposal for the search of extrasolar planets by occultation.
J. Schneider, M. Chevreton.
3. Symposium International de Bioastronomie, p. 33 (1991). Abstract. – See Abstr. 012.005 for the main entry.

036.111   Celestial seeing: three imaging technologies.
R. Berry.
Astronomy, Vol. 20, No. 2, p. 68 – 73 (Feb 1992).

Film, video, and CCD – all three are a part of modern amateur astronomy, each with its strengths and weaknesses.

036.112   Very high spatial resolution two–dimensional solar spectroscopy with video CCDs.
A. Johannesson, T. Bida, B. Lites, G. B. Scharmer.
Astron. Astrophys., Vol. 258, No. 2, p. 572 – 582 (May 1992).

The authors have developed techniques for recording and reducing spectra of solar fine structure with complete coverage of two–dimensional areas at very high spatial resolution and with a minimum of seeing–induced distortions. These new techniques permit one, for the first time, to place the quantitative measures of atmospheric structure that are afforded only by detailed spectral measurements into their proper context. The techniques comprise the simultaneous acquisition of digital spectra and slit–jaw images at video rates as the solar scene sweeps rapidly by the spectrograph slit. During data processing the slit–jaw images are used to monitor rigid and differential image motion during the scan, allowing measured spectrum properties to be remapped spatially. The resulting quality of maps of measured properties from the spectra is close to that of the best filtergrams. The authors present the techniques and show maps from scans over pores and small sunspots obtained at a resolution approaching 1/3 arcsec in the spectral region of the magnetically sensitive Fe I lines at 630.15 and 630.25 nm. The maps shown are of continuum intensity and calibrated Doppler velocity.

036.113   Radio–interferometric imaging of spectral lines. The problem of continuum subtraction.
T. J. Cornwell, J. M. Uson, N. Haddad.
Astron. Astrophys., Vol. 258, No. 2, p. 583 – 590 (May 1992).

The authors analyze two methods of continuum subtraction in radio interferometric imaging of spectral lines. These perform a linear fit to determine the continuum emission. In the two methods, UVLIN and IMLIN, the fitting is performed on the original visibilities and dirty images respectively. The authors find that both methods work very well when the continuum emission is spread over a small field of view and there are negligible instrumental errors. Simple expressions for the typical error level for both methods in this regime are given. Both remain robust for significant instrumental errors in the sense that the dynamic range is then proportional to the peak line strength rather than the peak continuum strength. Hybrid methods consisting of one of these methods preceded by the conventionally used UVSUB method should work well when imaging emission spread over a large field of view. The authors show examples of the application of these methods, including a demonstration of the dangers inherent in the UVBAS method in which linear fits to the visibility amplitude and phase are performed.

036.114   Systematic color transformation effects in Strömgren photometry.
J. Manfroid, C. Sterken.
Astron. Astrophys., Vol. 258, No. 2, p. 600 – 604 (May 1992).

A major cause of errors in photometry is to be attributed to the color transformation. A thorough analysis of numerous observing runs carried out in the Strömgren system shows how reliable the transformations are in the different indices. Conclusions are drawn about the reduction methods, the validity of homogenization procedures, and the choice of the standard stars which define the system.

036.115   Optical searches for ROSAT X–ray cluster sources.
R. G. Cruddace, H. T. MacGillivray.
Workshop on Traces of the Primordial Structure in the Universe, p. 205 – 221 (May 1991). – See Abstr. 012.020 for the main entry.

The authors describe the use of Schmidt telescope surveys, which have achieved the most detailed description of the whole sky to date. They desribe first the establishment of a digital data base summarizing the characteristics of some 500 million objects identified on UK Schmidt plates. This data base is interrogated routinely in identifying ROSAT sources. The authors describe automatic techniques used to search for clusters on a Schmidt plate at the location of a ROSAT source. They then describe a project, in which the digitised UK Schmidt data base is searched systematically for clusters of galaxies. The catalog resulting from this search will be published.

036.116   Bridge of ground–based laser enters space–laser guide stars.
Sun Jingwen.
Laser J., Vol. 12, No. 6, p. 281 – 284 (Dec 1991). In Chinese.

The principle and importance of laser guide stars in laser astronomy and adaptive imaging in astronomy is pointed out. The requirement of laser parameters for laser guide stars is discussed.

036.117   Interpretation of lightcurves of atmosphereless bodies. I. General theory and new inversion schemes.
M. Kaasalainen, L. Lamberg, K. Lumme, E. Bowell.
Astron. Astrophys., Vol. 259, No. 1, p. 318 – 332 (Jun 1992).

Theoretical methods of lightcurve inversion are presented that can be used in determination of the 3–D shape and/or the light–scattering behaviour of the surface of a body from disk–integrated photometry. They can be applied to atmosphereless bodies in the solar system. With no loss of generality, the objects of this study are referred to as asteroids. The inversion comprises three steps. First, a function (or functions) containing information about both the shape and the albedo variegation of an asteroid is determined. This step is feasible provided the surface is strictly convex and provided a sufficient number of lightcurves are available at different observing geometries and at nonzero phase angles. Also, the functional form of the surface light–scattering law must be known and it must be of a suitable type. This inversion problem is mathematically illposed; i.e., small errors in the data may have large effects on the results. Also, the number and range of the observing geometries have a significant effect on the inversion. In the second step, separate expressions for the inverse of the Gaussian curvature and the albedo distribution are derived from the information obtained in the first step. This is possible if the functional form of the scattering law as a function of albedo is known and is of a suitable form. In the third step, the non–trivial problem of determining the radius vector of the surface from the Gaussian curvature is solved by using iterative optimization procedures developed by the authors.

036.118   Interpretation of lightcurves of atmosphereless bodies. II. Practical aspects of inversion.
M. Kaasalainen, L. Lamberg, K. Lumme.
Astron. Astrophys., Vol. 259, No. 1, p. 333 – 340 (Jun 1992).

The authors have developed methods of inversion that can be used in the determination of the three–dimensional shape or the albedo distribution of the surface of a body from disk–integrated photometry, assuming the shape to be strictly convex. They call this problem field photomorphography. In this second paper they study the practical aspects of the inversion procedure. They also apply their methods to lightcurve data of 39 Laetitia and 16 Psyche. There are many observational factors having an influence on the outcome of inversion. Also, one must make some a priori assumptions that are used in the inversion process. The most important points are the number and range of the observing geometries, the assumed spin vector of the asteroid, possible large–scale nonconvexity of the surface, and the accuracy of the lightcurves. In an appendix, the authors also describe how the formalism they have presented can be used in generating arbitrary convex shapes conveniently and then in producing synthetic lightcurves efficiently with them.

036.119   The proximity parameter.
X. Luri, J. Torra, F. Figueras.
Astron. Astrophys., Vol. 259, No. 1, p. 382 – 385 (Jun 1992).

A statistical development is presented to use the positional information in photometric planes – the information contained in the position of a star with respect to a calibration line. This development is applied in the case of colour–colour planes, defining what the authors call the proximity parameter, which is used for luminosity classification purposes and for an improved use of calibrations.

036.120   Revision of the Theil fitting method.
A. Pegoraro.
Astron. Astrophys., Vol. 259, No. 1, p. 386 – 393 (Jun 1992).

The non parametric robust fitting method originally due to Theil is reviewed starting from the work of Sen (1968). In that paper only the estimate of the angular coefficient in a straight line fit is considered. Here it is shown that this method can be successfully employed to estimate also the intercept of a straight line, as well as to fit other functions such as polynomials and some non–linear curves: the Gaussian, the Lorentzian, the exponential and the power law. The general method scheme is retained, while a new formulation for the error estimator is proposed. The method proved to be unbiased and robust by the Monte–Carlo simulations, adopting the Gaussian and Cauchy error sources. Comparison with the standard methods are performed too. The suggested reliability of the Theil method to fit data with errors both in the dependent and independent variable is tested. The obtained results ar not fully satisfactory.

036.121   A new method for filling gaps in data.
T. Serre, M. Auvergne, M. J. Goupil.
Astron. Astrophys., Vol. 259, No. 1, p. 404 – 411 (Jun 1992).

A new method for filling gaps in time series is introduced which uses a prediction method for the evolution of a dynamical system. This method is based on the existence of an attractor for the system. Efficiency of the method is shown with a torus and a Rössler chaotic attractor. Application to an RV Tauri variable star light curve is made. The results show that this method allows to fill gaps of several periods, i.e. to identify and decrease aliases in frequency domain, if sufficiently long time sequences are used to faithfully reconstruct the attractor.

036.122   Analyzing Hubble Space Telescope data.
F. Murtagh.
Ir. Astron. J., Vol. 20, No. 3, p. 184 – 187 (Mar 1992). – See Abstr. 012.038 for the main entry.

The author briefly reviews image restoration work currently carried out at ST–ECF, in the context of the spherically–aberrated HST images.

036.123   Geometries of radar astronomy in future experiments.
V. R. Eshleman.
Bull. Am. Astron. Soc., Vol. 23, No. 3, p. 1154 (1991). Abstract. – See Abstr. 012.037 for the main entry.

036.124   Three–beam chopping: an efficient infrared observing technique.
R. Landau, G. Grasdalen, G. C. Sloan.
Astron. Astrophys., Vol. 259, No. 2, p. 696 – 700 (Jun 1992).

The authors describe a technique for canceling the background during infrared observations without moving the telescope (without beam–switching). The secondary mirror is chopped between three positions: one on the source and the other two on opposite sides of the source. Some advantages over the traditional two–beam chopping with beam–switching are: higher observing efficiency, continuous monitoring of the source, cancellation of more rapidly changing noise sources, and elimination of the need to maintain any relation between the chopper throw and the amplitude of telescope motion during beam–switching. The technique also greatly simplifies observing with infrared arrays. It is simple to implement the method, both for computer–controlled secondaries and for those controlled by analog signals. Observations between 2 and 20 μm from two observatories are discussed.

036.125   **Photometric analysis of astronomical images by the wavelet transform.**
G. Coupinot, J. Hecquet, M. Aurière, R. Futaully.
Astron. Astrophys., Vol. 259, No. 2, p. 701 – 710 (Jun 1992).

An application of the wavelet analysis to astronomical images is presented. The authors show that this new method is a powerful tool for the detection of features whose scales are known. The detection is easy and accurate because, in a wavelet analysis, the effect of the background is removed. If a model is fitted to the objects, photometric data can be deduced from the parameters of the wavelet transform. If Gaussian models fit to the objects and to the point–spread function, the process is analytical. To extend the method to other models, numerical simulations have to be performed. The method has been applied to CCD observations of the galaxies VV 523 and M31, performed at the Bernard Lyot 2 m telescope of the observatory Pic–du–Midi. In VV 523, 13 clumps have been detected and their sizes and B–I color indexes determined. The detection of 12 new globular clusters in the bulge of M31 (including photometry and core radius determination) enables to fill the faint part of the luminosity distribution of M31 globular clusters. Furthermore the new clusters are similar in magnitude and core radius to those classified as post core collapse clusters in the Milky Way.

036.126   **Possible application of circular polarization for remote sensing of cosmic bodies.**
V. S. Degtjarev (*V. S. Degtyarev*), L. O. Kolokolova.
Earth, Moon, Planets, Vol. 57, No. 3, p. 213 – 223 (Jun 1992).

Phase dependences of circular polarization were obtained with a precision Stokes polarimeter designed and constructed at the Main Astronomical Observatory of the Academy of Sciences of Ukraine. A study was made of dielectric and metallic powders with grains of diameter 10–100 μm. Metallic powders were found to produce an essential circular polarization – up to 3%, just as dielectric powders did not show circular polarization values more than 0.05%. Change of circular polarization with phase angle V greatly depends on surface structure. Loose powders give phase curves with the same sign of circular polarization everywhere and with maximum at large phase angles $V > 120°$. Measurements of compacted powders show curves which change the sign repeatedly and have additional maxima, including a maximum at small phase angles $V < 40°$. A theory was created which considers a circular polarization as a result of multiple reflections of light from particulate surface. The theory provides reasonable good fit to the experimental data. It was concluded that measurements of circular polarization can be used to find metals in surface material of cosmic bodies (especially asteroids) and to determine characteristics of surface structure, in particular, to establish presence of regolith on metal–rich bodies.

036.127   **Data archiving for the International Ultraviolet Explorer (IUE) satellite.**
W. Wamsteker, M. Barylak, C. Driessen.
ESA Bull., No. 70, p. 60 – 65 (May 1992).

This article examines the ways in which the International Ultraviolet Explorer (IUE) Project has solved the problems associated with the access and exploitation of vast volumes of archived astronomical data. The lessons learnt are relevant to the planning and implementation of other high–volume scientific data archives.

036.128   **La photo planétaire.**
L. Spede.
Ciel, Vol. 54, p. 59 – 63 (Mar 1992).

036.129   **La photographie du ciel profond.**
L. Spede.
Ciel, Vol. 54, p. 106 – 110 (Apr 1992).

036.130   **Removing cosmic rays and other randomly positioned spurious events from CCD images by taking the lesser image – statistical theory for the general case.**
L. Kay.
Meas. Sci. Technol., Vol. 3, No. 4, p. 400 – 405 (Apr 1992).

If two optical images of the same scene are obtained using a CCD, a third image (called the lesser image) may be formed in computer memory by taking the lesser of the two counts in each pixel. The process may be used to remove, or greatly reduce, the effect of spurious events such as cosmic rays. A complete statistical theory of the lesser image is given for the general case, thereby facilitating recovery of the true image from the lesser image.

036.131   **An approach to trend analysis in data with special reference to cometary magnitudes.**
J. R. Donnison, H. W. Peers.
Mon. Not. R. Astron. Soc., Vol. 256, No. 4, p. 647 – 654 (15 Jun 1992).

A recent approach to trend analysis in data is introduced and developed in the context of a Pareto power–law model. The way in which theory based on the normal distribution can be modified by suitable choice of parameter and sample size to deal with general parametric families of distributions is explained with numerical examples. The Pareto distribution used by Donnison to model the brightness (magnitude) distribution of long–period, ($P > 200$ yr) and short–period ($P < 15$ yr) comets is reanalysed in terms of this new trend approach. Overwhelming evidence is found that the brightness indices increase with discovery date for the long–period comets. A plateau in values is apparent in the comets discovered after about 1830; comets discovered prior to this date form a truncated sample due to observational selection and should be employed with care in any statistical analysis. No trends are found in the brightness distribution of these comets as a function of their perihelion distance. Similarly, no trends are apparent in the brightness distribution of the short–period comets with increased orbital period.

036.132   **Cross–correlation methods for surface photometry: two–component models.**
S. Phillipps, P. J. Boyce.
Mon. Not. R. Astron. Soc., Vol. 256, No. 4, p. 673 – 678 (15 Jun 1992).

The cross–correlation or matched filter technique for determining surface photometric parameters in low signal–to–noise ratio data is extended to two–component models (i.e. up to four model parameters). It is shown that the linearity and distributivity of the convolution process greatly reduce the number of templates with which the data must be convolved. Indeed, provided that we are dealing with a significant number of images, obtaining a full four–parameter maximization requires only two to three times as much work as for a simple single–component fit. The method is simple to implement and particularly suited to determining the component parameters for many images in a single frame.

036.133   **Physical parameters of VLBI structures.**
T. W. Jones.
7. IAP Meeting: Extragalactic radio sources – from beams to jets, p. 159 – 166 (1992). – See Abstr. 012.041 for the main entry.

036.134   **Investigtion of some methods for determining statistical weights of astrometric catalogues.**
V. I. Zhdanov, S. L. Parnovskij, V. V. Tel'nyuk–Adamchuk.
Visn. Kiiv. Univ., Fiz.–Mat. Nauki, Astron., Vip. 3, p. 49 – 53 (1992). In Ukrainian.

The iteration schemes to find the statistical weights of several individual catalogues for their unification into a combined one have been compared through computer simulations. There was determined the scheme which reproduces most adequately the prescribed weights of statistically generated catalogues. Only two iterations appeared to be sufficient to obtain the final result.

036.135   **On flux calibration of spectra.**
M. A. Fluks, P. S. Thé.
Messenger, No. 67, p. 42 – 46 (Mar 1992).

036.136   **Confidence intervals for the Lutz–Kelker correction.**
C. Koen.
Mon. Not. R. Astron. Soc., Vol. 256, No. 1, p. 65 – 68 (1 May 1992).

The distribution function of the bias in the absolute magnitude of a star, as determined from its trigonometric parallax, is calculated for the case where the error dispersion is imperfectly known. A general power law is assumed for the spatial distribution of stars. The distribution function is used to calculate most probable and mean bias corrections as well as confidence intervals for the correction, the latter providing an indication of the intrinsic accuracy of the Lutz–Kelker correction. The basic method may be used for bias correction in other situations also.

036.137   Gain calibration of CCD systems at VBO.
T. P. Prabhu, Y. D. Mayya, G. C. Anupama.
J. Astrophys. Astron., Vol. 13, No. 1, p. 129 – 144 (Mar 1992).
The system gain of two CCD systems in regular use at the Vainu Bappu Observatory, Kavalur, is determined at a few gain settings. The procedure used for the determination of system gain and base–level noise is described.

036.138   High spatial resolution techniques.
O. von der Lühe.
1. Canary Islands Winter School of Astrophysics: Solar observations – techniques and interpretation, p. 1 – 69 (1992). See also Abstr. 54.036.276. – See Abstr. 012.045 for the main entry.
Contents: (1) Introduction. (2) Basic considerations and concepts: diffraction and optical transfer; coherence, Van Cittert–Zernike theorem. (3) Wave propagation through the atmosphere: statistics of index of refraction fluctuations; mutual intensity of a wave disturbed by turbulence; structure functions; the instantaneous optical transfer function. (4) Single frame analysis: data collection and preparation; image selection; single picture restoration; time series analysis. (5) Interferometry: Michelson interferometry; interferometric arrays; speckle interferometry; the Labeyrie method; seeing calibration; noise calibration; speckle imaging, Knox–Thompson; speckle imaging, speckle masking; speckle interferometry and anisoplanatism. (6) Active wavefront compensation: image motion compensation; adaptive optics. (7) Other methods: radio observations; high resolution observations from space; solar optical universal polarimeter; orbiting solar laboratory; solar and heliospheric observatory.

036.139   Magnetic field measurements.
E. Landi Degl'Innocenti.
1. Canary Islands Winter School of Astrophysics: Solar observations – techniques and interpretation, p. 71 – 143 (1992). – See Abstr. 012.045 for the main entry.
Contents: (1) Description of polarized radiation. (2) A prototype polarimeter: setting of various devices at fixed angles; measurements with rotating wave plates; measurements with variable retarders. (3) Physical components of polarimeters: polarizers; retarders. (4) Generalities on polarization phenomena in spectral lines: Zeeman effect; impact and resonance polarization; resonance polarization and the Hanle effect; the role of collisions in resonance polarization; a classification scheme for polarimetric observations. (5) Radiative transfer for polarized radiation. (6) Line formation in a magnetic field. (7) Transfer equations for the Stokes parameters in a magnetized atmosphere. (8) Solutions of the transfer equations and magnetic field measurements: weak field solution; solution for a Milne–Eddington atmosphere; more general analytical solutions; numerical solutions; particular solutions. (9) Magnetic field measurements in unresolved structures: the line ratio technique. (19) Magnetic field measurements in prominences.

036.140   Conversion to the true optical center in the complete overlapping–plate method.
Z. A. Galieva, N. N. Matveev, E. M. Loginova.
Kinematics Phys. Celest. Bodies, Vol. 7, No. 4, p. 68 – 70 (1991). English translation of Kinematika Fiz. Nebesn. Tel., Tom 7, No. 4, p. 88 – 90 (1991).
Conversion to the true optical center for astronomical reduction makes it possible to use a six–constant (instead of an eight–constant) model. This possibility was theoretically validated by A.A. Kiselev (1989).

036.141   Comparison of Kalman and Wiener filtering techniques for processing Pioneer Venus radio occultation data.
J. M. Jenkins, P. G. Steffes.
Bull. Am. Astron. Soc., Vol. 23, No. 3, p. 1195 (1991). Abstract. – See Abstr. 012.037 for the main entry.

036.142   A robust algorithm for mapping minerals using spectroscopy.
R. N. Clark, G. A. Swayze, A. J. Gallagher.
Bull. Am. Astron. Soc., Vol. 23, No. 3, p. 1199 (1991). Abstract. – See Abstr. 012.037 for the main entry.

036.143   Fractal analysis: a new remote sensing tool.
B. C. Bruno, G. J. Taylor, S. K. Rowland, P. G. Lucey, S. Self.
Bull. Am. Astron. Soc., Vol. 23, No. 3, p. 1200 (1991). Abstract. – See Abstr. 012.037 for the main entry.

036.144   Real time wavefront reconstruction for a 512 subaperture adaptive optical system.
P. Johnson, R. Trissel, L. Cuellar, B. Arnold, D. Sandler.
SPIE Conference on Active and Adaptive Optical Components as Part of SPIE's International Symposium on Optical Applied Science and Engineering, p. 460 – 471 (1992). – See Abstr. 012.052 for the main entry.
This paper describes a system capable of real time wavefront reconstruction for a 512 subaperture shearing interferometer. The system was designed to interface with a 1536 channel (512 segment) deformable mirror for atmospheric compensation using an artificial beacon. The phase gradients were measured using a shearing interferometer operating at two distinct shear lengths with quantum limited performance at ~ 100 photoelectrons per subaperture. A 128 node parallel processor performed a sparse matrix multiply to reconstruct the phasefront in real time. The matrix truncation technique used allowed 90% of the elements to be removed with only minor penalty in wavefront accuracy.

036.145   Wavefront sensing in imaging through the atmosphere: a detector strategy.
M. Séchaud, G. Rousset, V. Michau, J. C. Fontanella, J. G. Cuby, F. Rigaut, J. C. Richard.
SPIE Conference on Active and Adaptive Optical Components as Part of SPIE's International Symposium on Optical Applied Science and Engineering, p. 479 – 490 (1992). – See Abstr. 012.052 for the main entry.
Wavefront sensing is a very powerful technique whose capability in the field of diffraction–limited imaging through turbulence has been demonstrated. The ultimate performance of a Hartmann–Shack wavefront sensor is analysed and used to define a detector choice strategy.

036.146   Recovery of atmospheric phase distortion from stellar images using an artificial neural network.
D. G. Sandler, T. K. Barrett, R. Q. Fugate.
SPIE Conference on Active and Adaptive Optical Components as Part of SPIE's International Symposium on Optical Applied Science and Engineering, p. 491 – 499 (1992). – See Abstr. 012.052 for the main entry.
The authors report recent experimental verification of a new method to determine atmospheric phase directly from focused images of starlight. An artificial neural network is used to infer the phase from two images of a star, one at the exact focus and another intentionally out of focus. The authors applied the network to images of Vega obtained on a 1.5-meter telescope. Neural network predictions agree well with phase reconstructions using a conventional Hartmann wavefront sensor. The network approach offers a simple, inexpensive way to implement adaptive optics on astronomical telescopes in the near term.

036.147 The MIT program for identifying occultations and appulses by planets.
S. W. McDonald, A. S. Bosh, C. B. Sybert, H. B. Hammel, J. L. Elliot.
Bull. Am. Astron. Soc., Vol. 23, No. 3, p. 1210 (1991). Abstract. – See Abstr. 012.037 for the main entry.

036.148 Eine neue Entfaltungsmethode für die Doppler–Kartographierung von Oberflächen kühler Sterne und ihre Anwendung auf AB Doradus.
M. Kürster.
Diss. (Dr.rer.nat.), München Univ. (Germany). Fakultät für Physik; Max–Planck–Institut für Physik und Astrophysik, Garching (Germany). Inst. für Extraterrestrische Physik. MPE—236, 281 p. (Mar 1992). ISSN 0178–0719.
Subject of this thesis is the development of methods for the mapping of stellar surface structures, in particular the mapping of photospheric star spot distributions on active cool stars. Besides an extensive discussion of those observational phenomena that contain information on the distribution of surface structures procedures of extracting this information from the observational data are studied. Major attention is given to a method called "Doppler imaging" which enables the observer to produce detailed 2–dimensional images of the photospheres of rapidly rotating stars. The newly developed method is applied to recent observations of the rapidly rotating pre–main–sequence star AB Dor. For the first time variations of the profile shapes of neutral metal lines (Ca I and Fe I) were found for this K0 V star. These are attributable to star spots. As a comparision a Doppler image was also produced with the maximum entropy method which is in good agreement with the CLEAN reconstruction with respect to the position of surface structures.

036.149 The selection of a sampling interval for digitisation: how fine is fine enough?
V. G. Laidler, B. M. Lasker, M. Postman.
2. Conference on Digitised Optical Sky Surveys, p. 95 – 101 (1992). – See Abstr. 012.059 for the main entry.
A critical property of digitized sky surveys prepared from photographic materials is the sample interval, δ, used by the microdensitometer. An issue is a tradeoff between coarse sample intervals that, while economical, do not allow the faithful reproduction of the information in photographic material, and fine intervals, which can record all of the information, but at a prohibitive cost. The authors investigate these issues by conducting Fourier analyses of images characteristic of the imaging process in the absence of noise and by examining the properties of object inventories for a test field digitized at a range of sampling intervals. Both analyses indicate that 15 μm is an adequate sampling interval for modern Schmidt plates on Type III emulsions.

036.150 Hardware and software aspects of CCD camera–based astrometric plate measurements.
L. Winter, C. de Vegt, M. Steinbach, N. Zacharias.
2. Conference on Digitised Optical Sky Surveys, p. 123 – 131 (1992). – See Abstr. 012.059 for the main entry.
This paper discusses the application of CCD cameras for digitisation of astrometric plate material. Details of the astrometric plate measuring systems and their design principles at Hamburg Observatory are discussed. First results concerning performance and obtainable measurement accuracies are presented.

036.151 Real–time data processing for SuperCOSMOS.
M. Paterson.
2. Conference on Digitised Optical Sky Surveys, p. 141 – 145 (1992). – See Abstr. 012.059 for the main entry.
The high data rates achieved on SuperCOSMOS and the large data volume on a Schmidt plate place significant demands on traditional computer systems. A multi–domain Transputer system which can process large volumes of data at the required rate has been designed for SuperCOSMOS and is currently under development. The processes which must be implemented on such a system are discussed, and the methodology which supports the design is outlined.

036.152 Developments in compression techniques for COSMOS/SuperCOSMOS data.
J. E. F. Baruch, A. J. Crolla, R. D. Boyle.
2. Conference on Digitised Optical Sky Surveys, p. 153 – 165 (1992). – See Abstr. 012.059 for the main entry.

036.153 The determination of the dead–time constant in photoelectric photometry.
E. Poretti.
Messenger, No. 68, p. 52 – 53 (Jun 1992).

036.154 Compression of the Guide Star digitised Schmidt plates.
R. L. White, M. Postman, M. G. Lattanzi.
2. Conference on Digitised Optical Sky Surveys, p. 167 – 175 (1992). – See Abstr. 012.059 for the main entry.
An effective technique for image compression may be based on the H–transform. The method developed can be used for either lossless or lossy compression. The Guide Star digitised sky survey images can be compressed by at least a factor of 10 with no major losses in the astrometric and photometric properties of the compressed images. The method has been designed to be computationally efficient: compression or decompression of a 512x512 image requires only 4 seconds on a Sun SPARCstation 1.

036.155 Proper motions of m.a.s./year accuracy with photographic Schmidt plates.
O. Bienaymé, C. Soubiran.
2. Conference on Digitised Optical Sky Surveys, p. 177 – 183 (1992). – See Abstr. 012.059 for the main entry.
The authors present two new photometric and astrometric (relative proper motion) surveys. The first survey was obtained in a 1.78 square degree field centered on $l = 3°$, $b = 47°$, near the globular cluster M5. The proper motion accuracy is below 3 mas/year. The other survey, near the NGP, was obtained in a 9 square degree field, the relative proper motion accuracy ranging from 1.4 to 2 mas/year.

036.156 Calibration of large batches of photographic images.
Z. Liu, C. Sterken, H. Hensberge, J.–P. De Cuyper.
2. Conference on Digitised Optical Sky Surveys, p. 193 – 198 (1992). – See Abstr. 012.059 for the main entry.
During the 1985 – 1986 appearance of comet Halley, hundreds of photographic images were taken by the authors. These images were obtained with several telescopes, using different photographic emulsions and filters. All photographic images have subsequently been digitised with a PDS microdensitometer. This paper describes the calibration procedure developed.

036.157 An automated image measuring system.
M. Doi, N. Kashikawa, S. Okamura, K. Tarusawa, M. Fukugita, M. Sekiguchi, H. Iwashita.
2. Conference on Digitised Optical Sky Surveys, p. 199 – 208 (1992). – See Abstr. 012.059 for the main entry.
An outline and the status of development are presented for a new Automated Image Measuring System (AIMS), which has been designed for data reduction of Schmidt plates and for a large format CCD camera. The image analysis programme developed for AIMS is currently being used for an extensive photometric study of galaxies in the Coma cluster region. First results for the central 5° × 5° region are presented.

036.158 Automated morphological classification of faint galaxies.
G. Spiekermann.
2. Conference on Digitised Optical Sky Surveys, p. 209 – 213 (1992). Also published in Astron. J., Vol. 103, No. 6, p. 2102 – 2110 (Jun 1992). – See Abstr. 012.059 for the main entry.
The two main procedures in automated morphological classification of galaxies are the analysis of the parameter structure of

prototype galaxies and the classification itself. Five morphological classes, equivalent to the Hubble–types, are used: E, S0, Sa, Sb, Sc/Ir. Prototype galaxies from digitized and segmented IIIa–J Schmidt plates are selected interactively using a colour workstation. Five intervals of apparent magnitudes and two groups of different inclinations are distinguished.

036.159   Automated star/galaxy discrimination with neural networks.
S. C. Odewahn, E. B. Stockwell, R. L. Pennington, R. M. Humphreys, W. A. Zumach.
2. Conference on Digitised Optical Sky Surveys, p. 215 – 224 (1992). Also published in Astron. J., Vol. 103, No. 1, p. 318 – 331 (Jan 1992). – See Abstr. 012.059 for the main entry.

036.160   A comparison of star/galaxy classification approaches on digitised POSS–II plates.
N. Weir, A. Picard.
2. Conference on Digitised Optical Sky Surveys, p. 225 – 230 (1992). – See Abstr. 012.059 for the main entry.
The authors have compared the use of so–called "parametric" and "resolution"–based object classifiers on digitized images from the Second Palomar Observatory Sky Survey. They find that the latter approach, involving the fitting of two–dimensional templates to each detected object, can provide reasonable discrimination at levels approximately a magnitude fainter than the former. The authors employ this technique to significantly increase the depth of a galaxy survey using Palomar Schmidt plates.

036.161   Automated stellar spectral classification via statistical moments.
R. J. Dodd, S. K. Leggett.
2. Conference on Digitised Optical Sky Surveys, p. 231 – 233 (1992). – See Abstr. 012.059 for the main entry.
A method for classifying stellar spectra using statistical image moments is described.

036.162   Mit Mondbedeckungen auf der Jagd nach jungen Doppelsternen im Taurus.
M. Simon, C. Leinert.
Sterne Weltraum, Jahrg. 31, Nr. 6, p. 380 – 385 (Jun 1992).

036.163   Optical identification of X–ray sources from digitised sky survey plates.
B. J. McLean, R. Burg.
2. Conference on Digitised Optical Sky Surveys, p. 465 – 469 (1992). – See Abstr. 012.059 for the main entry.
The authors describe the procedure that has been developed to identify the optical counterparts of the ROSAT X–ray sources in the northern hemisphere. This is performed by the identification and categorisation of all optical objects contained within the X–ray error circle using the digitised sky survey images that are available in the GSSS archive at STScI. Additional ground based observations can then be scheduled on particular objects of interest or to resolve any ambiguous identifications.

036.164   Past and future dark matter experiments.
J. Rich.
15. Texas Symposium on Relativistic Astrophysics and 4. ESO–CERN Symposium, p. 357 – 365 (1991). – See Abstr. 012.060 for the main entry.
Six years ago, Goodman and Witten suggested that hypothetical weakly interacting massive particles (wimps) in our galactic halo could be detected via wimp–nucleus elastic scattering. Wimps would orbit through the galaxy with velocities of order $10^{-3}c$, so gigaelectronvolt–mass wimps would produce nuclear recoils in the keV range. Such events could be detected if the nucleus is contained in a "calorimetric" detector of sufficiently low energy threshold. The rate would depend on the wimp's cross section and local wimp density, estimated to be 0.3 GeV cm^{-3}. While early discussion of experimental possibilities centered around exotic cryogenic detectors, the only existing limits come from experiments using well–established semiconductor technology, based on either germanium or silicon. Conventional scintillators or gas proportional chambers are also sensitive to dark matter. The primary purpose of this paper is to discuss the various possible wimp detectors and to evaluate their ability to improve on the existing limits.

036.165   Indirect detection of dark matter.
K. Freese.
15. Texas Symposium on Relativistic Astrophysics and 4. ESO–CERN Symposium, p. 368 – 381 (1991). – See Abstr. 012.060 for the main entry.
The three favorits candidates for nonbaryonic dark matter are light massive neutrinos (mass ~ 30 eV), axions (mass $\sim 10^{-5}$ eV), and weakly interacting massive particles, or wimps (mass few GeV – few TeV). In this paper the author focuses on wimp candidates for dark matter, which include supersymmetric particles in the gigaelectronvolt to teraelectronvolt mass range.

036.166   Direct detection of solar neutrinos.
D. Sinclair.
15. Texas Symposium on Relativistic Astrophysics and 4. ESO–CERN Symposium, p. 382 – 391 (1991). – See Abstr. 012.060 for the main entry.

036.167   Binary star observations with the HST Fine Guidance Sensors. II. Bright Hyades.
O. G. Franz, L. H. Wasserman, E. Nelan, M. G. Lattanzi, B. Bucciarelli, L. G. Taff.
Bull. Am. Astron. Soc., Vol. 23, No. 3, p. 1257 (1991). Abstract. – See Abstr. 012.065 for the main entry.

036.168   Observations de tavelures à l'observatoire Hoher List.
K. L. Bath.
Astronomie, p. 1 – 3 (Apr 1992).

036.169   A method of creating AAVSO observing charts by computer.
C. E. Scovil, R. A. Leitner.
J. Am. Assoc. Variable Star Obs., Vol. 20, No. 1, p. 1 – 5 (1991).
Computer scanning of photographs and a new program which makes all of the star images round promise new excellence in charts.

036.170   Computer image processing techniques for the production of AAVSO observing charts.
G. Weingarten.
J. Am. Assoc. Variable Star Obs., Vol. 20, No. 1, p. 6 – 13 (1991).
The author describes the computer image–analysis and chart–drawing programs used in the production of the new series of AAVSO observing charts made directly from astrophotographs as described by Scovil and Leitner. These programs reduce the time required from tens of hours to under five minutes. It is in the areas of defect–per–chart elimination and object classification, including the proper identification of "double" stars, that modern techniques of computer image processing are most valuable.

036.171   Estimate of the expected number of stars within the field of view of an astronomical instrument.
S. E. Zdor, V. S. Chernov, T. A. Baglai.
Sov. J. Opt. Technol., Vol. 59, No. 5, p. 312 – 313 (May 1992). Current Physics Microform No.: 9211X0056.
A method is proposed for a rapid estimate of the expected number of stars within the field of view of an instrument by means of a nomogram that takes into account the penetrating power of the instrument and the position of its line of sight on the celestial sphere.

036.172 BAO–CCD BVRI photometry.
Jiang Zhaoji, Li Yong, Chen Jiansheng.
Acta Astrophys. Sin., Vol. 12, No. 1, p. 47 – 53 (Jan 1992). In Chinese. English translation in Chin. Astron. Astrophys., Vol. 16, No. 2, p. 226 – 233 (Apr–Jun 1992).
The BAO Schmidt telescope with a Thomson CCD of 576×384 pixels was used for BVRI photometry. 9 secondary standard stars in 3 CCD frames were observed on October 31, 1989. The colour equations are presented for this photometric system. Some problems relevant to the capability of CCD photometry are discussed.

036.173 A photometric study of the BAO–CCD BVRI system.
Jiang Zhaoji, Li Yong, Chen Jiansheng.
Chin. Astron. Astrophys., Vol. 16, No. 2, p. 226 – 233 (Apr–Jun 1992). English translation of Acta Astrophys. Sin., Vol. 12, No. 1, p. 47 – 53 (Jan 1992). See Abstr. 036.172.
The BAO Schmidt telescope with a Thomson CCD of 576×384 pixels was used for BVRI photometry. 9 secondary photometric standard stars in 3 CCD frames were observed on Oct. 31, 1989, to establish the color equations. Some problems relevant to the capability of CCD photometry are discussed.

036.174 Optical ground–based network.
R. Pellinen, K. Kaila.
International Workshop on Cluster Dayside Polar Cusp, p. 147 – 158 (Dec 1991). – See Abstr. 012.071 for the main entry.
A short summary of the development of ground–based optical observations of auroras is given. Existing optical networks in the northern hemisphere are described emphasizing the Scandinavia–Svalbard sector. Different observational techniques covering all-sky photography, TV camers, photometers, spectrometers, and interferometers are described. The physical limits and usefulness of the different techniques are critically reviewed. Some data examples are presented to demonstrate the capabilities of the different methods. Need for improvement and extension of the present network are discussed. The specific problem of ground–based optical support to the forthcoming Cluster and EISCAT Svalbard Radar programs is discussed in more detail.

036.175 Imaging the Earth's magnetosphere using ground–magnetometer arrays.
K. H. Glassmeier, E. Friis–Christensen.
International Workshop on Cluster Dayside Polar Cusp, p. 159 – 166 (Dec 1991). – See Abstr. 012.071 for the main entry.
A magnetometer array may be regarded as a telescope allowing to image magnetospheric dynamics in the ultra–low–frequency, near DC–range of the electromagnetic spectrum. The field–of–view of such an array lies in the range of several thousand kilometers, depending on the array's size. The resolution is typically of the order of 50–100 km in the ionosphere, i.e. 1500–3000 km in the magnetosphere. Highly sophisticated image processing tools have been developed, which make magnetometer arrays a very important tool to complement in–situ measurements such as planned for the CLUSTER project. One of the aims of this project is the study of small–scale plasma processes in, for example, the dayside polar cusp region. As the archipelago of Svalbard lies within the region of the northern polar cusp the installation of an international magnetometer array on Spitsbergen is strongly suggested.

036.176 The requirements for data exchange between Cluster and ground–based observatories, using EISCAT as an illustrative example.
E. Dunford, D. M. Willis, R. Schmidt.
International Workshop on Cluster Dayside Polar Cusp, p. 183 – 190 (Dec 1991). – See Abstr. 012.071 for the main entry.
The formation of a dedicated working group to co–ordinate observations by Cluster and ground–based observatories implies the need for a ground data system that enables fast and reliable access to data bases. These data bases will not only hold science data derived from Cluster and ground observations, but also data sets required for the planning of joint measurement campaigns. These could include Cluster orbital information, schedules of ground observatories, access to plasma and field models, etc. There is a clear requirement that all participants follow standardized procedures when generating these data sets, in order to permit an efficient intercomparison of plasma quantities measured in space, in the ionosphere, and on the ground. In this paper the authors describe the Cluster data system and use EISCAT as an illustrative example for co–ordinated measurement campaigns.

036.177 Anisoplanatism and use of laser guide stars.
L. E. Goad.
SPIE Conference on Active and Adaptive Optical Systems as Part of SPIE's International Symposium on Optical Applied Science and Engineering, p. 100 – 109 (1991). – See Abstr. 012.074 for the main entry.
The effects of focal and angular anisoplanatism are computed in order to evaluate the utility of using a single laser–produced guide star for the correction of ground–based astronomical imaging. The equations for the calculation of the effects are derived and the performance of a sodium–layer laser guide star system is computed for a Hufnagel atmospheric turbulence model. These results are presented in scaled units and for selected telescope apertures from 2 – 8 meters in diameter operating at wavelengths from $0.5 - 2.0\ \mu m$. The limiting telescope aperture size which can be adequately corrected using a single sodium–layer guide star is shown to be much larger than previously estimated.

036.178 Laser guide stars for adaptive optics systems: Rayleigh scattering experiments.
L. Thompson, R. Castle, D. Carroll.
SPIE Conference on Active and Adaptive Optical Systems as Part of SPIE's International Symposium on Optical Applied Science and Engineering, p. 110 – 119 (1991). – See Abstr. 012.074 for the main entry.
Experiments have been conducted in the atmosphere above Mt. Laguna Observatory to measure the properties of laser guide stars. The experimental system consists of a high frame rate video camera which records the backscattered light from an Excimer laser working in the near–UV at 351 nm. The Mt. Laguna 1–meter telescope is used to both transmit the outgoing beam and to image the return beam. The outgoing laser pulse triggers a time–gated image intensifier within the video camera which, with an appropriately selected time delay, records a time slice of the backscattered return signal. Preliminary results from the experiment can be used to calibrate the laser power needed to operate a large ground–based adaptive optics telescope.

036.179 Algorithms for wavefront reconstruction out of curvature sensing data.
N. Roddier.
SPIE Conference on Active and Adaptive Optical Systems as Part of SPIE's International Symposium on Optical Applied Science and Engineering, p. 120 – 129 (1991). – See Abstr. 012.074 for the main entry.
The author describes numerical methods for reducing data taken with curvature sensing measurements. The Laplacian integration, spherical aberration estimate and removal, automatization of image parameter determination routines are explained. The author then describes the user instrument to do real time telescope testing.

036.180 Experimental search of gravitational waves.
G. Pizzella.
Symposium on Gravitation and Modern Cosmology – the Cosmological Constant Problem, p. 139 – 150 (1991). – See Abstr. 012.076 for the main entry.
The author presents the status of the experiments for the search of gravitational waves.

036.181   **Image restoration: method–independent limit of efficiency and its realization.**
V. Yu. Terebizh.
Astron. Astrophys. Trans., Vol. 1, No. 1, p. 3 – 29 (1991).
Due to radiational and external noise, any image forming system can be described only in a frame of a stochastic model, with probability f(N,S) of occurrence of the observed image N, for any searched object S. Thus, restoration problem must inevitably be considered as a statistical estimation of unknown parameters S. Proceeding from the appropriate definition of restoration efficiency, it is possible to show that the method-independent limit of efficiency and accuracy is set by the Rao–Cramer theorem. The most promising way to achieve the theoretical limit is based on a maximum likelihood image restoration (MLIR) method. Indeed, MLIR gives limiting accuracy when light intensity begins to exceed the mean level of external noise. The maximum entropy method includes non-necessary, and logically inconsistent, assumptions. The concrete formulae for the calculation of an object estimate, and its accuracy, are given. Test cases are given, and examples, of using the proposed approach to different inverse problems.

036.182   **Visibility limit of naked–eye sunspots.**
H. U. Keller, T. K. Friedli.
Q. J. R. Astron. Soc., Vol. 33, No. 2, p. 83 – 89 (Jun 1992).
Based on observations of 20 observers, a visibility limit for naked–eye sunspot observations has been calculated. A sunspot with a penumbral diameter of at least 41 arcsec and an umbral diameter of at least 15 arcsec can be detected by an average eye. An eyesight test was designed to solve the question whether the penumbral or the umbral diameter is conclusive for the visibility of a sunspot. The measured limit of 19.3 arcsec indicates that a combination of both is the determining factor.

036.183   **Adaptive optics system tests at the ESO 3.6–m telescope.**
F. Merkle, G. Gehring, F. Rigaut, P. Léna,
G. Rousett, J. C. Fontanella, J. P. Gaffard.
SPIE Conference on Active and Adaptive Optical Systems as Part of SPIE's International Symposium on Optical Applied Science and Engineering, p. 308 – 318 (1991). – See Abstr. 012.074 for the main entry.
This paper reports the results of the observations made with the VLT Adaptive Optics Prototype System "COME–ON" at the ESO 3.6 m telescope. The analysis of uncorrected and corrected images in the near infrared wavelengths range ($\leqslant 5 \, \mu m$) leads to a detailed assessment of the system performance in terms of improvement of angular resolution – reaching nearly the ideal diffraction profiles down to 1.7 μm wavelength –, and Strehl ratio – approaching 0.6 to 0.8 at 3.8 μm. A resolution of 0.12″ has been obtained with this system at 1.7 μm which is wavelength dependent on the temporal parameters of the observation. The current limiting magnitude for the reference source is $m_R = 11.5$ applying the full correction capabilities of the system, and $m_R = 13$ if only the wavefront tilt is corrected.

036.184   **Partially compensated speckle imaging: Fourier phase spectrum estimation.**
M. C. Roggemann, C. L. Matson.
SPIE Conference on Active and Adaptive Optical Systems as Part of SPIE's International Symposium on Optical Applied Science and Engineering, p. 477 – 487 (1991). – See Abstr. 012.074 for the main entry.
Predetection compensation combined with post detection image processing for the case of imaging through atmospheric turbulence is addressed. Full and partial predetection compensation using adaptive optics is combined with bispectrum speckle imaging post-processing, and performance improvements are assessed. Full compensation was found to provide a large improvement in the signal–to–noise ratio (SNR) of the power spectrum estimate compared to the uncompensated case. Lower degrees of correction provided smaller improvements in the power spectrum SNR, and a very low degree of compensation provided results indistinguishable from the uncompensated case.

036.185   **Measuring phase errors of an array or segmented mirror with a single far–field intensity distribution.**
R. K. Tyson.
SPIE Conference on Active and Adaptive Optical Systems as Part of SPIE's International Symposium on Optical Applied Science and Engineering, p. 62 – 75 (1991). – See Abstr. 012.074 for the main entry.
A technique for extracting the relative phase (piston) differences between telescopes of a phased array or segments of a mirror is examined. The analysis concentrates on examination of measuring the relative phase of three segments with a single far-field intensity distribution. The contour of the first minimum in the far–field diffraction spot is directly related to the relative phase differences between three segments or telescopes. The paper addresses various methods for extracting the information and using it to phase the segments or telescopes. Methods of extending the technique to real–time control are proposed.

036.186   **Atmospheric turbulence sensing for a multiconjugate adaptive optics system.**
D. C. Johnston, B. M. Welsh.
SPIE Conference on Active and Adaptive Optical Systems as Part of SPIE's International Symposium on Optical Applied Science and Engineering, p. 76 – 87 (1991). – See Abstr. 012.074 for the main entry.
Current adaptive optical telescope designs use a single deformable mirror (DM), usually conjugated to the aperture plane, to compensate for the cumulative effects of optical turbulence. The corrected field of view of an adaptive optics system could theoretically be increased through the use of multiple DMs conjugated to a number of corresponding planes which sample the turbulence region in altitude. To properly control each DM, one needs a method for determining the phase distortion contributed by atmospheric layers at the selected altitudes. This paper presents a theoretical analysis of a signal processing technique for determining these phase contributions.

036.187   **A proposal for the seeing measurement of the site testing.**
Tan Huisong.
Publ. Yunnan Obs., No. 1, p. 11 – 14 (1992). In Chinese.

036.188   **Surface map of the array of two–dimensional data.**
Wu Guangjie.
Publ. Yunnan Obs., No. 1, p. 63 – 68 (1992). In Chinese.
A surface map, with the pseudo–3D effect, made by means of the array of two dimensional data (e.g. various astronomical images) can demonstrate the variation in the distribution of the data. The relative variability of intensities can be described more effectively than both the contour map and the gray–scale one. A brief introduction of the task making a surface map is given.

036.189   **Long–baseline optical and infrared stellar interferometry.**
M. Shao, M. M. Colavita.
Annu. Rev. Astron. Astrophys., Vol. 30, p. 457 – 498 (1992).
Contents: 1. Introduction. 2. Interferometry on the ground. 3. Interferometry in space. 4. Summary.

036.190   **Two dimensional photon–counting techniques.**
Wang Chuan–jin.
Publ. Purple Mt. Obs., Vol. 11, No. 1, p. 37 – 49 (Mar 1992). In Chinese.

036.191   **Anomalies of residual of Beijing photoelectric astrolabe and earthquakes occurred in the region around Beijing.**
Han Yanben.
Publ. Beijing Astron. Obs., No. 18, p. 72 – 76 (Oct 1991).
The relationship between anomalies of astronomical time and latitude residuals obtained by Beijing Photoelectric Astrolabe and earthquakes occurred in the region near the instrument is disscused. The result shows that the relationship is close and classical astrometric instruments can make a certain contribution for a short–term earthquake prediction.

036.192   Gain measurement of small–sized antenna by means of radioastronomical methods.
Qin Zhi–hai, Zhou Yu–lan, Liang Zheng–you, Jiang Su–yin, Liu Yan, Wei Shuang–lin.
Publ. Purple Mt. Obs., Vol. 11, No. 1, p. 61 – 67 (Mar 1992). In Chinese.
This paper discusses the basic principles and methods about the gain measurement of small–sized antenna by means of radioastronomical methods with the Sun and the Moon as calibration beacons. The resultant measurements are given and the comparison is shown.

036.193   Numerical and parametric PSF extraction from CCD frames: application to ground–based and HST observations.
O. Bendinelli, M. G. Gatti.
Mem. Soc. Astron. Ital., Vol. 63, No. 1, p. 165 – 168 (1992). – See Abstr. 012.087 for the main entry.
The authors show how to get quickly a numerical or parametric determination of the intensity profile (i.e. of the PSF) and of the surrounding background dealing with CCD star images uncrowded and oversampled, so that off–centering can be neglected. The related programs, written in Fortran, require only few time seconds to run automatically on a PPC and can be easily incorporated as alternative routines in any extended image analysis package.

036.194   An identification technique for OB associations in unresolved galaxies.
P. Battinelli, R. Capuzzo–Dolcetta.
Mem. Soc. Astron. Ital., Vol. 63, No. 1, p. 169 – 172 (1992). – See Abstr. 012.087 for the main entry.
The authors wish to give a suitable technique to accomplish the task to select OB associations in unresolved fields on the basis of photometric data only. They identify a method based on a combined application of Principal Component Analysis (PCA) and Cluster Analysis (CA) to fluxes and colours. The aim is to obtain an artificial image carrying the global information of the photometric observations, where to perform a recognition of star forming regions.

036.195   Applied techniques of PDS in photographic astrometry. II. Measurements of the crowded field of stars.
Wang Jiaji, Chen Li.
Ann. Shanghai Obs., Acad. Sin., No. 13, p. 62 – 66 (1992). In Chinese.
The various problems in the formation of scan frame files and combination of grouping measurements for a plate with the measurement of crowded field of stars with PDS microdensitometer are discussed.

036.196   Use of DE303/LE303 ephemeris in the analysis of lunar laser ranging data.
Xu Huaguan, Jin Wenjing.
Ann. Shanghai Obs., Acad. Sin., No. 13, p. 96 – 103 (1992). In Chinese.
The procedure for creating the files DE303/LE303 ephemeris and lunar libration LLB303 on computer storage by JPL ephemeris tape is described.

036.197   Use of photon counts in the photoelectric astrolabe (Type–II).
Zhao Gang, Zhang Jianwei, Wang Rui.
Ann. Shanghai Obs., Acad. Sin., No. 13, p. 129 – 139 (1992). In Chinese.
This article describes the new improvement of the photoelectric astrolabe and primary results. This system can be used to observe 11.0 magnitude stars effectively with the photon–counting record.

036.198   The VLBI data acquisition system of Shanghai Observatory.
Wu Linda.
Ann. Shanghai Obs., Acad. Sin., No. 13, p. 159 – 166 (1992). In Chinese.
The procedure of setting up the VLBI data acquisition system of Shanghai Observatory at Zô–Sé is reviewed. The composition and specification of the system are described.

036.199   The PC control system of the SLR system at Shanghai Observatory.
Zhang Jianhua.
Ann. Shanghai Obs., Acad. Sin., No. 13, p. 170 – 174 (1992). In Chinese.

036.200   Data communication between HP1000F minicomputer and IBM PC/XT–286 microcomputer of Shanghai Observatory.
Xue Zhuhe, Chen Ying, Kong Yi.
Ann. Shanghai Obs., Acad. Sin., No. 13, p. 198 – 202 (1992). In Chinese.

036.201   Progress in CCD photometry.
P. B. Stetson.
J. R. Astron. Soc. Can., Vol. 86, No. 2, p. 71 – 88 (Apr 1992).
As we approach the end of CCD astronomy's first decade, the author presents a brief summary of a few recent triumphs in the field of stellar photometry. This is followed by a general discussion of the current state of the software for performing stellar photometry, including a few recent developments as specific examples.

036.202   Data analysis and algorithms for gravitational wave antennas.
G. V. Pallottino, G. Pizzella.
The detection of gravitational waves, p. 243 – 265 (1991). – See Abstr. 003.097 for the main entry.
Contents: 1. Introduction. 2. The antenna response to a gravitational wave. 3. The basic block diagram and the wide band electronic noise. 4. The narrow band noise and the total noise. 5. Data filtering: Detection of short bursts; Detection of longer bursts; Detection of periodic signals. 6. The cross–section and the antenna sensitivity. 7. Coincidence techniques.

036.203   Data processing, analysis and storage for interferometric antennas.
B. F. Schutz.
The detection of gravitational waves, p. 406 – 452 (1991). – See Abstr. 003.097 for the main entry.
Contents: 1. Introduction: Signal to look for. 2. Analysis of the data from individual detectors: Finding broad–band bursts; Extracting coalescing binary signals; Looking for pulsars and other fixed–frequency sources. 3. Combining lists of candidate events from different detectors: Threshold mode of data analysis; Deciding that a gravitational wave has been detected. 4. Using cross–correlation to discover unpredicted sources: The mathematics of cross–correlation: enhancing unexpected signals; Cross–correlating differently polarized detectors; Using cross–correlation to search for a stochastic background. 5. Reconstructing the signal: single bursts seen in several detectors. 6. Data storage and exchange: Storage requirements; Exchanges of data among sites. 7. Conclusions.

036.204   Gravitational wave detection at low and very low frequencies.
R. W. Hellings.
The detection of gravitational waves, p. 453 – 475 (1991). – See Abstr. 003.097 for the main entry.
Contents: 1. Introduction. 2. LF and VLF gravitational waves. 3. The effect of a gravitational wave on electromagnetically tracked free masses: One–way tracking; Two–way tracking; Interferometers. 4. Pulsar timing analysis. 5. Doppler spacecraft tracking. 6. Space interferometer gravitational wave experiments: Microwave interferometer; Laser interferometers.

036.205 **On the possibility of detection of small comets in Ly–α.**
M. Banaszkiewicz, S. Grzedzielski, D. Ruciński,
M. S. Staniucha.
1. COSPAR Colloquium: Physics of the outer heliosphere,
p. 97 – 100 (1990). – See Abstr. 012.100 for the main entry.

Possibility of detection of small comets in the heliosphere by
simple monitoring technique of the Ly–α interplanetary glow is
discussed. Estimates of the detection possibility of water–type
objects in the range of 0.1–1 km radius sublimating at a typical
cometary rate are provided for both periodic and non–periodic
populations of comets. A simple photometer with a 2° full FOV,
scanning the sky by spacecraft rotation could detect up to 30 such
small objects per year.

036.206 **Integrity of HRI images.**
D. E. Harris, C. P. Stern, J. A. Biretta.
Symposium on Imaging X–Ray Astronomy – a Decade of
Einstein Observatory Achievements, p. 299– 303 (1990). – See
Abstr. 012.097 for the main entry.

Image processing problems with the High Resolution Imager
(HRI) of the Einstein Observatory are stressed. Whenever source
structure on the scale of 3″ to 8″ is in doubt, an evaluation of the
LOCKED image should be made even if this sacrifices exposure
time. This is a relative straightforward procedure with current
software. However, there is no automatic method of selecting
data segments for which the target is clear of the gaps, and
evaluation of imperfect gapmap corrections is therefore much
more difficult.

036.207 **Axion searches.**
P. Sikivie.
15. Texas Symposium on Relativistic Astrophysics and
4. ESO–CERN Symposium, p. 366 – 367 (1991). – See Abstr.
012.060 for the main entry.

036.208 **Neutral solar wind experiment.**
M. A. Gruntman, S. Grzedzielski, V. B. Leonas.
1. COSPAR Colloquium: Physics of the outer heliosphere,
p. 355 – 358 (1990). – See Abstr. 012.100 for the main entry.

036.209 **An inverse method of determination of the interstellar
neutral gas distribution function.**
M. Banaszkiewicz, H. Rosenbauer, M. Witte.
1. COSPAR Colloquium: Physics of the outer heliosphere,
p. 359 – 362 (1990). – See Abstr. 012.100 for the main entry.

The Tarantola's method has been applied to solve an inverse
problem of determining parameters of the shifted Maxwellian
distribution function of interstellar helium from information on
specific cuts through velocity space.

**Updating the AGK3 reference star catalog with astrometrically
reduced Schmidt plates.**
See Abstr. 002.081.

**Optical identification of sources in the IRAS Faint Sources
database.**
See Abstr. 002.087.

**New microcomputer technologies to facilitate collaboration and
publication.**
See Abstr. 002.116.

Basics of interferometry.
See Abstr. 003.064.

Particle astronomy minisymposium: introductory remarks.
See Abstr. 011.016.

Solar observations: techniques and interpretation. Proceedings.
See Abstr. 012.045.

**Infrared technology XVII. Dedicated to the memory of Irving J.
Spiro. Proceedings.**
See Abstr. 012.051.

Desktop publishing in astronomy & space sciences. Proceedings.
See Abstr. 012.053.

Digitised optical sky surveys. Proceedings.
See Abstr. 012.059.

**New results on standard candles. Proceedings. Trani Workshop on
New Results on Standard Candles, Trani, Bari (Italy), 26 – 30 Aug
1991.**
See Abstr. 012.084.

Photometric calibration of the southern sky surveys.
See Abstr. 013.077.

Digitised sky surveys at the Space Telescope Science Institute.
See Abstr. 013.079.

**A colour/proper motion study of the southern sky – a planned
digitisation programme on SuperCOSMOS.**
See Abstr. 013.080.

Asteroid searches from UKST material.
See Abstr. 013.087.

**The ROSAT all–sky survey and first identifications of X–ray
sources using digitised optical plates.**
See Abstr. 013.090.

**Optical identification of the new PARKES/MIT/NRAO radio
sources from digitised sky survey data.**
See Abstr. 013.091.

Progress in extra–solar planet detection.
See Abstr. 013.121.

**Acquisizione digitale ed elaborazione numerica di immagini
planetarie (II).**
See Abstr. 014.055.

Video in der Planetenastronomie.
See Abstr. 014.069.

**Time series analysis in astronomy: an application to quasar
variability studies.**
See Abstr. 021.014.

Monte Carlo methods.
See Abstr. 021.021.

**Calculation of wave–front–tilt correlations associated with atmo-
spheric turbulence.**
See Abstr. 031.010.

Optical fibers in astronomical instruments.
See Abstr. 031.016.

MARTINI: sensing and control system design.
See Abstr. 031.046.

**Stray light issues for background–limited infrared telescope
operation.**
See Abstr. 031.047.

Neural network adaptive optics for the Multiple Mirror Telescope.
See Abstr. 031.054.

Solar astronomy with a 19–segment adaptive mirror.
See Abstr. 031.055.

The Johns Hopkins adaptive optics coronagraph.
See Abstr. 031.056.

Active optics system for a 3.5–meter structured mirror.
See Abstr. 031.057.

Alignment and focus control of a telescope using image sharpening.
See Abstr. 031.058.

The University of Hawaii adaptive system. I. General approach.
See Abstr. 031.060.

Optical performance of large ground–based telescopes.
See Abstr. 031.062.

A quick method of checking the pointing of the antenna pattern of the RATAN–600 telescope with respect to the angle of elevation.
See Abstr. 033.001.

A fast noise–immunity estimation of the noise–track level of a radiometer by means of the absolute median deviation.
See Abstr. 033.002.

Perspectives of advance of RATAN–600 solar observations.
See Abstr. 033.004.

Data transmission between the control computer and data processing computer of PMO 13.7 m radio telescope.
See Abstr. 033.016.

Astronomical observations at 10 and 20 μm with the NRL infrared camera.
See Abstr. 034.014.

A 5 – 18 μm array camera for high–background astronomical imaging.
See Abstr. 034.016.

Charge–coupled devices: frame adding as an alternative to long integration times and cooling.
See Abstr. 034.026.

A simple visual Cassegrain CCD camera for the Wyoming Infrared Observatory.
See Abstr. 034.032.

Zürich Imaging Stokes Polarimeter – ZIMPOL I. Design review.
See Abstr. 034.034.

Demodulation of all four Stokes parameters with a single CCD – ZIMPOL II. Conceptual design.
See Abstr. 034.035.

The University of Hawaii NICMOS–3 Near–Infrared Camera.
See Abstr. 034.036.

Visible and infrared wavefront sensing for astronomical adaptive optics.
See Abstr. 034.053.

MUSICOS: a fiber–fed spectrograph for multi–site observations.
See Abstr. 034.055.

Kryogener Shutter für die NIR–Kamera am VLT.
See Abstr. 034.073.

Die räumliche Verteilung von Nahinfrarotlinienstrahlung in galaktischen Kernen gemessen mit dem abbildenden Fabry–Pérot–Spektrometer FAST.
See Abstr. 034.074.

The MAMA facility: a survey of scientific programmes.
See Abstr. 034.077.

ASTROSCAN II: the next Leiden microdensitometer.
See Abstr. 034.078.

A new astrometric measuring machine: design and astronomical programmes.
See Abstr. 034.079.

Extragalactic supernova detector, neutrino mass, and the Supernova Watch.
See Abstr. 034.086.

Thermal detection of dark matter.
See Abstr. 034.089.

Prototype high speed optical delay line for stellar interferometry.
See Abstr. 034.095.

The University of Hawaii adaptive optics system: III. The wavefront curvature sensor.
See Abstr. 034.096.

Intensified CCD camera based remote guiding unit for VBT and observation of speckles with ICCD.
See Abstr. 034.098.

Fast photometry using CCD.
See Abstr. 034.099.

A rotating tomographic imager for solar extreme–ultraviolet / soft X–ray emission.
See Abstr. 035.042.

Operation and performance of the OSSE instrument.
See Abstr. 035.046.

Geometrical stability and evolution of the Hipparcos telescope.
See Abstr. 035.056.

COBE Differential Microwave Radiometers: calibration techniques.
See Abstr. 035.069.

Interferometric measurements and light aberration.
See Abstr. 041.003.

Positions and parallaxes from the Hipparcos satellite. A first attempt at a global astrometric solution.
See Abstr. 041.011.

Interferometric measurements and light aberration.
See Abstr. 041.012.

Hipparcos data reduction – construction of the Hipparcos star catalogue.
See Abstr. 051.060.

Prospects for relic neutrino detection.
See Abstr. 061.128.

Taking the pulse of white dwarfs.
See Abstr. 065.054.

Gravitational microlensing: powerful combination of ray–shooting and parametric representation of caustics.
See Abstr. 066.161.

Solar corona correction in VLBI observation.
See Abstr. 066.297.

Estimating the degradation of brightness power spectra of solar granulation from images outside the disk centre.
See Abstr. 071.003.

Photometric and height calibration of the spectra observed at the 1983 total solar eclipse.
See Abstr. 079.002.

On the ultimate accuracy of solar oscillation frequency measurements.
See Abstr. 080.034.

Does the solar neutrino flux vary?
See Abstr. 080.072.

First measurement of the integral solar neutrino flux by the Soviet/American Gallium Experiment.
See Abstr. 080.078.

Five–minute oscillations in the solar continuum.
See Abstr. 080.087.

Aperture–averaging factor for optical scintillations of plane and spherical waves in the atmosphere.
See Abstr. 082.002.

Sky spectra at a light–polluted site and the use of atomic and OH sky emission lines for wavelength calibration.
See Abstr. 082.010.

Optical seeing at La Palma Observatorty. I. General guidelines and preliminary results at the Nordic Optical Telescope.
See Abstr. 082.035.

Proposal to measure terrestrial Bradley aberration.
See Abstr. 082.053.

Empirical image motion spectrum. I. Seeing quality and the atmospheric limitation on the accuracy of meridian observations.
See Abstr. 082.068.

Simulation of frontal cloudiness in the atmospheres of giant planets. I. Method of calculations.
See Abstr. 091.012.

Occultation of the Pleiades star cluster by the Moon: a first analysis.
See Abstr. 096.004.

New channel for the photoionization of hydrogen atoms in the solar system.
See Abstr. 106.098.

Milliarcsecond proper motion measurements with MAMA.
See Abstr. 111.013.

Radial velocities from objective prism plates.
See Abstr. 111.016.

Radiation–driven winds of hot luminous stars. X. The determination of stellar masses, radii and distances from terminal velocities and mass–loss rates.
See Abstr. 112.057.

Mode identification of pulsating stars from line profile variations with the moment method.
See Abstr. 114.081.

Line profile asymmetries in chromospherically active stars.
See Abstr. 116.030.

Stellar magnetic–field measurement with a double–beam polarimeter fitted with a photoelastic modulator.
See Abstr. 116.033.

Modelling stellar photospheric spots using spectroscopy.
See Abstr. 116.042.

Comparisons between Doppler imaging and starspot modelling.
See Abstr. 116.043.

A new approach to Doppler imaging of late–type stars.
See Abstr. 116.046.

Mapping magnetic fields on rapidly rotating stars.
See Abstr. 116.049.

Surface mapping of slowly rotating, cool stars using line bisector variations.
See Abstr. 116.052.

Light curve analysis of stars with more than one spot.
See Abstr. 116.064.

Characterization of long–term X–ray variability in a sample of late–type stars.
See Abstr. 116.073.

Comparing restored HST and VLA imagery of R Aquarii.
See Abstr. 117.030.

Detection of RS CVn–type systems by doublet 2800 Mg II.
See Abstr. 117.047.

The identification of potential counterparts to X–ray binaries using COSMOS.
See Abstr. 117.256.

Preliminary observations of Be/X–ray binaries with the UBV photoelectric photometry system.
See Abstr. 117.267.

Computation of ephemerides for Long–Period Variable stars for the Hipparcos mission.
See Abstr. 122.058.

Period analysis of variable stars.
See Abstr. 122.091.

The use of nebular spectra of Type Ia supernovae for distance determinations. The distance to the Centaurus Group.
See Abstr. 125.066.

The projection of fractal objects.
See Abstr. 131.239.

Gamma–ray emission from pulsars.
See Abstr. 143.078.

Modelling the stellar intensity and radial velocity fields in triaxial galaxies by sums of Gaussian functions.
See Abstr. 151.001.

A deep proper motion survey of the Pleiades for very low mass stars and brown dwarfs.
See Abstr. 153.045.

Photometry of the star cluster R136 using the Faint Object Camera of HST.
See Abstr. 153.057.

Abundances in globular cluster stars: methods and CNO abundances.
See Abstr. 154.071.

Study of stellar populations using neural network techniques.
See Abstr. 155.064.

The spherical harmonics as an alternative tool for determining the kinematical parameters of the local Milky Way.
See Abstr. 155.065.

UBV star counts in Selected Area 54, and global structure of the Galaxy.
See Abstr. 155.141.

A search for macroscopic dark matter in the Galactic Halo through microlensing.
See Abstr. 155.142.

Optimal estimates of line–of–sight velocity distributions from absorption line spectra of galaxies: nuclear discs in elliptical galaxies.
See Abstr. 157.150.

Evidence for dwarf stars at D $\sim$ 100 kiloparsecs near the Sextans dwarf spheroidal galaxy.
See Abstr. 157.190.

Adaptive filtering of long slit spectra of extended objects.
See Abstr. 157.276.

The new gravitational lens candidate Q 1208 + 1011 and the importance of high quality data.
See Abstr. 159.080.

Wavelet analysis of subclustering: an illustration, Abell 754.
See Abstr. 160.040.

Statistics of pencil beams in Voronoi foams.
See Abstr. 160.095.

A general analytical solution to the problem of Malmquist bias due to lognormal distance errors.
See Abstr. 160.096.

On the error estimates of correlation functions.
See Abstr. 160.105.

An automated search for compact groups of galaxies.
See Abstr. 160.113.

Der Anteil aktiver Galaxien im Galaxienhaufen Cl 1409 + 52 (3C 295).
See Abstr. 160.118.

Positional Astronomy, Celestial Mechanics

041 Astrometry

041.001 Accurate radio and optical positions for southern radio sources.
B. R. Harvey, D. L. Jauncey, G. L. White, A. Nothnagel,
G. D. Nicolson, J. E. Reynolds, D. D. Morabito, N. Bartel.
Astron. J., Vol. 103, No. 1, p. 229 – 233 (Jan 1992). Current Physics Microform No.: 9202A2139.

Accurate radio positions with a precision of about 0.01 arcsec are reported for eight compact extragalactic radio sources south of –45° declination. The radio positions were determined using Very Long Baseline Interferometry (VLBI) at 8.4 GHz on the 9589 km Tidbinbilla (Australia) to Hartebeesthoek (South Africa) baseline. The sources were selected from the Parkes Catalogue to be strong, flat–spectrum radio sources with bright optical QSO counterparts. Optical positions of the QSOs were also measured from the ESO B Sky Survey plates with respect to stars from the Perth 70 Catalogue, to an accuracy of about 0.19 arcsec rms. These radio and optical positions are as precise as any presently available in the far southern sky. A comparison of the radio and optical positions confirms the estimated optical position errors and shows that there is overall agreement at the 0.1 arcsec level between the radio and Perth 70 optical reference frames in the far south.

041.002 A study of the astrometric accuracy of photographic plates obtained with standard astrographs.
J. Núñez, J. M. Codina, N. Torras.
Astron. J., Vol. 103, No. 5, p. 1687 – 1688 (May 1992). Current Physics Microform No.: 9205G0261.

Photographic plates taken for minor planet positions with standard astrographs (Carte du Ciel or longer focus) could be a useful source for positions and proper motions of stars in the zodiacal zone because such plates overlap this zone several times. The authors have measured a selection of plates obtained for minor planet positions with the astrograph of the Fabra Observatory (field = $2° \times 2°$, f = 4 m). The selection contains both old and modern plates, film and glass plates, and the AGK3 and SAO as reference catalogs. The influence of the material, age, and reference catalog on the astrometric accuracy of the plate is discussed.

041.003 Interferometric measurements and light aberration.
M. D. Kislik.
Sov. Astron., Vol. 36, No. 1, p. 109 – 111 (Jan 1992). Current Physics Microform No.: 9207Y0673. English translation of Astron. Zh., Tom 69, Vyp. 1, p. 214 – 218 (Jan–Feb 1992).

The problem of using the concept of light aberration in processing interferometric observations is discussed. It is concluded that one should not use that concept in this case. It is shown that applying Galilean transformations in processing interferometric observations leads to errors of the same order of magnitude as the aberration effect itself.

041.004 Internal and external criterion analysis of astrographic reduction models based on observations in Bolivia.
S. G. Valeev, D. D. Polozhentsev, R. F. Zal'es (R. F. Salles),
L. I. Yagudin.
Kinematics Phys. Celest. Bodies, Vol. 7, No. 3, p. 7 – 12 (1991). English translation of Kinematika Fiz. Nebesn. Tel, Tom 7, No. 3, p. 9 – 14 (1991).

The influence of the type of reduction formula and the method used to estimate the plate constants on the internal and external accuray of star–coordinate prediction (on the average over a series of plates, for a single plate, and for a specific star) is investigated. The external criteria indicate that use of a constant reduction model for a series of plates may result in significant errors in the star coordinates on the individual plates.

041.005 Making an astrometric standard in the region of Cygnus.
V. S. Kislyuk, E. V. Rel'ke.
Kinematika Fiz. Nebesn. Tel, Tom 8, No. 2, p. 56 – 64 (Mar–Apr 1992). In Russian. English translation in Kinematics Phys. Celest. Bodies, Vol. 8, No. 2.

The procedure of making an astrometric standard in the area centered at $\alpha = 21^h03^m$, $\delta = 52°$ is described. The work is based on photographic observations obtained with four wide–angle astrographs of the same type in accordance with the FON (Photographic Sky Survey) programme. Plates obtained with the 2–m Tautenburg Schmidt telescope were used for extension of the standard to faint stars (down to 22^m). Positions and magnitudes of 1429 stars down to 16^m in the field $4°x4°$ as well as of 699 stars from 15^m to 22^m in the field $3°x3°$ were determined. The catalogue has been punched on a magnetic tape at the Astronomical Data Centre of the Astronomical Institute, Academy of Sciences (Moscow).

041.006 Observations of solar–system bodies with the Belgrade Meridian Circle.
S. Sadžakov, M. Dačić, Z. Cvetković.
Astron. Astrophys., Suppl. Ser., Vol. 92, No. 3, p. 605 – 607 (Feb 1992).

Observations of the Sun and planets (Mercury, Venus, Mars) are described. The error estimate including both the systematic error and the random one is presented.

041.007 Optical astrometric positions of southern quasars.
M. Assafin, R. Vieira Martins.
Astron. Astrophys., Suppl. Ser., Vol. 93, No. 2, p. 247 – 253 (May 1992).

As the first result of an astrometric program of southern radiosources, optical positions of 10 quasars between –75° and –25° declination are given. They have been derived using different telescopes. The sources were measured on film copies of the ESO (B) Schmidt Survey. Each source has two independent positions, one referred to the Perth 70 Catalogue, the other to the PPM South Catalogue, preliminary version. The positions were reduced through an auxiliary secondary reference frame of stars of intermediate brightness $m_v = 12 - 14$, obtained in 1988 from an astrograph. All objects were scanned with a PDS microdensitometer, their centres determined by a Gaussian fit to the marginal distributions of the density arrays. The mean internal error of the film reductions is 0."23 when using the Perth 70 stars, and 0."21 when using the PPM Catalogue. The radio and optical positions on the average agreed to the tenth of an arcsecond.

041.008   Results of observations made with the seven–inch transit circle 1967 – 1973.
Observations of the Moon and minor planets.
Catalog of 23,001 stars for 1950.0.
Comparisons with FK4.
J. A. Hughes, C. A. Smith, R. L. Branham.
Publ. U.S. Nav. Obs., Second Ser., Vol. 26, Part 2.
Naval Observatory, Washington, DC (USA), U.S. Government Printing Office, Washington, DC (USA). p. 153 – 553 (1992).

041.009   High–precision VLBI astrometry of the radio–emitting star σ CrB – a step in linking the Hipparcos and extragalactic reference frames.
J.–F. Lestrade, R. B. Phillips, R. A. Preston, D. C. Gabuzda.
Astron. Astrophys., Vol. 258, No. 1, p. 112 – 115 (May 1992).

VLBI observations of the optically bright radio–emitting star σ CrB have yielded its position relative to an angularly nearby quasar, its annual proper motion and its trigonometric parallax with formal uncertainties slightly better than 0.2 milliarcsec. This is the first result of the VLBI astrometric program the authors are conducting on 11 radio stars similar to σ CrB. They plan to obtain comparable results for all stars when enough epochs of observations are acquired. These stars will be used to astrometrically link the Hipparcos and VLBI extragalactic reference frames.

041.010   Comparison of the first results from the Hipparcos star mappers with the Hipparcos Input Catalogue.
C. Turon, F. Arenou, D. W. Evans, F. van Leeuwen.
Astron. Astrophys., Vol. 258, No. 1, p. 125 – 133 (May 1992).

Preliminary positions and magnitudes derived from the analysis of 12 weeks of observations from the Hipparcos star mappers are systematically compared with the various sources of ground–based data used in the Hipparcos Input Catalogue. These comparisons flow to cross–check the accuracies claimed by the various sources of ground–based data and by the analysis method of star mapper data. The parameters obtained for double stars, relative position and orientation, are also compared with ground–based data.

041.011   Positions and parallaxes from the Hipparcos satellite. A first attempt at a global astrometric solution.
L. Lindegren, F. van Leeuwen, C. Petersen, M. A. C. Perryman, S. Söderhjelm.
Astron. Astrophys., Vol. 258, No. 1, p. 134 – 141 (May 1992).

The last stage of the basic astrometric reduction of Hipparcos data is to combine angular measurements along great–circle scans into a globally coherent system of positions, proper motions and trigonometric parallaxes. The authors report the results of a first attempt at such a "sphere solution" of positions and parallaxes using great–circle results calculated from a selection of the data collected between November 1989 and February 1991. The data, which were made available to the reduction consortia for test and validation purposes, constitute less than a fifth of the total data collected during that period. The solution included some 20000 stars, and for at least 10000 stars the authors have determined the positions at mid–epoch with an accuracy of about 5 milli–arcsec and the trigonometric parallaxes with an average external standard error of 7 milli–arcsec. The positions reveal systematic errors in the Hipparcos Input Catalogue of up to 0.2 arcsec, some of which can probably be traced back to zonal errors in the FK4 and/or FK5 system.

041.012   Interferometric measurements and light aberration.
M. D. Kislik.
Astron. Zh., Tom 69, Vyp. 1, p. 214 – 218 (Jan–Feb 1992). In Russian. English translation in Sov. Astron., Vol. 36, No. 1 (1992).

The problem of using the concept of light aberration in the treatment of interferometric observation data is considered. A conclusion is drawn that this concept ought not to be used in this case. It is shown that application of Galileo transformations in the treatment of interferometric observations results in errors, having the same order of magnitude as the value of the aberration effect itself.

041.013   Catalog–to–catalog reductions: results for the FK, N30, and GC catalogs.
· L. G. Taff, B. Bucciarelli, M. G. Lattanzi.
Astrophys. J., Vol. 392, No. 2, p. 746 – 759 (20 Jun 1992).

The authors show that previous computations of catalog–to–catalog differences, whether by the bin–and–average method or by least–squares adjustment combined with an expansion in an incomplete set of basis functions, have all been performed at inappropriate epoch(s) of place. As a consequence, such results are frequently much more informative regarding the systematic differences between the proper motion systems than they are with respect to the systematic differences between the positional systems. Moreover, and for the same reason, some sets of proper motion differences really do not portray the true state of the evolution of the proper motion systems. The authors correct this error by arguing for a more meaningful epoch of place for catalog–to–catalog comparisons. They apply their method of infinitely overlapping circles at these other epochs to the FK sequences of catalogs, the N30, and the GC. This demonstrates one of the main failings of the least–squares expansion in an incomplete set of basis functions procedure – namely its tendency to incorrectly retain a nonsignificant term in the artificially truncated series and then to globally broadcast it as part of the final differences. If true, the result is a bias in the computed differences, and the published FK5 – FK4 differences suffer from this defect.

041.014   The correction in right ascension of 508 stars determinated wiht PMO photoelectric transit instrument.
Yao Jinsheng.
Publ. Purple Mt. Obs., Vol. 10, No. 3, p. 214 – 228 (Sep 1991). In Chinese.

The correction in right ascension of 508 stars determined with Purple Mountain Observatory photoelectric transit instrument in 1980 – 1985 to FK5 system is given. The RMS error is 2.12 ms.

041.015   Positions of major planets from observations in the years 1988 – 1989.
A. Ya. Gregul', N. A. Chernega.
Visn. Kiiv. Univ., Fiz.–Mat. Nauki, Astron., Vip. 3, p. 57 – 59 (1992). In Ukrainian.

The coordinates of Mars and Jupiter obtained from observations with the meridian circle of the Kiev State University Astronomical Observatory are given. The observational method and treatment are briefly described.

041.016   Corrections to the luni–solar precession derived from a compilation catalogue of extragalactic radio sources.
H. G. Walter.
Celest. Mech. Dyn. Astron., Vol. 53, No. 1, p. 71 – 80 (1992).

Observation catalogues of extragalactic radio sources obtained by Very Long Baseline Interferometry during the last decade agree in the mean to a few milliarcseconds (mas). Within this range the position differences show constant, linear and periodic offsets. To reduce the influence of individual catalogue properties the construction of a compilation catalogue seems to be the appropriate procedure. In some detail the compilation method is described providing simultaneous adjustment of source positions and catalogue corrections. The compilation catalogue consists of 40 objects having positional errors of 0.2 mas in right ascension (RA) and 0.3 mas in declination (Dec). Comparing this catalogue with the IERS Celestial Reference Frame compiled by means of other precepts yields weighted root–mean–square differences of 0.7 mas in RA and 1.3 mas in Dec. Finally, the terms of general precession in RA and Dec are included in the adjustment process giving estimates of the correction to the luni–solar precession between –1 and –3 mas/yr, the latter figure applying when some early data are added.

041.017   Analysis of the General Catalogue of positions and proper motions of 4949 geodetic stars (CGS).
Li Zhengxing.
Publ. Purple Mt. Obs., Vol. 10, No. 3, p. 204 – 213 (Sep 1991). In Chinese.

A statistical analysis of the CGS has been done in detail. The main results are as follows: (1) The internal precision and some features of the error distribution of CGS are analysed. (2) The systematic differences between CGS and FK4Sup are listed. (3) The stars with large differences of position or proper motion between CGS and FK4 are listed. The sources of the differences are discussed.

CPC 2 – the second cape photographic catalog. II. Conventional plate adjustment and catalog construction.
See Abstr. 002.002.

The AGK3U: an updated version of the AGK3.
See Abstr. 002.021.

The astrometric data bank of the Pulkovo Observatory and some examples of its use.
See Abstr. 002.027.

Fifth Fundamental Catalogue (FK5). Part II. The FK5 extension – new fundamental stars.
See Abstr. 002.039.

Astrophysical supplement to the general catalogue of geodetic stars.
See Abstr. 002.040.

The Hipparcos Input Catalogue. I. Star selection.
See Abstr. 002.041.

The Hipparcos Input Catalogue. II. Astrometric data.
See Abstr. 002.042.

An astrometric catalogue of southern and equatorial dwarf novae.
See Abstr. 002.044.

Reliability of the Hipparcos Input Catalogue tested by the "First look".
See Abstr. 002.045.

The astrometric standard region in the Cyg constellation: catalog of 2197 stars.
See Abstr. 002.048.

The Hipparcos Input Catalogue. Volumes 1 – 7.
See Abstr. 002.055.

RRS2 – the list of reference stars in the field around 238 extragalactic radio sources.
See Abstr. 002.065.

Carlsberg Meridian Catalogue, La Palma. Number 6. Observations of positions of stars and planets 1990.
See Abstr. 002.071.

Determination of star coordinates and proper motion from FON program plates.
See Abstr. 002.074.

Updating the AGK3 reference star catalog with astrometrically reduced Schmidt plates.
See Abstr. 002.081.

Star catalogs: how good are they really.
See Abstr. 002.091.

Application of the PPM catalogue in the reduction of the Carte du Ciel plates.
See Abstr. 002.125.

Second general catalogue of stars observed with the photoelectric astrolabes (GCPA2)
See Abstr. 002.131.

Fourth preliminary catalogue of stars observed with the photoelectric astrolabe (PACP4).
See Abstr. 002.132.

The Tokyo PMC catalog 88: catalog of positions of 3800 stars observed in 1988 and planetary positions observed in 1986 to 1988 with Tokyo Photoelectric Meridian Circle.
See Abstr. 002.152.

Sizing up the solar system.
See Abstr. 014.018.

How far the star?
See Abstr. 014.019.

Automated instrument complex for determining positions of natural and artificial celestial bodies.
See Abstr. 032.009.

A new astrometric measuring machine: design and astronomical programmes.
See Abstr. 034.079.

SuperCOSMOS.
See Abstr. 034.080.

On the relationship between conventional and overlap reduction techniques in positional astronomy.
See Abstr. 036.001.

Statistical analysis and separation of systematic errors in astrometric data.
See Abstr. 036.021.

Automatic determination of the astronomical azimuth by observing a celestial body using the electronic theodolite Kern E2 and the laptop computer Toshiba T1600.
See Abstr. 036.030.

A study of polynomial models for astrographic plate reductions.
See Abstr. 036.057.

High–fidelity copying of large–area astronomical plates: principles and some practical results.
See Abstr. 036.064.

The FAST Hipparcos Data Reducation Consortium: overview of the reduction software.
See Abstr. 036.089.

The NDAC Hipparcos data analysis consortium. Overview of the reduction methods.
See Abstr. 036.090.

Attitude determination in the Hipparcos revised mission.
See Abstr. 036.091.

Method of comparison between determinations of the Hipparcos attitude.
See Abstr. 036.092.

Modelling the torques affecting the Hipparcos satellite.
See Abstr. 036.093.

Hipparcos great–circle reduction. Theory, results and intercomparisons.
See Abstr. 036.094.

Comparison of Hipparcos results obtained on different dates on the same great circle.
See Abstr. 036.095.

Early improvements to the Hipparcos Input Catalogue through the accumulation of data from the satellite. Including the NDAC attitude reconstruction description.
See Abstr. 036.096.

Detection and measurement of double stars with the Hipparcos satellite: NDAC reductions.
See Abstr. 036.099.

Hipparcos double star recognition and processing within the FAST consortium.
See Abstr. 036.100.

The treatment of Hipparcos observations of some peculiar double stars: anomalous cases.
See Abstr. 036.101.

Tycho data analysis. Overview of the adopted reduction software and first results.
See Abstr. 036.102.

Tycho star recognition.
See Abstr. 036.104.

Tycho astrometry calibration.
See Abstr. 036.105.

Investigtion of some methods for determining statistical weights of astrometric catalogues.
See Abstr. 036.134.

Conversion to the true optical center in the complete overlapping–plate method.
See Abstr. 036.140.

Long–baseline optical and infrared stellar interferometry.
See Abstr. 036.189.

Applied techniques of PDS in photographic astrometry. II. Measurements of the crowded field of stars.
See Abstr. 036.195.

A radio optical reference frame. III. Additional radio and optical positions in the southern hemisphere.
See Abstr. 043.002.

In–orbit performance of the Hipparcos astrometry satellite.
See Abstr. 051.021.

The Hipparcos observing programme: preparation of the input catalogue.
See Abstr. 051.059.

Hipparcos data reduction – construction of the Hipparcos star catalogue.
See Abstr. 051.060.

Early scientific results from Hipparcos and future expectations.
See Abstr. 051.061.

A mechanism for orbital period modulation in close binaries.
See Abstr. 065.019.

Empirical image motion spectrum. I. Seeing quality and the atmospheric limitation on the accuracy of meridian observations.
See Abstr. 082.068.

Astrometric observations of asteroid Hidalgo near its perihelion.
See Abstr. 098.142.

Astrometric observations of the faint outer satellites of Jupiter during the 1989–1990 opposition.
See Abstr. 099.015.

An analysis of photographic astrometric observations of the Galilean moons: USNO refractor, 1986 – 1990.
See Abstr. 099.104.

Astrometric observations of the irregular satellites of Jupiter.
See Abstr. 099.105.

New proper motion determination of Luyten catalogue stars (LTT) south of declination –40° and right ascension between 04 h 30 m and 16 h 00 m.
See Abstr. 111.012.

Milliarcsecond proper motion measurements with MAMA.
See Abstr. 111.013.

Radial velocities from objective prism plates.
See Abstr. 111.016.

Binary star observations with the Hubble Space Telescope fine guidance sensors. II. Bright Hyades.
See Abstr. 153.007.

A magnitude, colour, and proper–motion probe of the Galaxy at an intermediate galactic latitude.
See Abstr. 155.001.

A complete, multicolor survey of absolute proper motions to B ~ 22.5: Galactic structure and kinematics at the north Galactic pole.
See Abstr. 155.008.

Position and morphology of the compact non–thermal radio source at the Galactic Center.
See Abstr. 155.113.

042 Celestial Mechanics, Figures of Celestial Bodies

042.001 Equipotential surfaces in close binary systems: remarks on the time–dependent potential function.
I. Todoran.
Astrophys. Space Sci., Vol. 187, No. 1, p. 119 – 126 (Jan 1992).
By use of the mass–point model, the equations of the equipotential surfaces are reviewed. A difference between the time–dependent potential function and zero relative velocity surfaces is put in evidence. A drawback in the time–dependent transformation between (ξ,η,ζ) and (x,y,z) coordinate systems is underlined.

042.002 Periodic solutions of the motion of a rigid body about the center of mass in a central Newtonian field.
F. M. El–Sabaa.
Cosmic Res., Vol. 29, No. 2, p. 151 – 155 (Sep 1991). English translation of Kosm. Issled., Tom 29, Vyp. 2, p. 172 – 177 (1991).
The equations of motion of a rigid body about a fixed point in a central Newtonian field are reduced to the equation of plane motion under the action of potential and gyroscopic forces, using the isothermal coordinates on the inertia ellipsoid. The construction of periodic solutions in the neighbourhood of nearly equilibrium points, by using Lyapunov's theorem of the holomorphic integral is obtained and the necessary and sufficient conditions for the stability of the system are given.

042.003 Translational–rotary motion of a deformed axisymmetric body in a central force field.
A. V. Demin, Yu. G. Markov.
Cosmic Res., Vol. 29, No. 2, p. 156 – 159 (Sep 1991). English translation of Kosm. Issled., Tom 29, Vyp. 2, p. 178 – 182 (1991).
Approximate equations are obtained for the motion of a viscoelastic axisymmetric body in a central force field. Steady state motions are found and their stability is studied.

042.004 Planar quasistatic motions of a viscoelastic body in a gravitational field.
O. V. Kholostova.
Cosmic Res., Vol. 29, No. 2, p. 160 – 168 (Sep 1991). English translation of Kosm. Issled., Tom 29, Vyp. 2, p. 183 – 193 (1991).
Planar motions of a viscoelastic body in a central Newtonian gravitational field are considered. Elastic oscillations are assumed to become excited along one of the body–fixed axes. The quasistatic motion of the body is investigated. Relative equilibrium positions of the body in a circular orbit are found and their stability is investigated. Eccentricity oscillations of the body in nonresonance and resonance cases are considered.

042.005 Mechanism for rotation of the plane of a satellite orbit.
Yu. G. Markov, I. S. Minyaev.
Cosmic Res., Vol. 29, No. 2, p. 175 – 184 (Sep 1991). English translation of Kosm. Issled., Tom 29, Vyp. 2, p. 201 – 211 (1991).
Under consideration is a slow changing mechanism of a satellite's orbital plane of inclination due to the action of dissipative forces on the part of a deformable planet.

042.006 On the restricted three–body problem with generalized forces.
A. Elipe.
Astrophys. Space Sci., Vol. 188, No. 2, p. 257 – 269 (Feb 1992).
By introducing general functions which depend on distance, a general scheme which determines the equilibrium solutions for the generalized restricted three–body problem is given. Application to problems such as primaries considered as rigid bodies, influence of the radiation pressure of the primaries, and a combination of radiation pressure and rigid body are presented.

042.007 Three–body orbital stability criteria for circular orbits.
J. R. Donnison, D. F. Mikulskis.
Mon. Not. R. Astron. Soc., Vol. 254, No. 1, p. 21 – 26 (1 Jan 1992).
Previous investigations of the stability of hierarchical three–body systems by Harrington and by Black and his collaborators give conflicting results. A new numerical examination of such systems and the corresponding stability criteria indicate that the Black functions give the general qualitative behaviour for very large and very small values of the third component; while for the range close to the equal–mass case the Harrington criterion is the more appropriate. Neither criterion was found to be satisfactory but it was found that the new results obtained are, however, in very good quantitative agreement with those obtained from the analytical c^2H method. This suggests that this criterion has a wider range of validity than previously suspected. Boundaries of stability are derived for the satellite, inner planet and outer planet mass combinations. The numerical results show a larger range of stable orbits in the outer planet case where the restricted model gives inconclusive results.

042.008 General–relativistic celestial mechanics. II. Translational equations of motion.
T. Damour, M. Soffel, Xu Chongming.
Phys. Rev. D, Vol. 45, No. 4, p. 1017 – 1044 (15 Feb 1992). Current Physics Microform No.: 9204D1627.
The translational laws of motion for gravitationally interacting systems of N arbitrarily composed and shaped, weakly self–gravitating, rotating, deformable bodies are obtained at the first post–Newtonian approximation of general relativity. The derivation uses the recently introduced multi–reference system method and obtains the translational laws of motion by writing that, in the local center–of–mass frame of each body, relativistic inertial effects combine with post–Newtonian self– and externally generated gravitational forces to produce a global equilibrium (relativistic generalization of d'Alembert's principle). Within the first post–Newtonian approximation [i.e., neglecting terms of order $(v/c)^4$ in the equations of motion], this work is the first to obtain complete and explicit results, in the form of infinite series, for the laws of motion of arbitrarily composed and shaped bodies. The authors first derive the laws of motion of each body as an infinite series exhibiting the coupling of all the post–Newtonian multipole moments of this body to the post–Newtonian tidal moments felt by this body. Then, the explicit expression of these tidal moments in terms of post–Newtonian multipole moments of the other bodies is given.

042.009 Sums of coefficients in the main Keplerian motion series.
K. V. Kholshevnikov, O. K. Tublina.
Kinematics Phys. Celest. Bodies, Vol. 7, No. 3, p. 1 – 6 (1991). Abstract. English translation of Kinematika Fiz. Nebesn. Tel, Tom 7, No. 3, p. 3 – 8 (1991).

042.010 Estimation of the geopotential effect on the motion of a geostationary satellite.
Eh. D. Kuznetsov.
Kinematika Fiz. Nebesn. Tel, Tom 8, No. 2, p. 52 – 55 (Mar–Apr 1992). In Russian. English translation in Kinematics Phys. Celest. Bodies, Vol. 8, No. 2.
Formulae have been deduced allowing to estimate the errors of spherical coordinates of a geostationary satellite which arise when geopotential harmonics are neglected.

042.011 On determination of resonance dominant arguments in the motions of celestial bodies.
G. V. Stolyarov, S. A. Kanaev, N. G. Lisin.
Astron. Tsirk., No. 1550, p. 31 (Sep – Oct 1991). In Russian.
The dominant harmonics of the perturbation function in the motion of celestial bodies are determined.

042.012   On calculation of the coefficients of resonance equations.
G. V. Stolyarov, N. G. Lisin, S. A. Kanaev.
Astron. Tsirk., No. 1550, p. 32 (Sep – Oct 1991). In Russian.
An analytical representation of coefficients in the resonance equation of satellites was found in the case, when the resonant model might be characterised by the solitary harmonic perturbation function.

042.013   Eine einfache Methode der parabolischen Bahnbestimmung.
M. Dīrikis.
Sterne, Band 68, Heft 2, p. 94 – 97 (1992).

042.014   On the investigation of resonant equations of the motion of satellites.
G. V. Stolyarov.
Astron. Tsirk., No. 1551, p. 33 (Nov – Dec 1991). In Russian.
The analytical representation of perturbations in the resonance equations of satellites was found in the case, when the resonant model might be characterised by the solitary harmonic perturbation function.

042.015   Construction of resonance models in the motion of celestial bodies.
G. V. Stolyarov.
Astron. Tsirk., No. 1551, p. 32 (Nov – Dec 1991). In Russian.
Types of resonance models of the motion of celestial bodies are presented.

042.016   On perturbations of resonant motions of satellites.
G. V. Stolyarov.
Astron. Tsirk., No. 1551, p. 34 – 35 (Nov – Dec 1991). In Russian.
The results of a study of perturbations in resonant motions of satellites are presented.

042.017   A few points on the stability of the solar system.
J. Laskar.
IAU Symposium No. 152: Chaos, resonance and collective dynamical phenomena in the solar system, p. 1 – 16 (1992). – See Abstr. 012.025 for the main entry.
The secular equations which were used to exhibit the chaotic behaviour of the solar system (Laskar, 1989) are established here in a Hamiltonian framework. The integration of the former secular equations over 400 Myr showed that the two resonant arguments given in Laskar (1990) present both transitions from libration to circulation. During the circulation of the first argument, temporary libration of this argument is observed, revealing resonance overlap between these two resonances, which explains the existence of a large chaotic zone for the motion of the solar system.

042.018   Long term evolution of the solar system.
J. Wisdom.
IAU Symposium No. 152: Chaos, resonance and collective dynamical phenomena in the solar system, p. 17 – 24 (1992). – See Abstr. 012.025 for the main entry.
The mapping method of Wisdom (1982) has been generalized to encompass all n–body problems with a dominant central mass (Wisdom and Holman, 1991). The new mapping method is presented as well as a number of initial applications. These include billion year integrations of the outer planets, a number of 100 million year integrations of the whole solar system, and a systematic survey of test particle stability in the outer solar system.

042.019   Numerical experiments on the motion of the outer planets.
G. D. Quinlan.
IAU Symposium No. 152: Chaos, resonance and collective dynamical phenomena in the solar system, p. 25 – 32 (1992). – See Abstr. 012.025 for the main entry.
The author has integrated the motion of the four Jovian planets on Myr timescales in fictitious solar systems in which the

orbits differ from those of the real solar system. A change of ≲1% in the major axis of any one of the planets from its real value can lead to chaotic motion with a Lyapunov exponent larger than 10^{-5}yr^{-1}. A survey of fifty solar systems with initial conditions chosen at random from a reasonable probability distribution shows the majority of them to be chaotic.

042.020   Resonant orbital evolution in the putative planetary system of PSR 1257 + 12.
R. Malhotra, D. Black, A. Eck, A. Jackson.
Nature, Vol. 356, No. 6370, p. 583 – 585 (16 Apr 1992). Letter-to-the-editor.
Periodic variations in the arrival times of pulses from the millisecond pulsar PSR 1257 + 12 are most straightforwardly interpreted as indicating the presence of two planet–like companions orbiting the pulsar. The authors point out that if the masses of the two planets are more than ~10 times greater than the minimum values (3.4 and 2.8 Earth masses) allowed by the observations, then their orbits will be in an exact 3:2 resonance. The character of the predicted orbital parameter perturbations is then markedly different from the periodic perturbations that result from only a near–resonance. The amplitude of the perturbations is much greater, and is very sensitive to the planet masses.

042.021   Numerical simulations of planetary systems of the Jupiter–Saturn type.
R. A. Broucke.
IAU Symposium No. 152: Chaos, resonance and collective dynamical phenomena in the solar system, p. 33 – 36 (1992). – See Abstr. 012.025 for the main entry.
The author made a numerical study of the General Three-Body Problem in two dimensions, with the intention to obtain some statistical estimates of the outcome of the system after a long time. Two different sets of masses were used. In the first series of experiments masses in the ratio of 0.95, 0.04 and 0.01 are used. In the second series, masses are used that are exactly in the Sun–Jupiter–Saturn ratio. All integrations were performed for a maximum of 12500 canonical units of time, corresponding to about 2000 revolutions of Jupiter. The cause of termination or type of catastrophe for the system has been determined in all cases. In most cases, this is a close approach of Saturn with Jupiter, followed by ejection of Saturn from the system.

042.022   General theory for the outer planets.
P. Bretagnon, G. Francou.
IAU Symposium No. 152: Chaos, resonance and collective dynamical phenomena in the solar system, p. 37 – 42 (1992). – See Abstr. 012.025 for the main entry.
An iterative method for the construction of planetary theories has been developed in order to determine the high order perturbations with respect to the masses. These perturbations are indeed needed to enlarge the validity span of analytical theories up to some million years. The application to the simplified Sun–Jupiter–Saturn problem gives a solution accurate over several ten million years. Throughout the study of the four outer planets the authors meet with convergence difficulties especially in the determination of fundamental frequencies. One of the results of this study is it shows evidence of long period terms with large amplitude in the mean longitudes: 12000″ in Saturn longitude, 20000″ in that of Uranus.

042.023   A planetary theory with elliptic functions and elliptic integrals exhibiting no small divisors.
C. A. Williams.
IAU Symposium No. 152: Chaos, resonance and collective dynamical phenomena in the solar system, p. 43 – 48 (1992). – See Abstr. 012.025 for the main entry.
This paper develops a planetary theory in three dimensions with elliptic functions and elliptic integrals. In an earlier treatment, (Williams, Van Flandern, and Wright, 1987) presented a two dimensional planetary theory to the first order of a Picard iteration. The theory did avoid expansions in powers of the ratio of the semi–major axes and it contained only two explicit small

divisors, $n-n'$ and $2n-n'$. These advantages are retained in the new theory and in fact no small divisors appear explicitly. Secular terms are removed by adopting an averaging technique rather than continuing the Picard iteration. The Lie series method of Deprit (1969) is chosen for the averaging. In order to simplify the Lie operator, the framework for the problem is chosen to be the circular restricted three body problem written in the polar–nodal coordinates of Whittaker. The algorithm is described and a few representative terms are discussed.

042.024 **Puzzles and prospects in planetary ring dynamics.**
P. Goldreich.
IAU Symposium No. 152: Chaos, resonance and collective dynamical phenomena in the solar system, p. 65 – 73 (1992). – See Abstr. 012.025 for the main entry.
The author outlines some of the main processes that shape planetary rings. Then he focuses on two outstanding issues, the role of self–gravity in the precession of narrow rings and the dynamics of Neptune's arcs.

042.025 **Collisional, collective and resonance phenomena in planetary rings.**
A. M. Fridman, N. N. Gor'kavij (*N. N. Gor'kavyj*).
IAU Symposium No. 152: Chaos, resonance and collective dynamical phenomena in the solar system, p. 75 – 82 (1992). – See Abstr. 012.025 for the main entry.
Contents: 1. Particle collisions in the rings. 2. Collective phenomena. 3. Resonance nature of Uranian rings.

042.026 **Collisional simulations of isolated Lindblad resonances.**
J. Hänninen, H. Salo, J. Lukkari.
IAU Symposium No. 152: Chaos, resonance and collective dynamical phenomena in the solar system, p. 97 – 102 (1992). – See Abstr. 012.025 for the main entry.
The influence of the perturbing satellite on the planetary ring at isolated Lindblad resonances is studied with numerical computer simulations, combining the Aarseth's force polynomial method for orbit integrations with the calculation of particle–particle impacts. Observed angular momentum exchange between the satellite and the dissipative, non–selfgravitating ring agrees with the Goldreich–Tremaine formula for gravitating ring within 20%, verifying that the exerted torque is not sensitive to the details of the dominant physical processes. The theoretically predicted angular momentum luminosity reversal was also observed.

042.027 **Numerical simulations of dense collisional systems with extended distribution of particle sizes.**
H. Salo.
IAU Symposium No. 152: Chaos, resonance and collective dynamical phenomena in the solar system, p. 103 – 108 (1992). – See Abstr. 012.025 for the main entry.
The dynamical evolution of dense planetary rings is mainly governed by the mutual impacts between macroscopic icy particles. By assuming that the ring system possesses local azimuthal symmetry, one can limit the calculations to a local co–moving region, following the mean orbital motion of particles. This makes it possible to model dense portions of rings with reasonable number of particles.

042.028 **A collisional and self–gravitational model to simulate numerically the dynamics of planetary disks.**
F. P. Gama, J.–M. Petit, H. Scholl.
IAU Symposium No. 152: Chaos, resonance and collective dynamical phenomena in the solar system, p. 109 – 114 (1992). – See Abstr. 012.025 for the main entry.
The dynamical evolution of the planetary rings is simulated by means of a numerical model in which particles interact through mutual attraction and inelastic collisions. The authors use a mixed simulation: a deterministic integration of the N–body problem for large distances ("particle–mesh" method with an expansion of density and potential in spherical harmonics) and a Monte Carlo treatment for the close encounters. The implementation is done in the Connection Machine in order to be able to

make a detailed simulation using a greater number of particles (of the order of 10^5). The deterministic calculation of the action of a shepherding satellite on the particles will allow to study the effect of resonances on the formation and the evolution of the sharp edges of the rings.

042.029 **Stability of asteroid motions.**
Y. Kozai.
IAU Symposium No. 152: Chaos, resonance and collective dynamical phenomena in the solar system, p. 115 – 122 (1992). – See Abstr. 012.025 for the main entry.
The author presents evidences showing that for most of the asteroids the motions are stable in the sense that they never approach major planets very closely and explains about mechanisms to avoid very close approaches by investigating the variations due to the secular perturbations of the eccentricities as functions of the arguments of perihelion, particularly, for asteroids with high eccentricities and inclinations. It is believed that some kinds of dynamical evolution processes have made the asteroid motions stable. The author shows also that there were some kinds of collisions among asteroids in the past which produced families and present distribution of asteroids as there are very faint asteroids only near Kirkwood gaps.

042.030 **The locations of secular resonances and the evolution of small solar system bodies.**
C. Froeschlé, P. Farinella, C. Froeschlé, Z. Knežević, A. Milani.
IAU Symposium No. 152: Chaos, resonance and collective dynamical phenomena in the solar system, p. 123 – 132 (1992). – See Abstr. 012.025 for the main entry.
Generalizing the secular perturbation theory of Milani and Knežević (1990), the authors have determined in the a–e–I proper elements space the locations of the secular resonances between the precession rates of the longitudes of perihelion and node of a small body and the corresponding eigenfrequencies of the secular perturbations of the four outer planets. They discuss some implications of the results for the dynamical evolution of small solar system bodies. In particular, their findings include: (1) the fact that the $g = g_6$ resonance in the inner asteroid belt lies closer than previously assumed to the Flora region, providing a plausible dynamical route to inject asteroid fragments into planet–crossing orbits; (2) the possible presence of some low–inclination "stable islands" between the orbits of the outer planets; (3) the fact that none of the secular resonances considered in this work exists for semimajor axes > 50 AU, so that these resonances do not provide a mechanism for transporting inwards possible Kuiper–belt comets.

042.031 **Proper elements of the asteroids. A semi–analytical method.**
A. Lemaitre.
IAU Symposium No. 152: Chaos, resonance and collective dynamical phenomena in the solar system, p. 133 – 137 (1992). – See Abstr. 012.025 for the main entry.
The author proposes here the first results of the semi–analytical method of Henrard (1990) applied to the calculation of proper elements for the asteroids.

042.032 **The qualitative explanation of observed peculiarities of Hecuba and Hilda asteroids distribution by a common investigation.**
E. V. Alfimova, I. A. Gerasimov.
IAU Symposium No. 152: Chaos, resonance and collective dynamical phenomena in the solar system, p. 139 – 144 (1992). – See Abstr. 012.025 for the main entry.
The authors explain all peculiarities of the distribution of asteroids with commensurabilities 2:1 and 3:2.

042.033 **New results on the motions of asteroids in resonances.**
R. Dvorak.
IAU Symposium No. 152: Chaos, resonance and collective dynamical phenomena in the solar system, p. 145 – 152 (1992). – See Abstr. 012.025 for the main entry.
The author presents a numerical study of the motion of asteroids in the 2:1 and 3:1 resonance with Jupiter. He integrated

the equations of motion of the elliptic restricted 3–body problem for a great number of initial conditions within this 2 resonances for a time interval of 10^4 periods and for special cases even longer (which corresponds in the Sun–Jupiter system to time intervals up to 10^6 years). The comparison with recent results show quite a good agreement for the structure of the 3:1 resonance. For motions in the 2:1 resonance the numeric results are in contradiction to others: high eccentric orbits are also found which may lead to escapes and consequently to a depletion of this resonant regions.

042.034   Very–high–eccentricity librations at some higher order resonances.
J. C. Klafke, S. Ferraz–Mello, T. Michtchenko.
IAU Symposium No. 152: Chaos, resonance and collective dynamical phenomena in the solar system, p. 153 – 158 (1992). – See Abstr. 012.025 for the main entry.

Motions near the 3:1, 4:1 and 5:2 resonances with Jupiter are studied by means of numerical integrations of a semi–analytically averaged Sun–Jupiter–asteroid planar problem. In order to have a model including the very–high–eccentricity regions of the phase space, the authors adopted a set of local expansions of the disturbing potential, adequate to perform the numerical exploration of regions in the phase space with eccentricities higher than 0.9 (Ferraz–Mello and Klafke, 1991). Individual solutions and qualitative results thus obtained are completely reproduced by numerical integration of the complete equations by filtering off the short–period components of these solutions.

042.035   Application of Wisdom's perturbative method to the 5:2 and 7:3 resonances.
T. Yokoyama, J. M. Balthazar.
IAU Symposium No. 152: Chaos, resonance and collective dynamical phenomena in the solar system, p. 159 – 166 (1992). – See Abstr. 012.025 for the main entry.

Wisdom's perturbative method is applied to the 5:2 and 7:3 resonances. Some comparisons with Yoshikawa's model are performed: for values of eccentricity up to about 0.3 – 0.4, agreement exists and it is better for 5:2 resonance. A clear difference between the cases 5:2 and 7:3 is observed: the former one, like in the case 3:1, can show significant variations of eccentricity, even starting from very small values, close to zero, while the latter seems to undergo such variations, but with initial eccentricity not less than a value near 0.1.

042.036   Corotations in some higher–order resonances.
S. Ferraz–Mello, M. Tsuchida, J. C. Klafke.
IAU Symposium No. 152: Chaos, resonance and collective dynamical phenomena in the solar system, p. 167 – 170 (1992). – See Abstr. 012.025 for the main entry.

Some results concerning the resonances 3:1, 4:1 and 5:2 are given and compared to results published by other authors.

042.037   A catalogue of periodic orbits in the elliptic restricted 3–body problem.
R. Dvorak, J. Kribbel.
IAU Symposium No. 152: Chaos, resonance and collective dynamical phenomena in the solar system, p. 171 – 174 (1992). – See Abstr. 012.025 for the main entry.

Results of families of periodic orbits in the elliptic restricted problem are shown for some specific resonances. They are calculated for all mass ratios $0 \leqslant \mu \leqslant 1.0$ of the primary bodies and for all values of the eccentricity of the orbit of the primaries $e \leqslant 1.0$. The grid size is of 0.01 for both parameters. The classification of the stability is undertaken according to the usual one and the results are compared with the extensive studies by Contopoulos (1986) in different galactical models.

042.038   The possible orbital evolution of the near–Earth asteroids.
E. I. Timoshkova.
IAU Symposium No. 152: Chaos, resonance and collective dynamical phenomena in the solar system, p. 175 – 178 (1992). – See Abstr. 012.025 for the main entry.

The subject of this paper is a study of a possible orbital evolution for near–Earth asteroids. The investigation is fulfilled

in the frame of the restricted circular three body problem. It is based on the calculations of the Jacobi constant. The osculating elements of some real Apollo–Amor–Aten asteroids are used as the starting parameters. A comparison with the results of other authors is given.

042.039   Binary asteroids: secular perturbations.
R. Vilhena de Moraes, S. M. G. Winter.
IAU Symposium No. 152: Chaos, resonance and collective dynamical phenomena in the solar system, p. 183 – 184 (1992). – See Abstr. 012.025 for the main entry.

The motion of two small bodies orbiting each other whose barycenter is orbiting around a massive body is studied. The equations of motion are integrated considering the secular part of the disturbing function.

042.040   Chaotic layers in resonance problems.
J. Henrard, M. Moons, A. Morbidelli.
IAU Symposium No. 152: Chaos, resonance and collective dynamical phenomena in the solar system, p. 189 – 208 (1992). – See Abstr. 012.025 for the main entry.

The recent numerical simulations of Tittemore and Wisdom (1988, 1989, 1990) and Dermott et al. (1988), Malhotra and Dermott (1990) concerning the tidal evolution through resonances of some pairs of Uranian satellites have revealed interesting dynamical phenomena related to the interactions between close–by resonances. These interactions produce chaotic layers and strong secondary resonances. The slow evolution of the satellite orbits in this dynamical lanscape is responsible for temporary capture into resonance, enhancement of eccentricity or inclination and subsequent escape from resonance. The present contribution aims at developing analytical tools for predicting the location and size of chaotic layers and secondary resonances. The problem of the 1:3 inclination resonance between Miranda and Umbriel is analysed.

042.041   A synthetic theory of motion for Titan–Hyperion.
L. Duriez.
IAU Symposium No. 152: Chaos, resonance and collective dynamical phenomena in the solar system, p. 209 – 214 (1992). – See Abstr. 012.025 for the main entry.

The author presents an iterative method allowing to synthesize a semi–numerical solution for the equations of motion of the resonant Saturn's satellites Titan–Hyperion (limited now to the planar problem). The current theory of Hyperion by Taylor, Sinclair & Message (1987) gives the greatest terms of the long–period part of the solution (depending on two angles: the libration angle τ, and the angular distance of the pericenters ζ). Using it as a first approximation, this solution is substituted numerically in the exact Lagrange equations of motion for Titan and Hyperion, computed for many values of the three angles: τ, ζ and ϕ (the mean synodic longitude). Besides a complete determination of the short–period perturbations of Hyperion obtained here completely for the first time, some long–period perturbations of Titan by Hyperion are also found which would be non negligible at the 10 km level.

042.042   Cross–tidal effects and orbit–orbit resonances.
T. Pauwels.
IAU Symposium No. 152: Chaos, resonance and collective dynamical phenomena in the solar system, p. 215 – 218 (1992). – See Abstr. 012.025 for the main entry.

The author discusses the influence of the deformation of a planet caused by a first satellite on the orbit of a second satellite of the same planet, where both satellites are supposed to be involved in an orbit–orbit resonance. Numerical results are given for seven orbit–orbit resonances in the Solar System.

042.043   About the secular acceleration of Mimas.
A. Vienne, J. M. Sarlat, L. Duriez.
IAU Symposium No. 152: Chaos, resonance and collective dynamical phenomena in the solar system, p. 219 – 222 (1992). – See Abstr. 012.025 for the main entry.

The authors explain the high values of the acceleration ($\approx 2°\mathrm{cy}^{-2}$) found in the longitude of Mimas by Kozai and

Dourneau when they fit to observations their current theory of the Mimas' motion. In fact, the authors have found that very long–period terms are missing in these theories; their expansion in powers of t well agrees with the observed acceleration. Effects of tidal dissipation are far smaller and could be determined only after accounting of these long–period terms.

042.044   Hori auxiliary system for Mimas–Tethys.
W. Sessin.
IAU Symposium No. 152: Chaos, resonance and collective dynamical phenomena in the solar system, p. 223 – 226 (1992). – See Abstr. 012.025 for the main entry.
The problem of two natural satellites around a spheric planet is considered. Only gravitational forces act on the system. The satellites move in a central field disturbed by the mutual atraction between them. The disturbing function is developed in power series of the small parameter (ratio of mass of satellites and planet), of the eccentricities and inclinations (assumed small) of the satellites' orbits. The mean motions are supposed to be commensurable in the ratio 2:1. It is also assumed that the critical angles are of inclination type. The critical angles of eccentricity type circulates and may be eliminate as fast variables like the short–periodic terms. The hamiltonian is truncated up to the second order in the inclinations. The system of ordinary differential equations generated by this hamiltonian is called Hori Auxiliary System. Greenberg (1973) shows that with a suitable choice of the reference system the Hori Auxiliary System is solvable. Therefore, it is possible to construct a formal theory of motion for this type of resonance. This is the case of Mimas–Tethys system when the oblateness of Saturn is neglected. Here, the author presents the solutions of the Hori Auxiliary System for inclination type resonance using an arbitrary reference system.

042.045   An integrable model for Helene.
J. S. Bevilacqua, W. Sessin.
IAU Symposium No. 152: Chaos, resonance and collective dynamical phenomena in the solar system, p. 227 – 230 (1992). – See Abstr. 012.025 for the main entry.
The goal of this work is to determine the influence of Enceladus in the motion of Helene. The authors constructed a model considering an oblated central body (Saturn) and three satellites (Enceladus, Dione and Helene) under the action of central forces. The development of the potential was made assuming small eccentricities, null inclinations and two resonances present in this system (2:1 between Enceladus and Dione and 1:1 between Dione and Helene). The mean Hamiltonian preserves terms derived of the oblateness of the central body and also terms of both resonant arguments. The auxiliary Hori's system generated by this Hamiltonian is completely integrable. In the solution for Helene the authors included the perturbative effects of Dione and Enceladus which extend the usual treatment.

042.046   Periodic solutions for the eccentricity and inclination first order resonance.
M. A. Nitto, W. Sessin.
IAU Symposium No. 152: Chaos, resonance and collective dynamical phenomena in the solar system, p. 231 – 232 (1992). – See Abstr. 012.025 for the main entry.
For the first order resonance, the problem of the motion of two small masses around a primary body can be of three different types: eccentricity, inclination or eccentricity–inclination. In this paper the authors study a dynamical system that includes both eccentricity– and inclination–type of resonance. This study is based in the models developed by Sessin and Ferraz–Mello (1984) and Sessin (1991). The resulting system of differential equation is non–integrable; thus, the families of trivial periodic solutions are studied.

042.047   Resonance $p:p+q$ in satellite orbits.
P. R. Grosso, W. Sessin.
IAU Symposium No. 152: Chaos, resonance and collective dynamical phenomena in the solar system, p. 233 – 234 (1992). – See Abstr. 012.025 for the main entry.
In this paper the authors consider the motion of a satellite whose mean motion is commensurable with the frequency of the rotational motion of the planet in the ratio $p:p+q$.

042.048   Generalized canonical systems: Formal solutions and the main problem of satellite theory.
S. Da Silva Fernandes.
IAU Symposium No. 152: Chaos, resonance and collective dynamical phenomena in the solar system, p. 235 – 238 (1992). – See Abstr. 012.025 for the main entry.
The generalized canonical version of Hori's method is discussed using some properties of generalized canonical system and, then, applied in solving the main problem of satellite theory.

042.049   Corotation solutions in the elliptic asteroidal problem with Stokes drag.
C. Beaugé, S. Ferraz–Mello.
IAU Symposium No. 152: Chaos, resonance and collective dynamical phenomena in the solar system, p. 355 – 358 (1992). – See Abstr. 012.025 for the main entry.
The authors have searched for stable stationary solutions of the resonant elliptic restricted problem of three bodies with Stokes drag. Results for the 1/2 external Jovian resonance are shown as example, including a comparison with numerical results of an N–body simulation.

042.050   The effect of non–stationary action on the motion of particles in protoplanetary nebulae.
A. S. Baranov.
IAU Symposium No. 152: Chaos, resonance and collective dynamical phenomena in the solar system, p. 359 – 362 (1992). – See Abstr. 012.025 for the main entry.
The author considers various types of the non–stationary actions, which may be in certain cases strictly periodic, but it may be quasiperiodic in the sense that the amplitude and the phase change gradually from one period to the other. In the general form this phenomenon is well known in the theory of oscillations, for instance, when superpositioning the oscillations with close frequencies (beats). Various mechanisms of generating these oscillations are admissible. In case of nebulae, the non–linear effects or resonances can be a cause of such quasiperiodic disturbances.

042.051   New developments in dynamics: hyperbolicity and chaotic dynamics.
J. Palis.
IAU Symposium No. 152: Chaos, resonance and collective dynamical phenomena in the solar system, p. 363 – 368 (1992). – See Abstr. 012.025 for the main entry.
Two important theories in Dynamical Systems were constructed in the sixties: the hyperbolic theory for general systems and the KAM (after Kolmogorov, Arnold and Moser) theory for conservative systems as the ones that appear in celestial mechanics. Most of the discussions here concern dissipative (or locally dissipative) systems, although most questions are now being posed for area preserving maps (sympletic maps in higher dimensions). Moreover, one can argue that understanding dynamically small dissipative perturbations of conservative systems is of much importance: indeed it has been recently shown that a KAM curve (tori in higher dimension) can be destroyed and in fact engulfed in the basin of attraction of a Hénon–like strange attractor.

042.052   Krein stability in the disturbed two–body problem.
R. R. Cordeiro, R. V. Martins.
IAU Symposium No. 152: Chaos, resonance and collective dynamical phenomena in the solar system, p. 369 – 374 (1992). – See Abstr. 012.025 for the main entry.
The authors present a method for the study of the Krein signature in perturbed Hamiltonian integrable systems. The method is developed up to first order in the small parameter. This method is applied to a particular instance of the two–body problem in which the semi–major axis is not affected by the perturbation.

042.053   Mappings in astrodynamics.
C. Froeschlé.
IAU Symposium No. 152: Chaos, resonance and collective dynamical phenomena in the solar system, p. 375 – 390 (1992). – See Abstr. 012.025 for the main entry.

The author reviews mappings mainly devised for the study of the dynamics of comets and asteroids. An attempt of a typology according to the method used to devise the mapping and to its deterministic or stochastic character is made.

042.054   Mappings for the first order asteroidal resonance.
T. J. Stuchi, W. Sessin.
IAU Symposium No. 152: Chaos, resonance and collective dynamical phenomena in the solar system, p. 391 – 394 (1992). – See Abstr. 012.025 for the main entry.

The authors construct a two step algebraic mapping from Sessin's simplified model for the first order resonance. The orbits obtained with this mapping are compared to the ones calculated with the exact solution. The authors also derive a reduced Hamiltonian. A plane Poincaré mapping, using delta periodic function, is constructed and compared to the reduced Hamiltonian contour curves showing the splitting of the separatrix due to delta perturbation technique.

042.055   New methods for long–time numerical integration of planetary orbits.
H. Kinoshita, H. Nakai.
IAU Symposium No. 152: Chaos, resonance and collective dynamical phenomena in the solar system, p. 395 – 406 (1992). – See Abstr. 012.025 for the main entry.

When planetary orbits are numerically integrated for a long time by conventional integrators, the most serious problem is secular errors in the energy and the angular momentum of the planetary system due to discretization (truncation) errors. The secular errors in the energy and the angular momentum mean that the semi–major axes, the eccentricities, and the inclinations of planetary orbits have a secular error which grows linearly with time. Recently symplectic integrators and linear symmetric multistep integrators are found not to produce the secular errors in the energy and the angular momentum due to the discretization errors. Here the authors describe briefly both methods and discuss favorable properties of these integrators for a long–term integration of planetary orbits.

042.056   Mapping for the asteroidal resonances.
M. Šidlichovský.
Astron. Astrophys., Vol. 259, No. 1, p. 341 – 348 (Jun 1992).

A generalized approach is presented to obtaining a mapping for the main second, third and fourth–order asteroidal resonances. One computer program then works for all these resonances. The relevant coefficients in the disturbing function must be known. The mapping was determined for the planar elliptic restricted three–body problem. Terms up to the fourth degree in eccentricities were taken into account. The mapping calculations are used for surface of section calculations and for calculating of Lyapunov characteristic exponents. These methods are useful in studying chaotic regions connected with resonances.

042.057   Effects of dynamic tides on secular variations of orbital elements in close binary systems.
E. Ruymaekers.
Astron. Astrophys., Vol. 259, No. 1, p. 349 – 358 (Jun 1992).

The theory of the secular variations of the semi–major axis, the eccentricity, and the longitude of the periastron due to tidal action in close binary systems of stars is developed for dynamic tides displaying a time lag with respect to the motion of the tide-generating companion. Dynamic tides are treated as forced, linear, non–radial, isentropic oscillations of a non–rotating star. The weak–friction approximation is applied in which the time lag is considered to be small. In the limiting case of an infinite orbital period, the secular parts of the rates of change of the semi–major axis, the eccentricity, and the longitude of the periastron derived in the framework of the dynamic theory of tides reduce to those derived in the framework of the equilibrium theory of tides. As an example, the secular variations of the semi–major axis, the eccentricity, and the longitude of the periastron are determined for a $5\,M_\odot$ main sequence star in case of very short orbital periods. The use of the dynamic theory of tides leads to secular variations of the orbital elements different from those predicted in the framework of the equilibrium theory of tides. The differences may be appreciable for shorter orbital periods, orbits with larger eccentricities, and cases of resonances of dynamic tides with free oscillation modes of the tidally distorted star.

042.058   On the evolution of motion of a planet–satellite system in the field of an attracting centre.
Yu. G. Markov, I. S. Minyaev.
Astron. Zh., Tom 69, Vyp. 2, p. 416 – 427 (Mar–Apr 1992). In Russian. English translation in Sov. Astron., Vol. 36, No. 2 (1992).

Translational–rotational motion of a system, consisting of a deformable planet and a satellite, in the field of an attracting centre is considered. Approximate equations are obtained in canonical variables. It is shown that during the tidal evolution, the satellite orbit is tending toward a circular one, and the orbit's radius decreases monotonically.

042.059   Criterion for instability of the translational–rotational motion of absolutely rigid bodies.
V. V. Vidyakin.
Astron. Zh., Tom 69, Vyp. 2, p. 428 – 435 (Mar–Apr 1992). In Russian. English translation in Sov. Astron., Vol. 36, No. 2 (1992).

By means of Lyapunov's first method, conditions are found, in compliance with which the translational–rotational motion of two absolutely rigid bodies is instable. The structure and external form of the bodies are supposed to satisfy certain conditions.

042.060   A test particle survey of the outer solar system.
M. J. Holman, J. Wisdom.
Bull. Am. Astron. Soc., Vol. 23, No. 3, p. 1151 (1991). Abstract. – See Abstr. 012.037 for the main entry.

042.061   Evolution of the motion of a planet–satellite system in a central gravitational field.
Yu. G. Markov, I. S. Minyaev.
Sov. Astron., Vol. 36, No. 2, p. 209 – 214 (Mar 1992). Current Physics Microform No.: 9208X2045. English translation of Astron. Zh., Tom 69, Vyp. 2, p. 416 – 427 (1992).

The translational–rotational motion of a "deformable planet-satellite" system in a central gravitational field is analyzed. The approximate equations of motion are obtained in canonical variables. It is shown that in the course of tidal evolution, the satellite's orbit tends to become circular and its radius decreases monotonically.

042.062   Criterion for instability of translational–rotational motion of perfectly rigid bodies.
V. V. Vidyakin.
Sov. Astron., Vol. 36, No. 2, p. 214 – 217 (Mar 1992). Current Physics Microform No.: 9208X2050. English translation of Astron. Zh., Tom 69, Vyp. 2, p. 428 – 435 (1992).

Lyapunov's first method is used to find the conditions under which the translational–rotational motion of two absolutely rigid bodies is unstable. The structure and external shape of the bodies must satisfy certain conditions.

042.063   On basic families of three–dimensional periodic orbits of three massive bodies and their stability.
K. E. Papadakis, V. V. Markellos.
Astrophys. Space Sci., Vol. 191, No. 2, p. 223 – 229 (May 1992).

The basic families of three–dimensional periodic orbits of the general three–body problem are determined numerically in order to obtain a global view of the simpler patterns of periodic three-body motion in three dimensions. The stability of the orbits is also computed. It is found that most of the orbits are unstable but stability intervals do exist for some of the families.

042.064   A note on the Hori–Lie perturbation technique.
O. M. Kamel.
Earth, Moon, Planets, Vol. 56, No. 1, p. 53 – 56 (Jan 1992).
The author describes how to avoid the introduction of Hori's pseudo time in general planetary theory in the case of two planets.

042.065   Expression of the initial Poincaré canonical variables as functions of the new in a ninth order J–S theory.
O. M. Kamel, A. S. Soliman.
Earth, Moon, Planets, Vol. 56, No. 3, p. 209 – 231 (Mar 1992).
The authors establish the solution of the ninth order – in masses – canonical J–S equations of motion by Hori–Lie technique – i.e., by expressing the initial Poincaré canonical variables as functions of the new variables through the Hori–Lie canonical transformation. Terms of order higher than 9 in the masses are neglected.

042.066   Conformal geometry of the Kepler orbit space.
J. F. Cariñena, C. López, M. A. del Olmo, M. Santander.
Celest. Mech. Dyn. Astron., Vol. 52, No. 4, p. 307 – 343 (1991).
The authors present here a group theoretical analysis of the structure of the space Ω of orbits in the classical (plane) Kepler problem, and relate it to the description of the Kepler orbits as curves in configuration and in velocity spaces. A Minkowskian parametrization in Ω is introduced which allows a clear description of many aspects of this problem. In particular, this parametrization suggests the introduction in Ω of a Lorentzian metric, whose conformal group $SO(3,2)$ contains a seven-dimensional subgroup which is induced by point transformations in the configuration space X. A $SO(2,1)$ subgroup of this group still acts transitively on X, which is thus identified as a homogeneous space for $SO(2,1)$; each regular Kepler orbit is the trace of a one–dimensional subgroup whose canonical parameter automatically equals to the classical anomalies. These results are somehow a configuration space analogous of the geometrical structure of the Kepler problem in the velocity space previously known.

042.067   On a question of A. Wintner about Jacobi coordinates in the three–body problem.
H. E. Cabral.
Celest. Mech. Dyn. Astron., Vol. 52, No. 4, p. 375 – 379 (1991).
In his work on the elimination of the nodes in the three–body problem, Jacobi considered a certain straight line (defined in the Introduction of this paper) which lies always within the invariable plane. Here the author settles the question of Wintner on the classification of all motions for which this line fails to exist.

042.068   Construction of invariant tori for the spin–orbit problem in the Mercury–Sun system.
A. Celletti, C. Falcolini.
Celest. Mech. Dyn. Astron., Vol. 53, No. 2, p. 113 – 127 (1992).
The stability of spin–orbit resonances, namely commensurabilities between the periods of rotation and revolution of an oblate satellite orbiting around a primary body, is investigated using perturbation theory. The authors reduce the system to a model described by a one–dimensional, time–dependent Hamiltonian function. By means of KAM theory they rigorously construct bidimensional invariant surfaces, which separate the three dimensional phase space. In particular with a suitable choice of the rotation numbers of the invariant tori they are able to trap the periodic orbit associated with a given resonance in a finite region of the phase space. This technique is applied to the Mercury–Sun system. A connection with the probability of capture in a resonance is also provided.

042.069   Coordinates for perturbed Keplerian systems with axial . symmetry.
S. Ferrer, B. R. Miller.
Celest. Mech. Dyn. Astron., Vol. 53, No. 1, p. 3 – 10 (1992).
A coordinate system is defined on the phase space of a perturbed Keplerian system after the mean anomaly has been averaged out, for the purpose of explaining how eliminating the longitude of the ascending node reduces the orbital space to a two–dimensional sphere in case the system admits an axial symmetry. Concomitantly, on the submanifold of direct osculating ellipses, the CDM variables are replaced by functions which form the basis of a Poisson algebra isomorphic to the Lie algebra $so(3)$ of the rotation group $SO(3)$; furthermore, in these variables, the doubly reduced phase flow appears like a rotation of the reduced phase space.

042.070   Trajectories and envelopes in the repulsive two–fixed–centre problem and in the restricted three–body problem.
J. Kallrath.
Celest. Mech. Dyn. Astron., Vol. 53, No. 1, p. 37 – 57 (1992).
The envelope of iso–energetic trajectories in the (repulsive) two–fixed–centre problem is derived. The analytical calculations finally lead to a transcendental equation, only containing elliptic integrals and the Weierstraß function, from which the envelope is constructed. The results may serve as a simple model for the boundary layer between two colliding supersonic stellar wind flows in binary systems, in which at least one of the components has a strong radiation field. Beyond this, the effect of non-inertial forces (centrifugal and Coriolis force) due to the binary's orbital motion has been estimated by a numerical analysis within the scope of the (repulsive) restricted three–body problem. All calculations have been performed for a hot model (Wolf–Rayet/O–star) binary system with a set of parameters which might be appropriate for HD 152270. The envelope may be well approximated by a hyperboloid. The non–inertial forces slightly turn the envelope against the line connecting both stars.

042.071   On a class of variational equations transformable to the Gauss hypergeometric equation.
H. Yoshida.
Celest. Mech. Dyn. Astron., Vol. 53, No. 2, p. 145 – 150 (1992).
A new class of linear ordinary differential equations with periodic coefficients is found which can be transformed to the Gauss hypergeometric equation, and therefore the monodromy matrices are computable explicitly. These equations appear as the variational equations around a straight–line solution in Hamiltonian systems of the form $H = T(p) + V(q)$, where $T(p)$ and $V(q)$ are homogeneous functions of p and q, respectively.

042.072   The elliptic restricted problem at the 3:1 resonance.
J. D. Hadjidemetriou.
Celest. Mech. Dyn. Astron., Vol. 53, No. 2, p. 151 – 183 (1992).
Four 3:1 resonant families of periodic orbits of the planar elliptic restricted three–body problem, in the Sun–Jupiter–asteroid system, have been computed. These families bifurcate from known families of the circular problem, which are also presented. Two of them, I_c, II_c bifurcate from the unstable region of the family of periodic orbits of the first kind (circular orbits of the asteroid) and are unstable and the other two, I_e, II_e, from the stable resonant 3:1 family of periodic orbits of the second kind (elliptic orbits of the asteroid). One of them is stable and the other is unstable. All the families of periodic orbits of the circular and the elliptic problem are compared with the corresponding fixed points of the averaged model used by several authors. The coincidence is good for the fixed points of the circular averaged model and the two families of the fixed points of the elliptic model corresponding to the families I_c, II_c, but is poor for the families I_e, II_e. A simple correction term to the averaged Hamiltionian of the elliptic model is proposed in this latter case, which makes the coincidence good. This, in fact, is equivalent to the construction of a new dynamical system, very close to the original one, which is simple and whose phase space has all the basic features of the elliptic restricted three–body problem.

042.073   Explicit solutions of the three–dimensional inverse problem of dynamics, using the Frenet reference frame.
F. Puel.
Celest. Mech. Dyn. Astron., Vol. 53, No. 3, p. 207 – 218 (1992).
Given a two–parameter of three–dimensional orbits, the author constructs the unit tangent vector, the normal and the

binormal which define the Frenet reference frame. In this frame, by writing that the force is conservative, he explicitly obtains the potential as a function of the energy along the trajectories and of its derivatives.

042.074   **On the use and abuse of Newton's second law for variable mass problems.**
A. R. Plastino, J. C. Muzzio.
Celest. Mech. Dyn. Astron., Vol. 53, No. 3, p. 227 – 232 (1992).
The authors clarify some misunderstandings currently found in the literature that arise from improper application of Newton's second law to variable mass problems. In the particular case of isotropic mass loss, for example, several authors introduce a force that actually does not exist.

042.075   **An alternative deduction of the Hill–type surfaces of the spatial 3–body problem.**
Ge Yanchao, Leng Xiaoling.
Celest. Mech. Dyn. Astron., Vol. 53, No. 3, p. 233 – 254 (1992).
In the present paper, inequalities stronger than Sundman's and the best possible zero velocity surfaces of the spatial 3–body problem first obtained by Saari (1987) are deduced using a modified version of the transformation developed by Zare (1976). The notion of inertia ellipsoid is used to show the equivalence of the present authors' result to that of Saari's.

042.076   **A new analytic approach to the Sitnikov problem.**
J. Hagel.
Celest. Mech. Dyn. Astron., Vol. 53, No. 3, p. 267 – 292 (1992).
A new analytic approach to the solution of the Sitnikov problem is introduced. It is valid for bounded small amplitude solutions ($z_{max} = 0.20$ in dimensionless variables) and eccentricities of the primary bodies in the interval $-0.4 < e < 0.4$. First solutions are searched for the limiting case of very small amplitudes for which it is possible to linearize the problem. The solution for this linear equation with a time dependent periodic coefficient is written up to the third order in the primaries eccentricity. After that the lowest order nonlinear amplitude contribution (being of order z^3) is dealt with as perturbation to the linear solution. The author first introduces a transformation which reduces the linear part to a harmonic oscillator type equation. Then two near integrals for the nonlinear problem are derived in action angle notation and an analytic expression for the solution $z(t)$ is derived from them. The so found analytic solution is compared to results obtained from numeric integration of the exact equation of motion and is found to be in very good agreement.

042.077   **Optimal trajectories in gravitational fields allowing approximation by the central linear field.**
A. G. Azizov, N. A. Korshunova.
Cosmic Res., Vol. 29, No. 4, p. 452 – 458 (Jan 1992). English translation of Kosm. Issled., Tom 29, Vyp. 4, p. 525 – 531 (1991).
For motions allowing an approximation of the gravitational field by a central linear field, the equations of Mayer's variational problem for a point with a limited mass flow rate are shown to be reducible to an analytical solution. The case where the hodograph of the basis vector is an ellipse corresponding to an alternation of zero– and maximum–thrust segments is analyzed. The results are useful for the solution of the problem of the change in velocity of a point from local circular to elliptic in a central Newtonian force field.

042.078   **On the evolution of inclinations and rotations of celestial bodies.**
A. V. Demin, Yu. G. Markov, I. S. Minyaev.
Kosm. Issled., Tom 30, Vyp. 2, p. 157 – 164 (Mar–Apr 1992). In Russian. English translation in Cosmic Research, Vol. 30, No. 2.
The space version of the two–body problem is studied when the mass point moves in the attraction field of a viscous elastic planet. The approximate equations describing the general regularities of the evolution of inclinations and rotations of the system are obtained.

042.079   **On the tidal evolution of a planet – satellite system in the circular three–body problem.**
I. S. Minyaev.
Kosm. Issled., Tom 30, Vyp. 2, p. 278 – 281 (Mar–Apr 1992). In Russian. English translation in Cosmic Research, Vol. 30, No. 2.

042.080   **Orbital simulation of captured satellites: application to Triton.**
L. A. M. Benner, W. B. McKinnon.
Bull. Am. Astron. Soc., Vol. 23, No. 3, p. 1209 (1991). Abstract. – See Abstr. 012.037 for the main entry.

042.081   **New orbital elements useful for predicting a particle's behavior upon encounter with a planet.**
R. Greenberg, W. F. Bottke, A. Carusi, G. B. Valsecchi.
Bull. Am. Astron. Soc., Vol. 23, No. 3, p. 1224 (1991). Abstract. – See Abstr. 012.037 for the main entry.

042.082   **Averaging the elliptic asteroidal problem with a Stokes drag.**
S. Ferraz–Mello.
15. Ecole de Printemps d'Astrophysique de Goutelas: Interrelations between physics and dynamics for minor bodies in the solar system, p. 45 – 60 (1992). – See Abstr. 012.055 for the main entry.
Contents: 1. Introduction. 2. The heliocentric equations. 3. The transformation of the dissipative terms. 4. Transformation to action–angle variables. 5. Stokes drag in the protoplanetary nebula. 6. Expansion and averaging of the dissipative terms. 7. Resonance. 8. The equations of the averaged problem with drag. 9. Conclusion.

042.083   **Dissipative phenomena in resonance problems in the solar system or the "dei ex machina" of celestial mechanics.**
A. Lemaitre.
15. Ecole de Printemps d'Astrophysique de Goutelas: Interrelations between physics and dynamics for minor bodies in the solar system, p. 61 – 84 (1992). – See Abstr. 012.055 for the main entry.
Contents: 1. Introduction (Motivation. The first cosmogonical hypotheses. Possible dissipative effects). 2. The general formalism (The model. The variation of δ. Formal calculation of the probability of capture. Captures in the SFMR for δ decreasing). 3. Application to the Kirkwood gap. 4. The drag effect. 5. Poynting Robertson drag. 6. Other applications.

042.084   **Topological methods for the qualitative analysis of a numerical simulation close to a resonance.**
A. Morbidelli.
15. Ecole de Printemps d'Astrophysique de Goutelas: Interrelations between physics and dynamics for minor bodies in the solar system, p. 133 – 157 (1992). – See Abstr. 012.055 for the main entry.
Contents: Introduction. 2. Topological aspects of a numerical simulation (Local dynamics in the neighbourhood of a resonant orbit. Topological analysis of numerical integrations. Some more possible topological aspects of an orbit). 3. Conclusion.

042.085   **Introduction to stochastic modelling of cometary dynamics: Monte Carlo simulations and Markov process.**
C. Froeschle.
15. Ecole de Printemps d'Astrophysique de Goutelas: Interrelations between physics and dynamics for minor bodies in the solar system, p. 159 – 196 (1992). – See Abstr. 012.055 for the main entry.
Contents: 1. Introduction. 2. From the Oort Cloud to the Solar System: Monte Carlo simulations of stellar perturbations. The Remy–Mignard model. 3. Stochastic modelling of planetary perturbations (Monte Carlo simulations of planetary perturbations. Modelling in terms of Markov process). 4. Results on irreducible finite Markov chains with ergodic states, and application to the study of the orbital evolution of short period comets. 5. Results on finite Markov chain with recurrent and transient states, and application to the study of the dynamical behaviour of small bodies in the Kuiper belt.

042.086   Planetary ring dynamics: from Boltzmann's equation to celestial mechanics.
P.–Y. Longaretti.
15. Ecole de Printemps d'Astrophysique de Goutelas: Interrelations between physics and dynamics for minor bodies in the solar system, p. 453 – 586 (1992). – See Abstr. 012.055 for the main entry.
Contents: 1. Introduction. 2. Basic concepts and orders of magnitude (Angular momentum axis. Collisional quasi equilibrium. Angular momentum transport and the origin of the ring radial structures). 3. The Boltzmann equation and its moments. 4. The equation of motion and the continuity equations: streamlines, at last! (Fluid test particle motion. Epicyclic versus elliptic elements. Ring streamlines and kinematics. The surface density of the ring). 5. The pressure tensor (Dilute systems. Dense systems. Energy dissipation and viscous flux of angular momentum. Summary and parametrization of the pressure tensor). 6. Perturbation equations and the mass, energy and angular momentum budget of ring systems. 7. Applications to ring dynamics (The self–gravity model for elliptic rings. Density waves at Lindblad resonances with external satellites). 8. Conclusions.

042.087   Planetary ring dynamics: secular exchange of angular momentum and energy with a satellite.
B. Sicardy.
15. Ecole de Printemps d'Astrophysique de Goutelas: Interrelations between physics and dynamics for minor bodies in the solar system, p. 631 – 651 (1992). – See Abstr. 012.055 for the main entry.
Contents: 1. Introduction. 2. The secular torque in the impulse approximation (Heuristic remarks. Evolution of the complex eccentricity). 3. The secular torque at isolated Lindblad resonances (Torque derivation. Isolated resonances vs. overlapping resonances). 4. Conclusions.

042.088   The formalism of the translational–rotational motion of a rigid body in the field of a gravitating and radiating centre.
A. G. Mavraganis.
Astrophys. Space Sci., Vol. 192, No. 2, p. 257 – 262 (Jun 1992).
The author's aim is to formulate the problem of motion of a rigid body acted upon by the Newtonian attraction of a particle and the repulsive force of this particle radiation. These forces are taken to be proportional to each other in a way which extends this connection from the point–like description of a body to its actual picture.

042.089   Mass ratios of three bodies for stable low–inclination three–dimensional periodic motion.
E. Perdios, V. V. Markellos.
Astrophys. Space Sci., Vol. 192, No. 2, p. 291 – 297 (Jun 1992).
The intervals of possible stability, on the μ–axis, of the basic families of three–dimensional periodic motions of the restricted three–body problem (determined in an earlier paper) are extended into regions of the μ–m_3 parameter space of the general three–body problem. Sample three–dimensional periodic motions corresponding to these regions are computed and tested for stability. Six regions, corresponding to the vertical–critical orbits $l1v$, $m1v$, $m2v$, and $i1v$, survive this preliminary stability test – therefore, emerging as the mass parameters regions allowing the simplest types of stable low inclination three–dimensional motion of three massive bodies.

042.090   Perturbations with elliptic functions in a Hamiltonian context.
C. A. Williams.
Bull. Am. Astron. Soc., Vol. 23, No. 3, p. 1259 (1991). Abstract. – See Abstr. 012.065 for the main entry.

042.091   New applications of Fatou's problem.
V. V. Radzievskij, V. P. Tomanov.
Sol. Syst. Res., Vol. 25, No. 5, p. 445 – 449 (Mar 1992). English translation of Astron. Vestn., Tom 25, No 5, p. 593 – 598 (1991).
The problem of the motion of a small body m in the gravitational field of a massive body m_1, and several massive, homogeneous, coplanar rings m_i (the asteroid belt, planets with a uniform mass distribution) is investigated. It is shown that all parabolic comets are isolated from the Oort cloud: they move inside a closed surface of zero radial velocity with a radius of about 5000 AU. The influence of the asteroid belt on the motion of the perihelion of Mercury is investigated.

042.092   On a family of exact solutions of a celestial ballistics problem.
Yu. M. Kopnin.
Kosm. Issled., Tom 30, Vyp. 3, p. 301 – 304 (May–Jun 1992). In Russian. English translation in Cosmic Res., Vol. 30, No. 3.
The subject of the investigation is a family of spatial orbits in Newton's field under the effect of disturbing acceleration that varies in inverse proportion to the third power of the distance from the center of attraction.

042.093   Chaos in a restricted charged four–body problem.
J. Casasayas, A. Nunes.
NATO Advanced Study Institute on Predictability, Stability, and Chaos in N–Body Dynamical Systems, p. 3 – 9 (1991). – See Abstr. 012.089 for the main entry.
A restricted charged four body problem is considered, which reduces to a two degrees of freedom Hamiltonian system. It is shown that an appropriate restriction of a Poincaré map of the system is conjugate to the shift homeomorphism on a certain symbolic alphabet.

042.094   Chaos in the three body problem.
A. Milani.
NATO Advanced Study Institute on Predictability, Stability, and Chaos in N–Body Dynamical Systems, p. 11 – 33 (1991). – See Abstr. 012.089 for the main entry.
This paper gives an outline of the basic tools required to show the occurrence of chaotic motions in the simplest non–integrable problems in celestial mechanics, such as the circular three–body problem. First, the linear and local theory of ordinary differential equations in the neighborhood of a fixed point is described; the problems arising in the embedding of invariant stable and unstable manifolds are also considered. In the next section, periodic orbits are discussed; the topics treated include variational equations, surfaces of section, the continuation of periodic orbits in the restricted three–body problem, and bifurcation of hyperbolic periodic orbits from resonant periodic orbits. The following section covers fundamental models of resonance, and the global behaviour of separatrices and their intersections. Finally, an outline is presented of the proof of the fundamental result, by which homoclinic points must necessarily occur in the restricted problem.

042.095   A new route to chaos: generation of spiral characteristics.
G. Contopoulos.
NATO Advanced Study Institute on Predictability, Stability, and Chaos in N–Body Dynamical Systems, p. 35 – 46 (1991). – See Abstr. 012.089 for the main entry.

042.096   Chaos, stability and predictability in Newtonian dynamics.
V. Szebehely.
NATO Advanced Study Institute on Predictability, Stability, and Chaos in N–Body Dynamical Systems, p. 63 – 71 (1991). – See Abstr. 012.089 for the main entry.
The entrance of the subjects of limited predictability and of chaos into the fields of celestial mechanics and gravitational N–body dynamics is discussed here. The non–integrability of the gravitational many–body problem (for three or more participating masses), when combined with errors in modelling and with

the uncertain values of the initial conditions, leads to bundles of trajectories instead of single orbits for a given dynamical problem. The consequences of these realistic considerations are treated and their effects in celestial mechanics are discussed.

042.097 Predictability, stability and chaos in dynamical systems.
C. Marchal.
NATO Advanced Study Institute on Predictability, Stability, and Chaos in N–Body Dynamical Systems, p. 73 – 91 (1991). – See Abstr. 012.089 for the main entry.

Progress in the theory of stability is reviewed for the particular case of the Lagrangian motions of the three–body problem. The considerations are then generalized to N–body systems. The concept of predictability is discussed, and it is noted that, in some cases, chaotic motions allow better predictions than regular motions. The Arnold diffusion conjecture presents a general picture of Hamiltonian dynamical systems. Chaotic motions are an essential part of this picture.

042.098 Analytical framework in Poincaré variables for the motion of the solar system.
J. Laskar.
NATO Advanced Study Institute on Predictability, Stability, and Chaos in N–Body Dynamical Systems, p. 93 – 114 (1991). – See Abstr. 012.089 for the main entry.

The author presents a method for calculating the expansion of the Hamiltonian of the planetary system in Poincaré variables. The algebraic manipulator TRIP for explicitly determining the planetary Hamiltonian is described.

042.099 Modelling: an aim and a tool for the study of the chaotic behaviour of asteroidal and cometary orbits.
C. Froeschlé.
NATO Advanced Study Institute on Predictability, Stability, and Chaos in N–Body Dynamical Systems, p. 125 – 155 (1991). – See Abstr. 012.089 for the main entry.

Chaotic solutions of Newton equations are deeply rooted in both asteroidal and cometary dynamics. Great progress has been made in the last decade using tools and results of the theory of dynamical systems. Both the existence of the Kirkwood gaps and the transfer of comets into observable orbits are shown to be related to chaos. Mapping and massive parallel computers are the main tools which are discussed here.

042.100 Mapping models for Hamiltonian systems with application to resonant asteroid motion.
J. D. Hadjidemetriou.
NATO Advanced Study Institute on Predictability, Stability, and Chaos in N–Body Dynamical Systems, p. 157 – 175 (1991). – See Abstr. 012.089 for the main entry.

A systematic method is presented to construct a mapping model for a Hamiltonian system. The author starts with two degrees of freedom, and then extends the method to three degrees of freedom. The basic notions of the averaging method are presented first, since the method of constructing the mapping is based on the averaged Hamiltonian. A mechanism for the generation of chaos is described, based on the interaction of two degrees of freedom in systems with three degrees of freedom. Finally, the method is applied to the 3:1 resonance in asteroid motion for nonzero eccentricity of Jupiter, where chaotic behaviour of the eccentricity of the asteroid orbit is found.

042.101 A model for the study of very–high–eccentricity asteroidal motion: the 3:1 resonance.
S. Ferraz–Mello, J. C. Klafke.
NATO Advanced Study Institute on Predictability, Stability, and Chaos in N–Body Dynamical Systems, p. 177 – 184 (1991). – See Abstr. 012.089 for the main entry.

042.102 The location of secular resonances.
A. Morbidelli.
NATO Advanced Study Institute on Predictability, Stability, and Chaos in N–Body Dynamical Systems, p. 185 – 192 (1991). – See Abstr. 012.089 for the main entry.

New developments in the theory of secular resonances in the dynamics of the asteroidal belt are briefly outlined.

042.103 Temporary capture into resonance.
J. Henrard.
NATO Advanced Study Institute on Predictability, Stability, and Chaos in N–Body Dynamical Systems, p. 193 – 196 (1991). – See Abstr. 012.089 for the main entry.

It seems liekly that some pairs of satellites of Uranus have been temporarily captured into resonance in the past. In order to analyze these temporary captures, one must modify the model constructed for the capture into resonance of the satellites of Jupiter and Saturn. The key factor is the value of the oblateness of Uranus which is smaller than the corresponding value for Jupiter or Saturn. The smaller value allows some overlap of nearby resonances producing chaos and secondary resonances. The secondary resonances are instrumental in dragging the captured orbit back to the chaotic layer surrounding the primary resonance from which it can escape in the regular region outside the resonance.

042.104 Applications of the restricted many–body problem to binary asteroids.
V. Szebehely, J. R. Pojman.
NATO Advanced Study Institute on Predictability, Stability, and Chaos in N–Body Dynamical Systems, p. 197 – 204 (1991). – See Abstr. 012.089 for the main entry.

This paper presents a general approach to the problem of stability of binary configurations in solar system dynamics. As a special application of the analytical approach, the possible existence of binary asteroids is investigated by numerical techniques. The analytical results are based on the hierarchical model of the restricted four–body problem, in which the primaries are the Sun and Jupiter, and the "small masses" are the asteroid and its satellite. The numerical results offer a parametric study considering the effects of the initial orbital parameters of the binary on its stability.

042.105 The wavelet transform as clustering tool for the determination of asteroid families.
P. Bendjoya, E. Slezak, C. Froeschlé.
NATO Advanced Study Institute on Predictability, Stability, and Chaos in N–Body Dynamical Systems, p. 205 – 213 (1991). – See Abstr. 012.089 for the main entry.

This paper reports on the application of the wavelet transform technique to the identification of minor planet families in the main asteroidal belt.

042.106 Delivery of meteorites from the v_6 secular resonance region near 2 AU.
C. Froeschlé, H. Scholl.
NATO Advanced Study Institute on Predictability, Stability, and Chaos in N–Body Dynamical Systems, p. 215 – 223 (1991). – See Abstr. 012.089 for the main entry.

Numerical integrations in the frame of a Sun– Mars– Jupiter – Saturn model over 1 Myr have been performed in order to investigate the orbital evolution of asteroid fragments produced in the innermost asteroid belt (2.07 – 2.13 AU). Fragments injected in the vicinity of the v_6 secular resonance enhance their eccentricities and become Mars–crossers. Close encounters to Mars will then lead to a random walk in semimajor axes. Two different mechanisms may occur to produce Earth–crossers. In the first case, the fragment enters the 4:1 mean motion resonance and becomes an Earth–crosser within at least 2.6×10^5 years. In the second case, which involves only the secular resonance v_6, the shortest timescale for deriving meteorites is of the order of 5.6×10^5 years.

042.107 **Perturbation theory, resonance, librations, chaos, and Halley's comet.**
P. J. Message.
NATO Advanced Study Institute on Predictability, Stability, and Chaos in N–Body Dynamical Systems, p. 239 – 247 (1991). – See Abstr. 012.089 for the main entry.
This paper first presents a brief survey of quasi–ergodicity, wildness, and chaos–type phenomena in celestial mechanics, then continues with a very brief outline of the development of solar system perturbation theory. It is shown how resonance in orbital period leads to transitions between types of motion, so that, since rational values of the ratio of two orbital periods are everywhere dense, the motions in the solar system will show a complexity in which the eventual character of a particular orbit of the system may be expected to depend very finely on the initial conditions. Finally, some numerical investigations of resonant librations in the orbit of Halley's comet are described.

042.108 **Stability of satellites in spin–orbit resonances and capture probabilities.**
A. Celletti.
NATO Advanced Study Institute on Predictability, Stability, and Chaos in N–Body Dynamical Systems, p. 337 – 344 (1991). – See Abstr. 012.089 for the main entry.
The stability of satellites in spin–orbit resonances is investigated in the light of perturbation theory. Using KAM theory, the author constructs invariant surfaces trapping the periodic orbit associated with the resonance in a finite region of the phase space. In addition, the probability of capture into a resonance is studied, and applications to the Earth–Moon and the Sun–Mercury system are described.

042.109 **Statistical analysis of the effects of close encounters of particles in planetary rings.**
F. P. Gama, J.–M. Petit.
NATO Advanced Study Institute on Predictability, Stability, and Chaos in N–Body Dynamical Systems, p. 345 – 353 (1991). – See Abstr. 012.089 for the main entry.
Until now, simulations of planetary rings have been performed either as deterministic simulations – when gravitation between particles alone or collisions alone are taken into account – or Monte–Carlo simulations – when both effects are considered. However, in Monte Carlo simulations only one variable is kept for each particle orbit: its semi–major axis. Theoretical studies show that one cannot disregard the effects on the eccentricity. So one has to do simulations which are able to follow at least these two parameters and which include both inelastic collisions and gravitation between particles. The authors attempt this by a mixed simulation, that is, a deterministic integration (N–body model) for large distances, and a Monte–Carlo treatment of close encounters. Some preliminary results from this study are presented.

042.110 **Chaos in coorbital motion.**
F. Spirig, J. Waldvogel.
NATO Advanced Study Institute on Predictability, Stability, and Chaos in N–Body Dynamical Systems, p. 395 – 410 (1991). – See Abstr. 012.089 for the main entry.
The motion of two coorbital satellites – at their close encounters – can be adequately approximated by Hill's lunar problem. Therefore chaotic behaviour in Hill's problem implies chaos in coorbital motion. Although the nonintegrability of Hill's problem has not yet been proven, numerical evidence clearly shows complicated behaviour typical of chaotic systems. In this paper the family of solutions relevant for circular coorbital motion is explored in details, and an example of a homoclinic orbit is given.

042.111 **Remarkable termination orbits of the restricted problem.**
V. V. Markellos.
NATO Advanced Study Institute on Predictability, Stability, and Chaos in N–Body Dynamical Systems, p. 413 – 423 (1991). – See Abstr. 012.089 for the main entry.
Homoclinic orbits at $L_{4,5}$ are termination orbits of families of periodic orbits of the restricted three–body problem. Very few

such orbits are known in the literature. The author presents here a large number of remarkable new non–symmetric homoclinic orbits for a mass parameter $\mu = 0.45$.

042.112 **Periodic orbits in the isosceles three–body problem.**
C. C. Monleón, J. M. Alfaro.
NATO Advanced Study Institute on Predictability, Stability, and Chaos in N–Body Dynamical Systems, p. 425 – 431 (1991). – See Abstr. 012.089 for the main entry.

042.113 **Quasiperiodic orbits as a substitute of libration points in the solar system.**
G. Gómez, À. Jorba, J. Masdemont, C. Simó.
NATO Advanced Study Institute on Predictability, Stability, and Chaos in N–Body Dynamical Systems, p. 433 – 438 (1991). – See Abstr. 012.089 for the main entry.
This paper considers the Earth – Moon – particle system as a restricted three–body problem (RTBP). It is well known that there are two equilateral libration points. In the real life system, these points do not exist, due to the effect of the perturbations caused by the part of the solar system which is not taken into account. In this work, the full problem is presented as a perturbation of the RTBP, and the authors look for a dynamical equivalent of $L_{4,5}$, which seems to be a quasiperiodic orbit. A method for obtaining these orbits is presented, and their stability is discussed.

042.114 **Stability zones around the triangular Lagrangian points.**
R. Dvorak, E. Lohinger.
NATO Advanced Study Institute on Predictability, Stability, and Chaos in N–Body Dynamical Systems, p. 439 – 446 (1991). – See Abstr. 012.089 for the main entry.
The authors discuss the stability around the libration point L_4 (and L_5) in the circular restricted three–body problem. The dependence of the extent of the stability zones around these equilibrium points as a function of the mass parameter $\mu = m_2/(m_1 + m_2)$ is established by using numerical experiments. The results are compared with existing data for the Earth–Moon case, and are then extended to a large range of μ values. The shrinking and even disappearance of the stability zones is well explained by the existence of two additional critical mass parameters below the well known value of $\mu_{crit} = 0.03852$.

042.115 **Chaotic trajectories in the restricted problem of three bodies.**
R. H. Smith, V. Szebehely.
NATO Advanced Study Institute on Predictability, Stability, and Chaos in N–Body Dynamical Systems, p. 447 – 455 (1991). – See Abstr. 012.089 for the main entry.
A complete qualitative understanding of the solutions of a set of nonlinear differential equations requires an investigation of the chaotic properties of the system. The circular restricted problem of three bodies has a long history of qualitative analysis, yet few studies have examined the possible existence of chaos in this problem. This study concentrates on the advent of chaos for widely different points in the phase space which correspond to closed Jacobian curves. In addition to observing the non–periodicity of the trajectory, two methods, Liapounov character–istic numbers and Poincaré surface of sections, are used to classify an orbit as chaotic. Regularized and unregularized equations of motion were used in conjunction with a variable stepsize integrator to produce the orbits.

042.116 **New formulations of the Sitnikov problem.**
K. Wodnar.
NATO Advanced Study Institute on Predictability, Stability, and Chaos in N–Body Dynamical Systems, p. 457 – 466 (1991). – See Abstr. 012.089 for the main entry.
The Sitnikov problem assumes two primaries with equal, non–zero masses moving around each other on congruent coplanar ellipses, while a massless planet performs motion along an axis perpendicular to the primary orbit plane through the common barycenter of the primaries. The system thus represents a special

case of the elliptic restricted three–body problem. A review of new results from an analytical study of this problem is presented.

042.117 Periodic solutions for the elliptic planar restricted three-body problem: a variational approach.
M. L. Bertotti.
NATO Advanced Study Institute on Predictability, Stability, and Chaos in N–Body Dynamical Systems, p. 467 – 473 (1991). – See Abstr. 012.089 for the main entry.
 The author outlines a variational approach, developed to find periodic orbits of the satellite in the elliptic restricted three–body problem with any value of the masses of the primaries. This approach leads to a multiplicity of generalized periodic solutions. These solutions are uniform limits of classical periodic solutions, having a prescribed rotational behaviour with respect to the primaries.

042.118 Figures of equilibrium in close binary systems.
J. A. López Ortí, A. López García, R. López Machí.
Celest. Mech. Dyn. Astron., Vol. 53, No. 4, p. 311 – 322 (1992).
 The equilibrium configurations of close binary systems are analyzed. The autogravitational, centrifugal and tidal potentials are expanded in Clairaut's coordinates. From the set of the total potential angular terms an integral equations system is derived. The reduction of them to ordinary differential equations and the determination of the boundary conditions allow a formulation of the problem in terms of a single variable.

042.119 Hill–type stability and hierarchical stability of the general three–body problem.
Y.–C. Ge.
NATO Advanced Study Institute on Predictability, Stability, and Chaos in N–Body Dynamical Systems, p. 475 – 479 (1991). – See Abstr. 012.089 for the main entry.
 Systematic numerical experiments show that the Hill–type stability criterion does not indicate, in the case of the initially elliptic coplanar three–body problem, a practical hierarchical stability. In some situations, this analytical criterion seems to be too restrictive, so that empirical stability regions exist outside the Hill–type stability regions; in other situations, however, it can be shown that specific instabilities occur frequently inside the Hill–type stability regions.

042.120 Equilibrium connections on the triple collision manifold.
A. Susín, C. Simó.
NATO Advanced Study Institute on Predictability, Stability, and Chaos in N–Body Dynamical Systems, p. 481 – 491 (1991). – See Abstr. 012.089 for the main entry.
 In the three–body problem the triple collision manifold plays a fundamental role in describing passages near triple collision. To study the possible transitions from the approach to collision to the escape from it, the invariant submanifolds on that manifold are essential. Here, the authors study mainly the connections between the equilateral approaches and escapes.

042.121 Orbits asymptotic to the outermost KAM in the restricted three–body problem.
M. Sekiguchi, K. Tanikawa.
NATO Advanced Study Institute on Predictability, Stability, and Chaos in N–Body Dynamical Systems, p. 493 – 498 (1991). – See Abstr. 012.089 for the main entry.
 The authors investigate the conjecture: In the restricted three–body problem, there exist collision orbits with the planet in any neighborhood of the outermost KAM around the retrograde satellite orbit for certain values of the Jacobi constant C and the mass parameter μ. This conjecture is checked here by numerical integration for $C = 2.98$ and $\mu = 0.001$.

042.122 A new interpretation of collisions in the N–body problem.
J. G. Bryant.
NATO Advanced Study Institute on Predictability, Stability, and Chaos in N–Body Dynamical Systems, p. 501 – 507 (1991). – See Abstr. 012.089 for the main entry.

042.123 An impulsional method to estimate the long–term behaviour of a perturbed system: application to a case of planetary dynamics.
B. Chauvineau.
NATO Advanced Study Institute on Predictability, Stability, and Chaos in N–Body Dynamical Systems, p. 509 – 514 (1991). – See Abstr. 012.089 for the main entry.
 This paper presents a method for investigating the long–term evolution of a perturbed system. The unperturbed system is supposed to possess an integral of the form: $\dot{r}^2 + h(r)$ and the perturbation is assumed to be time–dependent. The results are compared to direct analytical and numerical computations in the case of a perturbed harmonic oscillator. As example, the lifetime of a binary asteroid, perturbed by Jupiter, is estimated.

042.124 Improved Bettis methods for long–term prediction.
J. M. Ferrándiz, S. Novo.
NATO Advanced Study Institute on Predictability, Stability, and Chaos in N–Body Dynamical Systems, p. 515 – 522 (1991). – See Abstr. 012.089 for the main entry.
 The authors introduce a modification of Bettis' method in order to improve the long–term numerical integration of perturbed oscillators. They give several examples, involving both single and coupled oscillators, to illustrate the efficiency of this approach with respect to the methods of Bettis and Adams – Bashforth.

042.125 Application of spherically exact algorithms to numerical predictability in two–body problems.
J. M. Ferrándiz, M. T. Pérez.
NATO Advanced Study Institute on Predictability, Stability, and Chaos in N–Body Dynamical Systems, p. 523 – 530 (1991). – See Abstr. 012.089 for the main entry.
 Numerical predictions are strongly dependent on the algorithms used in the integration, even in cases as simple as the two–body problem, perturbed or not. In this contribution, the authors show some numerical experiments comparing the results obtained by applying different codes. Among these, the authors include some with special preservation properties, such as being spherically exact.

042.126 Are there irregular families of characteristic curves?
J. Font, C. Simó.
NATO Advanced Study Institute on Predictability, Stability, and Chaos in N–Body Dynamical Systems, p. 531 – 540 (1991). – See Abstr. 012.089 for the main entry.
 For Hamiltonian systems of two degrees of freedom the symmetric periodic orbits of a given type appear in continuous families. Every periodic orbit can be represented by one point in some suitable plane of parameters, and the full family is represented by the so–called characteristic curve. Some of these curves have components which are isolated, and these are called irregular characteristic curves. Here, the authors consider one example of this kind of behaviour. They show that, if the given Hamiltonian is embedded in a one–parameter family, the components are no longer isolated. Furthermore, by using several invariant manifolds, a full explanation of the structure and evolution of the irregular characteristic curves is given.

042.127 Non–linearity in the angles–only initial orbit determination problem.
D. Kaya, D. Snow.
NATO Advanced Study Institute on Predictability, Stability, and Chaos in N–Body Dynamical Systems, p. 541 – 546 (1991). – See Abstr. 012.089 for the main entry.
 The authors examine the concept of determining an initial orbit using angles–only observations over a short arc from a single space based sensor. The methods of Laplace and Gauss for angles–only initial orbit determination are applied. To apply these methods, one has to solve a system of nonlinear equations which have multiple solutions. Typical solution schemes recommend using an iterative solution technique with a first guess to start. The authors show that this technique is not satisfactory for performing initial orbit determination over short arcs of data.

They discuss past methods used in solving these nonlinear equations, and then present a new approach for finding all the necessary solutions and for determining the correct solution.

042.128 Analytical expansions of torque–free motions for short and long axis modes.

H. Kinoshita.

Celest. Mech. Dyn. Astron., Vol. 53, No. 4, p. 365 – 375 (1992).

Torque–free motion of a rigid body is integrable and its solution is expressed in terms of elliptic functions and elliptic integrals. The conventional analytical expression of the solution, however, is complicated and not suitable for hand–calculation. Recently the rotational motions of small celestial bodies in the solar system are frequently investigated by numerically integrating the equations of motion instead of using the analytical solution, since the numerical evaluation of the analytical and exact solution is a little bit difficult. As the observational accuracy of the rotational motions of the small bodies in the solar system is quite low, what we need for the reduction of these observations are rough estimates of the period of Eulerian motion (or the free precession period) and the amplitudes of the main periodic terms. Here the author gives simple analytical expansions of torque–free motions for short– and long–axis modes, which are correct up to the second–order of a small parameter. These expressions include only trigonometric functions and are easily evaluated by hand calculation for estimates of the essential quantities from which one can determine a global rotational motion of the torque–free motion. They can also be used as the zero–th order solution in a perturbation method, when the motion is perturbed by external torques.

042.129 A perturbation of the relativistic Kepler problem.

A. Nunes, J. Casasayas, J. Llibre.

NATO Advanced Study Institute on Predictability, Stability, and Chaos in N–Body Dynamical Systems, p. 547 – 554 (1991). – See Abstr. 012.089 for the main entry.

The authors consider the Kepler problem with the first order relativistic correction and show that, for a suitable class of perturbations, "almost all" the invariant tori and cylinders of the unperturbed system persist and that the perturbed system shows strong evidence of non–integrability.

042.130 Integrable 3–dimensional dynamical systems and the Painlevé property.

C. Polymilis.

NATO Advanced Study Institute on Predictability, Stability, and Chaos in N–Body Dynamical Systems, p. 555 – 563 (1991). – See Abstr. 012.089 for the main entry.

The author investigates some classes of 3–dimensional potentials and finds all the cases that have a second integral of motion quadratic in the velocities. Whenever there is no third integral of motion the Painlevé or weak Painlevé property is not satisfied, because two of the resonances are complex.

042.131 Generic and nongeneric Hopf bifurcation.

F. Spierig.

NATO Advanced Study Institute on Predictability, Stability, and Chaos in N–Body Dynamical Systems, p. 565 – 572 (1991). – See Abstr. 012.089 for the main entry.

A system of ordinary differential equations depending on a parameter ε is considered. The origin is assumed to be an equilibrium point for all ε. Furthermore, the hypotheses are made that the linearized system admits a pair of complex conjugate eigenvalues $\alpha(\varepsilon) \pm i\beta(\varepsilon)$ with $\beta(\varepsilon) > 0$, and that for $\varepsilon = 0$ these eigenvalues are purely imaginary, whereas the other eigenvalues are not integer multiples of $i\beta(0)$. Generically $\alpha'(0) \neq 0$. In this case the well known Hopf bifurcation theorem states that there exists one family of periodic solutions. If, however, the genericity condition is violated, then more than one family of periodic solutions may branch off from the origin. In this paper, a procedure is proposed allowing to determine all small periodic solutions of the system and their stability.

042.132 The chaotic motion of a rigid body rotating about a fixed point.

F. El–Sabaa, M. El–Tarazi.

NATO Advanced Study Institute on Predictability, Stability, and Chaos in N–Body Dynamical Systems, p. 573 – 581 (1991). – See Abstr. 012.089 for the main entry.

The authors transform the equations of motion of a heavy rigid body into the dynamical system of two degrees of freedom. The new system has Jacobi's integral, and the authors investigate numerically the existence of the second integral. This integral occurs when the regular orbits lie on invariant tori in the phase space. Stochastic regions indicate that the system is not integrable.

042.133 Determination of the slopes of "zero relative velocity" curves from the elliptic restricted three–body problem with an averaging method.

A. Pál, T. Oproiu.

Rom. Astron. J., Vol. 1, No. 1 – 2, p. 97 – 102 (1991).

Starting from Rein's "semiaveraging" scheme of the elliptic restricted three–body problem (according to Mojseev's classification), there were determined the slopes of the "zero relative velocity" curves in the libration point L_1 for different values of the eccentricity and of the mass ratio. The results are plotted in figures.

042.134 Orbital motion with the Mücket–Treder post–Newtonian gravitational law.

V. Mioc, P. Blaga.

Rom. Astron. J., Vol. 1, No. 1 – 2, p. 103 – 108 (1991).

Adopoting a post–Newtonian gravitational law with supplementary logarithmic term, and a perturbative treatment, the difference between the nodal and Keplerian perids, as well as the changes of the orbital elements over a nodal period are obtained. The first order approximation shows only a rotation of the orbit in its plane, while the second order approximation points out an orbit deformation, too.

042.135 On adequate sources of the external potential of an ellipsoid.

R. Z. Muratov.

Astron. Zh., Tom 69, Vyp. 3, p. 604 – 616 (May – Jun 1992). In Russian. English translation in Sov. Astron., Vol. 36, No. 3 (May – Jun 1992).

The problem of substitution of distributed sources with polynomial volume density ϱ by an adequate surface sources distribution is considered with regard to an ellipsoidal region. The latter distribution (described by an unknown surface density σ) has to generate the same external potential as the former one. The method of multipole momenta is shown to provide the problem's solution with preservation of its symmetry and without using Lamé functions. By way of example, the σ expressions are found which substitute adequately the densities ϱ described by polynomials of zero, first and second order.

042.136 Sufficient conditions of stability of the regular motion of two absolutely rigid bodies.

V. V. Vidyakin.

Astron. Zh., Tom 69, Vyp. 3, p. 633 – 639 (May – Jun 1992). In Russian. English translation in Sov. Astron., Vol. 36, No. 3 (May – Jun 1992).

By means of the second Lyapunov method conditions are found, with which the translational–rotational motion of two absolutely rigid bodies is stable. The structure and external form of the bodies is supposed to obey certain conditions.

042.137 On the surfaces of zero velocity in the restricted three–body problem with variable masses.

L. G. Luk'yanov.

Astron. Zh., Tom 69, Vyp. 3, p. 640 – 648 (May – Jun 1992). In Russian. English translation in Sov. Astron., Vol. 36, No. 3 (May – Jun 1992).

The singular points of the surfaces of zero velocity are investigated. A construction of these surfaces for all possible values of two free dimensionless parameters is made.

042.138   The planetary few–body problem at three–frequency resonance.
V. N. Shinkin.
Astron. Zh., Tom 69, Vyp. 3, p. 649 – 654 (May – Jun 1992). In Russian. English translation in Sov. Astron., Vol. 36, No. 3 (May – Jun 1992).

The paper investigates the stability of motion in the general and restricted planetary few–body problems in the case of the three–frequency resonance. The equations of bodies' motion near the resonance surface are obtained. The stability of motion of bodies is investigated. The condition of libration of the resonance phase is obtained. The results of investigation are applied to the motion of asteroids in the Solar system and the satellites of Jupiter (Io – Europa – Ganymede) and Uranus (Miranda – Ariel – Umbriel).

Introduction to Hamiltonian dynamical systems and the N–body problem.
See Abstr. 003.019.

Himmelsmechanik. Band I: Grundlagen, Determinierung.
See Abstr. 003.023.

Newton's solution of the one–body problem.
See Abstr. 004.029.

The moon–test in Newton's *Principia*: accuracy of inverse–square law of universal gravitation.
See Abstr. 004.030.

Chaos, resonance and collective dynamical phenomena in the solar system. Proceedings.
See Abstr. 012.025.

Predictability, stability, and chaos in N–body dynamical systems. Proceedings.
See Abstr. 012.089.

Symplectic integrators for Hamiltonian systems: basic theory.
See Abstr. 021.008.

Numerical simulators of stability in astrophysical problems.
See Abstr. 021.013.

On the periodic solutions of a particular Lamé equation.
See Abstr. 021.016.

Numerical integration methods for orbital motion.
See Abstr. 021.017.

On a Hermite integrator with Ahmad–Cohen scheme for gravitational many–body problems.
See Abstr. 021.019.

Length of arc as independent argument for highly eccentric orbits.
See Abstr. 021.034.

On the efficiency of Runge–Kutta–Nystrom methods with interpolants for solving equations of the form $y'' = f(t,y,y')$ over short timespans.
See Abstr. 021.035.

Increased accuracy of computations in the main satellite problem through linearization methods.
See Abstr. 021.036.

Stochasticity of motions and the structure of directions field in the Henon – Heiles model.
See Abstr. 021.037.

Chaos studied beyond the limits of thermodynamics.
See Abstr. 022.020.

Cas particulier dans le mouvement d'un corps déformable avec un point fixe. Le cas de la terre.
See Abstr. 044.038.

Gravitation and celestial mechanics investigations with Galileo.
See Abstr. 051.033.

Relativistic geocentric satellite equations of motion in closed form.
See Abstr. 052.006.

Motion of artificial satellites in the set of Eulerian redundant parameters. III.
See Abstr. 052.013.

Relativistic effects in the critical inclination problem in artificial satellite theory.
See Abstr. 052.014.

On the instability of "folded" equilibria of a flexible nonstretchable thread attached to the satellite in a circular orbit.
See Abstr. 052.016.

Analytical solutions to the four post–Newtonian effects in a near–Earth satellite orbit.
See Abstr. 052.017.

On the initially circular motion of an orbiter in the oblate, rotating, Martian atmosphere.
See Abstr. 052.023.

Moon's influence on the transfer from the Earth to a halo orbit around L_1.
See Abstr. 052.032.

First order theory of perturbed circular motion: an application to artificial satellites.
See Abstr. 052.033.

Poincaré–similar variables including J_2–secular effects.
See Abstr. 052.034.

Measuring the lack of integrability of the J_2 problem for Earth's satellites.
See Abstr. 052.035.

The effects of the J_3–harmonic (pear shape) on the orbits of a satellite.
See Abstr. 052.036.

Long–time predictions of satellite orbits by numerical integration.
See Abstr. 052.037.

The dynamical evolution of tidal capture binaries.
See Abstr. 065.018.

Tidal capture of stars by a massive black hole.
See Abstr. 065.042.

The status of non–homogeneous Dedekind ellipsoids.
See Abstr. 065.056.

Note on the representation of many–particle dynamics as higher–order single–particle dynamics, with a means to relativistic elevation of Newtonian dynamics.
See Abstr. 066.016.

Binary systems: higher order gravitational radiation damping and wave emission.
See Abstr. 066.052.

Degrees of freedom in the two–body problem of general relativity.
See Abstr. 066.189.

Time–geostationary orbits in the solar system.
See Abstr. 066.232.

Relativistic motion of gyroscopes and space gradiometry.
See Abstr. 066.265.

The wave nature and dynamical quantization of the solar system.
See Abstr. 091.040.

The N–dipole problem and the rings of Saturn.
See Abstr. 091.058.

The three–dipole problem.
See Abstr. 091.059.

Planetary distance law and resonance.
See Abstr. 091.065.

Tidal deceleration of the Moon's mean motion.
See Abstr. 094.024.

Planetary and figure–figure effects on the Moon's rotational motion.
See Abstr. 094.061.

The Moon's physical librations. Part 1: Direct gravitational perturbations.
See Abstr. 094.101.

The Moon's physical librations. Part 2: Non–rigid Moon and direct non–gravitational perturbations.
See Abstr. 094.102.

Significant high number commensurabilities in the main lunar problem: a postscript to a discovery of the ancient Chaldeans.
See Abstr. 094.103.

Evolution of asteroidal orbits at the 5:2 resonance.
See Abstr. 098.044.

A determination of the mass of (704) Interamnia from observations of (993) Moultona.
See Abstr. 098.070.

Solar system: Wandering on a leash.
See Abstr. 098.071.

An example of stable chaos in the solar system.
See Abstr. 098.072.

The principle of equivalence and the Trojan asteroids. II.
See Abstr. 098.073.

Orbital stability zones about asteroids. II. The destabilizing effects of eccentric orbits and of solar radiation.
See Abstr. 098.155.

A general theory of motion for the eight major satellites of Saturn. III. Long–period perturbations.
See Abstr. 100.009.

Looking for changes in the Saturnian system between Voyager and Cassini.
See Abstr. 100.015.

On the ephemerides of Uranus.
See Abstr. 101.003.

The elimination of the 1:2 critical terms of a first order theory of Uranus perturbed by Neptune through Von Zeipel method.
See Abstr. 101.057.

On the unmodeled perturbations in the motion of Uranus.
See Abstr. 101.061.

Pulsating Hill surfaces and the origin of comets.
See Abstr. 102.013.

The evolution of Jupiter family comets over 2000 years.
See Abstr. 102.019.

The long–term dynamical behavior of small bodies in the Kuiper belt.
See Abstr. 102.020.

Stochastic motion of nearly–parabolic comets under perturbations by planets.
See Abstr. 102.046.

Rotational behaviour of comet nuclei.
See Abstr. 102.056.

The dynamics of meteoroid streams.
See Abstr. 104.096.

Collision and tidal interaction between planetesimals.
See Abstr. 107.019.

Survival of a captured satellite in a primary extended atmosphere.
See Abstr. 107.036.

Periodic motion in perturbed elliptic oscillators.
See Abstr. 151.012.

Theory and applications of radial orbit instability in collisionless gravitational systems.
See Abstr. 151.014.

On the sensitivity of the N–body problem to small changes in initial conditions. II.
See Abstr. 151.018.

Equilibrium figures of stellar systems with a needle–shaped velocity ellipsoid. Statement of the general problem.
See Abstr. 151.026.

Equilibrium figures of barred stellar systems with a needle–shaped velocity ellipsoid. Statement of the general problem.
See Abstr. 151.048.

Theory for the instability of radial orbits in the collisionless gravitating systems and its applications.
See Abstr. 151.066.

Chaos in the N–body problem of stellar dynamics.
See Abstr. 151.082.

043 Astronomical Constants, Reference Systems

043.001 On the use of supernovae as radio–optical astrometric fiducial points.
K. W. Weiler, K. J. Johnston, S. D. Van Dyk.
Publ. Astron. Soc. Pac., Vol. 104, No. 674, p. 246 – 250 (Apr 1992). Current Physics Microform No.: 9205C2202.

The high–accuracy radio astrometric reference frame is based on extragalactic radio sources which are optically faint ($m_V > 15$), while the most accurate optical astrometric reference frame is made up of bright galactic stars which are almost all radio quiet to current sensitivity limits. The HIPPARCOS mission is establishing positional accuracies for these bright stars at the several milliarcsecond level of precision, and advancements in space–based observations should eventually allow precisions at the submilliarcsecond level. However, the relationship between the two reference frames is difficult to establish directly to a level consistent with the internal precision of the two independent frames. The compact extragalactic radio sources are too faint for most optical astrometric techniques, and the bright stars which are active radio sources are often in binary or extended envelope systems where the radio and optical photocenters may be related at accuracies of only a few milliarcseconds. This difficulty may be alleviated by using the occasional optically and radio bright supernovae as pointlike fiducial marks for establishing relationships between the two reference frames. These independent reference–frame relationships can be accomplished from currently available ground–based platforms. The authors anticipate about two bright ($m_V \lesssim 12$) supernovae per decade which could serve as useful candidates for combined radio–optical astrometric observations.

043.002 A radio optical reference frame. III. Additional radio and optical positions in the southern hemisphere.
J. L. Russell, D. L. Jauncey, B. R. Harvey, G. L. White, J. E. Reynolds, C. Ma, K. J. Johnston, A. Nothnagel, G. Nicolson, K. Kingham, R. Hindsley, C. de Vegt, N. Zacharias, D. F. Malin.
Astron. J., Vol. 103, No. 6, p. 2090 – 2098 (Jun 1992). Current Physics Microform No.: 9207A0370.

Radio and optical positions are presented for southern hemisphere extragalactic sources from the Parkes 2.7 GHz survey. Sixty–one sources were observed with Mark III VLBI at 8.4 GHz (X band) between Tidbinbilla, Australia and Hartebeesthoek, South Africa. The results presented are part of the effort to establish a global reference frame of 400 extragalactic radio sources. Radio positions with ~10 milliarcsec errors have been estimated for 39 sources not previously stated in the authors' radio reference frame catalog, and provisional positions were obtained for two additional sources, bringing the total number of catalog sources to 276. The principal source of error is the uncalibrated ionosphere. Of the remaining sources five were completely undetected, six were either too faint or too resolved, and nine had previous catalog positions. Optical positions on the FK5 system have also been measured for four southern sources using prime focus plates from the Anglo–Australian 4 m telescope with an accuracy 0″.06. This raises the number to 40 radio sources with accurately measured positions for their optical counterparts.

043.003 The FK5 equator and equinox from observations of minor planets.
R. L. Branham Jr., J. G. Sanguin.
Astron. J., Vol. 103, No. 6, p. 2099 – 2101 (Jun 1992). Current Physics Microform No.: 9207A0379.

413 photographic observations on the FK4 system of 16 minor planets made at four observatories are used to estimate corrections to the FK5 equinox and equator. The observations are reduced to the system of the FK5 by use of procedures recommended by the International Astronomical Union. Equinox and equator corrections are calculated from the L1 method, standard least squares, and iteratively reweighted least squares

with three weight functions: Talwar, Biweight, and Welsch. Statistical tests indicate that the L1 solution, which calculates an equinox correction of –0″.153 ± 3″.157 and an equator correction of 0″.220 ± 1″.187, best represents the observations. Given the magnitude of the errors, however, the significance of the corrections may be questioned.

043.004 Discussion on the origin of right ascension.
H. J. Yan, E. Groten.
Manuscr. Geod., Vol. 17, No. 1, p. 65 – 86 (Feb 1992).

The definition, mathematical representation, properties and coordinate transformation of the Departure Point in a quasi–inertial geocentric equatorial coordinate system (QIGECS) defined on the moving true equator are discussed in detail, in this paper. A nutational right ascension precession introduced by the Departure Point might become a new term of the precession in right ascension. The application of the Departure Point to Universal Time and Terrestrial Reference System is discussed more deeply. But it is also found that the adoption of the Departure Point does not entirely eliminate precession and nutation uncertainties and this problem becomes serious with the development of time if we do not refer to the equinox to correct it. In a supplementary manner the Departure Point can find its usage in some fields of astronomical measurement specially in the Earth Orientation Parameters, but it cannot completely replace the equinox ultimately. A quite general discussion of this topic and a further detailed application of the Departure Point to the Earth Rotation are given in Yan and Groten (1992).

043.005 Relativistic reference frames including time scales: questions and answers.
M. H. Soffel, V. A. Brumberg.
Celest. Mech. Dyn. Astron., Vol. 52, No. 4, p. 355 – 373 (1991).

The subject of relativistic reference frames in astronomy is discussed with respect to the problems and needs of the various user groups. For didactical reasons the discussion is presented in form of a sequence of questions and answers.

043.006 Space physics coordinate transformations: a user guide.
M. A. Hapgood.
Planet. Space Sci., Vol. 40, No. 5, p. 711 – 717 (May 1992).

This report presents a comprehensive description of the transformations between the major coordinate systems in use in Space Physics. The work of Russell is extended by giving an improved specification of the transformation matrices, which is clearer in both conceptual and mathematical terms. In addition, use is made of modern formulae for the various rotation angles. The emphasis throughout has been to specify the transformations in a way which facilitates their implementation in software. The report includes additional coordinate systems such as heliocentric and boundary normal coordinates.

043.007 The radio reference frame – progress to data.
J. L. Russell, C. de Vegt, D. Jauncey, K. J. Johnston, C. Ma, J. E. Reynolds.
Bull. Am. Astron. Soc., Vol. 23, No. 3, p. 1259 (1991). Abstract. – See Abstr. 012.065 for the main entry.

043.008 On the celestial reference frame independent of the dynamical planes of the Earth.
Zhao Ming.
Chin. Astron. Astrophys., Vol. 16, No. 1, p. 103 – 108 (Jan–Mar 1992). English translation of Acta Astron. Sin., Vol. 32, No. 3, p. 297 – 303 (Sep 1991). See Abstr. 54.043.015.

Up to now, all the Conventional Celestial Reference Frames (CCRF), such as FK5 as well as available radio source catalogues, are traditionally based on the conceptions of the equator, ecliptic, precession and nutation. Therefore, any changes in the expressions of precession or nutation as well as in the definitions of the equator or ecliptic must require a reconstruction of the CCRF. Such work is so heavy and complicated that it can not be done strictly. In addition, the errors in the expressions of precession or nutation may cause some distortion in CCRF. In principle, a celestial coordinate triad can be defined by an arbitrary group of objects (stars or radio sources). Referring to it, a Directly defined Celestial Reference Frame (DCRF) can be constructed. In this paper, the author proves that the DCRF can be realized and can be applied to astrometry and geodesy. DCRF is independent of the dynamical planes of the Earth.

043.009   The IERS GPS Terrestrial Reference Frame.
C. Boucher, Z. Altamimi.
IERS Tech. Note, No. 10, p. 1 – 23 (Jan 1992).
This Technical Note intends to produce a global GPS Terrestrial Reference Frame expressed in the IERS Terrestrial Reference System (ITRS) to be used for GPS analysis. This product is based on a combination of the ITRF90 solution which is the up to date realization of the ITRS and local tie vectors between GPS tracking points and SLR or VLBI points.

043.010   The preliminary study of the GPS application in radio and optical reference system connection.
Chen Gang, Qian Zhihan.
Ann. Shanghai Obs., Acad. Sin., No. 13, p. 74 – 80 (1992). In Chinese.
The fundamental method about the use of GPS observations to connect the radio and optical reference systems by VLBI and photoelectronic measurements respectively is reviewed. Then the precision of the system connection with this method is discussed in detail.

The Hipparcos Input Catalogue. II. Astrometric data.
See Abstr. 002.042.

RRS2 – the list of reference stars in the field around 238 extragalactic radio sources.
See Abstr. 002.065.

The fundamental constants of physics and spectroscopy.
See Abstr. 022.004.

Strings, gravity, and the constants of nature.
See Abstr. 022.060.

High–precision VLBI astrometry of the radio–emitting star σ CrB – a step in linking the Hipparcos and extragalactic reference frames.
See Abstr. 041.009.

Catalog–to–catalog reductions: results for the FK, N30, and GC catalogs.
See Abstr. 041.013.

Determination of long terms of the systematic difference between the JYD and BIH polar coordinate systems.
See Abstr. 044.040.

In–orbit performance of the Hipparcos astrometry satellite.
See Abstr. 051.021.

An application of new relations for a determination of the Newtonian constant of gravitation.
See Abstr. 066.046.

On the theory of relativistic reference frames based on optical coordinates.
See Abstr. 066.188.

Secular variations in optical observations of planets.
See Abstr. 091.021.

044 Time and Latitude Determination, Earth Rotation, Polar Motion

044.001   Polar motion, atmospheric angular momentum excitation and earthquakes – correlations and significance.
J. R. Preisig.
Geophys. J. Int., Vol. 108, No. 1, p. 161 – 178 (Jan 1992).
Equatorial atmospheric angular momentum (AAM) excitation functions and polar motion excitation functions (derived by Kalman filtering Very Long Baseline Interferometry polar motion estimates) are compared with the times of 1984–mid–1988 large earthquakes (magnitude greater than or equal to 7.5). There is a moderate correlation between times of large earthquakes and peaks in polar motion excitation. A strong correlation exists between the times of large earthquakes and large peaks in equatorial AAM amplitude; such a correlation is evident for six out of the eight large earthquakes occurring over the studied time interval. The AAM results indicate potential for temporal prediction of large/great earthquakes.

044.002   Effect of melting glaciers on the Earth's rotation and gravitational field: 1965–1984.
A. S. Trupin, M. F. Meier, J. M. Wahr.
Geophys. J. Int., Vol. 108, No. 1, p. 1 – 15 (Jan 1992).

044.003   The Earth's angular momentum budget on subseasonal time scales.
J. O. Dickey, S. L. Marcus, J. A. Steppe, R. Hide.
Science, Vol. 255, No. 5042, p. 321 – 324 (17 Jan 1992).
Irregular length of day (LOD) fluctuations on time scales of less than a few years are largely produced by atmospheric torques on the underlying planet. Significant coherence is found between the respective time series of LOD and atmospheric angular momentum (AAM) determinations at periods down to 8 days, with lack of coherence at shorter periods caused by the declining signal–to–measurement noise ratios of both data types. Refinements to the currently accepted model of tidal Earth rotation variations are required, incorporating in particular the nonequilibrium effect of the oceans. The remaining discrepancies between LOD and AAM in the 100– to 10–day period range may be due to either a common error in the AAM data sets from different meteorological centers, or another component of the angular momentum budget.

044.004   Stability of the astronomical frequencies over the Earth's history for paleoclimate studies.
A. Berger, M. F. Loutre, J. Laskar.
Science, Vol. 255, No. 5044, p. 560 – 566 (31 Jan 1992).
The expected changes over the past 500 million years in the principal astronomical frequencies influencing the Earth's climate may be strong enough to be detectable in the geological

records, and such effects have been inferred in several cases. Calculations suggest that the shortening of the Earth–moon distance and of the length of the day back in time induced a shortening of the fundamental periods for the obliquity and climatic precession over the last half–billion years. At the same time, the precessional constant increased from 50 to 61 arc sec per year. The changes in the frequencies of the planetary system due to its chaotic motion are much smaller; their influence on the changes of the periods of climatic precession, obliquity, and eccentricity of the Earth's orbit around the sun can be neglected. Eccentricity periods used for Quaternary climate studies may therefore be considered to have been more or less constant for pre–Quaternary times.

044.005   **UTC time step.**
         G. M. R. Winkler.
Time Serv. Announcement, Ser. 14, No. 52.
Naval Observatory, Washington, DC (USA). 1 p. (24 Feb 1992).
     The International Earth Rotation Service has announced the introduction of a time step to occur at the end of June 1992.

044.006   **Daily time differences.**
         U.S. Nav. Obs., Ser. 4, No. 1300.
Naval Observatory, Washington, DC (USA). 7 p. (2 Jan 1992).
     Further information covering the period Jan – Jun 1992 is published in Nos. 1301 – 1325 (Jan – Jun 1992).

044.007   **IERS Bulletin–A. International Earth Rotation Service.**
         IERS Bull., A, Vol. 5, No. 1, p. 1 – 6 (2 Jan 1992).
Contents: 1. Observations – very long baseline interferometry, satellite laser ranging, University of Texas LLR. 2. Combined Earth orientation parameters – IERS rapid service, IERS final values. 3. Predictions – 90–day, long–term. 4. Celestial pole offset series – observations from VLBI, NEOS daily series, IERS final values, predictions.
Further information covering the period Jan – Jun 1992 is published in Nos. 2 – 26 (Jan – Jun 1992).

044.008   **Time and latitude (Jul – Sep 1991).**
         C. Ron, A. Veselý, J. Vondrák (comps.).
Circ. Czech. Obs., Time Latitude, Jul – Sep 1991.
Československá akademie věd, Ondřejov. Astronomický ústav. 22 p. (Jun 1992).

044.009   **BIPM. Circular T. Data for September, October 1991.**
         BIPM Circ. T, No. 47.
Bureau International des Poids et Mesures, 92312 – Sèvres (France). 6 p. (2 Jan 1992).
     Contents: 1. Coordinated Universal Time UTC. Computed values of UTC–UTC(k). 2. International Atomic Time TAI and local atomic time scales TA(k). 3. Notes on sections 1 and 2. 4. UTC–GPS time and TAI–GPS time. 5. UTC–GLONASS time. 6. Measurement of UTC(j)–UTC(k). 7. Duration of the TAI scale interval.
Further data for the period Nov 1991 – Apr 1992 are given in Nos. 48 – 52 (Feb – Jun 1992).

044.010   **International   Earth   Rotation   Service   (IERS). Bulletin B. Data for November, December 1991.**
IERS Bull., B, No. 47, p. 1 – 6 (3 Jan 1992).
     Contents: 1. Earth orientation parameters. 2. Daily interpolation of x, y, UT1, D, dψ, dε. 3. Normal values of the Earth orientation parameters at five–day intervals. 4. Duration of the day and angular velocity of the Earth. 5. Information on time scales. 6. Individual series of Earth orientation parameters.
Further data for the period Dec 1991 – May 1992 are given in Nos. 48 – 52 (Feb – Jun 1992).

044.011   **Deceleration in the Earth's rotation due to the Sun.**
         M. Burša.
Stud. Geophys. Geod., Vol. 35, No. 3, p. 145 – 150 (1991).
     The estimate of the tidal long–term decrease in the angular velocity ω of the Earth's rotation due to the Sun is given as $-(0.8 \pm 0.3) \times 10^{-22}$ rad s^{-2}. It was computed on the basis of the

observed total long–term decrease in ω, of the observed tidal deceleration of the Moon and the observed decrease in the second–degree zonal Stokes geopotential harmonic term. Adopting the estimate given, the product of the Love number and the tidal phase lag angle due to the Sun (in degrees) comes out as 0.53 ± 0.20.

044.012   **Hora legal en los diferentes países del globo.**
         H. Dávila S.
Bol. Obs. Astron. Quito, p. 70 – 84 (Jan 1992).

044.013   **Rapport annuel de la Section du temps du BIPM, 1991.**
         (Annual Report of the BIPM Time Section, 1991.)
Annu. Rep. BIPM Time Sect., Vol. 4.
Bureau International des Poids et Mesures, 92312 – Sèvres (France). vp. (1992). ISBN 92–822–2124–5.
     The establishment of International Atomic Time, TAI, and of Coordinated Universal Time, UTC (with the exception of the determination and the announcement of leap seconds of UTC) has been the responsibility of the BIPM.
Contents: Part A: Atomic time scales established by the BIPM. Part B: Tables of results. Part C: Time signals.

044.014   **Polar motion excitation by variations of the effective angular momentum function: considerations concerning deconvolution problem.**
         A. Brzeziński.
Manuscr. Geod., Vol. 17, No. 1, p. 3 – 20 (Feb 1992).
     In the recent investigations concerning atmospheric excitation of polar motion the formulation of Barnes et al. (1983) is usually applied, with excitation expressed by the effective angular momentum (EAM) function χ. The author addresses here two problems which are important in the analysis of rapid perturbations with periods down to a few days: (1) the EAM function χ differs from the excitation function ψ used in the classical polar motion equation of Munk and MacDonald (1960), (2) the changes in earth orientation are expressed in both these equations by the coordinates of the instantaneous rotation pole while most of the geodetic measurements refer to the Celestial Ephemeris Pole (CEP). The definition of the CEP is specified here to the form enabling the derivation of the equation describing polar motion of the CEP. The author discusses also how the deconvolution problem for polar motion can be solved by the Kalman filtering method combined with autoregressive modeling of the excitation process. The complete filter equations are computed and the observability conditions are derived.

044.015   **Determination of some geodynamical parameters based on reduction of LAGEOS and Etalon–1 observation data.**
         A. N. Marchenko.
Kinematika Fiz. Nebesn. Tel, Tom 8, No. 1, p. 81 – 88, 96 (Jan–Feb 1992). In Russian. English translation in Kinematics Phys. Celest. Bodies, Vol. 8, No. 1.
     The problem of computation of some geodynamical parameters on the basis of LAGEOS and Etalon–1 observations is considered. After modification, the program system GEORAN–2 was used for the determination of following parameters: station coordinates, Earth's rotation parameters, coefficients C_R and C_T, some fundamental constants of geodesy and astronomy, and secular variations $\dot{J}_2$, $\dot{J}_3$, $\dot{J}_4$ of zonal harmonic coefficients.

044.016   **IRIS Earth Orientation Bulletin.**
         IRIS Bull. A, No. 95, p. 1 – 7 (Jan 1992).
The Bulletin is the publication of record for the current determination of pole position, UT1, length of day, and nutation.
Further information is given in Nos. 96 – 100 (Feb – Jun 1992).

044.017   **Correspondence between theory and observations of polar motion.**
         R. S. Gross.
Geophys. J. Int., Vol. 109, No. 1, p. 162 – 170 (Apr 1992).
     The goal of this paper is to write in terms of reported values the standard theoretical equation describing the Earth's wobble,

namely, the linearized conservation of angular momentum equation known as the Liouville equation.

044.018   Sub–daily resolution of Earth rotation variations with Global Positioning System measurements.
S. M. Lichten, S. L. Marcus, J. O. Dickey.
Geophys. Res. Lett., Vol. 19, No. 6, p. 537 – 540 (20 Mar 1992).

Data from a worldwide Global Positioning System (GPS) tracking experiment have been used to determine variations in Earth rotation (UT1–UTC) over a time period of three weeks. Kalman filtering and smoothing enabled changes in UT1–UTC over intervals of 2 to 24 hrs to be detected with the GPS data. Internal consistency checks and comparisons with other solutions from very long baseline interferometry (VLBI) and satellite laser ranging (SLR) indicate that the GPS UT1–UTC estimates are accurate to about 2 cm. Comparison of GPS–estimated variations in UT1–UTC with 2–hr time resolution over 4 days with predicted variations computed from diurnal and semi–diurnal oceanic tidal contributions strongly suggests that the observed periodic sub–daily variations of ~ 0.1 msec (5 cm) are largely of tidal origin.

044.019   Durch GPS bestimmbare Aspekte der Erdrotation (im Rahmen von IGS).
I. Bauersima.
Satell.–Beobachtungsstn. Zimmerwald, Ber., Nr. 21.
Satelliten–Beobachtungsstation Zimmerwald (Switzerland), Universität Bern, Bern (Switzerland). 31 p. (1992).

044.020   Ephemeris time obtained from lunar occultation observations made in the USSR during 1981 – 1985.
A. K. Osipov, I. S. Balins'ka, A. A. Molotaj, V. I. Mazur, B. F. Sincheskul, L. V. Kazantseva.
Visn. Kiiv. Univ., Fiz.–Mat. Nauki, Astron., Vip. 3, p. 59 – 66 · (1992). In Ukrainian.

044.021   The estimate of the deceleration in the Earth's rotation due to the Sun.
M. Burša.
Earth, Moon, Planets, Vol. 56, No. 1, p. 57 – 60 (Jan 1992). Presented at the XXth General Assembly of the I.A.G., Vienna (Austria), 15 Aug 1991.

The tidal long–term decrease in the angular velocity of the Earth's rotation has been estimated on the basis of the angular momentum tidal balance in the Earth–Moon–Sun system. The observed (LLR) tidal long–term decrease in the Moon's mean motion, the apparent secular acceleration in the mean longitude of the Sun and the long–term decrease in the 2nd degree zonal geopotential parameter were used.

044.022   Time adjustment on 1992 June 30.
IAU Circ., No. 5463, p. 1 (2 Mar 1992).

044.023   A Hamiltonian theory for an elastic Earth: secular rotational acceleration.
J. Getino, J. M. Ferrándiz.
Celest. Mech. Dyn. Astron., Vol. 52, No. 4, p. 381 – 396 (1991).

In this article, the authors' previous Hamiltonian theory for the rotation of an Earth whose elastic mantle is deformed by rotation and lunisolar attraction is applied to the study of the secular acceleration of the Earth's rotation. Since it is a result of the inelasticity, the theory is extended to include a phase lag. So, the authors obtain, in a theoretical way, a value of -5.6×10^{-22} rad sec^{-2}, which agrees perfectly with the latest observational results.

044.024   UTC time step.
Yamamoto Circ., No. 2179, p. 1 (22 Feb 1992). In Japanese.

044.025   Elastic energy of a deformable Earth: general expression.
J. Getino.
Celest. Mech. Dyn. Astron., Vol. 53, No. 1, p. 11 – 36 (1992).

This work is the first in the second part of a project dedicated to elaborating a Hamiltonian theory for the rotational motion of a deformable Earth. In the four works which make up the first part the basis of this theory is laid down, studying the effects produced when the Earth's elastic mantle is deformed by lunisolar attraction. More specifically, in Getino and Ferrándiz (1991), the elastic energy which is produced on the deformation of the Earth's mantle is studied, considering solely the second order in the development in spherical harmonics of the perturbing potential (tidal potential). The present article can be considered as an amplification of the above mentioned work, obtaining, under the same hypotheses, but also very general, a general expression of the said elastic energy for any order of the development of the tidal potential. Although this expression, in its general form, is very complicated, the final result is extremely simple, and for the case $n = 2$, it coincides, obviously, with that already found by the above mentioned authors.

044.026   Wie lang ist ein platonisches Jahr?
H. Dierks.
Nachr. Olbers–Ges. Bremen, Nr. 157, p. 2 – 6 (Apr 1992).

044.027   Reduction of the infinite system of differential equations in the problem of the Earth's nutation.
V. V. Bykova.
Kinematics Phys. Celest. Bodies, Vol. 7, No. 6, p. 1 – 7 (1991). English translation of Kinematika Fiz. Nebesn. Tel., Tom 7, No. 6, p. 3 – 10 (Nov–Dec 1991).

The problem of the Earth's nutation is reduced to the solution of an infinite system of linear differential equations by expansion in spherical harmonics. The possibility of reduction of this system for two model Earths is investigated: one with a solid inner core and one without it. Numerical experiments showed that consideration of a finite number of equations can produce a solution to the problem only for the model without the inner core; the reduction is ill conditioned for the solid inner core model.

044.028   Theoretical derivation of relativistic precession and nutation.
Li Ling–huai, Huang Tian–yi.
Publ. Purple Mt. Obs., Vol. 11, No. 1, p. 27 – 36 (Mar 1992). In Chinese.

Beginning with the solution of the Einstein's field equation, this paper discusses the origin of the relativistic precession and nutation, and derives the theoretical expression of the geodetic, Lense–Thirring, Thomas precession, nutation and the relativistic advance of the ecliptic pole.

044.029   Inverse deduction of interannual variation of atmospheric circulation from historical observations of Earth rotation.
Zheng Dawei.
Ann. Shanghai Obs., Acad. Sin., No. 13, p. 1 – 6 (1992). In Chinese.

It is confirmed from the data series obtained by the modern astronomical and meteorological techniques that the variations in rotation of solid earth and in the atmospheric angular momentum are consistent with each other very well on time scales of up to several years. In this paper, the possibility of the observational datasets of Earth rotation indicated as a kind of data source of atmospheric circulation is discussed.

044.030   On the precise conception of the relation between UT0 and UT1.
Zhao Ming.
Ann. Shanghai Obs., Acad. Sin., No. 13, p. 7 – 12 (1992). In Chinese.

The conception of the relation between UT0 and UT1 mentioned in many papers are not clear and precise, even are wrong. In this paper, the author gives a precise conception about it and points out: (1) UT1, as one of the Earth Rotation Parameters (ERP), is in proportion to the rotation angle of the Earth round its instaneous ephemeris axis. (2) UT0 is not a parameter of the Earth rotation and has not an independent sense in astronomy.

044.031   The characteristic of the 30 – 60 day fluctuation in the Earth rotation, atmospheric angular momentum and solar activity.
Gu Hui, Zheng Dawai.
Ann. Shanghai Obs., Acad. Sin., No. 13, p. 13 – 25 (1992). In Chinese.
  The length–of–day variation, the atmospheric angular momentum and the sunspot relative number data series spanning 13 years are analyzed to demonstrate the 30 – 60 day fluctuation. From stastistical analysis and test, it is shown that the fluctuation on this time scale possesses the characteristic of stochastic moving.

044.032   A possible cause of the formation of some short periodic fluctuation in LOD.
Yang Zhigen.
Ann. Shanghai Obs., Acad. Sin., No. 13, p. 33 – 39 (1992). In Chinese.
  Short periodic fluctuation in both series of the length of day (LOD) and the relative sunspot numbers, with periodicities of about 95 days and 120 days was found. The same fluctuation was identified in the rate of change of the orbital angular momentum of the Sun's center in ecliptic barycenter coordinate of the solar system. The existence of these coincident periodic components in three different series suggests that the solar activities are much possibly modulated by the orbital motion of the major planets and the former influence the Earth's rotation.

044.033   The effects of pole tide on determination of baseline with VLBI.
Yang Zhigen.
Ann. Shanghai Obs., Acad. Sin., No. 13, p. 40 – 48 (1992). In Chinese.
  The effects of pole tide on the displacement of station are estimated by the author. The maximum radial displacement of sites caused by both the variations of rotation rate and the polar motion are about 0.4 mm and 16.1 mm respectively, and are 0.1 mm and 4.5 mm respectively for the horizontal displacements.

044.034   First years results from Chinese–German VLBI geodetic campaign (CGVGC).
Luo Shifang, Qian Zhihan.
Ann. Shanghai Obs., Acad. Sin., No. 13, p. 145 – 153 (1992). In Chinese.

044.035   Comparison and analysis of the effects of the receiving parameters on the precision of GPS time comparison.
Hu Jinlun, Wan Ningshan.
Ann. Shanghai Obs., Acad. Sin., No. 13, p. 193 – 197 (1992). In Chinese.

044.036   Nutations of the Earth.
P. M. Mathews, I. I. Shapiro.
Annu. Rev. Earth Planet. Sci., Vol. 20, p. 469 – 500 (1992).
  Contents: Earth rotation: general features. Gravitational perturbations. Precession, nutation, and wobble. Normal modes. Forced nutations: theoretical considerations for idealized earth models. Deviations from idealized models. Estimation of nutation amplitudes: the VLBI method. Observational results and theoretical interpretation. Discussion and outlook.

044.037   Periodical components in the polar motion of the Earth deduced from: "Results of the International Latitude Service in a homogeneous system 1899.9 – 1979.0".
L. Rusu.
Rom. Astron. J., Vol. 1, No. 1 – 2, p. 81 – 83 (1991).

044.038   Cas particulier dans le mouvement d'un corps déformable avec un point fixe. Le cas de la terre.
M. Ciobanu.
Rom. Astron. J., Vol. 1, No. 1 – 2, p. 85 – 89 (1991).
  The possible consequences of the centrifugal potential on the Euler period in the case of the Earth are investigated.

044.039   La durée de l'année tropique et l'unité fondamentale de temps.
I. Mihăilă.
Rom. Astron. J., Vol. 1, No. 1 – 2, p. 91 – 95 (1991).
  A more accurate formula for the length of the tropical year is derived in terms of the coefficients of the polynomial expression of the Sun's mean longitude. The improved length of the tropical year is shorter than the known length. The consequences regarding the definition of the second of ephemeris time and the calculation of the Besselian epoch corresponding to an Julian date are presented.

044.040   Determination of long terms of the systematic difference between the JYD and BIH polar coordinate systems.
Qian–Changxia, Li Zhengxin.
Ann. Shanghai Obs., Acad. Sin., No. 13, p. 56 – 61 (1992). In Chinese.

044.041   A prediction of UT series with auto–adaptive auto–regressive model.
Ding Yue–rong, Xiao Nai–yuan, Xia Yi–Fei.
Sci. China, Ser. A, Vol. 34, No. 12, p. 1484 – 1491 (Dec 1991).
  The prediction of UT can be separated into two parts, i.e. the prediction of a definitive component and that of a random component. In this paper, the first part is carried out with linear fitting extrapolation and periodic fitting extrapolation of NEOS UT1–UTC series of one–day interval with a span of two years, and the second part with an RLS recursive procedure of auto–adaptive ASR modeling. The combination of the two predicted values gives a satisfying result.

Activities of research work related to Earth rotation in China from January 1988 to June 1990.
See Abstr. 013.127.

Sensing the rotation of the Earth.
See Abstr. 014.002.

Discussion on the origin of right ascension.
See Abstr. 043.004.

The IERS GPS Terrestrial Reference Frame.
See Abstr. 043.009.

Mission objectives and scientific rationale for the magnetometer mission.
See Abstr. 051.051.

Relativistic geocentric satellite equations of motion in closed form.
See Abstr. 052.006.

The problem of clock synchronization: a relativistic approach.
See Abstr. 066.214.

Angle between the axes of rotation of the Earth's core and mantle.
See Abstr. 081.002.

Differential rotation of the liquid core of the Earth.
See Abstr. 081.007.

On the calculation of low–frequency oscillations of the Earth's core.
See Abstr. 081.016.

Some remarks about the rotations of a viscous planet and its homogeneous liquid core: linear theory.
See Abstr. 081.019.

Influence of viscoelastic coupling on the axial rotation of the Earth and its fluid core.
See Abstr. 081.020.

Crustal velocities from geodetic very long baseline interferometry.
See Abstr. 081.023.

045 Astronomical Geodesy, Satellite Geodesy, Navigation

045.001 Global geodesy using GPS without fiducial sites.
M. Heflin, W. Bertiger, G. Blewitt, A. Freedman, K. Hurst, S. Lichten, U. Lindqwister, Y. Vigue, F. Webb, T. Yunck, J. Zumberge.
Geophys. Res. Lett., Vol. 19, No. 2, p. 131 – 134 (24 Jan 1992).
   Baseline lengths and geocentric radii have been determined from GPS data without the use of fiducial sites. Data from the first GPS experiment for the IERS and Geodynamics (GIG '91) have been analyzed with a no–fiducial strategy. A baseline length daily repeatability of 2 mm + 4 parts per billion was obtained for baselines in the northern hemisphere. Comparison of baseline lengths from GPS and the global VLBI solution GLB659 (Caprette et al., 1990) show rms agreement of 2.1 parts per billion. The geocentric radius mean daily repeatability for all sites was 15 cm. Comparison of geocentric radii from GPS and SV5 (Murray et al., 1990) show rms agreement of 3.8 cm. Given n globally distributed stations, the n(n–1)/2 baseline lengths and n geocentric radii uniquely define a rigid closed polyhedron with a well–defined center of mass. Geodetic information can be obtained by examining the structure of the polyhedron and its change with time.

045.002 Submillimeter horizontal position determination using very long baseline interferometry.
T. A. Herring.
J. Geophys. Res., Vol. 97, No. B2, p. 1981 – 1990 (10 Feb 1992).
   An analysis of interferometric phase delays from 15 years of Mark I and Mark III very long baseline interferometry (VLBI) experiments carried out with two radio telescopes in Westford, Massachusetts, ~1.24 km apart, yields weighted root–mean–square (WRMS) scatters about the mean locally horizontal coordinates of 1.0 and 2.0 mm in the north and east directions, respectively. The vertical coordinate scatter over this same duration is 3.2 mm. The measurements made during the last decade using the more accurate Mark III VLBI system yield WRMS scatters of 0.7, 0.8, and 2.3 mm, for the north, east, and vertical coordinates, respectively. Analysis of this latter data set using the less accurate, but more commonly used, group delay measurements yields WRMS scatters of 1.8, 2.5, and 2.8 mm for the north, east, and height components, respectively. The estimated rates of change of the coordinates of the baselines from the phase delay solution are presented. From these results, the maximum admissible rate of change of this baseline, either from geophysical causes or from telescope deformation, can be bounded to be less than 0.5 mm/yr in all coordinate directions. The authors conclude from these studies that VLBI antennas of at least of the structural quality of the pair in Westford satisfy a necessary but not sufficient condition for being able to maintain a global reference system with submillimeter per year accuracy for intervals in excess of a decade. These data are also used to determine an error model for the VLBI group delay measurements.

045.003 Geodesia cósmica.
V. Yurevich, H. Dávila S.
Bol. Obs. Astron. Quito, p. 5 – 17 (Jan 1992).

045.004 Accurate determination of Cartesian coordinates at geodetic stations using the Global Positioning System.
T. Soler, W. E. Strange, L. D. Hothem.
Geophys. Res. Lett., Vol. 19, No. 6, p. 533 – 536 (20 Mar 1992).
   Comparison of Cartesian coordinates determined at collocated sites using two independent space techniques, very long baseline interferometry (VLBI) and Global Positioning System (GPS), shows remarkable agreement even when the points in question span transcontinental distances. The results corroborate the capabilities of commercial dual–frequency GPS receivers to perform geodetic work at the highest available accuracy. Adjusted geocentric coordinates of a configuration of GPS stations well distributed along the eastern half of the United States were accurately determined (better than 10^{-8}) in the rigorously defined International Earth Rotation Service (IERS) terrestrial reference frame ITRF 89.

045.005 A history of laser ranging at McDonald Observatory.
P. J. Shelus, A. L. Whipple, R. L. Ricklefs, J. G. Ries, J. R. Wiant.
Bull. Am. Astron. Soc., Vol. 23, No. 3, p. 1257 (1991). Abstract. – See Abstr. 012.065 for the main entry.

045.006 The potential use of fiducial ground networks.
G. Bianco.
International Workshop on The Solid–Earth Mission ARISTO-TELES, p. 107 – 109 (Dec 1991). – See Abstr. 012.082 for the main entry.
   Collocation of space geodetic techniques will play an important role for precision orbit determination of ARISTOTELES. The FLINN network concept is ideal for defining and maintaining an highly precise conventional terrestrial reference frame by means of collocated SLR, VLBI and GPS stations. The proposed, "on–line" ARISTOTELES GPS tracking network should be supported by an extended, "off–line" tracking network with several selected FLINN sites, in order to include the ARISTOTELES mission within a standard, high accuracy conventional terrestrial reference system.

045.007 Accurate estimation of baselines with quasi–simultaneous observations from two stations.
Feng Chugang, Zhu Wenyao, Teng Zhanming.
Ann. Shanghai Obs., Acad. Sin., No. 13, p. 90 – 95 (1992). In Chinese.
   The semi–dynamic and semi–geometric satellite method is introduced. In this method, quasi–simultaneous observations from two stations are transformed into simultaneous range differences. It can be anticipated for a baseline about 1000 km in

length to reduce the effects of orbital and observational residual biases by a factor of 10.

045.008   Satellite laser ranging observations at Shanghai Observatory in 1990.
SLR Group.
Ann. Shanghai Obs., Acad. Sin., No. 13, p. 104 – 115 (1992). In Chinese.

Automatic determination of the astronomical azimuth by observing a celestial body using the electronic theodolite Kern E2 and the laptop computer Toshiba T1600.
See Abstr. 036.030.

Pearson type VII distribution of errors of laser satellite observations.
See Abstr. 036.063.

The IERS GPS Terrestrial Reference Frame.
See Abstr. 043.009.

Durch GPS bestimmbare Aspekte der Erdrotation (im Rahmen von IGS).
See Abstr. 044.019.

First years results from Chinese–German VLBI geodetic campaign (CGVGC).
See Abstr. 044.034.

Determination of long terms of the systematic difference between the JYD and BIH polar coordinate systems.
See Abstr. 044.040.

Modeling radiation forces acting on TOPEX/Poseidon for precision orbit determination.
See Abstr. 052.012.

Non–Riemannian theories of gravity and lunar and satellite laser ranging.
See Abstr. 066.023.

An update of the LAGEOS III gravitomagnetic experiment.
See Abstr. 066.128.

Relativistic motion of gyroscopes and space gradiometry.
See Abstr. 066.265.

Crustal velocities from geodetic very long baseline interferometry.
See Abstr. 081.023.

The Earth's gravity field from satellite geodesy – a 30 year adventure.
See Abstr. 081.042.

Geopotential models of the Earth from satellite tracking, altimeter and surface gravity observations: GEM–T3 and GEM–T3S.
See Abstr. 081.043.

Measurement of crustal deformation using the Global Positioning System.
See Abstr. 081.046.

046 Ephemerides, Almanacs, Calendars, Chronology

046.001   Calculation of an arbitrary date A.D.
B. Novotný.
Říše hvězd, Vol. 72, No. 2, p. 30 – 31 (Feb 1991). In Czech.

046.002   Some words about the astronomical year–books.
V. Vanýsek.
Říše hvězd, Vol. 72, No. 5, p. 87 – 89 (May 1991). In Czech.

046.003   The Astronomical Almanac for the year 1993.
        Data for astronomy, space sciences, geodesy, surveying, navigation and other applications.
Nautical Almanac Office, Washington, DC (USA); HM Nautical Almanac Office, Cambridge (UK), US Government Printing Office, Washington, DC (USA). vp. (1992). ISBN 0–11–886943–4. UK ed.: Her Majesty's Stationery Office, London (UK).

046.004   Anuarul Astronomic 1992.
Academia Romana, Bucharest. Institutul Astronomic, Editura Academiei Romane, Bucuresti (Romania). 303 p. (1992). ISBN 973–27–0299–0. Price Lei 105.

046.005   Efemérides astronómicas 1992.
H. Dávila S.
Bol. Obs. Astron. Quito, p. 90 – 131 (Jan 1992).

046.006   Astronomische Grundlagen für den Kalender 1994.
R. Bien, R. Jährling (comps.).
Astronomisches Rechen–Institut, Heidelberg (Germany), Braun, Karlsruhe (Germany). 147 p. (1992). ISBN 3–7650–0193–7. ISSN 0067–0014. Price DM 98.00.

046.007   Ephémérides Nautiques pour l'An 1993.
        Ouvrage publié par le Bureau des Longitudes spécialement à l'usage des marins.
Gauthier–Villars, Paris (France). 500 p. (1992). ISBN 2–10–001008–5.

046.008   Ephémérides Astronomiques 1993.
J. Meeus (ed.).
Société Astronomique de France, 75 – Paris. 128 p. (1992). ISSN 0004–6302. Supplément à la revue l'Astronomie.

046.009   1993 Nautical Almanac.
Maritime Safety Agency, Tokyo (Japan). Hydrographic Dept., Japan Oceanographic Data Center, Pub. No. 681. 486 p. (1992). ISSN 0910–0407. In Japanese.

046.010   Annuaire 1992.
Ciel, Vol. 54, Numéro spécial, p. 1 – 44 (1992).

046.011   Das Himmelsjahr 1993. Sonne, Mond und Sterne im Jahreslauf.
H.–U. Keller (ed.).
Franckh–Kosmos, Stuttgart (Germany). 246 p. (1992). ISBN 3–440–06370–4. Unter Mitarbeit von E. Karkoschka. Price DM 19.80.

046.012   Ahnerts Kalender für Sternfreunde 1993. Kleines astronomisches Jahrbuch.
R. Luthardt (ed.).
Barth, Leipzig (Germany). 176 p. (1992). ISBN 3–335–00313–6. Price DM 19.80.

046.013   **Le ciel de 1993. Phénomènes célestes pour l'année 1993.**
J. Meeus, D. Savoie, L. Tartois.
Astronomie, Suppl., p. 1 – 52 (1992).

046.014   **Connaissance des Temps. Ephémérides Astronomiques pour l'An 1993.**
Bureau des Longitudes, Paris (France), Service Hydrographique et Océanographique de la Marine, Brest (France). 135 p. (Jun 1992). ISBN 2–11–080635–4. ISSN 0181–3048. Price FF 210.00 (France), FF 240.00 (Etranger).

046.015   **Muster–Kalender für das Jahr 1993.**
H. Haupt (comp.).
Fromme, Wien (Austria). 36 p. (1991).

046.016   **Tables of sunrise, sunset, twilight, moonrise, & moonset 1993.**
Philippine Atmospheric, Geophysical and Astronomical Services Administration (PAGASA), Manila. 70 p. (1992). ISSN 0115–3307. Prepared by the Astronomical Publication Unit, Astronomy Research and Development Section, Atmospheric, Geophysical and Space Sciences Branch, Quezon City (Philippines).

046.017   **Himmelskalender 1993. Ein kleines astronomisches Jahrbuch für Österreich.**
H. Mucke.
Österreichischer Astronomischer Verein, Wien, Astronomisches Büro, Wien (Austria). 128 p. (1992). 37. Jahrg.

046.018   **Annuaire de l'Observatoire Royal de Belgique 1993. (Jaarboek van de Koninklijke Sterrenwacht van België 1993).**
Observatoire Royal de Belgique, Brussels, Hayez, Bruxelles (Belgium). 227 p. (1992). ISSN 0373–4900. 160e année.

046.019   **Solstice determination at noon.**
D. A. Allen.
Archaeoastronomy (U.K.), No. 17, p. S21 – S31 (1992).

046.020   **The length of the lunar month.**
B. E. Schaefer.
Archaeoastronomy (U.K.), No. 17, p. S32 – S42 (1992).

046.021   **Der Sternenhimmel 1993. Astronomisches Jahrbuch für Sternfreunde.**
E. Hügli, H. Roth, K. Städeli (eds.).
Sauerländer, Aarau (Switzerland). 231 p. (1992). ISBN 3–7941–3542–3. Price SFr. 39.80. Salle, Frankfurt am Main (Germany). ISBN 3–7935–5023–0. 53. Jahrg.

046.022   **Almanac for Geodetic Engineers 1993.**
Philippine Atmospheric, Geophysical and Astronomical Services Administration (PAGASA), Manila. 42 p. (1992). ISSN 0569–0838. Prepared by the Astronomical Publication Unit, Astronomical Research and Development Section, Atmospheric, Geophysical and Space Sciences Branch, Quezon City (Philippines).

046.023   **Physische Ephemeriden der Planeten.**
K.-H. Bücke.
Astron. Raumfahrt, Jahrg. 29, Heft 2, p. 14 – 15 (1991).

046.024   **Astronomicheski kalendar za 1992 godina.** *(Astronomical calendar for the year 1992.)*
B'lgarska Akademiya na Naukite, Sofia. 122 p. (1992). ISSN 0861–1270. Godina 39. Price 8.50 Lv.

046.025   **Ein "Ewiger Kalender". Einfach und mit einer unübertroffenen Übersichtlichkeit.**
R. Montandon.
Orion, Jahrg. 50, Nr. 248, p. 18 – 24 (Feb 1992).

046.026   **Nautisches Jahrbuch oder Ephemeriden und Tafeln für das Jahr 1993 zur Bestimmung der Zeit, Länge und Breite auf See nach astronomischen Beobachtungen.**
Bundesamt für Seeschiffahrt und Hydrographie, Hamburg (Germany). vp. (1992). ISSN 0077–6211. 142. Jahrg.

046.027   **The Indian Astronomical Ephemeris for the year 1993.**
India Meteorological Dept., Calcutta. Positional Astronomy Centre, Controller of Publications, New Delhi (India). 574 p. (1992). Price Rs. 100.00 (India), £ 11.66, US$ 36.00 (foreign).

046.028   **Almanaque Nautico – 1993. Con Suplemento para la Navegacion Aerea.**
Real Instituto y Observatorio de la Armada, San Fernando, Cadiz (Spain). vp. (1992). ISBN 84–7469–067–6. ISSN 0210–735X.

046.029   **Observer's Handbook 1993.**
R. L. Bishop (ed.).
Royal Astronomical Society of Canada, Toronto, Ontario. 236 p. (1992). ISSN 0080–4193. Eighty-fifth year of publication.

Der französische Revolutionskalender (1792 – 1805). Planung, Durchführung und Scheitern einer politischen Zeitrechnung.
See Abstr. 003.078.

When have been the Mayan calendar started?
See Abstr. 004.019.

The history of the tropical year.
See Abstr. 004.033.

A study on the calculation of true new moon and almanac in Xuanming calendar.
See Abstr. 004.077.

A report on mediaeval calendars discovered at Syogonji–temple.
See Abstr. 004.082.

Ephemerides of the 48 Hipparcos minor planets for the year 1992.
See Abstr. 098.062.

Ephemerides of minor planets for 1993.
See Abstr. 098.171.

On the ephemerides of Uranus.
See Abstr. 101.003.

Space Research

051 Extraterrestrial Research Related to Astronomy and Astrophysics

051.001 **EUVE probes the local bubble.**
P. Chien.
Sky Telesc., Vol. 83, No. 2, p. 161 – 163 (Feb 1992).
   NASA's Extreme Ultraviolet Explorer satellite will open a new spectral window on the solar system, hot stars, and the gas in our neighborhood of the Milky Way.

051.002 **End of the space racing USA/USSR.**
M. Grün.
Kozmos, Vol. 22, No. 2, p. 39 – 41 (Mar 1991). In Czech.

051.003 **Voyager's journey across the Solar System.**
P. Koubský.
Říše hvězd, Vol. 72, No. 1, p. 3 – 10 (Jan 1991). In Czech.

051.004 **Space astronomy.**
B. Valníček.
Říše hvězd, Vol. 72, No. 2, p. 28 – 30 (Feb 1991). In Czech.

051.005 **Astronautics in 1990.**
M. Grün.
Říše hvězd, Vol. 72, No. 9, p. 161 – 165 (Sep 1991). In Czech.
   Further information is given in No. 10, p. 185 – 188 (Oct 1991).

051.006 **Gamma Ray Observatory in orbit.**
R. Hudec.
Říše hvězd, Vol. 72, No. 11, p. 204 – 207 (Nov 1991). In Czech.

051.007 **If these news will be confirmed (COBE).**
M. Plavec.
Vesmír, Vol. 71, No. 5, p. 234 – 235 (May 1992). In Czech.

051.008 **Astronomical observations from outside the atmosphere.**
K. Makishima.
Astron. Her., Vol. 85, No. 1, p. 6 – 12 (1992). In Japanese.

051.009 **Space at JPL.**
W. I. McLaughlin.
Spaceflight, Vol. 34, No. 2, p. 57 – 61 (Feb 1992).
   Contents: Navigation to Gaspra. A builder of missions. According to plan.

051.010 **Space at JPL.**
W. I. McLaughlin.
Spaceflight, Vol. 34, No. 5, p. 166 – 170 (May 1992).
   Contents: Ulysses swings by Jupiter. Radar tracking of asteroids. Lowering mission costs.

051.011 **Venus – a prime Soviet objective. Part I.**
D. F. Robertson.
Spaceflight, Vol. 34, No. 5, p. 158 – 161 (May 1992).

051.012 **Venus – a prime Soviet objective. Part II.**
D. F. Robertson.
Spaceflight, Vol. 34, No. 6, p. 202 – 205 (Jun 1992).

051.013 **NASA's space observatories get exciting results.**
Spaceflight, Vol. 34, No. 5, p. 162 – 165 (May 1992).
Hubble finds evidence of a black hole. Hubble discovers young star clusters. Compton Observatory makes new discoveries. Hubble snapshots probe the early universe.

051.014 **Extreme Ultraviolet Explorer: a view inside the bubble.**
Spaceflight, Vol. 34, No. 6, p. 190 – 193 (Jun 1992).

051.015 **The second coming of Giotto. Part one: Encounter with Halley.**
D. Burnham.
Spaceflight, Vol. 34, No. 6, p. 210 – 212 (Jun 1992).

051.016 **An overview of the cosmic background explorer (COBE) and its observations: new sky maps of the early universe.**
G. F. Smoot.
NATO Advanced Study Institute on The Infrared and Submillimetre Sky after COBE, p. 1 – 14 (1992). – See Abstr. 012.009 for the main entry.
   This introductory paper discusses the three instruments aboard NASA's Cosmic Background Explorer (COBE) satellite and presents early results obtained from the first six months of observations. The three instruments (FIRAS, DMR, and DIRBE) have operated well and produced significant new results. The FIRAS measurement of the CMB spectrum supports the standard Big Bang model. The maps made from the DMR instrument measurements show a spatially smooth early universe. The maps of galactic and zodiacal emission produced by the DIRBE instrument are needed to identify the foreground emissions from extragalactic and thus to interpret its and the other COBE results in terms of events in the early universe.

051.017 **Preliminary results from the FIRAS and DIRBE experiments on COBE.**
E. L. Wright.
NATO Advanced Study Institute on The Infrared and Submillimetre Sky after COBE, p. 231 – 248 (1992). – See Abstr. 012.009 for the main entry.
   The FIRAS and DIRBE experiments on COBE will provide data that strongly constrain models of Population III stars and dust. This paper describes the analysis procedures necessary to process COBE data, and compares simple Population III models to the preliminary COBE data.

051.018 **The Ulysses mission.**
K.-P. Wenzel, R. G. Marsden, D. E. Page,
E. J. Smith.
Astron. Astrophys., Suppl. Ser., Vol. 92, No. 2, p. 207 – 219 (Jan 1992). Special issue: Ulysses instruments.
   The Ulysses mission is unique in the history of the exploration of our solar system by spacecraft. The path followed by Ulysses will enable us, for the first time, to explore the heliosphere within a few astronomical units of the Sun over the full range of heliographic latitudes, thereby providing the first characterisation of the uncharted third heliospheric dimension. Advanced scientific instrumentation carried on board the spacecraft is

designed to measure the properties of the heliospheric magnetic field, the solar wind, the Sun/wind interface, solar radio bursts and plasma waves, solar energetic particles and galactic cosmic rays, solar X–rays, and interplanetary/interstellar neutral gas and dust. Ulysses will also be used to detect cosmic gamma–ray bursts and search for gravitational waves. The mission, a collaboration between ESA and NASA, was launched in October 1990 and employs a Jupiter gravity–assist to achieve the trajectory extending to high solar latitudes. Ulysses will spend a total of 234 days, equivalent to about 8 solar rotations, at latitudes in excess of 70°. The authors describe the characteristics of the Ulysses mission.

051.019 **Beyond a boundary of the heliosphere.**
M. Gola.
Urania, Rok 63, Nr. 3, p. 67 – 71 (Mar 1992). In Polish.

051.020 **Ulysses – the space Odyssey 1990 – 1995.**
L. M. Błecki.
Urania, Rok 63, Nr. 4, p. 101 – 105 (Apr 1992). In Polish.

051.021 **In–orbit performance of the Hipparcos astrometry satellite.**
M. A. C. Perryman, E. Høg, J. Kovalevsky, L. Lindegren, C. Turon, P. L. Bernacca, M. Crézé, F. Donati, M. Grenon, M. Grewing, F. van Leeuwen, H. van der Marel, C. A. Murray, R. S. Le Poole, H. Schrijver.
Astron. Astrophys., Vol. 258, No. 1, p. 1 – 6 (May 1992).
Launched in August 1989 into a geostationary transfer orbit, the Hipparcos astrometry satellite failed to reach its intended geostationary orbit through the failure of its apogee boost motor. Present indications are that the possible operational lifetime should nevertheless extend beyond the end of 1992, with extremely high–quality scientific data being returned for some 65% of the time. For an assumed 3–year operational lifetime, the original mission goals (positions, parallaxes, and annual proper motions with an accuracy of about 2 milli–arcsec) should be achievable (the pre–launch expectations of the planned 2.5–year mission would be achievable with a 3.5–year mission in the revised orbit). This paper describes the in–orbit performance of the satellite, summarising the system parameters (scanning law, observing strategy, data recovery fraction) affecting the scientific programme, the payload performances and their time–dependence, and the resulting expected mission accuracies.

051.022 **Mars Observer mission.**
A. L. Albee, R. E. Arvidson, F. D. Palluconi.
J. Geophys. Res., Vol. 97, No. E5, p. 7665 – 7680 (25 May 1992).
The Mars Observer mission will extend the exploration and characterization of Mars by providing new and systematic measurements of the atmosphere, surface, and interior of the planet. These measurements will be made from a low–altitude polar orbiter over a period of 1 Martian year, permitting repetitive observations of the surface and of the seasonal variations of the atmosphere. The mission will be conducted in a manner that will provide new and valuable scientific data using a distributed data system that minimizes operational complexity and cost.

051.023 **Radio science investigations with Mars Observer.**
G. L. Tyler, G. Balmino, D. P. Hinson, W. L. Sjogren, D. E. Smith, R. Woo, S. W. Asmar, M. J. Connally, C. L. Hamilton, R. A. Simpson.
J. Geophys. Res., Vol. 97, No. E5, p. 7759 – 7779 (25 May 1992).
Mars Observer radio science investigations focus on two major areas of study: the gravity field and the atmosphere of Mars. Measurement accuracies expressed as an equivalent spacecraft velocity are expected to be of the order of 100 μm/s (for both types of investigations) from use of an improved radio transponder for two–way spacecraft tracking and a highly stable on–board oscillator for atmospheric occultation measurements. Planned gravity investigations include a combination of classical and modern elements. A spherical harmonic (or equivalent) field model of degree and order in the range 30–50 will be obtained, while interpretation will be in terms of internal stress and density models for the planet, using the topography to be obtained from the Mars Observer laser altimeter. Atmospheric investigations will emphasize precision measurement of the thermal structure and dynamics in the polar regions, which are regularly accessible as a result of the highly inclined orbit. Studies based on the measurements will include polar processes, cycling of the atmosphere between the poles, traveling baroclinic disturbances, small–scale waves and turbulence, the planetary boundary layer, and (possibly) the variability and altitude of the ionosphere. As the radio occultation is insensitive to dust in the atmosphere per se and measures only the resulting change in thermal structure, it is expected that the radio technique can contribute to understanding of dust storm phenomena. Mutual observations of the atmosphere by means of radio occultation and by the pressure modulator infrared radiometer and the thermal emission spectrometer are expected to strengthen the reliability and accuracy of all three investigations.

051.024 **Giotto Radio–Science Experiment: drag deceleration and spacecraft attitude perturbations expected during the encounter with comet P/Grigg–Skjellerup in July 1992.**
M. Pätzold, H. Porsche, P. Edenhofer, M. K. Bird, H. Volland.
Astron. Astrophys., Vol. 259, No. 1, p. L15 – L18 (Jun 1992). Letter–to–the–editor.
The GIOTTO Radio–Science Experiment will repeat its observations performed during the GIOTTO–Halley fly by a the next encounter with comet P/Grigg–Skjellerup in July 1992. The change in velocity due to the influx of cometary gas and dust will be determined by X–band ranging and Doppler measurements. Due to the lower flyby velocity (five times smaller) and the lower dust generation rate compared to Halley (ten times smaller), the change in velocity is estimated to be between 4 mm/sec and 4 cm/sec for a closest approach between 1000 km and 100 km, respectively. Large dust particle impacts (masses greater than 10 mg) on the spacecraft body may change the spin rate, resulting in a degradation of the real–time data link to Earth or, in the worst case, loss of data. Nutation is not considered a problem for this encounter.

051.025 **Marsnet surface and atmosphere investigations.**
A. F. Chicarro.
Workshop on the Martian Surface and Atmosphere Through Time, p. 32 – 33 (1992). Abstract. – See Abstr. 012.031 for the main entry.

051.026 **Mars environmental survey (MESUR): science objectives and mission description.**
G. S. Hubbard, P. F. Wercinski, G. L. Sarver, R. P. Hanel, R. Ramos.
Workshop on the Martian Surface and Atmosphere Through Time, p. 74 – 75 (1992). Abstract. – See Abstr. 012.031 for the main entry.

051.027 **Discovery concepts for Mars.**
J. G. Luhmann, C. T. Russell, L. H. Brace, A. F. Nagy, B. M. Jakosky, C. A. Barth, J. H. Waite.
Workshop on the Martian Surface and Atmosphere Through Time, p. 93 (1992). Abstract. – See Abstr. 012.031 for the main entry.

051.028 **Planet–B: a Japanese Mars aeronomy observer.**
K. Tsuruda.
Workshop on the Martian Surface and Atmosphere Through Time, p. 161 (1992). Abstract. – See Abstr. 012.031 for the main entry.

051.029 **Discovery–class mission concepts for Mars.**
B. M. Jakosky, J. G. Luhmann, L. H. Brace, C. T. Russell, C. A. Barth, H. Waite, A. Nagy.
Bull. Am. Astron. Soc., Vol. 23, No. 3, p. 1176 – 1177 (1991). Abstract. – See Abstr. 012.037 for the main entry.

051.030   Die Astrometrie–Mission HIPPARCOS der ESA.
H. Mucke.
Sternenbote, Jahrg. 35, Nr. 3, p. 54 – 63 (1992). Nach Unterlagen der ESA.

051.031   Space Science Reviews volume on Galileo Mission: overview.
T. V. Johnson, C. M. Yeates, R. Young.
Space Sci. Rev., Vol. 60, No. 1–4, p. 3 – 21 (May 1992). Special issue: The Galileo Mission. C. T. Russell (ed.).
The Galileo Mission is an extremely complex undertaking. This paper provides a brief historical overview, a discussion of broad scientific objectives, and a description of the spacecraft and trajectory characteristics.

051.032   Galileo trajectory design.
L. A. d'Amario, L. E. Bright, A. A. Wolf.
Space Sci. Rev., Vol. 60, No. 1–4, p. 23 – 78 (May 1992). Special issue: The Galileo Mission. C. T. Russell (ed.).
The Galileo spacecraft was launched by the Space Shuttle Atlantis on October 18, 1989. A two–stage Inertial Upper Stage propelled Galileo out of Earth parking orbit to begin its 6–year interplanetary transfer to Jupiter. Galileo has already received two gravity assists: from Venus on February 10, 1990 and from Earth on December 8, 1990. After a second gravity–assist flyby of Earth on December 8, 1992, Galileo will have achieved the energy necessary to reach Jupiter. Galileo's interplanetary trajectory includes a close flyby of asteroid 951–Gaspra on October 29, 1991, and, depending on propellant availability and other factors, there may be a second asteroid flyby of 243–Ida on August 28, 1993. Upon arrival at Jupiter on December 7, 1995, the Galileo Orbiter will relay data back to Earth from an atmospheric Probe which is released five months earlier. For about 75 min, data is transmitted to the Orbiter from the Probe as it descends on a parachute to a pressure depth of 20–30 bars in the Jovian atmosphere. Shortly after the end of Probe relay, the Orbiter ignites its rocket motor to insert into orbit about Jupiter. The orbital phase of the mission, referred to as the satellite tour, lasts nearly two years, during which time Galileo will complete 10 orbits about Jupiter. On each of these orbits, there will be a close encounter with one of the three outermost Galilean satellites (Europa, Ganymede, and Callisto). The gravity assist from each satellite is designed to target the spacecraft to the next encounter with minimal expenditure of propellant. The nominal mission is scheduled to end in October 1997 when the Orbiter enters Jupiter's magnetotail.

051.033   Gravitation and celestial mechanics investigations with Galileo.
J. D. Anderson, J. W. Armstrong, J. K. Campbell, F. B. Estabrook, T. P. Krisher, E. L. Lau.
Space Sci. Rev., Vol. 60, No. 1–4, p. 591 – 610 (May 1992). Special issue: The Galileo Mission. C. T. Russell (ed.).
The gravitation and celestial mechanics investigations during the cruise phase and Orbiter phase of the Galileo mission depend on Doppler and ranging measurements generated by the Deep Space Network (DSN) at its three spacecraft tracking sites in California, Australia, and Spain. The authors group their investigations into four broad categories as follows: (1) the determination of the gravity fields of Jupiter and its four major satellites during the orbital tour, (2) a search for gravitational radiation as evidenced by perturbations to the coherent Doppler link between the spacecraft and Earth, (3) the mathematical modeling, and by implication tests, of general relativistic effects on the Doppler and ranging data during both cruise and orbiter phases, and (4) an improvement in the ephemeris of Jupiter by means of spacecraft ranging during the Orbiter phase. In addition to the primary objectives of their investigations, the authors discuss two secondary objectives: the determination of a range fix on Venus during the flyby on 10 February, 1990, and the determination of the Earth's mass (GM) from the two Earth gravity assists, EGA1 in December 1990 and EGA2 in December 1992.

051.034   Magellan: overview of science findings.
R. S. Saunders.
Bull. Am. Astron. Soc., Vol. 23, No. 3, p. 1204 (1991). Abstract. – See Abstr. 012.037 for the main entry.

051.035   ROSAT: early results.
J. E. Trümper.
15. Texas Symposium on Relativistic Astrophysics and 4. ESO–CERN Symposium, p. 141 – 150 (1991). – See Abstr. 012.060 for the main entry.
The general scientific objectives of ROSAT are to perform (a) the first all–sky surveys using imaging X–ray and extreme ultraviolet telescopes and (b) detailed investigations of interesting sources in a guest investigator program. The survey operations commenced in August 1990 and at the time of the 15th Texas/ESO–CERN Conference about 75 percent of the sky had been scanned. In February 1991 the first half year guest observer program was started for which in total 738 proposals were received. A glimpse of what can be achieved by pointed observations was obtained during the calibration and verification measurements in the early phase of the mission (June/July 1990). The author gives a brief summary of the results obtained with the X–ray telescope so far, and a few remarks on the scientific instruments.

051.036   COBE.
E. L. Wright.
15. Texas Symposium on Relativistic Astrophysics and 4. ESO–CERN Symposium, p. 190 – 198 (1991). – See Abstr. 012.060 for the main entry.
The Cosmic Background Explorer (COBE) was launched on November 18, 1989. NASA's first satellite designed primarily for observational cosmology, COBE is performing detailed studies of the cosmic microwave background radiation from the big bang, and is searching for the cosmic infrared background radiation from the first objects to form after the big bang. The author concentrates here on effects that will be important in the final analysis of COBE data.

051.037   The International Ultraviolet Explorer project (IUE).
W. Wamsteker.
AIP Conf. Proc., No. 245: Workshop on Basic Space Science, p. 253 – 265 (1992). – See Abstr. 012.064 for the main entry.
A summary of the definition and evolution of the International Ultraviolet Explorer Satellite (IUE) is given. Some examples highlighting the exceptional success of the project are presented. The specific nature of the IUE project which has made this success possible are equally applicable to the stimulation of the participation of scientists from the developing countries in the IUE project. This will stimulate the development of the capabilities needed, to make a larger participation of the developing countries in the basic space sciences a reality.

051.038   Early results from the Cosmic Background Explorer (COBE).
J. C. Mather, M. G. Hauser, C. L. Bennett, N. W. Boggess, E. S. Cheng, R. E. Eplee Jr., H. T. Freudenreich, R. B. Isaacman, T. Kelsall, C. M. Lisse, S. H. Moseley Jr., R. A. Shafer, R. F. Silverberg, W. J. Spiesman, G. N. Toller, J. L. Weiland, S. Gulkis, M. Janssen, P. M. Lubin, S. S. Meyer, R. Weiss, T. L. Murdock, G. F. Smoot, D. T. Wilkinson, E. L. Wright.
AIP Conf. Proc., No. 245: Workshop on Basic Space Science, p. 266 – 276 (1992). – See Abstr. 012.064 for the main entry.
The Cosmic Background Explorer, launched November 18, 1989, has nearly completed its first full mapping of the sky with all three of its instruments: a Far Infrared Absolute Spectrophotometer (FIRAS) covering 0.1 to 10 mm, a set of Differential Microwave Radiometers (DMR) operating at 3.3, 5.7, and 9.6 mm, and a Diffuse Infrared Background Experiment (DIRBE) spanning 1 to 300 μm in ten bands. A preliminary map of the sky derived from DIRBE data is presented. There are no significant anisotropies in the microwave sky detected. At shorter wavelengths, the sky spectrum and anisotropies are dominated

by emission from 'local' sources of emission within our Galaxy and Solar System. Preliminary comparison of IRAS and DIRBE sky brightnesses toward the ecliptic poles shows the IRAS values to be significantly higher than found by DIRBE at 100 μm. The spacecraft, instrument designs, and data reduction methods are described.

051.039   Gamma Ray Observatory/BATSE status.
S. Howard.
Bull. Am. Astron. Soc., Vol. 23, No. 3, p. 1257 (1991). Abstract. – See Abstr. 012.065 for the main entry.

051.040   The SIRTF reference mission.
P. Eisenhardt, M. Osmolovsky, M. W. Werner, P. Hacking, E. L. Wright.
Bull. Am. Astron. Soc., Vol. 23, No. 3, p. 1269 (1991). Abstract. – See Abstr. 012.067 for the main entry.

051.041   Study of cosmic dust particles on board LDEF and MIR space station.
J. C. Mandeville.
IAU Colloquium No. 126: Origin and evolution of interplanetary dust, p. 11 – 14 (1991). – See Abstr. 012.068 for the main entry.

Two French experiments partly devoted to the detection of cosmic dust have been flown recently in space: on the NASA Long Duration Exposure Facility (LDEF), and on the Soviet MIR Space Station. The experimental approach and preliminary results are described briefly.

051.042   The present status of the Munich Dust Counter experiment on board of the HITEN spacecraft.
E. Igenbergs, A. Hüdepohl, K. Uesugi, T. Hayashi, H. Svedhem, H. Iglseder, G. Koller, A. Glasmachers, E. Grün, G. Schwehm, H. Mizutani, T. Yamamoto, A. Fujimura, N. Ishii, H. Araki, K. Yamakoshi, K. Nogami.
IAU Colloquium No. 126: Origin and evolution of interplanetary dust, p. 15 – 20 (1991). – See Abstr. 012.068 for the main entry.

The Munich Dust Counter (MDC) is a scientific experiment on board the MUSES–A mission of Japan measuring cosmic dust. The satellite HITEN of this mission has been launched on January 24th, 1990 from Kagoshima Space Center. The present status of the MDC experiment is summarized. The number of dust particles measured so far is presented together with first and preliminary results of flux calculations and spatial as well as directional distributions of cosmic dust particles measured until July 25, 1990. A clear evidence of particles coming from the inner solar system (beta–meteoroids) already has been found. These are compared to particles coming from the apex direction.

051.043   In–situ exploration of dust in the solar system and initial results from the Galileo dust detector.
E. Grün, H. Fechtig, M. S. Hanner, J. Kissel, B.–A. Lindblad, D. Linkert, G. Morfill, H. A. Zook.
IAU Colloquium No. 126: Origin and evolution of interplanetary dust, p. 21 – 28 (1991). – See Abstr. 012.068 for the main entry.

In–situ measurements of interplanetary dust have been performed in the heliocentric distance range from 0.3 AU out to 18 AU. The Galileo dust detector combines the high mass sensitivity of impact ionization detectors (10^{-15}g) together with a large sensitive area (0.1 m^2). The Galileo spacecraft was launched on October 18, 1989 and is on its solar system cruise towards Jupiter. Initial measurements of the dust flux from 0.7 to 1.2 AU are presented.

051.044   The NASA Solar Probe mission: in situ determination of interplanetary out–of–the ecliptic and near–solar dust environments.
B. T. Tsurutani, J. E. Randolph.
IAU Colloquium No. 126: Origin and evolution of interplanetary dust, p. 29 – 32 (1991). – See Abstr. 012.068 for the main entry.

The NASA Solar Probe mission will be one of the most exciting dust missions ever flown and will lead to a revolutionary advance in our understanding of dust within our solar system. Solar Probe will map the dust environment from the orbit of Jupiter (5 AU), to within 4 solar radii of the Sun's center. Solar Probe will also reach heliographic latitudes as high as ~15° at 28° above (below) the ecliptic on its trajectory inbound (outbound) to (from) the Sun. This, in addition to the ESA/NASA Ulysses mission, will help determine the out–of–the–ecliptic dust environment.

051.045   Dynamic modelling transformations for the low Earth orbit satellite particulate environment.
J. A. M. McDonnell, K. Sullivan, S. F. Green, T. J. Stevenson, D. H. Niblett.
IAU Colloquium No. 126: Origin and evolution of interplanetary dust, p. 37 – 40 (1991). – See Abstr. 012.068 for the main entry.

A simple dynamic model to investigate the relative fluxes and particle velocities on a spacecraft's different faces is presented. The results for LDEF are consistent with a predominantly interplanetary origin for the larger particulates, but a sizable population of orbital particles with sizes capable of penetrating foils of thickness < 30 μm. Data from experiments over the last 30 years do not show the rise flux expected if these were space debris. The possibility of a population of natural orbital particulates awaits confirmation from chemical residue analysis.

051.046   Particle astrophysics on the moon.
K. Lande.
15. Texas Symposium on Relativistic Astrophysics and 4. ESO–CERN Symposium, p. 635 – 641 (1991). – See Abstr. 012.060 for the main entry.

The moon provides a site with rather special characteristics that permit to view the particle emission of the astronomical world and study properties of particles in an environment not accessible on the earth. In this review, the author describes some of the questions that will benefit from observatories on the moon, indicates why these cannot be carried out on either the earth or in earth orbit, and considers characteristics of a multipurpose detector with which some of these questions could be attacked.

051.047   Astrophysics from the moon.
H. J. Smith.
15. Texas Symposium on Relativistic Astrophysics and 4. ESO–CERN Symposium, p. 642 – 648 (1991). – See Abstr. 012.060 for the main entry.

Contents: 1. Introduction. 2. Why is the moon so good for astronomy? 3. Drawbacks of the moon for astronomy. 4. Some lunar–specific telescopes and observational goals. 5. How likely is all this to happen?

051.048   Future observation of the F–corona with the LASCO coronograph space experiment.
P. L. Lamy, A. Llebaria, A. Maucherat, S. Koutchmy, F. Giovane.
IAU Colloquium No. 126: Origin and evolution of interplanetary dust, p. 191 – 194 (1991). – See Abstr. 012.068 for the main entry.

The Wide–field White Light and Spectrometric Coronograph (LASCO) to be flown on SOHO in 1995 will observe the corona from just above the limb at 1.1 out to 30 solar radii (R$_\odot$). In addition to the fundamental problems of coronal physics (heating of the corona, acceleration of the solar wind, coronal transients), the scientific objectives incorporate the distribution and properties of dust particles including those released from sun–grazing comets, and interactions of coronal plasma with the dust.

051.049   Lesson learned from GEOS and ISEE.
A. Pedersen.
International Workshop on Cluster Dayside Polar Cusp, p. 47 – 51 (Dec 1991). – See Abstr. 012.071 for the main entry.

The GEOS and ISEE–1 and ISEE–2 spacecraft were central spacecraft in the International Magnetospheric Study (IMS) which was initiated in 1976. These spacecraft had new instruments based on the experience of the first years in magnetospheric research, and the IMS aimed at comparing data from many instruments on spacecraft and between many different spacecraft and ground observations. Valuable aims were not reached, partly because the technical means (computers, networks and storage media) were not sufficiently developed to deal with such a tremendous task. Certain organisational aspects could also have been improved. These missions are reviewed with the aim to propose improvements for the Cluster mission.

051.050   Balloons for conjugate cusp studies in the 1990's.
W. Riedler, K. M. Torkar, I. B. Iversen.
International Workshop on Cluster Dayside Polar Cusp, p. 171 – 178 (Dec 1991). – See Abstr. 012.071 for the main entry.
This paper reviews the advantages of balloon–borne measurements for magnetospheric studies and develops a scenario of possible balloon campaigns during the Cluster mission, with emphasis on the dayside polar cap region.

051.051   Mission objectives and scientific rationale for the magnetometer mission.
R. A. Langel.
International Workshop on The Solid–Earth Mission ARISTO-TELES, p. 17 – 26 (Dec 1991). – See Abstr. 012.082 for the main entry.
Based on a review of the characteristics of the geomagnetic field, objectives for the magnetic portion of the ARISTOTELES mission are: (1) To derive a description of the main magnetic field and its secular variation. (2) To investigate the correlation between the geomagnetic field and variations in the length of day. (3) To study properties of the fluid core. (4) To study the conductivity of the mantle. (5) To model the state and evolution of the crust and upper lithosphere. (6) To measure and characterize field aligned currents and ionospheric currents and to understand their generation mechanisms and their role in energy coupling in the interplanetary–magnetospheric–ionospheric systems. Procedures for these investigations are outlined.

051.052   Progress in the spacewise approach to ARISTOTELES data reduction.
M. Bassanino, F. Migliaccio, F. Sanso.
International Workshop on The Solid–Earth Mission ARISTO-TELES, p. 33 – 43 (Dec 1991). – See Abstr. 012.082 for the main entry.
ARISTOTELES is a satellite mission conceived to determine a global model of the Earth's gravity field, in a homogeneous way, with high resolution and high accuracy. In the paper it is shown, after some simplifications and idealizations, how this task can be accomplished by using in a boundary value problem mode the data available from the satellite, i.e. gradiometry, accelerometry and GPS data.

051.053   Gravity field data products from the ARISTOTELES mission.
G. Balmino.
International Workshop on The Solid–Earth Mission ARISTO-TELES, p. 111 – 113 (Dec 1991). – See Abstr. 012.082 for the main entry.
The ARISTOTELES mission will bring a wealth of homogeneous information about the Earth gravity field enabling new direct and inverse modeling of geophysical structures at various scales, yielding a reference geoid surface of great quality for oceanographic studies, leading to global models of high resolution for versatile applications and in particular precise orbit determination of artificial satellites. The author's purpose is to review the different types of measurements involved in these investigations, the various levels of processing and how they can be phased with the scientific activities, and the expected products. Also, some general schemes are proposed along which the different tasks can be undertaken.

051.054   X–ray astronomy missions.
H. V. D. Bradt, T. Ohashi, K. A. Pounds.
Annu. Rev. Astron. Astrophys., Vol. 30, p. 391 – 427 (1992).
Contents: 1. Introduction. 2. Early years (1962 – 1969). 3. Small satellite era (1967 – 1977). 4. The large HEAO missions (1977 – 81). 5. The 1980s. 6. The 1990s. 7. The future.

051.055   Three–component penetrator accelerometer for Mars exploration.
G. Kh. Mardirosyan, V. M. Fremd.
Aerospace Res. Bulg., Vol. 8, p. 39 – 46 (1991). In Bulgarian.

051.056   Implementation of the revised Hipparcos mission at ESOC.
J. Van der Ha.
ESA Bull., No. 69, p. 9 – 15 (Feb 1992).

051.057   ESOC's role in routine Hipparcos operations.
D. Heger, A. McDonald, A. Schütz, O. Ojanguren, C. Sollazzo.
ESA Bull., No. 69, p. 16 – 25 (Feb 1992).

051.058   The Hipparcos payload's in–orbit performance.
G. Ratier, K. van Katwijk, G. Fade, M. A. C. Perryman.
ESA Bull., No. 69, p. 27 – 32 (Feb 1992).

051.059   The Hipparcos observing programme: preparation of the input catalogue.
C. Turon.
ESA Bull., No. 69, p. 36 – 42 (Feb 1992).
Two years into the Hipparcos measurement programme is a convenient point at which to reflect on the observing programme of ESA's revolutionary astrometry mission, and to examine the extent to which it has fulfilled its scientific and operational objectives. The comprehensive scientific observing programme, and its complex operational implementation, have passed all post–launch tests supremely well and promise a scientific return unsurpassed in the history of astronomical positional measurements.

051.060   Hipparcos data reduction – construction of the Hipparcos star catalogue.
E. Høg, J. Kovalevsky, L. Lindegren.
ESA Bull., No. 69, p. 43 – 50 (Feb 1992).
The on–ground treatment of the Hipparcos satellite data is the largest and most complex data–analysis problem ever undertaken in the history of astrometry. Assembling a final and highly accurate catalogue of star positions, distances and motions presents a unique problem, both scientifically and from the data–management viewpoint. The development of the data–analysis techniques guided the development of the satellite design itself. Self–checking procedures, implicit in the satellite measurement and data–analysis techniques, demonstrate that the ultimate astrometric catalogue promised by the Hipparcos mission will shortly become a reality.

051.061   Early scientific results from Hipparcos and future expectations.
The Hipparcos Science Team.
ESA Bull., No. 69, p. 51 – 57 (Feb 1992).
The first results of the lengthy Hipparcos Star Catalogue construction process are now becoming available. The star positions and distance estimates emerging from the data analyses give a dramatic indication of the outstanding scientific results that ESA's Hipparcos mission can be expected to provide.

051.062   F–corona–experiment: requirements for remote sensing of interplanetary dust.
I. Mann, H. Hartwig.
10. ESA Symposium on European Rocket and Balloon Programmes and Related Research, p. 335 – 337 (Nov 1991). – See Abstr. 012.088 for the main entry.
The near–solar dust that produces the brightness of the E–corona represents the central region of the interplanetary dust cloud. The structure of the inner dust cloud and the properties of the meteoritic complex and its relation to other constituents of the interplanetary medium have to be considered for the concept of near solar missions. To investigate this complex a synoptical observation in different spectral ranges is needed. The concept of a dedicated rocket borne experiment is outlined.

051.063 **Magellan mission description and radar system.**
W. T. K. Johnson.
International Symposium on Radars and Lidars in Earth and Planetary Sciences, p. 33 – 38 (Dec 1991). – See Abstr. 012.093 for the main entry.

The Magellan radar mission to Venus has collected and processed radar data from the spacecraft in an elliptical orbit around the planet. This multimode radar is the only science instrument on the mission and had the objective of mapping at least 70% of the planet surface during the primary mission or first cycle of 243 days. The spacecraft was launched in May, 1989 and arrived at Venus in August, 1990. A few weeks later the mapping of the planet began. During this first cycle of mapping nearly 84% of the surface was mapped. The second cycle missed during the first cycle and also using the radar in some non–standard imaging modes. The radar has three modes: synthetic aperture radar (SAR), altimetry, and passive radiometry. The radar system has produced maps of almost all of the Venusian surface with a resolution better than 600 m equivalent optical line pair. The system design and operation required to produce the high resolution images was complex and innovative. This paper describes the mission and the radar system design.

051.064 **In–situ Doppler velocimeter of very large grains: an essential goal for future cometary investigations.**
J. F. Crifo.
International Symposium on Radars and Lidars in Earth and Planetary Sciences, p. 65 – 70 (Dec 1991). – See Abstr. 012.093 for the main entry.

The essential role played by the solid grain ejection velocity in several critical fields of cometary investigation is briefly recalled. Comparison between the various approximating expressions currently in use for this velocity already reveals severe discrepancies. Recent indirect determinations based on model fits to dust tail images yield values which conflict with all these expressions. The author argues that both these theoretical expressions and these indirect determinations yield unreliable results, especially for large (cm sized and larger) grains. Future efforts along the same lines are not expected to provide significant improvements, because key informations (the detailed structure of the nucleus surface and the dispersion in physical properties of the grains) are not accessible. Direct in–situ Doppler velocimetry of large grains from comet rendez–vous spacecrafts thus appears as a potentially essential goal for future comet rendez–vous missions.

051.065 **ASTRA: altimetry and sounding of Titan with a radar on a descending craft.**
P. Kamoun, J. C. Anne, P. Ford.
International Symposium on Radars and Lidars in Earth and Planetary Sciences, p. 71 – 78 (Dec 1991). – See Abstr. 012.093 for the main entry.

Titan, the largest satellite of Saturn is still very poorly known. Radar waves which can propagate almost undisturbed through its optically thick atmospheric haze offer one of the best tools to observe its surface. Here, a small altimetric and sounding radar instrument is proposed to observe and characterize this surface from a descending craft. It is a simple lightweight and cheap instrument with a high scientific return for the knowledge of Titan surface and subsurface.

051.066 **The Solar–A mission experiments and the targets.**
Y. Uchida, Y. Ogawara.
IAU Colloquium No. 133: Eruptive solar flares, p. 309 – 313 (1992). Short summary of the talk. – See Abstr. 012.096 for the main entry.

051.067 **Japanese lunar mission in the mid–1990s.**
H. Mizutani.
Proc. Astron. Soc. Aust., Vol. 9, No. 2, p. 332 – 333 (1991). – See Abstr. 012.090 for the main entry.

This paper outlines the plans for a Japanese lunar exploration mission in the mid–1990s.

051.068 **Pioneers 10 and 11 deep space missions.**
P. Dyal.
1. COSPAR Colloquium: Physics of the outer heliosphere, p. 373 – 382 (1990). – See Abstr. 012.100 for the main entry.

Since launch, the Pioneers have measured large–scale properties of the heliosphere during more than one complete 11–year solar sunspot cycle, and have measured the properties of the expanding solar atmosphere, the transport of cosmic rays into our heliosphere, and the high–energy trapped radiation belts and magnetic field asssociated with the planets Jupiter and Saturn.

051.069 **Ulysses: a status report.**
D. E. Page, R. G. Marsden, E. J. Smith, K.–P. Wenzel.
1. COSPAR Colloquium: Physics of the outer heliosphere, p. 383 – 388 (1990). – See Abstr. 012.100 for the main entry.

The mission, the spacecraft and the science instrumentation are outlined and references are given indicating where these ares are given in detail.

The Hipparcos Input Catalogue. I. Star selection.
See Abstr. 002.041.

The Hipparcos Input Catalogue. II. Astrometric data.
See Abstr. 002.042.

The Hipparcos Input Catalogue. III. Photometry.
See Abstr. 002.043.

Reliability of the Hipparcos Input Catalogue tested by the "First look".
See Abstr. 002.045.

The Tycho Input Catalogue. Cross–matching the Guide Star Catalog with the Hipparcos INCA Data Base.
See Abstr. 002.046.

The 1984 – 1987 Solar Maximum Mission event list.
See Abstr. 002.088.

The 1988 Solar Maximum Mission event list.
See Abstr. 002.089.

The 1989 Solar Maximum Mission event list.
See Abstr. 002.090.

Space sailing.
See Abstr. 003.025.

Far encounter. The Neptune system.
See Abstr. 003.031.

Rendezvous in space. The science of comets.
See Abstr. 003.034.

The Galileo mission.
See Abstr. 003.061.

Summary: miniworkshops on space–based astrophysics.
See Abstr. 011.017.

The Compton Observatory Science Workshop. Proceedings.
See Abstr. 012.018.

Basic space science. Proceedings.
See Abstr. 012.064.

Cluster dayside polar cusp. Planning and coordination of measurements from Cluster, ground stations, balloons and rockets in the dayside polar–cusp region. Proceedings.
See Abstr. 012.071.

Coordination of SOHO and ground–based observations. Proceedings.
See Abstr. 012.080.

Proceedings of the 10th ESA Symposium on European Rocket and Balloon Programmes and Related Research.
See Abstr. 012.088.

Radars and lidars in Earth and planetary sciences. Proceedings.
See Abstr. 012.093.

Imaging X–ray astronomy. A decade of Einstein Observatory achievements.
See Abstr. 012.097.

ESA astronomical projects.
See Abstr. 013.004.

Prospects of the development of ground–based and space astronomy in the USSR.
See Abstr. 013.005.

The NASA airborne astronomy program: a perspective on its contributions to science, technology, and education.
See Abstr. 013.008.

The current status and future focus of basic space science in India.
See Abstr. 013.093.

Space research: developing university level curricula and research projects.
See Abstr. 013.096.

A look inside the bubble.
See Abstr. 013.104.

Man behind the mission.
See Abstr. 013.106.

The need of coordinated observations in space and from ground.
See Abstr. 013.114.

Space astronomy in India.
See Abstr. 013.140.

Mössbauer spectroscopy on the surface of Mars. Why?
See Abstr. 022.008.

The magnetic field investigation on the Ulysses mission: instrumentation and preliminary scientific results.
See Abstr. 035.030.

The Ulysses solar wind plasma experiment.
See Abstr. 035.031.

The unified radio and plasma wave investigation.
See Abstr. 035.033.

The interstellar neutral–gas experiment on ULYSSES.
See Abstr. 035.035.

The Ulysses Cosmic Ray and Solar Particle Investigation.
See Abstr. 035.037.

Geometrical calibration and assessment of the stability of the Hipparcos payload.
See Abstr. 035.055.

Geometrical stability and evolution of the Hipparcos telescope.
See Abstr. 035.056.

Science applications of the Mars Observer gamma ray spectrometer.
See Abstr. 035.057.

Mars Observer camera.
See Abstr. 035.058.

Thermal emission spectrometer experiment: Mars Observer mission.
See Abstr. 035.059.

Atmosphere and climate studies of Mars using the Mars Observer pressure modulator infrared radiometer.
See Abstr. 035.060.

The Mars Observer laser altimeter investigation.
See Abstr. 035.061.

Mars Observer magnetic fields investigation.
See Abstr. 035.062.

Mars 96 subsurface radar.
See Abstr. 035.129.

Definition of a L–band SAR for a Mars Rover mission.
See Abstr. 035.130.

The Italian involvement in Cassini Radar.
See Abstr. 035.131.

Considerations of a solar mass ejection imager in a low–earth orbit.
See Abstr. 035.134.

The FAST Hipparcos Data Reducation Consortium: overview of the reduction software.
See Abstr. 036.089.

The NDAC Hipparcos data analysis consortium. Overview of the reduction methods.
See Abstr. 036.090.

Attitude determination in the Hipparcos revised mission.
See Abstr. 036.091.

Method of comparison between determinations of the Hipparcos attitude.
See Abstr. 036.092.

Modelling the torques affecting the Hipparcos satellite.
See Abstr. 036.093.

Hipparcos great–circle reduction. Theory, results and intercomparisons.
See Abstr. 036.094.

Comparison of Hipparcos results obtained on different dates on the same great circle.
See Abstr. 036.095.

Early improvements to the Hipparcos Input Catalogue through the accumulation of data from the satellite. Including the NDAC attitude reconstruction description.
See Abstr. 036.096.

Hipparcos photometry: FAST main mission reduction.
See Abstr. 036.097.

Hipparcos photometry: NDAC reductions.
See Abstr. 036.098.

Detection and measurement of double stars with the Hipparcos satellite: NDAC reductions.
See Abstr. 036.099.

Hipparcos double star recognition and processing within the FAST consortium.
See Abstr. 036.100.

The treatment of Hipparcos observations of some peculiar double stars: anomalous cases.
See Abstr. 036.101.

Tycho data analysis. Overview of the adopted reduction software and first results.
See Abstr. 036.102.

Tycho transit detection.
See Abstr. 036.103.

Tycho star recognition.
See Abstr. 036.104.

Tycho astrometry calibration.
See Abstr. 036.105.

Tycho background determination and monitoring.
See Abstr. 036.106.

Tycho photometry calibration and first results.
See Abstr. 036.107.

Comparison of the first results from the Hipparcos star mappers with the Hipparcos Input Catalogue.
See Abstr. 041.010.

Positions and parallaxes from the Hipparcos satellite. A first attempt at a global astrometric solution.
See Abstr. 041.011.

Experimental gravitation and the Columbus program.
See Abstr. 066.135.

Spacecraft searches for gravitational waves from massive coalescing binaries: detection and false–alarm probabilities.
See Abstr. 066.136.

Synthetic maps of the brightness and polarization of the F–corona.
See Abstr. 074.060.

The Earth's gravity field from satellite geodesy – a 30 year adventure.
See Abstr. 081.042.

NASA's never–ending mission.
See Abstr. 091.023.

Differential VLBI measurements of the Venus atmosphere dynamics by balloons: VEGA project.
See Abstr. 093.001.

The first results of the Magellan mission.
See Abstr. 093.016.

Venus geology, geochemistry, and geophysics. Introduction.
See Abstr. 093.024.

Magellan's global view of the Venusian surface.
See Abstr. 093.100.

Solar system objects observed by Hipparcos.
See Abstr. 098.065.

Near–Earth asteroids and the history of planetary formation.
See Abstr. 098.176.

Mission to Jupiter.
See Abstr. 099.011.

Titan and exobiological aspects of the Cassini–Huygens mission.
See Abstr. 100.005.

A magnetosphere of Neptune.
See Abstr. 101.021.

Giotto extended mission.
See Abstr. 103.017.

Interplanetary dust observed by Galileo and Ulysses.
See Abstr. 106.045.

Computation of ephemerides for Long–Period Variable stars for the Hipparcos mission.
See Abstr. 122.058.

Gamma–ray monitoring of AGN and galactic black hole candidates by the Gamma–Ray Observatory.
See Abstr. 143.088.

COBE DMR results and implications.
See Abstr. 161.173.

Mapping the sky with the COBE Differential Microwave Radiometers.
See Abstr. 161.176.

COBE detects ripples of Big Bang?
See Abstr. 161.443.

052 Astrodynamics, Navigation of Space Vehicles

052.001 Global Positioning System radiation force model for geodetic applications.
H. F. Fliegel, T. E. Gallini, E. R. Swift.
J. Geophys. Res., Vol. 97, No. B1, p. 559 – 568 (10 Jan 1992).
   To generate the highly precise ephemerides of Global Positioning System (GPS) satellites necessary for modern geodetic applications, one must have an accurate force model that includes the pressure of solar radiation and spacecraft thermal emission. The authors present the dimension and optical parameters of Block I and Block II GPS satellites, show how they are used to form the models of the solar force, and compare predictions of these models with values estimated from tracking data. Simple approximating functions are given for the solar/thermal radiation pressure, and the problem of estimating a smaller, unmodeled force called Y bias is discussed. A simple model is given for the effect of earthshine on GPS spacecraft.

052.002 TOPEX orbit determination and gravity recovery using Global Positioning System data from repeat orbits.
Wu Jiuntsong, T. P. Yunck.
J. Geophys. Res., Vol. 97, No. B2, p. 1973 – 1980 (10 Feb 1992).
   A covariance analysis is presented for satellite tracking and gravity recovery with a differential Global Positioning System-based technique to be demonstrated on TOPEX in the early 1990s. The technique employs data from an ensemble of repeat ground tracks to recover a unique satellite epoch state for each track and a set of invariant positional parameters common to all tracks. The positional parameters represent the effect of mismodeled gravitational field on the satellite orbit. At an altitude of 1336 km, where gravity modeling is the dominant systematic error, averaging of random error over many arcs and adjustment of the gravity model reduce the final satellite position error. The positional parameters can then be used to produce a refined global gravity model. The analysis indicates that errors ranging from 5 to 8 cm in TOPEX altitude and 0.05 to 0.2 mGal for the gravity field can be achieved, depending on the number of repeat arcs used.

052.003 Navigation from relative measurements during approach to Mars.
N. M. Ivanov, P. R. Ivankov.
Cosmic Res., Vol. 29, No. 2, p. 212 – 218 (Sep 1991). English translation of Kosm. Issled., Tom 29, Vyp. 2, p. 247 – 254 (1991).
   A method is suggested for a solution of the navigation problem during approach to Mars, which can be used for aerodynamical deceleration at the entry into the planet's atmosphere. The results of a numerical simulation for estimating the efficiency of the method under the action of perturbations are considered.

052.004 An investigation of trajectories into a halo orbit near the L_2 libration point in the Earth–Sun system by using the Moon's gravity.
M. L. Lidov, V. A. Lyakhova, N. M. Teslenko.
Sov. Astron. Lett., Vol. 17, No. 6, p. 465 – 470 (Nov 1991). Current Physics Microform No.: 9207X1903. English translation of Pis'ma Astron. Zh., Tom 17, No. 12, p. 1124 – 1134 (Dec 1991).
   The plan for flight into a halo orbit of the Earth–Sun system suggested by Farquhar (1991) is investigated in connection with the Relikt 2 mission. According to this plan, the spacecraft first executes two revolutions along an orbit with a large eccentricity. The orbital parameters are chosen such that on the first half of the third revolution a close approach to the Moon occurs, as a result of which the spacecraft goes over into a trajectory that is asymptotic to some halo orbit near the L_2 libration point. A transfer flight by using the Moon's gravity enables one to achieve a halo orbit with a significantly smaller size than for a direct flight. An economical algorithm was devised to approximately determine the desired trajectories. A preliminary analysis of such trajectories in 1994 has been conducted.

052.005 Analysis of orbital perturbations acting on objects in orbits near geosynchronous earth orbit.
L. J. Friesen, A. A. Jackson IV, H. A. Zook, D. J. Kessler.
J. Geophys. Res., Vol. 97, No. E3, p. 3845 – 3863 (25 Mar 1992).
   Orbital evolution has been numerically simulated for objects started in geosynchronous Earth orbit (GEO) or in orbits near GEO, during a project to study potential orbital debris problems in this region. Perturbations simulated include nonspherical terms in the Earth's geopotential field, lunar and solar gravity, and solar radiation pressure. Objects simulated include large satellites, for which solar radiation pressure is insignificant, and small particles (a few microns in diameter), for which solar radiation pressure is an important force.

052.006 Relativistic geocentric satellite equations of motion in closed form.
V. A. Brumberg.
Astron. Astrophys., Vol. 257, No. 2, p. 777 – 782 (Apr 1992).
   In extending the results of Brumberg and Kopejkin (1989) the relativistic (post–Newtonian) geocentric satellite equations of motion are given in closed form avoiding expansion in geocentric coordinates of a satellite. These equations may be applied for the description of motion of a distant Earth's satellite in DGRS or KGRS (dynamically or kinematically non–rotating geocentric reference system, respectively). As a by–product of the transformation between BRS and GRS (barycentric and geocentric reference system, respectively) one obtains the relationship between BRS and GRS angular velocity of rotation of the Earth.

052.007 Optical properties of the Earth's surface and long–term perturbations of LAGEOS's semimajor axis.
D. Lucchesi, P. Farinella.
J. Geophys. Res., Vol. 97, No. B5, p. 7121 – 7128 (10 May 1992).
   The authors have reproduced the numerical model of Rubincam et al. (1987) for estimating the maximum along–track orbit–averaged perturbative acceleration $\langle T \rangle$ on LAGEOS's orbit due to radiation pressure from sunlight anisotropically reflected by the oceans. For the two optical models discussed by Rubincam et al. the authors have obtained $\langle T \rangle = 1.17$ and $2.79 \times 10^{-12}\,\mathrm{m/s^2}$. The latter value is about 3 times larger than the corresponding result of Rubincam et al., who made an error, and shows that a reasonable reflection model can give acceleration peaks of the same order as those observed in the residuals of LAGEOS's orbit.

052.008 On the influence of the moon's gravitational field on the motion of the artificial satellites.
A. N. Marchenko.
Manuscr. Geod., Vol. 16, No. 6, p. 360 – 366 (1991).
   The problem of the computation of the influence of non-centricity of the Lunar gravitational field on the motion of geodynamical satellites is considered. This problem is solved on the basis of introduction of an instantaneous center of planet's attraction. The analytical expression is constructed for the disturbing function of the non–central Moon's gravitational field. The numerical algorithm is developed for computation of the considered effect. This algorithm is tested by means of prediction of the "Lageous" and "Etalon-1" orbits. Quantiative estimations of perturbations from the Moon's non–centricity are computed for different orbital arcs of satellites mentioned.

052.009 On phase constraints in the problem of estimation with unmodeled disturbances.
A. I. Matasov.
Kosm. Issled., Tom 30, Vyp. 1, p. 3 – 9 (Jan–Feb 1992). In Russian. English translation in Cosmic Res., Vol. 30, No. 1.
   Many problems of determination of the orbit of a spacecraft are reduced to the so–called estimation problem with unmodeled disturbances. It is necessary in some cases to allow for additional constraints for the derivatives of this disturbances, which generate the phase constraints in the equivalent problem of

estimation. It has been shown in the note that in the problem under consideration the presence of the phase constraints is equivalent to the additional measurements.

052.010 **Estimation of satellite orbit parameters on the basis of measurements of its angular position on board an orbital spacecraft.**
A. I. Daugavet, E. V. Postnikov.
Kosm. Issled., Tom 30, Vyp. 1, p. 45 – 51 (Jan–Feb 1992). In Russian. English translation in Cosmic Res., Vol. 30, No. 1.
A method of satellite orbit calculation on the basis of its angular position measurements on board a spacecraft with known orbit and orientation parameters is described. The method may be used for the case of short time observations. Formulae for calculation of rsm errors of estimated parameters are derived.

052.011 **Correction of an inertial navigation system by relative measurements of artificial Earth satellites.**
N. M. Ivanov, P. R. Ivankov.
Kosm. Issled., Tom 30, Vyp. 1, p. 60 – 66 (Jan–Feb 1992). In Russian. English translation in Cosmic Res., Vol. 30, No. 1.
A method of correction of an inertial navigation system during the transition from the Earth satellite orbit to reentry trajectory is given. As navigation information are considered non–gravity acceleration and measurements of the components of the spacecraft state vector made by accelerometers and by means of the satellite navigation system. The method permits to determine orientation of the axes of sensitivity of the elements of the inertial navigation system. The dependence of the accuracy of the method on statistical characteristics of perturbibng factors is studied.

052.012 **Modeling radiation forces acting on TOPEX/Poseidon for precision orbit determination.**
J. A. Marshall, S. B. Luthcke, P. G. Antreasian, G. W. Rosborough.
National Aeronautics and Space Administration, Washington, DC (USA); Goddard Space Flight Center, Greenbelt, MD (USA), NASA–TM—104564. 75 p. (Jun 1992).
Geodetic satellites such as TOPEX/Poseidon require accurate orbital computations to support the scientific data they collect. For this satellite, a "box–wing" satellite form has been investigated that models the satellite as a combination of flat plates arranged in a box shape with a connected solar array. The nonconservative forces acting on each of the eight surfaces are computed independently, yielding vector accelerations which are summed to compute the total aggregate effect on the satellite center–of–mass. In order to test the validity of this concept, "micro–models" based on finite element analysis of TOPEX/Poseidon have been used to generate acceleration histories in a wide variety of orbit orientations. These profiles are then compared to the box–wing model. The results of these simulations and their implication on the ability to precisely model the TOPEX/Poseidon orbit are discussed.

052.013 **Motion of artificial satellites in the set of Eulerian redundant parameters. III.**
M. A. Sharaf, M. E. Awad, S. A. A. Najmuldeen.
Earth, Moon, Planets, Vol. 56, No. 2, p. 141 – 164 (Feb 1992).
In this paper, the classical and generalized Sundman time transformations are used to establish new generating set of differential equations of motion in terms of the Eulerian redundant parameters. The implementation of this set on digital computers for the commonly used independent variables is developed once and for all. Motion prediction algorithms based on these equations are developed in a recursive manner of the motions in the Earth's gravitational field with axial symmetry whatever the number of the zonal harmonic terms may be. Applications for the two types of short and long term predictions are considered for the perturbed motion in the Earth's gravitational field with axial symmetry with zonal harmonic terms up to J_{36}. Numerical results proved the very high efficiency and flexibility of the developed equations.

052.014 **Relativistic effects in the critical inclination problem in artificial satellite theory.**
A. H. Jupp, V. A. Brumberg.
Celest. Mech. Dyn. Astron., Vol. 52, No. 4, p. 345 – 353 (1991).
It is well known that in artifical satellite theory special techniques must be employed to construct a formal solution whenever the orbital inclination is sufficiently close to the critical value $\cos^{-1}(1\sqrt{5})$. In this article the authors investigate the consequences of introducing certain relativistic effects into the motion of a satellite about an oblate primary. Particular attention is paid to the critical inclination(s), and for such critical motions an appropriate method of solution is formulated.

052.015 **Effects of small external forces on the planar oscillation of a cable connected satellites system.**
Manaziruddin, R. B. Singh.
Celest. Mech. Dyn. Astron., Vol. 53, No. 3, p. 219 – 226 (1992).
The effects of small external dissipative and disturbing forces on the non–linear planar oscillation of a cable connected satellites system in the central gravitational field of earth have been studied. Typical non–linear oscillation's phenomena arizing from the aforesaid external forces are shown to take place. The presence of these forces enables the application of asymptotic methods of the theory of non–linear oscillations due to Bogoliubov and Mitropolsky to the equation characterizing the non–linear oscillation of the system.

052.016 **On the instability of "folded" equilibria of a flexible nonstretchable thread attached to the satellite in a circular orbit.**
S. D. Furta.
Celest. Mech. Dyn. Astron., Vol. 53, No. 3, p. 255 – 266 (1992).
The author considers the problem of Lyapunov's stability of relative equilibria of a flexible nonstretchable thread attached to the satellite moving in a circular Keplerian orbit in the first approximation. When it is in the position of relative equilibrium, the thread is known to be situated either along the radius vector of the orbit (the "radia" equilibrium) or along the circular orbit (the "tangential" equilibrium) and in each case the thread can be in a "folded" state. The author shows that "folded radial" equilibria of the thread are always unstable while "tangential" ones are unstable if the thread is sufficiently short in comparison with the radius of the orbit. The generalized Chetaev functional has been constructed to prove the instability.

052.017 **Analytical solutions to the four post–Newtonian effects in a near–Earth satellite orbit.**
Huang Cheng, Liu Lin.
Celest. Mech. Dyn. Astron., Vol. 53, No. 3, p. 293 – 307 (1992).
On the basis of the results by Huang et al. (1990), this paper further discusses and analyses the four post–Newtonian effects in a near–Earth satellite orbit: the Schwarzschild solution, the post–Newtonian effects of the geodesic precession, the Lense–Thirring precession and the oblateness of the Earth. A full analytical solution to the effects including their direct perturbations and mixed perturbations due to the Newtonian oblateness (J_2) perturbation and the Schwarzschild solution is obtained using the quasi–mean orbital element method analogous to the Kozai's mean orbital element method. Some perturbation properties of the post–Newtonian effects are revealed. The results obtained not only can provide a sound scientific basis for the precise determination of a man–made satellite orbit but they are also suitable for similar mechanics systems, such as the motions of planets, asteroids and natural satellites.

052.018 **Re–entry aerodynamics derived from space debris trajectory analysis.**
R. Crowther.
Planet. Space Sci., Vol. 40, No. 5, p. 641 – 646 (May 1992).
This paper considers the technique of orbital analysis as a means of determining the ill–defined gas–surface interaction between spacecraft and atmospheric molecules in low Earth orbit. The interaction is a major uncertainty in trajectory predictions for a body moving within an atmosphere. The rate of

change of the orbital period of a debris object, the uncontrolled Salyut 7/Kosmos 1686 space station, is analysed in order to determine the free molecular drag coefficient. The results are compared with theoretical values for the drag coefficient calculated using a complex representation of the vehicle configuration and motion and applying the Monte Carlo Test Particle method. Results suggest a nature of re–emission very close to the classical diffuse, totally accommodated case was occurring at the surface of the debris object as it approached re–entry. However the determined drag coefficient and therefore the derived interaction are found to be very sensitive to the neutral density and therefore the atmospheric model used in the analysis.

052.019   **Determining the actual motion of the "Salyut–7" – "Kosmos–1686" orbital complex with respect to the center of mass in high orbit.**
V. A. Sarychev, V. V. Sazonov, M. Yu. Belyaev, N. I. Efimov, I. L. Lapshina, V. M. Stazhkov.
Kosm. Issled., Tom 30, Vyp. 2, p. 147 – 156 (Mar–Apr 1992). In Russian. English translation in Cosmic Research, Vol. 30, No. 2.

Dynamic effects due to the gravitational one–axis stabilization of the "Salyut–7" – "Kosmos–1686" orbital complex are described. The effects have been revealed in statistical data processing of onboard measurements of solar and magnetic sensors.

052.020   **Trajectory optimization for space flights from Earth to Mars by solar sail.**
B. Ya. Sapunkov, V. A. Egorov, V. V. Sazonov.
Kosm. Issled., Tom 30, Vyp. 2, p. 194 – 202 (Mar–Apr 1992). In Russian. English translation in Cosmic Research, Vol. 30, No. 2.

052.021   **Chromatic observations of stellar scintillations for autonomous satellite navigation.**
A. S. Gurvich, S. V. Sokolovskij.
Kosm. Issled., Tom 30, Vyp. 2, p. 226 – 230 (Mar–Apr 1992). In Russian. English translation in Cosmic Research, Vol. 30, No. 2.

A method for determination of the planetary refraction angles by observations of the stars' rises and sets from space is considered. Instead of direct angular measurements it is proposed to measure the temporal lag of the stars' scintillations at two wavelengths.

052.022   **Variation in eccentricity for the orbit of Cosmos 373, 1970–87A.**
C. J. Brookes.
Planet. Space Sci., Vol. 40, No. 6, p. 847 – 857 (Jun 1992).

The orbital parameters of the satellite Cosmos 373, 1970–87A, have been determined at 105 epochs during a 9 year period prior to its decay in March 1980. With an initial perigee height of 472 km and an orbital plane inclined at 62.9° to the equator, the satellite was regarded as emminently suitable for the study of the gravitational field of the Earth, since the amplitude of the oscillation in eccentricity becomes very large for inclinations close to the critical angle of 63.4°. The analysis has yielded extremely accurate values of eccentricity, with standard deviations (S.D.s) down to 0.000004. In particular, 80 values of eccentricity were examined, covering a complete cycle of the argument of perigee, and corrections incorporated for the effects due to lunisolar perturbations, solid–Earth tides, solar radiation pressure and air drag. The analysis of the modified data has indicated an amplitude of oscillation of 0.00836 ± 0.00001, equivalent to almost 60 km in perigee height – the largest yet recorded for any near–Earth orbit of high accuracy.

052.023   **On the initially circular motion of an orbiter in the oblate, rotating, Martian atmosphere.**
V. Mioc, E. Radu, C. Blaga.
Rev. Mex. Astron. Astrofis., Vol. 24, No. 1, p. 15 – 19 (Apr 1992).

Using the density distribution law proposed by Sehnal and Pospišilová (1988), the initially circular motion of an orbiter in Mars' atmosphere (considered as oblate and rotating) is studied. The difference between the nodal period and the corresponding Keplerian one, and the changes of five independent orbital elements, over a nodal period, are analytically estimated.

052.024   **Motion of a satellite with flexible viscoelastic booms in a noncentral gravitational field.**
A. V. Shatina.
Cosmic Res., Vol. 29, No. 6, p. 699 – 705 (May 1992). English translation of Kosm. Issled., Tom 29, Vyp. 6, p. 815 – 821 (1991).

In the present study approximation equations that describe the motion of an artificial satellite in the form of a planar disc with rigid and visoelastic booms travelling in the gravitational field of a nonsymmetrical planet (in its equatorial plane) are obtained in terms of Delaunay's canoncical variables by means of the averaging method.

052.025   **Motions of a spacecraft asymptotic to its regular precessions.**
B. S. Bardin.
Cosmic Res., Vol. 29, No. 6, p. 705 – 710 (May 1992). English translation of Kosm. Issled., Tom 29, Vyp. 6, p. 822 – 827 (1991).

Previously described methods are used to investigate solutions, asymptotic to the equilibrium position, of an autonomous Hamiltonian system with two degrees of freedom in second–order resonance. Necessary and sufficient conditions are determined for the existence of asymptotic solutions, and an analytical representation is obtained for those solutions. The results are applied to the problem of motions, asymptotic to regular precessions, of a satellite in a circular orbit.

052.026   **Spatial rotations of a satellite in the circular three–body problem with fractional resonances.**
P. S. Krasil'nikov.
Cosmic Res., Vol. 29, No. 6, p. 734 – 745 (May 1992). English translation of Kosm. Issled., Tom 29, Vyp. 6, p. 858 – 871 (1991).

Spatial rotations of a satellite about its center of mass are considered within the circular three–body problem for the fractional resonances $\Omega = \omega_0/2$, $\Omega = 2\omega_0$, where Ω is the angular velocity of undisturbed satellite rotation, ω_0 is the mean motion of the finite masses. It is assumed that the trajectory of motion of the center of mass of the solid body is described by arbitrary periodic functions of time, while its central ellipsoid of inertia is close to a sphere. It will be shown that the averaged equations of motion of an asymmetric satellite admit a family of integral manifolds upon which the solution of the problem is reducible to quadratures. Satellite rotations on these manifolds are described. Motions of an axisymmetric body are studied in detail, and a geometric interpretation of resonant satellite rotations is given.

052.027   **Generation of trajectories and choice of routes for a passive flyby of a group of celestial bodies moving in Keplerian orbits.**
M. Yu. Akhlebininskij, M. S. Konstantinov.
Cosmic Res., Vol. 29, No. 6, p. 759 – 773 (May 1992). English translation of Kosm. Issled., Tom 29, Vyp. 6, p. 899 – 904 (1991).

The problem of generating trajectories for a flyby of several given celestial bodies moving in Keplerian orbits is solved in a new formulation. It is demonstrated that there exist trajectories along which one vehicle can fly by at least three bodies. A theorem asserting the existence of a trajectory for the flyby of three bodies moving in circular orbits of common radius is formulated and probed. The uniqueness of such trajectories is analyzed. An effective numerical method is proposed for generating trajectories for the flyby of three bodies moving in arbitrary elliptical orbits. The possibilities of the method are illustrated with examples. Questions regarding the insertion of the vehicle into the flyby trajectory are not discussed here.

052.028   **On the regularization of equatorial orbits.**
V. A. Kuz'minykh.
Cosmic Res., Vol. 29, No. 6, p. 811 – 814 (May 1992). English translation of Kosm. Issled., Tom 29, Vyp. 6, p. 944 – 947 (1991).

052.029 Optimal launching of a spacecraft from the lunar surface to the fixed point of its artificial satellite circular orbit.
K. G. Grigor'ev, E. V. Zapletina, M. P. Zapletin.
Kosm. Issled., Tom 30, Vyp. 3, p. 321 – 332 (May–Jun 1992). In Russian. English translation in Cosmic Res., Vol. 30, No. 3.

052.030 Flights to asteroids from an earth satellite orbit with perturbation–aerodynamic–impulsive maneuver near Mars.
L. B. Livanov.
Kosm. Issled., Tom 30, Vyp. 3, p. 333 – 342 (May–Jun 1992). In Russian. English translation in Cosmic Res., Vol. 30, No. 3.
The author considers flights to asteroids from an earth satellit orbit with perturbation–aerodynamic–impulsive maneuver near Mars. The sum of the three impulses – at Earth, Mars, and asteriod – is as a rule by 2 to 3 km/sec less than the sum of the impulses required for an optimal trajectory of direct Earth-asteroid flights.

052.031 Optimal launching of a spacecraft from the surface of the Moon to the circular orbit of its satellite.
K. G. Grigor'ev, M. P. Zapletin, D. A. Silaev.
Cosmic Res., Vol. 29, No. 5, p. 595 – 603 (Mar 1992). English translation of Kosm. Issled., Tom 29, Vyp. 5, p. 695 – 704 (Sep–Oct 1991).

052.032 Moon's influence on the transfer from the Earth to a halo orbit around L_1.
G. Gómez, À. Jorba, J. Masdemont, C. Simó.
NATO Advanced Study Institute on Predictability, Stability, and Chaos in N–Body Dynamical Systems, p. 283 – 290 (1991). – See Abstr. 012.089 for the main entry.
The influence of the Moon in the transfer of a satellite from the Earth to a halo orbit around L_1 in the Earth + Moon – Sun system, is analysed by means of a simple bicircular model for the motion of the Earth and the Moon. The results suggest that using the stable manifold of the halo orbit, slightly bent by the Moon, it is possible to carry out the transfer avoiding the insertion manoeuvre in the halo orbit.

052.033 First order theory of perturbed circular motion: an application to artificial satellites.
E. Bois, I. Wytrzyszczak.
NATO Advanced Study Institute on Predictability, Stability, and Chaos in N–Body Dynamical Systems, p. 291 – 295 (1991). – See Abstr. 012.089 for the main entry.
This paper describes briefly the characteristics of an analytical theory of perturbed circular motion. The main advantage of the solution, expanded in Fourier series and in nonsingular variables, is the presence of iterative formation laws for its coefficients. An application to the case of a geosynchronous satellite and the comparison of the results with a numerical integration show the degree of accuracy of the first–order solution.

052.034 Poincaré–similar variables including J_2–secular effects.
L. Floría, J. M. Ferrándiz.
NATO Advanced Study Institute on Predictability, Stability, and Chaos in N–Body Dynamical Systems, p. 297 – 303 (1991). – See Abstr. 012.089 for the main entry.
In earlier work the authors have defined a set of eight generalized canonical Delaunay – similar (GDS) variables incorporating the first–order secular effects present in the main problem in the theory of artificial Earth satellites. The new GDS set was derived by means of a canonical transformation whose generating function is inspired by Deprit's radial intermediary. When applied to Deprit's intermediary, the proposed variables lead to a simple solution, the momenta being constant and the co-ordinates being either a constant or a linear function of the independent variable. As a further step, a set of generalized Poincaré – similar (PS) canonical variables is constructed here; the new GPS set also exhibits the feature of containing the whole first–order secular contribution of the J_2 zonal harmonic of the Earth's potential and is free from singularities.

052.035 Measuring the lack of integrability of the J_2 problem for Earth's satellites.
C. Simó.
NATO Advanced Study Institute on Predictability, Stability, and Chaos in N–Body Dynamical Systems, p. 305 – 309 (1991). – See Abstr. 012.089 for the main entry.
The author considers the motion of a satellite around an oblate primary, keeping only the J_2 term in the expansion of the potential in spherical harmonics. The problem has cylindrical symmetry. It has been proved recently by Irigoyen (1990) that the problem is not integrable. However, even if the system is non integrable, the size of the stochastic zones can be so small that they can be neglected for all practical purposes. This is demonstrated here, and it is shown that for the case of the Earth, and considering possible real orbits, i.e., non colliding with the Earth, the effect of the non–integrability can be completely neglected.

052.036 The effects of the J_3–harmonic (pear shape) on the orbits of a satellite.
R. A. Broucke.
NATO Advanced Study Institute on Predictability, Stability, and Chaos in N–Body Dynamical Systems, p. 311 – 335 (1991). – See Abstr. 012.089 for the main entry.
The object of the present article is a detailed numerical investigation of the perturbation on the orbit of a satellite, caused by the pear–shape or J_3–harmonic of the central body. The author uses in this study concepts from the general theory of periodic orbits, such as Poincaré surfaces of section, stability theory, characteristic exponents and bifurcations. He finds several new families of periodic orbits in the J_3–problem. The orbital characteristics of five principal families are decribed in detail. The orbits of these families are periodic only in the rotating meridian plane which contains the satellite. They are not symmetric with respect to the equator.

052.037 Long–time predictions of satellite orbits by numerical integration.
J. M. Ferrándiz, M. E. Sansaturio, J. Vigo.
NATO Advanced Study Institute on Predictability, Stability, and Chaos in N–Body Dynamical Systems, p. 387 – 394 (1991). – See Abstr. 012.089 for the main entry.
The authors study limits of predictability for analytical and/or numerical calculations of the orbital behaviour of artificial satellites. These limits depend, of course, on the accuracy required, on the specific dynamical models formulated, on the sets of variables chosen to describe them, on the numerical or analytical techniques used and, especially, on the specific trajectories to be established. In order to check the reliability of the predictions, first integrals, constraints among redundant variables, and backward integrations from the end points to the initial conditions have been used.

ISPP–11. Physics of charged bodies in space plasmas. Proceedings.
See Abstr. 012.049.

Spacecraft, orbits and slingshots.
See Abstr. 014.010.

Minimax property of a discrete Kalman filter for imprecisely specified variance matrices.
See Abstr. 021.003.

Numerical integration methods for orbital motion.
See Abstr. 021.017.

Increased accuracy of computations in the main satellite problem through linearization methods.
See Abstr. 021.036.

Chaos in coorbital motion.
See Abstr. 042.110.

The potential use of fiducial ground networks.
See Abstr. 045.006.

Galileo trajectory design.
See Abstr. 051.032.

Geopotential models of the Earth from satellite tracking, altimeter and surface gravity observations: GEM–T3 and GEM–T3S.
See Abstr. 081.043.

The upper atmosphere as sensed by satellite orbits.
See Abstr. 082.023.

053 Artificial Satellites, Space Probes

053.001 **Choice of shape of bodies with minimum aerodynamic heating during motion in the atmospheres of planets in the solar system.**
M. A. Korchagina, N. N. Pilyugin.
Cosmic Res., Vol. 29, No. 2, p. 257 – 266 (Sep 1991). English translation of Kosm. Issled., Tom 29, Vyp. 2, p. 298 – 309 (1991).

Expressions are derived for convective and radiative heat fluxes and the wavedrag and friction–drag coefficients for axisymmetric and flat bodies moving in the atmosphere of planets in the solar system. Calculations of the total heating and drag of blunted cones moving along a trajectory in the atmospheres of the Earth, Venus, Mars, and Jupiter were performed. Possible formulations of variational problems on lowering the aerodynamic heating of probes were established on the basis of analysis of these calculations.

053.002 **Satellite Digest – 243.**
Spaceflight, Vol. 34, No. 3, p. 80 – 81 (Mar 1992).
1992 March listing of satellite and spacecraft launches.

053.003 **Satellite Digest – 244.**
Spaceflight, Vol. 34, No. 5, p. 171 (May 1992).
1992 May listing of satellite and spacecraft launches.

053.004 **Ginga – ein äußerst erfolgreicher japanischer Röntgen-satellit.**
J. Greiner.
Ahnerts Kalender für Sternfreunde 1993.
p. 169 – 174 (1992). With plate 48. – See Abstr. 046.012 for the main entry.

053.005 **Optimal soft landing of a spacecraft on the lunar surface from the lunar satellite circular orbit.**
K. G. Grigor'ev, E. V. Zapletina, M. P. Zapletin.
Kosm. Issled., Tom 30, Vyp. 2, p. 203 – 211 (Mar–Apr 1992). In Russian. English translation in Cosmic Research, Vol. 30, No. 2.

053.006 **Observations of space debris at Goldstone.**
R. M. Goldstein, L. W. Randolph.
International Symposium on Radars and Lidars in Earth and Planetary Sciences, p. 15 – 18 (Dec 1991). – See Abstr. 012.093 for the main entry.

The authors have used the planetary radar at the Jet Propulsion Laboratory's Goldstone Tracking Station to monitor small particles of orbital debris. This radar can detect metallic objects as small as 1.8 mm in diameter at an altitude of 600 km. The results of a first set of observations show a flux (at 600 km) of 6.4 objects per sq km per day, of equivalent size of 1.8 mm or larger. Forty percent of the observed particles appear to be concentrated into one or two orbits. An orbital ring with the same inclination as the radar (35.1°) is suggested. However, an orbital band with the much higher inclination of 66° is also a possibility.

Satellites, space debris, aircraft and astronomy.
See Abstr. 013.033.

Low–cost, high–resolution, single–structure array telescopes for imaging of low–Earth–orbit satellites.
See Abstr. 032.001.

Extreme Ultraviolet Explorer: a view inside the bubble.
See Abstr. 051.014.

The Ulysses mission.
See Abstr. 051.018.

The Hipparcos payload's in–orbit performance.
See Abstr. 051.058.

Modélisation des radiations ionisantes de l'environnement spatial.
See Abstr. 084.045.

Mars Rover Sample Return Mission: systemic model and optimization of scientific results. A case for large valley outlets.
See Abstr. 097.025.

An efficient HZETRN (a galactic cosmic ray transport code).
See Abstr. 144.060.

Theoretical Astrophysics

061 General Aspects (Nucleosynthesis, Elementary Particles, Neutrino Astronomy, etc.)

061.001 Constraints on 17 keV neutrinos.
J. M. Cline, T. P. Walker.
Phys. Rev. Lett., Vol. 68, No. 3, p. 270 – 273 (20 Jan 1992).
Current Physics Microform No.: 9203A1967.
Recent experiments indicate that a 17 keV neutrino mixes with v_e with 1% probability. The authors discuss the constraints from neutrinoless double β decay, cosmology, and astrophysics that any model possessing a 17 keV neutrino must obey. Several recently proposed models are shown to be incompatible with these constraints.

061.002 Monopole annihilation and baryogenesis at the electroweak scale.
V. V. Dixit, M. Sher.
Phys. Rev. Lett., Vol. 68, No. 5, p. 560 – 563 (3 Feb 1992).
Current Physics Microform No.: 9203B2082.
If the standard model is extended to include two Higgs doublets and a singlet, there is a region of parameter space for which electromagnetism is broken in a narrow temperature range just below the electroweak transition temperature. In this range, the universe superconducts, and monopoles will rapidly annihilate. It is shown that as the universe passes through this temperature range, the abundance of monopoles will be depleted to acceptable levels, and, in addition, the process of annihilation could generate the observed baryon asymmetry.

061.003 Minimal electroweak model for monopole annihilation.
T. H. Farris, T. W. Kephart, T. J. Weiler, T. C. Yuan.
Phys. Rev. Lett., Vol. 68, No. 5, p. 564 – 567 (3 Feb 1992).
Current Physics Microform No.: 9203B2086.
The authors construct the minimal extension of the standard model implementing the Langacker–Pi mechanism for reducing the grand unified theory monopole cosmic density to an allowed level. The model contains just a single charged scalar field in addition to the standard Higgs doublet, and is easily embeddable in any GUT. The authors identify the region of parameter space where monopoles annihilate in the higher temperature early universe. A particularly alluring possibility is that the demise of monopoles at the electroweak scale is in fact the origin of the universe's net baryon number.

061.004 Single explanation for both baryon and dark matter densities.
D. B. Kaplan.
Phys. Rev. Lett., Vol. 68, No. 6, p. 741 – 743 (10 Feb 1992).
Current Physics Microform No.: 9203G1271.
It is shown that in a general class of models in which the baryon number of the universe is created by electroweak anomalies, the energy density in dark matter may be related to the energy density in baryons as $\Omega_b/\Omega_{dm} = c \cdot$(proton mass) / (weak scale), where the number c is of order unity and calculable from the anomaly equation. The scenario unambiguously predicts charged and neutral particles with weak–scale masses which carry a new conserved quantum number and can be pair produced via the weak interactions.

061.005 Nucleosynthesis confronts an unstable, inert 17 keV state.
K. Enqvist, K. Kainulainen, M. Thomson.
Phys. Rev. Lett., Vol. 68, No. 6, p. 744 – 747 (10 Feb 1992).
Current Physics Microform No.: 9203G1274.
The authors study the cosmological consequences of an inert 17 keV state mixing with the electron neutrino. They find that the nucleosynthesis upper bound on the primordial helium abundance prohibits the existence of such a state, unless its lifetime falls into the range 6×10^{-4}s $\lesssim \tau_{vac} \lesssim 2 \times 10^{-2}$s. In this range the decay occurs after the chemical decoupling of the electron neutrinos and before the beginning of the nucleosynthesis, with the result that the predicted helium abundance can be lower than what it would be in the standard scenario.

061.006 Energy dependence of solar neutrino – electron scattering as a test of neutral currents.
W. Kwong, S. P. Rosen.
Phys. Rev. Lett., Vol. 68, No. 6, p. 748 – 751 (10 Feb 1992).
Current Physics Microform No.: 9203G1278.
The energy dependence of $v - e$ scattering of solar neutrinos is investigated in the framework of neutrino oscillations and the nonadiabatic Mikheyev – Smirnov – Wolfenstein effect. It is shown that, with sufficient data, it will be possible to establish unambiguously whether neutrino oscillations are actually occurring and whether the electron neutrino oscillates into active or inactive (sterile) neutrino flavors.

061.007 Astrophysical implications of the direct measurement of the ^{13}N(p,γ)^{14}O cross section.
M. Arnould, G. Paulus, A. Jorissen.
Astron. Astrophys., Vol. 254, No. 1/2, p. L9 – L12 (Feb 1992).
Letter–to–the–editor.
The first direct measurement of the ^{13}N(p,γ)^{14}O cross section at energies of astrophysical interest has been successfully conducted at the Louvain–la–Neuve cyclotron facility using an intense and highly purified ^{13}N beam (Decrock et al. 1991). From this pioneering experiment, a partial γ–width $\Gamma_\gamma = 3.8 \pm 1.2$ eV has been deduced for the astrophysically dominant resonance in ^{14}O at a centre of mass energy of 0.545 MeV. The authors discuss in this paper the impact of the corresponding new reaction rate on the conditions of development of the hot CNO mode of hydrogen burning, on the nucleosynthesis in novae, and on the s–process of nucleosynthesis in a variety of stellar environments where the necessary neutrons can be produced via ^{13}C(α,n)^{16}O.

061.008 Subatomic astronomy.
B. Sadoulet, J. W. Cronin.
Sky Telesc., Vol. 83, No. 1, p. 25 – 28 (Jan 1992).
The sciences of the very large and the very small have joined forces to address fundamental questions about the origin of the universe.

061.009 A measurement of the stellar $^{14}C(n, \gamma)^{15}C$ reaction rate.
H. Beer, F. Käppeler, M. Wiescher, J. Görres,
P. E. Koehler.
Verh. Dtsch. Phys. Ges., Vol. 27, No. 1, p. 127 – 128 (1992).
Abstract. Paper presented at the Frühjahrstagung des Fachverbandes Physik der Hadronen und Kerne der Deutschen Physikalischen Gesellschaft e.V. (DPG) Gemeinsam mit der Österreichischen Physikalischen Gesellschaft (OePG), Salzburg (Austria), 24 – 28 Feb 1992.

061.010 Neutron capture in ^{148}Sm and ^{150}Sm: the neutron density of the s–process.
K. Guber, K. Wisshak, F. Voss, F. Käppeler.
Verh. Dtsch. Phys. Ges., Vol. 27, No. 1, p. 128 (1992). Abstract.
Paper presented at the Frühjahrstagung des Fachverbandes Physik der Hadronen und Kerne der Deutschen Physikalischen Gesellschaft e.V. (DPG) Gemeinsam mit der Österreichischen Physikalischen Gesellschaft (OePG), Salzburg (Austria), 24 – 28 Feb 1992.

061.011 Isotopic r–abundances and nuclear structure far from stability: implications for the r–process mechanism.
K. L. Kratz, P. Möller, B. Pfeiffer, A. Wöhr, J. P. Bitouzet,
F.-K. Thielemann.
Verh. Dtsch. Phys. Ges., Vol. 27, No. 1, p. 225 – 226 (1992).
Abstract. Paper presented at the Frühjahrstagung des Fachverbandes Physik der Hadronen und Kerne der Deutschen Physikalischen Gesellschaft e.V. (DPG) Gemeinsam mit der Österreichischen Physikalischen Gesellschaft (OePG), Salzburg (Austria), 24 – 28 Feb 1992.

061.012 The cross section of $^{19}F(p, \alpha_2)^{16}O$ at low energies and the hydrogen burning of ^{19}F.
S. Ambacher, A. Denker, H. W. Drotleff, J. W. Hammer,
H. Knee, U. Atzrott, G. Staudt, H. Herndl, H. Oberhummer,
H. Scsribany.
Verh. Dtsch. Phys. Ges., Vol. 27, No. 1, p. 226 (1992). Abstract.
Paper presented at the Frühjahrstagung des Fachverbandes Physik der Hadronen und Kerne der Deutschen Physikalischen Gesellschaft e.V. (DPG) Gemeinsam mit der Österreichischen Physikalischen Gesellschaft (OePG), Salzburg (Austria), 24 – 28 Feb 1992.

061.013 The $^{13}C(\alpha, n)^{16}O$–reaction at low energies.
M. Soiné, A. Denker, H. W. Drotleff, J. W. Hammer,
H. Knee, G. Wolf.
Verh. Dtsch. Phys. Ges., Vol. 27, No. 1, p. 226 (1992). Abstract.
Paper presented at the Frühjahrstagung des Fachverbandes Physik der Hadronen und Kerne der Deutschen Physikalischen Gesellschaft e.V. (DPG) Gemeinsam mit der Österreichischen Physikalischen Gesellschaft (OePG), Salzburg (Austria), 24 – 28 Feb 1992.

061.014 Does the 17 keV neutrino annihilate in the early universe?
R. Foot, S. F. King.
Z. Phys., C, Vol. 54, No. 2, p. 317 – 321 (May 1992).
Recent experiments indicate that the electron neutrino contains a heavy 17 keV component. If these experiments and their interpretation are correct then this will require a modification of the minimal standard model. The standard cosmological model gives significant constraints on the properties of a 17 keV neutrino. It is usually assumed that these constraints imply that the 17 keV neutrino must decay rapidly into Goldstone bosons. The authors construct a class of gauge models which describe the 17 keV neutrino but which do not involve Goldstone bosons. The 17 keV neutrino is long lived, but annihilates sufficiently in the early universe so that its present day abundance is cosmologically acceptable.

061.015 High energy neutrinos from galactic sources.
M. Treichel.
Z. Phys., C, Vol. 54, No. 3, p. 469 – 481 (Jun 1992).
Analytical expressions are derived which allow to calculate flux densites of energetic neutrinos from hypothetical galactic sources, consisting of a proton accelerator and a dilute gas beam dump. The same formalism is used to calculate atmospheric muon and μ–neutrino fluxes. From the results, rates of upward going muons, both from the atmosphere and galactic sources, are computed and detection limits for neutrino emitters in the sky are established. Finally, the background in a surface detector, caused by scattered muons and charm decays in the rock, is estimated for the case of a flat surrounding.

061.016 Is it possible to detect the gamma–ray line from neutralino–neutralino annihilation?
V. S. Berezinsky (*V. S. Berezinskij*), A. Bottino, V. de Alfaro.
Phys. Lett. B, Vol. 274, No. 1, p. 122 – 127 (2 Jan 1992).
The authors present the calculations of the gamma–ray flux produced by neutralino annihilation ($\chi + \chi \rightarrow \gamma + \gamma$) in our Galaxy. Most attention is paid to the case in which the dark matter in and outside our Galaxy consists of neutralinos. It is shown that in this case the flux is essentially determined by only two physical quantities: the critical density $\varrho_c 1.1 \times 10^{-29} g/cm^3$ and the local (solar neighbourhood) density of dark matter $\varrho_\odot \cong 0.3 \ GeV/cm^3$. A (marginally) detectable gamma–ray flux is found only for an extreme combination of assumptions: the neutralino is almost a pure photino, the lightest s fermion is a slepton with mass $\sim 50 \ GeV$, and the average density of neutralinos in the universe is $\Omega_\chi = 0.03$.

061.017 Diurnal modulation effects in cold dark matter experiments.
J. I. Collar, F. T. Avignone III.
Phys. Lett. B, Vol. 275, No. 1/2, p. 181 – 185 (23 Jan 1992).
The effects of elastic scattering of cold dark matter (CDM) candidate particles on the constituent nuclei of the earth are predicted. The geology, nuclear physics, the earth's orientation, rotation, and trajectory through the galactic halo are included, as well as the isotropic maxwellian velocity distribution of CDM in the rest–frame of the halo. Observable modulations of detection rates and energy spectra are predicted for some interesting theoretical and experimental scenarios.

061.018 Total cross sections and thermonuclear reaction rates for $^{13}C(d, n)$ and $^{14}C(d, n)$.
C. R. Brune, R. W. Kavanagh.
Phys. Rev. C, Vol. 45, No. 3, p. 1382 – 1388 (Mar 1992). Current Physics Microform No.: 9205A2136.
The $^{13}C(d, n)$ and $^{14}C(d, n)$ cross sections have been measured for $0.2 \leqslant E_{c.m.} \leqslant 2.1 \ MeV$ and $0.2 \leqslant E_{c.m.} \leqslant 1.3 \ MeV$, respectively, using a 4π neutron detector. The cross sections are used to calculate the thermonuclear reaction rates for temperatures below 10 GK. The implications of these and other new nuclear-physics results for inhomogeneous primordial nucleosynthesis are discussed.

061.019 Charged– and neutral–current solar–neutrino cross sections for heavy–water Cherenkov detectors.
S. Ying, W. C. Haxton, E. M. Henley.
Phys. Rev. C, Vol. 45, No. 4, p. 1982 – 1987 (Apr 1992). Current Physics Microform No.: 9205G1938.
Charged– and neutral–current neutrino cross sections for deuterium have been calculated for the Bonn, Paris, and Hamada–Johnson potentials in order to estimate event rates (and their uncertainties) for solar and supernova neutrino detection in the Sudbury Solar Neutrino Observatory. Tests of the wave functions are provided by calculations of the j = 1/2 hyperfine–state muon capture rate and of the total cross section for absorbing ν_es from stopped muon decay. Detailed tables of the Paris potential results are given, and comparisons are made to the work of Doi and Kubodera.

061.020   Analysis of ^{8}Li(α, n)^{11}B below the Coulomb barrier in the potential model.
T. Rauscher, K. Grün, H. Krauss, H. Oberhummer, E. Kwasniewicz.
Phys. Rev. C, Vol. 45, No. 4, p. 1996 – 2000 (Apr 1992). Current Physics Microform No.: 9205G1952.
The reaction ^{8}Li(α, n)^{11}B is of interest in inhomogeneous big bang nucleosynthesis. A distorted wave Born approximation calculation employing folding potentials is presented for energies below the Coulomb barrier. The recently observed resonance at about 540 keV center–of–mass energy can be reproduced. The astrophysical S factor is calculated for the ground–state transition as well as for the transitions to the first four excited states of ^{11}B. The reaction rate is derived and compared to literature data. The inclusion of the excited states increases the rate by a factor of 1.5 compared to the ground–state transition.

061.021   Neutron capture in ^{122,123,124}Te: critical test for s process studies.
K. Wisshak, F. Voss, F. Käppeler, G. Reffo.
Phys. Rev. C, Vol. 45, No. 5, p. 2470 – 2486 (May 1992). Current Physics Microform No.: 9207C0432.
The neutron capture cross sections of ^{122,123,124,125,126}Te were measured in the energy range from 10 to 200 keV at the Karlsruhe Van de Graaff accelerator using gold as a standard. Several sets of measurements were performed under different experimental conditions to study the systematic uncertainties in detail. Maxwellian–averaged neutron capture cross sections were calculated for thermal energies between kT = 10 and 100 keV by normalizing the cross section shape up to 600 keV neutron energy reported in literature to the present data. These stellar cross sections were used in an s process analysis. With the classical approach the abundances of the three s only isotopes ^{122,123,124}Te could be reproduced within the experimental uncertainties of ~1%. The accuracy of the present data also allowed to derive constraints for the existing stellar models with respect to the effective neutron density. Furthermore, the p process abundances for the tellurium isotopes are discussed.

061.022   Neutron cross sections of ^{122}Te, ^{123}Te, and ^{124}Te between 1 and 60 keV.
Y. Xia, T. W. Gerstenhöfer, S. Jaag, F. Käppeler, K. Wisshak.
Phys. Rev. C, Vol. 45, No. 5, p. 2487 – 2493 (May 1992). Current Physics Microform No.: 9207C0449.
The currently favored s process scenario of helium shell burning in low mass stars involves a range of thermal energies from kT = 12 to 25 keV with most of the neutron exposure taking place at low temperatures. Therefore, differential cross sections are required down to the region of resolved resonances for the reliable determination of the Maxwellian–averaged cross sections typical of the stellar plasma. This work deals with the neutron capture cross sections of the important s only isotopes ^{122}Te, ^{123}Te, and ^{124}Te, which were measured between 1 and 60 keV neutron energy. The results represent the first set of experimental data in this energy range.

061.023   Gravitational particle production by cosmic textures.
J. Pullin, E. Verdaguer.
Mod. Phys. Lett. A, Vol. 7, No. 3, p. 181 – 185 (30 Jan 1992).
Using a perturbative technique the authors compute the rate of particle production in the evolution of a cosmic texture, based on the stress energy tensor introduced by Turok and Spergel. They compare the results with those for other topological defects.

061.024   Neutrino dark matter with a galactic–range new force.
M. Kawasaki, H. Murayama, T. Yanagida.
Mod. Phys. Lett. A, Vol. 7, No. 7, p. 563 – 570 (7 Mar 1992).
The authors propose a new force of the galactic range acting among massive neutrinos to weaken the phase space constraint for the neutrino dark halo. As a consequence they find that the neutrino with mass 10 eV becomes a candidate for the galaxy dark halo to explain the observed flat rotation curves. Cosmological consequence of the new force is briefly discussed. It is,

furthermore, pointed out that the new force can be naturally incorporated into the see–saw model for the neutrino mass.

061.025   A high–resolution study of the ^{20}Ne(^{3}He, t)^{20}Na reaction and the ^{19}Ne(p, γ)^{20}Na reaction rate.
M. S. Smith, P. V. Magnus, K. I. Hahn, A. J. Howard, P. D. Parker, A. E. Champagne, Z. Q. Mao.
Nucl. Phys. A, Vol. 536, No. 2, p. 333 – 348 (13 Jan 1992).
A high–precision measurement of the ^{20}Ne(^{3}He, t)^{20}Na reaction has been made using implanted ^{20}Ne transmission targets to obtain pertinent information on the low–energy resonances in the ^{19}Ne(p, γ)^{20}Na reaction. Resonance energies (447 $\pm$ 5, 658 $\pm$ 5, 787 $\pm$ 5, and 857 $\pm$ 5 keV) and upper limits on total intrinsic widths (< 10, < 6, < 10, and < 16 keV) have been measured for four excited states above the 2.199 MeV proton threshold in ^{20}Na. The stellar ^{19}Ne(p, γ)^{20}Na reaction rate is calculated for temperatures between 1×10^8 and 1×10^9K. When combined with a recent study of the ^{15}O(α, γ)^{19}Ne reaction, a new estimate is made of the conditions required for breakout from the hot CNO cycle to the rapid proton capture process.

061.026   Level structure of ^{21}Mg and the ^{20}Na(p, γ)^{21}Mg stellar reaction rate.
S. Kubono, Y. Funatsu, N. Ikeda, M. Yasue, T. Nomura, Y. Fuchi, H. Kawashima, S. Kato, H. Miyatake, H. Orihara, T. Kajino.
Nucl. Phys. A, Vol. 537, No. 1/2, p. 153 – 166 (3 – 10 Feb 1992).
The nuclear level structure of ^{21}Mg has been studied by the ^{24}Mg(^{3}He, ^{6}He)^{21}Mg reaction at 74 MeV. Angular distributions of the three–nucleon transfer reaction (^{3}He, ^{6}He) are measured for the first time, and successfully analyzed. More than 20 states have been identified with excitation energy and spin–parity determinations, including a possible s–wave resonance just above the proton threshold. One of the s–wave resonances assumed in the previous stellar reaction rate estimates is found to be a bound state. The stellar reaction rate of the ^{20}Na(p, γ)^{21}Mg process is estimated using the experimental data. The results predict the ignition of the proton radiative–capture process at T = 1×10^8K under typical nova conditions. This temperature happens to be in agreement with the previous theoretical estimates. The results also suggest that the nucleosynthesis flow of the rapid–proton process will run up to ^{21}Mg immediately after breakout from the hot–CNO cycle.

061.027   The ^{48}Ti(α, n)^{51}Cr and ^{48}Ti(α, p)^{51}V cross sections.
A. J. Morton, S. G. Tims, A. F. Scott, V. Y. Hansper, C. I. W. Tingwell, D. G. Sargood.
Nucl. Phys. A, Vol. 537, No. 1/2, p. 167 – 182 (3 – 10 Feb 1992).
The cross sections of the reactions ^{48}Ti(α, n)^{51}Cr and ^{48}Ti(α, p)^{51}V have been measured over bombarding–energy ranges of 4.98 – 9.52 MeV and 5.52 – 8.98 MeV, respectively. The data, together with those for all other N = 28 compound nucleus reactions reported in the literature, have been compared with the results of statistical–model calculations. New formulae prescribing optical–model parameters for use exclusively with N = 28 compound nucleus reactions are presented. Thermonuclear reaction rates for ^{48}Ti(α, n), (α, p) have been calculated for temperatures appropriate for stellar nucleosynthesis, and have been compared with published rates derived both from experimental measurements and from statistical–model calculations.

061.028   Nuclear reactions in astrophysics: recent experimental and theoretical studies, and further quests.
M. Arnould.
Nucl. Phys. A, Vol. 538, p. 493c – 504c (2 – 9 Mar 1992). Paper presented at the 4. International Conference on Nucleus–Nucleus Collisions (NN–4), Kanazawa (Japan), 10 – 14 Jun 1991.
A brief review is presented of recent theoretical and experimental efforts that have led to an improvement in our knowledge of nuclear reaction rates of interest in astrophysics. Emphasis is also put on the still existing (sometimes very large) uncertainties that affect some important rates. This is especially the case when short–lived nuclei are involved in the entrance channel.

061.029   Experimental approach to explosive nucleosynthesis.
S. Kubono.
Nucl. Phys. A, Vol. 538, p. 505c – 514c (2 – 9 Mar 1992). Paper presented at the 4. International Conference on Nucleus–Nucleus Collisions (NN–4), Kanazawa (Japan), 10 – 14 Jun 1991.

Recent development of experimental studies on explosive nucleosynthesis, especially the rapid proton process and the primordial nucleosynthesis were discussed with a stress on unstable nuclei. New development in the experimental methods for the nuclear astrophysics is also discussed which use unstable nuclear beams.

061.030   Measurement of the ^{3}H(^{1}Li, n_0)^{9}Be cross section within the energy range of big–bang nucleosynthesis.
A. Coc, P. Aguer, S. Barhoumi, G. Bogaert, J. Kiener,
A. Lefebvre, J. P. Thibaud, F. M. Baumann, H. Freiesleben,
C. Rolfs, P. Delbourgo–Salvador.
Nucl. Phys. A, Vol. 538, p. 515c – 522c (2 – 9 Mar 1992). Paper presented at the 4. International Conference on Nucleus–Nucleus Collisions (NN–4), Kanazawa (Japan), 10 – 14 Jun 1991.

The differential cross sections for the ^{3}H(^{7}Li, n_0)^{9}Be reaction were measured at 5 angles in the energy range E_{CM} = 0.2 – 0.9 MeV using a pulsed ^{7}Li beam and time–of–flight technique. Absolute values of the cross section were obtained by comparison with the well known cross section of ^{3}H(d, n)^{4}He at E_d = 1.0 MeV. The resulting reaction rates are obtained at temperatures relevant to big–bang nucleosynthesis, and consequences for primordial ^{9}Be abundances are discussed.

061.031   Radiative capture of protons by light nuclei at low energies.
F. E. Cecil, D. Ferg, H. Liu, J. C. Scorby, J. A. McNeil,
P. D. Kunz.
Nucl. Phys. A, Vol. 539, No. 1, p. 75 – 96 (16 Mar 1992).

Gamma–ray–to–charged–particle branching ratios, and gamma–ray angular distributions have been measured for the radiative capture of protons by ^{6}Li, ^{7}Li, ^{9}Be and ^{11}B for proton bombarding energies between 40 and 180 keV. Except for the 163 keV resonance in the reaction ^{11}B(p, γ)^{12}C, the branching ratios are roughly independent of energy and the angular distributions are isotropic. These measurements are used to deduce reaction S–factors and infer thermonuclear reactivities. The measurements are compared to distorted–wave–Boron–approximation direct–capture cross–section calculations.

061.032   Direct proton capture on ^{32}S.
C. Iliadis, U. Giesen, J. Görres, M. Wiescher,
S. M. Graff, R. E. Azuma, C. A. Barnes.
Nucl. Phys. A, Vol. 539, No. 1, p. 97 – 111 (16 Mar 1992).

The ^{32}S(p, γ)^{33}Cl reaction has been measured in the proton–energy range E_p = 0.4 – 2.0 MeV. Non–resonant γ–transitions were observed to the final states in ^{33}Cl at E_x = 0, 811 and 2846 keV. The corresponding spectroscopic factors have been extracted from fits to the excitation functions and are compared to values from stripping data as well as theoretical model calculations. The astrophysical aspects of the ^{32}S(p, γ)^{33}Cl reaction are also discussed.

061.033   Implications of Majorana neutrino transition magnetic moments for neutrino signals from supernovae.
E. Kh. Akhmedov, Z. G. Berezhiani.
Nucl. Phys. B, Part. Phys., Vol. 373, No. 2, p. 479 – 497 (6 Apr 1992).

The authors study resonant neutrino oscillations (RNO) and resonant spin–flavor precession (RSFP) of Majorana neutrinos inside supernovae. The response of H_2O and D_2O detectors to supernova neutrinos is calculated for the following scenarios: (1) Standard model; (2) RNO; (3) RSEP, and (4) RNO + RSFP. For the H_2O detectors, event numbers for both the Kamiokande II and IMB detection efficiencies are calculated. It is shown that for sufficiently large H_2O detectors the combined analysis of the isotropic/directional event–number ratio and prompt neutronization burst signal will enable one to distinguish between all the considered possibilities provided the distance to

supernova is less than 10 kpc. Even more rich and conclusive information may be reached by using the D_2O detector since it will allow one to observe the neutrinos via two charged–current reaction modes and a neutral–current mode in addition to the ν–e scattering.

061.034   Production and distribution of ^{26}Al in the galaxy: The role of massive stars.
N. Prantzos.
AIP Conf. Proc., No. 232, p. 129 – 148 (1 Aug 1991). Current Physics Microform No.: 9110A0825. – See Abstr. 012.003 for the main entry.

The various sources of production of ^{26}Al in the Galaxy are discussed in the light of recent observational and theoretical data. It is found that with current stellar and nucleosynthesis models, all the candidate sources still fall short (by factors of > 5) from producing the 3 $M_\odot$/10^6y required by the observations. The corresponding flux distribution in the galactic plane is also discussed; contrary to previous claims, it is argued that this signature cannot easily help to discriminate between the various candidate sources. Possibilities of unambiguous identification of massive stars are discussed, and relevant flux profiles taking into account the spiral structure of the Galaxy are presented.

061.035   Radionuclides of interest for γ–ray line astronomy from novae and red giants.
G. Paulus, M. Forestini.
AIP Conf. Proc., No. 232, p. 183 – 189 (1 Aug 1991). Current Physics Microform No.: 9110A0879. – See Abstr. 012.003 for the main entry.

The contribution of asymptotic giant branch (AGB) stars to the ^{26}Al content of the galactic disk is discussed, as well as the role of novae in the nucleosynthesis of ^{22}Na and ^{26}Al. In particular, the effects of various uncertainties are stressed in some detail.

061.036   High energy γ–ray lines: a probe for a cold dark matter halo.
P. Salati, P. Chardonnet.
AIP Conf. Proc., No. 232, p. 458 – 463 (1 Aug 1991). Current Physics Microform No.: 9110A1154. – See Abstr. 012.003 for the main entry.

The authors study the possibility of detecting halo cold dark matter through the annihilation process $\chi\chi \to \gamma\gamma$. This process produces monoenergetic γ–rays, and many be a clear signature of particle dark matter. If there is a closure density of photino – so far a favoured candidate – the authors show that it will be very difficult to observe this annihilation line from a satellite or space station borne experiment. On the contrary, triplet neutrinos which have recently been suggested as a plausible solution of the dark matter conundrum, whilst elusive at accelerators, should produce a clear γ–ray line signal, well above background.

061.037   Spectroscopy with enriched detectors: double beta decay and perspectives in astrophysical γ–ray spectroscopy and in dark matter detection.
H.V. Klapdor–Kleingrothaus.
AIP Conf. Proc., No. 232, p. 464 – 476 (1 Aug 1991). – See Abstr. 012.003 for the main entry.

A new area, of second generation experiments, using detectors made from enriched material is starting at present in double beta research. This allows to strongly improve the present limits on the neutrino mass. The author reviews the present status, in particular of the Heidelberg–Moscow experiment, by which for the first time evidence has been given that the technology of enriched HP Ge detector production can be mastered nowadays.

061.038   Orthopositronium, mirror universe, and primordial nucleosynthesis.
Ya. M. Kramarovskij, B. M. Levin, V. P. Chechev.
Sov. J. Nucl. Phys., Vol. 55, No. 2, p. 243 – 244 (Feb 1992). Current Physics Microform No.: 9205X1811.

It is shown that the observational data on the abundance of primordial ^{4}He does not rule out the existence of a "mirror universe". Its existence is in agreement with new experimental data on the annihilation of orthopositronium.

061.039   The anthropic principle and nucleosynthesis.
V. Vanýsek.
Říše hvězd, Vol. 72, No. 10, p. 188 – 190 (Oct 1991). In Czech.

061.040   Neutrino properties in matter.
C. W. Kim.
4. Asia Pacific Physics Conference, p. 173 – 179 (1991). – See Abstr. 012.006 for the main entry.

When neutrinos pass through a medium, their basic properties such as mass, mixing angle, magnetic moment are drastically modified. One of the consequences is the well-known MSW effect which can explain the depletion of the solar neutrinos. The medium effects can also dramatically alter the neutrino decays, both radiative and Majoron emission modes. A general discussion on how medium effects can modify neutrino properties is presented. Several implications of the effects on the solar neutrinos and supernova neutrinos are also discussed.

061.041   Big Bang archeology: WIMP capture by the Earth at finite optical depth.
A. Gould.
Astrophys. J., Vol. 387, No. 1, p. 21 – 26 (1 Mar 1992).

Capture of weakly interacting massive particles (WIMPs) by the Earth may be dominated by multiple collisions, even when the "optical depth" of the Earth is small ($\tau \ll 1$), and hence multiple collisions constitute a small fraction of all interactions. Here $\tau \equiv N\sigma/\pi R_c^2$, where R_c is the radius of the Earth's core, σ is the WIMP cross section with iron, and N is the number of iron atoms in the Earth. For scalar neutrinos of mass $m_x = 75$ GeV, this effect increases the capture rate by a factor ~ 5, even though $\tau \sim 0.3$. A WIMP with $\tau \sim 1$ would have capture enhanced by up to a factor ~ 30. These results are primarily of interest in the search for cosmological relics, stable particles whose relic density is too small to make up the dark matter, and whose mass is too large for them to have been seen in accelerator experiments.

061.042   Constraints of the neutrino magnetic moment from white dwarf cooling.
Jin Wang.
Astrophys. Space Sci., Vol. 189, No. 1, p. 1 – 4 (Mar 1992).

The neutrino magnetic moment provides an additional energy emission in stars. It will accelerate the white dwarf cooling process and reduce the life time of the white dwarf, but it causes a conflict with the observation. The author uses observational constraints to derive an upper limit for the neutrino magnetic moment: $\mu_\nu \sim 4.0 \times 10^{-12}\mu_B$.

061.043   Electroweak bubbles: nucleation and growth.
N. Turok.
Phys. Rev. Lett., Vol. 68, No. 12, p. 1803 – 1806 (23 Mar 1992). Current Physics Microform No.: 9205F0017.

In the standard electroweak theory, if the Higgs boson mass is comparable with the weak scale, the electroweak transition is weakly first order, and proceeds via bubble nucleation. It is shown that the universe supercools beyond the point where phase equilibrium is possible. When true-vacuum bubbles nucleate, they expand at velocities of the order of the speed of light until they fill the universe. These considerations are important for the recently proposed electroweak baryogenesis mechanisms.

061.044   Constraints to the decays of Dirac neutrinos from SN 1987A.
S. Dodelson, J. A. Frieman, M. S. Turner.
Phys. Rev. Lett., Vol. 68, No. 17, p. 2572 – 2575 (27 Apr 1992). Current Physics Microform No.: 9206C2120.

The wrong-helicity states of keV mass Dirac neutrinos are emitted copiously from the cores of newly born neutron stars, including that associated with SN 1987A. If they decay into proper-helicity neutrinos – ν_e, ν_μ, $\bar{\nu}_e$, or $\bar{\nu}_\mu$ – they should have produced distinctive high-energy events in the Kamiokande II and Irvine – Michigan – Brookhaven detectors. The absence of

such events excludes these decay modes for a Dirac neutrino of mass between order 1 keV and 300 keV and lifetime $10^{-9}[m_\nu/(1 \text{ keV})] \sec \lesssim \tau \lesssim 5 \times 10^7[m_\nu/(1 \text{ keV})] \sec$. This places severe constraints on models for Simpson's 17 keV neutrino.

061.045   Comment on "Multiple–scattering suppression of the bremsstrahlung emission of neutrinos and axions in supernovae".
R. F. Sawyer.
Phys. Rev. Lett., Vol. 68, No. 20, p. 3115 (18 May 1992). Current Physics Microform No.: 9207E1467.

A comment on the Letter by G. Raffelt and D. Seckel, Phys. Rev. Lett., Vol. 67, No. 19, p. 2605 – 2608 (4 Nov 1991). See Abstr. 54.061.058. A reply by Raffelt and Seckel is included.

061.046   Decaying Dirac neutrinos.
A. Acker, S. Pakvasa, J. Pantaleone.
Phys. Rev. D, Vol. 45, No. 1, p. R1 – R4 (1 Jan 1992). Current Physics Microform No.: 9203E0677.

Constraints on Dirac neutrino decay into invisible particles are surveyed. Neutrino lifetimes short enough to explain the solar neutrino problem are allowed by present terrestrial and cosmological measurements. A model in which Dirac neutrinos can have such short lifetimes is proposed. The recently resurrected 17 keV neutrino is incorporated into this model.

061.047   17 keV neutrino, MSW mechanism, and supernova constraints.
K. S. Babu, R. N. Mohapatra, I. Z. Rothstein.
Phys. Rev. D, Vol. 45, No. 1, p. R5 – R9 (1 Jan 1992). Current Physics Microform No.: 9203E0681.

A simple form for the neutrino mass matrix describing ν_e, ν_μ, ν_τ, and a sterile state ν_s is proposed which accommodates the 17 keV neutrino as a $\nu_\mu - \nu_\tau$ pseudo Dirac pair and simultaneously resolves the solar neutrino puzzle via $\nu_e - \nu_s$ Mikheyev – Smirnov – Wolfenstein (MSW) oscillation. This model, which is a specific realization of a scheme proposed recently by Caldwell and Langacker, is automatically free of all supernova constraints. It is shown that the mass matrix follows in a technically natural manner in extensions of the standard model with spontaneously broken global $U(1)_{Le} - L_\mu \times U(1)_{L\tau}$ symmetry. The 17 keV neutrino decays to ν_e and a Majoron with a lifetime of order $10^{-1} - 10^{-2}$ sec, thus satisfying all cosmological and astrophysical constraints.

061.048   Baryogenesis via leptogenesis.
M. A. Luty.
Phys. Rev. D, Vol. 45, No. 2, p. 455 – 465 (15 Jan 1992). Current Physics Microform No.: 9203G0989.

If right-handed Majorana neutrinos are added to the standard model, then lepton–number violating out–of–equilibrium decays of right-handed neutrinos combined with anomalous electroweak processes can generate the baryon number of the universe. The author analyzes this mechanism in detail, and determines the ranges of parameters for which the correct baryon number is generated. The scenario works for a wide range of parameters in the neutrino sector, including right-handed neutrino masses ranging from ~ 1 TeV to $\sim 10^{19}$ GeV, depending on the assumptions made about the structure of the neutrino mass matrices.

061.049   Strong CP violation, electroweak baryogenesis, and axionic dark matter.
V. A. Kuz'min, M. E. Shaposhnikov, I. I. Tkachev.
Phys. Rev. D, Vol. 45, No. 2, p. 466 – 475 (15 Jan 1992). Current Physics Microform No.: 9203G1000.

The authors discuss the influence of strong CP-violating effects in invisible axion models on electroweak baryogenesis. It is found that they can be important only if the electroweak phase transition (EPT) occurs simultaneously with the chirality breaking phase transition in QCD. The predicted baryon asymmetry of the universe (BAU) is comparable to the observed one only if there is no entropy production in the EPT. At the same time the dark matter constraints put a lower bound on the temperature after the EPT, so that it is impossible to satisfy simultaneously both requirements (BAU and axion dark matter). The authors conclude that possible strong CP violation plays no role in electroweak baryogenesis, unless there exists some yet unknown mechanism for the BAU generation, producing baryons at a rate which has to be at least 3 orders of magnitude higher than the sphaleron–induced rate.

061.050   Grand unified gauge–boson condensation on the cosmic string.
G. Dvali, S. M. Mahajan.
Phys. Rev. D, Vol. 45, No. 2, p. 665 – 674 (15 Jan 1992). Current Physics Microform No.: 9203G1199.

Expectation values of grand unified Higgs scalars can be strongly changed in the core of a cosmic string. The authors show that in certain cases such unusual Higgs structures imply the existence of nonzero classical gauge currents in the lowest–energy state of the system. This automatically triggers the condensation of the grand unified gauge bosons interacting linearly with this current, which could be either trivial or nontrivial under the U(1) subgroup responsible for the string. For the former, the gauge–boson condensate accumulated in the core of the defect is strictly radial, while in the latter case it also acquires an azimuthal (and magnetic) component. Existence of such types of condensates on the boundaries of the expanding vacuum bubbles (which arise in high–temperature phase transitions) can play an important role in creating the present baryon asymmetry.

061.051   Class of models leading to depletions of solar ν_e and atmospheric ν_μ fluxes.
C. H. Albright.
Phys. Rev. D, Vol. 45, No. 3, p. R725 – R728 (1 Feb 1992). Current Physics Microform No.: 9204D1963.

A model incorporating three left–handed $SU(2)_L \times U(1)_Y$–doublet neutrinos and three right–handed singlet neutrinos, with leptonic Dirac mass submatrices similar to empirical forms which work well for quarks, is required to yield both the nonadiabatic Mikheyev – Smirnov – Wolfenstein solar neutrino effect and the depleted flux of atmospheric muon neutrinos observed by Kamiokande. The author finds that a small segment of the δm_{23}^2 versus $\sin^2(2\theta_{23})$ plane exists where such solutions can be achieved and identifies the class of right–handed Majorana submatrices required.

061.052   Solar and supernova neutrino physics with Sudbury Neutrino Observatory.
A. B. Balantekin, F. Loreti.
Phys. Rev. D, Vol. 45, No. 4, p. 1059 – 1065 (15 Feb 1992). Current Physics Microform No.: 9204D1669.

The salient features of solar and supernova neutrino physics at the Sudbury Neutrino Observatory are examined. The possibility of detecting a solar antineutrino flux, predicted as a result of resonant spin–flavor conversion, is emphasized. The antineutrino flux at Sudbury due to commercial power reactors is calculated. Simplified, analytical expressions for the expected Galactic supernova count rates are given.

061.053   Dirac neutrinos and SN 1987A.
M. S. Turner.
Phys. Rev. D, Vol. 45, No. 4, p. 1066 – 1075 (15 Feb 1992). Current Physics Microform No.: 9204D1676.

Within the standard electroweak theory, "wrong–helicity" neutrinos are produced in a nascent neutron star by "spin–flip" processes (at a rate proportional to m_ν^2). These particles can freely escape, and may lead to an excessively rapid cooling of the newly born neutron star. Previous work has shown that the observed cooling of the neutron star associated with SN 1987A excludes a Dirac neutrino mass greater than ~ 20 keV for either ν_e, ν_μ, or ν_τ. The author reexamines the emission of "wrong–helicity" Dirac neutrinos from SN 1987A and concludes that, due to neutrino degeneracy and additional emission processes, the effect of a Dirac neutrino on the cooling of SN 1987A has been *underestimated*. While a precise Dirac mass limit awaits the incorporation of the new rates into detailed numerical cooling models, the author believes that the limit that follows from the

cooling of SN 1987A is better, probably much better, than 10 keV. In particular, it is concluded that SN 1987A definitely excludes a 17 keV Dirac mass neutrino that mixes with the electron neutrino at the 1% level.

061.054   Pair creation and decay of a massive particle near and far away from a cosmic string.
J. Audretsch, A. Economou, D. Tsoubelis.
Phys. Rev. D, Vol. 45, No. 4, p. 1103 – 1112 (15 Feb 1992). Current Physics Microform No.: 9204D1713.

The authors study the total transition probabilities of the tree–level processes of the pair creation and decay of a massive particle for real Klein–Gordon fields in the spacetime of an infinite straight static cosmic string. Basing the discussion on cylindrical modes characterized by an approximate radius of closest approach r_{min}, it is possible to approximately localize the non–Minkowskian processes to cylindrical effective interaction regions around the cosmic string. A physical understanding of the space dependence of the transition probabilities is obtained on the basis of analytic expressions for different energy domains referring to regions close to and far away from the string. For pair creation the Compton wavelength λ_C of the created particles proves to be a crucial length scale. For $r_{min} \ll \lambda_C$ the creation probability is insensitive to a variation of r_{min}. For large r_{min} it falls off at least exponentially with r_{min}.

061.055   Axions and the QCD phase transition.
M. Hindmarsh.
Phys. Rev. D, Vol. 45, No. 4, p. 1130 – 1138 (15 Feb 1992). Current Physics Microform No.: 9204D1740.

Axion dynamics during a first–order QCD phase transition are investigated. If the mass of the axion is different on either side of the phase boundary, low–energy axions are trapped in the shrinking regions of the high–temperature phase. A number of different possibilities present themselves: the axions may escape, resulting in the generation of isocurvature density perturbations; if a substantial population of quark nuggets is produced the axions remain trapped for the lifetime of the nugget; or the pressure of the trapped axions may itself prevent the collapse of the bubble of the high–temperature phase, leading to the formation of "axion nuggets".

061.056   Neutrino coherent forward scattering and its index of refraction.
Liu Jiang.
Phys. Rev. D, Vol. 45, No. 4, p. 1428 – 1431 (15 Feb 1992). Current Physics Microform No.: 9204D2038.

It is pointed out that, if neutrinos are to maintain coherence over the required distance for the Mikheyev – Smirnov – Wolfenstein solutions to the solar neutrino problem, effects arising from neutrino multiple scattering must be considered. The author gives a simple derivation for the neutrino index of refraction that takes into account this effect. The same method is also shown to be useful for situations with varying matter densities and neutrino mixing. The author also examines the question whether the coherence of propagating neutrinos in matter will be affected by switching on an external magnetic field, assuming neutrinos have a large magnetic moment.

061.057   Majoron decay of neutrinos in matter.
C. Giunti, C. W. Kim, U. W. Lee, W. P. Lam.
Phys. Rev. D, Vol. 45, No. 5, p. 1557 – 1568 (1 Mar 1992). Current Physics Microform No.: 9205A1083.

Using an helicity formalism for Majorana neutrino fields, the authors calculate the decay rates for the helicity – flipping and helicity – conserving Majoron decays of neutrinos propagating in dense media. They discuss the subtlety involved in the neutrino mixing in matter in a quantum field theoretical approach. General formulas, derived in a model – independent approach, can be applied to any number of neutrino generations and to the decay with production of any massless pseudoscalar boson. In particular, the authors discuss the two generation case and show that in matter the helicity – flipping decays are dominant over the helicity – conserving decays. The implications of the Majoron decay for the neutrinos from astrophysical objects are also briefly discussed.

061.058 Lepton mass and mixing matrices involving a 17 keV Dirac τ neutrino.

C. H. Albright.
Phys. Rev. D, Vol. 45, No. 5, p. 1624 – 1627 (1 Mar 1992).
Current Physics Microform No.: 9205A1150.

The author extends his study of neutrino masses and mixings to the case of a 17 keV Dirac τ neutrino, which has reappeared in recent β–decay experiments. A special set of Dirac submatrices is applied which yields a top–quark mass near 135 GeV. However, the model cannot simultaneously explain all the data including the 17 keV neutrino, its weak coupling to the electron, the present accelerator neutrino oscillation bounds, and the preferred solar neutrino nonadiabatic Mikheyev – Smirnov – Wolfenstein effect interpretation.

061.059 17 keV neutrino, solar neutrino puzzle, and supernova 1987A.

J. M. Cline.
Phys. Rev. D, Vol. 45, No. 5, p. 1628 – 1635 (1 Mar 1992).
Current Physics Microform No.: 9205A1154.

The author presents a model of the 17 keV neutrino which naturally admits Mikheyev – Smirnov – Wolfenstein oscillations of v_e and v_μ in the Sun, while avoiding forbidden v_τ oscillations. Alternatively, the large mixing angle needed for vacuum oscillations to reduce the solar v_e flux can be explained, in which case the v_e mass might naturally be close to its laboratory limit. The model is consistent with all known laboratory, cosmological, and astrophysical constraints, including the supernova bound on Dirac neutrino masses.

061.060 Seesaw–model predictions for the τ neutrino mass.

S. A. Bludman, D. C. Kennedy, P. G. Langacker.
Phys. Rev. D, Vol. 45, No. 5, p. 1810 – 1813 (1 Mar 1992).
Current Physics Microform No.: 9205A1336.

The observed deficit of solar neutrinos cannot be explained by a cooler solar model, which would suppress high–energy neutrinos more than low–energy neutrinos, contrary to observation. The small neutrino masses and flavor mixing implied by matter – amplified neutrino oscillations in the Sun are most naturally interpreted in terms of minimal grand unification theories (GUTs) incorporating the seesaw mechanism. In two such theories, SO(10) GUT and supersymmetric GUT, that are consistent with all laboratory experiments, the neutrino masses are proportional to the squares of the up–quark masses. For the SO(10) GUT model, the μ neutrino mass is close to that observed in solar neutrino oscillations. Although the seesaw model mass predictions are less reliable than the mixing–angle predictions, the τ neutrino mass may lie in the cosmologically important range 4 – 28 eV and be accessible to laboratory neutrino oscillation experiments or to observation in a nearby supernova.

061.061 Artificial opacity – numerical implementation into flux–limited neutrino diffusion.

R. Dgani, H.–T. Janka.
Astron. Astrophys., Vol. 256, No. 2, p. 428 – 432 (Mar 1992).

Recently a modification to flux–limited neutrino diffusion schemes was suggested by introduction of a correction term, which acts like an additional non–physical opacity (Janka 1990, 1991). By this "artificial opacity" the evolution of the neutrino flow in the outer layers of compact stellar objects like newly formed neutron stars is determined. It allows to reduce or even eliminate the error in the momentum balance associated with solutions of the standard flux–limited diffusion schemes. The authors demonstrate the principle applicability of the modified approach. The corresponding technical changes can easily be incorporated in any standard flux–limiting method and lead to results in better agreement with neutrino transport by direct solution of the Boltzmann equation, when an appropriate flux–limiter is used. This is of immediate importance for reliable investigations of the neutrino–heated delayed explosion mechanism of type–II supernovae.

061.062 Flux–limited neutrino diffusion versus Monte–Carlo neutrino transport.

H.–T. Janka.
Astron. Astrophys., Vol. 256, No. 2, p. 452 – 458 (Mar 1992).

Monte Carlo methods are applied to the transport of neutrinos in supernovae and protoneutron stars to check the applicability of currently used flux–limited diffusion techniques. Significant and principle differences are found. Motivated by these findings improvements of flux–limited diffusion are worked out which allow for a better and more consistent approximation of neutrino transfer. The introduction of a correction function, which carries information about the geometry of the considered problem and acts like an opacity term ("artificial opacity"), allows to achieve consistency between the radiation energy and momentum equations. It controls and governs the evolution of the radiation field towards free streaming when the matter opacity becomes small. Furthermore, better agreement with solutions of the Boltzmann equation requires the use of a new flux–limiter which is deduced from the Monte Carlo results for neutrino transport in a variety of typical situations.

061.063 Cosmological density of WIMPs from solar and terrestrial annihilations.

A. Gould.
Astrophys. J., Vol. 388, No. 2, p. 338 – 344 (1 Apr 1992).

If weakly interacting massive particles (WIMPs) comprise the dark matter, then the ratio of their annihilation signals from the centers of the Sun and Earth can be used to measure the degree of which WIMP capture and annihilation have come into equilibrium in the Earth. This would provide a measure of, or a limit on, the WIMP annihilation cross section. The annihilation cross section is, in turn, inversely related by a standard Lee–Weinberg calculation to the cosmological density of WIMPs, Ω_x. If the next generation of detectors finds a signal from the Sun, one may infer an upper bound, lower bound, or measurement of Ω_x, depending on whether there is a comparable signal, no signal, or a small signal from the Earth. The chain of reasoning which leads to this conclusion is broken if the WIMPs have primarily spin–dependent interaction cross sections, but this possibility may be checked using direct–detection experiments. A very accurate analytic formula is given for the integrated WIMP capture by the entire Sun.

061.064 Physics with radioactive nuclear beams.

R. N. Boyd, I. Tanihata.
Phys. Today, Vol. 45, No. 6, p. 44 – 52 (Jun 1992).

Recently developed facilities allow a wide range of new investigations of the reactions and properties of short–lived nuclei. These studies may help to solve puzzles of nuclear structure and the Big Bang.

061.065 Simulation of anomalous extensive air showers initiated by strong neutrino–quark interactions.

S. Mrenna.
Phys. Rev. D, Vol. 45, No. 7, p. 2371 – 2377 (1 Apr 1992).
Current Physics Microform No.: 9205H0195.

The observation of extensive air showers (EAS) in the atmosphere initiated by ultrahigh–energy cosmic rays offers a test of new physics. In particular, some showers, initiated by neutral particles from point sources, contain a larger number of muons than can be explained by the standard model. A strong interaction between quarks and neutrinos, induced by some new physics, is presented as an explanation. For definiteness, the new physics is assumed to be the manifestation of a composite structure of quarks and leptons, though the general features of the interaction are common to many new physics scenarios. The consequences of such an interaction on the generation and development of EAS are studied with a phenomenological model. Properties of the electromagnetic, muonic, and hadronic components of simulated EAS for neutrino–induced and ordinary proton–induced showers are presented for the observation level of the CYGNUS experiment at Los Alamos.

061.066   Variant of the $S_3 \times Z_3$ model for the 17 keV neutrino.
T. V. Duong, E. Ma.
Phys. Rev. D, Vol. 45, No. 7, p. 2570 – 2573 (1 Apr 1992).
Current Physics Microform No.: 9205H0394.

In extending the recently proposed $S_3 \times Z_3$ model of quark mass matrices to include leptons to account for a possible 17 keV neutrino, a solution was already obtained with a certain choice for the Majorana neutrino mass matrix. The authors show here that there is another choice which results in a qualitatively new solution. There will be three pseudo Dirac neutrinos, one of which is formed mostly out of ν_e and $\bar{\nu}_\mu$. The 17 keV neutrino decays into $\nu_{e,\mu}$ and a Majoron with a lifetime of about 3×10^{-3}s.

061.067   Fermion masses, $SU(2)_L \times U(1)_Y$, and the solar neutrino problem.
R. Gandhi.
Phys. Rev. D, Vol. 45, No. 7, p. R2192 – R2195 (1 Apr 1992).
Current Physics Microform No.: 9205H0016.

The connection between fermion mass generation in the standard model and matter resonant oscillation [Mikheyev – Smirnov – Wolfenstein (MSW)] solutions to the solar and atmospheric neutrino deficits is examined. Given our lack of understanding of the wide range of values spanned by fermion masses, as emphasized by the heaviness of the top quark, it is not impossible that fermions obtained their masses in a manner unrelated to electroweak symmetry breaking. If this is the case, then, adding only right-handed neutrino fields, the author emphasizes that the MSW solution to both the above deficits can be naturally realized via active $\rightleftharpoons$ sterile oscillations, with both small and large vacuum mixing angles. In the large angle case, a flux in the vicinity of 40 – 50 neutrino units is expected in the gallium experiments, along with a suppressed neutral current event rate in the Sudbury Neutrino Observatory detector.

061.068   Cosmological constraints on pseudo Nambu–Goldstone bosons.
J. A. Frieman, A. H. Jaffe.
Phys. Rev. D, Vol. 45, No. 8, p. 2674 – 2684 (15 Apr 1992).
Current Physics Microform No.: 9206F1400.

Particle physics models with pseudo Nambu–Goldstone bosons (PNGBs) are characterized by two mass scales: a global spontaneous symmetry–breaking scale f and a soft (explicit) symmetry–breaking scale Λ. The authors investigate general model–insensitive constraints on this two–dimensional parameter space arising from the cosmological and astrophysical effects of PNGBs. In particular, the authors study constraints arising from vacuum misalignment and thermal production of PNGBs, topological defects, and the cosmological effects of PNGB decay products, as well as astrophysical constraints from stellar PNGB emission.

061.069   Electroweak phase transition and baryogenesis.
G. W. Anderson, L. J. Hall.
Phys. Rev. D, Vol. 45, No. 8, p. 2685 – 2698 (15 Apr 1992).
Current Physics Microform No.: 9206F1411.

The authors give an analytic treatment of the one–Higgs-doublet, electroweak phase transition which demonstrates that the phase transition is first order. The phase transition occurs by the nucleation of thin–walled bubbles and completes at a temperature where the order parameter $\langle \varphi \rangle_T$ is significantly smaller than it is when the origin becomes absolutely unstable. The rate of anomalous baryon number violation is an exponentially sensitive function of $\langle \varphi \rangle_T$. In very minimal extensions of the standard model it is quite easy to increase $\langle \varphi \rangle_T$ so that anomalous baryon number violation is suppressed after the completion of the phase transition. Hence, baryogenesis at the electroweak phase transition is tenable in minimal extensions of the standard model with one Higgs doublet.

061.070   QCD sphalerons at high temperature and baryogenesis at the electroweak scale.
R. N. Mohapatra, Zhang Xinmin.
Phys. Rev. D, Vol. 45, No. 8, p. 2699 – 2705 (15 Apr 1992).
Current Physics Microform No.: 9206F1425.

Effects of a QCD sphaleron–like transition on baryogenesis at the weak scale are examined. The authors explain how QCD sphaleron–like plus SU(2) sphaleron processes convert right–handed quark numbers into baryon number during phase transitions.

061.071   Chiral cosmic strings.
J. D. Bekenstein.
Phys. Rev. D, Vol. 45, No. 8, p. 2794 – 2801 (15 Apr 1992).
Current Physics Microform No.: 9206F1520.

The author considers all global cosmic–string solutions of U(1) scalar field theory with a cylindrically symmetric energy density. These can be characterized as extrema of the field's energy for given angular and linear momenta. This string class comprises, apart from the well–known ordinary and rotating strings, twisted string configurations whose isophase surfaces twist as one moves along the axis, and lightlike–phase strings in which the twisted phase propagates at the speed of light parallel to the string axis. The author rules out global strings whose isophase contours are spirals in a plane, and proves, in a unified way and for a broad class of scalar potentials, the stability of ordinary, rotating, and lightlike–phase strings with a unit winding number against small perturbations.

061.072   Effective potential at finite temperature in the standard model.
M. E. Carrington.
Phys. Rev. D, Vol. 45, No. 8, p. 2933 – 2944 (15 Apr 1992).
Current Physics Microform No.: 9206F1659.

There has been much recent interest in the nature of the electroweak phase transition. This information is of importance in the context of the sphaleron models that have recently been proposed to explain the observed net baryon number in the universe. The presence of a term that is cubic in the Higgs condensate in the one–loop effective potential appears to indicate a first–order phase transition. However, the infrared singularities inherent in massless models produce cubic terms that are of the same order in the coupling. In this paper, these terms are included, and it is shown that the standard model has a first-order phase transition.

061.073   Possibility of observing "Centauro" events at the BNL Relativistic Heavy Ion Collider.
A. D. Panagiotou, A. Petridis, M. Vassiliou.
Phys. Rev. D, Vol. 45, No. 9, p. 3134 – 3142 (1 May 1992).
Current Physics Microform No.: 9207C0606.

A phenomenological model for the production of "Centauro"–type cosmic ray events is developed, assuming a nucleus-nucleus interaction in the upper atmosphere. The model is used to estimate, in a self–consistent way, several thermodynamic and kinematical quantities, characterizing the observed "Centauro" events. On the basis of this model, the authors describe a typical "Centauro" event, possibly produced in the fragmentation rapidity of an A + A central collision at 200A GeV at the BNL Relativistic Heavy Ion Collider. The authors suggest several characteristic signatures for these events, as well as the possibility of observing "strangelets".

061.074   Electroweak phase transition in supersymmetry.
G. F. Giudice.
Phys. Rev. D, Vol. 45, No. 9, p. 3177 – 3182 (1 May 1992).
Current Physics Microform No.: 9207C0649.

The electroweak phase transition in supersymmetric models is studied, analyzing the constraint on the Higgs boson mass coming from the condition that the cosmic baryon asymmetry is not washed out soon after the phase transition. It is found that, in the minimal supersymmetric model, baryogenesis at the weak scale requires a Higgs boson lighter than about 50 – 55 GeV. On the other hand, in extended supersymmetric models, it is possible to have the lightest Higgs boson as heavy as 100 GeV and still satisfy the requirement of weak–scale baryogenesis.

061.075 **Constants and cosmology: the nature and origin of fundamental constants in astrophysics and particle physics.**
P. S. Wesson.
Space Sci. Rev., Vol. 59, No. 3/4, p. 365 – 406 (Feb 1992).
The author asks about the nature and origin of the fundamental constants of astrophysics and particle physics, notably the speed of light c, the gravitational constant G, Planck's constant h, and the magnitude of the electron charge e. The author considers general relativity and the theories of the electromagnetic, weak and strong interactions that make up the Standard Model; together with the Lagrangians of Einstein, Maxwell, Schrödinger, Klein–Gordon, Dirac, Proca, and Yang–Mills. Then he looks in a more qualitative way at how the equations of physics are set up, their dimensional content, and the removal of constants from them by a suitable choice of units. The author concludes with Hoyle and Narlikar, Jeffreys and McCrea that parameters like c, G, and h are merely manmade dimensional conversion constants. They arise because of our subjective view that mass, length, and time are different concepts. These constants can be removed in a manner analogous to the removal of the permittivity of free space ε_0 from electrodynamics, and none are really fundamental. The charge e is different, being the low–energy limit of a funtion related to properties of the vacuum, but because of this it is not a fundamental constant either. The author suggests there are no constants which truly deserve to be called fundamental, and that an aim of physics ought to be to write down laws in which no constants appear.

061.076 **Neutrino radiation from supernovae.**
J. A. Grifols.
Nucl. Instrum. Methods Phys. Res., Sect. A, Vol. 314, No. 2, p. 386 – 389 (15 Apr 1992). Paper presented at the 5. International Symposium on Radiation Physics (ISRP-5), Dubrovnik (Yugoslavia), 10 – 14 Jun 1991.
In a supernova event about 10^{53}erg of gravitational binding energy are released, the bulk of which is carried off by neutrinos. The neutrino burst detected in February 1987 recorded a supernova explosion that took place in the Large Magellanic Cloud and confirmed the basics of models of stellar collapse. The neutrino luminosity inferred from observation agrees with the luminosity predicted by detailed model calculations of neutrino transport. However, there is some room for extra exotic sources of energy drain. Among them, there are right–handed neutrinos, which appear in almost any extension of the minimal standard model of electroweak interactions. In particular, if neutrinos are massive Dirac particle, the right–handed degrees of freedom should be copiously emitted in a supernova collapse. Using SN 1987A data one can place bounds on neutrino masses or on the magnetic moment of the neutrino. Furthermore, independently of the energetics of stellar dynamics, the actual detection at IMB and Kamioka has led to relevant limits on neutrino mass, neutrino lifetime, and neutrino charge.

061.077 **Relativistic neutrons in active galactic nuclei. I. Energy transport from the core.**
A. M. Atoyan.
Astron. Astrophys., Vol. 257, No. 2, p. 465 – 475 (Apr 1992).
It is shown that in the accretion models of AGN assuming efficient acceleration of relativistic protons (RPs) up to ultrarelativistic energies in the vicinity of supermassive black hole (BH) at dimensionless distances $r \equiv R/R_g \leqslant 10$ (R_g being the gravitational radius) about a third of the protons acceleration power $\dot{W}_p$ is transferred to relativistic neutrons (RNs) which then escape from the acceleration region, producing secondary RPs at distances up to kpc scales. For the power–law differential spectra of accelerated protons with the index $\alpha \sim 2$ the resulting rate of the energy released by RNs at large r ($\leqslant 10^8$) may essentially exceed the power that can be provided by the local gravitational field. It is shown that for the AGN model under consideration to be self–consistent, the secondary RPs should have an effective lifetime much shorter than the RP cooling time in the ambient ordinary thermal accretion plasma. Two possibilities to reach high cooling rates are discussed, namely: (i) the cooling of RPs in

small and dense clouds or filaments responsible for the line emission of AGN (two–phase accretion models); (ii) the RP energy losses due to development of plasma turbulence.

061.078 **Relativistic neutrons in active galactic nuclei. II. Gamma–rays of high and very high energies.**
A. M. Atoyan.
Astron. Astrophys., Vol. 257, No. 2, p. 476 – 488 (Apr 1992).
In the frames of the AGN model proposed in Paper I, the problem of the electromagnetic cascade initiated in a spatially non–homogeneous field of background photons due to propagation of relativistic neutrons (RNs) from the proton acceleration region $r \leqslant r_0 \sim 10$ and their decay at large radii, $r_0 < r < 10^7$–10^8, is considered (the N–cascade). In contrast to the H–cascade initiated by the relativistic electrons and gamma–rays from the region of $r \leqslant r_0$, which is similar to the ordinary pair–photon cascades in the spatially homogeneous field of soft photons, the radiation spectra resulting from the N–cascade can be described by the power–law functions gradually steepening with increasing gamma–ray energies. The integral luminosity L_{mx} of the AGN radiation from the submillimeter to X–ray bands is the general parameter defining the N–cascade development. The features of the high–energy and very–high–energy gamma–radiation resulting from the N–cascade are investigated. It is shown that the model proposed can provide significant fluxes of very high energy ($E \geqslant 1$ TeV) gamma–rays from the bright extragalactic sources with relatively low luminosities, $L_{mx} < 10^{45}$erg s^{-1}.

061.079 **The role of the neutrino electromagnetic moments in the stellar energy loss rate.**
B. K. Kerimov, S. M. Zeinalov, V. N. Alizade, A. M. Mourão.
Phys. Lett. B, Vol. 274, No. 3/4, p. 477 – 482 (16 Jan 1992).
If a massive Dirac neutrino has magnetic $\mu_v = F_{2v}(0)\mu_B$ and/or electric $d_v = g_{2v}(0)\mu_B$ dipole moments, where $\mu_B = e\hbar/2m_ec$ is the Bohr magneton, it would take part in electromagnetic interactions. The authors investigate the electromagnetic production mechanism of neutrino pairs in processes which are relevant for the neutrino luminosity of stars. They also calculate the stellar energy loss rates due to these processes and compare them with those due to the weak production mechanism.

061.080 **Is axino dark matter possible in supergravity?**
T. Goto, M. Yamaguchi.
Phys. Lett. B, Vol. 276, No. 1/2, p. 103 – 107 (6 Feb 1992).
The mass spectrum of an axion supermultiplet is examined in the context of supergravity. It is shown that in a non–scale supergravity a light axino can be realized, which serves as an interesting candidate of dark matter. Cosmological implications of an associated light saxion are also discussed.

061.081 **Coulomb and nuclear effects in direct breakup of 54-MeV ^{7}Li + ^{12}C, ^{197}Au.**
J. E. Mason, S. B. Gazes, R. B. Roberts, S. G. Teichmann.
Phys. Rev. C, Vol. 45, No. 6, p. 2870 – 2878 (Jun 1992). Current Physics Microform No.: 9208E1846.
Direct breakup of ^{7}Li$\rightarrow\alpha$+t was measured for 54–MeV ^{7}Li+^{12}C, ^{197}Au reactions. The breakup products were detected in coincidence using a close–geometry detection system, and small scattering angles forward of grazing, were investigated. Three–body classical trajectory calculations are found to provide qualitative agreement with the observations, provided the projectile is treated as an extended object. These strong Coulomb and nuclear effects severely complicate the extraction of radiative–capture cross sections and low–energy astrophysical S factors.

061.082 **Reaction ^{36}Ar(p, γ)^{37}K in explosive hydrogen burning.**
C. Iliadis, J. G. Ross, J. Görres, M. Wiescher, S. M. Graff, R. E. Azuma.
Phys. Rev. C, Vol. 45, No. 6, p. 2989 – 2994 (Jun 1992). Current Physics Microform No.: 9208E1965.
The reaction ^{36}Ar(p, γ)^{37}K has been measured in the proton energy range of $E_p = 0.32 – 0.93$ MeV. A new resonance was

found at $E_R = 321$ keV. The authors have measured the branching ratios and the resonance strength. The stellar reaction rates which are dominated by this resonance for temperatures $T = 0.07 - 0.9$ GK are calculated. Network calculations have been performed to investigate the influence of the new stellar rates on the time evolution of the ^{36}Ar abundance during explosive H burning.

061.083   Closed form theory of elastic breakup and applications to astrophysically relevant heavy ion reactions.
C. A. Bertulani, L. F. Canto, M. S. Hussein.
Phys. Rev. C, Vol. 45, No. 6, p. 2995 – 2999 (Jun 1992). Current Physics Microform No.: 9208E1971.

The authors investigate the breakup process in light heavy ion reactions. With the help of a separable approximation proposed in a previous paper and using closed forms for radial integrals, they obtain simple expressions for the breakup cross section. The theory is applied to the reaction $^{16}O + ^{28}Si \rightarrow \alpha + ^{12}C + ^{28}Si$, and the results are shown to agree with the experimental data.

061.084   Implications of the $^{14}C(\alpha, \gamma)^{18}O$ reaction for nonstandard big bang nucleosynthesis.
M. Gai.
Phys. Rev. C, Vol. 45, No. 6, p. R2548 – R2551 (Jun 1992). Current Physics Microform No.: 9208E1524.

The thermonuclear burning rates for the $^{14}C(\alpha, \gamma)^{18}O$ radiative capture reaction are calculated at temperatures ($0.3 < T < 10$GK) of relevance to the early universe. These rates are particularly important for estimating the formation of heavy elements in an inhomogeneous big bang nucleosynthesis. The authors investigate the effect of a possible new broad ($\Gamma \approx 0.45$ MeV) 1^- state, at approximately 9.0 MeV in ^{18}O as would be deduced from the Yale–Michigan State University measurement of the beta-delayed alpha–particle emission of ^{18}N and suggested by the Notre Dame–Caltech measurement of the nonresonant $^{14}C(\alpha, \gamma)^{18}O$ cross section. The gamma widths of the proposed broad state is estimated using the Alhassid, Gai, and Bertsch sum rule, and an experimental study is proposed.

061.085   On the contribution of ^{22}Ne to the synthesis of ^{54}Fe and ^{58}Ni in thermonuclear supernovae.
E. Bravo, J. Isern, R. Canal, J. Labay.
Astron. Astrophys., Vol. 257, No. 2, p. 534 – 538 (Apr 1992).

The white dwarf explosion model accounts fairly well for the global characteristics of type Ia supernovae. There are, however, several problems concerning the dynamics of the burning front (deflagration or detonation regime), initial conditions and nucleosynthesis, that still deserve attention. In particular, the overproduction of neutron–rich nuclides such as ^{54}Fe and ^{58}Ni, due to electron captures and to the neutron excess trapped in ^{22}Ne nuclei, still poses a major problem. In this paper the authors examine the possibilities for ^{22}Ne to migrate towards the central regions of the star and they show that if the white dwarf has enough time to solidify, the contribution of ^{22}Ne to the neutron excess becomes negligible.

061.086   Can a closure mass neutrino help solve the supernova shock reheating problem?
G. M. Fuller, R. Mayle, B. S. Meyer, J. R. Wilson.
Astrophys. J., Vol. 389, No. 2, p. 517 – 526 (20 Apr 1992).

The authors point out that a ν_μ or ν_τ neutrino with a cosmologically significant mass (10 – 100 eV) and a small mixing angle with a light ν_e would result in a matter–enhanced Mikheyev– Smirnov– Wolfenstein resonant transformation between these species in a region above the neutrino sphere but below the stalled shock during the reheating phase of a Type II supernova. The neutrino heating behind the shock is due to charged–current ν_e and $\bar{\nu}_e$ captures. Since the ν_μ and ν_τ have considerably higher average energies than do the ν_e, neutrino flavor mixing would result in higher effective ν_e energies behind the shock and a concomitant increase in the heating rate. Numerical calculations suggest that this effect results in a 60% increase in the supernova explosion energy, possibly helping to solve the energy problem of delayed–mechanism supernova models.

061.087   Measurement of the ^{76}Se(n, γ) capture cross section and phenomenological s–process studies: the weak component.
H. Beer, G. Walter, F. Käppeler.
Astrophys. J., Vol. 389, No. 2, p. 784 – 790 (20 Apr 1992).

The authors have measured the excitation function of the reaction ^{76}Se(n, γ) in the energy range from 3 to 200 keV neutron energy. The Maxwellian – averaged capture cross section versus thermal energy kT has been calculated. For kT = 30 keV a value of 164 ± 8 mbarn has been derived. The weak component s–process in the mass range A = 56 to 90 has been studied. Two models for the weak component, an s–process with an exponential exposure distribution and an s–process with a distinct single exposure, have been discussed. For a consistent description of the abundances of the weak component s–process under realistic stellar conditions a considerable part of the material produced during core He–burning in massive stars must be reprocessed by subsequent C–burning. The solar Kr–abundance was determined to be $(47 \pm 5)/10^6$ Si.

061.088   Neutrino electromagnetic scattering in astrophysics.
A. W. Poon, C. E. Waltham.
Can. J. Phys., Vol. 70, No. 2/3, p. 140 – 142 (Feb–Mar 1992).

The authors consider the electromagnetic scattering of neutrinos via their magnetic moment in astrophysical systems. They find that in the Sun the scattering processes flipping the neutrino from its left–handed state to its right–handed state are unimportant for all allowable values of the neutrino magnetic moment. However, considering these processes in the supernova SN 1987A, they find that the upper limit of the neutrino magnetic moment to be in the order of $3 \times 10^{-11} \mu_B$.

061.089   Neutrino astronomy.
Y. Totsuka.
Rep. Prog. Phys., Vol. 55, No. 3, p. 377 – 430 (Mar 1992).

Observations of extraterrestrial neutrinos are reviewed. Solar neutrinos and supernova neutrinos have been successfully detected and provided valuable information for astronomy and particle physics. Observation of the 1.9K cosmic background neutrinos would be most challenging. It probably requires finite neutrino masses and ultrahigh–energy neutrinos of cosmological and perhaps of grand unified theory (GUT) related origin. Ultrahigh–energy neutrinos have not yet been observed, but they will provide a clue about the origin of cosmic rays, and probably exciting information about cosmology and GUT.

061.090   On the calculation of Maxwellian–averaged capture cross sections.
H. Beer, F. Voss, R. R. Winters.
Astrophys. J., Suppl. Ser., Vol. 80, No. 1, p. 403 – 424 (May 1992).

The authors discuss a recipe for the calculation of Maxwellian–averaged capture cross sections as a function of temperature kT from the existing excitation function. Detailed formulae are given to handle the various portions of this excitation function which in general contains both resolved resonance and an unresolved resonance parts. For the temperature region considered, the details of the thermal (0.0253 eV) capture cross section and a possible direct capture component are of importance. The authors discuss the treatment of resolved resonances depending on the characteristics of the resonance parameters. Particular attention is given to a commonly encountered situation in which the excitation function is known only over a portion of the relevant energy range. In addition to a compilation of calculated Maxwellian–averaged capture cross sections versus temperature, special examples are discussed to illustrate the various details of the recipe.

061.091   Relic densities of neutralinos.
J. McDonald, K. A. Olive, M. Srednicki.
Phys. Lett. B, Vol. 283, No. 1/2, p. 80 – 84 (4 Jun 1992).
The cosmological relic density of the lightest supersymmetric particle, assumed to be a linear combination of the four neutralinos, is calculated to high precision using the exact tree-level annihilation cross section for all two–body final states.

061.092   The s–process in massive stars of variable composition.
I. Baraffe, M. F. El Eid, N. Prantzos.
Astron. Astrophys., Vol. 258, No. 2, p. 357 – 367 (May 1992).
The authors investigate the s–process nucleosynthesis during the core helium burning phase of massive stars for various values of the initial metallicity (Z). Models of stars of masses 15, 20 and 30 $M_\odot$ with $10^{-8} \leqslant Z \leqslant 0.02$ are calculated from the zero–age main sequence to the end of core helium burning. An updated nuclear reaction network coupled to the stellar evolution code is used to follow the s–process nucleosynthesis. As in previous investigations, it is found that the production of most heavy elements does not depend linearly on metallicity; the relationship of Cu and Zn to the metallicity is, however, close to a linear one. As a new result, it is found that the $^{13}C(\alpha,n)^{16}O$ neutron source becomes effective prior to the onset of core helium ignition in extremely metal–poor stars and leads to a significant production of some "light" s–nuclei. Finally, the authors study the sensitivity of their results to the nuclear data and stellar conditions, and give yields of elements from Ni to Zr as a function of stellar mass and metallicity, to be used in models of chemical evolution of the Galaxy.

061.093   Capture reactions on ^{14}C in nonstandard big bang nucleosynthesis.
M. Wiescher, J. Görres, F.–K. Thielemann.
AIP Conf. Proc., No. 238, p. 840 – 849 (15 Oct 1991). Current Physics Microform No.: 9203B1585. – See Abstr. 012.036 for the main entry.
Several capture reactions on ^{14}C, important in a nonstandard big bang scenario have been studied, the experimental techniques and results are presented. Network calculations for big bang conditions have been performed and show that the production of heavy elements in the early universe may be possible.

061.094   Cross–section measurements on radioactive samples.
P. E. Koehler, H. A. O'Brien.
AIP Conf. Proc., No. 238, p. 892 – 899 (15 Oct 1991). Current Physics Microform No.: 9203B1637. – See Abstr. 012.036 for the main entry.
The authors have developed a system for making (n, p) and (n, α) measurements on (mainly) radioactive nuclei. They are assembling a 4π detector of barium fluoride for making (n, γ) measurements on radioactive nuclei with relatively short half lives. Once operational, this new detector should allow to expand the measurements to many more nuclei, and to a broader range of nuclear physics and nuclear astrophysics issues addressed. Results of recent measurements are given and future plans are discussed.

061.095   Cross section of the $^7Li(n, \gamma)^8Li$ reaction at stellar energy.
Y. Nagai, M. Igashira, N. Mukai, K. Takeda, F. Uesawa, T. Ohsaki, T. Ando, H. Kitazawa, S. Kubono, T. Fukudo.
AIP Conf. Proc., No. 238, p. 900 – 902 (15 Oct 1991). Current Physics Microform No.: 9203B1645. – See Abstr. 012.036 for the main entry.
The reaction $^7Li(n, \gamma)^8Li$ is a trigger reaction for the nucleosynthesis of intermediate–mass nuclei in inhomogeneous big–bang models. The reaction rate was measured at neutron energy of 30 keV by detecting prompt γ–ray from the reaction. The obtained value is two times larger than that measured recently and is consistent with the estimated value from the thermal neutron capture cross section by using 1/v law.

061.096   The 30 keV averaged neutron capture cross sections of ^{56}Fe and ^{60}Ni.
F. Corvi, G. Fioni, A. Mauri, T. Babeliowsky.
AIP Conf. Proc., No. 238, p. 906 – 908 (15 Oct 1991). Current Physics Microform No.: 9203B1651. – See Abstr. 012.036 for the main entry.
High resolution neutron capture measurements of ^{56}Fe and ^{60}Ni were performed with total energy detectors in the energy range 1 – 200 keV. The data have been used to derive the Maxwellian–averaged cross sections for thermal energies around kT = 30 keV, corresponding to the mean temperature of the s–process of stellar nucleosynthesis.

061.097   Beta decay of some fp shell nuclei for presupernova stars.
K. Kar, S. Sarkar, A. Ray.
AIP Conf. Proc., No. 238, p. 909 – 910 (15 Oct 1991). Current Physics Microform No.: 9203B1654. – See Abstr. 012.036 for the main entry.
The authors describe a method for the calculation of beta decay rates for some fp shell nuclei which have important bearings on the presupernova evolution of massive stars (M $\geqslant$ 10 M$_\odot$).

061.098   The $^{180m}Ta(\gamma, \gamma')^{180}Ta$ cross section at 1.33 and 4.0 MeV and its astrophysical consequences.
Z. Németh, F. Käppeler, G. Reffo.
AIP Conf. Proc., No. 238, p. 917 – 919 (15 Oct 1991). Current Physics Microform No.: 9203B1662. – See Abstr. 012.036 for the main entry.
Enriched ^{180m}Ta samples were irradiated by an intense ^{60}Co source of 1.5 PBq and with 4 MeV bremsstrahlung. The results mean that ^{180m}Ta can survive at temperatures lower than $\sim 5 \times 10^8$K, but that it is quickly destroyed via photoexcitation above 7×10^8K. Accordingly, an s–process production of ^{180m}Ta during stellar helium burning remains a plausible possibility. The respective consequences for the observed abundance of nature's rarest stable isotope are discussed by means of new cross section information.

061.099   Positronium in astrophysical condition.
V. V. Burdyuzha, V. L. Kauts, N. P. Yudin.
IAU Symposium No. 150: Astrochemistry of cosmic phenomena, p. 41 – 45 (1992). – See Abstr. 012.029 for the main entry.
For the full paper see 54.061.085.

061.100   The p–nuclei: abundances and origins.
D. L. Lambert.
Astron. Astrophys. Rev., Vol. 3, No. 3–4, p. 201 – 256 (Apr 1992).
The review discusses the solar system (meteoritic) abundances and the possible modes of nucleosynthesis of the 30–odd p–nuclei from ^{74}Se to ^{196}Hg. In addition to a discussion of the abundances for bulk meteorites, isotopic anomalies related to the p–nuclei are discussed; e.g., the Xe–HL associated with the 'interstellar' diamonds and the extinct radionuclides ^{146}Sm and ^{92}Nb. Various proposed schemes of synthesizing p–nuclei are reviewed. It is noted that the γ–process (i.e., photoerosion) operating in SN Ia (exploding C–O white dwarfs) appears capable of accounting for the relative and absolute abundances of all but one or two of the rarest of p–nuclei. Synthesis of these latter nuclei is also discussed.

061.101   A way to limit selectron mass using supernova neutrino data.
P. Ram Babu, T. Chhabra.
Nuovo Cimento A, Vol. 105, No. 3, p. 387 – 393 (Mar 1992).
A way to estimate photino contamination in neutrino flux from supernovae is suggested. By comparing the experimental and theoretically expected ratios of forward–peaked events in the direction of neutrino source and isotropic events one can set a limit on selectron mass. As an example the data from SN 1987A have been used to set a limit on selectron mass.

061.102 Supergravity dark matter.
J. Ellis, L. Roszkowski.
Phys. Lett. B, Vol. 283, No. 3/4, p. 252 – 260 (11 Jun 1992).
The authors discuss the abundance of supersymmetric neutralino dark matter incorporating renormalized supergravity relations between sparticle masses in terms of the parameters m_0 (scalar mass), $m_{1/2}$ (gaugino mass), μ (Higgs mixing parameter), m_A (pseudoscalar Higgs mass) and tan β (ratio of Higgs VEVs), in the minimal supersymmetric extension of the standard model. They include radiative corrections to the Higgs boson masses. They describe the regions of the (μ, M_2), $(m_0, \tan \beta)$, and $(m_A, \tan \beta)$ planes where the neutralino could have the critical density.

061.103 Survey of atmospheric neutrino data and implications for neutrino mass and mixing.
E. W. Beier, E. D. Frank, W. Frati, S. B. Kim, A. K. Mann, F. M. Newcomer, R. van Berg, W. Zhang, K. S. Hirata, K. Inoue, T. Ishida, T. Kajita, K. Kihara, M. Nakahata, K. Nakamura, S. Ohara, A. Sakai, N. Sato, Y. Suzuki, Y. Totsuka, Y. Yaginuma, M. Mori, Y. Oyama, A. Suzuki, K. Takahashi, M. Yamada, M. Koshiba, K. Nishijima, T. Kajimura, T. Suda, T. Tajima, K. Miyano, H. Miyata, H. Takei, Y. Fukuda, E. Kodera, Y. Nagashima, M. Takita, H. Yokoyama, K. Kaneyuki, Y. Takeuchi, T. Tanimori.
Phys. Lett. B, Vol. 283, No. 3/4, p. 446 – 453 (11 Jun 1992).
A detailed comparison is made of the atmospheric neutrino results obtained by the Fréjus, IMB–3, and Kamiokande detectors. The implications of these results for vacuum neutrino oscillations are presented, and juxtaposed with the results for matter neutrino oscillations from the solar neutrino data.

061.104 Nucleation of strange matter in dense stellar cores.
J. E. Horvath, O. G. Benvenuto, H. Vucetich.
Phys. Rev. D, Vol. 45, No. 10, p. 3865 – 3868 (15 May 1992).
Current Physics Microform No.: 9207D1447.
The authors investigate the nucleation of strange quark matter inside hot, dense nuclear matter. Applying Zel'dovich's kinetic theory of nucleation, they find a lower limit of the temperature T for strange–matter bubbles to appear, which happens to be satisfied inside the Kelvin–Helmholtz cooling era of a compact star life but not much after it. These bounds thus suggest that a prompt conversion could be achieved, giving support to earlier expectations for nonstandard type II supernova scenarios.

061.105 Pseudo Dirac neutrinos and the solar neutrino problem.
H. Minakata, H. Nunokawa.
Phys. Rev. D, Vol. 45, No. 10, p. R3316 – R3320 (15 May 1992).
Current Physics Microform No.: 9207D0898.
Observational consequences of the pseudo Dirac–neutrino hypothesis of Kobayashi, Lim, and Nojiri are examined in the context of the solar neutrino problem. Detailed calculations are performed to obtain expected rates in the ^{37}Cl, Kamiokande II, and ^{71}Ga detectors. In a region of parameters consistent with the former two experiments the ^{71}Ga rate is predicted to be 50 – 65 solar neutrino units.

061.106 Properties of high–density matter in the electroweak symmetric phase.
D. Chandra, A. Goyal.
Phys. Rev. D, Vol. 45, No. 12, p. 4392 – 4399 (15 Jun 1992).
Current Physics Microform No.: 9208C1920.
The authors examine the bulk properties of matter at high densities and finite temperatures in the phase where electroweak symmetry is exact and fermions are massless, by taking the strong interactions into account perturbatively to lowest order in the quark–gluon chromodynamic coupling constant α_c. The authors also discuss the possibility of a phase transition of strange quark matter into this high–density matter in the electroweak symmetric phase at densities likely to be present in the core of dense neutron stars or collapsing stars. Finally, the authors study the properties of finite–size chunks of this matter by taking surface effects into account and give an estimate of the surface tension.

061.107 Particle astrophysics: experiments and observations in solar and stellar physics.
Yu. N. Gnedin.
Vulcano Workshop 1990: Frontier objects in astrophysics and particle physics, p. 189 – 200 (1991). – See Abstr. 012.040 for the main entry.
The new experiments in astrophysics are suggested for the generation and detection of pseudoscalar Goldstone bosons (such as axion, arion etc.). The main topic of this paper is connected with new experiments and astrophysical observations on the search of the exotic long–range pseudoscalar interaction and especially with existence of so–called axions.

061.108 Signatures of dark matter in underground detectors.
F. Halzen, T. Stelzer, M. Kamionkowski.
Phys. Rev. D, Vol. 45, No. 12, p. 4439 – 4442 (15 Jun 1992).
Current Physics Microform No.: 9208C1967.
The neutralino, the lightest superpartner in many supersymmetric theories, is arguably the leading dark matter candidate from both the cosmological and particle physics points of view. Its mass is bracketed by a minimum value of tens of GeV, determined from unsuccessful accelerator searches, and a maximum value of several TeV, above which neutralinos "overclose" the universe. If neutralinos exist in the Galactic halo, they will be gravitationally captured by scattering off elements in the Sun. Annihilation of neutralinos in the Sun will produce a neutrino flux which can be detected on Earth and thus provide indirect evidence for Galactic dark matter. The authors show that a 1 km² area is the natural scale of a neutrino telescope capable of probing the GeV – TeV neutralino mass range by searching for high–energy neutrinos produced by their annihilation in the Sun.

061.109 Photon and graviton Green's functions on cosmic string space–times.
B. Allen, J. G. McLaughlin, A. C. Ottewill.
Phys. Rev. D, Vol. 45, No. 12, p. 4486 – 4503 (15 Jun 1992).
Current Physics Microform No.: 9208C2014.
The authors present the Green's functions for photons and gravitons in the vicinity of an idealized cosmic string. They stress the importance of the Ward identities involved and the necessity for "smoothing" the curvature singularity in the space–time in order to carry out the calculations. The Green's functions are employed to determine the renormalized vacuum expectation value of the stress–energy tensor for scalar, electromagnetic, and linearized gravitational fields propagating in the neighborhood of an idealized cosmic string.

061.110 Improved cosmological and radiative decay constraints on neutrino masses and lifetimes.
S. A. Bludman.
Phys. Rev. D, Vol. 45, No. 12, p. 4720 – 4723 (15 Jun 1992).
Current Physics Microform No.: 9208C2248.
The best upper bounds on the masses of stable and unstable light neutrinos derive from the upper bound on the total mass density, as inferred from the lower limit $t_0 > 13$ Gyr on the dynamical age of the universe: if the universe is matter dominated, $m_\nu < 35(23) \cdot \max[1, (t_0/\tau_\nu)^{1/2}]$ eV, accordingly as a cosmological constant is (is not) allowed. The best bounds on the radiative decay of light neutrinos derive from the failure to observe prompt γ–rays accompanying the neutrinos from supernova 1987A: for any $m_\nu > 630$ eV, this provides a stronger bound on the neutrino transition moment than that obtained from red giants or white dwarfs. These results improve on earlier cosmological and radiative decay constraints by an overall factor 20 and allow neutrinos more massive than 35 eV only if they decay overwhelmingly into singlet Majorons or other new particles with a lifetime less than one month. The author reviews the 17 keV neutrino situation and notes that (1) its existence may be resolved by modest improvements in neutrino oscillation probabilities, and (2) double β decay and nucleosynthesis constraints require that its massive partner be an active neutrino, but

they allow solar neutrinos to oscillate into low–mass sterile neutrinos.

061.111 Massive Dirac neutrinos and SN 1987A.
A. Burrows, R. Gandhi, M. S. Turner.
Phys. Rev. Lett., Vol. 68, No. 26, p. 3834 – 3837 (29 Jun 1992). Current Physics Microform No.: 9209D1493.

The wrong–helicity states of a Dirac neutrino can provide an important cooling mechanism for young neutron stars. Based on numerical models of the early cooling of the neutron star associated with SN 1987A which self–consistently incorporate wrong–helicity neutrino emission, the authors argue that a Dirac neutrino of mass greater than 30 keV (25 keV if it is degenerate) leads to shortening of the neutrino burst that is inconsistent with the Irvine – Michigan – Brookhaven and Kamiokande II data. If pions are as abundant as nucleons in the cores of neutron stars, this limit improves to 15 keV.

061.112 r–process abundances and nuclear properties far from stability.
K. L. Kratz, H. Gabelmann, B. Pfeiffer, A. Wöhr, P. Möller, F.-K. Thielemann.
International Conference on Nuclear Data for Science and Technology, p. 635 – 637 (1992). – See Abstr. 012.043 for the main entry.

Recent measurements of β–decay properties of the 'waiting-point' nuclei ^{79}Cu, ^{80}Zn and ^{130}Cd, together with new QRPA shell–model predictions of so far unknown N $\cong$ 50 and N $\cong$ 82 isotopes in the r–process path, have allowed to explain the detailed isotopic composition in the A $\cong$ 80 and A $\cong$ 130 r–abundance peaks. The correlation between nuclear data far from stability and r–abundances suggests that the r–process involves a high–neutron–density β–flow equilibrium environment. Based on these results, the r–process components of nuclei in the 90 $\leqslant$ A $\leqslant$ 100 mass range were predicted for freeze–out conditions and compared to the solar–system r–process abundances.

061.113 (n, p) and (n, α) cross–section measurements with astrophysical applications.
C. Wagemans, S. Druyts, H. Weigmann, R. Barthélémy, P. Schillebeeckx, P. Geltenbort.
International Conference on Nuclear Data for Science and Technology, p. 638 – 640 (1992). – See Abstr. 012.043 for the main entry.

(n, p) and (n, α)–cross–sections with astrophysical importance have been measured for light nuclei with masses between 15 and 50 u. Thermal neutron induced measurements were performed at the high flux reactor of the I.L.L. (Grenoble); the resonance and keV neutron energy region was covered at the Geel Linear Accelerator. The importance of these data is illustrated by means of a typical s–process nucleosynthesis reaction network.

061.114 The stellar (n, γ) cross sections for ^{87}Rb and ^{192}Pt.
D. Neuberger, M. Tepe, F. Käppeler.
International Conference on Nuclear Data for Science and Technology, p. 641 – 643 (1992). – See Abstr. 012.043 for the main entry.

The stellar neutron capture cross sections of ^{87}Rb and ^{19}Pt have been measured by the activation technique relative to the gold standard cross section. The energy distribution of the resulting neutron spectrum allows directly to determine the stellar cross section for a thermal energy of kT = 25 keV. For ^{87}Rb, previous measurements exhibit sizable discrepancies that were resolved in the present study. The impact of the improved accuracy of the ^{87}Rb cross section on the s–process nucleosynthesis in the mass region 84 < A < 90 is discussed. The ^{192}Pt cross section has been investigated for the first time at keV energies. This isotope is of pure s–process origin and defines the branchings of the neutron capture chain at ^{191}Os and ^{192}Ir. Analysis of these branchings yields information on the s–process neutron density during helium shell burning in red giant stars.

061.115 Neutrino masses in the standard model.
P. Ramond.
Symposium on Quarks, Symmetries and Strings in Honor of Bunji Sakita's 60th Birthday, p. 32 – 42 (1991). – See Abstr. 012.044 for the main entry.

The author presents a generic catalogue of extensions of the Standard Model which violate lepton number and thus allow for Majorana neutrino masses. These fall into two classes, those which introduce new fermions and those which introduce new Higgs to break lepton number. In the process, the author discusses the axion–Majoron connection, and emphasizes the relevance of these models to the generation of baryon asymmetry in the early universe.

061.116 Gravitational instantons revisited – a possible origin of the cosmological principle.
K. Kikkawa.
Symposium on Quarks, Symmetries and Strings in Honor of Bunji Sakita's 60th Birthday, p. 276 – 283 (1991). – See Abstr. 012.044 for the main entry.

The physical role of a certain class of gravitational instantons is discussed. In the space where the instantons dominate matter distribution tends to be homogenized.

061.117 Transient flow of a relativistic radiating gas past a horizontal plate.
A. R. Bestman.
Astrophys. Space Sci., Vol. 192, No. 1, p. 133 – 139 (Jun 1992).

The problem of unsteady flow of a relativistic radiating neutrino gas is studied by imposing a time–dependent perturbation on a basic flow. When the perturbation is small, the problem, which is ill–posed, is reduced to a well–posed spatial value problem for the transverse velocity and the temperature. Subsequently the axial velocity and number density may be obtained by straightforward integration with respect to time and imposition of the initial condition. The solution for the initial value problem is tackled by the Laplace transform technique and the results are discussed quantitatively.

061.118 A hypothetical estimated mass of quark–lepton boson and mass of electron neutrino.
V. Skalský, M. Súkeník.
Astrophys. Space Sci., Vol. 192, No. 1, p. 149 – 150 (Jun 1992). Letter–to–the–editor.

From the assumption of symmetry of ratio of Fermion masses and ratio of masses of bosons the hypothetical mass of quark–lepton boson $m_{w_{es}}$ and hypothetical mass of electron neutrino m_{ve} result.

061.119 Neutrino emissivity of an ultrarelativistic plasma from positron and plasmino annihilation.
E. Braaten.
Astrophys. J., Vol. 392, No. 1, p. 70 – 73 (10 Jun 1992).

In an ultrarelativistic electron plasma, the electron acquires a large effective mass and an extra propagating mode called the plasmino. The neutrino emissivity from the annihilation process is calculated taking these plasma effects into account. When the temperature is sufficiently small compared to the electron Fermi energy, the emissivity from the annihilation of the electron with the positron plasmino exceeds that from the annihilation with the positron. Unless neutrinos become degenerate, the effect of the plasmino becomes significant only when the total annihilation contribution to the emissivity is negligible compared to that from photon and plasmon decay.

061.120 Searching for most of the universe.
D. O. Caldwell.
AIP Conf. Proc., No. 243, p. 1066 – 1072 (1 Jan 1992). Current Physics Microform No.: 9203C1914. – See Abstr. 012.047 for the main entry.

The non–luminous matter constituting probably more than 90% of the mass of the universe cannot be any particle in the Standard Model of particle physics. Non–accelerator experiments have eliminated as dark matter wide classes of candidate particles, spanning 12 orders of magnitude in mass and 20 orders of magnitude in cross section. Examples are weak isodoublet neutrinos $\gtrsim 30$ eV/c^2, sneutrinos, technibaryons, microcharged shadow matter, and cosmions, which could be both dark matter and solve the solar neutrino problem. Isodoublet neutrinos of all masses are eliminated if the 17–keV neutrino exists. A remaining candidate, the lightest supersymmetric particle (LSP), can be searched for with cryogenic detectors having potentially two orders of magnitude greater sensitivity than ionization detectors, as both ionization and phonons can be measured. The nuclear recoil signal produces mainly phonons, whereas the main background produces mainly ionization. Enriched ^{73}Ge will be used to look for the LSP.

061.121   Origin of ^{180m}Ta and the temperature of the s–process.
Z. Németh, F. Käppeler, G. Reffo.
Astrophys. J., Vol. 392, No. 1, p. 277 – 283 (10 Jun 1992).
The origin of ^{180m}Ta, nature's rarest isotope, represents a persisting puzzle of nuclear astrophysics. Recent studies indicate that it may be produced at a variety of nucleosynthesis sites, that is, by the s–process during stellar helium burning, or by the p– and v–processes, which occur in supernova explosions. Among these scenarios, the s–process is, at present, best suited for reliable predictions. For the quantitative discussion of a possible s–process origin of ^{180m}Ta, the required input data were improved in two directions: (1) By photoexcitation experiments it was found that the decay of ^{180m}Ta is not affected at typical s–process temperatures. (2) The stellar (n, γ)–rates for ^{179}Ta and ^{180m}Ta were calculated with a refined statistical model approach. The results show that the ^{180m}Ta abundances provide a lower or an upper limit for the temperature at the s–process site, depending on whether it is produced in stellar helium burning or in supernovae.

061.122   Electron conduction opacity for dense stellar plasmas.
N. Itoh.
Rev. Mex. Astron. Astrofis., Vol. 23, p. 231 – 234 (Mar 1992). – See Abstr. 012.054 for the main entry.
Thermal conductivity of the dense matter is calculated for the liquid metal state as well as for the crystalline lattice state. The calculation for the liquid metal state takes account of the best knowledge available on the structure factor of the ions in the high–temperature, classical limit and the dielectric screening due to degenerate electrons. The calculation for the crystalline lattice state takes account of the accurate form of the Debye–Waller factor.

061.123   Gravity–spin coupling and neutrino helicity oscillations.
G. Papini, Y. Q. Cai.
4. Canadian Conference on General Relativity and Relativistic Astrophysics, p. 355 – 359 (1992). – See Abstr. 012.056 for the main entry.
The covariant Dirac equation is solved exactly to first order in the metric deviation $\gamma_{\mu\nu}$. The solution yields general gravity–total angular momentum coupling terms and provides independent support for the existence of the rotation–spin effects for neutrons predicted by Mashhoon. The spin part of the general effect for fermions also applies to gravitational fields of arbitrary strength. Under specific conditions, the coupling is capable of flipping the helicity of the particles. This has, for neutrinos, interesting astrophysical implications.

061.124   Cosmic nucleosynthesis with fluctuations.
C. J. Hogan.
15. Texas Symposium on Relativistic Astrophysics and 4. ESO–CERN Symposium, p. 76 – 85 (1991). – See Abstr. 012.060 for the main entry.
    Contents: 1. Introduction. 2. Production of fluctuations. 3. Destruction of fluctuations. 4. Nucleosynthesis with fluctuations. 5. Conclusions.

061.125   Primordial nucleosynthesis and observed light element abundances.
B. E. J. Pagel.
15. Texas Symposium on Relativistic Astrophysics and 4. ESO–CERN Symposium, p. 131 – 140 (1991). – See Abstr. 012.060 for the main entry.
    Contents: 1. Introduction. 2. Deuterium and helium–3. 3. Lithium–7. 4. Helium–4. 5. Implications of the primordial helium abundance.

061.126   The quark–hadron phase transition.
H. Satz.
15. Texas Symposium on Relativistic Astrophysics and 4. ESO–CERN Symposium, p. 262 – 282 (1991). – See Abstr. 012.060 for the main entry.
    Contents: 1. Results from statistical quantum chromodynamics. 2. Conditions in nuclear collisions. 3. Volumes and lifetimes. 4. The onset of thermalization. 5. Primordial features. 6. Conclusions.

061.127   Light neutralino dark matter is still ok.
K. Griest.
1. International Trends in Astroparticle Physics Conference together with the SuperNova Watch Workshop, p. 405 – 410 (1992). – See Abstr. 012.102 for the main entry.
The author disputes recent claims that neutralinos lighter than 20 – 30 GeV have been ruled out as dark matter by the recent LEP accelerator experiments.

061.128   Prospects for relic neutrino detection.
P. F. Smith.
15. Texas Symposium on Relativistic Astrophysics and 4. ESO–CERN Symposium, p. 425 – 435 (1991). – See Abstr. 012.060 for the main entry.
The standard big bang model predicts a universal background of relic neutrinos, comparable in number density to the background microwave photons. Since the latter are observed, one can be confident that the neutrino background will also be present. This neutrino background is undetectable at the present time. It is therefore of interest to ask what technical developments would be needed to make its detection feasible, and whether any realistic experimental possibilities can be foreseen for the future. The aim of this paper is to illustrate the difficulties by summarizing six detection ideas that have been previously considered. The most promising direction for further study would appear to be that of coherent interactions, from which a considerably increased cross section results for scattering from bulk matter, producing small macroscopic forces. So far, no investigations of this idea have resulted in a practical detection scheme, but in this paper one new variation is suggested that could in principle give an observable effect, if the necessary stringent experimental conditions could be created.

061.129   Detection of high–energy astronomical point sources of neutrinos.
P. J. Litchfield.
15. Texas Symposium on Relativistic Astrophysics and 4. ESO–CERN Symposium, p. 474 – 482 (1991). – See Abstr. 012.060 for the main entry.
On the Soudan 2 proton–decay experiment the author and co–workers are in the business of detecting neutrinos as the main background to proton decay. Also, if previous results are correct, it may be possible to detect astronomical point sources using the high–energy cosmic–ray muons passing through the detector. The combination of the two subjects to detect neutrino point sources is obvious.

061.130   A neutrino astronomy test of the AGN paradigm and the broad line region.
F. W. Stecker, C. Done, M. H. Salamon, P. Sommers.
AIP Conf. Proc., No. 254, p. 341 – 344 (1 May 1992). – See Abstr. 012.057 for the main entry.
The authors calculate the spectrum of high energy neutrinos from AGN based on the accretion disk paradigm. They predict

that AGN will be the strongest source of ultrahigh energy background neutrinos in the universe. AGN neutrinos should be observable with further, and possible present, neutrino detectors. The authors show that high energy neutrinos can create a sphere of stellar disruption around the AGN with a radius ~ 30 light days, possibly accounting for the broad line region.

061.131 Neutrino emission from protoneutron star with modified URCA and nucleon bremsstrahlung processes.
H. Suzuki.
Symposium of Supercomputing Astronomy and Astrophysics in Japan, p. 267 – 272 (1991). – See Abstr. 012.069 for the main entry.
The author has performed numerical simulations of the quasistatic evolution of protoneutron star with neutrino transfer using multigroup flux limited diffusion scheme. The second half of the supernova neutrinos are emitted in this stage. He has found that the nucleon bremsstrahlung process is important as the source of ν_μ's. It complements the shortage of the emission rate for ν_μ's in comparison with ν_e's and $\bar{\nu}_e$'s.

061.132 High energy particle astrophysics.
Lee Wonyong.
1. Yanbian International Workshop on Modern Physics: Particles, quantum groups, high Tc, phase transitions and all that, p. 263 – 273 (1991). – See Abstr. 012.081 for the main entry.
The author discusses ultrahigh energy cosmic ray physics. In particle physics, higher energy increases the ability to discover higher mass particles and to probe smaller distances. In high energy cosmic ray physics, one detects the high energy radiation emitted by astronomical sources. By looking at the highest energies, one hopes to study the early universe.

061.133 Present status of neutrino physics.
C. W. Kim.
1. Yanbian International Workshop on Modern Physics: Particles, quantum groups, high Tc, phase transitions and all that, p. 312 – 324 (1991). – See Abstr. 012.081 for the main entry.
A brief review of the neutrino properties in matter as well as in vacuum is presented, with special emphasis on how the basic properties of neutrinos are modified when they pass through dense media such as the solar core and collapsing stars and how these changes can reveal their true nature.

061.134 Die Einheit von Mikro– und Makrokosmos.
H. Schopper.
Atomwirtsch., Atomtech., Vol. 37, No. 5, p. 230 – 237 (May 1992).
Developments in elementary particle physics and astrophysics as well as cosmology, together with the introduction of new concepts and paradigms, are opening up new possibilities to achieve a synthesis of the infinitesimally small and the infinitesimally large. This synthesis will have a great impact not only on specific fields of science, but will change the entire concept of the world. Elementary particle physics is in the process of abandoning Democritus' philosophy of indivisible atoms and approaching the Platonic concept of symmetry. Investigating the composition of matter also involves the question about what happened before the Big Bang, and whether there was any Big Bang as a singular event.

061.135 From stellar structures to fundamental physics.
V. Castellani, S. Degl'Innocenti.
Mem. Soc. Astron. Ital., Vol. 63, No. 1, p. 17 – 24 (1992). – See Abstr. 012.087 for the main entry.
The authors discuss astrophysical parameters which can give information about the value of the neutrino magnetic moment (n.m.m.). They found that pulsational properties of RR Lyrae pulsators in galactic globular clusters put an upper limit to n.m.m. of the order of $\mu_\nu = 10^{-12}\mu_B$. A similar upper limit is derived from the luminosity of the He burning giants in old galactic clusters.

061.136 Understanding n–capture nucleosynthesis: a test for stellar and galactic evolution.
M. Busso, R. Gallino, C. M. Raiteri.
Mem. Soc. Astron. Ital., Vol. 63, No. 1, p. 57 – 60 (1992). – See Abstr. 012.087 for the main entry.
Contents: 1. s– and r–processes and stellar evolution. 2. Galactic evolution traced by n–captures.

061.137 Quelques liens entre astrophysique et physique nucléaire.
B. Pichon.
J. Astron. Fr., No. 41, p. 16 (Feb 1992). – See Abstr. 012.070 for the main entry.

061.138 Le problème du lithium.
F. Spite.
J. Astron. Fr., No. 41, p. 19 (Feb 1992). – See Abstr. 012.070 for the main entry.

061.139 Aggiornamento sulle problematiche astronomiche e fisiche legate alla fenomenologia neutrino e alle attuali insufficienti conoscenze delle fondamentali proprietà di tale particella.
A. Masani.
G. Astron., Vol. 18, N. 1, p. 12 – 19 (Mar 1992).

061.140 Synthesis of nuclei in astrophysical environments.
F.–K. Thielemann, K.–L. Kratz.
22. Masurian Lakes Summer School on Nuclear Physics: Frontier topics in nuclear and astrophysics, p. 187 – 226 (1992). – See Abstr. 012.094 for the main entry.
Nuclear physics and astrophysics have at least two major points of common interest. These are first a wealth of nuclear reactions which explain the synthesis and abundance patterns of nuclei in nature. Another aspect is the behavior of nuclear matter at and beyond nuclear densities, being important in heavy–ion collisions and for the equation of state in the supernova collapse and in neutron stars. This review concentrates on nucleosynthesis processes. A brief introduction is given to the physics in astrophysical plasmas which governs composition changes. A short summary of abundances resulting from the big bang is given.

061.141 Cosmochronology and nucleochronometry.
E. Sheldon.
22. Masurian Lakes Summer School on Nuclear Physics: Frontier topics in nuclear and astrophysics, p. 227 – 249 (1992). – See Abstr. 012.094 for the main entry.
Contents: (1) Preamble. (2) Astronomic dating methods and indicants of age. (3) Nuclear dating.

061.142 Unified description of quark and lepton mass matrices.
E. Ma.
2. Winter School of Physics: Cosmology and elementary particles, p. 267 – 274 (1992). – See Abstr. 012.092 for the main entry.
Prompted by the present empirical information regarding quark masses and mixing angles, the author proposes an $S_3 X Z_3$ discrete symmetry from which two successful relationships are derived. This model is then extended to include leptons, resulting in two qualitatively different solutions for a possible 17 keV neutrino.

061.143 Time and prediction in quantum cosmology.
J. B. Hartle.
Conceptual problems of quantum gravity, p. 172 – 203 (1991). – See Abstr. 003.049 for the main entry.
A generalized quantum mechanics for cosmological spacetimes is suggested in which no variable plays the special role of the time in familiar quantum mechanics. In this generalization, the central role of time in familiar quantum mechanics arises not as a fundamental aspect of the formalism, but rather as an approximation appropriate to those initial conditions of the universe that lead to classical spacetime when it is large.

061.144   Time in quantum cosmology.
    J. J. Halliwell.
Conceptual problems of quantum gravity, p. 204 – 210 (1991). – See Abstr. 003.049 for the main entry.

061.145   The role of time in the interpretation of the wave function of the universe.
    R. M. Wald.
Conceptual problems of quantum gravity, p. 211 (1991). Abstract. – See Abstr. 003.049 for the main entry.

061.146   Space and time in the quantum universe.
    L. Smolin.
Conceptual problems of quantum gravity, p. 228 – 291 (1991). – See Abstr. 003.049 for the main entry.

This paper is devoted to the problem of constructing a quantum theory that could describe a closed system – a quantum cosmology. The author argues that this problem is an aspect of a much older problem – that of how to eliminate from the physical theories "ideal elements", which are elements of the mathematical structure whose interpretation requires the existence of things outside the dynamical system described by the theory. This discussion is aimed at uncovering criteria that a theory of quantum cosmology must satisfy, if it is to give physically sensible predictions. The author proposes three such criteria and shows that conventional quantum cosmology can only satisfy them, if there is an intrinsic time coordinate on the phase space of the theory. It is shown that approaches based on correlations in the wave function, that do not use an inner product, cannot satisfy these criteria. As example, the author discusses the problem of quantizing a class of relational dynamical models invented by Barbour and Bertotti. The dynamical structure of these theories is closely analogous to general relativity, and the problem of their measurement theory is also similar. It is concluded that these theories can only be sensibly quantized if they contain an intrinsic time.

061.147   Perspectives in particle astrophysics.
    B. C. Barish.
1. International Symposium on Particles, Strings and Cosmology, p. 185 – 206 (1991). – See Abstr. 012.098 for the main entry.

The author discusses some selected topics of current interest in particle astrophysics. The review concentrates on two topics: GUT magnetic monopoles and neutrino astrophysics. Present large detector projects are briefly characterized.

061.148   Intrinsic neutrino properties: as deduced from cosmology, astrophysics, accelerator, and non–accelerator experiments.
S. P. Rosen.
1. International Symposium on Particles, Strings and Cosmology, p. 207 – 223 (1991). – See Abstr. 012.098 for the main entry.

This paper reviews the intrinsic properties of neutrinos as deduced from cosmological, astrophysical, and laboratory experiments. Bounds on magnetic moments and theoretical models which yield large magnetic moments but small masses are briefly discussed. The MSW solution to the solar neutrino problem is reviewed in the light of the existing data from the ^{37}Cl and Kamiokande II experiments. The combined data disfavor the adiabatic solution and tend to support either the large angle solution or the nonadiabatic one. In the former case, the ^{71}Ga signal will be suppressed by the same factor as for ^{37}Cl, and in the latter case the suppression factor could be as large as 10 or more.

061.149   Gravitino–induced baryogenesis.
    J. Cline.
1. International Trends in Astroparticle Physics Conference together with the SuperNova Watch Workshop, p. 128 – 137 (1992). – See Abstr. 012.102 for the main entry.

The author investigates cosmic baryon generation in the minimal low energy supergravity model with the addition of dimension–four baryon–violating interactions. It is shown that gravitinos, if heavier than squarks, can give rise to a baryon asymmetry as they decay. Assuming that gravitinos are not

diluted by inflation in the post–Planckian epoch, the author finds a realistic baryon–to–photon ratio for various ranges of gaugino and squark masses between 100 GeV and 10 TeV.

061.150   Neutrino flavor–spin oscillations in the Sun.
    S. A. Bludman.
1. International Symposium on Particles, Strings and Cosmology, p. 537 – 544 (1991). – See Abstr. 012.098 for the main entry.

The resonance theory for flavor – and flavor–spin – changing neutrino oscillations is reviewed, stressing the constraints on adiabaticity that limit ν_{e1} conversion in the Sun. Flavor–spin oscillations predict the wrong energy dependence for the detected neutrino flux and allow no more than another 30% reduction in neutrino persistance probability at solar maximum.

061.151   Seesaw model predictions for the τ–neutrino mass.
    S. A. Bludman, D. C. Kennedy, P. G. Langacker.
1. International Trends in Astroparticle Physics Conference together with the SuperNova Watch Workshop, p. 378 – 382 (1992). – See Abstr. 012.102 for the main entry.

The small neutrino masses and flavor–mixing apparently observed in the Sun are most naturally interpreted in terms of grand unification theories incorporating the seesaw mechanism. In two such theories, SO(10) GUT and SUSY GUT, that are consistent with all laboratory data, the neutrino mixing is like up–quark (CKM) and the neutrino masses are proportional to the squares of the up–quark masses. For the SO(10) GUTS model, the symmetry breaking scale is intermediate and the μ–neutrino mass is close to that observed in solar neutrino oscillations. Although the seesaw model mass predictions are less reliable than the mixing angle predictions, the τ–neutrino mass may lie in the cosmologically important range (4 – 28) eV and be accessible to laboratory neutrino oscillation experiments or to observation in a nearby supernova.

High energy astrophysics. Volume 1. Particles, photons and their detection.
See Abstr. 003.018.

Strukturen des Universums.
See Abstr. 003.030.

Dynamics of the Standard Model.
See Abstr. 003.044.

Physics of massive neutrinos.
See Abstr. 003.091.

From steam to stars to the early universe.
See Abstr. 005.036.

LEP, the Laboratory for Electrostrong Physics, one year later.
See Abstr. 009.010.

Particle astronomy minisymposium: introductory remarks.
See Abstr. 011.016.

Proceedings of the Fourth Asia Pacific Physics Conference.
See Abstr. 012.006.

Z⁰ physics. Cargèse 1990.
See Abstr. 012.035.

Capture gamma–ray spectroscopy. Proceedings.
See Abstr. 012.036.

Nuclear data for science and technology. Proceedings.
See Abstr. 012.043.

Intersections between particle and nuclear physics. Proceedings.
See Abstr. 012.047.

Texas/ESO–CERN Symposium on Relativistic Astrophysics, Cosmology, and Fundamental Physics. Proceedings.
See Abstr. 012.060.

Structure of hadrons and hadronic matter. Proceedings.
See Abstr. 012.062.

Testing the standard model. Proceedings.
See Abstr. 012.063.

Cosmology and elementary particles. Proceedings.
See Abstr. 012.092.

Frontier topics in nuclear and astrophysics – graduate lectures.
See Abstr. 012.094.

Particles, strings and cosmology. Proceedings.
See Abstr. 012.098.

Trends in astroparticle physics. Proceedings.
See Abstr. 012.102.

Can astrophysics rescue particle physics from the standard model impasse?
See Abstr. 013.043.

Neutrino and gravitational wave astronomy: grand visions.
See Abstr. 013.098.

From photons to hadrons to galaxies – how to analyze the texture of matter distributions.
See Abstr. 021.009.

Measurement of the ^{8}Li$(\alpha, n)^{11}$B reaction cross section at energies of astrophysical interest.
See Abstr. 022.018.

A measurement of the ^{14}C$(n, \gamma)^{15}$C cross section at a stellar temperature of $kT = 23.3$ keV.
See Abstr. 022.024.

Thermal detection of dark matter.
See Abstr. 022.026.

Solar–neutrino neutral–current detection methods in the Sudbury Neutrino Observatory.
See Abstr. 022.075.

Searches for dark matter particles.
See Abstr. 022.233.

A proposed search for dark–matter axions in the $0.6 - 16$ μeV range.
See Abstr. 022.234.

"Long–term" neutrino flux integrations.
See Abstr. 022.240.

Study of mixing of solar neutrinos with a 1000 ton ICARUS detector.
See Abstr. 022.241.

Greenland '90: a first step toward using the polar ice cap as a Cherenkov detector.
See Abstr. 022.242.

A double beta decay experiment using bolometric detectors.
See Abstr. 022.243.

A cosmic axion experiment.
See Abstr. 022.244.

The possibility of radiogeochemical limits on stellar collapse rates in the Galaxy.
See Abstr. 022.246.

A large low energy neutrino detector for oscillations and supernovae watch.
See Abstr. 022.247.

The LVD supernova detector.
See Abstr. 022.248.

Expected supernova neutrino signal in IMB–3 detector.
See Abstr. 022.249.

Sudbury Neutrino Observatory.
See Abstr. 022.250.

Extra galactic supernova detector, neutrino mass and the supernova watch.
See Abstr. 022.251.

Supernova detectors based on coherent nuclear recoil.
See Abstr. 022.252.

Calculation of the upward–going muon background for a surface neutrino detector.
See Abstr. 034.060.

DUMAND–II progress report.
See Abstr. 034.069.

Extragalactic supernova detector, neutrino mass, and the Supernova Watch.
See Abstr. 034.086.

Thermal detection of dark matter.
See Abstr. 034.089.

Recent results on cryogenic detector developments.
See Abstr. 034.090.

Neutrino astronomy with large Cerenkov detectors.
See Abstr. 034.091.

Low–temperature extragalactic supernova neutrino detector.
See Abstr. 034.126.

Past and future dark matter experiments.
See Abstr. 036.164.

Indirect detection of dark matter.
See Abstr. 036.165.

Axion searches.
See Abstr. 036.207.

On the thermal conductivity due to collisions between relativistic degenerate electrons.
See Abstr. 062.087.

Unsteady flow of a relativistic radiating gas in a gravitational field.
See Abstr. 062.180.

s–processing in massive stars as a function of metallicity and interpretation of observational trends.
See Abstr. 065.027.

Astration and production in chemical evolution.
See Abstr. 065.033.

Constraints of axions from white dwarf cooling.
See Abstr. 065.072.

SiC particles from asymptotic giant branch stars: Mg burning and the s–process.
See Abstr. 065.094.

Constraints of axions from white dwarf cooling.
See Abstr. 065.109.

Supernova neutrinos: life after SN 1987A.
See Abstr. 065.113.

Solar and supernova neutrino interactions.
See Abstr. 065.119.

Two Kaluza–Klein wormhole solutions.
See Abstr. 066.040.

Updating nucleosynthesis bounds on Jordan–Brans–Dicke theories of gravity.
See Abstr. 066.102.

A description of semidegenerate self–gravitating spheres of fermions.
See Abstr. 066.158.

Self–gravitating general–relativistic cosmic strings.
See Abstr. 066.173.

Causal horizons, accelerations and strings.
See Abstr. 066.175.

Internal structure of a classical spinning electron.
See Abstr. 066.194.

A new class of stationary solutions to the five–dimensional Kaluza–Klein field equations.
See Abstr. 066.242.

Equations of state of cosmic strings in the presence of charged particles.
See Abstr. 066.281.

Baryosynthesis, gravitation and hot big–bang cosmology.
See Abstr. 066.288.

Virasoro model space and 2D gravity.
See Abstr. 066.295.

String fields in two–dimensional gravity.
See Abstr. 066.296.

Yang–Mills–like renormalizability of gauge–affine gravity and indirect applications.
See Abstr. 066.301.

Nucleosynthesis in a thick accretion disk around a 10 $M_\odot$ black hole.
See Abstr. 067.020.

Neutrino production from accreting X–ray pulsars.
See Abstr. 067.021.

Synchrotron emission of neutrino pairs in neutron stars.
See Abstr. 067.033.

The synthesis of ^{26}Al during combined hydrogen and helium–burning reactions.
See Abstr. 067.043.

Rotational properties of strange stars.
See Abstr. 067.070.

Rapid cooling of neutron stars by hyperons and Δ isobars.
See Abstr. 067.113.

Quasi–elastic neutrino scattering by nuclei in superdense matter of a collapsing star.
See Abstr. 067.145.

Bulk viscosity of hot neutron star matter from direct Urca processes.
See Abstr. 067.151.

Nuclear physics of dense matter.
See Abstr. 067.193.

Neutrinos from hell.
See Abstr. 067.231.

The solar neutrino problem and the neutrino magnetic moment.
See Abstr. 080.008.

Solar neutrino puzzle, horizontal symmetry of electroweak interactions and fermion mass hierarchies.
See Abstr. 080.055.

The fate of ^{7}Be in the Sun.
See Abstr. 080.057.

The solar neutrino problem.
See Abstr. 080.074.

Non–MSW solutions to the solar neutrino problem.
See Abstr. 080.095.

Can the strongly interacting dark matter be a heating source of Jupiter?
See Abstr. 099.034.

Correlated Si isotope anomalies and large ^{13}C enrichments in a family of exotic SiC grains.
See Abstr. 105.016.

Is beryllium in metal–poor stars of galactic or cosmological origin?
See Abstr. 114.044.

HEAO 3 limits on the ^{44}Ti yield in type I supernovae.
See Abstr. 125.013.

HEAO 3 limits on the ^{44}Ti yield in Galactic supernovae.
See Abstr. 125.023.

Neutrino transport in supernovae.
See Abstr. 125.063.

Neutral and charged current mediated neutrino heating of supernova shocks and supernova 1987A.
See Abstr. 125.064.

The supernova 1987A.
See Abstr. 125.090.

Resonance transitions to inert neutrinos in supernovae.
See Abstr. 125.093.

Hard X–rays from supernova 1987A: results of MIR–KVANT and GRANAT in 1987 – 1990 and expectations.
See Abstr. 125.098.

Majoron emission from SN 1987A.
See Abstr. 125.104.

SN 1987A: observations of the later phases.
See Abstr. 125.140.

Pion condensates in the Vela pulsar and the energetics of its glitch events.
See Abstr. 126.014.

Axion cooling of white dwarfs.
See Abstr. 126.107.

The interstellar deuterium–to–hydrogen ratio: a reevaluation of Lyman absorption–line measurements.
See Abstr. 131.106.

Abundances in the interstellar medium.
See Abstr. 131.131.

Sources of enrichment.
See Abstr. 131.282.

Spatially resolved spectroscopy of WR ring nebulae. III. New results and chemical properties of the sample.
See Abstr. 132.027.

A distribution for the galactic ^{26}Al and e^+e^- line emissions.
See Abstr. 143.019.

Shifts of the ^{26}Al line due to galactic rotation.
See Abstr. 143.022.

Gamma–ray lines from classical novae.
See Abstr. 143.023.

Neutrino astronomy on the 1 km^2 scale.
See Abstr. 144.008.

Observation of muons and neutrinos at great depth underground.
See Abstr. 144.015.

Implications of Cp violation in hyperon decays.
See Abstr. 144.018.

Observation of a small atmospheric v_μ/v_e ratio in Kamiokande.
See Abstr. 144.027.

The isotopic composition of iron–group cosmic rays.
See Abstr. 144.031.

Experimental search for direct detection of WIMPs.
See Abstr. 144.047.

Production of Li, Be, and B in the early Galaxy.
See Abstr. 155.009.

Possible sources of the Population I lithium abundance and light–element evolution.
See Abstr. 155.103.

OSSE observations of galactic 511 keV annihilation radiation.
See Abstr. 155.106.

Evolution of heavy–element abundances as a constraint on sites for neutron–capture nucleosynthesis.
See Abstr. 155.132.

Detecting dark matter via microlensing.
See Abstr. 155.136.

Indirect detection of heavy supersymmetric dark matter.
See Abstr. 155.158.

The metal–poor H II galaxy SBS 0335–052 and the primordial helium abundance.
See Abstr. 157.001.

Dwarf spheroidal galaxies and the mass of the neutrino.
See Abstr. 157.199.

The evolution of the alpha–elements in galaxies.
See Abstr. 157.299.

Astrophysical bags: a new paradigm for active galactic nuclei?
See Abstr. 158.189.

An observational search for axions.
See Abstr. 160.125.

Cosmological QCD Z(3) phase transition in the 10 TeV temperature range?
See Abstr. 161.005.

Charged particle creation in the steady state universe.
See Abstr. 161.008.

A single explanation for both the baryon and dark matter densities.
See Abstr. 161.014.

Model for cold dark matter.
See Abstr. 161.029.

Generalized dilaton couplings to dark matter.
See Abstr. 161.042.

Direct versus indirect searches for neutralino dark matter.
See Abstr. 161.058.

SUSY GUTs dark matter.
See Abstr. 161.066.

Thermal evolution of phases during the cosmological quark–hadron transition.
See Abstr. 161.078.

The thermodynamics of massless particle creation in an anisotropic universe.
See Abstr. 161.084.

QCD phase transition in the early universe and primordial nucleosynthesis.
See Abstr. 161.091.

Deflagration instability in the quark–hadron phase transition.
See Abstr. 161.116.

Almost–standard big bang nucleosynthesis with $\Omega_B h_2{}^0 \geqslant 0.015$: a reexamination of neutrino chemical potentials and $\varDelta$G.
See Abstr. 161.130.

Cosmological γ–rays from the decay of fourth family neutrinos: possibility and constraints.
See Abstr. 161.140.

Particle abundances in our universe: deterministic, or randomly determined via quantum cosmology or inflationary quantum fluctuations?
See Abstr. 161.144.

The spectral distortions of the cosmic microwave background radiation, LEP and heavy neutrinos.
See Abstr. 161.165.

Dark matter candidates and methods for detecting them.
See Abstr. 161.166.

Late phase transition – induced fluctuations in the cosmic neutrino distribution and the formation of structure in the universe.
See Abstr. 161.195.

Dark matter in the universe.
See Abstr. 161.215.

Late baryogenesis faces primordial nucleosynthesis.
See Abstr. 161.216.

Apocalypse according to the theoretician.
See Abstr. 161.220.

Inflation from strings. II. Reheating and baryogenesis.
See Abstr. 161.259.

System of self–gravitating semidegenerate fermions with a cutoff of energy and angular momentum in their distribution function.
See Abstr. 161.278.

The particle–cosmology connection: neutrino counting, dark matter and large–scale structure.
See Abstr. 161.288.

Cosmological limits on stable particle production at high energy.
See Abstr. 161.303.

Nucleation and bubble growth in a first–order cosmological electroweak phase transition.
See Abstr. 161.312.

Parameters of the ultrastable expansive non–decelerative universe and hypothetical mass of the electron neutrino.
See Abstr. 161.330.

Primordial nucleosynthesis bounds on the Brans–Dicke theory.
See Abstr. 161.359.

Particle production during inflationary reheating.
See Abstr. 161.367.

Semilocal strings and monopoles.
See Abstr. 161.387.

Primordial helium: the third decimal place.
See Abstr. 161.388.

Background radiation from dark matter.
See Abstr. 161.416.

Dark matter: an overview of direct searches.
See Abstr. 161.420.

Astroparticle physics and superstrings.
See Abstr. 161.425.

Constraints on heavy particles decaying into neutrinos.
See Abstr. 161.431.

Weyl–Dirac geometry and dark matter.
See Abstr. 161.434.

The cold dark matter model: does it work?
See Abstr. 161.448.

Inflationary axion cosmology.
See Abstr. 161.460.

Testing the Big Bang: light elements, neutrinos, dark matter and large–scale structure.
See Abstr. 161.506.

$\Omega_B = 1$ universe and primordial nucleosynthesis.
See Abstr. 161.507.

Neutrinos from dark matter annihilation
See Abstr. 161.515.

Physics at the Planck scale.
See Abstr. 161.516.

Inflation and axions.
See Abstr. 161.517.

Natural inflation.
See Abstr. 161.519.

Cosmic strings and large–scale structure: a comparative review.
See Abstr. 161.525.

Probing high energy physics with primordial nucleosynthesis.
See Abstr. 161.526.

Creation of a universe in the laboratory.
See Abstr. 161.528.

Inflationary cosmology.
See Abstr. 161.534.

Tau neutrinos and cold dark matter.
See Abstr. 161.535.

Testing big bang nucleosynthesis.
See Abstr. 161.536.

062 Hydrodynamics, Magnetohydrodynamics, Plasma

062.001 On a new, one–dimensional, time–dependent model for turbulence and convection. I. A basic discussion of the mathematical model.
M. Gehmeyr, K.–H. A. Winkler.
Astron. Astrophys., Vol. 253, No. 1, p. 92 – 100 (Jan 1992).

A model for turbulence and convection in regimes of very large Reynolds numbers is introduced by coupling an equation for turbulent energy to the equations of radiation hydrodynamics. These in turn are modified by nonlinear terms due to interactions with turbulence. All appearing Reynolds correlations are consistently modeled using a diffusion approximation. The equations of the model are discussed for spherical symmetry. The related equations for stellar structure are divided. In the hydrostatic regime, the free parameters are determined in the limit of a fully developed convection zone by comparison with the usual mixing length formulations. The general properties of the turbulence equation are characterized. Its consistency within the framework of the equations of radiation hydrodynamics is highlighted. It is shown that familiar notions like "semi-convection", "overshooting", and the criterion for convective instability find a natural interpretation in the model.

062.002 On a new, one–dimensional, time–dependent model for turbulence and convection. II. An elementary comparison of the old and the new model.
M. Gehmeyr, K.–H. A. Winkler.
Astron. Astrophys., Vol. 253, No. 1, p. 101 – 112 (Jan 1992).

A new model for turbulent convective transport in spherical symmetry is compared with a related version suggested by Stellingwerf (1982). Both formulations are compared in detail and crucial differences are highlighted. For various idealized situations the pure temporal and spatial behavior of both models is analyzed.

062.003 Formation of electric–current sheets in the magnetostatic atmosphere.
B. C. Low.
Astron. Astrophys., Vol. 253, No. 1, p. 311 – 317 (Jan 1992).
This paper presents an analytical illustration of the theorem due to Parker that a magnetic field with an arbitrarily prescribed topology in an electrically, perfectly conducting medium would, in general, seek equilibrium states with embedded electric current sheets, or magnetic tangential discontinuities. Low and Wolfson had previously shown that current sheets must form in certain magnetic fields driven to evolve quasi–statically through force–free states by the continuous displacements of the magnetic footpoints at the boundary. In particular, this process can take place in the absence of magnetic neutral points. This paper extends their result to the case of an isothermal atmosphere in which the Lorentz force does not vanish but balances pressure and gravitational forces in static equilibrium. Using the analytic magnetostatic solutions of Zweibel and Hundhausen, it is shown that the quasi–static evolution of the atmosphere in response to a continuously changing distribution of the plasma pressure at the atmospheric base can bring an initially smooth state to one in which an electric current sheet forms. The sheet formation is unavoidable in the end state because the balance of force requires the expulsion of magnetized plasma from between two magnetic surfaces under the frozen–in condition. A discussion is given of the result to relate it to recent developments.

062.004 MHD flow past an obstacle: large–scale flow in the magnetosheath.
C. C. Wu.
Geophys. Res. Lett., Vol. 19, No. 2, p. 87 – 90 (24 Jan 1992).
As a step to the study of the large–scale flow in the magnetosheath, the MHD flow past an obstacle is investigated. A time asymptotic method is used to obtain 3D steady–state solutions. The results indicate the formation of a depletion layer near the obstacle due to the increase of the magnetic field. Along the earth–sun line the plasma density increases first and then decreases from the post bow shock to the magnetopause. The local density maximum in front of the magnetopause may correspond to what was recently observed. When the interplanetary magnetic field direction is tilted from the solar wind flow, the IMF influences the shape of the bow shock as well as the location of the stagnation point at the magnetopause in a way consistent with observation. In addition, the results show the existence of a magnetosheath current in the post parallel–shock region.

062.005 Self–organization of cosmic radiation pressure instability. II. One–dimensional simulations.
C. J. Hogan, J. Woods.
Astrophys. J., Vol. 384, No. 1, p. 111 – 114 (1 Jan 1992).
The clustering of statistically uniform discrete absorbing particles moving solely under the influence of radiation pressure from uniformly distributed emitters is studied in a simple one–dimensional model. Radiation pressure tends to amplify statistical clustering in the absorbers; the absorbing material is swept into empty bubbles, the biggest bubbles grow bigger almost as they would in a uniform medium, and the smaller ones get crushed and disappear. Numerical simulations of a one–dimensional system are used to support the conjecture that the system is self–organizing. Simple statistics indicate that a wide range of initial conditions produce structure approaching the same self–similar statistical distribution, whose scaling properties follow those of the attractor solution for an isolated bubble. The importance of the process for large–scale structuring of the interstellar medium is briefly discussed.

062.006 Potential of a charged body in a cold plasma.
V. M. Barsukov, Yu. P. Mal'tsev.
Geomagn. Aeron., Vol. 30, No. 5, p. 653 – 656 (Apr 1990). English translation of Geomagn. Aehron., Tom 30, No. 5, p. 771 – 775 (1990).
The distribution of electric potential is computed in the vicinity of a conducting sphere injecting energetic electrons into a cold plasma. It is shown that the background electrons begin to undergo Langmuir oscillations immediately after the start of the injection. Simultaneously, they take on an average velocity directed toward the sphere, thus neutralizing its charge. The Langmuir oscillations die out in a time equal to the collision time of electrons with neutrals or ions. If the background electrons are magnetized, the equipotential surfaces external to the sphere are strongly extended along the magnetic field. The potential of the sphere turns out to be independent of the electron collision frequency. The influence of the injected electrons on the potential of the sphere is examined. It is shown that the influence is not significant if the radius of the sphere is less than the ratio of the velocity of the injected electrons to the Langmuir background frequency.

062.007 The problem of exact interior solutions for rotating rigid bodies in general relativity.
H. D. Wahlquist.
J. Math. Phys., Vol. 33, No. 1, p. 304 – 335 (Jan 1992). Current Physics Microform No.: 9202D1098.
The $(3+1)$ dyadic formalism for timelike congruences is applied to derive interior solutions for stationary, axisymmetric, rigidly rotating bodies. In this approach the mathematics is formulated in terms of three–space–covariant, first–order, vector–dyadic, differential equations for a and Ω, the acceleration and angular velocity three–vectors of the rigid body; for T, the stress dyadic of the matter; and for A and B, the "electric" and "magnetic" Weyl curvature dyadics which describe the gravitational field. It is shown how an appropriate ansatz for the forms of these dyadics can be used to discover exact rotating interior solutions such as the perfect fluid solution. By incorporating anisotropic stresses, a generalization is found of that previous solution and, in addition, a very simple new solution that can only exist in toroidal configurations.

062.008 The theory of magnetohydrodynamic wave generation by localized sources. III. Efficiency of plasma heating by dissipation of far–field waves.
W. Collins.
Astrophys. J., Vol. 384, No. 1, p. 319 – 332 (1 Jan 1992).
The flux in Alfvén waves derived with Lighthill's method is compared with fast and slow wave emission from a localized source in an isothermal, homogeneous plasma with constant B field. Since a small fraction of the fast and slow flux generated in the solar photosphere is transported to the corona, the ratio of the Alfvén to the total flux sets an uppper bound e on the efficiency of coronal heating from waves emitted by photospheric motions. The scaling of the fluxes with $\beta \sim p_{gas}/B^2$ shows that in the photosphere, where $\beta \gg 1$, slow and Alfvén emission are comparable and exceed fast radiation by $O(\beta^{7/2})$. These results imply that e is bound by $e \leqslant 1/2$. In order to investigate the sensitivity of the efficiency to the structure, a method is developed for comparing MHD fluxes by decomposition of arbitrary source fields into vector harmonics and computing the flux associated with the coupling between individual harmonics. For a simple class of sources in a high–β plasma, this technique shows that slow and Alfvén emission have similar magnitudes, giving $e \leqslant 1/2$. If these spherical sources are adopted as a model for oscillating solar granules with a 5 min period, the Alfvén fluxes are comparable to those required to heat the active regions.

062.009 Magnetohydrodynamic equilibria and cusp formation at an X–type neutral line by footpoint shearing.
G. E. Vekstein, E. R. Priest.
Astrophys. J., Vol. 384, No. 1, p. 333 – 340 (1 Jan 1992).
Some results concerning two–dimensional magnetohydrodynamic equilibria in the presence of an X–type neutral line are

presented. It is shown that in the framework of ideal MHD the latter is structurally unstable: shearing of an initial potential poloidal magnetic field results in the splitting of an X–point into a current sheet with cusp–points (or Y–points for some particular cases) at its ends. At the same time, one finds the formation of current sheets all along the separatrices even for a footpoint shearing displacement that is finite as the separatrix is approached. Scale–invariant or similarity solutions for the poloidal magnetic field inside the cusp are found. They are valid in the vicinity of a cusp–point on lengths much smaller than the global scale. These solutions predict fractional power–law singularities in the current density at the separatrix as well as a universal behavior for the poloidal magnetic field strength B_p near a cusp point, namely $B_p \sim r^{1/3}$. The global geometry of the poloidal magnetic field produced by different types of footprint shearing motions is discussed.

062.010 Linear resistive magnetohydrodynamic computations of resonant absorption of acoustic oscillations in sunspots.
M. Goossens, S. Poedts.
Astrophys. J., Vol. 384, No. 1, p. 348 – 360 (1 Jan 1992).

A numerical study of the resonant absorption of p–modes by sunspots is carried out in linear resistive MHD. The sunspot is idealized as a cylindrical axisymmetric flux tube stratified only in the radial direction and surrounded by a uniform unmagnetized plasma. First, the results of Lou in viscous MHD are reproduced in resistive MHD, and, hence, it is shown that the absorption efficiency is independent of the actual dissipation mechanism. Next, the parameter domain investigated by Lou is substantially enlarged to higher m–values, larger sunspots, and sunspot models with a twisted magnetic field. A parametric study reveals that the efficiency of the absorption mechanism depends significantly on both the equilibrium model and the characteristics of the p–modes. The overall picture resulting from this numerical survey of the relevant parameter domain is that the resonant absorption of p–modes is more efficient in larger sunspots with twisted magnetic fields. This is particularly true for p–modes with higher azimuthal wave numbers.

062.011 Statistical mechanics, Euler's equation, and Jupiter's Red Spot.
J. Miller, P. B. Weichman, M. C. Cross.
Phys. Rev. A, Vol. 45, No. 4, p. 2328 – 2359 (15 Feb 1992). Current Physics Microform No.: 9204D1042.

The authors construct a statistical–mechanical treatment of equilibrium flows in the two–dimensional Euler fluid which respects all conservation laws. The vorticity field is fundamental, and its long–range Coulomb interactions lead to an exact set of nonlinear mean–field equations for the equilibrium state. The authors illustrate the equations by solving them numerically in simple cases. In more complicated cases they use Monte Carlo techniques, with the eventual aim of detailed comparison with the Red Spot dynamical simulations of Marcus: Preliminary efforts show good agreement. Their techniques may be generalized to a number of other Coulomb–like Hamiltonian systems with an infinite number of conservation laws, including some in higher dimensions. For example, they rederive Lynden–Bell's theory of stellar–cluster formation, as well as the Debye–Hückel theory of electrolytes.

062.012 On the analysis of collision strengths and rate coefficients.
A. Burgess, J. A. Tully.
Astron. Astrophys., Vol. 254, No. 1/2, p. 436 – 453 (Feb 1992).

This paper presents a new way of critically assessing and compacting data for electron impact excitation of positive ions. Collision strengths Ω are scaled and then plotted as functions of the colliding electron energy, the complete range of energies being mapped onto the interval (0,1). The scaled Ω can then be represented to good accuracy by a 5–point spline. Similar scaling and spline fitting techniques enable thermally averaged collision strengths Υ to be obtained at all temperatures. Whole isoelectronic sequences may then be treated by mapping the ion charge onto (0,1) and using a second 5–point spline; thus reducing Υ, for the

whole sequence and all temperatures, to a 5×5 array. Three main types of transition (optically allowed, forbidden and exchange) are discussed separately and illustrated by numerical examples. Interactive programs for analysing atomic data in this way have been developed.

062.013 Simulation of ion acceleration in a charged dust cloud.
U. Motschmann, K. Sauer, T. Roatsch.
Geophys. Res. Lett., Vol. 19, No. 3, p. 225 – 228 (7 Feb 1992).

The interaction of a plasma flow with a cloud of charged dust particles is studied using a one–dimensional hybrid simulation model. Depending on the dust charge density and the cloud diameter, this cloud is either a weak or a strong obstacle in the plasma flow. For a weak dust obstacle a super–magnetosonic flow is decelerated monotonically inside the cloud whereas a sub–magnetosonic flow is accelerated. For a strong dust obstacle the flow configuration in the cloud changes dramatically: Deceleration and acceleration may change across the cloud and strong proton heating and shock generation may occur with the stationary regime being discarded. An interpretation of the proton dynamics in weak clouds is given on the basis of a simple analytic multifluid model.

062.014 A comparison of the reduced and approximate systems for the time dependent computation of the polar wind and multiconstituent stellar winds.
G. L. Browning, T. E. Holzer.
J. Geophys. Res., Vol. 97, No. A2, p. 1289 – 1302 (1 Feb 1992).

The "reduced" system of equations commonly used to describe the time evolution of the polar wind and multiconstituent stellar winds is derived from the equations for a multispecies plasma with known temperature profiles by assuming that the electron thermal speed approaches infinity. The reduced system is proved to have unbounded growth near the sonic point of the protons for many of the standard parameter cases. For the same parameter cases, however, the unmodified system (from which the reduced system is derived) exhibits growth in some of the Fourier modes, but this growth is bounded. A physical explanation is provided for the unbounded growth in the reduced system and the bounded growth in the unmodified system. An alternate system (the "approximate" system) in which the electron thermal speed is slowed down is introduced. The approximate system retains the mathematical behavior of the unmodified system and can be shown to accurately describe the smooth solutions of the unmodified system. The approximate system has a number of other advantages over the reduced system. For example, when the proton speed approaches the electron sound speed, the reduced system becomes inaccurate. Also, for three–dimensional flows the correct reduced system requires the solution of an elliptic equation, while the approximate system is hyperbolic and only requires a time step approximately 1 order of magnitude less than the reduced system. Numerical solutions from models based on the two systems are compared with each other to illustrate these points.

062.015 Semikinetic and generalized transport models of the polar and solar winds.
H. G. Demars, R. W. Schunk.
J. Geophys. Res., Vol. 97, No. A2, p. 1581 – 1596 (1 Feb 1992).

A comparison has been made, in as consistent a manner as possible, of a transport (bi–Maxwellian based 16–moment equations) and a semikinetic description of supersonic flow in the solar wind and also of both supersonic and subsonic flows in the polar wind for "steady state" conditions. The study shows: (1) remarkable agreement between the two models for supersonic collisionless flows, even for the higher–order moments; (2) the inadequacy of the semikinetic approach for modeling subsonic flows; and (3) the superiority of the 16–moment transport over the semikinetic approach for modeling the solar wind. The study provides further evidence that the bi–Maxwellian based transport equations are a useful tool for studying "thermal" space plasmas that develop non–Maxwellian features.

062.016 Shock drift acceleration for a near–perpendicular shock in a turbulent astrophysical plasma.
P. L. Newman, X. Moussas, J. J. Quenby, J. F. Valdes–Galicia, Z. Theodossiou–Ekaterinidi.
Astron. Astrophys., Vol. 255, No. 1/2, p. 443 – 452 (Feb 1992).

Analytical theory of diffusive shock acceleration from the microscopic view–point depends upon a knowledge of the mean energy gain per cycle across the shock. A numerical study is carried out on the basic shock drift energisation in a situation where two important assumptions of the analytical theory are broken. One violation is due to the introduction of a high degree of turbulence in the plasma, allowing continuous scattering right up to the shock surface. The second involves employing a near perpendicular ($\psi = 89°$), shock and an injection energy such that transformation to the electric–field–free ($E = 0$) frame involves the introduction of a significant anisotropy. The turbulent field is modelled employing in situ measurements of an interplanetary travelling shock. This turbulence keeps some particles in the drift acceleration region longer than expected and allows enhanced acceleration for some pitch angles if the results are compared with a scatter–free model employing a single discontinuity between two homogeneous fields. In addition, upstream acceleration occurs in the pre–shock foot and ramp structure. However, the mean energisation for the turbulent model is about the same as the computed mean energy gain for the scatter–free case. Both these computed energy gains nevertheless exceed analytical theory estimates by 18% or more, the discrepancy probably being due to the large anisotropy introduced in the $E = 0$ frame. First adiabatic conservation is found to hold to wihin 20%.

062.017 Theory of field line resonances of standing shear Alfvén waves in three–dimensional inhomogeneous plasmas.
S. Schulze–Berge, S. Cowley, L. Chen.
J. Geophys. Res., Vol. 97, No. A3, p. 3219 – 3222 (1 Mar 1992).

The authors have analyzed field line resonances of Alfvén waves in a rectangular box model with a straight uniform magnetic field but three–dimensionally varying density. Field line resonances are shown to exist even with this three–dimensional nonuniformity. For a given wave frequency the authors can construct the surface on which the resonance occurs and derive the local form of the singular solution. Magnetic perturbations are found to lie predominantly in the resonant surface. In the presence of azimuthal inhomogeneities the present theory could explain why some satellite measurements show geomagnetic pulsations of comparable magnitude in radial and azimuthal components.

062.018 Acoustic instability in cosmic ray mediated shocks.
H. Kang, T. W. Jones, D. Ryu.
Astrophys. J., Vol. 385, No. 1, p. 193 – 204 (20 Jan 1992).

The authors have examined the acoustic instability in cosmic ray dominated media that can amplify sound waves shorter than the scale height of the cosmic–ray pressure. The effects of the instability on the particle distribution have been studied using a time–dependent numerical method in which the diffusion–advection transport equation for the particle distribution function is solved self–consistently with the hydrodynamic conservation equations. Incident sound waves can grow into shocks in the precursors of strong cosmic ray mediated shocks, so that the gas entropy is increased significantly before shock passage, and the postshock cosmic ray pressure is slightly decreased. Fresh particles can be injected from the gas at the small–scale shocks, even after the initial large–scale shock becomes smooth due to the development of a strong cosmic ray pressure precursor. However, even though the instability can amplify the sound waves into moderately strong gas shocks, the cosmic ray pressure and the particle distribution are not significantly affected by these shocks. This is because diffusion of the cosmic rays causes the perturbation on the particle distribution to be much smaller than those of gasdynamic variables. The authors suggest that the amplified small–scale density structures might produce ESE, a class of flux variation observed in some compact radio sources and thought to be caused by scattering or refraction in the intervening interstellar medium.

062.019 Propagation of cylindrical shock waves in conducting media.
V. K. Singh.
Astrophys. Space Sci., Vol. 187, No. 1, p. 1 – 8 (Jan 1992).

Non–similarity solutions of the equations governing the motion of a perfect gas behind a cylindrical shock wave of variable strength have been obtained. These solutions are applicable to both the weak and the strong shocks. The nature of flow and field variables are illustrated through graphs. The total energy of the wave is taken to be constant.

062.020 Magnetohydrodynamic oscillation of a gas jet of zero inertia dispersed in a resistive liquid with energy conservation.
A. E. Radwan.
Astrophys. Space Sci., Vol. 187, No. 1, p. 9 – 25 (Jan 1992).

The dynamical oscillation and instability of a gas cylinder of zero inertia immersed in a resistive liquid has been developed for symmetric perturbations. In the absence of the magnetic field the author has used the conservation of energy to study such problem for all symmetric and asymmetric perturbations. In the latter it is found that the temporal amplification is much lower than that of the full fluid jet. The model is capillary stable for all short and long wavelengths in the asymmetric perturbation while in the symmetric disturbances it is stabilizing or not according the perturbed wavelength is shorter than the gas cylinder circumference or not. The resistivity is stabilizing or destabilizing according to restrictions. The electromagnetic body force is stabilizing for all wavelengths in the rotationally–symmetric disturbances. The Lorentz body force, for high magnetic field intensity, could be suppressing the destabilizing character of the present model. This may be due to the fact that the acting magnetic field is uniform and that the fluid is considered to be incompressible.

062.021 Gravitational instability of a composite and rotating plasma in the presence of a variable magnetic field through a porous medium.
R. C. Sharma, A. Rajput.
Astrophys. Space Sci., Vol. 187, No. 1, p. 105 – 111 (Jan 1992).

The gravitational instability of an infinite homogeneous self–gravitating plasma through porous medium is considered to include, separately, the effects due to rotation and collisions between ionized and neutral components. The dispersion relations are obtained in both cases. It is found that the gravitational instability of a composite and rotating plasma in the presence of a variable horizontal magnetic field through porous medium is determined by the Jeans's criterion.

062.022 Numerical simulations of hydrodynamical jets crossing a galactic halo / intracluster medium interface.
P. J. Wiita, M. L. Norman.
Astrophys. J., Vol. 385, No. 2, p. 478 – 490 (1 Feb 1992).

The authors have performed a series of high resolution two–dimensional hydrodynamical simulations of expanding axisymmetric jets using parameters expected to be reasonable for extragalactic radio sources. These model beams are initially conical and propagate first through X–ray emitting isothermal ($T \approx 10^7$K) galactic halos with densities declining as a power–law of radius. These beams then cross pressure–matched interfaces into hotter but less dense media, typical of intracluster or intergalactic gases. The results confirm most of the conclusions drawn from preliminary low–resolution runs, particularly the slowing down in relatively shallow power–law atmospheres and the transition from conical to quasi–cylindrical beams after crossing the interface. Good agreement with analytical estimates for conical jet propagation is also found. However, these superior simulations indicate that sufficiently slow jets do become unstable after passing the interface. Beams whose inputs of mass and momentum are removed both before and after crossing the interface are also examined. The resulting rarefaction waves, smoothed internal pressures, and slowed forward velocities imply that minimal shock acceleration of synchrotron emitting electrons will occur once a jet is no longer being fed.

062.023 Prominence sheets supported by constant–current force–free fields. II. Imposition of normal photospheric field component and prominence surface current.
C. Ridgway, T. Amari, E. R. Priest.
Astrophys. J., Vol. 385, No. 2, p. 718 – 730 (1 Feb 1992).
An analytical model for the magnetohydrostatic equilibrium of a sheet of mass and current in a constant–current force–free field is presented. The formulation of the problem differs from that of Paper I of the series, since here the authors impose the value of the surface current at the prominence rather than the normal magnetic field. The field topology and conditions for equilibrium are discussed. The model is used to generate particular normal (N–type) and inverse (I–type) magnetic field configurations.

062.024 Cylindrically symmetric force–free magnetic fields.
L. J. Porter, J. A. Klimchuk, P. A. Sturrock.
Astrophys. J., Vol. 385, No. 2, p. 738 – 745 (1 Feb 1992).
The magneto–frictional method was used to study the energy buildup in stressed coronal fields possessing cylindrical symmetry. Four different nonlinear, force–free magnetic field configurations were examined. It is found that, in all cases, a reasonable amount of twist in the field lines can produce enough free magnetic energy to power a typical flare. Furthermore, the rate of energy buildup is enhanced if the greatest twist and/or the magnetic flux is concentrated closer to the neutral line. It appears that the open–field configuration (a configuration for which the field lines extend to infinity and the current is confined to a current sheet separating outgoing and incoming field lines) is the limiting state as one imposes infinite shear. The results of this work do not contradict this theory, once numerical errors are taken into account.

062.025 Equilibrium of a magnetic flux tube in a compressible flow.
R. Kerswell, S. Childress.
Astrophys. J., Vol. 385, No. 2, p. 746 – 757 (1 Feb 1992).
Three–dimensional cellular convection concentrates a magnetic field into thin ropes and sheets when the magnetic Reynolds number is large. The authors examine the equilibrium structure of an axisymmetric flux rope sustained by turbulent, cellular convection of a compressible, electrically conducting gas. The back–reaction of the magnetic field on the turbulent flow is modeled in both the momentum and energy equations. Using boundary–layer techniques, the authors show that compressibility of the fluid is responsible for a non–Boussinesq boundary-layer effect, which is termed here "cooling runaway". This is a thermal breakdown which realizes in this model the convective collapse associated with superadiabatic effects. Cooling runaway may limit the flux carried by basic magnetic structures which are formed in the upper layer of the solar convection zone.

062.026 Thermal instability in a two–component magnetized plasma.
Yu. V. Vandakurov.
Sov. Astron. Lett., Vol. 17, No. 6, p. 433 – 436 (Nov 1991). Current Physics Microform No.: 9207X1871. English translation of Pis'ma Astron. Zh., Tom 17, No. 11, p. 1031 – 1038 (Nov 1991).
The cause of the discrepancies arising when the problem of thermal instability in a radiating magnetized plasma with an arbitrary angle between the wave vector and the equilibrium magnetic field is solved in the magnetohydrodynamic or the two–fluid approximation is examined. The displacement current is disregarded in the MHD approximation, but this is inconsistent with the equations of the two–fluid approximation; this is the cause of the discrepancies. It is shown that when the plasma motion is approximately perpendicular to the magnetic field, the region of the oscillatory mode of thermal instability excitation may be broader than when the motions are strictly transverse.

062.027 Finite amplitude fast magnetosonic waves in a low–density plasma.
A. V. Danilov.
Sov. Phys. – JETP, Vol. 74, No. 1, p. 48 – 52 (Jan 1992). Current Physics Microform No.: 9202X1852.
The author considers exact solutions of the quasi–hydrodynamical equations which describe periodic fast magnetosonic waves propagating in a non–isothermal plasma ($T_e \gg T_i$) at an arbitrary angle (not too close to $\pi/2$) to the external magnetic field. In this considerations the author uses an effective potential depending on a single parameter. He finds the conditions for which the ion dispersion cannot stop the nonlinear steepening of the wave and for which there appears an internal rotational discontinuity in its structure. He compares the results with the observations of low–frequency waves in the region in front of the bow shock wave of planets and comets.

062.028 Magnetic interchange instability of accretion disks.
M. Kaisig, T. Tajima, R. V. E. Lovelace.
Astrophys. J., Vol. 386, No. 1, p. 83 – 89 (10 Feb 1992).
A study is made of the nonlinear evolution of the magnetic interchange or buoyancy instability of a differentially rotating disk threaded by an ordered vertical magnetic field. As a model for the disk, the authors consider a two–dimensional ideal fluid in the equatorial plane of a central mass in the corotating frame of reference. The evolution is studied numerically by solving the equations of ideal magnetohydrodynamics. If the rotation rate of the disk is Keplerian, the disk is found to be stable. If the vertical magnetic field is sufficiently strong, and the field strength decreases with distance from the central object, and thus the rotation of the disk deviates from Keplerian, an instability develops. The magnetic flux and disk matter expand outward in certain ranges of azimuth, while disk matter with less magnetic flux moves inward over the remaining range of azimuth, showing the characteristic development of an interchange instability. Saturation and eventually decay occur when the field enters the outer Keplerian part of the disk. When a slow but steady supply of vertical magnetic field is present in the accretion flow, the magnetic perturbation persists and gives rise to a steady state disk configuration in which there is an outward angular momentum transport which corresponds to a dimensionless viscosity (Shakura–Sunyaev) parameter of $\alpha \sim 0.1$.

062.029 Knots in stellar jets from time–dependent sources.
A. C. Raga, L. Kofman.
Astrophys. J., Vol. 386, No. 1, p. 222 – 228 (10 Feb 1992).
The authors present an analytical model that describes the time–evolution of a highly supersonic jet of small opening angle ejected from a time–dependent source. With this model, the authors explore the possibility that the aligned knots observed along astrophysical (and in particular, stellar) jets might be the result of "internal working surfaces" produced in the jet flow as a result of a time–dependent ejection velocity. This model allows detailed predictions of the shock velocities, Hα luminosities and characteristic masses of the knots in a stellar jet. The authors present a comparison between these predictions and existing observations of the HH 34 outflow (which has a well defined, jetlike structure), and find a surprisingly good agreement between theoretical predictions and observations.

062.030 The current sheet and Joule heating of a slender magnetic tube in the upper photosphere.
T. Hirayama.
Sol. Phys., Vol. 137, No. 1, p. 33 – 50 (Jan 1992).
Joule heating in a slender magnetic flux tube is investigated. The distribution of the magnetic field and electric sheet current encircling a vertical cylindrical magnetic tube is determined by equating the converging magnetic flux, which results from the converging and downward flow of the granulation, and the dissipative expanding magnetic flux due to Ohmic decay. Here, to ensure the mass flux conservation, an overshooting convective flow pattern resembling recent simulations was assumed. Even with the electrical resistivity from neutral hydrogen, the width of the current sheet was found to be ≈ 2 km, being much smaller

than the tube diameter of ≈ 150 km, either from an exact or approximate (Gaussian) field distribution. The resultant energy flux density due to Joule heating averaged over the cylindrical cross sectional area, is $\approx 1 \times 10^9 \mathrm{erg\, cm^{-2} s^{-1}}$ for an assumed photospheric magnetic field of 1500 G. This amount may supply enough energy to heat the temperature minimum region of the flux tube by $\Delta T = 300K$ in accord with observations, though our estimation of the excess radiation loss which should be supplied by the Joule heating to keep $\Delta T = 300K$ is rather uncertain. A possible role of the Joule heating on spicule formation is briefly discussed together with discussions on the slab geometry, general flow patterns, and non–constant field distributions inside the flux tube.

062.031 The equilibrium shape of slender flux tubes in a linear force–free magnetic field.
J. Juan, J. L. Ballester.
Sol. Phys., Vol. 137, No. 2, p. 257 – 271 (Feb 1992).

The authors extend previous work of Browning and Priest (1984, 1986) by studying the equilibrium path of twisted and untwisted thin flux tubes in a stratified, isothermal atmosphere using as the ambient field a linear force–free field. When an untwisted flux tube is considered, one finds that shearing the magnetic arcade provides a different form to change the parameter λ which characterizes the external atmosphere, but at the same time this introduces a limitation in the width allowed for the external arcade. Also, the critical width found for the different analytical cases considered is always greater than one arch of the ambient arcade which prevents an eruption inside the arcade. In the case of twisted flux tubes, an analytical solution can be found for the critical λ_c, which separates regimes of strong and weak gravity, and the shape of the flux tube is now dependent on β, a parameter which represents the magnetic field enhancement of the loop at the photosphere.

062.032 Instabilities in astrophysical jets. I. Linear analysis of body and surface waves.
J.-H. Zhao, J. O. Burns, P. E. Hardee, M. L. Norman.
Astrophys. J., Vol. 387, No. 1, p. 69 – 82 (1 Mar 1992).

The authors investigate the instabilities of astrophysical jets in a new and more generalized way than in previous studies. They solve the dispersion equations for a thermally confined slab jet in a complex (k, ω) domain and find a number of singularities related to certain wave modes in this system. The authors show that the internal jet flow in the system is characterized by two internal sound waves, propagating against and with the flow, respectively. These two body waves grow in amplitude. The one propagating against the flow is condensed, which is crucial to disruption of jets. The other, propagating with the flow, is rarefied and less important in jet disruption. In addition, there are a number of surface waves existing in the interface between the jet flow and the ambient medium. Two of them are identified to couple to their relevant acoustic body waves. They also grow in amplitude. The results of this study suggest that the growing internal body waves and abundant growing surface waves can potentially explain the wealth of observed phenomena in astrophysical jets, such as quasi–periodic wiggles, jet disruption, limb–brightened features, and surface filaments.

062.033 Instabilities in astrophysical jets. II. Numerical simulations of slab jets.
J.-H. Zhao, J. O. Burns, M. L. Norman, M. E. Sulkanen.
Astrophys. J., Vol. 387, No. 1, p. 83 – 94 (1 Mar 1992). With plate 1.

The authors describe numerical simulations of an unstable supersonic slab–symmetric jet. The instabilities within the jet are characterized by growing internal body waves and their coupled surface waves that are also predicted in linear perturbation theory. The characteristic theory of fluid dynamics is used to help interpret the wave morphologies. The authors demonstrate that these waves can be excited by imposing an arbitrary disturbance. From the numerical simulations, it is found that the sound waves propagating against the flow slow down as they propagate outward, and they grow in amplitude. These waves eventually disrupt the jet at a certain length. This disruption length is related to the jet Mach number and the perturbation intensity. Thus, the Mach number of a jet observed with a radio telescope can be estimated by measuring the disruption length and estimating the perturbation intensity. The jet Mach numbers in radio tailed sources determined in this way agree quite well with estimates from ram pressure bending arguments. The wiggles and flares observed in many extragalactic jets, especially in tailed radio sources, appear to be intimately related to instabilities and the jet disruption process.

062.034 Astrophysical jets: Squeezing gas through space.
P. J. Wiita.
Nature, Vol. 355, No. 6360, p. 499 – 500 (6 Feb 1992).

062.035 Collimation of astrophysical jets by inertial confinement.
V. Icke, G. Mellema, B. Balick, F. Eulderink, A. Frank.
Nature, Vol. 355, No. 6360, p. 524 – 526 (6 Feb 1992). Letter–to–the–editor.

Many astrophysical objects, from young stars, Herbig–Haro objects and planetary nebulae up to active galactic nuclei, can be very simply modelled as isotropic sources of high–energy tenuous gas embedded in dense toroidal clouds. The authors describe numerical simulations showing how such an arrangement can in general circumstances give rise to a well collimated jet, as is observed in many of these systems. Their model is a two–dimensional generalization of the interacting–winds description of planetary nebulae. Where the two winds come into contact, a discontinuity is formed, which is dragged out by the fast outflowing gas into a chimney along the polar axis. High–energy gas rushes up this channel and flows out around the top, creating a hot backflow which keeps the chimney in place. The inner shock, enclosing the source of the fast wind, also aids in collimation, and ionization cones such as those observed in active galactic nuclei may also form.

062.036 Chaotic Alfvén waves in multispecies plasmas.
B. Buti.
J. Geophys. Res., Vol. 97, No. A4, p. 4229 – 4234 (1 Apr 1992).

By means of the Hamiltonian formalism, it is shown that the nonlinear Alfvén waves in a multispecies plasma in the presence of a driver can be chaotic provided the amplitude of the driver exceeds a certain threshold. This threshold is larger in the presence of heavier ions, e.g., oxygen in the case of comets and helium in the case of solar wind, than for plasma with only two species. This clearly shows that heavier ions tend to reduce the chaos.

062.037 On the thermodynamics of diamagnetic plasma expansions.
G. R. Gisler, T. G. Onsager.
J. Geophys. Res., Vol. 97, No. A4, p. 4265 – 4274 (1 Apr 1992).

Particle heating in a diamagnetic plasma expansion is studied by means of well–diagnosed simulation with an axisymmetric particle–in–cell code. Moments of the particle distribution function are obtained for spatially distinct subsets of the particles to examine temperature and density histories for different regions of the expanding plasma. The simulation is followed through one expansion–contraction cycle. While adiabatic behavior is observed during much of the cycle, significant deviations from the adiabatic result in strong particle heating. Anomalous ion heating occurs throughout the plasma during the expansion phase. This is manifested earliest in the ion parallel temperature, which increases first in the expanding plasma's outer reaches and last in the center. This heating originates from the entropy that is generated at the barrel ends of the plasma as the initial expansion along the ambient field is essentially free. Later, after a diamagnetic cavity is formed, ions within the cavity are reflected by the magnetic mirrors at the necks and transport some of the generated entropy back into the center of the cavity. The ion heating that occurs can easily raise the bulk plasma temperature by an order of magnitude over the initial adiabat.

062.038   **Inhibition of electron thermal conduction by electromagnetic instabilities.**
A. Levinson, D. Eichler.
Astrophys. J., Vol. 387, No. 1, p. 212 – 218 (1 Mar 1992).
Heat flux inhibition by electromagnetic instabilities in a hot magnetized plasma is investigated. Low–frequency electromagnetic waves become unstable due to anisotropy of the electron distribution function. The chaotic magnetic field thus generated scatters the electrons with an effective mean free path λ_{eff}. Saturation of the instability due to wave–wave interaction, nonlinear scattering, wave propagation, and collisional damping is considered. The effective mean free path is found self–consistently, using a simple model to estimate saturation level and scattering, and is shown to decrease with the temperature gradient length L. The results are applied to astrophysical systems. For interstellar clouds with $\beta \gg 1$ and whose sizes are well below 10^{21}cm, the instability is found to be important, keeping the value of $\varepsilon \equiv \lambda_{eff}/L$ near 10^{-3}.

062.039   **Nonaxisymmetric magnetogravitational instability of a streaming fluid cylinder ambient with a tenuous medium pervaded by transverse varying fields.**
A. E. Radwan.
Astrophys. Space Sci., Vol. 187, No. 2, p. 241 – 260 (Jan 1992).
The stability of a self–gravitating streaming fluid cylinder acting upon the electromagnetic force ambient with a tenuous medium of negligible inertia but pervaded by a transverse varying fields, has been developed. The stability criterion is derived, discussed analytically and the results are verified numerically. The cylinder is purely self–gravitating unstable in small axisymmetric domain and stable in all the rest states. The axial magnetic field internal the fluid jet has a strong stabilizing influence for all possible perturbations modes while the transverse field exterior the cylinder is stabilizing or destabilizing according to restrictions in the asymmetric modes and purely destabilizing in the symmetric one. The streaming has a strong destabilizing influence and that influence is independent of the kind of the perturbation and wavelengths. Both the streaming and the electromagnetic influence increase the gravitational axisymmetric unstable domain and shrink those of stability in the axisymmetric and non–axisymmetric perturbations. Moreover, the stabilizing character of the Lorentz force of some states will not be able to suppress the gravitational instability because the gravitational instability of sufficiently long waves will persist.

062.040   **The effects of Alfvén waves on heating plasma in post–flare loops.**
Lin Jun, Zhang Zhenda.
Astrophys. Space Sci., Vol. 187, No. 2, p. 291 – 306 (Jan 1992).
In investigating the effects of collision Alfvén waves on the heating of a cool–type solar loop, like the post–flare loop, models are proposed, and the distributions of ion or electron density, temperature, pressure, and wave energy density are simulated. The authors have assumed the magnetic field strength in the loop is about 100 G and found that Alfvén waves can propagate through the whole loop, that is to say, the decay length of collision Alfvén waves considered can reach to the height or length of the loop. Thus, the Alfvén wave heating is a considerable heating mechanism in cool loops. The authors have also found that the variations of density, pressure, and wave energy density are more significant than those of the temperature. In the whole loop, the temperature is of the order of $\sim 10^4$K. In comparison with other parameters, the temperature can be considered as homogeneous; hence, the heat conductive flux in the simulations is omitted.

062.041   **Populations of highly excited atoms and ions in the optically thin plasma.**
N. I. Rovenskaya.
Astrophys. Space Sci., Vol. 188, No. 1, p. 89 – 107 (Feb 1992).
Treating it as a boundary–value problem, the equations of highly excited state populations of atoms and ions are theoretically studied in case of recombining plasma. Scattering and spontaneous transitions as well as those induced by background radiation, are taken into account in the kinetic equations. The kinetic coefficients for inelastic scattering of incident charged particles on highly excited atoms and ions have been calculated in the asymptotically exact case: $1.6 \times 10^5 (z/n)^2 T_e^{-1} \ll 1$. The distribution functions over the Rydberg states, analytically found, allow to determine amplification factor and optical depth of radio–recombination lines as functions of cosmic plasma parameters.

062.042   **Large electric fields in acoustic waves and the stimulation of lightning discharges.**
W. Pilipp, T. W. Hartquist, G. E. Morfill.
Astrophys. J., Vol. 387, No. 1, p. 364 – 371 (1 Mar 1992).
Fluid acceleration can play a role similar to that of gravity in the establishment of charge separation generated electric fields in weakly ionized plasma in which solid particles are the dominant carriers of both negative and positive charge. The authors point out that the properties of fluid motions, the ionization structures, and grain size and charge distributions necessary to produce electric fields strong enough to induce discharge in the absence of gravity can be inferred from calculations of the electric fields in linear acoustic waves. This approach is examined in the context of the protosolar nebula where lightning may have been important for the thermal evolution of chondrules, meteoritic inclusions which must have been flash heated. The authors find that electric field strengths strong enough to induce lightning discharges could be produced in waves in that environment, but that the electrical energy in such a structure was far too small for significant heating and ionization to occur in the discharge channels associated with the wave. This implies that lightning discharges in the protosolar nebula, if they were responsible for melting of chondrules, must have been powered by electrical energy stored in much larger structures. In this case gravity acting on large, highly charged grains must have been important. The authors argue that the strong electric fields in waves may have provided the "seed conditions" for nonlinear charging of those large grains.

062.043   **A shock hodograph in collisionless plasma.**
M. A. Alabraba, A. R. Bestman.
Astrophys. Space Sci., Vol. 189, No. 1, p. 57 – 62 (Mar 1992).
In this paper, the shock hodograph (polar) for a collisionless transverse shock has been developed in the plane of the flow deflection angle and total pressure jump. The sonic point on the hodograph lies closer to the characteristic than to the point of maximum flow deflection for an attached shock. This hodograph is particularly useful in the analysis of three shock confluences and refraction of shock waves at gas interfaces. The first analysis is fully described in this paper. It is observed that the third shock wave is forward facing. It is pertinent to note that the limited region of supersonic flow also restricts the occurrences of three shock interactions.

062.044   **Arbitrary amplitude double layers in a multi–species electron–positron plasma.**
R. Bharuthram.
Astrophys. Space Sci., Vol. 189, No. 2, p. 213 – 222 (Mar 1992).
The existence and properties of arbitrary amplitude double layers in a four–component electron–positron plasma, consisting of two species of hot electrons, a hot and a cold positron species, are investigated as functions of plasma properties such as density and temperature ratios. Their behaviour for other plasma models is also discussed. Applications to the polar–cusp region of pulsars is considered.

062.045   **Chemically reacting flow of a compressible thermally radiating two–component plasma.**
A. R. Bestman.
Astrophys. Space Sci., Vol. 189, No. 2, p. 223 – 235 (Mar 1992).
The paper studies the compressible flow of a hot two–component plasma in the presence of gravitation and chemical reaction in a vertical channel. For the optically thick gas approximation, closed form analytical solutions are possible.

Asymptotic solutions are also obtained for the general differential approximation when the temperatures of the two bonding walls are the same. In the general case the problem is reduced to the solution of standard nonlinear integral equations which can be tackled by iterative procedure. The results are discussed quantitatively. The problem may be applicable to the understanding of explosive hydrogen–burning model of solar flares.

062.046 Asymmetric morphology of the propagating jet. II. The effect of atmospheric gradients.
P. E. Hardee, R. E. White III, M. L. Norman, M. A. Cooper, D. A. Clarke.
Astrophys. J., Vol. 387, No. 2, p. 460 – 483 (10 Mar 1992).

Simulations of slab jets propagating in atmospheres containing pressure gradients, density gradients, and temperature gradients or temperature jumps have been performed for a range of jet velocities. The use of slab geometry allows the study of asymmetric jet and lobe morphologies. Symmetry is broken by perturbing the jet at the inlet. The subsequent growth of the initial perturbation is the result of amplification by the Kelvin – Helmholtz instability. The distance that a jet propagates before oscillation amplitudes become large and highly collimated flow is disrupted increases as the Mach number increases and as the atmosphere's density decreases more rapidly. The increase in jet length occurs primarily as a result of adiabatic jet expansion. In constant atmospheres, jets interact with a surrounding cocoon or lobe of processed jet material which extends back to the origin. However, the present simulations show that the suppression of backflows which accompanies atmospheric gradients can eliminate the cocoon and the jet can interact directly with the external atmosphere before the jet reaches the lobe. An analytical model of lobe morphology is developed which reproduces the gross morphology seen in the simulations. Different jet, cocoon, and lobe morphologies arise primarily from changes in the propagation speed of an expanding jet head, changes in the density gradient in the external medium, and changes in backflows behind the jet head.

062.047 Radiative heat transfer to hydromagnetic flow of a slightly rarefied binary gas in a vertical channel.
A. R. Bestman, M. A. Alabraba, A. Ogulu.
Astrophys. Space Sci., Vol. 189, No. 2, p. 303 – 308 (Mar 1992).

The paper considers the fully–developed slip flow in a vertical channel with radiative heat transfer and mass transfer in the presence of an externally applied magnetic field. The problem is modelled by the compressible Navier–Stokes equations, so that the gas is only slightly rarefied. Invoking the exact integral equation for radiation, the problem is reduced to a set of ordinary integro–differential equations. By realistic assumptions, the set is linearized and the temperature is reduced to a mixed Fredholm–Volterra integral equation which is solved by standard iterative procedure. Thereafter the concentration equation is solved by the WKB approximation while the velocity is obtained by the finite difference scheme. These solutions are discussed qualitatively.

062.048 Applications of Lie groups to the equilibrium theory of cylindrically symmetric magnetic flux tubes.
L. M. Zelenyj, A. V. Milovanov.
Sov. Astron., Vol. 36, No. 1, p. 74 – 80 (Jan 1992). Current Physics Microform No.: 9207Y0638. English translation of Astron. Zh., Tom 69, Vyp. 1, p. 147 – 158 (Jan–Feb 1992).

A group analysis of the differential equations describing self–consistent equilibrium of cylindrically symmetric magnetic flux tubes is given. The existence of non–Maxwellian distribution functions, making it possible to find analytical expressions for magnetic fields and plasma density, is shown. The distribution function is assumed to depend on the integrals of the motion, making it possible to formulate and solve the problem of group classification of third–order, nonlinear differential equations with respect to the admissible groups of point transformations. Analytical solutions of the equations with a nontrivial symmetry group are found. The results can be used to analyze the fractal properties of plasma configurations in the atmospheres of the Sun and stars for different distribution functions.

062.049 Dynamical regimes and the possibility of microflares in a prominence.
V. V. Zajtsev, M. L. Khodachenko.
Sov. Astron., Vol. 36, No. 1, p. 81 – 87 (Jan 1992). Current Physics Microform No.: 9207Y0645. English translation of Astron. Zh., Tom 69, Vyp. 1, p. 159 – 172 (Jan–Feb 1992).

On the basis of solutions of the Kippenhahn–Schlüter magnetohydrostatic model, a dynamical model of a prominence is constructed that allows for plasma motion, having the form of cumulative flow, in the vicinity of the current sheet. An important circumstance in the solution of this problem is the partial ionization of plasma in the prominence, allowance for which under the conditions of nonstationary flow has considerable influence on the dynamics and energetics of the prominence. The equilbrium state described by the Kippenhahn–Schlüter solution than turns out to be unstable. Depending on the initial conditions in the system, various dynamical regimes may occur, corresponding to the "collapse" (compression) of prominence matter or its monotonic or quasiperiodic expansion. The energetics of this dynamical model of a prominence is investigated. In connection with the occurrence, under nonstationary conditions, of dissipation due to ion–atom collisions in the partially ionized plasma, the amount of energy released in the system increases considerably, exceeding ordinary Joule heating by several orders of magnitude. An approximate equation is obtained for the rate and degree of compression for which energy release exceeds radiative loss, i.e., the prominence is heated, providing the conditions for the development of a flare.

062.050 Plasma diagnostics based on self–reversed lines. I. Model calculation and application to argon arc measurements in the near–infrared and vacuum ultraviolet regions.
E. Sohns, M. Kock.
J. Quant. Spectrosc. Radiat. Transfer, Vol. 47, No. 5, p. 325 – 334 (May 1992).

A numerical model is developed which describes the formation of spectral lines emitted by an arc plasma. Experimental data of several self–reversed lines are compared with calculated line profiles to obtain information about the plasma inhomogeneity and the spectral parameters of the observed lines. Measurements have been carried out in the near infrared (n.i.r.) and vacuum ultraviolet (v.u.v.) using a wall–stabilised arc as the radiation source. A consistent set of transition probabilities and Stark constants has been derived for the lines under investigation. Comparisons are made with literature data.

062.051 Plasma diagnostics based on self–reversed lines. II. Application to nitrogen, carbon and oxygen arc measurements in the vacuum ultraviolet.
E. Sohns, M. Kock.
J. Quant. Spectrosc. Radiat. Transfer, Vol. 47, No. 5, p. 335 – 343 (May 1992).

A numerical model, which describes the formation of spectral lines by solving the equation of radiative transfer, has been applied to several self–reversed lines of nitrogen, carbon and oxygen in the spectral range between 115 and 200 nm using a wall–stabilized arc as a radiation source. The model calculation takes into account diffusion and demixing of the plasma components. By comparing measured with calculated line shapes, information is obtained on both the temperature distribution of the plasma column and the transition probabilities and Stark constants of the lines. Quasistatic broadenings due to ion interaction and, especially, foreign–gas interaction are found to play a major role in line–wing formation.

062.052 Dynamics of wind bubbles and superbubbles. I. Slow winds and fast winds.
B.–C. Koo, C. F. McKee.
Astrophys. J., Vol. 388, No. 1, p. 93 – 102 (20 Mar 1992).

Steady injection of mass and energy into an ambient medium produces a wind–blown bubble. This paper describes the overall evolution of such bubbles in a uniform medium from the initial, free–expansion stage to the final stage in which the pressure of the ambient medium is significant. The authors introduce the concepts of slow and fast winds, which naturally arise from consideration of radiative losses in the free–expansion stage. Bubbles blown by slow winds are radiative; bubbles blown by young stellar objects are the prototype. Bubbles blown by fast winds, on the other hand, are adiabatic by the time the swept–up gas dominates the gas injected by the wind, and they remain adiabatic unless there is additional mass injection into the bubble; bubbles blown by winds from OB stars are the prototype. Bubbles blown by either slow winds or fast winds eventually expand to the point that the pressure of the ambient medium is significant, and become pressure–confined bubbles. The authors also consider the evolution of bubbles in a plane–parallel disk, and discuss when a bubble can break out of a thin galactic disk. After breakout, bubbles can evolve into jets. Steady, collimated jets can form only over a limited range of wind luminosity and Mach number. The results are applied to the neutral stellar wind in the HH 7–11 region, to the north polar spur, and to the galactic winds in starburst galaxies.

062.053   **Dynamics of wind bubbles and superbubbles. II. Analytic theory.**
B.–C. Koo, C. F. McKee.
Astrophys. J., Vol. 388, No. 1, p. 103 – 126 (20 Mar 1992).

The analysis of wind bubbles and superbubbles in Paper I is extended to the case of power–law energy injection [$L_{in}(t) \sim t^{\eta-1}$] in a medium with a power–law density distribution [$\varrho_a(r) \sim r^{-k}$]. As before, the wind velocity is assumed to be constant, so that the energy injection rate is proportional to the mass injection rate. The evolution is followed from the free–expansion stage, in which the mass of the wind dominates the swept–up mass, through the self–similar stage, to the stage in which the bubble is confined by the pressure of the ambient medium. As in Paper I, winds may be divided into slow and fast, depending on the importance of radiative losses at the transition from the free expansion to the self–similar stage. Slow winds generally follow the radiative sequence of bubble evolution. If the wind velocity is not too low, the bubble evolves from a radiative bubble to a partially radiative bubble, in which the cooling time of the hot shocked wind is less than the age of the bubble but longer than the crossing time. The partially radiative bubble further evolves to an adiabatic bubble. The evolution of a bubble blown by a fast wind generally follows the adiabatic sequence of bubble evolution. and the shocked wind remains adiabatic unless there is additional mass injection into the bubble. Bubbles blown by either slow winds or fast winds eventually expand to the point that the pressure of the ambient medium is significant, and become pressure–confined bubbles. The authors have obtained characteristic evolutionary time scales, as well as the equation of motion for both the swept–up gas and the wind shock in each evolutionary stage.

062.054   **Energetic particle beams in quasars and active galactic nuclei.**
G. Pelletier, H. Sol.
Mon. Not. R. Astron. Soc., Vol. 254, No. 4, p. 635 – 646 (15 Feb 1992).

New developments of a two–flow model for extragalactic jets are presented. The authors study the possibility of ascribing Very Long Baseline Interferometry (VLBI) jet features and superluminal motion to beams of relativistic particles streaming through an ambient plasma such as a wind from the accretion disc. It is shown that stability of the beams relative to the excitation of Langmuir and Alfvénic waves can be ensured in the case of electron–positron beams with bulk Lorentz factor γ smaller than 43 (the square root of proton to electron mass ratio) as long as the magnetic field parallel to the beams and jets is larger than a critical value $B_c = 3.2 \times 10^{-3} n_p^{1/2}$, where n_p is the density of the ambient plasma. Conversely, electron–positron beams with higher bulk Lorentz factor, and electron–proton beams are rapidly destroyed. Therefore the authors propose that electron–positron beams with $\gamma \leqslant 43$ are responsible for VLBI features. Two main models for VLBI structures are then expected according to the relative values of parallel magnetic field and

ambient plasma density close to the central engine, corresponding to two different regimes for the beams, namely propagation when $B > B_c$ and dissolution when $B < B_c$.

062.055   **The stability of current–carrying jets.**
S. Appl, M. Camenzind.
Astron. Astrophys., Vol. 256, No. 2, p. 354 – 370 (Mar 1992).

The stability of supermagnetosonic current–carrying jets is examined on the basis of ideal magnetohydrodynamics. The stability properties are compared to hydrodynamic and magneto-hydrodynamic flows that carry no poloidal current. This allows to separate the effects of a purely poloidal magnetic field and a helical configuration corresponding to an electric current. Two equilibria with different current profile representative of an early propagation phase and of a kpc–jet surrounded by a cocoon are considered. The currents are in both cases balanced by a return current of equal size. A normal mode analysis is performed and results are presented for the pinch and kink modes. Current-carrying jets prove to be more stable than their current–free counterparts. Short wavelength instabilities are suppressed for both the current–carrying and current–free magnetically domi-nated jet, in contrast to the purely hydrodynamic one. This permits an effective energy transport in accordance with the observed low surface brightness of jets in FR 2 radiosources. The jets in powerful radiosources are likely to be supermagnetosonic and carry large electric currents. The domain of maximal instability of current–carrying jets is shifted towards longer wavelengths compared to flows without currents. This provides them with an enhanced stiffness. The results give strong support to the hydromagnetic origin of jets.

062.056   **The influence of boundary conditions on the excitation of disk dynamo modes.**
D. Moss, A. Brandenburg.
Astron. Astrophys., Vol. 256, No. 2, p. 371 – 374 (Mar 1992).

Calculations of mean field dynamos for galaxies have largely been for two rather disparate models. The thin disk model treats the ratio of disk height to radius explicitly as a small parameter, and applies zero tangential field boundary conditions at the disk surface. In contrast, the embedded disk model calculates the magnetic field in a spherical volume, whose radius is the disk radius and with the magnetic field fitting smoothly on to a curl–free exterior field at the surface of the sphere. The disk geometry is imposed by a flat distribution of the α–effect (and maybe also of the diffusivity η). For computational reasons this model has not been applied to very thin disks, so the regions of validity of the two models are almost disjoint. Comparison between their predictions is therefore difficult. In this paper the authors calculate, in linear theory, galactic dynamo modes according to both thin and embedded (or "thick") disk models for a simple underlying distribution of α–effect and differential rotation, using a common numerical scheme. For the smallest attainable ratio of disk height to radius, they find the critical dynamo numbers are similar, but that there are some significant differ-ences in field topology.

062.057   **Self–similar solutions and the stability of collapsing isothermal filaments.**
S. Inutsuka, S. M. Miyama.
Astrophys. J., Vol. 388, No. 2, p. 392 – 399 (1 Apr 1992).

Self–similar solutions which describe collapsing isothermal cylinders with self–gravity are derived. The solutions are parame-terized by their line masses. Their stability is investigated by two different methods in the linear regime. One is the approximate separation of variables as an eigenvalue problem, and the other is direct numerical integration of the evolution of perturbations. It is found that a self–gravitating cylinder is unstable to axisymmet-ric perturbations with wavelengths greater than about 2 times the diameter, when its line mass is nearly the same as that for equilibrium. In this case fragmentation is expected with separa-tions of about 4 times the diameter. When the line mass of the cylinder greatly exceeds the value for equilibrium, perturbations do not grow much and the entire cylinder collapses toward the axis. Therefore fragmentation is not expected, as long as the collapse is isothermal. Subsequent evolution in this case is also discussed, and fragmentation is expected after or during a change in the equation of state.

062.058 On the origin, acceleration and collimation of bipolar outflows and cosmic radio jets.
S. K. Chakrabarti, P. Bhaskaran.
Mon. Not. R. Astron. Soc., Vol. 255, No. 2, p. 255 – 260 (15 Mar 1992).
The authors show in a self–consistent way that it is possible to obtain well–collimated bipolar outflows and radio jets from magnetized protostellar discs and accretion discs in active galactic nuclei. They analytically solve the complete set of Euler–Maxwell equations both inside and outside the disc and appropriately match the solutions on the disc surface. The authors consider discs with a power–law angular–momentum distribution and assume self–similarity in the radial direction. The solutions are obtained from the mid–plane of the disc until the outflow reaches the Alfvén point. The authors provide specific examples of such solutions.

062.059 A test suite for magnetohydrodynamical simulations.
J. M. Stone, J. F. Hawley, C. R. Evans, M. L. Norman.
Astrophys. J., Vol. 388, No. 2, p. 415 – 437 (1 Apr 1992).
The authors have drawn together in this paper a collection of magnetohydrodynamic (MHD) problems that will provide researchers who have interests in modeling MHD phenomena with a battery of tests (a "test suite") for calibrating their numerical algorithms. Use of these tests will provide a common reference for comparison of different numerical MHD algorithms. The test suite includes both one– and two–dimensional problems. Taken together, these problems test the abilities of a numerical method to propagate accurately all of the MHD wave families in moving and stationary media, to capture MHD shocks and contact discontinuities, and to model accurately Lorentz force terms in multiple dimensions. An example solution of each test problem is presented. The authors demonstrate diagnostic convergence – testing procedures that provide a quantitative evaluation of MHD algorithms.

062.060 Convective instability in differentially rotating disks.
D. Ryu, J. Goodman.
Astrophys. J., Vol. 388, No. 2, p. 438 – 450 (1 Apr 1992).
The authors perform a normal mode analysis for nonaxisymmetric perturbations in a thin, differentially rotating disk with a vertical structure that is isothermal and convectively unstable. The vertical gravity is assumed to be external and constant. The perturbation scale is assumed to be much shorter than the radius of the disk but comparable to or less than the thickness. The initial value problem is formulated in shearing coordinates. Dispersion relations are obtained for the three limiting cases of zero shear, axisymmetric perturbations, and small radial wavelengths. The full effects of shear are studied by integrating numerically the initial value problem. It is found that nonaxisymmetric local Fourier modes have a radial wavenumber that increases linearly with time in proportion to the shear times the azimuthal wavenumber. While Coriolis forces exert stabilizing effects on the convective modes, shear has destablizing effects inasmuch as it reduces the epicyclic frequency at a given angular velocity. In a Keplerian disk, perturbations with azimuthal wavelengths about 2 times smaller than vertical wavelengths grow exponentially. Otherwise, perturbations do not grow until radial wavelengths become several times smaller than vertical wavelengths but grow exponentially after that. The angular momentum flux of the linear modes is nonzero only for nonaxisymmetric disturbances, and for these the flux is predominantly inward, i.e., in the direction of increasing angular velocity.

062.061 Dynamics and spectra of magnetically cushioned radiative shocks.
D. E. Innes.
Astron. Astrophys., Vol. 256, No. 2, p. 660 – 672 (Mar 1992).
This paper investigates the effects of magnetic fields on the evolution and spectral appearance of unstable radiative shocks.

In this work, a detailed treatment of the ionization evolution, the photoionizing radiation and its transfer through the gas, is coupled to a one–dimensional magnetohydrodynamics scheme. The cushioning effect of the magnetic field on the formation of secondary shocks in the postshock cooling flow is clearly demonstrated. The field strengths required to suppress shock formation are at least a factor two greater than predicted by linear stability analysis. Thus a field strength, transverse to the flow, of 9 μG is required to suppress shock formation behind a 175 km s^{-1} shock, travelling into a density of 1 cm^{-3}. Inclusion of the magnetic field results in realistic pressures and densities in the cool, postshock photoionized shell. This allows to follow the dynamics of the evolving photoionization zone where much of the low excitation optical radiation is emitted. The author discusses characteristic features in the spectra of unstable shocks and describe spectral diagnostics, based on the optical and UV line ratios and spatial intensity distributions, for their identification.

062.062 Non–linear evolution of synchrotron thermal instabilities.
G. Bodo, A. Ferrari, S. Massaglia, P. Rossi, K. Shibata, Y. Uchida.
Astron. Astrophys., Vol. 256, No. 2, p. 689 – 700 (Mar 1992).
The authors present a time–dependent, non–linear study of thermal instability in a magnetized plasma whose relativistic component supports the pressure and undergoes synchrotron losses while the thermal matter provides the inertia of the medium. They follow the temporal evolution of the instability in the non–linear regime by means of a one–dimensional, finite–difference MHD numerical code employing either free or periodic boundary conditions. They find that, in a plasma subject to constant heating, after an initial phase in which the instability growth rate follows the linear model, the instability reaches a quasi equilibrium state on timescales of the order of several synchrotron timescales. An essential condition for this instability to be efficient is that the plasma is out of equipartition with the relativistic particle energy exceeding the magnetic field energy. The mechanism considered can interpret the formation of filaments of enhanced emission observed, at high resolution, in lobes and jets of several extragalactic radio sources.

062.063 Nonlinear dust–acoustic waves in multispecies dusty plasmas.
F. Verheest.
Planet. Space Sci., Vol. 40, No. 1, p. 1 – 6 (Jan 1992).
A study is made of nonlinear dust–acoustic waves in a dusty plasma, which consists of any number of cold negatively charged dust grain species, in addition to the presence of the more usual isothermal hot and cold electrons and isothermal positive ions. The Sagdeev potential is obtained in general, together with limits on rarefactive solutions for the dust–acoustic solitons. Weak dust–acoustic solitons can be described by a modified Korteweg – de Vries equation. An application is given to a dusty plasma with one kind of negative grains, in the presence of protons, all electrons having been accreted unto the dust grains. Such dusty plasmas can support rarefractive supersonic solitons, in contrast to the usual ion–acoustic solitons which are compressive. Compressive weak dust–acoustic solitons are less likely to occur. When some streaming is included, one finds a slight decrease in the amplitude of the solitons.

062.064 Large amplitude double layers in dusty plasmas.
R. Bharuthram, P. K. Shukla.
Planet. Space Sci., Vol. 40, No. 4, p. 465 – 471 (Apr 1992).
The existence of arbitrarily large amplitude electrostatic double layers is investigated in a four–component plasma consisting of electrons, two distinct positive ion species of different temperatures, and massive micrometre–sized, negatively-charged dust particles. In the authors' model, the number densities of the electrons and ions are given by the Boltzmann distribution, whereas the dynamics of charged dust particles is described by the hydrodynamic equations. Criteria for the existence of finite amplitude double layers are obtained. The

dependence of the double layer amplitude and Mach number on several plasma parameters is examined.

062.065   Generation of radiation by upper–hybrid waves in non–uniform plasmas.
L. Stenflo, P. K. Shukla.
Planet. Space Sci., Vol. 40, No. 4, p. 473 – 476 (Apr 1992).

The decay of an electrostatic upper–hybrid wave into a high–frequency transverse wave and a low–frequency longitudinal wave in an inhomogeneous magnetoplasma is considered. The growth rate and threshold of this parametric instability are calculated. The present results should be suited for the under-standing of non–thermal radiation arising in non–uniform plasmas such as those found in the terrestrial magnetosphere as well as in the F–region of the ionosphere.

062.066   Downflows and entropy gradient reversal in deep convection.
K. L. Chan, D. Gigas.
Astrophys. J., Lett., Vol. 389, No. 2, p. L87 – L90 (20 Apr 1992).

The authors have performed a three–dimensional numerical experiment to investigate the behavior of columnar downflows in a deep convection zone. The downflows pump low–entropy fluid from the top to the lower region and create an entropy gradient reversal inside the convection zone. The negative flux of kinetic energy carried by the downflows was found to decrease in this region.

062.067   Detonation waves in relativistic hydrodynamics.
M. Cissoko.
Phys. Rev. D, Vol. 45, No. 4, p. 1045 – 1052 (15 Feb 1992).
Current Physics Microform No.: 9204D1655.

This paper is concerned with an algebraic study of the equations of detonation waves in relativistic hydrodynamics taking into account the pressure and the energy of thermal radiation. A new approach to shock and detonation wavefronts is outlined. The fluid under consideration is assumed to be perfect (nonviscous and nonconducting) and to obey the following equation of state: $p = (\gamma-1)\varrho$ where p, ϱ, and γ are the pressure, the total energy density, and the adiabatic index, respectively. The solutions of the equations of detonation waves are reduced to the problem of finding physically acceptable roots of a quadratic polynomial $\Pi(X)$ where X is the ratio τ/τ_0 of dynamical volumes behind and ahead of the detonation wave. The existence and the locations of zeros of this polynomial allow it to be shown that if the equation of state of the burnt fluid is known, then the variables characterizing the unburnt fluid obey well–defined physical relations.

062.068   On the stability of mean–field models of the solar convection zone.
G. Rüdiger, F. Spahn.
Sol. Phys., Vol. 138, No. 1, p. 1 – 9 (Mar 1992).

The stability of the solutions of the mean–field theories of turbulent media is questioned. It is done here for the model equations for the solar convection zone which have been used, in particular, to explain the differential rotation. The authors present an approximation valid for axisymmetric, short–wave disturbances. A critical local Rayleigh number can be defined – involving eddy diffusivities – above which the stratification becomes unstable. For mixing–length models of the solar convection zone one always finds sub–critical Rayleigh numbers. One must be careful, however, with other theoretical models. Those do not reach sufficiently high surface presure values so that there the associated Rayleigh numbers exceed their critical limits. In the outermost layers in such models, therefore, the solutions could really be unstable.

062.069   Some properties of finite energy constant–α force–free magnetic fields in a half–space.
J. J. Aly.
Sol. Phys., Vol. 138, No. 1, p. 133 – 162 (Mar 1992).

Some useful properties of a finite energy, constant–α, force–free magnetic field $\mathbf{B}_\alpha$ occupying a half–space D are presented. In particular: (a) Fourier and Green representations of $\mathbf{B}_\alpha$ are obtained and used to derive conditions for the existence and uniqueness of a $\mathbf{B}_\alpha$ having a given normal component B_z on the boundary ∂D. (b) The asymptotic behaviour of $\mathbf{B}_\alpha$ at infinity as well as stability results against changes in the boundary condition on ∂D and in the value of α are established. (c) The energy of $\mathbf{B}_\alpha$ is shown to be smaller than the energy of the open field having the same B_z on ∂D, thus confirming an earlier conjecture (Aly, 1984). (d) $\mathbf{B}_\alpha$ is proved to not be a Taylor–Heyvaerts–Priest state, in spite of the fact that its relative helicity H is finite and that it is the only solution of the Lagrange–Euler equation associated with the problem of minimizing the energy among all the fields having the same value of H and the same B_z on ∂D.

062.070   Review of the CIV phenomenon.
N. Brenning.
Space Sci. Rev., Vol. 59, No. 3/4, p. 209 – 314 (Feb 1992).

Alfvén's Critical Ionization Velocity (CIV) phenomenon is reviewed, with the main emphasis on comparisons between experimental and theoretical results. The review covers (1) the velocity measurements in laboratory experiments, (2) the effect of wall interaction, (3) the experimental and theoretical limits to the magnetic field strength and the neutral density, (4) ionospheric release experiments, (5) theoretical models for electron energiza-tion in comparison to experimental results, and (6) CIV models.

062.071   Magnetosphere–ionosphere coupling in multi–species magneto plasmas.
S. Guha, M. Asthana.
4. Asia Pacific Physics Conference, p. 1377 – 1380 (1991). – See Abstr. 012.006 for the main entry.

The nonlinear decay of an electromagetic wave into a lowerhybrid and upperhybrid wave in a multi–species plasma has been analytically investigated. A hydrodynamical model of the plasma is used. The nonlinear dispersion relation and growth rates are calculated for parametric decay, modulational and filamentation instabilities. As an application of the investigation growth rates are calculated for typical parameters of both laboratory and space plasmas.

062.072   Resistive tearing–mode instability in a current sheet with equilibrium viscous stagnation–point flow.
T. D. Phan, B. U. Ö. Sonnerup.
J. Plasma Phys., Vol. 46, No. 3, p. 407 – 421 (Dec 1991).

An analysis is presented of linear stability against tearing modes of a current sheet formed between two oppositely magnetized plasmas forced towards each other in two–dimen-sional steady stagnation–point flow. The velocity vector in this flow is confined to planes perpendicular to the reversing component of the magnetic field. The unperturbed state is an exact resistive and viscous equilibrium in which the resistive diffusion outwards from the current sheet is exactly balanced by the inward motion associated with the stagnation–point flow. Thus the behaviour of the tearing mode can be examined even when the resistive diffusion is comparable to or smaller than the growth time of the instability. The linear ordinary differential equation describing the mode structure is integrated numerically. The influence of the stagnation–point flow on the tearing mode is given. Finally, application of the results from this study to the problem of solar–wind plasma flow past the earth's magneto-sphere is briefly discussed.

062.073   Taylor relaxation of a Gold–Hoyle flux tube.
G. J. Rickard.
Astrophys. J., Vol. 389, No. 1, p. 413 – 420 (10 Apr 1992).

Force–free magnetic field equilibria satisfy the condition $\nabla \times \mathbf{B} = \alpha\mathbf{B}$. The author considers the relaxation of a noncon-stant–α force–free equilibrium – the Gold–Hoyle field – to linear force–free equilibria where α is a constant. The relaxation proceeds assuming the global magnetic helicity to be an invariant, the basis of the so–called Taylor hypothesis. This approach circumvents the need for sophisticated time–dependent nonlinear simulations, but clearly relies on the applicability of the Taylor hypothesis to the solar corona. The Gold–Hoyle tube has a normal field component threading the photosphere at each end, so the helicity of the final equilibria can be determined by specifying the constant α, subject to the constraint of global helicity conservation. There exists more than one possible value of α that satisfies the helicity constraint. However, it is believed that the global minimum in stability terms occurs for the smallest α value. It is found that these α–minimum equilibria contain up to 30% less energy than that stored in the original Gold–Hoyle tube.

062.074 **Magnetohydrodynamic waves in sharply and smoothly bounded cylinders.**
B. J. Shulman, E. G. Zweibel.
Astrophys. J., Vol. 389, No. 1, p. 428 – 439 (10 Apr 1992).
The propagation of waves in solar coronal loops is discussed. The waves are usually studied in a sharp boundary model where the density changes discontinuously from inside to outside the loop. The authors consider what happens if, instead of the sharp boundary condition at r = a the density varies continuously. A solar coronal loop is modeled as a cylindrical region of high density embedded in a surrounding medium of constant magnetic field. The density inhomogeneity was treated first as a step function (sharp boundary model) and next as a continuous function with a transition region joining the internal and external values across the cylinder's radius. The spectra for both models were than computed employing analytical and numerical procedures. The sharp boundary case consisted of a purely discrete spectrum. A continuous profile introduced a continuous spectrum. For the specific continuous profile chosen, the complex discrete spectrum was perturbed from the sharp boundary case. This implies that there is greater damping (in time) of the modes for the continuously varying model.

062.075 **Stability of a thin gravitating ring and systems of rings.**
V. L. Polyachenko.
Astron. Tsirk., No. 1550, p. 9 – 10 (Sep – Oct 1991). In Russian.
Approximate dispersion relations for a thin gravitating ring and for paired rings are derived. The results can help to explain some clumped ringlets of Saturn rings.

062.076 **Chaotic structure of nonlinear magnetic fields. I. Theory.**
N. C. Lee, G. K. Parks.
Geophys. Res. Lett., Vol. 19, No. 7, p. 637 – 640 (3 Apr 1992).
Tidman and Krall (1971) have studied extensively the behavior of nonlinear magnetic fields in MHD plasmas. The authors here have closely followed their work but the emphasis is different. They demonstrate here that the nonlinear magnetic field equations can be cast in the form of Duffing's equation. A Duffing system includes chaotic behavior; hence this study provides a different perspective on magnetic field development not previously discussed.

062.077 **Chaotic structures of nonlinear magnetic fields. II. Numerical results.**
N. C. Lee, G. K. Parks.
Geophys. Res. Lett., Vol. 19, No. 7, p. 641 – 644 (3 Apr 1992).
The authors present in this paper numerical solutions of the nonlinear magnetic field equations that have been cast in the form of Duffing's equation. These results show features that resemble the behavior of magnetic fields observed in space. The authors' work suggests that some of the observed magnetic field structures could be chaotic in origin.

062.078 **Critical density layer as obstacle at solar wind – exospheric ion interaction.**
K. Sauer, T. Roatsch, K. Baumgärtel, J. F. McKenzie.
Geophys. Res. Lett., Vol. 19, No. 7, p. 645 – 648 (3 Apr 1992).
Recent spacecraft observations near Venus and Mars have shown that the position and shape of the planetary bow shock do not agree well with predictions of gas dynamic models in which the obstacle boundary is defined by ionopause – magnetopause pressure balance estimations. In order to simulate some aspects of this problem, the authors have considered plasma flow into a

"heavy" ion cloud, whose maximum density is comparable to that of the inflowing protons. A simple bi–ion fluid model is used which takes into account the electrostatic coupling between protons and (heavy) ions of exospheric origin; magnetic field effects are ignored. One–dimensional studies show that for subsonic flows, representing the post–shock solar wind, a critical cloud density exists at which the flow is accelerated up to the bi–ion sound speed. Above this critical value no continuous solutions exist. The flow behavior in three dimensions was studied in an axially–symmetric model. For overcritical clouds the "critical density layer" appears as an "obstacle" forming an extended proton cavity. The planetopause at Mars may find an explanation within the present model.

062.079 **Applications of Lie groups to the theory of equilibrium of cylindrically–symmetric magnetic tubes.**
L. M. Zelenyj, A. V. Milovanov.
Astron. Zh., Tom 69, Vyp. 1, p. 147 – 158 (Jan–Feb 1992). In Russian. English translation in Sov. Astron., Vol. 36, No. 1 (1992).
A group analysis of differential equations for self–consistent equilibrium of cylindrically–symmetric magnetic tubes is presented. The existence of nonmaxwellian distribution functions, which make it possible to find analytical expressions for magnetic field and plasma density, is proved. Distribution functions are assumed to depend only upon constants of motion, and this allows us to formulate and solve the group classification problem for nonlinear third–order differential equations. The analytical solutions to differential equations with nontrivial symmetry group are calculated. The results obtained can be used for analysis of fractal properties of plasma configurations in solar and stellar atmospheres when treating different distribution functions.

062.080 **Dynamical regimes and possibility of origin of micro-flares in a prominence.**
V. V. Zajtsev, M. L. Khodachenko.
Astron. Zh., Tom 69, Vyp. 1, p. 159 – 172 (Jan–Feb 1992). In Russian. English translation in Sov. Astron., Vol. 36, No. 1 (1992).
On the basis of solutions of the Kippenhahn–Schlüter magnetohydrostatic model, a dynamical model of a solar prominence, taking plasma motion in the form of a cumulative flux into account, was developed. Of great importance is the fact that the prominence plasma is partially ionized. Ion–neutral collisions in nonsteady flux plasma conditions radically change the energetics and dynamics of the prominence model. In this case the equilibium state described by Kippenhahn–Schlüter solutions becomes unstable, and various dynamical regimes of plasma compression, as well as monotonic or quasiperiodic expansion of plasma can exist in the prominence, depending on the initial conditions. The energetics of this model of the prominence was investigated. Due to dissipation caused by ion–neutral collisions under nonsteady conditions, heating of the prominence greatly increases in comparison with the usual Joule heating. The estimated relation between the speed and degree of compression, when the prominence heating exceeds radiation losses, providing conditions for the development of the flare process, was obtained.

062.081 **Turbulent relaxation of magnetic fields. 1. Coarse–grained dissipation and reconnection.**
D. Tetreault.
J. Geophys. Res., Vol. 97, No. A6, p. 8531 – 8540 (1 Jun 1992).
A nonlinear, stochastic MHD model is suggested as an explanation for magnetic reconnection in the nearly collisionless environment of space plasma. The magnetic field line stochasticity (diffusion) results from the existence of overlapping $k \cdot B_0(x) = 0$ resonances and provides the anomalous resistivity driving the reconnection. The field lines diffuse in bundles which derive directly, via a nonlinear MHD instability, from the turbulent mixing of the large–scale fields. The model is a purely self–consistent MHD model in which the dynamical invariants of

one fluid MHD are satisfied. No ad hoc or kinetic instability–driven anomalous resistivity is required. Anomalous dissipation and reconnection are generated by a coarse–graining of the stochastic fields. The model is an example of "topological dissipation" of magnetic fields (Parker, 1972). Reconnection rates at a fraction of the Alfvén speed are predicted, and are in accord with previous conjecture (Parker, 1973).

062.082 **Turbulent relaxation of magnetic fields. 2. Self–organization and intermittency.**
D. Tetreault.
J. Geophys. Res., Vol. 97, No. A6, p. 8541 – 8547 (1 Jun 1992).
The intermittent nature of the stochastic MHD reconnection model of magnetic fields presented by Tetreault (1992) is developed. It is shown that the large–scale magnetic field topology relaxes in spatially localized regions or "domains". This occurs where the magnetic fluctuations, initially generated nonlinearly in the form of field line bundles, self–organize into coherent structures. The self–organization is driven by the tendency for parallel currents flowing along field lines within a bundle to attract each other. The self–organized structures take the form of flux tubes or ropes which in cross section are localized, topologically closed, magnetic island structures. The emergence of the structures out of the underlying MHD turbulence is an intermittent, macroscopic signature of the reconnection process and of a phase transition in the magnetic field topology. The structures are derived from a coarse–grained variational principle satisfying the dynamical invariants of MHD, and as such, their emergence can be expected to occur in a wide variety of magnetic field topologies. The model may have geophysical and astrophysical application, including reconnection and associated flux transfer events (FTEs) in the Earth's dayside magnetopause, as well as flux emergence and the occurrence of solar flares on the Sun.

062.083 **Relativistic shock waves and the excitation of plerions.**
J. Arons, Y. A. Gallant, M. Hoshino, A. B. Langdon, C. E. Max.
IAU Colloquium No. 128: The magnetospheric structure and emission mechanisms of radio pulsars, p. 78 – 85 (1992). – See Abstr. 012.022 for the main entry.
The authors have studied the structure of relativistic magnetosonic shock waves in plasmas composed purely of electrons and positrons, as well as those whose composition includes heavy ions as a minority constituent by number. The authors find that such shocks are not good candidates for the mechanism which converts rotational energy lost from a pulsar into the nonthermal synchrotron emission observed in plerions. However, a specific model is presented in some detail. Possible applications to the models of plerions and to constraints on theories of energy loss from pulsars are briefly outlined.

062.084 **The three–dimensional interaction of a supernova remnant with an interstellar cloud.**
J. M. Stone, M. L. Norman.
Astrophys. J., Lett., Vol. 390, No. 1, p. L17 – L19 (1 May 1992). With plates L3 – L4.
The authors present the hydrodynamic evolution in three dimensions of a spherical bubble embedded in a less dense uniform ambient medium as it interacts with a Mach 10 planar shock. This scenario is an idealized model of the interaction of an interstellar cloud with a large supernova remnant. A third–order Godunov numerical algorithm and 5.2×10^6 grid points are used to resolve the complex flow field which results. The authors find that the vortex rings observed in two–dimensional simulations are unstable in three dimensions, and the cloud fragments in all directions. Turbulent mixing of the cloud and interstellar medium is complete and is characterized by the formation of macroscopic vortex filaments. The strongest vortex filaments observed in these simulation may be responsible for the radio emission peaks observed in young supernova remnants.

062.085 **The effect of cosmic rays on the instability of a tangential discontinuity.**
S. V. Chalov.
J. Plasma Phys., Vol. 46, No. 2, p. 309 – 317 (Oct 1991).
The Kelvin–Helmholtz instability of a tangential velocity discontinuity in a plasma flow when cosmic–ray pressure is taken into account is studied. It is shown that a new unstable mode can exist. This appears when an effective Mach number exceeds some critical value. In the case of strong flow mediation by cosmic rays the wave vectors of the most–unstable waves are directed along the plasma velocity vector for both subsonic and supersonic flows.

062.086 **Particle acceleration in spherical wave fields.**
K. O. Thielheim.
IAU Colloquium No. 128: The magnetospheric structure and emission mechanisms of radio pulsars, p. 109 – 111 (1992). – See Abstr. 012.022 for the main entry.
Contents: 1. Particle acceleration by rotating magnets. 2. Vacuum wave fields of a rotating magnet. 3. Equations of motion. 4. Constant latitude approximation and restriction to the equatorial plane of rotation. 5. Substitution of eigentime by phase. 6. Mean development of dynamic variables. 7. Speculations about cosmic acceleration.

062.087 **On the thermal conductivity due to collisions between relativistic degenerate electrons.**
F. X. Timmes.
Astrophys. J., Lett., Vol. 390, No. 2, p. L107 – L109 (10 May 1992). = UCO/Lick Obs. Bull., No. 1214.
The double integral present in the expression for the thermal conductivity due to collisions between degenerate relativistic electrons is calculated numerically. The regime covered by the integration is valid for all ratios of the temperature to the electron plasma frequency temperature that are bounded from below by the melting temperature and bounded from above by the Fermi temperature. Excellent agreement is found with the limiting case expressions derived by Urpin and Yakovlev. The results are presented both in tabular form for applications requiring a high accuracy and in the form of a convenient fit formula. In addition, the fitting formula reduces in the proper limit to the first–order terms of the Urpin and Yakovlev asymptotic expansions. Effects of electron–electron collisions on the propagation speed of steady state deflagrations are briefly discussed.

062.088 **On the instability of decelerating shock waves.**
R. Nishi.
Prog. Theor. Phys., Vol. 87, No. 2, p. 347 – 365 (Feb 1992).
The author examines the decelerating shock instability (DSI) with a linear perturbation theory. He also investigates the effect of the self–gravity on the instability of a decelerating shocked layer. He finds that the characteristics of the DSI can be obtained with a thin layer approximation and that the DSI is caused by the combination of the fluctuation of column density induced by the rippling of the shocked layer and the deceleration of the layer as a whole. There is a typical wavenumber in the dispersion relation of the DSI. This dispersion relation may be related to the hierarchical structure of clumps and cores in the star forming regions around H II regions.

062.089 **Vortices on accretion disks.**
M. A. Abramowicz, A. Lanza, E. A. Spiegel, E. Szuszkiewicz.
Nature, Vol. 356, No. 6364, p. 41 – 43 (5 Mar 1992). Letter–to–the–editor.
Every rotating cosmic fluid that can be observed sufficiently closely displays either vortices or magnetic flux tubes on its surface; examples are tornadoes in the Earth's atmosphere, the Great Red Spot and other vortices in Jupiter's atmosphere, and sunspots. The authors suggest that hot accretion disks also produce coherent objects, and that these vortices and magnetic flux tubes will cause significant dissipation and other observable physical effects. They will facilitate the escape of collimated radiation from deep within hot disks, producing spectral changes

and time variability in the radiation from the disk. In the case of active galactic nuclei, modification of X–ray spectra due to the presence of vortices on accretion disks permits to explain several observational puzzles, including short–term variability and the low degree of linear polarization.

062.090 Dissipative stagnation–point flows at a current sheet with shear in the plasma velocity.
B. P. Besser, H. K. Biernat, R. P. Rijnbeek.
J. Atmos. Terr. Phys., Vol. 53, No. 11/12, p. 1081 – 1084 (Nov–Dec 1991). – See Abstr. 012.024 for the main entry.
The process of magnetic field annihilation is described using the framework provided by the MHD equations. An incompressible plasma is considered, and anomalous transport parameters in the form of uniform resistivity and viscosity are included. The planar stagnation–point geometry which the authors present models a current sheet separating two counterstreaming plasmas. One possible application of this configuration is the magnetopause, where coupling processes give rise to an exchange of energy and momentum between the magnetosheath and magnetospheric plasma. The authors allow for the possibility of a generally inclined plasma flow towards the current sheet so that, for example, the flow in the neighbourhood of the Earth's magnetospheric cusp and flank regions can be modelled as well as the subsolar region. A distinguishing feature of the exact solutions of the MHD equations which the authors present is that the flow is not irrotational and therefore viscous forces play an important role in the plasma behaviour.

062.091 Alfvén wings in the vicinity of a conducting body in a magnetized plasma.
I. I. Alexeev, E. S. Belenkaya.
J. Atmos. Terr. Phys., Vol. 53, No. 11/12, p. 1099 – 1101 (Nov–Dec 1991). – See Abstr. 012.024 for the main entry.
A solution of magneto–hydrodynamic equations is obtained for noncompressible plasma flow, in which a source of disturbance is located. Magnetic and electric fields are calculated in a stationary Alfvén structure appearing in the plasma flow. It is shown that the disturbance field may be up to twice the background field value.

062.092 Resonant behaviour of MHD waves on magnetic flux tubes. III. Effect of equilibrium flow.
M. Goossens, J. V. Hollweg, T. Sakurai.
Sol. Phys., Vol. 138, No. 2, p. 233 – 255 (Apr 1992).
The resonances that appear in the linear compressible MHD formulation of waves are studied for equilibrium states with flow. The conservation laws and the jump conditions across the resonance point are determined for 1D cylindrical plasmas. For equilibrium states with straight magnetic field lines and flow along the field lines the conserved quantity is the Eulerian perturbation of total pressure. Curvature of the magnetic field lines and/or velocity field lines leads to more complicated conservation laws. Rewritten in terms of the displacement components in the magnetic surfaces parallel and perpendicular to the magnetic field lines, the conservation laws simply state that the waves are dominated by the parallel motions for the modified slow resonance and by the perpendicular motions for the modified Alfvén resonance. The conservation laws and the jump conditions are then used for studying surface waves in cylindrical plasmas. These waves are characterized by resonances and have complex eigenfrequencies when the classic true discontinuity is replaced by a nonuniform layer. A thin non–uniform layer is considered here in an attempt to obtain analytical results. An important result related to earlier work by Hollweg et al. (1990) for incompressible planar plasmas is found for equilibrium states with straight magnetic field lines and straight velocity field lines. For these equilibrium states the incompressible and compressible surface waves have the same frequencies at least in the long wavelength limit and there is an exact correspondence with the planar case. As a consequence, the conclusions formulated by Hollweg et al. still hold for the straight cylindrical case. The effects of curvature are subsequently considered.

062.093 Numerical calculations of fast dynamos in smooth velocity fields with realistic diffusion.
D. J. Galloway, M. R. E. Proctor.
Nature, Vol. 356, No. 6371, p. 691 – 693 (23 Apr 1992). Letter–to–the–editor.
Many astrophysical magnetic fields are thought to arise by dynamo action due to internal fluid motions, but the natural timescale for magnetic field growth is the diffusion timescale, which in realistic astrophysical applications is very large. A fast dynamo is one that operates on the much shorter turnover timescale of the generating fluid flow, and the analytical intractability of smooth flows with diffusion has prompted the use of many ingenious models, differing from the true problem in having a modified or time–dependent diffusion or singularities in the flow field. Here the authors adopt a straightforward approach and present numerical computations of linear kinematic dynamos associated with periodic smooth flows, with diffusion explicitly included. Examples of time–varying flows depending on two spatial coordinates give convincing evidence of fast dynamo action for diffusion times up to 10,000 times greater than the turnover time. A three–dimensional steady flow shows similar behaviour, although computations have not been carried out so far and the asymptotic behaviour is less clear. All these flows have large regions where particle paths are chaotic.

062.094 Modeling structures of knots in jet flows with the Burgers equation.
L. Kofman, A. C. Raga.
Astrophys. J., Vol. 390, No. 2, p. 359 – 364 (10 May 1992).
The structures of high proper motion knots in both stellar and extragalactic jets could be the result of a time dependence in the ejection mechanism. Supersonic source velocity variations result in the formation of "internal working surfaces" (corresponding to pairs of shocks) that produce regions of enhanced emission and excitation which travel downstream along the jet flow, in qualitative agreement with the observations of astrophysical jets. The authors model such structures in terms of the one–dimensional viscous Burgers equation, which admits an analytic treatment. These solutions are used to obtain universal relations between the observed characteristics of the jets (e.g., their velocity profiles, knot separations, and the proper motions and intensities of the knots) and the kinetic energy ejected between two successive pulses of the time–dependent source. In this way, one obtains easily a recipe to interpret the "archeological record" of the past history of the source which is provided by the jet structure observed at the present time.

062.095 Ultrarelativistic hydrodynamics: high–resolution shock–capturing methods.
A. Marquina, J. M. Martí, J. M. Ibáñez, J. A. Miralles, R. Donat.
Astron. Astrophys., Vol. 258, No. 2, p. 566 – 571 (May 1992).
Some high–resolution shock–capturing methods have been designed recently to solve nonlinear hyperbolic systems of conservation laws. The authors have extended them to the relativistic hydrodynamics system of equations via a local characteristic approach. They are presenting preliminary tests of their procedure in the ultrarelativistic case displaying the main difficulties of this regime and the way to overcome them.

062.096 Switch shocks in molecular clouds. I. Plane–parallel magnetohydrodynamic analysis.
M. D. Smith.
Astrophys. J., Vol. 390, No. 2, p. 447 – 453 (10 May 1992).
Magnetohydrodynamic shock waves in molecular clouds are discussed. Plane–parallel flows with oblique magnetic fields are analyzed with particular regard to switch shocks. Switch–on shocks occur in the limit as the flow and field directions coincide; a transverse field is then switched on by the shock. In molecular clouds these become continuous or C–type, here termed C–switches. Their existence rules out the alternative J–type hydrodynamic flows. J–switches are only possible at low Alfvén numbers in strongly radiative shocks. C–switches yield both strong compression and large swings in the magnetic field

direction. They arise under a wide range of general conditions. One requirement is that the Alfvén speed exceeds the sound speed. The other requirement, to keep the flow supersonic, is that A^2/φ_0 is small, where A is the Alfvén number and φ_0 is a ratio of heating to cooling time scales. The temperature profile across an oblique shock is derived using the cool C–shock method. A formula is presented which determines the transverse shock of equivalent excitation. C–switch behavior influences flows within an angle $\sim 1/A$ of the field direction. Switch–off C–shocks, in which the field pressure generated by the switch–on shock is converted into thermal pressure as the transverse field disappears, are also discussed.

062.097   Relativistic magnetosonic shock waves in synchrotron sources: shock structure and nonthermal acceleration of positrons.
M. Hoshino, J. Arons, Y. A. Gallant, A. B. Langdon.
Astrophys. J., Vol. 390, No. 2, p. 454 – 479 (10 May 1992).

The authors study the theoretical properties of relativistic, transverse, magnetosonic collisionless shock waves in electron – positron – heavy ion plasmas of relevance to astrophysical sources of synchrotron radiation. They use both one–dimensional electromagnetic particle–in–cell simulations and quasi–linear theory to examine the spatial and kinetic structure of these nonlinear flows. All the upstream ions are electromagnetically reflected from the shock front, causing the magnetic field strength in the shock to overshoot its final downstream value in a series of long compressional oscillations with wavelengths comparable to the Larmor radius of ions with the rigidity appropriate to the downstream flow. The authors describe a new process of shock acceleration of nonthermal positrons, in which the gyrating reflected heavy ions dissipate their energy in the form of collectively emitted, left–handed magnetosonic waves which are resonantly absorbed by the positrons immediately behind the ion reflection region. This absorption gives rise to an ultrarelativistic downstream positron spectrum. It is argued that the nonthermal acceleration of positrons found here also applies to the partial acceleration of electrons. The authors briefly outline applications of these results to the termination shocks of pulsar winds and of jets emanating from active galactic nuclei. In particular, the model presented here can successfully account for the full spectral range of synchrotron radiation observed in the Crab Nebula.

062.098   Chaotic Alfvén waves in solar and cometary plasmas.
B. Buti.
IAU Symposium No. 152: Chaos, resonance and collective dynamical phenomena in the solar system, p. 297 – 298 (1992). Abstract. – See Abstr. 012.025 for the main entry.

062.099   Nonresonant resistive dissipation of compressible magnetohydrodynamic waves.
F. Califano, C. Chiuderi, G. Einaudi.
Astrophys. J., Vol. 390, No. 2, p. 560 – 566 (10 May 1992).

The effect of compressibility on the propagation and resistive dissipation properties of magnetohydrodynamic waves in non-uniform media is studied by using a normal mode analysis. This study concentrates on wave models for which the resistivity acts over the entire system, thus excluding localized resonant absorption. The authors identify the wave modes in inhomogeneous situations in terms of their familiar homogeneous counterparts and show that only the shear Alfvén and slow magnetosonic modes survive in the resistive regime. The most promising heating agents in the astrophysical context are shown to be the shear Alfvén waves that behave almost incompressibly.

062.100   Self–similar magnetohydrodynamics. V. Gravitating spheres and spheroids.
B. C. Low.
Astrophys. J., Vol. 390, No. 2, p. 567 – 572 (10 May 1992).

This paper presents a family of explicit, time–dependent magnetohydrodynamic solutions describing gravitating, magnetized, $\gamma = 4/3$ polytropes in the shape of a spheroid undergoing self–similar expansion or contraction in vacuum. Included in this family are the solutions for a gravitating, magnetized sphere, which are mathematically akin to Prendergast's equilibrium solutions for a static magnetic star. The solutions presented here are of basic physical interest and may be useful for the testing of multidimensional, numerical magnetohydrodynamic codes which include self–gravity as an important effect.

062.101   Magnetic reconnection in incompressible fluids.
E. E. DeLuca, I. J. D. Craig.
Astrophys. J., Vol. 390, No. 2, p. 679 – 686 (10 May 1992). With plates 12 – 15.

The authors investigate the dynamical relaxation of a disturbed X–type magnetic neutral point in a periodic geometry, with an ignorable coordinate, for an incompressible fluid. They find that the properties of the current sheet cannot be understood in terms of steady state reconnection theory or more recent linear dynamical solutions. Accordingly, the authors present a new scaling law for magnetic reconnection consistent with fast energy dissipation, i.e., the dissipation rate at current maximum is approximately independent of magnetic diffusivity η. The flux annihilation rate, however, scales as $\eta^{1/4}$, faster than the Sweet–Parker rate of $\eta^{1/2}$, but asymptotically much slower than the dissipation rate. These results suggest a flux pile–up regime in which the bulk of the free magnetic energy is released as heat rather than as kinetic energy of mass motion. The implications of these results for reconnection in the solar atmosphere and interior are discussed.

062.102   An integral on the shape of isolated magnetic loops.
J. Cheng.
Astron. Astrophys., Vol. 259, No. 1, p. 296 – 300 (Jun 1992).

Stationary flows in isolated magnetic flux tubes are studied using the thin flux tube approximation method. Assuming that the external atmosphere is a plane–stratified fluid in hydrostatic equilibrium, then an important integral is found, $(1-M_A^2)\,B\cos\theta = \text{const.}$, where M_A is the local Alfvén Mach number, B is the magnetic field strength and θ is the inclination of the flux tube with the horizontal direction. The integral is independent of the energy transfer mechanism and parameters of the ambient gas. Applying the integral, the authors obtain some theoretical results about the shape of isolated magnetic flux tubes and examine some numerical results of previous studies in this field. The applications of the integral to the solar atmosphere magnetic field are discussed briefly.

062.103   Relaxation processes in magnetohydrodynamics: a triad–interaction model.
V. Carbone, P. Veltri.
Astron. Astrophys., Vol. 259, No. 1, p. 359 – 372 (Jun 1992).

A truncated set of nonlinear ordinary differential equations for the velocity and magnetic field Fourier amplitudes has been derived from the MHD equations. In this reduced system the interactions take place only between three wavevectors. The dynamical evolution of this model is restricted to a six-dimensional vector space. Some three–dimensional time–invariant subspaces, have been identified. These subspaces are shown to determine the self–organization observed during the relaxation processes which occur in MHD, producing longliving nontrivial solutions like the "Taylor's vortex" or the "Alvénic structures". The authors' set of equations then assumes the role of a Lorenz-like model to study the relaxation processes in MHD and allows to make some predictions in the expected evolution of both fusion devices and astrophysical structures (coronal loop, low–frequency solar wind fluctuations).

062.104   Asymmetric disk accretion onto magnetized rotating compact stars.
G. D. Chagelishvili, J. G. Lominadze (D. G. Lominadze), Z. A. Sokhadze.
Astrophys. Lett. Commun., Vol. 28, No. 1, p. 11 – 14 (Jul 1990).

A case of the disk accretion onto an aligned gravimagnetic rotator with sufficiently strong large–scale azimuthal magnetic field in the accretion disk ($B_\varphi^2/8\pi \approx$ the median thermal energy density) is considered. This field is generated in the disk under certain circumstances as a result of the turbulent dynamo action in $\alpha\omega$ approximation. It is shown, that the existence of two magnetic fields of different origin in the system leads to the asymmetric accretion of the matter – the accretion mainly takes place selectively onto one of the magnetic poles depending on the ratio of the corotation and Alfvén radii. At the same time the accretion physics differs essentially from the disk accretion physics, which is realized at $B_\varphi = 0$.

062.105 H_2O maser pumping by shock waves.
J. R. D. Lepine, A. Heske.
IAU Symposium No. 150: Astrochemistry of cosmic phenomena, p. 245 – 247 (1992). – See Abstr. 012.029 for the main entry.

The authors discuss a simple H_2O maser pumping mechanism in which the population inversion of the masing levels takes place during the quick cooling of the gas behind a shock wave. The population of the rotational energy levels in the initial hot state and final cool state of the molecular gas, and the decay paths between levels are analysed to calculate the average number of 22 GHz photons emitted per H_2O molecule in the cooling process.

062.106 Hydrodynamics of encounters between star clusters and molecular clouds. I. Code validation and preliminary results.
T. Theuns.
Astron. Astrophys., Vol. 259, No. 2, p. 493 – 502 (Jun 1992).

A computer programme designed to study mixed systems containing both stars and gas is described and tested. The hydrodynamics part of the code is based on a 3D Lagrangian algorithm, while the stellar dynamics and self gravity parts are based on a high order direct summation code (N–body2). Advantage is taken of the multiple length and time scales in the system. Some preliminary results are presented of a first example of such mixed systems: the interaction between star clusters and molecular clouds in the Galactic plane. Encounters with dense clouds are catastrophic for the cluster. Tidal effects deform the initially spherical cluster into cylindrically shaped cluster debris. The denser parts of this debris contain loosely bound groups of stars as well as a rather high fraction of long period binaries, with periods comparable to the initial crossing time of the cluster. Diffuse clouds do not affect the cluster in a catastrophic way.

062.107 Hydrodynamics of encounters between star clusters and molecular clouds. II. Limits on cluster lifetimes.
T. Theuns.
Astron. Astrophys., Vol. 259, No. 2, p. 503 – 509 (Jun 1992).

Close encounters between star clusters and initially spherical molecular clouds in the Galactic plane are studied numerically. The full hydrodynamics of the cloud as well as the stellar dynamics of the cluster are included. The simulation results are compared with an improved version of the impulse approximation which is also valid for close encounters. With a realistic spectrum of cloud and encounter parameters taken into account, it is found that cluster lifetimes are limited to ≈ 100 Megayears due to catastrophic encounters with the most massive clouds of the spectrum.

062.108 The C:O ratio in dark clouds with cyclic star formation.
L. A. M. Nejad, D. A. Williams.
IAU Symposium No. 150: Astrochemistry of cosmic phenomena, p. 249 – 250 (1992). – See Abstr. 012.029 for the main entry.

062.109 Relativistic, perpendicular shocks in electron–positron plasmas.
Y. A. Gallant, M. Hoshino, A. B. Langdon, J. Arons, C. E. Max.
Astrophys. J., Vol. 391, No. 1, p. 73 – 101 (20 May 1992).

The authors use one–dimensional particle–in–cell plasma simulations to study the mechanical structure and thermalization properties of collisionless relativistic shock waves in electron–positron plasmas. The authors consider shocks propagating perpendicularly to the magnetic field direction, and show that these shocks are completely parameterized by σ, the ratio of the upstream Poynting flux to the upstream kinetic energy flux. The mechanical properties of the shock structure are controlled by magnetic reflection of all the particles, leading to a strong magnetic overshoot in the leading edge of the shock wave. The shock waves radiate a strong, semicoherent precursor wave with extraordinary mode polarization into the upstream medium, and also create an approximately Rayleigh–Jeans spectrum of extraordinary modes in the downstream medium. The spectrum of these downstream waves is broadband, extending to hundreds of harmonics above the cyclotron frequency based on upstream parameters. The downstream plasma is completely thermalized in the plane perpendicular to the magnetic field direction, with electron and positron distribution functions very close to relativistic Maxwellians. A two–dimensional simulation demonstrates the generality of the results beyond the assumption of the one–dimensional case. The thermalization mechanism is probably the formation of a synchrotron maser by the coherently reflected particles in the shock front. The authors suggest that recently observed short time–scale variability of the centimeter wave emission in compact extragalactic radio sources might be attributable to the precursor radiation from relativistic shocks in the magnetized coronae of accretion disks around the central black holes in AGNs.

062.110 MHD equilibria with flows in uniform gravity. I. 1–D prominence– and arcade–type solutions.
K. Tsinganos, G. Surlantzis.
Astron. Astrophys., Vol. 259, No. 2, p. 585 – 594 (Jun 1992).

A class of analytical MHD equilibria in a uniform gravitational field with flows along the magnetic lines is presented. The fieldlines have prominence–like valleys and arcade–type summits where the fields are horizontal. At the valleys the density is maximum and the flow speed subsonic and sub–Alfvenic, while at the summits the density is minimum, and the flow speed supersonic and super–Alfvenic. These solutions may be regarded as an extension of the familiar Kippenhahn–Schlüter model for a quiescence solar prominence and models for solar magnetic arcades, by including flows along the magnetic field lines. The periodic fine–scale of the magnetic field lines and streamlines is similar to the observed fine–scale fibril structure of solar prominences.

062.111 Magneto–radiative shock wave propagation in a conducting plasma.
V. K. Singh, K. K. Srivastava.
Astrophys. Space Sci., Vol. 190, No. 2, p. 169 – 176 (Apr 1992).

Similarity solutions describing the flow of a perfect gas behind a spherical and cylindrical shock wave in a magnetic field with radiation heat flux have been investigated. The total energy of the expanding wave has been assumed to remain constant. The solutions, however, are only applicable to a gaseous medium where the undisturbed pressure falls as the inverse square of the distance from the line of explosion.

062.112 Instability of a gravitating partially–ionized plasma.
A. Ali, P. K. Bhatia.
Astrophys. Space Sci., Vol. 191, No. 1, p. 89 – 100 (May 1992).

The effect of Hall currents and collision with neutrals on the instability of a horizontal layer of a self–gravitating partially-ionized plasma of varying density have been studied. It is assumed that the plasma is permeated by a variable horizontal magnetic field stratified vertically. A variational principle is shown to characterize the problem. By making use of the existence of the variational principle, proper solutions have been obtained for a semi–infinite plasma in which density has a one–dimensional (exponential) vertical stratification. The dispersion relation has been derived and solved numerically. It is found that the collisions with neutrals have a stabilizing influence while Hall currents have a destabilizing influence.

062.113 Hall effects on heat and mass transfer flow with variable suction and heat generation.
H. S. Takhar, P. C. Ram, S. S. Singh.
Astrophys. Space Sci., Vol. 191, No. 1, p. 101 – 106 (May 1992).

A study of the combined buoyancy effects of thermal and mass diffusion on MHD convection flow in the presence of Hall currents with variable suction and heat generation has been carried out. Analytical expressions for the velocity and the temperature of the fluid are given. The effects of Hall currents, parameter m, and heat source parameter δ on the velocity are discussed.

062.114 Sound waves in an electron–positron plasma.
L. N. Tsintsadze.
Astrophys. Space Sci., Vol. 191, No. 1, p. 151 – 155 (May 1992). Letter–to–the–editor.

By means of a kinetic description, the author obtains the sound wave velocity in an electron–positron plasma at relativistic temperatures.

062.115 Nonlinear ion–acoustic solitons in a warm plasma with adiabatic positive and negative ions and hot non-isothermal electrons.
S. K. El–Labany.
Astrophys. Space Sci., Vol. 191, No. 2, p. 185 – 194 (May 1992).

The propagation of an ion–acoustic soliton in a collisionless plasma with adiabatic positive and negative ions (with equal ion temperature) and hot non–isothermal electrons is studied by use of the renormalization method introduced by Kodama and Taniuti in the reductive perturbation method. The basic set of fluid equations describing the system is reduced to a Korteweg–de Vries (K–dV)–type equation for the first–order perturbed potential and to a linear inhomogeneous differential equation to the second–order of the perturbed potential. A stationary solution of the coupled equations is obtained.

062.116 The effect of relativistic particle beams on the evolution of supernova envelopes.
J. H. Beall.
Vulcano Workshop 1990: Frontier objects in astrophysics and particle physics, p. 269 – 288 (1991). – See Abstr. 012.040 for the main entry.

The author considers the effect of a relativistic particle beam produced as part of the collapse process of a supernova core into a neutron star or black hole on the evolution of the expanding envelope. Relativistic bremsstrahlung is the dominant energy loss mechanism until (or unless) the material through which the beam propagates becomes ionized. After an ionized channel is formed, plasma processes and inverse Compton losses become the dominate loss mechanisms. The energy loss processes associated with the beam impart significant momentum to the irradiated segments of the shell. This suggests a natural explanation for the asymmetric expansion of some supernovae, including SN1987a, and may account for the early mixing seen in that objects. It also implies that some fraction of the X–ray light from very early in a supernova explosion originates in an inverse Compton emission process wherein relativistic electrons from the beam collide with optical photons from the expanding envelope. The author presents self–consistent solutions to the rate equations for the energy loss due to collective processes and calculates the momentum transferred to the envelope by the beam. He then comments on the expected X–ray emission from SN1987a and makes estimates of the associated γ–ray flux.

062.117 On instability of an inhomogeneous plasma in active regions of the solar corona.
A. K. Yukhimuk, Yu. M. Vojtenko, A. N. Kristal'.
Visn. Kiiv. Univ., Fiz.–Mat. Nauki, Astron., Vip. 3, p. 26 – 29 (1992). In Ukrainian.

Influence of drift motions and kinetic effects on the stability of Alfvén–type waves at the boundary of solar coronal loops is investiated. It is shown that waves with frequency $\omega \gtrsim 1c^{-1}$ can be excited. Instability appears because of combined action of effects connected with inhomogeneity of plasma and finite value of the proton gyroradius.

062.118 The parameters of plasma in gamma–ray burst sources.
W. Tkaczyk, S. Karukuła.
Vulcano Workshop 1990: Frontier objects in astrophysics and particle physics, p. 289 – 296 (1991). – See Abstr. 012.040 for the main entry.

The models for gamma–ray burst sources are examined. In standard accretion and thermonuclear explosion models can exist unthermalized pair dominant plasma ($T_e + \neq T_e -$). The parameters of sources for which the regions with different temperatures of plasma can exist are evaluated. The constraints on temperatures of electrons and positrons are discussed.

062.119 Unsteady mixed convection flow with thermal diffusion effect.
B. K. Jha.
Astrophys. Space Sci., Vol. 191, No. 2, p. 283 – 288 (May 1992).

An exact analysis of the unsteady free and forced convection flow of an incompressible viscous fluid is presented in the presence of thermal diffusion effect. The Laplace transform technique is used to obtain the expressions for velocity, leading edge effects, and skin–friction. During the course of the discussion, the effects of S (thermal diffusion parameter), Pr (Prandtl number), and t (time parameter) on velocity, leading edge, and skin–friction are extensively discussed.

062.120 Influence of viscous dissipation on a hydromagnetic field.
T. I. Lekas, G. A. Georgantopoulos.
Astrophys. Space Sci., Vol. 191, No. 2, p. 299 – 305 (May 1992).

The effect of a transverse magnetic field and of the viscosity diffusion on the free–convection flow of an electrically–conducting incompressible fluid past a uniformly accelerated vertical plate is discussed. A finite difference method has been used to obtain a numerical solution. The influence of the various parameters on the flow field is discussed.

062.121 A note on MHD instability of a compressible gravitational streaming fluid cylinder.
A. E. Radwan.
Astrophys. Space Sci., Vol. 191, No. 2, p. 307 – 312 (May 1992).

The fundamental equations are formulated using cylindrical polar coordinates and then solved in the unperturbed state. The perturbation equations are determined, simplified, integrated and the constants of integrations are identified by applying appropriate boundary conditions across the perturbed fluid interface. A cumbersome stability criterion for MHD inviscid compressible self–gravitating streaming fluid cylinder is derived. The magnetic field is stabilizing, the streaming is destabilizing while both the self–gravitating and compressibility are stabilizing or not according to restrictions and the gravitational instability of sufficiently long waves will persist. Several approximations are required to obtain Chandrasekhar's and Fermi's dispersion relation (Chandrasekhar and Fermi, 1953).

062.122 Thermal–diffusion effects on MHD free–convective and mass–transfer flow past a moving infinite vertical plate in a rotating fluid.
N. Nanousis.
Astrophys. Space Sci., Vol. 191, No. 2, p. 313 – 322 (May 1992).

An analytical study of MHD free–convective and mass–transfer flow past a moving infinite vertical plate, in a rotating fluid, is presented, taking into account the thermal diffusion effects. The solution of the problem is obtained with the help of the Laplace transform technique. Analytical experessions are given for the velocity field and for skin–friction for two different cases, e.g., when the plate is impulsively started, moving on its own plane (case I) and when it is uniformly accelerated (case II). The effects on the velocity field and skin–friction, of the various parameters entering into the problem, are discussed with the help of graphs.

062.123 **Free and forced convection flow through a porous medium near the leading edge.**
M. A. Sattar.
Astrophys. Space Sci., Vol. 191, No. 2, p. 323 – 328 (May 1992).
Locally similar solutions of the combined free and forced convection flow in a semi–infinite vertical porous medium are obtained by a perturbation method. The solutions, obtained near the leading edge of the plate, are discussed in comparison with the numerical results obtained by Raptis and Perdikis (1988).

062.124 **Resistive instability and the magnetostrophic approximation.**
D. R. Fearn, W. S. Weiglhofer.
Geophys. Astrophys. Fluid Dyn., Vol. 63, No. 1–4, p. 111 – 138 (1992).
The authors investigate resistive instability of the toroidal magnetic field $B_0{}^*$ permeating a conducting fluid confined in an infinite cylindrical annulus. With application to planetary cores in mind, the system is rapidly rotating with uniform angular velocity. Resistive instability is most often associated with critical levels $k \cdot B_0{}^* = 0$ (where k is the wave vector). For the choice of field, critical levels are located at zeros of $B_0{}^*$. In this paper, the main emphasis is on studying resitive instability when no critical levels are present and the authors find instabilities for certain choices of $B_0{}^*$ when the cylindrical container is electrically insulating. Asymptotic results are obtained in the limit of high conductivity and in the limit of small axial wavenumber.

062.125 **The inner structure of an accretion disc around a magnetic neutron star.**
C. G. Campbell.
Geophys. Astrophys. Fluid Dyn., Vol. 63, No. 1–4, p. 179 – 198 (1992).
The effect of a magnetic binary neutron star on the inner structure of its accretion disc is considered. The stellar field penetrates the disc and toroidal field is created as a result of the vertical shearing motions, its magnitude being diffusion limited. If the disc's magnetic diffusivity, η, is treated as a function of $\bar{\omega}$, the distance from the stellar centre, with a radial length scale of $\sim \bar{\omega}$, a self–consistent solution can be found for the inner region. In the orbital plane, the magnetic force has a small effect on the radial momentum, so the angular velocity is Keplerian. The vertical equilibrium is a balance between the gradients in azimuthal magnetic pressure and thermal pressure, while the inflow is caused by the transfer of angular momentum from the disc to the star, via magnetic stresses. Vertical equilibrium breaks down as the inflow speed becomes a significant fraction of the Keplerian speed, this occurring in the region of the Alfvén radius based on radial free–fall. However, if η is prescribed by a simple model of turbulence, or has a magnetic buoyancy origin, the disc becomes rapidly disrupted vertically where the azimuthal magnetic force starts to exceed that due to viscosity. This occurs at a distance from the star significantly larger than the free–fall Alfvén radius.

062.126 **Magnetically–controlled disc accretion.**
C. G. Campbell.
Geophys. Astrophys. Fluid Dyn., Vol. 63, No. 1–4, p. 197 – 213 (1992).
A magnetic alternative to the viscous accretion disc, around a non–magnetic star, is presented. The disc field is generated by an $\alpha\omega$–dynamo; only weak turbulence is required to produce the α–effect, so the viscous force is negligible. The magnetic force does not significantly affect the vertical equilibrium or the radial momentum, hence stellar gravity balances the thermal pressure gradient vertically and the angular velocity is Keplerian up to a boundary layer close to the stellar surface. The azimuthal magnetic force causes the outward advection of angular momentum necessary for the transfer of matter through the disc at the externally imposed rate. The magnetic field has a quadrupolar structure.

062.127 **Rotation of the plasmasphere and the formation of the boundary layer at its limit.**
P. F. Krymskij.
Geomagn. Aeron., Vol. 31, No. 1, p. 39 – 43 (Aug 1991). English translation of Geomagn. Aehron., Tom 31, No. 1, p. 53 – 59 (1991).
Plasma flow in the vicinity of planets is considered. It is shown that at times of enhancement of the magnetospheric convection field at the boundary of a synchronously rotating plasmasphere a viscous boundary layer with a flow shift is formed. In the MHD approximation, with anisotropic conductivity in a dipole magnetic field, the distributions of velocity and plasma concentration, electric fields, and currents are computed.

062.128 **Numerical investigations of converging cylindrical and spherical shock waves.**
R. C. Srivastava, D. Leutloff, K. G. Roesner, K. Kuwahara.
Astrophys. Space Sci., Vol. 192, No. 1, p. 1 – 9 (Jun 1992).
The converging cylindrical and spherical shock waves have been numerically simulated. The flow is created by rupturing the diaphragm. The behaviour of the solution in the focussing stage is closely investigated and compared with the other results. An invariant difference scheme of Rusanov is used to follow the propagation of the shock wave. The study includes not only the moderate initial pressure ratio but also pressure ratios amounting up to thousands. The same scheme has been used from the initial stage of the focussing stage near the axis.

062.129 **MHD thermal–diffusion effects on free–convective and mass–transfer flow over an infinite vertical moving plate.**
N. G. Kafoussias.
Astrophys. Space Sci., Vol. 192, No. 1, p. 11 – 19 (Jun 1992).
The Soret effect on MHD free–convective and mass–transfer flow of an incompressible, viscous, and electrically–conducting fluid, past a moving vertical infinite plate is studied. The flow is assumed to be at small Reynolds number so that the induced magnetic field is neglected. The problem is solved with the help of the Laplace transform method for two different values of the dimensionless function $f(t)$ signifying two different cases, e.g., (i) when the boundary surface, the flat plate, is impulsively started, moving in its own plane and (ii) when it is uniformly accelerated. The effects on the velocity field as well as on the skin–friction of the various dimensionless parameters occurring into the problem, especially the magnetic parameter M and Soret number So, are discussed with the help of graphs.

062.130 **Effects of Hall current on hydromagnetic free–convective flow through a porous medium.**
H. S. Takhar, P. C. Ram.
Astrophys. Space Sci., Vol. 192, No. 1, p. 45 – 51 (Jun 1992).
An analysis of the effects of Hall current on hydromagnetic free–convective flow through a porous medium bounded by a vertical plate is theoretically investigated when a strong magnetic field is imposed in a direction which is perpendicular to the free stream and makes an angle α to the vertical direction. The influence of Hall currents on the flow is studied for various values of α.

062.131 **Geometry of magnetohydrostatic equilibria.**
C. Thakur, R. A. Yadav.
Astrophys. Space Sci., Vol. 192, No. 1, p. 75 – 78 (Jun 1992).
A geometrical study is made of magnetohydrostatic equilibrium of a perfectly conducting fluid considering magnetic field vector, constant along each current line. The authors have discussed the geometrical behaviour of Lorentz surfaces, field lines and current lines.

062.132 **Pulsated hydromagnetic flow with free–convection currents and mass transfer along a vertical infinite porous plate.**
T. I. Lekas, G. A. Georgantopoulos.
Astrophys. Space Sci., Vol. 192, No. 1, p. 91 – 101 (Jun 1992).
The influence of free–convection currents and mass transfer in the case of laminar, non–steady boundary layer is studied here.

The fluid is the air, considered incompressible and electrically conductive. The free–stream velocity oscillates about a mean value. A pulsated suction is also taken into account. A second material lies in small concentration in the fluid and absorbs part of the radiation. The influence of various parameters on the flow is examined.

062.133   Importance of plasma physics processes in extragalactic radio sources.
V. Fedorenko.
7. IAP Meeting: Extragalactic radio sources – from beams to jets, p. 1 – 6 (1992). – See Abstr. 012.041 for the main entry.
The author discusses relativistic particle acceleration with emphasis to the diffuse shock acceleration mechanisms.

062.134   Models of central engines of AGNs with electron–positron pair production.
A. A. Zdziarski.
7. IAP Meeting: Extragalactic radio sources – from beams to jets, p. 32 – 44 (1992). – See Abstr. 012.041 for the main entry.
The author reviews selected theoretical models of the central regions of active galactic nuclei with the emphasis on the models and scenarios that predict copious production of $e^{\pm}$ pairs. The models are classified based on their assumed geometry. One–zone models (with and without reflection), spherical accretion, and disk accretion will be considered.

062.135   Particle acceleration at relativistic shocks.
J. G. Kirk.
7. IAP Meeting: Extragalactic radio sources – from beams to jets, p. 176 – 182 (1992). – See Abstr. 012.041 for the main entry.
A brief review is given of the status of the theory of particle acceleration at relativistic shocks, indicating the importance of the effects which arise in oblique shocks. In the case of ultra–relativistic shocks, it is suggested that the main acceleration mechanism is shock–drift acceleration, because even a slight obliquity of the magnetic field suffices to inhibit the ability of a particle to cross and recross the shock front. However, for mildly relativistic flows in which the magnetic field is almost parallel to the shock normal the first order Fermi mechanism can be effective. In this case it is possible to estimate the efficiency of acceleration by considering the back reaction of particles on the fluid. The results of an investigation of the stationary solutions of this nonlinear problem show interesting deviations from the nonrelativistic results even at modest shock speeds ($\lesssim c/3$).

062.136   Time dependent numerical simulations of diffusive particle acceleration in astrophysical environments.
T. W. Jones, H. Kang.
7. IAP Meeting: Extragalactic radio sources – from beams to jets, p. 183 – 190 (1992). – See Abstr. 012.041 for the main entry.
As part of an effort to understand better the relationships between dynamical processes in and electromagnetic radiation from cosmic radio sources the authors have developed a program to study particle acceleration and propagation through numerical simulations.

062.137   Particle acceleration and magnetic field evolution.
J. A. Eilek, P. A. Hughes.
Beams and jets in astrophysics, p. 428 – 483 (1991). – See Abstr. 003.099 for the main entry.
Contents: 1. Introduction. 2. Stochastic acceleration: history and fundamentals; MHD waves and their interaction with particles; wave spectra and energetics; evolution of the particle spectra. 3. Acceleration by shocks: the physics of collisionless shock waves; the shock–Fermi process; the non–linear theory of the Fermi process; Fermi acceleration at relativistic shocks; threshold energy for the Fermi process. 4. The injection problem: resonances with nonrelativistic particles; the need for injection. 5. Electrodynamic acceleration: acceleration and heating by a DC electric field; shock drift acceleration; relation of electrodynamic acceleration to the injection problem. 6. Structure and maintenance of magnetic fields: turbulent dynamos; Taylor relaxation, consequences and applications. 7. Conclusions.

062.138   The motion of vortices within a rotating, fluid shell.
P. Lanzano.
Earth, Moon, Planets, Vol. 56, No. 1, p. 75 – 82 (Jan 1992).
The author considers a spherical, solid planet surrounded by a thin layer of an incompressible, inviscid fluid. The planet rotates with constant angular velocity. Within the constraints of the geostrophic approximation of hydrodynamics, the author determines the equation that governs the motion of a vortex tube within this rotating ocean. This vorticity equation turns out to be a nonlinear partial differential equation of the third order for the stream function of the motion. The author next examines the existence of particular solutions to the vorticity equation that represent travelling waves of permanent form but decaying at infinity.

062.139   Particle acceleration and flux outbursts.
K.–D. Fritz.
7. IAP Meeting: Extragalactic radio sources – from beams to jets, p. 191 – 194 (1992). – See Abstr. 012.041 for the main entry.
Flux outbursts of Blazar objects in the infrared to optical regime offer an opportunity to study particle acceleration in a time dependent scenario. The outburst spectra which show variability on timescales smaller than a day are interpreted as synchrotron radiation of high relativistic electrons. It is evident that common steady–state first order Fermi theory is not applicable in this case. Intrigued by polarization measurements during outbursts the possibility of a two–component synchrotron source is investigated.

062.140   From ultrarelativistic beams to radio jets: the accretion disk connection.
A. Königl.
7. IAP Meeting: Extragalactic radio sources – from beams to jets, p. 195 – 205 (1992). – See Abstr. 012.041 for the main entry.

062.141   Why do all the extragalactic jets have Lorentz factors less than twenty?
M. A. Abramowicz.
7. IAP Meeting: Extragalactic radio sources – from beams to jets, p. 206 – 213 (1992). – See Abstr. 012.041 for the main entry.
The observed Lorentz factors of extragalactic jets are never greater than 20. It was recently demonstrated (Abramowicz, Ellis and Lanza, 1989, 1990) that this may be explained as the effect of relativistic Compton drag, which gives an upper limit for the asymptotic Lorentz gamma factors of extragalactic jets, $\gamma_{max} \approx (m_p/m_e)^{1/3} \approx 20$. The precise value of γ_{max} depends only very weakly on astrophysical details.

062.142   Electron–positron beam production in compact radio sources.
R. Schlickeiser, U. Achatz.
7. IAP Meeting: Extragalactic radio sources – from beams to jets, p. 214 – 220 (1992). – See Abstr. 012.041 for the main entry.
The study of the acceleration of energetic electrons and positrons in active galactic nuclei has shown that the concerted action of accretion, accretion shock, magnetic fields and MHD turbulence produces a nearly mono–energetic distribution of relativistic electrons and positrons (the "pile–up"–distribution). The time evolution of the initially isotropic pile–up is studied for the two extreme cases of presence or absence of Alfvén waves in the galactic medium. The latter case gives ready explanations both for the formation of superluminal VLBI jets and for the radio gap phenomenon.

062.143   Production of relativistic $e^+ - e^-$ beams.
G. Henri, G. Pelletier.
7. IAP Meeting: Extragalactic radio sources – from beams to jets, p. 221 – 227 (1992). – See Abstr. 012.041 for the main entry.

062.144   MHD accretion–ejection flow for AGN.
G. Pelletier, R. Pudritz, J. Ferreira, F. Rosso.
7. IAP Meeting: Extragalactic radio sources – from beams to jets, p. 228 – 234 (1992). – See Abstr. 012.041 for the main entry.

062.145   Magnetically driven jets and winds.
R. V. E. Lovelace, J. Contopoulos.
7. IAP Meeting: Extragalactic radio sources – from beams to jets, p. 235 – 240 (1992). – See Abstr. 012.041 for the main entry.

A simple but general theory for the formation and propagation of jets and winds from magnetized accretion disks is derived from the equations of ideal magnetohydrodynamics.

062.146   Magnetic fields and reconnection in extragalactic jets.
M. M. Romanova, R. V. E. Lovelace.
7. IAP Meeting: Extragalactic radio sources – from beams to jets, p. 241 – 247 (1992). – See Abstr. 012.041 for the main entry.

Extra–galactic radio jets are investigated theoretically taking into account that the magnetic field is dragged out from the central source by the jet flow. Thus, magnetohydrodynamic models of jets are considered with zero net poloidal current and zero net poloidal flux. A simple model for particle acceleration at neutral layers in electron/positron and electron/proton plasmas is discussed.

062.147   The stability of force–free MHD jets.
S. Appl, M. Camenzind.
7. IAP Meeting: Extragalactic radio sources – from beams to jets, p. 248 – 255 (1992). – See Abstr. 012.041 for the main entry.

The authors investigated the stability properties of a cylindrical force–free jet propagating at supermagnetosonic velocity.

062.148   Application of the phase–integral method to the trapping of acoustic waves in a gravitating fluid. I. Planar polytrope and turning–point behavior.
G. H. Price.
Astrophys. J., Vol. 391, No. 2, p. 845 – 853 (1 Jun 1992).

The use of asymptotically based phase–integral methods to describe the behavior at short surface wavelengths of acoustic oscillations trapped within stratified, gravitating fluids is examined. The method is applied here to a planar polytrope, for which exact results are well known, both to assess the overall picture of oscillation behavior that it provides and more specifically to determine the reflection phase shifts that occur at the trapping boundaries. The analysis yields significantly differing descriptions of the field behavior near the surface, depending on the level of approximation used. A local analysis, although usefully delineating the trapping region, recovers the exact modal dispersion relation without overly elaborate reflection coefficients at the trapping limits only in the asymptotic limit. Refinement of the analysis to include gradient terms in the determination of the vertical wavenumber component recovers the exact dispersion relation without such restriction. Further refinement to recover the correct amplitude behavior well within the trapping region entails the neglect of terms that become large near the surface, with the consequence that the shallow subsurface upper turning point is lost in this analysis, which has the wave reflecting at the surface.

062.149   Stability of electron–positron beams.
U. Achatz, R. Schlickeiser.
7. IAP Meeting: Extragalactic radio sources – from beams to jets, p. 256 – 262 (1992). – See Abstr. 012.041 for the main entry.

The instability of relativistic electron–positron beams, recently discussed as a component of extragalactic jets, against the excitation of electromagnetic waves moving along a common axis of beam and magnetic field is examined.

062.150   Streaming particle beams in active galactic nuclei.
H. Sol.
7. IAP Meeting: Extragalactic radio sources – from beams to jets, p. 263 – 265 (1992). – See Abstr. 012.041 for the main entry.

062.151   Nonlinear interaction of the relativistic wind with the ambient plasma.
Ya. N. Istomin.
7. IAP Meeting: Extragalactic radio sources – from beams to jets, p. 266 – 271 (1992). – See Abstr. 012.041 for the main entry.

062.152   Physics of bright jets and their velocities.
G. V. Bicknell.
7. IAP Meeting: Extragalactic radio sources – from beams to jets, p. 272 – 278 (1992). – See Abstr. 012.041 for the main entry.

062.153   Mixed shocks and the formation of hot spots.
R. Lehoucq, J. Roland.
7. IAP Meeting: Extragalactic radio sources – from beams to jets, p. 287 – 293 (1992). – See Abstr. 012.041 for the main entry.

The hot spots in the extended lobes of extragalactic radio sources can be explained by the interaction of a non–relativistic supersonic jet with the intergalactic medium. The authors investigate the influence of the Mach number and the Alfvenic Mach number on: 1) the acceleration of relativistic particles by mixed shocks. 2) the spectral index of the power law distribution of the relativistic particles.

062.154   ZEUS–2D: a radiation magnetohydrodynamics code for astrophysical flows in two space dimensions. I. The hydrodynamic algorithms and tests.
J. M. Stone, M. L. Norman.
Astrophys. J., Suppl. Ser., Vol. 80, No. 2, p. 753 – 790 (Jun 1992).

The authors begin a detailed description of ZEUS–2D, a numerical code for the simulation of fluid dynamical flows in astrophysics, including a self–consistent treatment of the effects of magnetic fields and radiation transfer. The algorithms in ZEUS–2D divide naturally into three areas: (1) hydrodynamics (HD), (2) magnetohydrodynamics (MHD), and (3) radiation hydrodynamics (RHD). In this first paper, the authors give a detailed description of the HD algorithms which form the foundation for the more complex MHD and RHD algorithms. The authors use simple, well developed Eulerian HD algorithms based on the method of finite–differences implemented in a new covariant formalism which allows simulation in any orthogonal coordinate system. The effect of self–gravity on the flow dynamics is accounted for by an iterative solution of the sparse–banded matrix resulting from discretizing the Poisson equation in multidimensions. The results of an extensive series of HD test problems are presented.

062.155   ZEUS–2D: a radiation magnetohydrodynamics code for astrophysical flows in two space dimensions. II. The magnetohydrodynamic algorithms and tests.
J. M. Stone, M. L. Norman.
Astrophys. J., Suppl. Ser., Vol. 80, No. 2, p. 791 – 818 (Jun 1992).

The authors continue a detailed description of ZEUS–2D, a numerical code for the simulation of fluid dynamical flows in astrophysics, including a self–consistent treatment of the effects of magnetic fields and radiation transfer. In this paper, the authors give a detailed description of the magnetohydrodynamical (MHD) algorithms in ZEUS–2D. The recently developed constrained transport (CT) algorithm is implemented for the numerical evolution of the components of the magnetic field for MHD simulations. This formalism guarantees the numerically evolved field components will satisfy the divergence–free constraint at all times. It is found, however, that the method used to compute the electromotive forces must be chosen carefully to propagate accurately all modes of MHD wave families (in particular shear Alfvén waves). A new method of computing the electromotive force is developed using the method of characteristics (MOC). It is demonstrated through the results of an extensive series of MHD test problems that the resulting hybrid MOC–CT method provides for the accurate evolution of all modes of MHD wave families.

062.156   ZEUS–2D: a radiation magnetohydrodynamics code for astrophysical flows in two space dimensions. III. The radiation hydrodynamic algorithms and tests.
J. M. Stone, D. Mihalas, M. L. Norman.
Astrophys. J., Suppl. Ser., Vol. 80, No. 2, p. 819 – 845 (Jun 1992).

The authors conclude a detailed description of ZEUS–2D, a numerical code for the simulation of fluid dynamical flows in astrophysics including a self–consistent treatment of the effects of magnetic fields and radiation transfer. In this paper, the authors describe the radiation hydrodynamical (RHD) algorithms in ZEUS–2D. They developed a two–dimensional full transport algorithm to evolve the radiation moment equations. The moment equations are closed with the tensor variable Eddington factor whose components are computed from angular quadratures of the specific intensity which, in turn, is computed from a formal solution of the two–dimensional transfer equation using the method of short characteristics. This algorithm for multidimensional RHD differs significantly from more commonly used methods based on the diffusion approximation. The results of a collection of test problems, developed for RHD algorithms, are presented.

062.157   Absorption of acoustic waves by sunspots. II. Resonance absorption in axisymmetric fibril models.
C. S. Rosenthal.
Sol. Phys., Vol. 139, No. 1, p. 25 – 45 (May 1992).

Observations of the scattering of acoustic waves by sunspots show a substantial deficit in scattered power relative to incident power. A number of calculations have attempted to model this process in terms of adsorption at the magnetohydrodynamic Alfvén resonance. The results presented here extend these calculations to the case of a highly structured axisymmetric translationally invariant flux–tube embedded in a uniform atmosphere. The fractional energy absorbed is calculated for models corresponding to flux–tubes of varying radius, mean flux–density and location below the photosphere. The effects of twist are also included. It is found that absorption can be very efficient even in models with low mean magnetic flux density, provided the flux is concentrated into intense slender annuli. Twist is found to increase the range of wave numbers over which absorption is efficient, but it does not remove the low absorption at low azimuthal orders which is a feature of resonance absorption calculations in axisymmetric geometry, and which is in conflict with observation. These results suggest that resonance absorption could be an efficient mechanism in plage fields and fibril sunspots as well as in monolithic sunspots.

062.158   The prominence–corona transition region in transverse magnetic fields.
F. Chiuderi Drago, O. Engvold, E. Jensen.
Sol. Phys., Vol. 139, No. 1, p. 47 – 64 (May 1992).

An emission measure analysis is performed for the prominence–corona transition region (PCTR) under the assumption that the cool matter of quiescent filaments is contained in long, thin magnetic flux loops imbedded in hot coronal cavity gas. Consequently, there is a transition region around each thread. Comparison of the model and observations implies that the temperature gradient is perpendicular to the magnetic lines of force in the lower part of the PCTR. It is shown that in this layer the heating given by the divergence of the transverse conduction fails to account for the observed UV and EUV emission by several orders of magnitude. It is, therefore, suggested that the heating of these layers could be due to dissipation of Alfvén waves. In the high–temperature layers, where the plasma $\beta \geqslant 1$, the temperature gradient is governed by radiative cooling balancing conductive heating from the surrounding hot coronal gas. Also in these outer layers the presence of magnetic fields reduces notably the thermal conduction relative to the ideal field–free case. Numerical modelling gives good agreement with observed differential emission measure; the inferred value of the flux carried by Alfvén waves, as well as that of the damping length, greatly support the suggested form of heating. The model assumes that about 1/3 of the volume is occupied by threads and the rest by hot coronal cavity matter. The brightness of the EUV emission will depend on the angle between the thread structure and the line of sight, which may lead to a difference in brightness from observations at the limb and on the disk.

062.159   Ram–pressure confinement of a hypersonic jet.
C. Loken, J. O. Burns, D. A. Clarke, M. L. Norman.
Astrophys. J., Vol. 392, No. 1, p. 54 – 64 (10 Jun 1992).

The authors present the results of a high–resolution, hydrodynamical simulation of an extremely supersonic jet with a Mach number (relative to the undisturbed ambient medium) of 4650. Both the Mach number and the final length of the jet are greater than those of any previously published simulation. The hypersonic jet is highly dynamic in nature with varying head velocity and shape; the importance of the back action of the cocoon upon the jet is also clearly seen. Such hypersonically moving jets can account for the observations of luminous extragalactic radio jets, associated with classical double radio sources, which appear to be overpressured with respect to the surrounding ambient medium. The pressure jump across the bow shock of a hypersonic jet can easily be a factor of tens or hundreds, resulting in an overpressured cocoon which confines the overpressured jet. In the model the jet remains well collimated while the region behind the bow shock is overpressured by a factor of ~1600 with respect to the quiescent ambient medium. New VLA data are used to constrain the physical parameters of 3C 208 within the confines of this hypersonic jet simulation.

062.160   Dynamics of fronts in thermally bistable fluids.
C. Elphick, O. Regev, N. Shaviv.
Astrophys. J., Vol. 392, No. 1, p. 106 – 117 (10 Jun 1992).

Fronts between different thermal phases of a fluid, when the cooling function allows two thermally stable phases around an unstable one (bistability), are investigated. Fluid motion is included, in addition to thermal conduction. For a one–dimensional case the authors investigate the front dynamics by introducing an appropriate Lyapunov functional. It is assumed that the coefficient of thermal conductivity is small, so that the front thickness is very small compared with front separations. Pairs of adjacent fronts define a cloud (or an intercloud region), and their motion gives rise to the growth of the cloud (condensation) or of the intercloud region (evaporation). The authors discuss the properties of various types of fronts separating the different thermal phases of the fluid. An ensemble of stationary fronts is used to model a quasi–stationary cloudy medium. Equations of motion for fronts separating the hot and cold stable phases are derived. It is found that the medium exhibits an inverse cascade, with increasingly larger clouds predominating. The effects of small pressure variations on the front motions (and thus on cloud condensation and evaporation) are also included in the formalism. An example utilizing a simplistic approximation for the cooling function and a sinusoidal pressure variation is solved in detail.

062.161   Solar p–modes oscillations and heating of the corona.
V. I. Zhukov.
Sol. Phys., Vol. 139, No. 1, p. 201 – 203 (May 1992). Letter–to–the–editor.

The properties of the resonator are considered for fast magnetoacoustic waves. It is shown that tunnel penetration of waves from the resonator leads either to heating of the medium in the Alfvén resonance vicinity (if the inclination angle of the magnetic field is smaller than the critical angle), or to excitation of Alfvén waves at the Alfvén resonance (if the inclination angle is larger than the critical angle). This suggests that non–radiative heating of the corona can be due to solar p–mode oscillations.

062.162   Direct numerical simulations of turbulent convection: Part 2. Variable gravity and differential rotation.
W. Cabot, J. B. Pollack.
Geophys. Astrophys. Fluid Dyn., Vol. 64, No. 1–4, p. 97 – 133 (Mar 1992).

Direct numerical simulations of incompressible channel flow have been performed that explore the effects of centrifugally stable differential rotation on thermal convection. In order to provide greater correspondence to the interior regions of astrophysical accretion disks, the authors consider a gravity that varies linearly with distance from midplane and Keplerian rotation. The results suggest that convection in accretion disks is characterized by very long azimuthal wavelengths, and that, in some circumstances, Reynolds stresses can feed turbulence kinetic energy to the mean flow in contradiction to the conventional eddy–viscosity ansatz.

062.163   **A simple dynamo caused by conductivity variations.**
F. H. Busse, J. Wicht.
Geophys. Astrophys. Fluid Dyn., Vol. 64, No. 1–4, p. 135 – 144 (Mar 1992).

The problem of magnetic field generation by a flow of electrically conducting fluid parallel to a rigid plate is considered. It is shown that growing magnetic fields occur when the plate exhibits variations of the electrical conductivity in the streamwise direction of sufficient strength. In particular the case of sinusoidal variations is studied. The magnetic Reynolds number for dynamo action increases with decreasing conductivity of the plate and with decreasing amplitude of its modulation. Possible applications to geophysical problems are discussed.

062.164   **Dynamos with ambipolar diffusion drifts.**
M. R. E. Proctor, E. G. Zweibel.
Geophys. Astrophys. Fluid Dyn., Vol. 64, No. 1–4, p. 145 – 161 (Mar 1992).

In a weakly ionized medium, there is diffusive transport of magnetic field relative to the neutral gas. The effective diffusion coefficient is quadratic in the magnetic field strength so that the diffusion is nonlinear. The authors have included this nonlinear diffusion in a simple model of an α–ω dynamo in a slab. A rich variety of solutions, including steady states and steady travelling waves, are found. The results may be relevant to the generation of magnetic fields in galaxies and in accretion disks around young stars.

062.165   **Thin disc kinematic $\alpha\omega$–dynamo models. Part 1. Long length scale modes.**
A. M. Soward.
Geophys. Astrophys. Fluid Dyn., Vol. 64, No. 1–4, p. 163 – 199 (Mar 1992).

The assumptions and approximations commonly used in thin disk kinematic dynamo theory are discussed and appraised. Here attention is restricted to those modes, which have a radial length scale long compared to the disk thickness but short compared to the disc radius, upon which the characteristics of the disc vary. One of the commonly employed approximations pivots on the assumption that the radial structure is determined by lateral diffusion in the disc, whereas on those long radial scales the role of the external potential fields is more potent. Stix's $\alpha\omega$–galactic dynamo in an oblate spheroid is adopted as an illustrative example. The perturbation methods are applied to three distinct modes, namely the steady dipole, the steady and unsteady quadrupole. Each has very different characteristics. Despite the limitations of the theory, the lowest order approximation, which determines the critical dynamo number and the nature of the dynamo itself, gives generally reliable results, while the higher order theory predicts trends.

062.166   **Thin disc kinematic $\alpha\omega$–dynamo models. Part 2. Short length scale modes.**
A. M. Soward.
Geophys. Astrophys. Fluid Dyn., Vol. 64, No. 1–4, p. 201 – 225 (Mar 1992).

In thin disc kinematic dynamo theory there are two classes of dynamo mode. In the first class, the magnetic field generated has a radial length scale long compared with the short transverse length scale defined by the disc thickness. In the second class the radial length scale is short and comparable to the disc width. The former was investigated in Part I and Stix's $\alpha\omega$–galactic dynamo in an oblate spheroid was adopted as an illustrative example. Here the author again investigates Stix's model but, in contrast, considers the most readily excited oscillatory dipole mode, which happens to have short radial length scale and so belongs to the latter class. It is shown that, in the limit $\varepsilon \rightarrow 0$, local theory, which ignores the radial structure of the disc, leads to a critical dynamo number which is smaller than the actual value by about 20%. The explanation of this descrepancy is the main feature of the analysis.

062.167   **Registration of macro– and micro–inhomogeneities of plasma density in astrophysical objects.**
L. N. Kurochka.
Kinematics Phys. Celest. Bodies, Vol. 7, No. 4, p. 25 – 29 (1991). English translation of Kinematika Fiz. Nebesn. Tel, Tom 7, No. 4, p. 31 – 37 (1991).

The advantages of setting up observations of extended astrophysical objects (such as solar formations) using a broad-band filter beyond the limit of the Balmer continuum are indicated. The filtergrams would give information on macroscopic plasma–density inhomogeneities. It is shown that a rather complete picture of plasma–density micro–inhomogeneities (at the subtelescopic level) and extended regions with contrasting electron densities can be obtained from data on the intensity distribution in the region of the Balmer precontinuum or the precontinuum of another series.

062.168   **Vortices in non–uniform dusty plasmas.**
R. Bharuthram, P. K. Shukla.
Planet. Space Sci., Vol. 40, No. 5, p. 647 – 654 (May 1992).

By means of a two fluid model, the authors derive a set of non-linear mode coupling equations for weakly interacting low-frequency electrostatic fluctuations that are driven by a sheared ion velocity flow in a non–uniform dusty plasma embedded in an external magnetic field. The linear instability is re–examined. Furthermore, dipolar vortices are shown to be possible stationary solutions of the non–linear equations governing mode coupling of the ion velocity gradient fluctuations in dusty plasmas. The existence regions in parameter space for the dipolar vortex formation are presented. The relevance of the authors' investigation to astrophysical and space plasmas is discussed.

062.169   **Velocity modulation of a dusty plasma.**
K. E. Lonngren.
Planet. Space Sci., Vol. 40, No. 6, p. 763 – 765 (Jun 1992).

A beam of negatively charged dust particles whose velocity is modulated in phase space and passes through a neutralizing positive ion background is examined. It is noted that a density compression followed by a density rarefaction ensues.

062.170   **The evolution of cocoons surrounding light, extragalactic jets.**
D. F. Cioffi, J. M. Blondin.
Astrophys. J., Vol. 392, No. 2, p. 458 – 464 (20 Jun 1992).

If the mass density of supersonic, collimated material is less than that of the surrounding medium, a so–called light jet will be enveloped by a cocoon of overpressured shocked gas. Hydrodynamical simulations are used to understand the evolution of the cocoon. The cocoon's evolution is also compared to a simple analytic theory. To reconcile the theory with the simulations, the growth of the jet head must be taken into account. The overpressured cocoon stage exists for a relatively short astronomical time, after which only the region of cocoon near the jet head remains overpressured. The spatial distribution of the optical emission often observed in distant extragalactic jet systems can be explained with this improved understanding of cocoon evolution.

062.171   **Self–similar evolution of magnetized plasmas. I. Quasi–static solution.**
W.–H. Yang.
Astrophys. J., Vol. 392, No. 2, p. 465 – 469 (20 Jun 1992).

The concept of linear expansion, suggested in earlier papers, describes the self–similar evolution of a magnetic structure. Linear expansion can be represented by a single function $\zeta(t)$, which connects the evolving physical parameters of the plasma with their initial values in explicit forms. A general self–similar dynamic equation is derived. Its quasi–static solution is investigated here. It is shown that a $\gamma = 4/3$ polytrope may evolve through consecutive equilibria, if its magnetic field expands self–similarly. The change of the energy everywhere inside the plasma equals the work done by the internal plasma pressure and magnetic field for the expansion. For the special case of an expanding force–free magnetic field, the self–similar expansion is a clean expansion. No free magnetic energy is left anywhere inside the magnetic structure. The approximation in quasi–static modeling of a pressure confined magnetized plasmoid is analyzed. It requires that the characteristic Alfvén traveling time τ_A of the plasmoid is negligible compared to the relative rate of change of the external pressure. If the finite relaxing time is taken into account, excess magnetic potential energy may accumulate.

062.172   **MHD instabilities in a stratified atmosphere with anisotropic pressure.**
V. D. Kuznetsov, V. N. Oraevskij.
Pis'ma Astron. Zh., Tom 18, No. 6, p. 547 – 559 (Jun 1992). In Russian. English translation in Sov. Astron. Lett., Vol. 18, No. 3.
    Instability of a stratified atmosphere with horizontal magnetic field is considered in the framework of anisotropic MHD. The stability boundaries and increments for various types of perturbations are defined for the case of an exponential atmosphere with spatially homogeneous anisotropic temperature. The results obtained can be applied to the atmosphere of the Sun, space plasma and insterstellar medium.

062.173   **On regulation of the jet speed in SS 433.**
J. Fukue.
Publ. Astron. Soc. Jpn., Vol. 44, No. 2, p. 153 – 154 (1992).
    The speed of SS 433 jets has been observed to be temporally constant. It is proposed that this constancy is regulated by a pair creation processes in which the temperature reaches its maximum. When the energy input at the base of the jet increases, the additional energy is consumed so as to produce particles while the temperature maintains a maximum value. Thus, the jet speed, which is supposed to be attained by the conversion of thermal energy, does not change. If the jet emanates from the two–temperature pair atmosphere near $5r_g$, the jet would have a terminal speed of 0.2 c according to the hydrodynamical picture.

062.174   **Knots in stellar jets: crossing shocks or internal working surfaces?**
A. C. Raga, A. Noriega–Crespo.
Rev. Mex. Astron. Astrofis., Vol. 24, No. 1, p. 9 – 13 (Apr 1992).
    The authors discuss the observational implications of theoretical models for the formation of knots in stellar jets. In particular, they present predictions of the spatial offsets between the Hα and the [S II] emission from two possible models: a "crossing shock" model (in which the knots correspond to incident/reflected shock pairs), and an "internal working surface" model (in which the knots correspond to shock pairs that result from a time-variability of the outflow source). A comparison of the theoretical predictions with observations of the jet–like Herbig–Haro object HH 34 appears to favour the second of these scenarios.

062.175   **Effects of dust scattering on interstellar jet line profiles.**
N. Calvet, J. Cantó, L. Binette, A. C. Raga.
Rev. Mex. Astron. Astrofis., Vol. 24, No. 1, p. 81 – 97 (Apr 1992).
    The effect of scattering by dust particles in a stationary medium surrounding a moving source is to spread in velocity the intrinsic gaussian profile of the source. The extent and strength of the wings depend on the dust properties and on the optical depth of the medium. Theoretical models can be used to estimate the inclination of the jet to the line of sight, the source velocity, and the optical depth towards the source. The authors propose that the differences between the [S II] and Hα line profiles in the jets associated with HH 34 and HH 47 are due to different optical depths to the line emitting regions, which could be due to different sizes for these regions. The effects of scattering in reddening determinations that assume only attenuation of the light of the source are discussed. It is found that the optical depth determined by assuming only attenuation is smaller than the actual optical depth of the region.

062.176   **Two–dimensional slow shock reconnection.**
Y. C. Whang.
Astrophys. J., Vol. 392, No. 2, p. 637 – 646 (20 Jun 1992).
    The author uses a steady, compressible, magnetohydrodynamic (MHD) model to study two–dimensional slow shock reconnection in low-β plasmas. The model uses the exact Rankine-

Hugoniot solutions to calculate the jump conditions at shock crossings. This steady, X–type reconnection model assumes that the flow and the magnetic field are uniform in the region between the separatrix and the shock, and in the region behind the shock. Solutions exist in a small domain of the parametric space where the half–angle of the shock is a few degrees, the angle between the flow velocity and the magnetic field on the upstream side of the shock is of the order of 80°, and the β–ratio on the upstream side of the shock (β_1) must be very small of the order of 0.05. Across the shock the magnetic field changes by a factor of ~ 0.4, the plasma density increases by a factor of ~ 2.5, the thermal pressure increases by a factor of $\sim 1/\beta_1$, and the magnetic field lines deflect sharply by an angle between 50° and 70°. Behind the shock the plasma and magnetic fields are expelled into two narrow outflow regions where the flow speed is super–Alfvénic and superfast. In the outflow region the Alfvén number is ~ 5, the fast Mach number is ~ 1.8, and the plasma has a β–ratio of the order of 10.

062.177   **Dynamo action in stratified convection with overshoot.**
Å. Nordlund, A. Brandenburg, R. L. Jennings, M. Rieutord, J. Ruokolainen, R. F. Stein, I. Tuominen.
Astrophys. J., Vol. 392, No. 2, p. 647 – 652 (20 Jun 1992). With plates 3 – 4.
    The authors present results from direct simulations of turbulent compressible hydromagnetic convection above a stable overshoot layer. Spontaneous dynamo action occurs followed by saturation, with most of the generated magnetic field appearing as coherent flux tubes in the vicinity of strong downdrafts, where both the generation and destruction of magnetic field is most vigorous. Whether or not this field is amplified depends on the sizes of the magnetic Reynolds and magnetic Prandtl numbers. Joule dissipation is balanced mainly by the work done against the magnetic curvature force. It is this curvature force which is also responsible for the saturation of the dynamo.

062.178   **Equation of state of partially–ionized plasmas in the physical picture.**
F. J. Rogers, C. A. Iglesias.
Rev. Mex. Astron. Astrofis., Vol. 23, p. 133 – 140 (Mar 1992). – See Abstr. 012.054 for the main entry.
    The equation of state used by the OPAL opacity code is based on a many–body activity expansion of the grand canonical partition function. Here the authors give an overview of the method. This approach considers, from the outset, the basic Coulomb interaction between the electrons and nuclei in the system. Consequently, there is no need to factorize the free energy, make assertions about the effect of the plasma environment on bound states, or invoke a mechanism for truncating the internal partition function; as in free energy minimization methods.

062.179   **Annihilation radiation from a power–law distributed electron–positron plasma on the ground Landau level: the case of low magnetic fields.**
A. D. Kaminker, G. G. Pavlov, P. G. Mamradze.
Astrophys. Space Sci., Vol. 192, No. 2, p. 175 – 200 (Jun 1992).
    Intensity, polarization, and cooling rate of the two–photon annihilation radiation are studied in detail in the case of one-dimensional power–law distributions of electrons and positrons, assuming that they occupy the ground Landau level in a strong magnetic field $B \sim 10^{10}$–10^{12}G. Simple analytical expressions for limiting cases are obtained and results of numerical calculations of radiation characteristics are presented. Annihilation features and power–law–like hard tails observed in many gamma–ray burst spectra may be associated with the annihilation radiation of the magnetized power–law distributed plasma near neutron stars. Comparison of the observed and theoretical spectra allows one to estimate the power–law index of the $e^- e^+$-distribution and the gravitational redshift factor in the radiating region.

062.180 **Unsteady flow of a relativistic radiating gas in a gravitational field.**
A. R. Bestman.
Astrophys. Space Sci., Vol. 192, No. 2, p. 201 – 207 (Jun 1992).
The unsteady flow of a relativistic radiating neutrino gas is studied in a gravitational field. The curved body is assumed to be a vertical flat plate on which a time–dependent perturbation is imposed on a basic flow. For small perturbations, the ill–posed problem is reduced to a well–posed one and analytical solutions are developed.

062.181 **Nonstationary Petschek reconnection. Convective Zone.**
V. S. Semenov, N. V. Erkaev, M. F. Heyn.
Geomagn. Aeron., Vol. 31, No. 2, p. 176 – 180 (Oct 1991). English translation of Geomagn. Aehron., Tom 31, No. 2, p. 233 – 239 (1991).
The problem of nonstationary reconnection of magnetic lines of force is solved in the context of the Petschek approach. A solution is found in quite simple form in Cartesian coordinates. Considered are the development phase, in which the nonstationary reconnection is much like the Petschek model, and the separation phase in which the FR regions are propagated as self-contained solitary objects. Discussed are the relations of the reconnection process with the "slingshot" model, with the electromagnetic "fire", the decay of an arbitrary discontinuity, the longitudinal current model, transport processes, and the change in flow topology.

062.182 **Nonstationary Petschek recombination. Diffusion region.**
N. V. Erkaev, V. S. Semenov, M. F. Heyn.
Geomagn. Aeron., Vol. 31, No. 2, p. 181 – 185 (Oct 1991). English translation of Geomagn. Aehron., Tom 31, No. 2, p. 240 – 246 (1991).
A solution for the problem of reconnection in the diffusion region is obtained and asymptotically spliced with the solution in the convective zone. As a result, the Petschek condition is derived, which relates the behavior of the plasma conductivity in the vicinity on the X line with the overall characteristics of the recombinations. It turned out, just as in the Petschek estimates, that the degree of the recombination is weak, depending logarithmically on the conductivity. From the viewpoint of the transformation of magnetic energy into plasma energy it is shown that the pulse mode of reconnection, with a pulse length of the order of the diffusion time, is most effective.

062.183 **Effects of nuclear collisions in extragalactic radio sources.**
P. N. Okeke.
AIP Conf. Proc., No. 245: Workshop on Basic Space Science, p. 250 – 252 (1992). – See Abstr. 012.064 for the main entry.
The author reviews briefly the consequences of proton–proton (pp) collisions in extragalactic radio jets. Effort was made to answer the following questions: Can the results of pp collisions in the jets account for the observed radio, optical and gamma ray emission from the jets of extragalactic radio sources (EGRS)? What is the nature of these relativistic protons? What is the origin of the central gaps which are observed in EGRS?

062.184 **Magnetostatic configuration of a confined magnetic flux tube.**
X. Zhang, W. R. Hu.
Astrophys. Space Sci., Vol. 192, No. 2, p. 247 – 256 (Jun 1992).
In this paper an isolated magnetic flux tube confined in stratified atmosphere is studied for slender and axisymmetric model. The functions of the pressure, density, and temperature are expanded as a Taylor series of magnetic surface function ψ. Several models of an isolated magnetic flux tube confined in a stratified atmosphere are constructed, and the external pressure of the stratified atmosphere decreases reasonably with increasing height. The distribution of thermal dynamic quantities and the magnetic pressure in the flux tube are also obtained.

062.185 **Pair production in two temperature soft photon comptonized discs.**
C. Meirelles Filho.
Astrophys. Lett. Commun., Vol. 28, No. 5, p. 253 – 269 (Mar 1992).
With the allowance for a density dependent Coulomb logarithm the solution, for the two temperature soft proton comptonized disc with pairs becomes double–valued. Keeping the electronic temperature constant and varying α, one obtains a critical electronic temperature, below which pair equilibrium is possible. For temperature greater than the critical value, even allowance for pairs to escape the disc with the velocity of light cannot restore equilibrium. Treating α as a free parameter, allowing for temperature variation along the disc, one obtains a critical accretion rate, above which pair equilibrium is no longer possible. Below, the solution keeps its double–valueness. The effects of advection are seen to be negligible. It is also shown that the assumptions of ionic temperature lesser than virial temperature and thin disc are compatible only for supersonic turbulent regimes.

062.186 **The stability of stratified shear flows of an inviscid compressible fluid in MHD.**
D. N. Vyas, K. M. Srivastava.
Astrophys. Space Sci., Vol. 192, No. 2, p. 309 – 316 (Jun 1992).
By the generalized progressing wave expansion method, the effect of a magnetic field on the linear stability of the stratified horizontal flow of an inviscid compressible fluid has been studied. It is found that the magnetic field stabilizes the system.

062.187 **Astronomical bubbles with clumpy cores and accelerating haloes.**
J. E. Dyson, T. W. Hartquist.
Astrophys. Lett. Commun., Vol. 28, No. 5, p. 301 – 306 (Mar 1992).
Flows initiated in clumpy media can be radically different from those in smooth media. A particular class of flows relevant to such disparate objects as planetary nebulae and starburst galaxies arises when the clumps are concentrated around the flow initiation focus. The authors describe general features of such flows and comment on various astrophysical areas where they are of importance.

062.188 **The three dimensional interaction of a shock with an interstellar cloud.**
J. M. Stone, M. L. Norman.
Bull. Am. Astron. Soc., Vol. 23, No. 3, p. 1268 (1991). Abstract. – See Abstr. 012.067 for the main entry.

062.189 **Compact jets and the AGN paradigm.**
A. P. Marscher.
AIP Conf. Proc., No. 254, p. 377 – 385 (1 May 1992). – See Abstr. 012.057 for the main entry.
The nature of compact, nonthermal jets in active galactic nuclei is reviewed from both observational and theoretical standpoints. Despite rapid variations in brightness, the smallest emission regions in a jet may be relatively far removed from the central engine. Certain models for the formation of a relativistic particle beam that evolves downstream into the observed radio–infrared jet do, however, predict substantial high energy radiation from regions in the beam close to the accretion disk. Observations of such emission, and the general requirement that the jets be well collimated and accelerated to relativistic bulk velocities by the point where they are observed in the radio, can provide indirect constraints on the massive black hole accretion paradigm. Observations of well collimated, semi–relativistic jets in the accreting stellar systems SS 433 and Cyg X–3, as well as slower jets emanating from star forming regions, encourage one to associate jet formation with accretion of gas onto a compact object.

062.190   Jets on large scales.
A. H. Bridle.
AIP Conf. Proc., No. 254, p. 386 – 397 (1 May 1992). – See Abstr. 012.057 for the main entry.

Jets tell us that the AGN problem has a symmetry axis, at least for the radio–loud systems. The jets come in two primary "flavors", whose properties resemble those of the two primary modes of propagation of confined, light supersonic flows. In strong sources, many of the jet–lobe relationships and asymmetries are consistent with bulk relativistic velocities ($\geqslant c/2$) in the outflows of many–kpc scales. Viable alternatives exist, however, and some brightness asymmetries of weak sources and geometrical asymmetries of strong sources are unlikely to be the results of bulk relativistic motions. There is little evidence that the outflows are strongly magnetized, and there is some evidence that they are not.

062.191   Relativistic jets in AGN.
G. Ghisellini, P. Padovani, A. Celotti, L. Maraschi.
AIP Conf. Proc., No. 254, p. 398 – 408 (1 May 1992). – See Abstr. 012.057 for the main entry.

062.192   On the radiation field in astrophysical jets.
M.–a. Kondo.
AIP Conf. Proc., No. 254, p. 451 – 454 (1 May 1992). – See Abstr. 012.057 for the main entry.

Free jets in a radiation dominant state are considered from the purely thermal aspect, where the radiation field is established around the outlet of an optically thick accretion disk. The difference points from gaseous jets are the nondiabatic effect of radiative diffusion and the lateral boundary condition. Nonadiabaticity in the energy process affects the expansion rate of free jets. In the radiative diffusion case, the upwelling radiative heat flux and the sideward one are balancing with each other to keep the jet slender. The characteristic feature appears that a slender and dense jet wears a tenuous mantle.

062.193   Anisotropies in relativistic jets: beaming patterns for synchrotron and inverse–Compton emission.
S. P. Reynolds, D. C. Ellison.
AIP Conf. Proc., No. 254, p. 455 – 458 (1 May 1992). – See Abstr. 012.057 for the main entry.

The authors consider various forms of anisotropy in the angular distributions of relativistic electrons and photons in a jet with bulk relativistic velocity. Even if electrons are isotropic in a frame moving with some velocity, at optically thin frequencies (which dominate the radiation energy density) the synchrotron radiation field is not isotropic in that or any other frame, for most jet models. The authors exhibit the angular dependence of the synchrotron radio emission and inverse–Compton X–ray emission for simple jet models, and show that even for these oversimplified cases the assumption that this angular dependence can be described as some power of the Doppler factor is rather poor, for any value of that power. In addition, the X–ray flux does not have the same angular distribution as the radio flux, but is somewhat broader.

062.194   Mixing in ejecta of supernovae. I. General properties of 2–D Rayleigh–Taylor instabilities and mixing depth in ejecta of supernovae.
I. Hachisu.
Symposium of Supercomputing Astronomy and Astrophysics in Japan, p. 43 – 73 (1991). With 1 plate. – See Abstr. 012.069 for the main entry.

Nonlinear growth of the Rayleigh–Taylor (R–T) instabilities has been numerically studied to apply to the mixing in ejecta of supernovae. The author has considered an ideal two–dimensional R–T instability of compressible gas with an adiabatic constant $\gamma = 4/3$. It is numerically confirmed that the linear growth rate of the instability of compressible gas ($\gamma = 4/3$) is well described by that for incompressible fluid. The author has checked the dependency of the mixing width on the mesh resolution (128×128 and 256×256 grid systems), the order of the numerical accuracy (second–order and third–order), density ratio (1.5, 3,

and 10), the initial amplitude of the perturbations ($\varepsilon = 10^{-6}$, 10^{-4}, 10^{-4}, 0.01, 0.04, and 0.16), and the mode of the initial perturbation (random from mesh to mesh and sinusoidal of 1, 2, 4, 8, 16, and 32 waves). They have calculated directly supernova explosions of SN 1987A with the same numerical code as that for the R–T instability.

062.195   Interaction of supernova ejecta with circumstellar ring.
T. Suzuki, T. Shigeyama, K. Nomoto.
Symposium of Supercomputing Astronomy and Astrophysics in Japan, p. 75 – 82 (1991). – See Abstr. 012.069 for the main entry.

The ejecta of SN 1987A will eventually collide with the surrounding ring discovered by the Hubble Space Telescope. To investigate the hydrodynamical interaction of the ejecta with the ring, the authors perform 2–dimensional hydrodynamical calculations and estimate the X–ray emission from the free–free transitions of electrons during the first two years since collision.

062.196   Star formation induced by cloud–cloud collisions: collisions between clouds with different sizes and densities.
A. Habe, K. Ohta.
Symposium of Supercomputing Astronomy and Astrophysics in Japan, p. 111 – 134 (1991). With 1 plate. – See Abstr. 012.069 for the main entry.

In order to investigate the star formation triggered by cloud–cloud collisions, the authors perform the axially symmetric numerical hydrodynamic calculations of supersonic, non–identical cloud–cloud collision and study the condition of the gravitational collapse of clouds as a result of cloud–cloud collisions.

062.197   Three dimensional MHD simulation of the Parker instability.
R. Matsumoto, K. Shibata.
Symposium of Supercomputing Astronomy and Astrophysics in Japan, p. 177 – 189 (1991). – See Abstr. 012.069 for the main entry.

Three–dimensional magnetohydrodynamic simulations are performed to study the nonlinear evolution of the Parker instability in galactic gas disks and/or accretion disks and in emerging flux regions of the Sun.

062.198   MHD reconnection model for optical jets, H–H objects and GGD objects.
H. Hanami, T. Tajima.
Symposium of Supercomputing Astronomy and Astrophysics in Japan, p. 191 – 192 (1991). – See Abstr. 012.069 for the main entry.

062.199   Dynamics and structure of stellar jets in the magnetized interstellar medium – 2.5–D and 3–D MHD simulations.
Y. Todo.
Symposium of Supercomputing Astronomy and Astrophysics in Japan, p. 193 – 218 (1991). With 1 plate. – See Abstr. 012.069 for the main entry.

The author investigates the behavior of the stellar jets in the magnetized interstellar medium. In his picture, the large scale interstellar magnetic field is bunched with the contracting gas in the process of the star formation, and squeezed further by the toroidal field component produced by the rotation of the protostar and protostellar disk and propagated along the large scale field in the form of a large amplitude torsional Alfvén wavepacket. The high velocity stellar jet ejected from the protostar impinges into the interstellar medium along the axis of this helical magnetic field. The author identifies Herbig–Haro objects with shocked regions ahead of the stellar jets though the part connecting them is, in many cases, not visible. He performs 2.5–D and 3–D MHD simulations for this process. He finds that the magnetic field has strong effects on the global structure of the flow and shocks, Kelvin–Helmholtz instability, propagation velocity of the bow shock, and MHD helical instability.

062.200 A magnetic mechanism for astrophysical jet production and enhancement of disk accretion as its reaction – 2.5 D and 3 D MHD simulation studies.
Y. Uchida.
Symposium of Supercomputing Astronomy and Astrophysics in Japan, p. 219 – 231 (1991). – See Abstr. 012.069 for the main entry.
A brief review of MHD simulation studies on magnetodynamical production of astrophysical jets, behavior of accretion disks in response to the production of jets, and related phenomena, performed by the author's group is given.

062.201 Numerical simulations of galactic outflow and inflow phenomena.
K. Tomisaka.
Symposium of Supercomputing Astronomy and Astrophysics in Japan, p. 359 – 369 (1991). – See Abstr. 012.069 for the main entry.
The author summarizes recent progress of numerical studies on outflow phenomena from the galactic disk to halo. First, a galactic–scale outflow phenomenon is considered. If the high velocity cloud is formed from the radiatively cooled gas which was originally ejected from the disk as a hot gas, the physical condition at the base of the halo should be $\sim 10^6$K and 10^{-3}cm^{-3}. Next, the author tries to review recent results of numerical simulations of evolution of superbubbles, through which hot gas flows out to the halo. In the case of thin disk whose density scale height is $H \cong 100$ pc, the shell begins to be accelerated upwardly after several dynamical time scales. After that, the polar cap of the shell is broken and the hot gas flows away into the halo. Contrarily in the case of thick $H \cong 500$ pc or magnetized disk with a magnetic field parallel to the disk $B \cong 5$ μG, the shell is not accelerated and never shows blow–out.

062.202 Fragmentation of isothermal sheet–like clouds with magnetic fields.
S. M. Miyama.
Symposium of Supercomputing Astronomy and Astrophysics in Japan, p. 371 – 380 (1991). With 1 plate. – See Abstr. 012.069 for the main entry.
The stability of magnetized isothermal sheet–like clouds is investigated. Using a three–dimensional MHD code, nonlinear growth of perturbations and fragmentation processes in interstellar clouds are computed. The initial magnetic fields are assumed to be uniform and be parallel or perpendicular to the sheet–like clouds. In the case of the perpendicular magnetic fields, the gravitational instability is weakened and stabilized as the initial magnetic strength becomes large. In the case of the parallel ones, the instability is not weakened but all the fragments have filamentary structures and are aligned in the same direction.

062.203 The structure of radiative slow–mode shocks.
P. Xu, T. G. Forbes.
Sol. Phys., Vol. 139, No. 2, p. 315 – 342 (Jun 1992).
The authors investigate the structure of slow–mode MHD shocks in a plasma where both radiation and thermal conduction are important. In such plasma a slow shock dissociates into an extended foreshock, an isothermal subshock, and a downstream radiative cooling region. The analysis, which is both numerical and analytical, focuses on the nearly switch–off shocks which are generated by magnetic reconnection in a strong magnetic field. These shocks convert magnetic energy into kinetic energy and heat, and it is found that for typical flare conditions about 2/3 of the conversion occurs in the subshock while the remaining 1/3 occurs in the foreshock. The authors also find that no stable, steady–state solutions exist for radiative slow shocks unless the temperature in the radiative region downstream of the subshock falls below 10^5K. These results suggest that about 2/3 of the magnetic energy released in flare loops is released at the top of the loop, while the remaining 1/3 is released in the legs of the loop.

062.204 MHD intermediate shock wave.
Hu Youqiu.
Advances in solar–terrestrial science of China, p. 96 – 107 (1992). – See Abstr. 003.069 for the main entry.
This paper briefly surveys some current progress in the study of MHD intermediate shock. First of all, the author reviews the progress of cognition concerning the intermediate shock, with emphasis on Wu's contributions and conclusions. Next, he elucidates the concept of hybrid shock and describes a general rule for the temporal evolution and spatial connection of its various components. The author concludes with a qualitative analysis of the relationship between the intermediate shock and solar–terrestrial physics.

062.205 Polarized synchrotron emission from shock accelerated particles.
K.–D. Fritz.
Conference on the Variability of Blazars, p. 278 – 283 (1992). – See Abstr. 012.077 for the main entry.
Physical models for the flux outbursts in the nonthermal spectra of blazar objects have exclusively used power law spectral components in the past. Power laws, however, evolve only in systems which are in steady–state in a certain sense. At least for the highly variable components this is certainly not a proper assumption. The results from time–dependent calculations of first order Fermi aceleration show a significantly different spectral behavior.

062.206 Alfvén–magnetosonic waves interaction.
K. Murawski.
Acta Phys. Pol., A, Vol. 81, No. 3, p. 335 – 351 (Mar 1992).
The nonlinear propagation of the Alfvén and magnetosonic waves in the solar corona is investigated in terms of model equations. Due to viscous effects taken into account the propagation of the Alfvén wave itself is governed by a Burgers–type equation. The Alfvén waves exhibit a tendency to drive both the slow and fast magnetosonic waves. For this process model equations are a generalization of the Zakharov equations. The propagation of the magnetosonic waves is described by linearized Boussinesq–type equations with ponderomotive terms due to the Alfvén wave. Both long and short Alfvén waves are considered. Also the limits of the slow and fast modes are investigated. An approximate shock wave solution has been found for a vertically propagating slow mode. Numerical results for the fast mode propagating perpendicular to the magnetic field show the effect of inhomogeneity and pumping on a shock as the solution of the homogeneous Burgers equation.

062.207 Resonant structures within incompressible ideal MHD.
C. Zorzan.
J. Plasma Phys., Vol. 47, No. 2, p. 321 – 347 (Apr 1992).
The resonant characteristics of an incompressible ideal MHD fluid are highly structured. To help expose this structure, an equivalent electrical analogue of the MHD system is developed. The model, in the form of a transmission line, makes it possible to identify a number of new and important concepts, one of which is the effective impedance. The model also provides a more consistent interpretation of the spectrum of ideal MHD. The discrete Alfvén modes are found to be highly degenerate, while the transition to a discontinuous profile is accompanied by a redistribution of an uncountably infinite number of 'poles' from the continuous spectrum and onto the Alfvén modes. In addition, the electrical analogue shows that within a continuously structured fluid the characteristic behaviour is not necessarily dominated by the 'surface mode' alone. This view is also supported by the results of a numerical simulation of the linear MHD equations.

062.208 Smoothed particle hydrodynamics.
J. J. Monaghan.
Annu. Rev. Astron. Astrophys., Vol. 30, p. 543 – 574 (1992).
Contents: 1. Introduction. 2. Fundamentals. 3. Simple equations of motion. 4. Viscosity and thermal conduction.

5. Spatially–varying resolution. 6. Kernels. 7. Magnetic fields. 8. Special relativity. 9. Implementation. 10. Applications.

062.209   Aspects of three–dimensional reconnection.
    K. Schindler, M. Hesse, J. Birn.
International Workshop on Reconnection in Space Plasma, p. 5 – 9 (Jan 1989). – See Abstr. 012.099 for the main entry.
    In the past, magnetic reconnection has been studied largely in systems with a symmetry such as translational invariance. Recent developments in both theory and observation have established the need for a three–dimensional description of realistic reconnection processes. This paper adresses the basic notion of magnetic reconnection.

062.210   Relativistic quantum response of a strongly magnetised plasma. I. Mildly relativistic electron gas.
W. E. P. Padden.
Aust. J. Phys., Vol. 45, No. 2, p. 131 – 163 (1992).
    Approximate analytic expressions are derived for the linear response 4–tensor of a strongly magnetised, mildly relativistic electron plasma. The response derived is valid for radiation with frequency up to about the cyclotron frequency and is of use in the theory of spectra formation in X–ray pulsars.

062.211   Relativistic quantum response of a strongly magnetised plasma. II. Ultrarelativistic pair plasma.
W. E. P. Padden.
Aust. J. Phys., Vol. 45, No. 2, p. 165 – 183 (1992).
    Approximate analytic expressions are derived for the linear response 4–tensor of a strongly magnetised, ultrarelativistic thermal pair plasma. The results obtained are valid in general for frequencies below the electron cyclotron frequency. It is believed that the results could be of importance in some models of radio pulsars and gamma–ray bursters.

062.212   Three–dimensional MHD simulation of the Parker instability in galactic gas disks and the solar atmosphere.
R. Matsumoto, K. Shibata.
Publ. Astron. Soc. Jpn., Vol. 44, No. 3, p. 167 – 175 (1992).
    Three–dimensional (3D) magnetohydrodynamic simulations were performed in a study of the nonlinear evolution of the Parker instability in galactic gas disks and/or accretion disks, as well as in emerging flux regions (EFR) of the Sun. The initial magnetic fields are parallel to one of the horizontal coordinates in magnetostatic equilibrium. The effect of coupling between the Parker (undular) instability and the interchange instability was mainly studied.

062.213   The evolution of a magnetized superbubble.
    K. Tomisaka.
Publ. Astron. Soc. Jpn., Vol. 44, No. 3, p. 177 – 191 (1992).
    Superbubbles expanding in a magnetized interstellar matter were studied using a two–dimensional axisymmetric magnetohydrodynamics code.

062.214   Collisions of giant stars with compact objects: hydrodynamical calculations.
F. A. Rasio, S. L. Shapiro.
News Lett. Astron. Soc. N.Y., Vol. 4, No. 1, p. 31 (Feb 1992). Abstract. – See Abstr. 012.078 for the main entry.

062.215   The MHD description of cosmic plasmas.
    K. Tsinganos.
NATO Advanced Study Institute on the Sun: a Laboratory for Astrophysics, p. 139 – 154 (1992). – See Abstr. 012.091 for the main entry.
    The main goal of this chapter is to provide a physical understanding of ideal magnetohydrodynamics: Its governing equations are derived from the full set of Maxwell's equations coupled with a kinetic model of the fully ionized gas, described by Boltzmann's equation for each of the two species of the plasma, electrons and ions. The author then discusses the main assumptions and the region of validity of ideal MHD, in particular in the context of astrophysical and space plasmas. Some basic concepts

in MHD which are useful in understanding the gross physical properties of solar hydromagnetic structures and magnetic activity are briefly presented, such as, magnetic field stresses – pressure and tension – the freezing condition of the magnetic fieldlines with the fluid, magnetic buoyancy, the virial theorem, and the expansive propertiy of the magnetic field.

062.216   Symmetric and nonsymmetric MHD equilibria.
    K. Tsinganos.
NATO Advanced Study Institute on the Sun: a Laboratory for Astrophysics, p. 155 – 172 (1992). – See Abstr. 012.091 for the main entry.
    The basic equations of magnetohydrodynamics appropriate for an inviscid fluid of high electrical conductivity are considered for the case that the physical quantities are independent of one Cartesian coordinate, the angle Φ in cylindrical coordinates or, in general for the case that there is one ignorable coordinate in a general curvilinear coordinate system. The result allows a unified and systematic approach to the solution of problems. Next, the question of the existence of nonsymmetric MHD equilibria is investigated. The Taylor–Proudman and Parker theorems, in hydrostatics and magnetostatics, respectively, are discussed, together with their equivalent in MHD. The author explores the formal analogy between the field lines of symmetric equilibria and the surfaces of section of integrable Hamiltonian systems, concluding that spatially symmetric ideal MHD equilibria are topologically unstable to finite amplitude perturbations that do not share their symmetry properties. The resulting ergodicity of perturbed symmetric equilibria may have important consequences in the transport properties of magnetically dominated plasmas.

062.217   Spectroscopic diagnostics.
    A. H. Gabriel.
NATO Advanced Study Institute on the Sun: a Laboratory for Astrophysics, p. 261 – 276 (1992). – See Abstr. 012.091 for the main entry.
    A survey is presented of some of the available techniques for the interpretation of spectral intensities, in terms of the physical parameters of the emitting plasma. Following a discussion of the differences between LTE and "coronal" excitation conditions, the review is limited to "coronal" optically thin plasmas. The techniques include line ratio measurement of temperature and density, as well as the differential emission measure analysis, including continuum emission channels.

062.218   Exact 2–D MHD solutions for astrophysical outflows.
    K. Tsinganos, E. Trussoni, C. Sauty.
NATO Advanced Study Institute on the Sun: a Laboratory for Astrophysics, p. 349 – 376 (1992). – See Abstr. 012.091 for the main entry.
    The systematic method for obtaining and studying analytical solutions of the full MHD equations is illustrated with a class of dynamical plasma equilibria, corresponding to the problem of steady outflows along the open magnetic fieldlines of nonpolytropic stellar atmospheres in the central gravitational field of stellar objects. Those solutions extend the familiar Parker model for a solar/stellar wind by including (i) a nonspherically symmetric magnetic field and outflow speed, (ii) nonsperically symmetric pressure and density distributions, (iii) a selfconsistent calculations of the heating required to drive the flow pattern, and (iv) several novel hydromagnetic critical points in the flow which select a unique wind–type solution. In particular, the role of the poloidal and azimuthal magnetic fields and the Poynting energy flux to drive the outflow as in fast magnetic rotators is discussed. The rich topologies of these exact solutions present an interesting example of the mathematical complexity encountered in nonlinear physical systems. Such studies may be useful in efforts to understand the initial acceleration of axisymmetric hydromagnetic stellar winds and jets and should be taken into account in numerical modelling of those systems.

062.219 Physics of flares in stars and accretion disks.
J. Kuijpers.
NATO Advanced Study Institute on the Sun: a Laboratory for Astrophysics, p. 535 – 597 (1992). – See Abstr. 012.091 for the main entry.

This is a review of the physics of magnetic flares in the Sun. The details of the energy release processes in solar flare are, at present, not clear. The author points out fundamental aspects of the conversion of kinetic into magnetic energy. He shows the physical conditions required for magnetic flares to appear. Particular attention is paid to force free equilibria, energy storage, the amount of liberated energy, resistivity, and magnetic helicity. These findings are then applied to other stars, in particular to accretion disks around magnetic neutron stars.

062.220 Formation of giant molecular clouds associated with spiral structure.
Song Guoxuan.
Proc. Astron. Soc. Aust., Vol. 9, No. 2, p. 200 – 202 (1991). – See Abstr. 012.090 for the main entry.

Molecular hydrogen in spiral galaxies is distributed in clumps, i.e., molecular clouds, which have mass between $10^3 M_\odot$ and $10^6 M_\odot$ and a mass spectrum of $n(m) \propto m^{-1.6}$. Molecular clouds with masses greater than $10^5 M_\odot$, are called giant molecular clouds (GMCs). It is generally accepted that GMCs are formed by the coalescence of molecular clouds through their collision. This process is studied by both numerical simulation and numerical integration. The observation with high resolution identified a great number of CO emission cores in galaxies. Based on this result, the aggregation or clustering formation of GMCs is numerically simulated. In the process of either coalescence or clustering, spiral perturbation plays an important role.

062.221 The evolution of disturbed neutral point equilibria.
I. J. D. Craig.
Proc. Astron. Soc. Aust., Vol. 9, No. 2, p. 225 – 228 (1991). – See Abstr. 012.090 for the main entry.

The author considers the linear and non-linear evolution of a perturbed X-type neutral point. A semi-analytic treatment is given for the case of small disturbances of the equilibrium field. This problem admits well defined azimuthal modes which allow a formally exact determination of the magnetic annihilation rate. It is shown that the longest lived modes are purely radial and decay of the timescale $\tau \approx |\ln \eta|$ where η defines the resistivity of the coronal plasma. Higher azimuthal modes decay much faster, generally on a fraction $(O(m^{-1}))$ of the Alfvén timescale for the outer field. The author goes on to perform finite amplitude calculations that demonstrate the implosive current build-up that precedes the reconnective phase of the relaxation. In general both linear and non-linear studies support the idea of an initial implosive stage which rapidly releases the bulk of the energy associated with arbitrary field disturbances.

062.222 A simple method for numerical simulation of energetic particle transport in weakly inhomogeneous magnetic fields.
M. Ostrowski.
1. COSPAR Colloquium: Physics of the outer heliosphere, p. 223 – 226 (1990). – See Abstr. 012.100 for the main entry.

The author presents a simple method for solving problems involving transport of energetic charged particles in magnetized tenuous plasmas by Monte Carlo particle simulations.

062.223 MHD non–equilibrium: a numerical experiment.
P. Martens, Sun Mingtsung, Wu Shitsan.
IAU Colloquium No. 133: Eruptive solar flares, p. 65 – 68 (1992). – See Abstr. 012.096 for the main entry.

062.224 Field opening and reconnection.
T. G. Forbes.
IAU Colloquium No. 133: Eruptive solar flares, p. 79 – 88 (1992). – See Abstr. 012.096 for the main entry.

During an eruptive flare a large magnetic loop or plasmoid is ejected into interplanetary space, and the closed magnetic field structure which exists prior to the flare becomes opened. One of the requirements of flare models is to explain how the field can be opened while decreasing the overall magnetic energy of the system. After the field is opened, reconnection must occur in order to restore the field to its pre-flare configuration. The strongest evidence for reconnection in flares comes from observations of the chromospheric ribbons and the coronal loops which form after the onset of a large flare. The ribbons and loops appear to propagate through the chromosphere and corona during the flare, but Doppler-shift measurements show conclusively that these apparent motions are not due to mass motions of the solar plasma. The motions can only be explained by the upward propagation of an energy source in the corona, and in the MHD-reconnection model of flares, the propagating energy source is an x-line accompanied by slow-mode shocks.

062.225 Energy release at Alfvénic fronts in a force–free magnetic flux tube.
J. Nicholls.
IAU Colloquium No. 133: Eruptive solar flares, p. 89 – 92 (1992). – See Abstr. 012.096 for the main entry.

Melrose's model (1992) for the energy propagation into the energy release site in a solar flare in terms of Alfvénic fronts is extended to a constant-α force-free flux tube, relaxing the constraint of a uniform current profile. The important features of the model depend on α and the flux tube radius, r_0, only in the combination αr_0.

062.226 Ion–atom complexes and the recombination in stellar plasma.
A. A. Mihajlov, N. N. Ljepojević, M. S. Dimitrijević.
Workshop on the Atmospheres of Early–Type Stars, p. 365 – 367 (1992). – See Abstr. 012.095 for the main entry.

The authors show that for the study of recombination of ions and electrons in weakly ionized low temperature hydrogen plasmas, the processess $H + H^+ + e \rightarrow H + H^*(n)$ and $H_2^+ + e \rightarrow H + H^*(n)$ must both be considered since their contributions are comparable. A simple method for the calculation of the corresponding rate coefficients is presented. They also present the results of calculations for $H^*(n)$ excited to the level of the principal quantum number $n = 4$.

062.227 Relation between global and local aspects of reconnection.
V. M. Vasyliunas.
International Workshop on Reconnection in Space Plasma, p. 11 – 14 (Jan 1989). – See Abstr. 012.099 for the main entry.

062.228 Noise–free neutral sheets.
J. W. Dungey.
International Workshop on Reconnection in Space Plasma, p. 15 – 19 (Jan 1989). – See Abstr. 012.099 for the main entry.

062.229 Reconnection instability in collisionless plasma.
J. Büchner, L. M. Zelenyi (*L. M. Zelenyj*).
International Workshop on Reconnection in Space Plasma, p. 21 – 28 (Jan 1989). – See Abstr. 012.099 for the main entry.

The authors show that both plasma species, electrons as far as ions, must interact irreversibly with the tearing mode to initialize collisionless reconnection.

062.230 Nonlinear dynamics of the drift–tearing mode.
M. M. Kuznetsova, L. M. Zelenyi (*L. M. Zelenyj*).
International Workshop on Reconnection in Space Plasma, p. 29 – 34 (Jan 1989). – See Abstr. 012.099 for the main entry.

062.231 Reconnection and microturbulence.
J. Büchner, C.-V. Meister, B. Nikutowski.
International Workshop on Reconnection in Space Plasma, p. 35 – 40 (Jan 1989). – See Abstr. 012.099 for the main entry.

Space plasma instabilities are reviewed and, if possible, the related anomalous collision frequencies, i.e. the anomalous resistivity.

062.232   Anomalous collision frequency of magnetoactive plasmas with two–dimensional electrostatic turbulence (kB = 0).
C.–V. Meister.
International Workshop on Reconnection in Space Plasma, p. 41 – 44 (Jan 1989). – See Abstr. 012.099 for the main entry.

The paper deals with the study of current–driven instabilities in inhomogeneous space plasma exciting lower–hybrid–drift waves. Within the nonlinear theory of electrostatic flute instabilities an expression for the anomalous collision frequency is obtained containing the energy spectrum of the waves as a parameter. In the special case of the plasma of the Earth's magnetopause a collision frequency of $(0.03 - 0.3)$ Hz is found which may cause substantial resistivity enhancement even in the vicinity of the neutral sheet with large plasma–beta.

062.233   Comparison of compressible and incompressible models of reconnection.
M. Jardine, E. R. Priest.
International Workshop on Reconnection in Space Plasma, p. 45 – 49 (Jan 1989). – See Abstr. 012.099 for the main entry.

The authors consider two recent families of reconnection models, one for a compressible and one for an incompressible plasma and examine the differences between them.

062.234   Magnetic reconnection near a stagnation point.
M. F. Heyn, H. K. Biernat, R. P. Rijnbeek, F. T. Gratton.
International Workshop on Reconnection in Space Plasma, p. 51 – 54 (Jan 1989). – See Abstr. 012.099 for the main entry.

Starting from an exact solution of the dissipative MHD equations for a magnetic neutral layer with intrinsic stagnation point flow a normal magnetic field component is introduced as a perturbation. The resulting set of coupled ordinary differential equations is solved numerically and the behaviour of the perturbed quantities with distance from the neutral layer is presented and discussed.

062.235   Magnetic reconnection simulation using the 2.5 D electromagnetic direct implicit code "AVANTI".
D. W. Hewett, G. E. Francis, C. E. Max.
International Workshop on Reconnection in Space Plasma, p. 55 – 60 (Jan 1989). – See Abstr. 012.099 for the main entry.

062.236   Kinetic simulation of magnetic reconnection in the presence of shear.
G. E. Francis, D. W. Hewett, C. E. Max.
International Workshop on Reconnection in Space Plasma, p. 61 – 66 (Jan 1989). – See Abstr. 012.099 for the main entry.

062.237   Particle simulation of magnetic reconnection in collision–free plasma sheets.
W. Zwingmann.
International Workshop on Reconnection in Space Plasma, p. 67 – 70 (Jan 1989). – See Abstr. 012.099 for the main entry.

The onset of geomagnetic activity is widely regarded as being initiated by a large–scale instability of the plasma sheet. The ion-tearing instability has been shown to be a promising candidate for the tail dynamics. The author examines the non–linear behavior, using particle simulation.

062.238   Three–dimensional reconnection from a global viewpoint.
M. A. Berger.
International Workshop on Reconnection in Space Plasma, p. 83 – 86 (Jan 1989). – See Abstr. 012.099 for the main entry.

Three–dimensional reconnection is defined in terms of changes in a magnetic field topology, even in the absence of true separatrices. Global magnetic helicity conservation at magneto-pause flux transfer events is also discussed.

062.239   The evolution of MHD–perturbations in the intersection region of two separatrix planes.
S. V. Bulanov, S. G. Shasharina, D. Sunder.
International Workshop on Reconnection in Space Plasma, p. 203 – 206 (Jan 1989). – See Abstr. 012.099 for the main entry.

On the basis of two simple models the problem of forced magnetic field–line reconnection in the vicinity of the intersection of two separatrix planes which corresponds e.g. to the interaction of two magnetic islands is considered.

062.240   On particle energization by magnetic field line reconnection in extragalactic jets.
T. M. Kirchner.
International Workshop on Reconnection in Space Plasma, p. 303 – 308 (Jan 1989). – See Abstr. 012.099 for the main entry.

In order to investigate whether a magnetic field line reconnection process can be operative in cosmic jets for in situ re–energization of synchrotron radiating electrons the author considers the Gold–Parker mechanism of coronal heating (originally devised for the sun) under the conditions of extragalactic jets. He estimates that the power dissipated by this process, for the parameter values of two of the most powerful radio galaxies (M87 and Cen A), amounts to some $10^{39} \div 10^{40}$erg/s. This fits the energetic requirements of the electrons for the radio part of the synchrotron spectrum, but is possibly not enough for its optical and X–ray branch.

062.241   MHD solar wind – interstellar plasma interaction: 3D formulation by the projected characteristics method and the stability analysis.
R. Ratkiewicz.
1. COSPAR Colloquium: Physics of the outer heliosphere, p. 313 – 316 (1990). – See Abstr. 012.100 for the main entry.

The method of projected characteristics is developed for solving the nonlinear MHD solar wind – interstellar gas interaction problem in 3D Cartesian geometry. The von Neumann stability analysis for a linearized and simplified version of the used numerical procedure is performed.

062.242   The stability of jets.
M. Birkinshaw.
Beams and jets in astrophysics, p. 278 – 341 (1991). – See Abstr. 003.099 for the main entry.

The methods of hydrodynamic linear stability analysis for astrophysical jets are reviewed. In particular, the properties of the Kelvin–Helmholtz instability in an astrophysical environment are discussed. Some observed jet flows of extragalactic radio sources are compared with the shapes into which the Kelvin–Helmholtz instability should deform beams.

Physics of the plasma universe.
See Abstr. 003.020.

Solitary waves in plasmas and in the atmosphere.
See Abstr. 003.026.

An introduction to astrophysical hydrodynamics.
See Abstr. 003.037.

Accretion power in astrophysics.
See Abstr. 003.041.

Plasmaphysik im Sonnensystem.
See Abstr. 003.052.

The physics of astrophysics. Volume II: Gas dynamics.
See Abstr. 003.082.

Beams and jets in astrophysics.
See Abstr. 003.099.

Dusty plasmas. Proceedings. Special session during the 4. week of the Spring College on Plasma Physics, Trieste (Italy), 14 – 21 Jun 1991.
See Abstr. 012.023.

Extragalactic radio sources – from beams to jets. Proceedings.
See Abstr. 012.041.

ISPP–11. Physics of charged bodies in space plasmas. Proceedings.
See Abstr. 012.049.

Supercomputing astronomy and astrophysics. Proceedings.
See Abstr. 012.069.

Reconnection in space plasma. Proceedings.
See Abstr. 012.099.

Numerical simulations of two–dimensional and three–dimensional accretion flows.
See Abstr. 021.026.

A universal solver both for compressible and incompressible fluid by cubic interpolation.
See Abstr. 021.030.

Observation of dust shedding from material bodies in a plasma.
See Abstr. 022.015.

Regularities in experimental Stark shifts.
See Abstr. 022.042.

Low–frequency modes in dusty plasmas.
See Abstr. 022.089.

Two–stream instabilities in unmagnetized dusty plasmas.
See Abstr. 022.090.

Calculation of detailed atomic data using parametric potentials.
See Abstr. 022.186.

Relativistic free–free Gaunt factors for high–temperature stellar plasmas.
See Abstr. 022.193.

Flare–type plasma processes and explosive disruption of pinch current sheets.
See Abstr. 022.235.

Current sheet structure at different stages of magnetic reconnection.
See Abstr. 022.236.

Impulsive heating, plasma turbulence and magnetic reconnection inside current sheets.
See Abstr. 022.237.

Spontaneous and forced reconnection in a magnetic type–X structure.
See Abstr. 022.238.

Relativistic neutrons in active galactic nuclei. I. Energy transport from the core.
See Abstr. 061.077.

Transient flow of a relativistic radiating gas past a horizontal plate.
See Abstr. 061.117.

Neutrino emissivity of an ultrarelativistic plasma from positron and plasmino annihilation.
See Abstr. 061.119.

Some aspects of the theory of polarized radiation scattering in a randomly–inhomogeneous magnetoplasma.
See Abstr. 063.007.

Resonance line radiation instability in fully ionized plasmas.
See Abstr. 063.016.

Covariant flux–limited diffusion theories.
See Abstr. 063.019.

Scattering of electromagnetic waves by a distribution of charged dust particles in space plasmas.
See Abstr. 063.026.

Calculation of the quasi–energies and resonance behavior of the hydrogen Lyman–α problem.
See Abstr. 063.042.

A fast operator perturbation method for the solution of the special relativistic equation of radiative transfer in spherical symmetry.
See Abstr. 063.053.

Diamagnetic effects in synchrotron sources.
See Abstr. 063.062.

Plasma maser effects in astrophysics.
See Abstr. 063.065.

Theory of scattering of polarized radiation in a randomly inhomogeneous magnetized plasma.
See Abstr. 063.067.

Relation between emitted and received powers from a moving radiating source in a medium.
See Abstr. 063.094.

X–ray emission–line spectra of photoionized plasmas: density sensitivity of the Fe L–shell series.
See Abstr. 063.097.

Stimulated scattering of electromagnetic waves in dusty plasmas.
See Abstr. 063.099.

The Fe L–shell spectrum in compact astrophysical X–ray sources.
See Abstr. 063.103.

Comptonization modifications to the gamma–ray spectrum of active galactic nuclei.
See Abstr. 063.135.

Instabilities and evolution of spectra – more discussion on the relation of plasma emission and other radiation mechanisms.
See Abstr. 063.148.

Quantum theory of Stokes parameters for Thomson scattering in a magnetised plasma.
See Abstr. 063.149.

The effect of coupling of a corona to the turbulence in an accretion disk.
See Abstr. 064.007.

The effects of radiation drag on radial, relativistic hydromagnetic winds.
See Abstr. 064.009.

Spherically symmetric, polytropic flow.
See Abstr. 064.010.

Radiative shocks in atomic and molecular stellar–like atmospheres. V. Influence of the excited level of the hydrogen atom: the precursor structure.
See Abstr. 064.015.

Nonradial and nonpolytropic astrophysical outflows. I. Hydrodynamic solutions with flaring streamlines.
See Abstr. 064.016.

Turbulence in differentially rotating thin disks: a multi–component cascade model.
See Abstr. 064.017.

Numerical simulations of two–dimensional and three–dimensional accretion flows.
See Abstr. 064.035.

The stability of accretion tori. IV. Fission and fragmentation of slender, self–gravitating annuli.
See Abstr. 064.040.

On the local stability of accretion disks.
See Abstr. 064.041.

On stellar accretion of matter through a disk and the ejection of bipolar flows.
See Abstr. 064.042.

On the stability of magnetized accretion discs.
See Abstr. 064.045.

Radiation–hydrodynamic waves in an optically gray atmosphere. I. Homogeneous model.
See Abstr. 064.048.

The magnetic collimation of bipolar outflows. I. Adiabatic simulations.
See Abstr. 064.050.

Nonradial and nonpolytropic astrophysical outflows. II. Topology of MHD solutions with flaring streamlines.
See Abstr. 064.052.

Wave action conservation, over–reflection and over–transmission of non–axisymmetric waves in differentially rotating thin discs with self–gravity.
See Abstr. 064.069.

A three component model for MagnetoHydroDynamic turbulence in accretion disks.
See Abstr. 064.073.

Is the Oort A–value a universal growth rate limit for accretion disk shear instabilities?
See Abstr. 064.074.

Models for stellar wind of early–type stars.
See Abstr. 064.076.

Limit cycles in pair dominated accretion disks.
See Abstr. 064.077.

A new mechanism for viscosity in cloudy accretion disks: anomalous (turbulent) viscosity.
See Abstr. 064.078.

On the dynamics of the emerging pre–flare magnetic configuration.
See Abstr. 064.086.

A semi–analytical model of stellar flares.
See Abstr. 064.087.

Ion–atom complexes and the absorption of radiation in stellar plasma.
See Abstr. 064.106.

Pinch effect and stellar flares physics.
See Abstr. 064.121.

Speculation of the origin of X–ray emission in early–type stars.
See Abstr. 064.122.

Numerical study of colliding astrospherical stellar wind flows in binary systems.
See Abstr. 064.123.

The rate of mixing in semiconvective zones.
See Abstr. 065.002.

Angular momentum transport in magnetized stellar radiative zones. I. Numerical solution to the core spin–up model problem.
See Abstr. 065.035.

Bisystem oscillation theory of stars. I. Linear theory.
See Abstr. 065.044.

Bisystem oscillation theory of stars. II. Excitation mechanisms.
See Abstr. 065.045.

Linear polarization as a consequence of rotation in exploding stars.
See Abstr. 065.047.

Further improvements of a new model for turbulent convection in stars.
See Abstr. 065.058.

Criterion for convection in an inhomogeneous star.
See Abstr. 065.059.

Mixing in ejecta of supernovae. I. General properties of two–dimensional Rayleigh–Taylor instabilities and mixing width in ejecta of supernovae.
See Abstr. 065.066.

Numerical experiments on the effects of horizontal turbulent diffusion on transport by meridional circulation.
See Abstr. 065.067.

Ekman circulation and the synchronization of binary stars.
See Abstr. 065.075.

A computation of the equation of state for a Fermi gas.
See Abstr. 065.080.

Turbulent convection with overshooting: Reynolds stress approach.
See Abstr. 065.092.

Comparison of a fine–mesh hydrodynamic stellar model with a dynamic rezoning model.
See Abstr. 065.105.

Evolution of stars and gas in galactic nuclei.
See Abstr. 065.116.

The initial value problem for a class of general relativistic fluid bodies.
See Abstr. 066.010.

Spherically symmetric radiating solution with heat flow in general relativity.
See Abstr. 066.014.

Astrophysical jets and theories of gravitation.
See Abstr. 066.047.

Dynamics and gravitational interaction of waves in nonuniform media.
See Abstr. 066.051.

On the gravitational field of an arbitrary axisymmetric mass endowed with magnetic dipole moment.
See Abstr. 066.268.

Magnetic acceleration of broad emission–line clouds in active galactic nuclei.
See Abstr. 067.027.

Accretion in a Kerr black hole magnetosphere: energy and angular momentum transport between the magnetic field and the matter.
See Abstr. 067.037.

A generalization of the concept of adiabatic index for non–adiabatic systems.
See Abstr. 067.041.

Photon bubbles: overstability in a magnetized atmosphere.
See Abstr. 067.074.

Superluminal jets and neutron star winds.
See Abstr. 067.097.

Intrinsically asymmetric astrophysical jets.
See Abstr. 067.130.

Radiation hydrodynamics in accretion columns of neutron stars.
See Abstr. 067.154.

Geometrically thin, hot accretion disks: topology of the thermal equilibrium curves.
See Abstr. 067.169.

Relativistic fluid flows in the magnetized Schwarzschild spacetime.
See Abstr. 067.174.

Diskoseismology: probing accretion disks. II. Damping and growth of modes due to viscosity and gravitational radiation.
See Abstr. 067.201.

Magnetohydrodynamics of black holes and the origin of jets.
See Abstr. 067.205.

α–disks and the precession of extragalactic jets.
See Abstr. 067.206.

α–disks and the precession of extragalactic jets.
See Abstr. 067.208.

Structure of sunspot penumbrae: fallen magnetic flux tubes.
See Abstr. 072.024.

Solar active regions as a percolation phenomenon.
See Abstr. 072.036.

Vortex attraction and the formation of sunspots.
See Abstr. 072.037.

Dynamics of flaring loops. II. Flare evolution in the density–temperature diagram.
See Abstr. 073.002.

Weighted current sheets supported in normal and inverse configurations: a model for prominence observations.
See Abstr. 073.012.

Energy propagation into a flare kernel during a solar flare.
See Abstr. 073.013.

The modes of oscillation of a prominence. I. The slab with longitudinal magnetic field.
See Abstr. 073.019.

On the efficiency of electron cyclotron maser instability in solar flares.
See Abstr. 073.023.

Alfvénically driven slow shocks in the solar chromosphere and corona.
See Abstr. 073.042.

High–energy gamma–ray emission from pion decay in a solar flare magnetic loop.
See Abstr. 073.044.

The production of ^{3}He and heavy ion enrichments in ^{3}He–rich flares by electromagnetic hydrogen cyclotron waves.
See Abstr. 073.054.

A numerical simulation for the cooling process of the solar flare loop.
See Abstr. 073.080.

Energy release and transport in flare plasmas.
See Abstr. 073.084.

Triggering of eruptive flares: destabilization of the preflare magnetic field configuration.
See Abstr. 073.094.

Flare evolution in the density–temperature diagram.
See Abstr. 073.102.

Models of normal and inverse polarity filament eruptions and coronal mass ejections.
See Abstr. 073.111.

Particle acceleration by magnetic reconnection and fast magnetosonic shock waves in solar flares.
See Abstr. 073.128.

Quasi–static evolution of coronal magnetic fields.
See Abstr. 074.004.

Decay instability of incoherent Alfvén waves in the solar wind.
See Abstr. 074.008.

Modifications of magnetohydrodynamics as applied to the solar wind.
See Abstr. 074.018.

Spatial profiles of lines in steady–state coronal loops.
See Abstr. 074.024.

A self–consistent turbulent model for solar coronal heating.
See Abstr. 074.041.

Magnetic reconnection associated with emerging magnetic flux.
See Abstr. 074.061.

Alfvén–magnetosonic waves interaction in the solar corona.
See Abstr. 074.063.

MHD shocks and simple waves in CMEs.
See Abstr. 074.074.

A numerical simulation of magnetically driven coronal mass ejections.
See Abstr. 074.075.

Stability of force–free magnetic fields and the problem of solar coronal heating.
See Abstr. 074.076.

A neutral current sheet with flows: tearing and stratification modes and modelling of coronal structures.
See Abstr. 074.077.

On the MHD stability of the m = 1 kink mode in finite length coronal loops.
See Abstr. 074.078.

Dynamic development of coronal current sheets.
See Abstr. 074.079.

Interaction of the solar wind with the external plasma.
See Abstr. 074.087.

The interactions of the solar wind discontinuities in the vicinity of the heliopause.
See Abstr. 074.090.

The practical application of the magnetic virial theorem.
See Abstr. 075.003.

Three–dimensional force–free magnetic fields and flare energy buildup.
See Abstr. 075.004.

Linear force–free magnetic field around quiescent solar prominences computed from observable boundary conditions.
See Abstr. 075.010.

The properties of sources and sinks of a linear force–free field.
See Abstr. 075.020.

Energy buildup in sheared force–free magnetic fields.
See Abstr. 075.022.

A regularization method for extrapolation of solar potential magnetic fields.
See Abstr. 075.027.

Magnetic reconnection associated with emerging magnetic flux.
See Abstr. 075.039.

Deformation of magnetic null points.
See Abstr. 075.041.

Magnetic energy conversion on the sun.
See Abstr. 075.042.

Generation and non–equilibrium of solar atmospheric magnetic fields.
See Abstr. 075.043.

Connection between ambient density fluctuations and clumpy Langmuir waves in type III radio sources.
See Abstr. 077.007.

Properties of type III bursts in a gasdynamic model of the propagation of an electron stream in plasma.
See Abstr. 077.010.

The influence of the large–scale interplanetary shock structure on a low–energy particle event.
See Abstr. 078.010.

Stochastic fluctuations of the solar dynamo.
See Abstr. 080.002.

Emergence of magnetic flux from the convection zone into the solar atmosphere. I. Linear and nonlinear adiabatic evolution of the convective – Parker instability.
See Abstr. 080.015.

Alfvén wave transmission through the solar atmosphere.
See Abstr. 080.035.

Gravity wave and convection interaction in the solar interior.
See Abstr. 080.043.

The conversion of p–modes to slow modes and the absorption of acoustic waves by sunspots.
See Abstr. 080.049.

Waves in solar magnetic flux tubes: the observational signature of undamped longitudinal tube waves.
See Abstr. 080.058.

Convection.
See Abstr. 080.092.

Mean field dynamo theory.
See Abstr. 080.093.

The origin of the global system of field–aligned currents on the dayside of the magnetosphere.
See Abstr. 084.046.

Magnetotail reconnection, MHD theory and simulations.
See Abstr. 084.081.

The role of magnetic reconnection in magnetotail plasmoid dynamics.
See Abstr. 084.082.

2–D and 3–D simulation study of multiple X line reconnection.
See Abstr. 084.089.

Impulsive reconnection of the skewed magnetic fields at the magnetopause as a MHD model.
See Abstr. 084.090.

The structure of the reconnection layer in the Petschek reconnection model: theory and application to the earth's magnetopause.
See Abstr. 084.091.

Rotating magnetic anomalies as a possible accelerator of charged particles.
See Abstr. 099.030.

Acceleration mechanism of particles in the type–I cometary plasma.
See Abstr. 102.044.

Numerical simulation of the cometary shocks.
See Abstr. 102.059.

An analytical solution for the heliopause boundary and its comparison with numerical solutions.
See Abstr. 106.008.

Nonlinear effects of cosmic ray interaction with solar wind in the outer heliosphere.
See Abstr. 106.104.

Outer heliosphere: eigen pulsations, cosmic rays and stream kinetic instability.
See Abstr. 106.106.

Nonlinear surface wave propagation of heliospheric current sheet.
See Abstr. 106.110.

Secular ring instability in the protoplanetary accretion disk.
See Abstr. 107.022.

The collision and tidal interaction between planetesimals.
See Abstr. 107.048.

Numerical two–dimensional calculations of the formation of the solar nebula.
See Abstr. 107.050.

Three–dimensional evolution of early solar nebula.
See Abstr. 107.051.

Formation and evolution of the protoplanetary disk.
See Abstr. 107.052.

Magnetohydrodynamic puzzles in the protoplanetary nebula.
See Abstr. 107.054.

A magnetohydrodynamic model for Herbig–Haro objects: magnetically guided shocked flows associated with optical jets from young stellar objects.
See Abstr. 121.063.

Behavior of jets from young stellar objects in large–scale interstellar magnetic fields: MHD model of Herbig–Haro objects in 2.5–D simulations.
See Abstr. 121.064.

Two–dimensional simulations of hydrodynamic instability in supernova explosion. Part 2.
See Abstr. 125.005.

Non–equilibrium, non–LTE ionization in supernova remnants.
See Abstr. 125.007.

Hydrodynamic instabilities in supernova remnants: self–similar driven waves.
See Abstr. 125.059.

Completing the evolution of supernova remnants and their bubbles.
See Abstr. 125.060.

Hydrodynamical instabilities and mixing in SN1987A: 2D simulations of the first 3 months.
See Abstr. 125.101.

Asymmetric explosion in SN 1987A.
See Abstr. 125.128.

Evolution of a self–gravitating protocloud with turbulence.
See Abstr. 131.016.

Numerical models for the collapse and fragmentation of centrally condensed molecular cloud cores.
See Abstr. 131.040.

Ambipolar diffusion, cloud cores, and star formation: two–dimensional, cylindrically symmetric contraction. II. Results and a length scale for protostellar cores.
See Abstr. 131.104.

Ambipolar diffusion, cloud cores, and star formation: two–dimensional cylindrically symmetric contraction. III. A further parameter study and magnetically controlled accretion rate.
See Abstr. 131.105.

Magnetic diffusion in clumpy molecular clouds.
See Abstr. 131.124.

Ambipolar diffusion and star formation: formation and contraction of axisymmetric cloud cores. I. Formulation of the problem and method of solution.
See Abstr. 131.171.

Criteria for the collapse and fragmentation of rotating clouds.
See Abstr. 131.278.

Gravitational instability induced by a cloud–cloud collision: the case of head–on collisions between clouds with different sizes and densities.
See Abstr. 131.279.

The hydrodynamics of aspherical planetary nebulae. II. Numerical modelling of the early evolution.
See Abstr. 134.001.

On non–thermal particle generation in superbubbles.
See Abstr. 144.022.

Magnetic–drift–driven instability and cosmic–ray acceleration.
See Abstr. 144.026.

The Fokker–Planck coefficients of cosmic ray transport in random electromagnetic fields.
See Abstr. 144.030.

A note on the hydrodynamical description of cosmic ray propagation.
See Abstr. 144.036.

Close encounters in Coulomb and gravitational scattering. I. Relaxation of isotropic test–particle distributions by like–particle collisions.
See Abstr. 151.046.

Effect of an ensemble of explosions on the Galactic dynamo. I. General formulation.
See Abstr. 155.104.

Magnetic fields and radiative shocks in protogalaxies and the origin of globular clusters.
See Abstr. 157.211.

Dynamos in discs and halos of galaxies.
See Abstr. 157.248.

Bowshocks and the formation of the narrow–line region of Seyfert galaxies.
See Abstr. 158.084.

The starburst model for active galactic nuclei: the broad–line region as supernova remnants evolving in a high–density medium.
See Abstr. 158.091.

Suche nach Signaturen von Paarerzeugung und Reflexion in Hochenergie–Röntgenspektren aktiver Galaxien.
See Abstr. 158.111.

On the formation of extragalactic radio sources.
See Abstr. 158.145.

Parsec–scale linear polarization structure of active galactic nuclei.
See Abstr. 158.208.

Time–dependent inhomogeneous jet models for BL Lac objects.
See Abstr. 158.213.

BLR chemistry redux.
See Abstr. 158.231.

Gas velocities in the inner NLR: virial with inflow?
See Abstr. 158.233.

Interpretation of large scale extragalactic jets.
See Abstr. 158.303.

Interpretation of parsec scale jets.
See Abstr. 158.304.

From nucleus to hotspot: nine powers of ten.
See Abstr. 158.305.

Numerical simulations of radio source structure.
See Abstr. 158.306.

The production of jets and their relation to active galactic nuclei.
See Abstr. 158.307.

The optical jet of 3C 273: an electron–positron plasma beam viewed by the light of its quasar?
See Abstr. 159.042.

The role of Compton and Raman scattering in the quasar continuum.
See Abstr. 159.083.

Turbulence in cooling flows.
See Abstr. 160.042.

Cosmological Yang–Mills hydrodynamics.
See Abstr. 161.020.

A hydrodynamic approach to cosmology: methodology.
See Abstr. 161.096.

Radiative shocks and hydrogen molecules in pregalactic gas: the effects of postshock radiation.
See Abstr. 161.102.

Dynamical equations of a magnetofluid Universe.
See Abstr. 161.110.

Numerical simulation for cosmological fluid flows.
See Abstr. 161.213.

On the origin of cosmological magnetic fields.
See Abstr. 161.282.

Relativistic static spheres filled with infinitely conducting charged fluids.
See Abstr. 161.329.

Viscous fluid cosmological models in the presence of the zero–mass scalar field.
See Abstr. 161.355.

Numerische Untersuchung zur Kelvin–Helmholtz–Instabilität mit Selbstgravitation und kosmologischer Expansion: Ein Beitrag zur Entstehung von Struktur im Universum.
See Abstr. 161.400.

063 Radiative Transfer, Scattering

063.001 Radiative transfer problems in the solar and sun–like atmospheres.
J. Trujillo Bueno.
11. European Regional Astronomy Meeting of the IAU: New windows to the universe, Vol. 1, p. 119 – 160 (1990). – See Abstr. 012.001 for the main entry.

In this review the author first provides a survey of certain fundamental radiative transfer (RT) problems, i.e. of problems which arise because of the need to develop improved RT theories capable of accounting for the complexity of radiation fields originating in sun–like atmospheres. The author then concentrates on studying both the diagnostic and energy balance problems of sun–like atmospheres by means of various multi–dimensional RT calculations.

063.002 The Zeeman effect in astrophysical water masers and the observation of strong magnetic fields in regions of star formation.
G. E. Nedoluha, W. D. Watson.
Astrophys. J., Vol. 384, No. 1, p. 185 – 196 (1 Jan 1992).

The transfer equations for the polarized radiation of astro-physical, 22 GHz water masers are solved in the presence of a magnetic field which causes a Zeeman splitting that is much smaller than the spectral line breadth. This study discusses the relationship between the recently detected circular polarization in this maser radiation and the strength of the magnetic field. Effects specific for the transfer of the 22 GHz maser radiation are emphasized. These include the narrowing and rebroadening of the spectral line with increasing optical depth, the merging of hyperfine components, deviations of the molecular velocities from a Maxwellian distribution, unequal populations of the magnetic substates, the influence of cross–relaxation, and the generation of circularly polarized radiation due to changes in

direction of the linear polarization. Major uncertainties in the inferred strengths of the magnetic fields due to these consider-ations tend to occur mostly when the flux of maser radiation is high and the maser is saturated. The observed spectral line breadths are found to be good indicators of whether the radiative flux is high enough that such effects are significant. Uncertainty in the strength of the magnetic field due to lack of knowledge about which hyperfine transition is the source of the 22 GHz masers is removed. The observed 22 GHz maser feature is found to be the result of a merger of the three strongest hyperfine components.

063.003 Effect of elastic collisions on the frequency distribution of astrophysical maser radiation.
Y. C. Wu, M. Elitzur.
Phys. Lett. A, Vol. 162, No. 2, p. 137 – 143 (3 Feb 1992).

The frequency distribution of astrophysical maser radiation is an important ingredient in analyzing the properties of astrophysi-cal maser sources. The frequency profile of the absorption coefficient is studied analytically and computationally including the effect of elastic collisions. The evolution of that profile is detailedly discussed. Two important features that result from elastic collisions are: (1) The frequency profile of the absorption coefficient in the saturated regime becomes narrower and sharper. (2) Line narrowing continues during early saturation. These may help explain some observed narrow maser features.

063.004 Monte Carlo simulations of Raman scattered O VI emission lines in symbiotic stars.
H. M. Schmid.
Astron. Astrophys., Vol. 254, No. 1/2, p. 224 – 240 (Feb 1992).

Symbiotic binaries consist of a cool giant and a hot radiation source, which is ionizing a nebula. Recently two emission lines at

λ6825 and λ7082, which are only observed in these systems, have been identified as Raman scattering of O VI λ1032 and λ1038 photons by neutral hydrogen. Strong O VI lines are produced in the compact H II region, probably near the ionizing component. The conversion into $\lambda\lambda$6825, 7082 photons will occur in the extended atmosphere of the cool giant. Because the scattering geometry in symbiotic stars rotates due to the binary motion, systematic changes in the intensity and polarization of the Raman scattered lines are expected. In this paper Monte Carlo simulations are presented in order to investigate the flux and polarization properties of the Raman scattered emission lines. Photon paths are calculated for an illuminated atmosphere with scatterings according to the Rayleigh phase matrix and for various scattering and absorption coefficients. Calculated flux and polarization phase curves for symbiotic binaries are given, and diagnostic possibilities of the Raman scattered emission lines explored. The results show that the $\lambda\lambda$6825, 7082 emission lines allow a determination of orbital parameters, such as period, inclination and orientation of the orbital plane. Additionally, information on the geometric structure of the nebular O VI region and on absorbing particles in the outer atmosphere of the redgiant can be obtained.

063.005   Line scattering in an expanding shell: interferometric and polarimetric profiles.
J. Lefèvre.
Astron. Astrophys., Vol. 254, No. 1/2, p. 274 – 279 (Feb 1992).
The stellar light scattered by an expanding circumstellar dust shell is redshifted. The modification of line profiles due to this redistribution in wavelength has already been studied by Romanik and Leung (1981). In this paper the author considers other observable consequences of this process. Since the ratio between the direct starlight and the scattered light varies accross spectral features, at least two quantities must be affected: the visibility of interferometric fringes and, when the object presents a departure from spherical symmetry, the percentage linear polarization. Some results obtained by numerical simulation are presented. The relative variations of these two quantities are large for shells presenting a moderate extinction opacity ($\tau_{ex} \sim 1$). For luminous late–type variable with low expansion velocity, observations require a high spectral resolution ($\Delta\lambda \lesssim 1$ Å). However it seems possible to coordinate for brightest sources interferometric and polarimetric observations.

063.006   A note on the enhancement of J values in optically thick scattering atmospheres.
J. W. Kaminski, J. C. McConnell.
Can. J. Phys., Vol. 69, No. 8/9, p. 1166 – 1174 (Aug–Sep 1991).
In a planetary atmosphere the J value is determined by the angular–averaged radiance, or the average density of photons in an element of volume. The average density may be enhanced by multiple scattering of photons in a conservative, or near–conservative scattering atmosphere. The authors show that in a conservative semi–infinite medium this enhancement will be a factor of 5, for optical depths greater than about 20 for coherent scattering. They investigate the modification of the J values owing to multiple scattering in an optically thick medium of various optical depths, various single–scattering albedos of the scattering medium, and a range of surface albedos. The authors have applied the results to the calculation of J values in clouds in the terrestrial atmosphere and in the Rayleigh–scattering atmosphere of Uranus. They note that J values in a realistic atmosphere may be enhanced by as much as a factor of 5 throughout a large fraction of the atmosphere over those calculated without multiple scattering and surface reflection.

063.007   Some aspects of the theory of polarized radiation scattering in a randomly–inhomogeneous magnetoplasma.
A. V. Kukushkin, M. R. Ol'yak.
Pis'ma Astron. Zh., Tom 18, No. 1, p. 87 – 95 (Jan 1992). In Russian. English translation in Sov. Astron. Lett., Vol. 18, No. 1.
A theoretical analysis of the spatial coherence of the polarized radiation from cosmic radio sources shows that in a randomly–

inhomogeneous magnetoplasma a slight circular polarization of the initially nonpolarized radiation arises when the observation points are separated. The presence of magnetic field fluctuations results in the opposite phase velocity shifts of clockwise and counterclockwise waves which, in its turn, causes a relative linear polarization growth in the case when interferometer baseline increases.

063.008   A note on positivity of equilibrium populations in radiative transfer theory.
P. H. Damgaard, P. G. Hjorth, P. A. Thejll.
Astron. Astrophys., Vol. 254, No. 1/2, p. 422 – 425 (Feb 1992).
The rate matrix that arises when considering statistical equilibrium is usually solved iteratively in radiative transfer theory. The authors discuss here a theorem stating conditions under which a certain version of the matrix is guaranteed positive solution. This may have applications to improving the iterative methods commonly used to solve the radiation transfer problem.

063.009   Multiple resonant scattering in the Compton upscatter model of gamma–ray bursts.
J. J. Brainerd.
Astrophys. J., Vol. 384, No. 2, p. 545 – 559 (10 Jan 1992).
Resonant Compton scattering, an increasingly popular mechanism for suppressing X–rays and producing gamma rays, must be treated as a multiple–scattering process for conditions thought characteristic of gamma–ray bursts. Photons that multiply scatter with a beamed power–law electron distribution in a uniform magnetic field produce a flat spectrum between the cyclotron frequency and an optical–depth – dependent critical energy; this critical energy ranges between several hundred keV and several MeV. Above this critical energy, the gamma–ray spectrum has a shape determined by the electron distribution and described by a single–scattering model. Only electron distributions that are nearly proportional to the electron momentum are able to simultaneously suppress X–rays and to produce a single–scattering spectrum. As the Thomson optical depth approaches unity, photons that experience multiple scatterings often spawn additional photons at a rate that makes the model unphysical.

063.010   Line–of–sight results from H II region models.
R. B. Gruenwald, S. M. Viegas.
Astrophys. J., Suppl. Ser., Vol. 78, No. 1, p. 153 – 178 (Jan 1992).
Point–to–point observations of a gaseous nebula correspond to an integration over regions with different physical conditions. Therefore, observational data may vary from point to point and may differ from data for the whole nebula. Thus, an analysis of high spatial resolution data requires line–of–sight results from a given set of photoionization models. This paper presents results for several useful line–of–sight quantities related to H II regions, such as line intensity ratios, ionic column densities, and the mean square temperature fluctuation $\langle t^2 \rangle$. The value of $\langle t^2 \rangle$ obtained from the models depends on the considered ion, gas abundance, gas density, and stellar temperature, as well as on the considered line of sight. Based on these results, a comparison between abundances obtained from empirical methods and those adopted in the models shows that the appropriate value of $\langle t^2 \rangle$ must be used in any empirical abundance determination, especially when discussing abundance variations across a given nebula.

063.011   Upgrading the Accelerated Lambda Iteration technique by means of "least change secant methods".
L. Koesterke, W.–R. Hamann, P. Kosmol.
Astron. Astrophys., Vol. 255, No. 1/2, p. 490 – 494 (Feb 1992).
Non–LTE radiation transfer calculations can be made much more efficient and powerful, when the Iteration with Approximate Lambda Operators (ALI) is combined with a modern method for solving the non–linear rate equations which occur within that framework. The authors explain in detail how Broyden's method from the class of so–called "least change secant methods" can be applied for that purpose, replacing the conventional Newton iteration. Very positive experiences were collected when they implemented Broyden's method into their

non–LTE code for modeling spherically expanding stellar atmo-spheres (Wolf–Rayet stars). Compared to the conventional Newton algorithm, the new technique offers a twofold advantage. First, the computational effort to solve the equations of statistical equilibrium is greatly reduced (e.g., by a factor of eight in the authors' case), and, second, the number of atomic levels which can be accounted for is increased far beyond the usual limit of 100 where the Newton algorithm becomes numerically unstable.

063.012   The inverse Planckian transform and temperature spec-tra.
L. Salas.
Astrophys. J., Vol. 385, No. 1, p. 288 – 293 (20 Jan 1992).

The author presents an iterative technique for obtaining the temperature spectrum underlying the observed spectral energy distribution of unresolved thermal sources. This problem can be identified as an inverse transformation, and the technique presented here to solve it is based on the reinterpretation of Planck's function as a conditional probability. This allows the use of Bayesian inference through the Richardson–Lucy algorithm to perform the inversion. To demonstrate this method, the author analyzes two synthetic objects, and tests the convergence and noise dependence of the technique. The technique is also applied to the spectral energy distributions of T Tauri stars, although the class of objects that can be analyzed with this technquie is much broader.

063.013   Transfer of γ rays: deterministic solutions by means of the discrete–ordinate–matrix–exponential method.
R. Wehrse, M. Hof.
AIP Conf. Proc., No. 232, p. 477 – 482 (1 Aug 1991). Current Physics Microform No.: 9110A1173. – See Abstr. 012.003 for the main entry.

The authors consider a medium of non–negligible optical depth in which high energy line photons are Compton scattered. First results for a plane–parallel medium, which consists of cold electrons and which is irradiated from one side by γ line photons, are presented and the accuracies of the derived radiation fields are discussed.

063.014   Inverse   problem   of   thermal   sounding. Pt. III. Determination of the vertical profile of the mixture ratio of a minor gas component.
E. A. Ustinov.
Cosmic Res., Vol. 29, No. 2, p. 249 – 256 (Sep 1991). English translation of Kosm. Issled., Tom 29, Vyp. 2, p. 289 – 297 (1991).

The inverse problem of constructing the vertical profile of the volume mixture ratio of a minor gaseous component from observations of the outgoing thermal radiation in the absorption band of this component is formulated starting from a general formulation obtained by the author. An analytical expression is derived for the kernel of the corresponding linearized inverse problem. A numerical experiment demonstrating the workability of the formulated inverse problem was performed.

063.015   Radiative processes in strong magnetic fields.
A. K. Harding.
Radiat. Eff. Defects Solids, Vol. 122–123, Part 2, p. 625 – 641 (Dec 1991). Paper presented at the International Conference on Coherent Radiation Processes in Strong Fields, Washington, DC (USA), 18 – 22 Jun 1990.

The behavior of electromagnetic processes in strong magnetic fields is currently of great interest in high–energy astrophysics. Strong magnetic fields affect the physics in several fundamental ways: energies perpendicular to the field are quantized, transverse momentum is not conserved and electron/positron spin is important. The relaxation of transverse momentum conservation allows first order processes and their inverses: one–photon pair production and annihilation, synchrotron/cyclotron radiation and absorption, which are kinematically forbidden under field–free conditions. The second–order processes: two–photon pair production and annihilation and Compton scattering, are also modified in strong fields. The discreteness of e^+–e^- pair states

causes resonant behavior in the cross sections and decreases the second–order rates from their free–space values. These processes play an important role in modelling high energy emission from pulsars and gamma–ray bursts.

063.016   Resonance line radiation instability in fully ionized plasmas.
W. G. Mathews.
Astrophys. J., Vol. 386, No. 1, p. 90 – 100 (10 Feb 1992). = UCO/Lick Obs. Bull., No. 1195.

An instability is expected in any photoionized plasma in which the pressure of strongly trapped Lyα radiation is an appreciable fraction of the total pressure. This instability arises as a result of the increased local escape probability accompanying small–amplitude velocity fluctuations of small spatial scales. Equilibrium clouds ionized by isotropic UV radiation and in which the central gas pressure is depressed due to the contribution of Lyα pressure are subject to this instability, and their global structure is significantly affected. If the ionizing radiation is nonisotropic, density fluctuations resulting from the Lyα instability are pushed through the plasma parallel to the radiation flux vector, further complicating the velocity fields. The turbulent energy arising from these instabilities, unlike normal turbulence, is introduced on very small spatial scales. The standard redistribution theory for the detailed transfer of Lyα and other optically thick emission lines must be considerably modified by the presence of this instability. The escape probability of all strongly trapped emission lines is significantly increased compared to the static case. The instability may be important in broad–line or electron–scattering clouds in quasars and intergalactic clouds in the Lyα forest.

063.017   Flash photoionization of gamma–ray burst environ-ments.
D. L. Band, D. H. Hartmann.
Astrophys. J., Vol. 386, No. 1, p. 299 – 307 (10 Feb 1992).

If a gamma–ray burst occurs in a neutral medium, any ionizing radiation associated with the burst will flash–photoionize the surrounding medium. The line emission from this region can be a powerful diagnostic of the unobservable ionizing radiation associated with the burst and of the nature of the surrounding medium. In particular, the recombination Hα line flux from the ionized region is $F_{H\alpha} = 3.6 \times 10^{-25} J \cdot n_H$ ergs cm^{-2}s^{-1}, where J is the fluence (in photons cm^{-2}) of ionizing radiation; a line flux of $F_{H\alpha} > 3 \times 10^{-17}$ ergs cm^{-2}s^{-1} should be observable. Archival optical transients extrapolated into the ultraviolet suggest J = $10^{3.5}$– 10^5 photons cm^{-2}, and different theoretical models for optical emission from gamma–ray bursts predict similar ionizing fluxes. Thus the line emission should be observable from a sufficiently dense medium. If $n_H > 10^4$cm^{-3}, the source will fade on a human time scale as a result of the recombination of the medium, aiding in source detection. However, dust absorption in a dense cloud could suppress the Hα line emission; the burst is unlikely to destroy the dust grains along the line of sight. To circumvent the extinction, other hydrogen recombination lines at lower frequencies and infrared forbidden lines can be used. Therefore, observations of the line emission from the error boxes of strong gamma–ray bursts can provide constraints on the environment of the burst source as well as the unobservable ultraviolet burst emission.

063.018   Resonant Compton cooling and annihilation line produc-tion in gamma–ray bursts.
R. D. Preece, A. K. Harding.
Astrophys. J., Vol. 386, No. 1, p. 308 – 324 (10 Feb 1992).

A synchrotron self–Compton emission model for gamma–ray bursts is presented, which produces narrow annihilation features for a variety of field strengths, primary electron injection energies, and injection rates. In this model, energetic primary electrons are injected and cooled by synchrotron emission in a strong, homogeneous magnetic field, resulting in a pair cascade. Multiple resonant scattering with cyclotron photons efficiently traps and cools pairs in the ground state to an average energy where the Compton energy loss rate is zero, which is in agreement with previous estimates of a Compton temperature. This Compton energy is determined by the shape of the photon spectrum, which itself changes as the particles cool, so a self-consistent solution must be found iteratively. The particle distributions in the ground state are determined by numerically solving the Fokker–Planck equation in the steady state. Annihilation between pairs in these cooled distributions can be very efficient. In the case of isotropic injection of primary electrons, a significant narrow–line feature appears in the overall emission. In the case of beamed injection, the annihilation line is broadened to the extent that it would not be observable.

063.019   Covariant flux–limited diffusion theories.
A. M. Anile, V. Romano.
Astrophys. J., Vol. 386, No. 1, p. 325 – 329 (10 Feb 1992).
A general relativistic covariant flux–limited diffusion theory is presented for radiation propagating through inhomogeneous and nonstationary media. Explicit flux–limited expressions are obtained for the radiation energy flux and stress tensor in the case of small shear.

063.020   Radiative transfer for polarized light: equivalence between Stokes parameters and coherency matrix formalisms.
J. Sánchez Almeida.
Sol. Phys., Vol. 137, No. 1, p. 1 – 14 (Jan 1992).
Two formal solutions of the radiative transfer equation for polarized light have been proposed. One uses the Stokes parameters to describe the polarization, while the other uses the coherency matrix. It is shown in the present work that they are equivalent. Both can be used to compute response and contribution functions for the Stokes parameters and both require the solution of systems of differential equations with similar numbers of independent variables. New equations to solve the radiative transfer problem using the Stokes parameters formalism are presented. In addition, a computer code which synthesizes the Stokes profiles by means of these equations is described.

063.021   The derivation of parent electron spectra from bremsstrahlung hard X–ray spectra.
C. M. Johns, R. P. Lin.
Sol. Phys., Vol. 137, No. 1, p. 121 – 140 (Jan 1992).
The authors formulate a numerical method to derive the spectrum of the parent electrons from the hard X–ray spectrum produced in optically thin bremsstrahlung. The method can utilize any form for the bremsstrahlung cross sections, and it provides accurate estimates of uncertainties in the derived electron spectrum based on uncertainties in the photon measurements. This method is applied to test photon spectra, as well as to hard X–ray spectra of the 27 June, 1980 solar flare which was observed by high spectral resolution detectors.

063.022   Analysis of McCoyd's mechanism of the negative polarization of light scattered by atmosphereless celestial bodies.
Yu. G. Shkuratov, M. A. Kreslavskij, N. V. Opanasenko.
Astron. Vestn., Tom 26, No. 1, p. 46 – 53 (Jan – Feb 1992). In Russian. English translation in Solar Syst. Res., Vol. 26, No. 1.
McCoyd's model of the negative polarization is generalized for the three–dimensional case that is realistically. Calculations show that the generalized mechanism does not describe the negative polarization of light scattered by atmosphereless celestial bodies. Whereas the mechanism is suitable for explanation of the polarization degree as a function of parameters characterizing geometry of illumination and observation. The results of the calculation are confirmed by laboratory measurements of a glass plate with rough sides.

063.023   Radiative instabilities and 1000 second fluctuations in astrophysical masers.
G. A. Scappaticci, W. D. Watson.
Astrophys. J., Lett., Vol. 387, No. 2, p. L73 – L76 (10 Mar 1992).
A stability analysis for small (linear) perturbations is presented for the radiation in astrophysical masers treated in the usual, linear maser approximation. Instabilities that oscillate with a period of $\sim L/c$, where L is the length of the maser, are found. They occur (1) when the maser is partly but not heavily saturated, (2) when the decay rate Γ for the molecular states is near c/L, and (3) when the product of the brightness temperature T_0 of the incident radiation and the angle or the beaming is less than a critical value that depends upon the particular masing transition. A fourth parameter – the fractional inversion in the pumping multiplied by $(T_0/\text{frequency})$ – determines the importance of spontaneous emission which can eliminate the instabilities. These instabilities are a likely cause for the fluctuations in the radiation from the 18 cm OH masers that have been reported to occur on time scales as short as 1000 s. The calculations are applicable to other types of astrophysical masers as well and suggest that spontaneous emission will prevent similar instabilities in the H_2O and SiO masers.

063.024   Partially coherent scattering in stellar chromospheres. I. Effects on resonance line thermalization.
K. G. Gayley.
Astrophys. J., Suppl. Ser., Vol. 78, No. 2, p. 549 – 564 (Feb 1992).
Photon scattering in the extended wings of many strong resonance lines is nearly coherent under typical stellar chromospheric conditions, with free–electron densities of about 10^{12}cm^{-3} or less. When photon transport in these partially coherent wings dominates escape from the chromosphere, the depth of line thermalization will be strongly affected. Such effects of partial redistribution (PRD) are not accounted for in standard two–level thermalization models involving complete redistribution (CRD). The author investigates the conditions necessary for partially coherent scattering to influence the thermalization depth, and approximates these effects in homogeneous slab atmospheres, using several common resonance lines as examples. For electron densities above 10^{10}cm^{-3}, it is found that even when coherent scattering dominates the escape process, the thermalization depths of the strongest resonance lines of H, Ca II, and Mg II agree roughly with the standard result for complete redistribution over a Doppler profile. This occurs because of the importance of Doppler diffusion in frequency. However, at lower densities, such as for giant star chromospheres and QSO models, the results deviate strongly from the Doppler CRD case.

063.025   The Sobolev approximation for line formation with partial frequency redistribution.
D. G. Hummer, G. B. Rybicki.
Astrophys. J., Vol. 387, No. 1, p. 248 – 257 (1 Mar 1992).
The formation of a spectral line in a uniformly expanding infinite medium is investigated in the Sobolev approximation with particular attention to frequency redistribution. Numerical and analytic solutions of the transfer equation are presented for a number of redistribution functions and their approximations, including type I and type II partial redistribution, coherent scattering and complete redistribution, and the Fokker–Planck and uncorrelated approximation to the R_{II} function. The solutions for the mean intensity are shown to depend very much on the type of redistribution mechanism, while for the frequency–weighted mean intensity J, which enters the rate equations, this dependence is weak. This implies that the use of Sobolev escape probabilities based on complete redistribution can be an adequate approximation for many calculations for which only the radiative excitation rates are needed. However, it is shown that the criteria for applicability of Sobolev theory may be difficult to meet when transfer occurs primarily in the Voigt wings, especially for complete redistribution.

063.026   Scattering of electromagnetic waves by a distribution of charged dust particles in space plasmas.
U. de Angelis, A. Forlani, V. N. Tsytovich, R. Bingham.
J. Geophys. Res., Vol. 97, No. A5, p. 6261 – 6267 (1 May 1992).
The authors analyze the scattering of electromagnetic waves by a distribution of charged dust particles, a common component of many space plasmas. For the case when the dust grains cannot be considered as independent scatterers (that is, when the intergrain separation is of the order or less than the Debye length), the

scattering cross section is calculated for a statistical distribution of dust particles correlated via their electrostatic coupling. The nonneutrality of the plasma is taken into account, due to the charging of the grains by plasma currents, secondary emission, and photoelectrons (the Sun being the source of photons). The grain charges and corresponding scattering cross section are calculated for values of the plasma dust parameters relevant to many space plasma environments in the solar system, showing that enhancement with respect to scattering by free electrons is possible.

063.027   The theoretical polarisation from the obliquely rotating envelopes of single stars.
G. K. Fox.
Astrophys. Space Sci., Vol. 187, No. 2, p. 219 – 239 (Jan 1992).

The polarimetric variability of stars possessing an obliquely rotating envelope is investigated in the optically thin, single electron scattering approximation. It is shown that in the point light source treatment, one is unable to distinguish between polarimetric variability occurring due to rotation about a body axis and (binary) orbital motion. Nor is it possible, from the polarimetric variability, to infer the physical geometry of the obliquely rotating envelope. It is shown that polarimetric discrimination between envelope geometries is possible (to some extent) when the light source is considered to be finite in extent, due to the possible occultation of some of the scatterers, resulting in polarimetric variability that cannot be explained by the canonical point light source models. In this paper, the author considers, as illustrative examples of the effects of scatterer occultation, two diametrically opposite spots and an obliquely rotating, near planar, disc. Finally, an account is made of the spectroscopic variability of the two geometries considered, which again indicate that geometry discrimination is possible.

063.028   The Rosseland mean opacity of interstellar grains.
A. Ali, M. S. El-Nawawy, M. A. El Shalaby.
Astrophys. Space Sci., Vol. 188, No. 1, p. 109 – 116 (Feb 1992).

The authors have calculated the opacity of interstellar grains in the temperature range 10–1500K. Two composite grain models have been considered. One of them consists of silicate coated with an ice mantle and the second has a graphite core coated also with an ice mantle. These models are compared with isolated grain models. An exact analytical and computational development of Güttler's formulae for composite grain models has been used to calculate the extinction coefficient. It has been found that the thickness of the mantle affects the opacity of the interstellar grains. The opacity of composite models differs from that of the isolated models. The effect of the different species (ice, silicate, and graphite) is also clear.

063.029   Diagnostic possibilities of circular polarization for investigating the surfaces of atmosphereless bodies.
V. S. Degtyarev, L. O. Kolokolova, A. V. Morozhenko, M. F. Tsurul'.
Pis'ma Astron. Zh., Tom 18, No. 3, p. 279 – 283 (Mar 1992). In Russian. English translation in Sov. Astron. Lett., Vol. 18, No. 2.

Phase dependences for circular polarization of light scattered from a dust layer were obtained with a high-precision Stokes polarimeter. The surfaces were formed by a layer of tiny powder particles. Transparent dielectric particles, the same particles, coated by metal (nickel), and metallic particles were investigated. The particles had similar dimensions ($\cong 50\ \mu m$), but different shapes. It is seen from the measurements that the circular polarization is near zero if particles are dielectric, but it is considerable if particles are obstacle (coated by metal or metallic). Especially large values of the circular polarization – up to 3% – were measured for obstacle particles at large phase angles. Therefore measurements of the circular polarization for atmosphereless bodies can be used for discovery of metal on their surfaces. Such investigations can ascertain the nature of M-asteroids.

063.030   Time–dependent scattering and transmission function in an anisotropic two–layered atmosphere.
T. K. Deb, S. Karanjai, G. Biswas.
Astrophys. Space Sci., Vol. 189, No. 1, p. 95 – 117 (Mar 1992).

The authors consider the time–dependent diffuse reflection and transmission problems for a homogeneous anisotropically-scattering atmosphere of finite optical depth and solve it by the principle of invariance. They also consider the time–dependent diffuse reflection and transmission of parallel rays by a slab consisting of two anisotropic homogeneous layers, whose scattering and transmission properties are known. It is shown how to express the time–dependent reflected and transmitted intensities in terms of their components. In a manner similar to that given by Tsujita (1968), the authors assumed that the upward–directed intensities of radiation at the boundary of the two layers are expressed by the sum of products of some auxiliary functions depending on only one argument. Then, after some analytical manipulations, three groups of systems of simultaneous integral equations governing the auxiliary functions are obtained.

063.031   An exact solution of the equation of transfer for coherent scattering in an exponential atmosphere.
S. Karanjai, T. K. Deb.
Astrophys. Space Sci., Vol. 189, No. 1, p. 119 – 122 (Mar 1992).

An exact solution of the transfer equation for coherent scattering in stellar atmospheres with Planck's function as a nonlinear function of optical depth, of the form $B_\nu(T) = b_0 + b_1 e^{-\beta\tau}$, is obtained by the method of the Laplace transform and Wiener–Hopf technique.

063.032   The amplitude of the opposition effect due to weak localization of photons in discrete disordered media.
M. I. Mishchenko, J. M. Dlugach (*Zh. M. Dlugach*).
Astrophys. Space Sci., Vol. 189, No. 1, p. 151 – 154 (Mar 1992). Letter–to–the–editor.

Weak localization of photons in discrete disordered media is considered as a possible physical mechanism of the opposition effect of some atmosphereless bodies. The amplitude of the opposition effect is calculated by using the rigorous vector multiple–scattering theory and the scalar approximation. It is shown that the scalar approximation can significantly overestimate the amplitude of the opposition effect. Thus, this approximation should not be used in interpreting the observational data, and some previous results obtained with this approximation may require substantial revision.

063.033   A new method of calculating multi–level non–LTE line formation.
G. Q. Wu.
Astrophys. Space Sci., Vol. 189, No. 2, p. 171 – 180 (Mar 1992).

This paper introduces a new method of solving the equation of multi–level non–LTE radiative transfer subject to constraints. This method is based on the combination of the advantages of the complete linearization method by Auer and Mihalas (1969) and the simple separated–iteration technique (Mihalas, 1978). First, the author linearizes the equation of radiative transfer and constraints, respectively, then he solves the linearized equation of the radiative transfer and linearized constraints, separately. He overcomes the disadvantages of requiring the simultaneous solution of the corresponding equations by the complete linear-ization method and the poor convergence of the simple separat-ed–iteration technique.

063.034   Radiation transfer in a diffuse and specular reflecting slab with Rayleigh scattering.
M. S. Abdel Krim, E. M. Abulwafa, S. M. Shouman.
Astrophys. Space Sci., Vol. 189, No. 2, p. 279 – 287 (Mar 1992).

A method of analysis is presented for solving the radiative transfer problem in an absorbing, emitting, inhomogeneous, and anisotropically scattering plane–parallel medium with specular and diffuse reflecting boundaries and internal source (problem 1). Exact relations for the radiation heat flux at the boundaries of problem 1 are obtained in terms of the radiation density and albedos of the corresponding source-free medium with specular reflecting boundaries (problem 2). Two coupled integral equations for the radiation density and the second moment of the radiation intensity for problem 2 with Rayleigh phase functions are obtained. The Galerkin method is used to solve these equations. Albedos of problem 2 are compared with the F_N method. Numerical results for radiation heat fluxes at the boundaries of problem 1 are tabulated for different forms of the internal source.

063.035   A new scheme for multidimensional line transfer. II. ETLA method in one dimension with application to iron Kα lines.
J. I. Castor, P. G. Dykema, R. I. Klein.
Astrophys. J., Vol. 387, No. 2, p. 561 – 571 (10 Mar 1992).
  The one–dimensional version of the radiation hydrodynamics computer code ALTAIR is described. It is an implementation of the equivalent two–level atom method, and also incorporates a variety of other iterative techniques to achieve a fast, accurate solution of the problem of coupled atomic kinetics and radiative transfer. All the techniques are described in sufficient detail to enable their use by others. The application of the program is illustrated by a set of calculations of the iron X–ray emission spectrum from intercloud hot gas that may exist in the broad–line region of active galactic nuclei. These calculations demonstrate that such spectra may be a rich lode to mine for information about the hot gas in AGNs, and also that the concept of "the iron Kα line" inadequately depicts the spectrum.

063.036   Numerical computations of Neumann expansion coefficients of Chandrasekhar's H–function for isotropic scattering.
K. Kawabata, T. Satoh.
J. Quant. Spectrosc. Radiat. Transfer, Vol. 47, No. 1, p. 1 – 8 (Jan 1992).
  The authors present two numerical schemes to compute the Neumann series expansion coefficients of Chandrasekhar's H–function for isotropic scattering. The first scheme involves direct evaluation of the closed–form expression for the expansion coefficients derived by Rutily and Bergeat, while the second scheme is to generate the coefficients by making use of the reflection functions for successive orders of scattering. It is shown that the second scheme is significantly faster than the first, and that its use enables one to extend the computation of the Neumann series coefficients of the H–function to arbitrarily high orders of light scattering. For convenience in practical use, an approximate formula is also given for the computation of the correction term required to represent the total H–function with the Neumann series terminated at the 16th term.

063.037   A multi–layer discrete–ordinate method for vector radiative transfer in a vertically–inhomogeneous, emitting and scattering atmosphere. I. Theory.
F. Weng.
J. Quant. Spectrosc. Radiat. Transfer, Vol. 47, No. 1, p. 19 – 33 (Jan 1992).
  The author establishes a theory for discretizing the vector integral–differential radiative transfer equation. In this theory, the phase matrix is derived from averaging the scattering matrix over polydisperse particles and then making a linear transformation of the averaged scattering matrix according to spherical trigonometry. The phase matrix and radiative vector in the vector radiative transfer equation are both expanded into Fourier cosine and sine series. The complete set of solutions for the discrete matrix equations for cosine and sine modes of the radiative vectors is obtained by solving for the eigenvalues and eigenvectors and particular solutions. The integration coefficients in the solutions are determined through the continuity conditions at the vertically–layered interface and the top and bottom boundaries. The eigenvalue characteristics are also explained. A numerical method to handle ill–conditioning in solving for integration coefficients is also introduced.

063.038   A multi–layer discrete–ordinate method for vector radiative transfer in a vertically–inhomogeneous, emitting and scattering atmosphere. II. Application.
F. Weng.
J. Quant. Spectrosc. Radiat. Transfer, Vol. 47, No. 1, p. 35 – 42 (Jan 1992).
  The upwelling radiance from the vector radiative transfer model established in Part I is compared with Chandrasekhar's analytic solutions for a conservative Rayleigh scattering atmosphere. The results from the model and the theory agree well for either an optically–thin or thick atmosphere. The polarized brightness temperatures at microwave frequencies from the model are compared with results from Stamnes' intensity model for a precipitating atmosphere.

063.039   A solution to the problem of radiation transfer in inhomogeneous media using the SHM.
S. Tiné, S. Aiello, A. Belleni, C. Cecchi Pestellini.
J. Quant. Spectrosc. Radiat. Transfer, Vol. 47, No. 2, p. 95 – 102 (Feb 1992).
  The authors extend the spherical harmonics method (SHM) to solve the radiative transfer equation for the case of radially-varying extinction and scattering coefficients. The formalism is first introduced and discussed. Some numerical results of astrophysical importance are then presented.

063.040   A relation between the Schrödinger and radiative transfer equations.
J. F. Geurdes.
J. Quant. Spectrosc. Radiat. Transfer, Vol. 47, No. 2, p. 121 – 126 (Feb 1992).
  It is demonstrated in this paper that the quantum–mechanical Schrödinger equation can be reformulated into a radiative transfer equation for the probability density of the wave function. The physical relevance of the formalism is established by applying it to hydrogen. In addition, it is demonstrated that the uncertainty relation for momentum and position can be derived from the Stokes equation. An extension of the derived one–particle radiative transfer equation to an N–particle equation is also discussed.

063.041   The correlated–k coefficients calculated by random band models.
X. Zhu.
J. Quant. Spectrosc. Radiat. Transfer, Vol. 47, No. 3, p. 159 – 170 (Mar 1992).
  The correlated–k coefficient for the cumulative distribution of the absorption coefficient in random band models is calculated with a computationally efficient algorithm based on a numerical inverse Laplace transform. A scaling transformation is introduced to partially eliminate the ill–conditioned behavior around the singularity point. In the region very close to the singularity, an analytic expression for the k coefficient derived for the Malkmus model is used to match the whole solution. The algorithm yields accurate k coefficients with a maximum error for a few percent. When applied to the random band model with an S^{-1}–β–tailed line intensity distribution and a Voigt line profile, the algorithm yields a maximum error in the escape function of $< 3\%$ in comparison with line–by–line integrations. The algorithm is sufficiently general to be applicable to a variety of radiative transfer problems in planetary atmospheres.

063.042   Calculation of the quasi–energies and resonance behavior of the hydrogen Lyman–α problem.
W. M. Ruyten.
J. Quant. Spectrosc. Radiat. Transfer, Vol. 47, No. 3, p. 179 – 184 (Mar 1992).
  Recently, Bakshi and Kalman presented numerical results for the quasi–energies of the $n = 2$ multiplet in the hydrogen Lyman–α transition for a plasma in which both strong static and oscillating electric fields are present. Here the author shows how recent work on related magnetic and optical resonance problems provides a simplified mathematical treatment, as well as greater insight into the complex resonance behavior of this interaction.

063.043 A comparison of solutions for light scattering and absorption by agglomerated or orbitrarily-shaped particles.
J. C. Ku, K.-H. Shim.
J. Quant. Spectrosc. Radiat. Transfer, Vol. 47, No. 3, p. 201 – 220 (Mar 1992).
Three approximate solutions for light scattering and absorption by agglomerated or arbitrarily-shaped particles have been investigated by comparing both their solution formulas and numerical results.

063.044 Line profile variations caused by low-frequency non-radial pulsations of rapidly rotating stars. II.
U. Lee, C. S. Jeffery, H. Saio.
Mon. Not. R. Astron. Soc., Vol. 254, No. 2, p. 185 – 191 (15 Jan 1992).
Line profile variations caused by low-frequency non-radial pulsations of rapidly rotating stars have been calculated, which take into account the temperature dependence of the absorption line equivalent width. Test calculations have been made for Si III 4553 Å and He I 6678 Å. The relative amplitude of temperature variation has been assumed to be comparable to the amplitude of velocity variation relative to the equatorial rotation velocity, which is reasonable for low-frequency oscillations because of the boundary condition. It is found that the effects of temperature variation exceed that of velocity variation in line profile variations. Consequently, travelling features have enough amplitude around the line centre to be consistent with observation.

063.045 Anisotropic radiation transfer in a plane medium with specularly-reflecting boundary conditions.
M. T. Attia, M. A. Madkour, E. M. Abulwafa, M. M. Abd-Elnaby.
J. Quant. Spectrosc. Radiat. Transfer, Vol. 47, No. 3, p. 221 – 227 (Mar 1992).
Anisotropic scattering in radiation transfer through an inhomogeneous planar medium with internal energy sources and diffusely- and specularly-reflecting boundary conditions is considered (problem 1). The partial heat fluxes for this problem are given in terms of the albedos of the source-free problem with specularly-reflecting boundaries (problem 2). The Galerkin technique is used to calculate first the albedos for problem 2 and then to calculate the partial heat fluxes for problem 1. Results are obtained for isotropic and anisotropic scattering for uniform and nonuniform internal sources.

063.046 Solution of the radiative transfer equation by the moment method using polynomials of special form.
V. P. Solovjov.
J. Quant. Spectrosc. Radiat. Transfer, Vol. 47, No. 3, p. 229 – 236 (Mar 1992).
A new version of the moment method, based on the use of Cutteridge-Devyatov polynomials, is described. The procedure is illustrated by a sample solution of the radiative transfer equation in an absorbing and isotropically-scattering medium. The reflectivity and transmissivity of an isotropically-scattering, plane-parallel slab are determined for isotropically-incident radiation. Numerical results are compared with exact solutions. Generalization to cases involving inhomogeneous, anisotropically-scattering and emitting media are discussed.

063.047 Stokes parameters of radiation propagating through an aligned gaseous-dust medium.
A. Z. Dolginov, V. I. Siklitsky (V. I. Siklitskij).
Mon. Not. R. Astron. Soc., Vol. 254, No. 3, p. 369 – 382 (1 Feb 1992).
The Stokes parameters of radiation propagating through a gaseous-dust medium are obtained in explicit analytical form. The dust grains are assumed to be aligned. These formulae are applicable to any type of alignment, any shape, chemical composition, and size of grain, if the grains are not very optically thick. The axis of predominant alignment is the external magnetic field or the direction of the flux of ambient gas. The important role of the Barnett effect for the rotating grain is taken into

account. The formulae obtained are applied to some specific sorts of grains (anisotropic graphite flakes, silicates spheroids, etc.) and conditions in the medium (interstellar and circumstellar clouds). Stokes parameters for the zodiacal light polarization are obtained. Interpretation of polarimetric observation of the light scattered by circumstellar dust around the red supergiant α Orionis is presented.

063.048 Analytical expressions for radiative properties of planar Rayleigh scattering media, including polarization contributions.
E. Vermote, D. Tanré.
J. Quant. Spectrosc. Radiat. Transfer, Vol. 47, No. 4, p. 305 – 314 (Apr 1992).
The objective of this paper is to provide a convenient and fast way for computing the radiative properties of planar Rayleigh scattering media. Analytical expressions are developed for the three molecular atmospheric functions which are required in remote sensing: the atmospheric reflectance, the transmission function and the spherical albedo. The expressions are adjusted by using accurate computations performed with successive orders of a scattering code. The accuracy of the code is first checked by using previously published tabulations. The required accuracy of 0.001 in the reflectance unit is achieved by numerical adjustments. The contribution of polarization is considered implicitly. The expressions are shown to be valid for a fairly large range of observational conditions.

063.049 The contribution of third order linear absorption to the water vapor continuum.
E. Hudis, Y. Ben-Aryeh, U. P. Oppenheim.
J. Quant. Spectrosc. Radiat. Transfer, Vol. 47, No. 5, p. 319 – 323 (May 1992).
An improved statistical theory for the third-order linear absorption of electromagnetic radiation propagating through a molecular gas is presented and applied to water vapor. Numerical calculations have been made of the contributions of third-order linear absorption to the water vapor continuum in the i.r. $(750 - 1200 \text{ cm}^{-1})$ and in the millimeter wave $(30 - 300 \text{ GHz})$ regions. The results are compared with the available experimental data and are found to show semi-quantitative agreement.

063.050 The role of electron scattering in the X-ray rotational light curves of Intermediate Polars.
S. R. Rosen.
Mon. Not. R. Astron. Soc., Vol. 254, No. 3, p. 493 – 500 (1 Feb 1992).
The role of electron scattering in explaining the high-energy rotational X-ray modulation of Intermediate Polar binaries is investigated within the framework of the accretion curtain model for such systems. The author considers the implications and how the assumptions behind the model might be modified to explain the observations.

063.051 Photodissociation in strong magnetic fields and application to pulsars.
V. B. Bhatia, N. Chopra, N. Panchapakesan.
Astrophys. J., Vol. 388, No. 1, p. 131 – 137 (20 Mar 1992).
The cross sections for the photoionization of positronium in strong magnetic fields $(10^{12} - 10^{13} \text{G})$ are derived within the Born approximation. Further, the total ionization rate of the positronium by thermal photons in strong magnetic fields is computed. The authors discuss the application of this process in light of the recent suggestions that in the vicinity of a pulsar the high-energy photons can be captured by the strong magnetic field and be transformed into bound positronia.

063.052 Multiple light scattering by polydispersions of randomly distributed, perfectly-aligned, infinite Mie cylinders illuminated perpendicularly to their axes.
M. I. Mishchenko, J. M. Dlugach, Eh. G. Yanovitskij.
J. Quant. Spectrosc. Radiat. Transfer, Vol. 47, No. 5, p. 401 – 410 (May 1992).
The radiative transfer theory has been applied to calculate multiple light scattering by ensembles of randomly distributed,

perfectly aligned, polydisperse, infinite Mie cylinders illuminated perpendicularly to their axes. An efficient method for computing the single–scattering Stokes matrix for polydispersions of Mie cylinders is described. The albedo problem for a homogeneous half–space of scatterers is considered. Illustrative numerical results for the backscattering coefficients are computed for polydispersions of cylinders with different refractive indices. An application to the problem of weak localization of photons is given. Specifically, enhanced backscattering of light by two–dimensional, discrete, disordered media is considered and back-scattering enhancement factors in exactly the backscattering direction are computed for a number of scattering models.

063.053   **A fast operator perturbation method for the solution of the special relativistic equation of radiative transfer in spherical symmetry.**
P. H. Hauschildt.
J. Quant. Spectrosc. Radiat. Transfer, Vol. 47, No. 6, p. 433 – 453 (Jun 1992).
A fast method for the solution of the radiative transfer equation in rapidly moving spherical media, based on an approximate Λ–operator iteration, is described. The method uses the short characteristic method and a tridiagonal approximate Λ–operator to achieve fast convergence. The convergence properties and the CPU time requirements of the method are discussed for the test problem of a two–level atom with background continuum absorption and Thomson scattering. Details of the actual implementation for fast vector and parallel computers are given. The method is accurate and fast enough to be incorporated in radiation–hydrodynamic calculations.

063.054   **Benchmark results for single scattering by spheroids.**
F. Kuik, J. F. de Haan, J. W. Hovenier.
J. Quant. Spectrosc. Radiat. Transfer, Vol. 47, No. 6, p. 477 – 489 (Jun 1992).
The authors present and discuss single–scattering results for three different kinds of spheroidal particles. Models 1 and 2 represent monodisperse, randomly–oriented prolate spheroids while Model 3 is an ensemble of randomly–oriented, prolate spheroids with a modified gamma size distribution for the semi–major axes and represents a model for an earth volcanic aerosol. The authors expect the results for all three models to have accuracies of six decimals. Because of this high accuracy, they are suitable for use as benchmarks. The authors present results for the scattering matrix elements in the form of expansion coefficients to make them easily accessible for further use.

063.055   **Polarized radiation of an atmosphere containing randomly–oriented spheroids.**
W. M. F. Wauben, J. W. Hovenier.
J. Quant. Spectrosc. Radiat. Transfer, Vol. 47, No. 6, p. 491 – 504 (Jun 1992).
The polarized internal and external radiation is calculated for a plane–parallel, homogeneous atmosphere and three types of randomly–oriented spheroids. The tabulated values are accurate and may serve as benchmark results. They have been obtained in two independent ways, namely, by using (1) a scheme based on the adding/doubling method and (2) the F_N method.

063.056   **Radiative pressure in natural masers.**
A. A. Sumin, V. S. Strel'nitskij.
Astron. Tsirk., No. 1550, p. 11 – 12 (Sep – Oct 1991).
The authors precise the qualitative analysis and give the results of a numerical simulation of the radiative pressure effects in the one–dimensional two–streams H_2O maser.

063.057   **Radiative atomic Rosseland mean opacity tables.**
F. J. Rogers, C. A. Iglesias.
Astrophys. J., Suppl. Ser., Vol. 79, No. 2, p. 507 – 568 (Apr 1992).
For more than two decades the astrophysics community has depended on opacity tables produced at Los Alamos. In the present work the authors offer new radiative Rosseland mean

opacity tables calculated with the OPAL code developed independently at LLNL. They give extensive results for the recent Anders – Grevesse mixture which allow accurate interpolation in temperature, density, hydrogen mass fraction, as well as metal mass fraction. The tables use temperature as parameter and follow tracks of constant R, where $R = density/(temperature)^3$. The ranges of R and temperature are such as to cover typical stellar conditions from the interior through the envelope and the hotter atmospheres. Cool atmospheres are not considered, since photoabsorption by molecules is neglected. Only radiative processes are taken into account, so that electron conduction is not included. For comparison purposes the authors also present some opacity tables for the Ross–Aller and Cox–Tabor metal abundances. Although in many regions the OPAL opacities are similar to previous work, large differences are reported. For example, factors of 2 – 3 opacity enhancements are found for stellar envelope conditions.

063.058   **Particle heated atmospheres of magnetic white dwarfs.**
U. Woelk, K. Beuermann.
Astron. Astrophys., Vol. 256, No. 2, p. 498 – 506 (Mar 1992).
The authors discuss the particle heating of magnetic atmospheres in AM Her stars. For small accretion rates cyclotron losses can balance the energy input due to Coulomb encounters within one mean free path of the infalling protons and no shock forms. The authors solve the radiative transfer in the target atmosphere and present for first time temperature structure and cylcotron spectra of a particle heated atmosphere as solution of an LTE stellar atmosphere code. They find temperature inversions between the heated outer layers and the photosphere up to three orders of magnitude depending on the accretion rate, the white dwarf mass and the magnetic field strength. The spectra show strong cyclotron line emission in the optical/IR, an optically thick continuum in the UV/XUV regime and a weak, optically thin bremsstrahlung component at X–ray energies. Passing through the colder outer parts of the atmosphere, the cyclotron lines develop strong self absorption in the first two or three harmonics generating strongly deformed line profiles, clearly different from those in previous isothermal models.

063.059   **Radiative transfer in rotating stars.**
P. Hadrava.
Astron. Astrophys., Vol. 256, No. 2, p. 519 – 524 (Mar 1992).
The radiative transfer in atmosphere of a rotationally oblated star is investigated. The horizontal diffusion of light is calculated for a gray model in radiative equilibrium with a power law for opacity. Its influence on the gravity darkening is discussed.

063.060   **On the origin of the H I line emission associated with massive young stellar objects.**
J. C. Bunn, J. E. Drew.
Mon. Not. R. Astron. Soc., Vol. 255, No. 3, p. 449 – 459 (1 Apr 1992).
The authors address the problem of the interpretation of the H I line emission associated with massive young stellar objects by comparing results of LTE calculations of H I Brα and Pfγ profiles with observations of the exciting source of Sh2–106 obtained by Garden and Geballe. The theoretical results are derived by formal solution of the transfer equation as applied to a spherically symmetric stellar wind. Free–free and bound–free opacity are taken into account. It is found that the observed Brα profile and flux can be reproduced by LTE stellar wind models that also satisfy the radio flux constraint upon $\dot{M}/v_\infty$ obtained by Felli et al. The best–fitting models are characterized by terminal velocities of around 100 km s^{-1} and relatively gradual velocity laws similar to the one derived for P Cygni. However the observed F(Pfγ)/F(Brα) flux ratio cannot be reproduced. The effect of departures from LTE upon the line fluxes and flux ratio are discussed, and it is proposed that the observed asymmetry in the Pfγ profile is due to the combined effect of free-free opacity and a line source function that exceeds its LTE value. The failure to reproduce the observed F(Pfγ)/F(Brα) ratio implies the simple spherically symmetric wind model is in difficulties and that it may

have to be replaced by one locating the IR line emission in an outflow further removed from the stellar source.

063.061   Estimating the effect of finite angular light source dimensions on the opposition brightness effect in atmosphereless bodies.
Yu. G. Shkuratov.
Sol. Syst. Res., Vol. 25, No. 1, p. 54 – 57 (Jul 1991). English translation of Astron. Vestn., Tom 25, No. 1, p. 71 – 75 (1991).
Within the framework of the diffraction model of formation of the opposition effect an approximate expression is obtained for the dependence of the brightness peak at the phase null on the angular radius α_0 of the light source (Sun), where η is the characteristic radius of the light scattering region, referenced to the wavelength of the light. For the sufficiently large η, characteristic $F(\alpha_0)$ is significant. Thus, the intense opposition effect observed with Saturn's satellite Enceladus, the satellites of Uranus, and some bright asteroids is not only a consequence of unique structure and surface albedo of these bodies, but also the fact that they are quite far removed from the Sun.

063.062   Diamagnetic effects in synchrotron sources.
G. Bodo, G. Ghisellini, E. Trussoni.
Mon. Not. R. Astron. Soc., Vol. 255, No. 4, p. 694 – 700 (15 Apr 1992).
A magnetized plasma behaves as a diamagnetic material: particles injected in a magnetized region can lower the applied magnetic field, this screening being almost complete if the energy density of the particles is high enough with respect to the energy density of the applied magnetic field. The authors find the equilibrium states of such systems assuming that the applied field is produced by "batteries" external to the region under consideration. Under the influence of synchrotron losses, the system evolves. The internal equilibrium magnetic field increases with time, inducing an electric field which slows down the cooling. As a result, one electron can radiate more than its internal energy, the extra work being done by the "batteries" maintaining the external magnetic field. The consequences on the evolution of synchrotron sources are discussed.

063.063   On the density and field sensitivities of dielectronic recombination.
D. B. Reisenfeld, J. C. Raymond, A. R. Young, J. L. Kohl.
Astrophys. J., Lett., Vol. 389, No. 1, p. L37 – L40 (10 Apr 1992).
Dielectronic recombination dominates the recombination rates of most ions in coronal plasmas at their temperatures of peak concentration. Because dielectronic recombination goes by way of high nl doubly excited levels, it is susceptible to collisional excitation and ionization, leading to a decreased rate. On the other hand, theoretical studies show that Stark mixing of the nl levels by a modest electric field enhances the dielectronic recombination rate severalfold. A companion paper presents a calculation of the combined effects of plasma microfields and collisional processes on the dielectronic recombination rates of carbon ions. Here, the authors compute the ionization balance as a function of density. They find that the new results require increased emission measures to match the C IV emission line intensities observed in the Sun and in late–type stars. The new data also make it more difficult to interpret the overall EUV emission line spectrum of the Sun.

063.064   Non–LTE radiative transfer with Lambda–acceleration: convergence properties using exact full and diagonal Lambda–operators.
J. J. MacFarlane.
Wis. Astrophys., No. 396.
Wisconsin Univ., Madison (USA). 16 p. (Aug 1991).
The author investigates the convergence properties of Λ–acceleration methods for non–LTE radiative transfer problems in planar and spherical geometry. Matrix elements of the "exact" Λ–operator are used to accelerate convergence to a solution in which both the radiative transfer and atomic rate equations are simultaneously satisfied.

063.065   Plasma maser effects in astrophysics.
I. M. Lal Das.
4. Asia Pacific Physics Conference, p. 797 – 800 (1991). – See Abstr. 012.006 for the main entry.
The theory of electron–cyclotron maser instability has been frequently invoked to explain extremely intense radio emissions from planets, stars or stellar bodies. The paper gives an overview of the plasma maser instability of astrophysical importance.

063.066   Interference model of negative polarization of light scattered by the solid surfaces of celestial bodies.
Yu. G. Shkuratov.
Sol. Syst. Res., Vol. 25, No. 2, p. 110 – 117 (Sep 1991). English translation of Astron. Vestn., Tom 25, No. 2, p. 152 – 161 (1991).
A theoretical model is presented for the negative polarization of light scattered by surfaces having a complicated structure. The model is based on analysis of the interference of rays which interact with the same scatterers in the process of double scattering. The minimum of negative polarization, the angle of inversion, and some other polarimetric characteristics are calculated as a function of the parameters of the model: the albedo and the packing density of particles of the surface as well as the size of the particles and the constant characterizing the polarizability. The calculations are in good agreement with the experimental data.

063.067   Theory of scattering of polarized radiation in a randomly inhomogeneous magnetized plasma.
A. V. Kukushkin, M. R. Ol'yak.
Sov. Astron. Lett., Vol. 18, No. 1, p. 37 – 40 (Jan 1992). Current Physics Microform No.: 9209X1255. English translation of Pis'ma Astron. Zh., Tom 18, No. 1, p. 87 – 95 (1992).
A theoretical investigation of spatial coherence of polarized radio emission from cosmic sources at spatially separate observing sites has revealed that weak circular polarization is produced in a randomly inhomogeneous magnetized plasma. Fluctuations of the longitudinal magnetic field in the medium lead to opposite shifts of the phase velocity of right– and left–handed polarized waves, which results, in turn, in relative enhancement of linear polarization for a longer interferometer baseline.

063.068   Collective electron–positron annihilation.
A. A. Belyanin, V. V. Kocharovskii,
Vl. V. Kocharovskii.
IAU Colloquium No. 128: The magnetospheric structure and emission mechanisms of radio pulsars, p. 117 – 122 (1992). – See Abstr. 012.022 for the main entry.
The phenomenon of collective spontaneous annihilation of a magnetized electron–positron plasma is predicted. Like the superradiance in systems with discrete energy spectra, collective annihilation leads to the generation of powerful coherent radiation with the rate of this process considerably exceeding the spontaneous annihilation and collisional relaxation rates.

063.069   Linear acceleration emission: a detailed analysis.
E. T. Rowe.
IAU Colloquium No. 128: The magnetospheric structure and emission mechanisms of radio pulsars, p. 123 – 129 (1992). – See Abstr. 012.022 for the main entry.
An exact expression is given for the orbit of a charged particle accelerated by a non–propagating electrostatic wave. The corresponding current is obtained, for untrapped particles, as an expansion in Bessel functions, and a Lorentz transform allows us to treat the case of a propagating wave. An appropriate form for the absorption coefficient is derived, and an interesting angular and frequency dependence is revealed when the propagating wave is superluminal. Finally, the possible application to pulsars, particularly in explaining the multicomponent pulse profiles, is discussed briefly.

063.070 The Josephson effect as a possible alternative mechanism for pulsar radio emission: spectra, pulse structure, polarization, and X–/γ–ray emission.
Y. A. Kovalev.
IAU Colloquium No. 128: The magnetospheric structure and emission mechanisms of radio pulsars, p. 130 – 131 (1992). – See Abstr. 012.022 for the main entry.
Results and problems of the model for Josephson pulsars are summarized and discussed. In this model coherent radio emission is generated in the mantle of a neutron star by the Josephson effect. The matter from the mantle may flow along the magnetic field through cracks in the crust, forming a magnetized plasma wave guide. Radio emission propagates in this plasma flow as polarized normal wave–guide modes. A line–of–sight section gives the observed polarized structure of pulses. In some cases polarized X– and γ–ray emission may be generated in this flow in or over cracks. Such an approach allows to avoid the typical problems and is in agreement with the main observational results.

063.071 An empirical theory of pulsar emission.
J. M. Rankin.
IAU Colloquium No. 128: The magnetospheric structure and emission mechanisms of radio pulsars, p. 133 – 139 (1992). – See Abstr. 012.022 for the main entry.
A system of pulsar profile classification is used as a starting point to study the emission characteristics of pulsars. Two types or mechanisms of pulsar radiation are identified which combine geometrically to produce five major species of profile. The core emission, which forms a pencil beam of radiation, is apparently produced close to the stellar surface throughout the entire polar cap region by low γ particles. The conal emission, which consists of a hollow conical beam, then seems to be emitted at heights of 10 to 20 stellar radii by currents of high γ particles travelling along some of the most peripheral of the "open" field lines.

063.072 On a possible mechanism of pulsar radiation.
A. Z. Kazbegi, G. Z. Machabeli, G. I. Melikidze.
IAU Colloquium No. 128: The magnetospheric structure and emission mechanisms of radio pulsars, p. 232 – 235 (1992). – See Abstr. 012.022 for the main entry.

063.073 Pulsar emission beams and profiles in an inverse– Compton scattering model.
G. J. Qiao.
IAU Colloquium No. 128: The magnetospheric structure and emission mechanisms of radio pulsars, p. 238 – 241 (1992). – See Abstr. 012.022 for the main entry.
The author presents a calculation for both the "core" and hollow "cone" emission beams, as well as model pulse profiles in an inverse–Compton scattering model. Both "core" and hollow "cone" emission beams are obtained naturally in the calculations. Examples of pulse profiles of pulsars at different radio frequencies are presented. The theoretical shapes of the pulse profiles agree very satisfactorily with actual observations.

063.074 The locations of the core and conal emission regions in an inverse–Compton scattering model.
G. J. Qiao, C. G. Li, M. Li.
IAU Colloquium No. 128: The magnetospheric structure and emission mechanisms of radio pulsars, p. 242 – 244 (1992). – See Abstr. 012.022 for the main entry.
The physical conditions and locations of the emission regions for core and hollow cone emissions are very important in understanding the mechanism of radio pulsars. The authors present two related methods in an inverse–Compton scattering model which give a clear scenario for determining the location of the emission regions.

063.075 A model for the drifting–subpulse phenomenon.
A. Z. Kazbegi, G. Z. Machabeli, G. I. Melikidze.
IAU Colloquium No. 128: The magnetospheric structure and emission mechanisms of radio pulsars, p. 296 – 298 (1992). – See Abstr. 012.022 for the main entry.
The aim of this paper is to explain the phenomenon of drifting subpulses observed in the pulsar radio emission within the framework of the theory developed by the authors.

063.076 Coherent radio–emission mechanisms for pulsars.
D. B. Melrose.
IAU Colloquium No. 128: The magnetospheric structure and emission mechanisms of radio pulsars, p. 306 – 315 (1992). – See Abstr. 012.022 for the main entry.
Pulsar radio emission is the brightest of all known coherent emission, and its brightness temperature is close to the maximum conceivable in terms of energy efficiency. Three possible pulsar radio emission mechanisms warrant serious consideration in polar cap models: coherent curvature emission, relativistic plasma emission, and free electron maser emission, respectively.

063.077 Is the pulsar emission produced by superluminally moving charged patterns?
H. Ardavan.
IAU Colloquium No. 128: The magnetospheric structure and emission mechanisms of radio pulsars, p. 316 – 318 (1992). – See Abstr. 012.022 for the main entry.
Charge–current distributions whose patterns rotate around a fixed axis rigidly, and so have a phase speed that is greater than the speed of light in vacuo outside the light cylinder, emit radiation–different from both Cherenkov and synchrotron radiation which characteristics are given. The author suggests that this radiation, here referred to as Schott radiation, may in fact be that received from pulsars.

063.078 Computer simulation of electron–flow bunching in the pulsar magnetosphere.
Yu. A. Rylov.
IAU Colloquium No. 128: The magnetospheric structure and emission mechanisms of radio pulsars, p. 319 – 321 (1992). – See Abstr. 012.022 for the main entry.

063.079 A non–linear emission mechanism for pulsar radio radiation.
E. Asseo, G. Pelletier, H. Sol.
IAU Colloquium No. 128: The magnetospheric structure and emission mechanisms of radio pulsars, p. 322 – 325 (1992). – See Abstr. 012.022 for the main entry.

063.080 Depth of formation of lines in the solar atmosphere.
L. Achmad.
Sol. Phys., Vol. 138, No. 2, p. 411 – 414 (Apr 1992). Letter–to– the–editor.
A simple and reliable method has been established for the definition of the depth of line formation in a stellar atmosphere. It is based on the determination of the level of emission of the reemitted line radiation. The new definition, which is applied in this paper to the solar case, does not suffer from the ambiguities of previously derived expressions. This method is applied to two artificial weak oxygen I lines as examples.

063.081 1612 MHz OH maser emission from axisymmetric circumstellar envelopes: Miras.
A. J. Collison, J. D. Fix.
Astrophys. J., Vol. 390, No. 1, p. 191 – 212 (1 May 1992).
The authors perform radiative transfer calculations using a modified form of the Sobolev approximation to determine the inversion of the 1612 MHz line of OH in axisymmetric circumstellar envelopes around Miras. The mass loss is assumed to be occurring in the form of a smooth wind. Line profiles and maps are presented for three models of varying degrees of asymmetry and for various orientations of the envelopes. The authors conclude that the axisymmetric models can reproduce many of the features of observed profiles and maps, which both the standard, spherically symmetric model and the discrete emission model cannot easily explain. The model profiles reproduce all of the general features seen in the line profiles of real sources.

063.082 The polarization of pulsar radiation.
V. Radhakrishnan.
IAU Colloquium No. 128: The magnetospheric structure and emission mechanisms of radio pulsars, p. 367 – 372 (1992). – See Abstr. 012.022 for the main entry.
In this review the author has summarized a number of the most important observations of polarized pulsar emission and showed where a number of problems exist in their interpretation.

063.083 A mechanism for circular polarization in pulsar radiation.
A. Z. Kazbegi, G. Z. Machabeli, G. I. Melikidze.
IAU Colloquium No. 128: The magnetospheric structure and emission mechanisms of radio pulsars, p. 373 – 374 (1992). – See Abstr. 012.022 for the main entry.
  The authors propose to outline a possible mechanism for the generation of circular polarization consistent with the theory of pulsar radio–wave excitation.

063.084 On the nature of the circularly polarized component of pulsar radio emission.
Ya. N. Istomin.
IAU Colloquium No. 128: The magnetospheric structure and emission mechanisms of radio pulsars, p. 375 – 377 (1992). – See Abstr. 012.022 for the main entry.
  It is shown that circular polarization occurs in the region of cyclotron resonance because the group velocities of right–hand and left–hand polarized waves are different with respect to the direction of the magnetic field. Due to the dependence of the intensity of radio emission on the coordinates across the polar cap, this difference in group velocities leads to noncompensated circular polarization proportional to the derivative of the total intensity as a function of longitude. The indicated dependence corresponds to observations of the so–called core component of pulsar radio emission.

063.085 Modeling of pulsar polarization properties.
  L. M. Shier, F. C. Michel.
IAU Colloquium No. 128: The magnetospheric structure and emission mechanisms of radio pulsars, p. 378 – 383 (1992). – See Abstr. 012.022 for the main entry.
  In the popularly accepted empirical model for pulsar emission, bunches of charged particles traveling along open field lines near the magnetic pole emit curvature radiation. Such radiation is linearly polarized. Most of the radiation is assumed to be emitted from a ring shaped region centered on the pole (the hollow cone model). The authors have calculated the expected average polarization using this model and find them to be in disagreement with observations. The addition of a second ring inside the first with orthogonal polarization solves this problem. This new model explains other observed features of pulsar emission including discontinuities in position–angle profiles and multi-component profiles. Plasma interactions might account for a second ring.

063.086 Radiation damping and the two mode behavior in pulsars.
C.–I. Björnsson.
IAU Colloquium No. 128: The magnetospheric structure and emission mechanisms of radio pulsars, p. 391 – 393 (1992). – See Abstr. 012.022 for the main entry.
  It is argued that both observations and theory indicate that radiation damping plays an important role in pulsar emission. The two–mode behavior as well as the observed value of the brightness temperature can both be understood as a result of radiation damping. In this context, a possible cause for the enhanced depolarization in the wings of pulsar profiles is discussed.

063.087 Curvature radiation and polarized emission from PSR 2303+30.
J. A. Gil.
IAU Colloquium No. 128: The magnetospheric structure and emission mechanisms of radio pulsars, p. 394 – 399 (1992). – See Abstr. 012.022 for the main entry.
  The author presents 430 MHz Arecibo single pulse polarization measurements for PSR 2303+30. Single pulses typically consist of one or two subpulses. Each subpulse is associated with circular polarization which reverses sense near the subpulse peak and with its own swing of the position angle which does not coincide with the average position–angle curve. The author attempts to explain these complicated characteristics within the

framework of a coherent curvature radiation model of pulsar emission.

063.088 A determination of pulsar emission geometry from polarization observations.
Wu Xinji, Xu Wen.
IAU Colloquium No. 128: The magnetospheric structure and emission mechanisms of radio pulsars, p. 400 – 403 (1992). – See Abstr. 012.022 for the main entry.
  One of the important problems in pulsar studies is to determine the magnetic inclination angle α, the intrinsic width of the radiation beam (2ϱ) and the angle $(\alpha + \beta)$ between the observer's direction and the rotation axis. The authors solve this problem for individual pulses by using the observed pulse width $(2\Delta\phi)$, the swing of polarization angle $(2\Delta\psi)$, and its central gradient $(d\psi/d\phi)_{max}$. The results are shown to be sensitively connected to the polarization–angle swing $(2\Delta\psi)$, which is not well measured for most pulsars. The authors discuss this method for the determination of pulsar geometry in comparison with other methods.

063.089 Theoretical profiles of Lyman α satellites and application to synthetic spectra of DA white dwarfs.
N. F. Allard, D. Koester.
Astron. Astrophys., Vol. 258, No. 2, p. 464 – 468 (May 1992).
  The authors present new theoretical calculations for the red wing of the Lyα profile. Close collisions with neutral and ionized hydrogen lead to the formation of the pseudo–molecules H–H and H–H$^+$ with the appearance of satellite features near 1600 and 1400 Å. The calculations include multiperturber effects, which are responsible for the formation of H_3^+ and H_3 with features near 1950 and 2600 Å. The theoretical absorption profiles are included in stellar atmosphere codes and used to predict synthetic spectra for DA white dwarfs of intermediate temperatures (20000 to 8000K). These new calculations offer a unique opportunity to determine accurate effective temperatures and surface gravities for the variable ZZ Ceti stars.

063.090 Hanle effect with partial frequency redistribution. II. Linear polarization of the solar Ca I 4227 Å line.
M. Faurobert–Scholl.
Astron. Astrophys., Vol. 258, No. 2, p. 521 – 534 (May 1992).
  This paper is devoted to the interpretation of observations of the linear polarization in the resonance line of Ca I at 4227 Å over solar active and non–active regions performed by Stenflo (1982) and Stenflo et al. (1980). Theoretical polarization profiles of the Ca I line are calculated by solving a non–LTE polarized transfer problem with a two–level atom interative method which takes into account partial frequency redistribution and the Hanle effect. Comparisons of the theoretical line wing polarization with observational data from non–magnetic regions suggest an empirical value for γ_{vw} (the van der Waals coefficient of neutral calcium) which is in agreement with the results of previous theoretical calculations. In magnetic regions, the observed polarization rates are found to be consistent with the presence of a magnetic canopy lying in the low solar chromosphere at an altitude of about 950 km above $\tau_{5000} = 1$. Low lying chromo-spheric canopies have recently been predicted by Solanki and Steiner (1990) from theoretical calculations on the magnetic equilibrium of hot flux tubes imbedded in a cooler chromosphere.

063.091 Partially coherent scattering in stellar chromospheres. II. The first–order escape probability method.
K. G. Gayley.
Astrophys. J., Vol. 390, No. 2, p. 573 – 582 (10 May 1992).
  The author derives approximate analytic expressions for resonance line wing diagnostics, accounting for frequency redistribution effects, for homogeneous slabs and slabs with a constant Planck function gradient. The objective is to describe resonance line emission profiles from a simplified conceptual standpoint, thereby allowing observers to obtain a rapid under-standing of the basic physical parameters of the line forming

layers prior to performing detailed numerical simulations. As an example of the utility of the approach, the author derives an approximate analytic expression for the dependence on stellar surface gravity of the location of the Ca II and Mg II resonance line profile peaks. This provides a helpful first step in quantifying the effects of partial redistribution on the overall scaling law known as the Wilson-Bappu effect.

063.092   Partially coherent scattering in stellar chromospheres. III. A second-order escape probability method.
K. G. Gayley.
Astrophys. J., Vol. 390, No. 2, p. 583 - 589 (10 May 1992).
   Radiative transfer employing "second-order" escape probability methods has only been applied in cases where photons undergo complete redistribution (CRD) in each scattering. The author develops an approximate radiative transfer equation using generalized escape probabilities, applicable even in the presence of nearly coherent scattering in the damping wings of resonance lines. Solving the derived equation produces so-called second-order accuracy, because it accounts for the interception of photons originating deep in the atmosphere by layers closer to the surface, analogous to probabilistic methods in CRD. Approximate analytic solutions are also derived, which can be applied in special regimes and achieve good agreement with accurate numerical results. Because the methods derived are extremely fast, applications to radiative hydrodynamics and multidimensional transfer are anticipated, although refinements to the present static plane-parallel approach will be required.

063.093   Gamma-ray deposition and nonthermal excitation in supernovae.
C. Kozma, C. Fransson.
Astrophys. J., Vol. 390, No. 2, p. 602 - 621 (10 May 1992).
   The authors calculate the γ-ray deposition in supernovae by solving the Spencer-Fano equation. Ionization, excitation, and heating rates are presented for the different chemical composition zones of a core collapse supernova, as well as for a solar composition applicable to, for example, active galactic nuclei. The authors also discuss the thermalization in pure helium, oxygen, and iron plasmas. The latter is of particular interest for Type Ia supernovae. Convenient analytical expressions are given to facilitate the use of these results for the calculation of the physical conditions and emission from supernovae. The authors also discuss the spectral characteristics of the emission following nonthermal excitation and ionization in the supernova. In particular, they calculate the fraction of the absorbed γ-ray energy which is reemitted as UV photons with energy above 3.4 eV, the n = 2 threshold of hydrogen. These results are applied to the ionization of hydrogen and the formation of the Hα line. Good qualitative and quantitative agreement between the model and observations of SN 1987A is found. In particular, absorption of UV photons in the Balmer continuum dominates the excitation before day 500, while direct nonthermal excitation from the ground state dominates thereafter. It is also discussed how the wing of the Hα line can be used as a probe of the density in the envelope.

063.094   Relation between emitted and received powers from a moving radiating source in a medium.
H. M. Lai, C. S. Ng.
Astrophys. Lett. Commun., Vol. 28, No. 1, p. 27 - 32 (Jul 1990).
   It is proposed that the ratio of the emitted power to the received power from a moving radiating source in a medium is to be considered and evaluated at constant ω_0, the proper frequency characteristic of the source, instead of at constant ω, the frequency of the radiation received. The general formula thus obtained differs from that given by Ko and Chuang, even in the case of an isotropic medium. Application to a field-free plasma is made, which includes the situation of the complex Doppler effect, and the implication of the results to the case of a magnetoplasma is discussed.

063.095   Influence of X-ray radiation on stellar atmospheres and the problem of helium abundance in B-type stars.
A. S. Mitskevich, V. V. Tsymbal.
Astron. Zh., Tom 69, Vyp. 2, p. 333 - 346 (Mar-Apr 1992). In Russian. English translation in Sov. Astron., Vol. 36, No. 2 (Mar-Apr 1992).
   A method of computation of model stellar atmospheres, irradiated by external emission with an arbitrary spectrum, is described. The computed equivalent widths and profiles of He I lines in B-type star atmospheres are presented; the effects of external X-ray radiation were accounted for. It is shown that hard coronal radiation considerably diminishes the equivalent widths of the lines discussed, due to the effect of "superionization" of He I. This effect takes place only for eefective stellar temperatures $T_{eff} \leqslant 25,000K$. The He I lines, computed with an account for X-ray radiation, were then used for a classical LTE analysis, aiming to evaluate the influence of hard radiation on the determination of helium abundance. As a result, a quasideficiency of helium of the order of 0,4 dex is obtained from the blue region lines, and as high as 2 dex from the red region lines, for the ratio of X-ray and bolometric luminosities $L_x/L_{bol} = 10^{-8} - 10^{-7}$. In the framework of the proposed model of a star irradiated by external coronal emission, an attempt is made to explain the equivalent widths of helium lines, observed in "He-weak" B-type stars and sdOB stars; so far, these lines have been interpreted in the framework of the model of chemical element diffusion within the stellar atmosphere. An experiment is suggested, which would allow, basing on optical-range observations, to choose between the two above-mentioned models.

063.096   Influence of X-rays on stellar atmospheres and the problem of helium abundance in B stars.
A. S. Mitskevich, V. V. Tsymbal.
Sov. Astron., Vol. 36, No. 2, p. 169 - 175 (Mar 1992). Current Physics Microform No.: 9208X2005. English translation of Astron. Zh., Tom 69, Vyp. 2, p. 333 - 346 (1992).
   A method of calculating stellar atmospheres irradiated from outside by radiation with an arbitrary spectrum is described. The equivalent widths and profiles of lines of neutral helium produced in the atmosphere of a B star, calculated with allowance for X-rays from outside, are presented. It is shown that hard coronal emission considerably reduces the equivalent widths of those lines due to "superionization" of He I. That effect occurs only for effective stellar temperatures $T_{eff} \leqslant 25,000K$. The He I lines calculated with allowance for X-rays are used for a classical LTE analysis to estimate the influence of hard X-rays on the determination of helium abundance. The result is a helium quasideficit of about 0.4 dex based on lines in the blue range and up to 2 dex based on red spectral lines for a ratio of X-ray to bolometric luminosity $L_x/L_{bol} = 10^{-8}-10^{-7}$. The proposed model (a star irradiated by external coronal radiation) is used to try to explain the observed equivalent widths of helium lines of "He-weak" B stars and sdOB stars, presently interpreted in a model of diffusion of chemical elements in the stellar atmosphere. An experiment is suggested for choosing between these two models on the basis of observations in the visible.

063.097   X-ray emission-line spectra of photoionized plasmas: density sensitivity of the Fe L-shell series.
D. A. Liedahl, S. M. Kahn, A. L. Osterheld, W. H. Goldstein.
Astrophys. J., Vol. 391, No. 1, p. 306 - 317 (20 May 1992).
   The circumsource environments of accretion-powered X-ray sources are likely to support relatively dense ($n > 10^{11}cm^{-3}$) photoionized X-ray emission-line regions. The Fe L-shell ions provide a versatile class of plasma diagnostics in this regime, their multielectron structures resulting in diverse spectral phenomena. In this paper, the authors concentrate on the spectral response of Fe L-shell ions to variations in electron density over the range $10^{11}cm^{-3} - 10^{16}cm^{-3}$. They find that density-sensitive line ratios exist in the wavelength interval 12 - 17 Å for the ions Fe XVII - XXI. The prominent role of radiative recombination in the population kinetics distinguishes the density-sensitive Fe lines in photoionized plasmas from those which operate in coronal equilibrium plasmas. The authors present the results of

detailed atomic modeling of these ions and discuss applications to spectroscopic observations of accretion–driven X–ray sources.

063.098   Coherent backscattering: a vector formulation for effects of polarization, absorption, and small or large scatterers.
K. J. Peters.
Bull. Am. Astron. Soc., Vol. 23, No. 3, p. 1171 (1991). Abstract. – See Abstr. 012.037 for the main entry.

063.099   Stimulated scattering of electromagnetic waves in dusty plasmas.
P. K. Shukla, L. Stenflo.
Astrophys. Space Sci., Vol. 190, No. 1, p. 23 – 32 (Apr 1992).

The nonlinear coupling between a large amplitude electromagnetic wave and the slow background motion in a dusty plasma is considered. Stimulated scattering instabilities are investigated. The relevance of this investigation to cometary and astrophysical plasmas is pointed out.

063.100   X–ray polarization by Compton scattering in Seyfert galaxies.
G. Matt, G. C. Perola, L. Piro.
Vulcano Workshop 1990: Frontier objects in astrophysics and particle physics, p. 99 – 106 (1991). – See Abstr. 012.040 for the main entry.

The authors have calculated, by means of Montecarlo simulations, the dependence of the intensity and of the degree of polarization on the inclination angle and on the energy of the radiation reflected by cold, optically thick accreting matter located close to the central X–ray source; they have also calculated the polarization of the radiation scattered by a cloud of free electrons possibly surrounding the X–ray source and the Broad Lines Region.

063.101   X and γ–ray emission from relativistic jets: the case of 3C 273.
Vulcano Workshop 1990: Frontier objects in astrophysics and particle physics, p. 175 – 181 (1991). – See Abstr. 012.040 for the main entry.

The angular dependence of the spectrum of secondary electrons and positrons which are produced in the interaction of a relativistic proton beam with ambient matter is first investigated. Results are then used to model the X and γ–ray emission from the quasar 3C 273. It is found that the core of 3C 273 must emit a power of $P \approx 1.5 \times 10^{48}$ ergs/s which corresponds to a mass loss of relativistic protons $\dot{M} \approx 1.1 \, M_{\odot}$/yr in the jet.

063.102   The synchro–Compton limit of the brightness temperature of nonstationary radio sources.
V. I. Slysh.
Astrophys. J., Vol. 391, No. 2, p. 453 – 455 (1 Jun 1992).

The brightness temperature of synchrotron emission from nonstationary radio sources during the cooling down by the inverse Compton effect is calculated. It is shown that brightness temperatures as high as 5×10^{15} K at 1 GHz are allowed during the first day after injection of relativistic electrons of sufficiently high energy. This is about four orders of magnitude higher than the canonical synchro–Compton limit introduced by Kellermann and Pauliny–Toth for stationary radio sources. A stationary situation with the in situ first–order Fermi acceleration will give a brightness temperature of about 10^{15} K at 1 GHz due to the compensation of the inverse Compton losses by particle acceleration. The high brightness temperature effect is most pronounced at low frequencies and is proposed as the explanation of the low–frequency variability phenomenon. Strong high–energy emission is predicted during phases of high brightness temperature.

063.103   The Fe L–shell spectrum in compact astrophysical X–ray sources.
D. A. Liedahl, S. M. Khan, W. H. Goldstein, A. L. Osterheld.
AIP Conf. Proc., No. 257, p. 181 – 189 (15 May 1992). Current Physics Microform No.: 9207G1901. Paper presented at the 8. Biennial APS Topical Conference on Atomic Processes in Plasmas, Portland, ME (USA), 25 – 29 Aug 1991.

Compact X–ray sources derive their luminosities from the conversion of gravitational potential energy through accretion. A centrally–produced hard X–ray continuum photoionizes the surrounding accretion flow, which can produce discrete X–ray emission. To date, as applied to compact X–ray sources, a detailed diagnostic approach to the analysis of spectroscopic data has been largely ineffective, owing to the limited quality of available line spectra and the lack of appropriate models of X–ray line emission in the presence of an ionizing continuum radiation field. The authors are involved in an atomic modeling program to investigate the discrete spectral response to variations in ambient plasma conditions, with the aim of establishing a new set of plasma diagnostics appropriate to this physical regime. They discuss the physical mechanisms and applications of several of the diagnostics which have been discovered in the Fe L–shell spectrum, under conditions appropriate to X–ray photoionized plasmas.

063.104   Collective radiation from jets.
G. Benford.
7. IAP Meeting: Extragalactic radio sources – from beams to jets, p. 85 – 91 (1992). – See Abstr. 012.041 for the main entry.

Collective emission can occur when plasma oscillations scatter relativistic electrons. For typical jet conditions the spectrum has the same polarization as ordinary synchrotron emission, but far greater power, because the electrons bunch together (antenna mechanism). This allows higher emissivity and relaxes the demand for jet energy. Flickering emerges naturally as coherence varies, without demanding highly relativistic jet motions.

063.105   Solution of the equation of transfer for coherent scattering in an exponential atmosphere by Busbridge's method.
T. K. Deb, S. Karanjai.
Astrophys. Space Sci., Vol. 192, No. 1, p. 127 – 132 (Jun 1992).

A solution of the transfer equation for coherent scattering in stellar atmosphere with Planck's function as a nonlinear function of optical depth, viz. $B_{\nu}(T) = b_0 + b_1 e^{-\beta \tau}$ is obtained by the method developed by Busbridge (1953).

063.106   Solution of the equation of transfer with Rayleigh's phase function in a thin atmosphere.
S. Karanjai, L. Biswas.
Astrophys. Space Sci., Vol. 192, No. 1, p. 151 – 156 (Jun 1992). Letter–to–the–editor.

The equation of radiative transfer with scattering according to Rayleigh's phase function has been solved in a thin atmosphere by use of a modification of the spherical–harmonic method suggested by Wan et al. (1986).

063.107   Atmospheric effect on the upwelling radiation at the top of the atmosphere over a stream.
T. Takashima, K. Masuda.
Earth, Moon, Planets, Vol. 57, No. 1, p. 75 – 84 (Apr 1992).

A new version is adopted for the evaluation of the upwelling radiation from atmosphere bounded by the surface, where the surface is composed of two half semi–infinite Lambert surfaces and a stream is inserted between them. The contrast of the stream is discussed with respect to the atmospheric effect. The width of the stream is considered to be 0.5, 1, and 3 km; the solar and observational direction is located in the normal plane to the stream. The observational site is located at altitude 30 km. The horizontal distance of observational site to the stream is fixed to $6°28$. The atmosphere is assumed to be homogeneous, which is composed of aerosol and molecules, where the model aerosol is of the oceanic type. In the computational procedure, a probability of radiation interacting with respective half surfaces and the stream is calculated based on the assumption of single scattering in the atmosphere, where isotropic scattering is undertaken. The numerical simulation exhibits the extraordinary effect near the stream. The contrast of the stream depends upon the albedo of the surrounding surfaces. It increases with the increase of the stream width and decreases with the optical thickness.

**063.108   Incorporation of density fluctuations into photoioniza-
tion calculations.**
R. E. Williams.
Astrophys. J., Vol. 392, No. 1, p. 99 – 105 (10 Jun 1992).
  A method is proposed whereby a distribution of density inhomogeneities can be incorporated into photoionization calculations. A continuous spectrum of densities is assumed for the gas by defining a density–dependent volume filling factor $\varepsilon_N \sim N^{-\gamma}$. The ionization in the condensations is affected by shadowing, and can be computed in terms of the density by dividing the condensations into distinct ionization zones of hydrogen and helium, and assuming each to have a constant continuum optical depth throughout the zone. Emission line fluxes are computed by transforming the standard emissivity volume integral into an integral over density, utilizing the filling factor. Applied to novae, it is shown that the density above which condensations become neutral remains constant in time during the early expansion, and that a considerable fraction of the ejecta is shadowed from the central ionizing source.

**063.109   A   spherical   circumstellar   dust   model   for
IRAS 09371 + 1212.**
G. Robinson, R. G. Smith, A. R. Hyland.
Mon. Not. R. Astron. Soc., Vol. 256, No. 3, p. 437 – 448 (1 Jun 1992).
  The authors present the results of fitting a spherically symmetric circumstellar dust model to the peculiar source IRAS 09371 + 1212, a bipolar nebula with an extremely cool dust shell which exhibits clear evidence for large quantities of water ice. The entire spectrum in the wavelength range 2–100 μm can be well represented by a model employing grains with a silicate core and a water ice mantle. However, a small quantity of an additional grain species, such as graphite, is necessary to provide an extra source of opacity in the near–infrared. In particular, the 3.1–μm ice absorption feature and the two ice emission features at 44 and 62 μm can simultaneously be well fitted, showing that it is the same dust which gives rise to both features. It is found that a good fit can be obtained using crystalline ice particles, in agreement with the results of earlier simple models. Furthermore in view of the sharply peaked nature of the observed far–infrared energy distribution, radiative transfer effects cannot compensate for a substantially incorrect opacity law, such as that appropriate to amorphous water ice. Core region limb brightening is predicted at wavelengths where the shell is optically thin. A consequence of this is that the 3.1–μm ice absorption feature should be deepest at the inner boundary of the dust shell. Thus, the location of the inner boundary of the shell could be determined by obtaining spectra of the 3.1–μm feature at an angular resolution of better than 0.3 arcsec. The evolutionary status is discussed and the possibility that IRAS 09371 + 1212 may be a weak or naked T Tauri star rather than an evolved object is examined.

**063.110   Derivation of the master equation for the atomic density
matrix for line polarization studies in the presence of
magnetic field and depolarizing collisions in astrophysics.**
V. Bommier, S. Sahal–Bréchot.
Ann. Phys. (Paris), Vol. 16, No. 5, p. 555 – 598 (Oct 1991).
  The authors derive in a coherent manner the master equation for the density matrix of an atom interacting with a bath of perturbers and photons, in the presence of a weak magnetic field. This paper has been inspired by astrophysical purposes: the interpretation of line polarization induced by anisotropic excitation of the levels, eventually modified by the local magnetic field (the Hanle effect), the polarization can be due to scattering of the incident anisotropic radiation, as in solar prominences, or to impact polarization, as in solar flares. The physical conditions are then those of numerous astrophysical media: any directions of polarization and magnetic field, two–level atom approximation not valid, weak radiation field, weak density of perturbers. The master equation for the atomic density matrix has been derived in the framework of the impact approximation.

**063.111   Derivation of the radiative transfer equation for line
polarization studies in the presence of magnetic field in
astrophysics.**
V. Bommier.
Ann. Phys. (Paris), Vol. 16, No. 5, p. 599 – 622 (Oct 1991).
  In the preceding paper (063.110), the master equation for the atomic density matrix has been derived in the framework of the impact approximation. In the present paper, the author obtains the spontaneous emission 4–vector and absorption and stimulated emission 4×4 matrices entering the transfer equation for polarized radiation, as functions of the absorbed and emitted radiation polarization tensors, themselves functions of the Zeeman coherences of the atomic density matrix. Line profiles have been ignored.

063.112   Line profiles emitted from an accretion torus.
Y. Kojima, J. Fukue.
Mon. Not. R. Astron. Soc., Vol. 256, No. 4, p. 679 – 684 (15 Jun 1992).
  The authors have calculated the emission–line profile from a geometrically thick accretion disc. Assuming the emission from large radii to dominate, all general relativistic effects are neglected in this paper. The line profile emitted from a geometrically thin disc is generally a double–peaked one at large enough radii. The line profile emitted from the torus is different, however. A flat–top profile can be observed if the angle θ_0 between the line of sight and the rotation axis is $\theta_* < \theta_0 < 180° - \theta_*$, while a double–peaked profile can be observed otherwise, where θ_* is a critical angle and larger than the inclination angle of the funnel wall of the torus to the rotation axis. The critical angle decreases with the increase of the thickness of the torus.

**063.113   Inverse problem of photometric observation of solar
radiation reflected by an optically dense planetary
atmosphere. Mathematical methods and weighting functions of
linearized inverse problem.**
E. A. Ustinov.
Cosmic Res., Vol. 29, No. 4, p. 519 – 532 (Jan 1992). English translation of Kosm. Issled., Tom 29, Vyp. 4, p. 604 – 620 (1991).
  A general approach to the solution of the inverse radiative transfer problem based on the investigation of solutions of the corresponding direct and inverses radiative transfer problems is considered. The approach is applied to the case of an optimally dense planetary atmosphere. The adjoint transfer equation and corresponding boundary conditions are derived. Numerical solutions are obtained using spherical harmonics. Weighting functions of the vertical profiles of the coefficients of the expansion of the atmospheric scattering phase function in Legendre polynomials are computed for a simulated Rayleigh atmosphere.

**063.114   Electron impact polarization of atomic spectral lines.
I. A general theoretical scheme.**
S. Fineschi, E. Landi Degl'Innocenti.
Astrophys. J., Vol. 392, No. 1, p. 337 – 352 (10 Jun 1992).
  A suitable theoretical scheme able to describe, in a wide variety of astrophysical situations, the phenomenon of atomic line polarization by electron impact is developed. Assuming the Born approximation, the rate equations for the density matrix elements of a multilevel atomic system, interacting with a nonrelativistic electron beam having any kind of angular distribution, are derived in full generality. The resulting theory generalizes the previous ones by accounting for the collisional rates and the cross sections concerning both inelastic and superelastic collisions, and by taking into account the coherences among Zeeman sublevels split by a magnetic field. The final equations for the collisional rates are made quite compact by the introduction of suitable spherical tensors connected with the components of the electric multipole moments of the atomic transitions. As an example of particular relevance, the general formulae derived here are applied to the case of the electric dipole interactions.

063.115 A modified Eddington–Barbier relation in highly coherent resonance–line wings.
K. G. Gayley.
Astrophys. J., Vol. 392, No. 1, p. 353 – 356 (10 Jun 1992).

The Eddington–Barbier relation in the presence of complete redistribution (CRD) allows the depth dependence of the frequency–independent line source function to be approximately inferred from the observed line profile intensity. For this reason, CRD line profiles are often considered easier to interpret than resonance–line profiles, which typically require partial redistribution (PRD). However, the authors show here that the Eddington–Barbier relation can also be generalized to a useful form when scattering in the wings is highly coherent. This allows the depth dependence of the frequency–averaged line source function to be approximated directly from observed resonance–line profiles, facilitating profile interpretation and guiding atmospheric modeling. The authors show that depth–dependent information about smoothly varying static atmospheres can be extracted from highly coherent resonance–line wings, although this information is compressed more toward line center than in the CRD case and so requires better spectral resolution to be extracted.

063.116 New Rosseland mean opacities for molecular clouds and the solar nebula accretion disk.
D. P. Simonelli, J. B. Pollack, D. J. Hollenbach, W. Fong.
Bull. Am. Astron. Soc., Vol. 23, No. 3, p. 1232 (1991). Abstract. – See Abstr. 012.037 for the main entry.

063.117 On the use of forbidden lines as density diagnostics in stratified media.
P. N. Safier.
Astrophys. J., Vol. 392, No. 2, p. 492 – 496 (20 Jun 1992).

The author investigates the influence of density and ionization gradients on the observed density–sensitive forbidden line ratios from a spatially unresolved source by means of a simple model. It is shown that these ratios can be used to estimate the spatial dependence of density and ionization as well as the source geometry.

063.118 Equivalent width of cyclotron lines in gamma–ray burst spectra.
O. Nishimura, T. Ebisuzaki.
Publ. Astron. Soc. Jpn., Vol. 44, No. 2, p. 109 – 115 (1992).

The X–ray Astronomy satellite, Ginga, detected first– and second–harmonic cyclotron lines around 20 and 40 keV in the energy spectra of gamma–ray bursts. These cyclotron scattering lines can be explained in terms of the transmission model, which consists of a magnetized slab illuminated from below by incident gamma–ray radiation. The authors performed multi–angle radiative transfer calculations which included the natural line width in a plane–parallel slab threaded by a uniform strong magnetic field oriented parallel to the slab normal. For large optical depth $\tau_T > 1.77 \times 10^{-3}$ ($N_e > 2.7 \times 10^{25}$ electrons m^{-2}) and small zenith angle $\mu = \cos\theta < 0.2$, the authors find that the ratio of the equivalent width of second–to–first harmonic line is as large as 3–30; the first line can't be detected. This large ratio is found to be caused by the effect of a Lorentzian wing due to the natural line width that the authors include in the present calculation.

063.119 Results obtained using the OPAL code.
C. A. Iglesias, F. J. Rogers.
Rev. Mex. Astron. Astrofis., Vol. 23, p. 161 – 170 (Mar 1992). – See Abstr. 012.054 for the main entry.

The Rosseland mean opacities computed using the OPAL codes show significant differences when compared to the Los Alamos results. The differences have been traced to improvements in both the equation of state and atomic physics in the OPAL code. Furthermore, recent work suggests that the OPAL opacities condiderably improve the agreement between observations and stellar models.

063.120 "Finding" the "missing" solar ultraviolet opacity.
R. L. Kurucz.
Rev. Mex. Astron. Astrofis., Vol. 23, p. 181 – 186 (Mar 1992). – See Abstr. 012.054 for the main entry.

The author has computed new opacities for model stellar atmospheres and envelopes using a large grant of Cray computer time at the San Diego Supercomputer Center. The opacities include 58,000,000 atomic and diatomic molecular lines. Twelve–step distribution functions are tabulated for 56 temperatures in the range from 2000 K to 200000 K, for 21 log pressures from –2 to 8, for 1212 wavelength intervals from 10 to 10000 nm, for microturbulent velocities 0, 1, 2, 4, and 8 km/s, for scaled solar abundances [+1.0], [+0.5], [+0.3], [+0.2], [+0.1], [+0.0], [–0.1], [–0.2], [–0.3], [–0.5], [–1.0], [–1.5], [–2.0], [–2.5], [–3.0], [–3.5], [–4.0], [–4.5], [–5.0], and [+0.0, no He] (log abundance of elements heavier than helium relative to solar). Rosseland means are also tabulated for each case. The final files for each abundance require two 6250 bpi VAX backup tapes. The author is now distributing tape copies. He hopes to have CD–ROMs available in the near future. A solar photospheric model computed with the new opacities matches the observed energy distribution.

063.121 Remaining line opacity problems for the solar spectrum.
R. L. Kurucz.
Rev. Mex. Astron. Astrofis., Vol. 23, p. 187 – 194 (Mar 1992). – See Abstr. 012.054 for the main entry.

We need high–resolution, high–signal–to–noise spectra of the sun with complete spectral coverage. The solar spectrum provides the insolation that controls the atmospheric chemistry of the earth and of all the solar system bodies. It is important for stellar astrophysics as the "standard" stellar spectrum because it can be observed better than that of any other star. It is important for understanding the sun, for it allows us to study the conditions and motions at its surface. It is an important high–temperature laboratory source for atomic and molecular spectroscopy. To interpret the spectrum we require accurate energy levels, accurate wavelengths, accurate gf values, accurate damping constants. We require hyperfine splitting, isotopic splitting, and Zeeman splitting. We require completeness in order to deconvolve blends. We need every level below the lowest ionization or dissociation energy. For molecules that is every vibrational and rotational level.

063.122 Sampling methods.
U. G. Jørgensen.
Rev. Mex. Astron. Astrofis., Vol. 23, p. 195 – 202 (Mar 1992). – See Abstr. 012.054 for the main entry.

A tremendous increase in computing capacity has allowed us to solve the radiative transfer equation at several thousand frequency points during the iterative solution of the classic model atmosphere problem and to construct a much more detailed and realistic sampling of the opacity than in earlier models. Whereas the harmonic mean and straight mean enable a very fast computation of the model structure, the errors in such simple opacity treatments may be bigger than the errors due to neglect of all improvements in the physics included in computations of model atmospheres of cool stars during the last 25 years. Yet the most simple harmonic mean sampling method – the Rosseland mean – is often still used to compute boundary conditions for evolutionary models of cool stars. Statistical transformation of measured straight mean absorption coefficients into individual lines, followed by careful sampling, has been a considerable improvement over the direct use of the straight mean absorption coefficient in the construction of model atmospheres.

063.123 Molecular opacities in M–star atmospheres.
F. Allard, M. Scholz, R. Wehrse.
Rev. Mex. Astron. Astrofis., Vol. 23, p. 203 – 215 (Mar 1992). – See Abstr. 012.054 for the main entry.

The general properties of M giant and dwarf atmospheres are briefly reviewed, and the problems of modeling are described. The authors show, by means of recently calculated dwarf atmospheres, that molecules play a key role in the equation of state and that their absorption determines in most cases the slope of the spectrum. A comparison of calculated and observed spectra of the M–dwarf system Gliese 866 demonstrates the limitations of present modeling and the need for improved data on molecular equilibria and opacities.

063.124 Molecular opacity and stellar structure.
T. R. Carson, G. Luo, C. M. Sharp.
Rev. Mex. Astron. Astrofis., Vol. 23, p. 217 – 221 (Mar 1992). – See Abstr. 012.054 for the main entry.

Models of low–mass stars on the Main Sequence have been constructed with the help of new molecular opacities. It is found that, in spite of the differences between the new and old molecular opacities, the models are virtually unchanged. This circumstance permits more easily the assessment of the importance of other aspects of the physics, such as equation of state and energy generation, used in the modelling.

063.125 Hydrogen opacities at high densities.
P. Lenzuni, D. Saumon.
Rev. Mex. Astron. Astrofis., Vol. 23, p. 223 – 230 (Mar 1992). – See Abstr. 012.054 for the main entry.

The authors discuss continuum opacities in a high–density ($\varrho \geqslant 10^{-3} \mathrm{g\,cm}^{-3}$) hydrogen gas, with special emphasis on collision–induced absorption by H_2 and photodetachment of H^-.

063.126 Opacity problems in accretion disks around young stellar objects.
N. Calvet.
Rev. Mex. Astron. Astrofis., Vol. 23, p. 235 – 240 (Mar 1992). – See Abstr. 012.054 for the main entry.

Accretion disks around young stellar objects are cool and dense; dust and molecules constitute the most important opacity sources. Calculations of internal and atmospheric structure and of synthetic spectra for these disks have been hindered by the lack of appropriate and reliable opacity data. Specific problems are discussed.

063.127 The effects of lines on the mean opacities in novae, supernovae, and accretion disks.
R. Wehrse, B. Baschek, G. Shaviv.
Rev. Mex. Astron. Astrofis., Vol. 23, p. 247 – 251 (Mar 1992). – See Abstr. 012.054 for the main entry.

For the solution of the hydrodynamic equations in differentially moving media the calculation of the radiation flux, the gradient of the radiation pressure, and the energy balance is required. After a brief discussion of the shortcomings of the approximations which are frequently made, the authors suggest a fast method for the accurate calculation of these quantities by means of three mean opacity coefficients which replace the Rosseland mean of the static case. This method allows a consistent inclusion of the important contribution of spectral lines to the radiation field.

063.128 Radiative opacities and stellar pulsations.
N. R. Simon, S. M. Kanbur.
Rev. Mex. Astron. Astrofis., Vol. 23, p. 253 – 258 (Mar 1992). – See Abstr. 012.054 for the main entry.

The authors review some current problems in stellar pulsations in which the radiative opacities play a crucial role. Particular attention is paid to the already–published OPAL opacities, which seem to go a long way toward resolving long–standing difficulties in pulsational models. A brief, first look is also given to a preliminary version of opacties from the Opacity Project.

063.129 Polarization shifts in nonsymmetric theories of gravity.
J. H. Palmer, R. B. Mann, M. D. Gabriel,
M. P. Haugan.
4. Canadian Conference on General Relativity and Relativistic Astrophysics, p. 40 – 44, 350 – 354 (1992). – See Abstr. 012.056 for the main entry.

Theories of gravity that include a totally antisymmetric component of the metric coupled to electromagnetism predict that light will propagate through a gravitational field with a polarization dependent velocity. The observed polarization depends upon the initial polarization and the path traveled by the light. For extended sources of polarized light within magnetically active regions on the Sun, this leads to an observed depolarization. Observations of the polarization of solar spectral lines near the limb of the Sun are used to place limits on the nonsymmetric charge of the Sun in Moffat's Nonsymmetric Gravitation Theory (NGT).

063.130 On the radiation forces.
F. Mignard.
15. Ecole de Printemps d'Astrophysique de Goutelas: Interrelations between physics and dynamics for minor bodies in the solar system, p. 419 – 451 (1992). – See Abstr. 012.055 for the main entry.

The author discusses few points related to the radiation forces exerted by the Sun on small particles together with their dynamical consequences. The four sections comprise first an historical introduction, indicating the highlights of the subject from the time of Kepler to the beginning of the twentieth century. The second section deals with the physics of radiation forces for particles at rest with respect to the radiation source. The author outlines there the methods used to determine the magnitude of the radiation pressure. The case of a moving particle is taken up with some detail in sec. 3, with the introduction of the tangential component of the radiation force, known as the Poynting–Robertson drag. The analysis is conducted both in the laboratory and particle frames and includes a new derivation. The orbital evolution of small particles in solar orbit is sketched out in the last section.

063.131 Solution of the equation of transfer for coherent scattering in an exponential atmosphere by the method of discrete ordinates.
S. Karanjai, T. K. Deb.
Astrophys. Space Sci., Vol. 192, No. 2, p. 209 – 217 (Jun 1992).

A solution of the transfer equation for coherent scattering in stellar atmosphere with Planck's function as a nonlinear function of optical depth, viz., $B_v(T) = b_0 + b_1 e^{-\beta\tau}$ is obtained by the method of discrete ordinates originally due to Chandrasekhar.

063.132 Inverse problem of photometry of solar radiation reflected by an optically thick planetary atmosphere. II. Numerical aspects and requirements on the observation geometry.
E. A. Ustinov.
Cosmic Res., Vol. 29, No. 6, p. 785 – 800 (May 1992). English translation of Kosm. Issled., Tom 29, Vyp. 6, p. 917 – 932 (1991).

The nonlinear inverse problem of radiation transfer is studied. This problem is formulated for the reconstruction of the angular and altitude variation of the atmospheric–scattering phase function over the planetary disk from photometric data on the solar radiation reflected by the planetary atmsophre. The interative solution includes, in turn, a solution of the corresponding direct and conjugate problems of radiation transfer. A numerical algorithm is formulated for solving the conjugate problem by the method of spherical harmonics. The angular oscillations introduced in the computed weighting functions of the linearized inverse problem by the use of this method are eliminated by integration of the complete source function of the transfer equation. The conditions imposed on the geometry of observation of the planetary disk are studied from the standpoint of the information content of these observations.

063.133 Multiple Compton scattering of electrons in a photon field.
I. P. Ivanenko, V. V. Sizov.
Astrophys. Space Sci., Vol. 192, No. 2, p. 219 – 245 (Jun 1992).

Consideration is given to the motion of electrons in a photon field of the monoenergetic or power–law spectrum under the conditions when the main mechanism of energy loss is the inverse Compton scattering by field photons. This process changes the primary spectrum of electrons and converts low–energy field photons to high–energy gamma–quanta for which the electron confinement region is assumed to be optically thin. The electron and gamma–ray spectra have been obtained in a wide energy interval including the Klein–Nishina and Thomson regions. A simple qualitative dependence of the solutions found on the field parameters and the primary spectrum of electrons has been established. The electron and gamma–ray spectra have been obtained by numerically solving the kinetic equation dependent on two variables: the energy of electrons and their path (or the time of motion) in a photon field. The results dramatically differ from the solution of the steady–state kinetic equation which depends only on the electron energy and is frequently used in the given problem.

063.134   The theoretical gamma–ray spectrum of quasars.
M. E. Dixson, W. A. Stein.
AIP Conf. Proc., No. 254, p. 321 – 324 (1 May 1992). – See Abstr. 012.057 for the main entry.
Preliminary results of calculations concerning the theoretical γ–ray spectrum of quasi–stellar objects is presented with special application to 3C 273. The calculated spectra extent (in γ–ray energy) from approximately 0.1 to 1000 MeV. In this simple model numerous γ–ray producing mechanisms resulting from inelastic pp interactions are included to contribute to the total output. These include the decay of neutral pions, electron and positron bremsstrahlung, positron annihilation and the Compton scattering of synchrotron photons. Although very preliminary, the resulting model spectrum seems able to account for the observed spectrum. The emission may arise in various regions in which relativistic protons encounter sufficient quantities of target nuclei in the ambient gas.

063.135   Comptonization modifications to the gamma–ray spectrum of active galactic nuclei.
M. G. Baring.
AIP Conf. Proc., No. 254, p. 325 – 328 (1 May 1992). – See Abstr. 012.057 for the main entry.
Pair cascade models predict the appearance of 512 keV pair annihilation lines above the gamma–ray continuum. The moderate optical depths associated with pair cascades suggests Comptonization of the line will contribute to the hard X–ray continuum. Here numerical calculations of the Comptonization of the line and analytic approximations to this are presented for mildly relativistic plasmas. Earlier calculations in the Thomson limit are inappropriate and here the time–independent solution of the photon kinetic equation yields a ghost line at about 170 keV due to Compton backscattering. This feature has been predicted for the Galactic Centre. A cold plasma assumption is employed in order to develop the analytic technique for approximating the spectrum.

063.136   Radiation from magnetized accretion disks in AGN.
G. B. Field, R. D. Rogers.
AIP Conf. Proc., No. 254, p. 329 – 332 (1 May 1992). – See Abstr. 012.057 for the main entry.
A model of AGNs based on accretion disks around $10^8 M_\odot$ black holes is presented. This model, developed to explain the X–ray background, is based upon a strong ($B \sim 10^4 G$) magnetic field in the disk, which forms loops above the disk in which $B \sim 200$ G, where shock waves initiated by magnetic reconnection accelerate electrons to relativistic energies. The electrons emit synchrotron and inverse Compton radiation. The spectrum calculated for the range 10^{-6} keV to 10^4 keV is presented.

063.137   High energy emission in accretion flows in AGN.
M. Kafatos.
AIP Conf. Proc., No. 254, p. 333 – 336 (1 May 1992). – See Abstr. 012.057 for the main entry.
Hot accretion disks developing in the inner accretion flows onto supermassive black holes produce copious amounts of relativistic electron–positron pairs, X–rays and γ–rays. The relativistic pairs upscatter the X–rays emanating from the disk and produce γ–rays up to ~ 300 MeV. These coupled with pion produced γ–rays are blueshifted in the disk and emerge with

energies of hundreds of MeV. Penrose pair–produced pairs emerge with energies up to 4–5 GeV. Hot accretion disks are discussed in the context of high energy extragalactic sources like 3C 279.

063.138   High energy spectra from the jet/disk interface.
J. F. Kartje, A. Königl, P. S. Coppi.
AIP Conf. Proc., No. 254, p. 337 – 340 (1 May 1992). – See Abstr. 012.057 for the main entry.
The authors investigated the dynamical interaction between ultrarelativistic particle beams and the local radiation fields of accretion disks in the central engines of AGNs. For electron–proton or electron–positron jets accelerated to initial Lorentz factors $\gamma_i \lesssim 10^4$, the particles mainly interact with the photons via Thomson scattering, which provides an effective means for both decelerating the relativistic beams and for producing the hard X–ray spectra seen in many AGNs. For larger γ_i, scattering by the beam leads to the development of an $e^- e^+$ cascade, resulting from photon–photon, photon–electron, and photon–photon pair production. The authors study the structure of these pair cascades and present preliminary calculations of the soft X–ray through γ–ray spectrum produced by the Comptonized disk photons.

063.139   The diagnostic potentialities of circular polarization for studying the surfaces of atmosphereless cosmic bodies.
V. S. Degtyarev, L. O. Kolokolova, A. V. Morozhenko, M. F. Tsurul'.
Sov. Astron. Lett., Vol. 18, No. 2, p. 113 – 115 (Mar 1992). Current Physics Microform No.: 9211X0752. English translation of Pis'ma Astron. Zh., Tom 18, No. 3, p. 279 – 283 (1992).
The phase dependences of the circular polarization of light scattered by surfaces formed by a layer of small particles are found by means of a high–precision laboratory Stokes polarimeter. Transparent dielectric particles, the same particles coated with a metal (nickel) layer, and purely metallic particles were studied. The particles were of similar sizes ($\approx 50\ \mu$m) but of different shapes. The measurements showed that, independent of particle shape, the circular polarization of light scattered by surfaces formed by particles with small absorption coefficients (dielectrics) is essentially zero, but is far from zero for strongly absorbing particles (metallic and metallized), and is especially large (up to 3%) at large phase angles. Consequently, the measurement of the circular polarization of light scattered by surfaces of atmosphereless cosmic bodies can be used to detect metals in their surface layer material which, in particular, can be used to more precisely determine the nature of the M–type asteroids.

063.140   Radiative transfer in axi–symmetric dust clouds around AGN.
E. A. Pier, J. H. Krolik.
AIP Conf. Proc., No. 254, p. 508 – 511 (1 May 1992). – See Abstr. 012.057 for the main entry.
The authors have developed a new radiative transfer code to study the thermally reradiated emission from an optically and geometrically thick torus of dust which partially obscures the central regions of an active galactic nucleus. The authors discuss the spectral and angular distributions of the radiated energy over a range of model parameters and viewing angles.

063.141   Scattering calculations on the basis of the Fredholm integral equation method.
M. Matsumura, M. Seki.
IAU Colloquium No. 126: Origin and evolution of interplanetary dust, p. 203 – 206 (1991). – See Abstr. 012.068 for the main entry.
The Fredholm integral equation method (FIM) is one of the solutions to the scattering of electromagnetic radiation by homogeneous and isotropic ellipsoidal particles. Some numerical calculations are performed with the FIM. The results for spherical particles are compared with those by the Mie theory. It is confirmed that the agreement between them is satisfactory for all the models calculated. On the basis of this method, the authors examine profiles of the absorption band around $\lambda = 10\ \mu$m for

spherical and ellipsoidal particles composed of crystalline olivine. It is found that the profile strongly depends on the shape of the particle. Even when the particle is moderately elongated (axial ratios are $2:\sqrt{2}:1$), the profile is significantly different from that of a sphere.

063.142 The scattering matrix of randomly oriented infinite cylinders.
P. Stammes.
IAU Colloquium No. 126: Origin and evolution of interplanetary dust, p. 207 – 210 (1991). – See Abstr. 012.068 for the main entry.

A method is outlined to compute the scattering matrix of an ensemble of infinite cylinders in random orientation, as an approximation to the scattering matrix of an ensemble of very long but finite cylinders. Numerical checks are presented, which show that the results for infinite cylinders agree with theoretical results for very thick and very thin cylinders, as well as with numerical data for prolate spheroids and short finite cylinders.

063.143 Asymptotic values of some scattering parameters from Mie theory.
G. A. Shah.
Kodaikanal Obs. Bull., Vol. 11, p. 9 – 17 (1991). – See Abstr. 012.072 for the main entry.

The scattering parameters such as extinction efficiency, albedo, asymmetry parameter, etc. have been studied on the basis of the Mie theory of scattering of electromagnetic radiation by a sphere as well as the classical geometrical optics and diffraction (GOD). The spheres are assumed to be composed of water– or ice–like dielectric and absorbing materials with index of refraction, $m = m^1 - im^{11}$; $m^1 = 1.33$ or 1.333 and $m^{11} = 0$, 0.0033, 0.033, 0.1, 0.33 or 3.3. The minimum value of the size–to–wavelength parameter (the circumference of the sphere divided by the wavelength) has been estimated for each scattering parameter so that the asymptotic value obtained from Mie theory calculations agree, at moderate accuracy, with the corresponding result based on GOD.

063.144 Absorption–line formation in an optically thick weakly absorbing planetary atmosphere. I. Homogeneous slab.
N. N. Fomin, Eh. G. Yanovitskij.
Kinematics Phys. Celest. Bodies, Vol. 7, No. 5, p. 24 – 33 (1991). English translation of Kinematika Fiz. Nebesn. Tel., Tom 7, No. 5, p. 29 – 38 (Sep–Oct 1991).

A rigorous asymptotic theory of absorption–line formation in a homogeneous weakly absorbing cloudy atmosphere of high optical thickness is developed. Basic formulas that determine the profile and equivalent widths of the line, for both diffusely reflected and diffusely transmitted radiation, are derived. In the latter case, the profile and equivalent widths of the line do not depend on the zenith distance of the Sun or the direction of arrival of the radiation at the observer. The corresponding growth curves are calculated. The problem of formation of a "superweak" absorption line when multiple light scattering does not deform the original profile of the absorption coefficient of the line or molecular band in any way is examined in detail.

063.145 Absorption–line formation in an optically thick weakly absorbing planetary atmosphere. II. Vertically inhomogeneous media.
N. N. Fomin, Eh. G. Yanovitskij.
Kinematics Phys. Celest. Bodies, Vol. 7, No. 5, p. 34 – 42 (1991). English translation of Kinematika Fiz. Nebesn. Tel., Tom 7, No. 5, p. 39 – 47 (Sep–Oct 1991).

Absorption–line formation in an optically thick inhomogeneous atmosphere is studied. It is assumed that the atmosphere is weakly absorbing, both at the line frequencies and in the continuous spectrum. Rigorous analytic solutions of the problem are given for the following cases: 1) an isothermal slab with barometric vertical pressure variation; 2) a polytropic model atmosphere (for the central line frequency). A procedure for calculating the profile of an absorption line is given for the general case of arbitrary dependence of the optical properties of the atmosphere on depth. This method is based on reduction of the corresponding boundary–value problem to a Cauchy problem.

063.146 Scattering cross section for randomly oriented particles of arbitrary shape.
M. I. Mishchenko.
Kinematics Phys. Celest. Bodies, Vol. 7, No. 5, p. 83 – 86 (1991). English translation of Kinematika Fiz. Nebesn. Tel., Tom 7, No. 5, p. 93 – 95 (Sep–Oct 1991).

The scattering of light by randomly oriented particles of arbitrary shape is discussed. Waterman's T–matrix approach is used to derive a simple analytic expression for the total scattering cross section.

063.147 Inverse problems of radiative transfer in sounding of planetary atmospheres.
E. A. Ustinov.
Diss. (Dr.phil.), Tartu Univ. (Estonia). Estonian Academy of Sciences, Tartu (Estonia)34 p. (1992).

Contents: 1. Introduction. 2. Statistical approach to regularization and solution of inverse problems. 3. In situ photometry of the optically thick planetary atmosphere. 4. Remote sounding of the purely absorbing atmosphere in the thermal IR region. 5. Remote sounding of the scattering planetary atmosphere. 6. Conclusion.

063.148 Instabilities and evolution of spectra – more discussion on the relation of plasma emission and other radiation mechanisms.
Huang Guang–li, Wang Ren–yin.
Publ. Purple Mt. Obs., Vol. 11, No. 1, p. 1 – 7 (Mar 1992). In Chinese.

A point of view is proposed that the growth of the electromagnetic waves excited by plasma instabilities has an important effect on the evolution of their spectra, which depends on wave–particle interaction in plasma radiation processes. Thus, the mechanism of plasma emission is based on the classical electrodynamics and has been developed as an unstable radiation theory, and with the equilibrium of different basic radiation processes, such as bremsstrahlung, cyclotron and synchrotron radiation, and so on, as the initial condition.

063.149 Quantum theory of Stokes parameters for Thomson scattering in a magnetised plasma.
Chou Chih–Kang, Chen Hui–Hwa.
Proc. Astron. Soc. Aust., Vol. 9, No. 2, p. 325 (1991). Extended abstract. – See Abstr. 012.090 for the main entry.

063.150 Introduction: Synchrotron and inverse–Compton radiation.
P. A. Hughes, L. Miller.
Beams and jets in astrophysics, p. 1 – 51 (1991). – See Abstr. 003.099 for the main entry.

Astronomical masers.
See Abstr. 003.004.

High energy astrophysics. Volume 1. Particles, photons and their detection.
See Abstr. 003.018.

Diffraction effects in semiclassical scattering.
See Abstr. 003.058.

The physics of astrophysics. Volume II: Gas dynamics.
See Abstr. 003.082.

Beams and jets in astrophysics.
See Abstr. 003.099.

The magnetospheric structure and emission mechanisms of radio pulsars. Proceedings.
See Abstr. 012.022.

Astrophysical opacities. Proceedings. Workshop on Astrophysical Opacities (WAO), Caracas (Venezuela), 15 – 19 Jul 1991.
See Abstr. 012.054.

Testing the AGN paradigm. Proceedings.
See Abstr. 012.057.

The Opacity Project – equation of state.
See Abstr. 013.084.

The Opacity Project – results for opacities.
See Abstr. 013.085.

The Opacity Project – a post–script.
See Abstr. 013.086.

Smoothed particle rendering for fluid visualization – three–dimensional accretion disk and jet formation.
See Abstr. 021.027.

Level populations for Fe III applicable to astrophysical plasmas and a comparison with planetary nebula observations.
See Abstr. 022.006.

Simple algorithms for remote determination of mineral abundances and particle sizes from reflectance spectra.
See Abstr. 022.033.

Quantitative subpixel spectral detection of targets in multispectral images.
See Abstr. 022.034.

Rate coefficients for the excitation of infrared and ultraviolet lines in C II, N III, and O IV.
See Abstr. 022.093.

Collision strengths and excitation rate coefficients for transitions in Ca XV.
See Abstr. 022.094.

Excitation rate coefficients for transitions among the n = 1, 2, and 3 levels of He$^+$.
See Abstr. 022.095.

L–shell X–ray opacity of many–electron atoms.
See Abstr. 022.160.

Relativistic free–free Gaunt factors for high–temperature stellar plasmas.
See Abstr. 022.193.

Interpretation of lightcurves of atmosphereless bodies. I. General theory and new inversion schemes.
See Abstr. 036.117.

Self–organization of cosmic radiation pressure instability. II. One–dimensional simulations.
See Abstr. 062.005.

Populations of highly excited atoms and ions in the optically thin plasma.
See Abstr. 062.041.

Plasma diagnostics based on self–reversed lines. I. Model calculation and application to argon arc measurements in the near–infrared and vacuum ultraviolet regions.
See Abstr. 062.050.

Plasma diagnostics based on self–reversed lines. II. Application to nitrogen, carbon and oxygen arc measurements in the vacuum ultraviolet.
See Abstr. 062.051.

Relativistic magnetosonic shock waves in synchrotron sources: shock structure and nonthermal acceleration of positrons.
See Abstr. 062.097.

Relativistic, perpendicular shocks in electron–positron plasmas.
See Abstr. 062.109.

Particle acceleration at relativistic shocks.
See Abstr. 062.135.

Time dependent numerical simulations of diffusive particle acceleration in astrophysical environments.
See Abstr. 062.136.

Particle acceleration and flux outbursts.
See Abstr. 062.139.

From ultrarelativistic beams to radio jets: the accretion disk connection.
See Abstr. 062.140.

ZEUS–2D: a radiation magnetohydrodynamics code for astrophysical flows in two space dimensions. III. The radiation hydrodynamic algorithms and tests.
See Abstr. 062.156.

Effects of dust scattering on interstellar jet line profiles.
See Abstr. 062.175.

Equation of state of partially–ionized plasmas in the physical picture.
See Abstr. 062.178.

On the radiation field in astrophysical jets.
See Abstr. 062.192.

Anisotropies in relativistic jets: beaming patterns for synchrotron and inverse–Compton emission.
See Abstr. 062.193.

Relativistic quantum response of a strongly magnetised plasma. I. Mildly relativistic electron gas.
See Abstr. 062.210.

Relativistic quantum response of a strongly magnetised plasma. II. Ultrarelativistic pair plasma.
See Abstr. 062.211.

Spectroscopic diagnostics.
See Abstr. 062.217.

The calcium infrared triplet lines in stellar spectra.
See Abstr. 064.006.

Cooling functions for hot dense gas: application to accretion discs in close binaries.
See Abstr. 064.008.

The effects of radiation drag on radial, relativistic hydromagnetic winds.
See Abstr. 064.009.

Dynamics of the envelopes of Be stars in the equatorial plane.
See Abstr. 064.011.

Fe II emission lines. II. Excitation mechanisms in cool stars.
See Abstr. 064.012.

Modelling the spectrum of WC–type Wolf–Rayet stars.
See Abstr. 064.014.

Radiative shocks in atomic and molecular stellar–like atmospheres. V. Influence of the excited level of the hydrogen atom: the precursor structure.
See Abstr. 064.015.

Winds from T Tauri stars. II. Balmer line profiles for inner disk winds.
See Abstr. 064.019.

Balmer line profiles for infalling T Tauri envelopes.
See Abstr. 064.020.

π^0–decay gamma–ray emission from winds of massive stars.
See Abstr. 064.022.

X–ray–heated models of stellar flare atmospheres: theory and comparison with observations.
See Abstr. 064.024.

On the influence of the convective efficiency on the determination of the atmospheric parameters of DA white dwarfs.
See Abstr. 064.025.

Spherical opacity sampling model atmospheres for M–giants. I. Techniques, data and discussion.
See Abstr. 064.038.

Ionization and excitation in cool giant stars. I. Hydrogen and helium.
See Abstr. 064.043.

Reflection with limb darkening.
See Abstr. 064.047.

Radiation–hydrodynamic waves in an optically gray atmosphere. I. Homogeneous model.
See Abstr. 064.048.

Stellar wind collision and X–ray generation in massive binaries.
See Abstr. 064.053.

Models of planetary nebula spectral evolution.
See Abstr. 064.055.

Dependence of Stokes profiles on the temperature of photospheric models.
See Abstr. 064.057.

Temperature dependence of Stokes profiles for model photospheres.
See Abstr. 064.058.

Shock waves and Hα profiles in the hydrodynamical model for RR Lyrae.
See Abstr. 064.066.

The outer atmospheres of the "hybrid" bright giants: the chromospheres of α TrA (K4 II), ι Aur (K3 II), γ Aql (K3 II) and θ Her (K1 II).
See Abstr. 064.067.

Three–dimensional structured shocks in AM Herculis–type systems. II. Cyclotron emission from ridge–shaped emission regions.
See Abstr. 064.068.

Irradiation of accretion disks around young objects. II. Continuum energy distribution.
See Abstr. 064.072.

Models for stellar wind of early–type stars.
See Abstr. 064.076.

Speculation of the origin of X–ray emission in early–type stars.
See Abstr. 064.122.

Type Ia supernovae: theoretical light curves with a slow pre–maximum rise.
See Abstr. 065.001.

Strongly nonadiabatic pulsations and strange modes.
See Abstr. 065.095.

Accretion flows near black holes mediated by radiative viscosity.
See Abstr. 067.005.

Compton scattering in a converging fluid flow: spherical near–critical accretion onto neutron stars.
See Abstr. 067.007.

The fate of accreted CNO elements in neutron star atmospheres: X–ray bursts and gamma–ray lines.
See Abstr. 067.008.

A possible explanation for intraday variability in active galactic nuclei. Magnetic reconnection and coherent plasma emission.
See Abstr. 067.011.

Magnetic acceleration of broad emission–line clouds in active galactic nuclei.
See Abstr. 067.027.

Radiation dynamics in X–ray binaries. I. Type 1 bursts.
See Abstr. 067.029.

Radiation dynamics in X–ray binaries. II. Type 2 bursts.
See Abstr. 067.030.

Radiation dynamics in X–ray binaries. III. Extremely compact objects.
See Abstr. 067.031.

Standing shocks in accretion disks and the spectra of active galactic nuclei.
See Abstr. 067.038.

Dynamical effects of radiation pressure due to synchrotron absorption in turbulent spherical accretion.
See Abstr. 067.069.

Energy dependence of normal branch quasi–periodic intensity oscillations in low–mass X–ray binaries.
See Abstr. 067.071.

Near–critical spherical accretion onto magnetized neutron stars: modified magnetospheric radius.
See Abstr. 067.095.

The effect of light bending and redshift on pulsar beaming the case of shorter rotation periods.
See Abstr. 067.118.

Nonstationary processes in the pulsar magnetosphere.
See Abstr. 067.119.

Steady, periodic gamma–ray emission from accreting X–ray pulsars.
See Abstr. 067.133.

Quasi–elastic neutrino scattering by nuclei in superdense matter of a collapsing star.
See Abstr. 067.145.

Radiation hydrodynamics in accretion columns of neutron stars.
See Abstr. 067.154.

Modeling of Lyman edge features in accretion disk spectra.
See Abstr. 067.178.

Neutron star crusts.
See Abstr. 067.194.

Nonthermal electron–positron pairs and cold matter in the central engines of active galactic nuclei.
See Abstr. 067.196.

The high energy AGN continuum: models without e^+e^-–pairs.
See Abstr. 067.197.

Structure and spectra of electron–positron pair cascade atmospheres.
See Abstr. 067.198.

Pair cascades triggered by relativistic protons in AGNs.
See Abstr. 067.199.

Relativistic electron production by inelastic and elastic collisions in active galactic nuclei.
See Abstr. 067.200.

Diurnal variation of the linear polarization across the Hβ Fraunhofer line of the terrestrial atmosphere. I.
See Abstr. 071.016.

Diurnal variation of the linear polarization across the Hα and Hβ Fraunhofer lines of the terrestrial atmosphere, and a detection of a "daylight flash". II.
See Abstr. 071.017.

The formation mechanism of the lines of the C I infrared multiplet at λ1069.5 nm in the spectrum of the Sun.
See Abstr. 071.020.

The depth of formation of absorption lines in the solar atmosphere.
See Abstr. 071.022.

What caused an unusually broad He I 10830 Å emission line in a solar limb flare?
See Abstr. 073.025.

High–energy gamma–ray emission from pion decay in a solar flare magnetic loop.
See Abstr. 073.044.

Spatial profiles of lines in steady–state coronal loops.
See Abstr. 074.024.

Nontraditional emission mechanism of type III bursts at twice the plasma frequency.
See Abstr. 077.011.

A thermal/nonthermal model for solar microwave bursts.
See Abstr. 077.028.

Interior opacities and the standard solar model.
See Abstr. 080.026.

Waves in solar magnetic flux tubes: the observational signature of undamped longitudinal tube waves.
See Abstr. 080.058.

Departure from local thermodynamic equilibrium and its effect on solar continuous absorption.
See Abstr. 080.059.

Line–by–line computation of the atmospheric absorption spectrum using the decomposed Voigt line shape.
See Abstr. 082.034.

Light scattering by nonspherical ice grains: an application to noctilucent cloud particles.
See Abstr. 082.078.

Auroral resonance line radiative transfer.
See Abstr. 084.010.

The shadow effect for a planetary surface with Gaussian mesorelief.
See Abstr. 091.010.

Estimating the behaviour of the radiation field inside a planetary atmosphere.
See Abstr. 091.048.

Opacities of the 1973 dust storm over the Solis Lacus, Hellas, and Syrtis Major areas of Mars.
See Abstr. 097.004.

An analysis of the influence of interstellar scattering on interplanetary scintillation observations.
See Abstr. 106.001.

Scattering of solar UV on local neutral gases.
See Abstr. 106.094.

Numerical two–dimensional calculations of the formation of the solar nebula.
See Abstr. 107.050.

Measure of the grain velocity structure in the circumstellar envelope "Frosty Leo".
See Abstr. 112.006.

A modelling of circumstellar Ba II lines for the hypergiant ϱ Cassiopeiae.
See Abstr. 112.011.

Infrared observations and thermal models of the β Pictoris disk.
See Abstr. 112.019.

Mass loss from OH/IR stars: models for the infrared emission of circumstellar dusts shells.
See Abstr. 112.059.

The origin of the far–ultraviolet continuum in solar and stellar flares.
See Abstr. 116.007.

Spectropolarimetry of the nova–like variable RW Trianguli.
See Abstr. 117.001.

A two–wind radiation–driven model for the atmospheric features of β Lyrae.
See Abstr. 117.011.

Fe II fluorescence and anomalous C IV doublet intensities in symbiotic novae.
See Abstr. 117.154.

X–ray scattering and fluorescence in the wind of a massive X–ray binary.
See Abstr. 117.155.

Spectral evolution of accretion disks of dwarf novae. III. Outburst cycles of SS Cygni.
See Abstr. 117.162.

Spiral shocks and subday variability in SS 433.
See Abstr. 117.180.

Disk instability and the time–dependent X–ray emission from the intermediate polar GK Persei.
See Abstr. 117.192.

The accretion halo in AM Herculis systems.
See Abstr. 117.221.

Model of supernova 1979c with radiative transfer in the envelope.
See Abstr. 125.084.

A model of Supernova 1979c including radiative transfer in the shell.
See Abstr. 125.085.

Hydrogen recombination at high optical depth and the spectrum of SN 1987A.
See Abstr. 125.106.

The [O I] $\lambda\lambda6300$, 6364 doublet of SN 1987A.
See Abstr. 125.115.

Line fluorescence from the ring around supernova 1987A.
See Abstr. 125.118.

New evidence on the shape of pulsar beams.
See Abstr. 126.049.

Radius–to–frequency mapping in the radio pulsar emission mechanism.
See Abstr. 126.050.

On the shape of the pulsar emission diagram.
See Abstr. 126.051.

Microstructure and the pulsar radio emission mechanism.
See Abstr. 126.085.

Photodissociation of H_2 and the H/H_2 transition in interstellar clouds.
See Abstr. 131.005.

On graphite and the variations in the ultraviolet extinction curve.
See Abstr. 131.043.

Grain cooling in collapsing clouds.
See Abstr. 131.058.

UV irradiated clumps in the Orion A molecular cloud: intepretation of low–J CO observations.
See Abstr. 131.126.

Astronomical masers.
See Abstr. 131.275.

The development of H I dissociation zones around new H II regions.
See Abstr. 132.004.

Millimeter recombination lines in the planetary nebula BD $+30°3639$.
See Abstr. 134.015.

N III line emission in planetary nebulae: continuum fluorescence.
See Abstr. 134.023.

Spectrophotometry of Bowen resonance fluorescence lines in three planetary nebulae.
See Abstr. 134.028.

The non–thermal continuum of compact sources.
See Abstr. 158.017.

Carbon monoxide in active galaxies.
See Abstr. 158.023.

Gas clouds from tidally disrupted stars in active galactic nuclei.
See Abstr. 158.027.

Anisotropic line emission and the geometry of the broad–line region in active galactic nuclei.
See Abstr. 158.051.

High–energy gamma radiation from extragalactic radio sources.
See Abstr. 158.076.

Bowshocks and the formation of the narrow–line region of Seyfert galaxies.
See Abstr. 158.084.

The response of the broad emission line region to ionizing continuum variations. II. Numerical simulations.
See Abstr. 158.088.

The starburst model for active galactic nuclei: the broad–line region as supernova remnants evolving in a high–density medium.
See Abstr. 158.091.

The response of the broad emission line region to ionizing continuum variations. III. An atlas of transfer functions.
See Abstr. 158.154.

Hard X–ray observations of AGN – do accretion disks exist?
See Abstr. 158.194.

Constraints on pairs and reflection in active galaxies.
See Abstr. 158.205.

Models for the gamma–ray emission of 3C 279.
See Abstr. 158.212.

Are the extended narrow line regions in AGN photoionized by the emission from thick accretion disks?
See Abstr. 158.221.

The nature of the broad line region: optical/UV/X–ray studies.
See Abstr. 158.224.

The contribution of stars to the AGN line emission.
See Abstr. 158.225.

Star–disk collisions in active galactic nuclei and the origin of the broad line region.
See Abstr. 158.226.

Observational constraints on stellar wind models of broad emission lines of AGNs.
See Abstr. 158.227.

BLR chemistry redux.
See Abstr. 158.231.

Gas velocities in the inner NLR: virial with inflow?
See Abstr. 158.233.

Far–infrared and radio signatures of starbursts and monsters.
See Abstr. 158.237.

The optical jet of 3C 273: an electron–positron plasma beam viewed by the light of its quasar?
See Abstr. 159.042.

Collective emission from rapidly variable quasars.
See Abstr. 159.063.

The role of Compton and Raman scattering in the quasar continuum.
See Abstr. 159.083.

Reflections: the optical jet of 3C 273.
See Abstr. 159.113.

Radiative shocks and hydrogen molecules in pregalactic gas: the effects of postshock radiation.
See Abstr. 161.102.

Numerical simulation for cosmological fluid flows.
See Abstr. 161.213.

064 Stellar Atmospheres, Stellar Envelopes, Mass Loss, Accretion

064.001 The effects of the velocity stratification of a Mira photosphere on line strengths and curves of growth.
M. Scholz.
Astron. Astrophys., Vol. 253, No. 1, p. 203 – 209 (Jan 1992).

The velocity stratification of a Mira photosphere resulting from matter outflow below and matter infall above the outgoing shock front may or may not strongly increase the equivalent width of lines in the transition region from the linear to the flat part and on the flat part of the curve of growth. Misinterpretation of this effect as microturbulent broadening leads to incorrect results. The real presence of microturbulence reduces the increase caused by outflow/infall velocities. The strengths of weak lines and blue–shifted lines with symmetric profiles formed below the shock front hardly depend on outflow/infall velocities and may be submitted to a classical coarse analysis. A fine analysis of such lines through a spherically extended or even plane hydrostatic model atmosphere is often justified but may fail in some cases.

064.002 Chromospheric phenomena in late–type stars.
P. Ulmschneider.
11. European Regional Astronomy Meeting of the IAU: New windows to the universe, Vol. 1, p. 45 – 64 (1990). – See Abstr. 012.001 for the main entry.

In this short overview first the characteristic NLTE thermodynamics of chromospheres is discussed. Then the observational evidence for magnetic fields and their relation to the chromospheric emission activity of the sun and of stars is summarized. A third unique feature of chromospheres and coronae is that they need persistent mechanical heating. Proposed heating mechanisms and in particular recent advances in acoustic heating are outlined. Finally the role of chromospheres and coronae in the generation of eruptive flows is discussed.

064.003 The solar wind and the winds from cool stars.
R. Hammer.
11. European Regional Astronomy Meeting of the IAU: New windows to the universe, Vol. 1, p. 77 – 97 (1990). Mitt. Kiepenheuer Inst., Nr. 319. – See Abstr. 012.001 for the main entry.

This review begins with a brief summary of observed characteristics of the winds from cool stars. The mechanisms that drive these winds are not yet known; but some of their basic properties can be inferred already from time–independent stellar wind theory. These properties include the size and spatial distribution of the energy input, and the question if the driving occurs via body forces or via heating the outer stellar atmosphere to high temperatures. It is shown that compressive waves are an unlikely mechanism to produce hot coronae, even for old, slowly rotating stars with inefficient dynamos. Recent time–dependent models of the winds from stars with and without coronae are reviewed.

064.004 The formation of coronal regions in accretion disks.
S. D. Murray, D. N. C. Lin.
Astrophys. J., Vol. 384, No. 1, p. 177 – 184 (1 Jan 1992).

While there is extensive evidence for the existence of hot, low-density gas or winds above accretion disks, the mechanism for heating the gas remains undetermined. The authors propose a mechanism similar to that used to model chromospheric heating in the Sun and other stars. Sound waves propagating through an accretion disk are refracted away from the central plane by the strong density gradient. As they move into regions of lower density, the sound waves accelerate to form shocks, which heat the gas, leading to the formation of a hot, low–density region. Results of hydrodynamical simulations show that waves with frequencies similar to the local Keplerian frequency, Ω, lead to the most efficient heating. For the optically thin region modeled here, the result is the formation of coronal regions with densities $n \lesssim 10^{10} \mathrm{cm}^{-3}$, and temperatures ranging from 10^4K to 10^6K over a few scale heights. Waves with frequencies several times Ω experience strong dissipation in the denser regions of the disk, and so have a weaker effect upon the tenuous outer regions. Waves with lower frequencies experience less dissipation, and result in sharper temperature transitions. The hot, low–density gas which results from the shock heating may be responsible for the observed ultraviolet lines from cataclysmic variables, as well as the spatially coincidental lines of H and He I.

064.005 Line shapes of rotating stars with application to Alpha Lyrae.
G. H. Elste.
Astrophys. J., Vol. 384, No. 1, p. 284 – 291 (1 Jan 1992).

Profiles of weak and medium strong lines in the high-resolution spectrum of Vega can in principle be understood by the center–to–limb variation of their equivalent widths, predicted by a standard model atmosphere. Finer details may require a different model structure depending on the latitude of the rotating star, and perhaps differential rotation.

064.006 The calcium infrared triplet lines in stellar spectra.
U. G. Jørgensen, M. Carlsson, H. R. Johnson.
Astron. Astrophys., Vol. 254, No. 1/2, p. 258 – 265 (Feb 1992).

Observations of the infrared triplet lines of ionized calcium are often used as diagnostics of surface gravity among the stars contributing to the integrated light of distant galaxies. The authors have calculated the equivalent widths of these lines for a series of models with a broad range of input parameters to test their sensitivity to surface gravity, temperature, and calcium abundance. Over a larger range of parameter space than in earlier investigations, they find the sensitivity to be more complex than previously thought. They derive theoretical relations between equivalent width and log(g) for different values of T_{eff} and metallicity and compare their results with observations. By actual NLTE calculations, it is shown that departures from LTE are small. Comparison of the results with observations of stars in the

Galaxy, suggest that the Ca/Fe ratio is an increasing function of metallicity.

064.007 The effect of coupling of a corona to the turbulence in an accretion disk.
G. T. Geertsema, M. Kuperus.
Astron. Astrophys., Vol. 254, No. 1/2, p. 426 – 435 (Feb 1992).

Assuming a preferred length scale at which energy flows from magnetohydrodynamic turbulence in a thin accretion disk to a magnetically coupled corona, the authors investigate the differences between the energy dissipated in the accretion disk and the energy flowing into the corona. The study shows that the turbulence is very stable even against a strong coronal energy sink. A weak coronal energy sink reduces the amplitude of the fluctuations in the turbulent energy. The temporal spectra of the disk turbulence i.e. the dissipated energy and of the energy flowing into the corona are compared. The slopes of the power spectra of the energy which flows into the corona are steeper than those for the dissipated energy. A case of strong coronal coupling is further examined. Remarkable peaks occur in the Fourier spectra, which the authors identify with energy bursts in the total turbulent energy.

064.008 Cooling functions for hot dense gas: application to accretion discs in close binaries.
A. M. Dumont.
Astron. Astrophys., Vol. 254, No. 1/2, p. 454 – 459 (Feb 1992).

The author computes cooling functions for a hot and dense collisionally heated gas taking into account optical thickness effects. He shows that these cooling functions depend strongly on the density and on the column density, and so are very different from the classical "optically thin" case (Cox and Tucker 1969). In particular the well known maximum of the cooling function at a temperature of the order 10^5K disappears, owing to line collisional quenching, and the cooling function is smaller than the optically thin ones, even at very high temperatures ($\sim 10^8$K), if the surface density is a few g cm^{-2}. Computations are limited to a temperature and a density such that the bremsstrahlung–compton process is negligible. These computations are applied to accretion discs in close binary stars, and they show that the "hot" solution is thermally unstable. A grid of cooling functions and Rosseland mean opacities are given, and could be used for other purposes.

064.009 The effects of radiation drag on radial, relativistic hydromagnetic winds.
Li Zhiyun, M. C. Begelman, Chiueh Tzihong.
Astrophys. J., Vol. 384, No. 2, p. 567 – 579 (10 Jan 1992).

The authors study the effects of drag on an idealized relativistic MHD wind of radial geometry. The astrophysical motivation is to understand the effects of radiation drag on the dynamics of a jet or wind passing through the intense radiation field of an accreting compact object. From a critical point analysis, it is found that a slow magnetosonic point can appear in a dragged flow even in the absence of gravitational force, as a result of a balance between the drag force and the combination of thermal pressure and centrifugal forces. As in the undragged case, the Alfvén point does not impose any constraints on the flow. Although it is formally possible for a dragged flow to possess more than one fast magnetosonic point, the authors show that this is unlikely in practice. In the limit of a "cold", centrifugally driven flow, the fast magnetosonic point moves to infinite radius, just as in the drag–free case. For a given mass flux, the total energy output carried to infinity, and the final partition between the kinetic energy and the Poynting flux, are the same for the dragged and the drag-free flows. The main effects of radiation drag are to increase the amount of energy and angular momentum extracted from the source and to redistribute the regions where acceleration occurs in the flow. For a relativistic wind, the dissipated energy can exceed the final kinetic energy of the flow and may be comparable to the total flow energy (which is dominated by Poynting flux).

064.010 Spherically symmetric, polytropic flow.
T. Theuns, M. David.
Astrophys. J., Vol. 384, No. 2, p. 587 – 604 (10 Jan 1992).

The authors determine, in closed form, the flow profile corresponding to stationary, spherically symmetric, polytropic flow around a gravitating and radiating point mass. The profile is a branch of a quartic equation, relating distance to the central object to, for example, the local sound speed. Two branches of this equation are complex. The other two describe three different types of flow: subsonic and transonic accretion flow, transonic wind, and supersonic, super–Eddington flow. At any position in this flow, a supersonic regime may be linked to a subsonic one, through an adiabatic shock. Given the profiles, the authors determine the position of this shock, and vice versa. They also derive a stability criterion for the continous and shocked flows.

064.011 Dynamics of the envelopes of Be stars in the equatorial plane.
H. Chen, J. M. Marlborough, L. B. F. M. Waters.
Astrophys. J., Vol. 384, No. 2, p. 605 – 612 (10 Jan 1992).

The authors have investigated the dynamics in the equatorial plane of models for the envelopes of Be stars first by inverting the equation of motion to solve for the unknown force, $F_x(r)$, in addition to those arising from gravitation, rotation, and gas pressure gradient, required to produce a radial component of velocity having an assumed functional form. The authors then determine $F_x(r)$ for both a beta velocity and power–law velocity dependence, and for each investigate the consequences for $F_x(r)$ of the assumptions of angular momentum conservation, Keplerian rotation, and the variation of kinetic temperature with r. The models of Waters and coworkers, deduced from an analysis of the continuous energy distributions of six Be stars, were also applied. The results indicate that in all cases $F_x(r)$ has a similar characteristic shape. Between the surface of the star and a turning point of order $10 – 100$ stellar radii, $F_x(r)$ decreases with increasing r, but less rapidly than does gravity. Beyond the turning point, $F_x(r)$ increases with r at least initially; however, it may again decrease with further increases in r.

064.012 Fe II emission lines. II. Excitation mechanisms in cool stars.
P. G. Judge, C. Jordan, U. Feldman.
Astrophys. J., Vol. 384, No. 2, p. 613 – 623 (10 Jan 1992).

The authors discuss excitation mechanisms for the "resonance" transitions (between the $3d^64s$, $3d^7$, and $3d^64p$ configurations) of Fe II observed in emission in the near–ultraviolet spectra of cool stars. The analysis is based upon (1) emission measure analysis of previously measured lines in IUE spectra of cool giants, (2) discussion of the behavior of Fe II lines observed above the solar limb from Skylab spectra, (3) approximate radiative transfer calculations in a 59 level Fe II model atom using mean escape probabilities and a parameterization of optical radiation fields, and (4) accurate radiative transfer calculations in a smaller atomic model. The solar spectra show unambiguous evidence that electron collisions are responsible for most of the Fe II emission observed above the white–light limb. For cool giants, the z^6D^0, z^6F^0 terms are also excited primarily by electron collisions. However, the z^6P^0, z^4D^0, z^4F^0 and z^4P^0 terms cannot be excited solely by electron collisions or by previously identified line fluorescence processes. These terms are excited by electron excitation of metastable quartet terms below ~ 4 eV, followed by photoexcitation in lines at optical wavelengths by photospheric radiation.

064.013 Non–LTE, line–blanketed model atmospheres for late O– and early B–type stars.
J. A. Grigsby, N. D. Morrison, L. S. Anderson.
Astrophys. J., Suppl. Ser., Vol. 78, No. 1, p. 205 – 237 (Jan 1992).

This paper reports the first use of non–LTE, line–blanketed model atmospheres to analyze the spectra of hot stars. The stars analyzed are members of clusters and associations, have spectral types in the range O9 – B2 and luminosity classes in the range III – V, have slow to moderate rotation, and are photometrically constant. The model–atmosphere code, PAM, has been described by Anderson in 1985 and 1987. In the present study, sampled line opacities of iron–group elements were incorporated into the radiative transfer solution. Solar abundances were assumed. Synthetic line profiles of Hα, Hβ, Hγ, and He I λλ4471 and 4922 were fitted to the stellar spectra in order to choose T_{eff}, log g, and v sin i for each star. In most of the program stars, the authors obtained good to excellent agreement between the computed profiles and essentially all the line profiles used to fix the model, and reliable stellar parameters could be derived. With other lines, however, the models demonstrate less success in reproducing observed line strengths, probably due to inadequate model ions for He II, N II, and C II. In Hα, the fit to the inner line wings is improved over what was achieved in previous studies, but the present study finds that the synthetic line cores are systematically too deep. The cause of this discrepancy is unknown. The authors constructed the relation between the observed Strömgren c_0 index and the derived values of T_{eff}. The result is a useful temperature calibration for c_0. The behavior of the observed equivalent widths of N II, N III, C II, and C III lines as a function of T_{eff} was also studied.

064.014 Modelling the spectrum of WC–type Wolf–Rayet stars.
W.–R. Hamann, U. Leuenhagen, L. Koesterke, U. Wessolowski.
Astron. Astrophys., Vol. 255, No. 1/2, p. 200 – 214 (Feb 1992).

Model calculations for carbon–rich Wolf–Rayet stars (WC) are presented which account for complex model atoms of helium and, especially, of carbon. The multi–level non–LTE radiation transfer in expanding atmospheres is simulated under the standard assumptions of spherical symmetry, homogeneity and stationarity. A small set of models is presented, and the dependence of the resulting spectra on the basic stellar and atmospheric parameters is studied. The synthetic spectra are compared with the observed spectrum of WR 111 (alias HD 165763, subtype WC5), identifying the numerous lines and blends and discussing their significance for the purpose of spectral analyses. The models seem capable of reproducing the observation with reasonable accuracy. Thus the tools for future quantitative analyses of WC spectra are prepared. The theoretical spectra are found to depend not on the stellar radius R_* and the mass–loss rate $\dot{M}$ individually, but (in good approximation) only on the ratio $R_*/\dot{M}^{2/3}$. This "transformation law" which is already known from pure–helium models is confirmed here for the carbon lines as well.

064.015 Radiative shocks in atomic and molecular stellar–like atmospheres. V. Influence of the excited level of the hydrogen atom: the precursor structure.
E. Huguet, D. Gillet, J.–P. J. Lafon.
Astron. Astrophys., Vol. 255, No. 1/2, p. 233 – 245 (Feb 1992).

This paper is concerned with the problem of modelization of a raditive shock in atomic and molecular gas of hydrogen. The temperature (500–5000K) and the density (10^{-8}–10^{-14}g cm^{-3}) lie in the ranges found in cool pulsating stars (W Virginis, RV Tauri and Mira stars). In this work the authors include the first excited level of the hydrogen atom and, consequently, both the Balmer continuum and the Lyman–α line. The consequences of the introduction of a three–level atom are investigated. In particular, it is shown that coupling between the Lyman–α line and the H_2 molecular spectra is negligible. Finally a three–photon precursor model for weak radiative shocks (i.e. with negligible recombination rates) is given. The results show a smaller temperature than in the two–level atom models. The precursor sensitivity to the parameter $\tilde{T}_{e,B}$, describing the shape of the Balmer spectrum, is also examinated.

064.016 Nonradial and nonpolytropic astrophysical outflows. I. Hydrodynamic solutions with flaring streamlines.
K. Tsinganos, C. Sauty.
Astron. Astrophys., Vol. 255, No. 1/2, p. 405 – 419 (Feb 1992).

An analytical expression is presented for the streamline shape of steadily but nonspherically expanding and hydrodynamically dominated but nonpolytropic and rotating outflows from a gravitating central object. Deviations from the original Parker model are parametrized with the introduction of appropriate parameters for rotation and deviations of the density, pressure and streamlines from the spherically symmetric case. With those parameters equal to zero, a nonrotating, bipolar, spherically symmetric in density, pressure and streamlines, Parker–type wind outflow is obtained as a special case. It is shown that a streamline flaring and a pressure increase toward the equator, as well as rotation, suppress the acceleration of the outflow, while a density decrease polewards strongly enhances the acceleration. On the meridional plane, the streamlines become asymptotically radial, the axisymmetric sonic surface moves away from the base as the polar angle increases, while isoheating surfaces are elongated along the axis of the flow. The study is illustrated by applying it to fast solar wind streams from polar coronal holes and the jets of the stellar object SS433.

064.017 Turbulence in differentially rotating thin disks: a multi–component cascade model.
G. T. Geertsema, A. Achterberg.
Astron. Astrophys., Vol. 255, No. 1/2, p. 427 – 442 (Feb 1992).

In this paper, a theoretical model is presented which models some of the aspects of the dynamics of three–dimensional magneto–hydrodynamic turbulence in a differentially rotating accretion disk. A set of simplified non–linear dynamical equations is derived. Numerical solutions of these equations show chaotic behavior. In particular, the turbulent shear stress shows large fluctuations on timescales of a few rotation periods. The relevance of these results for the mass–transport and structure of thin accretion disks is briefly discussed.

064.018 On the shaping of circumstellar envelopes.
G. Pascoli, J. Leclercq, B. Poulain.
Publ. Astron. Soc. Pac., Vol. 104, No. 673, p. 182 – 188 (Mar 1992). Current Physics Microform No.: 9204G1286.

The authors examine the evolutionary process of expansion of a magnetized ring of cool matter ejected from a red giant. The subsequent formation of a circumstellar shell or preplanetary nebula is also analyzed. The velocity, density, and magnetic field are calculated along a time sequence, on a large two–dimensional grid of points close to and far away from the star. Owing to long–range forces such as thermal pressure gradients and magnetic stresses, the velocity field becomes a linear function of the distance. The latitudinal dependence of this function is poor, while conversely the density distribution is clearly aspheric. The magnetic field intensity at a distance $\sim 10^{15} - 10^{16}$cm is in good agreement with the observational data. The isophotes are plotted and compared to observational contours obtained from the literature. The filamentary structure appearance and its stability within the envelope are also discussed.

064.019 Winds from T Tauri stars. II. Balmer line profiles for inner disk winds.
N. Calvet, L. Hartmann, R. Hewett.
Astrophys. J., Vol. 386, No. 1, p. 229 – 238 (10 Feb 1992).

Balmer line profiles have been calculated using escape probability methods for T Tauri wind models with non–spherically symmetric geometry. The wind is assumed to originate in the inner regions of an accretion disk surrounding the T Tauri star, and flows outward in a "cone" geometry. Two types of wind models are considered, both with monotonically increasing expansion velocities as a function of radial distance. For flows with large turbulent velocities, such as the high frequency Alfvén wave – driven wind models, the effect of cone geometry is to increase the blue wing emission, and to move the absorption reversal close to line center. This type of model can provide good agreement with observed Hα line profiles of some T Tauri stars. However, even in cone geometry turbulent wind models still tend to produce too much blueshifted absorption in Hβ and Hγ. The authors have also calculated line profiles for a wind model rotating with the same angular velocity as the inner disk. The Balmer lines of this model are significantly broader than observed in most objects, suggesting that the observed emission lines do not arise in a region rotating at Keplerian velocity. Both turbulent and rotating wind models have difficulties explaining the centrally peaked emission line profiles often observed.

064.020 Balmer line profiles for infalling T Tauri envelopes.
N. Calvet, L. Hartmann.
Astrophys. J., Vol. 386, No. 1, p. 239 – 247 (10 Feb 1992).

The authors consider the possibility that the Balmer emission lines of T Tauri stars arise in infalling envelopes rather than winds. Line profiles for the upper Balmer lines are presented for models with cone geometry, intended to simulate the basic features of magnetospheric accretion from a circumstellar disk. An escape probability treatment is used to determine line source functions in nonspherically symmetric geometry. The authors find that thermalization effects can produce nearly symmetric Hα line profiles, when the higher Balmer series lines exhibit inverse P Cygni profiles. The infall models produce centrally peaked emission line wings, in good agreement with observations of many T Tauri stars. Furthermore, if it is assumed that the flow follows a dipole pattern, infall models can produce narrow central absorption reversals that are slightly blueshifted when viewed from certain angles. It is suggested that the Balmer emission of many T Tauri stars may be produced in an infalling envelope, with blueshifted absorption contributed by an overlying wind. Some of the observed narrow absorption components with small blueshifts may also arise in the accretion column.

064.021 Colliding winds from early–type stars in binary systems.
I. R. Stevens, J. M. Blondin, A. M. T. Pollock.
Astrophys. J., Vol. 386, No. 1, p. 265 – 287 (10 Feb 1992).

The collision of massive, supersonic stellar winds in early–type binary systems will produce a substantial flux of soft X–rays from the shock–heated material in the wind interaction region. Here the authors study the dynamics of the wind and shock structure formed by the wind collision. They employ a two–dimensional hydrodynamics code, which self–consistently accounts for radiative cooling, and calculate the X–ray luminosity and spectra of the shock–heated region, accounting for wind attenuation and the influence of different abundances on the resultant level and spectra of X–ray emission. The authors also examine a variety of dynamical instabilities that are found to dominate the intershock region. These instabilities disrupt the postshock flow and add a time variability of order 10% to the X–ray luminosity. The X–ray spectrum of these systems is found to vary with the nuclear abundances of the winds, which can be very nonsolar in the case of Wolf–Rayet stars. The authors use these theoretical models to study several massive binary systems, in particular V444 Cyg and HD 193793.

064.022 π^0–decay gamma–ray emission from winds of massive stars.
R. L. White, W. Chen.
Astrophys. J., Lett., Vol. 387, No. 2, p. L81 – L84 (10 Mar 1992).

The authors propose that the chaotic stellar winds from massive stars are potential π^0–decay γ–ray sources. A small fraction of the thermal ions is first–order Fermi accelerated to high energies by the shocks embedded in the highly unstable radiatively driven winds. These particles interact with the thermal ions to produce γ–rays peaked around 0.1 GeV via π^0–decay. When the mass–loss rate of an early–type star is high enough, the wind becomes opaque to the ion–ion interactions and a substantial fraction of the nonthermal ions' energy is emitted in γ–rays. The calculated γ–ray flux from some nearby O stars is marginally detectable by the EGRET experiment on board the Compton Gamma–Ray Observatory. The γ–ray flux from W–R stars cold be much higher, although there are large uncertainties in the model for W–R winds.

064.023 On preplanetary–nebula formation.
G. Pascoli.
Publ. Astron. Soc. Pac., Vol. 104, No. 675, p. 350 – 361 (May 1992). Current Physics Microform No.: 9206D2104.

The author proposes a comprehensive model in which he examines the process of gas outflow by an evolved star (post–AGB stars) and the subsequent circumstellar shell formation.

First, he investigates the buoyant transport of an isolated toroidlike tube through the red–giant convective envelope, originating from the innermost regions of the star and composed of strongly magnetized gas. When carrying up to a position immediately above the stellar surface, such a tube is no longer confined by the gas pressure existing throughout the red–giant envelope and consequently expands. It is hypothesized that the confluence of such magnetized tubes can produce a disklike structure of cool matter which surrounds the star. The author has also analyzed the blowing process by a fast wind of the early ejected cool matter. The large– and small–scale structures of the preplanetary nebula shell such as the bipolarity, the filamentary pattern, and the multiple–shell appearance are then discussed. A critical comparison with other concurrent models are also summarily presented.

064.024 X–ray–heated models of stellar flare atmospheres: theory and comparison with observations.
S. L. Hawley, G. H. Fisher.
Astrophys. J., Suppl. Ser., Vol. 78, No. 2, p. 565 – 598 (Feb 1992).

The authors compute a sequence of five model atmospheres consisting of the photosphere, chromosphere, and transition region. The models represent the response of the gas in a magnetically confined loop to intense flare energy release. The authors assume that the energy release is confined to the corona, and include the effects of chromospheric evaporation and indirect heating of the lower atmosphere by X–rays emitted from the coronal plasma. The models are computed in hydrostatic and energetic equilibrium and incorporate a detailed non–LTE solution of the radiative transfer and statistical equilibrium equations for a 6 level plus continuum hydrogen atom, a 5 level plus continuum Ca II ion, and a 3 level plus continuum Mg II ion. Line and continuum surface fluxes are presented in the wavelength range 1000 – 9000 Å and are compared with those observed during a giant flare on the M dwarf star AD Leo. The observed transition–region line fluxes are well described by the model. The observed flare continuum, however, is much bluer than that computed from the models; the observations fit a blackbody spectrum with T $\sim$ 8500K – 9500K. The authors propose that the flare continuum is formed by photospheric reprocessing of intense ultraviolet to extreme ultraviolet (EUV) line emission from the upper chromosphere. It is suggested that if the UV/EUV line emission is formed in response to the deposition of a large flux of nonthermal electrons, the continuum luminosity and color temperature can be used to determine both the energy flux and and the flare area being bombarded by energetic electrons. The same reprocessing mechanism may be responsible for some solar "white light" flares.

064.025 On the influence of the convective efficiency on the determination of the atmospheric parameters of DA white dwarfs.
P. Bergeron, F. Wesemael, G. Fontaine.
Astrophys. J., Vol. 387, No. 1, p. 288 – 293 (1 Mar 1992).

The authors present a detailed calculation of model atmospheres for DA white dwarfs where several versions of the mixing–length theory, with different associated convective efficiencies, are used. The predicted emergent fluxes, color indices, and equivalent widths are most sensitive to the assumed parameterization of the theory in the range $T_{eff} \sim$ 8000K – 15,000K. This, it turns out, is also the region where the Balmer jump is a most useful gravity discriminant. The authors discuss the implications of the calculations for previous determinations of atmospheric parameters of DA white dwarfs, and show that these results are much more model–dependent than previously believed.

064.026 Dynamics of a wind driven shell in steady–state accretion. The self–similar case.
B. I. Gnatyk, V. A. Krol'.
Pis'ma Astron. Zh., Tom 18, No. 3, p. 228 – 233 (Mar 1992). In Russian. English translation in Sov. Astron. Lett., Vol. 18, No. 2.

The self–similar (power–like) solutions of the problem of motion of a stellar (galaxy) wind driven shell in steady–state accretion flow are found in the thin layer approximation taking into account gravity, counter–pressure and radiative losses. The property of the power–like solutions of equations describing the steady–state accretion and the dynamics of the wind driven envelope are analysed.

064.027 Model atmospheres for population synthesis.
R. L. Kurucz.
IAU Symposium No. 149: The stellar populations of galaxies, p. 225 – 232 (1992). – See Abstr. 012.007 for the main entry.

The author has used his newly calculated iron group line list together with his earlier atomic and molecular line data to compute new opacities for the temperature range 2000K to 200000K. He uses the new line opacities, additional continuous opacities, and an approximate treatment of convective over-shooting. Thus far he has completed a grid of 7000 model atmospheres at 2 km/s for all the abundances, for the tempera-ture range 3500K to 50000K, and for log g from 0.0 to 5.0. This grid will allow a consistent theoretical treatment of photometry from K stars to B stars. Fluxes are tabulated from .09 to 160 micrometers. Preliminary results are reported for many photo-metric systems. The models, fluxes, and colors are available on magnetic tape and will also be distributed on CD–ROMs.

064.028 Mass loss by active late–type stars.
O. G. Badalyan, M. A. Livshits.
Sov. Astron., Vol. 36, No. 1, p. 70 – 74 (Jan 1992). Current Physics Microform No.: 9207Y0634. English translation of Astron. Zh., Tom 69, Vyp. 1, p. 138 – 146 (Jan–Feb 1992).

Coronal arches have come to be considered the source of X–rays from late–type stars. Plasma outflow is hindered in such a model. There are reasons to think that the interaction between large–scale magnetic fields and stellar wind leads to the forma-tion of streamers in stellar coronae (just as on the Sun). Using an earlier solution to the problem of plasma flow in slowly diverging coronal rays, the rate of mass loss M by stars with surface activity is calculated under the assumption that streamers occupy one fourth of the stellar surface at the level where the stellar wind is initiated. It is found that $M \geqslant 10^{-11} M_\odot$/yr for active dwarfs, and M is even larger for K subgiants that are in RS CVn systems. These values exceeding the solar rate of mass loss by two or three orders of magnitude, pertain to late–type stars for which the observed magnetic flux is close to the maximum value. The consequences of these results are discussed briefly.

064.029 Injection of mass and energy into the ISM by massive stars.
C. Leitherer, L. Drissen, C. Robert.
IAU Symposium No. 149: The stellar populations of galaxies, p. 447 (1992). – See Abstr. 012.007 for the main entry.

064.030 Influence of ion–atom collisions on the absorption of radiation in white dwarfs.
A. A. Mihajlov, M. S. Dimitrijević.
Astron. Astrophys., Vol. 256, No. 1, p. 305 – 308 (Mar 1992).

In order to provide the relevant absorption coefficients for the interpretation of the continuum absorption spectra in a number of white dwarfs with helium dominated atmo-spheres, the processes $He_2^+ + \hbar\omega \rightarrow He + He^+$ and $He + He^+ + \hbar\omega \rightarrow He + He^+$ have been considered together. The authors present also the absorption coefficients for the conditions of DA white dwarf atmospheres calculated by taking into account together the processes $H_2^+ + \hbar\omega \rightarrow H + H^+$ and $H + H^+ + \hbar\omega \rightarrow H + H^+$.

064.031 Wind accretion by compact objects: the "flip–flop" instability.
M. Livio.
IAU Symposium No. 151: Evolutionary processes in interacting binary stars, p. 185 – 194 (1992). – See Abstr. 012.008 for the main entry.

The problem of the stability of wind accretion onto compact objects is examined. Recent analytical and numerical calculations show that in two dimensions, Bondi–Hoyle accretion flows are unstable to a "flip–flop" instability. The instability can manifest itself as bursts in the accretion rate and as a random walk–type spin–up, spin–down behaviour of the accreting compact object. The nature of the flow in three dimensions needs further clarification. Possible observational implications are reviewed.

064.032 Accretion disc instabilities.
A. R. King.
IAU Symposium No. 151: Evolutionary processes in interacting binary stars, p. 195 – 203 (1992). – See Abstr. 012.008 for the main entry.

The author gives a brief discussion of accretion disc instabili-ties, concentrating mainly on tidal instabilities caused by the presence of binary companion. The superhumps observed in superoutbursts of SU UMa dwarf novae probably result from the excitation of a resonance in the accretion disc near the 3:1 commensurability with the binary orbit. This resonance can only appear for mass ratios $q = M_2/M_1 < q_{crit} \cong 0.25 - 0.33$: for larger mass ratios the available resonances are considerably weaker. Application of this picture to other types of binary suggests that the condition $q < q_{crit}$ may be necessary but not sufficient. Further, some cataclysmic systems show phenomena which could be tidal in origin even though the condition evidently fails.

064.033 The viscous evolution of elliptical accretion discs.
D. Syer, C. J. Clarke.
Mon. Not. R. Astron. Soc., Vol. 255, No. 1, p. 92 – 104 (1 Mar 1992).

The authors examine critically the widely accepted assumption that an accretion flow should circularize on a time–scale much shorter than the viscous time–scale. They provide an analytic calculation, supplemented by a generalization of the circular discs of Shakura and Sunyaev, and some "sticky–particle" numerical simulations, both of which suggest that the converse might be the case (i.e. that initially elliptical discs could retain their shape over a viscous time–scale, or even become more elliptical). The authors argue that there are astrophysical implications of such a result for initially eccentric streams of gas, either in a Kepler potential, or where the accretion rate is high compared to the rate of differential precession of streamlines.

064.034 Three–dimensional hydrodynamic simulation of an ac-cretion flow in a close binary system.
K. Sawada, T. Matsuda.
Mon. Not. R. Astron. Soc., Vol. 255, No. 1, p. 17P – 20P (1 Mar 1992).

Three–dimensional hydrodynamic simulations of Roche–lobe overflow are performed by solving the Euler equations. The gas is assumed to be inviscid and adiabatic except in the case of shocks. Gas filling the critical Roche lobe of a mass–losing component flows through the L1 point towards a compact object to form an accretion disc. It is found that the accretion disc is not axisymmetric but characterized by a pair of spiral shocks, as was the case in two–dimensional calculations.

064.035 Numerical simulations of two–dimensional and three–dimensional accretion flows.
T. Matsuda, T. Ishii, N. Sekino, K. Sawada, E. Shima, M. Livio, U. Anzer.
Mon. Not. R. Astron. Soc., Vol. 255, No. 2, p. 183 – 191 (15 Mar 1992).

Numerical simulations of 2D and 3D accretion flows past a gravitating compact object from a uniform flow at a large distance upstream are performed by solving the Eulerian equations. The authors find that 2D flows exhibit a "flip–flop instability" if the central accreting body is small. If the central body is enlarged at some instance in the oscillating flow, then the accretion shock shows a rather periodic oscillation similar to the von Karman vortex street. In the case of 3D flows, it is found that the shock cone is much more robust than in 2D, and the flip–flop instability takes a different, probably less violent form. The causes for the instabilities are discussed.

064.036   Chemistry in a protoplanetary nebula.
D. A. Howe, T. J. Millar, D. A. Williams.
Mon. Not. R. Astron. Soc., Vol. 255, No. 2, p. 217 – 226 (15 Mar 1992).

The authors investigate gas–phase chemistry in the remnant "superwind" of a carbon–rich red giant star, during its transition to a planetary nebula. The interacting stellar winds model is used. It is found that during the first few hundred years of transition, significant abundances of a few small molecules and ions (e.g. CH^+, CH_2^+, CH_3^+, CH, CH_2, NH) may occur in the thin, dense, shocked shell of gas predicted by this model, but that most molecules observed in protoplanetary nebulae will be rapidly destroyed, through photodissociation by strong UV from the central star. If dense clumps are present during transition, they may allow the gas–phase formation and/or survival of small amounts of some molecules, such as HCN, CN, C_2H_2, and HC_3N, until about 2000 yr after termination of the superwind; and young, fully developed planetary nebulae may show observable amounts of polyatomic molecules by this means. Such clumping may explain the existence of, e.g., HCN in NGC 7027. These results support the view that the large abundances of heavy molecules observed in young transition objects are either survivors from the red–giant superwind, or produced by non–gas–phase processes, such as desorption from grains. If they are relics of the red–giant ear, the authors predict that the abundances of such species in CRL 618 will fall, perhaps on a time-scale of decades, and suggest that these be monitored.

064.037   A 3–D hydrodynamical simulation of wind accretion.
H. M. J. Boffin.
IAU Symposium No. 151: Evolutionary processes in interacting binary stars, p. 453 – 456 (1992). – See Abstr. 012.008 for the main entry.

064.038   Spherical opacity sampling model atmospheres for M–giants. I. Techniques, data and discussion.
B. Plez, J. M. Brett, Å. Nordlund.
Astron. Astrophys., Vol. 256, No. 2, p. 551 – 571 (Mar 1992).

The authors present a new code for the calculation of static, spherically symmetric, opacity sampling model atmospheres in local thermodynamic equilibrium for cool giants and supergiants of spectral type M. Up–to–data for line opacities have been included, with new calculated line lists for the TiO, VO and H_2O molecules. This is the first time both sphericity and a realistic description (opacity sampling) of all opacity sources are used simultaneously for the calculation of cool stellar atmospheres and this results in a better agreement between calculated and observed fluxes and colours. A grid of models covering $3000K \leqslant T_{eff} \leqslant 4000K$, $-0.5 \leqslant \log g \leqslant 1.5$ for $1 M_\odot \leqslant M \leqslant 5 M_\odot$ with solar composition will be published separately.

064.039   The effect of irradiation in Algol–type binaries: metallicity and position in the HR diagram.
A. Claret, A. Giménez.
Astron. Astrophys., Vol. 256, No. 2, p. 572 – 580 (Mar 1992).

The authors present a study of the most relevant effects of irradiation in the secondary components of Algol–type binaries. They have used the so–called Uppsala Model Atmosphere (UMA) code to simulate the cool secondary stars while the external fluxes of the hotter primaries were computed using the models by Kurucz interpolated for the characteristic wavelengths of the UMA program. Theoretical bolometric albedos have been found to be in good agreement with observations. A comparison between the spectral energy distribution of model atmospheres with and without irradiation, for the same T_{eff} and log g, denotes clear differences mainly in the depth of the absorption lines. Furthermore, the authors have found that the computed irradiated spectra are very similar, in some spectral regions, to the equivalent model atmosphere without irradiation but a lower metallicity. This means that, if observed atmospheric distributions of flux in close binaries are not conveniently corrected for the effect of irradiation, systematic errors may be introduced in their interpretation, in particular with respect to the metal

content and effective temperatures of the secondaries. On the other hand, irradiation produces an increase in log T_{eff} due to the heating of the upper layers. The authors argue that this shift could be responsible, at least partly, for the observed overluminosity of the secondaries of classical Algols.

064.040   The stability of accretion tori. IV. Fission and fragmentation of slender, self–gravitating annuli.
D. M. Christodoulou, R. Narayan.
Astrophys. J., Vol. 388, No. 2, p. 451 – 466 (1 Apr 1992).

The authors present results on the dynamical evolution of two–dimensional, geometrically slender, differentially rotating, self–gravitating annuli orbiting around a central mass. These are idealized models of thick (pressure supported) accretion disks. The initial axisymmetric equilibrium models studied here are compressible with polytropic index n = 1 and have constant specific angular momentum. The linear growth rates of the different modes are calculated through a solution of the linearized fluid equations. Some model evolutions have also been carried into the nonlinear regime using a two–dimensional, Eulerian hydro code. Three kinds of unstable modes are known to operate in these annuli, the P, I, and J modes. In the nonlinear regime, the P–mode causes a breakup of the slender annulus into "planets" that do not separate completely, but eventually merge to form a new, roughly axisymmetric configuration. The I–mode is shown to be due to a fission instability. Under the action of an I–mode with azimuthal wavenumber m = 2, the circular annulus initially deforms into an elliptical shape and later separates into symmetrical clumps that move out to a larger radius. If the central star is allowed to move in response to the developing nonaxisymmetry of the annulus, then power is shifted to the I–mode with m = 1. In this case, the annulus breaks up into one major clump with debris, forming a binary system with the original central star. The J–mode has been previously identified with the Jeans mechanism of instability. In the nonlinear regime, the behavior of this mode is analogous to the fragmentation process of collapsing clouds. The mode generates severe clumpiness in the annulus, ultimately causing it to break up into multiple blobs.

064.041   On the local stability of accretion disks.
B. Dubrulle, E. Knobloch.
Astron. Astrophys., Vol. 256, No. 2, p. 673 – 678 (Mar 1992).

The local stability of accretion disks is investigated through the study of an unbounded rotating compressible viscous plane Couette flow. Bulk viscosity is included. Such a flow is shown to be stable against all small amplitude disturbances. The implication of this result for the origin of the assumed turbulent viscosity is explored. In particular, the possible role of finite amplitude perturbations in the onset of turbulence is stressed.

064.042   On stellar accretion of matter through a disk and the ejection of bipolar flows.
M. V. Torbett, D. L. Gilden.
Astron. Astrophys., Vol. 256, No. 2, p. 686 – 688 (Mar 1992).

Numerical simulations in one dimension with radiative losses approximately taken into account were conducted to evaluate the effectiveness of a hydrodynamical mechanism proposed to account for the ejection of bipolar flows from systems undergoing disk accretion. When radiative cooling is taken into account, the mechanism fails to accelerate matter to escape velocity for reasonable values of protostellar disk densities. However, a significant fraction, $\sim 10^{-2}$, of the accreted matter can be accelerated to velocities in excess of 0.25–0.5 of the orbital velocity.

064.043   Ionization and excitation in cool giant stars. I. Hydrogen and helium.
D. G. Luttermoser, H. R. Johnson.
Astrophys. J., Vol. 388, No. 2, p. 579 – 594 (1 Apr 1992).

The authors demonstrate the influence that non–LTE radiative transfer has on the electron density, ionization equilibrium, and excitation equilibrium, in model atmospheres representative of both oxygen–rich and carbon–rich red giant stars. The radiative transfer and statistical equilibrium equations are solved self–consistently for H, H⁻, H₂, He I, C I, C II, Na I, Mg I, Mg II, Ca I, and Ca II in a plane–parallel static medium. Calculations are made for both (1) radiative equilibrium model photospheres alone, and (2) model photospheres with attached chromospheric models as determined semiempirically with IUE spectra of g Her (M6 III) and TX Psc (C6,2). This paper reports on the excitation and ionization results for hydrogen and helium. Although hydrogen is strongly overionized with respect to LTE throughout much of the atmosphere of the photospheric models, almost all the hydrogen is still in the ground state or associated in H₂, and the hydrogen contribution to the electron density is small. Thus this overionization has little impact on the atmospheric structure or the emergent spectrum. The calculations show that excitation and ionization for helium is negligible throughout the entire photosphere and lower chromosphere of these models. The models are presented in two subsections: the first discusses ionization and excitation in the "pure" photospheric models, and the second describes the effect that the chromospheric radiation field has on the level populations both in the chromosphere and the photosphere ("chromospheric backwarming").

064.044   Condensation of graphite and refractory carbides in stellar atmospheres.
B. Fegley Jr.
54. Annual Meeting of the Meteoritical Society, p. 60 (1991). Abstract. – See Abstr. 012.010 for the main entry.

064.045   On the stability of magnetized accretion discs.
E. Knobloch.
Mon. Not. R. Astron. Soc., Vol. 255, No. 3, p. 25P – 28P (1 Apr 1992).
The shear instability of differentially rotating magnetized accretion discs is formulated as an eigenvalue problem. A monotonic instability caused by the presence of weak vertical magnetic fields, claimed by Balbus and Hawley on the basis of a local analysis, is shown to be absent whenever an azimuthal magnetic field is present. Instead the instability may take the form of overstable oscillations. A necessary condition for such an instability is obtained.

064.046   The evolution of circumstellar matter around an isolated massive star during the red supergiant phase – the role of heat conduction.
A. D'Ercole.
Mon. Not. R. Astron. Soc., Vol. 255, No. 4, p. 572 – 580 (15 Apr 1992).
The evolution of circumstellar matter around a massive star during the red supergiant phase is investigated. It is shown that, if heat conduction is present, the cavity carved around the star during its main–sequence phase may deflate substantially and even disappear completely in extreme cases. The author also finds that thermal conduction may cause a shrinking of the cavity already during the main–sequence phase; in this case the bubble tends to settle at a smaller radius. The author considers how the evolution of Wolf–Rayet ring nebulae and supernova remnants is affected, and discusses particular astronomical objects such as SN 1987A, the Cygnus Loop, N132D and the WR nebula around HD 197406.

064.047   Reflection with limb darkening.
P. Hadrava.
Astron. Astrophys., Vol. 257, No. 1, p. 218 (Apr 1992).
Corrected analytical expression for the reflection effect in binaries is given for the case of reradiation with linear limb darkening.

064.048   Radiation–hydrodynamic waves in an optically gray atmosphere. I. Homogeneous model.
N. S. Dzhalilov, Y. D. Zhugzhda (*Yu. D. Zhugzhda*), J. Staude.
Astron. Astrophys., Vol. 257, No. 1, p. 359 – 365 (Apr 1992).
The authors consider the coupled set of equations of hydrodynamics and radiative transfer. Only radiative transfer is taken into account in the energy equation, while other mechanisms of

heat transfer are ignored. Disturbances of the tensor of radiative pressure are included in the linearized momentum equation. This term proves to be of great importance for radiation–acoustic waves. LTE as well as a gray approximation and homogeneity of the atmosphere are assumed. The hydrodynamic equations are solved by using an exact solution of the radiative transfer equation (no Eddington approximation is made). The dispersion equation for the nonadiabatic oscillations is derived. This equation is valid for arbitrary optical thickness and generalizes all hitherto known results. The analysis of the dispersion equation points out a strong coupling of thermal and acoustic waves on a level where $\chi_0/k \approx 1$ (χ_0 = opacity, k = wave number). The damping length and decrements and the decay time of temperature disturbances of the p–modes and radiative oscillations have been calculated as well. The results of the present analysis can be applied to the solar p–mode oscillations.

064.049   Classical novae as fast magnetic rotators.
M. Orio, E. Trussoni, H. Ögelman.
Astron. Astrophys., Vol. 257, No. 2, p. 548 – 556 (Apr 1992).
UV and X–ray observations of classical novae after the outburst suggest a short turn–off scale, of the order of a few years. This is reconciled with the post–nova nuclear burning stage of the remaining envelope only if the nova white dwarfs are very massive. The authors examine this issue considering that a relevant number of systems that undergo classical nova outburst can be intermediate polars and AM Her type, implying that they have sufficiently strong magnetic fields and high rotation rates to be fast magnetic rotators. Approximate post–nova magnetic wind solutions are compared with the optically thick winds studied by Kato in a series of papers from 1983 to 1989. The relationship of the envelope mass M_{env} to the mass–loss rate $\dot{M}$ changes for rotating white dwarfs with magnetic fields of the order of a few 10^6 Gauss; the mass loss phase is shorter than in the case of low magnetic field and less envelope mass is left on the white dwarf at the end. Intermediate polars nova systems can enter the regime of centrifugal magnetic rotators, in which the whole envelope can be ejected no matter under which conditions the thermonuclear runaway occurred. The authors can thus reconcile the models with the observations without having to resort to very high white dwarf masses in classical novae.

064.050   The magnetic collimation of bipolar outflows. I. Adiabatic simulations.
J. M. Stone, M. L. Norman.
Astrophys. J., Vol. 389, No. 1, p. 297 – 304 (10 Apr 1992).
The collimation of an isotropic, nonmagnetic protostellar wind by an ordered magnetic field in the ambient medium is investigated using self–consistent magnetohydrodynamic simulations in two dimensions. The strong protostellar wind inflates a low–density bubble in the ambient medium. The growth of this bubble is confined along the direction of the ambient magnetic field when the gas pressure in the interior drops to the level of the magnetic pressure at the surface. For typical molecular cloud conditions, collimation of the protostellar wind into a bipolar outflow occurs on time scales of $10^4 - 10^5$ yr. When a self–consistent model for a magnetically supported cloud is used as a more realistic initial condition, the wind–blown bubble can "blowout" along diverging field lines in the direction of the steepest density gradient.

064.051   Radiation–driven winds of hot stars: a simplified model.
M. Villata.
Astron. Astrophys., Vol. 257, No. 2, p. 677 – 680 (Apr 1992).
In the framework of radiation–driven–wind theory an approximated line force is derived from the classical model. This line force has the property of being conservative (dependent on the radial distance only) and this fact simplifies the integration of the equation of motion, which assumes the form of a Parker equation. The agreement between the results (wind velocity fields and mass loss rates) of the simplified and the classical models is very good and represents an "a posteriori" justification of the proposed approximation.

064.052 Nonradial and nonpolytropic astrophysical outflows. II. Topology of MHD solutions with flaring streamlines.
K. Tsinganos, C. Sauty.
Astron. Astrophys., Vol. 257, No. 2, p. 790 – 806 (Apr 1992).

The detailed dependence of the topology of a nonradial and nonpolytropic magnetized outflow from a gravitating central object on the degree of magnetization, density latitudinal inhomogeneity and deviations of the streamlines from radiality (flaring), is examined. The topology is governed by several novel saddle and nodal hydromagnetic critical points that select a characteristic wind–type outflow solution. This critical solution starts sub–Alfvenically at the base with flaring streamlines, becoming super–Alfvenic further downstream with the streamlines asymptotically radial. Other jet–type noncritical solutions also exist and correspond to the streamlines forming a nozzle above the base. The multiple hydromagnetic critical points and the involved nature of the topology that appear in this relatively simple analytical class of solutions are indicative of the complexity of the MHD wind–phenomenon; this fact may be taken into account in numerical modelling of magnetized nonradial outflows, such as in the hydromagnetic solar wind, or, other jet–type magnetized astrophysical flows.

064.053 Stellar wind collision and X–ray generation in massive binaries.
V. V. Usov.
Astrophys. J., Vol. 389, No. 2, p. 635 – 648 (20 Apr 1992).

The gas flow, stellar wind collision, and generation of X–ray emission in massive binaries are considered. X–ray emission of such binaries is caused by gas heating to temperatures of 10^7–10^8K behind the front of shock waves which are formed due to the collision of the stellar wind flowing out from the components of a binary system, either with the surface of the other component or with the stellar wind flowing out from the other component. The distributions of the hot gas temperature and density in the shock layers are obtained. Using these distributions, the bremsstrahlung X–ray emission from the region of stellar wind collision is calculated.

064.054 Mass loss from active late–type stars.
O. G. Badalyan, M. A. Livshits.
Astron. Zh., Tom 69, Vyp. 1, p. 138 – 146 (Jan–Feb 1992). In Russian. English translation in Sov. Astron., Vol. 36, No. 1 (1992).

Coronal arches are believed to be the source of X–ray emission of late–type stars. Plasma outflow is hindered in this model. Interaction between large–scale magnetic fields and stellar wind appears to result (as well as on the Sun) in formation of streamers in stellar coronae. Using the obtained solution for a plasma flow in slowly diverging coronal rays, the authors calculated the mass losses dM/dt from stars with surface activity under the assumption that streamers occupy 1/4 of the stellar surface on the level where the wind originates. It is shown that the dM/dt values are $\geqslant 10^{-11}$M$_\odot$/yr for active dwarfs and are still larger for K–type subgiants in RS CVn–type binary systems. These values of dM/dt which are larger by 2 – 3 orders of magnitude than the solar one, characterize late–type stars with the observed magnetic fluxes close to maximal ones. The implications of the obtained results are discussed.

064.055 Models of planetary nebula spectral evolution.
K. Volk.
Astrophys. J., Suppl. Ser., Vol. 80, No. 1, p. 347 – 368 (May 1992).

First results are presented for models of the overall spectral evolution of planetary nebulae, with emphasis on the far–infrared emission of the remnant asymptotic giant branch (AGB) dust shells. The models combine a stellar/nebular spectrum, either an approximate model assuming the nebula to be strongly ionization bounded or a proper photoionization model, with a dust radiative transfer calculation. A series of these calculations are used to follow the evolution of the spectrum from the end of the AGB through to the white dwarf cooling stage. Combining these spectral results with an assumed Galactic distribution of

planetary nebulae allows the simulation of a large number of observational quantities including the nebular V–magnitude, Hβ flux, the IRAS colors, and the 5 GHz radio flux density. Those models using the full photoionization calculation also yield simulated sky image profiles at radio wavelengths and in some optical forbidden lines. Comparison of the model results to the IRAS color–color diagram indicate that the Schönberner 0.64 M$_\odot$ central star model evolves too slowly to match the observations. The first results indicate, still with some uncertainty, that a range of initial 10 μm dust optical depths from 5 to 20 plus a faster evolving central star gives a reasonable match to the IRAS colors. The models indicate that sky images should be able to clearly distinguish between matter–bounded and ionization–bounded nebulae and between the interacting winds model and the superwind model of planetary nebula formation.

064.056 A new nonlinear approximation to the limb–darkening of hot stars.
J. Díaz-Cordovés, A. Giménez.
Astron. Astrophys., Vol. 259, No. 1, p. 227 – 231 (Jun 1992).

The authors suggest the use of a new expression for the emergent flux distribution in stellar atmosphere, in the form, $I(\mu) = I(1)[1-c(1-\mu)-d(1-\sqrt{\mu})]$, where μ is the cosine of the angle subtended by the emergent radiation and the direction perpendicular to the stellar surface. Previous analytical expressions, e.g. linear and quadratic approximations, are compared to the new one and it is shown that a much better fitting to theoretical values is obtained using the above equation for effective temperatures larger than 8500K. The authors, thus, propose to use in future works the new expression in the analysis of the light curves of eclipsing binaries, lunar occultations of stars, or determinations of stellar diameters by interferometric methods.

064.057 Dependence of Stokes profiles on the temperature of photospheric models.
S. G. Mozharovskij.
Astron. Zh., Tom 69, Vyp. 2, p. 368 – 376 (Mar–Apr 1992). In Russian. English translation in Sov. Astron., Vol. 36, No. 2 (Mar–Apr 1992).

The magnetoactive line profiles dependence on effective temperature of photospheric models has been studied by computer simulations. It is shown that most of Fe I lines have maximum intensification for models having a temperature lying between the values in the quiet photosphere and sunspot umbra. I, V and Q Stokes profiles have different temperature sensitivity; the causes of such differences are explained. The strongest I, V and Q Stokes profiles correspond to models that have successively higher temperature. The V Stokes profiles change weakly in a wide range of temperatures and therefore it is wrong to use them in all cases for a separate analysis of hot and cold components of a two–component umbra.

064.058 Temperature dependence of Stokes profiles for model photospheres.
S. G. Mozharovskij.
Sov. Astron., Vol. 36, No. 2, p. 185 – 189 (Mar 1992). Current Physics Microform No.: 9208X2021. English translation of Astron. Zh., Tom 69, Vyp. 2, p. 368 – 376 (1992).

The profiles of magnetoactive lines as a function of the effective temperature of model photospheres are studied in a computer simulation. It is shown that most lines of neutral iron are most enhanced in models intermediate in temperature between models of the quiet photosphere and of a sunspot umbra. The I, V, and Q Stokes profiles have different temperature sensitivities; the reasons for such a difference are explained. The strongest I, V, and Q profiles correspond to models with successively higher temperatures. The profiles of circular polarization vary quite weakly over a wide temperature range, so they may not always be suitable for a separate analysis of the cool and hot components of a two–component umbra.

064.059 Hydrogen non–LTE computations in M dwarf chromospheres.
P. M. Panagi.
Ir. Astron. J., Vol. 20, No. 3, p. 197 – 198 (Mar 1992). – See Abstr. 012.038 for the main entry.

064.060   Are molecules responsible for origin of cold giants mass loss?
I. K. Shmeld, V. S. Strelnitskij, A. V. Fedorova, O. V. Fedorova.
IAU Symposium No. 150: Astrochemistry of cosmic phenomena, p. 413 – 414 (1992). – See Abstr. 012.029 for the main entry.

It is shown that radiative pressure in vibronic transitions of molecules may play an important role in origin of mass outflow of cold giants and supergiants.

064.061   Circumstellar chemistry.
A. E. Glassgold, G. A. Mamon.
Symposium on Chemistry and Spectroscopy of Interstellar Molecules, p. 261 – 266 (1992). – See Abstr. 012.039 for the main entry.

Recent theoretical studies of circumstellar chemistry are discussed for both red–giant and protostellar winds. The generalized photochemical model is able to account for the recently discovered silicon–bearing molecules in the prototypical, C–rich, AGB star IRC + 10216. The surprising occurrence of CO in protostellar winds that are largely atomic is interpreted to be the result of the high density and the rapid decrease of temperature with distance that is expected for such winds.

064.062   The radiation from accretion discs.
G. Shaviv, R. Wehrse.
Vulcano Workshop 1990: Frontier objects in astrophysics and particle physics, p. 49 – 60 (1991). – See Abstr. 012.040 for the main entry.

The authors discuss the theory of the radiation from accretion discs concentrating on a particular accretion disc, namely accretion discs in nova–like variables. Nova–like variables are binary systems in which the accreting star, the primary, is a white dwarf and the secondary, the mass giving star is a low mass main sequence star. The mass transferred from the secondary onto the WD has angular momentum and hence cannot fall directly onto the WD but forms a disc. There are several sources of radiation in a general binary system with an accretion disc. These sources are the stars themselves, the point where the stream of mass from the secondary hits the disc (known as the hot spot) etc. Hence in order to get the spectra of the accretion disc one has to remove from the observed spectra all the other contributions. Nova–like variables possess a high accretion rate and for this reason the radiation from the accretion disc is the dominant radiation source and the above problem does not exist.

064.063   On accretion by compact objects from a stellar wind.
M. Livio.
Vulcano Workshop 1990: Frontier objects in astrophysics and particle physics, p. 67 – 80 (1991). – See Abstr. 012.040 for the main entry.

The problem of Bondi–Hoyle accretion by a compact object from a stellar wind is reviewed. In particular, the problem of accretion of angular momentum is examined. Recent results indicate that Bondi–Hoyle accretion is unstable to a "flip–flop" instability in which the wake oscillates from side to side. The instability is accompanied by large amplitude variations in the accretion rate and in the specific angular momentum of the accreted material. The implications of the instability for the observations of wind fed X–ray sources are examined.

064.064   Magnetic fields in accretion disks around protostars and fragmentation of the disks.
T. Nakano.
Mem. Soc. Astron. Ital., Vol. 62, No. 4, p. 841 – 846 (1991). – See Abstr. 012.042 for the main entry.

The author investigates the behavior of magnetic field in accretion disks around protostars. Using the alpha model for a steady accretion disk with a constant plasma beta he finds that the magnetic field is decoupled from the gas only at the radius $r \lesssim r_d \approx 10^2 (M_*/M_\odot)^{1/3}$AU, where M_* is the mass of the central star, for $\beta \approx 1$. Because the magnetic torque is inefficient at $r \lesssim r_d$, the efficiency of angular momentum transport is low, and then the density is high. The author finds that the disk becomes gravitationally unstable and breaks into fragments of

mass 0.01 $M_\odot$ at $r \approx r_d$. Phenomena caused by the fragments are discussed.

064.065   The evolution of disk accreting protostars and the origin of the birthline.
G. S. Stringfellow.
Mem. Soc. Astron. Ital., Vol. 62, No. 4, p. 847 – 850 (1991). – See Abstr. 012.042 for the main entry.

Theoretical evolutionary calculations for disk accreting protostars are presented and comparisons made with classical T Tauri stars in the H–R diagram. The total luminosity of the composite disk accreting system, given by the sum of the protostellar luminosity and the luminosity arising in the boundary layer, reproduces the stellar "birthline" when plotted against the core's T_{eff}.

064.066   Shock waves and Hα profiles in the hydrodynamical model for RR Lyrae.
A. B. Fokin.
Mon. Not. R. Astron. Soc., Vol. 256, No. 1, p. 26 – 36 (1 May 1992).

The shock phenomena and Hα formation in the atmosphere of RR Lyrae are investigated by means of numerical simulations. The full amplitude hydrodynamical model is generated adopting $M = 0.578\ M_\odot$, $L = 62\ L_\odot$, $T_e = 7175$K and population II abundance. The Hα profiles are obtained by solution of the non–LTE line transfer problem for the multilevel hydrogen atom. A close quantitative agreement between the predicted and observed profiles is found, confirming the adequacy of the model used. In the course of pulsations the model produces an extended atmosphere of about 16 static scaleheights, in which two shocks are successively generated during one period. These results generally confirm the earlier conclusions of Hill. It is suggested that the bump of the light curve is connected with the early shock generation, the mechanism for which is discussed. The full development of the main shock occurs very high in the atmosphere. This can explain both the lack of doubling of the weak metallic lines, formed at a relatively large depth, and the absence of prominent emission in Hα due to large deviations from LTE near the shock front. Strong emission in the Lyman lines is predicted, reaching maximum intensity near the light maximum.

064.067   The outer atmospheres of the "hybrid" bright giants: the chromospheres of α TrA (K4 II), ι Aur (K3 II), γ Aql (K3 II) and θ Her (K1 II).
G. M. Harper.
Mon. Not. R. Astron. Soc., Vol. 256, No. 1, p. 37 – 64 (1 May 1992).

Detailed models of the chromospheres of the "hybrid" bright giants α TrA (K4 II), ι Aur (K3 II), γ Aql (K3 II) and θ Her (K1 II) are presented. The models run from the base of the chromosphere and into the transition region. The chromospheric models are based on radiative transfer calculations for a range of atoms and emission lines, including: Mg II h and k; Al II] λ2669 Å; C II λ1335 Å; C II] λ2325 Å; Si II $\lambda\lambda$1808.0, 1816.9 and 1817.5 Å. The transition region models are made using standard emission measure techniques. The presence of steep turbulent velocity gradients leads to profiles of optically thin chromospheric emission lines that are strongly dependent on the distribution of emitting material with temperature. The widths of the optically thin emission line of Al II] at λ2669 Å and the optically thick lines of Fe I at $\lambda\lambda$2823 and 2844 Å provide valuable upper limits to the turbulence on the lower and middle chromosphere. These one–component models provide the starting point for studies of the chromospheric energy balance. They constrain conditions in the lower boundaries of magnetically driven winds and also give a basis for studying chromospheric inhomogeneities.

064.068   Three–dimensional structured shocks in AM Herculis–type systems. II. Cyclotron emission from ridge–shaped emission regions.
K. Wu, D. T. Wickramasinghe.
Mon. Not. R. Astron. Soc., Vol. 256, No. 2, p. 329 – 338 (15 May 1992).

The authors study the properties of cyclotron emission from accretion shocks in AM Herculis binaries which result from non–

uniform accretion on to linearly extended regions on the surface of the magnetic white dwarf. The authors show that the resulting ridge–shaped post–shock structures yield cyclotron spectra with prominent harmonic features over a wide range of viewing aspects, even for high total accretion rates, in contrast to what has been found previously for axisymmetric accretion shocks.

064.069 Wave action conservation, over–reflection and over–transmission of non–axisymmetric waves in differentially rotating thin discs with self–gravity.

Y. Nakagawa, M. Sekiya.
Mon. Not. R. Astron. Soc., Vol. 256, No. 4, p. 685 – 694 (15 Jun 1992).

The effects of self–gravity on the over–reflection phenomena of linear non–axisymmetric waves in differentially rotating (Keplerian) discs are examined. The authors use coordinates comoving with the unperturbed differential rotation. In the comoving coordinates, Poisson's equation for the perturbed gravitational potential can be readily solved and the perturbation equations reduce to a single second–order ordinary differential equation with respect to time t. Also an exact expression for the conserved wave action density can be found. Connecting the WKBJ solutions for $t \rightarrow -\infty$ and $t \rightarrow \infty$ numerically, the authors obtain the transmission coefficient T and the reflection coefficient R of waves as functions of the normalized azimuthal wavenumber K_y for various values of the self–gravity parameter Q^{-1} (the inverse of Toomre's parameter). The results show that self–gravity enhances wave amplification significantly; in addition to over–reflection, over–transmission (i.e., T > 1) occurs for $Q^{-1} \gtrsim 0.5$, and the waves are extremely amplified by orders of magnitude for $Q^{-1} \gtrsim 1$. The authors find T = 0 and R = 1 for some special discrete values of K_y. This phenomenon may be understood by analogy with the one–dimensional Schrödinger equation, i.e. wave interference in broadened or double–peaked potentials.

064.070 Line blanketed model atmospheres of Ap stars. VII. Miscellaneous models.

H. Muthsam, K. Stepień.
Acta Astron., Vol. 42, No. 2, p. 117 – 138 (1992).

Sets of model atmospheres of Ap stars with three different chemical compositions and a range of effective temperatures are presented. Additional models are available from the Acta Astronomica Archive through the Internet.

064.071 Time evolution of the accretion disk radius in a dwarf nova.

S. Ichikawa, Y. Osaki.
Publ. Astron. Soc. Jpn., Vol. 44, No. 1, p. 15 – 26 (1992).

The time evolution of the accretion disk radius in the outburst cycle of a dwarf nova has been studied. A numerical scheme for calculating the time evolution of accretion disks is presented, in which both the mass and the angular momentum of the system are exactly conserved during evolution, and in which the tidal torque and tidal dissipation exerted by the secondary star on the accretion disk are taken into account. By using this method, the authors have examined the time evolution of the disk radius in the dwarf nova outburst cycle based on both the mass transfer burst model and the disk instability model. The disk instability model is able to explain the basic observed features of the disk radius variation; i.e., expansion of the disk during outburst and its gradual shrinkage in quiescence. The mass–transfer burst model gives, at first sight, a similar variation in the disk radius with time. However, if examined closely, it is found that the mass–transfer burst model gives rise to a transient decrease in the disk radius at the onset of an outburst, and very little shrinkage in the disk radius in quiescence; those two features are inconsistent with observations.

064.072 Irradiation of accretion disks around young objects. II. Continuum energy distribution.

N. Calvet, G. Magris C., A. Patiño, P. D'Alessio.
Rev. Mex. Astron. Astrofis., Vol. 24, No. 1, p. 27 – 42 (Apr 1992).

The authors have calculated the emergent flux for systems composed of a young star and a steady, infinitely thin, optically thick accretion disk subject to irradiation from the central star.

064.073 A three component model for MagnetoHydroDynamic turbulence in accretion disks.

G. T. Geertsema.
Diss., Rijksuniversiteit Utrecht (Netherlands). 115 p. (1992).

In this thesis, the author concentrates on accretion disks around stellar binary systems and more precisely on the cataclysmic variables and the low mass X-ray binaries. Contents: 1. Introduction. 2. MHD turbulence and accretion disks. 3. Time evolution of a magnetic seed–field in a turbulent accretion disk. 4. Porting a 3D–turbulence simulation to a multi-processor system. 5. Turbulence in differentially rotating thin disks. 6. The effect of coupling of a corona to the turbulence in an accretion disk. 7. Nederlandse samenvatting.

064.074 Is the Oort A–value a universal growth rate limit for accretion disk shear instabilities?

S. A. Balbus, J. F. Hawley.
Astrophys. J., Vol. 392, No. 2, p. 662 – 666 (20 Jun 1992).

A weak–field local MHD instability that is of importance to accretion disks and has recently been studied by the authors is reexamined in more detail. The maximum growth rate of the instability is found to be not only independent of the magnetic field strength but independent of field geometry as well. In particular, all Keplerian disks are unstable in the presence of any weak poloidal field, with the ratio of the maximum growth rate to disk angular velocity given by 3/4. More generally, the maximum growth rate of any weak field configuration that is not purely toroidal is given by the local Oort A–value of the disk. The authors investigate the reasons behind this remarkably general behavior by using a form of the dynamical Hill equations. It is conjectured that the Oort A–value is an upper bound to the growth rate of any instability feeding upon the free energy of differential rotation.

064.075 Molecular equilibrium in stars.

T. R. Carson.
Rev. Mex. Astron. Astrofis., Vol. 23, p. 151 – 159 (Mar 1992). –
See Abstr. 012.054 for the main entry.

A review is given of the methods for the study of molecular equilibrium under stellar conditions. All the equilibrium constants for diatomic and triatomic molecules are given in terms of spectroscopic constants, allowing full account to be taken of isotopic variation. Numerical procedures are outlined for the solution of the equations for the equilibrium concentrations of an arbitrary number of elements (including ions) and molecules.

064.076 Models for stellar wind of early–type stars.

R. Blomme.
Rev. Mex. Astron. Astrofis., Vol. 23, p. 241 – 246 (Mar 1992). –
See Abstr. 012.054 for the main entry.

The radiatively driven wind theory does not correctly predict the observed values of the terminal velocity and mass loss rate for O and early B–type stars. Hoping to explain this discrepancy the author investigated how sensitive the theoretical predictions are to errors in the oscillator strengths. The results show that the stellar wind dynamics is quite sensitive to changes in the oscillator strengths for the iron–group elements. For some stars the mass loss rate increases by a factor of 2, bringing them much closer to the observed rates. Unfortunately at the same time the terminal velocity also increases, making the discrepancy of that parameter with the observations even larger.

064.077 Limit cycles in pair dominated accretion disks.

G. Björnsson, R. Svensson.
AIP Conf. Proc., No. 254, p. 260 – 263 (1 May 1992). – See Abstr. 012.057 for the main entry.

064.078 A new mechanism for viscosity in cloudy accretion disks: anomalous (turbulent) viscosity.

A. Fridman, L. Ozernoy.
AIP Conf. Proc., No. 254, p. 280 – 283 (1 May 1992). – See Abstr. 012.057 for the main entry.

The authors analyze the dynamics of a rotating cloudy gaseous medium relative to 2–D perturbations having wavelengths shorter than the disk thickness. The viscosity due to cloud–cloud collisions is confronted with the derived critical viscosity, below which a collective mode of perturbations in the disk is able to grow until going into the saturation regime where fully developed turbulence is established. It is shown that for the assumed parameters of AGN accretion disks consistent with the observational data, the viscosity due to collective modes usually dominates over that due to cloud–cloud collisions.

064.079 **The shock temperature and superionization in the wind compressed disk model for Be stars.**
J. E. Bjorkman, J. P. Cassinelli.
Bull. Am. Astron. Soc., Vol. 23, No. 3, p. 1265 (1991). Abstract. – See Abstr. 012.067 for the main entry.

064.080 **Protostellar accretion disks: a model for connecting star to disk.**
K. R. Bell, D. N. C. Lin, J. Papaloizou.
Bull. Am. Astron. Soc., Vol. 23, No. 3, p. 1267 – 1268 (1991). Abstract. – See Abstr. 012.067 for the main entry.

064.081 **Dynamics of a windblown shell in stationary accretion flow. Self–similar case.**
B. I. Gnatyk, V. A. Krol'.
Sov. Astron. Lett., Vol. 18, No. 2, p. 92 – 95 (Mar 1992). Current Physics Microform No.: 9211X0731. English translation of Pis'ma Astron. Zh., Tom 18, No. 3, p. 228 – 233 (1992).
The self–similar (power–law) solutions of the problem of the motion of a shell driven by stellar (galactic) wind in stationary accretion flow are found in the thin–layer approximation, taking into account gravity, counterpressure, and radiative cooling. The characteristics of the power–law solutions of the equations of stationary accretion and dynamics of a windblown shell are analyzed.

064.082 **A model analysis of the infrared data of late type stars surrounded by the circumstellar dust envelopes.**
O. Hashimoto.
Sci. Rep. Tôhoku Univ., Eighth Ser., Vol. 12, Nos. 2/3, p. 133 – 143 (Jan 1992). – See Abstr. 012.073 for the main entry.
A lot of model infrared spectral energy distributions of the oxygen–rich late type stars surrounded by the circumstellar dust envelopes are calculated by solving the radiative transport equations in the spherical dust envelope with various model input parameters. Using IRAS photometric data at 12, 25 and 60 μm and the strength of the silicate band feature at 10 μm from LRS spectra the best fit model is searched for each IRAS oxygen–rich evolved star with the circumstellar dust envelope.

064.083 **Physical models of solar and stellar spots.**
A. Skumanich.
Armagh Observatory Bicentenary Colloquium on Surface Inhomogeneities on Late–Type Stars, p. 94 – 107 (1992). – See Abstr. 012.085 for the main entry.
The present observational picture of the magnetic structure of sunspots is reviewed first. It is argued that the continued use of the Beckers–Schröter (1969) magnetic profile in spot models is untenable. The dipole representation of recent magnetic profile data, as well as the magnitude of the field aligned currents for "twisted" spots, is also reviewed. Next, the theoretical picture of both the subsurface and above–surface magnetic structure is reviewed for a variety of theoretical "conjectures" including both current sheet and distributed current models. It is found that lateral force balance leads to a fairly simple relation in which the product of the square–root of the pressure excess and the area of the flux tube is essentially constant with depth and equal to the emerged magnetic flux. Thus, the mean field strength (flux per unit area) should scale as the square–root of the pressure not only for thin flux tubes, but also for spots. Observational evidence for such a scaling is reviewed.

064.084 **Numerical simulations of flares on late–type stars: hydrodynamics and X–ray spectra.**
C.–C. Cheng, R. Pallavicini.
Armagh Observatory Bicentenary Colloquium on Surface Inhomogeneities on Late–Type Stars, p. 258 – 260 (1992). – See Abstr. 012.085 for the main entry.

064.085 **Non–local thermal conduction in solar and stellar coronal loops.**
A. Ciaravella, G. Peres, S. Serio.
Armagh Observatory Bicentenary Colloquium on Surface Inhomogeneities on Late–Type Stars, p. 261 – 263 (1992). – See Abstr. 012.085 for the main entry.

064.086 **On the dynamics of the emerging pre–flare magnetic configuration.**
H. A. Harutyunian (G. A. Arutyunyan), V. S. Hayrapetyan (V. S. Ajrapetyan).
Armagh Observatory Bicentenary Colloquium on Surface Inhomogeneities on Late–Type Stars, p. 284 – 285 (1992). – See Abstr. 012.085 for the main entry.
The theoretical description of flare activity in the Sun and dMe stars as it is generally accepted, is based on magnetic field reconnection. This paper shows that in the upper layers of stellar atmospheres closed magnetic configurations with low β can arise, which may exhibit pinching and pre–flare reconnection effects.

064.087 **A semi–analytical model of stellar flares.**
R. A. Kopp, G. Poletto.
Armagh Observatory Bicentenary Colloquium on Surface Inhomogeneities on Late–Type Stars, p. 295 – 297 (1992). – See Abstr. 012.085 for the main entry.
The authors propose a simple "point" model for flaring coronal plasma, which includes thermal conduction, chromospheric evaporation, radiative losses, and gravitational loop draining, and which gives the temporal evolution of temperature, density and velocity averaged over the flaring loop. The model allows a rapid survey to be made of flare–loop parameter space: it is shown that the high flare emission measures can be interpreted within the framework of the model, although it is not possible to evaluate uniquely the physical parameters of the flaring region.

064.088 **Accretion disc phenomena.**
J. E. Pringle.
Reviews in Modern Astronomy 5: Variabilities in stars and galaxies, p. 97 – 104 (1992). – See Abstr. 003.087 for the main entry.
Contents: Introduction. Accretion disc spectra. Disc dynamics. Viscosity.

064.089 **Strength of the O I triplet λ7771–5 and atmospheric microturbulence in A–F stars.**
Y. Takeda.
Publ. Astron. Soc. Jpn., Vol. 44, No. 3, p. 309 – 324 (1992).
The behavior of atmospheric microturbulence (ξ) on the H–R diagram with regard to normal A–F stars was investigated using the O I triplet λ7771–5, the strength (W) of which is sensitive to ξ as well as to a non–LTE effect. The latter effect was properly taken into account by performing detailed non–LTE calculations with a realistic atomic model of O I. A comparison of the theoretical W(7771–5) with the observed values is presented and discussed. As an application of the results of the non–LTE calculations, oxygen abundances from various O I lines of α CMi, α Per, α Cyg, and α Lyr were determined, showing that the abundances derived from different lines turn out to be mutually consistent if the non–LTE effect is taken into account: this results in the tentative conclusion that the abundances of oxygen for the former three stars are nearly normal, while α Lyr is slightly O–deficient.

064.090   Grain formation in stellar outflows: is classical nucleation theory valid?
M. P. Egan, C. M. Leung.
News Lett. Astron. Soc. N.Y., Vol. 4, No. 1, p. 33 (Feb 1992).
Abstract. – See Abstr. 012.078 for the main entry.

064.091   Ionizing radiation from early–type stars.
D. Kunze, R.–P. Kudritzki, J. Puls.
Workshop on the Atmospheres of Early–Type Stars, p. 45 – 47 (1992). – See Abstr. 012.095 for the main entry.
A grid of NLTE model atmospheres for early–type stars is presented. Model atoms of eleven different elements with various abundances and, for a subset of the models, effects of wind blanketing, are taken into account. These model atmospheres supply realistic emergent fluxes which can be used in ionization calculations for circum– and interstellar gas. The dependence of the ionizing radiation of these stars on spectral type, luminosity class, evolutionary state and stellar environment is discussed.

064.092   Analyses of B–type stars in the Magellanic Clouds.
A. Jüttner, B. Wolf.
Workshop on the Atmospheres of Early–Type Stars, p. 63 – 76 (1992). – See Abstr. 012.095 for the main entry.
Model atmosphere analysis abundance determinations from B–type stars in the Magellanic Clouds are reviewed.

064.093   Wolf–Rayet stars.
W. R. Hamann.
Workshop on the Atmospheres of Early–Type Stars, p. 87 – 100 (1992). – See Abstr. 012.095 for the main entry.
Based on the so–called standard model for expanding atmospheres, non–LTE model calculations have been developed during the last years. Their application for the quantitative spectral analysis of WR stars is in progress. Recent results about the parameters and chemical composition of WR atmospheres are reviewed. From the achieved consistency between model predictions and observations, including the test case V444 Cygni, the author concludes that the standard model is an adequate description of WR atmospheres to a satisfactory degree of accuracy. The empirical results of the spectral analyses obtained so far fit into the scenario of post–red–supergiant evolution, although there is not yet a detailed agreement with calculated tracks. The driving mechanism of WR winds remains an open question.

064.094   Diffusion, mass loss and accretion in stars.
G. Michaud.
Workshop on the Atmospheres of Early–Type Stars, p. 189 – 202 (1992). – See Abstr. 012.095 for the main entry.
Contents: Astrophysical context. Competition between mass loss, accretion ad diffusion. AmFm stars, λ Bootis stars and the Li gap. Magnetic stars. The abundance connection.

064.095   Radiative accelerations on Ga and Al ions in stable atmospheres of CP stars.
J. Budaj, M. Zboril, J. Zverko.
Workshop on the Atmospheres of Early–Type Stars, p. 210 – 212 (1992). – See Abstr. 012.095 for the main entry.
Radiative accelerations of ions (Ga I, II and Al I–V) are computed in the atmospheres of CP stars.

064.096   Si II autoionization lines in stratified atmosphere of Bp star.
M. Zboril, J. Budaj.
Workshop on the Atmospheres of Early–Type Stars, p. 213 – 215 (1992). – See Abstr. 012.095 for the main entry.
Synthetic spectra were computed under the assumption of atmosphere stratification for a Bp type star. At $\lambda5200$ depression both Si normal and autoionization lines strengthen at $\lambda5187$ and $\lambda5204$ and may contribute to $\lambda5175$ peaked contributor of $\lambda5200$ depression.

064.097   Non–LTE analysis of the Palomar Green subdwarf O stars.
P. Thejll, F. Bauer, D. Kunze, R. Saffer, J. Liebert, H. Shipman.
Workshop on the Atmospheres of Early–Type Stars, p. 261 – 263 (1992). – See Abstr. 012.095 for the main entry.
Using NLTE model spectra, the sdO subdwarfs found in the Palomar–Green survey, are analyzed in order to determine their relative contribution to the white dwarf cooling sequence. The authors confirm that only a small fraction of the WD's, and thus also of the DB WD's, can be related to the sdO channel.

064.098   The atmospheres of extreme helium stars.
C. S. Jeffery.
Workshop on the Atmospheres of Early–Type Stars, p. 293 – 297 (1992). – See Abstr. 012.095 for the main entry.
Extreme helium (EHe) stars are low–mass early–type supergiants with hydrogen–deficient and carbon–rich atmospheres. Progress and recent results in the quantitative analysis of their atmospheres are reviewed, including the first results to include the effects of line–blanketing in the model atmospheres. The hot RCrB star DY Cen has been found to be relatively hydrogen–rich, but extremely metal–poor. A comparison of EHe star abundances with other H–deficient stars is made.

064.099   Mass loss from v Sgr and other helium stars.
R. E. Dudley, C. S. Jeffery.
Workshop on the Atmospheres of Early–Type Stars, p. 298 – 300 (1992). – See Abstr. 012.095 for the main entry.
Mass–loss rates have been determined for 6 hydrogen–deficient stars by modelling their wind line profiles in the UV from high resoltuion spectra. Line profiles were calcualted with using the SEI method (Sobolev approximation with Exact Integration of the transfer equation: Lamers, Cerruti–Sola and Perinotto 1987). Optimum fits to the observations were obtained using a least squares procedure. Results for single helium stars are in good agreement with previous studies. The stellar wind from the binary star v Sgr has been modelled succesfully for the first time by adopting an extended two component wind.

064.100   Blue post–asymptotic giant branch stars at high galactic latitude.
R. J. H. McCausland, E. S. Conlon, P. L. Dufton, F. P. Keenan.
Workshop on the Atmospheres of Early–Type Stars, p. 301 – 304 (1992). – See Abstr. 012.095 for the main entry.
Model atmosphere analyses are presented for high resolution spectra of six stars at high galactic latitude. Although their derived atmospheric parameters are consistent with their previous classification as early B–type stars, their metal abundances are significantly different from those expected for Population I objects. However both their chemical composition and atmospheric parameters appear consistent with a Post Asymptotic Giant Branch evolutionary status. Additional evidence for this hypothesis is present in the spectra of one star (LS IV–12°111), where its higher effective temperature is sufficient to excite emission lines from the surrounding nebula.

064.101   Stellar flares: confined or eruptive events?
R. Pallavicini.
IAU Colloquium No. 133: Eruptive solar flares, p. 289 – 300 (1992). – See Abstr. 012.096 for the main entry.
Observations in different spectral bands have shown the existence for many similarities between solar and stellar flares, in spite of the far larger energies that are typically involved in the latter. The analogy may go as far as to include the occurrence on stars of both confined and eruptive flares similar to those observed on the Sun. The observational evidence for the existence of stellar eruptive flares is reviewed and it is shown that the data are still inconclusive in this respect. Models of stellar flares as either confined or eruptive magnetic structures are also discussed and it is concluded that models are unable at present to discriminate between the two cases.

064.102   Atmospheric models of flare stars.
     P. J. D. Mauas, A. Falchi, R. Falciani.
IAU Colloquium No. 133: Eruptive solar flares, p. 390 – 393 (1992). – See Abstr. 012.096 for the main entry.
    The authors report preliminary results on the computation of the first semiempirical atmospheric models for the star AD Leo, both in the quiescent and flaring state.

064.103   Carbon enrichment in the outer layer of hot helium–rich high gravity stars.
K. Unglaub, I. Bues.
Workshop on the Atmospheres of Early–Type Stars, p. 334 – 336 (1992). – See Abstr. 012.095 for the main entry.
    The atmospheres of many white dwarfs and subdwarfs O (sdO's) are helium–rich with traces of heavier elements. Since these stars are in an evolutionary stage after the core helium burning, at least carbon should be present with significant abundance in the innermost regions. This is why the authors investigated in detail which carbon abundance is to be expected in the atmospheres of the stars in an equilibrium state of sedimentation, ordinary diffusion and selective radiative forces.

064.104   The opacity project – a review.
     K. Butler.
Workshop on the Atmospheres of Early–Type Stars, p. 345 – 358 (1992). – See Abstr. 012.095 for the main entry.
    The aims and methods of the international collaboration known as the Opacity Project are described. Results from all aspects of the work are presented, culminating in pulsation models for Cepheid variables which are in agreement with observation.

064.105   The Lyman α line wing and application for synthetic spectra of DA white dwarfs.
N. F. Allard, D. Koester.
Workshop on the Atmospheres of Early–Type Stars, p. 359 – 361 (1992). – See Abstr. 012.095 for the main entry.
    New theoretical absorption profiles are included in stellar atmosphere codes and used to predict synthetic spectra for DA white dwarfs of intermediate temperatures (20000 to 8000K). These new calculations offer a unique opportunity to determine acccurate effective temperatures and surface gravities for the variable ZZ Ceti stars.

064.106   Ion–atom complexes and the absorption of radiation in stellar plasma.
A. A. Mihajlov, M. S. Dimitrijević.
Workshop on the Atmospheres of Early–Type Stars, p. 362 – 364 (1992). – See Abstr. 012.095 for the main entry.
    In order to provide the relevant absorption coefficients for the interpretation of the continuum absorption spectra in stellar atmospheres, the processes $H_2^+ + \hbar\omega \rightarrow H + H^+$ and $H + H + \hbar\omega \rightarrow H + H^+$ have been considered together, as well as the same processes in the helium case. The authors present the corresponding coefficients for the conditions of stellar atmospheres.

064.107   Stark broadening parameters for spectral lines of multicharged ions in stellar atmospheres: C IV, N V, O VI lines and regularities within an isoelectronic sequence.
M. S. Dimitrijević, S. Sahal–Bréchot.
Workshop on the Atmospheres of Early–Type Stars, p. 368 – 370 (1992). – See Abstr. 012.095 for the main entry.
    Using a semiclassical approach, the authors have calculated recently, electron–, proton–, and ionized helium–impact line widths and shifts for 39 C IV, 30 N V and 30 O VI multiplets. This comprehensive set of data has been used for the investigation of Stark broadening parameter regularities within isoelectronic sequences.

064.108   On Stark line shifts in spectra of very hot stars.
     V. Kršljanin, M. S. Dimitrijević.
Workshop on the Atmospheres of Early–Type Stars, p. 371 – 373 (1992). – See Abstr. 012.095 for the main entry.
    Starting from semiclassical calculations of Stark broadening parameters, Stark line shifts of several important UV lines of C IV, Si IV, N V and O VI in spectra of very hot main sequence and high–gravity stars have been estimated and possibilities for their observational confirmation have been discussed.

064.109   Accelerated Lambda Iteration.
     I. Hubeny.
Workshop on the Atmospheres of Early–Type Stars, p. 377 – 392 (1992). – See Abstr. 012.095 for the main entry.
    Accelerated Lambda Iteration, or ALI, methods are reviewed. An emphasis is given to the critical evaluation of various methods, analysing their physical and mathematical meaning, and recommending the most advantageous methods to interested non–specialists who consider applying these methods to solving actual line formation and model stellar atmosphere problems.

064.110   Instabilities in hot–star winds: basic physics and recent developments.
S. P. Owocki.
Workshop on the Atmospheres of Early–Type Stars, p. 393 – 408 (1992). – See Abstr. 012.095 for the main entry.
    The winds of the hot, luminous, OB stars are driven by the line–scattering of the star's continuum radiation flux. Several kinds of observational evidence indicate that such winds are highly structured and variable, and it seems likely that a root cause of this variability is the strong instability of the line–driving mechanism. This paper reviews the basic physics of the linear instability and summarizes results from numerical simulations of its nonlinear evolution. Particular emphasis is placed on the dynamical importance of the diffuse, scattered radiation field, and on recent methods for incorporating such scattering effects into the numerical simulations. The author also summarizes recent preliminary results on synthetic UV line, Hα, IR continuum spectra in dynamical wind models with extensive structure.

064.111   Application of the ETLA approach in the comoving frame to the study of winds in hot stars.
M. Perinotto, C. Catala.
Workshop on the Atmospheres of Early–Type Stars, p. 417 – 419 (1992). – See Abstr. 012.095 for the main entry.
    Among the various methods developed for studying the formation of line profiles in an expanding atmosphere, the one based on the direct solution of the coupled comoving–frame transfer and statistical equilibrium equations with multilevel, multi–ion model atoms in the equivalent–two–level–atom (ETLA) scheme, offers particular advantages over other methods. The authors have adapted a version of this code for application to hot stars. Appending a wind model to a photosphere calculated with the non–LTE TLUSTY code of Hubeny (1989), they have computed the C IV line profiles in a test case with the entirely different technique, called "unified atmosphere–wind theory". The effect of the turbulence across the winds appears to be important to reproduce observed profiles in hot stars.

064.112   Interactive LTE spectrum synthesis.
     A. E. Lynas–Gray.
Workshop on the Atmospheres of Early–Type Stars, p. 420 – 424 (1992). – See Abstr. 012.095 for the main entry.
    Local Thermodynamic Equilibrium (LTE) is known to be a useful simplifying assumption when computing line–blanketed model stellar atmospheres and synthetic spectra. Element abundance adjustment can optimise agreement between a synthetic and an observed spectrum, providing a useful technique for abundance determination when spectrum lines are blended. Interactive abundance adjustment techniques are proposed as a means of minimising the effort involved in using LTE spectrum synthesis to obtain abundances for several elements in a large sample of stars.

064.113   **Calculation of line positions in the presence of magnetic and electric fields in white dwarf spectra.**
S. Friedrich, H. Ruder, R. Östreicher.
Workshop on the Atmospheres of Early–Type Stars, p. 425 – 427 (1992). – See Abstr. 012.095 for the main entry.
The authors present first results of their line position calculations in the presence of strong magnetic and electric fields, as one can find in the dense atmospheres of magnetic White Dwarfs. The problem was solved by second order perturbation theory with hydrogenic wavefunctions calculated in a pure magnetic field and arbitrary orientations of the electric and magnetic fields.

064.114   **Treatment of strong magnetic fields in very hot stellar atmospheres.**
D. Engelhardt, I. Bues.
Workshop on the Atmospheres of Early–Type Stars, p. 428 – 430 (1992). – See Abstr. 012.095 for the main entry.
Polarized radiative transfer in a hydrogen–rich model atmosphere of an effective temperature of 50000K and a magnetic field strength of B = 1000 Tesla must be investigated in the NLTE scheme. The four transfer equations for each Zeemann transition $\Delta m = 0, +1, -1$ coupled by the interaction of the polarized beam and the electron, are connected to the rate equations in terms of energy. Within the framework of the Dirac equation as a transfer equation the four equations decouple including magnetooptical parameters. Application of the Green's function method leads to a formal solution, which is inverted numerically.

064.115   **Blends, frequency grids and the Scharmer scheme.**
M. J. Stift.
Workshop on the Atmospheres of Early–Type Stars, p. 431 – 434 (1992). – See Abstr. 012.095 for the main entry.
The use of operator perturbation methods such as the Scharmer schema makes it possible to solve problems involving complex atomic and atmopsheric models.

064.116   **Multilevel non–LTE radiative transfer using exact complete and diagonal operators.**
J. J. MacFarlane.
Workshop on the Atmospheres of Early–Type Stars, p. 435 (1992). Abstract. – See Abstr. 012.095 for the main entry.

064.117   **Iron line blanketing in NLTE model atmospheres for O stars: first results.**
S. Dreizler, K. Werner.
Workshop on the Atmospheres of Early–Type Stars, p. 436 – 439 (1992). – See Abstr. 012.095 for the main entry.
The authors constructed two NLTE model atmospheres for post–AGB stars to explore the effect of iron line blanketing in hot star atmospheres.

064.118   **Calculations of non–LTE radiative transfer in extended outflowing atmospheres using the Sobolev approximation for line transfer.**
A. de Koter, H. J. G. L. M. Lamers, W. Schmutz.
Workshop on the Atmospheres of Early–Type Stars, p. 440 – 444 (1992). – See Abstr. 012.095 for the main entry.
Contents: Introduction. Properties of the model code (Density structure, temperature structure, chemical composition). Computational method (Approximate lambda operator, preconditioning statistical equilibrium). Validity of the Sobolev approximation for LBVs. Conclusions.

064.119   **3D radiative line–transfer for disk–shaped Be stars envelopes.**
W. Hummel.
Workshop on the Atmospheres of Early–Type Stars, p. 445 – 447 (1992). – See Abstr. 012.095 for the main entry.
Be stars, defined as early–type emission–line stars of luminosity classes III – V, exhibit stellar absorption and circumstellar emission features in their spectra. A new interpretation of winebottle–type emission lines is presented.

064.120   **Thermal Comptonization in standard accretion disks.**
L. Maraschi, S. Molendi.
Symposium on Imaging X–Ray Astronomy – a Decade of Einstein Observatory Achievements, p. 283 – 288 (1990). – See Abstr. 012.097 for the main entry.
Contents: Accretion disk model. Comptonized bremsstrahlung flux. Results: temperature profiles and emission spectra. Conclusions.

064.121   **Pinch effect and stellar flares physics.**
V. S. Hayrapetyan, A. G. Nikhogossian, V. V. Vikhrev.
International Workshop on Reconnection in Space Plasma, p. 163 – 167 (Jan 1989). – See Abstr. 012.099 for the main entry.
A new pinch–model of primary energy release of solar and stellar flares is suggested.

064.122   **Speculation of the origin of X–ray emission in early–type stars.**
A. M. T. Pollock.
International Workshop on Reconnection in Space Plasma, p. 309 – 311 (Jan 1989). – See Abstr. 012.099 for the main entry.
In recent years observations of OB and Wolf–Rayet stars have shown them to be unexpectedly strong sources of X–ray and radio radiation. Nonthermal radio emission shows the presence of relativistic electrons and magntic fields far out in the powerful stellar winds that are characteristic of these stars. This suggests an alternative interpretation of the X–ray emission to Lucy and White's shock model as a result of the rapid Compton cooling that energetic electrons would suffer against photospheric radiation. It is proposed that it is the ubiquitous presence of relativistic electrons in early–type stellar winds that is the case of both radio and X–ray emission and that magnetic reconnection in single and colliding pairs of winds may be the agent of electron acceleration.

064.123   **Numerical study of colliding astrospherical stellar wind flows in binary systems.**
J. Kallrath.
1. COSPAR Colloquium: Physics of the outer heliosphere, p. 321 – 325 (1990). – See Abstr. 012.100 for the main entry.
The dynamical and geometrical aspects of two quasi radial supersonic counterstreaming gas flows are investigated. The results are used as initial data for the hyperbolic system of equations describing the conservation of mass, momentum and energy. In a model WR/O–binary system a subsonic region with an extension of a few solar radii is established. The procedure used here might also be appropriate to investigate a variety of phenomena near the heliopause.

The observation and analysis of stellar photospheres.
See Abstr. 003.028.

Accretion power in astrophysics.
See Abstr. 003.041.

Astrophysical opacities. Proceedings. Workshop on Astrophysical Opacities (WAO), Caracas (Venezuela), 15 – 19 Jul 1991.
See Abstr. 012.054.

The atmospheres of early–type stars. Proceedings.
See Abstr. 012.095.

Smoothed particle rendering for fluid visualization – three–dimensional accretion disk and jet formation.
See Abstr. 021.027.

Stark–broadening parameters of ionized mercury spectral lines of astrophysical interest.
See Abstr. 022.050.

Atomic and molecular data for opacity calculations.
See Abstr. 022.187.

Molecular opacity data for stellar atmospheres.
See Abstr. 022.188.

Calculation of transition frequencies and line strengths of water for cool star opacities.
See Abstr. 022.189.

Trajectories and envelopes in the repulsive two–fixed–centre problem and in the restricted three–body problem.
See Abstr. 042.070.

Formation of electric–current sheets in the magnetostatic atmosphere.
See Abstr. 062.003.

A comparison of the reduced and approximate systems for the time dependent computation of the polar wind and multiconstituent stellar winds.
See Abstr. 062.014.

Magnetic interchange instability of accretion disks.
See Abstr. 062.028.

Knots in stellar jets from time–dependent sources.
See Abstr. 062.029.

Applications of Lie groups to the equilibrium theory of cylindrically symmetric magnetic flux tubes.
See Abstr. 062.048.

Dynamics of wind bubbles and superbubbles. I. Slow winds and fast winds.
See Abstr. 062.052.

Dynamics of wind bubbles and superbubbles. II. Analytic theory.
See Abstr. 062.053.

Convective instability in differentially rotating disks.
See Abstr. 062.060.

Applications of Lie groups to the theory of equilibrium of cylindrically–symmetric magnetic tubes.
See Abstr. 062.079.

Asymmetric disk accretion onto magnetized rotating compact stars.
See Abstr. 062.104.

The effect of relativistic particle beams on the evolution of supernova envelopes.
See Abstr. 062.116.

Dynamos with ambipolar diffusion drifts.
See Abstr. 062.164.

A magnetic mechanism for astrophysical jet production and enhancement of disk accretion as its reaction – 2.5 D and 3 D MHD simulation studies.
See Abstr. 062.200.

Exact 2–D MHD solutions for astrophysical outflows.
See Abstr. 062.218.

Physics of flares in stars and accretion disks.
See Abstr. 062.219.

Ion–atom complexes and the recombination in stellar plasma.
See Abstr. 062.226.

Radiative transfer problems in the solar and sun–like atmospheres.
See Abstr. 063.001.

Monte Carlo simulations of Raman scattered O VI emission lines in symbiotic stars.
See Abstr. 063.004.

Line scattering in an expanding shell: interferometric and polarimetric profiles.
See Abstr. 063.005.

Upgrading the Accelerated Lambda Iteration technique by means of "least change secant methods".
See Abstr. 063.011.

The inverse Planckian transform and temperature spectra.
See Abstr. 063.012.

Partially coherent scattering in stellar chromospheres. I. Effects on resonance line thermalization.
See Abstr. 063.024.

The Sobolev approximation for line formation with partial frequency redistribution.
See Abstr. 063.025.

The theoretical polarisation from the obliquely rotating envelopes of single stars.
See Abstr. 063.027.

Time–dependent scattering and transmission function in an anisotropic two–layered atmosphere.
See Abstr. 063.030.

An exact solution of the equation of transfer for coherent scattering in an exponential atmosphere.
See Abstr. 063.031.

Line profile variations caused by low–frequency non–radial pulsations of rapidly rotating stars. II.
See Abstr. 063.044.

The role of electron scattering in the X–ray rotational light curves of Intermediate Polars.
See Abstr. 063.050.

A fast operator perturbation method for the solution of the special relativistic equation of radiative transfer in spherical symmetry.
See Abstr. 063.053.

Radiative atomic Rosseland mean opacity tables.
See Abstr. 063.057.

Particle heated atmospheres of magnetic white dwarfs.
See Abstr. 063.058.

Radiative transfer in rotating stars.
See Abstr. 063.059.

On the origin of the H I line emission associated with massive young stellar objects.
See Abstr. 063.060.

On the density and field sensitivities of dielectronic recombination.
See Abstr. 063.063.

1612 MHz OH maser emission from axisymmetric circumstellar envelopes: Miras.
See Abstr. 063.081.

Theoretical profiles of Lyman α satellites and application to synthetic spectra of DA white dwarfs.
See Abstr. 063.089.

Partially coherent scattering in stellar chromospheres. II. The first–order escape probability method.
See Abstr. 063.091.

Partially coherent scattering in stellar chromospheres. III. A second–order escape probability method.
See Abstr. 063.092.

Gamma–ray deposition and nonthermal excitation in supernovae.
See Abstr. 063.093.

X–ray emission–line spectra of photoionized plasmas: density sensitivity of the Fe L–shell series.
See Abstr. 063.097.

Solution of the equation of transfer for coherent scattering in an exponential atmosphere by Busbridge's method.
See Abstr. 063.105.

Solution of the equation of transfer with Rayleigh's phase function in a thin atmosphere.
See Abstr. 063.106.

Incorporation of density fluctuations into photoionization calculations.
See Abstr. 063.108.

A spherical circumstellar dust model for IRAS 09371 + 1212.
See Abstr. 063.109.

Line profiles emitted from an accretion torus.
See Abstr. 063.112.

A modified Eddington–Barbier relation in highly coherent resonance–line wings.
See Abstr. 063.115.

"Finding" the "missing" solar ultraviolet opacity.
See Abstr. 063.120.

Sampling methods.
See Abstr. 063.122.

Molecular opacities in M–star atmospheres.
See Abstr. 063.123.

Molecular opacity and stellar structure.
See Abstr. 063.124.

Hydrogen opacities at high densities.
See Abstr. 063.125.

Opacity problems in accretion disks around young stellar objects.
See Abstr. 063.126.

The effects of lines on the mean opacities in novae, supernovae, and accretion disks.
See Abstr. 063.127.

Solution of the equation of transfer for coherent scattering in an exponential atmosphere by the method of discrete ordinates.
See Abstr. 063.131.

Type Ia supernovae: theoretical light curves with a slow pre–maximum rise.
See Abstr. 065.001.

Excitation of pressure modes in common envelopes.
See Abstr. 065.023.

Low–mass red giants and masses of NPNs.
See Abstr. 065.030.

Winds from rotating, magnetic, hot stars: consequences for the rotational evolution of O and B stars.
See Abstr. 065.037.

Linear polarization as a consequence of rotation in exploding stars.
See Abstr. 065.047.

On the pulsations of relativistic accretion disks and rotating stars: the Cowling approximation.
See Abstr. 065.051.

Jet formation in the transition from the asymptotic giant branch to planetary nebulae.
See Abstr. 065.053.

Shock waves in stellar atmospheres and breaking waves on an ocean beach.
See Abstr. 065.057.

Weak chaos in long–period variables.
See Abstr. 065.063.

Turbulent convection with overshooting: Reynolds stress approach.
See Abstr. 065.092.

The evolution of intermediate–mass protostars. II. Influence of the accretion flow.
See Abstr. 065.099.

The equation of state for stellar envelopes: comparison of theoretical results.
See Abstr. 065.101.

Hydrodynamic studies of accretion onto ONeMg white dwarfs using a large nuclear reaction network.
See Abstr. 065.103.

Comparison of a fine–mesh hydrodynamic stellar model with a dynamic rezoning model.
See Abstr. 065.105.

Accretion flows near black holes mediated by radiative viscosity.
See Abstr. 067.005.

Compton scattering in a converging fluid flow: spherical near–critical accretion onto neutron stars.
See Abstr. 067.007.

The fate of accreted CNO elements in neutron star atmospheres: X–ray bursts and gamma–ray lines.
See Abstr. 067.008.

The X–ray spectrum of modified α–viscosity accretion disc.
See Abstr. 067.047.

Disc oscillation model for quasi–periodic light variations in cataclysmic variables.
See Abstr. 067.057.

Photon bubbles: overstability in a magnetized atmosphere.
See Abstr. 067.074.

Geometrically thin, hot accretion disks: topology of the thermal equilibrium curves.
See Abstr. 067.169.

Evolution of a Compton heated wind from an irradiated accretion disk.
See Abstr. 067.191.

The stability of a warped accretion disc.
See Abstr. 067.227.

Oscillations and seismological diagnostics of sunspots.
See Abstr. 072.059.

Formation and evolution of the protoplanetary disk.
See Abstr. 107.052.

Hydrogen infrared recombination line profiles in Be star winds.
See Abstr. 112.003.

Quiescent X–ray emission of cool stars.
See Abstr. 112.005.

A modelling of circumstellar Ba II lines for the hypergiant ϱ Cassiopeiae.
See Abstr. 112.011.

Infrared observations and thermal models of the β Pictoris disk.
See Abstr. 112.019.

Radio and infrared studies of mass–losing evolved stars.
See Abstr. 112.031.

Constraints on Be star wind geometry by linear polarisation and IR excess.
See Abstr. 112.039.

Mass loss in main–sequence A–type stars?
See Abstr. 112.058.

Mass loss from OH/IR stars: models for the infrared emission of circumstellar dusts shells.
See Abstr. 112.059.

Magnetically confined wind on the Ap star 53 Camelopardalis?
See Abstr. 112.071.

Hα observations of early–type stars.
See Abstr. 112.092.

Stellar chromospheres, coronae, and winds.
See Abstr. 112.109.

Radiation–driven wind theory: not (yet?) working.
See Abstr. 112.113.

Radiation–driven wind theory: the influence of turbulence.
See Abstr. 112.114.

Spectroscopic study of microturbulence in the atmosphere of Arcturus.
See Abstr. 114.008.

Atmospheric model and dynamical state of the atmosphere of the supergiant Eta Leonis (A0 Ib).
See Abstr. 114.024.

The energy distribution of 21 Comae Berenices (A3p): comparison to standard stars of similar spectral types and modeling.
See Abstr. 114.027.

Galactic abundance gradients from OB–type stars in young clusters and associations.
See Abstr. 114.067.

Abundance patterns in A stars: carbon and silicon.
See Abstr. 114.068.

Statistical equilibrium of Al I/II in A stars and the abundance of aluminium in Vega.
See Abstr. 114.070.

Effective temperature and models for early A stars: application to 78 Vir (A2p).
See Abstr. 114.083.

NLTE analysis of a sdO binary: HD128220.
See Abstr. 114.092.

NLTE analysis of helium rich subdwarf O stars.
See Abstr. 114.093.

Surface structures and flares in solar–like stars.
See Abstr. 116.001.

Spectroscopic signatures of active regions on main sequence stars.
See Abstr. 116.047.

Modelling inhomogeneities on B and A type magnetic peculiar stars.
See Abstr. 116.055.

A two–wind radiation–driven model for the atmospheric features of β Lyrae.
See Abstr. 117.011.

The X–ray hardness ratio variation in low–mass X–ray binaries.
See Abstr. 117.048.

6 Hz quasiperiodic oscillations from low–mass X–ray binaries: the sound of an accretion disk?
See Abstr. 117.150.

X–ray scattering and fluorescence in the wind of a massive X–ray binary.
See Abstr. 117.155.

Spectral evolution of accretion disks of dwarf novae. III. Outburst cycles of SS Cygni.
See Abstr. 117.162.

Might nondegenerate helium stars be the accretors in cataclysmic binary systems?
See Abstr. 117.194.

Might non–degenerate helium stars be the accreting components in cataclysmic binaries?
See Abstr. 117.199.

Quasi–periodic precession of disks in X–ray binaries, with possible application to Centaurus X–3.
See Abstr. 117.260.

Modelling of the large X–ray flare on II Peg observed with Ginga.
See Abstr. 117.272.

Properties of swept–up molecular outflows.
See Abstr. 121.015.

Dust around young stars. Model of envelope of the Ae Herbig star WW Vul.
See Abstr. 121.024.

Magnetic structure in cool stars. XVIII. UV–line emissions from T Tauri stars.
See Abstr. 121.027.

On the interpretation of the spectrum of FU Orions.
See Abstr. 121.052.

Surface inhomogeneity on T Tau stars and the structure of their stellar winds.
See Abstr. 121.062.

The atmospheric motion of the β Cephei star: 12 Lacertae.
See Abstr. 122.055.

Stellar flares: observations and modelling.
See Abstr. 122.157.

Luminous blue variables.
See Abstr. 122.158.

Abundances of classical novae.
See Abstr. 124.009.

The secondary outburst maximum of T Coronae Borealis: implications for the physics of accretion disks.
See Abstr. 124.057.

Limit cycle instabilities in a free accretion disk – a source of energy in SN 1987A.
See Abstr. 125.091.

Postexplosion hydrodynamics of SN 1987A.
See Abstr. 125.114.

The [O I] $\lambda\lambda6300$, 6364 doublet of SN 1987A.
See Abstr. 125.115.

Barium and other s–process elements in the early–time spectrum of SN 1987A.
See Abstr. 125.127.

The ring around SN 1987A and rotation of the progenitor.
See Abstr. 125.137.

The constitution of the atmospheric layers and the extreme ultraviolet spectrum of hot hydrogen–rich white dwarfs.
See Abstr. 126.097.

Iron abundance in the hot DA white dwarfs Feige 24 and G191–B2B.
See Abstr. 126.108.

Analysis of PG 1159 stars.
See Abstr. 126.119.

Infall in collapsing protostars.
See Abstr. 131.196.

Initiation of bipolar flows by magnetic field twisting in protostellar nebulae.
See Abstr. 131.262.

Heating the haloes of planetary nebulae.
See Abstr. 134.022.

NLTE analysis of the hydrogen–deficient central star of the planetary nebula Abell 78.
See Abstr. 134.044.

Observational constraints on stellar wind models of broad emission lines of AGNs.
See Abstr. 158.227.

065 Stellar Structure and Evolution

065.001   Type Ia supernovae: theoretical light curves with a slow pre–maximum rise.
A. Khokhlov (*A. M. Khokhlov*), E. Müller, P. Höflich.
Astron. Astrophys., Vol. 253, No. 1, p. L9 – L12 (Jan 1992). Letter–to–the–editor.
    The authors present theoretical light curve (LC) calculations which reproduce Type Ia supernovae with a slow pre–maximum rise (V maximum at ~ 20 days), like the recently observed SN1990N. They use the same input physics as in their previous work (Höflich et al. 1991). In addition, both Thomson and line scattering has been included, which is crucial for reproducing the monochromatic LC. For the radiation transport the moment equations are solved. The calculations are based on delayed detonation models in which the transition from a deflagration to a detonation occurs due to the pulsation of an exploding white dwarf. The interaction with an extended low density envelope created during the pulsation leads to a substantial redistribution of the kinetic energy inside the ejecta.

065.002   The rate of mixing in semiconvective zones.
H. C. Spruit.
Astron. Astrophys., Vol. 253, No. 1, p. 131 – 138 (Jan 1992).
    Mixing in semiconvective zones is treated as a double–diffusive process. The mixing takes place by overturning cells in horizontal layers, separated by stable interfaces across which transport takes place by diffusion. Effective diffusion coefficients are derived and tested against laboratory results. Easily implementable expressions for stellar evolution calculations are given. They imply a much smaller mixing rate than assumed in other formulations.

065.003   Effect of horizontal turbulent diffusion on transport by meridional circulation.
B. Chaboyer, J.–P. Zahn.
Astron. Astrophys., Vol. 253, No. 1, p. 173 – 177 (Jan 1992).
    The authors examine the effect of horizontal turbulent motions on the transport of chemicals by meridional circulation in a stellar radiation zone. They show that strong enough turbulent diffusion inhibits the advection of a chemical element, because it tends to homogenize horizontal layers. The vertical transport then reduces to a diffusion process, a phenomenon similar to the classical shear dispersion in a pipe flow (Taylor 1953). Likewise, the differential rotation is presumably smoothed out in latitude, but this will not greatly hinder the advection of angular momentum. The presence of such horizontal turbulence may thus explain the discrepancy between the transport of chemicals and that of angular momentum which is observed in the Sun (Law et al. 1984; Pinsonneault et al. 1989). Assuming that horizontal turbulence enforces nearly spherical rotation, the authors formulate the advection of angular momentum as a one–dimensional problem.

065.004   Evolution of solar and stellar rotation.
S. Catalano.
11. European Regional Astronomy Meeting of the IAU: New windows to the universe, Vol. 1, p. 161 – 178 (1990). – See Abstr. 012.001 for the main entry.
    The paper focuses on the observational aspects which help to delineate the rotational history of the sun and its behavior among stars of similar type.

065.005 On the uniqueness of static perfect–fluid solutions in general relativity.
R. Beig, W. Simon.
Commun. Math. Phys., Vol. 144, No. 2, p. 373 – 390 (Feb 1992).
  Following earlier work of Masood–ul–Alam, the authors consider a uniqueness problem for non–rotating stellar models. Given a static, asymptotically flat perfect–fluid spacetime with barotropic equation of state $\varrho(p)$, and given another such spacetime which is spherically symmetric and has the same $\varrho(p)$ and the same surface potential: it is proved that both are identical provided $\varrho(p)$ satisfies a certain differential inequality. This inequality is more natural and less restrictive than the conditions required by Masood–ul–Alam.

065.006 Turbulent shear flow and rotation.
J.–P. Zahn.
11. European Regional Astronomy Meeting of the IAU: New windows to the universe, Vol. 1, p. 291 – 300 (1990). – See Abstr. 012.001 for the main entry.
  Differential rotation shares most properties of plane parallel shear flow. In the absence of stabilizing effects due to vertical stratification, magnetic field, etc., it becomes unstable for sufficiently large Reynolds number. Depending on the profile of the flow, the instability is either linear, or of finite amplitude. In differentially rotating stars, those shear instabilities reach a turbulent regime. The properties of that turbulence are sketched out, and it is shown how to estimate the transport of chemicals and angular momentum.

065.007 Rotation, age and lithium.
E. Schatzman.
11. European Regional Astronomy Meeting of the IAU: New windows to the universe, Vol. 1, p. 375 (1990). Abstract. – See Abstr. 012.001 for the main entry.

065.008 White dwarfs.
J. Isern, E. Garcia–Berro, M. Hernanz, R. Mochkovitch.
11. European Regional Astronomy Meeting of the IAU: New windows to the universe, Vol. 1, p. 391 – 400 (1990). – See Abstr. 012.001 for the main entry.

065.009 Diagnostics of stellar evolution: the oxygen isotopes.
D. S. P. Dearborn.
Phys. Rep., Vol. 210, No. 6, p. 367 – 382 (Jan 1992).
  The oxygen isotope ratios provide a probe of the physics of stellar interiors. Among the phenomena that can be studied through the oxygen isotope ratios are the maximum depth of penetration of the convective envelopes on the giant branch, the efficiency of nonconvective mixing, as well as restrictions on the mass loss that occurs. The oxygen isotope ratios also provide an indicator of the nuclear processing that occurs during third dredge up and carbon star formation. In addition to the oxygen isotopes, the author indicates the expected behavior of nitrogen and the light elements.

065.010 Carbon detonations in rapidly rotating white dwarfs.
M. Steinmetz, E. Müller, W. Hillebrandt.
Astron. Astrophys., Vol. 254, No. 1/2, p. 177 – 190 (Feb 1992).
  The authors have performed a set of two–dimensional hydro-dynamic simulations of the propagation of detonation waves in rapidly rotating white dwarfs. The axisymmetric initial models used in the simulations are in rotational equilibrium and possess a density stratification similar to that of configurations predicted by the merging scenario of Type Ia supernovae. The energy release is approximated by a single exothermic nuclear reaction $14\ ^{12}C \rightarrow 3\ ^{56}Ni$ processing at the rate of the $^{12}C(^{12}C,^{4}He)^{20}Ne$ reaction. Central and off–center ignitions have been studied. When burning is ignited in the center the results show, that independently of the shape of the star, all matter is burned into iron group elements and the star is completely disrupted after typically 200 msec. Igniting the white dwarf in an extended central region having a more or less flattened spheroidal or cylindrical shape gives rise to an initially different behaviour, but

within about 100 msec the shape of the detonation wave approaches that of one being ignited only in the center. When burning is ignited off–center all carbon is consumed, too. Thus detonations in non–spherical massive white dwarfs cannot explain the observed intermediate mass elements seen in the spectra of Type Ia supernovae.

065.011 Advantages and limitations of the bipolytrope model for computing the secular evolution of cataclysmic binaries.
U. Kolb, H. Ritter.
Astron. Astrophys., Vol. 254, No. 1/2, p. 213 – 223 (Feb 1992).
  For the purpose of parameter studies of the secular evolution of cataclysmic binaries the authors improved and generalized the bipolytrope description for mass donors with masses $\lesssim 1\ M_\odot$. In this simplified model the secondary is treated as a polytrope with different indices in its radiative core and convective envelope, where in contrast to previous work the core index is now allowed to vary with the stellar mass. For quantitative investigations using this bipolytrope model it is necessary to gauge the 2 free parameters of the model against full stellar models. The specific problem to which the bipolytrope model is to be applied determines the way of calibrating. The authors present a detailed comparison between evolutionary sequences obtained with a full stellar evolution code and corresponding bipolytrope sequences in a calibration designed to give the best overall agreement. Apart from sequences turning on mass transfer with an initial secondary mass close to the mass where ZAMS models become fully convective the simplified bipolytrope sequences reproduce the full ones with an accuracy of better than 5%.

065.012 On mass–transfer rates in classical nova precursors.
I. Iben Jr., M. Y. Fujimoto, J. MacDonald.
Astrophys. J., Vol. 384, No. 2, p. 580 – 586 (10 Jan 1992).
  A simple model is presented to describe the evolution of cataclysmic variables (CVs). Mass transfer in long period CVs ($P_{orb} > 3$ hr) is assumed to be due to a magnetic stellar wind and in short–period CVs ($P_{orb} < 2$ hr) to be due to gravitational wave radiation. The critical accreted mass for a classical nova event is adopted to be that for which the thermal structure of the white dwarf is in steady state after many nova events. The frequency with which long–period CVs (accretion rate $10^{-8.2\pm0.6}M_\odot yr^{-1}$) experience classical nova outbursts is predicted to be 20 – 35 times the frequency with which short–period CVs (accretion rate less than $2 \times 10^{-10}M_\odot yr^{-1}$) experience such outbursts, whereas the number of observable CVs with small accretion rates is estimated to be comparable with the number of observable CVs with large accretion rates. Hence, known CVs with small accretion rates are not expected to have experienced a recent nova event and CVs with the highest accretion rates have the greatest chance of being associated with an historical nova. This may help understand why most known precursors and followers of historical novae show high accretion rates.

065.013 Evolutionary models of halo stars with rotation. II. Effects of metallicity on lithium depletion, and possible implications for the primordial lithium abundance.
M. H. Pinsonneault, C. P. Deliyannis, P. Demarque.
Astrophys. J., Suppl. Ser., Vol. 78, No. 1, p. 179 – 203 (Jan 1992).
  The authors have computed models of metal–poor stars with rotation and have compared their lithium depletion with observations of halo stars. The models that have turn–off ages compatible with the observations have a nearly flat $Li - T_{eff}$ relationship in the region of the Spite lithium "plateau". Depending on the initial angular momentum, the models have a depletion factor ranging between a factor of 5 and a factor of 10 at fixed T_{eff}, implying a maximum initial lithium abundance of 3.1. Both the dispersion and the overall depletion factor are much smaller for metal–poor models than for solar metallicity ones. The factors that determine lithium depletion in rotational models are discussed, and the different depletion patterns in solar metallicity and metal–poor models are traced to differences in their structure and evolution. The dependence of the lithium

depletion on age, mass, initial angular momentum, and metallicity is also discussed. The dispersion predicted from these models is not inconsistent with the observations.

065.014   **Low–mass X–ray binary evolution and the origin of milisecond pulsars.**
J. Frank, A. R. King, J.–P. Lasota.
Astrophys. J., Lett., Vol. 385, No. 2, p. L45 – L48 (1 Feb 1992).
    The authors consider the evolution of low–mass X–ray binaries (LMXBs). It is shown that X–ray irradiation of the companion stars causes these systems to undergo episodes of rapid mass transfer followed by detached phases. The systems are visible as bright X–ray binaries only for a short part of each cycle, so that their space density must be considerably larger than previously estimated. This removes the difficulty in regarding LMXBs as the progenitors of low–mass binary pulsars. The authors identify the low–accretion–rate phase of the cycle with the soft X–ray transients. They show that 3 hr is likely to be the minimum orbital period for LMXBs with main–sequence companions and suggest that the evolutionary endpoint for many LMXBs may be systems which are sources of gamma–ray bursts.

065.015   **Erratum: "Dynamic tides in stars as forced isentropic oscillations and their effects on free oscillations"**
[Astron. Astrophys., Vol. 237, No. 1, p. 110 – 124 (Oct 1990)].
R. Polfliet, P. Smeyers.
Astron. Astrophys., Vol. 255, No. 1/2, p. 495 (Feb 1992). See Abstr. 52.065.025.

065.016   **Convection zone of white–dwarf stars.**
Z. M. Atweh, D. Eryurt–Ezer.
Astrophys. Space Sci., Vol. 187, No. 1, p. 27 – 36 (Jan 1992).
    The boundary convection zones of hot helium white–dwarf stars in the range $17000K \leqslant T_e \leqslant 30000K$ are studied. Recently, an anisotropic mixing–length theory which determines the mixing–length parameter locally is applied for the convection zones calculation. Comparing with the calculations by using the mixing–length theory, it is found that maximum velocity decreases appreciably, and the other boundary conditions are affected.

065.017   **Frequency–spectra solutions for multiperiodic variable stars: analysis and interpretation of artificial data.**
R. Ventura.
Astrophys. Space Sci., Vol. 187, No. 1, p. 37 – 55 (Jan 1992).
    An accurate and extensive analysis on the present possibility of obtaining definite, unambiguous frequency–spectra solutions for multiperiodic δ–Scuti stars has been performed using artificial data. Some convenient photometric data sampling techniques have been adopted to produce very realistic light curves, with gaps and noise and subsequently critically analyzed. Useful suggestions about those to adopt in order to detect very low–amplitude components of the signal have been inferred. The results obtained point out the need of extensive devoted series of measurements, spread over a very long time span and also over a wide baseline of observers.

065.018   **The dynamical evolution of tidal capture binaries.**
C. S. Kochanek.
Astrophys. J., Vol. 385, No. 2, p. 604 – 620 (1 Feb 1992).
    The formation of a binary by the tidal capture of a main–sequence star (mass M_1, radius R_1) by a neutron star (mass M_2) is an extremely violent dynamical event. The star, which can be characterized as an irrotational Riemann ellipsoid in which the change in energy is proportional to the change in angular momentum, emerges from the first pericentric passage with both a large oscillation and a bulk rotation. If the energy in the oscillation exceeds approximately $0.15 GM_1{}^2/R_1$, the star becomes dynamically unstable during the encounter, and either disrupts or loses a substantial fraction of its mass. Viscosity is unimportant until late in the evolution (ellipticity $e \lesssim 0.5$), and in most cases nonlinear sources of viscosity such as mode–mode coupling in the atmosphere of the star are the dominant source of viscosity. The energy of the oscillations is stored in only a few

modes, and their amplitudes evolve in a slowly damped random walk, which leads to large fluctuations in the orbital binding energy. The time scale for the evolution of the system is dominated by the fluctuations into low–energy, long–period orbits. A small fraction (1% – 2%) of the binaries become unbound as a result of de–excitation or heating from encounters with other stars. Another 2% – 5% remain bound but are scattered into orbits with pericenters $R_p/R_1 > 50$, owing to angular momentum transfer during encounters with other stars. While these represent a small fraction of tidal encounters, they can be a large fraction of non–merging encounters, and they represent all of the objects which in the traditional picture of tidal encounters form binaries with orbital radii of $6R_1 – 10R_1$.

065.019   **A mechanism for orbital period modulation in close binaries.**
J. H. Applegate.
Astrophys. J., Vol. 385, No. 2, p. 621 – 629 (1 Feb 1992).
    Some eclipsing variables are observed to undergo orbital period modulations of amplitude $\Delta P/P \sim 10^{-5}$ over time scales of decades or longer. These modulations can be explained by the gravitational coupling of the orbit to variations in the shape of a magnetically active star in the system. The variable deformation of the active star is produced by variations in the distribution of angular momentum as the star goes through its activity cycle. This mechanism typically requires that the active star be variable at the $\Delta L/L \approx 0.1$ level, and be differentially rotating at the $\Delta\Omega/\Omega \approx 0.01$ level. The torque needed to redistribute the angular momentum can be exerted by a mean subsurface magnetic field of several kilogauss.

065.020   **Classical nova models with accretion heating at accretion rates of 10^{-9} and $10^{-10}M_\odot$ per year.**
D. Prialnik, A. Kovetz.
Astrophys. J., Vol. 385, No. 2, p. 665 – 669 (1 Feb 1992).
    Two evolutionary sequences of accretion onto a 1.25 $M_\odot$ C–O white dwarf are computed for $\dot{M} = 10^{-10}M_\odot yr^{-1}$ and $\dot{M} = 10^{-9}M_\odot yr^{-1}$ and are carried through the thermonuclear runaway and the dynamical phases of mass loss, followed by contraction of the remnant and decline. Improved algorithms are used for diffusion during the accretion phase and for the computation of mass loss. Accretional heating is taken into account. It is found that outbursts typical of classical novae may be achieved for the adopted rather high accretion rates that are compatible with observations. The heavy element mass fractions in the ejecta are found to be 0.31 and 0.19, the maximal velocities reached are 2300 and 1600 km s^{-1}, and mass loss lasts for ~ 12 and ~ 25 days, respectively, for the two models. The effect of diffusion and its dependence on the accretion rate are discussed.

065.021   **Evolution of a millisecond binary pulsar in the period gap.**
A. V. Fedorova, E. V. Ergma.
Sov. Astron. Lett., Vol. 17, No. 6, p. 419 – 422 (Nov 1991). Current Physics Microform No.: 9207X1857. English translation of Pis'ma Astron. Zh., Tom 17, No. 11, p. 999 – 1007 (Nov 1991).
    The authors investigated as a function of the evaporation efficiency f′ the evolution of a millisecond binary pulsar (P $\approx$ 1.8 msec, B = 5·10^{8}G) in the period gap. They found that evolution in the period gap is determined by the characteristic times τ_{gr}, τ_{nc}, τ_{evap}, and τ_{sd}, and a small change in f can significantly affect the evolution of the system. They show that for f = 0.015 – 0.017, the Roche lobe is filled when the orbital period P $\approx$ 50 min. The authors discuss the application of these results to the system MXB 1916–05.

065.022   **Can full convection explain the observed short–period limit of the W UMa–type binaries?**
S. M. Rucinski.
Astron. J., Vol. 103, No. 3, p. 960 – 966 (Mar 1992). Current Physics Microform No.: 9203E2168.
    Stars become fully convective at low effective temperatures. The unique mass–radius relation for fully convective stars (at a given effective temperature) cannot be reconciled with such a

relation for Roche geometry except for a unique combination of the orbital periods and effective temperatures: the "Hayashi line" for contact binaries. The dynamically stable contact systems cannot exist with effective temperatures lower than the full–convection point. It is shown that the more massive component has relatively thicker convective envelope and at low temperatures becomes fully convective first. Because of the energy transfer to the secondary, the full–convection point of the primary component is shifted to somewhat larger masses than for single stars (depending on the mass–ratio which determines how much radiating area provides the secondary). The full–convection limit for solar–abundance, main–sequence systems is located at about B–V ≈ 1.5, almost independently of the total mass or mass ratio of a system. This point is some distance from the colors of the currently known shortest–period systems. Thus, although the Hayashi line is a definite limit on parameters of contact binaries, something else produces the observed cutoff in the period distribution.

065.023   Excitation of pressure modes in common envelopes.
N. Soker.
Astrophys. J., Vol. 386, No. 1, p. 190 – 196 (10 Feb 1992).
The author studies the excitation of oscillatory modes by a low–mass star orbiting inside a common envelope with a more massive star. This paper studies adiabatic high spherical harmonic degree ($l \gg 1$) modes propagating outward in the envelope. The dominant oscillatory modes are those for which the spherical harmonic order is large with $m \sim l$, and thus the amplitudes are large close to the equatorial plane and small closer to the poles. A secondary of mass $\sim 1\%$ of the primary mass excites modes with relative surface amplitudes of a few times 10%. Even a Jupiter–like brown dwarf, when it is very deep in the envelope of an asymptotic giant branch star, can cause perturbations of relative surface amplitudes of $\sim 10\%$ near the equatorial plane. If mass loss is influenced by oscillations, then this mechanism can lead to higher mass loss in the equatorial plane. In the case of an asymptotic giant branch primary, this might lead to the formation of an elliptical planetary nebulae. Oscillation excitation by a low–mass secondary might also be relevant to the formation of the ring around SN 1987A.

065.024   The frequencies of supernovae in binaries.
A. V. Tutukov, L. R. Yungelson (*L. R. Yungel'son*), I. Iben Jr.
Astrophys. J., Vol. 386, No. 1, p. 197 – 205 (10 Feb 1992).
The formation frequencies of about two dozen different kinds of supernovae have been obtained with a numerical program which calculates such frequencies according to algorithms based on several observed properties of binary stars (distribution of initial main–sequence pairs with regard to primary mass, mass ratio q, and orbital separation A) and on rough estimates of the consequences of Roche lobe overflow (a transformation between initial mass and remnant mass; orbital expansion in mass–conservative events and orbital shrinkage due to angular momentum loss by a magnetic stellar wind; gravitational wave radiation; and braking in a common envelope). The total rate at which supernova explosions are calculated to occur in our Galaxy is between $0.032 \, \text{yr}^{-1}$ and $0.048 \, \text{yr}^{-1}$, depending on assumptions about the initial distribution over q and about the common envelope angular momentum loss parameter. From 25% to 45% of supernovae originate in initially close binaries. The fact that the number of theoretically predicted types of supernova and supernova–like events exceeds the number of types which have received the status of forming a class is consistent with the growing recognition that there is considerable variety among observed supernovae.

065.025   Constraints on the surface magnetic fields of hot stars with winds.
M. Maheswaran, J. P. Cassinelli.
Astrophys. J., Vol. 386, No. 2, p. 695 – 702 (20 Feb 1992).
Several constraints on the surface magnetic fields of rotating stars with winds are discussed. It is shown that there are two allowed ranges for the strengths of surface radial magnetic fields, which the authors call the "strong field" and "weak field" ranges. In a previous paper, it was shown that, in a fast magnetic rotator, the radial component must be strong with a lower bound determined by the speed of subsurface motions. In the present paper, rotating hot stars with winds and weaker surface magnetic fields are considered. If there is appreciable mass loss caused, say, by radiation forces, the surface magnetic field may have a weak radial component, which has an upper bound that varies directly as the mass–loss rate and the lower bound of the strong field. Since the strong–field lower bound increases with rotation speed, so does the weak–field upper bound. For the O–type main–sequence star 9 Sgr and the B supergiant star ζ^1 Sco, the upper bound for a weak field is found to be of order 1 G, which is consistent with recent interpretations of radio observations. For a Wolf–Rayet star, with stellar parameters similar to those of CV Ser, it is found to be $\sim 20 \, \text{G}$, at faster rotation speeds. The average strengths of a surface radial field, in hot star with moderate or rapid rotation, cannot lie between the lower bound for the "strong field" and the upper bound for the "weak field". There is some observational support for the claim that a surface radial field should belong to one of these two distinct intervals.

065.026   The evolution through H and He burning of Galactic cluster stars.
V. Castellani, A. Chieffi, O. Straniero.
Astrophys. J., Suppl. Ser., Vol. 78, No. 2, p. 517 – 536 (Feb 1992).
The authors extend previous evolution computations of intermediate–mass stars to lower masses in order to include the red giant branch (RGB) phase transition, i.e., the transition between models igniting He in a highly degenerate core and those igniting He quiescently. Selected stellar models down to 1 $M_\odot$ and having solar metallicity have been computed from the zero–age main sequence up to the onset of thermal pulses on the asymptotic giant branch. On this basis, the authors produce theoretical isochrones and synthetic clusters in the range of ages $30 \leqslant t \leqslant 11,000$ Myr, covering the range of age spanned by the large majority of Galactic clusters. It is found that the variations among the relative frequencies of subgiants, red giants, and clump (central He–burning) stars are not strictly connected to the RGB phase transition. Further, very few stars (if any) are expected to populate the bright part of the RGB of any known Galactic open cluster. It is also found that in the range of ages between 0.6 and 10 Gyr, the luminosity of the central He–burning models is almost insensitive to the age. The fit to the color–magnitude diagrams of a set of Galactic clusters, i.e., the Pleiades, Hyades, Praesepe, NGC 2420, and NGC 188, does show a good agreement between observational data and theoretical expectations. This is taken as evidence that claims postulating a core overshooting scenario are unfounded. Tables of selected isochrones are finally presented in a form suitable for comparisons with observed color – magnitude diagrams.

065.027   s–processing in massive stars as a function of metallicity and interpretation of observational trends.
C. M. Raiteri, R. Gallino, M. Busso.
Astrophys. J., Vol. 387, No. 1, p. 263 – 275 (1 Mar 1992).
The authors first analyze the s–process in massive stars as a function of metallicity. The nucleosynthesis occurring in both core helium and shell carbon burning is investigated by numerically modeling nuclear reaction networks, subject to the conditions provided by stellar models. The efficiency of the neutron–capture mechanism depends on the amount of enhancement that is assumed for the α–rich nuclei, that are observed to be overabundant with respect to iron in stars of metallicity lower than solar. The authors find that the s–process in massive stars is "secondary–like" in the Galactic disk, the amount of s–process matter ejected being roughly proportional to Fe. In the Galactic halo, the s–efficiency drops at low metallicities, below [Fe/H] $\sim$ –2. A critical neutron poison during He burning is ^{16}O; its effects are studied in detail. In the second part, the elemental s–contributions from massive stars to the solar abundances from iron to zirconium (the so–called weak component) are presented. It is found that the weak component can account for a consistent fraction of the solar Cu, Ga, Ge, and Se. For this atomic mass

range, the present analysis is able to separate the contributions to the solar system abundances coming from the various nucleosynthetic sources. The authors discuss the evolutionary trends of elements such as Co, Ni, Cu, Zn, Rb, Sr, Y, and Zr in comparison with observations of halo stars.

065.028   Meridional circulation in rotating stars.
          T. Nakakita.
Astrophys. Space Sci., Vol. 188, No. 1, p. 41 – 87 (Feb 1992).
    The redistribution of angular momentum caused by the meridional circulation in a rotating Cowling–type star is studied as a nonlinear initial–value problem, employing the first–order perturbation theory and Legendre expansion. The difficulty of the Eddington–Sweet theory, that is, the meridional velocity becomes infinitely large both on the free surface and on the interface between the radiative and convective regions, is removed following Osaki's (1972) or Sakurai's (1975) proposal. The present work shows that the star moves toward the almost circulation free state in the Eddington–Sweet time–scale. However, the resultant almost circulation free state is quite different from the Roxburgh (1964) solution; the random circulation argued by Kippenhahn (1974) never appears.

065.029   Constant–mass sequences of differentially–rotating po-
          lytropes.
          D. Galli.
Astrophys. Space Sci., Vol. 188, No. 2, p. 241 – 255 (Feb 1992).
    In this paper, differentially–rotating polytropic constant–mass sequences are computed by implementation of the so–called "constant units technique". Then, numerical results concerning such constant–mass sequences are compared with their respective values obtained when corresponding constant–central–density sequences are computed.

065.030   Low–mass red giants and masses of NPNs.
          A. Harpaz, G. Shaviv.
Astrophys. Space Sci., Vol. 189, No. 1, p. 11 – 21 (Mar 1992).
    The evolutionary track of low–mass red giant stars (0.7–0.9 $M_\odot$) is computed with the aim to demonstrate the conditions under which low–mass white dwarfs (WDs) can form through the evolution of single stars. Also, the influence of the mixing length to the scale height ratio on the radius of the star is calculated and the coupling between the mixing–length and the mass–loss rate parameters is investigated. The conclusions are that the uncertainties in mass–loss and mixing–length to scale-height ratio leave enough parameter space to allow the formation of low–mass WD via single star evolution. The authors also conclude that the gap between proto–WD stars without any nebula and stars with well–defined nebulae is bridged by stars which have a dilute gas cloud around them which cannot be observed as a nebula.

065.031   New form of dynamo equation for convective regions of
          rotating stars.
          G. S. Bisnovatyi–Kogan (*G. S. Bisnovatyj–Kogan*).
Astrophys. Space Sci., Vol. 189, No. 1, p. 147 – 149 (Mar 1992).
Letter–to–the–editor.
    Differential rotation generated by convection in rotating stars leads to amplification of the toroidal magnetic field. As there is a direct connection between rotational velocity Ω and its gradient $\nabla\Omega$, the equation of magnetic field generation contains terms, depending directly on Ω. It follows, therefore, that the mechanism of the α–ω dynamo is always realized in convective zones of stationary rotating stars making the action of the α^2–dynamo unimportant.

065.032   The pulsation of Delta–Scuti stars.
          H. Milligan, T. R. Carson.
Astrophys. Space Sci., Vol. 189, No. 2, p. 181 – 211 (Mar 1992).
    Evolution, linear pulsation, and nonlinear pulsation codes were combined to produce nonlinear models of Delta–Scuti stars in an instability region extending over $3.8 < \log T_e < 3.95$ and $0.6 < \log L/L_\odot < 2.0$. The linear analysis upheld the consensus that they are normal Population I stars of about 2 $M_\odot$, in stages

of evolution corresponding to central hydrogen burning and shell hydrogen burning. The growth rates were very slow; driving was due to an opacity mechanism in the second helium ionization region; periods and period ratios of the lowest modes of the models were in the same range as those observed. A wide range of nonlinear models was investigated. When eigenfunctions from the linear analysis were used as initial velocity profiles, it was found that the dominant peak in the periodogram of the light curve corresponded to the mode initiated. For a small subset of models, limiting amplitudes were identified, and were found to be in close agreement with observed light amplitudes.

065.033   Astration and production in chemical evolution.
          L. I. Arany–Prado, R. de la Reza.
IAU Symposium No. 149: The stellar populations of galaxies, p. 387 (1992). – See Abstr. 012.007 for the main entry.

065.034   New theoretical isochrones.
          A. Bressan, G. Bertelli, C. Chiosi, F. Fagotto,
          E. Nasi.
IAU Symposium No. 149: The stellar populations of galaxies, p. 395 (1992). – See Abstr. 012.007 for the main entry.
    Using evolutionary tracks computed by Alongi et al. (1991) for the two chemical compositions Z = 0.020, Y = 0.28 and Z = 0.008, Y = 0.25 the authors constructed theoretical isochrones from 20×10^9yr to 3×10^6yr.

065.035   Angular momentum transport in magnetized stellar
          radiative zones. I. Numerical solution to the core spin–up
          model problem.
          P. Charbonneau, K. B. MacGregor.
Astrophys. J., Vol. 387, No. 2, p. 639 – 661 (10 Mar 1992). With plates 3 – 6.
    The authors investigate the time evolution of the angular momentum and induced toroidal magnetic field distribution in an initially nonrotating radiative stellar envelope containing a large scale poloidal magnetic field, following the impulsive spin–up of the underlying core. The authors present a large set of numerical calculations pertaining to monopolar, dipolar and quadrupolar magnetic configurations, with and without density gradients across the envelope, as well as a set of solutions for which the poloidal field is only partially anchored to the core. The use of the Galerkin finite element method yields an extremely robust computational scheme, allowing the calculations to extend to Reynolds numbers up to 10^5 and to time intervals up to 10^4 Alfvén times. The most striking feature of the fully core–anchored solutions is that a state of solid–body rotation is always attained, independent of poloidal field geometry or choice of Reynolds numbers. However, the time required to enforce a state of strict solid–body rotation (or very mean solid–body rotation) is in general much longer than the core – surface Alfvén transit time, and increases rather slowly with increasing Reynolds numbers. Nevertheless, for field strengths of order 1 G, it still remains considerably shorter than main–sequence nuclear evolutionary time scales, and is one order of magnitude smaller than early main–sequence rotational evolution time scales. It is shown that the relatively rapid transition toward solid–body rotation depends critically on *all* field lines having at least one footpoint anchored on the rigidly rotating core.

065.036   Effects of Fe/C phase separation on the ages of white
          dwarfs.
          Z. W. Xu, H. M. Van Horn.
Astrophys. J., Vol. 387, No. 2, p. 662 – 672 (10 Mar 1992).
    The authors have calculated the energy release associated with the phase separation of Fe from C in a predominantly C white dwarf. Using an extension of the equation of state developed by Salpeter and Zapolsky, the authors have computed the total gravitational plus internal energy differences between models of homogeneous composition and those with Fe–enriched cores. In the unlikely case where the core is pure Fe, a substantial extension of the white dwarf cooling times is found, even with the small cosmic abundance of this element. For the more realistic core compositions that result if the Fe/C phase diagram is either

of the spindle or of the azeotropic type, the energy release is still sufficient to prolong the cooling times by ~ 0.6 Gyr. The authors have computed the luminosity function for Fe/C phase separation and find that it produces an appreciable "bump" in the luminosity function. The authors have also considered both the possiblity that C/O, Fe/C, and Ne/C phase separation may occur sequentially, yielding a maximum age ~ 13 Gyr, as well as the possibility that subsequent phase separation processes may be prevented by the first one to occur. The latter case seems the most probable, making the faintest white dwarf less than 10 Gyr old.

065.037   Winds from rotating, magnetic, hot stars: consequences for the rotational evolution of O and B stars.
K. B. MacGregor, D. B. Friend, R. L. Gilliland.
Astron. Astrophys., Vol. 256, No. 1, p. 141 – 147 (Mar 1992).
   In an effort to obtain estimates of magnetic field strengths in hot stars, the authors have computed the evolution of rigidly rotating 15 and 30 $M_\odot$ stars, including the effects of magnetically coupled, line–driven mass loss as described by the model of Friend and Mac–Gregor (1984). Using mean rotational velocities (i.e. $v \sin i$ values) derived from observations of main–sequence stars of these masses to specify the initial state, they followed the variations in time of the surface rotation rate due to internal structural changes and wind–related breaking. In each case the initial magnetic field strength was varied until the calculated rotational velocity at the onset of the blue supergiant phase was in resonable agreement with the corresponding mean $v \sin i$ value derived from observations. For both evolutionary models, it is found that the computed rotational velocity decrease in the presence of a magnetic field of even modest strength ($\lesssim 100$ G) exceeds the limits set by observations. The limitations of this study and its implications for both the magnetic fields and winds of hot stars are discussed.

065.038   The evolution of high–metallicity horizontal–branch stars and the origin of the ultraviolet light in elliptical galaxies.
E. Horch, P. Demarque, M. Pinsonneault.
Astrophys. J., Lett., Vol. 388, No. 2, p. L53 – L56 (1 Apr 1992).
   Evolutionary calculations of high–metallicity horizontal–branch stars show that for the relevant masses and helium abundances, post–HB evolution in the HR diagram does not proceed toward and along the AGB, but rather toward a "slow blue phase" in the vicinity of the helium–burning main sequence, following the extinction of the hydrogen shell energy source. For solar and twice solar metallicity, the blue phase begins during the helium shell burning phase; for 3 times solar metallicity, it begins earlier, during the helium core burning phase. This behavior differs from what takes place at lower metallicities. The implications for high–metallicity old stellar populations in the Galactic bulge and for the integrated colors of elliptical galaxies are discussed.

065.039   The effect of angular momentum exchange in the evolution of binary systems.
R. Q. Huang, K. N. Yu, Z. W. Han.
Astron. Astrophys., Vol. 256, No. 2, p. 438 – 441 (Mar 1992).
   The evolution of a binary system consisting of a 8 $M_\odot$ and 5.5 $M_\odot$ star has been studied in the conservative approximation of Case B mass transfer. Specific attention is given to the effect of exchange between the orbital angular momentum and the spin angular momentum as the primary component evolves to a red giant and in the phase of mass transfer. It has been found that the system considering the spin angular momentum will evolve to a longer orbital period and with a smaller endmass of the primary. It shows that the spin angular momentum cannot be neglected in the calculating of the evolution of binary systems.

065.040   Heavy element opacities and the pulsations of β Cepheid stars.
M. Kiriakidis, M. F. El Eid, W. Glatzel.
Mon. Not. R. Astron. Soc., Vol. 255, No. 1, p. 1P – 5P (1 Mar 1992).
   Using recently published opacity tables, the evolution and stability of a 15–$M_\odot$ star with initial metallicities $Z = 0.02$ and

0.03 is re–examined. For $Z = 0.03$, radial and non–radial pulsations are found to be unstable for $4.326 < \log T_{eff} < 4.43$. Implications of the results for the β Cepheid phenomenon are discussed.

065.041   Progenitor masses of Type Ib/c supernovae.
E. Baron.
Mon. Not. R. Astron. Soc., Vol. 255, No. 2, p. 267 – 268 (15 Mar 1992).
   Recent models for the origin of Type Ib/c supernovae have proposed that the progenitors consist of relatively low–mass ($M \sim 12$–18 $M_\odot$) stars in binary systems that have lost their hydrogen envelope. On the basis of the nickel mass, light curve and the amount of mixing, these models identify the low–mass end with Ic's and the higher mass end with Ib's. The author shows that the dependence of the nickel mass on the progenitor mass is highly uncertain and suggest that the helium lines in Type Ib's and the lack thereof in Type Ic's could be more naturally explained by identifying the low–mass end with Ib's and the high–mass end with Ic's.

065.042   Tidal capture of stars by a massive black hole.
I. D. Novikov, C. J. Pethick, A. G. Polnarev.
Mon. Not. R. Astron. Soc., Vol. 255, No. 2, p. 276 – 284 (15 Mar 1992).
   The processes leading to the tidal capture of stars by a massive black hole and the consequences of these processes in a dense stellar cluster are discussed in detail. When the amplitude of a tide and the subsequent oscillations are sufficiently large, the energy deposited in a star after periastron passage and formation of a bound orbit cannot be estimated directly using the linear theory of oscillations of a spherical star but rather numerical estimates must be used. The evolution of a star after tidal capture is discussed. The maximum ratio R of the cross–section for tidal capture to that for tidal disruption is about 3 for real systems. For the case of a stellar system with an empty capture loss cone, even in the case when the impact parameter for tidal capture only slightly exceeds the impact parameter for direct tidal disruption, tidal capture would be much more important than tidal disruption.

065.043   Diffusion and mixing in accreting white dwarfs.
I. Iben Jr., M. Y. Fujimoto, J. MacDonald.
Astrophys. J., Vol. 388, No. 2, p. 521 – 540 (1 Apr 1992).
   Numerical experiments have been conducted to determine the degree of enhancement of CNO elements in the envelope of a 1 $M_\odot$ carbon–oxygen white dwarf accreting hydrogen–rich material at rates of 10^{-10}, 10^{-9}, and $10^{-8} M_\odot yr^{-1}$. Three initial configurations have been adopted: (1) no initial surface helium layer, 10^9 yr of cooling prior to start of accretion; (2) no initial surface helium layer, a steady state interior thermal structure that is expected after many thermonuclear outbursts; and (3) an initial layer of $10^{-3} M_\odot$ of helium, 10^9 yr of cooling, with diffusion, before the start of accretion at the rate $10^{-10} M_\odot yr^{-1}$. Only ordinary particle diffusion and convective mixing are taken into account. In the first case, an order of magnitude enhancement in CNO occurs for the highest accretion rate, and the degree of enhancement increases with decreasing accretion rate. For case (2), enhancements are even larger than in the case of cold initial models, and the degree of enhancement increases with increasing accretion rate. In case (3) with a helium buffer layer, approximately 3% of the mass of the helium layer is incorporated into the convective shell during the runaway, and negligible surface enhancements of heavy elements are found. However, estimating the amount of mass lost in a nova explosion, the authors argue that, if $\dot{M} \leqslant 2 \times 10^{-10} M_\odot yr^{-1}$, the helium layer can be eroded, and, after 60 or so nova explosions, the mass which is dredged into the convective layer during a thermonuclear runaway is larger than the mass of the helium zone produced by continued hydrogen burning in the remnant. Hence, for a small enough mass accretion rate, mixing due to particle diffusion and convection alone is sufficient to produce a large enhancement of heavy elements in the nova ejectum, and the mass of the white dwarf decreases with time.

065.044   **Bisystem oscillation theory of stars. I. Linear theory.**
Li Yan.
Astron. Astrophys., Vol. 257, No. 1, p. 133 – 144 (Apr 1992).
   The linear oscillation theory of stars is improved to include the nonequilibrium effects of the gas and radiation. The nonequilibrium diffusion approximation is used to accurately describe the interaction of the gas with the radiation. The concept of two thermodynamic systems of a radiating fluid is introduced to show the physical proceses more clearly in the nonequilibrium diffusion, and it is completely compatible with the radiation-hydrodynamic description of the fluid. The linearized equations as well as the boundary conditions for the oscillation are derived. The characteristics of the oscillation are discussed on the assumption of the adiabatic oscillation. It is found that the equivalent adiabatic compression coefficient, Γ_{1e}, which partly determines the restoring forces of the oscillation, is some kind of weighted average of Γ_1's for the gas and radiation, and the weight functions depend on the process that the fluid undergoes. For the adiabatic oscillation, the weight functions are the pressures, pressure gradients and pressure perturbations of the gas and radiation. In the nonadiabatic case, a proper Γ_{1e} can also be defined. But by contrast to the adiabatic case, there is a phase delay, which determines whether the oscillation is overstable or damped, between the restoring force and the oscillation. The equivalent exponent Γ_{1e} in the equilibrium diffusion may be different from the real one by an order of $\sim 20\%$.

065.045   **Bisystem oscillation theory of stars. II. Excitation mechanisms.**
Li Yan.
Astron. Astrophys., Vol. 257, No. 1, p. 145 – 152 (Apr 1992).
   The excitation mechanisms of the stellar oscillation due to the interaction of the gas with the radiation are investigated according to the bisystem oscillation theory. The equation for the temporal variation in the kinetic energy of the oscillation and the expression for the work integral valid in the nonequilibrium diffusion are derived. It is found that the excitation mechanisms, which transform the radiation energy into the oscillation one, are of two types, i.e. the "modulation" mechanism and "frozen temperature" mechanism, which result from the nonequilibrium between the gas and radiation. The former is due to the modulation of the process that the gas absorbs or emits the radiation in the unperturbed state by the opacity perturbation, and the latter is due to the nonequilibrium of the gas and radiation produced by the oscillation. Through the careful examination of some previously proposed excitation mechanisms, it is found that the $\varkappa$–mechanism and γ–mechanism belong to the frozen temperature mechanism, and the convective blocking mechanism belongs to the modulation mechanism, though their operating processes were not correctly understood before. However, no mechanism has been found in the bisystem oscillation theory which corresponds to the r–mechanism or bump mechanism, which indicates that these two kinds of excitation mechanism are not real.

065.046   **Tests of the CM model for turbulent convection. I. Application to M67.**
F. D'Antona, I. Mazzitelli, R. G. Gratton.
Astron. Astrophys., Vol. 257, No. 2, p. 539 – 547 (Apr 1992).
   The mixing length theory (MLT) provides a rather approximate description of turbulent convection in stars. Recently, Canuto and Mazzitelli (CM) have suggested a new model, the main goal of which is to account for the wide spectrum of eddies. The new model, which has no adjustable parameters, has been found to fit the solar T_{eff} within 0.5%. The authors' purpose here is to test the performance of the CM model on stellar objects other than the Sun. They concentrate on the comparison of new evolutionary tracks and isochrones, computed in the CM framework, with the HR diagram features of the open cluster M67. The CM isochrones are adequate to reproduce the general features of the HR diagram morphology, at least as well as the MLT isochrones. The CM model appears to do even a better job of matching the observed red giants, though this may depend on the inferred temperatures of the few cluster giants. It is concluded

that the CM model has passed also the test for the red giants. Consideration of theoretical and observational error bars leads to choose as best theoretical fit for M67 the one obtained by increasing the distance modulus of the cluster, from the currently accepted value 9.7 to 9.9. This choice is based on the location of the clump (core helium burning stars), and on the existence and location of the turn–off gap. The global result of adopting the CM isochrones and of increasing the distance modulus, is to lower the age of M67 by $\sim 20\%$ with respect to the previous determinations.

065.047   **Linear polarization as a consequence of rotation in exploding stars.**
M. Steinmetz, P. Höflich.
Astron. Astrophys., Vol. 257, No. 2, p. 641 – 654 (Apr 1992).
   The authors have performed a set of two–dimensional hydrodynamic simulations of the propagation of spherically initiated shock waves in the hydrogen–rich envelope of a rotating massive star. Rigidly and differentially rotating polytropes have served as initial models. Mass, extension and rotational velocity of these models were selected to represent the observed quantities of the SN 1987A progenitor. The authors show that within the framework of their model the progenitor must rotate differentially to explain the observed polarization.

065.048   **Rare thermonuclear explosions in short–period cataclysmic variables with possible application to the nova–like red variable in the galaxy M31.**
I. Iben Jr., A. V. Tutukov.
Astrophys. J., Vol. 389, No. 1, p. 369 – 374 (10 Apr 1992).
   A very bright nova with a peak luminosity significantly larger than the relevant Eddington luminosity might be initiated by a hydrogen shell flash on a very cold low–mass degenerate dwarf accreting matter from a close low–mass companion which fills its Roche lobe in a system with orbital period 80 min to 2 hr. Under appropriate conditions, which are calculated here, the mass of accreted matter can be as large as 0.005 $M_\odot$ when the explosion occurs, and a luminosity larger than the Eddington limit may be generated by the frictional interaction between the matter in a dense common envelope and the interior stellar cores. The frequency of such explosions in our Galaxy may be about six events per millennium. The most likely precursor is a cataclysmic variable with period near 80 min, and the final result of the super nova explosion may be dissolution of the low–mass companion into the escaping high–velocity wind. An outburst that was detected in Andromeda in 1988 may be an example.

065.049   **On the interior structure of contact binaries and the light–curve paradox.**
J. L. Tassoul.
Astrophys. J., Vol. 389, No. 1, p. 375 – 385 (10 Apr 1992).
   The author presents a hydrodynamical model for energy transfer between the components of a contact binary. The dominant feature of the model is the nonuniform heating of the base of the common envelope, which is due to the unequal emergent luminosities at the inner Roche lobes. By continuity, temperature differences on each Roche equipotential above these lobes do exist, so that the common envelope must be treated as a three–dimensional barocline. By making use of expansions in Roche coordinates, the author shows how to obtain steady baroclinic solutions for both the large–scale flow and the temperature field. The velocity field is essential to satisfy the Navier–Stokes equations within the common envelope. The baroclinic temperature field is determined mainly by the three–dimensional radiative transfer within the common envelope – the lateral flux being steadily maintained by the temperature differences on each Roche equipotential. Since the surface effective gravities of the components of a Roche model are nearly equal, it follows that the effective temperatures should be also nearly equal. Although this new baroclinic model is based on many simplifying approximations, it provides a new way toward a resolution of the light–curve paradox for the contact binaries while providing consistent interior structures. Moreover, because the mechanism for the energy transfer within the common

envelope does not involve lateral convection along the Roche equipotentials, it applies to both the early–type and the late–type contact binaries.

065.050   Rotating stars: the angular momentum constraints.
F. de Felice, L. Di G. Sigalotti.
Astrophys. J., Vol. 389, No. 1, p. 386 – 391 (10 Apr 1992).
An important parameter to be considered in stellar evolution is the ratio cJ/GM^2, where J is the angular momentum of a star and M is its total mass. This is a pure number which must be $\leqslant 1$ for a neutron star or a black hole, but can take values much larger than one in a star at its birth. Here, the authors investigate the values of the ratio predicted by the numerical models of a stellar object during protostellar contraction as well as at the neutron star phase. For the latter case, relativistic models of both slowly and fast rotating configurations are considered.

065.051   On the pulsations of relativistic accretion disks and rotating stars: the Cowling approximation.
J. R. Ipser, L. Lindblom.
Astrophys. J., Vol. 389, No. 1, p. 392 – 399 (10 Apr 1992).
This paper shows that the hydrodynamic degrees of freedom of the adiabatic pulsation of relativistic fluids (e.g., accretion disks or rotating stars) can be described by a single scalar potential. When the gravitational perturbations are neglected – the Cowling approximation – this potential is determined by a second–order (typically elliptic) partial differential equation. A variational principle is developed from which the pulsation frequencies may be evaluated in this approximation. For objects like accretion disks in which self–gravitation effects are negligible, this approximation becomes an exact description of the pulsations.

065.052   Three–dimensional hydrodynamical simulations of colliding stars. III. Collisions and tidal captures of unequal–mass main–sequence stars.
W. Benz, J. G. Hills.
Astrophys. J., Vol. 389, No. 2, p. 546 – 557 (20 Apr 1992).
The authors used a 3D smooth–particle hydrodynamics (SPH) code with 7000 particles to simulate 30 collisions between lower main–sequence (MS) stars whose masses differ by a factor of 5. Encounters with point–mass intruders were also simulated. The two MS stars become gravitationally bound in physical collision if their relative velocity, V, at infinity is less than a critical velocity, V_d. The authors find that V_d decreases from 1000 km s^{-1} in a head–on collision to 150 km s^{-1} in a grazing one. If the less massive star is replaced by a point mass, V_d remains the same for grazing collisions and tidal encounters, but it drops to ~600 km s^{-1} in head–on collisions. If $V > V_d$, the lower mass MS star, which is the denser of the two, passes through the more massive MS star and escapes without becoming gravitationally bound to it. In collisions between two main–sequence stars, V_d increases with V due to increased shock dissipation. If the two MS stars coalesce in a physical collision, the denser (lower mass) MS star settles to the center of the other one. The violent mixing of the more massive star during the encounter and the settling of the low–mass star to its center should reset the nuclear clock of the coalesced star, so it contracts to the (helium–rich) MS. It is found that tidal captures of binary stars in globular clusters can only occur in encounters in which the closest approach of the stars to their center of mass is less than 2.0 times the sum of their radii. Coalescence of two MS stars in a globular cluster is more probable than their forming a binary by tidal capture.

065.053   Jet formation in the transition from the asymptotic giant branch to planetary nebulae.
N. Soker.
Astrophys. J., Vol. 389, No. 2, p. 628 – 634 (20 Apr 1992).
The author demonstrates the feasibility of a scenario by which a star in a common envelope blows a collimated wind as it evolves from the asymptotic giant branch (AGB) to the central star of a planetary nebula. The collimated flow turns into jets along the symmetry axis. These jets then form in their leading front the ansae – two opposite bright knots along the major axis which are observed in many elliptical planetary nebulae. An essential part of the model is that the evolving AGB star is in a common envelope with a low–mass companion. The low–mass companion, which orbits close to the giant core, spins up the envelope efficiently at late stages of the evolution. The author calculates the surface shape of the rotating envelope by assuming a simple angular momentum distribution. The deformed surface of the rotating envelope is assumed to blow a substantial fraction of the wind along the rotation axis. The proposed scenario leads to two types of jets and ansae. One type is slow, ~ 50 km s^{-1}, with a total mass of $\leqslant 10^{-3} M_\odot$. The second type, which can continue to be active during the planetary nebula stage, is fast, 100 – 300 km s^{-1}, with a total mass of ~$10^{-4} M_\odot$.

065.054   Taking the pulse of white dwarfs.
R. E. Nather, D. E. Winget.
Sky Telesc., Vol. 83, No. 4, p. 374 – 378 (Apr 1992).
A stellar seismology project may pin down the age of the universe.

065.055   Type I supernovae and accretion–induced collapses from cataclysmic variables?
M. Livio, J. W. Truran.
Astrophys. J., Vol. 389, No. 2, p. 695 – 703 (20 Apr 1992).
The authors examine the possibility that the white dwarfs in cataclysmic variables might grow in mass to the Chandrasekhar limit. First, it is demonstrated that the masses of the white dwarfs in classical nova systems generally decrease as a result of nova explosions. Then, the authors present observational and theoretical evidence in support of the view that the masses of the white dwarfs in recurrent nova systems increase. The critical parameters for which a mass increase is obtained are approximately $M_{WD} > 1.2\ M_\odot$ and $\dot{M} > 10^{-8} M_\odot yr^{-1}$. The system in which the white dwarfs are most likely to reach the Chandrasekhar limit are those which satisfy the above limits on the mass of the white dwarf and on the accretion rate and in which, in addition, the secondary stars are transferring helium–rich material. A subclass of the recurrent novae satisfies all these conditions. The cataclysmic variable systems in which the white dwarf components can grow to the Chandrasekhar limit are most likely to produce an accretion–induced collapse, rather than a Type Ia supernova. The authors calculate the rate of these events and find it to be consistent with the birth rate of low–mass X–ray binaries.

065.056   The status of non–homogeneous Dedekind ellipsoids.
J. C. Miller, A. Lanza.
9. Italian Conference on General Relativity and Gravitational Physics, p. 547 – 550 (1991). – See Abstr. 012.021 for the main entry.

065.057   Shock waves in stellar atmospheres and breaking waves on an ocean beach.
G. Wallerstein, S. Elgar.
Science, Vol. 256, No. 5063, p. 1531 – 1536 (12 Jun 1992).
The phenomenon of ocean waves breaking on a beach is analogous to shock waves in the atmosphere of a pulsating star. In both cases a velocity discontinuity is clearly present. In stars the upper, expanding layer halts and falls back so as to interact with the rising gas at a shock. Similarly, a bore on a beach reaches its maximum extension before sliding back onto the next coming wave. Analogous quantities such as the surface gravity of the star and the beach gradient in the ocean have similar effects on the flows and the nature of the discontinuity between them. Phenomena that are not analogous include the thermodynamic properties of the two media. Ocean observations may help solve some problems in shock phenomena associated with stellar pulsations.

065.058   Further improvements of a new model for turbulent convection in stars.
V. M. Canuto, I. Mazzitelli.
Astrophys. J., Vol. 389, No. 2, p. 724 – 730 (20 Apr 1992).
The authors analyze the effect of several improvements of the input physics of a recent model for stellar turbulent convection.

They first study the effect of (1) the inclusion of a variable molecular weight (Cox and Giuli's variable Q), and (2) the use of the latest opacities of Rogers and Iglesias. Using evolutionary tracks for the Sun, the authors conclude that the original model for turbulence with the mixing length $\Lambda = z$, together with $Q \neq 1$ and the new opacities, yields a fit to the solar T_{eff} within 0.5%. The model has no adjustable parameters. Second, the authors propose a formulation of the mixing length Λ that extends the purely nonlocal $\Lambda = z$ expression so as to include local effects. They derive the expression $\Lambda = \alpha(S, a) \cdot z$, where $S = 160A^2(\nabla - \nabla_{ad})$ and where $1-a$ represents the weight of local effects. The new expression generalizes both the mixing–length theory (MLT) phenomenological expression $\Lambda = \alpha \cdot H_p$, as well as the present model $\Lambda = z$. By adjusting the parameter a, an even better fit to the solar T_{eff} can be obtained.

065.059 **Criterion for convection in an inhomogeneous star.**
R. B. Stothers, C.–W. Chin.
Astrophys. J., Lett., Vol. 390, No. 1, p. L33 – L35 (1 May 1992).
To resolve the question of whether the Schwarzschild criterion or the Ledoux criterion should be used to test for convective instability in a star, a well–observed cluster of chemically inhomogeneous massive stars, in which the choice of the criterion for convection makes a crucial and easily observable difference, is required. NGC 330, a metal–poor cluster in the Small Magellanic Cloud, is ideal for this test. Its large evolved stellar population contains both blue and red supergiants. Its many red supergiants should be absent if a gradient of mean molecular weight did not choke off rapid convective motions in the inhomogeneous region connecting the envelope and core. Thus the Ledoux criterion for convection is strongly indicated as being correct.

065.060 **Adiabatic properties of pulsating DA white dwarfs.**
II. Mode trapping in compositionally stratified models.
P. Brassard, G. Fontaine, F. Wesemael, C. J. Hansen.
Astrophys. J., Suppl. Ser., Vol. 80, No. 1, p. 369 – 401 (May 1992).
As part of a study of the adiabatic gravity–mode period structure of models of pulsating DA white dwarfs, the authors examine qualitatively and quantitatively the phenomenon of mode trapping caused by compositional layering in these stars. This is motivated by the possibility that the pulsation modes detected in ZZ Ceti stars are actually trapped in the outer hydrogen layer. The most important effect of compositional layering is the formation of a nonuniform period distribution which bears the signature of the strength and location of the composition discontinuity. The trapped modes are conspicuous in that they show the largest deviations from the mean behavior in the period distribution, and are the most sensitive to the characteristic of the H/He composition transition zone. The authors derive a semianalytic formula relating the periods of trapped modes to model properties, which can be used to estimate the hydrogen layer mass in a ZZ Ceti star. They then investigate the effect of a true discontinuity (an idealization of the H/He transition zone in a DA white dwarf) in the Brunt–Väisälä frequency profile on the g–mode period spectrum of a stellar model. The authors derive an analytic expression which gives the full period spectrum of the model in terms of only four parameters. Comparisons with exact numerical calculations are presented for both the semianalytic approach and the second analytic formula for the period spectra. It is shown that the second analytic expression reproduces quite accurately the period spectrum of trapped modes.

065.061 **Testing theories of star formation.**
J. N. Evans II.
3. Symposium International de Bioastronomie, p. 17 – 18 (1991).
– See Abstr. 012.005 for the main entry.

065.062 **Stellar evolution in blue populous clusters of the Small Magellanic Cloud and the problems of envelope semicon-vection and convective core overshooting.**
R. B. Stothers, C.–w. Chin.
Astrophys. J., Vol. 390, No. 1, p. 136 – 143 (1 May 1992).
Two of the blue populous clusters in the Small Magellanic Cloud, NGC 330 and NGC 458, contain a considerably larger number of evolved stars than are found in comparably young cluster in our Galaxy. This richness makes them valuable tools for studying post–main–sequence evolution in the mass range $4 - 15$ M$_\odot$. New theoretical evolutionary sequences for stars with low metallicities, appropriate to the SMC, are derived here with both standard Cox–Stewart opacities and the new Rogers–Iglesias opacities. The authors find that only those sequences with little or no convective core overshooting can reproduce the two most critical observations: the maximum effective tempera-ture displayed by the hot evolved stars and the difference between the average bolometric magnitudes of the hot and cool evolved stars. An upper limit to the ratio of the mean overshoot distance beyond the classical Schwarzschild core boundary to the local pressure scale height can be set at $d/H_P < 0.2$. The frequency of cool supergiants in NGC 330 implies that the Ledoux criterion (rather than the Schwarzschild criterion) for convection and semiconvection in the envelopes of massive stars is strongly favored. NGC 330 and NGC 458 turn out to have ages of $\sim 3 \times 10^7$ yr and $\sim 1 \times 10^8$ yr, respectively.

065.063 **Weak chaos in long–period variables.**
V. Icke, A. Frank, A. Heske.
Astron. Astrophys., Vol. 258, No. 2, p. 341 – 356 (May 1992).
When low– to intermediate–mass stars reach the very late stages of their evolution, they show long period oscillations. There is an increasing body of observational evidence which indicates that these oscillations become more and more irregular as a star begins to leave the asymptotic giant branch (AGB); that mass loss accompanies these oscillations, and that the mass loss rate increases progressively as the oscillations become more erratic; and that the mass loss can be very aspherical. The authors have attempted to account for the increasing irregularity of the long–period variations by investigating how the outer layers ("mantle") of an evolved AGB star respond to pulsations which originate in the stellar interior. It is assumed that the oscillations arise in a zone of instability well below the stellar photosphere. The mantle of the star is driven periodically by pressure waves generated by the interior motion. This type of driven oscillator exhibits chaotic motions for a wide range of relevant parameters. The authors study their driven–oscillator equations in detail, in various approximations. Analytic estimates show why and where chaos ought to set in. When comparing the outcome of the analysis with observations, the authors note that the studied characteristics can be traced directly to the dynamics of a driven oscillator: motions are regularly periodic at first, then become multiperiodic, and finally chaotic, while the mass of the mantle decreases more and more rapidly through mass loss.

065.064 **A theoretical investigation of population II red giant clumps.**
G. Bono, V. Castellani.
Astron. Astrophys., Vol. 258, No. 2, p. 385 – 388 (May 1992).
Theoretical models predicting a clump of stars along the red giant branches of population II stellar clusters have been based on the assumption of a sharp discontinuity of the chemical composition at the bottom edge of external convection. The influence on the clumps of a smoothing of this chemical discontinuity is investigated by using numerical experiments. The authors show that for reasonable assumptions about the size of the partially mixed region the occurrence of red giant "bumps" remains a firm theoretical prediction. The structural behaviour of clump stars is briefly discussed.

065.065 **Core overshooting and stellar evolution.**
E. Brocato, V. Castellani.
Astron. Astrophys., Vol. 258, No. 2, p. 397 – 398 (May 1992).
The authors discuss recent estimates of surface gravities of blue stragglers which have been reported as evidence for the occur-rence of overshooting in H–burning stars. The adoption of updated evolutionary tracks makes such evidence much less conclusive. Moreover, any relation between gravity and over-shooting depends on the evolutionary status of blue stragglers in galactic open clusters, which is far from being clearly established.

065.066   Mixing in ejecta of supernovae. I. General properties of two–dimensional Rayleigh–Taylor instabilities and mixing width in ejecta of supernovae.
I. Hachisu, T. Matsuda, K. Nomoto, T. Shigeyama.
Astrophys. J., Vol. 390, No. 1, p. 230 – 252 (1 May 1992).

The nonlinear growth of two–dimensional Rayleigh–Taylor (R–T) instabilities is numerically simulated in a study of mixing in supernova ejecta. The authors first present much refined calculations of mixing in a realistic model of SN 1987A with an improved code and various mesh resolutions. The results show that the mixing width due to Rayleigh–Taylor instabilities is still too small to account for the observations even with relatively large initial perturbation. To clarify the basic properties of the R–T instabilities and the dependence of the mixing width on the initial density structure, initial perturbation, and numerical resolution, the authors consider simplified ideal models of R–T instabilities of a compressible gas with an adiabatic constant $\gamma = 4/3$. The ideal R–T instabilities are calculated for various mesh resolutions, numerical accuracy, density ratio, initial amplitude of the perturbations, and mode of the initial perturbation (random from mesh to mesh and sinusoidal waves). The mixing width is found to depend mainly on the initial amplitudes and the density ratio. This implies that the mixing width in supernova ejecta depends mainly on the initial amplitude of the perturbation and on the density structure of the presupernova models (density ratio).

065.067   Numerical experiments on the effects of horizontal turbulent diffusion on transport by meridional circulation.
P. Charbonneau.
Astron. Astrophys., Vol. 259, No. 1, p. 134 – 142 (Jun 1992).

It has recently been shown by Chaboyer and Zahn (1992) that in the presence of strong horizontal turbulence, the transport of angular momentum and chemicals normally mediated by rotationally–induced meridional circulation could be formally described by a one dimensional transport equation. They also have shown that in the limit of very strong horizontal tubulence, the net vertical transport of chemicals by meridional circulation is effectively shut off. This paper presents a set of numerical experiments aimed at quantifying various aspects of this horizontal turbulence ansatz. It is shown that the transition to 1–D behavior for both particle and angular momentum transport occurs for horizontal Reynolds numbers $R_H \lesssim 1$, and the effective inhibition of meridional circulation–mediated particle transport for $R_H \lesssim 10^{-2}$. These values are essentially independent of the vertical Reynolds number R_V.

065.068   Nonradial oscillations of white dwarfs – variational formulation revisited.
H. P. Singh.
Astron. Astrophys., Vol. 259, No. 1, p. 155 – 158 (Jun 1992).

The fundamental frequencies of nonradial (Kelvin) mode of oscillation of completely degenerate configurations have been evaluated in the first as well as the second approximation by a variational method. The computations have been made for different values of the central degeneracy parameter $1/y_0^2$ and for uniform temperatures of 0, 2×10^7K and 10^8K in the interior. The zero–temperature white dwarf calculations confirm the existence of a crossover in frequency between the fundamental radial (p) mode ($l = 0$) and the f or Kelvin ($l = 2$) mode at $\sim 0.45\,M_\odot$. For hot white dwarf models under consideration, this resonance condition between the two modes is seen to remain unchanged.

065.069   The subseismic approximation for low–frequency modes of the Earth applied to low–degree, low–frequency g–modes of non–rotating stars.
I. De Boeck, T. Van Hoolst, P. Smeyers.
Astron. Astrophys., Vol. 259, No. 1, p. 167 – 174 (Jun 1992).

The subseismic approximation proposed by Smylie and Rochester (1981) for low–frequency modes in the Earth is used in an asymptotic treatment of low–frequency g–modes belonging to low–degree spherical harmonics in the case of a non–rotating, spherically symmetric star in which the square of the Brunt–Väisälä frequency is different from zero everywhere inside the star except at the center and keeps the same sign. The approximation neglects the contribution of the Eulerian perturbation of the pressure to the Lagrangian perturbation of that quantity relative to the contribution stemming from the equilibrium pressure gradient. In contrast to the procedure adopted in the standard asymptotic theory, the Eulerian perturbation of the gravitational potential is not neglected. The subseismic approximation leads to an inadequate representation of the radial component of the Lagrangian displacement near the star's surface. This inadequate representation affects the equation determining the asymptotic eigenfrequencies. The eigenvalue equation is applied to second–degree g$^-$–modes of the compressible equilibrium configuration with uniform mass density and second–degree g$^+$–modes of the polytropic model with index $n = 3$. For the latter model, the asymptotic theory based on the subseismic approximation yields better approximations of the eigenfrequencies of second–degree g–modes up to the order 30 than does the standard asymptotic theory.

065.070   Equations of state and bump Cepheids. II. Non–linear results.
S. M. Kanbur.
Astron. Astrophys., Vol. 259, No. 1, p. 175 – 182 (Jun 1992).

The Hummer–Mihalas–Dappen (1988) and a simple Saha type equation of state are used to obtain non–linear pulsation characteristics of a grid of models spanning the Hertzsprung sequence (Cox 1974). The grid of models is taken from Simon and Davis (1983) who used the Los Alamos equation of state in their computations. The result is a sensitivity analysis of theoretical non–linear bump Cepheid models to the equation of state employed in the calculation. The results obtained with these equations of state are different, though not enough to resolve the Cepheid bump mass discrepancy (Stobie 1969; Simon and Schmidt 1976; Simon 1986).

065.071   The dynamical and physical evolution of close solar type binaries as a consequence of angular momentum loss.
F. van't Veer.
13. Journée de Strasbourg: Stades avancés dans l'évolution des étoiles binaires serrées, p. 1 – 17 (1992). – See Abstr. 012.034 for the main entry.

Angular momentum loss is the driving force for the dynamical evolution of close tidally interacting binaries with solar type components. A description of the different stages of this evolution is given, and the variation with time of the binary orbit is formulated. The final single star stage of these binaries is also described and a first tentative is made to distinguish between ordinary single stars and coalesced binaries, with particular attention to differences of the chemical composition of their atmospheres. A short review of the possible presence of disks of circumstellar material around very close angular momentum losing binaries is presented.

065.072   Constraints of axions from white dwarf cooling.
Jin Wang.
Mod. Phys. Lett. A, Vol. 7, No. 17, p. 1497 – 1502 (7 Jun 1992).

The author performs numerical calculations on the white dwarf cooling process. He found that it provides a constraint on the axion mass. The upper bound is 0.01 eV.

065.073   Chemical composition effects and double–mode RR Lyrae masses.
G. Kovács, J. R. Buchler, A. Marom, C. A. Iglesias, F. J. Rogers.
Astron. Astrophys., Vol. 259, No. 2, p. L46 – L48 (Jun 1992). Letter–to–the–editor.

On the basis of accumulating observational evidence that Pop. II stars are overabundant in oxygen and other heavy elements, the authors compute RR Lyrae models with new mixtures of heavy elements. Comparing these results with the ones obtained with a solar (Anders–Grevesse) composition they conclude that: (i) with the same total metal abundance Z the models with Pop. II mixtures yield masses smaller than the ones with a solar mixture, (ii) with the same absolute Fe abundance the new masses are higher, (iii) these two effects increase with Z; (iv) the masses of the double–mode RR Lyrae (especially Oosterhoff I) stars obtained with the period ratio method therefore are left uncertain up to $\approx 0.1\,M_\odot$ until accurate abundances for the dominant elements become available for individual stars.

065.074   Gamma–ray light curves and spectra for Type Ia supernovae.
P. Höflich, A. Khokhlov (*A. M. Khokhlov*), E. Müller.
Astron. Astrophys., Vol. 259, No. 2, p. 549 – 566 (Jun 1992).

The authors have computed the gamma–ray energy deposition functions, light curves and spectra for a set of theoretical Type Ia supernova models including deflagration, detonation, delayed detonation and tamped detonation models. The results have been obtained with a Monte Carlo gamma–ray deposition scheme which takes into account all relevant gamma–transitions and interaction processes.

065.075   Ekman circulation and the synchronization of binary stars.
M. Rieutord.
Astron. Astrophys., Vol. 259, No. 2, p. 581 – 584 (Jun 1992).

The author shows that large–scale flows driven by Ekman pumping in the spin–up/down of a tidally distorted star is not efficient enough to reduce the synchronization time. This latter time remains of the order of the viscous time, if the star is made of an incompressible viscous fluid. The computation of the synchronization time scale of early–type binaries should follow the approach proposed by Zahn and recently improved by Rocca, and Goldreich and Nicholson.

065.076   The collapse of white dwarfs to neutron stars.
S. E. Woosley, E. Baron.
Astrophys. J., Vol. 391, No. 1, p. 228 – 235 (20 May 1992). = UCO/Lick Obs. Bull., No. 1206.

The observable consequences of an accreting white dwarf collapsing directly to a neutron star are considered. The outcome depends critically upon the nature of the wind that is driven by neutrino absorption in the surface layers as the dwarf collapses. Unlike previous calculations which either ignored mass loss or employed inadequate zoning to resolve it, the authors find a characteristic mass–loss rate of about $0.005\,M_\odot s^{-1}$ and an energy input of $5 \times 10^{50} ergs\,s^{-1}$. Such a large mass–loss rate almost completely obscures any prompt electromagnetic display and certainly rules out the production by this model of γ–ray bursts situated at cosmological distances. The occurrence of such collapses within our own Galaxy might, however, be detected and limited by their nucleosynthesis and γ–ray line emission. To avoid the overproduction of rare neutron–rich isotopes heavier than iron, such events must be very infrequent, probably happening no more than once every thousand years.

065.077   Presupernova evolution in massive interacting binaries.
P. Podsiadlowski, P. C. Joss, J. J. L. Hsu.
Astrophys. J., Vol. 391, No. 1, p. 246 – 264 (20 May 1992).

The authors have systematically investigated how binary interaction affects the presupernova evolution of massive close binaries and the resulting supernova explosions, using a modified Henyey–type stellar evolution code. With the modified code, one is able to follow the effects of mass and angular momentum loss from the binary, as well as mass transfer within the binary system. The authors find that a large number of binary scenarios can be distinguished, depending on the type of binary interaction and the evolutionary stage of the supernova progenitor at the time of the interaction. In general, the structure of a massive star can be affected in three fundamentally different ways: by mass loss, mass accretion, or common–envelope evolution. As a result of mass loss by Roche lobe overflow, stars can lose all of their hydrogen–rich envelopes and become helium stars, which are potential candidates for the progenitors of Type Ib supernovae. Mass accretion can also significantly alter the structure of the

supernova progenitor, if it takes place after the main–sequence phase of the accreting star. The star may then end its life as a blue supergiant instead of a red supergiant. In this case, the resulting supernova explosion would resemble SN 1987A. In the most dramatic case of binary interaction, in which a supernova progenitor captures its companion in a common envelope, two different outcomes are possible, depending on whether the envelope is ejected during the spiral–in phase or remains bound. If the envelope is ejected, the progenitor will become a helium star and the subsequent supernova explosion may be of the Type Ib variety. If the binary components merge completely, the final outcome would be a single star with no trace of the original secondary. The authors performed Monte Carlo simulations to estimate the frequencies of occurrence of the individual scenarios. They find that, because of a previous binary interaction, 15% – 30% of all massive stars (with initial masses $\gtrsim 8\,M_\odot$) become helium stars, and another $\sim 5\%$ of all massive stars end their lives as blue supergiants rather than as red supergiants. The fate of the companion of the exploding star is considered in each scenario. A detailed discussion of the evolutionary history of the apparent progenitor Sk –69°202 of SN 1987A is included.

065.078   Implementation of a complex–plane strategy to the computation of rotating polytropic models.
V. S. Geroyannis.
Astrophys. Space Sci., Vol. 190, No. 1, p. 43 – 55 (Apr 1992).

The author implements a "complex–plane strategy" and a "multiple partition technique" to the computation of polytropic models distorted by very strong and very rapid differential rotation. He also verifies with his numerical results a "heuristic relation" between stability and virial theorem.

065.079   The effect of toroidal magnetic field and differential rotation on self–gravitating polytropic models.
V. S. Geroyannis, M. G. Sidiras.
Astrophys. Space Sci., Vol. 190, No. 1, p. 139 – 144 (Apr 1992).

The authors construct a polytropic model distorted by toroidal magnetic field and differential rotation. They then compute states of critical rotation of this model. In the computations they implement the so–called "complex–plane strategy" and "multiple partition technique" which are numerical methods deviced recently by the first author.

065.080   A computation of the equation of state for a Fermi gas.
M. Yildiz, D. Eryurt–Ezer.
Astrophys. Space Sci., Vol. 190, No. 2, p. 233 – 242 (Apr 1992).

Calculation of equation of state in stellar interiors becomes difficult as contained gas deviates from perfect gas. The authors present a method for the calculation of electron pressure in terms of density and temperature in the presence of degeneracy. The method is applicable for $T < 10^9 K$, and requires complete ionization.

065.081   Comparison of stellar opacities at low temperatures.
C. R. Acad. Sci., Sér. II, Tome 314, No. 13, p. 1435 – 1441 (18 Jun 1992).
C. Sharp, Y. Lebreton, A. Baglin.

This paper gives a critical evaluation of opacities in the intermediate domain of temperature where both molecules and atoms are contributing, a region which is very important for the structure of stars of intermediate type.

065.082   Dependence of tidal effects on the mass distribution in binary stellar systems.
P. V. Subrahmanyam.
Astrophys. Space Sci., Vol. 191, No. 1, p. 137 – 146 (May 1992).

Analytic expressions for tidal disruption and tidal coalescence for binary stellar systems have been derived, using impulsive approximation, as a function of the mass ratio and separation of the components. These expressions can be used for any mass distribution of the components for non–interpenetrating cases.

065.083 **The formation of intermediate–mass stars.**
F. Palla.
Mem. Soc. Astron. Ital., Vol. 62, No. 4, p. 873 – 880 (1991). – See Abstr. 012.042 for the main entry.

The formation and early evolution of stars in the mass range 2 to 10 $M_\odot$ is described. The results of detailed numerical calculations of accreting protostars are presented for a variety of initial conditions. It is shown that the transition from low–mass stars to more massive objects is marked by a sequence of events, the appearance of a radiative barrier, the ignition of deuterium in a shell, the rapid swelling of the star, gravitational construction to the Main–Sequence that do not depend on the assumptions of the computations. The derivation of a theoretical birthline shows an excellent agreement with the observed distribution of Herbig Ae/Be stars in the H–R diagram. Finally, the implications on the conditions for star formation in cloud cores are discussed.

065.084 **Oxygen–enhanced models for globular cluster stars. I. Evolutionary tracks.**
D. A. VandenBerg.
Astrophys. J., Vol. 391, No. 2, p. 685 – 709 (1 Jun 1992).

Evolutionary tracks for 13 – 16 mass values (depending on the assumed chemical composition) in the range $0.15 \leqslant M/M_\odot \leqslant 1.25$ have been computed for each of 10 metallicities between [Fe/H] = −2.26 and [Fe/H] = −0.47. The helium content was assumed to vary from Y = 0.2350 to 0.2454 over this range in [Fe/H]. Oxygen overabundances were chosen to obey [O/Fe] = −0.5·[Fe/H], for [Fe/H] $\geqslant$ −1.0, and [O/Fe] = −0.2·[Fe/H] + 0.3, for lower iron abundances. Solar number abundance ratios were adopted for all of the other heavy elements. The tracks were extended either to an age of 30 Gyr or, in the case of models with masses $\geqslant 0.7\ M_\odot$, to the tip of the red giant branch. Altogether, 46 tracks were evolved to the helium flash point. The T_{eff} and B–V scales of the models are normalized as accurately as possible to available observational constraints. The properties of selected models along the entire evolutionary sequences for 0.8 $M_\odot$ (or 0.85 $M_\odot$ in the case of the three highest [Fe/H] values) are tabulated in detail. In addition, for all of the computed giant branch tracks, such information as the luminosity and core mass at the helium flash, the final $^{12}C/^{13}C$ and C/N number abundance ratios obtained on completion of the first dredge–up phase, and the luminosity where the H–burning shell contacts the chemical composition discontinuity is also provided.

065.085 **Pulsational study of BL Herculis models. I. Radial velocities.**
J. R. Buchler, P. Moskalik.
Astrophys. J., Vol. 391, No. 2, p. 736 – 749 (1 Jun 1992).

The linear and nonlinear pulsational behavior of nine sequences of BL Herculis models is studied, and their radial velocity curves are discussed in detail. The pulsations of these stars, in analogy to the classical Cepheids, are strongly affected by internal resonances, most importantly the 2:1 resonance with the second overtone. This latter coupling causes a characteristic systematic progression of the Fourier phases and amplitude ratios as the period ratio P_2/P_0 is varied. In contrast to Cepheids, the strength of the resonance depends very sensitively on the stellar mass and luminosity, and the morphology of the Fourier progression changes significantly when M or L are varied. In most of the model sequences one finds narrow windows in which the pulsations exhibit periodic alternations of deep and shallow minima in the radial velocity and light curves. This behavior occurs for periods somewhere in the range from 2^d0 to 2^d6, depending on the sequence. It is caused by the 3:2 resonance between the fundamental mode and the first overtone. In the two most nonadiabatic sequences the same resonance causes windows of chaotic oscillations.

065.086 **The structure of horizontal–branch models. I. The zero–age horizontal branch.**
B. Dorman.
Astrophys. J., Suppl. Ser., Vol. 80, No. 2, p. 701 – 724 (Jun 1992).

The author presents a detailed study of the structure of zero–age horizontal–branch (HB) models. He first demonstrates explicitly the properties of composite polytropes on the homology–invariant (U, V)–plane and then uses the discussion as a framework for understanding the results from more detailed stellar models. The author reexamines the role of the CNO elements as nuclear catalysts and of envelope opacity sources, as well as the envelope helium abundance. General results are derived concerning the equilibrium of HB envelopes, which can predict qualitatively how structure and evolutionary calculations will vary with the input physics and assumptions about realistic stellar abundances in globular clusters. In particular, it is found that, for stars of a fixed range of mass arriving on the HB, the stellar distribution is determined mainly by CNO for low metallicities ([Fe/H] $\lesssim$ −1), but mainly by opacity sources for high metallicities. The value of [Fe/H], where CNO ceases to dominate, depends strongly on the adopted opacity. These results have relevance for the discrepancy between the masses derived for RR Lyrae stars from stellar evolution and pulsation calculations, for the observed relationship between HB luminosity and metallicity, and for the HB morphology of globular clusters.

065.087 **A note concerning the pulsations of Ap stars.**
A. Gautschy.
Mon. Not. R. Astron. Soc., Vol. 256, No. 1, p. 11P – 14P (1 May 1992).

Observations of oscillations of Ap stars show frequencies above the acoustic cut–off frequency in the atmosphere. Time–series analysis reveals several discrete eigenfrequencies to be present in the observational data and not a continuum of modes of different frequencies. Theoretically a discrete eigenspectrum can exist also for frequencies above the cut–off frequencies. Numerical calculations are presented demonstrating the discreteness of the eigensolutions and the development of the damping rate due to wave leakage at the surface of the star.

065.088 **Adiabatic properties of pulsating DA white dwarfs. III. A finite–element code for solving nonradial pulsation equations.**
P. Brassard, C. Pelletier, G. Fontaine, F. Wesemael.
Astrophys. J., Suppl. Ser., Vol. 80, No. 2, p. 725 – 752 (Jun 1992).

The authors present a new numerical code to solve the system of ordinary differential equations which describes the linear, adiabatic, nonradial pulsations of stellar models. This code, based on the Galerkin finite–element method of weighted residuals, is characterized by its stability, speed, accuracy, high–order convergence, and ease of use. Its performance is illustrated in tests carried out both with a homogeneous, polytropic stellar model and with a realistic stellar evolution model. It is also contrasted with the performance of two other codes, previously used in adiabatic studies of pulsating DA white dwarfs, which are based on either relaxation or shooting methods. The finite–element code outperforms both of them in terms of accuracy, effectiveness, and stability.

065.089 **The treatment of highly non–adiabatic, non–radial pulsations by application of the Riccati method to the example of hydrogen–deficient carbon stars.**
W. Glatzel, A. Gautschy.
Mon. Not. R. Astron. Soc., Vol. 256, No. 2, p. 209 – 218 (15 May 1992).

The Riccati method, developed to solve the linear boundary–value problem of radial stellar pulsations, is generalized and shown to be applicable to the calculation of non–radial, non–adiabatic oscillations. Its advantages are demonstrated for the extreme example of a highly non–adiabatic helium star. The acoustic spectrum of helium stars is studied in detail. Mode–coupling instabilities occur in the non–radial spectrum for a wide range of the harmonic degree l, their appearance being similar to the radial case studied previously. With increasing l, the width of the instability strip in the HRD – extending up to log $T_{eff} \approx$ 4.3 for low l – shrinks and the growth rates of the instabilities decrease. Stabilization is reached between l = 10 and 100.

065.090 **Spin–down of rapidly rotating, convective stars.**
C. A. Tout, J. E. Pringle.
Mon. Not. R. Astron. Soc., Vol. 256, No. 2, p. 269 – 276 (15 May 1992).
The authors construct a model for the spin–down of rapidly rotating, convective stars. They assume that rotation and convection cause a star to rotate differentially, that differential rotation and convection generate a magnetic dynamo, that a magnetic dynamo gives rise to mass loss and a magnetically controlled stellar wind, and that such a wind results in angular momentum loss and hence stellar spin–down. Using the model, the authors show that a protostar accreting from a circumstellar disc reaches an equilibrium rotation period significantly below break–up. The authors also show that the spin–down rates they derive are consistent with those required to drive the orbital evolution of cataclysmic variables.

065.091 **The production of surface carbon depletions among globular cluster giants by interior mixing.**
G. H. Smith, C. A. Tout.
Mon. Not. R. Astron. Soc., Vol. 256, No. 3, p. 449 – 456 (1 Jun 1992).
Published measurements of the carbon abundances of giants in several metal–poor globular clusters have indicated that [C/H] decreases progressively from the base to the tip of the giant branch by a total of 1.0 or more. Standard stellar models indicate that throughout much of giant branch evolution the convective envelope of a cluster star is not expected to be in contact with regions near the nuclear–burning shell wherein the C→N process of hydrogen burning is taking place. The observations indicate, therefore, that transport of material must take place between the base of the convective envelope and the hydrogen–burning shell throughout much of the giant branch phase of evolution. A carbon abundance gradient of $d[C/H]/dM_V \approx 0.25$ dex mag^{-1} is observed on the giant branch of the cluster M92 by Carbon et al. The authors find that this gradient can be accounted for if circulation currents having speeds $\sim 10^{-2}$cm s^{-1} within the M92 giants are cycling material from the convective envelope down into the proximity of the hydrogen–burning shell, where it is processed and subsequently, returned to the envelope. Giant branch mixing may be capable of producing significant modifications from the initial C and N abundance distribution of cluster stars. In order to account for the existence of CN–strong stars on the upper giant branches of globular clusters, it may be necessary to infer that the atmospheres of such stars have undergone substantial O→N processing.

065.092 **Turbulent convection with overshooting: Reynolds stress approach.**
V. M. Canuto.
Astrophys. J., Vol. 392, No. 1, p. 218 – 232 (10 Jun 1992).
Turbulent convection is a phenomenon relevant to both stellar structure and accretion disks. In the latter, a basic parameter such as the turbulent viscosity v_t is still treated phenomenologically; in the case of stellar structure, most of the work still relies on the mixing–length theory, which assumes homogeneity and thus lacks diffusion terms (divergence of third–order moments). To include them, one needs a new formalism. The author reviews and discusses the Reynolds stress approach (proven successful in other fields) which provides a set of coupled differential equations that yield all the turbulent quantities of interest. Although the system can only be solved numerically, some features are evident: Inclusion of the diffusion terms related to $\langle w^2\theta \rangle$ and $\langle w\theta^2 \rangle$ contributes a countergradient term, which may carry heat from cold to hot regions. Inclusion of the diffusion term related to $\langle 1/2q^2w \rangle$ (turbulent kinetic energy flux) contributes an additional term, which is responsible for overshooting. In addition to the convective flux, the author also derives a model expression for v_t as a function of both shear and buoyancy: it is needed in the numerical simulation of stellar convection and in accretion disks to replace the phenomenological expressions used thus far.

065.093 **The creation of superrich lithium giants.**
I.–J. Sackmann, A. I. Boothroyd.
Astrophys. J., Lett., Vol. 392, No. 2, p. L71 – L74 (20 Jun 1992).
A time–dependent "convective diffusion" algorithm has been coupled for the first time with a self–consistent full evolutionary computation in order to investigate the creation of superrich lithium stars on the asymptotic giant branch. The authors considered stars of mass 3, 4, 5, 6, and 7 M$_\odot$, for Z = 0.02 and Z = 0.001. Superrich lithium stars were produced when convective envelope base temperatures exceeded 50×10^6K. Abundances of log $\varepsilon(^7$Li$) \sim 4.5$ were obtained for stars of mass 4 – 6 M$_\odot$, lying in the luminosity range M$_{bol} \sim -6$ to -7, in excellent agreement with observations. High lithium abundances persist for $10^4 - 10^5$yr. The superrich lithium abundances are independent of a star's earlier lithium history.

065.094 **SiC particles from asymptotic giant branch stars: Mg burning and the s–process.**
L. E. Brown, D. D. Clayton.
Astrophys. J., Lett., Vol. 392, No. 2, p. L79 – L82 (20 Jun 1992).
The authors discuss whether isotopically anomalous SiC particles found in meteorites originate in AGB stars. It is shown that if the peak helium shell flash temperatures of massive (6 – 9 M$_\odot$) AGB stars are about 10% larger than they are normally assumed to be, alpha particle reactions with the magnesium will become significant. Then the ^{26}Mg(α, n)^{29}Si reaction produces a large excess of ^{29}Si. With a light element nuclear reaction network, the authors calculate the evolution of the silicon isotopic composition during AGB evolution and find that the experimentally determined correlation between excess ^{29}Si and excess ^{30}Si in SiC particles from carbonaceous chondrites can indeed be naturally produced in this way. This demonstration contrasts with previous negative conclusions for temperatures too low for magnesium burning to contribute. The authors therefore suggest that if the large isotopically anomalous SiC particles carrying nearly pure s–process krypton and xenon do indeed originate in AGB stars, those stars are massive and have peak shell flash temperatures near 450×10^6K.

065.095 **Strongly nonadiabatic pulsations and strange modes.**
J. Zalewski.
Publ. Astron. Soc. Jpn., Vol. 44, No. 1, p. 27 – 43 (1992).
The properties of linear nonadiabatic pulsations for radial and nonradial acoustic modes in low–mass supergiant stars in the diffusion approximation and with radiative transfer are discussed. Although strange modes are found both for radial and nonradial modes, their frequencies and excitation rates depend sensitively on the description of the outer stellar layers. An asymptotic WKB analysis which takes into account the strong nonlinearity in entropy perturbation is used to examine the influence of nonadiabaticity on the pulsation modes; it is found that for strong nonadiabaticity two types of modes are possible, viz. either ordinary or strange modes. Their frequencies are determined by the location of the H/He ionization zone.

065.096 **Possible scenario of a supernova explosion as a result of gravitational collapse of a massive stellar core.**
V. S. Imshennik.
Pis'ma Astron. Zh., Tom 18, No. 6, p. 489 – 504 (Jun 1992). In Russian. English translation in Sov. Astron. Lett., Vol. 18, No. 3.
The author presents an explosion scenario, in which the main role belongs to the rotation of the initial stellar core, while the collapse developes in two stages. The first stage, leading to the formation of a rotating preneutron star, has been studied previously by mathematical simulation (Imshennik, Nadezhin 1977, 1991). The second stage, commencing with the fragmentation of the preneutron star results in formation of a close binary star system, the parameters of which are determined by conservation of total mass and angular momentum law. The subsequent evolution of the close binary system is determined by gravitational radiation, binary components mutual approach, loss of mass by the low mass component and its explosion with energy release of $\sim 10^{51}$ergs, when the mass reaches a value of ~ 0.1 M$_\odot$, according to Blinnikov et al. (1984, 1990). All stages of this

evolution are described by corresponding evaluations which do not contradict to the SN 1987A explosion data. This scenario is proposed as an alternative to the previously formulated magneto-rotational mechanism of explosion (Bisnovatyj-Kogan, 1970; Imshennik, Nadezhin, 1991) and needs to be confirmed and developed in particular by mathematical simulation.

065.097 Coupling coefficients of pulsation for radiative stellar models.
M. Takeuti, F. Yamakawa, T. Ishida.
Publ. Astron. Soc. Jpn., Vol. 44, No. 2, p. 101 – 107 (1992).

The second-order theory of coupling is discussed regarding the radial pulsation of radiative stellar models, while taking nonadiabatic effects into consideration. Numerical values are given for model classical cepheids. Adiabatic coupling coefficients are changed slightly when convection is ignored. The entropy changes near the stellar-surface increase the strength of coupling.

065.098 Setting the clock of stellar models.
G. Bertelli, A. Bressan, C. Chiosi.
Astrophys. J., Vol. 392, No. 2, p. 522 – 529 (20 Jun 1992).

The study of old open clusters such as IC 4651 has brought into evidence that the age assigned to these clusters heavily depends on the type of stellar models in use, that is, standard models or models with a certain amount of convective overshoot from the core. Current stellar models indicate that the age of IC 4651 lies in the range 1.3×10^9 to 4×10^9yr, and that the turnoff mass falls in the range $1.5 - 1.7\,M_\odot$, depending on the adopted color excess, distance modulus, and type of models (classical versus overshoot). In this paper, the authors first show that a fundamental inconsistency affects the models with core overshoot and corresponding isochrones calculated by Maeder and Meynet. Second, they re-derive the color excess, distance modulus, and age of IC 4651 with the aid of new models for both the classical and the overshoot mixing scheme. Using new evolutionary tracks calculated by Fagotto and by Alongi et al., the authors derive with the standard models a color excess $E_{B-V} = 0.16$, a distance modulus $(m-M)_0 = 9.6$, and an age of 1.4×10^9yr. With the overshoot models, they obtain a color excess $E_{B-V} = 0.14$, a distance modulus $(m-M)_0 = 9.5$, and an age of 2.0×10^9yr.

065.099 The evolution of intermediate–mass protostars. II. Influence of the accretion flow.
F. Palla, S. W. Stahler.
Astrophys. J., Vol. 392, No. 2, p. 667 – 677 (20 Jun 1992).

The authors consider the growth of spherical protostars that condense from diffuse interstellar clouds. As in their previous study, they focus on the internal structure of protostars in the intermediate–mass range: $2\,M_\odot \lesssim M_* \lesssim 10\,M_\odot$. Three evolutionary sequences with mass accretion rates of $\dot{M} = 1 \times 10^{-5}$, 3×10^{-5} and $1 \times 10^{-4}M_\odot\mathrm{yr}^{-1}$are computed. In all cases, the protostar exhibits the same evolutionary phases found previously: full convection, appearance of a radiative barrier, shell ignition of deuterium, rapid swelling, gravitational contraction, and central hydrogen ignition. The change of surface conditions from the previously used accretion shock has little quantitative effect, while increasing $\dot{M}$ swells the radius and delays the final ignition of hydrogen. For $\dot{M} = 10^{-4}M_\odot\mathrm{yr}^{-1}$, the protostar joins the ZAMS at $M_* = 15\,M_\odot$. The stellar birthline constructed with $\dot{M} = 10^{-5}M_\odot\mathrm{yr}^{-1}$ shows good agreement with the observed distribution of pre–main–sequence stars in the H–R diagram. The authors therefore propose that all stars with $M_* \lesssim 10\,M_\odot$ form with an accretion rate of this order of magnitude, and from molecular cloud cores with similar density and temperature structures.

065.100 Upper limit to the mass of pulsationally stable stars with uniform chemical composition.
R. B. Stothers.
Astrophys. J., Vol. 392, No. 2, p. 706 – 709 (20 Jun 1992).

Nuclear–energized pulsational instability is a well–known feature of models of chemically homogeneous stars above a critical mass. With the new Rogers – Iglesias opacities, the

instability occurs above $120 - 150\,M_\odot$ for normal Galactic Population I chemical compositions, and above $\sim 90\,M_\odot$ for stars in metal–poor environments like the outer Galaxy and the Small Magellanic Cloud. Models of homogeneous helium burning stars are unstable above masses of 19 and $14\,M_\odot$, respectively. These significant increases of the critical masses, in the normal metallicity cases, over the values derived previously with the Los Alamos opacities can explain the stability of the brightest observed O type stars, but they do not exclude the possibility that the most luminous hydrogen–deficient Wolf–Rayet stars are experiencing this type of instability.

065.101 The equation of state for stellar envelopes: comparison of theoretical results.
W. Däppen.
Rev. Mex. Astron. Astrofis., Vol. 23, p. 141 – 149 (Mar 1992). – See Abstr. 012.054 for the main entry.

A previous comparison of thermodynamical quantities, computed in the chemical and physical pictures, revealed a remarkable agreement in the H and He ionization zones of the Sun, despite the radically different treatment of bound states in the two formalisms. This agreement was due to an unexpectedly dominating (classical) Coulomb pressure term. New comparisons, for higher temperatures and densities, and for a representative solar mixture (H, He and O), have demonstrated substantial differences in the O–ionization fractions. Also, the thermodynamic quantities reflect these differences to a degree that is within reach of helioseismology.

065.102 The effect of chemical abundance oscillations on the pulsational properties of a 5 $M_\odot$ hydrogen–helium star.
B. Uyaniker, H. Kirbiyik.
Astrophys. Space Sci., Vol. 192, No. 2, p. 275 – 289 (Jun 1992).

A model of a first generation intermediate star of 5 $M_\odot$, with $Z = 0$ has been considered. The model is at an advanced stage of its evolution and has a double shell burning. It burns helium in the inner shell, and hydrogen, via CNO cycle, in the outer shell. $\varepsilon_\varrho = (\partial\log \varepsilon/\partial\log \varrho)_T$ and $\varepsilon_T = (\partial\log \varepsilon/\partial\log T)_\varrho$ were computed allowing for the oscillations of the relative mass abundance of the reagents in nuclear reactions. Including $\mu_\varrho = (\partial\log \varepsilon/\partial\log \varrho)_T$ and $\mu_T = (\partial\log \varepsilon/\partial\log T)_\varrho$ of mean molecular weight and the effect of the oscillations of abundances due to nuclear reactions, stability has been studied. Contrary to the results of the static calculations, the authors have found that instability due to the excitation mechanism provided by the high temperature sensitivity of energy generation rate propagates up to the surface. Thus the model in question has been found to be unstable against radial adiabatic pulsations, in its fundamental mode.

065.103 Hydrodynamic studies of accretion onto ONeMg white dwarfs using a large nuclear reaction network.
S. Starrfield, M. Politano, J. W. Truran, W. M. Sparks, A. Weiss.
Bull. Am. Astron. Soc., Vol. 23, No. 3, p. 1265 (1991). Abstract. – See Abstr. 012.067 for the main entry.

065.104 Der innere Aufbau der Hauptreihensterne.
H. Zimmermann.
Astron. Sch., Jahrg. 28, Heft 6, p. 6 – 8 (Dec 1991).

065.105 Comparison of a fine–mesh hydrodynamic stellar model with a dynamic rezoning model.
T. Ishida, M. Takeuti.
Symposium of Supercomputing Astronomy and Astrophysics in Japan, p. 273 – 284 (1991). – See Abstr. 012.069 for the main entry.

The authors have compared a hydrodynamic model of classical cepheid containing 500 grid points with that containing 100 grid points. In linear model, growth rates of pulsation modes appear to grow as increase of grid points, and to saturate at about 200 meshes. In hydrodynamic model, some shock phenomena appear to be captured only in 500 grid model. It indicates that the 500 grid model is very useful to study the shock waves in pulsating envelope. The newly constructed 500 grid model is applied to study the ionization and recombination fronts and shock waves in pulsating envelope. The motion of the discontinuities over a whole pulsation phase is shown for the first time.

065.106 **Modelling supergiant pulsations.**
J. Zalewski.
Sci. Rep. Tôhoku Univ., Eighth Ser., Vol. 12, Nos. 2/3, p. 157 – 161 (Jan 1992). – See Abstr. 012.073 for the main entry.
The observational data on variable intermediate spectral type supergiants and pulsational models of supergiants are discussed.

065.107 **New developments in understanding the HR diagram.**
C. Chiosi, G. Bertelli, A. Bressan.
Annu. Rev. Astron. Astrophys., Vol. 30, p. 235 – 285 (1992).
Contents: 1. Introduction.   2. Basic stellar evolution. 3. Nuclear reactions, opacities, and stellar winds. 4. Semiconvection and convective overshoot. 5. Old star clusters. 6. Intermediate age clusters and Cepheid stars. 7. Supergiant stars in the Milky Way and LMC.

065.108 **Influence of starspots on internal stellar structure.**
H. C. Spruit.
Armagh Observatory Bicentenary Colloquium on Surface Inhomogeneities on Late–Type Stars, p. 78 – 84 (1992). – See Abstr. 012.085 for the main entry.

065.109 **Constraints of axions from white dwarf cooling.**
J. Wang.
1. International Trends in Astroparticle Physics Conference together with the SuperNova Watch Workshop, p. 146 – 153 (1992). – See Abstr. 012.102 for the main entry.
The author performs numerical calculations on the white dwarf cooling process. He finds that it provides a constraint on the axion mass. The upper bound is 0.01 eV.

065.110 **Effects of envelope overshoot in intermediate and massive stars.**
A. Bressan.
Mem. Soc. Astron. Ital., Vol. 63, No. 1, p. 25 – 28 (1992). – See Abstr. 012.087 for the main entry.

065.111 **Quelques aspects actuels de la physique stellaire.**
C. Catala.
J. Astron. Fr., No. 41, p. 3 – 8 (Feb 1992). – See Abstr. 012.070 for the main entry.

065.112 **Element mixing in a stellar interior by the Balbus–Hawley mechanism.**
S. Kato.
Publ. Astron. Soc. Jpn., Vol. 44, No. 3, p. L31 – L33 (1992).
Recently, a powerful local instability of differentially rotating magnetic accretion disks was presented by Balbus and Hawley. This should also be effective as a process of the element mixing which is required to explain the evolution of some kind of stars, like the progenitor of SN 1987A.

065.113 **Supernova neutrinos: life after SN 1987A.**
A. Burrows.
1. International Trends in Astroparticle Physics Conference together with the SuperNova Watch Workshop, p. 463 – 476 (1992). – See Abstr. 012.102 for the main entry.
The author concentrates in this paper solely on the theory of neutrino bursts. He provides an overview and preliminary answers to the following questions: what are the distinctive and diagnostic features of the neutrino emissions? What might we learn about supernova physics and the supernova mechanism from an abundance of events? Of what theoretical issues should the experimental teams be aware?

065.114 **Testing stellar evolution theory with oscillation frequency data.**
W. Dziembowski.
Reviews in Modern Astronomy 5: Variabilities in stars and galaxies, p. 142 – 160 (1992). – See Abstr. 003.087 for the main entry.
Contents: Introduction. Mixing in stellar interiors – an unresolved problem. Methods of asteroseismology. Mixing in the Sun's core? On the interface between convective envelope and radiative interior. What did we learn from helioseismology about the angular momentum evolution? Can δ Scuti star observations help us to solve the overshooting problem?

065.115 **Note on 5:7 phase–locking in the two–mode coupling of stellar pulsation.**
Y. Tanaka, K. Seya, M. Takeuti.
Publ. Astron. Soc. Jpn., Vol. 44, No. 3, p. 331 – 333 (1992).
The nonlinear theory of oscillation suggests phase–locking between moderately coupled modes. This is referred to as an explanation for the fact that the observed period–ratio of double–mode cepheids is close to 5:7. Numerical calculations of coupled model oscillators show the existence of 5:7 phase–locking for an appropriate period–ratio.

065.116 **Evolution of stars and gas in galactic nuclei.**
R. Spurzem.
Reviews in Modern Astronomy 5: Variabilities in stars and galaxies, p. 161 – 173 (1992). – See Abstr. 003.087 for the main entry.
Contents: Introduction. Stellar hydrodynamics. Dense star–gas systems. Multi–mass momentum models.

065.117 **Stellar evolution.**
J. Mitton.
Images of the universe, p. 110 – 120 (1991). – See Abstr. 003.090 for the main entry.

065.118 **B–type stars in young clusters.**
D. Schönberner, V. Wenske.
Workshop on the Atmospheres of Early–Type Stars, p. 27 – 29 (1992). – See Abstr. 012.095 for the main entry.
The authors studied B–type stars of the two young stellar clusters NGC 2264 and α Per (Melotte 20) for stellar evolution purposes.

065.119 **Solar and supernova neutrino interactions.**
W. C. Haxton.
1. International Trends in Astroparticle Physics Conference together with the SuperNova Watch Workshop, p. 483 – 495 (1992). – See Abstr. 012.102 for the main entry.
Two topics are addressed, the interactions of neutrinos during a type II supernova and the effect of current eddies on solar neutrino oscillations. The supernova discussion focuses on the nucleosynthesis that accompanies inelastic neutral current interactions of neutrinos in the mantle of a collapsing star, and on the effect of neutrino "down–scattering" and preheating on the explosion mechanism. The second half of the paper deals with the influence of solar turbulence (or density fluctuations) on the neutrino effective mass and the possibility that a time–varying neutrino flux could result. The effects of harmonic density or three–current perturbations on the oscillation probability are explored analytically and numerically.

065.120 **The mechanism of mirror–symmetry violation of the magnetic field in rotating stars and possible astrophysical applications.**
G. S. Bisnovatyj–Kogan, S. G. Moiseenko.
Astron. Zh., Tom 69, Vyp. 3, p. 563 – 571 (May – Jun 1992). In Russian. English translation in Sov. Astron., Vol. 36, No. 3 (May – Jun 1992).
A mirror–symmetry violation mechanism of the magnetic field by interaction of poloidal and toroidal field components in differentially rotating stars is suggested. The appearing mirror–

non–symmetrical field with differential rotation becomes stronger and can lead to one–side eruption and recoil effect. Concrete examples are given.

Introduction to stellar astrophysics. Volume 3. Stellar structure and evolution.
See Abstr. 003.016.

Stellar astrophysics.
See Abstr. 003.057.

Stellar evolution: Stars arriving two by two.
See Abstr. 011.011.

New windows to the universe. Volumes I, II. Invited review papers and general lectures.
See Abstr. 012.001.

Evolutionary processes in interacting binary stars. Proceedings.
See Abstr. 012.008.

Astrophysical opacities. Proceedings. Workshop on Astrophysical Opacities (WAO), Caracas (Venezuela), 15 – 19 Jul 1991.
See Abstr. 012.054.

Surface inhomogeneities on late–type stars. Proceedings.
See Abstr. 012.085.

The Opacity Project – equation of state.
See Abstr. 013.084.

The Opacity Project – results for opacities.
See Abstr. 013.085.

Das Riesenstadium der Sterne.
See Abstr. 014.051.

Numerical simulators of stability in astrophysical problems.
See Abstr. 021.013.

Basic research in mathematical and space sciences.
See Abstr. 021.022.

A measurement of the $^{14}C(n, \gamma)^{15}C$ cross section at a stellar temperature of kT = 23.3 keV.
See Abstr. 022.024.

Relativistic free–free Gaunt factors for high–temperature stellar plasmas.
See Abstr. 022.193.

Level structure of ^{21}Mg and the $^{20}Na(p, \gamma)^{21}Mg$ stellar reaction rate.
See Abstr. 061.026.

Radionuclides of interest for γ–ray line astronomy from novae and red giants.
See Abstr. 061.035.

Constraints of the neutrino magnetic moment from white dwarf cooling.
See Abstr. 061.042.

Dirac neutrinos and SN 1987A.
See Abstr. 061.053.

The role of the neutrino electromagnetic moments in the stellar energy loss rate.
See Abstr. 061.079.

Reaction $^{36}Ar(p, \gamma)^{37}K$ in explosive hydrogen burning.
See Abstr. 061.082.

On the contribution of ^{22}Ne to the synthesis of ^{54}Fe and ^{58}Ni in thermonuclear supernovae.
See Abstr. 061.085.

Can a closure mass neutrino help solve the supernova shock reheating problem?
See Abstr. 061.086.

Measurement of the $^{76}Se(n, \gamma)$ capture cross section and phenomenological s–process studies: the weak component.
See Abstr. 061.087.

On the calculation of Maxwellian–averaged capture cross sections.
See Abstr. 061.090.

The s–process in massive stars of variable composition.
See Abstr. 061.092.

Beta decay of some fp shell nuclei for presupernova stars.
See Abstr. 061.097.

The $^{180m}Ta(\gamma, \gamma')^{180}Ta$ cross section at 1.33 and 4.0 MeV and its astrophysical consequences.
See Abstr. 061.098.

Nucleation of strange matter in dense stellar cores.
See Abstr. 061.104.

Massive Dirac neutrinos and SN 1987A.
See Abstr. 061.111.

The stellar (n, γ) cross sections for ^{87}Rb and ^{192}Pt.
See Abstr. 061.114.

Origin of ^{180m}Ta and the temperature of the s–process.
See Abstr. 061.121.

From stellar structures to fundamental physics.
See Abstr. 061.135.

Understanding n–capture nucleosynthesis: a test for stellar and galactic evolution.
See Abstr. 061.136.

On a new, one–dimensional, time–dependent model for turbulence and convection. I. A basic discussion of the mathematical model.
See Abstr. 062.001.

Downflows and entropy gradient reversal in deep convection.
See Abstr. 062.066.

On the thermal conductivity due to collisions between relativistic degenerate electrons.
See Abstr. 062.087.

Self–similar magnetohydrodynamics. V. Gravitating spheres and spheroids.
See Abstr. 062.100.

Dynamo action in stratified convection with overshoot.
See Abstr. 062.177.

Line profile variations caused by low–frequency non–radial pulsations of rapidly rotating stars. II.
See Abstr. 063.044.

Radiative atomic Rosseland mean opacity tables.
See Abstr. 063.057.

Gamma–ray deposition and nonthermal excitation in supernovae.
See Abstr. 063.093.

Results obtained using the OPAL code.
See Abstr. 063.119.

Molecular opacity and stellar structure.
See Abstr. 063.124.

Radiative opacities and stellar pulsations.
See Abstr. 063.128.

On preplanetary–nebula formation.
See Abstr. 064.023.

The effect of irradiation in Algol–type binaries: metallicity and position in the HR diagram.
See Abstr. 064.039.

The stability of accretion tori. IV. Fission and fragmentation of slender, self–gravitating annuli.
See Abstr. 064.040.

Classical novae as fast magnetic rotators.
See Abstr. 064.049.

Models of planetary nebula spectral evolution.
See Abstr. 064.055.

Boson star in a gravitation theory with dilaton.
See Abstr. 066.061.

A description of semidegenerate self–gravitating spheres of fermions.
See Abstr. 066.158.

A generalization of the relativistic equilibrium equations for a non–rotating star.
See Abstr. 066.162.

Radiating soliton stars in the thin–wall approximation.
See Abstr. 066.164.

A new class of barotropic stellar models. III. Equations for the relativistic approximation.
See Abstr. 066.289.

Scenes from general relativistic astrophysics.
See Abstr. 066.306.

Nuclear astrophysics of dense matter.
See Abstr. 067.010.

Post–Newtonian frequencies for the pulsations of rapidly rotating neutron stars.
See Abstr. 067.028.

The structure and evolution of Thorne–Żytkow objects.
See Abstr. 067.034.

Supernovae: where and why do they break off?
See Abstr. 067.042.

Model for the completeness of quasinormal modes of relativistic stellar oscillations.
See Abstr. 067.044.

Gamma–ray bursts and cosmic rays from accretion–induced collapse.
See Abstr. 067.062.

W–modes: a new family of normal modes of pulsating relativistic stars.
See Abstr. 067.067.

Rotational properties of strange stars.
See Abstr. 067.070.

Application of the improved Hartle method for the construction of general relativistic rotating neutron star models.
See Abstr. 067.137.

X–ray binaries and their evolution.
See Abstr. 067.202.

Could final states of stellar evolution proceed towards naked singularities?
See Abstr. 067.211.

Interior opacities and the standard solar model.
See Abstr. 080.026.

Standard solar model.
See Abstr. 080.032.

The fate of ^{7}Be in the Sun.
See Abstr. 080.057.

Standard solar model. II. g–modes.
See Abstr. 080.065.

The role of the molecular – metallic transition of hydrogen in the evolution of Jupiter, Saturn, and brown dwarfs.
See Abstr. 091.038.

Surface abundances of light elements in stars.
See Abstr. 114.005.

Which evolution for an Am star?
See Abstr. 114.007.

The CH stars. I. Carbon isotope ratios.
See Abstr. 114.014.

Carbon, nitrogen, and oxygen abundances in early B–type stars.
See Abstr. 114.019.

WR 134 line–profile variability revisited.
See Abstr. 114.023.

The chemical composition of luminous stars: problems or opportunities?
See Abstr. 114.038.

Hot subluminous stars.
See Abstr. 114.089.

A flux–limited sample of galactic carbon stars.
See Abstr. 115.001.

The properties of post–Hayashi track evolutionary phases and the problem of the ultraviolet emission from elliptical galaxies.
See Abstr. 115.024.

The metallicity effect on the luminosity level of the AGB clump in low mass stars.
See Abstr. 115.025.

Observed evolutionary changes in the visual magnitude of the luminous blue variable P Cygni.
See Abstr. 115.026.

Asteroseismology as a probe of stellar structure and evolution.
See Abstr. 116.002.

Rotation and transition layer emission in cool giants.
See Abstr. 116.008.

Magnetic fields at the surfaces of stars.
See Abstr. 116.026.

Spectroscopic signatures of active regions on main sequence stars.
See Abstr. 116.047.

Is there a recycled old pulsar in Cygnus X–3?
See Abstr. 117.008.

Outbursts by low–mass white dwarfs in symbiotic variables.
See Abstr. 117.018.

Evolutionary state of W UMa–type systems.
See Abstr. 117.049.

Evolutionary sequences for binary stars in the mass range 9 to 40 M⊙.
See Abstr. 117.097.

Dynamical theory of viscous tides in binary systems.
See Abstr. 117.152.

DF Hydrae: a W UMa system with spotted components.
See Abstr. 117.164.

The interacting binary V367 Cygni. I. Photometric solution including the influence of an accretion disk.
See Abstr. 117.165.

Constraints on the evolution of AM Herculis stars derived from the observed distribution of orbital periods.
See Abstr. 117.173.

Astrophysical parameters of X Persei.
See Abstr. 117.202.

Pulsating X–ray sources: long–term period variations.
See Abstr. 117.247.

The synchronization of binary components from the motion of starspots.
See Abstr. 117.277.

On the mass discrepancy of the Cepheid stars.
See Abstr. 122.012.

Toward a resolution of the bump and beat Cepheid mass discrepancies.
See Abstr. 122.017.

The binary Cepheid BP Cir: a test of evolutionary tracks.
See Abstr. 122.024.

On the instability strip of the Cepheid stars.
See Abstr. 122.032.

New opacities and the origin of the β Cephei pulsation.
See Abstr. 122.035.

The pulsations of yellow semiregular variables. I. Double–mode behaviour of UU Herculis.
See Abstr. 122.036.

The solution of the light curve of 44 Tauri: insights into the pulsational behaviour of δ Scuti stars.
See Abstr. 122.037.

Variable stars and stellar evolution.
See Abstr. 122.090.

Intrinsic properties of Cepheid variables.
See Abstr. 122.143.

On pulsations of luminous stars.
See Abstr. 122.153.

Nonradial pulsations of O– and B–stars.
See Abstr. 122.159.

A prediction of the γ–ray flux from Nova Herculis 1991.
See Abstr. 124.025.

Disk instability and outburst properties of the intermediate polar GK Persei.
See Abstr. 124.041.

Thermonuclear detonations and deflagrations in supernovae.
See Abstr. 125.004.

Two–dimensional simulations of hydrodynamic instability in supernova explosion. Part 2.
See Abstr. 125.005.

Supernova 1987A and formation of rotating neutron stars.
See Abstr. 125.024.

SN 1987A and rotating neutron star formation.
See Abstr. 125.065.

The energy sources powering the the late–time bolometric evolution of SN 1987A.
See Abstr. 125.087.

The supernova 1987A.
See Abstr. 125.090.

Postexplosion hydrodynamics of SN 1987A.
See Abstr. 125.114.

The ring around SN 1987A and rotation of the progenitor.
See Abstr. 125.137.

Possible models for the Type Ia supernova 1990N.
See Abstr. 125.143.

Modeling the iron–dominated spectra of the Type Ia supernova SN 1991T at premaximum.
See Abstr. 125.149.

Break–up of quasi–periodic motion as an interpretation of white–dwarf variability.
See Abstr. 126.015.

Maximum rates of period change for DA white dwarf models with carbon and oxygen cores.
See Abstr. 126.100.

Axion cooling of white dwarfs.
See Abstr. 126.107.

More on the progenitors of white dwarfs.
See Abstr. 126.118.

Numerical models for the collapse and fragmentation of centrally condensed molecular cloud cores.
See Abstr. 131.040.

Numerical models of tidally interacting protostellar binary systems.
See Abstr. 131.051.

The implications of runaway OB stars for high–mass star formation.
See Abstr. 131.067.

Ambipolar diffusion, cloud cores, and star formation: two–dimensional, cylindrically symmetric contraction. II. Results and a length scale for protostellar cores.
See Abstr. 131.104.

Ambipolar diffusion, cloud cores, and star formation: two–dimensional cylindrically symmetric contraction. III. A further parameter study and magnetically controlled accretion rate.
See Abstr. 131.105.

Ambipolar diffusion and star formation: formation and contraction of axisymmetric cloud cores. I. Formulation of the problem and method of solution.
See Abstr. 131.171.

Infall in collapsing protostars.
See Abstr. 131.196.

Towards understanding the stellar initial mass function.
See Abstr. 131.247.

Initiation of bipolar flows by magnetic field twisting in protostellar nebulae.
See Abstr. 131.262.

Old planetary nuclei and their evolutionary connections.
See Abstr. 134.045.

Two–component gravitating systems and red giant – like structure.
See Abstr. 151.010.

Rotation–induced mixing and lithium depletion in galactic clusters.
See Abstr. 153.003.

Solar calibration and the ages of the old disk clusters M67, NGC 188, and NGC 6791.
See Abstr. 153.005.

The oscillating blue stragglers in the open cluster M67.
See Abstr. 153.017.

Old open clusters as a test for the mixing schemes.
See Abstr. 153.052.

Setting the clock of stellar models.
See Abstr. 153.056.

Evolved stars in ω Centauri. I. Radial distribution of blue sub-dwarfs.
See Abstr. 154.022.

Evolution versus pulsation along the horizontal branch of M15.
See Abstr. 154.025.

The effect of helium diffusion on the ages of globular clusters.
See Abstr. 154.039.

Possible observational consequences of primordial binaries in globular clusters.
See Abstr. 154.047.

Are there two kinds of blue stragglers in globular clusters?
See Abstr. 154.056.

Looking for gaps.
See Abstr. 154.061.

The evolution of low mass stars and the enhancement of the α–elements.
See Abstr. 154.062.

Constraints on the age and evolution of the Galaxy from the white dwarf luminosity function.
See Abstr. 155.023.

Evolution of spiral galaxies. I. Halo–disk connection for the evolution of the solar neighborhood.
See Abstr. 155.028.

Possible sources of the Population I lithium abundance and light–element evolution.
See Abstr. 155.103.

Evolution of heavy–element abundances as a constraint on sites for neutron–capture nucleosynthesis.
See Abstr. 155.132.

The star formation history of the Large Magellanic Cloud.
See Abstr. 156.026.

Stellar evolution and stellar populations in galaxies.
See Abstr. 157.099.

The contribution of advanced post–Hayashi track evolutionary phases to the ultraviolet light of elliptical galaxies.
See Abstr. 157.210.

Can the first stars formed be pre–galactic?
See Abstr. 161.128.

The Hubble constant from nickel radioactivity in Type Ia supernovae.
See Abstr. 161.381.

066 Relativistic Astrophysics, Gravitation Theory

066.001   Search for a coupling of the Earth's gravitational field to nuclear spins in atomic mercury.
B. J. Venema, P. K. Majumder, S. K. Lamoreaux, B. R. Heckel, E. N. Fortson.
Phys. Rev. Lett., Vol. 68, No. 2, p. 135 – 138 (13 Jan 1992). Current Physics Microform No.: 9202G2134.

The authors have measured the ratio of nuclear spin-precession frequencies of ^{199}Hg and ^{201}Hg atoms for two orientations of magnetic field relative to the Earth's gravitational field. They find that the spin–dependent component of gravitational energy is less than 2.2×10^{-21}eV, a substantial improvement over previous limits. This result provides a test of the equivalence principle for nuclear spins, and sets limits on the magnitude of possible scalar – pseudoscalar interactions which would couple to the spins.

066.002   An obstacle to building a time machine.
S. M. Carroll, E. Farhi, A. H. Guth.
Phys. Rev. Lett., Vol. 68, No. 3, p. 263 – 266 (20 Jan 1992). Current Physics Microform No.: 9203A1960.

Gott has shown that a spacetime with two infinite parallel cosmic strings passing each other with sufficient velocity contains closed timelike curves. The authors discuss an attempt to build such a time machine. Using the energy–momentum conservation laws in the equivalent $(2+1)$–dimensional theory, they explicitly construct the spacetime representing the decay of one gravitating particle into two. It is found that there is never enough mass in an open universe to build the time machine from the products of decays of stationary particles. More generally, the Gott time machine cannot exist in any open $(2+1)$–dimensional universe for which the total momentum is timelike.

066.003   Knot invariants as nondegenerate quantum geometries.
B. Brügmann, R. Gambini, J. Pullin.
Phys. Rev. Lett., Vol. 68, No. 4, p. 431 – 434 (27 Jan 1992). Current Physics Microform No.: 9203A2128.

The loop–space representation based on Ashtekar's new variables has allowed for the first time the construction of quantum states of the gravitational field. However, all states presently known were associated with spacetime metrics that were everywhere degenerate. Here, the authors present a new exact solution of the constraint equations of quantum gravity that is the first quantum state of the gravitational field known to be associated with a not–everywhere degenerate metric. The state is associated with the second coefficient of the Alexander–Conway polynomial of knot theory.

066.004   Toward a canonical formalism of non–perturbative two–dimensional gravity.
T. Yoneya.
Commun. Math. Phys., Vol. 144, No. 3, p. 623 – 639 (Mar 1992).

On the basis of the previously proposed action principle describing the theory space of 2D gravity in less than one–dimension, the authors develop a systematic canonical formalism for studying the properties of the string equation in the phase space of the cosmological constant and its canonical conjugate, the puncture operator. As a consequence, the geometrical origin of the generalized Virasoro condition on the partition function is understood to be the symmetry under the regular area–preserving diffeomorphisms ($w_{1+\infty}$ symmetry) in the deformed phase space. The deformed canonical formalism can be regarded as a quantization of a classical canonical formalism describing the sphere limit of the theory.

066.005   Semi–infinite homology and 2D gravity. I.
B. H. Lian, G. J. Zuckerman.
Commun. Math. Phys., Vol. 145, No. 3, p. 561 – 593 (Apr 1992).

The authors studied the constraint problem for two–dimensional quantum gravity in the conformal gauge. In this gauge, they proposed an ansatz for the gravitational sector. Using this

ansatz, the authors established a striking connection between the matrix models and continuum 2D gravity. They also announced several results on semi–infinite homology of the Virasoro algebra with coefficients in a suitable class of positive energy modules. In this article, the authors will provide details of the proof of the announced results.

066.006   The structure of naked singularity in self–similar gravitational collapse.
P. S. Joshi, I. H. Dwivedi.
Commun. Math. Phys., Vol. 146, No. 2, p. 333 – 342 (May 1992).

The authors study the structure and formation of naked singularities in self–similar gravitational collapse for an adiabatic perfect fluid. Conditions are obtained for the singularity to be either locally or globally naked and for the families of non–spacelike geodesics to terminate at the singularity in past. This is shown to be a strong curvature naked singularity in a powerful sense and an interesting relationship is pointed out between positivity of energy and occurrence of naked singularity.

066.007   A generating technique for Einstein gravity conformally coupled to a scalar field with Higgs potential.
D. V. Gal'tsov, B. C. Xanthopoulos.
J. Math. Phys., Vol. 33, No. 1, p. 273 – 277 (Jan 1992). Current Physics Microform No.: 9202D1067.

Starting from any solution of the Einstein equations, with cosmological term, coupled to a minimally coupled massless scalar field, a solution of the Einstein equations is constructed, conformally coupled to a massless self–interacting scalar field with the usual Higgs potential. When the cosmological constant vanishes, the Higgs term disappears and the transformation procedure reduces to that obtained by Bekenstein in 1974. As an example, a nonsingular cosmological solution is constructed that describes the restoration of spontaneously broken symmetry.

066.008   Global gravitational instantons and their degrees of symmetry.
R. A. Baadhio.
J. Math. Phys., Vol. 33, No. 2, p. 721 – 724 (Feb 1992). Current Physics Microform No.: 9202F1935.

It is shown that the n ten–dimensional very exotic spheres – or, equivalently, global gravitational instantons or solitons – admit only a finite number of involutions. They are realized as smooth SU(3) biaxial actions on Σ^{10} and are classified by the universal group manifold CP². A formula, referred to as the degree of symmetry, which quantizes the maximal symmetry that a global gravitational instanton could possess is derived. Namely, it is established that Σ^{10} admits no more than 20 SU(3) involutions.

066.009   N = 2 conformal supergravity and superconformal anomaly.
M. Kachkachi, T. Lhallabi.
J. Math. Phys., Vol. 33, No. 2, p. 725 – 734 (Feb 1992). Current Physics Microform No.: 9202F1939.

The differential geometry formalism for N = 2 conformal supergravity in harmonic superspace is developed. The N = 2 superdiffeomorphisms and local super–Weyl transformations and their BRST–like symmetries are derived. These lead to the formulation of the cohomology problem and the N = 2 superconformal anomaly.

066.010   The initial value problem for a class of general relativistic fluid bodies.
A. D. Rendall.
J. Math. Phys., Vol. 33, No. 3, p. 1047 – 1053 (Mar 1992). Current Physics Microform No.: 9203G0471.

A body or collection of bodies made of perfect fluid can be described in general relativity by a solution of the Einstein–Euler system where the mass density has spatially compact support. It is shown that for certain equations of state there exists a wide class of solutions of this type corresponding to appropriate initial data given on a spacelike hypersurface. This class is not constrained by any symmetry requirements. The key element of the proof is to write the equations as a symmetric hyperbolic system which is regular both for nonvanishing density and in vacuum.

066.011   Head–on collision of gravitational plane waves with noncollinear polarization: a new class of analytic models.
D. Tsoubelis, Wang Anzhong.
J. Math. Phys., Vol. 33, No. 3, p. 1054 – 1064 (Mar 1992). Current Physics Microform No.: 9203G0478.

A new four–parameter class of exact solutions of Einstein's field equations is obtained, using the inverse scattering method of Belinsky and Zakharov. Its members represent the head–on collision of a variably polarized gravitational plane wave with one having constant polarization and, in general, different profile, or with an infinitely thin shell of null dust. In some of these models no curvature singularity develops along the future boundary of the region of interaction. In certain cases the singularity avoidance is the direct result of the noncollinear polarization of the waves involved in the collision.

066.012   Gravitational interaction of plane gravitational waves and matter shells.
Wang Anzhong.
J. Math. Phys., Vol. 33, No. 3, p. 1065 – 1072 (Mar 1992). Current Physics Microform No.: 9203G0489.

The Weyl, Ricci scalars and the Bianchi identities are explicitly given in terms of distributions. Using the obtained Bianchi identities, the gravitational interaction of colliding shock or impulsive gravitational plane waves or matter shells is studied. It is found that, when two impulsive shells of null dust collide, they necessarily produce a tail. The tail could be a "Coulomb–like" gravitational field, or a matter current, or a mixture of both.

066.013   Minimal coupling of electromagnetic fields in Riemann–Cartan space–times for perfect fluids with spin density.
L. L. Smalley, J. P. Krisch.
J. Math. Phys., Vol. 33, No. 3, p. 1073 – 1081 (Mar 1992). Current Physics Microform No.: 9203G0497.

The electromagnetic field is minimally coupled to gravity in a Riemann–Cartan space–time containing a charged magnetized spinning fluid. It is required that the overall Lagrangian of the gravitational field, spinning matter, and the electromagnetic field be invariant under a gauge transformation of the vector potential. The theory preserves both charge conservation and particle number conservation. The electromagnetic field, via the vector potential, now interacts directly with the spin energy momentum. The spin transport equation, in addition to the usual Fermi–Walker transport term, contains a contribution due to the torque of the electromagnetic field acting on a magnetic dipole. In the absence of electromagnetism, the field equations reduce to those of the usual self–consistent Lagrangian formalism for a perfect fluid with spin density.

066.014   Spherically symmetric radiating solution with heat flow in general relativity.
D. Kramer.
J. Math. Phys., Vol. 33, No. 4, p. 1458 – 1462 (Apr 1992). Current Physics Microform No.: 9204G1636.

A model describing a collapsing sphere with radial heat flow is studied. Vaidya's pure radiation field is taken as the exterior solution. The motion of the boundary surface follows from the matching conditions. In the remote part, the evolution starts with the interior and exterior Schwarzschild solution.

066.015   Einstein gravity coupled to a massless conformal scalar field in arbitrary space–time dimensions.
B. C. Xanthopoulos, T. E. Dialynas.
J. Math. Phys., Vol. 33, No. 4, p. 1463 – 1471 (Apr 1992). Current Physics Microform No.: 9204G1641.

For space–times of arbitrary dimensionality two transformations are obtained that applied to any solution of the Einstein equations coupled to a minimally coupled scalar field construct a solution of the Einstein equations coupled to a conformal scalar field. The transformations are conformal for the metrics and algebraic for the reparametrization of the scalar fields. With these generating techniques static, spherically symmetric, asymptotically flat solutions of the Einstein–conformal scalar field equations in arbitrary space–time dimensions are obtained. Black hole solutions are systematically searched for and it is found that the only (nonvacuum) one exists in four space–time dimensions. This is the Bekenstein black hole for gravity coupled to a conformal scalar field.

066.016   Note on the representation of many–particle dynamics as higher–order single–particle dynamics, with a means to relativistic elevation of Newtonian dynamics.
E. H. Kerner.
J. Math. Phys., Vol. 33, No. 5, p. 1675 – 1684 (May 1992). Current Physics Microform No.: 9205H1195.

It is shown how (first in Newtonian physics), by processes of repeated differentiation of equations of motion and of algebraic elimination, the dynamics of many particles may be brought to another equivalent representation, that of a single higher–order equation of motion for a single particle. Here a higher–order Lagrangian followed by an Ostrogradsky Hamiltonian may be brought into play. The higher–order one–particle representation translates simply and directly (but not uniquely) into relativistic generalization (because only a single world line is being described), in which the Poincaré group is canonically represented in an Ostrogradsky Hamiltonian formulation. The examples of the Kepler problem and the harmonic oscillator are elaborated in detail.

066.017   A multipolar stationary object embedded in a gravitational field.
K. D. Krori, D. Goswami.
J. Math. Phys., Vol. 33, No. 5, p. 1780 – 1781 (May 1992). Current Physics Microform No.: 9205H1300.

The task of embedding a multipolar stationary object in a gravitational field is undertaken in this paper. It is believed that such an object will be astrophysically interesting.

066.018   Morse theory and gravitational microlensing.
A. O. Petters.
J. Math. Phys., Vol. 33, No. 5, p. 1915 – 1931 (May 1992). Current Physics Microform No.: 9205H1435.

Morse theory is used to rigorously obtain counting formulas and lower bounds for the total number of images of a background point source, not on a caustic, undergoing lensing by a single–plane microlens system having compact bodies plus either subcritical or supercritical continuously distributed matter. An image–counting formula is also found for the case when external shear is added.

066.019   Wormholes and time machines.
P. Davies.
Sky Telesc., Vol. 83, No. 1, p. 20 – 23 (Jan 1992).

066.020   On the strength of a gravitational field.
B. Mashhoon.
Phys. Lett. A, Vol. 163, No. 1/2, p. 7 – 14 (9 Mar 1992).

The notion of gravitational field strength is critically examined in this paper within the framework of the general theory of relativity. It is shown that this concept is strongly observer dependent in close analogy with electromagnetism. The results are applied to the linearized theory of gravitation in order to show that local Lorentz invariance in general precludes the possibility of a generally covariant analysis of gravitational perturbations.

066.021 New exact three–parameter solution of the Einstein–Maxwell equations for a charged spinning mass.
N. R. Sibgatullin, V. S. Manko.
Phys. Lett. A, Vol. 163, No. 5/6, p. 364 – 366 (30 Mar 1992).

An exact asymptotically flat solution of the Einstein–Maxwell equations referring to a charged rotating mass and different from the Kerr–Newman spacetime is presented in explicit form. The solution obtained has the Schwarzschild static pure vacuum limit.

066.022 Empirical and theoretical evidence for gravitational polarization of matter.
H. Essén.
Phys. Scr., Vol. 45, No. 1, p. 22 – 25 (Jan 1992).

Empirical evidence indicating modification of gravity inside matter is summarized. It is argued that four independent sets of facts point in the same direction: a weakening of gravity inside matter. Indirect evidence comes from discrepancies between the theoretical Standard Solar Model (SSM) and the observed neutrino flux, the observed seismic oscillations of the Sun, and geological evidence about the climate on the early Earth. The possible modifications of standard general relativity that might be compatible with the above and all other known facts are considered.

066.023 Non–Riemannian theories of gravity and lunar and satellite laser ranging.
I. Ciufolini, R. Matzner.
Int. J. Mod. Phys. A, Vol. 7, No. 4, p. 843 – 852 (10 Feb 1992).

Theories of gravity with a non–Riemannian manifold have been studied since the advent of Einstein's general relativity. In this paper, after an introduction on theories of gravity with a non–Riemannian spacetime and in particular on the nonsymmetric Moffat theory, the authors briefly describe the techniques of lunar and satellite laser ranging. Among the various applications of lunar and satellite laser ranging are several important measurements and tests of Einstein's general relativity as well as constraints on some alternative gravity theories. In particular, lunar and satellite laser ranging put strong validity limits on the 1983 nonsymmetry Moffat theory.

066.024 Backwards time–travel induced by combined magnetic and gravitational fields.
M. Novello, N. F. Svaiter, M. E. X. Guimarães.
Mod. Phys. Lett. A, Vol. 7, No. 5, p. 381 – 386 (20 Feb 1992).

The authors analyze the behavior of a macroscopic particle submitted to combined magnetic and gravitational fields on Gödel's universe. The examination is made in a local Gaussian system of coordinates.

066.025 Gravity waves in classical string theory.
P. M. Schwarz.
Nucl. Phys. B, Part. Phys., Vol. 373, No. 2, p. 529 – 556 (6 Apr 1992).

Gravity waves that serve as conformally invariant background for the string sigma model can create space–time singularities when they collide. The author describes some of the properties and pathologies of these waves, and computes the $O(\alpha')$ perturbation to the metric for the space–time with two colliding plane waves. The perturbative correction to the dilaton field indicates that the α' expansion becomes strongly coupled when the space–time curvature is large. The null convergence condition required for proving the singularity theorem for these space–times can be violated for certain choices of amplitude and polarization of the colliding waves.

066.026 Higher–order gravity in an $R^1 \times S^1 \times S^2$ space–time.
S. Chakraborty.
Astrophys. Space Sci., Vol. 187, No. 1, p. 135 – 141 (Jan 1992). Letter–to–the–editor.

The action with general Lagrangian is transformed to the canonical form. The classical field equations are analysed for a particular choice of the Lagrangian. A study of the Wheeler–Dewitt equation has also been performed.

066.027 Hyperbolic evolution system for numerical relativity.
C. Bona, J. Massó.
Phys. Rev. Lett., Vol. 68, No. 8, p. 1097 – 1099 (24 Feb 1992).
Current Physics Microform No.: 9205C2041.

Einstein evolution equations are written as a hyperbolic system of balance laws. A harmonic time coordinate is used with zero shift vector (harmonic slicing). The principal part of the evolution system reduces to a set of uncoupled wave equations in first order form. The relevance for three–dimensional numerical relativity of both the harmonic slicing and the resulting evolution system is stressed.

066.028 Fermat's principle in general relativity.
R. Nityananda, J. Samuel.
Phys. Rev. D, Vol. 45, No. 10, p. 3862 – 3864 (15 May 1992).
Current Physics Microform No.: 9207D1444.

Kovner has observed that Fermat's principle can be used to describe the motion of light rays in arbitrary gravitational fields, not just stationary ones. The authors give a simple demonstration of this fact.

066.029 Naked singularities: gravitationally collapsing configurations of dust or radiation in spherical symmetry, a unified treatment.
J. P. S. Lemos.
Phys. Rev. Lett., Vol. 68, No. 10, p. 1447 – 1450 (9 Mar 1992).
Current Physics Microform No.: 9205D0197.

It is shown that the naked singular solutions that form in the gravitational spherical collapse of radiation are of the same nature as the ones which form the collapse of dust matter. The stability, the strength, and other features of the singularities and Cauchy horizons are analyzed. This has implications in a possible formulation of the cosmic censorship hypothesis.

066.030 Matching of the Vaidya and Robertson–Walker metric.
F. Fayos, X. Jaen, E. Llanta, J. M. M. Senovilla.
Classical Quantum Gravity, Vol. 8, No. 11, p. 2057 – 2068 (Nov 1991).

The authors give the necessary conditions for the matching of a general Robertson–Walker geometry to a general spherically symmetric radiation metric. They also derive the conditions for the matching of a Vaidya metric to a general Robertson–Walker metric. The possible applications of the results to stellar collapse and to the study of local inhomogeneities in a cosmological context are considered. An alternative interpretation of the energy–momentum tensor of the Robertson–Walker part of spacetime is given in such a way that the physical processes can be better understood.

066.031 On a covariant Hamilton–Jacobi framework for the Einstein–Maxwell theory.
P. Hořava.
Classical Quantum Gravity, Vol. 8, No. 11, p. 2069 – 2084 (Nov 1991).

The author extends the previously developed covariant Hamilton–Jacobi description of pure gravity in any dimension greater than two, to coupled electromagnetic and gravitational fields. In the case of pure gravity, the framework is based on a first–order differential equation on the bundle of metrics over spacetime manifold. In the case of coupled gravi– and electrodynamics, the Hamilton–Jacobi equation is reduced naturally to a set of first–order equations on the bundle of metrics. This reduction indicates that the description of the Einstein–Maxwell theory is considerably simplified at by applying the covariant Hamilton–Jacobi framework.

066.032 Solutions of the vacuum Einstein equation having toroidal infinite red–shift surface.
D. A. Korotkin.
Classical Quantum Gravity, Vol. 8, No. 11, p. L219 – L222 (Nov 1991). Letter–to–the–editor.

The author presents a brief description of the simplest algebraic geometric solution of the stationary axisymmetric vacuum Einstein equation. This localized solution has a ring singularity surrounded by the infinite red–shift surface representing the infinite family of converging tori containing one another and the singular ring. So it seems natural to name this solution the "toron". It has zero mass and non–zero NUT parameter and is set by three real parameters: site on symmetry axis, diameter of the singular ring and an additional "NUT density" parameter responsible for the size of the largest torus of infinite red–shift surface.

066.033 A note on gravitational and SU(2) instantons with Ashtekar variables.
S. Uehara.
Classical Quantum Gravity, Vol. 8, No. 11, p. L229 – L234 (Nov 1991). Letter–to–the–editor.

The author examines the ansatz proposed by Samuel which leads to gravitational instanton solutions. He presents conditions which also lead to gravitational instanton solutions of the Einstein equations with a cosmological constant. Samuel's ansatz is one of the explicit solutions of these conditions.

066.034 Can string theory overcome deep problems in quantum gravity?
J. H. Schwarz.
Phys. Lett. B, Vol. 272, No. 3/4, p. 239 – 244 (5 Dec 1991).

Studies of the consequences of requiring a reconciliation of quantum mechanics and general relativity have led to a number of distasteful allegations. These include claims that pure quantum states evolve into mixed states and that wormhole contributions to the euclidean path integral render the parameters of particle physics stochastic. My point of view is that a sensible reconciliation requires replacing general relativity by string theory, and it should provide more satisfying answers to these problems. One possibility is that it provides many different types of 'quantum hair', perhaps enough to account for all the entropy associated with black holes. Some possible sources of quantum hair in string theory are considered.

066.035 Antimatter gravity and the weak equivalence principle.
M. H. Holzscheiter, R. E. Brown, J. Camp, T. Darling, P. Dyer, D. B. Holtkamp, N. Jarmie, N. S. P. King, M. M. Schauer, S. Cornford, K. Hosea, R. A. Kenefick, M. Midzor, D. Oakley, R. Ristinen, F. C. Witteborn.
AIP Conf. Proc., No. 233, p. 573 – 575 (5 Aug 1991). Current Physics Microform No.: 9110A1813. – See Abstr. 012.004 for the main entry.

Ideas are presented for an experiment to compare the acceleration, g, of antiprotons in the Earth's gravitational field with that of particles of normed matter, such as protons or hydrogen ions. The experiments will test whether antiprotons obey the weak equivalence principle.

066.036 Gravitational radiation associated with anticollapse and supernova explosions.
A. P. Trofimenko, V. S. Gurin.
Astrophysics, Vol. 34, No. 1, p. 52 – 55 (Jan – Feb 1991). English translation of Astrofizika, Tom 34, Vyp. 1, p. 83 – 89 (Feb 1991).

A model of D bodies as anticollapsing objects (white holes) is considered in an extended spacetime manifold. The field of gravitational radiation is taken into account. The spacetime of a spherical gravitational wave is described by means of a solution of Robinson–Trautman type, which exhibits an oscillatory nature of the global structure of the spacetime manifold of a gravitationally radiating white hole. The model is used to explain the energy balance of the gravitational burst from the supernova 1987A. It is noted that the production of D bodies is also accompanied by a burst of gravitational radiation.

066.037 Experimental constraints on strong–field relativistic gravity.
J. H. Taylor, A. Wolszczan, T. Damour, J. M. Weisberg.
Nature, Vol. 355, No. 6356, p. 132 – 136 (9 Jan 1992).

Experiments in our Solar System can test relativistic gravity only in the weak–field limit, but systems containing pulsars necessarily involve the effects of strong–field gravity. Timing observations of three binary pulsars yield tight constraints on the nature of gravity in the strong–field regime, allowing the measurement of velocity–dependent and nonlinear phenomena separately from the effects of gravitational radiation. General relativity passes these new experimental tests with complete success.

066.038 Conformally–invariant scalar field with trace–free energy–momentum tensor in Robertson–Walker models.
N. I. Singh, N. B. Singh.
Astrophys. Space Sci., Vol. 188, No. 1, p. 165 – 167 (Feb 1992). Letter–to–the–editor.

Exact solutions of Einstein's field equations for a conformally–invariant scalar field with trace–free energy–momentum tensor is presented for the Robertson–Walker models with $K = +1, -1$. The physical properties of the solution are also studied.

066.039 Three–dimensional gravity from the Turaev–Viro invariant.
S. Mizoguchi, T. Tada.
Phys. Rev. Lett., Vol. 68, No. 12, p. 1795 – 1798 (23 Mar 1992). Current Physics Microform No.: 9205F0009.

The authors study the q–deformed SU(2) spin network as a three–dimensional quantum gravity model. They show that in the semiclassical continuum limit the Turaev–Viro invariant obtained recently defines a naturally regularized path integral, in which a contribution from the cosmological term is effectively included. The regularization – dependent cosmological constant is found to be $4\pi^2/k^2 + O(k^{-4})$, where $q^{2k} = 1$. The authors also discuss the relation to the Euclidean Chern – Simons – Witten gravity in three dimensions.

066.040 Two Kaluza–Klein wormhole solutions.
Jin Wang.
Astrophys. Space Sci., Vol. 189, No. 1, p. 5 – 9 (Mar 1992).

The author examines models of ten–dimensional supergravity and monopoles. Wormhole solutions are obtained under the Freund–Rubin–type spontaneous compactification. The meaning of the solutions is discussed in terms of the tunnelling process.

066.041 Post–newtonian approximation of gravitation theory in flat space–time.
W. Petry.
Astrophys. Space Sci., Vol. 189, No. 1, p. 63 – 77 (Mar 1992).

The post–Newtonian approximation of the gravitational field of a perfect fluid for a previously stated theory of gravitation in flat space–time is studied. The conservation laws of energy–momentum and angular momentum are derived and the equivalence of the conservation law of energy–momentum and the equations of motion is shown to the studied accuracy. The equations of motion are stated. All the results of the post–Newtonian approximation of the gravitation theory in flat space–time and of the general theory of relativity, as considered by Will in his famous book, agree to the studied accuracy.

066.042 Fractal structure of two–dimensional gravity coupled to c = −2 matter.
N. Kawamoto, V. A. Kazakov, Y. Saeki, Y. Watabiki.
Phys. Rev. Lett., Vol. 68, No. 14, p. 2113 – 2116 (6 Apr 1992). Current Physics Microform No.: 9205F0327.

The authors present the results of numerical calculations of internal fractal dimensions of a dynamically triangulated two–dimensional gravity coupled to $c = -2$ matter. Internal fractal dimensions as a function of internal geodesic distance are measured from the number of boundaries and the total length of boundaries, and show clear scaling behavior, while the number of triangles starts to scale only around the maximum size. The authors apply a recursive sampling algorithm for $c = -2$, which makes it possible to generate the most probable triangulations of spherical topology.

066.043   Precursory singularities in spherical gravitational collapse.
K. Lake.
Phys. Rev. Lett., Vol. 68, No. 21, p. 3129 – 3132 (25 May 1992).
Current Physics Microform No.: 9208B1011.

General conditions are developed for the formation of naked precursory ("shell–focusing") singularities in spherical gravitational collapse. These singularities owe their nakedness to the fact that the gravitational potential fails to be single valued prior to the onset of a true gravitational singularity. It is argued that they do not violate the spirit of cosmic censorship. Rather, they may well be an essentially generic feature of relativistic gravitational collapse.

066.044   Reinterpretation of Jordan–Brans–Dicke theory and Kaluza–Klein cosmology.
Y. M. Cho.
Phys. Rev. Lett., Vol. 68, No. 21, p. 3133 – 3136 (25 May 1992).
Current Physics Microform No.: 9208B1015.

The author emphasizes that it is the Pauli metric, not the Jordan metric, which describes the massless spin–two graviton in the Brans–Dicke theory. Similarly in the "Jordan–Brans–Dicke theory" based on Kaluza–Klein unification, only the Pauli metric can correctly describe Einstein's theory of gravitation. This necessitates a completely new reinterpretation of the "old" Kaluza–Klein cosmology as well as the Brans–Dicke theory. More significantly, the analysis shows that the Kaluza–Klein dilaton must generate a fifth force which could violate the equivalence principle.

066.045   Superstrings in quantum cosmology.
Wang Jin.
Phys. Rev. D, Vol. 45, No. 2, p. 412 – 416 (15 Jan 1992). Current Physics Microform No.: 9203G0946.

The author considers the dynamics of a four–dimensional universe using the pointlike field–theory effective action of superstrings. He uses the boundary conditions of Hartle and Hawking and Vilenkin to solve the Wheeler – DeWitt equation. It is found that under certain conditions the universe will tunnel from a Euclidean regime to a Lorentzian regime. This process is studied in detail.

066.046   An application of new relations for a determination of the Newtonian constant of gravitation.
A. G. Gasanalizade.
Astrophys. Space Sci., Vol. 189, No. 1, p. 155 – 158 (Mar 1992).
Letter–to–the–editor.

The author has established an earlier–unknown relation between the Newtonian constant of gravitation G and other fundamental physical and astronomical constants. He has calculated the constant of gravitation from such a new relationship, and found $G(\text{cal}) = 6.67197926 \times 10^{-11} \text{m}^3 \text{kg}^{-1} \text{s}^{-2}$, which is in very good agreement with the value of G recommended by "CODATA–86" (Cohen and Taylor, 1987), $G = 6.67259(85) \times 10^{-11} \text{m}^3 \text{kg}^{-1} \text{s}^{-2}$ (128 ppm).

066.047   Astrophysical jets and theories of gravitation.
M. Livio, N. Rosen.
Astrophys. J., Vol. 387, No. 2, p. 458 – 459 (10 Mar 1992).

The authors point out that a recently discovered hydrodynamical jet formation mechanism (see work by Fryxell, Taam, and McMillan; and Matsuda et al.) provides us with a tool that is capable of distinguishing between general relativity and bimetric general relativity. It is for the first time that such a large–scale phenomenon as the mere presence or absence of relativistic jets can distinguish between theories of gravitation. The present observational situation seems to favor general relativity over bimetric general relativity.

066.048   Singularity–free space–time.
F. J. Chinea, L. Fernández–Jambrina,
J. M. M. Senovilla.
Phys. Rev. D, Vol. 45, No. 2, p. 481 – 486 (15 Jan 1992). Current Physics Microform No.: 9203G1015.

The authors show that the solution of Einstein's field equations published by Senovilla [51.066.043] is geodesically complete and singularity–free. The solution also satisfies the stronger energy and causality conditions, such as global hyperbolicity, the strong energy condition, causal symmetry, and causal stability.

066.049   Global structure of Gott's two–string spacetime.
C. Cutler.
Phys. Rev. D, Vol. 45, No. 2, p. 487 – 494 (15 Jan 1992). Current Physics Microform No.: 9203G1021.

Gott has recently obtained exact solutions to Einstein's equation representing two infinitely long, straight cosmic strings that gravitationally scatter off each other. A remarkable feature of these solutions is that they contain closed timelike curves when the relative velocity of the strings is sufficiently high. Here, the author elucidates the global structure of Gott's two–string spacetime. In particular, he proves that the closed timelike curves are confined to a certain region of the spacetime, and that the spacetime contains complete spacelike, edgeless, achronal hypersurfaces, from which the causality–violating regions may be said to evolve.

066.050   Gravitational wave bursts with memory: the Christodoulou effect.
K. S. Thorne.
Phys. Rev. D, Vol. 45, No. 2, p. 520 – 524 (15 Jan 1992). Current Physics Microform No.: 9203G1054.

The "memory" of a gravitational wave burst is the permanent relative displacement that it imposes on free test masses, or more precisely, the permanent change in the burst's gravitational wave field h_{jk}^{TT}. This memory, in general, is equal to the change in the transverse–traceless (TT) part of the "1/r, Coulomb–type" gravitational field generated by the four–momenta of the source's various independent pieces. Christodoulou has recently identified a contribution to a burst's memory that arises from nonlinearities in the vacuum Einstein field equation. This paper shows that the Christodoulou memory is precisely the TT part of the "1/r, Coulomb–type" gravitational field produced by the burst's gravitons, and it therefore gets built up over the same length of time as it takes for the source to emit the gravitons. The sensitivity of broad–band gravitational wave detectors such as LIGO to the Christodoulou memory is analyzed and discussed.

066.051   Dynamics and gravitational interaction of waves in nonuniform media.
R. Kulsrud, A. Loeb.
Phys. Rev. D, Vol. 45, No. 2, p. 525 – 531 (15 Jan 1992). Current Physics Microform No.: 9203G1059.

The authors derive the generally covariant equations describing the propagation of waves with an arbitrary dispersion relation in a nonuniform, nondissipative medium. The back–reaction of the waves on the medium is expressed in terms of the wave energy–momentum tensor. The formalism is based on variations of the Lagrangian of the system with respect to the wave amplitude and phase and the particle orbits. The Lagrangian approach is considered in detail in the context of a cold, unmagnetized plasma. It is shown that the "inertial" mass of a photon in a plasma, namely the plasma frequency, is also its gravitational mass. Extremely precise experiments are needed to measure the gravitational "free fall" of phonons, plasmons, or photons in laboratory media.

066.052   Binary systems: higher order gravitational radiation damping and wave emission.
W. Junker, G. Schäfer.
Mon. Not. R. Astron. Soc., Vol. 254, No. 1, p. 146 – 164 (1 Jan 1992).

The paper treats the motion of binary systems under the back–reaction of the gravitational radiation generated by the quasi–elliptic and quasi–hyperbolic post–Newtonian motions of the binaries. The angular momentum losses are calculated and, together with the already known energy losses, the changes of the eccentricities and semimajor axes are derived, allowing the determination of a detailed picture of the radiation damping in binary systems up to the 3.5 post–Newtonian order. The waveforms of the higher order gravitational radiation are presented. In particular, their dependencies on the periastron shifts are made explicit. Various cases of the motion of the binaries and of the waveforms are shown graphically.

066.053   Gravitational radiative corrections in N = 1 supergravity.

C. Chiou–Lahanas, A. Kapella–Economou, A. B. Lahanas, X. N. Maintas.

Phys. Rev. D, Vol. 45, No. 2, p. 534 – 544 (15 Jan 1992). Current Physics Microform No.: 9203G1068.

In an ungauged N = 1 supergravity theory, defined on an arbitrary Kahlerian manifold, the authors compute the divergent one–loop corrections to the bosonic part of the effective action. Although the theory is not renormalizable such a calculation may be of relevance in view of the fact that N = 1 supergravities emerge as effective nonrenormalizable theories in the low–energy limit of some superstring models. The authors pay special attention to the one–loop scalar potential of the theory. It is shown that, by a proper redefinition of the metric, geometric objects such as scalar curvature can be made not to interact with the scalars. The definition of the potential of the theory becomes in this way unambiguous.

066.054   Scaling solutions in cosmic–string networks.

E. Copeland, T. W. B. Kibble, D. Austin.

Phys. Rev. D, Vol. 45, No. 4, p. 1000 – 1004 (15 Feb 1992). Current Physics Microform No.: 9204D1610.

The evolution of a cosmic–string network is examined in terms of two length scales: ξ, related to the long–string density, and xi, the persistence length along the left– or right–moving string, respectively. Previous work is extended by allowing for the dependence of some of the parameters on these scales. The changes have some dramatic effects. As before, an important role is played by the parameter q describing the relative kinkiness of a loop as compared to a section of string of the same length. The authors show that scaling solutions, in which both ξ and xi are of similar length, both proportional to the horizon size, exist for all values of q. However, for small values of q these solutions are unstable, so the scaling solution will actually be reached only for q larger than some critical value of order 2. The results are compared with those of Allen and Caldwell, and the possibility that scaling has not in fact been reached is briefly discussed.

066.055   Evanescent black holes.

C. G. Callan, S. B. Giddings, J. A. Harvey, A. Strominger.

Phys. Rev. D, Vol. 45, No. 4, p. 1005 – 1009 (15 Feb 1992). Current Physics Microform No.: 9204D1615.

A renormalizable theory of quantum gravity coupled to a dilaton and conformal matter in two space–time dimensions is analyzed. The theory is shown to be exactly solvable classically. Included among the exact classical solutions are configurations describing the formation of a black hole by collapsing matter. The problem of Hawking radiation and back reaction of the metric is analyzed to leading order in a 1/N expansion, where N is the number of matter fields. The results suggest that the collapsing matter radiates away all of its energy before an event horizon has a chance to form, and black holes thereby disappear from the quantum mechanical spectrum. It is argued that the matter asymptotically approaches a zero–energy "bound state" which can carry global quantum numbers, and that a unitary S matrix including such states should exist.

066.056   Second–order parametrized–post–Newtonian Lagrangian.

M. J. Benacquista.

Phys. Rev. D, Vol. 45, No. 4, p. 1163 – 1173 (15 Feb 1992). Current Physics Microform No.: 9204D1773.

A many–body Lagrangian to second post–Newtonian order using an extension of the PPN formalism is introduced and the properties of new parameters are explored. A parametrized gauge transformation is developed to permit comparison with theories of gravity in a variety of different coordinate systems. A procedure to impose Lorentz invariance on a general second–order post–Newtonian Lagrangian is developed. The Lagrangian is then constrained to possess Lorentz invariance and a "Lorentz–invariant" gauge is introduced. The constrained Lagrangian is found to be described by ten new second–order PPN parameters. When the Lagrangian is further constrained to describe theories of gravity for which test particles move along geodesics, one of the ten new parameters is given entirely in terms of first–order PPN parameters, leaving only nine PPN parameters to describe the second–order gravitational interaction. A "metric" gauge is introduced which reduces to the gauge associated with the formalism of Arnowitt, Deser, and Misner, when the general relativity values of the PPN parameters are used.

066.057   Generating solutions of the Einstein–Maxwell equations with prescribed physical properties.

H. Quevedo.

Phys. Rev. D, Vol. 45, No. 4, p. 1174 – 1177 (15 Feb 1992). Current Physics Microform No.: 9204D1784.

A linear transformation for generating electrovacuum solutions is presented. The multipole moments of the new solutions can be expressed explicitly in terms of those of the seed solution. This provides a method of generating solutions with physical properties determined a priori. The author discusses the specific example of a solution representing the gravitational field of a deformed rotating source with zero charge and a vanishing magnetic–monopole moment but nonvanishing higher electric and magnetic multipoles.

066.058   Traveling waves on a magnetic universe.

D. Garfinkle, M. A. Melvin.

Phys. Rev. D, Vol. 45, No. 4, p. 1188 – 1191 (15 Feb 1992). Current Physics Microform No.: 9204D1798.

The authors find solutions of the Einstein–Maxwell equations representing gravitational and electromagnetic waves traveling along the axis direction in a cylindrical magnetic universe. The waves are strongly gravitating and can have an arbitrary profile.

066.059   Multidimensional quantum wormholes.

A. Zhuk.

Phys. Rev. D, Vol. 45, No. 4, p. 1192 – 1197 (15 Feb 1992). Current Physics Microform No.: 9204D1802.

The integrable cosmological model with n (n > 1) spaces of constant curvature is investigated in the case when only one of them is not Ricci flat. This spacetime is minimally coupled with a massless scalar field. Two types of quantum wormholes with a continuous and discrete spectrum are found. The analogous result takes place for pure gravity also when the scalar field is absent.

066.060   Strong–field tests of relativistic gravity and binary pulsars.

T. Damour, J. H. Taylor.

Phys. Rev. D, Vol. 45, No. 6, p. 1840 – 1868 (15 Mar 1992). Current Physics Microform No.: 9204H1226.

Observations of pulsars in binary systems provide a unique opportunity for testing the strong–field regime of relativistic gravity. The authors present a detailed account of the "parametrized post–Keplerian" (PPK) formalism, which allows analysis of pulsar timing and pulse structure data in a theory – independent way by using a certain number of fitted post–Keplerian parameters. It is shown that as many as 19 such parameters can be measured, under favorable conditions, giving access to 15 possible tests of relativistic gravity. The authors isolate and quantify the theoretical content of these tests by deriving expressions linking the phenomenological parameters to the inertial masses of the pulsar and its companion, and to the polar angles of the spin axis of the pulsar. It is shown that the recently discovered binary pulsar PSR 1534 + 12 should give access to two new strong–field tests of relativistic gravity. Moreover, in the long run, the first–discovered binary pulsar, PSR 1913 + 16, could give access to three strong–field tests, beyond the presently obtained omega – γ – P_b test. Finally, the authors show how, by combining the PPK approach with the predictions of a rather generic class of tensor biscalar theories, one can bring together tests based on observations of several different pulsars.

066.061   Boson star in a gravitation theory with dilaton.
Tao Zhijian, Xun Xue.
Phys. Rev. D, Vol. 45, No. 6, p. 1878 – 1883 (15 Mar 1992).
Current Physics Microform No.: 9204H1264.

The authors study the possible static spherically symmetric configuration of stellar objects comprised of bosons in a gravitation theory differing from Einstein's theory in the sense that the dilaton field appears in the theory as a nonlinear realization of dilatation invariance. Different scales of the dilaton interacting with a complex scalar field and different scales of the mass of the complex scalar field are considered.

066.062   Gravitational radiation from cosmic strings.
B. Allen, E. P. S. Shellard.
Phys. Rev. D, Vol. 45, No. 6, p. 1898 – 1912 (15 Mar 1992).
Current Physics Microform No.: 9204H1284.

A cosmic string network in an expanding universe evolves by losing energy to loops which in turn oscillate and emit gravitational radiation. The power radiated at a frequency corresponding to the nth fundamental mode of oscillation of the loop is characterized by a dimensionless constant P_n. The authors determine these constants for loops produced in a numerical simulation of the cosmic string network. The resulting numerical values of the P_n appear to show a linear dependence on loop size, indicating that small–scale structure on the loops is very important in determining the overall radiation power. Long–string radiation is also studied, confirming this conclusion. The power radiated by a horizon–length string increases with time, because in the current simulations the small–scale structure on the string does not yet scale relative to the horizon length.

066.063   Gauge–invariant cosmic structures – a dynamic systems approach.
A. Woszczyna.
Phys. Rev. D, Vol. 45, No. 6, p. 1982 – 1988 (15 Mar 1992).
Current Physics Microform No.: 9204H1368.

Gravitational instability is expressed in terms of dynamic systems theory. The gauge–invariant Ellis–Bruni equation and Bardeen's equation are discussed in detail. It is shown that in an open universe filled with matter of constant sound velocity the Jeans criterion does not adequately define the length scale of the gravitational structure.

066.064   Time machine and self–consistent evolution in problems with self–interaction.
I. D. Novikov.
Phys. Rev. D, Vol. 45, No. 6, p. 1989 – 1994 (15 Mar 1992).
Current Physics Microform No.: 9204H1375.

Physical processes with self–interactions in spacetimes with closed timelike curves are discussed. Examples of self–consistent solutions of the corresponding problems are obtained.

066.065   General class of inhomogeneous perfect–fluid solutions.
E. Ruiz, J. M. M. Senovilla.
Phys. Rev. D, Vol. 45, No. 6, p. 1995 – 2005 (15 Mar 1992).
Current Physics Microform No.: 9204H1381.

The authors present a general class of solutions to Einstein's field equations with two spacelike commuting Killing vectors by assuming the separation of variables of the metric components. The solutions can be interpreted as inhomogeneous cosmological models. It is shown that the singularity structure of the solutions varies depending on the different particular choices of the parameters and metric functions. There exist solutions with a universal big–bang singularity, solutions with timelike singularities in the Weyl tensor only, solutions with singularities in both the Ricci and the Weyl tensors, and also singularity–free solutions.

066.066   Gravitational collapse of rotating spheroids and the formation of naked singularities.
S. L. Shapiro, S. A. Teukolsky.
Phys. Rev. D, Vol. 45, No. 6, p. 2006 – 2012 (15 Mar 1992).
Current Physics Microform No.: 9204H1392.

The authors explore numerically the effect of rotation on the collapse of collisionless gas spheroids in full general relativity. The spheroids are initially prolate and consist of equal numbers of corotating and counterrotating particles. The authors have previously shown that in the absence of rotation the spheroids all collapse to spindle singularities. When the spheroids are sufficiently compact, the singularities are hidden inside black holes. However, when the spheroids are large enough, there are no apparent horizons. These nonrotating spheroids are strong candidates for naked singularities. The present simulations suggest that rotation significantly modifies the evolution when it is sufficiently large. Imploding configurations with appreciable rotation ultimately collapse to black holes. However, for small enough angular momentum, the simulations cannot at present distinguish rotating from nonrotating collapse: spindle singularities appear to arise without apparent horizons. Hence it is possible that even spheroids with some angular momentum may form naked singularities.

066.067   Four–dimensional quantum gravity in the conformal sector.
I. Antoniadis, E. Mottola.
Phys. Rev. D, Vol. 45, No. 6, p. 2013 – 2025 (15 Mar 1992).
Current Physics Microform No.: 9204H1399.

The authors study the trace–anomaly induced dynamics of the conformal factor of four–dimensional quantum gravity. The resulting effective scalar theory is ultraviolet renormalizable, and possesses a nontrivial, infrared stable fixed point. The exact anomalous scaling dimension of the conformal factor at the critical point is derived. The authors argue that this theory describes 4D gravity at large distances and provides a framework for a dynamical solution of the cosmological constant problem.

066.068   Naked singularities in non–selfsimilar gravitational collapse of radiation shells.
P. S. Joshi, I. H. Dwivedi.
Phys. Rev. D, Vol. 45, No. 6, p. 2147 – 2150 (15 Mar 1992).
Current Physics Microform No.: 9204H1533.

Non–selfsimilar gravitational collapse of imploding radiation is shown to give rise to a strong curvature naked singularity. The conditions are specified for the singularity to be globally naked and the strength of the same is examined along nonspacelike curves and along all the families of nonspacelike geodesics terminating at the singularity in the past.

066.069   Thermodynamics of event horizons in (2 + 1)–dimensional gravity.
B. Reznik.
Phys. Rev. D, Vol. 45, No. 6, p. 2151 – 2154 (15 Mar 1992).
Current Physics Microform No.: 9204H1537.

Although gravity in 2 + 1 dimensions is very different in nature from gravity in 3 + 1 dimensions, it is shown that the laws of thermodynamics for event horizons can be manifested also for (2 + 1)–dimensional gravity. The validity of the classical laws of horizon mechanics is verified in general and exemplified for the (2 + 1)–dimensional analogues of Reissner– Nordström and Schwarzschild – de Sitter spacetimes. The entropy is given by 1/4L, where L is the length of the horizon. A consequence of having consistent thermodynamics is that the second law fixes the sign of Newton's constant to be positive.

066.070   Capture cross–sections of photons and slow uncharged particles by spherically–symmetric charged compact body in relativistic theory of gravity.
A. F. Zhakharov.
Teor. Mat. Fiz., Vol. 90, No. 1, p. 148 – 154 (Jan 1992). In Russian. English translation in Theor. Math. Phys.

Capture cross–sections of photons and slow uncharged particles by spherically–symmetric charged compact body are obtained in the framework of relativistic theory of gravity. Comparison of the results with the ones from GTR is given.

066.071   Phase transition on the curvature in quantum R^2–gravity and induction of Einstein's gravity.
S. D. Odintsov, I. L. Shapiro.
Teor. Mat. Fiz., Vol. 90, No. 3, p. 469 – 480 (Mar 1992). In Russian. English translation in Theor. Math. Phys.
A phase transition on the curvature in an effective potential of quantum R^2–gravity with matter is studied. In the framework of the renormalization group approach the effective potential is calculated up to the terms linear on the curvature. An universal expression for the induced gravitational and cosmological constants is obtained. The effective potential and the induced gravitational and cosmological constants depend on relations between coupling constants of the theory and on gauge parameters. In the case when the matter is represented by a scalar field the coupling constants are fixed by the requirement of the asymptotic freedom. The gauge dependence is absent for a single parametrization and gauge invariant effective action.

066.072   Extended thermodynamics of Friedmann – Robertson – Walker models in the Landau – Lifshitz frame.
M. O. Calvão, J. M. Salim.
Classical Quantum Gravity, Vol. 9, No. 1, p. 127 – 135 (Jan 1992).
The authors study dissipative Friedmann – Robertson – Walker models with the help of the relativistic extended thermodynamics as formulated in the Landau – Lifshitz (energy) frame. The physically most relevant solutions are shown to tend asymptotically to models in which the initial material anisotropy due to particle drift (diffusion) is washed out. Some recent results obtained by Turok and Barrow in a different context can be recovered.

066.073   Straight cosmic strings in linearized $R + R^2$ gravity.
B. Linet, P. Teyssandier.
Classical Quantum Gravity, Vol. 9, No. 1, p. 159 – 171 (Jan 1992).
The gravitational field of a straight U(1)–gauge cosmic string is investigated within the linear approximation to the gravitational theories with Lagrangian densities. The metric of a straight infinitely thin string is obtained in closed form. The gravitational properties of this metric are explored. It is found that a short range gravitational force is exerted on a slowly moving test particle, a feature not present in general relativity. The deflection of a light ray travelling in a plane orthogonal to the string is also determined.

066.074   No–hair theorem for spherical monopoles and dyons in SU(2) Einstein Yang–Mills theory.
P. Bizon, O. T. Popp.
Classical Quantum Gravity, Vol. 9, No. 1, p. 193 – 205 (Jan 1992).
It is shown that spherically symmetric Einstein Yang–Mills (EYM) equations do not admit essentially non–Abelian asymptotically flat static solutions, which have non–zero YM charge and are non–singular, either globally or outside the horizon. This proves, under the assumption of spherical symmetry, that there are no globally regular static EYM monopoles and dyons, and that all static EYM black holes with non–zero YM charge belong to the Reissner–Nordstrøm family.

066.075   Colliding plane wave spacetimes.
J. B. Griffiths.
Classical Quantum Gravity, Vol. 9, No. 1, p. 207 – 215 (Jan 1992).
When two plane waves collide, it is shown that the metric in the interaction region is only the product of two 2–spaces in certain restricted cases which include gravitational and electromagnetic waves. The form of the metric in the completely general case is defined here, and is illustrated for the case of colliding plane neutrino waves. Two classes of exact solutions, including the general case of pure neutrino waves, are given and some general properties are deduced.

066.076   Interior axisymmetric stationary perfect fluid solution of Einstein's equations.
E. Kyriakopoulos.
Classical Quantum Gravity, Vol. 9, No. 1, p. 217 – 223 (Jan 1992).
A solution of Einstein's equations for the interior of a rigidly rotating axisymmetric perfect fluid is presented, with an equation of state $\mu + 3p = 0$. The equipressure surfaces are closed. The solution has a static limit. The metric functions of the line element of the solution, in the usual coordinates, are expressed in terms of the Weierstrass elliptic function.

066.077   Radiative Einstein–Maxwell spacetimes and "no–hair" theorems.
W. Simon.
Classical Quantum Gravity, Vol. 9, No. 1, p. 241 – 256 (Jan 1992).
Friedrich's construction of purely radiative solutions out of static, asymptotically flat solutions of Einstein's vacuum equations is generalized to the Einstein–Maxwell case. These radiative solutions can be considered as being conformal and analytic extensions, to neighbourhoods of infinity, of local solutions to a certain time–symmetric initial value problem. The author points out that the non–existence of non–trivial global solutions to the corresponding constraints provides the basis for known uniqueness proofs for static, asymptotically flat black hole and perfect fluid solutions.

066.078   Active and passive gravitational mass of a Schwarzschild sphere.
W. B. Bonnor.
Classical Quantum Gravity, Vol. 9, No. 1, p. 269 – 274 (Jan 1992).
It is shown that for a static sphere of uniform density the passive gravitational mass is greater than the active gravitational mass. Implications for dynamics are briefly discussed.

066.079   Quantum mechanics of a solid–state bar gravitational antenna.
L. P. Grishchuk.
Phys. Rev. D, Vol. 45, No. 8, p. 2601 – 2608 (15 Apr 1992). Current Physics Microform No.: 9206F1327.
A quantum mechanical treatment of a bar gravitational antenna is presented. The theory takes into account the crystalline structure of the bar and collective behavior of its mass elements. The low–frequency and high–frequency (Debye) modes of oscillations are considered. It is shown that the quantum mechanial derivation of the absorption cross section of the gravitational antenna agrees totally with the classical result. The recent claims of a significant (six orders of magnitude) quantum mechanical enhancement of the cross section are shown to be incorrect.

066.080   Gravity and the Poincaré group.
G. Grignani, G. Nardelli.
Phys. Rev. D, Vol. 45, No. 8, p. 2719 – 2731 (15 Apr 1992). Current Physics Microform No.: 9206F1445.
The authors discuss gravity as a gauge theory of the Poincaré group in three and four dimensions, i.e., in a metric–independent fashion. The fundamental fields of the theory are the gauge potentials, the matter fields, and the so–called Poincaré coordinates $q^a(x)$: a set of fields that are defined on the space–time manifold, but that transform as Poincaré vectors under gauge transformations. The authors discuss the procedure needed to connect this theory with the Einstein formulation of gravity, and they show that the field equations for the gauge potentials, for pointlike sources, and for scalar and spinor matter fields, reproduce the Einstein equations, the geodesics equations, and the Klein–Gordon and the Dirac equations in curved space–time, respectively.

066.081 Collisions of relativistic clusters and the formation of black holes.
S. L. Shapiro, S. A. Teukolsky.
Phys. Rev. D, Vol. 45, No. 8, p. 2739 – 2750 (15 Apr 1992).
Current Physics Microform No.: 9206F1465.

The authors perform numerical simulations of head–on collisions of relativistic clusters. The cluster particles interact only gravitationally, and so satisfy the collisionless Boltzmann equation in general relativity. The authors construct and follow the evolution of three classes of initial configurations: spheres of particles at rest; spheres of particles boosted towards each other; and spheres of particles in circular orbits about their respective centers. In the first two cases, the spheres implode towards their centers and may form black holes before colliding. These scenarios thus can be used to study the head–on collision of two black holes. In the third case the clusters are initially in equilibrium and cannot implode. In this case collision from rest leads either to coalescence and virialization, or collapse to a black hole. This scenario is the collisionless analog of colliding neutron stars in relativistic hydrodynamics.

066.082 Gravitational instantons with a scalar field.
S. Mukherjee, B. C. Paul, N. Dadhich, A. Kshirsagar.
Phys. Rev. D, Vol. 45, No. 8, p. 2772 – 2775 (15 Apr 1992).
Current Physics Microform No.: 9206F1498.

Analytic gravitational instanton solutions in a theory which includes a self-interacting scalar field are obtained. The relevant potential for the scalar field satisfying equations of motion is determined by developing a general technique. The corresponding actions are calculated exactly and the qualitative features of the solutions are discussed.

066.083 Gravitational wave forms at finite distances and at null infinity.
R. Gómez, J. Winicour.
Phys. Rev. D, Vol. 45, No. 8, p. 2776 – 2782 (15 Apr 1992).
Current Physics Microform No.: 9206F1502.

Motivated by numerical studies of gravitational radiation, the authors investigate the discrepancies that arise if wave forms are observed at a finite distance as opposed to infinity. This study considers scalar radiation from a spherically symmetric Einstein Klein–Gordon system. This allows to isolate the effects of backscattering and redshifting while avoiding more complicated effects arising in nonspherical systems. The authors show that discrepancies close to 100% can arise at large observation distances for sufficiently periodic systems. The effects are most pronounced for radiation losses between one–quarter and one–half of the initial mass. This falls within the expected regime of the spiral infall of a relativistic binary system. The predominant contribution to this discrepancy stems from a time–dependent redshift arising from radiative mass loss.

066.084 Strings falling into spacetime singularities.
H. J. de Vega, N. Sánchez.
Phys. Rev. D, Vol. 45, No. 8, p. 2783 – 2793 (15 Apr 1992).
Current Physics Microform No.: 9206F1509.

The authors study the dynamics of strings near spacetime singularities. They consider gravitational–wave backgrounds with a singularity of the type $|U|^{-\beta}$, U being a null coordinate. New features in the string behavior appear: when $\beta \geqslant 2$, the string does not propagate through the gravitational wave and it escapes to infinity grazing the singularity plane U = 0; one transverse coordinate does not oscillate in time (neither classically nor quantum mechanically) and the tunnel effect does not take place. The expectation value of the mass squared $\langle M_{>}{}^2 \rangle$ and mode number $\langle N_{>} \rangle$ operators and of the energy–momentum tensor are computed. When the transverse size (ϱ_0) of the gravitational–wave front is infinite, divergences in $\langle M_{>}{}^2 \rangle$ and $\langle N_{>} \rangle$ appear for $1 \leqslant \beta < 2$ and $3/2 \leqslant \beta < 2$, respectively. The authors argue that the short–distance spacetime singularity at U = 0 is not responsible for these divergences, but the infinite amount of energy carried by the gravitational wave when $\varrho_0 = \infty$.

066.085 Higher derivatives and renormalization in quantum cosmology.
F. D. Mazzitelli.
Phys. Rev. D, Vol. 45, No. 8, p. 2814 – 2822 (15 Apr 1992).
Current Physics Microform No.: 9206F1540.

In the framework of the canonical quantization of general relativity, quantum field theory on a fixed background formally arises in an expansion in powers of the Planck length. In order to renormalize the theory, quadratic terms in the curvature must be included in the gravitational action from the beginning. These terms contain higher derivatives which change completely the Hamiltonian structure of the theory. The author shows that it is possible to avoid this problem by replacing the higher–derivative theory by a second–order one. The classical solutions of the latter are also solutions of the former. The author quantizes the theory, renormalizes the infinities, and shows that there is a smooth limit between the classical and the renormalized theories. A Robertson–Walker minisuperspace with a quantum scalar field is applied.

066.086 Curved–space magnetic monopoles.
M. E. Ortiz.
Phys. Rev. D, Vol. 45, No. 8, p. R2586 – R2589 (15 Apr 1992).
Current Physics Microform No.: 9206F1312.

Explicit solutions of the coupled Einstein – Yang – Mills – Higgs field equations representing a magnetic monopole are constructed, both in and away from the Bogomolny – Prasad – Sommerfield limit. The solutions are seen to tend towards black hole solutions, as the strength of the gravitational coupling is increased. A careful analysis of solutions near the transition to a black hole shows that the monopole loses its non–Abelian hair as it develops a horizon. In certain cases, solutions without a horizon are seen to be unstable to gravitational collapse.

066.087 Tolman's formula in scalar–tensor theories of gravitation.
M. R. Avakyan, L. Sh. Grigoryan, A. A. Saaryan.
Astrophysics, Vol. 34, No. 2, p. 134 – 137 (Mar – Apr 1991).
English translation of Astrofizika, Tom 34, Vyp. 2, p. 265 – 270 (Apr 1991).

In the framework of scalar–tensor and scalar–tensor bimetric theories of gravitation, the total mass of matter and of a constant gravitational field is expressed in the form of an integral that is only over the space occupied by the matter.

066.088 The strength of Einstein's equations.
N. F. J. Matthews.
Gen. Relativ. Gravitation, Vol. 24, No. 1, p. 17 – 33 (Jan 1992).

The strength of Einstein's empty–space field equations is computed anew and shown to be equal to the amount of initial data required for a local solution of the equations. This same amount of initial data is shown to be precisely that required for a set of 16 unknown first–order differential equations containing 10 fields variables and having six identities of second order. The 10 field variables must be functions of second order in the metric coefficients. The 16 field equations $C^{\alpha\beta\mu\sigma}{}_{,\alpha} = 0$, where $C^{\alpha\beta\mu\sigma}$ is Weyl's conformal tensor, are shown to have the same properties as those of the unknown equations, suggesting that $C^{\alpha\beta\mu\sigma}{}_{,\alpha} = 0$ is a satisfactory local first–order formulation of Einstein's second–order empty–space field equations.

066.089 The exterior gravitational field of a static and stationary mass with an arbitrary set of multipole moments.
V. S. Manko.
Gen. Relativ. Gravitation, Vol. 24, No. 1, p. 35 – 45 (Jan 1992).

New exact asymptotically flat solutions of Einstein's vacuum equations for the description of the exterior gravitational field of a static and stationary mass with an arbitrary mass–multipole structure are presented.

066.090   Geodesic Bianchi type cosmological models.
D. Sklavenites.
Gen. Relativ. Gravitation, Vol. 24, No. 1, p. 47 – 58 (Jan 1992).

Some perfect fluid solutions of Einstein's field equations are obtained in spacetimes with two hypersurface orthogonal space-like commuting Killing vectors. The flow is assumed to be geodesic. The solutions depend on an arbitrary function of time which determines the equation of state. In the models derived, one additional Killing vector exists and the solutions are actually Bianchi–type cosmological models.

066.091   Structure of the generalized Friedmann problem.
R. T. Jantzen, C. Uggla.
Gen. Relativ. Gravitation, Vol. 24, No. 1, p. 59 – 85 (Jan 1992).

An investigation of those cases of the generalized Friedmann equation which are solvable in terms of elementary or elliptic functions is undertaken together with a study of the time gauges which allow this to occur. This is accomplished by examining the natural choices of independent and dependent variables in this problem using manipulations like those of the Kepler problem, which is shown to be equivalent to a generalized Friedmann problem, thus clarifying the similarities between the simplest solutions of each.

066.092   Type N twisting vacuum gravitational fields.
G. Ludwig, Y. B. Yu.
Gen. Relativ. Gravitation, Vol. 24, No. 1, p. 93 – 110 (Jan 1992).

For vacuum, type N, twisting gravitational fields in Einstein field equations reduce to partial differential equations for two functions, one real, the other complex, which may be regarded as initial data on a local T^+. If it is assumed that in some frame these initial data take on a certain product form, one factor involving only a spatial variable, the other only a retarded time variable, then these equations become relatively tractable and reduce further to two ordinary differential equations. Rejecting all solutions which lead to Minkowski space or to zero twist leaves just two possibilities. One corresponds to the only explicitly known spacetime of the kind, namely that of Hauser. The other leads to new type N twisting metrices. However, these metrices can be constructed explicitly only when a single nonlinear third–order ordinary differential equation has been solved.

066.093   A spinning string.
H. H. Soleng.
Gen. Relativ. Gravitation, Vol. 24, No. 1, p. 111 – 117 (Jan 1992).

The Einstein–Cartan field equations are solved for a string source with spin–polarisation along the axis of symmetry. The interior solution is matched to an exterior vacuum space–time using Arkuszewski – Kopczyński – Ponomariev junction conditions. The exterior solution is a four dimensional extension of the space–time outside a spinning point particle in three dimensional Einstein theory. It reduces to the geometry outside a conventional straight cosmic string in the case of vanishing spin.

066.094   A search for "fifth force": the tower gravity experiment.
Yang Xinshe, Liu Yicheng, Wang Qianshen, Zhou Wenhu.
4. Asia Pacific Physics Conference, p. 1251 – 1254 (1991). – See Abstr. 012.006 for the main entry.

In a test of Newton's inverse–square law of gravitation, the authors compared gravity measured on a 320 m tower with gravity estimates calculated from ground measurements. They found no significant departure from Newton's inverse–square law, and the accuracy of the experiment constrains the Yukawa-potential coupling constant α to be less than 0.001.

066.095   General relativistic textures and their interactions with matter and radiation.
R. Durrer, M. Heusler, P. Jetzer, N. Straumann.
Nucl. Phys., B, Vol. 368, No. 2, p. 527 – 553 (13 Jan 1992).

Time–dependent solutions of the coupled Einstein–σ–model equations are studied analytically and numerically. The authors analyse in detail a spherically symmetric self–similar solution which describes the general relativistic collapse of a global texture. In view of the cosmological implications of the texture scenario for large scale structure formation, the authors derive exact analytical formulae for the light deflection, photon redshift and density fluctuations in the gravitational field of this texture solution. They also discuss the behavior of collisionless particles. Numerical results of a parameter study of other texture solutions are presented and are compared with the special self–similar solution.

066.096   Gravitational microlensing by a single star plus external shear.
S. Mao.
Astrophys. J., Vol. 389, No. 1, p. 63 – 67 (10 Apr 1992).

Gravitational microlensing by a single star plus external shear is considered. It is shown that for a general cusp the magnification probability distribution follows $p_c(A)dA \sim A^{-7/2}dA$ for sufficiently large magnifications. An adaptive grid technique is developed to calculate the magnification probability distributions. The results could be useful for cases of microlensing where the surface mass density is low.

066.097   Elliptic–type motion in a Schwarzschild – de Sitter gravitational field.
P. Blaga, V. Mioc.
Europhys. Lett., Vol. 17, No. 3, p. 275 – 278 (14 Jan 1992).

The elliptic–type motion in a Schwarzschild – de Sitter gravitational field is being studied by a classic method, considering separately the contribution of the Schwarzschild space–time local deformation and that of the cosmological background as perturbing forces. The first–order variations of the orbital elements over one nodal period (and the second–order changes only for the Schwarzschild "force") as well as those of the nodal period itself are determined. The domain in which the used methods are applicable (estimated for a black hole of 10 solar masses) results to be large enough to allow investigations in realistic astronomical situations.

066.098   SL(2,R) coset constructions and two–dimensional gravity.
Hu Hongliang.
Phys. Lett. B, Vol. 273, No. 4, p. 445 – 449 (26 Dec 1991).

The relation between coset constructions of SL(2,R)/SO(1,1) and SL(2,R)/U(1) and the theories of two–dimensional gravity coupled to conformal matters is discussed, using the free field representations of the former. The author finds that both the topological gravity theory and the theory of c = 1 conformal matter coupled to 2D gravity can be viewed as coset constructions of SL(2,R) current algebra. He also compares these results with 2D black hole theory.

066.099   Regular gravitational lagrangians.
N. Dragon.
Phys. Lett. B, Vol. 276, No. 1/2, p. 31 – 35 (6 Feb 1992).

The Einstein action with vanishing cosmological constant is for appropriate field content the unique local action which is regular at the fixed point to affine coordinate transformations. Imposing this regularity requirement one excludes also Wess–Zumino counterterms which trade gravitational anomalies for Lorentz anomalies. One has to expect dilatational and SL(D) anomalies. If these anomalies are absent and if the regularity of the quantum vertex functional can be controlled then Einstein gravity is renormalizable.

066.100   Black p–brane solutions of D = 11 supergravity theory.
R. Güven.
Phys. Lett. B, Vol. 276, No. 1/2, p. 49 – 55 (6 Feb 1992).

Various classes of extended black hole solutions of D = 11 supergravity theory are presented. It is shown that D = 11 supergravity admits a class of 'electric' black p–brane solutions for p = 2, 4, 6 and a 'magnetic' type of black five–brane solution. Each of these solutions is characterized by a mass and a charge parameter. The only supersymmetric members of these families are the extreme cases where Bogomol'nyi type of bounds are saturated by the parameters. The extreme cases also allow multi-source generalizations. Upon double dimensional reduction these families give rise to two new solutions of Type IIA string theory and one of these is a black five–brane.

066.101   The vacuum in three–dimensional simplicial quantum gravity.
J. Ambjørn, S. Varsted, D. V. Boulatov, A. Krzywicki.
Phys. Lett. B, Vol. 276, No. 4, p. 432 – 436 (20 Feb 1992).

The authors present evidence for a first–order phase transition in simplicial quantum gravity in three dimensions. It implies the existence of a vacuum, which has at least certain of the characteristics needed for a physically acceptable continuum limit of the theory. The action is assumed to contain the cosmological and the Einstein terms only. The behaviour of the entropy of manifolds stabilizes the theory.

066.102   Updating nucleosynthesis bounds on Jordan–Brans–Dicke theories of gravity.
J. A. Casas, J. García–Bellido, M. Quirós.
Phys. Lett. B, Vol. 278, No. 1/2, p. 94 – 96 (19 Mar 1992).

Nucleosynthesis bounds on Jordan–Brans–Dicke theories of gravity are updated using improved determinations of ^{4}He, $D + {}^3$He and ^{7}Li primordial abundances and neutron life–time. The authors find, for $N_v = 3$ and most of the allowed $\Omega_0 h_0^2$ range, stronger bounds than those obtained from post–newtonian experiments, $|\omega| > 250$ (2σ). Furthermore, it is suggested that GR predictions with $N_v = 3$ are hardly consistent with nucleosynthesis observations at 90% CL, while GR and $N_v = 2$ seems to accommodate better the observations.

066.103   Gravity theories with action densities bounded from below.
E. I. Guendelman.
Phys. Lett. B, Vol. 279, No. 3/4, p. 254 – 258 (16 Apr 1992).

The author discusses ways of constructing theories of gravity where the action density is bounded from below. This property can make the euclidean path integral better defined than in the standard Einstein case. For a class of action densities which depend (non–linearly) on the scalar curvature, such models are at the classical level equivalent to scalar–tensor theories of gravity with scalar potentials that imply a stable vacuum and where there is the possibility of an inflationary phase, as well as other interesting features. Scalars with such types of potentials appear naturally in some Kaluza–Klein and other theories. When integrating out the scalars in such theories, one could end up with the models discussed.

066.104   Computational methods for vacuum spacetimes.
D. Hobill.
Banff Summer Institute on Gravitation, p. 98 – 112 (1991). – See Abstr. 012.019 for the main entry.

Computational relativity is beginning to make significant contributions to our understanding of important physical systems and the behavior of the Einstein equations. This article discusses methods for the construction of numerical solutions to the vacuum equations in a high performance computing environment. Simulations in zero, one and two spatial dimensions are discussed and serve to demonstrate the problems encountered in computational relativity and gravitation.

066.105   Integration contours in the path integral approach to quantum gravity.
J. Louko.
Banff Summer Institute on Gravitation, p. 171 – 185 (1991). – See Abstr. 012.019 for the main entry.

The author reviews some of the problems that would need to be confronted in order to give a consistent path integral quantisation of general relativity. He then concentrates on the contour of integration and describes some recent work on this topic within simple cosmological models. The relevance of cosmological model analyses for the full theory is discussed.

066.106   Canonical quantum gravity.
A. Ashtekar.
Banff Summer Institute on Gravitation, p. 189 – 220 (1991). – See Abstr. 012.019 for the main entry.

After a brief historical introduction to the subject of quantum gravity as a whole, a recently revived version of the canonical approach is outlined. The emphasis is on the conceptual framework and basic ideas involved in various constructions. The proofs and detailed arguments can be found in the original papers and monographs referred to in the text.

066.107   Unimodular relativity, general covariance, time, and the Ashtekar variables.
L. Bombelli.
Banff Summer Institute on Gravitation, p. 221 – 232 (1991). – See Abstr. 012.019 for the main entry.

The author considers the so–called unimodular theories of gravity, obtained by considering only metrics with a fixed local volume element, and shows that they are equivalent, in the spatially compact case, to generally covariant theories, obtained by fixing only the value of the total spacetime volume. The author develops the Hamiltonian formulation using Ashtekar's variables, discusses the structure of the gauge group, sets up the corresponding quantum theory, and discusses in what sense the known loop state solutions of quantum general relativity are also solutions in the present theory.

066.108   The gravitational interaction in 2 + 1 dimensions.
J. Gegenberg, G. Kunstatter, H. P. Leivo.
Banff Summer Institute on Gravitation, p. 233 – 245 (1991). – See Abstr. 012.019 for the main entry.

2 + 1 gravity coupled to a scalar and anti–symmetric tensor gauge field is examined. The quantum theory is finite–dimensional and exactly solvable, containing two phase space degrees of freedom in addition to those in the vacuum theory.

066.109   Time in 2 + 1 dimensional quantum gravity.
S. Carlip.
Banff Summer Institute on Gravitation, p. 246 – 259 (1991). – See Abstr. 012.019 for the main entry.

By examining the exact quantization of general relativity in 2 + 1 dimensions, one can investigate the nature of time in quantum gravity, while at the same time avoiding the difficult technical problems of 3 + 1 dimensions. It is shown that a manifestly gauge–invariant, time–independent quantization is possible, and is exactly equivalent – at least for simple spatial topologies – to a gauge–fixed quantization with an explicit choice of time. In particular, Hilbert space norms and inner products can be defined without any reference to time, and operators that commute with the super–Hamiltonian nevertheless permit a full dynamical description of 2 + 1 dimensional gravity. General relativity may thus need no fundamental revision in order to solve the "problem of time".

066.110   No time and quantum gravity.
W. G. Unruh.
Banff Summer Institute on Gravitation, p. 260 – 275 (1991). – See Abstr. 012.019 for the main entry.

The role of time in the interpretation of Canonical Quantum Gravity is examined. In particular, some common claims about how time enters into the theory are examined and the key problems are listed. Three proposals are examined with respect to their treatment of time in General Relativity. In the third, a minor change in General Relativity is suggested which introduces an explicit external time, but difficulties with this theory are also pointed out.

066.111   Some classical results of possible relevance for semi–classical relativity.
N. Van den Bergh, E. Gunzig, P. Nardone, M. Castagnino.
Banff Summer Institute on Gravitation, p. 276 – 283 (1991). – See Abstr. 012.019 for the main entry.

Within the context of semi–classical gravity the instability of Minkowski space–time is a consequence of the non–linear feedback between quantum fluctuations of the matter fields and their space–time geometric response. This leads to a cosmological history finding its roots in an instability rather than in a singularity. The authors present indicating the existence of a close relationship between the assumed initial conformal flatness and the spatial homogeneity of the universe.

066.112 Semi–classical gravity and the black hole singularity.
E. Poisson.
Banff Summer Institute on Gravitation, p. 284 – 292 (1991). – See Abstr. 012.019 for the main entry.

The author examines the effects on the Schwarschild singularity of a particular semi–classical model based on the Gauss–Bonnet quadratic combination, in a spacetime possessing more than four dimensions. In this model, the field equations are non-trivial and consist of a system of second–order differential equations for the metric tensor (any other combination would yield fourth–order equations). The results depend strongly on the overall sign of the quadratic term: for positive coupling, as dictated by string theory, the singularity occurs sooner than in the classical descriptions: for negative coupling, the model breaks down because the radius of the compactified space is forced to collapse to zero for some non–vanishing value of the radial coordinate.

066.113 Relativistic wave equations in 1 + 1 dimensions.
S. M. Morsink.
Banff Summer Institute on Gravitation, p. 293 – 302 (1991). – See Abstr. 012.019 for the main entry.

Semi–Classical gravity is explored in (1 + 1) dimensions by examining the solutions to the Klein–Gordon and Dirac wave equations in the presence of a point source. Solutions are found for both massless and massive particles for space–times containing either a naked or cloaked singularity. The non–relativistic limit of the Klein–Gordon equation has the same solutions found using Schrodinger's equation. It is possible to quantize the energy of particles in a space–time containing a naked singularity. In addition, the wavefunctions are quantized for a space–time containing a black hole.

066.114 Topics in two dimensional non–Riemannian gravity.
P. F. Kelly, R. McArthur, R. B. Mann,
G. Kunstatter, D. Vincent, J. Gegenberg.
Banff Summer Institute on Gravitation, p. 303 – 313 (1991). – See Abstr. 012.019 for the main entry.

A number of recent results from an investigation of non–Riemannian geometries in two spacetime dimensions are described. The authors have discovered a fully geometric dynamical theory of 1 + 1 classical gravity; and have constructed a variety of solutions. The usual bosonic sigma model on a 1 + 1 dimensional base space and N dimensional target space (both Riemannian) is often augmented by a torsion (Wess–Zumino–Witten) term. Constructing an ordinary bosonic sigma model on non–Riemannian base and target spaces provides a natural geometric origin for the torsion contribution. The non–Riemannian bosonic sigma model exhibits proper classical behavior, but an examination of the conformal anomaly in the quantum theory reveals that the model is fundamentally anomalous.

066.115 Geometrical approach to the effective action.
G. Kunstatter.
Banff Summer Institute on Gravitation, p. 365 – 400 (1991). – See Abstr. 012.019 for the main entry.

The geometrical approach to the effective action is discussed in the context of sigma models and gauge theories. The formalism provides an elegant geometrical interpretation for the Faddeev–Popov ansatz that clearly illustrates the analogy between gauge fixing in the functional integral and dimensional reduction in Kaluza–Klein theory. One can also use this geometrical approach to construct a class of effective actions that are invariant under field reparametrizations, generate reparametrization covariant Feynman rules, and in the case of gauge theories, are also gauge independent and gauge invariant. This construction, due originally to Volkovisky and DeWitt, is described in detail. The

advantages and shortcomings of the resulting class of effective actions are stressed.

066.116 Classification of gravitational anomalies.
F. Brandt, N. Dragon, M. Kreuzer.
Banff Summer Institute on Gravitation, p. 401 – 410 (1991). – See Abstr. 012.019 for the main entry.

The authors determine all solutions to the consistency equations for gravity and Yang–Mills theories in arbitrary dimensions. They point out the essential elements of a constructive method for a complete computation of the relevant cohomology. As gravity and higher dimensional Yang–Mills are not renormalizable, no restriction on mass dimensions apply. The authors prove that all solutions can be constructed in terms of forms and covariant tensors.

066.117 Particles, fields and general relativity.
F. I. Cooperstock.
Banff Summer Institute on Gravitation, p. 411 – 435 (1991). – See Abstr. 012.019 for the main entry.

The author's research program of modelling elementary particles in singularity–free gauge invariant field theory is described. Work in progress on the description of the electron as a quantum soliton in Dirac–Maxwell theory is presented. The problem of balancing separated charged masses in general relativity is discussed. The author shows how the Herlt class can be fully integrated to yield a new exact solution for separated sources. It is shown that there is a distance–dependent balance possible and its potential physical consequences are discussed.

066.118 The confrontation between general relativity and experiment: a 1990 update.
C. M. Will.
Banff Summer Institute on Gravitation, p. 439 – 498 (1991). – See Abstr. 012.019 for the main entry.

The status of experimental tests of general relativity and of theoretical frameworks for analysing them are reviewed. Einstein's equivalence principle is well supported by experiments such as the Eötvös experiment, tests of special relativity, and the gravitational redshift experiment. Tests of general relativity have reached high precision, including the light deflection, the Shapiro time delay, the perihelion advance of Mercury, and the Nordtvedt effect in lunar motion. Gravitational wave damping has been detected to one percent using the binary pulsar. The status of the 'fifth force' is discussed, along with the frontiers of experimetal relativity, including proposals for testing relativistic gravity advanced technology and spacecraft.

066.119 Low frequency strategy for VIRGO, the French–Italian interferometric gravitational wave antenna.
C. Bradaschia, E. Calloni, M. Cobal, R. Del Fabbro,
A. Di Virgilio, A. Giazotto, L. E. Holloway, H. Kautzky,
B. Michelozzi, V. Montelatici, D. Passuello, W. Velloso.
Banff Summer Institute on Gravitation, p. 499 – 514 (1991). – See Abstr. 012.019 for the main entry.

Long arms interferometric gravitational wave antennas are planned all over the world. The French–Italian antenna is designed in order to detect Gravitational Waves in a wide frequency range from 10 Hz to few kHz. The antenna is a high sensitivity interferometer with 3 km arms and the 10 Hz region is achievable by means of ad-hoc suspensions (called Super-Attenuator, SA), for the optical components, in order to reduce the seismic noise. The SA's measurements have shown that suspension is adequate to reduce the seism well below other noise sources.

066.120 Gravitationally induced birefringence in nonsymmetric theories of gravity.
J. H. Palmert, R. B. Mann, M. D. Gabriel, M. P. Haugan.
Banff Summer Institute on Gravitation, p. 515 – 522 (1991). – See Abstr. 012.019 for the main entry.

Theories of gravity which employ a nonsymmetric metric coupled to matter show interesting new physical predictions tied in with violation of the Einstein Equivalence Principle. The authors present two such effects: the polarization dependent deflection of light by a static, spherically symmetric gravitational source, and polarization dependent time delay in radar echo experiments. The polarization dependence is a result of coupling the antisymmetric part of the metric to electromagnetism. The magnitude of the effects is computed in the case of Moffat's nonsymmetric theory of gravity.

066.121   Review of the nonsymmetric gravitational theory.
J. W. Moffat.
Banff Summer Institute on Gravitation, p. 523 – 597 (1991). Abstract. – See Abstr. 012.019 for the main entry.

066.122   Torsion trace and variable G.
H. H. Soleng.
Banff Summer Institute on Gravitation, p. 598 – 602 (1991). – See Abstr. 012.019 for the main entry.
In the Einstein–Cartan theory only the traceless part of torsion can be coupled to spin. Here is is argued that the trace part of torsion singifies intrinsic dilation currents and a variable gravitational coupling.

066.123   Gravitational lensing.
R. L. Webster.
Banff Summer Institute on Gravitation, p. 603 – 624 (1991). – See Abstr. 012.019 for the main entry.
The first of these lectures describes the basic ideas and formalisms of gravitational lensing. In the second lecture, some specific applications of gravitational lensing are discussed, in particular those problems which relate to the determination of cosmological parameters in the most generally accepted cosmological models – the homogeneous, isotropic FRW models in General Relativity. In the final lecture, one example of gravitational lensing is discussed, the quasar 2237+0305, and in particular the observation of microlensing in some of its images. Implications of these observations are explored.

066.124   Do gravitational waves really exist?
M. Łysik.
Urania, Rok 63, Nr. 5, p. 141 – 144 (May 1992). In Polish.

066.125   Non–local variables for general relativity – a review.
C. Kozameh, E. T. Newman.
9. Italian Conference on General Relativity and Gravitational Physics, p. 47 – 53 (1991). – See Abstr. 012.021 for the main entry.
It is the purpose of this note to try to briefly survey and outline a new point of view towards the Einstein vacuum equations that has been developing and evolving over the past five or six years. The basic idea is that our space–time and all its metric properties, are to be derived concepts and that there is an underlying, more primitive structure from which the space–time, the metric and the Einstein equations are to be derived. This point of view is based on the use of two non–local variables; the holonomy operator (the parallel propagator around closed loops) and the ''light–cone'' cut function (the intersection of the light–cone of any spacetime point with a canonically chosen light–cone, usually taken as null infinity).

066.126   Projective Unified Field Theory and its relationship to Klein–Kaluza type theories.
E. Schmutzer.
9. Italian Conference on General Relativity and Gravitational Physics, p. 54 – 67 (1991). – See Abstr. 012.021 for the main entry.
After a short review of the Projective Unified Field Theory (PUFT), its physical interpretation and predictions the author presents some basic statements on its relationship to Klein–Kaluza–type theories.

066.127   Experimental gravitation and very low temperatures.
M. Cerdonio, M. Bonaldi, A. Cavalleri, R. Dolesi, G. Durin, P. Falferi, G. Fontana, P. Fortini, R. Macchietto, A. Maraner, R. Mezzena, A. Ortolan, G. A. Prodi, C. Ravanelli, B. Tiveron, R. Tommasini, C. Valentini, J. P. Zendri, S. Vitale.
9. Italian Conference on General Relativity and Gravitational Physics, p. 114 – 130 (1991). – See Abstr. 012.021 for the main entry.
The authors report on the progress in the feasibility study for a ground based cryogenic experiment aimed to measure the Lense–Thirring–Schiff dragging of intertial frames due to the rotating earth. They present the goals of the experiment AURIGA, to be set up at LNL: an ultracryogenic, $T \approx 0.1 K$, resonant antenna, $M \cong 2300$ kg, will be operated in continuous coincidence with the similar NAUTILUS at CERN to search for gravitational waves impulses from galaxies of the Local Group.

066.128   An update of the LAGEOS III gravitomagnetic experiment.
I. Ciufolini.
9. Italian Conference on General Relativity and Gravitational Physics, p. 131 – 138 (1991). – See Abstr. 012.021 for the main entry.
The author reports the latest results regarding the LAGEOS III experiment, in particular he describes the results of a joint ASI–NASA study on this experiment to measure the gravitomagnetic field.

066.129   Use of very low temperatures for detecting gravitational waves.
I. Modena, E. Coccia.
9. Italian Conference on General Relativity and Gravitational Physics, p. 139 – 146 (1991). – See Abstr. 012.021 for the main entry.
The challenging problem in the gravitational wave research is the very small size of the effect one wishes to measure. Three advantages emerge from cooling a resonant antenna to very low temperatures: one based on the general reduction of the kT thermal noise, one based on the use of superconducting devices and one due to the better acoustic properties of antenna materials. The main requirements and features of cryogenic systems for gravitational wave experiments above and below 1K temperature are reported. The problem of the acoustic isolation from the cooling sources is discussed.

066.130   Quadratic gravity: old and new.
F. Occhionero, M. Litterio, S. Capozziello, L. Amendola.
9. Italian Conference on General Relativity and Gravitational Physics, p. 365 – 374 (1991). – See Abstr. 012.021 for the main entry.
Corrections from quantum gravity suggest that the Einstein–Hilbert Lagrangian be enlarged to include quadratic and higher order terms in the Ricci scalar, R, when the latter is large. The ensuing field equations are then fourth order, rather than second. For the early stages of cosmological evolution the situation can be intuitively visualized if one projects the trajectories onto a two–dimensional phase space or, better yet, onto its Poincaré projection. The authors first review this method for the cases of a non–minimally coupled scalar field and of the compactification in a multidimensional cosmology of inner radius b(t). Then they apply it to fourth order gravity: they show the existence of inflationary attractors and forbidden regions of the $(R, \dot{R})$ phase space. In conclusion, as a natural follow–up of these models, the authors outline the possibility of a scalar coupling with the curvature squared term, which makes time dependent the so called Starobinsky scalaron mass. With this mechanism the latter may attain dynamically values of order of $10^{-5} - 10^{-6}$ as required by inflation down from the natural initial values of order unity without any fine tuning.

066.131   **On the causal structure of Tolman–Bondi spacetimes.**
G. Grillo.
9. Italian Conference on General Relativity and Gravitational Physics, p. 450 – 455 (1991). – See Abstr. 012.021 for the main entry.

The author studies the spherically symmetric, non–homogeneous gravitational collapse of dust (i.e. of a pressure–free perfect fluid), as described by the non–synchronous, recollapsing Tolman–Bondi metrics. A class of metrics is classified in which the corresponding "shell–focusing" singularity is at least locally naked, in terms of the behaviour in a neighbourhood of $r = 0$ (in comoving coordinates) of the arbitrary functions $m(r)$, $E(r)$ and $t_0(r)$; this happens whenever the local deviation from homogeneity is sufficiently high. Finally it is shown that there exist cases in which the corresponding locally naked singularity is a strong curvature singularity in the sense of Tipler, by checking the validity of the strong limiting focusing condition.

066.132   **Violation of Mach's principle by a time–dependent rotating metric.**
R. Bergamini, E. Pian.
9. Italian Conference on General Relativity and Gravitational Physics, p. 486 – 491 (1991). – See Abstr. 012.021 for the main entry.

066.133   **Cataclysmic variables as sources of gravitational waves.**
F. Barone, F. C. D'Ambrosio, L. Di Fiore, L. Milano, G. Russo.
9. Italian Conference on General Relativity and Gravitational Physics, p. 596 – 602 (1991). – See Abstr. 012.021 for the main entry.

The gravitational radiation background from a sample of 137 catalysmic variables has been evaluated. This analysis enlarges the scenarios of continuous gravitational waves sources. The authors estimate a GW flux at Earth of $10^{-12} \text{erg sec}^{-1} \text{cm}^{-2}$, and a dimensionless amplitude $h \approx 10^{-21}$.

066.134   **Conservation laws in general relativity and the determination of the PPN parameters.**
A. D. A. M. Spallicci.
9. Italian Conference on General Relativity and Gravitational Physics, p. 608 – 610 (1991). – See Abstr. 012.021 for the main entry.

066.135   **Experimental gravitation and the Columbus program.**
A. D. A. M. Spallicci.
9. Italian Conference on General Relativity and Gravitational Physics, p. 611 – 614 (1991). – See Abstr. 012.021 for the main entry.

The Attached Pressurised module and the Free Flyer of the Columbus program might offer the potential for experiments in gravitational physics. Three classes of experiments are outlined. The mechanical experiments (those involving measurements and control of very small accelerations, and determination of bodies position) which determine e.g. the fifth force and the constant of gravitation by using accelerometers. An other category of experiments might be performed by clocks to measure relativistic effects. Finally laser interferometers for gravitational waves are to be considered.

066.136   **Spacecraft searches for gravitational waves from massive coalescing binaries: detection and false–alarm probabilities.**
M. Tinto, J. W. Armstrong.
9. Italian Conference on General Relativity and Gravitational Physics, p. 615 – 619 (1991). – See Abstr. 012.021 for the main entry.

066.137   **Source location determination for gravitational wave pulses with a network of four earth–based laser–interferometric detectors.**
Y. Gürsel, M. Tinto.
9. Italian Conference on General Relativity and Gravitational Physics, p. 620 – 623 (1991). – See Abstr. 012.021 for the main entry.

The authors extend the method described in a previous paper to the case of four, widely separated, laser–interferometric gravitational wave detectors. They show that for conventional signal–to–noise ratios of about 10, it is possible to locate the source within an angular error of 1×10^{-6} steradians, which is about 10 times better than the case of a network consisting of three widely separated detectors.

066.138   **Post–Newtonian corrections in gravitation from superstring theory.**
G. Cristofano, M. Fabbrichesi, K. Roland.
9. Italian Conference on General Relativity and Gravitational Physics, p. 648 – 651 (1991). – See Abstr. 012.021 for the main entry.

At the Planckian regime of very high center–of–mass energies and small transferred momenta gravity is the dominant interaction. A decoupling of all states but the graviton would allow a direct comparison between superstring theory and classical gravitation, regardless of compactification details. The authors test the extent of such a decoupling by calculating the deflection angle of a graviton in the field of a massive state of the superstring and compare it to the corresponding result of general relativity up to the second order in Newton's constant.

066.139   **Negative energy fluxes and cosmic censorship.**
L. H. Ford, T. A. Roman.
9. Italian Conference on General Relativity and Gravitational Physics, p. 680 – 683 (1991). – See Abstr. 012.021 for the main entry.

Quantum field theory allows violations of the weak energy condition in the form of locally negative energy fluxes and densities. If there are no restrictions on such fluxes, then one could conceivably use the negative energy flux generated by a moving mirror to violate cosmic censorship. This might be accomplished by shining the negative energy flux from the mirror on an extreme Reissner–Nordström black hole, thus producing a naked singularity. It is shown, for physically reasonable trajectories of a mirror moving in a two–dimensional black hole background, that this is not possible. The authors find that the change in the mass of the black hole $|\Delta M|$, due to the absorption of the negative energy flux, and the effective lifetime ΔT of the naked singularity thus produced, are limited by an uncertainty principle–type inequality of the form $|\Delta M| \Delta T < 1$. The conclusion is that the resulting violation of cosmic censorship is classically unobservable, since it is below the scale of the normal quantum fluctuations of the mass of the black hole on this timescale.

066.140   **On the rotation of polarization by a gravitational lens.**
C. C. Dyer, E. G. Shaver.
Astrophys. J., Lett., Vol. 390, No. 1, p. L5 – L7 (1 May 1992).

It is shown that the polarization directions of photons in a beam propagating under the influence of a gravitational lens remain unaffected by the lens in most cases of astrophysical interest.

066.141   **String–modified four–dimensional cosmology.**
K. D. Krori, A. Das Purkayastha.
Can. J. Phys., Vol. 70, No. 2/3, p. 179 – 182 (Feb–Mar 1992).

Recently Gegenberg proposed a string–modified four–dimensional gravity theory comprising gravity, electromagnetism, a dilaton field, and a Kalb–Ramond field. In this paper, some general cosmological aspects of this theory are discussed.

066.142 Renormalization group equations in curved space–time with nontrivial topology.
E. Elizalde, S. D. Odintsov.
Europhys. Lett., Vol. 19, No. 4, p. 261 – 265 (15 Jun 1992).
Renormalization group equations for massless GUT's in curved space–time with nontrivial topology are formulated. The asymptotics of the effective action both at high and low energies are obtained. It is shown that the Casimir energy contribution at high curvature (early universe) becomes nonessential in the effective action.

066.143 Continuum and discretum – unified field theory and elementary constants.
H.–J. Treder.
Found. Phys., Vol. 22, No. 3, p. 395 – 420 (Mar 1992).
Unitary field theories and "SUPER–GUT" theories work with an universal continuum, the structurd spacetime of R. Descartes, B. Spinoza, B. Riemann, and A. Einstein, or a structured vacuum according the quantum theory of unitary fields. Planck's conception of the three elementary constants $\hbar$, c, and G may be the key to general relativistic quantum field theory like unitary theory. However, the elementary constants are a question of measurement–theory, also. According to Popper's theory of induction, such unitary theories are "universal explaining theories". The fundamental constants involve the complementarity between the universal statements in unitary theory and the "basic statements" in the language of classical observables.

066.144 Two–loop approach to the effective action in quantum gravity.
I. L. Buchbinder, S. D. Odintsov, O. A. Fonarev.
Int. J. Mod. Phys. A, Vol. 7, No. 14, p. 3203 – 3233 (10 Jun 1992).
For the first time the authors present the general formalism and results of calculation of the two–loop effective action in Einstein quantum gravity on the background $M_N \times T_k$, where M_N is Minkowski space and T_k is a k–dimensional torus. They discuss the case of a zero cosmological constant as well as of a nonzero one. The method of calculating variations of the action on a metric tensor and the technique of calculating momentum integrals in dimensional regularization are presented. Some applications to spontaneous compactification are discussed, as well as some prospects.

066.145 The one–way propagation of light near the surface of the Earth in metric theories of gravity.
L. Marchildon, A. F. Antippa.
Nuovo Cimento B, Vol. 107, No. 2, p. 153 – 166 (Feb 1992).
The synchronization of clocks at distant spatial points is a question of convention. If a synchronization not involving electromagnetic radiation is agreed upon, the one–way velocity of light becomes meaningful. The authors develop the prediction of general relativity and other metric theories of gravity for the one–way speed of light near the surface of the rotating Earth, distant clocks being synchronized by means of clock transport. This prediction may soon be checked against experimental measurements.

066.146 Holes in space–time.
D. K. Ross.
Nuovo Cimento B, Vol. 107, No. 2, p. 203 – 210 (Feb 1992).
The author explores the possibility that the topology of space–time may not be that of Minkowski's space–time but may include many topological holes. He shows that these holes should behave like massless scalar particles with Lorentz invariant cross–sections which depend upon their size. The author argues that they should have a thermal energy distribution with a temperature the same as the cosmic neutrino background. A photon interacting with this sea of holes is red–shifted in an energy–dependent way. This allows to put severe constraints on their cross–sections using radio and optical quasar data.

066.147 Thermal stress–energy tensor of a scalar field in Reissner–Nordström space–time.
Huang Chaoguang.
Phys. Lett. A, Vol. 164, No. 5/6, p. 384 – 388 (27 Apr 1992).
The approximate expression for the renormalised stress–energy tensor for a conformally invariant scalar field in the Reissner–Nordström space–time is obtained. At the event horizons, it is regular for $e^2 < M^2$, but not for $e^2 = M^2$.

066.148 Static plane–symmetric space–time with a conformally coupled massless scalar field.
O. Grøn, H. H. Soleng.
Phys. Lett. A, Vol. 165, No. 3, p. 191 – 193 (18 May 1992).
The authors investigate static plane–symmetric space–times. The source of curvature in the Taub space is identified, and the modification hereof due to the presence of a conformally coupled scalar field is calculated. For a specific value of the scalar charge, the scalar field has an energy–momentum tensor of the Casimir type. In this case the source is a domain wall.

066.149 Gravitational instantons and quantisation of the cosmological constant.
I. Moss.
Phys. Lett. B, Vol. 283, No. 1/2, p. 52 – 54 (4 Jun 1992).
Complex gravitational instantons, introduced recently into quantum cosmology, may be important in the small scale structure of spacetime. If the cosmological constant is negative, then there is a complex instanton which breaks the electromagnetic gauge symmetry unless the cosmological constant is quantised. $\Lambda = -3e^2/N^2$, where e is the unit of charge and N is a natural number.

066.150 Topology change by quantum tunneling in (2+1)–dimensional Einstein gravity.
Y. Fujiwara, S. Higuchi, A. Hosoya, T. Mishima, M. Siino.
Prog. Theor. Phys., Vol. 87, No. 2, p. 253 – 268 (Feb 1992).
The authors investigate possibilities of topology change and nucleation of universes in the (2+1)–dimensional Einstein gravity model with negative cosmological constant. They demonstrate that topology change and nucleation phenomena can occur by quantum tunneling by explicitly constructing Euclidean signature hyperbolic 3–geometry. The amplitude is explicitly calculated in the WKB approximation. Point particles can be accommodated in this scheme as topological defects in space–time. They are pair created and annihilated when the universe changes its topology.

066.151 Anisotropic cosmological models in N = 2, D = 5 supergravity.
L. O. Pimentel.
Classical Quantum Gravity, Vol. 9, No. 2, p. 377 – 381 (Feb 1992).
Exact solutions of N = 2 supergravity in five dimensions are found in a Bianchi type I spacetime. The singularity is not avoided, in contrast to the isotropic case.

066.152 The unbounded action and the "density of states" in non–perturbative quantum gravity.
E. Myers.
Classical Quantum Gravity, Vol. 9, No. 2, p. 405 – 411 (Feb 1992).
One of the problems which must be overcome before a non–perturbative theory of quantum gravity can be formulated is the unboundedness of the Euclidean action. The author presents a brief outline of the "density of states reconstruction" method of analysing data from Monte Carlo simulations and then proposes a way to use this method to overcome the unboundedness problem.

066.153 **The fluid–ray tetrad formulation of Einstein's field equations.**
W. R. Stoeger, S. D. Nel, R. Maartens, G. F. R. Ellis.
Classical Quantum Gravity, Vol. 9, No. 2, p. 493 – 507 (Feb 1992).
The authors describe the fluid–ray tetrad, which is based on the four-velocity u of the cosmological fluid and the null vector k which lies along the generators of the null geodesics in the past light cones $C^-(p_0)$ centred on our world line C. Then, they formulate the field equations in terms of this tetrad, giving also the Jacobi identities, the metric variable equations for observational coordinates, and the contracted Bianchi identities. The equivalent set of equations, involving the Ricci identities, the full Bianchi identities, along with the contracted Bianchi identities, is also given. This formulation facilitates the integration of the field equations in certain cosmological contexts.

066.154 **Inhomogeneous perfect fluid cosmologies.**
N. van den Bergh, J. Skea.
Classical Quantum Gravity, Vol. 9, No. 2, p. 527 – 532 (Feb 1992).
The authors analyse inhomogeneous perfect fluid models obeying a γ law equation of state and admitting an Abelian two-dimensional group of motions with two hypersurface – orthogonal Killing vectors. The models are asymptotically self–similar and generalize a recently found non–singular solution with $\gamma = 4/3$.

066.155 **Static charged fluid surrounded by a black anti–hole: an enlarged Klein solution.**
M. Cataldo, N. V. Mitskiévič.
Classical Quantum Gravity, Vol. 9, No. 2, p. 545 – 552 (Feb 1992).
A new Petrov type D exact solution of the Einstein–Maxwell equations with a charged perfect fluid is obtained from the seed Klein metric for a static spherically symmetric distribution of incoherent radiation. A special case of the new solution is studied in detail, its spacetime being found to possess one horizon inside which it is static. Since the fluid behaves acausally outside the horizon, the authors match the solution to a continuous set of electrovacuum spacetimes with cosmological terms (the cosmological constant Λ depending on the junction radius). In the new spacetime, synchronous coordinates are constructed and a redshift effect is evaluated.

066.156 **From the Weyl theory to a theory of locally anisotropic spacetime.**
G. Yu. Bogoslovsky.
Classical Quantum Gravity, Vol. 9, No. 2, p. 569 – 575 (Feb 1992).
It is shown that Weyl's ideas, pertaining to local conformal invariance, find natural embodiment within the framework of a relativistic theory based on a viable Finslerian model of spacetime. This is associated with the peculiar property of the Finslerian metric which describes a locally anisotropic space of events. Such a metric, in contrast to the Riemannian one, is conformally invariant, in which case the local conformal transformations of the Riemannian metric tensor, apart from spacetime intervals, leave invariant rest masses as well as all observables and thus appear as local gauge transformations. The corresponding Finslerian theory of gravitation turns out to be an Abelian gauge theory. It satisfies the principle of correspondence with Einstein's theory and predicts a number of non–trivial physical effects accessible for experimental test.

066.157 **Six years of the fifth force.**
E. Fischbach, C. Talmadge.
Nature, Vol. 356, No. 6366, p. 207 – 215 (19 Mar 1992).
The enunciation of the "fifth force" hypothesis in 1986 spawned a generation of experiments searching for deviations from newtonian gravity. Although no compelling evidence for any new weak forces has emerged in the past six years, the searches for anomalous gravitational effects have produced a large number of important experimental and theoretical results.

066.158 **A description of semidegenerate self–gravitating spheres of fermions.**
M. Membrado, A. F. Pacheco, J. Sañudo.
Astrophys. J., Vol. 390, No. 1, p. 88 – 95 (1 May 1992).
The authors analyze the structure of a sphere of self–gravitating fermions, at low constant temperature, substituting the usual Fermi–Dirac distribution function by a trapezoidal approximation. This model coincides with the exact result up to the first correction in the low–temperature expansion while leading to finite equilibrium configurations; thus, it represents an alternative to the use of the habitual energy – or density – cutoffs. Analytical expressions (up to terms in T^2) are obtained for magnitudes such as total energy, radius of the sphere, and local values for the density and pressure.

066.159 **Strong curvature naked singularities in non–self–similar gravitational collapse.**
P. S. Joshi, I. H. Dwivedi.
Gen. Relativ. Gravitation, Vol. 24, No. 2, p. 129 – 137 (Feb 1992).
It is shown that strong curvature naked singularities form in a non–self–similar gravitational collapse of radiation. The imploding radiation space–times with a general form of mass function are analyzed and it is shown that a strong curvature property holds along all families of non–spacelike geodesics terminating at the singularity in past. In view of the strength of the singularity and the non–self–similar nature of space–time, the authors believe this to be a very serious counter–example, which must be taken into account for any possible formulation of the cosmic censorship hypothesis.

066.160 **Gravitation: Time travel on a string.**
B. Allen, J. Simon.
Nature, Vol. 357, No. 6373, p. 19 – 21 (7 May 1992).

066.161 **Gravitational microlensing: powerful combination of ray–shooting and parametric representation of caustics.**
J. Wambsganss, H. J. Witt, P. Schneider.
Astron. Astrophys., Vol. 258, No. 2, p. 591 – 599 (May 1992).
The authors present a combination of two very different methods for numerically calculating the effects of gravitational microlensing: the backward–ray–tracing that results in two-dimensional magnification patterns, and the parametric representation of caustic lines; they are in a way complementary to each other. The combination of these methods is much more powerful than the sum of its parts. It allows to determine the total magnification and the number of microimages as a function of source position. The mean number of microimages is calculated analytically and compared to the numerical results. The peaks in the lightcurves, as obtained from one–dimensional tracks through the magnification pattern, can now be divided into two groups: those which correspond to a source crossing a caustic, and those which are due to sources passing outside cusps. The authors determine the frequencies of those two types of events as a function of the surface mass density, and the probability distributions of their magnitudes. They find that for low surface mass density as many as 40% of all events in a lightcurve are not due to caustic crossings, but rather due to passings outside cusps.

066.162 **A generalization of the relativistic equilibrium equations for a non–rotating star.**
G. Magli, J. Kijowski.
Gen. Relativ. Gravitation, Vol. 24, No. 2, p. 139 – 158 (Feb 1992).
The problem of elastomechanical equilibrium for a static, spherically symmetric star composed of an elastic material is analyzed. A suitable formulation of relativistic elasticity theory is used, and the second order equilibrium equations are found. It is shown that the equilibrium conditions with "anisotropic pressure" introduced ad hoc by some authors are in fact the dynamical conditions for a relativistic elastic material. The corresponding first order equations for the components of the metric and of the energy–momentum tensor reduce to the Tolman – Oppenheimer – Volkoff equations if the material

exhibits no shape–rigidity. Two interesting classes of solutions are discussed.

066.163 Spatial anisotropy in nonsymmetric gravitation theories.
Zhou Ziye, M. P. Haugan.
Phys. Rev. D, Vol. 45, No. 10, p. 3336 – 3340 (15 May 1992). Current Physics Microform No.: 9207D0918.

The authors show that the energies of atomic states can depend strongly on an atom's orientation in space according to theories of gravity which couple the antisymmetric part of a nonsymmetric tensor gravitational field to the electromagnetic field. Atomic physics experiments designed to test the isotropy of space are directly sensitive to such orientation dependence. The authors derive a new constraint that these experiments impose on nonsymmetric theories. This constraint implies that the magnitude of effects due to an antisymmetric tensor component of the gravitational field must be more than 1000 times smaller than suggested in the recent literature on nonsymmetric theories.

066.164 Radiating soliton stars in the thin–wall approximation.
M. Esculpi, L. Herrera.
Phys. Rev. D, Vol. 45, No. 10, p. 3341 – 3354 (15 May 1992). Current Physics Microform No.: 9207D0923.

The authors study the nonadiabatic evolution of nontopological soliton stars (scalar and fermion) in the thin–wall approximation. For the scalar case the space–time within the star is assumed to be self–similar and both (extreme) mechanisms of radiation transport (diffusion and streaming out) coexist. For the fermion star only the streaming–out regime will be present. In the scalar case, transitions to the static regime are allowed only for surface potentials equal to 0.25. For fermion solitons (once the external pulse has passed away) all models tend to the stationary regime, which is characterized by the asymptotic value of the surface gravitational potential equal to $4/7$.

066.165 Compact baby universe model in ten dimension and probability function of quantum gravity.
Yan Jun, Hu Shike.
High Energ. Phys. Nucl. Phys., Vol. 15, No. 10, p. 890 – 897 (Oct 1991). In Chinese.

In this paper, the quantum probability functions are calculated for a ten–dimensional compact baby universe model. It is found that the probability for the Yang–Mills baby universe to undergo a spontaneous compactification down to a four–dimensional spacetime is greater than that to remain in the original homogeneous multidimensional state. Some questions about large–wormhole catastrophe are also discussed.

066.166 A generalized metric of gravitation.
A. El–Tahir.
Int. J. Mod. Phys. A, Vol. 7, No. 13, p. 3133 – 3139 (20 May 1992).

Useful expressions relating the metric components g_{rr} and g_{tt} of the static isotropic space–time, and the scalar curvature R, with the general Lagrangian are obtained. A generalized metric of gravity is introduced, which is essentially nonsingular, and reducible to cosmological and hence Schwarzschild geometries by imposing a weak–field constraint.

066.167 An algorithm to generate classical solutions of string effective action.
S. K. Kar, S. P. Khastgir, A. Kumar.
Mod. Phys. Lett. A, Vol. 7, No. 17, p. 1545 – 1551 (7 Jun 1992).

It is shown explicitly that a number of solutions for the background field equations of the string effective action in space-time dimension D can be generated from any known lower dimensional solution when background fields have only time dependence. An application of the result to the two–dimensional charged black hole is presented. The case of background with more general coordinate dependence is also discussed.

066.168 Modified black holes in two dimensional gravity.
N. Mohammedi.
Phys. Lett. B, Vol. 281, No. 1/2, p. 36 – 42 (7 May 1992).

The SL(2, R)/U(1) gauged WZWN model is modified by a topological term and the accompanying change in the geometry of the two dimensional target space is determined. The possibility of this additional term arises from a symmetry in the general formalism of gauging an isometry subgroup of a non–linear sigma model with an antisymmetric tensor. It is shown, in particular, that the space–time exhibits some general singularities for which the recently found black hole is just a special case. From a conformal field theory point of view and for special values of the unitary representations of SL(2, R), this topological term can be interpreted as a small perturbation by a $(1, 1)$ conformal operator of the gauge WZWN action.

066.169 A charged Kerr metric solution in new general relativity.
T. Kawai, N. Toma.
Prog. Theor. Phys., Vol. 87, No. 3, p. 583 – 598 (Mar 1992).

The authors give an exact solution of the gravitational and electromagnetic field equations with a charged rotating source in new general relativity. The solution has three parameters Q, h and a, and it gives a charged Kerr metric space–time. The parallel vector fields and the electromagnetic vector potential are axially symmetric. In this space–time, one cannot discriminate new general relatively from general relativity, so far as scalar, the Dirac and the Yang–Mills fields and macroscopic bodies are used as probes. The space–time does not have singularities at all, although it has an "effective singularity". Two kinds of Reissner–Nordström metric solutions, one is the authors' solution with h = 0 and the other is a solution given by Hayashi and Shirafuji, are physically equivalent with each other. Nevertheless, these are markedly different from each other with regard to the asymptotic behavior of the torsion tensor for r→∞ and the space–time singularities.

066.170 String theory at nonzero temperature and two–dimensional gravity.
S. D. Odintsov.
Riv. Nuovo Cimento, Vol. 15, No. 2, p. 1 – 64 (1992).

Part I is devoted to bosonic string. The author calculates the free energy at any genus. The vacuum energy for torus–compactified bosonic string is found. In part II he briefly describes the free energy for open and closed superstrings. In part III he discusses the noncritical strings thermodynamics. The free energy in the closed noncritical bosonic string with dynamical Weyl mode is found. The same model of open string at nonzero temperature is considered in an external magnetic field. Finally, the thermodynamics of Chamseddine's noncritical string model based on Jackiw–Teitelboim two–dimensional gravity is investigated. The general description of two–dimensional gravity actions is given in part IV.

066.171 The relation between the relativistic 5D Wesson theory and the Newtonian variable mass theory.
J. C. Carvalho, J. A. S. Lima.
Gen. Relativ. Gravitation, Vol. 24, No. 2, p. 171 – 177 (Feb 1992).

The existence of Newtonian analogs to spatially homogeneous and isotropic cosmological models of the relativistic 5D Wesson variable mass theory is investigated. By treating the continuous universe "matter creation process" by the methods of standard hydrodynamics, it is shown that classical analogs are obtained only if the cosmological constant is null and the spatial curvature is positive.

066.172 A one–parameter family of cylindrically symmetric perfect fluid cosmologies.
W. Davidson.
Gen. Relativ. Gravitation, Vol. 24, No. 2, p. 179 – 185 (Feb 1992).

Non–stationary cylindrically symmetric one–parameter solutions to Einstein's equations are given for a perfect fluid. There is a time singularity (t = 0) at which the pressure p and density μ

are equal to $+\infty$ throughout the radial coordinate range $0 \leqslant r < \infty$, but the solutions are well behaved for $t > 0$, p and μ decreasing steadily to zero as r increases through the range $0 \leqslant r < \infty$, or as t increases through the range $0 < t < \infty$. The motion is irrotational with shear, expansion and acceleration. The family of solutions, of Petrov type I, are generally spatially inhomogeneous, of class B(ii), having two spacelike Killing vectors which are mutually orthogonal and hypersurface orthogonal, associated with an orthogonally transitive group G_2. The particular members for which there are equations of state $p = \mu/3$ and $p = \mu$ are specially considered.

066.173   Self–gravitating general–relativistic cosmic strings.
E. Shaver.
Gen. Relativ. Gravitation, Vol. 24, No. 2, p. 187 – 198 (Feb 1992).
The author examines the coupled Einstein – Euler – Lagrange equations for nonstationary cosmic strings. Self–consistent solutions to all the equations are found under the assumption that the energy–momentum tensor is of the form $T_t{}^t = T_z{}^z$, while all other components vanish. It is shown that the strings are necessarily static in this case and that the scalar field potential must be of the usual quartic form with the coupling constants satisfying $e^2 = 8\lambda$.

066.174   Closed spaces in cosmology.
H. V. Fagundes.
Gen. Relativ. Gravitation, Vol. 24, No. 2, p. 199 – 217 (Feb 1992).
This paper deals with two aspects of relativistic cosmologies with closed spatial sections. These spacetimes are based on the theory of general relativity, and admit a foliation into space sections S(t), which are spacelike hypersurfaces satisfying the postulate of the closure of space: each S(t) is a three–dimensional closed Riemannian manifold. The topics discussed are: (1) a comparison, previously obtained, between Thurston geometries and Bianchi – Kantowski – Sachs metrics for such three–manifolds is here clarified and developed; and (2) the implications of global inhomogeneity for locally homogeneous three–spaces of constant curvature are analyzed from an observational viewpoint.

066.175   Causal horizons, accelerations and strings.
M. Gasperini.
Gen. Relativ. Gravitation, Vol. 24, No. 2, p. 219 – 223 (Feb 1992).
All the points of a string are always causally connected provided their relative acceleration is smaller than the critical value $a_c = (m\alpha')^{-1}$, where m is the mass and $1/\alpha'$ the string tension. It is pointed out that this limiting acceleration characterizes the transition to an unstable regime, in which an approximate description of the string motion around the classical path of a particle is no longer consistent.

066.176   Erratum: "Null charts and naked singularities in spherically symmetric, homothetic spacetimes" [Gen. Relativ. Gravitation, Vol. 23, No. 5, p. 527 – 581 (May 1991)].
R. N. Henriksen, K. Patel.
Gen. Relativ. Gravitation, Vol. 24, No. 2, p. 231 (Feb 1992). See Abstr. 53.066.176.

066.177   General–relativistic model of a spinning cosmic string.
B. Jensen, H. H. Soleng.
Phys. Rev. D, Vol. 45, No. 10, p. 3528 – 3533 (15 May 1992).
Current Physics Microform No.: 9207D1110.
The authors investigate the infinite, straight, rotating cosmic string within the framework of Einstein's general theory of relativity. A class of exact interior solutions is derived for which the source satisfies the weak and the dominant energy conditions. The interior metric is matched smoothly to the exterior vacuum. A subclass of these solutions has closed timelike curves both in the interior and the exterior geometries.

066.178   Dynamics of plane–symmetric thin walls in general relativity.
Wang Anzhong.
Phys. Rev. D, Vol. 45, No. 10, p. 3534 – 3543 (15 May 1992).
Current Physics Microform No.: 9207D1116.
Plane walls (including plane domain walls) without reflection symmetry are studied in the framework of Einstein's general relativity. Using the distribution theory, all the Einstein field equations and Bianchi identities are split into two groups: one holding in the regions outside of the wall and the other holding at the wall. The Einstein field equations at the wall are found to take a very simple form, and are given explicitly in terms of the discontinuities of the metric coefficients and their derivatives. The Bianchi identities at the wall are also given explicitly. Using the latter, the interaction of a plane wall with gravitational waves and some specific matter fields is studied.

066.179   Breaking Weyl invariance in the interior of a bubble.
W. R. Wood, G. Papini.
Phys. Rev. D, Vol. 45, No. 10, p. 3617 – 3627 (15 May 1992).
Current Physics Microform No.: 9207D1199.
The basis on which Weyl's unified theory of gravitation and electromagnetism was rejected is reconsidered from a new perspective. It is argued that while Weyl's theory, as indeed any classical theory, is incapable of explaining atomic phenomena, this does not nullify the geometric interpretation of the exterior electromagnetic field; it simply reflects the fact that some form of quantization is needed to account for atomic standards of length. In support of this argument the Gauss – Mainardi – Codazzi formalism is employed to demonstrate that it is possible to construct a bubble in Weyl space where the exterior geometry is conformally invariant and the electromagnetic field can be given a geometric interpretation, while at the same time a standard of length can be introduced into the theory by breaking the conformal invariance in the interior of the bubble.

066.180   General analytic solution of R^2 gravity with dynamical torsion in two dimensions.
W. Kummer, D. J. Schwarz.
Phys. Rev. D, Vol. 45, No. 10, p. 3628 – 3635 (15 May 1992).
Current Physics Microform No.: 9207D1210.
Using light–cone variables, the authors show that R^2 gravity with dynamical torsion in two dimensions is one of the rare field theories whose complete classical solution in closed form can be obtained. It fulfils an invariant relation between the cosmological constant, the curvature scalar, and the scalar formed by the torsion tensor. The authors conjecture that this relation, interpreted as a local conservation law, is closely connected to the integrability of the theory. The solutions may possess a rich spectrum of singularities in curvature and torsion. Special cases, including one with nonvanishing torsion, can be used to elucidate some physical properties of the solution, where by "physical" the authors imply the validity of concepts from general relativity such as measurements of distances and times and of extremal trajectories of a scalar test particle.

066.181   A diagram illustrating the resolution of the twins paradox.
T. Kiang.
Ir. Astron. J., Vol. 20, No. 3, p. 201 – 206 (Mar 1992). – See Abstr. 012.038 for the main entry.
The so–called "twins paradox" in special relativity arises out of a confusion between clock rates and clock readings. However, to resolve the paradox properly, one has to recognize not only (1) that the travelling twin uses two coordinate systems, (x', t') on the way out and (x", t") on the way back, but also (2) that the Lorentz transformation between (x", t") and the Earth–bound twin's system (x, t) is not of the usual form given in texts but contains constants pertaining to the coordinates at the turn–around. A diagram is given showing how differential ageing obtains while the relative clock rates are symmetrical throughout.

066.182 **Linear and nonlinear gravidynamics: static field of a collapsar.**
V. V. Sokolov.
Astrophys. Space Sci., Vol. 191, No. 2, p. 231 – 258 (May 1992).

In the bounds of the consistent dynamic interpretation of gravitation (gravidynamics) a gravitational field has been divided into two components: scalar and tensor, each one interacting with its source by the same coupling constant. Consequently, a spherically–symmetrical gravitational field in vacuum generated by a massive object influences test bodies as an algebraic sum of attraction and repulsion. Field energy in vacuum around the source is also a sum of energies of two components – purely tensor and scalar components of gravitation. At distances from a gravitating object much greater than its gravitational radius, energies of each separate field component are equal to each other at the same point of space. In the bounds of gravidynamics based on the so–called Einstein's "linearized" equation and proceeding from general principles of theory of classical fields a statement (a theorem) has been formulated on the static gravitational field of a collapsar: a spherically–symmetric object generating a static field in vacuum may always only occupy a finite, nonzero volume.

066.183 **Qualitative analysis of a generalized Maxwell–Einstein system. Application for a cosmological model.**
J. Tossa, J. C. Fabris, C. Romero.
C. R. Acad. Sci., Sér. II, Tome 314, No. 4, p. 339 – 343 (13 Feb 1992).

The coupling of gravitation to a Maxwellian–type field, in even n dimensions, leads to Einstein equations, coupled to two scalar fields after reduction to four dimensions. The qualitative features of this system are analyzed in this note by a dynamical system method.

066.184 **The theory of relativity and super–luminal speeds. III. The catastrophe of the space–time on the Finsler metric.**
Cao Shenglin.
Astrophys. Space Sci., Vol. 190, No. 2, p. 303 – 315 (Apr 1992).

According to the different properties between the ds^2 and the ds^4, it is discussed that the space–time will have the catastrophic nature on the Finsler metric ds^4 (see Cao, 1990). The space–time transformations and the physical quantities will suddenly change at the catastrophic theory of the space–time. It is supposed that only the dual velocity of the super–luminal–speed could be observed (see Cao, 1988). If so, a particle with the super–luminal–speed $v > c$ could be regarded as its anti–particle with the dual velocity $v_1 = c^2/v < c$.

066.185 **Theoretical problems in nonsymmetric gravitational theory.**
T. Damour, S. Deser, J. McCarthy.
Phys. Rev. D, Vol. 45, No. 10, p. R3289 – R3291 (15 May 1992).
Current Physics Microform No.: 9207D0871.

It has recently been noted that the nonsymmetric metric model of gravity faces severe observational constraints. The authors show here that it is also subject to physically unacceptable formal difficulties even as an effective field theory: When expanded about a Riemannian background, the model exhibits curvature–coupled negative–energy (ghost) modes and unacceptable asymptotic behavior.

066.186 **Wormhole spectrum of a quantum Friedmann – Robertson – Walker cosmology minimally coupled to a power–law scalar field and the cosmological constant.**
S. P. Kim, D. N. Page.
Phys. Rev. D, Vol. 45, No. 10, p. R3296 – R3300 (15 May 1992).
Current Physics Microform No.: 9207D0878.

The expansion of the wave function of a quantum Friedmann – Robertson – Walker cosmology minimally coupled to a scalar field with a power–law potential by its scalar–field part decouples the gravitational–field part into an infinite system of linear homogeneous differential equations (equivalent to a matrix equation). The solutions for the gravitational–field part are found in the product integral formulation. It is shown that there exists a spectrum of the wave functions exponentially damped for large three–geometries under the condition that the cosmological constant should vanish. These are interpeted as the Hawking– Page wormholes.

066.187 **Relativity and space–time measurements.**
T. Grabińska.
Astrophys. Space Sci., Vol. 191, No. 1, p. 23 – 42 (May 1992).

A new reconstruction of special relativity (SR) is presented. The question of length and time measurements forms the basic plane of the author's search. The question why Milne's procedure, applied to clocks synchronization by means of light signal, supersedes the Einsteinian procedure by means of moving clock, is elaborated. According to H. E. Ives' hints the consistency of velocity notion in SR is reanalysed. The justification of the light postulate from the point of view of measurement procedures and anti–absolutistic heuristics of SR is discussed. The new correspondence relation between SR and Ives' space–time theory is presented. The fundamental problems related to operationistic interpretation of SR and to research programmes of kinematics are discussed in the new theoretical fashion.

066.188 **On the theory of relativistic reference frames based on optical coordinates.**
A. N. Aleksandrov, V. I. Zhdanov.
Visn. Kiiv. Univ., Fiz.–Mat. Nauki, Astron., Vip. 3, p. 6 – 11 (1992). In Ukrainian.

A relativistic approach to astrometric reference frames is developed on the basis of optical coordinates. The formulae for the metric tensor and other tensor fields as generalized Taylor expansions in powers of optical coordinates are given. The relation of optical and harmonic coordinates for weakly gravitating moving bodies is found in the case when the distance from them to the observer is comparable or exceeds a characteristic length scale of the gravitational field.

066.189 **Degrees of freedom in the two–body problem of general relativity.**
V. I. Zhdanov.
Visn. Kiiv. Univ., Fiz.–Mat. Nauki, Astron., Vip. 3, p. 12 – 17 (1992). In Ukrainian.

The qualitative behaviour of a gravitationally bound two–body system in infinite past is studied taking into account the effect of radiation damping. Under general assumptions the weakly relativistic trajectories of the system are shown to have the classical number of degrees of freedom.

066.190 **Spinor structural equations for Killing tensors in Einstein spaces of general relativity.**
Yu. M. Kudrya.
Visn. Kiiv. Univ., Fiz.–Mat. Nauki, Astron., Vip. 3, p. 21 – 26 (1992). In Ukrainian.

Structural equations and their first series of integrability conditions are obtained for Killing tensors of rank two associated with quadratic first integrals of motion of test bodies in Einstin space–times by the use of spinor formalism of general relativity. The complete sets of integrability conditions are derived in symmetric spaces of Petrov type N and D. The numbers of quadratic integrals are calculated.

066.191 **Cosmological models in the scalar–tetradic theory B.**
P. Chauvet, L. O. Pimentel.
Gen. Relativ. Gravitation, Vol. 24, No. 3, p. 243 – 258 (Mar 1992).

The authors present two methods for solving the cosmological equations of the scalar–tetradic theory B when a Friedmann – Robertson – Walker geometry is assumed. Among the many solutions found, there are several physically meaningful ones including inflationary universe solutions.

066.192 **Biframe bundle geometry and an extension of RMW theory: application to a charged perfect fluid.**
K. S. Hammon.
Gen. Relativ. Gravitation, Vol. 24, No. 3, p. 259 – 279 (Mar 1992).

An extension of the original Rainich – Misner – Wheeler (RMW) theorem to include Einstein–Maxwell spacetimes with geometrical sources has recently been accomplished by generalizing the geometrical arena from the linear frame bundle LM to the bundle of biframes L²M. The assumptions of a Riemannian connection one–form on LM and a general connection one–form on L²M necessarily implies the existence of a difference form K. The author provides new algebraic and differential conditions on an arbitrary triple (M, g, K), in addition to those already imposed by the generalization of the RMW theorem, which guarantee the form of the coupled Einstein–Maxwell field equations associated with a charged perfect fluid spacetime. All physical quantities associated with these field equations, namely the Maxwell field strength, the mass–energy density, the pressure, the electric and magnetic charge to mass ratios, and the unit four velocity of the fluid, can be recovered from the geometry.

066.193 **Black holes and Newtonian physics.**
A. K. Raychaudhuri.
Gen. Relativ. Gravitation, Vol. 24, No. 3, p. 281 – 283 (Mar 1992).

It is argued that one–way passage is inconsistent with Newtonian physics, and thus the dark bodies as thought of by Michell and Laplace cannot be considered as exact analogues of relativistic black holes.

066.194 **Internal structure of a classical spinning electron.**
C. A. López.
Gen. Relativ. Gravitation, Vol. 24, No. 3, p. 285 – 296 (Mar 1992).

A classical model of the spinning electron in general relativity consisting of a rotating charge distribution with Poincaré stresses is set up. It is made up of a continuous superposition of thin charged shells with differential rotation. Each elementary shell is maintained in stationary equilibrium in the gravitational field created by the others. A class of interior solutions of the Kerr–Newman field is thus obtained. The corresponding stress–energy tensor splits into the sum of two terms. The first one is the Maxwell tensor associated to a rotating charge distribution, and the second one corresponds to a material source having zero energy density everywhere, no radial pressure, and an isotropic transverse stress. These negative pressures or tensions are identified with the cohesive forces introduced by Poincaré to stabilize the Lorentz electron model. They are shown to be the source of a negative gravitational mass density and thereby of the violation of the energy conditions inside the electron.

066.195 **Limitations of the geometrodynamic clock.**
D. E. Brahm, R. P. Gruber.
Gen. Relativ. Gravitation, Vol. 24, No. 3, p. 297 – 303 (Mar 1992).

The authors examine sources of error in the geometrodynamic (or "Marzke – Wheeler") clock, and choose parameters to minimize the total error. In theories with time–varying masses, there is an unavoidable minimum error. For a human–scale clock, the dominant error is from quantum uncertainty in the photon location.

066.196 **Traversible wormholes in (2 + 1) dimensions.**
G. P. Perry, R. B. Mann.
Gen. Relativ. Gravitation, Vol. 24, No. 3, p. 305 – 321 (Mar 1992).

The authors investigate traversible wormhole solutions to the Einstein field equations in (2 + 1) dimensions. The constraints on the field equations to obtain a wormhole solution are presented and further constraints for traversibility of the wormhole are also given. The authors show that there is no analog of the (3 + 1)–dimensional Schwarzschild wormhole in (2 + 1) dimensions. For general wormholes, the radial tension and lateral pressure at the throat of the wormhole must be zero, and the energy density must be negative. Two specific wormhole solutions are presented, and a stability analysis of these solutions is given.

066.197 **Gravitational wave background from a sample of cataclysmic variables.**
F. Barone, L. Di Fiore, L. Milano, G. Russo.
Gen. Relativ. Gravitation, Vol. 24, No. 3, p. 323 – 341 (Mar 1992).

The authors analyze cataclysmic variables (CVs) as sources of gravitational waves (GW), basing their analysis only on known objects (168 CVs taken from the recent Ritter catalog). A similar analysis has already been performed for eccentric binaries and pulsars. The authors try to evaluate the gravitational wave background, outlining all the potentially interesting sources, in two different ways, showing the substantial agreement of the results obtained. Although not completely new, such results are based on real samples of data, plus some statistical analysis, and therefore constitute a solid basis for planning the construction of GW detectors (especially space–borne GW antennas). Moreover, they provide the possibility of experimentally proving the effectiveness of the mechanism of gravitational radiation on CV evolution. Furthermore, for the sake of completeness, the GW emission from known low–mass X–ray binaries and CV related objects has been evaluated.

066.198 **The phase of scalar field driven wormholes at one loop in the path integral formulation for Euclidean quantum gravity.**
A. Carlini, M. Martellini.
Classical Quantum Gravity, Vol. 9, No. 3, p. 629 – 640 (Mar 1992).

The authors calculate the one–loop approximation to the Euclidean quantum gravity coupled to a scalar field around the classical Carlini and Mijić wormhole solutions. The main result is that the Euclidean partition functional Z_{EQG} in the "little wormhole" limit is real. Extension of the CM solutions with the inclusion of a bare cosmological constant to the case of a sphere S^4 can lead to the elimination of the destabilizing effects of the scalar modes of gravity against those of the matter.

066.199 **The Poincaré limit 2 + 1 dimensional quantum de Sitter gravity.**
L. F. Urrutia, F. Zertuche.
Classical Quantum Gravity, Vol. 9, No. 3, p. 641 – 650 (Mar 1992).

The quantum traces algebra for the 2 + 1 Poincaré gravity in first oder formalism is explicitly constructed by contracting the corresponding traces algebra of the de Sitter gravity. Unbounded representations of the latter, in the case $\Lambda < 0$, are constructed in terms of an underlying SU(1, 1) algebra. Unfortunately, these representations do not possess a well defined Poincaré limit. Nevertheless an explicit realization of the Poincaré traces algebra is constructed in terms of two pairs of canonical variables.

066.200 **Multigrid in general relativity: II. Kerr spacetime.**
A. Lanza.
Classical Quantum Gravity, Vol. 9, No. 3, p. 677 – 696 (Mar 1992).

This is the second test problem of a series, aiming at the treatment of self–gravitating tori rotating around rapidly rotating black holes. The multigrid method is being applied to solve numerically the stationary and axisymmetric vacuum Einstein equations. Numerical results are presented for the model problem where the equations are restricted to the Kerr metric.

066.201 **A generalization of the Kerr–Schild ansatz.**
S. Bonanos.
Classical Quantum Gravity, Vol. 9, No. 3, p. 697 – 711 (Mar 1992).

The author introduces a class of spacetimes in which the metric tensor can be expressed covariantly in terms of the Minkowski metric and two vector fields, orthogonal to each other. When certain covariantly defined components of the Einstein tensor vanish ("main equations"), the remaining components satisfy a true conservation law. The author also obtains the conditions under which a solution of the linearized main equations is a solution of the full main equations. Several simple solutions to these equations – among them the Schwarzschild solution – are obtained.

066.202   A direct comparison of two codes in numerical relativity.
M. W. Choptuik, D. S. Goldwirth, T. Piran.
Classical Quantum Gravity, Vol. 9, No. 3, p. 721 – 750 (Mar 1992).

The authors discuss a detailed numerical comparison of the results of two codes which treat the same model problem in numerical relativity. The model consists of a single, massless scalar field minimally coupled to the gravitational field with a further restriction to spherical symmetry. The comparison was complicated by the fact that the codes were based on different formalisms, used different coordinate systems and employed different numerical solution techniques. The authors also describe some new algorithms which use Richardson extrapolation to significantly increase the accuracy of one of the codes at a given resolution. It is shown that both codes are convergent even in the regime where the field interactions are significantly non-linear and highly time-dependent.

066.203   Relativistic effects for Doppler measurements near solar conjunction.
B. Bertotti, G. Giampieri.
Classical Quantum Gravity, Vol. 9, No. 3, p. 777 – 793 (Mar 1992).

The authors consider the relativistic corrections to the Doppler effect in a metric theory of gravity, in the weak field and slow motion approximation, for a source and a receiver at great distances from, but near alignment with, the perturbing body (i.e. near conjunction). The formalism is applied to an ideal experiment, related to that of the deflection of light rays by the solar mass. The authors then introduce a differential Doppler technique, which allows some new possibilities in the field of experimental gravitation. With a spacecraft equipped with a hydrogen maser, one can measure the difference between the fractional frequency shift in both directions. Near conjunction, it is possible to obtain information on the angular momentum of the Sun or on a possible effect of a privileged cosmological frame of reference. This technique also allows a drastic reduction of the effect due to the plasma of the solar corona. The maximum fractional frequency change induced by the angular momentum of the Sun is about 7×10^{-16}, which is barely consistent with the stability of hydrogen masers currently available.

066.204   About the non-existence of perfect fluid bodies with the Kerr metric outside.
T. Wolf, G. Neugebauer.
Classical Quantum Gravity, Vol. 9, No. 3, p. L37 – L42 (Mar 1992). Letter-to-the-editor.

On the basis of conditions of Boyer for fluid surfaces in the Kerr spacetime, a one-parameter family of hypothetical perfect fluid bodies with the Kerr metric outside is investigated concerning properties of their shape. A comparison with the Newtonian series of MacLaurin ellipsoids of the same surface potential, mass, angular velocity and angular momentum shows clear qualitative differences. This goes together with the absence of a phase transition of the Lagrangian which describes the Kerr metric in contrast to a phase transition of the Lagrangian describing classical MacLaurin ellipsoids. In the classical case this transition represents the situation when angular velocity is starting to decrease for increasing angular momentum, because of flattening.

066.205   Numerical treatment of the spherically symmetric general relativistic Boltzmann equation for massless and massive particles.
H. Harleston, E. T. Vishniac.
Phys. Rev. D, Vol. 45, No. 12, p. 4458 – 4472 (15 Jun 1992). Current Physics Microform No.: 9208C1986.

The Arnowitt – Deser – Misner formalism is used to write the Einstein – Boltzmann coupled system of equations. The sources of gravitational field are represented by ordinary matter described by a perfect fluid approximation together with a particle gas described by a phase-space distribution function obeying the general relativistic Boltzmann transport equation. Through the use of the Liouville operator in phase space, the authors obtain a form of the Boltzmann equation that makes it very amenable for numerical treatment. The resulting system of equations can be used for the numerical study of either massless or massive particles interacting with ordinary matter.

066.206   Exakte Lösungen der Einsteinschen Feldgleichungen für rotierende Massenschalen. *(Exact solutions for the Einstein field equations for rotating mass shells.)*
W. Konrad.
Diss. (Dr.rer.nat.), Tübingen Univ. (Germany). Fakultät für Physik. 149 p. (26 Jun 1990).

066.207   Does general relativity allow an observer to view an eternity in a finite time?
M. L. Hogarth.
Found. Phys. Lett., Vol. 5, No. 2, p. 173 – 181 (Apr 1992).

The author investigates whether there are general relativistic spacetimes that allow an observer μ to collect in a finite time all the data from the worldline of another observer λ, where the proper length of λ's worldline is infinite. The existence of these spacetimes has a bearing on certain problems in computation theory. A theorem shows that most standard spacetimes cannot accommodate this scenario. There are however spacetimes which can: anti – de Sitter spacetimes is one example.

066.208   A minimal time and time-temperature uncertainty principle.
V. De Sabbata, C. Sivaram.
Found. Phys. Lett., Vol. 5, No. 2, p. 183 – 189 (Apr 1992).

The authors show that introducing torsion in general relativity, that is, physically, considering the effect of the spin and linking the torsion to defects in spacetimes topology, one can have a minimal unit of time. Also an uncertainty relation between time and temperature is suggested. The interesting thing is that with this minimal time one can eliminate the divergence of the self-energy integral without introducing any ad hoc cut-off, and it is also possible to understand black-hole evaporation as a process of quantum diffusion which leads directly to the Hawking formula. A minimal operationally definable temperature in a cosmological context is discussed.

066.209   On the interaction of bubbles with gravitational and matter fields.
Wang Anzhong.
Mod. Phys. Lett. A, Vol. 7, No. 20, p. 1779 – 1789 (28 Jun 1992).

The space-time containing a single spherically symmetric bubble is studied by using the distribution theory. The Einstein equations on the bubble wall are given explicitly in terms of the discontinuities of metric coefficients and their derivatives. The interaction of a bubble with surrounding gravitational and matter fields is also investigated by the "generalized" Bianchi identities. In particular, it is found that an electromagnetic field does not interact with any bubble, and is continuous across the bubble wall without reflecting or absorbing, while the interaction of a bubble produced with a scalar field or a perfect fluid is possible.

066.210   Planetary orbits as limit cycles.
J. Díaz Bejarano, C. Miró Rodríguez.
Nuovo Cimento B, Vol. 107, No. 5, p. 497 – 501 (May 1992).

General relativistic orbits in the Schwarzschild metric can be analysed in terms of the anharmonic asymmetric oscillator. A simple generalization of the usual Fourier series, suitable for this type of nonlinear system, is used with a harmonic-balance method to study approximate solutions of perturbed orbits. The existence of solutions of the limit cycle type is demonstrated quantitatively.

066.211 **Wormhole in the Einstein theory.**
Shen Yougen, Tan Zhenqiang.
Nuovo Cimento B, Vol. 107, No. 6, p. 653 – 656 (Jun 1992).
The authors discuss the wormhole model with the axion charge in the Einstein theory with the cosmological constant, and deduce the corresponding wormhole equation. By solving this equation, they get a wormhole which connects the corresponding points in two Euclidean – de Sitter space–times.

066.212 **The problem of expanding shell in relativistic gravitational theory.**
A. A. Vlasov, O. V. Monovskij.
Teor. Mat. Fiz., Vol. 91, No. 2, p. 334 – 345 (May 1992). In Russian.
The expanding shell gravitational field continuously–differentially sewed on the boundary is found in the linear approach. A second order analysis allows to establish the convergence of the decomposition used over the gravitational constant.

066.213 **On spherical gravitation waves in relativistic theory of gravity.**
A. V. Genk.
Teor. Mat. Fiz., Vol. 91, No. 2, p. 346 – 352 (May 1992). In Russian.
The dependence of the energy transferred by spherical gravitation waves (radiated by a central–symmetrical body during a transition between two static states) on the distance to the source is found with the help of conservation laws. The point of observation of these waves is discussed.

066.214 **The problem of clock synchronization: a relativistic approach.**
S. A. Klioner.
Celest. Mech. Dyn. Astron., Vol. 53, No. 1, p. 81 – 109 (1992).
The problem of synchronization of the Earth–based clocks has been discussed in the framework of General Relativity Theory. The synchronization is considered as the transformation of the observers' proper time scales to the coordinate time scale of local inertial geocentric reference system, which is single for all the observers. The formulas for the relativistic corrections occurring in some methods of Earth–based clock synchronization (transported clock, duplex communication via geostationary satellite and meteor–burst link, LASSO experiments) have been derived enabling one to attain the accuracy of 0.1 ns.

066.215 **Relativity of inwards and outwards: an example.**
M. A. Abramowicz.
Mon. Not. R. Astron. Soc., Vol. 256, No. 4, p. 710 – 718 (15 Jun 1992).
The meaning of the inward and outward directions in a curved space is not absolute and this may create paradoxes and confusions. The author discusses an interesting example of such a situation which occurs around a compact star.

066.216 **Completion of the ten–dimensional anomaly free supergravity programme: the field equations.**
I. Pesando.
Classical Quantum Gravity, Vol. 9, No. 4, p. 823 – 866 (Apr 1992).
The author gives details of the derivations of the equations of motion of the ten–dimensional anomaly free supergravity and prove that torsionless Ricci–flat manifolds with the spin connection embedded in the gauge group are solutions of the equations of motion.

066.217 **Point particles as defects in (2 + 1)–dimensional quantum gravity.**
Y. Fujiwara, S. Higuchi, A. Hosoya, T. Mishima, M. Siino.
Classical Quantum Gravity, Vol. 9, No. 4, p. 867 – 872 (Apr 1992).
Recently the authors have shown explicit examples of topology change of the universe by quantum tunnelling in the (2 + 1)–dimensional quantum gravity. In the present work it is shown that mass points can be created in pairs as spacetime defects in the course of the topology changing processes.

066.218 **General relativity as the low–energy limit in higher derivative quantum gravity.**
S. D. Odintsov, I. L. Shapiro.
Classical Quantum Gravity, Vol. 9, No. 4, p. 873 – 882 (Apr 1992).
The curvature induced phase transition in quantum R^2–gravity with matter is investigated. The renormalization group approach is used to calculate the effective potential. The universal expressions for induced gravitational and cosmological constants are obtained. The effective potential and the induced cosmological and gravitational constants depend on the coupling constants of the original theory and on the gauge parameters. When matter is described by a scalar field, coupling constants subject to asymptotic freedom are chosen. Then a curvature induced phase transition is possible. If the Vilkovisky – DeWitt effective action formalism is used, the induced gravitational and cosmological constants do not depend on gauge parameters.

066.219 **The $G_{Newton} \rightarrow 0$ limit of Euclidean quantum gravity.**
L. Smolin.
Classical Quantum Gravity, Vol. 9, No. 4, p. 883 – 893 (Apr 1992).
Using the Ashtekar formulation, it is shown that the $G_{Newton} \rightarrow 0$ limit of Euclidean or complexified general relativity is not a free field theory, but is a theory that describes a linearized self–dual connection propagating on an arbitrary anti–self–dual background. This theory is quantized in the loop representation and, as in the full theory, an infinite dimensional space of exact solutions to the constraints are found. An inner product is also proposed. The path integral is constructed from the Hamiltonian theory and the measure is explicitly computed non–perturbatively, without relying on a semiclassical expansion.

066.220 **Effective action in quantum gravity.**
G. A. Vilkovisky.
Classical Quantum Gravity, Vol. 9, No. 4, p. 894 – 903 (Apr 1992).
It is argued that the effective action theory may be regarded as a phenomenological theory describing a certain class of measurements irrespective of the nature of fundamental quantum objects. The effective action in quantum field theory is discussed in detail and used as a guide. A connection between the effective field and observables is established. The approach is applied to the gravitational collapse problem. A procedure for building a basis of non–local gravitational invariants is described and a result for the vacuum radiation in a spherically symmetric in–state is presented.

066.221 **Averaging Einstein's equations.**
N. V. Zotov, W. R. Stoeger.
Classical Quantum Gravity, Vol. 9, No. 4, p. 1023 – 1031 (Apr 1992).
Einstein's equations are averaged, first for a space–like distribution of stars, then for an expanding system of galaxies. The results indicate the possible nature of the extra terms introduced as in Ellis' conjecture, and also show that the rate of expansion will be different from that of a FLRW model with comparable density. The results also show that it is possible to recover the form of the FLRW metric by averaging over inhomogeneities on a scale on which Einstein's equations are known to hold.

066.222 **Infinitely many cosmological constants.**
I. Bengtsson, O. Boström.
Classical Quantum Gravity, Vol. 9, No. 4, p. L47 – L51 (Apr 1992). Letter–to–the–editor.
Using Ashtekar's formulation of general relativity an infinite parameter family of neighbours of Einstein's equations has been discovered. The authors give solutions to the equations to first order in the parameters assumed to be small. This shows that an infinite number of these parameters are physically different.

066.223   Gravitational collapse of massive stars, supernovae, and SN 1987A in the Large Magellanic Cloud.
V. S. Imshennik.
Astrophysics on the threshold of the 21st century, p. 167 – 188 (1992). – See Abstr. 003.048 for the main entry.
   The mechanism of a supernova explosion has a special significance in the theory of collapse which has not been solved up to now in both versions of the theory. The author presents an analysis of all existing attempts to solve this problem and indicates possible paths of their solution. He points out the unarguable indications in the SN 1987A burst of the presence of an effective explosion mechanism just in the center of the star, the progenitor Sk–69°202.

066.224   Flat FRW models with variable G and Λ.
D. Kalligas, P. Wesson, C. W. F. Everitt.
Gen. Relativ. Gravitation, Vol. 24, No. 4, p. 351 – 357 (Apr 1992).
   The authors consider Einstein's equations with variable gravitational coupling G and cosmological term Λ. For a power–law time dependence of G, the cosmological term varies in proportion to the inverse square of the time, provided the equation of state is not that of vacuum. There is then no dimensional constant associated with Λ. For a vacuum equation of state, the model is compatible with classical inflation for a wide class of functions $G(t)$ and $\Lambda(t)$. For non–power–law behaviour of $G(t)$, it is possible to have a scale factor that increases exponentially without a vacuum equation of state. For this case the energy density associated with Λ decreases exponentially, which at time zero it is equal with opposite sign to the regular energy density, so there is zero total energy initially.

066.225   Is there evidence for torsion?
Zhang Chengmin, Yang Guochen, Chen Fangpei, Wu Xinji.
Gen. Relativ. Gravitation, Vol. 24, No. 4, p. 359 – 371 (Apr 1992).
   Under the assumption that a pulsar's magnetic field originates from the net polarized spin of its neutrons, and by using the post–Newtonian approximated torsion in the fith order in (v/c), the spin precession arising from torsion is coordinate dependent, which influences the magnetic field of the pulsar and makes the magnetic inclination close to the rotation axis. Assuming the possibility of the magnetic inclination density to be in random alignment at the initial time, the calculations presented here show that most pulsars should have smaller inclinations at the age of $10^6 - 10^7$yr, and that the inclination decreases with increasing age of the pulsar. This is consistent with the astronomically observed distribution.

066.226   The early Weyl universe.
T. Bradfield.
Gen. Relativ. Gravitation, Vol. 24, No. 4, p. 373 – 387 (Apr 1992).
   The consequences of a period of Weyl invariance in the early universe are investigated. It is argued that the natural outcome of such a period is a Kaluza–Klein style compactification of an internal space in which any time variation of the scale factor of this space is absorbed (via a Weyl transformation) into the gravitational coupling. A five–dimensional test model is shown to undergo exponential inflation of the space–time sector due to a false vacuum state of the non–metric part of the connection.

066.227   Vacuum cosmological solution in a 6D universe.
T. Fukui.
Gen. Relativ. Gravitation, Vol. 24, No. 4, p. 389 – 395 (Apr 1992).
   A simple vacuum cosmological solution that is a function of ct, Gm/c^2 and $eG^{1/2}/c^2$ is obtained in the 6D space–time–mass–charge universe which is proposed by Wesson with the introduction of the sixth coordinate of charge in order to obtain a unified theory of gravity and electromagnetism along the line of his original 5D space–time–mass universe. It reduces to a solution similar to that of the radiation era in the 4D FRW universe

through the compactification of the extra dimensions. The trajectory of a "test particle" in the 6D universe is also studied by using the solution.

066.228   Some exact solutions of string cosmology in Bianchi III space–time.
R. Tikekar, L. K. Patel.
Gen. Relativ. Gravitation, Vol. 24, No. 4, p. 397 – 404 (Apr 1992).
   Following the techniques used by Letelier and Stachel, some exact Bianchi III cosmological solutions of massive strings in the presence of magnetic field are obtained and their physical features are discussed. Some string solutions in which magnetic fields are absent are also discussed.

066.229   Some canonical forms for the metric of spacetimes admitting a rational first integral of the geodesic equation.
E. G. L. R. Vaz, C. D. Collinson.
Gen. Relativ. Gravitation, Vol. 24, No. 4, p. 405 – 418 (Apr 1992).
   Canonical forms are obtained for the metrics of space–times admitting a surface generating Killing pair, one member of which is hypersurface orthogonal.

066.230   Post–Newtonian limit of Finsler space theories of gravity and solar system tests.
I. W. Roxburgh.
Gen. Relativ. Gravitation, Vol. 24, No. 4, p. 419 – 431 (Apr 1992).
   Finsler geometry is considered as a wider framework for analysing solar system tests of theories of gravity than is afforded by Riemannian geometry. The post–Newtonian limit for the spherically symmetric one–body problem is examined by expanding the Finsler metric about the Minkowski space of special relativity for those Finsler spaces whose null surface is Riemannian. In such a framework there are five PPN parameters instead of the three in Riemannian geometry. The classical solar system tests can readily be satisfied, leaving two arbitrary parameters. These parameters could be determined from measurements of the second order gravitational red–shift and periodic perturbations in particle orbits, thus providing a consistency check on the Riemannian metric hypothesis of general relativity. Such an experiment is possible on a satellite on an orbit with perihelion of a few solar radii.

066.231   A class of empty spacetimes admitting a rational first integral of the geodesic equation.
C. D. Collinson, P. J. O'Donnell.
Gen. Relativ. Gravitation, Vol. 24, No. 4, p. 451 – 455 (Apr 1992). With a correction in Vol. 24, No. 6, p. 691 (Jun 1992).
   The empty space field equations are solved for one of the canonical forms obtained previously by Vaz and Collinson, for the metrics of space–times admitting a surface generating Killing pair, one member of which is hypersurface orthogonal.

066.232   Time–geostationary orbits in the solar system.
M. Hosokawa, F. Takahashi.
Publ. Astron. Soc. Jpn., Vol. 44, No. 2, p. 159 – 162 (1992).
   According to general relativity, each object has its own proper time. The authors show here that in the solar system there are three free–fall circular orbits that possess the same proper time as that on the geoid surface. These are the orbits around the Earth, the Sun and Jupiter. The authors call these "time–geostationary orbits". In the future space age these orbits will play important roles in establishing interplanetary time standards and in allowing highly accurate astronomical measurements.

066.233   Gravitational waves and causality.
P. E. Ehrlich, G. G. Emch.
Rev. Math. Phys., Vol. 4, No. 2, p. 163 – 221 (Jun 1992).
   A strictly ordered hierarchy of eight causal properties encountered in general relativity is reviewed for the explicit case of the

gravitational plane waves. Illustrative proofs are given to the effect that the place of these space–times is precisely known in the hierarchy: they are causally continuous, but not causally simple. The other conditions of the hierarchy are also discussed separately, as are some causality conditions that belong outside the hierarchy.

066.234 Connections, loops and quantum general relativity.
A. Ashtekar, C. Rovelli.
Classical Quantum Gravity, Vol. 9, Suppl., p. S3 – S12 (1992). – See Abstr. 012.058 for the main entry.
The current status of a programme for non–perturbative canonical quantisation of general relativity is briefly summarized.

066.235 Fluctuations in the matter–gravity system.
G. Venturi.
Classical Quantum Gravity, Vol. 9, No. 5, p. 1217 – 1230 (May 1992).
The matter–gravity system is examined in a path integral approach for the case of matter (scalar fields) coupled to a FRW spacetime. The gravitational wavefunction including the back-reaction of matter, represented by the average of the matter Hamiltonian, is first constructed and subsequently the full wavefunction for the matter–gravity system is derived. The semiclassical limit for gravitation, the way time arises and the equation satisfied by matter are examined using the stationary phase approximation. It is shown how the presence of a large number of matter fields can lead to the "decoherence" of the gravitational wavefunction.

066.236 Gravitational effect of the quantum vacuum outside a cosmic string.
Ø. Grøn, H. H. Soleng.
Classical Quantum Gravity, Vol. 9, No. 5, p. 1231 – 1238 (May 1992).
The conical geometry of a static symmetric cosmic string gives a non–vanishing contribution to the vacuum expectation value of the energy–momentum tensor of a quantum field. The authors construct a model of a cosmic string taking into account the gravitational field of this energy density. The strongly curved interior of the string is modeled by a homogeneous fluid with vanishing tangential and radial stress. The total Tolman mass of the system is positive, so the string produces an attractive gravitational field.

066.237 A differential form approach for rotating perfect fluids in general relativity.
F. J. Chinea, L. M. González–Romero.
Classical Quantum Gravity, Vol. 9, No. 5, p. 1271 – 1302 (May 1992).
A compact formulation for general relativistic, axisymmetric perfect fluids in stationary rotation is introduced; the case of differential rotation is included. The basic variables are differential 1–forms with immediate physical relevance. As an illustration of the method, the authors present the derivation of an irrotational solution found recently. With the matching problem in mind, stationary, axially symmetric vacuum fields are considered as formal perfect fluids; it is shown that they can always be brought to a "rigid rotation" or to an "irrotational" form. Two discrete transformations that may be useful for solution-generating purposes are introduced. A formulation of the Ernst type is given for the "irrotational" vacuum case.

066.238 Causality in (2+1)–dimensional gravity.
G. t'Hooft.
Classical Quantum Gravity, Vol. 9, No. 5, p. 1335 – 1348 (May 1992).
A method is presented to characterize fully the evolution of an arbitrary set of spinless particles in (unquantized) (2+1)–dimensional gravity theory. The method produces a complete series of time ordered Cauchy surfaces, which are being triangulated. By construction, closed timelike curves never arise, even if the initial conditions contain a Gott pair. This construction shows that the configuration proposed by Carrol et al, in

which a Gott pair is formed in a closed universe, nevertheless does not admit closed timelike curves; this universe has a finite lifetime, ending in a "big crunch".

066.239 Spherical collapse of a non–rotating perfect fluid in the non–symmetric theory of gravitation.
P. Savaria.
Classical Quantum Gravity, Vol. 9, No. 5, p. 1349 – 1363 (May 1992).
A set of time–dependent differential equations governing the behaviour of a spherically symmetric non–rotating perfect fluid in the non–symmetric theory of gravitation (NGT), is derived in a form suitable for numerical computations. This is then applied to the pressureless collapse of such a fluid. If the Newtonian potential does not exceed a certain value, and if the NGT potential is large enough, the collapse stops and reverses itself into an expansion before the Schwarzschild horizon is reached. A solution is presented for the case where the attractive tensor component of the NGT force is negligible. Bounds are found on the Newtonian potential and the ratio of NGT to Newtonian potentials within which the bounce phenomenon must occur.

066.240 Scalar–tensor theories of gravity with Φ–dependent masses.
J. A. Casas, J. García–Bellido, M. Quirós.
Classical Quantum Gravity, Vol. 9, No. 5, p. 1371 – 1384 (May 1992).
The authors study new physical phenomena and constraints in generalized scalar–tensor theories of gravity with Φ–dependent masses. They investigate a scenario with two types of Φ–dependent masses which could correspond to visible and dark matter sectors. The parameters of this theory are constrained from post–Newtonian bounds, primordial nucleosynthesis and the age of the universe. The authors present a perfect fluid formalism for the dark matter sector with variable masses and find an entropy increase effect during the matter era and, in principle, a measurable effect on the motion of the halo of spiral galaxies. For the case of string effective theories, the constancy of gauge couplings provides new bounds which are orders of magnitude stronger than the previous ones.

066.241 Effect of a weak plane GW on a light beam.
J. A. Lobo.
Classical Quantum Gravity, Vol. 9, No. 5, p. 1385 – 1394 (May 1992).
The effect of a weak plane gravitational wave on a beam of light is analysed by solving Maxwell's electromagnetic equations in a curved background geometry. The results are applied to the consideration of the function of laser–interferometric GW antennae.

066.242 A new class of stationary solutions to the five–dimensional Kaluza–Klein field equations.
R. Becerril, T. Matos.
Gen. Relativ. Gravitation, Vol. 24, No. 5, p. 465 – 476 (May 1992).
Using the five–dimensional potential formalism, a set of six new stationary axisymmetric solutions to the Kaluza–Klein field equations are constructed from the one dimensional subspace of the potential space. All the solutions have scalar potential and magnetic field, two of them possess magnetic monopoles and the other describes a magnetic dipole.

066.243 Finding isometry groups in theory and practice.
M. E. Araujo, T. Dray, J. E. F. Skea.
Gen. Relativ. Gravitation, Vol. 24, No. 5, p. 477 – 500 (May 1992).
An algorithm is given for determining the isometry group of an arbitrary spacetime (in four dimensions). Numerous examples are given and the partial implementation of this algorithm using the symbolic manipulation package CLASSI is discussed.

066.244   A gravitating light–ball: stationary states and developing collapse.
S. N. Sokolov.
Gen. Relativ. Gravitation, Vol. 24, No. 5, p. 519 – 536 (May 1992).
A ball filled with a light–gas is studied in the frame of GR in the spherically symmetric case. Equations of motion suitable for numerical solution are developed and it is shown that the variables of the gravitational field can be excluded from the equations of motion, thus reducing the gravitational interaction into an equivalent direct interaction of the matter with itself. The stationary states, oscillation modes and instability modes are computed and the processes of decay of unstable states and of collapse are studied.

066.245   Already unified field theory with non–covariant geometrization.
H.–J. Treder.
Gen. Relativ. Gravitation, Vol. 24, No. 5, p. 537 – 541 (May 1992).
In general relativity the non–covariant ansatz $A^i = \delta^i_{\ 4}$ for the vector–potential A_k gives the general solution of the Maxwell equations as four coordinate conditions which are the conditions of integrability of the Einstein equations. In the some sense the ansatz $\Phi = X^4$ is a general solution of the scalar wave equation in a reference system given by one coordinate condition. The author discusses the meaning of the canonical quantization of the fields in such reference systems.

066.246   Physical interpretation of vacuum solutions of Einstein's equations. I. Time–independent solutions.
W. B. Bonnor.
Gen. Relativ. Gravitation, Vol. 24, No. 5, p. 551 – 574 (May 1992).
This article is a review of interpretations which have been given to some well known solutions of the vacuum equations. Special attention is paid to those of Schwarzschild, Curzon and Kerr, and it is argued that the bizarre topologies they have been endowed with are physically unrealistic. Among others discussed here are the two–centres solution of Bach and Weyl, the NUT solution, and solutions for an infinite line–mass, both static and rotating.

066.247   Study of the Robinson–Trautman metrics in the asymptotic future.
S. Frittelli, O. M. Moreschi.
Gen. Relativ. Gravitation, Vol. 24, No. 6, p. 575 – 597 (Jun 1992).
A systematic study of the Robinson–Trautman metrics in the asymptotic future is presented. As a by–product, another technique that can be used for determining the existence of solutions of the Robinson–Trautman equation, is found. All these metrics present an exponential asymptotic limit to the Schwarzschild metric in this regime.

066.248   Shear–free spherically symmetric perfect fluid solutions with conformal symmetry.
P. Havas.
Gen. Relativ. Gravitation, Vol. 24, No. 6, p. 599 – 615 (Jun 1992).
In 1987, Dyer, McVittie and Oattes determined the general relativistic field equations for a shear–free perfect fluid with spherical symmetry and a conformal Killing vector in the t – r plane, which depend on an arbitrary constant m. Two particular solutions of these equations were given recently by Maharaj, Leach and Maartens, as well as a partial solution thought to be valid for almost all m. In this paper, this solution is completed for four values of m, and it is shown that it cannot be completed for any others by currently available techniques. However, a new solution of a different form, but also depending on a Weierstrass elliptic function, is found for a further value of m.

066.249   General co–moving frames in stationary and axisymmetric metrics.
A. D. Rogava.
Gen. Relativ. Gravitation, Vol. 24, No. 6, p. 617 – 624 (Jun 1992).
The author derives analytical expressions for all non–zero components of the tetrads of general co–moving frames.

066.250   The generalized radiation gauge in curved space–time.
J. Tolksdorf.
Gen. Relativ. Gravitation, Vol. 24, No. 6, p. 625 – 639 (Jun 1992).
The author gives a necessary and sufficient criterion for the existence of the U(1) radiation gauge in a curved space–time with isometry. This criterion is purely geometric and leads to a (local) 3 + 1–split of spacetime with vanishing extrinsic curvature. If the symmetry is timelike and in absence of charges, the generalization of the Coulomb gauge leads to a time evolution of the Maxwell field which is analogous to that in flat space–time.

066.251   The harmonic–map structure of the axially symmetric stationary Einstein equations.
A. P. Whitman, W. R. Stoeger.
Gen. Relativ. Gravitation, Vol. 24, No. 6, p. 641 – 658 (Jun 1992).
The authors systematically review the solutions of the vacuum Einstein equations for the axially symmetric stationary case which are harmonic maps. In particular, it is shown that the interesting part of the Kerr solution is a composition of a harmonic map into $H_1^{\ 2}$ with a totally geodesic map from $H_1^{\ 2}$ into SS(1, 1). The authors also point out, relying on Sanchez' results, that there is an analogous structure for the Lorentz–domain cases involving cylindrical gravitational waves and colliding plane waves.

066.252   Generalised splitting of spacetime.
D. J. McManus.
Gen. Relativ. Gravitation, Vol. 24, No. 6, p. 659 – 677 (Jun 1992).
Normally, when a spacetime splitting is considered, the ADM 3 + 1 split is brought to mind. In this paper, the idea of spacetime splitting is extended to include an m + n splitting of spacetime. The global spacetime has dimension (m + n) and the foliating spaces have dimension m. There are n independent normals to each of these foliating spaces, thus giving n different extrinsic curvatures. The generalized Gauss–Weingarten and the generalised Gauss–Codazzi equations associated with this splitting are derived. These generalised equations reduce to the familar ADM equations when a 3 + 1 split is considered. The generalised equations are found to have a particularly elegant form when an orthogonal splitting of spacetime is examined.

066.253   Bianchi type V perfect fluid cosmologies.
K. Fišer, K. Rosquist, C. Uggla.
Gen. Relativ. Gravitation, Vol. 24, No. 6, p. 679 – 686 (Jun 1992).
Bianchi type V solutions of the Einstein equations are studied using the Hamiltonian approach. Explicit expressions depending on a single quadrature are given for the metric components in the general orthogonal perfect fluid case. It is shown that the quadrature can be evaluated in terms of elementary or elliptic integrals when the parameter γ in the equation of state $p = (\gamma{-}1)\varrho$ takes the values 1, 10/9, 4/3, 14/9, 5/3, and 2.

066.254   Some exact Bianchi solutions.
R. Maartens, M. F. Wolfaardt.
Classical Quantum Gravity, Vol. 9, No. 6, p. 1525 – 1533 (Jun 1992).
A technique for constructing first integrals is used to regain two orthogonal Bianchi perfect fluid solutions in a unified way and to find an apparently new Bianchi type V tilted solution with stiff fluid. The solution includes a big bang model which is asymptotically tilted.

066.255 **Gravitational lensing.**
R. Narayan, R. Blandford.
15. Texas Symposium on Relativistic Astrophysics and 4. ESO–CERN Symposium, p. 117 – 130 (1991). – See Abstr. 012.060 for the main entry.
Contents: 1. Introduction.   2. Observed lens candidates. 3. Lensing optics. 4. Application in cosmography. 5. Nature and distribution of dark matter. 6. Galaxies/galactic nuclei at high redshift. 7. Concluding remarks.

066.256 **Change of signature in classical relativity.**
G. F. R. Ellis, A. Sumeruk, D. Coule, C. Hellaby.
Classical Quantum Gravity, Vol. 9, No. 6, p. 1535 – 1554 (Jun 1992).
The authors point out that the classical Einstein field equations, suitably interpreted, allow a change of signature of spacetime. Specific examples of such changes are constructed in the case of Robertson–Walker geometries. The authors obtain classical solutions that have properties similar to those obtained in quantum cosmologies obeying the Hartle–Hawking "no boundary" condition: these singularity–free universes have no beginning, but they do have an origin of time. They can be regarded either as classical analogues of the quantum cosmology results, or as classical solutions where a quantum cosmology era is avoided.

066.257 **The derivation of Einstein equations from invariance principles.**
D. R. Grigore.
Classical Quantum Gravity, Vol. 9, No. 6, p. 1555 – 1571 (Jun 1992).
In a first–order Lagrangian theory of gravitation, the author proves rigorously that general coordinate transformations are Noetherian symmetries, if and only if the Lagrangian is a linear combination of a cosmological term and of a Lagrangian giving Einstein equations. Only the pure gravity case is considered.

066.258 **Quantum fluctuations of a spherical gravitational shock wave.**
M. Hortaçsu.
Classical Quantum Gravity, Vol. 9, No. 6, p. 1619 – 1629 (Jun 1992).
The author shows that quantum fluctuations, in particular vacuum polarization, vanish in the background of the spherical shock wave solution of the Einstein field equations, recently found by Nutku. This result is due to the smoothness properties of the new solution.

066.259 **Exact solution of the Einstein–Maxwell equations referring to a charged spinning mass.**
T. E. Denisova, V. S. Manko.
Classical Quantum Gravity, Vol. 9, No. 6, p. L57 – L60 (Jun 1992). Letter–to–the–editor.
The full metric representing a charged generalization of the Gutsunaev–Manko stationary vacuum solution is given in explicit form.

066.260 **Strings and gravity.**
G. Veneziano.
15. Texas Symposium on Relativistic Astrophysics and 4. ESO–CERN Symposium, p. 180 – 189 (1991). – See Abstr. 012.060 for the main entry.
The subject of this paper is the highly nontrivial relationship between strings and gravity. Such a relation is not just interesting, it is the very *raison d'être* of string theory as a (candidate) consistent theory of classical and quantum gravity. Two sides to this subject are discussed in this paper: (1) a classical side subtitled "Strings in gravity", and (2) a quantum side with the subtitle "Gravity from strings".

066.261 **The exact solution of Einstein's disc.**
J. F. Fan, G. Z. Xie, Z. M. Tang, Y. J. Wang.
Astrophys. Space Sci., Vol. 192, No. 2, p. 325 – 328 (Jun 1992). Letter–to–the–editor.
In this Letter, Einstein's equations of gravitational field have been used to calculate the metric of Einstein's disc. The known result has been obtained, and its physical significance is made very clear.

066.262 **What's new with gravity?**
T. Van Flandern.
Bull. Am. Astron. Soc., Vol. 23, No. 3, p. 1258 (1991). Abstract. – See Abstr. 012.065 for the main entry.

066.263 **General relativity and experiment: a brief review.**
T. Damour.
Classical Quantum Gravity, Vol. 9, Suppl., p. S55 – S59 (1992). – See Abstr. 012.058 for the main entry.

066.264 **Equivalence principle violation, antigravity and anyons induced by gravitational Chern–Simons couplings.**
S. Deser.
Classical Quantum Gravity, Vol. 9, Suppl., p. S61 – S72 (1992). – See Abstr. 012.058 for the main entry.
The interaction of point sources with topologically massive gravity in $2+1$ dimensions leads to a variety of novel effects. These include generation of gravimagnetic fields by a structureless point particle, which becomes an anyon. The latter has an arbitrary spin with corresponding statistics under particle exchange. An intrinsically spinning particle leads to solutions which violate the equivalence principle and can even generate a repulsive Newtonian potential.

066.265 **Relativistic motion of gyroscopes and space gradiometry.**
E. Gill, M. Soffel, H. Ruder, M. Schneider.
Dtsch. Geod. Komm. Bayer. Akad. Wiss., Reihe A, Nr. 107. Verlag der Bayerischen Akademie der Wissenschaften, München (Germany). 82 p. (1992). ISBN 3–7696–8189–4.
Relativistic effects in the motion of freely falling test particles moving along circular orbits in the exterior gravitational field of a spherically symmetric, slowly rotating or slightly oblate body are considered. To this end the authors solve the geodesic equation in the (parametrized) post–Newtonian metric, the linearized Kerr metric and the linearized Erez–Rosen metric, respectively. The main part of this article deals with the "pseudo–Newtonian resonances" found by Mashhoon and Theiss (1982 – 1986) in the motion of torque–free gyroscope axes, moving along inclined orbits in the gravitational field of a rotating mass. To investigate relativistic effects in space gradiometry the authors calculate the relativistic tidal forces in a local inertial system, using the previously derived results for geodesic and spin motion. According to the development of high precision superconducting satellite gradiometers the measurement of gravito–magnetic effects of the Earth's rotation may in principle be achieved in the near future. Analyzing the motion of the Earth–Moon system in the gravitational field of the rotating Sun, they find that gravito–magnetic effects can be neglected.

066.266 **Variations of constants and exact solutions in multidimensional gravity.**
S. B. Fadeev, V. D. Ivashchuk, V. N. Melnikov
(*V. N. Mel'nikov*).
Symposium on Gravitation and Modern Cosmology – the Cosmological Constant Problem, p. 37 – 49 (1991). – See Abstr. 012.076 for the main entry.
In this article the author considers the problem of stability of the gravitational constant.

066.267   Null surface canonical formalism.
J. N. Goldberg, D. C. Robinson, C. Soteriou.
Symposium on Gravitation and Modern Cosmology – the Cosmological Constant Problem, p. 59 – 64 (1991). – See Abstr. 012.076 for the main entry.
The problem of constructing the canonical formalism for general relativity on a null cone is resumed.

066.268   On the gravitational field of an arbitrary axisymmetric mass endowed with magnetic dipole moment.
I. D. Novikov, V. S. Manko.
Symposium on Gravitation and Modern Cosmology – the Cosmological Constant Problem, p. 121 – 128 (1991). – See Abstr. 012.076 for the main entry.

066.269   Twistors as spin 3/2 charges.
R. Penrose.
Symposium on Gravitation and Modern Cosmology – the Cosmological Constant Problem, p. 129 – 137 (1991). – See Abstr. 012.076 for the main entry.
It is pointed out that twistors play a role as the charges for helicity 3/2 massless fields. Since such fields can be defined consistently in general Ricci–flat 4–manifolds, a possible new approach to defining twistors in vacuum space–times is indicated.

066.270   Projective unified field theory in context with the cosmological term and the variability of the gravitational constant.
E. Schmutzer.
Symposium on Gravitation and Modern Cosmology – the Cosmological Constant Problem, p. 179 – 184 (1991). – See Abstr. 012.076 for the main entry.
The 5–dimensional variant "Projective Unified Field Theory" leads to a constant gravitational coupling factor (gravitational constant) and a variable cosmological coupling factor ("cosmological constant"). This last fact could be interesting for quantum field theory in connection with the quantum fluctuations of the vacuum.

066.271   The introduction of the cosmological constant.
E. L. Schucking.
Symposium on Gravitation and Modern Cosmology – the Cosmological Constant Problem, p. 185 – 187 (1991). – See Abstr. 012.076 for the main entry.

066.272   Pulsation of a supermassive star in tensor field gravitation theory.
Yu. V. Baryshev.
Conference on the Variability of Blazars, p. 52 – 54 (1992). – See Abstr. 012.077 for the main entry.
It is shown that unlike general relativity, the tensor field gravitation theory leads to the post–Newtonian stability of a supermassive star. Its gravitational binding energy can be a more effective power source than nuclear reactions. Within the field gravitation theory framework there might exist supermassive magnetoids, synchro–Compton caldrons, and plasma–turbulent reactors as power sources in active galactic nuclei. Pulsations of these objects could appear in the observed variability of their emission.

066.273   Velocity of propagation of gravitational radiation, mass of the gravitation, range of the gravitational force, and the cosmological constant.
J. Weber.
Symposium on Gravitation and Modern Cosmology – the Cosmological Constant Problem, p. 217 – 223 (1991). – See Abstr. 012.076 for the main entry.
Einstein's equations with cosmological constant are considered. With an appropriate set of coordinates, the vacuum equations have the same form as the Klein Gordon Equation. The range of the gravitational force is the Compton Wavelength of the graviton with rest mass m. The cosmological constant is one half the reciprocal of the squared Compton Wavelength. Limits on the graviton mass are obtained by considering the

observational data on the advance of the perihelion of Mercury, the observed gravitational radiation from Supernova 1987A, and the known gravitational binding of clusters of galaxies. The observed pulses from Supernova 1987A are in good agreement with the cross section theory published in 1984 and 1986, and reviewed here. As first discussed by F. Zwicky, the graviton mass, as deduced from known gravitational binding of clusters of galaxies, is less than 1.2×10^{-63}g. The cosmological constant is less than 6.4×10^{-52}cm^{-2}.

066.274   Lens effect in a Kerr field.
J. Pluta.
Rep. Math. Phys., Vol. 29, No. 3, p. 289 – 295 (Jun 1991).
A ray of light passing through a graviational field of a rotating black hole experiences a deflection. The equations describing an asymptotic shape of its trajectory are derived, first for a ray passing in the plane $\Theta = \pi/2$, next for the general case. An image of a light source seen by the distant observer (the gravitational lens effect) is described in both cases.

066.275   A note on the uniqueness of the Wyman solution.
A. Krasiński.
Rep. Math. Phys., Vol. 29, No. 3, p. 337 – 339 (Jun 1991).
It is pointed out that the barotropic equation of state splits the family of spherically symmetric shearfree expanding perfect fluid solutions of Einstein's equations into two distinct families: the Wyman solution and the Robertson–Walker solutions. The latter ones are not contained in the former as a limiting case. Based on this fact, it is argued that the barotropic equation of state is unnatural in inhomogeneous cosmological models.

066.276   Conservation laws in general relativity.
M. Ferraris, M. Francaviglia.
Classical Quantum Gravity, Vol. 9, Suppl., p. S79 – S95 (1992). – See Abstr. 012.058 for the main entry.
The general "Lagrangian" method to generate conserved currents in field theories starting from the so–called Poincaré–Cartan form is reviewed, with examples of application to general relativity.

066.277   Two–dimensional gravity: quantum group structure of the continuum theory.
J.–L. Gervais.
Classical Quantum Gravity, Vol. 9, Suppl., p. S97 – S116 (1992). – See Abstr. 012.058 for the main entry.
Current progress in understanding quantum gravity from the operator viewpoint is reviewed. It is based on the $U_q(sl(2))$ quantum–group structure recently put forward, for the chiral components of the metric in the conformal gauge.

066.278   How does quantum gravity modify the Schrödinger equation for matter fields?
C. Kiefer.
Classical Quantum Gravity, Vol. 9, Suppl., p. S147 – S156 (1992). – See Abstr. 012.058 for the main entry.
The author performs an expansion of the Wheeler – DeWitt equation of canonical quantum gravity with respect to powers of the Planck mass. The first three orders lead to the picture of a quantised matter field propagating on a fixed gravitational background. Its state obeys a functional Schrödinger equation with respect to a many–fingered time defined by this background. The next order yields quantum gravitational corrections to this equation. The meaning of these terms is discussed in detail.

066.279   Ashtekar variables and unification of gravitational and electromagnetic interactions.
A. Magnon.
Classical Quantum Gravity, Vol. 9, Suppl., p. S169 – S181 (1992). – See Abstr. 012.058 for the main entry.
It is shown that Ashtekar new canonically conjugate variables recently extracted from the space–time geometry, enable one to view the gravitational field as a (self–interacting) Yang–Mills field. The author proves that Einstein's field equations then arise from a single variational principle and that general relativity can

be interpreted as a gauge theory in the (anti) self–dual regime. The author then presents an interaction which could govern the large scale structure of the universe. This interaction is the result of an SU(2) × U(1) unification of gravitational and electromagnetic degrees of freedom, and is mediated by a field of the Yang–Mills type. This Yang–Mills field is shown to carry a charge which encompasses both gravitational and Maxwellian aspects.

066.280 Perturbations in Bianchi I universes.
P. G. Miedema, W. A. Van Leeuwen.
Classical Quantum Gravity, Vol. 9, Suppl., p. S183 – S185 (1992). – See Abstr. 012.058 for the main entry.

The evolution equations for small perturbations in the metric, energy density and material velocity for an anisotropic viscous Bianchi I universe are studied. It is found (whether or not viscosity is present) that, just as in the flat FRW universe, the general solution of the perturbation equations can be split up into three non–coupled perturbations, namely gravitational waves ("tensor perturbations"), vortex motions ("vector perturbations") and density enhancements ("scalar perturbations").

066.281 Equations of state of cosmic strings in the presence of charged particles.
P. Peter.
Classical Quantum Gravity, Vol. 9, Suppl., p. S197 – S206 (1992). – See Abstr. 012.058 for the main entry.

In a 4D field theory with U(1) broken invariance coupled with electromagnetism in which cosmic strings can form, the author considers the case of a straight vortex at a microscopic level. He solves the field equations numerically around the vortex, computes the total conserved 4–current and the stress energy tensor both in charged and current carrying situations. Integration then provides for the "charge", "current", tension and energy per unit length: this gives the equation of state and shows the existence of charged particle emission configurations.

066.282 Gravitational micro–lensing effects and astrophysical applications.
K. Chang.
Publ. Korean Astron. Soc., Vol. 7, No. 1, p. 97 – 105 (1992). – See Abstr. 012.079 for the main entry.

The most favourable possibilities to observe the phenomena of gravitational lensing are the high amplification events and the time delay between the images. These effects provide the information to determine the Hubble parameter and the matter distribution in the universe. The image properties due to micro–lensing also is of an importance to find out the size and the structure of the source.

066.283 Foundational problems in quantum gravity and quantum cosmology.
E. Prugovečki.
Found. Phys., Vol. 22, No. 6, p. 755 – 806 (Jun 1992).

The conventionalistically based instrumentalist epistemology and methodology underlying the various approaches to the quantization of gravity is contrasted with the operationally based logical analysis practiced by the founders of relativity theory and quantum mechanics in developing their respective disciplines. The foundational problems to which they give rise are described. Their origins are traced to instrumentalist practices which have been in the past the objects of criticisms by Dirac, Heisenberg, Born, and others, but which have nevertheless prevailed in relativistic quantum physics after the emergence of the conventional renormalization program. The operationally based premises of a recently developed geometro–stochastic approach to the quantization of gravity are analyzed. It is shown that their roots lie in the epistemology adopted by the founders of relativity theory and quantum mechanics, and that they reflect a conceptualization of quantum reality which offers the possibility of a resolution of the main foundational problems encountered by the other approaches to quantum gravity.

066.284 Cosmological applications of gravitational lensing.
R. D. Blandford, R. Narayan.
Annu. Rev. Astron. Astrophys., Vol. 30, p. 311 – 358 (1992).

Contents: 1. Introduction. 2. Secure lenses. 3. Gravitational lens optics. 4. Cosmography. 5. Dark matter. 6. Quasars and galaxies at high redshift. 7. Searches for gravitational lenses. 8. Outlook.

066.285 Some new solution of Einstein–Maxwell equations derived from Kaluza–Klein theory.
Shen Yougen.
Ann. Shanghai Obs., Acad. Sin., No. 13, p. 116 – 120 (1992). In Chinese.

066.286 The solution of electromagnetic wave equations in curved space–time.
Luo Jixiong, Cheng Zongyi.
Ann. Shanghai Obs., Acad. Sin., No. 13, p. 121 – 128 (1992). In Chinese.

066.287 Gravitational micro–lensing effects and astrophysical applications.
K. Chang.
Proc. Astron. Soc. Aust., Vol. 9, No. 2, p. 215 – 218 (1991). – See Abstr. 012.090 for the main entry.

The most favourable methods of observing the phenomenon of gravitational lensing are through high–amplification events and the time delay between the images. These effects provide us with the information to determine the Hubble parameter and the matter distribution in the universe. The image properties due to micro–lensing can be used to find the size and structure of the source.

066.288 Baryosynthesis, gravitation and hot big–bang cosmology.
J. D. Barrow.
2. Winter School of Physics: Cosmology and elementary particles, p. 171 – 214 (1992). – See Abstr. 012.092 for the main entry.

This paper deals with mathematical and formal aspects of Newtonian and general relativistic gravitation as well as some higher–order lagrangian theories of gravity. It introduces the Friedman metrics and places them in the context of all solutions to Einstein's equations. The emphasis is upon basic principles and ideas that do not lean too heavily upon specific models.

066.289 A new class of barotropic stellar models. III. Equations for the relativistic approximation.
V. Ureche.
Rom. Astron. J., Vol. 1, No. 1 – 2, p. 51 – 55 (1991).

In the previous two papers (Ureche, 1988, 1989) a new class of barotropic stellar models was proposed. The basic equations of equilibrium and the power series solution for the Newtonian approximation were obtained. A numerical analysis of these equations was performed, and a stability criterion was given. Basic equations for a relativistic approximation are given. With the equation of state in the non–dimensional form, the equations of hydrostatic equilibrium and the equations for the geometry of the space–time continuum are given (also in a non–dimensional form).

066.290 The Wheeler – DeWitt equation and the path integral in minisuperspace quantum cosmology.
J. J. Halliwell.
Conceptual problems of quantum gravity, p. 75 – 115 (1991). – See Abstr. 003.049 for the main entry.

066.291 Interpreting the density matrix of the universe.
D. N. Page.
Conceptual problems of quantum gravity, p. 116 – 121 (1991). – See Abstr. 003.049 for the main entry.

066.292   Gravitational waves in general relativity.
D. G. Blair.
The detection of gravitational waves, p. 3 – 15 (1991). – See
Abstr. 003.097 for the main entry.
  Contents: 1. Introduction to general relativity. 2. Stress energy
and curvature.   3. Non–linearity   and   wave   phenomena.
4. Introduction to gravitational waves. 5. The effects of gravitational waves.

066.293   Sources of gravitational waves.
D. G. Blair.
The detection of gravitational waves, p. 16 – 42 (1991). – See
Abstr. 003.097 for the main entry.
  Contents: 1. Gravitational waves and the quadrupole formula.
2. Strain amplitude, flux and luminosity. 3. Supernovae.
4. Binary coalescence. 5. Other sources of gravitational waves:
Black holes; Pulsars; Binary stars; Cosmological sources. 6. The
rate of burst events: Galactic high frequency sources; Massive
black hole events. 7. Thorne diagrams. 8. Conclusion.

066.294   Space–time topology and quantum gravity.
J. L. Friedman.
Conceptual problems of quantum gravity, p. 539 – 572 (1991). –
See Abstr. 003.049 for the main entry.
  Characteristic features are discussed of a theory of quantum
gravity that allows space–time with a non–Euclidean topology.
The review begins with a summary of the manifolds that can
occur as classical vacuum space–times and as space–times with
positive energy. Local structures with non–Euclidean topology –
topological geons – collapse, and one may conjecture that in
asymptotically flat space–times non–Euclidean topology is hidden from view. In the quantum theory, large diffeos can act
nontrivially on the space of states, leading to state vectors that
transform as representations of the corresponding symmetry
group π_0(Diff). In particular, in a quantum theory that, at
energies $E < E_{Planck}$, is a theory of the metric alone, there appear
to be ground states with half–integral spin, and in higher–
dimensional gravity, with the kinematical quantum numbers of
fundamental fermions.

066.295   Virasoro model space and 2D gravity.
P. Nelson.
1. International Symposium on Particles, Strings and Cosmology, p. 449 – 453 (1991). – See Abstr. 012.098 for the main entry.

066.296   String fields in two–dimensional gravity.
S. R. Das.
1. International Symposium on Particles, Strings and Cosmology, p. 475 – 485 (1991). – See Abstr. 012.098 for the main entry.
  The author discusses how the dynamical degree of freedom of
two–dimensional gravity results in an extra dimension in non–
critical string theory. It is shown that reparameterization
invariance determines the renormalization of couplings in 2D
gravity, and it is explained how this leads to the classical string
field theory equations in non–critical strings.

066.297   Solar corona correction in VLBI observation.
Zhu Jin.
Chin. Sci. Bull., Vol. 37, No. 10, p. 837 – 841 (May 1992).
  The author calculates the VLBI time delay due to corona
effects from the two methods, the direct method and the
deflection method, and compares the results.

066.298   Quantum string gravity near the Planck scale.
G. Veneziano.
1. International Symposium on Particles, Strings and Cosmology, p. 486 – 501 (1991). – See Abstr. 012.098 for the main entry.
  The structure of quantum string gravity, as it emerges from the
study of gedanken experiments at Planckian energies, is outlined.
According to the regime considered, either classical general
relativity phenomena or quantum string–size effects dominate. A
consistent picture emerges whereby the fundamental string–
length parameter shields the classical and quantum infinities of
the point–particle limit.

066.299   Topological gravity and supergravity.
A. H. Chamseddine.
1. International Symposium on Particles, Strings and Cosmology, p. 547 – 559 (1991). – See Abstr. 012.098 for the main entry.
  A gauge invariant action is constructed for gravity, in a metric
independent way. In space–times of dimensions $2n + 1$ this is
based on the $2n + 1$ Chern–Simons form, with the gauge groups
ISO($2n + 1$) or SO(1, $2n + 1$) or SO(2, 2n), depending on the sign
of the cosmological constant. In space–times of dimensions 2n, a
scalar field in the adjoint representation of the gauge group is also
needed. Supersymmetrization can be performed by gauging the
appropriate graded groups. These theories, although topological,
are non–trivial and allow for a propagating graviton.

066.300   Gravity and gauge theories in three dimensions.
S. Deser.
1. International Symposium on Particles, Strings and Cosmology, p. 560 – 572 (1991). – See Abstr. 012.098 for the main entry.
  Contents: 1. Introduction. 2. How is D = 3 different from all
other dimensions. 3. Chern–Simons terms and novel actions.
4. Vector CS theories – anyons. 5. Gravity, CS terms, anyons.
6. Quantum theory.

066.301   Yang–Mills–like renormalizability of gauge–affine gravity and indirect applications.
Y. Ne'eman.
1. International Symposium on Particles, Strings and Cosmology, p. 589 – 599 (1991). – See Abstr. 012.098 for the main entry.
  The author outlines a proof of the renormalizability in a non–
Riemannian model, based on gauging GL(4, R), in which
Einstein's gravity dominates the low–energy region through a
Goldstone–Higgs spontaneous symmetry breakdown mechanism.

066.302   Wormholes in dimensions 1 – 4.
S. W. Hawking.
1. International Symposium on Particles, Strings and Cosmology, p. 623 – 634 (1991). – See Abstr. 012.098 for the main entry.

066.303   Non–perturbative string theory.
D. J. Gross.
1. International Symposium on Particles, Strings and Cosmology, p. 657 – 671 (1991). – See Abstr. 012.098 for the main entry.
  The matrix model solution of c = 1 matter coupled to two
dimensional quantum gravity is reviewed, both in the case where
the target space is the real line and a circle of finite radius. The
role and physical significance of the nonsinglet states is analysed.
The meaning of this theory as a two dimensional string theory is
discussed and a fermionic field theory representation is constructed.

066.304   Self–duality in classical gravity.
J. Samuel.
Recent advances in general relativity. Essays in honor of Ted
Newman, p. 72 – 84 (1992). – See Abstr. 003.051 for the main
entry.

066.305   Holonomies in quantum gravity.
C. Rovelli.
Recent advances in general relativity. Essays in honor of Ted
Newman, p. 85 – 102 (1992). – See Abstr. 003.051 for the main
entry.
  The author reviews the present status of a new approach to the
problem of the quantization of general relativity called the loop
representation.

066.306   Scenes from general relativistic astrophysics.
J. L. Friedman.
Recent advances in general relativity. Essays in honor of Ted
Newman, p. 127 – 145 (1992). – See Abstr. 003.051 for the main
entry.
  Four topics from relativistic astrophysics are discussed here.
These include: (1) the upper limit on rotation of relativistic stars
and its relation to the equation of state of matter at high density;

(2) astrophysical evidence against Witten's hypothesis that, instead of iron, strange quark matter may be the true ground state of baryons; (3) the normal modes of oscillation of rotating stars (knowledge of these is needed to fix the upper limit on rotation of relativistic stars; (4) the normal modes of oscillation of black holes.

066.307   **Asymptotic structure of space–time.**
H. Friedrich.
Recent advances in general relativity. Essays in honor of Ted Newman, p. 146 – 181 (1992). – See Abstr. 003.051 for the main entry.

066.308   **Sources of gravitational waves and prospects for their detection.**
K. S. Thorne.
Recent advances in general relativity. Essays in honor of Ted Newman, p. 196 – 229 (1992). – See Abstr. 003.051 for the main entry.
Contents: 1. Introduction. 2. Aspects of gravitational waves that are important for their detection. 3. Present and future gravitational wave detectors. 4. Sources of gravitational waves and challenges for theorists: binary stars; coalescence of neutron star and black hole binaries and capture of stars by black holes; supernovae; rotating neutron stars; black hole births; cosmological studies; unexpected sources. 5. Conclusion.

066.309   **Supernova 1987A gravitational wave antenna observations, cross sections, correlations with six elementary particle detectors, and resolution of past controversies.**
J. Weber.
Recent advances in general relativity. Essays in honor of Ted Newman, p. 230 – 240 (1992). – See Abstr. 003.051 for the main entry.

066.310   **On Joseph Weber's new cross section for resonant–bar gravitational wave detectors.**
K. S. Thorne.
Recent advances in general relativity. Essays in honor of Ted Newman, p. 241 – 250 (1992). – See Abstr. 003.051 for the main entry.

Spacetime and gravitation.
See Abstr. 003.002.

Gauge gravitation theory.
See Abstr. 003.009.

Special relativity.
See Abstr. 003.010.

Spacetime physics. Introduction to special relativity.
See Abstr. 003.033.

Causality electromagnetic induction and gravitation. A different approach to the theory of electromagnetic and gravitional fields.
See Abstr. 003.042.

Conceptual problems of quantum gravity.
See Abstr. 003.049.

Studies in the history of general relativity.
See Abstr. 003.050.

Recent advances in general relativity. Essays in honor of Ted Newman.
See Abstr. 003.051.

Introducing Einstein's relativity.
See Abstr. 003.063.

Theoretical foundations of cosmology. Introduction to the global structure of space–time.
See Abstr. 003.066.

Relativity theory. Concepts and basic principles.
See Abstr. 003.071.

Gravitational lenses.
See Abstr. 003.088.

Relativity on curved manifolds.
See Abstr. 003.096.

The detection of gravitational waves.
See Abstr. 003.097.

Gravitation. A Banff Summer Institute.
See Abstr. 012.019.

9th Italian Conference on General Relativity and Gravitational Physics. Proceedings.
See Abstr. 012.021.

General relativity and relativistic astrophysics. Proceedings.
See Abstr. 012.056.

Les Journées Relativistes, Cargèse (France), 1 – 4 May 1991.
See Abstr. 012.058.

Texas/ESO–CERN Symposium on Relativistic Astrophysics, Cosmology, and Fundamental Physics. Proceedings.
See Abstr. 012.060.

Gravitation and modern cosmology. The cosmological constant problem. Volume in honor of Peter Gabriel Bergmann's 75th birthday. Proceedings.
See Abstr. 012.076.

Correlation between the Maryland and Rome gravitational wave detectors and the IMB detector during SN1987A.
See Abstr. 013.040.

Coincidences among Mont Blanc, Kamiokande, Baksan, IMB, Frejus, Homestake and Plateau Rosa detectors during SN1987A.
See Abstr. 013.041.

What effect has been detected by underground detectors and gravitational wave antennas a few hours before SN1987A observation?
See Abstr. 013.042.

Space radiointerferometry and gravitational waves.
See Abstr. 013.068.

Neutrino and gravitational wave astronomy: grand visions.
See Abstr. 013.098.

Tunnels through time.
See Abstr. 015.044.

Clock synchronization and isotropy of the one–way speed of light.
See Abstr. 022.029.

Multipump and quasistroboscopic back–action evasion measurements for resonant–bar gravitational wave antennas.
See Abstr. 022.057.

Strings, gravity, and the constants of nature.
See Abstr. 022.060.

Present and future of the Rome gravitational wave experiment.
See Abstr. 022.080.

Approaching the dc SQUID limit for a conventional cryogenic gravitational radiation detector.
See Abstr. 022.081.

Progress report on the development of the Gyromagnetic Electron Gyroscope.
See Abstr. 022.082.

Spectral analysis for gravitational antennas.
See Abstr. 022.083.

Fast estimation of the noise of a graviational wave antenna.
See Abstr. 022.084.

Fabry–Perot resonators with oscillating mirrors for interferometric GW antennas.
See Abstr. 022.085.

Report on the operation of the 389 kg cryogenic gravitational wave antenna ALTAIR at I.F.S.I.
See Abstr. 022.086.

Low noise ultra high frequency R.F.–SQUID.
See Abstr. 022.087.

Theoretical problems on gravitational–wave detectors.
See Abstr. 022.163.

Gravity wave astronomy.
See Abstr. 022.165.

Study of new fundamental forces in a microgravity environment.
See Abstr. 022.177.

New measurements with a torsion pendulum during the solar eclipse.
See Abstr. 022.197.

Time and interpretations of quantum gravity.
See Abstr. 022.198.

A Galilean experiment using a holographic technique.
See Abstr. 022.227.

Measurement of energy transfer in a five–mode gravitational wave bar detector.
See Abstr. 034.041.

A new–type antenna for continuous gravitational radiation.
See Abstr. 034.042.

Catching the wave.
See Abstr. 034.044.

Automatic control of a Michelson interferometer.
See Abstr. 034.052.

Ultralow temperatures gravitational wave detectors.
See Abstr. 034.063.

The gravitational wave experiment.
See Abstr. 035.041.

Experimental search of gravitational waves.
See Abstr. 036.180.

General–relativistic celestial mechanics. II. Translational equations of motion.
See Abstr. 042.008.

Orbital motion with the Mücket–Treder post–Newtonian gravitational law.
See Abstr. 042.134.

Relativistic reference frames including time scales: questions and answers.
See Abstr. 043.005.

Theoretical derivation of relativistic precession and nutation.
See Abstr. 044.028.

Gravitation and celestial mechanics investigations with Galileo.
See Abstr. 051.033.

Relativistic geocentric satellite equations of motion in closed form.
See Abstr. 052.006.

Relativistic effects in the critical inclination problem in artificial satellite theory.
See Abstr. 052.014.

Pair creation and decay of a massive particle near and far away from a cosmic string.
See Abstr. 061.054.

Chiral cosmic strings.
See Abstr. 061.071.

Constants and cosmology: the nature and origin of fundamental constants in astrophysics and particle physics.
See Abstr. 061.075.

Photon and graviton Green's functions on cosmic string space–times.
See Abstr. 061.109.

Transient flow of a relativistic radiating gas past a horizontal plate.
See Abstr. 061.117.

Time and prediction in quantum cosmology.
See Abstr. 061.143.

Time in quantum cosmology.
See Abstr. 061.144.

Space and time in the quantum universe.
See Abstr. 061.146.

Detonation waves in relativistic hydrodynamics.
See Abstr. 062.067.

Ultrarelativistic hydrodynamics: high–resolution shock–capturing methods.
See Abstr. 062.095.

Covariant flux–limited diffusion theories.
See Abstr. 063.019.

Polarization shifts in nonsymmetric theories of gravity.
See Abstr. 063.129.

On the uniqueness of static perfect–fluid solutions in general relativity.
See Abstr. 065.005.

Asymptotics of gravitational collapse of scalar waves.
See Abstr. 067.009.

Pulsar timing and relativisitic gravity.
See Abstr. 067.032.

A generalization of the concept of adiabatic index for non–adiabatic systems.
See Abstr. 067.041.

Model for the completeness of quasinormal modes of relativistic stellar oscillations.
See Abstr. 067.044.

Mechanics of apparent horizons.
See Abstr. 067.052.

Construction of three–dimensional black hole initial data via multiquadrics.
See Abstr. 067.065.

Cauchy horizon instability for Reissner–Nordström black holes in de Sitter space.
See Abstr. 067.077.

Black holes in magnetic monopoles.
See Abstr. 067.082.

Quantum hair and quantum gravity.
See Abstr. 067.085.

On the light curve of an orbiting spot.
See Abstr. 067.091.

Thermodynamics and quantum aspects of black holes in $(1+1)$ dimensions.
See Abstr. 067.124.

Numerically generated black–hole spacetimes: interaction with gravitational waves.
See Abstr. 067.142.

Are horned particles the end point of Hawking evaporation?
See Abstr. 067.143.

Lower dimensional black holes.
See Abstr. 067.167.

Relativistic fluid flows in the magnetized Schwarzschild spacetime.
See Abstr. 067.174.

Black hole evaporation: an open question.
See Abstr. 067.182.

Numerical calculations of black holes and naked singularities.
See Abstr. 067.183.

Gravitational radiation from coalescing binary neutron stars. IV. Tidal disruption.
See Abstr. 067.210.

Could final states of stellar evolution proceed towards naked singularities?
See Abstr. 067.211.

1D numerical relativity applied to neutron star collapse.
See Abstr. 067.215.

Black holes and partition functions.
See Abstr. 067.228.

Black hole thermodynamics and the space–time discontinuum.
See Abstr. 067.229.

Black holes: the inside story.
See Abstr. 067.232.

Radar and spacecraft ranging to Mercury between 1966 and 1988.
See Abstr. 092.014.

The principle of equivalence and the Trojan asteroids. II.
See Abstr. 098.073.

AS Camelopardalis – disagreement with the general theory of relativity?
See Abstr. 117.022.

Expected supernovae rates and gravitational wave antennas at ultralow temperatures.
See Abstr. 125.043.

General relativity: The good companions.
See Abstr. 126.011.

PSR 0655 + 64: an astrophysical laboratory for testing relativistic gravity theories.
See Abstr. 126.096.

Equilibrium stellar systems with spindle singularities.
See Abstr. 151.031.

The distinctive feature of relativistic stellar systems.
See Abstr. 151.067.

The interior gravitational field of a star cluster with a massive black hole at its center.
See Abstr. 151.078.

On the nature of the dark halo of our Galaxy.
See Abstr. 155.006.

Extending the Macho search to $\sim 10^6 M_\odot$.
See Abstr. 155.139.

PKS 0537–441: an elusive case of a gravitationally lensed blazar.
See Abstr. 158.020.

Q1208 + 1011: the most distant imaged quasar, or a binary?
See Abstr. 159.003.

Inversion of the amplification bias and the number excess of foreground galaxies around high–redshift QSOs.
See Abstr. 159.009.

A search for excess galaxies around distant flat–spectrum radio quasars.
See Abstr. 159.011.

Wavelet analysis of "double quasar" flux data.
See Abstr. 159.015.

Quasars gezien door de zwaartekrachts 'bril'.
See Abstr. 159.017.

An automated search for widely spaced gravitational lenses performed on 25 grens plates.
See Abstr. 159.023.

Probability distributions for the magnification of quasars due to microlensing.
See Abstr. 159.025.

Caustic–induced features in microlensing magnification probability distributions.
See Abstr. 159.026.

Der Mikrogravitationslinseneffekt – Theorie und Anwendungen.
See Abstr. 159.054.

Gravitational microlensing: the effect of a single lens and the impact of sparse sampling.
See Abstr. 159.091.

Resolving accretion discs with microlensing.
See Abstr. 159.103.

Redshift measurements of the brightest cluster galaxies of the gravitational lens 0957 + 561.
See Abstr. 160.041.

A new straight arc detected in a cluster of galaxies at $z = 0.423$.
See Abstr. 160.047.

Effects of Wesson's 5D–STM theory of gravity on the 4D Universe model.
See Abstr. 161.001.

Can galaxies exist within our particle horizon with Hubble recessional velocities greater than c?
See Abstr. 161.051.

Kinematics between comoving photon exchangers in a closed, matter–dominated universe.
See Abstr. 161.052.

The statistics of multiple gravitational lensing and the application of a single lensing approximation.
See Abstr. 161.075.

Exact Bianchi–type VIII and IX models in the presence of zero–mass scalar fields.
See Abstr. 161.076.

Torsion, minimum time, string tension and its physical implications in cosmology.
See Abstr. 161.080.

Dynamical equations of a magnetofluid Universe.
See Abstr. 161.110.

Singularity–free self–creation cosmology.
See Abstr. 161.111.

Exact Bianchi–type VIII and IX models in the presence of the self–creation theory of cosmology.
See Abstr. 161.112.

Slowly rotating cosmological viscous fluid universe in Brans–Dicke theory.
See Abstr. 161.113.

A nonsingular universe.
See Abstr. 161.114.

Exact solutions in string–motivated scalar–field cosmology.
See Abstr. 161.148.

No Starobinsky inflation from self–consistent semiclassical gravity.
See Abstr. 161.155.

Non–scale–invariant density perturbations from chaotic extended inflation.
See Abstr. 161.156.

Texture collapse.
See Abstr. 161.157.

On the stability of tension stars.
See Abstr. 161.159.

Gravitational lensing in a universe model with realistic mass distribution. II. Results.
See Abstr. 161.188.

Did the universe evolve?
See Abstr. 161.190.

Decoherence effects of gravitons in quantum cosmology.
See Abstr. 161.194.

Cosmological dispersion, the corrected redshift formula, and large–scale structure.
See Abstr. 161.200.

Coarse–graining approach to quantum cosmology.
See Abstr. 161.202.

A new creation cosmology.
See Abstr. 161.206.

Limits on a possible violation of the strong equivalence principle from primordial nucleosynthesis.
See Abstr. 161.207.

Apocalypse according to the theoretician.
See Abstr. 161.220.

Singularities in colliding plane–wave spacetimes.
See Abstr. 161.229.

Quantum strings in curved spacetimes.
See Abstr. 161.232.

Qualitative analysis of two–fluid Bianchi cosmologies.
See Abstr. 161.241.

Spacetime wormholes as analytic continuation of closed expanding universes.
See Abstr. 161.252.

Absorption and emission of radiation by a sourceless Abelian gauge wall in a Robertson–Walker space–time.
See Abstr. 161.262.

The isotropic singularity in cosmology.
See Abstr. 161.273.

Observational cosmology. III. Exact spherically symmetric dust solutions.
See Abstr. 161.274.

Schwarzschild – de Sitter type wormhole.
See Abstr. 161.294.

The dark matter problem and quantum gravity.
See Abstr. 161.297.

Using gamma–ray bursts to detect a cosmological density of compact objects.
See Abstr. 161.300.

Gravitational waves from global phase transitions.
See Abstr. 161.306.

A large–scale structure model for gravitational lensing.
See Abstr. 161.311.

Cosmological constraints on cosmic–string gravitational radiation.
See Abstr. 161.315.

Quantum fluctuations on domain walls, strings, and vacuum bubbles.
See Abstr. 161.316.

Structure of wiggly–cosmic–string wakes.
See Abstr. 161.317.

Global momentum loss in a non–expanding universe.
See Abstr. 161.328.

Relativistic static spheres filled with infinitely conducting charged fluids.
See Abstr. 161.329.

Obtaining the metric of our universe.
See Abstr. 161.335.

Bianchi type–II, VIII, and IX in certain new theories of gravitation.
See Abstr. 161.340.

Gravitational radiation from colliding vacuum bubbles.
See Abstr. 161.353.

Duration of inflation and possible remnants of the preinflationary universe.
See Abstr. 161.356.

Conformal gravity and the flatness problem.
See Abstr. 161.358.

Primordial nucleosynthesis bounds on the Brans–Dicke theory.
See Abstr. 161.359.

Gravity in the expansive nondecelerative Universe.
See Abstr. 161.363.

Some consequences of quadratic gravity for the early universe.
See Abstr. 161.372.

On the exact solutions and invariant rays of the phase plane in Brans–Dicke theory.
See Abstr. 161.373.

String theory and the quantization of gravity.
See Abstr. 161.374.

Trapped surfaces due to spherical inhomogeneities in expanding open universes.
See Abstr. 161.390.

Gauge–invariant perturbations in a scalar field dominated universe.
See Abstr. 161.391.

Gravitational waves without gravitons.
See Abstr. 161.401.

Faint galaxies and dark matter.
See Abstr. 161.424.

Cosmic strings in Lyra geometry.
See Abstr. 161.435.

Inflationary solution in higher dimensions.
See Abstr. 161.440.

Bianchi type–II string–dust universes.
See Abstr. 161.441.

Film of the extension of Schwarzschild space through the $r = 0$ singularity.
See Abstr. 161.446.

Is physics consistent with closed timelike curves?
See Abstr. 161.447.

Galactic gravitational lensing and cosmological models.
See Abstr. 161.472.

Torsion, quantum effects and the problem of cosmological constant.
See Abstr. 161.477.

Qualitative cosmology.
See Abstr. 161.479.

Third quantization of gravity and the cosmological constant problem.
See Abstr. 161.480.

Pre–post history of Tolman's cosmos.
See Abstr. 161.484.

Some ideas on the cosmological constant problem.
See Abstr. 161.485.

Gravitational lensing and the geometry of the universe.
See Abstr. 161.494.

The cosmological constant.
See Abstr. 161.499.

Physics at the Planck scale.
See Abstr. 161.516.

Creation of a universe in the laboratory.
See Abstr. 161.528.

067 Astrophysics of Compact Objects (Neutron Stars, Black Holes)

067.001   Models of self–gravitating accretion disks.
G. Bodo, A. Curir.
Astron. Astrophys., Vol. 253, No. 1, p. 318 – 328 (Jan 1992).

The authors compute the equilibrium structure of a self–gravitating thick accretion disk by an interative procedure which produces a final density distribution in equilibrium with the potential coming from it, and discuss the main differences in their properties with respect to the models computed without self-gravity.

067.002   How to find a black hole.
F. Flam.
Science, Vol. 255, No. 5046, p. 794 – 795 (14 Feb 1992).

First look for a distinctive X–ray signature, then try to measure its mass, say observers. That strategy has yielded the best black hole candidate yet.

067.003   Non–smoothness of event horizons of Robinson–Trautman black holes.
P. T. Chruściel, D. B. Singleton.
Commun. Math. Phys., Vol. 147, No. 1, p. 137 – 162 (Jun 1992).

It is shown that generic "small data" Robinson-Trautman space–times cannot be C^{123} extended beyond the "r = 2m Schwarzschild–like" event horizon. This implies that an observer living in such a space–time can determine by local measurements whether or not he has crossed the event–horizon of the black–hole.

067.004   On the instability of the n = 1 Einstein–Yang–Mills black holes and mathematically related systems.
R. M. Wald.
J. Math. Phys., Vol. 33, No. 1, p. 248 – 255 (Jan 1992). Current Physics Microform No.: 9202D1042.

The usual approach to analyze the linear stability of a static solution of some system of equations consists of searching for linearized solutions which satisfy suitable boundary conditions spatially and which grow exponentially in time. In the case of the n = 1 Einstein–Yang–Mills (EYM) black hole, an interesting situation occurs. There exists a perturbation which grows exponentially in time–and spatially decreases to zero at the horizon–but nevertheless is physically singular on the horizon. Thus, this unstable mode is unacceptable as initial data, and the question arises as to whether the n = 1 EYM black hole is stable. The author analyzes this issue here in the more general case. He proves that there exists smooth initial data of compact support in M which give rise to a solution which grows unboundedly with time. This implies that the n = 1 EYM black hole and other mathematically similar systems are unstable despite the nonexistence of physically acceptable exponentially growing modes.

067.005   **Accretion flows near black holes mediated by radiative viscosity.**
A. Loeb, A. Laor.
Astrophys. J., Vol. 384, No. 1, p. 115 – 128 (1 Jan 1992).

Angular momentum transport by photons is analyzed for thin and thick configurations of steady accretion flows near black holes. The radiative viscosity coefficient is derived accurately from kinetic theory for arbitrary photon spectra. Thin Keplerian disks cannot be supported by radiative viscosity. However, in a quasi–spherical accretion most of the initial angular momentum of the gas can be transported away by the photons it produces, allowing gas flow toward the central black hole. Nearly spherical accretion may therefore result from more general boundary conditions than usually assumed. Compton drag can also be effective in the outer optically thin part of the flow. The radiative transport of angular momentum may provide the needed accretion to fuel massive black holes in active galactic nuclei.

067.006   **Thermal structure of neutron stars with very low accretion rates.**
J. L. Zdunik, P. Haensel, B. Paczyński, J. Miralda–Escudé.
Astrophys. J., Vol. 384, No. 1, p. 129 – 135 (1 Jan 1992).

Steady state models of neutron stars accreting matter with the heavy element content $0.000002 \leqslant Z \leqslant 0.02$ at the rate $10^{-16} \leqslant \dot{M} \leqslant 10^{-11} M_{\odot} yr^{-1}$ are presented. The models have steady state envelopes, with stationary hydrogen and helium–burning shell sources. The capture of electrons by protons is taken into account; this process is important in models with the lowest heavy–element content. Two types of neutron star models are calculated: ordinary, and models with pion–condensed cores. The latter have strongly enhanced neutrino cooling, and hence their isothermal cores have a lower temperature. The hydrogen and helium shells burn in the pycnonuclear regime and are stable for accretion rates lower than $\sim 10^{-12.5} M_{\odot} yr^{-1}$ for ordinary neutron stars, and lower than $\sim 10^{-11} M_{\odot} yr^{-1}$ for stars with pion–condensed cores. This critical rate decreases with the increase of heavy element content of the accreted matter. Gravitational settling of heavy elements may increase this rate.

067.007   **Compton scattering in a converging fluid flow: spherical near–critical accretion onto neutron stars.**
A. Mastichiadis, N. D. Kylafis.
Astrophys. J., Vol. 384, No. 1, p. 136 – 142 (1 Jan 1992).

The authors study Compton scattering of low–frequency photons in a converging flow of cold plasma. They solve analytically the equation of radiative transfer in the case of spherical near–critical steady state accretion onto a neutron star. The inner boundary condition is that the neutron star surface is completely reflective, or that there is a magnetopause with an empty cavity inside it. The photons escape diffusively and electron scattering is the dominant source of opacity. The energy gain of the photons comes entirely from the bulk motion of the converging flow of the accreting gas. The spectrum observed at infinity is a power law at high frequencies with photon number

spectral index –1. This spectrum is significantly flatter than that found for accretion onto black holes.

067.008   **The fate of accreted CNO elements in neutron star atmospheres: X–ray bursts and gamma–ray lines.**
L. Bildsten, E. E. Salpeter, I. Wasserman.
Astrophys. J., Vol. 384, No. 1, p. 143 – 176 (1 Jan 1992).

The authors describe the fate of incident ^{12}C, ^{14}N, and ^{16}O in accreting neutron star atmospheres. When the accreting material is stopped by Coulomb collisions with atmospheric electrons, all incoming elements heavier than helium thermalize at higher altitudes in the atmosphere than the accreting protons. The incoming protons and helium then destroy the elements via nuclear spallation reactions. A small fraction of the nuclear reactions cause nuclear excitation and subsequent γ–ray emission. The probability for a nucleus to survive this bombardment depends on how long it spends in the hazardous region of the atmosphere. For typical accretion rates, the nucleus resides in the hazardous region for many destruction times, and therefore has a small survival probability. The authors calculate the fractions of incident ^{12}C, ^{14}N, and ^{16}O that survive proton bombardment as a function of the accretion rate and the mass and radius of the neutron star. The subsequent paucity of CNO nuclei decreases hydrogen–burning rates in the deep regions of the atmosphere, thereby reducing the amount of helium available for the unstable nuclear flashes that cause type I X–ray bursts. X–ray bursts still occur, but they are predominantly of the mixed hydrogen–helium type. The authors also determine the γ–ray line emission from this collisional deceleration scenario. It is shown that the incident electrons are crucial in determining the macroscopic electric field present in the atmosphere. The authors also briefly discuss how the results are modified in the high magnetic field environment (B $\gtrsim 10^{12}$G) of X–ray pulsars. Alternative stopping scenarios such as collisionless shocks and disk accretion are also discussed.

067.009   **Asymptotics of gravitational collapse of scalar waves.**
R. Gómez, J. Winicour.
J. Math. Phys., Vol. 33, No. 4, p. 1445 – 1457 (Apr 1992). Current Physics Microform No.: 9204G1623.

Through a combination of analytic and numerical techniques, the formation of a black hole by a self–gravitating, spherically symmetric, massless scalar field is investigated. The evolution algorithm incorporates a Penrose compactification so that the Bondi mass, the news function, and other radiation zone limits can be obtained numerically. The late time behavior recently established by Christodoulou is confirmed and new asymptotic relations for late time and for large amplitude limits are derived. It is found that the scalar monopole moment decays exponentially during black hole formation in contrast to the perturbation theory result for a power law decay rate in an Oppenheimer–Snyder background. It is demonstrated that the Newman–Penrose constant for the scalar field is globally well defined and has significant effects.

067.010   **Nuclear astrophysics of dense matter.**
D. Vautherin.
Paris–11 Univ., 91 Orsay (France). Inst. de Physique Nucléaire, IPNO/TH—91–90. 37 p. (Dec 1991). Lectures given at the Univ. degli Studi di Trento, Dipt. di Fisica, Povo (Italy), 5 – 21 Nov 1991.

Starting from the equation of state for a non–relativistic Fermi gas the author describes the equilibrium state of stars whereby the equation of state is generalized to the relativistic case for the description of white dwarfs. Then the evolution of massive stars is described in this framework regarding the thermonuclear burning phase, the gravitational collapse, the neutronization, and the neutrino diffusion. Then the equation of state of supernova matter and the cooling of neutron stars are considered. The author concludes that this approach is somewhat oversimplified in the case of neutron stars, while it is very useful in the case of white dwarfs, where residual interactions can be neglected.

067.011 A possible explanation for intraday variability in active galactic nuclei. Magnetic reconnection and coherent plasma emission.
H. Lesch, M. Pohl.
Astron. Astrophys., Vol. 254, No. 1/2, p. 29 – 38 (Feb 1992).
A model for rapid variability (less than one day) in active galactic nuclei is developed. At radii smaller than 5×10^{14}cm (which correspond to 10 Schwarzschild radii for a black hole with $10^8 M_\odot$) accretion disks contain strong turbulent magnetic fields ($B \cong 10^2$Gauss). Due to convective motions these fields reach the disk corona and are dissipated by magnetic reconnection. In the small reconnection zones (with a lenght of about 10^{11}cm) electrons are accelerated to relativistic energies. Then the interplay of acceleration and subsequent excitation of Langmuir waves leads to coherent emission of electromagnetic waves by inverse Compton scattering. This scenario may lead to an explanation of the observed brightness temperatures exceeding the Compton limit and the extremely small rising time of the variability (10^{46}erg s^{-1} during 1 s (Wagner 1991)). This model is not restricted to the optical band. A slight variation of the parameters allows e.g. for variable X–ray emission.

067.012 Impact of the nuclear equation of state on models of general relativistic, rotating neutron stars.
F. Weber, M. K. Weigel, N. K. Glendenning.
Verh. Dtsch. Phys. Ges., Vol. 27, No. 1, p. 230 – 231 (1992).
Abstract. Paper presented at the Frühjahrstagung des Fachverbandes Physik der Hadronen und Kerne der Deutschen Physikalischen Gesellschaft e.V. (DPG) Gemeinsam mit der Österreichischen Physikalischen Gesellschaft (OePG), Salzburg (Austria), 24 – 28 Feb 1992.

067.013 Pycnonuclear triple–alpha–fusion rate in hypernetted chain approximation.
H. M. Müller, K. Langanke.
Z. Phys., A, Vol. 342, No. 2, p. 133 – 140 (May 1992).
The authors have studied the helium plasma in the hypernetted–chain–approximation considering both short–ranged internuclear and long–ranged Coulomb interactions. The two-particle distribution functions are used to determine the pycnonuclear triple–alpha–fusion rate in the density regime 10^8g/cm$^3 \leqslant \varrho \leqslant 10^{10}$g/cm^3, in which the fusing of three alpha-particles to form a ^{12}C–nucleus plays an important role in the crust evolution of an accreting old neutron star. The calculation supports the idea that the helium liquid undergoes a phase transition to stable ^{8}Be matter at densities slightly lower than $\varrho \approx 3 \times 10^9$g/cm^3 as the plasma induced screening potential then becomes strong enough to bind the ^{8}Be ground state.

067.014 Neutron stars, hybrid stars and constraints on the equation of state from neutron stars and hybrid stars.
A. Rosenhauer, E. F. Staubo, L. P. Csernai.
Z. Phys., A, Vol. 342, No. 2, p. 235 – 238 (May 1992).
Hybrid stars composed of a strange matter core surrounded by neutron matter are investigated. The authors apply star models based on phenomenological equations of state (EOS) from nuclear reactions including a phase transition between the hadronic phase and the quark gluon plasma. For specific equations of state hybrid stars might exist. While the nuclear part of the EOS has only a minor influence on the properties of hybrid stars, the EOS for the quark gluon phase has a crucial impact on the existence of such objects.

067.015 Instability of Einstein–Yang–Mills black holes.
D. V. Gal'tsov, M. S. Volkov.
Phys. Lett. A, Vol. 162, No. 2, p. 144 – 148 (3 Feb 1992).
An analytic proof of the instability of magnetic Abelian and non–Abelian SU(2) Einstein–Yang–Mills black holes is given by functional methods.

067.016 Black holes with rings.
T. Azuma, T. Koikawa.
Phys. Lett. A, Vol. 162, No. 5, p. 365 – 369 (24 Feb 1992).
The authors present a peculiar feature of their soliton solution to the higher–dimensional Einstein equations. The solution describes a black hole surrounded by null rings.

067.017 Charged non–abelian SU(3) Einstein–Yang–Mills black holes.
D. V. Gal'tsov, M. S. Volkov.
Phys. Lett. B, Vol. 274, No. 2, p. 173 – 178 (9 Jan 1992).
It is found that, unlike the SU(2) case, for the SU(3) Einstein–Yang–Mills system non–linear superpositions of the Reissner–Nordstrøm and the non–abelian black hole may exist. These solutions are shown to admit the extreme black hole structures.

067.018 Q–stars at finite temperature.
Su Rukeng, Su Chenggang, Pan Rongshi.
Phys. Lett. B, Vol. 278, No. 3, p. 297 – 301 (26 Mar 1992).
The discussions on Q–stars are extended to finite temperature. The authors find that below the critical temperature T_c the radius, mass and the frequency of the Q–star will increase with temperature. The critical temperature at which the Q–star will disappear and the epoch in which the Q–star can exist are given.

067.019 Massive binary black holes and wiggling jets.
J. S. Kaastra, N. Roos.
Astron. Astrophys., Vol. 254, No. 1/2, p. 96 – 98 (Feb 1992).
Jets in extragalactic radio sources often show small, semi-periodic deviations from a straight line, like wiggles, knots etc. Here the authors investigate the possibility that at least a part of these features is caused by a modulation due to orbital motion in a binary black hole. The periodically changing orbit velocity direction of the black hole producing the jet leads to a modulation of the jet velocity as seen by a distant observer. Application of the model to 3C 273 and M87 leads to acceptable estimates of the orbital radius, velocity and period and of the minimum mass present in both black holes.

067.020 Nucleosynthesis in a thick accretion disk around a $10\,M_\odot$ black hole.
K. Arai, M. Hashimoto.
Astron. Astrophys., Vol. 254, No. 1/2, p. 191 – 197 (Feb 1992).
Nucleosynthesis in a geometrically thick accretion disk around a $10\,M_\odot$ black hole is calculated with use of a nuclear reaction network which contains the elements up to Ge. The constructed disk model is in hydrostatical equilibrium supported predominantly by radiation pressure and has a stationary flow with a supercritical accretion rate. The α parameter is used in taking into account viscosity inside the disk. When α becomes as low as 10^{-10}, the maximum temperature reaches 4.2×10^9K in the central region of the disk. A considerable amount of elements up to ^{56}Fe can be produced deep inside the disk.

067.021 Neutrino production from accreting X–ray pulsars.
K. S. Cheng, K. W. Ng, T. Cheung, M. M. Lau.
J. Phys. G, Vol. 18, No. 4, p. 725 – 738 (Apr 1992).
There exist electrostatic acceleration regions (accelerator gaps) above the inner part of a Keplerian accretion disc of an X–ray pulsar which can accelerate positively charged particles to extreme relativistic energies towards the accretion disc. The subsequent hadronic collisions between such ultrarelativistic protons (ions) and the disc material can produce numerous secondary mesons whose decay generates a luminous neutrino flux. Although the neutrinos can directly escape from the collision regions without any energy loss, the disc magnetic field can modify the neutrino spectrum because their unstable charged parent particles can lose energy via synchrotron radiation. This can provide direct information on the structure of the accretion disc. In calculating the meson production rate, the authors have used an improved scaling violation model which has the effect of enhancing the neutrino production rate. It appears that a large fraction of the accretion power of X–ray pulsars ($\approx 10^{37}$erg s^{-1}) could be carried away by the neutrino flux.

067.022   The SU(3) black hole.

V. D. Dzhunushaliev.

JETP Lett., Vol. 55, No. 3, p. 157 – 162 (10 Feb 1992). Current Physics Microform No.: 9204X1992.

The Einstein–Yang–Mills equations solution which is a SU(3) black hole was built in this article.

067.023   Accretion disks in active galactic nuclei: vertically averaged models.

J. K. Cannizzo, C. M. Reiff.

Astrophys. J., Vol. 385, No. 1, p. 87 – 93 (20 Jan 1992).

The authors construct simple models of accretion disks surrounding supermassive black holes in active galactic nuclei. By considering several different opacity laws, the authors show that some conclusions arrived at by previous investigators are model–dependent. They present a simple technique for solving the full vertically averaged disk equations taking into account both self–gravity and central object gravity, and gas pressure and radiation pressure. The authors also discuss the self–consistency of the standard non–self–gravitating, geometrically thin accretion disk models and confirm the findings of Lin and Shields that such models are valid only in a restricted parameter space.

067.024   Accretion disks in active galactic nuclei: vertically explicit models.

J. K. Cannizzo.

Astrophys. J., Vol. 385, No. 1, p. 94 – 107 (20 Jan 1992).

The author explores a region of the parameter space accessible to accretion disks in active galactic nuclei using vertically integrated models to describe the thin–disk structure. He examines models with $0.01 \leqslant \alpha \leqslant 3$ and radii near 10^{-3}pc (assuming a central black hole of mass $10^8\,M_\odot$). Particular attention is paid to the role of convection. It is found that it may be appropriate to take the mixing length to be roughly one–third of a pressure scale height when applying standard mixing–length theory to accretion disks. The author also presents scalings for various critical points in the locus of steady state solutions and provides crude estimates of time scales for behavior associated with the disk instability. As pointed out by other authors, the disk instability solves the problem of fueling quasars by requiring that they be "on" only a fraction of the time. In addition, the fact that this duty cycle is roughly equal to the ratio of quasar–like objects to all galaxies at the height of the quasar era implies that most or all galaxies once had active cores, and therefore most present–day galaxies must still have massive black holes at their centers.

067.025   Gamma–ray lines from accreting neutron stars.

L. Bildsten.

AIP Conf. Proc., No. 232, p. 401 – 406 (1 Aug 1991). Current Physics Microform No.: 9110A1097. – See Abstr. 012.003 for the main entry.

The current generation of γ–ray telescopes (SIGMA, GRO) provides a new opportunity for observing red–shifted γ–ray lines from the atmospheres of accreting neutron stars. A successful observation would provide important information about how the accretion stream settles onto the neutron star and might allow limits to be placed on the nuclear equation of state. Thus, the author has undertaken a theoretical re–analysis of different γ–ray emission mechanisms. This paper describes his results on the 4.438 MeV γ–ray line emission from ^{12}C and ^{16}O and outlines the ongoing calculation of the 2.2 MeV D–recombination line flux expected from the spallation of incident Helium. The author shows that a neutron star accreting material of solar abundances will produce a 4.438 MeV γ–ray line flux that is below the current observational limits.

067.026   Gamma–ray emission from black holes.

J. C. Ling.

AIP Conf. Proc., No. 232, p. 407 – 415 (1 Aug 1991). Current Physics Microform No.: 9110A1103. – See Abstr. 012.003 for the main entry.

Strong continuum gamma–ray emission at ~ 1 MeV possibly correlated with a narrow annihilation line at 511 keV has been observed from both Cyg X–1 and the Galactic Center. Such correlated emission has been interpreted as a unique gamma–ray signature for theoretically predicted relativistic, positron–electron pair–dominated plasma in regions surrounding the black holes. The author reviews primarily the Cyg X–1 results, which have provided important new insights about the source. Cyg X–1 may be considered a canonical reference stellar black hole whose spectral and temporal characteristics can be used for comparison with those of other black–hole candidates including the Galactic Center and AGN.

067.027   Magnetic acceleration of broad emission–line clouds in active galactic nuclei.

R. T. Emmering, R. D. Blandford, I. Shlosman.

Astrophys. J., Vol. 385, No. 2, p. 460 – 477 (1 Feb 1992).

It is proposed that broad emission lines in active galactic nuclei are formed by dense clouds in a molecular, hydromagnetic wind accelerated radiatively and centrifugally away from an accretion disk orbiting a massive black hole. These clouds are supposed to be photoionized by the UV continuum produced in the innermost radii of the accretion disk where the radiation flux is sufficient to evaporate the dust grains. If the wind is responsible for extracting most of the disk angular momentum, then the estimated ionizing flux, electron density, cloud filling factor, column density, and velocity are in accord with values for these quantities inferred from studies of the line widths, strengths, and variability. In order to reproduce typical line profiles with cusp shapes, blue asymmetry, and blueshifted peaks, it is found that the volume emissivity must scale roughly as $r^{-3.3}$, where r is the cylindrical radius; in this case each decade of radius contributes nearly equally to the line profile. Simple models of the cloud emissivity approximate this dependence. The authors study the possibility that the broad wings of the lines are formed through electron scattering by a $\sim 10^6$K warm intercloud medium. It is found that such a phase can be thermally stable, when magnetically confined, and that a Thomson optical depth $\tau_T \sim 1$ is automatically maintained.

067.028   Post–Newtonian frequencies for the pulsations of rapidly rotating neutron stars.

C. Cutler, L. Lindblom.

Astrophys. J., Vol. 385, No. 2, p. 630 – 641 (1 Feb 1992).

The formalism for computing the oscillation frequencies of rapidly rotating stars in the post–Newtonian approximation is reviewed and extended. Numerical results are presented for the frequencies of the l = m f–modes of rapidly rotating neutron stars. The ratios of the critical angular velocities (where the mode frequencies pass through zero) to $(\pi G \bar{\varrho}_0)^{1/2}$ (with $\bar{\varrho}_0$ the average density) are lower than their Newtonian counterparts by up to 10%. Thus post–Newtonian effects tend to enhance the gravitational radiation – induced instability in rotating stars.

067.029   Radiation dynamics in X–ray binaries. I. Type 1 bursts.

M. A. Walker.

Astrophys. J., Vol. 385, No. 2, p. 642 – 650 (1 Feb 1992).

The author presents equations describing the evolution of a thin, axisymmetric, viscous, relativistic, irradiated accretion disk, and numerical solutions of these equations in the case where irradiation results from a thermonuclear flash on the surface of the accreting neutron star. These calculations verify the notion that the radiation torque induces a substantial increase in accretion rate, during a type 1 X–ray burst, and provide insight into the factors which influence the dynamical response of the disk.

067.030   Radiation dynamics in X–ray binaries. II. Type 2 bursts.

M. A. Walker.

Astrophys. J., Vol. 385, No. 2, p. 651 – 660 (1 Feb 1992).

The author presents a new model for the source XBT 1730–335, the rapid burster. The model is based on the idea that radiation plays an important role in transporting angular momentum from relativistic accretion disks. The calculations presented are fully relativistic and cast in a Kerr geometry; they describe the evolution of a geometrically thin, axisymmetric, irradiated accretion disk around a nonmagnetic neutron star.

The model describes not only temporal, but also spectral properties, and addresses many of the significant features of the real system, including the burst–energy / waiting time relationship, the unique nature of this object, the source spectrum, the outburst cycle, the nonbursting state, and the unusual type 1 burst properties which are sometimes observed. By comparison with the observed properties, the author infers that the rapid burster is a non–magnetic, "critically compact", slowly rotating neutron star in a highly eccentric binary system with a period of 6 months; accretion proceeds via a massive, nearly dissipationless disk which is at an inclination of about 50° to the line of sight.

067.031 Radiation dynamics in X–ray binaries. III. Extremely compact objects.
M. A. Walker.
Astrophys. J., Vol. 385, No. 2, p. 661 – 664 (1 Feb 1992).

The author calculates the spectral modifications which arise from the scattering of photons by accretion disks around nonmagnetic neutron stars. It is found that extremely compact stars (which lie inside the marginally stable orbit) must exhibit a copious hard X–ray flux, and that the radiation field entirely dominates the dynamical behavior of disks around these objects. Based on these findings, the author reinterprets the "black hole candidates" as extremely compact stars. He also offers a simple explanation for the observed hardening of low mass X–ray binary spectra, at low accretion rates, in terms of nongeodesic infall onto mildly compact stars.

067.032 Pulsar timing and relativisitic gravity.
J. H. Taylor.
AIP Conf. Proc., No. 233, p. 540 – 548 (5 Aug 1991). Current Physics Microform No.: 9110A1780. – See Abstr. 012.004 for the main entry.

Radio pulsars provide unique opportunities for precision measurments of gravitational phenomena, including instances where relativistically strong fields are known to be important. The author summarizes the status of two continuing experiments involving detailed investigation of the gravitational interactions between a pulsar and its orbiting companion. One experiment yields high–accuracy mass measurements for two neutron stars, and a quantitative test which proves with better than 1% precision – that a gravitationally bound binary system loses energy at the rate expected from gravitational radiation. The second experiment, involving a different binary pulsar, allows measurement of two parameter characterizing the general relativistic "Shapiro delay" in the system, and from these the masses of the pulsar and its white dwarf companion. Both experiments also yield rigorous limits on the rate of change of Newton's constant, G. The limits are presently at the level of parts in 10^{11} per year.

067.033 Synchrotron emission of neutrino pairs in neutron stars.
A. D. Kaminker, K. P. Levenfish, D. G. Yakovlev.
Sov. Astron. Lett., Vol. 17, No. 6, p. 450 – 454 (Nov 1991). Current Physics Microform No.: 9207X1888. English translation of Pis'ma Astron. Zh., Tom 17, No. 12, p. 1090 – 1100 (Dec 1991).

The energy loss rate Q from synchrotron radiation of neutrino pairs emitted by a degenerate relativistic electron gas in a magnetic field **B** is calculated and approximated by a simple formula when the electrons populate many Landau levels. Over a broad range of conditions, the loss rate is independent of electron density. In neutron star envelopes the synchrotron losses are comparable to or exceed other neutrino energy losses.

067.034 The structure and evolution of Thorne–Żytkow objects.
R. C. Cannon, P. P. Eggleton, A. N. Żytkow, P. Podsiadlowski.
Astrophys. J., Vol. 386, No. 1, p. 206 – 214 (10 Feb 1992).

The authors construct models of spherically symmetric stars in which a central neutron star core is accreting material from an envelope (Thorne–Żytkow objects). Core masses are allowed to grow by accretion from ~0.5 to 2.0 $M_\odot$, and envelope masses range from ~0 to 30 $M_\odot$. The authors find models populating

the entire range of both parameters, and a limited range of parameter space, in which two distinct types of model can coexist, one where gravitational energy and one where nuclear energy is the dominant source of energy.

067.035 Dynamical model of the magnetic field of neutron stars.
F. W. Cummings, D. D. Dixon, P. E. Kaus.
Astrophys. J., Vol. 386, No. 1, p. 215 – 221 (10 Feb 1992).

A dynamical model of the magnetization of a neutron star is given. The model shows three distinct behaviors, as characterized in the parameter space of the two relevant parameters. In one region of the parameter space, the magnetization, and correspondingly the magnetic field, behaves erratically, non–periodically, occasionally producing large pulses of directional electric and magnetic field. It is suggested that this field is associated with the mysterious "bursters", very energetic gamma–rich pulses of the order of 1 s duration, and believed to emanate from single neutron stars. A second region of parameter space shows the magnetization precessing at a constant period around the spin or conserved angular momentum direction. This region is associated with the well–known pulsars. The third region is a "dead" region, where the magnetization is aligned with the spin axis, and the star is nonradiating. The model suggests a life history of a neutron star, in which the star evolves initially from a burster, later becoming a pulsar, and ending as a dead star; an alternative evolution is from burster directly to a dead star.

067.036 On the physical nature of bursters. I.
G. S. Saakyan, G. P. Alodzhants, A. V. Sarkisyan.
Astrophysics, Vol. 34, No. 1, p. 15 – 27 (Jan – Feb 1991). English translation of Astrofizika, Tom 34, Vyp. 1, p. 21 – 40 (Feb 1991).

The phenomenon of burster flashes is examined. In a compact binary system consisting of a neutron star and an ordinary star, a characteristic quantity is an accretion rate $\dot{M}_E \approx 1.3 \times 10^{18}$ g/sec, which corresponds to the Eddington luminosity limit L_E. If the accretion rate is $\dot{M} < \dot{M}_E$, then there is a source of soft X rays that in the presence of a strong magnetic field is manifested as an X–ray pulsar. Bursters are objects with $\dot{M} \gtrsim \dot{M}_E$ and apparently comparatively weak magnetic field. Finally, in objects with relativistic jets (for example, SS 433) $\dot{M} > \dot{M}_E$. It is shown that in bursters the temperature of the neutron star is practically the same from the center to its surface, where the accretion flow is stopped; the temperature is $T \approx 2 \times 10^7$.

067.037 Accretion in a Kerr black hole magnetosphere: energy and angular momentum transport between the magnetic field and the matter.
K. Hirotani, M. Takahashi, S. Nitta, A. Tomimatsu.
Astrophys. J., Vol. 386, No. 2, p. 455 – 463 (20 Feb 1992).

The authors study a MHD interaction between accreting matter and magnetic field in a stationary and axisymmetric magnetosphere surrounding a Kerr black hole, which may exist in active galactic nuclei. The critical condition that the MHD ingoing flows must pass through the fast magnetosonic point is analyzed in detail. It is found that this condition restricts the shape of the magnetic field lines threading the event horizon. For example, cylindrical field lines must be bent into radial or paraboloidal ones. The plasma is expected to be injected from a thin disk into the magnetosphere and accretes to the event horizon. The authors calculate the fluid's energy and angular momentum both at the injection region and at the event horizon to investigate their MHD transport along the field lines. It is shown that the fluid's energy and angular momentum at the event horizon do not depend so much on the initial conditions at the injection region, but rather vary with the field line geometry near the event horizon. Therefore, the transport process between the fluid and the magnetic field should be sufficiently effective in the accretion. Interestingly, the fluid's energy can become negative at the event horizon, even if it is positive at the injection region. This is a Penrose mechanism due to the MHD energy extraction from the gravitational field into the magnetic field.

067.038 Standing shocks in accretion disks and the spectra of active galactic nuclei.
S. K. Chakrabarti, P. J. Wiita.
Astrophys. J., Lett., Vol. 387, No. 1, p. L21 – L24 (1 Mar 1992).

Standing shocks in the inner regions of accretion disks around active galactic nuclei will modify the disks' spectra. The authors illustrate the types of spectral energy distributions that emerge from simple models of this kind and compare them with observational results. Good fits, particularly in the UV, are found for 1202 + 281 and 2130 + 099. The models also reproduce the IR region very well, when one adds an additional cool disk component of much larger size.

067.039 Black holes: Placing faith in the masses?
J. F. Dolan.
Nature, Vol. 355, No. 6361, p. 589 – 590 (13 Feb 1992).

Although massive compact objects in X–ray binaries, such as V404 Cyg and Cyg X–1, cannot be classical neutron stars, proving that they are black holes will require the detection of some phenomenon unique to a black hole other than its total mass.

067.040 Vortex drag and the spin–up time scale for pulsar glitches.
R. I. Epstein, G. Baym.
Astrophys. J., Vol. 387, No. 1, p. 276 – 287 (1 Mar 1992).

The authors investigate the coupling of the superfluid of a neutron star with the solid crust arising from the scattering of individual nuclei in the inner crust with dynamical vortex lines of the superfluid. Such interactions generate quantized Kelvin mode vortex oscillations (kelvons). The authors calculate the rate of kelvon production and the consequent drag produced on the superfluid. The coupling is sufficiently strong to permit glitch spin–up time scales $\lesssim 60$ times the rotation period. Catastrophic vortex–unpinning events are therefore capable of producing giant glitches with rapid spin–ups on the scale observed in the Vela pulsar.

067.041 A generalization of the concept of adiabatic index for non–adiabatic systems.
W. Barreto, L. Herrera, N. Santos.
Astrophys. Space Sci., Vol. 187, No. 2, p. 271 – 290 (Jan 1992).

The concept of adiabatic index, measuring the stiffness of the equation of state for adiabatic systems, and which plays a fundamental role in the study of gravitational collapse, is extended to systems emitting and/or absorbing energy. In the light of the new definition, the collapse of two models of radiating, general relativistic spheres is analyzed in detail, at different regimes of radiation transport. The conspicuous role played by the new variable is clearly exhibited.

067.042 Supernovae: where and why do they break off?
L. Herrera, L. A. Núñez.
Astrophys. Space Sci., Vol. 188, No. 1, p. 9 – 18 (Feb 1992).

The precise mechanism whereby gravitational collapse leads to a type II supernova event is one of the most controversial points in the understanding of the final stages of stellar evolution. The "bounce–shock" mechanism and the proposed "long–term neutrino mediated" processes compete to explain the ejection of the outer envelopes of the star. Despite their differences, both descriptions consider the shock as mainly responsible for the ejection of the outer mantle in a supernova burst. In this note the authors discuss results from collapse calculations in which the shock is considered as an interface separating the quasi–state stiff core from the outer mantle. In the models considered the shock fades out and becomes a new boundary surface delimiting the compact homogeneous remnant. The resulting pictures become intelligible in the light of a recently proposed generalization of the concept of adiabatic index for systems where radiation flux is present.

067.043 The synthesis of ^{26}Al during combined hydrogen and helium–burning reactions.
A. Goswami, S. Ramadurai, H. L. Duorah.
Astrophys. Space Sci., Vol. 188, No. 2, p. 233 – 239 (Feb 1992).

The authors have studied the synthesis of ^{26}Al during combined hydrogen and helium–burning processes in high temperature and density conditions. The possible sites for these processes are believed to be the neutron star surfaces where the density ranges from $\varrho = 10^4$–10^7g cm^{-3} and temperature range from 10^8–8×10^8K. The screening effect which leads to an enhancement of nuclear reaction rates is taken into account whenever necessary. A detailed calculation of the abundances of ^{26}Al and ^{27}Al isotopes is presented here. Finite amounts of ^{26}Al is found to be produced at $T = 2 \times 10^8$K and $\varrho = 10^6$g cm^{-3} due to these combined reactions. This situation is likely to be realized during the γ–ray burst events on neutron star surface. The amount of material processes in the burst sources is very little compared to the amount of material processed in novae or supernovae. Thus it is suggested that rather than contributing to the overall amount of ^{26}Al, γ–ray bursts are likely to contribute more significantly to the inhomogeneity of ^{26}Al distribution in interstellar medium.

067.044 Model for the completeness of quasinormal modes of relativistic stellar oscillations.
R. H. Price, V. Husain.
Phys. Rev. Lett., Vol. 68, No. 13, p. 1973 – 1976 (30 Mar 1992).
Current Physics Microform No.: 9205F0187.

Relativistic stellar oscillations are fluid and metric perturbations characterized by complex frequencies called quasinormal modes. This letter presents a model which clarifies the extent to which such modes can be complete, in the sense of normal modes. It is found, for a specific class of Cauchy data, that, describing purely outgoing radiation, an expansion of the dynamics in terms of quasinormal modes can be constructed. For more general Cauchy data, it is shown that no useful distinction is possible between data that do and do not stimulate quasinormal oscillations.

067.045 Structure of the singularity inside a realistic rotating black hole.
A. Ori.
Phys. Rev. Lett., Vol. 68, No. 14, p. 2117 – 2120 (6 Apr 1992).
Current Physics Microform No.: 9205F0331.

The author presents results of an analysis of the asymptotic behavior of nonlinear, asymmetric, metric perturbations near the Cauchy horizon inside a Kerr black hole. This analysis suggests that metric perturbations, to all orders in the perturbation expansion, are finite and small at the Cauchy horizon, even though their gradients (and the curvature) diverge there. Accordingly, objects which fall into a realistic rotating black hole a long time after the collapse will not be crushed by a tidal gravitational deformation as they approach the curvature singularity.

067.046 A classical instability of Reissner–Nordström solutions and the fate of magnetically charged black holes.
K. Lee, V. P. Nair, E. J. Weinberg.
Phys. Rev. Lett., Vol. 68, No. 8, p. 1100 – 1103 (24 Feb 1992).
Current Physics Microform No.: 9205C2044.

Working in the context of spontaneously broken gauge theories, the authors show that the magnetically charged Reissner–Nordström solution develops a classical instability if the horizon is sufficiently small. This instability has significant implications for the evolution of a magnetically charged black hole. In particular, it leads to the possibility that such a hole could evaporate completely, leaving in its place a nonsingular magnetic monopole.

067.047 The X–ray spectrum of modified α–viscosity accretion disc.
Cao Xinwu.
Astrophys. Space Sci., Vol. 189, No. 1, p. 141 – 145 (Mar 1992).
Letter–to–the–editor.

The X–ray spectrum of an accretion disc around a black hole is obtained under the modified α–viscosity law. Both electron scattering opacity and free–free absorption opacity are taken into account in the calculation. The author also finds that the requirement of a geometrically–thin disc forces a limit on the accretion rate $\dot{M} < 0.25\,\dot{M}_{cr}$ (i.e. $L < 0.25\,L_{edd}$). Several previous disc calculations violate this limit and their results are questionable.

067.048   **Remarks on the relative populations of neutron and strange stars.**
J. E. Horvath, G. A. Foglia.
Astrophys. Space Sci., Vol. 189, No. 1, p. 159 – 162 (Mar 1992). Letter–to–the–editor.
The authors consider a simple qualitative model to estimate the time–scale for neutron → strange matter decay in dense stellar environments. It is argued that a large mismatch between the former and the microscopic weak interaction time–scale suggests that a dual population of both types of compact objects is unlikely. Assuming the correctness of the strange matter hypothesis all of them should be strange stars. If one instead postulates accretion as the decisive feature for the conversion, a consideration of neutron stars structure indicates a fairly narrow range for the onset of the critical density before the corresponding Chandrasekhar mass is achieved.

067.049   **Formation of the pulse profile of the Hercules X–1 pulsar. Mathematical simulation.**
M. B. Averintsev, L. G. Titarchuk, E. K. Sheffer.
Sov. Astron., Vol. 36, No. 1, p. 35 – 40 (Jan 1992). Current Physics Microform No.: 9207Y0599. English translation of Astron. Zh., Tom 69, Vyp. 1, p. 71 – 81 (Jan–Feb 1992).
The eclipse of the emitting region on the surface of the neutron star by the nearby inner edge of the accretion disk is simulated mathematically. The results are compared with the light curve constructed from observations of the X–ray emission of the Her X–1 pulsar. Such a comparison makes it possible to narrow the range of a number of geometrical parameters of the Her X–1/HZ Her binary system.

067.050   **Equilibrium configurations of degenerate neutron stars with frozen–in superstrong magnetic fields.**
G. A. Shul'man.
Sov. Astron., Vol. 36, No. 1, p. 58 – 62 (Jan 1992). Current Physics Microform No.: 9207Y0622. English translation of Astron. Zh., Tom 69, Vyp. 1, p. 116 – 124 (Jan–Feb 1992).
The equilibrium mass of a degenerate, magnetized neutron star as a function of magnetic field strength is estimated by an approximate energy method for a number of values of the density at its center, $\varrho_c \gtrsim 1.67 \cdot 10^{14}\,\mathrm{g\cdot cm^{-3}}$. It is shown that for a magnetic field strength $H_c \gtrsim 10^{17}\,\mathrm{G}$ at its center, the mass of a magnetized neutron star exceeds that of a similar star with a weaker field. It is also shown that an equilibrium magnetized neutron star may be stable for a number of values of H_c. If the magnetic field strength H_c reaches values close to $10^{18}\,\mathrm{G}$, the contributions of the electron and neutron gases to the pressure in the stellar interior are about the same.

067.051   **Decay of weak poloidal magnetic fields in the liquid layer of neutron star envelopes.**
U. Geppert, H.–L. Wiebicke.
Astron. Astrophys., Vol. 256, No. 1, p. L9 – L10 (Mar 1992). Letter–to–the–editor.
If a strong temperature gradient exists in the liquid layer of neutron star envelopes the interaction of the heat flux with the magnetic field is very effective. For weak magnetic fields ($B \lesssim 10^9\,\mathrm{G}$) the induction and heat transport equations describing this interaction can be linearized. In this case the poloidal magnetic field component is not influenced by the evolution of both, the toroidal field and the temperature distribution. However, the strong radial temperature gradient inside the liquid layer acts on the poloidal field by the thermal drift velocity, which leads to a non–negligible convective term in the induction equation. The hotter the neutron star surface is, the larger is the temperature gradient and the more slowly the poloidal field in this region decays ($\tau_p \gtrsim 40000\,\mathrm{yr}$). During typical timescales of

small–scale toroidal field growth ($\tau_i \approx 1\,\mathrm{yr}$) due to thermoelectricity the poloidal field is nearly stationary.

067.052   **Mechanics of apparent horizons.**
W. Collins.
Phys. Rev. D, Vol. 45, No. 2, p. 495 – 498 (15 Jan 1992). Current Physics Microform No.: 9203G1029.
An equation for the variation in the surface area of an apparent horizon is derived which has the same form as the thermodynamic relation $T \cdot dS = dQ$. For a stationary vacuum black hole, the expression corresponding to a temperature equals the temperature of the event horizon. Also, if the black hole is perturbed infinitesimally by weak matter and gravitational fields, the area variation of the apparent horizon asymptotically approaches the Hartle–Hawking result for the event horizon. These results support the idea that a local version of black hole thermodynamics in nonstationary systems can be constructed for apparent horizons.

067.053   **Amplification of the black hole Hawking radiation by stimulated emission.**
J. Audretsch, R. Müller.
Phys. Rev. D, Vol. 45, No. 2, p. 513 – 519 (15 Jan 1992). Current Physics Microform No.: 9203G1047.
If bosonic particles or antiparticles are sent in during the collapse to a Schwarzschild black hole, they stimulate an additional creation of particle – antiparticle pairs out of the curved background. This results in an induced amplification of the thermal Hawking radiation. Using wave–packet modes, the authors show that the packet of incoming particles reappears in the outer region at a time and with an energy which corresponds to the propagation and redshift of classical ray optics. It is accompanied by additional particles of the same energy. Their intensity varies with energy according to the thermal Hawking spectrum and is proportional to the number of incoming particles. Another contribution of the same structure goes back to antiparticles coming in at a different trajectory. They stimulate the emission of additional pairs. The respective antiparticles fall through the horizon and their particle partners add to the outgoing particle content. A localization of the processes is obtained.

067.054   **The soliton bag model of the quark star.**
R. Mańka, I. Bednarek, J. Syska.
Mon. Not. R. Astron. Soc., Vol. 254, No. 1, p. 87 – 92 (1 Jan 1992).
Using an analogy to a soliton model of hadrons the existence of a hot quark star is examined.

067.055   **The evolution of a black hole's force–free magnetosphere.**
I. Okamoto.
Mon. Not. R. Astron. Soc., Vol. 254, No. 2, p. 192 – 220 (15 Jan 1992).
This paper elucidates the structure of a stationary axisymmetric force–free magnetosphere of a Kerr black hole, and the hole's evolution due to extraction of rotational energy by the Blandford–Znajek process.

067.056   **On the construction of a simple model pulsar magnetosphere.**
L. Mestel, M. H. L. Pryce.
Mon. Not. R. Astron. Soc., Vol. 254, No. 2, p. 355 – 360 (15 Jan 1992).
The simplest example illustrating the effect of magnetospheric charges on a pulsar magnetic field has the region within the light cylinder filled with the Goldreich–Julian charge density which corotates with the star. It is shown that two superficially different techniques for constructing the magnetic field are formally equivalent. The relative advantages of each representation are briefly discussed.

067.057   Disc oscillation model for quasi–periodic light variations in cataclysmic variables.
T. Okuda, K. Ono, M. Tabata, S. Mineshige.
Mon. Not. R. Astron. Soc., Vol. 254, No. 3, p. 427 – 434 (1 Feb 1992).

A radial oscillation model of accretion discs for quasi–periodic light variations in cataclysmic variables is examined by means of numerical simulations. It is known that when the α model is used for the viscosity, accretion discs around compact objects are overstable against radial axisymmetric oscillations under a variety of conditions. The authors' calculations demonstrate that radial oscillations of the accretion discs around white dwarfs yield quasi–periodic oscillations (QPOs) in the disc of luminosity L_d. The calculated QPO periods are ~ 80–400 s for $\alpha = 0.1$ and for the mass–transfer rate of $\sim 10^{17}$–10^{19} g s^{-1}. The oscillation amplitudes range from ~ 0.3 to 0.6% of L_d and are higher for high L_d. These characteristics are in good agreement with the observations.

067.058   The perturbations of a fully general relativistic and rapidly rotating neutron star. I. Equations of motion for the solid crust.
D. Priou.
Mon. Not. R. Astron. Soc., Vol. 254, No. 3, p. 435 – 452 (1 Feb 1992). With microfiche MN 254/1.

The general relativistic theory of elasticity developed by Carter and Quintana is applied to the study of the solid crust of a fully general relativistic and rapidly rotating neutron star, the equations of motion of which are set up in the case of a uniform angular velocity. This unperturbed configuration is then perturbed in the most general way in order to write explicitly the general equations of motion of the solid part of the neutron star. No choice of gauge is made: keeping this gauge freedom as long as possible will simplify the comparison with other investigations, for which different choices of gauge were made; moreover, writing the equations of motion before choosing any gauge will help one to find the one by means of which they simplify as much as possible. The perturbed stress–energy tensor of elasticity is then studied together with the Lagrangian strain tensor and the constant volume strain tensor.

067.059   Neutron stars and planet–mass companions.
I. R. Stevens, M. J. Rees, P. Podsiadlowski.
Mon. Not. R. Astron. Soc., Vol. 254, No. 3, p. 19P – 22P (1 Feb 1992).

The authors propose a formation mechanism for the recently discovered planet–mass companions to neutron stars, namely that the stellar companion of a millisecond pulsar is disrupted by rapid mass loss via an evaporative wind driven by the pulsar radiation. Because the star loses mass on a time–scale short compared to its thermal time–scale, it expands and overspills its Roche lobe. This process can result in the disruption of the stellar companion and the formation of a massive disc around the neutron star. This disc then evolves, in a manner analogous to the solar nebula, to form planet–mass objects around the neutron star.

067.060   Light curves of rotating, oscillating neutron stars.
T. E. Strohmayer.
Astrophys. J., Vol. 388, No. 1, p. 138 – 147 (20 Mar 1992).

The author has developed a technique for computing the light curve produced by a rotating, oscillating neutron star that emits radiation from circular polar cap regions. The model includes the effects of general relativity on the photon trajectories and allows for anisotropic (beamed) emission from the stellar surface. It also allows general orientations of the line of sight, neutron star rotation axis, and polar cap emission region. The author adopts a spherical harmonic decomposition to describe the angular dependence of the time–dependent intensity produced by the stellar oscillations. Several examples are given of light curves produced by single, low–order ($l = 1, 2$) oscillation modes. Using a Gaussian beaming function, the author has simulated typical radio pulsar beam widths in order to investigate a neutron star oscillation model for subpulse drift in pulsars. He has also simulated X–ray bursts and X–ray pulsars to assess the possibility of detecting such oscillations in these sources with XTE and AXAF.

067.061   Proton acceleration in neutron star magnetospheres.
I. A. Smith, J. I. Katz, P. H. Diamond.
Astrophys. J., Vol. 388, No. 1, p. 148 – 163 (20 Mar 1992).

To explain the emission of TeV and PeV gamma rays from accreting X–ray binary sources, protons must be accelerated to several times the gamma–ray energy. The authors proposed previously that protons could be accelerated to these energies in the outer regions of a neutron star's closed magnetosphere. They show here that, at certain times, the plasma in the accretion column of the neutron star may form a deep enough pool that the top portion becomes unstable to convective motions in spite of the strong magnetic field. The resulting turbulence produces fluctuations in the strength of the magnetic field that travel up the accretion column, taking energy out to the region of the energetic protons. The protons resonantly absorb this energy and are accelerated to high energies. Including the synchrotron radiation losses of the protons, the authors show that the protons can be accelerated to energies that are high enough to explain the gamma–ray observations.

067.062   Gamma–ray bursts and cosmic rays from accretion–induced collapse.
A. Dar, B. Z. Kozlovsky, S. Nussinov, R. Ramaty.
Astrophys. J., Vol. 388, No. 1, p. 164 – 170 (20 Mar 1992).

Accretion of matter onto the surface of a white dwarf in a binary system can push it over the Chandrasekhar mass limit and may cause it to collapse into a naked or nearly naked neutron star without detectable optical emission. Such an optically quiet neutron star birth should be accompanied by a neutrino burst which could be detected with underground neutrino detectors only if the collapse took place in our own Galaxy or in very close nearby galaxies. However, neutrino – antineutrino annihilation outside the neutron star into electron – positron pairs will produce a gamma–ray burst that can be observed out to distances of at least 300 Mpc, if the mass surrounding the newly formed neutron star is less than about $3 \times 10^{-4} M_\odot$. Part of this mass will be injected into the interstellar space with energy above 10 MeV per nucleon. Such nuclei can be further accelerated to cosmic–ray energies in the interstellar space. Thus, accretion–induced collapse may be an important source of cosmic rays and of cosmological gamma–ray bursts. Conversely, the observed rate of gamma–ray bursts and cosmic–ray data can be used to limit the birthrate of naked, or nearly naked, neutron stars to less than one per 10^3 yr in galaxies similar to ours. This rate is too small to contribute significantly to the birthrate of pulsars, and it implies that it is very unlikely that a neutrino burst unaccompanied by optical emission will be detected in the near future by underground neutrino detectors.

067.063   Gravitational radiation from supermassive black holes.
H. E. Kandrup, M. E. Mahon.
Phys. Rev. D, Vol. 45, No. 4, p. 1013 – 1016 (15 Feb 1992). Current Physics Microform No.: 9204D1623.

If the cores of most galaxies contain supermassive black holes, galaxy – galaxy collisions could lead to their coalescence and a consequent emission of gravitational radiation. For holes with masses $\geqslant 10^6 M_\odot$, the amplitude of this radiation should be sufficiently large (metric perturbation $\sim 10^{-18}$–10^{-17}) as to be detectable at the present horizon distance of $\sim 10^{10}$ light years, using current technology. It is shown here that there is solid evidence indicating that galaxy collisions could have been sufficiently frequent at early times (redshifts $z \sim 2 - 3$) to lead to a rate of potentially observable events as short as one every $\sim 1 - 100$ yr.

067.064   On radial oscillations in viscous accretion discs surrounding neutron stars.
X. Chen, R. E. Taam.
Mon. Not. R. Astron. Soc., Vol. 255, No. 1, p. 51 – 60 (1 Mar 1992).

Radial oscillations resulting from axisymmetric perturbations in viscous accretion discs surrounding neutron stars in X–ray binary systems have been investigated. Within the framework of the α–viscosity model a series of hydrodynamic calculations demonstrates that the oscillations are global for $\alpha \sim 1$. On the other hand, for $\alpha \lesssim 0.4$ the oscillations are local and are confined to the disc boundaries. If viscous stresses acting in the radial direction are included, however, it is found that the disc can be stabilized. The application of such instabilities in accretion discs, without reference to the boundary layer region between the neutron star (or magnetosphere) and the inner edge of the disc, to the phenomenology of quasi–periodic oscillations is brought into question.

067.065   Construction of three–dimensional black hole initial data via multiquadrics.
M. R. Dubal.
Phys. Rev. D, Vol. 45, No. 4, p. 1178 – 1187 (15 Feb 1992). Current Physics Microform No.: 9204D1788.
  Numerical solutions of the $3+1$ Hamiltonian constraint equation for single black hole initial data are presented. When expressed in Cartesian coordinates the solutions for the conformal factor are fully three–dimensional, and therefore the approach described here can straightforwardly produce initial data for two or more black holes with arbitrary positions, spins, and linear momenta. The numerical method used is the multiquadric approximation scheme.

067.066   General relativistic electric potential drops above pulsar polar caps.
A. G. Muslimov, A. I. Tsygan.
Mon. Not. R. Astron. Soc., Vol. 255, No. 1, p. 61 – 70 (1 Mar 1992).
  The authors study the general relativistic electrodynamics of an isolated, rotating, magnetic neutron star. They consider the region of a neutron star magnetosphere with steady, space charge limited flow along open magnetic field lines. The explicit solutions to the Maxwell equations are obtained. Being the simplest, this model enables one to carry out analytically a general relativistic treatment, and to demonstrate the influence of the effects of General Relativity on the creation of an electric field in the afore–mentioned region. Of particular importance is the effect of dragging of inertial frames of reference. The incorporation of this effect leads to a new result: the authors demonstrate the possibility of generation of an electric field component, which is purely general relativistic in origin and is proportional to $\cos \chi$ (where χ is the inclination angle of a pulsar). For an oblique rotator the characteristic magnitude of this electric field is, approximately, $4 \times 10^2 (r_g/a) P^{1/2}$ times stronger than that in a flat space–time limit. An interesting consequence of this analysis is that for typical pulsar parameters ($P \cong 1$ s, $B \cong 10^{12}$G) the development of an electron – positron avalanche is possible at a characteristic height of approximately a stellar radius above the surface. In addition, the authors point out that the heating of a polar cap by the return flow of positrons can be less efficient, because the power put into their acceleration is reduced by a factor of $6 \times 10^{-3} (r_g/a) P^{-1/2}$ compared to the total energy loss rate on the acceleration of electrons.

067.067   W–modes: a new family of normal modes of pulsating relativistic stars.
K. D. Kokkotas, B. F. Schutz.
Mon. Not. R. Astron. Soc., Vol. 255, No. 1, p. 119 – 128 (1 Mar 1992).
  The authors demonstrate explicitly the existence of a new family of outgoing–wave normal modes of pulsating relativistic stars, the first such family known that has no analogue in Newtonian stars. These modes were discovered earlier by the authors in a toy model, where they were called strongly damped normal modes. Kojima then found the first examples of these modes in realistic spherical polytropic stellar models. Here the authors give a number of arguments that demonstrate the existence of this family unequivocally, and they calculate a large number of eigenfrequencies. An interesting feature of w–modes is that the lowest order mode of each sequence has a frequency similar to that of the lowest order mode of a spherical black hole.

For higher modes, the spectrum diverges from the black–hole spectrum, but shows remarkable similarity to that of the strongly damped modes of the toy problem. As carriers of gravitational–wave information, w–modes may be important and observable in the burst of gravitational radiation that follows the formation of a neutron star. They should also be essential in solving the problem of the completeness of the outgoing–wave normal modes of radiating systems.

067.068   Model atmospheres for neutron stars.
M. C. Miller.
Mon. Not. R. Astron. Soc., Vol. 255, No. 1, p. 129 – 145 (1 Mar 1992).
  While the presence of an atmosphere on a neutron star will not significantly modify its total thermal emission, it may change the emission in the sensitivity bands of detectors such as Einstein or ROSAT so that the inferred surface temperature (from a blackbody curve) may be quite different from the actual surface temperature. This in turn may affect deduced cooling curves. Previous calculations of model atmospheres of neutron stars have used atomic data calculated for zero magnetic field. However, many neutron stars are expected to have extremely high magnetic fields, on the order of $B \geqslant 10^{12}$G, and it is important to take this into account. This paper uses atomic data in high magnetic fields computed using a multiconfigurational Hartree–Fock code, and the data were presented in Miller and Neuhauser. The effects of ionization and polarization in strong magnetic fields are discussed, and the prospects for observation by satellites are investigated.

067.069   Dynamical effects of radiation pressure due to synchrotron absorption in turbulent spherical accretion.
A. Mason.
Mon. Not. R. Astron. Soc., Vol. 255, No. 2, p. 203 – 209 (15 Mar 1992).
  Models of turbulent spherical accretion with magnetic dissipation for black hole masses of $M = 10$ and $10^8 M_\odot$ and accretion rates of $\dot{M} \sim \dot{M}_c$ ($\equiv L_{Edd}/c^2$) are considered. With magnetic fields $B \sim 10^4$–10^7G and temperatures of the order of 10^9–10^{11}K near the hole, emission and self–absorption of cyclo–synchrotron radiation is important. The dynamical effects of the radiation pressure due to cyclo–synchrotron absorption on the infalling gas are investigated. The large cross–section for synchrotron absorption processes ($\gg \sigma_T = 6.65 \times 10^{-25}$cm^2) produces a radiation pressure that might balance and even overcome the gravitational attraction. The author discusses a semi–analytical criterion for the existence of an unsteady flow for accretion models with M typical of a stellar black hole and a black hole in a galactic nucleus, and with $\dot{M} \sim \dot{M}_c$ and luminosities of $\sim 10^{-2} L_{Edd}$.

067.070   Rotational properties of strange stars.
M. Colpi, J. C. Miller.
Astrophys. J., Vol. 388, No. 2, p. 513 – 520 (1 Apr 1992).
  The authors present results from an investigation of the rotational properties of strange stars, using models with a canonical value of the bag constant. The changes in structure resulting from uniform rotation have been calculated within the slow rotation regime and the minimum rotation periods consistent with stability to nonaxisymmetric perturbations have also been calculated. The minimum period is found to be set by the onset of instability in either the $m = 2$ or $m = 3$ mode. The first of these modes, which is probably inaccessible to standard neutron stars, may be the critical one for old strange stars spun up by accretion and this could be of importance in giving an observational test for distinguishing between strange stars and standard neutron stars.

067.071   Energy dependence of normal branch quasi–periodic intensity oscillations in low–mass X–ray binaries.
G. S. Miller, F. K. Lamb.
Astrophys. J., Vol. 388, No. 2, p. 541 – 554 (1 Apr 1992).
  The properties of the ~ 6 Hz quasi–periodic X–ray intensity oscillations observed in the low–mass X–ray binary Cyg X–2 when it is on the normal spectral branch, are shown to be

consistent with a model in which protons from a central source with a fixed spectrum are Comptonized by an oscillating radial inflow. As the electron scattering optical depth of the flow varies, the spectrum of the escaping X–rays appears to rotate about a pivot energy E_p that depends mainly on the electron temperature in the flow. The temperature derived from the observed energy dependence of the Cyg X–2 normal branch oscillations is approximately 1 keV, in good agreement with the estimated Compton temperature of its X–ray spectrum. The mean optical depth τ of the Comptonizing flow is inferred to be about 10, while the change in τ over an oscillation is estimated to be about 1. The effect of induced scattering is investigated and used to place an approximate lower bound on the volume of the Comptonizing region. The authors conjecture that the 6 Hz normal branch oscillations observed in other Z–class sources are also produced largely by oscillations in the degree of Comptonization of a central source.

067.072   CNO destruction by spallation and type I X–ray bursts.
J. C. Tillett, J. MacDonald.
Astrophys. J., Vol. 388, No. 2, p. 555 – 560 (1 Apr 1992).
Recent work on the surface boundary conditions of accreting neutron stars (Bildsten) indicates that appreciable amounts of CNO elements can be destroyed by spallation. Previous authors assumed an accretion of near–solar abundance material which is not altered by boundary effects. Here the authors construct steady state models of accreting neutron stars assuming all but traces of metals are destroyed due to spallation in the surface layers. The authors also construct models with solar accretion for comparison. The models are compared with EXOSAT observations of 4U/MXB 1636–53. It is found that the simplifications introduced by assuming CNO destruction do not bring theory closer to describing what is observed.

067.073   Physical properties of a soliton black hole at finite temperature.
R. Pan, R. Su.
Phys. Rev. D, Vol. 45, No. 6, p. 2144 – 2146 (15 Mar 1992).
Current Physics Microform No.: 9204H1530.
It is shown that the nontopological scalar black hole suggested by Friedberg, Lee, and Pang is dynamically stable at finite temperature. The heat capacity of a scalar soliton black hole is positive. The physical properties of a scalar black hole at finite temperature are discussed.

067.074   Photon bubbles: overstability in a magnetized atmosphere.
J. Arons.
Astrophys. J., Vol. 388, No. 2, p. 561 – 578 (1 Apr 1992).
Linear stability theory is used to study the formation of "photon bubbles" in a convectively stable scattering atmosphere supported against gravity entirely by radiation pressure. A WKB analysis shows that internal waves in an isothermal magnetized atmosphere are overstable, when the magnetic pressure of a vertical magnetic field satisfies $B^2/8\pi > (M_0/2\gamma)p_{rad}$. Here $M_0 \ll 1$ is the ratio of the sound crossing time to the photon diffusion time in the atmosphere and γ is the ratio of specific heats. The unstable waves are buoyant, rising with a group velocity $\sim c_0 M_0$, where c_0 is the isothermal sound speed. The most unstable waves correspond to the formation of long vertical fingers of radiation where the density is depleted relative to the surroundings. A simple model is developed for the two–dimensional structure of a plasma mound formed by laminar accretion onto the magnetic poles of a neutron star, in which upward photon diffusion balances downward photon advection with the plasma. It is shown that photon bubbles would form in a polar accretion mound under the conditions expected in accretion–powered pulsars within a few tenths of a millisecond. Because long–wavelength modes have the largest rise speeds, eventual dominance by a few large bubbles is suggested, and possible connections between bubble formation and short–time variability in accretion–powered pulsars is briefly discussed, as well as a possible connection of the photon bubble phenomenon

to the rapid time variability observed in the Rapid Burster and in quasi–periodic oscillator sources.

067.075   The "inner–horizon thermodynamics" of Kerr black holes.
I. Okamoto, O. Kaburaki.
Mon. Not. R. Astron. Soc., Vol. 255, No. 3, p. 539 – 544 (1 Apr 1992).
For Kerr black holes, it is argued that the thermodynamics of the event horizons and of the inner horizons are described in a unified manner by using a dimensionless parameter h. It is defined by $h = a/r_H \equiv h_+$ on the event horizons and varies in the range of $0 \leqslant h \leqslant 1$, while it is also given by $h = r_H/a = a/r_- \equiv h_-$ on the inner horizons in the range of $1 \leqslant h \leqslant \infty$, where $a = J/Mc$ is the angular momentum radius and $r_H \equiv r_+$, r_- are the radii of the outer and inner horizons respectively. Every thermodynamic variable, other than the mass M and the angular momentum J, is also defined on both the outer and inner horizons, and its values form a pair of the roots of a quadratic equation with M and J contained in the coefficients, similar to the horizon radii $r_\pm$ and $h_\pm$. It is also shown that this pair, of each variable, constitute mirror images to each other with respect to the extreme Kerr state.

067.076   Constraints on the critical temperature of the phase transition to the quark–gluon plasma.
L. A. Kondratyuk, B. V. Martemyanov, M. I. Krivoruchenko.
Sov. J. Nucl. Phys., Vol. 55, No. 6, p. 914 – 916 (Jun 1992).
Current Physics Microform No.: 9208X0754.
Low temperatures of the phase transition to the quark–gluon plasma correspond to low values of the bag model constant and to absolutely stable strange quark matter. Some of the observed pulsars are identified quite reliably as neutron stars. If strange matter is stable, the central density of these pulsars has to be smaller than the critical density of the phase transition to nonstrange quark matter. The nonstrange quark matter being formed turns to more stable strange matter on a weak interaction timescale, converting neutron stars into strange stars. The requirement of stability of old and newly born neutron stars is used to constrain the bag model constant and the critical temperature of the phase transition to the quark–gluon plasma at zero chemical potential.

067.077   Cauchy horizon instability for Reissner–Nordström black holes in de Sitter space.
P. R. Brady, E. Poisson.
Classical Quantum Gravity, Vol. 9, No. 1, p. 121 – 125 (Jan 1992).
The radiative tail of gravitational and electromagnetic perturbations accumulates at the Cauchy horizon of a Reissner–Nordström black hole, producing an instability of that horizon. This instability is usually interpreted as an infinite proper–time compression effect: a free–falling observer crossing the Cauchy horizon sees an infinite number of waves within a finite proper time, hence measuring infinite energy densities. Here the authors show, using a simple model in which a Reissner–Nordström black hole immersed in de Sitter space is perturbed by a radial stream of infalling lightlike particles, that proper–time compression, though sufficient, is not a necessary condition for the instability of the Cauchy horizon. In this model, the Cauchy horizon is unstable whenever the surface gravity of the Cauchy horizon is greater than that of the cosmological horizon, despite the fact that there is no corresponding infinite proper–time compression.

067.078   Iron Kα line from X–ray illuminated relativistic disks.
G. Matt, G. C. Perola, L. Piro, L. Stella.
Astron. Astrophys., Vol. 257, No. 1, p. 63 – 68 (Apr 1992).
The intensity and profile of the iron Kα fluorescence line from a flat, optically thick accretion disk rotating around a Schwarzschild black hole and illuminated by a central X–ray source are computed using a fully relativistic treatment of the photon intensity and shifts. The X–ray source is modelled as an isotropic point source located on the symmetry axis at a height h in units of

the gravitational radius. These calculations represent a refinement and an extension of those presented by Matt et al. (1991) for $h = 20$, carried out using a weak field approximation and therefore of validity limited to inclination angles $\lesssim 70°$. Here it is shown that at high inclination angles purely relativistic effects lead to the growth of features between the two Doppler horns and that, as a consequence, the line equivalent width maintains a sizeable value, while the centroid energy and the line width go through a broad maximum at about $\sim 80°$. The statistical implications for the expected distribution of the line parameters in a sample of randomly oriented disks in Seyfert galaxies are briefly discussed.

067.079 **Black hole normal modes: phase–integral treatment.**
N. Fröman, P. O. Fröman, N. Andersson, A. Hökback.
Phys. Rev. D, Vol. 45, No. 8, p. 2609 – 2616 (15 Apr 1992).
Current Physics Microform No.: 9206F1335.
A simple phase–integral formula, valid in an arbitrary order of approximation, for the determination of the normal–mode frequencies of a Schwarzschild black hole is derived rigorously. Numerical results obtained from this formula show that the phase–integral method in general yields the normal–mode frequencies more accurately than previous approximate analytical treatments.

067.080 **Quasinormal modes of Schwarzschild black holes: defined and calculated via Laplace transformation.**
H. P. Nollert, B. G. Schmidt.
Phys. Rev. D, Vol. 45, No. 8, p. 2617 – 2627 (15 Apr 1992).
Current Physics Microform No.: 9206F1343.
Quasinormal modes play a prominent role in the literature when dealing with the propagation of linearized perturbations of the Schwarzschild geometry. The authors show that space–time properties of the solutions of the perturbation equation imply the existence of a unique Green's function of the Laplace transformed wave equation. This Green's function may be constructed from solutions of the homogeneous time–independent equation, which are uniquely characterized by the boundary conditions they satisfy. The authors show that quasinormal–mode frequencies can be defined as the poles of the Green's function for the Laplace transformed equation. On the basis of this definition a new technique for the numerical calculation of quasinormal frequencies is developed.

067.081 **Quantum–mechanical scattering of charged black holes.**
J. Traschen, R. Ferrell.
Phys. Rev. D, Vol. 45, No. 8, p. 2628 – 2635 (15 Apr 1992).
Current Physics Microform No.: 9206F1354.
The authors describe the quantum–mechanical scattering of slowly moving maximally charged black holes. They develop a canonical quantization procedure on the parameter space of possible static classical solutions. With this, the authors compute the capture cross sections for the scattering of two black holes. Finally, they discuss how quantization on this parameter space relates to quantization of the degrees of freedom of the gravitational field.

067.082 **Black holes in magnetic monopoles.**
K. Lee, V. P. Nair, E. J. Weinberg.
Phys. Rev. D, Vol. 45, No. 8, p. 2751 – 2761 (15 Apr 1992).
Current Physics Microform No.: 9206F1477.
The authors study magnetically charged classical solutions of a spontaneously broken gauge theory interacting with gravity. They show that nonsingular monopole solutions exist only if the Higgs–field vacuum expectation value v is less than or equal to a critical value v_{cr}, which is of the order of the Planck mass. In the limiting case, the monopole becomes a black hole, with the region outside the horizon described by the critical Reissner–Nordström solution. For $v < v_{cr}$, the authors find additional solutions which are singular at r = 0, but which have this singularity hidden within a horizon. These have nontrivial matter fields outside the horizon, and may be interpreted as small black holes lying within a magnetic monopole. The nature of these solutions as a function

of v and of the total mass M and their relation to the Reissner–Nordström solutions are discussed.

067.083 **Euclidean black–hole vortices.**
F. Dowker, R. Gregory, J. Traschen.
Phys. Rev. D, Vol. 45, No. 8, p. 2762 – 2771 (15 Apr 1992).
Current Physics Microform No.: 9206F1488.
The authors demonstrate the existence of solutions of the Euclidean Einstein equations that correspond to a vortex sitting at the horizon of a black hole. They find the asymptotic behaviors, at the horizon and at infinity, of vortex solutions for the gauge and scalar fields in an Abelian Higgs model on a Euclidean Schwarzschild background and interpolate between them by integrating the equations numerically. Calculating the back reaction shows that the effect of the vortex is to cut a slice out of the Euclidean Schwarzschild geometry. The consequences of these solutions for black hole thermodynamics are discussed.

067.084 **Gamma–ray bursts from planet – magnetosphere systems around neutron stars.**
H. Hanami.
Astrophys. J., Lett., Vol. 389, No. 2, p. L71 – L74 (20 Apr 1992).
The author studied a planet – magnetosphere interaction model for γ–ray bursts using the model of a current circuit which consists of the magnetosphere and the surface of a neutron star. The motion of planets or comets in the magnetosphere works as the battery in the circuit system. The physical condition on the surface of the neutron star is important to make a closed circuit inducing good conversion from the kinetic energy of a planet to that of the magnetosphere oscillation. An old and cooled neutron star with a temperature ≈ 10 eV can prepare the conditon for closing the current circuit in the surface of the neutron star. The planet – magnetosphere system is unstable to a feed–back instability, related to the variability of the γ–ray bursts.

067.085 **Quantum hair and quantum gravity.**
S. Coleman, L. M. Krauss, J. Preskill, F. Wilczek.
Gen. Relativ. Gravitation, Vol. 24, No. 1, p. 9 – 16 (Jan 1992).
A black hole may carry quantum numbers that are not associated with massless gauge fields, contrary to the spirit of the "no–hair" theorems. The "quantum hair" is invisible in the classical limit, but measurable via quantum interference experiments. Quantum hair alters the temperature of the radiation emitted by a black hole. It also induces non–zero expectation values for fields outside the event horizon; these expectation values are non–perturbative in $h/2\pi$, and decay exponentially far from the hole. The existence of quantum hair demonstrates that a black hole can have an intricate quantum–mechanical structure that is completely missed by standard semiclassical theory.

067.086 **The ferromagnetic transition and the neutron star models.**
Huang Wenhong, Yu Jiuwei, Gao Shanghui.
4. Asia Pacific Physics Conference, p. 1242 – 1246 (1991). – See Abstr. 012.006 for the main entry.
The effect of each component of a two–body central potential on the critical density for ferromagnetic transition is calculated for neutron matter by applying the lower order constraint variation method of Owen. The results show that the radius of the hard core play an important role in the critical density. For neutron matter with M–S interaction, the critical density of the ferromagnetic transition, the equation of state and the neutron star models are calculated. The maximum mass of the neutron star with the ferromagnetic phase is lower than that without the ferromagnetic phase.

067.087 **Self–gravitating thin disks around rapidly rotating black holes.**
A. Lanza.
Astrophys. J., Vol. 389, No. 1, p. 141 – 156 (10 Apr 1992).
Sequences of equilibrium numerical models have been constructed for self–gravitating thin disks encircling rapidly rotating black holes. The multigrid method was used for solving numerically the stationary, axisymmetric Einstein equations. The

black hole is described either by specifying its angular velocity ω_H and the coordinate radius of the horizon h/2 or, alternatively, by specifying the area of the horizon A_H and the angular momentum J_H.

067.088   Electrical conductivity of neutron star cores in the presence of strong magnetic fields: effects of interactions and superfluidity of nucleons.
E. Østgaard, D. G. Yakovlev.
Nucl. Phys. A, Vol. 540, No. 1/2, p. 211 – 226 (13–20 Apr 1992).

The electric resistivity tensor of npe–matter in the cores of neutron stars with magnetic field B is calculated for the case when the protons are nonsuperfluid and the neutrons are either normal or superfluid due to the singlet–state pairing. The tensor contains the longitudinal, transverse and Hall resistivities, $R_\parallel$, $R_\perp$, and R_H. Their explicit expressions are obtained for densities ϱ around the standard nuclear matter density $\varrho = 2.8 \times 10^{14}$ g cm^{-3} in terms of nucleon Fermi momenta and effective masses renormalized due to medium effects. $R_\parallel$ is equal to the field–free resistivity R_0, $R_H \sim B$; neither of them depends on the neutron superfluidity. $R_\perp = R_0 + R_B$ contains the term $R_B \sim B^2$ which greatly enhance $R_\perp$, if B is strong. When $T \geqslant T_c$, R_B scales as T^{-2}. If $0.2\, T_c \lesssim T < T_c$, R_B becomes larger than its nonsuperfluid value. At lower T, on the contrary, R_B is greatly suppressed by the superfluidity. These results can be used to study the cooling rates of neutron stars with large internal magnetic fields.

067.089   Cosmic string winding around a black hole.
A. L. Larsen.
Phys. Lett. B, Vol. 273, No. 4, p. 375 – 379 (26 Dec 1991).

The author considers a charged circular cosmic string of the 'degenerate' kind which can be described by the Nambu–Goto action, winding around a Kerr–Newman black hole in the equatorial plane, and look for stable stationary configurations in some special cases.

067.090   Supernova and neutron stars with relativistic equations of state.
K. Sumiyoshi, H. Toki, R. Brockmann.
Phys. Lett. B, Vol. 276, No. 4, p. 393 – 397 (20 Feb 1992).

The authors study the properties of supernova hot stars and neutron stars using equations of state at finite temperature with arbitrary proton–to–neutron ratios derived from the relativistic mean field theory. The coupling strengths and masses of the relativistic mean field theory are fixed by the known nuclear properties and describe also unstable nuclei. They calculate also the star profiles with equations of state obtained by the relativistic Brueckner–Hartree–Fock theory, which reproduces the nuclear matter saturation property.

067.091   On the light curve of an orbiting spot.
V. Karas, G. Bao.
Astron. Astrophys., Vol. 257, No. 2, p. 531 – 533 (Apr 1992).

The authors investigate the influence of eclipses due to the thickness of an accretion disc on the light curve of an X–ray emitting spot located on the disc surface. The disc is assumed to orbit a compact object which is described by the Kerr metric. General relativistic effects affecting photon paths and energy are treated within the framework of geometrical optics. The authors find that the eclipse curve on the disc surface has a "cusp" which is caused by gravitational lensing. They give an example of the light curve, which is significantly narrower compared to the case with no eclipse.

067.092   The signature of corotating spots in accretion disks.
G. Bao.
Astron. Astrophys., Vol. 257, No. 2, p. 594 – 598 (Apr 1992).

The author investigates the general relativistic effects of a spot co–moving with the accretion disk around a black hole. The observed flux increases when the spot is behind the black hole due to gravitational lensing (via an increase in the observed solid angle) and, to a lesser extent, due to the gravitational corrections to the Doppler shift. The Doppler effect and the gravitational lensing make the X–ray signal non-sinusoidal, which flattens the

variability power spectrum of many spots and weakens the predicted "cut–off" feature at the high–frequency end. The results may explain why so far the sharp roll–off phenomenon at high frequency in X–ray variability power spectrum of AGN has not been observed.

067.093   Soft gamma rays from black holes versus neutron stars.
E. P. Liang.
Compton Observatory Science Workshop, p. 173 – 184 (Feb 1992). – See Abstr. 012.018 for the main entry.

The recent launches of GRANAT and GRO provide unprecedented opportunities to study compact collapsed objects from their hard X–ray and gamma ray emissions. The spectral range above 100 keV can now be explored with much higher sensitivity and time resolution than before. The author reviews the soft gamma ray spectral data of black holes and neutron stars, radiation and particle energization mechanisms and potentially distinguishing gamma ray signatures. He also outlines some of the highest priority future observations that will shed much light on such systems.

067.094   A burst from a thermonuclear runaway on an ONeMg white dwarf.
S. Starrfield, M. Politano, J. W. Truran, W. M. Sparks.
Compton Observatory Science Workshop, p. 377 – 386 (Feb 1992). – See Abstr. 012.018 for the main entry.

The authors have performed studies which examine the consequences of accretion, at rates of $10^{-9} M_\odot yr^{-1}$ and $10^{-10} M_\odot yr^{-1}$, onto an ONeMg white dwarf with a mass of 1.35 $M_\odot$. The initial abundance distribution corresponded to a mixture that was enriched to either 25%, 50%, or 75% in products of carbon burning. The remaining material in each case is assumed to have a solar composition. The evolution of the thermonuclear runaway on the 1.35 $M_\odot$ white dwarf produced peak temperatures in the shell source exceeding 300 million degrees. The sequence produced significant amounts of ^{22}Na from proton captures onto ^{20}Ne and significant amounts of ^{26}Al from proton captures on ^{24}Mg. This sequence ejected $5.2 \times 10^{-6} M_\odot$. When the mass accretion rate was decreased to $10^{-10} M_\odot$, the resulting thermonuclear runaway produced a shock that moved through the outer envelope of the white dwarf and raised the surface luminosity to L > $10^7 L_\odot$ and the effective temperature to values exceeding $10^7 K$. The interaction of the material expanding from off of the white dwarf with the accretion disk should produce a burst of γ–rays.

067.095   Near–critical spherical accretion onto magnetized neutron stars: modified magnetospheric radius.
A. Mitra.
Astron. Astrophys., Vol. 257, No. 2, p. 807 – 810 (Apr 1992).

The author envisages here a new magnetospheric phenomenon in that, for a suitable accretion geometry, the magnetosphere of an accreting X–ray pulsar can start receding (rather than contracting) as the accretion rate exceeds some transition value $(\dot{M}_{cr})$. To this effect, he considers the repulsive effects of the feedback radiation pressure on the accretion flow around a magnetized neutron star. In a simplified approach to the problem, it is found that the canonical relationship between magnetospheric radius (r_m) and mass accretion rate $(\dot{M})$, $r_m \sim \dot{M}^{-2/7}$ gets modified to $r_m \sim M^{-2/7} \varepsilon^{-1/7}$, where $\varepsilon \equiv (1 - L_\infty / L_{ed})$; L_∞ is the luminosity at infinity, and L_{ed} is the Eddington luminosity of the neutron star. More accurate treatment further modifies this relationship – but the important point to note here is that as $\dot{M} > \dot{M}_{tr} \sim 0.66\, \dot{M}_{ed}$, the magnetosphere starts expanding because of relatively stronger radiation pressure.

067.096   Wave and stability properties of black holes.
J. D. Krige, J. F. McKenzie.
9. Italian Conference on General Relativity and Gravitational Physics, p. 456 – 462 (1991). – See Abstr. 012.021 for the main entry.

The authors transform the wave equation governing gravitational perturbations of a Schwarzschild black hole from its standard Schrödinger or Regge–Wheeler form to a Klein–Gordon type wave equation. This latter form reveals that incoming waves with frequencies $\omega \ll (\gg) \omega_{cml}$, a critical frequency, are completely reflected (transmitted). Moreover, those high frequency waves ($\omega > \omega_{cml}$) which penetrate through to the region near the Schwarzschild radius r_s are, on crossing this event horizon, attenuated by a factor $\exp(-\pi\omega r_s/c)$. Also, local instability in the vicinity of r_s indicates that the neighbourhood around r_s is dynamically active, and, as well as acting like a Hawking type particle creator, will behave as a wave emitter in order to relax the "stresses" on the metric.

067.097 Superluminal jets and neutron star winds.
A. Lanza.
9. Italian Conference on General Relativity and Gravitational Physics, p. 537 – 542 (1991). – See Abstr. 012.021 for the main entry.

067.098 Profile formation of Her X–1 main pulse. Numerical simulation of the process.
M. B. Averintsev, L. G. Titarchuk, E. K. Sheffer.
Astron. Zh., Tom 69, Vyp. 1, p. 71 – 81 (Jan–Feb 1992). In Russian. English translation in Sov. Astron., Vol. 36, No. 1 (1992).
A numerical simulation of the process of occultation of the emitting region on the surface of the neutron star by the inner edge of the accretion disc was made. Calculation results were compared with the observed X–ray emission of Her X–1. Such comparison allowed to find restrictions for a number of geometrical parameters of the Her X–1/HZ Her system.

067.099 Equilibrium configurations of degenerate neutron stars with frozen–in superstrong magnetic fields.
G. A. Shul'man.
Astron. Zh., Tom 69, Vyp. 1, p. 116 – 124 (Jan–Feb 1992). In Russian. English translation in Sov. Astron., Vol. 36, No. 1 (1992).
In the framework of the approximate energy method, the equilibrium mass of a degenerate magnetized neutron star is estimated as a function of the magnetic field intensity for a number of given values of the matter density in the centre of the star $\varrho_c \gtrsim 1{,}67 \cdot 10^{14} \mathrm{g \cdot cm^{-3}}$. It is shown that in the presence of a superstrong magnetic field in the centre of the star, $H_c \gtrsim 10^{17}\mathrm{G}$, the mass of the magnetic star increases in comparison with the mass of the analogous star in the absence of the field. It is shown also that for a number of H_c values, the equilibrium magnetized neutron star may be stable. If the magnetic field H_c is as high as $10^{18}\mathrm{G}$, the contributions of the electron and neutron gases to the pressure in the stellar interior are nearly equal.

067.100 The evolution of the magnetic fields of neutron stars.
D. Bhattacharya.
IAU Colloquium No. 128: The magnetospheric structure and emission mechanisms of radio pulsars, p. 27 – 34 (1992). – See Abstr. 012.022 for the main entry.
The evolution of the magnetic field strength plays a major role in the life history of a neutron star. In this article the observational evidence of field evolution, in particular that of field decay and magnetic alignment, are critically examined. It is concluded that the observed decay of the spindown torque on radio pulsars cannot be caused by a secular evolution of the "obliqueness" of the neutron star, as suggested by some authors. Recent observations provide a strong indication that the decay of the magnetic field strength of a neutron star may be closely related to its evolution in a binary system. Theoretical models for such an evolution are discussed.

067.101 Planet survival?
K. A. Postnov, M. E. Prokhorov.
Astron. Astrophys., Vol. 258, No. 2, p. L17 – L18 (May 1992). Letter–to–the–editor.
A planet around a radiopulsar could survive supernova explosion of the pulsar progenitor if the planetary orbit were eccentric prior the explosion. The probability for the planet to

settle in a low eccentric orbit after the explosion is found to be sufficiently large.

067.102 Evolution of thermally generated neutron–star magnetic fields.
R. W. Romani, L. E. Hernquist.
IAU Colloquium No. 128: The magnetospheric structure and emission mechanisms of radio pulsars, p. 46 – 48 (1992). – See Abstr. 012.022 for the main entry.
The authors describe a new model of neutron star magnetic moments, assuming that the fields are generated at birth and following their evolution to ages as large as the Hubble time. With realistic thermal evolution and conductivities, isolated neutron stars will maintain large magnetic dipole fields. As suggested elsewhere field modification under mass accretion might lead to torque decay. The authors identify an operative mechanism for this process; the results of this unified picture are in agreement with observations of a wide range of neutron star systems.

067.103 Magnetospheric structure of rotation–powered neutron stars.
J. Arons.
IAU Colloquium No. 128: The magnetospheric structure and emission mechanisms of radio pulsars, p. 56 – 77 (1992). – See Abstr. 012.022 for the main entry.
The author surveys recent theoretical work on the structure of the magnetospheres of rotation–powered pulsars, within the observational constraints set by their observed spindown, their ability to power synchrotron nebulae and their ability to produce beamed collective radio emission, while putting only a small fraction of their energy into incoherent X– and gamma radiation. The author finds no single theory has yet given a consistent description of the magnetosphere, but he concludes that models based on a dense outflow of pairs from the polar caps, permeated by a lower density flow of heavy ions, are the most promising avenue for future research.

067.104 Hawking radiation from a non–static black hole.
Zhao Zheng, Dai Xianxin.
Chin. Phys. Lett., Vol. 8, No. 10, p. 548 – 550 (1991).
A method exactly determining an event horizon and its temperature in a non–static space–time is proposed. Using the generalized Tortoise coordinate, the authors give an exact location of the event horizon and exact Hawking temperature for a general spherically symmetric evaporating black hole.

067.105 Hydrodynamics of relativistic pulsar wind.
J. G. Lominadze, J. I. Javakhishvili, E. G. Tsikarishvili.
IAU Colloquium No. 128: The magnetospheric structure and emission mechanisms of radio pulsars, p. 90 – 93 (1992). – See Abstr. 012.022 for the main entry.
In this paper the authors get a closed set of relativistic hydrodynamical equations, which describes relativistic strongly magnetized, collisionless plasma with an anisotropic pressure tensor.

067.106 A DC–circuit model of the pulsar magnetosphere.
S. Shibata.
IAU Colloquium No. 128: The magnetospheric structure and emission mechanisms of radio pulsars, p. 94 – 95 (1992). – See Abstr. 012.022 for the main entry.
The structure of the pulsar magnetosphere is studied in terms of a DC–circuit analogy. The author finds (1) the pulsar's death results from the "disappearance" of operating points of this DC–circuit, (2) the outer gap is necessary, (3) the kinetic energy flux of the pair plasma wind and the electromagnetic flux are comparable, and (4) for rapid pulsars only two types of structure (a wind dominated one and an outer–gap dominated one) are possible.

067.107 **The axisymmetric pulsar magnetosphere: a classical model.**
R. Fitzpatrick, L. Mestel.
IAU Colloquium No. 128: The magnetospheric structure and emission mechanisms of radio pulsars, p. 96 – 97 (1992). – See Abstr. 012.022 for the main entry.
The aim of this work is to elucidate how a neutron star with a dipolar magnetic field of axis k and an instantaneous angular velocity αk would spin down, within the framework of classical physics.

067.108 **Global structure of an axially symmetric pulsar magnetosphere.**
Yu. A. Rylov.
IAU Colloquium No. 128: The magnetospheric structure and emission mechanisms of radio pulsars, p. 98 – 100 (1992). – See Abstr. 012.022 for the main entry.
A self–consistent model of the pulsar magnetosphere is presented. The principal point of the model is its use of the massless approximation.

067.109 **Conditions on the capture–region boundary.**
Yu. A. Rylov.
IAU Colloquium No. 128: The magnetospheric structure and emission mechanisms of radio pulsars, p. 101 – 102 (1992). – See Abstr. 012.022 for the main entry.
Some expressions for discontinuities of the electromagnetic field and its derivatives on the capture–region boundary are obtained and investigated. The capture–region boundaries are classified by types.

067.110 **Self–consistent numerical modelling of pulsar magnetospheres.**
H. Herold, T. Ertl, B. Finkbeiner, H. Ruder.
IAU Colloquium No. 128: The magnetospheric structure and emission mechanisms of radio pulsars, p. 105 – 108 (1992). – See Abstr. 012.022 for the main entry.
The magnetosphere of a rapidly rotating, strongly magnetized neutron star with aligned magnetic and rotational axes (parallel rotator) is modelled numerically. Including the radiation of the particles accelerated to relativistic energies as an efficient damping mechanism, the authors obtain a quasi–stationary self–consistent solution to this classical problem. The numerical simulation, which was started from the well–known vacuum solution, yields a global magnetospheric structure that can be characterized by two regions of oppositely charged particles, which eventually produce a relativistic pulsar wind, separated by a vacuum gap of considerable extent.

067.111 **Equilibrium of the return–current sheet and structure of the pulsar magnetosphere.**
Yu. E. Lyubarskii (*Yu. Eh. Lyubarskij*).
IAU Colloquium No. 128: The magnetospheric structure and emission mechanisms of radio pulsars, p. 112 – 113 (1992). – See Abstr. 012.022 for the main entry.

067.112 **The energy emission of sheets in the pulsar magnetosphere.**
Yi Tong, Li Zhong Yuan.
IAU Colloquium No. 128: The magnetospheric structure and emission mechanisms of radio pulsars, p. 114 – 116 (1992). – See Abstr. 012.022 for the main entry.
The authors present a possible emission mechanism based on the idea of current sheets in magnetohydrodynamics. The current sheets are formed close to the light cylinder due to a relativistic effect involving partly frozen–in particles. The authors estimate that the energy emitted by the current sheets fits the observations fairly well.

067.113 **Rapid cooling of neutron stars by hyperons and Δ isobars.**
M. Prakash, M. Prakash, J. M. Lattimer, C. J. Pethick.
Astrophys. J., Lett., Vol. 390, No. 2, p. L77 – L80 (10 May 1992).
The authors show that direct Urca processes with hyperons and/or nucleon isobars can occur in dense matter as long as the concentration of Λ hyperons exceeds a critical value that is less than 3% and is typically about 0.1%. The neutrino luminosities from the hyperon Urca processes are about 5 – 100 times less than the typical luminosity from the nucleon direct Urca process, if the latter process is not forbidden, but they are larger than those expected from other sources. These new direct Urca processes provide avenues for rapid cooling of neutron stars, which invoke neither exotic states nor the large proton fraction (of order 0.11 – 0.15) required for the nucleon direct Urca process.

067.114 **Very high energy γ–ray generation near the light cylinder of an axisymmetric rotator: COS–B like γ–ray sources.**
S. V. Bogovalov, Yu. D. Kotov.
IAU Colloquium No. 128: The magnetospheric structure and emission mechanisms of radio pulsars, p. 207 – 208 (1992). – See Abstr. 012.022 for the main entry.
Super–hard γ–ray radiation spectra have been calculated. This radiation is generated near the velocity–of–light cylinder through the process of inverse–Compton scattering of relativistic electrons by thermal photons radiated by a neutron star. These calculations have been compared with observations of the Crab and Vela pulsars at 1000–GeV γ–ray energies. A correlation between γ–ray flares and those in soft ($E_X \cong 1$ keV) X–rays are predicted.

067.115 **Critical temperatures for superconducting quark matter existence in dense stellar cores.**
J. E. Horvath, O. G. Benvenuto, H. Vucetich.
Mod. Phys. Lett. A, Vol. 7, No. 11, p. 995 – 999 (10 Apr 1992).
If quark matter is actually a component of compact stars it can probably develop a superconducting phase as a result of QCD interactions. This effect may be harmless for (or dramatically affect) the properties of the star, depending on the actual value of the strong coupling constant α_c. Explicit expressions for the critical temperature Tc are derived by using some recent results on the long–range behavior of the gluon propagators. The consequences for the cooling histories of compact stars and possible trends are briefly discussed.

067.116 **Recycled pulsars and low mass X–ray binaries.**
G. S. Bisnovatyi–Kogan (*G. S. Bisnovatyj–Kogan*).
IAU Colloquium No. 128: The magnetospheric structure and emission mechanisms of radio pulsars, p. 209 – 212 (1992). – See Abstr. 012.022 for the main entry.
A magnetized neutron star may appear as a radio pulsar or an X–ray source. The latter is connected with a binary system where accretion from a normal star onto the neutron star produces X–ray emission. At the end of the evolution of a normal non–massive star, accretion stops and the neutron star becomes a recycled radio pulsar. Further evolution may lead to an additional transition from a radio pulsar to a low mass X–ray binary (LMXB). The formation of a single recycled pulsar is considered and a new mechanism of "enhanced evaporation" in globular clusters is analyzed.

067.117 **The maximum moment of intertia of neutron stars and its implications for pulsar observations.**
P. Haensel.
IAU Colloquium No. 128: The magnetospheric structure and emission mechanisms of radio pulsars, p. 217 – 219 (1992). – See Abstr. 012.022 for the main entry.
A simple approximate formula, expressing the maximum moment of inertia of a neutron star as a function of the mass and radius of the configuration with a maximum allowable mass, is shown to be a quite precise representation of the results obtained for a broad set of equations of state of dense matter. The resulting

possible observational constraints in the mass–radius plane for neutron star models are discussed.

067.118 The effect of light bending and redshift on pulsar beaming the case of shorter rotation periods.
R. C. Kapoor.
IAU Colloquium No. 128: The magnetospheric structure and emission mechanisms of radio pulsars, p. 225 – 227 (1992). – See Abstr. 012.022 for the main entry.

An estimate of the effect of light bending and redshift on pulsar beam characteristics has been made using a weak Kerr metric for the case of a $1.4\,M_\odot$ neutron star with a radius in the range 6 – 10 km and rotation periods of 1.56 ms and 33 ms, respectively.

067.119 Nonstationary processes in the pulsar magnetosphere.
A. V. Gurevich, Yu. N. Istomin.
IAU Colloquium No. 128: The magnetospheric structure and emission mechanisms of radio pulsars, p. 229 – 231 (1992). – See Abstr. 012.022 for the main entry.

The authors investigate several instabilities which give different characteristic times for nonstationary processes. Several instabilities are connected with the mechanism of plasma generation in the polar cap gap region. Another nonstationary process is due to the nonlinear phenomenon arising in the magnetosphere during the propagation of the flux of electromagnetic radiation.

067.120 Neutron stars, hybrid stars and the equation of state.
A. Rosenhauer, E. F. Staubo, L. P. Csernai,
T. Øvergård, E. Østgaard.
Nucl. Phys. A, Vol. 540, No. 3/4, p. 630 – 645 (27 Apr 1992).

Gross properties of hybrid stars consisting of a core of strange matter surrounded by ordinary neutron matter are investigated. The authors discuss star models based on phenomenological equations of state from nuclear reactions including a phase transition between the hadronic phase and the quark–gluon plasma. For certain parameters, such equations of state support the existence of hybrid stars. The identification of such objects could provide detailed information on the properties of strange quark matter.

067.121 Cracking of self–gravitating compact objects.
L. Herrera.
Phys. Lett. A, Vol. 165, No. 3, p. 206 – 210 (18 May 1992).

The author reports and discusses the "cracking" (breaking) of a family of self–gravitating spheres, which results from the appearance of total radial forces of different signs in different regions of the sphere, once the equilibrium configuration has been perturbed. The occurrence of such a "cracking" (in the context of the equation of state considered) is induced by either the local anisotropy of the fluid (independently of its origin) or by the emission of incoherent radiation in the streaming out limit. Prospective applications of this effect to some astrophysical scenarios are discussed.

067.122 Spinning a charged dilaton black hole.
K. Shiraishi.
Phys. Lett. A, Vol. 166, No. 5/6, p. 298 – 302 (29 Jun 1992).

A charged dilaton black hole which possesses infinitesimal angular momentum is studied. The author finds that the gyromagnetic ratio of the dilaton black hole depends not only on the parameter which appears the interaction between the dilaton and the electric field but also nonlinearly on the ratio of the charge to the mass of the black hole. The moment of inertia for the charged dilaton hole in the limit of infinitesimal angular momentum is also calculated.

067.123 Energy spectrum of a quantum black hole.
J. Louko, B. F. Whiting.
Classical Quantum Gravity, Vol. 9, No. 2, p. 457 – 473 (Feb 1992).

The authors discuss a "minisuperspace" path integral for the partition function of a Schwarzschild black hole in thermal equilibrium within a finite spherical box. They define and evaluate a partition function using a non–trivial complex integration contour. The partition function solves exactly the relevant differential equation related to the Wheeler–DeWitt equation, and it has the desired semiclassical behaviour indicating in particular thermodynamical stability. For a given size of the box, the density of states is non–vanishing only in a finite energy interval whose upper end is twice as high as would be classically expected without negative temperatures. When negative temperatures are included, this discrepancy is resolved, and the system is then analogous to certain systems in ordinary quantum statistical mechanics which admit negative temperatures.

067.124 Thermodynamics and quantum aspects of black holes in $(1 + 1)$ dimensions.
R. B. Mann, T. G. Steele.
Classical Quantum Gravity, Vol. 9, No. 2, p. 475 – 492 (Feb 1992).

The thermodynamic and quantum properties of black holes in two–dimensional spacetime are investigated. It is demonstrated that the relationships between mass, temperature and entropy of the black hole following from a standard Euclidean analysis agree with those obtained from a quantum field theoretic treatment of vacuum stress–energy. The authors show that such black holes tend to quantize in units of a fundamental mass parameter in order to maximize their entropy. Finally, estimates are made of the back–reaction on the black hole from quantum stress–energy, and a modified form of the model particle detector is used to demonstrate a quantum version of the equivalence principle.

067.125 The origin of planets orbiting millisecond pulsars.
M. Tavani, L. Brookshaw.
Nature, Vol. 356, No. 6367, p. 320 – 322 (26 Mar 1992). Letter–to–the–editor.

At least two Earth–sized planets have been discovered around the 6–ms pulsar PSR 1257 + 12, which, like millisecond pulsars in general, has probably been spun up by accretion of material from a companion star. In addition, two "star–vaporizing" millisecond pulsars (SVPs), 1957 + 20 and 1744–24A, show evidence of mass outflows from their low–mass companions, which are thought to be vaporized by pulsar radiation. The authors suggest a model for the formation of planets around millisecond pulsars such as 1257 + 12, which no longer have stellar companions. They present detailed hydrodynamical models which suggest that planet formation can occur either in a low–mass X–ray binary progenitor to a progenitor of an SVP when the neutron star is accreting material driven off its companion by X–ray irradiation, or after a pulsar has formed and is vaporizing its companion. In both cases a circum–binary disk is created in which planets can form on a timescale of $10^5 - 10^6$ years and the planets can survive a second phase in which the companion star moves towards the pulsar and is completely vaporized.

067.126 Pair–production avalanches revisited.
F. C. Michel.
IAU Colloquium No. 128: The magnetospheric structure and emission mechanisms of radio pulsars, p. 236 – 237 (1992). – See Abstr. 012.022 for the main entry.

067.127 An analytical solution of the problem of plasma ejection from the magnetosphere of an axisymmetric rotator.
S. V. Bogovalov.
IAU Colloquium No. 128: The magnetospheric structure and emission mechanisms of radio pulsars, p. 245 – 247 (1992). – See Abstr. 012.022 for the main entry.

The flow of e^+e^- plasma ejected by an axisymmetrically rotating magnetized neutron star is considered in a hydrodynamical approximation. It is shown that in the vicinity of the light cylinder a helical discontinuity is formed. The transformation of toroidal magnetic field energy into plasma energy takes place at this discontinuity. Particles are accelerated to an energy of

10 TeV for a neutron star with the characteristics of the Crab pulsar.

067.128   General relativistic electrodynamics of the magnetic polar regions of neutron stars.
A. G. Muslimov, A. I. Tsygan.
IAU Colloquium No. 128: The magnetospheric structure and emission mechanisms of radio pulsars, p. 248 – 251 (1992). – See Abstr. 012.022 for the main entry.

The induction of the electric fields near a rotating neutron star is considered within the framework of General Relativity. It is demonstrated that within the open magnetic field region, filled by relativistically moving charged particles, a sufficiently strong component of the electric field is generated. This component is due to the effect of dragging of inertial frames of reference and predominates in the case when a neutron star is not exactly an orthogonal rotator. Finally, some implications of these results on the theory of radio pulsars are discussed.

067.129   Probing magnetospheres of rotation–driven neutron stars.
J. M. Cordes.
IAU Colloquium No. 128: The magnetospheric structure and emission mechanisms of radio pulsars, p. 253 – 260 (1992). – See Abstr. 012.022 for the main entry.

Magnetospheric issues are discussed as they relate to radio intensity variations, spindown, rotational noise, and emission altitudes. Radio fluctuations trace the overall state(s) of the magnetosphere in spite of the fact that $L_{radio} \ll I\Omega\dot{\Omega}$. Five methods for estimating emission altitudes r_e are discussed. Four measure either the transverse width or radial depth of the emission region; assumption of a dipolar field and, in some cases, a radius to frequency mapping, yields estimates for the absolute radius of the emission region. The fifth method (v/c effects in Stokes parameter waveforms) provides direct altitude estimates. Most results suggest that $r_e \lesssim 0.02\,R_{LC}$ at $\nu = 0.4$ GHz. A radius to frequency mapping appears viable for some but not necessarily all objects. The mapping need not be one–to–one. Prospects are discussed for using simultaneous radio and high energy observations to make further progress in our understanding of magnetospheres.

067.130   Intrinsically asymmetric astrophysical jets.
J. C. L. Wang, M. E. Sulkanen, R. V. E. Lovelace.
Astrophys. J., Vol. 390, No. 1, p. 46 – 65 (1 May 1992).

The authors extend their previous treatment of the origin of self–collimated electromagnetic jets to the general case where there is no reflection symmetry of the magnetic field about the equatorial plane of the disk. They obtain the axisymmetric field structure inside the disk by solving for the magnetic flux function, $\Psi(r, z)$, and the toroidal magnetic field $B_\varphi(r, z)$, from the generalized thin–disk induction equation. The asymptotic (large–z) magnetic field structure outside the disk is obtained by solving the force–free Grad–Shafranov equation semianalytically. The authors find jet solutions in which the power flow is carried mainly by the Poynting flux of the electromagnetic field, and the angular momentum outflow from the disk is carried by the magnetic field. This power flow is different in general above and below the disk. The ratio of the jet luminosities (top/bottom) depends directly on the degree of asymmetry of the field and can easily be much greater than unity. In addition, a significant fraction of the accretion power can be carried by these jets. It is argued that the degree of field asymmetry in the disk is determined by the asymmetry of the weak galactic field fed into the disk at large distances over long periods of time. In addition to jet energetics, the authors study the dynamical effects of asymmetric jets and fields on the black hole – disk system.

067.131   Oscillations of rotating neutron stars.
T. E. Strohmayer.
IAU Colloquium No. 128: The magnetospheric structure and emission mechanisms of radio pulsars, p. 299 – 304 (1992). – See Abstr. 012.022 for the main entry.

The author uses a perturbation technique to compute the rotational corrections to the non–radial oscillation spectrum of a realistic neutron–star model. He finds that $l = l_0$ oscillations are coupled to $l = l_0 \pm 1$ oscillations by the Coriolis force. For the toroidal modes, this coupling introduces a non–zero radial component to the velocity field. The author computes the neutrino damping rates for several corrected toroidal modes. The neutrino damping time can approach the gravitational radiation damping time in rotating neutron stars if the central temperature is high enough. The rotationally induced coupling of spheroidal oscillations to toroidal modes can also produce significant displacements at the stellar surface. This may have interesting implications for channeling energy, e.g., that associated with a glitch, to the surface of the star. Perhaps this might produce observable effects in the pulsar emission process or a γ–ray burst event.

067.132   Nonlinear temporal model for formation of pulsar microstructures.
A. C.-L. Chian.
IAU Colloquium No. 128: The magnetospheric structure and emission mechanisms of radio pulsars, p. 356 – 361 (1992). – See Abstr. 012.022 for the main entry.

A nonlinear plasma model which may account for temporal modulation of pulsar radio pulses is presented. Envelope solitons and envelope nonlinear wave trains can result from the nonlinear interaction of the high–frequency coherent pulsar radiation with the pulsar magnetosphere. Theories of electromagnetic envelope solitons and electromagnetic envelope nonlinear wave trains in electron–positron plasmas are reviewed. The application of this model for observation of pulsar microstructures is discussed.

067.133   Steady, periodic gamma–ray emission from accreting X–ray pulsars.
K. S. Cheng, M. M. Lau, T. Cheung, P. P. Leung, K. Y. Ding.
Astrophys. J., Vol. 390, No. 2, p. 480 – 485 (10 May 1992).

The authors study the production mechanisms of medium–energy γ–rays from accreting X–ray pulsars. The models concern electrons accelerated in strongly shielded accelerators above the inner part of the accretion disk whose energies are limited by curvature radiation reaction. The curvature photons are sufficiently energetic to produce secondary $e^\pm$ pairs in collisions with the hard X–rays from the accretion shock front above the neutron star surface outside the acceleration regions, which are shielded from X–rays. A part of the synchrotron photons radiated by these secondary pairs is still energetic enough to produce a third generation of pairs which also emit synchrotron photons. These processes will continue until the energies of the synchrotron photons are low enough to avoid the pair production process in collision with the stellar X–rays. Summing over all generations of the surviving synchrotron photon spectra gives the predicted γ–ray spectrum. The authors suggest that those accreting X–ray pulsars with reported detection of transient very high energy γ–rays (e.g., Her X–1, Cen X–3, and Cyg X–3) should be detectable sources for the GRO in the energy range of 10^5– 10^7eV.

067.134   Can accretion onto isolated neutron stars produce γ–ray bursts?
A. K. Harding, M. Leventhal.
Nature, Vol. 357, No. 6377, p. 388 – 389 (4 Jun 1992). Letter–to–the–editor.

The isotropy and flat count spectrum of γ–ray bursts revealed by the BATSE detector on the Compton Gamma–Ray Observatory have led to suggestions that the burst sources are an extended galactic halo of high–velocity neutron stars. The authors show that if slow accretion onto these neutron stars from the interstellar medium is to be the origin of γ–ray bursts, the accretion physics is very different from what applies for local, low–velocity neutron stars. For halo neutron stars with high magnetic fields and velocities ($v > 190$ km s^{-1}), electromagnetic dipole radiation pressure prevents accretion unless the period is longer than tens of seconds; the centrifugal barrier will then prevent accretion until the period reaches several thousand seconds. For periods as long as this, accretion may proceed through Kelvin–Helmholtz instability at the magnetopause boundary. At interstellar densities and neutron–star magnetic fields of $\sim 10^{12}$G, the accretion rate by this process can be much larger than the Bondi–Hoyle accretion rate, but is still well below what is needed for slow–accretion burst models. The authors conclude that slow accretion onto high–velocity neutron stars in the halo cannot be the origin of γ–ray bursts.

067.135   Millisecond pulsars with extremely strong magnetic fields as a cosmological source of γ–ray bursts.
V. V. Usov.
Nature, Vol. 357, No. 6378, p. 472 – 474 (11 Jun 1992). Letter–to–the–editor.

The spatial and luminosity distribution of γ–ray bursts as observed by the BATSE instrument on the Compton Gamma Ray Observatory provides support for the revival of the idea that the burst sources are at cosmological distances. The author presents a new model for γ–ray bursts at cosmological distances, based on the formation of rapidly rotating neutron stars with surface magnetic fields of the order of 10^{15}G. Such objects could form by the gravitational collapse of accreting white dwarfs with anomalously high magnetic fields in binaries, as in magnetic cataclysmic binaries. Once formed, such rapidly rotating and strongly magnetized neutron stars would lose their rotational kinetic energy catastrophically, on a timescale of seconds or less: rotation of the magnetic field creates a strong electric field, and hence an electron–positron plasma, which the author shows to be optically thick and in quasi–thermodynamic equilibrium. This plasma flows away from the neutron star at relativistic speeds, and X–ray and γ–ray emission at the photosphere of this relativistic wind may then reproduce the observational characteristics of a γ–ray burst.

067.136   Black–hole disk coronae with e^+e^- pairs.
S. Mineshige, M. Kusunose.
Astron. Her., Vol. 85, No. 5, p. 191 – 195 (1992). In Japanese.

067.137   Application of the improved Hartle method for the construction of general relativistic rotating neutron star models.
F. Weber, N. K. Glendenning.
Astrophys. J., Vol. 390, No. 2, p. 541 – 549 (10 May 1992).

Models of general relativistic rotating neutron stars, constructed from Hartle's perturbative "slow" rotation formalism of massive relativistic objects, are compared with their counterparts obtained from the exact solution of Einstein's equations. It is found that both methods, perturbative and exact, lead to compatible results down to rotational Kepler periods $P_K \approx 0.5$ ms, a value which is by far smaller than the smallest yet observed pulsar period. This finding rests on the reinvestigation of Hartle's method, (1) supplementing it by a self-consistency condition inherent in the determination of the Kepler frequency, and (2) analyzing carefully sequences of star models near their end points. A collection of 17 representative neutron matter equations of state served as an input. Because of its simple structure, Hartle's method should prove to be a practical tool for testing models of the nuclear equation of state with data on pulsar periods. The form of an approximate empirical formula for the general relativistic Kepler frequency is obtained, and proportionality to the Newtonian expression arises in about equal parts from the equatorial flattening and the frame dragging.

067.138   Gauged Q–stars.
Chang Hongbo.
High Energ. Phys. Nucl. Phys., Vol. 15, No. 11, p. 991 – 996 (Nov 1991). In Chinese.

The author derives the theory about gauged non–topological soliton stars and their black holes, and finds that the gauged Q–stars with maximum particle number q_{max} in a definite range of mass are cold, stable and in coherent stars of very large mass. Their characteristics are similar to those of general soliton stars. When $q > q_{max}$, the gauged Q–stars are not stable.

067.139   Comments of no–hair theorems and stability of black holes.
S. P. de Alwis.
Phys. Lett. B, Vol. 281, No. 1/2, p. 43 – 48 (7 May 1992).

In the light of recent black hole solutions inspired by string theory, the author reviews some old statements on field theoretic hair on black holes. He also discusses some stability issues. In particular he argues that the two–dimensional string black hole solution is semi–classically stable while the naked singularity is unstable to tachyon fluctuations. Finally the author comments on the relation between the linear dilaton theory and the 2d black hole solution.

067.140   Supersymmetric black holes.
R. Kallosh.
Phys. Lett. B, Vol. 282, No. 1/2, p. 80 – 88 (21 May 1992).

The effective action of $d = 4$, $N = 2$ supergravity is shown to acquire no quantum corrections in background metrics admitting supercovariantly constant spinors. In particular, these metrics include the Robinson–Bertotti metric with all eight supersymmetries unbroken. Another example is a set of arbitrary number of extreme Reissner–Nordström black holes. These black holes break four of eight supersymmetries, leaving the other four unbroken. The author has found manifestly supersymmetric black holes, which are non–trivial solutions of the flatness condition $D^2 = 0$ of the corresponding (shortened) superspace. Their bosonic part describes a set of extreme Reissner–Nordström black holes. The super black hole solutions are exact even when all quantum supergravity corrections are taken into account.

067.141   Eigenfrequencies of radial pulsations of strange quark stars.
B. Datta, P. K. Sahu, J. D. Anand, A. Goyal.
Phys. Lett. B, Vol. 283, No. 3/4, p. 313 – 318 (11 Jun 1992).

The authors calculate the range of eigenfrequencies of radial pulsations of stable strange quark stars, using the general relativistic pulsation equation and adopting a realistic equation of state for degenerate strange quark matter.

067.142   Numerically generated black–hole spacetimes: interaction with gravitational waves.
A. Abrahams, D. Bernstein, D. Hobill, E. Seidel, L. Smarr.
Phys. Rev. D, Vol. 45, No. 10, p. 3544 – 3558 (15 May 1992). Current Physics Microform No.: 9207D1126.

The authors present results from a new two–dimensional numerical relativity code used to study the interaction of gravitational waves with a black hole. The initial data correspond to a single black hole superimposed with time–symmetric gravitational waves (Brill waves). A gauge–invariant method is presented for extracting the gravitational waves from the numerically generated spacetime. The authors show that the interaction between the gravitational wave and the black hole excites the quasinormal modes of the black hole. An extensive comparison of these results is made with black hole perturbation theory. For low amplitude initial gravitational waves, excellent agreement is found between the theoretically predicted scrl = 2 and scrl = 4 wave forms and the wave forms generated by the code.

067.143   Are horned particles the end point of Hawking evaporation?
T. Banks, A. Dabholkar, M. R. Douglas, M. O'Loughlin.
Phys. Rev. D, Vol. 45, No. 10, p. 3607 – 3616 (15 May 1992). Current Physics Microform No.: 9207D1189.

The authors investigate the proposal by Callan, Giddings, Harvey, and Strominger (CGHS) that two–dimensional quantum fluctuations can eliminate the singularities and horizons formed by matter collapsing on the nonsingular extremal black hole of dilaton gravity. The authors argue that this scenario could in principle resolve all of the paradoxes connected with Hawking evaporation of black holes. However, it is shown that the generic solution of the model of CGHS is singular. The authors propose modifications of this model which may allow the scenario to be realized in a consistent manner.

067.144   Obliquity and magnetic dipole radiation from collapsing, rotating, magnetized stars.
W. Y. Chau, J. L. Zhang.
Astrophys. Space Sci., Vol. 190, No. 1, p. 131 – 138 (Apr 1992).
The authors consider in this paper the evolution of a collapsing (or exploding), uniformly rotating, uniformly magnetized spheroidal star with non–aligned rotational and magnetic axes. Analytical expressions were obtained for the change in angle (obliquity) between the two axes (based on the frozen field condition), and the energy loss via magnetic dipole radiation. Numerical estimates with typical data show that the obliquity increases (asymptotically to $\pi/2$) with the collapse from white dwarf to neutron star, and the energy loss could be as much as 4×10^{39}erg, about twice the amount emitted when the two axes are aligned.

067.145   Quasi–elastic neutrino scattering by nuclei in superdense matter of a collapsing star.
L. B. Leinson.
Astrophys. Space Sci., Vol. 190, No. 2, p. 271 – 280 (Apr 1992).
The interaction of neutrinos with nuclei in the superdense matter of a collapsing star is studied, taking into account the collective modes and thermal fluctuations of the medium density. It is shown that the elastic neutrino scattering by nuclei with a momentum transfer less than or of the order of the inverse distance between the ions in the nonideal Coulomb plasma, differs considerably from the analogous scattering by a single nucleus. The weak νA interaction screening by medium electrons is taken into account. The collision integral and transport cross section of neutrino scattering by nuclei are calculated in terms of macroscopic medium parameters.

067.146   Evolution of close binary systems.
V. M. Lipunov.
Vulcano Workshop 1990: Frontier objects in astrophysics and particle physics, p. 29 – 47 (1991). – See Abstr. 012.040 for the main entry.
A review is given concerning the modern state of the theory of close binary system evolution. A special attention is given to formation and evolution of relativistic stars (neutron stars, black holes and white dwarfs) in binary systems. The account of the intrinsic evolution of the magnetized compact star is shown, both theoretically and numerically, to be the decisive factor in explaining observable properties and prophesing yet unknown properties of high–energy radiation sources in our and other galaxies. The main results are given for the modern evolutionary scenario simulations by the Monte–Carlo method.

067.147   A model of SS Cygni.
L. G. Filipov, M. Dimitrova.
Vulcano Workshop 1990: Frontier objects in astrophysics and particle physics, p. 61 – 66 (1991). – See Abstr. 012.040 for the main entry.
A model of the cataclysmic SS Cyg is discussed. Based on observational data it is shown that in such object with magnetic field of order of 10^6G two types of accretion onto a white dwarf are possible: disk accretion with formation of a "dead" disk with following cut through the magnitosphere and manifestation of short period bright outburst followed by the formation of hot corona from the rest of the disk. The newly formed configuration when corona and inflowing flux interact represents a hybrid between accretion disk and quasispheric nucleus where the magnetic dipole with the white dwarf is situated. The generation of condition for Rayleigh–Taylor instability in this envelope explains the appearence of the second type (continuous outbursts).

067.148   A numerical model of X–ray bursts – an attempt to interpret some behavioural aspects of the rapid burster.
L. G. Filipov, E. D. Georgieva.
Vulcano Workshop 1990: Frontier objects in astrophysics and particle physics, p. 95 – 98 (1991). – See Abstr. 012.040 for the main entry.
The authors present a numerical model of X–ray bursts which appear during the active phase of a transient source, obtained under the assumption of a "dead disk" around a strongly magnetized fast rotating neutron star. Accretion onto neutron stars is controlled by two conditions – switching on and switching off, which determine the maximum mass of the disk that may contribute to accretion and therefore the total energy emitted during a burst. The authors discuss the dependence of the burst profiles, duration and time intervals between them, the maximum luminosity and energy radiated during the burst on the parameters of the model. The model can be taken in account when studying the behaviour of the Rapid Burster as well as the 1.2 – 2.2 hours flares of EXO 2030 + 375.

067.149   Vacuum nonsingular black hole.
I. Dymnikova.
Gen. Relativ. Gravitation, Vol. 24, No. 3, p. 235 – 242 (Mar 1992).
The spherically symmetric vacuum stress–energy tensor with one assumption concerning its specific form generates the exact analytic solution of the Einstein equations which for large r coincides with the Schwarzschild solution, for small r behaves like the de Sitter solution, and describes a spherically symmetric black hole, which is everywhere singularity free.

067.150   On quasilevels in the gravitational field of a black hole.
A. B. Gaina, O. B. Zaslavskii.
Classical Quantum Gravity, Vol. 9, No. 3, p. 667 – 676 (Mar 1992).
Quantum tunnelling of scalar particles with energies near the top of an effective potential barrier which corresponds to a circular classically unstable orbit is investigated. The corresponding lifetime has a rather smooth logarithmic dependence on the parameters of the problem. For the Kerr metric the lifetime of a particle counter–rotating with respect to a black hole along a circular orbit is shown to depend weakly on the hole angular momentum, whereas in the co–rotating case it increases significantly for a rapidly rotating black hole.

067.151   Bulk viscosity of hot neutron star matter from direct Urca processes.
P. Haensel, R. Schaeffer.
Phys. Rev. D, Vol. 45, No. 12, p. 4708 – 4712 (15 Jun 1992).
Current Physics Microform No.: 9208C2236.
Direct Urca processes, occurring in neutron star matter with a proton fraction exceeding the critical value of (11 – 15)%, can strongly enhance the bulk viscosity of the matter.

067.152   Remarks on the continued fraction method for computing black hole quasinormal frequencies and modes.
E. W. Leaver.
Phys. Rev. D, Vol. 45, No. 12, p. 4713 – 4716 (15 Jun 1992).
Current Physics Microform No.: 9208C2241.
The author discusses the stability, accuracy, validity, and convergence of the continued fraction method for computing black hole quasinormal frequencies and modes.

067.153   Astrophysik in der Schwarzschildmetrik am Beispiel von Quasi–Normalmoden schwarzer Löcher und Lichtablenkung bei Röntgenpulsaren. *(Astrophysics in the Schwarzschild metric: The quasi–normal modes of black holes and the light deflection of X–ray pulsars.)*
H. P. Nollert.
Diss. (Dr.rer.nat.), Tübingen Univ. (Germany). Fakultät für Physik. 97 p. (18 Sep 1990).

067.154 Strahlungshydrodynamik in Akkretionssäulen auf Neutronensternen. (Radiation hydrodynamics in accretion columns of neutron stars.)
K. Wolf.
Diss. (Dr.rer.nat.), Tübingen Univ. (Germany). Fakultät für Physik. 117 p. (11 Sep 1990).

067.155 A new method dealing with Hawking effects of evaporating black holes.
Zhao Zheng, Dai Xianxin.
Mod. Phys. Lett. A, Vol. 7, No. 20, p. 1771 – 1778 (28 Jun 1992).
Both the location and the temperature of event horizons of evaporating black holes can be easily given if one proposes the Klein–Gordon equation approaches the standard form of the wave equation near event horizons by using tortoise–type coordinates.

067.156 T–symmetry violation as a result of virtual gravitational interaction.
G. S. Bisnovatyj–Kogan.
Nuovo Cimento B, Vol. 107, No. 3, p. 357 – 359 (Mar 1992).
T–symmetry violation observed in the weak K–decays is interpreted as a result of an influence of the virtual black holes, inside which the space and time coordinates change places. When this decay happens inside the black hole the P–symmetry violation inherent to (V–A)–type of weak Lagrangian is accepted by the observer as T–symmetry violation. CP–symmetry conservation corresponds to CT–symmetry conservation.

067.157 On pair creation in a black hole magnetosphere.
V. S. Beskin, Ya. N. Istomin, V. I. Pariev.
7. IAP Meeting: Extragalactic radio sources – from beams to jets, p. 45 – 51 (1992). – See Abstr. 012.041 for the main entry.

067.158 Above the accretion disc & black hole: a starburst stellar cluster?
J. J. Perry.
7. IAP Meeting: Extragalactic radio sources – from beams to jets, p. 52 – 58 (1992). – See Abstr. 012.041 for the main entry.
The author reviews the consequences of assuming that the AGN phenomenon manifests itself when a nuclear starburst stellar cluster coexists with a massive central object responsible for the continuum.

067.159 Formation of relativistic jets in quasars and the origin of quasi–periodicity.
M. Camenzind.
7. IAP Meeting: Extragalactic radio sources – from beams to jets, p. 71 – 77 (1992). – See Abstr. 012.041 for the main entry.
Relativistic jets are naturally formed in the rapidly rotating magnetospheres of geometrically thin accretion disks around rapidly rotating black holes. Disk fields generated by dynamo effects in the boundary layer between disk and rotating hole immerse the hole into a rotating magnetosphere which is loaded by plasma from the disk. Magnetized winds ejected from the surface of the disk will be collimated by magnetic effect on scales somewhat larger than the light cylinder of the magnetosphere. Non–axisymmetric perturbations dragged along by these jets will rotate and produce quasi–periodic emission by a kind of lighthouse effect.

067.160 Magnetic reconnection in active galactic nuclei.
H. Lesch.
7. IAP Meeting: Extragalactic radio sources – from beams to jets, p. 78 – 84 (1992). – See Abstr. 012.041 for the main entry.
A model for the emission processes causing rapid variability (less than one day) in active galactic nuclei is developed. This model explains the very rapid variability in the X–ray. According to this scenario the higher the variable frequency is, the closer to the central black hole it should originate.

067.161 Formation of very strongly magnetized neutron stars: implications for gamma–ray bursts.
R. C. Duncan, C. Thompson.
Astrophys. J., Lett., Vol. 392, No. 1, p. L9 – L13 (10 Jun 1992).
Neutron stars with unusually strong magnetic dipole fields, $B_{dipole} \sim 10^{14}G – 10^{15}G$, can form when conditions for efficient helical dynamo action are met during the first few seconds after gravitational collapse. Such high–field neutron stars, "magnetars", initially rotate with short periods ~ 1 ms, but quickly lose most of their rotational energy via magnetic braking, giving a large energy boost to the associated supernova explosion. Several mechanisms unique to magnetars can plausibly generate large (~ 1000 km s^{-1}) recoil velocities. These include magnetically induced anisotropic neutrino emission, core rotational instability and fragmentation, and/or anisotropic magnetic winds. Magnetars are relatively difficult to detect because they drop below the radio death line faster than ordinary pulsars, and because they probably do not remain bound in binary systems. The authors conjecture that their main observational signatures are gamma–ray bursts powered by their vast reservoirs of magnetic energy. If they acquire large recoils, most magnetars are unbound from the Galaxy or reside in an extended, weakly bound Galactic corona.

067.162 Crustal magnetic field decay and neutron star cooling.
V. A. Urpin, A. G. Muslimov.
Mon. Not. R. Astron. Soc., Vol. 256, No. 2, p. 261 – 268 (15 May 1992).
The ohmic decay of the magnetic field initially confined to the surface layers of the neutron star crust is considered. It is shown that the neutron star cooling can effectively increase the conductivity σ and the characteristic ohmic decay time τ_B. The scenario of the crustal field evolution is as follows. At the initial stage the field decays comparatively rapidly: it can decrease by a factor of 10–100 during the first $\sim 10^6$–10^7yr depending on initial field location. At this stage the conductivity is determined by electron–phonon scattering. After about 10^6–10^7yr, σ in the crust begins to be determined by the scattering on impurities and may be sufficiently high in pure crystals. At late stages the decay time τ_B is large due to the higher conductivity, and it can increase during the evolution because of magnetic field diffusion into the deep layers of the crust. The authors' calculations show that the crustal field decays rather slowly even in the case when the field is initially confined to the outer regions of the crust. For instance, if the electric currents were initially concentrated within the layer with $\varrho \leqslant (2$–$4) \times 10^{11}$g cm^{-3}, then the surface field weakens by a factor of $\sim 10^2$–10^3 after $\sim 10^{10}$yr.

067.163 A Newtonian description of the geometry around a rotating black hole.
S. K. Chakrabarti, R. Khanna.
Mon. Not. R. Astron. Soc., Vol. 256, No. 2, p. 300 – 306 (15 May 1992).
The authors suggest that most of the relevant properties of the Kerr space–time could be retrieved by suitably choosing a scalar potential which they present here. This potential can be used in the same manner as one uses the gravitational potential in Newtonian flat space–time. The authors present examples using both the general relativistic equations and their potential. The examples include test–particle motion and fluid motion around a Kerr black hole, as well as the evolution of the ellipticity of a self–gravitating, slowly rotating collapsing star.

067.164 Planetary systems around radiopulsars.
A. V. Tutukov.
Astron. Vestn., Tom 26, No. 3, p. 44 – 46 (May – Jun 1992). In Russian. English translation in Sol. Syst. Res., Vol. 26, No. 3.
Two planetary systems around radiopulsars have been discovered recently. Possible ways of their formation in the framework of the theory of close binary evolution are discussed. The dynamical disruption of the companion filling its Roche lobe leads to the formation of a gaseous disk around the neutron star or to an oxygen–neon dwarf. Depending on time, this dwarf accretes matter of the disk, and collapses into a neutron star. The disk size significantly increases by viscosity, its temperature

decreases and dust particles can be formed in it. Coagulation of dust particles leads finally to accumulation of planets on circular orbits around neutron stars.

067.165   Ensemble dependence of the stability of thermal black holes.
G. L. Comer.
Classical Quantum Gravity, Vol. 9, No. 4, p. 947 – 962 (Apr 1992).
    Within the Euclidean path integral approach to statistical mechanics the author examines the question of the ensemble dependence of the stability of thermal black holes. In both of the ensembles considered it is found that there is only one system configuration which can satisfy the given thermodynamic boundary conditions. It is found that throughout the parameter space of one ensemble black holes are never stable, whereas in the other ensemble they are nearly always stable. The effects of quantum fluctuations are included in the partition function for the stable ensemble. This gives a gravitational field entropy which has terms not included in the Bekenstein – Hawking formula.

067.166   Massive scalar quasi–normal modes of Schwarzschild and Kerr black holes.
L. E. Simone, C. M. Will.
Classical Quantum Gravity, Vol. 9, No. 4, p. 963 – 977 (Apr 1992).
    The frequencies of low–lying normal modes of Schwarzschild and Kerr black holes due to massive scalar perturbations are calculated using a higher–order WKB approximation. In the Schwarzschild case, for a range of scalar–field angular momenta, it is found that when the mass of the scalar field increases, the oscillation frequency increases, while the damping decreases. The convergence of the approximation is studied as a function of the mass. For the Kerr case, the same general variation with the mass of the field is found.

067.167   Lower dimensional black holes.
R. B. Mann.
Gen. Relativ. Gravitation, Vol. 24, No. 4, p. 433 – 449 (Apr 1992).
    A survey of black hole physics in two spacetime dimensions is presented. Basic properties, specific solutions and quantum aspects are considered in turn. The relationship between string theoretic black holes and those arising in other (1 + 1)–dimensional theories of gravity is discussed.

067.168   Static and stationary black holes with QCD hairs.
Yu Hongwei, Wang Yongjiu.
Sci. China, Ser. A, Vol. 34, No. 10, p. 1233 – 1242 (Oct 1991). In Chinese.
    The coupled SU(5) Einstein–Yang–Mills–Higgs system of fields is investigated. A family of static spherically symmetric and stationary axisymmetric black hole solutions is obtained. The result shows that the black hole can carry not only the Abelian electric and magnetic charges but also the non–Abelian SU(3)$_c$ color charge.

067.169   Geometrically thin, hot accretion disks: topology of the thermal equilibrium curves.
M. Kusunose, S. Mineshige.
Astrophys. J., Vol. 392, No. 2, p. 653 – 661 (20 Jun 1992).
    The authors present all the possible thermal equilibrium states of geometrically thin α–disks around stellar mass black holes. They employ a (vertically) one–zone disk model and assume that a main energy source is viscous heating of protons and that cooling is due to bremsstrahlung and Compton scattering. Results can be summarized in the (Σ = surface density, $\dot{M}$ = accretion rate) plane. There exist various branches of the thermal equilibrium solution, depending on whether disks are effectively optically thick or thin, radiation pressure–dominated or gas pressure–dominated, composed of one–temperature plasmas or of two–temperature plasmas, and with high concentration of e^+e^- pairs or without pairs. The thermal equilibrium curves at high temperatures ($T \gtrsim 10^8$K) are substantially modified by the

presence of e^+e^- pairs. The authors examine the thermal stability of these branches. There are no thermally stable branches at high temperatures, $T \gtrsim 10^7$K, if the standard α–model is adopted and pair balance is assumed. A hard X–ray emitting disk cannot therefore stay on any branches on time scales longer than the thermal time scales (of order 10 ms).

067.170   Hawking radiation of a quantum black hole in an inflationary universe.
Huang Wunghong.
Classical Quantum Gravity, Vol. 9, No. 5, p. 1199 – 1209 (May 1992).
    The quantum stress–energy tensor of a massless scalar field propagating in the two–dimensional Vaidya – de Sitter metric, which describes a classical model spacetime for a dynamical evaporating black hole in an inflationary universe, is analysed. The author presents a possible way to obtain the Hawking radiation terms for the model with arbitrary functions of mass. It is used to see how the expansion of universe will affect the dynamical process of black hole evaporation. The results show that the cosmological inflation has the tendency to depress the black hole evaporation. However, if the cosmological constant is sufficiently large, then the back–reaction effect tends to increase the black hole evaporation. The author also shows that the evaporation will always produce a divergent flux of outgoing radiation along the Cauchy horizon where the curvature has a finite value.

067.171   Uniqueness proof of static charged black holes revisited.
A. K. M. Masood-ul-Alam.
Classical Quantum Gravity, Vol. 9, No. 5, p. L53 – L55 (May 1992). Letter–to–the–editor.
    The author gives an alternative and rigorous proof of the uniqueness of non–degenerate static electrovac black holes.

067.172   Spinor particles and black holes in 1 + 1 dimensions.
S. M. Morsink.
4. Canadian Conference on General Relativity and Relativistic Astrophysics, p. 25 – 29 (1992). – See Abstr. 012.056 for the main entry.
    The Hawking radiation of massive spinor particles from a 1 + 1 dimensional black hole is examined so that the temperature of the black hole can be determined. Using the solution of Dirac's equation in the 1 + 1 black hole metric the vacuum expectation value of the spinor stress–energy tensor is calculated. The energy density is compared with the energy density of black body radiation. This allows an identification to the black hole of a temperature of $M/2\pi$. The temperature found here is the same found when a similar analysis is done on massive scalar particles.

067.173   More colored black holes.
H. P. Künzle.
4. Canadian Conference on General Relativity and Relativistic Astrophysics, p. 76 – 80 (1992). – See Abstr. 012.056 for the main entry.
    The discrete sequences of globally regular static spherically symmetric solutions of the SU(2)–Einstein–Yang–Mills equations and the similar sequences of black hole solutions are generalized to the gauge groups SU(n). Spherical symmetry is defined as invariance of the Yang–Mills connection and the space–time metric under any suitable action of the group SU(2) on the SU(n)–principal bundle. All these actions are classified. Many of them lead to equations that admit only a Reissner–Nordström like solution but there are some "irreducible" cases that appear to admit discrete sets of regular and black hole solutions.

067.174   Relativistic fluid flows in the magnetized Schwarzschild spacetime.
E. P. Esteban.
AIP Conf. Proc., No. 254, p. 52 – 55 (1 May 1992). – See Abstr. 012.057 for the main entry.
    Relativistic magnetohydrodynamics plays an important role in accretion processes. Unfortunately, a complete and rigorous

study of this subject is indeed very difficult. It involves, for example, hydromagnetic flows in evolving spacetimes as well as the interaction of charged fluids with electromagnetic fields. Nonetheless, by making a set of restrictive assumptions, analytical and numerical relativistic hydrodynamics solutions for the vacuum Schwarzschild and Kerr spacetimes were found. The authors propose an extension of this work by considering as a fixed background the magnetized Schwarzschild metric. Two cases (in the equatorial plane) are considered. The geodesic infall (pressure neglected), and the accretion (including pressure) of apolytropic gas.

067.175 **On the luminosity of black hole cluster model of active galactic nuclei.**
W. R. Stoeger, A. G. Pacholczyk, T. F. Stepinski.
AIP Conf. Proc., No. 254, p. 61 – 64 (1 May 1992). – See Abstr. 012.057 for the main entry.
The luminosity of a nuclear cluster of accreting black holes and other objects is discussed in terms of two accretion regimes; external supply of gas from outside the cluster and internal supply resulting from tidal disruption and capture of stars within the cluster. The external supply regime results in radiation being emitted from the innermost parts of the cluster while the internal supply can efficiently feed the holes in the outer parts of the cluster as long as it is not too compact a cluster and is embedded in a distribution of stars with a density larger than $10^7 M_\odot/pc^3$.

067.176 **Towards a "dark paradigm"? or an empirical approach to accretion disks in AGN.**
S. Collin–Souffrin.
AIP Conf. Proc., No. 254, p. 119 – 128 (1 May 1992). – See Abstr. 012.057 for the main entry.
The origin of the UV–bump is looked for in the framework of the model of a black hole fueled by accretion. It is emitted by a medium of large optical thickness and large spatial covering factor. Assuming that the X–ray source is located near the black hole, the absence of hard X–ray absorption constrains this medium to be a flat extended structure, most likely identified with a disk. The "cold blob model" is ruled out and it is shown that the UV–bump cannot be produced in a radial accretion flow. This does not exclude however that a large fraction of the accretion flow is radial. It is then argued that the disk does not only emit the optical–UV continuum, but should also contribute to the line emission, in particular to the linewings. This picture leads to strong constraints, since models based on fitting both the continuum and the lines can be used to determine the central mass and the accretion rate. They lead to the result that the disk does not necessarily draw the whole mass rate required to fuel the black hole at the bolometric luminosity.

067.177 **Comments on the nature of the Lyman edge from thin accretion disks.**
A. Laor.
AIP Conf. Proc., No. 254, p. 155 – 158 (1 May 1992). – See Abstr. 012.057 for the main entry.
The detection of an intrinsic Lyman edge in the continuum emission of AGN can serve as an important evidence for the thermal nature of the UV emission. Non LTE and electron scattering effects can significantly decrease the amplitude of the edge. These effects, combined with strong relativistic effects, expected if the emission originates from a thin accretion disk, can make it hard to detect the edge in existing data. High S/N observations over a large wavelength range are required in order to put interesting constraints on the presence of a thin disk edge, in particular, and on the thermal nature of the UV emission in general.

067.178 **Modeling of Lyman edge features in accretion disk spectra.**
G. Lee, G. A. Kriss, A. F. Davidsen.
AIP Conf. Proc., No. 254, p. 159 – 162 (1 May 1992). – See Abstr. 012.057 for the main entry.
The authors consider two possibilities which can change the spectral shape of the Lyman–limit region in the radiation emitted

by the accretion disks of Active Galactic Nuclei (AGN) and quasars. One is the effect produced by the strong gravitational field of the black hole, and the other is Comptonization of the emitted disk spectrum by a surrounding hot corona. The authors present synthesized spectra in the region of the Lyman limit for optically thick, geometrically thin accretion disk models incorporating the above two effects.

067.179 **Nuclear and particle astrophysics.**
N. K. Glendenning.
International Summer School on the Structure of Hadrons and Hadronic Matter – NATO Advanced Study Institute, p. 275 – 350 (1991). – See Abstr. 012.062 for the main entry.
The author discusses the physics of matter that is relevant to the structure of compact stars. This includes nuclear, neutron star matter and quark matter and phase transitions between them. Many aspects of neutron star structure and its dependance on a number of physical assumptions about nuclear matter properties and hyperon couplings are investigated. He also discusses the prospects for obtaining constraints on the equation of state from astrophysical sources. Neutron star masses although few are known at present, provide a very direct constraint in as much as the connection to the equation of state involves only the assumption that Einstein's general theory of relativity is correct at the macroscopic scale. The rapid rotation of pulsars is also discussed. It is shown that for periods below a certain limit it becomes increasingly difficult to reconcile them with neutron stars. Strange stars are possible if strange matter is the absolute ground state. The author discusses such stars and their compatibility with observation.

067.180 **Brick walls for black holes.**
R. B. Mann, L. Tarasov, A. Zelnikov (A. Zel'nikov).
Classical Quantum Gravity, Vol. 9, No. 6, p. 1487 – 1494 (Jun 1992).
The authors examine the prescription of t'Hooft for cutting off eigenmodes of particle wavefunctions in the vicinity of black hole event horizons in N dimensions. They find for any N > 3 that the cutoff occurs as a consequence of the causal structure of spacetime, independently of the strength of the source, in agreement with the four–dimensional case. The authors separately discuss the special case N = 2, showing why in this case the cutoff depends on the strength of the source.

067.181 **Inner–structure instability of a Reissner – Nordstrom black hole.**
N. Yu. Gnedin, M. L. Gnedina.
Astron. Zh., Tom 69, Vyp. 3, p. 584 – 592 (May – Jun 1992). In Russian. English translation in Sov. Astron., Vol. 36, No. 3 (May – Jun 1992).
The instability of the inner horizon of a Reissner – Nordstrom black hole with respect to scalar perturbations arising on the outer horizon is simulated numerically. It is shown that perturbations falling from the outside on to the black hole change the black hole structure and provoke a singularity. If the mass of perturbations is large enough, the singularity consists of two isotropic parts and a spacelike one.

067.182 **Black hole evaporation: an open question.**
T. Jacobson.
15. Texas Symposium on Relativistic Astrophysics and 4. ESO–CERN Symposium, p. 104 – 116 (1991). – See Abstr. 012.060 for the main entry.
Contents: 1. Introduction. 2. Hawking radiation and short distances. 3. Quantum fluid–flow model of a black hole. 4. Do black holes evaporate? 5. Discussion.

067.183 **Numerical calculations of black holes and naked singularities.**
S. L. Shapiro, S. A. Teukolsky.
15. Texas Symposium on Relativistic Astrophysics and 4. ESO–CERN Symposium, p. 158 – 163 (1991). – See Abstr. 012.060 for the main entry.
The authors have presented numerical evidence that the hoop conjecture is a valid criterion for the formation of black holes

during gravitational collapse and they have found the first numerical candidates for the formation of naked singularities from nonspherical collapse of well–behaved initial configurations.

067.184 Diskoseismology: signatures of black hole accretion disks.
M. Nowak, R. V. Wagoner.
AIP Conf. Proc., No. 254, p. 231 – 234 (1 May 1992). – See Abstr. 012.057 for the main entry.

General relativity requires the existence of a spectrum of oscillations which are trapped near the inner edge of accretion disks around black holes. The authors have developed a general formalism for analyzing the normal modes of such acoustic perturbations of arbitrary thin disk models, approximating the dominant relativistic effects via a modified Newtonian potential (these modes do not exist in Newtonian gravity). The eigenfunctions and eigenfrequencies of a variety of disk models are found to fall in to two main classes, which are analogous to the p–modes and g–modes in the sun. In this work, the authors compute the eigenfunctions and eigenfrequencies of isothermal disks. The (relatively small) rates of growth or damping of these oscillations due to gravitational radiation and parametrized model of viscosity are also computed.

067.185 Tidal interaction and brightening of accretion disks around supermassive black holes.
S.–W. Kim, I. Yi.
AIP Conf. Proc., .No. 254, p. 239 – 241 (1 May 1992). – See Abstr. 012.057 for the main entry.

The authors study the tidal interaction between a supermassive black hole and a star. The brightening of an accretion disk around a supermassive black hole as a consequence of the tidal disruption of the star is investigated with the Newtonian α–disk model, although relativistic effects are expected to be substantial. A large mass, 10^8–$10^9 M_\odot$, for the supermassive black hole is required to produce possible optical flares on time scales of a few years. The numerical results imply that only a few percent of the mass of a disrupted star forms a very thin layer of accretion disk outside the marginally stable orbit. The authors suggest that the formation of such a thin disk could be interesting for producing potentially observable optical thermal flares with characteristic time scales on the order of a few years.

067.186 Rapid variability in AGN and accretion disk hot–spots.
P. J. Wiita, A. V. Mangalam, S. K. Chakrabarti.
AIP Conf. Proc., No. 254, p. 251 – 254 (1 May 1992). – See Abstr. 012.057 for the main entry.

The authors present models for the optical microvariability (significant variations on timescales less than a day) that, over the past few years, has been conclusively established to be common in BL Lacs and some other AGN. The scenario invokes the excess emission produced by flares or "hot–spots" on the accretion disks around supermassive black holes. Phenomenological models, based on steady–state "β–disks" with flares that are assumed to turn on randomly in location and time, have been computed including gravitational and Doppler frequency shifts as well as eclipses; these can match the general features of the optical and ultraviolet observations, such as the power density spectrum. The relative variations in different spectral bands depend upon the type and parameters of the models, and thus future multi–color observations may allow us to distinguish between possible situations. Two–dimensional hydrodynamical simulations of the evolution of disks subject to non–axisymmetric perturbations have also been performed. Such disks develop spiral shocks which grow, fragment and reform on time–scales short enough to provide a physical basis for the random hot–spots.

067.187 A universal method determining Hawking effect in spherically symmetric or plane–symmetric non–static space–times.
Zhao Zheng.
Chin. Phys. Lett., Vol. 9, No. 9, p. 401 – 404 (1992).

A universal proof of the new method determining the location and the temperature of event horizons is given. The method is valid to every non–static spherically symmetric or plane–symmetric space–time. An additional term appears in the standard form of the wave equation near the event horizon in the spherically symmetric case. The coefficient of this term is just the relative change rate of the entropy of a black hole.

067.188 Accretion disk coronas in QSOs.
W. A. Stein.
AIP Conf. Proc., No. 254, p. 268 – 270 (1 May 1992). – See Abstr. 012.057 for the main entry.

Radiation pressure on gas in the hypothesized accretion disks of QSOs will result in a corona. Electrons of 340 keV, equivalent to $T_e \sim 3 \times 10^8$K, may result in a significant Compton optical depth for the implied mass loss rate. Comptonization of ultraviolet to 30 keV may result in contradiction to observations.

067.189 Do accretion disk instabilities drive a thermal wind in AGN?
M. Dopita, C. Coleman.
AIP Conf. Proc., No. 254, p. 271 – 275 (1 May 1992). – See Abstr. 012.057 for the main entry.

A linear stability analysis of a warped accretion disk beyond the Bardeen–Petterson radius has shown that this is violently unstable to shearing instabilities of a size comparable with the disk density scale height. The authors hypothesize that this instability will lead to fragmentation of the disk in the unstable zone, even in objects which have only a moderate misalignment between the angular momentum vectors of the outer accretion disk and the central black hole. Compton heating and evaporation of the clouds resulting from disk fragmentation is shown to be capable of driving a thermal Parker wind in these cases, which, if optically thick to electron scattering will result in an electron scattering photosphere. The authors propose that this is the source of the "big blue bump" UV emission seen in QSOs and other AGN. In such a model the broad–line region is caused by photoionization by this UV field of both the outer accretion disk and of cloud fragments entrained in the wind. The authors suggest that such a model is capable of describing the $\sim 90\%$ of QSOs which are radio–quiet. The radio–loud objects would be those in which the accretion rate is sufficiently low, and for which the disk and Black Hole angular momentum vectors are sufficiently well–aligned to allow the escape of the relativistic electron plasma from the vicinity of the Black Hole.

067.190 Black holes in the Galaxy.
J. McClintock.
15. Texas Symposium on Relativistic Astrophysics and 4. ESO–CERN Symposium, p. 495 – 502 (1991). – See Abstr. 012.060 for the main entry.

There is no decisive evidence that black holes exist. However, it is widely believed that they power active galactic nuclei and some X–ray binaries, and that they inhabit the nuclei of many normal galaxies. The compulsion to believe that black holes are physical objects is driven by our faith in general relativity, the idea that a black hole is a logical end point of stellar and galactic evolution, and the knowledge that the familiar neutron star is a small step away from a black hole. The clearest evidence for black holes comes from dynamical studies of four X–ray binaries, which are described and compared in this paper. Topics discussed here include nondynamical signatures of black holes, alternative models for Cyg X–1, and recent results obtained for Cyg X–1 and A 0620–00.

067.191 Evolution of a Compton heated wind from an irradiated accretion disk.
D. T. Woods, J. I. Castor, R. I. Klein, C. F. McKee, J. B. Bell.
AIP Conf. Proc., No. 254, p. 276 – 279 (1 May 1992). – See Abstr. 012.057 for the main entry.

The authors present 2–D hydrodynamical calculations of an axially symmetric Compton heated wind from an accretion disk surrounding a massive ($\approx 10^8 M_\odot$) black hole. The accretion disk is assumed to flare with radius, thus allowing the disk to be exposed to the hard ionizing radiation of the central source. The heating produces a hot corona above the disk with a temperature on the order of the Compton temperature, which, at sufficiently large distances (~ Compton radius) from the central object, leads to a thermally driven wind. The authors' code utilizes a second-order Godunov scheme with adaptive mesh refinement to which rotation, central gravity, and energy sources/sinks have been added. The authors present the results of a calculation which has been evolved from a nearly static thermal equilibrium for a luminosity of the central object of $L/L_{EDD} = 0.3$. The wind which develops appears to be hydrodynamically stable for their optically thin heating function. The authors show the structure of the wind and compare the mass flux densities to analytic estimates.

067.192 Black hole accretion with radiative viscosity.
A. Loeb, A. Laor.
AIP Conf. Proc., No. 254, p. 284 – 288 (1 May 1992). – See Abstr. 012.057 for the main entry.

Angular momentum transport by photons is analyzed for thin and thick configurations of steady accretion flows near black holes. The radiative viscosity coefficient is derived accurately from kinetic theory for arbitrary photon spectra. Thin Kelperian disks cannot be supported by radiative viscosity. However, in a quasi–spherical accretion most of the initial angular momentum of the gas can be transported away by the photons it produces, allowing gas flow towards the central black hole. Compton drag can also be effective in the outer optically–thin part of the flow. The radiative transport of angular momentum may provide the needed accretion to fuel massive black holes with gas in AGN.

067.193 Nuclear physics of dense matter.
C. J. Pethick, D. G. Ravenhall.
15. Texas Symposium on Relativistic Astrophysics and 4. ESO–CERN Symposium, p. 503 – 509 (1991). – See Abstr. 012.060 for the main entry.

Over the past few years there have been a number of important developments in applications of nuclear physics to astrophysical dense matter. The authors discuss the nuclear and hadronic constituents assumed in the nuclear models. In dense matter there are also electrons, muons, and neutrinos as the thermodynamic conditions call for them. They interact electromagnetically and/or via weak interactions, and their thermodynamic properties may be calculated by the standard methods for almost ideal relativistic Fermi gases.

067.194 Neutron star crusts.
L. Hernquist.
15. Texas Symposium on Relativistic Astrophysics and 4. ESO–CERN Symposium, p. 510 – 518 (1991). – See Abstr. 012.060 for the main entry.

This paper is a summary of recent developments in the understanding of the physics of neutron star crusts. It focuses on a limited number of topics; mainly those related to isolated neutron stars, that is, those not in binaries, and which involve effects occurring at low densities in the outer crust.

067.195 Evolution of the magnetic fields of neutron stars.
G. Srinivasan.
15. Texas Symposium on Relativistic Astrophysics and 4. ESO–CERN Symposium, p. 538 – 547 (1991). – See Abstr. 012.060 for the main entry.

Contents: 1. Do neutron star magnetic fields decay? 2. The low magnetic field binary pulsars. 3. Asymptotic fields. 4. Field decay due to mass accretion. 5. The nature of the core field. 6. The fishing rods. 7. Field evolution of solitary pulsars. 8. Neutron stars in binaries. 9. The residual field.

067.196 Nonthermal electron–positron pairs and cold matter in the central engines of active galactic nuclei.
A. A. Zdziarski.
AIP Conf. Proc., No. 254, p. 291 – 300 (1 May 1992). – See Abstr. 012.057 for the main entry.

The nonthermal $e^\pm$ pair model of the central engine of active galactic nuclei (AGNs) is discussed. The model assumes that nonthermal $e^\pm$ pairs are accelerated to highly relativistic energies

in a compact region close to the central black hole and in the vicinity of some cold matter. The model has a small number of free parameters and explains a large body of AGN observations from EUV to soft γ-rays. In particular, the model explains the existence of the UV bump, the soft X–rays excess, the canonical hard X–ray power law, the spectral hardening above ~ 10 keV, and some of the variability patterns in the soft and hard X–rays. In addition, the model explains the spectral steepening above ~ 50 keV seen in NGC 4151.

067.197 The high energy AGN continuum: models without e^+e^-–pairs.
D. Kazanas.
AIP Conf. Proc., No. 254, p. 301 – 310 (1 May 1992). – See Abstr. 012.057 for the main entry.

The general properties of the high energy AGN continuum are reviewed with particular emphasis on the connection between radio loud and radio quiet AGN in the broad context of the dynamics of accretion onto a black hole. Arguments are provided indicating the possibility that the emission from radio loud AGN originates at distances much larger than a few Schwarzschild radii from the black hole and hence the role of pairs in defining the observed spectrum is limited. A connection is made between the radio and high energy continuum of AGN.

067.198 Structure and spectra of electron–positron pair cascade atmospheres.
B. G. Tritz, S. Tsuruta.
AIP Conf. Proc., No. 254, p. 311 – 312 (1 May 1992). – See Abstr. 012.057 for the main entry.

The self–consistent spatial structure and radiation spectra of steady–state electron–positron pair cascade shower atmospheres surrounding AGN accretion flows are calculated, by developing computer codes to model geometry–dependent radiation transfer, including: Compton scattering, pair production, and pair annihilation. Disk and spherical accretion flow geometries are investigated, as well as a wide range of plausible primary accretion flow spectrum. It is found that substantial pair atmospheres can develop above accretion flows which emit even a modest fraction of their luminosity as gamma radiation. The radiation spectrum emitted by the flow can be significantly reprocessed during transit through the atmosphere, leading to observational constraints on model parameters.

067.199 Pair cascades triggered by relativistic protons in AGNs.
B. E. Stern, M. Sikora, R. Svensson.
AIP Conf. Proc., No. 254, p. 313 – 316 (1 May 1992). – See Abstr. 012.057 for the main entry.

The authors present results of Large Particle Monte Carlo simulations of relativistic proton supported hadronic and electromagnetic cascades in the central regions of AGNs. They assume that the relativistic protons are injected into a region filled with UV radiation and with rarified proton–electron thermal plasma (e.g., the corona above an accretion disk). The results show that electromagnetic spectra produced in this way can be much harder at 100 keV < hν < 10 MeV and can have much weaker annihilation lines than those produced by pure electromagnetic cascades. The authors also show the energy spectra of escaping relativistic neutrons and neutrinos.

067.200 Relativistic electron production by inelastic and elastic collisions in active galactic nuclei.
M. G. Baring.
AIP Conf. Proc., No. 254, p. 317 – 320 (1 May 1992). – See Abstr. 012.057 for the main entry.

The production of relativistic electrons by elastic and inelastic collisional mechanisms is studied, with the view to assessing whether they can provide suitable injections for AGN pair cascade models. A distribution of relativistic protons is injected into the emission region, perhaps by shock acceleration, and these collide with the protons and pairs in the ambient thermal gas. The ep collisions boost the cold electrons to relativistic energies by Coulomb scattering and bremsstrahlung. The pp collisions are inelastic and create electrons through pion decay modes. Here it is observed that significant electron injection luminosities can be generated via these processes. It is also suggested that the peak luminosity at $\gamma_e \sim 100$ that is required in pair cascade models can be produced via the decay of charged pions and also by pair production between X–rays and the photonic decay products of neutral pions. The inelastic processes generate high–energy power–law electron injections.

067.201   **Diskoseismology: probing accretion disks. II. Damping and growth of modes due to viscosity and gravitational radiation.**
M. A. Nowak, R. V. Wagoner.
Bull. Am. Astron. Soc., Vol. 23, No. 3, p. 1268 (1991). Abstract. – See Abstr. 012.067 for the main entry.

067.202   **X–ray binaries and their evolution.**
W. Kluźniak.
15. Texas Symposium on Relativistic Astrophysics and 4. ESO–CERN Symposium, p. 587 – 596 (1991). – See Abstr. 012.060 for the main entry.
  Contents: 1. Introduction. 2. Low mass X–ray binaries. 3. Neutron stars or black holes? 4. Conclusions.

067.203   **Coalescing binary neutron stars.**
T. Nakamura, K.–I. Oohara.
15. Texas Symposium on Relativistic Astrophysics and 4. ESO–CERN Symposium, p. 597 – 604 (1991). – See Abstr. 012.060 for the main entry.
  There are several motivations for the study of coalescing binary neutron stars. One is the real existence of binary neutron stars. In particular, PSR 1913 + 16 has been observed precisely and it is believed to consist of two neutron stars. Very recently it was found that two more binary neutron stars, PSR 2127 + 11 C and PSR 1534 + 12, exist. The orbital parameters of these binary neutron stars are very similar to those of PSR 1913 + 16. They look like twins. It seems that six neutron stars in these three binaries have almost the same mass $\sim 1.4 \, M_\odot$. Therefore two neutron stars in these systems will coalesce in $\sim 10^8$y because of the emission of gravitational waves.

067.204   **Line profiles of radially moving clouds in the presence of an accretion disk.**
Cao Xinwu, Zhang Jialü.
Chin. Astron. Astrophys., Vol. 16, No. 1, p. 17 – 23 (Jan–Mar 1992). English translation of Acta Astrophys. Sin., Vol. 11, No. 4, p. 321 – 326 (Oct 1991). See Abstr. 54.067.074.
  In the black hole model of active galactic nuclei, light rays emitted by ionized clouds moving in the vicinity of the black hole will be subjected to the effects of Doppler shift, gravitational redshift and bending, as well as shielding by an accretion disk, should this exist. The authors have taken all these effects into consideration in a rigorous solution of the photon transport equation for the line profile under the Schwarzschild metric. It is found that asymmetric double peak structure appears in some cases.

067.205   **Magnetohydrodynamics of black holes and the origin of jets.**
M. Camenzind.
15. Texas Symposium on Relativistic Astrophysics and 4. ESO–CERN Symposium, p. 610 – 619 (1991). – See Abstr. 012.060 for the main entry.
  Contents: 1. Introduction. 2. The magnetic structure of AGN disks. 3. Magnetohydrodynamics of rapidly rotating compact objects. 4. Magnetized accretion onto rapidly rotating black holes and the Blandford–Znajek process. 5. Axisymmetric magnetospheres of black holes and the origin of jets in AGNs.

067.206   **α–disks and the precession of extragalactic jets.**
Lu Jufu.
Acta Astrophys. Sin., Vol. 12, No. 1, p. 9 – 12 (Jan 1992). In Chinese. English translation in Chin. Astron. Astrophys., Vol. 16, No. 2, p. 133 – 136 (Apr–Jun 1992).
  This paper considers a mechanism in which a black hole is driven to precess by a surrounding accretion disk, and points out the dependence of the precession period on the accretion rate. The result is in good agreement with the observational data of extragalactic jets.

067.207   **Broad lines from tidally disrupted stars.**
N. Roos.
AIP Conf. Proc., No. 254, p. 556 – 559 (1 May 1992). – See Abstr. 012.057 for the main entry.
  Recent investigations of the tidal break–up of solar–type stars near massive black holes have shown that about half of the stellar debris mass is strongly bound to the hole and will quickly be swallowed while the rest flies away in large, radially elongated clouds with typical velocities of order 5000 $M_6^{1/6}$km s^{-1}, where $M_6 = M_{hole}/10^6 M_\odot$. Tidal disruptions of stars near a central massive black hole may therefore provide fuel for AGN and also be a source of cold, dense gas which will emit broad permitted lines when photoionized by the central continuum source (Roos, 1992). The structure and properties of freely expanding (unbound) remnant clouds irradiated by the central continuum source of an AGN are discussed.

067.208   **α–disks and the precession of extragalactic jets.**
Lu Jufu.
Chin. Astron. Astrophys., Vol. 16, No. 2, p. 133 – 136 (Apr–Jun 1992). English translation of Acta Astrophys. Sin., Vol. 12, No. 1, p. 9 – 12 (Jan 1992). See Abstr. 067.206.
  This paper considers a mechanism in which a black hole is driven to precess by a surrounding accretion disk, and points out the dependence of the precession period on the accretion rate. The result is in good agreement with the observational data of extragalactic jets.

067.209   **Coalescing binary neutron stars.**
K.–i. Oohara, T. Nakamura.
Symposium of Supercomputing Astronomy and Astrophysics in Japan, p. 83 – 91 (1991). – See Abstr. 012.069 for the main entry.

067.210   **Gravitational radiation from coalescing binary neutron stars. IV. Tidal disruption.**
T. Nakamura, K.–i. Oohara.
Symposium of Supercomputing Astronomy and Astrophysics in Japan, p. 93 – 110 (1991). – See Abstr. 012.069 for the main entry.

067.211   **Could final states of stellar evolution proceed towards naked singularities?**
N. Dallaporta.
Symposium on Gravitation and Modern Cosmology – the Cosmological Constant Problem, p. 11 – 18 (1991). – See Abstr. 012.076 for the main entry.

067.212   **Gamma lines nuclei ^{6,7}Li, ^{7}Be from accreting material on the neutron star.**
V. V. Burdyuzha, V. M. Lipunov, N. P. Yudin.
Astron. Astrophys. Trans., Vol. 1, No. 1, p. 51 – 54 (1991).
  The possibility of the observation of gamma lines from nuclei ^{6,7}Li, ^{7}Be, produced at the accreting material on a neutron star is discussed. The $\alpha\alpha$ collisions were considered. These lines with E ~ 0.429; 0.478; 3.56 MeV probably may be observed by an instrument with sensitivity better than $\sim 10^{-5}$phot/cm^{2}sec from X–ray sources with normal chemical abundance.

067.213 **The proto neutron star.**
J. Kim.
1. Yanbian International Workshop on Modern Physics: Particles, quantum groups, high Tc, phase transitions and all that, p. 274 – 283 (1991). – See Abstr. 012.081 for the main entry.
A description of the formation of a proto neutron star is presented. It is then compared with the observed data from SN 1987A including a constraint on L.E.P. (light exotic particle).

067.214 **Quark stars.**
C. H. Lee, H. K. Lee.
1. Yanbian International Workshop on Modern Physics: Particles, quantum groups, high Tc, phase transitions and all that, p. 284 – 292 (1991). – See Abstr. 012.081 for the main entry.
If the pressure inside a collapsing star becomes sufficiently large a phase transition from nuclear matter to quark matter may occur and the resulting configuration is called a quark star. Equilibrium characteristics of quark stars, such as the mass, radius and surface redshift, are calculated in two different cases where strange matter is self–bound and non–self–bound respectively.

067.215 **1D numerical relativity applied to neutron star collapse.**
E. Gourgoulhon.
Classical Quantum Gravity, Vol. 9, Suppl., p. S117 – S125 (1992). – See Abstr. 012.058 for the main entry.
A complete system of equations sufficient to handle the fully general relativistic hydrodynamics in the case of spherical symmetry is presented, as well as its numerical implementation in a pseudo–spectral code. This code is able to follow tiny radial perturbations of a neutron star near the maximum mass, and especially to show the growth of their fundamental mode as the central density increases. It is also able to describe a complete collapse towards a black hole when the instability threshold is passed. In other respects, the code is employed to address the problem of the minimum mass of a black hole formed via gravitational collapse.

067.216 **The dynamics of the solid crust of a general relativistic and rapidly rotating neutron star.**
D. Priou.
Classical Quantum Gravity, Vol. 9, Suppl., p. S207 – S212 (1992). – See Abstr. 012.058 for the main entry.
The oscillations of a neutron star's solid crust have so far mainly been investigated in the purely Newtonian framework. Here the author reports on his attempts to construct a general relativistic theory of the oscillations of a rapidly rotating neutron star.

067.217 **Magnetohydrodynamic wave propagation in the "ionosphere" of the central black hole in an active galactic nucleus.**
S. J. Park.
Publ. Korean Astron. Soc., Vol. 7, No. 1, p. 71 – 77 (1992). – See Abstr. 012.079 for the main entry.
An axisymmetric, stationary electrodynamic model of the central engine of an active galactic nucleus has been well formulated by Macdonald and Thorne. In this model the relativistic region around the central black hole must be filled by highly conducting plasma and the equations of magnetohydrodynamics are then satisfied. In this paper the author analyzes magnetohydrodynamic wave propagation in this region. He finds that there are three distinct types of waves – the Alfvén wave and two magnetosonic waves. The wave equations turn out to be not very different from those in nonrelativistic case except they are redshifted.

067.218 **Black holes in galactic nuclei: alternatives and implications.**
H. M. Lee.
Publ. Korean Astron. Soc., Vol. 7, No. 1, p. 89 – 96 (1992). – See Abstr. 012.079 for the main entry.
Recent spectroscopic observations indicate concentration of dark masses in the nuclei of nearby galaxies. This has been usually interpreted as the presence of massive black holes in these nuclei. Alternative explanations such as the dark cluster composed of low mass stars (brown dwarfs) or dark stellar remnants are possible provided that these systems can be stably maintained for the age of galaxies. The author considers the possible outcomes of interactions between the black hole and the surrounding stellar system. Under typical conditions of M31 or M32, tidal disruption will occur every 10^3 to 10^4 years. The author presents a simple scenario for the evolution of stellar debris based on basic principles. Finally the author outlines recent effort to simulate the process of tidal disruption and subsequent evolution of the stellar debris numerically using Smoothed Particle Hydrodynamics technique.

067.219 **Hawking radiation to an observer with variable acceleration.**
Zhao Zheng, Luo Zhiqiang, Huang Chaoguang.
Chin. Phys. Lett., Vol. 9, No. 5, p. 269 – 272 (1992).
There exists an event horizon and thermal radiation depending on time to a non–uniformly accelerating observer. The Hawking temperature is proportional to its instantaneous acceleration.

067.220 **Temperature of non–stationary Kerr–Newman black hole.**
Zhao Zheng, Huang Weihua.
Chin. Phys. Lett., Vol. 9, No. 6, p. 333 – 336 (1992).
Following a new method developed by the authors the temperature of a non–stationary Kerr–Newman black hole is given. The temperature is homogeneous everywhere at the event horizon and only depends on the time.

067.221 **Quantum temperature near the ring singularity in the Kerr space–time.**
Zhao Zheng.
Chin. Phys. Lett., Vol. 9, No. 7, p. 390 – 392 (1992).
There exists a local event horizon at the ring singularity in the equatorial plane of the Kerr black hole, whose surface gravity diverges. The Hawking temperature near the ring is infinite if the relation between the temperature and the surface gravity is still valid.

067.222 **Metric perturbations induced by a particle falling into a Schwarzschild black hole. I. Formulation.**
M. Shibata, T. Nakamura.
Prog. Theor. Phys., Vol. 87, No. 5, p. 1139 – 1157 (May 1992).
Metric perturbations induced by a test particle and a spheroidal dust shell falling straightly into a Schwarzschild black hole are examined in radial as well as isothermal gauge conditions. The authors calculate the wave forms at several radii. They find that the wave form of the quasi–normal mode is essentially the same, but the wave form of the precursor depends on the radius. For one particle case, the metric perturbation does not approach zero for $|t| \to \infty$ and it also depends on the gauge conditions, while for a spheroidal dust shell case, it approaches zero for $|t| \to \infty$ irrespective of gauges. As for the energy flux in the radial gauge, the authors find that to estimate the energy flux within 20% accuracy one needs $r \sim 40$ M for the dust shell case, while for one particle case one needs at least $r \sim 150$ M. This suggests that in the two body problem such as the collision of two black holes or coalescing binary neutron stars in numerical relativity, one must carefully treat the metric to estimate the energy flux.

067.223 **Boson bound states near a Kerr–Newman naked singularity.**
Li Yuanjie.
Aust. J. Phys., Vol. 45, No. 2, p. 127 – 130 (1992).
The author discusses boson bound states near a Kerr–Newman (KN) naked singularity by means of spectroscopic eigenvalue analysis. The results show that in the background of a KN naked singularity, the self–conjugate extension operator of a boson Hamiltonian has and only has discrete eigenvalues.

067.224   Accretion disk models for microvariability.
P. J. Wiita, H. R. Miller, N. Gupta,
S. K. Chakrabarti.
Conference on the Variability of Blazars, p. 311 – 319 (1992). –
See Abstr. 012.077 for the main entry.
The authors' models involve the random excess emission
produced by flares or "hot–spots" on the accretion disks around
supermassive black holes that are believed to provide power-
house for AGN.

067.225   Light curves of oscillating neutron stars.
T. Strohmayer.
News Lett. Astron. Soc. N.Y., Vol. 4, No. 1, p. 29 (Feb 1992).
Abstract. – See Abstr. 012.078 for the main entry.

**067.226   Equation of state of dense matter and astronomical
observations.**
P. Haensel.
22. Masurian Lakes Summer School on Nuclear Physics: Frontier topics in nuclear and astrophysics, p. 163 – 185 (1992). – See
Abstr. 012.094 for the main entry.
Neutron stars are the densest stable objects observed in the
present day Universe. The existence of neutron stars was
proposed immediately after discovery of the neutron. Neutron
stars were discovered in 1967 as radio pulsars. Then, they were
observed also as X–ray pulsars, X–ray bursters, and most
probably – as gamma–ray bursters. They consitute unique
cosmic laboratories, in which the theory of superdense matter can
be confronted, through astrophysical scenarios, with astronomical observations.

067.227   The stability of a warped accretion disc.
C. S. Coleman, S. Kumar.
Proc. Astron. Soc. Aust., Vol. 9, No. 2, p. 249 – 250 (1991). – See
Abstr. 012.090 for the main entry.
An accretion disc becomes warped when subjected to a torque
which is misaligned with the disc plane. Such torques may be
caused by Lense–Thirring precession near a spinning compact
object, or the quadrupole field of a binary star. Here the flow in
an adiabatic warped disc is modelled as a two–dimensional shear
layer with linear velocity profile and free surface boundary
conditions, and is investigated by means of a linear stability
analysis. The flow is found to be unstable whenever it contains a
critical layer, i.e., a level at which the shear velocity is equal to the
phase velocity. The instability occurs over a broad wavenumber
range and has a typical dimensionless growth rate ≈ 0.1 for both
the compressible and incompressible cases. These waves grow
with a time–scale of about one orbital period, and are likely to
have a major effect on the disc viscosity.

067.228   Black holes and partition functions.
J. W. Yorck Jr.
Conceptual problems of quantum gravity, p. 573 – 596 (1991). –
See Abstr. 003.049 for the main entry.
The train of developments leading from the early viewpoint on
black hole thermodynamics to the present understanding in terms
of statistical thermodynamics of the gravitational field in black
hole topologies is reviewed. The key idea is to take into strict
account the boundary conditions, that is, the information
supposed to be known to a given observer, in constructing
appropriate ensembles. Statistical mechanics of gravitational
fields describing the black hole topological sector, and the
correspondence to thermodynamics, are then discussed in the
canonical ensemble. The Euclidean action is evaluated on the
constraint hypersurface and a measure is obtained from a
heuristic argument, resulting in a path–integral form of the
canonical partition function. From this one obtains the usual
black–hole entropy, supplemented by statistical corrections when
the temperature and size of the system are appropriate. Under
other boundary conditions, it follows that a phase transition
(change of topology) must occur.

**067.229   Black hole thermodynamics and the space–time discon-
tinuum.**
T. Jacobson.
Conceptual problems of quantum gravity, p. 597 – 599 (1991). –
See Abstr. 003.049 for the main entry.

**067.230   A study of the temperature distribution of neutron stars
with magnetic charge and magnetic moment.**
Zhou Sanqing.
Chin. Sci. Bull., Vol. 37, No. 11, p. 894 – 898 (Jun 1992).

067.231   Neutrinos from hell.
F. W. Stecker, C. Done, M. H. Salamon, P. Sommers.
1. International Trends in Astroparticle Physics Conference
together with the SuperNova Watch Workshop, p. 326 – 349
(1992). – See Abstr. 012.102 for the main entry.
The authors calculate the spectrum and high energy
v background flux from photomeson production in Active
Galactic Nuclei (AGN), using the recent UV and X–ray
observations to define the photon fields and an accretion disk
shock acceleration model for producing high energy particles.
Collectively, AGN produce the dominant isotropic v background
between 10^4 and 10^{10}GeV, detectable with current instruments.
AGN v's should produce a sphere of stellar disruption which may
explain the "broad line region" seen in AGN.

067.232   Black holes: the inside story.
W. Israel.
Recent advances in general relativity. Essays in honor of Ted
Newman, p. 103 – 126 (1992). – See Abstr. 003.051 for the main
entry.
Contents: 1. Introduction. 2. Some recent developments.
3. Colored black holes. 4. Nature of the black hole interior.
5. Core inflation in spherical black holes. 6. Collision of light–
like shells and schematic mass–inflation. 7. Cosmological implications.

Updated catalogue of DC dwarfs (1987 variant).
See Abstr. 002.004.

White and black holes in the Universe.
See Abstr. 003.014.

Accretion power in astrophysics.
See Abstr. 003.041.

**Recent advances in general relativity. Essays in honor of Ted
Newman.**
See Abstr. 003.051.

Astrophysics of neutron stars.
See Abstr. 003.054.

Black holes.
See Abstr. 003.056.

Introducing Einstein's relativity.
See Abstr. 003.063.

Relativity theory. Concepts and basic principles.
See Abstr. 003.071.

**The magnetospheric structure and emission mechanisms of radio
pulsars. Proceedings.**
See Abstr. 012.022.

Frontier objects in astrophysics and particle physics. Proceedings.
See Abstr. 012.040.

General relativity and relativistic astrophysics. Proceedings.
See Abstr. 012.056.

Testing the AGN paradigm. Proceedings.
See Abstr. 012.057.

Texas/ESO–CERN Symposium on Relativistic Astrophysics, Cosmology, and Fundamental Physics. Proceedings.
See Abstr. 012.060.

Structure of hadrons and hadronic matter. Proceedings.
See Abstr. 012.062.

How has space astrophysics expanded the horizon of physics?
See Abstr. 013.020.

Numerical simulations of two–dimensional and three–dimensional accretion flows.
See Abstr. 021.026.

Smoothed particle rendering for fluid visualization – three–dimensional accretion disk and jet formation.
See Abstr. 021.027.

Hydrogen molecules and chains in a superstrong magnetic field.
See Abstr. 022.007.

Artificial opacity – numerical implementation into flux–limited neutrino diffusion.
See Abstr. 061.061.

Flux–limited neutrino diffusion versus Monte–Carlo neutrino transport.
See Abstr. 061.062.

Relativistic neutrons in active galactic nuclei. I. Energy transport from the core.
See Abstr. 061.077.

Nucleation of strange matter in dense stellar cores.
See Abstr. 061.104.

Properties of high–density matter in the electroweak symmetric phase.
See Abstr. 061.106.

Particle astrophysics: experiments and observations in solar and stellar physics.
See Abstr. 061.107.

Massive Dirac neutrinos and SN 1987A.
See Abstr. 061.111.

Electron conduction opacity for dense stellar plasmas.
See Abstr. 061.122.

A neutrino astronomy test of the AGN paradigm and the broad line region.
See Abstr. 061.130.

Neutrino emission from protoneutron star with modified URCA and nucleon bremsstrahlung processes.
See Abstr. 061.131.

Synthesis of nuclei in astrophysical environments.
See Abstr. 061.140.

Arbitrary amplitude double layers in a multi–species electron–positron plasma.
See Abstr. 062.044.

Particle acceleration in spherical wave fields.
See Abstr. 062.086.

Relativistic magnetosonic shock waves in synchrotron sources: shock structure and nonthermal acceleration of positrons.
See Abstr. 062.097.

Asymmetric disk accretion onto magnetized rotating compact stars.
See Abstr. 062.104.

The effect of relativistic particle beams on the evolution of supernova envelopes.
See Abstr. 062.116.

The inner structure of an accretion disc around a magnetic neutron star.
See Abstr. 062.125.

Magnetically–controlled disc accretion.
See Abstr. 062.126.

Models of central engines of AGNs with electron–positron pair production.
See Abstr. 062.134.

On regulation of the jet speed in SS 433.
See Abstr. 062.173.

Annihilation radiation from a power–law distributed electron–positron plasma on the ground Landau level: the case of low magnetic fields.
See Abstr. 062.179.

Compact jets and the AGN paradigm.
See Abstr. 062.189.

Jets on large scales.
See Abstr. 062.190.

Relativistic jets in AGN.
See Abstr. 062.191.

A magnetic mechanism for astrophysical jet production and enhancement of disk accretion as its reaction – 2.5 D and 3 D MHD simulation studies.
See Abstr. 062.200.

Relativistic quantum response of a strongly magnetised plasma. I. Mildly relativistic electron gas.
See Abstr. 062.210.

Relativistic quantum response of a strongly magnetised plasma. II. Ultrarelativistic pair plasma.
See Abstr. 062.211.

Collisions of giant stars with compact objects: hydrodynamical calculations.
See Abstr. 062.214.

Physics of flares in stars and accretion disks.
See Abstr. 062.219.

Multiple resonant scattering in the Compton upscatter model of gamma–ray bursts.
See Abstr. 063.009.

Radiative processes in strong magnetic fields.
See Abstr. 063.015.

Resonant Compton cooling and annihilation line production in gamma–ray bursts.
See Abstr. 063.018.

Photodissociation in strong magnetic fields and application to pulsars.
See Abstr. 063.051.

Collective electron–positron annihilation.
See Abstr. 063.068.

Linear acceleration emission: a detailed analysis.
See Abstr. 063.069.

The Josephson effect as a possible alternative mechanism for pulsar radio emission: spectra, pulse structure, polarization, and X-/γ-ray emission.
See Abstr. 063.070.

Is the pulsar emission produced by superluminally moving charged patterns?
See Abstr. 063.077.

On the nature of the circularly polarized component of pulsar radio emission.
See Abstr. 063.084.

Modeling of pulsar polarization properties.
See Abstr. 063.085.

A determination of pulsar emission geometry from polarization observations.
See Abstr. 063.088.

X-ray emission–line spectra of photoionized plasmas: density sensitivity of the Fe L–shell series.
See Abstr. 063.097.

The Fe L–shell spectrum in compact astrophysical X–ray sources.
See Abstr. 063.103.

Collective radiation from jets.
See Abstr. 063.104.

The theoretical gamma–ray spectrum of quasars.
See Abstr. 063.134.

Radiation from magnetized accretion disks in AGN.
See Abstr. 063.136.

High energy emission in accretion flows in AGN.
See Abstr. 063.137.

High energy spectra from the jet/disk interface.
See Abstr. 063.138.

The formation of coronal regions in accretion disks.
See Abstr. 064.004.

The effects of radiation drag on radial, relativistic hydromagnetic winds.
See Abstr. 064.009.

Wind accretion by compact objects: the "flip–flop" instability.
See Abstr. 064.031.

The viscous evolution of elliptical accretion discs.
See Abstr. 064.033.

Three–dimensional hydrodynamic simulation of an accretion flow in a close binary system.
See Abstr. 064.034.

Numerical simulations of two–dimensional and three–dimensional accretion flows.
See Abstr. 064.035.

The radiation from accretion discs.
See Abstr. 064.062.

On accretion by compact objects from a stellar wind.
See Abstr. 064.063.

A three component model for MagnetoHydroDynamic turbulence in accretion disks.
See Abstr. 064.073.

Is the Oort A–value a universal growth rate limit for accretion disk shear instabilities?
See Abstr. 064.074.

Limit cycles in pair dominated accretion disks.
See Abstr. 064.077.

A new mechanism for viscosity in cloudy accretion disks: anomalous (turbulent) viscosity.
See Abstr. 064.078.

Low–mass X–ray binary evolution and the origin of millisecond pulsars.
See Abstr. 065.014.

Evolution of a millisecond binary pulsar in the period gap.
See Abstr. 065.021.

Tidal capture of stars by a massive black hole.
See Abstr. 065.042.

Rotating stars: the angular momentum constraints.
See Abstr. 065.050.

On the pulsations of relativistic accretion disks and rotating stars: the Cowling approximation.
See Abstr. 065.051.

The status of non–homogeneous Dedekind ellipsoids.
See Abstr. 065.056.

The collapse of white dwarfs to neutron stars.
See Abstr. 065.076.

Possible scenario of a supernova explosion as a result of gravitational collapse of a massive stellar core.
See Abstr. 065.096.

Einstein gravity coupled to a massless conformal scalar field in arbitrary space–time dimensions.
See Abstr. 066.015.

Can string theory overcome deep problems in quantum gravity?
See Abstr. 066.034.

Gravitational radiation associated with anticollapse and supernova explosions.
See Abstr. 066.036.

Precursory singularities in spherical gravitational collapse.
See Abstr. 066.043.

Astrophysical jets and theories of gravitation.
See Abstr. 066.047.

Binary systems: higher order gravitational radiation damping and wave emission.
See Abstr. 066.052.

Evanescent black holes.
See Abstr. 066.055.

Strong–field tests of relativistic gravity and binary pulsars.
See Abstr. 066.060.

Boson star in a gravitation theory with dilaton.
See Abstr. 066.061.

Gravitational collapse of rotating spheroids and the formation of naked singularities.
See Abstr. 066.066.

Naked singularities in non–selfsimilar gravitational collapse of radiation shells.
See Abstr. 066.068.

Thermodynamics of event horizons in (2 + 1)–dimensional gravity.
See Abstr. 066.069.

Collisions of relativistic clusters and the formation of black holes.
See Abstr. 066.081.

Elliptic–type motion in a Schwarzschild – de Sitter gravitational field.
See Abstr. 066.097.

SL(2,R) coset constructions and two–dimensional gravity.
See Abstr. 066.098.

Black p–brane solutions of D = 11 supergravity theory.
See Abstr. 066.100.

Semi–classical gravity and the black hole singularity.
See Abstr. 066.112.

Relativistic wave equations in 1 + 1 dimensions.
See Abstr. 066.113.

Negative energy fluxes and cosmic censorship.
See Abstr. 066.139.

Thermal stress–energy tensor of a scalar field in Reissner–Nordström space–time.
See Abstr. 066.147.

Strong curvature naked singularities in non–self–similar gravitational collapse.
See Abstr. 066.159.

Radiating soliton stars in the thin–wall approximation.
See Abstr. 066.164.

An algorithm to generate classical solutions of string effective action.
See Abstr. 066.167.

Modified black holes in two dimensional gravity.
See Abstr. 066.168.

A charged Kerr metric solution in new general relativity.
See Abstr. 066.169.

Linear and nonlinear gravidynamics: static field of a collapsar.
See Abstr. 066.182.

Black holes and Newtonian physics.
See Abstr. 066.193.

Multigrid in general relativity: II. Kerr spacetime.
See Abstr. 066.200.

About the non–existence of perfect fluid bodies with the Kerr metric outside.
See Abstr. 066.204.

A minimal time and time–temperature uncertainty principle.
See Abstr. 066.208.

Relativity of inwards and outwards: an example.
See Abstr. 066.215.

Is there evidence for torsion?
See Abstr. 066.225.

Spherical collapse of a non–rotating perfect fluid in the non–symmetric theory of gravitation.
See Abstr. 066.239.

Pulsation of a supermassive star in tensor field gravitation theory.
See Abstr. 066.272.

Lens effect in a Kerr field.
See Abstr. 066.274.

Scenes from general relativistic astrophysics.
See Abstr. 066.306.

On the power spectra of the wind–fed X–ray binary pulsar GX 301–2.
See Abstr. 117.033.

The X–ray hardness ratio variation in low–mass X–ray binaries.
See Abstr. 117.048.

Synthesis of light curves of close binary systems containing a relativistic object. Asymmetric, geometrically thick disk around the relativistic object.
See Abstr. 117.051.

X–ray studies of the Hercules X–1 pulsar with the ASTRON satellite.
See Abstr. 117.052.

X–ray binaries and related systems.
See Abstr. 117.072.

Outburst phenomena in X–ray binaries.
See Abstr. 117.088.

Note on LMC X–4.
See Abstr. 117.124.

On the masses and on the mass transfer in the interactive binary SS 433.
See Abstr. 117.128.

Hard X–ray observations of Vela X–1 and A0535 + 26 with HEXE: discovery of cyclotron lines.
See Abstr. 117.149.

Synthesis of light curves of close binary systems with a relativistic object. Asymmetric geometrically thick disk around a relativistic object.
See Abstr. 117.159.

X–ray studies of Hercules X–1 from the Astron space station.
See Abstr. 117.160.

Spectral variability of He3–640, optical counterpart of the X–ray source A 1118–61.
See Abstr. 117.168.

Thermonuclear flash model for long X–ray tails from Aquila X–1.
See Abstr. 117.172.

Spiral shocks and subday variability in SS 433.
See Abstr. 117.180.

What drives the mass transfer in Cygnus X–3?
See Abstr. 117.187.

Canonical time variations of X–rays from black hole candidates in the low–intensity state.
See Abstr. 117.191.

Some results on the physics of accreting X–ray sources (Her X–1 and A 0535 + 26).
See Abstr. 117.209.

Optical light curve of HZ Herculis/Hercules X–1: look inside the system structure.
See Abstr. 117.210.

Pulsating X–ray sources: long–term period variations.
See Abstr. 117.247.

Black hole accretion disks exhibiting superhumps.
See Abstr. 117.249.

Problems for the standard blackhole/accretion disk models in Cygnus X–1?
See Abstr. 117.255.

Evidence for black holes in stellar binary systems.
See Abstr. 117.265.

A 6.5–day periodicity in the recurrent nova V404 Cygni implying the presence of a black hole.
See Abstr. 124.059.

Supernova 1987A and formation of rotating neutron stars.
See Abstr. 125.024.

SN 1987A and rotating neutron star formation.
See Abstr. 125.065.

Shall we observe a radio binary pulsar in SN 1987A?
See Abstr. 125.126.

On the decay of the magnetic fields of single radio pulsars.
See Abstr. 126.002.

Radio emission altitudes in the pulsar magnetosphere.
See Abstr. 126.006.

Gamma–ray bursts and radio pulsar glitches.
See Abstr. 126.016.

Orbital angular momentum loss in PSR 1957 + 20.
See Abstr. 126.017.

A magnetospheric heating model for the evaporation of the companion to PSR 1957 + 20.
See Abstr. 126.018.

Geometry of pulsar emission regions.
See Abstr. 126.022.

Determining the coherence of micropulses.
See Abstr. 126.033.

Pulsars, their evolution, winds and radiation.
See Abstr. 126.046.

Postglitch behavior of the Crab pulsar: evidence for external torque variations.
See Abstr. 126.047.

New evidence on the shape of pulsar beams.
See Abstr. 126.049.

Intensity dependence of the PSR 0329 + 54 pulse profile.
See Abstr. 126.064.

Detecting UV radiation from nearby pulsars with the Hubble Space Telescope.
See Abstr. 126.071.

Microstructure and the pulsar radio emission mechanism.
See Abstr. 126.085.

Observational evidence for persistent microstructure periodicities.
See Abstr. 126.088.

Remarks on triaxiality, spin–up, and magnetic field obliquity of some millisecond pulsars.
See Abstr. 126.094.

On the 440 keV line in the Crab Nebula pulsar.
See Abstr. 126.095.

PSR 0655 + 64: an astrophysical laboratory for testing relativistic gravity theories.
See Abstr. 126.096.

The pulsar radio–frequency emission problem.
See Abstr. 126.110.

Magnetic fields of degenerate stars.
See Abstr. 126.113.

The story of the pulsars.
See Abstr. 126.116.

Pulsars: the new celestial clocks.
See Abstr. 126.117.

Discovery and early X–ray lightcurve of the transient GRS 1124–68 (Nova Muscae 1991).
See Abstr. 142.001.

Hard X–ray observation of galactic center region.
See Abstr. 142.005.

SIGMA results on the fast variability of Cyg X–1 in the hard X–ray/gamma–ray energy band.
See Abstr. 142.007.

Broadband X–ray spectra of black–hole candidates, X–ray pulsars, and low–mass binary X–ray systems. Kvant module results.
See Abstr. 142.015.

Detection of quasiperiodic oscillations of X rays from the black–hole candidate GX 339–4.
See Abstr. 142.016.

The best black hole in the Galaxy.
See Abstr. 142.030.

X–rays from galactic black hole candidates and active galactic nuclei.
See Abstr. 142.063.

Soft gamma ray observations of Cygnus X–1 with the coded–aperture SIGMA telescope.
See Abstr. 143.001.

GRANAT images of the Galactic Center region in 4 – 1300 keV band: localization of the possible candidate for 511 keV source.
See Abstr. 143.010.

Compton backscattered 511 keV annihilation line and the 170 keV line from the Galactic Center.
See Abstr. 143.015.

On the two population model for gamma–ray bursts.
See Abstr. 143.027.

Cyclotron lines in gamma–ray bursts and magnetic field decay.
See Abstr. 143.030.

Gamma–ray bursts in the galactic halo.
See Abstr. 143.034.

Issues in the analysis and interpretation of cyclotron lines in gamma–ray bursts.
See Abstr. 143.050.

Gamma–ray monitoring of A.G.N. and galactic black hole candidates by the Compton Gamma–ray Observatory.
See Abstr. 143.054.

Two–population model for the sources of γ–ray bursts.
See Abstr. 143.057.

Cosmic gamma–ray bursts from black hole tidal disruption of stars?
See Abstr. 143.066.

Gamma–ray burst theory.
See Abstr. 143.086.

Gamma–ray monitoring of AGN and galactic black hole candidates by the Gamma–Ray Observatory.
See Abstr. 143.088.

Acceleration of ultra–high–energy cosmic rays in compact sources.
See Abstr. 144.040.

Cosmic rays from primordial black holes.
See Abstr. 144.083.

Angular momentum and black hole formation in stellar systems: a statistical approach.
See Abstr. 151.006.

The distinctive feature of relativistic stellar systems.
See Abstr. 151.067.

Tidal disruption of a star by a supermassive blackhole.
See Abstr. 151.075.

Rapid orbital decay of a black hole binary in merging galaxies.
See Abstr. 151.084.

An accreting black hole model for Sagittarius A*.
See Abstr. 155.025.

Gravitational lensing by a massive black hole at the Galactic center.
See Abstr. 155.026.

The neutron star population in the Galaxy.
See Abstr. 155.051.

The stellar populations of neutron and strange stars in the Galaxy.
See Abstr. 155.066.

The galactic center in the far–red.
See Abstr. 155.102.

Diagnostics of a putative black hole at the galactic center.
See Abstr. 155.143.

Testing the AGN paradigm for our home galaxy.
See Abstr. 155.144.

The a/m ratio in spiral galaxies: a consistency check for M31.
See Abstr. 157.215.

A search for black holes in galaxy nuclei.
See Abstr. 157.277.

The structure of M32 at red wavelengths.
See Abstr. 157.278.

Supermassive black holes in the centers of galaxies?
See Abstr. 157.279.

Probing the monster in luminous starburst galaxies.
See Abstr. 157.281.

Fitting the broad line spectrum and UV continuum by accretion discs in active galactic nuclei.
See Abstr. 158.002.

The non–thermal continuum of compact sources.
See Abstr. 158.017.

Optical and radio morphology of elliptical dust–lane galaxies. Comparison between CCD images and VLA maps.
See Abstr. 158.021.

The lighthouse effect of relativistic jets in blazars. A geometric origin of intraday variability.
See Abstr. 158.022.

Gas clouds from tidally disrupted stars in active galactic nuclei.
See Abstr. 158.027.

Origin of the central radio gaps in extragalactic radio sources.
See Abstr. 158.057.

Variability pattern from X–ray to IR wavelengths in the active nucleus of NGC 1566.
See Abstr. 158.078.

Universal energy spectrum from point sources.
See Abstr. 158.103.

On the X–ray variability power spectra of AGNs.
See Abstr. 158.110.

Does an orbiting star cause periodic modulation of X–rays from NGC 6814?
See Abstr. 158.113.

Active galactic nuclei. IV. Supplying black hole clusters by tidal disruption and by tidal capture of stars.
See Abstr. 158.127.

The standard model and some new directions.
See Abstr. 158.179.

Double–peaked line profiles in AGNs – testing for supermassive binary black holes.
See Abstr. 158.180.

Infrared angular size and the supermassive black hole hypothesis in AGN.
See Abstr. 158.181.

Orbital models for the periodic X–ray flares of NGC 6814.
See Abstr. 158.182.

NGC 7469. The perfect triptych: interaction, circumnuclear starburst and black hole?
See Abstr. 158.185.

Strong black holes as cores of AGN.
See Abstr. 158.187.

Red holes, not black holes, at AGN centers?
See Abstr. 158.190.

Do accretion disks exist? IR through radio observations.
See Abstr. 158.191.

Observational evidence for thin AGN disks.
See Abstr. 158.192.

Optical properties of ultra–soft X–ray AGN.
See Abstr. 158.198.

Double–peaked line profiles in AGNs – searching for reverberation from an accretion disk.
See Abstr. 158.200.

On the origin of double peaked broad line profiles in AGN: accretion disk with hot spots or biconical BLR?
See Abstr. 158.202.

UV variability and the nature of the continuum source in NGC 5548.
See Abstr. 158.203.

Parsec–scale linear polarization structure of active galactic nuclei.
See Abstr. 158.208.

Star–disk collisions in active galactic nuclei and the origin of the broad line region.
See Abstr. 158.226.

Observational constraints on stellar wind models of broad emission lines of AGNs.
See Abstr. 158.227.

Evidence for an extranuclear AGN fuel source.
See Abstr. 158.232.

Testing the AGN Paradigm: summary
See Abstr. 158.238.

Many black hole nuclei in blazars.
See Abstr. 158.241.

The production of jets and their relation to active galactic nuclei.
See Abstr. 158.307.

The jets of quasar 1928 + 738: superluminal motion and large–scale structure.
See Abstr. 159.050.

Lyman edges: signatures of accretion disks.
See Abstr. 159.094.

Using color–color diagrams to test models for the "blue bump".
See Abstr. 159.100.

Observations of diskoseismology.
See Abstr. 159.102.

Resolving accretion discs with microlensing.
See Abstr. 159.103.

A search for microvariability in five OVV quasars.
See Abstr. 159.104.

Gamma–rays from 3C 279 and universal energy spectrum.
See Abstr. 159.110.

Starburst–black hole QSO model. I: Broad emission lines and the central luminosity.
See Abstr. 159.115.

Starburst–black hole QSO model. II: Broad absorption lines.
See Abstr. 159.116.

Elements of string cosmology.
See Abstr. 161.069.

Torsion, minimum time, string tension and its physical implications in cosmology.
See Abstr. 161.080.

Charged cosmic string near a black hole.
See Abstr. 161.266.

The origin of the Big–Bang.
See Abstr. 161.332.

Contribution of gauge theories to baryon asymmetry.
See Abstr. 161.333.

A hypothetical end of creative–expansive evolution phase of the expansive non–decelerative Universe.
See Abstr. 161.364.

String theory and the quantization of gravity.
See Abstr. 161.374.

The origin of black holes in active galactic nuclei.
See Abstr. 161.418.

Astroparticle physics and superstrings.
See Abstr. 161.425.

Sun

071 Photosphere, Spectrum

071.001 Dynamics of the solar granulation. I. A phenomenological approach.
A. Nesis, A. Hanslmeier, R. Hammer, R. Komm, W. Mattig, J. Staiger.
Astron. Astrophys., Vol. 253, No. 2, p. 561 – 566 (Jan 1992).
= Mitt. Kiepenheuer–Inst. Nr. 347.

Based on time series of excellent spectrograms, taken with the vacuum tower telescope in Izaña, Tenerife, the authors present traces of line Doppler shifts which exhibit strong asymmetries within solar granules. They discuss this result in terms of different granulation flow models. By means of a power analysis they investigate the properties of the distribution of continuum intensity fluctuations as a function of spatial scale.

071.002 The formation of the Mg I emission features near 12 μm.
M. Carlsson, R. J. Rutten, N. G. Shchukina.
Astron. Astrophys., Vol. 253, No. 2, p. 567 – 585 (Jan 1992).

The authors explain the formation of the high–n emission lines of Mg I in the solar spectrum near 12 μm employing standard plane–parallel NLTE modeling with a radiative–equilibrium model atmosphere without chromosphere. The emission is a natural consequence of the replenishing of population depletion in a minority species from the population reservoir in the next higher ionization stage. The population depletion is primarily driven by lines with 6–7 eV excitation energy that become optically thin in the photosphere. The authors obtain excellent agreement with the observational constraints from a comprehensive but straightforward atomic model. They reproduce: (i) the observed 12 μm emission peaks; (ii) the observed wide absorption troughs; (iii) the observed peak–and–trough limb brightening; (iv) the observed strength ratios of different Mg I emission lines; (v) the observed absorption profiles of other Mg I Rydberg lines. The authors also explain why infrared emission features are present for Mg I, Si I and Al I, why they are absent for Na I and K I, and why corresponding H I Rydberg lines gain dominance only at longer wavelengths. The suggestion by Lemke and Holweger (1987) is confirmed that the 12 μm lines are formed in the photosphere and the claim by Zirin and Popp (1989) is disproved that the temperature minimum occurs much deeper than in standard models of the solar atmosphere.

071.003 Estimating the degradation of brightness power spectra of solar granulation from images outside the disk centre.
I. Rodríguez Hidalgo, M. Collados, M. Vázquez.
Astron. Astrophys., Vol. 254, No. 1/2, p. 371 – 380 (Feb 1992).

A method is presented to estimate the influence of the atmospheric turbulence and the telescope on the mean power spectra of the granulation brightness distribution at positions outside the disk centre, derived using the determination of Fried's parameter. The procedure is based on the differential degradation suffered by the directions perpendicular and parallel to the solar limb; the former one is more affected because it contains more power at high frequencies, due to the geometrical foreshortening effect. A certain spectral ratio is defined, which can be evaluated after the observed power spectra and can be described directly by a Korff's function for a given value of the Fried parameter r_0. The best fit of the measured attenuation to the theoretical one is calculated, allowing to obtain the adequate Modulation Transfer Function to reconstruct the original power

spectra. A test has been performed by evaluating the mentioned ratio after restored spectra showing that these present the required elongation up to a resolution of about 0″.4.

071.004 About spectroscopic measurements of the solar meridional motion.
F. Cavallini, G. Ceppatelli, A. Righini.
Astron. Astrophys., Vol. 254, No. 1/2, p. 381 – 386 (Feb 1992).

A large number of spectroscopic measurements of meridional mass motion on the Sun has been carried out in the past. The results are rather contradictory, although a 10 m s^{-1} poleward flow might be inferred. Some authors have pointed out that the so called "meridional motion" might be ascribed to a latitudinal dependence of the convective flux. In this paper the authors describe new observations carried out with the spectro–interferometer installed at the G. B. Donati Solar Tower in Arcetri: these measurements have been performed using lines having either similar or very different physical parameters in order to disentangle convective effects from true mass flows. The authors discuss the results and conclude that the great variety of "meridional motions" so far observed by spectroscopic means may be ascribed to an intrinsic variability of the meridional flow pattern and in part to the presence of a large scale velocity field.

071.005 What do red solar rays tell us?
E. V. Kononovič.
Říše hvězd, Vol. 72, No. 3, p. 43 – 47 (Mar 1991). In Czech.

071.006 Die turbulente Sonnenoberfläche.
A. Hanslmeier.
Sternenbote, Jahrg. 35, Nr. 6, p. 126 – 138 (1992).

071.007 Spherical harmonic analysis of steady photospheric flows. II.
D. H. Hathaway.
Sol. Phys., Vol. 137, No. 1, p. 15 – 32 (Jan 1992).

The use of the spherical harmonic functions to analyse the nearly steady flows in the solar photosphere is extended to situations in which B_0, the latitude at disk center, is nonzero and spurious velocities are present. The procedures for extracting the rotation profile and meridional circulation are altered to account for the seasonal tilt of the Sun's rotation axis toward and away from the observer. A more robust and accurate method for separating the limb shift and meridional circulation signals is described. The analysis procedures include the ability to mask out areas containing spurious velocities (velocity–like signals that do not represent true flow velocities in the photosphere). The procedures are shown to work well in extracting the various flow components from realistic artificial data with a broad, continuous spectrum for the supergranulation. The presence of this supergranulation signal introduces errors of a few m s^{-1} in the measurements of the rotation profile, meridional circulation, and limb shift from a single Doppler image. While averaging the results of 24 hourly measurements has little effect in reducing these errors, an average of 27 daily measurements reduces the errors to well under 1 m s^{-1}.

071.008 Is mesogranulation a distinct regime of convection?
T. Straus, F.-L. Deubner, B. Fleck.
Astron. Astrophys., Vol. 256, No. 2, p. 652 – 659 (Mar 1992).

The authors investigate the dynamics of the "mesogranulation" phenomenon by analyzing spatially two–dimensional spectral time series taken at disk center and cos θ = 0.8. After a three–dimensional Fourier transformation, they integrate the power and crosspower spectra in azimuth, and calculate k–ω power, phase difference, and crosspower spectra. The spatial power spectra are compared with those obtained from numerical simulations of stationary cell patterns. In the deepest atmospheric layers the V–I phase difference spectra reveal a uniform regime of convective motions ($\phi_{VI} \approx 0°$) at frequencies below the Lamb mode. The power spectra exhibit at all levels a significant distinction between supergranulation and convective patterns of smaller scales ($\lesssim 10$ Mm). On the other hand, the "mesogranulation" phenomenon cannot be identified as an independent convective regime in the deep photosphere, distinct from granulation. Rather, the mesostructures that appear to emerge in the middle photosphere and temperature minimum seem to be a product of the overshoot driving only the largest elements ($\gtrsim 4$ Mm) of an extended distribution of granular sizes while the smaller elements have already died out at a lower level.

071.009 Speckle observations of solar granulation.
C. R. de Boer, F. Kneer, A. Nesis.
Astron. Astrophys., Vol. 257, No. 2, p. L4 – L6 (Apr 1992). Letter–to–the–editor.

The authors present observations of solar granulation in a plage region near disc centre obtained with the Vacuum Tower Telescope at Observatorio del Teide, Tenerife. Speckle methods were employed for data acquisition and data reduction. The images show small–scale structures of the size near the telescopic diffraction limit of 0.2 arcsec. Attention is called to bright lanes at the borders between granules and intergranular areas. Conceivably, they are the intensity signature of strong upflows at the border of granules or of shocks in supersonic convection which are predicted by computer simulations of the granular phenomenon.

071.010 Convective motion on the Sun.
N. Weiss.
Nature, Vol. 356, No. 6367, p. 287 (26 Mar 1992).

071.011 Evolution and advection of solar mesogranulation.
R. Muller, H. Auffret, T. Roudier, J. Vigneau,
G. W. Simon, Z. Frank, R. A. Shine, A. M. Title.
Nature, Vol. 356, No. 6367, p. 322 – 325 (26 Mar 1992). Letter–to–the–editor.

Granular structure on the Sun's surface, with a typical scale of 1 – 2 Mm, has been known since 1800. More recently an intermediate "mesogranular" structure was found, with a characteristic scale of 3 – 10 Mm. The authors have obtained a three-hour sequence of observations at the Pic du Midi Observatory which shows the evolution of mesogranules from appearance to disappearance with unprecedented clarity. They see that the supergranules, which are known to carry along (advect) the granules with their convective motion, also advect the mesogranules to their boundaries. This process controls the evolution and disappearance of mesogranules.

071.012 Lifetimes in Fe II and the solar abundance of iron.
P. Hannaford, R. M. Lowe, N. Grevesse, A. Noels.
Astron. Astrophys., Vol. 259, No. 1, p. 301 – 306 (Jun 1992).

New atomic lifetimes have been determined for eight quartet (z^4D, z^4F) levels and ten sextet (z^6D, z^6F, z^6P) levels in Fe II and the results for seven of these levels have been combined with existing experimental branching fractions to obtain a revised set of log gf–values for 15 solar lines in Fe II. The new log gf data are used together with equivalent widths determined from the Liège solar atlas to derive a value for the iron photospheric abundance: $A_{Fe} = \log N_{Fe} = 7.48 \pm 0.04$ (relative to log N_H taken as 12.00), which is consistent with the currently accepted meteoritic result, 7.51 ± 0.01. The results are compared with the atomic lifetime

data, log gf data and solar analyses used in three other recent solar abundance determinations based on Fe II lines.

071.013 Solar hydrogen lines in the infrared.
M. Carlsson, R. J. Rutten.
Astron. Astrophys., Vol. 259, No. 2, p. L53 – L56 (Jun 1992). Letter–to–the–editor.

The authors study recently observed H I lines in the infrared solar spectrum, employing detailed NLTE modeling to explain their formation and to evaluate their diagnostic merits. The solar infrared H I lines vary much in character, depending on opacity and wavelength; the authors' computations reproduce the observations closely. The line wings are primarily set by Stark broadening due to metal ions and protons; the line cores are sensitive to NLTE population departure divergence which is driven by Balmer–continuum photoionization. The formation heights of the H I lines range from the deep photosphere for near–infrared line wings to the chromosphere for line cores with $\lambda > 10$ μm; these features provide valuable diagnostics of the thermal structure of the solar atmosphere.

071.014 Diamagnetic abundance differentiation in the solar system.
R. Steinitz, E. Kunoff.
IAU Symposium No. 150: Astrochemistry of cosmic phenomena, p. 425 – 426 (1992). – See Abstr. 012.029 for the main entry.

Chemical abundances in the solar corona or solar wind compared to those in the photosphere differentiate according to first ionization potential. The authors suggest that the effect is the result of diamagnetic diffusion pumps operating in the presence of gravitation and diverging magnetic structures. Implications are given concerning abundances in the solar system and chemically peculiar stars.

071.015 Spectral lines of the solar spectrum for investigation of magnetic fields by the line–ratio method.
N. A. Gladushina, V. G. Lozitskij, T. T. Tsap.
Visn. Kiiv. Univ., Fiz.–Mat. Nauki, Astron., Vip. 3, p. 29 – 34 (1992). In Ukrainian.

The line–ratio method's possibility has been discussed for additional information on subtelescopic magnetic fluxtubes. The fluxtubes' line profiles have been obtained by using three specially selected lines. So it is possible to measure strength of the magnetic fields and to define thermodynamical parameters. The lines of the multiplets Fe I 1, 2, 62, 816 and Cr I–multiplet 18 are proposed for that.

071.016 Diurnal variation of the linear polarization across the Hβ Fraunhofer line of the terrestrial atmosphere. I.
H. M. Basurah.
Astrophys. Space Sci., Vol. 192, No. 1, p. 21 – 33 (Jun 1992).

When spectra of the daytime sky are compared with those directly from the Sun, it is found that the depths of the Fraunhofer lines are reduced. This indicates the presence of an added light, the Ring effect. Most previous research on the Ring effect has been performed spectroscopically, with the notion that the added component is always unpolarized. Here the author has presented spectropolarimetric observations using the principle of the line depth method (compression between the line centre and its continuum) to investigate this effect. The general scattered light of the blue sky is polarized and as the additional component may or may not also be polarized, the filling–in effect should be detectable by performing spectropolarimetry, which indeed has been achieved. The observations have involved the use of high precision polarimetry with spectral resolution ≈ 2 Å at the Hβ Fraunhofer spectral line. The data indicate a variation of the linear polarization in the Hβ line centre as a function of the solar zenith angle. One possible advantage of the polarimetric technique is that the Ring effect detections can be made without recourse to solar measurements directly or indirectly using an attenuator. Finally, a model without knowledge of the filling–in mechanism, has yielded a good qualitative agreement with the observational behaviours and it can be taken as a first step to explain the variation of the linear polarization across the spectral line.

071.017   Diurnal variation of the linear polarization across the Hα and Hβ Fraunhofer lines of the terrestrial atmosphere, and a detection of a "daylight flash". II.
H. M. Basurah.
Astrophys. Space Sci., Vol. 192, No. 1, p. 35 – 44 (Jun 1992).

In Paper I, the ability of using the polarimetric technique to observe the filling-in of the Fraunhofer spectrum line profiles – the Ring effect – has been presented (using the Hβ Fraunhofer line only). It has also been found that the added light, which caused the filling-in, is polarized. Here, a description is given of some observations, which, hopefully, will lead to a better understanding of the Ring effect using the Hα and Hβ Fraunhofer spectral lines. The observations of the filling-in effect at the zenith have covered many conditions which might be considered as controlling its behaviour at these two spectral regions. The data reveal variations of polarization at the centres of those Fraunhofer lines as a function of solar zenith angle. The results show no uniformity for the added light intensity, its value was depending on the kind of day (clear and turbid) and/or on the solar zenith angle. Also the photometric observational studies reported here suggest that a new phenomenon behaving like airglow has been discovered. It is referred to as a "daylight flash" and was recorded (for short periods) on three days at both the Hα and Hβ Fraunhofer lines with different strengths. The cores of these lines were filled-in while the near continua were unaffected.

071.018   The observed limb effect in Fraunhofer lines, II.
A. K. Pierce.
Sol. Phys., Vol. 139, No. 1, p. 1 – 12 (May 1992).

The absolute limb effect for a number of Fraunhofer lines observed at the McMath Solar Observatory is given. Results, uncorrected for scattered light, are given for the following lines: Fe I λ3734.9, λ3735.3, λ5123.7, λ5250.2, λ5434.5, λ6678.0, and λ8886.6. Additional lines observed are five lines of CN λλ3876.3–3880.0, two lines of Ca I λ6161.3 and λ6162.2, one line of Na I, λ6160.7, and one CN line, λ7957.0, of the red system.

071.019   Telluric water vapor contamination of the Mount Wilson solar Doppler measurements.
C. S. Carter, H. B. Snodgrass, C. Bryja.
Sol. Phys., Vol. 139, No. 1, p. 13 – 24 (May 1992).

It has been shown that the solar line λ5250.2 (Fe I) is weakly blended with a telluric line in the water vapor spectrum, and that magnetograms taken using this line are therefore inaccurate. The authors investigate the effects of this contamination on the Mount Wilson synoptic magnetograph data, which is based on λ5250.2. Using spectrum scans taken at Kitt Peak, they model the contamination and develop a procedure that would correct for it, whenever the slant water vapor along the line of sight to the Sun is known. As this information is not available for the data collected thus far at Mount Wilson, the authors use the variation of determined quantities with airmass to obtain an average, or first-order, correction. Concentrating on the fitted coefficients for the solar rotation, the correction is found to be very slight, ∼0.5%, raising the value for the A coefficient, averaged over the period 3 December, 1985 to 22 July, 1990, from 2.8289 to 2.8422 μrad s^{-1}. The correction also removes a slight annual variation that has become dicernible in the data collected since 1986.

071.020   The formation mechanism of the lines of the C I infrared multiplet at λ1069.5 nm in the spectrum of the Sun.
N. G. Shchukina.
Kinematics Phys. Celest. Bodies, Vol. 7, No. 4, p. 30 – 38 (1991). English translation of Kinematika Fiz. Nebesn. Tel, Tom 7, No. 4, p. 38 – 47 (1991).

Mechanisms of formation of the lines of the C I infrared multiplet at λ1069.5 nm in the solar atmosphere are discussed. It is shown that the principal causes of departure from LTE of its level populations are "pumping" of the lower level in the upper photosphere and lower chromosphere by the UV radiation field

in the lines λλ165.75 and 258.29 nm and recombination–cascade processes due to an IR–photon deficit in subordinate bound–free continua and highly excited lines of C I. These latter processes not only compensate the drainage of atoms from muliplet levels on absorption of photons of highly excited lines in the visible region of the spectrum, but also result in additional population of these levels. Photons of the near UV continuum with energies of 2.4 – 3.8 eV are incapable of significantly disturbing the established distribution of the LTE departure coefficients.

071.021   A CAMAC–MERA 60 data–acquisition system applied to solar spectra and maps in the He I 10830 Å line.
A. B. Bukach, L. V. Didkovskij, N. N. Stepanyan, G. A. Sunitsa, Z. A. Shcherbakova.
Bull. Crimean Astrophys. Obs., Vol. 82, p. 158 – 168 (1992). English translation of Izv. Krym. Astrofiz. Obs., Tom 82, p. 172 – 184 (1990).

A universal spectrophotometer has been built around the BST–2 tower solar telescope with spectrograph and a CAMAC–MERA 60 data–acquisition system. The receiving devices consist of an infrared dissector, FEU–83 photomultiplier, and T–22 photodiode. Observations have been made on the Sun with all these detectors. The spectrophotometer with FEU–83 and photodiode gives spectral maps in two IR ranges simultaneously with a resolution of 2″ × 3″. The time required to map a single active region is 3 – 5 min, while the entire disk can be mapped with a resolution of 5″ × 5″ in 1 – 1.5 hr. In 1988, 223 maps were obtained of active regions and large parts of the Sun. Solar spectra have been recorded with a resolution of 70,000 in 70 – 90 sec with the scanning spectrometer and IR image dissector.

071.022   The depth of formation of absorption lines in the solar atmosphere.
V. A. Sheminova.
Kinematika Fiz. Nebesn. Tel, Tom 8, No. 3, p. 44 – 62 (May–Jun 1992). In Russian. English translation in Kinematics Phys. Celest. Bodies, Vol. 8, No. 3.

The characteristics of the depression contribution functions are studied for the Stokes profiles of the Fraunhofer lines formed in a magnetic field. The form of the depression functions depends mainly on the strength of splitting and the Zeeman component intensity, and is of a complicated character with a distinctly pronounced asymmetry. The depths of formation of magnetically sensitive lines are found by means of these contribution functions. The calculation revealed that, in a strong longitudinal magnetic field, the steep part of the line profile is formed higher than the centre of the line profile. The Stokes profiles that describe the polarization characteristics are formed only several kilometres higher than the Stokes profile that specifies the general depression of the unpolarized and polarized radiation. The averaged depth of formation of the whole line profile is practically independent of the magnetic field strength. The depths of formation of 17 photospheric lines usually used in magneto-spectroscopic observations are calculated for the models of the quiet photosphere, a flux tube, and the sunspot umbra.

071.023   On the effect of scintillations on the photosphere brightness distribution near the solar limb from eclipse observations.
L. A. Akimov, I. L. Belkina, N. P. Dyatel, G. P. Marchenko.
Kinematika Fiz. Nebesn. Tel, Tom 8, No. 3, p. 63 – 68 (May–Jun 1992). In Russian. English translation in Kinematics Phys. Celest. Bodies, Vol. 8, No. 3.

Photoelectric observations of the integrated fluxes of Mars, Jupiter, and stars have been made, the signal integration time being 20 ms for different telescope apertures. The quadratic mean flux fluctuation is found to be a function of the angular size of the source and aperture. The scintillation parameters for Mars obtained with an 8–cm aperture are used for simulating the effect of scintillations on the darkening function near the solar limb determined from eclipse observations. The scintillations are found to affect substantially the results being obtained. Averaging of several curves perturbed by scintillations gives nevertheless an almost monotonous fall of brightness at the extreme limb.

071.024 Variability with time of absorption lines in the solar spectrum.
R. I. Kostyk.
Kinematics Phys. Celest. Bodies, Vol. 7, No. 5, p. 43 – 45 (1991). English translation of Kinematika Fiz. Nebesn. Tel., Tom 7, No. 5, p. 48 – 50 (Sep–Oct 1991).
   B. T. Babii's conclusion (1991) that the Sun's photosphere is in a stationary state up to a certain level (in the formation region of weak absorption lines) and nonstationary above that level, which was obtained from analysis of the variations of Fraunhofer–line central depths observed in 1960 – 1985 is placed in doubt.

071.025 The Sun.
I. Nicolson.
Images of the universe, p. 93 – 109 (1991). – See Abstr. 003.090 for the main entry.
   Contents: (1) Inside the Sun. (2) The surface and beyond. (3) Solar physics.

071.026 The photosphere.
H. Zirin.
NATO Advanced Study Institute on the Sun: a Laboratory for Astrophysics, p. 175 – 190 (1992). – See Abstr. 012.091 for the main entry.
   The photosphere is the surface of the Sun that we see, and the direct source of its energy. It is dominated by granulation, supergranulation, and magnetic fields. The author discusses the significance of limb darkening, the opacity, and model structure. The magnetic fields are dominated by the network, but the weaker fields are also of great interest. The author discusses the Fraunhofer spectrum, and what it tells us about the surface, and finally, the newly–discovered emission lines at 12μ.

071.027 Photospheric lines redshift from balloon ultraviolet spectra of the quiet Sun.
D. Samain.
10. ESA Symposium on European Rocket and Balloon Programmes and Related Research, p. 363 – 366 (Nov 1991). – See Abstr. 012.088 for the main entry.
   Wavelength shifts between solar photospheric lines observed in two quiet areas of the Sun and telluric O_2 absorption lines of the Schuman–Runge bands have been measured in the 1950–2000 Å region. The results are deduced from high resolution spectra taken on board a balloon experiment at an altitude of 39 km. They show a systematic redshift of the solar lines relative to the reference telluric spectrum. After correction for the gravitational redshift and for all the known relative motions between Sun and observer, the average residual value is $+7$ mÅ, that corresponds in terms of velocity to an equivalent Doppler–Fizeau shift on the whole spectrum of about 1 km/s away from the observer. Has the redshift a solar origin, or are the O_2 telluric lines blueshifted in the terrestrial atmosphere?

The "Fraunhofer Solar Spectrum" data bank.
See Abstr. 002.032.

Sir William Herschel's notebooks: abstracts of solar observations.
See Abstr. 004.005.

The optically thick C III spectrum. I. Term populations and multiplet intensities at lower optical depths.
See Abstr. 022.051.

Molecular data from solar spectroscopy.
See Abstr. 022.190.

Neutral Ti line oscillator strengths.
See Abstr. 022.201.

A new instrument for high resolution, two–dimensional solar spectroscopy.
See Abstr. 034.051.

Very high spatial resolution two–dimensional solar spectroscopy with video CCDs.
See Abstr. 036.112.

The current sheet and Joule heating of a slender magnetic tube in the upper photosphere.
See Abstr. 062.030.

Remaining line opacity problems for the solar spectrum.
See Abstr. 063.121.

Detection of "invisible sunspots".
See Abstr. 072.008.

Magnetic fields and thermodynamic conditions in the solar flare of June 8, 1989.
See Abstr. 073.022.

Hydrogen photoionization rates for chromospheric and prominence plasmas.
See Abstr. 073.027.

The tangential discontinuous surface of velocity in solar chromosphere.
See Abstr. 073.065.

Filament eruptions, flaring arches and eruptive flares.
See Abstr. 073.093.

White–light flares.
See Abstr. 073.099.

Variations in the relative elemental abundances of oxygen, neon, magnesium, and iron in high–temperature solar active–region and flare plasmas.
See Abstr. 074.032.

Patterns in the photospheric magnetic field and percolation theory.
See Abstr. 075.001.

The effect of the anomalous dispersion in the solar atmosphere on results of magnetic field measurements using the line–ratio method.
See Abstr. 075.006.

Peculiar photospheric velocity fields and magnetic energy build–up.
See Abstr. 075.013.

A true–field magnetogram in a solar plage region.
See Abstr. 075.016.

Direct indication of magnetic reconnection in solar photosphere.
See Abstr. 075.031.

Small–scale photospheric magnetic fields.
See Abstr. 075.040.

On the evidence for mesogranules in solar power spectra.
See Abstr. 080.028.

A new solar irradiance calibration from 3295 Å to 8500 Å derived from absolute spectrophotometry of Vega.
See Abstr. 080.048.

Acoustic wave ducting along magnetic flux tubes in the sunspot regions.
See Abstr. 080.054.

072 Sunspots, Faculae, Activity Cycles, Solar Patrol

072.001 Superficial activity as a trace of the internal mechanisms of the solar cycle.
J. I. García de la Rosa.
11. European Regional Astronomy Meeting of the IAU: New windows to the universe, Vol. 1, p. 27 – 43 (1990). – See Abstr. 012.001 for the main entry.
The authors are approaching the 30th anniversary of Babcock's purely observational model of the Solar Cycle (Babcock, 1961). The occasion seems apt to check its health, after almost three decades of improved solar observations. Although the present review cannot accomplish this whole task it will attempt to give some hints in that direction.

072.002 American relative sunspot numbers for January 1992.
Sol. Bull. (AAVSO), Vol. 48, No. 1, p. 1 – 2 (Jan 1992).
Sunspot numbers for Feb – Jun 1992 are given in Vol. 48, Nos. 2 – 6 (Feb – Jun 1992).

072.003 Sonnenfleckenrelativzahlen der SONNE–Gruppe und des S.I.D.C. für Oktober 1991.
Sterne Weltraum, Jahrg. 31, Nr. 1, p. 71 (Jan 1992).
Further listings of sunspot numbers concerning the period Nov 1991 – Mar 1992 are published in Nr. 2, p. 117 (Feb 1992), Nr. 3, p. 183 (Mar 1992), Nr. 4, p. 251 (Apr 1992), Nr. 5, p. 350 (May 1992), and Nr. 6, p. 399 (Jun 1992), respectively.

072.004 Sunspot numbers.
Sky Telesc., Vol. 83, No. 1, p. 111 (Jan 1992).
Sunspot numbers for Sep 1991 are given. Further lists concerning the time span Oct 1991 to Feb 1992 are given in No. 2, p. 228 (Feb 1992), No. 3, p. 346 (Mar 1992), No. 4, p. 466 (Apr 1992), No. 5, p. 584 (May 1992), and No. 6, p. 711 (Jun 1992), respectively.

072.005 New information on solar activity, 1779 – 1818, from Sir William Herschel's unpublished notebooks.
D. V. Hoyt, K. H. Schatten.
Astrophys. J., Vol. 384, No. 1, p. 361 – 384 (1 Jan 1992).
Sir William Herschel observed the Sun from 1779 to 1818 with most of these solar observations made from 1799 to 1806. Wolf in reconstructing the long–term history of solar activity did not have Herschel's notebooks for analysis and thus was limited to scattered observations for the period 1801 – 1807. According to Wolf, the Wolf sunspot numbers for these years are very uncertain. Thus, by analyzing Herschel's notebooks, a better reconstruction of solar activity during cycle 5 is feasible. From Herschel's observations the authors find that the peak sunspot number occurred in the 1801 – 1803 time period rather than 1805 as Wolf deduced. Instead of a solar cycle of 17 yr in length, the authors find a cycle length of 14 yr. Second, the authors find that the peak yearly mean sunspot number is only about 38 rather than 45, as Wolf deduced. Thus, this solar cycle was the weakest cycle since the Maunder Minimum and has been called the Dalton or Modern Minimum. ^{10}Be, ^{14}C, aurora records, and magnetic declination observations all tend to support these conclusions. In the course of this study, a technique for making early solar observations homogeneous with modern sunspot observations has been developed. It is discussed here.

072.006 Model evaluation of the year–to–year variability in an 11–year sunspot cycle.
P. K. Pasricha, S. Aggarwal, B. M. Reddy.
Ann. Geophys., Vol. 9, No. 10, p. 696 – 702 (Oct 1991).
The peak values of the 11–year sunspot cycles constitute a "sunspot envelope", with a cyclic trend of about 80 years. This paper evaluates the variability of an 11–year sunspot cycle over different portions of this envelope. It is shown that the year–to–year fluctuations in a sunspot cycle may be described by a Yule autoregressive process. Useful predictions of the smoothed sunspot numbers 1 year in advance are thereby possible. The prediction basis in this statistical model approach is an average sunspot cycle, averaged over three cycles, separately at different portions of the sunspot envelope. By a variance analysis, the smoothed sunspot numbers in this prediction scheme are shown to be roughly equivalent to 12–month running averaged sunspot numbers. The predictions made by the present model are found to be satisfactory during the decay phase of an 11–year sunspot cycle, over different portions of the sunspot envelope. It probably implies that the sunspot envelope does not follow the peaks of sunspot numbers adequately with an 80–year period.

072.007 High spatial resolution magnetograms of solar active regions.
C. U. Keller, J. O. Stenflo, O. von der Lühe.
Astron. Astrophys., Vol. 254, No. 1/2, p. 355 – 361 (Feb 1992).
Using the Universal Birefringent Filter at the Sacramento Peak Vacuum Tower Telescope the authors have obtained simultaneous observations of left and right circular polarization in various solar magnetic features with a resulting spatial resolution of 0″.7 in the magnetograms. They describe the data reduction in some detail and discuss some instrumental effects. In particular they show that seeing can create features in magnetograms. A penumbra near disk center shows small–scale features in the magnetogram which are associated with the bright filaments. Bright features in the umbra of a small spot exhibit considerable polarization signals. In a pore region opposite polarities are found with a few seconds of arc.

072.008 Detection of "invisible sunspots".
H. Zirin, H. Wang.
Astrophys. J., Lett., Vol. 385, No. 1, p. L27 – L29 (20 Jan 1992). With plates L1 – L4.
Using a new CCD system, the authors have detected tiny sunspots, which they term "micropores", associated with elements of the magnetic network far from active regions. The smallest micropores detected are less than 1″ in diameter and about half the size of the associated magnetic feature. The pore size is systematically smaller than the size of the magnetic element seen in the magnetograms, but the ratio of pore size versus magnetic element size is independent of pore size, suggesting that the difference is real. For the smaller elements there is a good linear relation between the brightness deficit of the pore and the total flux of the magnetic element. The micropores are distinguishable from dark lanes in the granulation by their long life, large brightness deficit, and association with magnetic fields.

072.009 The periodic behaviour of solar activity: the near 155–day periodicity in sunspot areas.
M. Carbonell, J. L. Ballester.
Astron. Astrophys., Vol. 255, No. 1/2, p. 350 – 362 (Feb 1992).
The historical record of daily sunspot areas (1878–1982), covering cycles 12 to 21, has been analysed, looking for the periodicity around 155 d found in other indicators of solar activity, mainly solar flares. The results indicate that a periodicity between 150–160 d (77–72 nHz) seems to be significant during solar cycles 16–21, while it cannot be detected in solar cycles 12 to 15. However, a sliding–window analysis reveals some time intervals within each of most of the cycles considered, in which a strong periodicity appears at 155.6 d. This feature suggests an intermittent character. Also, some evidence is found suggesting that the periodicity appears or is more important in the hemisphere were sunspot areas have been dominant during the solar cycle.

072.010 Solar activity and brightness of Neptune.
V. Vanýsek.
Říše hvězd, Vol. 72, No. 7, p. 127 – 128 (Jul 1991). In Czech.

072.011 **How high will be the next solar maximum?**
M. Kopecký.
Vesmír, Vol. 70, No. 5, p. 266 – 267 (May 1991). In Czech.

072.012 **Fractal properties of sunspots.**
L. M. Zelenyj, A. V. Milovanov.
Sov. Astron. Lett., Vol. 17, No. 6, p. 425 – 427 (Nov 1991).
Current Physics Microform No.: 9207X1863. English translation
of Pis'ma Astron. Zh., Tom 17, No. 11, p. 1013 – 1019
(Nov 1991).
The authors present a fractal model of sunspots. They assume
that fractal structures are formed when magnetic force tubes
collect to a fractal cluster. They obtain expressions for magnetic
field distributions in sunspot umbras and penumbras, and find an
analytic expression for the fractal dimension assuming the
cluster's free energy is minimized. They discuss the relationship
between a spot's fractal dimension and the distribution function
of particles in a plasma. The distribution function corresponding
to an observed fractal dimension is found.

072.013 **Statistische Gesetzmäßigkeiten photosphärischer Fak-
keln.**
H. Stetter.
Sonne, Jahrg. 16, Nr. 61, p. 10 – 12 (Mar 1992).

072.014 **The growth and decay of sunspot groups.**
R. F. Howard.
Sol. Phys., Vol. 137, No. 1, p. 51 – 65 (Jan 1992).
Digitized Mount Wilson sunspot data from 1917 to 1985 are
analyzed to examine the growth and decay rates of sunspot group
umbral areas. These rates are distributed roughly symmetrically
about a median rate of decay of a few μhemisphere/day.
Percentage area change rates average $502\%\text{day}^{-1}$ for growing
groups and $-45\%\text{day}^{-1}$ for decaying groups. These values are
significantly higher than the comparable rates for plage magnetic
fields because spot groups have shorter lifetimes than do plages.
The distribution of percentage decay rates also differs from that
of plage magnetic fields. Small spot groups grow at faster rates on
average than they decay, and large spot groups decay on average
at faster rates than they grow. Near solar minimum there is a
marked decrease in daily percentage spot area growth rates. This
decrease is not related to group area, nor is it due to latitude
effects. Sunspot groups with rotation rates close to the average
(for each latitude) have markedly slower average rates of daily
group growth and decay than do those groups with rotation rates
faster or slower than the average. Similarly, sunspot groups with
latitude drift rates near zero have markedly slower average rates
of daily group growth and decay than do groups with significant
latitude drifts in either direction. Both of these findings are
similar to results for plage magnetic fields. These various
correlations are discussed in the light of our views of the
connection of the magnetic fields of spot groups to subsurface
magnetic flux tubes. It is suggested that a factor in the rates of
growth or decay of spot groups and plages may be the inclination
angle to the vertical of the magnetic fields of the spots or plages.
Larger inclination angles may result in faster growth and decay
rates.

072.015 **Intermediate–term periodicities in solar activity.**
R. Oliver, M. Carbonell, J. L. Ballester.
Sol. Phys., Vol. 137, No. 1, p. 141 – 153 (Jan 1992).
The presence of intermediate–term periodicities in solar
activity, at approximately 323 and 540 days, has been claimed by
different authors. In this paper, the authors have performed a
search for them in the historical records of two main indices of
solar activity, namely, the daily sunspot areas (cycles 12 – 21) and
the daily Zürich sunspot number (cycles 6 – 21). Two different
methods to compute power spectra have been used. The results
obtained for the periodicity near 323 days indicate that it has
only been present in cycle 21, while in previous cycles no
significant evidence for it has been found. On the other hand, a
significant periodicity at 350 days is found in sunspot areas and
Zürich sunspot number during cycles 12 – 21 considered all
together, also having been detected in some individual cycles.

However, this last periodicity must be looked into with care due
to the lack of confirmation for it coming from other features of
solar activity. The periodicity around 540 days is found in cycles
12, 14, and 17 in sunspot areas, while during cycles 18 and 19 it is
present, with a very high significance, in sunspot areas and
Zürich sunspot number. It also appears at 528 days in sunspot
areas during cycles 12 – 21. On the ohter hand, it is important to
note the coincidence between the asymmetry, favouring the
northern hemisphere, of sunspot areas and solar flares during
cycle 19, and the fact that the periodicity at 540 days was only
present, with high significance, in that hemisphere during that
solar cycle.

072.016 **Fitting the sunspot cycles 10 – 21 by a modified
F–distribution density function.**
W. Elling, H. Schwentek.
Sol. Phys., Vol. 137, No. 1, p. 155 – 165 (Jan 1992). With a
correction in Vol. 138, No. 2, p. 425 – 426 (Apr 1992).
It has been found that sunspot cycles 10 – 21, represented by
quarterly mean values of Zürich sunspot number, can be suitably
described by the F–distribution density function provided it is
modified by introducing five characteristic parameters, in order
to achieve an optimal fitting of each cycle. The average cycle
calculated from cycles 10 – 21 has been used as a basis to forecast
time and magnitude of the maximum of each cycle, as a function
of various numbers of the first quarterly mean values in the
beginning $N = 8$ to 16 quarters. The standard deviations at a
99% significance level calculated from the observed values
depend on N, and vary from 1.6 to 1.1 quarters and 65 to 16 units
of sunspot number. A rather sufficient forecast is obtained from
$N = 12$ quarters (with inaccuracy of ± 1.5 quarters and ± 24
units); the forecast for cycle 22 yielded, for $N = 12$, the values
$t_m = (15.4 \pm 1.5)$ quarters ($\sim 1990.\text{I}$) and $f(t_m) = (175 \pm 24$ units).

072.017 **Cyclic variation of the global magnetic field indices.**
V. N. Obridko, B. D. Shelting.
Sol. Phys., Vol. 137, No. 1, p. 167 – 177 (Jan 1992).
The energetical aspect of solar phenomena of different spatial
and time scales has been studied with special attention to global
magnetic fields. Cyclic regularities in the heliosphere are determined by energetics of global magnetic fields. The energy
variation of global fields consists of a number of maxima and
minima coinciding with reference points of the sunspot cycle. The
correlations of a number of well–known indices in the heliosphere
with Wolf numbers and with indices of energetics of the global
magnetic field have been investigated. The results can be used to
identify more exactly the reference points of the cycle.

072.018 **The east–west inclination of magnetic field lines in
sunspots.**
R. F. Howard.
Sol. Phys., Vol. 137, No. 2, p. 205 – 213 (Feb 1992).
Digitized Mount Wilson sunspot data covering the interval
from 1917 to 1985 are analyzed to examine the average areas of
individual sunspot umbrae over small zones of central meridian
distance. Assuming that systematic, east–west differences in these
quantities are due to the inclination of the magnetic fields of the
spots, one can calculate average east–west inclination angles for
all spots and for subsets of the full data set. It is found from such
an analysis that on average spot fields are inclined such as to trail
the rotation by a few deg. Leading and following spots may show
a tendency to be inclined slightly away from each other, in
contrast to the results of an earlier study of plage magnetic fields.
Growing spots tend to be inclined much more to the east than
decaying spots. This is in the opposite sense to the analogous
result derived from plage magnetic fields.

072.019 **Near–infrared CCD observations of umbral dots.**
M. W. Ewell Jr.
Sol. Phys., Vol. 137, No. 2, p. 215 – 223 (Feb 1992).
Umbral fine structures have been observed at 8500 Å using a
new CCD detector. Four frames with diffraction–limited seeing
were obtained. Between 68 and 91 umbral dots with a brightness
contrast greater than 2% were found in each frame, although no

dots were found in the darkest part of the umbra. The intrinsic flux of the umbral dots varies widely, indicating that their intrinsic brightness does as well. The mean dot lifetime is estimated as 15 min, although some dots were observed to live more than 2 h. Some of the umbral dots are flowing into the umbra at speeds up to 0.5 km s^{-1}. These dots have higher than average contrast and are associated with penumbral grains.

072.020 The latitude of filament bands at the sunspot minimum and the activity level in the two following 11–year solar cycles.
V. I. Makarov, V. P. Mikhailutsa (*V. P. Mikhajlutsa*).
Sol. Phys., Vol. 137, No. 2, p. 385 – 394 (Feb 1992).

The zonal structure of the distribution of filaments is considered. The mean latitudes of two filament bands are calculated in each solar hemisphere at the minima of the sunspot cycle in the period 1924 – 1986: middle latitude $\Phi_{2,m}$ and low latitude $\Phi_{1,m}$. It is shown that the mean latitude of the filament band $\Phi_{2,m}$ at the minimum – m of the cycle correlates, with $\varrho = 0.94$, with the maximum – M sunspot area S(M) and maximum Wolf number W(M) in the succeeding solar cycle M. It is shown that the mean latitude of the low–latitude filament band $\Phi_{1,m}$ is linearly dependent on the mean latitude filament band $\Phi_{2,m+1}$ at the succeeding minimum. The authors found a correlation of the latitude of the low–latitude filament band $\Phi_{1,m}$ with the maximum sunspot area in the M + 1 cycle. This enables to predict the power of two succeeding 11–year solar cycles on the basis of the latitude of filament bands at the minimum of activity, 1985 – 1986: W(22) $\cong$ 205 $\pm$ 10, W(23) $\cong$ 210 $\pm$ 10. The importance of the relationships found for theory and applied aspects is emphasized. An attempt is made to interpret the relationships physically.

072.021 Global solar cycle in the distribution of the green coronal emission period: 1940 – 1989.
V. V. Bortzov (*V. V. Borzov*), V. I. Makarov, V. P. Mikhailutsa (*V. P. Mikhajlutsa*).
Sol. Phys., Vol. 137, No. 2, p. 395 – 400 (Feb 1992).

The authors have studied the latitude–time distribution of the green (5303 Å) coronal line emission for 1940 – 1989 and compared these data with the distributions of the weak magnetic field, and of polar faculae and sunspots. They have found that a new cycle of coronal activity commences after the polar field reversal in the form of two components in each hemisphere. The global coronal activity cycle has a duration of 16 – 17 years and is described by two components that reflect the activity of polar faculae and sunspots.

072.022 On classification and systematization of solar activity indices.
M. Kopecký.
Geomagn. Aehron., Tom 32, No. 1, p. 15 – 18 (Jan – Feb 1992). In Russian. English translation in Geomagn. Aeron., Vol. 32, No. 1.

072.023 The 77 day periodicity in the flare rate of cycle 22.
T. Bai.
Astrophys. J., Lett., Vol. 388, No. 2, p. L69 – L72 (1 Apr 1992).

The occurrence times of major flares of solar cycle 22 are analyzed to detect periodicities. The author finds that a periodicity of 77 days was in operation in the 15 month interval from 1988 November to 1990 February for six cycles. During the rest of 1990 no periodicity was discernible; however, it seems that the 77 day periodicity has resumed in 1991. This 77 day period is interpreted as the third subharmonic of the fundamental period of ~25$^\text{d}$5. In this interpretation, the 154 day periodicity is the sixth subharmonic. It is also found that when the 77 day periodicity or the 154 day periodicity is in operation, the occurrence rate of major flares is much higher than is expected from the relative sunspot number.

072.024 Structure of sunspot penumbrae: fallen magnetic flux tubes.
D. G. Wentzel.
Astrophys. J., Vol. 388, No. 1, p. 211 – 217 (20 Mar 1992).

The author presents a model of a sunspot penumbra involving magnetic flux tubes that have fallen into the phosphere and float there. An upwelling at the inner end of a fallen tube continuously provides additional gas. This gas flows along and lengthens the tube and is ovservable as the Evershed flow. Fallen flux tubes may appear as bright streaks near the upwelling, but they become dark filaments further out. The model is corroborated by recent optical high resolution magnetic data for the penumbral filaments, by the 12 μm magnetic measurements relevant to the height of the temperature minimum, and by photographs of the umbra / penumbra boundary. If the boundaries of the penumbra are defined in terms of the location of penumbral filaments, then these boundaries should be rather ragged, as observed, because they depend on the energy of the upwellings relative to the energy needed for flux tubes to fall. The latter energy depends on the shape of the spot magnetic field and places the boundaries of the penumbra near those identified in the "return–flux" model of sunspots.

072.025 Dynamic phenomena in the chromospheric layer of a sunspot.
C. E. Alissandrakis, A. A. Georgakilas, D. Dialetis.
Sol. Phys., Vol. 138, No. 1, p. 93 – 105 (Mar 1992).

The authors have studied running penumbral waves, umbral oscillations, umbral flashes and their interrelations from Hα observations of a large isolated sunspot. Using a subtraction image processing technique they removed the sharp intensity gradient between the umbra and the penumbra and enhanced the low contrast, fine geatures. They observed running penumbral waves which started in umbral elements with a size of a few arcseconds, covered the umbra and subsequently propagated through the penumbra. The period of the waves was 190 s and the mean propagation velocity was about 15 km s^{-1}. The authors detected intense brightenings, located between umbral elements from where waves started, which had the characteristics of umbral flashes. There are indications that umbral flashes are related to the propagation of the waves through the umbra and their coupling. The subtraction images also show considerable fine structure in the chromospheric umbra, with size between 0".3 and 0".8.

072.026 Study of periodicities of solar nuclear gamma ray flares and sunspots.
V. K. Verma, G. C. Joshi, D. C. Paliwal.
Sol. Phys., Vol. 138, No. 1, p. 205 – 208 (Mar 1992). Letter–to–the–editor.

The authors have carried out a power–spectrum analysis of solar nuclear gamma–ray (NGR) flares observed by SMM and HINOTORI satellites. The solar NGR flares show a periodicity of 152 days, confirming the existence of a 152 – 158 days periodicity in the occurrence of solar activity phenomena and also indicating that the NGR flares are a separate class of solar flares. The power–spectrum analysis of the daily sunspot areas on the Sun for the period 1980 – 1982 shows a peak around 159 days while sunspot number data do not show any periodicity. Therefore, only sunspot area data should be treated as an indicator of solar activity and not the daily sunspot number data.

072.027 Description, analysis and impact of major solar activity during recent U.S. Shuttle missions.
M. J. Golightly, A. C. Hardy, W. Atwell, K. Hardy.
Adv. Space Res., Vol. 12, No. 2/3, p. 335 – 338 (1992). – See Abstr. 012.016 for the main entry.

Since STS–26, three large solar events have occurred during Shuttle missions; a geomagnetic storm during STS–29 and solar particle events (SPEs) during STS–28 and –34. The maximum dose to a crew attributed to an SPE was estimated to be 30 μGy. Time–resolved dosimetry measurements of the SPE dose during STS–28 were made using the Air Force Radiation Monitoring Equipment (RME)–III. Comparison of calculated and measured dose demonstrated a discrepancy, possibly a result of deficiencies in the geomagnetic cutoff model used. This experience demonstrates that dose from an SPE is strongly dependent on numerous factors such as orbit inclination, SPE start time, spectral parameters and geomagnetic field conditions; the exact combination of these factors is fortuitous. New sources of data and procedures are being investigated for incorporation into operational Shuttle radiation support practices.

072.028   Absorbed in the Sun.
A. Michałec.
Urania, Rok 63, Nr. 5, p. 136 – 141 (May 1992). In Polish.

072.029   On the relation between the intensities of bright features and the local background in sunspot umbrae.
M. Sobotka, J. A. Bonet, M. Vázquez.
Astron. Astrophys., Vol. 257, No. 2, p. 757 – 762 (Apr 1992).

A photometric study of umbral bright features (umbral dots, clusters of umbral dots, bright grains of light bridges) in sunspots is presented. It is based on white–light images taken with a CCD video camera driven by an automatic image selection system. A direct linear relation between the observed brightness of the features and that of the surrounding background areas has been found. To eliminate the influence of the seeing on this result, the radiative fluxes, which are much less sensitive to the image degradation, have been examined and an analogous relation has been found. Finally, a possible physical explanation is suggested.

072.030   Coordinated solar observations obtained during the GRO/Max'91 target–of–opportunity campaign of June 1991.
A. L. Kiplinger.
Compton Observatory Science Workshop, p. 489 (Feb 1992). Abstract. – See Abstr. 012.018 for the main entry.

072.031   Solar variability as a manifestation of the Sun's motion.
I. Charvátová, J. Střeštík.
J. Atmos. Terr. Phys., Vol. 53, No. 11/12, p. 1019 – 1025 (Nov–Dec 1991). – See Abstr. 012.024 for the main entry.

Based on records of solar activity available for the last two millenia, the relationship between the motion of the Sun around the barycentre, or centre of mass, of the solar system and the variability of solar activity was studied. The results indicate that the motion of the Sun could be the source of solar variability.

072.032   A new look at Wolf sunspot numbers in the late 1700's.
D. V. Hoyt, K. H. Schatten.
Sol. Phys., Vol. 138, No. 2, p. 387 – 397 (Apr 1992).

The authors show how the Wolf sunspot number can be derived from the number of sunspot groups alone. They utilize this approach to obtain a 'Group Wolf number'. This technique has advantages over the classical method of determining the Wolf number because corrections for observer differences are reduced and long–term self–consistent time series can be developed. The level of activity can be calculated to an accuracy of $\pm 5\%$ using this method. Applying the technique to Christian Horrebow's observations of solar cycles 1, 2, and 3 (1761 – 1777), the authors find that the standard Wolf numbers are nearly homogeneous with sunspot numbers measured from 1875 to 1976 except the peak of solar cycle 2. This result suggests that further analyses of early sunspot observations could lead to significant improvements in the uniformity of the measurements of solar activity. Such improvements could have important impacts upon our understanding of long–term variations in solar activity, such as the Gleissberg cycle, or secular variations in the Earths' climate.

072.033   Longitudinal position of sunspots and chromospheric filaments. Solar rotation No. 1829.
A. Garcia.
Coimbra Univ. (Portugal). Observatorio Astronomico. 8 p. (1990). In Portuguese and English.

This issue, Vol. 1, Fasc. 6, presents graphs on the longitudinal position of sunspots (diameter > 4000 km) and main chromospheric filaments derived from monochromatic images of the solar chromosphere (the K_{IV} line of ionized calcium at $\lambda = 3933$ Å and the Hα line of hydrogen at $\lambda = 6563$ Å). It concerns the solar rotation No. 1829 and has four groups of graphs for the northern and southern hemispheres, respectively. Photographic negatives of the spectroheliograms in Hα, Ca K_3 and Ca K_{IV} are enclosed.

072.034   Longitudinal position of sunspots and chromospheric filaments. Solar rotation No. 1830.
A. Garcia.
Coimbra Univ. (Portugal). Observatorio Astronomico. 8 p. (1990). In Portuguese and English.

This issue, Vol. 1, Fasc. 7, presents graphs on the longitudinal position of sunspots (diameter > 4000 km) and main chromospheric filaments derived from monochromatic images of the solar chromosphere (the K_{IV} line of ionized calcium at $\lambda = 3933$ Å and the Hα line of hydrogen at $\lambda = 6563$ Å). It concerns the solar rotation No. 1830 and has four groups of graphs for the northern and southern hemispheres, respectively. Photographic negatives of the spectroheliograms in Hα, Ca K_3 and Ca K_{IV} are enclosed.

072.035   Longitudinal position of sunspots and chromospheric filaments. Solar rotation No. 1831.
A. Garcia.
Coimbra Univ. (Portugal). Observatorio Astronomico. 8 p. (1990). In Portuguese and English.

This issue, Vol. 1, Fasc. 8, presents graphs on the longitudinal position of sunspots (diameter > 4000 km) and main chromospheric filaments derived from monochromatic images of the solar chromosphere (the K_{IV} line of ionized calcium at $\lambda = 3933$ Å and the Hα line of hydrogen at $\lambda = 6563$ Å). It concerns the solar rotation No. 1831 and has four groups of graphs for the northern and southern hemispheres, respectively. Photographic negatives of the spectroheliograms in Hα, Ca K_3 and Ca K_{IV} are enclosed.

072.036   Solar active regions as a percolation phenomenon.
D. G. Wentzel, P. E. Seiden.
Astrophys. J., Vol. 390, No. 1, p. 280 – 289 (1 May 1992).

The authors model the appearance of solar active regions using percolation theory. The motivation is: magnetic fields of active regions presumably are released and rise from some deep site of the solar dynamo. The authors attempt to bundle all the very complicated magnetic phenomena into two dimensionless parameters. The main parameter is the probability, P_{st}, that the release and rise of one flux tube stimulates the subsequent release and rise of a neighboring flux tube. A second parameter measures the lifetime of flux once it has arrived at the surface. With this hypothesis one can reproduce several properties of the distribution of active regions on the Sun: (1) The active regions persist for a long time. Magnetic flux emerges mostly where there is flux already. (2) There are persistent empty regions, reminiscent of coronal holes. (3) The dependence of P_{st} is that of a phase transition. For example, for P_{st} near a critical value, a change in P_{st} by merely 1% can change the area covered by active regions by a factor of 10. Surface activity provides a highly amplified signal of small changes at the site of the dynamo during the solar cycle. (4) The size distribution of the active regions is close to exponential, as observed.

072.037   Vortex attraction and the formation of sunspots.
E. N. Parker.
Astrophys. J., Vol. 390, No. 1, p. 290 – 296 (1 May 1992).

A downdraft vortex ring in a stratified atmosphere exhibits universal attraction for nearby vertical magnetic flux bundles. The author speculates that the magnetic fields emerging through the surface of the Sun are individually encircled by one or more subsurface vortex rings, providing an important part of the observed clustering of magnetic fibrils to form pores and sunspots.

072.038 **High resolution observations of the Evershed flow.**
P. Börner, F. Kneer.
Astron. Astrophys., Vol. 259, No. 1, p. 307 – 312 (Jun 1992).

Observations with high spatial resolution of the Evershed effect in a sunspot near the limb are presented. They were obtained with the Gregory Coudé Telescope at Observatorio del Teide, Tenerife, and consist of photographic spectrograms in the wavelength region 5160–5180 Å and corresponding slit–jaw images in white light at various positions of the sunspot image on the slit of the spectrograph. The measurement of the velocity field covers a height range of approximately 100 km (above $\tau_5 = 1$) to 700 km in the Mg b_2 line. The Evershed flow is inhomogeneous at all heights, though decreasing in amplitude with increasing height. The authors find maximum velocities of 4 km s^{-1} at the lowest layer. The flow goes clearly beyond the outer penumbra border. In Mg b_2 the Evershed effect is inverse, on average, while at the 500 km level it is not. The authors suggest that in one fortunate case, with slit orientation parallel to a flow tube, they have seen siphon flows outward, reaching to the height of the Mg b_2 layer and ending in photospheric faculae. The stationarity of the flow on small scales is questioned.

072.039 **Spatial structure of the Evershed effect.**
E. Wiehr, D. Degenhardt.
Astron. Astrophys., Vol. 259, No. 1, p. 313 – 317 (Jun 1992).

Spectra of a sunspot penumbra at $\vartheta = 27°$ are taken in four adjacent slit positions at exceptionally good seeing conditions, yielding smallest continuum structures of 0.35 arcsec width. The Doppler shifts of line–core and line–wings of the non–magnetic line Fe I 7090.4 are cospatial, however, their amplitudes are unrelated. The amount of line asymmetry is thus not related to the corresponding line–core shift. A good correlation with the continuum occurs only in spectra achieving highest spatial resolution. A slightly less resolved spectrum does not show a similar correlation, thus explaining the missing relation in former observations. The amplitudes of shift and asymmetry are not related to those of the continuum intensity. The lack of any amplitude relation in spite of an almost perfect spatial correlation among shift, asymmetry, and continuum might be explained by the different angles–of–view through fine–structures aligned along fluxtubes of individual inclination angles recently observed. In addition, actual penumbral structures being essentially smaller than 250 km as deduced from white light pictures might yield different influences of spatial smearing on continuum, line shift, and line asymmetry.

072.040 **Der 22. Sonnenfleckenzyklus. Die Aktivität in der Zeit 1990/1991.**
W. Diehl.
Ahnerts Kalender für Sternfreunde 1993.
p. 156 – 162 (1992). – See Abstr. 046.012 for the main entry.

072.041 **Sunspots. Jan – Dec 1989.**
Q. Bull. Sol. Act., Vol. 31, Part I, p. 1 – 7 (1989).
Sunspot relative–numbers and sunspot areas are given for Jan – Dec 1989.

072.042 **Provisional sunspot–numbers for December 1991.**
Yamamoto Circ., No. 2177, p. 2 (26 Jan 1992). In Japanese.
Provisional sunspot–numbers for January – May 1992 are given in Nos. 2178, 2181, 2183, 2185, 2186.

072.043 **Detailed comparison between sunspot activity in "hot spots" and galactic cosmic–ray intensity.**
M. Akioka, J. Kubota, M. Suzuki, I. Tohmura.
Sol. Phys., Vol. 139, No. 1, p. 177 – 187 (May 1992). = Contrib. Kwasan Hida Obs., Kyoto Univ., No. 303.

The influence of sunspot activity on the condition of the solar–terrestrial environment during cycle 21 was examined using the data of sunspots and the modulation of the galactic cosmic–ray intensity. The "hot spots" discussed by Ichimoto et al. (1985) and Bai (1987, 1988) were also found by analyzing the longitudinal distribution of sunspot groups. A detailed comparison between the time change of the sunspot activity in hot spots and that of the galactic cosmic rays observed by the neutron monitor reveals that several trasient diminutions of the GCR intensity (with much longer duration than a Forbush decrease) occur at nearly the same time as the sporadic enhancement of sunspot area in the hot spots.

072.044 **Sunspots, planetary alignments and solar magnetism: a progress review.**
P. A. H. Seymour, M. Willmott, A. Turner.
Vistas Astron., Vol. 35, Part 1, p. 39 – 71 (Jan 1992).

This paper discusses a new theory of the solar cycle which is based on the concept of resonant coupling between the tidal forces due to the planets and the evolving magnetic field of the Sun. The theory combines the advantages of two earlier classes of theories and also overcomes the disadvantages of these classes. In order to set the new proposals in context, the authors describe briefly the salient features of sunspots, the solar cycle and some of the theories relevant to the presentation of the new theory. They also introduce aspects of canal tidal theory since these form the basis of this proposal.

072.045 **An emerging current/magnetic flux system with delta structure in AR 2372.**
Ding Youji, Hong Qinfang.
Chin. Astron. Astrophys., Vol. 16, No. 1, p. 60 – 70 (Jan–Mar 1992). English translation of Acta Astron. Sin., Vol. 32, No. 3, p. 225 – 232 (Sep 1991). See Abstr. 54.072.130.

The characteristics of an emerging current/magnetic flux system (CMFS) with delta structure near the neutral line of the active region AR 2372 of April 1980 are described and interpreted with a simple MHD theory. The main results are: (1) It is easy to identify the CMFS and to estimate its current from measurements of the transverse magnetic field and the derived current density map. (2) The existence of old potential field and its Lorentz force on the CMFS can explain its rotation and the formation of the sheared configuration along the neutral line. (3) The CMFS looks like the current sheet in the neutral layer of the old longitudinal field, though it cannot be that. (4) The shearing motion of the spots along the neutral line manifests the separating motion of the footpoints of the CMFS. (5) The top of the CMFS shows a rapid rise. (6) Both the shearing motion and the rapid rise are caused by the m = 1 twist instability of the CMFS itself.

072.046 **Definitive American relative sunspot numbers for 1990.**
P. O. Taylor.
J. Am. Assoc. Variable Star Obs., Vol. 20, No. 1, p. 98 – 100 (1991).

072.047 **A long–lived active area on the Sun.**
P. O. Taylor, M. Alexescu.
J. Am. Assoc. Variable Star Obs., Vol. 20, No. 1, p. 101 – 104 (1991).

The development and major energetic phenomena associated with a long–lived solar active area are examined.

072.048 **The sunspot observations made in 1989.**
H. H. Esenoğlu.
Publ. Istanbul Univ. Obs., No. 157.
Istanbul Univ. (Turkey). Observatory. 22 p. (1992). = Istanbul Üniv. Astronomi ve Fizik Der., Vol. 55, p. 1 – 22 (1990).

This paper gives the heliographic coordinates for the sunspot groups observed in 1989.

072.049 **Study of possible subsurface influences on the emerging active regions.**
T. Baranyi, A. Ludmány.
Sol. Phys., Vol. 139, No. 2, p. 247 – 254 (Jun 1992).

The large–scale distribution of the orientations of emerging sunspot groups has been studied for the year 1977. It is probable that the declinations from the azimuthal directions are not entirely randomly scattered, but they can be governed also by subsurface velocity fields. If this assumption is correct, the most probable internal velocity distribution is a non–axisymmetric (columnar) giant convection pattern with a longitudinal wave number $l = 11$ rotating slightly slower than the Carrington system, on the basis of the given material.

072.050   Infrared array measurements of sunspot magnetic fields.
M. R. McPherson, H. Lin, J. R. Kuhn.
Sol. Phys., Vol. 139, No. 2, p. 255 – 266 (Jun 1992).

The authors have used a 128×128 format HgCdTl infrared array with the Sacramento Peak Observatory Vacuum Telescope (VTT) and Echelle spectrograph to obtain two–dimensional observations of the true magnetic field strength in a sunspot. The system retains all of the spectral information contained in the unpolarized IR Fraunhofer line profile with time resolution of about a minute (depending on the scan area and spatial resolution). Infrared observations allow direct field strength measurements out to the outer edge of the penumbra. The data suggest that the magnetic flux density in the outer penumbra is not well described by an extrapolation of the quadratic polynomial, in normalized central distance, that describes the umbral field. The authors measure a relatively high field strength of 800 G at the penumbra–quiet–Sun boundary, which is consistent with the "return–flux" model of Osherovich and Garcia (1989).

072.051   Determination of thermodynamical conditions in the sunspot chromosphere by solving an inverse problem. II. Numerical simulation of pressure and mass density distributions.
R. B. Teplitskaya, I. P. Turova, V. G. Skochilov.
Kinematika Fiz. Nebesn. Tel, Tom 8, No. 3, p. 27 – 37 (May–Jun 1992). In Russian. English translation in Kinematics Phys. Celest. Bodies, Vol. 8, No. 3.

In Paper I the accuracy was estimated to which the temperature and electron density distributions, $T(\tau_{14})$ and $n_e(\tau_{14})$ can be reconstructed (τ_{14} is the optical depth at the centre of the H line) on the basis of the inversion of the Ca II H and K central reversal profiles. In the present paper the derived relationships $T(\tau_{14})$ and $n_e(\tau_{14})$ are used for reconstructing total pressure P, mass density ϱ and geometric height h. The hypothesis of hydrostatic equilibrium is nowhere applied. The calculations are performed using a test model that simulates conditions in the chromosphere above the sunspot umbra. The accuracy of reconstruction of $P(\tau_{14})$ and $\varrho(\tau_{14})$ is inferior to that for $T(\tau_{14})$ and $n_e(\tau_{14})$, but this is not an obstacle in clearing up the main question: Does hydrostatic equilibrium exist in the umbra chromosphere? A criterion has been formulated which will allow to answer this question on the basis of the results of calculations. Application of the proposed algorithm is likely to allow also to evaluate the inclination of the sunspot flux tube axis within the framework of the magnetohydrostatic equilibrium.

072.052   Time variations of the tangential velocity component in the Evershed effect.
S. A. Druzhinin, A. A. Pevtsov, V. I. Levkovskij, M. V. Nikonova.
Kinematics Phys. Celest. Bodies, Vol. 7, No. 5, p. 46 – 53 (1991). English translation of Kinematika Fiz. Nebesn. Tel., Tom 7, No. 5, p. 51 – 60 (Sep–Oct 1991).

Observations of the tangential velocity component in the Evershed effect (torsional oscillations of sunspots) are reported. The spectral composition of the radial–velocity signals from two penumbral areas at positions symmetric about the umbra was investigated for six sunspots. The selected fields were on a line perpendicular to the direction to the center of the Sun's image, i.e., the rotational velocities of the gas in the sunspot were measured. The observations were made in the lines Fe I $\lambda543.45$ nm and H_β 486.13 nm. The results of the study confirm the existence of torsional oscillations of sunspots at photosphere level with a period of about an hour. These oscillations were measured for the first time at chromosphere level (period about 30 min). It is concluded from observations of two sunspots over three and four days that torsional oscillations with a period of several days occurred in these sunspots.

072.053   Dynamics in solar active regions.
Fang Cheng.
Advances in solar–terrestrial science of China, p. 52 – 59 (1992). – See Abstr. 003.069 for the main entry.

The recent progress in research in China on the evolution of solar active regions, energy storage of solar flares, the dynamics of solar flare atmosphere as well as the dynamics of prominences is briefly reviewed.

072.054   Fine structures in solar active regions.
Wu Mingchan, Li Zhikai.
Advances in solar–terrestrial science of China, p. 89 – 85 (1992). – See Abstr. 003.069 for the main entry.

This paper briefly describes recent work with the 26 cm Highresolution Vacuum Solar Telescope at Yunnan Observatory in solar cycle 22 observations. The spatial resolution of the optical data is 0.75 arcsec for photospheric observations and 1.2 arcsec for chromospheric observations. The AR 5395 in March 1989, the strongest active region of Cycle 22, was investigated together with the sunspot group, flare and mass ejections associated with flares. The analyses of intensity and velocity fields of the two–ribbon flare in AR 5470 on May 5, 1989, provided good evidence for chromopsheric evaporation. A report on the umbral dots first observed in China is introduced.

072.055   Morphological evolution and motion of the photospheric sunspot group in AR 5312.
Hong Qinfang, Wang Hongzhao, Li Weibao.
Publ. Yunnan Obs., No. 1, p. 36 – 41 (1992). In Chinese.

The morphology and evolutionary process of the active region AR 5312 are described. It is pointed out that (1) there is a long period δ–structure with the arrangement of the reversal of polarities in the active region and (2) the newly emerging flux and the rapid development of the sunspot motion have close relation to the eruption of a lot of flares in this region.

072.056   Sunspots: an observational overview.
P. Maltby.
Armagh Observatory Bicentenary Colloquium on Surface Inhomogeneities on Late–Type Stars, p. 124 – 138 (1992). – See Abstr. 012.085 for the main entry.

The author reviews some observational aspects of sunspots that may be relevant to other stars. Particular emphasis is on magnetic fields and large–scale gas flows in large sunspots.

072.057   UV observations of sunspots.
J. B. Gurman.
Armagh Observatory Bicentenary Colloquium on Surface Inhomogeneities on Late–Type Stars, p. 147 – 162 (1992). – See Abstr. 012.085 for the main entry.

Contents: 1. Introduction. 2. The temperature minimum and chromosphere. 3. The transition region: penumbral activity; nonthermal line broadening; emission measure; electron pressure; comparisons with starspots. 4. Mass flows. 5. Oscillations. 6. The magnetic field. 7. Conclusions.

072.058   Observation and study of the brightness and the longitudinal field in the penumbra of a sunspot (I).
Liu Jianqiang, Cao Ai, Chen Jimin, Ai Guoxiang.
Publ. Beijing Astron. Obs., No. 18, p. 54 – 61 (Oct 1991).

Using the Solar Magnetic Field Telescope in Huairou Solar Observing Station of Beijing Astronomical Observatory the authors have observed a series of longitudinal magnetic fields of a sunspot with a very short integration time and obtained monochrome images simultaneously. All of these magnetograms are aligned according to correlation analysis and add together. After comparing the resulting magnetograms to the monochrome images of the sunspot, one finds that magnetic filaments correspond to bright features of monochrome images in the penumbra of the sunspot.

072.059 Oscillations and seismological diagnostics of sunspots.
J. Staude.
Armagh Observatory Bicentenary Colloquium on Surface Inhomogeneities on Late–Type Stars, p. 181 – 191 (1992). – See Abstr. 012.085 for the main entry.

This paper reviews some fundamental aspects of observations and of the theory of sunspot oscillations with a view on the possible role of magneto–atmospheric waves in energy transport and heating in stellar atmospheres. Emphasis is placed on attempts to explain the observed features of sunspot umbral oscillations by comparison with detailed model calculations. The latter require us to consider waves under the simultaneous influence of compressibility, gravity, and magnetic field in a realistic model of the convective zone and the atmosphere of a large sunspot umbra. This model provides an important tool for sounding subphotospheric layers of sunspots and possibly of other magnetized stellar atmospheres (Ap stars).

072.060 The Sun in 1991.
H. Barnes.
South. Stars, Vol. 34, No. 7, p. 399 – 402 (Jun 1992).

072.061 The north–south asymmetry of sunspot distribution.
Wang Yi.
J. R. Astron. Soc. Can., Vol. 86, No. 2, p. 89 – 98 (Apr 1992).

The sunspot distribution in the northern and southern hemispheres of the Sun has been studied in detail. This paper is concerned with the sunspot N–S asymmetry during the period 1874 – 1987. Statistical analyses show that the distribution of sunspots in the northern and southern hemispheres is not random.

072.062 Sunspots: a laboratory for solar physics.
J. B. Gurman.
NATO Advanced Study Institute on the Sun: a Laboratory for Astrophysics, p. 221 – 243 (1992). – See Abstr. 012.091 for the main entry.

After a brief discussion of sunspot morphology and the better–known aspects of the solar activity cycle, somewhat more detail is given on sunspot–to–photospheric continuum contrast, as a function of both wavelength and phase of the solar cycle. Umbral dots are a poorly understood phenomenon which may show the real limits on convection in the presence of a strong magnetic field. Simple, magnetostatic, and one–dimensional radiative transfer models of the observable layers of umbrae are introduced. Ultraviolet observations are used to determine whether the upper reaches of the umbral atmosphere are close to hydrostatic equilibrium, and models of umbral oscillations are discussed, as a paradigm of sunspots as "controls" for the less ordered parts of the solar atmosphere.

072.063 Solar activity.
H. Zirin.
NATO Advanced Study Institute on the Sun: a Laboratory for Astrophysics, p. 449 – 463 (1992). – See Abstr. 012.091 for the main entry.

The magnetic cycle lasts 22 years, reversing polarity in each 11–year half. Spots reach the surface in emerging flux regions (EFR) and grow in complexity. Ephemeral regions bring considerable flux to the surface, but play no great role. Diffusion models of the cycle are discussed, but the low observed diffusion constant makes them a poor fit to reality. The author then turns to the properties of active regions, especially highly active regions producing many flares, and we discusses the properties of those flares.

072.064 Quasibiennial periodicity of solar and planetary phenomena.
I. Predeanu.
Rom. Astron. J., Vol. 1, No. 1 – 2, p. 69 – 79 (1991).

The quasibiennial oscillation (QBO) of various solar and geophysical parameters is analysed, taking some planetary configurations as temporal reference points. The incidence of the QBO minima in the proximity of Sun–Mars oppositions is discussed. The increase of this effect when Mars is near the perihelion or Jupiter is conjunct to the Sun is pointed out.

072.065 Active region classifications, complexity, and flare rates.
P. L. Bornmann, D. Shaw.
IAU Colloquium No. 133: Eruptive solar flares, p. 337 – 340 (1992). – See Abstr. 012.096 for the main entry.

072.066 The sunspot active regions for the 22nd solar cycle.
Zhao Ai–di.
Publ. Purple Mt. Obs., Vol. 11, No. 2, p. 69 – 74 (Jun 1992). In Chinese.

072.067 The downturn in solar activity during solar cycles 5 and 6.
J. O. Murphy.
Proc. Astron. Soc. Aust., Vol. 9, No. 2, p. 330 – 331 (1991). – See Abstr. 012.090 for the main entry.

The atmospheric ^{14}C record, the corresponding W_M values derived from a carbon reservoir model, auroral numbers and the Zurich relative annual sunspot numbers all demonstrate a substantial downturn in solar activity for the duration of solar cycles 5 and 6. This reduction is also imbedded in some dendrochronological proxy data sets, which describe an annual index radial growth rate for trees at high–altitude sites. A significant lagged correlation can exist between tree–ring indices and the 11–year solar cycle during periods of high solar activity, a feature which is not evident during quiescent periods.

072.068 Estimation of sunspot magnetic field parameters using the superpenumbral topology.
J. Linke, J. H. Hernández.
International Workshop on Reconnection in Space Plasma, p. 129 – 130 (Jan 1989). – See Abstr. 012.099 for the main entry.

Applying the model of Molodensky et al. (1987) the authors have investigated the fibril directions (H–alpha level) for the penumbral and superpenumbral region of a sunspot, namely the main, but following spot of the small, simple active region SD 135/1984 on June 24. To reduce the number of free parameters of the model the authors used vector magnetograms. Structures of H–alpha filtergrams and heliograms are compared with the theoretical field model.

072.069 Long–term evolution of coronal magnetic fields basing on noise storm continuum observations in the 20th and 21st cycle.
A. Böhme.
International Workshop on Reconnection in Space Plasma, p. 157 – 161 (Jan 1989). – See Abstr. 012.099 for the main entry.

Basing on the noise storms of the cycles No. 20 and 21 the paper deals with a study of the long–term variations of the radiation signatures of sunspot groups during a solar cycle. The results obtained lead to the conclusion that non–potential loops favouring the emission of a strong type I continuum at the low frequencies (< 100 MHz) tend to exist more frequently above the sunspot groups in the later phase of a solar cycle than above comparable groups during its first activity maximum.

072.070 Statistical analysis of solar flare, neutrino flux and sunspot data.
G. Chattopadhyay, P. Raychaudhuri.
1. COSPAR Colloquium: Physics of the outer heliosphere, p. 393 – 394 (1990). – See Abstr. 012.100 for the main entry.

Sir William Herschel's notebooks: abstracts of solar observations.
See Abstr. 004.005.

Historical reverse of solar activities.
See Abstr. 004.076.

Solar physics at Potsdam: vector magnetic field measurements, diagnostics and modelling of sunspot structure and dynamics.
See Abstr. 013.119.

Beobachtungen mit dem Vakuum–Turm–Teleskop auf Teneriffa.
See Abstr. 032.003.

Bestimmung des Wilson–Effektes an H–Flecken.
See Abstr. 036.037.

Visibility limit of naked–eye sunspots.
See Abstr. 036.182.

Linear resistive magnetohydrodynamic computations of resonant absorption of acoustic oscillations in sunspots.
See Abstr. 062.010.

Absorption of acoustic waves by sunspots. II. Resonance absorption in axisymmetric fibril models.
See Abstr. 062.157.

Physical models of solar and stellar spots.
See Abstr. 064.083.

The depth of formation of absorption lines in the solar atmosphere.
See Abstr. 071.022.

Periodicity in filamentary activity.
See Abstr. 073.016.

Determination of thermodynamical conditions in the chromosphere above the sunspot umbra by solving an inverse problem. I. Numerical simulation of temperature and electron density distributions.
See Abstr. 073.024.

Prominence activity during the ascending phase of solar cycle 22.
See Abstr. 073.050.

The basal and strong–field components of the solar atmosphere.
See Abstr. 073.051.

Motion along some chromospheric fibrils.
See Abstr. 073.066.

Flares above sunspots and magnetic fields. I.
See Abstr. 073.067.

The relation between solar flare and longitudinal current density on Jan. 14, 1989.
See Abstr. 073.069.

Basic magnetic configuration and energy supply processes for an interacting flux model of eruptive solar flares.
See Abstr. 073.087.

The role of cancelling magnetic fields in the buildup to erupting filaments and flares.
See Abstr. 073.088.

Variation of the vector magnetic field in an eruptive flare.
See Abstr. 073.089.

Intrinsically hot flares and possible connection to deep convective magnetic fields.
See Abstr. 073.090.

Interaction of large–scale magnetic structures in solar flares.
See Abstr. 073.091.

"Post" flare loops.
See Abstr. 073.100.

Flares activity during solar cycle 21.
See Abstr. 073.103.

Preliminary research on the hot spots of energy flares since Cycle 21.
See Abstr. 073.123.

Flare instability and driving mechanism.
See Abstr. 073.133.

Coronal Magnetic Structures Observing Campaign. IV. Multiwaveband observations of sunspot and plage–associated coronal emission.
See Abstr. 074.010.

Intermediate–term periodicities in the green corona brightness of the Sun.
See Abstr. 074.020.

Counterstreaming solar wind halo electron events: solar cycle variations.
See Abstr. 074.021.

Flare–associated high–speed solar plasma streams.
See Abstr. 074.040.

On neutralized currents in the solar corona.
See Abstr. 074.050.

Evidence for mass outflow in the low solar corona over a large sunspot.
See Abstr. 074.053.

The practical application of the magnetic virial theorem.
See Abstr. 075.003.

The reversal of the solar polar magnetic fields. III. The large–scale fields and the first major active regions of cycle 22.
See Abstr. 075.007.

On the rotation of large–scale background fields in the 21st cycle of solar activity.
See Abstr. 075.008.

Global modes constituting the solar magnetic cycle. I. Search for 'dispersion relations'.
See Abstr. 075.009.

Vector magnetogram and Dopplergram observation of magnetic flux emergence and its explanation.
See Abstr. 075.012.

Peculiar photospheric velocity fields and magnetic energy build–up.
See Abstr. 075.013.

Global modes constituting the solar magnetic cycle. II. Phases, "geometrical eigenmodes", and coupling of field behaviour in different latitudes.
See Abstr. 075.018.

Energy buildup in sheared force–free magnetic fields.
See Abstr. 075.022.

Spatially extended measurements of magnetic field strength in solar plages.
See Abstr. 075.024.

The cancellation of magnetic flux in the solar atmosphere.
See Abstr. 075.037.

Characteristics of the magnetic field of the active region of the flare of 1989 July 5.
See Abstr. 075.038.

Possibilities and problems of the interpretation of solar magneto-graph measurements and applications to flare–active regions.
See Abstr. 075.044.

Gamma–ray measurements from the Space Shuttle during a solar flare.
See Abstr. 076.009.

The Sun as a star: high spectral resolution solar data degraded to low–dispersion IUE resolution.
See Abstr. 076.019.

Large–scale patterns on the Sun observed in the millimetric wavelength range.
See Abstr. 077.004.

High dynamic range multifrequency radio observations of a solar active region.
See Abstr. 077.008.

Compact sources of suprathermal microwave emission detected in quiescent active regions during lunar occultations.
See Abstr. 077.020.

Millisecond radio spikes on long centimetre and short decimetre wavelengths and occurrence frequency.
See Abstr. 077.025.

Radio burst at 245 MHz and the forecast of solar activities.
See Abstr. 077.040.

Compact sources of suprathermal microwave emission detected in quiescent active regions during lunar occultations.
See Abstr. 077.044.

Solar proton events at the growth phase of the 22nd solar activity cycle.
See Abstr. 078.009.

Stochastic fluctuations of the solar dynamo.
See Abstr. 080.002.

Variability of the quiet atmosphere of the sun in the context of the activity cycle.
See Abstr. 080.003.

The low l solar p–mode spectrum at maximum and minimum solar activity.
See Abstr. 080.018.

The 22–year cycle of the Sun.
See Abstr. 080.021.

Interpretation of solar–cycle variability in high–degree p–mode frequencies.
See Abstr. 080.030.

Large–scale flows and solar luminosity variations.
See Abstr. 080.041.

Precise ground–based solar photometry and variations of total irradiance.
See Abstr. 080.045.

The conversion of p–modes to slow modes and the absorption of acoustic waves by sunspots.
See Abstr. 080.049.

Scattering of p–modes by a sunspot.
See Abstr. 080.050.

Acoustic wave ducting along magnetic flux tubes in the sunspot regions.
See Abstr. 080.054.

Local acoustic diagnostics of the solar interior.
See Abstr. 080.067.

Magnetic fields and motions in the solar atmosphere.
See Abstr. 080.086.

The effect of surface inhomogeneities on total solar irradiance.
See Abstr. 080.088.

Seismic investigation of the solar interior.
See Abstr. 080.091.

Mean field dynamo theory.
See Abstr. 080.093.

Effects of vibrationally excited molecular nitrogen on ionospheric–thermospheric coupling for different levels of solar activity.
See Abstr. 082.046.

Response of photoionization rates and electron densities to solar activity in the 100 – 200 km height region.
See Abstr. 083.005.

Solar activity control of ionospheric and thermospheric processes.
See Abstr. 083.016.

27–day fluctuations in the ionospheric D–region.
See Abstr. 083.017.

Geomagnetic and solar data. September – October 1991.
See Abstr. 084.013.

Long–term variations in geomagnetic and solar activities and secular variations of the geomagnetic field components.
See Abstr. 084.015.

Geomagnetic and solar data. November – December 1991.
See Abstr. 084.021.

Geomagnetic and solar data. January – February 1992.
See Abstr. 084.029.

A study of geomagnetic variations with periods of four years, six months and 27 days.
See Abstr. 084.079.

Interplanetary disturbance structures and geomagnetic storm.
See Abstr. 085.015.

The effect of solar activity on the light curves of comets Churyumov–Gerasimenko (1982 VIII) and Halley (1986 III).
See Abstr. 103.006.

Narrow–band IPD images of cometary CN and C_2: the effect of solar activity on coma scales.
See Abstr. 103.016.

Meteoritic evidence for an active early sun.
See Abstr. 105.064.

Intensity variations in the interplanetary magnetic field measured by Voyager 2 and the 11–year solar cycle modulation of Galactic cosmic rays.
See Abstr. 106.012.

On the large–scale effects of two interplanetary shocks on the associated particle events.
See Abstr. 106.039.

Unfolding mysteries of stellar cycles.
See Abstr. 116.028.

Surface inhomogeneities of late–type stars: Summary and conclusions.
See Abstr. 116.075.

The cosmic radiation in the heliosphere at successive solar minima.
See Abstr. 144.013.

Components of the 11– and 22–year variation of cosmic rays.
See Abstr. 144.058.

Bi–annual variations in solar activity and cosmic rays.
See Abstr. 144.061.

Neutron monitor investigations relating modulated cosmic ray spectra with heliospheric magnetic field polarity reversals.
See Abstr. 144.076.

A time–dependent drift model with a simulated wavy neutral sheet for the solar modulation of cosmic rays.
See Abstr. 144.078.

Solar effects on underground muons at 570 hg/cm².
See Abstr. 144.079.

The long term cosmic ray variation relevant to solar wind structure in the outer heliosphere.
See Abstr. 144.084.

073 Chromosphere, Flares, Prominences

073.001 **The effects of magnetic field geometry on the confinement of energetic electrons in solar flares.**
K. G. McClements.
Astron. Astrophys., Vol. 253, No. 1, p. 261 – 268 (Jan 1992).
   A Fokker–Planck code which includes the effects of a variable magnetic field is used to compute the steady–state distribution of energetic electrons in a flaring coronal loop. Results are obtained for several different functional forms of $B(s)$, the magnetic induction as a function of distance along a particular field line. The three cases studied are: (a) B varying exponentially with s; (b) B varying quadratically with s; and (c) $B(s)$ corresponding to a magnetic dipole located below the transition region. In each case, solutions of the Fokker–Planck equation are obtained for two different types of boundary condition in electron pitch angle space. From these results are computed the rate at which electrons are lost from the system (via the ends of the loop), and the corresponding fraction η of the total X–ray flux which will be produced by the loop footpoints. η is obtained for photon energies in the range 19–24 keV and 57–100 keV, corresponding to two of the channels on the Hard X–ray Telescope of the Solar–A spacecraft, due to be launched in August 1991. Generally speaking, the results are insensitive to both the functional form of $B(s)$ and the choice of pitch angle space boundary condition, and correspond fairly closely with crude analytical estimates. By comparing these results with Solar–A data, when this becomes available, it should be possible to obtain information about the plasma parameters of a flaring loop.

073.002 **Dynamics of flaring loops. II. Flare evolution in the density–temperature diagram.**
J. Jakimiec, B. Sylwester, J. Sylwester, S. Serio, G. Peres, F. Reale.
Astron. Astrophys., Vol. 253, No. 1, p. 269 – 276 (Jan 1992).
   In the present paper the evolution of basic thermodynamic parameters of a single flaring solar loop has been investigated in terms of density–temperature (N–T) diagram. A grid of hydrodynamic models has been calculated for this purpose using the Palermo–Harvard code. The calculated models differ in their initial conditions and the form of the energy input i.e. the heating rate value, the heating duration, the assumed time profile. The authors have considered the consequences of variation of these model parameters on the evolutionary paths in the density–temperature diagrams. They show that, over a substantial duration, the decay occurs along a $T \sim N^2$ trajectory when the impulsive flare heating function is switched–off abruptly. The results obtained in this paper can be very useful as diagnostics of the flare heating process based on soft X–ray observations.

073.003 **Eruptive phenomena on the sun.**
Z. Svestka.
11. European Regional Astronomy Meeting of the IAU: New windows to the universe, Vol. 1, p. 99 – 117 (1990). – See Abstr. 012.001 for the main entry.
   This paper reviews phenomena in which plasma is injected into preexisting coronal structures (surges, X–ray bright surges, flaring arches) and phenomena in which the magnetic field is disrupted (sprays, erupting filaments, eruptive (dynamic) flares). At some places the reader is referred to another recent review (Svestka, 1989) where dynamic flares were discussed in more detail.

073.004 **The morphological characteristics and cooling mechanisms of the post–flare loop system of April 28, 1980.**
J. Lin, Z. Zhang, Z. Wang, R. N. Smartt.
Astron. Astrophys., Vol. 253, No. 2, p. 557 – 560 (Jan 1992).
   The morphology and evolution of a post–flare loop system are described in the paper. The observational data show that this loop system is very complex, in which the loops are overlapped and inlaid. The whole loop system is in the relatively stable stage of the main phase, during which the variations of its form with time is slow and the variations of its brightness are also small. A simple method is used to compute the cooling of postflare loops by radiation, conduction and expansion. The relative importance of radiative, conductive and expansive losses are briefly discussed. Finally, the authors investigate the most likely resource which permits the loops to stay bright for a long time.

073.005 **The relationship between global X–ray luminosity and flaring on the Sun.**
G. Pearce, R. A. Harrison, B. J. I. Bromage, A. G. M. Pickering.
Astron. Astrophys., Vol. 253, No. 2, p. 601 – 603 (Jan 1992).
   The authors investigate the relationship between solar global flaring activity and X–ray luminosity, as part of a solar–stellar investigation. In keeping with previous suggestions, they find that the rate of flaring on the Sun is closely related to X–ray luminosity with a positive correlation. This is in contrast to relationships found on dMe stars, and can be interpreted in terms of flare activity both on a microscopic and macroscopic scale varying sympathetically. The microscopic activity has significance for outstanding problems such as coronal heating.

073.006   Energetics and dynamics in a large solar flare of 1989 March.
J.-P. Wülser, D. M. Zarro, R. C. Canfield.
Astrophys. J., Vol. 384, No. 1, p. 341 – 347 (1 Jan 1992).

For the first time in a large (X1.2) solar flare, the authors have combined SMM X–ray observation and NSO / Sacramento Peak Hα spectra to test predictions of chromospheric heating and evaporation by nonthermal thick–target electrons. It is demonstrated that the ratio of Hα flare energy flux to the energy flux deposited by thick–target electrons obeys a power–law dependence on electron heating flux, with a slope that is consistent with that predicted by the thick–target electron transport and heating model in a one–dimensional hydrostatic atmosphere. Comparison of the observed Hα emission with that predicted by the thick–target model implies a constant electron precipitation area of $(2.2 \pm 0.7) \times 10^{17} cm^2$ during the impulsive phase. The electron precipitation area deduced for this X–flare is consistent with the mean value of $\sim 10^{17} cm^2$ obtained from a similar thick–target analysis of Hα and extreme ultraviolet observations in weaker solar flares. Upflowing coronal material (as seen in blueshifted Ca XIX soft X–rays) and downflowing chromospheric material (as seen in redshifted Hα) appear simultaneously at the beginning of impulsive hard X–ray emission. It is concluded that the thick–target model satisfactorily accounts for the observed magnitude of chromospheric Hα emission, and for the timing and amplitudes of oppositely directed plasma motions during the impulsive phase of this X flare.

073.007   Fluid flow in a jet and the Ca XIX line profiles observed during solar flares.
P. L. Bornmann, J. R. Lemen.
Astrophys. J., Vol. 385, No. 1, p. 363 – 374 (20 Jan 1992).

Two types of fluid models were considered as methods for reproducing the blueshifts and line broadening observed in soft X–ray lines during the rise of solar flares. The fluid models representing the laminar flow of material in a jet and through a pipe were used to derive the velocity at each location in the flow. These velocities were then converted to velocity distributions and convolved with a thermal Maxwellian distribution to produce theoretical line profiles. The resulting theoretical profiles were then compared with Ca XIX line profiles observed by SMM during the 1980 May 21 flare. Reasonable agreement was found for the jet model, while the pipe model was less successful at reproducing the observed line profiles. The optimal values for the free parameters in the jet model are within the ranges expected for solar flare conditions. The jet is much smaller than typical flare volumes estimated from spatial images, but the resulting densities are consistent with the upper limit of densities derived with density–sensitive line ratios. This supports previous reports of small filling factors and is interpreted as evidence for a chromospheric origin for the flow.

073.008   Gamma–ray lines from solar flares.
E. Rieger.
AIP Conf. Proc., No. 232, p. 421 – 438 (1 Aug 1991). – See Abstr. 012.003 for the main entry.

The observation of γ–ray lines during solar flares is the most convincing evidence that nuclear reaction take place between high energy particles and the solar atmosphere. The application of data from the measurements of these lines is very wide spread. It ranges from information about the acceleration of charged particles to elemental composition determination of the various solar atmospheric layers. After a brief review of the production of the nuclear lines it is shown, how their measuremnts can be applied in different areas of the solar flare phenomenon.

073.009   Dynamical problems in Hα flare loops on the solar disc.
Xu Ao–ao, Tang Yu–hua.
Sci. China, Ser. A, Vol. 35, No. 4, p. 463 – 471 (Apr 1992). In Chinese.

The dynamical process in Hα flare loops on the solar disc is quantitatively discussed by using the observed data on the two–ribbon flare on May 16, 1981, taken with the Hα spectroheliograph and Hα chromospheric telescope at Yunnan Observatory, Kunming, as well as the X–ray data of the same flare from satellite "HINOTORI". It is firstly found that matter at the top of the Hα flare loops has initial velocity of several hundred km/s and the falling velocity of matter within the loops decelerates. The authors believe that these dynamical problems are related to the physical mechanism of flare loops.

073.010   A burst model for line emission in the solar atmosphere I. XUV lines of He I and He II in impulsive flares.
J. M. Laming, U. Feldman.
Astrophys. J., Vol. 386, No. 1, p. 364 – 370 (10 Feb 1992).

A model in which the solar chromosphere is heated by explosive events is developed and used to interpret XUV spectra of He I and He II in impulsive flares observed by the Skylab spectroheliograph. From a comparison of relative line intensities from He I and He II emitted within the flares, the model establishes sizes, durations, and frequencies for the individual events. By comparison with lines from other elements observed in the same flares, the model is shown to be consistent with a helium abundance relative to hydrogen of 0.1 for burst temperatures of 15 – 18 eV.

073.011   A theory of filament eruptions before the impulsive phase of solar flares.
J.-i. Sakai, S. Koide.
Sol. Phys., Vol. 137, No. 2, p. 293 – 306 (Feb 1992).

The authors present a theory of filament eruption before the impulsive phase of solar flares. They show that the upward motion of the magnetic X–point tracing the filament eruption begins several minutes before the impulsive phase of the flare, where the explosive magnetic reconnection starts of the X–point magnetic field configuration located under the filament. No change occurs in the character of the motion of the X–point during the onset of the explosive magnetic reconnection. The upward speed of the X–point is about $110 \ km \ s^{-1}$ at the onset of the impulsive phase. The authors give an important condition leading to filament eruptions, which relate to the state of the current sheet under the filament, where the magnetic energy can be released.

073.012   Weighted current sheets supported in normal and inverse configurations: a model for prominence observations.
P. Démoulin, T. G. Forbes.
Astrophys. J., Vol. 387, No. 1, p. 394 – 402 (1 Mar 1992).

The authors analyze the magnetic support of solar prominences in two dimensions by using a technique which incorporates both photospheric and prominence magnetic field observations. The prominence is modeled by a mass–loaded current sheet which is supported against gravity by magnetic fields from a bipolar source in the photosphere and a massless line current in the corona. With this system, prominence support can be achieved in three distinct kinds of configurations: (1) an arcade topology with a normal polarity, (2) a helical topology with a normal polarity, and (3) a helical topology with an inverse polarity. In all cases the important parameter is the variation of the horizontal component of the prominence field with height. In the arcade topology, only configurations where the field decreases with height can be supported, but in the helical topology, configurations can be supported even when the field does not decrease with height. Adding a line current external to the prominence eliminates the nonsupport problem which plagues virtually all previous prominence models with inverse polarity.

073.013   Energy propagation into a flare kernel during a solar flare.
D. B. Melrose.
Astrophys. J., Vol. 387, No. 1, p. 403 – 413 (1 Mar 1992).

Energy released in a flare kernel is stored everywhere around at least the coronal portion of the flaring flux tube, and this energy must propagate into the flare kernel during a flare. A model for the energy propagation into the energy release site is formulated in terms of Alvénic fronts where stored magnetic energy is partially converted into an energy flux. The fronts are assumed to be created by a sudden onset of dissipation at the start of a flare, modeled by a resistance, R_c, switched on at t = 0. The energy flux provides the power released in the resistance. The model implies that (1) the current flowing into R_c immediately after it is switched on is $I_1 = I_0/(1 + R_c/R_A)$, with $R_A = \mu_0 \cdot v_A/4\pi$, where v_A is the Alfvén speed; (2) the maximum rate of energy release occurs for $R_c = R_A$ when all the nonpotential magnetic energy stored inside the coronal portion of the flux tube is released in an Alfvén propagation time; (3) if a resistance R_0 is included at the photospheric boundary, then after an infinite number of reflections from R_0 and R_c the current approaches the value $I_\infty = I_0/(1 + R_c/R_0)$ implied by a circuit model. When the coronal resistance is switched off, energy continues to propagate from below the photosphere to resupply the magnetic energy initially stored in the corona.

073.014   **E–region effective electron loss rates and electron temperatures under different solar flare conditions.**
A. C. Balachandra Swamy.
Astrophys. Space Sci., Vol. 188, No. 2, p. 271 – 278 (Feb 1992).
Pressuming the total life–time for recombination of molecular ions with electrons remains constant at each altitude under different solar flare conditions, one can estimate the mean dissociative recombination coefficient and the electron temperatures, with a knowledge of the quiet time recombination coefficient and the production rate profiles under these conditions. To estimate the electron temperatures, the temperature dependence of the form $(300/T_e)^{1/2}$ for the electron loss coefficient has been assumed. The computations show very high electron temperatures below about 130 km for the representative solar events considered.

073.015   **Periodicities of the number of solar flares for the period 1966–1988.**
J. Xanthakis, C. Poulakos, B. Petropoulos.
Astrophys. Space Sci., Vol. 188, No. 2, p. 321 – 329 (Feb 1992).
In the present paper a study is made of the mean monthly number of grouped solar flares F for the time period 1966–1988. Corresponding data F were taken from the catalogue published by Coffey (1989). Periodicities of 140, 120, 48, 18, 12, and 11 months as well as shorter periodicities of the order of 6 and 3 months for the solar flares have been found. The emphasis is given as far as the period of 48 months is concerned, which is for the first time revealed by the present investigation.

073.016   **Periodicity in filamentary activity.**
M. Dizer.
Astrophys. Space Sci., Vol. 189, No. 2, p. 289 – 301 (Mar 1992).
This paper presents the results of the study on the periodicity in filament activity. The spectral analysis of the number of filaments shows a basic period at 141 (~ 10.5 yr), at 138 (~ 10.3 yr), and at 144 (~ 10.7 yr) Carrington rotation in the northern and southern hemisphere, respectively. The time series concerning the index of filament activity shows also a typical period at 135 Carrington rotation (~ 10.1 yr), at 144 Carrington rotation (~ 10.7 yr) and at 133 Carrington rotation (~ 9.9 yr), respectively, in the northern and southern hemisphere. The power spectrum analysis of the time series of the filamentary activity in the short–term also yields less pronounced but still noticeable peaks which are statistically significant.

073.017   **Rapid motion of filaments: eruptive prominences.**
M. M. Molodenskij, B. P. Filippov, N. S. Shilova.
Sov. Astron., Vol. 36, No. 1, p. 92 – 97 (Jan 1992). Current Physics Microform No.: 9207Y0656. English translation of Astron. Zh., Tom 69, Vyp. 1, p. 181 – 191 (Jan–Feb 1992).
Rapid motion of filaments is interpreted to be the result of the interaction of the filament current with the field of the active region and the field of the "reflected" current. Observations of the velocity of eruptive prominences departing from the limb are summarized. The intergral of the filament energy is considered. The most rapid filament motion arises if loss of equilibrium occurs near a "saddle" point on the potential surface. On the basis of the authors' observations at the large GAS coronagraph

of the Main Astronomical Observatory with an Opton interference–polarization–filter in June – July 1989, the parameters of eruptive prominences with bounded and unbounded motions are compared.

073.018   **Analysis of the optical spectra of the solar flares. VI. Velocity field in the 13 June 1980 flare area.**
A. Falchi, R. Falciani, L. A. Smaldone.
Astron. Astrophys., Vol. 256, No. 1, p. 255 – 263 (Mar 1992).
The 13 June 1980 flare area was observed at NSO–Sacramento Peak Observatory, simultaneously with the Universal Spectrograph and with the Universal Birefringent Filter in parallel with the Zeiss H_α filter. The authors consider the flare emission measured with the spectrograph in the H_β, H_γ, H_δ, Ca II–K and Na–D_2 lines to detect possible asymmetry in their profiles. A characteristic blue asymmetry, indicative of coronal upflows, is present in the Ca XIX spectrum obtained with the Bent Crystal Spectrometer on the Solar Maximum Mission. This is qualitatively consistent with the generally accepted scenario of a chromophseric evaporation sufficiently rapid to drive both coronal upflows and chromospheric downflows. The velocities obtained from chromospheric lines are compared with the ones predicted by numerical simulations of gas dynamics in flare loops (Fisher 1986, 1989). The results show that the chromospheric condensation, predicted to be moving downwards with constant velocity within the condensation, probably has a velocity gradient and that the layers ahead of it seem to be affected by the motion of the condensation.

073.019   **The modes of oscillation of a prominence. I. The slab with longitudinal magnetic field.**
P. S. Joarder, B. Roberts.
Astron. Astrophys., Vol. 256, No. 1, p. 264 – 272 (Mar 1992).
The authors consider theoretically the modes of oscillation of a solar quiescent prominence, treating the prominence as a three–dimensional, magnetized plasma slab embedded in a uniform magnetic field. They examine the normal modes of oscillation of such a system. The existence of fast MHD surface modes with periods of the order of an hour is of particular interest. These modes, which are closely analogous to the classical modes of vibration of an elastic membrane, may correspond to the observed long–period oscillations of quiescent prominences. Shorter periodicities, in the range 3–5 min, may be associated with the fundamental of the magnetic Pekeris mode and the first harmonic of the magnetic Love mode.

073.020   **γ–ray and X–ray time profiles expected from a trap–plus–precipitation model for the 7 June 1980 and 27 April 1981 solar flares.**
E. Hulot, N. Vilmer, E. L. Chupp, B. R. Dennis, S. R. Kane.
Astron. Astrophys., Vol. 256, No. 1, p. 273 – 285 (Mar 1992).
Hard X–ray and prompt γ–ray line emissions are the most direct signatures of, respectively, electron and ion acceleration during solar flares. The peak time of the γ–ray emission for some events is delayed with respect to the peak time of the hard X–ray flux. These delays are either interpreted as evidence of a two–step acceleration process or as the result of the partial trapping and/or propagation of the particles from the acceleration region to the emission sites. It was shown earlier that hard X–ray and prompt γ–ray line delays can be qualitatively reproduced in the frame of the latter hypothesis with models describing the time dependent transport of energetic electrons and ions between these two sites. Here the authors focus on the close examination of the temporal evolution of X–ray and γ–ray fluxes for the 7 June 1980 and 27 April 1981 events which exhibit delays between X–ray and γ–ray maxima. The parameters of the ambient medium and of the accelerated partices are deduced for the two events and it is shown that the relative timing of X–ray and γ–ray emissions is quantitatively reproduced in the present context. The implications of a coronal contribution to γ–ray line emission are also discussed for these two events.

073.021 Chromospheric dynamics based on infrared solar brightness variations.
G. Kopp, C. Lindsey, T. L. Roellig, M. W. Werner, E. E. Becklin, F. Q. Orrall, J. T. Jefferies.
Astrophys. J., Vol. 388, No. 1, p. 203 – 210 (20 Mar 1992).
The authors have used the Kuiper Airborne Observatory to observe far–infrared continuum brightness fluctuations in the lower chromosphere due to solar 5 min oscillations on the quiet Sun. Brightness measurements made at 50, 100, 200, and 400 μm show a strong correlation with visible–line Doppler measurements from photospheric and chromospheric altitudes. The motion of the chromosphere is nearly in phase over a large range of heights, while the infrared brightness lags the Doppler velocity by phases varying from significantly less than 90° at low altitudes to nearly 90° at higher altitudes. The authors propose that this is the result of a nonadiabatic reponse of the chromospheric gas to compression and may indicate an important mechanism for wave dissipation. The authors estimate thermal relaxation times ranging from about 40 s at 340 km above the $\tau_{5000} = 1$ photosphere to ~ 300 s at 600 km.

073.022 Magnetic fields and thermodynamic conditions in the solar flare of June 8, 1989.
Eh. A. Baranovskij, N. I. Lozitskaya, V. G. Lozitskij.
Kinematics Phys. Celest. Bodies, Vol. 7, No. 3, p. 49 – 54 (1991). English translation of Kinematika Fiz. Nebesn. Tel, Tom 7, No. 3, p. 52 – 58 (1991).
Spectral observations of the importance 1B solar flare made with a circular polarization analyzer are analyzed. An empirical model of the flare is constructed by joint interpretation of data on the lines Fe I $\lambda\lambda 513.15$ and 543.5 nm, Fe II $\lambda\lambda 429.39$ and 501.84 nm, and the Balmer lines H_α and H_γ. Effects of deviation from LTE were taken into account in the calculations. It was found that the temperature and density in the flare at photospheric optical depths $\tau_{0.5} > 4 \cdot 10^{-2}$ are the same as those in the undisturbed atmosphere, and that the magnetic field intensity is near zero. The temperature is higher in the range $10^{-4} < \tau_{0.5} < 4 \cdot 10^{-2}$, with the elevation maximum corresponding to $\tau_{0.5} \approx 4 \cdot 10^{-3}$, where T = 6100K. The intensity of the longitudinal magnetic field at this depth is 80 – 90 mT. The field is even stronger, around 120 mT, at the places where the Balmer lines are formed. The density in the chromosphere is 1 – 1.5 orders of magnitude higher, and the turbulent velocity is 3 – 4 km/sec.

073.023 On the efficiency of electron cyclotron maser instability in solar flares.
A. Mackinnon, L. Vlahos, N. Vilmer.
Astron. Astrophys., Vol. 256, No. 2, p. 613 – 617 (Mar 1992).
The authors describe a simple model, designed to estimate the efficiency of the electron cyclotron maser (ECM) instability inside a magnetic trap. The emphasis in this model is on the energy loss associated with the formation of the loss cone distribution, rather than on the microscopic plasma physics. The authors find that the net radiation efficiency is low. Repeated mirroring does not greatly enhance the intrinsic efficiency, because of the energy lost from the corona in precipitating electrons. In consequence, absorption of ECM radiation is unlikely to be an important factor in understanding soft X–ray observations.

073.024 Determination of thermodynamical conditions in the chromosphere above the sunspot umbra by solving an inverse problem. I. Numerical simulation of temperature and electron density distributions.
R. B. Teplitskaya, V. G. Skochilov, S. A. Grigor'eva.
Kinematika Fiz. Nebesn. Tel, Tom 8, No. 1, p. 3 – 11 (Jan–Feb 1992). In Russian. English translation in Kinematics Phys. Celest. Bodies, Vol. 8, No. 1.
A numerical simulation is performed to reconstruct the temperature and electron density distributions in the chromosphere above the sunspot umbra. It is supposed that the only source of information are profiles of central reversals of the H and K Ca II lines and source functions (derived through their inversion) dependent on the optical depth at the H–line centre. It is shown that one can reconstruct the temperature and electron density with a maximum deviation $\pm 340K$ and $\pm 2 \cdot 10^{10} cm^{-3}$ respectively.

073.025 What caused an unusually broad He I 10830 Å emission line in a solar limb flare?
J. Q. You, G. K. Oertel.
Astrophys. J., Lett., Vol. 389, No. 1, p. L33 – L35 (10 Apr 1992).
This letter discusses an observation of an unusually broad He I 10830 Å line from a flare on the solar limb. The line profile is close to a dispersion profile with full half–width of 4.2 Å. One interpretation of this observation could be a peculiar distribution of mass motions in the flare, that is just right to mimic a dispersion–like line profile. Pressure broadening is usually indicated by a dispersion profile, but is an unlikely cause of the broadening because it would imply an electron density in the flare near $4 \times 10^{17} cm^3$. The authors found no other credible explanations for the observed broadening. They propose a critical experiment to clearly distinguish between the two interpretations or to point the way toward another.

073.026 Spatio–temporal fluctuations in He I 10830 Å line parameters: evidence for spicule formation.
P. Venkatakrishnan, S. K. Jain, S. Singh, F. Recely, W. C. Livingston.
Sol. Phys., Vol. 138, No. 1, p. 107 – 121 (Mar 1992).
The equivalent width, line depth, line width, and Doppler shift of the He I 10930 Å line were extracted from two time series of spectra. Scatter plots of time–averaged line depth, line width, and Doppler shifts, as well as the root mean square temporal fluctuation of these quantities against the time–averaged equivalent width at a few hundred spatial locations were obtained. The statistical behaviour of these line parameters and their fluctuations was used to infer plausible reasons for the fluctuations. Examination of these results showed that the line parameter fluctuations could be caused by fluctuations in the coronal UV radiation (which could drive the spicules) or by the appearance of density inhomogeneities such as spicules within the line forming domain. In either case, the data can be interpreted as representing the initial phases of spicules.

073.027 Hydrogen photoionization rates for chromospheric and prominence plasmas.
P. Rudawy, P. Heinzel.
Sol. Phys., Vol. 138, No. 1, p. 123 – 131 (Mar 1992).
New values of hydrogen photoionization rates for subordinate continua arising from bound levels with the principal quantum number $i = 2 - 5$ have been evaluated numerically, using an extensive compilation of the observed photospheric radiation fields. These rates can be directly incorporated into the equations of statistical equilibrium as so–called fixed rates. The authors tabulate the photoionization rates and equivalent radiation temperatures for various height above the photosphere, which is particularly useful for chromospheric and prominence non–LTE modeling. The results are compared with those previously obtained by other authors.

073.028 Solar flares: an overview.
D. M. Rust.
Adv. Space Res., Vol. 12, No. 2/3, p. 289 – 301 (1992). – See Abstr. 012.016 for the main entry.
This is a survey of solar phenomena and physical models that may be useful for improving forecasts of solar flares and proton storms in interplanetary space. Knowledge of the physical processes that accelerate protons has advanced because of gamma–ray and X–ray observations from the Solar Maximum Mission telescopes. Several possibilities for improvements in the art of flare forecasting are presented, among them: the use of acoustic tomography to probe for subsurface magnetic fields; a satellite–borne solar magnetograph; and an X–ray telescope to monitor the corona for eruptions.

073.029   **Magnetic reconnection in magnetoplasma of solar flare.**
K. W. Min, J. Y. Shin, H. S. Yun.
Bull. Am. Astron. Soc., Vol. 23, No. 3, p. 1264 (1991). Abstract. – See Abstr. 012.067 for the main entry.

073.030   **BATSE flare observations in solar cycle 22.**
R. A. Schwartz, B. R. Dennis, G. J. Fishman, C. A. Meegan, R. B. Wilson, W. S. Paciesas.
Compton Observatory Science Workshop, p. 457 – 468 (Feb 1992). – See Abstr. 012.018 for the main entry.

073.031   **GRO solar flare observations.**
R. J. Murphy.
Compton Observatory Science Workshop, p. 469 (Feb 1992). Abstract. – See Abstr. 012.018 for the main entry.

073.032   **COMPTEL solar flare observations.**
J. M. Ryan, H. Aarts, K. Bennett, H. Debrunner, C. de Vries, J. W. den Herder, G. Eymann, D. J. Forrest, R. Diehl, W. Hermsen, R. Kippen, L. Kuiper, J. Lockwood, M. Loomis, G. Lichti, J. Macri, M. McConnell, D. Morris, V. Schönfelder, G. Simpson, M. Snelling, H. Steinle, A. Strong, B. N. Swanenburg, W. R. Webber, C. Winkler.
Compton Observatory Science Workshop, p. 470 – 479 (Feb 1992). – See Abstr. 012.018 for the main entry.
COMPTEL observed the sun during the period of high solar activity from 7 Jun to 15 Jun 1991. Major flares were observed on Jun 9 and 11. Although, both flares were large GOES events, they were not extraordinary in terms of γ–ray emission. Only the decay phase of the Jun 15 flare was observed by COMPTEL. The authors report the preliminary analysis of data from these flares, including the first spectroscopic measurement of solar flare neutrons. The deuterium formation line at 2.223 MeV was present in both events, and for at least the Jun 9 event, was comparable to the flux in the nuclear line region of 4 – 8 MeV, consistent with SMM observations. A clear neutron signal was present in the flare of 9 Jun with the spectrum extending up to 80 MeV and consistent in time with the emission of γ–rays, confirming the utility of COMPTEL in measuring the solar neutron flux at low energies. The neutron flux below 100 MeV appears to be lower than that of the 1982 Jun 3 flare by more than an order of magnitude.

073.033   **Excitation of the solar flare far–ultraviolet continuum by line irradiation.**
J. G. Doyle, K. J. H. Phillips.
Astron. Astrophys., Vol. 257, No. 2, p. 773 – 776 (Apr 1992).
Observations of the far–ultraviolet ($\lambda < 1682$ Å) continuum by Skylab during an intense solar flare confirm previous calculations that the excitation of this continuum is due to the ionization of neutral silicon atoms near the temperature-minimum region irradiated by ultraviolet line radiation emitted by the upper chromosphere or transition region. The evidence is an observed proportionality of the continuum intensity with the intensities of C IV (1548/51 Å) and C II (1335/36 Å) lines.

073.034   **Neutron and gamma ray production in the 1991 June X–class flares.**
R. Ramaty, X. M. Hua, B. Kozlovsky, R. E. Lingenfelter, N. Mandzhavidze.
Compton Observatory Science Workshop, p. 480 – 485 (Feb 1992). – See Abstr. 012.018 for the main entry.
The authors present new calculations of pion radiation and neutron emission from solar flares. They fit the recently reported high energy GAMMA–1 observations with pion radiation produced in a solar flare magnetic loop. The expected neutron emission in such a loop model is calculated and predictions are made of the neutron fluences expected from the 1991 Jun X–class flares.

073.035   **Stereoscopic observations of hard X–ray sources in solar flares made with GRO and other spacecraft.**
S. R. Kane, K. Hurley, J. M. McTiernan, J. G. Laros.
Compton Observatory Science Workshop, p. 486 – 488 (Feb 1992). – See Abstr. 012.018 for the main entry.
Since the launch of the Gamma Ray Observatory (GRO) in Apr 1991, the BATSE instrument on GRO has recorded a large number of solar flares. Some of these flares have also been observed by the Gamma–Ray Burst Detector on the Pioneer Venus Orbiter and/or by the Solar X–ray/Cosmic Gamma–Ray Burst Experiment on the Ulysses spacecraft. A preliminary list of common flares observed during the period May – Jun 1991 is presented and the possible joint studies are indicated.

073.036   **Millimeter and hard X–ray/γ–ray observations of solar flares during the June 91 GRO campaign.**
M. R. Kundu, S. M. White, N. Gopalswamy, J. Lim.
Compton Observatory Science Workshop, p. 502 – 513 (Feb 1992). – See Abstr. 012.018 for the main entry.

073.037   **Radio synthesis imaging during the GRO solar campaign.**
D. E. Gary.
Compton Observatory Science Workshop, p. 514 (Feb 1992). Abstract. – See Abstr. 012.018 for the main entry.

073.038   **VLA, PHOENIX and BATSE observations of an X1 flare.**
R. F. Willson, M. J. Aschwanden, A. O. Benz.
Compton Observatory Science Workshop, p. 515 – 521 (Feb 1992). – See Abstr. 012.018 for the main entry.
The authors present observations of an X1 flare (Jul 18, 1991) detected simultaneously with the VLA, the PHOENIX Digital Radio Spectrometer and the Burst and Transient Source Experiment (BATSE) aboard the Gamma Ray Observatory. The VLA was used to produce snapshot maps of the impulsive burst emission on timescales of 1.7 sec at both 20 and 91 cm. The results indicate electron acceleration in the higher corona several minutes before the onset of the hard X–ray burst detected by BATSE. Comparisons with high spectral and temporal observations by PHOENIX reveal a variety of radio bursts at 20 cm, such as type III. bursts, intermediate drift bursts, and quasi–periodic pulsations during different stages of the X1 flare. The described X1 flare is unique in the sense that it appeared at the east limb, providing the most accurate information on the vertical structure of different flare tracers visible in radio wavelengths.

073.039   **Prompt particle acceleration around moving X–point magnetic field during impulsive phase of solar flares.**
J. Sakai.
Compton Observatory Science Workshop, p. 528 – 535 (Feb 1992). – See Abstr. 012.018 for the main entry.
The author presents a model for high–energy solar flares to explain prompt proton and electron acceleration, which occurs around moving X–point magnetic field during the implosion phase of the current sheet. He derives the electromagnetic fields during the strong implosion of the current sheet, which is driven by the converging flow toward the center of the magnetic arcade. Test particle motion is investigated in the strong electromagnetic fields derived from the MHD equations. It is shown that both protons and electrons can be promptly (within 1 s) accelerated to ~ 70 MeV and ~ 200 MeV, respectively. This acceleration mechanism can be applicable for the impulsive phase of the gradual gamma ray and proton flares, which have been called two–ribbon flares.

073.040   **Search for evidence of low energy protons in solar flares.**
T. R. Metcalf, J.–P. Wülser, R. C. Canfield, H. S. Hudson.
Compton Observatory Science Workshop, p. 536 – 541 (Feb 1992). – See Abstr. 012.018 for the main entry.

073.041 **Rapid motions of filaments: eruptive prominences.**
M. M. Molodenskij, B. P. Filippov, N. S. Shilova.
Astron. Zh., Tom 69, Vyp. 1, p. 181 – 191 (Jan–Feb 1992). In Russian. English translation in Sov. Astron., Vol. 36, No. 1 (1992).

Rapid motion of filaments is interpreted as a result of the interaction between the filament current, the active region field and the "mirror" current field. The summary of eruptive prominence velocity observations during its way from the limb is listed. The filament energy integral is considered. The most rapid filament motion arises when the loss of equilibrium occurs near the "saddle" point of the potential surface. A comparison of parameters of eruptive prominences with infinite and finite motions is realized on the basis of the authors' observations.

073.042 **Alfvénically driven slow shocks in the solar chromosphere and corona.**
J. V. Hollweg.
Astrophys. J., Vol. 389, No. 2, p. 731 – 738 (20 Apr 1992).

The author considers, via a numerical simulation, the nonlinear evolution of a single torsional Alfvénic pulse launched from the solar photosphere on a thin vertical magnetic flux tube which extends into an open coronal region. The pulse steepens into a fast shock, and a slow shock is formed in the chromosphere behind the torsional pulse. Some of the Alfvénic energy is reflected downward by the transition region (TR) and by the steep rise of the Alfvén speed in the upper chromosphere. However, the slow shock reflects part of this energy upward; the slow shock is thus able to enhance the flux of Alfvénic energy into the corona and to enhance the dynamical effects of the Alfvén wave on the TR and upper chromosphere. The energy in the torsional pulse can lead to repeated upward ejections of the TR and underlying chromosphere. After three upward ejections there is a density plateau comparable to the densities found on solar spicules. It is also found that the nonlinear dynamics leads to very impulsive behavior, as manifested by the impulsive flux of energy into the corona and by the short–lived but large amplitude transverse velocities in the corona. Thus, at least some of the events that have been called microflares or explosive events could be the consequence of nonlinear wave dynamics.

073.043 **Excitation of Alfvén waves and local turbulence by energetic ion beams in solar flares.**
V. P. Meytlis, H. R. Strauss.
J. Geophys. Res., Vol. 97, No. A6, p. 8701 – 8705 (1 Jun 1992).

Solar flares release a large part of their energy in the form of beams of high–energy particles. Energetic ion beams can excite Alfvén waves when the beam velocity is greater than the Alfvén speed. In turn, the Alfvén waves can be unstable with respect to secondary Kelvin–Helmholtz instabilities, which produce turbulence and limit the Alfvén wave amplitude. In this way, the ion beam leads to the production of fluid turbulence. The turbulence is strong enough to limit the waves to an energy density which is small compared to the beam energy density. However, the turbulent resistivity produced by the waves and secondary instability is many orders of magnitude larger than molecular resistivity. The turbulence may enhance reconnection, triggering more flares.

073.044 **High–energy gamma–ray emission from pion decay in a solar flare magnetic loop.**
N. Mandzhavidze, R. Ramaty.
Astrophys. J., Vol. 389, No. 2, p. 739 – 755 (20 Apr 1992).

The authors have investigated the production of high–energy gamma rays resulting from pion decay in a solar flare magnetic loop. They took into account magnetic mirroring, MHD pitch–angle scattering, and all of the relevant loss processes and photon production mechanisms. The model treated the transport of both the primary ions and the secondary positrons resulting from the decay of the positive pions, as well as the transport of the produced gamma–ray emission. The distributions of the gamma rays were calculated as a function of atmospheric depth, time, emission angle, and photon energy. The obtained angular distributions are not sufficiently anisotropic to account for the

observed limb brightening of the flare emission at energies larger than 10 MeV, indicating that the bulk of this emission is bremsstrahlung from primary electrons. The calculations are compared with the available data for pion decay radiation from the 1982 June 3 flare.

073.045 **Characteristics of hard X–ray spectra of impulsive solar flares.**
G. A. Dulk, A. L. Kiplinger, R. M. Winglee.
Astrophys. J., Vol. 389, No. 2, p. 756 – 763 (20 Apr 1992).

The authors study the spectral characteristics of 93 impulsive, hard X–ray flares that were observed by the hard X–ray burst spectrometer on the SMM spacecraft. Major findings are: (1) During the initial few seconds after onset of rapidly rising bursts there is a "high–energy delay", where the rise in flux at energies $\gtrsim 150$ keV is delayed by a few seconds relative to that at energies $\lesssim 100$ keV. (2) At the times of peak flux, the power–law spectra almost always "break downward" at an energy of ~ 100 keV, i.e., the spectra at higher energies are steeper than those at lower energies. (3) During the decay phase of bursts, the slopes of the power–law distributions at high and low energies change or cross over in such a way that the overall spectrum assumes the form of either a single power law or a broken power law that breaks up. The break energies of the spectra are usually lower after the crossover, but in $\sim 30\%$ of the cases they are higher. The authors relate these observational results to models of solar flares that invoke electric fields as a particle acceleration mechanism.

073.046 **Structure and physics of solar faculae. V. Study of the roughness effect in the quiet chromosphere–corona transition zone.**
J.–C. Pecker, S. Dumont, Z. Mouradian.
Sol. Phys., Vol. 138, No. 2, p. 213 – 222 (Apr 1992).

Taking into account the effect of roughness (or local departures from sphericity) of the emitting layers in the chromosphere–corona transition zone allows one to determine the optical depths of layers responsible for resolved structures in C II, C III, O IV, and O VI lines. The obtained values correspond not too badly with determinations made in Paper III, by methods not exceedingly influenced by the spherical symmetry hypothesis.

073.047 **Formation of solar prominences by photospheric shearing motions.**
G. S. Choe, L. C. Lee.
Sol. Phys., Vol. 138, No. 2, p. 291 – 329 (Apr 1992).

A numerical simulation is performed to investigate the prominence formation in a magnetic arcade by photospheric shearing motions. A two–and–a–half–dimensional magnetohydrodynamic code is used, in which the gravitational force, radiative cooling, thermal conduction and a simplified form of coronal heating are included. It is found that a footpoint shear induces an expansion of the magnetic arcade and cooling of the plasma in it. Simultaneously the denser material from the lower part of the arcade is pulled up by the expanding field lines. A local enhancement of radiative cooling is thus effected, which leads to the onset of thermal instability and the condensation of coronal plasma. The condensed material grows vertically to form a sheet–like structure making dips on field lines, leading to the formation of the Kippenhahn–Schlüter type prominence. The mass of the prominence is found to be supplied not only by the condensation of the material in the vicinity but also by the siphon–type upflows. The upward growth of the vertical sheet–structure of the prominence is saturated at a certain stage and the newly condensed material is found to slide down from above the prominence along magnetic field lines. This drainage of material leads to the formation of an arc–shaped cavity of low density and low pressure around the prominence. The problem of force and heat balance is addressed and the prominence is found to be not in a static equilibrium but in a dynamic interaction with its environment.

073.048   The fibril structure of prominences.
A. W. Hood, E. R. Priest, U. Anzer.
Sol. Phys., Vol. 138, No. 2, p. 331 – 351 (Apr 1992).

The authors present several magneto–hydrostatic equilibrium models for prominences with fibril–like fine structure. For all the models ad hoc temperature profiles are used without discussing the energetics. The authors assume fine structure to occur either across the prominence axis or along it. This approach is intended as a first step towards more realistic models based upon a series of vertical fibril structures.

073.049   The Hα development of an intense limb flare and associated flaring arches.
A. Antalová, M. B. Ogir.
Sol. Phys., Vol. 138, No. 2, p. 361 – 378 (Apr 1992).

The Hα analysis of the development of the strong impulsive and faint gradual phase of the June 26, 1983 flare indicates the following: (1) The flare originated from two microprominences on the southeast border of NOAA 4227. (2) The main flare structure was a flare cone, which consisted of a bright surge–like stream, elevated above two flare ribbons (located in the cone's base). The flare cone had a height of about 40×10^3km and lasted 4 min in Hα. The upper part of the cone was terminated by a very fine loop, which was bent to the west, where later a chromospheric brightening occurred at the footpoint of a flaring arch. A 300 keV burst and radio spikes were observed during the maximum flare phase. (3) The flaring arch system, with its apex at a height of about 48×10^3km, formed the skeleton for the coronal helmet structure. The velocity of the plasma moving along the flaring arch was between 3500 km s^{-1} and 6900 km s^{-1} during the first brightening.

073.050   Prominence activity during the ascending phase of solar cycle 22.
V. N. Dermendjiev, K. Y. Stavrev, P. L. Andreeva.
Sol. Phys., Vol. 138, No. 2, p. 415 – 418 (Apr 1992). Letter–to–the–editor.

Longitude–latitude and time–latitude distributions of the number and area of prominences observed at Lomnický Stit coronal station in the years 1986 – 1990 are studied using the method of contour maps construction with different degree of smoothing. Special attention is paid to the bifurcation in the prominence distribution. Comparison with the ascending phase of solar cycle 21 is made.

073.051   The basal and strong–field components of the solar atmosphere.
C. J. Schrijver.
Astron. Astrophys., Vol. 258, No. 2, p. 507 – 520 (May 1992).

Spectroheliograms of quiet and active solar regions, observed in spectral lines originating in the upper chromosphere and transition region, are studied. Relationships between line intensities originating at different temperatures in the solar atmosphere are quantified presupposing a two–component model, comprising (i) a background basal emission and (ii) a magnetically controlled emission which shows power–law dependences between emissions in different spectral lines. The consistency of the results of the modelling yields strong evidence in favour of a basal emission component that is most likely non–magnetic in origin. The basal component dominates the emission from outside the magnetic network, but is also present in pixels of at least moderate activity in network and plage, at the resolution of $5'' \times 5''$. The inferred solar basal flux density in the C II line equals the basal flux found for solar–like dwarf stars. The distribution of intensities associated with the basal component is asymmetric, with a relatively strong high–intensity tail. This skewness appears related to the observed statistics of temporal variability. The relationships between upper–chromospheric and transition–region intensities in excess of the basal intensities are generally weakly but significantly non–linear with the power–law index deviating more strongly from unity with increasing difference of the temperatures of formation of the two compared emissions.

073.052   Flares!
R. E. Hill.
Astronomy, Vol. 20, No. 2, p. 74 – 78 (Feb 1992).

073.053   The simultaneous effects of collisions, reverse currents and magnetic trapping on the temporal evolution of energetic electrons in a flaring coronal loop.
K. G. McClements.
Astron. Astrophys., Vol. 258, No. 2, p. 542 – 548 (May 1992).

The temporal and spatial evolution of a population of high–energy electrons in a flaring coronal loop is simulated by solving numerically the appropriate kinetic equation. Coulomb collisions, reverse currents, magnetic field convergence, and precipitation, are all taken into account. Two scenarios are investigated: (1) electrons initially having a single–temperature Maxwellian distribution throughout the loop, the acceleration process having ceased; and (2) electrons accelerated isotropically close to the loop apex, the acceleration process having a linear rise and decay time profile. The spatial distribution of hard X–ray emission is determined, and in case (1) it is found that emission from the chromosphere (i.e. the loop footpoints) generally exceeds that from the corona, even when the magnetic field is strongly converging. Reverse current ohmic losses in the corona tend to reduce the chromospheric X–ray flux, while the coronal X–ray flux is unaffected. In case (2), X–ray emission is concentrated in the corona to a greater extent than in case (1), the spatial distribution of emission again depending on the magnitude of the reverse current. In agreement with previous authors, the authors find a positive correlation between ohmic losses and the timescale over which electrons are accelerated. They conclude that hard X–ray observations are best explained by pulsed, isotropic acceleration of electrons near the apex of a loop with a strongly converging magnetic field.

073.054   The production of ³He and heavy ion enrichments in ³He–rich flares by electromagnetic hydrogen cyclotron waves.
M. Temerin, I. Roth.
Astrophys. J., Lett., Vol. 391, No. 2, p. L105 – L108 (1 Jun 1992).

The authors present a new model for the production of ^{3}He and heavy ion enrichments in ^{3}He–rich flares using a direct single–stage mechanism. In analogy with the production of electromagnetic hydrogen cyclotron waves in Earth's aurora by electron beams, the authors suggest that such waves should exist in the electron acceleration region of impulsive solar flares. Both analytic and test–particle models of the effect of such waves in a nonuniform magnetic field show that these waves can selectively accelerate ^{3}He and heavy ions to MeV energies in a single–stage process, in contrast to other models which require a two–stage mechanism.

073.055   Electrodynamics of electron beams in solar flares.
G. H. J. van den Oord.
Ir. Astron. J., Vol. 20, No. 3, p. 188 – 190 (Mar 1992). – See Abstr. 012.038 for the main entry.

The electrodynamical aspects of beam/return current systems during the impulsive phase of solar flares are discussed. The relevant equations for the electrostatic and the inductive response are given and the solutions are briefly discussed. The conditions for beam propagation are derived and it is shown that under certain conditions beam propagation is inhibited leading to bulk plasma heating.

073.056   Time evolution of a two–ribbon flare: characteristics of post–flare loops.
Gu Xiaoma, Lin Jun, Luan Ti, B. Schmieder.
Astron. Astrophys., Vol. 259, No. 2, p. 649 – 662 (Jun 1992).

Observations of a two ribbon flare on 5 May, 1989 were performed at Yunnan Observatory and at Meudon. Magnetograms have been obtained in Huairou–Beijing Observatory. The topology of the flaring region and the slow evolution of post–flare loops (cool flare loops) can be derived from Yunnan filtergrams. The ribbons seem to be the footpoints of the loops. The analysis of Hα line profiles in the loops, obtained with the

Multichannel Subtractive Double Pass spectrograph of Meudon allows to derive some physical parameters of the loops. The importance of the reference background intensity is discussed. The cool flare loops are low ($h_{max} \sim 25000$ km) dense ($n_e \sim 10^{11} cm^{-3}$) and relatively inhomogeneous. They are rising with a velocity of 5 to 10 km s^{-1}.

073.057 Impulsive phase Fe Kα emission in a flare of 1989 March.

D. M. Zarro, B. R. Dennis, G. L. Slater.
Astrophys. J., Vol. 391, No. 2, p. 865 – 871 (1 Jun 1992).

Observations of the Fe Kα soft X–ray line made with the SMM bent crystal spectrometer on 0557 UT 1989 March 7 show evidence for enhanced Fe Kα line emission coincident with an intense hard X–ray (> 50 keV) burst. The authors study three different models for explaining the Kα enhancement: (1) photoexcitation by soft X–ray thermal bremsstrahlung radiation from an isothermal source; (2) collisional excitation by nonthermal thick–target electrons; (3) photoexcitation by a nonthermal hard X–ray flux distribution which extends with a power–law spectrum down to the Fe Kα ionization threshold at 7.1 keV. Within the limits of isothermal temperature and emission measure set by soft X–ray observations, the thermal photoexcitation model cannot reproduce satisfactorily the intensity of enhanced Kα emission during the hard X–ray impulsive phase. On the other hand, the electron–collisional model and the nonthermal power–law model are both equally capable of explaining the excess Kα flux (above that produced by thermal photoexcitation). The authors discuss the implications of this result for the nonthermal interpretation of impulsive hard X–ray bursts in solar flares.

073.058 The sudden disappearance of a dark filament observed on October 26, 1989.

J. Kubota, R. Kitai, I. Tohmura, A. Uesugi.
Sol. Phys., Vol. 139, No. 1, p. 65 – 79 (May 1992).

The authors present Hα monochromatic and spectroscopic observations of the sudden disappearance of a dark filament located near the center of the solar disk on October 26, 1989. The event was not associated with the flare activity. The dark filament first disintegrated into two loop–like components, and then each component successively showed ascending motion with a velocity greater than 30 km s^{-1}. Comparison of the Hα pictures taken before and after the start of this event suggests that the dark filament was originally composed of two magnetic flux loops.

073.059 Development of a topological model for solar flares.

P. Démoulin, J. C. Hénoux, C. H. Mandrini.
Sol. Phys., Vol. 139, No. 1, p. 105 – 123 (May 1992).

In order to understand the role of interacting large–scale structures in solar flares, the authors have analysed the topology of three–dimensional potential and linear force–free fields. The magnetic field has been modelled by a distribution of charges or dipols located below the photosphere. This modelling permits to define the field connectivity by the charges or the dipoles at both ends of every field line. The authors found that the appearance of a separator above the photosphere is more likely when a parasitic bipole emerges outside the axis that joins the main polarities and when the field lines are characteristic of a field created by dipoles. The separatrices derived in the potential and force–free hypothesis have different shapes. However, in the strong field regions where flares usually occur, the separatrices of the potential and force–free field models become closer. This property makes possible the use of the potential field, as a first estimate, for computing the location in the photosphere of the separatrices and for comparing this location with the position of observed Hα kernels. Displacements of the separatrices of a force–free field result from modifications of the free energy of the field. Then force–free fields have the further capability of predicting the kernel displacement. In all cases a configuration suitable for prominence support is found above the separator.

073.060 A solar flare model including the formation and destruction of the current sheet in the corona.

A. I. Podgornyj, I. M. Podgornyj.
Sol. Phys., Vol. 139, No. 1, p. 125 – 145 (May 1992).

A numerical simulation method is used to show the possibility of forming a current sheet in the solar corona in an active region with four magnetic poles. The evolution of the quasi–stationary current sheet can lead to its transfer to an unsteady state. The MHD instability of this sheet causes its decay, accompanied by a set of events which characterizes the solar flare. The electrodynamical model of a solar flare includes a system of field–aligned currents typical of a magnetospheric substorm. Several events in substorms and solar flares are explained by the generation of field–aligned currents.

073.061 The white–light solar flare of March 27, 1991.

Z. B. Korobova.
Sol. Phys., Vol. 139, No. 1, p. 205 – 207 (May 1992). Letter–to–the–editor.

A white–light flare was recorded on March 27, 1991 at Tashkent. It occurred at the penumbra of a large, complex sunspot group. The energy released per unit time was 2.4×10^{28} erg s^{-1}.

073.062 White–light flares of 1991 June in the NOAA region 6659.

T. Sakurai, K. Ichimoto, E. Hiei, M. Irie, K. Kumagai,
M. Miyashita, Y. Nishino, K. Yamaguchi, G. Fang,
M. A. Kambry, Z. Zhao, K. Shinoda.
Publ. Astron. Soc. Jpn., Vol. 44, No. 1, p. L7 – L13 (1992). Letter–to–the–editor.

The authors report on observations of flare activities in an active region NOAA 6659, which appeared on the sun in 1991 June. Among six X–class flares in this region, the authors observed three flares (June 4, 9 and 11), all of which were white–light flares. A detailed discussion is given concerning a particularly interesting white–light flare which occurred on June 11.

073.063 White–light flare observed at the solar limb.

E. Hiei, Y. Nakagomi, H. Takuma.
Publ. Astron. Soc. Jpn., Vol. 44, No. 1, p. 55 – 62 (1992).

A white–light flare occurring at the solar limb and its associated loop prominence system were observed in white light on 1989 August 16. Nine photographs of these phenomena were reduced. The brightening of the flare at the limb was explained by an increase in temperature, estimated to be of the order of 5250K; its total energy emitted in the WLF was inferred to be 10^{30} erg. The bright top of the flare loops is thought to be due to bound–free/free–free emission, and its electron density was estimated to be about $10^{12-13} cm^{-3}$.

073.064 Ca II K line diagnostics of the dynamics of the solar flare atmosphere.

C. Fang, E. Hiei, S.–y. Yin, W.–q. Gan.
Publ. Astron. Soc. Jpn., Vol. 44, No. 1, p. 63 – 72 (1992).

Observations of the Ca II K line profiles for 12 solar flares have been analyzed and some characteristics of the red asymmetry of the Ca II K line are given. Based on Non–LTE calculations, the influence of velocity fiels in the lower atmosphere of flares on the profiles of the Hα and Ca II K lines has been explored. The result indicates that a downward motion of plasma above the temperature minimum region (TMR) as well as a contracting motion of the plasma toward TMR can well explain the red asymmetry observed at the K$_1$ positions. The typical velocity is 10–30 km s^{-1}.

073.065 The tangential discontinuous surface of velocity in solar chromosphere.

Hu Wenrui, Zhang Hongqi, Chen Jiming.
Sci. China, Ser. A, Vol. 34, No. 11, p. 1354 – 1364 (Nov 1991). In Chinese.

The data of velocity and magnetic fields in the solar photosphere (5324 Å) and the chromosphere (4861 Å) clearly show the features of tangential discontinuity of velocity in the chromo-sphere. The velocity fields in and near the solar active region No. 88029 by the Huairou Station have been analyzed in detail. A lot of magnetohydrodynamic discontinuous surfaces, especially the tangential discontinuities, are shown from the observations. The calculations of the thickness of discontinuous layer and the evolution time of instability agree with the observational results.

073.066   Motion along some chromospheric fibrils.
L. G. Kartashova.
Bull. Crimean Astrophys. Obs., Vol. 82 , p. 107 – 113 (1992). English translation of Izv. Krym. Astrofiz. Obs., Tom 82, p. 116 – 124 (1990).

Radial velocities have been measured from the H_α lines for a series of chromospheric fibrils joining spots, pores, or flocculi having magnetic fields opposite in polarity and which also belong to superpenumbras. It is found that: 1. the fibrils are arched structures, since one finds mainly two types of motion: a) descent almost throughout the fibril, apart from a small part near the middle, where there is rise at a rate 1 – 6 km/sec. The sinking rate increases smoothly from the central part to the ends, where it attains 10 – 60 km/sec; b) flow of material from one fibril to another with a speed of 10 – 60 km/sec. 2. there is evident that the sinking rate at the ends varies in time synchronously and in antiphase with the rise speed at the middle. During a flare, there may be increased rise at the middle, while the sinking rates at the ends decrease.

073.067   Flares above sunspots and magnetic fields. I.
A. N. Babin, A. N. Koval'.
Bull. Crimean Astrophys. Obs., Vol. 82 , p. 119 – 126 (1992). English translation of Izv. Krym. Astrofiz. Obs., Tom 82, p. 129 – 138 (1990).

Magnetic fields have been examined for two flares of magnitude 2B in the penumbra of spots with δ–configurations by reference to the emission lines of iron, D_1 and D_2 lines of sodium, and D_3 for helium, the magnetic fields of the spots on which the flares occurred were also determined. The emission from the metals and helium occurred not more than 3″ from the points at which the fields changed sign in the spots; the field strengths indicated by the metal emission lines were over 2000 G, while those indicated by helium were about 1000 G, but with the polarity unaltered. The observations indicate that there are elements unresolved by the telescope having magnetic fields opposite in direction. The line-of-sight velocities have been determined from the emission and absorption lines. Material in the deeper layers rises and that in the upper layers sinks with velocities of 1 – 3 km/sec.

073.068   Helium radiation diffusion in prominences with variable electron density. I. Theoretical contours of the λ584 Å resonance line He I.
C. Lkhagvazhav.
AIP Conf. Proc., No. 245: Workshop on Basic Space Science, p. 193 – 200 (1992). – See Abstr. 012.064 for the main entry.

It is shown that on the basis of a solution of the integral diffusion equations for radiation in a multilevel helium atom under low–temperature plasma conditions in the "vertical slab" model with variable electron densities: (a) Neutral helium is ionized by coronal radiation mainly in the $\lambda\lambda 100 - 300$ Å spectral region. (b) The 2^3S level is destroyed by electron impacts, and the 2^1S level decays via the escape of the quanta of λ584 Å through the 2^1P level. (c) Emission in the resonance line λ584 Å $(2^1P \rightarrow 1^1S)$ occurs due to recombination to 2^1S with subsequent absorption of quanta of infrared radiation λ20581 Å. (d) The radiation of helium is generated in the vicinity of the boundary planes in the penetration of radiation with $\lambda = 200$ Å, where the density of matter decreases gradually down to the coronal value. In the subordinate lines, the radiation is conditioned by quasi-resonance scattering of photospheric radiation. (e) The calculated absolute values of the intensities of helium and hydrogen lines are in good agreement with the observations.

073.069   The relation between solar flare and longitudinal current density on Jan. 14, 1989.
Chen Jimin, Zhang Hongqi.
Acta Astrophys. Sin., Vol. 12, No. 1, p. 82 – 86 (Jan 1992). In Chinese.

The authors compare the methods of Krall and potential field in order to estimate the direction of the solar transverse magnetic field, and find the latter method is better than the former in estimating the directions. They analyse the relation between the flare and longitudinal current density in the AR 5312 active region on Jan. 14, 1989, and obtain the results that initial bright points in H_β correspond to the largest value of the longitudinal current density in the photosphere.

073.070   Observation and analysis of the asymmetry of Ca II K line of solar flares.
Yin Suying, Fang Cheng, Gan Weiqun.
Chin. Astron. Astrophys., Vol. 16, No. 2, p. 187 – 195 (Apr–Jun 1992). English translation of Acta Astron. Sin., Vol. 32, No. 4, p. 353 – 361 (Dec 1991). See Abstr. 54.073.168.

The authors study the Ca II K line spectra of eight flares observed with the Solar Tower Telescope at the Nanjing University. It is seen that the red asymmetry is common to all flares, i.e., that the intensity of the red wing at K_{1r} is stronger than that of the blue wing at K_{1b} and the distance between K_{1r} and the line center is larger than that for K_{1b}. Such red asymmetry is most obvious near the maximum of the K line intensity. However, the K line center does not display on obvious Doppler shift for all flares the authors study. By using the non-LTE theory when calculating the Ca II K line profiles, it is shown that the asymmetry may be explained by the movement of material around the temperature minimum region. Moreover, a statistical study shows that the temperature and the depth of the temperature minimum region are proportional to the maximum intensity of the flare emission.

073.071   An observational evidence of the flare model related to filament currents.
Xu Aoao.
Chin. Astron. Astrophys., Vol. 16, No. 2, p. 195 – 206 (Apr–Jun 1992). English translation of Acta Astron. Sin., Vol. 32, No. 4, p. 343 – 352 (Dec 1991). See Abstr. 54.073.167.

Using the observational data about the regularities of motion of four filaments related to two–ribbon flares before and after the beginning of the impulsive phase, the author has solved the momentum equation and the energy equation of evolution and movement of filament currents in terms of the Kuperus–Raadu prominence (filament) model, and obtained the curves of changes of the current intensity and total energy of filament currents with time before and after the beginning of the impulsive phase. He finds that the beginning moments of the impulsive phase of four flares all agree with the maximum moments of the filament energy. Moreover, it is also found that the importance of X-ray flares is nearly proportional to the rate of increase of filament energy before the occurrence of flares. These results reveal the intrinsic relation between the flare eruption and the evolution of filament currents: the explosion of filaments causes the appearance of the flares, and the energy of the flares is supplied by filament currents. This provides a strong observational evidence for the flare model related to filament currents.

073.072   Dynamic regimes of prominence evolution.
N. M. Bakhareva, V. V. Zaitsev (*V. V. Zajtsev*), M. L. Khodachenko.
Sol. Phys., Vol. 139, No. 2, p. 299 – 314 (Jun 1992).

The dynamical model of a solar prominence taking into account plasma motion in the form of a cumulative flux was developed. Of great importance is the fact that the prominence plasma is partially ionized. Ion–neutral collisions in nonsteady flux conditions radically change the energetics and dynamics of the prominence model. In this case the equilibrium state described by the Kippenhahn–Schlüter solutions becomes unstable, and various dynamic regimes of plasma compression, as well as monotonic or quasiperiodic expansion of the plasma can exist in the prominence.

073.073   Toward the circuit theory of solar flares.
V. V. Zaitsev (*V. V. Zajtsev*), A. V. Stepanov.
Sol. Phys., Vol. 139, No. 2, p. 343 – 356 (Jun 1992).
It has been shown that the main problems of the circuit theory of solar flares have been resolved considering the case of magnetic loop emergence and the correct application of Ohm's law. The generalized Ohm's law for solar flares is obtained. The conditions for flare energy release are as follows: large current value, $> 10^{11}$A, nonsteady–state character of the process, and the existence of a neutral component in a flare plasma. As an example, the coalescence of a flare loop and a filament is considered. It has been shown that the current dissipation has increased drastically as compared with that in a completely ionized plasma. The current dissipation provides effective Joule heating of the plasma and particle acceleration in a solar flare. The ion–atom collisions play the decisive role in the energy release process. As a result the flare loop resistance can grow by 8 – 10 orders of magnitude. For this one does not need the anomalous resistivity driven by small–scale plasma turbulence. The energy release emerging from the upper part of a flare loop stimulates powerful energy release from the chromospheric level.

073.074   On the saturation of electron–cyclotron masers in solar flares.
Yu. E. Charikov, G. D. Fleishman (*G. D. Flejshman*).
Sol. Phys., Vol. 139, No. 2, p. 387 – 399 (Jun 1992).
The authors consider the relaxation of an unstable distribution of fast non–relativistic electrons. Langmuir turbulence generated by the electrons is found to determine the saturation of an electron–cyclotron maser. The important role of nonlinear processes in Langmuir and electromagnetic waves is shown. The characteristic saturation time is about 1 ms. It is shown that both cyclotron maser emission and the transformation of plasma waves to transverse ones can be essential in the formation of observable radio spectra from solar flares.

073.075   Collisional interaction between an electron beam and partially ionized chromospheric plasma.
V. V. Zharkova, V. A. Kobylinskij.
Kinematika Fiz. Nebesn. Tel, Tom 8, No. 3, p. 38 – 43 (May–Jun 1992). In Russian. English translation in Kinematics Phys. Celest. Bodies, Vol. 8, No. 3.
Energy spectrum for a stationary electron beam injected from the solar corona into the chromosphere is obtained, as well as density variations with the depth of penetration into the chromospheric plasma of an arbitrary degree of ionization. The density of electrons in the beam is shown to decrease with decreasing degree of ionization more slowly than under the complete ionization of the plasma, this resulting in an increase of the rates of nonthermal excitation and ionization of hydrogen atoms. The stability of reverse current in the chromospheric layers is estimated on the basis of the initial energy flux of the beam at the upper boundary. For hard beams ($\gamma = 3$) the reverse current in the lower chromosphere is stable up to flux values of 10^4J cm^{-2}s^{-1}, and for the beams with a more soft spectrum ($\gamma = 5$) the current stability breaks when the fluxes become greater than 4×10^3J cm^{-2}s^{-1}.

073.076   Strong magnetic fields in solar flares: observational data and a theoretical model.
N. I. Lozitskaya, V. G. Lozitskij, A. A. Solov'ev.
Kinematics Phys. Celest. Bodies, Vol. 7, No. 6, p. 25 – 31 (1991). English translation of Kinematika Fiz. Nebesn. Tel., Tom 7, No. 6, p. 40 – 47 (Nov–Dec 1991).
Spectral observations of five solar flares made in 1981 and 1989 with a circular–polarization analyzer at the echelle spectrograph of the horizontal solar telescope of the Kiev University Astronomical Observatory are analyzed. Direct measurements of the Zeeman splitting of the emission of Fe I, Fe II, Na I, He I and H I lines indicated that the intensity of the longitudinal magnetic field H ∥ in the flares ranges from 10 – 20 to 350 mT, and that H^e∥ =

120 – 180 mT at the points at which He I emission is formed. The field H^p∥ at photospheric level, measured in the splitting of the Fraunhofer profiles, is systematically lower than the emission H^e∥. This may indicate the existence of a local (vertically) intensification of the magnetic field in flares. It is shown that a theoretical model according to which strong fields arise as a result of twisting of magnetic flux tubes in the upper levels of the solar atmosphere agrees satisfactorily with the observational data.

073.077   Research of solar events in optical, radio and X–ray regions.
Wang Jialong.
Advances in solar–terrestrial science of China, p. 67 – 76 (1992). – See Abstr. 003.069 for the main entry.
This paper describes some results of solar event research in optical, radio and X–ray regions. In the first section the time relationship between phenomena of a discrete solar event in the spectral region is shown. Spatial structure of solar events is discussed based on observations in the second section. In the third section, time sequences of solar and relevant geophysical events are investigated. Finally, a new classification of solar flares is suggested by using optical, radio and X–ray data of solar events.

073.078   Exploration of the variation in the source function within a loop prominence by means of the simplex method.
Li Kejun.
Publ. Yunnan Obs., No. 1, p. 47 – 54 (1992). In Chinese.
A method for exploring the variation in the source function in the line profiles of a loop prominence is proposed. Some symmetric lines of the loop prominence of 1984 Feb 18 on the solar limb are fitted by using this method with the simplex method. It is found that only consideration of a simple layer for the symmetrical lines is premitted, even if the source function changes with the optical depth of the line.

073.079   Solar flares and coronal mass ejections.
S. W. Kahler.
Annu. Rev. Astron. Astrophys., Vol. 30, p. 113 – 141 (1992).
    Contents: 1. Introduction.   2. Historical   perspective.
3. Flare/CME   relationships.   4. Interplanetary   effects.
5. Conclusions.

073.080   A numerical simulation for the cooling process of the solar flare loop.
W. Q. Gan, C. Fang, H. Q. Zhang.
Proc. Astron. Soc. Aust., Vol. 9, No. 2, p. 317 (1991). Abstract. – See Abstr. 012.090 for the main entry.

073.081   An update on high–speed studies of solar flares.
J. C. Krueger, D. Perry, G. D. Toot.
News Lett. Astron. Soc. N.Y., Vol. 4, No. 1, p. 12 (Feb 1992). Abstract. – See Abstr. 012.078 for the main entry.

073.082   Chromospheric structure.
J. B. Gurman.
NATO Advanced Study Institute on the Sun: a Laboratory for Astrophysics, p. 245 – 259 (1992). – See Abstr. 012.091 for the main entry.
A brief introduction is given to the variety of chromospheric features seen in Hα images. Despite their wealth of detail, images obtained in a single, narrow band in Hα also suffer from a surfeit of missing information. The most basic chromospheric structures are the supergranular network and spicules; these are visible in lines formed higher in the chromosphere, as well, which also show evidence for small–scale heating. The heating of even the lower chromosphere is still problematic, but may be explained by large–amplitude acoustic wave dissipation. A current challenge is understanding the apparent thermal bifurcation of the chromosphere into regimes of distinctly different scale heights. Finally, the possibility of chromospheric wave cavities is briefly discussed.

073.083   Overview of solar flares.
A. G. Emslie.
NATO Advanced Study Institute on the Sun: a Laboratory for Astrophysics, p. 465 – 474 (1992). – See Abstr. 012.091 for the main entry.

The author discusses the energetics of, and characteristic emissions associated with, solar flares, and various flare "classification schemes" that have been devised. It is argued that flares derive their energy from current–carrying, but force–free, magnetic fields; observations of the magnetic "shear" associated with these currents are presented. The author discusses the basic principles of flare modeling, both "forward" (i.e., starting with a prescribed initial configuration and solving the relevant equations) and "inverse" (i.e., fitting observed emissions to a self-consistent model of energy release).

073.084   Energy release and transport in flare plasmas.
A. G. Emslie.
NATO Advanced Study Institute on the Sun: a Laboratory for Astrophysics, p. 489 – 508 (1992). – See Abstr. 012.091 for the main entry.

The author discusses the fundamental issues involved in the dissipation of stored magnetic energy during solar flares, and the various magnetic field topologies and geometries that have been involed to enhance the energy release rate. Transport of the released energy, particularly by energetic electrons, is reviewed, noting especially the global electrodynamic consequences of such a scenario. Results from hydrodynamic simulations of electron-heated flare atmospheres are compared critically with recent observational data.

073.085   Temporal and spatial characteristics of solar flares from observations.
Wang Jialong.
Proc. Astron. Soc. Aust., Vol. 9, No. 2, p. 203 – 208 (1991). – See Abstr. 012.090 for the main entry.

Temporal and spatial characteristics of solar flares are briefly reviewed in this paper. The global, temporal and spatial behaviours of flares are given first. Besides the 154–day periodicity, an 80–day periodicity of occurrence rate of large hard X–ray bursts for the period 1980 February – 1985 December, and the delay of the peak occurrence rate of large flares are pointed out, then the gregariousness of major flares is shown. In the third section, the time process and spatial structure of individual flares are shown and described according to space and ground–based observations. In the last section two problems on flare properties are discussed. (i) Previous classifications of solar flares are based generally on observations in a single spectral region. A new classification of flares based on observations in multi–spectral regions is given. (ii) Energy released in part of a loop seems to be not enough for a whole flare, and a qualitative model in which the energy is supplied by the untwisting of magnetic fields is proposed.

073.086   History and basic characteristics of eruptive flares.
Z. Švestka, E. W. Cliver.
IAU Colloquium No. 133: Eruptive solar flares, p. 1 – 11 (1992). – See Abstr. 012.096 for the main entry.

The authors review the evolution of the knowledge and understanding of the eruptive (dynamic, two–ribbon) flare phenomenon.

073.087   Basic magnetic configuration and energy supply processes for an interacting flux model of eruptive solar flares.
E. R. Priest.
IAU Colloquium No. 133: Eruptive solar flares, p. 15 – 32 (1992). – See Abstr. 012.096 for the main entry.

A review is given of the current understanding of: the basic magnetic configuration of a flare, including loop and arcade structures, prominence models and interactions of separate flux systems; the process of preflare energy storage in excess of potential due to photospheric motions; the conditions for flare occurrence, including shear in the corona and complexity and the roles of spot motions, flux cancellation and prominences;

theories for eruption by magnetic nonequilibrium. In the course of this, the elements of a new Interacting Flux Model for flares are outlined.

073.088   The role of cancelling magnetic fields in the buildup to erupting filaments and flares.
S. F. Martin, S. H. B. Livi.
IAU Colloquium No. 133: Eruptive solar flares, p. 33 – 45 (1992). – See Abstr. 012.096 for the main entry.

The authors present a scenario for understanding the role of cancelling magnetic fields in the build–up to eruptive solar flares. The key intermediate step in this scenario involves the formation of a filament magnetic field in the corona above a photospheric polarity inversion where cancelling magnetic fields are observed. To illustrate the observable phases of this scenario, the authors describe the build–up to two simple eruptive flares in a small active region.

073.089   Variation of the vector magnetic field in an eruptive flare.
D. M. Rust, G. Cauzzi.
IAU Colloquium No. 133: Eruptive solar flares, p. 46 – 52 (1992). – See Abstr. 012.096 for the main entry.

Observation of a 3B, M6 flare on April 2, 1991 appear to confirm earlier evidence that eruptive flares are triggered by measurable magnetic field changes. In the eight hours before the flare, the shear in the magnetic fields increased. The development that likely triggered the flare was the emergence into the active region and rapid proper motion of new flux. One of the small spots marking the negative leg of the new flux pushed into an established positive field at 0.2 km/s. Data from the JHU/APL vector magnetograph show that this motion led to the development of a sheared field. The flare started near the newly–sheared fields and spread to engulf most of the spot region. A magnetogram taken 45 min after flare onset shows possible relaxation of the sheared fields.

073.090   Intrinsically hot flares and possible connection to deep convective magnetic fields.
H. A. Garcia, P. S. McIntosh.
IAU Colloquium No. 133: Eruptive solar flares, p. 53 (1992). – See Abstr. 012.096 for the main entry.

073.091   Interaction of large–scale magnetic structures in solar flares.
C. H. Mandrini, P. Démoulin, J. C. Hénoux.
IAU Colloquium No. 133: Eruptive solar flares, p. 54 (1992). Abstract. – See Abstr. 012.096 for the main entry.

073.092   The intrinsic relationship between flares and eruption of filament currents.
S. T. Wu, A. A. Xu.
IAU Colloquium No. 133: Eruptive solar flares, p. 55 – 58 (1992). – See Abstr. 012.096 for the main entry.

In this study, the authors have employed the data from four flares with filament eruptions given by Kahler et al. (1988) to deduce the filament current and total energy as functions of time according to the Kuperus–Raadu filament model (1974). From these results, they found that the impulsive phase of these four flares is at the maximum of the total energy contained in the filament, which shows that the total energy of the filament decreases at the initiation of the impulsive phase of the flare. The amount of the filament energy decreases as required by the flare energy. Further, the authors discovered that the importance of the X–ray flare is proportional to the rate of the increase of the total energy prior to the filament eruption. On the basis of these results, it is concluded that there is an intimate relationship between the flare and eruptive filament quantitatively.

073.093   Filament eruptions, flaring arches and eruptive flares.
A. Bhatnagar, A. Ambastha, N. Srivastava.
IAU Colloquium No. 133: Eruptive solar flares, p. 59 – 64 (1992). – See Abstr. 012.096 for the main entry.

Several cases of erupting filaments showing distinctly their "feet" have been studied. Role of the feet and their anchorage with the photosphere in maintaining filament stability is established, apart from the footpoint separation and height criteria. Further, a homologous series of more energetic events, namely, the flaring arches and eruptive flares of March 5–7, 1991, suggest a repetitive restoration of magnetic field conditions and energy build–up within a day. High resolution H–alpha observations of these events indicate that large amount of ejected material was "siphoned out" from the chromosphere through the top a low-lying compact emission loop within the active region.

073.094 Triggering of eruptive flares: destabilization of the preflare magnetic field configuration.
R. L. Moore, G. Roumeliotis.
IAU Colloquium No. 133: Eruptive solar flares, p. 69 – 78 (1992). – See Abstr. 012.096 for the main entry.

This paper takes the three–dimensional configuration of the magnetic field in and before eruptive flares as the main guide to how the preflare field comes to lose its stability and erupt. From observed characteristics (1) of the preflare magnetic field configuration, (2) of the onset and development of the eruption of this configuration before and during the flare, and (3) of the onset and development of the flare energy release within the erupting field, the typical erupting field configuration for two–ribbon eruptive flares is constructed. The observational centerpiece for this construction is the evidence from the Marshall Space Flight Center vector magnetograph that strong magnetic shear along the main magnetic inversion line is critical for large eruptive flares. From (a) the empirical field configuration and (b) the observation that the initial flare brightening typically stems from points where opposite–polarity flux is gradually merging and canceling at or near the main inversion line, it is argued (1) that eruptive flares are driven by the eruptiv expansion of the strongly sheared core of the preflare magnetic field, (2) that this eruption is triggered by preflare slow reconnection accompanying flux cancellation in the sheared core, and (3) that in some flares the triggering reconnection and flux cancellation is between opposite–polarity strands of the extant preflare sheared core field, while in other flares it is between the sheared core field and new emerging flux.

073.095 Energy transport in solar flares: implications for Ca XIX emission.
D. M. Zarro.
IAU Colloquium No. 133: Eruptive solar flares, p. 95 – 104 (1992). – See Abstr. 012.096 for the main entry.

Results of recent numerical calculations of the soft X–ray Ca XIX resonance line profile are reviewed. The calculations were based on three different energy transport models: nonthermal thick–target electron precipitation; thermal conduction; and electric–field acceleration. Predictions made by each of these models for the Ca XIX blueshift and nonthermal broadening are compared with observations obtained during the flare impulsive phase. Agreements and discrepancies between theory and observation are discussed.

073.096 Fluid flow in a jet and the Ca XIX line profiles observed during solar flares.
P. L. Bornmann, J. R. Lemen.
IAU Colloquium No. 133: Eruptive solar flares, p. 105 (1992). For the full paper see Astrophys. J., Vol. 385, No. 1, p. 363 – 374 (20 Jan 1992) – see Abstr. 55.073.007. – See Abstr. 012.096 for the main entry.

073.097 Characteristics of the impulsive phase of flares.
A. O. Benz, M. J. Aschwanden.
IAU Colloquium No. 133: Eruptive solar flares, p. 106 – 115 (1992). – See Abstr. 012.096 for the main entry.

The impulsive phase of flares is an observational concept, characterized by spiky emissions from γ–rays to radio waves. It is generally agreed that during this time a large fraction of the original flare energy resides in energetic particles which are manifested in these emissions. Here the authors concentrate on recent decimeter and microwave observations that indicate a high level of fragmentation of this energy release when related to hard X–ray flux. Recent attempts to characterize the flare and the distribution of the radio bursts in time and frequency by statistical methods are also reviewed.

073.098 Comparison of UV and X–ray solar flare observations and theoretical models.
J. E. Rodriguez, M. G. Rovira, M. E. Machado, E. J. Reichmann.
IAU Colloquium No. 133: Eruptive solar flares, p. 116 – 117 (1992). – See Abstr. 012.096 for the main entry.

The authors have studied SMM Hard X–ray Imaging Spectrometer (HXIS) and Ultraviolet Spectrometer and Polarimeter (UVSP) observations of a flare on April 10, 1980, where both instruments simultaneously imaged a hard X–ray footpoint area. The UVSP recorded the emission in the N V line at 1238.8 Å, with a spatial resolution of 3″ and a temporal resolution of 10 s. While the overall footpoint area as seen in the hard X–ray images (16–30 keV) is of the order of two HXIS $(8'')^2$ pixels $(6 \times 10^{17} cm^2)$, the UV observations show evidence of smaller scale structure $(\approx 5 \times 10^{16} cm^2)$ for individual peaks throughout the duration of the hard X–ray burst. The ultraviolet peaks are also characterized by redshifts in the line, which is formed at the 10^5K level within the transition region.

073.099 White–light flares.
J. C. Hénoux, J. Aboudarham.
IAU Colloquium No. 133: Eruptive solar flares, p. 118 – 123 (1992). – See Abstr. 012.096 for the main entry.

The observed good temporal correlation between white–light flares and hard X–ray bursts suggests that energetic electrons could be the cause of white–light flare emission. However, even if sufficient energy can be deposited in the chromosphere by electron bombardment, direct collisonal heating of the photosphere requires to high a flux of electrons of a few hundred keV and does not appear plausible. The authors show that non–thermal effects increae the opacity of the upper photosphere and temperature minimum region. Then these regions are consequently radiatively heated both by the beam–produced flare chromospheric emission and by the quiet photospheric emission. As a consequence, a temporary decrease of the continuum intensity (negative flare) is expected. After less than 20 s, the radiative heating of the upper photosphere and temperature minimum region produces a white–light flare.

073.100 "Post" flare loops.
B. Schmieder.
IAU Colloquium No. 133: Eruptive solar flares, p. 124 – 133 (1992). – See Abstr. 012.096 for the main entry.

"Post"–flare loops or cool flare loops when observed in the Hα line are magnetic structures within an active region which are clearly a fundamental part of the flare itself and not an external phenomenon or a consequence of flares. After a two–ribbon flare the field–line reconnection gives rise to flare loops of hot temperature which shrink and become cool within few minutes to an hour. The Hα flare loops appear generally as dark loops during the gradual phase of the flare; viewed on the limb they may appear as loops in emission, reaching 50000 km. They create a system of quasi–steady arches lasting up to several hours. Large downflows are observed along the legs of the loops with deceleration by comparing to free–fall motions. As derived from radiative transfer diagnostics, the gas pressure range of the loops is between 0.2 and 5 dyn cm^{-2} and the electron density is between 10^{10} and $10^{12} cm^{-3}$. The magnetic free energy needs to be continuously replenished from low levels during the flare. Different mechanisms may be considered, dynamical ones such as upward motion from the convection zone pushing the coronal field and twisting of the field lines, or thermal ones such as evaporation or ablation of chromospheric material. The author discusses some observational evidence supporting the Forbes and Malherbe (1986) reconnection model based on the Kopp and Pneuman configuration.

073.101 **Plasma parameters derived from MSDP observations of cool flare loops.**
P. Heinzel, B. Schmieder, P. Mein.
IAU Colloquium No. 133: Eruptive solar flares, p. 134 (1992).
Abstract. – See Abstr. 012.096 for the main entry.

073.102 **Flare evolution in the density–temperature diagram.**
S. Serio, F. Reale, G. Peres, J. Jakimiec, B. Sylwester, J. Sylwester.
IAU Colloquium No. 133: Eruptive solar flares, p. 135 – 138 (1992). – See Abstr. 012.096 for the main entry.

The scope of this work is to set up diagnostic tools for the X-ray flare decay phase in terms of the density–temperature (n–T) diagram. It originates from the observation of the interesting characteristics of this diagram, which were pointed out by Jakimiec et al. (1986, 1987) in the analysis of SMM flare data. Their basic observation was that many flares follow approximate power–law trajectories in this diagram, during their decay phase. The authors have used the Palermo–Harvard hydrodynamical code to compute the evolution of flare temperature and density for a grid of confined loop flares. The hydrodynamic code solves the conservation equations for mass, velocity and energy in a semicircular coronal loop anchored in the chromosphere, and subject to a stationary heating function, providing for steady state equilbrium, and to an impulsive heating function causing the flare (s. Peres and Serio, 1984).

073.103 **Flares activity during solar cycle 21.**
S. Dinulescu, V. Dinulescu.
Rom. Astron. J., Vol. 1, No. 1 – 2, p. 63 – 67 (1991).

The present paper investigates the statistical behaviour of the solar flares through cycle 21 (1976 – 1986). The total number of solar flares and the number of flares of various importance are given.

073.104 **The Neupert effect: what can it tell us about the impulsive and gradual phases of eruptive flares?**
B. R. Dennis, B. M. Uberall, D. M. Zarro.
IAU Colloquium No. 133: Eruptive solar flares, p. 139 – 143 (1992). – See Abstr. 012.096 for the main entry.

The Neupert effect is the name given to the correlation observed in many flares between the impulsive time profile of the microwave and hard X-ray emissions and the time derivative of the soft X-ray profile. The authors have used data collected between 1980 and 1989 from the Hard X-Ray Burst Spectrometer (HXRBS) on the Solar Maximum Mission (SMM) and the soft X-ray detector on GOES to determine which events show this correlation and which do not. They have found that of 66 HXRBS events observed in 1980 with a peak rate of > 1000 counts/s, 58 (80%) showed good correlation with peaks in the GOES time derivative plot corresponding to peaks in the HXR plots to within ± 20 s. The more gradually varying X-ray events that are commonly referred to as Type C flares and that are often accompanied by eruptive flares, tend to show poorer correlation between the SXR time derivative and the HXR time profile. In several cases studied, the later, more gradually varying, peaks either did not register at all in the SXR time derivative plots or resulted in very broad peaks that, in one case on 1981 April 26, was delayed by 13 min.

073.105 **Particle acceleration in the impulsive phase of solar flares.**
D. B. Melrose.
IAU Colloquium No. 133: Eruptive solar flares, p. 147 – 156 (1992). – See Abstr. 012.096 for the main entry.

Theoretical ideas on particle acceleration in solar flares are discussed with emphasis on bulk energization of electrons during the impulsive phase. Many localized, short–lived episodes of energy release must be involved. An analogy with the acceleration of auroral electrons by multiple weak double layers (WLDs) is suggested and explored. Based on the auroral analogy a mechanism for reflecting electrons by solitary waves (SWs) in proposed. The prompt acceleration of relativistic electrons is discussed briefly.

073.106 **Kinetic description of electron beams in the chromosphere.**
D. Gómez, P. J. D. Mauas.
IAU Colloquium No. 133: Eruptive solar flares, p. 157 – 160 (1992). – See Abstr. 012.096 for the main entry.

073.107 **Nuclear reactions in flares.**
E. Rieger.
IAU Colloquium No. 133: Eruptive solar flares, p. 161 – 170 (1992). – See Abstr. 012.096 for the main entry.

The explosive release of energy stored in sheared magnetic fields and its subsequent transformation to a large extent into kinetic energy of charged particles is a common phenomenon occurring in plasma throughout the universe from a place as close as the Earth's magnetosphere to objects at cosmological distances such as quasars. On this vast distance scale the Sun is of crucial importance for the understanding of these physical processes. Due to its proximity, flares can be investigated in the gamma–ray regime, which is the domain where the accelerated particles leave their fingerprints most clearly and flare–generated particles can be recorded in space and related to particular events. In this paper the author discusses the production of gamma–ray line radiation and neutrons and approach the problem of the acceleration and energy release of charged particles from the viewpoint of the observer.

073.108 **Radio emission of eruptive flares.**
M. Karlický.
IAU Colloquium No. 133: Eruptive solar flares, p. 171 – 176 (1992). – See Abstr. 012.096 for the main entry.

Radio spectra of some eruptive flares are described. Most of them do not conform to the classical spectrum schema. Both the type I burst chains associated with the filament activation phase of the May 16, 1981 flare, and the slow negative drift of the group of type III and U bursts in the July 12, 1982 and June 15, 1991 flares illustrate the upwards expansion of complete magnetic structures. In some eruptive flares, e.g. April 24, 1985, reverse drift bursts are observed prior to the upwards expansion. In addition, narrowband dm–spikes and a variety of positively drifting features are frequently observed. These are believed to be radio signatures of localized reconnections and of the spreading of flare dissipative processes. Results of numerical simulations supporting these ideas are presented.

073.109 **High energetic solar proton flares of 19 to 29 October 1989.**
M. A. Mosalam Shaltout.
IAU Colloquium No. 133: Eruptive solar flares, p. 190 – 193 (1992). – See Abstr. 012.096 for the main entry.

The evolution of solar flare activity was followed by the method of cumulative summation curves according to observed H–alpha flares and X–ray bursts (measured on satellites) in the active region No. 5747 during one rotation when the energetic solar flares of 19 October 1989 occurred. It was found that the steep trend of increased activity sets on several tens of hours prior to the occurrence of the energetic flare, which could be used, together with other methods, for forecasts of major flares.

073.110 **On the association between large scale X–ray brightenings and solar flares.**
C. H. Mandrini, M. E. Machado.
IAU Colloquium No. 133: Eruptive solar flares, p. 220 (1992). Abstract. – See Abstr. 012.096 for the main entry.

073.111 **Models of normal and inverse polarity filament eruptions and coronal mass ejections.**
D. F. Smith, E. Hildner, R. S. Steinolfson.
IAU Colloquium No. 133: Eruptive solar flares, p. 272 – 275 (1992). – See Abstr. 012.096 for the main entry.

073.112   An observational–conceptual model of the formation of filaments.
S. F. Martin.
IAU Colloquium No. 133: Eruptive solar flares, p. 331 – 332 (1992). Extended abstract. – See Abstr. 012.096 for the main entry.

073.113   Dynamics in the prominence–corona transition region from HRTS spectra.
B. Schmieder, K. P. Dere, J. E. Wiik.
IAU Colloquium No. 133: Eruptive solar flares, p. 333 – 336 (1992). – See Abstr. 012.096 for the main entry.
Line profiles of UV emission lines between 1206 Å and 1670 Å observed in two prominences with the High Resolution Telescope and Spectrograph (HRTS) are analysed. Microturbulent velocities from 0 to 25 km s⁻¹ are found in both prominences.

073.114   Velocity field in the 13 June 1980 flare area.
A. Falchi, R. Falciani, L. A. Smaldone.
IAU Colloquium No. 133: Eruptive solar flares, p. 342 (1992). – See Abstr. 012.096 for the main entry.

073.115   The X12 limb flare and spray of 01 June 1991.
V. Gaizauskas, C. R. Kerton.
IAU Colloquium No. 133: Eruptive solar flares, p. 347 – 350 (1992). – See Abstr. 012.096 for the main entry.

073.116   Analysis of X–ray flares observed by the SMM spacecraft.
A. M. Hernandez, M. G. Rovira, C. H. Mandrini, M. E. Machado.
IAU Colloquium No. 133: Eruptive solar flares, p. 351 – 354 (1992). – See Abstr. 012.096 for the main entry.
The authors show preliminary results of a study of hard and soft X–ray emission correlations in solar flares, using total fluxes as well as image information provided by the SMM spacecraft.

073.117   Distribution function for electron beams in the chromosphere.
P. J. D. Mauas, D. Gómez.
IAU Colloquium No. 133: Eruptive solar flares, p. 355 – 358 (1992). – See Abstr. 012.096 for the main entry.

073.118   Sub–second variations of HXR and H–alpha flare emission.
P. Heinzel, M. Karlický.
IAU Colloquium No. 133: Eruptive solar flares, p. 359 – 362 (1992). – See Abstr. 012.096 for the main entry.
For a series of electron beam pulses, the authors have computed the time–dependent chromospheric heating and the corresponding hard X–ray (HXR) flux. Moreover, by solving the time–dependent NLTE problem for hydrogen, they theoretically predict the Hα–line intensity variations on sub–second time scales. Both HXR–fluxes and Hα wing intensities do exhibit a spiky behaviour, consistent with short pulse–beam heating. However, the spikes in Hα are unexpectedly "inverse", i.e. the line intensity decreases during the beam heating. They correlate rather well with HXR emission peaks computed for 24 keV channel. The authors compare their theoretical results with recent observations of Kiplinger et al. (1991).

073.119   The role of protons in solar flares.
D. Heristchi, R. Boyer.
IAU Colloquium No. 133: Eruptive solar flares, p. 363 – 366 (1992). – See Abstr. 012.096 for the main entry.

073.120   Simultaneous Hα and microwave observations of a limb flare on June 20, 1989.
M. Graeter, T. A. Kucera.
IAU Colloquium No. 133: Eruptive solar flares, p. 372 (1992). Extended abstract. – See Abstr. 012.096 for the main entry.

073.121   Nonresonant ion–beam turbulence in solar flares.
F. Verheest.
IAU Colloquium No. 133: Eruptive solar flares, p. 373 – 376 (1992). – See Abstr. 012.096 for the main entry.
Special type III radio bursts are attributed to low–energy, super–Alfvénic proton beams. Such beams are unstable against scattering by resonant and nonresonant parallel waves. The latter may dominate and are of two types, with long wavelengths and linear polarization, or else with shorter wavelengths and right/left–hand polarization. Extensions of Fowler's theorem to unstable beam–plasmas indicate higher levels of magnetic field turbulence a shorter wavelengths, up to a sizeable fraction of the flare magnetic field. The beams become sub–Alfvénic before completely travelling down the loop, leading to considerable heating of the flare particles.

073.122   The emerging picture of eruptive solar flares.
P. A. Sturrock.
IAU Colloquium No. 133: Eruptive solar flares, p. 397 – 409 (1992). Also published in Comments Astrophys., Vol. 16, No. 2, p. 71 – 85 (May 1992) – see Abstr. 55.073.124. – See Abstr. 012.096 for the main entry.
This is a summary of the IAU Colloquium No. 133. The author describes the picture of eruptive solar flares as the result of the presentations at Iguazú.

073.123   Preliminary research on the hot spots of energy flares since Cycle 21.
Luo Baorong.
Proc. Astron. Soc. Aust., Vol. 9, No. 2, p. 318 – 319 (1991). – See Abstr. 012.090 for the main entry.
The time and spatial distributions of the "energy flare indices" which have been observed since Cycle 21 are analysed and 13 hot spots of energy flares during this period are given in this article. These active regions of the "hot spots" appear repeatedly where there erupted the energy flares accounting for 63.2 percent of the total indices. The characteristics of the hot spots of the energy flares and the relationship between the hot spots and the evolution of the large–scale magnetic fields are also further discussed in this paper.

073.124   The emerging picture of eruptive solar flares.
P. A. Sturrock.
Comments Astrophys., Vol. 16, No. 2, p. 71 – 85 (May 1992).
This is a reprint of a paper originally published as conference summary for IAU Colloquium No. 133 on "Eruptive Solar Flares", which took place on 2 – 6 August 1991 in Iguazu, Argentina (see abstract 073.122).

073.125   Unneutralised currents in solar flares.
D. B. Melrose.
Proc. Astron. Soc. Aust., Vol. 9, No. 2, p. 326 – 327 (1991). – See Abstr. 012.090 for the main entry.
Observations of the vector magnetic field imply that there is a net current flowing through the corona, and there is also evidence that this current is directly related to flares. It is argued that if the currents are unneutralised, then this requires a radical rethinking of several widely accepted ideas on solar magnetic fields and their role in solar flares.

073.126   The spectral observation of continuous emission of a solar flare on Jan. 18, 1989.
Xuan Jia–yu, Zhong Shu–hua, Luo Zhi, Lin Jun, Ye Hui–lian, Chen Xue–kun.
Chin. Sci. Bull., Vol. 37, No. 5, p. 391 – 396 (Mar 1992).

073.127   Spectral line asymmetries and Doppler shifts of the 1984 January 26 flare.
Ding Ming–de, Fang Cheng.
Chin. Sci. Bull., Vol. 37, No. 6, p. 479 – 483 (Mar 1992).

073.128 **Particle acceleration by magnetic reconnection and fast magnetosonic shock waves in solar flares.**
J. Sakai, Y. Ohsawa.
International Workshop on Reconnection in Space Plasma, p. 111 – 116 (Jan 1989). – See Abstr. 012.099 for the main entry.

The paper reviews recent development of the theory of current loop coalescence and fast magnetosonic shock waves, giving particular attention to particle acceleration caused by these processes.

073.129 **The relation between H–alpha and SXR emission of the flares.**
A. Antalová.
International Workshop on Reconnection in Space Plasma, p. 145 – 148 (Jan 1989). – See Abstr. 012.099 for the main entry.

An evolutionary coincidence between the occurence of the LDE flares and the formation of the evolving delta configuration, in a certain active region, was found.

073.130 **The precursors of the 9 July 1982 solar proton flare event.**
V. V. Zaitsev (*V. V. Zajtsev*), E. A. Averianikhina (*E. A. Aver'yanikhina*), H. Aurass, A. Krüger.
International Workshop on Reconnection in Space Plasma, p. 149 – 152 (Jan 1989). – See Abstr. 012.099 for the main entry.

The authors study some radio and X–ray precursor phenomena of the big 09 July 1982 flare and type IV burst event.

073.131 **The scales of solar microwave bursts and scenarios of flare energy release.**
A. Krüger, B. Kliem, J. Hildebrandt.
International Workshop on Reconnection in Space Plasma, p. 169 – 174 (Jan 1989). – See Abstr. 012.099 for the main entry.

Based on earlier observational evidence that characteristic time scales of different solar microwave burst types are distributed over a wide range ($10^{-3} – 10^4$sec), different mechanisms of energy release have been considered to account for the impulsive flux increase (time scale $<10^3$sec). Among different competing processes the coalescence instability is found to be a promising candidate to combine sufficiently short time scales with substantial energy release.

073.132 **An analogy between magnetospheric substorms and solar flares: new version.**
V. M. Mishin.
International Workshop on Reconnection in Space Plasma, p. 233 – 240 (Jan 1989). – See Abstr. 012.099 for the main entry.

The known analogy between magnetospheric substorms and solar flares is based on magnetic reconnection inside the active solar region as a main content of both phenomena. A new scenario takes into account, as an additional element, the reconnection on the active region boundary which promotes the energy input to this region from outside. So, a flare, like a substorm, is not only the unloading but the driven process too. It is emphasized that efficiency of the above energy input increases as $L_T^2 \approx h^2$ where L_T is the length of the closed magnetospheric tail, and h is height of arcades. Observational tests of the model are suggested.

073.133 **Flare instability and driving mechanism.**
P. Raychaudhuri.
1. COSPAR Colloquium: Physics of the outer heliosphere, p. 403 – 404 (1990). – See Abstr. 012.100 for the main entry.

The 1984 – 1987 Solar Maximum Mission event list.
See Abstr. 002.088.

The 1988 Solar Maximum Mission event list.
See Abstr. 002.089.

The 1989 Solar Maximum Mission event list.
See Abstr. 002.090.

Physical processes in solar flares.
See Abstr. 003.083.

Eruptive solar flares. Proceedings.
See Abstr. 012.096.

Reconnection in space plasma. Proceedings.
See Abstr. 012.099.

Preliminary calibration results for the BATSE instrument on CGRO.
See Abstr. 036.080.

Prominence sheets supported by constant–current force–free fields. II. Imposition of normal photospheric field component and prominence surface current.
See Abstr. 062.023.

Cylindrically symmetric force–free magnetic fields.
See Abstr. 062.024.

The effects of Alfvén waves on heating plasma in post–flare loops.
See Abstr. 062.040.

Chemically reacting flow of a compressible thermally radiating two–component plasma.
See Abstr. 062.045.

Dynamical regimes and the possibility of microflares in a prominence.
See Abstr. 062.049.

Dynamical regimes and possibility of origin of microflares in a prominence.
See Abstr. 062.080.

MHD equilibria with flows in uniform gravity. I. 1–D prominence– and arcade–type solutions.
See Abstr. 062.110.

The prominence–corona transition region in transverse magnetic fields.
See Abstr. 062.158.

Registration of macro– and micro–inhomogeneities of plasma density in astrophysical objects.
See Abstr. 062.167.

The structure of radiative slow–mode shocks.
See Abstr. 062.203.

Physics of flares in stars and accretion disks.
See Abstr. 062.219.

The evolution of disturbed neutral point equilibria.
See Abstr. 062.221.

Field opening and reconnection.
See Abstr. 062.224.

Energy release at Alfvénic fronts in a force–free magnetic flux tube.
See Abstr. 062.225.

Hanle effect with partial frequency redistribution. II. Linear polarization of the solar Ca I 4227 Å line.
See Abstr. 063.090.

Derivation of the master equation for the atomic density matrix for line polarization studies in the presence of magnetic field and depolarizing collisions in astrophysics.
See Abstr. 063.110.

Derivation of the radiative transfer equation for line polarization studies in the presence of magnetic field in astrophysics.
See Abstr. 063.111.

X-ray-heated models of stellar flare atmospheres: theory and comparison with observations.
See Abstr. 064.024.

On the dynamics of the emerging pre-flare magnetic configuration.
See Abstr. 064.086.

Pinch effect and stellar flares physics.
See Abstr. 064.121.

Solar hydrogen lines in the infrared.
See Abstr. 071.013.

The Sun.
See Abstr. 071.025.

The 77 day periodicity in the flare rate of cycle 22.
See Abstr. 072.023.

Longitudinal position of sunspots and chromospheric filaments. Solar rotation No. 1829.
See Abstr. 072.033.

Longitudinal position of sunspots and chromospheric filaments. Solar rotation No. 1830.
See Abstr. 072.034.

Longitudinal position of sunspots and chromospheric filaments. Solar rotation No. 1831.
See Abstr. 072.035.

Dynamics in solar active regions.
See Abstr. 072.053.

Fine structures in solar active regions.
See Abstr. 072.054.

Morphological evolution and motion of the photospheric sunspot group in AR 5312.
See Abstr. 072.055.

UV observations of sunspots.
See Abstr. 072.057.

Solar activity.
See Abstr. 072.063.

Active region classifications, complexity, and flare rates.
See Abstr. 072.065.

Statistical analysis of solar flare, neutrino flux and sunspot data.
See Abstr. 072.070.

Focused transport of energetic particles along magnetic field lines draped around a coronal mass ejection.
See Abstr. 074.006.

Martens-Kuin models of normal and inverse polarity filament eruptions and coronal mass ejections.
See Abstr. 074.015.

The X-ray corona, the coronal hole, and the heliosphere.
See Abstr. 074.019.

The birth of giant post-flare arches.
See Abstr. 074.028.

Variations in the relative elemental abundances of oxygen, neon, magnesium, and iron in high-temperature solar active-region and flare plasmas.
See Abstr. 074.032.

Steady siphon flows in closed coronal structures: comparison with extreme-ultraviolet observations.
See Abstr. 074.034.

Stereoscopic observations of a solar flare hard X-ray source in the high corona.
See Abstr. 074.042.

Magnetic reconnection associated with emerging magnetic flux.
See Abstr. 074.061.

Large-scale quasi-stationary X-ray coronal structures associated with eruptive solar flares.
See Abstr. 074.066.

Large scale structures associated with eruptive flares and radio waves.
See Abstr. 074.067.

Coronal millimeter sources associated with eruptive flares.
See Abstr. 074.068.

A giant post-flare coronal arch observed by Skylab.
See Abstr. 074.069.

The solar sources of coronal mass ejections.
See Abstr. 074.071.

Remote sensing observations of mass ejections and shocks in interplanetary space.
See Abstr. 074.072.

Microwave and soft X-ray emission of solar flare events associated with coronal mass ejections.
See Abstr. 074.089.

The practical application of the magnetic virial theorem.
See Abstr. 075.003.

Three-dimensional force-free magnetic fields and flare energy buildup.
See Abstr. 075.004.

Linear force-free magnetic field around quiescent solar prominences computed from observable boundary conditions.
See Abstr. 075.010.

Magnetic field configuration associated with solar gamma-ray flares in June 1991.
See Abstr. 075.014.

Magnetic shear in flaring regions. I. Quantitative evaluation of the change in shear.
See Abstr. 075.017.

Energy buildup in sheared force-free magnetic fields.
See Abstr. 075.022.

Electric currents and magnetic-field loops in solar active regions.
See Abstr. 075.028.

The cancellation of magnetic flux in the solar atmosphere.
See Abstr. 075.037.

Characteristics of the magnetic field of the active region of the flare of 1989 July 5.
See Abstr. 075.038.

Magnetic reconnection associated with emerging magnetic flux.
See Abstr. 075.039.

Magnetic energy conversion on the sun.
See Abstr. 075.042.

X–rays and inner–shell transitions in the solar atmosphere.
See Abstr. 076.001.

Solar flare gamma–ray line profiles.
See Abstr. 076.002.

C IV line ratios in the Sun.
See Abstr. 076.004.

Transitions from metastable levels emitted during short–duration bursts: how valid are their calculated intensities?
See Abstr. 076.005.

Mg IX line ratios in the Sun.
See Abstr. 076.006.

Solar Si II line ratios from the High–Resolution Telescope and Spectrograph.
See Abstr. 076.008.

COMPTEL neutron response at 17 MeV.
See Abstr. 076.010.

Ne V line ratios in the EUV spectra of solar flares.
See Abstr. 076.011.

The Sun as a star: high spectral resolution solar data degraded to low–dispersion IUE resolution.
See Abstr. 076.019.

Lower limits to the speed of transport of impulsive flare energy deduced from joint UVSP/HXRBS observations.
See Abstr. 076.021.

New statistical properties of solar X–ray and γ–ray bursts.
See Abstr. 076.023.

Multifrequency observations of a remarkable solar radio burst.
See Abstr. 077.003.

Two–dimensional model maps of flaring loops at cm–wavelengths.
See Abstr. 077.012.

Solar radio flares 1989 – 1991.
See Abstr. 077.022.

A thermal/nonthermal model for solar microwave bursts.
See Abstr. 077.028.

The quasi–periodic fine structures in a solar radio burst at 3.2 cm wavelength on 23 May 1990.
See Abstr. 077.032.

Observation and analysis of solar decimetric spike radiation.
See Abstr. 077.033.

The effect of the plasma background parameter on spike radiation.
See Abstr. 077.034.

Radio observations of the M8.1 solar flare of 23 June, 1988: evidence for energy transport by thermal processes.
See Abstr. 077.035.

Microwave and soft X–ray radiation during flares evolving in strong magnetic fields.
See Abstr. 077.036.

Rapid fluctuation in solar radio emission.
See Abstr. 077.039.

Meter–decameter radio emission associated with a coronal mass ejection.
See Abstr. 077.043.

Microwave flare characteristics in 8 and 3 mm Metsähovi measurements compared with optical and H–alpha data.
See Abstr. 077.046.

Nonlinear emission mechanism of type III solar radio bursts.
See Abstr. 077.047.

Diagnostics of energy release and magnetic fields on the sun by radio methods.
See Abstr. 077.049.

The solar flare event on 1990 May 24: evidence for two separate particle accelerations.
See Abstr. 078.003.

Energy spectra of ions from impulsive solar flares.
See Abstr. 078.006.

Composition and azimuthal spread of solar energetic particles from impulsive and gradual flares.
See Abstr. 078.011.

Trapping and escape of the high energy particles responsible for major proton events.
See Abstr. 078.020.

Coronal and interplanetary transport of solar flare protons from the ground level event of 29 September 1989.
See Abstr. 078.021.

Acoustic waves in the solar atmosphere. IX. Three minute pulsations driven by shock overtaking.
See Abstr. 080.004.

Observations of high–frequency and high–wavenumber solar oscillations.
See Abstr. 080.066.

Atmospheric electricity response to an isolated solar flare and to a series of flares.
See Abstr. 082.015.

An observing and analyzing for ionospheric effect during the period of the solar flare.
See Abstr. 083.019.

Nitric oxide and lower ionosphere quantities during solar particle events of October 1989 after rocket and ground–based measurements.
See Abstr. 083.025.

Ionospheric response of solar flares.
See Abstr. 083.042.

Terrestrial response to eruptive solar flares: geomagnetic storms.
See Abstr. 084.077.

Modeling the time–intensity profile of solar flare generated particle fluxes in the inner heliosphere.
See Abstr. 106.033.

Transient interplanetary and geomagnetic disturbances associated with solar flares.
See Abstr. 106.090.

Mg II absolute line profiles for late–type stars and for spatially–resolved solar regions.
See Abstr. 112.038.

The origin of the far–ultraviolet continuum in solar and stellar flares.
See Abstr. 116.007.

Electron densities and temperatures in solar and stellar atmospheres.
See Abstr. 116.067.

Stellar flares: observations and modelling.
See Abstr. 122.157.

Changes in the cosmic–ray power spectrum during solar flares.
See Abstr. 144.003.

074 Corona, Solar Wind

074.001   Solar radio corona.
M. Pick, G. Trottet.
11. European Regional Astronomy Meeting of the IAU: New windows to the universe, Vol. 1, p. 65 – 76 (1990). – See Abstr. 012.001 for the main entry.
This review is restricted to the structure of the corona including three topics: the quiet corona, the active corona and the topology of flaring sites. The interest of imaging radio observations with high spatial and time resolutions is emphasized.

074.002   Kosmos–900 satellite detection of high–energy electron fluxes associated with high–velocity recurrent flows of the solar wind.
E. V. Gorchakov, K. G. Afanas'ev, V. A. Iozenas, M. V. Ternovskaya.
Geomagn. Aeron., Vol. 30, No. 5, p. 721 – 722 (Apr 1990). English translation of Geomagn. Aehron., Tom 30, No. 5, p. 850 – 851 (1990).
For the first time the generation of electrons with energies of 15 – 30 MeV has been detected in a broad range of magnetic shells, including shells close to the boundary of the magnetosphere. The generation is linked with the entry of the Earth into a high–velocity recurrent flux.

074.003   Variability of coronal structures and ion components in the solar wind.
G. Zastenker, L. Avanov, Yu. Yermolaev (*Yu. Ermolaev*), P. Bochsler, Z. Němeček, J. Šafránková.
Czech. J. Phys., Vol. 41, No. 10, p. 1001 – 1008 (Oct 1991). Paper presented at the 7. STP Symposium, The Hague (Netherlands), 1990.
The variations of solar wind ion fluxes of protons and α–particles are studied in a wide timescale: from parts of a second to several months. A "persistence time" of about 60 hours was obtained for the large–scale variations of α–particles. Power density spectra of velocity, density and magnetic field were studied in the frequency range from 10^{-5} to 10^{-3}Hz. Middle–scale fluctuations of both protons and α–particles are close to each other and the spectrum for α–particles has a somewhat greater slope than that for protons. Estimates of the variations of the flux power density are given in the frequency range from 10^{-3} to 3 Hz.

074.004   Quasi–static evolution of coronal magnetic fields.
D. W. Longcope, R. N. Sudan.
Astrophys. J., Vol. 384, No. 1, p. 305 – 318 (1 Jan 1992).
A formalism is developed to describe the purely quasi–static part of the evolution of a coronal loop driven by its footprints. This is accomplished under the assumption of a long, thin loop. The quasi–static equations reveal the possibility for sudden "loss of equilibrium", at which time the system evolves dynamically rather than quasi–statically. Such quasi–static crises produce high–frequency Alfvén waves and, in conjunction with Alfvén wave dissipation models, form a viable coronal heating mechanism. Furthermore, an approximate solution to the quasi–static equations by perturbation methods verifies the development of small–scale spatial current structure.

074.005   Adiabatic cooling of solar wind electrons.
O. Sandbæk, E. Leer.
J. Geophys. Res., Vol. 97, No. A2, p. 1571 – 1580 (1 Feb 1992).
In thermally driven winds emanating from regions in the solar corona with base electron densities of $n_0 \geqslant 10^8 cm^{-3}$, a substantial fraction of the heat conductive flux from the base is transferred into flow energy by the pressure gradient force. The adiabatic cooling of the electrons causes the electron temperature profile to fall off more rapidly than in heat conduction dominated flows. Alfvén waves of solar origin, accelerating the basically thermally driven solar wind, lead to an increased mass flux and enhanced adiabatic cooling. The reduction in electron temperature may be significant also in the subsonic region of the flow and lead to a moderate increase of solar wind mass flux with increasing Alfvén wave amplitude. In the solar wind model presented here the Alfvén wave energy flux per unit mass is larger than in models where the temperature in the subsonic flow is not reduced by the wave, and consequently the asymptotic flow speed is higher.

074.006   Focused transport of energetic particles along magnetic field lines draped around a coronal mass ejection.
L. C. Tan, F. M. Ipavich, G. M. Mason, M. A. Lee, B. Klecker.
J. Geophys. Res., Vol. 97, No. A2, p. 1597 – 1608 (1 Feb 1992).
The transport of particles in interplanetary space can be dominated by adiabatic focusing of particle pitch angle distributions under conditions when the ratio of the particle scattering mean free path to the magnetic field focusing length, λ/L, is ~ 1. Since for the average Archimedean spiral field configuration, $L \sim 1$ AU at Earth orbit, the only reported events dominated by focusing are those nearly "scatter–free" events where $\lambda \sim 1$ AU. However, if the interplanetary magnetic field is distorted so that the focusing length L is small, then even if the interplanetary scattering mean free path were small, focused transport would be expected to dominate, since λ/L would still be ~ 1. The authors present evidence for such an event, obtained with observations of ~ 0.1–1 MeV nucleon^{-1} protons and alpha particles obtained with the University of Maryland/Max–Planck–Institut experiment on the ISEE 3 spacecraft during the decay phase of the June 6, 1979, solar particle event.

074.007   The magnetic topology of solar coronal structures following mass ejections.
S. W. Kahler, A. J. Hundhausen.
J. Geophys. Res., Vol. 97, No. A2, p. 1619 – 1632 (1 Feb 1992).
The bright radial structures observed in the solar corona for 1–2 days following a coronal mass ejection (CME) have traditionally been interpreted as unidirectional magnetic fields, commonly known as "legs", at the sides of the ejections. The

authors examine in detail the bright structures following 16 CMEs observed with the coronagraph on the Solar Maximum Mission (SMM) spacecraft and find that these structures can form anywhere within the lateral span of a CME, not only at the sides. The authors suggest that a more plausible interpretation is that the bright radial structures are the tops of coronal streamers containing magnetic neutral sheets across which the magnetic fields reverse direction. The observational support for this view is presented and discussed.

074.008   Decay instability of incoherent Alfvén waves in the solar wind.
H. Umeki, T. Terasawa.
J. Geophys. Res., Vol. 97, No. A3, p. 3113 – 3119 (1 Mar 1992).
   The nonlinear evolution of a large–amplitude incoherent Alfvén wave is studied via one–dimensional magnetohydrodynamic simulation. The initial wave magnetic field is given as a superposition of circularly polarized Alfvén waves with the same helicity and propagation direction but with different wave numbers. What one observes is contrary to the previous belief that incoherent Alfvén waves are stable against decay: In a low β plasma ($\beta = 0.2$) one can clearly see the growth of backscattered Alfvén waves, which are opposite in helicity and propagation direction from the original Alfvén waves. In a high β plasma ($\beta = 2.0$), on the other hand, no backscattered Alfvén waves are observed. These results are consistent with the expectation from the theory of the parametric decay instability developed for a coherent Alfvén wave. The authors also show that incoherent Alfvén waves in the solar wind can decay parametrically in the region of 4–20 R_s where β is sufficiently low. For the decay process to work, the period of Alfvén waves should be less than several minutes.

074.009   Proton and alpha particle fluxes in the solar wind: results of a three–fluid model.
A. Bürgi.
J. Geophys. Res., Vol. 97, No. A3, p. 3137 – 3150 (1 Mar 1992).
   A three–fluid model for the solar wind, with continuity, momentum, and energy equations for the three species protons, alpha particles, and electrons, including electron heat conduction and a parametrized coronal heat source, is used to study the behavior of the resulting proton and alpha fluxes as a function of coronal energy input and the proton–to–alpha density ratio at the coronal base.

074.010   Coronal Magnetic Structures Observing Campaign. IV. Multiwaveband observations of sunspot and plage–associated coronal emission.
J. W. Brosius, R. F. Willson, G. D. Holman, J. T. Schmelz.
Astrophys. J., Vol. 386, No. 1, p. 347 – 358 (10 Feb 1992).
   Simultaneous observations of an active region located near central meridian were obtained with the VLA, the SMM X–ray Polychromator, and the Beijing Observatory magnetograph on 1987 December 18, during the Coronal Magnetic Structures Observing Campaign. An asymmetric looplike structure connects the strong leading sunspot with a nearby region of opposite polarity. Both 6 and 20 cm emission lie along this structure, rather than over the sunspot, with higher frequency emission originating closer to the footpoint inside the sunspot. The 20 cm emission is due to a superposition of second – and third – harmonic gyroemission, where the field strength is 160 – 300 G, while the 6 cm emission is due to third – harmonic gyroemission from a region where the magnetic field strength ranges from 547 to 583 G. The X–ray data associated with an area of trailing plage were used to predict the brightness temperature structure due to thermal bremsstrahlung emission in the 6 and 20 cm wavebands. The predicted 6 cm brightness temperature in and around the location of the X–ray peak is low, consistent with the lack of observed 6 cm plage emission. The predicted 20 cm brightness temperature is consistent with that observed in the central portions of the plage, but the high 20 cm polarization requires the presence of cool ($T_e \leqslant 5 \times 10^5$K), absorbing plasma overlying the hot plasma observed in X–rays.

074.011   Loss of equilibrium in coronal loops.
R. M. Lothian, A. W. Hood.
Sol. Phys., Vol. 137, No. 1, p. 105 – 120 (Jan 1992).
   The loss of equilibrium in coronal magnetic field structures is a possible source of energy for coronal heating and solar flares. The authors investigate whether such a loss of equilibrium occurs when a coronal loop is progressively twisted by photospheric motions. In studies of 2–D cylindrical equilibria, long loops have been found to be of constant cross–sectional area along most of their length, with axial variations being confined to narrow boundary layers. The authors use this information to develop a 1–D line–tied model, for a 2–D coronal loop. They specify the twist in terms of the azimuthal field and more physically, in terms of the photospheric footpoint displacement. In the former case they find a loss of equilibrium, but not in the latter. The authors also examine a twisted loop with a non–zero plasma pressure. The loss of equilibrium is only found at high–plasma β. It is conjectured that such high–β can occur in flare loops and prior to a prominence eruption. However, when the plasma evolves adiabatically, there is no loss of equilibrium.

074.012   Relationships of quasi–stationary solar wind flows with their sources on the Sun.
V. G. Eselevich.
Sol. Phys., Vol. 137, No. 1, p. 179 – 197 (Jan 1992).
   By considering an example of four Carrington rotations (1671, 1672, 1681, and 1682), it is shown that there generally exists an exhaustive correspondence between quasi–stationary flows of "fast" and "slow" solar wind (SW), on the one hand, and their sources on the Sun: coronal holes (CHs) and the heliospheric current sheet (HCS), on the other. It is also shown that by knowing characteristics such as the coordinate of the center of gravity of CHs on the Sun, their areas S and the positions of the neutral line (NL) and of the HCS without the NL on the Sun, it becomes possible to calculate the time of appearance and the amplitude of three points on the SW velocity profile at the Earth's orbit. Calculated values agree with those observed.

074.013   Intensity oscillations of the Fe XIV ($\lambda = 530.3$ nm) solar corona: disturbances due to the wave processes in the Earth's atmosphere.
V. N. Dermendjiev, G. V. Kolarov, Ts. A. Mitsev.
Sol. Phys., Vol. 137, No. 1, p. 199 – 202 (Jan 1992). Letter–to–the–editor.
   The short period oscillations of the Earth's atmosphere transparency for the spectral line $\lambda = 510.6$ nm are studied using lidar data. The results obtained are discussed in connection with the problem of the reliability of short period intensity oscillations of the Fe XIV ($\lambda = 530.3$ nm) solar coronal emission line. The main conclusion is that the green coronal oscillations with periods shorter than 60 s are realistic.

074.014   The evolution of twisted coronal loops.
J. A. Robertson, A. W. Hood, R. M. Lothian.
Sol. Phys., Vol. 137, No. 2, p. 273 – 292 (Feb 1992).
   The evolution of coronal loops in response to slow photospheric twisting motions is investigated using a variety of methods. Firstly, by solving the time–dependent equations it is shown that the field essentially evolves through a sequence of 2–D equilibria with no evidence of rapid dynamic evolution. Secondly, a sequence of 1–D equilibria are shown to provide a remarkably good approximation to the 2–D time–dependent results using a fraction of the computer time. Thus, a substantial investigation of parameter space is now possible. Finally, simple bounds on the 3–D stability of coronal loops are obtained. Exact stability bounds can be found by using these bounds to reduce the region of parameter space requiring further investigation. Twisting the loop too much shows that a 3–D instability must be triggered.

074.015   Martens–Kuin models of normal and inverse polarity filament eruptions and coronal mass ejections.
D. F. Smith, E. Hildner, N. P. M. Kuin.
Sol. Phys., Vol. 137, No. 2, p. 317 – 328 (Feb 1992).
   An analysis is made of the Martens–Kuin filament eruption model in relation to observations of coronal mass ejections

(CMEs). The field lines of this model are plotted in the vacuum or infinite resistivity approximation with two background fields. The first is the dipole background field of the model and the second in the potential streamer model of Low. The assumption is made that magnetic field evolution dominates compression or other effects which is appropriate for a low–β coronal plasma. The Martens–Kuin model predicts that, as the filament erupts, the overlying coronal magnetic field lines rise in a manner inconsistent with observations of CMEs associated with eruptive filaments. An alternate case is considered in which the directions of currents in the Martens–Kuin model are reversed resulting in a so–called normal polarity configuration of the filament magnetic field. In this case, a neutral line occurs above the current–carrying filament. The background field lines now distort to support the filament and help eject it. While the vacuum field results make this configuration appear very promising, a full two– or more– dimensional MHD simulation is required to properly analyze the dynamics resulting from this configuration.

074.016 Concerning solar sources of large–scale heliospheric disturbances.
V. G. Fainshtein.
Sol. Phys., Vol. 137, No. 2, p. 329 – 343 (Feb 1992).

Initial observational data and their analysis made by Hewish and Bravo (1986) are examined critically. It is shown that the conclusion drawn by them that coronal holes are solar sources of all eruptive flows recorded in interplanetary space by the method of IPS–images using 900 radio sources, is ungrounded. At the same time it is argued that, under certain conditions, coronal holes are indeed able to emit eruptive flows which are capable of exciting interplanetary shock waves.

074.017 The influence of the anomalous cosmic–ray component on the dynamics of the solar wind.
H.-J. Fahr, H. Fichtner, S. Grzedzielski.
Sol. Phys., Vol. 137, No. 2, p. 355 – 383 (Feb 1992).

It is well known that both the galactic and anomalous cosmic rays show positive intensity gradients in the outer heliosphere which are connected with corresponding pressure gradients. Due to an efficient dynamical coupling between the solar wind plasma and these highly energetic media by means of convected MHD turbulences, there exists a mutual interaction between these media. As one consequence of this scenario the enforced pressure gradients influence the distant solar wind expansion. The authors concentrate on the interaction of the solar wind only with the anomalous cosmic–ray component. They use the standard two– fluid model in which the cosmic–ray fluid modifies the solar wind flow via the cosmic–ray pressure gradient. They derive numerical solutions and can show in their model, which fits the available observational data, radial decelerations of the distant solar wind by between 5 to 11% are to be expected.

074.018 Modifications of magnetohydrodynamics as applied to the solar wind.
D. Montgomery.
J. Geophys. Res., Vol. 97, No. A4, p. 4309 – 4310 (1 Apr 1992).

The effect of including the Braginskii viscous stress tensor in magnetohydrodynamics is remarked upon. It is shown that semiquantitative agreement with a recent observed anisotropy in the turbulent solar wind spectrum can be achieved in this way. The modifications of the dynamical equations are simple enough to permit their inclusion in numerical codes. The effects of large "ion parallel viscosities" also may be significant for plasmas in quite different regimes than the solar wind.

074.019 The X–ray corona, the coronal hole, and the heliosphere.
E. N. Parker.
J. Geophys. Res., Vol. 97, No. A4, p. 4311 – 4316 (1 Apr 1992).

The X–ray emission from the Sun arises primarily from the gas trapped in the bipolar magnetic fields of both small and large active regions. It appears that the trapped gas is heated by the intermittent dissipation of magnetic energy (nanoflares) at the current sheets that arise spontaneously in any magnetic field subject to continuous deformation. The solar wind issues from regions of weak field pushed open by the expanding corona. Most of the heat input is close to the Sun, in the first $1-2\,R_S$, raising the gas slowly out through the gravitational field and gradually accelerating it up through the speed of sound at a distance of perhaps $3-5\,R_S$. The waves generated by photospheric convection dissipate only at distances of $5\,R_S$ and beyond, where their heat input and momentum accelerate the wind to the high velocities of 600–800 km/s sometimes observed. The only source for the principle heat input close to the Sun appears to be the network activity, as suggested by Martin, Porter, and Moore. Thus the mass loss and the formation of the heliosphere are primarily a consequence of the smallest–scale activity supplemented by occasional flares and coronal mass ejections. The X–ray emission is largely a consequence of the smallest flares, the nanoflares, supplemented by occasional X–ray bursts from large flares. It is presumed that the mass loss and associated circumstellar spheres and the X–ray emission from most other stars arise in the same manner.

074.020 Intermediate–term periodicities in the green corona brightness of the Sun.
A. Özgüç.
Astrophys. Space Sci., Vol. 187, No. 2, p. 197 – 207 (Jan 1992).

The author has analysed intermediate–term periodicities in the green corona by dividing 10° latitudinal belts for the solar cycles, 18, 19, and 20 (1947–1976). Discrete Fourier transform technique was used and three noticeable periodicities (3.48, 2.57, and 2.27 years) were found. The physical origin of these periods is not known, but evidence in the results exclude the possibility that the observed periods are a harmonic due to the method of analyse. The period of 3.48–year is the strongest one. 17.6–month periodicity was found only on around +40° belt while 155–day periodicity was not found in the analysis.

074.021 Counterstreaming solar wind halo electron events: solar cycle variations.
J. T. Gosling, D. J. McComas, J. L. Phillips, S. J. Bame.
J. Geophys. Res., Vol. 97, No. A5, p. 6531 – 6535 (1 May 1992).

Previous work has suggested that counterstreaming solar wind halo electron events are a reliable signature of coronal mass ejections (CMEs) in the solar wind at 1 AU. Using observations of counterstreaming events made by the Los Alamos plasma experiment on ISEE 3/ICE, the authors have determined the percentage of time that counterstreaming was observed each year during an ~12.4 year interval extending from August 1978 through December 1990. These annual percentages vary roughly in phase and amplitude both with the 11–year sunspot cycle and with annual CME rates derived from coronagraph observations. The total variation in annual percentages over the solar cycle was approximately a factor of 21. Near solar activity maximum, counterstreaming events accounted for ~14.7% of all ISEE 3/ICE solar wind measurements, while near solar minimum they accounted for ~0.7% of all the measurements. Inferred rates of counterstreaming events at a single point in the ecliptic plane ranged from ~72/yr near solar maximum to less than 12/yr near solar minimum. These inferred rates are in reasonable quantitative accord with predicted rates for CME events in the ecliptic plane derived from SMM coronagraph observations, further supporting the thesis that most CMEs in the solar wind at 1 AU can be reliably identified by the counterstreaming electron signature.

074.022 The equilibrium of coronal flux tubes under toroidal forces.
J. Juan, J. L. Ballester.
Astrophys. Space Sci., Vol. 188, No. 2, p. 279 – 288 (Feb 1992).

The properties of slender isolated flux tubes, taking into account curvature effects, were investigated by Parker (1975, 1979) and Spruit (1981), and many studies have been made concerning the equilibrium of slender flux tubes in the solar corona. In this paper the authors use a different approach considering the coronal loop as a part of a circular torus and studying the position of its top when the loop is in equilibrium under toroidal forces. Toroidal forces were considered by

Shafranov (1966) for toroidal pinches and the equilibrium can be studied for different values of the toroidal current intensity and external magnetic field. The results show that it is possible to have a coronal flux tube in equilibrium without considering gravity and external magnetic field. Furthermore, the total twist of the flux tube and its variation with the toroidal intensity has been studied.

074.023   Corona and solar wind in a quasidipole magnetic field.
    I. V. Chashej.
Sov. Astron., Vol. 36, No. 1, p. 98 – 102 (Jan 1992). Current Physics Microform No.: 9207Y0662. English translation of Astron. Zh., Tom 69, Vyp. 1, p. 192 – 200 (Jan–Feb 1992).
    The formation of the corona and solar wind by an Alfvén wave energy source in a quasidipole magnetic field is considered. The coronal temperature and the height of the cusp line above closed equatorial regions and the coronal temperature and asymptotic flow velocity in polar regions are estimated. The main results of the model are in agreement with observations for a solar activity minimum.

074.024   Spatial profiles of lines in steady–state coronal loops.
    S. Chandra, L. Prasad.
J. Quant. Spectrosc. Radiat. Transfer, Vol. 47, No. 6, p. 533 – 537 (Jun 1992).
    A steady–state model of the magnetically confined plasma in coronal loops is presented by expressing the magnetic field and fluid velocity in terms of a single Chandrasekhar–Kendall function. For the Chandrasekhar–Kendall function with $n = 0 = m$, the authors found a radial variation of pressure. For constant electron density, the temperature increases radially outwards. The variations of the fluxes for lines from C II, C III, O IV, Ne VII, and Mg X along the radius of the loop are plotted. The hotter lines generated near the surface of the loop show larger spatial extents than the cooler lines emitted in the core of the loop, which is in agreement with observations.

074.025   Coronal density and temperature structure from coordi-
    nated observations associated with the total solar eclipse
    of 1988 March 18.
M. Guhathakurta, G. J. Rottman, R. R. Fisher, F. Q. Orrall, R. C. Altrock.
Astrophys. J., Vol. 388, No. 2, p. 633 – 643 (1 Apr 1992).
    This paper explores and compares diagnostics for temperature and density within large–scale structures of the inner corona, based on cospatial and cotemporal spectrophotometric observations made at the time of the total solar eclipse of 1988 March 17/18. The basic data consists of the following: coronal soft X–ray images near 173 Å observed from a sounding rocket; eclipse observations of the K–corona in white light and in the green coronal line (5303 Å, Fe XIV); and ground–based corona-graph observations of the green and red (6374 Å, Fe X) coronal lines. The K–coronal observations alone provide a measure of electron density and of the scale–height temperature T_s, both as functions of position angle. The plasma temperature T can be derived unambiguously from the intensity ratios Fe XIV/XUV or Fe XIV/Fe X, since all the emission lines come from the ionized state of Fe and the ratios are only weakly dependent on density. These temperatures and the densities found in well–defined large–scale coronal structures are discussed. The emission–line temperature is found to be high in the coronal structures with enhanced white–light emission and associated with new cycle high–latitude magnetic fields separated from the old cycle polar field of opposite polarity. Also the average of the ratio of T_s/T over the entire range of position angles is roughly unity, although the ratio is higher than unity (1.3 – 1.6) in the three most prominent streamers.

074.026   Current instability and Alfve'n waves in coronal loops.
    P. P. Malovichko, A. K. Yukhimuk.
Kinematika Fiz. Nebesn. Tel, Tom 8, No. 1, p. 20 – 23 (Jan–Feb 1992). In Russian. English translation in Kinematics Phys. Celest. Bodies, Vol. 8, No. 1.
    Current instability of MHD–waves in coronal loops is analysed. It is shown that the instability may develop effectively

only in the corona, where plasma density is rather low. Disturbance development can cause magnetic field reconnection to stimulate the flare creation.

074.027   Structure of the solar corona, and the heliospheric
    current sheet.
R. A. Gulyaev, B. P. Filippov.
Sov. Phys. – Dokl., Vol. 37, No. 1, p. 4 – 5 (Jan 1992). Abstract. Current Physics Microform No.: 9207X2088.

074.028   The birth of giant post–flare arches.
    G. Poletto, Z. Švestka.
Sol. Phys., Vol. 138, No. 1, p. 189 – 199 (Mar 1992).
    Using short accumulation times, the authors have succeeded in the detection in HXIS images of the initial growth of the giant post–flare arch of 6 November, 1980 and part of the initial growth of the giant arch of 7 November, 1980. These observa-tions are relevant to the problem of the origin of giant arches: the authors discuss some models.

074.029   Instabilities in the solar corona.
    A. W. Hood.
Plasma Phys. Controlled Fusion, Vol. 34, No. 4, p. 411 – 442 (Apr 1992).
    This review begins with an introduction to some of the various solar phenomena that may be investigated using MHD stability theory. The geometry of the coronal magnetic field for many models is either a loop or an arcade and it is the stability properties of these structures that are investigated. Section 2 presents a simple physical description of the basic MHD instabilities and describes the main difference between laboratory and coronal plasmas due to the presence of an extremely dense plasma at the footpoints of the coronal magnetic fields. The implications of this density interface are discussed. In Section 3 linear stability theory is applied to some solar situations and the more recent non–linear simulations are discussed in Section 4. For MHD instabilities, it is the presence of the cool, dense lower layers of the Sun's atmospheres that provides the main difference between laboratory and solar plasmas. By photospheric line-tying the Sun is able to stabilize the stressed coronal magnetic field in a stable configuration for periods between a day or so, to months on end.

074.030   Solar wind from a corona with a large helium abundance.
    E. Leer, T. E. Holzer, E. C. Shoub.
J. Geophys. Res., Vol. 97, No. A6, p. 8183 – 8201 (1 Jun 1992).
    Observations of quasi–steady high–speed solar wind streams show that the proton mass flux density at 1 AU is remarkably constant, varying by less than 10% over long time periods. The observations are problematic, for simple theoretical models predict that the proton mass flux density is a sensitive function of the coronal base temperature, which is not expected to be unvarying to the degree required by the observations. In this paper the authors investigate the possibility that the presence of alpha particles in the coronal base region can reduce the sensitivity of the proton mass flux to base temperature. The equations of mass and momentum conservation are solved for electrons, protons, and alpha particles using a variety of assumed temperature profiles for each species. A wide range of base conditions are considered. The authors find that for an alpha particle to proton density ratio at the base as small as 10%, alpha particles can reduce the sensitivity of the proton mass flux density to variations in the base temperature. The authors also study the effects of enhanced collisional coupling and of Alfvén waves on the flux of protons and alpha particles. As an aid to future observational determination of the alpha particle density in the corona, they present calculations of the intensities of the resonantly scattered lines He II λ304 and H I λ1216 for selected models.

074.031   Corona and solar wind in a quasidipole magnetic field.
I. V. Chashej.
Astron. Zh., Tom 69, Vyp. 1, p. 192 – 200 (Jan–Feb 1992). In Russian. English translation in Sov. Astron., Vol. 36, No. 1 (1992).

The corona and solar wind formation by an Alfven wave energy source in a quasidipole magnetic field is considered. The estimates of coronal temperature and cusp line altitude above the closed nearequatorial regions, coronal temperature and asymptotic flow speed in the open polar regions are obtained. The main consequences of the model are in agreement with the observational data for the minimum solar activity period.

074.032   Variations in the relative elemental abundances of oxygen, neon, magnesium, and iron in high–temperature solar active–region and flare plasmas.
D. L. McKenzie, U. Feldman.
Astrophys. J., Vol. 389, No. 2, p. 764 – 776 (20 Apr 1992).

The authors have used X–ray spectral line ratios that are weakly dependent upon temperature to search for evidence of elemental abundance variations in the solar corona. Using spectral lines from Ne IX, O VIII, Fe XVII, Fe XVIII, and Mg XI, the authors found evidence for Fe/Ne abundance ratio variations of more than a factor of 4, Fe/O abundance ratio variations of factors of 2.4 – 3.1, and O/Ne abundance ratio variations of approximately a factor of 2. These last variations are the first seen for the abundance ratio of two elements in the "high–FIP" group (those having first ionization potential (FIP) greater than 10 eV). No evidence was found for abundance ratio variations for Fe and Mg, two low–FIP elements, but the statistics are such that variations as large as a factor of 1.5 cannot be ruled out. The data suggest that a mechanism that favors the promotion of charged species to the corona is operative in the upper chromosphere at a temperature of $\sim 1.5 \times 10^4$ K. The authors argue that the minimum observed low–FIP / high–FIP abundance ratios can be no lower than the photospheric ratios. They suggest that the abundance of iron in the photosphere corresponds to that measured in C1 chondrites rather than the higher value obtained by spectroscopic techniques.

074.033   Structural changes in the solar corona during the July 1991 eclipse.
J. B. Zirker, S. Koutchmy, C. Nitschelm, G. Stellmacher, J. P. Zimmermann, P. Martinez, I. Kim, N. Dzubenko (*N. I. Dzyubenko*), L. Kurochka (*L. N. Kurochka*), V. Makarov (*V. I. Makarov*), M. Fatianov (*M. P. Fat'yanov*), V. Rusin, L. Klocok, O. T. Matsuura.
Astron. Astrophys., Vol. 258, No. 2, p. L1 – L4 (May 1992). Letter–to–the–editor.

Preliminary results of the analysis of radially filtered pictures obtained at several hours interval during the July 11, 1991 total solar eclipse are given. Instrumental parameters and method are outlined. Structural changes are discussed.

074.034   Steady siphon flows in closed coronal structures: comparison with extreme–ultraviolet observations.
G. Peres, D. Spadaro, G. Noci.
Astrophys. J., Vol. 389, No. 2, p. 777 – 783 (20 Apr 1992).

The authors have computed models of steady siphon flows in coronal loops, in order to compare the intensities of some EUV transition region emission lines synthesized from these models with observations of typical solar regions. In particular, the authors have examined the C III (977 Å), O IV (554 Å), O V (630 Å) and O VI (1032 Å) lines, performing a detailed fit of previously published data concerning a faint loop structure extending over a typical active region, loops in the core of an active region, and a large–scale loop structure interconnecting active regions. The purpose is to explore whether siphon flow loop models can improve the fitting of transition region lines with respect to static models, since static models can hardly fit the data for the C III line. The authors have calculated the line emission using ion densities determined by considering nonequilibrium ionization. Comparing the present fit with that relative to the same data previously obtained with static coronal loop models,

the authors find that siphon flow models of compact active region loops are in better agreement with observations of EUV transition region lines, while for large loops interconnecting different active regions static models work slightly, but not significantly, better.

074.035   Variable carbon and oxygen abundances in the solar wind as observed in Earth's magnetosheath by AMPTE/ CCE.
R. von Steiger, S. P. Christon, G. Gloeckler, F. M. Ipavich.
Astrophys. J., Vol. 389, No. 2, p. 791 – 799 (20 Apr 1992).

During periods of enhanced solar wind pressure, the dayside boundary of Earth's magnetosphere occasionally was compressed below the apogee of the AMPTE/CCE satellite. Then, while in the magnetosheath, the CHEM instrument on CCE sampled the charge state composition of the solar wind thermalized by the bow shock. The authors present a comprehensive list of the solar wind oxygen and carbon charge states and abundances measured in all magnetosheath periods during the lifetime of CCE, 1984 – 1988. Two different charge state ratios, C^{6+}/C^{5+} and O^{7+}/O^{6+}, are used to infer the coronal source temperature of the solar wind flow responsible for the magnetosphere compression. In most cases, the authors can also identify the solar source feature (coronal hole, flare, or filament) of the flow. The abundance ratio C/O is found to be highly variable between different flow intervals. Specifically, there is a strong correlation between the inferred coronal source temperature and C/O: While the flows emanating from coronal holes have an average value consistent with the photospheric C/O value, the flows from flares or disappearing filaments have a higher average C/O value than the photosphere. Also, the range of C/O values in the latter flows is much greater than the range of coronal hole C/O values.

074.036   Solar corona 22 July 1990. Preliminary information on photographic observations made by the Kiev University expedition in Markovo, Magadan region.
N. I. Dzyubenko, G. A. Rubo, V. V. Bondarchuk.
Astron. Tsirk., No. 1551, p. 24 – 25 (Nov – Dec 1991). In Russian.

The results of the solar corona photographic observations on July 22, 1990 in Markovo (Siberia) are presented. 10–meter coronograph, 3–meter coronograph with radial–gradient neutral density filter and camera "Salut" had been used. The quality of the negatives and colored positive coronal images is good. The general structure up to 5 $R_\odot$ of the quasi–maximum solar corona of the July 22, 1990 eclipse is discussed.

074.037   Estimation of the mass of a coronal mass ejection from radio observations.
N. Gopalswamy, M. R. Kundu.
Astrophys. J., Lett., Vol. 390, No. 1, p. L37 – L39 (1 May 1992). With plate L5.

The authors estimate the mass of a coronal mass ejection (CME) using meter–decametric observations obtained with the Clark Lake multifrequency radioheliograph. Mass estimates in the past have been made using coronagraph and white–light photometer observations. Since the radiation at radio and optical wavelength regimes has different physical origins, the radio method may provide an independent check on the mass estimates. The present estimate of the 1986 February 16 CME using the radio method is close to the average value of CME masses reported in the literature.

074.038   Role of periodic fluctuations in solar wind–magnetosphere coupling.
A. K. Murali, S. R. P. Nayar, V. V. Somayajulu.
J. Atmos. Terr. Phys., Vol. 53, No. 10, p. 881 – 887 (Oct 1991).

In this paper, the behaviour of the fluctuating component in the solar wind parameters (V, B_z), the auroral electrojet indices (AU, AL), the ring current index (D_{st}) and the interplanetary electric field (V × B_z) during 10 magnetic storms is analysed to understand the solar wind–magnetosphere coupling. It is found that during the moderate storms ($D_{st} > -100$ nT), the fluctuating component of 3–4 h periodicity is clearly discernible in all the parameters, and during the intense storms ($D_{st} < -100$ nT) the periodic fluctuations are not well defined.

074.039 The evolution of line–tied coronal arcades including a converging footpoint motion.
B. Inhester, J. Birn, M. Hesse.
Sol. Phys., Vol. 138, No. 2, p. 257 – 281 (Apr 1992).

It has been demonstrated in the past that single, two–dimensional coronal arcades are very unlikely driven unstable by a simple shear of the photospheric footpoints of the magnetic field lines. By means of two–dimensional, time–dependent MHD simulations, the authors present evidence that a resistive instability can result if in addition to the footpoint shear a slow motion of the footpoints towards the photospheric neutral line is included. Unlike the model recently proposed by van Ballegooijen and Martens (1989), the photospheric footpoint velocity in the model is nonsingular and the shear dominates everywhere. Starting from a planar potential field geometry for the arcade, the authors find that after some time a current sheet is formed which is unstable with respect to the tearing instability. The time of its onset scales with the logarithm of the magnetic diffusivity assumed in the calculation. In its nonlinear phase, a quasi–stationary situation arises in the vicinity of the x–line with an almost constant reconnection rate. The height of the x–line above the photosphere and the distance of the separatrix footpoints remain almost constant in this phase, while the helical flux tube, formed above the neutral line, continuously grows in size.

074.040 Flare–associated high–speed solar plasma streams.
H. O. Vats.
Sol. Phys., Vol. 138, No. 2, p. 379 – 386 (Apr 1992).

Characteristics of flare–associated high–speed solar plasma streams are investigated using measurements from space probes and Earth–orbiting spacecraft for the period 1964 – 1982. The maximum observed velocity (V_m) of these streams range from 400 to 850 km s^{-1} with peak probability for $\sim$600 km s^{-1}. These remained for a period of 1 – 10 days (peak $\sim$3 days). The difference between the pre–stream velocity (V_0) and the maximum velocity (V_m) of any high–speed stream serves as a measure of its intensity. The yearly percentage occurrence, total duration and the product of mean (V_m–V_0) with total duration of the high–speed streams during the year correlates well with solar activity, e.g., maximum during high solar activity period and minimum during low solar activity. The study suggests that the presence of sunspots plays a significant role in the generation of flare associated high–speed solar streams.

074.041 A self–consistent turbulent model for solar coronal heating.
J. Heyvaerts, E. R. Priest.
Astrophys. J., Vol. 390, No. 1, p. 297 – 308 (1 May 1992).

The rate of solar coronal heating induced by the slow random motions of the dense photosphere is calculated in the framework of an essentially parameter–free model. This model assumes that these motions maintain the corona in a state of small–scale MHD turbulence. The associated dissipative effects then allow a large–scale stationary state to be established. The solution for the macroscopic coronal flow and the heating flux is first obtained assuming the effective (turbulent) dissipation coefficients to be known. In a second step these coefficients are calculated by the self–consistency argument that they should result from the level of turbulence associated with this very heating flux. For the sake of tractability the derivation is restricted to a two–dimensional situation where boundary flows are translationally symmetric. The resulting value of the heating rate and the predicted level of microturbulent velocity compare satisfactorily with the observational data.

074.042 Stereoscopic observations of a solar flare hard X–ray source in the high corona.
S. R. Kane, J. McTiernan, J. Loran, E. E. Fenimore, R. W. Klebesadel, J. G. Laros.
Astrophys. J., Vol. 390, No. 2, p. 687 – 702 (10 May 1992).

The hard X–ray burst observed on 1984 February 16 by the 5 keV – 3.2 MeV X–ray spectrometer aboard the ICE/ISEE 3 spacecraft was associated with a flare which occurred $\sim$40° behind the west limb of the Sun. The X–ray source directly associated with the flare was partially occulted by the photosphere from the view of the ICE instrument. The occultation height above the solar limb was $\sim$1.6 × 10^{5}km. The occultation height for the GOES soft X–ray (0.5 – 8 Å) monitor, which also observed the flare, was $\sim$2.1 × 10^{5}km. The flare was in full view of the 100 keV – 2 MeV gamma–ray burst detector on the PVO spacecraft which was located at that time $\sim$16° behind the west limb of the Sun. The X–ray sources observed by the ICE and GOES instruments are identified as the unocculted coronal parts of the primary X–ray source associated with the flare. It is found that the sources of the impulsive hard X–ray ($\gtrsim$25 keV) and impulsive soft X–ray (2 – 5 keV) emissions in this flare extended to coronal altitudes $\gtrsim$2 × 10^{5}km above the photosphere. During the impulsive phase, the average rate of injection of electrons >5 keV was $\sim$6 × 10^{36} electrons s^{-1}, the total number injected being $\sim$10^{39}. The rate of energy injection and the total injected energy were $\sim$10^{29}ergs s^{-1} and $\sim$10^{31}ergs, respectively. The volume of the impulsive hard X–ray source in the corona is estimated to be $\gtrsim$10^{30}cm^3 with an average density of $\lesssim$10^9 ions cm^{-3}, indicating a large diffuse hard X–ray source in the corona.

074.043 On an inhomogeneous model of the inner corona of the Sun.
O. G. Badalyan.
Astron. Zh., Tom 69, Vyp. 2, p. 377 – 382 (Mar–Apr 1992). In Russian. English translation in Sov. Astron., Vol. 36, No. 2 (1992).

An analytical inhomogeneous model of the inner corona is proposed, consisting of a dense hot component with temperature $T = 2 \cdot 10^6$K and density $n_0 = (1-2) \cdot 10^9$cm^{-3} and of a low–density cool one (background component) with $T = 1.4 \cdot 10^6$K and $n_0 = (1-4) \cdot 10^8$cm^{-3}. The fraction of hot elements has been defined, using the data on white–light corona brightness, which proves to be equal to $\sim$10% and decreases inversely proportional to the square of the distance from the Sun's center. The model considered does not contradict observations of the green coronal line λ5305 Å, carried out in not very active regions of the corona.

074.044 Inhomogeneous model of the inner solar corona.
O. G. Badalyan.
Sov. Astron., Vol. 36, No. 2, p. 190 – 192 (Mar 1992). Current Physics Microform No.: 9208X2026. English translation of Astron. Zh., Tom 69, Vyp. 2, p. 377 – 382 (1992).

An analytic inhomogeneous model is suggested for the inner solar corona, which includes a dense hot component with temperature $T = 2 \cdot 10^6$K and density $n_0 = (1-2) \cdot 10^9$cm^{-3} and a tenuous cool (background) component with $T = 1.4 \cdot 10^6$K and $n_0 = (1-4) \cdot 10^8$cm^{-3}. From data on the brightness of the white–light corona, it is found that the fraction of the hot elements is $\sim$10% by volume at the base of the corona and is inversely proportional to the square of the distance from the center of the Sun. This model is consistent with observations of the 5303 Å green coronal line in low–activity regions of the corona.

074.045 Theoretical analysis of the Fe XVIII X–ray spectrum and application to solar coronal observations.
M. Cornille, J. Dubau, M. Loulergue, F. Bely–Dubau, P. Faucher.
Astron. Astrophys., Vol. 259, No. 2, p. 669 – 681 (Jun 1992).

A theoretical study of Fe XVIII X–ray spectrum is presented and a comparison of the corresponding atomic parameters with other calculations is done. Results are used to construct different theoretical spectra which are compared to a solar active–region spectrum in the 13–19 Å range. From this analysis, new Fe XVIII line identifications are proposed.

074.046 Intensité de la couronne solaire en lumière monochromatique, selon des angles de position variant de 5° en 5°.
Jan – Dec 1989.
Q. Bull. Sol. Act., Vol. 31, Part IV, p. 1 – 31 (1989).

Contributing observatories: Norikura, Kislovodsk, Lomnický Štit.

074.047 **The variability of elemental abundances in the upper solar atmosphere.**
U. Feldman.
AIP Conf. Proc., No. 257, p. 171 – 180 (15 May 1992). Current Physics Microform No.: 9207G1891. Paper presented at the 8. Biennial APS Topical Conference on Atomic Processes in Plasmas, Portland, ME (USA), 25 – 29 Aug 1991.
Coronal elemental abundances are found to change by as much as an order of magnitude relative to those present in the solar photosphere. Observations of modifications in coronal elemenal abundances are reviewed and a tentative model governing the changes is discussed.

074.048 **The hydrodynamic picture of the solar wind flow around the Earth's magnetosphere and a comet in the quasi-turbulent regime.**
N. G. Ptitsyna, Z. A. Kereselidze.
Geomagn. Aeron., Vol. 31, No. 1, p. 135 – 136 (Aug 1991). English translation of Geomgn. Aehron., Tom 31, No. 1, p. 185 – 187 (1991).
Long–period (T ≳ 300 seconds) geomagnetic field pulsations in the Earth's magnetosphere are modeled on the basis of a forced oscillation representation. The oscillation generator is considered to be a large–scale eddy formed during the flow of the solar wind around the magnetosphere. With this model representation the characteristic pulsation periods as a function of the solar wind parameters and criteria characterizing the quasi–turbulent flow regime can be found. To visualize the flow around the magnetosphere, use is made of the data on comets, for which rotations and oscillations of the tail relative to the Sun–Earth line are frequently observed. A joint analysis is made of specific events – the sudden rotation of the tail of the Bradfield comet and pulsations at the Earth, 6 February 1980.

074.049 **Structure and dynamics of cool flare loops.**
P. Heinzel, B. Schmieder, P. Mein.
Sol. Phys., Vol. 139, No. 1, p. 81 – 104 (May 1992).
MSDP observations of the 16 May, 1981 two–ribbon flare are used to study the physical structure and the dynamical behaviour of cool flare loops. Using the first–order differential cloud model (DCM) technique, the authors derive empirically some basic plasma parameters at 15 points along one loop leg. The flow velocities and the true heights have been reconstructed with respect to a geometrical projection. Subsequently, detailed non-LTE models of cool loops have been constructed in order to fit Hα source function values previously derived from DCM1 analysis. It is demonstrated that this source function is rather sensitive to the radial component of the flow velocity (the so–called Doppler brightening) and to enhanced irradiation of the loops from the underlying flare ribbons. The authors have been able to estimate quantitatively all plasma parameters which determine the physical structure of cool loops as well as the momentum–balance condition within the loops. For these dark loops they have arrived relatively low gas pressures of the order of $0.1 - 0.5$ dyne cm^{-2} with corresponding electron densities around 10^{11}cm^{-3}. Pressure–gradient forces have been found to be of small importance in the momentum–balance equation, and thus they cannot explain departures from a free–fall motion found in the MSDP data analysis. Three possible solutions to this problem are proposed.

074.050 **On neutralized currents in the solar corona.**
L. K. Wilkinson, A. G. Emslie, G. A. Gary.
Astrophys. J., Lett., Vol. 392, No. 1, p. L39 – L42 (10 Jun 1992).
Using Ampere's integral law applied to vector magnetograph transverse field data, the authors have analyzed the current pattern in an active region associated with flaring activity. The greater sensitivity of this integral (rather than differential) formalism of Ampere's law may allow one to identify regions of weak return current around the actual positive current region. The effects of Faraday rotation on the inferred magnetic field measurements, and hence on the deduced current pattern, are discussed. It is concluded that the current patterns in the region consist either of a neutralized (return) current system, well

modeled by a coaxial cable, or of a potential (current–free) field, with the apparent currents in the core an artifact induced by magneto–optical effects. In either case, the data do not convincingly show an unneutralized current system.

074.051 **Study of formation of the supersonic solar wind.**
N. A. Lotova, O. A. Korelov, Ya. V. Pisarenko.
Geomagn. Aehron., Tom 32, No. 3, p. 78 – 84 (May – Jun 1992). In Russian. English translation in Geomagn. Aeron., Vol. 32, No. 3.

074.052 **On potential field models of the solar corona.**
Y.-M. Wang, N. R. Sheeley Jr.
Astrophys. J., Vol. 392, No. 1, p. 310 – 319 (10 Jun 1992).
The current–free approximation has been widely used to infer the magnetic structure of the solar corona from magnetograph observations. In such calculations, the potential field is usually matched directly to the observed line–of–sight component of the photospheric field. However, this procedure is invalid because the magnetograph measurements apply to deep atmospheric layers where the magnetic field is nonpotential and nearly radial. A better approach is to match only the radial component of the potential field to the photospheric data (corrected for line–of–sight projection on the assumption that the field lines are radially oriented), while allowing a discontinuity in the tangential field components. The implied current sheet is a mathematical idealization of the finite "boundary layer" within which the rapidly diverging field lines make their transition to a potential configuration. This procedure yields much stronger polar fields than the conventional line–of–sight method and resolves a number of discrepancies between earlier potential field calculations and observations of polar coronal holes, the brightness distribution of the K–corona, and the shape of the interplanetary current sheet.

074.053 **Evidence for mass outflow in the low solar corona over a large sunspot.**
W. M. Neupert, J. W. Brosius, R. J. Thomas, W. T. Thompson.
Astrophys. J., Lett., Vol. 392, No. 2, p. L95 – L98 (20 Jun 1992). With plate L5.
Spatially resolved extreme ultraviolet (EUV) coronal emission–line profiles have been obtained in a solar active region, including a large sunspot, using an EUV imaging spectrograph. Relative Doppler velocities were measured in the lines of Mg IX, Fe XV, and Fe XVI with a sensitivity of $2 - 3$ km s^{-1} at 350 Å. The only significant Doppler shift occurred over the umbra of the large sunspot, in the emission line of Mg IX (at $T_e \approx 1.1 \times 10^6$K). The maximum shift corresponded to a peak velocity toward the observer of 14 ± 3 km s^{-1} relative to the mean of measurements in this emission line made elsewhere over the active region. The magnetic field in the low corona was aligned to within 10° of the line of sight at the location of maximum Doppler shift. Depending on the closure of the field, such a mass flow could either contribute to the solar wind or reappear as a downflow of material in distant regions on the solar surface. The site of the source, near a major photospheric field boundary, was consistent with origins of low–speed solar wind typically inferred from interplanetary plasma observations.

074.054 **Element abundances and plasma properties in a coronal polar plume.**
K. G. Widing, U. Feldman.
Astrophys. J., Vol. 392, No. 2, p. 715 – 721 (20 Jun 1992). With plates 5 – 6.
Element abundances have been determined in a coronal polar plume previously observed to have a low Ne/Mg abundance ratio associated with an open magnetic field. This study is based on images of the plume in Ne VI, Ne VII, Mg VI – Mg VIII, Na VIII, and Ca IX, Ca X photographed between 300 and 600 Å by the NRL spectroheliograph on Skylab. The density variation with height derived from Mg VIII 436.7 Å is satisfactorily fitted with a temperature of 843,000K in hydrostatic equilibrium, close to the temperature expected in ionization equilibrium. It is also consistent with a small departure from hydrostatic equilibrium

and outflow velocities in the range of $10-20$ km s^{-1}. The electron density derived from the Mg VIII doublet ratio is 1×10^9cm^{-3}. Element abundances relative to magnesium at several altitudes were derived by combining plots of the ion differential emission measures. It is found that the relative abundances of Na, Mg, and Ca in the plume are the same (within a factor of 2) as those in the photosphere, whereas the abundance of neon relative to magnesium is only 1/10 of this ratio in the photosphere.

074.055   The extended structure of the transition region of the solar wind.
N. A. Lotova, O. A. Korelov, Ya. V. Pisarenko, O. P. Medvedeva.
Geomagn. Aeron., Vol. 31, No. 2, p. 168 – 171 (Oct 1991). English translation of Geomagn. Aehron., Tom 31, No. 2, p. 223 – 227 (1991).
In June 1988 an extensive experiment was conducted on the translucence of the transition region of the solar wind from subsonic to supersonic flow. Five sources were used simultaneously to extend the sounding to more distant regions in circumsolar space. It was shown that the extent of the region of formation of supersonic flow is greater than previously assumed.

074.056   Fluctuations of the magnetic field and galactic cosmic rays in recurrent high–velocity flows of the solar wind.
N. S. Kaminer, A. E. Kuz'micheva, N. V. Mymrina.
Geomagn. Aeron., Vol. 31, No. 2, p. 172 – 175 (Oct 1991). English translation of Geomagn. Aehron., Tom 31, No. 2, p. 228 – 232 (1991).
Changes in the degree of irregularity of the modulus of the magnetic field and fluctuations of cosmic rays in recurrent high-velocity flows of the solar wind are considered. It is shown that the dispersion of the field is greater in the leading and lagging portions of the flow than in the central portion. The high regularity of the magnetic field in the central portion of the flow is maintained with an increase in its width. The dependence of the field dispersion on the solar wind–velocity within the flows is investigated. High correlation between the field dispersion and the plasma velocity is observed.

074.057   Solar wind interaction with Mars and Venus during solar activity cycle.
T. K. Breus.
Kosm. Issled., Tom 30, Vyp. 3, p. 396 – 419 (May–Jun 1992). In Russian. English translation in Cosmic Res., Vol. 30, No. 3.
Analysis and comparison of results of the Phobos 2 mission are carried out. The features of the boundaries observed in the interaction region of the SW flow with comet Halley and Venus and Mars during the periods of minimum and maximum of solar activity are presented. It is shown that all these bodies of the Solar system are similar in some aspects. They have one common feature – the cometopause like chemical boundary. However this cometopause like chemical boundary is prominent near the planets when there is high solar wind dynamic pressure during solar maximum. The boundary detected by Phobos 2 near the planet Mars and identified as the magnetopause or the planetopause seems to be not the real impenetrable obstacle where the force balance with the SW dynamic pressure takes place. It has some features which are similar to the cometopause of comet Halley and the cometopause like chemical boundary observed at Venus during the periods of high SW dynamic pressure in the vicinity of the solar maximum. Phobos 2 had revealed the features of the SW/Mars interaction which are even more cometary like than these observed at Venus. The obstacle to the SW flow at Mars can be both the magnetospheric like and the ionospheric like during the maximum of solar activity.

074.058   The density structure of polar plumes.
A. B. C. Walker Jr., J. F. Lindblom, C. E. DeForest, E. S. Paris, M. J. Allen, R. B. Hoover, T. W. Barbee Jr.
Bull. Am. Astron. Soc., Vol. 23, No. 3, p. 1264 (1991). Abstract. – See Abstr. 012.067 for the main entry.

074.059   Atmospheric responses to magnetic flux eruptions. III. Shock evolution near the Sun.
Hu Youqiu, Wang Chi, Zheng Huinan.
Acta Astrophys. Sin., Vol. 12, No. 1, p. 54 – 62 (Jan 1992). In Chinese. English translation in Chin. Astron. Astrophys., Vol. 16, No. 2, p. 215 – 225 (Apr–Jun 1992).
As the last one of the series, this paper discusses the shock evolution in the coronal atmosphere. It shows that the nonuniformity of the plasma and the magnetic field in the coronal background is determinative for the structure and evolution of shocks. A slow shock – fast magnetosonic wave system formed near the Sun evolves into a hybrid shock with an intermediate shock as its necessary component while propagating outward. The hybrid shock remains as it propagates along an electric current sheet but keeps evolving into a pure fast shock as it propagates along a unipolar open magnetic field.

074.060   Synthetic maps of the brightness and polarization of the F–corona.
Y. Fang, P. L. Lamy, A. Llebaria.
IAU Colloquium No. 126: Origin and evolution of interplanetary dust, p. 195 – 198 (1991). – See Abstr. 012.068 for the main entry.
The Wide–Field Light and Spectrometric Coronograph (LASCO) to be flown on SOHO in 1995 is designed to perform accurate photopolarimetric observation of the solar corona. For simulation purpose but also to have a two–dimensional model of the F–corona, the authors have realized synthetic maps of its brightness and polarization.

074.061   Magnetic reconnection associated with emerging magnetic flux.
K. Shibata, S. Nozawa, R. Matsumoto.
Symposium of Supercomputing Astronomy and Astrophysics in Japan, p. 169 – 176 (1991). With 1 plate. – See Abstr. 012.069 for the main entry.
The two–dimensional magnetohydrodynamic numerical simulations have been performed to study magnetic reconnection between an emerging flux and an overlying coronal magnetic field. The effect of gravity is properly taken into account. It is shown that the inclusion of gravity produces a number of new effects in the reconnection associated with the emerging flux, which have not been noted so far: The reconnection starts when most of chromospheric mass in the current sheet between the emerging flux and the coronal field drain down along the loop because of the gravity. This means that the start of a compact flare or an X–ray bright point follows the disappearance of an arch filament. It is also found that multiple magnetic islands are created in the sheet, which confine cool, dense chromospheric plasmas. These islands dynamically coalesce each other and are ejected along the sheet with Alfven speed.

074.062   Atmospheric responses to magnetic flux eruptions. III. Shock evolution near the Sun.
Hu Youqiu, Wang Chi, Zheng Huinan.
Chin. Astron. Astrophys., Vol. 16, No. 2, p. 215 – 225 (Apr–Jun 1992). English translation of Acta Astrophys. Sin., Vol. 12, No. 1, p. 54 – 62 (Jan 1992). See Abstr. 074.059.
As the last one of the series, this paper discusses the shock evolution in the coronal atmosphere. It shows that the nonuniformity of the plasma and the magnetic field in the coronal background is determinative for the structure and evolution of shocks. A slow shock–fast magnetosonic wave system formed near the Sun evolves into a hybrid shock with an intermediate shock as its necessary component while propagating outward. The hybrid shock remains as it propagates along an electric current sheet but keeps evolving into a pure fast shock as it propagates along a unipolar open magnetic field.

074.063   Alfvén–magnetosonic waves interaction in the solar corona.
K. Murawski.
Sol. Phys., Vol. 139, No. 2, p. 279 – 297 (Jun 1992).
The nonlinear propagation of the Alfvén and magnetosonic waves in the solar corona is investigated in terms of model

equations. Due to viscous effects taken into account the propagation of the fast wave itself is governed by Burgers type equations possessing both expansion and compression shock solutions. Numerical simulations show that both parallely and perpendicularly propagating fast waves can steepen into shocks if their amplitudes are in excess of some sizeable fraction of the Alfvén velocity. However, if the magnetic field changes linearly in the perpendicular direction, then formation of perpendicular shocks can be hindered. The Alfvén waves exhibit a tendency to drive both the slow and fast magnetosonic waves whose propagation is described by linearized Boussinesq type equations with ponderomotive terms due to the Alfvén wave. The limits of the slow and fast waves are investigated.

074.064   Solar wind acceleration
Hu Wenrui.
Advances in solar–terrestrial science of China, p. 108 – 120 (1992). – See Abstr. 003.069 for the main entry.
Features of solar wind acceleration in general and the energy addition of Alfvén fluctuations in detail are discussed. Studies of the interplanetary magnetic field and its influence on solar wind acceleration are reviewed.

074.065   The solar corona.
A. H. Gabriel.
NATO Advanced Study Institute on the Sun: a Laboratory for Astrophysics, p. 277 – 296 (1992). – See Abstr. 012.091 for the main entry.
The structure of the quiet solar corona is considered, excluding the effect of magnetic active regions. Starting with a simple one-dimensional geometry, models are derived both from ab initio theory and from analysis of spectral line intensities. To overcome the inconsistencies which arise, account is taken progressively of two–dimensional effects; the concentration of magnetic fields at the supergranule cell boundaries and the effect of coronal holes. The relationship is discussed between coronal holes, solar wind onset, the observation of fast streams, and the implication for other stars. The transition region is shown to be thicker than often assumed, thereby allowing the neglect of some of the more sophisticated non–Maxwellian effects which have been proposed in the past.

074.066   Large–scale quasi–stationary X–ray coronal structures associated with eruptive solar flares.
R. A. Kopp, G. Poletto.
IAU Colloquium No. 133: Eruptive solar flares, p. 197 – 206 (1992). – See Abstr. 012.096 for the main entry.
About ten years ago the Hard X–Ray Imaging Spectrometer aboard the Solar Maximum Mission detected for the first time large, faint coronal structures associated with dynamic flares. These have come to be known as "giant arches". Notwithstanding their extreme faintness, the energy content of these giant structures is of the order of 1–10% of the total energy released by large flares ($\cong 10^{32}$erg), thus representing a non–negligible term in the global flare energy balance. Analogously, their mass content ($\cong 10^{15}$g) is quite similar to that inferred for post–flare loops and coronal mass ejections. In spite of their obvious relevance, little is known about these giant features. Only about ten such arches have been identified to date, and the observations have both insufficient spatial resolution and inadequate time coverage. As a consequence, it is hard even to define their "typical" behavior and to ascertain their basic characteristics. Because of these difficulties, we still lack a generally accepted flare scenario which accounts for the presence of such structures. After reviewing the observational properties of giant arches the authors describe the hypotheses advanced thus far to explain their origin, evolution and energy supply. Uncertainties in the interpretations are emphasized and alternative models proposed.

074.067   Large scale structures associated with eruptive flares and radio waves.
N. Gopalswamy, M. R. Kundu.
IAU Colloquium No. 133: Eruptive solar flares, p. 207 – 213 (1992). – See Abstr. 012.096 for the main entry.
The authors review some recent results obtained from 2–dimensional imaging observations of the Sun using the Clark

Lake multifrequency radioheliograph. The radioheliograph produced images of the Sun's corona on a daily basis at several frequencies within the range 20–125 MHz during the period 1982–87. Using these images both large scale structures as well as transient phenomena such as bursts have been studied. In this paper the authors discuss the nature of radio emission associated with eruptive filaments and CMEs. It is possible to trace the structure of magnetic fields in the corona based on the multifrequency observations of moving type IV bursts at meter and decameter wavelengths. The authors illustrate this by discussing specific events. They discuss a rare case of the detection of thermal radio emission in association with a fast CME. They estimate the CME mass using spatially resolved radio data.

074.068   Coronal millimeter sources associated with eruptive flares.
A. Krüger, S. Urpo.
IAU Colloquium No. 133: Eruptive solar flares, p. 214 – 219 (1992). – See Abstr. 012.096 for the main entry.
Sporadically occurring coronal millimeter wave sources (CMMSs) detected as off–limb brightenings corresponding to a height of the order of 10^5km above the photosphere attract much interest. The lifetime of the phenomenon is several hours and it can be interpreted as a special kind of long–duration post–flare (and inter–flare) emission at coronal altitudes. There is a relationship to other solar radio phenomena, e.g. GRF–microwave bursts and noise storms, as well as to longduration events in soft X–ray coronal hard X–ray arches, and coronal mass ejections. The main results obtained until now on CMMSs are comprehensively summarized for the first time and new aspects included. The relevant emission processes have been analysed considering the whole microwave range. Open questions and implications for the study of eruptive flare dynamics and coronal structures by a new diagnostic tool are discussed. Finally, prospects of further investigations have been considered.

074.069   A giant post–flare coronal arch observed by Skylab.
Z. Švestka, S. Šimberová.
IAU Colloquium No. 133: Eruptive solar flares, p. 221 – 224 (1992). – See Abstr. 012.096 for the main entry.
The limb event of 13 August 1973, observed by Skylab in soft X–rays, exhibited typical characteristics of the giant post–flare arches observed by HXIS and FCS on board SMM in the 1980s. The authors present here examples of the processed Skylab images which yield 4 times better angular resolution than the SMM experiments and thus, for the first time, make it possible to distinguish the real fine structure of a giant post–flare arch.

074.070   Characteristics of coronal mass ejections.
E. Hildner.
IAU Colloquium No. 133: Eruptive solar flares, p. 227 – 233 (1992). – See Abstr. 012.096 for the main entry.
Coronal mass ejections (CMEs) are important and beautiful solar phenomena. Their frequency of occurrence, locations, speeds, and sizes are presented, measured over a significant fraction of a solar cycle. As yet, there is no consensus model which adequately explains these measurements, in part because the initiation and evolution of CMEs appear extremely complex and intractable.

074.071   The solar sources of coronal mass ejections.
D. F. Webb.
IAU Colloquium No. 133: Eruptive solar flares, p. 234 – 247 (1992). – See Abstr. 012.096 for the main entry.
Despite nearly two decades of observations and study of coronal mass ejections (CMEs), the question of the physical and phenomenological origins of CMEs remains unanswered. This question has been addressed in several different types of studies, each having important limitations. These include statistical analyses of the probabilities of association with CMEs of certain kinds of solar activity near CME onset, studies of the timing and location of activity associated with individual events, and hybrid

studies combining these methods. These studies suggest significant levels of association between CMEs and such near–surface activity as eruptive prominences, optical and X–ray flares, and gradual microwave and metric type II and IV radio bursts. The highest frequencies of association are between CMEs and prominence eruptions and long–enduring X–ray events. This relationship appears strong enough to suggest that prominence-related ejections may form a separate physical class of CMEs. On the other hand, about half of all CMEs appear to have no "good" associations. The author reviews the results of studies of CME origins, including recent results on the timing and location of CMEs and related activity.

074.072 Remote sensing observations of mass ejections and shocks in interplanetary space.

B. V. Jackson.
IAU Colloquium No. 133: Eruptive solar flares, p. 248 – 257 (1992). – See Abstr. 012.096 for the main entry.

It has long been known that disturbances can propagate from Sun to Earth with periods of a few days following large solar flares. These distrubances involve a significant portion of the lower solar corona and the energy of the flare. Several techniques have been used to remotely detect and follow different structures as they propagate outward from the Sun. These techniques include interplanetary scintillation (IPS), kilometric radio and HELIOS photometer observations. Structures in the interplanetary medium can generally be classed as those which propagate outward from the Sun and those which co–rotate with approximately the solar rotation rate. In this review the author concentrate on ejecta and shocks known to be associated with large solar flares observed on the surface of the Sun and their manifestations using various observational techniques. Examples of the data related to mass ejections from each of these techniques are shown and interpreted.

074.073 In situ observations of coronal mass ejections in interplanetary space.

J. T. Gosling.
IAU Colloquium No. 133: Eruptive solar flares, p. 258 – 267 (1992). – See Abstr. 012.096 for the main entry.

Coronal mass ejections, CMEs, in the solar wind at 1 AU generally have distinct plasma and field signatures by which they can be distinguished from the ordinary solar wind. These include one or more of the following: helium abundance enhancements, ion and electron temperature depressions, unusual ionization states, strong magnetic fields, low plasma beta, low magnetic field variance, coherent field rotations, counterstreaming (along the field) energetic protons, and counterstreaming suprathermal electrons. The msot reliable of these appears to be counterstreaming electrons, which indicates that CMEs at 1 AU typically are closed field structures either rooted at both ends in the Sun or entirely disconnected from it as plasmoids. About 1/3 of all CMEs have sufficiently high speeds to produce transient interplanetary shock disturbances at 1 AU; the remainder simply ride along with the solar wind. The frequency of occurrence of CMEs in the ecliptic plane, as distinguished by the counterstreaming electron signature, varies roughly in phase and amplitude with the 11–yr solar activity cycle. Near solar maximum they account for ~15% of all solar wind measurements, while near solar minimum they account for less than 1% of all the measurements. All but one of the 37 largest geomagnetic storms near the last solar maximum were associated with Earth–passage of interplanetary disturbances driven by fast CMEs; that is, CMEs are the prime link between solar and geomagnetic activity. However, more than half of all earthward directed CMEs are relatively ineffective in a geomagnetic sense.

074.074 MHD shocks and simple waves in CMEs.

R. S. Steinolfson.
IAU Colloquium No. 133: Eruptive solar flares, p. 276 (1992). – See Abstr. 012.096 for the main entry.

074.075 A numerical simulation of magnetically driven coronal mass ejections.

W. P. Guo, J. F. Wang, B. X. Liang, S. T. Wu.
IAU Colloquium No. 133: Eruptive solar flares, p. 381 – 384 (1992). – See Abstr. 012.096 for the main entry.

Using a quasi–steady, helmet–streamer as the initial corona, the authors simulate the dynamical evolution of the mass ejection due to magnetic eruption at the base of the streamer. The simulated ejections reproduce some of the observed features of loop–shaped coronal mass ejections. A comparison between the present magnetically driven model and previous thermal driven models is presented. It is shown that both the driving mechanism and the initial corona are important in the simulation of CMEs.

074.076 Stability of force–free magnetic fields and the problem of solar coronal heating.

G. E. Vekstein.
International Workshop on Reconnection in Space Plasma, p. 93 – 97 (Jan 1989). – See Abstr. 012.099 for the main entry.

Stability of force–free magnetic fields is considered in connection with the problem of magnetic coronal heating.

074.077 A neutral current sheet with flows: tearing and stratification modes and modelling of coronal structures.

V. M. Gubchenko, V. V. Zaitsev (*V. V. Zajtsev*).
International Workshop on Reconnection in Space Plasma, p. 99 – 104 (Jan 1989). – See Abstr. 012.099 for the main entry.

The nature of the solar corona fine structure (its r, θ, and φ variations) is studied using a model of a neutral current sheet submerged into the solar wind plasma flow.

074.078 On the MHD stability of the m = 1 kink mode in finite length coronal loops.

M. Velli, A. W. Hood, G. Einaudi.
International Workshop on Reconnection in Space Plasma, p. 105 – 109 (Jan 1989). – See Abstr. 012.099 for the main entry.

A general method for studying the ideal and resistive MHD stability of plasma configurations with line–tying is presented and applied to the case of the m = 1 kink mode in coronal loops.

074.079 Dynamic development of coronal current sheets.

B. Kliem.
International Workshop on Reconnection in Space Plasma, p. 117 – 122 (Jan 1989). – See Abstr. 012.099 for the main entry.

The possible interplay of tearing, coalescence, and current-driven kinetic instabilities in current sheets is considered on the basis of order of magnitude estimates of instability thresholds and scalings. Arguments are given that favour the development of internal dynamics for current sheets with stationary inflow and appropriate combination of parameters (the "impulsive bursty regime of reconnection"). The model relates newly emerging flux to flare energy release and provides the basis for an understanding of some moving type IV radio bursts associated with coronal mass ejections.

074.080 Cosmic rays and magnetosonic instabilities of solar wind flow near the heliospheric shock wave.

S. V. Chalov.
1. COSPAR Colloquium: Physics of the outer heliosphere, p. 219 – 221 (1990). – See Abstr. 012.100 for the main entry.

It is known that the flow of thermal plasma interacting with cosmic rays in front of a sufficiently strong shock wave is unstable with respect to short–wavelength magnetosonic oscillations. The results of the general theory of this instability are applied to the solar wind flow in front of a heliospheric shock wave close to the equator.

074.081 Plasma observations in the distant heliosphere: a view from Voyager.

A. J. Lazarus, R. L. McNutt Jr.
1. COSPAR Colloquium: Physics of the outer heliosphere, p. 229 – 234 (1990). – See Abstr. 012.100 for the main entry.

The authors summarize solar wind proton observations from 1 AU to a position of Voyager 2 near 30 AU.

074.082 Distant solar wind plasma – view from the Pioneers.
A. Barnes.
1. COSPAR Colloquium: Physics of the outer heliosphere,
p. 235 – 240 (1990). – See Abstr. 012.100 for the main entry.
Contents: 1. Introduction. 2. Variation of the solar wind with
heliocentric distance. 3. Changes in global heliospheric morphology near sunspot minimum. 4. Coronal changes and the global
heliosphere near sunspot minimum.

**074.083 Coronal magnetic field strengths determined from fiber
bursts.**
G. Mann, K. Baumgärtel.
International Workshop on Reconnection in Space Plasma,
p. 153 – 156 (Jan 1989). – See Abstr. 012.099 for the main entry.
A method of estimation of coronal magnetic field strengths
above active regions is presented by means of fiber bursts
appearing as a special fine structure in solar type–IV radio bursts.

074.084 Shock heating of the solar wind plasma.
Y. C. Whang, Liu Shouliang, L. F. Burlaga.
1. COSPAR Colloquium: Physics of the outer heliosphere,
p. 241 – 244 (1990). – See Abstr. 012.100 for the main entry.
Contents: 1. Radial increase in the entropy of the solar wind
plasma. 2. Shocks observed in 1973–1982. 3. MHD simulation
for shock heating of the solar wind.

**074.085 The role of the magnetic reconnection in the solar
wind/magnetosphere interaction.**
M. I. Pudovkin, V. S. Semenov.
International Workshop on Reconnection in Space Plasma,
p. 281 – 286 (Jan 1989). – See Abstr. 012.099 for the main entry.
Processes of the solar wind flow around the magnetosphere
and of the magnetic field reconnection at the magnetopause are
considered in a self–consistent manner.

074.086 Solar wind vortex flow in the outer heliosphere.
I. S. Veselovsky (*I. S. Veselovskij*).
1. COSPAR Colloquium: Physics of the outer heliosphere,
p. 277 – 280 (1990). – See Abstr. 012.100 for the main entry.
The structure of the T. von Karman vorticity wake type near
the heliomagnetic equator is considered. Linearized equations are
used for polytropic gas on the background of axially symmetric
supersonic flow with the velocity minimum at the equatrial plane
in the heliosphere.

074.087 Interaction of the solar wind with the external plasma.
V. B. Baranov.
1. COSPAR Colloquium: Physics of the outer heliosphere,
p. 287 – 297 (1990). – See Abstr. 012.100 for the main entry.
A critical survey of the present–day situation in gasdynamical
models of solar wind interaction with the local interstellar
medium is presented.

074.088 Signature of a viscous interaction at the heliopause.
H. Pérez–de–Tejada.
1. COSPAR Colloquium: Physics of the outer heliosphere,
p. 299 – 305 (1990). – See Abstr. 012.100 for the main entry.
Observations of the solar wind interaction with weakly
magnetized bodies in the solar system (Venus, Mars, comets) are
used to define guidelines to the region of interaction between the
local interstellar plasma and the solar wind.

**074.089 Microwave and soft X–ray emission of solar flare events
associated with coronal mass ejections.**
I. M. Chertok, A. A. Gnezdilov, E. P. Zaborova.
Astron. Zh., Tom 69, Vyp. 3, p. 593 – 603 (May – Jun 1992). In
Russian. English translation in Sov. Astron., Vol. 36, No. 3
(May – Jun 1992).
The features of microwave and soft X–ray bursts accompanying the Coronal Mass Ejections (CMEs) observed with P78–1
satellite are analysed. It is shown that the distribution of the
events on the burst "intensity–duration" plots allows not only to
separate flares with CMEs and the ones without CMEs, but to
determine the relations between the burst characteristics and

main parameters of CMEs (angular sizes, speed, mass, shapes).
The large and fast CMEs of complex shapes are observed, as a
rule, in combination with the most intensive and long–duration
bursts. The CMEs having middle parameters and simple shapes
are mainly identified with moderate non–impulsive soft X–ray
and microwave bursts, as well as with the "gradual rise and fall"
radio bursts. The majority of impulsive bursts are not accompanied by CMEs in general, but the most intense of them may be
sometimes associated with small and simple CMEs. The relationships between CMEs and flare energy release are discussed.

**074.090 The interactions of the solar wind discontinuities in the
vicinity of the heliopause.**
S. A. Grib, E. A. Pushkar (*E. A. Pushkar'*), A. A. Barmin.
1. COSPAR Colloquium: Physics of the outer heliosphere,
p. 317 – 320 (1990). – See Abstr. 012.100 for the main entry.
The normal and oblique interactions of different solar wind
discontinuities such as double shocks and tangential discontinuities with the heliospheric shock, heliopause and hypothetical
interstellar shock are considered.

**074.091 Investigation of the solar wind transonic region at meter
wavelength.**
N. A. Lotova.
1. COSPAR Colloquium: Physics of the outer heliosphere,
p. 397 – 398 (1990). – See Abstr. 012.100 for the main entry.

Eruptive solar flares. Proceedings.
See Abstr. 012.096.

Reconnection in space plasma. Proceedings.
See Abstr. 012.099.

Physics of the outer heliosphere. Proceedings.
See Abstr. 012.100.

The Ulysses solar wind plasma experiment.
See Abstr. 035.031.

The Solar Wind Ion Composition Spectrometer.
See Abstr. 035.032.

The coronal–sounding experiment.
See Abstr. 035.040.

**A SPAN MCP detector for the SOHO Coronal Diagnostic
Spectrometer.**
See Abstr. 035.043.

Considerations of a solar mass ejection imager in a low–earth orbit.
See Abstr. 035.134.

Neutral solar wind experiment.
See Abstr. 036.208.

**Future observation of the F–corona with the LASCO coronograph
space experiment.**
See Abstr. 051.048.

**F–corona–experiment: requirements for remote sensing of inter-
planetary dust.**
See Abstr. 051.062.

**The theory of magnetohydrodynamic wave generation by localized
sources. III. Efficiency of plasma heating by dissipation of far–field
waves.**
See Abstr. 062.008.

**Magnetohydrodynamic equilibria and cusp formation at an X–type
neutral line by footpoint shearing.**
See Abstr. 062.009.

Simulation of ion acceleration in a charged dust cloud.
See Abstr. 062.013.

Semikinetic and generalized transport models of the polar and solar winds.
See Abstr. 062.015.

Cylindrically symmetric force–free magnetic fields.
See Abstr. 062.024.

The equilibrium shape of slender flux tubes in a linear force–free magnetic field.
See Abstr. 062.031.

Chaotic Alfvén waves in multispecies plasmas.
See Abstr. 062.036.

Resistive tearing–mode instability in a current sheet with equilibrium viscous stagnation–point flow.
See Abstr. 062.072.

Taylor relaxation of a Gold–Hoyle flux tube.
See Abstr. 062.073.

Magnetohydrodynamic waves in sharply and smoothly bounded cylinders.
See Abstr. 062.074.

Critical density layer as obstacle at solar wind – exospheric ion interaction.
See Abstr. 062.078.

Relaxation processes in magnetohydrodynamics: a triad–interaction model.
See Abstr. 062.103.

On instability of an inhomogeneous plasma in active regions of the solar corona.
See Abstr. 062.117.

The prominence–corona transition region in transverse magnetic fields.
See Abstr. 062.158.

Solar p–modes oscillations and heating of the corona.
See Abstr. 062.161.

Nonstationary Petschek reconnection. Convective Zone.
See Abstr. 062.181.

Nonstationary Petschek recombination. Diffusion region.
See Abstr. 062.182.

The structure of radiative slow–mode shocks.
See Abstr. 062.203.

Alfvén–magnetosonic waves interaction.
See Abstr. 062.206.

Spectroscopic diagnostics.
See Abstr. 062.217.

MHD solar wind – interstellar plasma interaction: 3D formulation by the projected characteristics method and the stability analysis.
See Abstr. 062.241.

On the density and field sensitivities of dielectronic recombination.
See Abstr. 063.063.

The solar wind and the winds from cool stars.
See Abstr. 064.003.

Nonradial and nonpolytropic astrophysical outflows. I. Hydrodynamic solutions with flaring streamlines.
See Abstr. 064.016.

Nonradial and nonpolytropic astrophysical outflows. II. Topology of MHD solutions with flaring streamlines.
See Abstr. 064.052.

Non–local thermal conduction in solar and stellar coronal loops.
See Abstr. 064.085.

Solar corona correction in VLBI observation.
See Abstr. 066.297.

The Sun.
See Abstr. 071.025.

Detailed comparison between sunspot activity in "hot spots" and galactic cosmic–ray intensity.
See Abstr. 072.043.

Long–term evolution of coronal magnetic fields basing on noise storm continuum observations in the 20th and 21st cycle.
See Abstr. 072.069.

The effects of magnetic field geometry on the confinement of energetic electrons in solar flares.
See Abstr. 073.001.

Dynamics of flaring loops. II. Flare evolution in the density–temperature diagram.
See Abstr. 073.002.

Eruptive phenomena on the sun.
See Abstr. 073.003.

Energetics and dynamics in a large solar flare of 1989 March.
See Abstr. 073.006.

Fluid flow in a jet and the Ca XIX line profiles observed during solar flares.
See Abstr. 073.007.

Energy propagation into a flare kernel during a solar flare.
See Abstr. 073.013.

Analysis of the optical spectra of the solar flares. VI. Velocity field in the 13 June 1980 flare area.
See Abstr. 073.018.

On the efficiency of electron cyclotron maser instability in solar flares.
See Abstr. 073.023.

Alfvénically driven slow shocks in the solar chromosphere and corona.
See Abstr. 073.042.

Structure and physics of solar faculae. V. Study of the roughness effect in the quiet chromosphere–corona transition zone.
See Abstr. 073.046.

The Hα development of an intense limb flare and associated flaring arches.
See Abstr. 073.049.

The basal and strong–field components of the solar atmosphere.
See Abstr. 073.051.

The simultaneous effects of collisions, reverse currents and magnetic trapping on the temporal evolution of energetic electrons in a flaring coronal loop.
See Abstr. 073.053.

A solar flare model including the formation and destruction of the current sheet in the corona.
See Abstr. 073.060.

White–light flare observed at the solar limb.
See Abstr. 073.063.

Toward the circuit theory of solar flares.
See Abstr. 073.073.

Exploration of the variation in the source function within a loop prominence by means of the simplex method.
See Abstr. 073.078.

Solar flares and coronal mass ejections.
See Abstr. 073.079.

Fluid flow in a jet and the Ca XIX line profiles observed during solar flares.
See Abstr. 073.096.

Flare evolution in the density–temperature diagram.
See Abstr. 073.102.

Models of normal and inverse polarity filament eruptions and coronal mass ejections.
See Abstr. 073.111.

Dynamics in the prominence–corona transition region from HRTS spectra.
See Abstr. 073.113.

The large–scale magnetic field of the Sun and the structure of the solar wind in 1973 – 1974.
See Abstr. 075.002.

Three–dimensional force–free magnetic fields and flare energy buildup.
See Abstr. 075.004.

Energy buildup in sheared force–free magnetic fields.
See Abstr. 075.022.

Peculiarities of quasi–periodic variations of large–scale solar magnetic fields and of solar wind properties.
See Abstr. 075.026.

A regularization method for extrapolation of solar potential magnetic fields.
See Abstr. 075.027.

Electric currents and magnetic–field loops in solar active regions.
See Abstr. 075.028.

On the method of calculating two–dimensional potential magnetic fields with current sheets.
See Abstr. 075.035.

Magnetic energy conversion on the sun.
See Abstr. 075.042.

Fe XVIII emission–line intensities in the Sun.
See Abstr. 076.003.

C IV line ratios in the Sun.
See Abstr. 076.004.

Transitions from metastable levels emitted during short–duration bursts: how valid are their calculated intensities?
See Abstr. 076.005.

X–ray observations of limb flare loops and post–flare coronal arch.
See Abstr. 076.022.

A computer simulation study of Type III radio burst propagation through the solar corona.
See Abstr. 077.001.

On the effect of radio waves group delay in the solar corona.
See Abstr. 077.002.

Multifrequency observations of a remarkable solar radio burst.
See Abstr. 077.003.

Quiet–Sun emission and local sources at meter and decimeter wavelengths and their relationship with the coronal neutral sheet.
See Abstr. 077.005.

Connection between ambient density fluctuations and clumpy Langmuir waves in type III radio sources.
See Abstr. 077.007.

High dynamic range multifrequency radio observations of a solar active region.
See Abstr. 077.008.

Observations of mode coupling in the solar corona and bipolar noise storms.
See Abstr. 077.013.

Millisecond microwave spikes at 8 GHz during solar flares.
See Abstr. 077.017.

The role of signal group delay in the solar corona.
See Abstr. 077.018.

Heliolongitudinal variations of parameters of type III burst radiation and its harmonic structure.
See Abstr. 077.023.

Solar longitude variations in the parameters of type III radio emission and its harmonic structure.
See Abstr. 077.024.

Decimetric solar type U bursts: VLA and PHOENIX observations.
See Abstr. 077.026.

Meter–decameter radio emission associated with a coronal mass ejection.
See Abstr. 077.043.

Chaotic radiopulsations and coronal magnetic field estimates.
See Abstr. 077.050.

Superevents: their origin and propagation through the heliosphere from 0.3 to 35 AU.
See Abstr. 078.004.

The great solar energetic particle events of 1989 observed from geosynchronous orbit.
See Abstr. 078.005.

Coronal and interplanetary transport of solar flare protons from the ground level event of 29 September 1989.
See Abstr. 078.021.

One day on the sun.
See Abstr. 079.006.

The 22–year cycle of the Sun.
See Abstr. 080.021.

Alfvén wave transmission through the solar atmosphere.
See Abstr. 080.035.

The secrets of the solar atmosphere.
See Abstr. 080.060.

The present understanding of the cusp.
See Abstr. 083.034.

The magnetic field strength of currents in the magnetopause and the dynamic pressure of the solar wind.
See Abstr. 084.004.

The quiet geomagnetic field at geosynchronous orbit and its dependence on solar wind dynamic pressure.
See Abstr. 084.008.

On impulsive penetration of solar wind plasmoids into the geomagnetic field.
See Abstr. 084.022.

The semiannual variation of great geomagnetic storms and the postshock Russell–McPherron effect preceding coronal mass ejecta.
See Abstr. 084.030.

The Earth's bow shock and magnetopause position as a result of the solar wind – magnetosphere interaction.
See Abstr. 084.034.

Solar wind–terrestrial magnetosphere coupling: application of linear prediction theory.
See Abstr. 084.036.

Solar control of the Earth's emission of energetic O^+.
See Abstr. 084.038.

Antarctic studies of the cusp.
See Abstr. 084.067.

Solar wind–magnetosphere coupling processes observed near the dayside cusp/cleft.
See Abstr. 084.068.

Mathematical model of physical processes responsible for generation of magnetospheric electric fields and their penetration into the Earth's magnetosphere.
See Abstr. 084.071.

Terrestrial response to eruptive solar flares: geomagnetic storms.
See Abstr. 084.077.

Coronal mass ejections: the link between solar and geomagnetic activity.
See Abstr. 084.078.

Solar wind and magnetospheric processes and various coupling mechanisms including those observed on other planets.
See Abstr. 085.014.

Turbulent pick–up of new–born ions near Venus and Mars and problems of numerical modelling of the solar wind interaction with these planets. I. Features of the solar wind interaction with planets.
See Abstr. 093.014.

Turbulent pick–up of new–born ions near Venus and Mars and problems of numerical modelling of the solar wind interaction with these planets. II. Two–fluid HD model.
See Abstr. 093.015.

Systematic variations in solar wind fluence with lunar location: implications for resource utilization.
See Abstr. 094.074.

Observations of plasma boundaries and phenomena around Mars with Phobos 2.
See Abstr. 097.013.

Interaction of the solar wind with Mars. The Phobos 2 results.
See Abstr. 097.187.

Plasma–beam instabilities in cometary ionospheres.
See Abstr. 102.001.

Further evolution of velocity shell distribution of cometary and interstellar pickup ions and excitation of oblique Alfvén waves.
See Abstr. 102.004.

A self–consistent model for the particles and fields upstream of an outgassing comet. 2. A time–dependent description.
See Abstr. 102.012.

Plasma–beam instabilities in cometary ionospheres.
See Abstr. 102.014.

The role of the interplanetary magnetic field reorientation in the mechanism of the comet's brightness outburst occurrence.
See Abstr. 102.023.

Acceleration mechanism of particles in the type–I cometary plasma.
See Abstr. 102.044.

Comets: pre– and post–Halley.
See Abstr. 102.058.

The effect of solar activity on the light curves of comets Churyumov–Gerasimenko (1982 VIII) and Halley (1986 III).
See Abstr. 103.006.

Interplanetary magnetic flux: measurement and balance.
See Abstr. 106.005.

An analytical solution for the heliopause boundary and its comparison with numerical solutions.
See Abstr. 106.008.

Nonstationary propagation of flare–generated particles in the heliosphere.
See Abstr. 106.009.

A search during the 1991 solar eclipse for the infrared signature of circumsolar dust.
See Abstr. 106.013.

Large–scale structure of the circumsolar medium based on scintillations.
See Abstr. 106.017.

On the existence of the heliospheric current sheet without a neutral line (HCS without NL).
See Abstr. 106.023.

Anisotropy of the energetic neutral atom flux in the heliosphere.
See Abstr. 106.025.

Interplanetary magnetic field connection to the Sun during electron heat flux dropouts in the solar wind.
See Abstr. 106.036.

Large–scale configuration of the circumsolar plasma as determined by scintillations.
See Abstr. 106.037.

Polarization parameters of large–scale turbulence of the solar wind and fluctuations of the cosmic ray intensity.
See Abstr. 106.052.

Interplanetary medium disturbances generated by a slow isolated magnetic cloud.
See Abstr. 106.056.

Introductory lecture – the heliosphere.
See Abstr. 106.092.

Model predictions and remote observations of the hydrogen density profile in the distant heliosphere.
See Abstr. 106.095.

Comparison of Ly–α and Ly–β interplanetary glows observed by the Voyager ultra–violet spectrometer.
See Abstr. 106.096.

Ion acceleration to cosmic ray energies.
See Abstr. 106.101.

Nonlinear effects of cosmic ray interaction with solar wind in the outer heliosphere.
See Abstr. 106.104.

Evolution of the ideas about the heliosphere and cosmic ray modulation in interplanetary space.
See Abstr. 106.105.

Outer heliosphere: eigen pulsations, cosmic rays and stream kinetic instability.
See Abstr. 106.106.

Outer heliosphere as a many–component medium for cosmic ray propagation.
See Abstr. 106.107.

Magnetic fields in the heliosphere: Pioneer observations.
See Abstr. 106.111.

Radio noise in the heliospheric cavity.
See Abstr. 106.112.

Trapped radiation in the outer heliosphere.
See Abstr. 106.113.

A model ot the reconnection pattern at the heliopause.
See Abstr. 106.114.

Filtration of the interstellar neutrals at the heliospheric interface and their coupling to the solar wind.
See Abstr. 106.115.

LISM–heliosphere interaction mediated by suprathermal particles.
See Abstr. 106.116.

Expected beams of energetic neutral atoms in the outer heliosphere.
See Abstr. 106.117.

Motion of the strong disturbances in the interplanetary medium.
See Abstr. 106.118.

Reconnection pattern at the heliopause.
See Abstr. 106.119.

Off–disk implantation of early solar wind into a planetesimal–dust cloud.
See Abstr. 107.049.

The interaction of interstellar pick–up ions with the solar wind – probing the interstellar medium by in–situ measurements.
See Abstr. 131.288.

The simulated features of heliospheric cosmic–ray modulation with a time–dependent drift model. I. General effects of the changing neutral sheet over the peiod 1985 – 1990.
See Abstr. 144.016.

The simulated features of heliospheric cosmic–ray modulation with a time–dependent drift model. III. General energy dependence.
See Abstr. 144.057.

Diffusion, drifts, and modulation of galactic cosmic rays in the heliosphere.
See Abstr. 144.075.

The predictions of a time–dependent drift model compared with cosmic–ray intensity observations from 1976 to 1989.
See Abstr. 144.077.

A time–dependent drift model with a simulated wavy neutral sheet for the solar modulation of cosmic rays.
See Abstr. 144.078.

The anomalous component of cosmic rays.
See Abstr. 144.081.

The long term cosmic ray variation relevant to solar wind structure in the outer heliosphere.
See Abstr. 144.084.

On using the ACR to probe the LISM/heliosphere interface.
See Abstr. 144.086.

075 Magnetic Fields

075.001   **Patterns in the photospheric magnetic field and percolation theory.**
C. J. Schrijver, C. Zwaan, A. C. Balke, T. D. Tarbell, J. K. Lawrence.
Astron. Astrophys., Vol. 253, No. 1, p. L1 – L4 (Jan 1992).
Letter–to–the–editor.
    The magnetic field in solar plages forms a highly structured pattern with no apparent characteristic length scale. This pattern appears to be a fractal with a dimension between 1.45 and 1.60. Small–scale displacements of concentrations of magnetic flux in the network are consistent with a random walk on a fractal with a similar dimension. The authors argue that percolation theory offers an effective explanation for observed geometric properties of small–scale flux concentrations in the solar photosphere, by demonstrating the close correspondence with clusters formed by randomly placed tracers on a two–dimensional (irregular) lattice. Percolation theory also offers a model for the subdiffusive behaviour of tracers performing a random walk on clusters formed by bonded sites. The authors demonstrate that the geometry of flux concentrations and of the displacement of magnetic flux as a function of time are equivalent to situations in percolation theory below a critical value called the percolation threshold.

075.002 **The large–scale magnetic field of the Sun and the structure of the solar wind in 1973 – 1974.**
N. M. Rudneva, P. M. Svidskij, Ya. V. Nagelis, B. I. Ryabov.
Geomagn. Aeron., Vol. 30, No. 5, p. 615 – 619 (Apr 1990). English translation of Geomagn. Aehron., Tom 30, No. 5, p. 723 – 729 (1990).

It has been shown on the basis of a combined analysis of data on the large–scale magnetic field of the Sun, manifestations of solar activity, and the parameters of the solar wind and the interplanetary magnetic field that a high–velocity solar wind flows predominantly out of the near–polar regions of the equivalent equatorial central dipoles. Activity complexes are characterized by a strong large–scale magnetic field at the photosphere and in the lower corona and are sources of discrete increases of the modulus of the interplanetary magnetic field. In some activity complexes coronal holes, which are sources of high–velocity streams of solar wind, can be formed as a result of readjustment of the magnetic field.

075.003 **The practical application of the magnetic virial theorem.**
J. A. Klimchuk, R. C. Canfield, J. E. Rhoads.
Astrophys. J., Vol. 385, No. 1, p. 327 – 343 (20 Jan 1992).

The magnetic virial theorem states that the magnetic energy contained in a coronal force–free magnetic field is given by a surface integral at the photospheric boundary involving the three vector magnetic field components. It is possible to use this theorem together with vector magnetograph data to compute the magnetic energy of solar active regions. In particular, one might compute the energy of an active region both before and after a flare occurs to determine whether the energy observed to be released during the flare is actually extracted from the magnetic field, as is commonly believed. In order to attach any significance to such a determination, one must understand how errors in the vector magnetograph measurements produce errors in the virial theorem energy. The authors have numerically simulated the effects of realistic errors on known magnetic fields. These include errors due to random polarization noise, crosstalk between different polarization signals, systematic polarization bias, and seeing–induced crosstalk. Analytical expressions for the energy errors were also derived, which apply under certain idealized conditions. The results serve as a useful tool for evaluating the ability of vector magnetographs to provide suitable data for the accurate determination of magnetic energies using the virial theorem.

075.004 **Three–dimensional force–free magnetic fields and flare energy buildup.**
J. A. Klimchuk, P. A. Sturrock.
Astrophys. J., Vol. 385, No. 1, p. 344 – 353 (20 Jan 1992).

The authors have used the "magneto–frictional" method to compute fully three–dimensional models of force–free magnetic fields. Beginning with a potential field produced by a point dipole buried below the solar surface, they displaced the magnetic footpoints at the photosphere to investigate the buildup of magnetic energy. It is found that reasonable footpoint shearing displacements can increase the total magnetic energy by at least one–third. The energy buildup is greater when the shearing displacements are concentrated closer to the magnetic neutral line. Roughly half of the energy buildup is free magnetic energy. The absolute quantity of free magnetic energy (10^{30}ergs – 10^{33}ergs, depending upon the scaling of the models) is sufficient to explain solar flares. There is no evidence for "loss of equilibrium".

075.005 **Sign reversal of the high latitude solar magnetic field.**
E. E. Benevolenskaya, V. I. Makarov.
Pis'ma Astron. Zh., Tom 18, No. 3, p. 266 – 270 (Mar 1992). In Russian. English translation in Sov. Astron. Lett., Vol. 18, No. 2.

It is supposed that the existence of 3–fold magnetic field reversal in one of the solar hemispheres is a result of two types of variations in the background magnetic field. The first type is the Hale 22–year cycle, the second is a quasi–biannual cycle. In the first approximation, the background magnetic field evolution may be described by diffusion equations corrected for meridional circulation and with a periodically changing source. Numerical modelling shows that in the case of multiplicity of the frequencies and with a certain relation between the amplitudes of these periods it appears that the existence of 3–fold polarity reversals may take place in the even cycles (according to Zurich numbering).

075.006 **The effect of the anomalous dispersion in the solar atmosphere on results of magnetic field measurements using the line–ratio method.**
V. G. Lozitskij, V. A. Sheminova.
Kinematika Fiz. Nebesn. Tel, Tom 8, No. 1, p. 12 – 19 (Jan–Feb 1992). In Russian. English translation in Kinematics Phys. Celest. Bodies, Vol. 8, No. 1.

On the basis of calculations of the Stokes parameters for the Fe I $\lambda\lambda$524.7 and 525.0 nm lines and the Holweger–Müller model atmosphere, the effect of the anomalous dispersion on the line–ratio method results is analysed. It is shown that with present–day accuracy of observational data, the anomalous dispersion should be taken into account only when the following conditions are simultaneously fulfilled: a) the magnetic field slope angle exceeds 20°; b) the magnetic field strength is larger than 100 mT; c) the fluxtube magnetic profile is rectangular; and d) the parts of the spectral line profile close to the line centre ($\Delta\lambda < 4$ pm) are used.

075.007 **The reversal of the solar polar magnetic fields. III. The large–scale fields and the first major active regions of cycle 22.**
P. R. Wilson.
Sol. Phys., Vol. 138, No. 1, p. 11 – 21 (Mar 1992).

It is shown that the first new cycle active regions of the longitude range emerged across the neutral lines of a cell, which continued to grow and expand across the equator for several rotations. The development of a parallel trans–equatorial band of flux of opposite (negative) polarity and the emergence of both new and old cycle active regions across a neutral line of this cell are also described. Simulations show that, while the growth of the positive region could, in part, be explained by the decay of flux from these new regions, there were significant differences between synoptic contour charts based on the simulations and those constructed from the observed fields. They also show that the development of the negative region cannot reasonably be explained by the decay of the observed active regions. A further example of the counter rotation of decaying active region fields is reported. Here the initial tilt of the negative–positive magnetic axes of two adjacent regions is normal, and simulations based on these data show their combined follower flux moving preferentially polewards. However, the observations show that, after three rotations, the decaying leader flux is entirely poleward of the follower flux.

075.008 **On the rotation of large–scale background fields in the 21st cycle of solar activity.**
V. I. Mordvinov, E. M. Tikhomolov.
Sol. Phys., Vol. 138, No. 1, p. 23 – 33 (Mar 1992).

The authors study some peculiarities of the time variation of dipole components in the longitudinal field distribution in individual low–latitude belts of the Sun. For analyzing the horizontal dipole rotation and variations of amplitudes we used magnetic and Hα data. From 1979 to 1981 the rotation of the dipoles of the northern and southern low–latitude belts occurs with periods of about 26.8 days (N) and 28.2 days (S), in agreement with the results reported by other authors. A uniform rotation of the low–latitude dipoles of these belts continued until the end of 1981. Following the next coincidence of the magnetic poles in longitude the dipoles change in their rotation character. During about 15 – 20 rotations the low–latitude dipoles co–rotate with a new period close to the Carrington period. This is followed by a rapid (in 3 – 5 rotations) transition of the poles to a new stable state, also with the Carrington rotation period. Unlike the trajectories of the poles, the dipole amplitudes of the low–latitude belts showed a significant variability.

075.009 Global modes constituting the solar magnetic cycle. I. Search for 'dispersion relations'.
M. H. Gokhale, J. Javaraiah, K. Narayanan Kutty, B. A. Varghese.
Sol. Phys., Vol. 138, No. 1, p. 35 – 47 (Mar 1992).

The spherical–harmonic–Fourier (SHF) analysis of the Sun's magnetic field inferred from the Greenwich sunspot data is refined and extended to include the full length (1874 – 1976) of the data on the magnetic tape provided by H. Balthasar. Perspective plots and grey level diagrams of the SHF power spectra for the odd and the even degree axisymmetric modes are presented. Comparing these with spectra obtained from two simulated data sets with random redistribution within the wings in the butterfly diagrams, the authors conclude that there is no clear evidence for the existence of any relation between the harmonic degree and the temporal frequency of the power concentrations of the inferred field. It is suggested that the solar magnetic cycle consists of some global oscillations of the Sun "forced" at a frequency $\sim 1/21.4\,\mathrm{y}^{-1}$ and, perhaps, weak resonances at its odd harmonics. The band width of the forcing frequency seems to be much less than $1/107\,\mathrm{y}^{-1}$. In case the global oscillations are torsional MHD, the significance of their parity and power peak is pointed out.

075.010 Linear force–free magnetic field around quiescent solar prominences computed from observable boundary conditions.
P. Démoulin, M. A. Raadu, J. M. Malherbe.
Astron. Astrophys., Vol. 257, No. 1, p. 278 – 286 (Apr 1992).

The authors analyse the magnetic support of solar prominences in two–dimensional linear force–free fields. The prominence is modeled as a vertical current sheet with mass in equilibrium between gravity and magnetic forces. A finite difference numerical technique is used which incorporates both vertical photospheric and horizontal prominence magnetic field observations.

075.011 A comparison of vector magnetograms from the Marshall Space Flight Center and Mees Solar Observatory.
R. S. Ronan, F. Q. Orrall, D. L. Mickey, E. A. West, M. J. Hagyard, K. S. Balasubramaniam.
Sol. Phys., Vol. 138, No. 1, p. 49 – 68 (Mar 1992).

The authors compare completely independent vector magnetic field measurements from two very different polarimetric instruments. They obtained active region magnetic field data with both the Marshall Space Flight Center (MSFC) and MSO system on five days during June 1985. They conclude: (1) the spatially–averaged line–of–sight components agree quite well; (2) although the MSO spatial grid is coarser, the quality of the MSO image is better than that of the MSFC data because of better seeing conditions; (3) the agreement between the transverse magnitudes is affected by the poor image quality of the MSFC data; and (4) if the effects of Faraday rotation caused by including line–center linear polarization in the method of analysis are taken into account, the azimuths show good agreement within the scatter in the data caused by the averaging process.

075.012 Vector magnetogram and Dopplergram observation of magnetic flux emergence and its explanation.
Zhang Hongqi, Song Mutao.
Sol. Phys., Vol. 138, No. 1, p. 69 – 92 (Mar 1992).

During 23 – 28 August 1988, at the Huairou Solar Observation Station of Beijing Observatory, the full development process of the region HR 88059 was observed. It emerged near the center of the solar disk and formed a medium active region. A complete series of vector magnetograms and photospheric and chromospheric Dopplergrams was obtained. From an analysis of these data, combined with some numerical simulations, the following conclusions can be drawn. (1) The emergence of new magnetic flux from enhanced networks followed by sunspot formation can be simply described by MHD numerical simulation. (2) New opposite bipolar features emerge within the former bipolar field with an identical strength which will develop a sunspot group complex. Also, arch filament systems appear there located in the

position of flux emergence. The neutral line is often pushed aside and curved, leading to faculae heating and the formation of a current sheet.

075.013 Peculiar photospheric velocity fields and magnetic energy build–up.
F. Zuccarello.
Astron. Astrophys., Vol. 257, No. 1, p. 298 – 306 (Apr 1992).

The in situ storage of magnetic energy is studied in a linear force–free arcade sheared by a relative velocity between its footpoints. The photospheric velocity fields used in the analysis are related to sunspots proper motions and to the solar differential rotation pattern. As observations have shown, the flare productivity is enhanced in sites where anomalies in the solar differential rotation pattern are present, where with anomalies, rigidly rotating structures, like pivot points in long–lived filaments, and parasitic polarities are indicated. Aim of this work is to verify whether these anomalies may play a significative role in the pre–flare stage of magnetic energy storage. The author has calculated the amount of energies involved in the magnetic energy build–up phase. The results show that sunspots proper motions are the most efficient to build–up magnetic energy ($\sim 2.7 \times 10^{33}\mathrm{erg}$ after 100 h with a velocity of $1.5 \times 10^4\mathrm{cm\,s}^{-1}$), but that also the values of the stored magnetic energy in a region characterized by rigidly rotating structures coupled with new born sunspots ($\sim 10^{33}\mathrm{erg}$ after 100 h), may be sufficient to allow flaring activity. Therefore it may be concluded that the anomalies in the solar angular velocity pattern, as deduced by observations, may have an important role in the magnetic energy build–up.

075.014 Magnetic field configuration associated with solar gamma–ray flares in June 1991.
M. J. Hagyard, E. A. West, J. E. Smith, F.–M. Trussart, E. G. Kenney.
Compton Observatory Science Workshop, p. 490 – 501 (Feb 1992). – See Abstr. 012.018 for the main entry.

The authors describe the vector magnetic field configuration of the solar active region AR 6659 that produced very high levels of flare activity in Jun 1991. The morphology and evolution of the photospheric fields are described for the period Jun 7 to Jun 10, and the flares taking place around these dates and their locations relative to the photospheric fields are indicated. By comparing the observed vector field with the potential field calculated from the observed line–of–sight flux, the authors identify the nonpotential characteristics of the fields along the magnetic neutral lines where the flares were observed. These results are compared with those from an earlier study of γ–ray flares.

075.015 On inversion polarity lines of large–scale magnetic fields.
G. V. Kuklin, M. V. Nikonova.
Astron. Tsirk., No. 1551, p. 22 – 23 (Nov – Dec 1991).

075.016 A true–field magnetogram in a solar plage region.
D. Rabin.
Astrophys. J., Lett., Vol. 390, No. 2, p. L103 – L106 (10 May 1992). With plates L10 – L11.

A new instrument, the Near–Infrared Magnetograph, has been used to make the first two–dimensional image of true magnetic field strength in the solar photosphere. The magnitude of the magnetic field vector is derived with a typical formal precision of $\pm 75\,\mathrm{G}$ (2σ) from circularly polarized spectra of a highly Zeeman–sensitive iron line at $6388.6\,\mathrm{cm}^{-1}$ ($1.565\,\mu\mathrm{m}$). The true–field map demonstrates that the properties of "kilogauss" flux tubes vary coherently on a variety of spatial scales within the 1' field of view. The measured fields span the range $1000\,\mathrm{G}$ – $1700\,\mathrm{G}$. The amplitude of the polarized signal implies that the spatial filling factor of the flux tubes can approach 0.3 at the seeing–limited resolution of 2″. Magnetic field strength and magnetic flux are statistically related in the sense that weak–field areas are weak–flux areas, but strong fields are present in both strong–flux and weak–flux areas. This implies a degree of independence in the relationship between the filling factor of flux tubes and their individual properties, such as field strength, pressure, and temperature.

075.017 Magnetic shear in flaring regions. I. Quantitative evaluation of the change in shear.
K. R. Sivaraman, R. R. Rausaria, S. M. Aleem.
Sol. Phys., Vol. 138, No. 2, p. 353 – 360 (Apr 1992).

The authors have evaluated the shear angle of the neutral line of the non–potential magnetic field for one or two days prior to and after the flare event for 10 cases. They have used the Hα filament positions to evaluate the shear in the neutral line. They find from the samples studied that it is a change in the shear that occurs a day prior to the flare that can lead to the event. This change can be in either direction, i.e., it can be a large increase from a small value or a decrease from a large initial value. Thus it is the change in the shear angle that seems to be deciding criterion for a flare to occur and not a large value for the shear angle itself. The authors have one instance where there was no significant change in the shear angle over a period of a few days and this region, although similar to other active regions studied, did not produce any flare activity.

075.018 Global modes constituting the solar magnetic cycle. II. Phases, "geometrical eigenmodes", and coupling of field behaviour in different latitudes.
M. H. Gokhale, J. Javaraiah.
Sol. Phys., Vol. 138, No. 2, p. 399 – 410 (Apr 1992).

The authors show that the axisymmetric odd degree SHF modes of 21.4–yr periodicity and degrees $l \leqslant 29$ in the solar magnetic field (as inferred from sunspot data during 1874 – 1976), are at least approximately stationary. Among the sine and cosine components of these SHF modes four groups are found, each defining the geometry of a coherent global oscillation characterized by a distinct power hump and its own level of variation. The first two of these "geometrical eigenmodes" (viz., B_1 and B_2), define the large–scale structure of the butterfly diagrams. Remaining SHF modes define the orderliness of the field distribution even within the wings of the butterflies down to scales $l \approx 29$. These include the "geometrical eigenmodes" B_3 and B_4, which are not present in simulated data sets in which the latitudes of the sunspot groups are randomly redistributed within the wings of the butterflies. Superposition of B_1, B_2, B_3, and B_4 is necessary and sufficient to reproduce important observed properties of the latitude–time distribution of the real field, not only in the sunspot zone, but also in the middle (35°–75°) and the high ($\gtrsim 75°$) latitudes, with appropriate relative orders of magnitude and phases. Thus, B_1, B_2, B_3, and B_4 seem to represent really existing global oscillations in the Sun's internal magnetic field. The geometrical form of B_1 may also be the form of the forcing oscillation.

075.019 High latitude solar magnetic fields.
N. Murray.
Sol. Phys., Vol. 138, No. 2, p. 419 – 422 (Apr 1992). Letter–to–the–editor.

The author uses Kitt Peak magnetograms to measure polar magnetic fields. The polar mean absolute field increases at the same time as the polar mean field decreases. That is, the polar mean absolute field varies in phase with solar activity, in contrast to the out of phase variation of the mean polar field. It is found that the polar fields have a large bipolar component even at solar minimum, with a magnitude equal to that found at low latitudes outside the active latitude bands.

075.020 The properties of sources and sinks of a linear force–free field.
P. Démoulin, E. R. Priest.
Astron. Astrophys., Vol. 258, No. 2, p. 535 – 541 (May 1992).

In a highly conducting plasma, the magnetic field topology determines where, e.g., current sheets can form, which is of great importance as a potential coronal heating source. With the classical extrapolation of a continuous weak photospheric field, the determination of topology is in general a difficult challenge. Because of the concentration of the photospheric field at intense

flux tubes in supergranulation boundaries a more realistic field representation may be a description in terms of magnetic singularities located just below the photosphere. The authors analyse in detail the generalization to linear force–free fields of the standard multipole expansion for singular potential fields. Solutions are presented in spherical coordinates with the constraint that all singularities are located in the half–space $z < 0$ below the solar photospheric plane ($z = 0$). A great variety of solutions is shown to exist depending on two continuous and one discrete parameter. The properties of monopole and dipole solutions in particular are discussed and it is shown that isolated magnetic charges exist only in the potential limit and not in a linear force–free field.

075.021 Synoptic charts of solar magnetic fields, Mount Wilson Observatory. Jan – Dec 1989.
R. K. Ulrich.
Q. Bull. Sol. Act., Vol. 31, Part II, p. 1 – 8 (1989).

The synoptic charts are constructed from digital data of the daily magnetograms obtained at the 150–foot Tower Telescope at Mount Wilson and concern solar rotations 1810 – 1823. The spectrum line employed is 5250.2, Fe I. Each synoptic chart is made up of computer–drawn segments from individual day's observation.

075.022 Energy buildup in sheared force–free magnetic fields.
R. Wolfson, B. C. Low.
Astrophys. J., Vol. 391, No. 1, p. 353 – 358 (20 May 1992).

Photospheric displacement of the footpoints of solar magnetic field lines results in shearing and twisting of the field, and consequently in the buildup of electric currents and magnetic free energy in the corona. The sudden release of this free energy may be the orgin of eruptive events like coronal mass ejections, prominence eruptions, and flares. An important question is whether such an energy release may be accompanied by the opening of magnetic field lines that were previously closed, for such open field lines can provide a route for matter frozen into the field to escape the Sun altogether. This paper presents the results of numerical calculations showing that opening of the magnetic field is permitted energetically, in that it is possible to build up more free energy in a sheared, closed, force–free magnetic field than is in a related magnetic configuration having both closed and open field lines. Whether or not the closed force–free field attains enough energy to become partially open depends on the form of the shear profile; the results presented here compare the energy buildup for different shear profiles. Implications for solar activity are discussed briefly.

075.023 Indices for the large–scale solar magnetic field.
L. Z. Shukhova, M. I. Pudovkin, V. D. Zholudev.
Geomagn. Aeron., Vol. 31, No. 1, p. 122 – 125 (Aug 1991). English translation of Geomgn. Aehron., Tom 31, No. 1, p. 172 – 176 (1991).

The principal–component method in statistics was applied to a time series of solar magnetograms in order to calculate integrated indices for the large–scale solar magnetic field. The two indices obtained are used to calculate the solar wind velocity at the earth's orbit via multivariate linear regression using solar data for the period from 31 March to 27 August 1979.

075.024 Spatially extended measurements of magnetic field strength in solar plages.
D. Rabin.
Astrophys. J., Vol. 391, No. 2, p. 832 – 844 (1 Jun 1992). With plates 2 – 4.

Magnetic field strengths along one spatial dimension of a plage region have been determined from circularly polarized (Stokes V) spectra of a highly Zeeman–sensitive iron line at 6388.6 cm^{-1} (1.565 μm). The measured fields are found to lie primarily in the range 1200 G – 1700 G. The mean formal precision for a single determination is ± 65 G (2σ). The field strength is coherently organized on spatial scales from 1′ to the limit of angular resolution (2″). The amplitude of the V signal implies that the spatial filling factor of the strong–field elements can approach 0.5

within a 2″ resolution element. Magnetic field strength and amplitude are correlated in the sense that locations with stronger mean fields have larger V amplitudes, but the relationship shows more scatter than can be explained by errors in measurement. The individual σ–components of the V profile are broader than an average quiet–Sun line profile would produce; the likely explanation is Zeeman broadening due to a range of magnetic field strength within the resolution element. The shapes of the V profiles become complex as the slit crosses the main polarity inversion line of the active region, where inclined fields, flows, and multiple field components must be considered.

075.025 **Generation of a strong toroidal magnetic field near the bottom of the convective zone.**
V. N. Krivodubskij.
Visn. Kiiv. Univ., Fiz.–Mat. Nauki, Astron., Vip. 3, p. 85 – 87 (1992). In Ukrainian.
The generation mechanism of the toroidal magnetic field by the angular velocity radial gradient acting on the relict poloidal magnetic field at the boundary between the convective and radiative zone is proposed. The magnetic induction magnitude of the toroidal field reaches about 2×10^2 T, the limiting effect of the magnetic buoyancy being taken into account. This value agrees with the estimation of the toroidal field obtained from helioseismological data.

075.026 **Peculiarities of quasi–periodic variations of large–scale solar magnetic fields and of solar wind properties.**
V. A. Kovalenko, S. I. Molodykh.
Planet. Space Sci., Vol. 40, No. 5, p. 741 – 747 (May 1992).
An analysis is made of variations in length of quiescent solar filaments and of solar wind parameters. The study revealed some peculiarities of annual and quasi–biennial variations of these characteristics; in particular, it was found that in March–April of the odd years there is a significant increase in the number of days occupied by high speed streams, which triggers a substantial enhancement of geomagnetic activity. It is shown that the dynamics of large–scale solar magnetic fields are quite well manifested in the particle flux density of the solar wind for differnt velocity regimes and the occurrence frequency of high ($V > 500$ km s^{-1}) and low ($V < 400$ km s^{-1}) solar wind velocities. The authors discuss the possible changes of solar magnetic fields which could be responsible for the observed variations of these parametes. They analyse, in terms of a solar wind formation model developed by the present authors, the relationship between the variability of large–scale solar magnetic fields and properties of the solar wind for its different formation regimes.

075.027 **A regularization method for extrapolation of solar potential magnetic fields.**
G. A. Gary, Z. E. Musielak.
Astrophys. J., Vol. 392, No. 2, p. 722 – 735 (20 Jun 1992).
This paper discusses the mathematical basis of a Tikhonov regularization method for extrapolating the chromospheric – coronal magnetic field using photospheric vector magnetograms. This analysis is the first step in the direction of mathematical rigor needed for the direct numerical integration techniques for the general force–free magnetic field Cauchy problem. The authors describe the mathematical formalism and develop the stability criteria and the accuracy level of extrapolated solutions as a function of the extrapolation height. The basic result of the paper is that, by introducing an appropriate smoothing of the initial data of the Cauchy potential problem, an approximate Fourier integral solution is found and an upper bound to the error in the solution is derived. The noise in the data would cause the solutions to diverge, but the smoothing decreases exponentially the contributions of the high spatial frequencies. This specific regularization technique, which is a function of magnetograph measurement sensitivities, provides a method to extrapolate the potential magnetic field above an active region into the chromosphere and low corona.

075.028 **Electric currents and magnetic–field loops in solar active regions.**
V. I. Abramenko, S. I. Gopasyuk, M. B. Ogir'.
Bull. Crimean Astrophys. Obs., Vol. 82 , p. 99 – 106 (1992). English translation of Izv. Krym. Astrofiz. Obs., Tom 82, p. 108 – 116 (1990).
Observations have been used on the magnetic–field vector from the 5250 Å Fe I line with H$_\alpha$ pictures for two active regions to examine the magnetic–field loop structures and the relationships between the electric currents and the fields. Much of the photospheric transverse field in an active region is due to currents flowing in the photosphere. Most of the filamentary structures in H$_\alpha$ do not coincide with projections of the field lines calculated in the potential approximation from the vertical component H$_2$. Those few that do coincide with line projections form three systems of loop structures differing in height. The numbers of these filaments are dependent on the active–region evolution. Currents flow in the upper chromosphere and adjacent corona. The self–induction coefficient for unit length of a current loop is 0.1 – 0.4, and it is larger where the transverse field is stronger. The energy stored in the local–current field is exceptionally high and is sufficient to provide the energy input to the most powerful flares.

075.029 **Convection in magnetic elements outside active solar regions.**
T. T. Tsap.
Bull. Crimean Astrophys. Obs., Vol. 82 , p. 114 – 118 (1992). English translation of Izv. Krym. Astrofiz. Obs., Tom 82, p. 124 – 129 (1990).
Studies have been made on the correlation between longitudinal magnetic fields and line–of–sight–velocities near the center of the solar disk. The fields and velocities have been observed with the double magnetograph at the Crimean Astrophysical Observatory on the lines Fe I 5250 Å, Fe I 5253 Å, Fe I 5233 Å, Mg I 5184 Å with high spatial resolutions (1″ × 1″ and 1″ × 2″). There is a close correlation between the velocities measured on those lines. Convective motions in the magnetic field elements are very small and do not exceed about 20 m/sec.

075.030 **Sign reversal of the high–latitude solar magnetic field.**
E. E. Benevolenskaya, V. I. Makarov.
Sov. Astron. Lett., Vol. 18, No. 2, p. 108 – 110 (Mar 1992). Current Physics Microform No.: 9211X0747. English translation of Pis'ma Astron. Zh., Tom 18, No. 3, p. 266 – 270 (1992).
The authors propose that triple sign reversals of the solar magnetic field are the result of two types of variation of the background magnetic field. The first is the Hale 22–year cycle, and the second is a quasi–biannual cycle. The evolution of the background magnetic field can be described to first order by a diffusion equation with allowance for meridional circulation and a two–frequency source of radial field. From numerical calculations it follows that in the case of multiplicity of frequencies, and for a certain ratio of amplitudes of these periods, zones of alternating polarity may be formed during the maxima of even–numbered 11–year cycles (Zurich numbering).

075.031 **Direct indication of magnetic reconnection in solar photosphere.**
Wang Jingxiu, Shi Zhongxian.
Chin. Astron. Astrophys., Vol. 16, No. 1, p. 71 – 77 (Jan–Mar 1992). English translation of Acta Astrophys. Sin., Vol. 11, No. 4, p. 389 – 393 (1991). See Abstr. 54.075.046.
A direct indication of magnetic reconnection in the solar photosphere has been found for the first time with the aid of the time sequence of vector magnetograms. The reconnection takes place in the interface between one pole of an emerging flux region and the old flux of opposite polarity. It occurs well after a subflare with X–ray classification of C2.9 when flux cancellation in the interface had lasted for several hours. It is suggested that the reconnection in the photosphere would be a common phenomenon on the sun.

075.032 An approach to the development of magnetic shear.
Wang Jingxiu.
Acta Astrophys. Sin., Vol. 12, No. 1, p. 75 – 81 (Jan 1992). In Chinese. English translation in Chin. Astron. Astrophys., Vol. 16, No. 2, p. 207 – 214 (Apr–Jun 1992).

In view of the approximate description of magnetic shear configuration by a force–free magnetic field, the force–free factor α would be a measure of magnetic shear, and its time changes would be analytically described by a differential equation. It is shown by this equation that the magnetic shear is generated by local dynamo action resulting from the interaction between magnetic field and plasma motion. It is also illustrated that the emergence and submergence of magnetic flux and the squeezing and pressing of the opposite polarity fields would be as effective as shear motion in producing the magnetic shear.

075.033 An approach to the development of magnetic shear.
Wang Jingxiu.
Chin. Astron. Astrophys., Vol. 16, No. 2, p. 207 – 214 (Apr–Jun 1992). English translation of Acta Astrophys. Sin., Vol. 12, No. 1, p. 75 – 81 (Jan 1992). See Abstr. 075.032.

In view of an approximate description of magnetic shear by non–potential character of a force–free magnetic field, the force–free factor would be a measure of magnetic shear, and the shear development would be analytically described by a differential equation. It is clearly shown by this equation that the magnetic shear is generated by local dynamo action resulting from the interaction between magnetic field and plasma motion. It is also illustrated that the squeezing and pressing of opposite polarity fields, the flux emergence and submergence would be as effective as shear motion in producing magnetic shear.

075.034 Observational evidence for various models of moving magnetic features.
J. W. Lee.
Sol. Phys., Vol. 139, No. 2, p. 267 – 273 (Jun 1992).

The author presents new measurements of moving magnetic features (MMFs) based on the observations of the active region NOAA 5612 made at Big Bear Solar Observatory on 2 August, 1989. He checks the existing theoretical models against the new observations and discusses the origin of MMFs conjectured from the deduced observational constraints.

075.035 On the method of calculating two–dimensional potential magnetic fields with current sheets.
V. S. Titov.
Sol. Phys., Vol. 139, No. 2, p. 401 – 404 (Jun 1992). Letter–to–the–editor.

The method of calculating two–dimensional potential magnetic configurations with current sheets (CS) is proposed, the number of CS in corona and the degree of asymmetry being both arbitrary. As a given boundary value the component of magnetic field normal to the photosphere is considered.

075.036 Observations and studies of solar magnetic fields and velocity fields.
Ai Guoxiang, Zhang Hongqi.
Advances in solar–terrestrial science of China, p. 25 – 37 (1992). – See Abstr. 003.069 for the main entry.

The authors look back on the development of instruments for measuring the solar magnetic field and velocity field and introduce the video vector magnetograph (Solar Magnetic Field Telescope) at Huairou Solar Observation Station of Beijing Astronomical Observatory. The authors also briefly summarize the important results of their research at Huairou in the past four years.

075.037 The cancellation of magnetic flux in the solar atmosphere.
Wang Jingxiu.
Advances in solar–terrestrial science of China, p. 77 – 88 (1992). – See Abstr. 003.069 for the main entry.

As one of the basic phenomenological modes of solar flux evolution, cancellation has been described as the gradual and mutual flux loss in closely spaced magnetic fields of opposite polarities. The apparent bipole region made up by these two cancelling components is referred to as one cancelling magnetic feature. Cancellation is the most common mode of the disappearance of magnetic flux on the sun. In this review paper, the author first presents a complete description of the observed mode of flux cancellation with emphasis on recent progress in clarifying its physical nature. A review is given of the apparent negatives and affirmatives of the association of flares to cancelling magnetic features and the possible explanation.

075.038 Characteristics of the magnetic field of the active region of the flare of 1989 July 5.
Zhong Shuhua, Luo Zhi, Xuan Jiayu, Lin Jun.
Publ. Yunnan Obs., No. 1, p. 42 – 46 (1992). In Chinese.

The magnetic field of the active region corresponding to the flare with the continuum emission, which erupted on July 5, 1989, is analysed. The results show that the magnetic intensity is stronger and the sunspot area is larger when the preceding sunspot appears, and the former is weaker and the latter is smaller when the following sunspot occurs.

075.039 Magnetic reconnection associated with emerging magnetic flux.
K. Shibata, S. Nozawa, R. Matsumoto.
Publ. Astron. Soc. Jpn., Vol. 44, No. 3, p. 265 – 272 (1992).

Two–dimensional (2D) magnetohydrodynamic (MHD) numerical simulations have been performed in order to study magnetic reconnection between an emerging flux and an overlying coronal magnetic field, while taking into account the effect of gravity, high spatial resolution, and a sufficient time span.

075.040 Small–scale photospheric magnetic fields.
M. Schüssler.
NATO Advanced Study Institute on the Sun: a Laboratory for Astrophysics, p. 191 – 220 (1992). – See Abstr. 012.091 for the main entry.

The author starts with a description of observational methods and results concerning non–spot magnetic fields with particular emphasis on Stokes profile analysis and the line ratio method as well as on recent results from helioseismology. It follows a discussion of the theoretical approaches which have been used to describe small–scale solar magnetic fields, i.e. magnetoconvection, thin flux tubes, and numerical model calculations and simulations. Then the author goes through the whole life cycle of magnetic elements and discusses observational and theoretical results concerning the various phases: formation by flux expulsion and thermal effects, magnetic and thermal structure in quasi–stationary equilibrium, dynamical processes and, eventually, instabilities, destruction, and decay.

075.041 Deformation of magnetic null points.
K. Galsgaard, Å. Nordlund.
IAU Colloquium No. 133: Eruptive solar flares, p. 343 – 346 (1992). – See Abstr. 012.096 for the main entry.

075.042 Magnetic energy conversion on the sun.
E. R. Priest.
International Workshop on Reconnection in Space Plasma, p. 73 – 81 (Jan 1989). – See Abstr. 012.099 for the main entry.

075.043 Generation and non–equilibrium of solar atmospheric magnetic fields.
N. Seehafer.
International Workshop on Reconnection in Space Plasma, p. 87 – 92 (Jan 1989). – See Abstr. 012.099 for the main entry.

075.044 **Possibilities and problems of the interpretation of solar magnetograph measurements and applications to flare–active regions.**
J. Staude, A. Hofmann.
International Workshop on Reconnection in Space Plasma, p. 123 – 128 (Jan 1989). – See Abstr. 012.099 for the main entry.
  The authors review the basic uncertainties encountered in the interpretation of magnetograph data, but also possibilities for deriving more reliable information. The problems are illustrated by describing the data handling of the Potsdam vector magnetograph in more detail, moreover, examples of observed flare–active regions demonstrate the state of information which nowadays can be obtained.

075.045 **Note on an apparent highly–sheared magnetic field structure.**
J. Linke, G. Bachmann.
International Workshop on Reconnection in Space Plasma, p. 131 – 132 (Jan 1989). – See Abstr. 012.099 for the main entry.
  Using Potsdam vector magnetograms and a force–free model, the authors have investigated the magnetic field structure of the active region BBR 18474 of 1982 July 15, and 1982 July 16. By comparison of the azimuths of the current–free model of the magnetic field structure in the central part of this active region with the map of the azimuth values estimated from the observation the authors find a good qualitative agreement. Additional earlier results have shown, that the appearance of highly sheared magnetic field structure near the neutral line in the central part of this active region may be a simulated effect generated by a subphotospheric current system.

Reconnection in space plasma. Proceedings.
See Abstr. 012.099.

Solar physics at Potsdam: vector magnetic field measurements, diagnostics and modelling of sunspot structure and dynamics.
See Abstr. 013.119.

Beobachtungen mit dem Vakuum–Turm–Teleskop auf Teneriffa.
See Abstr. 032.003.

Zürich Imaging Stokes Polarimeter – ZIMPOL I. Design review.
See Abstr. 034.034.

Magnetic field measurements.
See Abstr. 036.139.

Prominence sheets supported by constant–current force–free fields. II. Imposition of normal photospheric field component and prominence surface current.
See Abstr. 062.023.

Cylindrically symmetric force–free magnetic fields.
See Abstr. 062.024.

Equilibrium of a magnetic flux tube in a compressible flow.
See Abstr. 062.025.

The current sheet and Joule heating of a slender magnetic tube in the upper photosphere.
See Abstr. 062.030.

The equilibrium shape of slender flux tubes in a linear force–free magnetic field.
See Abstr. 062.031.

Some properties of finite energy constant–α force–free magnetic fields in a half–space.
See Abstr. 062.069.

Taylor relaxation of a Gold–Hoyle flux tube.
See Abstr. 062.073.

Magnetic reconnection in incompressible fluids.
See Abstr. 062.101.

An integral on the shape of isolated magnetic loops.
See Abstr. 062.102.

MHD equilibria with flows in uniform gravity. I. 1–D prominence– and arcade–type solutions.
See Abstr. 062.110.

Three dimensional MHD simulation of the Parker instability.
See Abstr. 062.197.

Field opening and reconnection.
See Abstr. 062.224.

Energy release at Alfvénic fronts in a force–free magnetic flux tube.
See Abstr. 062.225.

Spectral lines of the solar spectrum for investigation of magnetic fields by the line–ratio method.
See Abstr. 071.015.

High spatial resolution magnetograms of solar active regions.
See Abstr. 072.007.

Detection of "invisible sunspots".
See Abstr. 072.008.

Cyclic variation of the global magnetic field indices.
See Abstr. 072.017.

The east–west inclination of magnetic field lines in sunspots.
See Abstr. 072.018.

Structure of sunspot penumbrae: fallen magnetic flux tubes.
See Abstr. 072.024.

Solar active regions as a percolation phenomenon.
See Abstr. 072.036.

Vortex attraction and the formation of sunspots.
See Abstr. 072.037.

An emerging current/magnetic flux system with delta structure in AR 2372.
See Abstr. 072.045.

Infrared array measurements of sunspot magnetic fields.
See Abstr. 072.050.

Sunspots: an observational overview.
See Abstr. 072.056.

Observation and study of the brightness and the longitudinal field in the penumbra of a sunspot (I).
See Abstr. 072.058.

Estimation of sunspot magnetic field parameters using the super-penumbral topology.
See Abstr. 072.068.

The effects of magnetic field geometry on the confinement of energetic electrons in solar flares.
See Abstr. 073.001.

The modes of oscillation of a prominence. I. The slab with longitudinal magnetic field.
See Abstr. 073.019.

Magnetic fields and thermodynamic conditions in the solar flare of June 8, 1989.
See Abstr. 073.022.

The basal and strong–field components of the solar atmosphere.
See Abstr. 073.051.

Time evolution of a two–ribbon flare: characteristics of post–flare loops.
See Abstr. 073.056.

The tangential discontinuous surface of velocity in solar chromosphere.
See Abstr. 073.065.

Flares above sunspots and magnetic fields. I.
See Abstr. 073.067.

The relation between solar flare and longitudinal current density on Jan. 14, 1989.
See Abstr. 073.069.

Strong magnetic fields in solar flares: observational data and a theoretical model.
See Abstr. 073.076.

Overview of solar flares.
See Abstr. 073.083.

Basic magnetic configuration and energy supply processes for an interacting flux model of eruptive solar flares.
See Abstr. 073.087.

The role of cancelling magnetic fields in the buildup to erupting filaments and flares.
See Abstr. 073.088.

Variation of the vector magnetic field in an eruptive flare.
See Abstr. 073.089.

Intrinsically hot flares and possible connection to deep convective magnetic fields.
See Abstr. 073.090.

Interaction of large–scale magnetic structures in solar flares.
See Abstr. 073.091.

Triggering of eruptive flares: destabilization of the preflare magnetic field configuration.
See Abstr. 073.094.

"Post" flare loops.
See Abstr. 073.100.

Radio emission of eruptive flares.
See Abstr. 073.108.

Models of normal and inverse polarity filament eruptions and coronal mass ejections.
See Abstr. 073.111.

Preliminary research on the hot spots of energy flares since Cycle 21.
See Abstr. 073.123.

Unneutralised currents in solar flares.
See Abstr. 073.125.

The equilibrium of coronal flux tubes under toroidal forces.
See Abstr. 074.022.

On neutralized currents in the solar corona.
See Abstr. 074.050.

On potential field models of the solar corona.
See Abstr. 074.052.

Atmospheric responses to magnetic flux eruptions. III. Shock evolution near the Sun.
See Abstr. 074.059.

Magnetic reconnection associated with emerging magnetic flux.
See Abstr. 074.061.

Atmospheric responses to magnetic flux eruptions. III. Shock evolution near the Sun.
See Abstr. 074.062.

A numerical simulation of magnetically driven coronal mass ejections.
See Abstr. 074.075.

Coronal magnetic field strengths determined from fiber bursts.
See Abstr. 074.083.

Large–scale patterns on the Sun observed in the millimetric wavelength range.
See Abstr. 077.004.

Decimetric solar type U bursts: VLA and PHOENIX observations.
See Abstr. 077.026.

Diagnostics of energy release and magnetic fields on the sun by radio methods.
See Abstr. 077.049.

Stochastic fluctuations of the solar dynamo.
See Abstr. 080.002.

A nonlocal convection model of the solar convection zone.
See Abstr. 080.011.

Emergence of magnetic flux from the convection zone into the solar atmosphere. I. Linear and nonlinear adiabatic evolution of the convective – Parker instability.
See Abstr. 080.015.

The 22–year cycle of the Sun.
See Abstr. 080.021.

Generation of low–frequency pulsations in magnetic waveguides in the Sun's atmosphere.
See Abstr. 080.025.

Solar differential rotation due to magnetic stresses.
See Abstr. 080.053.

Acoustic wave ducting along magnetic flux tubes in the sunspot regions.
See Abstr. 080.054.

Waves in solar magnetic flux tubes: the observational signature of undamped longitudinal tube waves.
See Abstr. 080.058.

Doppler–Oszillationen unter dem Einfluss solarer Magnetfelder.
See Abstr. 080.068.

Short–period variations in the Sun's global magnetic field.
See Abstr. 080.071.

Numerical solution of magnetic flux emergence in gravity–stratified solar atmosphere.
See Abstr. 080.081.

Numerical solution of magnetic flux emergence in gravity–stratified solar atmosphere.
See Abstr. 080.083.

Magnetic fields and motions in the solar atmosphere.
See Abstr. 080.086.

Mean field dynamo theory.
See Abstr. 080.093.

Mg II absolute line profiles for late–type stars and for spatially–resolved solar regions.
See Abstr. 112.038.

The general magnetic field of the Sun and biennial variations of cosmic rays.
See Abstr. 144.051.

Frequency spectra of quasi–periodic variations in the general magnetic field of the Sun, in the parameters of the interplanetary magnetic field and cosmic rays.
See Abstr. 144.052.

076 UV, X, Gamma Radiation

076.001   X–rays and inner–shell transitions in the solar atmosphere.
G. A. Doschek.
AIP Conf. Proc., No. 215, p. 603 – 621 (6 Dec 1990). Current Physics Microform No.: 9103C1143. Paper presented at the 15. International Conference on X–Ray and Inner–Shell Processes, Knoxville, TN (USA), 9 – 13 Jul 1990. T. A. Carlson, M. O. Krause, S. T. Manson (eds.). ISBN 0–88318–790–6.
During the 1980's, very high spectral resolution solar X–ray spectra were obtained from a number of Bragg crystal spectrometer experiments on orbiting spacecraft. Taken together, these instruments covered the solar X–ray spectrum from about 1.8 Å up to about 25 Å. Inner–shell transitions in highly ionized ions were observed for several solar abundant elements, such as iron and calcium. Most of the spectra were obtained from solar flare plasmas at temperatures of about 1 – 3.5 keV. It was possible to study the time–behavior of line intensities and line profiles for all phases of a solar flare. From these spectra, the accuracy of certain atomic physics calculations can be determined, and parameters in the solar flare plasma such as electron temperature and density can be measured. The most significant results from these space missions are reviewed, and a new Bragg crystal spectrometer experiment is described.

076.002   Solar flare gamma–ray line profiles.
F. L. Lang, C. W. Werntz.
AIP Conf. Proc., No. 232, p. 445 – 452 (1 Aug 1991). Current Physics Microform No.: 9110A1141. – See Abstr. 012.003 for the main entry.
Solar gamma–ray lines are produced through collisions of pairs of positive ions whose center of mass energies are above the relevant thresholds for excitation of gamma–ray emitting states. Because of the low density of the solar plasma, prompt gamma rays are emitted by recoiling ions before significant energy loss has occurred. Thus, the lines are expected to be Doppler broadened to widths of the order of a hundred keV. Solar flare gamma–ray line profiles of the carbon and oxygen lines from heavy ions in the ambient solar medium interacting with representative high–energy proton and alpha particle populations will be presented.

076.003   Fe XVIII emission–line intensities in the Sun.
D. L. McKenzie, F. P. Keenan, S. M. McCann, K. A. Berrington, A. Hibbert, M. Mohan.
Astrophys. J., Vol. 385, No. 1, p. 378 – 380 (20 Jan 1992).
Recently calculated electron impact excitation rates among the $2s^2 2p^5$, $2s 2p^6$, and $2s^2 2p^4 nl$ levels of Fe XVIII are used to derive theoretical emission line ratios applicable to solar X–ray spectra. Oscillator strengths for all levels with $n \leqslant 3$ are used to take cascade involving the levels fully into account. Overall, the agreement between the theory and spectra measured by crystal spectrometers aboard the OV 1–17, P78–1, and SMM satellites is good. The effects of cascade from levels with $n \geqslant 4$ are small, at least for the strongest lines. The labor required to take into account these higher levels is not justified by the current discrepancies between theory and observations.

076.004   C IV line ratios in the Sun.
F. P. Keenan, E. S. Conlon, L. K. Harra, V. M. Burke, K. G. Widing.
Astrophys. J., Vol. 385, No. 1, p. 381 – 383 (20 Jan 1992).
Recent R–matrix calculations of electron impact excitation rates for transitions in C IV are used to derive the theoretical electron temperature–sensitive emission–line ratios $R_1 = I(312.43 \text{ Å})/I(419.74 \text{ Å})$, $R_2 = I(384.14 \text{ Å})/I(419.74 \text{ Å})$, $R_3 = I(312.43 \text{ Å})/I(419.49 \text{ Å})$, and $R_4 = I(384.14 \text{ Å})/I(419.49 \text{ Å})$. A comparison of these ratios with solar observational data obtained with the NRL S082A slitless spectrograph on board Skylab reveals good agreement between theory and observation for R_2 and R_4, which provides experimental support for the accuracy of the atomic data adopted in the analysis. However, most of the observed values of R_1 and R_3 lie above the theoretical high–temperature limit, which is probably due to blending of the 312.43 Å line.

076.005   Transitions from metastable levels emitted during short–duration bursts: how valid are their calculated intensities?
U. Feldman.
Astrophys. J., Vol. 385, No. 2, p. 758 – 762 (1 Feb 1992).
Recent spectroscopic studies of the quiescent and transient phenomena that take place in the Sun's atmosphere suggest that short–duration bursts may play a significant role in the heating processes of the upper solar atmosphere. Accepting bursts as working processes, concepts of electron density diagnostics in the upper solar atmosphere, and element abundance determinations, methods which are based mostly on line ratios formed in steady state coronal equilibrium are being reexamined. A review of the Fe IX density diagnostics and the effects of high–temperature short–lived bursts on this unique atomic system is given. An account of some of the metastable emission lines used in solar plasma diagnostics, and the electron densities at which they may become affected as a result of short–lived bursts, is also given.

076.006   Mg IX line ratios in the Sun.
F. P. Keenan, E. S. Conlon, L. K. Harra, K. G. Widing.
Astrophys. J., Vol. 386, No. 1, p. 371 – 374 (10 Feb 1992).
Theoretical Mg IX electron density sensitive emission line ratios, derived using electron impact excitation rates interpolated from accurate R–matrix calculations, are presented for $R_1 = I(443.97 \text{ Å})/I(368.07 \text{ Å})$, $R_2 = I(439.17 \text{ Å})/I(368.07 \text{ Å})$, $R_3 = I(368.07 \text{ Å})/I(443.07 \text{ Å})$, and $R_4 = I(441.20 \text{ Å})/I(368.07 \text{ Å})$. A comparison of these with observational data for solar flares, obtained with the NRL S082A spectrograph on board Skylab, reveals excellent agreement between theory and observations for R_1 and R_2, which confirms the usefulness of these ratios as N_e diagnostics for solar flares, as well as providing experimental

support for the accuracy of the atomic data adopted in the line ratio calculations. However, the observed values of both R_3 and R_4 generally imply unrealistically high electron densities, which is probably due to blending in the 443.40 and 441.20 Å lines, probably with Ar IV 443.44 Å, and Mg VI/Mg VII 441.22 Å, respectively.

076.007 Registration of high energy gamma–rays with the "Gamma–1" telescope from the solar flares on March 26 and June 15, 1991.
V. V. Akimov, V. G. Afanas'ev, A. S. Belousov,
I. D. Blokhintsev, V. A. Volzhenskaya, L. F. Kalinkin,
N. G. Lejkov, V. E. Nesterov, A. M. Gal'per, S. A. Voronov,
V. M. Zemskov, V. G. Kirillov–Ugryumov, B. I. Luchkov,
Yu. V. Ozerov, A. V. Popov, V. A. Rud'ko, M. F. Runtso,
V. Yu. Chesnokov, L. V. Kurnosova, M. A. Rusakovich,
N. P. Topchiev, M. I. Fradkin, E. I. Chujkin, V. Yu. Tugaenko,
T. N. Tyan, V. N. Ishkov, M. Gros, I. Grenier, E. Barouch,
P. Wallin, A. R. Baser–Bachi, J.–M. Lavigne, J. F. Olive,
J. Juchniewicz.
Pis'ma Astron. Zh., Tom 18, No. 2, p. 167–172 (Feb 1992). In Russian. English translation in Sov. Astron. Lett., Vol. 18, No. 1.

Gamma–radiation with energy range extending up to ≈2 GeV from the solar flares of March 26 and June 15, 1991 was registered with the "Gamma–1" telescope on board the astrophysical observatory "Gamma". The values of the fluxes and energy spectra of the gamma–rays were determined.

076.008 Solar Si II line ratios from the High–Resolution Telescope and Spectrograph.
F. P. Keenan, J. W. Cook, P. L. Dufton, A. E. Kingston.
Astrophys. J., Vol. 387, No. 2, p. 726–731 (10 Mar 1992).

New calculations of electron impact excitations rates for allowed transitions in Si II are used to derive theoretical emission–line ratios involving the $3s^23p\ ^2P^0 - 3s^23d\ ^2D$, $3s^23p\ ^2P^0 - 3s3p^2\ ^2S$, and $3s^23p\ ^2P^0 - 3s^24s\ ^2S$ multiplets near 1262, 1306, and 1530 Å respectively. A comparison of these line ratios with observational data from a quiet solar region, a sunspot, and an active region, obtained with the High–resolution Telescope and Spectrograph on board a sounding rocket flight reveals that the 1530 Å multiplet is optically thick, which is consistent with a calculation of the optical depth of these lines through a model atmosphere. The 1262 and 1306 Å multiplets appear to be effectively optically thin. The average discrepancy between the theoretical and observed ratios is ∼40%, which may not be significant, since the estimated uncertainties in both the calculated and experimental data are approximately 30%.

076.009 Gamma–ray measurements from the Space Shuttle during a solar flare.
P. S. Haskins, J. E. McKisson, A. G. Weisenberger, D. W. Ely,
T. A. Ballard, C. S. Dyer, P. R. Truscott, R. B. Piercey,
A. V. Ramayya.
Adv. Space Res., Vol. 12, No. 2/3, p. 331–334 (1992). – See Abstr. 012.016 for the main entry.

An X2/2B level solar flare occurred on 12 Aug, 1989, during the last day of the flight of the Space Shuttle Columbia. Detectors on the GOES 7 satellite observed increased X-ray fluxes at approximately 1400 GMT and a solar particle event at approximately 1600 GMT. Measurements with the bismuth germanate detector of the Shuttle Activation Monitor experiment showed factors of two to three increases in count rates at high latitudes comparable to those seen during South Atlantic Anomaly passages beginning at about 1100 GMT. That increased activity was observed at both north and south high latitudes in the 57°, 300 km orbit and continued until the detector was turned off at 1800 GMT. Measurements made earlier in the flight over the same geographic coordinates did not produce the same levels of activity. This increase in activity may not be entirely accounted for by observed geomagnetic phenomena which were not related to the solar flare.

076.010 COMPTEL neutron response at 17 MeV.
T. J. O'Neill, F. Ait–Ouamer, J. Morris, O. T. Tumer,
R. S. White, A. D. Zych.
Compton Observatory Science Workshop, p. 109–115 (Feb 1992). – See Abstr. 012.018 for the main entry.

The COMPTEL instrument was exposed to 17 MeV d, t neutrons prior to launch. These data have been analyzed and are compared with Monte Carlo calculations. Energy and angular resolutions are compared and absolute efficiencies are calculated at 0° and 30° incident angle. The COMPTEL neutron responses at 17 MeV and higher energies are needed to understand solar flare neutron data.

076.011 Ne V line ratios in the EUV spectra of solar flares.
F. P. Keenan, E. S. Conlon, L. K. Harra,
K. M. Aggarwal, K. G. Widing.
Astrophys. J., Vol. 389, No. 1, p. 440–442 (10 Apr 1992).

Theortical line ratios involving $2s^22p^2 - 2s2p^3$ transitions in Ne V between 359 and 572 Å are presented. A comparison of these with solar flare observational data from the NRL S082A spectrograph on board Skylab reveals excellent agreement between theory and experiment, with discrepancies that average only 8%. This provides experimental support for the accuracy of the atomic data adopted in the line ratio calculations, and in addition resolves discrepancies between theory and observations previously found for this species. The potential usefulness of the N V lines ratios as electron temperature diagnostics for the solar transition region is briefly discussed.

076.012 Flare γ–ray continuum emission from neutral pion decay.
D. Alexander, A. L. MacKinnon.
Compton Observatory Science Workshop, p. 523–527 (Feb 1992). – See Abstr. 012.018 for the main entry.

The authors investigate in detail the production of solar flare gamma ray emission above 10 MeV via the interaction of high energy protons with the ambient solar atmosphere. The considerations are restricted to the broad–band gamma ray spectrum resulting from the decay of neutral pions produced in p–H reactions. Inferences about the form of the proton spectrum at 10–100 MeV have already been drawn from de–excitaion gamma ray lines. The aim is to constrain the proton spectrum at higher energies. The detailed shape of the gamma ray spectra around 100 MeV is found to have a strong dependence on the spectral index of the power–law and on the turnover energy (from Bessel function to power–law). As would be expected the harder the proton spectrum the wider the 100 MeV feature. The photon spectra are to be compared with observations and used to place limits upon the number of particles accelerated and to constrain acceleration models.

076.013 High–energy gamma rays recorded by the Gamma 1 telescope from the 26 March and 15 June 1991 solar flares.
V. V. Akimov, V. G. Afanas'ev, A. S. Belousov,
I. D. Blokhintsev, V. A. Volzhenskaya, L. F. Kalinkin,
N. G. Lejkov, V. E. Nesterov, A. M. Gal'per, S. A. Voronov,
V. M. Zemskov, V. G. Kirillov–Ugryumov, B. I. Luchkov,
Yu. V. Ozerov, A. V. Popov, V. A. Rud'ko, M. F. Runtso,
V. Yu. Chesnokov, L. V. Kurnosova, M. A. Rusakovich,
N. P. Topchiev, M. I. Fradkin, E. I. Chujkin, V. Yu. Tugaenko,
T. N. Tyan, V. N. Ishkov, M. Gros, I. Grenier, E. Barouch,
P. Wallin, A. R. Baser–Bachi, J.–M. Lavigne, J. F. Olive,
J. Juchniewicz.
Sov. Astron. Lett., Vol. 18, No. 1, p. 69–71 (Jan 1992). Current Physics Microform No.: 9209X1287. English translation of Pis'ma Astron. Zh., Tom 18, No. 2, p. 167–172 (1992).

The Gamma 1 gamma–ray telescope aboard the Gamma astrophysical observatory recorded gamma emission extending to energies ∼2 GeV from the solar flares of 26 March and 15 June 1991. The flux and energy spectra of these gamma rays have been determined.

076.014 Solar EUV/UV and equatorial airglow measurements from San Marco–5.
G. Schmidtke, H. Doll, C. Wita, S. Chakrabarti.
J. Atmos. Terr. Phys., Vol. 53, No. 8, p. 781 – 785 (Aug 1991).
Paper presented at the 8. International Symposium on Equatorial Aeronomy (ISEA–8), San Miguel de Tucuman (Argentina), 21–27 March 1990.

Equatorial airglow and solar radiation measurements from 20 to 700 nm have been conducted with two spectrophotometric instruments aboard the San Marco–5 satellite from March to December 1988. The in–flight performance of the experiment and its in–flight calibration aspects are described. Preliminary results of the solar flux measurements and the planned scheme of the airglow modelling are presented.

076.015 Intercomparisons of the solar irradiance measurements from the Nimbus–7 SBUV, the NOAA–9 and NOAA–11 SBUV/2, and the STS–34 SSBUV instruments: a preliminary study.
R. P. Cebula, M. T. DeLand, D. F. Heath, E. Hilsenrath, R. D. Hudson, B. M. Schlesinger.
J. Atmos. Terr. Phys., Vol. 53, No. 11/12, p. 993 – 997 (Nov–Dec 1991). – See Abstr. 012.024 for the main entry.

Solar irradiance measurements in the spectral range 160–400 nm at approximately 0.15–0.20 nm intervals and at 1 nm resolution have been made by Solar Backscatter Ultraviolet (SBUV) series instruments continually since November 1978. Solar irradiance data from the Nimbus–7 SBUV satellite instrument, the SBUV/2 instruments on the NOAA–9 and NOAA–11 satellites, and the October 1989 flight (STS–34) of the Shuttle SBUV (SSBUV) instrument are presented and intercompared. Uncertainties in the instruments' absolute and long–term radiometric calibrations, which vary among the four instruments, are discussed. Comparisons of the initial, or "day 1", solar spectra from the four instruments show agreements to within approximately 10%, with spectral biases on the order of ±4%. Irradiances measured by the two NOAA instruments and SSBUV agree to within about 5% overall from 270 to 360 nm, with spectral biases on the order of about ±2%; the Nimbus–7 SBUV irradiances are an additional 5–10% lower in this region than those measured by the other three instruments.

076.016 Variability of solar ultraviolet irradiance.
J. M. Pap, R. F. Donnelly, H. S. Hudson, G. J. Rottman, R. C. Willson.
J. Atmos. Terr. Phys., Vol. 53, No. 11/12, p. 999 – 1003 (Nov–Dec 1991). – See Abstr. 012.024 for the main entry.

A model of solar Lyman alpha irradiance developed by multiple linear regression analysis, including the daily values and 81–day running means of the full disk equivalent width of the Helium line at 1083 nm, predicts reasonably well both the short– and long–term variations observed in Lyman alpha. In contrast, Lyman alpha models calculated from the 10.7 cm radio flux overestimate the observed variations in the rising portion and maximum period of solar cycle, and underestimates them during solar minimum. The authors show models of Lyman alpha based on the He line equivalent width and 10.7 cm radio flux for those time intervals when no satellite observations exist, namely back to 1974 and after April 1989, when the measurements of the Solar Mesosphere Satellite were terminated.

076.017 Revised solar extreme ultraviolet flux model.
W. K. Tobiska.
J. Atmos. Terr. Phys., Vol. 53, No. 11/12, p. 1005 – 1018 (Nov–Dec 1991). – See Abstr. 012.024 for the main entry.

An extended and revised solar extreme ultraviolet irradiance model for aeronomical use during the 1990s has been developed. The extensions significantly increase the application of the SERF2 solar EUV model beyong the October 1981 – April 1989 time–frame. The model can be used from 1947 to the present for coronal EUV full–disk irradiances and from 1976 to the present for chromospheric EUV full–disk irradiances. Substantial revisions to SERF2 were made which significantly improve the ability of the model to reproduce observed 27–day and solar cycle

EUV temporal variations. A multiple linear regression method is used to obtain coefficients for modelled EUV photon flux. This method allows for the inclusion of new rocket and satellite datasets into the model as they become available. The solar H Lyman–α and He I 10,830 Å equivalent width measurements are used as the independent model parameters for the chromospheric irradiances while the 10.7–cm radio emission daily and 81–day running mean values are the independent parameters for the coronal and transition region irradiances. The results of the model give full–disk photon fluxes at 1 AU for 39 EUV wavelength groups and discrete lines between 1.8 and 105.0 nm for a given date. The OSO, AEROS, AE satellite datasets and six rocket datasets used in the model development are summarized, the modelling technique is described in detail, the model formulation is presented, and the comparisons of the model to the datasets are discussed.

076.018 Solar variation 1979–1987 estimated from an empirical model for changes with time in the sensitivity of the solar backscatter ultraviolet instrument.
B. M. Schlesinger, R. P. Cebula.
J. Geophys. Res., Vol. 97, No. D9, p. 10119 – 10134 (20 Jun 1992).

Variations with time in the solar irradiance have been derived from a new empirical model for change in the sensitivity of the solar backscatter ultraviolet (SBUV) instrument aboard Nimbus 7. The SBUV sensitivity change is modeled as the product of two separate exponential decays: one with instrument exposure to the Sun, the other with time. Time constants for these exponentials are calculated from functional fits to the measured UV irradiances for three time intervals, short compared with the instrument lifetime. At nearly all wavelengths the degradation coefficients for exposure dependence (r_λ) do not vary significantly from interval to interval, but those for time dependence (s_λ) vary significantly with time at the shorter wavelengths. A new instrument sensitivity function is defined to be the product of an exponential decay with exposure with constant r_λ and an exponential decay with time with time–varying s_λ. The time dependence of s_λ is constrained by the values derived for the three fitting intervals. Use of this functional form to correct the measured irradiances for instrument changes yields solar irradiance variations between 175 and 400 nm within 3% of those predicted by the Mg core–to–wing ratio, appropriately scaled with wavelength. This result strongly suggests that the instrument parameters derived during the short fit intervals can be used to improve the representation of instrument behavior during those intervals and can serve as constraints on any function designed to characterize long–term changes in the SBUV instrument. The new characterization yields an estimate for the variation of solar irradiance between maximum and minimum of 5–8% at 205 nm and 1–4% between 210 and 260 nm, consistent with the 8–9 and 3–4% predicted using the Mg index.

076.019 The Sun as a star: high spectral resolution solar data degraded to low–dispersion IUE resolution.
J. G. Doyle, J. W. Cook.
Astrophys. J., Vol. 391, No. 1, p. 393 – 402 (20 May 1992).

The authors present high spectral resolution (~0.06 Å) solar data for an active region and a large two–ribbon flare degraded to the typical resolution of low–dispersion (~5 Å) spectra of the IUE satellite. This clearly shows the amount of detail yet to be acquired in stellar spectra and indicates the main spectral regions where line blending is a problem.

076.020 High resolution extreme ultraviolet (EUV) studies of the Sun.
S. K. Jain.
AIP Conf. Proc., No. 245: Workshop on Basic Space Science, p. 159 – 169 (1992). – See Abstr. 012.064 for the main entry.

The author briefly discusses some of the currently unanswered problems in solar astronomy which lend themselves to investigations particularly in the EUV region of the spectrum. In this context, need of high spatial, spectral and temporal resolution is emphasized for such studies. Finally, after a brief discussion of the EUV instrumentation, a high resolution stigmatic EUV spectroheliometer is described which is currently under development for high resolution studies of the solar chromosphere, transition region and corona. The spectroheliometer will be flown aboard a sounding rocket.

076.021 Lower limits to the speed of transport of impulsive flare energy deduced from joint UVSP/HXRBS observations.
R. A. Schwartz, S. A. Drake.
Bull. Am. Astron. Soc., Vol. 23, No. 3, p. 1264 (1991). Abstract. – See Abstr. 012.067 for the main entry.

076.022 X–ray observations of limb flare loops and post–flare coronal arch.
Z. Švestka, K. L. Smith, K. T. Strong.
Sol. Phys., Vol. 139, No. 2, p. 405 – 408 (Jun 1992). Letter–to–the–editor.

The authors present observations of a post–flare arch following an eruptive flare, detected in X–ray lines above the western solar limb on 2 May 1985.

076.023 New statistical properties of solar X–ray and γ–ray bursts.
Zhang Heqi.
Advances in solar–terrestrial science of China, p. 60 – 66 (1992). – See Abstr. 003.069 for the main entry.

In this paper, the SMM/GRS data obtained during the solar maximum are studied. Some new statistical properties pointed out here will be significant in establishing models of particle acceleration in solar flares. SMM/GRS data have been extensively studied and some interesting results of data analysis have been published in the past five years. In this work, taking a new point of view, the author makes a detailed study of over one hundred solar X–ray and γ–ray bursts obtained from SMM/GRS during the 1980 – 1982 solar maximum. A preliminary analysis leads to some important statistical results, which point out several statistical properties of GMS data and some possible astrophysical implications for charged particle acceleration in solar flares.

076.024 High–energy flare emissions.
A. G. Emslie.
NATO Advanced Study Institute on the Sun: a Laboratory for Astrophysics, p. 475 – 488 (1992). – See Abstr. 012.091 for the main entry.

The author reviews the various components of high energy emission and the mechanisms responsible for them. A disproportionately large part of the discussion is concerned with hard X–ray emission – its spectrum, polarization, spatial structure, and temporal characteristics. Both "nonthermal" and "thermal" models for hard X–ray bursts are reviewed, with the main thrust of the argument aimed at laying this historically significant, but physically questionable, dichotomy to rest. The current status of our understanding, together with a prognosis for the future, concludes this chapter.

The 1984 – 1987 Solar Maximum Mission event list.
See Abstr. 002.088.

The 1988 Solar Maximum Mission event list.
See Abstr. 002.089.

The 1989 Solar Maximum Mission event list.
See Abstr. 002.090.

7. Quadrennial Symposium on Solar–Terrestrial Physics (SCOSTEP–7), The Hague (Netherlands), 25 – 30 Jun 1990.
See Abstr. 012.024.

Emission lines from O IV as a plasma diagnostic.
See Abstr. 022.097.

An EUV imaging spectrograph for high–resolution observations of the solar corona.
See Abstr. 035.023.

The solar X–ray/cosmic gamma–ray burst experiment aboard Ulysses.
See Abstr. 035.038.

A rotating tomographic imager for solar extreme–ultraviolet / soft X–ray emission.
See Abstr. 035.042.

Radiometric calibration of solar space telescopes – the development of a vacuum–ultraviolet transfer source standard.
See Abstr. 035.067.

Solar ultraviolet instrumentation.
See Abstr. 035.123.

Soft X–ray instrumentation.
See Abstr. 035.124.

X–ray instrumentation.
See Abstr. 035.125.

Observation of the solar Lyman–α line.
See Abstr. 035.127.

The derivation of parent electron spectra from bremsstrahlung hard X–ray spectra.
See Abstr. 063.021.

Photospheric lines redshift from balloon ultraviolet spectra of the quiet Sun.
See Abstr. 071.027.

UV observations of sunspots.
See Abstr. 072.057.

The sunspot active regions for the 22nd solar cycle.
See Abstr. 072.066.

The relationship between global X–ray luminosity and flaring on the Sun.
See Abstr. 073.005.

Energetics and dynamics in a large solar flare of 1989 March.
See Abstr. 073.006.

Fluid flow in a jet and the Ca XIX line profiles observed during solar flares.
See Abstr. 073.007.

Gamma–ray lines from solar flares.
See Abstr. 073.008.

A burst model for line emission in the solar atmosphere I. XUV lines of He I and He II in impulsive flares.
See Abstr. 073.010.

γ–ray and X–ray time profiles expected from a trap–plus–precipitation model for the 7 June 1980 and 27 April 1981 solar flares.
See Abstr. 073.020.

Excitation of the solar flare far–ultraviolet continuum by line irradiation.
See Abstr. 073.033.

High–energy gamma–ray emission from pion decay in a solar flare magnetic loop.
See Abstr. 073.044.

Characteristics of hard X–ray spectra of impulsive solar flares.
See Abstr. 073.045.

Structure and physics of solar faculae. V. Study of the roughness effect in the quiet chromosphere–corona transition zone.
See Abstr. 073.046.

The Hα development of an intense limb flare and associated flaring arches.
See Abstr. 073.049.

The simultaneous effects of collisions, reverse currents and magnetic trapping on the temporal evolution of energetic electrons in a flaring coronal loop.
See Abstr. 073.053.

Impulsive phase Fe Kα emission in a flare of 1989 March.
See Abstr. 073.057.

Research of solar events in optical, radio and X–ray regions.
See Abstr. 073.077.

Temporal and spatial characteristics of solar flares from observations.
See Abstr. 073.085.

Energy transport in solar flares: implications for Ca XIX emission.
See Abstr. 073.095.

Fluid flow in a jet and the Ca XIX line profiles observed during solar flares.
See Abstr. 073.096.

Characteristics of the impulsive phase of flares.
See Abstr. 073.097.

Comparison of UV and X–ray solar flare observations and theoretical models.
See Abstr. 073.098.

White–light flares.
See Abstr. 073.099.

Flare evolution in the density–temperature diagram.
See Abstr. 073.102.

The Neupert effect: what can it tell us about the impulsive and gradual phases of eruptive flares?
See Abstr. 073.104.

Nuclear reactions in flares.
See Abstr. 073.107.

On the association between large scale X–ray brightenings and solar flares.
See Abstr. 073.110.

The X12 limb flare and spray of 01 June 1991.
See Abstr. 073.115.

Analysis of X–ray flares observed by the SMM spacecraft.
See Abstr. 073.116.

Sub–second variations of HXR and H–alpha flare emission.
See Abstr. 073.118.

The precursors of the 9 July 1982 solar proton flare event.
See Abstr. 073.130.

Coronal Magnetic Structures Observing Campaign. IV. Multiwaveband observations of sunspot and plage–associated coronal emission.
See Abstr. 074.010.

The X–ray corona, the coronal hole, and the heliosphere.
See Abstr. 074.019.

The birth of giant post–flare arches.
See Abstr. 074.028.

Steady siphon flows in closed coronal structures: comparison with extreme–ultraviolet observations.
See Abstr. 074.034.

Stereoscopic observations of a solar flare hard X–ray source in the high corona.
See Abstr. 074.042.

Theoretical analysis of the Fe XVIII X–ray spectrum and application to solar coronal observations.
See Abstr. 074.045.

Evidence for mass outflow in the low solar corona over a large sunspot.
See Abstr. 074.053.

Element abundances and plasma properties in a coronal polar plume.
See Abstr. 074.054.

Large–scale quasi–stationary X–ray coronal structures associated with eruptive solar flares.
See Abstr. 074.066.

A giant post–flare coronal arch observed by Skylab.
See Abstr. 074.069.

Microwave and soft X–ray emission of solar flare events associated with coronal mass ejections.
See Abstr. 074.089.

Gamma–ray and microwave emission from 1991 June events.
See Abstr. 077.016.

Microwave and soft X–ray radiation during flares evolving in strong magnetic fields.
See Abstr. 077.036.

Launch–times of MHD shocks observed as type II bursts.
See Abstr. 077.037.

Solar abundances from gamma–ray spectroscopy: comparisons with energetic particle, photospheric, and coronal abundances.
See Abstr. 080.022.

27–day fluctuations in the ionospheric D–region.
See Abstr. 083.017.

077 Radio, Infrared Radiation

077.001 A computer simulation study of Type III radio burst propagation through the solar corona.
M. A. Itkina, B. N. Levin.
Astron. Astrophys., Vol. 253, No. 2, p. 521 – 524 (Jan 1992).
Type III solar radio burst propagation through large–scale coronal structure is numerically simulated. It is shown that radio wave refraction in an overdense streamer results in an increase of the apparent radial distance of the Type III fundamental source and produces a broadening of the radiation polar diagram in agreement with the observations. It is also verified that the well known fine frequency structure of Type IIIb emission can be due to the fibrous character of the streamer.

077.002 On the effect of radio waves group delay in the solar corona.
A. B. Eremin.
Pis'ma Astron. Zh., Tom 18, No. 1, p. 74 – 80 (Jan 1992). In Russian. English translation in Sov. Astron. Lett., Vol. 18, No. 1.
A calculation of the radio waves group delay has been made for solar type III radio emission. As a model for the electron density in the corona the Newkirk formula is used. It is shown that neglecting the effect considered may cause an error about 40% in the evaluation of the mean velocity of fast electron beams exciting solar type III radio bursts. The calculation has been made in the frame of the type III fundamental generation scheme proposed by Eremin and Zajtsev (1985).

077.003 Multifrequency observations of a remarkable solar radio burst.
S. M. White, M. R. Kundu, T. S. Bastian, D. E. Gary,
G. J. Hurford, T. Kucera, J. H. Bieging.
Astrophys. J., Vol. 384, No. 2, p. 656 – 664 (10 Jan 1992).
Observations of an impulsive solar radio burst from three observatories are presented. The striking observational aspects of this flare are that the time profile was identical throughout at 8.6, 15, and 86 GHz, that the spectrum was apparently flat from 15 to 86 GHz, and that there was a sharp cutoff in the spectrum between 5.0 and 8.6 GHz. The simplest interpretation of the cutoff, a plasma frequency effect, leads to the conclusion tht there was exceptionally high–density material in the solar corona ($\sim 5 \times 10^{11}$cm). VLA images at 15 GHz show a single loop structure which brightened uniformly and showed little change in size during the whole impulsive phase. The flat spectrum is consistent with optically thin thermal bremsstrahlung emission, but the lack of observed soft X–ray emission and other properties of the flare cannot easily be accommodated by this mechanism. The authors explore the possibility that the emission is optically thick due to thermal absorption of nonthermal gyrosynchrotron emission, or optically thin gyrosynchrotron emission absorbed by high–density material intervening along the line of sight. Both of these explanations also face difficulties.

077.004 Large–scale patterns on the Sun observed in the millimetric wavelength range.
B. Vršnak, S. Pohjolainen, S. Urpo, H. Teräsranta, R. Brajša,
V. Ruždjak, Z. Mouradian, S. Jurač.
Sol. Phys., Vol. 137, No. 1, p. 67 – 86 (Jan 1992).
The nature and behaviour of large–scale patterns on the solar surface, indicated by the areas of brightness–temperature depressions in the millimetric wavelength range, is studied. A large sample of 346 individual, low–temperature regions (LTRs) was employed to provide reliable statistical evidence. An association of 99% was found between the locations of LTRs and the large–scale magnetic field inversion lines, and 60% of the LTRs were associated with the inversion line filaments. A tentative physical association with filaments is reconsidered, and one particularly well–observed case is presented. The heights of the perturbers causing brightness–temperature depressions are discussed. The long–term evolution of the latitudinal distribution of LTRs is presented in a butterfly diagram. Two belts of low–temperature regions outline the active region belts, shifting with them towards the equator during the solar activity cycle. The low–temperature region belts of the forthcoming cycle appear already at the maximum of the actual cycle at latitudes of about 55°. The superpositions of the temperature minima distributions in the synoptic maps show patterns appearing as "giant cells" and compatible with indications inferred from magnetographic data. The reliability of the inferred cells is considered, and a statistical analysis reveals a negligible probability for an accidental distribution appearing in the form of giant cells.

077.005 Quiet–Sun emission and local sources at meter and decimeter wavelengths and their relationship with the coronal neutral sheet.
P. Lantos, C. E. Alissandrakis, D. Rigaud.
Sol. Phys., Vol. 137, No. 2, p. 225 – 256 (Feb 1992).
The authors analysed multifrequency 2–dimensional maps of the solar corona obtained with the Nançay radioheliograph during two solar rotations in 1986. They discuss the emission of the quiet Sun, coronal holes and local sources and its association with chromospheric and coronal features as well as with large–scale magnetic fields. They give statistics of source brightness temperatures, as well as distributions in longitude and latitude. Although we found no significant center–to–limb effect in the brightness temperature, the sources were not visible far from the central meridian (apparently a refraction effect). The brightest sources at 164 MHz were near, but not directly above active regions and had characteristics of faint type I continua. At 408 MHz some sources were observed directly above active regions and one was unambiguously a type I continuum. The majority of the fainter sources showed no association with chromospheric features seem on Hα synoptic charts, including filaments. Most of them were detected at one frequency only. Sources identified at three frequencies (164, 327, and 408 MHz) were located in regions of enhanced large–scale magnetic field, some of them at the same location as decayed active regions visible one rotation before on synoptic Hα charts. Multifrequency sources are associated with maxima of the green line corona. The comparison with K–corona synoptic charts shows a striking association of the radio sources with dense coronal regions, associated with the coronal neutral sheet. Furthermore, the authors detected an enhanced brightness region which surrounds the local sources and is stable over at least one solar rotation. They identify it with the radio counterpart of the coronal neutral sheet.

077.006 Radial variation of type–III source parameters.
P. A. Robinson.
Sol. Phys., Vol. 137, No. 2, p. 307 – 315 (Feb 1992).
The parameters of type–III sources have been observed to vary as powers of the distance of the source from the Sun. Here, the values of the observed exponents are reviewed and theoretical relationships between them are discussed and extended. It is shown that 11 observed exponents can be derived from a four–element subset. A least–squares fit is carried out by varying these four exponents and it is shown that the results are consistent with observation to within the observational uncertainties. Best–fit expressions are given for the plasma density and temperature, the solar wind speed, beam velocity and density, frequency drift rate, peak Langmuir fields, brightness temperature, volume emissivity, beam duration and burst decay time.

077.007 Connection between ambient density fluctuations and clumpy Langmuir waves in type III radio sources.
P. A. Robinson, I. H. Cairns, D. A. Gurnett.
Astrophys. J., Lett., Vol. 387, No. 2, p. L101 – L104 (10 Mar 1992).
A recent stochastic – growth theory of clumpy Langmuir waves in type III sources is shown to imply that the clumps will have the same size distribution as the ambient low–frequency density fluctuations in the solar wind. Spectral analysis of Langmuir–wave time series from the ISEE 3 plasma wave

instrument confirms this prediction to within the uncertainties in the spectra. The smallest Langmuir clump size is inferred to be in the range 0.4 – 30 km in general, and 2 – 30 km for beam-resonant waves. It is concluded that the diffusion of waves in the source is anomalous.

077.008 High dynamic range multifrequency radio observations of a solar active region.
S. M. White, M. R. Kundu, N. Gopalswamy.
Astrophys. J., Suppl. Ser., Vol. 78, No. 2, p. 599 – 617 (Feb 1992). With plate 15.

The authors present high range, multifrequency radio observations of a solar active region. The evolution of the region is followed at 5 GHz as it rotates from the limb to disk center, and when it is at disk center, observations at 0.33, 1.5, 5, 8.4, and 15 GHz are used to analyze the distribution of density and magnetic field within the active region. A dynamic range of up to 1500 (at 8.4 GHz) was achieved. This high dynamic range allows to image material at brightness temperatures ranging from 4000K to 1.7×10^6K in the same map. The authors can thus unambiguously identify the signatures of both optically thick gyroresonance emission, outlining magnetic fields, and optically thin thermal free–free emission, indicating density contrast, simultaneously. By comparing images at 5 and 8.4 GHz, the authors identify regions in the trailing part of the active region where optically thin fourth–harmonic gyroresonance emission is contributing to the observed brightness temperatures at 5 GHz, indicating the presence of 450 G fields. An important result is that in comparing higher–frequency emission with the 1.5 GHz image, the authors find that the 1.5 GHz flux cannot be completely explained by free–free emission and conclude that there is a component of gyroresonance emission at the edges of the active region, where the corona is not opaque due to free–free opacity, and that there are 140 G fields in the low corona.

077.009 Some characteristics of the S–component and outbursts of solar emission at millimeter wavelengths.
I. G. Moiseev, N. S. Nesterov, P. S. Nikitin.
Pis'ma Astron. Zh., Tom 18, No. 2, p. 173 – 182 (Feb 1992). In Russian. English translation in Sov. Astron. Lett., Vol. 18, No. 1.

Simultaneous observations at wavelengths of 13 mm (parameters I and V) and 8 mm (I) resulted in detection of a correlation of burst emission spectral index α_b with α of the parent local source and with burst rise time. The shorter is the burst and the faster is its development the steeper is the emission spectrum of the event. It is found that the circular polarization degree does not exceed 35%, spectral indices are in the range between –0.5 and –6.5. The observed characteristics of burst may be explained by models of inhomogeneous thermal (free–free) sources or by combination both of nonthermal and of thermal radiation.

077.010 Properties of type III bursts in a gasdynamic model of the propagation of an electron stream in plasma.
V. N. Mel'nik.
Kinematics Phys. Celest. Bodies, Vol. 7, No. 3, p. 55 – 63 (1991). English translation of Kinematika Fiz. Nebesn. Tel, Tom 7, No. 3, p. 59 – 68 (1991).

A model of the type III burst is based on a gasdynamic theory of the scattering of an electron stream in plasma. In this model, the burst consists of a large number of microbursts of short duration (1 – 100 msec at decameter wavelengths) with high brightness temperature. The microbursts are beam–plasma structures that move at constant and approximately equal velocities ($v_{pl} = 0.3$ s). The velocity scatter explains the shape of the bursts, and the value found for the velocity is close to the average observed velocity of the sources of type III bursts.

077.011 Nontraditional emission mechanism of type III bursts at twice the plasma frequency.
V. N. Mel'nik.
Kinematics Phys. Celest. Bodies, Vol. 7, No. 3, p. 64 – 66 (1991). English translation of Kinematika Fiz. Nebesn. Tel, Tom 7, No. 3, p. 69 – 71 (1991).

A nontraditional radioemission mechanisms of type III bursts at twice the plasma frequency that consists in merging of

longitudinal and transverse waves (l + t→t) at frequency $\omega = \omega_p$ is discussed. The former are excited by a stream of fast electrons, while the latter are formed by Rayleigh scattering of the longitudinal waves on plasma ions. It is shown that this mechanism is of threshold nature. The emission brightness temperatures of the bursts at the second harmonic are found in a gasdynamic model of the type III burst to be $T_{br}^{III} = 10^9 - 10^{10}$K at decameter wavelengths.

077.012 Two–dimensional model maps of flaring loops at cm–wavelengths.
P. Preka–Papadema, C. E. Alissandrakis.
Astron. Astrophys., Vol. 257, No. 1, p. 307 – 314 (Apr 1992).

The authors present a complete set of model computations of the microwave emission from a flaring loop. Two–dimensional maps in total intensity and circular polarization are given, as a function of wavelength, heliocentric distance and the orientation of the loop with respect to the direction of the limb. In agreement with the authors' previous one–dimensional computations, the emission in the optically thin case comes from the one or both feet of the loop, with the primary maximum usually at the diskward side; in the optically thick case the emission comes from the entire loop, with one or two maxima near the top. The authors discuss in detail the effects of the orientation of the loop, as well as the polarization structure produced by self–absorption and gyro-resonance absorption. The latter produce a patchy structure in the V map, with regions polarized both in the extraordinary sense (in the upper part of the loop) and in the ordinary sense (in the lower part of the loop). Finally the authors compare their computations with high spatial resolution observations of simple bursts.

077.013 Observations of mode coupling in the solar corona and bipolar noise storms.
S. M. White, G. Thejappa, M. R. Kundu.
Sol. Phys., Vol. 138, No. 1, p. 163 – 187 (Mar 1992).

The authors review high–spatial–resolution observations of the Sun which reflect on the role of mode coupling in the solar corona, and present a number of new observations. They show that typically polarization inversion is seen at 5 GHz in active region sources near the solar limb, but not at 1.5 GHz. Although this is apparently in contradiction to the simplest form of mode coupling theory, in fact it remains consistent with current models for the active region emission. Microwave bursts show no strong evidence for polarization inversion. The authors discuss bipolar noise storm continuum emission in some detail, utilizing recent VLA observations at 327 MHz. They show that bipolar sources are common at 327 MHz. Further, the trailing component of the bipole is frequently stronger than the leading component, in apparent conflict with the "leading–spot" hypothesis. The observations indicate that at 327 MHz mode coupling is apparently strong at all mode–coupling layers in the solar corona. The 327 MHz observations require a much weaker magnetic field strength in the solar corona to explain this result than did earlier lower–frequency observations: maximum fields are 0.2 G. This is a much weaker field than is consistent with current coronal models.

077.014 Spectral structures of type "line" in the microwave burst emission at the Sun.
A. P. Klassen.
Sol. Phys., Vol. 138, No. 1, p. 201 – 203 (Mar 1992). Letter–to–the–editor.

Observational results of radioburst spectra in the range of 2.0 – 4.0 GHz are presented. It is shown that the narrow–band spectral structures of the type "line" with line widths of $Df/f \le 0.1$ and intensity $DF/F > 10\%$ are observed in the impulsive phase of the bursts.

077.015 Significance of inverse Cherenkov effect and its relevance to interpret the structures of type IV radio emissions.
N. L. Kotcherlakota.
4. Asia Pacific Physics Conference, p. 801 – 804 (1991). – See Abstr. 012.006 for the main entry.

The inverse Cherenkov effect followed by the refractive enhanced amplification of p–mode waves within clouds of active plasma regions of solar bursts is suggested to lead to newer instabilities. A modulation of these waves by modified cyclotron waves is noted as responsible for the richness of detail delivered by solar type IV radio bursts. Refractive indices, drift velocities and growth times characterizing patterns such as zebra–tadpole, rope–like, and spikes etc. structures occurrence are given.

077.016 Gamma–ray and microwave emission from 1991 June events.
S. Enome, H. Nakajima, H. S. Hudson, R. Schwartz.
Compton Observatory Science Workshop, p. 522 (Feb 1992). Abstract. – See Abstr. 012.018 for the main entry.

077.017 Millisecond microwave spikes at 8 GHz during solar flares.
A. O. Benz, H. Su, A. Magun, W. Stehling.
Astron. Astrophys., Suppl. Ser., Vol. 93, No. 3, p. 539 – 544 (Jun 1992).
High circularly polarized spikes have been observed during solar flares up to 8 GHz. The typical half–power bandwidth was 120 MHz, and the duration was less than the time resolution of 100 ms. A group of 46 spikes in the 6.5 – 8.0 GHz range has been observed at the maximum of an $H\alpha$ flare and within 3 s of the peak of microwave emission observed at 2.7 GHz by other observatories. The peak flux of the spikes reached 60 sfu above background. These emissions closely resemble the spikes previously reported at lower frequency. If interpreted as second harmonic of the electron gyrofrequency, a magnetic field exceeding 1400 G would be requested in the corona.

077.018 The role of signal group delay in the solar corona.
A. B. Eremin.
Sov. Astron. Lett., Vol. 18, No. 1, p. 32 – 34 (Jan 1992). Current Physics Microform No.: 9209X1250. English translation of Pis'ma Astron. Zh., Tom 18, No. 1, p. 74 – 80 (1992).
The group delay of signals in the coronal plasma is calculated for type III solar radio emission using the Newkirk coronal model. It is shown that neglect of this effect leads to an error of 40% in determining the mean velocity of the fast electron fluxes which generate type III solar radio radiation. The calculation was performed using the fundamental–mode generation scheme for type III flares proposed by Eremin and Zajtsev (1985).

077.019 Some characteristics of S–component emission and solar flares at millimeter wavelengths.
I. G. Moiseev, N. S. Nesterov, P. S. Nikitin.
Sov. Astron. Lett., Vol. 18, No. 1, p. 71 – 75 (Jan 1992). Current Physics Microform No.: 9209X1289. English translation of Pis'ma Astron. Zh., Tom 18, No. 2, p. 173 – 182 (1992).
Simultaneous observations at 13 mm (I and V parameters) and 8 mm (I parameter) have revealed a correlation between the slope of the S–component source spectrum and the duration of the flare rise time, with more rapid flares having steeper spectra. The degree of circular polarization in flares does not exceed 35%. The spectral index is in the range between –0.5 to –6.5. The flare characteristics observed at mm wavelengths can be explained both by models with thermally inhomogeneous sources and by nonthermal emission.

077.020 Compact sources of suprathermal microwave emission detected in quiescent active regions during lunar occultations.
E. Correia, P. Kaufmann, F. M. Strauss.
Sol. Phys., Vol. 138, No. 2, p. 223 – 231 (Apr 1992).
Solar quiescent active regions are known to exhibit radio emission from discrete structures. The knowledge of their dimensions and brightness temperatures is essential for understanding the physics of quiescent, confined plasma regions. Solar eclipses of 10 August, 1980 and 26 January, 1990, observed with high sensitivity and high time resolution at 22 GHz, allowed an unprecedented opportunity to identify Fresnel diffraction effects during lunar occultations of active regions. The results indicate the presence of quiescent discrete sources smaller than one arcsecond in one dimension. Assuming symmetrical sources,

their brightness temperatures were larger than 2×10^7K and 8×10^7K, for the 1980 and 1990 observations, respectively.

077.021 Solar microwave radiation maps measured at Metsähovi Radio Research Station in 1991.
S. Urpo, K. Karlamaa, S. Pohjolainen, H. Teräsranta.
Helsinki Univ. Technol., Metsähovi Radio Res. Stn., Rep. Ser. A, No. 10.
Helsinki University of Technology, Espoo (Finland). Metsähovi Radio Research Station. 122 p. (1992). ISBN 951–22–0938–7.
Solar measurements with the 14 m radio telescope at Metsähovi, include maps of the whole or part of the Sun, tracking of active regions and solar oscillation monitoring. Most of the solar maps have been measured using receivers operating at 22.2 GHz or at 36.8 GHz. This volume contains selected maps measured in 1991. A list of all measured maps during this period is included.

077.022 Solar radio flares 1989 – 1991.
S. Urpo, S. Pohjolainen, H. Teräsranta.
Helsinki Univ. Technol., Metsähovi Radio Res. Stn., Rep. Ser. A, No. 11.
Helsinki University of Technology, Espoo (Finland). Metsähovi Radio Research Station. 119 p. (1992). ISBN 951–22–1183–1.
This volume contains a list of radio flares observed at Metsähovi Radio Research Station in 1989 – 1991 and a selection of time profiles of the bursts.

077.023 Heliolongitudinal variations of parameters of type III burst radiation and its harmonic structure.
Ya. G. Tsybko.
Astron. Zh., Tom 69, Vyp. 2, p. 383 – 390 (Mar–Apr 1992). In Russian. English translation in Sov. Astron., Vol. 36, No. 2 (1992).
During a noise storm of simple type III radio bursts with diffuse dynamic spectra which can be observed at frequencies $f \lesssim 30$ MHz, emission with two–component harmonic structure is generated in the sources. Bursts, related with the fundamental plasma frequencies f_p are observed when the active region is located in the central heliolongitudinal range, whereas at other (limb) stages of the storm the second harmonics with lower drift rates $|\Delta t / \Delta f|^{-1}$ are registered. The initial interpretation of the parameter $\Delta t(\Delta f)$ variation as a stereo effect of observation of subrelativistic sources appears to be incorrect, together with the statement on the harmonic homogeneity ($f = 2f_p$) of hectometric bursts registered during the period of half a solar rotation. This leads to an overestimation by several times of the values of type III source average speed, as well as of the radial scales $\Delta r(\Delta f_p)$ of the solar corona. The normalized amplitude of the heliolongitudinal variation of $\Delta t(\Delta f)$ is in good agreement with the power law index of the standard optical model $N_e(r)$ of the corona.

077.024 Solar longitude variations in the parameters of type III radio emission and its harmonic structure.
Ya. G. Tsybko.
Sov. Astron., Vol. 36, No. 2, p. 193 – 196 (Mar 1992). Current Physics Microform No.: 9208X2029. English translation of Astron. Zh., Tom 69, Vyp. 2, p. 383 – 390 (1992).
During a noise storm of simple type III radio bursts at frequencies $f \lesssim 30$ MHz, radiation with a two–component harmonic structure is produced in the sources. Bursts at the plasma frequency fundamental f_p are seen when the active region lies in the central sector of solar longitude, while at other (limb) stages of a storm bursts are detected at the second harmonic ($2f_p$) with a reduced drift rate $|\Delta f / \Delta t|$. The initial explanation for variations of $\Delta t(\Delta f)$ as a stereoscopic effect of observation of subrelativistic sources is incorrect, as is the statement that hectometer–wave bursts detected in the course of half a solar rotation are harmonically uniform ($f = 2f_p$). That led to severalfold overestimates of the average velocity of type II sources and of the radial scale $\Delta r(\Delta f_p)$ of the solar corona. The normalized amplitude of the solar longitude dependence of $\Delta t(\Delta f)$ agrees well with the power–law exponent of the standard optical $N_e(r)$ model of the corona.

077.025 Millisecond radio spikes on long centimetre and short decimetre wavelengths and occurrence frequency.
Ji Shuchen, Xie Ruixiang, Wang Min.
Astrophys. Space Sci., Vol. 190, No. 1, p. 33 – 41 (Apr 1992).
High time–resolution data observed in two periods, respectively, by three frequencies (1.42, 2.84, and 3.67 GHz) or four frequencies (1.42, 2.00, 2.84, and 4.00 GHz) of fast sampling radiotelescopes were processed. Obtained were some significant results showing that during the obviously rising or maximum phases of solar cycle 22, the occurrence frequency of millisecond radio spikes at three or four frequencies decreased with the frequency increase and the highest occurrence frequency was at 1.42 GHz. Provided that the second x–mode is pre–dominant in the growth rate of ECM instability, the authors calculate the magnetic intensity of source regions with spike bursts at the four frequencies and interpret the occurrence frequency of millisecond radio spikes on long centimetre and short decimetre wavelengths. Finally, it is suggested that, owing to the Razing effect, the occurrence frequency of millisecond radio spikes starts to decrease when $f \leqslant 126$ MHz.

077.026 Decimetric solar type U bursts: VLA and PHOENIX observations.
M. J. Aschwanden, T. S. Bastian, A. O. Benz, J. W. Brosius.
Astrophys. J., Vol. 391, No. 1, p. 380 – 392 (20 May 1992).
The authors report on observations of type U bursts, simultaneously detected by the VLA at 1.446 GHz and by the broad-band spectrometer PHOENIX (ETH Zürich) in the 1.1 – 1.7 GHz frequency band. The VLA was operated with a time resolution of 1.66 s, while PHOENIX used a time resolution of 40 ms. A sequence of ~6 inverted–U bursts with similar morphology occurred between 1355 – 1359 UT, on 1989 August 13, accompanied by a soft X–ray flare of GOES class of C6.6. The peak flux was 7 SFU above background. The narrow bandwidth of ≈ 10% and the weak polarization in the sense of the ordinary mode is consistent with harmonic plasma emission. The turnover frequency of 1.4 GHz at the top of the U bursts then corresponds to a plasma density of $6 \times 10^9 \mathrm{cm}^{-3}$, while the footpoints at more than 1.7 GHz are rooted in densities greater than ~$10^{10}\mathrm{cm}^{-3}$. The source covers a wide angle of diverging magnetic field lines whose footpoints originate close to a magnetic intrusion of negative polarity into the main sunspot group of the active region (NOAA/USAF 5629) with dominant positive polarity. The authors interpret the U bursts in terms of electron beams propagating in closed coronal loops.

077.027 The radio image of the Sun at the wavelengths 3.5, 2.8, 2.25 and 1.95 cm at the period previous to solar eclipse on July 11, 1991.
N. V. Baranov, L. I. Tsvetkov.
Pis'ma Astron. Zh., Tom 18, No. 5, p. 467 – 476 (May 1992). In Russian. English translation in Sov. Astron. Lett., Vol. 18, No. 3.
The results of solar emission observations on July 10 – 11, 1991 made with the Crimean Astrophysical Observatory 22–m radio telescope using a four–wavelength polarimeter are given. A comparison of the radio sources and the optical formations on the Sun was made. Some physical parameters were estimated for the regions of increased radio brightness.

077.028 A thermal/nonthermal model for solar microwave bursts.
S. G. Benka, G. D. Holman.
Astrophys. J., Vol. 391, No. 2, p. 854 – 864 (1 Jun 1992).
High–resolution spectra of microwave bursts from Owens Valley Radio Observatory show numerous departures from expectations based on simple thermal or nonthermal models. In particular, (1) ~80% of the events show more than one spectral peak; (2) many bursts have a low–side spectral index steeper than the maximum expected slope; and (3) the peak frequency stays relatively constant, and changes intensity in concert with the secondary peaks throughout a given event's evolution. The authors develop a theoretical formalism allowing both thermal and nonthermal particles to coexist in the flaring plasma. Within this formalism, they show that the observed spectral features can be accounted for by gyrosynchrotron radiation. The authors

present the model distribution function, use it to calculate gyrosynchrotron spectra, and systematically analyze these spectra. Finally, the physical interpretation of the thermal/nonthermal distribution function, as it relates to solar flares, is discussed .

077.029 Clumpy Langmuir waves in type III radio sources.
P. A. Robinson.
Sol. Phys., Vol. 139, No. 1, p. 147 – 163 (May 1992).
A model is developed for the clumpy Langmuir waves observed in type III source regions. In this model the waves are generated by instability of a beam which propagates outward from the Sun in a state close to marginal stability. Ambient density perturbations cause fluctuations about the marginally stable state, leading to nonuniformities in both beam and waves and, hence, to spatially inhomogeneous growth. High damping rates and high wave levels are strongly anti–correlated, leading to suppression of the net damping. Below saturation stochastic growth causes the waves to follow a random walk in the logarithm of their energy density and the resulting probability of observing a field of magnitude E is approximately proportional to E^{-1}. Comparison with observations shows that this model can account for the levels and clumpiness of the Langmuir waves, the small net dissipation required for the beams to propagate to 1 AU, the characteristic decay time of type III electromagnetic emission, and the negative mean growth rate observed in situ in type III sources. At 1 AU only the very highest fields approach the threshold for nonlinear wave collapse, but this threshold may be more commonly exceeded closer to the Sun.

077.030 Solar observations at Metsähovi in January – June 1992.
S. Urpo, S. Pohjolainen, H. Teräsranta.
Helsinki Univ. Technol., Metsähovi Radio Res. Stn., Rep. Ser. A, No. 12.
Helsinki University of Technology, Espoo (Finland). Metsähovi Radio Research Station. 65 p. (1992). ISBN 951–22–1237–4.
This volume contains a selection of solar maps measured during the first half of 1992, a list of observed radio flares, and a selection of their time profiles.

077.031 The relation between the spectrum and directivity of solar radio emission bursts.
K. Shibasaki, V. M. Moskovkina, I. E. Pogodin.
Geomagn. Aeron., Vol. 31, No. 2, p. 275 – 277 (Oct 1991). English translation of Geomagn. Aehron., Tom 31, No. 2, p. 357 – 359 (1991).
It is demonstrated that the character of the spatial directivity of solar–radio bursts is determined by the frequency of the spectral maximum, which is due to the difference in the position of the source and the orientation of the corresponding magnetic fields.

077.032 The quasi–periodic fine structures in a solar radio burst at 3.2 cm wavelength on 23 May 1990.
Qin Zhihai, Li Chunsheng, Wei Shuangling, Chen Yunxia.
Chin. Astron. Astrophys., Vol. 16, No. 1, p. 78 – 83 (Jan–Mar 1992). English translation of Acta Astron. Sin., Vol. 32, No. 3, p. 247 – 251 (Sep 1991). See Abstr. 54.077.071.
This paper describes the quasi–periodic fine structures (FS) in a solar radio burst which was observed with time resolution of 10 ms at 3.2 cm wavelength on 23 May 1990. The radio burst was associated with a M8.7/1B flare occurred in the active region AR 6063. The average period of these fine structures is about 1.5 s and the amplitude modulation range from 2% to 5%. A preliminary analysis shows that there is a linear relation between the mean flux S of the radio burst and the repetition rate R of the quasi–periodic fine structures at 3.2 cm wavelength. This relation is similar to that obtained by Kaufmann et al. in millimeter wavelength region. The authors suggest that the quasi–periodic fine structures probably reflect the modulated gyro–synchrotron emission due to MHD waves oscillating along magnetic field lines.

077.033 **Observation and analysis of solar decimetric spike radiation.**
Wang Min, Sie Ruixiang, Wang Congyun.
Chin. Astron. Astrophys., Vol. 16, No. 1, p. 84 – 90 (Jan–Mar 1992). English translation of Acta Astron. Sin., Vol. 32, No. 3, p. 252 – 257 (Sep 1991). See Abstr. 54.077.072.

From January, 1988 to June, 1989 the solar radio synchronous observational system with high temporal resolution of the Yunnan Observatory was used to carry out observations at a decimetric wavelength (1420 MHz), and the data are analyzed in this paper. The result shows a relation between the millisecond spike radiation of the solar decimetric radio emission and the radio bursts recorded with a time constant of one second. It also reveals the coexistence of decimetric radio millisecond spike radiations and Hα flares, type III radio bursts as well as soft X–ray bursts. A solar activity event with abundant radio millisecond spike radiations is investigated preliminarily and the relation between the duration and flux of individual spikes is discussed.

077.034 **The effect of the plasma background parameter on spike radiation.**
Huang Guangli, Wang Deyu.
Chin. Astron. Astrophys., Vol. 16, No. 1, p. 91 – 102 (Jan–Mar 1992). English translation of Acta Astrophys. Sin., Vol. 11, No. 4, p. 352 – 361 (Oct 1991). See Abstr. 54.077.042.

The authors discuss the effect of the plasma background parameter ω_{pe}/Ω_e on spike radiation on the basis of the theory of electron cyclotron maser instability. They discuss the physical meaning of the parameter, its numerical range and the polarization and harmonics structure of the radiation that are especially sensitive to changes in this parameter.

077.035 **Radio observations of the M8.1 solar flare of 23 June, 1988: evidence for energy transport by thermal processes.**
T. S. Bastian, D. E. Gary.
Sol. Phys., Vol. 139, No. 2, p. 357 – 385 (Jun 1992).

The Very Large Array and the frequency agile interferometer at the Owens Valley Radio Observatory were used to observe the M8.1 flare of 23 June, 1988. The observations were supplemented by radiometer measurements made by the USAF RSTN network site at Palehua, HI, by GOES soft X–ray observations, by USAF SOON Hα filtergrams, and by a KPNO photospheric magnetogram. The radio data reveal a wide variety of phenomena, including: (i) multiply impulsive microwave burst that is essentially thermal in character; (ii) stationary discrete components at 1.5 GHz, associated temporally and spatially with distant brightenings in Hα; (iii) a dynamical component at 1.5 GHz associated with hot plasma moving subsonically into the corona; (V) the appearance of intense, short–lived, decimetric burst activity near the lead sunspot in the active region at 1.5 GHz, indicative of a high degree of inhomogeneity in the source. The unusually complete radio coverage allows to investigate the transport of energy from the initial site to sites of distant Hα brightenings. The transport of energy appears to be most consistent with slow, thermal processes, rather than rapid transport by nonthermal electron beams.

077.036 **Microwave and soft X–ray radiation during flares evolving in strong magnetic fields.**
B. Vršnak, V. Ruždjak, P. Zlobec, S. Jurač.
Hvar Obs. Bull., Vol. 15, No. 1, p. 1 – 9 (1991).

Microwave and soft X–ray bursts associated with Hα flares protruding over major sunspot umbrae are studied. The probability for a detectable microwave radiation is larger in flares protruding over umbrae with a stronger magnetic field. The peak fluxes of microwave radiation measured at about 3 GHz are exponentially larger for stronger umbral fields. The slope of the established relation is independent on the importance of the flares in the sample. The soft X–ray flux data measured onboard the GOES–satellite were used to estimate the effective plasma temperature in the flaring volume. Statistically, the temperature is higher in flares occurring in stronger magnetic fields.

077.037 **Launch–times of MHD shocks observed as type II bursts.**
B. Vršnak, P. Zlobec, V. Ruždjak.
Hvar Obs. Bull., Vol. 15, No. 1, p. 11 – 20 (1991).

28 "high–frequency" type II bursts characterized by the starting frequency higher than 237 MHz have been selected to estimate accurately the "launch–time" of the flare initiated MHD shock. The time association between the shock ignition and the development of the flare as observed in microwave, Hα and X–ray emission was established. The MHD shock is launched around the time of the hard X–ray burst peak and during the fast rise of the microwave emission at 3 GHz, which favors the energy pulse mechanism of the shock ignition.

077.038 **Intensity variations and short time evolution of solar microwave low temperature regions.**
S. Pohjolainen, S. Urpo, H. Teräsranta, B. Vršnak, R. Brajša, V. Ruždjak, S. Jurač.
Hvar Obs. Bull., Vol. 15, No. 1, p. 21 – 29 (1991).

Solar microwave low temperature regions (LTR), i.e. areas with brightness temperature considerably below the quiet Sun level, were observed during solar mapping at the Metsähovi Radio Research Station, Finland. It was possible to observe some of the areas as they passed across the solar disc. The authors present the evolution (intensities, rotation rates, and latitudinal motions) of the four selected LTRs and two nearby active regions observed on May 20 – 26, and August 26 – September 2, 1990, April 10, 1990 was chosen to study the short time evolution of LTRs in the vicinity of an active region producing a burst.

077.039 **Rapid fluctuation in solar radio emission.**
Fu Qijun.
Advances in solar–terrestrial science of China, p. 38 – 51 (1992). – See Abstr. 003.069 for the main entry.

Rapid variations of solar radio emission in solar flares have been found in different wavebands for more than two decades. Millisecond time scale radio spikes and fast fine structures superimposed on radio bursts are a quickly developing area of solar radio physics. The Solar Radio Astronomy Group of Beijing Astronomy Observatory found spike emission of microwave bursts on msec time scales at 2840 MHz in 1981. Since 1983, a comprehensive project for setting up a joint–observation network for observing solar radio emission with high time resolution has been jointly carried out by the Chinese solar radio astronomy community. The equipments in the network cover a wide frequency domain from meter to short centimeter wavelengths. At present, the network has been set up. Considerable successes have been achieved not only in the instrumental and observational aspects but also in the theoretical aspect. In this paper, a summary of the efforts in the 22nd solar activity maximum period is presented.

077.040 **Radio burst at 245 MHz and the forecast of solar activities.**
Xia Zhiguo, Chen Jingying.
Publ. Yunnan Obs., No. 1, p. 55 – 62 (1992). In Chinese.

The data of the radio burst at 245 MHz are used for the real–time alert of solar activities. Some characteristics of the radio burst at the frequency of 245 MHz and the relation between the narrow–band bursts at meter wavelengths and other solar activities are introduced.

077.041 **Solar radio observations.**
G. J. Hurford.
NATO Advanced Study Institute on the Sun: a Laboratory for Astrophysics, p. 297 – 312 (1992). – See Abstr. 012.091 for the main entry.

Solar radio observations of the quiet and active Sun are reviewed in the context of the properties of the emission processes. For plasma radiation, the emphasis is on interpretation of the frequency of the emission in terms of height. The roles of free–free bremsstrahlung in the quiet Sun, gyroresonance emission in active regions and thermal and nonthermal gyrosynchrotron emission in flares are discussed. Observational examples emphasize cases where the straightforward physics seems to work.

077.042   Arcsecond determination of solar burst centers of emission simultaneous to high time resolution and high sensitivity at 48 GHz.
J. E. R. Costa, E. Correia, P. Kaufmann, R. Herrmann, A. Magun.
IAU Colloquium No. 133: Eruptive solar flares, p. 177 – 179 (1992). Extended abstract. – See Abstr. 012.096 for the main entry.

077.043   Meter–decameter radio emission associated with a coronal mass ejection.
M. R. Kundu, N. Gopalswamy.
IAU Colloquium No. 133: Eruptive solar flares, p. 268 – 271 (1992). – See Abstr. 012.096 for the main entry.
A study of meter–decameter radio emission associated with the 1986 Feb 10 coronal mass ejection event is presented here. The event was accompanied by a major flare (optical importance 1B and X–ray importance C9.6), preceded by a filament disappearance.

077.044   Compact sources of suprathermal microwave emission detected in quiescent active regions during lunar occultations.
E. Correia, P. Kaufmann, F. M. Strauss.
IAU Colloquium No. 133: Eruptive solar flares, p. 341 (1992). Abstract. For the full paper see Sol. Phys., Vol. 138, No. 2, p. 223 – 231 (Apr 1992). – See Abstr. 012.096 for the main entry.

077.045   High spectral resolution of mm–wavelength (23–18 GHz) solar bursts.
H. S. Sawant, R. R. Rosa, J. R. Cecatto, F. C. R. Fernandes.
IAU Colloquium No. 133: Eruptive solar flares, p. 367 (1992). – See Abstr. 012.096 for the main entry.

077.046   Microwave flare characteristics in 8 and 3 mm Metsähovi measurements compared with optical and H–alpha data.
S. Pohjolainen, S. Urpo, H. Teräsranta, M. Tornikoski.
IAU Colloquium No. 133: Eruptive solar flares, p. 368 – 371 (1992). – See Abstr. 012.096 for the main entry.
In 1989-1990 several microwave bursts were detected at 37 and 90 GHz (8 and 3 mm wavelengths, respectively) during solar mapping or source tracking at the Metsähovi Radio Research Station, Finland. Nine of the events are analysed and discussed in this paper.

077.047   Nonlinear emission mechanism of type III solar radio bursts.
A. C.–L. Chian, F. B. Rizzato.
IAU Colloquium No. 133: Eruptive solar flares, p. 377 – 380 (1992). – See Abstr. 012.096 for the main entry.

077.048   The quasi–periodic pulsations in solar microwave bursts on May 3, 1989.
Zhang Heng, Gong Yuan–fang, Shang Qiong–zhen, Hu Han–min, Lü Song–quan, Yang Rong–bang.
Chin. Sci. Bull., Vol. 37, No. 6, p. 473 – 478 (Mar 1992).

077.049   Diagnostics of energy release and magnetic fields on the sun by radio methods.
A. Krüger.
International Workshop on Reconnection in Space Plasma, p. 133 – 138 (Jan 1989). – See Abstr. 012.099 for the main entry.
The impact of radio astronomy on the knowledge of energy release processes and magnetic fields in the solar atmosphere is briefly reviewed.

077.050   Chaotic radiopulsations and coronal magnetic field estimates.
J. Kurths, M. Karlicky.
International Workshop on Reconnection in Space Plasma, p. 175 – 177 (Jan 1989). – See Abstr. 012.099 for the main entry.
Analyzing the time structure of a coronal pulsation event of April 3, 1980 provides an evolution from a regular periodic phase to an irregular behaviour. The irregular stage can be described by a low–dimensional chaotic attractor. The nature of this structural change reveals that this pulsation event evolves along the Ruelle–Takens–Newhouse route to chaos. This event is discussed in terms of MHD waves in coronal loops yielding estimates of the plasma parameters in the solar corona, particularly of the magnetic field strength.

Tremsdorf solar radio astronomy observatory – the scientific programme and an interesting observation.
See Abstr. 009.024.

Molecular data from solar spectroscopy.
See Abstr. 022.190.

A solar decimeter radio dynamic spectrometer.
See Abstr. 033.018.

The unified radio and plasma wave investigation.
See Abstr. 035.033.

The formation mechanism of the lines of the C I infrared multiplet at $\lambda 1069.5$ nm in the spectrum of the Sun.
See Abstr. 071.020.

Long–term evolution of coronal magnetic fields basing on noise storm continuum observations in the 20th and 21st cycle.
See Abstr. 072.069.

Chromospheric dynamics based on infrared solar brightness variations.
See Abstr. 073.021.

Radio synthesis imaging during the GRO solar campaign.
See Abstr. 073.037.

On the saturation of electron–cyclotron masers in solar flares.
See Abstr. 073.074.

Research of solar events in optical, radio and X–ray regions.
See Abstr. 073.077.

Characteristics of the impulsive phase of flares.
See Abstr. 073.097.

The Neupert effect: what can it tell us about the impulsive and gradual phases of eruptive flares?
See Abstr. 073.104.

Radio emission of eruptive flares.
See Abstr. 073.108.

Simultaneous Hα and microwave observations of a limb flare on June 20, 1989.
See Abstr. 073.120.

Nonresonant ion–beam turbulence in solar flares.
See Abstr. 073.121.

The precursors of the 9 July 1982 solar proton flare event.
See Abstr. 073.130.

The scales of solar microwave bursts and scenarios of flare energy release.
See Abstr. 073.131.

Solar radio corona.
See Abstr. 074.001.

Coronal Magnetic Structures Observing Campaign. IV. Multi-waveband observations of sunspot and plage–associated coronal emission.
See Abstr. 074.010.

Estimation of the mass of a coronal mass ejection from radio observations.
See Abstr. 074.037.

Large scale structures associated with eruptive flares and radio waves.
See Abstr. 074.067.

Coronal millimeter sources associated with eruptive flares.
See Abstr. 074.068.

Dynamic development of coronal current sheets.
See Abstr. 074.079.

Coronal magnetic field strengths determined from fiber bursts.
See Abstr. 074.083.

Microwave and soft X–ray emission of solar flare events associated with coronal mass ejections.
See Abstr. 074.089.

High–energy flare emissions.
See Abstr. 076.024.

On a relation of solar proton characteristics with some parameters of solar microwave bursts.
See Abstr. 078.016.

A search during the 1991 solar eclipse for the infrared signature of circumsolar dust.
See Abstr. 106.013.

078 Cosmic Radiation

078.001   Features of the focused diffusion of solar cosmic rays.
G. A. Bazilevskaya, R. M. Golynskaya.
Geomagn. Aeron., Vol. 30, No. 5, p. 725 – 727 (Apr 1990). English translation of Geomagn. Aehron., Tom 30, No. 5, p. 853 – 855 (1990).
Time profiles of the intensity and anisotropy of solar cosmic rays at various distances from the source are considered as a function of the ratio of the mean free path to the focusing length of the interplanetary field. The role of solar–wind velocity, radial change in transport path, and the nature of pitch–angle scattering is examined.

078.002   Observation of anomalously high flux densities of low-energy heavy nuclei on the Salyut–6, Salyut–7, and Mir orbital stations.
Yu. F. Gagarin, Ya. V. Dvoryanchikov, V. A. Dergachev, A. P. Lobakov, V. I. Lyagushin, A. Yu. Ovchinnikova, A. V. Solov'ev, I. G. Khilyuto, E. A. Yakubovskij.
JETP Lett., Vol. 55, No. 2, p. 88 – 91 (25 Jan 1992). Current Physics Microform No.: 9204X0448.
Significant flux densities of Ca–Fe nuclei with energies below 50 – 100 MeV/nucleon have been detected by a track detector at an altitude ~300 – 350 km. The flux densities increase with decreasing energy. The uniquely high flux densities of nuclei in the energy range 140 – 5 MeV/nucleon in 1988 – 1990 are greater than those in previous exposures by one to four orders of magnitude may be correlated with intense solar proton flares in August – October 1989.

078.003   The solar flare event on 1990 May 24: evidence for two separate particle accelerations.
H. Debrunner, J. A. Lockwood, J. M. Ryan.
Astrophys. J., Lett., Vol. 387, No. 1, p. L51 – L54 (1 Mar 1992).
It has been reported that direct solar neutrons were measured by cosmic–ray neutron monitors (NMs) in the first increase of the solar cosmic–ray event on 1990 May 24. A careful study of this first increase led the authors to conclude that there were no solar neutrons detected by the North American NMs. A semiquantitative analysis of the records of a few NM stations with good response to the first and/or second increase suggests that the first increase beginning during the interval 2045 – 2050 UT was produced by solar protons presumably accelerated at the Sun in the impulsive phase of the flare; and that the second larger increase which started no earlier than 2105 UT was probably due to solar protons accelerated by an extended coronal shock. The authors report preliminary results of an analysis of this solar cosmic–ray event which provides evidence for two entirely separate accelerations of protons from two separate particle populations at the Sun. The optical class 1B solar flare was at heliographic coordinates 36°N, 76°W in NOAA region 6063.

078.004   Superevents: their origin and propagation through the heliosphere from 0.3 to 35 AU.
W. Dröge, R. Müller–Mellin, E. W. Cliver.
Astrophys. J., Lett., Vol. 387, No. 2, p. L97 – L100 (10 Mar 1992).
Superevents are long–lasting enhancements of the interplanetary particle population that are observed initially in the inner heliosphere ($\leqslant 1$ AU) and that propagate to the outer heliosphere ($\geqslant 35$ AU) with speeds of ≈ 800 km s^{-1}. Superevents are observed in nuclei at energies up to tens of MeV as well as in MeV electrons and differ distinctly from known intensity increases due to single solar flare particle events, corotating events, and energetic storm particle events. They remain one or more orders of magnitude above background at 1 AU for at least one solar rotation, and their intensity variation with heliolongitude is small. Superevents are associated with local minima of the Galactic cosmic–ray intensity. Between 1974 and 1986, the authors identified 16 superevents. They examined the solar activity associated with the most prominent superevents and find in general that these events originate in extended (0.5 – 2 month) episodes of coronal mass ejection (CME) activity from single active regions or narrow ranges of active longitudes. Analyses of the variation of particle peak intensity as a function of time, ecliptic longitude, and radial distance suggests that superevents result when systems of CMEs, with their associated shocks and particle events, create an outward propagating shell encompassing the Sun. The relatively weak negative radial intensity gradients of superevents indicate that local acceleration and trapping, as well as flare – accelerated particles, are key factors in their formation.

078.005   The great solar energetic particle events of 1989 observed from geosynchronous orbit.
G. D. Reeves, T. E. Cayton, S. P. Gary, R. D. Belian.
J. Geophys. Res., Vol. 97, No. A5, p. 6219 – 6226 (1 May 1992).
Los Alamos energetic proton instruments at geosynchronous orbit observed more major solar energetic particle events during

1989 than any other year since this series of detectors began observations in 1976. The temporal flux profiles of four intervals, which contain six distinct events, are compared illustrating the uniqueness of each event. Characteristic risetime and decay time are computed for each event. During two of these events, brief order–of–magnitude increases of the proton flux are observed. They are associated with sudden commencement events and dramatic changes in the solar wind. The authors conclude that these two brief events are likely the result of shock acceleration in the solar wind. They have fit the measured count rates to a spectral form which is exponential in rigidity, and they have examined the changes in spectral slope with time for each of the four intervals. In general, harder spectra are measured near the onset of an event followed by a softening of the spectrum as the fluxes decay. The authors have also investigated the effects of these events on geomagnetic activity by comparing the fluxes of > 30–keV electrons at geosynchronous orbit and Kp geomagnetic index during the early part of two of the solar energetic particle events.

078.006   Energy spectra of ions from impulsive solar flares.
D. V. Reames, I. G. Richardson, K.–P. Wenzel.
Astrophys. J., Vol. 387, No. 2, p. 715 – 725 (10 Mar 1992).
   The authors report on a study of the energy spectra of ions from impulsive solar flares in the 0.1 – 100 MeV region obtained from the combined observations of three experiments on the ISEE 3 and IMP 8 spacecraft. Most of the events studied are dominated by He and these He spectra show a persistent steepening or break above ~ 10 MeV resulting in an increase in the power–law spectral indices from ~ 2 to ~ 3.5 or more. Spectra of H, ^{3}He, ^{4}He, O and Fe have spectral indices that are consistent with a value of ~ 3.5 above ~ 2 MeV amu^{-1}. One event, dominated by protons, shows a clear maximum in the spectrum near 1 MeV. If the roll–over in the spectrum below 1 MeV is interpreted as a consequence of matter traversal in the solar atmosphere, then the source of the acceleration would lie only ~ 800 km above the photosphere, well below the corona. Alternative interpretations are that trapping in the acceleration region directly causes a peak in the resulting ion spectrum or that low–energy particles encounter significant additional scattering during transport from the flare. In two of the events, the normal velocity dispersion in the arrival of the ions is modulated by the passage of spatial structures in the interplanetary plasma.

078.007   Prediction and evaluation of solar particle events based on precursor information.
G. R. Heckman, J. M. Kunches, J. H. Allen.
Adv. Space Res., Vol. 12, No. 2/3, p. 313 – 320 (1992). – See Abstr. 012.016 for the main entry.
   Protection from the radiation effects of solar particle events for deep space mission crews requires a warning system to observe solar flares and predict subsequent charged particle fluxes. Such a system relates precursor information observed in each flare to the intensity, delay, and duration of the subsequent Solar Particle Event (SPE) at other locations in the solar system. A warning system of this type is now in operation at the NOAA Space Environment Services Center in Boulder, Colorado for support of space missions. It has been used to predict flare particle fluxes at the earth for flares of Solar Cycle 22. The flare parameters used and the effectiveness of the current warning system are presented, with an examination of the shortcomings. Needed improvements to the system include more complete observations of solar activity, especially information on the occurrences of solar mass ejections; and consideration of the effects of propagation conditions in the solar corona and interplanetary medium.

078.008   "MIR" radiation dosimetry results during the solar proton events in September–October 1989.
T. P. Dachev, Yu. N. Matviichuk, N. G. Bankov, J. V. Semkova, R. T. Koleva, Ya. J. Ivanov, B. T. Tomov, V. M. Petrov, V. A. Shurshakov, V. V. Bengin, V. S. Machmutov,

N. A. Panova, T. A. Kostereva, V. V. Temny (*V. V. Temnyj*), Yu. N. Ponomarev, R. Tykva.
Adv. Space Res., Vol. 12, No. 2/3, p. 321 – 324 (1992). – See Abstr. 012.016 for the main entry.
   Using data from dosimetry – radiometry system "Liulin" on board of "Mir"–space station the particle flux and doserate during Sep–Oct, 1989 has been studied. Special attention has been payed to the flux and doserate changes inside the station after intensive solar proton events (SPE) on 29 Sep, 1989. The comparison between the doses before and after the solar flares shows increase of the calculated mean dose per day by factor of 10 to 200. During the SPE on 29 Sep the additional dose was 310 mrad. The results of the experiment are compared with the data for the solar proton fluxes obtained on the GOES–7 satellite.

078.009   Solar proton events at the growth phase of the 22nd solar activity cycle.
S. I. Avdjushin (*S. I. Avdyushin*), M. N. Nazarova, N. K. Perejaslova, I. E. Petrenko, S. G. Frolov.
Adv. Space Res., Vol. 12, No. 2/3, p. 325 – 329 (1992). – See Abstr. 012.016 for the main entry.
   Meteor satellite observations in March, August, September and October 1989 recorded intensive solar proton events which caused a disturbed radiation situation in the near–Earth space. The paper presents the results of analyzing flux and spectral characteristics of the events and their relation to heliogeophysical situation.

078.010   The influence of the large–scale interplanetary shock structure on a low–energy particle event.
A. M. Heras, B. Sanahuja, Z. K. Smith, T. Detman, M. Dryer.
Astrophys. J., Vol. 391, No. 1, p. 359 – 369 (20 May 1992).
   The authors have developed a numerical model to study the influence of the large–scale shock topology on the associated low–energy (less than 2 MeV) particle event upstream of the shock. It includes particle injection at the solar corona, two–dimensional MHD simulation of the propagation of the shock, modeling of particle propagation through the interplanetary medium, and particle injection at the shock. The injection at the shock is represented by a numerical source function in the equation that describes particle propagation. The MHD simulation provides the time when the shock intersects the interplanetary magnetic field line which connects with the spacecraft, the evolution of this connection point over the shock front, and its plasma characteristics as it propagates out into the heliosphere. By fitting the observed particle flux and anisotropy profiles, one can determine the parameters which characterize particle propagation and the rate of particle injection at the shock front. The model is used to study the particle event associated with the interplanetary shock observed on 1979 April 24 by ISEE–3 in the energy range 35 – 1600 keV.

078.011   Composition and azimuthal spread of solar energetic particles from impulsive and gradual flares.
M.–B. Kallenrode, E. W. Cliver, G. Wibberenz.
Astrophys. J., Vol. 391, No. 1, p. 370 – 379 (20 May 1992).
   The authors compiled a list of 77 flare–associated solar energetic particle (SEP) events observed from 1974 to 1985 by at least one of the two Helios space probes. They classified the SEP parent flares as impulsive (25 cases) or gradual (52 cases) on the basis of their soft X–ray durations. The authors then compared the intensities of the prompt component of ~ 0.5 MeV electrons, ~ 10 MeV protons, and ~ 10 MeV per nucleon helium for the two classes of SEP flares. It is found that SEPs from gradual flares have higher intensities than SEPs from impulsive flares. These differences are most pronounced for protons (about two orders of magnitude) and less for electrons (about one order of magnitude), and helium (about a factor of 5). The SEPs from impulsive flares have a "cone of emission" of $\pm 50°$ versus $\pm 120°$ for gradual flares. These results are discussed in the context of recent works on particle acceleration in solar flares.

078.012   Spectrum of accelerated particles in solar proton events with fast component.
J. Perez–Pereza, A. Gallegos–Cruz, Eh. V. Vashenyuk,
L. I. Miroshnichenko.
Geomagn. Aehron., Tom 32, No. 2, p. 1 – 12 (Mar – Apr 1992).
In Russian. English translation in Geomagn. Aeron., Vol. 32,
No. 2.

078.013   The effect of statistical acceleration on propagation of particles from a solar flare in the interplanetary medium.
M. F. Bakhareva, I. V. Moskalenko.
Geomagn. Aehron., Tom 32, No. 2, p. 13 – 18 (Mar – Apr 1992).
In Russian. English translation in Geomagn. Aeron., Vol. 32,
No. 2.

078.014   Temporal variations of decay protons for the solar events of 21.06.1980, 03.06.1982 and 24.04.1984.
I. G. Kurganov, V. M. Ostryakov.
Geomagn. Aehron., Tom 32, No. 3, p. 149 – 153 (May – Jun 1992). In Russian. English translation in Geomagn. Aeron., Vol. 32, No. 3.

078.015   Observation of shock–accelerated protons by Giotto and IMP–8 under solar minimum conditions in February 1986.
E. Kirsch, P. C. Trochoutsos, E. T. Sarris,
S. McKenna–Lawlor.
Sol. Phys., Vol. 139, No. 1, p. 165 – 175 (May 1992).
Solar and shock–accelerated protons were observed by the interplanetary S/C GIOTTO and aboard the Earth orbiting satellite IMP–8 during solar minimum conditions in February 1986. Forward and reverse shock configurations developed on 9 – 11 February, 1986 and could be recognized by sunward and anti–sunward proton propagation. The results are consistent with a general east–west asymmetry in particle fluxes observed within a distance of 1 AU from the Sun and caused by quasi–perpendicular and quasi–parallel shock configurations in the interplanetary magnetic field. The high–energy protons (> 1 GeV) measured simultaneously by ground–based neutron monitors showed a complementary enhanced amplitude in the diurnal variation on 9 February, 1986.

078.016   On a relation of solar proton characteristics with some parameters of solar microwave bursts.
L. V. Yasnov.
Geomagn. Aehron., Tom 32, No. 3, p. 159 – 163 (May – Jun 1992). In Russian. English translation in Geomagn. Aeron., Vol. 32, No. 3.

078.017   A system of loop like interplanetary bottles containing solar cosmic rays in June 1974.
G. P. Lyubimov, E. A. Chuchkov.
Cosmic Res., Vol. 29, No. 6, p. 778 – 784 (May 1992). English translation of Kosm. Issled., Tom 29, Vyp. 6, p. 910 – 916 (1991).
The authors consider an increase in solar protons with energies 1 – 15 MeV which occurred on June 8 – 12, 1974. They show that the source of this increase was a solar flare on June 6 which injected solid cosmic rays into a system of quasistationary interplanetary looplike bottles. They analyze in detail the solar and interplanetary data, and approximate the solar cosmic ray data.

078.018   Recurrent enhancements of energetic particle intensity during the decreasing phase of the 21st solar activity cycle.
Yu. Logachev, M. Zel'dovich, V. Stolpovskij, K. Gringauz,
M. Verigin, I. Klimenko, A. Shomodi, A. Vagra,
K. Kecskemety.
Kosm. Issled., Tom 30, Vyp. 3, p. 368 – 377 (May–Jun 1992). In Russian. English translation in Cosmic Res., Vol. 30, No. 3.
Recurrent enhancements of energetic particle fluxes recorded on board space probes VEGA–1 and –2 during the first half of 1985 formed a long–lasting series of events in the decreasing phase of the 21st solar activity cycle. On the basis of intensity data of protons with energies 1 – 2 MeV measured by IMP–7 and –8 spacecraft it is found that the series spans 26 solar rotations. The characteristic features of these events are outlined. The change of intensity and energy spectrum is compared with ESP events. The last ten enhancements of the series were associated with a coronal hole on the Sun.

078.019   Statistical features of solar particle energy spectra with heliolongitude.
Yu Xing–feng, Yao Jin–xing.
Publ. Purple Mt. Obs., Vol. 11, No. 1, p. 21 – 26 (Mar 1992). In Chinese.
A statistical study of proton, GLE (Ground Level Effect), γ–ray and X–ray events has been carried out. It is found that there is a "preferred region" of proton and GLE events between 20°W and 80°W in the active regions, but there are not "preferred regions" of electron events derived from γ–ray and X–ray events.

078.020   Trapping and escape of the high energy particles responsible for major proton events.
D. V. Reames.
IAU Colloquium No. 133: Eruptive solar flares, p. 180 – 185 (1992). – See Abstr. 012.096 for the main entry.
Energetic particles from impulsive flares are characterized as rich in electrons, ^{3}He, and Fe and high ionization states of Fe indicate that the material has been heated in the flare. Particles from these events are only seen from magnetically well–connected flares, and the particles reach maximum intensity in a few hours. Particles from gradual events have heavy element abundances and ionization states that are near their coronal values but protons are strongly enhanced. Gradual events come from a wide longitude region and particle intensities remain high for several days, much longer than the associated phenomena at the Sun. Most major proton events are gradual events, but some are impulsive or have both impulsive and gradual phases. The extended evolution of the major proton events in space and time can no longer be understood in terms of slow diffusive transport of particles through the corona and interplanetary medium, since particles from the impulsive events are found to behave much differently. The observations can be understood only if the major gradual events involve a large interplanetary shock wave that accelerates particles over an extended region of longitude for a long time. Wave–particle interactions play a major role in the trapping and acceleration of particles in large events. Diffusive containment allows particle acceleration to occur on open as well as closed field lines in both impulsive and gradual events.

078.021   Coronal and interplanetary transport of solar flare protons from the ground level event of 29 September 1989.
P. H. Stoker.
IAU Colloquium No. 133: Eruptive solar flares, p. 186 – 189 (1992). – See Abstr. 012.096 for the main entry.
This behind–the–limb (assumed West 105 degrees) solar proton event was recorded by all global neutron monitors with an onset time at 1148 UT ± 2 min irrespective of viewing direction and cutoff rigidity. Neutron monitors with viewing directions of asymptotic longitude between 180 and 340 degrees East recorded a much faster rise time (~ 10% per minute) in count rate from onset to maximum than the ~2% rise time between 0 to 100 degrees East, irrespective of cutoff rigidity. The exponential decay time changed from ~40 min for $P_c \geqslant 11$ GV to ~200 min for $P_c \lesssim 1$ GV. These observations are compared with computations from models of coronal and interplanetary transport.

The ULYSSES energetic particle composition experiment EPAC.
See Abstr. 035.034.

The Ulysses Cosmic Ray and Solar Particle Investigation.
See Abstr. 035.037.

Study of periodicities of solar nuclear gamma ray flares and sunspots.
See Abstr. 072.026.

The production of ^{3}He and heavy ion enrichments in ^{3}He–rich flares by electromagnetic hydrogen cyclotron waves.
See Abstr. 073.054.

High energetic solar proton flares of 19 to 29 October 1989.
See Abstr. 073.109.

The role of protons in solar flares.
See Abstr. 073.119.

The precursors of the 9 July 1982 solar proton flare event.
See Abstr. 073.130.

Focused transport of energetic particles along magnetic field lines draped around a coronal mass ejection.
See Abstr. 074.006.

Strong increase of neutron flux during 22 July, 1990 eclipse.
See Abstr. 079.004.

Seasonal variations in the data of a chlorine–argon detector of solar neutrinos.
See Abstr. 080.046.

Nitric oxide and lower ionosphere quantities during solar particle events of October 1989 after rocket and ground–based measurements.
See Abstr. 083.025.

Voyager energetic particle observations at interplanetary shocks and upstream of planetary bow shocks: 1977–1990.
See Abstr. 106.027.

Solar and interplanetary shock effects on the flux and anisotropy profiles of an Energetic Storm Particle (ESP) event observed on 23–25 April 1979.
See Abstr. 106.038.

Radiations from space: swift charged particles and neutrons.
See Abstr. 144.005.

The simulated features of heliospheric cosmic–ray modulation with a time–dependent drift model. II. On the energy dependence of the onset of new modulation in 1987.
See Abstr. 144.035.

The additional fluxes of cosmic rays in the stratosphere in the various half–periods of the 22–year solar magnetic cycle.
See Abstr. 144.080.

079 Solar Eclipses

079.001 **Solar eclipses in the years 1990 – 2000.**
         P. Kotrč.
Říše hvězd, Vol. 72, No. 2, p. 25 – 27 (Feb 1991). In Czech.

Solar eclipse 1983 June 11

079.002 **Photometric and height calibration of the spectra observed at the 1983 total solar eclipse.**
You Jianqi, Fan Zhongyu, Ji Haisheng, Qiao Qiyuan.
Acta Astrophys. Sin., Vol. 12, No. 1, p. 93 – 100 (Jan 1992). In Chinese.
    The photometric and height calibration procedures of the spectra obtained at the 1983 total solar eclipse are given in detail. The accuracy of the results is discussed. For the absolute calibration of the characteristic curves, two methods are introduced. In both cases, the Sun serves as a standard source. In the first case, the atmospheric extinction curve is derived from the slit spectra of the disk center at different zenith distances before and after the eclipse, thereby the absolute calibration being immediately carried out from the intrinsic intensity of the Sun and the slit width, etc. On the other hand, the slitless spectra around the second contact are used to plot the $E_\lambda(h) \sim h$ curve, and the base of the chromosphere is determined as the height corresponding to the inflection point of the $E_{5000\,\text{Å}}(h) \sim h$ curve. One can calibrate the characteristic curves in a way different from the first one. There is indication that the results of both methods are quite consistent, implying feasibility of the calibration procedures. Finally, the temperature at the base of the chromosphere is calculated to be 4385 ± 50K, in accordance with the data of the slitless spectra.

Solar eclipse 1990 July 22

079.003 **Observation of the total solar eclipse 1990 July 22.**
         M. I. Dzyubenko, V. M. Efimenko, L. M. Kurochka, S. I. Musatenko, G. A. Rubo, V. V. Tel'nyuk–Adamchuk, K. I. Churyumov.
Visn. Kiiv. Univ., Fiz.–Mat. Nauki, Astron., Vip. 3, p. 35 – 38 (1992). In Ukrainian.
    The Kiev University expedition observed the total solar eclipse on July 22, 1990 at Chukotka, south–eastern Yakutiya and Taimyr. The largescale pictures of the solar corona, its spectra were obtained. The geophysical effects of the solar eclipse were registered. Some preliminary results of the observations are given.

079.004 **Strong increase of neutron flux during 22 July, 1990 eclipse.**
N. N. Volodichev, B. M. Kuzhevskij, O. Yu. Nechaev, M. I. Panasyuk.
Kosm. Issled., Tom 30, Vyp. 3, p. 422 – 424 (May–Jun 1992). In Russian. English translation in Cosmic Res., Vol. 30, No. 3.

Solar eclipse 1991 July 11

079.005 **Observation of solar eclipse on 11 July 1991 in Mexico for solar diameter determination.**
Eh. A. Gurtovenko, V. V. Tel'nyuk–Adamchuk, S. M. Okulov, Yu. A. Buzdugan, P. A. Olijnyk.
Astron. Tsirk., No. 1550, p. 29 – 30 (Sep – Oct 1991). In Russian.
    Within the framework of investigation of the long period solar diameter variations the observations were made during the solar eclipse at 11 July 1991 in South California, Mexico. A preliminary analysis of the data obtained gives the possibility to get a

detailed information of the flux changes. These data enable to determine the maximum totality moments, totality durations in each of the points for purposes of solar diameter determinations.

079.006 One day on the sun.
D. Bruning.
Astronomy, Vol. 20, No. 1, p. 48 – 54 (Jan 1992).

079.007 Eclipse totale de soleil du 11 juillet 1991: les observations photographiques et CCD de la polarisation coronale.
F. Clette.
Ciel Terre, Vol. 108, No. 3, p. 85 – 94 (May – Jun 1992).

079.008 The great eclipse.
J. M. Pasachoff.
Natl. Geogr., Vol. 181, No. 5, p. 30 – 51 (May 1992).

079.009 Die totale Sonnenfinsternis vom 11. Juli 1991.
M. Richert.
Ahnerts Kalender für Sternfreunde 1993.
p. 163 – 165 (1992). With plates 36, 37. – See Abstr. 046.012 for the main entry.

079.010 Photographing the solar eclipse of 11 July 1991, La Paz, Mexico.
F. Diego.
Vistas Astron., Vol. 35, Part 1, p. 1 – 2 (Jan 1992).

079.011 The 1991 eclipse: "Telling it like it is".
P. Beer.
Vistas Astron., Vol. 35, Part 1, p. 3 – 4 (Jan 1992).

079.012 The 11 July 1991 eclipse: The view from Hawaii.
N. Henbest.
Vistas Astron., Vol. 35, Part 1, p. 5 – 7 (Jan 1992).

079.013 All at sea ... The eclipse'n me.
R. E. White.
Vistas Astron., Vol. 35, Part 1, p. 9 – 10 (Jan 1992).

Solar eclipse 1992 January 4

079.014 Ring of fire.
R. Talcott.
Astronomy, Vol. 20, No. 4, p. 68 – 71 (Apr 1992).
Amateur photographs of the 1992 annular eclipse.

Solar eclipse 1992 June 30

079.015 The total solar eclipse of Tuesday 30 June 1992.
J. Knight.
Mon. Notes Astron. Soc. S. Afr., Vol. 51, Nos. 5/6, p. 40 – 44 (Jun 1992).

Totality. Eclipses of the Sun.
See Abstr. 003.005.

Astronomical records in the *Ch'un–ch'iu* chronicle.
See Abstr. 004.008.

The 1816 solar eclipse and comet 1811 I in John Linnell's astronomical album.
See Abstr. 004.014.

A re–investigation of the "double dawn" event recorded in the Bamboo Annals.
See Abstr. 004.074.

Die totale Sonnenfinsternis vom 11. Juli 1991 – Beobachtungsort San Pedrito.
See Abstr. 014.040.

Abnormalities of the time comparisons of atomic clocks during the solar eclipses.
See Abstr. 034.087.

On the effect of scintillations on the photosphere brightness distribution near the solar limb from eclipse observations.
See Abstr. 071.023.

Coronal density and temperature structure from coordinated observations associated with the total solar eclipse of 1988 March 18.
See Abstr. 074.025.

Structural changes in the solar corona during the July 1991 eclipse.
See Abstr. 074.033.

Solar corona 22 July 1990. Preliminary information on photographic observations made by the Kiev University expedition in Markovo, Magadan region.
See Abstr. 074.036.

Compact sources of suprathermal microwave emission detected in quiescent active regions during lunar occultations.
See Abstr. 077.020.

The radio image of the Sun at the wavelengths 3.5, 2.8, 2.25 and 1.95 cm at the period previous to solar eclipse on July 11, 1991.
See Abstr. 077.027.

Compact sources of suprathermal microwave emission detected in quiescent active regions during lunar occultations.
See Abstr. 077.044.

A possible atmospheric pressure wave from the total solar eclipse of 22 July 1990.
See Abstr. 082.055.

Ionospheric effects of the solar eclipse of September 23, 1987, around the equatorial anomaly crest region.
See Abstr. 083.002.

On a possible use of total solar eclipse below the horizon for observations of the inner zodiacal light (as applied to the eclipse of 30 June, 1992).
See Abstr. 106.028.

080 Atmosphere, Figure, Internal Constitution, Neutrinos, Rotation, etc.

080.001 Amplitude modulation of low–degree solar p–modes.
S. Ehgamberdiev (*Sh. A. Ehgamberdiev*), S. Khalikov
(*Sh. Khalikov*), M. Lazrek, E. Fossat.
Astron. Astrophys., Vol. 253, No. 1, p. 252 – 260 (Jan 1992).
The appearance of the power spectrum obtained with one day of full–disk helioseismic data is strongly changing from day to day. This is a combined effect of the changing interferences between unresolved individual p–modes and the real amplitude modulation of these p–modes. The statistical distribution of the power inside one given peak has been studied as a function of the magnitude of the amplitude modulation. The result depends on the level of interdependence of the fluctuations of amplitude displayed by different individual modes. Using 99 days of data provided by the IRIS instrument of Kumbel (Uzbekistan) and 18 different spectral peaks, it has been possible to obtain, within some assumptions regarding the interdependence or independence of the individual modes, two independent and converging values of the amplitude modulation rate. The result is $0.20 < \sigma_a < 0.37$, with the most probable value being 0.23 ± 0.02. A comparison is then made of the amplitude modulation, possibly provided by modelling one p–mode as a superposition of damped sine waves impulsively excited at random times with random amplitudes. A somewhat surprising result is that the range of possibilities covered by this model does not bracket the measured values.

080.002 Stochastic fluctuations of the solar dynamo.
A. R. Choudhuri.
Astron. Astrophys., Vol. 253, No. 1, p. 277 – 285 (Jan 1992).
Attempts in the past to model the irregularities of the solar cycle were based on studies of the nonlinear feedback of magnetic fields on the dynamo source terms. Since the α–coefficient is obtained by averaging over the turbulence, it is expected to have stochastic fluctuations, and the author shows that these fluctuations can explain the irregularities of the solar cycle in a more satisfactory way. He solves the dynamo equations in a slab with a single mode, taking the α–coefficient to be constant in space but fluctuating stochastically in time with some given amplitude and given correlation time. The same level of percentile fluctuations (about 10%) produces no effect on an $\alpha\omega$ dynamo, but makes an α^2 dynamo completely chaotic. The level of irregularities in an $\alpha^2\omega$ dynamo qualitatively agrees with the solar behavior, reinforcing the conclusion of Choudhuri (1990) that the solar dynamo is of the $\alpha^2\omega$–type.

080.003 Variability of the quiet atmosphere of the sun in the context of the activity cycle.
R. Muller.
11. European Regional Astronomy Meeting of the IAU: New windows to the universe, Vol. 1, p. 1 – 25 (1990). – See Abstr. 012.001 for the main entry.
During the last few years, it has been found that the structure of the solar photosphere, outside active regions, varies over the 11–year activity cycle. The structure of the photosphere and of the convection zone varies as indicated by the observed variation of the granulation size, the shape of photospheric lines as well as the solar radius, luminosity and effective temperature; however there is not a good agreement between these various results as yet. Outside active regions, the variation of the magnetic flux, which is an important parameter necessary to understand the solar variability, is not well known yet: while no variation is revealed by magnetograph observations, a variation in antiphase with the sunspot number is suggested by indirect indicators like NBPs, XBPs, spicules. The differential rotation as well as the meridional circulation are variable with a period of 11 years. All those variable phenomena have to be taken into account to understand the origin of the solar activity.

080.004 Acoustic waves in the solar atmosphere. IX. Three minute pulsations driven by shock overtaking.
W. Rammacher, P. Ulmschneider.
Astron. Astrophys., Vol. 253, No. 2, p. 586 – 600 (Jan 1992).
The authors show that short–period acoustic waves with periods less than 40 s, by the process of shock overtaking, are able to drive three minute type first overtone pulsations of the solar chromosphere. Waves with such periods are the main driving mechanism for these pulsations which receive of the order of 50% of the available wave energy. Simulations of the Mg II k and Ca II K line profiles show a very similar emission core evolution as is observed for Ca II K_{2V} cell grains.

080.005 Die Sonne und die magnetischen Eigenschaften der Neutrinos.
R. Plaga.
Sterne Weltraum, Jahrg. 31, Nr. 1, p. 20 – 23 (Jan 1992).

080.006 Neutrino oscillations in the Sun and the ^{205}Tl solar neutrino experiment.
S. Neumaier, A. Urban, E. Nolte.
Z. Phys., A, Vol. 341, No. 2, p. 239 – 242 (Jan 1992).
Neutrino induced transition rates from ^{205}Tl to excited states in ^{205}Pb were calculated for neutrino fluxes from the different hydrogen burning reactions in the Sun. Suppression factors for electron neutrinos due to flavor oscillations in the Sun were obtained. The influence of neutrino oscillations on the neutrino capture rate of ^{205}Tl in dependence of the mixing angle and neutrino mass difference is discussed.

080.007 Describing analytically the MSW effect for solar neutrinos in the presence of solar density perturbations.
A. Abada, S. T. Petcov.
Phys. Lett. B, Vol. 279, No. 1/2, p. 153 – 160 (9 Apr 1992).
The authors derive a simple analytic expression for the average probability of two–neutrino matter–enhanced transitions of solar neutrinos in the sun in the case of existence of a thin layer in the interior of the sun in which the density changes much faster (but continuously) than is predicted by the standard solar model.

080.008 The solar neutrino problem and the neutrino magnetic moment.
J. Pulido.
Phys. Rep., Vol. 211, No. 4, p. 167 – 199 (Feb 1992).
The physics of the proposed solution to the solar neutrino puzzle based on the neutrino magnetic moment is reviewed. The magnetic moment transition mechanism from active to sterile neutrinos can be either resonant or non–resonant and its kinship to matter enhanced oscillations is shown. The transition probability in the adiabatic approximation is calculated and the limits to adiabaticity are discussed. The full probability incorporating both the adiabatic and non–adiabatic regimes is derived using the Landau–Zener approximation for the non–adiabatic regimes. The available experimental data from the three existing solar neutrino experiments (Davis, Kamiokande II and SAGE) are compared with the results of the theory. The uncertainties in the solar magnetic field are considerable and the ansatz used takes a value of 10^5G along the solar core and the radiation zone, decreasing then linearly along the convection zone. An anticorrelation between neutrino flux and solar activity, although consistent with the theory, cannot be clearly predicted.

080.009 On the coherence condition for resonant spin–flavor precession of Majorana solar neutrinos.
R. Horvat.
Int. J. Mod. Phys. A, Vol. 7, No. 6, p. 1309 – 1314 (10 Mar 1992).
One of the most attractive solutions to the solar–neutrino problem (including an anticorrelation of the solar–neutrino flux

with sunspot activity) incorporates a Majorana neutrino having a flavor–changing transition moment as large as $(0.1 - 1) \times 10^{-10}$ Bohr magnetons. This solution is compatible with all known laboratory, astrophysical and cosmological bounds. The author shows the consistency of the solution with the coherence condition for effective–mass eigenstates inside the sun.

080.010   **Solar radial velocity and oscillations as measured by sodium and potassium resonant scattering spectrometers.**
P. L. Pallé, C. Régulo, T. Roca Cortés, L. Sánchez Duarte, F. X. Schmider.
Astron. Astrophys., Vol. 254, No. 1/2, p. 348 – 354 (Feb 1992).
Since the beginning of 1990, a sodium based resonance scattering spectrometer belonging to the IRIS Network has operated at Observatorio del Teide side by side with the MkI instrument (a resonance scattering spectrometer based on potassium) from the University of Birmingham and operated since 1975. A 3 month series of simultaneous data obtained in summer 1990 is the basis for this study that compares the performance of both instruments in measuring the solar radial velocity and oscillations. Interesting features of the p–modes as seen at two different levels of the solar atmosphere such as the ratio of energies per unit mass and the difference of phases are obtained. These results demonstrate that the p–modes are truly standing waves. Crosscorrelation of the spectra of the series obtained with both instruments shows the existence of signals well above the cutoff frequency of the solar atmosphere, being interpreted as travelling waves, also called pseudomodes. An estimation of the cutoff frequency gives a value of 5.6 ± 0.1 mHz, higher than theoretically predicted.

080.011   **A nonlocal convection model of the solar convection zone.**
D. R. Xiong, Q. L. Chen.
Astron. Astrophys., Vol. 254, No. 1/2, p. 362 – 370 (Feb 1992).
The models of the solar convection zone with various convective parameters c_1 and c_2 have been constructed based on Xiong's nonlocal convection theory (Xiong 1979, 1981). The departure form radiative equilibrium is taken into account by means of the generalized Eddington approximation (Xiong 1989). The nonlocal convection model has a large temperature gradient in the surface superadiabatic convection zone and a rather shallow convectively instable zone in comparison with the local convection model with the same c_1. The departures from radiative equilibrium are negligible except at the most upper layer of the convection zone, where the maximum relative departure is about 3%. A comparison of the model based on the Eddington approximation with the one based on the radiative diffusion approximation shows that they are very close to each other. A comparison of the theoretical eigenfrequencies of the adiabatic p–mode oscillations and the surface lithium depletion with the observed ones shows that the model with $c_1 = 0.75$ and $c_2 = 0.25$ seems to be optimal. The theoretical eigenfrequencies of the modes of degree $l < 60$ are systematically smaller than the observed ones and the overshooting seems slightly deeper in views of the surface lithium depletion. A reasonable explanation may be that there is an about 10^4G magnetic field in the lower convection zone. The magnetic fields will brake extensive overshooting and block the convective heat transport, which will increase the temperature gradient in the lower convection zone.

080.012   **Neutrino decay solution of the solar neutrino problem revisited.**
Z. G. Berezhiani, G. Fiorentini, M. Moretti, A. Rossi.
JETP Lett., Vol. 55, No. 3, p. 151 – 156 (10 Feb 1992). Current Physics Microform No.: 9204X1986.
The neutrino decay solution of the solar neutrino problem is revisited in the context of majoron models. It is shown that for a particular range of parameters this scenario reconciles both the Homestake data and the Kamiokande data. The prediction for gallium detectors is also given. It is shown that the sensitivity of Borexino is sufficient to observe the solar $\bar{\nu}_e$ signal, which is the crucial prediction of this scenario, and to distinguish it from the alternative $\bar{\nu}_e$ signal provided by the hybrid models of neutrino oscillation and magnetic moment transitions.

080.013   **The Nimbus 7 solar total irradiance: a new algorithm for its derivation.**
D. V. Hoyt, H. L. Kyle, J. R. Hickey, R. H. Maschhoff.
J. Geophys. Res., Vol. 97, No. A1, p. 51 – 64 (1 Jan 1992).
The Nimbus 7 satellite has measured the solar total irradiance from November 1978 to July 1991 (153 months). These measurements are important both in solar physics and for climate change. To insure that the Nimbus 7 measurements are capturing the true behavior of the Sun, it is essential that the properties of the radiometer and its changes over time be understood. The calibration of the radiometer can be viewed as a process of removing instrumental influences from the raw measurements, leaving the experimeter with an estimate of solar variability. In this paper the changing radiometer pointing, the zero offsets, the stability of the gain, the temperature sensitivity, and the influences of other platform instuments are all examined and their effects on the measurements considered. Only the question of relative accuracy (not absolute) is examined. The resulting derived solar irradiances are compared to previous analyses of the Nimbus 7 radiometer and to the Solar Maximum Mission (SMM) measurements.

080.014   **The resolving power of current helioseismic inversions for the Sun's internal rotation.**
J. Schou, J. Christensen–Dalsgaard, M. J. Thompson.
Astrophys. J., Lett., Vol. 385, No. 2, p. L59 – L62 (1 Feb 1992).
The inferred internal solar rotation rate obtained from helioseismic inversion is a nonlocal average of the true rotation rate; hence some care is required in the interpretation of the results of such inversions. The authors present kernels which describe the weighting of the average, for an inversion that is representative of recent work. Of particular interest is the fact that the so–called polar rate is actually an extrapolation from lower latitudes: the current helioseismic data do not permit a solution whose averaging kernel is localized at high solar latitude. Nonetheless, the results do not change the basic conclusion that the average radial gradient of the rotation rate in the solar convection zone is small. Finally, the authors demonstrate that more localized measures of the solar rotation are possible with the more complete data that should become available from new helioseismology projects.

080.015   **Emergence of magnetic flux from the convection zone into the solar atmosphere. I. Linear and nonlinear adiabatic evolution of the convective – Parker instability.**
S. Nozawa, K. Shibata, R. Matsumoto, A. C. Sterling, T. Tajima, Y. Uchida, A. Ferrari, R. Rosner.
Astrophys. J., Suppl. Ser., Vol. 78, No. 1, p. 267 – 282 (Jan 1992).
The authors study the linear and nonlinear properties of the evolution of emerging magnetic flux from the solar convection zone into the photosphere, chromosphere, and corona. An isolated flux sheet initially embedded in a simple model convection zone below a model solar atmosphere is susceptibile to both convective instability and the Parker instability. The authors perform a linear stability analysis of the partially magnetized convection zone. They find that the growth rate of this combined convective – Parker instability differs significantly from that of the Parker instability in the absence of convection. Furthermore, the authors use a two–dimensional MHD code to study the nonlinear evolution of the convective – Parker instability and the subsequent expansion of magnetic flux into the overlying atmosphere. It is found that, in spite of the superadiabaticity of the unstable layer, the nonlinear development of the magnetic field is similar to that of the Parker instability in a convectively stable layer.

080.016   **Tests of the detection and mode classifications of low–degree solar gravity modes with 1978 solar diameter observations.**
H. A. Hill.
Astrophys. J., Suppl. Ser., Vol. 78, No. 1, p. 283 – 300 (Jan 1992).
The 1978 solar diameter observations of Caudell et al. have been analyzed for evidence of the low–degree gravity modes

classified by Hill and by Hill and Gu, and to furnish tests of the classifications with respect to angular order m. These tests are based on correlations of the power densities, the mean power densities, and the mean–square deviations of the power densities found in the 1978 power spectrum with properties of the classified spectrum and properties of the 1979 differential radius observations. The degrees of $l = 1,...,5$ are included in this work. In the series of independent correlation tests designed to test for evidence of the classified modes, positive results were obtained. The results of the present study give further independent evidence consistent with the hypothesis that gravity–mode signals have been detected. The combined findings strongly support the position that the 1978 power spectrum contains evidence of gravity modes and that a significant fraction of the mode classifications used in the analyses are correct with respect to m.

080.017   **Seesaw model solutions of the solar neutrino problem.**
S. A. Bludman, D. C. Kennedy, P. G. Langacker.
Nucl. Phys. B, Part. Phys., Vol. 374, No. 2, p. 373 – 391 (27 Apr 1992).
The authors re–examine solutions to the solar neutrino problem involving neutrino oscillations, decaying neutrinos and a cooler center of the sun. Comparison of the Homestake and Kamiokande II observations implies that low–energy neutrinos are more suppressed than high–energy, disfavoring the large–mass adiabatic MSW resonance and allowing to exclude non-standard solar models with a simple reduction of core temperature. The authors also present two specific seesaw models compatible with electroweak neutral current and proton decay constraints, including radiative corrections to the seesaw predictions.

080.018   **The low *l* solar p–mode spectrum at maximum and minimum solar activity.**
M. Anguera Gubau, P. L. Pallé, F. Pérez Hernández, C. Régulo, T. Roca Cortés.
Astron. Astrophys., Vol. 255, No. 1/2, p. 363 – 372 (Feb 1992).
Velocity measurements of disc integrated sunlight obtained at the Observatorio del Teide from 1980 throughout 1989 are used to find the frequencies and amplitudes of each p–mode with $l < 4$ and $5 < n < 33$. The 32 best monthly spectra obtained are averaged in two separate groups corresponding to periods with solar activity maximum and minimum. The observational p–mode frequencies differ from those predicted by standard solar models by more than their errors; however the frequency separations such as $\delta\nu_{02}$, and $\delta\nu_{13}$ are in reasonable agreement. Such parameters, which give information about the core of the Sun, can be determined for the maximum and the minimum of solar activity cycle. Significant differences between maximum and minimum of these parameters have not been found and they agree well with the predictions of standard models, although the small observational errors achieved allow to distinguish amongst them. The relative variations of the frequencies between minimum and maximum are in qualitative agreement with other investigations, although some significant differences found suggest possible changes in layers other than the most external ones. Furthermore, these variations are different for modes of different *l* value, suggesting a shape of the perturbation other than spherically symmetric. It is also found that the energy of the modes is smaller at the maximum than during the minimum of solar activity, telling us a bit more about the structure and efficiency of the sunlayers in exciting these modes.

080.019   **Seismological tests of standard solar models calculated with new opacities.**
W. A. Dziembowski, A. A. Pamyatnykh, R. Sienkiewicz.
Acta Astron., Vol. 42, No. 1, p. 5 – 15 (1992).
The authors calculated models of the Sun adopting opacity data of Iglesias and Rogers (1991) for three heavy element mixtures. Calculations were made with a standard stellar evolution code ignoring effects of gravitational settling and the convective overshooting. Using nearly 2300 measured frequencies of solar p–modes the authors determined corrections to the sound speed and density distribution in these models. The

corrections were found to be significantly smaller than those in models calculated with earlier opacities. The model calculated with the Anders and Grevesse (1989) mixture of heavy elements shows a remarkable agreement with the helioseismic data. There is, however, a contradiction between this consistency and a sizable correction to the surface helium abundance, $\Delta Y \approx -0.04$. The authors argue that this large value may be a spurious result caused by inadequacies in the MHD (Mihalas, Däppen and Hummer, 1988) thermodynamics used in the models. The authors found new evidence for such inadequacies in relative large corrections to the sound speed in the fractional radius range 0.85–0.95.

080.020   **Solar models with enhanced energy transport in the core.**
J. Christensen-Dalsgaard.
Astrophys. J., Vol. 385, No. 1, p. 354 – 362 (20 Jan 1992).
The discrepancy between the observed and computed flux of solar neutrinos can be eliminated if the temperature gradient in the core of the model, and hence its central temperature, is reduced. This could in principle be accomplished by reducing the core opacity; alternatively, it has been proposed that a fraction of the energy transport in the solar core results from the motion of hypothetical particles (the so–called WIMPs). The resulting changes in solar structure have measurable effects on the frequencies of the solar 5 minute oscillations. Here the author considers models where the opacity in the solar core has been decreased in a manner which roughly simulates the effects of the WIMPs. The analysis of the models and their frequencies provides insight into the consequences of modifications to the physics of solar models. In particular, it is found that models with the observed neutrino flux are inconsistent with observations of low–degree solar oscillations.

080.021   **The 22–year cycle of the Sun.**
W. Kundt.
Astrophys. Space Sci., Vol. 187, No. 1, p. 75 – 85 (Jan 1992).
The 22–yr solar cycle is being explained as due to a spin–aligned magnetic quadrupole frozen into the (radiative) core. Differential rotation of the poorly conducting convection zone gives rise to flux winding and to an alternating dominance, near the surface, of one of the two constituent dipoles. Equatorial superrotation and polar subrotation are stabilized by the same magnetic flux that transfers angular momentum from the (radiative) core to the (escaping) solar wind. The time– and radius–dependent magnetic torque is also responsible for the rigid and twisting oscillations of a thin surface layer, and for the relative velocities of different tracers.

080.022   **Solar abundances from gamma–ray spectroscopy: comparisons with energetic particle, photospheric, and coronal abundances.**
R. J. Murphy, R. Ramaty, B. Kozlovsky.
AIP Conf. Proc., No. 232, p. 439 – 444 (1 Aug 1991). Current Physics Microform No.: 9110A1135. – See Abstr. 012.003 for the main entry.
The authors have derived accelerated particle and ambient gas abundances using solar flare gamma–ray spectroscopy. They find that the derived accelerated particle composition is different from the composition observed in large proton flares; rather, it resemble that observed in ^{3}He–rich flares. The analysis also suggests that the ambient composition differs from the composition of both the photosphere and the corona.

080.023   **The journey into the centre of the Sun.**
L. Neslušán, J. Rybák.
Kozmos, Vol. 23, No. 1, p. 8 – 14 (Feb 1992). In Slovak.

080.024   **Can physics explain the mystery of missing solar neutrinos?**
J. Grygar.
Říše hvězd, Vol. 72, No. 2, p. 31 – 32 (Feb 1991). In Czech.

080.025 Generation of low–frequency pulsations in magnetic waveguides in the Sun's atmosphere.
V. M. Nakaryakov, N. S. Petrukhin, S. M. Fajnshtejn.
Sov. Astron. Lett., Vol. 17, No. 6, p. 423 – 424 (Nov 1991). Current Physics Microform No.: 9207X1861. English translation of Pis'ma Astron. Zh., Tom 17, No. 11, p. 1008 – 1012 (Nov 1991).

In this paper the authors derive analytically the nonlinear Schrödinger equation which describes the nonlinear dynamics of the amplitude of eigenmodes in a magnetic flux tube in the solar atmosphere. The derivation uses the approximation of an ideal magnetic layer. They investigate the generation of low–frequency pulsations which arise as a result of modulation instability in packets of trapped modes. They present estimates of the period of the low–frequency pulsations for photospheric flux tubes.

080.026 Interior opacities and the standard solar model.
J. Faulkner, F. J. Swenson.
Astrophys. J., Lett., Vol. 386, No. 2, p. L55 – L58 (20 Feb 1992). = UCO/Lick Obs. Bull., No. 1208.

The authors investigate the sensitivity to changes in the interior opacities of a number of solar parameters commonly inferred from standard solar models. In particular, they discuss the sensitivity of the central temperature, the radius at the base of the convective zone, and the degree of lithium depletion to changes in the opacity prescription. The authors examine the impact of the new OPAL opacities on these quantities, and find that further interior opacity increases concentrated around 2×10^6K are indicated from the solar oscillation data. The authors provide some additional OPAL opacities needed to produce accurate standard solar models.

080.027 Messung der Randverdunklung der Sonne im Weißlicht.
H. Joppich.
Sonne, Jahrg. 16, Nr. 61, p. 15 (Mar 1992).

080.028 On the evidence for mesogranules in solar power spectra.
G. P. Ginet, G. W. Simon.
Astrophys. J., Vol. 386, No. 1, p. 359 – 363 (10 Feb 1992).

The power spectrum of the horizontal component of the solar convective velocity field has recently been estimated from observations of the Doppler shifts of surface flows at and near disk center (Chou et al). Those authors assert that "there is no evidence of apparent energy excess at the scale of mesogranulation". It is shown in this paper that their conclusion is incorrect and that the shape of the observational spectrum does indeed confirm the presence of both supergranules and mesogranules in the solar convective flow. To establish this claim, the authors have extended existing kinematic models of convection at the solar surface and have introduced power spectra diagnostics. It is found that models with supergranule cells alone do not produce spectra that match the observations, but if mesogranules are included, then there is excellent agreement between the model and observational spectra, when the model parameters are chosen to be consistent with proper motion and Doppler measurements.

080.029 On the sensitivity of high–degree p–mode frequencies to the solar convection zone helium abundance.
J. A. Guzik, A. N. Cox.
Astrophys. J., Vol. 386, No. 2, p. 729 – 733 (20 Feb 1992).

For the standard solar model of Guzik and Cox, incorporating the latest opacities and the Mihalas et al. (MHD) equation of state (EOS), the calculated nonadiabatic p–mode frequencies of degree l = 1 – 200 differed from observed frequencies by at most ~4 μHz. For degrees l = 300 – 1000, which sample the upper convection zone, larger discrepancies of 5 – 30 μHz were found, which might be attributed to a decrease in convection zone helium abundance due to gravitational and thermal diffusion. Here the authors compare nonadiabatic frequencies for modes of degree l = 200 – 1000 for the MHD EOS model, described in Guzik and Cox, with Y = 0.27, and a model with convection zone Y = 0.24, corresponding to the decrease expected due to diffusion. The frequencies of the l = 300 – 600 modes, which

sample best the helium ionization region, show the greatest sensitivity to this helium abundance change. The frequency sensitivity to helium abundance is not affected by changes in the mixing–length / pressure scale height ratio or the inclusion of turbulent pressure, although these effects can alter considerably the upper convection zone structure and the calculated frequencies. The O – C differences for the accurately observed modes favor a reduction in convection zone helium mass fraction of about 0.03, consistent with that expected due to diffusion.

080.030 Interpretation of solar–cycle variability in high–degree p–mode frequencies.
D. J. Evans, B. Roberts.
Nature, Vol. 355, No. 6357, p. 230 – 232 (16 Jan 1992). Letter–to–the–editor.

Recently Libbrecht and Woodard and Elsworth et al. have demonstrated that the frequencies of solar acoustic p–mode oscillations vary significantly over the solar cycle. The authors have previously suggested that cyclic variations in the magnetic activity of the Sun could modulate the p–mode frequencies in a similar way. In particular, they investigated simple models of the "magnetic canopy", which permeates the solar atmosphere and overlies all of the Sun's surface, to determine its influence on p–mode frequencies. Here the authors make a comparison of their model predictions with the observations of Libbrecht and Woodard. They find that, despite the simplicity of their model, they are able to obtain good agreement with the observed frequency shifts for modes of frequency less than 4 mHz, through a mechanism in which an increasing magnetic field induces "stiffening" of the Sun's chromosphere. Above this frequency there is clearly something missing from their model. The authors speculate that the behaviour above 4 mHz is related to a cutoff frequency in the solar atmosphere, above which waves are trapped only partially.

080.031 Solar neutrinos: Is no SNUs good news?
L. M. Krauss.
Nature, Vol. 355, No. 6359, p. 399 – 400 (30 Jan 1992).

080.032 Standard solar model.
D. B. Guenther, P. Demarque, Y.–C. Kim, M. H. Pinsonneault.
Astrophys. J., Vol. 387, No. 1, p. 372 – 393 (1 Mar 1992).

A set of solar models has been constructed, each based on a single modification to the physics of a reference solar model. In addition, a model combining several of the improvements has been calculated to provide a "best" solar model. Improvements were made to the nuclear reaction rates, the equation of state, the opacities, and the treatment of the atmosphere. The impact on both the structure and the frequencies of the low–l p–modes of the model to these improvements are discussed. The authors find that the combined solar model, which is based on the best physics available ar present (and does not contain any ad hoc assumptions), reproduces the observed oscillation spectrum (for low–l) within the errors associated with the uncertainties in the model physics (primarily opacities).

080.033 Localized excitation of solar oscillations.
P. R. Goode, D. Gough, A. Kosovichev.
Astrophys. J., Vol. 387, No. 2, p. 707 – 711 (10 Mar 1992).

Solar oscillation data are well described in terms of waves produced by isolated expansive events occurring less than 200 km below the base of the photosphere. The events last about 5 minutes.

080.034 On the ultimate accuracy of solar oscillation frequency measurements.
K. G. Libbrecht.
Astrophys. J., Vol. 387, No. 2, p. 712 – 714 (10 Mar 1992).

If one assumes that solar p– and f–mode oscillations are stochastically excited, then the measured mode properties and solar background noise can be used to calculate maximum–likelihood uncertainties in mode frequency measurements, σ_ν, for a given observation time. Such calculations agree quite well with current data, if one uses the measured background noise, which includes instrumental as well as solar contributions. Assuming negligible instrumental background, the author finds that it should be possible with a 3 yr continuous observation to measure individual mode frequencies v_{nlm} to accuracies as high as $\sigma_v/v \approx 1.4 \times 10^{-5}$, which is a factor of ~ 5 times better than our current best one–season measurements. The most precise measurements should be for low–l modes in the $1-2$ mHz range, and the longest periods observable in a 3 yr observation would be approximately 20 min. These fundamental limitations in the eventual accuracy of p–mode frequency measurements are set by the solar background noise and the stochastic nature of the driving mechanism.

080.035   Alfvén wave transmission through the solar atmosphere.
P. L. Similon, S. Zargham.
Astrophys. J., Vol. 388, No. 2, p. 644 – 647 (1 Apr 1992).
The transmission and reflection coefficients of shear Alfvén waves in an open magnetic structure emerging from the Sun are determined. Both two–dimensional slab and axisymmetric static equilibrium configurations are solved numerically on a curvilinear coordinate grid adapted to the magnetic field geometry and to the density stratification. The equilibrium structure obtained numerically extends from the photosphere up to an altitude of 20,000 km, spanning 14 density scale heights and allowing for a magnetic flux tube expansion by a factor $40 - 400$, corresponding to a filling factor of 2.5% to 0.25%. From 20,000 km up to 4 R$_\odot$ a magnetic monopole model is adopted. The Alfvén wave equation is solved for wave periods ranging from 10 s to 10^4s for which the WKB approximation is not valid. It is shown that, because of magnetic expansion, the coefficient of transmission of Alfvén waves from the photosphere to 4 R$_\odot$ has values about $5\% - 30\%$ for wave periods in the minutes range. Thus magnetic expansion allows a significant Alfvén energy flux to propagate from the photosphere to the corona, an effect especially important for open magnetic structures, where other heating mechanisms such as nanoflares are not operative.

080.036   Solar pulsational stability. I. Pulsation–mode thermodynamics.
N. J. Balmforth.
Mon. Not. R. Astron. Soc., Vol. 255, No. 4, p. 603 – 631 (15 Apr 1992).
It is not currently known what excites solar five–minute oscillations. Of the two most plausible possibilities, thermal overstability and stochastic excitation by turbulent convection, the single most important discriminating factor is the intrinsic stability of the pulsation modes. In view of this fact, the author addresses the problem of the linear stability of model solar envelopes. He employs a time–dependent, non–local mixing–length prescription for convection, and the Eddington approximation to radiative transfer. The calculations reveal that low–degree acoustic modes are damped. Moreover, the theoretical damping rates compare well with measurements of solar oscillation line widths. Turbulent pressure fluctuations play a critical role in stabilizing the pulsations. Finally the author relates his results to those of previous investigations.

080.037   Solar pulsational stability. II. Pulsation frequencies.
N. J. Balmforth.
Mon. Not. R. Astron. Soc., Vol. 255, No. 4, p. 632 – 638 (15 Apr 1992).
The processes that damp solar pulsations also influence pulsation frequencies. This creates deviations from frequencies computed assuming adiabatic motion without turbulent stress and is one possible reason why the observed frequencies of the five–minute oscillations do not agree with theoretical eigenfrequencies. Another possible cause is that the commonly used local mixing–length theory fails to reproduce the mean structure of the Sun in the turbulent surface boundary layer. The computational machinery developed for linear stability analysis in an earlier paper is used to gauge the degree of error incurred in employing these two standard approximations. The results indicate that the assumption of adiabatic stress–free motion is less severe than

errors likely to occur in modelling the solar superadiabatic boundary layer by conventional formulations of mixing–length theory.

080.038   Solar pulsational stability. III. Acoustical excitation by turbulent convection.
N. J. Balmforth.
Mon. Not. R. Astron. Soc., Vol. 255, No. 4, p. 639 – 649 (15 Apr 1992).
Current opinion presumes that solar five–minute oscillations are intrinsically damped and excited as a consequence of the emission of acoustical radiation by turbulent convective flows. Prescriptions modelling this process are employed to estimate modal mean amplitudes. The resulting oscillation power spectra can be made to agree with observation by adjusting various theoretical parameters. Whilst this ruins the predictive power of the theory, it does demonstrate that the hypothesis of stochastic excitation is plausible. Uncertainties in the theory are shown to have substantial repercussions on the mean amplitudes.

080.039   Characteristics of solar p–modes: results from the IPHIR experiment.
T. Toutain, C. Fröhlich.
Astron. Astrophys., Vol. 257, No. 1, p. 287 – 297 (Apr 1992).
Solar p–modes were observed in irradiance during more than 160 d by the IPHIR experiment on the USSR PHOBOS Mission in 1988. They are characterized by their frequency, splitting, linewidth and amplitudes, determined by fitting Lorentzians to the lines. Because of the long uninterrupted time series the frequencies are probably the most accurate available at present. They are compared with results from other observations and theoretical models. Very good agreement is observed with a recent standard model, MHD–S2, of Christensen–Dalsgaard (1991), for both the absolute frequencies ($<3~\mu$Hz) and the difference $\delta_{02} = v_{n,0} - v_{n-1,2}$ ($<0.15~\mu$Hz), which means that the standard solar model is a good approximation to the real Sun and that the solution to the "neutrino puzzle" has to be sought from particle physics. From the splittings of the $l = 1$ and 2 modes the rotation of the core ($0.0 < r < 0.2$) is inferred to about 4.6 times the surface rate. The damping of p–modes is determined from the linewidths; lifetimes between 24 and 2 d are found for $n = 16,...,26$. For the first time the long uninterrupted time series allows a detailed analysis of the temporal evolution of the amplitudes of p–modes. The statistical behavior for the energy distribution in time confirms the assumption of excitation by turbulent motions at the top of the convection zone. The results are in good agreement with the findings from the analysis of the linewidths and amplitudes of the whole time series.

080.040   How does the Sun become lean?
J. Mergentaler.
Urania, Rok 63, Nr. 3, p. 66 – 67 (Mar 1992). In Polish.

080.041   Large–scale flows and solar luminosity variations.
S. Arendt.
Astrophys. J., Vol. 389, No. 1, p. 421 – 427 (10 Apr 1992).
Recent measurements with radiometers have shown that the solar irradiance varies by about 0.1% peak–to–peak over the 11 yr solar cycle. The irradiance is in phase with the level of solar activity, reaching a maximum during solar maximum and a minimum during solar minimum. A potential contributing cause of this irradiance variability is a variation in large–scale flow patterns present in the solar convection zone. In this paper, the influence of a large–scale flow variation on solar luminosity is considered. The convection zone is modeled as a plane–parallel fluid layer through which heat is transported by small–scale convection as described by the mixing–length theory, and the large–scale flow is treated as a small perturbation superposed on the layer. The excess energy transported by the flow is calculated and is found to depend critically on the mixing length. The excess luminosity is found to be carried as small–scale convective heat flow, rather than bulk enthalpy transport, through all but the top few percent of the convection zone. The results are applied to the specific examples of supergranules and giant cells, and it is found

that the effect may be important, although the uncertainties involved prevent a firm conclusion.

080.042   Preliminary results of a balloon flight of the solar disk sextant.
E. Maier, L. Twigg, S. Sofia.
Astrophys. J., Vol. 389, No. 1, p. 447 – 452 (10 Apr 1992).

This paper reports preliminary results of a ballon flight on 1990 October 11 of the solar disk sextant (SDS) experiment. The SDS is an instrument which measures the solar diameter at different orientations with respect to the solar polar axis. The flight yielded useful data for ~6 hr, which included 26 full rotations of the instrument about its axis in addition to three 20 min periods at fixed angles. For the two fixed angle data sets reduced, the probable error is of the order of 15 mas per 13 ms measurement. Fitting straight lines through these data sets, with time as the independent variable, yields slopes of $(7.1 \pm 1.5) \times 10^{-3}$ and $(6.7 \pm 1.6) \times 10^{-3}$mas s^{-1}, consistent with the value of 6.47×10^{-3}mas s^{-1} expected from the Earth's approach to the Sun due to the orbital motion toward perihelion. Upon rotating the instrument on its axis a sinusoidal component of the diameter measurements was observed in each rotation cycle, with a (variable) amplitude of ~150 mas. This variation, due to instrumental inadequacies, greatly complicates the determination of the solar oblateness for the present instrument configuration, resulting in a relatively large error. The present result is $\varepsilon = (5.6 \pm 6.3) \times 10^{-6}$, ~30° offset from the polar-equator position. The absolute diameter obtained by means of the FFT definition is found to be $(1919.269 \pm 0.240)''$ or $(1919.131 \pm 0.240)''$, depending on the orientation mode of the measurement.

080.043   Gravity wave and convection interaction in the solar interior.
Ø. Andreassen, B. N. Andersen, C. E. Wasberg.
Astron. Astrophys., Vol. 257, No. 2, p. 763 – 769 (Apr 1992).

Methods developed to numerically simulate hydrodynamic waves in the terrestrial atmosphere have been utilized to investigate the similar phenomena in the solar interior. The spectral collocation method with open horizontal boundaries used is well suited for solar type studies. The current study is the start of a program to investigate the degree of penetration of gravity waves through the solar convection zone and to investigate the possible excitation of gravity waves in the solar interior by convection. The preliminary results indicate that a significant fraction of the wave energy in a gravity type wave in the convectively stable region in the solar interior may tunnel through the solar convective zone to the surface. For a wave with a horizontal extent equivalent to a global mode with degree l about 15–16 the energy transmission is of the order of 0.02% in the current model.

080.044   Cosmic ray albedo γ–rays from the quiet Sun.
D. Seckel, T. Stanev, T. K. Gaisser.
Compton Observatory Science Workshop, p. 542 – 549 (Feb 1992). – See Abstr. 012.018 for the main entry.

The authors estimate the flux of gamma–rays that result from collisions of high energy galactic cosmic rays with the solar atmosphere. An important aspect of their model is the propagation of cosmic rays through the magnetic fields of the inner solar system. The authors use diffusion to model propagation down to the bottom of the corona. Below the corona they trace particle orbits through the photospheric fields to determine the location of cosmic ray interactions in the solar atmosphere and evolve the resultant cascades. The author predict an integrated flux of gamma rays (at 1 AU) of $F \approx 5 \times 10^{-8}$cm^{-2}sec^{-1}. This can be an order of magnitude above the galactic background, and should be observable by EGRET.

080.045   Precise ground–based solar photometry and variations of total irradiance.
G. A. Chapman, A. D. Herzog, J. K. Lawrence, S. R. Walton, H. S. Hudson, B. M. Fisher.
J. Geophys. Res., Vol. 97, No. A6, p. 8211 – 8219 (1 Jun 1992).

Variations in the total solar irradiance measured by the active cavity radiometer irradiance monitor (ACRIM) on SMM have been correlated with measures of magnetic activity on the solar disk. Quantitative indices of magnetic activity were derived from ground–based, full–disk, photometric images of the Sun at red (6723 Å) and violet (3934–Å K line) wavelengths. The red images have been obtained on a daily basis at the San Fernando Observatory since 1985, and the K line images since 1988. Sunspot irradiance deficits are calculated directly from the red images while proxy measures of facular irradiance excesses are derived from the K line images. The images analyzed here were made during 21 days between June 20 and July 14, 1988, a period centered on the disk passage of a large sunspot group. The best two–parameter multiple correlation coefficient between the ACRIM data and the photometric data is $R^2 = 0.97$ (21 data points, 18 degrees of freedom). The zero point $S_0 = 1367.27$ W m^{-2} agrees well with the solar irradiance measured by ACRIM/SMM during the 1986 activity minimum: the residual standard deviation was 0.13 W m^{-2} (about 100 ppm). The multiple correlations were extended to include measures of the irradiance contribution of "network" magnetic fields, unassociated with active regions. NOAA 9 spacecraft observations of UV Mg II lines at 2800 Å gave $R^2 = 0.99$ (17 degrees of freedom) with $S_0 = 1366.68 \pm 0.08$ W m^{-2}. The index of 10.7–cm microwave flux gave $R^2 = 0.98$, with $S_0 = 1366.43 \pm 0.11$ W m^{-2}. The authors can thus model short–term irradiance changes to within 100 ppm relative precision from ground–based data.

080.046   Seasonal variations in the data of a chlorine–argon detector of solar neutrinos.
Yu. R. Rivin.
Astron. Tsirk., No. 1551, p. 26 – 27 (Nov – Dec 1991).

The data obtained with the chlorine–argon detector (South Dakota) show a seasonal variation. The variation of argon–37 can be approximated by an annual wave with a period of ~12 months and an amplitude of ~0.3 atom/day, with maximum in spring and minimum in autumn. The author investigates the asymmetry in measurements of this wave and the modulation of its amplitude.

080.047   Reopening the solar neutrino question.
L. M. Krauss.
Nature, Vol. 357, No. 6378, p. 437 (11 Jun 1992).

080.048   A new solar irradiance calibration from 3295 Å to 8500 Å derived from absolute spectrophotometry of Vega.
G. W. Lockwood, H. Tüg, N. M. White.
Astrophys. J., Vol. 390, No. 2, p. 668 – 678 (10 May 1992).

By imaging sunlight diffracted by 20 and 30 μm diameter pinholes onto the entrance aperture of a photoelectric grating scanner, the authors determined the solar spectral irradiance relative to the spectrophotometric standard star Vega, observed at night with the same instrument. Solar irradiances are tabulated at 4 Å increments from 3295 Å to 8500 Å. Over most of the visible spectrum, the internal error of measurement is less than 2%. The authors compare their calibration with earlier irradiance measurements by Neckel and Labs, by Arvesen, Griffin, and Pearson, and with the high–resolution solar atlas by Kurucz et al. The three calibrations agree well in visible light but differ by as much as 10% in the ultraviolet.

080.049   The conversion of p–modes to slow modes and the absorption of acoustic waves by sunspots.
H. C. Spruit, T. J. Bogdan.
Astrophys. J., Lett., Vol. 391, No. 2, p. L109 – L112 (1 Jun 1992). With plates L9 – L10.

The possibility that the conversion of p–modes to flux tube – guided slow modes (s–modes) is responsible for the reported absorption of the solar acoustic oscillations in sunspots and neighboring plage, is discussed in the context of a simple model atmosphere – a plane–parallel stratified adiabatic polytrope threaded by a uniform vertical magnetic field. The normal modes of oscillation of this atmosphere have complex eigenfrequencies. The authors argue that the acoustic absorption suffered by an incident p–mode scales like the imaginary part of these complex eigenfrequencies of the internal sunspot atmosphere. It is found that (1) for the f–mode, the absorption coefficient increases monotonically from small to large horizontal wavenumbers, and (2) along the nth p–mode ridge, this same general trend is modulated by the presence of n localized absorption minima. These characteristic signatures of acoustic absorption by p–mode / s–mode conversion afford the diagnostic possibility of determining the sunspot magnetic field strength from the location in wavenumber of the predicted absorption minima.

080.050   Scattering of p–modes by a sunspot.
D. C. Braun, T. L. Duvall Jr., B. J. LaBonte,
S. M. Jefferies, J. W. Harvey, M. A. Pomerantz.
Astrophys. J., Lett., Vol. 391, No. 2, p. L113 – L116 (1 Jun 1992).
The acoustic scattering properties of a large sunspot are determined from a Fourier–Hankel decomposition of p–mode amplitudes as measured from a 68 hr subset of a larger set of observations made at the south pole in 1988. The authors show that significant improvement in the measurement of p–mode scattering amplitudes results from the increased temporal frequency resolution provided by this data. Scattering phase shifts are unambiguously determined for the first time, and the dependence of the p–mode phase shift and absorption with wavenumber and frequency is presented.

080.051   Helioseismology.
S. V. Vorontsov.
Astron. Zh., Tom 69, Vyp. 2, p. 347 – 367 (Mar–Apr 1992). In Russian. English translation in Sov. Astron., Vol. 36, No. 2 (Mar–Apr 1992).
This is a short review on the relatively young and rapidly developing branch of solar physics – the study of the solar interior using the observational frequencies of its acoustic oscillations. The main properties of the acoustic oscillations are described, together with some new results obtained during the last two or three years.

080.052   Helioseismology.
S. V. Vorontsov.
Sov. Astron., Vol. 36, No. 2, p. 175 – 185 (Mar 1992). English translation of Astron. Zh., Tom 69, Vyp. 2, p. 347 – 367 (1992).
This short review is devoted to a relatively young, actively developing field of solar physics: the study of the solar interior based on the observed frequencies of acoustic oscillations. The main properties of the acoustic oscillations are presented briefly and some new results obtained in the last two years are described.

080.053   Solar differential rotation due to magnetic stresses.
H. Volland.
Astron. Astrophys., Vol. 259, No. 2, p. 663 – 668 (Jun 1992).
It is argued that the components of differential rotation of the convection zone (defined as the deviation from rigid rotation) may be driven by Maxwell stresses. The zonally averaged flow within the upper convection zone is described by Rossby–Haurwitz waves of zonal wavenumber $m = 0$, and an estimate is given of the configuration of the poloidal and the toroidal magnetic fields necessary to drive the observed differential rotation. The magnetic field components derived for minimum and maximum solar activity are not inconsistent with the solar magnetic fields measured at the solar surface.

080.054   Acoustic wave ducting along magnetic flux tubes in the sunspot regions.
R. N. Singh.
Astrophys. Space Sci., Vol. 191, No. 1, p. 125 – 130 (May 1992).
The acoustic waves generated in the solar atmosphere propagate globally as well as upwards. These waves interact with the solar magnetic field structures and are ducted upwards. The velocity of these modified acoustic waves is shown to vary in a modelled solar atmosphere. The solar plasma propagating upwards with these waves are likely to alter the observed features of spicules, granules, and supergranules during changing phases of sunspot regions.

080.055   Solar neutrino puzzle, horizontal symmetry of electro-weak interactions and fermion mass hierarchies.
R. N. Mohapatra.
Symposium on Quarks, Symmetries and Strings in Honor of Bunji Sakita's 60th Birthday, p. 43 – 53 (1991). – See Abstr. 012.044 for the main entry.
The author discusses the possibility that the apparent anti-correlation between the solar neutrino data of Davis and the number of sunspots may be an indication of a possible local horizontal symmetry operating among leptons of the first and second generation. The horizontal symmetry enables to reconcile the possible large magnetic moment of neutrino required to understand the solar neutrino data with small upper limits on neutrino mass. Extending this symmetry to the quark sector, leads to a mass formula for quarks and leptons where the fermions of the third generation pick up mass at the three level and induce the mass of the second and first generation in higher loop level, thereby explaining the observed mass and mixing hierarchy among charged fermions qualitatively.

080.056   Rotational evolution of the Sun.
N. Kiziloğlu.
Astrophys. Space Sci., Vol. 192, No. 1, p. 83 – 89 (Jun 1992).
The evolutionary behaviour of rotating solar models with different initial angular–momentum distributions has been investigated through the pre–main–sequence and main–sequence phases. The angular momentum was removed from the convective envelope of the solar models according to the Kawaler's model of magnetic stellar wind (Kawaler, 1988). The models show that (i) the surface rotational velocities of the solar mass stars are independent of initial angular momentum for ages greater than 10^8 yr and (ii) it is not possible to explain the neutrino problem and the sufficient depletion of lithium in the Sun.

080.057   The fate of ^{7}Be in the Sun.
C. W. Johnson, E. Kolbe, S. E. Koonin, K. Langanke.
Astrophys. J., Vol. 392, No. 1, p. 320 – 327 (10 Jun 1992).
The authors reexamine the electron– and proton–capture rates of ^{7}Be important to the solar neutrino "problem". Although the assumptions implied by the traditional Debye approximation for plasma screening are not valid, a careful numerical study changes the electron capture rate by less than 2%. The authors extrapolate experimental data on the proton capture reaction to astrophysically relevant energies using an energy dependence that includes d–wave scattering. The proton capture rate is shown to be relatively independent of the model space and interaction used. The authors find that the solar proton capture rate is lowered by approximately 7% from the currently accepted value.

080.058   Waves in solar magnetic flux tubes: the observational signature of undamped longitudinal tube waves.
S. K. Solanki, B. Roberts.
Mon. Not. R. Astron. Soc., Vol. 256, No. 1, p. 13 – 25 (1 May 1992).
Linear calculations of undamped longitudinal waves in thin solar magnetic flux tubes are presented. The influence of such waves, having a variety of parameters, on the Stokes I and V profiles of eight photospheric spectral lines is studied. Diagnostics based on the Stokes parameters of the properties of flux tube waves, in particular of the amount of energy transported by them into the upper solar atmosphere, are developed. The importance of radiative transfer effects and of the thermodynamic changes associated with the waves are pointed out. Some qualitative comparisons with the observational data are considered. The influence of flux tube waves on spatially and temporally unresolved observations is also considered.

080.059   Departure from local thermodynamic equilibrium and its effect on solar continuous absorption.
T. L. John.
Mon. Not. R. Astron. Soc., Vol. 256, No. 1, p. 69 – 79 (1 May 1992).
Limb darkening observations are reanalysed allowing for departures from local thermodynamic equilibrium (LTE). Comparisons of resulting "observed" absorption coefficients with

theoretical absorption cross–sections indicate that departures, larger than 2% in the NLTE parameter B from unity, are unlikely. Direct computations based on radiative equilibrium considerations support larger departures.

080.060   The secrets of the solar atmosphere.
H. Zirin.
Astrophysics on the threshold of the 21st century, p. 53 – 62 (1992). – See Abstr. 003.048 for the main entry.
Contents: 1. A little history. 2. Whither the chromosphere? 3. The corona and the solar wind. 4. The problem of coronal heating. 5. The stars.

080.061   Status of the Soviet–American gallium experiment.
A. I. Abazov, O. L. Anosov, E. L. Faizov,
V. N. Gavrin, A. V. Kalikhov, T. V. Knodel, I. I. Knyshenko,
V. N. Kornoukhov (*V. N. Kornokhov*), S. A. Mezentseva,
I. N. Mirmov, A. I. Ostrinsky, A. M. Pshukov, N. Ye. Revzin,
A. A. Shikhin, P. V. Timofeyev (*P. V. Timofeev*),
E. P. Veretenkin, V. M. Vermul, G. T. Zatsepin, T. J. Bowles,
B. T. Cleveland, S. R. Elliott, H. A. O'Brien, D. L. Wark,
J. F. Wilkerson, R. Davis Jr., K. Lande, M. L. Cherry,
R. T. Kouzes.
AIP Conf. Proc., No. 243, p. 1116 – 1121 (1 Jan 1992). Current Physics Microform No.: 9203C1964. – See Abstr. 012.047 for the main entry.
A radiochemical ^{71}Ga–^{71}Ge experiment to determine the primary flux of neutrinos from the Sun has begun operation at the Baksan Neutrino Observatory in the USSR. The number of ^{71}Ge atoms extracted from thirty tons of gallium was measured in five runs during the period of January to July 1990. Assuming that the extraction efficiency for ^{71}Ge atoms produced by solar neutrinos is the same as natural Ge carrier, this corresponds to a limit on the product of the neutrino flux and the cross section for all sources of neutrinos of less than 72 NU (90% CL). This is to be compared with the flux of 132 SNU predicted by the Standard Solar Model.

080.062   Solar neutrino observations with the Homestake ^{37}Cl detector.
K. Lande, B. Cleveland, T. Daily, R. Davis, J. Distel,
C. K. Lee, A. Weinberger, P. Wildenhain, J. Ullman.
AIP Conf. Proc., No. 243, p. 1122 – 1133 (1 Jan 1992). Current Physics Microform No.: 9203C1970. – See Abstr. 012.047 for the main entry.
The continuous twenty year record of these measurements of the solar neutrino flux with the Homestake chlorine detector, indicates that the average solar neutrino flux is 2.2 ± 0.3 SNU and that this ν_e flux appears to vary with the 11 year solar activity cycle. Higher ν_e fluxes are observed during solar quiet periods and lower ν_e fluxes during solar active periods. When the Homestake data is combined with the Kamiokande results, the region of overlap between the two experiments is for an observed to predicted ^8B neutrino flux ratio of about 0.4 and very little low energy neutrino flux. If the Kamiokande results are corrected for MSW effects, neutral current scatterings by non–electron neutrinos, then both the ^8B and the ^7Be fluxes are about 1/3 of the Standard Solar Model predictions.

080.063   Response of ^{127}I to solar neutrinos.
J. Engel.
AIP Conf. Proc., No. 243, p. 1134 – 1136 (1 Jan 1992). Current Physics Microform No.: 9203C1982. – See Abstr. 012.047 for the main entry.
The author describes a calculation of the expected event rate in an ^{127}I solar neutrino detector. The cross section for ^8B neutrinos is considerably larger than the corresponding cross section in ^{37}Cl. The ^7Be neutrino cross section is even more enhanced, and the total event rate in a 1000–ton iodine detector (assuming Standard Solar Model fluxes) should be 7 – 10 times that in Homestake ^{37}Cl experiment.

080.064   Dynamics of the solar atmosphere.
P. Mein.
1. Canary Islands Winter School of Astrophysics: Solar observations – techniques and interpretation, p. 179 – 246 (1992). – See Abstr. 012.045 for the main entry.
Contents: (1) Introduction. (2) Outline of the solar structure: internal sun; solar atmosphere; solar activity. (3) One–dimension non–magnetic model atmosphere continuum and spectral line formation: model atmosphere; spectral intensity and source function; formation of the continuous spectrum; formation of spectral lines. (4) Diagnostic methods for velocity measurements: in situ measurements; indirect measurements; intensity measurements (continuum and lines); Dopplershifts of line profiles; averaging effects in line Doppler–shifts; instrumental aspects: velocity measurements. (5) Rotation and convection: solar rotation from Doppler measurements; solar rotation from tracers; large–scale meridional circulation; convection (intermediate and small scales). (6) Waves in the non–magnetic atmosphere: the five minutes oscillations; the k–ω diagram; power spectrum in the k–ω diagram; phase–lags between two lines or between intensity and Dopplershift; mechanical energy flux; coronal heating. (7) Motions in magnetic flux tubes and spots. (8) Velocity fields in prominences and filaments: classification, structure; velocity measurements; steady flows; oscillations; instabilities ("disparitions brusques", eruptions). (9) Mass ejections – instabilities.

080.065   Standard solar model. II. g–modes.
D. B. Guenther, P. Demarque, M. H. Pinsonneault,
Y.–C. Kim.
Astrophys. J., Vol. 392, No. 1, p. 328 – 336 (10 Jun 1992).
The authors present the g–mode oscillation for a set of modern solar models. Each model is based on a single modification or improvement to the physics of a reference solar model. Improvements were made to the nuclear reaction rates, the equation of state, the opacities, and the treatment of the atmosphere. The authors estimate the error in the predicted g–mode periods associated with the uncertainties in the model physics and describe the specific sensitivities of the g–mode periods and their period spacings to the different model structures. In addition, the models are compared to a sample of published observations. Remarkably good agreement is found between the "best" solar model and the observations of Hill and Gu.

080.066   Observations of high–frequency and high–wavenumber solar oscillations.
D. N. Fernandes, P. H. Scherrer, T. D. Tarbell, A. M. Title.
Astrophys. J., Vol. 392, No. 2, p. 736 – 738 (20 Jun 1992). With plate 7.
Doppler shift measurements of the Na D1 absorption line have revealed solar oscillations in a new regime of frequency and wavenumber. Oscillations of vertical velocities in the temperature minimum and low chromosphere of the Sun are observed with frequencies ranging up to 9.5 mHz. The fundamental modes appear with wavenumbers up to 5.33 Mm^{-1} (equivalent spherical harmonic degree 3710). The authors find no evidence for chromospheric modes of 3 min period.

080.067   Local acoustic diagnostics of the solar interior.
D. C. Braun, C. Lindsey, Y. Fan, S. M. Jefferies.
Astrophys. J., Vol. 392, No. 2, p. 739 – 745 (20 Jun 1992). With plates 8 – 12.
The observed absorption of p–modes by sunspots and solar magnetic fields has raised the possibility of developing holographic techniques to probe local magnetic features within the solar interior. Two simple diagnostic utilities are developed and tested to search for evidence of subsurface p–mode absorption. These consist of maps of acoustic power and maps of the surface acoustic flux vector. Maps of acoustic power for a 50 hr sequence of solar Ca II K–line images show power deficits at 3 mHz corresponding to surface magnetic flux and power enhancements surrounding active regions (halos) at 6 mHz. The 6 mHz halos are observed to extend well beyond the active regions into the quiet Sun and it is likely that they represent a true emission of waves. Faint 3 mHz power deficits are also seen, extending in

long fingers from active regions far into areas of the quiet Sun. These fingers rotate with the Sun and probably represent subsurface absorption of p–mode power. Acoustic power maps alone do not unambiguously determine the presence of absorbing or emitting regions. A vector quantity, termed "surface acoustic flux vector" by the authors, is used to circumvent this problem. This flux vector is essentially a measure of the time–reversal variance of the wave motion and its divergence is an indicator of wave emission or absorption.

080.068 Doppler–Oszillationen unter dem Einfluß solarer Magnetfelder.
H. Balthasar.
Sterne Weltraum, Jahrg. 31, Nr. 5, p. 304 – 305 (May 1992).

080.069 Solar neutrinos: new physics?
J. N. Bahcall.
15. Texas Symposium on Relativistic Astrophysics and 4. ESO–CERN Symposium, p. 11 – 13 (1991). – See Abstr. 012.060 for the main entry.
The author concentrates in this report on two theoretical arguments that suggest – independent of particular solar or particle physics models – that solar neutrino experiments reveal new physics beyond the standard electroweak model.

080.070 Solar pulsations: effects due to the 22–year activity cycle?
V. A. Kotov, T. T. Tsap, L. V. Didkovskij.
Bull. Crimean Astrophys. Obs., Vol. 82 , p. 127 – 134 (1992). English translation of Izv. Krym. Astrofiz. Obs., Tom 82, p. 138 – 146 (1990).
Regular measurements have been made on the differential Doppler velocity in the Crimea between 1974 and 1987 (in all, 902 days, 5612 hr of observation), which confirm the long–term phase–coherent pulsation with a period of 160.01 min. The new data also suggest that the 160–min pulsation may have a multiplet fine structure. In particular, there have been large changes in the amplitude and phase of the pulsation in the period 1983 – 1987, which may mean that after 1982 – 1983 one has observed the 160 min oscillations related to the second half of the 22–year magnetic activity cycle. This new and unexpected feature opens up scope for probing the solar interior, and also for researching the internal rotation and the 11 (22)–year solar cycle.

080.071 Short–period variations in the Sun's global magnetic field.
M. L. Demidov, V. A. Kotov, V. M. Grigor'ev.
Bull. Crimean Astrophys. Obs., Vol. 82 , p. 135 – 140 (1992). English translation of Izv. Krym. Astrofiz. Obs., Tom 82, p. 147 – 153 (1990).
Measurements have been made on the Sun's general magnetic field in 1975 – 1978 and in 1987 at the Crimean and Sayan observatories and at Mount Wilson; these have shown that there are more or less stable oscillations with periods of about 47, 60, 85, and 160 min with mean amplitudes of about 1 μT. The new Sayan observations of 1987 confirm the previous conclusion that there is long–time coherence in the oscillation with period 160.0101 min.

080.072 Does the solar neutrino flux vary?
B. M. Vladimirskij, L. D. Kislovskij.
Bull. Crimean Astrophys. Obs., Vol. 82 , p. 141 – 147 (1992). English translation of Izv. Krym. Astrofiz. Obs., Tom 82, p. 153 – 161 (1990).
The variations in solar neutrino flux at the Brookhaven detector may be due to variations in the performance in extracting Ar^{37} from the perchloroethylene. Such changes could be due to the Ar^{37+} being trapped in clathrate structures that give rise to metastable molecular complexes. If so, the measurements should show macroscopic fluctuations. Some such features are, in fact, observed: in particular, the measurements tend to represent a set of discrete states, which vary with the magnetic activity at the end of the observation time. The interpretation suggests that the observed flux is probably close to the one predicted by the

standard model. Possible ways of checking this by experiment are briefly considered.

080.073 Seismic constraints on the solar neutrino problem.
D. O. Gough.
15. Texas Symposium on Relativistic Astrophysics and 4. ESO–CERN Symposium, p. 199 – 217 (1991). – See Abstr. 012.060 for the main entry.
Contents: 1. Introduction. 2. Standard solar models and the reactions of the pp chain. 3. The solar neutrino problem. 4. Helioseismic inversion. 5. Results of the inversion. 6. Modified standard models, and WIMP accretion. 7. Macroscopic motion in the core. 8. Conclusion.

080.074 The solar neutrino problem.
D. C. Kennedy.
Theoretical Advanced Study Institute (TASI) in Elementary Particle Physics: Testing the standard model, p. 807 – 860 (1991). – See Abstr. 012.063 for the main entry.
The theoretical and experimental status of the solar neutrino problem is reviewed, with major emphasis on particle theory solutions: neutrino masses and magnetic moments, resonant flavor and spin conversion in the Sun, and adiabatic vs. non-adiabatic conversions. The properties of solar model calculations, thermonuclear fusion cycles and solar magnetic fields are discussed. Current experimental results and expectations from future measurements are interpreted in light of astrophysical and particle theory.

080.075 Radiochemical solar neutrino experiments.
T. Kirsten.
15. Texas Symposium on Relativistic Astrophysics and 4. ESO–CERN Symposium, p. 392 – 393 (1991). – See Abstr. 012.060 for the main entry.

080.076 Long–term solar variability and solar seismology: I.
H. A. Hill, R. J. Kroll.
AIP Conf. Proc., No. 245: Workshop on Basic Space Science, p. 170 – 180 (1992). – See Abstr. 012.064 for the main entry.
Solar seismology and solar variability studies are paradigms of multi–disciplinary efforts. They offer a wide range of educational and research experiences and opportunities to students and researchers in many fields. Solar variability research is many-faceted, and a variety of techniques are employed to study it. Some of the more prominent efforts are surveyed, ranging from direct solar measurement to observations of planets and stars. Emphasis is given to the SCLERA program, which has detected changes on the Sun which may be relevant to the solar energy output and to the long–term climate of the Earth. This emphasis is given in part to project an example where education is an important consideration. The future of solar variability research is outlined, and the SCLERA International Network is described. This network is based on the idea that collaboration among scientists is the foundation on which are built opportunities for scientific discoveries, educational development, and technological progress.

080.077 Long–term solar variability and solar seismology: II.
H. A. Hill, P. Oglesby, Gu Yeming.
AIP Conf. Proc., No. 245: Workshop on Basic Space Science, p. 181 – 192 (1992). – See Abstr. 012.064 for the main entry.
There are a number of interesting aspects of the Sun's internal and surface structure that can be seen in the observed properties of the Sun's normal modes of oscillation. Two examples of these are the speed of sound and the internal rotation. There are other manifestations that are reported but not yet confirmed; or predicted but not yet observed. Examples here are the abundance of He in the convection zone and the detection of the internal gravity modes of oscillation. The current situation regarding both classes of works is discussed and exciting opportunities are indicated for the next generation of researchers. These opportunities are open to both space and ground-level observing programs.

080.078   First measurement of the integral solar neutrino flux by the Soviet/American Gallium Experiment.
V. N. Gavrin.
15. Texas Symposium on Relativistic Astrophysics and 4. ESO–CERN Symposium, p. 394 – 398 (1991). – See Abstr. 012.060 for the main entry.

080.079   Electron screening effects on thermonuclear reactions in the Sun.
J. Arafune, M. Fukugita.
Prog. Theor. Phys., Vol. 87, No. 6, p. 1467 – 1471 (Jun 1992).

The electron screening effect in the plasma is calculated for the thermonuclear pp reaction in the Sun, with a particular attention to the effect at the short distance, where the conventional Debye–Hückel approximation fails. It is shown that the correct treatment for the short distance effect modifies the estimate with the Debye–Hückel approximation at most by 3%, which would cause the change in the conventional estimate of the ^8B solar neutrino flux by < 8%. This modification is much smaller than the effect argued recently by Kurucz.

080.080   Solar neutrino flux variations.
V. G. Gavryusev, E. A. Gavryuseva.
15. Texas Symposium on Relativistic Astrophysics and 4. ESO–CERN Symposium, p. 483 – 494 (1991). – See Abstr. 012.060 for the main entry.

Almost uninterrupted measurements of the ^{37}Ar production rate using 610 tons of perchlorethylene, which have been conducted over the past 20 years by R. Davis and his collaborators, offers a unique possibility for the investigation of the processes in the deep interior of the Sun. The first results obtained by this group have already shown an intriguing discrepancy between the high–energy neutrino fluxes predicted for the Sun by the stellar evolution theory and the measured values of the ^{37}Ar production rate. Approximately at the end of the first ten years of measurements, a number of investigators asked whether there were time variations in the data of the counting rate of neutrinos in the Cl–Ar detector. The investigations showed that there is a noticeable anticorrelation between the ^{37}Ar production rate and the solar activity. Time variations with other periods were also revealed.

080.081   Numerical solution of magnetic flux emergence in gravity–stratified solar atmosphere.
Song Mutao, Zhang Hongqi.
Acta Astrophys. Sin., Vol. 12, No. 1, p. 63 – 74 (Jan 1992). In Chinese. English translation in Chin. Astron. Astrophys., Vol. 16, No. 2, p. 173 – 186 (Apr–Jun 1992).

Numerical simulation in MHD Lagrangian scheme is adopted to compute the dynamic evloution of solar magnetic bipolar–field in a gravity–stratified atmosphere when a parallel or an opposite magnetic dipole emerges from the subphotosphere. It is shown that the emergence of an opposite dipole would lead to converging and descending motion of plasma, and that an augmented pressure has to be used in order to led this dipole float. The floating evolution process results in the formation of a current sheet in the interface between the old and the new emerging fields. The computation reveals that the gas at the arch's top is rising with a small velocity while the gases at the arch's legs are falling down with a large speed. This peculiar property stems from the nonhomogeneity of density in the gravity–stratified atmosphere. The authors simulate the formation of an usual solar active region by making a strong magnetic dipole float into a weak background field in the solar photosphere and chromosphere. Finally, the computation results are applied to explain magnetic emerging flux data obtained at Huairou, Beijing Astronomical Observatory.

080.082   The preliminary results for five–minute oscillations detected from the solar continuum intensity.
Zhang Heng.
Acta Astrophys. Sin., Vol. 12, No. 1, p. 87 – 92 (Jan 1992). In Chinese.

The data of time sequence for the solar intensity continuum spectra are analysed by using the Fourier analysis, after the instrumental response profile and the Earth's atmospheric effect in the data are removed. Low degree five–minute oscillation is detected by investigating the superposed power and time coherence in the spectrum. As the noises have not been completely removed the results are preliminary.

080.083   Numerical solution of magnetic flux emergence in gravity–stratified solar atmosphere.
Song Mutao, Zhang Hongqi.
Chin. Astron. Astrophys., Vol. 16, No. 2, p. 173 – 186 (Apr–Jun 1992). English translation of Acta Astrophys. Sin., Vol. 12, No. 1, p. 63 – 74 (Jan 1992). See Abstr. 080.081.

Numerical MHD simulation in Lagrangian scheme is made of the dynamical evolution of the solar magnetic bipolar field in a gravity–stratified atmosphere when a parallel or an antiparallel dipole emerges from the subphotosphere. It is shown that the emergence will lead to converging and descending motion of the plasma, and that a pressure has to be added for the dipole to float. The floating results in the formation of a current sheet in the interface between the old and the new fields. The sheet may bring about explosive phenomena such as the Ellerman bombs. The computation reveals that the gas at the top of the arch is rising with a small velocity while the gas at the feet is falling with a large velocity. This peculiar property stems from the nonhomogeneity of the physical parameters in the gravity–stratified atmosphere. The authors also simulate a usual solar active region by making a strong magnetic dipole of 1500 G float into a weak background field of 100 G. It is indicated that a diminished pressure should be added on the boundary in order to avoid strong shocks. The emerging process should last at least several hours. After emergence a strong current sheet is formed in the chromosphere which, on arriving in the lower corona, will become a potential source of solar flares. Finally, the results of computation will be used to interpret the magnetic emerging flux data obtained at Huairou Station.

080.084   The peculiarities of rotation of the solar polar regions.
Yu. A. Solonsky (*Yu. A. Solonskij*), V. V. Makarova.
Sol. Phys., Vol. 139, No. 2, p. 233 – 245 (Jun 1992).

The sidereal rotation rate of the high–latitude solar regions is examined using long–lived photospheric polar faculae. The observations were carried out from 1982 to 1986. The following facts have been established: (a) There is a differential rotation of the polar faculae close to the maximum of solar activity, while the amount of latitude gradient of solar rotation decreases towards the sunspot minimum; (b) small differences of rotation in the northern and southern hemispheres of the Sun are observed; (c) some deviations of differential rotation curves constructed for each Carrington rotation from the mean curve of differential rotation are revealed. The total amplitude of the maximum positive and negative excesses is about $40 - 50$ m s^{-1}. The positive surplus velocities of solar rotation (the amplitude of which is about $20 - 25$ m s^{-1}) move in the form of a wave from heliographic latitudes $\approx 40°$ with a velocity of 1.6 m s^{-1}. The latitude width of this flow is $\Delta B \approx 15°$. This wave of abnormally high velocity starts in the year of minimum solar activity and reaches the pole 11 years later. The picture is symmetrical relative to the equator.

080.085   On the recording of regions with increased plasma density on the Sun.
L. N. Kurochka, I. P. Kryachko, E. Markova.
Sol. Phys., Vol. 139, No. 2, p. 275 – 277 (Jun 1992).

For the purpose of studying the statistics, dynamics, and morphology of parts of the solar surface with enhanced emission measure and considerable inhomogeneities of electron concentration it is suggested to introduce a new type of solar observation: taking filtergrams by means of interference filters, calculated for the Balmer continuum limit and the region of the blue continuum. To determine accurately the inhomogeneity of the electron concentration in an object, it is suggested to scan the surface of the object and record the spectrum of the blue continuum by means of a spectrochronograph.

080.086   **Magnetic fields and motions in the solar atmosphere.**
S. I. Gopasyuk.
Kinematics Phys. Celest. Bodies, Vol. 7, No. 5, p. 1 – 7 (1991). English translation of Kinematika Fiz. Nebesn. Tel., Tom 7, No. 5, p. 3 – 10 (Sep–Oct 1991).

This review is devoted to analysis of observations of magnetic–field structure and motions in an active region on the Sun. Characteristic types of motions and their relations to magnetic structures and electric currents are designated. A role of magnetic–field structure in the formation of certain plasma–motion types is indicated. Similarities between active and undisturbed regions are noted in magnetic–field structure and the relation of plasma motions to the magnetic field.

080.087   **Five–minute oscillations in the solar continuum.**
Zhang Heng.
Publ. Yunnan Obs., No. 1, p. 20 – 35 (1992). In Chinese.

The five–minute oscillations in the solar continuum are analyzed. The principle for the application the solar continuum to the study of the oscillations is introduced and the analysis of the observed data is given. The instrumental response profile is removed by taking the ratio of the intensity observed from the visible continuum to the infrared one at an arbitrary instant to that at a given instant. The effect of the Earth's atmosphere on the data is computed by taking the integration over the wavelengths. The fast Fourier transform is used to get the power spectrum.

080.088   **The effect of surface inhomogeneities on total solar irradiance.**
J. Lean.
Armagh Observatory Bicentenary Colloquium on Surface Inhomogeneities on Late–Type Stars, p. 167 – 180 (1992). – See Abstr. 012.085 for the main entry.

Simultaneous observations from the Active Cavity Radiometer (ACRIM) of the SMM satellite and the Earth Radiation Budget (ERB) experiment on Nimbus 7 have demonstrated conclusively that the Sun's spectrally integrated emission from its entire disk is indeed variable. These observations, together with data from the ERBS and NOAA–9 experiments, are reviewed here. It is shown that variations in the Sun's total irradiance reflect the growth and evolution of magnetic active regions throughout the Sun's activity cycle, with the relative contributions of dark sunspots and bright faculae/plages dependent on wavelength. Over the 27 day time scale of solar rotation, enhanced emission from bright faculae competes with the emission deficit in sunspots. The 11 year cycle in total solar irradiance reflects the increase, from minimum to maximum activity, in both the sunspots and faculae on the solar disk. The decrease in total irradiance, which has been observed by satellites during the descending phase of solar cycle 21, can be explained by a brightness source which more than compensates the sunspot deficit. This same source of brightness variations is responsible for changes in the ultraviolet range of the Sun's spectrum, suggesting that its origin is related to magnetic flux tubes.

080.089   **The structure and evolution of the Sun.**
J. Christensen–Dalsgaard.
NATO Advanced Study Institute on the Sun: a Laboratory for Astrophysics, p. 11 – 28 (1992). – See Abstr. 012.091 for the main entry.

Computations of "standard" solar models are based on assumptions of equilibrium between the forces acting on different parts of the Sun and between the amount of energy produced in the core and radiated from the surface. The structure of the resulting models depends on the assumed physics of the solar interior (equation of state, opacity, energy generation). Although accurate calculations are required to obtain detailed information about solar structure, simple estimates can give a feel for the general properties of the solar interior. A firm prediction of the calculation is that the solar luminosity has increased somewhat since the formation of the Sun. All standard models predict a neutrino flux which is substantially larger than the observed value.

080.090   **L'esperimento Gallex del Gran Sasso ed il problema dei neutrini solari.**
V. Castellani, S. Degl'Innocenti.
G. Astron., Vol. 18, N. 2, p. 23 – 30 (Jun 1992).

080.091   **Seismic investigation of the solar interior.**
J. Christensen–Dalsgaard.
NATO Advanced Study Institute on the Sun: a Laboratory for Astrophysics, p. 29 – 80 (1992). – See Abstr. 012.091 for the main entry.

Observation of a large number of modes of solar oscillation has permitted detailed investigation of the solar interior. To illustrate the diagnostic potential of the frequencies, some properties of the observed p–modes are discussed in terms of simple ray picture of the oscillations. Solar models and their frequencies are used to illustrate how the frequencies depend on the physics of the solar interior. From inverse analyses of the frequencies, one may determine, e.g., the variation of sound speed with position. Solar rotation causes fine structure in the frequencies. By inverting the observations, it is possible to infer the angular velocity as a function of depth and latitude in much of the Sun. Recent observations have given detailed information about frequency changes during the solar cycle.

080.092   **Convection.**
M. Schüssler.
NATO Advanced Study Institute on the Sun: a Laboratory for Astrophysics, p. 81 – 98 (1992). – See Abstr. 012.091 for the main entry.

Convection in stars is discussed on basic level, with emphasis on the convection zone of the Sun. The treatment starts with the classical criteria for convective instability and the role of temperature and molecular weight gradients. This gives a basis for introducing the mixing length formalism as a means for describing the large–scale properties of a convective region. It follows a discussion of convective overshoot into the stably stratified layers adjacent to a convection zone. With aid of a non–local extension of the mixing length formalism, the depth and structure of an overshoot layer below the solar convection zone can be calculated. Observed solar convective flow patterns are characterized briefly. Finally, numerical simulations of convection are discussed in some detail. After presenting the basic equations and some general results from simulations, the author describes the new picture of stellar convection which emerges from these calculations. The lecture concludes with a comparison of these new insights with the assumptions which form the basis of the mixing length formalism.

080.093   **Mean field dynamo theory.**
P. Hoyng.
NATO Advanced Study Institute on the Sun: a Laboratory for Astrophysics, p. 99 – 138 (1992). – See Abstr. 012.091 for the main entry.

Mean field dynamo theory has witnessed a very rapid development during the last 25 years, resulting in many models which reproduce most of the basic features of the large scale field of the Sun and the Earth. In this review the application to the Sun features prominently. After a brief discussion of the relevant observations and some basic results from laminar dynamo theory, the traditional picture of linear mean field theory is outlined, including an analysis of the properties of the plane wave solutions of the dynamo equation. The emphasis is on explaining the physical mechanisms and theoretical concepts in an elementary fashion, and not on a detailed comparison of various solar mean field models. A rigorous derivation of the dynamo equation under various conditions is also presented (isotropic/anisotropic turbulence, short/long correlation time and the two–scale approximation). Finally, the author discusses some of the major problems and recent developments such as the structure of the magnetic field, nonlinear dynamos, boundary layer dynamos, internal and external forcing, and subcritical dynamo operation.

080.094   An explanation for the solar neutrino problem – a new mechanism of thermonuclear reaction.
Wang Hongzhang.
Proc. Astron. Soc. Aust., Vol. 9, No. 2, p. 313 – 314 (1991). – See Abstr. 012.090 for the main entry.

A new mechanism of thermonuclear reaction is briefly introduced. It shows that a certain amount of thermonuclear reaction can take place in dense, low temperature ($T < 1 \times 10^5$K) plasmas. As most regions in the Sun are at moderate and low temperature, a sufficient amount of fusion energy is generated there. Therefore, the current standard solar models, in which the solar central temperature must be slightly lower than 15×10^6K, must be modified, and this would make the flux of high–energy neutrinos conform with the observational results.

080.095   Non–MSW solutions to the solar neutrino problem.
J. Pantaleone.
1. International Trends in Astroparticle Physics Conference together with the SuperNova Watch Workshop, p. 356 – 368 (1992). – See Abstr. 012.102 for the main entry.

Recent experimental data have reaffirmed the solar neutrino problem. However the present ^{37}Cl, Kamiokande–II and ^{71}Ga data can be brought into agreement with the Standard Solar Model predictions by any of the several different modifications of neutrino propagation. Some of the less discussed possibilities are reviewed here, along with possible experimental tests.

Plasmaphysik im Sonnensystem.
See Abstr. 003.052.

Unsere Sonne – ein rätselhafter Stern? Erkenntnisse und Spekulationen der Astrophysik.
See Abstr. 003.053.

New windows to the universe. Volumes I, II. Invited review papers and general lectures.
See Abstr. 012.001.

Astrophysical opacities. Proceedings. Workshop on Astrophysical Opacities (WAO), Caracas (Venezuela), 15 – 19 Jul 1991.
See Abstr. 012.054.

The Sun: a laboratory for astrophysics. Proceedings.
See Abstr. 012.091.

Can astrophysics rescue particle physics from the standard model impasse?
See Abstr. 013.043.

Basic research in mathematical and space sciences.
See Abstr. 021.022.

Solar–neutrino neutral–current detection methods in the Sudbury Neutrino Observatory.
See Abstr. 022.075.

Molecular data from solar spectroscopy.
See Abstr. 022.190.

"Long–term" neutrino flux integrations.
See Abstr. 022.240.

Study of mixing of solar neutrinos with a 1000 ton ICARUS detector.
See Abstr. 022.241.

The possibility of radiogeochemical limits on stellar collapse rates in the Galaxy.
See Abstr. 022.246.

Direct detection of solar neutrinos.
See Abstr. 036.166.

Energy dependence of solar neutrino – electron scattering as a test of neutral currents.
See Abstr. 061.006.

Charged– and neutral–current solar–neutrino cross sections for heavy–water Cherenkov detectors.
See Abstr. 061.019.

Neutrino properties in matter.
See Abstr. 061.040.

17 keV neutrino, MSW mechanism, and supernova constraints.
See Abstr. 061.047.

Class of models leading to depletions of solar ν_e and atmospheric ν_μ fluxes.
See Abstr. 061.051.

Solar and supernova neutrino physics with Sudbury Neutrino Observatory.
See Abstr. 061.052.

Seesaw–model predictions for the τ neutrino mass.
See Abstr. 061.060.

Fermion masses, $SU(2)_L \times U(1)_Y$, and the solar neutrino problem.
See Abstr. 061.067.

Neutrino electromagnetic scattering in astrophysics.
See Abstr. 061.088.

Neutrino astronomy.
See Abstr. 061.089.

Survey of atmospheric neutrino data and implications for neutrino mass and mixing.
See Abstr. 061.103.

Pseudo Dirac neutrinos and the solar neutrino problem.
See Abstr. 061.105.

Searching for most of the universe.
See Abstr. 061.120.

Present status of neutrino physics.
See Abstr. 061.133.

Aggiornamento sulle problematiche astronomiche e fisiche legate alla fenomenologia neutrino e alle attuali insufficienti conoscenze delle fondamentali proprietà di tale particella.
See Abstr. 061.139.

Neutrino flavor–spin oscillations in the Sun.
See Abstr. 061.150.

Seesaw model predictions for the τ–neutrino mass.
See Abstr. 061.151.

Linear resistive magnetohydrodynamic computations of resonant absorption of acoustic oscillations in sunspots.
See Abstr. 062.010.

Equilibrium of a magnetic flux tube in a compressible flow.
See Abstr. 062.025.

On the stability of mean–field models of the solar convection zone.
See Abstr. 062.068.

Chaotic Alfvén waves in solar and cometary plasmas.
See Abstr. 062.098.

Magnetic reconnection in incompressible fluids.
See Abstr. 062.101.

Application of the phase–integral method to the trapping of acoustic waves in a gravitating fluid. I. Planar polytrope and turning–point behavior.
See Abstr. 062.148.

Solar p–modes oscillations and heating of the corona.
See Abstr. 062.161.

MHD instabilities in a stratified atmosphere with anisotropic pressure.
See Abstr. 062.172.

Three–dimensional MHD simulation of the Parker instability in galactic gas disks and the solar atmosphere.
See Abstr. 062.212.

Radiative transfer problems in the solar and sun–like atmospheres.
See Abstr. 063.001.

Depth of formation of lines in the solar atmosphere.
See Abstr. 063.080.

Results obtained using the OPAL code.
See Abstr. 063.119.

"Finding" the "missing" solar ultraviolet opacity.
See Abstr. 063.120.

Remaining line opacity problems for the solar spectrum.
See Abstr. 063.121.

Radiation–hydrodynamic waves in an optically gray atmosphere. I. Homogeneous model.
See Abstr. 064.048.

Evolution of solar and stellar rotation.
See Abstr. 065.004.

Further improvements of a new model for turbulent convection in stars.
See Abstr. 065.058.

The equation of state for stellar envelopes: comparison of theoretical results.
See Abstr. 065.101.

Solar and supernova neutrino interactions.
See Abstr. 065.119.

Empirical and theoretical evidence for gravitational polarization of matter.
See Abstr. 066.022.

About spectroscopic measurements of the solar meridional motion.
See Abstr. 071.004.

Die turbulente Sonnenoberfläche.
See Abstr. 071.006.

Convective motion on the Sun.
See Abstr. 071.010.

Evolution and advection of solar mesogranulation.
See Abstr. 071.011.

Solar hydrogen lines in the infrared.
See Abstr. 071.013.

Solar active regions as a percolation phenomenon.
See Abstr. 072.036.

Oscillations and seismological diagnostics of sunspots.
See Abstr. 072.059.

Statistical analysis of solar flare, neutrino flux and sunspot data.
See Abstr. 072.070.

Chromospheric dynamics based on infrared solar brightness variations.
See Abstr. 073.021.

Flare instability and driving mechanism.
See Abstr. 073.133.

Intensity oscillations of the Fe XIV ($\lambda = 530.3$ nm) solar corona: disturbances due to the wave processes in the Earth's atmosphere.
See Abstr. 074.013.

Peculiar photospheric velocity fields and magnetic energy build–up.
See Abstr. 075.013.

Generation and non–equilibrium of solar atmospheric magnetic fields.
See Abstr. 075.043.

Observation of solar eclipse on 11 July 1991 in Mexico for solar diameter determination.
See Abstr. 079.005.

Effect of geomagnetic disturbances on the flux intensity of direct solar radiation.
See Abstr. 085.004.

Cosmic ray physics with underground detectors.
See Abstr. 144.043.

Earth

081 Structure, Figure, Gravity, Orbit, etc.

081.001 Geoid anomalies and dynamic topography from convection in cylindrical geometry: applications to mantle plumes on Earth and Venus.
W. S. Kiefer, B. H. Hager.
Geophys. J. Int., Vol. 108, No. 1, p. 198 – 214 (Jan 1992).

A variety of evidence suggests that at least some hotspots are formed by quasi–cylindrical mantle plumes upwelling from deep in the mantle. The authors model such plumes in cylindrical, axisymmetric geometry with depth–dependent, Newtonian viscosity. Cylindrical and sheet–like, Cartesian upwellings have significantly different geoid and topography signatures. However, Rayleigh number–Nusselt number systematics in the two geometries are quite similar. The geoid anomaly and topographic uplift over a plume are insensitive to the viscosity of the surface layer, provided that it is at least 1000 times the interior viscosity. Increasing the Rayleigh number or including a low–viscosity asthenosphere decreases the geoid anomaly and the topographic uplift associated with an upwelling plume. Increasing the aspect ratio increases both the geoid anomaly and the topographic uplift of a plume. The Nusselt number is a weak function of the aspect ratio, with its maximum value occurring at an aspect ratio of slightly less than 1.

081.002 Angle between the axes of rotation of the Earth's core and mantle.
L. V. Nikitina.
Geomagn. Aeron., Vol. 30, No. 5, p. 702 – 705 (Apr 1990). English translation of Geomagn. Aehron., Tom 30, No. 5, p. 832 – 836 (1990).

The magnetic field of the Earth is determined by currents in its liquid core. The joint precession of the core and mantle, acted upon by tidal and cohesive forces, is considered and the angle between their axes of rotation (it is of the order of 10^{-6}rad) is found. With the angle taken into account the velocity field in the outer core is derived, and it is shown that the main rotation of the currents occurs in the boundary layer.

081.003 Gravity fields of the southern ocean from Geosat data.
D. C. McAdoo, K. M. Marks.
J. Geophys. Res., Vol. 97, No. B3, p. 3247 – 3260 (10 Mar 1992).

In August 1990, the U.S. Navy declassified all Geodetic Mission (GM) radar altimeter data acquired by the Geosat satellite over oceanic regions south of 60°S. The authors have used these GM data in conjunction with the unclassified, lower–resolution Geosat Exact Repeat Mission (ERM) altimeter data to construct high–resolution gravity fields on a 5–km grid covering the annular region of the southern ocean, which lies between 60°S and 72°S and encircles Antarctica.

081.004 The inner core translational triplet and the density near Earth's center.
D. E. Smylie.
Science, Vol. 255, No. 5052, p. 1678 – 1682 (27 Mar 1992).

Four long records from superconducting gravimeters yield evidence of the triplet of translational oscillations of the solid inner core about its central position. Calculations of core oscillation modes allow identification of the three translational resonances. Each resonance is defined by approximately 20

successive spectral estimates. A new Earth model brings the computed periods into agreement with observation.

081.005 The constant part of the tidal field in the theory of heights.
A. Zeman.
Stud. Geophys. Geod., Vol. 35, No. 2, p. 75 – 80 (1991).

The problem of the constant part of the tidal field is still topical in view of the recommendations of IAG to eliminate the tidal effect of external masses from all geodetic measurements under preservation of the effect of the time–constant tidal deformation of the Earth. The paper discusses the consequences of accepting this recommendation for normal heights, and suggests a solution based on the new definition of the normal gravity field.

081.006 On the possible influence of Solar System onto the planet Earth.
I. Charvátová.
Vesmír, Vol. 70, No. 5, p. 270 – 273 (May 1991). In Czech.

081.007 Differential rotation of the liquid core of the Earth.
L. V. Nikitina, A. A. Ruzmajkin.
Geomagn. Aehron., Tom 32, No. 1, p. 140 – 144 (Jan – Feb 1992). In Russian. English translation in Geomagn. Aeron., Vol. 32, No. 1.

081.008 Formation of spinels in cosmic objects during atmospheric entry: a clue to the Cretaceous–Tertiary boundary event.
E. Robin, P. Bonté, L. Froget, C. Jéhanno, R. Rocchia.
Earth Planet. Sci. Lett., Vol. 108, No. 4, p. 181 – 190 (Feb 1992).

Magnetic spinels produced by oxidation of extraterrestrial objects in the atmosphere have a composition distinct from terrestrial spinels. They are characterized by a high iron oxidation state, arising from crystallization under high oxygen fugacities, and a high nickel concentration due to the relatively high abundance of this element in extraterrestrial material. The iron oxidation state increases from micrometeorites, to meteoroid ablation material and to impact–generated products. This reflects a progressive increase of the oxygen fugacity, corresponding to decreasing altitudes of crystallization. Spinels found at the Cretaceous–Tertiary boundary are similar to those that crystallized from meteoroid ablation material and impact–generated products, supporting the view that a collisional event did occur at the end of the Cretaceous.

081.009 Iodine abundances in oceanic basalts: implications for Earth dynamics.
B. Déruelle, A. Jambon, G. Dreibus.
Earth Planet. Sci. Lett., Vol. 108, No. 4, p. 217 – 227 (Feb 1992).

081.010 Neodymium and strontium isotopic study of Australasian tektites: new constraints on the provenance and age of target materials.
J. D. Blum, D. A. Papanastassiou, C. Koeberl, G. J. Wasserburg.
Geochim. Cosmochim. Acta, Vol. 56, No. 1, p. 483 – 492 (Jan 1992).

Nd and Sr isotopic studies of Australasian tektites provide information on the age and provenance of the target materials and allow one to characterize the target area and the impact process leading to tektite formation. The Nd and Sr isotopic data provide evidence that all Australasian tektites were derived from a single sedimentary formation with a narrow range of stratigraphic ages close to 170 Ma. The authors suggest that all of the Australasian tektites were derived from a single impact event, and that the australites represent the upper part of a melt sheet ejected at high velocity, whereas the indochinites represent melts formed at a lower level in the target material which were distributed closer to the area of impact. The impact site is inferred to be within an area of Jurassic sedimentary bedrock, which spans the geopolitical boundaries between northern Cambodia, southern Laos, and southeastern Thailand.

081.011 Approximation of the height of the geoid.
K. V. Pishchukhina.
Kinematics Phys. Celest. Bodies, Vol. 7, No. 3, p. 13 – 20 (1991). English translation of Kinematika Fiz. Nebesn. Tel, Tom 7, No. 3, p. 15 – 21 (1991).
A modification of Molodenskij's root–mean–square approximation of the geoid height is developed on the basis of the simple–layer potential instead of the Stokes formula. The compactness of the derived formulas makes it easy to include harmonics up to any desired degree in the expansion of the far–zone component and to conduct a qualitative analysis of the coefficients of the harmonics as functions of their degree and the radius of the near zone. Numerical experiments indicate that the method is most efficient in approximation of the far–zone effect outside of a spherical cap of radius 5°.

081.012 Planetary evolution and global tectonics.
A. S. Monin.
Tectonophysics, Vol. 199, No. 2–4, p. 149 – 164 (10 Dec 1991).
Using simplified quantitative models of the Earth's composition, its core formation (including discussion of the mega–impact hypothesis), the fate of iron, mobile substances and water, the release of radiogenic heat, gravitational and tidal energy, mantle convection and polar wandering are considered. Plate tectonics is treated as a consequence of mantle convection caused by gravitational differentiation of heavy and light substances (essentially by growth of the iron core).

081.013 Potassium, rubidium, and cesium in the Earth and Moon and the evolution of the mantle of the Earth.
W. F. McDonough, S.–S. Sun, A. E. Ringwood, E. Jagoutz, A. W. Hofmann.
Geochim. Cosmochim. Acta, Vol. 56, No. 3, p. 1001 – 1012 (Mar 1992). – See Abstr. 012.013 for the main entry.
The aim of this paper is to examine the data base available to constrain the Rb/Cs ratio of the Earth and its bulk K, Rb, and Cs contents. In addition, the authors have reviewed the analogous data on lunar rocks in order to compare the Rb/Cs ratio of the Silicate Earth and the Moon. This information is used to test whether Rb and Cs abundances can be used to constrain models of lunar origin. An additional discussion on Rb–Cs mantle geochemistry and its secular evolution is included.

081.014 Geochemistry and origin of Muong Nong–type tektites.
C. Koeberl.
Geochim. Cosmochim. Acta, Vol. 56, No. 3, p. 1033 – 1064 (Mar 1992). – See Abstr. 012.013 for the main entry.
The present report is based on the detailed study of nineteen Muong Nong–type tektite specimens. The petrographical characteristics and the chemical composition of these samples were studied in great detail, using a variety of techniques. Electron microprobe studies have been performed to investigatre the chemical differences between layers and within layers. Furthermore, light and dark layers have been separated from some samples for trace element analyses. Here the author presents and discusses the complete data set, and tries to put it in context with related observations to arrive at some conclusions regarding the origin of the Muong Nong tektites.

081.015 Oxygen isotopic homogeneity of the Earth: new evidence.
F. Robert, A. Rejou–Michel, M. Javoy.
Earth Planet. Sci. Lett., Vol. 108, No. 1–3, p. 1 – 9 (Jan 1992).
The two oxygen isotope ratios $^{18}O/^{16}O$ and $^{17}O/^{16}O$ were analyzed in Precambrian (3.5 Gy) and modern cherts and in mantle–derived lavas. All samples exhibit oxygen isotopic compositions consistent with mass–dependent isotopic fractionation from a single reservoir and thus suggest that 3.5 Gy ago the Earth was already a well–mixed body. These results would therefore not support models of ocean accretion by cometary impact later than 3.7 ± 0.1 Gy. New measurements on lunar rocks confirm oxygen isotopic homogeneity of the Earth–Moon system and thus suggests that the internal mixing of the two bodies pre–dates their accretion. The recent giant impact model used to explain how the Moon was formed is also compatible with the authors' results.

081.016 On the calculation of low–frequency oscillations of the Earth's core.
S. Diakonov.
Geophys. J. Int., Vol. 107, No. 2, p. 291 – 296 (Nov 1991).
The problem of calculation of low–frequency oscillations of an ideal rotating compressible fluid is investigated. An original method of solving such a problem based on using characteristic functions of the Poincaré operator is proposed. An efficient scheme of calculation of characteristic numbers and functions of the Poincaré operator is derived. A high rate of convergence of the method is shown. The liquid compressibility is found to have an essential influence on the theoretical nutation amplitude.

081.017 Numerical calculation of modes of oscillation of the Earth's core.
D. E. Smylie, X. Jiang, B. J. Brennan, K. Sato.
Geophys. J. Int., Vol. 108, No. 2, p. 465 – 490 (Feb 1992).
This paper describes the numerical implementation of a variational principle for the calculation of the very low–frequency ($< 300\ \mu$Hz) modes of oscillation of the fluid outer core using realistic models of Earth structure.

081.018 The subseismic approximation in core dynamics.
D. J. Crossley, M. G. Rochester.
Geophys. J. Int., Vol. 108, No. 2, p. 502 – 506 (Feb 1992).
Scaling arguments have been used to justify the subseismic approximation (SSA) for analysing long–period oscillations of the Earth's liquid core. This approximation neglects the contribution of flow pressure to elastic compression relative to that of transport along the equilibrium pressure gradient. Here the authors present numerical tests of the SSA for a suite of 90 core undertones (gravity modes) in non–rotating Earth models with uniformly stable liquid cores.

081.019 Some remarks about the rotations of a viscous planet and its homogeneous liquid core: linear theory.
M. Lefftz, H. Legros.
Geophys. J. Int., Vol. 108, No. 3, p. 705 – 724 (Mar 1992).
Linear equations governing the rotation of the Earth are developed for a model with a Maxwell homogeneous mantle and a homogeneous inviscid fluid core having a differential rotation relative to the mantle. The authors find four eigenfrequencies for the equatorial perturbations in rotation. Two are well known: the rotational nearly diurnal frequency and the Chandlerian frequency with a damping related to the relaxation time of the Earth. The other two frequencies, one being a heavily damped long–period oscillation and the other one zero, are related to the relaxation modes, but are nevertheless coupled with the rotational eigenfrequencies. The authors investigate the kinetic and deformation energy resulting from both impulsive and time–constant geophysical sources. Using a generalized notation, they derive an analytical solution for the rotations of the Earth and its fluid core due to various excitation sources at the Earth's surface and at the core–mantle boundary. The authors obtain some results concerning phenomena acting at the CMB which are able to produce a significant shift of the rotation axis.

081.020 Influence of viscoelastic coupling on the axial rotation of the Earth and its fluid core.
M. Lefftz, H. Legros.
Geophys. J. Int., Vol. 108, No. 3, p. 725 – 739 (Mar 1992).
  Linear equations governing the axial rotation of the Earth are developed for a model with a Maxwell homogeneous mantle and a homogeneous inviscid fluid core having a differential rotation relative to the mantle. The authors find three eigenfrequencies for the axial perturbations in rotation; the first one is relative to friction acting at the core–mantle boundary and the others are relaxation modes associated with viscoelastic Love numbers. The authors point out the differential rotation of the core when the Earth is submitted to zonal geophysical excitations like glaciation and deglaciation, tidal torque or other processes occurring at the core–mantle boundary (CMB). They show that the tidal deceleration involves an eastward drift of the core with respect to the mantle. However, for precise values of the mantle viscosity and of the frictional constant at the CMB, the authors find that the last deglaciation involves a westward drift of the core with respect to the mantle which may be correlated with the observed westward drift of the geomagnetic field of the Earth.

081.021 Geosat–derived geoid anomalies at medium wavelength.
A. Cazenave, S. Houry, B. Lago, K. Dominh.
J. Geophys. Res., Vol. 97, No. B5, p. 7081 – 7096 (10 May 1992).
  Geosat profiles of the Exact Repeat Mission have been averaged over a 1–year period and high–pass–filtered using inverse method techniques. The geoid surface constructed with both ascending and descending profiles shows at medium wavelengths band–shaped anomalies preferentially elongated in the east–west direction. These anomalies have an average amplitude of ∼30 cm and dominant wavelengths of 750 km and 1100 km. The authors have performed numerous tests to show that the lineations are not artefacts created by the filtering process. Moreover, two–dimensional (2–D) filtering with the inverse method applied on a regional basis over the Pacific gives essentially similar results, indicating that the filtered geoid is not affected by directional bias. Seafloor topography in the Pacific filtered by 2–D inverse method also shows east–west trending depth anomalies positively correlated to medium–wavelength geoid lineations.

081.022 Gravity field approximation using airborne gravity gradiometer data.
D. Arabelos, I. N. Tziavos.
J. Geophys. Res., Vol. 97, No. B5, p. 7097 – 7108 (10 May 1992).
  A set of airborne Gravity Gradiometer Survey System (GGSS) data was used in combination with the OSU89B geopotential model to predict gravity anomalies and deflections of the vertical. The gradiometer data were collected during an airborne survey of the GGSS in the Texas/Oklahoma area. Nineteen flight tracks were in the authors' disposal for further analysis. These tracks results from 54 initial distinct tracks from which a processing was performed by the Analytic Sciences Corporation. In order to assess the quality of the GGSS measurements, a first crossover analysis was carried out. The results showed large discrepancies at the crossing points between East–West (EW) and North–South (NS) tracks. The least squares collocation method and the fast Fourier transform technique were used to produce gravity anomalies and deflections of the vertical referred to geoid surface from second–order derivatives measured at the flight level. The predicted gravity values were then compared to "truth–based" point free air gravity anomalies with a standard deviation of the differences at the level of 3 mGal in the best cases.

081.023 Crustal velocities from geodetic very long baseline interferometry.
F. W. Fallon, W. H. Dillinger.
J. Geophys. Res., Vol. 97, No. B5, p. 7129 – 7136 (10 May 1992).
  The authors have used very long baseline interferometry observations from the International Radio Interferometric Surveying and Crustal Dynamics Project programs taken over a time span of 5–8 years (through August 1990) to derive relative velocities of 16 sites on the North American, Eurasian, Pacific,

and African plates. The data reduction scheme simultaneously estimates Earth orientation parameters and nutation for each session, local atmosphere and clock correction terms, source positions, and initial site positions, as well as the site velocities. No a priori geophysical crustal model whatsoever is imposed to obtain the velocities. Instead the authors introduce a minimal set of geometric constraints to obtain the solution. Two alternative constraint formulations are considered; they are shown to be equivalent in that they yield equivalent velocity sets with allowance for translation and rotation. These are (1) setting the secular motion of the pole and mean length of day to fixed values and (2) fixing the net rotation of the sites. The resulting velocities have formal standard errors typically <0.2 cm/yr, and most velocities are significantly different from zero. They agree closely to within 0.5 cm/yr, with velocities predicted by the Minster–Jordan and NUVEL–1 plate motion models.

081.024 Progress in the determination of the gravitational coefficient of the Earth.
J. C. Ries, R. J. Eanes, C. K. Shum, M. M. Watkins.
Geophys. Res. Lett., Vol. 19, No. 6, p. 529 – 531 (20 Mar 1992).
  In most of the recent determinations of the geocentric gravitational coefficient (GM) of the Earth, the laser ranging data to the Lageos satellite have had the greatest influence on the solution. These data, however, have generally been processed with a small but significant error in one of the range corrections. In a new determination of GM using the corrected center–of–mass offset, a value of 398600.4415 km^3/sec^2 (including the mass of the atmosphere) has been obtained, with an estimated uncertainty (1 σ) of 0.0008 km^3/sec^2.

081.025 The Earth's core in a nutshell.
J. A. Jacobs.
Nature, Vol. 356, No. 6367, p. 286 – 287 (26 Mar 1992).

081.026 Analytical model for solidification of the Earth's core.
B. A. Buffett, H. E. Huppert, J. R. Lister, A. W. Woods.
Nature, Vol. 356, No. 6367, p. 329 – 331 (26 Mar 1992). Letter–to–the–editor.
  The Earth's solid inner core is generally thought to have formed by gradual solidification of the liquid core as the Earth cooled. To elucidate the relative importance of the various physical effects on the thermal evolution of the core, the authors have developed an analytical model based on global heat conservation, which describes the cooling of the vigorously convecting, fluid outer core and the concomitant growth of the inner core. They obtain a simple form for the evolution of the inner–core radius which allows the consequences of changes to the model's input parameters to be readily assessed. For most of this evolution, inner–core growth is controlled primarily by the heat capacity of the outer core and the history of the heat flux into the base of the mantle.

081.027 Internally heated mantle convection and the thermal and degassing history of the Earth.
D. R. Williams, V. Pan.
J. Geophys. Res., Vol. 97, No. B6, p. 8937 – 8950 (10 Jun 1992).
  A parameterized internally heated convection model for the Earth has been developed. The mantle viscosity is temperature and volatile–content dependent. A heat flow/Rayleigh number relationship appropriate for an internally heated mantle is assumed. For each model an initial homologous temperature and bulk water content are assumed and the initial radiogenic heat production is constrained to give a present day heat flow of 0.07 W m^{-2}. The model is run for 4.6 Gyr, and temperature, heat flow, degassing and regassing rates, stress, and viscosity are calculated. A nominal case is established which shows good agreement with accepted mantle values. The effects of changing various parameters have also been tested.

081.028 **A study of the astronomical theory of ice ages in a two–dimensional nonlinear climate model.**
R. Q. Lin, R. X. Huang, J. R. Aprel.
J. Geophys. Res., Vol. 97, No. D9, p. 10029 – 10036 (20 Jun 1992).

The authors introduce a new one–level nonlinear seasonal energy balance climate model with a two–dimensional land–sea geography to study the astronomical theory of ice ages. The new model is more physically consistent and mathematically reliable than those previously reported. The model reproduces the current climate state and explains how the Earth's orbit causes ice ages. The authors have found a set of bifurcation points of warm and cold orbit for both backward and forward processes. Moreover, they found that because the Earth's orbit varies slowly, gradual increases (or decreases) in the concentration of CO_2 will also cause sudden climate transitions. Thus, variations in the Earth's orbit and the concentration of CO_2 appear to be the two major mechanisms that cause ice ages or climate changes; furthermore, these changes are usually sudden. Finally, the authors predict the near–future climate resulting from a continual increase in CO_2 concentration. The model does not yet include slower components such as the deep–ocean circulation. Therefore, variations occur more rapidly than in the real world.

081.029 **Thermal evolution and chemical differentiation of the terrestrial magma ocean.**
Y. Abe.
Workshop on the Physics and Chemistry of Magma Oceans from 1 bar to 4 Mbar, p. 9 – 10 (1992). Abstract. – See Abstr. 012.032 for the main entry.

081.030 **A magma ocean and the Earth's internal water budget.**
T. J. Ahrens.
Workshop on the Physics and Chemistry of Magma Oceans from 1 bar to 4 Mbar, p. 13 – 14 (1992). Abstract. – See Abstr. 012.032 for the main entry.

081.031 **The role of hard turbulent thermal convection in the Earth's early thermal evolution.**
U. Hansen, D. A. Yuen, Zhao Wuling, A. V. Malevsky.
Workshop on the Physics and Chemistry of Magma Oceans from 1 bar to 4 Mbar, p. 19 (1992). Abstract. – See Abstr. 012.032 for the main entry.

081.032 **Magma ocean: mechanisms of formation.**
W. M. Kaula.
Workshop on the Physics and Chemistry of Magma Oceans from 1 bar to 4 Mbar, p. 33 (1992). Abstract. – See Abstr. 012.032 for the main entry.

081.033 **Pressure regimes and core formation in the accreting Earth.**
H. E. Newsom.
Workshop on the Physics and Chemistry of Magma Oceans from 1 bar to 4 Mbar, p. 42 – 43 (1992). Abstract. – See Abstr. 012.032 for the main entry.

081.034 **Terrestrial magma ocean and core segregation in the Earth.**
E. Ohtani, N. Yurimoto.
Workshop on the Physics and Chemistry of Magma Oceans from 1 bar to 4 Mbar, p. 44 – 45 (1992). Abstract. – See Abstr. 012.032 for the main entry.

081.035 **Evolution of a terrestrial magma ocean: thermodynamics, kinetics, rheology, convection, differentiation.**
V. S. Solomatov, D. J. Stevenson.
Workshop on the Physics and Chemistry of Magma Oceans from 1 bar to 4 Mbar, p. 53 – 54 (1992). Abstract. – See Abstr. 012.032 for the main entry.

081.036 **Dynamics and evolution of a magma ocean.**
D. J. Stevenson.
Workshop on the Physics and Chemistry of Magma Oceans from 1 bar to 4 Mbar, p. 55 (1992). Abstract. – See Abstr. 012.032 for the main entry.

081.037 **The effect of stochastic variation on the size distribution of bodies impacting the Earth.**
M. C. Nolan, R. Greenberg.
Bull. Am. Astron. Soc., Vol. 23, No. 3, p. 1151 (1991). Abstract. – See Abstr. 012.037 for the main entry.

081.038 **On a relationship between earthquake centers and regions of high–energy particle precipitations from the radiation belts.**
M. E. Aleshina, S. A. Voronov, A. M. Gal'per, S. V. Koldashov, L. V. Maslennikov.
Kosm. Issled., Tom 30, Vyp. 1, p. 79 – 83 (Jan–Feb 1992). In Russian. English translation in Cosmic Res., Vol. 30, No. 1.

Experimental data about a relationship between high–energy charged particle streams in the radiation belts and the Earth's seismic activity by interaction of these particles with ULF emission of seismic origin in the ionosphere are analysed. The data show that this interaction may lead to particle precipitation from the radiation belts observed in experiments as an abrupt increase of the particle count rates. A spatial correlation between the regions of particle precipitations in the near–Earth space and earthquake centers is found which allows to predict the latitude of new earthquake centers.

081.039 **A preliminary study of the relationship between large earthquakes and precipitation for the region of Athens, Greece.**
Y. Liritzis, B. Petropoulos.
Earth, Moon, Planets, Vol. 57, No. 1, p. 13 – 21 (Apr 1992).

The annual rainfall data of Athens rain gauge stations, for the last 119 years, is compared with the occurence of large ($M \geqslant 6$) earthquakes along the fault and thrust systems in the vicinity of Athens and interesting correlations have been observed. This preliminary investigation reveals the possible occurrence of a large earthquake in the region of Athens till 1993.

081.040 **Multiband imaging by Galileo: new views of the Earth.**
W. R. Thompson.
Bull. Am. Astron. Soc., Vol. 23, No. 3, p. 1204 (1991). Abstract. – See Abstr. 012.037 for the main entry.

081.041 **Degassing.**
J. C. G. Walker.
US–USSR Workshop on Planetary Sciences, p. 191 – 202 (1991). – See Abstr. 012.075 for the main entry.

Important new information has become available in recent years concerning the release of gases from the interior of the Earth. The most fruitful source of information has been the measurement of rare gas concentrations in sea floor basalts. The results set important constraints that need to be incorporated into any comprehensive understanding of the early history of the planets. In his review the author describes some of the hightlights of these results and gives an indication of how they are derived.

081.042 **The Earth's gravity field from satellite geodesy – a 30 year adventure.**
R. H. Rapp.
International Workshop on The Solid–Earth Mission ARISTOTELES, p. 29 – 32 (Dec 1991). – See Abstr. 012.082 for the main entry.

The first information on the Earth's gravitational field from artificial satellite observations was published in 1958. The next years have seen a dramatic improvement in the resolution and accuracy of the series representation of the Earth's gravity field. The improvements have taken place slowly taking advantage of improved measurement accuracy and the increasing number of satellites. The proposed ARISTOTELES mission would provide the opportunity to take a significant leap in improving our knowledge of the Earth's gravity field.

081.043   Geopotential models of the Earth from satellite tracking, altimeter and surface gravity observations: GEM–T3 and GEM–T3S.
F. J. Lerch, R. S. Nerem, B. H. Putney, T. L. Felsentreger, B. V. Sanchez, S. M. Klosko, G. B. Patel, R. G. Williamson, D. S. Chinn, J. C. Chan, K. E. Rachlin, N. L. Chandler, J. J. McCarthy, J. A. Marshall, S. B. Luthcke, D. W. Pavlis, J. W. Robbins, S. Kapoor, E. C. Pavlis.
National Aeronautics and Space Administration, Washington, DC (USA), NASA–TM–104555. 124 p. (Jan 1992).

Improved models of the Earth's gravitational field have been developed from conventional tracking data (GEM–T3S) and from a combination of satellite tracking, satellite altimeter and surface gravimetric data (GEM–T3). This combination model represents a significant improvement in the modeling of the gravity field at half–wavelengths of 300 km and longer. Both models are complete to degree and order 50.

081.044   The Milankovitch's astronomical theory of climatic variation and the prediction of temperature variations.
Shi Guang–cheng, Yao Jin–sheng, Yang Ben–you, Wang Shun–zhen.
Publ. Purple Mt. Obs., Vol. 11, No. 2, p. 97 – 110 (Jun 1992). In Chinese.

This paper deals with the Milankovitch's astronomical theory of the climatic variation. The method of prediction of secular variations on the earth surface temperature is explored, and the secular variations of the temperature for the future 5×10^5 years in the latitudes $\varphi = 65°$ (N,S) are given.

081.045   Seismic tomography of the Earth's mantle.
B. Romanowicz.
Annu. Rev. Earth Planet. Sci., Vol. 19, p. 77 – 99 (1991).

The author reviews the most recent progress in the retrieval of the three–dimensional (3D) structure of the Earth's mantle by tomographic methods.

081.046   Measurement of crustal deformation using the Global Positioning System.
B. H. Hager, R. W. King, M. H. Murray.
Annu. Rev. Earth Planet. Sci., Vol. 19, p. 351 – 382 (1991).

Geophysicists have classically been interested in interpreting the least significant digits provided by geodesy. It is instructive to have a historical perspective on this matter, so the authors start with a brief history of the marriage between geodesy and geophysics. They then discuss the motivations for the surge of activity using this new tool. Next, they describe the basic principles behind relative positioning using GPS and document how well the system works. The authors provide specific examples of results relevant to tectonics that have already been obtained, as well as an overview of work in progress. They end by discussing new developments on the horizon.

Chaotic processes in the geological sciences.
See Abstr. 003.068.

Annual Review of Earth and Planetary Sciences. Volume 19.
See Abstr. 003.092.

Annual Review of Earth and Planetary Sciences. Volume 20.
See Abstr. 003.093.

Planetary geology: Killer acid at the K/T boundary.
See Abstr. 011.010.

The Taylor Colloquium: Origin and evolution of planetary crusts, Canberra (Australia), 1 – 2 Oct 1990.
See Abstr. 012.013.

Physics and chemistry of magma oceans from 1 bar to 4 Mbar. Abstracts of presented papers.
See Abstr. 012.032.

The Solid–Earth Mission ARISTOTELES. Proceedings.
See Abstr. 012.082.

Geophysics news 1991.
See Abstr. 013.132.

Endogenous production, exogenous delivery and impact–shock synthesis of organic molecules: an inventory for the origins of life.
See Abstr. 015.011.

Comets and meteorites were a minor source of prebiotic organic compounds on the early Earth.
See Abstr. 015.042.

Sources of organic material for the origin of life on Earth.
See Abstr. 015.043.

Evidences for the terrestrial magma ocean from high–pressure melting experiments.
See Abstr. 022.133.

Fractal analysis: a new remote sensing tool.
See Abstr. 036.143.

Polar motion, atmospheric angular momentum excitation and earthquakes – correlations and significance.
See Abstr. 044.001.

Effect of melting glaciers on the Earth's rotation and gravitational field: 1965–1984.
See Abstr. 044.002.

Stability of the astronomical frequencies over the Earth's history for paleoclimate studies.
See Abstr. 044.004.

Determination of some geodynamical parameters based on reduction of LAGEOS and Etalon–1 observation data.
See Abstr. 044.015.

Elastic energy of a deformable Earth: general expression.
See Abstr. 044.025.

Progress in the spacewise approach to ARISTOTELES data reduction.
See Abstr. 051.052.

Gravity field data products from the ARISTOTELES mission.
See Abstr. 051.053.

TOPEX orbit determination and gravity recovery using Global Positioning System data from repeat orbits.
See Abstr. 052.002.

A simple dynamo caused by conductivity variations.
See Abstr. 062.163.

The character of the field during geomagnetic reversals.
See Abstr. 084.076.

Periodic system of multi–ring planetary structures as result of interference of variously oriented lithospheric waves.
See Abstr. 091.015.

Extra–long lithospheric waves forming the morphotectonic face of planets.
See Abstr. 091.016.

Recent grazing impacts on the Earth recorded in the Rio Cuarto crater field, Argentina.
See Abstr. 105.018.

High–pressure melting of carbonaceous chondrite.
See Abstr. 105.222.

^{135}Cs–^{135}Ba: a new cosmochronometric constraint on the origin of the earth and the astrophysical site of the origin of the solar system.
See Abstr. 107.007.

Origin of the biosphere of the Earth.
See Abstr. 107.011.

Cooling of the magma ocean due to accretional disruption of the surface insulating layer.
See Abstr. 107.015.

Chemical composition of the Earth after the giant impact.
See Abstr. 107.047.

Comet impacts and chemical evolution on the bombarded Earth.
See Abstr. 107.064.

082 Atmosphere (Refraction, Scintillation, Extinction, Airglow, Site Testing)

082.001 Lightning induced brightening in the airglow layer.
W. L. Boeck, O. H. Vaughn Jr., R. Blakeslee,
B. Vonnegut, M. Brook.
Geophys. Res. Lett., Vol. 19, No. 2, p. 99 – 102 (24 Jan 1992).
   This report describes a transient luminosity observed at the altitude of the airglow layer (about 95 km) in coincidence with a lightning flash in a tropical oceanic thunderstorm directly beneath it. This event provides new evidence of direct coupling between lightning and ionospheric events. This luminous event in the ionosphere was the only one of its kind observed during an examination of several thousand images of lightning recorded under suitable viewing conditions with Space Shuttle cameras. Several possible mechanisms and interpretations are discussed briefly.

082.002 Aperture–averaging factor for optical scintillations of plane and spherical waves in the atmosphere.
L. C. Andrews.
J. Opt. Soc. Am. A, Vol. 9, No. 4, p. 597 – 600 (Apr 1992).
Current Physics Microform No.: 9206D2021.
   Previous analyses of aperture averaging of optical scintillations in the turbulent atmosphere have generally involved either numerical integrations or approximation formulas based on asymptotic results for large apertures. The author develops the exact expressions for the aperture–averaging factor in the weak-turbulence regime for both plane and spherical waves. For computational ease, accurate approximation or interpolation formulas that can be applied in most cases of interest with greater accuracy than previous approximations are also developed.

082.003 Differential geometric approach to atmospheric refraction.
W. C. Kropla, W. H. Lehn.
J. Opt. Soc. Am. A, Vol. 9, No. 4, p. 601 – 608 (Apr 1992).
Current Physics Microform No.: 9206D2025.
   Differential geometric techniques are presented and used to model the optical properties of the atmosphere under conditions that produce superior mirages. Optical path length replaces the usual Euclidean metric as a distance–measuring function and is used to construct a surface on which the paths of light rays are geodesics. The geodesic equations are shown to be equivalent to the ray equation in the plane. A differential equation that relates the Gaussian curvature of the surface and the refractive index of the atmosphere is derived. This equation is solved for the cases in which the curvature vanishes or is constant. Illustrative examples based on observation demonstrate the use of geometric techniques in the analysis of mirage images.

082.004 E.S.O site evaluation for the V.L.T.
M. Sarazin.
11. European Regional Astronomy Meeting of the IAU: New windows to the universe, Vol. 1, p. 435 – 449 (1990). – See Abstr. 012.001 for the main entry.
   The methodology and instrumentation of the site evaluation campaign for the V.L.T innovate in several ways. In particular, seeing monitors based on the measurement of differential image motion have been developed for that purpose. The availability of such instruments opens a new era in the understanding of site related parameters, and permits to plan a more flexible use of modern telescopes in the future.

082.005 Die 10– bis 12jährige Schwingung in der Stratosphäre.
K. Labitzke, H. van Loon.
Sterne Weltraum, Jahrg. 31, Nr. 2, p. 98 – 102 (Feb 1992).

082.006 Polar airglow and aurora.
D. J. McEwen, D. A. Harrington.
Can. J. Phys., Vol. 69, No. 8/9, p. 1055 – 1058 (Aug–Sep 1991).
   A survey of night airglow emissions in the polar cap shows stable emission intensities during quiet periods through the winter solstice. Those affected by particle precipitation, O I $\lambda5577$ and $\lambda6300$, show great variability with solar activity and the state of the interplanetary magnetic field. A statistical study of electron–precipitation occurrence above 78° geomagnetic latitude shows events sufficient to result in observable enhancements in O I emission intensities in about 40% of the satellite passes in the magnetic latitude range from 79° to 83° and in about 15% of the passes for latitudes above 85°.

082.007 High spectral resolution measurement of gamma ray lines from the Earth's atmosphere.
J. B. Willett, W. A. Mahoney.
J. Geophys. Res., Vol. 97, No. A1, p. 131 – 140 (1 Jan 1992).
   A search for gamma ray line features from the Earth's atmosphere has been conducted using data from the third High Energy Astronomy Observatory (HEAO 3) high spectral resolution gamma ray spectrometer. In addition to the strong line at 0.511 MeV, other intrinsically broadened line features have been observed at 1.63, 2.31, 3.67, 4.43, 5.09, and 6.13 MeV. Since the spectral resolution of the instrument is much finer than the width of the observed line features, the intrinsic width as well as the energy and intensity of each of these lines are reported. Several other predicted lines have also been observed. The characteristics of the lines seen by HEAO 3 are generally consistent with theoretical predictions as well as with previous measurements.

082.008   Measurement of atmospheric and diffuse radiation using a time–of–flight telescope.
J. A. da Costa Ferreira Neri, B. Agrinier, J. M. Lavigne.
J. Geophys. Res., Vol. 97, No. A2, p. 1541 – 1548 (1 Feb 1992).

A time–of–flight telescope constructed by the Commissariat à l'Energie Atomique (Saclay) and the Centre d'Etude Spatiale des Rayonnements (Toulouse) was used for gamma ray flux measurements. The atmospheric component was studied from balloon as a function of the pressure and the energy. Gamma ray fluxes were also measured with the telescope axis oriented in various directions and with the experiment inverted. The atmospheric and diffuse components were determined in various energy ranges.

082.009   Upper limits on spacecraft–induced ultraviolet emissions from the Space Shuttle (STS–61C).
D. Morrison, P. D. Feldman, R. C. Henry.
J. Geophys. Res., Vol. 97, No. A2, p. 1633 – 1638 (1 Feb 1992).

Ultraviolet spacecraft–induced emissions from low Earth–orbiting satellites have been reported by several investigators. Several R/Å of ultraviolet emission were obtained from the S3–4 satellite at altitudes between 180 and 250 km and from the Spacelab 1 shuttle mission at an altitude of 250 km. Conway et al. (1987) showed that N_2 Lyman–Birge–Hopfield (LBH) emissions observed by S3–4 at night are probably the result of spacecraft interaction with the atmosphere. The authors have searched for band emission of N_2, OH, O_2, and NO in nightglow spectra obtained in January 1986 with the Johns Hopkins ultraviolet background experiment (UVX) flown on the space shuttle Columbia (STS–61C) at an altitude of 330 km.

082.010   Sky spectra at a light–polluted site and the use of atomic and OH sky emission lines for wavelength calibration.
D. E. Osterbrock, A. Martel.
Publ. Astron. Soc. Pac., Vol. 104, No. 671, p. 76 – 82 (Jan 1992). = UCO, Lick Obs. Bull., No. 1205. Current Physics Microform No.: 9202F2102.

Spectra of the night sky, taken at Lick Observatory in 1988 and 1989 as byproducts of nebular spectra, show the great increase of light pollution by sodium high– and low–pressure lamps in comparison with previous spectra taken in 1975. The usefulness of the emission lines of the night sky spectrum for wavelength calibration is mentioned. In the far–red and near–infrared regions, where there are only few atomic night–sky lines, the OH vibration–rotation spectrum may be used for this purpose. Accurate rest wavelengths for these lines, calculated from the best laboratory determinations, are tabulated, and the special suitability of the P_1 (and to a lesser extent P_2) lines is discussed.

082.011   Effects of gravity waves on complex airglow chemistries. 1. $O_2(b^1\Sigma^+_g)$ emission.
D. W. Tarasick, G. G. Shepherd.
J. Geophys. Res., Vol. 97, No. A3, p. 3185 – 3193 (1 Mar 1992).

The theory of Hines and Tarasick (1987) for the effects of gravity waves on airglow emissions is extended to consider more complex airglow chemistries, including multiple and multiple–step production mechanisms, quenching, and other loss processes. Relations for the dependence of η, the ratio of brightness to temperature fluctuation, on emission chemistry are presented in a generalized form which is readily applicable to other emissions. The specific case of $O_2(b^1\Sigma^+_g)$ airglow is examined in detail for three different proposed chemical production mechanisms. Loss mechanisms are also considered; quenching is found to be important to predictions of η. The results obtained are compared with the very limited set of published observations of gravity waves in $O_2(b^1\Sigma^+_g)$ airglow. It is shown that gravity wave observations may place new constraints on possible mechanisms for airglow production, constraints that are not obtainable by other means.

082.012   Effects of gravity waves on complex airglow chemistries. 2. OH emission.
D. W. Tarasick, G. G. Shepherd.
J. Geophys. Res., Vol. 97, No. A3, p. 3195 – 3208 (1 Mar 1992).

The theory of Hines and Tarasick (1987) for the effects of gravity waves on airglow emissions is extended to consider more complex airglow chemistries, including multiple and multiple–step production mechanisms, quenching, and other loss processes. Attention is given to nonsteady state chemistry, since the time constants of some species involved in OH* production are comparable with gravity wave periods. Relations for the dependence of η, the ratio of brightness to temperature fluctuations, on emission chemistry are presented in a generalized form which is readily applicable to other emissions. The specific case of OH airglow is examined in detail for the quenching rates of Llewellyn et al. (1978) and for those of Lowe (1991). Quenching is found to be important to predictions of η. The results obtained are compared with the limited set of published observations of gravity waves in OH airglow. Although the theory predicts significant differences in the behavior of η for the two sets of rates, the existing data are not adequate to draw firm conclusions.

082.013   Seeing measurements and observing statistics at the U.S. Naval Observatory, Flagstaff Station.
H. C. Harris, F. J. Vrba.
Publ. Astron. Soc. Pac., Vol. 104, No. 672, p. 140 – 145 (Feb 1992). Current Physics Microform No.: 9203B2057.

Direct measurements of the seeing for the U.S. Naval Observatory, Flagstaff Station are presented for the years 1986 – 1991 from the USNO CCD parallax program at the 1.55–m telescope. Cloud–cover and seeing–related statistics are presented for all programs from the same telescope for the years 1969 – 1990. The data show that seeing at the zenith has a median of 1.3 arcsec and a mode of 1.0 arcsec FWHM; the distribution of observed seeing probably includes a significant contribution from local dome seeing and/or telescope effects. Cloud–cover statistics show that a yearly average of approximately 45% of the nighttime hours are clear and 65% usable. Both the cloud–cover and seeing statistics show significant seasonal variations. The Flagstaff data are also compared with published statistics from Kitt Peak National Observatory.

082.014   Observing conditions at the Wise Observatory.
N. Brosch.
Q. J. R. Astron. Soc., Vol. 33, No. 1, p. 27 – 32 (Mar 1992).

Data pertaining to the observational qualities of the site of the Wise Observatory and the efficiency of the observations were collected during 17 years and are presented.

082.015   Atmospheric electricity response to an isolated solar flare and to a series of flares.
V. M. Sheftel', A. K. Chernyshev.
Geomagn. Aehron., Tom 32, No. 1, p. 111 – 117 (Jan – Feb 1992). In Russian. English translation in Geomagn. Aeron., Vol. 32, No. 1.

082.016   Aerosol coagulation model in the middle atmosphere at minimal meteor matter influx.
M. Begkhanov, O. Kurbanmuradov, V. N. Lebedinets.
Astron. Vestn., Tom 26, No. 2, p. 102 – 111 (Mar – Apr 1992). In Russian. English translation in Solar Syst. Res., Vol. 26, No. 2.

The coagulation model of cosmic aerosols in the middle atmosphere is developed. The influx of meteor vapours and of micrometeorites, the processes of condensation, coagulation, sedimentation and diffusion are taken into account. The physical mechanism of the formation of the light scattering layer in the upper stratosphere is revealed for the first time.

082.017   Geotechnical survey of the LEST site on La Palma.
P. Søndergaard, T. B. Hansen.
LEST Found., Tech. Rep., No. 52.
Oslo Univ. (Norway). Inst. for Teoretisk Astrofysikk. 17 p. (1992).

This technical report presents a summary of the results of the site investigations together with recommendations to the foundation of the LEST Solar Telescope at Roque de los Muchachos, La Palma, The Canary Islands. The results of the geotechnical survey suggest some modifications of the setup below the telescope in order to minimize the height of the subsurface structures, which could result in a much more favourable foundation in the upper basalt formation.

082.018 The role of cometary and meteor matter in noctilucent clouds genesis.

V. N. Lebedinets, O. Kurbanmuradov.
Astron. Vestn., Tom 26, No. 1, p. 83 – 92 (Jan – Feb 1992). In Russian. English translation in Solar Syst. Res., Vol. 26, No. 1.

A global balance water model of cometary origin in the atmosphere at 60 – 140 km heights is developed. If the mini-comets bring into the Earth's atmosphere $(0.5 - 1) \times 10^6 t\, d^{-1}$ molecules of H_2O, their relative concentration in the mesopause is about 10^{-4} what is sufficient for water vapour condensation at the temperature of 160 – 170K. Fine particles of interplanetary dust or the dust particles brought by mini-comets can be considered as condensation nuclei. The main sources of water vapour are different in the troposphere, stratosphere and mesopause.

082.019 Wind regime of the meteor zone and its interaction with processes in the strato– and mesosphere.

R. B. Bekbasarov, K. A. Karimov, Kh. N. Nabotov, R. Shminder.
Astron. Vestn., Tom 26, No. 1, p. 98 – 102 (Jan – Feb 1992). In Russian. English translation in Solar Syst. Res., Vol. 26, No. 1.

Measurements of major wind components by methods D_1 and D_2 in 1984 are presented. The time variations of wind components in the main seasons are considered. Dates of beginning, end and duration of spring variations of the circulation regime are compared. The time variation of wind regime in the low thermosphere in winter is analysed in detail. A close correlation of the time variation of circulation regime of M–zone with the processes in the mesosphere and stratosphere in the cold half-year is shown.

082.020 Observations of noctilucent clouds and aerosol layers in the strato–mesosphere from orbital stations "Salyut–7" and "Mir".

A. I. Lazarev, V. N. Lebedinets, L. A. Mirzoeva, V. P. Savinykh, V. G. Titov.
Astron. Vestn., Tom 26, No. 1, p. 115 – 125 (Jan – Feb 1992). In Russian. English translation in Solar Syst. Res., Vol. 26, No. 1.

Informations from the cosmonauts V. P. Savinykh and V. G. Titov's log-book on observations of noctilucent and mesospheric clouds from space in 1985 and 1988 are presented. The equatorial mesospheric clouds (EMC) discovered earlier by V. V. Kovalenko and A. S. Ivanchenkov differ significantly from polar mesospheric clouds (PMC). The problems of PMC and EMC genesis in the light of the hypothesis on cometary water origin in the upper atmosphere are considered.

082.021 A mixed method for measuring the inner scale of atmospheric turbulence.

A. Consortini, Sun Yi Yi, Li Zhi Ping, G. Conforti.
J. Mod. Opt., Vol. 37, No. 10, p. 1555 – 1560 (Oct 1990). Letter-to-the-editor.

A new method for measuring the inner scale and structure constant of atmospheric turbulence is proposed and tested. It utilizes both phase and amplitude fluctuations of laser radiation propagating in the medium. The measured quantities are the intensity variance of a spherical wave and the transverse–displacement variance (wandering) of a narrow laser beam. The method is developed within the limits of geometrical optics. Experimental laboratory results are presented. Values of inner scale of a few millimetres are measured.

082.022 Nightglow emissions from oxygen in the lower thermosphere.

D. R. Bates.
Planet. Space Sci., Vol. 40, No. 2/3, p. 211 – 221 (Feb–Mar 1992).

Several independent trains of reasoning lead to the conclusion that $O_2(c^1\Sigma_u^-)$ is the precursor for the emission of the 557.7 nm line of atomic oxygen. It is shown that the measured altitude profile of the volume emission rate of the Herzberg I system is reproduced by using the expected rate coefficient for the production of $O_2(A^3\Sigma_u^+)$ by termolecular association and rate coefficients for quenching by atomic and molecular oxygen that accord with laboratory evidence. Termolecular association does not contribute directly to the atmospheric system because the $O_2(b^1\Sigma_g^+)$ it generates is mainly in high vibrational levels and vibrational deactivation of these by interchange of electronic energy with normal oxygen molecules is not rapid enough to prevent deactivation to $O_2(a^1\Delta_g$ or $X^3\Sigma_g^-)$ from occurring. The prime source of the $O_2(b^1\Sigma_g^+,\ v=0)$ is excitation of the $O_2(X^3\Sigma_g^-,\ v=0)$ collision partner involved in $O_2(A^3\Sigma_u^+, A'^3\Delta_u$ and $c^1\Sigma_u^-)$ deactivation. This source can account for the measured altitude profile of the volume emission rate of the atmospheric system.

082.023 The upper atmosphere as sensed by satellite orbits.

D. G. King–Hele.
Planet. Space Sci., Vol. 40, No. 2/3, p. 223 – 233 (Feb–Mar 1992).

This paper offers a personal (and therefore biased) survey of the advances in knowledge of the thermosphere made possible by measuring the changes in the orbits of satellites, particularly in the early years of the space era. Reliable numerical values of air density at heights above 200 km were obtained for the first time, and many different types of variation in density were discovered and codified. Subsequently the orbital changes were analysed to determine other atmospheric parameters, such as scale height and zonal winds.

082.024 Airglow hydroxyl emissions.

G. G. Sivjee.
Planet. Space Sci., Vol. 40, No. 2/3, p. 235 – 242 (Feb–Mar 1992).

Airglow OH emissions originate from the mesopause and provide optical signatures of various physical and chemical processes occurring in this region of the upper atmosphere. Emissions from different vibrational levels of the mesopause OH (in the ground electronic state) may peak at different heights, roughly between 80 and 90 km. The thermalized rotational distribution of these band emissions is widely employed to infer the kinetic temperature (and a small temperature gradient) of the mesopause. Variations in airglow OH band intensities and rotational temperature are interpreted as signatures of various disturbances propagating in the mesopause. These and allied topics of airglow OH emissions are reviewed to identify some unanswered questions in this field which form the bases for current aeronomic investigations of the mesosphere.

082.025 Sodium chemistry: a brief review and two new mechanisms for sudden sodium layers.

W. Swider.
Planet. Space Sci., Vol. 40, No. 2/3, p. 247 – 253 (Feb–Mar 1992).

The aeronomy of sodium has improved immensely since the rate coefficient of the key three–body loss process for Na was determined a decade ago. Models for Na in the mesosphere compare reasonably well with observations, including seasonal and diurnal variations. The computed Na nightglow also agrees with data. Less well understood are the circumstances responsible for the sporadic occurrence of narrow intense layers of Na. Dynamical processes undoubtedly are involved in the formation of these sudden sodium layers (SSLs). Two new SSL mechanisms tentatively are proposed. One involves ion–ion recombination between coincident layers of Na^+, and O^- ions formed by $e + O_2$. This reaction, endothermic at room temperatures, may be important during auroras when energetic electrons are present and/or if sufficiently large electric fields are present. A less likely conjecture is that layers of atomic oxygen may mix with layers of sodium compounds to produce SSLs.

082.026   **Stratospheric ion chemistry: present understanding and outstanding problems.**
E. Arijs.
Planet.   Space   Sci.,   Vol. 40,   No. 2/3,   p. 255 – 270 (Feb–Mar 1992).
Our present knowledge on stratospheric ion chemistry is reviewed. Available experimental data as well as modelling efforts are discussed and the needs for further research are pointed out.

082.027   **The polar lower thermosphere.**
R. G. Roble.
Planet.   Space   Sci.,   Vol. 40,   No. 2/3,   p. 271 – 297 (Feb–Mar 1992).
The Earth's mesosphere and lower thermosphere are the least explored regions of the Earth's atmosphere. The observations that have been made in this region, however, indicate that it is a dynamically active region, especially the polar lower thermosphere where most of the auroral energy that impacts the Earth's atmosphere is deposited. This energy is redistributed globally by the thermospheric wind system and there are important dynamic, chemical, radiational, and electrodynamic couplings that occur between the lower thermosphere and the middle atmosphere. There are also couplings with the upper thermosphere, ionosphere, and magnetosphere. A brief review of the available observations important for understanding global dynamic processes in the lower thermosphere is given. Results of simulations made with the NCAR thermosphere/ionosphere general circulation model (TIGCM) are presented to illustrate interactions of polar lower thermospheric chemistry and dynamics.

082.028   **Nitric oxide in the lower thermosphere.**
C. A. Barth.
Planet.   Space   Sci.,   Vol. 40,   No. 2/3,   p. 315 – 336 (Feb–Mar 1992).

082.029   **Thermospheric odd nitrogen.**
J.–C. Gérard.
Planet.   Space   Sci.,   Vol. 40,   No. 2/3,   p. 337 – 353 (Feb–Mar 1992).
The photochemistry of NO, $N(^2D)$ and $N(^4S)$ and results of recent space measurements of their density distribution are discussed. In particular, the role of the reaction between metastable $N(^2D)$ atoms and O_2 as a source of $O(^1D)$ is discussed in the light of laboratory and aeronomical observations. Global satellite measurements are compared with results of two- and three–dimensional models including transport. The possibility of explaining the odd nitrogen observations gathered in the Venusian and Martian thermospheres with the current understanding of the terrestrial models adapted to CO_2– rich atmospheres is examined. It is concluded that the understanding of the processes governing the distribution of odd nitrogen in terrestrial planets is generally satisfactory, although several aspects require further quantitative investigation.

082.030   **The changing stratosphere.**
M. B. McElroy, R. J. Salawitch, K. Minschwaner.
Planet.   Space   Sci.,   Vol. 40,   No. 2/3,   p. 373 – 401 (Feb–Mar 1992).

082.031   **Natural and anthropogenic perturbations of the stratospheric ozone layer.**
G. P. Brasseur.
Planet.   Space   Sci.,   Vol. 40,   No. 2/3,   p. 403 – 412 (Feb–Mar 1992).
The paper reviews potential causes for reduction in the ozone abundance. The response of stratospheric ozone to solar activity is discussed. Ozone changes are simulated in relation with the potential development of a fleet of high–speed stratospheric aircraft and the release in the atmosphere of chlorofluorocarbons. The calculations are performed by a two–dimensional chemical–radiative–dynamical model. The importance of heterogeneous chemistry in polar stratospheric clouds and in the Junge layer (sulfate aerosol) is emphasized. The recently reported ozone trend over the last decade is shown to have been largely caused by

the simultaneous effects of increasing concentrations of chlorofluorocarbons and heterogeneous chemistry. The possibility for a reduction in stratospheric ozone following a large volcanic eruption such as that of Mount Pinatubo in 1991 is discussed.

082.032   **Stratospheric evidence of relativistic electron precipitation.**
A. C. Aikin.
Planet.   Space   Sci.,   Vol. 40,   No. 2/3,   p. 413 – 431 (Feb–Mar 1992).
The hypothesis that relativistic electron precipitation is modifying the high–latitude southern hemisphere ozone distribution is tested by examining simultaneous electron density data as measured with a ground–based partial reflections sounder and ozone mixing ratios data in the 40 to 50 km region obtained from the satellite–borne SBUV instrument.

082.033   **NO and O_2 ultraviolet nightglow and spacecraft glow from the S3–4 satellite.**
R. W. Eastes, R. E. Huffman, F. J. Leblanc.
Planet. Space Sci., Vol. 40, No. 4, p. 481 – 493 (Apr 1992).
Observations (1600–2950 Å) from the S3–4 satellite of the Earth's nightglow at ~ 6 Å resolution have been analyzed. These data indicate that the only significant emissions from NO are the δ and γ bands. The relative brightness of these bands was consistent with previous observations at lower resolution. The data support the presence of O_2 emissions in the Herzberg I, II, and III bands in ratios ($\sim 4.5{:}1{:}1.4$), consistent with analysis of previous ground–based observations of these band systems. Spacecraft glow is seen from the N_2 Lyman–Birge–Hopfield bands, as previously reported; however, no spacecraft–related emissions were seen from NO or O_2.

082.034   **Line–by–line computation of the atmospheric absorption spectrum using the decomposed Voigt line shape.**
A. Uchiyama.
J. Quant. Spectrosc. Radiat. Transfer, Vol. 47, No. 6, p. 521 – 532 (Jun 1992).
A new line–by–line method has been developed to calculate the spectral absorption coefficients of absorption lines with a Voigt line shape in the terrestrial atmosphere. This method has also been applied to several problems in the field of atmospheric radiation. In the present method, the Voigt line shape is decomposed into several sub–functions, with even quadratic functions used as sub–functions used as sub–functions. This functional form allows for easy estimation of the half–width of the sub–function. The final spectral absorption is obtained by the superposition of independent spectral absorption coefficients. The finer details of the absorption spectrum have then been examined to determine if they are zero or not at every grid point. This procedure eliminates the need for unnecessary calculations in interpolations of absorption coefficients from coarser to finer resolution and also saves storage capacity.

082.035   **Optical seeing at La Palma Observatorty. I. General guidelines and preliminary results at the Nordic Optical Telescope.**
J. Vernin, C. Muñoz–Tuñón.
Astron. Astrophys., Vol. 257, No. 2, p. 811 – 816 (Apr 1992).
The results of one night of observations corresponding to an intensive site assessment campaign, held at the Observatorio del Roque de los Muchachos (La Palma) on 22 July 1990, are presented and discussed. The goal of the campaign was to evaluate the contribution of free atmosphere, boundary layer, surface layer and dome plus mirror seeing to image degradation at the Nordic Optical Telescope (NOT). To fulfill this objective the Scidar instrument was adapted to the NOT focus, whilst launched balloons and an instrumented mast were used to evaluate the exra–dome seeing. Although the seeing measurements reported here result from a single night of observations only, and thus have no statistical bearing on the seeing quality at La Palma, they nevertheless revealed a potentially excellent site; seeing contributions of respectively 0".08, 0".50, and 0".40 were measured for the surface layer, boundary layer and free

atmosphere, with a resulting net effect of 0"69. At the beginning of the night, extremely good seeing conditions (about 0"5) where encountered at the NOT focus, demonstrating that, compared to the atmospheric conditions, there was virtually no significant dome and mirror seeing contributing to the image degradation.

082.036 **Neutral atomic oxygen density from nighttime radar and optical wind measurements at Millstone Hill.**
M. J. Buonsanto, Y.-K. Tung, D. P. Sipler.
J. Geophys. Res., Vol. 97, No. A6, p. 8673 – 8679 (1 Jun 1992).
Neutral atomic oxygen densities [O] in the thermosphere are calculated from analysis of coincident Millstone Hill incoherent scatter radar and Fabry Perot interferometer measurements taken during 14 nights with widely varying geomagnetic conditions in 1990 and 1991, and the results are compared with the mass spectrometer/incoherent scatter 1986 (MSIS–86) model.

082.037 **Stratospheric clouds at South Pole during 1988. 1. Results of lidar observations and their relationship to temperature.**
G. Fiocco, M. Cacciani, P. di Girolamo, D. Fuà, J. DeLuisi.
J. Geophys. Res., Vol. 97, No. D5, p. 5939 – 5946 (20 Apr 1992).
An optical radar–lidar–has been operational at the Amundsen–Scott South Pole Station since summer 1987–1988. The observations were specially directed to the detection of aerosol layers and polar stratospheric clouds (PSCs). The lidar utilized a Nd–YAG laser followed by a second harmonic generator, and a 0.5–m diameter Cassegrain receiving telescope. Results obtained during the period May–October 1988 are summarized. Some 10,000 profiles of the lidar echoes, each the result of 1–min averaging, were obtained. Data sets consisting of profiles of the scattering ratio and of the backscattering cross section B_a, based on half–hour averaging, are presented. The data can be related to profiles of the atmospheric temperature T, usually obtained on a daily basis at South Pole. Stratifications appear to have two distinct types of structures: one structure shows only a modest variation with height; the other is characterized by sharp features, with large changes of the cross section with height. The basic results, the relationship between B_a, and T, and their statistical relevance are considered in this paper.

082.038 **Stratospheric clouds at South Pole during 1988. 2. Their evolution in relation to atmospheric structure and composition.**
D. Fuà, M. Cacciani, P. di Girolamo, G. Fiocco, A. di Sarra.
J. Geophys. Res., Vol. 97, No. D5, p. 5947 – 5952 (20 Apr 1992).
A lidar was installed at the Amundsen–Scott South Pole Station in the austral summer 1987–1988: aerosol layers, generically identified as polar stratospheric clouds, were frequently observed in the period May–October 1988 through the altitude range 8–20 km. On the basis of previous work (Fiocco et al., 1991) the behavior of the aerosol backscattering cross section B_a has been related to the temperature T through linear fits between the two variables (Fiocco et al., 1992). The resulting coefficients, namely the slope $b = dB_a/dT$ and T_f, the temperature at which the onset of condensation occurs, help to classify these clouds as Type I or II, in view of the thermodynamic properties of the condensing species. A strong dependence of B_a on T, represented by a large value of b, is interpreted as evidence of ice condensation and leads the authors to identify the clouds as Type II, while a lower value of b, characteristic of diffuse structures, is taken as evidence for Type I clouds, composed of nitric acid trihydrate. The evolution of these features is reported.

082.039 **Seasonal variations of mesospheric hydrogen and ozone concentrations derived from ground–based airglow and lidar observations.**
H. Takahashi, B. R. Clemesha, Y. Sahai, P. P. Batista, D. M. Simonich.
J. Geophys. Res., Vol. 97, No. D5, p. 5987 – 5994 (20 Apr 1992).
Upper atmospheric airglow emissions in the mesopause region, O I 557.7 nm, NaD at 589.3 nm and OH(9,4) band at 774.6 nm have been observed since 1977 at Cachoeira Paulista (22.7°S, 45.0°W), Brazil. The seasonal dependence of the intensity variations of these emissions was studied using the data from a total of 560 nights, during the period June 1977 to November 1986. Strong semiannual variations in the NaD and O I 557.7 nm emissions were observed with maxima at the equinoxes, in contrast to the OH(9,4) emission, which showed a much smaller seasonal variation. Hydrogen and ozone concentrations at around 87–89 km were calculated using the observed NaD and OH(9,4) band intensities and the sodium concentration observed by lidar at a nearby station. The obtained seasonal variations of the two species, opposite in phase with each other, are in good agreement with current atmospheric dynamics models, which show a seasonal variation of the vertical transport of atomic oxygen and hydrogen.

082.040 **On observations of upper atmosphere night airglow in the O I 557.7 nm line in Abastumani during high seismic activity in Georgia in April – June 1991.**
T. I. Toroshelidze, L. M. Fishkova, S. P. Chilingarashvili.
Astron. Tsirk., No. 1551, p. 38 – 39 (Nov – Dec 1991). In Russian.
Observations made in Abastumani have shown that night airglow O I 557.7 nm emission intensity was disturbed in April – June 1991. These disturbances coincided with the catastrophic earthquake on April 29, 1991 and the following seismic activity.

082.041 **Global ozone depletion and the Antarctic ozone hole.**
G. Pitari, G. Visconti, M. Verdecchia.
J. Geophys. Res., Vol. 97, No. D8, p. 8075 – 8082 (30 May 1992).
The secular trend of the Antarctic ozone hole has been studied with a two–dimensional model which can simulate formation of polar stratospheric clouds and includes heterogeneous chemical reactions. Results from the numerical simulation have been validated by comparison with available experimental data. Trends up to the year 2010 using standard (i.e., homogeneous) and heterogeneous chemistry have been compared and show that global ozone depletion reached 5–6% in the last 30 years and will average 8% for the next 20 years. Subtracting a 2% loss due to standard chemistry in the presence of trace gas increase in the last 30 years, the authors find a 3–4% global ozone loss due to heterogeneous chemistry. The depletion is evident even outside the southern hemisphere spring season and at mid–latitudes, pointing to an increase in global ozone sink.

082.042 **The spectrum of the tropical oxygen nightglow observed at 3 Å resolution with the Hopkins Ultraviolet Telescope.**
P. D. Feldman, A. F. Davidsen, W. P. Blair, C. W. Bowers, S. T. Durrance, G. A. Kriss, H. C. Ferguson, R. A. Kimble, K. S. Long.
Geophys. Res. Lett., Vol. 19, No. 5, p. 453 – 456 (3 Mar 1992).
Ultraviolet spectra of the tropical oxygen night–glow in the range of 830 to 1850 Å (in first order) at 3 Å resolution were obtained with the Hopkins Ultraviolet Telescope during the Astro–1 space shuttle mission in December 1990. The authors present here the data obtained on a setting celestial target as the zenith angle of the line–of–sight varied from 77 to 95°C. The dominant features in the spectrum (other than geocoronal hydrogen) are O I $\lambda\lambda$1304 and 1356 and the radiative recombination continuum near 911 Å. The continuum is resolved and found to be consistent with an electron temperature in the range 1000–1250K. The observed ratio of the brightness of O I λ1356 to the continuum suggests that O^+–O^- mutual neutralization contributes about 40% to the 1356 Å emission. The dependence of the optically thin emissions on zenith angle is consistent with a simple ionospheric model. Weak O I λ989 emission is also detected, but there is no evidence for any similarly produced atomic nitrogen emissions.

082.043 **Observation of high–N hydroxyl pure rotation lines in atmospheric emission spectra by the CIRRIS 1A Space Shuttle experiment.**
D. R. Smith, W. A. M. Blumberg, R. M. Nadile, S. J. Lipson, E. R. Huppi, N. B. Wheeler, J. A. Dodd.
Geophys. Res. Lett., Vol. 19, No. 6, p. 593 – 596 (20 Mar 1992).
Pure rotation line emissions from highly rotationally excited OH have been observed between 80 and 110 km tangent height under both nighttime and daytime quiescent conditions. Data were obtained using the cryogenic CIRRIS 1A interferometer,

operated on the Space Shuttle. Transitions from OH ($v = 0$–2, N' $\leqslant$ 33) were identified between 400 and 1000 cm^{-1}, corresponding to states with energies as high as 23000 cm^{-1}. These are the first definitive observations of OH pure rotation transitions in the airglow, and by far the highest N levels observed in any type of OH airglow emission spectrum. The present observations of highly excited rotational states of OH parallel those made during recent field studies of NO and laboratory studies of OH, NO, and CO.

082.044   **Mesospheric OH airglow temperature fluctuations: a spectral analysis.**
E. M. Dewan, W. Pendleton, N. Grossbard, P. Espy.
Geophys. Res. Lett., Vol. 19, No. 6, p. 597 – 600 (20 Mar 1992).

A field campaign under Project MAPSTAR was conducted in Colorado during May–July 1988. As part of this effort ground based measurements of OH airglow rotational temperatures at 85 km were made by means of an infrared Fourier spectrometer (IRFWI of USU). These measurements employed a least squares spectral fitting technique involving the 3–1 Meinel band. These data and their estimated PSD's are presented. The latter are interpreted by means of the relations between gravity wave temperature fluctuations and vertical displacement fluctuations given by Makhlouf et al. (1990). The results are compared with the predictions of the gravity wave model of Dewan (1990, 1991) and shown to be in reasonable agreement thus lending further support to the local wave–cascade hypothesis.

082.045   **A thermosphere/ionosphere general circulation model with coupled electrodynamics.**
A. D. Richmond, E. C. Ridley, R. G. Roble.
Geophys. Res. Lett., Vol. 19, No. 6, p. 601 – 604 (20 Mar 1992).

A new simulation model of upper atmospheric dynamics is presented that includes self–consistent electrodynamic interactions between the thermosphere and ionosphere. This model, which the authors call the National Center for Atmospheric Research thermosphere – ionosphere – electrodynamic general circulation model (NCAR/TIE–GCM), calculates the dynamo effects of thermospheric winds, and uses the resultant electric fields and currents in calculating the neutral and plasma dynamics. A realistic geomagnetic field geometry is used. Sample simulations for solar maximum equinox conditions illustrate two previously predicted effects of the feedback. Near the magnetic equator, the afternoon uplift of the ionosphere by an eastward electric field reduces ion drag on the neutral wind, so that relatively strong eastward winds can occur in the evening. In addition, a vertical electric field is generated by the low–latitude wind, which produces east–west plasma drifts in the same direction as the wind, further reducing the ion drag and resulting in stronger zonal winds.

082.046   **Effects of vibrationally excited molecular nitrogen on ionospheric–thermospheric coupling for different levels of solar activity.**
K. Serafimov, M. Serafimova.
J. Atmos. Terr. Phys., Vol. 53, No. 11/12, p. 1139 – 1143 (Nov–Dec 1991). – See Abstr. 012.024 for the main entry.

An analysis has been made of the interaction between neutral and ionized thermospheric components and processes of the generation, transfer and dissipation of vibrationally excited nitrogen molecules. Analytical expressions for these interactions with the topside thermosphere have been developed. It has been shown that the collision frequency v_{ei} is a direct measure of the excited components $N_{2,v}$. Naturally, a complete analysis of $N_{2,v}$ and its effect is possible only when it is rationally combined with airglow measurements. As an example, the saturation effects of $N_{2,v}$ have been considered for different levels of solar activity, in particular the considerable increase of $[NO^+]/[O_2^+]$ during a very active Sun and its role in restricting electron density growth rates.

082.047   **Mean temperature fields in the upper stratosphere.**
Yu. P. Koshelkov.
J. Atmos. Terr. Phys., Vol. 53, No. 11/12, p. 1195 – 1202 (Nov–Dec 1991). – See Abstr. 012.024 for the main entry.

Long–term temperature fields based on satellite data from CIRA–1986 and NMC charts are considered. The amplitudes

and phases of stationary waves are determined for the 3–month winter period, and the wave characteristics in the two hemispheres are compared.

082.048   **Nature of the optical anomalies of summer of 1908.**
V. A. Romejko.
Sol. Syst. Res., Vol. 25, No. 4, p. 362 – 368 (Jan 1992). English translation of Astron. Vestn., Tom 25, No. 4, p. 482 – 489 (1991).

Analysis of optical phenomena occurring in the Earth's atmosphere in the summer of 1908 indicates that those phenomena were caused by the presence of a strong field of noctilucent clouds initiated by the explosion of the Tungus meteorite, as well as volcanic activity of 1907 – 1908.

082.049   **Nature of the anomalous sky luminescence connected with the Tungus event.**
V. A. Bronshtehn.
Sol. Syst. Res., Vol. 25, No. 4, p. 369 – 381 (Jan 1992). English translation of Astron. Vestn., Tom 25, No. 4, p. 490 – 504 (1991).

A quantitative explanation is suggested for the anomalous sky luminescence which was observed on the night of June 30 – July 1, 1908 (and on subsequent nigths) over a vast region from western Siberia to the British Isles. An estimate of the sky brightness at this time has been made from five observations: 10^{-6} to 10^{-7}sb. The mechanism of secondary scattering of sunlight by dust which was contained in the head of the Tungus comet and entered the Earth's atmosphere at the same time as the Tungus body has been considered. The densities and optical properties of the dust particles, and their size distribution have been assumed to be the same as in the head of Comet Halley. Transport of these dust particles (in the mass range from 10^{-13}g to 10^{-5}g) is not done by winds, but in the Earth's gravity field. Calculation of their motion enables one to explain the position of the western boundary of the luminescence region, and the position of its eastern boundary speaks in favor of low–angle entry of the Tungus body into the atmosphere. With allowance for Stokes settling after braking of the dust particles, the height range of the luminescent dust was 50 km to 70 km. Here its density was increased by $\sim$3000 times due to braking and settling. Calculation of the luminescence brightness of a dust cloud that is illuminated both by primary scattering by the dust itself and also by molecular scattering gave a value of 10^{-6}sb, in good agreement with the observational data.

082.050   **Generalized scale invariance and differential rotation in cloud radiances.**
S. Lovejoy, K. Pflug, D. Schertzer.
Physica A, Vol. 185, No. 1–4, p. 121 – 128 (15 Jun 1992). Paper presented at the International Conference on Complex Systems: Fractals, Spin Glasses and Neural Networks, Miramare, Trieste (Italy), 2 – 6 Jul 1991.

Scale invariant symmetries are usually restricted to systems involving extremely special types of scale changes; self–similar and self–affine fractals involving isotropy and differential stratification respectively. In contrast, geophysical and astrophysical systems can be scale invariant but display complex anisotropy including differential rotation. The formalism required to handle such symmetries is generalized scale invariance; using the new Monte Carlo differential rotation technique the authors test it on satellite cloud radiances over the range 1 – 1200 km. The results underscore the limited usefulness of self–similar and self–affine scaling ideas in atmospheric dynamics since one finds substantial differential rotation which is, nevertheless, scaling.

082.051   **Nonlinear Schroedinger equation for low frequency acoustic–gravity waves.**
Huang Chaosong, Li Jun.
Acta Geophys. Sin., Vol. 34, No. 5, p. 531 – 537 (Sep 1991). In Chinese.

Nonlinear behaviours of low frequency acoustic–gravity waves are studied. The atmosphere is considered to be imcompressible and isothermal, and the gravitational acceleration is independent of the height. The nonlinear wave equations are obtained by using the technique of multiple time and space scales. It is shown that the acoustic–gravity waves propagating parallel to the earth's surface satisfy the nonlinear Schroedinger equation which has solutions of soliton–envelopes of wave trains.

082.052   Global climatic effects of atmospheric dust from large asteroid or comet impacts on Earth.
C. Covey, S. L. Thompson.
Bull. Am. Astron. Soc., Vol. 23, No. 3, p. 1150 – 1151 (1991). Abstract. – See Abstr. 012.037 for the main entry.

082.053   Proposal to measure terrestrial Bradley aberration.
J. P. Wesley.
Found. Phys. Lett., Vol. 5, No. 1, p. 77 – 82 (Feb 1992).
Because parallax exactly masks Bradley aberration when ordinary terrestrial sources are used; it is proposed to measure the angle of parallax, and thus, the angle of aberration, by observing telescopically the appearance of a three dimensional object used as a source. For a setup rigidly fixed to the Earth's surface at a northern latitude the variation of the appearance of the object as a function of the time of day can then yield the magnitude and direction of the absolute velocity of the Earth.

082.054   Waves in the atmosphere caused by the solar terminator. A review.
V. M. Somsikov.
Geomagn. Aeron., Vol. 31, No. 1, p. 1 – 8 (Aug 1991). English translation of Geomagn. Aehron., Tom 31, No. 1, p. 1 – 8 (1991).
A systematic review is proposed of the theoretical and experimental information of investigations of the solar terminator (ST) which have accumulated over the approximately 20 years since the start of investigations of it as a regular global source of disturbances of the entire thickness of the atmosphere, including the thermosphere. The existing models of the ST and possible mechanisms of formation of disturbances in the neutral atmosphere and the ionospheric plasma are discussed. The results of calculations and analysis of the characteristics and structures and the waves generated by the ST throughout the entire atmospheric and ionospheric plasma are outlined. The results of experimental investigations of the effects created by the ST in the atmosphere and ionosphere, an appreciable part of which has been carried out within the framework of the All–Union "Terminator" program are given.

082.055   A possible atmospheric pressure wave from the total solar eclipse of 22 July 1990.
B. W. Jones, G. J. Miseldine, R. J. A. Lambourne.
J. Atmos. Terr. Phys., Vol. 54, No. 2, p. 113 – 115 (Feb 1992).
Total solar eclipses are of importance in studies of pressure waves in the Earth's atmosphere, because they provide relatively well defined forcing functions. Over the south east of the U.K. the authors have observed an atmospheric pressure disturbance that might have been a wave caused by the total solar eclipse of 22 July 1990. At each of three microbarometers, sited at the apices of a triangle of sides 21.5, 27.9 and 35.8 km, the authors observed the order of a 30 Pa rise and fall in pressure over a period of about 2 h. The direction and speed of travel are weakly constrained but are consistent with an eclipse origin. The nature of the wave is uncertain.

082.056   Quasi–biennial and quasi–annual ozone variations.
M. A. Nuzhdina.
Visn. Kiiv. Univ., Fiz.–Mat. Nauki, Astron., Vip. 3, p. 80 – 84 (1992). In Ukrainian.
The total ozone anomalies for the Northern hemisphere stations' data (1961 – 1985 years) have been analysed by statistical methods. The quasi–biennial and the quasi–annual cycles have been investigated in detail. These cycles are concluded to be independent one from another but have similar nature. It is supposed that these cycles are a consequence of galactic cosmic ray variations.

082.057   Coagulation model of the middle atmosphere aerosols. Cosmic dust influx optimization.
V. N. Lebedinets, O. Kurbanmuradov.
Astron. Vestn., Tom 26, No. 3, p. 53 – 60 (May – Jun 1992). In Russian. English translation in Sol. Syst. Res., Vol. 26, No.3.
Modern evaluations of the general cosmic dust influx into the Earth's atmosphere M and of the mean dust particles density δ are analyzed. The model height profile of the atmospheric turbidity σ_a/σ_m (where σ_a, σ_m are aerosol and molecular light scattering coefficients) approximates to the observed one at $M \approx 1000$ t × day^{-1} and $\delta \approx 1$ g × cm^{-3}. The known light scattering layer at the height 50 km is formed by optimization of the mean aerosol dimensions for light scattering at this height.

082.058   On possible cometary origin of a background sulphate layer in the stratosphere.
V. N. Lebedinets.
Astron. Vestn., Tom 26, No. 3, p. 61 – 64 (May – Jun 1992). In Russian. English translation in Sol. Syst. Res., Vol. 26, No. 3.
A global balanced model is developed of sulphur brought by mini–comets into the middle atmosphere. It is shown that in the intervals between powerful volcano eruptions the sulphate aerosol Junge layer can be formed from mini–cometary sulphur. Required influx of cometary sulphur corresponds to a sulphur depletion relative to oxygen content (as compared to their space abundances) by a factor of 30.

082.059   Equatorial atomic oxygen profiles derived from rocket observations of O I 557.7 nm airglow emission.
D. Gobbi, H. Takahashi, B. R. Clemesha, P. P. Batista.
Planet. Space Sci., Vol. 40, No. 6, p. 775 – 781 (Jun 1992).
Two rocket observations of the O I λ557.7 nm nightglow emission have been carried out from the equatorial region at Natal (5.8°S, 35.2°W), one on 9 December 1985 and the other on 31 October 1986. The volume emission rate obtained from these observations showed a maximum at a height of 97 ± 1 km with a half–width of 6 km for the first experiment and 98 ± 1 km with a half–width of 7 km for the second experiment. The observed emission profiles showed large variability, suggesting dynamical control of the atomic oxygen. The atomic oxygen density profiles deduced from the present data are compared with atmospheric models and currently available experimental data.

082.060   Inverse problem of the photometry of solar radiation reflected by an optically thick planetary atmosphere. III. Remote sensing of minor gaseous constituents and atmospheric aerosol.
E. A. Ustinov.
Kosm. Issled., Tom 30, Vyp. 2, p. 212 – 225 (Mar–Apr 1992). In Russian. English translation in Cosmic Research, Vol. 30, No. 2.
Applications are considered of the inverse problem formulated earlier in a general form to problems of remote sensing of minor gaseous constituents and aerosol in planetary atmospheres using data of observations of reflected solar radiation. The radiative transfer equation is considered in a two–component medium consisting of the background molecular atmosphere and an atmospheric constituent being sounded. General expressions are obtained which describe variations of intensity of the outgoing radiation due to variations of the extinction coefficient and scattering phase function of the sounded constituent. These expressions are used to formulate the inverse problems of retrieval of minor gaseous constituent mixing ratio, volume extinction coefficient of scattering aerosol and microphysical parameters of aerosol particles. Expressions for weighting functions of corresponding linearized inverse problems are obtained.

082.061   Altitude profiles of the atmospheric system of O$_2$ and of the green line emission.
M. J. López–González, J. J. López–Moreno, R. Rodrigo.
Planet. Space Sci., Vol. 40, No. 6, p. 783 – 795 (Jun 1992).
Measurements of the Atmospheric System (AS) of the O$_2$ and green line of O(^{1}S) emission profiles have been analysed and

photochemical schemes chosen to explain these measurements. During night–time, both emissions can be mainly explained by energy transfer mechanisms following atomic oxygen recombination. The analysis of the emissions shows that the $O_2(b^1\Sigma_g^+)$ state could be produced via a precursor having the ratio of the rate coefficient of quenching by atomic oxygen to that by molecular oxygen equal to three, while the precursor of the $O(^1S)$ state would have a quenching ratio greater than five. The only available observation of the AS dayglow has been used to derive an altitude profile of the mesospheric ozone by the use of a photochemical scheme that includes O_2 photoabsorption and transfer from $O(^1D)$ as the main sources of $O_2(b^1\Sigma_g^+)$ during the daytime.

082.062   **Atmospheric fluctuations: empirical structure functions and projected performance of future instruments.**
M. Bester, W. C. Danchi, C. G. Degiacomi, L. J. Greenhill, C. H. Townes.
Astrophys. J., Vol. 392, No. 1, p. 357 – 374 (10 Jun 1992).

The extension of high quality astronomical observations towards larger apertures, adaptive optics, and infrared wavelengths leads to extrapolation of present knowledge of astronomical "seeing" by means of theoretical models, such as Kolmogorov turbulence combined with Taylor's "frozen atmosphere" swept past the observer by winds. Observations of path length fluctuations from a star to a two–telescope spatial interferometer at 11 μm wavelength, and also measurements of path length fluctuations 3 m above the ground by laser distance interferometers, show substantial deviations from such a model. Path lengths 3 m above the ground show for short times ($0.1\,s \leqslant t \leqslant 1\,s$) that under good seeing conditions fluctuations are approximated by a random walk, with a structure function of time proportional to t rather than $t^{5/3}$ as expected from the asymptotic approximation of a Kolmogorov–Taylor model. The implications of these measurements for current theoretical models of atmospheric turbulence are reviewed in this paper. In general, the results indicate that large–aperture telescopes or long baseline interferometry, particularly for IR wavelengths, will often provide better imaging than is expected on the basis of the common assumption that relative fluctuations in path lengths through the atmosphere increase with the 5/3 power of their separation. The results are also favorable for adaptive optics.

082.063   **The influence of the Pinatubo eruption on the atmospheric extinction at La Silla.**
H.–G. Grothues, J. Gochermann.
Messenger, No. 68, p. 43 – 44 (Jun 1992).

082.064   **Airglow – was ist das eigentlich?**
P. Riepe, S. Binnewies.
Sterne Weltraum, Jahrg. 31, Nr. 6, p. 410 – 412 (Jun 1992).

082.065   **A fresh analysis of some recent data on atmospheric refraction near the horizon with implications in archaeoastronomy.**
H. Exton.
Archaeoastronomy (U.K.), No. 17, p. S57 – S58 (1992).

082.066   **Optically resolved emission features produced in the atmosphere by a pulsed source of ultraviolet radiation.**
N. V. Eliseev, V. A. Kiselev, S. I. Kozlov.
Cosmic Res., Vol. 29, No. 6, p. 816 – 818 (May 1992). English translation of Kosm. Issled., Tom 29, Vyp. 6, p. 950 – 951 (1991).

082.067   **Temporal and spatial variations of the atmospheric diffuse light.**
S. M. Kwon, S. S. Hong, J. L. Weinberg.
IAU Colloquium No. 126: Origin and evolution of interplanetary dust, p. 179 – 182 (1991). – See Abstr. 012.068 for the main entry.

The Barbier's relation for the diffusely scattered airglow has been modified in such a way that it may describe, with simple changes of two parameter values, the dependence on zenith distance of the atmospheric diffuse light at any time of the night.

082.068   **Empirical image motion spectrum. I. Seeing quality and the atmospheric limitation on the accuracy of meridian observations.**
P. F. Lazorenko.
Kinematika Fiz. Nebesn. Tel, Tom 8, No. 3, p. 78 – 91, 99 (May–Jun 1992). In Russian. English translation in Kinematics Phys. Celest. Bodies, Vol. 8, No. 3.

The model presented in this paper conforms in general with the present–day knowledge on the turbulence spectrum and experimental data on the wind velocity and temperature fluctuations. The value of the constant c is proposed to be a measure of the seeing quality at an astronomical site, its value being defined by the intensity of turbulence. Some expressions are given that relate c to the measured image motion dispersion, telescope aperture, and frequency characteristics of image motion monitors. The estimates of c obtained for several plain sites lie within the interval $0.16 – 0.18''$, while for the best sites in highlands $c \approx 0.08''$. The model gives a possibility to evaluate the contribution of the atmosphere to the measurement error for large angles when the integration time $T > 40\,s$. It is possible to reach an observational error $\sigma(T) = 0.01 – 0.03''$ when the telescope operates in optimum seeing conditions and $T \approx 2$ min.

082.069   **Some aspects of Rayleigh and Mie scattering in the atmosphere over Pune.**
P. C. S. Devara, P. Ernest Raj.
Kodaikanal Obs. Bull., Vol. 11, p. 1 – 7 (1991). – See Abstr. 012.072 for the main entry.

Computations of aerosol mixing ratio $[(N_a + N_m)/N_m]$ have been made using the aerosol number density (N_a) estimated from the laser radar (lidar) observations carried out at the HTM, Pune and air molecular number density (N_m) derived from the radiometersonde data of temperature and pressure obtained from the IMD, Pune for the days of lidar observations. The results of the study of the seasonal variation in the vertical distribution (up to 5 km AGL) of aerosol mixing ratio obtained from the above observations collected for the two–year period, October 1986 – September 1988 are presented in this paper.

082.070   **Runaway greenhouse atmospheres: applications to Earth and Venus.**
J. F. Kasting.
US–USSR Workshop on Planetary Sciences, p. 234 – 245 (1991). – See Abstr. 012.075 for the main entry.

Runaway greenhouse atmospheres are discussed. The critical solar flux required to trigger a runaway greenhouse is at least 1.4 times the solar flux at Earth's orbit (S_o). Rapid water loss may occur, however, at as little as 1.1 S_0, from a type of atmosphere termed a "moist greenhouse." The moist greenhouse model provides the best explanation for loss of water from Venus, if Venus did indeed start out with a large amount of water. The present enrichment in the D/H ratio on Venus provides no unambiguous answer as to whether or not it did. A runaway greenhouse (or "steam") atmosphere may have been present on the Earth during much of the accretion process. Evidence from neon isotopes supports this hypothesis and provides some indication for how long a steam atmosphere may have lasted. Finally, the theory of runaway and moist greenhouse atmospheres can be used to estimate the position of the inner edge of the continuously habitable zone around the Sun. Current models place this limit at about 0.95 AU.

082.071   **On the accuracy of correction for atmospheric extinction in spectrophotometric observations of stars.**
G. A. Terez, Eh. I. Terez.
Kinematics Phys. Celest. Bodies, Vol. 7, No. 5, p. 61 – 66 (1991). English translation of Kinematika Fiz. Nebesn. Tel., Tom 7, No. 5, p. 68 – 74 (Sep–Oct 1991).

Numerical modeling is used to analyze the possible errors of determination of atmospheric extinction in spectrophotometric star observations. A relation is obtained for the final rms errors of extraatmospheric magnitudes as function of the number of stars observed during a night and their relative positions.

082.072   **Astroclimate features of an observing station in Bolivia.**
G. A. Alekseeva, A. A. Arkharov, V. D. Galkin,
V. V. Novikov.
Kinematics Phys. Celest. Bodies, Vol. 7, No. 6, p. 40 – 41 (1991).
English translation of Kinematika Fiz. Nebesn. Tel., Tom 7,
No. 6, p. 71 – 73 (Nov–Dec 1991).
    Analysis of spectrophotometer data showed that an observing station near Tarija (the Bolivian–Soviet Observatory) has a moderate astroclimate. Clear nights dominate in the fall and winter. The values of the extinction coefficient for the wavelength $\lambda 550$ nm vary in the range $0.13 - 0.46^m$. The amount of water vapor varies from 5 to 30 mm of precipitable water, and reaches 15 – 20 mm even in the winter. A sharp increase in the extinction coefficient with increasing amount of water in the atmosphere is noted.

082.073   **Ground–based and balloon–borne measurements of atmospheric ozone.**
Wang Gengchen, Wang Yingjian.
Advances in solar–terrestrial science of China, p. 223 – 234 (1992). – See Abstr. 003.069 for the main entry.
    A comprehensive review on the present status of detection techniques for atmospheric ozone in China is given in this paper.

082.074   **20 years with electron beam experiments on rockets and satellites.**
B. N. Maehlum.
10. ESA Symposium on European Rocket and Balloon Programmes and Related Research, p. 55 – 61 (Nov 1991). – See Abstr. 012.088 for the main entry.
    Rocket– and satellite–borne electron accelerators have been utilized for some 20 years for conducting "controlled" simulation studies of processes in the upper atmosphere and in the magnetosphere. Investigations of the "natural" processes in space have been conducted by this method as well as basic studies of the interaction between beams of fast electrons and the plasma environment. In this paper highlights from these studies and some of the experimental difficulties are discussed.

082.075   **The excitation of the Chamberlain and Herzberg II bands in the terrestrial nightglow.**
D. P. Murtagh, J. Stegman, G. Witt.
10. ESA Symposium on European Rocket and Balloon Programmes and Related Research, p. 127 – 130 (Nov 1991). – See Abstr. 012.088 for the main entry.
    The authors' previous nightglow model has now been augmented to include the Chamberlain and Herzberg II emissions. The Chamberlain data derive from two rocket measurements of the volume emission profile of the 5–2 band combined with ground based spectroscopic measurements of the nightglow in the near UV. The Herzberg II data come solely from the ground based measurements. The excitation mechanism that is consistent with the data involves only direct excitation through the three body association of oxygen atoms. This is in contrast to the Atmospheric band and O I green line emissions. Empirical parameters for the quenching of the two upper states have been derived under the assumption that only O and O_2 are responsible.

082.076   **Night sky brightness from visual observations. II. A visual photometer.**
A. R. Upgren.
News Lett. Astron. Soc. N.Y., Vol. 4, No. 1, p. 11 (Feb 1992). Abstract. – See Abstr. 012.078 for the main entry.

082.077   **Night sky spectrum at Okayama Astrophysical Observatory.**
M. Iye, E. Nishihara, H. Sugai.
Rep. Natl. Astron. Obs. Jpn., Vol. 1, No. 3, p. 221 – 228 (1992). In Japanese.
    Spectroscopic CCD observations of night sky emission at Okayama Astrophysical Observatory were carried out in December 1988 and in April 1991 using the Cassegrain spectrograph mounted on the 188 cm telescope. It is emphasized that although the site is heavily contaminated by city light at the optical band

below 7000 Å, the dominant light source at the near infrared band beyond 7000 Å is the atmospheric emission lines of OH radicals. This fact shows that the observation at Okayama Astrophysical Observatory in the near infrared band is not so heavily handicapped in terms of the sky brightness as the observation in the optical wavelength band. The spectra in the 9000 Å band of a Seyfert galaxy I Zw 1 and a star forming galaxy Mrk 1259 are shown to illustrate the results of elimination of the atmospheric emission lines.

082.078   **Light scattering by nonspherical ice grains: an application to noctilucent cloud particles.**
M. I. Mishchenko.
Earth, Moon, Planets, Vol. 57, No. 3, p. 203 – 211 (Jun 1992).
    The light scattered by noctilucent cloud particles is nearly fully polarized at scattering angles in the vicinity of 90°. This was one of the reasons to conclude that the upper limit of their sizes is not larger than about 0.12 μm. Nevertheless, this estimate was made on the basis of the Mie scattering theory for spherical particles, whereas many investigators noted usefulness of highly aspherical shapes of noctilucent cloud particles. In this paper, the author used rigorous light scattering theory for randomly oriented nonspherical particles to calculate the degree of linear polarization of the scattered light for ice grains of different shape. By comparing these calculations with rocket polarization measurements of noctilucent clouds, he shows that, as for spherical particles, the upper limit of particle equal–volume radii for slightly flattened and elongated grains is of about 0.12 μm, while for highly aspherical plate–like and needle–like particles this upper limit is substantially larger and is of about 0.18–0.20 μm. The author also reports calculations of the volumetric scattering cross–section for particles of different shape and shows that randomly oriented spheroids have (slightly) smaller scattering cross section per unit particle mass than equal–volume spherical grains. Nevertheless, if in noctilucent clouds plate–like and needle–like grains grow to much larger sizes than spherical particles, their scattering efficiency may be much greater.

082.079   **Coupling stratosphere–lower ionosphere at middle latitudes.**
B. A. de la Morena, F. Caballero, A. Giménez.
10. ESA Symposium on European Rocket and Balloon Programmes and Related Research, p. 419 – 422 (Nov 1991). – See Abstr. 012.088 for the main entry.
    A study is presented about the coupling of the stratosphere and the lower ionosphere during winter periods, supported by analysis of wind, stratospheric temperature and ionospheric absorption in the D region, obtained at El Arenosillo in Southern Spain (37.1 N, 6.7 W). Superloki datasondes with rocket launchings and A–3 absorption measuring equipments were used. A clear difference between the coupling phenomena shown in soutwest and central Europe is revealed by this study.

082.080   **Rocket–borne infrared measurements in the arctic upper atmosphere.**
K. U. Grossmann, D. Homann, J. Schulz.
10. ESA Symposium on European Rocket and Balloon Programmes and Related Research, p. 423 – 427 (Nov 1991). – See Abstr. 012.088 for the main entry.
    In spring 1990 and 1991 and in summer 1990 the SISSI program (Spectroscopic Infrared Structure Signatures Investigation) was carried out over Esrange, Kiruna. Four Skylark payloads were successfully launched. The main experiment of each payload was a liquid helium cooled infrared spectrometer which measured the thermal infrared emission from the arctic mesosphere and lower thermosphere. The recorded spectra will be used to determine trace gas densities in the mesosphere and to study excitation processes in the mesopause region and above.

082.081   **The atmospheric extinction at Yonsei University Observatory in 1982 – 89.**
J. H. Jeong.
Proc. Astron. Soc. Aust., Vol. 9, No. 2, p. 320 (1991). – See Abstr. 012.090 for the main entry.
    More than 1000 coefficients of atmospheric extinction in the U, B and V passbands were obtained as a secondary result of long–

term photometric monitoring of stellar objects at Yonsei University Observatory during 1982 – 1989. The long–term variation of the extinction coefficients of each passband are presented in this paper.

Solar–terrestrial models and application software.
See Abstr. 002.033.

Diffraction effects in semiclassical scattering.
See Abstr. 003.058.

7. Quadrennial Symposium on Solar–Terrestrial Physics (SCOSTEP–7), The Hague (Netherlands), 25 – 30 Jun 1990.
See Abstr. 012.024.

18. Optical Society of India Symposium on Optical Science and Engineering, Bangalore (India), 21 – 23 Mar 1990.
See Abstr. 012.072.

Proceedings of the 10th ESA Symposium on European Rocket and Balloon Programmes and Related Research.
See Abstr. 012.088.

The protection of astronomical and geophysical sites: general introduction.
See Abstr. 013.028.

Light pollution.
See Abstr. 013.029.

Radio–interference.
See Abstr. 013.030.

Millimeter band radio astronomy.
See Abstr. 013.031.

Pollution of geophysical sites.
See Abstr. 013.032.

U.S. ground–based space research programs in Svalbard.
See Abstr. 013.109.

Cusp and cleft studies in Greenland.
See Abstr. 013.112.

Airglow observation program carried out by Tokyo Astronomical Observatory.
See Abstr. 013.139.

Laboratory studies of water vapor absorption in the atmospheric window at 213 GHz.
See Abstr. 022.049.

High resolution absorption cross sections in the transmission window region of the Schumann–Runge bands and Herzberg continuum of O_2.
See Abstr. 022.072.

Optical manifestation of microbursts of electron fluxes.
See Abstr. 022.091.

Polynomial coefficients for calculating O_2 Schumann–Runge cross sections at 0.5 cm^{-1} resolution.
See Abstr. 022.099.

$O(^1S)$ and $O(^1D)$ quantum yields from rocket measurements of electron densities and 557.7 and 630.0 nm emissions in the nocturnal F–region.
See Abstr. 022.173.

$N(^2D) + O_2$: a source of thermospheric 6300 Å emission?
See Abstr. 022.175.

Prediction of atmospherically induced wave–front degradations.
See Abstr. 031.011.

De atmosfeer overwonnen.
See Abstr. 031.013.

Adaptive optics using curvature sensing.
See Abstr. 031.053.

Optical performance of large ground–based telescopes.
See Abstr. 031.062.

Water tunnel test of telescope enclosure models.
See Abstr. 032.027.

Application of the intensified CCD to airglow and auroral measurements.
See Abstr. 034.037.

A cloud detector for automated telescopes.
See Abstr. 034.124.

Balloon–borne solar occultation Fourier transform spectrometry for measurements of stratospheric trace species.
See Abstr. 035.120.

Remote sensing of trace gases with a balloon borne version of the Michelson interferometer for passive atmospheric sounding (MIPAS).
See Abstr. 035.121.

A light UV–visible spectrometer for atmospheric composition measurements by solar occultation.
See Abstr. 035.122.

On the correction of stellar spectra for the loss of radiation during its passage through the Earth's atmosphere and through the spectrograph–slit.
See Abstr. 036.023.

Observations at (sub)millimetre wavelengths: effect of atmosphere and telescope sidelobes.
See Abstr. 036.067.

On flux calibration of spectra.
See Abstr. 036.135.

Anisoplanatism and use of laser guide stars.
See Abstr. 036.177.

Laser guide stars for adaptive optics systems: Rayleigh scattering experiments.
See Abstr. 036.178.

Partially compensated speckle imaging: Fourier phase spectrum estimation.
See Abstr. 036.184.

Atmospheric turbulence sensing for a multiconjugate adaptive optics system.
See Abstr. 036.186.

A proposal for the seeing measurement of the site testing.
See Abstr. 036.187.

Polar motion, atmospheric angular momentum excitation and earthquakes – correlations and significance.
See Abstr. 044.001.

The Earth's angular momentum budget on subseasonal time scales.
See Abstr. 044.003.

The characteristic of the 30 – 60 day fluctuation in the Earth rotation, atmospheric angular momentum and solar activity.
See Abstr. 044.031.

Re–entry aerodynamics derived from space debris trajectory analysis.
See Abstr. 052.018.

A note on the enhancement of J values in optically thick scattering atmospheres.
See Abstr. 063.006.

Analytical expressions for radiative properties of planar Rayleigh scattering media, including polarization contributions.
See Abstr. 063.048.

The contribution of third order linear absorption to the water vapor continuum.
See Abstr. 063.049.

Atmospheric effect on the upwelling radiation at the top of the atmosphere over a stream.
See Abstr. 063.107.

Diurnal variation of the linear polarization across the Hβ Fraunhofer line of the terrestrial atmosphere. I.
See Abstr. 071.016.

Diurnal variation of the linear polarization across the Hα and Hβ Fraunhofer lines of the terrestrial atmosphere, and a detection of a "daylight flash". II.
See Abstr. 071.017.

Photospheric lines redshift from balloon ultraviolet spectra of the quiet Sun.
See Abstr. 071.027.

Intensity oscillations of the Fe XIV ($\lambda = 530.3$ nm) solar corona: disturbances due to the wave processes in the Earth's atmosphere.
See Abstr. 074.013.

Solar EUV/UV and equatorial airglow measurements from San Marco–5.
See Abstr. 076.014.

A study of the astronomical theory of ice ages in a two–dimensional nonlinear climate model.
See Abstr. 081.028.

Interactive ionosphere modeling: a comparison between TIGCM and ionosonde data.
See Abstr. 083.014.

Numerical modelling of the thermosphere–ionosphere–protonosphere system.
See Abstr. 083.015.

Solar activity control of ionospheric and thermospheric processes.
See Abstr. 083.016.

Quasi–periodic fluctuations in ionospheric absorption in relation to planetary activity in the stratosphere.
See Abstr. 083.018.

The ionospheric response to the atmospheric quasi–biennial oscillation.
See Abstr. 083.038.

Auroral energy deposition rate.
See Abstr. 084.023.

Earth's atmosphere: terrestrial or extraterrestrial?
See Abstr. 091.022.

Isolation of major Venus thermospheric cooling mechanism and implications for Earth and Mars.
See Abstr. 093.009.

Influence of aerodynamic roughness length on aeolian processes: Earth, Mars, Venus.
See Abstr. 097.037.

Fine resolution brightness distribution of the visible zodiacal light.
See Abstr. 106.079.

The role of impacting processes in the chemical evolution of the atmosphere of primordial Earth.
See Abstr. 107.060.

Atmospheric gamma–ray spectrum between 10 and 100 MeV.
See Abstr. 143.070.

The geophysical effects of cosmic rays.
See Abstr. 144.073.

The additional fluxes of cosmic rays in the stratosphere in the various half–periods of the 22–year solar magnetic cycle.
See Abstr. 144.080.

083 Ionosphere

083.001 **The role of ion drift in the formation of ionisation troughs in the mid– and high–latitude ionosphere – a review.**
A. S. Rodger, R. J. Moffett, S. Quegan.
J. Atmos. Terr. Phys., Vol. 54, No. 1, p. 1 – 30 (Jan 1992).
F–layer ionisation troughs are frequently observed in the sub–auroral and high–latitude ionospheres. The authors define the mid–latitude trough as the region of low plasma concentration of F–region altitudes that occurs near the equatorward side of the low latitude edge of the energetic electron precipitation boundary of the auroral oval. High latitude troughs are simply defined as troughs that occur in the auroral oval and polar cap. The authors review the progress that has been made in describing the

phenology and morphology of the mid–latitude trough since the review by Moffett and Quegan.

083.002 **Ionospheric effects of the solar eclipse of September 23, 1987, around the equatorial anomaly crest region.**
K. Cheng, Y.–N. Huang, S.–W. Chen.
J. Geophys. Res., Vol. 97, No. A1, p. 103 – 112 (1 Jan 1992).
The ionospheric responses to the solar eclipse of September 23, 1987, in the equatorial anomaly crest region have been investigated by using ionospheric vertical sounding, VLF propagation delay time, and differential Doppler shift data observed at Chungli, which is located near the northern equatorial anomaly crest region.

083.003 **Alfvén waves in the auroral ionsophere: a numerical model compared with measurements.**
D. J. Knudsen, M. C. Kelley, J. F. Vickrey.
J. Geophys. Res., Vol. 97, No. A1, p. 77 – 90 (1 Jan 1992).

The authors solve a linear numerical model of Alfvén waves reflecting from the high–latitude ionosphere, both to better understand the role of the ionosphere in the magnetosphere/ ionosphere coupling process and to compare model results with in situ measurements. They use the model to compute the frequency–dependent amplitude and phase relations between the meridional electric and the zonal magnetic fields due to Alfvén waves. These relations are compared with measurements taken by an auroral sounding rocket flown in the morningside oval and by the HILAT satellite traversing the oval at local noon.

083.004 **Ionospheric simulation compared with Dynamics Explorer observations for November 22, 1981.**
J. J. Sojaka, M. Bowline, R. W. Schunk, J. D. Craven,
L. A. Frank, J. R. Sharber, J. D. Winningham, L. H. Brace.
J. Geophys. Res., Vol. 97, No. A2, p. 1245 – 1256 (1 Feb 1992).

Dynamics Explorer (DE) 2 electric field and particle data have been used to constrain the inputs of a time–dependent ionospheric model (TDIM) for a simulation of the ionosphere on November 22, 1981. The simulated densities have then been critically compared with the DE 2 electron density observations. This comparison uncovers a model – data disagreement in the morning sector trough, generally good agreement of the background density in the polar cap and evening sector trough, and a difficulty in modelling the observed polar F layer patches. From this comparison, the consequences of structure in the electric field and precipitation inputs can be seen. This is further highlighted during a substorm period for which DE 1 auroral images were available. Using these images, a revised dynamic particle precipitation pattern was used in the ionosphere model; the resulting densities were different from the original simulation. With this revised dynamic precipitation model, improved density agreement is obtained in the auroral/polar regions where the plasma convection is not stagnant.

083.005 **Response of photoionization rates and electron densities to solar activity in the 100 – 200 km height region.**
Chen Yao–wu, Xu Xiu–juan, R. J. Hung.
Acta Geophys. Sin., Vol. 35, No. 2, p. 142 – 149 (Mar 1992). In Chinese.

By using the current knowledge about solar EUV radiation fluxes, neutral thermospheric structure, absorption and ionization cross sections, the response of photoionization rate and electron density varying with geometric height and solar zenith angle to solar activity in the 100 – 200 km height region of the atmosphere were calculated. The characteristics of the depth and width of the $E–F_1$ valley were analysed.

083.006 **Estimation of the height of the main ionospheric maximum by the analytical extrapolation method.**
G. I. Ostrovskij.
Geomagn. Aehron., Tom 32, No. 1, p. 159 – 160 (Jan – Feb 1992). In Russian. English translation in Geomagn. Aeron., Vol. 32, No. 1.

083.007 **Incoherent scatter observations of the E and F regions.**
H. Rishbeth, B. S. Lanchester.
Planet. Space Sci., Vol. 40, No. 2/3, p. 355 – 372 (Feb–Mar 1992).

This paper outlines the principles of the incoherent scatter technique, and briefly reviews the ways in which it has been used to study the ionospheric E and F regions. Examples are given of the important scientific advances that have been made over the last thirty years.

083.008 **A study of rocket measurements of ionospheric currents. I. General setting and night–time ionospheric currents.**
C. A. Onwumechili.
Geophys. J. Int., Vol. 108, No. 2, p. 633 – 640 (Feb 1992).

Some 76 results of rocket measurements of ionospheric currents worldwide from 1948 to 1973, have been collected and arranged into seven groups suggested by the characteristic of geomagnetic variations which they cause. Their organization and planned study in five parts are outlined. This first part also presents the study of the group of rocket measurements of night–time ionospheric currents. The finding from nine rocket flight results is that the attempts to detect night–time ionospheric currents with rocket–borne instruments in the low and mid–latitudes have not found distinctly clear indication of such currents. The remnant E–region electron density at night measured aboard two rockets led to calculated currents so small that they support the above finding. Discussion of the rocket and other relevant studies led to the conclusion that nocturnal ionospheric currents do not seem to exist at any part of the solar cycle.

083.009 **A study of rocket measurements of ionospheric currents. II. Ionospheric currents outside the dip equatorial zone.**
C. A. Onwumechili.
Geophys. J. Int., Vol. 108, No. 2, p. 641 – 646 (Feb 1992).

This second part of the series has studied ionospheric currents outside the dip equatorial zone in the three groups of: high–latitude currents beyond 70° latitude (Group 6), daytime currents within 35° to 70° dip latitude (Group 5), and daytime currents with 7° to 35° dip latitude (Group 4). Ionospheric currents have been found and measured with rockets in all three groups, and their directions are in conformity with expectations from geomagnetic variations: eastwards equatorward of the Sq focus (Group 4), westwards between the Sq focus and high latitudes (Group 5), and complex in the higher and polar latitudes (Group 6). Of great interest is the discovery that the currents occur in the two eastward layers in Group 4 but only in one layer in Groups 5 and 6.

083.010 **A study of rocket measurements of ionospheric currents. III. Ionospheric currents at the magnetic dip equator.**
C. A. Onwumechili.
Geophys. J. Int., Vol. 108, No. 2, p. 647 – 659 (Feb 1992).

The 31 daytime rocket flight results within 0° or 2° dip latitude are presented and discussed.

083.011 **A study of rocket measurements of ionospheric currents. IV. Ionospheric currents in the transition zone and the overview of the study.**
C. A. Onwumechili.
Geophys. J. Int., Vol. 108, No. 2, p. 660 – 672 (Feb 1992).

The study of rocket measurements suggests the following conclusions. (1) Ionospheric currents are basically in two layers. (2) The upper current layer with a steady altitude extent of 18 ± 3 km is global and should be regarded as the worldwide part of the Sq current system. (3) The intense eastward lower current layer at the magnetic dip equator that terminates in a focus within 2° to 4° dip latitude, together with the reverse westward lower current layer that peaks around $5.3 \pm 0.7°$ dip latitude, should be regarded as the equatorial electrojet (EEJ). (4) The EEJ and the worldwide part of the Sq current system are coupled, sometimes overlapping in varying degrees, especially within 0° to 0.5° dip latitude where the overlap often leads to a hybrid current system. (5) The lunar geomagnetic variations result from the modulations of the above current layers and not from a separate current layer.

083.012 **A study of rocket measurements of ionospheric currents. V. Modelling rocket profiles of low–latitude ionospheric currents.**
C. A. Onwumechili.
Geophys. J. Int., Vol. 108, No. 2, p. 673 – 682 (Feb 1992).

Some 25 rocket profiles of the current density of the eastward ionospheric lower current layer close to the magnetic dip equator, one profile of the reverse westward lower current layer at about 5° dip latitude and five profiles of the eastward upper current layer within 0° to about 8° dip latitude have been modelled and are discussed in detail.

083.013 A confirmation of the validity of the electric current distribution determined by a ground–based magnetometer network.
S.-I. Akasofu.
Geophys. J. Int., Vol. 109, No. 1, p. 191 – 196 (Apr 1992).

Determination of the distribution of electric currents at the ionospheric level by a ground–based magnetometer network has been sought for by our pioneers, K. Birkeland and S. Chapman, from the beginning of this century. Both recognized difficulties associated with the determination. However, during the last decade, we have overcome the difficulties by establishing several meridian chains of magnetic observatories and developing several computer codes to invert the magnetic data. On the other hand, it has been difficult to confirm the current distribution thus obtained by a completely independent method. It is fortunate that an opportunity for the independent determination has arisen by a recent satellite–based observation. It is shown that both the ground–based and satellite–based results agree fairly well with each other. Thus, a ground–based magnetometer network has been proven to be able to determine the 3–D current distribution which couples the magnetosphere and the ionosphere.

083.014 Interactive ionosphere modeling: a comparison between TIGCM and ionosonde data.
M. V. Codrescu, R. G. Roble, J. M. Forbes.
J. Geophys. Res., Vol. 97, No. A6, p. 8591 – 8600 (1 Jun 1992).

Results from a time dependent geomagnetic simulation of the coupled thermosphere and ionosphere using the new interactive thermosphere – ionosphere general circulation model (TIGCM) (Roble et al., 1988) of the National Center for Atmospheric Research are compared with F2–layer data obtained from a latitudinal chain of East Asian ionosonde Stations situated close to the –165° magnetic meridian and separated by about 5° in magnetic latitude. This is among the first extended comparisons (10 days) between the TIGCM modeled ionosphere and data, where the effects of neutral dynamics on the ionosphere are studied using a global, fully interactive thermosphere–ionosphere model.

083.015 Numerical modelling of the thermosphere–ionosphere–protonosphere system.
A. A. Namgaladze, Yu. N. Korenkov, V. V. Klimenko, I. V. Karpov, V. A. Surotkin, N. M. Naumova.
J. Atmos. Terr. Phys., Vol. 53, No. 11/12, p. 1113 – 1124 (Nov–Dec 1991). – See Abstr. 012.024 for the main entry.

The results of a numerical calculation of the thermospheric circulation, the ion and electron concentrations and temperatures as well as electric fields of magnetospheric and dynamo origin are presented for solar minimum, low geomagnetic activity and spring (equinox) conditions. The results are obtained from a numerical hydrodynamical model of the Earth's ionosphere and protonosphere for the height range from 80 km to a geocentric distance of 15 R_E. The concentrations and temperature of the neutral atmosphere are computed from the MSIS–83 empirical model of the thermosphere. The effects of small electric fields on the global distributions of the thermospheric circulation, concentrations and temperatures of the charged particles in the ionosphere and protonosphere are discussed. It is shown that electric fields smaller than even $20 \, mV \, m^{-1}$ form all the main structures of the global plasma distribution at F–region altitudes and in the protonosphere, such as the main ionospheric trough, the light ion trough, the plasmapause and "hot spots" in the ion and electron temperature distributions.

083.016 Solar activity control of ionospheric and thermospheric processes.
N. Jakowski, B. Fichtelmann, A. Jungstand.
J. Atmos. Terr. Phys., Vol. 53, No. 11/12, p. 1125 – 1130 (Nov–Dec 1991). – See Abstr. 012.024 for the main entry.

Ionospheric electron content observations carried out in Neustrelitz (53.3°N, 13.1°E) and Havana, Cuba (23.1°N, 82.5°W) are analyzed with respect to their solar activity dependence. Cross correlation studies with the solar radio flux at 10.7 cm wavelength indicate a response time of the ionosphere to the

27–day solar radiation cycle of the order of 1–2 days. It is assumed that this time shift is caused by solar radiation control of the atomic oxygen concentration in the thermosphere which follows the change of the solar radiation due to photodissociation of molecular oxygen, but with a time lag of several days. This hypothesis is discussed using a one–dimensional numerical model which includes photochemistry and diffusion processes. To a first approximation, temperature changes have been neglected and use is made of a constant temperature profile according to CIRA 72. The results demonstrate a time lag of the atomic oxygen concentration variation of 2 days with respect to the solar radiation variation. The 27–day run of the model indicates changes in the concentration of atomic oxygen by about 7% at 180 km height.

083.017 27–day fluctuations in the ionospheric D–region.
D. Pancheva, R. Schminder, J. Laštovička.
J. Atmos. Terr. Phys., Vol. 53, No. 11/12, p. 1145 – 1150 (Nov–Dec 1991). – See Abstr. 012.024 for the main entry.

The 27–day fluctuations in ionospheric absorption of radio waves along three radio paths in Central and Southeastern Europe are investigated over the interval 1982–1988 together with analogous fluctuations in the solar Lyman–α flux and in the prevailing wind in the lower thermosphere. The 27–day fluctuations in the lower ionosphere are of direct solar origin only if the Lyman–α flux exhibits a very well expressed solar rotation variation. The absorption fluctuations are largest in winter near solar activity minimum, in fair coincidence with the maxima of corresponding fluctuations in zonal and particularly meridional winds. This indicates a dynamical forcing (maybe of solar origin).

083.018 Quasi–periodic fluctuations in ionospheric absorption in relation to planetary activity in the stratosphere.
D. Pancheva, J. Laštovička, B. A. de la Morena.
J. Atmos. Terr. Phys., Vol. 53, No. 11/12, p. 1151 – 1155 (Nov–Dec 1991). – See Abstr. 012.024 for the main entry.

Quasi–periodic fluctuations (2–15 days) in ionospheric absorption along three radio paths in Central and Southern Europe are investigated for three winter periods 1985/86, 1986/87 and 1987/88. The periods of dominant fluctuations in absorption and the time variations of their amplitudes are similar for all radio paths and every winter period. The shorter–period fluctuations are found to be associated with enhancements of planetary wave to activity in the stratosphere (30 hPa, 60°N), while longer period fluctuations appear to be associated with intensification of planetary wave on activity in the stratosphere.

083.019 An observing and analyzing for ionospheric effect during the period of the solar flare.
Liu Wantong, Huang Zhen, Zheng Ruimin.
Acta Geophys. Sin., Vol. 34, No. 5, p. 651 – 656 (Sep 1991). In Chinese.

From 31 Jul to 9 Sep, 1989, VLF Omega signals have been received at the following three sites: Xinxiang, Lintong and Urumqi respectively. At the same time, the chromosphere observation and the ionosphere surveying were also doing. A total of 36 activities were recorded simultaneously at three sites. The solar flare of 16 Aug resulted in interrupt of radio telecommunication via ionosphere, which lasted for more than 4 h. The paper analyzes the ionosphere effect of these flare events by synthetically utilizing VLF observation data of the three sites, the chromosphere data, ionosphere surviving data and X–ray data of satellite GOES–7.

083.020 Increases of electron density in the upper auroral ionosphere observed by the "Prognoz–5" spacecraft.
V. P. Grigor'eva, V. V. Pisareva.
Kosm. Issled., Tom 30, Vyp. 1, p. 84 – 88 (Jan–Feb 1992). In Russian. English translation in Cosmic Res., Vol. 30, No. 1.

Regions of increased electron densities in the upper ionosphere and magnetosphere are fixed by the spectrum analyser installed on board the "Prognoz–5" spacecraft. The occurrence of these regions is associated mainly with the ionospheric electron jet appeared under the action of ultraviolet solar radio emission.

083.021   Equatorial electrojet parameters and the relevance of electromagnetic drifts (EMD) over Thumba.
A. C. Balachandra Swamy.
Astrophys. Space Sci., Vol. 191, No. 2, p. 203 – 211 (May 1992).

The thermal imbalance in the E–region heights is inescapable and an undisputed fact. The equatorial electrojet parameters are being evaluated for the Indian Equatorial Station of Thumba, making use of the observed electron density, electron temperature, and current density profiles for two rocket flights on 3 March, 1973 and 7 April, 1972 around local noon, corresponding to low and medium solar active conditions. The computed Joule heating due to equatorial electrojet current (EEC) does not account for the observed difference between the electron and neutral gas temperatures. The discrepancy of about 6 km in the peaks of the observed and computed current density profiles may be attributed to the presence of the electromagnetic drifts (EMD). In order to see whether or not EMD plays an important role, the photoionization balance between production and loss rates have been computed by making use of the latest available solar flux and cross sections and chemical reaction rate constants for the appropriate solar epoch conditions including the transport term due to EMD. There is an excellent agreement between the observed and computed electron density profiles indicating its relevance.

083.022   Capabilities of a phase–difference method for analysis of ionospheric structural inhomogeneities.
A. I. Agaryshev, Yu. B. Ivanov, I. A. Shemetov.
Geomagn. Aeron., Vol. 31, No. 1, p. 86 – 89 (Aug 1991). English translation of Geomagn. Aehron., Tom 31, No. 1, p. 120 – 125 (1991).

083.023   Modeling of the electron temperature in the E–region of the auroral ionosphere.
I. O. Voronkov, A. V. Shirochkov.
Geomagn. Aeron., Vol. 31, No. 1, p. 95 – 97 (Aug 1991). English translation of Geomagn. Aehron., Tom 31, No. 1, p. 133 – 136 (1991).

A nonstationary model of the space–time distribution of the electron temperature at heights of 95 – 145 km is described; the model takes into account the processes of collisional energy release of the accelerating electron flux, Joule heating, heating resulting from the development of double–flow instability, and energy exchange of electrons with neutral molecules. The limits of applicability of the approximation of thermodynamic equilibrium are pointed out and the relative contributions of the various sources of heating are analyzed. A comparison is made of the calculations and the data obtained from the EISCAT incoherent scattering installation.

083.024   Thin walls of ionization inhomogeneities in the polar ionosphere observed by satellite radio–sounding.
N. P. Danilkin, S. V. Zhuravlev, L. P. Morozova,
V. I. Pogorelov, K. L. Tol'skij.
Geomagn. Aeron., Vol. 31, No. 1, p. 98 – 101 (Aug 1991). English translation of Geomagn. Aehron., Tom 31, No. 1, p. 137– 142 (1991).

Experimental radio–sounding data on irregular ionosphere inhomogeneities of a specific type ("walls") from the Kosmos–1809 satellite are analyzed. Conclusions are reached regarding their geometrical dimensions, internal structure, and indicatrix of radio–scattering by plasma elements forming the inhomogeneities.

083.025   Nitric oxide and lower ionosphere quantities during solar particle events of October 1989 after rocket and ground–based measurements.
A. M. Zadorozhny, G. A. Tuchkov, V. N. Kikhtenko,
J. Laštovička, J. Boška, A. Novák.
J. Atmos. Terr. Phys., Vol. 54, No. 2, p. 183 – 192 (Feb 1992).

The most dramatic demonstrations of solar activity are solar proton flares. One such very strong flare, accompanied by a solar proton event (SPE) and a large ground level enhancement of cosmic rays on Earth, was observed in October 1989. During this SPE, ion density and nitric oxide concentration profiles were measured by rockets launched from the Soviet research vessel 'Akademik Shirshov' in the southern part of the Indian Ocean. The rocket experiment yielded the first in–situ measurement of NO concentration increased by SPE. The NO concentrations estimated from ion–pair production rates due to measured fluxes of high energy particles agree fairly well with the observed NO concentrations in the stratopause region. The results of rocket measurements are compared with measurements of the radio wave absorption in the lower ionosphere performed at similar latitudes in central Europe. Model calculations of absorption show that while the night–time enhancement of absorption can be explained by increased electron density related to the measured increase of ion density as a consequence of enhanced penetration of high energy particles, the daytime increase of absorption needs to be explained mainly in terms of the observed increase of nitric oxide concentration.

083.026   The appropriate dyadic applicable to the lower region of the ionospheric plasma.
S. S. De, S. Bandyopadhyay, S. K. Dubey, B. K. Sarkar,
A. C. Sen, S. K. Adhikari.
Earth, Moon, Planets, Vol. 56, No. 1, p. 1 – 6 (Jan 1992).

An expression for the susceptibility dyadic appropriate to the lower regions of the atmospheric plasma is derived using Maxwell's field equations and the equation of conservation of momentum. The contributions due to viscous effect and convection current density are incorporated in the physical processes within the stated medium. Utilizing the approximation of linearized equations, second order coupled wave equations have been derived through the dyadic.

083.027   Oscillations of the alternating magnetic field of the Earth with a period of about 1.5^h.
A. V. Aleksandrov, V. S. Bychkov, I. A. Larin, I. V. Komkov.
Geomagn. Aehron., Tom 32, No. 3, p. 119 – 124 (May – Jun 1992). In Russian. English translation in Geomagn. Aeron., Vol. 32, No. 3.

083.028   Field–aligned flows of H$^+$ and He$^+$ in the mid–latitude topside ionosphere at solar maximum.
G. J. Bailey, R. Seller.
Planet. Space Sci., Vol. 40, No. 6, p. 751 – 762 (Jun 1992).

A time–dependent mathematical model of the Earth's ionosphere and plasmasphere has been used to investigate the field–aligned flows of H$^+$ and He$^+$ in the topside ionosphere at L = 3 during solar maximum.

083.029   Thermosphere–ionosphere interaction in a period of ionosphere storms.
A. D. Danilov, D. D. Belik.
Geomagn. Aeron., Vol. 31, No. 2, p. 157 – 167 (Oct 1991). English translation of Geomagn. Aehron., Tom 31, No. 2, p. 209 – 222 (1991).

083.030   Modeling of the dynamics of electron density profiles in the E–region of the auroral ionosphere.
I. O. Voronkov, A. V. Shirochkov, L. N. Makarova,
K. Schlegel.
Geomagn. Aeron., Vol. 31, No. 2, p. 228 – 230 (Oct 1991). English translation of Geomagn. Aehron., Tom 31, No. 2, p. 303 – 307 (1991).

It is shown, on the basis of data obtained with the EISCAT incoherent scattering station and the nonstationary model of the auroral E–region of the ionosphere, which makes it possible to calculate the electron and ion density profiles and the changes occurring in these profiles during disturbances, that the model reproduces the experimentally observed changes in the altitude of the maximum of the layer and the change in the electron density and shape of the electron density profile in the layer. The basic characteristics of the change occurring in the electron density profiles during auroral disturbances are examined.

083.031 Interpretation of Vertikal–6 and –7 rocket electron density measurements.
A. V. Pavlov.
Geomagn. Aeron., Vol. 31, No. 2, p. 231 – 234 (Oct 1991). English translation of Geomagn. Aehron., Tom 31, No. 2, p. 308 – 311 (1991).

The electron density N_e measured on the Vertikal–6 (low solar activity) and Vertikal–7 (elevated solar activity) rockets at altitudes of 180 – 1500 km is compared with the theoretical values on N_e. On the basis of this comparison it is hypothesized that the elevated, as compared with the MSIS–86 model, value of the atomic oxygen density measured in the Vertikal–6 experiment lasts for a brief time and does not affect N_e. For both experiments the best agreement betweeen the computed and measured values of N_e is achieved with the MSIS–86 model atmospheric pressure and density for the calculation of the plasma drift velocity. The IRI–86/87 empirical model gives a value of N_e that agrees fairly well with the rocket measurements of N_e.

083.032 Calculation of the heating of the electronic component of the ionosphere and plasmasphere by Alfvén and fast magnetosonic waves.
S. V. Tanygin, G. V. Khazanov, A. A. Chernov.
Geomagn. Aeron., Vol. 31, No. 2, p. 245 – 248 (Oct 1991). English translation of Geomagn. Aehron., Tom 31, No. 2, p. 324 – 328 (1991).

An algorithm for calculating the heat fluxes and local sources of heat arising in the Earth's magnetosphere owing to dissipation of the energy of the ring current as the ring current interacts with the outer plasmasphere is presented. It is shown for a simple example that heating occurs far from the geomagnetic equator and is more efficient when the R mode is excited. The calculations showed that the heat fluxes can reach values of $\sim 3 \cdot 10^{11} \mathrm{eV \cdot s^{-1} \cdot cm^{-2}}$. Such heat fluxes could significantly increase the temperature of the outer plasmasphere ($\sim 8500\mathrm{K}$).

083.033 Correction to a mathematical model of a high–latitude ionosphere based on oblique–incidence sounding data.
G. A. Alad'ev, V. M. Lukashkin, V. S. Mingalev, M. I. Orlova.
Geomagn. Aeron., Vol. 31, No. 2, p. 292 – 293 (Oct 1991). English translation of Geomagn. Aehron., Tom 31, No. 2, p. 373 – 375 (1991).

Oblique incidence sounding data on radio routes through high latitudes are used to refine the control parameters of a multi–ion two–dimensional mathematical model of a high–latitude ionosphere.

083.034 The present understanding of the cusp.
R. Lundin, J. Woch, M. Yamauchi.
International Workshop on Cluster Dayside Polar Cusp, p. 83 – 95 (Dec 1991). – See Abstr. 012.071 for the main entry.

The polar cusp, first identified from low–altitude orbiting satellites, is the only region in the topside terrestrial ionosphere that maintains continuous contact with the solar wind plasma. The continuous inflow of plasma through the cusp leads to a direct transfer of solar wind energy and momentum to the ionosphere and atmosphere. Thus, the cups represents a particular "hot spot" in the solar–terrestrial relationship. This report reviews what has been learnt on the dynamics and topology of the polar cusp from in situ plasma measurements primarily on board low and medium altitude satellites. New improved magnetic field mapping models have also helped elucidating the connection of the low–altitude cusp to the magnetospheric boundary region.

083.035 The ionospheric signature of flux transfer events.
S. W. H. Cowley, M. P. Freeman, M. Lockwood, M. F. Smith.
International Workshop on Cluster Dayside Polar Cusp, p. 105 – 112 (Dec 1991). – See Abstr. 012.071 for the main entry.

The authors consider the effects at ionospheric heights which take place when transient reconnection events (i.e. flux transfer events (FTEs)) occur at the dayside magnetopause. They discuss the nature of the FTE–related ionospheric flows, the associated current systems, and the plasma precipitation. In particular, the authors outline the nature of the time–dependent cusp precipitation which occurs in this case, and compare expectations with those based on steady magnetopause reconnection.

083.036 Ground–based studies of sensing magnetosphere/ionosphere interactions: convection and substorms.
Y. Kamide, A. D. Richmond.
International Workshop on Cluster Dayside Polar Cusp, p. 129 – 137 (Dec 1991). – See Abstr. 012.071 for the main entry.

The importance of coordinated satellite and ground–based observations in studies of ionosphere/magnetosphere interactions is stressed by showing examples of ionospheric electrodynamic parameters from most recent efforts.

083.037 Relationships between ionospheric and tail phenomena.
S. Perraut, J. A. Sauvaud.
International Workshop on Cluster Dayside Polar Cusp, p. 191 – 201 (Dec 1991). – See Abstr. 012.071 for the main entry.

This review deals with large scale processes taking place in the magnetotail and on their consequences upon the ionosphere. The topics have been selected in the light of the new information that can be expected from the CLUSTER mission and from the associated ground based measurements. Plasma convection is discussed, in particular the authors try to establish the link between the classical picture of the convection at the ionospheric level and its magnetospheric projection. It is shown that Cluster can help determining the location and the shape of the Harang discontinuity in the magnetosphere provided that the S/C are close enough to the Earth. The existence of a steady state convection is questioned in light of recent models. The relation between convection and magnetic activity is also discussed. Then the mapping between auroral arcs and the various magnetospheric boundaries is investigated. Plasma dynamics in the mid tail and in the inner plasma sheet is described with particular emphasis on some of the models proposed for substorms. Suggestions are made about complementary measurements to be performed by Cluster and ground based instruments to understand plasma circulation in the magnetosphere and its relation to substorms.

083.038 The ionospheric response to the atmospheric quasi–biennial oscillation.
Chen Pei–ren.
Acta Geophys. Sin., Vol. 35, No. 3, p. 288 – 294 (May 1992). In Chinese.

083.039 Regional features of the ionosphere over China.
Huang Xinyu, Li Jun.
Advances in solar–terrestrial science of China, p. 173 – 181 (1992). – See Abstr. 003.069 for the main entry.

The morphology of both ionospheric profile and transient phenomena in the East Asia area has been studied to reveal the regional features and their formation mechanism in the present paper. Based on the systematic and recent data, the results describe in detail the "Far Eastern Anomaly" and regional features of "Equatorial Anomaly".

083.040 The day–to–day variability of the equatorial ionization anomaly.
Chen Peiren.
Advances in solar–terrestrial science of China, p. 182 – 196 (1992). – See Abstr. 003.069 for the main entry.

Recently the day–to–day variability of the equatorial ionization anomaly (EIA) has been studied in China. The main progress made so far is reviewed in this paper.

083.041 Ionospheric observations by DGS–256 at Hainan station.
Xu Chufu.
Advances in solar–terrestrial science of China, p. 197 – 211 (1992). – See Abstr. 003.069 for the main entry.

A brief description of the Hainan synthetic experimental site, and some initial results of ionospheric observations are presented. Based on data analyses and relating to main problems on the equatorial ionosphere, discussions on the ionospheric phenomenology and morphology near the equator region are given.

083.042   Ionospheric response of solar flares.
Pan Liande.
Advances in solar–terrestrial science of China, p. 292 – 302 (1992). – See Abstr. 003.069 for the main entry.
Studies of the effect of solar activity on the ionosphere carried out by Chinese scientists in recent years are outlined and the main topic of the effects of solar flares on the ionosphere is discussed in this paper.

083.043   Comparison of auroral emission and EISCAT–derived plasma parameters during the flight of NEED–II.
F. Søraas, K. Aarsnes, C. Hall, A. Brekke, M. Rietveld,
U. P. Løvhaug.
10. ESA Symposium on European Rocket and Balloon Programmes and Related Research, p. 75 – 81 (Nov 1991). – See Abstr. 012.088 for the main entry.
During the flight of the NEED–II rocket from Andøya Rocket Range in 1990 the EISCAT radar monitored the state of the ionosphere at apogee. Simultaneously, all–sky TV registered time development of auroral forms and hence the intensity of emissions along the magnetic field line passing through the EISCAT scattering volume. Here, the intensities of the emissions are compared with the plasma parameters determined by EISCAT. A clear correlation between the F–region electron temperature and the upper E–region optical emission intensity was observed.

Solar–terrestrial models and application software.
See Abstr. 002.033.

7. Quadrennial Symposium on Solar–Terrestrial Physics (SCOSTEP–7), The Hague (Netherlands), 25 – 30 Jun 1990.
See Abstr. 012.024.

Cluster dayside polar cusp. Planning and coordination of measurements from Cluster, ground stations, balloons and rockets in the dayside polar–cusp region. Proceedings.
See Abstr. 012.071.

Proceedings of the 10th ESA Symposium on European Rocket and Balloon Programmes and Related Research.
See Abstr. 012.088.

Research programmes at Svalbard.
See Abstr. 013.110.

The Polar Geophysical Institute program of the theoretical and experimental investigation on cusp.
See Abstr. 013.111.

China meridian chain of magnetometers and the global electric current system.
See Abstr. 013.126.

Optical manifestation of microbursts of electron fluxes.
See Abstr. 022.096.

A method for characterizing transient ionospheric disturbances using a large radiotelescope array.
See Abstr. 036.076.

Generation of radiation by upper–hybrid waves in non–uniform plasmas.
See Abstr. 062.065.

Review of the CIV phenomenon.
See Abstr. 062.070.

Lightning induced brightening in the airglow layer.
See Abstr. 082.001.

A thermosphere/ionosphere general circulation model with coupled electrodynamics.
See Abstr. 082.045.

Effects of vibrationally excited molecular nitrogen on ionospheric–thermospheric coupling for different levels of solar activity.
See Abstr. 082.046.

Coupling stratosphere–lower ionosphere at middle latitudes.
See Abstr. 082.079.

Auroral energy deposition rate.
See Abstr. 084.023.

The origin of the global system of field–aligned currents on the dayside of the magnetosphere.
See Abstr. 084.046.

Antarctic studies of the cusp.
See Abstr. 084.067.

Solar wind–magnetosphere coupling processes observed near the dayside cusp/cleft.
See Abstr. 084.068.

A model of the dayside magnetosphere and breakups in the cusp region.
See Abstr. 084.070.

Mathematical model of physical processes responsible for generation of magnetospheric electric fields and their penetration into the Earth's magnetosphere.
See Abstr. 084.071.

On the theory of natural oscillations of planetary ionospheres.
See Abstr. 091.055.

The question of the luminosity of meteor trails.
See Abstr. 104.001.

The geophysical effects of cosmic rays.
See Abstr. 144.073.

084 Aurorae, Geomagnetic Field, Magnetosphere

084.001 Geomagnetic disturbance fields: an analysis of observatory monthly means.
D. N. Stewart, K. A. Whaler.
Geophys. J. Int., Vol. 108, No. 1, p. 215 – 223 (Jan 1992).
This work quantifies the extent to which disturbance phenomena contribute to the observed geomagnetic field on the timescale of months to years.

084.002 High latitude pulsating aurorae revisited.
Q. Wu, T. J. Rosenberg.
Geophys. Res. Lett., Vol. 19, No. 1, p. 69 – 72 (3 Jan 1992).
Dayside auroral pulsations (10–40 s periods) have been studied for different levels of geomagnetic disturbance with emission data obtained at South Pole station, Antarctica ($-74.2°$ MLAT).

084.003 Great magnetic storms.
B. T. Tsurutani, W. D. Gonzalez, F. Tang, Y. T. Lee.
Geophys. Res. Lett., Vol. 19, No. 1, p. 73 – 76 (3 Jan 1992).
The five largest magnetic storms that occurred between 1971 to 1986 are studied to determine their solar and interplanetary causes. All of the events are found to be associated with high speed solar wind streams led by collisionless shocks. The high speed streams are clearly related to identifiable solar flares. It is found that: (1) it is the extreme values of the southward interplanetary magnetic fields rather than solar wind speeds that are the primary causes of great magnetic storms, (2) shocked and draped sheath fields preceding the driver gas (magnetic cloud) are at least as effective in causing the onset of great magnetic storms (3 of 5 events) as the strong fields within the driver gas itself, and (3) precursor southward fields ahead of the high speed streams allow the shock compression mechanism (item 2) to be particularly geoeffective.

084.004 The magnetic field strength of currents in the magnetopause and the dynamic pressure of the solar wind.
Ts. D. Porchkhidze, Ya. I. Fel'dshtejn.
Geomagn. Aeron., Vol. 30, No. 5, p. 638 – 641 (Apr 1990). English translation of Geomagn. Aehron., Tom 30, No. 5, p. 753 – 756 (1990).

084.005 Asymmetry of the disturbed geomagnetic field at middle– and low–latitudes.
A. Grafe, Ya. I. Fel'dshtejn, A. Prigancova, P. V. Sumaruk.
Geomagn. Aeron., Vol. 30, No. 5, p. 732 – 735 (Apr 1990). English translation of Geomagn. Aehron., Tom 30, No. 5, p. 859 – 862 (1990).
During the intense magnetic storm of March 23 – 24, 1969, an investigation was made of features of the latitudinal asymmetry of the field variations at middle and low latitudes from data on the geomagnetic field variations. For this storm it is shown that the asymmetry of the long–period component (>1 hour) is greater at the lower latitudes than at middle latitudes. The field asymmetry for the short–period variations grows with increasing latitude. The long–period component of the asymmetry correlates better with the activity of the western electrojet than with the eastern electrojet. This correlation is higher at low than at middle latitudes. The observed features of the magnetic field asymmetry are attributable to the nonuniform longitudinal distribution of ions composing the ring current and to the contribution of the field of the longitudinal currents.

084.006 The ultraviolet imager experiment on the Swedish Viking satellite: contributions to auroral physics.
L. L. Cogger, J. S. Murphree, R. D. Elphinstone, D. J. Hearn, R. A. King.
Can. J. Phys., Vol. 69, No. 8/9, p. 1032 – 1039 (Aug–Sep 1991).
The ultraviolet imager on board the Swedish Viking satellite was designed to provide real–time monitoring of the auroral distribution from space. This objective was achieved over the nominal lifetime of the satellite, March – December, 1986 during which period approximately 45000 auroral images were acquired. A number of technical and operational innovations have resulted in a rich data base for studies of auroral and magnetospheric processes. Some of the significant scientific advances that have resulted from the investigation of the temporal and spatial development of the auroral distribution include observations of rapid changes of dayside aurora, the effects of this distribution due to the interplanetary magnetic field, and more detailed knowledge of the substorm process.

084.007 Spatial relationship of visual arcs to auroral electrojets.
Q. Pao, L. L. Cogger, D. D. Wallis, A. G. McNamara.
Can. J. Phys., Vol. 69, No. 8/9, p. 1047 – 1054 (Aug–Sep 1991).
Observations of the 5577 Å emission detected with an all–sky imager are superimposed onto maps of auroral E–region coherent backscatter at 50 MHz made by the bistatic auroral radar system (BARS) to obtain the spatial and temporal relationships of stable visual arcs to auroral electrojets and the Harang discontinuity.

084.008 The quiet geomagnetic field at geosynchronous orbit and its dependence on solar wind dynamic pressure.
C. L. Rufenach, R. L. McPherron, J. Schaper.
J. Geophys. Res., Vol. 97, No. A1, p. 25 – 42 (1 Jan 1992).

084.009 Comparison of Echo 7 field line length measurements to magnetospheric model predictions.
R. J. Nemzek, P. R. Malcolm, J. R. Winckler.
J. Geophys. Res., Vol. 97, No. A2, p. 1279 – 1288 (1 Feb 1992).
The Echo 7 sounding rocket experiment injected electron beams on central tail field lines near $L = 6.5$. Numerous injections returned to the payload as "conjugate echoes" after mirroring in the southern hemisphere. The authors compare field line lengths calculated from measured conjugate echo bounce times and energies to predictions made by integrating electron trajectories through various magnetospheric models: the Olson–Pfitzer Quiet and Dynamic models and the Tsyganenko–Usmanov model.

084.010 Auroral resonance line radiative transfer.
G. R. Gladstone.
J. Geophys. Res., Vol. 97, No. A2, p. 1377 – 1388 (1 Feb 1992).
A model is developed for simulating the two–dimensional radiative transfer of resonance line emissions in auroras. The method of solution utilizes Fourier decomposition of the horizontal dependence in the intensity field so that the two–dimensional problem becomes a set of one–dimensional problems having different horizontal wavenumbers. The individual one–dimensional problems are solved for using a Feautrier–type solution of the differential–integral form of the radiative transfer equation. In the limit as the horizontal wavenumber becomes much larger than the local line–center extinction coefficient, the scattering integral becomes considerably simplified, and the final source function is evaluated in closed form. The two–dimensional aspects of the model are tested against results for nonresonance radiative transfer studies, and the resonance line part of the model is tested against results of existing plane–parallel resonance line radiative transfer codes. Finally, the model is used to simulate the intensity field of O I 1304 Å for hard and soft auroras of various Gaussian horizontal widths. The results demonstrate the importance of considering the effects of two–dimensional radiative transfer when analyzing auroral resonance line data.

084.011 On the possibility of auroral remote sensing with the Viking Ultraviolet Imager.
D. P. Steele, D. J. McEwen, J. S. Murphree.
J. Geophys. Res., Vol. 97, No. A3, p. 2845 – 2862 (1 Mar 1992).
An investigation was carried out to assess the value of ultraviolet auroral images for remote sensing of electron precipitation. The authors compared auroral images, obtained by both

cameras of the Viking Ultraviolet Imager during April and early May 1986, with simultaneous measurements of electron precipitation from the DMSP F7 and HiLat satellites at low altitudes above the auroral zone.

084.012   Coordinated measurements made by the Sondrestrom radar and the Polar Bear Ultraviolet Imager.
R. Robinson, T. Dabbs, J. Vickrey, R. Eastes, F. Del Greco, R. Huffman, C. Meng, R. Daniell, D. Strickland, R. Vondrak.
J. Geophys. Res., Vol. 97, No. A3, p. 2863 – 2872 (1 Mar 1992).

In 1986 and 1987 the Sondrestrom incoherent scatter radar in Greenland was operated routinely in coordination with selected overpasses of the Polar Bear satellite. For these experiments the auroral imaging remote sensor on Polar Bear obtained images of auroral emissions in two far ultraviolet wavelength bands centered at approximately 136 and 160 nm and one visible band centered at 391.4 nm. Measurements at these three wavelengths were extracted from the images for comparison with the coincident radar measurements.

084.013   Geomagnetic and solar data. September – October 1991.
H. E. Coffey.
J. Geophys. Res., Vol. 97, No. A3, p. 3229 – 3230 (1 Mar 1992).

084.014   On diffusion of the tangential fluctuations of the geomagnetic field through the mantle.
D. D. Sokoloff (*D. D. Sokolov*), B. G. Zinchenko.
Astron. Nachr., Vol. 313, No. 2, p. 115 – 123 (1992).

The authors develop a statistical approach to resolve the transport problem for the tangential fluctuations of the geomagnetic field in the mantle. For the sake of simplicity they treat the mantle as a thick layer of vacuum and assume in addition that only a radial component of the magnetic field of the core penetrates through the core–mantle boundary. These assumptions allow to find exact expressions for the tangential field components throughout the mantle. By using such expressions the authors construct a correlation tensor of tangential components and then, since the mantle is thick enough, study its asymptotic properties on the Earth surface. Incidentally, the correlation tensor trace happens to be equal to the correlation function of the radial component that was obtained by Pilipenko and Sokoloff (1992). Indeed, the authors provide a simple boundary problem which initially describes the diffusion functions. They also pay a special attention to transformation properties of the correlation tensor and find here some interesting analogies with secular variation data of the geomagnetic field.

084.015   Long–term variations in geomagnetic and solar activities and secular variations of the geomagnetic field components.
J. Střeštík.
Stud. Geophys. Geod., Vol. 35, No. 1, p. 1 – 6 (1991).

After the removal of the eleven–year periodicity, long–term patterns of the aa indices of geomagnetic activity and of Wolf's sunspot numbers are defined. The positions of maxima and minima exhibit the same regularities as the secular variations of the geomagnetic field components. This result is associated with the motion of the Sun round the barycentre of the solar system.

084.016   The 15 m.y. geomagnetic reversal periodicity: a quantitative test.
A. Mazaud, C. Laj.
Earth Planet. Sci. Lett., Vol. 107, No. 3/4, p. 689 – 696 (Dec 1991).

The authors have analyzed the three most commonly used geomagnetic polarity time scales that cover the past 100 m.y.. In the three cases, different spectral analyses show a 13–16 m.y. periodicity in the rate of reversal occurence. To test whether this periodicity is real or simply arises from a random generator the authors have compared these polarity time scales with a large number of synthetic sequences produced by a random process, characterized by a linear time variation of its mean activity. Geomagnetic and generated sequences were regularly sampled by

using sliding windows, and then Fourier spectra of the obtained frequency signals were compared. This test shows that the detected periodicity is presumably not a simple statistical fluctuation of an aperiodic generator, and consequently that a long–term periodicity in the geodynamo must be seriously considered.

084.017   A global simulation of the magnetosphere with a long tail: no interplanetary magnetic field.
A. Kageyama, K. Watanabe, T. Sato.
J. Geophys. Res., Vol. 97, No. A4, p. 3929 – 3944 (1 Apr 1992).

A global simulation of the magnetosphere with a long tail ($\sim 100 \, R_E$) is performed. A magnetosphere with a neutral sheet is constructed from a dipole field by solar wind dynamic pressure (no interplanetary magnetic field, IMF). Concentration of the plasma sheet current occurs preferentially at $14–18 \, R_E$ on the tail side of the Earth, which is an indication that, magnetically, this is the most fragile region of the tail structure. It is the demarcation region between the dipolar and streaming field line structures. Therefore, in the presence of resistivity, magnetic reconnection can be preferentially driven here by a compressional disturbance of some sort. In the present initial value problem, reconnection occurs at $15–20 \, R_E$ in the tail and develops into a large lump of plasma surrounded by reconnected field lines, a plasmoid, which is ejected tailward. The time scale of the plasmoid formation and ejection process is very slow, of the order of several hours, when no IMF exists. After ejecting one plasmoid, reconnection occurs again in the plasma sheet, and a second plasmoid is formed and ejected. This result shows that a magnetosphere that has a sufficiently long tail and a neutral sheet is fragile and subject to plasmoid formation. The authors also show that the hot plasma, when mapped down to the ionosphere along the field lines, encompasses the auroral oval in agreement with the DE satellite observations.

084.018   Electron acceleration by Alfvén waves in the magnetosphere.
C.–H. Hui, C. E. Seyler.
J. Geophys. Res., Vol. 97, No. A4, p. 3953 – 3963 (1 Apr 1992).

The self–consistent electron kinetics of Alfvén waves on the electron inertial scale is studied using a two–dimensional hybrid–kinetic description. The ions follow a fluid description for Alfvén waves at frequencies below the ion cyclotron frequency. The parallel electron dynamics are treated kinetically using particle–in–cell techniques. In this model the electron plasma mode is eliminated and only the physics of the Alfvén waves is retained. At sufficiently large amplitudes, it is found that oblique Alfvén waves break due to finite electron inertia in a cold plasma. The consequence of wave breaking is the formation of an electron beam which can be unstable to the beam–plasma instability. The electrons supporting the parallel current thermalize into a non-Maxwellian distribution with an energetic tail up to several keV, assuming a reasonable magnetospheric Alfvén speed. In hot plasma simulations, electron trapping is the principal mechanism of electron acceleration. It is proposed that wave breaking or electron trapping of oblique Alfvén waves at $1 \, R_E$ can result in electron acceleration and may explain some observed auroral phenomena.

084.019   Three–dimensional MHD modeling of magnetotail dynamics for different polytropic indices.
M. Hesse, J. Birn.
J. Geophys. Res., Vol. 97, No. A4, p. 3965 – 3976 (1 Apr 1992).

The evolution of the three–dimensional resistive tearing instability in a magnetotail configuration is analyzed by means of a resistive MHD code for various forms of the energy equation including ohmic and compressional heating.

084.020   A quasi–static magnetospheric convection model in two dimensions.
G. M. Erickson.
J. Geophys. Res., Vol. 97, No. A5, p. 6505 – 6522 (1 May 1992).

A self–consistent, two–dimensional model of subsonic, plasma sheet convection is presented. Specifically, time sequences of static equilibrium solutions for the two–dimensional magneto-spheric magnetic field are constructed consistent with adiabatic convection. This model self–consistently includes a dipole field and a reasonable accounting for the effects of inner magneto-spheric shielding. Starting from a relaxed magnetospheric equi-librium, the earthward convection of plasma sheet flux tubes results in the stretching of inner plasma sheet field lines, the development of a local minimum in the equatorial magnetic field B_e in the near–Earth plasma sheet, and an increasing lobe magnetic field. This evolution in time occurs generally, indepen-dent of the specific magnetopause or far–tail boundary condi-tions, provided the plasma sheet is not at marginal interchange stability. This behavior results solely from the convection of flux tubes of increasing entropy into the near–Earth plasma sheet. The author discusses these results in the general context of earthward convection in Earth's plasma sheet and in the specific context of magnetospheric substorms.

084.021 **Geomagnetic and solar data. November – Decem-ber 1991.**
H. E. Coffey.
J. Geophys. Res., Vol. 97, No. A5, p. 6549 – 6550 (1 May 1992).

084.022 **On impulsive penetration of solar wind plasmoids into the geomagnetic field.**
M. Roth.
Planet. Space Sci., Vol. 40, No. 2/3, p. 193 – 201 (Feb–Mar 1992).

084.023 **Auroral energy deposition rate.**
M. H. Rees.
Planet. Space Sci., Vol. 40, No. 2/3, p. 299 – 313 (Feb–Mar 1992).
Processes that lead to energy deposition by externally imposed electric fields and by energetic particle precipitation are described and the consequences on the atmosphere and ionosphere are examined. The importance of global observations is emphasized and current shortcomings of available measurements are noted.

084.024 **Modelling of the magnetic field of magnetospheric ring current as a function of interplanetary medium parame-ters.**
Y. I. Feldstein (*Ya. I. Fel'dshtejn*).
Space Sci. Rev., Vol. 59, No. 1/2, p. 83 – 165 (Jan 1992).
The models are examined which are proposed elsewhere for describing the magnetic field dynamics in ring–current DR during magnetic storms on the basis of the magnetospheric energy balance equation. The equation parameters, the functions of injection F and decay τ, are assumed to depend on interplanetary medium parameters (F and τ during the storm main phase) and on ring–current intensity (τ during the recovery phase). The present–day models are shown to be able of describing the DR variations to within a good accuracy.

084.025 **Intensity of the Earth's magnetic field since Precambrian from Thellier–type paleointensity data and interferences on the thermal history of the core.**
M. Prévot, M. Perrin.
Geophys. J. Int., Vol. 108, No. 2, p. 613 – 620 (Feb 1992).
The authors present a compilation of palaeointensity data obtained by the Thellier method from magnetic rocks up to 3.5 Gyr old. No apparent very long–term variation occurs from present to Early Precambrian. The overall data are compatible with a gravitationally powered dynamo model in which the inner core is some 4.0 Gyr old. However, palaeointensity data are still too incomplete to dismiss the possibility of a more complex field history with a geodynamo first thermally driven then gravitation-ally powered. Despite the limited number of data currently available, large differences in the dipole field moment are clearly documented with a time–scale of the order of 50–100 Myr. The changes in intensity can be due to repetitive instabilities of the thermal boundary layer at the base of the mantle.

084.026 **Absence of upstream energetic ions under turbulent radial interplanetary magnetic field.**
E. T. Sarris, G. C. Anagnostopoulos, S. M. Krimigis.
J. Geophys. Res., Vol. 97, No. A6, p. 8231 – 8237 (1 Jun 1992).
According to Fermi models as applied to the Earth's bow shock energetic particles are accelerated most efficiently under radial interplanetary magnetic field (IMF) conditions. In an earlier paper (Sarris and Krimigis, 1988) the author tested the Fermi mechanism for cases of radial IMF when no detectable ambient energetic particle fluxes were present. They concluded that the above mechanism could not account for the observations in the vicinity of the bow shock. In this work the authors extend the previous test by examining cases observed by the IMP 8 spacecraft where, in addition to the radial IMF, the following particle and field conditions were present: (1) in situ cyclotron – resonant wave activity, (2) a seed energetic particle population, and (3) small ($\lesssim 25°$) θ_{Bn} at the points of connection of the spacecraft to the bow shock. Examination of data from days 67, 1979, and 303, 1980, show that despite the fact that all of the above conditions were satisfied, no ion enhancements ($\gtrsim 50$ keV) attributable to the Fermi process could be discerned. The authors conclude that, even with the addition of criteria much more stringent than those applied in all previously published upstream events, which provided the observational underpinning for the development of the Fermi model, the effects of the Fermi process in accelerating ions to energies greater than ~ 50 keV are essentially undetectable ($\varepsilon < 10^{-3}$) in the region upstream of Earth's bow shock.

084.027 **Ground and satellite observations of an auroral event at the cusp/cleft equatorward boundary.**
P. E. Sandholt, P. T. Newell.
J. Geophys. Res., Vol. 97, No. A6, p. 8685 – 8691 (1 Jun 1992).
Detailed observations by ground–based and satellite instru-ments of a transient auroral event in the ≈ 10–11 MLT sector during negative IMF B_z are reported.

084.028 **Regions of negative Bz in the Tsyganenko 1989 model neutral sheet.**
E. F. Donovan, G. Rostoker, C. Y. Huang.
J. Geophys. Res., Vol. 97, No. A6, p. 8697 – 8700 (1 Jun 1992).
The authors point out a disturbing feature of the Tsyganenko (1989) model magnetic field, namely the occurrence of negative Bz in the model neutral sheet. On the basis of observations of Bz in the neutral sheet the authors conclude that this is an artifact of the model and not a real effect. This feature of the model should be considered when the model is used either to infer mappings from the ionosphere to the vicinity of the neutral sheet or as a tool in theoretical studies. The authors propose that in the develop-ment of future models, it would be useful for the distribution of Bz in the neutral sheet to be imposed as a constraint on the model.

084.029 **Geomagnetic and solar data. January – February 1992.**
H. E. Coffey.
J. Geophys. Res., Vol. 97, No. A6, p. 8723 – 8726 (1 Jun 1992).

084.030 **The semiannual variation of great geomagnetic storms and the postshock Russell–McPherron effect preceding coronal mass ejecta.**
N. U. Crooker, E. W. Cliver, B. T. Tsurutani.
Geophys. Res. Lett., Vol. 19, No. 5, p. 429 – 432 (3 Mar 1992).
The occurrence rate of great geomagnetic storms displays a pronounced semiannual variation. Of the forty–two great storms during the period 1940–1990, none occurred during the solstitial months of June and December, and 40% (17) occurred during the equinoctial months of March and September. This suggests that the semiannual variation found by averaging indices is not the result of some statistical effect superposed on the effects of random storm occurrence but rather is dominated by the storms themselves. Recent results indicate that the intense southward interplanetary magnetic fields (IMFs) responsible for great storms can reside in the postshock plasma preceding the driver gas of coronal mass ejections (CMEs) as well as in the driver gas

itself. Here the authors propose that strong southward fields in the postshock flow result from a major increase in the Russell–McPherron polarity effect through a systematic pattern of compression and draping within the ecliptic plane. Differential compression at the shock increases the Parker spiral angle and, consequently, the azimuthal field component that projects as a southward component onto Earth's dipole axis. The resulting prediction is that southward fields in the postshock plasma maximize at the spring (fall) equinox in CMEs emerging from toward (away) sectors. This pattern produces a strong semiannual variation in postshock IMF orientation and may account at least in part for the observed semiannual variation of the occurrence of great geomagnetic storms.

084.031 The relationship of periodic structures in auroral luminosity in the afternoon sector of ULF pulsations.
G. Rostoker, B. Jackel, R. L. Arnoldy.
Geophys. Res. Lett., Vol. 19, No. 6, p. 613 – 616 (20 Mar 1992).
Preliminary studies of Viking imager data by Lui et al. (1987) have revealed the presence, on occasion, of bright auroral structures periodically arrayed along the auroral oval in the post noon quadrant. It was suggested that these auroral forms reflect the action of a Kelvin–Helmholtz instability. In this paper the authors report the results of such a study of periodically arrayed auroral structures with concurrent ULF pulsations. Assuming a common origin of the two phenomena to be the Kelvin–Helmholtz instability, the authors make quantitative estimates of some key physical properties of the source region for the ULF and auroral disturbances. They further report the observation of a westward traveling surge in the vicinity of 18 MLT, and argue on the basis of the mapping factor from the ionosphere to the magnetotail why surges closer to midnight do not appear to exhibit a significant westward propagation velocity.

084.032 Geomagnetic quiet daily variations in the Australian region – information from a new station at Charters Towers (20.1°S).
R. J. Stening, P. A. Hopgood.
J. Atmos. Terr. Phys., Vol. 53, No. 10, p. 959 – 964 (Oct 1991).
The establishment of the new magnetic observatory at Charters Towers is described. Hourly values of the magnetic elements are analysed from two data sets (1984–1985 and 1986–1987) to provide solar diurnal variations during different months. These are compared with other stations in Australia and Papua New Guinea to study the behaviour of the focus of the Sq current systems. The lunar geomagnetic variations are consistent with other Australian stations. An oceanic lunar tide is detected in the vertical element Z.

084.033 Palaeomagnetic constraints on the geometry of the geomagnetic field during reversals.
J.-P. Valet, P. Tucholka, V. Courtillot, L. Meynadier.
Nature, Vol. 356, No. 6368, p. 400 – 407 (2 Apr 1992).
Palaeomagnetic records of the path of the pole during reversals of the Earth's magnetic field provide a test of the hypothesis that dipolar or low–order axisymmetric components of the field dominate during reversals. Multiple records of reversals during the past 12 Myr show no simple or consistent geographical pattern. Although a more robust analysis of the transitional field awaits a greater number of well–distributed sampling sites, the present data are not inconsistent with the simplest models, in which a field reminiscent of the non–dipole component of the present–day field becomes dominant.

084.034 The Earth's bow shock and magnetopause position as a result of the solar wind – magnetosphere interaction.
Z. Němeček, J. Šafránková.
J. Atmos. Terr. Phys., Vol. 53, No. 11/12, p. 1049 – 1054 (Nov–Dec 1991). – See Abstr. 012.024 for the main entry.
The paper deals with a study of the variations of the Earth's magnetopause and the bow shock position based on simultaneous measurements by the Prognoz 10 and IMP–8 spacecraft. The real boundary position determined by one spacecraft has been compared with the boundary position calculated from Formisano's model. The solar wind parameters, required for the calculation, have been taken from measurements of the other spacecraft. The differences between the calculated and observed boundary positions are discussed from the point of view of possible influences of different solar wind parameters on these deviations. From the authors' discussion it follows that the magnetopause shape calculated by Olson, Fairfield or Choe et al. for a dipole magnetic field seems to be a better approximation than Formisano's three–dimensional fit. On the other hand, Formisano's fit to the bow shock shape can be used for determination of the bow shock position if the additional influence of the interplanetary magnetic field strength is taken into account.

084.035 Geomagnetic storms categorized by varieties of interplanetary structures.
G.-l. Zhang.
J. Atmos. Terr. Phys., Vol. 53, No. 11/12, p. 1055 – 1067 (Nov–Dec 1991). – See Abstr. 012.024 for the main entry.
The morphology of geomagnetic storms is a sensitive function of the forcing interplanetary structures. The initial phase of a geomagnetic storm is related to the interplanetary high density structure in front of the magnetic cloud. The onset, development, and recovery of the strong main phase of the storm are controlled by the magnetic field orientation of the cloud with respect to the ecliptic plane. For a negative magnetic cloud, with the magnetic field vector first being southward, the main phase of the storm will be weakened or delayed with the decrease of both the speeds of the magnetic cloud and the ambient solar wind streams. A storm with a long delayed main phase is observed with the arrival of a positive magnetic cloud, and in the low–speed case the weak main phase can be separated from the sudden commencement by a relatively quiet period. A reduction in the recovery rate of a strong main phase may be a signature of the rear boundary crossing of the cloud. The solar wind streams behind the magnetic cloud keep affecting the slow recovery of the main phase.

084.036 Solar wind–terrestrial magnetosphere coupling: application of linear prediction theory.
H. O. Rucker, K. J. Trattner.
J. Atmos. Terr. Phys., Vol. 53, No. 11/12, p. 1069 – 1072 (Nov–Dec 1991). – See Abstr. 012.024 for the main entry.
By using high time resolution ISEE 3 solar wind data and geomagnetic indices, in particular AL, AU and D_{st}, the linear filtering technique was applied to model the magnetospheric response. The computed impulse response function exhibits a specific low pass filter characteristic which indicates that the magnetosphere cannot follow fast fluctuations. The time delay of the magnetospheric response depends strongly on the strength of the energy input. However, the response of the magnetosphere depends not only on a single solar wind parameter. Comparing the results of the single channel analysis with the results of the more sensitive technique of multi channel filtering the authors get a better look into the solar wind–magnetosphere coupling. The most predictable geomagnetic index is the ring current index D_{st} where the authors can predict over 70% of the D_{st} variances by using VB_s and pV^2 as input signals.

084.037 Magnetic flux rope type structures in the geomagnetic tail.
A. E. Antonova, A. P. Kropotkin.
J. Atmos. Terr. Phys., Vol. 53, No. 11/12, p. 1073 – 1079 (Nov–Dec 1991). – See Abstr. 012.024 for the main entry.
The authors identify some structures in the geomagnetic tail observed by the Prognoz 9 and ISEE spacecraft as "magnetic flux ropes". The following new features are emphasized: (1) The structures are associated with considerable fluxes of energetic ions and electrons. (2) Particles are effectively energized at magnetic field discontinuities, resulting in the generation of spectra extending up to MeV energies. (3) An external field source (i.e. the interplanetary magnetic field) may be of essential importance for the generation of the flux ropes whose axes lie in the cross–tail direction.

084.038 Solar control of the Earth's emission of energetic O⁺.
W. Lennartsson.
J. Atmos. Terr. Phys., Vol. 53, No. 11/12, p. 1103–1111 (Nov–Dec 1991). – See Abstr. 012.024 for the main entry.

Energetic (0.1–16 keV/e) O⁺ data obtained in the Earth's plasma sheet (between 10 and 23 R_E) by an ion mass spectrometer on the ISEE-1 spacecraft are compared statistically with published data on the concurrent solar wind and IMF. The most strongly variable parameter of the plasma sheet O⁺ is its density, which is found to be well correlated with certain solar wind parameters, especially with the solar wind flow speed and the IMF component perpendicular to the flow vector. When those two solar wind parameters are combined to form an electric field (–vxB), both the number density and the energy density of the O⁺ are found to vary in proportion to the square of that electric field, on average, suggesting that the emission of energetic O⁺ ions from the Earth may be powered by that same field. Based on this and on the previously published correlation with solar activity, it is argued that the emission of O⁺ is controlled by a combination of high–frequency (ionizing) and quasi–static (accelerating) solar electromagnetic fields.

084.039 Paleointensity of the geomagnetic field during the last 80,000 years.
E. Tric, J. P. Valet, P. Tucholka, M. Paterne, L. Labeyrie, F. Guichard, L. Tauxe, M. Fontugne.
J. Geophys. Res., Vol. 97, No. B6, p. 9337–9352 (10 Jun 1992).

High–resolution records of the relative paleointensity of the geomagnetic field have been obtained from five marine cores. Three duplicate records were used to estimate the regional coherency of the data within a single area (Tyrrhenian Sea) while the two others document the field variations in the eastern Mediterranean and the southern Indian Ocean. Careful investigations of distinct rock magnetic parameters have established the downcore uniformity of the sediments in terms of magnetic mineralogy and grain sizes. The time–depth control was provided by oxygen isotopes, and small–scale variations in the deposition rates were constrained by means of tephrachronology. The synthetic curve calculated from the Mediterranean records provides a continuous record of the intensity variations during the last 80,000 years (80 kyr), which correlates well with the sparse volcanic data available for the period 0–40 kyr. The fact that identical behavior is seen in both data sets and that they also compare quite well with results from a core collected in the Pacific Ocean establishes the truly dipolar character of these variations. The dipole field moment is characterized by large–scale changes as shown by the existence of pronounced drops (at 39 and 60 kyr) alternating with periods of higher intensity. The record suggests a periodic nature for these intensity variations. These results demonstrate the potential of sediments for such studies and constitute a first step towards obtaining a global paleointensity record over a long period of time.

084.040 Fire in the sky.
R. Sampson.
Astronomy, Vol. 20, No. 3, p. 38–43 (Mar 1992).

084.041 Simple radio telescopes and aurora.
K. Maeda.
Astron. Her., Vol. 85, No. 2, p. 58–62 (1992). In Japanese.

084.042 ULF waves – their relationship to the structure of the Earth's magnetosphere.
W. Allan, E. M. Poulter.
Rep. Prog. Phys., Vol. 55, No. 5, p. 533–598 (May 1992).

The Earth's magnetosphere is highly structured, in terms of both magnetic field and plasma characteristics. This structure has a profound influence on the propagation of plasma waves, especially ultra–low–frequency (ULF) waves with mHz frequencies, which have wavelengths comparable with typical magnetospheric dimensions. In this review the authors illustrate how the basic theory of ULF hydromagnetic wave propagation in an infinite, homogeneous, uniformly magnetized plasma has been extensively modified to cope with the requirements of applying it

to the magnetosphere, a natural laboratory for the physics of ULF waves. They consider the field–line–guided Alfvén wave modes and the isotropic fast hydromagnetic wave modes, and show how the existence of magnetospheric boundaries can affect the structure of both modes.

084.043 A gallery of aurora photos – night of the great aurora (8 November 1991).
Astronomy, Vol. 20, No. 3, p. 89–91 (Mar 1992).

084.044 Discovery and investigation of the nonthermal continuum at the frequencies 1486 and 992 kHz in the northern polar region of the terrestrial magnetosphere.
V. N. Kuril'chik, V. P. Grigor'eva, A. Tirpak, S. V. Mironov, L. Fisher, A. Ya. Yaroshevich.
Kosm. Issled., Tom 30, Vyp. 1, p. 107–119 (Jan–Feb 1992). In Russian. English translation in Cosmic Res., Vol. 30, No. 1.

Nonthermal continuum radiation at the unusually high frequencies for this type of emission, namely 1486 and 992 kHz, was discovered by Prognoz 10 satellite observations in the nightside of the northern polar region of the terrestrial magnetosphere. The emission propagates away from the Earth in form of beams and shows the diurnal and seasonal variations of the intensity with its appearance at the time of the displacement of the North Magnetic Pole to the magnetospheric tail. The continuum emission has a pulse–noise character with pronounced limit of maximal intensity at saturation level 10⁻¹⁷ W/m²·Hz. The region of generation of the continuum is apparently located at the geocentric distance <2 R_E in the night mid–latitude ionospheric trough.

084.045 Modélisation des radiations ionisantes de l'environnement spatial.
J. Lemaire, L. Bossy.
Ciel Terre, Vol. 108, No. 2, p. 41–46 (Mar–Apr 1992).

Accurate and up–to–date models for the distribution of energetic particle fluxes in the Earth's magnetosphere are required by space scientists and by aerospace engineers to evaluate the radiation doses to be experienced by future space systems during their orbital mission.

084.046 The origin of the global system of field–aligned currents on the dayside of the magnetosphere.
M. V. Samokhin.
Astrophys. Space Sci., Vol. 191, No. 2, p. 195–202 (May 1992).

There is a component of the current normal to the boundary near the tangential discontinuity (the magnetopause) if the plasma is frozen in the magnetic field. On the assumption that the plasma density obeys the model of Gold's distribution $n \sim r^{-4}$, one finds that if one closes the component of the current in the ionosphere, the global system of field–aligned currents is created which is consistent with the Triad data on the value, direction, and the distribution with the local time.

084.047 Facular structures of polar auroras.
V. R. Tagirov, M. V. Mal'kov.
Geomagn. Aeron., Vol. 31, No. 1, p. 76–79 (Aug 1991). English translation of Geomagn. Aehron., Tom 31, No. 1, p. 106–110 (1991).

A relation of the facular structures of polar auroras to geomagnetic pulsations has been shown on the basis of auroral and magnetic data. According to the results of model calculations it turned out that the facular structures are projected onto the equatorial plane of the magnetosphere at a distance of 5–10 R_E from the earth. The existing explanations of the mechanisms of formation of facular structures are discussed.

084.048 Sudden changes in the secular variations of the geomagnetic field during the late 1970s.
V. P. Golovkov, A. O. Simonyan.
Geomagn. Aeron., Vol. 31, No. 1, p. 117–121 (Aug 1991). English translation of Geomgn. Aehron., Tom 31, No. 1, p. 165–171 (1991).

A sudden change in the series describing the first derivatives of the mean annual values of the geomagnetic elements was observed between 1977 and 1980. Treating these jerks as smooth stepwise increases in the constant acceleration enables us to interpret this phenomenon as a typical jerk. Spherical harmonic analysis is used to construct a global constant–acceleration field relative to 1985. Comparison of the pre–jerk and post–jerk maps of the global distribution of the constant acceleration revealed the presence of quantitative errors in the post–jerk maps. The large error in the post–jerk maps can be explained by the fact that the initial data are worse because of a lack of current data and non-uniformity of the observatory network used. The statistical correlation between jerks and years of maximum solar activity was also valid for this jerk.

084.049   Feasibility of ground–based diagnostics of the location of the plasmapause according to electron precipitations.
S. N. Samsonov, V. D. Sokolov.
Geomagn. Aeron., Vol. 31, No. 1, p. 140–141 (Aug 1991). English translation of Geomagn. Aehron., Tom 31, No. 1, p. 191–192 (1991).

The feasibility of ground–based diagnostics of the position of the plasmapause according to measurements of electrons precipitating into the Earth's atmosphere is demonstrated. The method proposed is based on the difference of the minimal energies of the electrons precipitating from the outer and inner sides of the plasmapause. With the goal of improving the reliability of establishing the location of the plasmapause, the modulation characteristics of the precipitating electron fluxes are compared with the VLF–radiation intensities, using spectral and correlation analyses. The experimental results pointing to the feasibility of using the proposed method are presented.

084.050   The Aurora 1990.
R. J. Livesey.
J. Br. Astron. Assoc., Vol. 102, No. 3, p. 151–157 (Jun 1992).

084.051   The effect of lateral restriction of the electron flux on the formation of radiation forms of aurorae.
N. Ya. Kotsarenko, V. P. Pas'ko.
Geomagn. Aehron., Tom 32, No. 3, p. 40–46 (May–Jun 1992). In Russian. English translation in Geomagn. Aeron., Vol. 32, No. 3.

084.052   Analytical study of the dynamics of charged particles of the plasma sheet of the Earth's magnetosphere taking into account their pitch–angle distribution.
S. V. Smolin.
Geomagn. Aehron., Tom 32, No. 3, p. 98–104 (May–Jun 1992). In Russian. English translation in Geomagn. Aeron., Vol. 32, No. 3.

084.053   Current instability and generation of Alfvén waves in the Earth's magnetosphere.
P. P. Malovichko, A. K. Yukhimuk.
Geomagn. Aehron., Tom 32, No. 3, p. 163–167 (May–Jun 1992). In Russian. English translation in Geomagn. Aeron., Vol. 32, No. 3.

084.054   International reference geomagnetic field updated in 1991.
R. A. Langel.
Geomagn. Aehron., Tom 32, No. 3, p. 183–188 (May–Jun 1992). In Russian. English translation in Geomagn. Aeron., Vol. 32, No. 3.

084.055   Generation of VLF and ELF radiation in inhomogeneous plasma of the plasmapause region.
N. I. Izhovkina, V. I. Larkina.
Cosmic Res., Vol. 29, No. 4, p. 509–512 (Jan 1992). English translation of Kosm. Issled., Tom 29, Vyp. 4, p. 593–596 (1991).

Experimental data and simulation results are presented on the generation of very low frequency (VLF) and extremely low frequency (ELF) radiation in the inhomogeneous plasma of the plasmapause region. It is assumed that inhomogeneities of plasma density in that region can be represented in the form of tubes aligned along the geomagnetic field lines.

084.056   Dayside aurorae and their relation to other geophysical phenomena.
S. V. Leontyev (S. V. Leon'tev), G. V. Starkov, V. G. Vorobjev (V. G. Vorob'ev), V. L. Zverev, Ya. I. Feldstein (Ya. I. Fel'dshtejn).
Planet. Space Sci., Vol. 40, No. 5, p. 621–639 (May 1992).

Principal morphological peculiarities of auroral luminosity are investigated on the basis of the data from multi–year aurorae observations in day hours at Spitzbergen and Franz Josef Land. It is shown that in this region the typical forms of aurorae are moving poleward rayed arcs appearing at the equatorward boundary of the auroral oval and disappearing at its pole boundary. Discrete forms of aurorae are located inside a much broader red luminosity band in its equatorward part. Auroral pulsations with a period of 10–50 s are observed in the prenoon sector in a region of much harder precipitations found more equatorward with respect to the daytime red luminosity band. The influence of a B_zIMF component upon daytime aurorae is exercised both directly through an equatorward (poleward) shift of daytime aurorae upon decreasing (increasing) B_z and via an increase in a planetary geomagnetic activity related to the appearance of substorms during which the whole region of the daytime luminosity is shifted to much lower latitudes. A decrease of intensity of daytime aurorae with duration of 5–10 min before the beginning of an expansive phase of a substorm on the night side is detected. The peculiarities of the daytime aurorae dynamics during substorms are also investigated. A scheme of the daytime auroral luminosity distribution is presented. Analytical expressions of the dependence of the daytime aurorae position on IMF are provided. Certain physical mechanisms that can explain the peculiarities of daytime aurorae dynamics are also discussed.

084.057   Magnetospheric chorus emissions: a review.
S. S. Sazhin, M. Hayakawa.
Planet. Space Sci., Vol. 40, No. 5, p. 681–697 (May 1992).

A review of chorus emissions observed on ground–based stations and in the Earth's magnetosphere is presented. Different approaches to modelling these emissions are discussed. It is pointed out that the most likely energy source of chorus emissions lies in the anisotropic hot electrons in the equatorial magnetosphere. The energy of these electrons is transferred to waves via the electron cyclotron instability. Then the non–linear deformation of the hot electron distribution function under the influence of these waves produces almost monochromatic wavelets which, in their turn, generate a chorus element in a manner similar to the generation of artificially stimulated emissions by ground–based transmitters.

084.058   Dynamics of aurorae in the cusp region and characteristics of magnetic reconnection at the magnetopause.
M. I. Pudovkin, S. A. Zaitseva (S. A. Zajtseva), P. E. Sandholt, A. Egeland.
Planet. Space Sci., Vol. 40, No. 6, p. 879–887 (Jun 1992).

The development of dayside auroral breakups is studied, and results are compared with the magnetic field reconnection model. It is shown that periods of poleward motion of auroral arcs are preceded by relatively short (T ~ 100 s) intervals of equatorward auroral expansion. These equatorward motions are analogous in some aspects to poleward jumps of aurorae at the night side of the oval during auroral substorms, and they are thought to be the direct signatures of magnetic field reconnection pulses at the magnetopause. FTEs associated with dayside breakups are shown to be rather large–scale phenomena, the lengths of the reconnection line at the magnetopause equals 4–6 R_E, and "the meridional" extent of the reconnected tube along the magnetopause is also about 4 R_E. The rate of energy input into the reconnection region is estimated.

084.059   Plasmamessungen im Magnetosphärenschweif.
W. Baumjohann.
Habil.-Schrift, München Univ. (Germany). Fakultät für Geowissenschaften; Max-Planck-Institut für Physik und Astrophysik, Garching (Germany). Inst. für Extraterrestrische Physik. MPE-233, 32 p. (Jan 1992). ISSN 0178-0718.

084.060   Modeling the pitch–angle distribution of charged particles in the terrestrial magnetosphere.
S. V. Smolin.
Geomagn. Aeron., Vol. 31, No. 2, p. 186 – 190 (Oct 1991). English translation of Geomagn. Aehron., Tom 31, No. 2, p. 247 – 252 (1991).
Approximate analytical equations are derived from an analysis of experimental data that make it possible in two bands of equatorial pitch angles of $0 - 30°$ and $30 - 90°$ to more realistically predict for energetic charged particles (protons and electrons) their pitch–angle distribution in the terrestrial magnetosphere. It is proposed that on a qualitative level the origin and existence of the Earth's radiation belts can also be attributed to the increasing rise (experimentally confirmed) in the pitch–angle distribution of charged particles with large pitch angles as the plasma approaches Earth.

084.061   Regions of the development of a Kelvin–Helmholtz instability and the electrical potential distribution at the magnetosphere boundary.
A. V. Mezentsev.
Geomagn. Aeron., Vol. 31, No. 2, p. 269 – 271 (Oct 1991). English translation of Geomagn. Aehron., Tom 31, No. 2, p. 351 – 353 (1991).
The configurations of the Kelvin–Helmholtz instability regions at the magnetosphere boundary are obtained in the incompressible case as a function of the angle θ between the interplanetary magnetic field and the geomagnetic field for different density ratios at the boundary based on a numerical solution to the problem of streamline flow around the magnetosphere near the magnetic barrier. For $\theta = 0$, the instability regions lie near the equatorial plane. This region shrinks and shifts towards higher latitudes with increasing angle θ. Growth of the shock front at the discontinuity serves to reduce the size of the instability regions. Using these results it was possible to estimate the thickness of the viscous boundary layer and the distribution of the electrical potential at the magnetosphere boundary near the equatorial plane.

084.062   MHD wave generation from plasma motion along the geomagnetic field.
B. M. Chistoserdov.
Geomagn. Aeron., Vol. 31, No. 2, p. 282 – 283 (Oct 1991). English translation of Geomagn. Aehron., Tom 31, No. 2, p. 364 – 366 (1991).
The stability of MHD waves from magnetospheric plasma motion along a longitudinally inhomogeneous geomagnetic field is examined. Plasma motion in the stronger field direction is demonstrated to lead to generation of MHD oscillation. The instability mechanism examined in this paper may be responsible for generation of MHD oscillations in the region of open magnetic field lines of the terrestrial magnetosphere.

084.063   Formation of the periodic structures of auroras.
N. Ya. Kotsarenko, G. V. Lizunov, V. P. Pas'ko.
Geomagn. Aeron., Vol. 31, No. 2, p. 309 – 311 (Oct 1991). English translation of Geomagn. Aehron., Tom 31, No. 2, p. 388 – 390 (1991).
It is shown that the stratification of electron streams precipitating into the atmosphere into small–scale forms (rays, laminae, etc.) may result from the development of filamentary (aperiodic) instability. The ray structure of the polar aurora, in both nature and external form, is similar to the picture of the filamentary nature of relativistic electron beams observed in laboratory experiments. The instability is dissipative in nature; it requires the presence of collisions of the precipitating electrons with the

surrounding medium, and in general does not require the presence of a background plasma.

084.064   The interaction of regular and random components in the analysis of a time series of geomagnetic reversals.
V. Yu. Varygin, V. P. Aparin.
Geomagn. Aeron., Vol. 31, No. 2, p. 312 – 313 (Oct 1991). English translation of Geomagn. Aehron., Tom 31, No. 2, p. 390 – 392 (1991).
Some aspects of the use of spectrum analysis in the investigation of the time series of geomagnetic reversals within the recent superchron of mixed polarity are considered. The hypothesis of a random distribution of reversals is analyzed, taking into account the nonstationary nature of the geomagnetic field reversal process.

084.065   The greatest light show in the north.
N. Bone.
Astron. Now, Vol. 6, No. 6, p. 19 – 22 (Jun 1992).
Over the last few years, spectacular auroral displays corresponding to high solar activity have stretched down from their northern polar latitudes. One can expect a few more light shows before solar activity dies away.

084.066   Polar cap boundary and structure of dayside cusp as determined by ion precipitation.
O. A. Troshichev.
International Workshop on Cluster Dayside Polar Cusp, p. 31 – 34 (Dec 1991). – See Abstr. 012.071 for the main entry.
The results of particle measurements on board DMSP F6 and F7 spacecraft show that the ion precipitation features in the auroral oval and in the polar cap are systematically diverse: the ion precipitation in the oval is of a smoothed character, whereas that in the polar cap is of a patchy type. The boundary between these two types of ion precipitation can be usually detected by a quick fall in the ion total number flux below some definite level. The polar cap identified in such a way has a shape of a roughly sun-aligned ellipse when IMF is northward. Under the influence of the azimuthal IMF the northern polar cap is shifted toward the dusk (dawn) when $B_y > 0$ ($B_y < 0$). In the southern polar cap the effect is opposite. The following structural zones in the dayside oval can be separated using the ion data: cusp, cusp core (near the noon meridian), poleward edge of the cusp, equatorward edge of the cusp.

084.067   Antarctic studies of the cusp.
A. S. Rodger.
International Workshop on Cluster Dayside Polar Cusp, p. 35 – 41 (Dec 1991). – See Abstr. 012.071 for the main entry.
Antarctica offers many unique advantages for studies of the ionospheric signatures of processes associated with the magnetospheric cusp. These result from the very large displacement of the geographic and geomagnetic poles, there being a land mass upon which to deploy experiments, and a pollution–free environment in which to make observations. In recent years, important new results concerning the coupling of solar wind processes through the magnetosphere and into the ionosphere have been determined using Antarctic data–sets. These new results are briefly summarised in this paper. The present and planned observing programmes that are likely to be in place during the Cluster mission are also described.

084.068   Solar wind–magnetosphere coupling processes observed near the dayside cusp/cleft.
E. Friis–Christensen, K. H. Glassmeier.
International Workshop on Cluster Dayside Polar Cusp, p. 97 – 103 (Dec 1991). – See Abstr. 012.071 for the main entry.
The physical processes involved in solar wind–magnetosphere coupling are rather complex and not possible to describe adequately by means of single point measurements. Even in case of multiple spacecraft missions like CLUSTER, future investigations will have to rely on the full spectrum of available observations, including ground–based measurements in the cusp

and cleft region. In this paper the authors identify some of the unsolved problems in solar wind–magnetosphere coupling and demonstrate the capabilities of using modern dedicated ground-based research facilities to obtain "maps" of highly dynamic ionospheric signatures of magnetosphere boundary layer processes. These "maps" may provide crucial information about the distribution and motion of the sources of the processes.

084.069 **Mapping of the auroral oval to the tail.**
G. Gustafsson.
International Workshop on Cluster Dayside Polar Cusp, p. 113 – 117 (Dec 1991). – See Abstr. 012.071 for the main entry.
During the last decade it has become possible to use multisatellite measurements to study auroral substorms. Global satellite images of the aurora can be combined with simultaneous particle and field measurements at geosynchronous altitude and in the magnetotail. The main tool to relate observations in different regions of the magnetosphere is a magnetic field model and it is, therefore, of crucial importance in studies of substorm phenomena. Three different cases of mapping between the ionosphere and the magnetotail are discussed. These studies indicate the development and the need to further improvements for the auroral source regions in the tail.

084.070 **A model of the dayside magnetosphere and breakups in the cusp region.**
M. I. Pudovkin.
International Workshop on Cluster Dayside Polar Cusp, p. 139 – 144 (Dec 1991). – See Abstr. 012.071 for the main entry.
A model of the geomagnetic field in the dayside magnetosphere is discussed. The model is based on the supposition on the partial penetration of the magnetosheath's magnetic field into the magnetosphere in the vicinity of the reconnection line at the magnetopause. Configuration of the plasma in the vicinity of that line is analyzed, and electric field equipotentials are projected onto the polar ionosphere along the field lines of the model geomagnetic field. Effects of the potential and vortex components of the magnetopause electric field on the dynamics of the aurorae in the cusp region are discussed. On the base of the model proposed, an interpretation of the auroral dynamics in the cusp region is given.

084.071 **Mathematical model of physical processes responsible for generation of magnetospheric electric fields and their penetration into the Earth's magnetosphere.**
V. V. Denisenko, N. V. Erkaev, S. S. Zamay, A. V. Kitaev, A. V. Mezentsev, I. T. Matveenkov, V. G. Pivovarov.
International Workshop on Cluster Dayside Polar Cusp, p. 211 – 216 (Dec 1991). – See Abstr. 012.071 for the main entry.
A program for theoretical studies of the generation of electric fields and currents in the ionosphere and magnetosphere connected with the interaction of the solar wind and the Earth's magnetic field is presented. Because of the problem's complexity, it is divided into subsequent steps every one of which is studied in detail. The starting point of the program is a problem of formation of a magnetic barrier near the magnetopause that is the key for the whole magneospheric physics. The important step of the program is generation of electric field in the magnetosphere due to viscous interaction and a process of reconnection. The proposed program can serve as a theoretical basis for planned global satellite and ground projects.

084.072 **Dynamics of high energy captured radiation in the Earth's inner radiation belt.**
N. N. Volodichev, A. A. Gusev, Yu. V. Mineev, G. I. Pugacheva, P. I. Shavrin.
Cosmic Res., Vol. 29, No. 5, p. 678 – 682 (Mar 1992). English translation of Kosm. Issled., Tom 29, Vyp. 5, p. 790 – 794 (Sep–Oct 1991).

084.073 **Dynamics of the neutral sheet in the magnetotail during substorm.**
Xu Ronglan.
Advances in solar–terrestrial science of China, p. 121 – 139 (1992). – See Abstr. 003.069 for the main entry.
This paper is a brief summary of the author's studies on magnetotail phenomena, based on the dynamnics of the neutral sheet in the magnetotail during substorms.

084.074 **The monitoring and studying of the near earth electromagnetic environment by VLF and ULF waves.**
Yang Shaofeng.
Advances in solar–terrestrial science of China, p. 159 – 164 (1992). – See Abstr. 003.069 for the main entry.
This paper introduces the study of the near earth electromagnetic environment by means of the observation and analyses of VLF and ULF waves during the 22th solar maximum cycle. It presents the importance of geomagnetic pulsations and whistlers in the solar maximum years. Some results are published for the first time. The observation and study of geomagnetic pulsations and whistlers have become an important means of monitoring the electromagnetic environment of the near earth space.

084.075 **The magnetosphere.**
C. T. Russell.
Annu. Rev. Earth Planet. Sci., Vol. 19, p. 169 – 182 (1991).
Contents: Introduction. The solar wind. The magnetospheric cavity. The energization of the magnetosphere. The interior of the magnetosphere. The aurora. Concluding remarks.

084.076 **The character of the field during geomagnetic reversals.**
S. W. Bogue, R. T. Merrill.
Annu. Rev. Earth Planet. Sci., Vol. 20, p. 181 – 219 (1992).
In this review, the authors begin with some general background on the geodynamo and rock magnetism for the reader unfamiliar with these topics. They then summarize current knowledge of the transitional field. This discussion centers on those studies that have been especially influential in shaping ideas. The authors touch on problems related to the fidelity of the various paleomagnetic recorders, an issue that is at the forefront of much current research. Finally, the authors show how knowledge of the transitional geomagnetic field might lead to a better understanding of the processes occurring in the earth's outer core.

084.077 **Terrestrial response to eruptive solar flares: geomagnetic storms.**
W. D. Gonzalez, B. T. Tsurutani.
IAU Colloquium No. 133: Eruptive solar flares, p. 277 – 286 (1992). – See Abstr. 012.096 for the main entry.
During the interval of August 1978 – December 1979, 56 unambiguous fast forward shocks were indentified using magnetic field and plasma data collected by the ISEE–3 spacecraft. Because this interval is at a solar maximum the authors assume the streams causing these shocks are associated with coronal mass ejections and eruptive solar flares. For these shocks they describe the shock–storm relationship for the level of intense storms. Then, they discuss the interplanetary structures that are associated with the large–amplitude and long–duration negative B_z fields, which are found in the sheath field and/or driver gas regions of the shock and are thought to be the main cause of the intense storms. The authors also present for the solar physicist a summary of the interplanetary/magnetosphere coupling functions, based on the magnetopause reconnection process. They end by giving an overview of the long–term evolution of geomagnetic storms such as those associated with the seasonal and solar cycle distributions.

084.078 **Coronal mass ejections: the link between solar and geomagnetic activity.**
J. T. Gosling, D. J. McComas, J. L. Phillips.
IAU Colloquium No. 133: Eruptive solar flares, p. 385 (1992). – See Abstr. 012.096 for the main entry.

084.079 **A study of geomagnetic variations with periods of four years, six months and 27 days.**
A. L. Clúa de Gonzalez, W. D. Gonzalez, S. L. G. Dutra, B. T. Tsurutani.
IAU Colloquium No. 133: Eruptive solar flares, p. 386 – 389 (1992). – See Abstr. 012.096 for the main entry.

The monthly and daily samples of the *Ap* geomagnetic index for 51 years, in the 1932–1982 interval, were investigated by means of the power spectrum technique. Although in general the results confirm previous findings about possible periodicities in the geomagnetic activity, some aspects are either new or they are now interpreted somewhat differently than other authors have done. The period around 4 years in the monthly *Ap* power spectrum is associated to the dual peak–structure observed in the geomagnetic activity variation (Gonzales et al., 1990). Several of the peaks shown by the daily *Ap* spectrum are interpreted as harmonics of the six–months period, and others as caused by the solar rotation periodicity, in such a way that the two series of Fourier sequences are considered to be juxtaposed. A strong solar cycle modulation is observed in these series, particularly in that related to the solar rotation period, which almost disappears for the solar maximum phase. Furthermore, a statistic analysis of the geomagnetic storm occurrence has confirmed the findings related to the dual–peak distribution as well to the seasonal variation.

084.080 **Generation mechanism of magnetic noise bursts in the Earth's magnetotail.**
Zhou Guo–Cheng, Cao Jin–Bn.
Sci. China, Ser. A, Vol. 34, No. 12, p. 1492 – 1499 (Dec 1991).

In this paper the authors study the electromagnetic instability driven by ion beams and plasma inhomogeneity in the plasma sheet boundary layer and for the first time obtain the growth rate spectrum of electromagnetic waves in a broad frequency range. The fundamental characteristics of the growth rate spectrum are in agreement with the satellite observations of the magnetic noise bursts in the magnetotail.

084.081 **Magnetotail reconnection, MHD theory and simulations.**
J. Birn, M. Hesse, K. Schindler.
International Workshop on Reconnection in Space Plasma, p. 217 – 221 (Jan 1989). – See Abstr. 012.099 for the main entry.

The authors discuss the present understanding of magnetotail reconnection leading to plasmoid formation and ejection. They emphasize three–dimensional structures and deviations from earlier imposed symmetries, based on MHD simulations and topological considerations. The authors find that, in general, the separation of the plasmoid will take a finite amount of time. During this state the plasmoid is characterized by filamentary structures of interwoven flux tubes with different topological connections.

084.082 **The role of magnetic reconnection in magnetotail plasmoid dynamics.**
A. Otto.
International Workshop on Reconnection in Space Plasma, p. 223 – 226 (Jan 1989). – See Abstr. 012.099 for the main entry.

In taillike configurations magnetic reconnection necessarily leads to the formation of plasmoids. The paper analyzes the dynamical evolution of the developing plasmoids and the influence of the magnetic reconnection on the properties of plasmoids.

084.083 **Broad electrostatic waves as a consequence of magnetotail reconnection.**
M. Ashour–Abdalla, D. Schriver.
International Workshop on Reconnection in Space Plasma, p. 241 – 248 (Jan 1989). – See Abstr. 012.099 for the main entry.

Using one– and two–dimensional particle simulations the authors examine ion beam driven instabilities within the plasma sheet boundary layer of the earth's magnetotail and the consequences these instabilities have in terms of plasma heating.

084.084 **The role of cold electrons in the generation of the high–frequency part of broadband electrostatic noise.**
T. M. Burinskaya, C.–V. Meister.
International Workshop on Reconnection in Space Plasma, p. 249 – 252 (Jan 1989). – See Abstr. 012.099 for the main entry.

084.085 **Electromagnetic electron temperature anisotropy instability in the earth's magnetotail plasma sheet.**
B. Nikutowski.
International Workshop on Reconnection in Space Plasma, p. 253 – 258 (Jan 1989). – See Abstr. 012.099 for the main entry.

The electromagnetic fire house instability driven by an electron temperature anisotropy is considered for parallel propagating waves as a possible explanation for the observations of magnetic noise bursts in the central plasma sheet of the geomagnetic tail.

084.086 **ELF plasma turbulence observed by Prognoz–8 in plasma jets in the magnetospheric tail.**
J. Blecki, K. Kossacki, B. Popielawska, S. A. Romanov, S. P. Savin.
International Workshop on Reconnection in Space Plasma, p. 259 – 262 (Jan 1989). – See Abstr. 012.099 for the main entry.

084.087 **Flux transfer events: reconnection without separators?**
M. Hesse, J. Birn, K. Schindler.
International Workshop on Reconnection in Space Plasma, p. 263 – 268 (Jan 1989). – See Abstr. 012.099 for the main entry.

The authors present a topological analysis of a simple model magnetic field of a perturbation at the magnetopause modeling an apparent flux transfer event.

084.088 **Multiple X line reconnection at the dayside magnetopause.**
L. C. Lee.
International Workshop on Reconnection in Space Plasma, p. 269 – 274 (Jan 1989). – See Abstr. 012.099 for the main entry.

A brief review of the recent studies on magnetic reconnection by the Alaskan group is presented.

084.089 **2–D and 3–D simulation study of multiple X line reconnection.**
Z. F. Fu.
International Workshop on Reconnection in Space Plasma, p. 275 – 279 (Jan 1989). – See Abstr. 012.099 for the main entry.

An incompressible MHD simulation is used to verify the proposed multiple X line reconnection process at the dayside magnetosphere. The 2–D simulation results are briefly reviewed and the preliminary 3–D simulation results are reported. The condition for the occurrence and the dynamical property of the multiple X line reconnection are discussed. The 3–D topology of the reconnected flux tubes is discussed with special attention directed to the ends of the tubes.

084.090 **Impulsive reconnection of the skewed magnetic fields at the magnetopause as a MHD model.**
V. S. Semenov.
International Workshop on Reconnection in Space Plasma, p. 287 – 289 (Jan 1989). – See Abstr. 012.099 for the main entry.

The unsteady reconnection problem in skewed fields under the finite length of the X line is considered in the frame of an ideal incompressible MHD approximation. The propagation of the discontinuities as well as the evolution of the magnetic field, plasma flow and pressure are computed analytically for the arbitrary given function which describes the time behaviour of the reconnection rate.

084.091 **The structure of the reconnection layer in the Petschek reconnection model: theory and application to the earth's magnetopause.**
R. P. Rijnbeek, M. F. Heyn, H. K. Biernat, V. S. Semenov.
International Workshop on Reconnection in Space Plasma, p. 291 – 296 (Jan 1989). – See Abstr. 012.099 for the main entry.

Reconnection is formulated in terms of a Riemann problem and the structure of the reconnection layer, which corresponds to the layer of MHD wave modes in Petschek's model, is analysed as a function of the parameters in the inflow regions. The influence of asymmetric inflow regions and the relevance to the earth's magnetopause is discussed.

Solar–terrestrial models and application software.
See Abstr. 002.033.

7. Quadrennial Symposium on Solar–Terrestrial Physics (SCOSTEP–7), The Hague (Netherlands), 25 – 30 Jun 1990.
See Abstr. 012.024.

Cluster dayside polar cusp. Planning and coordination of measurements from Cluster, ground stations, balloons and rockets in the dayside polar–cusp region. Proceedings.
See Abstr. 012.071.

Proceedings of the 10th ESA Symposium on European Rocket and Balloon Programmes and Related Research.
See Abstr. 012.088.

Reconnection in space plasma. Proceedings.
See Abstr. 012.099.

Space plasma physics at the Applied Physics Laboratory over the past half–century.
See Abstr. 013.037.

The Norwegian upper atmosphere programme in Svalbard.
See Abstr. 013.108.

U.S. ground–based space research programs in Svalbard.
See Abstr. 013.109.

Research programmes at Svalbard.
See Abstr. 013.110.

The Polar Geophysical Institute program of the theoretical and experimental investigation on cusp.
See Abstr. 013.111.

Cusp and cleft studies in Greenland.
See Abstr. 013.112.

Plans for the EISCAT Svalbard radar.
See Abstr. 013.113.

Requirements for coordinated Cluster and ground–based observations of the cusp.
See Abstr. 013.115.

China meridian chain of magnetometers and the global electric current system.
See Abstr. 013.126.

Airglow observation program carried out by Tokyo Astronomical Observatory.
See Abstr. 013.139.

$N(^2D) + O_2$: a source of thermospheric 6300 Å emission?
See Abstr. 022.175.

Magnetic and radio detection of aurorae.
See Abstr. 034.054.

Optical ground–based network.
See Abstr. 036.174.

Imaging the Earth's magnetosphere using ground–magnetometer arrays.
See Abstr. 036.175.

The requirements for data exchange between Cluster and ground–based observatories, using EISCAT as an illustrative example.
See Abstr. 036.176.

Lesson learned from GEOS and ISEE.
See Abstr. 051.049.

Balloons for conjugate cusp studies in the 1990's.
See Abstr. 051.050.

Mission objectives and scientific rationale for the magnetometer mission.
See Abstr. 051.051.

MHD flow past an obstacle: large–scale flow in the magnetosheath.
See Abstr. 062.004.

A comparison of the reduced and approximate systems for the time dependent computation of the polar wind and multiconstituent stellar winds.
See Abstr. 062.014.

Semikinetic and generalized transport models of the polar and solar winds.
See Abstr. 062.015.

Theory of field line resonances of standing shear Alfvén waves in three–dimensional inhomogeneous plasmas.
See Abstr. 062.017.

Generation of radiation by upper–hybrid waves in non–uniform plasmas.
See Abstr. 062.065.

Magnetosphere–ionosphere coupling in multi–species magneto plasmas.
See Abstr. 062.071.

Resistive tearing–mode instability in a current sheet with equilibrium viscous stagnation–point flow.
See Abstr. 062.072.

Dissipative stagnation–point flows at a current sheet with shear in the plasma velocity.
See Abstr. 062.090.

Noise–free neutral sheets.
See Abstr. 062.228.

Anomalous collision frequency of magnetoactive plasmas with two–dimensional electrostatic turbulence (kB = 0).
See Abstr. 062.232.

Particle simulation of magnetic reconnection in collision–free plasma sheets.
See Abstr. 062.237.

Three–dimensional reconnection from a global viewpoint.
See Abstr. 062.238.

An analogy between magnetospheric substorms and solar flares: new version.
See Abstr. 073.132.

Role of periodic fluctuations in solar wind–magnetosphere coupling.
See Abstr. 074.038.

In situ observations of coronal mass ejections in interplanetary space.
See Abstr. 074.073.

The role of the magnetic reconnection in the solar wind/magnetosphere interaction.
See Abstr. 074.085.

On a relationship between earthquake centers and regions of high–energy particle precipitations from the radiation belts.
See Abstr. 081.038.

Polar airglow and aurora.
See Abstr. 082.006.

20 years with electron beam experiments on rockets and satellites.
See Abstr. 082.074.

Alfvén waves in the auroral ionsophere: a numerical model compared with measurements.
See Abstr. 083.003.

A confirmation of the validity of the electric current distribution determined by a ground–based magnetometer network.
See Abstr. 083.013.

Calculation of the heating of the electronic component of the ionosphere and plasmasphere by Alfvén and fast magnetosonic waves.
See Abstr. 083.032.

The present understanding of the cusp.
See Abstr. 083.034.

The ionospheric signature of flux transfer events.
See Abstr. 083.035.

Ground–based studies of sensing magnetosphere/ionosphere interactions: convection and substorms.
See Abstr. 083.036.

Relationships between ionospheric and tail phenomena.
See Abstr. 083.037.

Comparison of auroral emission and EISCAT–derived plasma parameters during the flight of NEED–II.
See Abstr. 083.043.

Solar wind and magnetospheric processes and various coupling mechanisms including those observed on other planets.
See Abstr. 085.014.

Interplanetary disturbance structures and geomagnetic storm.
See Abstr. 085.015.

Parameter identification of geomagnetic disturbances initiated by the solar wind.
See Abstr. 085.021.

Inductive acceleration of protons and electrons in planetary magnetotails.
See Abstr. 091.066.

The response of the quiet–time auroral configuration to short– and long–term interplanetary magnetic field variations.
See Abstr. 106.004.

Transient interplanetary and geomagnetic disturbances associated with solar flares.
See Abstr. 106.090.

Particle acceleration at the front of a head shock–wave.
See Abstr. 144.055.

The geophysical effects of cosmic rays.
See Abstr. 144.073.

085 Solar-terrestrial Relations

085.001 **The geoeffectiveness of the primary indices of spot formation on the Sun.**
V. A. Dergachev, A. V. Shershnev.
Geomagn. Aeron., Vol. 30, No. 5, p. 625 – 627 (Apr 1990). English translation of Geomagn. Aehron., Tom 30, No. 5, p. 736 – 739 (1990).
    The connection between global geomagnetic disturbances and variations of the primary indices of spot formation has been analyzed. A close correlation has been shown to exist between the strength of solar magnetic phenomena and the unrest of the geomagnetic field. The delay of the aa indices relative to changes in the lifetime of groups of spots T_0 is evidently related to drift towards the pole of the solar magnetic field, manifested in migration towards the pole of high–latitude proturberances, polar faculae, and phenomena of coronal activity. It has been concluded that the high–latitude magnetic field of the Sun influences the lifetime of spot groups in high–latitude regions of the spot formation zone.

085.002 **The spatial distribution of the association between total ozone and the 11–year solar cycle.**
K. Labitzke, H. van Loon.
Geophys. Res. Lett., Vol. 19, No. 4, p. 401 – 404 (21 Feb 1992).
    The pattern of correlation between the 11–year solar cycle and heights and temperatures in the lower stratosphere is in all months shaped as a crescent with its axis in the subtropics. The change of total ozone from the solar maximum in 1979–1980 to the minimum in 1985–1986 has the same shape. Although the effect of the solar cycle is said to have been removed from the ozone data, two thirds of the stations which have been used for this purpose lie outside the regions where the stratosphere is significantly correlated with the solar cycle. For this reason it is unlikely that the influence of the cycle has been completely eliminated.

085.003 **No dependence of the temperature of the troposphere at Berlin on the solar activity cycle.**
W. Elling, H. Schwentek.
Sol. Phys., Vol. 137, No. 2, p. 401 – 402 (Feb 1992). Letter–to–the–editor.
    The application of statistical analysis on the set of tropospheric temperature data obtained from daily radiosonde launchings at Berlin–Tempelhof from June 1958 to June 1986 does not confirm a dependence of tropospheric temperature on sunspot number previously considered to be inside the bounds of possibility.

085.004 **Effect of geomagnetic disturbances on the flux intensity of direct solar radiation.**
M. I. Pudovkin, S. V. Veretenko.
Geomagn. Aehron., Tom 32, No. 1, p. 148 – 150 (Jan – Feb 1992). In Russian. English translation in Geomagn. Aeron., Vol. 32, No. 1.

085.005 Neutral networks and predictions of solar–terrestrial effects.
H. Lundstedt.
Planet. Space Sci., Vol. 40, No. 4, p. 457 – 464 (Apr 1992).

Neural networks for predictions of solar–terrestrial effects, such as geomagnetic induced currents (GICs), are presented. The following assumptions are made: the geomagnetic activity is mainly controlled by the southward B_z-component of the solar wind. There are three major solar causes of the southward B_z-components, namely the solar sector boundaries (SSB), the coronal mass ejections (CME) and the coronal holes (CH). The mean GIC size has an exponential relation to the geomagnetic activity index K_p. The neural networks were trained with solar input data from various U.S. data bases: SSB–data from CSSA, Stanford, California, CME–data and solar wind–data from NOAA/SEL, Boulder, Colorado and finally CH–data from SacPeak/AFGL.

085.006 Sun–controlled spatial and time–dependent cycles in the climatic/weather system.
E. C. Njau.
Nuovo Cimento C, Vol. 15, N. 1, p. 17 – 23 (Jan–Feb 1992).

The author shows, on the basis of meteorological records, that certain spatial and time–dependent cycles exist in the earth–atmosphere system. These cycles seem to be associated with sunspot cycles.

085.007 States and interstate switching in meteorological parameters.
E. C. Njau.
Nuovo Cimento C, Vol. 15, N. 1, p. 25 – 35 (Jan–Feb 1992).

The author shows theoretically and also through actual records that variations in air/surface temperature and associated meteorological parameters take place in three different states. Systematic switching from one state to another does take place at several places. Available evidence indicates that interstate switching between states 1 and 2 is primarily caused by the 11 year solar cycle and some of those other cycles responsible for variations in the solar energy incident onto the earth–atmosphere system. Finally it is shown that one of the possible causes of double sunspot cycle variations in meteorological parameters is interstate switching between state 1 and state 2.

085.008 Solar and geomagnetic variability and changes of weather and climate.
V. Bucha.
J. Atmos. Terr. Phys., Vol. 53, No. 11/12, p. 1161 – 1172 (Nov–Dec 1991). – See Abstr. 012.024 for the main entry.

Short–term and long–term changes in solar activity, the geomagnetic field, weather and climate show very similar quasi–periodic variations. The author's results that are substantiated by statistical tests are used to investigate the hypotesis that the changes in solar activity can influence processes in the auroral oval which modulate the alternation of meridional and zonal types of atmospheric circulation, leading to changes in temperature and pressure. Long–term climate fluctuations can be explained as being due to the wander of the geomagnetic poles influenced by processes in the Earth's interior (which are modified by the Earths orbital characteristics). These poles represent a centre of the auroral zone, the shift of which could lead to crucial changes of atmospheric flow and thus to the occurrence of glacial and interglacial periods. The increase of global temperature in the past may participate also in changes in the CO_2 concentration. The results can contribute to the solution of the problem of how responses to anthropogenic impacts on the Earth can be distinguished from the variability of the natural system, mainly of the natural forcing mechanisms and of the climate.

085.009 Solar activity, geomagnetic variations, and climate changes.
D. K. Nurgaliev.
Geomagn. Aeron., Vol. 31, No. 1, p. 14 – 18 (Aug 1991). English translation of Geomagn. Aehron., Tom 31, No. 1, p. 20 – 26 (1991).

Relations between changes of the climate and the magnetic field of the earth are established on the basis of a comparison of paleoclimatic and geomagnetic data for the last ~25 thousand years and during one Late Permian reversal. Possible mechanisms of the influence of the variations and geomagnetic field reversal on the climate are discussed.

085.010 Centers of flare activity and their link with geomagnetic activity.
A. V. Borovik, L. V. Borovik, V. A. Parkhomov.
Geomagn. Aeron., Vol. 31, No. 1, p. 137 – 139 (Aug 1991). English translation of Geomagn. Aehron., Tom 31, No. 1, p. 187 – 190 (1991).

The space–time characteristics of the flare activity of 130 groups of spots are examined according to the present Solar–Geophysical Data. It is established that flares of importance S and time series of flares originate in certain sites of active regions–centers of flare activity. Is is found that after periods of active operation of centers of flare activity a growth in geomagnetic activity and sudden–onset geomagnetic storms follow.

085.011 Monthly variations of the Caspian sea level and solar activity.
P. R. Romanchuk, M. N. Pasechnik.
Visn. Kiiv. Univ., Fiz.–Mat. Nauki, Astron., Vip. 3, p. 74 – 79 (1992). In Ukrainian.

The connection between 11–year cycle of solar activity and the Caspian sea level is investigated. Seasonal changes of the Caspian sea level and annual variations of the sea level with variations of solar activity are studied. The results of the verifications of the sea level forecasts obtained with application of the rules discovered by the authors are given.

085.012 Response of the middle atmosphere to solar proton events in October 1989.
A. M. Zadorozhnyj, V. N. Kikhtenko, G. A. Kokin, O. M. Raspovov, O. I. Shumilov, G. A. Tuchkov, M. I. Tyasto, A. F. Chizhov, O. V. Shtrykov, E. A. Kasatkina, Eh. V. Vashenyuk.
Geomagn. Aehron., Tom 32, No. 2, p. 32 – 40 (Mar – Apr 1992). In Russian. English translation in Geomagn. Aeron., Vol. 32, No. 2.

085.013 Macroscopic fluctuations, Sun–Earth links, and methodological aspects of excat measurements.
B. M. Vladimirskij.
Bull. Crimean Astrophys. Obs., Vol. 82, p. 148 – 157 (1992). English translation of Izv. Krym. Astrofiz. Obs., Tom 82, p. 161 – 172 (1990).

Major phenomenological features are briefly described for macroscopic fluctuations. Some such fluctuation parameters are dependent on solar activity, in which the agent directly affecting the system is the background electromagnetic field at low and infralow frequencies. Macroscopic fluctuations occur very widely, and in many fairly precise measurements, there may be a correlation between the measures quantity and the solar–activity parameters, e.g., in precision measurements on physical constants such as the velocity of light. Sometimes, these fluctuations can be a source of serious errors in interpreting measurements. Arguments are presented to show that the solar–activity dependence can affect the recording performance in some systems. This is, for example, the source of the microvariations in intensity in the cover system at the Baksan neutrino observatory, or fluctuations in brightness in certain Seyfert galaxies with the period of the solar pulsations, 160 min. It is suggested that such fluctuations are associated with the triggering of gravitational–wave detectors. The scope for rigorous experimental test is discussed.

085.014 **Solar wind and magnetospheric processes and various coupling mechanisms including those observed on other planets.**
C. A. Reddy.
AIP Conf. Proc., No. 245: Workshop on Basic Space Science, p. 84 – 117 (1992). – See Abstr. 012.064 for the main entry.

This overview paper presents a brief description of the basic physical processes involved in the interactions between the solar wind and the magnetosphere, the ionosphere and the thermosphere of the Earth's space environment. Sources for information on the magnetospheres of the other planets of the solar system are given.

085.015 **Interplanetary disturbance structures and geomagnetic storm.**
Zhang Gongliang.
AIP Conf. Proc., No. 245: Workshop on Basic Space Science, p. 118 – 124 (1992). – See Abstr. 012.064 for the main entry.

This article summarizes main results obtained by the author and his colleagues through recent studies of solar disturbance plasma outputs. The discussions involve the basic characteristics of corotating and transient disturbances and their solar cycle variations. He presents evidences for quasi–steady structure of corotating disturbances, and stresses the idea about magnetic expansion of magnetic clouds as the basic character of strong transient disturbance. Finally, the morphology of geomagnetic storms is discussed in response to interplanetary structures.

085.016 **On research of solar activities–climate/weather relationship.**
Lu Daren, Huang Zhen, Shen Mei.
Advances in solar–terrestrial science of China, p. 212 – 222 (1992). – See Abstr. 003.069 for the main entry.

In this paper, a short review on the present status of the research on the solar activity–climate/weather relationship is presented, including statistical research, different points of view on the relationship and some possible mechanisms relevant to this subject. Emphasis is on the newly–founded Solar Cycle–QBO–Climate Statistics. Comments on problems to be further studied are discussed.

085.017 **Solar activities and atmospheric electric processes.**
Yan Muhong, Zhang Yijun.
Advances in solar–terrestrial science of China, p. 235 – 247 (1992). – See Abstr. 003.069 for the main entry.

The probable physical relationship between solar activities and thunderstorm activities has been studied in this paper by using correlation analysis of data and theoretical model research.

085.018 **Climate response to solar activity.**
Wang Shaowu.
Advances in solar–terrestrial science of China, p. 248 – 257 (1992). – See Abstr. 003.069 for the main entry.

Studies on solar–climatic connections are reviewed. Significant correlations are found between the atmospheric circulation indices, temperatures and precipitations on the one hand, and different scale of solar variabilities on the other. The important impacts of the secular cycle, Hale cycle and 5 – 6 year cycle on the climate of the earth are emphasized.

085.019 **Analysis of solar–terrestrial general effects.**
Gao Meiqing, Kong Nan, Xiang Jingtian, Wen Xiaolei, Lu Wensong.
Advances in solar–terrestrial science of China, p. 303 – 311 (1992). – See Abstr. 003.069 for the main entry.

This paper reviews the study and analysis of solar–terrestrial general effects. Particularly it introduces the authors' study in China on general solar–terrestrial effects of the major solar events in February, 1986 and March, 1989 and the prediction of the geomagnetic activity maximum and peak time in the 22nd solar cycle. The expert system for geomagnetic activity prediction is introduced. The analysis of the relationship between satellite anomaly with geomagnetic storms, substorms and the environment is made.

085.020 **Synergetic analysis of the solar–terrestrial relations.**
D. G. Gochev, P. I. Nenovski.
Aerospace Res. Bulg., Vol. 8, p. 3 – 12 (1991). In Bulgarian.

085.021 **Parameter identification of geomagnetic disturbances initiated by the solar wind.**
Zhou Xiao–yan, Tschu Kung–kun.
Acta Geophys. Sin., Vol. 35, No. 3, p. 278 – 287 (May 1992). In Chinese.

Solar–terrestrial models and application software.
See Abstr. 002.033.

Advances in solar–terrestrial science of China.
See Abstr. 003.069.

7. Quadrennial Symposium on Solar–Terrestrial Physics (SCOSTEP–7), The Hague (Netherlands), 25 – 30 Jun 1990.
See Abstr. 012.024.

Solar–terrestrial models at the National Space Science Data Center.
See Abstr. 013.027.

The need of coordinated observations in space and from ground.
See Abstr. 013.114.

The program on global character research of solar–terrestrial system in the maximum period of the 22nd solar cycle.
See Abstr. 013.123.

Coordinated observations and measurements of solar–terrestrial system.
See Abstr. 013.124.

The progress of solar–terrestrial sciences in China.
See Abstr. 013.125.

The characteristic of the 30 – 60 day fluctuation in the Earth rotation, atmospheric angular momentum and solar activity.
See Abstr. 044.031.

A possible cause of the formation of some short periodic fluctuation in LOD.
See Abstr. 044.032.

MHD intermediate shock wave.
See Abstr. 062.204.

On classification and systematization of solar activity indices.
See Abstr. 072.022.

The downturn in solar activity during solar cycles 5 and 6.
See Abstr. 072.067.

Research of solar events in optical, radio and X–ray regions.
See Abstr. 073.077.

Long–term solar variability and solar seismology: I.
See Abstr. 080.076.

Terrestrial response to eruptive solar flares: geomagnetic storms.
See Abstr. 084.077.

Coronal mass ejections: the link between solar and geomagnetic activity.
See Abstr. 084.078.

A study of geomagnetic variations with periods of four years, six months and 27 days.
See Abstr. 084.079.

Planetary System

091 Physics and Dynamics of the Planetary System

091.001 Nuclear fission reactors as energy sources for the giant outer planets.
J. M. Herndon.
Naturwissenschaften, Jahrg. 79, Heft 1, p. 7 – 14 (Jan 1992).
Of the giant planets, Jupiter, Saturn, and Neptune presently radiate into space approximately twice as much energy as they receive from the Sun; Uranus, however, emits little, if any, energy other than absorbed solar energy. The purpose of the present paper is to suggest the possibility of naturally occurring nuclear fission reactors in the giant outer planets. The discovery of a naturally occurring, terrestrial, nuclear fission, 'breeder' reactor is reviewed. Quantitative estimates are made of the planetary energy release by nuclear fission and of the duration that present planetary power output levels could be sustained by nuclear fission energy.

091.002 The greenhouse effect within the solar system.
R. Courtin, C. P. McKay, J. Pollack.
Recherche, Vol. 23, No. 243, p. 542 – 549 (May 1992). In French.

091.003 Planetary atmospheres.
Z. Pokorný.
Kozmos, Vol. 22, No. 4, p. 111 – 117 (Aug 1991). In Slovak.

091.004 Venus, Earth, Mars... – the alchemy of planetary climates.
Z. Pokorný.
Kozmos, Vol. 22, No. 4, p. 120 – 122 (Aug 1991). In Slovak.

091.005 Three–dimensional perturbations of particles in a narrow planetary ring.
R. A. Kolvoord, J. A. Burns.
Icarus, Vol. 95, No. 2, p. 253 – 264 (Feb 1992).
In an effort to understand the observed complex form of Saturn's F ring, the authors have used Gauss' perturbation equations to numerically model the short–term, three–dimensional dynamics of narrow rings. They consider ring particles that are perturbed by local moonlets orbiting with small ($i \leqslant 0°.1$) inclination; collisions are ignored. It is confirmed that, as expected, the distance of closest approach determines the strength of the interaction; this distance depends on the orientation of the orbits, as well as the orbital eccentricity and the separation in semimajor axes of the ring particle from the perturbing satellite. It is found that, with an appropriate choice of parameters, ring particles can be perturbed to substantial inclinations. The authors present simulations of a model narrow ring consisting initially of three strands typically with 800 particles apiece, which show the short–term effects of the neighboring moons and the influence of different orientations at encounter.

091.006 Equilibrium velocities in planetary rings with low optical depth.
K. Ohtsuki.
Icarus, Vol. 95, No. 2, p. 265 – 282 (Feb 1992).
The author examines the equilibrium random velocity in planetary rings of one and two particle size components with low optical depth, where the epicyclic approximation for particle

motions can be applied. First, assuming that the random velocity is large enough to allow particle gravity to be neglected, the rate of random velocity evolution due to collisions is evaluated. In the case that the restitution coefficient of the particles is independent of impact velocity, the author derives analytic expressions for the rate of evolution. He also estimates the equilibrium random velocity, using the restitution coefficient of icy particles obtained by impact experiments. Next, he includes particle gravity with the aid of orbital calculations of the three–body problem. Finally, he examines the equilibrium velocities of two–component systems for arbitrary mass ratios. It is found that the smaller particles will have equilibrium velocity determined by the velocity–dependent restitution coefficient, while velocity of larger particles seems to tend toward energy balance with smaller particles if the mass ratio is not large and the mass of large particles does not exceed a certain value. If these conditions are violated, the velocity of large particles is also determined by the velocity–dependent restitution coefficient.

091.007 Analysis of stellar occultation data for planetary atmospheres. I. Model fitting, with application to Pluto.
J. L. Elliot, L. A. Young.
Astron. J., Vol. 103, No. 3, p. 991 – 1015 (Mar 1992). Current Physics Microform No.: 9203E2199.
An analytic model for a stellar–occultation light curve has been developed for a small, spherically symmetric planetary atmosphere that includes thermal and molecular weight gradients in a region that overlies an extinction layer. This work applies to the thermal structure of the upper part of Pluto's atmosphere probed by current stellar occultation data, so the issue of whether the lower part should be modeled as an extinction layer or sharp thermal gradient is not addressed. The model can be described by two equivalent sets of parameters. One set specifies the occultation light curve in terms of signal levels, times, and time intervals. Consequently, it is the more suitable set to use for fitting the light curve. The other set specifies physical parameters of the planetary atmosphere. Equations are given for the transforming between the sets of parameters, including their errors and correlation coefficients. Detailed numerical calculations are presented for a benchmark case. In order to establish the formal errors in the model parameters expected for datasets of different quality, least–squares fitting tests are carried out on synthetic datasets with different noise levels. This model has also been fit to the KAO data from the 1988 June 9 stellar occultation by Pluto.

091.008 Aerosol component of the atmospheres of major planets.
V. G. Tejfel'.
Astron. Vestn., Tom 26, No. 1, p. 3 – 27 (Jan – Feb 1992). In Russian. English translation in Solar Syst. Res., Vol. 26, No. 1.
The main results of a study of the aerosol component (clouds and haze) in the atmospheres of Jupiter, Saturn, Uranus and Neptune are briefly reviewed. The albedo and color characteristics of clouds, their microphysical structure and the altitude differences, derived from photometry, polarimetry and the molecular absorption bands intensity, physical–chemical composition of the clouds and the properties of stratospheric aerosol are considered.

091.009   **Volcanism and tectonics on planets and satellites of solar system: dependence on the body size and on orbital period.**
A. T. Bazilevskij, M. A. Kreslavskij.
Astron. Vestn., Tom 26, No. 2, p. 66 – 76 (Mar – Apr 1992). In Russian. English translation in Solar Syst. Res., Vol. 26, No. 2.

28 planets and satellites, for which good enough images are available, are classified according to the degree of their late (after heavy bombardment) endogenic activity. This activity occurred at different degree on 16 of the studied bodies and did not occur in a noticeable form on 12 bodies. It was found that, as a rule, the larger the studied body and the less the period of its rotation around the central body, the higher is its endogenic activity. Physical phenomena controlling this dependence are briefly considered. It has been shown that the body radius is a parameter which is proportional to the amount of heat generated by early and late sources and to the degree of preservation of early heat at late geologic epochs. The orbital period is a parameter of effectiveness of tidal heating of satellites. For some poorly studied bodies of the solar system a prognosis of their late endogenic activity has been made. The most chances for that has Titan, a Saturn satellite, less chances have Pluto and its satellite Charon. On other poorly studied bodies, which are some satellites of giant planets and all asteroids, the late endogenic activity was evidently absent.

091.010   **The shadow effect for a planetary surface with Gaussian mesorelief.**
Yu. G. Shkuratov, D. G. Stankevich.
Astron. Vestn., Tom 26, No. 2, p. 89 – 101 (Mar – Apr 1992). In Russian. English translation in Solar Syst. Res., Vol. 26, No. 2.

The Smith – Fuks method of calculation of the shadow effect for a rough surface is generalized for the case which takes into account a correlation for propagation of incident and emergent light rays. The probability that surface points are both illuminated and visible are obtained for arbitrary azimuth and incident and emergent angles. For large angles of incidence and emergence the azimuth dependence has a sharp increase at the zero azimuth angle, that resembles the opposition effect of atmosphereless celestial bodies. Calculations are conformed to computer modelling.

091.011   **Confirmation of resonant structure in the Solar System.**
J. Laskar, T. Quinn, S. Tremaine.
Icarus, Vol. 95, No. 1, p. 148 – 152 (Jan 1992).

Using a semianalytical secular theory, Laskar computed the orbits of the planets over 200 million years and found that their motion, and especially the motion of the inner planets, is chaotic, with a Liapunov exponent of $1/(5 Myr)$. It was found that this chaotic behavior originated in the existence of the resonances between the main secular frequencies associated with the perihelia and nodes of the planets. The recent numerical integration of the Solar System over 6 Myr by Quinn et al. is in excellent agreement with the results of the secular theory and confirms the validity of the use of both solutions for orbital calculations related to Milankovich climate theory. The 6–Myr time span is too short to confirm the chaotic behavior directly, but the existence of the secular resonance between the Earth and Mars is confirmed, and the behavior of the resonant argument shows very good quantitative agreement, thus providing strong indirect support that chaos is actually present.

091.012   **Simulation of frontal cloudiness in the atmospheres of giant planets. I. Method of calculations.**
K. Yu. Ibragimov.
Kinematika Fiz. Nebesn. Tel, Tom 8, No. 2, p. 15 – 24, 35 (Mar–Apr 1992). In Russian. English translation in Kinematics Phys. Celest. Bodies, Vol. 8, No. 2.

A system of equations describing the formation and evolution of two–component frontal cloud systems in the atmospheres of giant planets is considered. Horizontal motions of atmospheric masses are supposed to be absent at the boundaries of the region and the velocities inside the region are supposed to change according to the parabolic law. Space distributions of waterance

and ammoniance in the atmosphere of Jupiter are calculated for such a simplest dynamical model. The following parameters were assumed as variables: concentrations x_i; diffusion coefficient k supposed to be constant within the region; maximum vertical velocity w_m; and maximum values of horizontal velocities of the wind u_m and v_m. Variation intervals for them are as follows: $x_i = 10^5 - 10^{-3}$, $k = 10^6 - 10^7 cm^2/s$, $w_m = 20 - 40 cm/s$, $u_m = 50 - 100 m/s$, $v_m = 50 - 100 m/s$.

091.013   **Radiative association in planetary atmospheres.**
A. Dalgarno, J. F. Babb, Y. Sun.
Planet. Space Sci., Vol. 40, No. 2/3, p. 243 – 246 (Feb–Mar 1992).

A summary account is presented of radiative association processes that may be significant in producing continuum emission in the atmospheres of the planets. The importance of radiative association of N and O and of NO+O has been established by the presence in the airglow of the emission spectra. The possible contributions of $O+O$, $O+O_2$ and $O^+ +O$ are discussed. The radiative association of He^+ and H may be an important source of HeH^+ in the Jovian ionospheres.

091.014   **Here and there on a topography of the largest moons of the solar system.**
S. R. Brzostkiewicz.
Urania, Rok 63, Nr. 6, p. 168 – 179 (Jun 1992). In Polish.

091.015   **Periodic system of multi–ring planetary structures as result of interference of variously oriented lithospheric waves.**
G. G. Kochemasov.
Astron. Tsirk., No. 1550, p. 35 – 36 (Sep – Oct 1991). In Russian.

Regular positions of giant ring planetary structures on the Moon and the Earth were noted earlier. An equatorial belt of the Earth's giant rings consists of 8 regularly spaced superstructures. Characteristic size of each is $2\pi R/8 \approx 3/4R$. The regular net of such planetary structures is linked to interference of lithospheric waves of four main directions.

091.016   **Extra–long lithospheric waves forming the morphotectonic face of planets.**
G. G. Kochemasov.
Astron. Tsirk., No. 1550, p. 37 – 38 (Sep – Oct 1991). In Russian.

Giant ring superstructures are linked to lithospheric waves with length about a planet radius or so. Extra–long waves (more than a radius) are responsible for a general morphotectonic appearance of a solid planet. The Eastern hemisphere of the Earth can be modelled by interference of lithospheric waves of 4 directions with wave lengths about 3 radii.

091.017   **Shock drift acceleration of energetic protons at a planetary bow shock.**
J. Giacalone.
J. Geophys. Res., Vol. 97, No. A6, p. 8307 – 8318 (1 Jun 1992).

The author presents the results of numerical orbit integrations of the interaction of suprathermal charged particles (protons) with a planetary bow shock. The primary goal of this study is to analyze the effect of the changing geometry of the shock, due to its curvature, on the kinematics of the particle/shock interaction.

091.018   **Transformation of the orbits of small bodies by close encounters with the terrestrial planets.**
G. V. Andreev, A. K. Terent'eva, O. A. Bayuk.
Sol. Syst. Res., Vol. 25, No. 2, p. 129 – 131 (Sep 1991). English translation of Astron. Vestn., Tom 25, No. 2, p. 177 – 180 (1991).

The Laplace method of the unperturbed two–body problem in the spheres of influence of the planets and Sun is used to show that close encounters of small bodies with the terrestrial planets could be one of the sources of short–period orbits. At least some of the Amur and Apollo asteroids and meteor swarms like the Geminids and Arietids could be of cometary origin.

091.019 **Collisionless braking of dust particles in the electrostatic field of planetary dust rings.**
O. Havnes, T. Aslaksen, F. Melandsø, T. Nitter.
Phys. Scr., Vol. 45, No. 5, p. 491 – 496 (May 1992). – See Abstr. 012.023 for the main entry.

The authors show that a dust test particle moving in a periodic orbit in electrostatic fields, as those due to planetary rings, will experience a net deceleration also in the absence of any dust–dust collisions. The varying charge on the moving dust, as it moves in and out of regions of differing electron and ion densities, will be out of phase with the equilibrium charge in a way which leads to a net braking of the particle. This effect has been shown to damp the coherent oscillations of electrostatically supported dust rings and to damp the oscillations of levitated dust particles in plasma sheaths at surfaces of solid bodies. It is shown that this effect will lead to a damping of internal random velocities in a planetary ring, at a rate which in many cases can be much faster than that due to dust–dust collisions.

091.020 **Collisionless damping of oscillations in electrically supported dust rings.**
F. Melandsø.
Phys. Scr., Vol. 45, No. 5, p. 515 – 520 (May 1992). – See Abstr. 012.023 for the main entry.

Low frequency density waves in a planetary ring, will be damped if there is a significant time delay in charging of dust particles. This damping can in many cases be more important than damping due to dust–dust and dust–plasma collisions. The author We has derived the response in plasma potential and dust particle charge when the dust density is varied periodically with frequency. This response is then applied to oscillations of dust particles in Jupiter's and Saturn's tenuous rings in the direction normal to the central plane of the rings. For tenuous rings one finds analytically expressions which approximate the eigenfrequencies and timescales of damping for the oscillation. These expressions are compared with numerically solutions of the original set of equations.

091.021 **Secular variations in optical observations of planets.**
P. K. Seidelmann.
IAU Symposium No. 152: Chaos, resonance and collective dynamical phenomena in the solar system, p. 49 – 51 (1992). Abstract. – See Abstr. 012.025 for the main entry.

091.022 **Earth's atmosphere: terrestrial or extraterrestrial?**
L. Broadhurst.
Astronomy, Vol. 20, No. 1, p. 38 – 45 (Jan 1992).

New studies indicate that the atmosphere of Earth – like that of the other terrestrial planets – may have been born with the planet itself and not simply outgassed in volcanic eruptions.

091.023 **NASA's never–ending mission.**
J. Williams.
Astronomy, Vol. 20, No. 2, p. 38 – 43 (Feb 1992).

091.024 **Elemental mapping of planetary surfaces using gamma–ray spectroscopy.**
R. C. Reedy.
AIP Conf. Proc., No. 238, p. 994 – 1002 (15 Oct 1991). Current Physics Microform No.: 9203B1739. – See Abstr. 012.036 for the main entry.

The gamma rays escaping from a planet can be used to map the concentrations of various elements in its surface. In a planet, the high–energy particles in the galactic cosmic rays induce a cascade of particles that includes many neutrons. The γ rays are made by nuclear excitations induced by these cosmic–ray particles and their secondaries and by the decay of the naturally–occurring radioelements. After a short history of planetary γ–ray spectroscopy and its applications, the γ–ray spectrometer planned for the Mars Observer mission is presented. Laboratory experiments that simulate the cosmic–ray bombardments of planetary surfaces or measure cross sections for the production of γ rays are reviewed. Theoretical calculations for the processes that make and transport neutrons and γ rays are discussed. The emphasis is on studies of Mars and on new ideas, concepts, and problems that have arisen over the last decade, such as Doppler broadening and peaks from neutron scattering with germanium nuclei in a high–resolution γ–ray spectrometer.

091.025 **Deuterium in the solar system.**
T. Owen.
IAU Symposium No. 150: Astrochemistry of cosmic phenomena, p. 97 – 101 (1992). – See Abstr. 012.029 for the main entry.

Values of D/H measured in the methane on the giant planets and Titan indicate the presence of two distinct reservoirs of deuterium in the outer solar system. The dominant reservoir is in hydrogen gas, the second, multi–component reservoir is found in the hydrogen that is bound in condensed compounds. Both reservoirs appear to have originated in the interstellar medium. In contrast, the values of D/H in water vapor on Mars and Venus (especially) exhibit a large enrichment from the "condensed matter" starting value. Interpretation of this enrichment may illuminate the history of water on these two planets.

091.026 **On the importance of differential rotation in planetary dynamos.**
H. Houben.
Bull. Am. Astron. Soc., Vol. 23, No. 3, p. 1134 – 1135 (1991). Abstract. – See Abstr. 012.037 for the main entry.

091.027 **Planetary neutral clouds.**
J. D. Richardson.
Bull. Am. Astron. Soc., Vol. 23, No. 3, p. 1148 (1991). Abstract. – See Abstr. 012.037 for the main entry.

091.028 **Evidence for magma oceans on asteroids, the Moon, and Earth.**
G. J. Taylor, M. D. Norman.
Workshop on the Physics and Chemistry of Magma Oceans from 1 bar to 4 Mbar, p. 58 – 65 (1992). Abstract. – See Abstr. 012.032 for the main entry.

091.029 **Magma ocean formation due to giant impacts.**
W. B. Tonks, H. J. Melosh.
Workshop on the Physics and Chemistry of Magma Oceans from 1 bar to 4 Mbar, p. 66 – 67 (1992). Abstract. – See Abstr. 012.032 for the main entry.

091.030 **Giant planet magnetospheres.**
F. Bagenal.
Bull. Am. Astron. Soc., Vol. 23, No. 3, p. 1152 (1991). Abstract. – See Abstr. 012.037 for the main entry.

091.031 **A model for the mean zonal velocities and alternating jets in planetary atmospheres.**
H. G. Mayr, K. L. Chan, I. Harris.
Bull. Am. Astron. Soc., Vol. 23, No. 3, p. 1165 – 1166 (1991). Abstract. – See Abstr. 012.037 for the main entry.

091.032 **Inconsistencies in Voyager satellite interpretations: new evidence for endogenic modification of crater size distribution.**
W. K. Hartmann.
Bull. Am. Astron. Soc., Vol. 23, No. 3, p. 1169 (1991). Abstract. – See Abstr. 012.037 for the main entry.

091.033 **Surface frosting: an important aspect of the behavior of thermally buffered frosts and ices.**
D. A. Paige, H. H. Kieffer.
Bull. Am. Astron. Soc., Vol. 23, No. 3, p. 1169 (1991). Abstract. – See Abstr. 012.037 for the main entry.

091.034 **Bistatic radar studies of planetary surfaces.**
R. A. Simpson.
Bull. Am. Astron. Soc., Vol. 23, No. 3, p. 1177 (1991). Abstract. – See Abstr. 012.037 for the main entry.

091.035   **Non–linear Lorentz resonances.**
J. A. Burns, D. P. Hamilton.
Bull. Am. Astron. Soc., Vol. 23, No. 3, p. 1180 (1991). Abstract. –
See Abstr. 012.037 for the main entry.

091.036   **Dynamics of dust in a plasma sheath and injection of dust into the plasma sheath above Moon and asteroidal surfaces.**
T. Nitter, O. Havnes.
Earth, Moon, Planets, Vol. 56, No. 1, p. 7 – 34 (Jan 1992).
The authors have examined single dust particle dynamics in a plasma sheath near the surface of solid bodies in space, considering conditions which resemble those of planetary system bodies, when photoelectric effect can be neglected. The forces on the dust particles are assumed to be from the electric field in the sheath and from gravitation only.

091.037   **The molecular – metallic transition of hydrogen and the structure of Jupiter and Saturn.**
G. Chabrier, D. Saumon, W. B. Hubbard, J. I. Lunine.
Astrophys. J., Vol. 391, No. 2, p. 817 – 826 (1 Jun 1992).
Recently, a new equation of state for hydrogen which predicts a molecular – metallic phase transition at finite temperature has become available. It is combined with a helium equation of state, and the resulting thermodynamic description of H/He mixtures is used to compute interior models of Jupiter and Saturn, subject to the constraints of the measured gravitational harmonics of both planets. The authors discuss the inferred heavy element abundance distribution in their interiors and the possible consequences on their formation. In particular, the Z–element enhancement and smaller core in Saturn relative to Jupiter, a conclusion of this study, may indicate a depletion of water ice in the Jupiter formation zone.

091.038   **The role of the molecular – metallic transition of hydrogen in the evolution of Jupiter, Saturn, and brown dwarfs.**
D. Saumon, W. B. Hubbard, G. Chabrier, H. M. Van Horn.
Astrophys. J., Vol. 391, No. 2, p. 827 – 831 (1 Jun 1992).
An equation of state for hydrogen which predicts a molecular – metallic phase transition at finite temperatures has become available recently. A companion paper addresses the issue of the internal structures of Jupiter and Saturn, as derived with this new equation of state. Here the authors study the effect of this phase transition on the cooling histories of these two giant planets and of substellar brown dwarfs. The phase transition alters the present age of Jupiter and of Saturn by a few percent. Interestingly, the cooling of brown dwarfs is most strongly affected at the time when the interior adiabat crosses the critical point of the phase transition.

091.039   **Love numbers of the Moon and of the terrestrial planets.**
C. Z. Zhang.
Earth, Moon, Planets, Vol. 56, No. 3, p. 193 – 207 (Mar 1992).
In the IERS Standards (1989), for the Moon the adopted value of the tide Love number, k_2, is equal to 0.0222. In this paper using the latest geodetic parameters of the Moon a group of internal structure models are constructed for this celestial body, then the dependence of the Moon's core size on calculated value of k_2 is explored. The obtained results indicate that the second degree Love number, $k_2 = 0.02664$, of the lunar model 91–04 is near its observed value (0.027 ± 0.006). This implies that the Moon may possess an outer core of 660 km radius and of 300 kbar mean rigidity. With the same method the static Love numbers from degree 2 to 30 are computed for the terrestrial planets – Mercury, Venus, and Mars, and the influence of some parameters (such as the rigidity) of the outer core on low degree Love numbers is discussed. Finally, the likely range of the second degree Love numbers is determined for the terrestrial planets. It seems that if low degree Love numbers of a terrestrial planet can be detected in the future space explorations, there is some possibility to improve the planetary internal structure model. For example, as soon as space techniques yield an observed value of $k_2 > 0.10$ for

Mercury, there will be reason to anticipate that a partly melted iron core exists in this planet.

091.040   **The wave nature and dynamical quantization of the solar system.**
V. N. Damgov, D. B. Douboshinsky.
Earth, Moon, Planets, Vol. 56, No. 3, p. 233 – 242 (Mar 1992).
A heuristic model is proposed of the mean distances between the solar–system planets, their satellites and the primaries. The model is based on: (i) the concept of the solar system structure wave nature; (ii) the micro–mega analogy of the micro– and megasystem structures, and (iii) the oscillator amplitude "quantization" phenomenon, occuring under wave action, discovered on the basis of the classical oscillations theory (Damgov et al., 1990, 1991). From the equation, describing the charge rotation under the action of an electromagnetic wave, an expression is obtained for the discrete set of probable stationary motion amplitudes. The discrete amplitude values – the "quantization" phenomenon – are defined by the argument values at the extreme points of the N–order Bessel functions. Using this expression, the mean related distances are computed from the solar system planets and the Saturn, Uranian and Jovian satellites to the primaries.

091.041   **Escape processes in planetary atmospheres.**
V. I. Moroz.
Astrophysics on the threshold of the 21st century, p. 63 – 80 (1992). – See Abstr. 003.048 for the main entry.
A significant contribution to the development of escape theory was made by S. B. Pikel'ner and I. S. Shklovskii who are the inspiration for this article. The author presents here another viewpoint and introduces new facts.

091.042   **Numerical simulations of isolated Lindblad resonances in collisional ring.**
J. Hänninen, H. Salo.
Bull. Am. Astron. Soc., Vol. 23, No. 3, p. 1181 (1991). Abstract. – See Abstr. 012.037 for the main entry.

091.043   **Observational and theoretical evidence that satellite–ring torques are much smaller than the "standard model" predicts.**
Y. Nakagawa, T. G. Brophy, P. A. Rosen.
Bull. Am. Astron. Soc., Vol. 23, No. 3, p. 1182 (1991). Abstract. – See Abstr. 012.037 for the main entry.

091.044   **Equatorial superrotation in a slowly rotating GCM: implications for Titan and Venus.**
A. D. Del Genio, W. B. Rossow, T. P. Eichler.
Bull. Am. Astron. Soc., Vol. 23, No. 3, p. 1187 (1991). Abstract. – See Abstr. 012.037 for the main entry.

091.045   **Temperature dependence of the intensities, shifts, and widths of the thermal infrared lines of trace gases in the atmospheres of the outer planets and Titan.**
P. Varanasi.
Bull. Am. Astron. Soc., Vol. 23, No. 3, p. 1187 (1991). Abstract. – See Abstr. 012.037 for the main entry.

091.046   **Modeling stratiform clouds in the atmospheres of giant planets with mutual solubility of condensing components. II.**
M. V. Bujkov, K. Yu. Ibragimov, G. A. Kirienko, A. M. Pirnach.
Kinematics Phys. Celest. Bodies, Vol. 7, No. 4, p. 18 – 24 (1991). English translation of Kinematika Fiz. Nebesn. Tel, Tom 7, No. 4, p. 23 – 30 (1991).
The conditions of condensation of water-ammonia clouds under the conditions of strong departure from the Raoul–Henry law are discussed. Approximate expressions derived for the saturation vapor pressure above a drop of aqueous ammonia solution at various concentrations are derived on the basis of Antoine's law and available experimental phase diagrams of the water–ammonia solution and give more accurate results than the

Clausius law. They were used in modeling Jovian clouds. A total of 49 models were computed, and analysis indicates significant variation of both the macro– and the micro–structure of the clouds. Cases of strong and weak mutual penetration of the clouds are presented by way of illustration.

091.047   The internal structure of the giant planets.
V. N. Zharkov.
Sol. Syst. Res., Vol. 25, No. 6, p. 465 – 483 (May 1992). English translation of Astron. Vestn., Tom 25, No. 6, p. 627 – 649 (1991).

The basic ideas and data which are relevant to the problem of constructing models of the giant planets are discussed in this review. The following questions are considered: the establishment of the concept of Jupiter and Saturn as hydrogen planets, the theory of the figures for rotating planets in hydrostatic equilibrium, the gas–liquid adiabatic model of the giant planets, observational data, the abundances of elements and of the group of cosmochemical substances, the equations of state, models of the internal structure of Jupiter, Saturn, Uranus, and Neptune, the question of the time of formation of Jupiter, a two–component model of the formation of the terrestrial planets and the role of Jupiter in the formation of the Earth, and the role of Jupiter in the formation of the giant planets.

091.048   Estimating the behaviour of the radiation field inside a planetary atmosphere.
J. W. Hovenier, W. M. Wauben.
Bull. Am. Astron. Soc., Vol. 23, No. 3, p. 1193 (1991). Abstract. – See Abstr. 012.037 for the main entry.

091.049   Stochastic behavior of the orbits of planets during their accretion.
I. N. Ziglina.
Sol. Syst. Res., Vol. 25, No. 6, p. 526 – 543 (May 1992). English translation of Astron. Vestn., Tom 25, No. 6, p. 703 – 722 (1991).

The evolution of the orbital elements of a growing planet during the accretion process is examined. The planetary orbit is subject to perturbations from random ecounters and collisions with bodies of its accretion zone and to gravitational perturbations from an already formed massive planet (Jupiter). The Fokker–Planck equation describing the behavior of the distribution function of the eccentricity and inclination of the planetary orbit as a function of time is formulated and solved. The present mean value of the orbital eccentricities and inclinations of the terrestrial planets is consistent with their accretion at the final stage from a swarm of bodies with a mean mass of $\sim 10^{-2} M_\oplus$ and mean eccentricities and inclinations of ~ 0.2.

091.050   Collisional simulations of satellite Lindblad resonances.
J. Hänninen, H. Salo.
Icarus, Vol. 97, No. 2, p. 228 – 247 (Jun 1992).

The influence of a perturbing satellite on a planetary ring at isolated Lindblad resonances is studied with numerical computer simulations, combining Aarseth's force polynomial method for orbit integrations with the calculation of particle–particle impacts. Observed torque exchange between the satellite and the dissipative non–self–gravitating ring agrees with the theoretical torque for a gravitating ring to within 20%. This verifies that angular momentum exchange is not very sensitive to the details of the dominant physical processes. The theoretical estimate for the formation of a gap at a 2:1 inner Lindblad resonance is also shown to be of the correct order of magnitude. The theoretically predicted flux–reversal phenomenon due to streamline distortion is also verified.

091.051   Strukturen in planetaren Ringen.
F. Spahn.
Phys. Unserer Zeit, Jahrg. 23, Nr. 3, p. 121 – 128 (May 1992).

091.052   Lorentz resonances and the vertical structure of dusty rings: analytical and numerical results.
L. Schaffer, J. A. Burns.
Icarus, Vol. 96, No. 1, p. 65 – 84 (Mar 1992).

Dust grains orbiting the giant outer planets carry a weak and nearly constant electrical charge. Through interaction with the planet's irregular magnetic field, the inclinations and eccentricities of these charged grains can be pumped up significantly. This occurs especially when a period of the Lorentz force matches the radial or out–of–plane epicyclic periods. Since orbiting particles sample the spatial structure of the planetary magnetic field at differing rates depending on their mean orbital radius, these Lorentz resonances (LR) occur at specific orbital radii. Several of the strongest LR have been related to structural features of the Jovian dust ring system. In this paper the authors use a combination of analytical and numerical techniques in order to understand the nature of these resonances. A simple extension of the perturbation theory for the LRs yields the charge–to–mass ratio corrections to the periods of in–plane and out–of–plane motion, thus allowing for an accurate determination of the resonance locations.

091.053   Numerical simulations of dense collisional systems. II. Extended distribution of particle sizes.
H. Salo.
Icarus, Vol. 96, No. 1, p. 85 – 106 (Mar 1992).

A local simulation method with N up to 4000 particles is applied to the collisional dynamics of dense planetary rings with power–law distribution of particle sizes. The equilibrium state is found to depend mainly on the maximum particle size, while the slope and the width of the distribution have less significance. In all the cases studied, the maximum ratio between velocity dispersions of different sizes stays below a factor of about five. The derived effective geometric thickness for Saturn's rings is about 25 m for the layer of cm–sized particles, and about 10 m for the largest particles. Simulations with bimodal distribution of sizes show that systems are far from energy equipartition for mass ratios larger than about 10, even if the optical thicknesses of the two populations are the same. The possibility of selective viscous instability among small particles is also addressed but found not possible for the standard elastic model of icy particles.

091.054   Dust in planetary ring systems.
M. R. Showalter.
IAU Colloquium No. 126: Origin and evolution of interplanetary dust, p. 349 – 356 (1991). – See Abstr. 012.068 for the main entry.

Each of the outer gas giants, Jupiter, Saturn, Uranus and Neptune, is now known to be encirled by a system of rings. Some of these, such as the A, B, and C rings of Saturn and the nine narrow Uranian rings, are rather optically thick and are composed primarily of large bodies (1 cm to 10 m). However, every other system has been found to contain a large population of micron–sized dust. Such rings reveal the effects of a variety of physical processes that are also acting on interplanetary and interstellar grains. When such rings are examined as members of a general class, recurring patterns begin to emerge.

091.055   On the theory of natural oscillations of planetary ionospheres.
N. Ya. Kotsarenko, A. A. Shvydkij.
Kinematika Fiz. Nebesn. Tel, Tom 8, No. 3, p. 106 – 108 (May–Jun 1992). In Russian. English translation in Kinematics Phys. Celest. Bodies, Vol. 8, No. 3.

The spectrum of natural low frequency oscillations of an ionosphere is found, the ionosphere being simply simulated as a homogeneous plasma bounded by two spherical surfaces. The role of ion collisions and geomagnetic field is discussed.

091.056   Dark matter in the solar system: hydrogen cyanide polymers.
C. N. Matthews.
Origins Life Evol. Biosphere, Vol. 21, No. 5 – 6, p. 421 – 434 (1991 – 1992). – See Abstr. 012.083 for the main entry.

091.057   The inner planets.
R. Baum.
Images of the universe, p. 20 – 39 (1991). – See Abstr. 003.090 for the main entry.

Of the solar system's nine primary planets, the four nearest to the Sun are known as the inner or terrestrial planets. All move in orbits within 1.52 AU of the Sun, and are similar in size and density. All have rocky surfaces and can be studied by geologic techniques. Unlike the jovian or outer planets, the terrestrial planets have few, if any, satellites. Until Dec 1962, when Mariner 2 flew past Venus, our knowledge of the terrestrial planets was very limited. Centuries of direct visual observation had established a sound understanding of their orbital theory, sizes and densities, and the rotation of Mars, but virtually nothing about their surfaces. This paper is a brief exposition of these worlds as they are perceived in the aftermath of the first phase of space exploration.

091.058   The N–dipole problem and the rings of Saturn.
C. L. Goudas.
NATO Advanced Study Institute on Predictability, Stability, and Chaos in N–Body Dynamical Systems, p. 371 – 385 (1991). – See Abstr. 012.089 for the main entry.

N–magnetic dipoles each located on a star–member of an N–body star system, are assumed to move with their carrier stars and control the motions of charged grains in their vicinity. The case N = 5, in a special configuration, where four dipoles perform rigid rotation about the fifth, while all have magnetic moments parallel to the angular velocity vector, is used as a test case to show that the "spaces of trapping" found to exist in the two and three dipole problems, receive a form similar to the rings of Saturn and that pairs of "spaces of trapping" are separated by gaps similar to the Cassini division. The effect of gravity of a rotating planet within which the five dipoles, of internal "dynamo" origin exist and corotate are taken into account.

091.059   The three–dipole problem.
C. L. Goudas, E. G. Petsagourakis.
NATO Advanced Study Institute on Predictability, Stability, and Chaos in N–Body Dynamical Systems, p. 355 – 370 (1991). – See Abstr. 012.089 for the main entry.

As a further generalization of Störmer's problem of one magnetic dipole, the authors investigate the three–dipole problem. Each of the three magnetic dipoles is assumed to be located on one member of a three–star system that performs Newtonian motions. Charged particles, positive or negative, moving in the vicinity of the three moving dipoles perform motions which are the object of this study. The authors show that if the three stars perform the Lagrangean circular solution of the three–body problem and if the magnetic moments of their dipoles are perpendicular to their plane of motion, then three, or two, or one, closed space regions exist, where charged particles of appropriate energy are permanently trapped. These regions of trappings can be considered as generalized Van Allen zones.

091.060   Late stage of accumulation and chemical composition of the terrestrial planets.
S. V. Kozlovskaya, G. V. Pechernikova.
Earth, Moon, Planets, Vol. 57, No. 3, p. 225 – 230 (Jun 1992).

The model in which the differences of chemical composition of the terrestrial planets are determined by special conditions at the later stage accumulation is discussed. Impact heating would rapidly lead to differentiation of Mercury's interiors. Subsequent high–velocity of Mercury with planetesimals of a comparable size would erode away much of the silicate crust and mantle; such silicates would be accumulated by Venus and fall into the Sun. This model is in agreement with the current models of terrestrial planets internal constitution.

091.061   VLA/Goldstone Planetary Radar results.
A. W. Grossman, D. O. Muhleman, B. J. Butler, M. A. Slade.
International Symposium on Radars and Lidars in Earth and Planetary Sciences, p. 19 – 22 (Dec 1991). – See Abstr. 012.093 for the main entry.

Recent results from an entirely new technique of planetary radar astronomy are presented. The VLA/Goldstone Planetary Radar combines the transmitter of the Goldstone 70–meter antenna and the receivers of the VLA interferometer to create a

synthesis imaging radar instrument with unprecedented capabilities. The technique yields improved sensitivity and produces a direct sky–map of radar flux density while avoiding the ambiguities associated with conventional range–doppler mapping. The method is illustrated by application to radar mapping of Mars and radar detection of Titan.

091.062   Bistatic radar studies of the planets.
I. R. Linscott, G. L. Tyler, E. A. Marouf, R. A. Simpson.
International Symposium on Radars and Lidars in Earth and Planetary Sciences, p. 23 – 29 (Dec 1991). – See Abstr. 012.093 for the main entry.

The use of bistatic radar for planetary studies is discussed with the objective of presenting important, new capabilities for space borne systems. In this presentation, the success of bistatic radar, as a probe of the solar system's planets and moons, is first reviewed. Then the relevance of new, micro–power and signal processing technologies is assessed, and the scientific opportunities for extending the study of planetary surfaces, atmospheres and rings are presented.

091.063   Giant planet magnetospheres.
F. Bagenal.
Annu. Rev. Earth Planet. Sci., Vol. 20, p. 289 – 328 (1992).

The orientations of the planet and its magnetic field control the morphology and dynamics of a planet's magnetosphere. The author first considers how the giant planet magnetospheres fall into two categories – the large, symmetric magnetospheres of Jupiter and Saturn and the smaller, irregular magnetospheres of Uranus and Neptune. The author then discusses the characteristics of the plasma and the current understanding of the magnetospheric processes for each planet in turn. Finally, the author compares the energetic particle populations, radio emissions, and remote sensing of magnetospheric processes in these giant planet magnetospheres.

091.064   Origin of noble gases in the terrestrial planets.
R. O. Pepin.
Annu. Rev. Earth Planet. Sci., Vol. 20, p. 389 – 430 (1992).

Contents: Introduction. Primary solar–system volatile sources and processes. The current data base and its implications for evolutionary processing. Modeling approaches to the problem. Early astrophysical environments and planetary accretion histories. Models of atmospheric evolution by hydrodynamic escape. Summary.

091.065   Planetary distance law and resonance.
J. J. Rawal.
Proc. Astron. Soc. Aust., Vol. 9, No. 2, p. 329 (1991). – See Abstr. 012.090 for the main entry.

In this paper the relation between the planetary distance law and the resonant structures is shown, in that the resonance relation has been expressed in terms of Roche's constant (Rawal 1984, 1986, 1989). This brings forth a coherent, elegant and unified picture of the Solar System and satellite systems.

091.066   Inductive acceleration of protons and electrons in planetary magnetotails.
A. L. Taktakishvili, L. M. Zeleny (L .M. Zelenyj).
International Workshop on Reconnection in Space Plasma, p. 227 – 232 (Jan 1989). – See Abstr. 012.099 for the main entry.

The paper deals with the unified model of particle acceleration (both electrons and ions) in a planetary magnetotail due to explosive spontaneous reconnection. It is shown that the inductive electric field is able to generate energetic particle bursts with special velocity dispersion features. The proposed mechanism accelerates both species up to MeV–energies in the Earth's magnetosphere (but with the electron flux lower than ion) and vice versa practically only electrons in the case of the compact Mercury magnetosphere. The theory conforms with the experimental data obtained thus far in terrestrial and Hermean magnetotails.

Moons of the solar system. An illustrated encyclopedia.
See Abstr. 002.059.

Bibliography (1990 Part 1).
See Abstr. 002.093.

The Tokyo PMC catalog 88: catalog of positions of 3800 stars observed in 1988 and planetary positions observed in 1986 to 1988 with Tokyo Photoelectric Meridian Circle.
See Abstr. 002.152.

Bibliography.
See Abstr. 002.153.

Astrophysical data. Planets and stars.
See Abstr. 003.017.

Das neue Bild vom Sonnensystem.
See Abstr. 003.032.

Plasmaphysik im Sonnensystem.
See Abstr. 003.052.

Worlds in the sky. Planetary discovery from earliest times through Voyager and Magellan.
See Abstr. 003.072.

Satellites of the outer planets. Worlds in their own right.
See Abstr. 003.081.

Annual Review of Earth and Planetary Sciences. Volume 19.
See Abstr. 003.092.

Annual Review of Earth and Planetary Sciences. Volume 20.
See Abstr. 003.093.

Ground–based planetary science at Lick Observatory 1888 – 1938.
See Abstr. 004.050.

Freezing fire: measuring planetary heat, 1900 – 1930.
See Abstr. 004.051.

Planetary geology: Killer acid at the K/T boundary.
See Abstr. 011.010.

Chaos, resonance and collective dynamical phenomena in the solar system. Proceedings.
See Abstr. 012.025.

Physics and chemistry of magma oceans from 1 bar to 4 Mbar. Abstracts of presented papers.
See Abstr. 012.032.

23. Annual Meeting of the Division for Planetary Sciences (DPS) of the American Astronomical Society (AAS), Palo Alto, CA (USA), 4 – 8 Nov 1991. Abstracts of presented papers.
See Abstr. 012.037.

Lunar and planetary science, Volume 22. Proceedings.
See Abstr. 012.046.

Interrelations between physics and dynamics for minor bodies in the solar system. Proceedings.
See Abstr. 012.055.

Radars and lidars in Earth and planetary sciences. Proceedings.
See Abstr. 012.093.

Reconnection in space plasma. Proceedings.
See Abstr. 012.099.

The role of solar system photometry.
See Abstr. 013.018.

Geophysics news 1991.
See Abstr. 013.132.

Report of the IAU/IAG/COSPAR Working Group on Cartographic Coordinates and Rotational Elements of the Planets and Satellites: 1991.
See Abstr. 013.141.

Sizing up the solar system.
See Abstr. 014.018.

What makes a planet habitable, and how to search for habitable planets in other solar systems.
See Abstr. 015.009.

Exobiological habitats: an overview.
See Abstr. 015.016.

Monte Carlo methods.
See Abstr. 021.021.

Observation of dust shedding from material bodies in a plasma.
See Abstr. 022.015.

Atmospheric effects on cratering efficiency.
See Abstr. 022.030.

Secondary electron yields of solar system ices.
See Abstr. 022.031.

Simple algorithms for remote determination of mineral abundances and particle sizes from reflectance spectra.
See Abstr. 022.033.

Quantitative subpixel spectral detection of targets in multispectral images.
See Abstr. 022.034.

Application of the Stogryn–Hirschfelder treatment of weak dimers to planetary atmospheres.
See Abstr. 022.038.

Tunable diode laser measurements on the 951.7393 cm^{-1} line of $^{12}C_2H_4$ at planetary atmospheric temperatures.
See Abstr. 022.044.

Intensity and linewidth measurements in the 13.7 μm fundamental bands of $^{12}C_2H_2$ and $^{12}C^{13}CH_2$ at planetary atmospheric temperatures.
See Abstr. 022.048.

Circular polarization as an instrument for investigation of surfaces of atmosphereless celestial bodies. I. Laboratory measurements of highly absorptive substances.
See Abstr. 022.068.

Effect of temperature on shock metamorphism of single–crystal quartz.
See Abstr. 022.098.

Computational model of the collision induced absorption spectra of H_2–He pairs in the fundamental band.
See Abstr. 022.124.

Density shifts and line strengths for 4–0 and 5–0 quadrupole transitions in molecular hydrogen: implications for the spectra of the Jovian planets.
See Abstr. 022.126.

Light scattering by a randomly oriented cluster of spheres.
See Abstr. 022.131.

An expanded program of laboratory measurements on ammonia's microwave absorption spectrum.
See Abstr. 022.134.

Optical constants of poly–HCN for astronomical applications.
See Abstr. 022.172.

Laboratory and theoretical studies of thermal emission spectroscopy.
See Abstr. 022.178.

The physics and dynamics of charged dust grains.
See Abstr. 022.195.

The role of volume scattering in reducing spectral contrast of reststrahlen bands in spectra of powdered minerals.
See Abstr. 022.202.

Laboratory polarimeter for measurement of the Stokes vector of light scattered from surfaces.
See Abstr. 034.039.

Extraterrestrial Mössbauer spectrometry.
See Abstr. 035.071.

Planetary imaging with a small CCD camera.
See Abstr. 036.049.

Possibility of outer space X–ray diffractometry.
See Abstr. 036.068.

Statistical analysis of data in planetary observations.
See Abstr. 036.108.

Interpretation of lightcurves of atmosphereless bodies. I. General theory and new inversion schemes.
See Abstr. 036.117.

Geometries of radar astronomy in future experiments.
See Abstr. 036.123.

Fractal analysis: a new remote sensing tool.
See Abstr. 036.143.

A few points on the stability of the solar system.
See Abstr. 042.017.

Long term evolution of the solar system.
See Abstr. 042.018.

Numerical experiments on the motion of the outer planets.
See Abstr. 042.019.

Numerical simulations of planetary systems of the Jupiter–Saturn type.
See Abstr. 042.021.

Cross–tidal effects and orbit–orbit resonances.
See Abstr. 042.042.

New methods for long–time numerical integration of planetary orbits.
See Abstr. 042.055.

Planetary ring dynamics: from Boltzmann's equation to celestial mechanics.
See Abstr. 042.086.

Planetary ring dynamics: secular exchange of angular momentum and energy with a satellite.
See Abstr. 042.087.

Analytical framework in Poincaré variables for the motion of the solar system.
See Abstr. 042.098.

Temporary capture into resonance.
See Abstr. 042.103.

Perturbation theory, resonance, librations, chaos, and Halley's comet.
See Abstr. 042.107.

Stability of satellites in spin–orbit resonances and capture probabilities.
See Abstr. 042.108.

Statistical analysis of the effects of close encounters of particles in planetary rings.
See Abstr. 042.109.

Quasiperiodic orbits as a substitute of libration points in the solar system.
See Abstr. 042.113.

An impulsional method to estimate the long–term behaviour of a perturbed system: application to a case of planetary dynamics.
See Abstr. 042.123.

The planetary few–body problem at three–frequency resonance.
See Abstr. 042.138.

Choice of shape of bodies with minimum aerodynamic heating during motion in the atmospheres of planets in the solar system.
See Abstr. 053.001.

r–process abundances and nuclear properties far from stability.
See Abstr. 061.112.

Simulation of ion acceleration in a charged dust cloud.
See Abstr. 062.013.

Finite amplitude fast magnetosonic waves in a low–density plasma.
See Abstr. 062.027.

Large electric fields in acoustic waves and the stimulation of lightning discharges.
See Abstr. 062.042.

Nonlinear dust–acoustic waves in multispecies dusty plasmas.
See Abstr. 062.063.

Large amplitude double layers in dusty plasmas.
See Abstr. 062.064.

Resistive instability and the magnetostrophic approximation.
See Abstr. 062.124.

Rotation of the plasmasphere and the formation of the boundary layer at its limit.
See Abstr. 062.127.

The motion of vortices within a rotating, fluid shell.
See Abstr. 062.138.

Vortices in non–uniform dusty plasmas.
See Abstr. 062.168.

Velocity modulation of a dusty plasma.
See Abstr. 062.169.

A note on the enhancement of J values in optically thick scattering atmospheres.
See Abstr. 063.006.

Inverse problem of thermal sounding. Pt. III. Determination of the vertical profile of the mixture ratio of a minor gas component.
See Abstr. 063.014.

Analysis of McCoyd's mechanism of the negative polarization of light scattered by atmosphereless celestial bodies.
See Abstr. 063.022.

Scattering of electromagnetic waves by a distribution of charged dust particles in space plasmas.
See Abstr. 063.026.

Diagnostic possibilities of circular polarization for investigating the surfaces of atmosphereless bodies.
See Abstr. 063.029.

Benchmark results for single scattering by spheroids.
See Abstr. 063.054.

Estimating the effect of finite angular light source dimensions on the opposition brightness effect in atmosphereless bodies.
See Abstr. 063.061.

Interference model of negative polarization of light scattered by the solid surfaces of celestial bodies.
See Abstr. 063.066.

Coherent backscattering: a vector formulation for effects of polarization, absorption, and small or large scatterers.
See Abstr. 063.098.

Inverse problem of photometric observation of solar radiation reflected by an optically dense planetary atmosphere. Mathematical methods and weighting functions of linearized inverse problem.
See Abstr. 063.113.

Inverse problem of photometry of solar radiation reflected by an optically thick planetary atmosphere. II. Numerical aspects and requirements on the observation geometry.
See Abstr. 063.132.

The diagnostic potentialities of circular polarization for studying the surfaces of atmosphereless cosmic bodies.
See Abstr. 063.139.

Absorption–line formation in an optically thick weakly absorbing planetary atmosphere. I. Homogeneous slab.
See Abstr. 063.144.

Absorption–line formation in an optically thick weakly absorbing planetary atmosphere. II. Vertically inhomogeneous media.
See Abstr. 063.145.

Inverse problems of radiative transfer in sounding of planetary atmospheres.
See Abstr. 063.147.

Planetary orbits as limit cycles.
See Abstr. 066.210.

Sunspots, planetary alignments and solar magnetism: a progress review.
See Abstr. 072.044.

Quasibiennial periodicity of solar and planetary phenomena.
See Abstr. 072.064.

Some remarks about the rotations of a viscous planet and its homogeneous liquid core: linear theory.
See Abstr. 081.019.

Solar wind and magnetospheric processes and various coupling mechanisms including those observed on other planets.
See Abstr. 085.014.

Magnitudes of selected stellar occultation candidates for Pluto and other planets, with new predictions for Mars and Jupiter.
See Abstr. 096.008.

Occultations of stars by solar system objects. IX. Occultations of catalog stars by asteroids, planets, and major satellites in 1992 and 1993.
See Abstr. 096.009.

Rotating magnetic anomalies as a possible accelerator of charged particles.
See Abstr. 099.030.

Theory of the Neptunian arcs. I. Stability of an individual epiton.
See Abstr. 101.008.

The Oort cloud.
See Abstr. 102.053.

The solar motion, the Galaxy and the aphelia distribution of long period comets.
See Abstr. 103.019.

The sporadic meteoroid flux near Venus, Earth, and Mars.
See Abstr. 104.064.

An estimation of meteoroid flux at outer Martian space for steady meteor streams.
See Abstr. 104.088.

Voyager energetic particle observations at interplanetary shocks and upstream of planetary bow shocks: 1977–1990.
See Abstr. 106.027.

Physical processes on circumplanetary dust.
See Abstr. 106.084.

Accretion in the inner nebula: the relationship between terrestrial planetary compositions and meteorites.
See Abstr. 107.013.

Collision and tidal interaction between planetesimals.
See Abstr. 107.019.

Secular ring instability in the protoplanetary accretion disk.
See Abstr. 107.022.

The rate of planet formation and the solar system's small bodies.
See Abstr. 107.057.

Advancement of photoionization and photodissociation rates relevant to astrochemistry.
See Abstr. 131.129.

092 Mercury

092.001 Mercury's cool surprise.
J. K. Beatty.
Sky Telesc., Vol. 83, No. 1, p. 35 – 36 (Jan 1992).
Radar probings of the innermost planet may have turned up something about as likely as a snowball in hell.

092.002 Subsurface emissions from Mercury: VLA radio observations at 2 and 6 centimeters.
M. J. Ledlow, J. O. Burns, G. R. Gisler, J.–H. Zhao, M. Zeilik, D. N. Baker.
Astrophys. J., Vol. 384, No. 2, p. 640 – 655 (10 Jan 1992).
The authors present radio observations of Mercury made with the VLA: once in 1986, and on two dates in February of 1988. These observations are the first to spatially map both hot regions associated with the theoretical "hot poles". These "hot poles" are separated by 180° and are a result of the unusual diurnal heating from Mercury's 3/2 spin–orbit resonance and eccentric orbit. The highest resolution maps reveal details of the planet as small as 330 km. The authors include maps of total intensity, brightness temperature, polarized intensity, fractional polarization, depolarization, and spectral index. The subsurface thermal emissions from Mercury are characteristic of blackbody reradiation from the solar insolation over a diurnal cycle. The authors use these observations to produce full–disk thermophysical models. The one–dimensional, time–dependent heat–diffusion equation has been solved for all observed disk elements at each epoch in order to constrain thermophysical parameters and properties of the subsurface material. Using typical lunar values for several of the parameters, the authors reproduce the temperature morphology and most of the observed temperature values. The best–fit models require a substantial contribution of the heat transport in the subsurface to be radiative in nature.

092.003 Why Mercury and Venus have no satellites?
Z. Mikolášek.
Říše hvězd, Vol. 72, No. 1, p. 13 – 14 (Jan 1991). In Czech.

092.004 Mercury: an interim report of the evening elongation 1991 March – April.
D. Graham.
J. Br. Astron. Assoc., Vol. 102, No. 1, p. 36 (Feb 1992).

092.005 Mercury: a review of recent (1989 – 1991) results from observations and theory.
A. Sprague.
Bull. Am. Astron. Soc., Vol. 23, No. 3, p. 1192 (1991). Abstract. – See Abstr. 012.037 for the main entry.

092.006 An S–band radar anomaly at the north pole of Mercury.
J. K. Harmon, M. A. Slade.
Bull. Am. Astron. Soc., Vol. 23, No. 3, p. 1197 (1991). Abstract. – See Abstr. 012.037 for the main entry.

092.007 Mercury Goldstone/VLA radar: Part 1.
M. A. Slade, R. Jurgens, B. Butler, D. Muhleman.
Bull. Am. Astron. Soc., Vol. 23, No. 3, p. 1197 (1991). Abstract. – See Abstr. 012.037 for the main entry.

092.008 Radio imaging of Mercury's subsurface.
D. L. Mitchell, I. de Pater.
Bull. Am. Astron. Soc., Vol. 23, No. 3, p. 1197 (1991). Abstract. – See Abstr. 012.037 for the main entry.

092.009 Impact processes as an explanation of the spin resonance state of Mercury.
J. Boyce, D. J. Stevenson.
Bull. Am. Astron. Soc., Vol. 23, No. 3, p. 1197 – 1198 (1991). Abstract. – See Abstr. 012.037 for the main entry.

092.010 Mercury Goldstone/VLA radar: Part 2.
B. Butler, D. Muhleman, M. Slade, R. Jurgens.
Bull. Am. Astron. Soc., Vol. 23, No. 3, p. 1200 (1991). Abstract. – See Abstr. 012.037 for the main entry.

092.011 A Monte Carlo simulation of Mercury's atmosphere.
K. R. Garlow.
Bull. Am. Astron. Soc., Vol. 23, No. 3, p. 1200 (1991). Abstract. – See Abstr. 012.037 for the main entry.

092.012 Terrace width variations in complex Mercurian craters, and the transient strength of cratered Mercurian and lunar crust.
A. C. Leith, W. B. McKinnon.
Bull. Am. Astron. Soc., Vol. 23, No. 3, p. 1202 (1991). Abstract. – See Abstr. 012.037 for the main entry.

092.013 Mercury's high metal content and super–adiabaticity within the proto–solar cloud.
A. J. R. Prentice.
Bull. Am. Astron. Soc., Vol. 23, No. 3, p. 1232 (1991). Abstract. – See Abstr. 012.037 for the main entry.

092.014 Radar and spacecraft ranging to Mercury between 1966 and 1988.
J. D. Anderson, M. A. Slade, R. F. Jurgens, E. L. Lau, X. X. Newhall, E. M. Standish Jr.
Proc. Astron. Soc. Aust., Vol. 9, No. 2, p. 324 (1991). – See Abstr. 012.090 for the main entry.
Improved solutions have been obtained for the orbit and equatorial cross–section of Mercury using radar ranging data spanning 22 years. These data have yielded new results on the precession of Mercury's perihelion and better limits on a possible time variation in the gravitational constant G.

Passage de Mercure devant le Soleil, observé par Gassendi (1592 – 1655), le 7 novembre 1631.
See Abstr. 004.066.

Observations of solar–system bodies with the Belgrade Meridian Circle.
See Abstr. 041.006.

Construction of invariant tori for the spin–orbit problem in the Mercury–Sun system.
See Abstr. 042.068.

New applications of Fatou's problem.
See Abstr. 042.091.

Love numbers of the Moon and of the terrestrial planets.
See Abstr. 091.039.

Inductive acceleration of protons and electrons in planetary magnetotails.
See Abstr. 091.066.

Impact–induced thermal effects in the lunar and Mercurian regoliths.
See Abstr. 094.008.

093 Venus

093.001 Differential VLBI measurements of the Venus atmosphere dynamics by balloons: VEGA project.
R. Z. Sagdeyev (*R. Z. Sagdeev*), V. V. Kerzhanovitch
(*V. V. Kerzhanovich*), L. R. Kogan, V. I. Kostenko,
V. M. Linkin, L. I. Matveyenko (*L. I. Matveenko*),
R. R. Nazirov, S. V. Pogrebenko, I. A. Struckov (*I. A. Strukov*),
R. A. Preston, J. Purcel, C. E. Hildebrand,
V. A. Grishmanovskiy (*V. A. Grishmanovskij*), A. N. Kozlov,
E. P. Molotov, J. E. Blamont, L. Boloh, G. Laurans,
P. Kaufmann, J. Galt, F. Biraud, A. Boischot,
A. Ortega–Molina, C. Rosolen, G. Petit, P. G. Mezger,
R. Schwartz, B. O. Rönnäng, R. E. Spencer, G. Nicolson,
A. E. E. Rogers, M. H. Cohen, R. M. Martirosyan,
I. G. Moiseyev (*I. G. Moiseev*), J. S. Jatskiv (*Ya. S. Yatskiv*).
Astron. Astrophys., Vol. 254, No. 1/2, p. 387 – 392 (Feb 1992).

Results are reported of the measurement of balloon motion in the upper layers of the Venus atmosphere. The measurement of the balloon trajectories was carried out by differential very–long–baseline interferometry observations of the balloons relative to the fly–by spacecrafts (VEGA–1 and VEGA–2). Balloon–2, launched in the southern hemisphere of the planet led to the discovery of gas motion with average velocities of 65.3 m s^{-1} and 3.4 m s^{-1} in longitude and latitude, respectively. Balloon–1, launched in the northern hemisphere, showed that gas motions occur with an average velocity of 68.7 m s^{-1} in the longitudinal direction, but no wind was detected in the latitudinal direction. Pendulum motion of the balloon baskets with a period of 7.4 s was detected on many occasions. This occurred 10 times more frequently in the southern hemisphere than in the northern one. The nature of these phenomena is discussed.

093.002 Venus unveiled.
S. J. Goldman.
Sky Telesc., Vol. 83, No. 3, p. 258 – 262 (Mar 1992).

After beaming back trillions of bits of data, the Magellan spacecraft has delivered a detailed, global view of our cloud–shrouded neighbor.

093.003 Theory of small–scale density and electric field fluctuations in the nightside Venus ionosphere.
J. D. Huba.
J. Geophys. Res., Vol. 97, No. A1, p. 43 – 50 (1 Jan 1992).

Recently, it has been reported that small–scale ($\lambda \sim 0.1$–2 km) density irregularities occur during 100–Hz electric field bursts in the nightside ionosphere of Venus. The correlation of field and plasma fluctuations suggests that a local plasma instability may be responsible for the turbulence. In this paper the author provides a detailed analysis of the lower–hybrid–drift instability as a mechanism to generate the observed irregularities. He develops a fully electromagnetic theory that is relevant to the finite β plasma in Venus' ionosphere and includes collisional effect (e.g., electron–ion electron–neutral, and ion–neutral collisions). The key features of the analysis that favor this instability are (1) it is a flute mode and propagates orthogonal to the ambient magnetic field, (2) it is a relatively short wavelength mode and the Doppler–shifted frequency can be $\gtrsim 100$ Hz, (3) it can produce both electric field and density fluctuations, as well as magnetic field fluctuations in a finite β plasma, and (4) it is most unstable in low–β plasma so that it is likely to occur in the low–density, high–magnetic–field ionospheric holes. These features are consistent with observational results.

093.004 Venus after Magellan.
M. Eliáš.
Kozmos, Vol. 23, No. 2, p. 4 – 12 (Mar 1992). In Czech.

093.005 A global traveling wave on Venus.
M. D. Smith, P. J. Gierasch, P. J. Schinder.
Science, Vol. 256, No. 5057, p. 652 – 655 (1 May 1992).

The dominant large–scale pattern in the clouds of Venus has been described as a "Y" or "Ψ" and tentatively identified by earlier workers as a Kelvin wave. A detailed calculation of linear wave modes in the Venus atmosphere verifies this identification. Cloud feedback by infrared heating fluctuations is a plausible excitation mechanism. Modulation of the large–scale pattern by the wave is a possible explanation for the "Y". Momentum transfer by the wave could contribute to sustaining the general circulation.

093.006 An experimental study of aeolian structures on Venus.
J. R. Marshall, R. Greeley.
J. Geophys. Res., Vol. 97, No. E1, p. 1007 – 1016 (25 Jan 1992).

Experiments to simulate the formation of aeolian bed forms on Venus show that a high–density atmosphere produces small transverse bed forms with dimensionless similarities to terrestrial dunes but with both dimensional and behavioral similarities to subaqueous current ripples. Their development is influenced by wind speed, particle size, and atmospheric density. Although aeolian bed forms should be observed at all elevations on Venus, their optimum expression is compatible with the lowest elevations where atmospheric pressure is greatest. Their development is relatively unhindered by the presence of dense grains, the lack of sorting in source sediment from which they form, or the addition of cohesive dust. Small (~ 10 cm) bed forms are efficient in sorting materials either by density or particle size. Bed forms developed in the limited size of the wind tunnel are probably representative of small bed forms on Venus; considerations suggest that bed forms on Venus may grow to larger sizes. Discovery of dune fields on Magellan images of Venus support this prediction.

093.007 Chemistry of the surface and lower atmosphere of Venus.
B. Fegley Jr., A. Treiman.
Astron. Vestn., Tom 26, No. 2, p. 3 – 65 (Mar – Apr 1992). In Russian. English translation in Solar Syst. Res., Vol. 26, No. 2.

The authors give a comprehensive review of the chemical interactions between the atmosphere and surface of Venus. Earth–based, Earth–orbital, and spacecraft data on the composition of the atmosphere and surface of Venus are presented and applied to quantitative discussions of the chemical interactions between carbon, hydrogen, sulfur, chlorine, fluorine, and nitrogen gases and plausible minerals on the surface of Venus. The results of these calculations are used to predict stable minerals and mineral assemblages on the surface of Venus, to determine which, if any, atmospheric gases are buffered by mineral assemblages on the surface, and to critically review and assess the prior work on atmosphere–surface chemistry on Venus. Several specific conclusions are presented.

093.008 Venusian highlands: geoid to topography ratios and their implications.
S. E. Smrekar, R. J. Phillips.
Earth Planet. Sci. Lett., Vol. 107, No. 3/4, p. 582 – 597 (Dec 1991).

Using Pioneer Venus line–of–sight gravity data and orbit simulation procedures, the authors have estimated apparent depths of isostatic compensation (ADCs) for twelve Venusian highland features: Asteria, Atla, Bell, Beta, Ovda, Phoebe, Tellus, Thetis and Ulfrun Regiones, and Nokomis, Gula and Sappho Montes. ADCs range from 50 km to 270 km; half of the values are less than 100 km. Using these ADCs, the authors estimate geoid to topography ratios (GTRs) for each area to allow comparison with convection calculations and with terrestrial data for oceanic hot spots, swells and plateaus. The difference between terrestrial and Venusian GTR ranges can be explained largely by the lack of a low viscosity zone on Venus.

093.009 Isolation of major Venus thermospheric cooling mechanism and implications for Earth and Mars.
G. M. Keating, S. W. Bougher.
J. Geophys. Res., Vol. 97, No. A4, p. 4189 – 4197 (1 Apr 1992).

A study of Pioneer Venus orbiter atmospheric drag (OAD) data establishes that there is a weak but clear response of the atmosphere to short–term variations associated with the 27–day rotation of the Sun. Thermospheric temperature residuals relative to the mean diurnal variation have been found to be associated with variations in solar activity after correction for the Earth–Sun–Venus angle. All of the dayside OAD thermospheric data are found to exhibit this characteristic with peak–to–peak temperature variations of approximately 25K. These results are compared with National Center for Atmospheric Research theoretical models of the Venus thermosphere that include processes not in local thermodynamic equilibrium. It is found that the low amplitude of the 27–day oscillations, combined with the cooling necessary for observed 300–K dayside temperatures, may not be explained by eddy conduction cooling and may only be explained by very strong 15–μm cooling. Implications concerning the decrease in amplitude of the 11–year Venus thermospheric variability due to the 15–μm thermostat effect are also examined. In addition, the cooling mechanism must apply elsewhere, including Earth and Mars.

093.010   Erratum: "Global hybrid simulation of the solar wind interaction with the dayside of Venus" [J. Geophys. Res., Vol. 96, No. A5, p. 7779 – 7791 (1 May 1991)].
K. R. Moore, V. A. Thomas, D. J. McComas.
J. Geophys. Res., Vol. 97, No. A4, p. 4317 (1 Apr 1992). See Abstr. 53.093.047.

093.011   Cellular convection in the atmosphere of Venus.
R. D. Baker II, G. Schubert.
Nature, Vol. 355, No. 6362, p. 710 – 712 (20 Feb 1992). Letter–to–the–editor.
Among the most intriguing features of the atmosphere of Venus is the presence of cellular structures near and downwind of the subsolar point. It has been suggested that the structures are atmospheric convection cells, and, in fact, numerical simulations of three–dimensional compressible convection produce features reminiscent of the cellular structures found in the subsolar Venus atmosphere. There has been some difficulty, however, in accounting for the sizes of these structures. Here the authors propose that strongly penetrative convection into the stable regions above and below the neutrally stable cloud layer coupled with penetrative convection from the surface increases the vertical dimensions of the cells, thereby helping to explain their large horizontal extent.

093.012   Magma reservoirs and neutral buoyancy zones on Venus: implications for the formation and evolution of volcanic landforms.
J. W. Head, L. Wilson.
J. Geophys. Res., Vol. 97, No. E3, p. 3877 – 3903 (25 Mar 1992).
The authors examine the production of magma reservoirs and neutral buoyancy zones on Venus and the implications of their development for the formation and evolution of volcanic landforms.

093.013   Evidence for enhanced dynamic flow in ionospheric holes from the Pioneer Venus Orbiter Neutral Mass Spectrometer.
W. T. Kasprzak, H. B. Niemann.
Planet. Space Sci., Vol. 40, No. 1, p. 33 – 45 (Jan 1992).
The Pioneer Venus Orbiter Neutral Mass Spectrometer (ONMS) was operated in ion mode during the third nightside cycle of the Orbiter around Venus. The subject of current interest in this paper is measurements of one component of the ion drift obtained from the ONMS instrument for the outbound holes of orbit numbers 530 and 531. One component ion drifts are also obtained for orbit number 500 which is symmetrically placed with respect to the midnight meridian but without a hole. The ONMS data suggest that enhanced dynamic flow is taking place on the nightside in the hole regions, in regions without a hole, and near the ionopause. The direction of the ion flow cannot be determined. However, the minimum speed of the ion flow can be estimated and bounds on its direction in the ecliptic plane established.

093.014   Turbulent pick–up of new–born ions near Venus and Mars and problems of numerical modelling of the solar wind interaction with these planets. I. Features of the solar wind interaction with planets.
T. K. Breus, A. M. Krymskii (*A. M. Krymskij*).
Planet. Space Sci., Vol. 40, No. 1, p. 121 – 130 (Jan 1992).
This paper deals with mass–loading near Venus. It is shown that heavy ions born upstream of the Venusian shockfront do not significantly change the solar wind (SW) parameters (in particular, Mach number). In the Venusian magnetosheath the number of heavy ions undergoing acceleration in the largescale field, which can be source of the asymmetry and non–hydrodynamic properties of the plasma, is a few percent of the total ion flux from the dayside to the downstream mantle. The most intensive mass–loading of the SW flow is near the ionopause. Pick–up instabilities are possible there and plasma with two ion species will have hydrodynamical features due to turbulence resulting from instabilities.

093.015   Turbulent pick–up of new–born ions near Venus and Mars and problems of numerical modelling of the solar wind interaction with these planets. II. Two–fluid HD model.
T. K. Breus, A. M. Krymskii (*A. M. Krymskij*), V. Ya. Mitnitskii (*V. Ya. Mitnitskij*).
Planet. Space Sci., Vol. 40, No. 1, p. 131 – 138 (Jan 1992).
A two–fluid hydrodynamic (HD) model with anomalous friction between ion species (protons and O^+ ions) due to turbulence in the magnetosheath is suggested for solar wind (SW) interactions with Venus and Mars. The method of calculation and results are given for both planets. The results are compared with experimental data from the Pioneer–Venus and PHOBOS–2 spacecraft and also with the results of the one–fluid HD model with mass–loading. The results of the two–fluid HD model are in better agreement with the experimental data than those obtained earlier.

093.016   The first results of the Magellan mission.
S. R. Brzostkiewicz.
Urania, Rok 63, Nr. 13, p. 9 – 13 (1992). In Polish.

093.017   Past and present water budget of Venus.
T. M. Donahue, R. R. Hodges Jr.
J. Geophys. Res., Vol. 97, No. E4, p. 6083 – 6091 (25 Apr 1992).
A detailed analysis of Pioneer Venus large–probe neutral mass spectrometer (LNMS) data confirms an earlier report that the abundance of Venus deuterium relative to hydrogen is 2 orders of magnitude larger than that of terrestrial deuterium. These results have recently received confirmation from ground–based spectroscopic observations. The LNMS data can be analyzed to obtain height profiles for hydrogen compounds because of the very different isotopic signatures of Venus hydrogen and terrestrial hydrogen. The mixing ratio for water increases by a factor of 4 between 1 and 10 km to a value of 2.2 times the ^{36}Ar mixing ratio (67 ppm) and drops to 10 ppm above 50 km. Preliminary analysis has not revealed the presence of any hydrogen constituent that can compensate for the decrease in water vapor mixing ratio below 10 km. No combination of gas phase and surface chemistry has been found to account for the strange behavior, and vertical transport alone is not feasible. These results are inexplicable unless they can be interpreted in terms of large–scale temporal and spatial variations (for which other evidence also exists). It is argued that present day deuterium and hydrogen cannot be in a steady state with stochastic cometary inputs but that these results call for an early Venus endowed with at least 2 orders of magnitude more water above the surface than is presently there. From the isotope ratio measurement alone, however, how the water got there or how much was originally present cannot be discerned.

093.018   Pervasive large–scale magnetic fields in the Venus nightside ionosphere and their implications.
J. G. Luhmann.
J. Geophys. Res., Vol. 97, No. E4, p. 6103 – 6121 (25 Apr 1992).
When the solar wind dynamic pressure at Venus was extraordinarily high during the primary mission of the Pioneer Venus

Orbiter (PVO), "disappearing ionospheres" occurred on the nightside, with accompanying pervasive near–periapsis magnetic fields of tens of nanoteslas. These nightside counterparts of the generally horizontal large–scale magnetic fields in the dayside ionosphere are found to exhibit some dependence of field magnitude on the solar wind pressure but not on solar zenith angle. Their statistical behavior suggests a global configuration in which the low–altitude field wraps around the planet, while the field at higher altitudes is draped like the induced magnetotail field. The toroidal low–altitude field geometry implies the possible existence of magnetic x points in the low–altitude wake.

093.019 Wind interaction with falling ejecta: origin of the parabolic features on Venus.
R. J. Vervack Jr., H. J. Melosh.
Geophys. Res. Lett., Vol. 19, No. 6, p. 525 – 528 (20 Mar 1992).

Unusual parabolic features associated with impact craters have been observed by Magellan on Venus. A strong correlation exists between the orientation of the features and the zonal winds on Venus. The authors propose a quantitative model in which the parabolic features are produced by the interaction of the zonal winds with material ejected ballistically from the impact crater. As the ejecta particles fall through the atmosphere, the winds transport them downwind from their entry point, smaller particles being transported a greater distance. Since the ejecta distribution is initially axially symmetric and smaller particles are thrown farther from the crater, the winds blow the particles on the upwind side back upon one another, leading to a pile–up of material. On the downwind side, the winds disperse the ejecta particles and no pile–up occurs. The resulting thickness distribution on the Venusian surface matches the observed parabolic features closely. The dual parabolic feature associated with the crater Carson is also explained by this model.

093.020 First Magellan results from Venus.
J. Raitala.
Annual meeting of the Finnish Astronomical Society, p. 15 – 16 (Apr 1992). – See Abstr. 012.033 for the main entry.

093.021 Airburst origin of dark shadows on Venus.
K. J. Zahnle.
J. Geophys. Res., Vol. 97, No. E6, p. 10243 – 10255 (25 Jun 1992).

A simple analytic model for the catastrophic disruption and deceleration of impactors in a thick atmosphere is used to (1) reproduce observed Venusian cratering statistics and (2) generate radar dark disks by the impact of atmospheric shock waves with the surface. When used as input to Monte Carlo simulations of Venusian cratering, the model nicely reproduces the observed low diameter cutoff. Venusian craters are found to be more consistent with an asteroidal rather than a cometary source. The radar–dark "shadows" of the title are surface features, usually circular, that have been attributed to airbursting impactors. A typical craterless airburst is the equivalent of a $\sim 10^6$ megaton explosion. The airburst is treated as a massive, extended explosion using a thin–shell, isobaric cavity approximation. The strong atmospheric shock waves excited by the airburst are then coupled to surface rock using the usual impedance matching conditions. Peak shock pressures experienced by surface rock typically exceed 0.2 GPa for distances 15–30 km from ground zero (the place on the surface immediately beneath the site of the airburst), and 1 GPa for 10–20 km. These high shock pressures are felt to considerable depth, often more than a kilometer. Beneath the airburst the shock could reduce surface rocks to fine rubble, while at greater distance the weaker shock would leave fields of broken blocks, perhaps in part accounting for radar–bright halos that often surround the dark shadows.

093.022 Magnetic fields in Venus nightside ionospheric holes: collected Pioneer Venus Orbiter magnetometer observations.
J. G. Luhmann, D. S. Russell.
J. Geophys. Res., Vol. 97, No. E6, p. 10267 – 10282 (25 Jun 1992).

The magnetic fields detected by the Pioneer Venus Orbiter (PVO) magnetometer within the electron density depletions called "holes" in the nightside ionosphere are typically larger and more organized than the fields in the surrounding ionosphere. Moreover, they have substantial sunward/antisunward components which cause them to appear as near–radial fields near the antisolar point. The collection of observations presented here illustrates the variety of appearances of the fields in holes. Some new results which summarize their average properties, their dependence on solar wind conditions, and their lack of geographical control are also presented. These results are potentially pertinent to the interpretation of data from the PVO entry at the end of 1992 and from the impending Mars Observer mission, which will probe the magnetic fields in the low–altitude wake of weakly magnetized Mars.

093.023 Estimates of Venusian atmospheric torque.
J. McCue, J. R. Dormand, A. M. Gadian.
Earth, Moon, Planets, Vol. 57, No. 1, p. 1 – 11 (Apr 1992).

An estimate is derived of the solar gravitational torque on the thermal atmospheric tide of Venus. The value obtained is compared with the computed torque on the body of the planet itself caused by viscous coupling between it and the superrotating atmosphere. The comparison suggests that the solar thermal torque and the viscous torque are effective in the maintenance of the four–day superrotation of the Venusian atmosphere.

093.024 Venus geology, geochemistry, and geophysics. Introduction.
A. T. Basilevsky (*A. T. Bazilevskij*), O. V. Nikolayeva (*O. V. Nikolaeva*), V. P. Volkov.
Venus geology, geochemistry, and geophysics, p. 1 – 7 (1992). – See Abstr. 003.047 for the main entry.

093.025 Volcanism.
E. N. Slyuta, O. V. Nikolayeva (*O. V. Nikolaeva*).
Venus geology, geochemistry, and geophysics, p. 13 – 30 (1992). – See Abstr. 003.047 for the main entry.

The Venera 15/16 radar survey of the northern quarter of the planet has given reliable morphologic evidence of various forms of basaltic volcanism represented by the plains–forming areal eruptions and edifices of various size.

093.026 Hot–spot structures.
A. M. Nikishin, A. A. Pronin, A. T. Basilevsky (*A. T. Bazilevskij*).
Venus geology, geochemistry, and geophysics, p. 31 – 67 (1992). – See Abstr. 003.047 for the main entry.

Within the area surveyed by Venera 15/16 many uplands and highlands with associated tectonism and volcanism are observed. Their diameters vary from 100 to 3000 km and the height reaches 4 – 5 km. The endogenic origin of these features is obvious. They may have formed as a result of the activity of hot nonlinear, more or less planimetrically equidimensional areas in the venusian mantle. For them the term hot spots is appropriate. The features of this group include dome–like uplands, exemplified by Beta Regio, and smaller features of this type as well as so–called coronae and arachnoids.

093.027 The Lakshmi phenomenon.
A. A. Pronin.
Venus geology, geochemistry, and geophysics, p. 68 – 81 (1992). – See Abstr. 003.047 for the main entry.

The structure of Lakshmi and its surrounding mountains occupying the western part of Ishtar Terra highland is a unique phenomenon in many respects. The conformity of the pattern of the Lakshmi ridge–and–grove structures surrounding the plateau indicates the unity of the process of formation of this giant structure more than 2000 km across. Its central part, Laskshmi Planum, the plateau itself, is represented by a vast lava plain about 1500 km across and elevated over the mean level of venusian plains by 3 – 4 km. This lava plateau is unusual for Venus. The mountains surrounding the plateau are elevated over the plateau by 3 – 6 km.

093.028 **Tesserae.**
A. L. Sukhanov.
Venus geology, geochemistry, and geophysics, p. 82 – 95 (1992). – See Abstr. 003.047 for the main entry.

The areas of regional deformations were called parquet at first, because their typical patterns resemble a sort of rhombical or orthogonal parquetry (Barsukov et al., 1985). Later the name tessera (tiles in Greek) was accepted, but the terms do not quite coincide. Both words mean peculiar types of ridge–and–groove terrain, evidently, due to deformations. But the word tessera(e) is also used as a generic term to designate rather large massifs with parquet surface (usually in combination with some name).

093.029 **Ridge belts on plains.**
V. P. Kryuchkov.
Venus geology, geochemistry, and geophysics, p. 96 – 112 (1992). – See Abstr. 003.047 for the main entry.

Within the area surveyed by Venera 15/16 the volcanic plains contain many ridge belts. They resemble the mountain ridge systems surrounding the Lakshmi structure and the ridge systems encircling typical coronae. At first they were interpreted as belts of compressional deformation. Later structures have been found favoring their origin by extensional deformation.

093.030 **Impact craters.**
B. A. Ivanov.
Venus geology, geochemistry, and geophysics, p. 113 – 128 (1992). – See Abstr. 003.047 for the main entry.

Identification of impact craters on Venus has been made on radar images processed from Venera 15/16 data. They are subdivided into morphological classes. Crater density and dimensions are given.

093.031 **Evidence on the crustal dichotomy.**
O. V. Nikolayeva (*O. V. Niokolaeva*), M. A. Ivanov, V. K. Borozdin.
Venus geology, geochemistry, and geophysics, p. 129 – 139 (1992). – See Abstr. 003.047 for the main entry.

093.032 **Global tectonic style.**
A. T. Basilevsky (*A. T. Bazilevskij*).
Venus geology, geochemistry, and geophysics, p. 140 – 152 (1992). – See Abstr. 003.047 for the main entry.

The current data on Venus topography, geophysics, and geology are not adequate for reliable understanding of the global tectonic style of this planet. But the desire to arrange the data in the form of models forces us to take new steps in this direction. This attempt, based mainly on the results of the Venera 15/16 radar survey, is given in this paper. It aims not to solve but to define the problems, some of which may be resolved by the Magellan mission.

093.033 **Resurfacing.**
A. T. Basilevsky (*A. T. Bazilevskij*), O. V. Nikolayeva (*O. V. Nikolaeva*), R. O. Kuzmin.
Venus geology, geochemistry, and geophysics, p. 153 – 160 (1992). – See Abstr. 003.047 for the main entry.

In the case of Venus, the only way now available of absolute dating of geologic formations is the estimation of the areal density of the impact craters. So the parameter of absolute age of the given geologic formation is the crater retention age.

093.034 **Venusian igneous rocks.**
V. L. Barsukov.
Venus geology, geochemistry, and geophysics, p. 165 – 176 (1992). – See Abstr. 003.047 for the main entry.

A minimum of detailed information on rocks on the surface of Venus is available obtained by Venera and Vega landing sites. A classification of Venusian surface rocks can be made by comparing U/K_2O and Th/K_2O ratios with those characteristic for various types of the terrestrial and lunar magmatic rocks.

093.035 **Chemical processes on the planetary surface.**
M. Yu. Zolotov, V. P. Volkov.
Venus geology, geochemistry, and geophysics, p. 177 – 199 (1992). – See Abstr. 003.047 for the main entry.

The understanding of chemical processes on the Venus surface depends on knowledge of the chemical composition of the lower atmosphere, especially the redox conditions at the surface–atmosphere interface.

093.036 **Volatiles in atmosphere and crust.**
V. P. Volkov.
Venus geology, geochemistry, and geophysics, p. 200 – 207 (1992). – See Abstr. 003.047 for the main entry.

The author presents comments on generalized tables of Venus' volatile inventories and their isotopic composition in the outer shells in relation to other inner planets.

093.037 **Expansion of topography into spherical harmonics.**
V. N. Zharkov.
Venus geology, geochemistry, and geophysics, p. 214 – 217 (1992). – See Abstr. 003.047 for the main entry.

093.038 **Gravity field, loading coefficients, anomalous density waves, and the case of long waves.**
V. N. Zharkov.
Venus geology, geochemistry, and geophysics, p. 218 – 227 (1992). – See Abstr. 003.047 for the main entry.

093.039 **Rotation.**
V. N. Zharkov.
Venus geology, geochemistry, and geophysics, p. 228 – 230 (1992). – See Abstr. 003.047 for the main entry.

The problem of the rotation of Venus is not yet fully understood. The author gives a short summary.

093.040 **Statistical properties of topography and gravity field.**
V. N. Zharkov.
Venus geology, geochemistry, and geophysics, p. 231 – 232 (1992). – See Abstr. 003.047 for the main entry.

093.041 **Model of the interior structure: Earth–like models.**
V. N. Zharkov.
Venus geology, geochemistry, and geophysics, p. 233 – 240 (1992). – See Abstr. 003.047 for the main entry.

The author uses and constructs Earth–like models of Venus.

093.042 **A physical model of Venus.**
V. N. Zharkov.
Venus geology, geochemistry, and geophysics, p. 241 – 258 (1992). – See Abstr. 003.047 for the main entry.

The author explains the thermodynamics of the mantle and core and estimates the coefficients of thermal and electrical conductivity, and the mechanical quality Q. A rheological model of the crust and mantle is also considered.

093.043 **The stress state of Venusian crust and variations of its thickness: implications for tectonics and geodynamics.**
V. N. Zharkov, K. I. Marchenkov.
Venus geology, geochemistry, and geophysics, p. 259 – 273 (1992). – See Abstr. 003.047 for the main entry.

The data for the heights of the relief and the external gravitational field of Venus for spherical harmonics with degree and order up to 18, obtained from spacecraft, allow to start theoretical investigations of the crust–mantle boundary and of the stress state of the planetary interior. The authors make a preliminary comparative analysis of the tectonic structure and the geodynamic style of the planet.

093.044 **Convection: a simplified version.**
V. N. Zharkov, V. S. Solomatov.
Venus geology, geochemistry, and geophysics, p. 274 – 279 (1992). – See Abstr. 003.047 for the main entry.

The authors maintain a theory, in which convection takes place independently in the upper and lower mantles of the Earth (and

Venus) (in first approximation) and both convective systems interact on their boundary. Detailed argumentation is given.

093.045   Models of the thermal evolution of Venus.
V. N. Zharkov, V. S. Solomatov.
Venus geology, geochemistry, and geophysics, p. 280 – 319 (1992). – See Abstr. 003.047 for the main entry.

The thermal evolution of Venus can be divided into three periods: adjustment of the upper mantle to the thermal regime of the lower mantle (~ 0.5 by), transition of the whole mantle to the asymptotic regime ($\sim 3 - 4$ by), and the asymptotic regime. The present thermal regime of Venus is close to the asymptotic one. The present temperature of the upper layers of the Venusian mantle is near 1700K; the temperature at the core–mantle boundary is 3500 – 4000K and is close to that of the Earth's. A preliminary comparison of the models with one– and two–layer convection in the mantle shows that the heat flux due to the cooling of the planet is almost the same for both cases. The essential difference lies in the temperature distribution in the mantle. The upper mantle of Venus is partially melted. This can suppress the thermal convection and result in chemical convection. Reasons for the lack of the intrinsic magnetic field of Venus are considered. The thermal regime of the crust is characterized by heat and mass transfer: conductive heat transport, circulation of melts in the crust, exchange of basalt between the upper mantle and the crust, formation of the mantle plumes, convection in the lower part of the crust, involving of the crust into the mantle convection. It must have regional preculiarities due to unstationarity of the convection in the crust and the mantle and has a characteristic time of $10^8 - 10^9$y.

093.046   Atlas of Venus surface images.
A. I. Sidorenko, O. N. Rzhiga, Yu. N. Alexandrov (*Yu. N. Aleksandrov*), A. I. Zakharov, Yu. S. Tyuflin, A. A. Pronin, G. A. Burba.
Venus geology, geochemistry, and geophysics, p. 325 – 381 (1992). – See Abstr. 003.047 for the main entry.

This atlas presents radar images of the surface of Venus obtained by Venera 15 and Venera 16 orbiters.

093.047   Names of topographic features on Venus.
G. A. Burba.
Venus geology, geochemistry, and geophysics, p. 383 – 400 (1992). – See Abstr. 003.047 for the main entry.

093.048   The structures of gravitational outflow in the massive Maxwell Mt. on Venus.
A. A. Pronin, M. A. Kreslavskij.
Astron. Vestn., Tom 26, No. 3, p. 26 – 43 (May – Jun 1992). In Russian. English translation in Sol. Syst. Res., Vol. 26, No. 3.

On the basis of an analysis of Venera–15/16 and Magellan radar images and altimetry data the surface features of Maxwell Mt. on Venus are interpreted in terms of stress signs. Spatial distribution of the features observed is compared with a theoretical model of viscous gravitational relaxation. A satisfactory agreement between model and real stress distributions points to the efficiency of gravitational tectonics on Venus.

093.049   Correlation of Earth–based NIR imagery and Pioneer–Venus Orbiter imagery and data.
B. Ragent, L. Travis, D. Crisp, D. Allen, P. Steffes, J. Jenkins, G. Deardorff, Y. Hung.
Bull. Am. Astron. Soc., Vol. 23, No. 3, p. 1192 (1991). Abstract. – See Abstr. 012.037 for the main entry.

093.050   High–resolution spectroscopy of Venus's night side in the 2.3, 1.7 and 1.1 – 1.3 μm windows.
B. Bézard, C. de Bergh, J. P. Maillard, D. Crisp, J. Pollack, D. Grinspoon.
Bull. Am. Astron. Soc., Vol. 23, No. 3, p. 1192 (1991). Abstract. – See Abstr. 012.037 for the main entry.

093.051   The flight of the Galileo spacecraft past Venus, Earth, and the Moon.
A. T. Bazilevskij.
Sol. Syst. Res., Vol. 25, No. 6, p. 507 – 512 (May 1992). English translation of Astron. Vestn., Tom 25, No. 6, p. 677 – 685 (1991).

A short review is given of the results of scientific observations of Venus, Earth, and the Moon by the Galileo spacecraft on its way to Jupiter via a complex trajectory using gravitational maneuvers around Venus and Earth. Television and infrared observations of Venus yielded new information on the structure and dynamics of its atmosphere. Observations of the atmosphere and surface of Earth, the properties of which are well–known, will be used to calibrate Galileo's on–board instruments. During the flyby of the Moon images of its western hemisphere were obtained, including the well–studied regions of its visible face as well as poorly studied regions of its far side. Spectrozonal maps can be used to study the distributions of basic types of lunar rocks. Spectra of the Mare Orientale basin do not show significant amounts of the mafic material, which sets a limit on the excavation depth of the initial crater of this basin at less than 70 km. The existence of the vast South Pole–Aitken basin (D = 2000 km) was confirmed. Spectra of the edges of this basin indicate an enhanced content of mafic material, which supports the presence of buried mare lava or excavation by this basin of underlying material. Galileo's next object of investigation is the asteroid 951 Gaspra, scheduled for flyby in October 1991.

093.052   Simulations of near infrared spectra of Venus' nightside with high temperature databases.
J. B. Pollack, J. B. Dalton, D. Grinspoon, R. B. Wattson, D. Crisp, D. Allen, B. Bezard, C. de Bergh, R. Freedman.
Bull. Am. Astron. Soc., Vol. 23, No. 3, p. 1193 (1991). Abstract. – See Abstr. 012.037 for the main entry.

093.053   Characteristics of gravity waves generated in the lower Venus atmosphere.
R. E. Young, H. Houben, R. L. Walterscheid, G. Schubert.
Bull. Am. Astron. Soc., Vol. 23, No. 3, p. 1193 (1991). Abstract. – See Abstr. 012.037 for the main entry.

093.054   Venus lower thermosphere: 3D model implications.
S. W. Bougher, M. J. Alexander.
Bull. Am. Astron. Soc., Vol. 23, No. 3, p. 1193 – 1194 (1991). Abstract. – See Abstr. 012.037 for the main entry.

093.055   Solar–locked features in Venus thermospheric oxygen.
M. J. Alexander, A. I. F. Stewart, S. C. Solomon, S. W. Bougher.
Bull. Am. Astron. Soc., Vol. 23, No. 3, p. 1194 (1991). Abstract. – See Abstr. 012.037 for the main entry.

093.056   Fractionation of hydrogen and deuterium due to non–thermal escape on Venus.
M. Gurwell, Y. L. Yung.
Bull. Am. Astron. Soc., Vol. 23, No. 3, p. 1194 (1991). Abstract. – See Abstr. 012.037 for the main entry.

093.057   Laboratory measurements of the millimeter–wave (3 mm) opacity of gaseous SO_2 under simulated conditions of the middle atmosphere of Venus.
A. K. Fahd, P. G. Steffes.
Bull. Am. Astron. Soc., Vol. 23, No. 3, p. 1194 (1991). Abstract. – See Abstr. 012.037 for the main entry.

093.058   Intense localized near–infrared oxygen airglow on the Venus night side.
D. Crisp, D. A. Allen, B. Bezard, C. de Bergh, J.–P. Maillard.
Bull. Am. Astron. Soc., Vol. 23, No. 3, p. 1194 – 1195 (1991). Abstract. – See Abstr. 012.037 for the main entry.

093.059 **Galileo/NIMS at Venus: middle–atmosphere zonal and meridional winds, and implications for the observed mid–level cloud morphology.**
K. H. Baines, R. W. Carlson.
Bull. Am. Astron. Soc., Vol. 23, No. 3, p. 1195 (1991). Abstract. – See Abstr. 012.037 for the main entry.

093.060 **Dimers of carbon dioxide in the atmospheres of Venus and Mars.**
K. Fox, S. J. Kim, Z. Slanina.
Bull. Am. Astron. Soc., Vol. 23, No. 3, p. 1195 (1991). Abstract. – See Abstr. 012.037 for the main entry.

093.061 **Magnetic fields in Venus nightside ionospheric holes: collected Pioneer Venus Orbiter magnetometer observations.**
J. G. Luhmann, D. S. Russell.
Bull. Am. Astron. Soc., Vol. 23, No. 3, p. 1195 – 1196 (1991). Abstract. The full paper is published in J. Geophys. Res., Vol. 97E, p. 10267 – 10282 (25 Jun 1992). – See Abstr. 012.037 for the main entry.

093.062 **Thermal observations of Venus during Galileo Venus encounter.**
T. Z. Martin, G. S. Orton, L. D. Travis, A. A. Lacis.
Bull. Am. Astron. Soc., Vol. 23, No. 3, p. 1196 (1991). Abstract. – See Abstr. 012.037 for the main entry.

093.063 **Rocket observations of Venus SO_2 and SO.**
C. Y. Na, L. W. Esposito, W. E. McClintock, C. A. Barth.
Bull. Am. Astron. Soc., Vol. 23, No. 3, p. 1196 (1991). Abstract. – See Abstr. 012.037 for the main entry.

093.064 **Waves in the Venus atmosphere: idealized Cartesian model.**
P. J. Schinder, M. D. Smith, P. J. Gierasch.
Bull. Am. Astron. Soc., Vol. 23, No. 3, p. 1196 (1991). Abstract. – See Abstr. 012.037 for the main entry.

093.065 **Waves in the Venus atmosphere: spherical model.**
M. D. Smith, P. J. Schinder, P. J. Gierasch.
Bull. Am. Astron. Soc., Vol. 23, No. 3, p. 1196 (1991). Abstract. – See Abstr. 012.037 for the main entry.

093.066 **Radio occultation studies of the Venus atmosphere with the Magellan spacecraft.**
P. G. Steffes, J. M. Jenkins, R. S. Austin, G. L. Tyler, E. H. Seale.
Bull. Am. Astron. Soc., Vol. 23, No. 3, p. 1196 – 1197 (1991). Abstract. – See Abstr. 012.037 for the main entry.

093.067 **Images of the Venus cloud deck from Galileo.**
P. J. Gierasch, M. Smith, P. Helfenstein, P. J. Schinder, J. Veverka, M. J. S. Belton, K. P. Klaasen, J. B. Pollack, K. A. Rages, D. Morrison, A. P. Ingersoll, C. D. Anger, M. H. Carr, C. R. Chapman, M. E. Davies, F. P. Fanale, R. Greeley, R. Greenberg, J. W. Head III, G. Neukum, C. B. Pilcher.
Bull. Am. Astron. Soc., Vol. 23, No. 3, p. 1203 – 1204 (1991). Abstract. – See Abstr. 012.037 for the main entry.

093.068 **Galileo near infrared mapping spectroscopy of Venus: composition and cloud physics.**
R. W. Carlson, K. H. Baines, L. W. Kamp, T. Encrenaz, P. Drossart, E. Lellouch, F. W. Taylor, A. D. Collard, S. B. Calcutt, J. B. Pollack, D. Grinspoon.
Bull. Am. Astron. Soc., Vol. 23, No. 3, p. 1204 (1991). Abstract. – See Abstr. 012.037 for the main entry.

093.069 **Galileo observations of radio signals from lightning at Venus.**
D. A. Gurnett, W. S. Kurth, A. Roux, R. Gendrin, C. F. Kennel, S. J. Bolton.
Bull. Am. Astron. Soc., Vol. 23, No. 3, p. 1204 (1991). Abstract. – See Abstr. 012.037 for the main entry.

093.070 **Tectonic processes on Venus: a global perspective.**
S. C. Solomon.
Bull. Am. Astron. Soc., Vol. 23, No. 3, p. 1205 (1991). Abstract. – See Abstr. 012.037 for the main entry.

093.071 **Global distribution and styles of volcanism on Venus and implications for resurfacing: a synthesis of Magellan results.**
J. W. Head.
Bull. Am. Astron. Soc., Vol. 23, No. 3, p. 1205 (1991). Abstract. – See Abstr. 012.037 for the main entry.

093.072 **Impact craters on Venus: What are they telling us?**
G. G. Schaber.
Bull. Am. Astron. Soc., Vol. 23, No. 3, p. 1205 (1991). Abstract. – See Abstr. 012.037 for the main entry.

093.073 **Channels on Venus.**
V. R. Baker, G. Komatsu, V. C. Gulick, J. S. Kargel, R. G. Strom, T. J. Parker.
Bull. Am. Astron. Soc., Vol. 23, No. 3, p. 1205 (1991). Abstract. – See Abstr. 012.037 for the main entry.

093.074 **Magellan: measurements of the electrical and physical properties of the surface of Venus.**
D. B. Campbell, N. J. S. Stacy, P. G. Ford, G. H. Pettengill, R. E. Arvidsen, B. A. Campbell.
Bull. Am. Astron. Soc., Vol. 23, No. 3, p. 1206 (1991). Abstract. – See Abstr. 012.037 for the main entry.

093.075 **Constraints on the interior dynamics of Venus from Magellan data.**
R. J. Phillips, R. E. Grimm, R. R. Herrick, B. Parsons, S. E. Smrekar.
Bull. Am. Astron. Soc., Vol. 23, No. 3, p. 1206 (1991). Abstract. – See Abstr. 012.037 for the main entry.

093.076 **The origin and evolution of coronae on Venus.**
S. Squyres, D. Janes, G. Baer, D. Bindschadler, G. Schubert, V. Sharpton, E. Stofan.
Bull. Am. Astron. Soc., Vol. 23, No. 3, p. 1219 – 1220 (1991). Abstract. – See Abstr. 012.037 for the main entry.

093.077 **Global characteristics of coronae on Venus: a preliminary assessment from Magellan data.**
G. Schubert, G. Baer, D. L. Bindschadler, E. R. Stofan, D. M. Janes, S. Squyres, D. Sandwell, V. L. Sharpton.
Bull. Am. Astron. Soc., Vol. 23, No. 3, p. 1220 (1991). Abstract. – See Abstr. 012.037 for the main entry.

093.078 **Constraints on the thermal structure of Venus mountain belts from Magellan observations of volcanism and deformation.**
N. Namiki, S. C. Solomon.
Bull. Am. Astron. Soc., Vol. 23, No. 3, p. 1220 (1991). Abstract. – See Abstr. 012.037 for the main entry.

093.079 **Formation mechanism of Venusian channels.**
G. Komatsu, V. R. Baker, R. G. Strom, J. S. Kargel.
Bull. Am. Astron. Soc., Vol. 23, No. 3, p. 1220 (1991). Abstract. – See Abstr. 012.037 for the main entry.

093.080 **Composition and petrogenesis of Venusian channel–forming lavas.**
J. S. Kargel, J. S. Lewis, G. Komatsu.
Bull. Am. Astron. Soc., Vol. 23, No. 3, p. 1220 – 1221 (1991). Abstract. – See Abstr. 012.037 for the main entry.

093.081   **Wind streaks on Venus: Magellan results.**
R. Greeley, M. A. Geringer, R. E. Arvidson,
C. Elachi, J. J. Plaut, R. S. Saunders, E. R. Stofan,
E. J. P. Thouvenot, S. D. Wall, C. M. Weitz, G. Schubert.
Bull. Am. Astron. Soc., Vol. 23, No. 3, p. 1221 (1991). Abstract. –
See Abstr. 012.037 for the main entry.

093.082   **Venus: detailed topography of the Artemis, Diana and Dali Chasmata.**
P. G. Ford, Liu Fang, G. H. Pettengill.
Bull. Am. Astron. Soc., Vol. 23, No. 3, p. 1221 (1991). Abstract. –
See Abstr. 012.037 for the main entry.

093.083   **Volumetric analysis of large impact craters on Venus.**
J. B. Garvin, G. G. Schaber.
Bull. Am. Astron. Soc., Vol. 23, No. 3, p. 1221 (1991). Abstract. –
See Abstr. 012.037 for the main entry.

093.084   **Areas of low emissivity in Alpha Regio and other locations.**
K. A. Tryka, D. O. Muhleman.
Bull. Am. Astron. Soc., Vol. 23, No. 3, p. 1221 (1991). Abstract. –
See Abstr. 012.037 for the main entry.

093.085   **Magellan preliminary report on the rotation period, the direction of the north pole, and the geodetic control network of Venus.**
M. E. Davies, T. R. Colvin, P. G. Rogers, P. W. Chodas,
W. L. Sjogren.
Bull. Am. Astron. Soc., Vol. 23, No. 3, p. 1222 (1991). Abstract. –
See Abstr. 012.037 for the main entry.

093.086   **Distribution and classification of fluidized ejecta blankets (FEBs) associated with Venusian impact craters.**
J. R. Johnson, R. G. Strom, T. S. Roessler, G. Komatsu,
V. R. Baker.
Bull. Am. Astron. Soc., Vol. 23, No. 3, p. 1222 (1991). Abstract. –
See Abstr. 012.037 for the main entry.

093.087   **Global distribution and geological settings of Venusian channels.**
G. Komatsu, V. R. Baker, R. G. Strom, T. J. Parker.
Bull. Am. Astron. Soc., Vol. 23, No. 3, p. 1222 (1991). Abstract. –
See Abstr. 012.037 for the main entry.

093.088   **Inversion of Magellan altimeter data to empirical scattering functions.**
M. J. Maurer, G. L. Tyler, R. A. Simpson.
Bull. Am. Astron. Soc., Vol. 23, No. 3, p. 1222 (1991). Abstract. –
See Abstr. 012.037 for the main entry.

093.089   **Thick lava flows on Venus.**
H. J. Moore, J. J. Plaut, P. M. Schenk, J. W. Head.
Bull. Am. Astron. Soc., Vol. 23, No. 3, p. 1222 – 1223 (1991).
Abstract. – See Abstr. 012.037 for the main entry.

093.090   **Geology of Cochran Crater, Venus.**
H. J. Moore, G. G. Schaber.
Bull. Am. Astron. Soc., Vol. 23, No. 3, p. 1223 (1991). Abstract. –
See Abstr. 012.037 for the main entry.

093.091   **Magellan observations of extended impact crater related deposits on the surface of Venus.**
N. J. S. Stacy, D. B. Campbell, G. S. Musser, R. E. Arvidson,
W. Newman, C. Schaller.
Bull. Am. Astron. Soc., Vol. 23, No. 3, p. 1223 (1991). Abstract. –
See Abstr. 012.037 for the main entry.

093.092   **Venus surface mineralogy: observational and theoretical constraints.**
B. Fegley Jr., A. H. Treiman, V. L. Sharpton.
22. Lunar and Planetary Science Conference, p. 3 – 20 (1992). –
See Abstr. 012.046 for the main entry.

The authors utilize Earth–based, Earth–orbital, and spacecraft observations of the atmosphere and surface of Venus, thermodynamic models of atmosphere–lithosphere interactions, and where available kinetic data on relevant gas–solid reactions to place constraints on the mineralogy of the surface of Venus. They discuss which minerals and mineral assemblages are stable on the surface of Venus and which, if any, of these minerals are involved in controlling the abundances of reactive gases in the atmosphere of Venus. The authors conclude by identifying key issues facing us today about the mineralogy and geochemistry of the surface of Venus and suggest experimental, observational, and theoretical studies that can improve our knowledge of these important questions.

093.093   **Laboratory measurements of the microwave and millimeter–wave opacity of gaseous sulfur dioxide (SO_2) under simulated conditions for the Venus atmosphere.**
A. K. Fahd, P. G. Steffes.
Icarus, Vol. 97, No. 2, p. 200 – 210 (Jun 1992).

Gaseous sulfur dioxide (SO_2) has long been recognized as one of the primary absorbers in the Venus atmosphere at microwave frequencies. However, the effects of gaseous SO_2 on the millimeter–wave emission of Venus are not fully understood. This is mainly due to the lack of measurements of opacity under Venus–like conditions at millimeter wavelengths. As a result, we have made laboratory measurements of the opacity of gaseous SO_2 in a CO_2 atmosphere at 13.3 cm, 1.32 cm, and 0.32 cm. The results of the measurements show a close agreement with the absorptivity predicted from a Van Vleck–Weisskopf formalism at the two shortest wavelengths but not at the longest wavelength. In addition, the results show a frequency dependence that is slower than the f^2 dependence of absorptivity proposed by Janssen and Poynter and Steffes and Eshleman. The results are incorporated into a radiative transfer model to infer a new abundance profile for gaseous SO_2 in the middle atmosphere of Venus.

093.094   **Rapporto osservativo del programma Venere elongazioni est 1989 e ovest 1990.**
G. Quarra Sacco, D. Sarocchi.
Astronomia UAI, N. 3, p. 15 – 22 (May – Jun 1992).

093.095   **First results of radar survey of Venus by Magellan.**
A. T. Bazilevskij.
Sol. Syst. Res., Vol. 25, No. 5, p. 412 – 427 (Mar 1992). English translation of Astron. Vestn., Tom 25, No 5, p. 548 – 568 (1991).

A status review of the high–resolution radar survey, as of May 1991, is given. High–quality images, altimetric maps, and maps of radiophysical surface properties have been obtained. Analysis of these materials has resulted in discovery of a number of new phenomena, including: 1) pancake features 20 – 25 km across, evidently formed by eruptions of viscous nonbasaltic lava; 2) long "channels", probably formed by flows of low–viscous liquids, possibly comatiite, carbonatite, or sulfur lavas; 3) areas of polygonal, sometimes very regular gridlike fracturing on volcanic plains; 4) enigmatic flows originating within ejected blankets of impact craters; 5) traces of wind erosion/accumulation of loose material. The western part of Aphrodite Terra, which was previously viewed by some authors as a possible analog of the Earth's mid–oceanic spreading zones, has turned out to be tessera and does not look like a spreading zone. No reliable evidence of plate tectonics on Venus has been found. The survey is continuing.

093.096   **The global and local correlation between the Venus gravity field and its topography.**
M. Burša, Z. Šíma, J. Kostelecký.
Earth, Moon, Planets, Vol. 57, No. 2, p. 123 – 138 (May 1992).

The correlation between the undulations of the equipotential and topography surface of Venus has been investigated over 16 meridional sections as well as in the global scale. The correlation is in general high. However, the figure parameters of the ellipsoids best fitting the surface in question differ significantly. The global features like figures of the best–fitting ellipsoids are mutually strange.

093.097   Zonal wind in the south polar regions of Venus from data of a radio transillumination.
V. N. Gubenko, S. S. Matyugov, O. I. Yakovlev, I. R. Vaganov.
Kosm. Issled., Tom 30, Vyp. 3, p. 390 – 395 (May–Jun 1992). In Russian. English translation in Cosmic Res., Vol. 30, No. 3.

The temperature and pressure data in the southern hemisphere of Venus obtained by the method of radio transillumination using the Venera 15 and 16 satellites are used to determine the altitude profiles of zonal wind velocity at the altitudes from 50 to 80 km. The altitude and latitude dependences of zonal wind velocity in the south latitude interval from 66 to 83° are given. A yet in the circumpolar atmosphere of the southern hemisphere at the latitudes 70 – 72° is discovered, at which wind velocity achieves the maximum value 115 m s^{-1} at the altitude 62 km.

093.098   The thermal conditions of Venus.
V. N. Zharkov, V. S. Solomatov.
US–USSR Workshop on Planetary Sciences, p. 174 – 190 (1991). – See Abstr. 012.075 for the main entry.

This paper examines models of Venus' thermal evolution. The models include the core which is capable of solidifying when the core's temperature drops below the liquidus curve, the mantle which is proposed as divided into two, independent of the convecting layers (upper and lower mantle), and the cold crust which maintains a temperature on the surface of the convective mantle close to 1200°C. The models are based on the approximation of parametrized convection, modified here to account for new investigations of convection in a medium with complex rheology.

093.099   Lithospheric and atmospheric interaction on the planet Venus.
V. P. Volkov.
US–USSR Workshop on Planetary Sciences, p. 218 – 233 (1991). – See Abstr. 012.075 for the main entry.

A host of interesting problems related to the probability of a global process of chemical interaction of the Venusian atmosphere with that planet's surface material has emerged in the wake of flights by the Soviet space probes, "Venera-4, -5, -6, and -7" (1967 – 70). It was disclosed during these flights that the temperature of Venus' surface attains 750K, pressure is approximately 90 atm., and CO_2 constitutes 97% of the atmosphere. The author explores several of these issues which were discussed in the pioneering works of Mueller (1963, 1969) and Lewis (1968, 1970): (1) Is Venus' troposphere in a state of chemical equilibrium? (2) Can one assume that the chemical composition of the troposphere is buffered by the minerals of surface rock? (3) What are the scales and mechanisms involved as exogenic processes take place? (4) To what degree is the composition of cloud particles tied to the process of lithospheric–atmospheric interaction?

093.100   Magellan's global view of the Venusian surface.
R. E. Arvidson, R. J. Phillips, N. Izenberg.
Earth Space, Vol. 4, No. 7, p. 5 – 10 (Mar 1992).

093.101   Magellan onthult Venus.
T. van der Meij.
Zenit, Jaarg. 19, Nr. 3, p. 106 – 109 (Mar 1992).

093.102   Results from the Magellan altimeter.
P. G. Ford, G. H. Pettengill, Liu Fang.
International Symposium on Radars and Lidars in Earth and Planetary Sciences, p. 39 – 44 (Dec 1991). – See Abstr. 012.093 for the main entry.

During the nominal Magellan mission, the altimeter has made some 3 million measurements of the Venus surface, arranged into a set of "footprints" covering the latitude range from 85°N to 80°S. Range correlation, Doppler filtering, multi–burst summation, and range migration are used in order to focus the data. Maps have been prepared showing the global distribution of topography, meter–scale slope, and power reflection coefficient. The results are similar to those reported in previous experiments – the surface radius exhibits a uni–modal distribution with more than 70% of the surface lying within 1 km of the mean radius, but the high resolution of the Magellan altimeter has disclosed several surprisingly steep features, e.g. the west face of Maxwell Montes, the southern face of Danu Montes, and the chasmata to the east of Thetis Regio, where average kilometer–scale slopes of greater than 30° are not uncommon. This conclusion is corroborated by close inspection of SAR imagery.

093.103   Venus: surface dielectric properties.
G. H. Pettengill, P. G. Ford.
International Symposium on Radars and Lidars in Earth and Planetary Sciences, p. 45 – 48 (Dec 1991). – See Abstr. 012.093 for the main entry.

Observations of radiothermal emission, as well as of radar reflectivity, made both from the Earth and from Venus–orbiting spacecraft, have disclosed localized areas of high effective dielectric permittivity on the surface of Venus. In parts of Maxwell Mons and in some highland portions of Aphrodite Terra, the 12–cm–wavelength surface radioemissivity drops to values as low as 0.30, accompanied by a radar power reflection coefficient approaching 0.65. These internally consistent, but unusual, results imply an effective dielectric permittivity in excess of 80, a value which stands in contrast to approximately 5 for most lowland regions of Venus. In the absence of liquid water (known not to exist on the hot surface of Venus), such high values of permittivity are unexpected, and most likely arise from the presence of small, electrically conducting inclusions in a host dielectric, producing a "loaded" dielectric.

093.104   Dynamics of the rotation of Venus with account for the superrotation of its atmosphere.
A. A. Khentov.
Astron. Zh., Tom 69, Vyp. 3, p. 655 – 659 (May – Jun 1992). In Russian. English translation in Sov. Astron., Vol. 36, No. 3 (May – Jun 1992).

It is proved that the quick retrograde rotation of the atmosphere can stabilize the retrograde rotation of Venus.

Venus geology, geochemistry, and geophysics. Research results from the USSR.
See Abstr. 003.047.

SO_2 absorption cross–section measurements from 197 nm to 240 nm.
See Abstr. 022.010.

Laboratory measurements of weak carbon dioxide bands relevant to Venus' nightside emission spectrum at 2.2 microns.
See Abstr. 022.176.

Comparison of Kalman and Wiener filtering techniques for processing Pioneer Venus radio occultation data.
See Abstr. 036.141.

Observations of solar–system bodies with the Belgrade Meridian Circle.
See Abstr. 041.006.

Venus – a prime Soviet objective. Part I.
See Abstr. 051.011.

Venus – a prime Soviet objective. Part II.
See Abstr. 051.012.

Magellan: overview of science findings.
See Abstr. 051.034.

Magellan mission description and radar system.
See Abstr. 051.063.

Critical density layer as obstacle at solar wind – exospheric ion interaction.
See Abstr. 062.078.

Solar wind interaction with Mars and Venus during solar activity cycle.
See Abstr. 074.057.

Geoid anomalies and dynamic topography from convection in cylindrical geometry: applications to mantle plumes on Earth and Venus.
See Abstr. 081.001.

Magma ocean: mechanisms of formation.
See Abstr. 081.032.

Thermospheric odd nitrogen.
See Abstr. 082.029.

Runaway greenhouse atmospheres: applications to Earth and Venus.
See Abstr. 082.070.

Love numbers of the Moon and of the terrestrial planets.
See Abstr. 091.039.

Equatorial superrotation in a slowly rotating GCM: implications for Titan and Venus.
See Abstr. 091.044.

Why Mercury and Venus have no satellites?
See Abstr. 092.003.

Comparisons of peak ionosphere pressures at Mars and Venus with incident solar wind dynamic pressure.
See Abstr. 097.016.

Influence of aerodynamic roughness length on aeolian processes: Earth, Mars, Venus.
See Abstr. 097.037.

Possible discrete annular dust formations around Mars and Venus.
See Abstr. 097.186.

Imaging of Saturn and Venus in the thermal infrared.
See Abstr. 100.019.

Numerical simulation of impact crater gravity collapse.
See Abstr. 105.015.

094 Moon

094.001   **Lunar impact basins and crustal heterogeneity: new western limb and far side data from Galileo.**
M. J. S. Belton, J. W. Head III, C. M. Pieters, R. Greeley, A. S. McEwen, G. Neukum, K. P. Klaasen, C. D. Anger, M. H. Carr, C. R. Chapman, M. E. Davies, F. P. Fanale, P. J. Gierasch, R. Greenberg, A. P. Ingersoll, T. Johnson, P. Paczkowski, C. B. Pilcher, J. Veverka.
Science, Vol. 255, No. 5044, p. 570 – 576 (31 Jan 1992).

Multispectral images of the lunar western limb and far side obtained from Galileo reveal the compositional nature of several prominent lunar features and provide new information on lunar evolution. The data reveal that the ejecta from the Orientale impact basin lying outside the Cordillera Mountains was excavated from the crust, not the mantle, and covers pre-Orientale terrain that consisted of both highland materials and relatively large expanses of ancient mare basalts. The inside of the far side South Pole–Aiken basin has low albedo, red color, and a relatively high abundance of iron– and magnesium–rich materials. These features suggest that the impact may have penetrated into the deep crust or lunar mantle or that the basin contains ancient mare basalts that were later covered by highlands ejecta.

094.002   **First X–ray portrait of the Moon.**
R. Hudec.
Říše hvězd, Vol. 72, No. 6, p. 111 – 112 (Jun 1991). In Czech.

094.003   **Theory of the physical libration of the Moon.**
V. I. Zubkov.
Sov. Phys. – Dokl., Vol. 36, No. 12, p. 809 – 811 (Dec 1991). Abstract. Current Physics Microform No.: 9207X0141.

094.004   **Polarimetric and photometric properties of the Moon: telescopic observations and laboratory simulations. 1. The negative polarization.**
Yu. G. Shkuratov, N. V. Opanasenko, M. A. Kreslavsky (*M. A. Kreslavskij*).
Icarus, Vol. 95, No. 2, p. 283 – 299 (Feb 1992).

In 1985 – 1990 a wide program of photometric and polarimetric observations of the Moon was carried out by means of a new spectropolarimeter which made possible an accuracy of polarimetric measurements up to 0.04%. In order to interpret these and other observations, laboratory photometric and polarimetric measurements of some natural and artificial samples have been made. Results of linear and parabolic regression analysis for parameters of the negative polarization and photometric characteristics of the Moon are presented. A principal component analysis was also carried out. Only two statistically significant principal components were found: the first is predominantly determined by albedo and the second is controlled by some parameters of the negative polarization. Some relationships between polarimetric and photometric parameters are studied in detail.

094.005   **The Moon from Flamsteed to La Palma.**
D. Jones.
Gemini, No. 35, p. 8 – 10 (Mar 1992).

094.006   **Getting started:moonlighting.**
J. E. Westfall.
Strolling Astron., Vol. 36, No. 1, p. 23 – 25 (Mar 1992).

094.007   **Rb–Sr and Sm–Nd chronology of an Apollo 17 KREEP basalt.**
C. Y. Shih, B. M. Bansal, H. Wiesmann, L. E. Nyquist.
Earth Planet. Sci. Lett., Vol. 108, No. 4, p. 203 – 215 (Feb 1992).

Sm–Nd and Rb–Sr mineral isochrons were determined for an Apollo 17 KREEP (pigeonite) basalt clast (,543) from breccia

72275 collected from Boulder 1, Station 2 in the Valley of Taurus–Littrow. Sm–Nd analyses of the basalt yield a precise mineral isochron age of 4.08 ± 0.07 Ga for $\lambda(^{147}Sm) = 0.00654$ Ga^{-1}. The Rb–Sr mineral isochron age of 4.13 ± 0.08 Ga for $\lambda(^{87}Rb) = 0.0139$ Ga^{-1} is concordant with the Sm–Nd age. However, the Rb–Sr isotopic system has been slightly disturbed, probably by the Serenitatis impact event ~ 3.95 Ga ago. The Sm–Nd age is interpreted as the crystallization age of this Apollo 17 KREEP basalt. Ages and initial $^{87}Sr/^{86}Sr$ and ε_{Nd} values for this basalt and Apollo 14 and 15 KREEP basalts suggest that these two types of KREEP basalts could be produced from sources having similar Rb/Sr and Sm/Nd ratios. These new age data suggest that KREEP basaltic volcanism started ~ 200 Ma earlier than the ~ 3.85 Ga age generally accepted from studies of Apollo 15 pristine KREEP basalts. Apollo 17 KREEP basalts probably pre–date the Serenitatis impact, strongly indicating that not all KREEP basaltic volcanism was necessarily related to basin–forming impacts.

094.008 Impact–induced thermal effects in the lunar and Mercurian regoliths.
M. J. Cintala.
J. Geophys. Res., Vol. 97, No. E1, p. 947 – 974 (25 Jan 1992).
This paper investigates thermal effects of micrometeoroid impact into the regoliths of the Moon and Mercury and offers some comparisons between the regoliths of the two planets.

094.009 Noble gases in lunar anorthositic rocks 60018 and 65315: acquisition of terrestrial krypton and xenon indicating an irreversible adsorption process.
S. Niedermann, O. Eugster.
Geochim. Cosmochim. Acta, Vol. 56, No. 1, p. 493 – 509 (Jan 1992).
The authors present results of gas analyses of rocks 60018 and 65315 and of experiments designed to elucidate whether atmospheric contamination is the only source of terrestrial–like Xe in lunar samples, and if so, what the nature of the contamination process is and whether it affects other noble gases as well.

094.010 Determination of the position of the selenodetic initial point from Greenwich meridian observations of the crater Mösting A.
V. S. Kislyuk, V. I. Belan, R. L. Semerenko.
Kinematics Phys. Celest. Bodies, Vol. 7, No. 3, p. 21 – 27 (1991). English translation of Kinematika Fiz. Nebesn. Tel, Tom 7, No. 3, p. 22 – 28 (1991).
Quasi–dynamic coordinates of the crater Mösting A are determined from reduction of Greenwich meridian observations made of the crater in January 1960 – April 1982. The lunar motion theories j = 2 and DE200/LE200 are used. The results indicate that the Moon's center of mass is 1.5 km north of the center of the Moon's figure according to Koziel (1967) and 4.8 km closer to the Earth. The results are compared with those of other authors.

094.011 Gravitational field and certain surface properties of the Moon.
L. A. Savrov, E. K. Kuchik.
Sol. Syst. Res., Vol. 25, No. 1, p. 19 – 23 (Jul 1991). English translation of Astron. Vestn., Tom 25, No. 1, p. 27 – 33 (1991).
From covariational and harmonic analyses, a weak linear dependence is found between the Moon's gravitational field and certain of its surface properties, such as albedo, degree of polarization, and crater density distribution.

094.012 Determination of the chemical composition of the lunar surface from combined γ–ray, X–ray, and optical remote measurements.
N. N. Evsyukov, Yu. A. Surkov, Eh. I. Chumak.
Sol. Syst. Res., Vol. 25, No. 1, p. 24 – 32 (Jul 1991). English translation of Astron. Vestn., Tom 25, No. 1, p. 34 – 44 (1991).
By combined X–ray, optical, and γ–surveys one can map the chemical composition of the lunar surface in terms of the principal components with a space resolution accessible to optical instruments and a reliability typical of γ– and X–ray measurements.

094.013 Calcalong Creek: a KREEPy lunar meteorite.
D. H. Hill, W. V. Boynton, R. A. Haag.
54. Annual Meeting of the Meteoritical Society, p. 91 (1991). Abstract. – See Abstr. 012.010 for the main entry.

094.014 FeO and MgO trends in plagioclases of two Apollo 15 mare basalts.
O. B. James, J. J. McGee.
54. Annual Meeting of the Meteoritical Society, p. 106 (1991). Abstract. – See Abstr. 012.010 for the main entry.

094.015 The origin of amorphous rims on lunar plagioclase grains: solar wind damage or vapor condensates?
L. P. Keller, D. S. McKay.
54. Annual Meeting of the Meteoritical Society, p. 114 (1991). Abstract. – See Abstr. 012.010 for the main entry.

094.016 There are too many kinds of mafic impact melt breccias at Apollo 16 for them all to be basin products.
R. L. Korotev.
54. Annual Meeting of the Meteoritical Society, p. 122 (1991). Abstract. – See Abstr. 012.010 for the main entry.

094.017 Relationships among basaltic lunar meteorites.
M. M. Lindstrom.
54. Annual Meeting of the Meteoritical Society, p. 138 (1991). Abstract. – See Abstr. 012.010 for the main entry.

094.018 Plagioclase, ilmenite, lunar magma oceans and mare basalt.
G. Ryder.
54. Annual Meeting of the Meteoritical Society, p. 204 (1991). Abstract. – See Abstr. 012.010 for the main entry.

094.019 Natural thermoluminescence and anomalous fading: terrestrial age, transit times and perihelia of lunar meteorites.
S. Symes, P. H. Benoit, H. Sears, D. W. G. Sears.
54. Annual Meeting of the Meteoritical Society, p. 225 (1991). Abstract. – See Abstr. 012.010 for the main entry.

094.020 Remnant of water–rock interaction on the lunar surface.
K. Takahashi, A. Masuda.
54. Annual Meeting of the Meteoritical Society, p. 226 (1991). Abstract. – See Abstr. 012.010 for the main entry.

094.021 Possible inheritance of silicate differentiation during lunar origin by giant impact.
P. H. Warren.
54. Annual Meeting of the Meteoritical Society, p. 239 (1991). Abstract. – See Abstr. 012.010 for the main entry.

094.022 Geochemistry of lunar crustal rocks from breccia 67016 and the composition of the Moon.
M. D. Norman, S. R. Taylor.
Geochim. Cosmochim. Acta, Vol. 56, No. 3, p. 1013 – 1024 (Mar 1992). – See Abstr. 012.013 for the main entry.
Global differentiation of the Moon has produced a plagioclase–rich crust overlying an ultramafic mantle. The authors have conducted a major and trace element study of anorthositic clasts from an Apollo 16 breccia to investigate the geochemical features of these highlands lithologies and their role in lunar crustal evolution.

094.023 Magnesium isotope fractionation in lunar soils.
T. M. Esat, S. R. Taylor.
Geochim. Cosmochim. Acta, Vol. 56, No. 3, p. 1025 – 1031 (Mar 1992). – See Abstr. 012.013 for the main entry.
Oxygen and silicon extracted by partial fluorination from lunar soils show large, percent–level, mass–dependent enrichments in heavy isotopes. Similar isotope fractionations occur in

sulphur in grain size separates and possibly in potassium in bulk soils. In contrast calcium, extracted using similar methods, exhibits only marginal isotope effects. The authors have measured magnesium in water and acid leaches and in surface layers of individual soil grains. They have not been able to find any large isotope fractionation in magnesium. The magnesium data are in agreement with the calcium data within errors and in sharp contrast to the large effects in O, Si, S, and K. The results for calcium and magnesium place severe constraints on any theory attempting to account for the large mass fractionation effects in O, Si, S, and K in lunar soils. The authors propose that the large effects in oxygen and silicon are probably artifacts of analytical procedures.

094.024   Tidal deceleration of the Moon's mean motion.
M. K. Cheng, R. J. Eanes, B. D. Tapley.
Geophys. J. Int., Vol. 108, No. 2, p. 401 – 409 (Feb 1992).
The secular change in the mean motion of the Moon, dn/dt, caused by the tidal dissipation in the ocean and solid Earth is due primarily to the effect of the diurnal and semidiurnal tides. The long–period ocean tides produce an increase in dn/dt, but the effects are only 1% of the diurnal and semidiurnal ocean tides. In this investigation, expressions for these effects are obtained by developing the tidal potential in the ecliptic reference system. The computation of the amplitude of equilibrium tide and the phase corrections is also discussed. The averaged tidal deceleration of the Moon's mean motion, dn/dt, from the most recent satellite ocean tide solutions is -25.25 ± 0.4 arcseconds/century^2. The value for dn/dt inferred from the satellite–determined ocean tide solution is in good agreement with the value obtained from the analysis of 20 years of lunar laser ranging observations.

094.025   An investigation of the depth of excavation and thickness of basalt fill for the lunar mascon basins.
A. M. Mullis.
Geophys. J. Int., Vol. 109, No. 1, p. 233 – 239 (Apr 1992).
In a previous paper (Mullis, 1991), a model for the formation of the Imbrium gravity anomaly was described, which yielded estimates of both the maximum depth of excavation of the basin and depth of basalt fill in the resulting topographic depression. This model has now been extended to the Serenitatis, Humorum, Crisium, Nectaris and Orientale basins. In each case, depth/diameter ratios for the original excavation close to 0.08 have been determined. This is much shallower than would be expected from extrapolation of depth/diameter ratios for small fresh lunar craters, and would suggest that none of the impact structures considered excavated below the level of the crust/mantle boundary. Mare basalt thicknesses for the structures considered range from 1.8 km for Orientale to 4.9 km for Serenitatis.

094.026   The violent side of mare volcanism.
C. R. Coombs, D. S. McKay, B. R. Hawke.
Workshop on Mare Volcanism and Basalt Petrogenesis: "Astounding Fundamental Concepts (AFC)" Developed Over the Last Fifteen Years, p. 9 – 10 (1991). Abstract. – See Abstr. 012.028 for the main entry.

094.027   Formation of Apollo 14 aluminous mare basalts by replenishment fractional crystallization and assimilation or precursor crust.
T. L. Dickinson, D. O. Nelson.
Workshop on Mare Volcanism and Basalt Petrogenesis: "Astounding Fundamental Concepts (AFC)" Developed Over the Last Fifteen Years, p. 11 – 12 (1991). Abstract. – See Abstr. 012.028 for the main entry.

094.028   Ancient mare volcanism.
B. R. Hawke, P. G. Lucey, J. F. Bell, P. D. Spudis.
Workshop on Mare Volcanism and Basalt Petrogenesis: "Astounding Fundamental Concepts (AFC)" Developed Over the Last Fifteen Years, p. 13 – 14 (1991). Abstract. – See Abstr. 012.028 for the main entry.

094.029   Geological setting and morphology of mare volcanic deposits: implications for chronology, petrogenesis, and eruption conditions.
J. W. Head.
Workshop on Mare Volcanism and Basalt Petrogenesis: "Astounding Fundamental Concepts (AFC)" Developed Over the Last Fifteen Years, p. 15 – 16 (1991). Abstract. – See Abstr. 012.028 for the main entry.

094.030   Pristine mare glasses: primary magmas?
P. C. Hess.
Workshop on Mare Volcanism and Basalt Petrogenesis: "Astounding Fundamental Concepts (AFC)" Developed Over the Last Fifteen Years, p. 17 – 18 (1991). Abstract. – See Abstr. 012.028 for the main entry.

094.031   Mare Serenitatis/mare Tranquillitatis shelf region: identification of basalt types from multispectral reflectance measurements.
R. Jaumann, G. Neukum, H. Hoffmann.
Workshop on Mare Volcanism and Basalt Petrogenesis: "Astounding Fundamental Concepts (AFC)" Developed Over the Last Fifteen Years, p. 19 – 20 (1991). Abstract. – See Abstr. 012.028 for the main entry.

094.032   Exploration of relationships between low–Ti and high–Ti pristine lunar glasses using an armalcolite assimilation model.
J. H. Jones, J. W. Delano.
Workshop on Mare Volcanism and Basalt Petrogenesis: "Astounding Fundamental Concepts (AFC)" Developed Over the Last Fifteen Years, p. 21 – 22 (1991). Abstract. – See Abstr. 012.028 for the main entry.

094.033   Dynamical melting models of mare basalts.
J. Longhi.
Workshop on Mare Volcanism and Basalt Petrogenesis: "Astounding Fundamental Concepts (AFC)" Developed Over the Last Fifteen Years, p. 23 – 24 (1991). Abstract. – See Abstr. 012.028 for the main entry.

094.034   Remelting mechanisms for shallow source regions of mare basalts.
M. Manga, J. Arkani–Hamed.
Workshop on Mare Volcanism and Basalt Petrogenesis: "Astounding Fundamental Concepts (AFC)" Developed Over the Last Fifteen Years, p. 25 – 26 (1991). Abstract. – See Abstr. 012.028 for the main entry.

094.035   REE distribution coefficients for pigeonite: constraints on the origin of the mare basalt europium anomaly, III.
G. McKay, J. Wagstaff, L. Le.
Workshop on Mare Volcanism and Basalt Petrogenesis: "Astounding Fundamental Concepts (AFC)" Developed Over the Last Fifteen Years, p. 27 – 28 (1991). Abstract. – See Abstr. 012.028 for the main entry.

094.036   Cyclical AFC at Apollo 14: Sr isotope evidence from high–alumina basalts.
C. R. Neal, L. A. Taylor.
Workshop on Mare Volcanism and Basalt Petrogenesis: "Astounding Fundamental Concepts (AFC)" Developed Over the Last Fifteen Years, p. 29 – 30 (1991). Abstract. – See Abstr. 012.028 for the main entry.

094.037   Models for mare basalt petrogenesis developed over the last 20 years.
C. R. Neal, L. A. Taylor.
Workshop on Mare Volcanism and Basalt Petrogenesis: "Astounding Fundamental Concepts (AFC)" Developed Over the Last Fifteen Years, p. 31 – 35 (1991). Abstract. – See Abstr. 012.028 for the main entry.

094.038   **Identification and origin of source heterogeneities for Apollo 17 basalts using isotopic tracers.**
C. R. Neal, L. A. Taylor, J. B. Paces, A. N. Halliday.
Workshop on Mare Volcanism and Basalt Petrogenesis: "Astounding Fundamental Concepts (AFC)" Developed Over the Last Fifteen Years, p. 36 – 38 (1991). Abstract. – See Abstr. 012.028 for the main entry.

094.039   **The history of lunar volcanism.**
L. E. Nyquist, C.-Y. Shih.
Workshop on Mare Volcanism and Basalt Petrogenesis: "Astounding Fundamental Concepts (AFC)" Developed Over the Last Fifteen Years, p. 39 – 42 (1991). Abstract. – See Abstr. 012.028 for the main entry.

094.040   **The probable continuum between emplacement of plutons and mare volcanism in lunar crustal evolution.**
C. M. Pieters.
Workshop on Mare Volcanism and Basalt Petrogenesis: "Astounding Fundamental Concepts (AFC)" Developed Over the Last Fifteen Years, p. 43 – 44 (1991). Abstract. – See Abstr. 012.028 for the main entry.

094.041   **The moon: dead or alive.**
P. H. Schultz.
Workshop on Mare Volcanism and Basalt Petrogenesis: "Astounding Fundamental Concepts (AFC)" Developed Over the Last Fifteen Years, p. 45 – 46 (1991). Abstract. – See Abstr. 012.028 for the main entry.

094.042   **The relevance of picritic glasses to mare basalt volcanism.**
C. K. Shearer, J. J. Papike, N. Shimizu.
Workshop on Mare Volcanism and Basalt Petrogenesis: "Astounding Fundamental Concepts (AFC)" Developed Over the Last Fifteen Years, p. 47 – 48 (1991). Abstract. – See Abstr. 012.028 for the main entry.

094.043   **Lunar mare volcanism: recent advances in petrogenesis.**
J. W. Shervais, S. K. Vetter.
Workshop on Mare Volcanism and Basalt Petrogenesis: "Astounding Fundamental Concepts (AFC)" Developed Over the Last Fifteen Years, p. 49 – 50 (1991). Abstract. – See Abstr. 012.028 for the main entry.

094.044   **Geochemical constraints (and pitfalls) on remelting of lunar magma ocean cumulates for the generation of high–Ti mare basalts.**
G. A. Snyder, L. A. Taylor, C. R. Neal.
Workshop on Mare Volcanism and Basalt Petrogenesis: "Astounding Fundamental Concepts (AFC)" Developed Over the Last Fifteen Years, p. 51 – 52 (1991). Abstract. – See Abstr. 012.028 for the main entry.

094.045   **The sources of mare basalts: a model involving lunar magma ocean crystallization, plagioclase flotation, and trapped instantaneous residual liquid.**
G. A. Snyder, L. A. Taylor, C. R. Neal.
Workshop on Mare Volcanism and Basalt Petrogenesis: "Astounding Fundamental Concepts (AFC)" Developed Over the Last Fifteen Years, p. 53 – 54 (1991). Abstract. – See Abstr. 012.028 for the main entry.

094.046   **Lunar magma transport phenomena.**
F. J. Spera.
Workshop on Mare Volcanism and Basalt Petrogenesis: "Astounding Fundamental Concepts (AFC)" Developed Over the Last Fifteen Years, p. 55 – 58 (1991). Abstract. – See Abstr. 012.028 for the main entry.

094.047   **Titanium–rich basalts within the Flamsteed–P region of Oceanus Procellarum: new results from high spatial resolution CCD data.**
J. M. Sunshine, C. M. Pieters.
Workshop on Mare Volcanism and Basalt Petrogenesis: "Astounding Fundamental Concepts (AFC)" Developed Over the Last Fifteen Years, p. 59 – 60 (1991). Abstract. – See Abstr. 012.028 for the main entry.

094.048   **Apollo 15 olivine–normative and quartz–normative mare basalts: a common origin by dynamic melting.**
S. K. Vetter, J. W. Shervais.
Workshop on Mare Volcanism and Basalt Petrogenesis: "Astounding Fundamental Concepts (AFC)" Developed Over the Last Fifteen Years, p. 61 – 62 (1991). Abstract. – See Abstr. 012.028 for the main entry.

094.049   **A spinifex–textured mare basalt: comparison with komatiites.**
P. H. Warren, E. A. Jerde, G. W. Kallemeyn.
Workshop on Mare Volcanism and Basalt Petrogenesis: "Astounding Fundamental Concepts (AFC)" Developed Over the Last Fifteen Years, p. 65 – 66 (1991). Abstract. – See Abstr. 012.028 for the main entry.

094.050   **The whole Earth–facing side of the Moon.**
P. Moore.
Astron. Now, Vol. 6, No. 2, p. 43 (Feb 1992).
Moon walking. For parts 1 – 4 see:
Part 1: July 1991 – North–western quadrant.
Part 2: September 1991 – South–western quadrant.
Part 3: November 1991 – South–eastern quadrant.
Part 4: January 1992 – North–eastern quadrant.
Pictorial: February 1992 – A selection of readers photographs and slides.

094.051   **Models of the structure and the origin of the Moon.**
E. L. Ruskol.
Sol. Syst. Res., Vol. 25, No. 4, p. 308 – 317 (Jan 1992). English translation of Astron. Vestn., Tom 25, No. 4, p. 408 – 421 (1991).
Fundamental data on the inner structure of the Moon obtained from space and earth–based investigations are briefly summarized. The early thermal history and the origin of the Moon still remain largely obscure. Two opposite models of the formation of the Moon from a circumterrestrial disk are compared: (1) an instant formation of the disk due to a megaimpact of a large protoplanetary body into the mantle of a growing Earth, and (2) a continuously replenished disk from a feeding zone of the Earth (co–accretion). Both models of the Moon's origin need a careful quantitative elaboration, including a tying–in with the thermal history of the Earth, particularly for the megaimpact mechanism.

094.052   **Geological structure of the Moon.**
A. L. Sukhanov.
Sol. Syst. Res., Vol. 25, No. 4, p. 318 – 324 (Jan 1992). English translation of Astron. Vestn., Tom 25, No. 4, p. 422 – 430 (1991).
A short summary is given of available data on the main structures of the lunar surface highlands, maria, basins, impact craters, and volcanic structures, and data on their internal structure and the covering regolith. The morphological features are described, which shed some light on their genesis and time of formation and allow reconstructing of a stratigraphic succession from the crust formation to the present time.

094.053   **Selenodesy and lunar cartography.**
V. S. Kislyuk.
Sol. Syst. Res., Vol. 25, No. 4, p. 324 – 330 (Jan 1992). English translation of Astron. Vestn., Tom 25, No. 4, p. 431 – 438 (1991).
The most recent studies on the gravitational field and geometric and dynamic figures of the Moon are reviewed. Selenodetic reference systems and catalogs are considered. Generalized selenodetic and selenodynamic parameters are

presented, together with information on the basic studies of lunar surface cartography.

094.054   The distribution of anorthosite on the nearside of the Moon.
B. R. Hawke, P. G. Lucey, G. J. Taylor, P. D. Spudis.
Workshop on the Physics and Chemistry of Magma Oceans from 1 bar to 4 Mbar, p. 20 – 21 (1992). Abstract. – See Abstr. 012.032 for the main entry.

094.055   Origin of the moon and lunar core formation.
V. J. Hillgren.
Workshop on the Physics and Chemistry of Magma Oceans from 1 bar to 4 Mbar, p. 24 – 25 (1992). Abstract. – See Abstr. 012.032 for the main entry.

094.056   A new angle on lunar ferroan–suite differentiation.
B. L. Jolliff.
Workshop on the Physics and Chemistry of Magma Oceans from 1 bar to 4 Mbar, p. 28 – 29 (1992). Abstract. – See Abstr. 012.032 for the main entry.

094.057   Imperfect fractional crystallization of the lunar magma ocean and formation of the lunar mantle: a "realistic" chemical approach.
G. A. Snyder, L. A. Taylor.
Workshop on the Physics and Chemistry of Magma Oceans from 1 bar to 4 Mbar, p. 51 – 52 (1992). Abstract. – See Abstr. 012.032 for the main entry.

094.058   Inheritance of magma ocean differentiation during lunar origin by giant impact.
P. H. Warren.
Workshop on the Physics and Chemistry of Magma Oceans from 1 bar to 4 Mbar, p. 68 – 69 (1992). Abstract. – See Abstr. 012.032 for the main entry.

094.059   The ferroan–anorthositic suite and the extent of primordial lunar melting.
P. H. Warren, G. W. Kallemeyn.
Workshop on the Physics and Chemistry of Magma Oceans from 1 bar to 4 Mbar, p. 72 – 73 (1992). Abstract. – See Abstr. 012.032 for the main entry.

094.060   Domes associated with the south–eastern dark floor area in the lunar crater Schickard.
K. W. Abineri.
J. Br. Astron. Assoc., Vol. 102, No. 3, p. 163 – 166 (Jun 1992).
Continuing the microscopic examination of Orbiter IV HR160/2 microfilm, the programme has been extended to include detailed drawing of three suspected domes associated with the south–eastern dark floor area in Schickard. In order to confirm the larger features in Schickard, (shown only under very restricted lighting conditions on the microfilm images), the aid of experienced lunar observers is requested. Visual, photographic, or CCD imaging techniques could be used by those who have moderate or large aperture telescopes.

094.061   Planetary and figure–figure effects on the Moon's rotational motion.
E. Bois, I. Wytrzyszczak, A. Journet.
Celest. Mech. Dyn. Astron., Vol. 53, No. 2, p. 185 – 201 (1992).
An accurate theory of the rotation of the Moon has been constructed by numerical integration. All direct perturbations on the Moon's rotational motion have been analysed. The requirements of the current observational accuracy are such that some improvements had to be added to the theoretical models. First, the gravitational figure of the Moon has been developed up to the fifth degree harmonics. Second, mutual potential effects between the figure of the Moon and the figure of the Earth have been expanded farther up. The direct action of planets must be taken into account, its effects being very small but not always negligible. The physical librations resulting of planetary effects and Earth–Moon figure–figure interactions are presented in this paper.

094.062   Dividing the lunar disk into districts according to albedo, color index and polarization degree. First quarter phase.
O. I. Kvaratskheliya, V. V. Novikov, Kh. G. Tadzhidinov.
Astron. Vestn., Tom 26, No. 3, p. 14 – 25 (May – Jun 1992). In Russian. English translation in Sol. Syst. Res., Vol. 26, No. 3.
Optically typical districts of the lunar surface were found as a result of calculative treatment of telescopic lunar images obtained by means of polarization and spectral–zone filters. The representation of seven landing places was studied also using optical characteristics of the eastern part of the lunar disk. In case of using the relations albedo–color or albedo–polarization degree the landing places correspond to 40% of the area studied, while the division into districts according to all three optical parameters gives a representation of 15%. A visualization method is developed to find the most probable areas where are lunar rocks corresponding to terrestrial mineralogical collections. A possibility of mineralogical mapping of the lunar surface is discussed using the relations between the optical and chemcial properties of lunar regolith.

094.063   Separation of optically typical districts on the lunar disk according to the albedo and polarization degree.
O. I. Kvaratskheliya, V. V. Novikov, Kh. G. Tadzhidinov.
Astron. Vestn., Tom 26, No. 3, p. 99 – 110 (May – Jun 1992). In Russian. English translation in Sol. Syst. Res., Vol. 26, No. 3.
Here are discussed two possible variants of two–parameter separation (districting) of the lunar surface on the basis of the dependence of albedo and polarization degree. Seven lunar landing places (mean space of 800 km^2) representing 40% of the eastern part of the lunar disc are found.

094.064   Petrology of lunar rocks.
M. I. Korina.
Sol. Syst. Res., Vol. 25, No. 6, p. 513 – 525 (May 1992). English translation of Astron. Vestn., Tom 25, No. 6, p. 686 – 702 (1991).
Lunar rock data from various sources are compiled. Mineralogy, chemistry, and petrology of highland and maria rocks are discussed. A classification of these rocks is constructed. Principal features of lunar geochemistry are discussed. Petrological hypotheses of lunar rock origin are outlined, with their mutual inconsistencies pointed out. The importance of accretional impact processes for the highland material formation is noted.

094.065   Lunar and asteroid surface composition from solar wind–sputtered secondary ions.
R. C. Elphic, H. O. Funsten, B. L. Barraclough, D. J. McComas.
Bull. Am. Astron. Soc., Vol. 23, No. 3, p. 1198 (1991). Abstract. – See Abstr. 012.037 for the main entry.

094.066   Application of Chamberlain's exospheric theory to the lunar atmosphere.
D. M. Hunten.
Bull. Am. Astron. Soc., Vol. 23, No. 3, p. 1198 (1991). Abstract. – See Abstr. 012.037 for the main entry.

094.067   Sodium in the lunar atmosphere: dependence of release mechanisms on local solar zenith angle.
R. W. H. Kozlowski, D. M. Hunten, W. K. Wells, A. L. Sprague, F. A. Grosse.
Bull. Am. Astron. Soc., Vol. 23, No. 3, p. 1198 (1991). Abstract. – See Abstr. 012.037 for the main entry.

094.068   Multicolor photometry of Apollo 14 and 16 landing sites: first results.
H. Rebhan, G. Neukum.
Bull. Am. Astron. Soc., Vol. 23, No. 3, p. 1198 (1991). Abstract. – See Abstr. 012.037 for the main entry.

094.069   Compositional variability of the southern Mare Sereni-tatis and northern Mare Tranquillitatis regions of the Moon from CCD imaging.
J. F. Bell III, J. B. Adams, B. R. Hawke, K. A. Horton.
Bull. Am. Astron. Soc., Vol. 23, No. 3, p. 1199 (1991). Abstract. –
See Abstr. 012.037 for the main entry.

094.070   The terrain northwest of Humorum Basin.
C. A. Peterson, B. R. Hawke, P. G. Lucey,
G. J. Taylor, J. F. Bell, D. T. Blewett, B. A. Campbell,
P. D. Spudis.
Bull. Am. Astron. Soc., Vol. 23, No. 3, p. 1199 (1991). Abstract. –
See Abstr. 012.037 for the main entry.

094.071   A detailed spectral study of the Schiller–Schickard region of the Moon.
D. T. Blewett, B. R. Hawke, J. F. Bell, P. G. Lucey,
C. A. Peterson, G. J. Taylor, P. D. Spudis.
Bull. Am. Astron. Soc., Vol. 23, No. 3, p. 1200 (1991). Abstract. –
See Abstr. 012.037 for the main entry.

094.072   Mapping lunar titanium abundances at high spatial resolution.
D. E. Melendrez, S. M. Larson, J. R. Johnson, R. B. Singer.
Bull. Am. Astron. Soc., Vol. 23, No. 3, p. 1201 (1991). Abstract. –
See Abstr. 012.037 for the main entry.

094.073   Western Oceanus Procellarum basalts: preliminary analysis of Galileo multi–spectral images.
J. Sunshine, C. Pieters, J. Head, A. McEwen, R. Greeley,
M. Belton.
Bull. Am. Astron. Soc., Vol. 23, No. 3, p. 1201 (1991). Abstract. –
See Abstr. 012.037 for the main entry.

094.074   Systematic variations in solar wind fluence with lunar location: implications for resource utilization.
T. D. Swindle, J. R. Johnson, S. M. Larson, R. B. Singer.
Bull. Am. Astron. Soc., Vol. 23, No. 3, p. 1201 (1991). Abstract. –
See Abstr. 012.037 for the main entry.

094.075   Orientale and South Pole–Aitken Basins: Galileo Solid–State Imaging System results.
J. Head, C. Pieters, S. Murchie, J. Mustard, E. Fischer,
J. Plutchak, R. Greeley, C. Pilcher, G. Neukum, H. Hoffmann,
A. McEwen.
Bull. Am. Astron. Soc., Vol. 23, No. 3, p. 1203 (1991). Abstract. –
See Abstr. 012.037 for the main entry.

094.076   Compositional diversity of the lunar limb and farside from Galileo SSI images.
C. Pieters, J. Head, J. Sunshine, E. Fischer, S. Murchie,
S. Pratt, A. McEwen, R. Greeley, G. Neukum, H. Hoffmann.
Bull. Am. Astron. Soc., Vol. 23, No. 3, p. 1203 (1991). Abstract. –
See Abstr. 012.037 for the main entry.

094.077   Galileo SSI observation of lunar maria.
R. Greeley, S. D. Kadel, D. A. Williams,
M. J. S. Belton, J. W. Head, S. L. Murchie, C. M. Pieters,
J. M. Sunshine, L. R. Gaddis, A. S. McEwen, G. Neukum.
Bull. Am. Astron. Soc., Vol. 23, No. 3, p. 1203 (1991). Abstract. –
See Abstr. 012.037 for the main entry.

094.078   Micrometer–sized glass spheres in Apollo 16 soil 61181: implications for impact volatilization and condensation.
L. P. Keller, D. S. McKay.
22. Lunar and Planetary Science Conference, p. 137 – 141 (1992).
– See Abstr. 012.046 for the main entry.
Micrometer–sized glass spheres from a mature highland lunar soil (61181) were analyzed for major elements with a transmission electron microscope.

094.079   Evolution of isotopic signatures in lunar–regolith nitro-gen: noble gases and nitrogen in grain–size fractions from regolith breccia 79035.
J. F. Kerridge, J. S. Kim, Y. Kim, K. Marti.
22. Lunar and Planetary Science Conference, p. 215 – 224 (1992).
– See Abstr. 012.046 for the main entry.

094.080   Modeling the evolution of N and $^{15}N/^{14}N$ in the lunar regolith: mixing models involving two components.
J. F. Kerridge, P. Bochsler, O. Eugster, J. Geiss.
22. Lunar and Planetary Science Conference, p. 239 – 248 (1992).
– See Abstr. 012.046 for the main entry.
A computer simulation of N buildup on the lunar surface, employing two independent sources of N, generates a good match with most, but not all, aspects of N abundance and isotope systematics observed in regolith samples. This suggests that a two–component model, at least as tested here, is insufficient to explain the lunar data.

094.081   Revisited geology of Gassendi Crater from earth–based near–infrared multispectral solid state imaging.
S. Chevrel, P. C. Pinet.
22. Lunar and Planetary Science Conference, p. 249 – 258 (1992).
– See Abstr. 012.046 for the main entry.
The authors present an extended high–spatial–resolution mapping of Gassendi Crater in 10 bands in the visible to near–infrared domain (0.40 – 1.05 μm). Several spectral units have been identified in relation to morphological structures of the crater, such as the center peaks, and low–albedo units either associated with fractures on the floor or with the mare–like material unit in the southern portion of the crater.

094.082   A high–resolution radar and CCD imaging study of crater rays in Mare Serenitatis and Mare Nectaris.
B. A. Campbell, J. F. Bell III, S. H. Zisk, B. R. Hawke,
K. A. Horton.
22. Lunar and Planetary Science Conference, p. 259 – 274 (1992).
– See Abstr. 012.046 for the main entry.
High–resolution 3.0–cm and 70–cm radar images, along with recently acquired CCD imaging data, are used to characterize crater ray deposits in Mare Serenitatis and Mare Nectaris.

094.083   Compositional variations in Apollo 17 soils and their relationship to the geology of the Taurus–Littrow site.
R. L. Korotev, D. T. Kremser.
22. Lunar and Planetary Science Conference, p. 275 – 301 (1992).
– See Abstr. 012.046 for the main entry.
New compositional data for major and trace elements in Apollo 17 soils are combined with literature data to constrain a mass–balance model that accounts for the compositions of most soils as mixtures of a small number of lithologic components observed at the site.

094.084   Pyroclastic deposits on the western limb of the Moon.
C. T. Coombs, B. R. Hawke.
22. Lunar and Planetary Science Conference, p. 303 – 312 (1992).
– See Abstr. 012.046 for the main entry.
Localized pyroclastic deposits are more common on the western limb of the Moon than once thought. The authors have conducted geologic and remote–sensing studies of 26 individual localized dark–mantle deposits in an elongated zone on the western limb.

094.085   The origin of the Moon flash of May 23, 1985.
G. Kolovos, J. H. Seiradakis, H. Varvoglis,
S. Avgoloupis.
Icarus, Vol. 97, No. 1, p. 142 – 144 (May 1992).
The authors present further evidence that the bright spot on the Moon observed on May 23, 1985, was a natural phenomenon that occurred slightly above the lunar surface. The interpretation that the spot could be attributed to an instantaneous reflection from a satellite's solar panel fails to explain several important characteristics of the observation.

094.086 Apollo 15 green glass: relationships between texture and composition.
A. M. Steele.
22. Lunar and Planetary Science Conference, p. 329 – 341 (1992).
– See Abstr. 012.046 for the main entry.
A suite of 365 Apollo 15 green–glass particles was analyzed by INAA and then described petrographically so that comparisons between composition and physical characteristics could be made.

094.087 Origin of picritic green glass magmas by polybaric fractional fusion.
J. Longhi.
22. Lunar and Planetary Science Conference, p. 343 – 353 (1992).
– See Abstr. 012.046 for the main entry.

094.088 Origin of yellow glasses associated with Apollo 15 KREEP basalt fragments.
A. Basu, B. B. Holmberg, E. Molinaroli.
22. Lunar and Planetary Science Conference, p. 365 – 372 (1992).
– See Abstr. 012.046 for the main entry.
An unusual yellowish glass is found in the interstices, in fracture fillings, and as encrustation of a few uncommon, mildly fragmented KREEP basalt particles in Apollo 15 soils. Chemical compositions of these glasses are analyzed.

094.089 Chemical variation and zoning of olivine in lunar dunite 72415: near–surface accumulation.
G. Ryder.
22. Lunar and Planetary Science Conference, p. 373 – 380 (1992).
– See Abstr. 012.046 for the main entry.

094.090 U–Th–Pb, Rb–Sr, and Sm–Nd isotopic systematics of lunar troctolitic cumulate 76535: implications on the age and origin of this early lunar, deep–seated cumulate.
W. R. Premo, M. Tatsumoto.
22. Lunar and Planetary Science Conference, p. 381 – 397 (1992).
– See Abstr. 012.046 for the main entry.

094.091 Petrogenesis of the western highlands of the Moon: evidence from a diverse group of whitlockite–rich rocks from the Fra Mauro formation.
G. A. Snyder, L. A. Taylor, Yun–Gang Liu, R. A. Schmitt.
22. Lunar and Planetary Science Conference, p. 399 – 416 (1992).
– See Abstr. 012.046 for the main entry.
A suite of twenty–seven "new" 2 – 10 mm rocklets has been separated from Apollo 14 soils and analyzed.

094.092 Trace elements in 59 mostly highland Moon rocks.
M. Ebihara, R. Wolf, P. H. Warren, E. Anders.
22. Lunar and Planetary Science Conference, p. 417 – 426 (1992).
– See Abstr. 012.046 for the main entry.
The authors report new chemical analyses for up to 26 trace elements, including seldom–determined highly siderophile elements Ir, Os, Re, Au, Pd, and Ge, for 59 lunar samples.

094.093 Mineralization on the Moon?: Theoretical considerations of Apollo 16 "rusty rocks", sulfide replacement in 67016, and surface–correlated volatiles on lunar volcanic glass.
R. O. Colson.
22. Lunar and Planetary Science Conference, p. 427 – 436 (1992).
– See Abstr. 012.046 for the main entry.
Theoretical considerations of vapor–rock interactions in the lunar environment are a useful supplement to petrologic studies of mineralization or alteration in rocks from the Moon. They also provide insights into the potential for the existence of more extensive mineralization on the Moon than is found in our limited sample set. Discussed in this paper are the coexistence and textural association in 66095 of the phases lawrencite, troilite, schreibersite, iron metal, and sphalerite; the replacement of olivine in certain clasts of 67016 by troilite and enstatite; and the existence of $Zn + S$ deposits on the surfaces of volcanic glass beads.

094.094 The sodium and potassium atmosphere of the Moon and its interaction with the surface.
A. L. Sprague, R. W. H. Kozlowski, D. M. Hunten, W. K. Wells, F. A. Grosse.
Icarus, Vol. 96, No. 1, p. 27 – 42 (Mar 1992).
Observations of lunar atmospheric sodium and potassium from May 1988 to Jul 1991 are reported and analyzed. Densities at 80° north and south are less than equatorial ones by a factor of 2 – 3. For the observations the apparent scale heights for the intensity are 119 – 611 km for Na, and 85 and 154 km for K; most of these are much larger than would be expected for atoms thermalized to the surface temperature. However, the intensity drops off with increasing radius at a much greater rate than would be observed for an atmosphere that is mostly escaping. The authors interpret their data using both single– and two–component analyses. They amplify an earlier suggestion that source atoms are quickly redistributed into thermal and supra–thermal populations by "competing release mechanisms" acting at the surface.

094.095 Lunar surface from remote sensing data.
V. V. Shevchenko, Yu. G. Shkuratov, N. V. Opanasenko.
Sol. Syst. Res., Vol. 25, No. 5, p. 428 – 434 (Mar 1992). English translation of Astron. Vestn., Tom 25, No 5, p. 569 – 577 (1991).
The review covers remote–sensing investigations of the lunar surface on the basis of earth telescope and spacecraft data. Remote sensing is a system of measurements of space distributions of static fields, particle fluxes, and electromagnetic radiation. An analysis of these lunar surface characteristics is given. Magnetic fields are studied by direct and indirect methods. The chemical composition of the surface rocks is determined by registration of α, γ, and X–rays. Data on mineral composition are based on studies of visual, ultraviolet, and infrared albedo by using colorimetric and spectroscopic methods. Surface layer structure is studied by radar methods and by photometric and polarimetric means. Data on the characteristic space distributions have been used for thematic lunar maps.

094.096 Lunar mare volcanism: stratigraphy, eruption conditions, and the evolution of secondary crusts.
J. W. Head III, L. Wilson.
Geochim. Cosmochim. Acta, Vol. 56, No. 6, p. 2155 – 2175 (Jun 1992). – See Abstr. 012.086 for the main entry.
Developments and trends in the last fifteen years of geological analysis of lunar mare volcanism are highlighted by (1) documentation of the distribution and stratigraphy of mare units, (2) a more thorough understanding of the principles of ascent and eruption of lunar magmas, (3) increased knowledge of the implications of volcanic landform and deposit morphology for eruption conditions, and (4) convergence of sample analysis research and the understanding of processes of ascent and eruption of magma.

094.097 Petrogenesis of mare basalts: a record of lunar volcanism.
C. R. Neal, L. A. Taylor.
Geochim. Cosmochim. Acta, Vol. 56, No. 6, p. 2177 – 2211 (Jun 1992). – See Abstr. 012.086 for the main entry.
Returned rock and soil samples from our nearest planetary neighbor have provided the basis for much of our understanding of the origin and evolution of the Moon. Of particular importance are the mare basalts, which have revealed considerable information about lunar volcanism and the nature of the mantle, as well as post–magma–generation processes. This paper is a critical review of the petrogenetic models for the generation of mare basalts formulated over the last twenty years.

094.098 The isotopic record of lunar volcanism.
L. E. Nyquist, C.–Y. Shih.
Geochim. Cosmochim. Acta, Vol. 56, No. 6, p. 2213 – 2234 (Jun 1992). – See Abstr. 012.086 for the main entry.
The timing and extent of lunar volcanism can be assumed to have reflected the internal thermal evolution of the Moon. The

authors summarize some fundamental aspects of lunar history which have been established as a result of the Apollo program of lunar exploration and contemporaneous Soviet lunar missions. They emphasize those aspects of the isotopic data which record lunar global evolution most directly and so give few details of the local geology of the sites where the samples were collected.

094.099   Experimental petrology and petrogenesis of mare volcanics.
J. Longhi.
Geochim. Cosmochim. Acta, Vol. 56, No. 6, p. 2235 – 2251 (Jun 1992). - See Abstr. 012.086 for the main entry.

Mare volcanics consist of basalts and picritic pyroclastic glasses spanning a wide range of TiO_2 concentration. The more primitive low–Ti basalts and picritic glasses have olivine along on their low–pressure liquidi. Most of the chemical variation among the low–Ti basalts is the result of olivine fractionation in a series of parental MgO–rich liquids differing in TiO_2 concentration. Controlled–cooling–rate crystallization studies on a variety of mare compositions have provided the basis for reconstructing the size and, in some cases, stratigraphy of mare flows. Groundmass textures, crystal size, crystal morphology, nucleation density, and zoning patterns have all been employed to quantify cooling histories of mare basalts.

094.100   Lunar magma transport phenomena.
F. J. Spera.
Geochim. Cosmochim. Acta, Vol. 56, No. 6, p. 2253 – 2265 (Jun 1992). - See Abstr. 012.086 for the main entry.

The fluid dynamics of the generation, segregation, ascent and emplacement, or eruption, of lunar magma is intimately coupled to the interpretation of their geochemical and petrological characteristics. An accurate understanding of lunar petrogenesis is necessarily an iterative conceptual process where "facts" based on magma dynamics. When the "facts" disagree, an opportunity for substative new insight is offered.

094.101   The Moon's physical librations. Part 1: Direct gravitational perturbations.
I. Wytrzyszczak, E. Bois.
NATO Advanced Study Institute on Predictability, Stability, and Chaos in N–Body Dynamical Systems, p. 257 – 264 (1991). - See Abstr. 012.089 for the main entry.

An accurate model of the Moon's rotation has been derived by numerical integration. Direct gravitational perturbations of the Moon's rotational motion have been analysed. The resulting librations are presented in this paper. These include complete physical librations, planetary effects, and Earth–Moon figure – figure interactions.

094.102   The Moon's physical librations. Part 2: Non–rigid Moon and direct non–gravitational perturbations.
E. Bois, I. Wytrzyszczak.
NATO Advanced Study Institute on Predictability, Stability, and Chaos in N–Body Dynamical Systems, p. 265 – 271 (1991). - See Abstr. 012.089 for the main entry.

Some relations between the non–rigidity of the Moon and its physical librations are described here, including librations due to the tides and others due to the rotational motion. Starting from their nature, their cause and their behaviour, the different families of physical librations are presented here in a compact classification scheme.

094.103   Significant high number commensurabilities in the main lunar problem: a postscript to a discovery of the ancient Chaldeans.
A. E. Roy, B. A. Steves, G. B. Valsecchi, E. Perozzi.
NATO Advanced Study Institute on Predictability, Stability, and Chaos in N–Body Dynamical Systems, p. 273 – 282 (1991). - See Abstr. 012.089 for the main entry.

The existence of the Saros implies a near repetition of the orbital elements of the Moon not only at eclipses, but at any other time during the Saros; both the JPL ephemeris and a numerical integration of the elliptic restricted 3–body problem confirm this finding. As a consequence, the Moon moves in a nearly periodic orbit of period equal to the Saros. Moreover, in the circular restricted 3–body problem it is possible to find periodic orbits, with period equal to the Saros, for which the behaviour of the osculating orbital elements in time is strikingly similar to that of the real Moon.

Moons of the solar system. An illustrated encyclopedia.
See Abstr. 002.059.

An automated information system for the Lunar Nomenclature data base processing.
See Abstr. 002.092.

Historical review of a long–overlooked paper by R. A. Daly concerning the origin and early history of the Moon.
See Abstr. 004.024.

The moon–test in Newton's *Principia*: accuracy of inverse–square law of universal gravitation.
See Abstr. 004.030.

The man who found a city in the Moon.
See Abstr. 004.047.

Conference: The scientific problems of creating a Lunar base. Moscow, 5 – 8 Feb 1991.
See Abstr. 011.018.

Production and uses of simulated lunar materials.
See Abstr. 012.026.

Mare volcanism and basalt petrogenesis: "Astounding Fundamental Concepts (AFC)" developed over the last fifteen years. Abstracts of presented papers.
See Abstr. 012.028.

Physics and chemistry of magma oceans from 1 bar to 4 Mbar. Abstracts of presented papers.
See Abstr. 012.032.

Mare volcanism and basalt petrogenesis. Papers presented at a workshop at the Annual Meeting of the Geological Society of America, Dallas, TX (USA), 27 – 28 Oct 1990.
See Abstr. 012.086.

Xylan: a potential contaminant for lunar samples and antarctic meteorites.
See Abstr. 022.191.

Partition coefficients for iron between plagioclase and basalt as a function of oxygen fugacity: implications for Archean and lunar anorthosites.
See Abstr. 022.228.

A history of laser ranging at McDonald Observatory.
See Abstr. 045.005.

On the influence of the moon's gravitational field on the motion of the artificial satellites.
See Abstr. 052.008.

Optimal launching of a spacecraft from the lunar surface to the fixed point of its artificial satellite circular orbit.
See Abstr. 052.029.

Optimal soft landing of a spacecraft on the lunar surface from the lunar satellite circular orbit.
See Abstr. 053.005.

Potassium, rubidium, and cesium in the Earth and Moon and the evolution of the mantle of the Earth.
See Abstr. 081.013.

Oxygen isotopic homogeneity of the Earth: new evidence.
See Abstr. 081.015.

Periodic system of multi–ring planetary structures as result of interference of variously oriented lithospheric waves.
See Abstr. 091.015.

Dynamics of dust in a plasma sheath and injection of dust into the plasma sheath above Moon and asteroidal surfaces.
See Abstr. 091.036.

Love numbers of the Moon and of the terrestrial planets.
See Abstr. 091.039.

Terrace width variations in complex Mercurian craters, and the transient strength of cratered Mercurian and lunar crust.
See Abstr. 092.012.

The flight of the Galileo spacecraft past Venus, Earth, and the Moon.
See Abstr. 093.051.

Light scattering by rough surfaces on asteroidal/lunar regoliths.
See Abstr. 098.058.

Mid–infrared (7.5 – 12.8 μm) spectra of 4 Vesta, 16 Psyche, 24 Themis, 113 Amalthea and the Moon.
See Abstr. 098.074.

Ejecta from lunar impacts: where is it on earth?
See Abstr. 105.058.

Lunar mare meteorites.
See Abstr. 105.206.

Nitrogen, noble gases, and nuclear tracks in lunar meteorites MAC 88104/105.
See Abstr. 105.252.

Mineralogical studies of lunar mare meteorites EET 87521 and Y 793274.
See Abstr. 105.253.

Chemical composition of the Earth after the giant impact.
See Abstr. 107.047.

095 Lunar Eclipses

095.001 Video image analysis – partial lunar eclipse 1991 December 21.
B. W. Soulsby.
Aust. J. Astron., Vol. 4, No. 3, p. 143 – 148 (Apr 1992).
    Video recordings of the 1991 Dec 21 partial lunar eclipse have provided permanent records for image analysis to test the improved lunar eclipse ephemerides predictions for fourth contact and to add to the data base for estimates of the oblateness of the umbra and Earth's upper atmosphere.

095.002 Zur Mondfinsternis vom 9./10. Dezember 1992. Simulationen im Computerprogramm "Voyager".
J. Alean.
Orion, Jahrg. 50, Nr. 250, p. 100 – 102 (Jun 1992).

095.003 Wie dunkel wird die Dezember–Mondfinsternis?
T. Baer.
Orion, Jahrg. 50, Nr. 250, p. 103 – 107 (Jun 1992).

Significant high number commensurabilities in the main lunar problem: a postscript to a discovery of the ancient Chaldeans.
See Abstr. 094.103.

096 Lunar and Planetary Occultations

096.001 Lunar occultations of southern near–infrared stellar sources.
A. Richichi, F. Lisi, A. Di Giacomo.
Astron. Astrophys., Vol. 254, No. 1/2, p. 149 – 166 (Feb 1992).
    Fifteen lunar occultation events have been observed in the course of a program aimed at the measurement of angular diameters of southern near–infrared sources and the detection of possible circumstellar shells. As a result, angular diameters have been determined for the first time for nine stars with spectra cooler than K5. The authors present the first practical implementation of a method for the removal of low–frequency fluctuations induced by atmospheric turbulence in occultation traces. Also, a detailed analysis is given of the limiting resolution set by the instrumentation and of the errors in the final results. The angular diameters are discussed in connection with available multi–wavelength photometry, obtained in part in the course of the observations, leading to direct estimates of the effective temperatures. The authors also suggest and discuss the presence of circumstellar shells around four sources with low effective temperatures and infrared excesses.

096.002 Stellar occultation candidates from the Guide Star Catalog. I. Saturn, 1991–1999.
A. S. Bosh, S. W. McDonald.
Astron. J., Vol. 103, No. 3, p. 983 – 990 (Mar 1992). Current Physics Microform No.: 9203E2191.
    The authors present a list of 203 potential occultations by Saturn and its rings of stars from the Hubble Space Telescope Guide Star Catalog (GSC), during the years 1991–1999. Because the GSC is not a complete catalog, this is not an exhaustive list of Saturn occultations. In particular, stars brighter than magnitude 8 are not included. However, this list does include many fainter candidates than do current occultation candidate lists for Saturn; these fainter stars also can provide a high signal–to–noise ratio if observed with a large telescope or in the infrared where Saturn

and its rings have absorption bands. The authors list the occultation circumstances, as well as star information found in the GSC.

096.003   Grazing occultation observations.
D. Stockbauer.
Occultation Newsl., Vol. 5, No. 7, p. 157 (May 1992).

096.004   Occultation of the Pleiades star cluster by the Moon: a first analysis.
A. Gerritsen, T. Schoenmaker.
Occultation Newsl., Vol. 5, No. 7, p. 159 – 162, 164 (May 1992).

096.005   Solar system occultations during 1992.
D. W. Dunham.
Occultation Newsl., Vol. 5, No. 7, p. 167 – 169, 176 (May 1992).

096.006   Lunar occultation prediction and software news.
D. W. Dunham.
Occultation Newsl., Vol. 5, No. 7, p. 171 – 174 (May 1992).

096.007   New double stars.
T. Murray.
Occultation Newsl., Vol. 5, No. 7, p. 184 – 186 (May 1992).

096.008   Magnitudes of selected stellar occultation candidates for Pluto and other planets, with new predictions for Mars and Jupiter.
C. B. Sybert, A. S. Bosh, L. M. Sauter, J. L. Elliot, L. H. Wasserman.
Astron. J., Vol. 103, No. 4, p. 1395 – 1398 (Apr 1992). Current Physics Microform No.: 9204G0349.
Occultation predictions for the planets Mars and Jupiter are presented along with BVRI magnitudes of 45 occultation candidates for Mars, Jupiter, Saturn, Uranus, and Pluto. Observers can use these magnitudes to plan observations of occultation events. The optical depth of the Jovian ring can be probed by a nearly central occultation on 1992 July 8. Mars occults an unusually red star in early 1993, and the occultations for Pluto involving the brightest candidates would possibly occur in the spring of 1992 and the fall of 1993.

096.009   Occultations of stars by solar system objects. IX. Occultations of catalog stars by asteroids, planets, and major satellites in 1992 and 1993.
L. H. Wasserman, E. Bowell, R. L. Millis.
Astron. J., Vol. 103, No. 6, p. 2079 – 2089 (Jun 1992). Current Physics Microform No.: 9207A0359.
Predictions are given for occultations of catalog stars by asteroids, planets, and major satellites for 1992 and 1993. The predictions are based on a computerized comparison of the occulting bodies' ephemerides and nine major star catalogs. The asteroid search is complete for all numbered asteroids whose angular diameters are thought to exceed 0.08 arcsec during the search years. Preliminary ground tracks are shown for the more favorable occultations by asteroids. No specially favorable occultations were found involving planets, and no occultations by major satellites were found.

096.010   Occultation observations in 1990.
Data Rep. Hydrogr. Obs., Ser. Astron. Geod., No. 26, p. 1 – 39 (Mar 1992).
In 1990, timing data of 798 lunar occultations, including 432 photoelectric observations, of reliable quality were obtained at four astronomical stations of JHD. Reduction and analysis give the following results for moon's longitude and latitude: $\Delta L = +0\overset{s}{.}41 \pm 0\overset{s}{.}03$(m.e.); $\Delta B = -0\overset{s}{.}19 \pm 0\overset{s}{.}05$(m.e.) for 1990.5 on the FK5 system.

096.011   The eclipse of star 28 Sgr by Titan.
A. N. Rudenko, A. I. Movchan, Yu. S. Romanov, N. R. Burlak, N. I. Koshkin, I. S. Bryukhanov.
Astron. Tsirk., No. 1550, p. 33 – 34 (Sep – Oct 1991). In Russian.
The results of observations of the eclipse of 28 Sgr by Titan at the Odessa Astronomical Observatory on the 4th July, 1989 are given. The light curve is represented in the instrumental system from observations with the 50–cm reflector and electrophotometer at the Astronomical Station in Mayaki settlement. The moments are determined of the eclipse beginning, reaching the region of the light minimum, the commencement of going out of the full phase and the eclipse end from observations with four telescopes.

096.012   Occultations of stars by the moon observed at the Kiev University Astronomical Observatory in 1989.
A. K. Osipov, V. I. Mazur, N. I. Buromskij, A. A. Zhitetskij, L. V. Kazantseva, S. S. Tryashin, A. A. Cheshkov, Yu. A. Moskalenko.
Visn. Kiiv. Univ., Fiz.–Mat. Nauki, Astron., Vip. 3, p. 66 – 73 (1992). In Ukrainian.
Timing date of total and grazing observations of lunar occultations in 1989 were reported.

096.013   The 18 August 1991 stellar occultation by the Neptune system, and upcoming occultations by Neptune's rings.
R. G. French, S. Maene, J. D. Goguen, K. J. Meech, R. L. Baron.
Bull. Am. Astron. Soc., Vol. 23, No. 3, p. 1181 (1991). Abstract. – See Abstr. 012.037 for the main entry.

096.014   The photoelectric occultations of two double stars.
Qian Bochen, Fan Qingyuan, Wang Yi.
Ann. Shanghai Obs., Acad. Sin., No. 13, p. 81 – 83 (1992). In Chinese.
The photoelectric occultation observations of two double stars are presented.

Quelques occultations chez Képler.
See Abstr. 004.065.

A high speed photometer in the optical region for lunar occultation studies.
See Abstr. 034.105.

The MIT program for identifying occultations and appulses by planets.
See Abstr. 036.147.

Mit Mondbedeckungen auf der Jagd nach jungen Doppelsternen im Taurus.
See Abstr. 036.162.

Ephemeris time obtained from lunar occultation observations made in the USSR during 1981 – 1985.
See Abstr. 044.020.

Analysis of stellar occultation data for planetary atmospheres. I. Model fitting, with application to Pluto.
See Abstr. 091.007.

Saturn's rings: optical depth profiles at $\lambda 3.9$ μm from the occultation of 28 Sgr.
See Abstr. 100.033.

Saturn ring masses and lightcurve morphology from IRTF observations of the occultation of 28 Sgr.
See Abstr. 100.034.

The kinematics of eccentric features in Saturn's Cassini Division from combined Voyager and groundbased data.
See Abstr. 100.035.

Detailed study of Prometheus' and Pandora's density waves.
See Abstr. 100.042.

Shape and opacity of Titan's stratosphere from the 28 Sgr occultation.
See Abstr. 100.044.

A global Titan upper atmosphere model from the occultation of 28 Sgr.
See Abstr. 100.059.

The 25 and 28 June 1991 stellar occultations by the Uranian rings.
See Abstr. 101.063.

Occultation constraints on atmospheric models for Pluto.
See Abstr. 101.082.

The radius of Pluto from the 9 June 1988 occultation.
See Abstr. 101.083.

Multiplicity among the young stars in Taurus.
See Abstr. 121.004.

097 Mars, Mars Satellites

097.001 **The atmospheric composition of Mars: ISM and ground–based observational data.**
T. Encrenaz, E. Lellouch, J. Rosenqvist, P. Drossart,
M. Combes, F. Billebaud, I. de Pater, S. Gulkis, J. P. Maillard,
G. Paubert.
Ann. Geophys., Vol. 9, No. 12, p. 797 – 803 (Dec 1991).
   Monitoring the abundances of Martian minor constituents and searching for local variations can provide valuable information about the aeronomy of the Martian atmosphere. Recent results from the PHOBOS ISM experiment suggest possible local variations of the CO abundance. Additional constraints have been given by ground–based observations of CO in the infrared and millimeter range, a search for minor species in the millimeter range, and the millimeter detection of deuterated water. From these data, information on the thermal profile and the vertical distribution low CO and H_2O has been retrieved.

097.002 **Atomic oxygen in the Martian thermosphere.**
A. I. F. Stewart, M. J. Alexander, R. R. Meier,
L. J. Paxton, S. W. Bougher, C. G. Fesen.
J. Geophys. Res., Vol. 97, No. A1, p. 91 – 102 (1 Jan 1992).
   Modern models of thermospheric composition and temperature and of excitation and radiative transfer processes are used to simulate the O I 130–nm emission from Mars measured by the Mariner 9 ultraviolet spectrometer.

097.003 **De Mars–oppositie van 1990.**
J. Koet.
Zenit, Jaarg. 19, Nr. 1, p. 34 – 37 (Jan 1992).

097.004 **Opacities of the 1973 dust storm over the Solis Lacus, Hellas, and Syrtis Major areas of Mars.**
T. Akabane, K. Iwasaki, Y. Saito, Y. Narumi.
Astron. Astrophys., Vol. 255, No. 1/2, p. 377 – 382 (Feb 1992).
   The opacities of the October 1973 dust storm, occurred on Mars, were estimated by analyzing negatives exposed at Hida Observatory. The authors solved the equation of radiative transfer by the discrete ordinate method, applying the Minnaert formula to the ground reflectivity which constrains one of the boundary conditions of the equation. The opacity over Solis Lacus area was 3–5 in the initial phase of the storm, and 3–4 over Hellas in the early decay phase, when the dust cloud over Hellas was still active. The bright and dark contrast of Syrtis Major was extremely low in the early decay phase. The decrease of contrast suggests that the opacity over Syrtis Major was about 2.

097.005 **The Martian bow shock: wave observations in the upstream region.**
A. Skalsky, R. Grard, S. Klimov, C. M. C. Nairn,
J. G. Trotignon, K. Schwingenschuh.
J. Geophys. Res., Vol. 97, No. A3, p. 2927 – 2933 (1 Mar 1992).
   Distinct electric field oscillations in the frequency range from 4 to 40 kHz are detected by the plasma wave system in the upstream region of the bow shock during the operation of the

Phobos 2 spacecraft around Mars. The upstream electron foreshock boundary approximately coincides with the magnetic field line tangential to the bow shock surface. A polarization analysis shows clearly that the electric field vector of the waves is aligned with the local magnetic field. The features of these high-frequency oscillations appear to be very similar to those of the emissions seen near the electron plasma frequency in the electron foreshock regions upstream of the bow shocks of Venus and Earth.

097.006 **Solar wind erosion of the Mars early atmosphere.**
H. Perez–de–Tejada.
J. Geophys. Res., Vol. 97, No. A3, p. 3159 – 3167 (1 Mar 1992).
   A calculation of the amount of volatiles that Mars has lost in the past through solar wind erosion is presented. The analysis is based on the examination of the ionospheric plasma flow produced by the transfer of momentum of the solar wind to the planet's upper ionosphere and is carried out by using loss rates suitable to a dense Venus–like Martian early ionosphere. The results indicate that an amount of mass equivalent to that of a global water ocean at least 10 m deep was removed by this process over the planet's lifetime. Further calculations show that depending on the strength of the Martian early ionosphere the total mass eroded could have been the equivalent of a global ocean up to ~ 30 m deep. These numbers are much larger than those derived from current planetary escape rates inferred from the recent Phobos measurements and represent up to about nearly one half of the amount of water which is believed was delivered to the Martian atmosphere through early volcanic activity.

097.007 **The motion of Mars' pole. II. The effect of an elastic mantle and a liquid core.**
J. L. Hilton.
Astron. J., Vol. 103, No. 2, p. 619 – 637 (Feb 1992). Current Physics Microform No.: 9202D1949.
   A first–order approximation of the effects of an elastic mantle and liquid core on the motion of Mars' pole are explored. The effect on Mars' Chandler wobble (Eulerian free nutation) is much less dependent on Mars' structure than the Earth's Chandler wobble depends on the Earth's structure. The period of the liquid core free–core nutation (FCN), however, is found to be very sensitive to the mean core radius; if the FCN period is known with an uncertainty of 2 days, then the mean core radius can be inferred with an uncertainty of only 6 km. The amplitude of the forced nutation in the liquid core models is also sensitive to the mean core radius. The sensitivity is high enough that measuring of the amplitudes of the three largest nutation components with an accuracy of a milliarcsecond will produce measures of the mean core radius with uncertainties of 32, 38, and 67 km, respectively. Elastic mantle, solid core models, however, are found to produce no significant difference in the motion of the pole compared to the rigid solid core model. Evidence for some sort of nonrigid polar motion is shown to exist from the Viking lander radar ranges of Mars. Methods of obtaining higher quality

observations of Mars' orientation in space, and the applicability of the methods derived for Mars to other planets in the solar system are discussed.

097.008 **On the possibility of chemosynthetic ecosystems in subsurface habitats on Mars.**
P. J. Boston, M. V. Ivanov, C. P. McKay.
Icarus, Vol. 95, No. 2, p. 300 – 308 (Feb 1992).
The authors have reexamined the question of extant microbial life on Mars in light of the most recent information about the planet and recently discovered nonphotosynthetic microbial ecosystems on Earth–deep sea hydrothermal vent communities and deep subsurface aquifer communities. On Mars, protected subsurface niches associated with hydrothermal activity could have continued to support life even after surface conditions became inhospitable. Geochemical evidence from the SNC meteorites and geomorphological evidence for recent volcanism suggest that such habitats could persist to the present time. The authors suggest a possible deep subsurface microbial ecology similar to those discovered to depths of several kilometers below the surface of the Earth. They focus on anaerobic systems utilizing CO_2 as the primary source of carbon. The hypothetical ecosystem is neither supported, nor excluded, by current observations of Mars. Tests for such a subsurface system involve locating active geothermal areas associated with ground ice or detecting trace quantities of reduced atmospheric gases that would leak from such a system.

097.009 **The density of Martian craters as a function of elevation and its application to the identification of ancient sea beds.**
T. Y. Winarski.
Strolling Astron., Vol. 36, No. 1, p. 9 – 11 (Mar 1992).
This study uses a global mapping of the density of Martian impact craters as a function of the surface elevation in order to trace the oulines of ancient seas on that planet.

097.010 **Is the analysis of Viking–1 and –2 observational data on the optical properties of the Martian atmosphere reliable?**
A. V. Morozhenko.
Astron. Vestn., Tom 26, No. 1, p. 28 – 38 (Jan – Feb 1992). In Russian. English translation in Solar Syst. Res., Vol. 26, No. 1.
The analysis of the optical properties data for the Martian atmosphere in the periods of its high transparency, obtained by Viking–1, –2 landers, permits to conclude that: 1) the derived values of the effective radius of particles do not agree with the results of polarimetric observations; 2) in the time when the images of the Sun were obtained, the optical properties of the planetary atmosphere were unstable. The last fact was not taken into account in treating the Viking observations what permits to state that the data about the optical properties of the atmosphere, mentioned above, are not true.

097.011 **What is known about aerosols in the Mars atmosphere?**
V. I. Moroz.
Astron. Vestn., Tom 26, No. 1, p. 39 – 45 (Jan – Feb 1992). In Russian. English translation in Solar Syst. Res., Vol. 26, No. 1.
A short review of main properties of Martian aerosols is given, including composition, optical depth, size distribution. Characteristics of so called "constant haze" are described in more detail. New information about constant haze was obtained in some experiments of the Phobos mission. The height distribution of the extinction coefficient was defined for the first time. Full optical depth was about 0.2 twice less than measured by Viking landers at the same season and size particle distribution was found more narrow than from Viking data.

097.012 **Once more on the state of the Martian atmosphere in 1988.**
V. M. Mikhajlets.
Astron. Vestn., Tom 26, No. 2, p. 127 (Mar – Apr 1992). In Russian. Letter–to–the–editor. English translation in Solar Syst. Res., Vol. 26, No. 2.

097.013 **Observations of plasma boundaries and phenomena around Mars with Phobos 2.**
K. Sauer, T. Roatsch, U. Motschmann, K. Schwingenschuh, R. Lundin, H. Rosenbauer, S. Livi.
J. Geophys. Res., Vol. 97, No. A5, p. 6227 – 6233 (1 May 1992).
Magnetic field and plasma measurements on board the Soviet spacecraft Phobos 2 have been analyzed during five elliptical orbits around Mars. The existence of at least one separate plasma boundary and an adjacent plasma layer, called the planetopause and the transition region, between the bow shock and the ionopause seems to be a characteristic feature of the solar wind interaction with an almost nonmagnetized planetary ionosphere. It is suggested that the planetopause is a multiple–ion discontinuity, where a large number of solar wind protons are deflected at an exospheric density ramp. Strong changes in magnetic field and plasma flow direction within the transition region are interpreted as signatures of current sheets or internal shocks. The detected eclipse boundary in the tail is perhaps an elongation of the ionopause found at Venus. New ideas concerning the formation of multiple–ion flow boundaries by electrostatic plasma–plasma interaction are discussed. Finally, a remarkable Deimos event has been detected during the fifth orbit. This is explained as an interaction of the subsonic solar wind with a thin cloud of charged dust particles.

097.014 **Viking 2 electron observations at Mars.**
F. S. Johnson, W. B. Hanson.
J. Geophys. Res., Vol. 97, No. A5, p. 6523 – 6530 (1 May 1992).
The Viking retarding potential analyzer flown to Mars in 1976 had negative retarding potential sweeps in addition to the positive ion sweeps. This paper presents an analysis of the electron mode sweeps made in Viking 2 above the ionosphere.

097.015 **Experiment of constructing a photometric map of the normal albedo of the Martian surface.**
V. G. Tejfel', N. V. Sinyaeva, A. N. Aksenov, G. A. Kharitonova.
Pis'ma Astron. Zh., Tom 18, No. 3, p. 271 – 278 (Mar 1992). In Russian. English translation in Sov. Astron. Lett., Vol. 18, No. 2.
The technique and results of photometric and computer reduction of Mars photographs obtained near the 1990 Mars opposition are described. These measurements were used to construct a map of the normal albedo in red light. Measurements of the intensity distribution are carried out as well as the determination of the local and global limb darkening coefficients and reduction of the Mars images for limb darkening. From these data the fragments of the albedo map and general map were constructed in isophotes, semitone and three–dimensional presentation.

097.016 **Comparisons of peak ionosphere pressures at Mars and Venus with incident solar wind dynamic pressure.**
M. H. G. Zhang, J. G. Luhmann.
J. Geophys. Res., Vol. 97, No. E1, p. 1017 – 1025 (25 Jan 1992).
Previous calculations of the potential for pressure between the solar wind and the ionospheres of the weakly magnetized planets, Mars and Venus, indicated that the maximum or peak ionospheric thermal pressure is sufficient to stand off the solar wind at Venus but not at Mars. In this study the authors used radio occultation measurements of electron density profiles from Mariner 6 and 7, the Mariner 9 extended mission, and the U.S. Viking orbiters, together with model ion and electron temperature profiles, to derive thermal pressure profiles in the Mars ionosphere. Similarly, Pioneer Venus Orbiter (PVO) radio occultation data and temperature models were used to obtain ionospheric pressure profiles at Venus. Because the radio occultation data give information for both active and quiet phases of the solar cycle, this method allows one to consider how the balance changes between solar minimum and maximum.

097.017 **The ionospheric effects of a weak intrinsic magnetic field at Mars.**
H. Shinagawa, T. E. Cravens.
J. Geophys. Res., Vol. 97, No. E1, p. 1027 – 1035 (25 Jan 1992).
The existence of an intrinsic magnetic field at Mars has yet to be ascertained despite a number of missions to Mars over the last

few decades. It is now widely accepted that the thermal pressure of the Martian ionosphere is not large enough to balance the average solar wind dynamic pressure, therefore the ionosphere must be magnetized, either by a solar wind induced field or by an intrinsic field. Shinagawa and Cravens (1980) demonstrated that the behavior of the Martian ionosphere depends on the strength of the intrinsic field. The authors have improved their earlier model of the Martian ionosphere by allowing the magnetic field to have any direction in the horizontal plane, and they present results of calculations for several different intrinsic magnetic field strengths and directions.

097.018   Mars secular obliquity change due to the seasonal polar caps.
D. P. Rubincam.
J. Geophys. Res., Vol. 97, No. E2, p. 2629 – 2632 (25 Feb 1992).

There is a weak positive feedback mechanism between the astronomy and meteorology of Mars. The mechanism is this: the seasonal waxing and waning polar caps cause small changes in Mars' dynamical flattening. Because the changes in flattening are out of phase with the Sun, there is a net annual solar torque on the planet which increases the angle between the equatorial and orbital planes. On the basis of Viking observations of the present climate and simple atmospheric models of past climates these seasonal shifts of mass between the atmosphere and polar caps are capable of secularly increasing Mars' obliquity by about 1° or 2° since the origin of the solar system. Thus the climate, driven largely by the axial tilt, reacts back on the planet and slightly enhances the seasons on Mars as time progresses. More sophisticated models will probably not change this result much; therefore, this mechanism probably produced only minor changes in Mars' climate. It causes negligible changes in the axial tilt and climate of the Earth.

097.019   Origin of giant Martian polygons.
G. E. McGill, L. S. Hills.
J. Geophys. Res., Vol. 97, No. E2, p. 2633 – 2647 (25 Feb 1992).

Extensive areas of the Martian northern plains in Utopia and Acidalia planitiae are characterized by "polygonal terrane". Polygonal terrane consists of material cut by complex troughs defining a pattern resembling mudcracks, columnar joints, or frost–wedge polygons on Earth. However, the Martian polygons are orders of magnitudes larger than these potential Earth analogues, leading to severe mechanical difficulties for genetic models based on simple analogy arguments. Plate–bending and finite element models indicate that shrinkage of desiccating sediment or cooling volcanics accompanied by differential compaction over buried topography can account for the stresses responsible for polygon troughs as well as the large size of the polygons. Although trough widths and depths relate primarily to shrinkage, the large scale of the polygonal pattern relates to the spacing between topographic elevations on the surface buried beneath polygonal terrane material. Geological relationships favor a sedimentary origin for polygonal terrane material, but the authors' model is not dependent on the specific genesis. The authors' analysis also suggests that the polygons must have formed at a geologically rapid rate.

097.020   Mars radar mapping: strong backscatter from the Elysium basin and outflow channel.
J. K. Harmon, M. P. Sulzer, P. J. Perillat, J. F. Chandler.
Icarus, Vol. 95, No. 1, p. 153 – 156 (Jan 1992).

Radar reflectivity maps of Mars, obtained at Arecibo in 1990 using a modified delay–Doppler technique, have revealed strong depolarized echoes coming from the Elysium flood basin and outflow channel as well as from Elysium Mons. This result is consistent with recent studies which show that much of the basin/channel floor is covered with lava flows.

097.021   Degradation studies of Martian impact craters.
N. G. Barlow.
54. Annual Meeting of the Meteoritical Society, p. 12 (1991).
Abstract. – See Abstr. 012.010 for the main entry.

097.022 $^{142}Nd/^{144}Nd$ in SNCs and early differentiation of a heterogeneous Martian (?) mantle.
L. E. Nyquist, C. L. Harper, H. Wiesmann, B. Bansal, C. Y. Shih.
54. Annual Meeting of the Meteoritical Society, p. 178 (1991).
Abstract. – See Abstr. 012.010 for the main entry.

097.023   Radiative fluxes on a dustfree Mars.
H. Savijärvi.
Contrib. Atmos. Phys., Vol. 64, No. 2, p. 103 – 112 (May 1991).

Several longwave and shortwave radiation parameterization schemes were tested as for dustfree Mars. Band models (BM) with the Curtis–Godson approximation served as references for the CO_2 longwave spectrum. An emissivity scheme with pressure scaling was tuned with the help of the BM's. A broadband scheme, used previously both in a Martian GCM and a mesoscale model, was found to give slightly different results. It was also quite sensitive to its upper boundary condition. Water vapour effects on the thermal radiation were found to be negligible in winter but not in summer. The daytime heating of the Martian lower atmosphere by absorption of solar radiation was very small regarding water vapour and weak as to CO_2, when compared to the much stronger longwave daytime heating and nighttime cooling in a simulated summertime diurnal cycle.

097.024   Dust storms on Mars.
S. R. Brzostkiewicz.
Urania, Rok 63, Nr. 4, p. 106 – 111 (Apr 1992). In Polish.

097.025   Mars Rover Sample Return Mission: systemic model and optimization of scientific results. A case for large valley outlets.
N. A. Cabrol, E. A. Grin, A. Dollfus.
Sol. Syst. Res., Vol. 25, No. 2, p. 105 – 110 (Sep 1991). English translation of Astron. Vestn., Tom 25, No. 2, p. 145 – 151 (1991).

This paper is a proposal for a projective systemic model to assess the reliability level of each operation of a Mars Sample Return Mission in order to achieve optimum mission productivity. The instrumentation efficiency is analyzed in relation to the adequate site criteria. Candidate sites are tested by this systemic approach. These sites include the principal characteristics of the mission scientific goals and technical constraints.

097.026   The thermal stability of near–surface ground ice on Mars.
D. A. Paige.
Nature, Vol. 356, No. 6364, p. 43 – 45 (5 Mar 1992). Letter–to–the–editor.

The existence of subsurface water ice on Mars has been predicted in several theoretical studies, but there are no definitive observations of its present distribution. Geomorphic features on the surface of Mars have been widely interpreted as evidence for the presence of ground ice, but many of these features are found at near–equatorial latitudes, where thermal models have predicted that near–surface water ice should not be stable under present climate conditions. The author presents the results of thermal calculations which show that observed geographic variations in the thermal and reflectance properties of martian soils significantly affect subsurface temperatures. His results indicate that in certain regions, ground–ice deposits could exist much closer to the surface, and much closer to the equator, than previously thought. In the future, these deposits could be a valuable resource for human exploration.

097.027   Liquid water and life on early Mars.
C. P. McKay, L. R. Doyle, W. L. Davis, R. A. Wharton.
3. Symposium International de Bioastronomie, p. 190 – 192 (1991). – See Abstr. 012.005 for the main entry.

097.028 **New interpretation of crustal extension evidences on Mars.**
E. A. Grin.
3. Symposium International de Bioastronomie, p. 193 (1991). Abstract. – See Abstr. 012.005 for the main entry.

097.029 **Martian paleohydrology and its implications for exobiology science.**
N. A. Cabrol, E. A. Grin.
3. Symposium International de Bioastronomie, p. 194 – 198 (1991). – See Abstr. 012.005 for the main entry.

097.030 **Iron Mössbauer spectroscopy: superparamagnetism in hydrothermal vents and the search for evidence of past life on Mars.**
D. G. Agresti, T. J. Wdowiak.
Workshop on the Martian Surface and Atmosphere Through Time, p. 9 – 10 (1992). Abstract. – See Abstr. 012.031 for the main entry.

097.031 **The nanophase iron mineral(s) in Mars soil.**
A. Banin, T. Ben-Shlomo, L. Margulies, D. F. Blake, A. U. Gehring.
Workshop on the Martian Surface and Atmosphere Through Time, p. 11 – 12 (1992). Abstract. – See Abstr. 012.031 for the main entry.

097.032 **Martian impact crater degradation studies: implications for localized obliteration episodes.**
N. G. Barlow.
Workshop on the Martian Surface and Atmosphere Through Time, p. 13 – 14 (1992). Abstract. – See Abstr. 012.031 for the main entry.

097.033 **Midlatitude weather systems on Mars: is there a hemispheric asymmetry?**
J. R. Barnes.
Workshop on the Martian Surface and Atmosphere Through Time, p. 15 – 16 (1992). Abstract. – See Abstr. 012.031 for the main entry.

097.034 **Mars: compositional variability of ferric/ferrous minerals and polar volatiles from groundbased imaging spectroscopy.**
J. F. Bell III.
Workshop on the Martian Surface and Atmosphere Through Time, p. 17 – 18 (1992). Abstract. – See Abstr. 012.031 for the main entry.

097.035 **Thermally distinct ejecta blankets from martian craters.**
B. H. Betts, B. C. Murray.
Workshop on the Martian Surface and Atmosphere Through Time, p. 19 – 20 (1992). Abstract. – See Abstr. 012.031 for the main entry.

097.036 **Infrared imaging of Mars for volatile distribution and seasonal variability between 2.4 and 5.1 μm.**
D. L. Blaney.
Workshop on the Martian Surface and Atmosphere Through Time, p. 21 (1992). Abstract. – See Abstr. 012.031 for the main entry.

097.037 **Influence of aerodynamic roughness length on aeolian processes: Earth, Mars, Venus.**
D. G. Blumberg, R. Greeley.
Workshop on the Martian Surface and Atmosphere Through Time, p. 22 (1992). Abstract. – See Abstr. 012.031 for the main entry.

097.038 **Dust storm driven variations of the Mars thermosphere and exosphere: coupling of atmospheric regions.**
S. W. Bougher, C. G. Fesen, R. W. Zurek.
Workshop on the Martian Surface and Atmosphere Through Time, p. 23 (1992). Abstract. – See Abstr. 012.031 for the main entry.

097.039 **Rates of oxidative weathering on the surface of Mars.**
R. G. Burns.
Workshop on the Martian Surface and Atmosphere Through Time, p. 26 – 27 (1992). Abstract. – See Abstr. 012.031 for the main entry.

097.040 **Martian channel networks: a revised Strahler approach for quantitative morphometry.**
N. A. Cabrol, E. A. Grin.
Workshop on the Martian Surface and Atmosphere Through Time, p. 28 – 29 (1992). Abstract. – See Abstr. 012.031 for the main entry.

097.041 **Ice in the northern lowlands and southern highlands of Mars and its enrichment beneath the Elysium lavas.**
J. A. Cave.
Workshop on the Martian Surface and Atmosphere Through Time, p. 30 – 31 (1992). Abstract. – See Abstr. 012.031 for the main entry.

097.042 **Mars dust and cloud opacities and scattering properties.**
R. T. Clancy, S. W. Lee.
Workshop on the Martian Surface and Atmosphere Through Time, p. 34 (1992). Abstract. – See Abstr. 012.031 for the main entry.

097.043 **VLA mapping of 1.35 cm water emission from the Mars atmospheric limb.**
R. T. Clancy, A. W. Grossman, D. O. Muhleman.
Workshop on the Martian Surface and Atmosphere Through Time, p. 35 – 36 (1992). Abstract. – See Abstr. 012.031 for the main entry.

097.044 **The subsurface hydrologic response of Mars to the thermal evolution of its early crust.**
S. M. Clifford, M. H. Carr.
Workshop on the Martian Surface and Atmosphere Through Time, p. 37 – 38 (1992). Abstract. – See Abstr. 012.031 for the main entry.

097.045 **Analysis of Martian atmospheric and surface optical properties between 4.4 and 5.1 μm.**
D. Crisp, D. L. Blaney.
Workshop on the Martian Surface and Atmosphere Through Time, p. 39 – 40 (1992). Abstract. – See Abstr. 012.031 for the main entry.

097.046 **Micro weather stations for in situ measurements in the Martian planetary boundary layer.**
D. Crisp, W. J. Kaiser, T. W. Kenny, T. R. VanZandt, J. E. Tillman.
Workshop on the Martian Surface and Atmosphere Through Time, p. 41 – 42 (1992). Abstract. – See Abstr. 012.031 for the main entry.

097.047 **Soil texture and granulometry at the surface of Mars.**
A. Dollfus, M. Deschamps, J. Zimbelman.
Workshop on the Martian Surface and Atmosphere Through Time, p. 43 – 44 (1992). Abstract. – See Abstr. 012.031 for the main entry.

097.048 **The composition of Martian aeolian sands: thermal emissivity from Viking IRTM observations.**
K. S. Edgett, P. R. Christensen.
Workshop on the Martian Surface and Atmosphere Through Time, p. 45 – 46 (1992). Abstract. – See Abstr. 012.031 for the main entry.

097.049   Infrared photometric behavior and opposition effect of Mars.
S. Erard, J.–P. Bibring, P. Drossart.
Workshop on the Martian Surface and Atmosphere Through Time, p. 47 – 48 (1992). Abstract. – See Abstr. 012.031 for the main entry.

097.050   Atmosphere–surface interactions and atmospheric evolution on Mars.
B. Fegley Jr.
Workshop on the Martian Surface and Atmosphere Through Time, p. 49 – 50 (1992). Abstract. – See Abstr. 012.031 for the main entry.

097.051   The meteoritic contribution to dust and aerosols in the atmosphere of Mars.
G. J. Flynn.
Workshop on the Martian Surface and Atmosphere Through Time, p. 51 – 52 (1992). Abstract. – See Abstr. 012.031 for the main entry.

097.052   Nitrogen escape from Mars.
J. L. Fox.
Workshop on the Martian Surface and Atmosphere Through Time, p. 53 – 54 (1992). Abstract. – See Abstr. 012.031 for the main entry.

097.053   Implications of Early Hesperian ages for presumed Noachian age volcanic flows on Mars.
H. V. Frey.
Workshop on the Martian Surface and Atmosphere Through Time, p. 55 – 56 (1992). Abstract. – See Abstr. 012.031 for the main entry.

097.054   Mars surface weathering products and spectral analogs: palagonites and synthetic iron minerals.
D. C. Golden, D. W. Ming, R. V. Morris, H. V. Lauer Jr.
Workshop on the Martian Surface and Atmosphere Through Time, p. 59 – 60 (1992). Abstract. – See Abstr. 012.031 for the main entry.

097.055   Styles of crater gradation in southern Ismenius Lacus, Mars: clues from Meteor Crater, Arizona.
J. A. Grant, P. H. Schultz.
Workshop on the Martian Surface and Atmosphere Through Time, p. 61 – 62 (1992). Abstract. – See Abstr. 012.031 for the main entry.

097.056   Magmatic intrusions and hydrothermal systems on Mars.
V. C. Gulick.
Workshop on the Martian Surface and Atmosphere Through Time, p. 63 – 64 (1992). Abstract. – See Abstr. 012.031 for the main entry.

097.057   Mars: wavelength–dependent dual polarization global scattering.
J. K. Harmon, M. A. Slade, R. S. Hudson.
Workshop on the Martian Surface and Atmosphere Through Time, p. 65 – 66 (1992). Abstract. – See Abstr. 012.031 for the main entry.

097.058   The stable isotopic compositions of indigenous carbon–bearing components in EETA 79001.
C. P. Hartmetz, I. P. Wright, C. T. Pillinger.
Workshop on the Martian Surface and Atmosphere Through Time, p. 67 – 68 (1992). Abstract. – See Abstr. 012.031 for the main entry.

097.059   Greenhouse warming by minor gases on early Mars.
M. N. Heinrich, W. R. Thompson, C. Sagan.
Workshop on the Martian Surface and Atmosphere Through Time, p. 69 (1992). Abstract. – See Abstr. 012.031 for the main entry.

097.060   The Martian polar caps: stability and water transport at low obliquities.
B. G. Henderson, B. M. Jakosky.
Workshop on the Martian Surface and Atmosphere Through Time, p. 70 (1992). Abstract. – See Abstr. 012.031 for the main entry.

097.061   Mars polar caps at low obliquity.
B. G. Henderson, B. M. Jakosky.
Workshop on the Martian Surface and Atmosphere Through Time, p. 71 (1992). Abstract. – See Abstr. 012.031 for the main entry.

097.062   Dark material in the polar layered deposits on Mars.
K. Herkenhoff.
Workshop on the Martian Surface and Atmosphere Through Time, p. 72 – 73 (1992). Abstract. – See Abstr. 012.031 for the main entry.

097.063   Observations of Mars using Hubble Space Telescope.
P. B. James, R. T. Clancy, S. W. Lee, R. Kahn, R. Zurek, L. Martin, R. Singer.
Workshop on the Martian Surface and Atmosphere Through Time, p. 76 – 77 (1992). Abstract. – See Abstr. 012.031 for the main entry.

097.064   A liquidus phase diagram for the groundmass of EETA 79001A (Eg), a primitive shergottite composition.
J. H. Jones, A. J. G. Jurewicz, L. L. Le.
Workshop on the Martian Surface and Atmosphere Through Time, p. 80 – 81 (1992). Abstract. – See Abstr. 012.031 for the main entry.

097.065   Glacial geomorphic evidence for la late climatic change on Mars.
J. S. Kargel, R. G. Strom.
Workshop on the Martian Surface and Atmosphere Through Time, p. 82 – 83 (1992). Abstract. – See Abstr. 012.031 for the main entry.

097.066   Was early Mars warmed by ammonia?
J. F. Kasting, L. L. Brown, J. M. Acord, J. B. Pollack.
Workshop on the Martian Surface and Atmosphere Through Time, p. 84 – 85 (1992). Abstract. – See Abstr. 012.031 for the main entry.

097.067   Erosional landforms on the layered terrains in Valles Marineris.
G. Komatsu, R. G. Strom, V. C. Gulick, T. J. Parker.
Workshop on the Martian Surface and Atmosphere Through Time, p. 86 – 87 (1992). Abstract. – See Abstr. 012.031 for the main entry.

097.068   Mars: correcting surface albedo observations for effects of atmospheric dust loading.
S. W. Lee, R. T. Clancy.
Workshop on the Martian Surface and Atmosphere Through Time, p. 88 (1992). Abstract. – See Abstr. 012.031 for the main entry.

097.069   Simulations of the seasonal polar caps on Mars.
B. L. Lindner.
Workshop on the Martian Surface and Atmosphere Through Time, p. 89 – 90 (1992). Abstract. – See Abstr. 012.031 for the main entry.

097.070   Mars atmosphere evolution: escape to space.
J. G. Luhmann.
Workshop on the Martian Surface and Atmosphere Through Time, p. 91 – 92 (1992). Abstract. – See Abstr. 012.031 for the main entry.

097.071 **Nonlinear stratified flow over localized topographic obstacles on Mars.**
J. A. Magalhães, R. E. Young.
Workshop on the Martian Surface and Atmosphere Through Time, p. 94 – 95 (1992). Abstract. – See Abstr. 012.031 for the main entry.

097.072 **Volatile tracers of Martian atmospheric evolution: present measurement status and requirements for future investigations.**
P. Mahaffy, K. Mauersberger.
Workshop on the Martian Surface and Atmosphere Through Time, p. 96 (1992). Abstract. – See Abstr. 012.031 for the main entry.

097.073 **An ejection model for SNC meteorites: an indication for recent volcanism on Mars.**
J. P. Manker.
Workshop on the Martian Surface and Atmosphere Through Time, p. 97 – 98 (1992). Abstract. – See Abstr. 012.031 for the main entry.

097.074 **Observed changes in limb clouds immediately prior to the onset of planet–encircling dust storms.**
L. J. Martin, P. B. James, R. W. Zurek.
Workshop on the Martian Surface and Atmosphere Through Time, p. 99 – 100 (1992). Abstract. – See Abstr. 012.031 for the main entry.

097.075 **New dust opacity maps from Viking IR thermal mapper data.**
T. Z. Martin, M. I. Richardson.
Workshop on the Martian Surface and Atmosphere Through Time, p. 101 – 102 (1992). Abstract. – See Abstr. 012.031 for the main entry.

097.076 **Temporal variability of the surface and atmosphere of Mars: Viking orbiter color observations.**
A. S. McEwen.
Workshop on the Martian Surface and Atmosphere Through Time, p. 103 – 104 (1992). Abstract. – See Abstr. 012.031 for the main entry.

097.077 **Regional variations in the stability and diffusion of water–ice in the Martian regolith.**
M. T. Mellon, B. M. Jakosky.
Workshop on the Martian Surface and Atmosphere Through Time, p. 105 – 106 (1992). Abstract. – See Abstr. 012.031 for the main entry.

097.078 **Exploring compositional variations on the surface of Mars applying mixing modeling to a telescopic spectral image.**
E. Merényi, J. S. Miller, R. B. Singer.
Workshop on the Martian Surface and Atmosphere Through Time, p. 107 – 108 (1992). Abstract. – See Abstr. 012.031 for the main entry.

097.079 **Simulation of Martian surface–atmosphere interaction in a space–simulator: technical considerations and feasibility.**
D. Möhlmann, H. Kochan.
Workshop on the Martian Surface and Atmosphere Through Time, p. 109 (1992). Abstract. – See Abstr. 012.031 for the main entry.

097.080 **Mars dust storm simulations: analysis of surface stress.**
J. R. Murphy, C. B. Leovy.
Workshop on the Martian Surface and Atmosphere Through Time, p. 110 (1992). Abstract. – See Abstr. 012.031 for the main entry.

097.081 **Is ground ice stable near the Martian equator?**
D. A. Paige.
Workshop on the Martian Surface and Atmosphere Through Time, p. 111 – 112 (1992). Abstract. – See Abstr. 012.031 for the main entry.

097.082 **Distribution of coastal morphology in the Martian northern lowlands.**
T. J. Parker, D. S. Gorsline.
Workshop on the Martian Surface and Atmosphere Through Time, p. 113 – 114 (1992). Abstract. – See Abstr. 012.031 for the main entry.

097.083 **Recent Elysium volcanism–effects on the Martian atmosphere.**
J. B. Plescia, J. Crisp.
Workshop on the Martian Surface and Atmosphere Through Time, p. 115 – 116 (1992). Abstract. – See Abstr. 012.031 for the main entry.

097.084 **Determining the pH of Mars from the Viking labelled release reabsorption effect.**
R. C. Plumb.
Workshop on the Martian Surface and Atmosphere Through Time, p. 117 (1992). Abstract. – See Abstr. 012.031 for the main entry.

097.085 **Chemical reaction path modeling of hydrothermal processes on Mars: preliminary results.**
G. S. Plumlee, W. I. Ridley.
Workshop on the Martian Surface and Atmosphere Through Time, p. 118 – 119 (1992). Abstract. – See Abstr. 012.031 for the main entry.

097.086 **Short– and long–term climate changes on Mars.**
J. B. Pollack.
Workshop on the Martian Surface and Atmosphere Through Time, p. 120 (1992). Abstract. – See Abstr. 012.031 for the main entry.

097.087 **Influence of heat flow on early Martian climate.**
S. Postawko, F. P. Fanale.
Workshop on the Martian Surface and Atmosphere Through Time, p. 121 (1992). Abstract. – See Abstr. 012.031 for the main entry.

097.088 **Mid–infrared spectra of Martian komatiite.**
D. P. Reyes.
Workshop on the Martian Surface and Atmosphere Through Time, p. 122 – 123 (1992). Abstract. – See Abstr. 012.031 for the main entry.

097.089 **Chryse Planitia region, Mars: channeling history, flood–volume estimates, and scenarios for bodies of water in the northern plains.**
S. L. Rotto, K. L. Tanaka.
Workshop on the Martian Surface and Atmosphere Through Time, p. 124 – 125 (1992). Abstract. – See Abstr. 012.031 for the main entry.

097.090 **Dorsa Argentea type sinuous ridges, Mars: evidence for linear dune hypothesis.**
S. W. Ruff.
Workshop on the Martian Surface and Atmosphere Through Time, p. 126 – 127 (1992). Abstract. – See Abstr. 012.031 for the main entry.

097.091 **Atmospheric and surface temperatures and airborne dust amounts during late southern summer from Mariner 9 IRIS data.**
M. Santee, D. Crisp.
Workshop on the Martian Surface and Atmosphere Through Time, p. 128 – 129 (1992). Abstract. – See Abstr. 012.031 for the main entry.

097.092   **A carbonate–silicate aqueous geochemical cycle model for Mars.**
M. W. Schaefer, H. Leidecker.
Workshop on the Martian Surface and Atmosphere Through Time, p. 130 – 131 (1992). Abstract. – See Abstr. 012.031 for the main entry.

097.093   **Amazonis and Utopia Planitiae: Martian lacustrine basins.**
D. H. Scott, J. W. Rice Jr., J. M. Dohm, M. G. Chapman.
Workshop on the Martian Surface and Atmosphere Through Time, p. 132 – 133 (1992). Abstract. – See Abstr. 012.031 for the main entry.

097.094   **Evidence for crystalline hematite as an accessory phase in Martian soils.**
R. B. Singer, J. S. Miller.
Workshop on the Martian Surface and Atmosphere Through Time, p. 134 – 135 (1992). Abstract. – See Abstr. 012.031 for the main entry.

097.095   **Carbonate formation on Mars: history of the CO_2 atmosphere from models of diffusion–limited growth in non–aqueous environments.**
S. K. Stephens, D. J. Stevenson.
Workshop on the Martian Surface and Atmosphere Through Time, p. 136 – 137 (1992). Abstract. – See Abstr. 012.031 for the main entry.

097.096   **Atomic oxygen in the Martian thermosphere.**
A. I. F. Stewart, M. J. Alexander, R. R. Meier, L. J. Paxton, S. W. Bougher, C. G. Fesen.
Workshop on the Martian Surface and Atmosphere Through Time, p. 138 – 139 (1992). Abstract. – See Abstr. 012.031 for the main entry.

097.097   **Physical interpretation of thermal and reflected data on Martian surface units.**
E. L. Strickland III.
Workshop on the Martian Surface and Atmosphere Through Time, p. 140 – 141 (1992). Abstract. – See Abstr. 012.031 for the main entry.

097.098   **Physical properties of Deucalionis, Eos, and Xanthe–type units in the central equatorial region of Mars.**
E. L. Strickland III.
Workshop on the Martian Surface and Atmosphere Through Time, p. 142 – 143 (1992). Abstract. – See Abstr. 012.031 for the main entry.

097.099   **Physical properties of Meridiani Sinus–type units in the central equatorial region of Mars.**
E. L. Strickland III.
Workshop on the Martian Surface and Atmosphere Through Time, p. 144 – 145 (1992). Abstract. – See Abstr. 012.031 for the main entry.

097.100   **Physical properties of Oxia/Lunae Planum and Arabia–type units in the central equatorial region of Mars.**
E. L. Strickland III.
Workshop on the Martian Surface and Atmosphere Through Time, p. 146 – 147 (1992). Abstract. – See Abstr. 012.031 for the main entry.

097.101   **Surface photometric properties and albedo changes in the central equatorial region of Mars.**
E. L. Strickland III.
Workshop on the Martian Surface and Atmosphere Through Time, p. 148 – 149 (1992). Abstract. – See Abstr. 012.031 for the main entry.

097.102   **Glacial and marine chronology of Mars.**
R. G. Strom, J. S. Kargel, N. Johnson, C. Knight.
Workshop on the Martian Surface and Atmosphere Through Time, p. 150 – 151 (1992). Abstract. – See Abstr. 012.031 for the main entry.

097.103   **Characterization of Martian near–subsurface materials by determination of cohesion and angle of internal friction.**
R. J. Sullivan.
Workshop on the Martian Surface and Atmosphere Through Time, p. 152 – 153 (1992). Abstract. – See Abstr. 012.031 for the main entry.

097.104   **Topography of Apollinaris Patera and Ma'adim Vallis.**
G. D. Thornhill, D. A. Rothery, J. B. Murray, T. Day, A. Cook, J.–P. Muller, J. C. Iliffe.
Workshop on the Martian Surface and Atmosphere Through Time, p. 154 – 155 (1992). Abstract. – See Abstr. 012.031 for the main entry.

097.105   **Turbulent spectra, fluxes, stability and growth of the mixed layer in the boundary layer of Mars.**
J. E. Tillman, L. Landberg, S. E. Larsen.
Workshop on the Martian Surface and Atmosphere Through Time, p. 156 – 158 (1992). Abstract. – See Abstr. 012.031 for the main entry.

097.106   **Aqueous–alteration products in S–N–C meteorites and implications for volatile/regolith interactions on Mars.**
A. H. Treiman, J. L. Gooding.
Workshop on the Martian Surface and Atmosphere Through Time, p. 159 – 160 (1992). Abstract. – See Abstr. 012.031 for the main entry.

097.107   **Microcraters on Mars: evidence of past climatic variations.**
A. R. Vasavada, T. J. Milavec, D. A. Paige.
Workshop on the Martian Surface and Atmosphere Through Time, p. 162 – 163 (1992). Abstract. – See Abstr. 012.031 for the main entry.

097.108   **The ultraviolet albedo of Mars.**
R. Wagener.
Workshop on the Martian Surface and Atmosphere Through Time, p. 164 (1992). Abstract. – See Abstr. 012.031 for the main entry.

097.109   **Modelling the seasonal cycle of CO_2 on Mars: a fit to the Viking lander pressure curves.**
S. E. Wood, D. A. Paige.
Workshop on the Martian Surface and Atmosphere Through Time, p. 167 – 168 (1992). Abstract. – See Abstr. 012.031 for the main entry.

097.110   **Comparison of drift potential derived from Mars GCM (*General Circulation Model*) with rock abundance from IRTM.**
Pengyan Xu, R. Greeley.
Workshop on the Martian Surface and Atmosphere Through Time, p. 171 – 172 (1992). Abstract. – See Abstr. 012.031 for the main entry.

097.111   **Climatic implications of the simultaneous presence of CO_2 and H_2O in the Martian regolith.**
A. P. Zent.
Workshop on the Martian Surface and Atmosphere Through Time, p. 173 – 174 (1992). Abstract. – See Abstr. 012.031 for the main entry.

097.112   **The ancient oxygen exosphere of Mars: implications for atmosphere evolution.**
M. H. G. Zhang, J. G. Luhmann, A. F. Nagy, S. W. Bougher.
Workshop on the Martian Surface and Atmosphere Through Time, p. 175 (1992). Abstract. – See Abstr. 012.031 for the main entry.

097.113   **Forsterite/melt partitioning of argon and iodine: implications for atmosphere formation by outgassing of an early Martian magma ocean.**
D. S. Musselwhite, M. J. Drake, T. D. Swindle.
Workshop on the Physics and Chemistry of Magma Oceans from 1 bar to 4 Mbar, p. 40 – 41 (1992). Abstract. – See Abstr. 012.032 for the main entry.

097.114   **Ancient ice sheets on Mars.**
R. G. Strom, J. S. Kargel.
Bull. Am. Astron. Soc., Vol. 23, No. 3, p. 1172 – 1173 (1991). Abstract. – See Abstr. 012.037 for the main entry.

097.115   **Glaciation on Mars: When did it snow and for how long?**
J. S. Kargel, R. G. Strom, N. Johnson, C. Knight.
Bull. Am. Astron. Soc., Vol. 23, No. 3, p. 1173 (1991). Abstract. – See Abstr. 012.037 for the main entry.

097.116   **A regional example of a hydrologic cycle on Mars.**
J. M. Moore, D. R. Janke, G. D. Clow, W. L. Davis, R. M. Haberle, C. P. McKay, C. R. Stoker.
Bull. Am. Astron. Soc., Vol. 23, No. 3, p. 1173 (1991). Abstract. – See Abstr. 012.037 for the main entry.

097.117   **Microcraters on Mars: evidence of past climatic variations.**
A. R. Vasavada, T. J. Milavec, D. A. Paige.
Bull. Am. Astron. Soc., Vol. 23, No. 3, p. 1173 (1991). Abstract. – See Abstr. 012.037 for the main entry.

097.118   **Crater degradation in Arabia and Maia Valles, Mars.**
N. G. Barlow.
Bull. Am. Astron. Soc., Vol. 23, No. 3, p. 1173 – 1174 (1991). Abstract. – See Abstr. 012.037 for the main entry.

097.119   **Electrochemistry of the Martian soil.**
A. P. Zent, C. P. McKay, D. Bass.
Bull. Am. Astron. Soc., Vol. 23, No. 3, p. 1174 (1991). Abstract. – See Abstr. 012.037 for the main entry.

097.120   **Sand on Mars: composition from Viking IRTM thermal emission measurements.**
K. S. Edgett, P. R. Christensen.
Bull. Am. Astron. Soc., Vol. 23, No. 3, p. 1174 (1991). Abstract. – See Abstr. 012.037 for the main entry.

097.121   **The southern hemisphere of Mars in 1977: temporal observations of the dust storm initiation and surface interactions.**
A. S. McEwen, J. D. Swann, A. P. Ingersoll, E. DeJong.
Bull. Am. Astron. Soc., Vol. 23, No. 3, p. 1174 (1991). Abstract. – See Abstr. 012.037 for the main entry.

097.122   **Evidence for crystalline hematite as an accessory phase in Martian soils.**
R. B. Singer, J. S. Miller.
Bull. Am. Astron. Soc., Vol. 23, No. 3, p. 1174 – 1175 (1991). Abstract. – See Abstr. 012.037 for the main entry.

097.123   **Scattering behavior of the Valles Marineris dark sands.**
P. E. Geissler, R. B. Singer.
Bull. Am. Astron. Soc., Vol. 23, No. 3, p. 1175 (1991). Abstract. – See Abstr. 012.037 for the main entry.

097.124   **Mars: absolute calibration of 1988 visible and near–IR spectral images.**
J. S. Miller, R. B. Singer, W. K. Wells, L. Weller.
Bull. Am. Astron. Soc., Vol. 23, No. 3, p. 1175 (1991). Abstract. – See Abstr. 012.037 for the main entry.

097.125   **Mars surface and atmospheric compositional variability from groundbased imaging spectroscopy during 1988 and 1990.**
J. F. Bell III, P. G. Lucey, T. B. McCord, D. Crisp.
Bull. Am. Astron. Soc., Vol. 23, No. 3, p. 1175 (1991). Abstract. – See Abstr. 012.037 for the main entry.

097.126   **Martian spectroscopy between 4.4 and 5.1 μm: preliminary surface and atmospheric modeling results.**
D. L. Blaney, D. Crisp.
Bull. Am. Astron. Soc., Vol. 23, No. 3, p. 1175 – 1176 (1991). Abstract. – See Abstr. 012.037 for the main entry.

097.127   **Color classification of the Valles Marineris, Mars.**
E. Hauber, G. Neukum.
Bull. Am. Astron. Soc., Vol. 23, No. 3, p. 1176 (1991). Abstract. – See Abstr. 012.037 for the main entry.

097.128   **Multispectral study of Cerberus dark materials.**
J. N. Head, R. B. Singer, P. E. Geissler.
Bull. Am. Astron. Soc., Vol. 23, No. 3, p. 1176 (1991). Abstract. – See Abstr. 012.037 for the main entry.

097.129   **Geological applications of high resolution ground–based radar imaging of Mars.**
R. F. Jurgens, R. E. Arvidson, J. J. Plaut.
Bull. Am. Astron. Soc., Vol. 23, No. 3, p. 1177 (1991). Abstract. – See Abstr. 012.037 for the main entry.

097.130   **Exploring compositional variations on Mars: mixture modeling from a telescopic spectral image.**
E. Merényi, J. S. Miller, R. B. Singer.
Bull. Am. Astron. Soc., Vol. 23, No. 3, p. 1177 (1991). Abstract. – See Abstr. 012.037 for the main entry.

097.131   **Mars radar echoes at 3.5–cm.**
T. W. Thompson, R. F. Jurgens, M. A. Slade, T. C. O'Brien, H. J. Moore.
Bull. Am. Astron. Soc., Vol. 23, No. 3, p. 1177 – 1178 (1991). Abstract. – See Abstr. 012.037 for the main entry.

097.132   **Martian parent craters for the SNC meteorites.**
P. J. Mouginis–Mark, T. J. McCoy, G. J. Taylor, K. Keil.
J. Geophys. Res., Vol. 97, No. E6, p. 10213 – 10225 (25 Jun 1992).
The young ages (~ 1.3 Ga) and the basaltic to ultramafic compositions of the shergottites, nakhlites, and chassignites meteorites severely restrict their potential source regions on Mars. The authors have used this age and compositional information, together with geologic data derived from Viking Orbiter images, to identify 25 candidate impact craters in the Tharsis region of Mars that could be the source crater for these meteorites. None of these craters are close to the size (~ 100 km diameter) implied by the dynamical study of SNC ejection developed by Vickery and Melosh (1987). The craters in the authors' study were selected because they are > 10 km in diameter, have morphologies indicative of young craters, and satisfy both the petrologic criteria of the SNCs and the proposed 1.3 Ga crystallization ages. Of these 25 craters, only nine are found on geologic units believed to be young. No crater exists to satisfy well the criteria of sampling both a 1.3 Ga surface (nakhlites and Chassigny) and a 180 Ma surface (shergottites) without at the same time imposing significant constraints on the chronology of Mars as inferred from the cumulative crater curves. The relatively young age (based on their inferred position in the stratigraphic column of Tharsis (Scott et al., 1981) of the

SNCs implies that volcanic activity on the plains of the Tharsis region extended well past 1.3 Ga.

097.133 Phase transitions in the Mars mantle.
E. Severova.
Earth, Moon, Planets, Vol. 56, No. 1, p. 83 – 91 (Jan 1992). Letter–to–the–editor.

A number of the models of Mars was constructed on the basis of different mass concentration of chemical elements and equation of state, proposed by different authors. It was shown that in the majority of cases the pressure, necessary for the second phase transition, is reached in the mantle.

097.134 Martian lake basins and lacustrine plains.
R. A. De Hon.
Earth, Moon, Planets, Vol. 56, No. 2, p. 95 – 122 (Feb 1992).

Outflow channels and valley systems are evidence of water flow on the surface of Mars. Whenever there is a consequent flow of water on an irregular surface, temporary impoundment in surface depressions will form lakes. A classification of Martian lake basins based on the location of the basin in respect to water sources is proposed. The classes are Type 1: Valley–head basins, Type 2: Intravalley basins, Type 3: Valley–terminal basins, and Type 4: Isolated basins. Martian lakes are ephemeral features. Many craters and irregular depressions impounded water only until the basins filled and overflowed. Water escaping by spillover rapidly cut crevasses in the downstream side of basins and drained the ponds. Clastic lacustrine sediments collected in the lakes as flowing water lost velocity and turbulences. Evaporitic deposits may be significant in those basins that were not rapidly drained. Sediments deposited in lake basins from smooth, featureless plains. Lacustrine plains are potentially candidate sites for Mars landings and for the search for evidence of ancient life.

097.135 Infrared imaging of Mars between 2.4 and 5.1 μm.
D. L. Blaney.
Bull. Am. Astron. Soc., Vol. 23, No. 3, p. 1183 (1991). Abstract. – See Abstr. 012.037 for the main entry.

097.136 Thermal infrared spectra (5.5 – 9.2 μm) of Mars obtained from the Kuiper Airborne Observatory during the 1990 opposition.
T. L. Roush, F. Witteborn, J. Bregman, J. B. Pollack, D. Rank, A. Graps.
Bull. Am. Astron. Soc., Vol. 23, No. 3, p. 1183 (1991). Abstract. – See Abstr. 012.037 for the main entry.

097.137 Analysis of Mariner 7 thermal infrared spectra and comparison to recent airborne observations.
T. L. Roush, J. B. Pollack, T. Z. Martin.
Bull. Am. Astron. Soc., Vol. 23, No. 3, p. 1183 (1991). Abstract. – See Abstr. 012.037 for the main entry.

097.138 PHOBOS KRFM observations of Martian dust: evidence for inter–annual variability in dust loading.
L. W. Esposito, M. T. Mellon, M. Jones, J. R. Corn, V. I. Moroz, E. V. Petrova, L. V. Ksanfomaliti.
Bull. Am. Astron. Soc., Vol. 23, No. 3, p. 1184 (1991). Abstract. – See Abstr. 012.037 for the main entry.

097.139 Mapping the potato: a digital image map mosaic of Phobos.
D. P. Simonelli, B. T. Carcich, P. C. Thomas.
Bull. Am. Astron. Soc., Vol. 23, No. 3, p. 1184 (1991). Abstract. – See Abstr. 012.037 for the main entry.

097.140 Physical properties of regolith on Phobos.
L. Ksanfomaliti, V. Moroz, S. Murchie, D. Britt, N. Goroshkova, T. Duxbury, B. Zhukov, E. Kuehrt, B. Murray, G. Nikitin, E. Petrova, K. Pieters, A. Soufflot, P. Fisher, J. W. Head.
Cosmic Res., Vol. 29, No. 4, p. 533 – 550 (Jan 1992). English translation of Kosm. Issled., Tom 29, Vyp. 4, p. 621 – 640 (1991).

The present article presents the results of two experiments performed onboard the Phobos spacecraft, spectrophotometry of albedo in the range 300 – 600 nm and radiometry of intrinsic thermal emission in the range 6 – 50 μm. The thermophysical properties of fine–grained material (regolith) are similar to those of the moon. The reflectance properties of regolith on Phobos are largely heterogenous along the tracks that were studied and were most often associated with topographical features, mainly craters and the age of the craters. In every case the albedo, which is low in absolute value, grows from 350 to 600 nm. Optically modified Martian rocks are the closest analogs to this surface. Similarity to the reflectance spectra of carbonaceous chondrites, which had been expected, was not confirmed.

097.141 Methods and results of study of the aerosol component of the Martian atmosphere during periods of high transparency.
A. V. Morozhenko.
Kinematics Phys. Celest. Bodies, Vol. 7, No. 4, p. 1 – 17 (1991). English translation of Kinematika Fiz. Nebesn. Tel, Tom 7, No. 4, p. 3 – 22 (1991).

Existing methods of determination of the optical properties of the Martian atmosphere during periods of high atmospheric transparency, and the results obtained, are the subject of a critical analysis. It is found that evaluations of direct (Viking–1 and –2 spacecraft measurements directly at the surface of the planet) and remote methods do not agree. Analysis of optical–thickness estimates made at the landing sites of these vehicles pointed to the conclusion that the optical properties of the Martian atmosphere were probably subject to significant variations in time while the photometric experiments were being conducted, possibly as a result, for example of, fog dissipation (morning measurements) and condensation (evening measurements). It is shown that if this is not taken into account in workup of the observational data, the values obtained for the optical thickness will be high and in some cases even fictitious. It is concluded that the problem of the optical properties of the Martian atmosphere during periods of high transparency is still open, and that this must be taken into account in the design of new space experiments.

097.142 Photometric properties of Phobos regolith from data gathered on the Phobos mission.
L. V. Ksanfomaliti.
Sol. Syst. Res., Vol. 25, No. 6, p. 484 – 506 (May 1992). English translation of Astron. Vestn., Tom 25, No. 6, p. 650 – 676 (1991).

In the present article the reflectance properties of Phobos regolith are reviewed. The study is based on data gathered in 1989 by means of the KRFM spectrophotometer installed onboard the Phobos spacecraft. Improved spectrophotometry of Phobos in the 300 – 600 nm band, complemented with data on albedo in the shorter wavelengths of the ultraviolet and near infrared bands are presented. On the basis of the surface properties, Phobos would appear to possess an unambiguously nonhomogenous composition, which suggests a complex history of development. Judging from the spectrophotometric properties of Phobos regolith discovered in the experiment, its true reflectance properties, on the one hand, have little in common with data presented in previously published studies, while on the other hand, do not agree very well with the properties of carbonaceous chondrites or yield clearcut analogies to other meteorite materials. The significant inhomogeneity of the properties of regolith in the 315 – 600 nm band is associated with particular topographic features.

097.143 Limitations of spectral analysis of the Phobos magnetometer data in the search for an intrinsic Martian magnetic field.
C. T. Russell, J. G. Luhmann, K. Schwingenschuh.
Planet. Space Sci., Vol. 40, No. 5, p. 707 – 710 (May 1992).

The authors analyze both observed and synthetic time series of the magnetic field obtained in circular orbit around Mars by the Phobos spacecraft. Of the three reported spectral peaks at 8, 12 and 24 h only the 24 h peak could be due to intrinsic sources. However, the authors can also produce 24 h spectral peaks in synthetic time series with no intrinsic field effects included. Hence, they conclude that present spectral analyses of time series obtained with the Phobos magnetometers provide no constraints on the size or the existence of an intrinsic magnetic field at Mars.

097.144   Intensity and position of the Martian magnetic dipole, calculated from the observations of the satellite Phobos 2.
A. Grafe.
Planet. Space Sci., Vol. 40, No. 5, p. 719 – 730 (May 1992).
By comparing a model of the magnetosphere of Mars and Phobos 2 observations of the magnetic field, an attempt is made to determine whether the planet has an intrinsic magnetic field. The 3–D geomagnetosphere model of Voigt is used. This model entails a given magnetopause geometry which is represented on the day side by a hemisphere and on the night side by a semi-finite cylinder with constant radius. Voigt's model is adapted to Martian conditions. For comparison with Phobos 2 observations near Mars in the evening sector data are available only from three elliptical orbits: orbit 1, orbit 2 and orbit 3. Unfortunately due to the uncontrolled spin of the satellite the orientation errors for orbits 1 and 2 are too great. Therefore the comparison with the observation can only be done with the data from orbit 3 on 8 February 1989. Assuming that on this day the magnetopause crossing was about 05:48 U.T. (R = 4.554×10^6m) the best agreement between the model and observation is provided by a dipole having a magnetic moment of about 0.7×10^{12}T m^3 and a tilt to the equator plane of nearly $20°$ where its North Pole is directed northwards. This leads to the conclusion that the intrinsic magnetic field of Mars is weak.

097.145   Gamma–radiation of Mars as an indicator of Martian rock element composition (based on "Phobos–2" data).
Yu. A. Surkov, L. P. Moskaleva, V. P. Kharyukova,
O. S. Manvelyan, S. E. Zajtseva, G. G. Smirnov.
Kosm. Issled., Tom 30, Vyp. 2, p. 262 – 274 (Mar–Apr 1992). In Russian. English translation in Cosmic Research, Vol. 30, No. 2.
The element composition of Martian rocks was investigated by means of the scintillation gamma–spectrometer on the spacecraft "Phobos–2". The characteristic gamma–radiation of the rocks contains information about their element composition. The data analysis includes the following: the background correction, the decomposition of the spectra into the continuous and characteristic components, the calculation of the element concentration by means of the monoelement response functions. At present the element composition data are obtained for some equatorial areas of the planet, in particular, for the Lunar plateau and Tharsis.

097.146   Ancient oceans on Mars.
V. R. Baker.
Bull. Am. Astron. Soc., Vol. 23, No. 3, p. 1206 (1991). Abstract. – See Abstr. 012.037 for the main entry.

097.147   The ancient ocean hypothesis: a critique.
M. H. Carr.
Bull. Am. Astron. Soc., Vol. 23, No. 3, p. 1206 (1991). Abstract. – See Abstr. 012.037 for the main entry.

097.148   Martian xenology. III. The untold story.
K. Zahnle.
Bull. Am. Astron. Soc., Vol. 23, No. 3, p. 1211 (1991). Abstract. – See Abstr. 012.037 for the main entry.

097.149   Gas–phase organic chemistry of early Mars: laboratory yields and implications.
M. N. Heinrich, W. R. Thompson, C. Sagan.
Bull. Am. Astron. Soc., Vol. 23, No. 3, p. 1211 (1991). Abstract. – See Abstr. 012.037 for the main entry.

097.150   A 2–D annually–averaged study of potential early Martian atmospheres.
R. B. Schmunk, J. W. Chamberlain.
Bull. Am. Astron. Soc., Vol. 23, No. 3, p. 1212 (1991). Abstract. – See Abstr. 012.037 for the main entry.

097.151   Photochemistry of the Martian atmosphere (mean conditions).
V. A. Krasnopolsky (*V. A. Krasnopol'skij*).
Bull. Am. Astron. Soc., Vol. 23, No. 3, p. 1212 (1991). Abstract. – See Abstr. 012.037 for the main entry.

097.152   Solar zenith angle dependence of the Martian hot oxygen exosphere.
W. B. Colwell, A. I. F. Stewart.
Bull. Am. Astron. Soc., Vol. 23, No. 3, p. 1212 (1991). Abstract. – See Abstr. 012.037 for the main entry.

097.153   Regulation of CO and O$_2$ abundances by the escape of oxygen and hydrogen in the atmosphere of Mars.
H. Nair, M. Allen, Y. L. Yung.
Bull. Am. Astron. Soc., Vol. 23, No. 3, p. 1213 (1991). Abstract. – See Abstr. 012.037 for the main entry.

097.154   Vertical, latitudinal, and diurnal distributions of water in the Mars atmosphere from VLA 1.35 cm spectral line mapping.
R. T. Clancy, A. W. Grossman, D. O. Muhleman.
Bull. Am. Astron. Soc., Vol. 23, No. 3, p. 1213 (1991). Abstract. – See Abstr. 012.037 for the main entry.

097.155   Modelling Mars' water vapor and ozone budget during 1988 – 89.
B. Rizk, D. M. Hunten, R. M. Haberle, J. B. Pollack,
F. Espenak, M. J. Mumma.
Bull. Am. Astron. Soc., Vol. 23, No. 3, p. 1213 (1991). Abstract. – See Abstr. 012.037 for the main entry.

097.156   Modeling the seasonal CO$_2$ cycle on Mars: a fit to the Viking lander pressure curves.
S. E. Wood, D. A. Paige.
Bull. Am. Astron. Soc., Vol. 23, No. 3, p. 1213 (1991). Abstract. – See Abstr. 012.037 for the main entry.

097.157   Simulations of the seasonal pressure changes on Mars with a general circulation model.
J. B. Pollack, R. M. Haberle, J. Schaeffer, H. Lee.
Bull. Am. Astron. Soc., Vol. 23, No. 3, p. 1214 (1991). Abstract. – See Abstr. 012.037 for the main entry.

097.158   Search for spatial variations in the CO abundance on Mars at 2.3 micron.
F. Billebaud, J. Rosenqvist, E. Lellouch, T. Encrenaz,
J.–P. Maillard.
Bull. Am. Astron. Soc., Vol. 23, No. 3, p. 1214 (1991). Abstract. – See Abstr. 012.037 for the main entry.

097.159   Vertical distribution and granulometry of Martian dust particles from the Phobos/ISM and Auguste experiments.
E. Chassefière, J. Blamont, P. Drossart, J. Rosenqvist,
M. Combes, S. Erard, Y. Langevin, J.–P. Bibring.
Bull. Am. Astron. Soc., Vol. 23, No. 3, p. 1214 (1991). Abstract. – See Abstr. 012.037 for the main entry.

097.160   Submillimeter observations of the (3–2) ^{12}CO and ^{13}CO transitions in the atmosphere of Mars.
T. Encrenaz, E. Lellouch, S. Gulkis, I. de Pater.
Bull. Am. Astron. Soc., Vol. 23, No. 3, p. 1214 (1991). Abstract. – See Abstr. 012.037 for the main entry.

097.161   The upper limit to the Martian magnetic moment.
C. T. Russell, J. G. Luhmann, T. L. Zhang,
K. Schwingenschuh, W. Riedler, Ye. Yeroshenko
(*E. G. Eroshenko*).
Bull. Am. Astron. Soc., Vol. 23, No. 3, p. 1214 – 1215 (1991). Abstract. – See Abstr. 012.037 for the main entry.

097.162   Martian slope winds.
H. Savijärvi, T. Siili.
Bull. Am. Astron. Soc., Vol. 23, No. 3, p. 1215 (1991). Abstract. –
See Abstr. 012.037 for the main entry.

097.163   The ancient oxygen exosphere of Mars: implications for atmosphere evolution.
M. H. G. Zhang, J. G. Luhmann, A. F. Nagy, S. W. Bougher.
Bull. Am. Astron. Soc., Vol. 23, No. 3, p. 1215 (1991). Abstract. –
See Abstr. 012.037 for the main entry.

097.164   The LMD Martian general circulation model: results about the annual pressure cycle.
O. Talagrand, F. Hourdin, F. Forget.
Bull. Am. Astron. Soc., Vol. 23, No. 3, p. 1217 (1991). Abstract. –
See Abstr. 012.037 for the main entry.

097.165   Martian dust storms in 1990.
L. J. Martin, J. D. Beish, D. C. Parker.
Bull. Am. Astron. Soc., Vol. 23, No. 3, p. 1217 (1991). Abstract. –
See Abstr. 012.037 for the main entry.

097.166   Mars great dust storms as a chaotic phenomenon.
A. P. Ingersoll, J. R. Lyons.
Bull. Am. Astron. Soc., Vol. 23, No. 3, p. 1217 – 1218 (1991).
Abstract. – See Abstr. 012.037 for the main entry.

097.167   Monitoring Martian dust storms by speckle imaging – preliminary results.
J. W. Beletic, R. W. Zurek, R. M. Goody.
Bull. Am. Astron. Soc., Vol. 23, No. 3, p. 1218 (1991). Abstract. –
See Abstr. 012.037 for the main entry.

097.168   Martian atmospheric temperatures, dust amounts, and net radiative heating rates from Mariner 9 IRIS data.
M. L. Santee, D. Crisp.
Bull. Am. Astron. Soc., Vol. 23, No. 3, p. 1218 (1991). Abstract. –
See Abstr. 012.037 for the main entry.

097.169   First absolute wind measurements in the middle atmosphere of Mars.
J. Rosenqvist, E. Lellouch, J. J. Goldstein, S. W. Bougher,
G. Paubert.
Bull. Am. Astron. Soc., Vol. 23, No. 3, p. 1218 (1991). Abstract.
The full paper is published in Astrophys. J., Vol. 383, No. 1,
p. 401 – 406 (10 Dec 1991). – See Abstr. 012.037 for the main
entry.

097.170   Martian northern middle latitude atmospheric response to a simulated evolving Martian global dust storm.
J. R. Murphy, C. B. Leovy.
Bull. Am. Astron. Soc., Vol. 23, No. 3, p. 1218 – 1219 (1991).
Abstract. – See Abstr. 012.037 for the main entry.

097.171   Possible resonance of forced stationary waves in the Martian atmosphere.
J. R. Barnes, J. L. Hollingsworth.
Bull. Am. Astron. Soc., Vol. 23, No. 3, p. 1219 (1991). Abstract. –
See Abstr. 012.037 for the main entry.

097.172   Nonlinear stratified flow over localized topographic obstacles on Mars.
J. A. Magalhães, R. E. Young.
Bull. Am. Astron. Soc., Vol. 23, No. 3, p. 1219 (1991). Abstract. –
See Abstr. 012.037 for the main entry.

097.173   Interpretation of the KRFM–infrared measurements of Phobos.
E. Kührt, B. Giese, H. U. Keller, L. V. Ksanfomality
(*L. V. Ksanfomaliti*).
Icarus, Vol. 96, No. 2, p. 213 – 218 (Apr 1992).
The multichannel radiometer KRFM on board the Phobos–2
spacecraft measured the thermal flux from Phobos' surface along
two different ground tracks on Mar 25, 1989. The results for the

first track are interpreted within a thermal model that takes into
account the ellipsoidal shape and the surface roughness of
Phobos. Information about the thermal properties and the grain
size of the regolith is deduced by comparing measured with
calculated thermal fluxes.

097.174   Reflectance spectroscopy of palagonite and iron–rich montmorillonite clay mixtures: implications for the surface composition of Mars.
J. Orenberg, J. Handy.
Icarus, Vol. 96, No. 2, p. 219 – 225 (Apr 1992).
Mixtures of a Hawaiian palagonite and an iron–rich, montmorillonite clay were evaluated as Mars surface spectral analogs
from their diffuse reflectance spectra. The presence of the 2.2-μm
absorption band in the reflectance spectrum of clays and its
absence in the Mars spectrum have been interpreted as indicating
that highly crystalline aluminous hydroxylated clays cannot be a
major mineral component of the soil on Mars.

097.175   Valley systems on Tyrrhena Patera, Mars: earth–based radar measurements of slopes.
S. H. Zisk, P. J. Mouginis–Mark, J. M. Goldspiel, M. A. Slade,
R. F. Jurgens.
Icarus, Vol. 96, No. 2, p. 226 – 233 (Apr 1992).
Eight new topographic profiles across the Martian volcano
Tyrrhena Patera have been obtained from radar data collected by
the JPL Goldstone Radar System in 1988. These profiles, which
have a reproducible accuracy of better than 150 m, show the
volcano to rise ~ 1.5 km above the plains of Hesperia Planum to
the east, and to have an average height–to–diameter ratio of
~ 1:340. The slopes on the northern flanks of Tyrrhena Patera
bear little correlation with the width or depth of valley systems
found in that area. This suggests that erosion by gravity–driven
flows was not responsible for valley formation and that other
factors, such as spatial or temporal variations in the volume of
ground water released by sapping, or strength differences in the
materials comprising the surface units of the volcano, controlled
the geometry and locations of the valleys.

097.176   Analysis of Phobos mission gamma ray spectra from Mars.
J. I. Trombka, L. G. Evans, R. Starr, S. R. Floyd,
S. W. Squyres, J. T. Whelan, G. J. Bamford, R. L. Coldwell,
A. C. Rester, Yu. A. Surkov, L. P. Moskaleva,
V. P. Kharyukova, O. S. Manvelyan, S. Ye. Zaitseva
(*S. E. Zajtseva*), G. G. Smirnov.
22. Lunar and Planetary Science Conference, p. 23 – 29 (1992). –
See Abstr. 012.046 for the main entry.
The determination of the elemental composition of the surface
of a planetary body can be achieved, in many cases, by remote-
sensing gamma ray spectroscopy. A gamma ray spectrometer
was carried on the Soviet spacecraft Phobos–2, and obtained
data while in an elliptical orbit around Mars. The authors report
on the results of two independent approaches to data analysis,
one by the Soviet group and one by an American group. The
results for five elements are given for two different orbits of Mars.
Major geologic units that contribute to the signal for each orbit
have been identified. The results from the two techniques are in
general agreement and there appear to be no geologically
significant differences between the results for each orbit.

097.177   The Tharsis Montes, Mars: comparison of volcanic and modified landforms.
J. R. Zimbelman, K. S. Edgett.
22. Lunar and Planetary Science Conference, p. 31 – 44 (1992). –
See Abstr. 012.046 for the main entry.
The three Tharsis Montes shield volcanos, Arsia Mons,
Pavonis Mons, and Ascraeus Mons, have broad similarities that
have been recognized since the Mariner 9 reconnaissance in 1972.
Upon closer examination the volcanos are seen to have significant differences that are due to individual volcanic histories. All
three volcanos exhibit the following characteristics: gentle (< 5°)
flank slopes, entrants in the northwestern and southeastern
flanks that were the source for lavas extending away from each

shield, summit caldera(s), and enigmatic lobe–shaped features extending over the plains to the west of each volcano. The three volcanos display different degrees of circumferential graben and trough development in the summit regions, complexity of preserved caldera collapse events, secondary summit–region volcanic construction, and erosion on the lower western flanks due to mass wasting and the processes that formed the large lobe–shaped features. All three lobe–shaped features start at elevations of 10 to 11 km and terminate at 6 km. The complex morphology of the lobe deposits appear to involve some form of catastrophic mass movement followed by effusive and perhaps pyroclastic volcanism.

097.178   Polygenetic origin of Hrad Vallis region of Mars.
R. A. De Hon.
22. Lunar and Planetary Science Conference, p. 45 – 51 (1992). – See Abstr. 012.046 for the main entry.

Hrad Vallis is located in the transition zone between Elysium Mons and Utopia Planitia. Near its origin, at the northern edge of Elysium lavas, Hrad Vallis is characterized by a low–sinuosity channel within a north–northwest–trending, broad, flat–floored valley. A nearby flat–floored valley is parallel to the Hrad trend and parallel to elongate depressions, fissures, and faults in the region. An apparent hierarchy of landforms provides insight into the origin of the features associated with Hrad Vallis. An extended period of time is indicated during which freely circulating water existed on and beneath the surface of Mars. Karst and thermokarst processes imply very different climatic regimes and different host materials.

097.179   New evidence of lacustrine basins on Mars: Amazonis and Utopia Planitiae.
D. H. Scott, M. G. Chapman, J. W. Rice Jr., J. M. Dohm.
22. Lunar and Planetary Science Conference, p. 53 – 62 (1992). – See Abstr. 012.046 for the main entry.

Amazonis and Utopia Planitiae are two large basins on Mars that have morphologic features commonly associated with former standing bodies of water. The basins exhibit terraces and lineations resembling shorelines, etched and infilled floors marked by sinuous channels in places, inflow channels along their borders, and other geomorphic indicators believed to be related to the presence of water and ice.

097.180   Flood surge through the Lunae Planum outflow complex, Mars.
R. A. De Hon, E. A. Pani.
22. Lunar and Planetary Science Conference, p. 63 – 71 (1992). – See Abstr. 012.046 for the main entry.

The object of this study is to trace a single flood surge through the Lunae Planum outflow complex and to evaluate the effects of multiple ponding and multiple channel routing on the discharge onto Chryse Planitia. In addition, it examines the duration of flow from a large ponded source. The Lunae Planum outflow complex provides a convenient model for such a routing study because it has one primary source and a well–defined channel system. A study of the drainage system across rugged terrain provides new insights into the erosional and depositional regimes associated with the outflow event and provides realistic limits to the time span during which water was actively present in the drainage system.

097.181   Kasei Valles, Mars: interpretation of canyon materials and flood sources.
K. L. Tanaka, M. G. Chapman.
22. Lunar and Planetary Science Conference, p. 73 – 83 (1992). – See Abstr. 012.046 for the main entry.

What sorts of materials make up the canyon walls and floors? What were the sources of the flood waters? The authors propose answers for these questions by synthesizing previous work, available data, and new observations and interpretations based on their mapping of the north–central Kasei Valles.

097.182   Vertical structure and size distributions of Martian aerosols from solar occultation measurements.
E. Chassefière, J. E. Blamont, V. A. Krasnopolsky (V. A. Krasnopol'skij), O. I. Korablev, S. K. Atreya, R. A. West.
Icarus, Vol. 97, No. 1, p. 46 – 69 (May 1992).

Solar occultations performed with a spectrometer on board the Soviet spacecraft Phobos 2 provided data on the vertical structure of the Martian aerosols in the equatorial region near the northern spring equinox. Five clouds were detected above 45 km altitude and their vertical structure recorded at six wavelengths between 0.28 and 3.7 μm. Since the clouds seen from Phobos 2 are observed at twilight, which coincides with the diurnal maximum of the ambient temperature, they can be assumed to be in a steady state. If their thermodynamic state were to vary quickly during the day, the optical thickness at twilight would correspond to unrealistic values in earlier hours when the temperature is lower. Clouds are well fitted by theoretical profiles obtained assuming the steady state.

097.183   The composition, structure, and gravitational field of Mars.
V. N. Zharkov, E. M. Koshlyakov, K. I. Marchenkov.
Sol. Syst. Res., Vol. 25, No. 5, p. 387 – 411 (Mar 1992). English translation of Astron. Vestn., Tom 25, No 5, p. 515 – 547 (1991).

Questions connected with the problem of the internal structure of Mars, including its chemical composition and petrological model are considered in this review. The model of Mars is divided into three main regions: the crust, mantle, and core. The thickness of the crust may reach ~ 150 km to 200 km. The mantle adjoins the core at a depth $\sim 1600 \pm 200$ km, and is divided into an olivine zone, a phase transition zone, and a spinel zone. Sulfur and possibly hydrogen are the main impurities in the core. The core of Mars may be in a liquid state. A correlation of temperature with effective viscosity leads to a "low" temperature distribution in the interior of Mars. The lithosphere thickness is $\sim 500 \pm 200$ km. The question of the interpretation of the gravitational field by means of a Green's function technique is considered in detail. For this purpose, an expansion of the field and topography in spherical harmonics to degree and order 18 is used. The stress level in the Martian lithosphere is high and amounts to several hundred bars. The estimated value of the stresses in the mantle is ~ 30 bars.

097.184   Secular variations of the Martian magnetic field?
Sh. Sh. Dolginov.
Kosm. Issled., Tom 30, Vyp. 3, p. 425 – 428 (May–Jun 1992). In Russian. English translation in Cosmic Res., Vol. 30, No. 3.

097.185   Experiment to construct a photometric map of the normal albedo of the surface of Mars.
V. G. Tejfel, N. V. Sinyaeva, A. N. Aksenov, G. A. Kharitonova.
Sov. Astron. Lett., Vol. 18, No. 2, p. 110 – 113 (Mar 1992). Current Physics Microform No.: 9211X0749. English translation of Pis'ma Astron. Zh., Tom 18, No. 3, p. 271 – 278 (1992).

The authors describe the method and the results of photometric and computer processing of photographs of Mars obtained near the opposition of 1990, which have been used to construct a map of the normal albedo in the red part of the spectrum. They have measured the intensity distribution and determined the limb–darkening coefficients, both global and local. Based on these measurements they have reduced the photographs of Mars for limb darkening, and constructed parts of an albedo map and a general map in isophote, half–tone and three–dimensional representation.

097.186   Possible discrete annular dust formations around Mars and Venus.
Yu. K. Gulak, I. A. Dychko.
Kinematics Phys. Celest. Bodies, Vol. 7, No. 5, p. 22 – 23 (1991). English translation of Kinematika Fiz. Nebesn. Tel., Tom 7, No. 5, p. 27 – 28 (Sep–Oct 1991).

The positions of possible dust belts around Mars and Venus are calculated on the basis of a statistical theory of quasi–isoenergetic complexes in the form of radial standing waves.

097.187   Interaction of the solar wind with Mars. The Phobos 2 results.
T. K. Breus, A. M. Krymskij, Eh. M. Dubinin,
E. G. Eroshenko, V. Ya. Mitnitskij, N. F. Pisarenko,
V. A. Styazhkin, S. V. Barabash.
Cosmic Res., Vol. 29, No. 5, p. 635 – 645 (Mar 1992). English translation of Kosm. Issled., Tom 29, Vyp. 5, p. 741 – 753 (Sep–Oct 1991).

The properties of the boundary separating the region dominated by the solar wind and that of the plasma of planetary origin, the planetopause that has been detected in experiments on the Phobos 2 spacecraft, have been investigated. It is shown that the properties of this boundary are similar to those of analogous chemical boundaries; the cometopause near comet Halley and the chemical boundary which exists at Venus at times of high solar wind dynamic pressure. Numerical simulation of the flow past Mars shows that, at a maximum of solar activity, the interaction of the dense neutral atmosphere of Mars with the solar wind can lead to charge exchange between the solar wind protons and the neutral atmosphere, and to a significant decrease of the concentration of solar wind protons in the transition region near the planetopause.

097.188   Magnetic field and magnetosphere of Mars.
Sh. Sh. Dolginov.
Cosmic Res., Vol. 29, No. 5, p. 646 – 677 (Mar 1992). English translation of Kosm. Issled., Tom 29, Vyp. 5, p. 754 – 789 (Sep–Oct 1991).

097.189   Observation infrarouge du CO dans l'atmosphère de Mars.
F. Billebaud, E. Lellouch, T. Encrenaz, J. Crovisier,
J. P. Maillard.
J. Astron. Fr., No. 41, p. 12 (Feb 1992). – See Abstr. 012.070 for the main entry.

097.190   L'atmosphere martienne vue par ISM/Phobos.
J. Rosenqvist.
J. Astron. Fr., No. 41, p. 18 (Feb 1992). – See Abstr. 012.070 for the main entry.

097.191   Observations of Mars in 1990–91.
R. W. Schmude Jr.
J. R. Astron. Soc. Can., Vol. 86, No. 3, p. 117 – 129 (Jun 1992).
An intensive visual and photometric study of Mars was carried out during the 1990–91 opposition and the results are presented.

Mars. Unser geheimnisvoller Nachbar. Vom antiken Mythos zur bemannten Mission.
See Abstr. 003.024.

Scientific rationale and requirements for a global seismic network on Mars. Report of a workshop.
See Abstr. 012.027.

Mars surface and atmosphere through time (MSATT). Abstracts of presented papers.
See Abstr. 012.031.

Mars – erkundet und interpretiert.
See Abstr. 014.067.

Mössbauer spectroscopy on the surface of Mars. Why?
See Abstr. 022.008.

Adsorption of CO on oxide and water ice surfaces: implications for the Martian atmosphere.
See Abstr. 022.032.

Experimental simulation of Martian neutron leakage spectra.
See Abstr. 022.076.

Simulations of surface winds at the Viking Lander sites using a one–level model.
See Abstr. 022.102.

Simulations of surface winds at the Viking Lander sites.
See Abstr. 022.103.

Martian surface simulations.
See Abstr. 022.104.

The spectral absorption of CO_2 ice in the ultraviolet, visible, and near–infrared.
See Abstr. 022.136.

Reflectivity (visible and near IR), Mössbauer, static magnetic, and X ray diffraction properties of aluminum–substituted hematites.
See Abstr. 022.166.

Modelling particle size effects on the emissivity spectra of minerals in the thermal infrared.
See Abstr. 022.170.

The adsorption of HO_x on surfaces: implications for the stability of CO_2 in the atmosphere of Mars.
See Abstr. 022.181.

Shock modification and chemistry and planetary geologic processes.
See Abstr. 022.230.

Possible magnetic experiments on the surface of Mars.
See Abstr. 035.026.

Science applications of the Mars Observer gamma ray spectrometer.
See Abstr. 035.057.

Mars Observer camera.
See Abstr. 035.058.

Thermal emission spectrometer experiment: Mars Observer mission.
See Abstr. 035.059.

Atmosphere and climate studies of Mars using the Mars Observer pressure modulator infrared radiometer.
See Abstr. 035.060.

The Mars Observer laser altimeter investigation.
See Abstr. 035.061.

Mars Observer magnetic fields investigation.
See Abstr. 035.062.

Mössbauer backscattering spectrometer for mineralogical analysis of the Mars surface.
See Abstr. 035.064.

Measurement of the magnetic field vector from a rotating spacecraft.
See Abstr. 035.072.

Scintillation gamma–ray spectrometer for determining the element composition of the rocks on Mars from the Phobos spacecraft.
See Abstr. 035.112.

Mars 96 subsurface radar.
See Abstr. 035.129.

Definition of a L–band SAR for a Mars Rover mission.
See Abstr. 035.130.

Observations of solar–system bodies with the Belgrade Meridian Circle.
See Abstr. 041.006.

Positions of major planets from observations in the years 1988 – 1989.
See Abstr. 041.015.

Mars Observer mission.
See Abstr. 051.022.

Radio science investigations with Mars Observer.
See Abstr. 051.023.

Marsnet surface and atmosphere investigations.
See Abstr. 051.025.

Mars environmental survey (MESUR): science objectives and mission description.
See Abstr. 051.026.

Discovery concepts for Mars.
See Abstr. 051.027.

Planet–B: a Japanese Mars aeronomy observer.
See Abstr. 051.028.

Discovery–class mission concepts for Mars.
See Abstr. 051.029.

Three–component penetrator accelerometer for Mars exploration.
See Abstr. 051.055.

Navigation from relative measurements during approach to Mars.
See Abstr. 052.003.

On the initially circular motion of an orbiter in the oblate, rotating, Martian atmosphere.
See Abstr. 052.023.

Critical density layer as obstacle at solar wind – exospheric ion interaction.
See Abstr. 062.078.

Solar wind interaction with Mars and Venus during solar activity cycle.
See Abstr. 074.057.

Thermospheric odd nitrogen.
See Abstr. 082.029.

Elemental mapping of planetary surfaces using gamma–ray spectroscopy.
See Abstr. 091.024.

Love numbers of the Moon and of the terrestrial planets.
See Abstr. 091.039.

VLA/Goldstone Planetary Radar results.
See Abstr. 091.061.

Isolation of major Venus thermospheric cooling mechanism and implications for Earth and Mars.
See Abstr. 093.009.

Turbulent pick–up of new–born ions near Venus and Mars and problems of numerical modelling of the solar wind interaction with these planets. I. Features of the solar wind interaction with planets.
See Abstr. 093.014.

Turbulent pick–up of new–born ions near Venus and Mars and problems of numerical modelling of the solar wind interaction with these planets. II. Two–fluid HD model.
See Abstr. 093.015.

Dimers of carbon dioxide in the atmospheres of Venus and Mars.
See Abstr. 093.060.

Water in SNC meteorites: evidence for a martian hydrosphere.
See Abstr. 105.006.

A liquidus phase diagram for a primitive shergottite.
See Abstr. 105.086.

Extraterrestrial water of possible Martian origin in SNC meteorites: constraints from oxygen isotopes.
See Abstr. 105.090.

Microprobe studies of microtomed particles of "white druse" salts in shergottite EETA 79001.
See Abstr. 105.105.

Iddingsite in the Nakhla meteorite: TEM study of mineralogy and texture of pre–terrestrial (Martian?) alterations.
See Abstr. 105.167.

Chassigny and the nakhlites: carbon–bearing components and their relationship to martian environmental conditions.
See Abstr. 105.195.

Distribution of water on Mars: implications from SNC meteorites.
See Abstr. 105.208.

Hydrogen and carbon isotopic composition of volatiles in Nakhla: implications for weathering on Mars.
See Abstr. 105.220.

On the isotopic composition of magmatic carbon in SNC meteorites.
See Abstr. 105.221.

Fall days of the SNC meteorites: evidence for and SNC meteorid stream, and a common site of origin.
See Abstr. 105.230.

Ther early faint Sun 'paradox' revisited: massive greenhouse effects on early Earth and Mars?
See Abstr. 107.024.

098 Minor Planets

098.001 Orbital elements of numbered minor planets.
N. K. Sumzina, E. Bowell, E. Goffin, L. L. Filenko,
G. V. Williams.
Minor Planet Circ., Nos. 19347 – 20366 (Jan – Jun 1992).

The numbered minor planets are listed according to their
definitive number. Newly numbered objects are indicated by an
asterisk. The names of the orbit computers are given behind the
respective M.P.C. numbers:

(33), (60), (86) 19471 Sumzina; (91) 20313 Bowell; (142), (155),
(156) 19472 Sumzina; (160) 20125 Bowell; (177), (178), (180),
(182) 19472 Sumzina; (203), (209) 19983 Goffin; (210), (221),
(226), (229), (236) 19472, (248), (251), (255), (258) 19473
Sumzina; (264) 19983 Goffin; (266), (268) 19473 Sumzina; (284),
(285), (292), (293), (301), (306) 19473, (332) 19474 Filenko; (357),
(358), (368), (392), (395) 19474 Sumzina; (395) 19983 Goffin;
(396), (421), (445), (450) 19474 Sumzina; (454) 20125 Bowell;
(457) 19474, (463), (464), (468), (469), (508), (523), (528), (542),
(547), (553), (564), (576) 19475, (577), (602), (603) 19476
Sumzina; (609) 19983 Goffin; (613) 19476 Sumzina; (629), (630),
(641), (642), (643), (648), (660), (670) 19476, (676), (682), (684),
(686), (691), (694), (699), (706) 19477 Filenko; (741) 20125
Bowell; (744) 19983 Williams; (747), (764), (766), (768) 19477,
(773), (774), (776), (783), (786), (788), (791), (796), (802), (804),
(808), (816) 19478, (829), (832), (845), (854) 19479 Filenko; (855)
19983 Goffin; (859), (861), (862), (877), (881), (882) 19479
Filenko; (889) 20126 Bowell; (913), (916) 19479, (919), (925),
(938), (944) 19480 Sumzina; (951) 19983 Williams; (960), (967),
(983), (984), (985), (1002), (1011) 19480, (1013), (1014), (1019),
(1020), (1022), (1036), (1037) 19481 Sumzina; (1050), (1051),
(1058), (1059), (1062) 19481, (1065), (1070), (1077), (1082),
(1086), (1090), (1091), (1092), (1094), (1096), (1097), (1100)
19482, (1101), (1104), (1107), (1113), (1115), (1120), (1123),
(1126), (1129), (1132), (1133), (1135) 19483, (1137), (1139),
(1141), (1145) 19484, (1146), (1149) 19983, (1150), (1153), (1154),
(1162) 19984 Filenko; (1166) 19484, (1174), (1180), (1182),
(1184), (1185) 19984 Sumzina; (1190) 20126 Bowell; (1191),
(1193), (1194) 19984, (1196), (1199), (1203), (1204), (1207),
(1210), (1212), (1217), (1218), (1220), (1224), (1231) 19985
Sumzina; (1243), (1252), (1261), (1266), (1268) 19986 Filenko;
(1269) 19658 Williams; (1270), (1288), (1292), (1296), (1297),
(1301), (1305) 19986 Filenko; (1308), (1310) 19822 Williams;
(1311) 19987 Filenko; (1312) 19822 Williams; (1313), (1335)
19987 Filenko; (1335) 20126, (1375) 20313 Williams; (1399),
(1405), (1414), (1418), (1426), (1427) 19987 Sumzina; (1438)
20126 Bowell; (1439), (1440) 19987 Sumzina; (1453), (1495)
20126 Bowell; (1498) 19987 Sumzina; (1627) 19988 Goffin; (1653)
19658 Williams; (1709) 20126, (1863) 20313, (1873) 20126, (2070),
(2076) 20313, (2142) 20126 Bowell, (2210) 19988 Williams;
(2247), (2283) 20126 Bowell; (2327) 19822 Williams; (2346)
19484, (2355) 20126 Bowell; (2368) 19988 Goffin; (2404), (2450)
20127, (2455) 20313 Bowell; (2464) 19658 Williams; (2471) 20314,
(2502) 20127 Bowell; (2529), (2572) 19988 Williams; (2593) 20127
Bowell; (2596) 19988 Williams; (2620) 20127 Bowell; (2629)
19988 Goffin; (2788), (2826), (2846), (2851) 20127, (2866) 20314,
(2911) 20127 Bowell; (2960) 20127 Williams; (2997) 20127
Bowell; (3025), (3073) 19988 Williams; (3088), (3090) 20128,
(3099) 20314.
For a continuation of this list see Abstr. 098.002.

098.002 Orbital elements of numbered minor planets.
E. Bowell, G. V. Williams, E. Goffin, B. G. Marsden,
C. M. Bardwell, S. Nakano, D. W. E. Green, H. Kaneda,
T. Kobayashi.
Minor Planet Circ., Nos. 19347 – 20366 (Jan – Jun 1992).

This list is a continuation of Abstr. 098.001. The numbered
minor planets are listed according to their definitive number.
Newly numbered objects are indicated by an asterisk. The names
of the orbit computers are given behind the respective M.P.C.
numbers:

(3111) 20128 Bowell; (3119) 19988 Williams; (3131) 20128
Bowell; (3175), (3176) 20128 Williams; (3200) 19988 Goffin;

(3223) 20128, (3225) 20314, (3226) 20128 Williams; (3227), (3228)
20128, (3239) 20314, (3243) 20128 Bowell; (3245) 19988, (3266)
19484 Williams; (3268) 20314, (3279) 20128 Bowell; (3289) 19989,
(3292) 20129, (3294), (3307) 19989, (3332) 20129 Williams; (3347)
20129 Bowell; (3353) 19989 Williams; (3357) 20314 Bowell;
(3361), (3362) 19989 Goffin; (3368) 20314 Bowell; (3373) 20129
Williams; (3381) 20314 Bowell; (3393) 20129, (3408) 20314,
(3411) 19989 Williams; (3414) 20129, (3415) 20314 Bowell;
(3461), (3462), (3463), (3465) 20129, (3473), (3476), (3489), (3532)
19989 Williams; (3551) 20314, (3565) 20129 Bowell; (3577) 19484
Williams; (3597) 20130 Bowell; (3619), (3624), (3625) 20130,
(3629) 19989 Williams; (3703) 20130 Bowell; (3712) 19990, (3732)
20315, (3753) 19990 Williams; (3795), (3820), (3848) 20130
Bowell; (3913) 19990 Williams; (3963), (3984), (3989) 20130,
(4005) 19990, (4042) 20130, (4076) 20315 Bowell; (4144) 20131
Williams; (4189) 19990 Marsden; (4230), (4239), (4316) 20131
Bowell; (4341) 20315 Williams; (4432), (4507), (4514) 20131,
(4518), (4535), (4541), (4569), (4576), (4579) 20315 Bowell; (4599)
20315, (4658) 19990 Williams; (4778) 19822, (4781) 20131 Bowell;
(4846) 19822, (5013)* 19484–19485 Williams; (5014)* 19485
Marsden; (5015)* 19485 Bardwell; (5016)* 19485–19486 Williams; (5017)* 19486 Bardwell; (5018)* 19486–19487 Nakano;
(5019)* 19487 Green; (5020)* 19487 Marsden; (5021)* 19487–
19488, (5022)* 19488 Green; (5023)* 19488–19489, (5024)* 19489
Williams; (5025)* 19489 Bardwell; (5026)* 19489–19490 Nakano;
(5027)*, (5028)* 19490 Williams; (5029)* 19490–19491 Bardwell;
(5030)* 19491 Nakano; (5031)* 19491 Kaneda; (5032)* 19491–
19492 Williams; (5033)* 19492 Kaneda; (5034)* 19492 Williams;
(5035)*, (5036)* 19493 Kaneda; (5037)* 19493–19494 Williams;
(5038)* 19658, (5039)* 19658–19659, (5040)* 19659 Marsden;
(5041)* 19659–19660 Bardwell; (5042)* 19660 Kaneda; (5043)*
19660 Williams; (5044)*, (5045)* 19661 Bardwell; (5046)* 19661–
19662 Kaneda; (5047)* 19662 Williams; (5048)* 19662, (5049)*
19663 Marsden; (5050)* 19663 Kaneda; (5051)* 19663–19664
Williams; (5052)* 19664 Kaneda; (5053)* 19664 Marsden;
(5054)* 19664–19665 Williams; (5055)* 19665 Green; (5056)
19665 Williams; (5057)* 19665–19666, (5058)* 19666 Marsden;
(5059)* 19666–19667 Williams; (5060)* 19667 Kaneda; (5061)*
19667–19668 Williams; (5062)*, (5063)* 19668 Marsden; (5064)*
19668–19669, (5065)* 19669 Kaneda; (5066)* 19669 Williams;
(5067)* 19669–19670 Marsden; (5068)*, (5069)* 19670, (5070)*
19671 Kaneda; (5071)* 19671 Williams; (5072)* 19822 Nakano;
(5073)* 19823 Kobayashi; (5074)* 19823 Bardwell; (5075)*
19823, (5076)* 19823–19824 Kobayashi; (5077)*, (5078)* 19824
Nakano; (5079)* 19824–19825 Bardwell; (5080)* 19825 Williams;
(5081)* 19825–19826 Marsden; (5082)* 19826 Williams; (5083)*
19826 Nakano; (5084)* 19826–19827 Bardwell; (5085)* 19827
Nakano; (5086)* 19827 Kobayashi; (5087)* 19827–19828,
(5088)* 19828, (5089)*, (5090)* 19829, (5091)* 19829–19830
Nakano; (5092)* 19830 Williams; (5093)* 19830, (5094)* 19830–
19831 Nakano; (5095)*, (5096)* 19831 Williams; (5097)* 19832
Kobayashi; (5098)* 19832–19833 Williams; (5099)* 19833 Nakano.
For a continuation of this list see Abstr. 098.003.

098.003 Orbital elements of numbered minor planets.
G. V. Williams, C. M. Bardwell, S. Nakano,
B. G. Marsden, H. Kaneda, T. Kobayashi, D. W. E. Green,
K. Ichikawa, T. Urata.
Minor Planet Circ., Nos. 19347 – 20366 (Jan – Jun 1992).

This list is a continuation of Abstr. 098.002. The numbered
minor planets are listed according to their definitive number.
Newly numbered objects are indicated by an asterisk. The names
of the orbit computers are given behind the respective M.P.C.
numbers:

(5100)* 19833 Williams; (5101)* 19833–19834 Bardwell; (5102)*,
(5103)* 19834 Nakano; (5104)+ 19834–19835 Williams; (5105)*
19835 Marsden; (5106)* 19835 Nakano; (5107)* 19836 Bardwell;
(5108)* 19836, (5109)*, (5110)* 19837, (5111)* 19837–19838
Nakano; (5112)* 19838 Kaneda; (5113)* 19838 Nakano; (5114)*

19838–19839 Kobayashi; (5115)* 19839 Marsden; (5116)* 19839–19840, (5117)* 19840 Nakano; (5118)* 19840 Marsden; (5119)* 19840–19841, (5120)* 19841 Williams; (5121)* 19841, (5122)* 19841–19842, (5123)* 19842 Nakano; (5124)* 19842–19843 Williams; (5125)* 19843 Kaneda; (5126)* 19843 Marsden; (5127)*, (5128)* 19844 Nakano; (5129)* 19844–19845 Marsden; (5130)* 19845 Williams; (5131)* 19845–19846 Bardwell; (5132)*, (5133)* 19846 Williams; (5134)* 19846–19847 Nakano; (5135)* 19847 Kaneda; (5136)* 19847 Williams; (5137)* 19847–19848 Marsden; (5138)* 19848 Williams; (5139)* 19848–19849, (5140)* 19849 Kaneda; (5141)* 19849, (5142)* 19849–19850 Nakano; (5143)*, (5144)* 19850, (5145)* 19850–19851 Williams; (5146)* 19851, (5147)* 19851–19852 Kaneda; (5148)* 19852 Nakano; (5149)* 19852 Williams; (5150)* 19852–19853, (5151)* 19853 Nakano; (5152)* 19990 Kobayashi; (5153)* 19990–19991 Marsden; (5154)*, (5155)* 19991 Kobayashi; (5156)* 19991–19992 Williams; (5157)* 19992 Nakano; (5158)* 19992–19993 Kobayashi; (5159)* 19993 Williams; (5160)* 19993 Nakano; (5161)* 19993–19994, (5162)* 19994 Kobayashi; (5163)* 19994 Nakano; (5164)* 19994–19995 Williams; (5165)* 19995 Kobayashi; (5166)* 19995 Williams; (5167)* 19996 Kobayashi; (5168)* 19996 Williams; (5169)* 19996–19997 Kobayashi; (5170)* 19997 Nakano; (5171)* 19997–19998 Green; (5172)*, (5173)* 19998, (5174)* 19998–19999 Kaneda; (5175)* 19999 Bardwell; (5176)* 19999 Kaneda; (5177)* 20000 Williams; (5178)* 20000, (5179)* 20001 Green; (5180)* 20001, (5181)* 20001–20002, (5182)*, (5183)* 20002 Williams; (5184)* 20003 Kaneda; (5185)* 20003 Nakano; (5186)* 20003–20004 Williams; (5187)* 20004 Nakano; (5188)* 20004, (5189)* 20004–20005, (5189) 20315 Williams; (5190)*, (5191)* 20005, (5192)* 20005–20006, (5193)* 20006 Kaneda; (5194)* 20006 Nakano; (5195)* 20006–20007 Green; (5196)* 20007, (5197)* 20007–20008 Kobayashi; (5198)* 20131 Ichikawa; (5199)*, (5200)* 20132 Nakano; (5201)* 20132–20133 Williams; (5202)* 20133 Nakano; (5203)* 20133 Bardwell; (5204)* 20133–20134 Nakano; (5205)* 20134 Kaneda; (5206)* 20134–20135 Nakano; (5207)* 20135 Marsden; (5208)* 20135 Williams; (5209)*, (5210)* 20136 Nakano; (5211)* 20136–20137 Marsden; (5212)* 20137 Kaneda; (5213)* 20137 Williams; (5214)* 20137–20138, (5215)* 20138 Kaneda; (5216)* 20315–20316 Williams; (5217)* 20316 Kobayashi; (5218)* 20316, (5219)* 20316–20317 Nakano; (5220)*, (5221)* 20317 Marsden; (5222)* 20318 Nakano; (5223)* 20318 Marsden; (5224)* 20318–20319 Williams; (5225)* 20319 Bardwell; (5226)* 20319–20320 Nakano; (5227)* 20320 Williams; (5228)* 20320–20321 Nakano; (5229)* 20391 Marsden; (5230)* 20321 Nakano; (5231)* 20321–20322 Bardwell; (5232)* 20322 Marsden; (5233)* 20323 Nakano; (5234)* 20323 Bardwell; (5235)* 20323–20324 Williams; (5236)* 20324 Nakano; (5237)* 20324 Urata; (5238)* 20324–20325 Kaneda; (5239)* 20325 Nakano; (5240)* 20325–20326, (5241)* 20326 Kaneda; (5242)* 20326 Urata; (5243)* 20326–20327 Bardwell; (5243)* 20326 Bardwell.

098.004 Orbital elements of unnumbered minor planets.

G. V. Williams, H. Kaneda, T. Nagata, T. Urata, S. Nakano, S. Kobayashi, C. M. Bardwell, E. Bowell, B. G. Marsden, K. Ichikawa.
Minor Planet Circ., Nos. 19347 – 20366 (Jan – Jun 1992).

The unnumbered minor planets are sorted according to their provisional designations. The names of the orbit computers are given behind the respective M.P.C. numbers:
[A920 TA] 19853, [1931 FC] 19854 Williams; [1936 QE$_1$] 19854 Kaneda; [1936 UG] 19671, [1943 DL] 20008 Williams; [1953 GH] 20138 Kaneda; [1953 TD$_1$], [1955 UN$_1$] 19494 Nagata; [1967 JN] 20327 Urata; [1969 TQ$_1$] 19854, [1972 RU$_1$] 19854–19855 Nakano; [1973 SS$_4$] 20327, [1974 OE] 19855 Williams; [1974 SK$_1$] 19494–19495 Kobayashi; [1974 SD$_3$] 19672 Bardwell; [1975 SK] 19855, [1975 VN$_5$] 20138–20139, [1975 XP$_3$] 19672, [1976 GO$_3$] 19855 Williams; [1976 GY$_3$] 20008 Kaneda; [1976 UG$_2$] 20008 Bowell; [1976 YE$_1$] 19495, [1976 YR$_1$] 20009 Kaneda; [1977 DU] 19495 Kobayashi; [1977 DY$_3$] 20009, [1977 EX] 19856, [1977 EC$_2$] 20139 Williams; [1977 RD] 20139 Kaneda; [1977 RW$_6$] 19856 Nakano; [1977 RZ$_8$] 19495 Williams; [1977 TO$_6$] 20139–20140 Bowell; [1978 PT$_4$] 19495–19496 Kobayashi; [1978 RK], [1978 RZ] 20140 Williams; [1978 RR$_8$] 20327 Kobayashi; [1978 SD$_3$] 20009 Nakano; [1978 TH$_6$] 20009,

[1978 VP$_1$] 19856 Williams; [1978 VN$_3$] 19856 Nakano; [1978 VX$_3$] 20328 Williams; [1978 VT$_6$] 20140 Marsden; [1978 XW] 20140–20141, [1978 YM] 20010 Ichikawa; [1979 KQ] 20141 Williams; [1979 MF] 20010 Kaneda; [1979 MC$_2$] 20141 Bowell; [1979 MZ$_2$] 20010 Nakano; [1979 MA$_5$] 20141 Marsden; [1979 QW$_3$] 20328 Williams; [1980 TT$_3$] 19857 Kaneda; [1980 TA$_4$] 20010 Ichikawa; [1980 VG] 20010–20011 Bowell; [1981 CB$_1$] 20141–20142 Williams; [1981 DF] 19857 Kaneda; [1981 EH$_1$] 19857 Urata; [1981 ES$_5$] 19857–19858, [1981 EJ$_7$] 20328 Williams; [1981 EK$_7$] 19858, [1981 ET$_{10}$] 20011 Bowell; [1981 EY$_{13}$] 20328–20329, [1981 EZ$_{18}$] 19858, [1981 EA$_{19}$] 20329 Williams; [1981 EW$_{20}$] 19858–19859 Bowell; [1981 EZ$_{25}$], [1981 ES$_{27}$] 19859 Nakano; [1981 EF$_{28}$] 19672 Williams; [1981 FR] 20329 Bowell; [1981 JB$_2$] 20142 Kaneda; 1981 QK] 20011, [1981 QW$_2$] 20329–20330, [1981 QY$_2$] 19496, [1981 RO$_1$] 20330 Williams; [1981 UO$_{11}$] 19673 Bowell; [1981 WA$_1$] 19859 Williams; [1981 WS$_1$] 19496 Kaneda; [1982 BA] 20330 Williams; [1982 BD$_{13}$] 20142, [1982 QY$_1$], [1982 TT$_2$] 19497 Nakano; [1982 UC$_6$] 20011 Williams; [1982 UR$_6$] 19860 Bowell; [1982 UQ$_{10}$] 19497 Nakano; [1982 UE$_{12}$] 20330–20331 Williams; [1983 CY$_2$] 19497 Nakano; [1983 CQ$_3$] 19673 Urata; [1983 RV$_3$] 20331, [1984 QY$_1$] 20142 Williams; [1984 SU] 20012 Bowell; [1984 SQ$_2$] 20012, [1984 SV$_5$] 20331, [1984 SZ$_5$] 19673 Williams; [1984 WA$_1$] 19497–19498 Urata; [1984 YE$_4$] 19498 Kobayashi; [1985 CT], [1985 CJ$_1$] 19860 Williams; [1985 CM$_1$] 19860–19861 Kaneda; [1985 CU$_1$] 19673–19674 Williams; [1985 CZ$_1$] 19498 Nakano; [1985 FC$_2$] 20142–20143 Bowell; [1985 FE$_3$] 20331–20332 Williams; [1985 HS$_1$] 20012 Kaneda; [1985 JN$_1$] 20143 Nakano; [1985 PE] 20143 Bowell; [1985 PC$_2$] 19499 Williams; [1985 RD] 20012 Nakano; [1985 RP$_2$] 20143–20144 Bowell; [1985 TB$_1$] 20012–20013 Marsden; [1985 TD$_3$] 19674, [1985 UQ$_4$] 19499 Nakano; [1985 VF$_2$] 19674 Williams. For a continuation of this list see Abstr. 098.005.

098.005 Orbital elements of unnumbered minor planets.

S. Nakano, H. Kaneda, E. Bowell, G. V. Williams, B. G. Marsden, T. Nagata, T. Kobayashi, C. M. Bardwell, T. Urata, K. Ichikawa.
Minor Planet Circ., Nos. 19347 – 20366 (Jan – Jun 1992).

This list is a continuation of Abstr. 098.004. The unnumbered minor planets are sorted according to their provisional designations. The names of the orbit computers are given behind the respective M.P.C. numbers:
[1986 CE$_2$] 19499 Nakano; [1986 EE$_2$] 20144 Kaneda; [1986 EF$_5$] 20013, [1986 QO$_1$] 19674–19675 Bowell; [1986 QR$_3$] 20013 Williams; [1986 RA] 20013 Marsden; [1986 RQ] 20013–20014, [1986 RC$_1$] 20144 Williams; [1986 RE$_2$] 19499 Nakano; [1986 RT$_2$] 20332, [1986 RX$_2$] 19861, [1986 RT$_5$] 20332 Williams; [1986 RF$_7$] 20332 Nakano; [1986 RH$_{12}$], [1986 SZ$_1$] 19675, [1986 TL] 20144 Williams; [1986 TQ] 20144–20145 Kaneda; [1986 TR$_3$] 19500 Nagata; [1986 TR$_4$] 19861 Nakano; [1986 WM$_5$] 19861 Williams; [1986 XR$_5$] 19861 Bowell; [1987 BC] 20145, [1987 DY$_4$] 20014 Williams; [1987 DK$_6$] 19862 Nakano; [1987 HK] 19500 Williams; [1987 HK] 20014 Bowell; [1987 QZ$_1$] 19862 Kobayashi; [1987 RE$_1$] 19675 Kaneda; [1987 RQ$_2$] 20014 Williams; [1987 RA$_3$] 19500 Nagata; [1987 SL] 19676 Bardwell; [1987 SL] 20145, [1987 SR$_1$] 19862 Williams; [1987 SO$_9$] 20014–20015 Kaneda; [1987 UQ$_3$] 19863 Williams; [1987 VQ] 19676 Nakano; [1987 VB$_1$] 20332–20333 Bowell; [1987 WY] 19500 Williams; [1987 WO$_1$] 19863 Kaneda; [1987 XC] 19501 Nakano; [1987 YD] 20015 Kaneda; [1988 AO$_1$] 19676 Nakano; [1988 AV$_1$] 20145 Bowell; [1988 AL$_3$] 19677 Marsden; [1988 BC] 19676 Nakano; [1988 BV] 20333, [1988 BK$_2$], [1988 BG$_4$], [1988 BL$_5$] 19501 Williams; [1988 CF] 19677 Kaneda; [1988 CP$_2$] 19677 Nakano; [1988 DE$_2$] 20015 Kaneda; [1988 DD$_5$] 20015 Williams; [1988 EN] 20333 Kobayashi; [1988 FM] 19863 Nakano; [1988 GL] 20146 Urata; [1988 RA] 20146 Williams; [1988 RW$_3$] 19863 Bowell; [1988 RN$_{11}$] 20146 Williams; [1988 ST$_2$] 19864 Bowell; [1988 TL] 20146 Williams; [1988 TM$_1$] 20016, [1988 WC] 20147, [1988 XZ] 20016 Bowell; [1988 XV$_2$] 19502 Nagata; [1989 AF$_1$] 19502 Nakano; [1989 CH$_1$] 19678 Bardwell; [1989 EW$_1$] 19678, [1989 EC$_2$] 19502, [1989 EL$_2$] 19864, [1989 EN$_2$] 20016 Williams; [1989 FA] 19864 Nakano; [1989 GF$_1$] 19864–19865 Kaneda; [1989 GO$_4$] 20016–20017 Williams; [1989 GP$_4$] 20147 Marsden; [1989 GR$_4$]

20334 Nakano; [1989 JF] 20017 Kaneda; [1989 NO] 20334 Williams; [1989 SW$_2$], [1989 SU$_3$] 20017, [1989 TC$_3$] 19502–19503 Kaneda; [1989 UX$_5$] 19503, [1989 XD$_2$] 20017–20018 Bowell; [1989 YK$_8$] 20334 Nakano; [1990 BZ] 20018 Bowell; [1990 BF$_2$] 19865 Williams; [1990 DA$_1$] 19503 Nakano; [1990 DM$_2$] 19865 Ichikawa; [1990 DL$_3$] 19678 Bowell; [1990 EA$_5$] 20018, [1990 FQ$_1$] 19503 Williams; [1990 KE] 19504 Bowell; [1990 OA$_1$] 19679, [1990 OF$_2$] 20018 Williams; [1990 OO$_3$] 19679 Kaneda; [1990 OT$_3$] 19679 Nakano; [1990 OB$_4$] 20334 Bardwell; [1990 QW$_1$] 19865 Nakano; [1990 QZ$_1$] 19679–19680 Kaneda; [1990 QL$_2$] 20019 Nakano; [1990 QM$_2$] 20147 Bardwell; [1990 QK$_3$] 20335, [1990 QO$_3$] 19866, [1990 QP$_3$] 20019 Williams; [1990 QP$_5$] 20148 Marsden; [1990 QT$_9$] 20335, [1990 QC$_{19}$] 20148 Williams; [1990 RB] 19504 Urata; [1990 RH$_2$] 20335 Williams; [1990 RR$_2$] 19504, [1990 RE$_6$], [1990 RD$_9$] 19505 Kaneda; [1990 SB] 20148 Williams; [1990 SK$_6$], [1990 SZ$_7$] 19866 Kaneda; [1990 SL$_9$] 20019 Bowell; [1990 ST$_{10}$] 20020 Nakano; [1990 SD$_{14}$] 19867.
For a continuation of this list see Abstr. 098.006.

098.006   Orbital elements of unnumbered minor planets.
G. V. Williams, B. G. Marsden, H. Kaneda, S. Nakano, C. M. Bardwell, E. Bowell, K. Ichikawa, T. Kobayashi, T. Urata.
Minor Planet Circ., Nos. 19347 – 20366 (Jan – Jun 1992).
This list is a continuation of Abstr. 098.005. The unnumbered minor planets are sorted according to their provisional designations. The names of the orbit computers are given behind the respective M.P.C. numbers:
[1990 TJ] 20148–20149 Williams; [1990 TR] 20020 Marsden; [1990 TZ] 20020–20021, [1990 TB$_1$] 20336 Williams; [1990 TK$_1$] 19867, [1990 TK$_3$] 20021 Kaneda; [1990 TM$_5$] 20149 Marsden; [1990 TW$_7$] 20336 Williams; [1990 TR$_{12}$] 20336 Nakano; [1990 TV$_{12}$] 19867 Williams; [1990 UQ] 19680 Bardwell; [1990 UJ$_1$], [1990 UR$_1$] 20149 Williams; [1990 UF$_2$] 20149–20150 Nakano; [1990 UP$_3$] 20021 Kaneda; [1990 UQ$_{11}$] 19680, [1990 VL$_2$] 20150 Nakano; [1990 VN$_2$] 20150 Bowell; [1990 VS$_2$] 20021–20022 Nakano; [1990 VR$_3$] 20150 Williams; [1990 VQ$_5$] 20022 Nakano; [1990 VR$_8$] 20336 Kaneda; [1990 VU$_{14}$] 19867–19868 Williams; [1990 VB$_{15}$] 19505, [1990 VC$_{15}$] 19505–19506 Kaneda; [1990 WE] 19868, [1990 WB$_2$] 20336–20337, [1990 WN$_2$] 20022 Nakano; [1990 WY$_3$] 20022 Kaneda; [1990 XB] 20150–20151 Nakano; [1990 XK] 20337, [1990 YH] 20151 Urata; [1991 JW] 20023 Marsden; [1991 JY] 19680, [1991 JY] 20151, [1991 LE$_1$] 20023 Williams; [1991 NP] 20337 Marsden; [1991 NR$_2$] 19506 Bardwell; [1991 NE$_3$] 20023 Ichikawa; [1991 NT$_3$] 19868 Williams; [1991 NM$_6$] 20023 Ichikawa; [1991 PY$_5$] 19868 Nakano; [1991 PC$_6$] 20024 Bowell; [1991 PH$_8$] 19506 Williams; []1991 PO$_8$] 19868 Nakano; [1991 PE$_{10}$] 20337–20338 Ichikawa; [1991 PF$_{10}$] 19869, [1991 PO$_{10}$] 19506, [1991 PO$_{10}$] 19506–19507 Williams; [1991 PT$_{10}$] 19869 Kaneda; [1991 PG$_{11}$], [1991 PQ$_{11}$] 19507, [1991 PT$_{11}$] 20024 Williams; [1991 PA$_{12}$] 20338 Ichikawa; [1991 PR$_{12}$], [1991 PZ$_{12}$] 20338 Nakano; [1991 PB$_{13}$] 20151–20152 Bowell; [1991 PC$_{13}$] 20024 Williams; [1991 PO$_{14}$] 19869 Nakano; [1991 PH$_{15}$] 20024–20025, [1991 PK$_{15}$] 20025 Williams; [1991 PG$_{16}$] 20025 Bowell; [1991 PV$_{17}$], [1991 PW$_{17}$] 20025, [1991 PF$_{18}$], [1991 PN$_{18}$] 20026, [1991 QF] 19507 Williams; [1991 QG] 19507 Bardwell; [1991 RC] 19680 Marsden; [1991 RC] 20152 Bardwell; [1991 RV$_1$] 19507, [1991 RK$_2$] 20026 Williams; [1991 RO$_2$] 19508 Marsden; [1991 RD$_4$] 20338 Ichikawa; [1991 RY$_4$] 19869 Nakano; [1991 RM$_6$] 20026 Williams; [1991 RH$_7$] 19508 Marsden; [1991 RA$_{10}$], [1991 RE$_{11}$] 20152 Ichikawa; [1991 RF$_{14}$] 19870, [1991 RE$_{15}$] 19680–19681 Williams; [1991 RM$_{15}$], [1991 RP$_{15}$], [1991 RA$_{16}$] 20027, [1991 RY$_{16}$] 20339 Ichikawa; [1991 RQ$_{21}$] 19870 Williams; [1991 RD$_{24}$] 20152, [1991 RB$_{25}$] 20339 Ichikawa; [1991 SG$_1$] 19681 Kaneda; [1991 SS$_1$] 20027 Bowell; [1991 TY] 19681 Williams; [1991 TB$_1$] 19508 Bardwell; [1991 TB$_1$] 20028, [1991 TD$_1$] 19508 Williams; [1991 TL$_1$] 19509 Marsden; [1991 TV$_1$] 19509 Nakano; [1991 TW$_1$] 19509 Kobayashi; [1991 TH$_2$] 19509–19510, [1991 TJ$_2$] 19510 Nakano; [1991 TK$_2$] 19510 Marsden; [1991 TS$_4$] 19510 Nakano; [1991 TR$_6$] 19681 Marsden; [1991 UK] 19510, [1991 UM], [1991 UU], [1991 UV] 19511 Kaneda; [1991 UY] 19511–19512 Nakano; [1991 UG$_1$] 19681 Williams; [1991 UM$_1$] 19681–19682

Kaneda; [1991 UP$_1$] 19512 Urata; [1991 UB$_2$], [1991 UC$_2$] 19512 Kaneda; [1991 UK$_2$] 19512 Nakano; [1991 UL$_2$] 19513 Kaneda; [1991 UO$_2$] 19513 Nakano; [1991 UT$_2$], [1991 UV$_2$] 19513, [1991 UZ$_2$].
For a continuation of this list see Abstr. 098.007.

098.007   Orbital elements of unnumbered minor planets.
H. Kaneda, G. V. Williams, S. Nakano, B. G. Marsden, D. K. Yeomans, T. Urata, C. M. Bardwell, T. Kobayashi.
Minor Planet Circ., Nos. 19347 – 20366 (Jan – Jun 1992).
This list is a continuation of Abstr. 098.006. The unnumbered minor planets are sorted according to their provisional designations. The names of the orbit computers are given behind the respective M.P.C. numbers:
[1991 UC$_3$], [1991 UD$_3$] 19514, [1991 UE$_3$] 19514–19515, [1991 UG$_3$] 19515 Kaneda; [1991 UG$_3$] 20028 Williams; [1991 UU$_3$] 19515 Kaneda; [1991 UW$_3$] 19515 Nakano; [1991 UY$_3$] 19515 Kaneda; [1991 UJ$_4$] 20028 Williams; [1991 UL$_4$] 20028 Kaneda; [1991 VA], [1991 VB] 19516 Marsden; [1991 VB] 19870 Williams; [1991 VE] 19516 Marsden; [1991 VG] 19516 Yeomans; [1991 VH] 19516, [1991 VH] 19682, [1991 VK] 19516, [1991 VK] 20028–20029 Williams; [1991 VL] 19517 Marsden; [1991 VL] 19682 Williams; [1991 VN], [1991 VO] 19517 Kaneda; [1991 VP] 19517 Nakano; [1991 VR] 19517–19518, [1991 VS], [1991 VE$_1$], [1991 VM$_1$] 19518 Kaneda; [1991 VR$_1$] 19518–19519 Nakano; [1991 VZ$_1$], [1991 VD$_2$], [1991 VF$_2$] 19519, [1991 VG$_2$] 19519–19520, [1991 VH$_2$] 19520 Kaneda; [1991 VN$_2$], [1991 VB$_3$] 19520 Urata; [1991 VR$_3$] 19520–19521, [1991 VV$_3$] 19521 Kaneda; [1991 VX$_3$] 20339 Williams; [1991 VY$_3$] 19521, [1991 VF$_4$] 19682 Nakano; [1991 VK$_4$], [1991 VM$_4$] 20029, [1991 VP$_4$] 20339 Williams; [1991 VF$_5$] 19682–19683, [1991 VK$_5$] 19683 Nakano; [1991 WA] 19521 Marsden; [1991 WA] 19683, [1991 WB] 20030 Bardwell; [1991 WC] 19683 Nakano; [1991 XA] 19521, [1991 XB] 19522 Marsden; [1991 XB] 19683, [1991 XC] 20030 Williams; [1991 XZ] 19684 Urata; [1991 XC$_1$] 19684 Nakano; [1991 XO$_1$] 20030 Williams; [1991 YA] 19870 Bardwell; [1991 YC] 19684 Urata; [1991 YF] 19684–19685 Kaneda; [1991 YG], [1991 YH] 19685, [1991 YX] 19870–19871 Nakano; [1991 YZ] 19685 Kaneda; [1992 AA] 19522 Nakano; [1992 AA] 19685 Williams; [1992 AA] 19871, [1992 AA] 20030 Marsden; [1992 AB] 19686 Williams; [1992 AB] 19871 Bardwell; [1992 AB] 20030 Marsden; [1992 AC] 19522 Kobayashi;[1992 AD], [1992 AE] 19686, [1992 AE] 19871 Williams; [1992 AE] 20030 Marsden; [1992 AF] 19686 Urata; [1992 AJ] 20031 Williams; [1992 AL] 19686 Kobayashi; [1992 AX] 19871, [1992 AX] 20031, [1992 AX] 20339–20340 Williams; [1992 AB$_1$] 20340 Marsden; [1992 AD$_1$] 19687 Nakano; [1992 AF$_1$] 19687 Kobayashi; [1992 AH$_1$], [1992 AK$_1$] 19687 Nakano; [1992 AS$_1$] 20031 Kobayashi; [1992 AT$_1$] 19871 Kaneda; [1992 BA] 19688 Williams; [1992 BA] 19872 Marsden; [1992 BB] 19688, [1992 BB] 19872, [1992 BB] 20153 Williams; [1992 BC] 19688 Marsden; [1992 BC] 19872 Williams; [1992 BF] 19688 Marsden; [1992 BF] 19872 Bardwell; [1992 BF] 20031 Marsden; [1992 BK] 20031–20032 Nakano; [1992 BM] 19872 Kaneda; [1992 BW] 20032 Williams; [1992 BZ] 19688 Nakano; [1992 BX$_1$], [1992 BL$_2$] 20340 Marsden; [1992 BF$_4$] 20340–20341, [1992 CD] 20341 Williams; [1992 CE] 19872–19873 Kaneda; [1992 CT], [1992 CU] 19873 Nakano; [1992 CC$_1$] 19873, [1992 CC$_1$] 20032 Williams; [1992 CE$_1$] 20153, [1992 CG$_1$] 19873 Kaneda; [1992 CH$_1$] 19874 Marsden; [1992 CH$_1$] 20153 Williams; [1992 CQ$_2$] 20341 Nakano; [1992 CT$_2$] 20341 Marsden.
For a continuation of this list see Abstr. 098.008.

098.008   Orbital elements of unnumbered minor planets.
S. Nakano, G. V. Williams, B. G. Marsden, H. Kaneda, C. M. Bardwell, K. Ichikawa, E. Bowell, T. Kobayashi, D. W. E. Green.
Minor Planet Circ., Nos. 19347 – 20366 (Jan – Jun 1992).
This list is a continuation of Abstr. 098.007. The unnumbered minor planets are sorted according to their provisional designations. The names of the orbit computers are given behind the respective M.P.C. numbers:
[1992 DA] 19874 Nakano; [1992 DB] 20032–20033 Williams; [1992 DC] 19874 Marsden; [1992 DC] 20033, [1992 DC] 20153

Williams; [1992 DK] 20033 Nakano; [1992 DU] 19874 Marsden; [1992 DG$_1$] 20033 Nakano; [1992 DZ$_2$] 20341–20342, [1992 DL$_4$] 20342 Williams; [1992 EB] 20033–20034, [1992 EF], [1992 EL], [1992 EM], [1992 EP] 20034, [1992 ER] 20035 Kaneda; [1992 EU] 20342 Bardwell; [1992 EB$_1$] 20035, [1992 EB$_1$] 20153, [1992 EC$_1$] 20342, [1992 ED$_1$] 20153–20154 Williams; [1992 EL$_1$] 20035 Nakano; [1992 ES$_1$] 20154 Kaneda; [1992 FB] 20035 Nakano; [1992 FD] 20154, [1992 FE] 20036, [1992 FE] 20154 Williams; [1992 FE] 20342 Bardwell; [1992 FF] 20036, [1992 FJ] 20154, [1992 FN] 20155 Kaneda; [1992 FP] 20155 Ichikawa; [1992 FR] 20342–20343 Nakano; [1992 FS] 20155, [1992 FT], [1992 FV] 20036, [1992 FA$_1$] 20155 Kaneda; [1992 FB$_1$] 20343 Ichikawa; [1992 FJ$_1$] 20156, [1992 FL$_1$] 20036, [1992 FL$_1$] 20156 Williams; [1992 FP$_1$], [1992 FS$_1$] 20156 Kaneda; [1992 FY$_1$] 20343 Ichikawa; [1992 FZ$_1$] 20156–20157 Kaneda; [1992 GH] 20343 Williams; [1992 GZ] 20343–20344 Nakano; [1992 GA$_1$] 20344 Williams; [1992 HD] 20344 Nakano; [1992 HE] 20157, [1992 HE] 20344 Williams; [1992 HF] 20157, [1992 HF] 20344 Marsden; [1992 HH], [1992 HJ] 20345 Nakano; [1992 HK] 20345 Kaneda; [1992 JB] 20157, [1992 JB] 20345 Williams; [1992 JD] 20157, [1992 JD] 20345 Marsden; [1992 JE] 20157, [1992 JE] 20346 Williams; [1992 JF] 20346 Nakano; [1992 JG] 20157, [1992 JG] 20346 Marsden; [1992 KD] 20346 Williams; [2530 P–L] 19874, [2536 P–L] 19689 Bowell; [3086 P–L] 20037, [4319 P–L] 19875 Williams; [4556 P–L], [4559 P–L] 19875 Bowell; [4577 P–L] 19689 Williams; [4722 P–L] 20158, [6058 P–L] 19875, [6328 P–L] 19875–19876 Nakano; [6571 P–L] 20346, [6588 P–L], [6615 P–L] 19876, [6742 P–L] 20347 Williams; [7068 P–L] 19876 Kobayashi; [7075 P–L] 19689, [9508 P–L] 19876–19877, [9575 P–L] 19877 Bowell; [1081 T–1] 19877, [1171 T–1] 19522, [1181 T–1] 19877–19878, [1198 T–1], [1293 T–1], [1295 T–1] 19878, [2151 T–1] 19878–19879, [3100 T–1] 19879 Nakano; [3105 T–1] 20037 Bowell; [3163 T–1] 19879 Williams; [3196 T–1] 19523 Green; [3332 T–1] 19879, [4195 T–1] 19879–19880, [4214 T–1], [4232 T–1] 19880 Nakano; [4272 T–1] 19523 Williams; [4277 T–1] 19880, [4349 T–1], [4854 T–1] 19881 Nakano; [1079 T–2] 19881 Williams; [1210 T–2] 19881 Nakano; [1274 T–2] 19882 Bowell; [1335 T–2] 20037, [1617 T–2] 19882, [2087 T–2] 19689–19690 Williams; [2142 T–2], [2287 T–2], [4135 T–2] 19690 Bowell; [4234 T–2] 20037–20038, [4253 T–2], [4293 T–2] 20038 Williams; [1017 T–3] 19882 Nakano; [1194 T–3] 19883 Williams; [2157 T–3] 19691 Kaneda; [2247 T–3] 19883 Williams; [2327 T–3] 19883 Marsden; [2370 T–3] 19691, [3006 T–3] 20158 Williams; [3101 T–3] 20347, [3220 T–3] 19883 Bowell; [3398 T–3] 19691 Williams; [4032 T–3] 19691 Bowell; [4045 T–3], [4157 T–3] 19884 Nakano; [4310 T–3] 20347 Williams; [4391 T–3] 20038.

098.009   Observations of minor planets.

Minor Planet Circ., Nos. 19347 – 20366 (Jan – Jun 1992).

Positional data are published in Nos. 19371 – 19466 (19 Jan 1992), 19566 – 19653 (18 Feb 1992), 19731 – 19817 (18 Mar 1992), 19917 – 19979 (17 Apr 1992), 20073 – 20120 (16 May 1992), and 20195 – 20307 (15 Jun 1992), respectively. Observations made at the following observatories are published: Astron. Obs. Campo Imperatore, Aurec-sur-Loire, Bassano Bresciano, Burlington remote site, Calar Alto, Cavriana, Chorzów, Colleverde di Guidonia, Crimean Astrophys. Obs., Dynic Astron. Obs., El Leoncito, Eldagsen, ESO, Farra d'Isonzo, Geisei, Flagstaff (U.S. Naval Obs.), Foggy Bottom Obs., Fujieda, Goethe Link Obs., Haute Provence, JCPM Kagoshima Stn., JCPM Yakiimo Stn., Kannabe Obs. Albireo Stn., Kani, Karasuyama, Kitami, Kitt Peak, Kitt Peak (Steward Obs.), Kiyosato, Kleť, Kushiro, Kvistaberg, La Palma, Linz, Ljubljana, Lowell Obs., Lowell Obs. Anderson Mesa Stn., Mauna Kea Obs., Merate, Mérida, Minami–Oda, Miyasaka Obs., Mount John Obs., Nihondaira Obs. Oohira Stn., Nyukasa, Oak Ridge Obs., Oizumi, Ojima, Okutama, Ootake, Oss. Chaonis, Otomo, Palomar, San Marcello Pistoiese, San Pedro Martir, Santa Lucia Stroncone, Sendai Obs. Ayashi Stn., Siding Spring, Skalnaté Pleso, Sormano, Springe, Stakenbridge, Stefanik Obs., Susono, Tajimi, Tautenburg, Toyota, Uccle, Uenohara, Victoria (Climenhaga Obs.), Victoria (Dominion Astrophys. Obs.), Yatsugatake South Base Obs., Yatsuka, YGCO Chiyoda Stn.

098.010   Ephemerides of minor planets and comets.

Minor Planet Circ., Nos. 19347 – 20366 (Jan – Jun 1992).

The ephemerides are published in Nos. 19524 – 19558 (19 Jan 1992), 19700 – 19722 (18 Feb 1992), 19884 – 19908 (18 Mar 1992), 20039 – 20060 (17 Apr 1992), 20164 – 20190 (16 May 1992), and 20347 – 20366 (15 Jun 1992), respectively.

098.011   Identifications and identification changes, corrected and deleted observations of minor planets.

Minor Planet Circ., Nos. 19347 – 20366 (Jan – Jun 1992).

The data are published in Nos. 19356 – 19357 (19 Jan 1992), 19559 (18 Feb 1992), 19723 – 19726 (18 Mar 1992), 19909 – 19911 (17 Apr 1992), 20061 – 20062 (16 May 1992), and 20191 (15 Jun 1992), respectively.

098.012   Index to orbital elements of comets and minor planets.

Minor Planet Circ., Nos. 20062 – 20070 (16 May 1992).

The index refers to both comet and minor planet orbits published in MPC 19058 – 20061. The list is divided into the parts (1) comets, (2) numbered minor planets, (3) unnumbered minor planets, and (4) planets from the P–L, T–1, T–2, and T–3 surveys.

098.013   Observatory codes.

Minor Planet Circ., Nos. 19348 – 19356 (19 Jan 1992).

This list gives the coordinates of observatories engaged in minor planet and comet astrometric work. It contains the longitudes in degrees eastward from Greenwich and the parallax constants as products of the geocentric distance and the cosine and sine, respectively, of the geocentric latitude.

098.014   Critical list of minor planets.

Minor Planet Circ., Nos. 19357 (19 Jan 1992).

This lists updates the one published in MPC 16772 (see Abstr. 52.098.152) and contains all objects observed in only one, two, or three oppositions. Objects observed at four or more apparitions are listed, too, if the last observed opposition was before 1982.

098.015   Orbital elements of one–opposition minor planets.

Minor Planet Circ., Nos. 19347 – 20366 (Jan – Jun 1992).

The orbits are published in Nos. 19468 – 19471 (19 Jan 1992), 19655 – 19658 (18 Feb 1992), 19818 – 19822 (18 May 1992), 19980 – 19983 (17 Apr 1992), 20124 – 20125 (16 May 1992), and 20310 – 20313 (15 Jun 1992), respectively.

098.016   New names of minor planets.

Minor Planet Circ., Nos. 19347 – 20366 (Jan – Jun 1992).

Citations of new minor planet names are published in Nos. 19692 – 19699 (18 Feb 1992), and 20158 – 20164 (16 May 1992), respectively.

098.017   Asteroid collisional evolution: an integrated model for the evolution of asteroid rotation rates.

P. Farinella, D. R. Davis, P. Paolicchi, A. Cellino, V. Zappalà. Astron. Astrophys., Vol. 253, No. 2, p. 604 – 614 (Jan 1992).

The authors combine experimental data and theoretical results to develop a model for the changes in the rotation rates of fragments produced by large collisions that break up and disperse asteroidal targets. Collisions that partially disperse the target can also produce a despinning of the reaccumulated core due to the angular momentum "splash" effect arising from the preferential escape of material with higher than average angular momentum (Cellino et al. 1990). Combining these results with previously published work on spin rate changes for cratering impacts (Harris 1979; Dobrovolskis and Burns 1984), the authors present a comprehensive model for the changes in asteroid spin rates due to collisions. This spin change algorithm, when incorporated into an existing simulation of collisional effects on asteroid sizes, produces an integrated model for studying the simultaneous evolution of asteroid sizes and spin rates over solar system history.

098.018 A picture–perfect asteroid.
   J. K. Beatty.
Sky Telesc., Vol. 83, No. 2, p. 134 – 135 (Feb 1992).
   Pinpoint targeting allowed Galileo scientists to get a glimpse of the asteroid Gaspra soon after the spacecraft's historic flyby.

098.019 Galileo besucht Gaspra. Die erste Begegnung mit einem Kleinplaneten.
   D. Fischer.
Sterne Weltraum, Jahrg. 31, Nr. 2, p. 103 – 105 (Feb 1992).

098.020 Physical study of the asteroid 243 Ida, second fly–by target of the Galileo spacecraft.
   M. Gonano–Beurer, M. Di Martino, S. Mottola, G. Neukum.
Astron. Astrophys., Vol. 254, No. 1/2, p. 393 – 396 (Feb 1992).
   According to the present mission planning, 243 Ida will be the second main belt asteroid to be encountered by the Galileo spacecraft, the fly–by being scheduled for August 1993. In the framework of the observation program of space mission targets, the authors have undertaken CCD and photoelectric measurements of 243 Ida in the B and V Johnson bands from the observatories of Loiano (Italy) and ESO (La Silla, Chile) at the time of its 1990 apparition. From seven observations which spanned a period of three months in time and a phase angle range from about 4 to 20 degrees, they obtain three composite lightcurves of this asteroid. The good quality of these data allows a substantial improvement in the determination of the synodic rotational period of 243 Ida, which results to by $P_{syn} = 4.6330 \pm 0.0005$ hr. The authors derive for the first time the magnitude–phase relationship and an estimation for the amplitude–phase relationship (APR) of this asteroid. The computation of the parameters of the Bowell–Harris–Lumme relation from the measurements gives: $H = 10.042 \pm 0.008$ mag and $G = 0.15 \pm 0.02$ mag deg^{-1}. From the lightcurve amplitude–phase relation the authors estimates (APR) = 0.010 mag deg^{-1} and m = 0.016 deg^{-1}. They also derive the color index B–V = 0.87 ± 0.03.

098.021 Hoogtijdagen voor de planetoiden. Meer dan 5000 kleine planeten ontdekt.
   F. Börngen.
Zenit, Jaarg. 19, Nr. 2, p. 78 – 81 (Feb 1992).

098.022 Photographic positions of minor planets.
   H. Debehogne, M. Scardia.
Astron. Nachr., Vol. 313, No. 2, p. 107 – 114 (1992).
   Photographic precise positions of minor planets, obtained with the G.P.O. astrograph at ESO (La Silla – Chile) and with the ZEN astrograph at Merate, are given.

098.023 Asteroid 126 Velleda: rotation period and magnitude–phase curve.
   A. N. Dovgopol, Yu. N. Krugly, V. G. Shevchenko.
Acta Astron., Vol. 42, No. 1, p. 67 – 72 (1992).
   Photoelectric observations of the asteroid 126 Velleda were carried out during the 1990 opposition. A rotation period P = 5^{h}364 $\pm$ 0.003 and a light curve amplitude of 0.22 mag were determined. The magnitude–phase relation has been obtained in the range 0°14 ⩽ α ⩽ 23°8 (H = 9.28 mag, G = 0.29). A comparative analysis of the opposition effects of Velleda and other S–type asteroids was made.

098.024 Asteroid and comet orbits using radar data.
   D. K. Yeomans, P. W. Chodas, M. S. Keesey,
S. J. Ostro, J. F. Chandler, I. I. Shapiro.
Astron. J., Vol. 103, No. 1, p. 303 – 317 (Jan 1992). Current Physics Microform No.: 9202A2213.
   For the 30 asteroids and 4 comets for which radar astrometric data were given by Ostro (1991), orbits have been computed using both the radar and the existing optical measurements. The techniques required to process radar data in orbit determination solutions are outlined and future radar observation opportunities for asteroids and comets are identified. For asteroids and comets that have only short intervals of optical astrometric data, the additional use of only a few radar observations allows a far more accurate extrapolation of their future motions. The use of radar data can often ensure an object's successful recovery at future Earth returns and greatly assist efforts in monitoring the motions of the rapidly growing population of known near–Earth objects, including their future close–Earth approaches.

098.025 We are a moving target.
   L. Kresák.
Kozmos, Vol. 23, No. 3, p. 4 – 7 (Jun 1992). In Slovak.

098.026 Evidence for ammonium–bearing minerals on Ceres.
   T. V. V. King, R. N. Clark, W. M. Calvin,
D. M. Sherman, R. H. Brown.
Science, Vol. 255, No. 5051, p. 1551 – 1553 (20 Mar 1992).
   Spectra obtained from recent telescopic observation of (1) Ceres and laboratory measurements and theoretical calculations of three component mixtures of Ceres analog material suggest that an ammoniated phyllosilicate is present on the surface of the asteroid, rather than H_2O frost as had been previously reported. The presence of an ammoniated phyllosilicate, most likely ammoniated saponite, on the surface of Ceres implies that secondary temperatures could not have exceeded 400K.

098.027 Minor planets at the Kleť Observatory.
   A. Mrkos.
Říše hvězd, Vol. 72, No. 5, p. 89 – 91 (May 1991). In Czech.

098.028 Asteroid 1990 MB.
   P. Andrle.
Říše hvězd, Vol. 72, No. 5, p. 91 – 92 (May 1991). In Czech.

098.029 Asteroid (1834) Palach.
   J. Grygar.
Říše hvězd, Vol. 72, No. 12, p. 231 – 232 (Dec 1991). In Czech.

098.030 Trojan, Hilda, and Cybele asteroids: new lightcurve observations and analysis.
   R. P. Binzel, L. M. Sauter.
Icarus, Vol. 95, No. 2, p. 222 – 238 (Feb 1992).
   The authors present lightcurve observations for 23 Trojan, Hilda, and Cybele asteroids, representing nearly one–half of the currently published data set for these objects. Previous researchers have suggested that Trojan and Hilda asteroids display significantly higher mean lightcurve amplitudes than their comparable diameter main–belt counterparts. The authors present a correction procedure for the use of multiple aspect asteroid lightcurve observations and perform a quantitative bias–corrected analysis of lightcurve amplitude distributions for all published data on Trojan, Hilda, and Cybele asteroids. Lightcurve samples are now in hand for all Trojan asteroids larger than 65 km. The authors report a new finding that the largest Trojans (D > 90 km) have a higher mean lightcurve amplitude than their low albedo main–belt counterparts. On the other hand, smaller Trojans, plus all Hildas and Cybeles, display lightcurve properties similar to comparable main–belt objects. A possible explanation for the Trojan and main–belt difference may be that the lower relative velocities in the Trojan regions allowed more irregular aggregates to form from planetesimals. Only the largest Trojans have been able to substantially retain their initial forms after subsequent collisional evolution. Thus among the Trojans, 90 km may represent a transition size between primordial objects and collision fragments.

098.031 Reports of asteroidal appulses and occultations.
   J. Stamm.
Occultation Newsl., Vol. 5, No. 7, p. 165 – 166 (May 1992).

098.032 Statistik von 5000 Kleinplaneten.
   J. Jahn.
KPM, Jahrg. 7, No. 19, p. 24 – 27 (Apr 1992).
   A statistic about discovery circumstances and orbital elements of the first 5000 numbered minor planets.

098.033 MPC–Format astrometrischer Positionen.
J. Jahn.
KPM, Jahrg. 7, No. 19, p. 29 – 32 (Apr 1992).
The author presents the new MPC format of astrometric positions for the Minor Planet and Comet Circular.

098.034 Neue Namen von Kleinplaneten.
J. Jahn.
KPM, Jahrg. 7, No. 19, p. 40 – 45 (Apr 1992).
New names of minor planets are summarized.

098.035 Positionsbeobachtungen von Kleinen Planeten und Kometen.
J. Jahn.
KPM, Jahrg. 7, No. 19, p. 46 – 47 (Apr 1992).
Positions of minor planets and comets are given.

098.036 Liste générale des objets Aten, Apollo et Amor (classés par a croissant).
M.–A. Combes, J. Meeus.
Obs. Trav., No. 29, p. 20 – 28 (1992).

098.037 Chronique des objets A.A.A. (No. 1).
M.–A. Combes, J. Meeus.
Obs. Trav., No. 30, p. 16 – 23 (1992).

098.038 Icarus, 1991 RC and the daytime Arietids.
D. Steel.
WGN, Vol. 20, No. 1, p. 20 – 22 (Feb 1992).

098.039 Radio observations regarding Earth–grazing asteroids.
D. Artoos.
WGN, Vol. 20, No. 2, p. 96 – 98 (Apr 1992).

098.040 Rotational properties of small asteroids: photoelectric observations.
M. A. Barucci, M. Di Martino, M. Fulchignoni.
Astron. J., Vol. 103, No. 5, p. 1679 – 1686 (May 1992). Current Physics Microform No.: 9205G0253.
In 1984 the authors started an observational program on small asteroids (diameter lower than about 50 km) with the aim to enlarge the available dataset of the rotational periods of this size range objects. In this paper they report the results obtained from photometric observations of ten small asteroids from August 1984 to January 1989 at the European Southern Observatory (La Silla, Chile). The authors have determined reliable synodic rotational periods for the asteroids 269 Justitia (P = 16^h545), 289 Nenetta (P = 6^h902), 417 Suevia (P = 7^h034), 435 Ella (P = 4^h623), 537 Pauly (P = 16^h250), 995 Sternberga (P = 16^h406), 1186 Turnera (P = 12^h010), and 1693 Hertzsprung (P = 8^h825), while for 504 Cora (P = 24^h) and 1392 Pierre (18^h), they obtained only an estimate of the possible rotational period.

098.041 On pyroxene types of near–Earth and near–Mars asteroids.
D. I. Shestopalov, L. F. Golubeva.
Astron. Vestn., Tom 26, No. 2, p. 77 – 88 (Mar – Apr 1992). In Russian. English translation in Solar Syst. Res., Vol. 26, No. 2.
The colorimetric data of bright near–Earth and near–Mars asteroids from TRIAD and ECAS have been analysed. Composition fields of pyroxenes obtained for these asteroids by the value of $(u - x)$ and 505 nm ferrous absorption band position within the pyroxenes quadrangle. Pyroxenes of the S–asteroids from Apollo–Amor which have spectral parameters similar to achondrites may be presented by diopside–augite series. AA–asteroids (S–type) which spectral parameters are likely to L–chondrites have either chondritic composition or Fe–rich ortho– and clinopyroxenes, not being met in meteorites mineral. On S–asteroids surface from Mars–crossers (MC) and Mars–approachers (MA) may be pyroxenes with high content Ca and Fe (as of ferrosalite and hedenbergite) and pyroxenes of E–asteroids are typically for stone and ironstone meteorites (enstatite, hypersthene, pigeonite). It is found the average color-

index $<u - x>$ increases with increasing average perihelion $<q>$ from 1 to 1.8 AU that indicates the trend of pyroxenes chemical composition on the bright asteroids surface.

098.042 Comparative analysis of phase relations in the light of asteroids.
F. A. Tupieva.
Kinematics Phys. Celest. Bodies, Vol. 7, No. 3, p. 39 – 48 (1991). English translation of Kinematika Fiz. Nebesn. Tel, Tom 7, No. 3, p. 42 – 51 (1991).
It is shown on the basis of comparison of available measurements of the phase dependences of the light of various types of asteroids that: 1) no differences are observed between the light phase curves of different types of asteroids within the limits of measurement error in the range of the opposition effect ($\alpha < 7°$); 2) the variation of light with phase angle is practically the same for S and M asteroids on the linear parts of the phase curves ($\alpha > 7°$). It is also shown that the surface layers of asteroids have the same high degree of porosity as the surface of the Moon.

098.043 Asteroid surface materials.
D. F. Lupishko, I. N. Bel'skaya.
Sol. Syst. Res., Vol. 25, No. 1, p. 1 – 18 (Jul 1991). English translation of Astron. Vestn., Tom 25, No. 1, p. 5 – 26 (1991).
This survey is based on research results published primarily during the past three years and reflects the current point of view of the problem of asteroid materials. It examines the initial observational data for study of asteroid materials, the mechanisms for the formation of absorption spectra, the spectral features of the principal mineral phases, the asteroid taxonomic classes, and their general mineralogical characteristics. The nature of the materials of C (and other low–albedo classes), S, and M asteroids and of the asteroid 4 Vesta is then examined. The evolution of the dominating types of material with heliocentric distance is discussed, including the presently most plausible heating mechanism for the early thermal evolution of the asteroids.

098.044 Evolution of asteroidal orbits at the 5:2 resonance.
S. I. Ipatov.
Icarus, Vol. 95, No. 1, p. 100 – 114 (Jan 1992). With corrections in Icarus, Vol. 97, No. 2, p. 309 (Jun 1992).
The authors numerically solved the full equations of the three–body problem (Sun–Jupiter–asteroid) in order to investigate the time dependence of eccentricities for fictitious asteroids initially located near the 5:2 Jovian commensurability. The runs covered a time span of $T \geqslant 5 \times 10^3 t_j$ (t_j is the heliocentric orbital period of Jupiter) for the planar model, of $T \geqslant 10^4 t_j$ for cases with the initial inclination $5° \leqslant i_0 \leqslant 20°$ and of $T \geqslant 10^5 t_j$ for cases at $i_0 = 40°$. The authors investigated regions of the initial values of semimajor axes and eccentricities for which, at some starting orbital orientations and initial positions, the fictitious asteroids were Mars– and Earth–crossers. It is found that, for initial eccentricities $e_0 \leqslant 0.2$ and $i_0 \leqslant 20°$, these ranges were almost the same. The range in which the authors found asteroids to be Mars–crossers is close to that free of real asteroids. Close encounters of asteroids with Mars and Earth might be one of the causes of the 5:2 Kirkwood gap.

098.045 Asteroid lightcurve observations from 1981.
A. W. Harris, J. W. Young, T. Dockweiler, J. Gibson, M. Poutanen, E. Bowell.
Icarus, Vol. 95, No. 1, p. 115 – 147 (Jan 1992).
Results of photoelectric lightcurve observations from Table Mountain and Lowell Observatories are reported. The observations were made from May through Dec, 1981. Included are also observations of 230 Athamantis made in 1974 from TMO and of 2100 Ra–Shalom made in 1978 from Lowell Observatory, not previously reported. Observations of 59 different asteroids are reported. About 15 new or significantly revised periods are reported. All lightcurve observations are in the V band; B–V color indices are included for a few objects previously lacking this information. Estimates of the mean and maximum reduced magnitudes are given for each object. Sufficient phase angle

coverage was obtained for 28 objects to fit those data with the H–G magnitude relation. Fits are generally good for objects of moderate albedo (moderate slope of the phase relation), but one continues to see poor quality fits among the highest and lowest albedo objects (shallowest and steepest sloping phase curves, respectively).

098.046   The S–class asteroid debate: historical outline.
J. F. Bell.
54. Annual Meeting of the Meteoritical Society, p. 13 – 14 (1991).
Abstract. – See Abstr. 012.010 for the main entry.

098.047   S–type asteroids and the Gaspra fly–by.
C. R. Chapman.
54. Annual Meeting of the Meteoritical Society, p. 44 (1991).
Abstract. – See Abstr. 012.010 for the main entry.

098.048   The mineralogy of S–type asteroids: why doesn't spectroscopy find ordinary chondrites in the asteroid belt?
M. J. Gaffey.
54. Annual Meeting of the Meteoritical Society, p. 65 – 66 (1991).
Abstract. – See Abstr. 012.010 for the main entry.

098.049   Using neural networks to classify asteroid spectra.
E. S. Howell, E. Merényi, L. A. Lebofsky.
54. Annual Meeting of the Meteoritical Society, p. 96 (1991).
Abstract. – See Abstr. 012.010 for the main entry.

098.050   Light–curves of asteroid 51 Nemausa during its 1989 apparition.
A. N. Dovgopol, L. R. Lisina.
Kinematika Fiz. Nebesn. Tel, Tom 8, No. 2, p. 36 – 39 (Mar–Apr 1992). In Russian. English translation in Kinematics Phys. Celest. Bodies, Vol. 8, No. 2.
Light–curves of asteroid 51 Nemausa were obtained during 5 nights in 1989. The shape of the curves is found to depend strongly on the solar phase angle. The maximum amplitude increases from $0^m.11$ to $0^m.13$ while the phase angle changes from $5.4°$ to $26.5°$. The average value of the phase coefficient is 0.045 magnitude per degree.

098.051   Characterization of low albedo asteroids.
L. A. Lebofsky, E. S. Howell, D. T. Britt.
54. Annual Meeting of the Meteoritical Society, p. 128 (1991).
Abstract. – See Abstr. 012.010 for the main entry.

098.052   Oxidation during metamorphism: another argument against S–asteroids having chondritic compositions.
H. Y. McSween Jr.
54. Annual Meeting of the Meteoritical Society, p. 152 (1991).
Abstract. – See Abstr. 012.010 for the main entry.

098.053   Doublet craters and the tidal disruption of binary asteroids.
H. J. Melosh, J. Stansberry.
54. Annual Meeting of the Meteoritical Society, p. 154 (1991).
Abstract. – See Abstr. 012.010 for the main entry.

098.054   Radar constraints on asteroid metal abundances and meteorite associations.
S. J. Ostro.
54. Annual Meeting of the Meteoritical Society, p. 180 (1991).
Abstract. – See Abstr. 012.010 for the main entry.

098.055   A new model for the formation of the asteroids – the parent bodies of the meteorites.
G. W. Wetherill.
54. Annual Meeting of the Meteoritical Society, p. 248 (1991).
Abstract. – See Abstr. 012.010 for the main entry.

098.056   S asteroids: evidence from astronomy and orbital mechanics.
G. W. Wetherill.
54. Annual Meeting of the Meteoritical Society, p. 249 (1991).
Abstract. – See Abstr. 012.010 for the main entry.

098.057   The collisional lifetime of asteroid 951 Gaspra.
P. Farinella, D. R. Davis, A. Cellino, V. Zappalà.
Astron. Astrophys., Vol. 257, No. 1, p. 329 – 330 (Apr 1992).
The authors estimate the collisional lifetime of asteroid 951 Gaspra, the first target of an asteroid encounter by an interplanetary space probe. Plausible values range between 10^8 and a few times 10^9yr, mainly depending on the impact strength of the body and the existing population of sub–km sized projectiles – both quantities for which only indirect observational evidence is currently available. Thus, Gaspra is most likely to be a fragment from a larger asteroid that was shattered in the past. Moreover, unless its material is very resistant to impact break–up (such as iron), Gaspra's surface is among the youngest extraterrestrial surfaces seen in the solar system (apart from active comet nuclei). However, better data on the abundance of very small asteroids are needed to determine its age in a more reliable way.

098.058   Light scattering by rough surfaces on asteroidal/lunar regoliths.
T. Mukai, S. Mukai.
Adv. Space Res., Vol. 11, No. 12, p. 137 – 140 (1991). – See Abstr. 012.014 for the main entry.
The light scattering properties of a rough surface with feature size comparable to the wavelength of interest, which is similar to the surface consisting of coarse regolith, are estimated from Mie theory for representing the roughness as an ensemble of noninteracting spheres. It is found that a disk–integrated linear polarization deduced from singly scattered rays by Mie theory is smaller than that resulting from Fresnel reflection for a rough surface of fine regolith. In addition, negative polarization in the back scattering region observed in asteroids is well described by a mixing model of coarse and fine regoliths.

098.059   Asteroid taxonomy types.
M. A. Barucci.
Adv. Space Res., Vol. 11, No. 12, p. 183 – 191 (1991). – See Abstr. 012.014 for the main entry.
In recent years several works have been carried out with the aim of understanding some of the physical and compositional properties of asteroid populations. Three recent works are based on statistical analyses of those asteroids for which a complete set of selected parameters were available: namely, a set of 589 asteroids described by seven colors indices, a subset of that one, of 438 asteroids for which IRAS albedo values were available and a set of 357 asteroids described by three variables: two reflected light color indices and high–quality IRAS albedo. The author compares the different methods used and discusses the differences in the results obtained: some differences result from the grouping technique chosen, and some on the quality of the data sets used. Particularly the classifications of some peculiar asteroids is discussed: Earth–crossing objects, and asteroids that may be extinct cometary nuclei.

098.060   Photoelectric and CCD photometry of 951 Gaspra.
C. Blanco, M. Di Martino, W. Ferreri, M. Gonano, S. Mottola, G. Neukum.
Adv. Space Res., Vol. 11, No. 12, p. 193 – 196 (1991). – See Abstr. 012.014 for the main entry.
The asteroid 951 Gaspra is a fly–by target for the Galileo space mission. This encounter will represent the first possibility to show an asteroid in close–up. In preparation to this close encounter an earth–based international observing campaign started in 1988, with the goal to provide a substantial data base on the photometric and rotational properties of this object. This will allow on one side to support and optimize the planning of the fly–by and, on the other side, to verify the accuracy of current models for the determination of shapes, surface textures, spin axis orientation and composition. In this paper the authors present

the composite lightcurve of 951 Gaspra obtained from photoelectric and CCD observations, carried out during the 1990 apparition at the observatories of Asiago, Catania and Loiano (Italy). The authors have determined the sidereal rotational period, and prograde sense of rotation, the H and G parameters and the B–V and V–R color indices.

098.061 Physical study of outer belt asteroids.
M. Gonano, M. Di Martino, S. Mottola,
G. Neukum.
Adv. Space Res., Vol. 11, No. 12, p. 197–200 (1991). – See Abstr. 012.014 for the main entry.

Located in the proximity of Jovian resonances, the outer belt asteroids are kept from interacting with other asteroids. In the last decade the information on the spectral and photometric properties of distant asteroids has strongly increased, leading to the formulation of specific questions, the answer to which will enable a comprehensive picture of this class of bodies. Since 1988 the authors are carrying out a program devoted to the physical study of the Trojans and outer belt asteroids to characterize their rotational properties, composition and shapes. During several observing campaigns, carried out at different observations, reliable rotational periods and light curve amplitudes have been determined for eight distant asteroids using both CCD and photoelectric photometry. The authors will here present some preliminary results of their campaigns.

098.062 Ephemerides of the 48 Hipparcos minor planets for the year 1992.
A. Bec–Borsenberger.
Astron. Astrophys., Suppl. Ser., Vol. 93, No. 1, p. 11–60 (Apr 1992).

To be observed by Hipparcos, the position of each object has to be known with an accuracy of 1 arcsecond and its magnitude with an accuracy of 0.5 mag. Therefore, the orbital elements of the 48 selected minor planets of the Hipparcos mission, have been improved in order to extrapolate their ephemerides with the accuracy required. Updated ephemerides are given here for the year 1992, referred to the two reference frames: J2000 and 1950.0. The orbital elements which have been used to compute these ephemerides are also given.

098.063 Chiron – an unusual asteroid or a big comet.
K. Ziołkowski.
Urania, Rok 63, Nr. 3, p. 71–78 (Mar 1992). In Polish.

098.064 A study of the polarimetric lightcurve of the asteroid 16 Psyche.
P. Broglia, A. Manara.
Astron. Astrophys., Vol. 257, No. 2, p. 770–772 (Apr 1992).

A search for rotational modulation of the linear polarization of 16 Psyche is presented. This object has been found to show a polarimetric variation with rotation, suggesting that the surface of the asteroid is variegated. A period $P = 0.1753 \pm 0.0004$ d and a total amplitude of 0.12% for the polarization change were derived.

098.065 Solar system objects observed by Hipparcos.
A. Bec–Borsenberger.
Astron. Astrophys., Vol. 258, No. 1, p. 94–98 (May 1992).

While the vast majority of objects contained in the Hipparcos observing programme are stars, 48 minor planets and two satellites of the major planets are also observable and included within the observing programme. These observations have very particular scientific values, and the inclusion of the objects in the Input Catalogue have posed very specific problems: both in the preparatory astrometric and photometric work needed to prepare for the observations, and in the considerations applied to their selection.

098.066 Photometry of the asteroids 22 Kalliope and 79 Eurynome: magnitude–phase relation and rotation and shape parameters.
F. P. Velichko, T. Michalowski, Yu. N. Kruglyj, D. F. Lupishko.
Sol. Syst. Res., Vol. 25, No. 2, p. 118–124 (Sep 1991). English translation of Astron. Vestn., Tom 25, No. 2, p. 162–170 (1991).

The light curves of the asteroids 22 Kalliope and 79 Eurynome obtained at the oppositions of 1986 and 1989, respectively, are presented. All available data and the method combining photometric astrometry and the amplitude–aspect relation for Kalliope were used to determine the sidereal period $P_{sid} = 4^h08^m53^s521 \pm 0^s002$, pole coordinates $\lambda_0 = 198° \pm 6°$ and $\beta_0 = 23° \pm 3°$, retrograde rotation, and the axial ratios of the approximating ellipsoid a:b:c = 1,56:1,18:1,00. Two equally valid solutions, both prograde, are obtained for Eurynome: 1) $5^h58^m39^s70 \pm 0^s02$, $64° \pm 10°$ and $45° \pm 15°$, and 2.68:2.09:1.00 and 2) $5^h58^m39^s71 \pm 0^s02$, $226°10'$ and $52° \pm 15°$, and 2.41:1.90:1.00. The magnitude–phase relation constructed from observations of Eurynome in the phase-angle interval $4°8 - 21°2$ is described by the parameters $H = 8,02 \pm 0.02$ and $G = 0,27 \pm 0.03$.

098.067 Orbital evolution of the Aten asteroids over a period of 11,550 years (9300 BC to 2250 AD).
A. F. Zausaev, A. N. Pushkarev.
Sol. Syst. Res., Vol. 25, No. 2, p. 125–129 (Sep 1991). English translation of Astron. Vestn., Tom 25, No. 2, p. 171–176 (1991).

The orbital evolution of the five asteroids belonging to the Aten group is traced by the Everhart method over the period 9300 BC to 2250 AD. Minimum distances of the asteroids to the terrestrial planets during evolution are calculated. Stable resonances with the Earth and Venus over this period are found for four of the five asteroids.

098.068 Origin of the asteroid belt.
V. S. Safronov, I. N. Ziglina.
Sol. Syst. Res., Vol. 25, No. 2, p. 139–146 (Sep 1991). English translation of Astron. Vestn., Tom 25, No. 2, p. 190–199 (1991).

The reasons for the absence of a normal planet in the asteroid region are discussed. According to present concepts, at the early-stage of evolution of the solar nebula, the growth of bodies (planetesimals) in the asteroid region by the merger of small particles proceeded in the same manner as in the region of the terrestrial planets. However, the more rapid formation of the massive Jupiter in the adjacent region halted this process as collisions of the bodies began to fission, not fusion, because their velocities were increased by external perturbations. Possible mechanisms for the increase in velocities of the asteroids and their removal from the asteroid region is discussed in this article: a) gravitational perturbations by Jupiter of the resonance asteroids increased their orbital eccentricities, making possible their ejection from the asteroid region. This mechanism could have been effective if the position of the resonances shifted (scanned) radially over the asteroid region and thereby involved a significant number of bodies of the region. This scan would have occurred with the variation of the distance of Jupiter from the Sun as it accreted gas and as it ejected bodies from the solar system; b) bodies of the Jupiter region with more eccentric orbits penetrated into the asteroid region and, in being more massive than the asteroids, they swept them out of the asteroid region with collisions. The larger bodies increased the asteroid velocities at close encounters; c) particles and small bodies interacted effectively with gas. The small particles moved with it, while the larger particles and bodies moved radially toward the region of higher gas pressure. The removal of solid material from the asteroid region because of the gas could have been effective if most of the bodies of kilometer size and larger were fragmented by collisions to sizes less than a few decameters. Bodies of the Jupiter region evidently played the principal role in sweeping bodies from the asteroid region. But the removal of more than 99.9% of the solid material from the region would hardly have been possible without the combined action of all three mechanisms.

098.069   Galileo views Gaspra.
R. Talcott.
Astronomy, Vol. 20, No. 2, p. 52 – 54 (Feb 1992).

098.070   A determination of the mass of (704) Interamnia from observations of (993) Moultona.
W. Landgraf.
IAU Symposium No. 152: Chaos, resonance and collective dynamical phenomena in the solar system, p. 179 – 182 (1992). – See Abstr. 012.025 for the main entry.
On 1973 November 23, there was a close approach (0.013 AU) between (704) Interamnia as the fifth largest minor planet, and (993) Moultona. It was made an attempt to determine the mass of Interamnia from its perturbations on the motion of Moultona. For the mass of Interamnia a result of $(0.37 \pm 0.17) \times 10^{-10}$ solar masses was obtained. This result gives an independent information about the mass of Interamnia besides the estimation from the diameter and the suspected density.

098.071   Solar system: Wandering on a leash.
C. D. Murray.
Nature, Vol. 357, No. 6379, p. 542 – 543 (18 Jun 1992).
The author reports on stable chaotic motion, identified in a number of orbits of bodies in the solar system. An example of such a chaotic orbit is that of the minor planet (522) Helga.

098.072   An example of stable chaos in the solar system.
A. Milani, A. M. Nobili.
Nature, Vol. 357, No. 6379, p. 569 – 571 (18 Jun 1992). Letter–to–the–editor.
Many planets have been shown to have chaotic instabilities in their orbital motions, but the long–term significance of this is not fully understood. The authors show that the orbit of the near–Jupiter asteroid 522 Helga is chaotic, with an unusually short Lyapunov time of 6,900 yr. They integrate its motion, including perturbations from the outer giant planets, over a period, 1,000 times longer than this, and find no significant instability. Chaos in the orbit of 522 Helga is caused by a 7:12 resonance with the orbit of Jupiter, but the size of the chaotic region in phase space is small; stability is ensured because the eccentricity and precession of the orbit are such that it avoids close encounters with Jupiter. Asteroid orbits with larger proper eccentricity would, they suggest, be genuinely unstable, consistent with the sparse asteroid population near Helga. Although Helga is the first clear–cut example of a stable chaotic orbit, the authors argue that "stable chaos" may be a rather common feature of solar system dynamics.

098.073   The principle of equivalence and the Trojan asteroids. II.
R. B. Orellana, H. Vucetich.
IAU Symposium No. 152: Chaos, resonance and collective dynamical phenomena in the solar system, p. 185 – 188 (1992). – See Abstr. 012.025 for the main entry.
A new value for the Nordtvedt parameter and the mass of Saturn are computed using the ten first Trojan asteroids. From 1262 observations, the authors find the inverse mass of Saturn 3498.17 ± 0.51 and the Nordtvedt parameter 0.0 ± 0.3.

098.074   Mid–infrared (7.5 – 12.8 μm) spectra of 4 Vesta, 16 Psyche, 24 Themis, 113 Amalthea and the Moon.
A. Sprague, F. Witteborn, D. Cruikshank, M. Bartholomew, R. Kozlowski.
Bull. Am. Astron. Soc., Vol. 23, No. 3, p. 1138 (1991). Abstract. – See Abstr. 012.037 for the main entry.

098.075   A 0.43–μm absorption feature in C–class asteroid reflectance spectra.
E. Hatch, F. Vilas, S. Larson, M. Gaffey.
Bull. Am. Astron. Soc., Vol. 23, No. 3, p. 1138 (1991). Abstract. – See Abstr. 012.037 for the main entry.

098.076   Effect of assumed grain size on abundance determinations for Vesta using Hapke theory.
M. L. Nelson.
Bull. Am. Astron. Soc., Vol. 23, No. 3, p. 1138 – 1139 (1991). Abstract. – See Abstr. 012.037 for the main entry.

098.077   Ammonium–bearing minerals on Ceres: potential mineral mixtures.
T. V. V. King, R. N. Clark, W. M. Calvin, D. M. Sherman, R. H. Brown.
Bull. Am. Astron. Soc., Vol. 23, No. 3, p. 1139 (1991). Abstract. – See Abstr. 012.037 for the main entry.

098.078   Aqueous alteration products in the reflectance spectra of primitive asteroids.
F. Vilas, E. Hatch, M. Gaffey, S. Larson.
Bull. Am. Astron. Soc., Vol. 23, No. 3, p. 1139 (1991). Abstract. – See Abstr. 012.037 for the main entry.

098.079   Near–infrared reflectance spectrum and lightcurve of the E–type Apollo asteroid (3103) 1982 BB.
M. J. Gaffey, K. L. Reed, M. S. Kelley.
Bull. Am. Astron. Soc., Vol. 23, No. 3, p. 1140 (1991). Abstract. – See Abstr. 012.037 for the main entry.

098.080   Characterization of low albedo asteroids.
L. A. Lebofsky, E. S. Howell, D. T. Britt.
Bull. Am. Astron. Soc., Vol. 23, No. 3, p. 1140 (1991). Abstract. – See Abstr. 012.037 for the main entry.

098.081   Using neural networks to classify asteroid spectra.
E. S. Howell, E. Merényi, L. A. Lebofsky.
Bull. Am. Astron. Soc., Vol. 23, No. 3, p. 1140 – 1141 (1991). Abstract. – See Abstr. 012.037 for the main entry.

098.082   On the ages of asteroid dynamical families.
F. Marzari, D. R. Davis.
Bull. Am. Astron. Soc., Vol. 23, No. 3, p. 1141 (1991). Abstract. – See Abstr. 012.037 for the main entry.

098.083   Will Gaspra data solve the S–asteroid controversy?
C. R. Chapman.
Bull. Am. Astron. Soc., Vol. 23, No. 3, p. 1141 (1991). Abstract. – See Abstr. 012.037 for the main entry.

098.084   Recreating the ancient asteroid belt: reassembling asteroid family parent bodies.
J. C. Granahan, J. F. Bell.
Bull. Am. Astron. Soc., Vol. 23, No. 3, p. 1141 (1991). Abstract. – See Abstr. 012.037 for the main entry.

098.085   Geminid meteoroids and the probability for cometary activity on Phaethon.
L. Adolfsson, B. Å. S. Gustafson.
Bull. Am. Astron. Soc., Vol. 23, No. 3, p. 1141 (1991). Abstract. – See Abstr. 012.037 for the main entry.

098.086   S asteroids 387 Aquitania and 980 Anacostia: possible fragments of the breakup of a spinel–rich parent body.
T. H. Burbine, M. J. Gaffey, J. F. Bell.
Bull. Am. Astron. Soc., Vol. 23, No. 3, p. 1142 (1991). Abstract. – See Abstr. 012.037 for the main entry.

098.087   New results on C $\equiv$ N–bearing solid organics on asteroids and comets.
D. P. Cruikshank, W. K. Hartmann, D. J. Tholen.
Bull. Am. Astron. Soc., Vol. 23, No. 3, p. 1142 (1991). Abstract. – See Abstr. 012.037 for the main entry.

098.088 **951 Gaspra: results from observations spanning two oppositions.**
J. D. Goldader, D. J. Tholen, T. Herbst, W. Golisch,
J. R. Spencer, D. P. Cruikshank, W. K. Hartmann.
Bull. Am. Astron. Soc., Vol. 23, No. 3, p. 1142 (1991). Abstract. –
See Abstr. 012.037 for the main entry.

098.089 **Asteroid size/albedo calibration.**
S. T. Holfeltz, L. G. Taff.
Bull. Am. Astron. Soc., Vol. 23, No. 3, p. 1143 (1991). Abstract. –
See Abstr. 012.037 for the main entry.

098.090 **Pole and asymmetric shape of 39 Laetitia.**
K. Lumme, E. Bowell.
Bull. Am. Astron. Soc., Vol. 23, No. 3, p. 1143 (1991). Abstract. –
See Abstr. 012.037 for the main entry.

098.091 **Time–resolved spectrophotometry of 951 Gaspra with the Galileo filters: results for the 1991 opposition.**
S. Mottola, M. Di Martino, H. Hoffmann, M. Gonano–Beurer,
G. Neukum.
Bull. Am. Astron. Soc., Vol. 23, No. 3, p. 1143 (1991). Abstract. –
See Abstr. 012.037 for the main entry.

098.092 **Status of the IRAS minor planet survey (IMPS).**
E. F. Tedesco, G. J. Veeder, D. L. Matson,
J. R. Chillemi, J. W. Fowler.
Bull. Am. Astron. Soc., Vol. 23, No. 3, p. 1144 (1991). Abstract. –
See Abstr. 012.037 for the main entry.

098.093 **High–resolution radar ranging to near–Earth asteroids.**
S. J. Ostro, J. K. Harmon, A. A. Hine, P. Perillat,
D. B. Campbell, J. F. Chandler, I. I. Shapiro, R. F. Jurgens,
D. K. Yeomans.
Bull. Am. Astron. Soc., Vol. 23, No. 3, p. 1144 (1991). Abstract. –
See Abstr. 012.037 for the main entry.

098.094 **Infrared (JHK) photometry of family asteroids.**
G. J. Veeder, E. F. Tedesco.
Bull. Am. Astron. Soc., Vol. 23, No. 3, p. 1144 (1991). Abstract. –
See Abstr. 012.037 for the main entry.

098.095 **Discovery strategy for Earth–crossing asteroids.**
K. Muinonen, E. Bowell.
Bull. Am. Astron. Soc., Vol. 23, No. 3, p. 1151 (1991). Abstract. –
See Abstr. 012.037 for the main entry.

098.096 **Follow–up and recovery of asteroids using orbital error estimation.**
E. Bowell, K. Muinonen.
Bull. Am. Astron. Soc., Vol. 23, No. 3, p. 1151 (1991). Abstract. –
See Abstr. 012.037 for the main entry.

098.097 **The captive asteroids.**
C. J. Cunningham.
Astronomy, Vol. 20, No. 6, p. 40 – 44 (Jun 1992).

098.098 **Der Brocken – ein neuer Tautenburger Planetoid.**
F. Börngen.
Zeiss Inf. Jenaer Rundsch., Jahrg. 1, Nr. 1, p. 11 – 12 (1992).

098.099 **Evidence against dusty regoliths on small main belt asteroids.**
G. J. Veeder.
Bull. Am. Astron. Soc., Vol. 23, No. 3, p. 1155 (1991). Abstract. –
See Abstr. 012.037 for the main entry.

098.100 **CCD photometry of 2060 Chiron during 1985 and 1991.**
R. L. Marcialis, B. J. Buratti.
Bull. Am. Astron. Soc., Vol. 23, No. 3, p. 1155 (1991). Abstract. –
See Abstr. 012.037 for the main entry.

098.101 **A model of 951 Gaspra.**
M. A. Barucci, M. Fulchignoni, C. De Sanctis.
Bull. Am. Astron. Soc., Vol. 23, No. 3, p. 1156 (1991). Abstract. –
See Abstr. 012.037 for the main entry.

098.102 **M type asteroids: new observations and statistics.**
M. Fulchignoni, E. Dotto, M. A. Barucci.
Bull. Am. Astron. Soc., Vol. 23, No. 3, p. 1156 (1991). Abstract. –
See Abstr. 012.037 for the main entry.

098.103 **Near–Earth and small main–belt asteroids: a comparison of physical properties.**
R. P. Binzel, S. Xu, S. J. Bus, E. Bowell.
Bull. Am. Astron. Soc., Vol. 23, No. 3, p. 1156 (1991). Abstract. –
See Abstr. 012.037 for the main entry.

098.104 **The Palomar Planet–Crossing Asteroid Survey (PCAS) discoveries: the last three years.**
E. F. Helin.
Bull. Am. Astron. Soc., Vol. 23, No. 3, p. 1156 (1991). Abstract. –
See Abstr. 012.037 for the main entry.

098.105 **International search program for NEAs.**
D. Morrison.
Bull. Am. Astron. Soc., Vol. 23, No. 3, p. 1157 (1991). Abstract. –
See Abstr. 012.037 for the main entry.

098.106 **The first workshop on 2060 Chiron: What manner of beast is the Centaur?**
R. L. Marcialis, S. J. Bus.
Bull. Am. Astron. Soc., Vol. 23, No. 3, p. 1157 (1991). Abstract. –
See Abstr. 012.037 for the main entry.

098.107 **Submillimeter photometry of 2060 Chiron.**
D. Jewitt, J. Luu.
Bull. Am. Astron. Soc., Vol. 23, No. 3, p. 1158 (1991). Abstract. –
See Abstr. 012.037 for the main entry.

098.108 **Are extreme seasonal variations controlling Chiron's activity?**
M. V. Sykes, P. R. Weissman.
Bull. Am. Astron. Soc., Vol. 23, No. 3, p. 1158 – 1159 (1991).
Abstract. – See Abstr. 012.037 for the main entry.

098.109 **A diagnostic spectral indicator of the exposure age of an asteroidal surface.**
J. A. Nuth III.
IAU Symposium No. 150: Astrochemistry of cosmic phenomena,
p. 461 – 462 (1992). – See Abstr. 012.029 for the main entry.

098.110 **Kritische Kleinplaneten**
J. Jahn.
KPM, Jahrg. 7, No. 19, p. 13 – 17 (Apr 1992).
The author presents the latest critical list of minor planets from various MPCs with orbital elements and selected ephemerides of the brightest objects.

098.111 **1992 AD: une comète de plus ou bien le plus éloigné des astéroides?**
O. Hainaut.
Ciel, Vol. 54, p. 65 – 67 (Mar 1992).

098.112 **1991 VK, 1991 VL, 1991 WA and 1991 XB.**
IAU Circ., No. 5422, p. 1 (3 Jan 1992).

098.113 **1991 YA.**
IAU Circ., No. 5423, p. 1 (8 Jan 1992).

098.114 **1992 AA.**
IAU Circ., No. 5424, p. 1 (8 Jan 1992).

098.115 **1992 AB.**
IAU Circ., No. 5425, p. 1 (8 Jan 1992).

098.116   1992 AC.
IAU Circ., No. 5426, p. 1 (8 Jan 1992).
Further information is given in Nos. 5442, 5474.

098.117   Mass distribution in the asteroid belt.
J. Klačka.
Earth, Moon, Planets, Vol. 56, No. 1, p. 47 – 52 (Jan 1992).
The dependence of the cumulative number of numbered asteroids (up to 3720) on their absolute magnitude is investigated. The differential mass index k is derived from these relations for fainter asteroids. A steeper slope ($2.2 < k < 2.4$) is found in the four most populous asteroid families (Flora, Koronis, Eos and Themis) and a flatter slope ($1.3 < k < 1.6$) for non–family asteroids. This indicates that there are two different asteroid polulations in the asteorid belt. Total masses of the asteroid families may be greater than it is commonly accepted.

098.118   1992 AD = (5145) Pholus.
IAU Circ., No. 5434, p. 1 (23 Jan 1992).
Further information is given in Nos. 5435, 5449, 5450, 5451, 5458, 5462, 5480.

098.119   1992 AE.
IAU Circ., No. 5436, p. 1 (23 Jan 1992).

098.120   Another Chiron–type object.
R. M. West.
Messenger, No. 67, p. 34 – 35 (Mar 1992).

098.121   1992 BC.
IAU Circ., No. 5441, p. 1 (31 Jan 1992).
Further information is given in No. 5448.

098.122   1992 BF.
IAU Circ., No. 5443, p. 1 (3 Feb 1992).
Further information is given in No. 5466.

098.123   (2060) Chiron.
IAU Circ., No. 5457, p. 1 (24 Feb 1992).

098.124   1992 CC$_1$.
IAU Circ., No. 5459, p. 1 (27 Feb 1992).
Further information is given in No. 5483.

098.125   1984 WE$_1$.
IAU Circ., No. 5484, p. 1 (30 Mar 1992).

098.126   1992 FE.
IAU Circ., No. 5488, p. 1 (2 Apr 1992).

098.127   Additions to the Taurid Complex.
D. Steel.
Observatory, Vol. 112, No. 1108, p. 120 – 122 (Jun 1992).
Three recently–discovered Apollo–type asteroids (1991 GO, 1991 TB2, and 1991 VL) appear likely to be members of the Taurid Complex of interplanetary objects. The other macroscopic objects in this complex include P/Comet Encke, 2201 Oljato, 5025 P–L, 1982 TA, and 1984 KB, quite apart from the long–known meteor showers which give the complex its name. The study of these objects may elucidate many facets of the recent impact history of the Earth.

098.128   1992 HE.
IAU Circ., No. 5508, p. 1 (30 Apr 1992).
Further information is given in Nos. 5509, 5543.

098.129   1992 JB.
IAU Circ., No. 5510, p. 1 (4 May 1992).

098.130   1992 JD.
IAU Circ., No. 5511, p. 1 (4 May 1992).
Further information is given in No. 5512.

098.131   1992 JE.
IAU Circ., No. 5515, p. 1 (7 May 1992).
Further information is given in No. 5542.

098.132   1992 KD.
IAU Circ., No. 5531, p. 1 (30 May 1992).
Further information is given in No. 5550.

098.133   1992 LC.
IAU Circ., No. 5538, p. 1 (10 Jun 1992).

098.134   1992 LR.
IAU Circ., No. 5548, p. 1 (22 Jun 1992).

098.135   1992 AC.
Yamamoto Circ., No. 2177, p. 1 – 2 (26 Jan 1992). In Japanese.

098.136   1992 AD = (5145) Pholus.
Yamamoto Circ., No. 2177, p. 2 (26 Jan 1992). In Japanese.
Further information is given in No. 2180.

098.137   1984 WE$_1$.
Yamamoto Circ., No. 2182, p. 2 (4 Apr 1992). In Japanese.

098.138   Galileo fly–by near asteroid Gaspra.
A. T. Bazilevskij.
Astron. Vestn., Tom 26, No. 3, p. 3 – 7 (May – Jun 1992). In Russian. English translation in Sol. Syst. Res., Vol. 26, No. 3.
October 29, 1991 Galileo spacecraft has flown nearby asteroid 951 Gaspra, received TV images of the asteroid and made other observations. One of the images was delivered to Earth and is described in this paper. The imaged part of the asteroid is 12×16 km across. Shape of Gaspra is rounded–angular what is typical for fragmentation forms smoothed by small meteoroids blastering. This morphology is evidently typical for all small bodies of the Solar system, may be, besides comets the material of which evaporates when they pass near the Sun thus competing with impact phenomena.

098.139   Highly variable objects in the solar system.
W. Z. Wisniewski.
1. European Meeting of the American Association of Variable Star Observers (AAVSO): International Cooperation and Coordination in Variable Star Research, p. 159 – 168 (1992). – See Abstr. 012.048 for the main entry.

098.140   Determination of the mass of (1) Ceres from perturbations on (203) Pompeja and (348) May.
G. Sitarski, B. Todorovic–Juchniewicz.
Acta Astron., Vol. 42, No. 2, p. 139 – 144 (1992).
The authors collected 223 observations of Pompeja from 1879–1990 and 88 observations of May from 1892–1991 to use them for improvement of orbits of the asteroids and for determination of the mass of Ceres. They applied the recurrent power series method to integrate numerically equations of motion of Pompeja and of May as disturbed by Ceres; in the observational interval both asteroids approached Ceres closer than to 0.1 a.u. twice. The authors created thirteen normal equations corresponding to 589 observational equations, to correct by the least squares method the mass of Ceres along with orbital elements of Pompeja and of May. The authors found the mass of Ceres equal to $(4.796 \pm 0.085) \times 10^{-10}$ in solar unit mass.

098.141   The collision frequencies between Aten–Apollo–Amor objects and the Earth.
Lu Jianhua.
Publ. Purple Mt. Obs., Vol. 10, No. 3, p. 188 – 195 (Sep 1991). In Chinese.
An improvement to Kessler's formulae on the collision frequency of orbiting objects is given. The improved formulae are available to the objects of lower orbital inclination. Furthermore,

the collision frequencies between 41 Aten–Apollo–Amor objects and the Earth were calculated. The results show that the average collision frequency is 4.7×10^{-11}/year; the average collision lifetime is 2.1×10^{10} year.

098.142 Astrometric observations of asteroid Hidalgo near its perihelion.
T. Nakamura, G. Sasaki, S. Okamura, M. Hamabe, S. Yoshida, Y. Taniguchi.
Publ. Astron. Soc. Jpn., Vol. 44, No. 2, p. L19 – L21 (1992). Letter–to–the–editor.
The astrometric positions of asteroid Hidalgo have been obtained with the Kiso 105–cm Schmidt telescope near its perihelion passage of 1991. The observations cover from 1990 October through 1991 March.

098.143 A high resolution CCD spectroscopic survey of low albedo main belt asteroids – results and analysis.
S. R. Sawyer.
Bull. Am. Astron. Soc., Vol. 23, No. 3, p. 1235 (1991). Abstract. – See Abstr. 012.037 for the main entry.

098.144 On the original distribution of the asteroids. IV. Numerical experiments in the outer asteroid belt.
M. Lecar, F. Franklin, P. Soper.
Icarus, Vol. 96, No. 2, p. 234 – 250 (Apr 1992).
The authors assumed, in 1973, that the asteroids were the building blocks of the planets and initially were distributed approximately uniformly. They suggested that the present distribution was carved out by the perturbations of Jupiter. To test this hypothesis further, the authors have now integrated the orbits of 140 asteroids with 14 semimajor axes from 0.63 to 0.76, and at 10 values of the eccentricity from 0.01 to 0.19. The integrations extended for 1 million Jovian years unless the asteroid crossed Jupiter's orbit before then. The integrations were confined to a plane. In the main survey, the authors chose the most "unstable" phases of the asteroid relative to Jupiter. There are only 8 of the 8032 asteroids with reliable orbits with semimajor axes between 0.68 and 0.74. In the simulation, 71% of the asteroids with semimajor axes equal to or larger than 0.680 were removed. Of the 26 asteroids that remained, 16 had initial eccentricities equal to or less than 0.03. Ten of the remaining asteroids with $a \geqslant 0.68$ were integrated for 10 million Jovian years; 5 became Jupiter crossers and 5 remained. Including these longer integration, 77% of the asteroids with $a \geqslant 0.68$ escaped. However, the depletion was much reduced in those cases when the initial phases caused conjunctions to occur at the asteroid's perihelion. The authors found a strong correlation between the Lyapunov exponent and the crossing time for orbits with crossing times less than 1 million Jovian years. If this result extrapolates to 4.5 billion years, it would indicate that with Jupiter the only perturber, there are stable orbits where there are no asteroids today.

098.145 Asteroid families – an initial search.
J. G. Williams.
Icarus, Vol. 96, No. 2, p. 251 – 280 (Apr 1992).
A sample of both numbered and faint Palomar–Leiden Survey (PLS) asteroids was searched for clusters in three–dimensional proper element space. These clusters, selected by stereo examination, were then filtered objectively by requiring that their Poisson probability of chance occurrence be small. 104 clusters clusters were accepted as families. These families have populations from 4 to 102 members, with a median of 8 members. Well–populated families in uncrowded regions are the most certain. The Themis, Eos, and Koronis families are obvious; major families include Eunomia and Alexandra. Families are interpreted as generated by impact. The Eos and Koronis families are examples of total breakups and the Themis family is a partial breakup with all or most of the material excavated from one side of the parent body.

098.146 Evolution of Earth–crossing binary asteroids due to gravitational encounters with the Earth.
P. Farinella.
Icarus, Vol. 96, No. 2, p. 284 – 285 (Apr 1992).
During close encounters with the Earth, binary Earth–approaching asteroids have their orbital energy changed by Earth tidal forces. The total amount of this change, over the binary's lifetime, is of the order of the energy itself. As a consequence, an initial population of binaries with separations of a few times the sum of the radii of the components evolves to a broader orbital distribution, which includes both increased separations, leading to formation of "loose binaries" and possibly to escape of the components, and smaller orbital distances, eventually creating contact binary systems. The existence of well–separated systems is suggested by the terrestrial cratering record, while contact systems have been recently observed by radar.

098.147 Photoelectric photometry of asteroid 394 Arduina.
R. G. Hutton.
Rev. Mex. Astron. Astrofis., Vol. 24, No. 1, p. 43 – 44 (Apr 1992).
The lightcurve of the asteroid 394 Arduina has been studied photometrically in B and V colors. From four fragments of the curve, observed during the September 1990 opposition, a rotation period of 16.53 ± 0.01 h, a maximum amplitude of 0.54 ± 0.01 mag and a (B–V) color index of 0.840 ± 0.014 were obtained.

098.148 Collision rates and impact velocities in the main asteroid belt.
P. Farinella, D. R. Davis.
Icarus, Vol. 97, No. 1, p. 111 – 123 (May 1992).
The authors have computed mutual collision probabilities and impact velocities for a set of 682 asteroids of diameter > 50 km, intended to represent a bias–free sample of asteroid orbits. For every asteroid, they have obtained the intrinsic collision probability, P_i, the average collision velocity, V, and the number of projectile orbits which can intersect the target asteroid's orbit, N_{cross}, using the proper orbital elements. No significant differences were found in the average values of P_i, V, or N_{cross} using osculating elements instead of proper elements, although results for individual asteroids could change by $\approx 10\%$. Collision probabilities are nearly independent of eccentricities but show a significant decrease with larger inclinations. As expected, collisional velocities grow rapidly with increasing orbital eccentricities and inclinations, but they show surprisingly little variation across the asteroid belt.

098.149 Asteroid 951 Gaspra: pre–Galileo physical model.
P. Magnusson, M. A. Barucci, R. P. Binzel, C. Blanco, M. Di Martino, J. D. Goldader, M. Gonano–Beurer, A. W. Harris, T. Michałowski, S. Mottola, D. J. Tholen, W. Z. Wisniewski.
Icarus, Vol. 97, No. 1, p. 124 – 129 (May 1992).
Prior to the Oct 1991 encounter of 951 Gaspra, asteroids were the only class of Solar System objects not resolved by spacecraft images. Analysis of the data returned by the Galileo spacecraft will apply critical tests to many techniques developed during the past 50 years for deriving physical properties for several hundred asteroids. As a benchmark for comparison, the authors have applied these techniques to establish in advance the best estimates for Gaspra's properties. The authors estimate it has an elongated shape with a ratio between equatorial axes of about 1.5 – 1.7. Further, they derive its sidereal rotation period to be 0.2934197 days.

098.150 The steep red spectrum of 1992 AD: an asteroid covered with organic material?
U. Fink, M. Hoffmann, W. Grundy, M. Hicks, W. Sears.
Icarus, Vol. 97, No. 1, p. 145 – 149 (May 1992).
A spectrum of the newly discovered asteroid (5145) 1992 AD was obtained with a CCD camera and spectrometer 1992 Feb 01.23. The reflection spectrum of 1992 AD displays a very steep and constant red slope between 0.5 and 1.0 μm and exhibits

no absorption nor emission features. The red slope is steeper than that of any presently known Solar System object. The reflectivity ratio between 1.0 and 0.55 μm is a factor of 3.5, or using a slight extrapolation, a factor of 4.90 for the wavelength octave 1.0 to 0.5 μm. The steep red slope is difficult to match with conventional silicate or meteoritic materials. While allotropes of sulfur may give a partial match, the best match is provided by the steep red spectra of mixtures of tholins, the residues left after subjecting organic molecules to an energetic radiation environment.

098.151 **Extraordinary colors of asteroidal object (5145) 1992 AD.**
B. E. A. Mueller, D. J. Tholen, W. K. Hartmann, D. P. Cruikshank.
Icarus, Vol. 97, No. 1, p. 150 – 154 (May 1992).
The recently discovered outer Solar System object, (5145) 1992 AD, in a somewhat Chiron–like orbit, has colors far redder than any other known asteroids or comets, and represents a hitherto–unknown spectral class. The red color may be associated with exposure of organics that are purer or more pristine than those found on the surfaces of C, P, and D asteroids, and comets, and such materials are likely to show diagnostic spectral features in the infrared.

098.152 **High resolution surface brightness profiles of near–earth asteroids.**
J. X. Luu, D. C. Jewitt.
Icarus, Vol. 97, No. 2, p. 276 – 287 (Jun 1992).
The authors present a new method to search for and estimate mass loss in near–Earth asteroids (NEAs) using high resolution surface photometry. The method was applied to 11 NEAs observed with a CCD at the University of Hawaii 2.2–m telescope. The method yields limiting mass loss rates $\dot{M} \leqslant 0.1 \text{ kg sec}^{-1}$, 1 – 2 orders of magnitude smaller than the typical rates of weakly active comets. However, these mass loss rate upper limits imply fractional active areas in NEAs that are comparable to cometary fractional active areas. Because of the small sizes of the NEAs, the mass loss rates produced by these fractional active areas are below the detection limit of current techniques; thus there may exist low–level cometary activity amongst the NEAs which goes unnoticed.

098.153 **Asteroid families identified by two different methods.**
P. Bendjoya, A. Cellino.
15. Ecole de Printemps d'Astrophysique de Goutelas: Interrelations between physics and dynamics for minor bodies in the solar system, p. 19 – 43 (1992). – See Abstr. 012.055 for the main entry.
Contents: 1. Introduction. 2. The hierarchical clustering method. 3. The wavelet analysis method. 4. Discussion.

098.154 **On the search for asteroid families.**
A. Milani, P. Farinella, Z. Knežević.
15. Ecole de Printemps d'Astrophysique de Goutelas: Interrelations between physics and dynamics for minor bodies in the solar system, p. 85 – 132 (1992). – See Abstr. 012.055 for the main entry.
Contents: 1. Proper elements (Input catalogues. Mean elements and mean motion resonances. Secular perturbations and proper elements. Iterative procedure and convergence problems. Tests for stability of the proper elements). 2. Clustering (Metric. Clustering methods. Background. Rejection criteria and significance parameters. Robustness and noise. Asteroid families and secular resonances). 3. Physical evidence (Collisional theory. Mass distributions. Velocities, anisotropies, structures. Rotations and shapes. Taxonomy and compositions. Ages and evolution).

098.155 **Orbital stability zones about asteroids. II. The destabilizing effects of eccentric orbits and of solar radiation.**
D. P. Hamilton, J. A. Burns.
Icarus, Vol. 96, No. 1, p. 43 – 64 (Mar 1992).
The gravitational effects of the Sun on a particle orbiting another massive body which itself moves on a circular path around the Sun have been studied extensively. The authors now consider two effects analytically and numerically: the asteroid's

nonzero heliocentric eccentricity and solar radiation pressure. In both of these cases, the numerical intergrations apply directly to a spherical asteroid, "Amphitrite". For an asteroid on an eccentric orbit it is argued that the stability zone scales roughly as the size of the Hill sphere calculated at the asteroid's pericenter. This scaling holds for large values of eccentricity and allows results for one asteroid with a given mass, semimajor axis, and eccentricity to be used for another with different values of these parameters. The authors compare predictions of the scaling law to numerical integrations for an "Amphitrite" with various orbital eccentricities and find good agreement for prograde orbits and for those with orbital planes nearly normal to the asteroid's heliocentric path, but not for retrograde orbits. The authors apply their results to the minor planet 951 Gaspra.

098.156 **Minor planets at unusually favorable oppositions in 1992.**
F. Pilcher.
Minor Planet Bull., Vol. 19, No. 1, p. 2 – 3 (Jan – Mar 1992).

098.157 **Close mutual approaches of minor planets in 1992.**
E. Goffin.
Minor Planet Bull., Vol. 19, No. 1, p. 4 (Jan – Mar 1992).

098.158 **Photoelectric photometry opportunities, February – April.**
A. W. Harris, V. Zappalà.
Minor Planet Bull., Vol. 19, No. 1, p. 5 (Jan – Mar 1992).

098.159 **Close approaches of minor planets to naked eye stars in 1992.**
E. Goffin.
Minor Planet Bull., Vol. 19, No. 1, p. 5 – 7 (Jan – Mar 1992).

098.160 **Photoelectric photometry opportunities, May – July.**
A. W. Harris, V. Zappalà.
Minor Planet Bull., Vol. 19, No. 2, p. 16 (Apr – Jun 1992).

098.161 **Future earth approaches of 4179 Toutatis.**
P. Sicoli, M. Cavagna.
Minor Planet Bull., Vol. 19, No. 2, p. 17 (Apr – Jun 1992).
The authors report an investigation of near–Earth passages for the asteroid 4179 Toutatis during the years 1900 – 2100. The only approach closer than that during 1992 occurs in 2004 at a distance of 0.010 AU.

098.162 **The temperature of an asteroid.**
D. B. Henry.
Ann. Inst. Henri Poincaré, Phys. Théor., Vol. 55, No. 2, p. 719 – 750 (Oct 1991). Paper presented at the Workshop on Multiscale Phenomena, Sao Paulo (Brazil), 16 – 17 Aug 1990.
The heat equation in a solid body with radiation boundary conditions and periodic heating at the boundary – modeling a rotating asteroid – is shown to have a unique time–periodic solution and it is globally attracting. The author finds approximations to this solution for various cases relevant to the possible existence of ice in the asteroid belt.

098.163 **CCD astrometric observations of the asteroid 951 Gaspra.**
A. K. B. Monet, R. C. Stone, C. C. Dahn.
Bull. Am. Astron. Soc., Vol. 23, No. 3, p. 1258 (1991). Abstract. – See Abstr. 012.065 for the main entry.

098.164 **The ephemeris development effort for minor planets 951 Gaspra and 243 Ida.**
D. K. Yeomans.
Bull. Am. Astron. Soc., Vol. 23, No. 3, p. 1258 (1991). Abstract. – See Abstr. 012.065 for the main entry.

098.165   **The effect of crossing–point observations on ephemeris uncertainties for asteroid 243 Ida.**
P. W. Chodas, G. W. Null.
Bull. Am. Astron. Soc., Vol. 23, No. 3, p. 1259 (1991). Abstract. – See Abstr. 012.065 for the main entry.

098.166   **Rapid–response radar observation of earth–approaching asteroids.**
J. F. Chandler, I. I. Shapiro, S. J. Ostro.
Bull. Am. Astron. Soc., Vol. 23, No. 3, p. 1259 (1991). Abstract. – See Abstr. 012.065 for the main entry.

098.167   **5.000 astéroides numérotés.**
M.–A. Combes, J. Meeus.
Astronomie, p. 1 – 4 (Mar 1992).

098.168   **Method for position observations of fast–moving asteroids and the practice of its use.**
M. R. Nesteruk.
Kinematika Fiz. Nebesn. Tel, Tom 8, No. 3, p. 97 – 99 (May–Jun 1992). In Russian. English translation in Kinematics Phys. Celest. Bodies, Vol. 8, No. 3.
A method is discussed for position observations of fast-moving asteroids that gives a possibility to decrease the error of coordinate determination by a factor of 1.5 and to observe faint objects. Seven positions of asteroid 1036 Ganymed are given.

098.169   **Photometry and polarimetry of the asteroid 47 Aglaja.**
G. P. Chernova, D. F. Lupishko, V. G. Shevchenko, N. N. Kiselev, R. Salles.
Kinematics Phys. Celest. Bodies, Vol. 7, No. 5, p. 15 – 21 (1991). English translation of Kinematika Fiz. Nebesn. Tel., Tom 7, No. 5, p. 20 – 26 (Sep–Oct 1991).
The light curves and the phase curves of light, color, and polarization of the C–type asteroid 47 Aglaja were observed on 33 nights in September – November 1989. The light phase curve found in the phase-angle range $\alpha = 0.1 – 12.9°$ indicates that asteroid 47 does not show the sudden brightness increase at opposition, in the range $\alpha = 0 – 2°$, that had been detected previously in high–albedo E–type asteroids. Note is taken of the difference between the formation mechanisms of the opposition light curves of high– and low–albedo asteroids and a possible explanation. This was the first polarimetry and the first derivation of polarization–curve parameters for 47 Aglaja. The rotation period was adjusted to $P = 13.178 \pm 0.005$ hour; the albedo was estimated at $p_v = 0.073$ and the diameter at $D = 122.0$ km.

098.170   **Polarimetry of CMEU asteroids. II. A peculiarity of M–type asteroids.**
I. N. Bel'skaya, N. N. Kiselev, D. F. Lupishko, G. P. Chernova.
Kinematics Phys. Celest. Bodies, Vol. 7, No. 6, p. 8 – 11 (1991). English translation of Kinematika Fiz. Nebesn. Tel., Tom 7, No. 6, p. 11 – 14 (Nov–Dec 1991).
Measurements of the depth P_{min} of the negative branch of the polarization phase curves are reported for asteroids 147, 217, 325, 347, and 796. Four of them are classified as M–type on the basis of P_{min} and U–V color, and one – 147 Protogeneia – as a C– or other low–albedo type. It is shown that M asteroids differ from other types of asteroids in the range of values of the parameter $P_{min}p_v$, forming a certain sequence. It is suggested that this may be due to varying metal contents in the surface layers of M asteroids.

098.171   **Ehfemeridy malykh planet na 1993 god. (*Ephemerides of minor planets for 1993.*)**
Yu. V. Batrakov (ed.).
Rossijskaya Akademiya Nauk, Sankt-Peterburg. Inst. Teoreticheskoj Astronomii. 512 p. (1992).
Contents: 1. Introduction. 2. Information on new elements. 3. Elements. 4. Osculating elements and inverse masses of perturbing planets. 5. Minor planet lightcurve parameters. 6. Opposition dates. 7. Ephemerides. 8. Ephemerides of some unusual planets. 9. Status of minor planet observations.

10. Antisun and Moon. 11. Information on computer version of EMP (STAMP).

098.172   **Asteroids and comets – the relationship between near–earth asteroids and main–belt asteroids and comets.**
Wang Qi.
Publ. Purple Mt. Obs., Vol. 11, No. 2, p. 75 – 84 (Jun 1992). In Chinese.
This paper discusses: (1) the relation between asteroids and short period comets. (2) The relation of near–Earth asteroids (NEAs) to main–belt asteroids and to short period comets. Recent observations and studies indicate that some of the asteroids have originally relation with comets and some of the NEA originate probably from main–belt asteroids and extinct cometary nuclei.

098.173   **Asteroidal occultations results.**
R. Boninsegna (ed.).
AOR (Asteroidal Occultation Results), No. 16.
European Asteroidal Observation Network (EAON). 2 p. (1992).
Contents: 1. (83) Beatrix and $+1°1465$ on 1990 March 13.
2. (19) Fortuna and AGK3 $+22°623$ on 1990 September 24.
3. (139) Juewa and AGK3 $+7°61$ on 1990 October 22.
4. (50) Virginia and AGK3 $+16°373$ on 1991 December 31.

098.174   **Asteroidal occultations (May to December 1991).**
R. Boninsegna (ed.).
EAON Inf., No. 11.
European Asteroidal Observation Network (EAON). 10 p. (1992).

098.175   **Predictions of stellar occultations by asteroids (for the last months of the year 1992).**
E. Goffin, R. Boninsegna.
European Asteroidal Observation Network (EAON). 10 p. (1992).
Several graphs containing the details of occultations of stars by minor planets are given.

098.176   **Near–Earth asteroids and the history of planetary formation.**
T. D. Swindle, J. S. Lewis, L.–A. A. McFadden.
Earth Space, Vol. 4, No. 6, p. 11 – 14 (Feb 1992).
A future sample return mission to an asteroid would answer questions about the formation of the solar system and, possibly, lead to a new source of valuable materials.

098.177   **Les familles d'astéroides et la transformée en ondelette.**
P. Bendjoya.
J. Astron. Fr., No. 41, p. 11 (Feb 1992). – See Abstr. 012.070 for the main entry.

098.178   **Radar reconnaissance of near–earth asteroids.**
S. J. Ostro.
International Symposium on Radars and Lidars in Earth and Planetary Sciences, p. 9 – 13 (Dec 1991). – See Abstr. 012.093 for the main entry.
During the past decade, echoes from 29 near–Earth asteroids have provided new information about these objects' physical and dynamical properties. In the investigation of a near–Earth asteroid, the strategy is to detect the echo, refine the delay–Doppler ephemeris, and then determine the dual–polarization delay–Doppler signature at the finest time–frequency resolution possible, as a function of the target's rotation phase and sky position. In principle, with instrumentation expected to be available by the mid–1990s, such an experiment should be able to yield a physical characterization of an asteroid that is as informative as Mariner 13's images of Phobos and Deimos. The near–Earth asteroid population probably contains many tens of thousands of radar detectable objects, most of which can be discovered optically within a few decades.

098.179   Did Icarus have a twin brother?
D. Steel, R. H. Mcnaught, D. Asher.
Minor Planet Bull., Vol. 19, No. 1, p. 9 – 11 (Apr – Jun 1992).

098.180   Observations of minor planets in 1982 – 1985 at the Bucharest Astronomical Institute.
A. Alexiu, G. Bocsa, M. Stanescu.
Rom. Astron. J., Vol. 1, No. 1 – 2, p. 109 – 112 (1991).
   Minor planets precise positions observed at Bucharest Observatory in 1982 – 1985 are presented. For the observations, a 380/6000 mm astrograph was used. The plate measurements were carried out with a ASCORECORD machine.

098.181   Ecliptical orthogonal motion and perihelion distribution of the AAA asteroids.
S. Siregar.
Proc. Astron. Soc. Aust., Vol. 9, No. 2, p. 315 – 316 (1991). – See Abstr. 012.090 for the main entry.
   This paper briefly discusses the Aten, Apollo and Amor asteroid types (collectively referred to as AAAs), their perihelion distribution and orthogonal motion. There are 90 objects in these classes: 6 Aten, 46 Amors and 38 Apollos. The maximum distance to the ecliptic plane of each class follows an exponential function.

Dictionary of minor planet names.
See Abstr. 002.001.

Asteroid photometric database.
See Abstr. 002.084.

Asteroid Photometric Catalogue. Second update.
See Abstr. 002.128.

A brief history of the Minor Planet Center.
See Abstr. 004.078.

Cuno Hoffmeister und die Planetoiden.
See Abstr. 005.007.

Space dust and debris. Proceedings. Topical Meeting of the COSPAR Interdisciplinary Scientific Commission B (Meetings B2, B3 and B5) of the COSPAR 28. Plenary Meeting, The Hague (Netherlands), 25 Jun – 6 Jul 1990.
See Abstr. 012.014.

Asteroid searches from UKST material.
See Abstr. 013.087.

Orbital evolution of dust particles from comets and asteroids.
See Abstr. 021.020.

Circular polarization as an instrument for investigation of surfaces of atmosphereless celestial bodies. I. Laboratory measurements of highly absorptive substances.
See Abstr. 022.068.

Iron and chromium absorption bands in the spectra of terrestrial pyroxenes: application to mineralogic remote sensing of asteroid surfaces.
See Abstr. 022.119.

Laboratory study of the opposition effect.
See Abstr. 022.128.

Light scattering by a randomly oriented cluster of spheres.
See Abstr. 022.131.

Interpretation of lightcurves of atmosphereless bodies. I. General theory and new inversion schemes.
See Abstr. 036.117.

Interpretation of lightcurves of atmosphereless bodies. II. Practical aspects of inversion.
See Abstr. 036.118.

Eine einfache Methode der parabolischen Bahnbestimmung.
See Abstr. 042.013.

Stability of asteroid motions.
See Abstr. 042.029.

The locations of secular resonances and the evolution of small solar system bodies.
See Abstr. 042.030.

Proper elements of the asteroids. A semi–analytical method.
See Abstr. 042.031.

The qualitative explanation of observed peculiarities of Hecuba and Hilda asteroids distribution by a common investigation.
See Abstr. 042.032.

New results on the motions of asteroids in resonances.
See Abstr. 042.033.

Very–high–eccentricity librations at some higher order resonances.
See Abstr. 042.034.

Application of Wisdom's perturbative method to the 5:2 and 7:3 resonances.
See Abstr. 042.035.

Corotations in some higher–order resonances.
See Abstr. 042.036.

The possible orbital evolution of the near–Earth asteroids.
See Abstr. 042.038.

Binary asteroids: secular perturbations.
See Abstr. 042.039.

Corotation solutions in the elliptic asteroidal problem with Stokes drag.
See Abstr. 042.049.

Mappings in astrodynamics.
See Abstr. 042.053.

Mappings for the first order asteroidal resonance.
See Abstr. 042.054.

Mapping for the asteroidal resonances.
See Abstr. 042.056.

The elliptic restricted problem at the 3:1 resonance.
See Abstr. 042.072.

Averaging the elliptic asteroidal problem with a Stokes drag.
See Abstr. 042.082.

Dissipative phenomena in resonance problems in the solar system or the "dei ex machina" of celestial mechanics.
See Abstr. 042.083.

Topological methods for the qualitative analysis of a numerical simulation close to a resonance.
See Abstr. 042.084.

New applications of Fatou's problem.
See Abstr. 042.091.

Modelling: an aim and a tool for the study of the chaotic behaviour of asteroidal and cometary orbits.
See Abstr. 042.099.

Mapping models for Hamiltonian systems with application to resonant asteroid motion.
See Abstr. 042.100.

A model for the study of very–high–eccentricity asteroidal motion: the 3:1 resonance.
See Abstr. 042.101.

The location of secular resonances.
See Abstr. 042.102.

Applications of the restricted many–body problem to binary asteroids.
See Abstr. 042.104.

The wavelet transform as clustering tool for the determination of asteroid families.
See Abstr. 042.105.

Delivery of meteorites from the v_6 secular resonance region near 2 AU.
See Abstr. 042.106.

The planetary few–body problem at three–frequency resonance.
See Abstr. 042.138.

Space Science Reviews volume on Galileo Mission: overview.
See Abstr. 051.031.

Generation of trajectories and choice of routes for a passive flyby of a group of celestial bodies moving in Keplerian orbits.
See Abstr. 052.027.

Diagnostic possibilities of circular polarization for investigating the surfaces of atmosphereless bodies.
See Abstr. 063.029.

The diagnostic potentialities of circular polarization for studying the surfaces of atmosphereless cosmic bodies.
See Abstr. 063.139.

Dynamics of dust in a plasma sheath and injection of dust into the plasma sheath above Moon and asteroidal surfaces.
See Abstr. 091.036.

Lunar and asteroid surface composition from solar wind–sputtered secondary ions.
See Abstr. 094.065.

Occultations of stars by solar system objects. IX. Occultations of catalog stars by asteroids, planets, and major satellites in 1992 and 1993.
See Abstr. 096.009.

Neues aus der Kometen–, Planetoiden– und Meteorszene.
See Abstr. 103.033.

Size–dependent composition in the meteoroid/asteroid population: probable causes and possible implications.
See Abstr. 104.055.

Meteoroid swarms: formation, evolution, and relationship to comets and asteroids.
See Abstr. 104.071.

The Geminids and the object 3200 Phaethon.
See Abstr. 104.079.

Predictions of the meteor radiant point associated with an Earth–approaching minor planet.
See Abstr. 104.082.

Orbits of meteorite producing fireballs. The Glanerbrug – a case study.
See Abstr. 105.003.

Redox effects in ordinary chondrites and implications for asteroid spectrophotometry.
See Abstr. 105.013.

Delivery of meteorites from the asteroid belt.
See Abstr. 105.131.

Mineralogical study of metals in MAC88177 with reference to S–type asteroids.
See Abstr. 105.150.

A possible link between melted micrometeorites from Greenland and Antartica with an asteroidal origin: evidence from carbon stable isotopes.
See Abstr. 105.189.

Silicate darkening in ordinary chondrite parent body regoliths: evidence from gas–rich and shock–blackened ordinary chondrites.
See Abstr. 105.215.

Size–dependent composition in interplanetary material.
See Abstr. 105.216.

Where are the ordinary chondrite parent bodies? A review of the 1991 Meteoritical Society special session.
See Abstr. 105.223.

Meteorite–asteroid spectral comparison: the effects of comminution, melting, and recrystallization.
See Abstr. 105.254.

Asteroid dust reaching the Earth: an information source on the composition of main belt asteroids.
See Abstr. 106.046.

Modelling of asteroidal dust production rates.
See Abstr. 106.048.

The fate of small grains about asteroids.
See Abstr. 106.049.

Interior resonance trapping of dust particles from comets and asteroids.
See Abstr. 106.050.

Asteroidal dust and the zodiacal emission.
See Abstr. 106.082.

Collision and tidal interaction between planetesimals.
See Abstr. 107.019.

The rate of planet formation and the solar system's small bodies.
See Abstr. 107.057.

099 Jupiter, Jupiter Satellites

099.001 The complete polarization state of a storm of millisecond bursts from Jupiter.
G. A. Dulk, A. Lecacheux, Y. Leblanc.
Astron. Astrophys., Vol. 253, No. 1, p. 292–306 (Jan 1992).
The authors report on the complete polarization state (four Stokes parameters) of an extended storm of decametric radiation from Jupiter from the Io–related source "Io–B". Three kinds of bursts were observed, (i) normal ("Γ") bursts with right–hand (RH) elliptical polarization, (ii) millisecond ("S") bursts with RH elliptical polarization, and (iii) S bursts with LH elliptical polarization.

099.002 The abundance of O^{++} in the Jovian magnetosphere.
F. Bagenal, D. E. Shemansky, R. L. McNutt, R. Schreier, A. Eviatar.
Geophys. Res. Lett., Vol. 19, No. 2, p. 79–82 (24 Jan 1992).
From a synthesis of data from the Plasma Science and Ultraviolet Science instruments on the Voyager 1 spacecraft the authors present a radial profile of O^{++} abundance between 4.9 and 42 R_J. They observe a sharp rise in O^{++} mixing ratio near 7.5 R_J, coincident with a sharp rise in effective electron temperature at the outer boundary of the Io plasma torus. Beyond 8.5 R_J the O^{++} mixing ratio is found to be roughly constant which indicates freezing of the ionization prevailing at the outer edge of the hot torus.

099.003 Jovian bremsstrahlung X rays: a Ulysses prediction.
J. H. Waite Jr., D. C. Boice, K. C. Hurley, S. A. Stern, M. Sommer.
Geophys. Res. Lett., Vol. 19, No. 2, p. 83–86 (24 Jan 1992).
The Jovian aurora is the most powerful planetary aurora in the solar system; to date, however, it has not been possible to establish conclusively which mechanisms are involved in the excitation of the auroral emissions that have been observed at ultraviolet, infrared, and soft X–ray wavelengths. Precipitation of Iogenic heavy sulfur and oxygen ions, downward acceleration of electrons along Birkeland currents, and a combination of both of these mechanisms have all been proposed to account for the observed auroral emissions. Modeling results reported here show that precipitating auroral electrons with sufficient energy to be consistent with the Voyager UVS observations will produce bremsstrahlung X rays with sufficient energy and intensity to be detected by the Solar Flare X–Ray and Cosmic Ray Burst Instrument (GRB) on board the Ulysses spacecraft. The detection of such bremsstrahlung X rays at Jupiter would provide strong evidence for the electron precipitation mechanism, although it would not rule out the possibility of some heavy ion involvement, and would thus make a significant contribution toward solving the mystery of the Jovian aurora.

099.004 Methane band photometry of the faded South Equatorial Belt of Jupiter.
T. Satoh, K. Kawabata.
Astrophys. J., Vol. 384, No. 1, p. 298–304 (1 Jan 1992). With plate 1.
A preliminary analysis of the limb darkening curves along the unusually faint Jovian South Equatorial Belt (SEB), observed by the authors early in 1990 using a CCD camera mounted on the 188 cm reflector at Okayama, is presented: three limb–darkening curves extracted from the CCD images obtained in two strong methane bands at $0.725\,\mu m$ and $0.890\,\mu m$ plus one nearby continuum region $(0.750\,\mu m)$ have been compared with the theoretical computations taking into account the effect of multiple light scattering. It is found that the single–scattering albedo for the particles comprising the visible cloud layer is almost unity, in contrast to its typical value of 0.991 deduced by Tomasko et al. for the typical dark state of the SEB. It is suggested that a large–scale lifting of the entire lower cloud layer is likely to have taken place in this region. The cause for these changes in the SEB may be an unusual updraft motion from the deeper level of the atomsphere.

099.005 The Jovian hectometric radiation: an overview after the Voyager mission.
H. P. Ladreiter, Y. Leblanc.
Ann. Geophys., Vol. 9, No. 12, p. 784–796 (Dec 1991).
The authors review the main characteristics and the inferred properties of the hectometric (HOM) radiation which was observed by IMP–6, RAE–1 and Voyager spacecraft. This includes the occurrence of HOM in Jovian longitude, the strong beaming in latitude, the polarization properties, the local time effect, and the solar wind control of the emission. Like many radio emissions observed at the other magnetized planets, the hectometric component is likely to be generated in the R–X mode from sources in the northern and southern auroral zones. The emission is beamed along the surface of hollow cones and is possibly produced by the cyclotron maser mechanism. The HOM sources extend up to 7 R_J along field lines near $L = 20$ in the auroral zone; they are distinct from the Io–controlled decametric (DAM) sources but are probably the extension of the non–Io DAM sources. Finally, the authors discuss certain questions of still unresolved problems that could be answered when Ulysses passes around Jupiter.

099.006 A model of convection and corotation in Jupiter's magnetosphere: Ulysses predictions.
A. F. Cheng.
Geophys. Res. Lett., Vol. 19, No. 3, p. 221–224 (7 Feb 1992).
The Ulysses Jupiter encounter will include the first spacecraft pass through the dusk magnetosphere and will allow new tests of Jovian convection and corotation models. The Cheng and Krimigis (1989) model is extended to suggest that corotation lag occurs principally within the magnetodisk proper, which extends out to ∼60–70 R_J. It is predicted that outside this radius, in the pre–midnight sector, spin–up to near corotation may occur owing to a local reduction in the outward mass transport rate. The mass loss may occur mainly in the dawn sector, to the magnetosphere wind and the dawn magnetosheath.

099.007 Nonadiabatic particle motion and corotation lag in the Jovian magnetodisk.
A. F. Cheng, R. B. Decker.
J. Geophys. Res., Vol. 97, No. A2, p. 1397–1402 (1 Feb 1992).
In the sharply curved magnetic field lines of Jupiter's magnetodisk, energetic ions follow nonadiabatic orbits violating the first and second adiabatic invariants. Even in the corotating reference frame, the energetic ion distribution function is in general anisotropic, as is shown by numerically computed orbits in the Goertz et al. (1976) magnetodisk model. Pitch angle scattering that can be induced by magnetic field irregularities or plasma waves is included in the calculations. This numerically calculated anisotropy of nonadiabatic ions can account for the anisotropy of energetic ions measured by Voyager in the Jovian magnetodisk, depending strongly on the radial distribution of particles injected into the current sheet but not strongly on the scattering rate. The measured anisotropy has previously been interpreted as indicating approximate corotation in the outer Jovian magnetodisk, but the present calculations show that the measured anisotropy can be consistent with a substantial corotation lag.

099.008 Measurement of Jovian decametric Io–related source location and beam shape.
K. Maeda, T. D. Carr.
J. Geophys. Res., Vol. 97, No. A2, p. 1549–1556 (1 Feb 1992).
From measurements of the Jovian decametric activity that was recorded by Voyager 1 and 2 the authors have obtained new information on the locations of the Io–related sources A and C (i.e., Io–A and Io–C) and on the shapes of their emission beams.

099.009 Eternal storms.
G. Hunt.
Astron. Now, Vol. 6, No. 3, p. 41 – 43 (Mar 1992). Special issue: "Focus: Jupiter".
Planetary probes revealed a planet that may be Earth–like on the outside and star–like within.

099.010 Bright lights on Jupiter.
S. Miller.
Astron. Now, Vol. 6, No. 3, p. 46 – 48 (Mar 1992). Special issue: "Focus: Jupiter".

099.011 Mission to Jupiter.
D. Whitehouse.
Astron. Now, Vol. 6, No. 3, p. 49 – 51 (Mar 1992). Special issue: "Focus: Jupiter".

099.012 Jupiter's polar auroras.
P. Drossart, J.–P. Maillard.
Recherche, Vol. 23, No. 240, p. 236 – 238 (Feb 1992). In French.

099.013 Occurrence of global–scale emissions on Jupiter: proposed identification of Jovian dimer H_2 emission.
L. M. Trafton, J. K. G. Watson.
Astrophys. J., Vol. 385, No. 1, p. 320 – 326 (20 Jan 1992).
Two occasions of exceptionally widespread but distinct emission activity have been observed in Jupiter's near–infrared K–band spectrum during September and November of 1988. Two different sets of emission features were involved on the two dates of observation. During these occasions, these normally absent emission features extended from the South polar limb to at least the equator, over a large range of longitudes. Meanwhile, Jupiter's auroral H_2 and H_3^+ emissions remained confined to their usual magnetic polar domains. The global–scale emission features observed during those periods appear to have originated from the H_2 dimer, $(H_2)_2$, during two different excitation modes. Inverse predissociation may have driven the November event. The September event may have originated deeper within the Jovian atmosphere, where excited H_2 is more likely to combine with an unexcited H_2 before radiating. Unusual magnetospheric loading may have precipitated these events.

099.014 Tidal contribution of the satellites to removing the angular momentum of Jupiter.
M. Burša.
Stud. Geophys. Geod., Vol. 35, No. 2, p. 61 – 74 (1991).
It has been shown that dynamically, on the basis of the distribution of angular momenta, the Jovian system cannot be considered an analogue within the solar system. The total tidal decrease in the angular momentum of Jupiter and in its angular velocity of rotation have been estimated, as well as the loss of mechanical energy due to tidal dissipation. It has been concluded that there are no dynamical contradictions with the hypothesis of the common cosmogonic origin of Jupiter and of its eight close satellites.

099.015 Astrometric observations of the faint outer satellites of Jupiter during the 1989–1990 opposition.
A. L. Whipple, P. J. Shelus, G. F. Benedict.
Astron. J., Vol. 103, No. 2, p. 617 – 618 (Feb 1992). Current Physics Microform No.: 9202D1947.
Astrometric positions for the faint outer Jovian satellites VI-XIII during the 1989–1990 opposition have been obtained from the measurement of plates taken with the 2.1 m Otto Struve reflector at McDonald Observatory.

099.016 Astrometry of the Galilean satellites from mutual eclipses and occultations.
A. Mallama.
Icarus, Vol. 95, No. 2, p. 309 – 318 (Feb 1992).
Twelve eclipses and occultations of Io were recorded with a CCD camera during the mutual eclipse and occultation season of 1991. These observations give high quality astrometric information that should be useful in connection with infrared determinations of the locations of Io's volcanoes made by other investigators during the occultations. Differential astrometric positions were determined by fitting the photometry to a model light curve that is based on the photometric properties of the eclipsed or occulted satellite, as well as the apparent motion of the satellites during the event. The astrometric results agree with Lieske's E–3 ephemeris predictions at a 1σ level of about 13 milliarcsec in orbital latitude. However, an 80–milliarcsec longitude residual, seen in the Europa–Io events, is many times larger than the uncertainty. This mean longitude residual can be applied to the E–3 ephemeris in order to use it for Europa–Io events that were not observed photometrically but that need astrometric calibration.

099.017 High resolution observations of the 2.125–μm feature in Io's spectrum during 1975 and 1976.
H. P. Larson, R. Timmermann, U. Fink.
Icarus, Vol. 95, No. 2, p. 325 – 328 (Feb 1992).
The authors confirm the presence of a weak absorption feature in Io's spectrum at 2.125 μm using high resolution observations acquired in 1975 and 1976. If this feature is due to a new class of material on Io, it appears to be the only diagnostic spectral feature of it anywhere in those regions of Io's near–IR spectrum that are accessible at groundbased telescopes. The authors cannot definitively assign the feature to any atom or molecule.

099.018 Périodes de rotation de Jupiter.
M. Jacquesson.
Obs. Trav., No. 29, p. 45 – 46 (1992).

099.019 Jupiter update.
J. Olivarez, P. W. Budine, I. Miyazaki.
Strolling Astron., Vol. 36, No. 1, p. 12 – 14 (Mar 1992).

099.020 Changes of physical parameters of the atmosphere of Jupiter by time.
A. A. Atai, Sh. M. Namazov.
Astron. Vestn., Tom 26, No. 1, p. 54 – 61 (Jan – Feb 1992). In Russian. English translation in Solar Syst. Res., Vol. 26, No. 1.
The diffuse–reflected radiation coefficients have been obtained on the basis of spectrophotoelectric observations of Jupiter carried out in 1982 in the spectral region $\lambda\lambda 5500 – 7500$ Å for the central part of the disk (diaphragm 3".5) and a part of disk covered by the diaphragm 28". Some parameters of the Jovian external atmosphere and the cloud layer in the limits of a two–layer model for the formation of the absorption bands were determined from the data of intensity of methane and ammonia absorption bands. The analysis of the observational and calculated results, comparison of the results of previous years shows that the physical conditions of the formation of the absorption bands in the atmosphere of the Jupiter have been changed by time.

099.021 A generalized hinged–magnetodisc model of Jupiter's nightside current sheet.
K. K. Khurana.
J. Geophys. Res., Vol. 97, No. A5, p. 6269 – 6276 (1 May 1992).
For each of the three spacecraft that have visited the magnetotail of Jupiter, independent hinged–magnetodisc models have been constructed for Jupiter's nightside current sheet. Model parameters (especially the hinge point distance) derived from any one encounter were found to be inapplicable to data from other encounters. Khurana and Kivelson (1989) in a study that evaluated several models of Jupiter's current sheet pointed out that if the hinging of the magnetotail is caused by solar wind forcing then the hinging distance should be parameterized in terms of Jupiter–Sun–magnetospheric (JSM) x coordinate instead of the cylindrical (or planetocentric) radial distance. In this paper the authors develop a quantitative hinged–magnetodisc generalized model that incorporates this suggestion and explains the current sheet crossing data from all three of the flybys with the introduction of only three fitting parameters. The present model is comparable to or slightly superior than the models

published previously in terms of rms error of fit between the modeled and observed current sheet crossing locations. The added novelty is that unlike the previous models the generalized model does not require a different set of parameters for each of the three encounters. A comparison of the authors' results with the earlier models is presented.

099.022 High resolution spectra of Io's neutral sodium cloud.
G. Cremonese, N. Thomas, C. Barbieri, C. Pernechele.
Astron. Astrophys., Vol. 256, No. 1, p. 286 – 298 (Mar 1992).

117 echelle spectra of Io's neutral sodium cloud have been taken during the 1988/89 Jupiter opposition. The spectra, which are catalogued herein, were taken at D–line wavelengths (5890 and 5896 Å) on 15 nights over a 5 month period and were centred on Io. In addition to the "normal" slow sodium cloud (line of sight velocities of 2–3 km s^{-1}), high velocity neutral sodium atoms have been observed on several occasions showing maximum line of sight velocities of around 80 km s^{-1}. Double peak structures in the velocity distribution have also been observed with peak differential column abundances at around 0 and 40 km s^{-1}. Spectra separated by a period of 31.86 d (a beat period between Jupiter's rotational period and Io's orbital period) showed almost identical velocity distributions and similar column abundances. Observations taken with the same observing geometry but at a different magnetic field configuration were substantially different. Fast neutral production rates were highly variable but at times of high fast sodium abundances estimates of around 2–5 × 10^{25} atom s^{-1} were derived.

099.023 Equivalent widths of the CH$_4$–6190 Å and NH$_3$–6450 Å bands across Jupiter's disk: variations from 1985 to 1989.
A. Molina, F. Moreno.
Astron. Astrophys., Vol. 256, No. 1, p. 299 – 304 (Mar 1992).

Jovian spectra in the red region obtained in the 1985–1989 period are studied globally. All of the observations were carried out using the coudé spectrograph at 1.52 m telescope at Calar Alto Observatory (Spain). The center–to–limb behaviour of the equivalent widths of the CH$_4$–6190 Å and NH$_3$–6450 Å bands are analyzed as well as the absorption of these bands along the north–south direction at central meridian of the planet. The temporal variations obtained along the period are discussed. A high center–to–limb ratio for the ammonia band was found, which increased over the period of the observations. The absorption of these bands is slightly greater at belts than at zones. Also, a theoretical model for interpretation is proposed.

099.024 Erratum: "Jupiter's atmospheric parameters derived from spectroscopic observations in the red region during the 1988 opposition" [Astron. Astrophys., Vol. 241, No. 1, p. 243 – 250 (Jan 1991)].
F. Moreno, A. Molina.
Astron. Astrophys., Vol. 256, No. 1, p. 321 (Mar 1992). See Abstr. 53.099.002.

099.025 Valhalla basin on Callisto: an analog of a lunar mascon mare?
G. A. Lejkin, A. N. Sanovich.
Sol. Syst. Res., Vol. 25, No. 1, p. 33 – 39 (Jul 1991). English translation of Astron. Vestn., Tom 25, No. 1, p. 45 – 52 (1991).

Improvement of the morphology of this multiring structure enabled us to obtain new data on the origin of the basin. A new hypothesis is discussed for the formation of this structure, which is based on the numerical estimation of a number of parameters. In the opinion of the authors, the basin is an analog of a lunar mascon which was formed on an icy satellite. A system of standing waves was formed in the process of outflow of matter from the watery mantle, and this produced the multiring structure when the water froze.

099.026 The abundance and distribution of water vapor in the Jovian troposphere as inferred from Voyager IRIS observations.
B. E. Carlson, A. A. Lacis, W. B. Rossow.
Astrophys. J., Vol. 388, No. 2, p. 648 – 668 (1 Apr 1992).

The authors have reanalyzed the Voyager IRIS spectra of the Jovian North Equatorial Belt (NEB) hot spots using a radiative model which includes the full effects of anisotropic multiple scattering by clouds. The atmospheric model includes the three thermochemically predicted cloud layers, NH$_3$, NH$_4$SH, and H$_2$O. Spectrally dependent cloud extinction is modeled using Mie theory. The upper tropospheric temperature profile, gas abundances, height–dependent parahydrogen profile, and vertical distribution of NH$_3$ cloud opacity are retrieved from analysis of the far–infrared (180 – 1200 cm^{-1}) IRIS observations. With these properties constrained, the 5 μm (1800 – 2300 cm^{-1}) observations are analyzed to determine the atmospheric and cloud structure of the deeper atmosphere. Since the NEB hot spots correspond to regions of minimum cloud opacity, these observations can be used to probe the atmosphere down to the ≈ 5 bar level. The results show that the abundance of water is at least 1.5 times solar, with 2 times solar (2.76 × 10^{-3} mixing ratio relative to H$_2$) providing the best–fit to the hot spot data. IRIS observations of the Tropical and Equatorial regions are then used to examine spatial variations in relative humidity.

099.027 A catalogue of the observations of the mutual phenomena of the Galilean satellites of Jupiter made in 1985 during the PHEMU85 campaign.
J. E. Arlot, W. Thuillot, J. Barroso Jr., L. Bergeal, C. Blanco, R. Boninsegna, P. Bouchet, J. Bourgeois, D. Briot, H. Bulder, R. Burchi, J. A. Cano, F. Colas, V. D'Ambrosio, A. Di Paolantonio, G. Dourneau, M. Dumont, S. Ferrand, A. Figer, G. Francou, M. Froeschlé, J. M. Gomez–Forrellad, C. Gouiffes, G. Helmer, F. J. Jablonsky, P. Laques, J. F. Le Campion, J. Lecacheux, J. M. Lecontel, J. Manfroid, C. Meyer, B. Morando, G. R. Quast, J. Rémis, J. Renaudineau, D. Rouan, C. Ruatti, J. P. Sareyan, F. X. Schmieder, F. Sèvre, J. Souchay, J. C. Valtier, D. T. Vu, J. D. Wahiche.
Astron. Astrophys., Suppl. Ser., Vol. 92, No. 1, p. 151 – 205 (Jan 1992).

In this paper, all the usable light curves obtained during the PHEMU85 campaign of observations of the mutual phenomena of the Galilean satellites are presented. These observations will provide accurate astrometric positions of major interest for dynamical studies of the motion of the Galilean satellites. The aim of this work is to give observational data directly usable for theoretical studies. The authors made 166 observations of 64 mutual events from 28 sites. The corresponding data are given and compared with the theoretical predictions. The accuracy of each observation has been determined. For each observation, information is given about the telescope, the receptor, the site and the observational conditions.

099.028 On acceleration and heating of charged particles in the Jovian ionosphere.
V. V. Zajtsev, V. E. Shaposhnikov.
Pis'ma Astron. Zh., Tom 18, No. 4, p. 380 – 390 (Apr 1992). In Russian. English translation in Sov. Astron. Lett., Vol. 18, No. 2.

The authors propose an acceleration mechanism of charged particles which is connected with the existence of an acceleration potential along the Jovian magnetic field. This model is developed on the basis of the known phenomenon of increase of the resistance of partially ionized plasma when the current is instable. The region of high resistance is formed in the upper ionosphere of the planet above the height of maximum electron density at the bottom of the magnetic flux tube of Io. For typical Jovian ionosphere conditions the resistance may be increased by about seven orders of magnitude. The potential difference accelerats particles up to several MeV in this region. The plasma is heated due to the Joule dissipation and the Buneman turbulence up to ten keV.

099.029 Ulysses observations of escaping VLF emissions from Jupiter.
M. L. Kaiser, M. D. Desch, W. M. Farrell, R. J. MacDowall, R. G. Stone, A. Lecacheux, B.-M. Pedersen, P. Zarka.
Geophys. Res. Lett., Vol. 19, No. 7, p. 649 – 652 (3 Apr 1992).

The Ulysses URAP experiment has detected Jovian radio emissions in the VLF range at distances from Jupiter in excess of 1.5 A.U. The URAP observations represent the first synoptic observations of Jupiter in the VLF band, 3 to 30 kHz. In this band lie the low–frequency extent of the bKOM emission, the escaping continuum emission, and the Jovian type IIIs. Initial results indicate that the continuum varies in frequency with the solar wind ram pressure at Jupiter, whereas, the Jovian type IIIs appear to be controlled to some extent by the planetary rotation, often appearing when system III longitude 100° faces the spacecraft.

099.030 Rotating magnetic anomalies as a possible accelerator of charged particles.
W. W. Liu.
J. Geophys. Res., Vol. 97, No. A6, p. 8145 – 8155 (1 Jun 1992).

The author shows, through the formulation of a simple theoretical model, that there exists a possibility that a significant number of a energetic ions in the Jovian magnetosphere, and presumably in other astrophysical environments beyond our direct access, owe their origin to a large–scale longitudinal magnetic anomaly corotating with the central body anchoring the magnetosphere, if the invoked magnetic anomaly has a typical boundary thickness comparable to an ion gyroradius, which for a preheated 10–keV heavy ion in the outer Jovian magnetosphere may be of the order of a few Jovian radii. The energy source for the acceleration is a torque from the central planet arising from a differential rotation between the ionosphere and the magnetosphere. The premise and outstanding issues pertaining to the present proposal are further discussed as motivating points for future studies.

099.031 Polarization of the Io–C radio emission from Jupiter.
C. H. Barrow.
J. Geophys. Res., Vol. 97, No. A6, p. 8169 – 8172 (1 Jun 1992).

Old measurements of the Stokes polarization parameters for the Jovian decametric radio emission are reexamined for comparison with the broadband measuremens for Io–A and Io–B emission, published recently by the Decametric Radio Astronomy Group at the Meudon Observatory. Records of an Io–C source event, recorded in 1966 at 18 MHz, indicate average values of 0.63 and 0.72 for the degrees of circular and linear polarization, respectively, with a mean value of 0.92 for the total polarization fraction and 0.40 for the average axial ratio. These values, which have not previously been published for the Io–C emission, complement and support the Meudon results. Burst-by–burst sequences of the polarization parameters reveal much diversity in the polarization and allow partial circular polarization to be distinguished from pure elliptical.

099.032 Densities and vibrational distribution of H_3^+ in the Jovian auroral ionosphere.
Y. H. Kim, J. L. Fox, H. S. Porter.
J. Geophys. Res., Vol. 97, No. E4, p. 6093 – 6102 (25 Apr 1992). With a correction in Vol. 97, No. E6, p. 10283 (25 Jun 1992).

Observations of the H_3^+ infrared emission at 2 and 4 μm have suggested that H_3^+ is in local thermodynamic equilibrium (LTE) in the region of the Jovian ionosphere from which the emissions originate. The authors have tested this assumption by calculating the vibrational distribution of H_3^+ over the altitude range of 350 to 1500 km above the methane cloud tops (1 to $4 \times 10^{-3} \mu bar$). They have constructed a model of the Jovian auroral ionosphere in which the neutral temperatures are enhanced over those of the mid–latitude ionosphere, as suggested by observations and models of the auroral region. The authors have modeled the precipitation of 10–keV electrons with an energy flux of 1 erg cm^{-2}s^{-1}. Both the energy and energy flux are less than those that are implicated in the production of the UV aurora. The authors have computed the densities and vibrational distribution

of H_3^+ and find that the distribution of the six lowest states of H_3^+ can be determined fairly well in spite of uncertainties in the atomic and molecular data. Since the nearly resonant transfer of vibration from $H_2(v=1)$ is an important process in populating the $H_3^+(v_1=0,v_2=2)$ state, it is necessary to model the vibrational distribution of H_2 as well. The computed altitude profiles and vibrational distributions of H_3^+ and H_2 are consistent with the observations of infrared emission in the 2– and 4–μm regions. The H_3^+ is not in LTE near and above the H_3^+ peak, since loss of the $H_3^+(v_1=0,v_2=1)$ and $H_3^+(v_1=0,v_2=2)$ states by radiation is approximately equal to the collisional loss rate.

099.033 Ion–cyclotron waves at Jupiter: possibility of detection by Ulysses.
Y. Mei, R. M. Thorne, R. B. Horne.
Geophys. Res. Lett., Vol. 19, No. 6, p. 629 – 632 (20 Mar 1992).

The authors evaluate the convective growth of ion–cyclotron waves in the Io plasma torus using realistic plasma parameters. Significant wave amplification is restricted to two dominant frequency bands. As originally proposed by Thorne and Moses (1983, 1985), waves between the O^+ and H^+ gyrofrequencies may be excited at higher latitudes ($\lambda > 15°$) in the region where H^+ becomes the dominant ion. Strong cyclotron resonant damping should prevent wave propagation to lower latitude. Even under optimum conditions the path integrated gain of such waves is modest and extremely sensitive to the properties of both the thermal plasma and the cyclotron resonant energetic ions. Consequently, these waves should be confined to a limited region of the torus ($6 < L < 9$). The equatorial region of the torus can be unstable to L–mode waves below the O^+ gyro–frequency. Rapid amplification should drive waves to non linear amplitudes for any reasonable choice of the plasma properties. While the orbit of Ulysses passes through the Io torus close to the optimum location (near $L = 8$) to detect both the high frequency (10 Hz) wave at $\lambda > 15°$ and the low frequency waves (< 0.5 Hz) near the equator, the sensitivity of the magnetometer and the lower frequency threshold of the plasma wave detector offer only a marginal possibility to clearly identify either of these important waves.

099.034 Can the strongly interacting dark matter be a heating source of Jupiter?
M. Kawasaki, H. Murayama, T. Yanagida.
Prog. Theor. Phys., Vol. 87, No. 3, p. 685 – 692 (Mar 1992).

The authors show that the strongly–interacting massive particle (SIMP) with mass $3 \times 10^6 - 10^7 GeV$ is astrophysically interesting as a dark matter candidate in the galactic halo. The annihilation of SIMPs inside Jupiter naturally explains the intrinsic heat flux irrespective of details of the planetary models. The authors discuss its effect in all Jovian planets as well as in the Sun and the Earth. They also comment that such a SIMP is accommodated in a class of hadronic axion models.

099.035 The tropospheric abundances of NH_3 and PH_3 in Jupiter's GRS from Voyager IRIS observations.
C. A. Griffith, B. Bézard, D. Gauthier, T. Owen.
Bull. Am. Astron. Soc., Vol. 23, No. 3, p. 1131 (1991). Abstract. – See Abstr. 012.037 for the main entry.

099.036 Ortho–para hydrogen equilibration on Jupiter.
B. E. Carlson, A. A. Lacis, W. B. Rossow.
Bull. Am. Astron. Soc., Vol. 23, No. 3, p. 1131 (1991). Abstract. – See Abstr. 012.037 for the main entry.

099.037 High–resolution spectroscopic thermal infrared images of Jupiter in 1989 October.
G. Orton, J. Lacy, J. Achtermann, P. Parmar, A. Castillo.
Bull. Am. Astron. Soc., Vol. 23, No. 3, p. 1131 (1991). Abstract. – See Abstr. 012.037 for the main entry.

099.038 **The variability of Jovian atmospheric features at 890 nm.**
D. M. Kuehn.
Bull. Am. Astron. Soc., Vol. 23, No. 3, p. 1132 (1991). Abstract. – See Abstr. 012.037 for the main entry.

099.039 **Jovian seismology.**
B. Mosser, D. Gautier, J. Gay, D. Mekarnia, P. Delache, J.–P. Maillard.
Bull. Am. Astron. Soc., Vol. 23, No. 3, p. 1132 (1991). Abstract. – See Abstr. 012.037 for the main entry.

099.040 **Low latitude lightning storms on Jupiter.**
W. J. Borucki.
Bull. Am. Astron. Soc., Vol. 23, No. 3, p. 1132 (1991). Abstract. – See Abstr. 012.037 for the main entry.

099.041 **3.3 micron methane emission from the Jovian auroral zone: non–LTE aspect.**
D. P. Kratz, S. J. Kim, R. N. Halthore.
Bull. Am. Astron. Soc., Vol. 23, No. 3, p. 1132 (1991). Abstract. – See Abstr. 012.037 for the main entry.

099.042 **Observations of Jupiter's aurora, airglow and albedo, 830 – 1850 Å, with the Hopkins Ultraviolet Telescope.**
P. D. Feldman, H. W. Moos, S. T. Durrance, M. A. McGrath, A. F. Davidsen, W. P. Blair, C. W. Bowers, W. V. Dixon, R. C. Henry, G. A. Kriss, J. Kruk, O. Vancura, H. C. Ferguson, R. A. Kimble, K. S. Long.
Bull. Am. Astron. Soc., Vol. 23, No. 3, p. 1132 – 1133 (1991). Abstract. – See Abstr. 012.037 for the main entry.

099.043 **A picture of the auroral regions of Jupiter in H_3^+ emission.**
P. Drossart, S. Bouffiès, J.–P. Maillard, S. Kim, J. Caldwell, T. Herbst, M. Shure.
Bull. Am. Astron. Soc., Vol. 23, No. 3, p. 1133 (1991). Abstract. – See Abstr. 012.037 for the main entry.

099.044 **Correlative observations of the H_2 ultraviolet and the H_3^+ infrared aurorae on Jupiter.**
G. E. Ballester, R. Prangé, S. Kim, T. Livengood, H. W. Moos, J. Caldwell.
Bull. Am. Astron. Soc., Vol. 23, No. 3, p. 1133 (1991). Abstract. – See Abstr. 012.037 for the main entry.

099.045 **Oxygen emission in the extended Io atmosphere.**
M. E. Brown.
Bull. Am. Astron. Soc., Vol. 23, No. 3, p. 1134 (1991). Abstract. – See Abstr. 012.037 for the main entry.

099.046 **Towards numerical weather prediction for Jupiter.**
T. Dowling, A. Fischer, J. Harrington.
Bull. Am. Astron. Soc., Vol. 23, No. 3, p. 1134 (1991). Abstract. – See Abstr. 012.037 for the main entry.

099.047 **Hydrocarbon photochemistry in the upper atmosphere of Jupiter.**
G. R. Gladstone, M. Allen, Y. L. Yung, J. I. Moses.
Bull. Am. Astron. Soc., Vol. 23, No. 3, p. 1134 (1991). Abstract. – See Abstr. 012.037 for the main entry.

099.048 **Early documentation of the history of Jupiter's south equatorial belt disturbance.**
T. A. Hockey.
Bull. Am. Astron. Soc., Vol. 23, No. 3, p. 1134 (1991). Abstract. – See Abstr. 012.037 for the main entry.

099.049 **The Lyman alpha bulge of Jupiter: effects of non–thermal velocity field.**
L. B. Jaffel, J. T. Clarke, G. R. Gladstone, R. Prangé, B. R. Sandel, A. Vidal–Madjar, R. V. Yelle.
Bull. Am. Astron. Soc., Vol. 23, No. 3, p. 1135 (1991). Abstract. – See Abstr. 012.037 for the main entry.

099.050 **Search for H_2S on Jupiter at millimeter wavelengths: observations and laboratory measurements.**
J. Joiner, P. G. Steffes.
Bull. Am. Astron. Soc., Vol. 23, No. 3, p. 1135 (1991). Abstract. – See Abstr. 012.037 for the main entry.

099.051 **Densities and vibrational distribution of H_3^+ in the Jovian auroral ionosphere.**
Y. H. Kim, J. L. Fox, H. Porter.
Bull. Am. Astron. Soc., Vol. 23, No. 3, p. 1135 (1991). Abstract. The full paper is published in J. Geophys. Res., E, Vol. 97, No. 4, p. 6093 – 6102 (25 Apr 1992). – See Abstr. 012.037 for the main entry.

099.052 **Long period Earth–like oscillations of temperature and zonal wind in Jupiter's equatorial stratosphere.**
C. B. Leovy, A. J. Friedson, G. S. Orton.
Bull. Am. Astron. Soc., Vol. 23, No. 3, p. 1135 (1991). Abstract. – See Abstr. 012.037 for the main entry.

099.053 **Spectroscopy of H_3^+ in the Jovian atmosphere.**
S. Miller, H. Lam, J. Tennyson, S. Ridgway, R. D. Joseph.
Bull. Am. Astron. Soc., Vol. 23, No. 3, p. 1136 (1991). Abstract. – See Abstr. 012.037 for the main entry.

099.054 **Time–dependence and spatial correlations between temperatures and visual and infrared properties of clouds in Jupiter: 1984 – 1991.**
G. Orton, J. Friedson, T. Kanamori, T. Thaller, R. Beebe, L. Huber, J. Caldwell.
Bull. Am. Astron. Soc., Vol. 23, No. 3, p. 1136 (1991). Abstract. – See Abstr. 012.037 for the main entry.

099.055 **Horizontal motions around and within the Great Red Spot.**
P. V. Sada, R. Beebe.
Bull. Am. Astron. Soc., Vol. 23, No. 3, p. 1137 (1991). Abstract. – See Abstr. 012.037 for the main entry.

099.056 **Improved MAGPAC computer images of the Io plasma torus.**
M. H. Taylor, F. Bagenal, D. Balcom, M. Harel, D. E. Shemansky.
Bull. Am. Astron. Soc., Vol. 23, No. 3, p. 1137 (1991). Abstract. – See Abstr. 012.037 for the main entry.

099.057 **Imaging of [O II] in the Jovian plasma torus.**
J. T. Trauger, D. I. Brown, N. M. Schneider.
Bull. Am. Astron. Soc., Vol. 23, No. 3, p. 1137 (1991). Abstract. – See Abstr. 012.037 for the main entry.

099.058 **Jovian large–scale stratospheric circulation.**
R. A. West, A. J. Friedson, J. F. Appleby.
Bull. Am. Astron. Soc., Vol. 23, No. 3, p. 1137 – 1138 (1991). Abstract. – See Abstr. 012.037 for the main entry.

099.059 **Ion temperature of the Jupiter plasma torus in 1988.**
R. C. Woodward, F. Scherb, F. L. Roesler, R. J. Oliversen.
Bull. Am. Astron. Soc., Vol. 23, No. 3, p. 1138 (1991). Abstract. – See Abstr. 012.037 for the main entry.

099.060 **Infrared imaging of Jupiter's aurorae.**
G. E. Ballester, S. Miller, J. Tennyson, R. D. Joseph, R. Baron, T. Owen.
Bull. Am. Astron. Soc., Vol. 23, No. 3, p. 1145 (1991). Abstract. – See Abstr. 012.037 for the main entry.

099.061 **Multi–wavelength Jovian auroral modeling.**
J. H. Waite Jr.
Bull. Am. Astron. Soc., Vol. 23, No. 3, p. 1145 (1991). Abstract. – See Abstr. 012.037 for the main entry.

099.062 An unusual change in the Jovian Ly–α bulge.
M. A. McGrath.
Bull. Am. Astron. Soc., Vol. 23, No. 3, p. 1145 (1991). Abstract. – See Abstr. 012.037 for the main entry.

099.063 Jupiter's H Lyα emission line profiles.
J. T. Clarke, G. R. Gladstone, L. B. Jaffel.
Bull. Am. Astron. Soc., Vol. 23, No. 3, p. 1145 (1991). Abstract. – See Abstr. 012.037 for the main entry.

099.064 A system III dependence in the apparent jovicentric distance of the Io plasma torus.
A. J. Dessler, B. R. Sandel.
Bull. Am. Astron. Soc., Vol. 23, No. 3, p. 1145 – 1146 (1991). Abstract. – See Abstr. 012.037 for the main entry.

099.065 The role of the molecular–metallic transition of hydrogen in Jupiter and Saturn.
D. Saumon, G. Chabrier, W. B. Hubbard, J. I. Lunine, H. M. Van Horn.
Bull. Am. Astron. Soc., Vol. 23, No. 3, p. 1148 (1991). Abstract. – See Abstr. 012.037 for the main entry.

099.066 Germane photochemistry in the atmosphere of Jupiter and Saturn: the reaction of atomic hydrogen with germane.
L. J. Stief, D. F. Nava, W. A. Payne, G. Marston.
Bull. Am. Astron. Soc., Vol. 23, No. 3, p. 1148 – 1149 (1991). Abstract. – See Abstr. 012.037 for the main entry.

099.067 The Galilean moons of Jupiter.
Astron. Now, Vol. 6, No. 3, p. 44 – 45 (Mar 1992). Special issue: "Focus: Jupiter".

099.068 Comparisons of the icy regoliths of Europa, Callisto, and Rhea.
D. L. Domingue, G. W. Lockwood, D. T. Thompson.
Bull. Am. Astron. Soc., Vol. 23, No. 3, p. 1169 – 1170 (1991). Abstract. – See Abstr. 012.037 for the main entry.

099.069 The icy Galilean satellites: new radar results from Arecibo and Goldstone.
S. J. Ostro, D. B. Campbell, R. A. Simpson, R. S. Hudson, R. Velez, R. Winkler, K. D. Rosema, D. K. Yeomans, E. M. Standish, J. F. Chandler, I. I. Shapiro.
Bull. Am. Astron. Soc., Vol. 23, No. 3, p. 1170 (1991). Abstract. – See Abstr. 012.037 for the main entry.

099.070 Tidal origin of the Laplace resonance and the resurfacing of Ganymede.
R. Malhotra.
Bull. Am. Astron. Soc., Vol. 23, No. 3, p. 1170 (1991). Abstract. The full paper is published in Icarus, Vol. 94, No. 2, p. 399 – 412 (Dec 1991). – See Abstr. 012.037 for the main entry.

099.071 CCD photometry and spectroscopy of outer Jovian satellites.
J. X. Luu.
Bull. Am. Astron. Soc., Vol. 23, No. 3, p. 1171 (1991). Abstract. – See Abstr. 012.037 for the main entry.

099.072 Laboratory studies of planetary molecules and ices: the case of Io.
F. Salama, S. A. Sandford, L. J. Allamandola.
IAU Symposium No. 150: Astrochemistry of cosmic phenomena, p. 435 – 436 (1992). – See Abstr. 012.029 for the main entry.
The techniques of low temperature spectroscopy are applied to analyze infrared observational data of Io in the 2.0 – 5.0 μm range. The presence of solid H_2S and traces of H_2O in the SO_2– dominant surface ices are derived from this analysis and it is suggested that CO_2 clusters may as well be present near the surface of Io.

099.073 Die Bewegungen der Galileischen Jupitermonde.
H. Smutek.
Ahnerts Kalender für Sternfreunde 1993. p. 165 – 168 (1992). – See Abstr. 046.012 for the main entry.

099.074 A model and map of Amalthea.
P. J. Stooke.
Earth, Moon, Planets, Vol. 56, No. 2, p. 123 – 139 (Feb 1992).
A topographic model of Amalthea (JV) was derived from the shapes of limbs and terminators in Voyager images, locally to accomodate large craters and ridges. The model is presented in tabular and graphic form, including the first detailed shaded relief maps of the satellite. The shape is very irregular, with radii varying between about 53 and 151 ± 5 km. The minimum value occurs in a deep crater at the south pole. The volume is estimated to be $(2.5 \pm 0.5) \times 10^6 km^3$. A prominent groove or valley extends some 150 km across the trailing side. High albedo, spectrally distinct markings are mapped and found to have a less obvious relationship with relief than previously suggested.

099.075 Jupiter in 1989 – 90.
J. H. Rogers.
J. Br. Astron. Assoc., Vol. 102, No. 3, p. 135 – 150 (Jun 1992).

099.076 High–temperature volcanic thermal emission: from Io to the Andes.
J. R. Spencer, W. Sinton, C. Oppenheimer, C. Kaminski.
Bull. Am. Astron. Soc., Vol. 23, No. 3, p. 1227 (1991). Abstract. – See Abstr. 012.037 for the main entry.

099.077 Spatial resolution of Io's hot spots: simultaneous 3.8 and 10 μm photometry of occultations of Io by Europa.
J. D. Goguen, G. J. Veeder, D. L. Matson, T. V. Johnson, R. H. Brown, D. Toomey, W. M. Sinton, H. A. Nair, S. Wang, J. Gradie.
Bull. Am. Astron. Soc., Vol. 23, No. 3, p. 1227 (1991). Abstract. – See Abstr. 012.037 for the main entry.

099.078 Mutual occultation photometry of Io during 1991.
R. R. Howell, B. P. Uberuaga.
Bull. Am. Astron. Soc., Vol. 23, No. 3, p. 1227 (1991). Abstract. – See Abstr. 012.037 for the main entry.

099.079 Jupiter's satellite Io: occultation high speed photometry and compositional mapping.
R. M. Nelson, B. D. Wallis, L. J. Horn, A. L. Lane, W. D. Smythe, E. S. Barker, B. W. Hapke.
Bull. Am. Astron. Soc., Vol. 23, No. 3, p. 1227 – 1228 (1991). Abstract. – See Abstr. 012.037 for the main entry.

099.080 Mutual event observations of Io's 2.125 μm (4705 cm⁻¹) absorption feature: the latitudinal distribution of the source material.
L. M. Trafton, D. F. Lester, T. F. Ramseyer, A. L. Whipple.
Bull. Am. Astron. Soc., Vol. 23, No. 3, p. 1228 (1991). Abstract. – See Abstr. 012.037 for the main entry.

099.081 New FTS spectra of Io near 2 and 4 micrometers.
C. de Bergh, E. Lellouch, J.–P. Maillard, B. Schmitt.
Bull. Am. Astron. Soc., Vol. 23, No. 3, p. 1228 (1991). Abstract. – See Abstr. 012.037 for the main entry.

099.082 SO_2 weathering on Io.
D. S. Burnett, M. L. Johnson.
Bull. Am. Astron. Soc., Vol. 23, No. 3, p. 1228 (1991). Abstract. – See Abstr. 012.037 for the main entry.

099.083 How hot is Io's lower atmosphere?
E. Lellouch, T. Encrenaz, M. Belton, I. de Pater, S. Gulkis, G. Paubert.
Bull. Am. Astron. Soc., Vol. 23, No. 3, p. 1228 – 1229 (1991). Abstract. – See Abstr. 012.037 for the main entry.

099.084   Io's fast sodium. I. Molecular ion source.
J. K. Wilson, N. M. Schneider, J. T. Trauger,
D. I. Brown, R. E. Evans, D. E. Shemansky.
Bull. Am. Astron. Soc., Vol. 23, No. 3, p. 1229 (1991). Abstract. –
See Abstr. 012.037 for the main entry.

099.085   A chemical flow model of the Io plasma torus using the adiabatic equations of motion.
P. L. Matheson, D. E. Shemansky.
Bull. Am. Astron. Soc., Vol. 23, No. 3, p. 1229 (1991). Abstract. –
See Abstr. 012.037 for the main entry.

099.086   Measurement of emissions from the Io torus with the Hopkins Ultraviolet Telescope.
H. M. Moos, P. D. Feldman, S. T. Durrance, J. Blanchette,
A. F. Davidsen, W. P. Blair, C. W. Bowers, W. V. Dixon,
R. C. Henry, G. A. Kriss, J. Kruk, O. Vancura, H. C. Ferguson,
R. A. Kimble, K. S. Long.
Bull. Am. Astron. Soc., Vol. 23, No. 3, p. 1229 (1991). Abstract. –
See Abstr. 012.037 for the main entry.

099.087   Plasma heating of Io's atmosphere.
R. P. LeBeau, M. Pospieszalska, R. E. Johnson.
Bull. Am. Astron. Soc., Vol. 23, No. 3, p. 1229 – 1230 (1991).
Abstract. – See Abstr. 012.037 for the main entry.

099.088   Stability in the supply of the Io plasma torus by ejection of atmospheric molecules.
R. E. Johnson.
Bull. Am. Astron. Soc., Vol. 23, No. 3, p. 1230 (1991). Abstract. –
See Abstr. 012.037 for the main entry.

099.089   Eclipse spectroscopy of Io in the 4–μm region.
D. R. Klassen, R. R. Howell, D. P. Cruikshank.
Bull. Am. Astron. Soc., Vol. 23, No. 3, p. 1230 (1991). Abstract. –
See Abstr. 012.037 for the main entry.

099.090   Io's fast sodium. II. Plasma slowdown at Io.
N. M. Schneider, J. K. Wilson, J. T. Trauger.
Bull. Am. Astron. Soc., Vol. 23, No. 3, p. 1230 (1991). Abstract. –
See Abstr. 012.037 for the main entry.

099.091   Implication of the sodium zenocorona.
W. H. Smyth, M. R. Combi.
Bull. Am. Astron. Soc., Vol. 23, No. 3, p. 1230 (1991). Abstract. –
See Abstr. 012.037 for the main entry.

099.092   Another unidentified near–IR absorption feature on Io.
D. F. Lester, L. M. Trafton, T. F. Ramseyer.
Bull. Am. Astron. Soc., Vol. 23, No. 3, p. 1230 – 1231 (1991).
Abstract. – See Abstr. 012.037 for the main entry.

099.093   KPNO 4 m FTS observations of the infrared band at 2.125 μm (4705 cm^{-1}) in the spectrum of Io.
L. M. Trafton, D. F. Lester, T. F. Ramseyer, K. H. Hinkle.
Bull. Am. Astron. Soc., Vol. 23, No. 3, p. 1231 (1991). Abstract. –
See Abstr. 012.037 for the main entry.

099.094   Instability of the zonal jets and longitudinal thermal waves in a Jovian atmosphere.
J. A. Pirraglia.
Icarus, Vol. 96, No. 2, p. 161 – 168 (Apr 1992).
   Analysis of infrared observations of Jupiter's atmosphere made during the Voyagers 1 and 2 encounters by Magalhães et al., implies the existence of large scale features which appear to move very slowly. Groundbased measurements with the NASA Infrared Telescope Facility by Deming et al. show large scale waves possibly fixed, or at most moving very slowly, with respect to system III. These observations suggest that the observed features may originate in the deep atmosphere. The author has an alternative source. A model of the jets implies that the waves observed are the result of the quasigeostrophic instability of the jets in the upper troposphere, and are not long lived but do last long enough to give the impression of being so under the

conditions of the observations. Calculations with the model are consistent with the observations.

099.095   Spatial variation of 2 μm H$_3^+$ emission in the southern auroral region of Jupiter.
F. Billebaud, P. Drossart, J.–P. Maillard, J. Caldwell, S. Kim.
Icarus, Vol. 96, No. 2, p. 281 – 283 (Apr 1992).
   Observations of the southern auroral spot of Jupiter, made at the CFHT in Sep 1988, show evidence for spatial variation of the intensities of the 2–μm lines of H$_3^+$. A sequence of observations around the auroral region shows variation not only in the absolute intensities, but also in the relative intensities, of H$_3^+$ lines. Possible causes for these variations include temperature fluctuations or differing H$_3^+$ ortho/para ratio, possibly caused by spatial inhomogeneity in the ionosphere; precipitation of different kinds of magnetospheric particles; and different particle flux values within the auroral zone. Information on the systematic variation in the H$_3^+$ parameters is an interesting constraint for future modeling of the infrared aurorae.

099.096   Morphology of infrared H$_3^+$ emissions in the auroral regions of Jupiter.
P. Drossart, R. Prangé, J. P. Maillard.
Icarus, Vol. 97, No. 1, p. 10 – 25 (May 1992).
   From a careful analysis of the 4–μm emission of Jupiter, it has been possible to describe the morphology of the auroral emission in H$_3^+$, which takes place above a background of both thermal emission and reflection of sunlight by the clouds and hazes of Jupiter. The structure of the auroral emission is found to be strongly peaked in auroral spots, two of which can be observed simulaneously in one hemisphere. The position of the emission peak in the south exhibits a complex apparent motion, probably constrained by both the System III longitude and the local time. The northern spots are apparently more fixed in longitude.

099.097   Jovian ultraviolet auroral activity, 1981 – 1991.
T. A. Livengood, H. W. Moos, G. E. Ballester,
R. M. Prangé.
Icarus, Vol. 97, No. 1, p. 26 – 45 (May 1992).
   The distribution of auroral brightness on Jupiter's surface as a function of magnetic longitude (System III) is studied using IUE observations of H$_2$ ultraviolet emissions over the years 1981 to 1991. The authors develop a simple model brightness distribution with two peaks which is fitted to the IUE data for some observations. The effectiveness of the double–peaked models is compared with the single–peaked models for the same observations.

099.098   Diurnal variations on Jupiter and Saturn?
E. Karkoschka.
Icarus, Vol. 97, No. 2, p. 182 – 186 (Jun 1992).
   The author observed four hydrogen quadrupole absorption lines in the Equatorial Zones of Jupiter and Saturn. No asymmetry between morning and evening side could be detected. Upper limits for an asymmetry of equivalent widths are 2 and 3% for Jupiter and Saturn, respectively. This is in contradiction to the 5 – 15% asymmetry seen by Cunningham et al. (see Abstr. 46.099.019) on Jupiter. The different results can be fully explained by the use of an improved solar spectrum for the reduction of the data reported here. The observations indicate that the ortho to para ratio of hydrogen is near equilibrium, in agreement with Cunningham et al. Also, Doppler shifts in the spectra imply a wind speed for Saturn's equatorial jet which is consistent with that obtained by tracking cloud features in Voyager images.

099.099   CCD photometry for Jovian eclipses of the Galilean satellites.
A. Mallama.
Icarus, Vol. 97, No. 2, p. 298 – 302 (Jun 1992).
   Timings of nine Jovian eclipses of the Galilean satellites observed during 1990 and 1991 are compared to predictions from modern ephemerides and are shown to exhibit very little internal scatter. These data can be used to supplement astrometry from

the mutual occultations and eclipses of the satellites that occurred during the same time period.

099.100   Analysis of Voyager 2 images of Jovian lightning.
W. J. Borucki, J. A. Magalhães.
Icarus, Vol. 96, No. 1, p. 1 – 14 (Mar 1992).

Voyager 2 images have been examined and found to contain bright spots due to lightning activity that is confined to two narrow latitude bands centered at 49°N and 13.5°N latitude and to a single region near 60°N latitude. No lightning activity is observed in the southern hemisphere. The brightest band of lightning activity occurs at 49°N, but no unusual cloud features can be associated with this activity. This result is consistent with the depth of the lightning activity being 80 km below the ammonia clouds and thus not producing readily observed disturbances in the visible clouds. The secondary band of lightning activity that appears at 13.5° N is restricted in longitude to the "disturbed" zone. The energy dissipation rate is approximately 50% greater than that found in the Voyager 1 images. The ratio of the energy dissipated by Jovian lightning to the thermal flux available to drive convection motions is found to be three decades larger than the terrestrial ratio.

099.101   On the detectability of Jovian oscillations with infrared heterodyne measurements.
B. Mosser, D. Gautier, T. Kostiuk.
Icarus, Vol. 96, No. 1, p. 15 – 26 (Mar 1992).

The possibility of detecting global oscillations of Jupiter by measuring the Doppler shift of far infrared emission lines formed in the lower stratosphere of the planet was investigated. The pure rotational quadrupole lines S(0) and S(1) of molecular hydrogen, located near 28 and 17 μm, present attractive characteristics for this research. The authors have calculated the expected line profiles due to these transitions on Jupiter and used them to evaluate the feasibility of the method. Measurements of the S(0) and S(1) lines over a period of a few nights would permit the detection of modes with periods of 5 to 20 min and Doppler shifts corresponding to velocities as low as 1 m sec^{-1}. Possible limitations of the method resulting from instrumental considerations, the fluctuations of the telescope guiding system and from the atmospheric transmission and seeing are discussed.

099.102   Near–infrared CVF spectrophotometry of selected areas of Jupiter during the 1991 apparition.
F. Moreno, A. Molina, J. L. Ortiz.
Icarus, Vol. 96, No. 1, p. 129 – 142 (Mar 1992).

Near–infrared spectra of selected areas of Jupiter obtained during Feb 1991 are presented. The absolutely calibrated data cover the belt and zone regions, the equatorial limbs and the polar regions in the spectral ranges $\lambda\lambda$ 1.4 – 2.4 and 2.9 – 4.1 μm with a resolving power ranging from 50 to 130. Limb darkening is present at every wavelength observed in the range 1.4 – 2.4 μm. The spectra are in good agreement with those obtained at the center and the east limb by Clark and McCord except at the bottom of the strong 2.3–μm methane band. Strong north and south polar brightenings are observed in the vicinity of the core of that band. The flux from Jupiter in the 3.3–μm region is found to be about 10^{-18}W cm$^{-2}\mu$m^{-1}. These data can be used for modeling purposes to compute infrared heating rates in the Jovian stratosphere, as well as to infer the vertical distribution of clouds and determine the dependence of cloud properties with wavelength.

099.103   Locations of 4–μm hot spots on the poles of Jupiter.
S. J. Kim, J. Caldwell, T. M. Herbst.
Icarus, Vol. 96, No. 1, p. 143 – 148 (Mar 1992).

The authors present a list of apparent positions of 4–μm polar bright spots in the auroral regions of Jupiter observed with the Protocam at the Infrared Telescope Facility on Mar 7 and 8, 1991. Although gross background auroral features on the north and south poles seem to be fixed on the surface of the planet, isolated bright spots over the gross auroral features showed rapid displacement especially on the south pole. The positions are compared with theoretical magnetic field lines and Voyager UV observations and find no significant coincidence is found. The authors discuss the possible cause of the time variability of the positions of the bright spots, and compare the apparent rates of motion with those of the 8–μm bright spots reported by J. Caldwell et al.

099.104   An analysis of photographic astrometric observations of the Galilean moons: USNO refractor, 1986 – 1990.
D. Pascu, C. A. Adler, J. F. Bloomfield.
Bull. Am. Astron. Soc., Vol. 23, No. 3, p. 1255 (1991). Abstract. – See Abstr. 012.065 for the main entry.

099.105   Astrometric observations of the irregular satellites of Jupiter.
A. L. Whipple, P. J. Shelus, G. F. Benedict.
Bull. Am. Astron. Soc., Vol. 23, No. 3, p. 1255 (1991). Abstract. – See Abstr. 012.065 for the main entry.

099.106   Statistical analysis of the Great Red Spot relative intensities for the time period 1963–1967.
J. Xanthakis, B. Petropoulos, C. Banos, E. Sarris.
Earth, Moon, Planets, Vol. 57, No. 2, p. 99 – 113 (May 1992).

The authors have analysed the Great Red Spot relative intensities for the time period 1963-1967, at 4300, 5500 and 6400 Å and found periods of 6, 4 and 3 months. Analytical relations that represent these intensities have been calculated.

099.107   Charged particle acceleration and heating in the Jovian ionosphere.
V. V. Zajtsev, V. E. Shaposhnikov.
Sov. Astron. Lett., Vol. 18, No. 2, p. 144 – 148 (Mar 1992). Current Physics Microform No.: 9211X0783. English translation of Pis'ma Astron. Zh., Tom 18, No. 4, p. 380 – 390 (1992).

The authors propose a charged particle acceleration mechanism associated with the potential drop along Jovian magnetic field lines. The mechanism is based on the well–known rise in resistance of a partially ionized plasma when the current through is nonstationary. A high–resistance region that provides the requisite potential drop along the magnetic field arises in the Jovian upper ionosphere, above the electron density maximum at the base of the Io magnetic flux tube. For parameters typical of the upper ionosphere, the resistance rises by seven orders of magnitude, and the corresponding potential drop can accelerate particles to several MeV. Joule heating and heating due to Bunemann turbulence can then heat the plasma to temperatures of some tens of keV.

099.108   Refraction of the Jovian decameter radioemission. II. Distortion of L–component directional pattern.
A. V. Arkhipov.
Kinematics Phys. Celest. Bodies, Vol. 7, No. 5, p. 8 – 14 (1991). English translation of Kinematika Fiz. Nebesn. Tel., Tom 7, No. 5, p. 11 – 19 (Sep–Oct 1991).

It is shown that when a realistic plasma model is used, the observed geometry of the L–emission directional pattern can be explained by refraction, as can the flux–density and symmetry in that diagram. The influence of the electric currents of the Io magnetic tube on the refraction of the Jovian decameter radioemission is analyzed for the first time.

099.109   Jupiter.
J. Rogers.
Images of the universe, p. 40 – 51 (1991). – See Abstr. 003.090 for the main entry.

Contents: (1) The structure of the atmosphere. (2) Large–scale disturbances in the atmosphere. (3) The moons of Jupiter. (4) Magnetic field and radiation belts.

099.110   Inertial modes of Jupiter.
U. Lee.
News Lett. Astron. Soc. N.Y., Vol. 4, No. 1, p. 25 (Feb 1992). Abstract. – See Abstr. 012.078 for the main entry.

The Galileo mission.
See Abstr. 003.061.

Seeing red: observations of colour in Jupiter's Equatorial Zone on the eve of the modern discovery of the Great Red Spot.
See Abstr. 004.012.

Focus: Jupiter. Introduction.
See Abstr. 014.015.

Jupiter – a failed star?
See Abstr. 014.016.

Seeing the most on Jupiter.
See Abstr. 014.042.

The photolysis of NH_3 in the presence of substituted acetylenes: a possible source of oligomers and HCN on Jupiter.
See Abstr. 022.059.

Dusty plasmas.
See Abstr. 022.088.

Quasi–random narrow band model fits to $1.6 - 2.5$ μm laboratory methane spectra and application to Jupiter.
See Abstr. 022.123.

Laboratory studies of radiation chemistry in the Jovian atmosphere.
See Abstr. 022.125.

Laboratory reflectance in the UV/VIS of ion bombarded ices: application to the Jovian satellites.
See Abstr. 022.135.

Application of ion irradiation experiments to planetary surfaces in the outer solar system.
See Abstr. 022.167.

The unified radio and plasma wave investigation.
See Abstr. 035.033.

Energetic Particles Investigation (EPI).
See Abstr. 035.075.

The lightning and Radio Emission Detector (LRD) instrument.
See Abstr. 035.076.

Galileo Probe Mass Spectrometer experiment.
See Abstr. 035.077.

Retrieval of a wind profile from the Galileo Probe telemetry signal.
See Abstr. 035.078.

Galileo Probe Nephelometer Experiment.
See Abstr. 035.079.

The Galileo Probe Atmosphere Structure Instrument.
See Abstr. 035.080.

Galileo Net Flux Radiometer Experiment.
See Abstr. 035.081.

The Jupiter Helium Interferometer Experiment on the Galileo entry probe.
See Abstr. 035.082.

The plasma instrumentation for the Galileo Mission.
See Abstr. 035.083.

The Galileo heavy element monitor.
See Abstr. 035.084.

The Galileo Dust Detector.
See Abstr. 035.085.

The Galileo plasma wave investigation.
See Abstr. 035.086.

The Galileo magnetic field investigation.
See Abstr. 035.087.

The Galileo Energetic Particles Detector.
See Abstr. 035.088.

The Galileo Solid–State Imaging experiment.
See Abstr. 035.089.

Near–Infrared Mapping Spectrometer experiment on Galileo.
See Abstr. 035.090.

Galileo ultraviolet spectrometer experiment.
See Abstr. 035.091.

Galileo Photopolarimeter/Radiometer experiment.
See Abstr. 035.092.

Galileo radio science investigations.
See Abstr. 035.093.

Positions of major planets from observations in the years $1988 - 1989$.
See Abstr. 041.015.

General theory for the outer planets.
See Abstr. 042.022.

Puzzles and prospects in planetary ring dynamics.
See Abstr. 042.024.

Collisional, collective and resonance phenomena in planetary rings.
See Abstr. 042.025.

Space Science Reviews volume on Galileo Mission: overview.
See Abstr. 051.031.

Gravitation and celestial mechanics investigations with Galileo.
See Abstr. 051.033.

Statistical mechanics, Euler's equation, and Jupiter's Red Spot.
See Abstr. 062.011.

Simulation of frontal cloudiness in the atmospheres of giant planets. I. Method of calculations.
See Abstr. 091.012.

Radiative association in planetary atmospheres.
See Abstr. 091.013.

The molecular – metallic transition of hydrogen and the structure of Jupiter and Saturn.
See Abstr. 091.037.

The role of the molecular – metallic transition of hydrogen in the evolution of Jupiter, Saturn, and brown dwarfs.
See Abstr. 091.038.

Modeling stratiform clouds in the atmospheres of giant planets with mutual solubility of condensing components. II.
See Abstr. 091.046.

The internal structure of the giant planets.
See Abstr. 091.047.

Solar system objects observed by Hipparcos.
See Abstr. 098.065.

Impact–generated atmospheres over Titan, Ganymede, and Callisto.
See Abstr. 100.007.

Planetarty rings: observational constraints and collision dynamics.
See Abstr. 100.016.

Giant planets and their satellites: what are the relationships between their properties and how they formed?
See Abstr. 107.059.

100 Saturn, Saturn Satellites

100.001   The action of the satellite 1981 S13, and the optical depth profile of the Encke–gap ringlet in the Saturn A ring.
F. Spahn, A. Saar, S. Schmidt, U. Schwarz.
Astron. Nachr., Vol. 313, No. 2, p. 101 – 105 (1992).
The radial optical depth profile of the Encke ringlet obtained by the occultation experiment of the Voyager photopolarimeter is explained to be caused by the gravitational action of the recently discovered satellite 1981 S13 and a second smaller moonlet orbiting in the vicinity of one of the triangular libration points. To this aim the results of previous and new numerical particle simulations as well as an extension of the scattering theory concerning a single moonlet to a pair of satellites have been used leading to a triple–peaked ringlet near the orbits of the moonlets. The width and the shape of that ringlet and its separate peaks depend on the mass ratio of both moonlets and on their orbital eccentricities. Furthermore, the authors' results yield a size of 1981 S13 of < 15 km in diameter.

100.002   Great spot on Saturn.
T. Starecký.
Říše hvězd, Vol. 72, No. 3, p. 52 – 53 (Mar 1991). In Czech.

100.003   The onset and growth of the 1990 equatorial disturbance on Saturn.
R. F. Beebe, C. Barnet, P. V. Sada, A. S. Murrell.
Icarus, Vol. 95, No. 2, p. 163 – 172 (Feb 1992).
The development of cloud structures at low Saturnian latitudes in Sep 1990 can be divided into three phases: the onset and expansion of the initial disturbance (popularly known as the Great White Spot), the eastward and westward expansion of the bright cloud, and the eventual formation of wave–like structures encircling the equator. The three phases of this storm are consistent with a single convective event. This event is evaluated within the historical context of previous occurrences and general seasonal mechanisms are considered.

100.004   The motion of Hyperion, Saturn's seventh satellite.
D. Taylor, J. Message.
Gemini, No. 36, p. 8 – 9 (Jun 1992).

100.005   Titan and exobiological aspects of the Cassini–Huygens mission.
F. Raulin, C. Frère, P. Paillous, E. de Vanssay, L. Do, M. Khlifi.
J. Br. Interplanet. Soc., Vol. 45, No. 6, p. 257 – 271 (Jun 1992).
Special issue: Exobiology. S. Santoli (ed.).

100.006   Can weak localization of photons explain the opposition effect of Saturn's rings?
M. I. Mishchenko, J. M. Dlugach.
Mon. Not. R. Astron. Soc., Vol. 254, No. 2, p. 15P – 18P (15 Jan 1992).
Weak localization of photons in discrete disordered media (or the coherent backscattering mechanism) is shown to be a likely explanation of the opposition effect exhibited by Saturn's rings. Specifically, the authors assume that the particles of Saturn's rings are covered with small H_2O ice grains and compute theoretically the opposition effect produced by these grains via the coherent backscattering mechanism. Both the width and amplitude of the observed opposition effect at visible wavelengths are consistent with theoretical calculations for effective grain radii of about $0.1–1$ μm. Such grains are known to be present in the outer B ring of Saturn and give rise to the so–called "spokes". Thus, the authors demonstrate that the opposition effect of Saturn's rings may be due to the surface properties of the individual ring particles rather than to interparticle shadowing, as is usually assumed.

100.007   Impact–generated atmospheres over Titan, Ganymede, and Callisto.
K. Zahnle, J. B. Pollack, D. Grinspoon.
Icarus, Vol. 95, No. 1, p. 1 – 23 (Jan 1992).
The competition between impact erosion and impact supply of volatiles to planetary atmospheres can determine whether a planet or satellite accumulates an atmosphere. In the absence of other processes (e.g., outgasing), one finds either that a planetary atmosphere should be thick, or that there should be no atmosphere at all. The boundary between the two extreme cases is set by the mass and velocity distributions and intrinsic volatile content of the impactors. The authors apply teir model specifically to Titan, Callisto, and Ganymede. The impacting population is identified with comets, either in the form of stray Uranus–Neptune planetisimals or as dislodged Kuiper belt comets. Systematically lower impact velocities on Titan allow it to retain a thick atmosphere, while Callisto and Ganymede get nothing. Titan's atmosphere may therefore be an expression of a late-accreting, volatile-rich veneer. An impact origin for Titan's atmosphere naturally accounts for the high D/H ratio it shares with Earth, the carbonaceous meteorites, and Halley.

100.008   A physical model of Titan's aerosols.
O. B. Toon, C. P. McKay, C. A. Griffith, R. P. Turco.
Icarus, Vol. 95, No. 1, p. 24 – 53 (Jan 1992).
Microphysical simulations of Titan's stratospheric haze show that aerosol microphysics is linked to organized dynamical processes. The detached haze layer may be a manifestation of 1 cm sec⁻¹ vertical velocities at altitudes above 300 km. Tomasko and Smith's model, in which a layer of large particles above 220 km altitude is responsible for the high forward scattering

observed by Rages and Pollack, is a natural outcome of the detached haze layer being produced by rising motions if aerosol mass production occurs primarily below the detached haze layer. The aerosol's electrical charge is critical for the particle size and optical depth of the haze. Dynamical processes control the haze particles below about 150 km. The optical depths of hydrocarbon clouds are probably less than one, requiring that abundant gases such as ethane condense on a subset of the haze particles to create relatively large, rapidly removed particles. The lower atmosphere and surface should be visible outside of regions of methane absorption in the near infrared. Limb scans at 2.0 μm wavelength sould be possible down to about 75 km altitude.

100.009   A general theory of motion for the eight major satellites of Saturn. III. Long–period perturbations.
A. Vienne, L. Duriez.
Astron. Astrophys., Vol. 257, No. 1, p. 331 – 352 (Apr 1992).

The authors give a representation of the long–period perturbations of the following satellites of Saturn: Mimas, Enceladus, Tethys, Dione, Rhea, Titan and Iapetus. The first order partial derivatives of this representation with respect to the physical parameters and the initial conditions have been also computed. The internal precision, obtained by comparison with numerical integration, has been estimated to a few kilometers for all satellites except Iapetus for which discrepancies reach 100 km over 100 years. Important new terms due to the orbital precession of Iapetus, and some terms coming from indirect perturbations by Jupiter have been found in the solution of Rhea, Titan and Iapetus. The solutions obtained for Mimas and Tethys suggest existence of secular resonance; the corresponding terms strongly depends on the value of the Tethys' eccentricity.

100.010   Comment on: "Evidence of Saturn's magnetic field anomaly from Saturnian kilometric radiation high–frequency limit" [J. Geophys. Res., Vol. 96, No. A8, p. 14129 – 14140 (1 Aug 1991)] by P. Galopeau, A. Ortega–Molina, and P. Zarka.
J. E. P. Connerney, M. D. Desch.
J. Geophys. Res., Vol. 97, No. A6, p. 8713 – 8717 (1 Jun 1992). See Abstr. 54.100.012.

100.011   Titan's atmosphere probed by stellar occultation.
A. Brahic, B. Sicardy, C. Ferrari, D. Gautier.
3. Symposium International de Bioastronomie, p. 133 (1991). Abstract. – See Abstr. 012.005 for the main entry.

100.012   Study of transmitted light through Titan's atmosphere.
D. Toublanc, J. P. Parisot, J. Brillet.
3. Symposium International de Bioastronomie, p. 134 – 135 (1991). – See Abstr. 012.005 for the main entry.

100.013   Prebiotic chemistry in planetary environments.
F. Raulin.
3. Symposium International de Bioastronomie, p. 141 – 148 (1991). – See Abstr. 012.005 for the main entry.

The formation of reactive organic compounds, such as HCN or HCHO, followed by their evolution in solution is one of the earliest steps in chemical evolution which might have led to the emergence of life on the Earth. Such organics are key ingredients of the prebiotic chemistry, since, in the presence of liquid water, they can give rise to the building blocks of living systems. Similar processes are going on in present planetary environments, especially on Titan, but in the absence of liquid water. With a dense reduced atmosphere mainly composed of N_2 and CH_4, rich in organic compounds in the gas and aerosol phases, and with the likely presence of an ocean of liquid methane and ethane, this moon appears as a natural laboratory for studying prebiotic organic chemistry at a planetary scale.

100.014   Titan's atmosphere from Voyager infrared observations: parallels and differences with the primitive earth.
A. Coustenis.
3. Symposium International de Bioastronomie, p. 179 – 189 (1991). – See Abstr. 012.005 for the main entry.

100.015   Looking for changes in the Saturnian system between Voyager and Cassini.
N. Borderies.
IAU Symposium No. 152: Chaos, resonance and collective dynamical phenomena in the solar system, p. 53 – 64 (1992). – See Abstr. 012.025 for the main entry.

This paper reviews a number of time–dependent phenomena that are relevant to our understanding of the dynamics of planetary rings and that will be investigated using Voyager and Cassini data. A long time baseline may help to decipher the physics of the spokes, understand better the morphology of the F ring and the rigid precession of non–circular ringlets, measure more precisely than has been done so far the satellites' torques and the viscosity of the A ring, and discover small satellites in the Saturnian ring system. Two exciting possibilities are those of determining the recession rates of the small satellites that border the rings, and of oberserving changes due to viscous diffusion in the irregular structures of the B ring.

100.016   Planetarty rings: observational constraints and collision dynamics.
A. Brahic, C. Ferrari.
IAU Symposium No. 152: Chaos, resonance and collective dynamical phenomena in the solar system, p. 83 – 96 (1992). – See Abstr. 012.025 for the main entry.

Collisions between ring's particles and gravitational perturbations of nearby satellites should explain most of the ring's structures. However, important questions are still unanswered. It is not understood why the rings are so dissimilar. The rings' origin and their stability over billions of years, most of their complex structures, the existence of arcs, and color and optical depth variations are not yet explained. Among all the ring mysteries, the uniform precession of narrow ringlets and the azimuthal brightness asymmetries should receive a high priority.

100.017   Hubble Space Telescope imaging of Saturn: the north pole.
J. Caldwell, C. C. Cunningham, X.–M. Hua, B. Turgeon, J. Westphal, C. Barnet.
Bull. Am. Astron. Soc., Vol. 23, No. 3, p. 1146 (1991). Abstract. – See Abstr. 012.037 for the main entry.

100.018   Hot spots on Saturn at 5 μm.
T. Owen, R. Baron.
Bull. Am. Astron. Soc., Vol. 23, No. 3, p. 1146 (1991). Abstract. – See Abstr. 012.037 for the main entry.

100.019   Imaging of Saturn and Venus in the thermal infrared.
F. Espenak, M. J. Mumma, D. Gezari, D. Deming, T. Kostiuk, G. Bjoraker, J. P. Connerney.
Bull. Am. Astron. Soc., Vol. 23, No. 3, p. 1146 (1991). Abstract. – See Abstr. 012.037 for the main entry.

100.020   Zonal variation of ammonia abundance in the Saturnian atmosphere.
A. L. Weir, B. J. Conrath, P. J. Gierasch.
Bull. Am. Astron. Soc., Vol. 23, No. 3, p. 1146 (1991). Abstract. – See Abstr. 012.037 for the main entry.

100.021   Ethane on Saturn.
T. Kostiuk, F. Espenak, P. Romani, D. Zipoy, J. Goldstein.
Bull. Am. Astron. Soc., Vol. 23, No. 3, p. 1146 – 1147 (1991). Abstract. – See Abstr. 012.037 for the main entry.

100.022   Meridional variations of albedo on Saturn: comparison of Voyager and Hubble Space Telescope observations.
C. D. Barnet, R. F. Beebe, W. A. Baum, G. E. Danielson, J. A. Westphal.
Bull. Am. Astron. Soc., Vol. 23, No. 3, p. 1147 (1991). Abstract. – See Abstr. 012.037 for the main entry.

100.023 **The nature of Saturn's 1990 equatorial storm.**
R. F. Beebe, L. Huber, C. D. Barnet, J. A. Westphal,
W. A. Baum, R. Light.
Bull. Am. Astron. Soc., Vol. 23, No. 3, p. 1147 – 1148 (1991).
Abstract. – See Abstr. 012.037 for the main entry.

100.024 **A new model for the disc Lyman alpha emission of Saturn.**
L. B. Jaffel, D. Feng, R. V. Yelle, D. T. Hall, F. Herbert.
Bull. Am. Astron. Soc., Vol. 23, No. 3, p. 1148 (1991). Abstract. –
See Abstr. 012.037 for the main entry.

100.025 **Saturn's toroidal water and hydrogen atmosphere.**
C. Crosby, M. K. Pospieszalska, R. E. Johnson.
Bull. Am. Astron. Soc., Vol. 23, No. 3, p. 1148 (1991). Abstract. –
See Abstr. 012.037 for the main entry.

100.026 **Hydrocarbon number density profiles in Saturn's upper atmosphere from the Voyager 2 stellar occultation experiment.**
D. Feng, B. Herman, R. V. Yelle.
Bull. Am. Astron. Soc., Vol. 23, No. 3, p. 1148 (1991). Abstract. –
See Abstr. 012.037 for the main entry.

100.027 **Enhanced stratospheric temperatures over Saturn's great equatorial storm of 1990.**
G. Orton, J. Friedson, T. Thaller, D. Gezari, F. Varosi,
C. Barnet, P. Drossart, J. Lecacheux, P. Laques, F. Colas,
J. Caldwell.
Bull. Am. Astron. Soc., Vol. 23, No. 3, p. 1149 (1991). Abstract. –
See Abstr. 012.037 for the main entry.

100.028 **The occultation of 28 Sagittarii by Saturn.**
W. H. Allen.
South. Stars, Vol. 34, No. 6, p. 341 – 346 (Mar 1992).
This paper reports on photoelectric observations made during the occultation of 28 Sgr by Saturn on 3 Jul 1989 at the Adams Lane Observatory, Blenheim, New Zealand. A light curve for the visible portion of the event was obtained and portions of this light curve are presented. From the data obtained the dimensions of the A and B rings of Saturn may be compared to values obtained by the Voyager 2 spacecraft, when it recorded δ Sco passing behind the rings during the Satrun flyby in Aug 1981.

100.029 **Detection of albedo markings on Enceladus.**
A. Verbiscer, J. Veverka.
Bull. Am. Astron. Soc., Vol. 23, No. 3, p. 1168 (1991). Abstract. –
See Abstr. 012.037 for the main entry.

100.030 **The shape of Tethys.**
P. C. Thomas, S. F. Dermott.
Bull. Am. Astron. Soc., Vol. 23, No. 3, p. 1168 (1991). Abstract.
The full paper is published in Icarus, Vol. 94, No. 2, p. 391 – 398
(Dec 1991). – See Abstr. 012.037 for the main entry.

100.031 **Characterization of tectonic structures on Iapetus.**
J. N. Head, S. K. Croft.
Bull. Am. Astron. Soc., Vol. 23, No. 3, p. 1168 – 1169 (1991).
Abstract. – See Abstr. 012.037 for the main entry.

100.032 **Is Hyperion's rotation only mildly chaotic?: Evidence for a dense spherical core.**
J. D. Scargle.
Bull. Am. Astron. Soc., Vol. 23, No. 3, p. 1171 – 1172 (1991).
Abstract. – See Abstr. 012.037 for the main entry.

100.033 **Saturn's rings: optical depth profiles at $\lambda 3.9$ μm from the occultation of 28 Sgr.**
P. D. Nicholson, O. Perkovic, K. Matthews, R. G. French.
Bull. Am. Astron. Soc., Vol. 23, No. 3, p. 1178 (1991). Abstract. –
See Abstr. 012.037 for the main entry.

100.034 **Saturn ring masses and lightcurve morphology from IRTF observations of the occultation of 28 Sgr.**
J. Harrington, M. L. Cooke, E. W. Dunham, W. J. Forrest,
J. L. Pipher, J. L. Elliot.
Bull. Am. Astron. Soc., Vol. 23, No. 3, p. 1178 – 1179 (1991).
Abstract. – See Abstr. 012.037 for the main entry.

100.035 **The kinematics of eccentric features in Saturn's Cassini Division from combined Voyager and groundbased data.**
E. Turtle, C. Porco, V. Haemmerle, W. Hubbard, R. Clark.
Bull. Am. Astron. Soc., Vol. 23, No. 3, p. 1179 (1991). Abstract. –
See Abstr. 012.037 for the main entry.

100.036 **Meteoroid bombardment and the structure of Saturn's A and B–ring inner edges.**
R. H. Durisen, P. W. Bode, S. Dyck, J. N. Cuzzi.
Bull. Am. Astron. Soc., Vol. 23, No. 3, p. 1179 (1991). Abstract. –
See Abstr. 012.037 for the main entry.

100.037 **Saturn A ring surface mass densities.**
L. J. Horn, J. Hui.
Bull. Am. Astron. Soc., Vol. 23, No. 3, p. 1179 (1991). Abstract. –
See Abstr. 012.037 for the main entry.

100.038 **Density wave launch zone analysis: discrepancies between theory and observation.**
P. A. Rosen, T. G. Brophy.
Bull. Am. Astron. Soc., Vol. 23, No. 3, p. 1179 (1991). Abstract. –
See Abstr. 012.037 for the main entry.

100.039 **Photometric studies of Saturn's C ring.**
M. L. Cooke, P. D. Nicholson, M. R. Showalter.
Bull. Am. Astron. Soc., Vol. 23, No. 3, p. 1180 (1991). Abstract. –
See Abstr. 012.037 for the main entry.

100.040 **Color variation and material transport in Saturn's rings.**
J. Cuzzi, P. Estrada, P. Bode.
Bull. Am. Astron. Soc., Vol. 23, No. 3, p. 1180 (1991). Abstract. –
See Abstr. 012.037 for the main entry.

100.041 **Three–dimensional perturbations in Saturn's F ring.**
R. A. Kolvoord, J. A. Burns.
Bull. Am. Astron. Soc., Vol. 23, No. 3, p. 1180 (1991). Abstract. –
See Abstr. 012.037 for the main entry.

100.042 **Detailed study of Prometheus' and Pandora's density waves.**
C. C. Harris, L. W. Esposito.
Bull. Am. Astron. Soc., Vol. 23, No. 3, p. 1181 (1991). Abstract. –
See Abstr. 012.037 for the main entry.

100.043 **Solar–heating efficiencies in the thermosphere of Titan.**
J. L. Fox, R. V. Yelle.
Bull. Am. Astron. Soc., Vol. 23, No. 3, p. 1184 (1991). Abstract. –
See Abstr. 012.037 for the main entry.

100.044 **Shape and opacity of Titan's stratosphere from the 28 Sgr occultation.**
B. Sicardy, W. B. Hubbard, D. Toublanc.
Bull. Am. Astron. Soc., Vol. 23, No. 3, p. 1184 (1991). Abstract. –
See Abstr. 012.037 for the main entry.

100.045 **Properties of Titan's stratospheric C_4N_2 cloud.**
R. E. Samuelson.
Bull. Am. Astron. Soc., Vol. 23, No. 3, p. 1185 (1991). Abstract. –
See Abstr. 012.037 for the main entry.

100.046 **A physical model of Titan's aerosols.**
O. B. Toon.
Bull. Am. Astron. Soc., Vol. 23, No. 3, p. 1185 (1991). Abstract.
The full paper is published in Icarus, Vol. 95, No. 1, p. 24 – 53
(Jan 1992). – See Abstr. 012.037 for the main entry.

100.047   Seasonal variations in Titan's stratospheric haze and albedo.
W. T. Hutzell, W. L. Chameides, C. P. McKay.
Bull. Am. Astron. Soc., Vol. 23, No. 3, p. 1185 (1991). Abstract. – See Abstr. 012.037 for the main entry.

100.048   Modelling of the diurnal variations in the atmosphere of Titan.
D. Toublanc, J. P. Parisot, J. Brillet, D. Gautier.
Bull. Am. Astron. Soc., Vol. 23, No. 3, p. 1185 (1991). Abstract. – See Abstr. 012.037 for the main entry.

100.049   Detectability of minor species on Titan from spectroscopic measurements in the 160 – 300 nm range.
R. Courtin.
Bull. Am. Astron. Soc., Vol. 23, No. 3, p. 1185 – 1186 (1991). Abstract. – See Abstr. 012.037 for the main entry.

100.050   Windows through Titan's atmosphere?
W. Grundy, M. Lemmon, U. Fink, P. Smith, M. Tomasko.
Bull. Am. Astron. Soc., Vol. 23, No. 3, p. 1186 (1991). Abstract. – See Abstr. 012.037 for the main entry.

100.051   High-resolution observations of Titan in the 1 to 1.3 micron region: preliminary study.
A. Coustenis, E. Lellouch, J. P. Maillard, K. Strong, B. Schmitt, C. Griffith.
Bull. Am. Astron. Soc., Vol. 23, No. 3, p. 1186 (1991). Abstract. – See Abstr. 012.037 for the main entry.

100.052   Titan's condensates and tholins: surface interactions.
W. R. Thompson, C. Sagan.
Bull. Am. Astron. Soc., Vol. 23, No. 3, p. 1186 (1991). Abstract. – See Abstr. 012.037 for the main entry.

100.053   Numerical simulations of the circulation of Titan's atmosphere: study of a possible Venus–like superrotation.
F. Hourdin, O. Talagrand, P. Le Van, R. Courtin, D. Gautier, C. P. McKay.
Bull. Am. Astron. Soc., Vol. 23, No. 3, p. 1187 (1991). Abstract. – See Abstr. 012.037 for the main entry.

100.054   Fractal aerosols in Titan's atmosphere.
M. Cabane, E. Chassefière, G. Israel.
Bull. Am. Astron. Soc., Vol. 23, No. 3, p. 1187 (1991). Abstract. – See Abstr. 012.037 for the main entry.

100.055   Evidence for seasonal change on Titan.
C. C. Cunningham, J. J. Caldwell, H. P. White, D. M. Anthony.
Bull. Am. Astron. Soc., Vol. 23, No. 3, p. 1187 – 1188 (1991). Abstract. – See Abstr. 012.037 for the main entry.

100.056   Properties and distribution of the aerosols in the atmosphere of Titan.
M. Dobrijevic, L. Goutoulli, D. Toublanc, J. P. Parisot, J. Brillet.
Bull. Am. Astron. Soc., Vol. 23, No. 3, p. 1188 (1991). Abstract. – See Abstr. 012.037 for the main entry.

100.057   Gravity waves in Titan's troposphere.
J. Friedson.
Bull. Am. Astron. Soc., Vol. 23, No. 3, p. 1188 (1991). Abstract. – See Abstr. 012.037 for the main entry.

100.058   Voyager 1 EUV airglow observations of Titan.
D. T. Hall, T. M. Tripp, D. E. Shemansky.
Bull. Am. Astron. Soc., Vol. 23, No. 3, p. 1188 (1991). Abstract. – See Abstr. 012.037 for the main entry.

100.059   A global Titan upper atmosphere model from the occultation of 28 Sgr.
W. B. Hubbard, D. M. Hunten, C. C. Porco, B. Sicardy.
Bull. Am. Astron. Soc., Vol. 23, No. 3, p. 1188 (1991). Abstract. – See Abstr. 012.037 for the main entry.

100.060   Titan hydrocarbons photochemical model.
L. M. Lara, R. Rodrigo, J. J. López–Moreno, A. Coustenis, E. Chassefière.
Bull. Am. Astron. Soc., Vol. 23, No. 3, p. 1189 (1991). Abstract. – See Abstr. 012.037 for the main entry.

100.061   Spectrophotometric observations of Titan from 1.2 to 5.1 μm.
K. S. Noll, R. F. Knacke.
Bull. Am. Astron. Soc., Vol. 23, No. 3, p. 1189 (1991). Abstract. – See Abstr. 012.037 for the main entry.

100.062   Laboratory investigations of the aerosols in the stratosphere of Titan.
T. W. Scattergood, B. Stone, E. Y. Lau.
Bull. Am. Astron. Soc., Vol. 23, No. 3, p. 1189 (1991). Abstract. – See Abstr. 012.037 for the main entry.

100.063   Study of transmitted light through the atmosphere of Titan.
D. Toublanc, J. P. Parisot, J. Brillet, D. Gautier.
Bull. Am. Astron. Soc., Vol. 23, No. 3, p. 1189 – 1190 (1991). Abstract. – See Abstr. 012.037 for the main entry.

100.064   Non–LTE models of Titan's upper atmosphere.
R. V. Yelle.
Bull. Am. Astron. Soc., Vol. 23, No. 3, p. 1190 (1991). Abstract. The full paper is published in Astrophys. J., Vol. 383, No. 1, p. 380 – 400 (10 Dec 1991). – See Abstr. 012.037 for the main entry.

100.065   Formation and growth of photochemical aerosols in Titan's atmosphere.
M. Cabane, E. Chassefière, G. Israel.
Icarus, Vol. 96, No. 2, p. 176 – 189 (Apr 1992).
Recent developments in the understanding of the morphology (shape and size) of haze aerosols in Titan's atmosphere, aggregate particles and their associated optical properties, have been considered in light of a microphysical modeling of aerosols. The classical assumption of spherical particles is shown to be valid during the first growth stage. The mean ratio between the masses of two particles colliding together is quantified as a function of altitude. From this criterion, the formation region inside which monomers are generated and the settling region where aggregates build up may be clearly separated and the boundary between them precisely defined and located. The strong dependence of the monomer radius on the pressure (thus the altitude) is calculated using different sets of parameters in order to cover the widest range of possibilities. Using parameters favored by West in his analysis of the polarizing properties of Titan's haze, the formation altitude of aerosols is found to lie in the range from 350 to 400 km.

100.066   Titan: evidence for seasonal change – a comparison of Hubble Space Telescope and Voyager images.
J. Caldwell, D. Anthony, H. P. White, C. C. Cunningham, E. J. Groth, H. Hasan, K. Noll, H. A. Weaver, P. H. Smith, M. G. Tomasko.
Icarus, Vol. 97, No. 1, p. 1 – 9 (May 1992).
Images of Titan were obtained by the Hubble Space Telescope (HST) on 26 Aug 1990. Comparison with Voyager 1 and Voyager 2 images obtained 10 and 9 years earlier shows that the seasonal hemispheric brightness asymmetry has reversed near 440 and 550 nm wavelengths, with the northern hemisphere now being brighter. An additional, noisy HST image at 889 nm wavelength, for which there are no analogous Voyager data, suggests that the southern hemisphere may have been brighter than the northern at that wavelength in 1990.

100.067   Saturn's upper troposphere 1986 – 1989.
E. Karkoschka, M. G. Tomasko.
Icarus, Vol. 97, No. 2, p. 161 – 181 (Jun 1992).

This work describes observations of Saturn's atmosphere in the visible and near–infrared (460 – 940 nm) including 4 hydrogen quadrupole lines, 17 methane absorption bands ranging over 3 orders of magnitude in absorption strength, an ammonia absorption band, and the absolute calibrated continuum spectrum. All observations have complete coverage of Saturn's disk, in latitude as well as in center–to–limb position. A new method describing center–to–limb information is presented. This data set gives a quite complete description of Saturn's atmosphere in the visible and near infrared at the spatial resolution of ground–based observations. Weak absorption features of hydrogen, methane, and ammonia show a significant enhancement in the North Polar Region compared to the rest of the planet. An atmospheric model is given which fits all observations within estimated errors.

100.068   Vapor–liquid   equilibrium   thermodynamics   of $N_2 + CH_4$: model and Titan applications.
W. R. Thompson, J. A. Zollweg, D. H. Gabis.
Icarus, Vol. 97, No. 2, p. 187 – 199 (Jun 1992).

Calculations of the vapor–liquid equilibrium thermodynamics of the $N_2 + CH_4$ system show that the tropospheric clouds of Titan are not pure CH_4, but solutions of CH_4 containing substantial quantities of N_2. The conditions for saturation, latent heat of condensation, and droplet composition all depend on this equilibrium. The authors present a thermodynamic model for vapor–liquid equilibrium in the $N_2 + CH_4$ system which, by its structure, places strong constraints on the consistency of experimental equilibium data, and confidently embodies temperature effects by also including enthalpy (heat of mixing) data. Selected equilibrium and enthalpy data are used in a maximum likelihood determination of model parameters. The model can be readily evaluated to compute the saturation criteria, composition of condensate, and latent heat in Titan's atmosphere for a given pressure–temperature (p–T) profile.

100.069   The dynamics of Saturn's E ring particles.
M. Horanyi, J. A. Burns, D. P. Hamilton.
Icarus, Vol. 97, No. 2, p. 248 – 259 (Jun 1992).

Saturn's tenuous E ring, located between 3 and 8 Saturnian radii, peaks sharply near Enceladus' orbit and has recently been found to be composed predominantly of grains 1 μm in radius. The authors study analytically and numerically the motion of such grains launched from Enceladus as they evolve under the action of Saturn's oblate gravity field, solar radiation pressure, and electromagnetic forces.

100.070   Das System der Saturnringe und die Enckesche Teilung.
J. Gürtler.
Sterne, Band 68, Heft 3, p. 188 – 197 (1992).

100.071   Numerical integration of the orbit of Phoebe.
J. R. Rohde.
Bull. Am. Astron. Soc., Vol. 23, No. 3, p. 1255 (1991). Abstract. – See Abstr. 012.065 for the main entry.

100.072   Reading Saturn's ring spokes.
L. R. Doyle, E. Grün.
IAU Colloquium No. 126: Origin and evolution of interplanetary dust, p. 357 – 360 (1991). – See Abstr. 012.068 for the main entry.

The micron–sized dust forming the radial spoke–like features in Saturn's rings are studied using radiative transfer analysis. Theories for their likely origin and evolution are discussed in light of these results, and future work is outlined.

100.073   Saturn encountered.
R. McKim.
Images of the universe, p. 52 – 62 (1991). – See Abstr. 003.090 for the main entry.

Contents: (1) Setting the scene. (2) The globe of Saturn. (3) The magnetosphere. (4) The rings. (5) Amongst the moons. (6) The future.

Dusty plasmas.
See Abstr. 022.088.

Application of ion irradiation experiments to planetary surfaces in the outer solar system.
See Abstr. 022.167.

Optical properties of tholin from H_2O/C_2H_6(6:1) ice, and comparison with Titan tholin, kerogen and meteoritic organics.
See Abstr. 022.171.

The Italian involvement in Cassini Radar.
See Abstr. 035.131.

General theory for the outer planets.
See Abstr. 042.022.

Puzzles and prospects in planetary ring dynamics.
See Abstr. 042.024.

Collisional, collective and resonance phenomena in planetary rings.
See Abstr. 042.025.

A synthetic theory of motion for Titan–Hyperion.
See Abstr. 042.041.

About the secular acceleration of Mimas.
See Abstr. 042.043.

Hori auxiliary system for Mimas–Tethys.
See Abstr. 042.044.

An integrable model for Helene.
See Abstr. 042.045.

ASTRA: altimetry and sounding of Titan with a radar on a descending craft.
See Abstr. 051.065.

Stability of a thin gravitating ring and systems of rings.
See Abstr. 062.075.

Three–dimensional perturbations of particles in a narrow planetary ring.
See Abstr. 091.005.

Equilibrium velocities in planetary rings with low optical depth.
See Abstr. 091.006.

The molecular – metallic transition of hydrogen and the structure of Jupiter and Saturn.
See Abstr. 091.037.

The role of the molecular – metallic transition of hydrogen in the evolution of Jupiter, Saturn, and brown dwarfs.
See Abstr. 091.038.

Equatorial superrotation in a slowly rotating GCM: implications for Titan and Venus.
See Abstr. 091.044.

Temperature dependence of the intensities, shifts, and widths of the thermal infrared lines of trace gases in the atmospheres of the outer planets and Titan.
See Abstr. 091.045.

The internal structure of the giant planets.
See Abstr. 091.047.

Numerical simulations of dense collisional systems. II. Extended distribution of particle sizes.
See Abstr. 091.053.

VLA/Goldstone Planetary Radar results.
See Abstr. 091.061.

Stellar occultation candidates from the Guide Star Catalog. I. Saturn, 1991–1999.
See Abstr. 096.002.

The eclipse of star 28 Sgr by Titan.
See Abstr. 096.011.

Solar system objects observed by Hipparcos.
See Abstr. 098.065.

The principle of equivalence and the Trojan asteroids. II.
See Abstr. 098.073.

The role of the molecular–metallic transition of hydrogen in Jupiter and Saturn.
See Abstr. 099.065.

Germane photochemistry in the atmosphere of Jupiter and Saturn: the reaction of atomic hydrogen with germane.
See Abstr. 099.066.

Comparisons of the icy regoliths of Europa, Callisto, and Rhea.
See Abstr. 099.068.

Diurnal variations on Jupiter and Saturn?
See Abstr. 099.098.

Millimeter–wave observations of Saturn, Uranus, and Neptune: CO and HCN on Neptune.
See Abstr. 101.070.

Does the Saturn family of periodic comets exist?
See Abstr. 102.006.

Giant planets and their satellites: what are the relationships between their properties and how they formed?
See Abstr. 107.059.

101 Uranus, Neptune, Pluto, Transplutonian Planets

101.001 **Large quasi–circular features beneath frost on Triton.**
P. Helfenstein, J. Veverka, D. McCarthy, P. Lee, J. Hillier.
Science, Vol. 255, No. 5046, p. 824 – 826 (14 Feb 1992).
Specially processed Voyager 2 images of Neptune's largest moon, Triton, reveal three large quasi–circular features ranging in diameter from 280 to 935 km within Triton's equatorial region. The largest of these features contains a central, irregularly shaped area of comparatively low albedo about 380 km in diameter, surrounded by crudely concentric annuli of higher albedo materials. None of the features exhibit significant topographic expression, and all appear to be primarily albedo markings. The features are located within a broad equatorial band of anomalously transparent frost that renders them nearly invisible at the large phase angles at which Voyager obtained its highest resolution coverage of Triton. The features can be discerned at smaller phase angles at which the frost only partially masks underlying albedo contrasts. The origin of the features is uncertain but may have involved regional cryovolcanic activity.

101.002 **The effect of magnetic topography on high–latitude radio emission at Neptune.**
C. B. Sawyer, J. W. Warwick, J. H. Romig.
J. Geophys. Res., Vol. 97, No. A1, p. 1 – 10 (1 Jan 1992).
Occultation by a local elevation on the surface of constant magnetic field is proposed as a new interpretation for the unusual properties of Neptune high–latitude emission. Abrupt changes in intensity and polarization of this broadband smooth radio emission were observed as the Voyager 2 spacecraft passed near the north magnetic pole before closest approach. The observed sequence of cutoffs with polarization reversal would not occur during descent of the spacecraft through regular surfaces of increasing magnetic field. The sequence can be understood in terms of constant–frequency (constant–field) surfaces that are not only offset from the planet center but are locally highly distorted by an elevation that occults the outgoing extraordinary–mode beam. The required occulter is similar to the field enhancement observed directly by the magnetometer team when Voyager reached lower altitude farther to the west. The authors present evidence that the sources of the high–latitude emission are located near the longitude of the minimum–B anomaly associated with the dipole offset and that the local elevation of constant–B surfaces extends eastward from the longitude where it is directly measured by the magnetometer to the longitude where occultation of the remote radio source is observed. Together, the radio and magnetometer experiments indicate that the constant–frequency surfaces are distorted by an elevation that extends 0.3 rad in the longitudinal direction. On the 462 kHz surface, about 1.1 R_N from the dipole center, the local elevation must be at least 0.45 R_N above the undisturbed surface.

101.003 **On the ephemerides of Uranus.**
A. Brunini.
Astron. Astrophys., Vol. 255, No. 1/2, p. 401 – 404 (Feb 1992).
The author computes a new set of Uranus's initial conditions in order to investigate the nature of its observed systematic residuals. In addition, he pays special attention to the masses of the known perturbing planets. However, he finds that the only way to fit accurately the observations is through the partition of the whole period of observations into two: one from the discovery up to 1897 and another from this last date up to the present. The problem is then the existence of a discontinuity in the orbital elements. Some speculations are made about the impossibility to fit accurately the whole orbit.

101.004 **Neptune.**
V. Pohánka.
Kozmos, Vol. 22, No. 2, p. 57 – 61 (Mar 1991). In Slovak.

101.005 **The effect of surface roughness on Triton's volatile distribution.**
R. V. Yelle.
Science, Vol. 255, No. 5051, p. 1553 – 1555 (20 Mar 1992).
Calculations of radiative equilibrium temperatures on Triton's rough surface suggest that significant condensation of N_2 may be occurring in the northern equatorial regions, despite their relatively dark appearance. The bright frost is not apparent in the

Voyager images because it tends to be concentrated in relatively unilluminated facets of the surface. This patchwork of bright frost–covered regions and darker bare ground may be distributed on scales smaller than that of the Voyager resolution; as a result the northern equatorial regions may appear relatively dark. This hypothesis also accounts for the observed wind direction in the southern hemisphere because it implies that the equatorial regions are warmer than the south polar regions.

101.006   Solar control of the upper atmosphere of Triton.
J. R. Lyons, Y. L. Yung, M. Allen.
Science, Vol. 256, No. 5054, p. 204 – 206 (10 Apr 1992).
If the upper atmosphere and ionosphere of Triton are controlled by precipitation of electrons from Neptune's magnetosphere as previously proposed, Triton could have the only ionosphere in the solar system not controlled by solar radiation. However, a new model of Triton's atmosphere, in which only solar radiation is present, predicts a large column of carbon atoms. With an assumed, but reasonable, rate of charge transfer between N_2^+ and C, a peak C^+ abundance results that is close to the peak electron densities measured by Voyager in Triton's ionosphere. These results suggest that Triton's upper atmospheric chemistry may thus be solar–controlled.

101.007   The arcs near Neptune as chains of epitons in a continuous transparent ring.
N. N. Gor'kavyj.
Sov. Astron. Lett., Vol. 17, No. 6, p. 428 – 433 (Nov 1991).
Current Physics Microform No.: 9207X1866. English translation of Pis'ma Astron. Zh., Tom 17, No. 11, p. 1020 – 1030 (Nov 1991).
Gor'kavyj (1989) suggested that the existence of an arc is possible if it is a collection of epicyclic vortices (epitons) inside the dust ring. The particles of such an elliptical vortex revolve around the center of mass along epicycles. The exact equality of the semimajor axes of the particle orbits produces equality of their periods of orbital revolution, consequently, a vortex is stable and does not spread out along the orbit. A continuous narrow ring leads to the equality of the semimajor axes for the epitons, which guarantees the stability of many dozens of vortices. Voyager 2 photographed a regular chain of compact condensations situated in the arcs near Neptune. The existence of such structure is a direct confirmation of the epiton model: not one of the other hypotheses provides for the existence of internal structure for an arc.

101.008   Theory of the Neptunian arcs. I. Stability of an individual epiton.
N. N. Gor'kavyj, T. A. Tajdakova, N. M. Gaftonyuk.
Sov. Astron. Lett., Vol. 17, No. 6, p. 457 – 461 (Nov 1991).
Current Physics Microform No.: 9207X1895. English translation of Pis'ma Astron. Zh., Tom 17, No. 12, p. 1105 – 1115 (Dec 1991).
The principal question of arc dynamics is why they do not diffuse along the orbit. Gor'kavyj (1989) has suggested that an arc is a chain of epicyclic vortices in a continuous ring. This model is tested by direct numerical simulation taking into account inelastic collisions between identical spherical particles. It is shown that the vortex (epiton) may be considered a stable formation if there is a natural mechanism which arranges the vortex particles in a collisionless order.

101.009   Theory of the Neptunian arcs. II. Dynamics of epitons in the ring.
N. N. Gor'kavyj, T. A. Tajdakova.
Sov. Astron. Lett., Vol. 17, No. 6, p. 462 – 465 (Nov 1991).
Current Physics Microform No.: 9207X1900. English translation of Pis'ma Astron. Zh., Tom 17, No. 12, p. 1116 – 1123 (Dec 1991).
The dynamics of an epicyclic vortex (epiton) immersed in a dust ring is investigated. Numerical calculations taking vortex self–gravitation into account show that the epiton is compressed in the dust stream and drifts radially toward the densest part of the ring. This implies position stability of the vortex in the dust

ring. It is shown that the formation of an epiton chain is due to the interaction of the self–gravitating vortices produced by the ring dust particles.

101.010   Pluto's extended atmosphere: an escape model and initial observations.
J. T. Clarke, S. A. Stern, L. M. Trafton.
Icarus, Vol. 95, No. 2, p. 173 – 179 (Feb 1992).
The authors have calculated the rates of production and hydrodynamic outflow of atomic hydrogen resulting from the photodissociation of methane in the upper atmosphere of Pluto. Under the present near–perihelion conditions this yields an extended cloud of H around Pluto which is likely to be the most easily observable signature of Pluto's extended atmosphere, and thereby provide information on the extent, escape rate, and composition of Pluto's upper atmosphere. The authors have also performed initial observations with the IUE attempting to detect the H Ly α emission from the extended H cloud, which is used to derive upper limits to the cloud properties as a function of the cloud extent.

101.011   Capture probabilities for secondary resonances.
J. Henrard, M. Moons.
Icarus, Vol. 95, No. 2, p. 244 – 252 (Feb 1992).
The authors analyze the role of secondary resonances during tidal evolution within a j:j + 2 orbit–orbit resonance and apply it to the tidal evolution of Miranda and Umbriel in the 1:3, i^2_M– resonance. They improve the result of Malhotra (1990) concerning the probability of capture into secondary resonances both analytically by modifying her model of a perturbed pendulum and numerically by using a seminumerical perturbation method which enables one to bypass this modelization.

101.012   Photometric variability of Charon at 2.2 μm.
A. S. Bosh, L. A. Young, J. L. Elliot, H. B. Hammel.
Icarus, Vol. 95, No. 2, p. 319 – 324 (Feb 1992).
Several dozen Pluto–Charon images were obtained on each of 4 nights with the ProtoCAM on the IRTF, mostly at K but also at J and H. A two–source image model was fit to the blended images of Pluto and Charon, with the position of Charon and the ratio of its signal to that of Pluto as free parametes. At K, the authors find Charon to be fainter than Pluto by 1.80 ± 0.09, 2.39 ± 0.05, and 2.09 ± 0.05 mag at lightcurve phases 0.06, 0.42, and 0.95. Combining these magnitudes with combined photometry of the Pluto–Charon system one finds apparent K magnitudes for Charon of 15.01 ± 0.08 at lightcurve phase 0.06 and 15.46 ± 0.05 at lightcurve phase 0.42. It is concludeed that Charon is variable in this filter bandpass. The variation is most likely due to changes in its geometric albedo as a function of longitude.

101.013   The atmosphere of Neptune: an analysis of radio occultation data acquired with Voyager 2.
G. F. Lindal.
Astron. J., Vol. 103, No. 3, p. 967 – 982 (Mar 1992). Current Physics Microform No.: 9203E2175.
Recordings of the tracking signals received from Voyager 2 during its occultation by Neptune have been used to study the vertical structure of Neptune's atmosphere. The measurements, which began at a planetographic latitude of 62° north and ended near 45° south, cover an altitude interval of about 5000 km. Inversion and interpretation of the occultation data have provided new information on the distribution of free electrons in Neptune's ionosphere, the thermal structure and composition of its troposphere and stratosphere, and the zonal winds below the 1.7 bar level in the troposphere.

101.014   The albedos of Pluto and Charon: wavelength dependence.
R. L. Marcialis, L. A. Lebofsky, M. A. DiSanti, U. Fink, E. F. Tedesco, J. Africano.
Astron. J., Vol. 103, No. 4, p. 1389 – 1394 (Apr 1992). Current Physics Microform No.: 9204G0343.
The 1987 March 3 occultation of Charon by Pluto was monitored simultaneously with three telescopes in the vicinity of Tucson, Arizona. Each site covered a distinct wavelength

interval, with the total range spanning 0.44–2.4 μm. Observing the same event ensures an identical Sun–Pluto–Earth geometry for all three sites, and minimizes the assumptions which must be made to combine results. The authors have used this spectrophotometry to derive the individual geometric albedos of Pluto and Charon over a factor of $\geqslant 5$ in wavelength. Combining their results with those of Binzel (1988), the authors obtain improved (B–V) color estimates (on the "Johnson Pluto" system) for the components of the system at rotational phase 0.75: (Pluto + Charon) = 0.843 ± 0.006; Pluto alone = 0.866 ± 0.007; and Charon alone = 0.702 ± 0.010.

101.015   The 1991 apparition of Uranus.
R. W. Schmude Jr.
Strolling Astron., Vol. 36, No. 1, p. 20 – 22 (Mar 1992).

101.016   The masses of Uranus and its major satellites from Voyager tracking data and Earth–based Uranian satellite data.
R. A. Jacobson, J. K. Campbell, A. H. Taylor, S. P. Synnott.
Astron. J., Vol. 103, No. 6, p. 2068 – 2078 (Jun 1992). Current Physics Microform No.: 9207A0348.

Analysis of the Doppler–tracking data and star–satellite imaging from the Voyager 2 spacecraft combined with Earth–based astrometric satellite observations has yielded improved values for the masses of the Uranian system and the satellites Ariel, Umbriel, Titania, Oberon, and Miranda. Masses are expressed as the product GM, the universal gravitational constant G times the mass M of the body, in units of $(km^3 s^{-2})$. The satellite masses are (4.4 ± 0.5) for Miranda, (90.3 ± 8.0) for Ariel, (78.2 ± 9.0) for Umbriel, (235.3 ± 6.0) for Titania, and (201.1 ± 5.0) for Oberon. The mass of the Uranian system is $(5,794,548.6 \pm 7.0)$ and the ratio of the mass of the Sun to the mass of the Uranian system is $(22,902.982 \pm 0.028)$. Quoted errors are standard errors and are the authors' assessment of the true rather than the formal errors. The Uranus rotational pole orientation angles and gravity harmonic coefficients were fixed at the values determined by French et al. (1988) from stellar occultations of the Uranian rings observed from both the Earth and Voyager 2 and from the occultation of the spacecraft radio signal.

101.017   A theory for narrow–banded radio bursts at Uranus: MHD surface waves as an energy driver.
W. M. Farrell, S. A. Curtis, M. D. Desch, R. P. Lepping.
J. Geophys. Res., Vol. 97, No. A4, p. 4133 – 4142 (1 Apr 1992).

The authors describe a possible scenario for the generation of the narrow – banded radio bursts (n bursts) detected at Uranus by the Voyager 2 planetary radio astronomy experiment. This particular radio emission is suspected to originate within the Uranian northern polar cusp region. In order to account for the emission burstiness which occurs on time scales of hundreds of milliseconds, the authors propose that ULF magnetic surface turbulence, generated at the frontside magnetopause, propagates down the open/closed field line boundary and mode converts to kinetic Alfvén waves (KAW) deep within the polar cusp. The oscillating KAW potentials then drive a transient electron stream (beam or loss cone) that creates the bursty radio emission. To substantiate these ideas, the authors show Voyager 2 magnetometer measurements by enhanced ULF magnetic activity at the frontside magnetopause. They then demonstrate analytically that such magnetic turbulence should mode convert deep in the cusp at a radial distance of $3~R_U$. A condition for mode conversion from magnetic surface waves to KAW is $f_{pc}/f_{ce} < 0.2$, which is also a condition favorable for the generation of the R–X mode radio bursts. The fact that a similar plasma condition is required for both processes lends strong support that the active regions for mode conversion and R–X mode wave generation are one and the same.

101.018   The distribution of atomic hydrogen in the magnetosphere of Saturn.
D. E. Shemansky, D. T. Hall.
J. Geophys. Res., Vol. 97, No. A4, p. 4143 – 4161 (1 Apr 1992).

Three sets of previously unpublished Voyager ultraviolet spectrometer (UVS) observations in H Ly α emission reveals a complex three–dimensional distribution of atomic hydrogen in the Saturn system. The authors examine the impact of the measured amounts of the atomic hydrogen on the plasma sheet in the inner magnetosphere. They discuss the reactions controlling the plasma assuming a source derived from the icy satellites.

101.019   On the possibility of the existence of two Transplutonian planets.
A. S. Guliev.
Pis'ma Astron. Zh., Tom 18, No. 2, p. 183 – 189 (Feb 1992). In Russian. English translation in Sov. Astron. Lett., Vol. 18, No. 1.

The author considers the distribution of node distances of orbits of long–period comets relative to the plane of motion of two hypothetical Transplutonian planets. Comparison with the analogous distributions in other planes shows that there are the nodes overbalances in the plane with the parameters $\Omega_p = 272.9°$; $i_p = 29.6°$ at the interval $48.5 - 56.6$ a.u. and in the plane with $\Omega_p = 341°$; $i_p = 30.5°$ at the interval $102 - 112$ a.u. These overbalances are statistically significant and may be explained only by the existence of hypothetical planets.

101.020   Evaluation of relative aerosol abundance and effective radius of particles in the atmosphere of Uranus.
M. S. Dement'ev.
Kinematika Fiz. Nebesn. Tel, Tom 8, No. 2, p. 25 – 35 (Mar–Apr 1992). In Russian. English translation in Kinematics Phys. Celest. Bodies, Vol. 8, No. 2.

Residual intensities of the Raman lines at $\lambda\lambda 398.9$, 402,6, and 406.3 nm in the spectrum of Uranus were analysed within the framework of the model of a homogeneous semi–infinite gas–aerosol layer. The relative aerosol abundance was obtained as a function of the asymmetry parameter of the Henyey–Greenstein phase function, and of the effective radius of particles (in the case of real phase functions). The aerosol abundance in the atmosphere of Uranus is shown to be significantly lower in 1961 – 1973 than in 1981 – 1983. Comparison of relative aerosol abundances found with the Raman lines and with the absorption bands of methane at $\lambda\lambda 509$ and 576 nm gives the value of the effective radius of particles $r_{eff} = 0.30 \pm 0910~\mu m$. The ratio of the volume scattering coefficients for gas and gas + aerosol had values of $0.98 - 0.99$ and $0.10 - 0.30$ near $\lambda 400$ nm for the first and second periods of observations, respectiveley, and the specific abundances of methane were 0.008 ± 0.001 and 0.010 ± 0.003 for the same periods.

101.021   A magnetosphere of Neptune.
R. Schreiber.
Urania, Rok 63, Nr. 13, p. 6 – 9 (1992). In Polish.

101.022   Unlucky Triton.
J. Kuczyński.
Urania, Rok 63, Nr. 5, p. 130 – 133 (May 1992). In Polish.

101.023   On the thermal structure of Triton's thermosphere.
M. H. Stevens, D. F. Strobel, M. E. Summers, R. V. Yelle.
Geophys. Res. Lett., Vol. 19, No. 7, p. 669 – 672 (3 Apr 1992).

The analysis of the Voyager 2 Ultraviolet Spectrometer (UVS) solar occultation data obtained at Triton is consistent with a spherically symmetric, isothermal thermosphere above 400 km at $T_\infty = 96K$. A detailed calculation of energy loss processes in a pure N_2 atmosphere, heating and cooling rates, and resultant thermal structure associated with solar UV irradiance and magnetospheric electron precipitation indicates that solar heating, with calculated $T_\infty = 70K$, is insufficient to account for the inferred $T_\infty = 96K$. The magnetosphere must deposit twice as much power as the sun ($\lambda \leqslant 800$ Å) to heat the thermosphere to 96K and generate the observed N_2 tangential column densities above 450 km. The thermal escape of H and N atoms and the downward diffusion of N atoms to recombine below 130 km results in local ionospheric heating efficiency of 24%. An upper limit on the tropopause CO mixing ratio of $2x10^{-4}$ is inferred in the absence of aerosol heating to balance its efficient cooling by LTE rotational line emission.

101.024 Scintillations of the Uranian kilometric radiation: implications for the downstream magnetopause.
B. M. Pedersen, M. G. Aubier, M. D. Desch.
J. Geophys. Res., Vol. 97, No. A6, p. 8127 – 8133 (1 Jun 1992).

From January 27 to 30, 1986, while Voyager 2 was outbound from Uranus, the planetary radio astronomy experiment recorded strong (~ 3–6 dB) modulations of the broadband smooth Uranian radio emission (Warwick et al., 1987) at frequencies below 500 kHz. These fluctuations exhibited simultaneously two different time scales (~ 100 s and ~ 10 s) and were exclusively observed outside the Uranian magnetopause during six planetary rotations. The modulations were only present during a part of the total duration of each broadband smooth episode. This phenomenon is interpreted as being due to propagation effects through the magnetopause. The long–period modulation is due to the influence of ~ 8000-km wavelength surface waves on Uranus' downstream magnetopause. Such waves have only been studied in detail for the inbound passage (Lepping et al., 1987). The short–period oscillation might be the signature of a second magnetopause ripple period, similar to that measured by Aubry et al. (1971) for the Earth's magnetopause. There is strong evidence that the surface wave fluctuations causing the scintillations are induced by the solar wind interaction with the magnetosphere. The authors discuss why these modulations are confined to certain periods of time and to a certain frequency range.

101.025 Neptune's polar cusp region: observations and magnetic field analysis.
R. P. Lepping, L. F. Burlaga, A. J. Lazarus, V. M. Vasyliunas, A. Szabó, J. Steinberg, N. F. Ness, S. M. Krimigis.
J. Geophys. Res., Vol. 97, No. A6, p. 8135 – 8144 (1 Jun 1992).

This paper complements the recent paper by Szabo et al. (1991) which stresses the variation in Voyager 2 (thermal) plasma parameters and argues in favor of one of two models of the cusp. These models are (1) stagnant plasma in a region (called cusp) separated from the magnetosheath by a tangential discontinuity or (2) a mantlelike layer of dynamic plasma (cusp) separated from the sheath by a rotational discontinuity. They argue that model 2 is better satisfied by their observations. They also demonstrate some similarities to Earth's cusp. The authors here confirm and extend their principal results using a different approach requiring plasma and vector magnetic field quantities. And further, they obtain various MHD properties of the cusp–magnetopause boundary, which separates the cusp from the magnetosheath allowing thermal anisotropy. Some of these properties are the magnetopause normal, mass and normal momentum flux, boundary speed (and thickness), and their relationships. The complex nature of the cusp is then revealed by demonstrating that the magnetopause normal velocity is composed of two components: a propagation speed and the other consistent with the rotational motion of the magnetosphere.

101.026 The possible existence of two Transplutonian planets.
A. S. Guliev.
Sov. Astron. Lett., Vol. 18, No. 1, p. 75 – 78 (Jan 1992). Current Physics Microform No.: 9209X1293. English translation of Pis'ma Astron. Zh., Tom 18, No. 2, p. 183 – 189 (1992).

The author has examined the distribution of orbit node distances of the long–period comets relative to the plane of motion of two postulated Transplutonian planets. A comparison with analogous distributions in other planes shows an excess of nodes in the plane with parameters $\Omega_p = 272.9°$; $i_p = 29.6°$ in the interval 48.5 – 56.6 AU and in the plane $\Omega_p = 341°$; $i_p = 30.5°$ in the interval 102 – 112 AU. These excesses are statistically significant and can be explained only by the existence of the postulated planets.

101.027 Two–color photometry of Pluto.
R. I. Kiladze, V. D. Kukhianidze.
Sol. Syst. Res., Vol. 25, No. 4, p. 330 – 332 (Jan 1992). English translation of Astron. Vestn., Tom 25, No. 4, p. 439 – 441 (1991).

Results are presented of two–color photelectric photometry of Pluto performed at the Abastumanskaya Astrophysical Observatory from 1986 – 1990. A decrease in the planets color index

during 1990 was discovered. The albedo of Pluto (at least in the blue) reached a minimum in 1988. This contradicts the point of view of Buie et al. (1987), Cruikshank et al. (1980) and Stern et al. (1988) that Pluto's mean brightness decreased due to changes in surface physical conditions, for example by ice sublimation occurring as a result of changes in the distance from the Sun.

101.028 Uranian aurora – UVS observations.
F. Herbert, B. R. Sandel.
Bull. Am. Astron. Soc., Vol. 23, No. 3, p. 1147 (1991). Abstract. – See Abstr. 012.037 for the main entry.

101.029 Inferring heat sources and model atmospheres from Voyager 2 Ultraviolet Spectrometer occultation data at Uranus.
M. H. Stevens, D. F. Strobel, F. Herbert.
Bull. Am. Astron. Soc., Vol. 23, No. 3, p. 1147 (1991). Abstract. – See Abstr. 012.037 for the main entry.

101.030 Two dimensional photochemical transport models of the stratosphere of Uranus.
W. W. McMillan, D. F. Strobel.
Bull. Am. Astron. Soc., Vol. 23, No. 3, p. 1152 (1991). Abstract. – See Abstr. 012.037 for the main entry.

101.031 The thermal structure of Uranus' atmosphere.
M. S. Marley, J. B. Pollack, C. P. McKay.
Bull. Am. Astron. Soc., Vol. 23, No. 3, p. 1152 – 1153 (1991). Abstract. – See Abstr. 012.037 for the main entry.

101.032 Source of the diffuse UV emission from Neptune's night side.
R. J. Vervack Jr., B. R. Sandel, A. J. Dessler.
Bull. Am. Astron. Soc., Vol. 23, No. 3, p. 1153 (1991). Abstract. – See Abstr. 012.037 for the main entry.

101.033 Models of the thermal structure of Neptune's upper atmosphere.
Y. Wang, R. V. Yelle.
Bull. Am. Astron. Soc., Vol. 23, No. 3, p. 1153 (1991). Abstract. – See Abstr. 012.037 for the main entry.

101.034 Nucleation and hydrocarbon aerosol formation in Neptune's atmosphere.
J. I. Moses, Y. L. Yung, M. Allen.
Bull. Am. Astron. Soc., Vol. 23, No. 3, p. 1153 (1991). Abstract. – See Abstr. 012.037 for the main entry.

101.035 3– to 13–μm spectra of Neptune in 1990 and 1991.
H. B. Hammel, L. A. Young, J. Hackwell, R. Russell, D. Lynch, G. Orton.
Bull. Am. Astron. Soc., Vol. 23, No. 3, p. 1153 (1991). Abstract. – See Abstr. 012.037 for the main entry.

101.036 H_2S in Uranus deep atmosphere.
R. M. Killen, F. M. Flasar.
Bull. Am. Astron. Soc., Vol. 23, No. 3, p. 1154 (1991). Abstract. – See Abstr. 012.037 for the main entry.

101.037 Coherent signal arraying of Voyager Neptune/Triton occultation data.
E. Mizuno, P. A. Rosen, N. Kawashima.
Bull. Am. Astron. Soc., Vol. 23, No. 3, p. 1154 (1991). Abstract. – See Abstr. 012.037 for the main entry.

101.038 The albedo, effective temperature, and energy balance of Neptune from Voyager data.
J. C. Pearl, B. J. Conrath.
Bull. Am. Astron. Soc., Vol. 23, No. 3, p. 1154 (1991). Abstract. The full paper is published in J. Geophys. Res., A, Vol. 96, No. 11, p. 18921 – 18930 (30 Oct 1991). – See Abstr. 012.037 for the main entry.

101.039   Ethane ice haze in the upper atmosphere of Uranus: a coupled diffusion/microphysical model with a cosmic ray vapor sink.
T. M. Schulz.
Bull. Am. Astron. Soc., Vol. 23, No. 3, p. 1154 – 1155 (1991). Abstract. – See Abstr. 012.037 for the main entry.

101.040   To the edge: missions to Pluto and Neptune.
K. Croswell.
Astronomy, Vol. 20, No. 5, p. 34 – 41 (May 1992).

101.041   The Uranian satellites: geological histories and comparisons with other moons.
R. J. Stevenson.
South. Stars, Vol. 34, No. 6, p. 347 – 357 (Mar 1992).

Since the Voyager fly-by of Uranus, the moons of the outer Solar System have fascinated planetary scientists with their diversity of surfaces. In this paper, the geological history of the Uranian moons is determined using principles of relative stratigraphy and crater counting. Ariel and Miranda are the most geologically advanced despite their small sizes. A thermal drive model is extended to these satellites, and the moons compared and contrasted with others in the outer Solar System.

101.042   Temperature profiles of Neptune's stratosphere between 1983–1990: long term evolution and non isothermal features.
F. Roques, B. Sicardy.
Bull. Am. Astron. Soc., Vol. 23, No. 3, p. 1155 (1991). Abstract. – See Abstr. 012.037 for the main entry.

101.043   First detections of CO and HCN in the atmosphere of Neptune.
A. Marten, D. Gautier, T. Owen, D. Sanders, R. T. Tilanus, J. Deane, H. Matthews.
Bull. Am. Astron. Soc., Vol. 23, No. 3, p. 1164 (1991). Abstract. – See Abstr. 012.037 for the main entry.

101.044   Interior model constraints on super–abundances of volatiles in the atmosphere of Neptune.
M. Podolak, M. S. Marley.
Bull. Am. Astron. Soc., Vol. 23, No. 3, p. 1164 (1991). Abstract. – See Abstr. 012.037 for the main entry.

101.045   A search for planetary–scale waves in Neptune's atmosphere.
F. M. Flasar, B. J. Conrath.
Bull. Am. Astron. Soc., Vol. 23, No. 3, p. 1164 (1991). Abstract. – See Abstr. 012.037 for the main entry.

101.046   A wave observed in the atmosphere of Neptune.
D. P. Hinson, J. A. Magalhães.
Bull. Am. Astron. Soc., Vol. 23, No. 3, p. 1164 – 1165 (1991). Abstract. – See Abstr. 012.037 for the main entry.

101.047   Oscillations in the motions of Neptune's major cloud features.
L. A. Sromovsky.
Bull. Am. Astron. Soc., Vol. 23, No. 3, p. 1165 (1991). Abstract. – See Abstr. 012.037 for the main entry.

101.048   Neptune's atmosphere: zonal mean circulation.
B. J. Conrath, F. M. Flasar, P. J. Gierasch.
Bull. Am. Astron. Soc., Vol. 23, No. 3, p. 1165 (1991). Abstract. – See Abstr. 012.037 for the main entry.

101.049   Winds of Neptune: Voyager results.
S. S. Limaye, L. A. Sromovsky.
Bull. Am. Astron. Soc., Vol. 23, No. 3, p. 1165 (1991). Abstract. – See Abstr. 012.037 for the main entry.

101.050   The unusual photometric properties of Umbriel: What do they mean?
B. Buratti, J. Gibson, J. Mosher, B. Hapke.
Bull. Am. Astron. Soc., Vol. 23, No. 3, p. 1169 (1991). Abstract. – See Abstr. 012.037 for the main entry.

101.051   Statistics and histories of satellite disruptions.
J. E. Colwell, L. W. Esposito.
Bull. Am. Astron. Soc., Vol. 23, No. 3, p. 1170 (1991). Abstract. – See Abstr. 012.037 for the main entry.

101.052   A dynamical history of the inner Neptunian satellites.
D. Banfield, N. Murray.
Bull. Am. Astron. Soc., Vol. 23, No. 3, p. 1170 (1991). Abstract. – See Abstr. 012.037 for the main entry.

101.053   The Laplace planes of Uranus and Pluto.
A. R. Dobrovolskis.
Bull. Am. Astron. Soc., Vol. 23, No. 3, p. 1171 (1991). Abstract. – See Abstr. 012.037 for the main entry.

101.054   Neptune's arcs and narrow rings: morphology & nature.
C. Ferrari, A. Brahic.
Bull. Am. Astron. Soc., Vol. 23, No. 3, p. 1178 (1991). Abstract. – See Abstr. 012.037 for the main entry.

101.055   A closer look at the Uranian rings.
M. R. Showalter.
Bull. Am. Astron. Soc., Vol. 23, No. 3, p. 1178 (1991). Abstract. – See Abstr. 012.037 for the main entry.

101.056   Origins of the rings of Uranus and Neptune. 1. Statistics of satellite disruptions.
J. E. Colwell, L. W. Esposito.
J. Geophys. Res., Vol. 97, No. E6, p. 10227 – 10241 (25 Jun 1992).

Stochastic simulations of the collisional fragmentation of the small moons of Neptune and Uranus confirm the conclusions of Smith et al. (1986, 1989) that many of these moons cannot have survived intact since the end of planetary formation. The authors perform two types of stochastic simulations of the collisonal history of small moons. Monte Carlo simulations in which only the largest surviving fragment from each disruption is followed show bimodal probability distributions for the size of the largest fragment. Once the moon is destroyed the first time, the collisional cascade to smaller sizes proceeds relatively quickly. A Markov chain approach allows the authors to follow the size distribution from each disruption to arbitrarily small sizes. The results and consequences of these simulations are discussed in detail.

101.057   The elimination of the 1:2 critical terms of a first order theory of Uranus perturbed by Neptune through Von Zeipel method.
O. M. Kamel.
Earth, Moon, Planets, Vol. 57, No. 1, p. 23 – 64 (Apr 1992).

In this paper the author eliminates in a first order U–N theory the 1:2 critical terms up to the third degree with respect to eccentricity – inclination in both parts, main and indirect of the U–N planetary Hamiltonian. He operates with the Von Zeipel technique. He adopts, in this theory, the Jacobi–Radau coordinates, and the Poincaré canonical variables. Powers higher than the third in the eccentricity – inclination are neglected.

101.058   Uranus.
IAU Circ., No. 5492, p. 1 (9 Apr 1992).

101.059   No occultation by Pluto on 1992 May 21.
IAU Circ., No. 5500, p. 1 (20 Apr 1992).

101.060   Pluto.
IAU Circ., No. 5532, p. 1 (30 May 1992).

101.061 **On the unmodeled perturbations in the motion of Uranus.**
A. Brunini.
Celest. Mech. Dyn. Astron., Vol. 53, No. 2, p. 129 – 143 (1992).
The author applies a numerical method to determine unmodeled perturbations in an attempt to explain the observed discrepancies in the motion of Uranus. He finds that the estimated perturbation shows some significant periods that could be attributed to insufficient knowledge of the perturbations from some of the known planets. On the assumption that the gravitational attraction of an unknown planet is the origin of the deviations, the best planar solution of the inverse problem is a planet of 0.6 Earth masses, with true longitude of $133°$ (1990.5), semi major axis $a = 44$ AU and eccentricity $e = 0.007$.

101.062 **Gravito–electrodynamics: recent results on Uranus and Neptune's rings.**
M. Horanyi.
Bull. Am. Astron. Soc., Vol. 23, No. 3, p. 1181 (1991). Abstract. – See Abstr. 012.037 for the main entry.

101.063 **The 25 and 28 June 1991 stellar occultations by the Uranian rings.**
R. G. French, N. Chanover, E. Mason, K. J. Meech.
Bull. Am. Astron. Soc., Vol. 23, No. 3, p. 1182 (1991). Abstract. – See Abstr. 012.037 for the main entry.

101.064 **Voyager radio occultation of Triton: surface topography and radius.**
E. A. Marouf, G. L. Tyler, V. R. Eshleman, P. A. Rosen.
Bull. Am. Astron. Soc., Vol. 23, No. 3, p. 1207 (1991). Abstract. – See Abstr. 012.037 for the main entry.

101.065 **Voyager radio occultation observations of Triton's neutral atmosphere.**
E. M. Gurrola, V. R. Eshleman, G. L. Tyler, E. A. Marouf.
Bull. Am. Astron. Soc., Vol. 23, No. 3, p. 1207 (1991). Abstract. – See Abstr. 012.037 for the main entry.

101.066 **Carbon atoms and ions in the atmosphere of Triton.**
J. R. Lyons, Y. L. Yung, M. Allen.
Bull. Am. Astron. Soc., Vol. 23, No. 3, p. 1207 (1991). Abstract. – See Abstr. 012.037 for the main entry.

101.067 **Hazes and clouds in Triton's atmosphere.**
K. A. Rages, J. B. Pollack.
Bull. Am. Astron. Soc., Vol. 23, No. 3, p. 1207 (1991). Abstract. – See Abstr. 012.037 for the main entry.

101.068 **Triton: regional variations in haze opacity.**
J. Hillier, P. Helfenstein, J. Veverka.
Bull. Am. Astron. Soc., Vol. 23, No. 3, p. 1207 (1991). Abstract. – See Abstr. 012.037 for the main entry.

101.069 **Anomalous light scattering on Triton.**
P. Lee, P. Helfenstein, J. Veverka, D. McCarthy.
Bull. Am. Astron. Soc., Vol. 23, No. 3, p. 1208 (1991). Abstract. – See Abstr. 012.037 for the main entry.

101.070 **Millimeter–wave observations of Saturn, Uranus, and Neptune: CO and HCN on Neptune.**
J. Rosenqvist, E. Lellouch, P. N. Romani, G. Paubert, T. Encrenaz.
Astrophys. J., Lett., Vol. 392, No. 2, p. L99 – L102 (20 Jun 1992).
Saturn, Uranus, and Neptune were observed at millimeter wavelengths with the IRAM 30 m telescope. The major result is the detection of CO and HCN in Neptune's stratosphere, with respective mixing ratios of $(6.5 \pm 3.5) \times 10^{-7}$ and $(3 \pm 1.5) \times 10^{-10}$. CO seems to be present in Neptune's troposphere as well and appears to decrease slowly with altitude (scale height ~ 200 km). HCN is probably formed from reactions between CH_3 and N, which can be supplied in sufficient amounts by escape from Triton's atmosphere. The origin of CO, however, is more problematic, because (1) thermochemical models fail to reproduce the observed abundance by a factor ~ 1000, and (2) an external source would require a very large flux of oxygen. CO appears to be at least 15 times less abundant on Uranus than on Neptune. Finally, an upper limit of 10^{-7} for CO in Saturn's stratosphere suggests an internal origin for Saturnian CO.

101.071 **Rotationally resolved UV studies of Triton: IUE's preview and HST's promise.**
S. A. Stern, E. S. Barker, G. R. Gladstone.
Bull. Am. Astron. Soc., Vol. 23, No. 3, p. 1208 (1991). Abstract. – See Abstr. 012.037 for the main entry.

101.072 **Tentative detection of CO and CO_2 ices on Triton.**
D. P. Cruikshank, T. C. Owen, T. R. Geballe, B. Schmitt, C. DeBergh, J.–P. Maillard, B. L. Lutz, R. H. Brown.
Bull. Am. Astron. Soc., Vol. 23, No. 3, p. 1208 (1991). Abstract. – See Abstr. 012.037 for the main entry.

101.073 **A Triton thermal model.**
C. J. Hansen, D. A. Paige.
Bull. Am. Astron. Soc., Vol. 23, No. 3, p. 1208 (1991). Abstract. – See Abstr. 012.037 for the main entry.

101.074 **Energy transport in Triton's surface–atmosphere system.**
J. A. Stansberry, R. V. Yelle, J. I. Lunine, A. McEwen.
Bull. Am. Astron. Soc., Vol. 23, No. 3, p. 1209 (1991). Abstract. – See Abstr. 012.037 for the main entry.

101.075 **Models of the viscous spreading of Triton's permanent polar caps.**
R. L. Kirk, R. H. Brown.
Bull. Am. Astron. Soc., Vol. 23, No. 3, p. 1209 (1991). Abstract. – See Abstr. 012.037 for the main entry.

101.076 **Rotationally resolved, ground–based studies of the near UV reflectivity of the Pluto–Charon system.**
E. S. Barker, S. A. Stern, L. M. Trafton.
Bull. Am. Astron. Soc., Vol. 23, No. 3, p. 1209 (1991). Abstract. – See Abstr. 012.037 for the main entry.

101.077 **Coupling of internal heat to volatile transport on Triton.**
R. H. Brown, R. L. Kirk.
Bull. Am. Astron. Soc., Vol. 23, No. 3, p. 1210 (1991). Abstract. – See Abstr. 012.037 for the main entry.

101.078 **Pluto's observations in the 2.15 to 2.35 μm region: preliminary study.**
A. Coustenis, E. Lellouch, B. Bézard, B. Schmitt.
Bull. Am. Astron. Soc., Vol. 23, No. 3, p. 1210 (1991). Abstract. – See Abstr. 012.037 for the main entry.

101.079 **Seeing through frost on Triton: large–scale quasi–circular features.**
P. Helfenstein, J. Veverka, P. Lee, D. McCarthy.
Bull. Am. Astron. Soc., Vol. 23, No. 3, p. 1210 (1991). Abstract. – See Abstr. 012.037 for the main entry.

101.080 **A search for distant satellites of Pluto.**
S. A. Stern, J. W. Parker, E. S. Barker, L. M. Trafton, R. A. Fesen.
Bull. Am. Astron. Soc., Vol. 23, No. 3, p. 1210 – 1211 (1991). Abstract. The full paper is published in Icarus, Vol. 94, No. 1, p. 246 – 249 (Nov 1991). – See Abstr. 012.037 for the main entry.

101.081 **Resolved photometry of Pluto–Charon at several light-curve phases.**
L. A. Young, J. L. Elliot, A. S. Bosh, H. B. Hammel, R. L. Baron.
Bull. Am. Astron. Soc., Vol. 23, No. 3, p. 1211 (1991). Abstract. – See Abstr. 012.037 for the main entry.

101.082   Occultation constraints on atmospheric models for Pluto.
J. L. Elliot, L. A. Young.
Bull. Am. Astron. Soc., Vol. 23, No. 3, p. 1216 (1991). Abstract. – See Abstr. 012.037 for the main entry.

101.083   The radius of Pluto from the 9 June 1988 occultation.
L. H. Wasserman, R. L. Millis, J. L. Elliot, L. A. Young.
Bull. Am. Astron. Soc., Vol. 23, No. 3, p. 1216 (1991). Abstract. – See Abstr. 012.037 for the main entry.

101.084   How big is Pluto?
D. J. Tholen, M. W. Buie.
Bull. Am. Astron. Soc., Vol. 23, No. 3, p. 1216 (1991). Abstract. – See Abstr. 012.037 for the main entry.

101.085   Mapping Pluto's surface from mutual event lightcurves: a comparison of techniques and physical implications.
E. F. Young, R. P. Binzel.
Bull. Am. Astron. Soc., Vol. 23, No. 3, p. 1216 (1991). Abstract. – See Abstr. 012.037 for the main entry.

101.086   Insolation history on Pluto: implications for frost models.
R. P. Binzel, E. F. Young, E. S. Ditchburn.
Bull. Am. Astron. Soc., Vol. 23, No. 3, p. 1216 – 1217 (1991). Abstract. – See Abstr. 012.037 for the main entry.

101.087   Photometric variability of Charon at 2.2 μm.
A. S. Bosh, L. A. Young, J. L. Elliot, H. B. Hammel, R. L. Baron.
Bull. Am. Astron. Soc., Vol. 23, No. 3, p. 1217 (1991). Abstract. The full paper is published in Icarus, Vol. 95, No. 2, p. 319 – 324 (Feb 1992). – See Abstr. 012.037 for the main entry.

101.088   Triton and Pluto – Are they the same?
W. B. McKinnon.
Bull. Am. Astron. Soc., Vol. 23, No. 3, p. 1219 (1991). Abstract. – See Abstr. 012.037 for the main entry.

101.089   Shepherding satellites and dynamical structure of the rings of Uranus.
Y. Kozai.
Publ. Astron. Soc. Jpn., Vol. 44, No. 2, p. 135 – 139 (1992).
The structure of the α–, β– and ε–rings of Uranus, with non-circular orbits, the eccentricities increasing as the semi–major axes and the apsidal motions common in each ring, is explained by the gravitational action of unknown shepherding satellites, which produce appreciable eccentricities to the rings. Their orbits are much closer to the rings than any of the shepherding satellites discovered by the Voyager spacecraft. Since the eccentricities of the ring particles are shown to be produced due to actions of the satellite in this paper, their apsidal motions are synchronized in each ring and are equal to that of the shepherding satellite. Since the eccentricity is increased towards the outer edge of the ring, the satellite must be outside of the ring by 10–100 km apart from the outer edge of the ring. Their masses are estimated to be of the order of 10^{-11}–10^{-10} of that of Uranus, that is, 10^{18}–10^{19}g, and are less than that of any known satellite.

101.090   Albedo maps of Pluto and Charon: initial mutual event results.
M. W. Buie, D. J. Tholen, K. Horne.
Icarus, Vol. 97, No. 2, p. 211 – 227 (Jun 1992).
The authors present single–scattering albedo maps of the surfaces of Pluto and Charon based primarily on mutual event observations. The dataset contains 3374 photometric observations that cover 15 different satellite transit events, 14 satellite eclipse events, and other out–of–eclipse photometry spanning 1954 to 1986. The authors applied the technique of maximum entorpy image reconstruction to invert the light–curves, thus revealing surface maps of single–scattering albedo. The surface of Pluto is seen to have albedo features similar to the authors'

previous spot model maps. In particular, a south polar cap is evident in the map of Pluto. The north polar region is brighter than the equatorial regions but is not as bright as the south pole. The map of Charon is somewhat darker.

101.091   The masses of Uranus and its satellites from earthbased astrometric observations and Voyager navigation data.
R. A. Jacobson, J. K. Campbell, A. H. Taylor, S. P. Synnott.
Bull. Am. Astron. Soc., Vol. 23, No. 3, p. 1255 – 1256 (1991). Abstract. – See Abstr. 012.065 for the main entry.

101.092   The Pluto–Charon system.
S. A. Stern.
Annu. Rev. Astron. Astrophys., Vol. 30, p. 185 – 233 (1992).
Contents: 1. Overview. 2. Historical context. 3. Pluto's orbit, rotation, and pole position. 4. Charon's discovery, the mutual events, and the basic attributes of the Pluto–Charon system. 5. Pluto's surface properties. 6. Pluto's density, bulk composition, and interior structure. 7. Pluto's atmosphere. 8. Charon: surface, interior, and atmosphere. 9. Origin of Pluto, Charon, and the binary system. 10. Concluding remarks.

101.093   Uranus, Neptune and Pluto.
P. Moore.
Images of the universe, p. 63 – 76 (1991). – See Abstr. 003.090 for the main entry.

101.094   Ground–based observations of the rings of Neptune.
I. Mosqueira.
News Lett. Astron. Soc. N.Y., Vol. 4, No. 1, p. 24 (Feb 1992). Abstract. – See Abstr. 012.078 for the main entry.

Far encounter. The Neptune system.
See Abstr. 003.031.

210 years since the discovery of Uranus.
See Abstr. 004.017.

Die Entdeckung des Planeten Neptun an Enckes 55. Geburtstag. Vorgeschichte, Ablauf und Wirkung einer "geplanten" Entdeckung.
See Abstr. 004.053.

Application of ion irradiation experiments to planetary surfaces in the outer solar system.
See Abstr. 022.167.

The absorption coefficient of nitrogen with application to Triton.
See Abstr. 022.180.

The MIT program for identifying occultations and appulses by planets.
See Abstr. 036.147.

General theory for the outer planets.
See Abstr. 042.022.

Puzzles and prospects in planetary ring dynamics.
See Abstr. 042.024.

Collisional, collective and resonance phenomena in planetary rings.
See Abstr. 042.025.

Chaotic layers in resonance problems.
See Abstr. 042.040.

Orbital simulation of captured satellites: application to Triton.
See Abstr. 042.080.

A note on the enhancement of J values in optically thick scattering atmospheres.
See Abstr. 063.006.

Solar activity and brightness of Neptune.
See Abstr. 072.010.

Analysis of stellar occultation data for planetary atmospheres. I. Model fitting, with application to Pluto.
See Abstr. 091.007.

The 18 August 1991 stellar occultation by the Neptune system, and upcoming occultations by Neptune's rings.
See Abstr. 096.013.

Planetarty rings: observational constraints and collision dynamics.
See Abstr. 100.016.

Giant planets and their satellites: what are the relationships between their properties and how they formed?
See Abstr. 107.059.

102 Comets (Origin, Structure, Atmospheres, Dynamics)

102.001 **Plasma–beam instabilities in cometary ionospheres.**
O. P. Verkhoglyadova, N. Ya. Kotsarenko,
G. V. Lizunov, K. I. Churyumov.
Pis'ma Astron. Zh., Tom 18, No. 1, p. 81 – 86 (Jan 1992). In Russian. English translation in Sov. Astron. Lett., Vol. 18, No. 1.
It is shown that the solar wind interaction with cometary plasma occurs in the mixing region and excites a wide plasma waves spectrum, namely, ion acoustic waves with the frequency of the ion Langmuir order, electron–cyclotron ones with the intensity maximum on a low hybrid frequency, whistler and magnetoacoustic waves. The frequency spectrum is investigated. The linear increasing increment and lengths of excited waves are found. It is also shown that under certain conditions a filamentary instability yielding magnetic field space fluctuations and plasma density fluctuations with a characteristic scale by order of tens of kilometers may develop.

102.002 **Wo kommen die Kometen her? Ein altes Problem – neu durchdacht.**
E. Litzroth.
Sternenbote, Jahrg. 35, Nr. 5, p. 102 – 114 (1992).

102.003 **The contribution of methanol to the 3.4 micron emission feature in comets.**
D. C. Reuter.
Astrophys. J., Vol. 386, No. 1, p. 330 – 335 (10 Feb 1992).
In this paper a fluorescence model for cometary emission in the v_2, v_3, and v_9 bands of methanol is described. This model is used in conjunction with observed spectra of several comets to determine the possible contribution of CH_3OH to the frequently seen 3.4 μm "cometary organic" feature. For comets P/Halley and Wilson, methanol production rates are retrieved from the flux in the v_3 band centered at ~ 3.52 μm.

102.004 **Further evolution of velocity shell distribution of cometary and interstellar pickup ions and excitation of oblique Alfvén waves.**
Peter H. Yoon.
J. Geophys. Res., Vol. 97, No. A5, p. 6467 – 6478 (1 May 1992).
This paper shows how an initially isotropic spherical shell velocity distribution of cometary or interstellar pickup ions can become thermalized. The basic instability mechanism responsible for the thermalization is based on papers by Wu and Yoon (1990) and Yoon (1990). The unstable mode is in the Alfvén mode branch and propagates in an oblique direction, with the mode stable for exactly parallel and perpendicular propagations. The present paper discusses the nonlinear evolution of the ion distribution as well as the unstable waves. The method is based on the self–consistent quasi–linear equations. It is found that the thermalization can take place in 6.4×10^3 to 1.6×10^4 ion gyroperiods for the parameters used in the present analysis, while the unstable waves can grow to an appreciable level of the initial fluctuation.

102.005 **Crystallization, sublimation, and gas release in the interior of a porous comet nucleus.**
D. Prialnik.
Astrophys. J., Vol. 388, No. 1, p. 196 – 202 (20 Mar 1992).
A numerical code has been developed for evolutionary calculations of the thermal structure and composition of a porous comet nucleus made of water ice, in amorphous or crystalline form, other volatiles, dust, and gases trapped in amorphous ice. Bulk evaporation, crystallization, gas release, and free flow of gases through the pores are taken into account. The numerical scheme yields exact conservation laws for mass and energy. The code is used for studying the effect of bulk evaporation of ice in the interior of a comet nucleus during crystallization. It is found that evaporation controls the temperature distribution. The vapor prevents cooling of the crystallized layer of ice, by recondensation and release of latent heat. Thus high temperatures ($\sim 170K$) are maintained below the surface of the nucleus and down to depths of tens or hundreds of meters, even at large heliocentric distances, as long as crystallization goes on. Gas trapped in the ice and released during the phase transition flows both toward the interior and toward the surface and out of the nucleus. The progress of crystallization is largely determined by the contribution of gas fluxes to heat transfer.

102.006 **Does the Saturn family of periodic comets exist?**
A. S. Guliev.
Kinematika Fiz. Nebesn. Tel, Tom 8, No. 1, p. 55 – 61 (Jan–Feb 1992). In Russian. English translation in Kinematics Phys. Celest. Bodies, Vol. 8, No. 1.
An analysis of the characteristics of comets with the aphelia from 7.46 to 14.2 AU has been carried out. The statistical criteria used show that they differ considerably from the Jupiter family. This can be particularly seen in the parameters q, i, J (the Jacobi constant), L (the perihelion longitude), in the dependence $H_{10}(q)$, in the dynamics of increase of number of comets as well as in comets reaction on solar activity. The analysis of the J values with reference to Jupiter and Saturn shows that comets belong to the family of the latter. The assumption that this group of comets is an intermediate one has been rejected. The results of calculations of cometary orbits evolution are taken into account.

102.007 **Analysis of Tisserand's constant for periodic comets.**
A. S. Guliev.
Kinematika Fiz. Nebesn. Tel, Tom 8, No. 2, p. 40 – 46 (Mar–Apr 1992). In Russian. English translation in Kinematics Phys. Celest. Bodies, Vol. 8, No. 2.
The investigation of Tisserand's constant (J) brought to light several new effects. 1. The maximum of the N(J) distribution does not lie in the interval predicted by the capture theory. 2. The values of J follow an exponential law with a discontinuity near 2.670. 3. Minimum values of J are observed more often among comets with longitudes of perihelion close to 90°. 4. There is no relationship between the values of e and cos i that appear in

Tisserand's formula; this relationship manifests itself only in the group of comets that have their longitudes of perihelia in the interval 43 – 103°. 5. These comets are characterized by relatively large values of the declination i. 6. Correlation between e and i of "primordial" long–period comets appears most clearly for periodic comets with their longitudes of perihelia close to 90°. 7. Periodic comets which according to the Tisserand criterion meet at a larger scale the requirements of the capture mechanism are characterized by small q and large e. The author comes thus to the onclusion that only a part of periodic comets has been captured from the overall set of long–period comets.

102.008   Three–dimensional statistics of cometary aphelia.
S. Ya. Mashchenko.
Kinematika Fiz. Nebesn. Tel, Tom 8, No. 2, p. 47 – 51 (Mar–Apr 1992). In Russian. English translation in Kinematics Phys. Celest. Bodies, Vol. 8, No. 2.

Eight cometary families that obey a modified Bode's law for giant planets have been revealed as a result of three–dimensional statistical analysis of cometary aphelia distribution in the nearest outskirts of the solar system (at distances from 50 to 4000 a.u.). Three of the families (at the distances of 93, 320 and 1100 a.u.) have been studied in detail. An assumption is made about the origin of these cometary families.

102.009   Dust investigation of insolated comet nucleus analogues.
K. Thiel, G. Kölzer.
54. Annual Meeting of the Meteoritical Society, p. 230 (1991). Abstract. – See Abstr. 012.010 for the main entry.

102.010   Origin of sungrazers: a frequent cometary end–state.
M. E. Bailey, J. E. Chambers, G. Hahn.
Astron. Astrophys., Vol. 257, No. 1, p. 315 – 322 (Apr 1992).

The orbits of sungrazing comets originally have inclinations near 90° and perihelion distances in the approximate range 0–2 AU. Long–term secular perturbations cause correlated changes in the orbital elements, especially the perihelion distance, eccentricity and inclination, which eventually lead to a temporary sungrazing state of extremely small perihelion distance. At such a time the comet may suffer tidal disruption or destruction by solar heating. The minimum perihelion distance occurs at intervals on the order of 10^3 revolutions, at least ten times shorter than the dynamical ejection timescale. Long–period comets with initially high–inclination orbits ($i \cong 90° \pm 15°$) and moderately small perihelion distance ($q \lesssim 2 \, AU$) frequently become sungrazers; their destruction during episodes of small perihelion distance is an important cometary end–state and source of interplanetary matter.

102.011   Dust optical properties: a comparison between cometary and interplanetary grains.
A. C. Levasseur–Regourd, J. B. Renard, R. Dumont.
Adv. Space Res., Vol. 11, No. 12, p. 175 – 182 (1991). – See Abstr. 012.014 for the main entry.

A better understanding of cometary dust optical properties has been derived from extensive observations of comet Halley, complemented by other cometary observations at large phase angles and/or in the infrared. Also, further analysis of IRAS observations and improvements in inversion techniques for zodiacal light have led to some progress in the knowledge of interplanetary dust. Synthetic curves for phase angle dependence of intensity and polarization are presented, together with typical albedo values. The results obtained for interplanetary dust are quite reminiscent of those found for comets. However, the heterogeneity of the interplanetary dust cloud is demonstrated by the radial dependence of its local polarization and albedo.

102.012   A self–consistent model for the particles and fields upstream of an outgassing comet. 2. A time–dependent description.
K. Flammer, T. E. Birmingham, D. A. Mendis, T. G. Northrop.
J. Geophys. Res., Vol. 97, No. A6, p. 8173 – 8181 (1 Jun 1992).

In a previous paper (Flammer et al., 1991), the authors determined the global variation of the magnetic field and solar wind flow parameters in the unshocked region upstream of an outgassing comet using a kinetic treatment for the cometary ions. Two different assumptions were made concerning the cometary ion distribution function: the pickup cometary ions formed either a velocity space gyrotropic ring distribution or a velocity space isotropic shell distribution in the solar wind frame of reference. In the present paper the authors consider the general case wherein the newly picked up ions are elastically pitch angle scattered from the initial ring distribution to a shell with some characteristic time scale τ. Using theoretically determined parameters which reflect the conditions expected at comet Kopff (a typical short–period comet and a possible target for the CRAF/Cassini mission) for various heliocentric distances, the authors determine how the pitch angle scattering rate affects the global morphology.

102.013   Pulsating Hill surfaces and the origin of comets.
V. V. Radzievskij.
Sol. Syst. Res., Vol. 25, No. 2, p. 132 – 138 (Sep 1991). English translation of Astron. Vestn., Tom 25, No. 2, p. 181 – 189 (1991).

The phenomenon of tides in the atmosphere of a body m_2 moving on an elliptical or hyperbolic heliocentric orbit is examined. It is assumed that the atmosphere of the body m_2 consists of a large number of small bodies m_3 (cometary nuclei). It is shown that as the distance between the Sun (m_1) and the body m_2 decreases, a "throat" of the pulsating Hill sphere forms opposite the apex of the tidal bulge. The bodies m_3 pass through this throat into the heliocentric Hill sphere. It is emphasized that the suggested hypothesis for the origin of comets agrees well with comet statistical data.

102.014   Plasma–beam instabilities in cometary ionospheres.
O. P. Verkhoglyadova, N. Ya. Kotsarenko,
G. V. Lizunov, K. I. Churyumov.
Sov. Astron. Lett., Vol. 18, No. 1, p. 35 – 37 (Jan 1992). Current Physics Microform No.: 9209X1253. English translation of Pis'ma Astron. Zh., Tom 18, No. 1, p. 81 – 86 (1992).

It is shown that the interaction of the solar wind with cometary plasma which occurs in a mixing region leads to the excitation of a wide spectrum of plasma waves – specifically, ion–acoustic waves with frequency of the order of ion–Langmuir waves, electron cyclotron waves with maximum intensity at the lower hybrid frequency, whistler–mode waves, and fast magnetosonic waves. The frequency spectrum is studied, and the linear growth rates and lengths of the excited waves are found. It is also shown that under certain conditions it is possible for a filamentary instability to develop, which can lead to spatial fluctuations in the magnetic field and plasma density with characteristic scales of tens of kilometers.

102.015   Energetic hydrogen atoms in cometary comae: production of the metallic component at large distances from the Sun.
S. Ibadov.
Astron. Tsirk., No. 1551, p. 28 – 29 (Nov – Dec 1991). In Russian.

Production of metallic atoms and ions of refractory metals (Fe, Ni, Si, etc.) in the comae of Halley–type comets due to the action of energetic (1–10 keV) hydrogen atoms (from the charge exchange of the quiet and perturbed solar wind protons) onto the cometary dust is investigated. It is found that the concentrations of Fe^+–type ions detected by in situ measurements of VEGA and GIOTTO space missions may be explained by action of this mechanism only at conditions of passing the comet through the corpuscular flow from strong solar flares, giving at the heliocentric distance 1 AU the fluxes of the order of 10^{10} proton/(cm^{2}s).

102.016   Cometary studies: bioastronomical perspectives.
A. C. Levasseur–Regourd.
3. Symposium International de Bioastronomie, p. 109 – 116 (1991). – See Abstr. 012.005 for the main entry.

The main points of interest for bioastronomers are presented, with special emphasis on recent results, on physico–chemical properties of the nucleus, on deuterium enrichment, and on cometary dust structure and composition. The question of the possible supply, during the period of heavy bombardment of the young Earth, with water and complex organic molecules (from cometary or interplanetary origin) is addressed.

102.017   Comet showers.
J. A. Fernández.
IAU Symposium No. 152: Chaos, resonance and collective dynamical phenomena in the solar system, p. 239 – 254 (1992). – See Abstr. 012.025 for the main entry.
Variations in the influx rate of incoming Oort cloud comets, leading to the occurrence of comet showers, are reviewed with special emphasis on the dynamical processes that produce them.

102.018   Dynamics of comets.
A. Carusi, G. B. Valsecchi.
IAU Symposium No. 152: Chaos, resonance and collective dynamical phenomena in the solar system, p. 255 – 268 (1992). – See Abstr. 012.025 for the main entry.
The gravitational processes affecting the dynamics of comets are reviewed. At great distances from the Sun the motion of comets is primarily affected by the vertical component of the galactic field, as well as by encounter with stars and giant molecular clouds. When comets move in the region of the planets, encounters with these can strongly affect their motion. A good fraction of all periodic comets spend some time in temporary libration about mean motion resonances with Jupiter; some comets can be captured by this planet as temporary satellites. Finally, there is a small number of objects with orbital characteristics quite different from those of all other short–period comets.

102.019   The evolution of Jupiter family comets over 2000 years.
G. Tancredi, H. Rickman.
IAU Symposium No. 152: Chaos, resonance and collective dynamical phenomena in the solar system, p. 269 – 274 (1992). – See Abstr. 012.025 for the main entry.
The orbital evolution of the whole sample of short–period comets was computed by numerical integrations for a time interval of 2000 yr centered on the present epoch. This data base is intended to serve in various studies involving the statistics of orbital evolution and correlation with physical parameters or discovery circumstances. The authors present some results concerning the following aspects: the evolution of the orbital elements and their past–future asymmetry, statistics on the discovery of comets and on the encounters of comets with Jupiter.

102.020   The long–term dynamical behavior of small bodies in the Kuiper belt.
H. F. Levison.
IAU Symposium No. 152: Chaos, resonance and collective dynamical phenomena in the solar system, p. 275 – 279 (1992). – See Abstr. 012.025 for the main entry.
In this paper the author calculates the timescales of the evolution of objects in the Kuiper belt using a new technique that treats the evolution of orbits in integral space as a diffusion problem.

102.021   Splitting of periodic comets.
G. Forti.
IAU Symposium No. 152: Chaos, resonance and collective dynamical phenomena in the solar system, p. 281 – 285 (1992). – See Abstr. 012.025 for the main entry.
Among the comets that were observed to break in two or more fragments, only a few of them are periodic. So far the dynamic study of the relative motion of a secondary nucleus with respect to the primary has supposed that a cometary fragment is subject to a small and continuous radial nongravitational force after separation at rest. This force acts against the solar attraction and varies according to an inverse square law. A small impulse at break up may also be invoked in some case. A different approach is followed in this paper when dealing with a fragment of a periodic comet: after separation the motion of a secondary nucleus is characterized by nongravitational forces which vary according to the same $g(r)$ law currently used for the primary. Results of the study of comets P/Biela and P/du Toit–Hartley show that the motion of their fragments after separation is characterized by nongravitational parameters which are larger than those of the parent bodies. Both fragments lasted for about 2 full revolutions and three returns.

102.022   Rotational behavior of cometary–type bodies.
E. Bois, P. Oberti, C. Froeschlé.
IAU Symposium No. 152: Chaos, resonance and collective dynamical phenomena in the solar system, p. 291 – 296 (1992). – See Abstr. 012.025 for the main entry.
The present paper deals with a general dynamical qualitative study of the rotational motion for cometary–type bodies submitted to gravitational torques. Numerical experiments of the evolution of comet nucleus attitude have been then performed, including the Sun and Jupiter's disturbing torques in the model. Results show small effects of the solar gravitational perturbation for Halley–type orbits. Only a very close–approach with Jupiter induces notable effects. The latter configuration presents some interesting sensitivity to initial conditions.

102.023   The role of the interplanetary magnetic field reorientation in the mechanism of the comet's brightness outburst occurrence.
I. M. Podgorny (*I. M. Podgornyj*), D. A. Andrienko, V. V. Kleshchenok, I. I. Mischishina (*I. I. Mishchishina*).
Astrophys. Lett. Commun., Vol. 28, No. 1, p. 33 – 37 (Jul 1990).
Outbursts of the comet brightness occur while passing boundaries of the sectorial structure of the interplanetary magnetic field. Cometary brightness variations are explained by disintegration of the comet's induced magnetosphere during reorientation of the interplanetary field. Due to disintegration of the magnetosphere the energy accumulated in the magnetic field is released and simultaneously fast particles get a possibility to penetrate to the near–cometary space.

102.024   Polarization by irregular particles.
M. S. Hanner, P. Yanamandra–Fisher, R. A. West.
Bull. Am. Astron. Soc., Vol. 23, No. 3, p. 1143 (1991). Abstract. – See Abstr. 012.037 for the main entry.

102.025   Mysterious sungrazers.
G. L. Verschuur.
Astronomy, Vol. 20, No. 4, p. 46 – 49 (Apr 1992).

102.026   Star passages through the Oort cloud.
P. R. Weissman.
Bull. Am. Astron. Soc., Vol. 23, No. 3, p. 1158 (1991). Abstract. – See Abstr. 012.037 for the main entry.

102.027   A search for the Kuiper Disk of comets.
A. L. Cochran, W. D. Cochran, M. V. Torbett.
Bull. Am. Astron. Soc., Vol. 23, No. 3, p. 1158 (1991). Abstract. – See Abstr. 012.037 for the main entry.

102.028   Inferring the size of a comet nucleus from spectrophotometric observations.
M. Allen, B. Bochner.
Bull. Am. Astron. Soc., Vol. 23, No. 3, p. 1158 (1991). Abstract. – See Abstr. 012.037 for the main entry.

102.029   Where does the outgassing of comet nuclei occur?
M. C. Festou.
Bull. Am. Astron. Soc., Vol. 23, No. 3, p. 1159 (1991). Abstract. – See Abstr. 012.037 for the main entry.

102.030   Multiple scattering of light in a coma with an axisymmetric dust jet.
K. M. Chick, T. I. Gombosi.
Bull. Am. Astron. Soc., Vol. 23, No. 3, p. 1159 (1991). Abstract. – See Abstr. 012.037 for the main entry.

102.031 Light scattering by comet dust models.
B. Å. S. Gustafson, R. H. Zerull, K. Schulz,
E. Corbach.
Bull. Am. Astron. Soc., Vol. 23, No. 3, p. 1160 (1991). Abstract. –
See Abstr. 012.037 for the main entry.

102.032 Multispecies gas flows in the interior of comets.
C. J. Alexander, T. I. Gombosi.
Bull. Am. Astron. Soc., Vol. 23, No. 3, p. 1160 (1991). Abstract. –
See Abstr. 012.037 for the main entry.

102.033 Modeling the observed water production rate in comets.
M. R. Combi, B. J. Bos.
Bull. Am. Astron. Soc., Vol. 23, No. 3, p. 1161 (1991). Abstract. –
See Abstr. 012.037 for the main entry.

102.034 The determination of outflow velocity and photochemical lifetime from cometary coma observations.
H.–Y. Hu, D. E. Shemansky, H. P. Larson.
Bull. Am. Astron. Soc., Vol. 23, No. 3, p. 1162 (1991). Abstract. –
See Abstr. 012.037 for the main entry.

102.035 Prospects for observing sulfur–bearing cometary molecules at infrared wavelengths.
E. E. Roettger, M. J. Mumma.
Bull. Am. Astron. Soc., Vol. 23, No. 3, p. 1163 (1991). Abstract. –
See Abstr. 012.037 for the main entry.

102.036 How dirty are comet nuclei?
M. V. Sykes, R. G. Walker.
Bull. Am. Astron. Soc., Vol. 23, No. 3, p. 1163 (1991). Abstract. –
See Abstr. 012.037 for the main entry.

102.037 The "Little Bang" as the origin of comets.
T. Van Flandern.
Bull. Am. Astron. Soc., Vol. 23, No. 3, p. 1163 (1991). Abstract. –
See Abstr. 012.037 for the main entry.

102.038 The effect of electron collisions on rotational populations of cometary water.
X. Xie, M. J. Mumma.
Bull. Am. Astron. Soc., Vol. 23, No. 3, p. 1164 (1991). Abstract. The full paper is published in Astrophys. J., Vol. 386, No. 2, p. 720 – 728 (20 Feb 1992). – See Abstr. 012.037 for the main entry.

102.039 The rotational lines of methanol in comets.
D. Bockelée–Morvan, J. Crovisier, P. Colom,
D. Despois.
Bull. Am. Astron. Soc., Vol. 23, No. 3, p. 1167 (1991). Abstract. –
See Abstr. 012.037 for the main entry.

102.040 Observational evidence for heterogeneity of cometary nuclei: the ratio of methanol–to–formaldehyde as an indicator of cometary origins.
M. J. Mumma, D. C. Reuter, S. Hoban, M. A. DiSanti.
Bull. Am. Astron. Soc., Vol. 23, No. 3, p. 1167 (1991). Abstract. –
See Abstr. 012.037 for the main entry.

102.041 Cometary gas–phase chemistry taking into account homogeneous and ion–induced water recondensation.
J. F. Crifo.
Astrophys. J., Vol. 391, No. 1, p. 336 – 352 (20 May 1992).
A chemical model applicable to the circumnuclear region of water–dominated active comets is presented. The model self-consistently considers the vacuum outflow of coma gas and dust, the development of gas–phase photochemical reactions, and the formation of neutral and positively charged water clusters. Solutions are presented using simple boundary conditions consistent with P/Halley in situ gathered data. The results illustrate the strong propensity of such a medium toward partial recondensation, either by homogeneous or by ion–induced clustering. In particular, the dominance of heavy proton hydrates in the innermost coma with positive charge density is predicted.

The results are compared to similar effects observed in a closely related natural medium: the terrestrial mesopause region.

102.042 Dust tail streamers of a comet with rotating nucleus.
K. Beisser, H. Drechsel.
Astrophys. Space Sci., Vol. 191, No. 1, p. 1 – 22 (May 1992).
Synthetic images of the dust tail are presented for a comet which has a rotating nucleus with one predominant dust source fixed to it. The images have been generated using a new computer model which, unlike similar models, allows for the study of dust tails caused by a rotating nucleus with an anisotropic distribution of sources. The dust tail is studied in the post–perihelion phase of a parabolic comet with a perihelion distance of 0.5 AU. One finds that in the case of a rotating nucleus with anisotropic emission characteristics streamers caused solely by the dynamics of the dust particles are forming in the dust tail even if there is no dependence between the solar irradiation angle of the source and the amount of dust emitted. If the dust emission depends on the solar irradiation angle of the dust source, then the brightest tail regions do not necessarily coincide with the synchrones for the times of maximum dust emission. As a consequence, a thorough analysis of streamer patterns in a cometary dust tail requires assumptions on the rotational state and the dust source distribution of the nucleus. Otherwise, it seems not possible to discern between streamers which are caused dynamically by nucleus rotation and others which reflect variations in the emission activity.

102.043 Boulder complex model.
B. S. Sandhu.
Earth, Moon, Planets, Vol. 56, No. 2, p. 165 – 171 (Feb 1992).
Whipple's icy conglomerate model of cometary nucleus enjoyed wide acceptance and a long successful life. Considerable changes were brought to it based on the guidance of observational evidences and theoretical considerations. After the fractal model and the rubble pile model, icy glue model brought major alterations. But the concept of porous refractory boulders of the icy glue model seems to be quite unrealistic and unexplicable. Viewing minutely the process of the formation of comets a new model is proposed in which in the outskirts of the solar accretion disc ice and dust mixture formed small particles, which agglomerated to give large particles. These large particles further agglomerated to form tens of meters sized boulders. In this model it is assumed that up to the formation of boulders, ice and dust mix glue was consumed almost completely. The boulders collide with one another and get glued with the help of the glue formed due to the breakage of small particles at the interface, and form a boulder complex of the size of hundreds of meters. These complexes of boulders along with small boulders come together to form the cometary nucleus.

102.044 Acceleration mechanism of particles in the type–I cometary plasma.
Li Zhongyuan, Guo Sheyu.
Earth, Moon, Planets, Vol. 56, No. 3, p. 243 – 250 (Mar 1992).
In this paper, the accelerated effect of ions has been discussed. The transversal magnetic disturbance is able to bring about the magnetic annihilation and merge in some cometary area. The non–steady–state reconnection process can transform the magnetic energy of some cometary area into the kinetic energy of plasma. In addition, the "two stream instability" caused by both solar wind and cometary plasmas exists in Type–I tail, it can also lead the particles to be accelerated and heated in the plasma tail.

102.045 Search for the rendez–vous effect by means of a cometary globe.
V. V. Radzievskij, A. B. Artem'ev, S. B. Levakova, G. N. Levin, V. P. Tomanov, L. A. Shushnaeva.
Astron. Vestn., Tom 26, No. 3, p. 79 – 84 (May – Jun 1992). In Russian. English translation in Sol. Syst. Res., Vol. 26, No. 3.
As a result of rendez–vous of comets and planets there appears a specific distribution of poles of cometary orbits – (R–effect). An usual globe (black, stellar, geographical) is suggested for the

investigation of the distribution of poles of cometary orbits on the celestial sphere. Two effects have been found.

102.046 Stochastic motion of nearly–parabolic comets under perturbations by planets.
V. V. Emel'yanenko.
Pis'ma Astron. Zh., Tom 18, No. 6, p. 528 – 536 (Jun 1992). In Russian. English translation in Sov. Astron. Lett., Vol. 18, No. 3.

An analytical expression is obtained for the diffusion rate of any nearly–parabolic orbits in the restricted circular three–body problem. The convenient formulae and recurrence relations for calculations are given. The author presents the results of calculations of the diffusion rate for different perihelion distances and inclinations, the perturbations from all the major planets were taken into account.

102.047 Temperature evolution of porous ice samples covered by a dust mantle.
N. I. Kömle, G. Steiner.
Icarus, Vol. 96, No. 2, p. 204 – 212 (Apr 1992).

The existence of nonvolatile dust mantles covering the ices of cometary nuclei is suggested both from the behavior of many cometary lightcurves and from the direct observations of comet Halley's nucleus by the Giotto and VEGA spacecraft in Mar 1986. The authors present a systematic study which sheds some light on the evolution of such dust–ice systems when they are irradiated by the sun. Hereby it is assumed that the dust mantle as well as the underlying ice are grainy, porous structures permeable to gases. It is found that both the surface temperature of the dust mantle and the temperature at the dust–ice interface are strong functions of the average pore radius of the materials and of the thermal conductivity of the mantle.

102.048 Comments on "Gas release from comets" and related trapped–gas experiments.
R. L. Hudson, B. Donn.
Icarus, Vol. 96, No. 2, p. 286 – 288 (Apr 1992). Concerning Abstract 53.102.006.

A recent paper on Kr trapping and release by H_2O ice is used to emphasize the many variables that influence the results of trapped–gas experiments with cometary ice analogs. A proposed linear relationship between ice thickness and gas retained beyond the amorphous–to–cubic phase change of water is questioned. The many variables involved, combined with the uncertainty of cometary structure and composition, imply that great care is needed in the application of laboratory results to cometary phenomena and that definitive conclusions are difficult to reach.

102.049 Physico–dynamical evolution of aging comets.
H. Rickman.
15. Ecole de Printemps d'Astrophysique de Goutelas: Interrelations between physics and dynamics for minor bodies in the solar system, p. 197 – 263 (1992). – See Abstr. 012.055 for the main entry.

Contents: 1. Physical properties of comets (Tentative picture of cometary nuclei. Gas production rates and activity levels. Delsemme–Rud analysis and nongravitational force. Cometary magnitudes. Gas production curves and nongravitational effects). 2. Orbital and dynamical properties of comets (Structure of the Jupiter family. New and old long–period comets. Oort cloud dynamics. The capture of comets). 3. Physico–dynamical evolution (Dynamical evolution in the Jupiter family. Physical evolution in the Jupiter family. Cometary lifetimes and end states).

102.050 Thermal and physico–chemical processes in cometary nuclei.
B. Schmitt.
15. Ecole de Printemps d'Astrophysique de Goutelas: Interrelations between physics and dynamics for minor bodies in the solar system, p. 265 – 307 (1992). – See Abstr. 012.055 for the main entry.

This chapter is intended to give the basic idea and equations about the thermal and physico–chemical processes which govern the evolutions of comet nuclei during their penetration in the inner solar system. The implications of our knowledge of the structure and composition of comets on the evolution sketch of the nucleus will be first introduced. The author then describes the different physico–chemical and thermal processes which may or should control the propagation of heat inside the nucleus and the release of gases out of the nucleus. Finally some typical results are presented on the physical and chemical differentiations and on the thermal evolution of a comet nucleus on Halley's orbit. They have been obtained by one of the more advanced numerical models published today.

102.051 The change in matter concentration of a cometary head vs. heliocentric distance.
R. S. Osherov, K. I. Churyumov.
Sol. Syst. Res., Vol. 25, No. 5, p. 452 – 456 (Mar 1992). English translation of Astron. Vestn., Tom 25, No 5, p. 603 – 608 (1991).

Visual observations of Comet Halley support the earlier discovered relation between the matter concentration in the comet's head and heliocentric distance.

102.052 Hunting for comets and planets.
F. J. Dyson.
Q. J. R. Astron. Soc., Vol. 33, No. 2, p. 45 – 57 (Jun 1992).

102.053 The Oort cloud.
L. S. Marochnik, L. M. Mukhin, R. Z. Sagdeev.
US–USSR Workshop on Planetary Sciences, p. 246 – 258 (1991). – See Abstr. 012.075 for the main entry.

The original estimate of the Oort cloud's mass was made on the hypothesis that the nuclei of all comets are spherical. This produced an Oort cloud mass of 0.1 M_δ (Oort 1950). Therefore, the comet cloud that occupies the outer edge of the solar system appeared to be in a dynamically zero–gravity state, having no effect on the mass and angular momentum distribution in it. Since then the estimate of the Oort cloud's mass has gradually increased. The possibility cannot be ruled out that the Oort cloud has a concentration of mass comparable to the aggregate mass of the planets, in which the bulk of the solar system's angular momentum is concentrated (Marochnik et al. 1988).

102.054 The chaotic dynamics of comets and the problems of the Oort cloud.
R. Z. Sagdeev, G. M. Zaslavskij.
US–USSR Workshop on Planetary Sciences, p. 259 – 269 (1991). – See Abstr. 012.075 for the main entry.

This paper discusses the dynamic properties of comets entering the planetary zone from the Oort cloud. Even a very slight influence of the large planets (Jupiter and Saturn) can trigger stochastic cometary dynamics. Multiple interactions of comets with the large planets produce diffusion of the parameters of cometary orbits and a mean increase in the semi–major axis of comets. Comets are lifted towards the Oort cloud, where collisions with stars begin to play a substantial role. The transport of comets differs greatly from the customary law of diffusion and noticeably decelerates the average comet flow. The vertical tidal effect of the Galaxy in this region of motion is adiabatic and cannot noticeably alter cometary distribution.

102.055 On the origin of nearly parabolic comets.
V. V. Radzievskij.
Kinematics Phys. Celest. Bodies, Vol. 7, No. 6, p. 18 – 24 (1991). English translation of Kinematika Fiz. Nebesn. Tel., Tom 7, No. 6, p. 21 – 29 (Nov–Dec 1991).

The correlation $E(p|a^{-1})$ between the parameters p and reciprocal semimajor axes a^{-1} of nearly parabolic comets is investigated. It is established that this correlation is negative for elliptical orbits with p > 2.7 a.u. (correlation coefficient K = −0.94) and positive for orbits with p < 1.7 a.u. (K = +0.99). The sign of the correlation is reversed for hyperbolic orbits. It is shown that all nuances of the $E(p|a^{-1})$ correlation can be explained equally well either under the author's hypothesis of ejection of the comets by transplutonian planets or under the Safronov–Guseinov hypothesis of in situ comet formation.

102.056   Rotational behaviour of comet nuclei.
P. Oberti, E. Bois, C. Froeschlé.
NATO Advanced Study Institute on Predictability, Stability, and Chaos in N–Body Dynamical Systems, p. 249 – 254 (1991). – See Abstr. 012.089 for the main entry.
Numerical experiments of the rotational behaviour of comet nuclei have been performed, including the Sun and Jupiter's disturbing torques in the models. In a stable configuration, the solar torque induces great librations that remain unchanged along the orbit. A close approach with Jupiter can result in great changes of the rotational pattern. The unstable configuration is characterized by great librations of the nutation angle, and by the existence of a possibly large chaotic zone in the phase space.

102.057   Comets and meteors.
D. W. Hughes.
Images of the universe, p. 77 – 92 (1991). – See Abstr. 003.090 for the main entry.
Contents: (1) Introduction. (2) Comets from the ground. (3) Monitoring. (4) Imaging. (5) Spectroscopy. (6) Comets from space. (7) Meteors.

102.058   Comets: pre– and post–Halley.
K. S. Krishna Swamy.
Proc. Astron. Soc. Aust., Vol. 9, No. 2, p. 219 – 224 (1991). – See Abstr. 012.090 for the main entry.
The recent intensive study of Comet Halley based on in situ measurements, observations carried out with rockets and satellites and supplemented with co–ordinated Earth based observations has not only confirmed pre–Halley results, but also has given new insight into the nature of the nucleus, dust, gas and the interaction of cometary plasma with the solar wind. These observations also have raised many new questions and problems. Several of these aspects are discussed. For a better understanding of these problems, the planned future missions to comets are also discussed.

102.059   Numerical simulation of the cometary shocks.
A. S. Lipatov.
1. COSPAR Colloquium: Physics of the outer heliosphere, p. 395 – 396 (1990). – See Abstr. 012.100 for the main entry.

102.060   Computational study of radiation chemical processing in comet nuclei.
R. Navarro–González, C. Ponnamperuma, R. K. Khanna.
Origins Life Evol. Biosphere, Vol. 21, No. 5 – 6, p. 359 – 374 (1991 – 1992). – See Abstr. 012.083 for the main entry.
Cometary nuclei have been exposed to high levels of ionizing radiation since their formation. The authors present some results of a computer model calculation of the effect of ionizing radiation on cometary material. The external (cosmic rays) and internal (embedded radionuclides) contributions in the processing of cometary nuclei are considered. The data suggest that massive radiation chemical processing due to cosmic rays may have taken place only in the outer layers of comets. The internal contribution of radionuclides to the radiation processing of comet cores seems to be modest. Therefore, comets could be carriers of intact homochiral biomolecules.

Comets and the origin and evolution of life. Proceedings. Conference on Comets and the Origin and Evolution of Life, Eau Claire, WI (USA), 30 Sep – 2 Oct 1991.
See Abstr. 012.083.

Comets and meteorites were a minor source of prebiotic organic compounds on the early Earth.
See Abstr. 015.042.

Orbital evolution of dust particles from comets and asteroids.
See Abstr. 021.020.

Extremely low thermal conductivity of amorphous ice: relevance to comet evolution.
See Abstr. 022.047.

Dust emission phenomena of cometary analogues.
See Abstr. 022.073.

Low–frequency modes in dusty plasmas.
See Abstr. 022.089.

Type II clathrate hydrate formation in cometary ice analogs in vacuo.
See Abstr. 022.156.

The physics and dynamics of charged dust grains.
See Abstr. 022.195.

An approach to trend analysis in data with special reference to cometary magnitudes.
See Abstr. 036.131.

Eine einfache Methode der parabolischen Bahnbestimmung.
See Abstr. 042.013.

Mappings in astrodynamics.
See Abstr. 042.053.

Introduction to stochastic modelling of cometary dynamics: Monte Carlo simulations and Markov process.
See Abstr. 042.085.

New applications of Fatou's problem.
See Abstr. 042.091.

Modelling: an aim and a tool for the study of the chaotic behaviour of asteroidal and cometary orbits.
See Abstr. 042.099.

Perturbation theory, resonance, librations, chaos, and Halley's comet.
See Abstr. 042.107.

In–situ Doppler velocimeter of very large grains: an essential goal for future cometary investigations.
See Abstr. 051.064.

Chaotic Alfvén waves in multispecies plasmas.
See Abstr. 062.036.

Chaotic Alfvén waves in solar and cometary plasmas.
See Abstr. 062.098.

The role of cometary and meteor matter in noctilucent clouds genesis.
See Abstr. 082.018.

On possible cometary origin of a background sulphate layer in the stratosphere.
See Abstr. 082.058.

Asteroid taxonomy types.
See Abstr. 098.059.

Asteroids and comets – the relationship between near–earth asteroids and main–belt asteroids and comets.
See Abstr. 098.172.

The solar motion, the Galaxy and the aphelia distribution of long period comets.
See Abstr. 103.019.

Are there any comets coming from interstellar space?
See Abstr. 103.021.

Formation of comets: constraints from the abundance of hydrogen sulfide and other sulfur species.
See Abstr. 103.032.

The effect of electron collisions on rotational populations of cometary water.
See Abstr. 103.109.

Stability of the cometary ionopause.
See Abstr. 103.111.

Triaxiality of Halley's comet.
See Abstr. 103.129.

Organic particles of cometary dust in space.
See Abstr. 104.059.

Meteoroid swarms: formation, evolution, and relationship to comets and asteroids.
See Abstr. 104.071.

Interior resonance trapping of dust particles from comets and asteroids.
See Abstr. 106.050.

Astrophysical dust grains in stars, the interstellar medium, and the solar system.
See Abstr. 106.089.

Solar system – interstellar medium. A chemical memory of the origins.
See Abstr. 107.021.

Cometary diagnostics of solar nebula chemistry.
See Abstr. 107.043.

The rate of planet formation and the solar system's small bodies.
See Abstr. 107.057.

Cometary origin of carbon, nitrogen and water on the Earth.
See Abstr. 107.061.

Comets and the formation of biochemical compounds on the primitive Earth – a review.
See Abstr. 107.062.

Comets as a possible source of prebiotic molecules.
See Abstr. 107.063.

Comet impacts and chemical evolution on the bombarded Earth.
See Abstr. 107.064.

Cometary supply of terrestrial organics: lessons from the K/T and the present epoch.
See Abstr. 107.065.

Relative abundance of ices in disks around T Tauri stars: implications for comet formation.
See Abstr. 121.057.

The physics of dusty plasmas.
See Abstr. 131.093.

103 Comets (Individual Objects)

103.001 Roman numeral designations of comets in 1990.
Minor Planet Circ., Nos. 19357 – 19358 (19 Jan 1992).

103.002 Die periodischen Kometen des Jahres 1992.
J. Jahn.
Sterne Weltraum, Jahrg. 31, Nr. 1, p. 52 – 54 (Jan 1992).

103.003 Ephemerides of comets.
Minor Planet Circ., Nos. 19347 – 20366 (Jan – Jun 1992).
Ephemerides for the following comets are given in Nos. 19524 – 19530 (19 Ja 1992), 19700 – 19708 (18 Feb 1992), 19887 – 19889 (18 Mar 1992), 20041 – 20043 (17 Apr 1992), 20168 – 20171 (16 May 1992), and 20347 – 20354 (15 Jun 1992), respectively:
1978 I P/Schuster, 1984 XVIII P/Shoemaker 2, 1985 XVI P/Ciffréo, 1986 XI P/Singer Brewster, 1988 XIII P/Helin–Roman–Crockett, 1989h1 P/Van Biesbroeck, 1990 XXII McNaught–Russell, 1991d Shoemaker–Levy, 1991l Helin–Lawrence, 1991o P/Chernykh, 1991p P/Shoemaker 1, 1991q P/Levy, 1991r Helin–Alu, 1991t P/Hartley 2, 1991v McNaught–Russell, 1991a1 Shoemaker–Levy, 1991b1 P/Shoemaker–Levy 6, 1991d1 P/Shoemaker–Levy 7, 1991f1 P/Kowal 2, 1991g1 Zanotta–Brewington, 1991h1 Mueller, 1992a Helin–Alu, 1992b Bradfield, 1992d Tanaka–Machholz, 1992f P/Shoemaker–Levy 8, 1992h Spacewatch, P/Ashbrook–Jackson, P/Daniel, P/Encke, P/Forbes, P/Gehrels 3, P/Giclas, P/Gunn, P/Holmes, P/Howell, P/Schaumasse, P/Schwassmann–Wachmann 1, P/Schwassmann–Wachmann 2, P/Slaughter–Burnham, P/Wolf.

103.004 Orbital elements of comets.
S. Nakano, B. G. Marsden, T. Kobayashi, R. H. McNaught, G. Forti, K. Muraoka, D. K. Yeomans.
Minor Planet Circ., Nos. 19347 – 20366 (Jan – Jun 1992).
The comets are listed according to their Roman numeral designation or their preliminary designation. The names of the orbit computers are given behind the respective M.P.C. numbers:
1984 XV Shoemaker 20121 (Marsden); 1989 XXI 20308 (Muraoka); 1990 VI Skorichenko–George 20308 (Muraoka); 1990 XIX McNaught–Russell 20308 (Marsden); 1990 XXII 19654 (Marsden); 1990 XXIX P/Spacewatch 20308 (Muraoka); 1991d Shoemaker–Levy 19467 (Nakano); 1991l Helin–Lawrence 19654 (Marsden); 1991q P/Levy 20308 (Nakano); 1991r Helin–Alu 19468, 20121 (Nakano); 1991v McNaught–Russell 19468 (Nakano), 20309 (Marsden); 1991a1 Shoemaker–Levy 19468 (Kobayashi), 19655 (Marsden), 20309 (Nakano); 1991b1 P/Shoemaker–Levy 6 19467 (Marsden); 1991d1 P/Shoemaker–Levy 7 19467 (Nakano); 1991f1 P/Kowal 2 19467 (Nakano); 1991g1 Zanotta–Brewington 19467, 19654, 19818 (Nakano); 1991h1 Mueller 19468, 19654 (Marsden), 19818 (Nakano); 1992a Helin–Alu 19654, 19818 (Marsden); 1992b Bradfield 19654 McNaught, 19818 (Marsden); 1992d Tanaka–Machholz 19980, 20121, 20309 (Marsden); 1992f P/Shoemaker–Levy 8 19980, 20121 (Nakano), 20309 (Marsden); 1992g P/Mueller 4 20121 (Nakano), 20309 (Marsden); 1992h Spacewatch 20121, 20310 (Marsden); 1992i Bradfield 20309 (Marsden); P/Clark 20122 (Nakano); P/d'Arrest 20122 (Yeomans); P/de Vico–Swift 20122 (Nakano); P/Finlay 20122 (Nakano); P/Honda–Mrkos–Pajdušáková 20124 (Marsden); P/Jackson–Neujmin 20123 (Forti); P/Longmore 20123 (Muraoka); P/Perrine–Mrkos 20123 (Nakano); P/Tuttle–

Giacobini–Kresák 20122 (Nakano); P/Reinmuth 1 20123 (Marsden); P/Schwassmann–Wachmann 3 20123 (Marsden).

103.005   Observations of comets.
      Minor Planet Circ., Nos. 19347 – 20366 (Jan – Jun 1992).
Observations made at the following stations are published in Nos. 19358 – 19370 (19 Jan 1992), 19559 – 19566 (18 Feb 1992), 19726 – 19731 (18 Mar 1992), 19911 – 19917 (17 Apr 1992), 20070 – 20073 (16 May 1992), and 20191 – 20195 (15 Jun 1992), respectively:
Burlington remote site, Cambridge, Cerro Tololo, Chiyoda, Cima Ekar, Colchester, Colleverde di Guidonia, Crimea (Sternberg Astron. Inst. Stn.), Crni vrh (Slovenia), Dynic Astron. Obs., Eastfield Obs., ESO, Fabra Obs., Farra d'Isonzo, Fukaya, Geisei, JCPM Hamatonbetsu Stn., JCPM Kagoshima Stn., Kambah (nr Canberra), Kiso, Kitt Peak, Kitt Peak (Steward Obs.), Klet̆, Kushiro, La Palma, Lamalou–les–Bains, Linz, Loiano, Mauna Kea Obs., Mount John Obs., Nihondaira Obs., Oak Ridge Obs., Oohira Stn., Ohtsu, Oishi, Oizumi, Ojima, Otomo, Palomar, Perth Obs., Pian dei Termini, Prague, Santa Lucia Stroncone, Sengamine, Siding Spring, Sormano, Uenohara, Victoria (Climenhaga Obs.), Victoria (Dominion Astrophys. Obs.), Yatsugatake South Base Obs., Yatsuka, YGCO Chiyoda Stn., YGCO Nagano Stn., Yodoe. ·
Observations of the following comets – sorted according to their provisional designation and their Roman numeral designation – are published:
1983 XII Lovas, 1984 V P/Smirnova–Chernykh, 1984 XV Shoemaker, 1984 XXI P/Arend–Rigaux, 1986 III P/Halley, 1987 IV Shoemaker, 1987 X P/Grigg–Skjellerup, 1988 XIV P/Tempel 2, 1990 XIX McNaught–Russell, 1989 XV Schwassmann–Wachmann 1, 1990 XX Levy, 1990 XXII McNaught–Russell, 1990 XXV P/Kearns–Kwee, 1990 XXIX P/Spacewatch, 1991d Shoemaker–Levy, 1991i P/Kowal 1, 1991l Helin–Lawrence, 1991n P/Faye, 1991o P/Chernykh, 1991p P/Shoemaker 1, 1991q P/Levy, 1991s P/Wirtanen, 1991t P/Hartley 2, 1991v McNaught–Russell, 1991y P/McNaught–Hughes, 1991z P/Shoemaker–Levy 5, 1991a1 Shoemaker–Levy, 1991b1 P/Shoemaker–Levy 6, 1991c1 P/Tsuchinshan 1, 1991d1 P/Shoemaker–Levy 7, 1991e1 P/Tsuchinshan 2, 1991f1 P/Kowal 2, 1991g1 Zanotta–Brewington, 1991h1 Mueller, 1992a Helin–Alu, 1992b Bradfield, 1992c P/Howell, 1992d Tanaka–Machholz, 1992e P/Singer Brewster, 1992f P/Shoemaker–Levy 8, 1992g Mueller, 1992h Spacewatch, 1992i Bradfield.

103.006   The effect of solar activity on the light curves of comets Churyumov–Gerasimenko (1982 VIII) and Halley (1986 III).
K. I. Churyumov, V. S. Filonenko.
Sov. Astron. Lett., Vol. 17, No. 6, p. 470 – 473 (Nov 1991). Current Physics Microform No.: 9207X1908. English translation of Pis'ma Astron. Zh., Tom 17, No. 12, p. 1135 – 1142 (Dec 1991).
A statistically significant correlation has been detected between brightness outbursts and variations of the new short-period comet Churyumov–Gerasimenko with the level of solar activity. The authors show that, contrary to the earlier results of Orlov (1923), the integrated brightness of Comet Halley correlates with changes in solar activity indices and solar wind velocity.

103.007   Cometary dust trails. I. Survey.
M. V. Sykes, R. G. Walker.
Icarus, Vol. 95, No. 2, p. 180 – 210 (Feb 1992).
Cometary dust trails were first observed by IRAS and consist of large refractory particles ejected from short–period comets at low velocities. Consequently, they tend to be found near the orbital paths of their parent bodies, their long and narrow appearance reminiscent of airplane contrails. An examination of the entire sky as seen by IRAS has resulted in the detection of a total of eight trails associated with known short–period comets as well as many faint trails having no known parents. Trails tended

to be associated with objects having low perihelion distances that were observed near perihelion. Trail comets are found to lose the bulk of their mass in the large refractory trail particles, and are found to have a median refractory/volatile mass ratio of ~ 3. This suggests that comets in general may be more like "frozen mudballs" than the canonical "dirty snowballs".

103.008   An infrared search for formaldehyde in several comets.
D. C. Reuter, S. Hoban, M. J. Mumma.
Icarus, Vol. 95, No. 2, p. 329 – 332 (Feb 1992).
The authors describe a search for cometary infrated emission from the v_1 and v_5 bands of formaldehyde in the spectral region from 3.55 to 3.64 μm. For several comets they present 3σ upper limits to the production rate of H_2CO retrieved using models which assume either a parent or a distributed source for this species. The 3σ upper limit for the production rate of H_2CO relative to water ranges from $<0.15\%$ to $<0.80\%$ assuming a parent source and $<1.7\%$ to $<5.8\%$ assuming a distributed source.

103.009   Tabulation of comet observations.
Int. Comet Q., Vol. 14, No. 1, p. 3 – 27 (Jan 1992).
Concerning comets: 1974 III Bradfield, 1983 V Sugano–Saigusa–Fujikawa, 1983 VII IRAS–Araki–Alcock, 1987 XXIX Bradfield, 1988 V Liller, 1988 XIV P/Tempel 2, 1989 X P/Brorsen–Metcalf, 1989t P/Wild 2, 1989c1 Austin, 1989d1 P/Schwassmann–Wachmann 3, 1990c Levy, 1990f P/Honda–Mrkos–Pajdušáková, 1990i Tsuchiya–Kiuchi, 1991a P/Metcalf–Brewington, 1991b Arai, 1991d Shoemaker–Levy, 1991l Helin–Lawrence, 1991n P/Faye, 1991o P/Chernykh, 1991q P/Levy, 1991s P/Wirtanen, 1991t P/Hartley 2, 1991a1 Shoemaker–Levy, 1991b1 P/Shoemaker–Levy 6, 1991g1 Zanotta–Brewington, P/Encke, P/Arend–Rigaux, P/Schwassmann–Wachmann 1.

103.010   Recent news and research concerning comets.
D. W. E. Green.
Int. Comet Q., Vol. 14, No. 2, p. 52 (Apr 1992).
Recent news and research concerning comets.

103.011   Tabulation of comet observations.
Int. Comet Q., Vol. 14, No. 2, p. 31 – 52 (Apr 1992).
Concerning comets: 1980 XV Bradfield, 1987 XXIX Bradfield, 1988 V Liller, 1989 X P/Brorsen–Metcalf, 1989 XV P/Schwassmann–Wachmann 1, 1989 XIX Okazaki–Levy–Rudenko, 1989 XXII Aarseth–Brewington, 1989h1 P/Van Biesbroeck, 1990 V Austin, 1990 VI Skorichenko–George, 1990 XX Levy, 1990 XXVIII P/Wild 2, 1990g McNaught–Hughes, 1991a P/Metcalf–Brewington, 1991d Shoemaker–Levy, 1991l Helin–Lawrence, 1991n P/Faye, 1991o P/Chernykh, 1991p P/Shoemaker 1, 1991q P/Levy, 1991s P/Wirtanen, 1991t P/Hartley 2, 1991a1 Shoemaker–Levy, 1991b1 P/Shoemaker–Levy 6, 1991f1 P/Kowal 2, 1991g1 Zanotta–Brewington, 1991h1 Mueller, 1992b Bradfield, 1992d Tanaka–Machholz, P/Machholz, P/Arend–Rigaux.

103.012   Die Kometenentdeckungen des Jahres 1991.
M. Meyer.
KPM, Jahrg. 7, No. 19, p. 9 – 12 (Apr 1992).
The author presents the most interesting comet discoveries of the year 1991 with some news on the discovery circumstances.

103.013   VdS–Archiv Kometen.
A. Kammerer, J. Jahn.
KPM, Jahrg. 7, No. 19, p. 34 – 40 (Apr 1992).
Observations of comets P/Levy (1991q), P/Hartley 2 (1991t), P/Wirtanen (1991s), P/Faye (1991n), and Zanotta–Brewington (1991g1) are listed.

103.014   Comet corner.
D. E. Machholz.
Strolling Astron., Vol. 36, No. 1, p. 30 – 32 (Mar 1992).

103.015 **Detection of comet nuclei at large heliocentric distances.**
M. E. Bailey, J. E. Chambers, G. Hahn.
Mon. Not. R. Astron. Soc., Vol. 254, No. 4, p. 581 – 588 (15 Feb 1992).

The largest long–period comets which have passed perihelion during the past halfcentury should still be observable. Fourteen candidate comets with well–determined orbits are presented, including data on their absolute magnitudes, distances and predicted positions in the sky. Detection of these comets would provide: (1) absolute magnitudes and diameters for a significant number of large long–period comets uncontaminated by outgassing; (2) improved orbital elements for these comets; (3) a dynamical probe of "dark matter" in the outer planetary system, whether comet disc, an inner core of the Oort cloud or "Planet X", and (4) a sample of distant comets suitable for monitoring short–term brightness variations such as those observed in Chiron and P/Halley.

103.016 **Narrow–band IPD images of cometary CN and C_2: the effect of solar activity on coma scales.**
N. P. Meredith, M. K. Wallis, D. Rees.
Mon. Not. R. Astron. Soc., Vol. 254, No. 4, p. 693 – 704 (15 Feb 1992).

Comet Halley and other cometary targets have been observed over a five–year period extending from 1985 August to 1990 May. 1–m and 24–inch telescopes have been used for near–nucleus studies and a selection of wide–angle Nikon camera lenses for large–scale structures. The imaging systems employed an Imaging Photon Detector in conjunction with narrow–band interference filters to isolate the emissions of various cometary species. Integration over these images give profiles extending over a large range in radius, facilitating accurate determinations of the CN and C_2 parent scale–lengths, as well as the CN destruction scale. Dependence on solar activity is found and the results are analysed separately for solar maximum and minimum conditions.

103.017 **Giotto extended mission.**
G. Schwehm.
Adv. Space Res., Vol. 11, No. 12, p. 127 – 131 (1991). – See Abstr. 012.014 for the main entry.

The navigation of the ESA spacecraft Giotto to its encounter with comet P/Halley on 14 Mar 1986 required just 10% of the fuel available. The spacecraft was retargeted to return close to Earth to maintain the option to extend the mission to encounter another comet, P/Grigg–Skjellerup on 10 Jul 1992. On 2 Apr 1986 the spacecraft was put into hibernation configuration and had been orbiting the Sun in the ecliptic. On 19 Feb 1990 it was reactivated, spacecraft subsystems and the payload checked out to determine its health status. On 2 Jul 1990 Giotto performed successfully the first–ever Earth gravity assist manoeuvre of a spacecraft approaching the Earth from deep space and was retargeted for comet P/Grigg–Skjellerup.

103.018 **Relation of the secular decrease of brightness to the orbital elements of short–period comets.**
J. Svoreň.
Adv. Space Res., Vol. 11, No. 12, p. 141 – 148 (1991). – See Abstr. 012.014 for the main entry.

Secular variations in the absolute brightness of short–period comets were derived on the basis of their maximum apparent magnitudes in individual returns. The present paper deals with all the comets of Jupiter family observed by 31 Dec 1987 in three returns at least. P/Encke has not been included. The sample contains 405 observed returns of 61 short–period comets. Relations of the secular fading to the orbital period, perihelion distance and also to the numerical eccentricity were found. Two independent methods of determination of aging of short–period comets, i.e. the secular fading and nongravitational parameters, are compared. The distribution of the values of the secular decrease of brightness is not in agreement with the trends in nongravitational effects. More probably, the observed secular decrease is connected with the apparent magnitude of the comet and with the influence of instrumental effects upon it.

103.019 **The solar motion, the Galaxy and the aphelia distribution of long period comets.**
A. Brunini, D. M. Canosa.
IAU Symposium No. 152: Chaos, resonance and collective dynamical phenomena in the solar system, p. 287 – 290 (1992). – See Abstr. 012.025 for the main entry.

The authors analyze the incoming direction of 180 long period comets, taken into account the gravitational influence of the planets and the Galaxy as a whole. They have made an improvement of some previous works on the matter, by inclusion of comets observed up to 1989. An asymmetry of aphelia is detected that cannot be explained as due to observational selection. The authors conclude that the so called Dynamical Friction induced by the solar motion through the interstellar medium could be the proper answer to this observed fact.

103.020 **The plasma tails of comets Halley, Okazaki–Levy–Rudenko, Austin, and Levy in high time resolution images.**
T. L. Farnham, K. J. Meech, W. G. Weller.
Bull. Am. Astron. Soc., Vol. 23, No. 3, p. 1161 (1991). Abstract. – See Abstr. 012.037 for the main entry.

103.021 **Are there any comets coming from interstellar space?**
Ľ. Kresák.
Astron. Astrophys., Vol. 259, No. 2, p. 682 – 691 (Jun 1992).

The current list of 264 well–determined original orbits of long–period comets (Marsden 1989) includes 18 hyperbolic orbits with negative binding energies $-1/a_0$ exceeding 4×10^{-5} AU. According to Yabushita (1991) this is the upper limit compatible with the nongravitatinal accelerations of outgassing cometary nuclei, and higher values may be indicative of interstellar origin. Analysis of the published orbit computations shows that for 10 of the 18 suspect comets (1895 IV, 1898 VIII, 1904 II, 1911 IV, 1932 VII, 1940 III, 1959 III, 1968 VI, 1975 XI, 1986 XVII) the hyperbolic excess is within the range of computing uncertainty even for a purely gravitational solution. In another five cases (1953 II, 1957 III, 1960 II, 1971 V, 1989 XIX) nongravitational solutions yield positive values of $1/a_0$, and in two cases (1899 I and 1955 V) the remaining excess is smaller than its mean error. A difficult exception is comet 1976 I, but a detailed analysis of its observations has shown that they can be satisfied by an elliptic original orbit without the inclusion of nongravitational effects (Kresák 1992). It is concluded that there is no evidence for any comet coming from the interstellar space.

103.022 **2–D photometry of comets Austin and P/Brorsen–Metcalf.**
L. C. Ho, M. Dickinson, H. Spinrad, R. L. Newburn.
Bull. Am. Astron. Soc., Vol. 23, No. 3, p. 1162 (1991). Abstract. – See Abstr. 012.037 for the main entry.

103.023 **Searching for formaldehyde in comets at infrared wavelengths.**
S. Hoban, D. C. Reuter, M. J. Mumma.
Bull. Am. Astron. Soc., Vol. 23, No. 3, p. 1162 (1991). Abstract. – See Abstr. 012.037 for the main entry.

103.024 **The carbon isotope abundance ratios in comets.**
M. Kleine, S. Wyckoff, P. A. Wehinger,
B. A. Peterson.
Bull. Am. Astron. Soc., Vol. 23, No. 3, p. 1166 (1991). Abstract. – See Abstr. 012.037 for the main entry.

103.025 **Analysis of ultraviolet OH measured by IUE in 14 comets.**
D. G. Schleicher.
Bull. Am. Astron. Soc., Vol. 23, No. 3, p. 1168 (1991). Abstract. – See Abstr. 012.037 for the main entry.

103.026 Cometary chemistry.
     M. F. A'Hearn.
IAU Symposium No. 150: Astrochemistry of cosmic phenomena,
p. 415 – 420 (1992). – See Abstr. 012.029 for the main entry.
     The key problem in cometary chemistry is to observe abun-
dances of species in the coma and to reassemble those species into
the species that are present in the nucleus. The limitations are
primarily due to poorly constrained models and the lack of
uniqueness in reassembling parent species from the fragments
when not all of the fragments are observable. These problems will
be illustrated with several examples.

103.027 Cometary molecules.
     L. E. Snyder.
IAU Symposium No. 150: Astrochemistry of cosmic phenomena,
p. 427 – 434 (1992). – See Abstr. 012.029 for the main entry.
     Which molecular species are firmly identified as cometary
species and which are also interstellar? How may excitation
effects bias the data interpretation?

103.028 Observations of parent molecules in comets at radio
     wavelengths: HCN, H_2S, H_2CO and CH_3OH.
P. Colom, D. Bockelee–Morvan, J. Crovisier, D. Despois,
G. Paubert.
IAU Symposium No. 150: Astrochemistry of cosmic phenomena,
p. 439 – 440 (1992). – See Abstr. 012.029 for the main entry.
     The authors present observations of cometary parent mole-
cules at the IRAM radio telescope which led to the first
detections of H_2S and CH_3OH in comets, and confirmed the
presence of H_2CO and HCN. Production rates and abundances
relative to H_2O are given.

103.029 Radio interferometric observations of cometary mole-
     cules.
P. Palmer, L. E. Snyder, I. de Pater.
IAU Symposium No. 150: Astrochemistry of cosmic phenomena,
p. 441 – 442 (1992). – See Abstr. 012.029 for the main entry.
     A useful method for extracting cometary signals is demonstrat-
ed using VLA observations of comet Brorsen–Metcalf (1989o).

103.030 Gas production rates in comets.
     A. A. de Almeida.
IAU Symposium No. 150: Astrochemistry of cosmic phenomena,
p. 443 – 445 (1992). – See Abstr. 012.029 for the main entry.
     Emission fluxes of CN, C_2 and C_3 carbon–bearing molecular
species observed in the coma of comets Bennett (1969i ≡
1970 II), West (1975n ≡ 1976 VI), P/Halley (1982i), Hartley–
Good (1985l) and Bradfield (1987s) are analysed. CN, C_2 and C_3
production rates their dependence on the heliocentric distance,
and the possible correlations among these radicals are studied
and briefly discussed.

103.031 Mass loss rates of three comets.
     P. D. Singh, W. F. Huebner, D. C. Boice, I. Konno,
E. Scalise Jr.
IAU Symposium No. 150: Astrochemistry of cosmic phenomena,
p. 447 – 448 (1992). – See Abstr. 012.029 for the main entry.
     Emission features of C_2, C_3, CN, and dust in comets Thiele
(1985m), Hartley–Good (1985l), and Giacobini–Zinner (1984e)
have been analyzed and their mass loss rates of about 0.5, 1.1,
and 0.8 Mg s^{-1} have been determined.

103.032 Formation of comets: constraints from the abundance of
     hydrogen sulfide and other sulfur species.
D. Despois, J. Crovisier, D. Bockelee–Morvan, P. Colom.
IAU Symposium No. 150: Astrochemistry of cosmic phenomena,
p. 459 – 460 (1992). – See Abstr. 012.029 for the main entry.
     Recent determinations of H_2S and other sulfur compounds
abundances in comets and in Orion KL bring new tests of the
origin of cometary matter.

103.033 Neues aus der Kometen–, Planetoiden– und Meteor-
     szene.
J. Jahn.
KPM, Jahrg. 7, No. 19, p. 48 – 64 (Apr 1992).
     Various results of latest comet and asteroid observations are
presented. Some lightcurves and ephemerides are given. Hotline
news from different sources are announced.

103.034 Evaluation of gas production rates in some comets.
     A. A. de Almeida.
Earth, Moon, Planets, Vol. 56, No. 1, p. 61 – 74 (Jan 1992).
     Emission fluxes of CN, C_2 and C_3 carbon–bearing molecular
species observed in the coma of comets Bennett (1970 II), West
(1976 VI), P/Halley (1986 III), Hartley–Good (1985 XVII) and
Bradfield (1987 XXIX) are analysed in the framework of Haser
model. CN, C_2 and C_3 production rates are determined using
recently derived fluorescence efficiencies. The dependence of CN,
C_2 and C_3 production rates on the heliocentric distance and the
possible correlations among these radicals is studied and briefly
discussed.

103.035 Probable comets.
     IAU Circ., No. 5471, p. 1 (10 Mar 1992).

103.036 Probable comets.
     Yamamoto Circ., No. 2181, p. 1 – 2 (30 Mar 1992). In
Japanese.

103.037 Modeling dust fragmentation in comets.
     I. Konno, W. F. Huebner.
IAU Colloquium No. 126: Origin and evolution of interplanetary
dust, p. 221 – 224 (1991). – See Abstr. 012.068 for the main entry.
     The authors developed a 1–D hydrodynamic model of dusty
gas flow with dust fragmentation in a cometary atmosphere and
performed calculations for a dust–size distribution with radii a =
10^{-4} – 10 cm and densities variable with dust size. A comparison
was made with Giotto observations of dust jet intensities within
100 km of the nucleus comet Halley. The authors found that dust
fragmentation cannot be solely responsible for the flattening of
the dust intensity near the nucleus with respect to the 1/R law.
They conclude that a combination of geometric effects and grain
fragmentation may explain the observed intensity profiles.

103.038 The contribution of long period comets to the interplane-
     tary dust cloud.
M. Fulle, G. Cremonese.
IAU Colloquium No. 126: Origin and evolution of interplanetary
dust, p. 225 – 228 (1991). – See Abstr. 012.068 for the main entry.
     The numerical analysis of cometary dust tails (Fulle 1989) is
applied to three long period comets, namely C/Bennett 1970 II,
C/Bradfield 1987 XXIX and C/Liller 1988 V. Therefrom the
authors obtained that each long period comet injects in bound
orbits at least half of the total produced mass. Considering that a
typical long period comet can produce more than 10^{14}g of dust
along each perihelion passage, one obtains that the considered
long period comets alone injected an input mass rate of about
10^{16}g s^{-1} of meteoroids in bound orbits during the last 20 years,
a contribution which is very close to that from all short period
comets.

103.039 Long dust trails of short periodic comets.
     H. U. Keller, K. Richter.
IAU Colloquium No. 126: Origin and evolution of interplanetary
dust, p. 229 – 234 (1991). – See Abstr. 012.068 for the main entry.
     Radar, infrared and visible observations of dust trails along
cometary orbits are considered.

103.040   **Comets as a source of interplanetary and interstellar grains.**
F. Hoyle, N. C. Wickramasinghe.
IAU Colloquium No. 126: Origin and evolution of interplanetary dust, p. 235 – 240 (1991). – See Abstr. 012.068 for the main entry.

Properties of cometary dust with regard to bulk density, optical characteristics and sizes, derived from recent observations, are used to model scattering properties of cometary and interstellar grains. A wide range of astronomical observations are shown to be explained if cometary objects are hypothesised as a major source of dust grains in the Galaxy.

103.041   **Scattering properties of cometary dust based on polarimetric data.**
S. Mukai, T. Mukai, S. Kikuchi.
IAU Colloquium No. 126: Origin and evolution of interplanetary dust, p. 249 – 252 (1991). – See Abstr. 012.068 for the main entry.

A phase function of linear polarization of several comets is examined, especially in a region of phase angles near a maximum polarization. A lower maximum polarization observed in comet Austin(1989c1) than those in comets West(1975n) and P/Halley leads to the speculation that a mixing ratio of rough scattering to Mie scattering in comet Austin increases from a sun–comet distance r of 0.6 AU to 1.2 AU. This implies a shortage of large particles in comet Austin in r < 1 AU.

103.042   **Synchronic band and its implication in the cometary dust.**
J.-i. Watanabe, K. Nishioka.
IAU Colloquium No. 126: Origin and evolution of interplanetary dust, p. 253 – 256 (1991). – See Abstr. 012.068 for the main entry.

The unique morphology of the synchronic band in the cometary dust tail is explained by a finite–lifetime fragment model. However, this model needs a severe restriction on the lifetime of the dust fragments: 25 – 70 days (r = 1 AU). This implies that detailed analysis of the synchronic band may reveal physical properties of the cometary dust particles. The authors suggest that the fragments in the synchronic band are relatively pure ice if they are not organic grains.

103.043   **Komeetontdekkingen in 1991.**
A. Scholten.
Zenit, Jaarg. 19, Nr. 4, p. 170 – 171 (Apr 1992).

Comet 1976 I Sato

103.044   **The peculiar orbit of comet 1976 I Sato.**
Ĺ. Kresák.
Astron. Astrophys., Vol. 259, No. 2, p. 692 – 695 (Jun 1992).

1976 I Sato is being held for the most convincing example of a comet which entered the solar system on an hyperbolic orbit from the interstellar space. In the present paper it is shown that its observations can be adequately represented by an elliptic orbit coming from the Oort cloud, even without the inclusion of nongravitational accelerations. The new original orbit leaves only 10 to 15% larger mean O–C residuals of the positional measurements than the hyperbolic solutions. It is shown that it was peculiar observing geometry, appearance, and distribution of observations over the orbital arc which has made the hyperbolic excess of the previous orbit determinations much larger than its formal computing uncertainty.

Comet 1987 VII Wilson

103.045   **Postperihelion spectra and images of comet Wilson (1986l) obtained with a focal reducer.**
C. D. Prasad, K. Jockers, E. H. Geyer.
Icarus, Vol. 95, No. 2, p. 211 – 221 (Feb 1992).

Comet Wilson (1986l) was observed with the MPAE/Hoher List focal reducer–based CCD camera and grism spectrometer having spectral resolution of 4.9 Å per pixel, coupled to the ESO

1–m reflector. The data base consists of six long slit spectra, one of them is taken on the nucleus and the others at 1.5×10^5km in the tail, and images with filters centered at 6600 and 4250 Å. A "ray" structured tail longer than 400,000 km is detected in the emissions of CO^+. The continuum profile extracted from the spectrum centered on the nucleus shows a distinct sunward asymmetry. From the imaging data it is seen that most of the dust ejection occurs in a "beam" on the sunward side, producing a "fan" like structure. The fan is about 70° wide and originates at about 12° east of the subsolar point. The dust column density derived from the spectrum centered on the nucleus is found to be asymmetric in the sun–antisun direction.

103.046   **Spatial profiles of unidentified molecules observed in the high resolution spectra of comet Wilson.**
J. M. Hahn, T. W. Rettig, S. C. Tegler, S. Wyckoff.
Bull. Am. Astron. Soc., Vol. 23, No. 3, p. 1162 (1991). Abstract. – See Abstr. 012.037 for the main entry.

Comet 1988 V Liller

103.047   **The dust tail of Comet Liller 1988 V.**
M. Fulle, G. Cremonese, K. Jockers, H. Rauer.
Astron. Astrophys., Vol. 253, No. 2, p. 615 – 624 (Jan 1992).

The authors discuss the analysis of three CCD frames secured with the 1.06–m Hoher List telescope concerning the dust tail of Comet Liller 1988 V, which were used as input for the inverse numerical approach to the interpretation of comet dust tails (Fulle, 1989). The authors consider dust grains of diameters between 15 μm and 10 cm ejected during the time interval $-200 < t < +40$ (days related to perihelion) and obtain a lower limit of the dust mass production of $(4 \pm 1) \times 10^{13}$g (for an assumed albedo for the phase function $Ap(\alpha) = 0.06$, $\alpha = 50°$) and a power index of the time–averaged size distribution of -3.5 ± 0.2. The observing geometry leads to the detection of distinct subsets of the total dust distribution: large and slow pre–perihelion grains and smaller and faster post perihelion grains. The measured lower limit to the total dust mass loss rate reaches a broad maximum of $\approx 5 \times 10^6$g s^{-1} a week before perihelion, whereas the dust velocity reaches the highest values two weeks after. At least half of the total dust mass was injected into orbits bound in the Solar System, supplying a significant source of interplanetary meteoroids of sizes larger than 0.4 mm.

Comet 1988 XXIV Yanaka

103.048   **Comet Yanaka (1988r): a new class of carbon poor comet.**
U. Fink.
Bull. Am. Astron. Soc., Vol. 23, No. 3, p. 1160 (1991). Abstract. – See Abstr. 012.037 for the main entry.

Comet 1989 XIX Okazaki–Levy–Rudenko

103.049   **Observation of the OH radio lines in comet Okazaki–Levy–Rudenko 1989 XIX during the occultation of a point radio source.**
J. Crovisier, D. Bockelée–Morvan, G. Bourgois, E. Gérard.
Astron. Astrophys., Vol. 253, No. 1, p. 286 – 291 (Jan 1992).

While monitoring the OH radio lines in comet Okazaki–Levy–Rudenko 1989 XIX with the Nançay radio telescope, the authors observed the occultation of a point radio source (B2 1426 + 295) at about 1′ from the comet centre (corresponding to about 60000 km in the comet frame). The lines appear to be twice stronger than the average of the preceding and following days. It is suggested that this enhancement is not due to cometary variability, but to the increase of the continuum background, in agreement with the maser theory of the OH line excitation and with the current cometary models of OH density distribution. The variation of the signal intensity during the occultation, while the line–of–sight to the continuum source swept a path of about

50000 km in the comet frame, suggests that either the OH distribution or the excitation of the OH radicals is not homogeneous in the coma.

Comet 1990 V Austin

103.050   **Helium and argon abundance constraints and the thermal evolution of comet Austin (1989c$_1$).**
S. A. Stern, J. C. Green, W. Cash, T. A. Cook.
Icarus, Vol. 95, No. 1, p. 157 – 161 (Jan 1992).

The authors report results from the flight of a rocket–borne, far ultraviolet (FUV) spectrometer observation to establish upper limits on He and Ar abundances in the bright, dynamically "new", comet Austin (1989c$_1$). Previous to comet Austin, no comet had been spectroscopically observed to set helium and argon abundance constraints. Relative to solar abundance, the upper limits imply comet Austin is at least 1.5×10^4 depleted in He/O, and no more than 30 times enriched in Ar/O. These upper limits allow one to discuss the thermal conditions experienced by this comet during its formation and during its subsequent evolution. The upper limits achieved by this 258–sec suborbital rocket observation indicate the usefulness of future, longer-integrated or otherwise more sensitive observations of comets below 1200 Å, in the FUV.

103.051   **Near–ultraviolet spectroscopy of comet Austin (1989c1).**
J. H. Valk, C. R. O'Dell, A. L. Cochran,
W. D. Cochran, C. B. Opal, E. S. Barker.
Astrophys. J., Vol. 388, No. 2, p. 621 – 632 (1 Apr 1992).

Comet Austin (1989c1) was observed post–perihelion at a heliocentric distance near 1.25 AU. The wavelength range was from 3000 Å to 4000 Å. The coma spectra were calibrated into flux units and the contaminating sky spectrum and solar scattered light continuum were subtracted, leaving an UV spectrum of about 1.5 Å resolution and excellent S/N ratio. The spectrum is dominated by emissions from OH, NH, CH, C$_3$, and CN, some of the weaker emissions of which are seen here for the first time. More bands of CO$_2^+$ were found than in any previous study and several intensity anomalies were noted; H$_2$CO, OH$^+$, NCN, N$_2^+$, and CN$^+$ may be present. Several emission features well above the noise level remain unidentified. The relative intensities of the OH and CN bands agree with the predictions of resonance fluorescence, when one considers the potential effects of contamination by other molecules.

103.052   **Three–dimensional coma simulation of C/Austin.**
A. J. Ferro, S. Wyckoff, T. Boroson, P. Mack.
Bull. Am. Astron. Soc., Vol. 23, No. 3, p. 1161 (1991). Abstract. – See Abstr. 012.037 for the main entry.

103.053   **Comet Austin (1989c$_1$): analysis of narrowband photometry.**
D. J. Osip, D. G. Schleicher, P. V. Birch.
Bull. Am. Astron. Soc., Vol. 23, No. 3, p. 1163 (1991). Abstract. – See Abstr. 012.037 for the main entry.

103.054   **Fabry–Perot observations of [O I] 6300 emission from Comet Austin (1989c$_1$).**
D. Schultz, F. Scherb, F. L. Roesler, G. S. H. Li.
Bull. Am. Astron. Soc., Vol. 23, No. 3, p. 1163 (1991). Abstract. – See Abstr. 012.037 for the main entry.

103.055   **An infrared search for parent volatiles in comet Austin (1989c$_1$): detection of CO and upper limits to OCS.**
M. A. DiSanti, M. J. Mumma, J. Lacy, P. Parmar.
Bull. Am. Astron. Soc., Vol. 23, No. 3, p. 1166 (1991). Abstract. – See Abstr. 012.037 for the main entry.

103.056   **High resolution spectroscopy of the A–X and B–X systems of CH in comet Austin (1989c$_1$).**
S. J. Kim, M. F. A'Hearn, M. Brown, H. Spinrad.
Bull. Am. Astron. Soc., Vol. 23, No. 3, p. 1166 (1991). Abstract. – See Abstr. 012.037 for the main entry.

103.057   **Violet and red CN emission in comet Austin (1989c$_1$).**
H. Campins, S. C. Tegler, S. M. Larson, M. Kleine, M. J. Rieke, D. Kelly.
Bull. Am. Astron. Soc., Vol. 23, No. 3, p. 1167 (1991). Abstract. – See Abstr. 012.037 for the main entry.

103.058   **A possible detection of infrared emission from carbon monoxide in comet Austin (1989c1).**
M. A. DiSanti, M. J. Mumma, J. H. Lacy, P. Parmar.
Icarus, Vol. 96, No. 2, p. 151 – 160 (Apr 1992).

This paper reports the first search for and tentative detection of cometary CO emission using a modern two–dimensional infrared array detector, which was employed with a high dispersion echelle spectrometer to probe the $v = 1$–0 band in comet Austin.

103.059   **Comet Austin (1989c1) O(^{1}D) and H$_2$O production rates.**
D. Schultz, G. S. H. Li, F. Scherb, F. L. Roesler.
Icarus, Vol. 96, No. 2, p. 190 – 197 (Apr 1992).

Groundbased observations of comet Austin (1989c1) were carried out using the Wisconsin dual–etalon Fabry–Perot spectrometer at the McMath solar telescope on Kitt Peak. The field of view diameter on the sky was approximately 10.5. Spectral scans were acquired at a resolution of 0.21 Å, which was sufficient to resolve cometary [OI]6300 emission from nearby NH$_2$ and telluric [OI]6300 emissions. From these measurements, the authors obtained the O(^{1}D) production rate Q(O(^{1}D)), which is nearly model independent. By taking into account photodissociation of H$_2$O and OH as sources of O(^{1}D), they determine Q(H$_2$O) from the Q(O(^{1}D)) results.

103.060   **A sensitive upper limit to OCS in comet Austin (1989c1) from a search for v_3 emission at 4.85 μm.**
M. A. DiSanti, M. J. Mumma, J. H. Lacy.
Icarus, Vol. 97, No. 1, p. 155 – 158 (May 1992).

The authors observed the dynamically new Comet Austin (1989c1) UT 1990 May 18, at $\lambda/\Delta\lambda \sim 10^4$, using a cryogenic echelle grating spectrometer. The observations encompassed the P(1) through P(14) line frequencies of the v_3 band of carbonyl sulfide (OCS). Although OCS was not detected in the data, the authors derive a upper limit to the OCS abundance (relative to H$_2$O) of 5.5×10^{-3} for an assumed coma temperature of 50K. This value falls between previously reported upper limits of 2×10^{-3}, obtained from IRAM observations of comet Levy (1990c), and $\sim 8 \times 10^{-3}$ for Comet P/Halley, based on IKS measurements aboard Vega 1.

103.061   **Spectrophotometric study of comet Austin (1989c1).**
B. S. Rautela, B. B. Sanwal.
Earth, Moon, Planets, Vol. 57, No. 2, p. 115 – 121 (May 1992).

Spectrophotometric observations of the head of comet Austin (1990 V) during five nights in May 1990 are presented. Emission bands due to CN, C$_2$ and C$_3$ molecules have been identified. An estimate of their column densities and production rates have been made.

Comet 1990 VI Skorichenko–George

103.062   **The bands of Asundi $a'^3\Sigma^+ - a^3\pi r$ and triplet $d^3\Delta - a^3\pi$ in the spectrum of comet Skorichenko–George (1989e$_1$).**
K. I. Churyumov.
Visn. Kiiv. Univ., Fiz.–Mat. Nauki, Astron., Vip. 3, p. 46 – 49 (1992). In Ukrainian.

The comet Skorichenko–George (1989e$_1$) spectra obtained by V. L. Afanas'ev, A. I. Shapovalova and the author show to have emission bands of Asundi $a'^3\Sigma^+ - a^3\Pi r$ and triplet $d^3\Delta - a^3\Pi$ systems of neutral CO. This leads to suggest the comet nucleus includes the formaldehyde H$_2$CO or polyformaldehyde (H$_2$CO)$_5$ which give birth to HCO$^{\overline{+}}$–ions which if recombined and photodissociated result into CO–radicals at $a'^3\Sigma$ and $d^3\Delta$ levels necessary to excite the CO–emissions in the Asundi and triplet system bands.

Comet 1990 XX Levy

103.063 Imaging the 3.4–μm feature in comet Levy (1990c).
J. J. Klavetter, S. Hoban.
Icarus, Vol. 95, No. 1, p. 60 – 64 (Jan 1992).
The authors observed the organic emission feature which peaks near 3.36 μm in comet Levy (1990c) on 1990 Aug 8 and 10 with the ProtoCAM at the Infrared Telescope Facility. In this paper they present the first two–dimensional images of the spatial distribution of the 3.4–μm organic emission in a comet. It is found that the brightness profile of the emission is statistically indistinguishable from that of the dust continuum within 1225 km from the nucleus.

103.064 Ultraviolet and visible variability of the coma of comet Levy (1990c).
P. D. Feldman, S. A. Budzien, M. C. Festou, M. F. A'Hearn, G. P. Tozzi.
Icarus, Vol. 95, No. 1, p. 65 – 72 (Jan 1992).
Short–term variability of the coma of comet Levy (1990c) was detected and monitored with the IUE satellite observatory during Aug and Sep 1990 including 24 hr of continuous observation on 18 Sep. The visible lightcurve obtained on this date with the IUE Fine Error Sensor (FES) shows two distinct maxima separated by 17.0 ± 0.1 hr. However, this period cannot properly match in phase the FES data obtained during 8–hr shifts on 11 and 13 Sep, and suggests a decrease in apparent period of $\sim 1.3\%$ per day. A similar decrease is derived from a comparison with the period derived from ground–based data taken in late August. The variation in the ultraviolet emissions of OH, CS, and $CO_2{}^+$, as well as that of the ultraviolet continuum, was also determined from 18 consecutive long–wavelength IUE spectra taken on 18 Sep. From these data it is possible to model a production rate source function based on a nucleus whose activity exhibits a hemispherical asymmetry. Moreover, the ratio of gas to dust production rates is found to show a similar asymmetry.

103.065 A model to explain the activity of Comet Levy (1990c).
N. H. Samarasinha, M. F. A'Hearn, H. A. Weaver, P. D. Feldman, C. Arpigny, D. Hutsemekers.
Bull. Am. Astron. Soc., Vol. 23, No. 3, p. 1159 (1991). Abstract. – See Abstr. 012.037 for the main entry.

103.066 Submillimeter molecular line observations of comet Levy (1990c).
F. P. Schloerb, W. Ge.
Bull. Am. Astron. Soc., Vol. 23, No. 3, p. 1166 (1991). Abstract. – See Abstr. 012.037 for the main entry.

103.067 Ice–skater model for the nucleus of Comet Levy 1990c: spin–up by a shrinking nucleus.
J.–i. Watanabe.
Publ. Astron. Soc. Jpn., Vol. 44, No. 2, p. 163 – 166 (1992).
The activity period of Comet Levy 1990c diminished by about 10% during a period of about one half month. This suggests a gradual increase in the angular velocity of the rotating nucleus. The Ice–Skater model is proposed for this spin–up phenomenon as a result of a shrinking of the nucleus. Nucleus shrinking along with the formation of a dense mantle changes the moments of inertia of the nucleus, and produces a spin–up of the rotation. The observed mass–loss rate indicates an extremely low density ($\leqslant 0.14 \text{ g cm}^{-3}$), which suggests that primitive comets have a porous amorphous ice surface.

103.068 Inner coma imaging of comet Levy (1990c) with the Hubble Space Telescope.
H. A. Weaver, M. F. A'Hearn, P. D. Feldman, C. Arpigny, W. A. Baum, J. C. Brandt, R. M. Light, J. A. Westphal.
Icarus, Vol. 97, No. 1, p. 85 – 98 (May 1992).
Comet Levy (1990c) was observed with the Hubble Space Telescope (HST) on 27 Sep 1990 when both the heliocentric and geocentric distances were ~ 1 AU. A single Wide–Field Camera (WFC) pixel projected to a distance of ~ 78 km at the comet, and deconvolution techniques were used to recover virtually the full spatial resolution capability of the HST. The images show a highly asymmetrical coma in which the sunward–facing hemisphere is more than a factor of 2 brighter than the tailward hemisphere, consistent with volatile sublimation occurring primarily on the dayside of the nucleus. The azimuthal dependence of the spatial brightness distribution on the sunward side is roughly Gaussian and an axis of symmetry that is nearly coincident with the projected Sun–comet line. Radial brightness profiles perpendicular to the Sun–comet line are very symmetric about the nucleus and follow approximately a ϱ^{-1} law (where ϱ is the projected distance to the nucleus). The coma of comet Levy is definetely not in steady state.

Comet 1991a₁ Shoemaker–Levy

103.069 Comet Shoemaker–Levy (1991a₁).
IAU Circ., No. 5501, p. 1 (20 Apr 1992).
Further information is given in Nos. 5519, 5529.

103.070 Comet Shoemaker–Levy (1991a₁).
Yamamoto Circ., No. 2176, p. 2 (4 Jan 1992). In Japanese.
Further information is given in No. 2185.

103.071 La comète Shoemaker–Levy 1991a₁.
S. Thebault.
Astronomie, p. 8 – 9 (May 1992).

Comet 1991d Shoemaker–Levy

103.072 Comet Shoemaker–Levy (1991d).
IAU Circ., No. 5444, p. 1 (4 Feb 1992).
Further information is given in Nos. 5507, 5540.

103.073 Comet Shoemaker–Levy (1991d).
Yamamoto Circ., No. 2176, p. 2 (4 Jan 1992). In Japanese.
Further information is given in Nos. 2184, 2186.

Comet 1991g₁ Zanotta–Brewington

103.074 Comet Zanotta–Brewington (1991g₁).
IAU Circ., No. 5427, p. 1 (9 Jan 1992).
Further information is given in Nos. 5432, 5439, 5452, 5465, 5514.

103.075 Comet Zanotta–Brewington (1991g₁).
Yamamoto Circ., No. 2176, p. 1 (4 Jan 1992). In Japanese.
Further information is given in No. 2177.

Comet 1991h₁ Mueller

103.076 Comet Mueller (1991h₁).
IAU Circ., No. 5420, p. 1 (1 Jan 1992).
Further information is given in Nos. 5421, 5438, 5464, 5482, 5496.

103.077 Comet Mueller (1991h₁).
Yamamoto Circ., No. 2176, p. 1 – 2 (4 Jan 1992). In Japanese.
Further information is given in Nos. 2178, 2180, 2181, 2183.

Comet 1991l Helin–Lawrence

103.078 Comet Helin–Lawrence (1991l).
IAU Circ., No. 5455, p. 1 (21 Feb 1992).
Further information is given in No. 5521.

Comet 1991r Helin–Alu

103.079   **Comet Helin–Alu (1991r).**
IAU Circ., No. 5433, p. 1 (22 Jan 1992).
Further information is given in No. 5500.

Comet 1992a Helin–Alu

103.080   **Comet Helin–Alu (1992a).**
IAU Circ., No. 5432, p. 1 (21 Jan 1992).
Further information is given in Nos. 5439, 5486.

103.081   **Comet Helin–Alu (1992a).**
Yamamoto Circ., No. 2177, p. 1   (26 Jan 1992). In
Japanese.
Further information is given in No. 2178.

Comet 1992b Bradfield

103.082   **Comet Bradfield (1992b).**
IAU Circ., No. 5442, p. 1 (2 Feb 1992).
Further information is given in Nos. 5444, 5445, 5469.

103.083   **Comet Bradfield (1992b).**
Yamamoto Circ., No. 2178, p. 1   (16 Feb 1992). In
Japanese.
Further information is given in No. 2181.

Comet 1992d Tanaka–Machholz

103.084   **Comet Tanaka–Machholz (1992d).**
IAU Circ., No. 5487, p. 1 (1 Apr 1992).
Further information is given in Nos. 5488, 5489, 5491, 5506,
5507, 5518, 5524, 5528, 5531, 5544.

103.085   **Comet Tanaka–Machholz (1992d).**
Yamamoto Circ., No. 2182, p. 1 – 2 (4 Apr 1992). In
Japanese.
Further information is given in Nos. 2184 – 2186.

Comet 1992h Spacewatch

103.086   **Comet Spacewatch (1992h).**
IAU Circ., No. 5509, p. 1 (2 May 1992).
Further information is given in No. 5513.

103.087   **Comet Spacewatch (1992h).**
Yamamoto Circ., No. 2185, p. 1   (23 May 1992). In
Japanese.

Comet 1992i Bradfield

103.088   **Comet Bradfield (1992i).**
IAU Circ., No. 5514, p. 1 (4 May 1992).
Further information is given in Nos. 5516, 5524, 5530.

103.089   **Comet Bradfield (1992i).**
Yamamoto Circ., No. 2185, p. 1   (23 May 1992). In
Japanese.

Periodic comet Ashbrook–Jackson

103.090   **Periodic comet Ashbrook–Jackson (1992j).**
IAU Circ., No. 5546, p. 1 (16 Jun 1992).

Periodic comet Bradfield

103.091   **On the anti–tail of comet Bradfield (1987 XXIX).**
H. Akisawa, T. Oka, K. Sugawara.
IAU Colloquium No. 126: Origin and evolution of interplanetary
dust, p. 269 – 272 (1991). – See Abstr. 012.068 for the main entry.
The anti–tail of comet Bradfield 1987 XXIX was observed
when the Earth passed through the orbital plane of this comet on
December 20, 1987. The time variation of this phenomenon was
monitored continuously by Japanese amateur astronomers. The
authors analyzed these photographs by using the Bessel–
Bredkhin theory.

103.092   **Formation mechanisms of the split tail of comet Bradfield 1987 XXIX.**
K. Sugawara, J.–i. Watanabe.
IAU Colloquium No. 126: Origin and evolution of interplanetary
dust, p. 273 – 276 (1991). – See Abstr. 012.068 for the main entry.
Comet Bradfield 1987 XXIX showed a well–developed split
tail, which is a dust tail feature divided into two branches by a
dark gap. Similar structures have been observed in several
comets. The authors suggested in previous works that the trend
of variation of position angle of the dark gap coincides well with
that of the plasma tail or projected radial vector. They discuss the
formation mechanisms of the split tail from the view point of
dust–plasma interactions.

Periodic comet Brorsen–Metcalf

103.093   **NH_3 and NH_2 in the coma of comet Brorsen–Metcalf.**
S. C. Tegler, L. F. Burke, S. Wyckoff, M. Womack,
U. Fink, M. DiSanti.
Astrophys. J., Vol. 384, No. 1, p. 292 – 297 (1 Jan 1992).
Narrow–band CCD images of comet Brorsen–Metcalf have
been obtained using interference filters (FWHM = 20 Å) centered at 6250 Å and 6338 Å to isolate continuum and NH_2 (8–0)
$\tilde{A}^2A_1 – \tilde{X}^2B_1$ emission, respectively. The 6338 Å images corrected for background sky and dust–scattered solar continuum
isolate the NH_2 coma in the comet. The distribution of NH_2 is
symmetric and shows no evidence for jet structure at the 3σ level
above background emission. An azimuthal average of the NH_2
image produces an NH_2 surface brightness profile for comet
Brorsen–Metcalf, which provides a significant constraint on the
NH_2 photodissociation time scale in comets. A Monte Carlo
simulation of the comet coma assuming that NH_3 is the primary
source of NH_2 is described and compared with the observations.
The effects on the surface brightness distribution of NH_2 due to
(1) collisions in the inner coma and (2) non–steady state production rates were investigated with the Monte Carlo model. The
authors find for steady state conditions and a recently revised
NH_2 photodissociation time scale, $\tau \sim 3.3 \times 10^4$s at 1 AU, a
satisfactory match between the observed and computed NH_2
surface brightness profiles. These results support the assumption
that NH_3 is the dominant source of NH_2 in the coma of comet
Brorsen–Metcalf.

103.094   **8 – 13 μm spectroscopy and IR photometry of comet
P/Brorsen–Metcalf (1989o) near perihelion.**
D. K. Lynch, M. S. Hanner, R. W. Russell.
Icarus, Vol. 97, No. 2, p. 269 – 275 (Jun 1992).
CVF spectroscopy from 8 to 13 μm and 3.5 to 20 μm
photometry of comet P/Brorsen–Metcalf were obtained between
Aug 28 and Sep 6, 1989, using the NASA Infrared Telescope
Fecility (IRTF). No silicate emission was observed during late
Aug and early Sep 1989 when the comet was 0.5 – 0.6 AU from
the Sun. The spectra were consistent with grey body emission at
around 400 – 430K, 6 – 12% above the radiative equilibrium
blackbody temperature. The absence of silicate emission along
with only a small color temperature excess suggests that the
grains were larger than those typically found in new comets or
P/Halley. An upper limit of 5–km radius for the size of the
nucleus is obtained from the infrared flux on Aug 13.

103.095 g–factors of the SH (0–0) band and SH upper limit in comet P/Brorsen–Metcalf (1989o).
S. J. Kim, M. F. A'Hearn.
Icarus, Vol. 97, No. 2, p. 303 – 306 (Jun 1992).
Since H_2S was detected in comets Austin (1989c1) and Levy (1990c) in the microwave range, there has been increasing interest in searching for SH, which is the prime dissociative product of H_2S. The authors present g–factors for the A–X (0–0) band of SH as a function of heliocentric velocity at r = 1.0 AU. They derive an upper limit production rate for comet Brorsen–Metcalf (1989o) and calculate a dissociative lifetime of 105 sec.

Periodic comet Churyumov–Gerasimenko

103.096 Features of the light curve of the short–period comet Churyumov–Gerasimenko (1982 VIII or 1982f).
K. I. Churyumov, V. S. Filonenko.
Sol. Syst. Res., Vol. 25, No. 1, p. 84 – 89 (Jul 1991). English translation of Astron. Vestn., Tom 25, No. 1, p. 109 – 115 (1991).
The light curve of the periodic comet Churyumov–Gerasimenko has been constructed and investigated for its third observed apparition in 1982. The comet's brightness reached a maximum 40 to 45 days after perihelion. Photometric parameters, absolute stellar magnitude, and the parameters of brightness flares have been determined. A significant relation of the comet's brightness to its phase angle has been recognized. The phase coefficient $\beta = 0.031 \pm 0.004$ stellar mag/deg has been determined for the comet. Breaks in the light curve that are situated symmetrically with respect to perihelion have been detected. It is shown that their appearance is connected with the effect of the phase dependence of the comet's brightness on its light curve.

Periodic comet Encke

103.097 P/Encke – der schnellste unter den Kometen.
W. Pfau.
Sterne, Band 68, Heft 3, p. 179 – 187 (1992).

Periodic comet Faye

103.098 Periodic comet Faye (1991n).
IAU Circ., No. 5440, p. 1 (30 Jan 1992).

Periodic comet Giacobini–Zinner

103.099 Spectral observations of periodic comet Giacobini–Zinner.
V. F. Esipov, G. A. Lukina, O. Mamadov.
Astron. Vestn., Tom 26, No. 1, p. 109 – 111 (Jan – Feb 1992). In Russian. English translation in Solar Syst. Res., Vol. 26, No. 1.
A set of spectrograms of comet Giacobini–Zinner was obtained at the Crimean Station of the Sternberg State Astronomical Institute. Swan bands ($C_2(\Delta v = +2, +1, 0, -1)$), as well as CN (0,1) and CH (0,0) are identified according to these spectrograms. There is a lot of bands CO, H_2O^+ and H_2. Photometric measurements showed that the spectrophotometric gradient is equal to 5.1 and corresponds to the spectral class gK3 and colour temperature 3674K.

103.100 Some physical characteristics of periodic comet Giacobini–Zinner 1979 III and 1985 XIII.
M. Z. Markovich.
Astron. Vestn., Tom 26, No. 3, p. 47 – 52 (May – Jun 1992). In Russian. English translation in Sol. Syst. Res., Vol. 26, No. 3.
Photometeric parameters of the comet 1979 III have been calculated with the help of a mathematical model, determining the dependence of absolute brightness of the comet on the conditions of visibility and the level of solar activity:

$H_\gamma = 10^m1 \pm 3.6$; y = 11.2 ± 12.2 ($n = 4,5$). An optimal law of brightness change and corresponding photometric parameters have been obtained also for the reapparition of this comet 1985 XIII: $H_\gamma = 8^m8 \pm 0.2$; y = 10.2 ± 1.1 ($n = 4.08$). A prognosis of comet brightness in its forthcoming appearance in 1992: $H_0 = 9^m5$ has been given. Relative dust content in the cometary nucleus is $f_v = 0.3$.

103.101 IUE observations of H Lyman–α in comet P/Giacobini–Zinner.
M. R. Combi, P. D. Feldman.
Icarus, Vol. 97, No. 2, p. 260 – 268 (Jun 1992).
Comet P/Giacobini–Zinner (1985 XIII) was observed during the flyby of the ICE satellite and during the 3 preceding months by the IUE satellite. Models for the hydrogen abundance and spatial distribution, which were previously used to reproduce the observed spatial distribution of the wide–field Lyman–α comae of comets Kohoutek and Halley measured by rocket and spacecraft borne instruments, have now been successfully applied to these IUE observations. Vectorial models, which implicitly assume production of cometary OH by dissociation of water molecules, have been routinely used to infer global water production rates from various nucleus–centered IUE observations of comets. A spherical radiative transfer model, adapted for application to the H coma, has been used to analyze the IUE observations. This analysis yields water production rates in very good agreement with those calculated from vectorial model analysis of OH observations. The H model makes essentially the same physical assumptions as the vectorial model for OH except for optical density.

Periodic comet Grigg–Skjellerup

103.102 On the rotating nucleus of Comet P/Grigg–Skjellerup.
G. Sitarski.
Acta Astron., Vol. 42, No. 1, p. 59 – 65 (1992).
The author applied Sekanina's model for the forced precession of the spin–axis of a rotating cometary nucleus to investigate the nongravitational motion of Comet P/Grigg–Skjellerup during 1922–1991. Formulae for the forced precession were adapted to use them in orbital computations. The author used 437 astrometric observations of the comet to correct the orbit and determine six parameters connected with a nongravitational force and with the rotating and precessing nucleus of the comet. The author found that the nucleus of Comet P/Grigg–Skjellerup should be a prolate spheroid rotating around its longer axis, and $R_a:R_b \approx 10:14$ where R_a and R_b are the equatorial and polar radii of the nucleus.

103.103 Periodic comet Grigg–Skjellerup.
IAU Circ., No. 5478, p. 1 (18 Mar 1992).

103.104 Periodic comet Grigg–Skjellerup.
Yamamoto Circ., No. 2181, p. 2 (30 Mar 1992). In Japanese.

103.105 Observations of the GIOTTO target comet P/Grigg–Skjellerup at the Calar Alto Observatory.
K. Birkle, H. Boehnhardt.
Earth, Moon, Planets, Vol. 57, No. 3, p. 191 – 201 (Jun 1992).
CCD images of comet P/Grigg–Skjellerup, obtained for astrometric purposes with the 3.5 m telescope at the Calar Alto Observatory, were used for an analysis of the activity status of the nucleus and for a search of faint coma structures. The nucleus was found essentially inactive beyond 2.7 AU solar distance both inbound and outbound (observations on 12–13 August 1986, 21–23 October 1986, 22 August 1988, 18 October 1988, 9 and 12 September 1991 and 3 December 1991). The coma of the comet was well developed in May and July 1987 with a diameter of at least 190000 km on 24 May 1987 and of at least 80000 km on 24 July 1987. The coma showed a cone of diffuse brightness enhancement in the sunward hemisphere. The orientation of the cone axis changed from the Sun direction in May 1987 towards

about North in July 1987, i.e., it was almost perpendicular to the projected Sun–nucleus line on the sky. The cone opening angle became smaller from about 100° in May to about 50° in July 1987. A weak and narrow plasma tail was found in the images of May 1987.

Periodic comet Halley

103.106   Accelerated cometary ions observed downstream of the comet Halley bow shock.
K. Kecskeméty, T. E. Cravens.
J. Geophys. Res., Vol. 97, No. A3, p. 2891 – 2906 (1 Mar 1992).

Fluxes of energetic ions with energies exceeding 100 keV were observed upstream of the bow shock of comet Halley by the Tunde instrument which was on board the VEGA 1 spacecraft. Downstream of the shock, ion fluxes in the energy range 100 to 800 keV were observed. Cometary ions, such as O^+, newly picked up by the solar wind have energies of about 15 keV in the solar wind frame of reference; hence the measured ion fluxes indicate that acceleration processes must have been operating near comet Halley. The measured ion fluxes were transformed into distribution functions in the solar wind frame using a variety of assumptions concerning the energy dependence of the distribution function and the identity of the ion species. The distribution functions are compared with those obtained by similar instruments on Giotto and the International Cometary Explorer as well as with the predictions of several theoretical models that employ different acceleration mechanisms.

103.107   Electron distributions upstream of the comet Halley bow shock: evidence for adiabatic heating.
D. E. Larson, K. A. Anderson, R. P. Lin, C. W. Carlson, H. Rème, K.–H. Glassmeier, F. M. Neubauer.
J. Geophys. Res., Vol. 97, No. A3, p. 2907 – 2916 (1 Mar 1992).

The authors report on three–dimensional plasma electron (22 eV to 30 keV) observations upstream of the comet Halley bow shock, obtained by the RPA–1 COPERNIC (Rème Plasma Analyzer–Complete Positive Ion, Electron and Ram Negative Ion Measurements near Comet Halley) experiment on the Giotto spacecraft. Besides electron distributions typical of the undisturbed solar wind and backstreaming electrons observed when the magnetic field line intersects the cometary bow shock, the authors find a new type of distribution characterized by enhanced low energy (< 100 eV) flux which peaks at 90° pitch angles. These are most prominent when the spacecraft is on field lines which pass close to but are not connected to the bow shock. The 90° pitch angle electrons appear to have been adiabatically heated by the increase in the magnetic field strength resulting from the compression of the upstream solar wind plasma by the cometary mass loading. The authors present a model calculation of this effect which agrees qualitatively with the observed 90° flux enhancements.

103.108   Halley's comet still observable.
J. Bouška.
Říše hvězd, Vol. 72, No. 9, p. 169 – 170 (Sep 1991). In Czech.

103.109   The effect of electron collisions on rotational populations of cometary water.
Xie Xingfa, M. J. Mumma.
Astrophys. J., Vol. 386, No. 2, p. 720 – 728 (20 Feb 1992).

The e – H_2O collisional rate for exciting rotational transitions in cometary water is evaluated for conditions found in comet Halley during the Giotto spacecraft encounter. In the case of the $O_{00} \rightarrow 1_{11}$ rotational transition, the e – H_2O collisional rate exceeds that for excitation by neutral – neutral collisions at distances exceeding 3000 km from the cometary nucleus. The estimates are based on theoretical and experimental studies of e – H_2O collisions, on ion and electron parameters acquired in situ by instruments on the Giotto and Vega spacecraft, and on results obtained from models of the cometary ionosphere. Thus, the rotational temperature of the water molecule in the intermediate coma may be controlled by collisions with electrons rather than with neutral molecules, and the rotational temperature retrieved from high–resolution infrared spectra of water in comet Halley may reflect electron temperatures, rather than neutral gas temperatures in the intermediate coma. The contribution of electron collisions may explain the need for large H_2O – H_2O cross sections in models which neglect the effect of electrons. The importance of electron collisions is enhanced for populations of water molecules in regions where their rotational lines are optically thick.

103.110   Observations of plasma dynamics in the coma of P/Halley by the Giotto ion mass spectrometer.
B. E. Goldstein, R. Goldstein, M. Neugebauer, S. A. Fuselier, E. G. Shelley, H. Balsiger, G. Kettmann, W.–H. Ip, H. Rosenbauer, R. Schwenn.
J. Geophys. Res., Vol. 97, No. A4, p. 4121 – 4132 (1 Apr 1992).

Observations in the coma of P/Halley by the Giotto ion mass spectrometer (IMS) are reported. The high–energy range spectrometer (HERS) of the IMS obtained measurements of protons and alpha particles from the far upstream region to the near ionopause region and of ions of mass 12–32 at distances of about 250,000 to 40,000 km from the nucleus. Plasma parameters from the high–intensity spectrometer (HIS) of the IMS obtained between 150,000 and 5000 km from the nucleus are also discussed.

103.111   Stability of the cometary ionopause.
Chen Daohan, Liu Linzhong.
Astrophys. Space Sci., Vol. 189, No. 1, p. 45 – 55 (Mar 1992).

Taking into account the effect of ion–neutral drag, the stability of sunlit cometary ionopause is rediscussed. The authors relate their analysis to the spacecraft findings of P/Halley and then assume that the density of the cometary plasma varies as r^{-1}. For simplicity, they also assume that the plasma is approximately incompressible and the plasma velocity has only a finite, albeit constant, radial component. On these assumptions, they derive the dispersion equation which allows an analytic solution.

103.112   A note on the very small grains (VSGs) observed at Halley's comet.
M. N. Fomenkova, D. A. Mendis.
Astrophys. Space Sci., Vol. 189, No. 2, p. 327 – 331 (Mar 1992). Letter–to–the–editor.

The most striking feature in the spatial distribution of the smallest dust grains observed at Halley's comet by the VEGA–1 spacecraft is the sharp glitch at a cometocentric distance of about 180000 km, which approximately corresponds to the so–called cometopause inside which the contaminated solar wind plasma was rapidly cooled. The authors propose that this glitch was caused by the electrostatic disruption of larger composite grains which rapidly charged up as they traversed the cometopause. The clear asymmetry in the distribution between the inbound and outbound portion of the spacecraft trajectory is also consistent with the dynamical effects of grain charging although other causes are not excluded.

103.113   CN jet velocity in comet P/Halley.
J. J. Klavetter, M. F. A'Hearn.
Icarus, Vol. 95, No. 1, p. 73 – 85 (Jan 1992).

CN jet images provide a direct means of measuring the projected expansion velocity in the coma of comet P/Halley. After processing of the CN images with a radial profile subtraction technique, the authors precisely determined the positions of the jets by Gaussian fitting in r–θ space. By linear fitting to this portion, the distance from the nucleus was measured for all jets that overlapped in azimuth in a time series, both within a night and from night to night. Following specific features demonstrated the flow was approximately radial. The authors also found the velocity by measuring specific radial features in jets. The velocity found by both methods was consistent. Due to projection effects, only a lower limit to the projected velocity could be found. From this large dataset (34 time series of jets), a wide range of projected velocities is measured. Weak evidence of an acceleration is found in the region of the coma ranging from r = 20,000 to 80,000 km. The findings are compared with other observations and theoretical models.

103.114 Erratum: "Plasma flow inside comet P/Halley" [Astron. Astrophys., Vol. 238, No. 1/2, p. 401 – 412 (Nov 1990)].
V. Formisano, E. Amata, M. B. Cattaneo, P. Torrente,
A. Johnstone, A. Coates, B. Wilken, K. Jockers, H. Borg.
Astron. Astrophys., Vol. 256, No. 2, p. 723 (Mar 1992). See Abstr. 52.103.039.

103.115 A possible mechanism for outbursts of comet P/Halley large heliocentric distances.
B. Schmitt, S. Espinasse, J. Klinger.
54. Annual Meeting of the Meteoritical Society, p. 208 (1991). Abstract. – See Abstr. 012.010 for the main entry.

103.116 Lower and upper limits to comet P/Halley solid material loss rate.
J. F. Crifo.
Adv. Space Res., Vol. 11, No. 12, p. 155 – 160 (1991). – See Abstr. 012.014 for the main entry.
A radiative hydrodynamic model of comet P/Halley is used to investigate quantitatively the contraints imposed onto the comet loss rate in solids by both the in–situ data and the remote sensing data. In particular, for the first time, the in–situ data are used to fit composite spectra extending from near–IR wavelengths to the microwave region. Also, for the first time the uncertainty affecting the ejection velocity of large grains is taken into account. The results suggest than only future rendez–vous studies of the very large grain density and velocity in the vicinity of a comet would be able to provide definitely reliable values of the comet loss rate.

103.117 Crystallization of amorphous ice as the cause of comet P/Halley's outburst at 14 AU.
D. Prialnik, A. Bar-Nun.
Astron. Astrophys., Vol. 258, No. 2, p. L9 – L12 (May 1992). Letter–to–the–editor.
The post–perihelion eruption of comet P/Halley, detected in Feb. 1991 and believed to have started 3 months earlier, can be explained by crystallization of amorphous ice taking place in the interior of the porous nucleus, at depths of a few tens of meters, accompanied by the release of trapped gases. Numerical calculations show that for a bulk density of 0.5 g cm^{-3} and a pore size of 1 mμ crystallization occurs on the outbound leg of comet P/Halley's orbit, at heliocentric distances between 5 and 17 AU. The trapped gas is released and flows to the surface through the porous medium. It may also open wider channels, as the internal pressures obtained surpass the tensile strength of cometary ice. The outflowing gas carries with it grains of ice and dust, and thus can explain the large amounts of dust observed in the coma at 14.3 AU and beyond. The typical decline time of the process is found to be on the order of months, in agreement with observations. The rate of outgassing is two to three orders of magnitude higher than in quiescence. In an asymmetric, non-uniform nucleus – in contrast to the one–dimensional spherical model – the process should occur intermittently, such as was observed for comet P/Halley beyond 5 AU.

103.118 A different view of plasma flow inside P/Halley.
M. Neugebauer, B. E. Goldstein, R. Goldstein,
S. A. Fuselier, F. M. Neubauer, H. Balsiger, W.–H. Ip.
Astron. Astrophys., Vol. 258, No. 2, p. 549 – 554 (May 1992).
The Giotto spacecraft carried two different instruments – the JPA and the IMS – for the observation of hot ions in the coma of P/Halley. Although there are many similarities in the time and distance profiles of the plasma flow parameters (bulk velocity, number density, and temperature) computed from the two data sets, there are also some significant differences, especially at cometocentric distances $<5 \times 10^5$km. The principal discrepancies between the JPA results presented by Formisano et al. (1990) and the IMS observations are: (1) the IMS did not detect the levelling off of the speed and temperature profiles that Formisano et al. interpreted as flow stabilization; (2) the IMS

detected differential north–south flow between the solar wind and cometary ions for only a brief interval when the magnetic field was oriented nearly southward, whereas Formisano et al. reported more extensive differential north–south flow that was independent of the direction of the field; (3) the JPA ion densities were factors of 2 to 4 higher than the IMS ion densities which, in turn, were an order of magnitude greater than theoretical values. Although some additional source of ionization is required in the theoretical models, the authors cannot support Formisano et al.'s contention that the critical ionization phenomenon was an important source of ionization in the coma of P/Halley.

103.119 The hydrogen coma of comet P/Halley observed in Lyman α using sounding rockets.
R. P. McCoy, R. R. Meier, H. U. Keller, C. B. Opal,
G. R. Carruthers.
Astron. Astrophys., Vol. 258, No. 2, p. 555 – 565 (May 1992).
Hydrogen Lyman α (121.6 nm) images of comet P/Halley were obtained using sounding rockets launched from White Sands Missile Range on 24.5 February and 13.5 March 1986. The second rocket was launched 13 h before the fly–by of the Giotto spacecraft. An electrographic camera on both flights provided Lyman α images covering a 20° field of view with 3′ resolution. The data from both flights have been compared with a time-dependent model of hydrogen kinetics. To match the measured isophote contours, hydrogen sources with velocity components of 8 km s^{-1} and 20 km s^{-1} (from OH and H$_2$O, respectively) as well as a low velocity component (~ 2 km s^{-1}), are required. This low velocity component is thought to result from thermalization of fast hydrogen atoms within the collision zone, providing an important diagnostic of temperature and density near the nucleus. Hydrogen production rates of 3.8×10^{30}s^{-1} and 1.7×10^{30}s^{-1} have been obtained for the two observations.

103.120 Outburst of comet P/Halley.
K. J. Meech.
Bull. Am. Astron. Soc., Vol. 23, No. 3, p. 1159 (1991). Abstract. – See Abstr. 012.037 for the main entry.

103.121 Dust distribution and particle fragmentation near the nucleus of comet Halley.
H. U. Keller.
Bull. Am. Astron. Soc., Vol. 23, No. 3, p. 1160 (1991). Abstract. – See Abstr. 012.037 for the main entry.

103.122 Organic vs. mineral components of dust from comet Halley.
M. E. Lawler.
Bull. Am. Astron. Soc., Vol. 23, No. 3, p. 1160 (1991). Abstract. – See Abstr. 012.037 for the main entry.

103.123 Visualization of Halley's spin state.
M. J. S. Belton, S. M. Vail, T. Thompson.
Bull. Am. Astron. Soc., Vol. 23, No. 3, p. 1160 (1991). Abstract. – See Abstr. 012.037 for the main entry.

103.124 Methanol abundance in comet P/Halley from in–situ measurements.
P. Eberhardt, R. Meier, D. Krankowsky, R. R. Hodges.
Bull. Am. Astron. Soc., Vol. 23, No. 3, p. 1161 (1991). Abstract. – See Abstr. 012.037 for the main entry.

103.125 An extended source for CN near the nucleus of comet P/Halley.
J. J. Klavetter, M. F. A'Hearn.
Bull. Am. Astron. Soc., Vol. 23, No. 3, p. 1167 (1991). Abstract. – See Abstr. 012.037 for the main entry.

103.126 The spatial distribution of the hydrogen sulfide and formaldehyde sources in comet P/Halley.
R. Meier, P. Eberhardt, D. Krankowsky, R. R. Hodges.
Bull. Am. Astron. Soc., Vol. 23, No. 3, p. 1167 – 1168 (1991). Abstract. – See Abstr. 012.037 for the main entry.

103.127   The production of H₂O by Halley close to perihelion.
M. L. Marconi, W. H. Smyth.
Bull. Am. Astron. Soc., Vol. 23, No. 3, p. 1168 (1991). Abstract. –
See Abstr. 012.037 for the main entry.

**103.128   Quantitative Strukturanalyse der Zyan–Koma des Ko-
meten Halley. (Quantitative structural analysis of comet
Halley's cyanide coma.)**
R. Schulz.
Diss. (Dr.rer.nat.), Bochum Univ. (Germany). Fakultät für
Physik und Astronomie. 113 p. (1 Feb 1991).

103.129   Triaxiality of Halley's comet.
M. Burša, Z. Kopal, V. Vanýsek.
Earth, Moon, Planets, Vol. 57, No. 1, p. 65 – 73 (Apr 1992).
There are many aspects of observational evidence that
cometary nuclei have irregular or nonspherical shape. The
triaxial figure of the Halley's comet nucleus is a well known fact.
Therefore, the nucleus shape plays a significant role in consideration of the formation and evolution of comets and several
attempts have been made to explain their nonsphericity. These
studies were mainly based on the random–walk schemes for the
aggregation processes. Although some results indeed lead to
irregularities and deviation from sphericity, the spherical or
irregular shape seem to be prevailing results. On the other hand
the triaxial figure can be formed by the tidal and rotational
forces. Thus, the assumption that the shape of the cometary
nucleus due to some of these effects is in principle acceptable. In
here assumed scenario an already evolved cometary nucleus is
situated as a satellite in the gravitation field of a planetary–like
body. Since the rigidity of the nucleus is low, it may be easily
transferred in the state of a synchronous satellite and in its shape
could be imprinted the dynamical effects from this epoch. Here
presented results indicate that such a possibility should be
seriously considered. The theory of this process is applied to the
nucleus of comet Halley. It is shown, that the nucleus might be
synchronously orbiting around a planetary–like hypothetical
body with a period of 0.7 d. The minimal bulk tensile strength of
the cometary material of about $10^2 N\,m^{-2}$ is estimated.

103.130   Periodic comet Halley (1986 III).
IAU Circ., No. 5535, p. 1 (5 Jun 1992).

**103.131   Mass of comet Halley dust particles from results of the
PUMA experiment.**
E. I. Evlanov, O. F. Prilutskij, M. N. Fomenkova.
Cosmic Res., Vol. 29, No. 4, p. 551 – 555 (Jan 1992). English
translation of Kosm. Issled., Tom 29, Vyp. 4, p. 641 – 646 (1991).
The authors estimate the mass of dust particles whose spectra
were measured by the PUMA-1 and –2 instruments. The total set
of spectra of each instrument was divided into groups according
to the analog signals measured at the moment of dust impact on
the target. Absolute mass for each group was determined by
comparing the distribution obtained with data from the
SP–2 dust counter. It was determined that particles recorded by
the PUMA-1 instrument have mass in the range $5 \cdot 10^{17}$ to
$5 \cdot 10^{-12}$g, while particles recorded by the PUMA–2 instrument
have mass in the range $2 \cdot 10^{-16}$ to $5 \cdot 10^{-12}$g. The dependence of
dust properties on mass has also been examined.

**103.132   The connection between CN jets and CN shells in the
coma of comet P/Halley.**
R. Schulz.
Icarus, Vol. 96, No. 2, p. 198 – 203 (Apr 1992).
The evolution of CN jets and CN shells, discovered in the coma
of comet P/Halley in 1986, is morphologically investigated. A
detailed structural analysis of digitized and intensity–calibrated
photographs of comet P/Halley CN coma yields strong evidence
for a connection between both structures. The CN jets and CN
shells are simultaneously existent in each image and there are
strong hints for a transition between the two structures. This
leads to constraints on the formation mechanism of CN jets. The
results indicate that the CN radicals in the jets and shells

originate in secondary sources and there is further evidence of
these sources being organic "CHON" particles.

**103.133   A comparison of modeled and observed intensity profiles
for C₂, C₃, CN, and the continuum for P/Halley.**
T. A. Ellis, J. S. Neff.
Icarus, Vol. 97, No. 1, p. 99 – 110 (May 1992).
Intensity profiles were obtained for C_2, C_3, and CN emission
and the continuum of P/Halley. The observations were made on
14 Dec 1985 and 6 and 8 Jan 1986, when the comet was between
0.90 and 1.28 AU from the Sun. Model intensity profiles were
compared with the observations, and used to constrain the dust
and gas parameters. Most of these parameters were consistent
with expected values; however, the lifetimes of C_3 and its parent
were much smaller than expected on each night, and the dust
ejection velocity on 14 Dec was only 60% of the expected value.
The day:night production rate ratio was found to be approximately 1:1 for the gas, and ranged from 4:3 – 1:0 for the dust.

**103.134   Relative spectrophotometry of Halley's comet in the
near IR range.**
V. F. Esipov, P. P. Korsun, O. Mamadov, V. G. Parusimov.
Sol. Syst. Res., Vol. 25, No. 5, p. 449 – 451 (Mar 1992). English
translation of Astron. Vestn., Tom 25, No 5, p. 599 – 602 (1991).
Five examples were selected from a series of spectrograms of
Halley's comet in the near IR in a wavelength interval of
6000 – 9000 Å. Within those examples the following bands were
identified: CO(3–0), (4–1), (10–2), (11–3); H_2O^+ H_2O^+(0,7,0),
(0,6,0); NH_3(0,6,0), (0,5,0); the red system CN(2–0), (3–1), 4–2),
(5–3) and the Phillips system band C_2(2–0). An asymmetric
distribution explicable as the effect of jets in the near–core region
is found in CN cross sections.

**103.135   Spectroscopic evidence of organic molecules released by
the dust of Halley's inner coma.**
J. Clairemidi, P. Rousselot, G. Moreels.
IAU Colloquium No. 126: Origin and evolution of interplanetary
dust, p. 217 – 220 (1991). – See Abstr. 012.068 for the main entry.
New spectroscopic arguments supporting the probable presence of organic molecules in the material released by comet
Halley are deduced from the data obtained by the Vega three–
channel spectrometer. Two excesses of emission on the UV side of
the OH and CN bands at 305 and 383 nm are interpreted as being
due to "prompt" radiation emitted by electronically excited OH
and CN radicals directly produced by the photolysis of water
vapor and an organic X–CN molecule. A broad–band emission
progressively appears between 342 and 375 nm when the solar
scattered continuum has been substracted. This emission increases approximately as the inverse of the projected distance to
the nucleus. It is interpreted as a fluorescence emission of organic
molecules, possibly condensed polycyclic hydrocarbons. Present
observations support the hypothesis of grains coated with
organic material and give arguments in favor of a probable
interstellar origin for cometary dust.

103.136   Spectropolarimetry of comet Halley.
N. Visvanathan, Z. Meglick, D. T. Wickramasinghe.
IAU Colloquium No. 126: Origin and evolution of interplanetary
dust, p. 245 – 248 (1991). – See Abstr. 012.068 for the main entry.
Spectropolarimetric observations from 3800 to 7000 Å were
obtained for the nucleus of comet Halley for nine nights during
1985 – 86. The observations were spaced over phase angle of 2
to 66°. The continuum polarization without molecular–line
contamination as well as the polarization of the molecular lines
were evaluated. The plot of polarization versus the phase angle is
analyzed.

103.137   The dust in the coma of comet Halley.
J. I. Hage, J. M. Greenberg.
IAU Colloquium No. 126: Origin and evolution of interplanetary
dust, p. 261 – 264 (1991). – See Abstr. 012.068 for the main entry.
The interstellar dust model of comets is numerically worked
out to satisfy several basic constraints provided by observations
of comet Halley and to derive the porosity of coma dust. The

observational constraints are: (1) the strengths of the 3.4 μm and 9.7 μm emission bands; (2) the relative amount of silicates to organic materials; (3) the mass distribution of the dust. The results indicate that coma dust has a porosity in the range $0.93 < P < 0.975$. Preliminary calculations concerning the observed linear polarization of comet Halley are presented.

103.138 Spatial distribution and color of dust in Halley's inner coma.

J. Clairemidi, E. Brandon, P. Rousselot, G. Moreels.
IAU Colloquium No. 126: Origin and evolution of interplanetary dust, p. 277 – 280 (1991). – See Abstr. 012.068 for the main entry.

Composite images of the intensity of solar radiation scattered by dust in Halley's coma are constructed by using the three-channel spectra obtained during the approach phase of the Vega 2 spacecraft. They cover a sector centered on the nucleus that has radius of 40000 km and an angular extent of 50°.

103.139 Polarimetric properties of Halley's dust.

A. K. Sen, M. R. Deshpande, U. C. Joshi.
IAU Colloquium No. 126: Origin and evolution of interplanetary dust, p. 285 – 288 (1991). – See Abstr. 012.068 for the main entry.

Comet P/Halley was observed polarimetrically for seven nights in IHW and other continuum filters, during the pre and post pherihelion passages. The polarimetric observations have been combined with observations taken by other investigators, to get a complete picture of phase angle and wavelength dependence of polarization of comet P/Halley. Assuming Mie type scattering by cometary grains, the observed polarization data were fitted for a set of complex refractive indices.

103.140 Swan bands in the spectrum of comet Halley.

A. A. Atai, D. I. Shestopalov.
Kinematics Phys. Celest. Bodies, Vol. 7, No. 6, p. 12 – 17 (1991). English translation of Kinematika Fiz. Nebesn. Tel., Tom 7, No. 6, p. 15 – 20 (Nov–Dec 1991).

Spectrophotometric observations of comet Halley made at the Shemakha Astrophysical Observatory (December 1985 – January 1986) are analyzed. The Swan emission bands of $C_2(2,0)$ $\lambda438.2$ nm and CN ($B^2\Sigma^+$–$X^2\Sigma^+$) are identified in the range around $\lambda420.0$ nm. It is shown that the ratios of the illuminances in the Swan bands lg [F(0,1)/F(0,0)] and lg [F(1,0)/F(0,0)] depend on heliocentric distance: before perihelion they are proportional to $r^{-0.43\pm0.06}$ and $r^{-0.26\pm0.30}$, and after perihelion to $r^{0.33\pm0.40}$ and $r^{1.43\pm0.38}$. The excitation temperature of the C_2 molecules ranges from 5900 to 6500K (band (1,0) at $\lambda473.7$ nm) and 5900 to 6700K (band (0,1) at $\lambda563.5$ nm). It is established that C_2 molecules form at different rates in the atmosphere of comet Halley before and after perihelion.

103.141 Power spectral analysis of enhanced scintillation of quasar 3C 459 due to Comet Halley.

P. Janardhan, S. K. Alurkar, A. D. Bobra, O. B. Slee, D. Waldron.
Aust. J. Phys., Vol. 45, No. 1, p. 115 – 126 (1992).

The radio source 2314+038 (3C 459) showed enhanced scintillations on three days at a solar elongation of about 90° as the plasma tail of Halley's Comet swept across it on six days during 16 – 21 December 1985. If one assumes that the plasma velocities in the tail were not constant everywhere, but increased linearly from about 50 km s^{-1} at the tail axis to the normal average solar wind velocity of 400 km s^{-1} at the edges where the tail merged with the solar wind, a power spectral analysis of the scintillations shows two ranges of the rms electron density variation ΔN and scale size a. In particular, these are a fine scale zone near the axis where a is in the range 9 to 27 km and ΔN in the range 2 to 5 cm^{-3} and a zone near the edges with a and ΔN in the ranges 100 to 265 km and 0.4 to 0.8 cm^{-3} respectively. The assumption of a single velocity of 100 km s^{-1} throughout the tail shows similar fine scales near the tail axis and large scales near the edges.

103.142 Observations of comet Halley in 1985 – 1986 at the Bucharest Astronomical Institute.

A. Alexiu, G. Bocsa, M. Stanescu.
Rom. Astron. J., Vol. 1, No. 1 – 2, p. 113 (1991).

103.143 Spectrophotometry of comet P/Halley.

S. C. Joshi, B. B. Sanwal, B. S. Rautela.
Uttar Pradesh State Obs., Repr., No. 422.
Uttar Pradesh State Observatory, Naini Tal (India). 3 p. (1992).

Photoelectric spectrophotometry of the coma of comet P/Halley on two nights during the post–perihelion period has been presented. The emission features of C_2, CN, C_3 and CH have been identified. The abundance and the production rates of C_2 and CN have been derived.

Periodic comet Hartley 2

103.144 Hartley 2 – ein interessanter Komet.

V. Kasten.
Nachr. Olbers–Ges. Bremen, Nr. 157, p. 7 – 9 (Apr 1992).

Periodic comet Howell

103.145 Periodic comet Howell (1992c).

IAU Circ., No. 5472, p. 1 (11 Mar 1992).

103.146 Periodic comet Howell (1992c).

Yamamoto Circ., No. 2181, p. 1 (30 Mar 1992). In Japanese.

Periodic comet Kopff

103.147 A model of comet P/Kopff with dust and detailed chemistry.

D. C. Boice, W. F. Huebner, I. Konno.
Bull. Am. Astron. Soc., Vol. 23, No. 3, p. 1160 – 1161 (1991). Abstract. – See Abstr. 012.037 for the main entry.

Periodic comet Kowal 2

103.148 Periodic comet Kowal 2 (1991f₁).

Yamamoto Circ., No. 2176, p. 1 (4 Jan 1992). In Japanese.

Periodic comet Machholz

103.149 The connection between comet P/Machholz and the Quadrantid meteor stream.

R. Gonczi, H. Rickman, C. Froeschlé.
Mon. Not. R. Astron. Soc., Vol. 254, No. 4, p. 627 – 634 (15 Feb 1992).

The parent comet of the Quadrantid meteors was not observed until, perhaps, quite recently. Comet P/Machholz (1986 VIII) currently has orbital elements that differ drastically from those of the meteor stream, but its secular perturbations involve large-scale oscillations in inclination and perihelion distance similar to those earlier found for the Quadrantids. Whether there may have been an epoch in the past, when the Quadrantids were shed by comet P/Machholz depends upon whether or not there has been enough time since that epoch for differential Jovian perturbations to cause the 180° discrepancy in nodal longitudes presently observed. The authors have investigated the variation of the period of q–i oscillation between different meteor particles in relation to the hypothesis that the shedding of meteors occurred nearly 4000 yr ago, when comet P/Machholz last had a very small perihelion distance. The authors show that this hypothesis is viable in view of the ejection velocities typically expected and the resulting spread in the period of q–i oscillations. The most promising range of semimajor axes is just inside the 2/1 resonance, and detailed study reveals many cases of chaotic behaviour due to close encounters with Jupiter. The dynamics of the Quadrantid stream thus appears even more complex than earlier studies have indicated.

Periodic comet Metcalf–Brewington

103.150   **Motion of Comet P/Metcalf–Brewington (1906 VI = 1991a).**
G. Sitarski.
Acta Astron., Vol. 42, No. 1, p. 49 – 57 (1992).
   The comet was observed during two apparitions and the author collected 166 observations for the orbit improvement: 99 of 1906/07 and 67 of 1991. The observations were selected and weighted according to Bielicki's method. To link both apparitions the author had to take into account a nongravitational deceleration in the comet's motion. He considered the nongravitational effects in two ways: either as a daily change da/dt of the semi–major axis a of the comet's orbit or as the Marsden's parameters A_1, A_2, A_3 connected with angular parameters η, I, φ of the rotating comet's nucleus. It appeared that it was possible to determine reasonable values of the nongravitational parameters linking only two apparitions of the comet. Four sets of orbital elements were computed in order to predict the next return of the comet to perihelion in 2001. The prediction was difficult because of the approach of the comet to Jupiter to within 0.11 a.u. in March 1993. Ephemerides for 1999 and 2000 were calculated.

Periodic comet Mueller 4

103.151   **Periodic comet Mueller 4 (1992g).**
      IAU Circ., No. 5495, p. 1 (13 Apr 1992).
Further information is given in Nos. 5497, 5498, 5505.

103.152   **Periodic comet Mueller 4 (1992g).**
      Yamamoto Circ., No. 2183, p. 1 – 2 (18 Apr 1992). In Japanese.
      Further information is given in No. 2184.

Periodic comet Schwassmann–Wachmann 1

103.153   **Periodic comet Schwassmann–Wachmann 1.**
      IAU Circ., No. 5446, p. 1 (7 Feb 1992).
Further information is given in No. 5451.

Periodic comet Shoemaker–Levy 8

103.154   **Periodic comet Shoemaker–Levy 8 (1992f).**
      IAU Circ., No. 5493, p. 1 (9 Apr 1992).
Further information is given in Nos. 5495, 5506, 5540.

103.155   **Periodic comet Shoemaker–Levy 8 (1992f).**
      Yamamoto Circ., No. 2183, p. 1 (18 Apr 1992). In Japanese.
      Further information is given in No. 2184.

Periodic comet Singer Brewster

103.156   **Periodic comet Singer Brewster (1992e).**
      IAU Circ., No. 5490, p. 1 (4 Apr 1992).

103.157   **Periodic comet Singer Brewster (1992e).**
      Yamamoto Circ., No. 2183, p. 1 (18 Apr 1992). In Japanese.

Periodic comet Tempel 2

103.158   **A model of P/Tempel 2 with dust and detailed chemistry.**
      W. F. Huebner, D. C. Boice, I. Konno, P. D. Singh.
IAU Symposium No. 150: Astrochemistry of cosmic phenomena, p. 449 – 450 (1992). – See Abstr. 012.029 for the main entry.
   The authors apply their fluid dynamic model with chemical kinetics of dusty comet comae to P/Tempel 2. A brief summary of results concerning gas/dust dynamics and chemistry is given.

The comet light curve atlas. (The comet light curve catalogue/atlas. III. The atlas).
See Abstr. 002.029.

Rendezvous in space. The science of comets.
See Abstr. 003.034.

The 1816 solar eclipse and comet 1811 I in John Linnell's astronomical album.
See Abstr. 004.014.

Eine Medaille von der größten Kometenerscheinung des 17. Jahrhunderts.
See Abstr. 004.028.

The two comets of August AD 1165.
See Abstr. 004.032.

Space dust and debris. Proceedings. Topical Meeting of the COSPAR Interdisciplinary Scientific Commission B (Meetings B2, B3 and B5) of the COSPAR 28. Plenary Meeting, The Hague (Netherlands), 25 Jun – 6 Jul 1990.
See Abstr. 012.014.

Simulation and alteration for amorphous silicates with very broad bands in infrared spectra.
See Abstr. 022.058.

Laboratory spectra of amorphous and crystalline olivine: an application to comet Halley IR spectrum.
See Abstr. 022.215.

Chemical composition of an emanation from comets: identification of the 3 micron comet feature.
See Abstr. 022.216.

Ice particle emission from cometary analogues.
See Abstr. 022.217.

Polyoxymethylene in cometary dust: laboratory tests.
See Abstr. 022.218.

Electrostatic fragmentation of irregularly shaped particles.
See Abstr. 022.222.

Giotto – the second encounter.
See Abstr. 035.068.

On the possibility of detection of small comets in Ly–α.
See Abstr. 036.205.

Perturbation theory, resonance, librations, chaos, and Halley's comet.
See Abstr. 042.107.

The second coming of Giotto. Part one: Encounter with Halley.
See Abstr. 051.015.

Giotto Radio–Science Experiment: drag deceleration and spacecraft attitude perturbations expected during the encounter with comet P/Grigg–Skjellerup in July 1992.
See Abstr. 051.024.

Finite amplitude fast magnetosonic waves in a low–density plasma.
See Abstr. 062.027.

Chaotic Alfvén waves in solar and cometary plasmas.
See Abstr. 062.098.

Solar wind interaction with Mars and Venus during solar activity cycle.
See Abstr. 074.057.

Asteroid and comet orbits using radar data.
See Abstr. 098.024.

Positionsbeobachtungen von Kleinen Planeten und Kometen.
See Abstr. 098.035.

Radio observations regarding Earth–grazing asteroids.
See Abstr. 098.039.

New results on C≡N–bearing solid organics on asteroids and comets.
See Abstr. 098.087.

Highly variable objects in the solar system.
See Abstr. 098.139.

Origin of sungrazers: a frequent cometary end–state.
See Abstr. 102.010.

A self–consistent model for the particles and fields upstream of an outgassing comet. 2. A time–dependent description.
See Abstr. 102.012.

Energetic hydrogen atoms in cometary comae: production of the metallic component at large distances from the Sun.
See Abstr. 102.015.

Splitting of periodic comets.
See Abstr. 102.021.

The effect of electron collisions on rotational populations of cometary water.
See Abstr. 102.038.

Observational evidence for heterogeneity of cometary nuclei: the ratio of methanol–to–formaldehyde as an indicator of cometary origins.
See Abstr. 102.040.

The change in matter concentration of a cometary head vs. heliocentric distance.
See Abstr. 102.051.

Comets and meteors.
See Abstr. 102.057.

Comets: pre– and post–Halley.
See Abstr. 102.058.

The strong meteor display of November 5, 1991.
See Abstr. 104.015.

Evolution of the meteoroid swarm of comet Halley.
See Abstr. 104.063.

On ther relationship between comet P/Machholz and the Quadrantid meteor stream.
See Abstr. 104.069.

Comet Machholz 1986 VIII and Quadrantid meteoroid stream. Orbital evolution and relationship.
See Abstr. 104.078.

η Lyrid meteor stream associated with comet IRAS–Araki–Alcock, 1983 VII.
See Abstr. 104.089.

Lifetime of meteor streams associated with comet Halley.
See Abstr. 104.091.

Chemical composition of dust expected from condensation models.
See Abstr. 106.087.

Comet Halley and interstellar chemistry.
See Abstr. 131.202.

Comet Halley as an aggregate of interstellar dust and further evidence for the photochemical formation of organics in the interstellar medium.
See Abstr. 131.276.

104 Meteors, Meteor Streams

104.001   The question of the luminosity of meteor trails.
    A. Yu. Ol'khovatov.
Geomagn. Aeron., Vol. 30, No. 5, p. 714 – 716 (Apr 1990). English translation of Geomagn. Aehron., Tom 30, No. 5, p. 844 – 846 (1990).

    Questions connected with the luminosity of meteor trails are considered. It is shown that the heights at which meteor trails are most often observed are characterized by a number of special features. It is concluded that the luminosity of a meteor trail is attributable to the formation of above–thermal electrons with subsequent excitation of trail components (products of abrasion of the meteor and the atmospheric components) by electron collision. The electron temperature in the trail required to produce luminosity is determined to be 6000K or more. It is concluded as a result of comparison with rocket sounding data of the ionosphere that the mechanism of electron heating is linked with the mechanism of the formation of ionosphere inhomogeneities.

104.002   Geminiden 1991: een fraai schouwspel.
    N. de Kort, U. Poerink, S. van Leverink, B. Apeldoorn.
Zenit, Jaarg. 19, Nr. 6, p. 248 – 251 (Jun 1992).

104.003   Electrophonic bolids.
    C. Keay.
Kozmos, Vol. 23, No. 3, p. 26 – 27 (Jun 1992). In Slovak.

104.004   Geminiden 1991: een geslaagde aktie!
    C. Johannink, C. ter Kuile, K. Miskotte, H. Betlem, K. Jobse, M. de Lignie, M. van Vliet, A. Scholten, P. Jenniskens.
Radiant, Jaarg. 14, Nr. 1, p. 4 – 18 (Feb 1992).

104.005 **Perseiden 1991 in Japan.**
M. Koseki.
Radiant, Jaarg. 14, Nr. 1, p. 21 – 23 (Feb 1992).

104.006 **Winter 1991: Geminiden, Monocerotiden en snelle meteoren uit de Leeuw.**
P. Jenniskens.
Radiant, Jaarg. 14, Nr. 2, p. 28 – 33 (Apr 1992).
Excellent clear skies allowed an unprecedented view on the Geminids and associated minor streams. The results of 28 observers, about 8200 meteors, allow a presentation of a detailed activity curve of the Geminids and the first actvity curve ever of the Monocerotids.

104.007 **Post buurse: Geminidenaktie 1991.**
R. Schievink, J. de Jong van Lier.
Radiant, Jaarg. 14, Nr. 2, p. 40 – 42 (Apr 1992).

104.008 **De Perseiden van 1980: een gewone terugkeer ...**
P. Jenniskens.
Radiant, Jaarg. 14, Nr. 3, p. 55 – 58 (Jun 1992).
A reanalysis of meteor counts obtained in Switzerland.

104.009 **Eerste simultane TV meteoren verwerkt.**
M. de Lignie.
Radiant, Jaarg. 14, Nr. 3, p. 59 – 61 (Jun 1992).
For 5 out of 17 meteors preliminary data are presented.

104.010 **Wow ...**
P. Spurny, J. Borovicka, Z. Ceplecha.
Radiant, Jaarg. 14, Nr. 3, p. 64 – 65 (Jun 1992).
A very bright fireball of −18 maximum absolute magnitude was photographed by 3 Czech stations of the European network. The firefall travelled a 83 km luminous trajectory in 5.2 sec and terminated its light at the extremely low height of 16 km.

104.011 **Visual meteor data base statistics for 1989 and 1990.**
P. Roggemans.
WGN, Vol. 20, No. 1, p. 6 – 10 (Feb 1992).

104.012 **The importance of the Taurids.**
D. Steel.
WGN, Vol. 20, No. 1, p. 17 – 20 (Feb 1992).

104.013 **Taurid fireball proportions.**
A. McBeath.
WGN, Vol. 20, No. 1, p. 22 – 26 (Feb 1992).

104.014 **Daylight fireball, Czechoslovakia, September 22, 1991, 16^h48^m UT.**
J. Borovicka, P. Spurný.
WGN, Vol. 20, No. 1, p. 27 (Feb 1992).

104.015 **The strong meteor display of November 5, 1991.**
P. Brown, D. Asher, D. Steel.
WGN, Vol. 20, No. 1, p. 28 – 31 (Feb 1992).
Reports of unusually strong meteor activity viewed from Mauna Kea associated with a radiant in Pegasus on November 5, 1991, are presented and discussed. The activity occurred close to the time expected for meteors associated with P/Hartley 2 or possibly the Taurid complex, but the radiants are widely separated so that the source of the activity cannot be ascertained at this stage.

104.016 **The 1992 Quadrantids.**
R. Koschack, J. Rendtel, M. Gyssens, E. Hillestad, P. Brown.
WGN, Vol. 20, No. 1, p. 31 – 36 (Feb 1992).

104.017 **Possible α– and δ–Aurigid activity.**
A. McBeath.
WGN, Vol. 20, No. 1, p. 36 – 39 (Feb 1992).

104.018 **Telescopic observations of the 1991 Perseids in Czechoslovakia.**
P. Pravec.
WGN, Vol. 20, No. 1, p. 46 – 49 (Feb 1992).

104.019 **Telescopic Orionids in the night of October 22 – 23, 1990.**
T. Hansen.
WGN, Vol. 20, No. 1, p. 49 – 50 (Feb 1992).

104.020 **Fireball: Czechoslovakia, December 13, 1991, $3^h55^m22^s$ UT.**
P. Spurný, Z. Ceplecha.
WGN, Vol. 20, No. 2, p. 84 (Apr 1992).

104.021 **Fireball: Czechoslovakia, January 2, 1992, $20^h10^m01^s$ UT.**
P. Spurný.
WGN, Vol. 20, No. 2, p. 85 (Apr 1992).

104.022 **Fireball: Czechoslovakia, February 2, 1992, $19^h18^m04^s$ UT.**
P. Spurný, V. Porubčan.
WGN, Vol. 20, No. 2, p. 86 (Apr 1992).

104.023 **1991 – 92 fall and winter U.K. visual results.**
A. McBeath.
WGN, Vol. 20, No. 2, p. 89 – 92 (Apr 1992).

104.024 **Spanish visual observations in the winter 1991 – 92.**
J. M. Trigo.
WGN, Vol. 20, No. 2, p. 92 (Apr 1992).

104.025 **1992 Quadrantid and Coma Berenicid activity in Spain.**
L. R. Bellot Rubio.
WGN, Vol. 20, No. 2, p. 93 – 94 (Apr 1992).

104.026 **The 1992 Quadrantids in Crimea.**
A. I. Grishchenyuk.
WGN, Vol. 20, No. 2, p. 95 (Apr 1992).

104.027 **Radio observations of the 1990 Quadrantids.**
J. Van Wassenhove.
WGN, Vol. 20, No. 2, p. 95 – 96 (Apr 1992).

104.028 **Bright radio signals from Geminids and Quadrantids.**
G. M. Kristensen.
WGN, Vol. 20, No. 2, p. 98 – 101 (Apr 1992).

104.029 **Results of the IMO Aquarid project.**
R. Arlt, R. Koschack, J. Rendtel.
WGN, Vol. 20, No. 3, p. 114 – 135 (Jun 1992).
The IMO Aquarid Project was set up in 1989 to find out to which extent visual observations could contribute to our knowledge of the radiant structure of the minor summer showers in Aquarius and Capricornus. During this three–year project, 4989 visual meteor plots were obtained, mainly from mid–northern latitudes. The method used for radiant determination is presented and discussed.

104.030 **PosDat – the positional meteor database of the IMO.**
D. Koschny.
WGN, Vol. 20, No. 3, p. 136 – 139 (Jun 1992).

104.031 **On the presence of trains in meteor showers.**
L. R. Bellot Rubio.
WGN, Vol. 20, No. 3, p. 140 – 144 (Jun 1992).
Different mechanisms for meteor train generation are reviewed. Train percentages for different showers are calculated and compared. An attempt is made to correlate numbers of trains with train duration. Finally, fireball trains are considered.

104.032 **Fireball, Austria, January 17, 1992, $21^h21^m20^s$ UT.**
P. Spurný, Z. Ceplecha.
WGN, Vol. 20, No. 3, p. 147 (Jun 1992).

104.033 **Fireball, Germany, March 4, 1992, $19^h34^m52^s$ UT.**
P. Spurný.
WGN, Vol. 20, No. 3, p. 148 (Jun 1992).

104.034 **Fireball, Austria, March 9, 1992, $4^h06^m00^s$ UT.**
P. Spurný.
WGN, Vol. 20, No. 3, p. 149 (Jun 1992).

104.035 **A southern Taurid fireball over Japan.**
Y. Shiba, K. Ohtsuka.
WGN, Vol. 20, No. 3, p. 150 (Jun 1992).
The results of orbital calculations of a fireball photographed over Japan on November 3, 1991, is presented.

104.036 **The 1991 Perseids in Malta.**
F. Gatt.
WGN, Vol. 20, No. 3, p. 151 – 152 (Jun 1992).

104.037 **The 1990 and 1991 Geminids in Rumania.**
V. Grigore.
WGN, Vol. 20, No. 3, p. 152 – 153 (Jun 1992).

104.038 **The 1992 Quadrantids in Bulgaria.**
V. Velkov.
WGN, Vol. 20, No. 3, p. 153 – 154 (Jun 1992).
An overview is given of Bulgarian observations of the Quadrantids in the night of January 3 – 4, 1992. A very high activity was noticed.

104.039 **Quadrantid observations from Halifax, Nova Scotia.**
P. Gray.
WGN, Vol. 20, No. 3, p. 155 (Jun 1992).

104.040 **The 1992 Quadrantids in Czechoslovakia.**
P. Pravec.
WGN, Vol. 20, No. 3, p. 156 – 158 (Jun 1992).

104.041 **Hungarian observations of the 1991 Leonids.**
I. Tepliczky, P. Spányi.
WGN, Vol. 20, No. 3, p. 158 – 160 (Jun 1992).

104.042 **Meteors section news.**
R. D. Lunsford.
Strolling Astron., Vol. 36, No. 1, p. 16 – 19 (Mar 1992).

104.043 **Some physical characteristics of the Draconid meteor stream according to radar observations in 1985.**
K. K. Kostylev.
Astron. Vestn., Tom 26, No. 2, p. 112 – 115 (Mar – Apr 1992). In Russian. English translation in Solar Syst. Res., Vol. 26, No. 2.
The paper presents the processing results of the individual amplitude–temporal characteristics of meteor echo–signals of the Draconid stream in 1985 from the view–point of optimization methodics. The particles density ($0.25\,\text{g cm}^{-3}$), mean heights (95 km), braking coefficient ($30\,\text{km s}^{-2}$) and initial radius (2.1 m) are evaluated. The conclusion on a loose structure of the meteor stream particles is made.

104.044 **On radiation intensity in meteor spectra.**
V. A. Smirnov.
Astron. Vestn., Tom 26, No. 2, p. 116 – 120 (Mar – Apr 1992). In Russian. English translation in Solar Syst. Res., Vol. 26, No. 2.
Various mechanisms of radiation of spectral lines in meteor spectra have been considered. The theoretical intensities of the meteor spectral lines have been compared with the experimental values of intensity. The influence of different external physical conditions on the radiation in the meteor spectra formation process was estimated. The solution of the kinetic equation permits to obtain the concentration of radiating atoms in meteor plasma and the coefficient of radiation efficiency.

104.045 **Diffusion of meteor trails.**
U. Shodiev.
Astron. Vestn., Tom 26, No. 2, p. 121 – 125 (Mar – Apr 1992). In Russian. English translation in Solar Syst. Res., Vol. 26, No. 2.
Analysis of altitude and temporal dependence of the expansion velocity and the diffusion coefficient of meteor trails is given. The exponential increase of the expansion velocity with height, the decrease of the coefficient of turbulent diffusion with height and also the decrease of $\lg D$ at morning time are noted.

104.046 **Some structural features of the Geminid meteor shower.**
P. B. Babadzhanov, Sh. O. Isamutdinov, R. P. Chebotarev.
Astron. Vestn., Tom 26, No. 1, p. 93 – 97 (Jan – Feb 1992). In Russian. English translation in Solar Syst. Res., Vol. 26, No. 1.
The results of investigation of transversal and longitudinal structures of the Geminids are given. Density of the particle flux and variation of the parameter s of the integral law of meteoroid mass distribution of the whole period of observations (1965 – 1983) are shown. The average value of the parameter s for the Geminid meteor shower is equal to 1.67.

104.047 **The mass distribution of sporadic meteoroids as determined by radar observations of overdense meteor trails.**
P. B. Babadzhanov, R. Sh. Bibarsov.
Astron. Vestn., Tom 26, No. 1, p. 103 – 108 (Jan – Feb 1992). In Russian. English translation in Solar Syst. Res., Vol. 26, No. 1.
The diurnal and seasonal variations in the mass distribution of sporadic meteors was examined on the basis of more than 200,000 meteor echoes observed in 1965 – 1970. The diurnal variation of s index has a minimal value ($s = 1.96$) at $10 - 12^h$ Dushanbe Decrete Time (DDT), the maximal values ($s = 2.15 - 2.28$) – at $18 - 22^h$ DDT. The mean values of s is 2.1 ± 0.1 in the mass range of $10^{-3} - 10^2$g. A definite regularity in seasonal variation of index s is not observed.

104.048 **Television observations of meteors in Dushanbe in 1979.**
I. F. Malyshev.
Astron. Vestn., Tom 26, No. 1, p. 112 – 114 (Jan – Feb 1992). In Russian. English translation in Solar Syst. Res., Vol. 26, No. 1.
The television system and method of one–station observations of meteors is briefly described. Results are given for 21 meteors observed in 1979, the brightness of whose is approximately –1.9 to +2.4 of absolute magnitude.

104.049 **Certain properties of statistical distributions of the average dimensions of sporadic–meteor orbits.**
E. N. Kramer, Yu. M. Gorbanev.
Kinematics Phys. Celest. Bodies, Vol. 7, No. 3, p. 35 – 38 (1991). English translation of Kinematika Fiz. Nebesn. Tel, Tom 7, No. 3, p. 37 – 41 (1991).
Statistical analysis of the mean dimensions of the orbits of photographic and radar meteors does not support the existence of "stable" and "unstable" quantum levels in the gravitational field of the Sun.

104.050 **CHON–particles in interplanetary space.**
V. N. Lebedinets.
Sol. Syst. Res., Vol. 25, No. 1, p. 49 – 53 (Jul 1991). English translation of Astron. Vestn., Tom 25, No. 1, p. 65 – 70 (1991).
The hypothesis that meteors with anomalously high appearance heights are generated by organic meteor bodies similar to CHON–particles is justified. Estimates are made of their evaporation energy, $Q = (0.4...2.5) \cdot 10^{10} \text{erg} \cdot \text{g}^{-1}$, as well as the temperature at which intense evaporation (or decomposition) commences, $T_i < 1000$K. About one half of the meteor bodies in interplanetary space should be organic, and about 90% of those in the Draconid shower. The total influx of organic material from space into the atmosphere is estimated at $10^2 - 10^5 \text{ton} \cdot \text{day}^{-1}$.

104.051 **Modeling the formation of the initial radius of a meteor trail.**
G. V. Grusha.
Sol. Syst. Res., Vol. 25, No. 1, p. 58 – 62 (Jul 1991). English translation of Astron. Vestn., Tom 25, No. 1, p. 76 – 81 (1991).

For an interpretation of a weak observable correlation between the initial radius of an ionized meteor trail and the velocity of the meteoric body and atmospheric density, the influence of the fracturing of small meteoric bodies and random variations of the atmospheric density is evaluated. The model of the initial radius of the meteor trail is compared with two–frequency radar measurements of the intial radii of 1008 meteor trails.

104.052 **Evolution of the Quadrantid meteoroid swarm.**
P. B. Babadzhanov, Yu. V. Obrubov, A. N. Pushkarev.
Sol. Syst. Res., Vol. 25, No. 1, p. 63 – 70 (Jul 1991). English translation of Astron. Vestn., Tom 25, No. 1, p. 82 – 92 (1991).

The Everhart method has been used to investigate the orbital evolution of 36 model meteoroids of the Quadrantid swarm over the period from 3500 BC to 2250 AD. The initial values of the semimajor axis were taken in the interval 2.7 – 3.2 AU. The mathematical model of the Quadrantid evolution has confirmed the possible formation of eight related meteor streams from this swarm: the Quadrantids, Ursids, Northern and Southern δ–Aquarids, daytime Arietids, α–Cetids, Carinids, and ϰ–Velids. New theoretical data on the radiants and activity duration of these showers have been obtained, which agree satisfactorily with the observations.

104.053 **Simulations of motions of large and fast meteoric bodies.**
Yu. N. Kiselev, I. B. Kosarev, I. V. Nemchinov, V. B. Rozhdestvenskij, B. D. Kristoforov, V. L. Yur'ev.
Sol. Syst. Res., Vol. 25, No. 1, p. 71 – 79 (Jul 1991). English translation of Astron. Vestn., Tom 25, No. 1, p. 93 – 103 (1991).

For experimental modeling of the motion of a large and fast meteoritic body under conditions where irradiation has a key role, a method is proposed, where the body model is placed in a hypersonic jet of gas of a high atomic weight. On the basis of calculations with the characteristic equation of lead vapors and a comparison with the characteristic equation for air, the relationship is defined linking the principal parameters of the flow of an air jet around a body and a flow of lead vapors. The results of experiments are described, which simulated the motion in the Earth's atmosphere at altitudes of 23 – 40 km/sec. In front of the body, a luminescent region of the thermal wave was registered, which was several times as large as the body. A method for investigation of the luminescence of such meteors is proposed, which is based on injection of a high–velocity plasma jet into the real atmosphere.

104.054 **Brillant meteor of June 1, 1937 and the Simuna crater.**
V. A. Bronsthehn.
Sol. Syst. Res., Vol. 25, No. 1, p. 80 – 83 (Jul 1991). English translation of Astron. Vestn., Tom 25, No. 1, p. 104 – 108 (1991).

Data on the bright meteor observed over Estonia on June 1, 1937, and the results of a study of the Simuna blast crater 8.5 m in diameter formed before 1938 according to statements by local inhabitants, are examined. It is shown that this meteor was an overtaking one and entered the Earth's atmosphere with an initial velocity $v_0 = 11 – 20$ km/sec. An analysis of its motion in the atmosphere and also its maximum brightness indicate that the body's initial mass was $M_0 \geqslant 50$ tons, whereas the Simuna crater was formed by the impact of a body having a final mass $M_f \leqslant 67$ kg. It is suggested that the Simuna crater was formed by a large fragment separated from the main body after the meteor's explosion at an altitude of 28 km as observed by many witnesses.

104.055 **Size–dependent composition in the meteoroid/asteroid population: probable causes and possible implications.**
J. F. Bell.
54. Annual Meeting of the Meteoritical Society, p. 15 (1991). Abstract. – See Abstr. 012.010 for the main entry.

104.056 **Depth and size dependence of cosmogenic nuclide production rates in meteoroids.**
N. Bhandari, K. J. Mathew, M. N. Rao, U. Herpers, K. Bremer, S. Vogt, W. Wölfli, H. J. Hofmann, R. Michel, R. Bodemann.
54. Annual Meeting of the Meteoritical Society, p. 20 (1991). Abstract. – See Abstr. 012.010 for the main entry.

104.057 **The composition of meteoroids impacting LDEF.**
D. E. Brownlee, M. Laurance, F. Horz, R. P. Bernhard, J. Warren, J. P. Bradley.
54. Annual Meeting of the Meteoritical Society, p. 38 (1991). Abstract. – See Abstr. 012.010 for the main entry.

104.058 **Earth–grazing fireball of October 13, 1990.**
J. Borovička, Z. Ceplecha.
Astron. Astrophys., Vol. 257, No. 1, p. 323 – 328 (Apr 1992).

A fireball of –6 absolute magnitude, which left the atmosphere again after appearing at heights of around 100 km above Czechoslovakia and Poland was photographed at two Czech stations of the European Fireball Network. The body travelled a 409 km luminous trajectory in 9.8 s with initial velocity of 41.7 km/s. The type I fireball was produced by a meteoroid mass of 44 kg, from which only 0.35 kg was ablated. The meteoroid left the Earth in a changed orbit and with solidified fusion crust on its surface. Detailed data on the fireball trajectory and both the encounter and outcast orbits are given. A special method for long trajectory determination of nearly horizontal motion was invented. This method is based on angular velocity measurements from the excellent record of one station combined with one direction derived from the not–so–good record of the other station and computes pericenter position of Keplerian motion from observations very close to this point.

104.059 **Organic particles of cometary dust in space.**
V. N. Lebedinets.
Adv. Space Res., Vol. 11, No. 12, p. 149 – 153 (1991). – See Abstr. 012.014 for the main entry.

Main components of all stone and iron meteorites begin to evaporate intensively ar $T_f \geqslant 2300$K. Mathematical simulations of observed heights and meteor decelerations with allowance for meteor fragmentation shows that all types of meteorites can be found among meteoroids. However, some meteoroids can not have primnary composition similar with any type of meteorites, as their matter evaporates at temperatures much lower than 2300K. This manifests in anomalous high altitudes of such meteors.

104.060 **Observations of sporadic meteors.**
R. Kamiński.
Urania, Rok 63, Nr. 1, p. 8 – 11 (Jan 1992). In Polish.

104.061 **Electrical and electromagnetic phenomena associated with meteor flight.**
V. A. Bronshtehn.
Sol. Syst. Res., Vol. 25, No. 2, p. 93 – 104 (Sep 1991). English translation of Astron. Vestn., Tom 25, No. 2, p. 131 – 144 (1991).

The problem of electrical and electromagnetic phenomena in meteor trains is far from a solution at present, and requires serious theoretical analysis and collecting experimental (observational) data. Observations of electrophonic fireballs and meteor head echoes, and also individual magnetometer observations during high meteor shower activity and large meteorite falls are the only such data at present. One may consider as fairly well founded the conclusion of Tokhtas'ev that, unlike an artificial Earth satellite, a meteor body during ablation acquires a stabilized positive charge with a potential of several volts (due to the thermal emission of electrons). Allowance for the mechanical emission of electrons during the body's fragmentation reinforces this conclusion, but the mechanical emission phenomenon itself needs (as applied to meteor bodies) additional research. The situation is considerably worse for the problem of the occurrence of electrical currents in a meteor train. Several mechanisms for generating them were suggested in a paper in the 1960s, but they are all based on crude approximations and have not had further development. The use of a quasihydrodynamic method for solving the kinetic equations of motion of the electrons and ions at the first (initial) stage of train formation may enable one to shed light on this problem.

104.062 Diffraction of interplanetary solid bodies potentially generating meteorites.
V. N. Lebedinets.
Sol. Syst. Res., Vol. 25, No. 2, p. 147 – 153 (Sep 1991). English translation of Astron. Vestn., Vol. 25, No. 2, p. 200 – 207 (1991).

A method for determining meteorite types by semiempirical simulations of observed meteor heights with allowance for fragmentation is developed. Dense stone and iron meteorites amount to about 12% of the total. Other meteorites are very rich in volatile compounds, consist mainly of organics, or are very porous. An analysis of their composition and velocity distribution indicates that only about 1% of bright bolides may be followed by a meteorite fall. This significantly limits the possibility of studying the distribution of the characteristics of space dust particle by conventional methods.

104.063 Evolution of the meteoroid swarm of comet Halley.
P. B. Babadzhanov, A. Hajduk, Yu. V. Obrubov, A. N. Pushkarev.
Sol. Syst. Res., Vol. 25, No. 2, p. 153 – 160 (Sep 1991). English translation of Astron. Vestn., Vol. 25, No. 2, p. 208 – 216 (1991).

The orbital evolution of meteoroids ejected from the nucleus of comet Halley over an interval of 11 millennia has been investigated. The dependence of the formation of the spatial shape of the swarm on time and meteoroid ejection velocity is shown. A possible explanation is given for the irregular mass distribution of the swarm materials, which is observed as irregular variations in the particle flux density at the Earth. The genetic relationship of the Orionid and η Aquarid meteor streams with comet Halley is confirmed.

104.064 The sporadic meteoroid flux near Venus, Earth, and Mars.
V. M. Kolmakov.
Sol. Syst. Res., Vol. 25, No. 2, p. 160 – 166 (Sep 1991). English translation of Astron. Vestn., Vol. 25, No. 2, p. 217 – 224 (1991).

The possibility is examined of using meteor–produced ionization of the night ionosphere to determine the flux of sporadic meteoroids. This method is used to estimate the sporadic meteoroid flux near Venus, Earth, and Mars. The fluxes for Venus and Mars turn out to be higher than for the Earth, by factors of 4 and 10, respectively.

104.065 Determination of meteor radiant distribution from single radar observations with arrival angle measurements.
O. I. Bel'kovich, V. V. Sidorov, T. K. Filimonova.
Sol. Syst. Res., Vol. 25, No. 2, p. 166 – 172 (Sep 1991). English translation of Astron. Vestn., Vol. 25, No. 2, p. 225 – 232 (1991).

A method of determination of a meteor stream density distribution over the celestial sphere from observations by radar arrival angles measurements is described. Stability of the solution is discussed. Examples of the distributions are given.

104.066 On the periods of oscillations of the perihelion and eccentricity of the Geminids' swarm orbits.
A. K. Markina, V. I. Musij, I. S. Shestaka.
Astron. Tsirk., No. 1551, p. 30 – 31 (Nov – Dec 1991). In Russian.

The periods of oscillations of the argument of perihelion and eccentricity of the Geminids' orbits as functions of their semimajor axes are obtained. It is shown that the conclusion made by P. B. Babadzhanov and Yu. V. Obrubov on 4 meteor showers originated from the same swarm and occurring simultaneously during the same year is not confirmed.

104.067 The dynamics of meteoroid streams.
I. P. Williams.
IAU Symposium No. 152: Chaos, resonance and collective dynamical phenomena in the solar system, p. 299 – 313 (1992). – See Abstr. 012.025 for the main entry.

Meteor showers are seen at regular and frequent intervals on Earth. They are caused by meteoroids (that is small dust grains) in a coherent stream, all moving on similar heliocentric orbits, burning up on encountering the atmosphere of the Earth. The main evolutionary effect on such streams is gravitational perturbations by the planets. At a basic level, meteoroid streams represent a collective dynamical phenomenon in which all members display roughly the same behavior. One of the fundamental questions which can be investigated is whether the behavior of the mean orbit of the whole stream represents the mean behavior of the stream members. Within the boundaries of some meteor streams lie regions where the orbits are in high order resonance with Jupiter. This also represents a phenomenon of interest. Finally, the possibility exists that some streams are in chaotic regions and it is interesting to investigate whether or not meteoroids in such regions do display chaotic behavior.

104.068 Dynamical aspects of the Taurid meteor complex.
J. Štohl, V. Porubčan.
IAU Symposium No. 152: Chaos, resonance and collective dynamical phenomena in the solar system, p. 315 – 324 (1992). – See Abstr. 012.025 for the main entry.

Unusually long activity of the Taurid meteor complex, extending over 3 – 5 months according to new estimations based on various orbital similarity criteria, has been evoking controversies about the possible origin and dynamical evolution of this unique complex of meteor streams. It even casts doubts on the reality of the extension of the complex. In the present paper orbital elements and the extension of the Taurid meteor complex are re–examined on the bases of the most precise photographic meteor orbits available from the IAU Meteor Data Center in Lund. The results are evaluated and discussed from the viewpoint of various proposals on the origin and dynamical evolution of the complex.

104.069 On ther relationship between comet P/Machholz and the Quadrantid meteor stream.
R. Gonczi, H. Rickman, C. Froeschlé.
IAU Symposium No. 152: Chaos, resonance and collective dynamical phenomena in the solar system, p. 325 – 327 (1992). – See Abstr. 012.025 for the main entry.

104.070 The Quadrantid stream, chaos or not?
Wu Zidian, I. P. Williams.
IAU Symposium No. 152: Chaos, resonance and collective dynamical phenomena in the solar system, p. 329 – 332 (1992). – See Abstr. 012.025 for the main entry.

The Quadrantid stream covers a region of space which contains many strong resonances and commensurabilities with the Jovian orbit. The authors have numerically integrated the orbital evolution of over one hundred actual meteoroids backwards to BC 5000. The evolution is quite complex, but most of the meteoroids are quite well behaved with rapid but smooth changes in the orbital elements. One meteoroid however shows sharp sudden changes in its orbital parameters and these changes are generally indicative of the presence of chaos.

104.071 Meteoroid swarms: formation, evolution, and relationship to comets and asteroids.
P. B. Babadzhanov, Yu. V. Obrubov.
Sol. Syst. Res., Vol. 25, No. 4, p. 289 – 307 (Jan 1992). English translation of Astron. Vestn., Tom 25, No. 4, p. 387 – 407 (1991).

Current ideas about the formation and evolution of meteoroid swarms and their relationship to comets and Apollo–Aten asteroids are reviewed. Results of investigations of the dynamics of the Quadrantid, Geminid, Taurid, η Aquarid, and Orionid meteoroid swarms are given. It is shown that two to eight related meteor showers could be formed from one meteoroid swarm.

Five new small–body complexes of the solar system have been discovered.

104.072   Influx of meteoritic material onto the Earth.
Yu. I. Voloshchuk, B. L. Kashcheev.
Sol. Syst. Res., Vol. 25, No. 4, p. 341 – 350 (Jan 1992). English translation of Astron. Vestn., Tom 25, No. 4, p. 453 – 465 (1991).

Estimates of the influx of meteoritic material onto the Earth, and an analysis and comparison of the distributions of the orbital elements of meteoritic bodies over a broad range of masses from 10^{-6} g to 10 g calculated from photographic and radar observations of meteor trains in the Earth's atmosphere, are shown in this paper. It is shown that, of the five hypothetical types of meteoritic bodies connected with their origins, four can be identified from the distributions of their orbits. A noticeable contribution to the meteoritic system of bodies of asteroidal origin has been identified. The analysis of the collections of data on meteors recorded by different methods, different researchers, and at different times showed that a discrete component which conforms to integer values of certain quantum numbers is contained in the distributions of the orbital elements of meteoritic bodies.

104.073   Velocity distribution and flux density of sporadic meteors.
P. B. Babadzhanov, V. M. Kolmakov.
Sol. Syst. Res., Vol. 25, No. 4, p. 358 – 362 (Jan 1992). English translation of Astron. Vestn., Tom 25, No. 4, p. 476 – 481 (1991).

The authors present a new method to determine the velocity distribution of meteors from observations. This method can be used to calculate the flux density of meteors in any velocity interval.

104.074   Watching Halley's debris.
P. M. Bagnall.
Astronomy, Vol. 20, No. 5, p. 78 – 81 (May 1992).

104.075   Liste détaillée des essaims diurnes d'étoiles filantes.
A. Koeckelenbergh.
Ciel Terre, Vol. 108, No. 2, p. 54 – 55 (Mar – Apr 1992).

104.076   The errors of determination of meteor radiants by photographs from three stations.
L. M. Sherbaum, A. M. Kazantsev.
Visn. Kiiv. Univ., Fiz.–Mat. Nauki, Astron., Vip. 3, p. 38 – 46 (1992). In Ukrainian.

The errors of radiants for eleven meteors are determined. Each of them is photographed at three stations. Reduction of this observed material has permitted to calculate the mean square error ΔR from three independent values. Comparison with the error ΔR_1 which is found by the standard method for "double" meteors shows that ΔR is 5 – 6 times larger than ΔR_1. In both cases the errors do not increase 1°.

104.077   On the estimation of meteoroid density by photographic observation of meteors.
G. G. Novikov, V. V. Kayumov.
Astron. Vestn., Tom 26, No. 3, p. 65 – 69 (May – Jun 1992). In Russian. English translation in Sol. Syst. Res., Vol. 26, No. 3.

On the basis of the theory of luminosity of meteors taking into account quasi–continuous fragmentation, the results of mathematical modelling of the observed decelerations and heights of 19 faint photographic meteors (12 sporadic and 7 stream meteors) the densities δ_0 and energy of fragmentation Q_a of meteoroids were established. The coincidence of δ_0 and Q_a is obtained according to the deceleration and luminosity only for 10 meteors (6 sporadic and 4 stream meteors). Thus the application of composite approach to processing for determination of densities and fragmentation energies greatly increases the reliability of parameters obtained.

104.078   Comet Machholz 1986 VIII and Quadrantid meteoroid stream. Orbital evolution and relationship.
P. B. Babadzhanov, Yu. V. Obrubov.
Astron. Vestn., Tom 26, No. 3, p. 70 – 78 (May – Jun 1992). In Russian. English translation in Sol. Syst. Res., Vol. 26, No. 3.

The evolutionary model of the P/Machholz meteoroids stream was simulated. It shows that this comet may produce eight meteor showers. There are such well known major meteor showers as the Quadrantids, Ursids, Northern and Southern δ–Aquarids, Daytime Arietids and α–Cetids. The possible age of the stream is about 7.5 millennia.

104.079   The Geminids and the object 3200 Phaethon.
E. N. Kramer, I. S. Shestaka.
Astron. Vestn., Tom 26, No. 3, p. 85 – 90 (May – Jun 1992). In Russian. English translation in Sol. Syst. Res., Vol. 26, No. 3.

A checking of the validity of the hypothesis on the genetic relations of Geminids and three more meteor swarms with asteroid 3200 Phaethon was carried out. Based upon the estimations of the ejection velocity of meteor particles from Phaethon's surface by suggesting it be a comet nucleus and taking into account the real sizes of Phaethon and the meteor particles, the conclusion has been drawn on the limitation of ejection velocities and perihelion argument variations due to secular perturbations and consequently, on the impossibility of four meteor swarms' formation. The supposition itself on the cometary nature of Phaethon is subjected to criticizm.

104.080   A method for determination of the radiant coordinates of meteor streams.
Sh. O. Isamutdinov.
Astron. Vestn., Tom 26, No. 3, p. 91 – 94 (May – Jun 1992). In Russian. English translation in Sol. Syst. Res., Vol. 26, No. 3.

A method of determination of the radiant coordinates of meteor streams is considered.

104.081   On a class of observed light curves of meteors and on terminal flares.
G. G. Novikov, N. A. Konovalova.
Astron. Vestn., Tom 26, No. 3, p. 95 – 98 (May – Jun 1992). In Russian. English translation in Sol. Syst. Res., Vol. 26, No. 3.

With account of quasicontinuous fragmentation and variability of the heat transfer coefficient the effect of terminal flares and one class of the observed light curves with negligible variation of brightness along a long part of the meteor trajectory are explained.

104.082   Predictions of the meteor radiant point associated with an Earth–approaching minor planet.
I. Hasegawa, Y. Ueyama, K. Ohtsuka.
Publ. Astron. Soc. Jpn., Vol. 44, No. 1, p. 45 – 54 (1992).

Predictions of meteor orbits and radiant points are presented for earth–approaching minor planets discovered before the end of 1989. All meteor orbits available from the IAU Meteor Data Center (Lund Observatory) are compared to the predictions, and, on the basis of orbital similarity, possible identifications are found. Besides (3200) Phaethon, the parent body of the Geminids, (2201) Oljato is likely to be another most probable candidate for an extinct comet, belonging to the Taurid complex. In addition, several meteors possibly associated with (4450) Pan = 1987 SY and 1988 TA are also found.

104.083   Meteors.
Z. Ceplecha, J. Borovička.
15. Ecole de Printemps d'Astrophysique de Goutelas: Interrelations between physics and dynamics for minor bodies in the solar system, p. 309 – 367 (1992). – See Abstr. 012.055 for the main entry.

Contents: 1. Orbiting meteoroids encounter the Earth. After defining the subject and terminology, survey of 4 possible types of the atmospheric interaction at encounter of meteoroids with the Earth are explained: (1) meteor or fireball, (2) meteorite–dropping fireball, (3) cratering event, (4) dust particle.

2. Physics of meteoric phenomena. Mathematical modeling of meteoroid interaction with the Earth's atmosphere: preheating, ablation, dynamics and light emission. Applications to meteor records.

3. Meteoroid populations and orbits. Structure and density of meteoroids of different types (groups), their orbits, relation to other bodies of the solar system and their influx to the Earth.

104.084 Determination of the velocity of a meteor from photographic observations.
A. F. Zausaev, A. N. Pushkarev.
Sol. Syst. Res., Vol. 25, No. 5, p. 456 – 459 (Mar 1992). English translation of Astron. Vestn., Tom 25, No 5, p. 609 – 612 (1991).

An algorithm is proposed for the calculation of the preatmospheric velocity of reference meteors photographed through a rotating shutter. The preatmospheric velocity is found from the solution of a system of equations for the time dependence of the path traversed by the meteor.

104.085 The orbital distribution and origin of meteoroids.
D. Steel.
IAU Colloquium No. 126: Origin and evolution of interplanetary dust, p. 291 – 298 (1991). – See Abstr. 012.068 for the main entry.

Approximately 68,000 orbits of meteoroids, ranging from sizes of 10 cm and more down to microgram masses, are now available through the IAU Meteor Data Center. It is found that quite different distributions result in different mass regimes, with implications for the origin and evolution of these particles: for example the larger bodies, observed as fireballs, are associated with meteorites incoming from the region of the asteroid belt with low–inclination orbits, whereas the smaller meteoroids have more comet–like orbits. There is also evidence for several meteoroid streams associated with specific Apollo asteroids. The data may additionally be viewed as a suitable source function in investigations of the production of interplanetary dust from the fragmentation of larger meteoroids in mutual collisions. However, inspection of the data raises many questions: for instance there seem to be many meteoroids on small retrograde paths, but no possible parent objects are known to exist on such orbits.

104.086 A study of meteor orbits obtained in Japan.
B. A. Lindblad.
IAU Colloquium No. 126: Origin and evolution of interplanetary dust, p. 299 – 302 (1991). – See Abstr. 012.068 for the main entry.

A list of 325 two–station photographic meteor orbits obtained with 35 mm cameras has recently been published by Japanese amateur groups. The present study analyzes the data and concludes that the orbits are of high quality and very useful for scientific purposes.

104.087 The micrometeoroid in the upper atmosphere.
F. Kamijo.
IAU Colloquium No. 126: Origin and evolution of interplanetary dust, p. 303 – 306 (1991). – See Abstr. 012.068 for the main entry.

The temperature and the radius variation of micrometeoroids in the thermosphere and the mesosphere are calculated theoretically. If the radius and the initial velocity are $100\,\mu m$ and 30 km/sec respectively, the evaporation height and the velocity coincide almost exactly with those of the Capricornids and the Virginids from the meteor stream observation. Moreover, it is shown that the not evaporated debris till the end of the sublimation may become spherules in the bottom of deep sea; and that fluffy micrometeoroids ($10\,\mu m$ size) floating in the stratosphere are also consistent with our calculation. The recondensation and the coagulation of the evaporated gas molecules from the meteoroid are also calculated, and it is shown that these secondary particles are very small and few.

104.088 An estimation of meteoroid flux at outer Martian space for steady meteor streams.
K. Nagasawa.
IAU Colloquium No. 126: Origin and evolution of interplanetary dust, p. 307 – 310 (1991). – See Abstr. 012.068 for the main entry.

This work was done to estimate meteoroid fluxes of steady meteor streams at places far away from the Sun, for example,

beyond Martian orbit. For the estimation, a normal distribution model is introduced. This model assumes that the flow of meteoroids is steady and the meteoroids are distributed in the form of anormal distribution around the orbit of their parent comet. Keplerian motion of each meteoroid is also assumed. Some parameters, necessary for determining a practical model, can be obtained through observations. Calculations show that fluxes of streams sharply decrease as the solar distance increases and that meteoroids scatter chiefly along the orbital planes of the parent comets. The model seems to be useful to deduce rough structure of meteor streams.

104.089 η Lyrid meteor stream associated with comet IRAS–Araki–Alcock, 1983 VII.
K. Ohtsuka.
IAU Colloquium No. 126: Origin and evolution of interplanetary dust, p. 315 – 318 (1991). – See Abstr. 012.068 for the main entry.

The probable association of comet IRAS–Araki–Alcock with the η Lyrid meteor stream is suggested, and the possible relation of the radio meteor shower on 1983 May 10 is also discussed.

104.090 The annual variation of radio meteor echoes observed from 1981 to 1985.
K. Suzuki.
IAU Colloquium No. 126: Origin and evolution of interplanetary dust, p. 319 – 322 (1991). – See Abstr. 012.068 for the main entry.

Radio observations of meteor echoes using FM broadcasting were carried out from January, 1981 to June, 1985. Annual variation of meteor rates in the 5 years was obtained. Some features in this annual variation can be explained by the occurrence of the major meteor showers. After the effects of these major showers are removed there remains a significant annual variation which shows higher rates in the winter half of the year (October to March). This seems to be caused by the annual variation of non–shower meteors radiating from the apex region.

104.091 Lifetime of meteor streams associated with comet Halley.
M. Hajdukova, A. Hajduk.
IAU Colloquium No. 126: Origin and evolution of interplanetary dust, p. 323 – 326 (1991). – See Abstr. 012.068 for the main entry.

Critical examination of the orbital parameters of particles ejected from comet Halley rejects the low age hypotheses for meteor showers associated with the comet. The diffusion of the orbits of large particles is too slow for explaining the observed structural features of the stream. The mass–loss process as derived from space observations compared with the mass of the stream of particles deduced from flux data lead to comet lifetimes of the order of 10^5 years.

104.092 The Taurid Complex: giant comet origin?
D. I. Steel, D. J. Asher, S. V. M. Clube.
IAU Colloquium No. 126: Origin and evolution of interplanetary dust, p. 327 – 330 (1991). – See Abstr. 012.068 for the main entry.

The formation and evolution of the Taurid Complex of interplanetary objects is modelled on the basis of the parent being a giant comet which entered the inner solar system some time in the past 10,000 – 20,000 years. The orbital element distributions for the presently–observed meteor showers are discussed in terms of how these can constrain any model for the origin of the overall complex. The authors present results from numerical integrations of fictitious meteoroids released from a comet over ten millenia, this comet having initial elements similar to those derived from a backwards integration of P/Encke.

104.093 Stability and inherent precision of two methods for solution of motion and ablation equations for fireball forming bodies in the Earth atmosphere.
V. V. Kalenichenko.
Kinematika Fiz. Nebesn. Tel, Tom 8, No. 3, p. 69 – 77 (May–Jun 1992). In Russian. English translation in Kinematics Phys. Celest. Bodies, Vol. 8, No. 3.

The stability of two methods for solution of motion and ablation equations for a fireball forming body in the Earth atmosphere is investigated. The inherent precision of solution is estimated on the basis of the Prairie network catalogue. The ratio of initial and terminal velocities of a fireball on the visible trajectory is shown to be a reliable criterion of the fireball fitness for the determination of physical characteristics of its cosmic body. The optimum value of this parameter can be evaluated for every fireball catalogue and it characterizes the catalogue quality to a certain extent, the nearer the ratio to unity the higher the quality. For the Prairie network catalogue, the solution of the motion and ablation equations of a fireball forming body is stable for those fireballs whose ratio of initial and terminal velocities exceeds 1.7.

104.094   Preparatory calculation to double–stationed observation of meteor showers.
Xu Pin–xin.
Publ. Purple Mt. Obs., Vol. 11, No. 2, p. 85 – 91 (Jun 1992). In Chinese.

104.095   Determination of population index r of Orionids.
Feng Zhan–liang, Xue Jun, Hu Ling–mei.
Publ. Purple Mt. Obs., Vol. 11, No. 2, p. 92 – 96 (Jun 1992). In Chinese.

104.096   The dynamics of meteoroid streams.
I. P. Williams, Z. Wu.
NATO Advanced Study Institute on Predictability, Stability, and Chaos in N–Body Dynamical Systems, p. 225 – 238 (1991). – See Abstr. 012.089 for the main entry.
Meteor showers, seen at regular and frequent intervals on Earth, are caused by the interaction of meteoroids (that is small dust grains) in a coherent stream, all moving on similar heliocentric orbits, with the atmosphere of the Earth. The formation and dynamical evolution of such streams is discussed and techniques, including numerical integration, for following their evolution is described. In some cases the evolution is very critically dependent on initial conditions and the evolution may be chaotic. In addition to general considerations, some specific streams, all with individual areas of interest, are discussed.

104.097   DMS voorjaarsakties 1991.
P. Jenniskens.
Radiant, Jaarg. 14, Nr. 1, p. 19 – 20 (Feb 1992).
During spring 1991 the occasional ecliptic shower meteors of the Virginid–Scorpiid–Sagittariid complex as well as some Lyrid activity was observed.

104.098   De kurkdroge zomer van 1991: veel waarnemingen.
P. Jenniskens.
Radiant, Jaarg. 14, Nr. 3, p. 49 – 54 (Jun 1992).
42 observers logged 6137 meteors of the Perseids.

A catalogue of meteor showers in medieval Arab chronicles.
See Abstr. 002.013.

The IAU Meteor Data Center in Lund.
See Abstr. 002.094.

The 1095 AD meteor event as described in the Anglo Saxon Chronicle.
See Abstr. 004.025.

Feuerwerk aus dem All.
See Abstr. 014.053.

Een batterij Canon camera's geautomatiseerd.
See Abstr. 034.029.

Monitoring meteors.
See Abstr. 036.024.

Possibilities of RDS in meteor back–scatter.
See Abstr. 036.044.

The software "Radiant"
See Abstr. 036.045.

Delivery of meteorites from the v_6 secular resonance region near 2 AU.
See Abstr. 042.106.

Formation of spinels in cosmic objects during atmospheric entry: a clue to the Cretaceous–Tertiary boundary event.
See Abstr. 081.008.

The role of cometary and meteor matter in noctilucent clouds genesis.
See Abstr. 082.018.

Wind regime of the meteor zone and its interaction with processes in the strato– and mesosphere.
See Abstr. 082.019.

Icarus, 1991 RC and the daytime Arietids.
See Abstr. 098.038.

Geminid meteoroids and the probability for cometary activity on Phaethon.
See Abstr. 098.085.

Additions to the Taurid Complex.
See Abstr. 098.127.

Comets and meteors.
See Abstr. 102.057.

The connection between comet P/Machholz and the Quadrantid meteor stream.
See Abstr. 103.149.

Orbits of meteorite producing fireballs. The Glanerbrug – a case study.
See Abstr. 105.003.

Face–dependent impact probabilities upon LDEF for heliocentric particle orbits.
See Abstr. 106.066.

105 Meteorites, Meteorite Craters

105.001 Rhenium–osmium isotope constraints on the age of iron meteorites.
M. F. Horan, J. W. Morgan, R. J. Walker, J. N. Grossman.
Science, Vol. 255, No. 5048, p. 1118 – 1121 (28 Feb 1992).
Rhenium and osmium concentrations and the osmium isotopic compositions of iron meteorites were determined by negative thermal ionization mass spectrometry. Data for the IIA iron meteorites define an isochron with an uncertainty of approximately ± 31 million years for meteorites ~ 4500 million years old. Although an absolute rhenium–osmium closure age for this iron group cannot be as precisely constrained because of uncertainty in the decay constant of ^{187}Re, an age of 4460 million years ago is the minimum permitted by combined uncertainties. These age constraints imply that the parent body of the IIAB magmatic irons melted and subsequently cooled within 100 million years after the formation of the oldest portions of chondrites. Other iron meteorites plot above the IIA isochron, indicating that the planetary bodies represented by these iron groups may have cooled significantly later than the parent body of the IIA irons.

105.002 Dodelijke rotsblokken uit de ruimte.
E. Echternach.
Zenit, Jaarg. 19, Nr. 4, p. 150 – 155 (Apr 1992).

105.003 Orbits of meteorite producing fireballs. The Glanerbrug – a case study.
P. Jenniskens, J. Borovička, H. Betlem, C. ter Kuile, F. Bettonvil, D. Heinlein.
Astron. Astrophys., Vol. 255, No. 1/2, p. 373 – 376 (Feb 1992).
At 18:32:38 UT on April 7, 1990, a breccious L–LL type chondrite fell near Glanerbrug in the Netherlands. From visual observations of the meteor by 200 occasional observers, a heliocentric orbit is derived by several independent methods, including a new method using the slope of the meteor on the sky as seen from different locations. The orbit found has a relatively high inclination of 23 ± 5 degrees, adding weight to the high inclination tail of meteorite producing fireballs. The average value of i for this population matches that of the population of near Earth asteroids, but is significantly higher than that found for the possible meteorite producing fireballs registered in the Prairie Network (Wetherill and ReVelle 1981) and the Meteorite Observation and Recovery Project (Halliday et al. 1989).

105.004 Icy cometary meteorite or trash from airplanes?
J. Šilhán.
Říše hvězd, Vol. 72, No. 3, p. 49 – 50 (Mar 1991). In Czech.

105.005 Der Kandidat. Der Saurier–Killer – die Spur führt nach Mexico.
D. Fischer.
Sterne Weltraum, Jahrg. 31, Nr. 4, p. 230 – 232 (Apr 1992).

105.006 Water in SNC meteorites: evidence for a martian hydrosphere.
H. R. Karlsson, R. N. Clayton, E. K. Gibson Jr., T. K. Mayeda.
Science, Vol. 255, No. 5050, p. 1409 – 1411 (13 Mar 1992).
The Shergotty–Nakhla–Chassigny (SNC) meteorites, purportedly of martian origin, contain 0.04 to 0.4 percent water by weight. Oxygen isotopic analysis can be used to determine whether this water is extraterrestrial or terrestrial. Such analysis reveals that a portion of the water is extraterrestrial and furthermore was not in oxygen isotopic equilibrium with the host rock. Lack of equilibrium between water and host rock implies that the lithosphere and hydrosphere of the SNC parent body formed two distinct oxygen isotopic reservoirs. If Mars was the parent body, the maintenance of two distinct reservoirs may result from the absence of plate tectonics on the planet.

105.007 The breakup of a meteorite parent body and the delivery of meteorites to Earth.
P. H. Benoit, D. W. G. Sears.
Science, Vol. 255, No. 5052, p. 1685 – 1687 (27 Mar 1992).
Whether many of the 10,000 meteorites collected in the Antarctic are unlike those falling elsewhere is contentious. The Antarctic H chondrites, one of the major classes of stony meteorites, include a number of individuals with higher induced thermoluminescence peak temperatures than observed among non–Antarctic H chondrites. The proportion of such individuals decreases with the mean terrestrial age of the meteorites at the various ice fields. These H chondrites have cosmic–ray exposure ages of about 8 million years, experienced little cosmic–ray shielding, and suffered rapid postmetamorphic cooling. Breakup of the H chondrite parent body, 8 million years ago, may have produced two types of material with different size distributions and thermal histories. The smaller objects reached Earth more rapidly through more rapid orbital evolution.

105.008 New fall of meteorite in Bohemia?
T. Starecký.
Říše hvězd, Vol. 72, No. 8, p. 157 – 158 (Aug 1991). In Czech.

105.009 The meteorite craters in Australia.
S. Vrána.
Říše hvězd, Vol. 72, No. 11, p. 207 – 208 (Nov 1991). In Czech.

105.010 A new type of meteoritic diamond in the enstatite chondrite Abee.
S. S. Russell, C. T. Pillinger, J. W. Arden, M. R. Lee, U. Ott.
Science, Vol. 256, No. 5054, p. 206 – 209 (10 Apr 1992).
Diamonds have been isolated from the Abee enstatite chondrite by the same procedure used for concentrating Cδ, the putative interstellar diamond found ubiquitously in primitive meteorities. Because the Abee diamonds have typical solar system isotopic compositions for carbon, nitrogen, and xenon, they are presumably nebular in origin rather than presolar. Their discovery in an unshocked meteorite eliminates the possibility of origins normally invoked to account for diamonds in ureilites and iron meteorites and suggests a low–pressure synthesis.

105.011 Antarctic meteoritic fever.
P. Rajlich.
Vesmír, Vol. 70, No. 1, p. 10 – 12 (Jan 1991). In Czech.

105.012 Tektites.
J. Bouška.
Vesmír, Vol. 71, No. 3, p. 137 – 143 (Mar 1992). In Czech.

105.013 Redox effects in ordinary chondrites and implications for asteroid spectrophotometry.
H. Y. McSween Jr.
Icarus, Vol. 95, No. 2, p. 239 – 243 (Feb 1992).
Ordinary chondrites have been oxidized during metamorphism, resulting in progressive increases in the mean ferrous iron content of pyroxenes and in the relative proportions of olivine and pyroxene. Reflectance spectra are sensitive to both of these mineralogic changes. Asteroids with rubble–pile structures would presumably expose samples with different metamorphic histories on their surfaces. This study examines whether the metamorphically induced ranges in mineralogy and corresponding spectral parameters might explain the observed variations in S–asteroid rotational spectra. The calculated spectral ranges for any one chondrite class are too limited to account for the large excursions in rotational spectra, and their slopes have opposite sign to the slopes of asteroid data. This study thus provides a further argument against the idea that ordinary chondrites were derived from S–asteroids with variable rotational spectra.

105.014   Meteorite falls in Denmark.
G. M. Kristensen.
WGN, Vol. 20, No. 3, p. 151 (Jun 1992).

105.015   Numerical simulation of impact crater gravity collapse.
A. V. Potapov.
Astron. Vestn., Tom 26, No. 1, p. 77 – 82 (Jan – Feb 1992). In Russian. English translation in Solar Syst. Res., Vol. 26, No. 1.
A numerical model of the impact crater gravity collapse is developed. The model consists of a numerical method of solving viscous incompressible fluid flows (method markers–and–cells (MAC)) and a numerical analog of the Bingham rheological model–biviscous model. The model has been used to estimate deformations of rocks underlying complex impact craters and some other characteristic values. Ways of further improvement of the model are demonstrated.

105.016   Correlated Si isotope anomalies and large ^{13}C enrichments in a family of exotic SiC grains.
J. Stone, S. Epstein, I. D. Hutcheon, G. J. Wasserburg.
Earth Planet. Sci. Lett., Vol. 107, No. 3/4, p. 570 – 581 (Dec 1991).
A suite of morphologically distinctive silicon carbide (SiC) grains from the Orgueil and Murchison carbonaceous chondrite meteorites contains Si and C of highly anomalous isotopic composition. All of the SiC grains in this suite are characterized by a distinctive platy morphology and roughly developed hexagonal crystal forms that allow them to be distinguished from other types of SiC found in the host meteorites. The authors suggest that the distinctive morphological characteristics and comparatively simple Si isotope systematics identify the platy SiC crystals as a genetically related family, formed around a single, isotopically heterogeneous presolar star or an association of related stars. The enrichments in ^{13}C and the Si isotope systematics of the platy SiC are broadly consistent with theoretical models of nucleosynthesis in low–mass, carbon stars on the asymptotic giant branch.

105.017   Rhenium–osmium isotope systematics in meteorites. I. Magmatic iron meteorite groups IIAB and IIIAB.
J. W. Morgan, R. J. Walker, J. N. Grossman.
Earth Planet. Sci. Lett., Vol. 108, No. 4, p. 191 – 202 (Feb 1992).
Using resonance ionization mass spectrometry (RIMS), Re and Os abundances were determined by isotope dilution (ID) and ^{187}Os/^{186}Os ratios measured in nineteen iron meteorites: eight from group IIAB, ten from group IIIAB, and Treysa (IIIB anomalous). Abundances range from 1.4 to 4800 ppb Re, and from 13 to 65000 ppb Os, and generally agree well with previous ID and neutron activation (NAA) results. The Re and Os data suggest that abundance trends in these iron groups may be entirely explained by fractional crystallization. Addition of late–formed metal to produce Re–Os variation in the B subgroups is not essential but cannot be excluded. Whole–rock isochrons for the IIAB and IIIAB groups are statistically indistinguishable. Pooled data yield an initial ^{187}Os/^{186}Os of 0.794 ± 0.010, with a slope of $(7.92 \pm 0.20) \times 10^{-2}$ corresponding to a magmatic iron meteorite age of 4.65 ± 0.11 Ga (using a decay constant of $1.64 \times 10^{-11} a^{-1}$). Given the errors in the slope and half life, this age does not differ significantly from the canonical chondrite age of 4.56 Ga, but could be as young as 4.46 Ga.

105.018   Recent grazing impacts on the Earth recorded in the Rio Cuarto crater field, Argentina.
P. H. Schultz, R. E. Lianza.
Nature, Vol. 355, No. 6357, p. 234 – 237 (16 Jan 1992). Letter–to–the–editor.
The most probable angle of a meteoroid impact on a planet is 45°, and an impact at 15° from the horizontal or lower is as likely as at 75° or higher. Yet little direct evidence for oblique impacts exists on the Earth, for two reasons. Unless the impact angle is very low, any asymmetry created during the initial transfer of energy from impactor to target is lost as the crater is formed; moreover, the shallow craters formed by oblique impact are more easily obscured by subsequent erosion. During routine flights two

years ago, however, one of the authors (R.E.L.) noticed an anomalous alignment of oblong rimmed depressions ($4 \text{ km} \times 1 \text{ km}$) on the otherwise featureless farmland of the Pampas in Argentina. The authors argue here, from sample analysis and by analogy with laboratory experiments, that these structures resulted from a low–angle impact and ricochet of a chondritic body originally 150 – 300 m in diameter.

105.019   The natural thermoluminescence of meteorites. 4. Ordinary chondrites at the Lewis Cliff ice field.
P. H. Benoit, H. Sears, D. W. G. Sears.
J. Geophys. Res., Vol. 97, No. B4, p. 4629 – 4648 (10 Apr 1992).
Natural thermoluminescence (TL) measurements have been made on 302 meteorites from the vicinity of the Lewis Cliff in the Beardmore region of Antarctica. The data provide information on terrestrial age and unusual radiation and thermal histories, which, in turn, are helpful in identifying fragments of a single fall and in understanding ice sheet movements and the mechanisms by which meteorite concentration occurs at this site.

105.020   On the relationship between isolated and chondrule olivine grains in the carbonaceous chondrite ALHA77307.
R. H. Jones.
Geochim. Cosmochim. Acta, Vol. 56, No. 1, p. 467 – 482 (Jan 1992).
The origin of isolated olivine grains in carbonaceous chondrites has been suggested to be either by direct condensation or by fragmentation of chondrules. In an attempt to resolve this debate, isolated olivine grains in the carbonaceous chondrite ALHA77307 have been studied in detail. Zoning characteristics and minor element compositions of olivines from all isolated and chondrule occurrences are described.

105.021   Magnesium isotopes in particles separated from carbonaceous chondrites.
M. Adriaens, F. Adams, M. Hyman, M. W. Rowe, D. E. Brownlee.
54. Annual Meeting of the Meteoritical Society, p. 1 (1991). Abstract. – See Abstr. 012.010 for the main entry.

105.022   Determination of trace elements in extraterrestrial materials by ICP/MS.
A. Albrecht, G. S. Hall, G. F. Herzog, D. E. Brownlee.
54. Annual Meeting of the Meteoritical Society, p. 2 (1991). Abstract. – See Abstr. 012.010 for the main entry.

105.023 ^{26}Al and ^{10}Be contents of the Murchison (C2) chondrite.
A. Albrecht, S. Vogt, G. F. Herzog, J. Klein, D. Fink, R. Middleton.
54. Annual Meeting of the Meteoritical Society, p. 3 (1991). Abstract. – See Abstr. 012.010 for the main entry.

105.024   The origin of matrix and rims in Bishunpur (L/LL3); an ion probe study.
C. M. O'D. Alexander.
54. Annual Meeting of the Meteoritical Society, p. 4 (1991). Abstract. – See Abstr. 012.010 for the main entry.

105.025   Interstellar organics and possible connections with the carbonaceous components of meteorites and IDPs.
L. J. Allamandola.
54. Annual Meeting of the Meteoritical Society, p. 5 (1991). Abstract. – See Abstr. 012.010 for the main entry.

105.026   SIMS analysis of micrometeoroid impacts on LDEF.
S. Amari, J. Foote, C. Simon, P. Swan, R. M. Walker, E. Zinner, E. K. Jessberger, F. Stadermann.
54. Annual Meeting of the Meteoritical Society, p. 7 (1991). Abstract. – See Abstr. 012.010 for the main entry.

105.027 **Renazzo–like chondrites: a light element stable isotope study.**
R. D. Ash, M. M. Grady, A. D. Morse, C. T. Pillinger.
54. Annual Meeting of the Meteoritical Society, p. 9 (1991).
Abstract. – See Abstr. 012.010 for the main entry.

105.028 **PIXE measurements on the iron meteorite Mundrabilla.**
S. Bajt, K. L. Rasmussen.
54. Annual Meeting of the Meteoritical Society, p. 11 (1991).
Abstract. – See Abstr. 012.010 for the main entry.

105.029 **Thermoluminescence and C–14 of non–Antarctic meteorites: terrestrial ages of Prairie State finds.**
P. H. Benoit, Lu Jie, D. W. G. Sears, A. J. T. Jull.
54. Annual Meeting of the Meteoritical Society, p. 16 (1991).
Abstract. – See Abstr. 012.010 for the main entry.

105.030 **Ice movement, pairing and meteorite showers of ordinary chondrites from the Allan Hills.**
P. H. Benoit, H. Sears, D. W. G. Sears.
54. Annual Meeting of the Meteoritical Society, p. 17 (1991).
Abstract. – See Abstr. 012.010 for the main entry.

105.031 **Thermoluminescence of meteorites from the Lewis cliff: ice movements, pairing, orbit, and antarctic/non–antarctic comparisons.**
P. H. Benoit, H. Sears, D. W. G. Sears.
54. Annual Meeting of the Meteoritical Society, p. 18 (1991).
Abstract. – See Abstr. 012.010 for the main entry.

105.032 **TEM cathodoluminescence spectra of meteorite minerals.**
E. J. Benstock, P. R. Buseck.
54. Annual Meeting of the Meteoritical Society, p. 19 (1991).
Abstract. – See Abstr. 012.010 for the main entry.

105.033 **The rhenium osmium chronometer: the iron meteorites revisited.**
J. L. Birck, M. Roy–Barman, C. J. Allegre.
54. Annual Meeting of the Meteoritical Society, p. 21 (1991).
Abstract. – See Abstr. 012.010 for the main entry.

105.034 **New carbonaceous and type 3 ordinary chondrites from the Sahara desert.**
A. Bischoff, T. Grund, T. Geiger, M. Endreß, W. Beckerling, K. Metzler, H. Palme, B. Spettel, R. N. Clayton, T. K. Mayeda.
54. Annual Meeting of the Meteoritical Society, p. 22 (1991).
Abstract. – See Abstr. 012.010 for the main entry.

105.035 **Mid–IR spectroscopy of Antarctic Consortium meteorites: B–7904, Y–82162 and Y–86720.**
J. L. Bishop, C. M. Pieters.
54. Annual Meeting of the Meteoritical Society, p. 23 (1991).
Abstract. – See Abstr. 012.010 for the main entry.

105.036 **A possible origin of EII chondrites from a high temperature–high pressure solar gas.**
M. Blander, L. Unger, A. Pelton, G. Eriksson.
54. Annual Meeting of the Meteoritical Society, p. 25 (1991).
Abstract. – See Abstr. 012.010 for the main entry.

105.037 **^{39}Ar–^{40}Ar ages of achondrites: evidence for a lunar–like cataclysm?**
D. D. Bogard, D. H. Garrison.
54. Annual Meeting of the Meteoritical Society, p. 26 (1991).
Abstract. – See Abstr. 012.010 for the main entry.

105.038 **K/T spherules are altered microtektites.**
B. F. Bohor, W. J. Betterton.
54. Annual Meeting of the Meteoritical Society, p. 27 (1991).
Abstract. – See Abstr. 012.010 for the main entry.

105.039 **Aqueous alteration on the parent bodies of carbonaceous chondrites: computer simulations of late–stage oxidation.**
W. L. Bourcier, M. E. Zolensky.
54. Annual Meeting of the Meteoritical Society, p. 30 (1991).
Abstract. – See Abstr. 012.010 for the main entry.

105.040 **Discriminant analysis of the tektite chemical data.**
V. Bouška, H. Maslowská.
54. Annual Meeting of the Meteoritical Society, p. 31 (1991).
Abstract. – See Abstr. 012.010 for the main entry.

105.041 **Mineralogical and chemical studies bearing on the origin of accretionary rims in the Murchison CM2 carbonaceous chondrite.**
A. J. Brearley, T. Geiger.
54. Annual Meeting of the Meteoritical Society, p. 33 (1991).
Abstract. – See Abstr. 012.010 for the main entry.

105.042 **Mineralogy of an unusual Cr–rich inclusion in the Los Martinez (L6) chondrite breccia.**
A. J. Brearley, M. L. Miller, I. Casanova, K. Keil.
54. Annual Meeting of the Meteoritical Society, p. 34 (1991).
Abstract. – See Abstr. 012.010 for the main entry.

105.043 **Bidirectional reflectance spectra of the Divnoe anomalous achondrite.**
D. T. Britt, C. M. Pieters, M. I. Petaev, N. I. Zaslavskaya.
54. Annual Meeting of the Meteoritical Society, p. 37 (1991).
Abstract. – See Abstr. 012.010 for the main entry.

105.044 **Zircons in Padvarninkai brecciated eucrite.**
M. Bukovanská, T. R. Ireland, A. El Goresy.
54. Annual Meeting of the Meteoritical Society, p. 39 (1991).
Abstract. – See Abstr. 012.010 for the main entry.

105.045 **Could chondrule rims be formed or modified by parent body accretion events?**
T. Bunch, P. Cassen, R. Reynolds, S. Chang, M. Podolak, P. Schultz.
54. Annual Meeting of the Meteoritical Society, p. 40 (1991).
Abstract. – See Abstr. 012.010 for the main entry.

105.046 **Al–Mg isotopic record of the recrystallization of a refractory inclusion during accretion into the Leoville parent body.**
C. Caillet, G. J. MacPherson, E. K. Zinner.
54. Annual Meeting of the Meteoritical Society, p. 41 (1991).
Abstract. – See Abstr. 012.010 for the main entry.

105.047 **Distribution of silicon between kamacite and taenite.**
I. Casanova.
54. Annual Meeting of the Meteoritical Society, p. 42 (1991).
Abstract. – See Abstr. 012.010 for the main entry.

105.048 **The positive Eu anomaly and Sc enrichment of minerals A and B in enstatite meteorites.**
Chen Yongheng, Lin Yangting, Wang Daode, E. Pernicka.
54. Annual Meeting of the Meteoritical Society, p. 45 (1991).
Abstract. – See Abstr. 012.010 for the main entry.

105.049 **The trace element chemistry and composition of niningerite in enstatite meteorites.**
Chen Yongheng, Wang Daode, E. Pernicka.
54. Annual Meeting of the Meteoritical Society, p. 46 (1991).
Abstract. – See Abstr. 012.010 for the main entry.

105.050 **The effect of precursor grain size on chondrule textures.**
H. C. Connolly Jr., B. D. Jones, R. H. Hewins.
54. Annual Meeting of the Meteoritical Society, p. 49 (1991).
Abstract. – See Abstr. 012.010 for the main entry.

105.051 **Low temperature annealing and cathodoluminescence of type I chondrule compositions.**
J. M. DeHart, G. E. Lofgren.
54. Annual Meeting of the Meteoritical Society, p. 52 (1991).
Abstract. – See Abstr. 012.010 for the main entry.

105.052 **Fe/Mn ratios in basaltic achondrites and primitive meteorites.**
J. S. Delaney.
54. Annual Meeting of the Meteoritical Society, p. 53 (1991).
Abstract. – See Abstr. 012.010 for the main entry.

105.053 **Nanophase metals in Fremdlinge from Allende: "smokes" from the early solar system?**
D. D. Eisenhour, P. R. Buseck.
54. Annual Meeting of the Meteoritical Society, p. 58 (1991).
Abstract. – See Abstr. 012.010 for the main entry.

105.054 **Negative thermal ionization and isotope dilution applied to the determination of Re and Os concentrations and Os isotopic compositions in iron meteorites.**
T. Esat, V. Bennett.
54. Annual Meeting of the Meteoritical Society, p. 59 (1991).
Abstract. – See Abstr. 012.010 for the main entry.

105.055 **REE variations in old hamite from aubrites and EL6 chondrites.**
C. Floss, G. Crozaz.
54. Annual Meeting of the Meteoritical Society, p. 61 (1991).
Abstract. – See Abstr. 012.010 for the main entry.

105.056 **Volatile elements in large micrometeorites from Greenland.**
G. J. Flynn, S. R. Sutton, W. Klock.
54. Annual Meeting of the Meteoritical Society, p. 63 (1991).
Abstract. – See Abstr. 012.010 for the main entry.

105.057 **Cosmogenic ^{36}Ar from neutron capture by ^{35}Cl in the Chico L6 chondrite: additional evidence for large shielding.**
D. H. Garrison, D. D. Bogard, G. F. Herzog.
54. Annual Meeting of the Meteoritical Society, p. 67 (1991).
Abstract. – See Abstr. 012.010 for the main entry.

105.058 **Ejecta from lunar impacts: where is it on earth?**
D. E. Gault, P. H. Schultz.
54. Annual Meeting of the Meteoritical Society, p. 68 (1991).
Abstract. – See Abstr. 012.010 for the main entry.

105.059 **The CK chondrites – conditions of parent body metamorphism.**
T. Geiger, A. Bischoff.
54. Annual Meeting of the Meteoritical Society, p. 69 (1991).
Abstract. – See Abstr. 012.010 for the main entry.

105.060 **A noble gas study in St. Severin core AIII and Knyahinya samples.**
E. Gilabert, B. Lavielle.
54. Annual Meeting of the Meteoritical Society, p. 70 (1991).
Abstract. – See Abstr. 012.010 for the main entry.

105.061 **Compound specific isotope analysis of polycyclic aromatic hydrocarbons in carbonaceous chondrites.**
I. Gilmour, I. A. Franchi, C. T. Pillinger, P. Eakin, A. Fallick.
54. Annual Meeting of the Meteoritical Society, p. 71 (1991).
Abstract. – See Abstr. 012.010 for the main entry.

105.062 **The plessite structure in iron meteorites.**
J. I. Goldstein, D. B. Williams, J. Zhang.
54. Annual Meeting of the Meteoritical Society, p. 72 (1991).
Abstract. – See Abstr. 012.010 for the main entry.

105.063 **Constraints on the time of accretion and thermal evolution of chondrite parent bodies by precise U–Pb dating of phosphates.**
C. Göpel, G. Manhes, C. J. Allegre.
54. Annual Meeting of the Meteoritical Society, p. 73 (1991).
Abstract. – See Abstr. 012.010 for the main entry.

105.064 **Meteoritic evidence for an active early sun.**
J. N. Goswami.
54. Annual Meeting of the Meteoritical Society, p. 74 (1991).
Abstract. – See Abstr. 012.010 for the main entry.

105.065 **Titanium, calcium and magnesium isotopic compositions in hibonite–rich inclusion from Efremovka.**
J. N. Goswami, G. Srinivasan, A. A. Ulyanov.
54. Annual Meeting of the Meteoritical Society, p. 75 (1991).
Abstract. – See Abstr. 012.010 for the main entry.

105.066 **Acfer 182: an unusual chondrite with affinities to ALH 85085.**
M. M. Grady, R. D. Ash, A. D. Morse, C. T. Pillinger.
54. Annual Meeting of the Meteoritical Society, p. 76 (1991).
Abstract. – See Abstr. 012.010 for the main entry.

105.067 **Exposure ages of LL– and L/LL–chondrites and implications for parent body histories.**
T. Graf, K. Marti.
54. Annual Meeting of the Meteoritical Society, p. 77 (1991).
Abstract. – See Abstr. 012.010 for the main entry.

105.068 **Spinel–bearing refractory inclusions in Cold Bokkeveld (CM2).**
R. C. Greenwood, R. Hutchison, G. Cressey.
54. Annual Meeting of the Meteoritical Society, p. 78 (1991).
Abstract. – See Abstr. 012.010 for the main entry.

105.069 **The co–evolution of chondrules and matrix in ordinary chondrites.**
J. N. Grossman.
54. Annual Meeting of the Meteoritical Society, p. 79 (1991).
Abstract. – See Abstr. 012.010 for the main entry.

105.070 **Attempts to constrain the carbon isotopic composition of dispersed carbonate in EETA 79001.**
C. P. Hartmetz, I. P. Wright, C. T. Pillinger.
54. Annual Meeting of the Meteoritical Society, p. 83 (1991).
Abstract. – See Abstr. 012.010 for the main entry.

105.071 **Roter Kamm crater age: 3.5 to 4.0 Ma.**
J. Hartung, M. Kunk, W. Reimold, R. Miller,
R. Grieve.
54. Annual Meeting of the Meteoritical Society, p. 84 (1991).
Abstract. – See Abstr. 012.010 for the main entry.

105.072 **Proximal ejecta of the Chicxulub crater, Yucatán peninsula, Mexico.**
A. R. Hildebrand, W. V. Boynton.
54. Annual Meeting of the Meteoritical Society, p. 90 (1991).
Abstract. – See Abstr. 012.010 for the main entry.

105.073 **Size and composition effects to cosmogenic nuclides in meteorites.**
M. Honda, H. Nagai, M. Imamura, K. Kobayashi.
54. Annual Meeting of the Meteoritical Society, p. 92 (1991).
Abstract. – See Abstr. 012.010 for the main entry.

105.074 **Carbon, nitrogen and silicon isotopes in small interstellar SiC grains from the Murchison C2 chondrite.**
P. Hoppe, J. Geiss, F. Bühler, J. Neuenschwander, S. Amari,
R. S. Lewis.
54. Annual Meeting of the Meteoritical Society, p. 93 (1991).
Abstract. – See Abstr. 012.010 for the main entry.

105.075 **Dissemination and fractionation of projectile material in impact melts from the Wabar crater, Saudi Arabia.**
F. Horz, D. W. Mittlefehldt, T. H. See.
54. Annual Meeting of the Meteoritical Society, p. 94 (1991).
Abstract. – See Abstr. 012.010 for the main entry.

105.076 **Fayalitic halos around FeNi inclusions in forsterite in the Kaba carbonaceous chondrite.**
X. Hua, P. R. Buseck, A. El Goresy.
54. Annual Meeting of the Meteoritical Society, p. 97 (1991).
Abstract. – See Abstr. 012.010 for the main entry.

105.077 **Two distinct 'normal planetary' noble gases carriers in chondrites.**
G. R. Huss, R. S. Lewis.
54. Annual Meeting of the Meteoritical Society, p. 98 (1991).
Abstract. – See Abstr. 012.010 for the main entry.

105.078 **Cr isotopic composition of sulfides in the Qingzhen enstatite chondrite.**
I. D. Hutcheon, P. K. Carpenter, M. Bar–Matthews.
54. Annual Meeting of the Meteoritical Society, p. 99 (1991).
Abstract. – See Abstr. 012.010 for the main entry.

105.079 **The Sterlitamak downfall.**
M. I. Petaev, Eh. Z. Gareev.
Priroda, No. 5, p. 52 – 55 (May 1992). In Russian.
The event which took place about two years ago could be called unique without any exaggeration: this is the second craterforming downfall of a meteorite known in history of meteoritics which was witnessed by people.

105.080 **The L6 chondrite fall at Glatton, England, 1991 May 5.**
R. Hutchison, J. C. Barton, C. T. Pillinger.
54. Annual Meeting of the Meteoritical Society, p. 100 (1991).
Abstract. – See Abstr. 012.010 for the main entry.

105.081 **ADRAR 003: a new extraordinary unequilibrated ordinary chondrite.**
R. Hutchison, S. J. B. Reed, R. D. Ash, C. T. Pillinger.
54. Annual Meeting of the Meteoritical Society, p. 101 (1991).
Abstract. – See Abstr. 012.010 for the main entry.

105.082 **Na–Cr sulfide phases in the Indarch (EH4) chondrite.**
M. L. Hutson.
54. Annual Meeting of the Meteoritical Society, p. 102 (1991).
Abstract. – See Abstr. 012.010 for the main entry.

105.083 **Oxygen isotopic compositions of individual meteoritic magnetite grains from carbonaceous chondrites.**
M. Hyman, M. W. Rowe, E. K. Zinner.
54. Annual Meeting of the Meteoritical Society, p. 103 (1991).
Abstract. – See Abstr. 012.010 for the main entry.

105.084 **More Ti isotopic compositions of presolar SiC from the Murchison meteorite.**
T. R. Ireland, E. K. Zinner, S. Amari, E. Anders.
54. Annual Meeting of the Meteoritical Society, p. 104 (1991).
Abstract. – See Abstr. 012.010 for the main entry.

105.085 **Large Nb–Ta fractionations in Allende Ca, Al–rich inclusions.**
K. P. Jochum, H. Palme, B. Spettel.
54. Annual Meeting of the Meteoritical Society, p. 108 (1991).
Abstract. – See Abstr. 012.010 for the main entry.

105.086 **A liquidus phase diagram for a primitive shergottite.**
J. H. Jones, A. J. G. Jurewicz, L. Le.
54. Annual Meeting of the Meteoritical Society, p. 109 (1991).
Abstract. – See Abstr. 012.010 for the main entry.

105.087 **Effect of metamorphism on isolated olivine grains in CO3 chondrites.**
R. H. Jones.
54. Annual Meeting of the Meteoritical Society, p. 110 (1991).
Abstract. – See Abstr. 012.010 for the main entry.

105.088 **Raman scattering and laser–induced luminescence from micro diamonds in ureilites.**
H. Kagi, A. Masuda, K. Takahashi.
54. Annual Meeting of the Meteoritical Society, p. 111 (1991).
Abstract. – See Abstr. 012.010 for the main entry.

105.089 **Compositional studies of Antarctic carbonaceous chondrites possibly related to Al Rais and Renazzo.**
G. W. Kallemeyn.
54. Annual Meeting of the Meteoritical Society, p. 112 (1991).
Abstract. – See Abstr. 012.010 for the main entry.

105.090 **Extraterrestrial water of possible Martian origin in SNC meteorites: constraints from oxygen isotopes.**
H. R. Karlsson, E. K. Gibson, R. N. Clayton, T. K. Mayeda, R. A. Socki.
54. Annual Meeting of the Meteoritical Society, p. 113 (1991).
Abstract. – See Abstr. 012.010 for the main entry.

105.091 **Trace element partitioning within mesosiderite clasts.**
A. K. Kennedy, B. W. Stewart, I. D. Hutcheon, G. J. Wasserburg.
54. Annual Meeting of the Meteoritical Society, p. 116 (1991).
Abstract. – See Abstr. 012.010 for the main entry.

105.092 **Interstellar precursors in synthesis of meteoritic organic matter.**
J. F. Kerridge.
54. Annual Meeting of the Meteoritical Society, p. 117 (1991).
Abstract. – See Abstr. 012.010 for the main entry.

105.093 **Xenon in chondritic metals.**
J. S. Kim, K. Marti, C. Perron, P. Pellas.
54. Annual Meeting of the Meteoritical Society, p. 118 (1991).
Abstract. – See Abstr. 012.010 for the main entry.

105.094 **Muong–Nong–type and splashform–type tektites from Hainan, China.**
E. A. King, C. Koeberl.
54. Annual Meeting of the Meteoritical Society, p. 119 (1991).
Abstract. – See Abstr. 012.010 for the main entry.

105.095 **^{41}Ca in the Jilin (H5) chondrite: a matter of size.**
J. Klein, D. Fink, R. Middleton, S. Vogt, G. F. Herzog.
54. Annual Meeting of the Meteoritical Society, p. 120 (1991).
Abstract. – See Abstr. 012.010 for the main entry.

105.096 **Beaverhead impact structure, Montana: geochemistry of impactites and country rock samples.**
C. Koeberl, P. S. Fiske.
54. Annual Meeting of the Meteoritical Society, p. 121 (1991).
Abstract. – See Abstr. 012.010 for the main entry.

105.097 **Microdistribution of chromium in metal and sulfide of IAB silicate inclusions and winonaites.**
A. Kracher.
54. Annual Meeting of the Meteoritical Society, p. 123 (1991).
Abstract. – See Abstr. 012.010 for the main entry.

105.098 **Petrologic description of Eagles Nest: a new olivine achondrite.**
D. A. Kring, W. V. Boynton, D. H. Hill, R. A. Haag.
54. Annual Meeting of the Meteoritical Society, p. 124 (1991).
Abstract. – See Abstr. 012.010 for the main entry.

105.099   Maralinga (CK4): record of highly oxidizing nebular conditions.
G. Kurat, F. Brandstätter, H. Palme, B. Spettel, M. Prinz.
54. Annual Meeting of the Meteoritical Society, p. 125 (1991).
Abstract. – See Abstr. 012.010 for the main entry.

105.100   Formation and alteration of refractory inclusions within the CM chondrites cold Bokkeveld, Murchison and Murray.
M. R. Lee, D. J. Barber.
54. Annual Meeting of the Meteoritical Society, p. 129 (1991).
Abstract. – See Abstr. 012.010 for the main entry.

105.101   Compound chondrules in ordinary chondrites.
M. S. Lee, A. E. Rubin, J. T. Wasson.
54. Annual Meeting of the Meteoritical Society, p. 130 (1991).
Abstract. – See Abstr. 012.010 for the main entry.

105.102   Shape–differentiated size distributions of chondrules in type 3 ordinary chondrites.
J. M. Leenhouts, W. R. Skinner.
54. Annual Meeting of the Meteoritical Society, p. 131 (1991).
Abstract. – See Abstr. 012.010 for the main entry.

105.103   Evaluation of the Strecker synthesis as a source of amino acids on carbonaceous chondrites.
N. R. Lerner, E. Peterson, S. Chang.
54. Annual Meeting of the Meteoritical Society, p. 132 (1991).
Abstract. – See Abstr. 012.010 for the main entry.

105.104   Ca–Al–rich inclusions in Ningqian (CV3) chondrite: evidence for primordial high enrichment in Re in Pt–group element nuggets.
Y. T. Lin, A. El Goresy, H. Fang.
54. Annual Meeting of the Meteoritical Society, p. 136 (1991).
Abstract. – See Abstr. 012.010 for the main entry.

105.105   Microprobe studies of microtomed particles of "white druse" salts in shergottite EETA 79001.
D. J. Lindstrom.
54. Annual Meeting of the Meteoritical Society, p. 137 (1991).
Abstract. – See Abstr. 012.010 for the main entry.

105.106   Ordinary chondrite classification and meteoritic evidence regarding parent bodies.
M. E. Lipschutz.
54. Annual Meeting of the Meteoritical Society, p. 139 (1991).
Abstract. – See Abstr. 012.010 for the main entry.

105.107   Dynamic crystallization characteristics of enstatite chondrite chondrules.
G. E. Lofgren, J. M. DeHart, A. B. Lanier.
54. Annual Meeting of the Meteoritical Society, p. 142 (1991).
Abstract. – See Abstr. 012.010 for the main entry.

105.108   Related compositional and cathodoluminescence trends in chondrules from Semarkona.
Lu Jie, D. W. G. Sears, P. H. Benoit, M. Prinz, M. K. Weisberg.
54. Annual Meeting of the Meteoritical Society, p. 143 (1991).
Abstract. – See Abstr. 012.010 for the main entry.

105.109   Abundance ratios of molybdenum isotopes in some iron meteorites.
Qi Lu, A. Masuda.
54. Annual Meeting of the Meteoritical Society, p. 144 (1991).
Abstract. – See Abstr. 012.010 for the main entry.

105.110   Isotope systematics of cumulate eucrite EET–87520.
G. W. Lugmair, S. J. G. Galer, R. W. Carlson.
54. Annual Meeting of the Meteoritical Society, p. 145 (1991).
Abstract. – See Abstr. 012.010 for the main entry.

105.111   Impact craters: are they useful?
V. L. Masaitis.
54. Annual Meeting of the Meteoritical Society, p. 149 (1991).
Abstract. – See Abstr. 012.010 for the main entry.

105.112   Spinel–bearing, Al–rich chondrules: modified by metamorphism or unchanged since crystallization?
T. J. McCoy, A. Pun, K. Keil.
54. Annual Meeting of the Meteoritical Society, p. 150 (1991).
Abstract. – See Abstr. 012.010 for the main entry.

105.113   Olivines in angrite LEW87051: phenos or xenos?
G. McKay, L. Le, J. Wagstaff.
54. Annual Meeting of the Meteoritical Society, p. 151 (1991).
Abstract. – See Abstr. 012.010 for the main entry.

105.114   Ablation of Australian tektites supportive of a terrestrial origin.
W. L. Melnik.
54. Annual Meeting of the Meteoritical Society, p. 153 (1991).
Abstract. – See Abstr. 012.010 for the main entry.

105.115   Accretionary dust mantles in CM chondrites: chemical variations and calculated time scales of formation.
K. Metzler, A. Bischoff, G. Morfill.
54. Annual Meeting of the Meteoritical Society, p. 155 (1991).
Abstract. – See Abstr. 012.010 for the main entry.

105.116   Determination of the ^{81}Kr saturation activity and Kr production rates for various meteorite classes; application to exposure ages and terrestrial ages.
T. Michel, O. Eugster, S. Niedermann.
54. Annual Meeting of the Meteoritical Society, p. 157 (1991).
Abstract. – See Abstr. 012.010 for the main entry.

105.117   Petrology and geochemistry of the EETA79002 diogenite.
D. W. Mittlefehldt, B. Myers.
54. Annual Meeting of the Meteoritical Society, p. 158 (1991).
Abstract. – See Abstr. 012.010 for the main entry.

105.118   Anomalous shocked quartz in Australian impact craters.
Y. Miura, T. Kato.
54. Annual Meeting of the Meteoritical Society, p. 159 (1991).
Abstract. – See Abstr. 012.010 for the main entry.

105.119   Anomalous quartz from possible impact craters in Japan.
Y. Miura, T. Kato, M. Okamoto.
54. Annual Meeting of the Meteoritical Society, p. 160 (1991).
Abstract. – See Abstr. 012.010 for the main entry.

105.120   Cooling histories of primitive achondrites Yamato 74357 and MAC88177.
M. Miyamoto, H. Takeda.
54. Annual Meeting of the Meteoritical Society, p. 161 (1991).
Abstract. – See Abstr. 012.010 for the main entry.

105.121   A new mechanism for the formation of meteoritic kerogen–like material.
W. A. Morgan Jr., E. D. Feigelson, H. Wang, M. Frenklach.
54. Annual Meeting of the Meteoritical Society, p. 162 (1991).
Abstract. – See Abstr. 012.010 for the main entry.

105.122   High–temperature mass spectrometric degassing of enstatite chondrites: implications for pyroclastic volcanism on the aubrite parent body.
D. M. Muenow, K. Keil, L. Wilson.
54. Annual Meeting of the Meteoritical Society, p. 163 (1991).
Abstract. – See Abstr. 012.010 for the main entry.

105.123 Gas–solid phase diagram of olivine and its application to chondrites.
H. Nagahara, B. O. Mysen, I. Kushiro.
54. Annual Meeting of the Meteoritical Society, p. 166 (1991).
Abstract. – See Abstr. 012.010 for the main entry.

105.124 Highly fractionated REE in chondrules and mineral fragments from Murchison (CM2): alteration or igneous?
N. Nakamura, M. Inoue.
54. Annual Meeting of the Meteoritical Society, p. 167 (1991).
Abstract. – See Abstr. 012.010 for the main entry.

105.125 Shock–induced deformation recorded in the Leoville CV carbonaceous chondrite.
T. Nakamura, K. Tomeoka, H. Takeda.
54. Annual Meeting of the Meteoritical Society, p. 168 (1991).
Abstract. – See Abstr. 012.010 for the main entry.

105.126 Matrix lumps in Dhajala and Mezo–Madaras: implications for chondrule–matrix relationships in ordinary chondrites.
C. E. Nehru, M. K. Weisberg, M. Prinz, R. N. Clayton, T. K. Mayeda.
54. Annual Meeting of the Meteoritical Society, p. 169 (1991).
Abstract. – See Abstr. 012.010 for the main entry.

105.127 Re–Os chronology of IAB, IIE, and IIIAB iron meteorites.
S. Niemeyer, B. K. Esser.
54. Annual Meeting of the Meteoritical Society, p. 172 (1991).
Abstract. – See Abstr. 012.010 for the main entry.

105.128 ^{41}Ca production profile in the Allende meteorite.
K. Nishiizumi, J. R. Arnold, D. Fink, J. Klein, R. Middleton.
54. Annual Meeting of the Meteoritical Society, p. 174 (1991).
Abstract. – See Abstr. 012.010 for the main entry.

105.129 ^{10}Be and ^{53}Mn in non–Antarctic iron meteorites.
K. Nishiizumi, J. R. Arnold, M. W. Caffee, R. C. Finkel.
54. Annual Meeting of the Meteoritical Society, p. 175 (1991).
Abstract. – See Abstr. 012.010 for the main entry.

105.130 ^{36}Cl terrestrial ages of Antarctic meteorites.
K. Nishiizumi, J. R. Arnold, P. Sharma, P. W. Kubik.
54. Annual Meeting of the Meteoritical Society, p. 176 (1991).
Abstract. – See Abstr. 012.010 for the main entry.

105.131 Delivery of meteorites from the asteroid belt.
M. Nolan, R. Greenberg.
54. Annual Meeting of the Meteoritical Society, p. 177 (1991).
Abstract. – See Abstr. 012.010 for the main entry.

105.132 Radiation–induced diamond (carbonado): a possible mechanism for the origin of diamond in some meteorites.
M. Ozima, S. Zashu.
54. Annual Meeting of the Meteoritical Society, p. 183 (1991).
Abstract. – See Abstr. 012.010 for the main entry.

105.133 Plasma chemical inert gas release from the Allende meteorite.
R. L. Palma, S. Chaffee, M. Hyman, M. W. Rowe.
54. Annual Meeting of the Meteoritical Society, p. 184 (1991).
Abstract. – See Abstr. 012.010 for the main entry.

105.134 Reheating of Allende components before accretion.
H. Palme, S. Weinbruch, A. El Goresy.
54. Annual Meeting of the Meteoritical Society, p. 185 (1991).
Abstract. – See Abstr. 012.010 for the main entry.

105.135 Exceptionally unfractionated solar noble gases in the H3–H6 chondrite ACFER.
A. Pedroni, H. W. Weber.
54. Annual Meeting of the Meteoritical Society, p. 186 (1991).
Abstract. – See Abstr. 012.010 for the main entry.

105.136 Exotic clasts in meteoritic breccias.
P. Pellas.
54. Annual Meeting of the Meteoritical Society, p. 187 (1991).
Abstract. – See Abstr. 012.010 for the main entry.

105.137 Vis/near IR reflectance spectra of CI/CM Antarctic consortium meteorites: B7904, Y82162, Y86720.
C. M. Pieters, D. Britt, J. Bishop.
54. Annual Meeting of the Meteoritical Society, p. 189 (1991).
Abstract. – See Abstr. 012.010 for the main entry.

105.138 Chromium isotopic compositions of individual spinel crystals from the Murchison meteorite.
F. A. Podosek, C. A. Prombo, E. K. Zinner, L. Grossman.
54. Annual Meeting of the Meteoritical Society, p. 190 (1991).
Abstract. – See Abstr. 012.010 for the main entry.

105.139 Are chondrules precursors of some cosmic spherules?
T. Presper, H. Palme.
54. Annual Meeting of the Meteoritical Society, p. 191 (1991).
Abstract. – See Abstr. 012.010 for the main entry.

105.140 LEW88055: aubritic inclusions in a Si–free iron meteorite.
M. Prinz, M. K. Weisberg, N. Chatterjee.
54. Annual Meeting of the Meteoritical Society, p. 192 (1991).
Abstract. – See Abstr. 012.010 for the main entry.

105.141 S–process Ba in SiC from Murchison series KJ.
C. A. Prombo, F. A. Podosek, S. Amari, E. Anders, R. S. Lewis.
54. Annual Meeting of the Meteoritical Society, p. 193 (1991).
Abstract. – See Abstr. 012.010 for the main entry.

105.142 Two relatively young impact craters near Waupun, Wisconsin.
W. F. Read.
54. Annual Meeting of the Meteoritical Society, p. 195 (1991).
Abstract. – See Abstr. 012.010 for the main entry.

105.143 Cosmogenic–nuclide production in very large meteorites.
R. C. Reedy.
54. Annual Meeting of the Meteoritical Society, p. 196 (1991).
Abstract. – See Abstr. 012.010 for the main entry.

105.144 Chromium isotopic composition in the enstatite chondrite Qingzhen and in magnetite of Orgueil.
M. Rotaru, J. L. Birck, C. J. Allegre.
54. Annual Meeting of the Meteoritical Society, p. 198 (1991).
Abstract. – See Abstr. 012.010 for the main entry.

105.145 Silicate darkening and heterogeneous plagioclase in CK and ordinary chondrites.
A. E. Rubin.
54. Annual Meeting of the Meteoritical Society, p. 199 (1991).
Abstract. – See Abstr. 012.010 for the main entry.

105.146 A new kind of meteoritic diamond in Abee.
S. S. Russell, C. T. Pillinger, J. W. Arden.
54. Annual Meeting of the Meteoritical Society, p. 200 (1991).
Abstract. – See Abstr. 012.010 for the main entry.

105.147 Meteoritic silicon carbide – separate grain populations and multiple components revealed by stepped combustion.
S. S. Russell, R. D. Ash, C. T. Pillinger, J. W. Arden.
54. Annual Meeting of the Meteoritical Society, p. 201 (1991).
Abstract. – See Abstr. 012.010 for the main entry.

105.148   **A survey of CAIs in Leoville and Vigarano: RIM layers, brecciation, metamorphism, and alteration.**
A. Ruzicka, W. V. Boynton.
54. Annual Meeting of the Meteoritical Society, p. 202 (1991).
Abstract. – See Abstr. 012.010 for the main entry.

105.149   **Zone sequences, widths and compositions of olivine coronas in mesosiderites.**
A. Ruzicka, W. V. Boynton.
54. Annual Meeting of the Meteoritical Society, p. 203 (1991).
Abstract. – See Abstr. 012.010 for the main entry.

105.150   **Mineralogical study of metals in MAC88177 with reference to S–type asteroids.**
J. Saito, H. Takeda.
54. Annual Meeting of the Meteoritical Society, p. 205 (1991).
Abstract. – See Abstr. 012.010 for the main entry.

105.151   **The determination of platinum group elements (PGE) in target rocks and fall–back material of the Nördlinger Ries impact crater (Germany).**
G. Schmidt, E. Pernicka.
54. Annual Meeting of the Meteoritical Society, p. 207 (1991).
Abstract. – See Abstr. 012.010 for the main entry.

105.152   **Are twin craters caused by double impactors?**
P. H. Schultz, D. E. Gault.
54. Annual Meeting of the Meteoritical Society, p. 209 (1991).
Abstract. – See Abstr. 012.010 for the main entry.

105.153   **Impact heating of shocked chondrites.**
E. R. D. Scott, K. Keil, D. Stöffler.
54. Annual Meeting of the Meteoritical Society, p. 210 (1991).
Abstract. – See Abstr. 012.010 for the main entry.

105.154   **Volatile loss during chondrule formation.**
D. W. G. Sears, Lu Jie, P. H. Benoit.
54. Annual Meeting of the Meteoritical Society, p. 211 (1991).
Abstract. – See Abstr. 012.010 for the main entry.

105.155   **Profiles of Ti^{3+}/Ti^{tot} ratios in zoned fassaite in Allende refractory inclusions.**
S. B. Simon, L. Grossman.
54. Annual Meeting of the Meteoritical Society, p. 213 (1991).
Abstract. – See Abstr. 012.010 for the main entry.

105.156   **Implications of chondrule sorting and low matrix of type 3 ordinary chondrites.**
W. R. Skinner, J. M. Leenhouts.
54. Annual Meeting of the Meteoritical Society, p. 215 (1991).
Abstract. – See Abstr. 012.010 for the main entry.

105.157   **High sensitivity survey chemical analysis of metal–rich meteorites by secondary ion mass spectrometry and glow discharge mass spectrometry.**
S. P. Smith, J. C. Huneke.
54. Annual Meeting of the Meteoritical Society, p. 216 (1991).
Abstract. – See Abstr. 012.010 for the main entry.

105.158   **Carbon and oxygen isotope composition of carbonates from an L6 chondrite: evidence for terrestrial weathering from the Holbrook meteorite.**
R. A. Socki, E. K. Gibson, H. R. Karlsson, A. J. T. Jull.
54. Annual Meeting of the Meteoritical Society, p. 217 (1991).
Abstract. – See Abstr. 012.010 for the main entry.

105.159   **Magnesium isotopic fractionation in refractory inclusions: indications for a mineralogic control.**
G. Srinivasan, J. N. Goswami, A. A. Ulyanov.
54. Annual Meeting of the Meteoritical Society, p. 218 (1991).
Abstract. – See Abstr. 012.010 for the main entry.

105.160   **Magnetization of meteorites by dynamo–generated magnetic fields in the solar nebula.**
T. F. Stepinski.
54. Annual Meeting of the Meteoritical Society, p. 220 (1991).
Abstract. – See Abstr. 012.010 for the main entry.

105.161   **New shock classification of chondrites: implications for parent body impact histories.**
D. Stöffler, K. Keil, E. R. D. Scott.
54. Annual Meeting of the Meteoritical Society, p. 221 (1991).
Abstract. – See Abstr. 012.010 for the main entry.

105.162   **Proposal for a revised petrographic shock classification of chondrites.**
D. Stöffler, K. Keil, E. R. D. Scott.
54. Annual Meeting of the Meteoritical Society, p. 222 (1991).
Abstract. – See Abstr. 012.010 for the main entry.

105.163   **Nitrogen isotope in eucrite Yamato–792510.**
N. Sugiura, K. Hashizume.
54. Annual Meeting of the Meteoritical Society, p. 223 (1991).
Abstract. – See Abstr. 012.010 for the main entry.

105.164   **Noble gases in the Monticello howardite.**
T. D. Swindle, M. K. Burkland.
54. Annual Meeting of the Meteoritical Society, p. 224 (1991).
Abstract. – See Abstr. 012.010 for the main entry.

105.165   **Noble gases in Y–74063(unique).**
N. Takaoka, K. Nagao, Y. Miura.
54. Annual Meeting of the Meteoritical Society, p. 227 (1991).
Abstract. – See Abstr. 012.010 for the main entry.

105.166   **Thermal release pattern of Hg isotopes in chondrites.**
A. N. Thakur.
54. Annual Meeting of the Meteoritical Society, p. 229 (1991).
Abstract. – See Abstr. 012.010 for the main entry.

105.167   **Iddingsite in the Nakhla meteorite: TEM study of mineralogy and texture of pre–terrestrial (Martian?) alterations.**
A. H. Treiman, J. L. Gooding.
54. Annual Meeting of the Meteoritical Society, p. 231 (1991).
Abstract. – See Abstr. 012.010 for the main entry.

105.168   **A new Xe component in diamond–rich acid residues from Efremovka CV3 carbonaceous chondrite.**
A. Verchovsky, U. Ott.
54. Annual Meeting of the Meteoritical Society, p. 232 (1991).
Abstract. – See Abstr. 012.010 for the main entry.

105.169   **Water depletion in tektites.**
A. M. Vickery, L. Browning.
54. Annual Meeting of the Meteoritical Society, p. 233 (1991).
Abstract. – See Abstr. 012.010 for the main entry.

105.170   **Cosmogenic nuclides in short–lived meteorites.**
S. Vogt, A. Albrecht, G. F. Herzog, J. Klein, D. Fink, R. Middleton, H. Weber, L. Schultz.
54. Annual Meeting of the Meteoritical Society, p. 234 (1991).
Abstract. – See Abstr. 012.010 for the main entry.

105.171   **^{41}Ca and ^{36}Cl depth profiles in the iron meteorite grant.**
S. Vogt, G. F. Herzog, D. Fink, J. Klein, R. Middleton, G. Korschinek.
54. Annual Meeting of the Meteoritical Society, p. 235 (1991).
Abstract. – See Abstr. 012.010 for the main entry.

105.172   **Cosmogenic ^{26}Al activities in Antarctic and non–Antarctic meteorites.**
J. F. Wacker.
54. Annual Meeting of the Meteoritical Society, p. 236 (1991).
Abstract. – See Abstr. 012.010 for the main entry.

105.193 **An APFIM investigation of a weathered region of the Santa Catharina meteorite.**
M. K. Miller, K. F. Russell.
Surf. Sci., Vol. 266, No. 1–3, p. 441 – 445 (15 Apr 1992). Paper presented at the 38. International Field Emission Symposium (IFES–38), Vienna (Austria), 5 – 9 Aug 1991.

An atom probe field ion microscope characterization has been performed on a weathered region of the Santa Catharina meteorite (USNM No. 3043). This meteorite consists primarily of a metallic matrix, oxides, and Schreibersite (FeNi)$_3$P. The metallic matrix was found to have decomposed into a random distribution of ∼ 12 nm diameter precipitates with compositions ranging from 38.6 – 54 at.% Ni in a matrix with lower nickel levels. Atom probe composition profiles between precipitates revealed a characteristic W–shaped nickel profile ranging from approximately 15% nickel adjacent to the precipitates to approximately 25% nickel in the central region away from the precipitates.

105.194 **Samarium–neodymium evolution of meteorites.**
A. Prinzhofer, D. A. Papanastassiou, G. J. Wasserburg.
Geochim. Cosmochim. Acta, Vol. 56, No. 2, p. 797 – 815 (Feb 1992).

The authors have obtained Sm–Nd data on two differentiated meteorites, Ibitira, a eucrite with distinct basaltic texture and with evidence of crystallization; and Morristown, a group 3A mesosiderite; as well as on Acapulco, an unclassified meteorite with chondritic chemical composition and a highly recrystallized texture. The presence of in situ decay of short–lived ^{146}Sm is demonstrated in these meteorites with initial abundance of ^{146}Sm/^{144}Sm from 0.009 to 0.007 for the different meteorites. The results indicate that three meteorites studied, some with very low REE concentrations including a mesosiderite, are relatively ancient objects, formed within the first 50 to 100 m.y. of the solar system, by planetary differentiation and impact processes, and were subjected to late metamorphism.

105.195   Chassigny and the nakhlites: carbon–bearing components and their relationship to martian environmental conditions.
I. P. Wright, M. M. Grady, C. T. Pillinger.
Geochim. Cosmochim. Acta, Vol. 56, No. 2, p. 817 – 826 (Feb 1992).

The carbon and nitrogen inventories of Chassigny and the nakhlites have been investigated by low–resolution (100°C temperature increment) stepped combustion; in addition, the contents and isotopic compositions of carbonate minerals have been assessed by the use of an acid–dissolution technique. The meteorites investigated were found to contain 2.5 – 30 ppm carbon as carbonate. Variation in δ^{13}C and δ^{18}O of the carbonates indicates either a change in conditions during formation of the carbonate minerals or that there may be two distinct carbon sources.

105.196   Classification of mafic clasts from mesosiderites: implications for endogenous igneous processes.
A. E. Rubin, D. W. Mittlefehldt.
Geochim. Cosmochim. Acta, Vol. 56, No. 2, p. 827 – 840 (Feb 1992).

The authors have analyzed thirteen igneous pebbles from the Vaca Muerta, EET87500, and Bondoc mesosiderites by electron microprobe and instrumental neutron activation and combined these data with literature data for forty–three analyzed mesosiderite clasts. They classifly these well–characterized clasts into the five principal groups.

105.197   Refractory inclusions with unusual chemical compositions from the Vigarano carbonaceous chondrite.
P. J. Sylvester, L. Grossman, G. J. MacPherson.
Geochim. Cosmochim. Acta, Vol. 56, No. 3, p. 1343 – 1363 (Mar 1992).

Ten inclusions, nine from Vigarano and one from Leoville, both members of the reduced subgroup of C3V chondrites, were analyzed for major and trace elements by neutron activation. Most have some refractory element characteristics that are common in refractory inclusions from Allende, a member of the oxidized subgroup of C3V chondrites. Six of the Vigarano inclusions, however, have refractory element fractionations that are unusual in Allende inclusions. Vigarano apparently sampled a different population of refractory inclusions from Allende, presumably because refractory nebular materials were not well mixed where and when C3V chondrites accreted.

105.198   CI chondrite–like clasts in the Nilpena polymict ureilite: implications for aqueous alteration processes in CI chondrites.
A. J. Brearley, M. Prinz.
Geochim. Cosmochim. Acta, Vol. 56, No. 3, p. 1373 – 1386 (Mar 1992).

The authors have carried out a detailed petrographic and mineralogical study of carbonaceous chondrite matrix clasts in the Nilpena polymict ureilite. The bulk compositions of a number of clasts, determined by electron microprobe, show that they have close affinities to CI chondrite matrix and differ substantially from CM matrix material. Transmission electron microscopy studies show that the mineralogy of the clasts is also consistent with CI chondrite matrix. Based on the oxygen isotopic composition of the matrix, the phase chemistry, and the bulk clast chemistry, the authors suggest that the matrix clasts in Nilpena represent material which is less altered than Orgueil. The difference on oxygen isotopic composition of the Nilpena matrix

clasts from CI chondrites may be because the clasts underwent alteration at a lower water/rock ratio.

105.199   The oldest zircons in the solar system.
T. R. Ireland, F. Wlotzka.
Earth Planet. Sci. Lett., Vol. 109, No. 1/2, p. 1 – 10 (Mar 1992).

The authors report the occurrence, chemistry, and U–Th–Pb isotopic systematics of three meteoritic zircon assemblages, two from the Vaca Muerta mesosiderite and one from the Simmern H5 chondrite.

105.200   The implications of the magnetism of ordinary chondrite meteorites.
S. J. Morden, D. W. Collinson.
Earth Planet. Sci. Lett., Vol. 109, No. 1/2, p. 185 – 204 (Mar 1992).

The magnetic properties of eleven ordinary chondrites (eight LL–chondrites and three L–chondrites) have been analysed. The samples were repeatedly fragmented and oriented to a common reference direction, natural remanent magnetisation (NRM), susceptibility (χ), and anisotropy of susceptibility measured at each stage. The response of some fragments to alternating field demagnetisation, thermal demagnetisation, isothermal remanent magnetisation (IRM), and magnetic hysteresis has been measured. It was found that the orientation of the stable NRM was random, down to a scale of ~ 1 mm^3. The dominant magnetic carrier in types 4–6 was the ordered iron–nickel mineral γ"–tetrataenite. The type 3's measured contained no tetrataenite and the NRM was carried by fine–grained taenite. A magnetic fabric, analogous to a physical fabric, was found in all samples. In all but two, the fabric was foliated. Tuxtuac and Wold Cottage showed lineation. Thermal demagnetisations of NRM showed a trend towards higher blocking temperatures for type 6 chondrites over types 3–5. Because the fabric was continuous in the majority of samples, and the NRM randomly orientated, it was concluded that the magnetic carriers were magnetised before emplacement in the meteorite, and that the meteorites are not finescale breccias. The preservation of the random NRM leads to the preference of hot accretion as the mechanism for producing chondritic textures, as opposed to metamorphic reheating, as this would tend to erase the random magnetisation.

105.201   Measurement of cosmogenic radionuclides in meteorites with a sensitive gamma–ray spectrometer.
G. Bonino, G. Cini Castagnoli, N. Bhandari.
Nuovo Cimento C, Vol. 15, N. 1, p. 99 – 104 (Jan–Feb 1992).

A large–volume HPGe gamma–ray spectrometer in a NaI(Tl) well has been set up underground at 70 m.w.e. depth for whole body counting of cosmogenic radionuclides in meteorites. The detectors are housed in a 20 cm thick lead shield with a lining of cadmium and OFHC copper. The scintillator is simultaneously operated in anticoincidence as well as in coincidence in selected energy channels to achieve low background levels (in the range of counts per day) and high specificity. In this way a large number of radionuclides such as ^{26}Al, ^{44}Ti, ^{60}Co, ^{22}Na, ^{54}Mn, and shorter–lived nuclides produced in extraterrestrial materials like meteorites and lunar rocks can be analysed. Results on Bouvante and Bereba achondrites and Dhajala and Torino chondrites are presented.

105.202   Teardrops on the Pampas.
P. H. Schultz, J. K. Beatty.
Sky Telesc., Vol. 83, No. 4, p. 387 – 392 (Apr 1992).

A curious pilot's photographs have revealed a chain of unique impact craters gouged into the verdant plains of central Argentina.

105.203   Radiation history of Antarctic and non–Antarctic meteorites.
V. A. Alekseev.
Sol. Syst. Res., Vol. 25, No. 2, p. 172 – 181 (Sep 1991). English translation of Astron. Vestn., Vol. 25, No. 2, p. 233 – 244 (1991).

Radiation age distributions of non–Antarctic and Antarctic H–chondrites are studied. Peaks are found in the distributions corresponding to ages of 6.3 ± 0.2 Myr for non–Antarctic meteorites and 6.4 ± 0.4 Myr for Antarctic. The similarity in radiation ages most probably indicates origin of both non–Antarctic and Antarctic H–chondrites from a single parent body. The content of long–lived cosmogenic radionuclides ^{26}Al ($T_{1/2} = 0.705$ Myr) and ^{53}Mn (3.7 Myr) was studied in 880 non–Antarctic and Antarctic H–chondrites. In non–Antarctic H– and L–, LL–chondrites the mean ^{26}Al contents were identical (within standard deviation), while in Antarctic H–chondrites the mean ^{26}Al content was $15 \pm 3\%$ higher than in Antarctic L– and LL–chondrites. The data obtained can be explained by the proposal of approximately constant frequency of incidence on Earth of L– and LL–chondrites, while the frequency of incidence of H–chondrites changed over time, being at a maximum $\sim n \cdot 10^4$yr ago. This is the cause of the lower mean terrestrial age of Antarctic H–chondrites (~ 40 kyr) as compared to the greater mean terrestrial age of Antarctic L– and LL–chondrites (~ 230 kyr).

105.204 Carbon–rich micrometeorites and prebiotic synthesis.
M. Maurette, P. Bonny, A. Brack, C. Jouret, M. Pourchet, P. Siry.
3. Symposium International de Bioastronomie, p. 124 – 132 (1991). – See Abstr. 012.005 for the main entry.

About 5000 unmelted and well preserved "giant" chondritic micrometeorites with size $\sim 50 – 200$ μm have been extracted from ~ 100 tons of antarctica blue ice. They have been unexpectedly well shielded against both terrestrial weathering and frictional heating in the atmosphere. Mineralogical studies indicate that they are all related to primitive "unequilibrated" meteorites (mostly carbonaceous chondrites). About 50% of them are made of friable and porous aggregates of submicron–sized grains, that represent a highly desequilibrated assemblage of minerals, metal oxides and sulfides, and some carbonaceous material related to the broad family of "hydrogenated refractory carbon". Each carbon–rich micrometeorite might have behaved as a "minicenter" of prebiotic synthesis on the early Earth, through the "in–situ" catalyzed hydrolysis of this carbonaceous material.

105.205 A compositional classification scheme for meteoritic chondrules.
D. W. G. Sears, Lu Jie, P. H. Benoit, J. M. DeHart, G. E. Lofgren.
Nature, Vol. 357, No. 6375, p. 207 – 210 (21 May 1992).

A taxonomic scheme is proposed for the main component of the primitive chondrite meteorites – the chondrules. The scheme provides insight into the variety of chondrules originally produced in the primordial solar nebula, and the effects on them of secondary processes on meteorite parent bodies. It may also contribute to understanding the origin of compositional diveristy in the chondrites.

105.206 Lunar mare meteorites.
P. H. Warren, G. W. Kallemeyn.
Workshop on Mare Volcanism and Basalt Petrogenesis: "Astounding Fundamental Concepts (AFC)" Developed Over the Last Fifteen Years, p. 63 – 64 (1991). Abstract. – See Abstr. 012.028 for the main entry.

105.207 Alteration of tektite to form weathering products.
J. J. Mazer, J. K. Bates, J. P. Bradley, C. R. Bradley, C. M. Stevenson.
Nature, Vol. 357, No. 6379, p. 573 – 576 (18 Jun 1992). Letter–to–the–editor.

Recent use of tektites as evidence for a bolide impact at the Cretaceous/Tertiary (K/T) boundary has focused attention on their long–term stability. It was proposed in these studies that residual clay features with the spherical tektite morphology result from in situ alteration of the original glassy material. By contrast, examination of tektite alteration as an analogue for the long–term degradation of nuclear waste glass has revealed no evidence of alteration, hydration or devitrification either for samples found in nature or for those reacted in the laboratory: no residual clay minerals were observed, and therefore the glass was interpreted as having reacted by a complete dissolution or etching process. The authors show that these apparently incongruent observations can be reconciled through understanding the relationship between the environment in which the glass reacts and the chemical processes that control the reaction rate. Alteration of tektites to clays, as observed at the K/T boundary, can proceed only under conditions of limited water contact.

105.208 Distribution of water on Mars: implications from SNC meteorites.
J. H. Jones.
Workshop on the Martian Surface and Atmosphere Through Time, p. 78 – 79 (1992). Abstract. – See Abstr. 012.031 for the main entry.

105.209 Time of transit and trajectory of the Tungus meteorite, from data collected in 1908.
A. A. Yavnel'.
Sol. Syst. Res., Vol. 25, No. 4, p. 381 – 386 (Jan 1992). English translation of Astron. Vestn., Tom 25, No. 4, p. 505 – 511 (1991).

By comparing data on observation of the Tungus bolide of June 30, 1908, collected in various years, it is shown that the most precise results regarding the moment of its transit, close to seismic data, are given by information collected in 1908 (7 – 8 a.m.) while evidence of eye–witnesses collected in 1959 – 1974 contains large errors and may even refer to other bolides. On the basis of data obtained in 1908, the azimuth of the trajectory is estimated ($70 – 130°$ to the east of the meridian) together with the inclination ($\geqslant 25°$ above the horizon). These results do not contradict calculations based on the forest damage data. In the final outcome it is shown that on June 30, 1908 a single bolide was observed in the morning, moving in the general direction from east to west.

105.210 Petrogenesis of the nakhlite meteorites: evidence from cumulate mineral zoning.
R. P. Harvey, H. Y. McSween Jr.
Geochim. Cosmochim. Acta, Vol. 56, No. 4, p. 1655 – 1663 (Apr 1992).

A simple igneous petrogenesis for the meteorite Nakhla has previously been called into question because Mg/Fe ratios in olivine indicate substantial disequilibrium between the predominant cumulus minerals (olivine and augite). Comparative analyses of simulated diffusive zoning and the observed cumulus mineral zoning for all three nakhlites (Nakhla, Governador Valadares, and Lafayette) show that their current compositions do not necessarily reflect parental magma compositions. The nakhlites appear to be a series of relatively simple cumulate rocks which have undergone various amounts of late–magmatic and subsolidus diffusion, possibly reflecting their relative positions in a cooling cumulate pile.

105.211 Pregraphitic and poorly graphitised carbons in porous chondritic micrometeorites.
F. J. M. Rietmeijer.
Geochim. Cosmochim. Acta, Vol. 56, No. 4, p. 1665 – 1671 (Apr 1992).

Two forms of crystalline carbon in porous chondritic micrometeorites W7029E5, U2011C2, and U2022C7/C8 are mixed layered, pregraphitic carbons. Mixed layered carbons represent incomplete carbonisation and graphitisation of precursor material. In U2022C7/C8, carbonisation mostly involved volatile loss. The formation of pregraphitic carbons indicates a sustained thermal regime in parent bodies of these micrometeorites (i.e., short–period comets, outer–belt asteroids, or protocomet nuclei). Temperatures of the sustained thermal regime remain unspecified, but carbon reactions were probably facilitated by catalytic support from layer silicates in these samples. Poorly graphitised carbon in U2022C7/C8 formed during a transient thermal event which is most likely flash–heating during micrometeoroid deceleration in the Earth's atmosphere.

105.212   Age and isotopic relationships among the angrites Lewis Cliff 86010 and Angra dos Reis.

G. W. Lugmair, S. J. G. Galer.

Geochim. Cosmochim. Acta, Vol. 56, No. 4, p. 1673 – 1694 (Apr 1992).

Results of a wide–ranging isotopic investigation of the unique Antarctican angrite LEW–86010 (LEW) are presented, together with a reassessment of the type angrite Angra dos Reis (ADOR). The principal objectives of this study are to obtain precise radiometric ages, initial Sr isotopic compositions, and to search for the erstwhile presence of the short–lived nuclei ^{146}Sm and ^{26}Al via their daughter products. The isotopic compositions of Sm, U, Ca, and Ti were also measured. This allows a detailed appraisal to be made of the relations between, and the geneology of, these two angrites. The age and isotopic constraints are discussed with respect to current collapse, condensation, and accretion timescales calculated for the solar nebula.

105.213   A shock–metamorphic model for silicate darkening and compositionally variable plagioclase in CK and ordinary chondrites.

A. E. Rubin.

Geochim. Cosmochim. Acta, Vol. 56, No. 4, p. 1705 – 1714 (Apr 1992).

Silicate darkening in ordinary chondrites (OC) is caused by tiny grains of metallic Fe–Ni and troilite occurring mainly within curvilinear trails that traverse silicate interiors and decorate or, in some cases, cut across silicate grain boundaries. Highly shocked OC (characterized by olivine grains with undulose to mosaic extinction) tend to have greater degrees of silicate darkening than lightly shocked OC; this indicates that silicate darkening is probably a result of shock metamorphism. A few OC also contain thin melt veins of chromite; this implies that localized shock temperatures reached $\geqslant 1635°C$ in these meteorites. Silicate darkening is also evident in CK carbonaceous chondrites, where magnetite and pentlandite grains form analogous curvilinear trails. It is possible that magnetite and pentlandite in CK chondrites were mobilized during shock metamorphism and dispersed through silicate interiors; alternatively, metal and troilite may have been dispersed and then transformed into magnetite and pentlandite during subsequent oxidation of these chondrites.

105.214   Isotopic, optical, and trace element properties of large single SiC grains from the Murchison meteorite.

A. Virag, B. Wopenka, S. Amari, E. Zinner, E. Anders, R. S. Lewis.

Geochim. Cosmochim. Acta, Vol. 56, No. 4, p. 1715 – 1733 (Apr 1992).

Forty–one large SiC grains from the Murchison CM2 chondrite were analyzed by ion probe mass spectrometry for the isotopic compositions of C, N, Mg, and Si, and the concentration of Al, Ti, V, Fe, Zr, and Ba. Most grains were also examined by Raman spectroscopy. The majority have large isotopic anomalies, with $^{13}C/^{12}C$ and $^{14}N/^{15}N$ up to $30 \times$ and $9 \times$ solar, and ^{29,30}Si. Only two grains, characterized by extremely heavy carbon give evidence for fossil ^{26}Mg, with $(^{26}Al/^{27}Al)_0$ ratios of 2.1×10^{-3} and 3.9×10^{-3}. On the basis of C and Si isotopic composition, twenty–nine of the grains fall into three compact clusters, presumably from three discrete sources. Two of these clusters are anomalous and comprise only grains of cubic structure (according to their Raman spectra). The third, isotopically, normal cluster contains only anhedral, noncubic grains; and although contamination cannot be categorically excluded, an origin in a reducing environment in the early solar system is a viable possibility.

105.215   Silicate darkening in ordinary chondrite parent body regoliths: evidence from gas–rich and shock–blackened ordinary chondrites.

D. T. Britt, L. A. Lebofsky.

Bull. Am. Astron. Soc., Vol. 23, No. 3, p. 1139 – 1140 (1991). Abstract. – See Abstr. 012.037 for the main entry.

105.216   Size–dependent composition in interplanetary material.

J. F. Bell.

Bull. Am. Astron. Soc., Vol. 23, No. 3, p. 1140 (1991). Abstract. – See Abstr. 012.037 for the main entry.

105.217   Black ordinary chondrites: an analyis of abundance and fall frequency.

D. T. Britt, C. M. Pieters.

Meteoritics, Vol. 26, No. 4, p. 279 – 285 (Dec 1991).

Black ordinary chondrite meteorites sample the spectral effects of shock on ordinary chondrite material in the space environment. Since shock is an important regolith process, these meteorites may provide insight into the spectral properties of the regoliths on ordinary chondrite parent bodies. To determine how common black chondrites are in the meteorite collection and, by analogy, the frequency of shock–alteration in ordinary chondrites, several of the world's major meteorite collections were examined to identify black chondrites. Over 80% of all catalogued ordinary chondrites were examined and, using an optical definition, 61 black chondrites were identified. Black chondrites account for approximately 13.7% of ordinary chondrite falls.

105.218   Mineralogy and possible origin of an unusual Cr–rich inclusion in the Los Martinez (L6) chondrite.

A. J. Brearley, I. Casanova, M. L. Miller, K. Keil.

Meteoritics, Vol. 26, No. 4, p. 287 – 300 (Dec 1991).

During a petrological study of the previously unclassified ordinary chondrite Los Martínez, we discovered a highly unusual Cr–rich inclusion which the authors believe is unique in both extraterrestrial and terrestrial mineralogy. The inclusion is highly zoned both compositionally and optically, with a Ca–Al rich, cloudy core and an opaque, Cr–Na–rich rim. Detailed SEM and TEM studies show that the inclusion now consists of a highly zoned, single crystal of plagioclase intergrown with chromium–rich spinel. The spinel has a well–developed crystallographic orientation relationship with the host plagioclase, which indicates that it is the product of exsolution. Although superficially similar to a plagioclase feldspar in composition, in detail the inclusion is Si–deficient and Al–enriched relative to a stoichiometric feldspar. The authors have not been able to identify a viable precursor mineral phase to the plagioclase–chromite intergrowth and suggest that it may be an unknown metastable phase.

105.219   Spinel–bearing, Al–rich chondrules in two chondrite finds from Roosevelt County, New Mexico: indicators of nebular and parent body processes.

T. J. McCoy, A. Pun, K. Keil.

Meteoritics, Vol. 26, No. 4, p. 301 – 309 (Dec 1991).

Two rare, spinel–bearing, Al–rich chondrules have been identified in new chondrite finds from Roosevelt County, New Mexico – RC 071(L4) and RC 072(L5). These chondrules have unusual mineralogies, dominated by highly and asymmetrically zoned, Al–rich spinels. Two alternatives exist to explain the origin of this zoning–fractional crystallization or metamorphism. These two chondrules cooled rapidly from near liquidus, as indicated by the zoning, occurrence and sizes of spinels, radiating chondrule textures and localized chromite depletions. The range of mineralogies in other Al–rich chondrules of similar composition reflect a range of peak temperatures and cooling rates. The authors see no reason to believe that this range is fundamentally different from the range of thermal histories experienced by "normal" Fe–Mg–rich chondrules.

105.220   Hydrogen and carbon isotopic composition of volatiles in Nakhla: implications for weathering on Mars.

L. L. Watson, S. Epstein, E. M. Stolper.

Workshop on the Martian Surface and Atmosphere Through Time, p. 165 – 166 (1992). Abstract. – See Abstr. 012.031 for the main entry.

105.221 **On the isotopic composition of magmatic carbon in SNC meteorites.**
I. P. Wright, M. M. Grady, C. T. Pillinger.
Workshop on the Martian Surface and Atmosphere Through Time, p. 169 – 170 (1992). Abstract. – See Abstr. 012.031 for the main entry.

105.222 **High–pressure melting of carbonaceous chondrite.**
C. B. Agee.
Workshop on the Physics and Chemistry of Magma Oceans from 1 bar to 4 Mbar, p. 11 – 12 (1992). Abstract. – See Abstr. 012.032 for the main entry.

105.223 **Where are the ordinary chondrite parent bodies? A review of the 1991 Meteoritical Society special session.**
M. J. Gaffey.
Bull. Am. Astron. Soc., Vol. 23, No. 3, p. 1152 (1991). Abstract. – See Abstr. 012.037 for the main entry.

105.224 **Mössbauer, X–rays and SEM analysis of meteorites from the mineralogy museum of the University of Parma (Italy).**
A. Bonazzi, I. Ortalli, G. Pedrazzi, K. Jiang, X. Zhang.
Hyperfine Interact., Vol. 70, No. 1–4, p. 953 – 956 (Apr 1992).
Paper presented at the International Conference on the Applications of the Mössbauer Effect (ICAME'91), Nanjing (People's Republic of China), 16 – 20 Sep 1991.
Five meteorites, belonging to the private collection of the Mineralogy Museum of the University of Parma have been analyzed by Mössbauer spectroscopy, X–rays diffraction, X–rays fluorescence and by scanning electron microscope. Following standard classification they have been assigned to the ordinary chondrites class, L type, with different minor compositions.

105.225 **Temperature dependence of the Mössbauer parameters of the Fe–Ni phases in the Santa Catharina meteorite.**
E. de Grave, R. E. Vandenberghe, P. M. A. de Bakker, A. van Alboom, R. Vochten, R. van Tassel.
Hyperfine Interact., Vol. 70, No. 1–4, p. 1009 – 1012 (Apr 1992).
Paper presented at the International Conference on the Applications of the Mössbauer Effect (ICAME'91), Nanjing (People's Republic of China), 16 – 20 Sep 1991.
The temperature variation in the range 8 – 760K of the hyperfine parameters of the Fe–Ni phases in the Santa Catharina meteorite has been determined. It is suggested that the disordered 50–50 Fe–Ni phase actually consists of two distinct fractions, i.e. a completely disordered phase and one with intermediate long–range ordering parameter. The single–line subspectrum of the 28%–Ni phase was found to display magnetic ordering below approximately 25K.

105.226 **Against all odds – meteorites that have struck home.**
C. Spratt, S. Stephens.
Mercury, Vol. 21, No. 2, p. 50 – 56 (Mar – Apr 1992).

105.227 **Meteoritics and the origins of atomic nuclei.**
D. D. Clayton.
Meteoritics, Vol. 27, No. 1, p. 5 – 17 (Mar 1992). The Leonard Medal address, presented 25 Jul 1991 at Monterey, CA (USA).
The science of nucleosynthesis was substantially inspired by chemical analyses of meteorites. As if in repayment, that theory now imbues meteoritics with enlarged meaning. The author recounts the emergence of four great issues for nucleosynthesis – issues that received decades of the author's own attention; and he describes unexpected abundance patterns within meteorites that were suggested by the resolution of those issues. The latter have altered the information content of meteoritic science. The issues are: (1) a quantitative s–process theory, (2) cosmoradiogenic chronology, (3) explosive nucleosynthesis and gamma–ray astronomy, (4) cosmic chemical memory. Starting from historical origins for each issue, the author comments upon both the broad cultural canvas in which they lie and his own work in their establishment. Examples of predicted (or rationalized) meteoritic

measurements illustrate the surprised delight at the expansion of the range and power of meteoritic science.

105.228 **Mineral compositions in Antarctic and Greenland micrometeorites.**
M. C. Michel–Levy, M. Bourot–Denise.
Meteoritics, Vol. 27, No. 1, p. 73 – 80 (Mar 1992).
The mineral compositions of 250 micrometeorites have been studied and olivines and low–calcium pyroxenes with crystals larger than 5 μm have been analysed. While magnesium rich grains dominate, the Fa content of olivine may reach 50% and the Fs content of pyroxene may reach 26%. The Ca and Mn of the olivine show no consistent trends with increasing Fe, but Cr shows a negative correlation. For low–Ca pyroxene, Al and Cr contents are generally higher than in pyroxenes of equilibrated chondrites but similar to those of highly unequilibrated chondrites. All these minerals are found as coarse–grained particles often with adhering iron–rich scoria or as clasts in fine–grained or scoriaceous micrometeorites. Apart from a few particles which could be the debris of ordinary chondrites, most micrometeorites probably come from a common source similar, but not identical to carbonaceous chondrites, as shown by their lower Ni and S content and their different oxygen isotopic composition assuming two measurements performed on olivine grains prove to be typical.

105.229 **Maralinga, a metamorphosed carbonaceous chondrite found in Australia.**
L. P. Keller, J. C. Clark, C. F. Lewis, C. B. Moore.
Meteoritics, Vol. 27, No. 1, p. 87 – 91 (Mar 1992).
The Maralinga meteorite was found near the village Maralinga, South Australia in 1974, but was not recognized as a meteorite until 1989. One weathered individual was recovered with a total mas of 3.38 kg. The bulk composition and petrography of Maralinga indicate that it is a metamorphosed (petrographic type 4) carbonaceous chondrite with major similarities to the Vigarano–subtype. However, recent trace element data from the literature suggest that Maralinga should be included with the CK (Karoonda–type) carbonaceous chondrites. The authors classify Maralinga as an anomalous CK4 chondrite because of its abundant chondrules and refractory inclusions to other known members of the CK group.

105.230 **Fall days of the SNC meteorites: evidence for and SNC meteorid stream, and a common site of origin.**
A. H. Treiman.
Meteoritics, Vol. 27, No. 1, p. 93 – 95 (Mar 1992).
Four of the SNC meteorites of putative Martian origin are falls. Two of these fell on Oct 3: Chassigny in 1815 and Zagami in 1962. The probability of this coincidence arising from random fall days is approximately 1 in 60. If this coincidence is not the result of chance, it suggests that some of the SNC meteorites are derived from a meteoroid stream. In that Chassigny and Zagami span nearly the full range of SNC lithologies and histories, the coincidence of fall days is consistent with suggestions that all of the SNCs came from a single site (impact crater) on their parent planet.

105.231 **Tabbita: an L6c chondrite from New South Wales, Australia.**
A. W. R. Bevan, B. Griffin, R. E. Pogson, F. L. Sutherland.
Meteoritics, Vol. 27, No. 1, p. 97 – 98 (Mar 1992).
A crusted stone weighing 3.10 kg was found in 1983 near Tabbita in south central New South Wales, Australia. Compositions of the ferro–magnesian silicates show that the meteorite belongs to the L–group of chondrites. Uniformity of silicate compositions and the presence of abundant crystalline plagioclase feldspar show that the meteorite belongs to petrologic type 6. Silicates that display undulose extinction, and the absence of any thermal effects induced by shock indicate that Tabbita is shock facies c. Tabbita is distinct from several other L6 chondrites found in the same general area.

105.232   Tektite–like bodies at Lonar Crater, India? Very unlikely.
R. F. Fudali, K. Fredriksson.
Meteoritics, Vol. 27, No. 1, p. 99 – 100 (Mar 1992).

The "tektite–like bodies" reported recently from Lonar Crater, India are, in fact, high sodium artificial glasses and so need to be explained by unrealistic, natural mixing models. These bodies have no bearing on the problem of impactite chemistry or tektite generation.

105.233   Uranium accumulation during weathering of Cañon Diabolo meteoritic iron.
B. A. Hofmann.
Meteoritics, Vol. 27, No. 1, p. 101 – 103 (Mar 1992).

Cañon Diablo meteoritic iron oxide consists mainly of goethite and maghemite and contains 209 to 630 ppb uranium compared to < 0.02 ppb in the unweathered octahedrite. Significant radioactive disequilibria between ^{238}U, ^{234}U and ^{238}Th indicate that uranium was sorbed from soil porewater during terrestrial weathering after meteorite impact 50 ka ago. Depending on the model assumed for U uptake, corrected ^{230}Th –^{234}U ages of 24 to 48 ka were obtained. While the data presented may not allow an unambiguous interpretation, the potential of this approach in obtaining minimum terrestrial ages for weathered meteorites is demonstrated.

105.234   The Meteoritical Bulletin, No. 72.
F. Wlotzka.
Meteoritics, Vol. 27, No. 1, p. 109 – 117 (Mar 1992).

Place of fall/find, class and type of individual specimens, total weight and circumstances of fall/find of some meteorites are given.

105.235   Labile trace elements in carbonaceous chondrites: a survey.
X.–y. Xiao, M. E. Lipschutz.
J. Geophys. Res., Vol. 97, No. E6, p. 10199 – 10212 (25 Jun 1992).

The authors report radiochemical neutron activation analysis data for Co, Au, Ga, Rb, Sb, Ag, Se, Cs, Te, Zn, Cd, Bi, Tl, and In (ordered by increasing putative volatility in primary nebular processes) in 42 C2–C6 chondrites, all but three from Antarctica. From these and literature data for 19 additional chondrites, C1–normalized concentrations of the nine most volatile elements (Ag→In) are quite constant in most meteorites. Trace element trends in 39 Antarctic and 22 non–Antarctic carbonaceous chondrites are similar: no evidence exists for substantial alteration by weathering of samples in Antarctica, nor do the data reflect modification by open–system, parent body metamorphism at $\geqslant 500$ °C. Volatile element concentrations and siderophile ratios (Au/Co and Ga/Co) define continua which correlate at statistically significant levels. Carbonaceous chondrites sample not a few, compositionally distinct parents but rather a compositional continuum in which parent materials forming under more oxidizing conditions incorporated lesser complements of volatiles, essentially unfractionated from cosmic composition. This may well reflect the range of formation conditions (temperature, duration, and water/rock ratios) represented by oxygen isotope variations during preterrestrial aqueous alteration of parent materials.

105.236   Analysis of impact–induced Fe^{2+} disorder in the pyroxene of the Ibitira meteorite.
T. V. V. Costa, V. W. Vieira, M. A. B. de Araujo.
Hyperfine Interact., Vol. 67, No. 1–4, p. 463 – 466 (Nov 1991). Paper presented at the Latin American Conference on the Applications of the Mössbauer Effect (LACAME '90), Havanna (Cuba), 29 Oct – 2 Nov 1992.

Mössbauer spectroscopy and X–ray diffraction analysis indicate that the only iron compound present in the Ibitira meteorite is a pyroxene known as pigeonite. The Mössbauer spectrum shows Fe^{2+} in the two different crystallographic sites and the population ratio indicates a reasonable degree of cation order. Comparison between these results and results obtained from pyroxene shocked under controlled conditions in the laboratory suggests that, the meteorite was subjected to a impact of low intensity.

105.237   Vaca Muerta mesosiderite strewnfield.
H. Pedersen, C. C. de Bon, H. Lindgren.
Meteoritics, Vol. 27, No. 2, p. 126 – 135 (Jun 1992).

A field investigation is presented of the strewnfield of the mesosiderite Vaca Muerta, originally found in 1861. The area, 11.5 km long, 2.1 km wide, is located about 60 km southeast of Taltal, Chile, in the Atacama Desert. It has yielded 80 meteorites with a total mass exceeding 3782 kg. Most fragments were found in an undisturbed state, but some had been broken by prospectors. The present studies, in connection with historical records, indicate that the original mass of Vaca Muerta exceeded 6 metric tons. One impact feature, somewhat modified by man, consists of a 10.5–m diameter, 1.7–m deep hole, without an uplifted rim. Small masses were scattered up to 85 m from the hole.

105.238   Na–bearing Ca–Al–rich inclusions in the Yamato–791717 CO carbonaceous chondrite.
K. Tomeoka, K. Nomura, H. Takeda.
Meteoritics, Vol. 27, No. 2, p. 136 – 143 (Jun 1992).

Ca–Al–rich inclusions (CAIs) in the Yamato–791717 CO carbonaceous chondrite contain 5 to 80 vol% of nepheline, along with minor sodalite, and thus are among the most nepheline–rich CAIs known. The primary phases in inclusions are mainly spine, fassaite, aluminous diopside, perovskite, and hibonite. In contrast to many CO chondrites, melilite is rare. The majority of inclusions are single concentric objects or aggregates of concentric objects. Lightly altered inclusions have cores of spinel surrounded by bands of nepheline (replacing fassaite), fassaite, and diopside. In moderately altered inclusions, spinel cores are replaced by nepheline. In heavily altered inclusions, the major part of internal areas are replaced by nepheline. In some moderately and heavily altered inclusions, only diopside rims remain unaltered. The degree of alteration in Y791717 CAIs appears to be much higher than those in CAIs in other reported meteorites.

105.239   Electrophonic sounds from large meteor fireballs.
C. S. L. Keay.
Meteoritics, Vol. 27, No. 2, p. 144 – 148 (Jun 1992).

Anomalous sounds from large meteor fireballs, anomalous because they are audible simultaneously with the sighting, have been a matter for debate for over two centuries. Only a minority of observers perceive them. Ten years ago a viable physical explanation was developed which accounts for the phenomenon in terms of ELF/VLF radiation from the fireball plasma being transduced into acoustic waves whenever appropriate objects happen to be in the vicinity of an observer. This explanation has now been verified observationally and supported by other evidence including the study of meteor fireball light curves reported here.

105.240   A unique, (almost) unaltered spinel–rich fine–grained inclusion in Kainsaz.
B. B. Holmberg, A. Hashimoto.
Meteoritics, Vol. 27, No. 2, p. 149 – 153 (Jun 1992).

The authors report a unique, spinel–rich, extremely porous fine–grained inclusion in the Kainsaz (CO3) meteorite. This inclusion is the least altered fine–grained inclusion yet discovered, having escaped almost entirely the second alterations experienced by Allende fine–grained inclusions. The inclusion is comprised of loosely packed 5 – 30 μm spinel grains mantled by thin layers of melilite, anorthite, and diopsidic pyroxene. The inclusion is one of the most spinel–rich, most porous fine–grained inclusions seen to date. The mineralogy of the inclusion matches that which has been predicted for a precursor of the altered mineral assemblages of Allende fine–grained inclusions, though a lack of interstitial material in the Kainsaz inclusion reduces the likelihood of a direct genetic relationship between the two. Its mineralogical composition confirms that the precursors of other, more altered, fine–grained inclusions were assemblages of refractory minerals exclusively.

105.241 Mechanism of Muong Nong–type tektite formation and speculation on the source of Australasian tektites.

C. C. Schnetzler.
Meteoritics, Vol. 27, No. 2, p. 154 – 165 (Jun 1992).

The source crater of the youngest and largest of the tektite strewnfields, the Australasian strewnfild, has not been located. A number of lines of evidence indicate that the Muong Nong–type tektites, primarily found in Indochina, are more primitive than the much more abundant and widespread splash–form tektites, and are proximal to the source. In this study the spatial distribution of Muong Nong–type tektite sites and chemical character have been used to indicate the approximate location of the source. The variation of Muong Nong–type tektite chemical composition appears to be caused by mixing of two silicate rock end–members and a small amount of limestone, and not by vapor fractionation. The variation in compostion is not random, and does not support in–situ melting or multiple impact theories. The distribution of both Muong Nong and splash–form tektite sites suggest the source is in a limited area near the southern part of the Thailand–Laos border.

105.242 On the thermal history of heavily shocked Yanzhuang H–chondrite.

F. Begemann, H. Palme, B. Spettel, H. W. Weber.
Meteoritics, Vol. 27, No. 2, p. 174 – 178 (Jun 1992).

Partly shock–melted Yanzhuang H–chondrite, now classified as petrological type H6 but before the shock event to type H4, was subjected to the shock–heating 2.6 Ma ago at the time it was spalled of its parent body and came into being as a meteoroid of ca. 30 cm radius. At that time the unmelted portion of the meteoroid suffered an almost complete loss of its radiogenic ^{4}He and ^{40}Ar while the contents of the most volatile non–noble gas elements Zn and Se were not measurably affected. The melted portion of Yanzhuang is also essentially void of radiogenic ^{4}He but it has retained some 80% of its radiogenic ^{40}Ar, presumably because in the melt the increase of the diffusion length more than compensated the increase of the diffusion constant.

105.243 The Hashima, Japan H4 chondrite: a newly reported meteorite.

M. Hoshino, K. Suwa.
Meteoritics, Vol. 27, No. 2, p. 179 – 181 (Jun 1992).

The authors report a new chondrite that fell in Hashima City in central Japan sometime during the period 1868 – 1912. The chondrite weighs 1110.64 g and exhibits distinct structure. Chondrules occupy 24 vol% of the stone and consist of olivine, low–Ca pyroxene, devitrified glass and lesser amounts of oligoclase, kamacite, taenite, troilite and chrominan spinel. Matrix occupying 76 vol% of the stone consists of olivine, low–Ca pryoxene, kamacite, taenite, troilite, cryptocrystalline minerals and lesser amounts of chromian spinel and chlorapatite. Matrix minerals have the same compositions as those in chondrules. Mineral chemistry, bulk chemistry and magentic properties indicate that Hashima is an H–group chondrite.

105.244 The Melnikovo LL6 chondrite: a new find from Ukraine.

A. N. Krot, N. I. Zaslavskaya, M. I. Petaev,
N. N. Kononkova, L. D. Barsukova, G. M. Kolesov.
Meteoritics, Vol. 27, No. 2, p. 182 – 183 (Jun 1992).

Melnikovo is a relatively unweathered 545.6–g LL6 chondrite that was found in 1983. Only a few poorly defined chondrules are discernable in the examined sections; two of these are enriched in chromite. The meteorite contains olivine, low–Ca pyroxene, plagioclase, rare clinopyroxene, chlorapatite, merrillite and opaque minerals, which have a modal abundance (in wt%) of troilite (3.9%), kamacite (0.4%), taenite plus tetrataenite (0.7%), chromite (0.8%), and trace amounts of ilemenite and Mn–ilemenite. The meteorite appears unbrecciated on a centimeter scale.

105.245 Dahmani, a highly oxidised LL6 chondrite bearing Ni–rich taenite.

M. C. Michel–Levy, M. B. Denise.
Meteoritics, Vol. 27, No. 2, p. 184 – 185 (Jun 1992).

Dahmani is a shocked LL6 fragmental breccia. According to the composition of the silicates and of the metal it is one of the most oxidised known.

105.246 Elemental abundance data for the Manitouwabing iron meteorite.

R. R. Brooks, X. Guo, M. Hoashi, R. D. Reeves, D. E. Ryan,
J. Holzbecher, G. S. Henderson.
Meteoritics, Vol. 27, No. 2, p. 186 (Jun 1992).

The Manitouwabing meteorite whose trace constituents have not been previously quantified was analysed for Au, As, Ga, Ge, Ir, Ni, Os, Pd, Pt, Th and Ru. The data confirm that it belongs to subgroup IIIA of the IIIAB group and on the basis of the much higher concentrations of As, Ir, Os, Pt, Rh and Ru, it is not paired with Madoc as had previously been proposed.

105.247 Inventory of the meteorite collection of Muséum d'Histoire Naturelle, Geneva, Switzerland.

B. Dominik, J. Deferne.
Meteoritics, Vol. 27, No. 2, p. 187 – 188 (Jun 1992).

The first inventory of the meteorite collection of the Muséum d'Histoire Naturelle of Geneva is given. The collection numbers at present 164 fragments of 102 individual meteorites.

105.248 "Tektites" and microkrystites at the Cretaceous Tertiary boundary: two strewn fields, one crater?

J. Smit, W. Alvarez, A. Montanari, N. Swinburne,
T. M. Van Kempen, G. T. Klaver, W. J. Lustenhouwer.
22. Lunar and Planetary Science Conference, p. 87 – 100 (1992).
– See Abstr. 012.046 for the main entry.

Two different types of spherules, probably impact derived, occur at the Cretaceous–Tertiary (K/T) boundary. It is concluded that the K/T "tektites" were once molten ejecta, solidified in flight. The microkrystites, on the other hand, are smaller, do not show splash forms, and may represent recondensed material from the ejected vapor cloud from the same large impact.

105.249 The Manson impact structure; its contribution to impact materials observed at the Cretaceous/Tertiary boundary.

R. R. Anderson, J. B. Hartung.
22. Lunar and Planetary Science Conference, p. 101 – 110 (1992).
– See Abstr. 012.046 for the main entry.

The Manson impact structure (Iowa) has a diameter of about 35 km. Studies of water well cuttings from the area of the structure, three shallow cores, and a recently obtained seismic profile allowed development of a model for the formation of the structure. The model suggests that the impact of a chondritic bolide with a diameter of about 2.1 km produced the Manson structure. It is suggested that the Manson impact could have produced all the exotic materials, including the shocked quartz, observed in K/T boundary sections throughout the world.

105.250 Geochemistry of Manson impact structure rocks: target rocks, impact glasses, and microbreccias.

C. Koeberl, J. B. Hartung.
22. Lunar and Planetary Science Conference, p. 111 – 126 (1992).
– See Abstr. 012.046 for the main entry.

Fifteen samples assumed to be representative of the target rock stratigraphy at the Manson Crater were analyzed for major– and trace–element composition as well as mineralogical characterization.

105.251 Solar–type xenon: isotopic abundances in Pesyanoe.

J. S. Kim, K. Marti.
22. Lunar and Planetary Science Conference, p. 145 – 151 (1992).
– See Abstr. 012.046 for the main entry.

The authors have measured elemental and isotopic abundances of Ar and Xe in three grain–size separates in the dark phase of the enstatite achondrite Pesyanoe. A comparison of Pesyanoe and lunar data shows that isotopic signatures of solar wind Xe as sampled at two different points in solar system space and time are identical within experimental error.

105.252 Nitrogen, noble gases, and nuclear tracks in lunar meteorites MAC 88104/105.
S. V. S. Murty, J. N. Goswami.
22. Lunar and Planetary Science Conference, p. 225 – 237 (1992).
– See Abstr. 012.046 for the main entry.
Eleven meteorites of lunar origin have been identified so far from the Antarctic meteorite collection. Noble gas, N, and nuclear track records have been measured in the lunar meteorites MAC 88104 and MAC 88105 to decipher the exposure history of these meteorites and to understand the source of the various N components in them.

105.253 Mineralogical studies of lunar mare meteorites EET 87521 and Y 793274.
H. Takeda, H. Mori, J. Saito, M. Miyamoto.
22. Lunar and Planetary Science Conference, p. 355 – 364 (1992).
– See Abstr. 012.046 for the main entry.
Mineralogical comparisons of lunar meteorites EET 87521 and Y 793274, possibly derived from a mare region of the Moon, have been performed to find quickly cooled basaltic components and slowly cooled plutonic materials.

105.254 Meteorite–asteroid spectral comparison: the effects of comminution, melting, and recrystallization.
B. E. Clark, F. P. Fanale, J. W. Salisbury.
Icarus, Vol. 97, No. 2, p. 288 – 297 (Jun 1992).
Laboratory results from a simulation of the possible effects of spectral alteration on reflectance of the optical surface of ordinary chondrite parent bodies is presented. Diffuse reflectance spectra from 0.3 to 2.6 μm were obtained for three chondritic meteorites. To simulate possible regolith processes the samples were comminuted to finer grain sizes, and the effect of comminution on their reflectance spectra was measured. Following comminution, the samples were melted, recrystallized, recomminuted, and remeasured. These laboratory alterations produced a decrease in absorption band depths at 0.95 μm, and melting and recrystallization produced a significant drop in albedo. Thus, although it was found that spectral characteristics could each be significantly changed by these procedures, no set of procedures was able to simultaneously affect all relevant parameters in such a way as to improve the match between ordinary chondritic meteorites and S–class asteroids.

105.255 The relation between diogenite cumulates and eucrite magmas.
T. L. Grove, K. S. Bartels.
22. Lunar and Planetary Science Conference, p. 437 – 445 (1992).
– See Abstr. 012.046 for the main entry.
The purpose of this paper is to test whether or not a calculated model cumulate derived by fractional crystallization of a magnesian eucrite parent magma approximates the analyzed chemical compositions of diogenite meteorites.

105.256 Föreslagna impaktstrukturer i Norden och närliggande områden.
F. E. Wickman.
Astron. Tidsskr., Årg. 25, Nr. 2, p. 49 – 62 (Jun 1992).

105.257 Comparative thermochemical studies of carbonaceous chondrites.
H. G. Wiedemann, A. Reller.
Naturwissenschaften, Jahrg. 79, Heft 4, p. 172 – 175 (Apr 1992).

105.258 Radiant of the Tunguska meteorite from visual observations.
I. T. Zotkin, A. N. Chigorin.
Sol. Syst. Res., Vol. 25, No. 5, p. 459 – 464 (Mar 1992). English translation of Astron. Vestn., Tom 25, No 5, p. 613 – 620 (1991).
Eyewitness accounts containing astrometric data are analyzed. Methods for determining the radiant of a fireball graphically and by direct calculation of observation discrepancies are discussed.

The position of the radiant is found: azimuth 126°, elevation 20°, with an error of $\pm 12°$. The east alternative of the trajectory is confirmed.

105.259 Preliminary study on neogene microtektites in the core collected from North Pacific.
Peng Hanchang, Liu Zhenkun, Zhuang Shijie, Mao Xueying, Chai Zhifang.
IAU Colloquium No. 126: Origin and evolution of interplanetary dust, p. 57 – 60 (1991). – See Abstr. 012.068 for the main entry.
A great number of microtektites were found in core collected from North Pacific. Because abundant microtektites are restricted to a 20–30 cm thick zone of core, we called this zone its microtektite layer. The age of sediments concentrated microtektites is from the Pliocene to the Pleistocene epoch. Research results indicate that these microtektites are similar to the North American tektites and the Australasian tektites.

105.260 Katastrofa Tungusaka.
K. Włodarczyk.
Postepy Astron., Tom 40, Zesz. 1, p. 11 – 14 (Jan–Mar 1992).

105.261 The origin of the polycyclic aromatic hydrocarbons in meteorites.
M. R. Wing, J. L. Bada.
Origins Life Evol. Biosphere, Vol. 21, No. 5 – 6, p. 375 – 383 (1991 – 1992). – See Abstr. 012.083 for the main entry.
Polycyclic aromatic hydrocarbons (PAGs) in C1 and C2 carbonaceous chondrites appear to be the product of a high–temperature synthesis. This observation counters a prevailing view that PAHs in meteorites are a thermal alternation product of preexisting aliphatic compounds, which in turn required the presence of low–temperature mineral phases such as magnetite and hydrated phyllosilicates for their formation. Such a process would necessarily lead to a more low–temperature assemblage of PAHs, as many low–temperature minerals and compounds are extant in meteorites. The presence of indigenous PAHs and absence of indigenous amino acids in the H4 ordinary chondrite Forest Vale provides support of the contention that different processes and environments contributed to the synthesis of the organic matter in the solar system.

105.262 The fate of organic matter during planetary accretion: preliminary studies of the organic chemistry of experimentally shocked Murchison meteorite.
T. N. Tingle, J. A. Tyburczy, T. J. Ahrens, C. H. Becker.
Origins Life Evol. Biosphere, Vol. 21, No. 5 – 6, p. 385 – 397 (1991 – 1992). – See Abstr. 012.083 for the main entry.
The present study focuses on the behavior of organic matter in carbonaceous meteorites during hypervelocity impact.

105.263 Das Steinheimer Becken. Eine geologische Wanderung durch einen Meteoritenkrater.
W. Lüthi.
Orion, Jahrg. 50, Nr. 248, p. 15 – 17 (Feb 1992).

105.264 Evidence for distillation in the formation of HAL and related hibonite inclusions.
T. R. Ireland, E. K. Zinner, A. J. Fahey, T. M. Esat.
Geochim. Cosmochim. Acta, Vol. 56, No. 6, p. 2503 – 2520 (Jun 1992).
Four hibonite–bearing refractory inclusions, HAL from Allende, DH–H1 from the Dhajala H3 chondrite, 7–404 and 7–971 from the Murchison CM2 chondrite, have related chemical and isotopic systematics. These chemical and isotopic characteristics are consistent with the formation of HAL–type inclusions as distillation residues. A distillation origin is supported by chemical and isotopic measurements of a hibonite–bearing distillation residue produced in the laboratory by evaporating terrestrial kaersutite.

105.265 **Origin of metallic Fe–Ni in Renazzo and related chondrites.**
M. S. Lee, A. E. Rubin, J. T. Wasson.
Geochim. Cosmochim. Acta, Vol. 56, No. 6, p. 2521 – 2533 (Jun 1992).
Metal data on Renazzo and Al Rais show an intriguing positive correlation between Co and Ni in the metal and systematic compositional differences among metal grains in matrix and chondrules. It was suggested that these trends reflected formation by condensation from nebular gas and could be used as evidence of a nebular origin for a similar trend in the anomalous meteorite Allan Hills 85085. Because the authors were skeptical that nebular processes could produce all aspects of the trend, they embarked on the present detailed study of metal in the RAR chondrites.

105.266 **Barred olivine chondrule in the Allende meteorite.**
A. E. Rubin.
J. R. Astron. Soc. Can., Vol. 86, No. 1, p. 1 – 4 (Feb 1992).

105.267 **The fall of the Abee meteorite and its probable orbit.**
A. A. Griffin, P. M. Millman, I. Halliday.
J. R. Astron. Soc. Can., Vol. 86, No. 1, p. 5 – 14 (Feb 1992).
The Abee meteorite fell at 23^h05^m Mountain Standard Time on 9 June, 1952, about 80 km north of Edmonton. A 107–kg enstatite chondrite (E4) was recovered. The authors describe details of the luminous fireball, its path and the recovery of the meteorite. The apparent radiant derived from the visual data shows that the meteorite approached the Earth from approximately the antapex direction. If one accepts the conclusion from recent studies of meteorite orbits that meteorites do not cross the orbit of Jupiter, then one may specify certain details of the Abee orbit.

105.268 **Recent field research on potential meteorite falls from the Meteorite Observation and Recovery Project.**
P. Brown, M. Zalcik.
J. R. Astron. Soc. Can., Vol. 86, No. 3, p. 130 – 139 (Jun 1992).
Results of seven years of field investigation of fifteen locales where the Meteorite Observation and Recovery Project camera system observed possible meteorite falls are presented and discussed. The fall areas where meteorites are most likely to be recovered based on terrain and end mass are identified. No new meteorites have been recovered to date.

105.269 **Cosmic–ray exposure history of ordinary chondrites.**
K. Marti, T. Graf.
Annu. Rev. Earth Planet. Sci., Vol. 20, p. 221 – 243 (1992).
Contents: Introduction. The cosmic–ray record. Thermal history of chondrites. Exposure age (T_e) histograms. Dynamical considerations. Observations on the regolith history. Conclusions.

The meteorite of Ensisheim: 1492 to 1992.
See Abstr. 004.046.

Stardust memories.
See Abstr. 011.009.

Abstracts for the 54th Annual Meeting of the Meteoritical Society.
See Abstr. 012.010.

Impact craters: are they useful?
See Abstr. 013.046.

Canadian Arctic Meteorite Project (CAMP): 1990.
See Abstr. 013.047.

Comets and meteorites were a minor source of prebiotic organic compounds on the early Earth.
See Abstr. 015.042.

Shock–induced transformations in the system $NaAlSiO_4$–SiO_2: a new interpretation.
See Abstr. 022.005.

Spectrum of particles' size formed in the course of meteorites ablation under model conditions.
See Abstr. 022.022.

Meteoritics and the origins of atomic nuclei.
See Abstr. 022.063.

Simulation of the interaction of galactic protons with meteoroids: on the production of ^7Be, ^{10}Be and ^{22}Na in an artificial meteoroid irradiated isotropically with 1.6 GeV protons.
See Abstr. 022.066.

Simulation of the interaction of galactic protons with meteoroids: isotropic irradiation of an artificial meteoroid with 1.6 GeV protons.
See Abstr. 022.071.

Effect of temperature on shock metamorphism of single–crystal quartz.
See Abstr. 022.098.

Xylan: a potential contaminant for lunar samples and antarctic meteorites.
See Abstr. 022.191.

The effect of total pressure on vaporization of alkalis from partially molten chondritic material.
See Abstr. 022.203.

Optical constants of kerogen from 0.15 to 40 μm: comparision with meteoritic organics.
See Abstr. 022.208.

Noble metal enrichments in cosmic spherules.
See Abstr. 022.210.

Studies on isotopic ratios of osmium and iridium in cosmic spherules using instrumental neutron activation analysis.
See Abstr. 022.211.

The p–nuclei: abundances and origins.
See Abstr. 061.100.

SiC particles from asymptotic giant branch stars: Mg burning and the s–process.
See Abstr. 065.094.

Formation of spinels in cosmic objects during atmospheric entry: a clue to the Cretaceous–Tertiary boundary event.
See Abstr. 081.008.

Neodymium and strontium isotopic study of Australasian tektites: new constraints on the provenance and age of target materials.
See Abstr. 081.010.

Geochemistry and origin of Muong Nong–type tektites.
See Abstr. 081.014.

Nature of the optical anomalies of summer of 1908.
See Abstr. 082.048.

Nature of the anomalous sky luminescence connected with the Tungus event.
See Abstr. 082.049.

Natural thermoluminescence and anomalous fading: terrestrial age, transit times and perihelia of lunar meteorites.
See Abstr. 094.019.

A spinifex–textured mare basalt: comparison with komatiites.
See Abstr. 094.049.

^{142}Nd/^{144}Nd in SNCs and early differentiation of a heterogeneous Martian (?) mantle.
See Abstr. 097.022.

Styles of crater gradation in southern Ismenius Lacus, Mars: clues from Meteor Crater, Arizona.
See Abstr. 097.055.

The stable isotopic compositions of indigenous carbon–bearing components in EETA 79001.
See Abstr. 097.058.

A liquidus phase diagram for the groundmass of EETA 79001A (Eg), a primitive shergottite composition.
See Abstr. 097.064.

An ejection model for SNC meteorites: an indication for recent volcanism on Mars.
See Abstr. 097.073.

Aqueous–alteration products in S–N–C meteorites and implications for volatile/regolith interactions on Mars.
See Abstr. 097.106.

Martian parent craters for the SNC meteorites.
See Abstr. 097.132.

The mineralogy of S–type asteroids: why doesn't spectroscopy find ordinary chondrites in the asteroid belt?
See Abstr. 098.048.

Doublet craters and the tidal disruption of binary asteroids.
See Abstr. 098.053.

A new model for the formation of the asteroids – the parent bodies of the meteorites.
See Abstr. 098.055.

Additions to the Taurid Complex.
See Abstr. 098.127.

Diffraction of interplanetary solid bodies potentially generating meteorites.
See Abstr. 104.062.

Stardust and planet dust.
See Abstr. 106.010.

Transmission electron microscopy of an interplanetary dust particle with links to CI chondrites.
See Abstr. 106.022.

Transmission atmosphérique des micrométéorites polaires et implications.
See Abstr. 106.026.

An interplanetary dust particle with links to CI chondrites.
See Abstr. 106.034.

A cosmic matter accretion event around 660,000 years before present found in two dated, Central Pacific cores.
See Abstr. 106.068.

Aqueous alteration in hydrated interplanetary dust particles.
See Abstr. 106.070.

Constraints on the parent bodies of collected interplanetary dust particles.
See Abstr. 106.086.

Effects of nuclear reactions with accelerated particles during a supernova explosion.
See Abstr. 107.002.

Accretion in the inner nebula: the relationship between terrestrial planetary compositions and meteorites.
See Abstr. 107.013.

Interstellar and meteoritic organic matter at 3.4 μm.
See Abstr. 131.209.

106 Interplanetary Matter, Interplanetary Magnetic Field, Zodiacal Light

106.001　An analysis of the influence of interstellar scattering on interplanetary scintillation observations.
C. A. Hajivassiliou.
Astron. Astrophys., Vol. 253, No. 1, p. 244 – 251 (Jan 1992).

Interstellar scattering (ISS) causes the observed angular sizes of radio sources to appear larger, and its modelling is therefore particularly important in the interpretation of low–frequency–high–resolution observations such as interplanetary scintillation (IPS) surveys. These have revealed a marked deficit of sources with small angular sizes which, if not entirely due to ISS, might have cosmological significance. Using a simple mathematical model of galactic scattering, the author shows how ISS affects the number–angular–diameter relation observed in an IPS survey as a function of galactic latitude. He deduces that the average value for the scattering angle (in the polar direction) required to account for the deficit of compact sources in the Cambridge Interplanetary Scintillation Survey at 81.5 MHz is $\Theta_p = 0\rlap{.}''25 \pm 0\rlap{.}''03$.

106.002　The dynamics of isolated flare magnetic clouds in the quiet regularly nonuniform inner heliosphere.
K. G. Ivanov, A. F. Kharshiladze, V. V. Petrochenko.
Geomagn. Aeron., Vol. 30, No. 5, p. 620 – 624 (Apr 1990). English translation of Geomagn. Aehron., Tom 30, No. 5, p. 730 – 735 (1990).

A theoretical model of the motion of an isolated flare magnetic cloud in the regularly nonuniform inner heliosphere has been developed. Calculations are given of the velocities and times of flight of the center of mass of the cloud away from the Sun towards the Earth's orbit for different positions of the flares relative to the heliospheric current layer and for different inclinations of the heliospheric current layer to the solar equatorial plane. The calculations indicate a strong deceleration of the clouds upon their propagation through a heliospheric streamer and the eastern neighborhoods of the heliospheric current layer. The appearance (absence) of shock waves (SC) at the Earth is determined both by the specific oblateness of the shock waves and by the conditions of their evolution, which are disrupted upon deceleration of the clouds or their movement far away from the heliospheric current layer.

106.003   Forschung an den Grenzen des Sonnensystems.
H. J. Fahr.
Sterne Weltraum, Jahrg. 31, Nr. 1, p. 24 – 28 (Jan 1992).

106.004   The response of the quiet–time auroral configuration to short– and long–term interplanetary magnetic field variations.
J. S. Murphree, L. L. Cogger, R. D. Elphinstone, D. Hearn.
Can. J. Phys., Vol. 69, No. 8/9, p. 1040 – 1046 (Aug–Sep 1991).
Observations from the IMP–8 satellite of the interplanetary magnetic field (IMF) are compared with areas of the polar region bounded by the aurora as observed by the Viking spacecraft during quiet–time conditions. A variety of energy–coupling functions are investigated and it is determined that the auroral distribution can be best described by the inclusion of azimuthal terms in addition to standard energy–coupling functions. Observations by the Viking spacecraft indicate a dominance of dusk sector polar arcs in the spring time and dawn sector arcs in the fall. Two alternative mechanisms can explain the observations. One involves the ordering of the IMF in a solar equatorial coordinate system while the other involves the Sun's polarity and the traversal of the Earth's orbit through different heliographic latitudes. A test is proposed whereby the two hypotheses can be investigated during the next solar cycle.

106.005   Interplanetary magnetic flux: measurement and balance.
D. J. McComas, J. T. Cosling, J. L. Phillips.
J. Geophys. Res., Vol. 97, No. A1, p. 171 – 178 (1 Jan 1992).
The authors have developed a new method for determining the approximate magnetic flux content of various solar wind structures in the ecliptic plane, using single–spacecraft measurements. The two–dimensional magnetic flux in a region of the solar wind is given by the integral of the radial magnetic field component over an arc perpendicular to the radial. Unfortunately, such measurements cannot be achieved with single (or even several) spacecraft in the solar wind. The authors show that the desired two–dimensional, ecliptic plane magnetic flux integral, at least for regions with simple magnetic topologies, is equivalent to $\varphi = \int B_y |v| dt$, where B_y is the ecliptic plane field component perpendicular to the solar wind velocity vector v. Thus φ can be determined entirely from measured quantities. In this study the authors examine variations in the magnetic flux in the ecliptic plane over a 16–year interval. In addition, they address the question of the opening and closing of interplanetary magnetic flux by comparing the ecliptic plane flux content of both coronal mass ejections and heat flux dropouts.

106.006   Anisotropy of nonrelativistic particles deduced from analysis of their bulk flow speed.
L. C. Tan, G. M. Mason, G. Gloeckler, B. Klecker.
J. Geophys. Res., Vol. 97, No. A1, p. 179 – 184 (1 Jan 1992).
The authors present a new method for deducing the anisotropy vector δ' of nonrelativistic particles in the solar wind frame. To order V_{sw}/v, $\delta' = -v^{-1} [d \ln F_0(v)/d \ln v] (V_F - V_{sw})$, where v and $F_0(v)$ are the velocity and the mean velocity phase space distribution function, respectivly, of the particles in the spacecraft frame, V_F is the bulk flow speed of the energetic particles in the spacecraft frame, and V_{sw} is the solar wind speed. An illustration of the use of this equation is given from an interval during the upstream period before the April 5, 1979, interplanetary shock.

106.007   Location of the radio emitting regions of interplanetary shocks.
D. Lengyel–Frey.
J. Geophys. Res., Vol. 97, No. A2, p. 1609 – 1618 (1 Feb 1992).
The author analyzes 20 interplanetary type II radio bursts to determine the location of the type II source region relative to the interplanetary shock. He also reports the first determination of a

density–distance relationship (density model) appropriate for interplanetary type II source regions.

106.008   An analytical solution for the heliopause boundary and its comparison with numerical solutions.
R. Ratkiewicz.
Astron. Astrophys., Vol. 255, No. 1/2, p. 383 – 387 (Feb 1992).
Different theoretical approaches to describe the plasma interface ahead of the solar system are available in the literature. The comparison of different heliopause configurations obtained by alternatively using the Newtonian approximation (NA) or the well–known Parker's solution for a plasma–plasma type interface with the complete hydrodynamic calculations carried out by Baranov et al. (1979) shows that there exists a domain in which first, the Parker solution is very close to that obtained by NA and second, in which both are close to the solution for the inner bow shock worked out by Baranov et al. (1979). This leads to the idea of using the analytic form of Parker's solution (for subsonic flows) as the approximate solution for just the opposite case, namely, for the interaction between two hypersonic streams. By this means, treating the compressible layers between the two bow shocks as a thin layer and assuming the plasma velocity as a constant throughout the cross dimension of the shock layer a third order ordinary differential equation describing the heliopause is obtained. This equation is then solved numerically. An analytical solution of this equation on the basis of a Parker type solution is also found. With the help of this solution a simple procedure of describing the heliospheric interface is proposed.

106.009   Nonstationary propagation of flare–generated particles in the heliosphere.
M. F. Bakhareva.
Geomagn. Aehron., Tom 32, No. 1, p. 145 – 147 (Jan – Feb 1992). In Russian. English translation in Geomagn. Aeron., Vol. 32, No. 1.

106.010   Stardust and planet dust.
T. A. McGlynn.
Nature, Vol. 355, No. 6355, p. 20 – 21 (2 Jan 1992).
Meteoroid impacts on the Long Duration Exposure Facility satellite and radar tracking of meteoroids indicate that some interplanetary dust travels at speeds greater than the escape velocity of the solar system. Because these meteoroids with "hyperbolic" orbits are larger than interstellar grains are expected to be, it is tempting to suppose that they come from other solar systems. The author reports on this topic.

106.011   Multifractal structure of the magnetic field and plasma in recurrent streams at 1 AU.
L. F. Burlaga.
J. Geophys. Res., Vol. 97, No. A4, p. 4283 – 4294 (1 Apr 1992).
This paper demonstrates the existence of multifractal structure in the fluctuations of the magnetic field strength, temperature, and density in the recurrent flows at 1 AU. The multifractal scaling is observed over the range of periods from 2 hours to 32 hours for the magnetic field and density and from 2 hours to 16 hours for the temperature. The lower cutoff of 2 hours is related to the use of hour averages; it does not have physical significance. The upper cutoff of 32 hours (16 hours) represents the transition to the mesoscale regime in which the fluctuations are determined by the stream structure. Thus the multifractal structure represents fluctuations superimposed on the corotating stream profile and the heliospheric plasma sheet profile. One may think of the multifractal structure as a texture superimposed on the mesoscale profiles, the structure of the texture being related to the mesoscale structure itself. The multifractal structure of the magnetic field strength fluctuations might be produced by a multiplicative process involving the velocity fluctuations.

106.012   Intensity variations in the interplanetary magnetic field measured by Voyager 2 and the 11–year solar cycle modulation of Galactic cosmic rays.
J. S. Perko, L. F. Burlaga.
J. Geophys. Res., Vol. 97, No. A4, p. 4305 – 4308 (1 Apr 1992).
The authors present new evidence to support the hypothesis that the 11–year solar cycle modulation of galactic cosmic rays is

caused by strong particle diffusion inside long–lived, merged interaction regions. These regions are represented by local enhancements in the heliospheric magnetic field strength. To test this hypothesis, the authors solved the one–dimensional, force field approximation of the cosmic ray modulation equation. The only variables were the strength of the local magnetic field and the position of the spacecraft, both taken directly from Voyager 2 data. The authors assume that a constant solar wind speed convects magnetic field compressions and rarefactions unchanged through a model heliosphere. The result is a reasonable simulation of the integrated, high–energy cosmic ray intensity profile from about 1982 to mid–1989. This period encompasses both the full recovery portion of the last 11–year cosmic ray cycle and the first year and a half of the new cycle. In particular, this model responds to the Voyager 2 magnetic field data by correctly timing the beginning of the new modulation cycle in late 1987. The authors conclude that their hypothesis is consistent with the results of this simulation.

106.013   **A search during the 1991 solar eclipse for the infrared signature of circumsolar dust.**
K.–W. Hodapp, R. M. MacQueen, D. N. B. Hall.
Nature, Vol. 355, No. 6362, p. 707 – 710 (20 Feb 1992). Letter–to–the–editor.
Theoretical suggestions that there should be a ring of dust in near–ecliptic orbit about the Sun were supported by observations, during a total solar eclipse on 12 November 1966, of enhanced infrared emission from the solar corona at a distance of $4 R_\odot$ from the Sun's centre. The infrared emission was attributed to the sublimation of dust grains as they spiral into the Sun. Two months after the 1966 eclipse, the feature at $4 R_\odot$ was seen again in observations from a stratospheric balloon–borne coronagraph, as were additional features at 3.5, 8.7 and $9.2 R_\odot$. Observations since then, however, have failed unambiguously to corroborate the earlier observations. The authors searched for excess infrared emission in the solar equatorial plane during the 11 July 1991 eclipse, using a wide–angle infrared camera on Mauna Kea, but failed to find any signature of dust evaporation. They argue that the earlier observations were credible, and therefore that the circumsolar dust ring is a transient feature, perhaps due to the injection of dust into near–solar space by a Sun–grazing comet.

106.014   **Outer heliospheric radio emissions. 1. Constraints on emission processes and the source region.**
I. H. Cairns, D. A. Gurnett.
J. Geophys. Res., Vol. 97, No. A5, p. 6235 – 6244 (1 May 1992).
The Voyager 1 and 2 spacecraft observed low–frequency radio emissions near 2 and 3 kHz during the interval 1983–1987 while at heliocentric distances from 15 to 27 AU and 11 to 20 AU, respectively. The authors consider the detailed theoretical and observational requirements for this radiation to be produced near multiples of the plasma frequency f_p by nonlinear, weak turbulence, wave–wave processes involving electrostatic Langmuir waves. Constraints on the emission processes and source characteristics are discussed.

106.015   **Outer heliospheric radio emissions. 2. Foreshock source models.**
I. H. Cairns, W. S. Kurth, D. A. Gurnett.
J. Geophys. Res., Vol. 97, No. A5, p. 6245 – 6259 (1 May 1992).
Low–frequency radio emissions in the range 2–3 kHz have been observed by the Voyager spacecraft during the intervals 1983–1987 and 1989 to the present while at heliocentric distances greater than 11 AU. New analyses of the wave data are presented, and the characteristics of the radiation are reviewed and discussed.

106.016   **Volatiles in interplanetary dust particles: a review.**
E. K. Gibson Jr.
J. Geophys. Res., Vol. 97, No. E3, p. 3865 – 3875 (25 Mar 1992).
The paper presents a review of the volatiles found within interplanetary dust particles. These particles have been shown to represent primitive material from early in the solar system's

formation and also may contain records of stellar processes. The organogenic elements (i.e., H, C, N, O, and S) are among the most abundant elements in our solar system, and their abundances, distributions, and isotopic compositions in early solar system materials permit workers to better understand the processes operating early in the evolutionary history of solar system materials. Interplanetary dust particles have a range of elemental compositions, but generally they have been shown to be similar to carbonaceous chondrites, the solar photosphere, Comet Halley's chondritic cores, and matrix materials of chondritic chondrites. Recovery and analysis of interplanetary dust particles have opened new opportunities for analysis of primitive materials, although interplanetary dust particles represent major challenges to the analyst because of their small size.

106.017   **Large–scale structure of the circumsolar medium based on scintillations.**
N. A. Lotova, I. Yu. Yurovskaya, Ya. V. Pisarenko, M. K. Bird, M. Pätzold, R. Güsten, W. Sieber.
Sov. Astron., Vol. 36, No. 1, p. 88 – 91 (Jan 1992). Current Physics Microform No.: 9207Y0652. English translation of Astron. Zh., Tom 69, Vyp. 1, p. 173 – 180 (Jan–Feb 1992).
An experiment on transillumination of the circumsolar medium by water vapor maser sources was carried out in December 1987, using five sources simultaneously. The circumsolar plasma was probed in the vicinity of the transition of the solar wind from subsonic to supersonic flow. The large–scale structure of the medium has been reconstructed from a study of the radial dependence of the scintillation index, m(R). The relationship between the structure of the medium in the transition region of the solar wind and the structure of the solar corona is investigated.

106.018   **Robust estimation of interplanetary scintillation.**
G. Woan.
Mon. Not. R. Astron. Soc., Vol. 254, No. 2, p. 273 – 276 (15 Jan 1992).
Interplanetary scintillation (IPS) is a random process, so measurements of its power are affected by self–noise. The noise process is therefore non–stationary whenever the mean IPS power changes, such as when performing drift scans across a radio source. Here the author presents a new estimator for IPS which takes proper account of the noise statistics, and shows it to perform significantly better than a simple "matched" filter. The probabilistic basis of this estimator also makes it a powerful discriminator against interference, capable of delivering good estimates of scintillating power even when many of the data are badly corrupted.

106.019   **Electron energy loss spectroscopy of the fine grained matrices of interplanetary dust particles.**
J. P. Bradley.
54. Annual Meeting of the Meteoritical Society, p. 32 (1991). Abstract. – See Abstr. 012.010 for the main entry.

106.020   **Average minor and trace element contents in seventeen "chondritic" IDPs suggest a volatile enrichment.**
G. J. Flynn, S. R. Sutton.
54. Annual Meeting of the Meteoritical Society, p. 62 (1991). Abstract. – See Abstr. 012.010 for the main entry.

106.021   **New PIXE analyses of interplanetary dust particles.**
E. K. Jessberger, J. Bohsung, S. Chakaveh, K. Traxel.
54. Annual Meeting of the Meteoritical Society, p. 107 (1991). Abstract. – See Abstr. 012.010 for the main entry.

106.022   **Transmission electron microscopy of an interplanetary dust particle with links to CI chondrites.**
L. P. Keller, D. S. McKay, K. L. Thomas.
54. Annual Meeting of the Meteoritical Society, p. 115 (1991). Abstract. – See Abstr. 012.010 for the main entry.

106.023   On the existence of the heliospheric current sheet without a neutral line (HCS without NL).
V. G. Eselevich, V. G. Fainshtein (*V. G. Fajnshtejn*).
Planet. Space Sci., Vol. 40, No. 1, p. 105 – 119 (Jan 1992).

By investigating quasi–stationary solar wind (SW) streams, it has been shown that there exists, along with the heliospheric current sheet (HCS) with a neutral line (NL), the HCS without NL. Chains of streamers which in the solar corona separate regions having the same directions of the radial magnetic field B_r, correspond to it near the Sun. In the presence of a sporadic stream, SW structures involving a change of sign of B_r, can be interpreted as a bend of the HCS with NL.

106.024   Extraction of ^{4}He from IDPs by step–heating.
A. O. Nier, D. J. Schlutter.
54. Annual Meeting of the Meteoritical Society, p. 173 (1991).
Abstract. – See Abstr. 012.010 for the main entry.

106.025   Anisotropy of the energetic neutral atom flux in the heliosphere.
M. A. Gruntman.
Planet. Space Sci., Vol. 40, No. 4, p. 439 – 445 (Apr 1992).

Characteristics of the energetic neutral atoms born at the heliosphere interface are considered for plasma flow structure resulting from a two–shock model of the interaction between the solar wind and the interstellar medium. The energy distributions of heliospheric energetic neutral atoms (HELENAs) are calculated and it is shown that the HELENA flux is highly anisotropic at the Earth's orbit. The characteristics of the HELENA flux are highly sensitive to the size of the heliosphere. This supports the conclusion that measurements of HELENAs from the Earth's orbit would give us an efficient tool to remotely study the heliosphere.

106.026   Transmission atmosphérique des micrométéorites polaires et implications.
P. Bonny.
Office National d'Etudes et de Recherches Aerospatiales (ONERA), 92 – Chatillon (France), ONERA–NT—1991–5. 226 p. (1991). ISSN 0078–3781.

Three main characteristics of polar micrometeorites are incompatible with predictions of previous models of micrometeorite atmospheric entry: the high fraction of unmelted particles, the low mass loss of those that are melted, and a still partly preserved organic matter as suggested by laboratory analysis. The purpose of this paper is to understand how these cosmic grains make it through so unexpectedly "well". Firstly, the current theory is reviewed. Moreover, the effect an organic component has on heating and the pyrolysis is analysed accurately with a computational code of ONERA that simulates the entire process of ablation of a charforming material. Secondly, a code has been written to simulate the complete micrometeorite atmospheric entry trajectory, including melting, porosity change, loss of an organic component and vaporization of the major elements of the mineral component. After this, the code is applied on the entire domain of micrometeorite velocity and angle of entry. The corridor of entry required for each of the three characteristics of the polar micrometeorites is then evaluated. After weighting the angle and velocity allowed for each corridor, an entry velocity distribution compatible with actual polar micrometeorites characteristics is found by comparing computer–generated data with "experimental". Finally, the origin of the polar micrometeorites is discussed, along with the part they may have played in bringing organic matter to the early Earth.

106.027   Voyager energetic particle observations at interplanetary shocks and upstream of planetary bow shocks: 1977–1990.
S. M. Krimigis.
Space Sci. Rev., Vol. 59, No. 1/2, p. 167 – 201 (Jan 1992).

The Voyager 1 and 2 spacecraft include instrumentation that makes comprehensive ion ($E \gtrsim 28$ keV) and electron ($E \gtrsim 22$ keV) measurements in several energy channels with good temporal, energy, and compositional resolution. Data collected over the past decade (1977–1988), including observations upstream and downstream of four planetary bow shocks (Earth, Jupiter, Saturn, Uranus) and numerous interplanetary shocks to ∼ 30 AU, are reviewed and analyzed in the context of the Fermi and shock drift acceleration (SDA) models.

106.028   On a possible use of total solar eclipse below the horizon for observations of the inner zodiacal light (as applied to the eclipse of 30 June, 1992).
R. A. Gulyaev.
Sol. Phys., Vol. 138, No. 1, p. 209 – 211 (Mar 1992). Letter–to–the–editor.

The Moon's umbral shadow, tangentially penetrating the Earth's atmosphere, appreciably reduces the brightness of the twilight sky at points located under the shadow axis. This should yield favourable conditions for observation of the zodiacal light for small altitudes of the Sun below the horizon. The location of the projection of the Moon's shadow axis at the Earth's surface under the above conditions is calculated for the case of the total solar eclipse of 30 June, 1992.

106.029   Cosmic dust and orbital debris: collection on MIR space station.
J. C. Mandeville.
Adv. Space Res., Vol. 11, No. 12, p. 93 – 96 (1991). – See Abstr. 012.014 for the main entry.

Upon the last joint Soviet–French mission on the MIR Space Station, on Dec 1988, an experiment devoted to the collection and detection of cosmic dust and space debris has been deployed in space during 13 months. A variety of sensors and collecting devices has make possible the study of effects and distribution of cosmic particles after recovery of exposed material. Remnants of particles, suitable for chemical identification are expected to be found within the stacked foil detectors. Discrimination between true cosmic particles and man–made orbital debris is expected. Some preliminary results are presented here.

106.030   Study of cosmic dust particles on board LDEF: the FRECOPA experiment.
J. C. Mandeville.
Adv. Space Res., Vol. 11, No. 12, p. 101 – 107 (1991). – See Abstr. 012.014 for the main entry.

A French experiment partly devoted to the detection of cosmic dust has been flown on the Long Duration Exposure Facility (LDEF), launched in Apr 1984, and retrieved in Jan 1990. A variety of sensors and collecting devices will make possible the study of cosmic particles after recovery of exposed material. Remnants of particles, suitable for chemical identification are expected to be found within the stacked foil detectors. Discrimination between true cosmic particles and man–made orbital debris is expected.

106.031   First spatio–temporal results from the LDEF interplanetary dust experiment.
S. F. Singer, J. E. Stanley, P. C. Kassel, W. H. Kinard, J. J. Wortman, J. L. Weinberg, J. D. Mulholland, G. Eichhorn, W. J. Cooke, N. L. Montague.
Adv. Space Res., Vol. 11, No. 12, p. 115 – 122 (1991). – See Abstr. 012.014 for the main entry.

The LDEF Interplanetary Dust Experiment was unique in providing a time history of impacts of micron–sized particles on six orthogonal faces of the vehicle over a span of nearly a full year. Over 15000 hits were recorded, representing a mix of zodiacal dust, meteor stream grains, orbital debris, perhaps beta–meteoroids, and possibly interstellar matter. Although the total number was higher than predicted, the relative panel activity distribution was near expectations. Detailed deconvolution of the impact record with orbital data is underway, to examine each of these populations. Very preliminary results of the fairly crude "first look" analysis suggest that debris is the major particle component at 500 km. The data show clear evidence of some known meteor streams as sharp, tightly–focused events, unlike their visible counterparts.

106.032 Out–of–ecliptic distribution of interplanetary dust derived from near Earth flux.
B. Kneißel, I. Mann.
Adv. Space Res., Vol. 11, No. 12, p. 123 – 126 (1991). – See Abstr. 012.014 for the main entry.

The out–of–ecliptic distribution of interplanetary dust, i.e. its number density, mainly has been subject to optical or infrared remote sensing techniques. As the population in interplanetary space is made up of orbiting particles which will cross the ecliptic plane, determination of their orbital properties there gives a possibility also to derive their out–of–ecliptic distribution. Determination of orbital elements is provided by advanced detectors capable of measuring the vector of impact velocity. In a simple model, which applies for advanced detectors in near earth orbit, the feasibility of the method to determine the out–of–ecliptic spatial distribution of dust has been tested.

106.033 Modeling the time–intensity profile of solar flare generated particle fluxes in the inner heliosphere.
D. F. Smart, M. A. Shea.
Adv. Space Res., Vol. 12, No. 2/3, p. 303 – 312 (1992). – See Abstr. 012.016 for the main entry.

It is possible to model the time–intensity profile of solar particles expected in space after the occurrence of a significant solar flare on the sun. After the particles are accelerated in the flare process they may be released into the solar corona and then into space. The heliolongitudinal gradients observed in the inner heliosphere are extremely variable, reflecting the major magnetic structures in the solar corona which extend into space. These magnetic structures control the particle gradients in the inner heliosphere. The most extensive solar particle measurements are those observed by earth–orbiting space–craft, and forecast and prediction procedures are best for the position of the earth. There is no consensus of how to extend the earth–based models to other locations in space.

106.034 An interplanetary dust particle with links to CI chondrites.
L. P. Keller, K. L. Thomas, D. S. McKay.
Geochim. Cosmochim. Acta, Vol. 56, No. 3, p. 1409 – 1412 (Mar 1992).

W7013F5 is a chondritic, hydrated interplanetary dust particle whose composition and mineralogy is nearly identical to that found in the CI chondrites. Transmission electron microscope observations show that the phyllosilicates in W7013F5 consist largely of a coherent intergrowth of Mg–Fe serpentine and Fe–bearing saponite on the unitcell scale. This distinctive intergrowth of phyllosilicates has only been observed previously in the CI chondrites. The presence of kamacite in W7013F5 indicates that the particle is extraterrestrial, and a thin amorphous rim surrounding the particle provides evidence that it is not a piece of a meteorite that fragmented during transit through the atmosphere.

106.035 Particulate detection in the near Earth space environment aboard the long duration exposure facility LDEF: cosmic or terrestrial?
J. A. M. McDonnell, K. Sullivan, T. J. Stevenson, D. H. Niblett.
IAU Colloquium No. 126: Origin and evolution of interplanetary dust, p. 3 – 10 (1991). – See Abstr. 012.068 for the main entry.

Examination of surfaces exposed for more than five and a half years, from detectors with unique attitude stabilisation relative to the orbital velocity vector, offers scope for examining definitively the sources of hypervelocity space particulates. Surfaces reveal discrete crater morphologies, crater size distributions and incident flux distributions. Discrete crater studies will later also reveal the chemistry of residues which can, especially via the capture cell principle, lead to elemental analysis of micron dimensioned particles. First analyses of the flux data from the thin foil perforation experiments (MAP) involve a study of the statistics of the forward (ram) direction, the rear (trailing) direction and the space pointing direction. Modelling of the dynamics of geocentrically bound and unbound orbits yields

evidence that the characteristics of the particles, and hence probably their source, change over the particle size range measured by the experiment. Smaller particles (< 1 μm diameter) have lower velocities which could include geocentrically bound particulates, whereas the larger particles (5 – 10 μm diameter) can be identified with "cosmic" particles of interplanetary or interstellar origin.

106.036 Interplanetary magnetic field connection to the Sun during electron heat flux dropouts in the solar wind.
R. P. Lin, S. W. Kahler.
J. Geophys. Res., Vol. 97, No. A6, p. 8203 – 8209 (1 Jun 1992).

Gosling (1975) and MacQueen (1980) pointed out that coronal mass ejections (CMEs) add significant magnetic flux which is connected to the Sun, to the interplanetary medium. Since the interplanetary magnetic field is observed to stay at a roughly constant level, some compensating removal of flux is necessary. McComas et al. (1989) suggested that magnetic reconnection in current sheets near the Sun removes flux by producing U–shaped magnetic structures in the interplanetary medium which are disconnected from the Sun at both ends. Since solar wind halo electrons normally carry heat flux outward from the Sun's hot corona to the cool interplanetary medium, McComas et al. (1989) interpreted observations of dropouts of the solar wind electron heat flux as evidence for such disconnnections. Here the authors use 2– to 8.5–keV solar electrons observed by ISEE 3 as tracers of the magnetic topology for these heat flux dropout (HFD) events. They conclude that in many HFDs the interplanetary field is still connected to the Sun, and that some energy–dependent process may produce HFDs without significantly perturbing electrons of higher energies. For two of the 25 HFDs, however, the $\geqslant$ 2–keV electron observations exhibit all the characteristics of real magnetic disconnection events, including a depletion in the total $\geqslant$ 2–keV fluxes.

106.037 Large–scale configuration of the circumsolar plasma as determined by scintillations.
N. A. Lotova, I. Yu. Yurovskaya, Ya. V. Pisarenko,
M. K. Berd, M. Pätzold, R. Güsten, W. Sieber.
Astron. Zh., Tom 69, Vyp. 1, p. 173 – 180 (Jan–Feb 1992). In Russian. English translation in Sov. Astron., Vol. 36, No. 1 (1992).

An occultation experiment, using water vapour maser sources, has been carried out in December 1987. Five occulted sources were used simultaneously. The circumsolar plasma sounding was performed at the transition zone of the solar wind, where the subsonic outflow becomes supersonic. The large–scale configuration of the medium was reconstructed from studies of the radial dependence of the scintillation index. The relation between configuration of the solar wind transsonic region and that of the solar corona has been investigated.

106.038 Solar and interplanetary shock effects on the flux and anisotropy profiles of an Energetic Storm Particle (ESP) event observed on 23–25 April 1979.
A. M. Heras, B. Sanahuja, Z. K. Smith, T. Detman, M. Dryer.
J. Atmos. Terr. Phys., Vol. 53, No. 11/12, p. 1027 – 1032 (Nov–Dec 1991). – See Abstr. 012.024 for the main entry.

A large Energetic Storm Particle event was observed by ISEE–3 between 23 and 25 April 1979. This event was associated with an interplanetary shock which was generated by a solar disappearing filament. Flux and anisotropy profiles in the upstream region of the event have been reproduced by means of a particle propagation model which includes particle injection both at the Sun and at the shock front. Injection at the front takes into account the fact that the IMF connection between the shock and the spacecraft is disrupted by the shock itself. The model, coupled to a 2D MHD time–dependent model, makes it possible for the authors to study the parameters for the interplanetary propagation of the particles, the injection rates both at the Sun and at the shock front, from near the Sun out to 1 AU.

106.039   On the large–scale effects of two interplanetary shocks on the associated particle events.
A. M. Heras, B. Sanahuja, Z. K. Smith, T. Detman, M. Dryer.
J. Atmos. Terr. Phys., Vol. 53, No. 11/12, p. 1033–1038 (Nov–Dec 1991). – See Abstr. 012.024 for the main entry.

An evolutionary model, including particle and shock propagation through the interplanetary medium, has been used to reproduce the evolution of the flux and anisotropy in the upstream region of two low–energy particle events observed by ISEE-3. These events, on 24 April 1979 and 18 February 1979, originated at solar (helio)longitudes ≈ 50° apart. By fitting the observed particle fluxes and anisotropies, the conditions for the propagation of the particles through the interplanetary medium and the injection rates at the shock have been determined as a function of time. The results are discussed in terms of the interplanetary magnetic field connection between the observer and the shock front and they are related to the heliolongitude of the parent solar activity.

106.040   Multipoint observations of planar interplanetary magnetic field structures.
C. J. Farrugia, R. P. Lepping, M. W. Dunlop, S. Elliott, A. Balogh, S. W. H. Cowley, M. P. Freeman, D. G. Sibeck.
J. Atmos. Terr. Phys., Vol. 53, No. 11/12, p. 1039–1047 (Nov–Dec 1991). – See Abstr. 012.024 for the main entry.

The authors present interplanetary magnetic field (IMF) data made on 1 November 1984 by three spatially well–separated spacecraft in the solar wind: the IMP 8, AMPTE–UKS and –CCE spacecraft. The IMF measured by each of the spacecraft is found to consist of a multiplicity of structures within which the magnetic field varies in parallel planes. The orientations of these planes at the three spacecraft locations are similar. The planes are inclined at a large angle to the ecliptic, and they lie almost perpendicular to the nominal Parker spiral direction in the ecliptic. Intercomparisons of the measurements at the various spacecraft show that the IMF features at one spacecraft are clearly reproduced at another with, however, time delays required for the propagation of signals. Using these time delays and the mutual separations of the spacecraft, the authors infer that the structures are convecting with the ambient flow. Simultaneous observations made downstream of the bow shock in the magnetosheath reveal that the magnetosheath magnetic field, too, is planar. However, the plane of maximum variation of the magnetosheath magnetic field is sheared with respect to the corresponding plane in the solar wind, and the field now lies approximately tangential to the magnetopause. Finally, the authors' observations are contrasted to another IMF configuration showing gross departures from the nominal Parker spiral orientation: magnetic clouds.

106.041   Dynamics of the zodiacal cloud.
S. F. Dermott, R. S. Gomes, D. D. Durda, B. A. S. Gustafson, S. Jayaraman, Y. L. Xu, P. D. Nicholson.
IAU Symposium No. 152: Chaos, resonance and collective dynamical phenomena in the solar system, p. 333–347 (1992). – See Abstr. 012.025 for the main entry.

Contents: 1. Introduction. 2. Orbital evolution due to Poynting–Robertson light drag. 3. Modeling the background zodiacal cloud. 4. Modeling the solar system dust bands.

106.042   Some aspects of particle's dynamics.
R. S. Gomes.
IAU Symposium No. 152: Chaos, resonance and collective dynamical phenomena in the solar system, p. 349–354 (1992). – See Abstr. 012.025 for the main entry.

Some features of the dynamics of particles affected by drag, in the field of the Sun and planets, are presented here. In particular mean motion and secular resonances are investigated. When dust particles are considered as a whole in the zodiacal cloud, a simple secular theory can explain much of its geometry. Dynamics of particles near an inner planet is mostly dispersive, but an average behavior can be deduced from some analytical and numerical considerations.

106.043   Asteroidal and meteoric components of the zodiacal light.
N. B. Divari.
Sol. Syst. Res., Vol. 25, No. 4, p. 351–357 (Jan 1992). English translation of Astron. Vestn., Tom 25, No. 4, p. 466–475 (1991).

A review of the existing hypotheses of the origin of the interplanetary dust is presented. Two components seem to exist in the zodiacal light: asteroidal (or ecliptic) and meteoric (or interplanetary). The first component resulting from the dust in asteroidal rings is seen in the visible region as the zodiacal band. The meteoric component responsible for the light diffused throughout the sky is provided by the dust in the inner part of the solar system. These particles have the same origin as the meteoroids. The necessity to reconsider the time of dust particles spiraling to the Sun under the Poynting–Robertson effect with regard to perturbations from planets is emphasized.

106.044   Beta meteoroids: LDEF interplanetary dust experiment provides confirmation observations with first detection from low Earth orbit.
W. J. Cooke, N. L. Montague, J. D. Mulholland, J. P. Oliver, C. G. Simon, S. F. Singer, J. L. Weinberg, P. C. Kassel, J. J. Wortman.
Bull. Am. Astron. Soc., Vol. 23, No. 3, p. 1142 (1991). Abstract. – See Abstr. 012.037 for the main entry.

106.045   Interplanetary dust observed by Galileo and Ulysses.
E. Grün, M. Baguhl, H. Fechtig, J. Kissel, D. Linkert, G. Linkert, N. Siddique, M. S. Hanner, B.–A. Lindblad, J. A. M. McDonnell, G. E. Morfill, G. Schwehm, H. A. Zook.
Bull. Am. Astron. Soc., Vol. 23, No. 3, p. 1149 (1991). Abstract. – See Abstr. 012.037 for the main entry.

106.046   Asteroid dust reaching the Earth: an information source on the composition of main belt asteroids.
D. E. Brownlee, S. G. Love.
Bull. Am. Astron. Soc., Vol. 23, No. 3, p. 1149 (1991). Abstract. – See Abstr. 012.037 for the main entry.

106.047   Modelling the zodiacal dust cloud.
Y.–L. Xu, S. F. Dermott, B. Å. S. Gustafson, S. Jayaraman, D. D. Durda.
Bull. Am. Astron. Soc., Vol. 23, No. 3, p. 1149–1150 (1991). Abstract. – See Abstr. 012.037 for the main entry.

106.048   Modelling of asteroidal dust production rates.
D. D. Durda, S. F. Dermott.
Bull. Am. Astron. Soc., Vol. 23, No. 3, p. 1150 (1991). Abstract. – See Abstr. 012.037 for the main entry.

106.049   The fate of small grains about asteroids.
D. P. Hamilton, J. A. Burns.
Bull. Am. Astron. Soc., Vol. 23, No. 3, p. 1150 (1991). Abstract. – See Abstr. 012.037 for the main entry.

106.050   Interior resonance trapping of dust particles from comets and asteroids.
A. A. Jackson, H. A. Zook.
Bull. Am. Astron. Soc., Vol. 23, No. 3, p. 1150 (1991). Abstract. – See Abstr. 012.037 for the main entry.

106.051   Measurement of low–energy charged particles by means of the SF–3M spectrometer onboard the satellite "Cosmos-1809".
N. V. Baranets, V. A. Gladyshev, T. M. Mulyarchik, V. A. Notkin, R. N. Suncheleev, Yu. P. Sharko.
Kosm. Issled., Tom 30, Vyp. 1, p. 67–78 (Jan–Feb 1992). In Russian. English translation in Cosmic Res., Vol. 30, No. 1.

A spectrometer of electrons and ions of low energies was installed onboard the satellite "Cosmos-1809" launched 18.XII.86. The description of the instrument and of its calibration is given. Some results obtained by this instrument are described: the electron and ion fluxes above the auroral latitudes, escaped and conjugate photoelectron fluxes, electron and ion bursts caused by powerful radiopulses abord the satellite.

106.052 Polarization parameters of large–scale turbulence of the solar wind and fluctuations of the cosmic ray intensity.
I. A. Transkij, S. A. Starodubtsev.
Geomagn. Aeron., Vol. 31, No. 1, p. 19 – 23 (Aug 1991). English translation of Geomagn. Aehron., Tom 31, No. 1, p. 27 – 33 (1991).

The method of covariant matrices is used to investigate the polarization properties of fluctuations of the interplanetary magnetic field. For three arbitrarily chosen time intervals the effective polarization parameters: the degree of polarization, the polarization angle, and the ellipticity and coherence are determined in the $10^{-4} \leqslant f < 2 \cdot 10^{-3} \text{sec}^{-1}$ frequency band. It was assumed that the observed fluctuations of the IMF are representable in the form of a superposition of quasi–monochromatic waves. Analysis shows that magnetosonic waves predominate in to intervals in the polarization component of the IMF and Alfven waves in the third. It follows from a comparision of the power spectra of the fluctuations of the cosmic–ray intensity and the polarization parameters that statistically significant bursts appear in the power spectrum of fluctuations of cosmic rays with energy E > 1 GeV in the frequency band of increasing degree of polarization of magnetosonic waves.

106.053 A thermomagnetic instability in the interplanetary plasma and the formation of "hot diamagnetic cavities".
L. G. Genkin, O. Yu. Gol'dshmidt, L. M. Erukhimov.
Geomagn. Aeron., Vol. 31, No. 1, p. 24 – 27 (Aug 1991). English translation of Geomagn. Aehron., Tom 31, No. 1, p. 34 – 39 (1991).

An instability of the thermomagnetic type, which may develop in the weakly collisional magnetoactive cosmic plasma when developed turbulence and drift of ions relative to electrons orthogonal to the magnetic field is present, is discussed. An instability leads to an increase of perturbations of the plasma temperature and the magnetic field. The proposed mechanism may be responsible for the formation of hot diamagnetic cavities which have been detected in the solar wind near the terrestrial shock wave, as well as the formation of the non–uniform structure of the interplanetary plasma and the interstellar medium.

106.054 Background proton fluxes with energies from 20 to 500 keV in the interplanetary medium during the period December 1975 to April 1977.
R. N. Basilova, S. P. Ryumin.
Geomagn. Aehron., Tom 32, No. 2, p. 19 – 22 (Mar – Apr 1992). In Russian. English translation in Geomagn. Aeron., Vol. 32, No. 2.

106.055 Direction of the 23 – 50 keV proton flux in the transition region and interplanetary space at geocentric distances of 20 – 35 radii of the Earth.
R. N. Basilova, S. P. Ryumin.
Geomagn. Aehron., Tom 32, No. 2, p. 23 – 27 (Mar – Apr 1992). In Russian. English translation in Geomagn. Aeron., Vol. 32, No. 2.

106.056 Interplanetary medium disturbances generated by a slow isolated magnetic cloud.
K. G. Ivanov, E. P. Romashets, A. F. Kharshiladze.
Geomagn. Aehron., Tom 32, No. 3, p. 85 – 91 (May – Jun 1992). In Russian. English translation in Geomagn. Aeron., Vol. 32, No. 3.

106.057 Cosmic–ray perpendicular diffusion coefficient computed using Voyager data at 15 AU.
J. F. Valdes–Galacia, J. J. Quenby, X. Moussas.
Sol. Phys., Vol. 139, No. 1, p. 189 – 199 (May 1992).

Perpendicular diffusion due to either field line wandering or random gradient and curvature effects makes a substantial contribution to the radial transport of particles at distances of about 5 AU and beyond when lines of the interplanetary magnetic field (IMF) become almost azimuthal. Here test particle

trajectories are followed in a field model which uses a magnetic field sample taken by Voyager 2 at 14.8 AU. Techniques previously developed are employed to calculate the perpendicular diffusion coefficient for 100 MeV protons and arising from random gradient and curvature effects. The result $K_{\perp} = (2.8 \pm 0.9) \times 10^{21} \text{cm}^2 \text{s}^{-1}$ shows a substantial increase above the value determined previously at 5 AU and, assuming a power law radial dependence $K \sim r^{\beta}$, implies $\beta \cong 0.8 - 1.1$. This result is consistent with observations of the cosmic–ray radial gradient.

106.058 Zodiacal emission. III. Dust near the asteroid belt.
W. T. Reach.
Astrophys. J., Vol. 392, No. 1, p. 289 – 299 (10 Jun 1992).

Properties of the zodiacal dust bands are derived from fits to IRAS profiles of the ecliptic. Three observations lead to the conclusion that the dust–band material is spread over a range of heliocentric distances between the asteroid belt and the Sun: parallax, color temperature, and wavelength dependence of band latitudes. The orientations of the midplanes of the bands are found to be typical of asteroids. A model of "migrating bands", wherein dust is produced near the asteroid belt and spirals into the Sun under the influence of Poynting–Robertson drag, is used to explain the range of heliocentric distances of dust–band material.

106.059 Extraction of helium from individual interplanetary dust particles by step–heating.
A. O. Nier, D. J. Schlutter.
Meteoritics, Vol. 27, No. 2, p. 166 – 173 (Jun 1992).

Fragments from 20 individual particles, collected in the Earth's stratosphere and believed to be interplanetary dust particles (IDPs), were subjected to step–heating to see if differences in the release pattern for ^{4}He could be observed which might provide clues to the origin of the particles. Comparisons were made to the release pattern for 18 individual lunar surface grains heated in the same manner. Twelve of the IDP fragments contained an appreciable amount of ^{4}He, 50% of which was released by the time the particles were heated to approximately 630°C. For the 18 individual lunar grains the corresponding average temperature was 660°C. Four of the IDP fragments contained appreciaibly less ^{4}He, and this was released at a higher temperature. From Flynn's analyses of the problem of the heating of IDPs in their descent in the atmosphere, the present results suggest that the parent IDPs of the 12 particles which contained an appreciable amount of ^{4}He suffered very little heating in their descent and are likely of asteroidal origin, although one cannot rule out the possibility that at least some of them had a cometary origin and entered the earth's atmosphere at a grazing angle.

106.060 Mineralogy of 12 large "chondritic" interplanetary dust particles.
M. E. Zolensky, D. J. Lindstrom.
22. Lunar and Planetary Science Conference, p. 161 – 169 (1992). – See Abstr. 012.046 for the main entry.

The authors report mineralogical analyses of 12 large "chondritic" interplanetary dust particles (IDPs) and discuss the probable origin of these large IDPs, comparing hydrous to anhydrous particles. They note that it is commonly difficult to classify certain IDPs by the current scheme and suggest that the degree of compositional heterogeneity exhibited by anhydrous silicates could form the basis of an alternate IDP classification scheme.

106.061 Trace elements in chondritic stratospheric particles: zinc depletion as a possible indicator of atmospheric entry heating.
G. J. Flynn, S. R. Sutton.
22. Lunar and Planetary Science Conference, p. 171 – 184 (1992). – See Abstr. 012.046 for the main entry.

106.062 Impact cratering from LDEF's 5.75–year exposure: decoding of the interplanetary and earth–orbital populations.
J. A. M. McDonnell.
22. Lunar and Planetary Science Conference, p. 185 – 193 (1992). – See Abstr. 012.046 for the main entry.

106.063   A detailed petrological analysis of hydrated, low–nickel, nonchondritic stratospheric dust particles.
F. J. M. Rietmeijer.
22. Lunar and Planetary Science Conference, p. 195 – 201 (1992).
– See Abstr. 012.046 for the main entry.

106.064   Suitability of silica aerogel as a capture medium for interplanetary dust.
R. A. Barrett, M. E. Zolensky, F. Hörz, D. J. Lindstrom, E. K. Gibson.
22. Lunar and Planetary Science Conference, p. 203 – 212 (1992).
– See Abstr. 012.046 for the main entry.

106.065   Interplanetary dust: cometary or asteroidal origin?
A. C. Levasseur–Regourd.
15. Ecole de Printemps d'Astrophysique de Goutelas: Interrelations between physics and dynamics for minor bodies in the solar system, p. 401 – 418 (1992). – See Abstr. 012.055 for the main entry.
Contents: 1. Introduction. 2. Dynamical evolution of the dust (Cometary dust. Asteroidal dust. Interplanetary dust). 3. In situ studies approach (Meteor studies. IDP studies. Impact studies). 4. Remote observations approach (Zodiacal light and zodiacal emission. Evolution of optical properties). 5. Conclusion.

106.066   Face–dependent impact probabilities upon LDEF for heliocentric particle orbits.
D. Steel.
IAU Colloquium No. 126: Origin and evolution of interplanetary dust, p. 41 – 44 (1991). – See Abstr. 012.068 for the main entry.
If the impact record upon LDEF is to be interpreted so as to determine the flux, orbits, sizes and compositions of natural meteoroids and dust, and space debris, then it is necessary to relate the microcraters and perforations recorded to the likely source orbit of the particle in each case. Here a single–particle approach is used to calculate the relative impact probabilities upon six orthogonal faces of LDEF for particles coming from heliocentric orbits confined to the ecliptic; the results are presented as functions of impact velocity and impact angle for each face. The cratering ratios for the East (or leading) face compared to the West (or trailing) and the Earth–directed faces are strongly dependent upon the velocities of the particles and can therefore indicate of the velocity distribution of meteoroids and interplanetary dust.

106.067   Collection of stratospheric microparticles above the sulfate layer using balloon–borne collectors.
J. R. Stephens, Y. Nakada, T. Onaka, F. J. M. Rietmeijer.
IAU Colloquium No. 126: Origin and evolution of interplanetary dust, p. 49 – 52 (1991). – See Abstr. 012.068 for the main entry.
The authors report preliminary analytical electron microscope (AEM) analysis of nearly 300 stratospheric particles collected using balloon–borne collectors at 34 – 36 km altitude. The particles are predominantly silica, plagioclase feldspar, Mg, Fe–silicates and rare barite, metal oxides, and unidentified Fe, Ni, Zn and Pb particles. The majority of these generally submicron–sized particles are comparable to volcanic particles collected at 20 km altitude from the 1982 eruption of the El Chichon volcano. Because of the uniqueness in altitude and collected particle sizes the collection may also contain interplanetary dust particles of types poorly represented in present collections.

106.068   A cosmic matter accretion event around 660,000 years before present found in two dated, Central Pacific cores.
K. Yamakoshi, K.'i. Nogami, R. Omori, Ma Jianguo, Ma Shulan.
IAU Colloquium No. 126: Origin and evolution of interplanetary dust, p. 53 – 56 (1991). – See Abstr. 012.068 for the main entry.
Iridium contents and the ratios of (Co/Fe) in two dated, respective layers of the cores are determined. These samples were dated fortunately with the paleo–magnetic and also with the cosmogenic Be–10 methods. Ir enrichments are found at (0.660 ± 0.030) My before present.

106.069   Physical and mineralogical properties of anhydrous interplanetary dust particles in the analytical electron microscope.
J. P. Bradley.
IAU Colloquium No. 126: Origin and evolution of interplanetary dust, p. 63 – 70 (1991). – See Abstr. 012.068 for the main entry.
The fine grained mineralogy and petrography of anhydrous "pyroxene" and "olivine" classes of chondritic interplanetary dust have been investigated by numerous electronmicroscopic studies. The "pyroxene" interplanetary dust particles (IDPs) are porous, unequilibrated assemblages of mineral grains, metal, glass, and carbonaceous material. They contain enstatite whiskers, FeNi carbides, and high–Mn olivines and pyroxenes, all of which are likely to be well preserved products of nebular gas reactions. Solar flare tracks are prominent in most "pyroxene" IDPs, indicating that they were not strongly heated during atmospheric entry. The "olivine" IDPs are coarse grained, equilibrated mineral assemblages that have probably experienced strong heating. Since most "olivine" IDPs do not contain tracks, it is possible that this heating occurred during atmospheric entry.

106.070   Aqueous alteration in hydrated interplanetary dust particles.
K. Tomeoka.
IAU Colloquium No. 126: Origin and evolution of interplanetary dust, p. 71 – 78 (1991). – See Abstr. 012.068 for the main entry.
Interplanetary dust particles (IDPs) characterized by chondritic composition can be divided into two principal groups, anhydrous and hydrated. This paper summarizes recent results of mineralogical and petrological studies dealing with the IDPs of hydrated type. Studies on mineralogical characteristics, infrared absorption spectra, and isotopic properties of the hydrated particles have suggested that they are primitive in major part of layer silicates and resemble CI and CM carbonaceous chondrites. Mineralogical and chemical data of both IDPs and carbonaceous chodrites have accumulated, and it is now possible to compare the mineralogies of the IDPs and the meteorites in considerable detail. Evidence was found that a significant proportion of the hydrated IDPs have been processed by aqueous alteration, and the nature of the alteration resembles provide important clues to the possible origins of IDPs.

106.071   The zodiacal cloud complex.
A. C. Levasseur–Regourd, J. B. Renard, R. Dumont.
IAU Colloquium No. 126: Origin and evolution of interplanetary dust, p. 131 – 138 (1991). – See Abstr. 012.068 for the main entry.
The physical properties of the interplanetary dust grains are, out of the ecliptic plane, mainly derived from observations of zodiacal light in the visual or infrared domains. The bulk optical properties (polarization, albedo) of the grains are demonstrated to depend upon their distance to the Sun (at least in a 0.1 AU to 1.7 AU range in the symmetry plane) and upon the inclination of their orbits (at least up to 22°). Classical models assuming the homogeneity of the zodiacal cloud are no longer acceptable. A hybrid model, with a mixture of two populations, is proposed. It suggests that various sources (periodic comets, asteroids, non periodic comets...) play an improtant role in the replenishment of the zodiacal cloud complex.

106.072   Spatial distribution and orbital properties of zodiacal dust.
B. Kneißel, I. Mann.
IAU Colloquium No. 126: Origin and evolution of interplanetary dust, p. 139 – 146 (1991). – See Abstr. 012.068 for the main entry.
Within the recent years the spatial distribution of zodiacal dust has been subject to a variety of modelling approaches. Whereas models derived from observations in the visual range tend to demand for an increase of interplanetary matter above the solar poles (bulges), models based on infrared measurements and extended to small r seem to favor a decrease there (holes). The models are reviewed, and the dynamical structure implicated in the models is outlined.

106.073   On the Gegenschein and the symmetry plane.
S. S. Hong, S. M. Kwon.
IAU Colloquium No. 126: Origin and evolution of interplanetary dust, p. 147 – 150 (1991). – See Abstr. 012.068 for the main entry.

Using 3–dim density models of the zodiacal cloud, the authors have calculated brightness of the zodiacal light over an extended region around the anti–solar point. The isophotal contours of the model Gegenscheins differ from each other, morphologically, to the degree that they can differentiate the competing density models. The recently reduced Gegenschein observations of 2° resolution clearly favour the ellipsoid–type models to the fan–types, and also suggest that the surface of the densest dust concentration in the outer part of the cloud has its ascending node at longitude $100 \pm 20°$ and is inclined $2 \pm 0°.5$ with respect to the ecliptic plane.

106.074   Ultraviolet observations of the zodiacal light and the origin of interplanetary dust grains.
C. F. Lillie.
IAU Colloquium No. 126: Origin and evolution of interplanetary dust, p. 151 – 154 (1991). – See Abstr. 012.068 for the main entry.

Surface brightness photometry of the night sky from rocket and satellite experiments shows an increase in the scattering efficiency of interplanetary dust grains in the 1500 to 3000 Å region of the spectrum. This increase is best explained by the presence of small dielectric particles with a mean radius of 0.04 microns. The most likely source of these grains is the dissolution of agglomerates of these particles which are released by comets during their perihelion passage. Many of these agglomerates have been collected in the Earth's atmosphere by high flying aircraft. Submicron particles swept up from interplanetary space may be responsible for the high altitude haze observed in planetary atmospheres.

106.075   Light scattering by dust particles in the outer solar system.
J. W. Hovenier, P. B. Bosma.
IAU Colloquium No. 126: Origin and evolution of interplanetary dust, p. 155 – 158 (1991). – See Abstr. 012.068 for the main entry.

Photometric observations of the zodiacal light performed by Pioneer 10 indicated that there may be very little scattering by dust in the outer solar system. The authors formulate explicit expressions for interpreting the brightness observed by a spacecraft travelling inside or outside a finite homogeneous cloud of scattering particles. An application is made to the ecliptic zodiacal light brightness as observed by Pioneer 10 and tabulated by Toller and Weinberg (1985). A satisfactory interpretation of these data as well as earthbound observations can be given by means of a model having a particle density distribution or mean scattering cross section which vanishes beyond 2.8 – 3.7 AU. Some implications for the nature and spatial distribution of the interplanetary dust are discussed.

106.076   Light scattering by solar system dust: the opposition effect and the reversal of polarization.
K. Muinonen, K. Lumme.
IAU Colloquium No. 126: Origin and evolution of interplanetary dust, p. 159 – 162 (1991). – See Abstr. 012.068 for the main entry.

The opposition effect and the reversal of linear polarization, or negative polarization, at small phase angles have been almost universally observed in light scattered from atmosphereless solar system bodies. Recent investigations have indicated that both phenomena can be qualitatively understood as resulting from a common physical mechanism: coherent multiple backscattering. As for interplanetary dust, the coherent backscattering mechanism contributes both to the Gegenschein and to the almost certainly existing negative polarization branch. Theoretical results supporting the coherent backscattering explanation are briefly presented.

106.077   The optical properties of interplanetary dust.
P. L. Lamy, J. M. Perrin.
IAU Colloquium No. 126: Origin and evolution of interplanetary dust, p. 163 – 170 (1991). – See Abstr. 012.068 for the main entry.

After briefly evaluating the observations of the zodiacal light and F–corona, the authors review laboratory results on the light scattering by dust particles and the various theories which have been recently proposed. They then discuss the optical properties of the dust with emphasis on the phase function, the polarization, the color, the albedo and the local enhancement in the Gegenschein.

106.078   The infrared zodiacal light.
M. S. Hanner.
IAU Colloquium No. 126: Origin and evolution of interplanetary dust, p. 171 – 178 (1991). – See Abstr. 012.068 for the main entry.

Thermal emission from interplanetary dust is the main source of diffuse radiation at $\lambda 5 - 50~\mu m$. Analysis of infrared sky maps from IRAS and ZIP lead to the result that the average optical properties of the dust change with heliocentric distance. The present uncertainties in calibration should be resolved by COBE. Existence of a dust sublimation zone at 4 solar radii awaits confirmation at the next solar eclipse.

106.079   Fine resolution brightness distribution of the visible zodiacal light.
S. M. Kwon, S. S. Hong, J. L. Weinberg, N. Y. Misconi.
IAU Colloquium No. 126: Origin and evolution of interplanetary dust, p. 183 – 186 (1991). – See Abstr. 012.068 for the main entry.

Applying time–dependent corrections of the atmospheric diffuse light to the observed night sky brightness, the authors have determined brightness of the zodiacal light over the region $40° \leqslant \lambda - \lambda_\odot \leqslant 320°$ and $-20° \leqslant \beta \leqslant 20°$. The resulting map of equal brightness contours has an angular resolution of two degrees, and exhibits east–west and north–south asymmetries.

106.080   Interplanetary dust close to the Sun.
I. Mann.
IAU Colloquium No. 126: Origin and evolution of interplanetary dust, p. 187 – 190 (1991). – See Abstr. 012.068 for the main entry.

The optical and infrared brightness of the Fraunhofer–corona is produced by light scattering at the zodiacal dust particles and by their thermal emission. It is modelled within the ecliptic ($4~R_\odot \leqslant \varepsilon \leqslant 15~R_\odot$) taking into account investigations of the global zodiacal dust cloud due to remote sensing and in situ experiments. The input of near solar dust to the corona brightness is discussed.

106.081   Optical properties of interplanetary dust in the tangential plane.
J. B. Renard, A. C. Levasseur–Regourd, R. Dumont.
IAU Colloquium No. 126: Origin and evolution of interplanetary dust, p. 199 – 202 (1991). – See Abstr. 012.068 for the main entry.

Local intensity and emissivity, and consequently local polarization degree, temperature and albedo, can be retrieved from optical and thermal observations of zodiacal light. The local polarization degree (normalized at constant solar distance and phase angle) is found to decrease with elevation above the symmetry plane of the zodiacal cloud. The heterogeneity of the cloud, established towards the symmetry pole, is here demonstrated in the tangential plane (almost perpendicular to the ecliptic plane at 1 AU). The authors present a map of the local polarization degree in this plane.

106.082   Asteroidal dust and the zodiacal emission.
W. T. Reach.
IAU Colloquium No. 126: Origin and evolution of interplanetary dust, p. 211 – 214 (1991). – See Abstr. 012.068 for the main entry.

The contribution to the brightness of the infrared background by asteroidal dust, distinguished both by lower color temperature and 'band–pair' morphology, is determined using IRAS observations. Dust band pairs are associated with at least 7 asteroid families and groups, but very little is detected from the remainder of the asteroid belt, indicating that asteroid families and groups are the source of asteroidal dust.

106.083 Mass distribution and bulk density distribution of interplanetary dust.
A. Hajduk.
IAU Colloquium No. 126: Origin and evolution of interplanetary dust, p. 331 – 334 (1991). – See Abstr. 012.068 for the main entry.
Mass distribution of the interplanetary dust is reexamined taking into account bulk density distribution of the dust and larger particles. It can be shown that the mass index of particles depends on the evolutionary stage of the population and changes along the mass scale. The flattening of the mass distribution at the higher mass range may explain the problem of the equilibrium between the source and sink of the interplanetary dust.

106.084 Physical processes on circumplanetary dust.
J. A. Burns.
IAU Colloquium No. 126: Origin and evolution of interplanetary dust, p. 341 – 348 (1991). – See Abstr. 012.068 for the main entry.
The life cycles of grains in circumplanetary space are governed by various physical processes that alter sizes and modify orbits. Lifetimes are quite short, perhaps $10^2 - 10^4$ years for typical circumplanetary grains of 1 micron radius. Thus particles must be continually supplied to the circumplanetary complex, probably by the grinding down of larger parent bodies in collisions. Dust is eroded gradually through sublimation and through sputtering by the magnetospheric plasma but also is catastrophically destroyed through hypervelocity impacts with interplanetary micrometeoroids. Orbits evolve through momentum transfer (light drag, plasma or Coulomb drag and atmospheric drag), and through resonant gravitational and electromagnetic forces. Plasma drag is generally the most effective evolution mechanism, with the possible exceptions of exospheric drag at Uranus and of electromagnetic schemes for some conditions. Since grains become charged (with typical electric potentials of a few volts), they undergo associated orbital perturbations: variable electromagnetic forces can cause the systematic drain of energy (orbital collapse) or, at specific (resonant) orbital locations can force large orbital inclinations/eccentricities. Solar rediation induces a periodic orbital eccentricity that can reach substantial values for 1 micron particles distant from the giant planets.

106.085 Cometary and asteroidal sources of interplanetary dust.
M. V. Sykes.
IAU Colloquium No. 126: Origin and evolution of interplanetary dust, p. 389 – 396 (1991). – See Abstr. 012.068 for the main entry.
The Infrared Astronomical Satellite has provided extensive observations of the zodiacal cloud at high spatial resolution which will not be matched in the forseeable future. Whithin the zodiacal cloud, IRAS discovered extended dust structures providing the link between the interplanetary dust complex and the asteroids and comets which are its source. These are the asteroid dust bands and the cometary dust trails.

106.086 Constraints on the parent bodies of collected interplanetary dust particles.
S. A. Sandford.
IAU Colloquium No. 126: Origin and evolution of interplanetary dust, p. 397 – 404 (1991). – See Abstr. 012.068 for the main entry.
Samples of interplanetary dust particles (IDPs) have now been collected from the stratosphere, from the Earth's ocean beds, and from the ice caps of Greenland and Antarctica. The most likely candidates for the sources of these particles are comets and asteroids. Comparison of the infrared spectra, elemental compositions, and mineralogy of the collected dust with atmospheric entry models and data obtained from cometary probes and telescopic observations has provided important constraints on the possible sources of the various types of collected dust. These constraints lead to the following conclusions. First, most of the deep sea, Greenland, and Antarctic spherules larger than 100 μm are derived from asteroids. Second, the stratospheric IDPs dominated by hydrated layer–lattice silicate minerals are also most likely derived from asteroids. Finally, the stratospheric IDPs dominated by the anhydrous minerals olivine and pyroxene are most likely from comets. The consequences of these parent body assignments are discussed.

106.087 Chemical composition of dust expected from condensation models.
T. Yamamoto.
IAU Colloquium No. 126: Origin and evolution of interplanetary dust, p. 413 – 420 (1991). – See Abstr. 012.068 for the main entry.
This review examines to what degrees the present chemical equilibrium condensation models are effective in predicting chemical composition of grains observed in a variety of cosmic environments. The composition expected from the equilibrium calculations is reviewed separately for refractory (rocky and metallic) and volatile (icy) components. Comments are given on the limitation of the equilibrium calculations in predicting the grain composition. By taking cometary ice as a typical cosmic volatile condensate, it is pointed out that its composition is far from that expected from the equilibrium models. Theories on the formation of cometary volatiles are reviewed, and an observational clue helpful to testing the theories is pointed out. Discussion is given on the advantage for formation of organic materials from volatile solids.

106.088 The interplanetary medium is thriving.
J. M. Greenberg.
IAU Colloquium No. 126: Origin and evolution of interplanetary dust, p. 443 – 451 (1991). – See Abstr. 012.068 for the main entry.
Colloquium summary.

106.089 Astrophysical dust grains in stars, the interstellar medium, and the solar system.
R. D. Gehrz.
US–USSR Workshop on Planetary Sciences, p. 126 – 142 (1991). – See Abstr. 012.075 for the main entry.
The possible connections between extra–solar–system astrophysical dust grains and the grains in the solar system are explored. A recent suggestion that grains are rapidly destroyed in the interstellar medium by supernova shocks is discussed. Experiments to establish the relationships between extra–solar–system astrophysical grains and solar system grains, and between cometary dust and the zodiacal dust are suggested. Among the most promising are sample return missions and improved high–resolution infrared spectroscopic information.

106.090 Transient interplanetary and geomagnetic disturbances associated with solar flares.
Zhang Gongliang.
Advances in solar–terrestrial science of China, p. 268 – 277 (1992). – See Abstr. 003.069 for the main entry.
This review article summarizes the main results obtained by the author and his colleagues through recent systematic studies of interplanetary and geomagnetic disturbances associated with flare activities, with emphasis on the source and propagation characteristics of the solar transient disturbed plasma output. The discussions include the long–duration geo–effective flare, the abnormal increment in magnetic energy, azimuthal asymmetry, the corotating partner of a transient stream, and magnetic storms in groups.

106.091 The problem of the oxygen presence in the heliosphere and observational implications.
H. J. Fahr.
10. ESA Symposium on European Rocket and Balloon Programmes and Related Research, p. 349 – 354 (Nov 1991). – See Abstr. 012.088 for the main entry.
Due to the relative motion of the solar system with respect to the ambient interstellar medium (VLISM) neutral atomic VLISM gases can pass over the plasma interface region ahead of the solar system and can enter into the inner heliosphere, however only, after having suffered changes in their dynamic and thermodynamic properties. VLISM oxygen especially is subject to strong charge exchange loss processes, uncompensated by corresponding gain processes, before it enters the inner heliosphere. Thus the presence of VLISM oxygen in interplanetary space is crucial for the type of plasma interface ahead of the solar system which sensitively determines the net oxygen inflow. The author herewith proposes to observe the interplanetary oxygen

resonance glow intensity at 1304 Å with a rocket– or Space–Shuttle–borne experiment in order to clarify on the nature of this interface. For this purpose detector sensitivities down to 10^{-3} Rayleighs are required. These sensitivities are reached by an instrument called Hitch–Helly which is developed at present for a forthcoming Hitchhiker flight.

106.092   Introductory lecture – the heliosphere.
W. I. Axdorf.
1. COSPAR Colloquium: Physics of the outer heliosphere, p. 7 – 15 (1990). – See Abstr. 012.100 for the main entry.
A brief review is given of the historical ideas dealing with the development of the present concept of the heliosphere, together with a summary of the present status, particularly with regard to the distance to the solar wind shock termination.

106.093   A new observational approach to investigate the heliospheric interstellar wind interface: the study of extreme and far ultraviolet resonantly scattered solar radiation from neon, oxygen, carbon and nitrogen.
S. Bowyer, H. J. Fahr.
1. COSPAR Colloquium: Physics of the outer heliosphere, p. 29 – 36 (1990). – See Abstr. 012.100 for the main entry.
A new method to investigate the heliospheric–interstellar wind interface is suggested. This interface affects inflowing LISM carbon, nitrogen, oxygen, and neon atoms differently, depending upon the character of the interaction at the interface. A preliminary evaluation of this effect has been carried out, and the results indicate that the distribution of these species within the inner heliosphere varies with different model interface regions. The study of resonantly scattered solar radiation from these species will then provide a means to discriminate among these models.

106.094   Scattering of solar UV on local neutral gases.
R. Lallement.
1. COSPAR Colloquium: Physics of the outer heliosphere, p. 49 – 59 (1990). – See Abstr. 012.100 for the main entry.
The solar system is embedded in in a glow of the resonance lines of H and He atoms, through scattering of solar photons by flowing interstellar neutral gas. This flow is modified by solar gravitation, ionization, radiation pressure, and possibly by interaction with ionized species around the heliopause. The study of intensity distribution allows the derivation of estimates of densities, velocity and temperature of neutral H and He in the very local interstellar medium.

106.095   Model predictions and remote observations of the hydrogen density profile in the distant heliosphere.
D. L. Judge, P. Gangopadhyay, S. Grzedzielski.
1. COSPAR Colloquium: Physics of the outer heliosphere, p. 61 – 64 (1990). – See Abstr. 012.100 for the main entry.
Observations of the hydrogen Ly–α glow by the deep space UV photometers and spectrometers on board Pioneer 10 and Voyager 2 have been utilized to infer the presence of a nearby solar wind shock. The observational and theoretical basis for the current understanding of the heliospheric neutral density profile is presented.

106.096   Comparison of Ly–α and Ly–β interplanetary glows observed by the Voyager ultra–violet spectrometer.
E. Chassefière, J. C. Vial, R. Lallement, J. L. Bertaux.
1. COSPAR Colloquium: Physics of the outer heliosphere, p. 65 – 72 (1990). – See Abstr. 012.100 for the main entry.
The comparison between Ly–α and Ly–β glows exhibits significant discrepancies which can not be explained by radiation transfer in the Ly–α interplanetary line. They might be due, at least partially, to the temporal variability of the solar flux, larger at Ly–α center than at Ly–β center.

106.097   Lyman–alpha observations from Voyager (1–18 AU).
R. Lallement, J. L. Bertaux, E. Chassefière, B. R. Sandel.
1. COSPAR Colloquium: Physics of the outer heliosphere, p. 73 – 82 (1990). – See Abstr. 012.100 for the main entry.
A fraction of Voyager 1 and 2 UVS measurements of the interplanetary Lyman–α background during the 1977–1983 period is presented and compared with results from current models of the interaction between the sun and the neutral interstellar gas.

106.098   New channel for the photoionization of hydrogen atoms in the solar system.
M. A. Gruntman.
1. COSPAR Colloquium: Physics of the outer heliosphere, p. 83 – 86 (1990). – See Abstr. 012.100 for the main entry.
A new two step channel for the photoionization of hydrogen atoms in the interplanetary space is proposed. Hydrogen atoms are excited by photons, then they decay to the metastable H(2S) state, where they can be photoionized. Competing processes are considered, and the photoionization rate through the proposed indirect channel is calculated. This rate becomes higher than that of direct photoionization for the region closer than 0.4 AU to the Sun.

106.099   General description of time–dependent density fluctuations in the interplanetary neutral gas distribution by means of Fourier transform.
H. J. Fahr, K. Scherer.
1. COSPAR Colloquium: Physics of the outer heliosphere, p. 87 – 92 (1990). – See Abstr. 012.100 for the main entry.

106.100   Three–dimensional models of the global zodiacal dust cloud – a bimodal model of the solar dust heliosphere.
B. Kneissel, I. Mann.
1. COSPAR Colloquium: Physics of the outer heliosphere, p. 93 – 95 (1990). – See Abstr. 012.100 for the main entry.
The interplanetary dust cloud or dust heliosphere may be regarded as a superposition of one component mainly concentrated to the ecliptic plane with regular properties known from the zodiacal dust and a second one that is isotropically distributed and may result from long–period comets. This approach promises to give a consistent and more physical explanation of results from in–situ masurements as well as optical and infrared brightness observations.

106.101   Ion acceleration to cosmic ray energies.
M. A. Lee.
1. COSPAR Colloquium: Physics of the outer heliosphere, p. 157 – 168 (1990). – See Abstr. 012.100 for the main entry.
The acceleration and transport environment of the outer heliosphere is schematically described. Acceleration occurs where the divergence of the solar wind flow is negative, that is at shocks, and where second–order Fermi acceleration is possible in the solar wind turbulence. Accelerations of ions injected at the shock up to energies $\leqslant 300$ MeV/charge is expected to occur and to create the anomalous cosmic ray component.

106.102   Characteristics of large Forbush decreases associated with interplanetary magnetic clouds.
Badruddin.
1. COSPAR Colloquium: Physics of the outer heliosphere, p. 183 – 186 (1990). – See Abstr. 012.100 for the main entry.

106.103   Cosmic ray, energetic ion and magnetic field characteristics of magnetic clouds.
T. R. Sanderson, J. Beeck, R. G. Marsden, C. Tranquille, K.–P. Wenzel, R. B. McKibben.
1. COSPAR Colloquium: Physics of the outer heliosphere, p. 187 – 190 (1990). – See Abstr. 012.100 for the main entry.
The authors present preliminary results of a survey of the relation between magnetic clouds and Forbush decreases during the period August 1978 to May 1982, using energetic ion and magnetic field observations from ISEE–3 to identify the magnetic

clouds, and the ground–based observations to identify the Forbush decreases.

106.104 Nonlinear effects of cosmic ray interaction with solar wind in the outer heliosphere.
V. Kh. Babayan, L. I. Dorman.
1. COSPAR Colloquium: Physics of the outer heliosphere, p. 191 – 193 (1990). – See Abstr. 012.100 for the main entry.
A set of equations is solved which describe the nonlinear interactions of galactic cosmic rays with the moving magnetized solar wind plasma including the inverse effect of cosmic rays on plasma streams.

106.105 Evolution of the ideas about the heliosphere and cosmic ray modulation in interplanetary space.
I. V. Dorman, L. I. Dorman.
1. COSPAR Colloquium: Physics of the outer heliosphere, p. 195 – 198 (1990). – See Abstr. 012.100 for the main entry.
The history of the evolution of ideas, models, and the theories concerning modulation of galactic cosmic rays in the interplanetary space is described in brief, including the effects of the Sun's dipolar field, the steady–state diffusion, the corpuscular streams with induced electric fields, the dynamic solar wind, the magnetic inhomogeneity spectrum, the particle energy variations, the isotropic and anisotropic diffusion, the nonlinear and drift effects, the role of shock waves, the hysteresis and estimation of heliospheric dimensions, the neutral current sheet, and the terminal shock wave.

106.106 Outer heliosphere: eigen pulsations, cosmic rays and stream kinetic instability.
L. I. Dorman, V. S. Ptuskin, V. N. Zirakashvili.
1. COSPAR Colloquium: Physics of the outer heliosphere, p. 205 – 209 (1990). – See Abstr. 012.100 for the main entry.
The possible large–scale eigen oscillations of the solar wind cavity and the relevant variations of cosmic density are discussed. The stream kinetic instability of galactic cosmic rays entring the heliosphere is examined. The energy density of excited MHD waves is found. The modes of cosmic ray propagation in the outer heliosphere are discussed.

106.107 Outer heliosphere as a many–component medium for cosmic ray propagation.
L. I. Dorman, V. Kh. Shogenov.
1. COSPAR Colloquium: Physics of the outer heliosphere, p. 211 – 213 (1990). – See Abstr. 012.100 for the main entry.
The theory developed earlier by the authors for cosmic rays propagation is applied to the outer heliosphere where steady solar wind flow with frozen–in magnetic field occurs together with high velocity fluxes, shock waves and other formations moving at velocities other than the solar wind velocity.

106.108 On the propagation velocity of cosmic ray modulation wave in the outer heliosphere.
L. I. Dorman, A. G. Zusmanovich, O. A. Kryakunova.
1. COSPAR Colloquium: Physics of the outer heliosphere, p. 215 – 218 (1990). – See Abstr. 012.100 for the main entry.
The expected propagation velocity of cosmic ray modulation effects in the outer heliosphere is calculated numerically and analytically. The results are compared with Voyager-1,2 and Pioneer-10,11 data. A good agreement has been obtained.

106.109 Radial evolution of interaction regions.
Y. C. Whang, L. F. Burlaga.
1. COSPAR Colloquium: Physics of the outer heliosphere, p. 245 – 248 (1990). – See Abstr. 012.100 for the main entry.
Near and outside 1 AU, corotating interaction regions bounded by shocks form at the leading edges of high–speed streams. This paper describes the radial evolution of interaction regions, the merging of interaction regions belonging to neighboring streams, and the coalescence of interaction regions belonging to successive solar rotations.

106.110 Nonlinear surface wave propagation of heliospheric current sheet.
M. S. Ruderman.
1. COSPAR Colloquium: Physics of the outer heliosphere, p. 249 – 252 (1990). – See Abstr. 012.100 for the main entry.

106.111 Magnetic fields in the heliosphere: Pioneer observations.
E. J. Smith.
1. COSPAR Colloquium: Physics of the outer heliosphere, p. 253 – 265 (1990). – See Abstr. 012.100 for the main entry.
Major physical processes are reviewed that involve the magnetic field as measured by Pioneer 10 and 11. Topics discussed are spatial gradients out to 20 AU, the evolution of individual solar wind streams from 1 to 5 AU, the merging of solar wind structures from 5 to 20 AU, the search for waves near 10 AU associated with interstellar ions, the acceleration of energetic particles to MeV energies and both short duration decreases and solar cycle modulation of galactic cosmic rays. Attention is given to departures from expectation and phenomena that were not anticipated.

106.112 Radio noise in the heliospheric cavity.
W. S. Kurth.
1. COSPAR Colloquium: Physics of the outer heliosphere, p. 267 – 275 (1990). – See Abstr. 012.100 for the main entry.
The heliosphere is the source of a wide variety of radio emissions. These emissions are generated at the sun, in planetary magnetospheres, and in the heliospheric cavity, itself. The focus of this paper is the latter category, specifically, low frequency interplanetary emissions since these might possibly be generated as a result of the interaction of the solar wind with the interstellar medium. The author summarizes the observations by the Voyager spacecraft of these low frequency emissions and discusses the various possible sources of the emission.

106.113 Trapped radiation in the outer heliosphere.
A. Czechowski, S. Grzedzielski.
1. COSPAR Colloquium: Physics of the outer heliosphere, p. 281 – 284 (1990). – See Abstr. 012.100 for the main entry.
Time evolution of the electromagnetic radiation trapped in the outer heliosphere and interacting with the solar wind of fluctuating intensity is studied within linear theory in the geometrical optics approximation. The authors discuss the implications of the results for understanding the origin of the 2–3 kHz signals observed by the Voyager missions.

106.114 A model ot the reconnection pattern at the heliopause.
W. Macek, J. Bysiek.
International Workshop on Reconnection in Space Plasma, p. 299 – 302 (Jan 1989). – See Abstr. 012.099 for the main entry.
Plasma mixing between the solar wind and the ambient ionized component of the local interstellar medium (LISM) at the heliopause, the boundary separating both media, is discussed. By analogy with the case of planetary magnetospheres it is argued that field reconnection processes may play an important role for the plasma mixing at the heliopause. The reconnection areas at the frontal heliopause as viewed from the sun depending on the assumed direction of the external LISM magnetic field are computed.

106.115 Filtration of the interstellar neutrals at the heliospheric interface and their coupling to the solar wind.
H. J. Fahr.
1. COSPAR Colloquium: Physics of the outer heliosphere, p. 327 – 343 (1990). – See Abstr. 012.100 for the main entry.
Contents: 1. Penetration of LISM neutrals into the heliosphere. 2. Neutral hydrogen and oxygen presence in the heliosphere. 3. Interstellar neutrals in the heliosphere and the origin of the anomalous cosmic ray component.

106.116 LISM–heliosphere interaction mediated by suprathermal particles.
S. Grzedzielski, J. Ziemkiewicz.
1. COSPAR Colloquium: Physics of the outer heliosphere, p. 363 – 366 (1990). – See Abstr. 012.100 for the main entry.
Problems of the deceleration and termination of the solar wind are discussed taking into account the interactions with neutral interstellar H and He, galactic and anomalous cosmic ray populations, interstellar magnetic field and effects due to pickup ions in the LISM.

106.117 Expected beams of energetic neutral atoms in the outer heliosphere.
S. Grzedzielski, D. Ruciński.
1. COSPAR Colloquium: Physics of the outer heliosphere, p. 367 – 370 (1990). – See Abstr. 012.100 for the main entry.
Estimates of the fluxes of energetic neutral atoms in the distant heliosphere are calculated taking into account all important interactions between solar wind plasmas and neutral gases in the heliosphere.

106.118 Motion of the strong disturbances in the interplanetary medium.
N. L. Borodkova, Yu. I. Yermolaev (*Yu. I. Ermolaev*), G. N. Zastenker.
1. COSPAR Colloquium: Physics of the outer heliosphere, p. 391 – 392 (1990). – See Abstr. 012.100 for the main entry.
Features of the strong disturbances, mainly shock waves, moving through the interplanetary medium are reviewed on the basis of the solar wind measurements onboard the Soviet highapogee satellites Prognoz–7,8 in 1978–81.

106.119 Reconnection pattern at the heliopause.
W. M. Macek.
1. COSPAR Colloquium: Physics of the outer heliosphere, p. 399 – 402 (1990). – See Abstr. 012.100 for the main entry.
Views of the front side heliopause and different areas, where reconnection can take place, are provided for various assumed orientations of the magnetic field of the LISM. Within these reconnection patches the lines of constant angle between the magnetic field of unperturbed solar wind and the (uniform) LISM field, both projected onto the heliopause surface, are computed. Low frequency (2–3 kHz) interplanetary radio emission detected by Voyager mission and implications of possible observations of the heliospheric boundary are discussed.

The fullness of space. Nebulae, stardust, and the interstellar medium.
See Abstr. 003.085.

Space dust and debris. Proceedings. Topical Meeting of the COSPAR Interdisciplinary Scientific Commission B (Meetings B2, B3 and B5) of the COSPAR 28. Plenary Meeting, The Hague (Netherlands), 25 Jun – 6 Jul 1990.
See Abstr. 012.014.

7. Quadrennial Symposium on Solar–Terrestrial Physics (SCOSTEP–7), The Hague (Netherlands), 25 – 30 Jun 1990.
See Abstr. 012.024.

Origin and evolution of interplanetary dust. Proceedings.
See Abstr. 012.068.

Proceedings of the 10th ESA Symposium on European Rocket and Balloon Programmes and Related Research.
See Abstr. 012.088.

Reconnection in space plasma. Proceedings.
See Abstr. 012.099.

Orbital evolution of dust particles from comets and asteroids.
See Abstr. 021.020.

Hydrogen cyanide polymerization: a preferred cosmochemical pathway.
See Abstr. 022.101.

The physics and dynamics of charged dust grains.
See Abstr. 022.195.

Penetration of hypervelocity projectiles into low density materials.
See Abstr. 022.219.

Catastrophic disruption of solid bodies by collision – experimental approach.
See Abstr. 022.220.

Methods, difficulties, and first results in laboratory simulation of cosmic dust electric charging.
See Abstr. 022.221.

Electrostatic fragmentation of irregularly shaped particles.
See Abstr. 022.222.

Plasma emission from high velocity impacts of microparticles onto water ice.
See Abstr. 022.223.

Velocity distribution of fragments in collisional breakup.
See Abstr. 022.224.

Jets of fragments from catastrophic break–up and their astrophysical implications.
See Abstr. 022.225.

The magnetic field investigation on the Ulysses mission: instrumentation and preliminary scientific results.
See Abstr. 035.030.

The Solar Wind Ion Composition Spectrometer.
See Abstr. 035.032.

The unified radio and plasma wave investigation.
See Abstr. 035.033.

The ULYSSES energetic particle composition experiment EPAC.
See Abstr. 035.034.

The interstellar neutral–gas experiment on ULYSSES.
See Abstr. 035.035.

Heliosphere Instrument for Spectra, Composition and Anisotropy at Low Energies.
See Abstr. 035.036.

The Ulysses dust experiment.
See Abstr. 035.039.

The gravitational wave experiment.
See Abstr. 035.041.

The Galileo Dust Detector.
See Abstr. 035.085.

A balloon–borne detector for stratospheric cosmic dust detection.
See Abstr. 035.115.

The Munich Dust Counter – a cosmic dust experiment on board of the MUSES–A mission of Japan.
See Abstr. 035.116.

Contamination of terrestrial EUV observations by energetic particles.
See Abstr. 035.126.

On the possibility of detection of small comets in Ly–α.
See Abstr. 036.205.

Neutral solar wind experiment.
See Abstr. 036.208.

The Ulysses mission.
See Abstr. 051.018.

Beyond a boundary of the heliosphere.
See Abstr. 051.019.

Study of cosmic dust particles on board LDEF and MIR space station.
See Abstr. 051.041.

The present status of the Munich Dust Counter experiment on board of the HITEN spacecraft.
See Abstr. 051.042.

In–situ exploration of dust in the solar system and initial results from the Galileo dust detector.
See Abstr. 051.043.

The NASA Solar Probe mission: in situ determination of interplanetary out–of–the ecliptic and near–solar dust environments.
See Abstr. 051.044.

Future observation of the F–corona with the LASCO coronograph space experiment.
See Abstr. 051.048.

F–corona–experiment: requirements for remote sensing of interplanetary dust.
See Abstr. 051.062.

Pioneers 10 and 11 deep space missions.
See Abstr. 051.068.

Ulysses: a status report.
See Abstr. 051.069.

Shock drift acceleration for a near–perpendicular shock in a turbulent astrophysical plasma.
See Abstr. 062.016.

Nonstationary Petschek reconnection. Convective zone.
See Abstr. 062.181.

Nonstationary Petschek recombination. Diffusion region.
See Abstr. 062.182.

MHD solar wind – interstellar plasma interaction: 3D formulation by the projected characteristics method and the stability analysis.
See Abstr. 062.241.

Stokes parameters of radiation propagating through an aligned gaseous–dust medium.
See Abstr. 063.047.

On the radiation forces.
See Abstr. 063.130.

Detailed comparison between sunspot activity in "hot spots" and galactic cosmic–ray intensity.
See Abstr. 072.043.

Solar flares and coronal mass ejections.
See Abstr. 073.079.

Focused transport of energetic particles along magnetic field lines draped around a coronal mass ejection.
See Abstr. 074.006.

Concerning solar sources of large–scale heliospheric disturbances.
See Abstr. 074.016.

The X–ray corona, the coronal hole, and the heliosphere.
See Abstr. 074.019.

Structure of the solar corona, and the heliospheric current sheet.
See Abstr. 074.027.

Variable carbon and oxygen abundances in the solar wind as observed in Earth's magnetosheath by AMPTE/CCE.
See Abstr. 074.035.

The hydrodynamic picture of the solar wind flow around the Earth's magnetosphere and a comet in the quasi–turbulent regime.
See Abstr. 074.048.

Study of formation of the supersonic solar wind.
See Abstr. 074.051.

On potential field models of the solar corona.
See Abstr. 074.052.

Fluctuations of the magnetic field and galactic cosmic rays in recurrent high–velocity flows of the solar wind.
See Abstr. 074.056.

Solar wind acceleration
See Abstr. 074.064.

Remote sensing observations of mass ejections and shocks in interplanetary space.
See Abstr. 074.072.

In situ observations of coronal mass ejections in interplanetary space.
See Abstr. 074.073.

Cosmic rays and magnetosonic instabilities of solar wind flow near the heliospheric shock wave.
See Abstr. 074.080.

Plasma observations in the distant heliosphere: a view from Voyager.
See Abstr. 074.081.

Distant solar wind plasma – view from the Pioneers.
See Abstr. 074.082.

Solar wind vortex flow in the outer heliosphere.
See Abstr. 074.086.

Interaction of the solar wind with the external plasma.
See Abstr. 074.087.

Signature of a viscous interaction at the heliopause.
See Abstr. 074.088.

The interactions of the solar wind discontinuities in the vicinity of the heliopause.
See Abstr. 074.090.

Superevents: their origin and propagation through the heliosphere from 0.3 to 35 AU.
See Abstr. 078.004.

The influence of the large–scale interplanetary shock structure on a low–energy particle event.
See Abstr. 078.010.

The effect of statistical acceleration on propagation of particles from a solar flare in the interplanetary medium.
See Abstr. 078.013.

Observation of shock–accelerated protons by Giotto and IMP–8 under solar minimum conditions in February 1986.
See Abstr. 078.015.

A system of loop like interplanetary bottles containing solar cosmic rays in June 1974.
See Abstr. 078.017.

Recurrent enhancements of energetic particle intensity during the decreasing phase of the 21st solar activity cycle.
See Abstr. 078.018.

Modelling of the magnetic field of magnetospheric ring current as a function of interplanetary medium parameters.
See Abstr. 084.024.

Absence of upstream energetic ions under turbulent radial interplanetary magnetic field.
See Abstr. 084.026.

The semiannual variation of great geomagnetic storms and the postshock Russell–McPherron effect preceding coronal mass ejecta.
See Abstr. 084.030.

Geomagnetic storms categorized by varieties of interplanetary structures.
See Abstr. 084.035.

Solar control of the Earth's emission of energetic O^+.
See Abstr. 084.038.

Interplanetary disturbance structures and geomagnetic storm.
See Abstr. 085.015.

Dust in planetary ring systems.
See Abstr. 091.054.

Meteoroid bombardment and the structure of Saturn's A and B–ring inner edges.
See Abstr. 100.036.

Reading Saturn's ring spokes.
See Abstr. 100.072.

Origin of sungrazers: a frequent cometary end–state.
See Abstr. 102.010.

Dust optical properties: a comparison between cometary and interplanetary grains.
See Abstr. 102.011.

The role of the interplanetary magnetic field reorientation in the mechanism of the comet's brightness outburst occurrence.
See Abstr. 102.023.

The contribution of long period comets to the interplanetary dust cloud.
See Abstr. 103.038.

Long dust trails of short periodic comets.
See Abstr. 103.039.

Comets as a source of interplanetary and interstellar grains.
See Abstr. 103.040.

The dust tail of Comet Liller 1988 V.
See Abstr. 103.047.

Spectroscopic evidence of organic molecules released by the dust of Halley's inner coma.
See Abstr. 103.135.

Power spectral analysis of enhanced scintillation of quasar 3C 459 due to Comet Halley.
See Abstr. 103.141.

Diffraction of interplanetary solid bodies potentially generating meteorites.
See Abstr. 104.062.

The dynamics of meteoroid streams.
See Abstr. 104.067.

Dynamical aspects of the Taurid meteor complex.
See Abstr. 104.068.

The Quadrantid stream, chaos or not?
See Abstr. 104.070.

The orbital distribution and origin of meteoroids.
See Abstr. 104.085.

Off–disk implantation of early solar wind into a planetesimal–dust cloud.
See Abstr. 107.049.

Physical–chemical processes in a protoplanetary cloud.
See Abstr. 107.053.

The physics of dusty plasmas.
See Abstr. 131.093.

Characteristics of interstellar and circumstellar dust.
See Abstr. 131.269.

Characteristics of the local interstellar medium.
See Abstr. 131.284.

A lower limit on the ionization fraction of the very local interstellar medium.
See Abstr. 131.287.

The interaction of interstellar pick–up ions with the solar wind – probing the interstellar medium by in–situ measurements.
See Abstr. 131.288.

Rocket observation of the near–infrared spectrum of the sky.
See Abstr. 133.009.

The cosmic radiation in the heliosphere at successive solar minima.
See Abstr. 144.013.

The simulated features of heliospheric cosmic–ray modulation with a time–dependent drift model. I. General effects of the changing neutral sheet over the peiod 1985 – 1990.
See Abstr. 144.016.

Significance of the turbulent sheath following the interplanetary shocks in producing Forbush decreases.
See Abstr. 144.017.

On the interplanetary cosmic ray latitudinal gradient.
See Abstr. 144.028.

A correlation between IMF and the limiting primary rigidity for the cosmic ray diurnal anisotropy.
See Abstr. 144.032.

The simulated features of heliospheric cosmic–ray modulation with a time–dependent drift model. II. On the energy dependence of the onset of new modulation in 1987.
See Abstr. 144.035.

Frequency spectra of quasi–periodic variations in the general magnetic field of the Sun, in the parameters of the interplanetary magnetic field and cosmic rays.
See Abstr. 144.052.

On the form of the 1987 hydrogen spectrum in the outer heliosphere.
See Abstr. 144.056.

The simulated features of heliospheric cosmic–ray modulation with a time–dependent drift model. III. General energy dependence.
See Abstr. 144.057.

Studies of the low–energy Galactic cosmic–ray composition near 28 AU at sunspot minimum: the primary–to–primary ratios.
See Abstr. 144.059.

Cosmic rays in the local interstellar medium.
See Abstr. 144.074.

Diffusion, drifts, and modulation of galactic cosmic rays in the heliosphere.
See Abstr. 144.075.

Neutron monitor investigations relating modulated cosmic ray spectra with heliospheric magnetic field polarity reversals.
See Abstr. 144.076.

The predictions of a time–dependent drift model compared with cosmic–ray intensity observations from 1976 to 1989.
See Abstr. 144.077.

A time–dependent drift model with a simulated wavy neutral sheet for the solar modulation of cosmic rays.
See Abstr. 144.078.

Solar effects on underground muons at 570 hg/cm^2.
See Abstr. 144.079.

The additional fluxes of cosmic rays in the stratosphere in the various half–periods of the 22–year solar magnetic cycle.
See Abstr. 144.080.

The anomalous component of cosmic rays.
See Abstr. 144.081.

The long term cosmic ray variation relevant to solar wind structure in the outer heliosphere.
See Abstr. 144.084.

On using the ACR to probe the LISM/heliosphere interface.
See Abstr. 144.086.

107 Cosmogony

107.001   Formation of planets from planetesimals.
S. Ida.
Astron. Her., Vol. 85, No. 5, p. 186 – 190 (1992). In Japanese.

107.002   Effects of nuclear reactions with accelerated particles during a supernova explosion.
A. K. Lavrukhina, G. K. Ustinova.
Astron. Vestn., Tom 26, No. 1, p. 62 – 71 (Jan – Feb 1992). In Russian. English translation in Solar Syst. Res., Vol. 26, No. 1.
The production of the anomalous component of neon, Ne–E, in nuclear reactions with fast particles, accelerated during a local supernova explosion 4,700 Myr ago, has been considered. The irradiation parameters have been determined according to the observed abundances of Li, Be and B isotopes. The dependence of Ne–E(L) and Ne–E(H) contents in graphitic grains and silicon carbide on the cut–off energy and the form of the spectrum of accelerated particles has been analysed. The comparison of the results of the calculation with the available measurements in chondrites unambiguously testifies in favour of the origin of ^{22}Na, the radioactive precursor of Ne–E, in nuclear reactions at the front of shock waves in the expanding supernova shell. The supernova, the explosion of which had led to the collapse of the protosolar cloud and gave rise to a great number of isotopic anomalies in meteorites, is supposed to be an O–, B–star of the second generation.

107.003   Tidal disruption of viscous bodies.
S. Sridhar, S. Tremaine.
Icarus, Vol. 95, No. 1, p. 86 – 99 (Jan 1992).
The authors examine the tidal disruption of homogeneous, incompressible bodies ("planetesimals") of viscous fluid, passing by a planet on a parabolic orbit. If the planetesimal radius is small compared to the distance of closest approach, the surface of the planetesimal is always ellipsoidal, and one can therefore reduce the equations of fluid mechanics and Poisson's equation to a set of coupled ordinary differential equations that can be solved with high accuracy. Disrupted planetesimals evolve into needle–like ellipsoids but their density does not decrease. The viscous fluid treatment is approximately valid for solid planetesimals if tidal stresses exceed the strength while the planetesimal is still held together by self–gravity; this condition requires radii exceeding 100 – 200 km for ice or rock planetesimals.

107.004   Influx of interstellar material onto the protoplanetary disk.
I. N. Ziglina, T. V. Ruzmajkina.
Sol. Syst. Res., Vol. 25, No. 1, p. 40 – 45 (Jul 1991). English translation of Astron. Vestn., Tom 25, No. 1, p. 53 – 60 (1991).
The influx of interstellar material onto the circumsolar disk during the collapse of the slowly rotating nebula is examined. The dependence of the disk parameters on the heliocentric distance is assumed to be a power law for the surface density and a proportionality for the thickness. It is found that the ratio of the mass flux onto the peripheral parts of the lateral surface of the disk of mass M_d to the flux at the edge is of the order $(M_d/M) \times (a_2/a)^{1/2}$, where M is the mass of the protosun. Organic compounds and ices could have survived the process of material accretion onto the disk, in spite of heating in the shock wave.

107.005   A guide to the use of theoretical models of the solar nebula for the interpretation of the meteoritic record.
P. Cassen.
54. Annual Meeting of the Meteoritical Society, p. 43 (1991). Abstract. – See Abstr. 012.010 for the main entry.

107.006   Particle–gas dynamics in the protoplanetary nebula.
J. N. Cuzzi, J. M. Champney, A. R. Dobrovolskis.
54. Annual Meeting of the Meteoritical Society, p. 50 (1991). Abstract. – See Abstr. 012.010 for the main entry.

107.007 ^{135}Cs–^{135}Ba: a new cosmochronometric constraint on the origin of the earth and the astrophysical site of the origin of the solar system.
C. L. Harper, H. Wiesmann, L. E. Nyquist.
54. Annual Meeting of the Meteoritical Society, p. 81 (1991). Abstract. – See Abstr. 012.010 for the main entry.

107.008   Silicate melt structures, physical properties, and planetary accretion.
P. Jakes, S. Sen, K. Matsuishi.
54. Annual Meeting of the Meteoritical Society, p. 105 (1991). Abstract. – See Abstr. 012.010 for the main entry.

107.009   Elemental abundance patterns in presolar diamonds.
R. S. Lewis, G. R. Huss, E. Anders, Y. G. Liu, R. A. Schmitt.
54. Annual Meeting of the Meteoritical Society, p. 133 (1991). Abstract. – See Abstr. 012.010 for the main entry.

107.010   The role of sulfur in planetary core formation.
K. Lodders, H. Palme.
54. Annual Meeting of the Meteoritical Society, p. 141 (1991). Abstract. – See Abstr. 012.010 for the main entry.

107.011   Origin of the biosphere of the Earth.
A. H. Delsemme.
3. Symposium International de Bioastronomie, p. 117 – 123 (1991). – See Abstr. 012.005 for the main entry.
The paradigm that has emerged to describe the origin of the solar system excludes the presence of water and of carbon in the planetesimals that agglomerated to form the proto–Earth. An unlikely but possible primary atmosphere of solar composition was transient enough not to play any significant role in the retention of water or carbon. However, the latter evolution of the planetesimals formed in the zone of the Jovian planets, brings a large number of objects made at cooler temperatures into the zone of the terrestrial planets. These objects are mainly the comets, that are going to bring to the Earth more water than needed to explain the oceans, and more carbon than needed to explain the carbonates and the biosphere. This general mechanism seems to work in the later evolution of numerous accretion disks around young stars, and promises to bring enough water and volatile compounds on rocky planets that would have otherwise remained barren.

107.012   Terrestrial accretion of prebiotic volatiles and organic molecules during the heavy bombardment.
C. F. Chyba, C. Sagan, L. Brookshaw, P. J. Thomas.
3. Symposium International de Bioastronomie, p. 149 – 154 (1991). – See Abstr. 012.005 for the main entry.

107.013   Accretion in the inner nebula: the relationship between terrestrial planetary compositions and meteorites.
S. R. Taylor.
Meteoritics, Vol. 26, No. 4, p. 267 – 277 (Dec 1991).
The bulk compositions of the terrestrial planets are assessed. Venus and Earth probably have similar bulk compositions, but Mars is enriched in volatile elements. The inner planets are all depleted in volatile elements, as shown by K/U ratios, relative to most meteorites and the CI primordial values. The CI meteorite abundances, despite aqueous alteration, match the solar data and provide the best estimate for the composition of the solar nebula, including the iron abundance. The variation in composition among the meteorites and the apparent lack of mixing among the groups indicates accretion from narrow feeding zones. There appears to have been little mixing between meteorite and planetary formation zones, as shown by the oxygen isotope variations, lack of mixing of meteorite groups, and differences in K/U ratios. In summary, it appears that the final accretion of planets did not result in widespread homogenization, and that mixing zones were not more than about 0.3 A.U. wide.

107.014   Superheat in magma oceans.
P. Jakes.
Workshop on the Physics and Chemistry of Magma Oceans from 1 bar to 4 Mbar, p. 26 – 27 (1992). Abstract. – See Abstr. 012.032 for the main entry.

107.015   Cooling of the magma ocean due to accretional disruption of the surface insulating layer.
S. Sasaki.
Workshop on the Physics and Chemistry of Magma Oceans from 1 bar to 4 Mbar, p. 47 – 48 (1992). Abstract. – See Abstr. 012.032 for the main entry.

107.016   Early planetary differentiation: geophysical consequences.
G. Schubert.
Workshop on the Physics and Chemistry of Magma Oceans from 1 bar to 4 Mbar, p. 49 – 50 (1992). Abstract. – See Abstr. 012.032 for the main entry.

107.017   The three stages of magma ocean cooling.
P. H. Warren.
Workshop on the Physics and Chemistry of Magma Oceans from 1 bar to 4 Mbar, p. 70 – 71 (1992). Abstract. – See Abstr. 012.032 for the main entry.

107.018   Giant and large impacts in the context of planetary formation theory.
G. W. Wetherill.
Workshop on the Physics and Chemistry of Magma Oceans from 1 bar to 4 Mbar, p. 74 – 75 (1992). Abstract. – See Abstr. 012.032 for the main entry.

107.019   Collision and tidal interaction between planetesimals.
S. Watanabe, S. M. Miyama.
Astrophys. J., Vol. 391, No. 1, p. 318 – 335 (20 May 1992).
The evolution of two encountering planetesimals in the solar gravitational field has been simulated using a three–dimensional smoothed particle hydrodynamics code, in order to obtain the precise coalescence probability, including the effects of tidal forces and shock compression on two–body interaction, one of the elementary processes in planetary accumulation. The tidal forces are considered to be important because they may enlarge the coalescence rate of planetesimals. To study this, the authors have first simulated close encounters of self–gravitating fluid bodies and found that the enlargement factor of the coalescence radius takes its maximum value of 1.7, when the two bodies have uniform density distributions and zero initial relative velocity. The factor decreases with increases in central mass concentration and/or in initial relative velocity. In the cases with initial velocity exceeding $0.5v_e$, where v_e is their escape velocity, a tidal enhancement of the coalescence radius cannot be expected. For those cases with initial velocities lower than the escape velocities, the amount of escaping mass from the merger after collision is less than a few percent of the total mass.

107.020   Cometary origin of carbon and water on the terrestrial planets.
A. H. Delsemme.
IAU Symposium No. 150: Astrochemistry of cosmic phenomena, p. 421 – 422 (1992). – See Abstr. 012.029 for the main entry.

107.021   Solar system – interstellar medium. A chemical memory of the origins.
D. Despois.
IAU Symposium No. 150: Astrochemistry of cosmic phenomena, p. 451 – 458 (1992). – See Abstr. 012.029 for the main entry.
The author emphasizes the role played by the study of comets through the properties of the dust, the chemical composition of volatiles and the elemental abundances. These data inform on cometary matter formation, and hence on conditions in the protosolar nebula.

107.022 Secular ring instability in the protoplanetary accretion disk.
E. Willerding.
Earth, Moon, Planets, Vol. 56, No. 2, p. 173 – 192 (Feb 1992).
The aim of the present paper is to show that an accretion disk, the material of which satisfies the Navier–Stokes equations of a compressible fluid, is secularly unstable to axisymmetric density distrubances even if Toomre's parameter Q is greater than one. This instability process, which can also be interpreted as negative diffusion, leads immediately to the formation of rings in the accretion disk as an intermediate step in the formation of planets, at least for the outer gaseous ones. The author believes that the same process is also responsible for the ringlet–structure of planetary rings. In the global theory, he formulates a complex linear integral equation describing the diffusion instability in the accretion disk.

107.023 Numerical models of the formation of the solar nebula.
A. P. Boss.
Astron. Vestn., Tom 26, No. 3, p. 8 – 13 (May – Jun 1992). In Russian. English translation in Sol. Syst. Res., Vol. 26, No. 3.
A new numerical code has been developed for studying the radiative hydrodynamics of protostellar formation. This short paper describes the recent results of two ongoing projects of interest for the formation of the solar nebula: the determination of initial conditions leading to single protostellar objects, and calculation of the physical conditions in a solar nebula formed by collapse onto a solar–mass protostellar core.

107.024 Ther early faint Sun 'paradox' revisited: massive greenhouse effects on early Earth and Mars?
C. Sagan, C. F. Chyba.
Bull. Am. Astron. Soc., Vol. 23, No. 3, p. 1211 – 1212 (1991). Abstract. – See Abstr. 012.037 for the main entry.

107.025 Growth and destruction of proto–planetary dust particles.
J. Blum.
Bull. Am. Astron. Soc., Vol. 23, No. 3, p. 1223 (1991). Abstract. – See Abstr. 012.037 for the main entry.

107.026 Perturbation theory for moderatley fast gravitational encounters between planetesimals.
G. R. Stewart.
Bull. Am. Astron. Soc., Vol. 23, No. 3, p. 1224 (1991). Abstract. – See Abstr. 012.037 for the main entry.

107.027 Accretional evolution of a planetesimal swarm: the terrestrial zone.
S. J. Weidenschilling, D. R. Davis, F. Marzari.
Bull. Am. Astron. Soc., Vol. 23, No. 3, p. 1224 (1991). Abstract. – See Abstr. 012.037 for the main entry.

107.028 On the origin of planetary spin.
L. Dones, S. D. Tremaine.
Bull. Am. Astron. Soc., Vol. 23, No. 3, p. 1224 (1991). Abstract. – See Abstr. 012.037 for the main entry.

107.029 On the origin of the systematic component of planetary rotation. I. Analytic development.
J. J. Lissauer, D. M. Kary.
Bull. Am. Astron. Soc., Vol. 23, No. 3, p. 1224 – 1225 (1991). Abstract. – See Abstr. 012.037 for the main entry.

107.030 On the origin of the systematic component of planetary rotation. II. Numerical experiments.
D. M. Kary, J. J. Lissauer, Y. Greenzweig.
Bull. Am. Astron. Soc., Vol. 23, No. 3, p. 1225 (1991). Abstract. – See Abstr. 012.037 for the main entry.

107.031 Despin mechanism for proto giant planets.
T. Takata, D. J. Stevenson.
Bull. Am. Astron. Soc., Vol. 23, No. 3, p. 1225 (1991). Abstract. – See Abstr. 012.037 for the main entry.

107.032 Formation of the giant planets.
J. B. Pollack, O. Hubickyj, P. Bodenheimer, J. J. Lissauer, Y. Greenzweig, M. Podolak.
Bull. Am. Astron. Soc., Vol. 23, No. 3, p. 1225 (1991). Abstract. – See Abstr. 012.037 for the main entry.

107.033 Mapping arrival geometries for "beams" of small particles approaching a planet.
W. F. Bottke, R. Greenberg, A. Carusi, G. B. Valsecchi.
Bull. Am. Astron. Soc., Vol. 23, No. 3, p. 1225 – 1226 (1991). Abstract. – See Abstr. 012.037 for the main entry.

107.034 Linear disk response to a protoplanet: numerical integrations.
D. G. Korycansky, J. Pollack.
Bull. Am. Astron. Soc., Vol. 23, No. 3, p. 1226 (1991). Abstract. – See Abstr. 012.037 for the main entry.

107.035 Dynamo magnetic field generation in the solar nebula at the location of the present–day asteroid belt.
E. H. Levy, T. F. Stepinski.
Bull. Am. Astron. Soc., Vol. 23, No. 3, p. 1226 (1991). Abstract. – See Abstr. 012.037 for the main entry.

107.036 Survival of a captured satellite in a primary extended atmosphere.
S. Sasaki.
Bull. Am. Astron. Soc., Vol. 23, No. 3, p. 1226 (1991). Abstract. – See Abstr. 012.037 for the main entry.

107.037 Planetesimal formation in laminar and turbulent nebular disks: additional results.
S. J. Weidenschilling.
Bull. Am. Astron. Soc., Vol. 23, No. 3, p. 1226 (1991). Abstract. – See Abstr. 012.037 for the main entry.

107.038 Giant planet formation using shallow water simulations of a gaseous protostellar disk.
B. Zajac, A. P. Ingersoll, T. E. Dowling.
Bull. Am. Astron. Soc., Vol. 23, No. 3, p. 1231 (1991). Abstract. – See Abstr. 012.037 for the main entry.

107.039 Chaotic evolution of the solar system.
J. Wisdom, G. J. Sussman.
Bull. Am. Astron. Soc., Vol. 23, No. 3, p. 1231 (1991). Abstract. The full paper is published in Science, Vol. 257, No. 5066, p. 56 – 62 (3 Jul 1992). – See Abstr. 012.037 for the main entry.

107.040 Criteria for dynamo magnetic field generation in the primordial solar nebula.
T. F. Stepinski.
Bull. Am. Astron. Soc., Vol. 23, No. 3, p. 1231 – 1232 (1991). Abstract. – See Abstr. 012.037 for the main entry.

107.041 Gas dynamic heating of chondrule precursor grains in nebular shock waves.
L. L. Hood, M. Horanyi.
Bull. Am. Astron. Soc., Vol. 23, No. 3, p. 1233 (1991). Abstract. – See Abstr. 012.037 for the main entry.

107.042 Non–equilibrium deuterium fractionation in the early solar nebula.
R. W. Dissly, Y. L. Yung.
Bull. Am. Astron. Soc., Vol. 23, No. 3, p. 1233 (1991). Abstract. – See Abstr. 012.037 for the main entry.

107.043 Cometary diagnostics of solar nebula chemistry.
S. Wyckoff, M. Womack, L. Ziurys.
Bull. Am. Astron. Soc., Vol. 23, No. 3, p. 1234 (1991). Abstract. – See Abstr. 012.037 for the main entry.

107.044 A comparison of solar wind and estimated solar system xenon abundances: a test for solid/gas fractionation in the solar nebula.
R. C. Wiens, D. S. Burnett, M. Neugebauer, R. O. Pepin.
22. Lunar and Planetary Science Conference, p. 153 – 159 (1992). – See Abstr. 012.046 for the main entry.

The solar Xe elemental abundance is determined via solar wind measurements from lunar ilmenites and normalized to Si by spacecraft data. The results are compared with estimated abundances assuming no fractionation, which are relatively well constrained for Xe by s–process calculations, odd–mass abundance interpolations, and odd–even abundance systematics.

107.045 Generation of dynamo magnetic fields in the primordial solar nebula.
T. F. Stepinski.
Icarus, Vol. 97, No. 1, p. 130 – 141 (May 1992).

The author studies the problem of the existence of dynamo–generated magnetic fields in the primordial solar nebula. The combined action of Keplerian rotation and helical convection enables an $\alpha\omega$ dynamo to generate large–scale magnetic fields in parts of the nebula where levels of electrical conductivity are high enough to provide coupling between the gas and the magnetic field. The aim of this paper is to identify those regions of the nebula where an $\alpha\omega$ dynamo is able to maintain a magnetic field for a relatively long period of time. The author calculates the electrical conductivity of nebular gas and subsequently the radial distribution of the local dynamo number for two specific nebular models – a viscous accretion disk model, and the quiescent minimum – mass nebula. The calculations show that magnetic fields can be easily generated and maintained by an $\alpha\omega$ dynamo in the inner and outer parts of the nebula.

107.046 N–body simulation of gravitational interaction between planetesimals and a protoplanet. I. Velocity distribution of planetesimals.
S. Ida, J. Makino.
Icarus, Vol. 96, No. 1, p. 107 – 120 (Mar 1992).

The distribution of eccentricities (e) and inclinations (i) of planetesimals is investigated by direct N–body simulations. This distribution governs the strength of the dynamical friction to a protoplanet embedded in the planetesimals, and consequently influences how planetary accretion proceeds, since the dynamical friction drives runaway growth of a protoplanet. The authors examined the e, i–distribution obtained by three–dimensional simulation of identical 400 particles and found that it is expressed by Gaussian distribution with high accuracy. Theoretical argument shows that the dynamical friction under the Gaussian distribution is expected to be strong enough to depress e and i of a protoplanet to values satisfying the energy equipartition with those of planetesimals. This result supports the existence of the runaway growth.

107.047 Chemical composition of the Earth after the giant impact.
Liu Lingun.
Earth, Moon, Planets, Vol. 57, No. 2, p. 85 – 97 (May 1992).

The giant impact hypothesis for the origin of the Moon has been widely accepted. One of the most important features of this hypothesis is that the impactor's metallic core was incorporated in the Earth after impact. If the mass of the impactor is 0.82×10^{27}g, the mass of the impactor core was estimated to be 0.19×10^{27}g, which is about 1/10 of present Earth's core. Liu (1982) derived the bulk composition of the Earth from Cl chondrites, and concluded that the Fe content of his model appears to be low in comparison with the present Earth, which, however, can be rationalized by the addition of impactor core into the proto–Earth developed by Liu (1982). If the impactor's mantle contains 14 wt% FeO as suggested, the mass ratio of impactor/proto–Earth should not exceed 0.22. The same ratio is not likely to exceed 0.30, if a giant blowoff did not occur during impact.

107.048 The collision and tidal interaction between planetesimals.
S.–i. Watanabe, S. M. Miyama.
Symposium of Supercomputing Astronomy and Astrophysics in Japan, p. 135 – 168 (1991). – See Abstr. 012.069 for the main entry.

Evolution of two encountering planetesimals in the solar gravitational field has been simulated using a three–dimensional smoothed particle hydrodynamics code, in order to obtain the precise coalescence probability including the effects of tidal forces and shock compression on two–body interaction, one of the elementary processes in planetary accumulation.

107.049 Off–disk implantation of early solar wind into a planetesimal–dust cloud.
S. Sasaki.
IAU Colloquium No. 126: Origin and evolution of interplanetary dust, p. 425 – 428 (1991). – See Abstr. 012.068 for the main entry.

Off–disk implantation of ancient solar wind into a protoplanetary dust cloud can explain the present amounts of solar–type noble gases in gas–rich meteorites and Venus, even if the dust cloud is very opaque along its midplane.

107.050 Numerical two–dimensional calculations of the formation of the solar nebula.
P. H. Bodenheimer.
US–USSR Workshop on Planetary Sciences, p. 17 – 30 (1991). – See Abstr. 012.075 for the main entry.

There has been considerable recent interest in two–dimensional numerical hydrodynamical calculations with radiative transfer, applied to the inner regions of collapsing, rotating protostellar clouds of about 1 $M_\odot$. The calculations start at a density that is high enough so that the gas is decoupled from the magnetic field. During the collapse, mechanisms for angular momentum transport are too slow to be effective, so that an axisymmetric approximation is sufficiently accurate to give useful results. Until the disk has formed, the calculations can be performed under the assumption of conservation of angular momentum of each mass element. In a numerical calculation, a detailed study of the region of disk formation can be performed only if the central protostar is left unresolved.

107.051 Three–dimensional evolution of early solar nebula.
A. P. Boss.
US–USSR Workshop on Planetary Sciences, p. 31 – 43 (1991). – See Abstr. 012.075 for the main entry.

Contents: 1. Initial conditions for protostellar collapse. 2. Single versus binary star formation. 3. Angular momentum transport mechanisms. 4. Three–dimensional solar nebula models. 5. Implications for planetary formation.

107.052 Formation and evolution of the protoplanetary disk.
T. V. Ruzmaikina (T. V. Ruzmajkina), A. B. Makalkin.
US–USSR Workshop on Planetary Sciences, p. 44 – 60 (1991). – See Abstr. 012.075 for the main entry.

This paper discusses a disk formation model during collapse of the protosolar nebula with $J \sim 10^{52}$g cm^{2}s^{-1}, yielding a low–mass protoplanetary disk. The disk begins to form at the growth stage of the stellar–like core and expands during accretion to the present dimensions of the solar system. Accretion at the edge of the disk significantly affects the nature of matter fluxes in the disk and its thermal evolution.

107.053 Physical–chemical processes in a protoplanetary cloud.
A. K. Lavrukhina.
US–USSR Workshop on Planetary Sciences, p. 61 – 69 (1991). – See Abstr. 012.075 for the main entry.

According to current views, the protosun and protoplanetary disk were formed during the collapse of a fragment of the cold, dense molecular interstellar cloud and subsequent accretion of its matter to a disk. One of the most critical cosmochemical issues in this regard is the identification of relics of such matter in the least altered bodies of the solar system: chondrites, comets, and interplanetary dust. The presence of deuterium–enriched, carbon–containing components in certain chondrites (Pillinger 1984) and radicals and ions in comets (Shulman 1987) is promising. If a relationship is established between solar nebula and interstellar matter, one can then identify certain details, such as the interstellar cloud from which the Sun and the planets were formed, and also come to a deeper understanding of the nature of physico–chemical processes in the protoplanetary cloud.

107.054 **Magnetohydrodynamic puzzles in the protoplanetary nebula.**
E. H. Levy.
US–USSR Workshop on Planetary Sciences, p. 70 – 81 (1991). – See Abstr. 012.075 for the main entry.
Our knowledge of the basic physical processes that governed the dynamical state and behavior of the protoplanetary accretion disk remains incomplete. Many large–scale astrophysical systems are strongly magnetized and exhibit phenomena that are shaped by the dynamical behaviors of magnetic fields. Evidence and theoretical ideas point to the possibility that the protoplanetary nebula also might have had a strong magnetic field. This paper summarizes some of the evidence, some of the ideas, some of the implications, and some of the problems raised by the possible existence of a nebular magnetic field. The aim of this paper is to provoke consideration and speculation, rather than to try to present a balance, complete analysis of all of the possibilities or to imagine that firm answers are yet in hand.

107.055 **Formation of planetesimals.**
S. J. Weidenschilling.
US–USSR Workshop on Planetary Sciences, p. 82 – 97 (1991). – See Abstr. 012.075 for the main entry.
A widely accepted model for the formation of planetesimals is by gravitational instability of a dust layer in the central plane of the solar nebula. Such a dust layer is extremely sensitive to turbulence, which would prevent gravitational instability unless coagulation forms bodies large enough to decouple from the gas. Collisional accretion driven by differential motions due to gas drag may bypass gravitational instability completely. Previous models of coagulation assumed that aggregates were compact bodies with uniform density, but it is likely that early stages of grain coagulation produced fractal aggregates having densities that decreased with increasing size. Fractal structure greatly slows the rate of coagulation due to differential settling and delays the concentration of solid matter to the central plane. Low–density aggregates also maintain higher opacity in the nebula than would result from compact particles.

107.056 **Formation of the terrestrial planets from planetesimals.**
G. W. Wetherill.
US–USSR Workshop on Planetary Sciences, p. 98 – 115 (1991). – See Abstr. 012.075 for the main entry.
This article describes recent and current development of theories in which the terrestrial planets formed by the accumulation of much smaller (one– to 10–kilometer diameter) planetesimals. The alternative of forming these planets from massive gaseous instabilities in the solar nebula is not reviewed.

107.057 **The rate of planet formation and the solar system's small bodies.**
V. S. Safronov.
US–USSR Workshop on Planetary Sciences, p. 116 – 125 (1991). – See Abstr. 012.075 for the main entry.
The evolution of random velocities and the mass distribution of pre–planetary body at the early stage of accumulation are currently under review. Arguments have been presented for and against the view of an extremely rapid, runaway growth of the largest bodies at this stage with parameter values of $\Theta \gtrsim 10^3$. Difficulties are encountered assuming such a large Θ: (a) bodies of the Jovian zone penetrate the asteroid zone too late and do not have time to hinder the formation of a normal–sized planet in the asteroidal zone and thereby remove a significant portion of the mass of solid matter and (b) Uranus and Neptune cannot eject

bodies from the solar system into the cometary cloud. Therefore, the values $\Theta < 10^2$ appear to be preferable.

107.058 **Late stages of accumulation and early evolution of the planets.**
A. V. Vityazev, G. V. Pechernikova.
US–USSR Workshop on Planetary Sciences, p. 143 – 162 (1991). – See Abstr. 012.075 for the main entry.
This article briefly discusses recently developed solutions of problems that were traditionally considered fundamental in classical solar system cosmogony: determination of planetary orbit distribution patterns, values for mean eccentricity and orbital inclinations of the planets, and rotation periods and rotation axis inclinations of the planets. The authors examine two important cosmochemical aspects of accumulation: the time scale for gas loss from the terrestrial planet zone, and the composition of the planets in terms of isotope data. They conclude that the early beginning of planet differentiation is a function of the heating of protoplanets during collisions with large (thousands of kilometers) bodies. This paper considers energetics, heat mass transfer processes, and characteristic time scales of these processes at the early stage of planetary evolution.

107.059 **Giant planets and their satellites: what are the relationships between their properties and how they formed?**
D. J. Stevenson.
US–USSR Workshop on Planetary Sciences, p. 163 – 173 (1991). – See Abstr. 012.075 for the main entry.
The giant planets region in our solar system appears to be bounded inside by the limit of water condensation, suggesting that the most abundant astrophysical condensate plays an important role in giant planet formation. Despite some interesting systematics among the four major planets and their satellites, no simple picture emerges for the temperature structure of the solar nebula from observations alone. However, it seems likely that Jupiter is the key to our planetary system and a similar planet could be expected for other systems. It is further argued that we should expect a gradual transition from solar nebula dominance to interstellar dominance in the gas phase chemistry of the source material in the outer solar system because of the inefficiency of diffusion in the solar nebula.

107.060 **The role of impacting processes in the chemical evolution of the atmosphere of primordial Earth.**
L. M. Mukhin, M. V. Gerasimov.
US–USSR Workshop on Planetary Sciences, p. 203 – 217 (1991). – See Abstr. 012.075 for the main entry.
Existing and observed data provide evidence of the very early formation of a dense atmosphere on Earth. The authors consider possible scenarios for the formation of the Earth's early atmosphere and its initial chemical composition.

107.061 **Cometary origin of carbon, nitrogen and water on the Earth.**
A. H. Delsemme.
Origins Life Evol. Biosphere, Vol. 21, No. 5 – 6, p. 279 – 298 (1991 – 1992). – See Abstr. 012.083 for the main entry.
Two independent assumptions are substantiated; firstly, that the Earth accreted from dust particles that were hot enough not to contain any volatiles; secondly, that after the accretion was finished, all the volatiles of the biosphere, including the atmosphere and the oceans, were brought by a cometary bombardment.

107.062 **Comets and the formation of biochemical compounds on the primitive Earth – a review.**
J. Oró, T. Mills.
Origins Life Evol. Biosphere, Vol. 21, No. 5 – 6, p. 267 – 277 (1991 – 1992). – See Abstr. 012.083 for the main entry.
Thirty years ago it was suggested that comets impacting on the primitive Earth may have represented a significant source of terrestrial volatiles, including some important precursors for prebiotic synthesis. This possibility is strongly supported not only by models of the collisional history of the early Earth, but also by

astronomical evidence that suggests that frequent collisions of comet–like bodies from the circumstellar disk around the star β Pictoris are taking place. Although a significant fraction of the complex organic compounds that appear to be present in cometary nuclei were probably destroyed during impact, it is argued that cometary collisions with the primitive Earth represented an important source of both free–energy and volatiles, and may have created transient, gaseous environments in which prebiotic synthesis may have taken place.

107.063 **Comets as a possible source of prebiotic molecules.**
W. F. Huebner, D. C. Boice.
Origins Life Evol. Biosphere, Vol. 21, No. 5–6, p. 299–315 (1991–1992). – See Abstr. 012.083 for the main entry.
Prebiotic molecules derive from abiotic organic molecules, radicals, and ions that pervade the universe at temperatures as high as several 1000K. Here the authors review the role of organic molecules that condensed at low temperatures before or during comet formation in the early history of the Solar System.

107.064 **Comet impacts and chemical evolution on the bombarded Earth.**
V. R. Oberbeck, H. Aggarwal.
Origins Life Evol. Biosphere, Vol. 21, No. 5–6, p. 317–338 (1991–1992). – See Abstr. 012.083 for the main entry.
Amino acids yields for previously published shock tube experiments are used with minimum Cretaceous–Tertiary (K/T) impactor mass and comet composition to predict AIB amino acid K/T boundary sediment column density. Sites favorable for chemical evolution of amino acids are examined and it is concluded that chemical evolution could have occurred at or above the surface even during periods of intense bombardment of Earth before 3.8 billion years ago.

107.065 **Cometary supply of terrestrial organics: lessons from the K/T and the present epoch.**
D. Steel.
Origins Life Evol. Biosphere, Vol. 21, No. 5–6, p. 339–357 (1991–1992). – See Abstr. 012.083 for the main entry.
The time–scales of import with respect to the physical survival of planet–crossing bodies (asteroids, comets, meteoroids, dust) in the inner solar system are considered, and characteristic times for different masses reviewed.

107.066 **Terrestrial and extraterrestrial sources of molecular homochirality.**
W. A. Bonner.
Origins Life Evol. Biosphere, Vol. 21, No. 5–6, p. 407–420 (1991–1992). – See Abstr. 012.083 for the main entry.

107.067 **Flaws in the Modern Laplacian Theory of the solar system.**
J. J. Monaghan.
Proc. Astron. Soc. Aust., Vol. 9, No. 2, p. 240 (1991). – See Abstr. 012.090 for the main entry.
The Modern Laplacian Theory (Prentice 1978) of the origin of the solar system assumes a non–dissipative model of supersonic turbulence and the existence of stable rings left behind during the contraction of the proto–Sun. The author shows by numerical simulation that the turbulence is highly dissipative and the rings are unstable. As a result of the instability the rings spread and interact with the proto–Sun. The rings therefore cannot form in the way Prentice has proposed.

107.068 **Chemical fractionation in gas rings and the formation of the solar system.**
A. J. R. Prentice.
Proc. Astron. Soc. Aust., Vol. 9, No. 2, p. 321–323 (1991). – See Abstr. 012.090 for the main entry.
The anomalously high density of the planet Mercury and the higher–than–solar rock–to–ice ratio in Jupiter's moons Ganymede and Callisto cannot be explained by the conventional disc models of solar system formation. It is shown here that the unusual chemical signature of these bodies is the outcome of a process of chemical fractionation and orbital focussing which is peculiar to gas ring models of planet and regular satellite formation. Good numerical agreement with the observational data is obtained if the temperature of condensation of the planetary system closely followed the law $T_n \cong 680[R_\delta/R_n]^{0.9}$K, where R_n denotes the radial distance from the Sun.

Planetary sciences. American and Soviet research. Proceedings.
See Abstr. 012.075.

Comets and the origin and evolution of life. Proceedings. Conference on Comets and the Origin and Evolution of Life, Eau Claire, WI (USA), 30 Sep–2 Oct 1991.
See Abstr. 012.083.

What makes a planet habitable, and how to search for habitable planets in other solar systems.
See Abstr. 015.009.

Endogenous production, exogenous delivery and impact–shock synthesis of organic molecules: an inventory for the origins of life.
See Abstr. 015.011.

Chemical studies on the existence of extraterrestrial life.
See Abstr. 015.040.

The effect of H_2O gas on volatilities of planet–forming major elements: I. Experimental determination of thermodynamic properties of Ca–, Al–, and Si–hydroxide gas molecules and its application to the solar nebula.
See Abstr. 022.056.

Hot shock experiments: simulation of an important process in the early solar system and in multi–ring cratering.
See Abstr. 022.070.

Structure, viscosity, and changes of silicate melts at impact sites.
See Abstr. 022.179.

The physics and dynamics of charged dust grains.
See Abstr. 022.195.

Laboratory study of early solar nebula condensed object analogs.
See Abstr. 022.229.

New orbital elements useful for predicting a particle's behavior upon encounter with a planet.
See Abstr. 042.081.

Averaging the elliptic asteroidal problem with a Stokes drag.
See Abstr. 042.082.

Dissipative phenomena in resonance problems in the solar system or the "dei ex machina" of celestial mechanics.
See Abstr. 042.083.

Large electric fields in acoustic waves and the stimulation of lightning discharges.
See Abstr. 062.042.

New Rosseland mean opacities for molecular clouds and the solar nebula accretion disk.
See Abstr. 063.116.

Diamagnetic abundance differentiation in the solar system.
See Abstr. 071.014.

Oxygen isotopic homogeneity of the Earth: new evidence.
See Abstr. 081.015.

Degassing.
See Abstr. 081.041.

Deuterium in the solar system.
See Abstr. 091.025.

Stochastic behavior of the orbits of planets during their accretion.
See Abstr. 091.049.

Origin of noble gases in the terrestrial planets.
See Abstr. 091.064.

Mercury's high metal content and super–adiabaticity within the proto–solar cloud.
See Abstr. 092.013.

The thermal conditions of Venus.
See Abstr. 093.098.

Origin of the asteroid belt.
See Abstr. 098.068.

Tidal contribution of the satellites to removing the angular momentum of Jupiter.
See Abstr. 099.014.

Three–dimensional statistics of cometary aphelia.
See Abstr. 102.008.

Boulder complex model.
See Abstr. 102.043.

Physico–dynamical evolution of aging comets.
See Abstr. 102.049.

Computational study of radiation chemical processing in comet nuclei.
See Abstr. 102.060.

CHON–particles in interplanetary space.
See Abstr. 104.050.

Magnetization of meteorites by dynamo–generated magnetic fields in the solar nebula.
See Abstr. 105.160.

The origin of the polycyclic aromatic hydrocarbons in meteorites.
See Abstr. 105.261.

The fate of organic matter during planetary accretion: preliminary studies of the organic chemistry of experimentally shocked Murchison meteorite.
See Abstr. 105.262.

Stardust and planet dust.
See Abstr. 106.010.

The expected flux of debris from the formation in extra solar planets to Earth: Have we detected it?
See Abstr. 118.031.

The properties and environment of primitive solar nebulae as deduced from observations of solar–type pre–main sequence stars.
See Abstr. 121.061.

Molecular cloud diagnostics of solar nebula chemistry.
See Abstr. 131.253.